WITHDRAWN
FROM
UNIVERSITY OF PLYMOUTH
LIBRARY SERVICES

D1756776

HANDBOOK OF VERTEBRATE IMMUNOLOGY

HANDBOOK OF VERTEBRATE IMMUNOLOGY

Edited by

Paul-Pierre Pastoret
Faculty of Veterinary Medicine, University of Liège, Belgium

Philip Griebel
Veterinary Infectious Disease Organization, Saskatoon, Canada

Hervé Bazin
European Commission and Faculty of Medicine, University of Louvain, Belgium

André Govaerts
Faculty of Medicine, University of Brussels, Belgium

Academic Press
San Diego London Boston
New York Sydney Toyko Toronto

This book is printed on acid-free paper

Academic Press
525 B Street, Suite 1900, San Diego, California 92101-4495, USA
http://www.apnet.com

Academic Press Limited
24–28 Oval Road, London NW1 7DX, UK
http://www.hbuk.co.uk/ap/

ISBN 0-12-546401-0

Library of Congress Cataloging-in-Publication Data
Handbook of vertebrate immunology/edited by Paul Pierre Pastoret . . .
[et al.].
p. cm.
Includes index.
ISBN 0-12-546401-0 (alk. paper)
1. Vertebrates—Immunology—Handbooks, manuals, etc.
2. Veterinary immunology—Handbooks, manuals, etc. I. Pastoret, Paul-Pierre.
QR181.H277 1996
591.9′6196—dc21 97-45873
CIP

A catalogue record for this book is available from the British Library

Typeset by Paston Press Ltd, Loddon, Norfolk
Printed in Great Britain by The Bath Press, Bath,

98 99 00 01 02 03 BP 9 8 7 6 5 4 3 2 1

CONTENTS

List of Section Editors xi

List of Contributors xiii

Preface xix

I **INTRODUCTION** 1
P. P. Pastoret, P. Griebel, H. Bazin, A. Govaerts

II **IMMUNOLOGY OF FISHES** 3
C. McL. Press

1. Introduction 3
C. McL. Press, T. Jørgensen
2. Lymphoid organs and their anatomical distribution 3
C. McL. Press
3. Leukocytes and their markers 6
N. W. Miller
4. Leukocyte traffic and associated molecules 9
A. E. Ellis
5. Cytokines 11
B. Robertsen
6. T-cell receptor 13
I. Hordvik, J. Charlemagne
7. Immunoglobulins 15
L. Pilström
8. Major histocompatibility complex (MHC) antigens 23
Ø. Lie, U. Grimholt
9. Red blood cell antigens 25
D. C. Morizot, T. J. McConnell
10. Ontogeny of the immune system 26
A. E. Ellis
11. Passive transfer of immunity 30
M. F. Tatner
12. Nonspecific immunity 31
G. A. Ingram
13. Complement system 36
C. Koch
14. Mucosal immunity 39
J. H. W. M. Rombout, E. P. H. M. Joosten
15. Acquired immunodeficiencies 40
J. S. Lumsden
16. Tumors of the immune system 41
J. S. Lumsden
17. Conclusions 41
C. McL. Press, T. Jørgensen
18. References 43

III **IMMUNOLOGY OF AMPHIBIANS** 63
J. Charlemagne, A. Tournefier

1. Introduction 63
2. Lymphoid organs and lymphocytes 63
3. Lymphocyte antigen-specific receptors 65
4. The major histocompatibility complex (MHC) 67
5. Immunobiology of T and B cells 68
6. Ontogeny of the immune response 69
7. Tumors of the immune system 70
8. Conclusion 70
9. References 70

IV **AVIAN IMMUNOLOGY** 73
J. M. Sharma

1. Introduction 73
J. M. Sharma
2. Lymphoid organs and their anatomical distribution 73
O. J. Fletcher, H. J. Barnes
3. Leukocyte markers in the chicken 81
S. H. M. Jeurissen, O. Vainio, M. J. H. Ratcliffe
4. T-cell receptors 87
W. T. McCormack
5. Cell surface and secreted immunoglobulins in B cell development 89
S. L. Demaries, M. J. H. Ratcliffe
6. Major histocompatibility complex (MHC) antigens 92
K. Hala, J. Plachy, J. Kaufman
7. Cytokines 95
S. Rautenschlein, J. M. Sharma
8. Complement 99
T. L. Koppenheffer
9. Ontogeny of the immune system 101
T. P. Arstila, P. Toivanen
10. Immunocompetence of the embryo and the newly-hatched chick 104
P. S. Wakenell
11. Mucosal gut immunity 105
H. S. Lillehoj, S. B. Jakowlew
12. Tumors of the immune system 108
K. A. Schat
13. Autoimmunity in avian model systems 111
M. H. Kaplan

14. Effect of viruses on immune functions 113
T. L. Pertile, J. M. Sharma
15. Concluding remarks 113
J. M. Sharma
16. References 117

V IMMUNOLOGY OF THE RAT 137
F. G. M. Kroese
1 Introduction 137
F. G. M. Kroese
2. Lymphoid organs: their anatomical distribution and cell composition 137
P. Nieuwenhuis
3. Rat cluster of differentiation (CD) antigens 142
F. G. M. Kroese, N. A. Bos
4. Leukocyte traffic and associated molecules 142
R. Pabst, J. Westermann
5. Cytokines and their receptors 150
P. van der Meide, M. N. Hylkema
6. T-cell development 150
H. Groen
7. The T-cell receptor 162
N. E. Torres-Nagel, T. Hermann, T. Hünig
8. B-cell development 168
F. G. M. Kroese, G. J. Deenen
9. Rat immunoglobulins 171
H. Bazin
10. Rat Ig variable region genes 173
N. A. Bos, P. M. Dammers
11. B lymphoid tumors and immunocytomas 177
H. Bazin
12. The major histocompatibility complex 179
J. Rozing, H. Groen
13. Nonspecific immunity 183
C. D. Dijkstra, T. K. van den Berg
14. Components of the mucosal immune system 187
E. P. van Rees
15. The complement system 190
M. R. Daha, R. H. van den Berg
16. Models of immunodeficiency in the rat 192
J. Rozing
17. Autoimmunity 194
D. L. Greiner, B. J. Whalen
18. References 198

VI IMMUNOLOGY OF LAGOMORPHS 223
R. G. Mage
1. Introduction 223
R. G. Mage
2. Lymphoid organs, their anatomical distribution and ontogeny of the immune system 223
P. D. Weinstein, A. O. Anderson
3. Rabbit leukocyte markers 226
K. L. Knight, D. Lanning
4. Leukocyte traffic and associated molecules 230
S. Sharar, C. Ramamoorthy, J. Harlan, R. Winn
5. Cytokines 233
R. G. Mage
6. T-cell receptor (TCR) 233
A. Seto
7. Immunoglobulins 235
R. G. Mage
8. MHC antigens 238
N. Sittisombut
9. Rabbit blood groups 242
L. S. Rodkey, R. G. Tissot
10. Passive transfer of immunity 243
R. G. Mage
11. Nonspecific immunity 243
R. G. Mage
12. Complement system 243
M. Komatsu
13. Mucosal immunity 245
P. D. Weinstein, A. O. Anderson
14. Immunodeficiencies 248
R. G. Mage, B. Hague
15. Tumors of the immune system 248
B. Hague
16. Autoimmune diseases 248
B. Albini
17. Conclusions 250
R. G. Mage
18. References 251

VII IMMUNOLOGY OF THE DOG 261
P. J. Felsburg
1. Introduction 261
P. Felsburg
2. Lymphoid organs and their anatomical distribution 261
H. Hogenesch
3. Leukocytes and their antigens 263
P. J. Felsburg
4. Leukocyte traffic and associated molecules 264
P. J. Felsburg
5. Cytokines 267
H. Hogenesch
6. T-cell receptor 268
P. J. Felsburg
7. Immunoglobulins 268
P. J. Felsburg
8. Major histocompatibility complex (MHC) antigens 268
J. L. Wagner
9. Red blood cell antigens 270
U. Giger
10. Ontogeny of the immune system 271
G. S. Krakowka
11. Passive transfer of immunity 271
P. J. Felsburg

12. Neonatal immune responses 272
P. J. Felsburg
13. Nonspecific immunity 273
P. J. Felsburg
14. Complement system 274
J. A. Winkelstein, R. Ameratunga
15. Mucosal immunity 275
H. Hogenesch
16. Immunodeficiencies 277
P. J. Felsburg
17. Tumors of the immune system 278
J. F. Modiano, S. C. Helfand
18. Autoimmunity 278
P. J. Felsburg
19. Conclusions 282
P. J. Felsburg
20. References 282

VIII IMMUNOLOGY OF THE CAT 290
H. Lutz
1. Introduction 290
H. Lutz
2. Lymphoid organs and their anatomical position 290
S. W. Hüttner
3. Leukocyte differentiation antigens 294
B. J. Willett
4. Cytokines 297
H. Lutz, C. Leutenegger
5. The major histocompatibility complex 301
B. J. Willett
6. Red blood cell antigens 303
U. Giger
7. Immunoglobulins 304
M. Suter
8. Passive transfer of maternal immunity 305
R. Pu, J. K. Yamamoto
9. Neonatal immune responses 308
J. K. Levy, M. B. Tompkins
10. Nonspecific immunity in the cat 309
J. W. Ritchey, W. A. F. Tompkins
11. The complement system 311
C. K. Grant
12. Ontogeny of the immune system 313
C. R. Stokes, D. A. Harbour
13. Mucosal immunity 313
C. R. Stokes, D. A. Harbour
14. Immunological diseases 315
N. C. Pedersen
15. Immunodeficiency diseases 321
N. C. Pedersen
16. Tumors of the immune system 325
M. Reinacher
17. Conclusions 327
H. Lutz
18. References 327

IX IMMUNOLOGY OF MUSTELIDAE 337
H. Tabel
1. Introduction 337
2. Badger 337
3. Ferret 337
4. River otter 337
5. Sea otter 338
6. Skunk 338
7. Mink 338
8. Conclusions 340
9. References 340

X IMMUNOLOGY OF HORSES AND DONKEYS 343
D. P. Lunn, D. Hannant, D. W. Horohov
1. Introduction 343
2. Leukocytes and their antigens 343
3. Immunoglobulins 347
4. Cytokines 350
5. Major histocompatibility complex (MHC) antigens 353
6. Innate immunity 354
7. Development of equine immunity 355
8. Disorders of the immune system 356
9. Tumors of the immune system 359
10. Reproductive immunology 359
11. Conclusions 361
12. References 361

XI IMMUNOLOGY OF THE PIG 373
M. D. Pescovitz
1. Introduction 373
M. D. Pescovitz
2. Lymphoid organs and their anatomical distribution 373
R. Pabst, H. J. Rothkötter
3. Leukocytes and their markers 374
M. D. Pescovitz
4. Leukocyte traffic and associated molecules 377
R. Pabst, H. J. Rothkötter
5. Cytokines and interferons 379
M. P. Murtaugh, D. L. Foss
6. T-cell receptor 383
M. D. Pescovitz
7. Immunoglobulins 384
J. E. Butler
8. Major histocompatibility complex (MHC) antigens 387
J. Lunney
9. Red blood cell antigens 389
P. Vögeli, D. Llanes
10. Ontogeny of the immune system 394
I. Trebichavsky, H. Tlaskalová-Hogenová, B. Cukroswska, J. Sinkora
11. Passive transfer of immunity 395
C. Stokes, M. Bailey

12. Neonatal immune responses 397
I. Trebichavsky, H. Tlaskalová-Hogenova
13. Nonspecific immunity 398
M. Bailey, C. R. Stokes
14. Complement system 400
K. Høgåsen, B. P. Morgan
15. Mucosal immunity 402
M. Bailey, C. R. Stokes
16. Immunodeficiencies 405
M. D. Pescovitz
17. Tumors of the immune system 405
M. D. Pescovitz
18. Autoimmunity 406
M. D. Pescovitz
19. Conclusions 406
M. D. Pescovitz
20. References 407

XII IMMUNOLOGY OF CAMELS AND LLAMAS 421
R. Hamers, S. Muyldermans
1. Introduction 421
2. Lymphoid organs and their anatomical distribution 422
3. Leukocytes and their markers 422
4. Immunoglobulins 423
5. Phylogeny of the camelid immune system 429
6. Ontogeny of the immune system 430
7. Passive transfer of immunity 430
8. Blood groups 432
9. Nonspecific immunity 432
10. Complement system 432
11. Immunodeficiencies 432
12. Tumors of the immune system 434
13. Conclusions 434
14. Acknowledgements 435
15. References 435

XIII IMMUNOLOGY OF CATTLE 439
B. Goddeeris
1. Introduction 439
B. Goddeeris
2. Lymphoid organs 439
D. McKeever, N. MacHugh, B. Goddeeris
3. The leukocytes and their markers 444
C. Howard, P. Sopp
4. Cytokines 447
B. Mertens
5. The T-cell receptor 454
N. Ishiguro
6. Immunoglobulins 456
J. Naessens
7. Major histocompatibility complex (MHC) antigens 459
H. Lewin
8. Red blood cell antigens 462
E. Tucker
9. Passive transfer of immunity 464
A. Husband
10. Nonspecific immunity 466
J. Roth, D. Desmecht
11. Complement system 469
H. Tabel
12. References 471

XIV SHEEP IMMUNOLOGY AND GOAT PECULIARITIES 485
P. J. Griebel
1. Introduction 485
P. J. Griebel
2. Lymphoid organs and their anatomical distribution 485
T. Landsverk
3. Leukocytes and their markers 489
P. J. Griebel
4. Leukocyte migration 493
A. J. Young
5. Sheep cytokines 496
A. D. Nash, R. J. Hawken, A. E. Andrews, J. F. Maddox, H. M. Martin
6. T-cell receptors 500
W. R. Hein, E. Peterhans
7. Immunoglobulins 503
W. R. Hein
8. Ovine major histocompatibility complex antigens (Ovar) 505
J. F. Maddox
9. Sheep red blood cell antigens 509
T. C. Nguyen
10. Ontogeny of the immune system 510
W. R. Hein
11. Transmission of passive immunity in the sheep 513
D. Watson
12. The neonatal immune system 515
W. G. Kimpton, C. Cunningham, R. N. P. Cahill
13. Innate immune mechnisms in the sheep 518
D. Haig, E. Peterhans
14. Complement 522
P. J. Griebel
15. Mucosal immunity 523
S. McClure
16. Immunodeficiencies 528
B. Blacklaws
17. Tumors of the immune system 530
A. van den Broeke
18. Experimental investigation of autoimmunity in the sheep 531
P. P. McCullagh
19. Conclusions 533
20. References 533

XV IMMUNOLOGY OF THE BUFFALO 555
S. D. Carter
1. Introduction 555
2. Lymphoid organs 555
3. Leukocytes and their markers 556
4. Leukocyte traffic 557
5. Cytokines 557
6. T-cell receptor 558
7. Major histocompatibility complex (MHC) antigens 558
8. Immunoglobulins 558
9. Passive transfer of immunity 559
10. Neonatal immmunity 560
11. Nonspecific immunity 560
12. Complement 560
13. Immunomodulation 560
14. Tumors of the immune system 561
15. References 561

XVI THE MOUSE MODEL 563
M. P. Defresne
1. Introduction 563
M. P. Defresne
2. The lymphoid organs 563
E. Heinen, M. P. Defresne
3. Immunoglobulins 570
M. Brait, J. Urbain
4. Cellular immunology 586
M. P. Defresne
5. Murine models of immunodeficiency 589
M. P. Moutschen
6. References 594

XVII THE SCID MOUSE MUTANT: DEFINITION AND POTENTIAL USE AS A MODEL FOR IMMUNE DISORDERS 603
J. Y. Cesbron, V. Leblond, N. Delhem
1. Introduction 603
2. The SCID mouse 603
3. The SCID model of human immune responses 604
4. The SCID 'leaky' phenotype and new immune deficient murine strain 607
5. SCID model of normal and neoplastic human haematopoiesis 608
6. Use of the SCID model without reconstitution 610
7. Acknowledgements 611
8. References 611

XVIII XENOTRANSPLANTATION 619
P. Gianello, M. Soares, D. Latinne, H. Bazin
1. Introduction 619
2. Hyperacute and delayed discordant xenograft rejection 620
3. Human circulating xenoreactive natural antibodies 620
4. The xenoantigens 623
5. Complement activation 623
6. Endothelial cell activation 624
7. Tolerance and endothelial cell accommodation 626
8. Therapeutic approaches: future trends 626
9. Conclusions 629
10. References 629

XIX CONCLUSIONS 635

GLOSSARY 637

INDEX 661

A four page colour plate section appears between pages 428 and 429

List of Section Editors

Hervé Bazin
European Commission and Experimental Immunology, Faculty of Medicine, University of Louvain, Belgium

Stuart D. Carter
Department of Veterinary Pathology, University of Liverpool, United Kingdom

Jean-Yves Cesbron
Institut Pasteur, Faculty of Medicine, Lille, France

Jacques Charlemagne
Groupe d'Immunologie Comparée, Université Pierre et Marie Curie, Paris, France

Marie-Paule Defresne
Laboratory of Pathology, Faculty of Medicine, University of Liège, Belgium

Nadirah Delhem
Institut Pasteur, Lille, France

Peter J. Felsburg
Department of Clinical Studies, School of Veterinary Medicine, University of Pennsylvania, Philadelphia, PA, USA

Pierre Gianello
Experimental Surgery, Faculty of Medicine, University of Louvain, Belgium

Bruno Goddeeris
Laboratory of Animal Physiology and Immunology, Catholic University of Leuven, Belgium

André Govaerts
Immunology–Immunopathology, Faculty of Medicine, University of Brussels, Belgium

Philip J. Griebel
Veterinary Infectious Disease Organization, Saskatoon, Canada

Raymond Hamers
Institute of Molecular Biology, University of Brussels, Belgium

Duncan Hannant
Animal Health Trust,.Newmarket, United Kingdom

Frans G. M. Kroese
Department of Histology and Cell Biology, Immunology Section, University of Groningen, The Netherlands

Dominique Latinne
Transplantation Immunology, Faculty of Medicine, University of Louvain, Belgium

Véronique Leblond
Department of Haematology, Hôpital Pitié-Salpétrière, CNRS, Paris, France

Hans Lutz
Department of Internal Veterinary Medicine, University of Zürich, Switzerland

Rose G. Mage
Laboratory Immunology, Molecular Immunogenetics Section, NIH, Bethesda, MD, USA

Serge Muyldermans
Institute of Molecular Biology, University of Brussels, Belgium

Paul-Pierre Pastoret
Department of Immunology–Vaccinology, Faculty of Veterinary Medicine, University of Liège, Belgium

Mark D. Pescovitz
Departments of Surgery and Microbiology/Immunology, Indiana University, Indianapolis, IN, USA

Charles McL. Press
Department of Morphology, Genetics and Aquatic Biology, Norwegian College of Veterinary Medicine, Oslo, Norway

Jagdev M. Sharma
Department of Veterinary PathoBiology, College of Veterinary Medicine, University of Minnesota, St Paul, MN, USA

Miguel Soares
Experimental Immunology, Faculty of Medicine, University of Louvain, Belgium

Henry Tabel
Department of Veterinary Microbiology, University of Saskatchewan, Saskatoon, Canada

A. Tournefier
Laboratoire d'Immunologie Comparée, Université de Bourgogne, Dijon, France

List of Contributors

B. Albini
Department of Microbiology, State University of New York, Buffalo, NY, USA

B. Adler
Institute of Veterinary Virology, University of Berne, Switzerland

H. Adler
Institute of Veterinary Virology, University of Berne, Switzerland

R. Ameratunga
Department of Pediatrics, School of Medicine, Johns Hopkins University, Baltimore, MD, USA

A. O. Anderson
US AMRIID, Frederick, MD, USA

A. E. Andrews
Centre for Animal Biotechnology, School of Veterinary Science, University of Melbourne, Parkville, Australia

T. P. Arstila
Department of Medical Microbiology, Turku University, Finland

M. Bailey
Division of Molecular and Cellular Biology, University of Bristol, United Kingdom

H. J. Barnes
College of Veterinary Medicine, North Carolina State University, Raleigh, NC, USA

H. Bazin
European Commission and Experimental Immunology, Faculty of Medicine, University of Louvain, Belgium

G. Bertoni
Institute of Veterinary Virology, University of Berne, Switzerland

B. Blacklaws
Department of Clinical Veterinary Medicine, University of Cambridge, United Kingdom

N. A. Bos
Department of Histology and Cell Biology, University of Groningen, The Netherlands

M. Brait
Laboratory of Animal Physiology, Faculty of Sciences, University of Brussels, Belgium

J. E. Butler
Department of Microbiology, University of Iowa, Iowa City, IA, USA

R. N. P. Cahill
Laboratory for Foetal and Neonatal Immunology, School of Veterinary Science, The University of Melbourne, Cnr. Park Dr, Parkville, Victoria, Australia

S. D. Carter
Department of Veterinary Pathology, University of Liverpool, United Kingdom

J.-Y. Cesbron
Institut Pasteur, Faculty of Medicine, Lille, France

J. Charlemagne
Groupe d'Immunologie Comparée, Université Pierre et Marie Curie, Paris, France

B. Charley
Institut National de la Recherche Agronomique, Virologie et Immunologie Moléculaires, Centre de Recherches de Jouy-en-Josas, France

C. Cunningham
Laboratory for Foetal and Neonatal Immunology, School of Veterinary Science, The University of Melbourne, Flemington Road, Parkville, Victoria, Australia

B. Cukrowska
Division of Immunology and Gnotobiology, Institute of Microbiology, Czech Academy of Sciences, Praha, Czech Republic

M. R. Daha
Department of Renal Diseases, University of Leiden, The Netherlands

P. M. Dammers
Department of Histology and Cell Biology, University of Groningen, The Netherlands

G. J. Deenen
Department of Histology and Cell Biology, University of Groningen, The Netherlands

M. P. Defresne
Laboratory of Pathology, Faculty of Medicine, University of Liège, Belgium

N. Delhem
Institut Pasteur, Lille, France

S. L. Demaries
Department of Microbiology and Immunology, McGill University, Montréal, Canada

D. Desmecht
Department of Pathology, Faculty of Veterinary Medicine, University of Liège, Belgium

C. D. Dijkstra
Department of Cell Biology and Immunology, Free University, Amsterdam, The Netherlands

A. E. Ellis
DAF Marine Laboratory, Aberdeen, United Kingdom

R. Fatzer
Institute of Animal Neurology, University of Berne, Switzerland

P. J. Felsburg
Department of Clinical Studies, School of Veterinary Medicine, University of Pennsylvania, Philadelphia, PA, USA

O. J. Fletcher
College of Veterinary Medicine, North Carolina State University, Raleigh, NC, USA

A. Fluri
Institute for Animal Breeding, University of Berne, Switzerland

D. L. Foss
Department of Veterinary Pathobiology, University of Minnesota, St Paul, MN, USA

P. Gianello
Experimental Surgery, Faculty of Medicine, University of Louvain, Belgium

U. Giger
The School of Veterinary Medicine, University of Pennsylvania, Philadelphia, PA, USA

B. Goddeeris
Laboratory of Animal Physiology and Immunology, Catholic University of Leuven, Belgium

A. Govaerts
Immunology–Immunopathology, Faculty of Medicine, University of Brussels, Belgium

C. K. Grant
Custom Monoclonals International, Sacramento, CA, USA

D. L. Greiner
Department of Medicine, University of Massachusetts, Worcester, MA, USA

P. J. Griebel
Veterinary Infectious Disease Organization, Saskatoon, Canada

U. Grimholt
Department of Morphology, Genetics and Aquatic Biology, Norwegian College of Veterinary Medicine, Oslo, Norway

H. Groen
Department of Histology and Cell Biology, University of Groningen, The Netherlands

B. Hague
Laboratory of Immunogenetics, NIAID, Rockville, MD, USA

D. Haig
Moredun Research Institute, Edinburgh, United Kingdom

K. Hala
Institute for General and Experimental Pathology, University of Innsbruck Medical School, Innsbruck, Austria

R. Hamers
Institute of Molecular Biology, University of Brussels, Belgium

R. J. Hanken
Centre for Animal Biotechnology, School of Veterinary Science, The University of Melbourne, Parkville, Australia

D. Hannant
Animal Health Trust, Newmarket, United Kingdom

D. A. Harbour
Division of Molecular and Cellular Biology, University of Bristol, Bristol, United Kingdom

J. Harlan
Harborview Medical Center, Anesthesiology, Seattle, WA, USA

R. J. Hawken
Centre for Animal Biotechnology, School of Veterinary Science, University of Melbourne, Australia

W. R. Hein
Basel Institute for Immunology, Basel, Switzerland

E. Heinen
Faculty of Medicine, University of Liège, Belgium

S. C. Helfand
Department of Medical Sciences, School of Veterinary Medicine, University of Wisconsin, Madison, WI, USA

T. Hermann
Institute of Virology and Immunobiology, University of Würzburg, Germany

K. Høgåsen
Institute of Immunology and Rheumatology, University of Oslo, Norway

H. Hogenesch
Department of Veterinary Pathobiology, School of Veterinary Medicine, Purdue University, West Lafayette, GA, USA

I. Hordvik
Department of Fish and Marine Biology, High Technology Centre, Bergen University, Norway

D. W. Horohov
Department of Veterinary Microbiology and Parasitology, School of Veterinary Medicine, Louisiana State University, Baton Rouge, LA, USA

C. Howard
AFRC Institute for Animal Health, Compton Laboratory, Compton, Newbury, Berkshire, United Kingdom

T. Hünig
Institute of Virology and Immunobiology, University of Würzburg, Germany

A. Husband
Department of Veterinary Pathology, The University of Sydney, Australia

S. W. Hüttner
Department of Internal Veterinary Medicine, University of Zürich, Switzerland

M. N. Hylkema
Department of Histology and Cell Biology, University of Groningen, The Netherlands

G. A. Ingram
Department of Biological Sciences, University of Salford, Salford, United Kingdom

N. Ishiguro
Department of Veterinary Public Health, Obihiro University of Agriculture and Veterinary Medicine, Japan

S. B. Jakowlew
Immunology and Disease Resistance Laboratory, US Department of Agriculture, Beltsville, MD, USA

S. H. M. Jeurissen
Department of Immunology, ID-DLO, Lelystad, The Netherlands

E. Joosten
Wageningen Agricultural University, Wageningen, The Netherlands

T. Jørgensen
Norwegian Fisheries College, Tromsø University, Tromsø, Norway

T. W. Jungi
Institute of Biochemistry and Medical Chemistry, University of Innsbruck, Austria

M. H. Kaplan
The Walther Oncology Center, Indiana University School of Medicine, Indianapolis, IN, USA

J. Kaufman
AFRC Institute for Animal Health, Compton Laboratory, Compton, Newbury, Berkshire, United Kingdom

W. G. Kimpton
Centre for Animal Biotechnology, School of Veterinary Science, University of Melbourne, Australia

K. L. Knight
Department of Microbiology, Loyola University Medical Center, Maywood, IL, USA

C. Koch
Statens Seruminstitut, Copenhagen, Denmark

M. Komatsu
Department of Animal Breeding and Genetics, National Institute of Animal Industry, Ibaraki, Japan

T. L. Koppenheffer
Department of Biology, Trinity University, San Antonio, TX, USA

G. S. Krakowka
Department of Veterinary Biosciences, College of Veterinary Medicine, The Ohio State University, Columbus, OH, USA

F. G. M. Kroese
Department of Histology and Cell Biology, Immunology Section, University of Groningen, The Netherlands

T. Landsverk
Department of Pathology, Norwegian College of Veterinary Medicine, Oslo, Norway

D. Lanning
Department of Microbiology, Loyola University Medical Center, Maywood, CA, USA

D. Latinne
Transplantation Immunology, Faculty of Medicine, University of Louvain, Belgium

V. Leblond
Department of Haematology, Hôpital Pitié-Salpétrière, CNRS, Paris, France

C. Leutenegger
Department of Internal Veterinary Medicine, University of Zürich, Switzerland

J. K. Levy
Department of Microbiology, Pathology and Parasitology, College of Veterinary Medicine, North Carolina State University, NC, USA

H. Lewin
Department of Animal Sciences, University of Illinois, IL, USA

Ø. Lie
Department of Morphology, Genetics and Aquatic Biology, Norwegian College of Veterinary Medicine, Oslo, Norway

H. S. Lillehoj
US Department of Agriculture, Protozoan Disease Laboratory, Poultry Science Institute, Beltsville, MD, USA

D. Llanes
Departmento de Genetica, Universidad de Cordoba, Spain

J. S. Lumsden
Department of Veterinary Pathology and Public Health, Faculty of Veterinary Science, Massey University, Palmerston North, New Zealand

D. P. Lunn
School of Veterinary Medicine, University of Wisconsin, Madison, WI, USA

J. Lunney
US Department of Agriculture, Immunology and Disease Resistance, Beltsville, MD, USA

H. Lutz
Department of Internal Veterinary Medicine, University of Zürich, Switzerland

N. MacHugh
ILRI, Nairobi, Kenya

J. F. Maddox
Centre for Animal Biotechnology, School of Veterinary Science, University of Melbourne, Australia

R. G. Mage
Laboratory Immunology, Molecular Immunogenetics Section, NIH, Bethesda, MD, USA

H. M. Martin
Centre for Animal Biotechnology, School of Veterinary Science, The University of Melbourne, Parkville, Australia

S. McClure
CSIRO, McMaster Laboratory, Blacktown, Australia

T. J. McConnell
Department of Biology, East Carolina University, Greenville, NC, USA

W. T. McCormack
Department of Pathology, Immunology and Laboratory Medicine, Center for Mammalian Genetics, University of Florida College of Medicine, Gainesville, FL, USA

P. P. McCullagh
Development Physiology Group, Australian National University, Canberra City, Australia

D. McKeever
ILRI, Nairobi, Kenya

B. Mertens
ILRI, Nairobi, Kenya

N. W. Miller
Department of Microbiology, University of Mississippi, Jackson, MS, USA

J. F. Modiano
Department of Veterinary Pathobiology, College of Veterinary Medicine, Texas A and M University, College Station, TX, USA

B. P. Morgan
Department of Medical Biochemistry, University of Wales College of Medicine, Cardiff, United Kingdom

D. C. Morizot
M. D. Anderson Cancer Center, University of Texas, Smithville, TX, USA

M. P. Moutschen
Faculty of Medicine, University of Liège, Belgium

M. P. Murtaugh
Department of Veterinary Pathobiology, University of Minnesota, St Paul, MN, USA

S. Muyldermans
Institute of Molecular Biology, University of Brussels, Belgium

J. Naessens
ILRI, Nairobi, Kenya

A. D. Nash
Centre for Animal Biotechnology, School of Veterinary Science, University of Melbourne, Australia

T. C. Nguyen
Laboratoire de Génétique Biochimique, INRA, Jouy-en-Josas, France

P. Nieuwenhuis
Department of Histology and Cell Biology, University of Groningen, The Netherlands

G. Obexer-Ruff
Institute for Animal Breeding, University of Berne, Switzerland

R. Pabst
Center of Anatomy, Medical School of Hannover, Germany

P. P. Pastoret
Department of Immunology–Vaccinology, Faculty of Veterinary Medicine, University of Liège, Belgium

N. C. Pedersen
Department of Medicine and Epidemiology, School of Veterinary Medicine, University of California, CA, USA

T. L. Pertile
Department of Veterinary Pathobiology, University of Minnesota, College of Veterinary Medicine, St Paul, MN, USA

M. D. Pescovitz
Departments of Surgery and Microbiology/Immunology, Indiana University, Indianapolis, IN, USA

E. Peterhans
Institute of Veterinary Virology, University of Berne, Switzerland

L. Pilstöm
Department of Medical Immunology and Microbiology, Uppsala University, Uppsala, Sweden

J. Plachy
Institute of Molecular Genetics, Academy of Sciences of the Czech Republic, Praha, Czech Republic

C. McL. Press
Department of Morphology, Genetics and Aquatic Biology, Norwegian College of Veterinary Medicine, Oslo, Norway

R. Pu
Department of Pathobiology, College of Veterinary Medicine, University of Florida, FL, USA

C. Ramamoorthy
Harborview Medical Center, Anesthesiology, Seattle, WA, USA

M. J. H. Ratcliffe
Department of Microbiology and Immunology, McGill University, Montreal, Quebec, Canada

S. Rautenschlein
Department of Veterinary PathoBiology, University of Minnesota, College of Veterinary Medicine, St Paul, MN, USA

M. Reinacher
Institute for Veterinary Pathology, University of Leipzig, Germany

J. W. Ritchey
Department of Microbiology, Pathology and Parasitology, College of Veterinary Medicine, North Carolina State University, NC, USA

B. Robertsen
Norwegian Fisheries College, Tromsø University, Tromsø, Norway

L. S. Rodkey
Department of Pathology, University of Texas Medicine School, Houston, TX, USA

J. H. W. M. Rombout
Wageningen Agricultural University, Wageningen, The Netherlands

J. Roth
Department of Microbiology, Immunology and Preventive Medicine, College of Veterinary Medicine, Iowa State University, IA, USA

H. J. Rothkötter
Center of Anatomy, Medical School of Hannover, Germany

J. Rozing
Department of Histology and Cell Biology, University of Groningen, The Netherlands

K. A. Schat
Department of Avian and Aquatic Animal Medicine, College of Veterinary Medicine, Cornell University, Ithaca, NY, USA

A. Seto
Department of Microbiology, Shiga University of Medical Science, Japan

S. Sharar
Department of Anesthesiology, Harborview Medical Center, Seattle, WA, USA

J. M. Sharma
Department of Veterinary PathoBiology, College of Veterinary Medicine, University of Minnesota, St Paul, MN, USA

J. Sinkora
Division of Immunology and Gnotobiology, Institute of Microbiology, Czech Academy of Sciences, Praha, Czech Republic

N. Sittisombut
Department of Microbiology, Faculty of Medicine, Chiang Mai University, Chiang Mai, Thailand

M. Soares
Experimental Immunology, Faculty of Medicine, University of Louvain, Belgium

P. Sopp
AFRC Institute for Animal Health, Compton Laboratory, Compton, Newbury, Berkshire, United Kingdom

C. R. Stokes
Division of Molecular and Cellular Biology, University of Bristol, United Kingdom

M. Suter
Institute for Veterinary Virology, University of Zürich, Switzerland

H. Tabel
Department of Veterinary Microbiology, University of Saskatchewan, Saskatoon, Canada

M. F. Tatner
Division of Infection and Immunity, IBLS, University of Glasgow, Glasgow, United Kingdom

R. G. Tissot
Department of Pathology, Monroe, WI, USA

H. Tlaskalová-Hogenová
Division of Immunology and Gnotobiology, Institute of Microbiology, Czech Academy of Sciences, Praha, Czech Republic

P. Toivanen
Department of Medical Microbiology, Turku University, Finland

M. B. Tompkins
Department of Microbiology, Pathology and Parasitology, College of Veterinary Medicine, North Carolina State University, NC, USA

W. A. F. Tompkins
Department of Microbiology, Pathology and Parasitology, College of Veterinary Medicine, North Carolina State University, NC, USA

N. E. Torres-Nagel
Institute of Virology and Immunobiology, University of Würzburg, Germany

A. Tournefier
Laboratoire d'Immunologie Comparée, Université de Bourgogne, Dijon, France

I. Trebichavsky
Division of Immunology and Gnotobiology, Institute of Microbiology, Czech Academy of Sciences, Praha, Czech Republic

E. Tucker
Lucy Cavendish College, Cambridge, United Kingdom

J. Urbain
Laboratory of Animal Physiology, Faculty of Sciences, University of Brussels, Belgium

O. Vainio
Department of Medical Microbiology, Turku University, Finland

R. H. van den Berg
Department of Renal Diseases, University of Leiden, The Netherlands

T. K. van den Berg
Department of Cell Biology and Immunology, Free University, Amsterdam, The Netherlands

A. van den Broeke
Department of Molecular Biology, University of Brussels, Belgium

P. van der Meide
Biomedical Primate Research Center, Rijswijk, The Netherlands

M. Vandevelde
Institute of Animal Neurology, University of Berne, Switzerland

E. P. van Rees
Department of Cell Biology and Immunology, Free University, Amsterdam, The Netherlands

P. Vögeli
Department of Animal Science, Swiss Federal Institute of Technology, Zürich, Switzerland

J. L. Wagner
Fred Hutchinson Cancer Research Center, University of Washington School of Medicine, Seattle, WA, USA

P. S. Wakenell
Department of Preventive Medicine, University of California, Davis, CA, USA

D. Watson
CSIRO, Pastoral Research Laboratory, Armidale, Australia

P. D. Weinstein
Dedham, MA, USA

E. R. Werner
Institute of Biochemistry and Medical Chemistry, University of Innsbruck, Austria

J. Westermann
Department of Anatomy, School of Medicine, University of Hannover, Germany

B. J. Whalen
Department of Medicine, University of Massachussetts, Worcester, MA, USA

B. J. Willett
Department of Veterinary Pathology, University of Glasgow, United Kingdom

J. A. Winkelstein
Department of Pediatrics, School of Medicine, Johns Hopkins University, Baltimore, MD, USA

R. Winn
Harborview Medical Center, Anesthesiology, Seattle, WA, USA

J. K. Yamamoto
Department of Pathobiology, College of Veterinary Medicine, University of Florida, FL, USA

A. J. Young
Basel Institute for Immunology, Switzerland

A. Zurbriggen
Institute of Animal Neurology, University of Berne, Switzerland

PREFACE

Immunology is not restricted to the study of mouse and human immune systems. Historically, immunological studies on other animal species have greatly contributed to the development of immunology. Rabbits have been largely used for studies on immunoglobulin isotypes, allotypes and idiotypes, sheep for *in vivo* demonstration of lymphocyte recirculation and poultry for bursal differentiation of immunoglobulin secreting plasma cells. However, there has been no single comprehensive book available until now with a description of the immune system of various vertebrate species.

The *Handbook of Vertebrate Immunology* is the first comprehensive source of comparative immunology in which fully updated data about the immunology of more than 20 vertebrate animal species is available.

An understanding of the immune systems of animals other than mice and primates has several essential aspects: (1) in domestic animals it has obviously important applications in veterinary medicine (diagnosis, vaccinations, immune diseases or immunodeficiencies); (2) animal health and animal products quality, which benefit from immunology, are now of primary importance for human health; (3) animals also represent unique models for medical studies; (4) comparative immunology constitutes another way to revisit and to increase our knowledge of immune systems; (5) the possible future use of xenografts clearly implies a thorough immunological description of the animal tissues to be grafted (Charley and Wilkie, 1994; Hein, 1995; Charley, 1996).

For all these reasons, it has been necessary to summarize in a comprehensive way all available data, to identify gaps in our knowledge, to provide researchers, teachers and students, in biology, medicine and veterinary science, with both basic and practical information. These objectives have been fulfilled by this volume. The editors have succeeded in organizing the book following a constant format, each chapter including many of the same sections. In addition, useful basic and practical information is presented in concise, tabular form, including available reagents, accession numbers of sequences, species-specific data, etc.

B. Charley
Veterinary Immunology Committee of the International Union of Immunological Societies, INRA, Jouy-en-Josas, France

References

Charley, B. (1996). The immunology of domestic animals – its present and future. *Vet. Immunol. Immunopathol.* 54, 3–6.
Charley, B. and Wilkie, B. N. (1994). Why study the immunology of domestic animals. *The Immunologist* 2, 103–105.
Hein, W. R. (1995). Sheep as experimental animals for immunological research. *The Immunologist* 3, 12–18.

I INTRODUCTION

Rabbits come out of the brush to sit on the sand in the evening, and the damp flats are covered with the night tracks of "coons" and with the spread pads of the dogs from the ranches, and with the split-wedge tracks of the deer that come to drink in the dark.

John Steinbeck
Of Mice and Men

The recent development of modern immunology has been primarily a story of '*Mice and Men*'. Nevertheless, other species have played a crucial role in illuminating our understanding of the basic features of the structure and function of the immune system. The rabbit has contributed much to studies of humoral immune responses, sheep have provided an essential model for the study of lymphocyte trafficking, and birds have contributed to an understanding of lymphocyte lineages and B-cell differentiation in the Bursa of Fabricius. In 1986, Bruce Wilkie organized the First International Veterinary Immunology Symposium at the University of Guelph, Canada, to gather together scientists interested in the immune system of vertebrates other than mouse and man. This Symposium represented a real milestone in the development of comparative and veterinary immunology.

There is a tremendous wealth of knowledge on the immune systems of both wild and domestic animals and it was timely to gather this information, which is scattered throughout numerous scientific journals, in the format of a Handbook. This Handbook is intended to provide the first comprehensive source of comparative immunology for students and researchers in comparative medicine, animal health, and biology.

The choice of species and subjects to be included in the Handbook was based on three main criteria: first, the importance of a species in public and animal health sectors; second, the importance of a species for comparative research; and third, the level of available immunological knowledge. For comparative purposes, chapters on selected features of the mouse immune system and the SCID mouse model were also included. A growing interest in xenografts also justified a separate chapter on this topic.

All chapters devoted to descriptions of the immune system of individual species have been organized with a similar format. This format includes a broad range of topics that are presented in a consistent order and style. Whenever possible, the available knowledge for a topic is presented in concise tabular format and primary references provided. This organization is intended to facilitate interspecies comparisons, allow rapid identification of information gaps, and reveal species-specific aspects of immune system structure or function. Furthermore, the careful referencing of information will enable interested researchers to pursue knowledge to a much greater depth.

Gathering this wealth of information would not have been possible without the commitment and dedication of Section editors and the generous participation of a large number of contributors. We are greatly indebted to their massive amount of work and their willingness to set aside other priorities for this project.

The staff at Academic Press have shown enthusiasm and support from the beginning of this project and they frequently provided useful guidance. We are especially grateful to Tessa Picknett, Sarah Stafford and Duncan Fatz. Christina Espert Sanchez and Lucie Karelle Bui Thi coordinated communication among Section editors and the Handbook editors, and worked diligently to integrate the edited manuscripts into a coherent and comprehensive reference book.

The editors hope that the *Handbook of Vertebrate Immunology* provides a valuable tool for all interested in immunology. Our greatest wish is that this Handbook not only facilitates research through the knowledge presented but that it also stimulates further research in those species and areas of immunology about which little is presently known. Mouse and Man have been the source of much immunological knowledge during the last 20 years. However, we believe that a full understanding of the immune system will be acquired only through a comparative analysis of its structure, function, and physiology in a multitude of species.

Handbook of Vertebrate Immunology
ISBN 0-12-546401-0

II Immunology of Fishes

1. Introduction

The fishes are the largest vertebrate class and constitute over 20 000 species (Nelson, 1994). Broadly, the fishes can be divided into the jawless fish, represented by the lamprey and hagfish, and the jawed fish that can be further divided into the cartilaginous fishes, such as the sharks and rays, and the bony fishes. Of the modern bony fishes, the teleosts are by far the major grouping (for overview of taxa see Figure II.7.1).

The evolution of fishes and the tetrapods diverged from each other about 300 million years ago and it is natural that fishes should be the subject of investigation of the evolution of lymphoid tissues and the development of the immune system. Considerable research has been performed on unravelling the phylogenetics of immunoglobulin and major histocompatibility genes and mapping the evolution of immunologically important traits associated with the appearance of cell mediated and humoral immunity. However, in addition to comparative investigations, fish also represent potential experimental models for vertebrate immunology as a whole. The zebrafish is, for example, increasingly being used as a vertebrate model for the study of the activation and regulation of gene expression in general and recently for the expression of immune genes. Extensive studies of various fish species have shown that the basic mechanisms of acquired immunity in fish and mammals are surprisingly similar (Faisal and Hetrick, 1992). Differences exist but there are similarities in macrophage function, lymphocyte stimulation and characterized humoral factors such as antibodies. A body of information has emerged over the last 10 years on the structure and organization of immune-related genes in fish. The genes coding for immunoglobulins, T-cell receptors and major histocompatibility complex (MHC) molecules have been cloned and sequenced and these data provide a tantalizing insight into the evolution of vertebrate genes.

Fish and their immune system may also represent an important scientific tool in the monitoring of environmental quality, particularly of immunotoxic environmental pollutants. Fish occupy a variety of ecological niches in the aquatic environment and so changes in the immune parameters of fish have the potential to be a sensitive gauge of environmental deterioration (Wester *et al.*, 1994).

Finally, the rapid expansion of aquaculture as an industry and the disease problems that the industry has encountered have intensified the study of immunology of fishes. The practical need to produce effective vaccines against a number of bacterial diseases has led to a better understanding of the piscine immune system and the factors that influence its function. It can be claimed, at least for part of the international aquaculture industry, that without effective vaccines, fish farming would not be feasible, either biologically or economically.

The object of this chapter is to provide an overview of the active areas of research on immunology of fishes. Many of the contributing authors have drawn attention to the meagerness of data available on some aspects of the piscine immune system, and in highlighting the hindrances and challenges to our understanding, they all emphasize the need for more research on this vast, diverse group of vertebrates.

2. Lymphoid Organs and their Anatomical Distribution

Studies on the immune system of fishes have focused on a relatively small number of species, principally among the teleosts. This summary will be confined mainly to the description of the major lymphoid organs in the teleosts, while the lymphoid organs in some other fishes will be mentioned briefly.

Fish lack a bone marrow and lymph nodes

Fish do not possess lymph nodes or bone marrow and instead the major lymphoid organs in the teleosts are the thymus, kidney, spleen and gut-associated lymphoid tissues (Figure II.2.1). A brief summary of the cell populations, life history and immune functions of the lymphoid organs is presented in Table II.2.1. The ontogeny of the lymphoid organs and immune responsiveness and detailed descriptions of leukocytes are presented elsewhere in this chapter.

In teleosts, it is widely accepted that a lymphatic system is differentiated from the blood vascular system. The cartilaginous fishes possess a hemolymph system, in which vessels contain blood and lymph in various proportions.

Handbook of Vertebrate Immunology
ISBN 0-12-546401-0

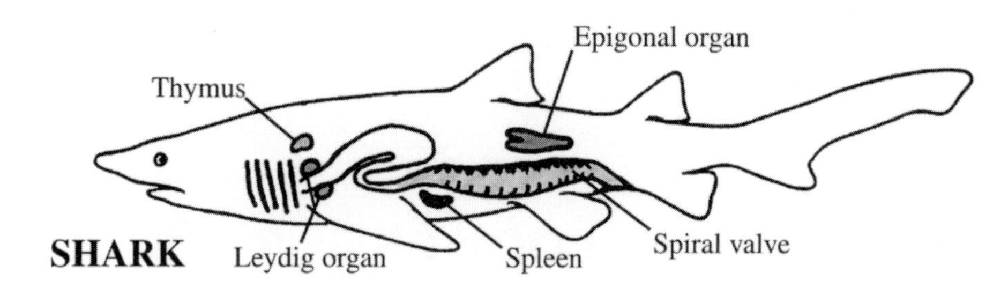

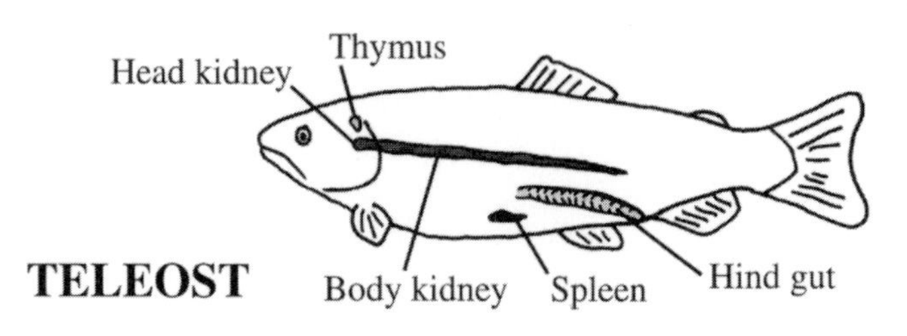

Figure II.2.1 The major lymphoid tissues of the elasmobranchs (shark) and the teleosts.

Table II.2.1 The major lymphoid organs of fishes[a]

Organ	*Cells*	*Life history*	*Immune function*
Epigonal organ[b] Cartilaginous fishes (sharks, rays)	Predominantly granulocytes, also lymphocytes	Granulopoiesis	—
Intestines, hind gut	Lymphocytes, macrophages	—	Mucosal immunity
Intestines, spiral valve Cartilaginous fishes (sharks, rays)	Lymphoid cells, macrophages	Early development, before spleen, Leydig's or epigonal organ	—
Kidney (lymphoid tissue absent from the kidney of most elasmobranchs)	Head kidney[b]: lymphocytes, granulocytes, macrophages, melanomacrophages, reticular cells Body kidney: lymphomyeloid and excretory tissue	Possesses pluripotent hemopoietic stem cells; lymphohemopoiesis	Blood-filtering organ Antigen trapping and immune responsive organ
Leydig's organ[b] Cartilaginous fishes (sharks, rays)	Predominantly granulocytes, also lymphocytes	Granulopoiesis	—
Meninges, cranial cavity[b] *Chondrostei* (sturgeons)	Granulocytes, lymphocytes	Predominantly granulopoiesis	Immune responsive?
Pericardial tissue[b] *Chondrostei* (sturgeons)	Granulocytes, lymphocytes	Lymphohemopoiesis	Immune responsive?
Periorbital and subcranial tissue[b] Holocephali (Chimaeras)	Predominantly granulocytes, also lymphocytes	Lymphohemopoiesis	Immune responsive?
Spleen All fishes except the jawless fish?	Often predominantly red pulp White pulp: lymphocytes, macrophages, melanomacrophages	Lymphoid populations appear late in development; lymphohemopoiesis	Blood-filtering organ Antigen trapping and immune responsive organ
Thymus All fishes except the jawless fish	Lymphocytes, macrophages, myoid cells, mast cells, epithelial cells	Site of first appearance of lymphocytes; involutes with age (not in some species)	Lymphopoiesis Direct involvement in immune response?

[a] Data from Zapata and Cooper (1990) and applicable mainly for the teleosts.
[b] Compared with mammalian bone marrow.

Wardle (1971) described a well-developed lymph system in teleosts and lymph collected from these vessels was devoid of erythrocytes and contained leukocytes in about the same proportion as in blood but in smaller numbers (Ellis and de Sousa, 1974). However, in some teleost species, the existence of a true lymphatic system has now been challenged and a secondary vascular system has been described (Vogel and Claviez, 1981; Steffensen and Lomholt, 1992). This system constitutes a separate, parallel circulatory system that originates from systemic arteries via tiny arterioarterial anastomoses, which form secondary arteries and supply their own capillary networks, and return to the systemic venous system. The secondary capillary beds have been demonstrated in the skin of the body surface, the fins, the surfaces of the mouth and pharynx and the peritoneum. It is proposed that a process of plasma skimming accounts for the paucity of cells in the secondary vessels (Steffensen and Lomholt, 1992). The tendency for cells to concentrate in the central stream of a vessel allows the relatively cell-free plasma flowing along the vessel wall to drain off into small side branches. Plasma skimming thus results in a lower hematocrit in the side branches than in the parent vessel.

Thymus

The thymus is well developed in both cartilaginous fish and bony fish but is absent in the jawless fish. No uniformity exists in thymic histology of the various species, but in general, the thymus is a paired organ present in the dorsolateral region of the gill chamber and is delineated by a connective tissue capsule that projects trabecula within the thymic parenchyma. In the cartilaginous fish, the thymus is a multilobulate bilateral organ located near the gills but the teleostean thymus consists of a pair of lobes located on each side of the gill cavity at the superior edge of the gill cover towards the opercular cavity.

The thymus is intraepithelial in teleosts (see Figure II.10.1) and the stroma is composed of epithelial-reticular cells that form a three-dimensional framework for the thymocytes and other free mesenchymal cells such as macrophages and myoid cells. A well-differentiated cortex and medulla are lacking in most teleosts but up to four layers have been reported in some species. Lymphocytes are the major cellular component but there are limited data on intrathymic cellular kinetics and migratory patterns. There is also only limited information on the possible functions of epithelial cells in the maturation of thymocytes and the acquisition of T cell specificities via the production of thymic factors (Chilmonczyk, 1992).

Blood vessels are the only system known to connect the thymus with other fish organs and in rainbow trout there is some evidence for a blood–thymus barrier of variable permeability (Chilmonczyk, 1983). Thymic involution is not a general rule in fish and the thymus persists in many older fish. Involution may occur in association with aging and sexual maturity and may be induced by stress, season and hormones.

A direct involvement of the thymus in defense mechanisms is suggested by several lines of evidence, including the appearance of plasma cells and plaque-forming cells after immunization (Ortiz-Muniz and Sigel, 1971; Sailendri and Muthukkaruppan, 1975; Chilmonczyk, 1992).

Kidney

In teleosts, the kidney is a paired retroperitoneal organ that extends as a sheet of tissue dorsal in the body cavity in close contact with the osseous vertebral spine (Zapata, 1979). The kidney consists of two segments: an anterior segment (cephalic or head kidney) that contains predominantly hemopoietic tissue (renal activity is absent); and a middle and posterior segment (body or trunk kidney) that is dominated by renal tissue but also contains lymphohemopoietic tissue (Zapata and Cooper, 1990). The kidney parenchyma, which is supported by a connective tissue capsule and a framework of reticular cells, is dispersed between an extensive system of sinusoids. The sinusoidal cells, which include endothelial cells and adventitial cells, form a barrier between the hemopoietic tissue and blood received from the renal portal vein (Ellis *et al.*, 1989). Macrophages are prominent, particularly melanomacrophages, and lymphocytes and plasma cells tend to be dispersed, although small clusters can occur (Meseguer *et al.*, 1991; Press *et al.*, 1994). The head kidney also contains granulocytes and nonspecific cytotoxic cells (Graves *et al.*, 1984; Bayne, 1986; Evans and Jaso-Friedmann, 1992). Endocrine tissues are present in the kidney including the corpuscles of Stannius and interrenal (adrenal) tissue.

In addition to its role as a major site of production of erythroid, lymphoid and myeloid cells (Smith *et al.*, 1970; Zapata, 1979; Zapata, 1981b), the kidney is important in the trapping of antigens and the production of antibody (Ellis, 1980; Zapata and Cooper, 1990).

Spleen

The spleen is a dark red to black organ located usually ventral and caudal to the stomach. The teleost spleen has a fibrous capsule and small trabeculae extend into the parenchyma, which can be divided into a red and white pulp. The red pulp, which may occupy the majority of the organ (Grace and Manning, 1980; Secombes and Manning, 1980), consists of a reticular cell network supporting blood-filled sinusoids that hold diverse cell populations including macrophages and lymphocytes. The white pulp is often poorly developed but may be divided into two

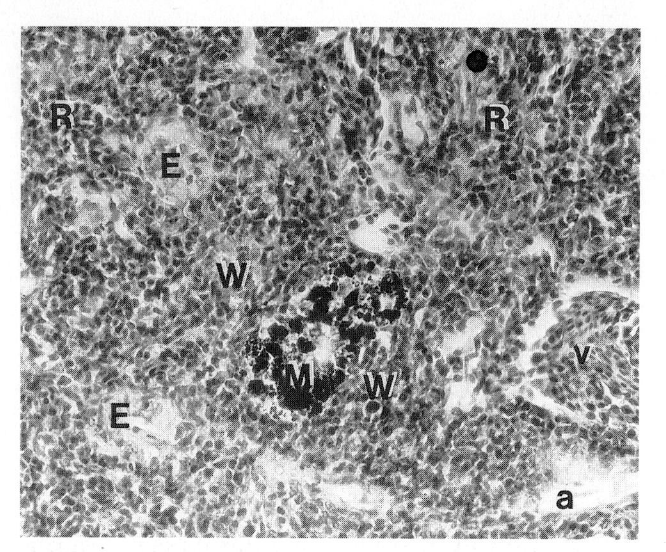

Figure II.2.2 Lymphoid tissue in the teleost spleen (Halibut, *Hippoglossus hippoglossus* L.). The white pulp (W) contains melanomacrophage centers (M) and ellipsoids (E) and is surrounded by extensive areas of red pulp (R). The melanomacrophage centers and lymphoid tissue are associated with arterioles (a) and venous sinuses (v).

compartments: the melanomacrophage centers and the ellipsoids (Figure II.2.2).

Melanomacrophage centers

Melanomacrophage centers are aggregations of closely packed macrophages that contain heterogeneous inclusions, the most frequent of which are melanin, hemosiderin and lipofuscin (Agius and Agbede, 1984). These aggregations of pigment-containing cells are present in the hemopoietic tissues of spleen and kidney, and in the periportal areas of the liver (Roberts, 1975) but their degree of organization varies between species. In many species, the centers are bound by a thin argyrophilic capsule, surrounded by white pulp and associated with thin-walled, narrow blood vessels (Agius, 1980; Herraez and Zapata, 1986; Ellis *et al.*, 1989). However, in salmonids, the accumulations of melanomacrophages are less well-defined and lack a capsule, but the association with blood vessels and lymphocytes is retained (Press *et al.*, 1994).

Melanomacrophage centers have been considered metabolic dumps but their ability to retain antigens for long periods, possibly in the form of immune complexes, has drawn comparisons with germinal centers in higher vertebrates (Ellis and de Sousa, 1974; Ferguson, 1976; Agius, 1980). However, follicular dendritic cells have not been demonstrated in fish, nor have splenic memory B cells in rainbow trout shown evidence of features associated with memory cells derived from germinal centers in higher vertebrates (Arkoosh and Kaattari, 1991).

Ellipsoids

The ellipsoids are terminations of arterioles with a narrow lumen that run through a sheath of reticular fibers, reticular cells and macrophages (Zapata and Cooper, 1990; Espenes *et al.*, 1995a). Ellipsoids appear to have a specialized function for the trapping of blood-borne substances, particularly immune complexes (Ellis, 1980; Secombes and Manning, 1980; Espenes *et al.*, 1995b), which are taken up by the rich population of macrophages in the ellipsoidal wall. The subsequent migration of laden macrophages to melanomacrophage centers has been described (Ellis *et al.*, 1976; Ferguson, 1976; Ellis, 1980).

Gut-associated lymphoid tissue

The teleosts lack organized gut-associated lymphoid tissue such as the Peyer's patches of mammals but, in species such as the carp, the lamina propria and epithelium of the hind gut contain significant populations of leukocytes. The participation of the hind gut and other organs in a mucosal immune system is discussed elsewhere in this chapter.

Elasmobranchs – sharks and rays

The main lymphoid organs in the elasmobranchs are the thymus and the spleen, which are well developed and possess many structural similarities to these organs in higher vertebrates (Zapata, 1980). Lymphomyeloid structures, which are mainly granulopoietic, are located in the esophagus (Leydig's organ) and in the gonads (epigonal organ) (Zapata, 1981a) (Figure II.2.1). Leydig's organ occupies a dorsal and ventral position in the gut submucosa extending from the oral cavity to the stomach. The epigonal organs occupy the parenchyma of the ovary and testis. Many elasmobranchs possess both Leydig's and epigonal organs but others have only one of these structures. Diffuse infiltrations with lymphocytes may also be found in the gut, especially in the spiral valve (Fänge, 1982; Hart *et al.*, 1986).

3. Leukocytes and their Markers

Introduction

All vertebrate immune responses are mediated by leukocytes and, therefore, an understanding of immunology must be based upon an understanding of leukocyte function. Hence, to assess function unambiguously, it is important not only to identify but also to isolate the cells of interest. In recent years, the identification and functional characterizations of subpopulations of human and murine leukocytes have been greatly facilitated by the development and use of numerous monoclonal antibodies (mAbs) specific for differentially expressed cell surface markers. These mAb-defined markers on human and murine leukocytes are known as clusters of differentiation (CD) antigens; there are currently more than 130 known CD antigens for human leukocytes, with the list growing continuously (Janeway and Travers, 1996). Fish, like mammals, possess leukocytes that can be classified as lymphocytes, monocytes/macrophages, and granulocytes based on morphology, ultrastructure, and cytochemical staining (Rowley *et al.*, 1988). However, in contrast to mammals, there is currently a paucity of cell surface markers that can be used to identify and to study fish leukocyte subpopulations functionally. As a result, little is currently known about fish leukocyte development and function. The purpose of this section is to highlight the known leukocyte markers in fish (see Tables II.3.1 and II.3.2) and to summarize how these have been used to define functionally various leukocyte subpopulations.

Lymphocytes

The existence of B cells has been directly demonstrated in many teleost species through the use of monoclonal antibodies specific for fish immunoglobulin (Ig) (Table II.3.1) and by the identification of Ig heavy and light chain genes (reviewed by Wilson and Warr, 1992; Warr, 1995; Pilström and Bengten, 1996). Much of our current understanding of

Table II.3.1 Monoclonal antibodies to fish immunoglobulin

Fish species	*mAb*	*Reactivity*	*References*
Atlantic cod	1F1	H chain	Pilström and Petersson (1991)
	Several	H chain	Israelsson *et al.* (1991)
	2A2	L chain	Pilström and Petersson (1991)
Atlantic salmon	Several	H chain	Killie *et al.* (1991)
	1206, 202	H chain	Magnadottir *et al.* (1996)
	G2H3	H chain	Pettersen *et al.* (1995)
Carp	WCI 4, WCI 12	H chain	Secombes *et al.* (1983)
	WCI M	H chain	Rombout *et al.* (1993b)
Channel catfish	Several	H chain	Lobb and Clem (1982), Lobb and Olson (1988)
	9E1	H chain	Sizemore *et al.* (1984), Miller *et al.* (1987)
	1G7, 3F12	L chain	Lobb *et al.* (1984)
European eel	WE11	H chain	Van der Heijden *et al.* (1995)
	WE12	L chain	Van der Heijden *et al.* (1995)
Rainbow trout	1.14	H chain	DeLuca *et al.* (1983)
	Several	H chain	Thuvander *et al.* (1990)
	Several	H chain	Sanchez *et al.* (1993a)
	2A1, 2H9	L chain	Sanchez and Dominguez (1991)
Red drum	RDG013, RDG048	H chain	MacDougal *et al.* (1995)
Sea bass	3B5, 6E11	H chain	Romestand *et al.* (1995)
	DLIg3	L chain	Scapigliati *et al.* (1996)
Sea bream	WSI 5	L chain	Navarro *et al.* (1993)
Turbot	Several	H chain	Navarro *et al.* (1993)

Table II.3.2 Monoclonal antibodies for fish non-B cells

Fish species	*mAb*	*Reactivity*	*References*
Atlantic salmon	E3D9	PMN leukocytes (70 kDa?)	Pettersen *et al.* (1995)
Catfish	13C10	Ig^- leukocytes (150 kDa)	Miller *et al.* (1987)
	13C5	Neutrophils	Bly *et al.* (1990)
	C3-1, 51A	Neutrophils	Ainsworth *et al.* (1990)
	5C6	NCC (40–41 kDa)	Evans *et al.* (1988)
	CfT1	T cells (35 kDa)	Passer *et al.* (1997)
	1H5	Leukocytes (180 and 90 kDa)	Yoshida *et al.* (1995)
Sea bass	DLT15	T cells? (170 kDa)	Scapigliati *et al.* (1995)

fish lymphocyte function has been obtained through the coupling of cell separation techniques, employing anti-Ig mAb, with immunologically relevant *in vitro* assay systems (reviewed by Clem *et al.*, 1991; Vallejo *et al.*, 1992). Studies, using mitogens, have revealed that Ig-positive cells from a variety of fish are responsive to lipopolysaccharide (LPS, a known mammalian B-cell mitogen) but not Concanavalin A (Con A, a known mammalian T cell mitogen); in contrast, fish Ig negative cells are responsive to Con A but not LPS (DeLuca *et al.*, 1983; Sizemore *et al.*, 1984; Miller *et al.*, 1987; Ainsworth *et al.*, 1990). The results of *in vitro* antibody studies have demonstrated that the generation of anti-hapten antibody responses to a T-independent antigen (as defined in mammals) in the channel catfish requires only surface Ig-positive cells and monocytes, whereas anti-hapten responses to a T-dependent antigen require not only surface Ig-positive cells and monocytes but also surface Ig-negative cells (Miller *et al.*, 1985, 1987). Other *in vitro* studies have shown that fish Ig-negative lymphocytes: (1) are the responding population in a mixed leukocyte reaction (Miller *et al.*, 1986); (2) produce a macrophage activating-like factor(s) upon stimulation with Con A and phorbol myristate acetate (Graham and Secombes, 1990a); (3) specifically proliferate in response to processed and presented antigen (Vallejo *et al.*, 1991a). All of these cell separation studies clearly show that fish have the functional equivalents of B and T cells. However in contrast to B cells, fish T cells have been indirectly defined as surface Ig-negative cells due to the lack of unambiguous cell surface markers. As a result, very few studies that give direct evidence for functional T cell markers in fish have been conducted. For example, in the channel catfish mAb13C10 was found to react with a large percentage of Ig-negative, but not Ig-positive leukocytes and the 13C10 reactive cells were shown to have T cell functions. However, this mAb reacted not only with catfish T cells, but also with other cell types including neutrophils and thrombocytes, and hence could not be considered T cell specific (Miller *et al.*, 1987). In the sea bass, mAb DLT15 was described that reacted with a majority of thymocytes and a low percentage of lymphoid cells from the blood, spleen and kidney (Scapigliati *et al.*, 1995). Based upon this differential pattern of reactivity, the authors suggested that this mAb reacted with a subpopulation of T cells. However, no functional data were given to support this contention. A recent study in the channel catfish describes a mAb (CfT1) that defines a single chain protein of M_r 35 000 expressed on most thymocytes and a subpopulation of lymphoid cells in the blood, spleen and head kidney, but not expressed on B cells, granulocytes, macrophages, thrombocytes or erythrocytes (Passer *et al.*, 1997). In addition, *in vitro* stimulation of PBLs with Con A resulted in an increased presence of CfT1 positive cells. Based on these findings, it appears that CfT1 defines a T lineage-specific marker in the channel catfish. However, definitive evidence will be obtained only when CfT1-positive cells are shown to express T-cell receptor gene products. In this regard, the recent identification of T-cell receptor genes in rainbow trout (Partula *et al.*, 1995) and channel catfish (Wilson *et al.*, 1998) should allow for the development of mAb to the T-cell receptor through use of recombinant proteins. It is also important to note that immortal clonal T and B cell lines, which express T cell receptor α/β and Ig genes, respectively, have been developed in the channel catfish (Miller *et al.*, 1994; Wilson *et al.*, 1998). The use of such cell lines should be of great value for developing reagents to other functionally important lymphocyte markers.

Natural cytotoxic cells (NCC)

Fish possess natural cytotoxic cells (NCC) which are functionally similar to mammalian natural killer (NK) cells, i.e. they kill certain types of xenogeneic target cells, including a number of pathogenic protozoa, without the requirement for previous exposure (reviewed by Evans and McKinney, 1991; Evans and Jaso-Friedmann, 1992). Fish NCC, unlike mammalian NK cells, are small agranular lymphocytes and exhibit diminished capacity to recycle lytic functions (Evans *et al.*, 1984a; Greenlee *et al.*, 1991). A mAb (5C6) has been developed that defines a cell surface marker of 40–41 kDa on catfish NCC (Evans *et al.*, 1988; Evans and Jaso-Friedmann, 1992). This marker appears to be an important function-associated antigen in that it is involved in cell target recognition, conjugate formation and cellular activation. Functional reactivity of this monoclonal antibody with human and rat NK cells has led to the speculation that NCC may be the teleost equivalent of mammalian NK cells. It should be noted that NCC are not the only type of NK-like effector cells present in fish. For example, recent studies have shown that non-immune channel catfish peripheral blood possesses potent cytotoxic effector cells for both allogeneic and virus-infected autologous cells, and these cytotoxic cell do not express the NCC 5C6 surface marker (Stuge *et al.*, 1995; Yoshida *et al.*, 1995; Hogan *et al.*, 1996). Currently there are no markers available that functionally define these non-NCC, NK-like effector cells from the catfish. However, cytotoxic activity of these effector cells can be inhibited by incubation with mAb 1H5 that recognizes putative LFA-1-like (CD11a/CD18) molecules on the surface of a majority of catfish peripheral blood leukocytes (Yoshida *et al.*, 1995).

Monocytes/macrophages

Macrophages are large agranular phagocytic cells distributed widely in body tissues, and are the mature form of monocytes which circulate in the blood. The cytochemical features of these cells from a variety of different fish species have been reviewed (Rowley *et al.*, 1988). In general, these cells are similar to mammalian monocytes/macrophages in that they are for the most part nonspecific esterase, acid phosphatase, periodic acid–Schiff, and peroxidase positive. In addition to the more phenotypically conventional macrophages, fish also possess pigmented macrophages (melanomacrophages) found as aggregates in many lymphoid tissues (Wolke, 1992). At present, no mAbs are available that functionally define monocyte/macrophage lineage cells. However, these cells can be isolated or depleted by a variety of methods, including plastic, fibronectin, or Sephadex G10 adherence, as well as density centrifugation on Percoll gradients (Sizemore *et al.*, 1984; Clem *et al.*, 1985; Chung and Secombes, 1987). The isolation of these cells has allowed for their functional characterization, which includes phagocytosis, chemotaxis, bactericidal activity via respiratory burst, production of IL-1-like activity, and antigen processing and presentation (Secombes and Fletcher, 1992; Vallejo *et al.*, 1992; Ellsaesser and Clem, 1994; Verburg-van Kemenade *et al.*, 1995). In addition, many of these functional studies suggest that fish monocytes/macrophages possess lectin, complement and Fc receptors in addition to MHC class II-like antigens. However, at present no reagents are available to assay directly for these functionally important molecules. The development of monocyte/macrophage long-term cell lines from both goldfish and channel catfish may serve as useful tools for identifying the expression of such markers (Vallejo *et al.*, 1991b; Wang *et al.*, 1995).

Granulocytes

The identification of fish granulocytes is based almost solely upon morphological, ultrastructural and cytochemical staining similarities with mammalian granulocytes (Rowley *et al.*, 1988; Ainsworth, 1992; Hine, 1992). As in mammals, three types of granulocytes have been identified in fish: neutrophils, eosinophils, and basophils. However, not all granulocyte types are found in every fish species studied. For example, only neutrophils have been identified in channel catfish (Ellsaesser *et al.*, 1985; Ainsworth and Dexiang, 1990), whereas, all three types are present in carp (Cenini, 1984; Tremmink and Bayne, 1987). Variations in morphology and histochemical staining have led to considerable confusion in the classification of fish granulocytes in a number of fish species (Ainsworth, 1992; Hine, 1992). Fish neutrophils are for the most part phagocytic and peroxidase positive; in some fish they are also Sudan black B positive (Ellsaesser *et al.*, 1985; Ainsworth, 1992; Hine, 1992). In contrast to monocytes/macrophages, fish neutrophils are not particularly adherent to coated or uncoated plastic surfaces although they can be isolated in relatively pure form from head kidney tissues by density gradient centrifugation (Waterstrat *et al.*, 1988; Lamas and Ellis, 1994a). The functions of isolated fish neutrophils include phagocytosis, chemotaxis, and bactericidal activity via the respiratory burst mode (Ainsworth, 1992; Hine, 1992; Secombes and Fletcher, 1992; Lamas and Ellis, 1994a). Some of these functional studies indirectly suggest the presence of lectin, complement, and Fc receptors on these cells, however, formal evidence is lacking. Nonspecific stress has been shown to cause neutrophilia in the channel catfish as a possible result of demargination of neutrophils from capillary beds in body tissues (Ellsaesser and Clem, 1986; Bly *et al.*, 1990). Several mAbs have been developed that identify neutrophils in catfish (Ainsworth *et al.*, 1990; Bly *et al.*, 1990) and salmon (Pettersen *et al.*, 1995), however the functional importance of these markers has not been fully determined. Fish eosinophils are peroxidase positive whereas basophils are not. However, neither is particularly phagocytic, albeit both are thought to be important in parasitic infections (Ainsworth, 1992; Hine, 1992). The lack of cell isolation techniques and cell markers for eosinophils and basophils has made these cells extremely difficult to study in an unambiguous fashion.

4. Leukocyte Traffic and Associated Molecules

Little is known about leukocyte traffic in fish and, with the exception of certain chemoattractants, nothing is known about the nature of the molecules that govern this phenomenon. The available information concerns aspects of *in vivo* migration within and between organs of lymphocytes and phagocytes under normal or inflammatory conditions and *in vitro* assays of chemotaxis.

Traffic within and between organs

Lymphocytes

Lymphocytes (and other leukocytes) can be recovered from the neural lymphatic duct of the plaice, *Pleuronectes platessa* (Wardle, 1971), and the lymph is returned to the blood vascular system via connection with the Duct of Cuvier. Radiolabelling experiments have shown that the neural lymphatic duct lymphocytes, on intravenous reinjection, settled preferentially in the lymphoid tissues of the kidney and spleen where they appeared to migrate to melanomacrophage centers (aggregates of macrophages containing pigment granules, pyroninophilic cells and lymphocytes) and where they were active in synthesizing RNA (Ellis and de Sousa, 1974). Fewer cells migrated through the gills, gut and liver, where they were not metabolically active. No circulating lymphocytes appeared to enter the thymus. Conversely, there is good evidence that thymic lymphocytes emigrate from the thymus. Tatner (1985a) labelled thymocytes of rainbow trout *in*

Table II.4.1 Stress and pathological conditions affecting fish leukocyte migration *in vivo*

Stimulant	*Effect*	*Reference*
Costiasis	Increased lymphocytes in gills	Ellis and Wooten (1978)
Inflammatory agents (bacteria, mineral oils)	Neutrophil and macrophage exudates	Finn and Nielson (1971)
Sexual maturation	Increased lymphocytes in skin	Peleteiro and Richards (1985)
Stress and corticosteroids	Lymphopenia	Ellis (1981a)
	Neutrophilia	Ellis (1981a)

situ by intrathymic injection of tritiated thymidine. Over the following 3 days the label accumulated mainly in the spleen and kidney, with a greater amount in the spleen. Much lower levels of label were detected in the liver, muscle and gut.

Macrophages of the reticulo-endothelial system

Macrophages in fish are present as reticulo-endothelial cells lining kidney sinusoids, within the walls of splenic ellipsoids and, in some species, such as the plaice, lining the atrial cavity of the heart. Following injection of carbon particles, these cells phagocytose the carbon, round up and migrate through the tissues or circulatory system and form aggregates within the kidney and spleen (Ellis *et al.*, 1976).

Stress and pathological conditions affecting leukocyte traffic

There are several reports describing increased numbers of lymphocytes in the gills, skin and gut of fish especially in response to irritation or infection (Ellis and Wooten, 1978) (Table II.4.1). Increased numbers of lymphocytes in the skin have been described in mature (both male and female) rainbow trout in comparison with immature trout (Peleteiro and Richards, 1985). Thus, it appears that lymphocyte traffic in fish may be modulated by physiological (endocrine) mechanisms. During stress or following administration of corticosteroids, there is a lymphocytopenia (Ellis, 1981a) suggesting that recirculatory traffic is affected by stress hormones.

Following tissue trauma induced by wounding, or injection of noxious substances (e.g. bacteria, mineral oils), there is an initial influx and accumulation of neutrophil granulocytes at the injured site lasting from about 6 h to 4 days, followed by a more sustained accumulation of macrophages detectable after about 24 h (Finn and Nielson, 1971). The molecular basis for inflammatory traffic in fish is not known but chemotactic stimuli are likely to be involved.

Chemotactic factors; *in vitro* migration

Fish leukocytes show an increased motility either randomly (chemokinesis) or directionally (chemotaxis) when exposed *in vitro* to a wide variety of substances (Table II.4.2) especially substances derived from parasites (Sharp *et al.*, 1991) and bacteria (Weeks-Perkins and Ellis, 1995). The nature of endogenously produced chemoattractants in fish is not clearly understood but activated complement components, eicosanoids and cytokines are believed to play a role.

Complement factors

Fresh trout serum has a chemotactic effect on fish blood leukocytes and the rate of migration is increased in the presence of antigen–antibody complexes (Griffin, 1984) and by zymosan-activated serum (Zelikoff *et al.*, 1991). It is known that fish serum contains the C5 component of complement (Nonaka *et al.*, 1981b) and it is believed that,

Table II.4.2 Endogenous factors modulating fish leukocyte migration *in vitro*

Increased migration by	*Nature of factor*	*Reference*
Activated serum components	Activated complement factors (C5?)	Griffin (1984), Zelikoff *et al.* (1991)
Decreased migration by antigen-induced leukocyte products	Migration inhibition factor (MIF)	Smith and Braun-Nesje (1982)
Human TNF-α	Evidence for TNF in fish	Jang *et al.* (1995a)
Leukocyte products	Leukotriene B_4 (LTB_4) Lipoxins	Hunt and Rowley (1986), Sharp *et al.* (1992)
Mitogen or antigen-induced leukocyte products	Unknown	Howell (1987), Bridges and Manning (1991)

as in mammals, the C5a cleavage product is responsible for the chemotactic responses.

Eicosanoids

Isolated blood granulocytes of dogfish show enhanced migration to leukotriene B_4 (LTB_4) and the leukocytes themselves released LTB_4 when stimulated with the calcium ionophore A23187 (Hunt and Rowley, 1986). Another class of eicosanoids, the lipoxins, are synthesized by rainbow trout macrophages in larger amounts than LTB_4 (Rowley *et al.*, 1991) and these are very potent chemoattractants for fish phagocytes (Sharp *et al.*, 1992). While isolated gut leukocytes of rainbow trout produced chemoattractant factors for kidney leukocytes in response to stimulation by the calcium ionophore A23187, gut leukocytes themselves were unresponsive to these factors (Davidson *et al.*, 1991).

Cytokines

TNF-α
Human tumor necrosis factor alpha (TNF-α) increases the migration rate of rainbow trout neutrophils and this response is abrogated by prior incubation of the cells with mAb to human TNF receptor (TNF-R1) (Jang *et al.*, 1995a). These data provide evidence that TNF exists in fish although no direct evidence for this is yet available.

Chemokines
Mitogen (PHA) or antigen stimulation of fish leukocytes can produce supernatants that can increase the migration of leukocytes (Howell, 1987; Bridges and Manning, 1991) or diminish the migration rate (migration inhibition factor, MIF) (Smith and Braun-Nesje, 1982) but such factors have not been otherwise characterized.

5. Cytokines

Fish cytokines have been discussed in a phylogenetic context by Cohen and Haynes (1991) and Secombes (1994a) and an excellent update of research on fish cytokines has recently been presented (Secombes *et al.*, 1996). As illustrated in Table II.5.1, several cytokines and one cytokine receptor have been demonstrated by biological activity, antigenic cross-reactivity or genetic techniques in fish. Some mammalian cytokines, IL-1 (Hamby *et al.*, 1986; Sigel *et al.*, 1986; Balm *et al.*, 1995), TGF-β (Jang *et al.*, 1994), TNF-α (Hardie *et al.*, 1994a), show biological activity in fish systems, suggesting the presence of receptors for these molecules in fish. Whilst little progress has been made with respect to purification of fish cytokines, there has recently appeared some promising reports of cDNA cloning and sequencing of fish cytokine genes. These include an IL-2-like gene (Tamai *et al.*, 1992) and an IFN-like gene (Tamai *et al.*, 1993a) from Japanese flatfish, and an IL-1β (Zou *et al.*, 1997; C. J. Secombes, personal communication) and a TGF-β gene from rainbow trout (Secombes *et al.*, 1996).

Interleukins

IL-1

Recent work by Ellsaesser and Clem (1994) has demonstrated that culture supernatants from channel catfish monocytes exert IL-1-like activity on mouse and catfish T cells. IL-1 was assayed by the proliferation of catfish lymphocytes or thymocytes in the presence of Con A, or by an IL-2 conversion assay using LBRM 33–1A5 cells. Gel filtration analyses indicated two forms of IL-1-like activity: a 70 kDa form that was active on catfish, but not mouse T cells and a 15 kDa form that was active on mouse, but not catfish, T cells. Both sizes of catfish IL-1 exhibited α and β determinants as shown by Western blot analyses and antibody inhibition assays using polyclonal antisera against human IL-1α and IL-1β.

Secretion of an IL-1-like factor by carp macrophages and neutrophilic granulocytes was shown by Verburg-van Kemenade *et al.* (1995). Culture supernatants costimulated PHA-induced proliferation of carp lymphocytes and stimulated the proliferation of the murine T cell line D10.G4.1. The latter is used as an IL-1 specific bioassay for human and murine IL-1 (Hopkins and Humphreys, 1989). Relatedness to IL-1 was further suggested by the inhibition of bioactivity in the supernatants by polyclonal antibodies against human recombinant IL-1α and IL-1β (Verburg-van Kemenade *et al.*, 1995). Western blot analysis of supernatants with polyclonal antisera against human IL-1 revealed protein bands at 22 kDa and 15 kDa.

In a mouse thymocyte proliferation assay, Atlantic salmon macrophages were shown to produce IL-1-like factors *in vitro* and the yeast β-glucan enhanced the production of such compounds (Robertsen *et al.*, 1994).

Although catfish and carp IL-1 induces reactivity from mouse T-cells, mammalian IL-1 does not appear to elicit biological activity from fish lymphocytes. Recombinant forms of mouse and human IL-1 failed to stimulate proliferation of catfish peripheral blood leukocytes which indicates either that the catfish cells require a high molecular weight IL-1 or that differences in bioactive structures exist between catfish and mammalian IL-1 (Ellsaesser and Clem, 1994). Likewise, recombinant human IL-1α and IL-1β failed to stimulate proliferation of carp lymphocytes (Verburg-van Kemenade *et al.*, 1995). The observations of Hamby *et al.* (1986) that purified human IL-1 costimulated Con A-induced proliferation of channel catfish blood leukocytes, may possibly be the result of IL-1 production by monocytes in the leukocyte population (Clem *et al.*, 1991). The recent report of cDNA cloning of a rainbow

Table II.5.1 Evidence for fish cytokines and cytokine receptors as suggested by bioactivity, antigenic cross-reactivity or genetic techniques[a]

Putative cytokine or receptor	*Species*	*Cell source*	*Detection method*	*Reference*
CSF	Carp	(Serum)	Colony stimulation	Moritoma *et al.* (1992)
	Rainbow trout	(Serum)	Colony stimulation	Kodama *et al.* (1994)
IFN	Japanese flatfish	Leukocytes	Antiviral, cloning	Tamai *et al.* (1993a,b)
IFN-α/β	Blue striped grunt	Fin	Antiviral	Beasley and Sigel (1967)
	Fathead minnow	Epithelial	Antiviral	Gravell and Malsberger (1965)
	Red swordtail	Embryonic	Antiviral	Kelly and Loh (1973)
	Rainbow trout	Gonad	Antiviral	de Sena and Rio (1975)
		Leukocytes	Antiviral	Rogel-Gaillard *et al.* (1993)
		(Serum)	Antiviral	Dorson *et al.* (1975), Renault *et al.* (1991)
	Sea bass	(Serum)	Antiviral	Pinto *et al.* (1993)
IFN-γ/MAF	Rainbow trout	Leukocytes	Macrophage activation, antiviral	Graham and Secombes (1988, 1990a,b)
IL-1α/β	Atlantic salmon	Macrophages	ACRI	Robertsen *et al.* (1994)
	Carp	Macrophages, neutrophils	Comitogenic in homologous and heterologous system	Verburg-van Kemenade *et al.* (1995)
	Channel catfish	Epidermal	Comitogenic in homologous and heterologous system	Sigel *et al.* (1986)
		Monocytes	Comitogenic in homologous and heterologous system, ACRW	Clem *et al.* (1991), Ellsaesser and Clem (1994)
	Rainbow trout	Macrophages	PCR/cloning	Secombes (1996), Zou *et al.* (1997)
IL-2	Carp	Optic nerves	ACRI, ACRW	Eitan *et al.* (1992)
	Japanese flatfish	Leukocytes	PCR/cloning	Tamai *et al.* (1992)
TGF-β	Rainbow trout	—	PCR/cloning	Secombes *et al.* (1996)
TNF-α	Atlantic salmon	Macrophages	ACRI	Robertsen *et al.* (1994)
TNF-α receptor	Rainbow trout	Macrophages	ACRI	Jang *et al.* (1995a)

[a] Abbreviations: ACRI = antigenic cross reactivity where inhibition of biological activity in an homologous system was demonstrated using antibodies against the mammalian cytokine equivalent, ACRW = antigenic cross reactivity where fish cytokine-like compounds were demonstrated in Western blots using antibodies against the mammalian cytokine equivalent, CSF = colony-stimulating factor, IL = interleukin, IFN = interferon, MAF = macrophage activating factor, TGF = transforming growth factor, TNF = tumor necrosis factor.

trout IL-1β gene and its expression in *E. coli* (Zou *et al.*, 1997; C. J. Secombes, personal communication), may represent a major breakthrough in fish interleukin research. The gene is deduced to encode a polypeptide of 260 amino acids that gives a mature peptide of 150 amino acids with a predicted molecular weight of 17 kDa. Amino acid similarities of the whole molecule to mammalian IL-1βs range from 49 to 55% (C. J. Secombes, personal communication).

IL-2

An IL-2-like lymphocyte growth-promoting factor was reported in supernatants from carp leukocyte cultures stimulated by mitogen or alloantigen (Caspi and Avtalion, 1984; Grondel and Harmsen, 1984). A 28 kDa factor cytotoxic for oligodendrocytes from carp was suggested to be related to IL-2 based on its reactivity with antibodies against human IL-2 (Eitan *et al.*, 1992).

Recently, an IL-2-like gene from Japanese flatfish was cloned by direct PCR amplification using primers from conserved nucleotide sequences of IL-2 genes from various mammals (Tamai *et al.*, 1992). However, the classification of this gene as an IL-2 gene based on sequence homology with the mammalian IL-2 gene is questionable (Secombes *et al.*, 1996). Nevertheless, expression of this IL-2 gene in COS-1 cells resulted in the production of a compound that stimulated proliferation of flatfish leukocytes (Tamai *et al.*, 1992). Biological cross-reactivity of mammalian IL-2 in fish systems is uncertain. Caspi and Avtalion (1984) showed that proliferation of carp T-like lymphoblasts was promoted in the presence of IL-2 containing supernatants of mammalian origin. Conversely, mouse recombinant IL-2 does not exhibit *in vitro* functional activity with catfish cells (Clem *et al.*, 1991).

Tumor necrosis factor

At present, there is apparently no cell cytotoxicity assay for TNF-α from fish. However, several lines of indirect evidence suggest the existence of TNF-α in salmonids. Human TNF-α was shown to be synergistic with rainbow trout macrophage-activating factor in elevating respiratory burst activity of trout macrophages and with mitogenic stimuli in heightening the proliferation responses of trout head kidney leukocytes (Hardie *et al.*, 1994a). The existence of a receptor for TNF-α in rainbow trout was later suggested by the observation that preincubation of trout phagocytes with monoclonal antibodies against the human TNF-α receptor, TNF-R1, inhibited the enhancement of macrophage respiratory burst and neutrophil migration by human TNF-α (Jang *et al.*, 1995a). Involvement of TNF-α in the glucan-induced enhancement of respiratory burst in Atlantic salmon macrophages is indicated by inhibition of the response with polyclonal antibodies against human TNF-α (Robertsen *et al.*, 1994).

Colony stimulating factors

Bioactivity for a macrophage colony-stimulating factor was observed in the serum of rainbow trout after injection of bacterial LPS (Kodama *et al.*, 1994). Colony-stimulating activity for head kidney granulocytes was also observed in the serum from carp and was shown to be enhanced in carp following injection with Freund's complete adjuvant (Moritoma *et al.*, 1992).

Interferons

All the type I IFN (IFN-α and IFN-β) genes are thought to have evolved from a single ancestral gene by gene duplication about 310 million years ago. This was around the time of the divergence of the tetrapods and suggests that fish may have only one type I IFN gene (Miyata *et al.*, 1985). In accordance with this suggestion, IFN-β-like sequences, but not IFN-α-like sequences were detected in teleost fish by hybridization experiments (Wilson *et al.*, 1983).

As shown in Table II.5.1, IFN activity has been detected in cell supernatants and serum from a wide range of fish species. Renault *et al.* (1991) developed an excellent spectrophotometric method for titration of trout IFN that can be used to measure IFN activity even in fry homogenates.

A 94 kDa IFN was partially purified from virus-infected rainbow trout gonad cells (RTG-2) that showed heat and pH stability characteristic of type I IFN (de Sena and Rio, 1975). Dorson *et al.* (1975) reported an IFN with an apparent molecular mass of 26 kDa in the serum of virus-infected rainbow trout.

Graham and Secombes (1988) isolated a macrophage activating factor (MAF) from mitogen-stimulated rainbow trout leukocytes that was shown to enhance respiratory burst and bactericidal activity in trout macrophages. Separation of trout leukocytes into sIg^- and sIg^+ lymphocytes by panning, showed that only sIg^- cells, most likely T cells, could produce MAF and that macrophages were necessary accessory cells (Graham and Secombes, 1990a). Further studies demonstrated that the MAF conferred viral resistance on a rainbow trout epithelial cell line and that MAF and IFN activity coeluted during size exclusion chromatography with a major peak at 19 kDa and a minor peak at 32 kDa (Graham and Secombes, 1990b). The mode of induction of MAF, its IFN activity and its acid and temperature sensitivity, suggest that it is an IFN-γ-like molecule (Graham and Secombes, 1990b).

Recently, an immortalized leukocyte cell line from Japanese flatfish was shown to secrete an IFN-like glycoprotein with a molecular weight of 16 kDa (Tamai *et al.*, 1993b). The IFN activity was relatively stable at 60°C, unstable at pH 2 and was species nonspecific in its action. The flatfish IFN cDNA was cloned by these workers (Tamai *et al.*, 1993a) and shown to encode a 138 amino acid polypeptide (11.6 kDa) that had little homology with mammalian IFNs or chicken IFN (Tamai *et al.*, 1993a; Sekellick *et al.*, 1994). On the basis of these properties, the flatfish IFN cannot be classified as either type I or type II IFN and may possibly represent a new IFN. Whilst type I IFN genes in fish have yet to be cloned, antiviral Mx genes that are specifically induced by type I IFNs in mammals, have been identified in salmonids. Three Mx genes have been cDNA cloned from rainbow trout (Trobridge and Leong, 1995; Trobridge *et al.*, 1997) and Atlantic salmon (Robertsen *et al.*, 1997).

Transforming growth factor

Jang *et al.* (1994) showed that natural bovine TGF-β1, depending on the concentration used, had both enhancing and suppressing effects on the respiratory burst activity of rainbow trout head kidney macrophages. TGF-β1 also inhibited the activation of trout macrophages by MAF or TNF-α. A TGF-β1-like gene has recently been cloned from rainbow trout (Secombes *et al.*, 1996). The translated amino acid sequence showed 53–68% homology to other vertebrate TGF-βs and, as with other known forms of TGF-β, the mature peptide consisted of 112 amino acids. These reports further confirm that TGF-βs cross-react and have highly conserved gene sequences among vertebrate species (Haynes and Cohen, 1993; Secombes *et al.*, 1996).

6. T-cell Receptor

Cell-mediated immunity, such as allograft rejection, occurs in all vertebrates and, at least in tetrapods, this reaction depends on thymus-derived lymphocytes (T cells). The participation of peripheral T-like lymphocytes (indirectly defined as surface Ig-negative: sIg^--cells) in fish immunity is now well documented, at least in some well-studied species such as the channel catfish where the sIg^- lymphoid population has most of the physiological properties of mammalian T cells (reviewed by Vallejo *et al.*, 1992). However, the difficulty in performing thymectomy in fish at early stages of development and the lack of antibody probes specific for T cell surface antigens have hampered the physiological analysis of T cell immunity in these species.

Polymerase chain reaction (PCR) strategies and cDNA library screenings have recently led to the successful cloning of cDNAs encoding TCR-like chains in several cold-blooded species, including fish. TCR_β-chain genes were successfully cloned from four fish species: the horned shark (Rast and Litman, 1994; Hawke *et al.*, 1996), the rainbow trout (Partula *et al.*, 1994, 1995), the Atlantic salmon (Hordvik *et al.*, 1996) and the channel catfish (Zhou *et al.*, 1997). Key residues of the V_β and C_β regions are in good agreement with those of higher vertebrates and in all four species, typical D_β and J_β regions are present. In other nonmammalian vertebrates (axolotl, chicken), the C_β connective peptide is 8–14 amino acids shorter than in mammals. A three-dimensional comparison indicates that the extended fragment of the β-chain in mammals represents a solvent exposed insertion with respect to other IgC domains (Bentley *et al.*, 1995), and consequently the fish TCR may be close in shape to the ancestral molecule. The most conserved region of the C_β chain is the transmembrane (TM) domain that maintains, in all species, the lysine residue which is thought, in mammals, to interact with CD3, and most of the residues that form the mammalian CART (conserved antigen receptor transmembrane) motif (Campbell *et al.*, 1994). In contrast to mammals, birds and amphibians, the C_β hinge region of trout, Atlantic salmon and channel catfish has no cysteine residue, suggesting that fish TCR_α- and TCR_β-chains are not associated covalently. A single C_β isotype has been described in salmon and trout, while two C_β isotypes were detected in channel catfish at the cDNA level.

TCR_α-chain gene sequences have been extensively described in rainbow trout (Partula *et al.*, 1996a) and cloned from salmon (Hordvik and coworkers, unpublished), channel catfish (Zhou *et al.*, 1997) and pufferfish (Rast *et al.*, 1995b). Key residues are conserved in the V_α-like regions, but the extracellular part of the C_α regions, which are not well conserved, are considerably smaller than the mammalian counterpart. However, the TM regions of fish C_α chains conserve the characteristic CART residues.

A TCR constant isotype, clearly different from the C_α or C_β isotypes was isolated in the horned shark (Rast *et al.*, 1995b) and tentatively assigned to C_δ, although this must be confirmed, as no $TCR\gamma$ chain equivalent has been isolated in the shark.

The variability of the rainbow trout TCR_α-chain has been extensively studied (Partula *et al.*, 1996a). Six V_α families were defined ($V_\alpha1$–$V_\alpha6$) on the basis of a minimum of 75% amino acid identity, and three different members of the $V_{\alpha2}$ family ($V_\alpha2.1$, $V_\alpha2.2$, $V_\alpha2.3$) were found. Sequence identity among families varied from 27% ($V_\alpha2.1$ vs $V_\alpha5$) to 55.5% ($V_\alpha4$ vs $V_\alpha6$). A phylogenetic analysis revealed that the six trout V_α segments were best matched to mammalian V_α and V_δ sequences, and the most identical regions were located around the conserved cysteine and tryptophan residues. The analysis of 40 V_α–J_α rearrangements led to the identification of 32 different J_α sequences, suggesting a potentially larger repertoire of genomic J_α segments. Several positions of the trout J_α segments are completely conserved with mammalian J_α, such as Phe106, Gly107, Gly109 and Thr110. However, residue 116, which is a proline in most mammals J_α was not conserved in the trout. The size of the trout CDRα3 region varies from 8 to 13 amino acids, and thus is not significantly different from mammalian CDRα3. Most of the structural properties were conserved between trout and human CDRα3 loops, suggesting that this region, which is implicated in Ag/MHC interaction has been under considerable selective pressure during vertebrate evolution.

Three V_α families and four J_α segments have been characterized in the channel catfish and Southern blot analysis suggested that each V_α family contains a large number of members.

The three trout V_β segments characterized (Partula *et al.*, 1995) represent different V_β families that are separated by considerable genetic distances: $V_\beta2$ and $V_\beta3$ are the more related (38.4% identical amino acids), but have only 30% identical residues with $V_\beta1$. $V_\beta2$ and $V_\beta3$ are not clearly related to any other V_β, but $V_\beta1$ has significant degrees of similarity with members of the human $V_\beta20$ family and is structurally related to the Kabat's subgroup II V_β sequences. The comparison of a large number of trout V_β–D_β–J_β junctions and the partial sequence analysis of the trout TCR_β locus (de Guerra and Charlemagne, 1997), led to the identification of 10 J_β and one D_β segments. The D_β gene is located about 600 bp 5′ to the J_β genes and the first C_β exon is located 3′ to the last 3′ J_β gene, suggesting that the trout and mammalian TCR_β loci have a similar structure. The J_β segments have very similar amino acid sequences, including eight completely conserved residues among the last 12 C-terminal positions, and the characteristic structure of the mammalian J_βs is conserved. The genomic sequence of the trout D_β is identical to those of the mouse, rat and human $D_\beta1$ genes (GGGACAGGGGGC) and their recombination signal sequences (RSS) are also very well conserved. However, although the RSS heptamers that flank the J_β genes are well conserved, the corresponding nonamers are not conserved and some cannot be distinguished. Interestingly, a large number of repetitive DNA sequences are located in the trout D_β–J_β–C_β locus, and some include many cryptic RSS sequences. The biological significance of these DNA repeats is unknown, but they may reflect some important events in TCR evolution. Among the eight J_β segments described in the Atlantic salmon (Hordvik *et al.*, 1996), six are clearly homologous to rainbow trout J_βs. The salmon D_β-like consensus octamer (GGGACAGGG) is identical to the core sequence of the trout D_β. From the known genomic sequence of the trout D_β–J_β–C_β locus, it was possible to provide an accurate interpretation of the V_β–D_β–J_β rearrangements and to confirm that, as in mammals, significant trimming operates before joining of the coding segments and that P- and N-nucleotides significantly contribute to the structural diversity of the junctions and to the corresponding $CDR_\beta3$ loops (de Guerra and Charlemagne, 1997). A significant proportion (32%) of the trout and salmon $V_\beta D_\beta J_\beta$ junctions are out of frame, suggesting that, as in mammals, allelic exclusion operates at the fish TCR_β loci.

A single V_β and three V_α families were detected in channel catfish. Southern blot analysis indicated that the V_α families are much more diversified than the V_β family. Four different J_αs and two J_βs were also identified, and the structure of the putative D_β core element (GGGACAGGGGG) is similar to those of the salmonid fish and mammalian D_β/$D_\beta1$ elements (Zhou *et al.*, 1997).

Seven distinct V_β families were defined in a cartilaginous fish, the horned shark (Hawke *et al.*, 1996), and these V_β chains contain most of the residues typically found in vertebrate TCR V_βs (Q6, C23, WY35, Y90 and C92). As in the rainbow trout, the shark V_β families are separated by significant genetic distance (they are less than 32% identical at the amino acid level), but a member of the $V_\beta4$ family (Hf89) has a relatively high sequence identity (45% amino acids) with human $V_\beta20$. The analysis of a large number of $V_\beta D_\beta J_\beta$ junctions led to the identification of a family of 18 different J_β segments and of a putative core D_β sequence (GGGACAAC). A clear diversity occurred at the junctions, with potential *N*-nucleotide additions. The lengths of the shark $CDR_\beta3$ loops (5–12 amino acids) are consistent with those of mammals. Thus, the functional constraints that operate in the contact interaction between TCR and MHC molecules may be similar from sharks to mammals. Pulse-field gel electrophoresis and genomic mapping suggested that shark TCR_β genes should be organized, as with the IgH and IgL genes, in multiple $V_\beta(D_\beta)J_\beta$ clusters (Rast and Litman, 1994). However, a limited number of TCR C_β genes or alleles appear to be expressed and the fact that different V_βs can associate with a similar J_β, and that different J_βs can associate with a similar V_β, suggested a combinatorial form of segmental rearrangements which is not fully compatible with a multiclustered organization of the TCR_β genes (Hawke *et al.*, 1996). One of the several explanations for this apparent contradiction is that TCR expression in shark operates only at a limited number of clusters.

In conclusion, it is now clear that fish possess highly complex and diverse molecules that are the structural homologues to the α and β chains, which in birds and mammals, form the $TCR_{\alpha\beta}$ receptors. However, it must be emphasized that the physiological role of these TCR-like chains in primitive vertebrates remains uncertain. Thus, the production of antibodies to TCR molecules will be the next step in the analysis of T-cell function in fish.

7. Immunoglobulins

In this section on immunoglobulins (Ig) in fish, the current state of knowledge is briefly summarized, and to improve the readability of this summary the bulk of the original literature dealing with fish Ig has been cited in Table II.7.1 rather than cited in the text. In Table II.7.1, the references upon which the present summary is based have been arranged according to the species and theme studied such as immunoglobulin biochemistry, antibody response and immunogenetics. Several more detailed reviews are recommended for a more complete picture of the different aspects of fish immunoglobulins (Hsu and Steiner, 1992; Wilson and Warr, 1992; Marchalonis *et al.*, 1993a,b; Du Pasquier, 1993a,b; Rast *et al.*, 1995a; Warr, 1995; Pilström and Bengten, 1996).

The fishes are a heterogeneous group of vertebrates with several branches that separated 400–500 million years ago. The systematics and phylogeny of fishes are complicated subjects and the reader is referred to G. Nelson (1989) or to J. S. Nelson (1994) for more detailed information (Figure II.7.1). It should be noted that even a taxon such as the bony fish (teleosts), which is considered as a homogeneous group by many biologists, is more heterogeneous than the mammals.

Agnatha

Of the still-living fish species, the jawless (agnathan) fish are the most primitive. This group comprises the lampreys and the hagfish and the immunoglobulin molecules in these species have been the focus of much research. In the late 1960s and early 1970s, the antigen binding and biochemical characterization of antibodies were described. However, subsequent studies in the hagfish have suggested that these 'antibodies' were most probably molecules of the complement system binding antigen (Fujii *et al.*, 1992). Nevertheless, these early observations, particularly in the lamprey, are intriguing and the immune system of the agnathans should be investigated further.

Table II.7.1 References to fish immunoglobulin that deal with immunoglobulin structure, activity and genes in fishes, arranged according to the species. The present summary of the current state of knowledge on immunoglobulins in fish has been drawn from these references

Species	*Common name*	*Immunoglobulin structure*	*Antibody activity*	*Genes*
Eptatretus stoutii and *E. burgeri*	Pacific hagfish	Raison *et al.* (1978a,b), Kobayashi *et al.* (1985a), Hanley *et al.* (1990), Varner *et al.* (1991), Fujii *et al.* (1992)		
Petromyzon marinus	Sea lamprey	Boffa *et al.* (1967), Marchalonis and Edelman (1968)		
Callorhinchus callorhincus	Ratfish	Sanchez *et al.* (1980), Garrido and De Ioannes (1981), De Ioannes and Aguila (1989)		
Hydrolagus collei				Rast *et al.* (1994)
Chlamydoselachus anguineus	Frill shark	Kobayashi *et al.* (1992)		
Heterodontus francisci	Horned shark	Frommel *et al.* (1971), Litman *et al.* (1971b)	Litman *et al.* (1980a,b, 1982, 1992), Mäkelä and Litman (1980)	Litman *et al.* (1985a,b), Hinds and Litman (1986), Kokubu *et al.* (1987, 1988a,b), Shamblott and Litman (1989a,b), Hinds-Frey *et al.* (1993), Rast *et al.* (1994)
Ginglymostoma cirratum	Nurse shark	Clem *et al.* (1967), Fidler *et al.* (1969), Small *et al.* (1970), Weinheimer *et al.* (1971), Gitlin *et al.* (1973), McCumber and Clem (1976), Klapper and Clem (1977), Fuller *et al.* (1978)	Rudikoff *et al.* (1970), Clem and Leslie (1971, 1982), Voss and Sigel (1972), Sigel *et al.* (1972), Shankey and Clem (1980a,b)	Vazquez *et al.* (1992), Greenberg *et al.* (1993, 1995, 1996)
Scyliorhinus canicula	Dogfish	Morrow *et al.* (1983)		
Mustelus canis	Smooth dogfish	Marchalonis and Edelman (1966)		
Carcharhinus sp.	Sandbar shark	Gitlin *et al.* (1973), Rosenshein *et al.* (1985), Rosenshein and Marchalonis (1987), Vazquez-Moreno *et al.* (1992)		Schluter *et al.* (1989, 1990), Hohman *et al.* (1992, 1993, 1995), Bernstein *et al.* (1996)
Negaprion brevirostris	Lemon shark	Gitlin *et al.* (1973)	Sigel *et al.* (1972)	
Raja sp.	Spiny rasp skate Little skate	Ellis and Parkhouse (1975), Kobayashi *et al.* (1984), Tomonaga and Kobayashi (1985), Hagiwara *et al.* (1985) Kobayashi and Tomonaga (1988)		Harding *et al.* (1990a,b), Anderson *et al.* (1994, 1995), Rast *et al.* (1994)
Bathyraja aleutica	Aleutian skate	Kobayashi *et al.* (1985b)		
Daysatis centroura	Sting ray	Marchalonis and Schonfeld (1970), Johnston *et al.* (1971)		

(continued)

Table II.7.1 *Continued*

Species	*Common name*	*Immunoglobulin structure*	*Antibody activity*	*Genes*
Latimeria chalumnae	Gombessa			Amemiya *et al.* (1993)
Neoceratodus forsteri	Australian lungfish	Marchalonis (1969)		
Protopterus aetiopicus	African lungfish	Litman *et al.* (1971c,d)		
Acipencer baeri	Siberian sturgeon	Partula and Charlemagne (1993)		Lundqvist *et al.* (1996)
Polyodon spathula	Paddlefish	Acton *et al.* (1971b,c), Weinheimer *et al.* (1971)	Legler *et al.* (1971)	
Lepisosteus osseus	Gar	Acton *et al.* (1971a,c), Bradshaw *et al.* (1971a,b), Weinheimer *et al.* (1971)		Wilson *et al.* (1995a,b)
Amia calva	Bowfin	Litman *et al.* (1971a,c)	Bradshaw and Sigel (1972)	Wilson *et al.* (1995a,b)
Elops saurus	Ladyfish			Amemiya and Litman (1990)
Cyprinus carpio	Common carp	Shelton and Smith (1970), Marchalonis (1971), Richter *et al.* (1973), Vilain *et al.* (1984), Atanassov and Botev (1988)	Lamers *et al.* (1985), Wetzel and Charlemagne (1985), Rombout *et al.* (1986), Kusuda *et al.* (1987), Wiegertjes *et al.* (1995)	
Carassius auratus	Goldfish	Marchalonis (1971), Warr and Marchalonis (1977), Vilain *et al.* (1984), Wilson *et al.* (1985)	Uhr *et al.* (1962), Trump (1970), Ruben *et al.* (1977), Desvaux and Charlemagne (1981), Wetzel and Charlemagne (1985)	Wilson *et al.* (1991)
Carassius carassius	Crucian carp		Nakanishi (1987b)	
Tinca tinca	Tench	Vilain *et al.* (1984)	Wetzel and Charlemagne (1985)	
Ictalurus punctatus	Channel catfish	Acton *et al.* (1971c), Weinheimer *et al.* (1971), Mestecky *et al.* (1975), Lobb and Clem (1981c), Lobb *et al.* (1984), Lobb (1986), Ourth (1986), Phillips and Ourth (1986), Lobb and Olson (1988), Klesius (1990), Hayman and Lobb (1993), Ledford *et al.* (1993)	Lobb and Clem (1983), Miller and Clem (1984), Lobb (1985, 1987), Vallejo *et al.* (1991a)	Ghaffari and Lobb (1989a,b, 1991, 1992, 1993), Wilson *et al.* (1990), Warr *et al.* (1991, 1992), Hayman *et al.* (1993), Jones *et al.* (1993), Ventura-Hollman *et al.* (1994, 1996), Magor *et al.* (1994a)
Ictalurus nebulosus	Brown bullhead	Isbell and Pauley (1983)		
Tachusurus australis	Sea catfish		Di Conza and Halliday (1971)	
Esox lucius	Northern pike	Clerx *et al.* (1980)		

(*continued*)

Table II.7.1 *Continued*

Species	*Common name*	*Immunoglobulin structure*	*Antibody activity*	*Genes*
Oncorhynchus mykiss (previously *Salmo gairdneri*)	Rainbow trout	Yamaga *et al.* (1978), Warr (1982), Elcombe *et al.* (1985), Olesen and Vestergård Jørgensen (1986), Sanchez *et al.* (1989, 1993), Sanchez and Dominguez (1991)	Etlinger *et al.* (1979), Tatner and Horne (1983), Bortz *et al.* (1984), St. Louis-Cormier *et al.* (1984), Cossarini-Dunier (1986), Desvaux *et al.* (1987), Thuvander *et al.* (1987), Thorburn and Jansson (1988), Gonzalez *et al.* (1989), Michel *et al.* (1990), Grayson *et al.* (1991), Arkoosh and Kaattari (1991), Kaattari and Shapiro (1994)	Matsunaga *et al.* (1990), Andersson and Matsunaga (1993, 1995), Daggfeldt *et al.* (1993), Lee *et al.* (1993), Hansen *et al.* (1994), Michard-Vanhee *et al.* (1994), Roman and Charlemagne (1994), Andersson *et al.* (1995), Roman *et al.* (1995, 1996), Partula *et al.* (1996)
Oncorhynchus kisutch	Coho salmon		Voss *et al.* (1978b), Olivier *et al.* (1985)	
Oncorhynchus keta	Chum salmon	Kobayashi *et al.* (1982), Fuda *et al.* (1992), Nagae *et al.* (1993)		
Oncorhynchus masou	Masu salmon	Fuda *et al.* (1991)		
Oncorhynchus tshawytscha	Chinook salmon		Wongtavatchai *et al.* (1995)	
Salmo salar	Atlantic salmon	Håverstein *et al.* (1988), Magnadottir (1990)	Killie *et al.* (1991), Magnadottir and Gudmundsdottir (1992), Lillehaug *et al.* (1993), Magnadottir *et al.* (1995)	Hordvik *et al.* (1992)
Salmo trutta	Brown trout	Ingram and Alexander (1979)	Ingram and Alexander (1976, 1980, 1981), O'Neill (1979)	
Gadus morhua	Atlantic cod	Pilström and Petersson (1991), Israelsson *et al.* (1991)	Bjørgan Schrøder *et al.* (1992)	Bengten *et al.* (1991, 1994), Daggfeldt *et al.* (1993), Bengten (1994)
Dicentrarchus labrax	Sea bass	Palenzuela *et al.* (1996)		
Epinephelus itaira	Giant grouper	Clem (1971)		
Perca fluviatilis	Perch	Whittington (1993)	Richter and Ambrosius (1972)	
Seriola quinqueradiata	Yellowtail	Matsubara *et al.* (1985)		
Lujtanus griseus	Gray snapper		Russell *et al.* (1970)	
Haemulon album	Margate	Clem and McLean (1975)		
Archosargus probatocephalus	Sheepshead	Lobb and Clem (1981a,b,d)		
Sparus aurata	Gilthead seabream	Navarro *et al.* (1993), Palenzuela *et al.* (1996)		
Sarotherodon galilaeus	Tilapia	Avtalion and Mor (1992)		
Scophthalmus maximus	Turbot	Estevez *et al.* (1994, 1995)		

(continued)

Table II.7.1 *Continued*

Species	*Common name*	*Immunoglobulin structure*	*Antibody activity*	*Genes*
Paralichthys dentatus	Summer flounder		Stolen *et al.* (1984), Laudan *et al.* (1986)	
Platichthys flesus	Flounder	Glynn and Pulsford (1993)	Pulsford *et al.* (1994)	
Pleuronectes platessa	Plaice	Fletcher and Grant (1969)	Fletcher and White (1973b)	
Pseudopleuronectus americanus	Winter flounder		Litman *et al.* (1982), Laudan *et al.* (1986, 1989)	
Sphoeroides glaber	Toadfish	Warr (1983)		

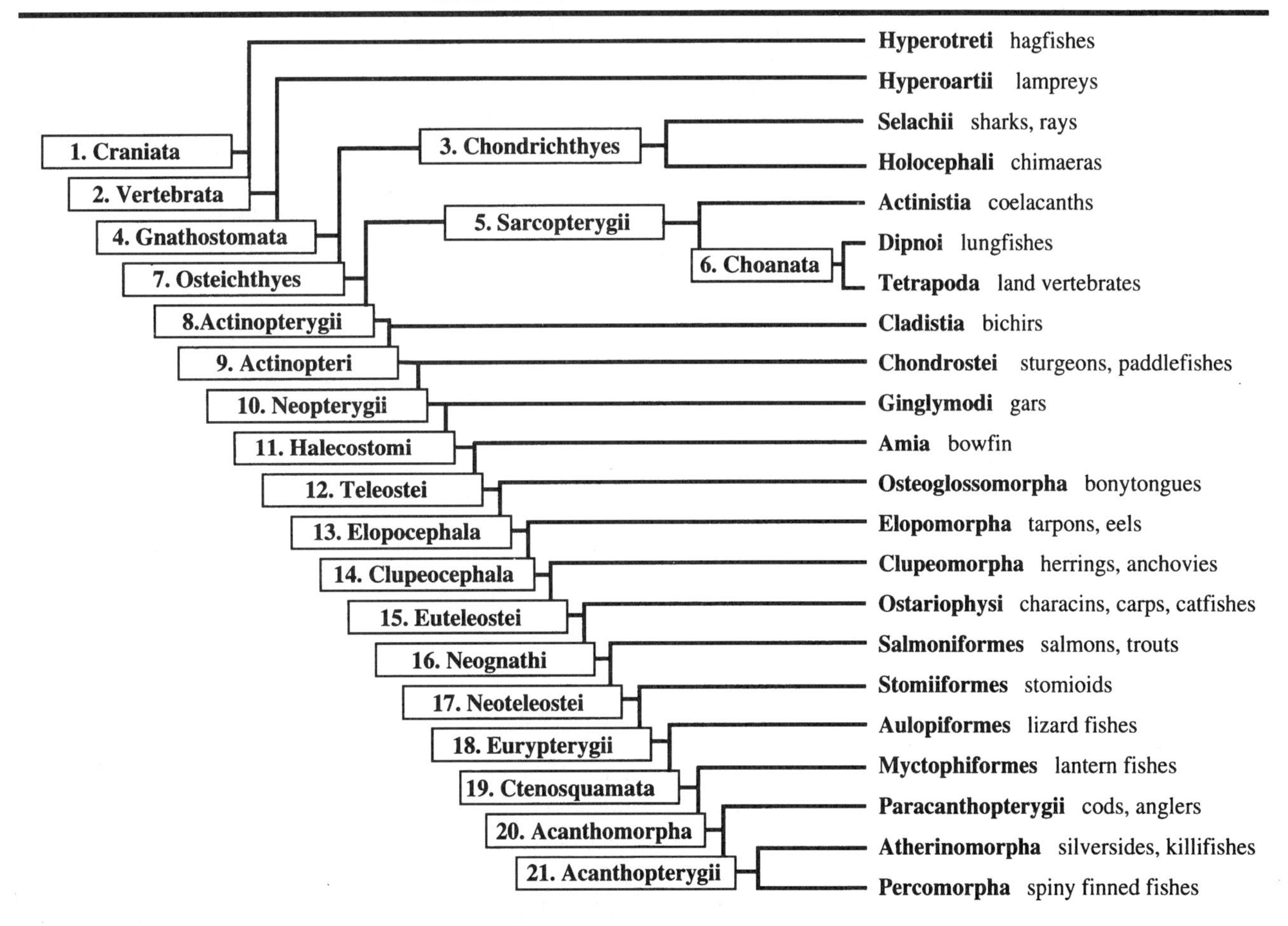

Figure II.7.1 Interrelationships of major fish groups. Inclusive taxa are numbered in each box. Reproduced with permission from Lundqvist, M., Bengtén, E., Strömberg, S. and Pilström, L. (1996) *J. Immunol.* **157**, 2031–2038. Copyright 1996. The American Association of Immunologists.

Biochemical characterization of Ig

Chondrichthyes (chimeras, sharks, rays and skates)

The most primitive gnathostoman group includes the cartilaginous fish (chondrichthyes) with the elasmobranchs (sharks and rays) and the holocephalans (chimeras). The immune system appears to have made a huge jump in evolutionary development from the agnatha to the chondrichthyes, because the latter possess a well-developed immune system with all the central immune organs and functions. IgM appears to be the first and most ancient Ig class because it is found in all gnathostoman vertebrate groups, including the cartilaginous fish. The designation of the immunoglobulin as IgM is justified by its similarity to the mammalian counterpart – a pentamer of the basic structure H_2L_2 and a heavy chain with four constant domains. There are reports indicating the J-chain is associated with the molecule. Comparative analyses with IgE also show that the amino acid sequences of the shark (and other fish) Ig are more similar to the μ than to the ε chain of mammals. The molecular mass for the whole molecule is in the range of 800–980 kDa and the carbohydrate content is around 5–10% (Acton *et al.*, 1972). The two chains are around 70–75 kDa and 20–25 kDa, respectively. The serum of sharks also contains a monomer of 160–200 kDa, which is serologically indistinguishable from the IgM, indicating a monomeric form. Nurse shark serum contains 11–14 mg Ig/ml corresponding to 17–21% of the total protein concentration.

Fuller *et al.* (1978) reported the isolation from the nurse shark of a 7S Ig with an H-chain of 50 kDa and a normal L-chain. In the frill shark, a similar Ig forms monomers and dimers of 150 and 300 kDa, respectively. A 45–50 kDa H-chain is antigenically different from the 68 kDa μ-chain and the two chains are produced by different plasma cells. Recently, PCR has been used to detect two non-IgM isotypes in sharks. In nurse shark, these isotypes are called IgNAR and IgNARC, the latter of which appears to correspond to IgW in the sandbar shark. Whether IgNAR is, in a strict sense, an Ig is debatable since it lacks light chains, although the molecule expresses different V-domains and is rearranged and diversified in a conventional manner. The molecule is less than 200 kDa and reduction yields one band around 80 kDa indicating that IgNAR has the form H_2. Interestingly, the H-chain contains six domains – one variable and five constant and a C-terminal peptide. The IgNARC/IgW has been less studied than IgNAR, but it is a true Ig in the sense that it binds an L-chain. The heavy chain has one variable and six constant domains and the amino acid sequence shows great similarities to IgNAR. IgNARC is reported to be expressed in very low amounts in nurse shark serum. In skates and rays, another isotype has been described called IgR. This non-covalent 320 kDa dimer separates upon reduction into a 45–50 kDa H-chain and a 23–27 kDa L-chain. The concentration of IgR is 0.73 mg/ml serum compared with 3.90 mg/ml for IgM. This isotype has only been found in rays and skates and does not correspond to the 50 kDa H-chain reported in the nurse shark. Different isotypes of Ig light chain have been described at the gene level in sharks and chimeras.

When the Ig functions as a B-cell receptor, the C-terminal region carries a hydrophobic peptide instead of the hydrophilic domain of the secreted form. Processing of the pre-mRNA determines which form should be synthesized. In the horned shark (as in mammals), a cryptic donor site in the CH4 exon is used for splicing the transmembrane exon, and this is probably valid for all Chondrichthyes.

Actinopterygii (rayfinned fish)

This group contains the largest number of species of all the vertebrates and the most primitive taxa are the sturgeons and paddlefish (Chondrostei), gars (Ginglymodi) and the bowfin (Amia), while the true bony fish (Teleostei) are considered more advanced. The Ig of these fishes is also of the IgM type but forms tetramers instead of pentamers and in some actinopterygian species monomers and dimers also occur. It has been hypothesized that the tetramer structure results from the number of amino acids downstream from the cysteine residue that forms the disulfide bridges of the polymer. A number higher than one should cause space problems, so the basic H_2L_2 units form a tetrahedron allowing the carboxy amino acids to protrude out from the disulfide ring structure (Pilström and Bengten, 1996). The molecular mass of these Ig is of course less than for IgM in the other vertebrates and values ranging from 650–850 kDa are generally reported with the two chains being 70–75 kDa and 20–25 kDa, respectively. The carbohydrate content varies (up to 16% has been reported) and the L-fucose content and the galactose : mannose ratios are higher than in mammals but very little, if anything, is known of the detailed structure of glycosyl moieties of fish Ig. The Ig concentration in serum is lower than that reported for the nurse shark, except in the holostean fish, the gar and the paddlefish, which have 14 and 17 mg Ig/ml serum, respectively, representing 40–50% of the total serum protein concentration. Salmonids appear to have very low amounts of Ig, usually around 1 mg/ml serum, which correspond to 2–6% of the total serum protein. Higher values are reported for other actinopterygian species, e.g. channel catfish (2–11 mg/ml, up to 21%), goldfish and cod (around 5 mg/ml, 17% of serum protein). In some species, it has been observed that the serum Ig concentration is higher in large fish than in smaller ones, but the reason for this phenomenon is unknown. The concentration of Ig in mucus and bile is much lower than the concentration in serum of the same fish. Ig is also found in the eggs of teleost fish but the concentrations are again very low compared with the levels in serum. The embryo of rainbow trout appears to start its own Ig production at a very early stage as transcripts of RAG1 and RAG2 have been demonstrated as early as 48 h after fertilization (Hansen and Kaattari, 1995, 1996).

In actinopterygian fish few isotypes other than IgM have been described, although different degrees of polymerization are often reported. In the giant grouper, an extremely small Ig-molecule has been found that reacts with anti-IgM antibodies. It was concluded that the molecule most probably represented a truncated form of IgM. Truncated forms of Ig have been found in ducks (Magor *et al.*, 1994b) and in the Australian lungfish. The Atlantic salmon has two types of tetrameric IgM encoded by two different loci that are expressed in equivalent amounts in serum. Recently, a peculiar form of IgD has been found in the channel catfish (Wilson *et al.*, personal communication). This molecule has a variable domain and is able to form a disulfide bridge with a light chain and seven constant domains, the first of which is $C_{\mu}1$ and the other six are similar to human and murine C_{δ} but the sequence similarities range from 9 to 29% for the different domains. Expression is achieved by an alternative RNA processing and not by class-switch, which is another similarity to mammalian IgD. Based on the difference in size and reactivity with monoclonal antibodies, two isotypes of Ig light chain have been described in rainbow trout and channel catfish. Recently, the existence of two different loci has been reported in rainbow trout.

Membrane bound Ig as the B-cell receptor lacks the CH4 domain in teleost fish and the transmembrane peptide is spliced to CH3. This arrangement was first demonstrated in goldfish (Warr and Marchalonis, 1977) but has subsequently been demonstrated in channel catfish, Atlantic cod, rainbow trout and salmon, so it is probably valid for all teleosts. The truncated B cell receptor appears to function normally as an antigen receptor and to mediate allelic exclusion (Ledford *et al.*, 1993; Miller *et al.*, 1994). The evolutionary transition from CH4 splicing to CH3 splicing appears to have occurred in the primitive actinopterygian bowfin and gar, which have both CH4 and CH3 splicing. The bowfin also has another cryptic splice donor site in the middle of the CH3 exon which is used for the production of membrane-mRNA.

Sarcopterygii

The Ig of the lobed finned fish has been little studied. The small Ig molecule of the Australian lungfish may be a truncated IgM (see above).

Antibody response

Chondrichthyes and Actinopterygii

After immunization, fish most often produce antibodies with specificity and measurable affinity for the immunizing antigen and these antibodies have biological properties such as agglutination, precipitation, complement fixation, opsonization, and skin sensitization. The Atlantic cod is unusual in that it does not produce any substantial amounts of antibodies upon immunization but is protected against a pathogenic *Vibrio* strain after vaccination with formalin-killed bacteria of the same strain. The antibody response of fish is generally affected by the environmental temperature so that warm temperatures produce higher titers than cold ones. Anamnestic responses characterized by rapid increases in serum titers, affinity, maturation and isotype switching are not found in fish but some reports exist claiming that a typical secondary response is present. Whether rays switch to IgR in an manner similar to mammalian IgG is unknown but no class switch in a strict sense can take place as the different Ig classes are encoded from different loci. Fish and other ectothermic vertebrates show an antibody response with a lower diversity of binding sites compared with the response of mammals. This lower diversity has been studied by means of isoelectric focusing on a population of antibodies specific for a given antigen such as bovine serum albumin, trinitrophenol and penicillin. These studies show that the antibody spectrotypes are very similar between individuals in a given species, so the restricted heterogeneity occurs within the given population or perhaps the whole species. One reason for the limited heterogeneity could be deficiencies in the genetic mechanisms for generation of diversity, but at least for teleost fish this does not appear to be so (see below).

There are high titers of anti-hapten antibodies, so called 'natural antibodies', in fish and many studies focus on these molecules. The most commonly studied hapten is trinitrophenol (TNP). The reason(s) for the high titers of the 'natural antibodies' has been thought by some to reflect similarities of epitopes with common bacteria in water, while others have speculated that 'natural antibodies' are the result of a response to environmental agents; no satisfactory explanation yet exists.

Immunoglobulin genes

The use of PCR with degenerate primers has led to the discovery of several new molecules of the immune system of fish including IgNAR, IgW and IgD, a second isotype of Ig light chain and the two chains of the T-cell receptor. However, cross-hybridization with mammalian probes and screening with expression vectors and antibodies resulted in the description of the genes of the two Ig chains before the PCR revolution.

Chondrichthyes

The Ig heavy chain locus of sharks and rays is organized in a manner that has become known as the 'multicluster type' where clusters of one V_H, two D, one J_H and one C_H segments are repeated as entities in the genome approximately 100 times. Some 50% of the clusters show joining of the V_H–D–J_H segments in the germ-line and appear to be functional, which limits the possibilities for antibody diversity. There are few reports on the number of different D and J_H segments but as there are two D segments per cluster the total number exceeds that of the V_H segments. The rearrangement takes place within a cluster but not between clusters, which also limits the diversification of antibodies. However, in the rearrangement either or both of the two D segments can be involved, which will increase the diversity; but as a D–J_H or D–D–J_H segment will always use the same V_H, the total repertoire of the heavy chain will be limited. Even so, the mechanisms for junctional imprecision and somatic mutations appear to be present and should contribute to the diversification of the antibodies.

The loci for the Ig light chain isotypes in sharks have a similar type of multicluster organization with V_L–J_L–C_L segments repeated in the genome. In the sandbar shark there are about 200 V_L segments and a similar number has been reported for the horned shark. There is no information to date on the germ-line repertoire of the new isotypes found in sharks but these additional loci will, of course, add to the final repertoire of antibodies.

The locus for IgNAR is small with a few genes and the V and C segments appear to be linked, suggesting a cluster organization. V segments have been classified into two groups based on criteria other than family. There are three D segments between the V and J segments and CDR3 has 11–23 amino acids. The IgNARC locus is also small but the V and C genes are not closely linked. There is no information on the IgW locus as yet and the 12 described cDNA clones of the V segment all belong to the same family with a CDR3 of 12–13 amino acids (range 8–21).

Actinopterygii

The Ig heavy chain locus of actinopterygian fish has an organization of the 'translocon' type, which is also found in amphibia and mammals. In this type, the V_H, D, and J_H segments have duplicated separately and form clusters separately in the germ-line configuration. Two teleost species have been studied thoroughly for their V_H repertoire – rainbow trout and channel catfish. These species, which represent two different branches on the phylogenetic tree, have more than 100 V_H genes each in their genome. V_H genes are generally grouped into families based on the similarities between their sequences, and at least 80% nucleotide identity must be shared between two V_H gene segments to belong to the same family. A high number of families indicates variability between the V segments. The rainbow trout has at least 11 different V_H families and channel catfish seven. In comparison, the mouse has 14 families (Tutter and Riblet, 1989; Tutter *et al.*, 1991) and human seven (Matsuda *et al.*, 1993; van Dijk *et al.*, 1993). V_H families have also been studied in other teleost species: the Atlantic cod has at least three families with 10–20 members in each family, the goldfish has also at least three families and the ladyfish two. The number of V_H segments of each family has usually been studied using the Southern blot technique, so it is impossible to judge how many are nonfunctional pseudogenes, as these have been shown to occur in some cases. With the translocon organization, the numbers of D_H and J_H segments will contribute significantly to the variability of CDR3. D-segments from fish have not been sequenced at the genomic level. Based on cDNA sequences, D-segments from rainbow trout have been sorted into 10 groups based on common motifs. Six families of J_H segments have been found in rainbow trout but they are not all used with the same frequency. The channel catfish has nine J_H segments in its genome all varying primarily at their 5′ end. The number of J_H segments in these two species corresponds to the number found in mammals (Höchtl *et al.*, 1982).

The rearrangement of V_H–D–J_H segments is not a precise process with nucleotides typically removed or added before ligation of the DNA. This imprecision adds to the variability of the CDR3. This imprecise joining (called junctional diversity) occurs in Atlantic cod and rainbow trout. The overall length of CDR3 in these species is shorter than that of mammals. The length in rainbow trout peaks at 8–9 (range 4–11) amino acids and in Atlantic cod at 10 (range 8–12) compared with 10–14 (range 2–26) in human. The length of CDR1 and 2 is constant in most species (Hsu and Steiner, 1992). Somatic mutation as a mechanism for affinity maturation has not been investigated in actinopterygian fish.

The organization of the Ig light chain loci of teleost fish is similar to the clustered type discussed above. Both channel catfish and rainbow trout have two isotypes of Ig light chain and, for at least one of the isotypes, the locus is (V_L–V_L–J_L–C_L) clusters. In Atlantic cod, only one isotype has been demonstrated with the locus organized as (V_L–J_L–C_L) clusters. A closer analysis reveals that the V_L segments have an opposite transcriptional orientation to the J_L and C_L segments but the effect of this organization on the generation of diversity is unknown. Not all actinopterygian fish have this type of organization of the Ig light chain loci. The Siberian sturgeon, which is a primitive actinopterygian species, has a translocon organization of its Ig light chain locus, as occurs with the κ locus in mouse and man. Thus, the clustered organization of the Ig light chain loci in sharks and teleosts does not represent the same evolutionary event.

In teleosts there are only a few families of V_L segments described in each of the three species studied and Southern blot analysis reveals that there are approximately 20 V_L-segments of the given family in the genome of each species. Because the V_L segments are clustered with the C_L segments and Southern blot analysis using V_L and C_L probes shows similar patterns, this is likely to be the true number of V_L segments. In Atlantic cod, five J_L segments have been described but they vary only at the 5′ end which contributes to CDR3 with two amino acids and cannot be considered to belong to different families. The nonteleost Siberian sturgeon has seven J_L segments upstream of the C_L, as seen in a PCR analysis of a genomic clone, but some segments may be pseudogenes. The few V_L described in this species all belong to the same family.

Sarcopterygii

Little is known about the Ig genes in this group of fish. One investigation in the gombessa shows that the V_H and D segments are clustered adjacent to each other, indicating a fourth type of organization of the heavy chain gene. The localization of the J_H and C_H segments is unknown.

Regulation of immunoglobulin genes

Very little is known about the transcriptional regulation of the Ig genes in fish. In the teleost heavy chain locus, sequence comparisons have shown a conserved promoter structure with an octamer and TATA box. In elasmobranchs, the octamer is missing but there is a decamer/nonamer motif similar to the promoter region of the T-cell receptor in mammals. The promoter from goldfish is the only one that has been used in functional studies and it induces a tissue-specific expression in both mouse and channel catfish cells. In elasmobranchs and teleosts, sequence homology has been used to identify several potential regulatory motifs in the J_H–C_H intron. Surprisingly, the 1.9 kb J_H–C_H intron of channel catfish does not contain any enhancer activity but instead an enhancer has been localized in a 1.8 kb segment 3′ of the constant segment that includes the TM2 exon. When paired with the mouse or goldfish V_H promoter, this enhancer generated tissue specific expression in both mouse and channel catfish cell lines. However, it is not similar to the mouse J_H–$C_\mu 1$ enhancer since it is not organized in a small core region and it lacks several of the E-box motifs present in the mouse. The channel catfish enhancer contains dispersed elements rich in octamer and μE5-related motifs. A strong evolutionary conservation in the function of nuclear transcription factors is supported by the findings that (1) the κ light chain promoter and IgH enhancer from the mouse promote a B-cell specific expression in transgenic rainbow trout; (2) the heavy chain enhancer from channel catfish can induce transcription together with a mouse promoter in a mouse cell line; and (3) the intronic enhancer from the mouse has a strong activity in a catfish B-cell line. No information about the regulation of L-chain transcription in fish is available. Large amounts of sterile transcripts of nonspliced and nonrearranged Ig light chain genes have been observed in rainbow trout, Atlantic cod and channel catfish.

Conclusions

The immune system of fish has immunoglobulins with many similarities to the mammalian counterpart but there are also deviations that reflect partly the evolutionary time that has elapsed since branching from a common ancestor and partly the adaptations to the different aquatic environments. Our knowledge of the field is limited because there are so many different species and the biology of the species is so different.

8. Major Histocompatibility Complex (MHC) Antigens

The fishes are in a phylogenetic position to provide important new insights into the evolution of vertebrate immunity. Individual fish species may also be of benefit to medical science by serving as efficient model systems for the investigation of basic immuno-developmental and immunopathological mechanisms (infections, neoplasms). More specifically, detailed knowledge of the major histocompatibility complex (MHC) in fish may be commercially utilized in aquaculture, in vaccine development, in selective breeding, and/or through molecular typing, to survey the structure and dynamics of wild populations, and to monitor and ensure the integrity of MHC diversity in fish stocks subjected to intense selective breeding.

Functional studies

Over the last two decades, functional studies of mixed lymphocyte reactivity (MLR), *in vitro* assays with alloantisera, allorecognition by graft rejection, *in vitro* and *in vivo* antibody responsiveness and biochemical analyses in many fish species have revealed typical features attributable to MHC, thus strongly indicating the existence of functional MHC molecules in teleostei.

Cell populations attributable to T cells, B cells and macrophages have been identified by structural and functional studies (DeLuca *et al.*, 1983; Secombes *et al.*, 1983; Miller *et al.*, 1987; Davidson *et al.*, 1991). Furthermore, studies on *in vitro* responsiveness to thymus-dependent hapten–carrier complexes have demonstrated cell cooperation between T cells, B cells and macrophages (Clem *et al.*, 1985; Miller *et al.*, 1985). Evidence for immune restriction has also been provided through the demonstration of strong MLRs in the absence of a detectable *in vitro* antibody response (Vallejo *et al.*, 1990). Moreover, a recent *in vivo* antibody response study in gynogenetic carps has revealed response patterns with major gene effects (Wiegertjes *et al.*, 1994).

Allograft recognition and rejection, and their kinetics,

are a classical test system that together with serology were employed in both mice (mid 1930s) and man (late 1950s) to define the MHC. Also, in several teleost fish species such as the common carp (*Cyprinus carpio*) (Komen *et al.*, 1990), goldfish (*Carrassius auratus*) (Nakanishi, 1987a), poecilids (*Poecilia formosa*) and platyfish (*Xiphophorus couchianus*) (Kallman, 1970), a series of transplantation studies have been carried out demonstrating the effect of major and minor loci. Employment of naturally occurring or artificially generated gynogenetic fish has, in recent years, also provided sound evidence for the existence of a few (one or two) major histocompatibility loci (Matsuzaki and Shima, 1989; Komen *et al.*, 1990; see also the review by Stet and Egberts, 1991).

To define a putative MHC in a new species using serological tools requires certain prerequisites. The challenge provided by the potentially vast number of polymorphic gene products of both MHC and non MHC origin capable of eliciting alloantisera, is obvious. Strategies employing inbred or gynogenetic lines and family material together with analysis of cosegregation of several MHC-associated features are thus preferable. Such an approach was used by Du Pasquier *et al.* (1975, 1977) to define an MHC in *Xenopus laevis*. Kaastrup *et al.* (1989) employed a similar strategy in fish providing convincing data on the cosegregation of serologically determined specificities using graft rejection kinetics in gynogenetic families of the common carp.

As reviewed by Stet and Egberts (1991), MLR has been observed in a considerable number of teleost fish species, such as Atlantic salmon (*Salmo salar*), rainbow trout (*Oncorhynchus mykiss*), channel catfish (*Ictalurus punctatus*) and common carp – all of which most likely reflect the involvement of MHC class II. A one-way matrix of MLR analysis in one single carp family resulting from a cross between two individuals of different geographical origin showed stimulation indices consistent with one locus and three alleles (one shared between parents). However, there are a number of obstacles to the interpretation of MLR responses in the context of detailed immunogenetic dissection. These obstacles include the variability in height and kinetics of both allogeneic and autologous responses, overall repeatability of the test and the surprising results in salmonids (e.g. rainbow trout). Thus, Kaastrup *et al.* (1988) indicated in a two-way MLR matrix in full sibs from two rainbow trout families that each of the two progenies displayed more genotypes than predicted, the reason for which could be the pseudotetraploid status of salmonids.

Until the successful molecular cloning of both genomic and expressed MHC in fish, a quite extensive series of experiments had been carried out with cross-reactive xenoantibodies, alloantisera and two-dimensional electrophoresis aimed at isolating products with expected characteristics. Several of these studies indicated the presence of both classes of MHC, either directly or indirectly in many teleost species, e.g. via β_2-microglobulin, as demonstrated in goldfish and carp (Shalev *et al.*, 1981, 1984) and in cod (*Gadus morhua*) (Lögdberg and Björck, 1983), but with no firm evidence (Kaufman *et al.*, 1990a,b). MHC class I-like signals (40–44 kDa) have also been detected using two-dimensional gel electrophoresis, with the molecular weight range possibly corresponding to the different alleles of the K locus present in the cell lysate of the carp material (Stet and Egberts, 1991). Products in both rainbow trout and Atlantic salmon with size and pI values typical for both MHC classes, have also been identified using two-dimensional electrophoresis of PBL lysates, labeled during preincubation with ^{35}S (P. Jones, personal communication).

Molecular cloning

Hashimoto *et al.* (1990) achieved the first successful isolation of MHC molecules in fish through the advantages of the polymerase chain reaction (PCR) method. Primers based on conserved motifs in man, mouse and chicken MHC sequences flanking the invariant cysteines of the MHC class II β_2 and MHC class I α_3 domains, were used to identify the analogous molecules in the common carp. This approach, using PCR with primers based on conserved regions, has also resulted in the successful isolation of MHC molecules from other lower vertebrates including amphibians, reptiles, various teleostei and cartilaginous fishes.

Partial cDNA and genomic MHC fragments are now being identified in a rapidly growing number of different fish species, and full-length MHC class I α chain cDNA molecules are described only in Atlantic salmon (Grimholt *et al.*, 1993), the common carp (van Erp *et al.*, 1996a), zebrafish (*Brachydanio rerio*) (Takeuchi *et al.*, 1995), in cichlid fishes (Sato *et al.*, 1997) and in rainbow trout (Hansen *et al.*, 1997). The β_2-microglobulin molecules, required for a functional MHC class I protein, have been identified in zebrafish (Ono *et al.*, 1993a), carp and tilapia (*Oreochromis niloticus* L.) (Dixon *et al.*, 1993) and rainbow trout (*Oncorhynchus mykiss*) (Shum *et al.*, 1996), thus strengthening the hypothesis that the fish MHC class I α molecules identified are translated into functional proteins. Full-length MHC class II β cDNA chain molecules have been identified in Atlantic salmon (Hordvik *et al.*, 1993), rainbow trout (Glamann, 1995), carp (Ono *et al.*, 1993b; van Erp *et al.*, 1996b), zebrafish (Ono *et al.*, 1992), an African great lake cichlid (*Cyphotilapia frontosa*) (Ono *et al.*, 1993c) and striped bass (*Morone saxatilis*) (Walker and McConnell, 1994). Full-length MHC class II α chain cDNA molecules have been identified in Atlantic salmon (U. Grimholt, E. Arnet, Ø. Lie, unpublished data), rainbow trout (R. J. M. Stet, personal communication), the common carp (van Erp *et al.*, 1996b), zebrafish (Sultmann *et al.*, 1993), striped bass (Hardee *et al.*, 1995) and the nurse shark *(Ginglymostoma cirratum*) (Kasahara *et al.*, 1992). The identification of expressed MHC molecules in sharks (*Triakis scyllia*, Hashimoto *et al.*, 1992; *G. cirratum*, Kasahara *et al.*, 1992), one of the lowest vertebrate groups studied so far, suggests that the MHC genes are older than 400 million years.

The MHC expressed molecules have many sequence similarities to those of higher vertebrates, probably reflecting both functional and structural requirements (reviewed in Kaufman *et al.*, 1994). The intradomain disulfide bridges, glycosylation sites and salt bridges found in higher vertebrate MHC molecules are generally also present in the fish cDNA sequences. Once the actual protein structure of fish MHC molecules has been resolved, important individual residues will be easier to identify.

Expression

Investigation of MHC class II β chain expression in the carp revealed high expression in the thymus, intermediate expression in peripheral blood, spleen, head kidney and gut, and no detectable expression in muscle and erythrocytes (Rodrigues *et al.*, 1995). A similar study in Atlantic salmon showed a high MHC class II β expression in gills, head kidney and spleen and low or undetectable expression in brain and muscle (Koppang *et al.*, 1998).

Polymorphism and genetic organization

Several groups have reported extensive fish MHC variability (Ono *et al.*, 1992, 1993b,c; Klein *et al.*, 1993; Sultmann *et al.*, 1993; Grimholt *et al.*, 1994; Miller and Withler, 1996a). However, since the number of loci and their sequence similarities are uncertain, defining alleles is difficult and polymorphism, i.e. defining the frequency of alleles in a population, has therefore not been established.

Restriction fragment length polymorphism (RFLP) studies have been performed using probes from the conserved class I α_3 and class II β_2 domain regions to identify the number of genes. In the common carp (Stet *et al.*, 1993), a total of 9–12 MHC class I and 3–5 MHC class II strongly hybridizing fragments were identified. In rainbow trout (Juul-Madsen *et al.*, 1992), hybridization with a class II β_2 probe in a family material resulted in two strongly hybridizing fragments with an allelic segregation pattern suggesting one loci and two alleles. In Atlantic salmon, preliminary pulse-field data indicate a minimum of five MHC class I loci clustered in a genomic region of approximately 1 Mb (M. Lundin, personal communication).

The chromosomal location of the MHC class I and class II genes is so far only established in zebrafish where the two classes seem to be located on different chromosomes (Postlewait *et al.*, 1994, and personal communication; Bingulac-Popovic *et al.*, 1997). In Atlantic salmon, a linkage analysis using satellite markers closely linked to the *Mhc-Sasa* class I and class II α genes shows no linkage, indicating that the two classes also reside on different chromosomes (A. Slettan, U. Grimholt, Ø. Lie, unpublished). Thus, the MHC class I and class II regions may well be located on two different chromosomes in all fish species.

Conclusions

Considering the sequence similarities, tissue distributions and variability, the MHC molecules in fish likely have a function similar to that of MHC molecules in higher vertebrates (for a more extensive review see Lie and Grimholt, 1996). The potential for application is especially dependent upon two major features of MHC, for which there are still very little or no data available in fish: the extent of polymorphism and the linkage or linkage disequilibria with disease. Thus, further studies on these crucial topics are needed before the practical implications of understanding fish MHC can be realized.

9. Red Blood Cell Antigens

The primary purpose of any review of red blood cell antigens of fishes must be to emphasize the paucity of information for any fish species and to reinforce caveats about generalizations for fishes, as a whole, due to the extraordinary diversity of genomic structure and genic differentiation in what is by far the largest class of vertebrates (Nelson, 1994). It will be apparent from this brief review that opportunities for significant research abound in even the most fundamental studies of comparative biochemistry, physiology, immunology, and genetics.

Erythrocytes have often been used as a plentiful and easily obtainable source to identify, characterize, and isolate molecules on a cell surface. Identification of fish erythrocyte antigens, however, has progressed quite slowly except for physiological investigations of membrane transport molecules. Phylogenetic variability in the structure and morphology of fish erythrocytes is extensive, as recently reviewed by Glomski *et al.* (1992). A pronounced difficulty in studies of teleost fishes is the molecular characterization of proteins coded by genes that are often part of gene families with unknown numbers of members. In many gene families, fishes have retained expression of genes presumably derived from chromosome duplications in vertebrate ancestors leading, in many instances, to more than double the number of expressed genes found in mammals (Morizot, 1994; Walter and Morizot, 1996). While this trend has been particularly well studied for isozyme genes (Fisher *et al.*, 1980), other examples include cellular oncogenes (Wittbrodt *et al.*, 1989; Harless *et al.*,

1995) and neurotrophins (Goetz *et al.*, 1994). Morizot *et al.* (1990) summarized gene expression in erythrocytes of isozymes in two teleosts, largemouth bass (*Micropterus salmoides*, order Perciformes) and platyfish (*Xiphophorus maculatus*, order Cyprinodontiformes). Even though relatively few enzymes were studied in small red cell samples from the latter species, striking differences in isozyme patterns were evident for about one-third of the enzymes compared. No simple pattern explained differences between taxa: in some cases, neural-specific isozymes were expressed in red cells of one species but not another, and in other instances multiple isozymes differentially expressed in many tissues all were expressed in erythrocytes. Because almost no comparative studies of fishes have been reported for erythrocyte surface proteins, it cannot be predicted whether such differences among taxa will be encountered, but differences may well be expected based upon intracellular protein profiles.

For the remainder of this review fish erythrocyte antigens will be defined as molecules that bind to B-cell receptors (cell membrane-associated or soluble forms). Molecules binding the T-cell receptor potentially would include all cellular proteins, including isozymes and hemoglobins, which have been extensively studied in fishes but cannot be covered in a brief summary. As will be seen, molecular characterization and sequence determination of most erythrocyte antigens studied in fishes lag far behind that accrued in other vertebrates.

Oligosaccharide erythrocyte antigens, which generally elicit a T-cell-independent IgM antibody response from B cells, have been well characterized in humans (Watkins, 1987; Yamamoto *et al.*, 1990) but not in fishes. The major human alloantigen system, the ABO histo-blood group, consists of three carbohydrate antigens (ABH), for which the molecular genetic basis has recently been determined (Yamamoto *et al.*, 1990). Glycosyltransferases convert the H antigen (O phenotype) into A or B antigens with the addition of terminal sugars to the carbohydrate of the glycoprotein. These antigens, therefore, are found covalently attached to cell surface proteins such as anion transport proteins. While fish erythrocyte surface membrane proteins are almost certainly glycosylated, as in most eukaryotic cell surface proteins, equivalent carbohydrate antigen systems have not yet been identified. The lack of a driving force such as evident transfusion reactions in relation to human health may be the primary reason for the limited understanding of fish blood group antigens (Sakai *et al.*, 1987).

Another set of cell surface antigens that play a prominent role in transplant reactions is the major histocompatibility complex (MHC) class I and class II antigens. Fish MHC genes are reviewed by Dixon *et al.* (1995) and in this chapter. The presence of MHC class I proteins on fish erythrocytes has not been demonstrated; these molecules are usually found on nucleated cells and thus are absent on mammalian erythrocytes, but not necessarily on fish red blood cells. Expression of MHC class II antigens is generally restricted to certain cell populations activated by cytokines and/or to professional antigen-presenting cells, and it thus also appears unlikely that these molecules will be found on fish erythrocyte cell surfaces.

Variability in the classical erythrocyte ABO system, the MHC, or virtually any other cell surface antigen system is typically characterized by antibody or T-cell reactions after blood transfusions. In a study by Sakai *et al.* (1987), erythrocytes were transfused between different fish species or among outbred members of the same species. Destruction or loss of foreign erythrocytes was assayed periodically using size differences as species-specific markers (or as allogeneic markers in diploid vs. triploid rainbow trout). Hemolytic and agglutination assays were also performed. All of these reactions were presumably due to the detection of nonself erythrocyte surface molecules. A striking result was that rainbow trout (*Oncorhynchus mykiss*) would not reject conspecific transfusions for more than 4 weeks. In one set of experiments, the speed of destruction of foreign erythrocytes increased with increasing evolutionary distances between donor rainbow trout and recipient species. The role of antibodies in erythrocyte lysis was not clear, as hemolysis did not always correlate with the presence of agglutination. In fact, hemolysis by serum (without previous exposure to foreign erythrocytes) occurred following transfusions between more distantly related taxa such as Salmoniformes to Cypriniformes. Although detailed biochemical or serological characterizations of erythrocyte surface molecules were not attempted by Sakai *et al.* (1987), the recent isolation of the immune response genes from several species of fishes has now made possible a more detailed analysis of tissue rejection.

10. Ontogeny of the Immune System

Differentiation of lymphoid organs

The larvae of different fish species hatch at different stages of development of their various organ systems.

In different species, lymphocytes differentiate within the lymphoid organs at different times relative to hatching. For example, in Atlantic salmon, the thymus and kidney are fully lymphoid at the time of hatching, while lymphocytes do not appear in lymphoid organs of sea bream until 1.5 months after hatching. Nevertheless, the order in which lymphocytes first appear in the lymphoid organs is the same for all species studied (Table II.10.1). Lymphocytes first appear in the thymus followed by the blood and head kidney and finally, after some considerable delay, in the spleen.

The origin of the thymic lymphocytes is not known, but most workers have identified hemopoietic blast cells in the kidney prior to differentiation of lymphocytes in the

Table II.10.1 Appearance of lymphocytes and the first appearance of expression of thymocyte (T)-specific or immunoglobulin (Ig)-specific markers in the developing lymphoid organs in fish. Days prior (−) or post (+) hatching

	Thymus	*Blood*	*Head kidney*	*Spleen*	*Size of fish when lymphocytes first appear in thymus (mm)*	*References*
Carp	+5 Ig^-T^+ +5 Ig^+T^- absent	+7–8	+7–8 Ig^-T^+ +10 Ig^+T^- +14	+8–9	4	Botham and Manning (1981) Secombes *et al.* (1983)
Plaice	+28	NK	+49	NK	8	Lele (1933)
Rainbow trout	+1–3 Ig^+ scarce +30	+5	+4 Ig^+ +4	+14–25 Ig^+ +30	17	Grace and Manning (1980) Razquin *et al.* (1990)
Rockfish[a]	+21	NK	+30	+44	NK	Nakanishi (1991)
Salmon	−22	−14	−14	+42	NK	Ellis (1977a)
Sea bass	+27 T^+ +30	NK	+10 T^+ +35	+44 T^+ +59	NK	Abelli *et al.* (1996)
Sea bream	+47	+54	+54	>+77	6	Josefsson and Tatner (1993)

[a] Days post-birth.
NK, Not known.

thymus and it is believed the thymic rudiment is populated initially by stem cells from the head kidney.

Following proliferation of lymphocytes in the thymus, many lymphocytes can be observed in the connective tissue separating the thymus from the head kidney, leading to the suggestion that lymphocytes migrate from the thymus to populate the head kidney.

The thymus in fish first develops as separate buds situated dorsal to several gill arches. These later coalesce into a single organ in each branchial cavity. The thymic buds develop as collections of lymphocytes between the pharyngeal epithelial cells and the epidermal basement membrane (Figure II.10.1). In young fish, the fully differentiated thymus is separated from the external environment by only a single layer of simple epithelial cells. In older fish, the epithelium thickens and the organ progressively becomes encapsulated in connective tissue.

During the first few months there is intense mitotic activity of lymphocytes in the thymus and then a decrease in activity. It is believed that many thymocytes migrate to peripheral organs during the first 2–3 months post-hatching and signs of involution of the thymus appear with the onset of sexual maturation.

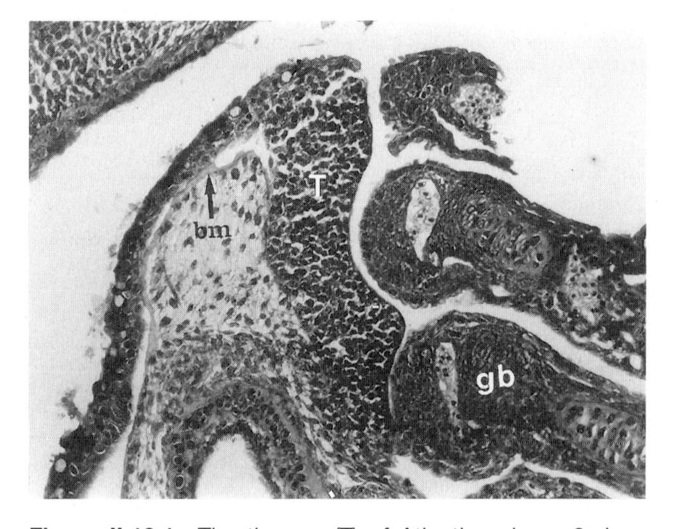

Figure II.10.1 The thymus (T) of Atlantic salmon 8 days prehatch. The organ develops within the pharyngeal epithelium, external to the basement membrane (bm) and is covered externally by a single layer of squamous epithelium. gb, primary gill bars.

Differentiation of the phagocytic system

The development of this system has been studied in rainbow trout that have received intraperitoneal injection of carbon particles (Tatner and Manning, 1985). In rainbow trout fry at 4 days post-hatching, phagocytic cells are mainly present in the integument, including the skin and gut and particularly in the gills. As the fish ages (18 days to 8 months), the kidney and spleen become the major sites of phagocytic cells.

Development of T and B lymphocytes

At present, no markers specific for mature T lymphocytes have been identified for fish. However, monoclonal antibodies (mAbs) have been produced that are considered to be specific for early T-cell markers and these provide evidence for initial differentiation of T lymphocytes within the thymus followed by their migration to the kidney and spleen (Table II.10.1).

Monoclonal antibodies specific for B lymphocytes are available for fish. These mAbs react with immunoglobulin (Ig) determinants expressed on the surface of B cells. Studies using these mAbs to determine the appearance of surface immunoglobulin (sIg) positive cells in the lymphoid tissues clearly demonstrate the importance of the kidney in their development (Table II.10.1).

The proportion of surface Ig-positive (sIg^+) B cells and plasma cells within the lymphoid organs in the developing carp has been studied using flow cytometry and immunohistochemistry of cytospin preparations, respectively (Koumans-van Diepen *et al.*, 1994a) (Table II.10.2). B cells were barely detected in fry 2 weeks post-hatching but their presence increased after 1 month. Plasma cells were first detected 1 month after hatching. The percentage of B cells in all organs continued to increase but did not reach adult levels until after 8 months old.

Ontogeny of B lymphocyte subpopulations

Using two mAbs (WC 14 and WC 12), which react with the heavy chain of serum Ig, three subpopulations of B cells have been identified in carp (Koumans-van Diepen *et al.*, 1995). The proportion of these cells in the head kidney and spleen changed during ontogeny but remained constant in the blood (Figure II.10.2). The functional significance of this is not known.

A polyclonal antiserum to Atlantic salmon serum Ig, which stained all blood lymphocytes of adult salmon, was used to study the differentiation of sIg^+ cells in salmon fry (Ellis, 1977a). Although small lymphocytes were present in the thymus and kidney prior to hatching, sIg positive lymphocytes were not detected until day 35 post-hatching. By day 48 post-hatching, the majority of lymphocytes stained with this antiserum. It is thus apparent, that while lymphocytes are present in the lymphoid organs of fish from an early age they do not initially express the surface

Table II.10.2 Percentages of lymphoid cells in blood and lymphoid organs of developing carp which are surface Ig^+ (B cells) or cytoplasmic Ig^+ (plasma cells). Data from Koumans-van Diepen *et al.* (1994a)

	Head kidney		*Body kidney*				
Age	*sIg*$^+$	*Plasma cells*	*sIg*$^+$	*Plasma cells*	*Spleen sIg*$^+$	*Thymus sIg*$^+$	*Blood sIg*$^+$
2 weeks	0.6	0.0	0.9	0.0	0.8	ND	0.0
1 month	7	0.2	4	0.3	6	0.5	ND
3 months	10	1.5	7	1	6	1.5	16
8 months	15	1	10	1	14	1.8	34
16 months	22	ND	17	ND	22	4	48

ND, not done.

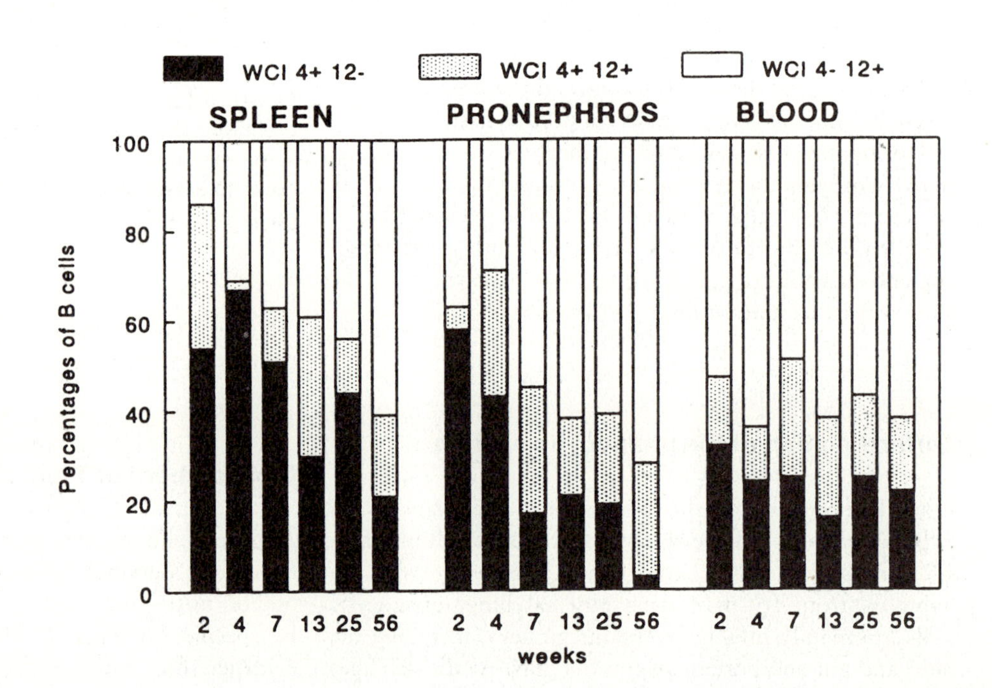

Figure II.10.2 B lymphocyte subpopulations ($WC14^+$ 12^-, $WC14^+$ 12^+ and $WC14^-$ 12^+) of carp in spleen, pronephros, and blood during ontogeny (age in weeks) expressed as percentages of the total B lymphocyte population; n = 6–8. Reprinted with permission from Koumans-van Diepen *et al.*, *Developmental and Comparative Immunology*, 1995, **19**(1), pp. 97–108, with kind permission of Elsevier Science Ltd.

markers characteristic of mature lymphocytes and a further period of time is necessary for functional differentiation.

Ontogeny of immune responsiveness

Antibody responses

The specific antibody titers induced by immunization increase with the age of the fish. The serum antibody titers in carp intramuscularly immunized with *Vibrio anguillarum* at 85, 99 and 128 days old significantly increased with age (Joosten *et al.*, 1995). These results are in agreement with the finding of Koumans-van Diepen *et al.* (1994a) who suggested that the development of the immune system of carp was completed between 3 and 8 months of age, based on the percentages of B cells and plasma cells found in the lymphoid tissues (see Table II.10.2). In addition, van Loon *et al.* (1981) stated that the adult level of serum Ig was attained when carp were 5–8 months old.

Ontogeny of immune protection

The earliest time to vaccinate fish is an important issue in aquaculture. Comparative studies with salmonid species indicate that immersion vaccination with *V. anguillarum* vaccines is ineffective in fish under 1 g (Johnson *et al.*, 1982a). As fish grow more slowly at lower temperatures, development correlates better with size rather than age. Accordingly, the onset of vaccine-induced protection correlates with the weight of the fish rather than time after hatch.

Furthermore, duration of protection increased with age. The maximum duration of protection was achieved in rainbow trout when they were immunized at about 4 g (Johnson *et al.*, 1982b) (Figure II.10.3).

Ontogeny of memory versus tolerance

The age of onset of specific antibody production appears to depend upon the nature of the antigen, probably pertaining to whether it is T-independent or T-dependent. As shown in Table II.10.3, early exposure (about 3–4 weeks post-hatching) of rainbow trout and carp to bacterial (*Aeromonas salmonicida*) antigens induced an antibody response and the development of memory (Manning *et al.*, 1982; Manning and Mughal, 1985). However, immunization of trout and carp up to 4 weeks post-hatching with human γ globulin (HGG) or sheep red blood cells (Srbc) resulted in the induction of tolerance that persisted for up to 23 weeks (van Loon *et al.*, 1981; Manning *et al.*, 1982; Manning and Mughal, 1985; van Muiswinkel *et al.*, 1985). A positive response and development of memory was not detected to these antigens until the fish were 8 weeks old.

A similar effect on suppression and priming of the antibody response has been reported in juvenile carp following oral immunization with *Vibrio* antigens. Oral priming of carp at 2 and 4 weeks post-hatching suppressed antibody production in response to injection of the antigen 10 weeks later compared with nonorally primed controls. However, oral priming at 8 weeks old resulted in enhanced responses to injected antigen 10 weeks later compared with controls (Joosten *et al.*, 1995).

These data suggest that B cells and T suppressor cells become functionally mature at about 4 weeks but T helper functions and memory cells mature later at about 8 weeks.

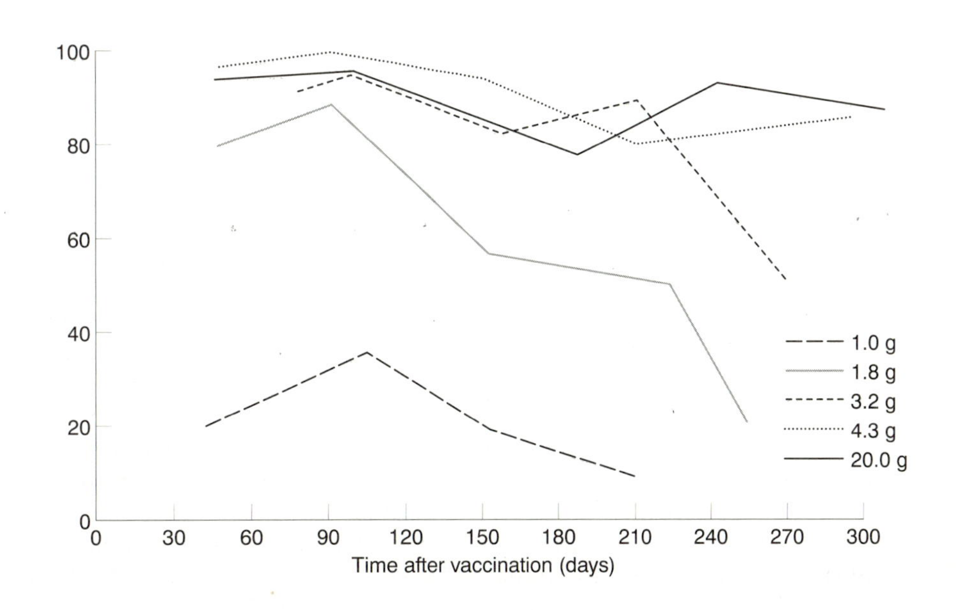

Figure II.10.3 Duration of protective immunity of rainbow trout at various sizes following a 20-s immersion in *Yersinia ruckeri* bacterin diluted 10^1. The 20-g fish were vaccinated by the shower method. Reprinted with permission from Johnson *et al.* (1982b).

Table II.10.3 Ontogeny of immune responsiveness in fish

Age (weeks) when first immunized, grafted or assayed	*Rainbow trout* Antibody responses to A.s.	*HGG*	*Graft rejection*	*Carp* Antibody response to A.s.	*HGG*	*SRBC*	Vibrio *(following oral priming)*[d]	*Graft rejection*	*Salmon* *MLR*[f]
1	–[e]	–(T)[e]	Incomplete[g]						–
2	–[e]	–(T)[e]	+[g]				S	+(M)[h,i]	
3	+[e]	–(T)[e]	+(M)[g]						–
4	+(M)[a]			+(M)[a]	–T[a]	–T[b]	S		–
6									+
8		+[e]			+(M)[a]	+[c]	P		

–, No response; +, positive response; M, memory induced; T, tolerance induced; S,P, suppressed or enhanced response to subsequent injection respectively; MLR, mixed leukocyte reaction; HGG, human gamma globulin; SRBC, sheep erythrocytes; *A.s.*, *Aeromonas salmonicida*.
Data from [a]Manning and Mughal (1985); [b]van Loon *et al.* (1981); [c]van Muiswinkel *et al.* (1985); [d]Joosten *et al.* (1995); [e]Manning *et al.* (1982); [f]Ellis (1977a); [g]Tatner and Manning (1983); [h]Botham and Manning (1981); [i]Botham *et al.* (1980).

Ontogeny of cell-mediated immunity (Table II.10.3)

The development of cell-mediated immunity (CMI) in young fish has been investigated using two tests: the mixed leukocyte response (MLR) and the allograft rejection response. In the Atlantic salmon, the MLR develops at day 42 post-hatching, coincident with first feeding (Ellis, 1977a). Skin rejection in carp and rainbow trout fry is developed by 16 days (Botham and Manning, 1981) and 14 days post-hatching (Tatner and Manning, 1983), respectively. In the trout, grafting at 5 days post-hatching resulted in incomplete rejection by day 30 post-grafting but the experiment was not continued to establish whether allograft application at this age eventually results in rejection or tolerance.

The ontogeny of cell-mediated immune memory has been investigated following the application of first set skin grafts to 26-day-old rainbow trout and 16-day-old carp (Botham *et al.*, 1980): in both cases, memory developed. It is thus apparent that in trout and carp, the CMI system has fully matured by 2–4 weeks post-hatching, suggesting that this system, with the production of cytotoxic T-like cells, matures a little earlier than the humoral immune response, particularly to the T-dependent antigen, HGG.

11. Passive Transfer of Immunity

Fish are not fully immunocompetent when they first hatch or are born, but may well have received some antigenic exposure before this time. Whether there is any immune protection afforded by the hen fish during this vulnerable period has been studied in only a few species. Nothing is known as to its mechanism, and very little as to its duration in relation to the onset of immunocompetence.

Nonspecific humoral factors, such as C-reactive protein and lectin-like agglutinins have been detected in the eggs of carp, *Cyprinus carpio* (van Loon *et al.*, 1981), plaice, *Pleuronectes platessa* (Bly *et al.*, 1986), tilapia, *Oreochromis aureus* (Mor and Avtalion, 1988), channel catfish, *Ictalurus punctatus* (Hayman and Lobb, 1993), chum salmon, *Oncorhynchus keta* (Fuda *et al.*, 1992) and coho salmon, *Oncorhynchus nerka* (Brown *et al.*, 1994).

Castillo *et al.* (1993) quantified the amount of IgM in unfertilized eggs of rainbow trout, *Oncorhynchus mykiss*, as 11.2–2.6 mg/g egg weight, increasing to a peak at the time of hatching, and declining slowly to initial values by 2 months post-hatching. Takemura (1993) studied changes in a maternally derived IgM-like protein in tilapia. The levels declined during prelarval stages when the yolk sac was being absorbed, and reached the lowest value at the time of first feeding (12 days after hatching, when all the yolk sac had been consumed).

Passive transfer of specific immunity would obviously be beneficial in the case of pathogens that are transmitted vertically and to which fry are particularly susceptible, e.g. infectious pancreatic necrosis virus. Mor and Avtalion (1990) demonstrated antibody activity specific to three protein antigens, used to immunize hen fish, in extracts of their embryos from 0–9 days old. The level of antibody activity in the embryos varied according to the time interval between the immunization of the hen fish and the spawning. It peaked between 14 and 37 days post-immunization and then rapidly declined. Mor and Avtalion (1988) have suggested that the immunity may be passed on via the yolk sac, as in birds.

Specific protective immunity against a pathogen was transferred in the ovoviviparous guppy, *Lebistes reticulatus*, from mother to fry (Takahashi and Kawahara, 1987).

Sin *et al.* (1994) have demonstrated passive transfer of protective immunity against ichthyopthiriasis in the mouth-brooding tilapia. Hen fish were vaccinated twice 1 month before spawning with live tomites of *Ichthyopthiriasis multifillis*. The fertilized eggs were collected from the mouths of vaccinated fish, and after brooding, had a survival after challenge of 95%, compared with unvaccinated mouth-brooded fry of 37%. The protective immunity was correlated with the anti-*Ichthyopthiriasis* titers in extracts of fry tissue and in the mother's plasma, and came from two sources: in the eggs via the mother and from the mother's mouth cavity.

Passive transfer of immunity against furunculosis, caused by *Aeromonas salmonicida*, has been shown in the white spotted char, *Salvelinus leucomaenis* (Kawahara *et al.*, 1993). Hen fish were immunized on three occasions with the detoxified extracellular products of the pathogen by intramuscular injection. Three weeks after the final injection, the eggs were collected and fertilized, and 28 days after hatching, the fry were challenged. Vaccinated fry had a cumulative mortality of 4%, compared with 28% in the unvaccinated controls.

The failure of passive immunity transfer may contribute to the very high mortality rates suffered by newly hatched fish fry in aquaculture systems; however, hen fish are not routinely immunized prior to spawning owing to a lack of knowledge on this subject.

12. Nonspecific Immunity

Nonspecific defense reactions in fish represent both humoral and nonlymphoid, cellular components of the immune system. Several nonimmunological humoral substances and cell secretions are considered to contribute to the natural resistance of fish to pathogenic and infectious agents. These include transferrin, toxins, lectins, (agglutinins of a nonimmunoglobulin nature), C-reactive protein, various lytic enzymes (e.g. lysozyme), interferon, nonenzymatic lysins, enzyme inhibitors and complement. This arsenal of innate mechanisms to combat disease is well documented in fish (see reviews by Ingram, 1980; Fletcher, 1981, 1982; Ellis, 1981b; Rijkers, 1982a; Alexander, 1985; Alexander and Ingram, 1992; Sakai, 1992; Manning, 1994; van Muiswinkel, 1995). In many instances, tentative defense roles have been ascribed to naturally occurring molecules identified in fish, although a limited number of functions has been determined (Table II.12.1).

Initial barriers to pathogen invasion in fish are epithelial surfaces (e.g. skin and gut) onto which is constantly secreted a layer of mucus from goblet cells. The close contact between the easily renewable mucus coat and the epidermal surface provides a primary physical and indeed chemical barrier against invasion and attachment of fungi, bacteria and various parasites. Many defense molecules, including immunoglobulins, are present in mucus (Tachibana, 1988; Alexander and Ingram, 1992; Shephard, 1994). Table II.12.1 indicates some of the known defense properties of mucus.

Complement (C) activation can occur by the classical pathway, initiated by specific immunoglobulin–antigen immune complexes, or via the alternative, antibody-independent route. Both pathways once activated cause opsonization and/or lysis of cells including bacteria (Sakai, 1992). Normal, nonimmune fish serum possesses spontaneous lytic activity (possibly alternative pathway mediated) that occurs independently of the complement system (Sakai, 1992). Examples include natural lysins in sea lamprey, *Petromyzon marinus* (Gewurz *et al.*, 1966), rainbow trout, *Oncorhynchus mykiss* (Chiller *et al.*, 1969), and other fish species (Sakai *et al.*, 1987).

Fish possess leukocytes, including granulocytes (eosinophils, neutrophils/heterophils, basophils) and monocytes/macrophages, that are involved in nonspecific cellular immune reactions (Ainsworth, 1992; Hine, 1992; Secombes and Fletcher, 1992). In most fish species, the structure and sites of differentiation of the various leukocyte types remain largely uninvestigated. Granulocyte classification has led to much confusion in the fish literature because the various types have been categorized on the basis of criteria applied upon comparison with mammalian leukocyte equivalents. Few functional investigations have been performed to aid granulocyte identification. Variations in fish granulocyte histological staining, ultrastructure, morphology and nomenclature among others have resulted in classification problems. Fish neutrophils appear to be the only determined functional equivalent to those of mammals. Table II.12.2 summarizes the types and function(s) of granulocytes in fish. Although morphological heterogeneity of granulocytes is well established, the roles of the various cell populations is less clear and may vary from one fish species to another. By comparison, the presence of circulating monocytes and fixed/tissue macrophages (mononuclear phagocytes) is well demonstrated in fish; these cells are structurally and functionally similar between species (Ellis, 1977b; Rowley *et al.*, 1988; Wolke, 1992). The macrophages appear to be derived from monocytes or pluripotential tissue cells and occur in both lymphomyeloid tissue and nonlymphomyeloid tissues (e.g. blood). Fish have pigments (melanin, lipofuscin, hemosiderin and ceroid) associated with macrophages and these cells, termed melanomacrophages, tend to occur in aggregates in the spleen, liver, kidney and gonads.

Phagocytosis, as the initial line of cellular defense, is universally found in all fish studied as a component of nonspecific immunity. This process, the stages of which are similar to those present in mammals, is performed largely by macrophages and neutrophils and is enhanced by opsonins and complement components. Little is known about how fish granulocytes function in nonspecific defense mechanisms. The characteristic inflammatory

Table II.12.1 Occurrence, types, source and known role(s) of substances involved in nonspecific defense mechanisms in fish

Compound	*Species*	*Source*	*Function*	*Reference*
Lysozyme	*Salmo gairdneri* (Rainbow trout)	Kidney	Anti-gram negative bacterial fish pathogens	Grinde (1989)
	Scophthalmus maximus (Turbot)	Lymphocytes, thrombocytes and plasma	Lysozyme and chitinase: anti-fungal (*Mucor mucedo*)	Manson *et al.* (1992)
	Pleuronectes platessa (Plaice)	Serum, mucus, plasma, lymph, kidney, spleen, stomach, neutrophils and skin epidermal cells	Not stated	Fletcher and White (1973a)
	Freshwater and marine teleosts, including *S. gairdneri*	As per *P. platessa* plus leukocytes, liver, gills, muscle and alimentary tract	Bacteriolytic	Lindsay (1986), Lie *et al.* (1989)
	Marine teleosts	Kidney and various tissues; erythrocytes	Bacteriolytic – intraspecies variation	Fänge *et al.* (1976), Grinde *et al.* (1988)
	Holocephalans and elasmobranchs	Epigonal organ; Leydig's organ; spleen; cranial LMT; granulocytes	Bacteriolytic	Fänge *et al.* (1976), Lundblad *et al.* (1979)
	Ictalurus punctatus (Channel catfish)	Serum	Anti-bacterial (*Salmonella paratyphi*)	Ourth (1980)
Natural antibodies	*Oncorhynchus mykiss*[a] (Rainbow trout)	Serum	Opsonic for *Aeromonas salmonicida* cells – increase kidney phagocytic cell activity	Michel *et al.* (1990)
			Neutralization activity of several viral fish pathogens	Gonzalez *et al.* (1989)
Lectin	*O. rhodurus* (Amago trout)	Ova	Opsonin role in inflammatory response for peritoneal exudate and kidney macrophages; phagocytes expressed membrane lectin receptor during nonspecific stimulation	Ozaki *et al.* (1983)
	O. mykiss	Eggs	Anti-fungal – agglutination of *Saprolegnia* or related fungi	Balakhnin and Dudka (1990), Balakhnin *et al.* (1990)
	Salmo salar (Atlantic salmon)	Serum	Mannan-binding protein MBP; opsonized bacteria and yeast phagocytosed by salmon phagocytes; strong yeast agglutination	Arason *et al.* (1994)
Lectin/agglutinin	*O. tshawytscha* (Chinook salmon)	Ova	Antibacterial – growth inhibition of 4 fish pathogenic bacteria	Voss *et al.* (1978a)
	O. keta (Chum salmon)	Ova	*Vibrio anguillarum* agglutination	Kamiya *et al.* (1990)
Agglutinin	*Conger myriaster* (Conger eel)	Skin mucus	Lysis of starfish *Asterina pectinifera* eggs; agglutination of *V. anguillarum* bacteria	Kamiya *et al.* (1988)
Natural precipitins and other factors	Salmon and trout species	Skin mucus	Genetic effects – furunculosis (*A. salmonicida*) resistant salmonid strains possess high mucus precipitating activity for *A. salmonicida* extracellular antigens and lack of pathogens in mucus	Cipriano and Heartwell (1986), Cipriano *et al.* (1994)

(continued)

Table II.12.1 *Continued*

Compound	*Species*	*Source*	*Function*	*Reference*
Bactericidal factors	Various fish species	Serum	Non-Ig, complement-dependent molecule; bactericidal towards various chemotype strains of enterobacteria (e.g. *Salmonella*)	Kawakami *et al.* (1984)
Protease (trypsin-like)	*O. mykiss*	Skin mucus	Vibriolytic action (*V. anguillarum*)	Hjelmeland *et al.* (1983)
Proteinaceous toxins	*Anguilla anguilla*, *A. japonica* and *Muraenesox cinerus* (Eels)	Skin mucus	Lethal to mice	Shiomi *et al.* (1994)
Toxic indole-derived vibrindole	*Ostracion cubicus* (Boxfish)	Skin mucus	Antibacterial but secreted as a metabolite into mucus by symbiotic *V. parahaemolyticus*	Bell and Carmeli (1994)
C-reactive protein (CRP)	*A. japonica* (Japanese eel), *O. mykiss*	Serum	Bacterial agglutination (*Streptococcus pneumoniae*)	Winkelhake and Chang (1982), Nunomura (1991)
	O. mykiss	Serum	Activates complement system; suppresses growth of *V. anguillarum* pathogen; CRP opsonic and enhanced phagocytosis of bacteria by peritoneal exudate cells	Nakanishi *et al.* (1991)
			Opsonic effect and macrophage activation for enhanced bacterial phagocytosis	Kodama *et al.* (1989)
			Lymphocyte surface CRP – nonspecific cytotoxicity for myeloma cells	Edagawa *et al.* (1993)
Complement (AP)	Salmonids	Serum	Bactericidal for *A. salmonicida*	Sakai (1983, 1992)
	O. nerka (Sockeye salmon)	Plasma (antibody-free)	Lysis of parasite *Crytobia salmositica*	Bower and Evelyn (1988)
	I. punctatus	Serum	Bactericidal for several bacterial species	Ourth and Wilson (1982a,b), Jenkins and Ourth (1990)
	O. mykiss	Serum	Inactivation of toxic bacterial extracellular products (antiprotease activity)	Sakai (1984)
Complement component (C5a-like activity)	*A. japonica*	Plasma, PB	Chemotactic for PB neutrophil migration	Suzuki (1986), Iida and Wakabayashi (1988)

AP, alternative pathway; LMT, lymphomyeloid tissue; PB, peripheral blood.
[a] *Oncorhynchus mykiss* (= *Salmo gairdneri*).

reactions against various infectious agents closely resemble those found in mammals and award some degree of protection by isolation (via encapsulation in some cases) of the infected area from the rest of the body (Suzuki and Iida, 1992). Acute inflammatory reactions in fish involve migration of neutrophils and/or eosinophils, or basophils in some species (Roberts, 1989), macrophages and lymphocytes (Suzuki and Iida, 1992). In chronic inflammation, granulomata, giant cells and epitheliod cells are present (Secombes, 1985; Ellis, 1986). The role of lymphocytes in fish inflammatory responses remains unknown. Leukocyte migration (by chemotaxis and/or chemokinesis) to inflammatory sites is enhanced by various host chemoattractant molecules that include

Table II.12.2 Occurrence, sites of granulopoiesis, types and functions of granulocytes in fish[a]

Group	*Common name (examples)*	*Granulopoietic sites*	*Granulocyte types*	*Function*
Agnathans	Hagfish	Pronephros, *lamina propria* of intestinal wall	NG, HG, T	Phagocytosis (N)
	Lamprey	Spleen; supraneural fat column. Prolarval blood islands (L); protospleen (L) and opisthonephric kidney (L)	NG, HG, EG, BG, T, N	Phagocytosis (LE, HG, N), Coagulation (T)
Holocephalans	Ghost sharks (Ratfish)	Orbital, subcranial (meningeal), shoulder cartilage, cardiac epithelium and intestinal spiral fold LMT	EG	Phagocytosis
Elasmobranchs	Sharks	Kidney LMT; epigonal organs; Leydig's organ; spleen	T, BG, B, EG, E, H, HG, NG, N	Phagocytosis (including T), Inflammation (NG), Coagulation (T), Migration (E)
	Dogfish	Spleen; kidney; epigonal and Leydig's organs; pancreas and intestinal submucosa	T, EG, HG, N, H, E	Phagocytosis (E), Coagulation (T)
	Rays	Spleen, meningeal LMT	T, E, H, HG, EG, A, NG	Phagocytosis (E)
Dipnoids	Lungfish	Kidney, liver, pancreas, gonads and intestine LMT	H, N, B, E	Phagocytosis (E)
Chondrosteans	Sturgeons and paddlefish	Kidney; CNS (spinal cord, brain associated tissue); heart (pericardium) associated LMT; spleen	EG, N, H, B, E	Phagocytosis (E)
Holosteans	Bowfins and gars	As per sturgeons	EG, H, N	Not determined
Teleosts	Modern bony fish (salmonids, cyprinids, anguillids etc.)	Head kidney, spleen	N, H, EG, BG, NG, HG, B, mast cells?, T	Phagocytosis (N), Inflammation (E and N), Anaphylactic-like response (EG/ mast cells?), Coagulation (T)

NG, neutrophilic granulocyte; HG, heterophilic granulocyte; EG, eosinophilic granulocyte; BG, basophilic granulocyte; L, larval; T, thrombocyte; N, neutrophil; B, basophil; E, eosinophil; H, heterophil; A, acidophil; LMT, lymphomyeloid tissue; CNS, central nervous system.
[a] Table compiled from Ainsworth (1992), Hine (1992), and Rowley *et al.* (1988).

eicosanoids (leukotrienes and lipoxins), cytokines, complement components (C5a-like) and other serum factors (Suzuki and Iida, 1992; Manning, 1994). Furthermore, fish neutrophils and macrophages exhibit respiratory burst activity, RBA (i.e. generation of reactive oxygen species including superoxide anion and H_2O_2), an indicator of bactericidal activity and oxygen-dependent toxic killing of pathogenic organisms. RBA is detected by measurement of the oxygen derivatives generated in fish leukocytes using chemiluminescence or reduction of nitroblue tetrazolium, ferricytochrome *c* or similar molecules. Fish macrophages and leukocyte cultures secrete cytokines upon activation by mitogen or antigen, but to date, the role of cytokine involvement in nonspecific immune reactions remains speculative (Secombes, 1994b).

Certain leukocytes of bony fish display cytotoxic/cytolytic activity towards a range of cells and parasites (Manning, 1994). These monocytes/macrophage-like nonspecific, nonphagocytic effector cytotoxic cells (or NCC) display a functional similarity to mammalian natural killer (NK) cells although they possess differences in cytoplasmic inclusions and temporal kinetics of cell killing (Evans and Jaso-Friedmann, 1992). Table II.12.3 gives examples of fish species possessing NCC and NK-like activities.

Selected examples of cells and mediators involved in nonspecific immune responses are given in Table II.12.3 and readers are referred to in-depth reviews dealing with granulocytes and macrophages in fish together with their potential functions in inflammation and other aspects of nonspecific cell-mediated immunity (Rowley *et al.*, 1988; Evans and Gratzek, 1989; Ainsworth, 1992; Hine, 1992; Secombes and Fletcher, 1992; Suzuki and Iida, 1992; Wolke, 1992).

Table II.12.3 Selected examples of fish leukocytes involved in nonspecific cellular immunity that exhibit functions comparable to mammalian types

Type	*Species*	*Source*	*Function(s)*	*Reference*
Neutrophil (N)	*Ictalurus punctatus* (Channel catfish)	Liver, K	Phagocytic recognition via surface galactosyl/mannosyl receptors of bacterium *Edwardsiella ictaluri*	Ainsworth (1993)
		HK	C3, mac-1-like protein membrane receptor involved in phagocytosis	Ainsworth (1994)
	Anguilla japonica (Japanese eel)	PB	Chemotaxis exhibited towards several bacterial species and enhancement of leukocytosis inducing activities	Iida and Wakabayashi (1988)
		PEC	Phagocytosis of *V. vulnificus* and other bacteria; RBA	Miyazaki and Kurata (1987), Moritomo *et al.* (1988)
	Pleuronectes platessa (Plaice)	K	Chemokinetic migration only for *Vibrio alginolyticus*	Nash *et al.* (1986)
	Salmo salar (Atlantic salmon)	PB	RBA, migration towards *A. salmonicida*, phagocytic and bactericidal	Lamas and Ellis (1994a,b)
Nonspecific cytotoxic cell (NCC)	*I. punctatus*	K, PB	Lytic killing of protozoans *Tetrahymena pyriformis* and *Ichthophthirius multifilliis*	Graves *et al.* (1985a,b)
		K, PB, spleen	Cytolysis of transformed mammalian cells	Evans *et al.* (1984b)
NK-like cytotoxic cells	*I. punctatus*	PB	Cytotoxicity towards uninfected allogeneic, autologous and channel catfish virus-infected allogeneic cells	Yoshida *et al.* (1995), Hogan *et al.* (1996)
Cytotoxic cells	*Pomacentrus partitus* (Damselfish)	K, spleen	Lysis of tumor cells	McKinney and Schmale (1994)
NK-like cells	*Cyprinus carpio* (Common carp)	K, PB	Cytotoxic against tumor cells	Suzumura *et al.* (1994)
Granular neutrophilic granulocyte (NG)	*C. carpio*	K	Spontaneous cytotoxic activity involving H_2O_2 causing lysis of tumor cells	Kurata *et al.* (1995)
Granulocyte (G) and lymphocyte-like cells	*Ginglymostoma cirratum* (Nurse shark)	PB	N/H regulate M-mediated spontaneous cytotoxicity	Haynes and McKinney (1991)
Eosinophilic granulocyte (EG)	*Scyllorhinus canicula* (Dogfish)	PB	Leukotriene B_4 enhances leukocyte migration	Hunt and Rowley (1986)
	P. platessa	Skin	Immediate cutaneous hypersensitivity reaction to fungal extracts of *Epidermophyton floccosum*; prostaglandins including PGE_2 involved in inflammatory response	Anderson *et al.* (1981), Jurd (1987)
Agranular mononuclear leukocyte (L) types	*Oncorhynchus mykiss* (Rainbow trout)	PB, HK, spleen	Nonspecific cytotoxic effector cells (NCC) causing apoptosis and necrosis of lymphoma target cells	Greenlee *et al.* (1991)
Leukocyte (L)	*O. mykiss* and *S. salar*	K	IPN-virus infected fish cells lysed spontaneously by nonspecific cytolytic activity	Moody *et al.* (1985)
	O. mykiss	K	Leukocyte response chemotactic for *O. mykiss* (host) chemoattractant but chemokinetic for parasite *Diphyllobothrium dendriticum* plerocercoid antigen extract	Sharp *et al.* (1991)
		K	MAF/INF-γ confer resistance to IPN virus-infected trout cells	Graham and Secombes (1990b)
	Several fresh water species	K	L (M?): cytotoxic for mammalian cell lines	Hinuma *et al.* (1980)
	G. cirratum	PB	Chemotactic and chemokinetic migration towards several chemoattractants	Obenauf and Smith (1985)

(continued)

Table II.12.3 *Continued*

Type	*Species*	*Source*	*Function(s)*	*Reference*
Macrophage (M)	*O. mykiss*	K	Larvicidal for digean *Diplostromum spathaceum* cercariae and diplostomules; RBA elicited	Whyte *et al.* (1989)
		K, PB	M and N: produce eicosanoids (lipoxins and leukotrienes); N lipoxin A4 and M lipoxins chemokinetic, and N leukotriene B4 chemotactic for leukocytes	Pettitt *et al.* (1989, 1991), Rowley *et al.* (1991), Sharp *et al.* (1992)
		HK	Phagocytosis of yeast cells reduced by inhibitor of eicosanoid biosynthesis	Rainger *et al.* (1992)
		K	MAF (=INF-γ?), RBA	Hardie *et al.* (1994a), Jang *et al.* (1995b,c)
	S. salar	K, PEC, spleen	Chemotactic for *Aeromonas salmonicida* cells but chemokinesis unaffected	Weeks-Perkins and Ellis (1995)
		K	MAF – activity; chemotactic migration; eicosanoid production; bactericidal (*A. salmonicida*) and RBA	Francis and Ellis (1994), Thompson *et al.* (1994)
		K	*V. salmonicida* phagocytosis; RBA and bactericidal (*A. salmonicida*)	Espelid and Jørgensen (1992), Jorgensen and Robertsen (1995)
	C. carpio	K	M and NG: IL-1-like factor produced	Verburg-van Kemenade *et al.* (1995)
	G. cirratum	PB	Cytotoxic activity	McKinney (1990)
	Leistomus xanthurus (Spot)	K	Chemotactic migration of bacterium *Legionella pneumophila*	Weeks *et al.* (1988)
	L. xanthurus and *Trinectes maculatus* (Hogchoker)	K	Chemotactic for *Escherichia coli* bacteria; phagocytosis occurred	Weeks and Warinner (1986)
	Tinca tinca (Tench)	Spleen	N and M: phagocytosis of yeast *Candida albicans*; RBA; chemotactic migration	Pedrera *et al.* (1992)
	Three marine teleost species	K	RBA; phagocytosis and bactericidal towards *Pasteurella piscicida*	Skarmeta *et al.* (1995)
	Limanda limanda (Dab)	K	RBA enhanced by recombinant TNF-α, β-glucan and lipopolysaccharide	Tahir and Secombes (1996)
	P. platessa	K	N and M: RBA, phagocytic and bactericidal towards *A. salmonicida*	Secombes *et al.* (1995)
Thrombocyte (T)	*Dicentrarchus labrax* (Sea bass)	PB	Phagocytosis of three bacterial species	Esteban *et al.* (1994)

N, neutrophil; NCC, nonspecific cytotoxic cell; NK, natural killer; M, macrophage; K, kidney; HK, head kidney; PB, peripheral blood; NG, neutrophilic granulocyte; G, granulocyte; L, leukocyte; PEC, peritoneal exudate cells; EG, eosinophilic granulocyte; T, thrombocyte; C3, third component of complement; PGE_2, prostaglandin E_2; MAF, macrophage activating factor; RBA, respiratory burst activity; IFN-γ, interferon gamma; IL-1, interleukin-1; TNF-α, tumor necrosis factor alpha

13. Complement System

The presence of the complement system has been demonstrated in all vertebrate classes, and in higher vertebrates it is possible to distinguish between a classical, antibody-dependent pathway, an alternative, nonantibody-dependent pathway, and the lytic pathway (Law and Reid, 1995). Recently, a lectin pathway has been described that has features in common with the classical pathway, but does not depend on antibodies and utilizes mannan-binding lectin (MBL) and its two serine proteases MASP 1 and 2, corresponding to the C_1-complex. The lectin pathway is initiated when MBL recognizes carbohydrate structures such as mannan on, for example, the surface of a bacteria (Holmskov *et al.*, 1994). The three initiation pathways are shown in Figure II.13.1.

Interest in fish complement has centered mainly on understanding the functional and structural evolution of the complement system, but the significance of nonspecific immune reactions especially in lower vertebrates has also led to the complement system in fish receiving increased attention. This is particularly relevant in species of commercial value as nonspecific defense mechanisms are

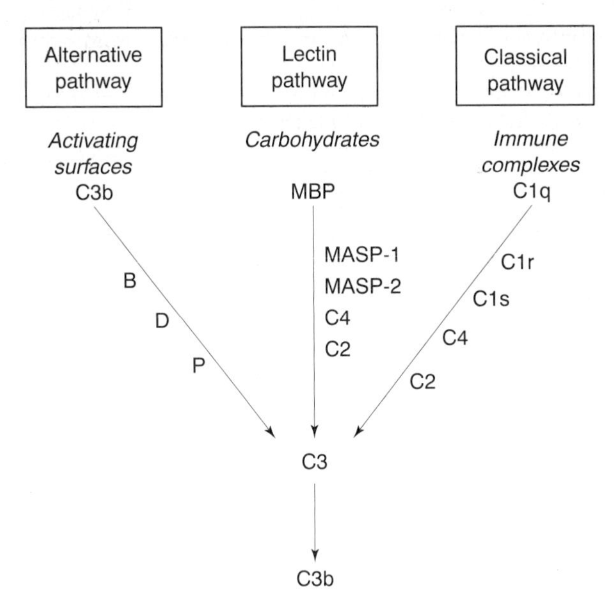

Figure II.13.1 The three activation pathways for the complement system. The complement proteins and factors associated with each pathway are shown adjacent to the arrows. Activation of C3 is followed by the common membrane activation complex, C5b–C9 (not shown on the figure). Abbreviations: B: factor B, D: factor D, P: properdin, MBP: mannan binding lectin, MASP: MBP-associated serine proteases. Reprinted with permission from Holmskov, U., Malhotra, R., Sim, R. B., Jensenius, J. C. (1994). Collectins: collagenous C-type lectins of the innate immune defense system. *Immunol. Today* **15**, 67–74.

important in protection against pathogenic microorganisms.

Methods used in mammalian complement analysis have often been used when examining fish complement, e.g. buffer systems known to distinguish between classical and alternative pathway, or activators of complement that are known to activate mammalian complement. Such generalizations are, however, seldom sustainable. At 4°C, fish complement shows significant activity, which makes the removal of antibodies by absorption difficult (Rijkers, 1982a). Mammalian complement is incubated at 56°C for heat inactivation, whereas temperatures around 45°C completely destroy complement activity in fish serum (Sakai, 1981). Individual complement components in mammals show functional compatibility, however, this is very seldom true between mammalian and fish components.

Jawless fish: Agnatha

Serum from jawless fish shows hemolytic activity that in some respects resembles complement activity, although the molecules involved do not appear to be complement related (Fujii and Murakawa, 1981; Nonaka *et al.*, 1984). The jawless fish do, however, possess complement proteins, the functions of which appear to be opsonic. A C3-related protein has been isolated from the lamprey (Nonaka *et al.*, 1984), and the cDNA cloned and sequenced (Nonaka and Takahashi, 1992). The lamprey protein has three chains, similar to those of mammalian C4. The sequence data reveal greater homology to C3 than to C4, and a sequence of the cleavage peptide more closely resembles C5a than C3a or C4a. This finding implies that the anaphylactic function borne mainly by mammalian C5 may be present in the lamprey C3-like protein. Similar evidence has been obtained from the hagfish (Ishiguro *et al.*, 1992). The evidence from these two agnathan species thus indicates that in these species there may be a single thioester-containing protein which has evolved from (or is) the common ancestor of C3, C4 and C5 in higher vertebrates.

Furthermore, only a single C2/factor B-like cDNA has been cloned from the lamprey. As in the case of the agnathan C3-like protein, this may indicate, that the gene duplication, leading to separate C2 and factor B proteins in mammals has occurred after the divergence of the agnathans from the line leading to the higher vertebrates (Nonaka *et al.*, 1984). Immunoglobulins have not been demonstrated in Agnathae (Fujii *et al.*, 1992), and complement-mediated reactions leading to opsonic activity may therefore represent an important humoral defense mechanism. These fish do possess lytic activity but it is unrelated to complement and there is no evidence of the existence of the mammalian C6–C9 homologues.

Cartilaginous fish: Chondrichthyes

Nurse shark (*Ginglymostoma cirratum*) serum shows antibody-dependent lytic activity (Jensen and Festa, 1980) and six individual components have been described (Jensen *et al.*, 1981). Structural and functional data are, however, lacking, and only a C1q, a C4 and a C2-like component appear to have been unequivocally defined. It is still disputed whether an alternative pathway exists in the nurse shark (Jensen *et al.*, 1981) and more data are needed before it can be stated to be nonexistent.

Bony fish: Teleosts

Serum from bony fish contains a strong lytic activity against mammalian red blood cells, and also a bactericidal effect (Nonaka *et al.*, 1981a; Ourth and Wilson, 1982b). Part of both these activities can be ascribed to the complement system. Very few attempts have, however, been made to analyze the complement system in detail (Ingram, 1980). Most information is available from the rainbow trout, but the carp has also been analyzed in some detail.

Nonaka *et al.* (1981a) immunized rainbow trout with SRBC and demonstrated both an antibody-dependent and

Table II.13.1 Calculated regional and total amino acid identity between trout factor B and factor B/C2 from other vertebrates[a]

	SCR-domains	*von Willebrand domain*	*Serine protease domain*	*Total*
Trout factor B versus human factor B	47%	34%	38%	38%
Trout factor B vs human C2	47%	32%	33%	34%
Trout factor B vs lamprey factor B	37%	27%	26%	28%
Trout factor B vs *Xenopus* factor B	43%	25%	36%	34%
Human factor B vs human C2	46%	38%	37%	

SCR, Short consensus repeats.
[a] Per cent amino acid identity was calculated using the program ALIGN within the DNASTAR software package.

an antibody-independent activation pathway, presumably corresponding to the mammalian classical and alternative complement pathways. When added to fish serum, LPS, zymosan or inulin, depleted all complement activity. A complex of six proteins, as visualized by sodium dodecyl sulfate-polyacrylamide gel electrophoresis (SDS-PAGE), were isolated from the surface of membranes from lysed red cells, and one of these proteins showed homology with mammalian C5 (Nonaka *et al.*, 1981b). This C5-like protein was necessary in both antibody-dependent and antibody-independent lysis of red cells. cDNA for a trout C9-like protein has been cloned and sequenced (Stanley *et al.*, 1987; Tomlinson *et al.*, 1993). The homology to mammalian C9 is clear, although it encodes an extra thrombospondin-domain at the C-terminus, giving an overall architecture similar to that of mammalian C8 α and β chains.

Rainbow trout C3 has been purified and characterized (Nonaka *et al.*, 1984; Jensen and Koch, 1991) and shows striking similarity to mammalian C3, having a two-chain structure. C3 cDNA clones have also been isolated and sequenced. The cloning revealed two isoforms, C3-1 and C3-2, of which C3-1 appears to be the functional gene, whereas C3-2 appears hemolytically inactive (Nonaka *et al.*, 1985; Lambris *et al.*, 1993). Rainbow trout C3 shows genetic polymorphism from electrophoretic separation in agarose gels, followed by immunoblotting onto nitrocellulose and detection with a monoclonal antibody to trout C3 (Jensen and Koch, 1991). Three allotypes have been demonstrated in a Danish population of farmed rainbow trout (Slierendrecht *et al.*, 1996).

Recently, a full-length clone corresponding to a putative trout factor B/C2 protein has been sequenced. Table II.13.1 lists the calculated regional and total amino acid identity between trout factor B, lamprey factor B, *Xenopus* factor B and human factor B and C2. Trout factor B shows 38% homology with human factor B and 34% homology with human C2, and comparisons with the other known sequences from mouse, amphibian (*Xenopus*) and lamprey have led to the phylogenetic tree of factor B and C2, shown in Figure II.13.2.

A C1 complex has been demonstrated in the carp (Yano *et al.*, 1988), and an analogue of factor D has been isolated (Yano and Nakao, 1994a). It is notable, that removal of carp factor D leads to a complete loss of both alternative and classical complement activity (Kaastrup and Koch, 1983; Yano and Nakao, 1994b).

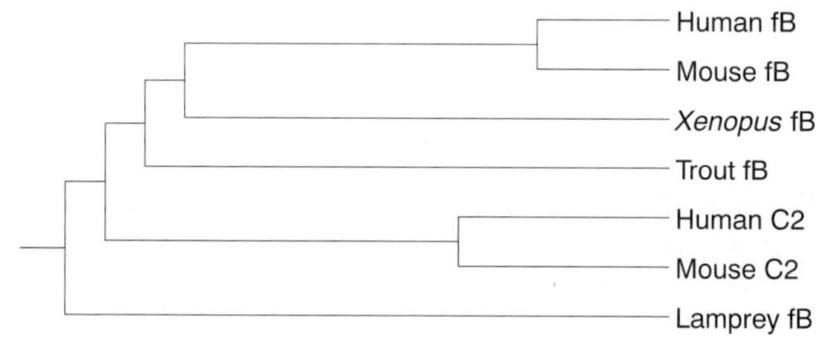

Figure II.13.2 Phylogenetic tree for factor B and C2. The relationships were obtained by the clustal method using the entire amino acid sequences.

Conclusions

It must be concluded that there is still relatively little known about the complement system in fish. The mammalian complement system is regulated by several control proteins, and complement receptors are of paramount importance when carrying out complement functions. The only known control protein belonging to the fish complement system is a putative factor H analogue, isolated from the barred sand bass (Kaidoh and Gigli, 1989) and sequenced from a cDNA clone (Dahmen *et al.*, 1994). The presence of complement receptors in fish is only indicated from functional studies.

The complement components as we know them in mammals are the results of a series of gene duplications (Factor D/C1s/C1r, Factor B/C2, C3/C4/C5, C6/C7, C8a/C8b). One of the fascinating aspects of fish complement is that fish appear to be at an evolutionary level where some of these duplications have not yet occurred, meaning that one component may carry functions later allocated to separate proteins.

Much remains before our understanding of fish complement will allow an appreciation of its functional importance in fish and yield additional insights into evolution.

14. Mucosal Immunity

Mucosal immunity in fish has been a poorly explored field of research. However, recent evidence indicates that this type of immunity exists in fish, as would be appropriate for an animal living in a pathogen-rich aquatic environment. In contrast to higher vertebrates, the skin of fish forms an extra mucosal barrier. In this section, data indicating the presence of a mucosal immune system in fish will be summarized and particular emphasis will be given to antigen transport and processing by mucosal surfaces, mucosal immune responses and the induction of oral tolerance against protein antigens.

The mucosal immune system of fish

Although fish lack clear lymphoid accumulations in their mucosae, many lymphoid cells and macrophages can be found in their intestine (Davina *et al.*, 1980; Temkin and McMillan, 1986; Doggett and Harris, 1991; Rombout *et al.*, 1993a) and skin (St. Louis-Cormier *et al.*, 1984; Lobb, 1987). As most research on mucosal immunity has been undertaken on the intestine of carp, data will be drawn principally from this species, but when necessary, important differences observed in other species will be cited. In carp, many Ig^+ cells (B cells and plasma cells) can be found in the lamina propria throughout the intestine, while most Ig^- cells and large Ig-binding macrophages were observed in the intestinal epithelium (Rombout *et al.*, 1993a). This situation is comparable with that in higher vertebrates, although Peyer's patch-like structures are absent, which is not surprising because fish lack lymph nodes or germinal center-like structures (van Muiswinkel *et al.*, 1991).

Therefore, it can be stated that fish have a more diffuse intestinal immune system. As far as is known, all fish have a second segment in their hindgut (20–25% of the gut length), which contains epithelial cells with a much higher endocytotic capacity and with more and larger vacuoles compared with the first segment (50–75% of the gut length, depending on the species (Rombout *et al.*, 1985). The second segment appears to be more adapted for the absorption of digested molecules. Enterocytes in this segment can transport antigens from the lumen to the lymphoid cells and macrophages in the mucosal tissue (carp, Rombout *et al.*, 1985; Rombout and van den Berg, 1989; trout, Georgopoulou *et al.*, 1986; tilapia, Doggett and Harris, 1991). In carp, many more and much larger intraepithelial macrophages can be found in this part of the gut compared with the first segment (Rombout *et al.*, 1989a). After oral or anal administration of antigens, intestinal epithelial cells appear to take up macromolecules by endocytosis and transport them to these large macrophages (Georgopoulou and Vernier, 1986; Rombout *et al.*, 1989a), which at least in carp appeared to have an Ig-binding capacity (Koumans-van Diepen *et al.*, 1994b) and an antigen-presenting function (Rombout *et al.*, 1986; Rombout and van den Berg, 1989). However, it should be noted that the abundance of these macrophages could not be confirmed for several other species such as sea bream (Joosten *et al.*, 1995) and trout (Joosten, 1997). Even so, the combination of an antigen-transporting epithelium and the presence of antigen-processing macrophages, next to abundant Ig^+ (B) and Ig^- (T) cells makes the second segment an important candidate site for the induction of a mucosal immune response. Although not much is known about antigen processing in gills and skin, it would appear that leukocytes are present and antigen can be taken up by these epithelia (Lobb, 1987; Iger and Wendelaar Bonga, 1994) and that immunocompetent cells are present at these locations (Rombout *et al.*, 1993a).

Mucosal immune responses

An important criterion for the existence of a mucosal immune system is the secretion of antigen-specific antibodies at mucosal surfaces. Antigen-specific antibodies could be detected in skin mucus, bile or the intestine after oral administration of a variety of antigens to several fish species such as plaice (Fletcher and White, 1973b), ayu (Kawai *et al.*, 1981), carp (Rombout *et al.*, 1986, 1989b), while small amounts of antigen-specific antibodies were found in serum. Similar results were obtained after bath vaccination of channel catfish, (Lobb, 1987) suggesting that mucosal responses can also be induced through the skin and/or gills. In contrast, intraperitoneal or intramuscular immunizations produced high titers of antibodies only in serum and not at the mucosal surfaces. However, these results are difficult to explain given, as stated by most fish immunologists, that fish have only one IgM-like molecule. In carp, the majority of skin mucus and serum IgM is tetrameric, but dimeric and monomeric forms are also found. Both IgM types possess similar H and L chains and are reactive with monoclonal antibodies (mAbs) against serum IgM. Recently, mAbs have been produced against skin mucus that react with mucus IgM and not with serum IgM (Rombout *et al.*, 1993b). These mAbs reacted specifically with a subpopulation of B cells and with skin epithelium and bile capillaries, the sites where mucosal Ig can be expected. In addition, after oral administration with a *Vibrio anguillarum* vaccine, specific Ig-secreting cells could be demonstrated in gills and gut with an enzyme-linked immunospot (ELISPOT) assay using these mAb, while these cells were not present in lymphoid organs involved in systemic responses after intramuscular injection (Joosten, 1997). Indications for another type of mucus Ig (including a secretory component) are also available for the sheepshead (Lobb and Clem, 1981c). These data clearly suggest that fish have an IgA homologue, although, at least in carp, it has to be considered as an IgM isotype. To consolidate these data, further investigation is needed of the molecular differences between both IgM molecules and the mechanisms behind the homing of mucus IgM-producing B cells and plasma cells.

Oral tolerance

Another important aspect of a mucosal immune system is the induction of tolerance after repeated oral administration of protein antigens, resulting in immunosuppression when the same protein is administered systemically. Indications are available that a similar mechanism can be found in fish. In carp (Rombout *et al.*, 1989b), trout (Davidson *et al.*, 1994) and salmon (Piganelli *et al.*, 1994) induction of oral tolerance has been indicated when fish were fed repeatedly with protein antigens. For mammals, it has been suggested that T cells migrating from the gut to other lymphoid organs may suppress the systemic immune responses by producing inhibitory cytokines (Friedman *et al.*, 1994). These intraepithelial T cells possess a number of membrane molecules that are distinct from T cells present in other lymphoid tissues. Recently, it has been demonstrated that carp intraepithelial Ig^- (T) cells also represent a distinct T-cell population (Joosten, 1997), which may play a role in the induction of oral tolerance.

Conclusions

This section summarizes the evidence available for a mucosal immune system in fish. The antigen-transporting capacity of the hind gut of fish, the abundant but diffuse presence of immuno-competent cells in this part of the gut, the detection of specific antibodies in skin mucus, bile or intestine only after intestinal or bath immunization, the indications for a distinct mucosal IgM isotype and the presence of cells secreting this isotype in gills and intestine, together strongly indicate that fish indeed have a mucosal immune system. Moreover, fish can also be tolerized by repeated feeding of protein antigens, which can be considered as another aspect of the mucosal immune system.

15. Acquired Immunodeficiencies

There have been no reports of congenital immunodeficiencies affecting phagocytes or lymphocytes in fish, and while natural antibodies exist (Table II.12.1), there are no reports of naturally occurring autoimmune disease in fish. There are numerous examples, however, of acquired immunodeficiencies. Three major causes of acquired immunodeficiency or immunosuppression that have been identified in fish are chemicals, temperature and stress.

Chemicals

A large number of chemicals produce a deficit in one or more immune function. Numerous heavy metals have been shown *in vitro* to reduce either antibody production (chromium in brown trout, O'Neill, 1981), cell-mediated immunity (cadmium-induced reduction in lymphocyte proliferation, Thuvander, 1989) and leukocyte numbers (Murad and Houston, 1988). Many other examples are documented (reviewed by Zelikoff, 1993) but the evidence is not clear-cut because results vary with the dose, length of exposure, species and immune assay employed. For example, exposure to cadmium has produced increased antibody production in rainbow trout (Thuvander, 1989) and striped bass (Robohm, 1986) but decreased antibody production in rainbow trout (Viale and Calamari, 1984) and cunners (Robohm, 1986). *In vivo* experiments have also demonstrated that exposure to metals, particularly copper, can increase the susceptibility of fish to infectious disease (Knittel, 1981; Baker *et al.*, 1983; Cossarini-Dunier *et al.*, 1988). Numerous pesticides and organic pollutants also produce a reduction in immune function but the significance of these effects requires further investigation (reviewed by Dunier and Siwicki, 1993). Of note are the observations that oxytetracycline has also been demonstrated to be immunosuppressive in fish, producing a reduction in plaque-forming cells after immunization (Grondel *et al.*, 1987; Siwicki *et al.*, 1989), reducing neutrophil activity (Siwicki *et al.*, 1989) and following long-term feeding, reduced both cellular and humoral immunity (Rijkers *et al.*, 1981).

Temperature

Temperature, not surprisingly, influences the immune response of fish, and the lower limits of a species temperature range are immunosuppressive. The effects of low temperature can be summarized as follows: prolonged allograft rejection (Rijkers, 1982b), suppression or inhibition of *in vivo* primary antibody response (Rijkers, 1982b; Bly and Clem, 1992), antigenic tolerance (Wishkovsky and Avtalion, 1982), suppression of *in vitro* anti-hapten antibody response to T-dependent antigens (Miller and Clem, 1984), reduced responsiveness to T-cell mitogens at low *in vitro* temperatures (Clem *et al.*, 1984), and reduced production of macrophage activating factor by leukocytes following low temperature exposure *in vivo* and *in vitro* (Hardie *et al.*, 1994b). The effect on T cells has been postulated to be a suppression of activation or generation of virgin T cells (Bly and Clem, 1992). Low temperature is also associated with a reduction in hemolytic complement in channel catfish (Hayman *et al.*, 1992).

Stress

Stress and corticosteroid administration increases the susceptibility of fish to disease (Kent and Hedrick, 1987) and can induce shedding of infectious agents from carriers (McCarthy, 1978). Administration of corticosteroids has been shown to reduce antibody production when administered before or at the same time as antigen (Houghton and Matthews, 1990), reduce splenic lymphocytes and circulating leukocytes (Maule *et al.*, 1987) and reduce phagocytic chemiluminescence (Stave and Roberson, 1985).

16. Tumors of the Immune System

Tumors of lymphoid and hemopoietic tissues have been reported in numerous fish species (Table II.16.1). The investigation of the pathogenesis and epidemiology of these tumors has been instigated in the northern pike and the muskellunge, but beyond these species, the pathogenesis and effects on the immune system of these tumors has received little attention.

17. Conclusions

For immunologists, the fishes are an interesting and increasingly useful group for the investigation of vertebrate immunology. Indeed, any consideration of the vertebrate immune system cannot overlook such a vast and diverse collection of species. With over 20 000 members and an evolutionary span of 400–500 million years, the immune system of fishes provides an insight into the innovation, elaboration and conservation that has occurred in vertebrate immunology.

The structure and organization of the vertebrate immune system becomes established within the fishes. A thymus and spleen appear as well-defined organs, conventional antibodies are recognized, as are lymphatics, and a distinction can be made between a systemic and a mucosal immune system. These standard elements of a vertebrate immune system make their appearance in different species across the spectrum of the fishes, from the jawless fishes, through the sharks and rays to the major grouping in the bony fishes, the teleosts. The form and presence of these features is still under active investigation, as indicated by the contentious status of the spleen in the jawless fish (Zapata and Cooper, 1990), of thymic involution in sharks (McKinney, 1992), and of the secondary vascular system in teleost fish (Steffensen and Lomholt, 1992).

Fish do not possess all the features of the modern mammalian immune system and fish lack a bone marrow and lymph nodes, as well as germinal centers. Yet, at present, the similarities are more striking than the differences. The functions of fish leukocytes are remarkably similar to their mammalian equivalents and abilities such as mitogen stimulation, antigen processing and presentation, and Ig gene expression are comparable in fishes and mammals. Convincing evidence of the conservation of Ig gene regulation and transcription has recently been provided by the demonstration that fish Ig genes can be expressed in mammalian B-cell lines and mouse Ig in transgenic rainbow trout (Section 7, this chapter).

The similarities within the vertebrate immune system are even more apparent at the gene level. The extensive series of studies in fishes that have cloned and sequenced the genes for TCR, Ig and MHC has exploited the high degree of homology that exists in gene organization and regulation across the vertebrates. For example, the genomic sequence of trout D_β is identical to those of the mouse, rat and human $D_\beta1$ genes and their recombination signal sequences are also very well conserved (Section 6, this chapter). In the teleost Ig heavy chain locus, sequence comparisons have shown a conserved promoter structure with an octamer and TATA box (Section 7, this chapter). While PCR primers based on conserved motifs in man, mouse and chicken MHC sequences have been used in the successful isolation of MHC molecules from various teleosts and cartilaginous fishes (Section 8, this chapter).

A direct benefit of this conservation has been the success of vaccines in controlling diseases in the aquaculture industry. The general principles and methods of vaccination have been used to create highly effective preparations, particularly against the major Gram negative pathogenic bacteria affecting the salmon industry (Press and Lillehaug, 1995). However, the challenge of this success for immunologists will be to extend the present understanding of the piscine immune system, so that other bacterial and viral diseases may also be controlled. Advances in the recognition of a mucosal immune system (Section 14, this chapter), hold particular promise for combating viral diseases in young fish and for improving the delivery of vaccines. Successful regimens for oral and immersion vaccination would be of considerable practical benefit to the aquaculture industry.

A further challenge posed by the aquaculture industry is its readiness to utilize new species. The introduction of new species into intensive aquaculture will most likely be accompanied by 'new' pathogens and diseases. While vaccination has shown itself to be the most effective and sustainable form of disease control in fish, the diversity that exists within the immune systems of fishes should encourage the investigation of the immune system of a new species prior to its adoption into commercial aquaculture.

The optimal starting point for the investigation of a new species is from a thorough knowledge of the immune system in the 'established' species such as the channel catfish and rainbow trout. Continued research in the species that have traditionally been used in immunological investigations will need to range from the gene level and the sequencing of immune-related genes to the definition of leukocyte markers through the use of monoclonal

Table II.16.1 Sporadic tumors of the immune system of fish

Species	*Neoplasm and site*	*Reference*
Adrianicthyidae		
Oryzias latipes	Lymphoma, cutaneous lymphoblastic	Harada *et al.* (1990)
	Lymphoma, thymic with metastasis	Battalora *et al.* (1990)
Amiidae[a]		
Amia calva	Granuloplastic leukemia	Anderson and Luther (1987)
Characidae		
Astyanax mexicanus	Lymphoma, thymic with metastasis	Nigrelli (1947)[b]
Pristella riddelei	Undifferentiated hematopoeitic neoplasm	Gross (1983)[c]
Hybrid characins	Lymphoma, thymic with metastasis	Rasquin and Hafner (1951)[b]
Cichlidae		
Aequidens maronii	Lymphoma, generalized	Harshbarger and Dawe (1973)
Clupeidae		
Clupea harengus harengus	Lymphoma	Johnstone (1926)[b]
Congridae		
Conger conger	Lymphoma, renal	Williams (1931)[b]
Cyprinidae		
Carassius auratus	Lymphoma, renal with metastasis	Plehn (1924)[b]
Rasbora lateristriata	Lymphoma, ceolomic with metastasis	Smith *et al.* (1936)[b]
Esocidae		
Esox lucius	Lymphoma[d], cutaneous with eventual metastasis	Ljungberg (1976), Mulchay (1976), Sonstegard (1976), Thompson (1982)[c]
Esox masquinongy	Lymphoma, cutaneous with eventual metastasis	Sonstegard (1975, 1976)[c]
Gadidae		
Gadus morhua	Lymphoma, orbit	Wolke and Wyand (1969)[b]
Ictaluridae		
Ictalurus punctatus	Lymphoma, renal with metastasis	Bowser *et al.* (1985)
Ictalurus nebulosus	Plasma cell leukemia	Gross (1983)[c]
Mugilidae		
Mugil curema	Leukemia	Pitombeira *et al.* (1975)[e]
Osmeridae		
Plecoglossus altivelis	Lymphoma, ovarian	Honma (1966)[b]
Percidae		
Stizostedion lucioperca	Lymphoma, cutaneous	Bekesi and Kovacs-Gayer (1986)
Pleuronectidae		
Platichthys flesus	Lymphoma, orbit	Johnstone (1911–12)[b]
Pomatomidae		
Pomatomus saltatrix	Lymphangioepithelioma	Harshbarger and Dawe (1973)
Salmonidae		
Oncorhynchus keta	Lymphoma	Honma and Hirosaki (1966)[b]
Oncorhynchus kisutch	Lymphoma, gastric	Ashley (1969)[b]
	Lymphoma (lymphoblastic), thymic, epitheliotropic	Kieser *et al.* (1991)
Oncorhynchus mykiss	Lymphoma, thymic with metastasis and leukemia	Warr *et al.* (1984)
	Lymphoma with metastasis and leukemia	Bernstein (1984)
Oncorhynchus tshawytscha	Plasmacytoid leukemia	Kent *et al.* (1990)[c]
Salmo salar	Lymphoma, renal	Haddow and Blake (1933)[b]
	Lymphoma, muscle	Roald and Håstein (1979)
	Lymphoma, (lymphoblastic) thymic with metastasis	Bowser *et al.* (1987)
Salvelinus namaycush	Lymphoma, kidney	Ehlinger (1963)[b]
S. namaycush × *S. fontinalis*	Lymphoma, thymic	Dunbar (1969)[b]
Salvelinus fontinalis	Lymphoma, thymic with metatasis and leukemia	Dunbar (1969)[b]
Thymallus thymallus	Lymphoma, kidney with metastasis	Hoffmann *et al.* (1988)

[a] Fish grouped as per Nelson (1994).
[b] Cited by Mawdesley-Thomas (1975).
[c] A viral etiology has been demonstrated or is suspected.
[d] Recently characterized as a diverse neoplasm and either a histiocytic diffuse malignant lymphoma or a lymphocytic diffuse malignant lymphoma (Bogovski *et al.*, 1994).
[e] Cited by Anderson and Luther (1987).

antibodies or gene probes. The establishment of a CD system for classification as in mammals should be a goal. A beneficial consequence of establishing such a system would be the more systematic characterization of fish leukocytes, particularly lymphocytes but also macrophages and granulocytes (Sections 3 and 12, this chapter). A major area of the piscine immune system awaiting detailed investigation, both at the functional and genetic level, is the cytokines. The preliminary nature of much of the present data available on cytokines in fish shows that this process has begun (Section 5, this chapter).

The study of immunology of fishes is at an exciting stage. As the piscine immune system becomes better understood, it is increasingly being adopted as a research model for medical science. The genetics and biology of fish species such as the zebrafish hold obvious advantages for the immunological researcher. The widening realization of the relevance and utility of fish not just to questions of phylogeny will lead more laboratories to adopt fish as the experimental animal of choice. Accordingly, the immunology of fishes should continue to be an important contributor to the understanding of vertebrate immunology as a whole.

18. References

Abelli, L., Picchietti, S., Romano, N., Mastrolia, L., Scapigliati, G. (1996). Immunocytochemical detection of a thymocyte antigenic determinant in developing lymphoid organs of sea bass, *Dicentrarchus labrax* (L). *Fish Shellfish Immunol.* **6**, 493–505.

Acton, R. T., Weinheimer, P. F., Dupree, H. K., Evans, E. E., Bennett, J. C. (1971a). Phylogeny of immunoglobulins. Characterization of a 14S immunoglobulin from the gar, *Lepisosteus osseus*. *Biochemistry* **10**, 2028–2036.

Acton, R. T., Weinheimer, P. F., Dupree, H. K., Russell, T. R., Wolcott, M., Evans, E. E., Schrohenloher, R. E., Bennett, J. C. (1971b). Isolation and characterization of the immune macroglobulin from the paddlefish, *Polyodon spathula*. *J. Biol. Chem.* **246**, 6760–6769.

Acton, R. T., Weinheimer, P. F., Hall, S. J., Niedermeier, W., Shelton, E., Bennett, J. C. (1971c). Tetrameric immune macroglobulins in three orders of bony fish. *Proc. Natl Acad. Sci. USA* **68**, 107–111.

Acton, R. T., Niedermeier, W., Weinheimer, P. F., Clem, L. W., Leslie, G. A., Bennett, J. C. (1972). The carbohydrate composition of immunoglobulins from diverse species of vertebrates. *J. Immunol.* **109**, 371–381.

Agius, C. (1980). Phylogenetic development of melano-macrophage centres in fish. *J. Zool. Lond.* **191**, 11–31.

Agius, C., Agbede, S. A. (1984). An electron microscopical study on the genesis of lipofuscin, melanin and haemosiderin in the haemopoietic tissues of fish. *J. Fish Biol.* **24**, 471–488.

Ainsworth, A. J. (1992). Fish granulocytes: morphology, distribution and function. *Annu. Rev. Fish Dis.* **2**, 123–148.

Ainsworth, A. J. (1993). Carbohydrate and lectin interaction with *Edwardsiella ictaluri* and channel catfish, *Ictalurus punctatus* (Rafinesque), anterior kidney leucocytes and hepatocytes. *J. Fish Dis.* **16**, 449–459.

Ainsworth, A. J. (1994). A β-glucan inhibitable zymosan receptor on channel catfish neutrophils. *Vet. Immunol. Immunopathol.* **41**, 141–152.

Ainsworth, A. J., Dexiang, C. (1990). Differences in the phagocytosis of four bacteria by channel catfish neutrophils. *Dev. Comp. Immunol.* **14**, 201–209.

Ainsworth, A. J., Dexiang, C., Greenway, T. (1990). Characterization of monoclonal antibodies to channel catfish, *Ictalurus punctatus*, leukocytes. *Vet. Immunol. Immunopathol.* **26**, 81–92.

Alexander, J. B. (1985). Non-immunoglobulin humoral defence mechanisms in fish. In *Fish Immunology* (eds Manning, M. J., Tatner, M. F.), pp. 133–140. Academic Press, London, UK.

Alexander, J. B., Ingram, G. A. (1992). Noncellular nonspecific defence mechanisms of fish. *Annu. Rev. Fish Dis.* **2**, 249–279.

Amemiya, C. T., Litman, G. W. (1990). Complete nucleotide sequence of an immunoglobulin heavy-chain gene and analysis of immunoglobulin gene organization in a primitive teleost species. *Proc. Natl Acad. Sci. USA* **87**, 811–815.

Amemiya, C. T., Ohta, Y., Litman, R. T., Rast, J. P., Haire, R. N., Litman, G. W. (1993). VH gene organization in a relict species, the coelacanth *Latimeria chalumnae*: Evolutionary implications. *Proc. Natl Acad. Sci. USA* **90**, 6661–6665.

Anderson, A. A., Fletcher, T. C., Smith, G. M. (1981). Prostaglandin biosynthesis in the skin of the plaice *Pleuronectes platessa* L. *Comp. Biochem. Physiol.* **70**, 195–199.

Anderson, M., Amemiya, C., Luer, C., Litman, R., Rast, J., Niimura, Y., Litman, G. (1994). Complete genomic sequence and patterns of transcription of a member of an unusual family of closely related, chromosomally dispersed Ig gene clusters in *Raja*. *Int. Immunol.* **6**, 1661–1670.

Anderson, M. K., Shamblott, M. J., Litman, R. T., Litman, G. W. (1995). Generation of immunoglobulin light chain gene diversity in *Raja erinacea* is not associated with somatic rearrangement, an exception to a central paradigm of B cell immunity. *J. Exp. Med.* **182**, 109–119.

Anderson, W. I., Luther, P. B. (1987). Poorly differentiated granuloplastic leukemia in a bowfin, *Amia calva* L. *J. Fish Dis.* **10**, 411–413.

Andersson, E., Matsunaga, T. (1993). Complete cDNA sequence of a rainbow trout IgM gene and evolution of vertebrate IgM constant domains. *Immunogenetics* **38**, 243–250.

Andersson, E., Matsunaga, T. (1995). Evolution of immunoglobulin heavy chain variable region genes: a V_H family can last for 150–200 million years or longer. *Immunogenetics* **41**, 18–28.

Andersson, E., Peixoto, B., Törmänen, V., Matsunaga, T. (1995). Evolution of the immunoglobulin M region genes of salmonid fish, rainbow trout (*Oncorhynchus mykiss*) and Arctic char (*Salvelinus alpinus*): implications concerning divergence time of species. *Immunogenetics* **41**, 312–315.

Arason, G. J., Gudmundsdottir, S., Elgaard, L., Thiel, S., Jensenius, J. (1994). An opsonin with mannan-binding protein (MBP)-like activity in the Atlantic salmon, *Salmo salar*. *Scand. J. Immunol.* **40**, 692.

Arkoosh, M. R., Kaattari, S. L. (1991). Development of immunological memory in rainbow trout (*Oncorhynchus mykiss*). I. An immunochemical and cellular analysis of the B cell response. *Dev. Comp. Immunol.* **15**, 279–293.

Ashley, L. M. (1969). Experimental fish neoplasia. In *Fish in*

Research (eds Neuhaus, O. W., Halver, J. E.), pp. 23–43. Academic Press, New York, USA.

Atanassov, C. L., Botev, B. A. (1988). Isolation and partial characterization of IgM-like immunoglobulins from the serum of carp (*Cyprinus carpio* L.), frog (*Rana ridibunda* Pall.) and tortoise (*Testudo graeca* Pall.). *Comp. Biochem. Physiol.* **89**B, 737–741.

Avtalion, R. R., Mor, A. (1992). Monomeric IgM is transferred from mother to egg in tilapias. *Israeli J. Aqua. Bamidgeh* **44**, 93–98.

Baker, R. J., Knittel, M. D., Fryer, J. L. (1983). Susceptibility of chinook salmon, *Oncorhynchus tshawytscha* (Walbaum), and rainbow trout, *Salmo gairdneri* Richardson, to infection with *Vibrio anguillarum* following sublethal copper exposure. *J. Fish Dis.* **6**, 267–275.

Balakhnin, I. A., Dudka, I. A. (1990). The interaction of fish mycopathogens with anti-B lectin of *Salmo gairdneri* eggs. *Mikol. Fitopatol.* **24**, 224–228.

Balakhnin, I. A., Dudka, I. A., Isaeva, N. M. (1990). Testing of fungi on their specific interaction with fish egg lectins. *Mikol. Fitopatol.* **24**, 416–420.

Balm, P. H. M., van Lieshout, E., Lokate, J., Wendelaar Bonga, S. E. (1995). Bacterial lipopolysaccharide (LPS) and interleukin (IL-1) exert multiple physiological effects in the tilapia *Oreochromis mossambicus* (Teleostei). *J. Comp. Physiol. B* **165**, 85–92.

Battalora, M. S. J., Hawkins, W. E., Walker, W. W., Overstreet, R. M. (1990). Occurrence of thymic lymphoma in carcinogenesis bioassay specimens of the Japanese medaka (*Oryzias latipes*). *Canc. Res. Suppl.* **50**, 5675s–5678s.

Bayne, C. J. (1986). Pronephric leucocytes of *Cyprinus carpio*: isolation, separation and characterization. *Vet. Immunol. Immunopathol.* **12**, 141–151.

Beasley, A. R., Sigel, M. M. (1967). Interferon production in cold-blooded vertebrates. *In Vitro* **3**, 154–165.

Bekesi, L., Kovacs-Gayer, E. (1986). Lymphosarcoma in a pike-perch (*Stizostedion lucioperca* L.): a case report. *Acta Vet. Hungarica* **34**, 101–102.

Bell, R., Carmeli, S. (1994). Vibrindole A, a metabolite of the marine bacterium, *Vibrio parahaemolyticus*, isolated from the toxic mucus of the boxfish *Ostracion cubicus*. *J. Nat. Prod.* **57**, 1587–1590.

Bengten, E. (1994). The immunoglobulin genes in Atlantic cod (*Gadus morhua*) and rainbow trout (*Oncorhynchus mykiss*). Ph.D., Department of Immunology, Uppsala University, Uppsala, Sweden.

Bengten, E., Leanderson, T., Pilström, L. (1991). Immunoglobulin heavy chain cDNA from the teleost Atlantic cod (*Gadus morhua* L.): nucleotide sequences of secretory and membrane form show an unusual splicing pattern. *Eur. J. Immunol.* **21**, 3027–3033.

Bengten, E., Strömberg, S., Pilström, L. (1994). Immunoglobulin VH regions in Atlantic cod (*Gadus morhua* L.): their diversity and relationship to VH families from other species. *Dev. Comp. Immunol.* **18**, 109–122.

Bentley, G. A., Boulot, G., Karjalainen, K., Mariuzza, R. A. (1995). Crystal structure of the β chain of a T cell antigen receptor. *Science* **267**, 1984–1987.

Bernstein, J. W. (1984). Leukaemic lymphosarcoma in a hatchery-reared rainbow trout, *Salmo gairdneri* Richardson. *J. Fish Dis.* **7**, 83–86.

Bernstein, R. M., Schluter, S. F., Shen, S., Marchalonis, J. J. (1996). A new high molecular weight immunoglobulin class from the carcharhine shark: implications for the properties of the primordial immunoglobulin. *Proc. Natl Acad. Sci. USA* **93**, 3289–3293.

Bingulac-Popovic, J., Figueroa, F., Sato, A., Talbot, W. S., Johnson, S. L., Gates, M., Postlethwait, J. H., Klein, J. (1997). Mapping of MHC class I and class II regions to different linkage groups in the zebrafish, *Danio rerio*. *Immunogenetics* **46**, 129–134.

Bjørgan Schrøder, M., Espelid, S., Jørgensen, T. O. (1992). Two serotypes of *Vibrio salmonicida* isolated from diseased cod (*Gadus morhua* L.): virulence, immunological studies and vaccination experiments. *Fish Shellfish Immunol.* **2**, 211–221.

Bly, J. E., Clem, L. W. (1992). Temperature and teleost immune functions. *Fish Shellfish Immunol.* **2**, 159–171.

Bly, J. E., Grimm, A. S., Morris, I. G. (1986). Transfer of passive immunity from mother to young in a teleost fish: haemagglutinating activity in the serum and eggs of plaice, *Pleuronectes platessa* L. *Comp. Biochem. Physiol.* **84**A, 309–313.

Bly, J. E., Miller, N. W., Clem, L. W. (1990). A monoclonal antibody specific for neutrophils in normal and stressed channel catfish. *Dev. Comp. Immunol.* **14**, 211–221.

Boffa, G. A., Fine, J. M., Drilhon, A., Amouch, P. (1967). Immunoglobulins and transferrin in marine lamprey sera. *Nature* **214**, 700–702.

Bogovski, S., Rossi, L., Bocchini, V., Lastraioli, S., Aiello, C., Santi, L. (1994). Immunohistochemical characterization of malignant lymphoma in northern pike, *Esox lucius* L., from Estonian Baltic waters. *J. Fish Dis.* **17**, 557–566.

Bortz, B. M., Kenny, G. E., Pauley, G. B., Garcia-Ortigoza, E., Anderson, D. P. (1984). The immune response in immunized and naturally infected rainbow trout (*Salmo gairdneri*) to *Diplostomum spathaceum* as detected by enzyme-linked immunosorbent assay (ELISA). *Dev. Comp. Immunol.* **8**, 813–822.

Botham, J. W., Manning, M. J. (1981). Histogenesis of the lymphoid organs in the carp, *Cyprinus carpio* L. and the ontogenic development of allograft reactivity. *J. Fish Biol.* **49**, 403–414.

Botham, J. W., Grace, M. F., Manning, M. J. (1980). Ontogeny of first set and second set alloimmune reactivity in fishes. In *Phylogeny of Immunological Memory* (ed. Manning, M. J.), pp. 83–92. Biomedical Press, Elsevier/North Holland, Amsterdam, The Netherlands.

Bower, S. M., Evelyn, T. P. T. (1988). Acquired and innate resistance to the haemoflagellate *Cryptobia salmositica* in sockeye salmon (*Oncorhynchus nerka*). *Dev. Comp. Immunol.* **12**, 749–760.

Bowser, P. R., McCoy, C. P., MacMillan, J. R. (1985). A lymphoproliferative disorder in a channel catfish, *Ictalurus punctatus* (Rafinesque). *J. Fish Dis.* **8**, 465–469.

Bowser, P. R., Wolfe, M. J., Wallbridge, T. (1987). A lymphosarcoma in an Atlantic salmon (*Salmo salar*). *J. Wildl. Dis.* **23**, 698–701.

Bradshaw, C., Sigel, M. M. (1972). Dinitrophenyl-reactive immunoglobulins in the serum of normal bowfin, *Amia calva*. *Immunology* **22**, 1029–1036.

Bradshaw, C. M., Clem, L. W., Sigel, M. M. (1971a). Immunologic and immunochemical studies of the gar, *Lepisosteus platyrhincus*. II. Purification and characterization of immunoglobulin. *J. Immunol.* **106**, 1480–1487.

Bradshaw, C. M., Richard, A. S., Sigel, M. M. (1971b). IgM

antibodies in fish mucus. *Proc. Soc. Exp. Biol. Med.* **136**, 1122–1124.

Bridges, A. F., Manning, M. J. (1991). The effects of priming immersions in various human gamma globulin (HGG) vaccines on humoral and cell mediated immune responses after intraperitoneal HGG challenge in the carp, *Cyprinus carpio* L. *Fish Shellfish Immunol.* **1**, 119–129.

Brown, L. L., Evelyn, T. P. T., Iwama, G. K. (1994). On the egg-mediated transfer of passive immunity from coho salmon to their progeny. *Dev. Comp. Immunol.* **18** (Suppl. 1), S96.

Campbell, K. S., Backstrom, B. T., Tiefenthaler, G., Palmer, E. (1994). CART: a conserved antigen receptor transmembrane motif. *Sem. Immunol.* **6**, 393–410.

Caspi, R. R., Avtalion, R. R. (1984). Evidence for the existence of an IL-2-like lymphocyte growth promoting factor in a bony fish, *Cyprinus carpio*. *Dev. Comp. Immunol.* **8**, 51–60.

Castillo, A., Sanchez, C., Domiguez, J., Kaattari, S. L., Villena, A. J. (1993). Ontogeny of sIgM and cIgM-bearing cells in rainbow trout. *Dev. Comp. Immunol.* **17**, 419–424.

Cenini, P. (1984). The ultrastructure of leukocytes in carp (*Cyprinus carpio*). *J. Zool.* **204**, 509–520.

Chiller, J. M., Hodgins, H. O., Weiser, R. S. (1969). Antibody response in rainbow trout (*Salmo gairdneri*). II. Studies on the kinetics of antibody-producing cells and on complement and natural hemolysin. *J. Immunol.* **102**, 1202–1207.

Chilmonczyk, S. (1983). The thymus of the rainbow trout (*Salmo gairdneri*). Light and electron microscopic study. *Dev. Comp. Immunol.* **7**, 59–68.

Chilmonczyk, S. (1992). The thymus in fish: development and possible function in the immune response. *Annu. Rev. Fish Dis.* **2**, 181–200.

Chung, S., Secombes, C. J. (1987). Activation of rainbow trout macrophages. *J. Fish Biol.* **31A**, 51–56.

Cipriano, R. C., Ford, L. A., Jones, T. E. (1994). Relationships between resistance of salmonids to furunculosis and recovery of *Aeromonas salmonicida* from external mucus. *J. Wildl. Dis.* **30**, 577–580.

Cipriano, R. C., Heartwell, C. M. (1986). Susceptibility of salmonids to furunculosis: differences between serum and mucus responses against *Aeromonas salmonicida*. *Trans. Am. Fish Soc.* **115**, 83–88.

Clem, L. W. (1971). Phylogeny of immunoglobulin structure and function. IV. Immunoglobulins of the giant grouper, *Epinephelus itaira*. *J. Biol. Chem.* **246**, 9–15.

Clem, L. W., De Boutaud, F., Sigel, M. M. (1967). Phylogeny of immunoglobulin structure and function. II. Immunoglobulins of the nurse shark. *J. Immunol.* **99**, 1226–1235.

Clem, L. W., Leslie, G. A. (1971). Production of 19S IgM antibodies with restricted heterogeneity from sharks. *Proc. Natl Acad. Sci. USA* **68**, 139–141.

Clem, L. W., Leslie, G. A. (1982). Phylogeny of immunoglobulin structure and function. XV. Idiotypic analysis of shark antibodies. *Dev. Comp. Immunol.* **6**, 463–472.

Clem, L. W., McLean, W. E. (1975). Phylogeny of immunoglobulin structure and function. VII. Monomeric and tetrameric immunoglobulins of the margate, a marine teleost. *Immunology* **29**, 791–799.

Clem, L. W., Sizemore, R. C., Ellsaesser, C. F., Miller, N. W. (1985). Monocytes as accessory cells in fish immune responses. *Dev. Comp. Immunol.* **9**, 803–809.

Clem, L. W., Miller, N. W., Bly, J. E. (1991). Evolution of lymphocyte subpopulations, their interactions and temperature sensitivities. In *Phylogenesis of Immune Function* (eds Warr, G. W., Cohen, N.), pp. 191–214. CRC Press, Boca Raton, Florida, USA.

Clem, L. W., Faulmann, E., Miller, N. W., Ellsaesser, C., Lobb, C. J., Cuchens, M. A. (1984). Temperature-mediated processes in teleost immunity: differential effects of *in vitro* and *in vivo* temperatures on mitogenic responses of channel catfish lymphocytes. *Dev. Comp. Immunol.* **8**, 313–322.

Clerx, J. P. M., Castel, A., Bol, J. F., Gerwig, G. J. (1980). Isolation and characterization of the immunoglobulin of pike (*Esox lucius* L.). *Vet. Immunol. Immunopathol.* **1**, 125–144.

Cohen, N., Haynes, L. (1991). The phylogenetic conservation of cytokines. In *Phylogenesis of Immune Functions* (eds Warr, G. W., Cohen, N.), pp. 241–268. CRC Press, Boca Raton, Florida, USA.

Cossarini-Dunier, M. (1986). Secondary response of rainbow trout (*Salmo gairdneri* Richardson) to DNP-haemocyanin and *Yersinia ruckeri*. *Aquaculture* **52**, 81–86.

Cossarini-Dunier, M., Desvaux, F.-X., Dorson, M. (1986). Variability in humoral responses to DNP-KLH of rainbow trout (*Salmo gairdneri*) comparison of antibody kinetics and immunoglobulins spectrotypes between normal trouts and trouts obtained by gynogenesis or self-fertilization. *Dev. Comp. Immunol.* **10**, 207–217.

Cossarini-Dunier, M., Demael, A., Lepot, D., Guerin, V. (1988). Effect of manganese ions on the immune response of carp (*Cyprinus carpio*) against *Yersinia ruckeri*. *Dev. Comp. Immunol.* **12**, 573–579.

Daggfeldt, A., Bengten, E., Pilström, L. (1993). A cluster type organization of the loci of the immunoglobulin light chain in Atlantic cod (*Gadus morhua* L.) and rainbow trout (*Oncorhynchus mykiss* Walbaum) indicated by nucleotide sequences of cDNA and hybridization analysis. *Immunogenetics* **38**, 199–209.

Dahmen, A., Kaidoh, T., Zipfel, P. F., Gigli, I. (1994). Cloning and characterisation of a cDNA representing a putative complement regulatory plasma protein from barred sand bass (*Parablax neblifer*). *Biochem. J.* **301**, 391–397.

Davidson, G. A., Ellis, A. E., Secombes, C. J. (1991). Cellular responses of leucocytes isolated from the gut of rainbow trout, *Oncorhynchus mykiss* (Walbaum). *J. Fish Dis.* **14**, 651–659.

Davidson, G. A., Ellis, A. E., Secombes, C. J. (1994). A preliminary investigation into the phenomenon of oral tolerance in rainbow trout (*Oncorhynchus mykiss* Walbaum, 1792). *Fish Shellfish Immunol.* **4**, 141–151.

Davina, J. H. M., Rijkers, G. T., Rombout, J. H. W. M., Timmermans, L. P. M., van Muiswinkel, W. B. (1980). Lymphoid and non-lymphoid cells in the intestine of cyprinid fish. In *Development and Differentiation of Vertebrate Lymphocytes* (ed. Horton, J. D.), pp. 129–140. Elsevier/North Holland Biomedical Press, Amsterdam, The Netherlands.

de Guerra, A., Charlemagne, J. (1997). Genomic oganization of the TcR β-chain diversity (Dβ) and joining (Jβ) segments in the rainbow trout: presence of many repeated sequences. *Mol. Immunol.* **34**, 653–662.

De Ioannes, A. E., Aguila, H. L. (1989). Amino terminal sequence of heavy and light chains from ratfish immunoglobulin. *Immunogenetics* **30**, 175–180.

de Sena, J., Rio, G. J. (1975). Partial purification and characterization of RTG-2 fish cell interferon. *Infect. Immun.* **11**, 815–822.

DeLuca, D., Wilson, M., Warr, G. W. (1983). Lymphocyte heterogeneity in the trout, *Salmo gairdneri*, defined with monoclonal antibodies to IgM. *Eur. J. Immunol.* **13**, 546–551.

Desvaux, F.-X., Charlemagne, J. (1981). The goldfish immune response. I. Characterization of the humoral response to particulate antigens. *Immunology* **43**, 755–762.

Desvaux, F.-X., Cossarini-Dunier, M., Chilmozcyk, S., Charlemagne, J. (1987). Antibody diversity in trout obtained by gynogenesis or self-fertilization. Comparative analysis of the heavy chain spectrotypes. *Dev. Comp. Immunol.* **11**, 577–584.

Di Conza, J. J., Halliday, W. J. (1971). Relationship of catfish antibodies to immunoglobulin in mucus secretions. *Aust. J. Exp. Biol. Med. Sci.* **49**, 517–519.

Dixon, B., Stet, R. J., van Erp, H. M., Pohajdak, B. (1993). Characterization of β_2-microglobulin transcripts from two teleost species. *Immunogenetics* **38**, 27–34.

Dixon, B., van Erp, S. H. M., Rodrigues, P. N. S., Egberts, E., Stet, R. J. M. (1995). Fish major histocompatibility complex genes: an expansion. *Dev. Comp. Immunol.* **19**, 109–133.

Doggett, T. A., Harris, J. E. (1991). Morphology of the gut associated lymphoid tissue of *Oreochromis mossambicus* and its role in antigen absorption. *Fish Shellfish Immunol.* **1**, 213–227.

Dorson, M., Barde, A., de Kinikelin, P. (1975). Egtved virus induced rainbow trout serum interferon: some physicochemical properties. *Ann. Microbiol. Paris* **126**, 485–489.

Du Pasquier, L. (1993a). Evolution of the immune system. In *Fundamental Immunology* (ed. Paul, W. E.). Raven Press Ltd., New York, USA.

Du Pasquier, L. (1993b). Phylogeny of B-cell development. *Curr. Op. Immunol.* **5**, 185–193.

Du Pasquier, L., Chardonnes, X., Miggiano, V. C. (1975). A major histocompatibility complex in the toad *Xenopus laevis* (Daudin). *Immunogenetics* **1**, 482–494.

Du Pasquier, L., Miggiano, V. C., Kobel, H. R., Fischberg, M. (1977). The genetic control of histocompatibility reactions in natural and laboratory-made polyploid individuals of the clawed toad *Xenopus*. *Immunogenetics* **5**, 129–141.

Dunbar, C. E. (1969). Lymphosarcoma of possible thymus origin in salmonid fishes. *Nat. Cancer Inst. Monogr.* **31**, 167–171.

Dunier, M., Siwicki, A. K. (1993). Effects of pesticides and other organic pollutants in the aquatic environment on immunity in fish: a review. *Fish Shellfish Immunol.* **3**, 423–438.

Edagawa, T., Murata, M., Hattori, O., Onuma, M., Kodama, H. (1993). Cell surface C-reactive protein of rainbow trout lymphocytes. *Dev. Comp. Immunol.* **17**, 119–127.

Ehlinger, N. F. (1963). Kidney disease in lake trout complicated by lymphosarcoma. *Prog. Fish Cult.* **25**, 3–7.

Eitan, S., Zisling, R., Cohen, A., Belkin, M., Hirschberg, D. L., Lotan, M., Schwartz, M. (1992). Identification of an interleukin 2-like substance as a factor cytotoxic to oligodendrocytes and associated with central nervous system regeneration. *Proc. Natl Acad. Sci. USA* **89**, 5442–5446.

Elcombe, B. M., Chang, R. J., Taves, C. J., Winkelhake, J. L. (1985). Evolution of antibody structure and effector functions: comparative hemolytic activities of monomeric and tetrameric IgM from rainbow trout, *Salmo gairdneri*. *Comp. Biochem. Physiol.* **80B**, 697–706.

Ellis, A. E. (1977a). Ontogeny of the immune response in *Salmo salar*. Histogenesis of the lymphoid organs and appearance of membrane immunoglobulin and mixed leucocyte reactivity. In *Developmental Immunobiology* (eds Solomon, J. B., Horton, J. D.), pp. 225–231. Biomedical Press, Elsevier/North Holland, Amsterdam, The Netherlands.

Ellis, A. E. (1977b). The leucocytes of fish: a review. *J. Fish Biol.* **11**, 453–491.

Ellis, A. E. (1980). Antigen-trapping in the spleen and kidney of the plaice *Pleuronectes platessa* L. *J. Fish Dis.* **3**, 413–426.

Ellis, A. E. (1981a). Stress and modulation of the immune response in fish. In *Stress and Fish* (ed. Pickering, A. D.), pp. 147–169. Academic Press, London, UK.

Ellis, A. E. (1981b). Non-specific defense mechanisms in fish and their role in disease processes. *Develop. Biol. Standard.* **49**, 337–352.

Ellis, A. E. (1986). The function of teleost fish lymphocytes in relation to inflammation. *Int. J. Tiss. Reac.* **8**, 263–270.

Ellis, A. E., Parkhouse, R. M. E. (1975). Surface immunoglobulins on the lymphocytes of the skate *Raja naevus*. *Eur. J. Immunol.* **5**, 726–728.

Ellis, A. E., de Sousa, M. A. B. (1974). Phylogeny of the lymphoid system. I. A study of the fate of circulating lymphocytes in plaice. *Eur. J. Immunol.* **4**, 338–343.

Ellis, A. E., Wooten, R. (1978). Costiasis of Atlantic salmon, *Salmo salar* L. smolts in sea water. *J. Fish Dis.* **1**, 389–393.

Ellis, A. E., Munro, A. L. S., Roberts, R. J. (1976). Defence mechanisms in fish. I. A study of the phagocytic system and the fate of intraperitoneally injected particulate material in the plaice (*Pleuronectes platessa*). *J. Fish Biol.* **8**, 67–78.

Ellis, A. E., Roberts, R. J., Tytler, P. (1989). The anatomy and physiology of teleosts. In *Fish Pathology* (ed. Roberts, R. J.), pp. 13–55. Baillière Tindall, London, UK.

Ellsaesser, C. F., Clem, L. W. (1986). Haematological and immunological changes in channel catfish stressed by handling and transport. *J. Fish Biol.* **28**, 511–521.

Ellsaesser, C. F., Clem, L. W. (1994). Functionally distinct high and low molecular weight species of channel catfish and mouse IL-1. *Cytokine* **6**, 10–20.

Ellsaesser, C. F., Miller, N. W., Cuchens, M. A., Lobb, C. J., Clem, L. W. (1985). Analysis of channel catfish peripheral blood leukocytes by bright-field microscopy and flow cytometry. *Trans. Am. Fish. Soc.* **114**, 279–285.

Espelid, S., Jørgensen, T. Ø. (1992). Antigen processing of *Vibrio salmonicida* by fish (*Salmo salar*) *in vitro*. *Fish Shellfish Immunol.* **2**, 131–141.

Espenes, A., Press, C. M., Dannevig, B. H., Landsverk, T. (1995a). Investigation of the structural and functional features of splenic ellipsoids in rainbow trout (*Oncorhynchus mykiss*). *Cell Tissue Res.* **279**, 469–474.

Espenes, A., Press, C. M., Dannevig, B. H., Landsverk, T. (1995b). Immune-complex trapping in the splenic ellipsoids of rainbow trout (*Oncorhynchus mykiss*). *Cell Tissue Res.* **282**, 41–48.

Esteban, M. A., Lopez-Ruiz, A., Garcia-Ayala, A., Mulero, V., Meseguer, J. (1994). Inhibition and quantification of bacterial uptake by blood thrombocytes of sea bass (*Dicentrarchus labrax* L.). *Dev. Comp. Immunol.* **18** (Suppl. 1), S86.

Estevez, J., Leiro, J., Santamarina, M. T., Dominguez, J., Ubeira, F. M. (1994). Monoclonal antibodies of turbot (*Scophthalmus maximus*) immunoglobulins: characterization and applicability in immunoassays. *Vet. Immunol. Immunopathol.* **41**, 353–366.

Estevez, J., Leiro, J., Santamarina, M. T., Ubeira, F. M. (1995). A sandwich immunoassay to quantify low levels of turbot

(*Scophthalmus maximus*) immunoglobulins. *Vet. Immunol. Immunopathol.* **45**, 165–174.

Etlinger, H. M., Chiller, J. M., Hodgins, H. O. (1979). Evolution of the lymphoid system. IV. Murine T-independent but not T-dependent antigens are very immunogenic in rainbow trout (*Salmo gairdneri*). *Cell. Immunol.* **47**, 400–406.

Evans, D. L., Jaso-Friedmann, L. (1992). Nonspecific cytotoxic cells as effectors of immunity in fish. *Annu. Rev. Fish Dis.* **2**, 109–121.

Evans, D. L., Gratzek, J. B. (1989). Immune defense mechanisms in fish to protozoan and helminth infections. *Am. Zool.* **29**, 409–418.

Evans, D. L., McKinney, E. C. (1991). Phylogeny of cytotoxic cells. In *Phylogenesis of Immune Function* (eds Warr, G. W., Cohen, N.), pp. 215–239. CRC Press, Boca Raton, Florida, USA.

Evans, D. L., Carlson, R. L., Graves, S. S., Hogan, K. T. (1984a). Nonspecific cytotoxic cells in fish (*Ictalurus punctatus*). IV: Target cell binding and recycling capacity. *J. Immunol.* **141**, 324–332.

Evans, D. L., Graves, S. S., Cobb, D., Dawe, D. L. (1984b). Nonspecific cytotoxic cells in fish (*Ictalurus punctatus*). II. Parameters of target cell lysis and specificity. *Dev. Comp. Immunol.* **8**, 303–312.

Evans, D. L., Jaso-Friedmann, L., Smith, E. E., St. John, A., Koren, H. S., Harris, D. T. (1988). Identification of a putative antigen receptor on fish nonspecific cytotoxic cells with monoclonal antibodies. *J. Immunol.* **141**, 324–332.

Faisal, M., Hetrick, F. M. (eds) (1992). *Annual Review of Fish Diseases*. Pergamon Press, New York, USA.

Fänge, R. (1982). A comparative study of lymphomyeloid tissue in fish. *Dev. Comp. Immunol.* **6** (Suppl. 2), 22–33.

Fänge, R., Lundblad, G., Lind, J. (1976). Lysozyme and chitinase in blood and lymphomyeloid tissues of marine fish. *Mar. Biol.* **36**, 277–282.

Ferguson, H. W. (1976). The relationship between ellipsoids and melano-macrophage centres in the spleen of turbot (*Scophthalmus maximus*). *J. Comp. Pathol.* **86**, 377–380.

Fidler, J. E., Clem, L. W., Small Jr, P. A. (1969). Immunoglobulin synthesis in neonatal nurse sharks (*Ginglymostoma cirratum*). *Comp. Biochem. Physiol.* **31**, 365–371.

Finn, J. P., Nielson, N. O. (1971). The inflammatory response of rainbow trout. *J. Fish Biol.* **3**, 463–478.

Fisher, S. E., Shaklee, J. B., Ferris, S. D., Whitt, G. S. (1980). Evolution of five multilocus isozyme systems in the chordates. *Genetica* **52/53**, 73–85.

Fletcher, T. C. (1981). Non-antibody molecules and the defence mechanisms of fish. In *Stress and Fish* (ed. Pickering, A. D.), pp. 171–183. Academic Press, London, UK.

Fletcher, T. C. (1982). Non-specific defence mechanisms of fish. *Dev. Comp. Immunol.* **6** (Suppl. 2), 123–132.

Fletcher, T. C., Grant, P. T. (1969). Immunoglobulins in serum and mucus of the plaice (*Pleuronectes platessa*). *Biochem. J.* **115**, 65P.

Fletcher, T. C., White, A. (1973a). Lysozyme activity in the plaice (*Pleuronectes platessa* L.). *Experienta* **29**, 1283–1285.

Fletcher, T. C., White, A. (1973b). Antibody production in the plaice *Pleuronectes platessa* after oral and parenteral immunization with *Vibrio anguillarum* antigens. *Aquaculture* **1**, 417–428.

Francis, C. H., Ellis, A. E. (1994). Production of a lymphokine (macrophage activating factor) by salmon (*Salmo salar*) leucocytes stimulated with outer membrane protein antigens of *Aeromonas salmonicida*. *Fish Shellfish Immunol.* **4**, 489–497.

Friedman, A., Al-Sabbagh, A., Santos, L. M. B., Fishman-Lobell, J., Polanski, M., Prabhu Das, M., Khoury, S. J., Weiner, H. L. (1994). Oral tolerance: a biological relevant pathway to generate peripheral tolerance against external and self antigens. In *Mechanisms of Immune Regulation* (ed. Granstein, R. D.), pp. 259–290. Karger, Basel, Switzerland.

Frommel, D., Litman, G. W., Finstad, J., Good, R. A. (1971). The evolution of the immune system. XI. The immunoglobulins of the horned shark, *Heterodontus francisci*: purification, characterization and structural requirements for antibody activity. *J. Immunol.* **106**, 1234–1243.

Fuda, H., Soyano, K., Yamazaki, F., Hara, A. (1991). Serum immunoglobulin M (IgM) during early development of masu salmon (*Oncorhynchus masou*). *Comp. Biochem. Physiol.* **99**A, 637–643.

Fuda, H., Hara, A., Yamazaki, F., Kobayashi, K. (1992). A peculiar immunoglobulin M (IgM) identified in eggs of chum salmon (*Oncorhynchus keta*). *Dev. Comp. Immunol.* **16**, 415–423.

Fujii, L., Murray, J., Sekizawa, A., Tomonaga, S. (1992). Isolation and characterization of a protein from hagfish serum that is homologous to the third component of the mammalian complement system. *J. Immunol.* **148**, 117–123.

Fujii, T., Murakawa, S. (1981). Immunity in lamprey. III. Occurrence of complement-like activity. *Dev. Comp. Immunol.* **5**, 251–259.

Fuller, L., Murray, J., Jensen, J. A. (1978). Isolation from nurse shark serum of immune 7S antibodies with two different molecular weight H-chains. *Immunochemistry* **15**, 251–259.

Garrido, J., De Ioannes, A. E. (1981). Electron microscopy of the natural IgM-like hemagglutinin of the ratfish (*Callorhynchus callorhynchus*). *Dev. Comp. Immunol.* **5**, 691–696.

Georgopoulou, U., Vernier, J. M. (1986). Local immunological response in the posterior intestinal segment of the rainbow trout after oral administration of macromolecules. *Dev. Comp. Immunol.* **10**, 529–537.

Georgopoulou, U., Sire, M. F., Vernier, J. M. (1986). Immunological demonstration of intestinal absorption and digestion of protein macromolecules in the trout (*Salmo gairdneri*). *Cell Tissue Res.* **245**, 387–395.

Gewurz, H., Finstad, J., Muschel, L. H., Good, R. A. (1966). Phylogenetic inquiry into the origin of the complement system. In *Phylogeny of Immunity* (eds Smith, R. T., Miescher, P. A., Good, R. A.), pp. 105–117. University of Florida Press, Gainesville, FL, USA.

Ghaffari, S. H., Lobb, C. J. (1989a). Cloning and sequence analysis of channel catfish heavy chain cDNA indicate phylogenetic diversity within IgM immunoglobulin family. *J. Immunol.* **142**, 1356–1365.

Ghaffari, S. H., Lobb, C. J. (1989b). Nucleotide sequence of channel catfish Ig heavy chain cDNA and genomic blot analysis. *J. Immunol.* **143**, 2730–2739.

Ghaffari, S. H., Lobb, C. J. (1991). Heavy chain variable region gene families evolved early in phylogeny. Ig complexity in fish. *J. Immunol.* **146**, 1037–1046.

Ghaffari, S. H., Lobb, C. J. (1992). Organization of immunoglobulin heavy chain constant and joining region genes in the channel catfish. *Mol. Immunol.* **29**, 151–159.

Ghaffari, S. H., Lobb, C. J. (1993). Structure and genomic

organization of immunoglobulin light chain in the channel catfish. An unusual genomic organizational pattern of segmental genes. *J. Immunol.* **151**, 6900–6912.

Gitlin, D., Perricelli, A., Gitlin, J. D. (1973). Multiple immunoglobulin classes among sharks and their evolution. *Comp. Biochem. Physiol.* **44B**, 225–239.

Glamann, J. (1995). Complete coding sequence of rainbow trout MHC II β chain. *Scand. J. Immunol.* **41**, 365–372.

Glomski, C. A., Tamnurlin, J., Chainani, M. (1992). The phylogenetic odyssey of the erythrocyte. III. Fish, the lower vertebrate experience. *Histol. Histopath.* 7, 501–528.

Glynn, P. J., Pulsford, A. L. (1993). Tryptic digestion of the serum immunoglobulin of the flounder, *Platichthys flesus. J. Mar. Biol. Ass. UK* **73**, 425–436.

Goetz, R., Koester, R., Winkler, C., Raulf, F., Lottspeich, F., Schartl, M., Thoenen, H. (1994). Neurotrophin 6 is a new member of the nerve growth factor family. *Nature* **372**, 266–269.

Gonzalez, R., Matsiota, P., Torchy, C., de Kinkelin, P., Avrameas, S. (1989). Natural anti-TNP antibodies from rainbow trout interfere with viral infection *in vitro*. *Res. Immunol.* **140**, 675–684.

Grace, M. F., Manning, M. J. (1980). Histogenesis of the lymphoid organs in rainbow trout, *Salmo gairdneri*. *Dev. Comp. Immunol.* **4**, 255–264.

Graham, S., Secombes, C. J. (1988). The production of a macrophage-activating factor from rainbow trout *Salmo gairdneri* leucocytes. *Immunology* **65**, 293–297.

Graham, S., Secombes, C. J. (1990a). Cellular requirements for lymphokine secretion by rainbow trout *Salmo gairdneri* leukocytes. *Dev. Comp. Immunol.* **14**, 59–68.

Graham, S., Secombes, C. J. (1990b). Do fish lymphocytes secrete interferon-γ? *J. Fish Biol.* **36**, 563–573.

Gravell, M., Malsberger, R. G. (1965). A permanent cell line from the fathead minnow (*Pimephales promelas*). *Ann. NY Acad. Sci.* **126**, 555–565.

Graves, S. S., Evans, D. L., Cobb, D., Dawe, D. L. (1984). Nonspecific cytotoxic cells in fish (*Ictalurus punctatus*). I. Optimum requirements for target cell lysis. *Dev. Comp. Immunol.* **8**, 293–302.

Graves, S. S., Evans, D. L., Dawe, D. L. (1985a). Antiprotozoan activity of non-specific cytotoxic cells (NCC) from the channel catfish (*Ictalurus punctatus*). *J. Immunol.* **134**, 78–85.

Graves, S. S., Evans, D. L., Dawe, D. L. (1985b). Mobilization and activation of non-specific cytotoxic cells (NCC) in the channel catfish (*Ictalurus punctatus*) infected with *Ichthyophthirius multifiliis*. *Comp. Immun. Microbiol. Infect. Dis.* **8**, 43–51.

Grayson, T. H., Jenkins, P. G., Wrathmell, A. B., Harris, J. E. (1991). Serum responses to the salmon louse, *Lepeophtheirus salmonis* (Kroyer, 1838), in naturally infected salmonids and immunised rainbow trout, *Oncorhynchus mykiss* (Walbaum), and rabbits. *Fish Shellfish Immunol.* **1**, 141–155.

Greenberg, A. S., Steiner, L., Kasahara, M., Flajnik, M. F. (1993). Isolation of a shark immunoglobulin light chain cDNA clone encoding a protein resembling mammalian κ light chains: Implications for the evolution of light chains. *Proc. Natl Acad. Sci. USA* **90**, 10603–10607.

Greenberg, A. S., Avial, D., Hughes, M., Hughes, A., McKinney, E. C., Flajnik, M. F. (1995). A new antigen receptor gene family that undergoes rearrangement and extensive somatic diversification in sharks. *Nature* **374**, 168–173.

Greenberg, A. S., Hughes, A. L., Guo, J., Avila, D., McKinney, E. C., Flajnik, M. F. (1996). A novel 'chimeric' antibody class in cartilaginous fish: IgM may not be the primordial immunoglobulin. *Eur. J. Immunol.* **26**, 1123–1129.

Greenlee, A. R., Brown, R. A., Ristow, S. S. (1991). Nonspecific cytotoxic cells of rainbow trout (*Oncorhynchus mykiss*) kill YAC-1 targets by both necrotic and apoptic mechanisms. *Dev. Comp. Immunol.* **15**, 153–164.

Griffin, B. R. (1984). Random and directed migration of trout (*Salmo gairdneri*) leucocytes: activation by antibody, complement and normal serum components. *Dev. Comp. Immunol.* **8**, 589–597.

Grimholt, U., Hordvik, I., Fosse, V. M., Olsaker, I., Endresen, C., Lie, Ø. (1993). Molecular cloning of major histocompatibility complex class I cDNAs from Atlantic salmon (*Salmo salar*). *Immunogenetics* **37**, 469–473.

Grimholt, U., Olsaker, I., de Vries Lindstrøm, C., Lie, Ø. (1994). A study of variability in the MHC class II β1 and class I α2 domain exons of Atlantic salmon (*Salmo salar*). *Anim. Genet.* **25**, 147–153.

Grinde, B. (1989). Lysozyme from rainbow trout, *Salmo gairdneri* Richardson, as an antibacterial agent against fish pathogens. *J. Fish Dis.* **12**, 95–104.

Grinde, B., Lie, Ø., Poppe, T., Salte, R. (1988). Species and individual variation in lysozyme activity in fish of interest in aquaculture. *Aquaculture* **68**, 299–304.

Grondel, J. L., Harmsen, E. G. M. (1984). Phylogeny of interleukins: growth factors produced by leucocytes of the cyprinid fish, *Cyprinus carpio* L. *Immunology* **52**, 477–482.

Grondel, J. L., Nouws, J. F. M., van Muiswinkel, W. B. (1987). The influence of antibiotics on the immune system: immunopharmokinetic investigations on the primary anti-SRBC response in carp, *Cyprinus carpio* L., after oxytetracycline injection. *J. Fish Dis.* **10**, 35–43.

Gross, L. (1983). Tumors, leukemia and lymphosarcoma in a fish. In *Oncogenic Viruses* (ed. Gross, L.), pp. 103–116. Pergamon Press, Oxford, UK.

Haddow, A., Blake, I. (1933). Neoplasms in fish: a report of six cases with a summary of the literature. *J. Path. Bact.* **36**, 41–47.

Hagiwara, K., Kobayashi, K., Kajii, T., Tomonaga, S. (1985). J-chain-like component in 18S immunoglobulin of the skate *Raja kenojei*, a cartilaginous fish. *Mol. Immunol.* **22**, 775–778.

Hamby, B. A., Huggins, E. M., Lachman, L. B., Dinarello, C. A., Sigel, M. M. (1986). Fish lymphocytes respond to human IL-1. *Lymphokine Res.* **5**, 157–162.

Hanley, P. J., Seppelt, I. M., Gooley, A. A., Hook, J. W., Raison, R. L. (1990). Distinct Ig H chains in a primitive vertebrate, *Eptatretus stoutii*. *J. Immunol.* **145**, 3823–3828.

Hansen, J. D., Kaattari, S. L. (1995). The recombination activating gene 1 (RAG1) of rainbow trout (*Oncorhynchus mykiss*). *Immunogenetics* **44**, 188–195.

Hansen, J., Kaattari, S. L. (1996). The recombination activating gene 2 (RAG2) of the rainbow trout *Oncorhynchus mykiss*. *Immunogenetics* **44**, 203–211.

Hansen, J., Leong, J.-A., Kaattari, S. (1994). Complete nucleotide sequence of a rainbow trout cDNA encoding a membrane-bound form of immunoglobulin heavy chain. *Mol. Immunol.* **31**, 499–501.

Hansen, J. D., Strassburger, P., Du Pasquier, L. (1997). Conservation of an alpha 2 domain within the teleostean world,

MHC class I from the rainbow trout *Onchorhynchus mykiss*. *Dev. Comp. Immunol.* 20, 417–425.

Harada, T., Hatanaka, J., Kubota, S. S., Enomoto, M. (1990). Lymphoblastic lymphoma in medaka, *Oryzias latipes* (Temminck et Schlegel). *J. Fish Dis.* 13, 169–173.

Hardee, J. J., Godwin, U., Benedetto, R., McConnell, T. J. (1995). Major histocompatibility complex class I A gene polymorphism in the striped bass. *Immunogenetics* 41, 229–238.

Hardie, L. J., Chappell, L. H., Secombes, C. J. (1994a). Human tumor necrosis factor α influences rainbow trout *Oncorhynchus mykiss* leucocyte responses. *Vet. Immunol. Immunopathol.* 40, 73–84.

Hardie, L. J., Fletcher, T. C., Secombes, C. J. (1994b). Effect of temperature on macrophage activation and the production of macrophage activating factor by rainbow trout (*Oncorhynchus mykiss*) leucocytes. *Dev. Comp. Immunol.* 18, 57–66.

Harding, F. A., Amemiya, C. T., Litman, R. T., Cohen, N., Litman, G. W. (1990a). Two distinct immunoglobulin heavy chain isotypes in a primitive, cartilaginous fish, *Raja erinacea*. *Nucl. Acids Res.* 18, 6369–6376.

Harding, F. A., Cohen, N., Litman, G. W. (1990b). Immunoglobulin heavy chain gene organization and complexity in the skate, *Raja erinacea*. *Nucl. Acids Res.* 18, 1015–1020.

Harless, J., Obermoeller, R. D., Walter, R. B., Svennson, R., Nairn, R. S., Vielkind, J. R., Kallman, K. D., Morizot, D. C. (1995). Characterization of a tyrosine kinase-peptidase A synteny in linkage group XIII of *Xiphophorus* fishes (Teleostei: Poeciliidae): implications for vertebrate chromosome evolution. *J. Hered.* 82, 256–259.

Harshbarger, J. C., Dawe, C. J. (1973). Hematopoietic neoplasms in invertebrate and poikilothermic vertebrate animals. In *Unifying Concepts of Leukemia* (eds Dutcher, R. M., Chieco-Bianchi, L.), pp. 1–25. Karger, Basel, Switzerland.

Hart, S., Wrathmell, A. B., Harris, J. E. (1986). Ontogeny of gut-associated lymphoid tissue (GALT) in the dogfish *Scyliorhinus canicula* L. *Vet. Immunol. Immunopathol.* 12, 107–116.

Hashimoto, K., Nakanishi, T., Kurosawa, Y. (1990). Isolation of carp genes encoding major histocompatibility complex antigens. *Proc. Natl Acad. Sci. USA* 87, 6863–6867.

Hashimoto, K., Nakanishi, T., Kurosawa, Y. (1992). Identification of a shark sequence resembling the major histocompatibility complex class I $\alpha 3$ domain. *Proc. Natl Acad. Sci. USA* 89, 2209–2212.

Håverstein, L. S., Aasjord, P. M., Ness, S., Endresen, C. (1988). Purification and partial characterization of an IgM-like serum immunoglobulin from Atlantic salmon (*Salmo salar*). *Dev. Comp. Immunol.* 12, 773–785.

Hawke, N. A., Rast, J. P., Litman, G. W. (1996). Extensive diversity of transcribed TCR-β in a phylogenetically primitive vertebrate. *J. Immunol.* 156, 2458–2464.

Hayman, J. R., Lobb, C. J. (1993). Immunoglobulin in the eggs of the channel catfish (*Ictalurus punctata*). *Dev. Comp. Immunol.* 17, 241–248.

Hayman, J. R., Bly, J. E., Levine, R. P., Lobb, C. J. (1992). Complement deficiencies in channel catfish (*Ictalurus punctatus*) associated with temperature and seasonal mortality. *Fish Shellfish Immunol.* 2, 183–192.

Hayman, J. R., Ghaffari, S. H., Lobb, C. J. (1993). Heavy chain joining region segments of the channel catfish. *J. Immunol.* 151, 3587–3596.

Haynes, L., Cohen, N. (1993). Transforming growth factor beta (TGFβ) is produced by and influences the proliferative response of *Xenopus laevis* lymphocytes. *Dev. Immunol.* 3, 223–230.

Haynes, L., McKinney, E. C. (1991). Shark spontaneous cytotoxicity: characterization of the regulatory cell. *Dev. Comp. Immunol.* 15, 123–134.

Herraez, M. P., Zapata, A. G. (1986). Structure and function of the melano-macrophage centres of the goldfish *Carassius auratus*. *Vet. Immunol. Immunopathol.* 12, 117–126.

Hinds, K. R., Litman, G. W. (1986). Major reorganization of immunoglobulin V_H-segmental elements during vertebrate evolution. *Nature* 320, 546–549.

Hinds-Frey, K. R., Nishikata, H., Litman, R. T., Litman, G. W. (1993). Somatic variation precedes extensive diversification of germline sequences and combinatorial joining in evolution of immunoglobulin heavy chain diversity. *J. Exp. Med.* 178, 815–824.

Hine, P. M. (1992). The granulocytes of fish. *Fish Shellfish Immunol.* 2, 79–98.

Hinuma, S., Abo, T., Kumaga, K., Hata, M. (1980). The potent activity of freshwater fish kidney cells in cell-killing. I. Characterization and species-distribution of cytotoxicity. *Dev. Comp. Immunol.* 4, 653–666.

Hjelmeland, K., Christie, M., Raa, J. (1983). Skin mucus protease from rainbow trout, *Salmo gairdneri* Richardson, and its biological significance. *J. Fish Biol.* 23, 13–22.

Höchtl, J., Müller, C. R., Zachau, H. G. (1982). Recombined flanks of the variable and joining segments of immunoglobulin genes. *Proc. Natl Acad. Sci. USA* 79, 1383–1387.

Hoffmann, R. W., Fischer-Scherl, T., Pfeil-Putzien, C. (1988). Lymphosarcoma in a wild grayling, *Thymallus thymallus* L.: a case report. *J. Fish Dis.* 11, 267–270.

Hogan, R. J., Stuge, T. B., Clem, L. W., Miller, N. W., Chinchar, V. G. (1996). Anti-viral cytotoxic cells in the channel catfish (*Ictalurus punctatus*). *Dev. Comp. Immunol.* 20, 115–127.

Hohman, V. S., Schluter, S. F., Marchalonis, J. J. (1992). Complete sequence of a cDNA clone specifying sandbar shark immunoglobulin light chain: gene organisation and implications for the evolution of light chains. *Proc. Natl Acad. Sci. USA* 89, 276–280.

Hohman, V. S., Schuchman, D. B., Schluter, S. F., Marchalonis, J. J. (1993). Genomic clone for sandbar shark λ light chain: generation of diversity in the absence of gene rearrangement. *Proc. Natl Acad. Sci. USA* 90, 9882–9886.

Hohman, V. S., Schluter, S. F., Marchalonis, J. J. (1995). Diversity of Ig light chain clusters in the sandbar shark (*Carcharhinus plumbeus*). *J. Immunol.* 155, 3922–3928.

Holmskov, U., Malhotra, R., Sim, R. B., Jensenius, J. C. (1994). Collectins: collagenous C-type lectins of the innate immune defense system. *Immunol. Today* 15, 67–74.

Honma, Y. (1966). Studies on the endocrine glands of the salmonid fish, the ayu, *Plecoglossus altevelis* Temminck et Schlegel. VI. Effect of artificially controlled light on the endocrines of the pond-cultured fish. *Bull. Jap. Soc. Sci. Fish.* 32, 32–40.

Honma, Y., Hirosaki, Y. (1966). Histopathology on the tumors and endocrine glands of the immature chum salmon, *Oncorhynchus keta*, reared in the Enoshima aquarium. *Jap. J. Icthyol.* 14, 74–83.

Hopkins, S. J., Humphreys, M. (1989). Simple, sensitive and

specific bioassay of interleukin-1. *J. Immunol. Methods* **120**, 271–276.

Hordvik, I., Grimholt, U., Fosse, V. M., Lie, Ø., Endresen, C. (1993). Cloning and sequence analysis of cDNAs encoding the MHC class II β chain in Atlantic salmon (*Salmo salar*). *Immunogenetics* **37**, 437–441.

Hordvik, I., Jacob, A. L. J., Charlemagne, J., Endresen, C. (1996). Cloning of T-cell antigen receptor beta chain cDNAs from Atlantic salmon, *Salmo salar*. *Immunogenetics* **45**, 9–14.

Hordvik, I., Voie, A. M., Glette, J., Male, R., Endresen, C. (1992). Cloning and sequence analysis of two isotypic IgM heavy chain genes from Atlantic salmon, *Salmo salar* L. *Eur. J. Immunol.* **22**, 2957–2962.

Houghton, G., Matthews, R. A. (1990). Immunosuppression in juvenile carp, *Cyprinus carpio* L.: the effects of the corticosteriods triamcinolone acetonide and hydrocortisone 21-hemisuccinate (cortisol) on acquired immunity and the humoral antibody response to *Ichthyophthirius multifiliis* Fouquet. *J. Fish Dis.* **13**, 269–280.

Howell, C. J. G. (1987). A chemotactic factor in the carp *Cyprinus carpio*. *Dev. Comp. Immunol.* **11**, 134–146.

Hsu, E., Steiner, L. A. (1992). Primary structure of immunoglobulin through evolution. *Curr. Opinion Struct. Biol.* **2**, 422–431.

Hunt, T. C., Rowley, A. F. (1986). Leukotriene B_4 induces enhanced migration of fish leucocytes *in vitro*. *Immunology* **59**, 563–568.

Iger, Y., Wendelaar Bonga, S. E. (1994). Cellular aspects of the skin of carp exposed to acidified water. *Cell Tissue Res.* **275**, 481–492.

Iida, T., Wakabayashi, H. (1988). Chemotactic and leukocytosis-inducing activities of eel complement. *Fish Pathol.* **23**, 55–58.

Ingram, G. A. (1980). Substances involved in the natural resistance of fish to infection – a review. *J. Fish Biol.* **16**, 23–60.

Ingram, G. A., Alexander, J. B. (1976). The immune response of brown trout (*Salmo trutta* L.) to injection with soluble antigens. *Acta Biol. Med. Germ.* **35**, 1561–1570.

Ingram, G. A., Alexander, J. B. (1979). The immunoglobulin of the brown trout, *Salmo trutta* and its concentration in the serum of antigen-stimulated and non-stimulated fish. *J. Fish Biol.* **14**, 249–260.

Ingram, G. A., Alexander, J. B. (1980). The immune response of brown trout, *Salmo trutta* to lipopolysaccharide. *J. Fish Biol.* **16**, 181–197.

Ingram, G. A., Alexander, J. B. (1981). The primary immune response of brown trout (*Salmo trutta*) to cellular and soluble antigens: enumeration of antibody-secreting and antibody-binding cells and the production of antibody. *Acta Biol. Med. Germ.* **40**, 317–330.

Isbell, G. H., Pauley, G. B. (1983). Characterization of immunoglobulins from the brown bullhead (*Ictalurus nebulosus*) produced against a naturally occurring bacterial pathogen, *Aeromonas hydrophila*. *Dev. Comp. Immunol.* **7**, 473–782.

Ishiguro, H., Kobayashi, K., Suzuki, M. (1992). Isolation of a hagfish gene that encodes a complement component. *EMBO J.* **11**, 829–837.

Israelsson, O., Petersson, A., Bengten, E., Wiersma, E. J., Andersson, B. S., Gezelius, G., Pilström, L. (1991). Immunoglobulin concentration in Atlantic cod, *Gadus morhua* L., serum and cross reactivity between anti-cod-antibodies and immunoglobulin from other species. *J. Fish Biol.* **39**, 265–278.

Janeway, C. A., Travers, P. (1996). *Immunobiology: The Immune System in Health and Disease*, 2nd edn. Current Biology Ltd., San Francisco and Philadelphia, USA.

Jang, S. I., Hardie, L. J., Secombes, C. J. (1994). Effects of transforming growth factor β1 on rainbow trout *Oncorhynchus mykiss* macrophage respiratory burst activity. *Dev. Comp. Immunol.* **18**, 315–323.

Jang, S. I., Mulero, V., Hardie, L. J., Secombes, C. J. (1995a). Inhibition of rainbow trout phagocyte responsiveness to human tumour necrosis factor α (hTNFα) with monoclonal antibodies to the hTNFα 55 kDa receptor. *Fish Shellfish Immunol.* **5**, 61–69.

Jang, S. I., Hardie, L. J., Secombes, C. J. (1995b). Elevation of rainbow trout *Oncorhynchus mykiss* macrophage respiratory burst activity with macrophage-derived supernatants. *J. Leuk. Biol.* **57**, 943–947.

Jang, S. I., Marsden, M. J., Kim, Y. G., Choid, M. S., Secombes, C. J. (1995c). The effect of glycyrrhizin on rainbow trout, *Oncorhynchus mykiss* (Walbaum), leucocyte responses. *J. Fish Dis.* **18**, 307–315.

Jenkins, J. A., Ourth, D. D. (1990). Membrane damage to *Escherichia coli* and bactericidal kinetics by the alternative complement pathway of channel catfish. *Comp. Biochem. Physiol.* **97**, 477–481.

Jensen, J. A., Festa, E. (1980). The 6-component complement system of the nurse shark (*Ginglymostoma cirratum*). In *Aspects of Developmental and Comparative Immunology* (ed. Solomon, J. B.), pp. 485–486. Pergamon Press, Oxford, UK.

Jensen, J. A., Festa, E., Smith, D. S., Cayer, M. (1981). The complement system of the nurse shark: hemolytic and comparative characteristics. *Science* **214**, 566–569.

Jensen, L. B., Koch, C. (1991). Genetic polymorphism of component C3 of rainbow trout (*Oncorhynchus mykiss*) complement. *Fish Shellfish Immunol.* **1**, 237–242.

Johnson, K. A., Flynn, J. K., Amend, D. F. (1982a). Onset of immunity in salmonid fry vaccinated by direct immersion in *Vibrio anguillarum* and *Yersinia ruckeri* bacterins. *J. Fish Dis.* **5**, 197–205.

Johnson, K. A., Flynn, J. K., Amend, D. F. (1982b). Duration of immunity in salmonids vaccinated by direct immersion with *Yersinia ruckeri* and *Vibrio anguillarum* bacterins. *J. Fish Dis.* **5**, 205–213.

Johnston Jr, W. H., Acton, R. T., Weinheimer, P. F., Niedermeier, W., Evans, E. E., Shelton, E., Bennett, J. C. (1971). Isolation and physico-chemical characterization of the IgM-like immunoglobulin from the stingray, *Dasyatis americana*. *J. Immunol.* **107**, 782–783.

Johnstone, J. (1911–12). Internal parasites and diseased conditions of fishes. *Proc. Trans. Lpool Biol. Soc.* **26**, 103–144.

Johnstone, J. (1926). Malignant and other tumors in marine fish. *Proc. Trans. Lpool Biol. Soc.* **40**, 75–98.

Jones, J. C., Ghaffari, S. H., Lobb, C. J. (1993). Immunoglobulin heavy chain constant and heavy chain variable region genes in phylogenetically diverse species of bony fish. *J. Mol. Evol.* **36**, 417–428.

Joosten, P. H. M., Aviles-Trigueros, M., Sorgeloos, P., Rombout, J. H. W. M. (1995). Oral vaccination of juvenile carp (*Cyprinus carpio*) and gilthead seabream (*Sparus aurata*) with bioencapsulated *Vibrio anguillarum* bacterin. *Fish Shellfish Immunol.* **5**, 289–299.

Joosten, P. H. M. (1997). Immunological aspects of oral vaccination in fish (*Cyprinus carpio* L.). Ph.D., Agricultural University, Wageningen, The Netherlands.

Jorgensen, J. B., Robertsen, B. (1995). Yeast β-glucan stimulates respiratory burst activity of Atlantic salmon (*Salmo salar* L.) macrophages. *Dev. Comp. Immunol.* **19**, 43–57.

Josefsson, S., Tatner, M. F. (1993). Histogenesis of the lymphoid organs in sea bream (*Sparus aurata* L.). *Fish Shellfish Immunol.* **3**, 35–49.

Jurd, R. D. (1987). Hypersensitivity in fishes: a review. *J. Fish Biol.* **31** (Suppl. A), 1–7.

Juul-Madsen, H., Glamann, J., Madsen, H. O., Simonsen, M. (1992). MHC class II beta-chain expression in the rainbow trout. *Scand. J. Immunol.* **35**, 687–694.

Kaastrup, P., Koch, C. (1983). Complement in the carp fish. *Dev. Comp. Immunol.* **7**, 781–782.

Kaastrup, P., Nielsen, B., Hørlyck, V., Simonsen, M. (1988). Mixed lymphocyte reactions (MLR) in rainbow trout (*Salmo gairdneri*) sibling. *Dev. Comp. Immunol.* **12**, 801–808.

Kaastrup, P., Stet, R. J. M., Tigchelaar, A. J., Egberts, E., van Muiswinkel, W. B. (1989). A major histocompatibility locus in fish: serological identification and segregation of transplantation antigens in the common carp (*Cyprinus carpio* L.). *Immunogenetics* **30**, 284–290.

Kaattari, S. L., Shapiro, D. A. (1994). *Determination of Serum Antibody Affinity Distributions Using ELISA-based Technology*. SOS Publications, Fair Haven, New York, USA.

Kaidoh, T., Gigli, I. (1989). Phylogeny of regulatory proteins of the complement system. Isolation and characterisation of a C4b/C3b inactivator and cofactor from sand bass plasma. *J. Immunol.* **142**, 1605–1613.

Kallman, K. D. (1970). Genetics of tissue transplantation in teleostei. *Transpl. Proc.* **2**, 263–271.

Kamiya, H., Muramoto, K., Goto, R. (1988). Purification and properties of agglutinins from conger eel, *Conger myriaster*, skin mucus. *Dev. Comp. Immunol.* **12**, 309–318.

Kamiya, H., Muramoto, K., Goto, R., Sakai, M., Ida, H. (1990). Properties of a lectin in chum salmon ova. *Nipp. Suis. Gakk.* **56**, 1139–1144.

Kasahara, M., Vazquez, M., Sato, K., Churchill McKinney, E. (1992). Evolution of the major histocompatibility complex: isolation of class II cDNA clones from the cartilaginous fish. *Proc. Natl Acad. Sci. USA* **89**, 6688–6692.

Kaufman, J., Salomonsen, J., Flajnik, M. (1994). Evolutionary conservation of MHC class I and class II molecules – different yet the same. *Sem. Immunol.* **6**, 411–424.

Kaufman, J. F., Ferrone, S., Flajnik, M. F., Kilb, M., Völk, H., Parisot, R. (1990a). MHC-like molecules in some nonmammalian vertebrates can be detected by some cross-reactive monoclonal antibodies. *J. Immunol.* **144**, 2273–2280.

Kaufman, J. F., Skjoedt, K., Salomonsen, J., Simonsen, M., Du Pasquier, L., Parisot, R., Riegert, P. (1990b). MHC-like molecules in some nonmammalian vertebrates can be detected by some cross-reactive xenoantisera. *J. Immunol.* **144**, 2258–2272.

Kawahara, E., Inarimori, T., Urano, K., Nomura, S., Takahashi, Y. (1993). Transfer of maternal immunity of white spotted char *Salvelinus leucomaenis* against furunculosis. *Nipp. Suis. Gakk.* **59**, 567.

Kawai, K., Kusuda, R., Itami, T. (1981). Mechanisms of protection in ayu orally vaccinated for vibriosis. *Fish Pathol.* **15**, 257–262.

Kawakami, M., Ihara, I., Ihara, S., Suzuki, A., Fukui, K. (1984). A group of bactericidal factors conserved by vertebrates for more than 300 million years. *J. Immunol.* **132**, 2578–2581.

Kelly, R. K., Loh, P. C. (1973). Some properties of an established fish cell line from *Xiphophorus helleri* (red swordtail). *In Vitro* **9**, 73–80.

Kent, M., Groff, J. M., Traxler, G. S., Zinkl, J. G., Bagshaw, J. W. (1990). Plasmacytoid leukemia in seawater reared chinook salmon, *Oncorhynchus tshawytscha*. *Dis. Aquat. Org.* **8**, 199–209.

Kent, M. L., Hedrick, R. P. (1987). Effects of cortisol implants on the PKX myxosporean causing proliferative kidney disease in rainbow trout, *Salmo gairdneri*. *J. Parasitol.* **73**, 455–461.

Kieser, D., Kent, M. L., Groff, J. M., McLean, W. E., Bagshaw, J. (1991). An epizootic of an epitheliotropic lymphoblastic lymphoma in coho salmon *Oncorhynchus kisutch*. *Dis. Aquat. Org.* **11**, 1–8.

Killie, J. E., Espelid, S., Jørgensen, T. Ø. (1991). The humoral immune response in Atlantic salmon (*Salmo salar* L.) against the hapten carrier antigen NIP-LPH: the effects of determinant (NIP) density and the isotype profiles of anti-NIP antibodies. *Fish Shellfish Immunol.* **1**, 33–46.

Klapper, D. G., Clem, L. W. (1977). Phylogeny of immunoglobulin structure and function: characterization of the cysteine-containing peptide involved in the pentamerization of shark IgM. *Dev. Comp. Immunol.* **1**, 81–92.

Klein, D., Ono, H., O'hUigin, C., Vincek, V., Goldsmidt, T., Kein, J. (1993). Extensive MHC variability in cichlid fishes of Lake Malawi. *Nature* **364**, 330–334.

Klesius, P. H. (1990). Effect of size and temperature on the quantity of immunoglobulin in channel catfish, *Ictalurus punctatus*. *Vet. Immunol. Immunopathol.* **24**, 187–195.

Knittel, M. D. (1981). Susceptibility of steelhead trout *Salmo gairdneri* Richardson to redmouth infection *Yersinia ruckeri* following exposure to copper. *J. Fish Dis.* **4**, 33–40.

Kobayashi, K., Tomonaga, S. (1988). The second immunoglobulin class is commonly present in cartilaginous fish belonging to the order Rajiformes. *Mol. Immunol.* **25**, 115–120.

Kobayashi, K., Hara, A., Takano, K., Hira, H. (1982). Studies on subunit components of immunoglobulin M from a bony fish, the chum salmon (*Oncorhynchus keta*). *Mol. Immunol.* **19**, 95–103.

Kobayashi, K., Tomonaga, S., Kajii, T. (1984). A second class of immunoglobulin other than IgM present in the serum of a cartilaginous fish, the skate *Raja kenojei*: isolation and characterization. *Mol. Immunol.* **21**, 397–404.

Kobayashi, K., Tomonaga, S., Hagiwara, K. (1985a). Isolation and characterization of immunoglobulin of hagfish, *Eptatretus burgeri*, a primitive vertebrate. *Mol. Immunol.* **22**, 1091–1097.

Kobayashi, K., Tomonaga, S., Teshima, K., Kajii, T. (1985b). Ontogenic studies on the appearance of two classes of immunoglobulin-forming cells in the spleen of the Aleutian skate, *Bathyraja aleutica*, a cartilaginous fish. *Eur. J. Immunol.* **15**, 952–956.

Kobayashi, K., Tomonaga, S., Tanaka, S. (1992). Identification of a second immunoglobulin in the most primitive shark, the frill shark, *Chlamydoselachus anguineus*. *Dev. Comp. Immunol.* **16**, 295–299.

Kodama, H., Yamada, F., Murai, T., Nakanishi, Y., Mikami, M., Izawa, H. (1989). Activation of trout macrophages and production of CRP after immunization with *Vibrio anguillarum*. *Dev. Comp. Immunol.* **13**, 123–132.

Kodama, H., Mukamoto, M., Baba, T., Mule, D. M. (1994). Macrophage-colony stimulating activity in rainbow trout (*Oncorhynchus mykiss*) serum. *Modulators Fish Immune Responses* **1**, 59–66.

Kokubu, F., Hinds, K., Litman, R., Shamblott, M. J., Litman, G. W. (1987). Extensive families of constant region genes in a phylogenetically primitive vertebrate indicate an additional level of immunoglobulin complexity. *Proc. Natl Acad. Sci. USA* **84**, 5868–5872.

Kokubu, F., Hinds, K., Litman, R., Shamblott, M. J., Litman, G. W. (1988a). Complete structure and organization of immunoglobulin heavy chain constant region genes in a phylogenetically primitive vertebrate. *EMBO J.* **7**, 1979–1988.

Kokubu, F., Litman, R., Shamblott, M. J., Hinds, K., Litman, G. W. (1988b). Diverse organization of immunoglobulin VH gene loci in a primitive vertebrate. *EMBO J.* **7**, 3413–3422.

Komen, J., van den Dobbelsteen, P. J. M., Slierendrecht, W. J., van Muiswinkel, W. B. (1990). Skin grafting in gynogenetic common carp (*Cyprinus carpio* L.). *Transplantation* **49**, 788–793.

Koppang, E. O., Lundin, M., Press, C. M., Rønningen, K., Lie, Ø. (1998). Differing levels of 8 MHC class II β chain expression in a range of tissues from Atlantic salmon, *Salmo salar* L. *Fish Shellfish Immunol.*, in press.

Koumans-van Diepen, J. C. E., Egberts, E., Peixoto, B. R., Taverne, N., Rombout, J. H. W. M. (1995). B cell and immunoglobulin heterogeneity in carp (*Cyprinus carpio* L.): an immuno(cyto)chemical study. *Dev. Comp. Immunol.* **19**, 97–108.

Koumans-van Diepen, J. C. E., Taverne-Thiele, J. J., van Rens, B. T. T. M., Rombout, J. H. W. M. (1994a). Immunocytochemical and flow cytometric analysis of B cells and plasma cells in carp (*Cyprinus carpio* L.): an ontogenetic study. *Fish Shellfish Immunol.* **4**, 19–28.

Koumans-van Diepen, J. C. E., van de Lisdonk, M. H. M., Taverne-Thiele, J. J., Verburg-van Kemenade, B. M. L., Rombout, J. H. W. M. (1994b). Characterisation of immunoglobulin-binding leucocytes in carp (*Cyprinus carpio* L.). *Dev. Comp. Immunol.* **18**, 45–56.

Kurata, O., Okamato, N., Ikeda, Y. (1995). Neutrophilic granulocytes in carp, *Cyprinus carpio*, possess a spontaneous cytotoxic activity. *Dev. Comp. Immunol.* **19**, 315–325.

Kusuda, R., Chen, C.-F., Kawai, K. (1987). The immunoglobulin of colored carp, *Cyprinus carpio*, against *Aeromonas hydrophila*. *Fish Pathol.* **22**, 179–183.

Lamas, J., Ellis, A. E. (1994a). Atlantic salmon (*Salmo salar*) neutrophil responses to *Aeromonas salmonicida*. *Fish Shellfish Immunol.* **4**, 201–219.

Lamas, J., Ellis, A. E. (1994b). Electron microscope observations of the phagocytosis and subsequent fate of *Aeromonas salmonicida* by Atlantic salmon neutrophils *in vitro*. *Fish Shellfish Immunol.* **4**, 539–546.

Lambris, J. D., Lao, Z., Pang, J., Alsenz, J. (1993). The component of trout complement cDNA cloning and conservation of functional sites. *J. Immunol.* **151**, 6123–6134.

Lamers, C. H. J., De Haas, M. J. H., van Muiswinkel, W. B. (1985). Humoral response and memory formation in carp after injection of *Aeromonas hydrophila* bacterin. *Dev. Comp. Immunol.* **9**, 65–75.

Laudan, R., Stolen, J. S., Cali, A. (1986). Immunoglobulin levels of the winter flounder (*Pseudopleuronectes americanus*) and the summer flounder (*Paralichthys dentatus*) injected with the microsporidan parasite *Glugea stephani*. *Dev. Comp. Immunol.* **10**, 331–340.

Laudan, R., Stolen, J. S., Cali, A. (1989). The effect of the microsporidia *Glugea stephani* on the immunoglobulin levels of juvenile and adult winter flounder (*Pseudopleuronectes americanus*). *Dev. Comp. Immunol.* **13**, 35–41.

Law, S. K. A., Reid, K. B. M. (1995). *Complement: In Focus*. IRL Press, Oxford, UK.

Ledford, B. E., Magor, B. G., Middleton, D. L., Miller, R. L., Wilson, M. R., Miller, N. W., Clem, L. W., Warr, G. W. (1993). Expression of a mouse-channel catfish chimeric IgM molecule in mouse myeloma cell. *Mol. Immunol.* **30**, 1405–1417.

Lee, M. A., Bengten, E., Daggfeldt, A., Rytting, A.-S., Pilström, L. (1993). Characterisation of rainbow trout cDNAs encoding a secreted and membrane-bound Ig heavy chain and the genomic intron upstream of the first constant exon. *Mol. Immunol.* **30**, 641–648.

Legler, D. W., Weinheimer, P. F., Acton, R. T., Dupree, H. K., Russell, T. R. (1971). Humoral immune factors in the paddlefish, *Polyodon spathula*. *Comp. Biochem. Physiol.* **38B**, 523–527.

Lele, S. H. (1933). On the phasical history of the thymus gland in plaice of various ages with note on the involution of the organ, including also notes on the other ductless glands in this species. *J. Univ. Bombay* **1**, 37–53.

Lie, Ø., Grimholt, U. (1996). The major histocompatibility complex of fish: genetic structure and function of the MHC of teleost species. In *The Major Histocompatibility Complex of Domestic Animal Species* (eds Schook, L., Lamont, S.), pp. 17–34. CRC Press, Boca Raton, Florida, FL, USA.

Lie, Ø., Evensen, Ø., Sorensen, A., Froysadal, E. (1989). Study on lysozyme activity in some fish species. *Dis. Aquat. Org.* **6**, 1–5.

Lillehaug, A., Ramstad, A., Bækken, K., Reitan, L. J. (1993). Protective immunity in Atlantic salmon (*Salmo salar* L.) vaccinated at different water temperatures. *Fish Shellfish Immunol.* **3**, 143–156.

Lindsay, G. J. H. (1986). The significance of chitinolytic enzymes and lysozyme in rainbow trout (*Salmo gairdneri*) defence. *Aquaculture* **51**, 169–173.

Litman, G. W., Frommel, D., Finstad, J., Good, R. A. (1971a). The evolution of the immune response. IX. Immunoglobulins of the bowfin: purification and characterization. *J. Immunol.* **106**, 747–754.

Litman, G. W., Frommel, D., Rosenberg, A., Good, R. A. (1971b). Circular dichroic analysis of immunoglobulins in phylogenetic perspective. *Biochim. Biophys. Acta* **236**, 647–654.

Litman, G. W., Rosenberg, A., Frommel, D., Pollara, B., Finstad, J., Good, R. A. (1971c). Biophysical studies of the immunoglobulins. The circular dichroic spectra of the immunoglobulins – a phylogenetic comparison. *Int. Arch. Allergy* **40**, 551–575.

Litman, G. W., Wang, A. C., Fudenberg, H. H., Good, R. A. (1971d). *N*-terminal amino-acid sequence of African lungfish immunoglobulin light chains. *Proc. Natl Acad. Sci. USA* **68**, 2321–2324.

Litman, G. W., Scheffel, C., Gerber-Jensen, B. (1980a). Immunoglobulin diversity in the phylogenetically primitive shark, *Heterodontus francisci*. *J. Immunogenet.* 7, 197–206.

Litman, G. W., Scheffel, C., Mäkelä, O. (1980b). Immunoglobulin diversity in the phylogenetically primitive shark, *Heterodontus francisci*: comparison of fine specificity in hapten binding by antibody to p-azobenzenearsonate. *Immunol. Lett.* **1**, 213–215.

Litman, G. W., Stolen, J. S., Sarvas, H. O., Mäkelä, O. (1982). The range and fine specificity of the anti-hapten immune response: phylogenetic studies. *J. Immunogenet.* **9**, 465–474.

Litman, G. W., Berger, L., Murphy, K., Litman, R., Hinds, K., Erickson, B. W. (1985a). Immunoglobulin VH-gene structure and diversity in *Heterodontus*, a phylogenetically primitive shark. *Proc. Natl Acad. Sci. USA* **82**, 2082–2086.

Litman, G. W., Hinds, K., Berger, L., Murphy, K., Litman, R. (1985b). Structure and organization of immunoglobulin V_H genes in *Heterodontus*, a phylogenetically primitive shark. *Dev. Comp. Immunol.* **9**, 749–758.

Litman, G. W., Haire, R. N., Hinds, K. R., Amemiya, C. T., Rast, J. P., Hulst, M. (1992). Evolutionary development of the B-cell repertoire. *Ann. NY Acad. Sci.* **351**, 360–368.

Ljungberg, O. (1976). Epizootical and experimental studies of skin tumors in northern pike (Esox lucius L.) in the Baltic Sea. In *Progress in Experimental Tumor Research* (eds Dawe, C. J., Scarpelli, D. G., Wellings, S. R.), pp. 156–165. S. Karger, Basel, Switzerland.

Lobb, C. J. (1985). Covalent structure and affinity of channel catfish anti-dinitrophenyl antibodies. *Mol. Immunol.* **22**, 993–999.

Lobb, C. J. (1986). Structural diversity of channel catfish immunoglobulins. *Vet. Immunol. Immunopathol.* **12**, 7–12.

Lobb, C. J. (1987). Secretory immunity induced in catfish, *Ictalurus punctatus*, following bath immunization. *Dev. Comp. Immunol.* **11**, 727–738.

Lobb, C. J., Clem, L. W. (1981a). The metabolic relationships of the immunoglobulins in fish serum, cutaneous mucus, and bile. *J. Immunol.* **127**, 1525–1529.

Lobb, C. J., Clem, L. W. (1981b). Phylogeny of immunoglobulin structure and function. X. Humoral immunoglobulins of the sheephead, *Archosargus probatocephalus*. *Dev. Comp. Immunol.* **5**, 271–282.

Lobb, C. J., Clem, L. W. (1981c). Phylogeny of immunoglobulin structure and function. XI. Secretory immunoglobulins in the cutaneous mucus of the sheepshead, *Archosargus probatocephalus*. *Dev. Comp. Immunol.* **5**, 587–596.

Lobb, C. J., Clem, L. W. (1981d). Phylogeny of immunoglobulin structure and function. XII. Secretory immunoglobulins in the bile of the marine teleost *Archosargus probatocephalus*. *Mol. Immunol.* **20**, 811–818.

Lobb, C. J., Clem, L. W. (1982). Fish lymphocytes differ in the expression of surface immunoglobulin. *Dev. Comp. Immunol.* **6**, 473–479.

Lobb, C. J., Clem, L. W. (1983). Distinctive subpopulations of catfish serum antibody and immunoglobulin. *Mol. Immunol.* **20**, 811–818.

Lobb, C. J., Olson, M. O. J. (1988). Immunoglobulin heavy H chain isotypes in a teleost fish. *J. Immunol.* **141**, 1236–1245.

Lobb, C. J., Olson, M. O. J., Clem, L. W. (1984). Immunoglobulin light chain classes in a teleost fish. *J. Immunol.* **132**, 1917–1923.

Lögdberg, L., Björck, L. (1983). Binding of mammalian β_2-microglobulin by glycoproteins in fish serum. *Mol. Immunol.* **20**, 885–891.

Lundblad, G., Fange, R., Slettengren, K., Lind, K. (1979). Lysozyme, chitinase and exo-*N*-acetyl-B-D-glucosaminidase (NAGase) in lymphomyeloid tissue of marine fishes. *Mar. Biol.* **53**, 311–315.

Lundqvist, M., Bengten, E., Strömberg, S., Pilström, L. (1996). The Ig light chain gene in the Siberian sturgeon (*Acipenser baeri*). Implications for the evolution of the immune system. *J. Immunol.* **157**, 2031–2038.

MacDougal, K. C., Johnstone, M. B., Burnett, K. G. (1995). Monoclonal antibodies reactive with the immunoglobulin heavy chain in Sciaenid fishes. *Fish Shellfish Immunol.* **5**, 389–391.

Magnadottir, B. (1990). Purification of immunoglobulin from the serum of Atlantic salmon (*Salmo salar* L.). *Icel. Agr. Sci.* **4**, 49–54.

Magnadottir, B., Gudmundsdottir, B. K. (1992). A comparison of total and specific immunoglobulin levels in healthy Atlantic salmon (*Salmo salar* L.) and in salmon naturally infected with *Aeromonas salmonicida* subsp. *achromogenes*. *Vet. Immunol. Immunopathol.* **32**, 179–189.

Magnadottir, B., Gudmundsdottir, S., Gudmundsdottir, B. K. (1995). Study of the humoral response of Atlantic salmon (*Salmo salar* L.), naturally infected with *Aeromonas salmonicida* subsp. *achromogenes*. *Vet. Immunol. Immunopathol.* **49**, 127–142.

Magnadottir, B., Kristjansdottir, H., Gudmundsdottir, S. (1996). Characterisation of monoclonal antibodies to separate epitopes on salmon IgM heavy chain. *Fish Shellfish Immunol.* **6**, 185–198.

Magor, K. E., Wilson, M. R., Miller, N. W., Clem, L. W., Middleton, D. L., Warr, G. W. (1994a). An Ig heavy chain enhancer of the channel catfish *Ictalurus punctatus*: evolutionary conservation of function but not structure. *J. Immunol.* **153**, 5556–5563.

Magor, K. E., Higgins, D. A., Middleton, D. L., Warr, G. W. (1994b). One gene encodes the heavy chain for three different forms of IgY in the duck. *J. Immunol.* **153**, 5549–5555.

Mäkelä, O., Litman, G. W. (1980). Lack of heterogeneity in antihapten antibodies of a phylogenetically primitive shark. *Nature* **287**, 639–640.

Manning, M. J. (1994). Fishes. In *Immunology: A Comparative Approach* (ed. Turner, R. J.), pp. 69–100. J. Wiley and Sons, Chichester, UK.

Manning, M. J., Mughal, M. S. (1985). Factors affecting the immune responses of immature fish. In *Fish and Shellfish Pathology* (ed. Ellis, A. E.), pp. 27–40. Academic Press, London, UK.

Manning, M. J., Grace, M. F., Secombes, C. J. (1982). Ontogenetic aspects of tolerance and immunity in carp and rainbow trout: studies on the role of the thymus. *Dev. Comp. Immunol.* **6** (Suppl. 2), 75–82.

Manson, F. D., Fletcher, T. C., Gooday, G. W. (1992). Localization of chitinolytic enzymes in blood of turbot, *Scophthalmus maximus*, and their possible roles in defence. *J. Fish Biol.* **40**, 919–927.

Marchalonis, J. J. (1969). Isolation and characterization of immunoglobulin-like proteins of the Australian lungfish (*Neoceratodus forsteri*). *Aust. J. Exp. Biol. Med. Sci.* **47**, 405–419.

Marchalonis, J. J. (1971). Isolation and partial characterization of immunoglobulins of goldfish (*Carassius auratus*) and carp (*Cyprinus carpio*). *Immunology* **20**, 161–173.

Marchalonis, J. J., Edelman, G. M. (1966). Polypeptide chains of immunoglobulins from the smooth dogfish (*Mustelus canis*). *Science* **154**, 1567–1568.

Marchalonis, J. J., Edelman, G. M. (1968). Phylogenetic origins of antibody structure. III. Antibodies in the primary immune response of the sea lamprey, *Petromyzon marinus*. *J. Exp. Med.* **127**, 891–914.

Marchalonis, J. J., Schonfeld, S. A. (1970). Polypeptide chain structure of sting ray immunoglobulin. *Biochim. Biophys. Acta* **221**, 604–611.

Marchalonis, J. J., Hohman, V. S., Kaymaz, H., Schluter, S. F. (1993a). Shared antigenic determinants of immunoglobulins in phylogeny and in comparison with T-cell receptors. *Comp. Biochem. Physiol.* **105B**, 423–441.

Marchalonis, J. J., Hohman, V. S., Schluter, S. F. (1993b). Antibodies of sharks. Novel methods of generation of diversity. *The Immunologist* **1**, 115–120.

Matsubara, A., Mihara, S., Kusuda, R. (1985). Quantitation of yellowtail immunoglobulin by enzyme-linked immunosorbent assay (ELISA). *Bull. Jap. Soc. Sci. Fish.* **51**, 921–925.

Matsuda, F., Shin, E. K., Nagaoka, H., Matsumura, R., Haino, M., Fukita, Y., Taka-ishi, S., Imai, T., Riley, J. H., Anand, R., Soeda, E., Honjo, T. (1993). Structure and physical map of 64 variable segments in the 3′ 0.8 megabase region of the human immunoglobulin heavy-chain locus. *Nature Genetics* **3**, 88–94.

Matsunaga, T., Chen, T., Törmänen, V. (1990). Characterization of a complete immunoglobulin heavy-chain variable region germline gene of rainbow trout. *Proc. Natl Acad. Sci. USA* **87**, 7767–7771.

Matsuzaki, T., Shima, A. (1989). Number of major histocompatibility loci in inbred strains of the fish *Oryzias latipes*. *Immunogenetics* **30**, 226–228.

Maule, A. G., Screck, C. B., Kaattari, S. L. (1987). Changes in the immune system of coho salmon (*Oncorhynchus kisutch*) during the parr-to-smolt transformation and after implantation of cortisol. *Can. J. Fish. Aquat. Sci.* **44**, 161–166.

Mawdesley-Thomas, L. E. (1975). Neoplasia in fish. In *The Pathology of Fishes* (eds Ribelin, W. E., Migaki, G.), pp. 805–870. University of Wisconsin Press, Madison, WI, USA.

McCarthy, D. H. (1978). Some ecological aspects of the bacterial fish pathogen *Aeromonas salmonicida*. In *Aquatic Microbiology* (eds Skinner, F. A., Shewen, J. M.), pp. 299–324. The Society for Applied Bacteriology Symposium Series No. 6, Academic Press Ltd., London, UK.

McCumber, L. J., Clem, L. W. (1976). Esterification of J chain and its effect on electrophoretic mobility in sodium dodecyl sulfate polyacrylamide gels. *Biochim. Biophys. Acta* **446**, 536–541.

McKinney, E. C. (1990). Shark cytotoxic macrophages interact with target membrane amino groups. *Cell. Immunol.* **127**, 506–513.

McKinney, E. C. (1992). Shark lymphocytes: primitive antigen reactive cells. *Annu. Rev. Fish Dis.* **2**, 43–51.

McKinney, E. C., Schmale, M. C. (1994). Damselfish with neurofibromatosis exhibit cytotoxicity toward tumour targets. *Dev. Comp. Immunol.* **18**, 305–313.

Meseguer, J., Esteban, M. A., Agulleiro, B. (1991). Stromal cells, macrophages and lymphoid cells in the head-kidney of sea bass (*Dicentrarchus labrax* L.). An ultrastructural study. *Arch. Histol. Cytol.* **54**, 299–309.

Mestecky, J., Kulhavy, R., Schrohenloher, R. E., Tomana, M., Wright, G. P. (1975). Identification and properties of J chain isolated from catfish macroglobulin. *J. Immunol.* **115**, 993–997.

Michard-Vanhee, C., Chourrout, D., Strömberg, S., Thuvander, A., Pilström, L. (1994). Lymphocyte expression in transgenic trout by mouse immunoglobulin promoter/enhancer. *Immunogenetics* **40**, 1–8.

Michel, G., Gonzalez, R., Avrameas, S. (1990). Opsonizing properties of natural antibodies of rainbow trout, *Oncorhynchus mykiss* (Walbaum). *J. Fish Biol.* **37**, 617–622.

Miller, K. M., Withler, R. E. (1996). Sequence analysis of a polymorphic MHC class II gene in Pacific salmon. *Immunogenetics* **43**, 337–351.

Miller, N. W., Clem, L. W. (1984). Temperature-mediated processes in teleost immunity: differential effects of temperature on catfish *in vitro* antibody responses to thymus-dependent and thymus-independent antigens. *J. Immunol.* **133**, 2356–2359.

Miller, N. W., Sizemore, R. C., Clem, L. W. (1985). The phylogeny of lymphocyte heterogeneity: the cellular response for *in vitro* antibody responses of channel catfish leukocytes. *J. Immunol.* **134**, 2884–2888.

Miller, N. W., Deuter, A., Clem, L. W. (1986). Phylogeny of lymphocyte heterogeneity: the cellular requirements for the mixed leukocyte reaction in channel catfish. *Immunology* **59**, 123–128.

Miller, N. W., Bly, J. E., van Ginkel, F., Ellsaesser, C. F., Clem, L. W. (1987). Phylogeny of lymphocyte heterogeneity: identification and separation of functionally distinct subpopulations of channel catfish lymphocytes with monoclonal antibodies. *Dev. Comp. Immunol.* **11**, 739–747.

Miller, N. W., Rycyzyn, M. A., Wilson, M. R., Warr, G. W., Naftel, J. P., Clem, L. W. (1994). Development and characterization of channel catfish longterm B cell lines. *J. Immunol.* **152**, 2180–2189.

Miyata, T., Hayashida, H., Kikuno, R., Toh, H., Kawade, Y. (1985). Evolution of interferon genes. *Interferon* **6**, 1–30.

Miyazaki, T., Kurata, K. (1987). Phagocytic response of peritoneal exudate cells of the Japanese eel against *Vibrio vulnificus*. *Bull. Fac. Fish. Mie. Univ.* **14**, 33–40.

Moody, C. E., Serreze, D. V., Reno, P. W. (1985). Non-specific cytotoxic activity of teleost leukocytes. *Dev. Comp. Immunol.* **9**, 51–64.

Mor, A., Avtalion, R. R. (1988). Evidence of transfer of immunity from mother to eggs in tilapias. *Israeli J. Aquaculture* **40**, 22–28.

Mor, A., Avtalion, R. R. (1990). Transfer of antibody activity from immunised mother to embryo in tilapias. *J. Fish Biol.* **37**, 249–255.

Moritoma, T., Noda, H., Yamaguchi, I., Watanabe, T. (1992). Enhancement of granulocyte-poiesis by Freund's complete adjuvant. *Nipp. Suis. Gakk.* **58**, 1145–1149.

Moritomo, T., Iida, T., Wakabayashi, H. (1988). Chemiluminescence of neutrophils isolated from peripheral blood of eel. *Fish Pathol.* **23**, 49–53.

Morizot, D. C. (1994). Reconstructing the gene map of the vertebrate ancestor. *Anim. Biotech.* **5**, 113–122.

Morizot, D. C., Schmidt, M. E., Carmichael, G. J., Stock, D. W., Williamson, J. H. (1990). Minimally invasive tissue sampling.

In *Electrophoretic and Isoelectric Focusing Techniques in Fisheries Management* (ed. Whitmore, D. H.), pp. 143–156. CRC Press, Boca Raton, FL, USA.

Morrow, W. J. W., Harris, J. E., Davies, D., Pulsford, A. (1983). Isolation and partial characterization of dogfish (*Scyliorhinus canicula*) antibody. *J. Mar. Biol. Ass. UK* **63**, 409–418.

Mulchay, M. F. (1976). Epizootiological studies of lymphomas in northern pike in Ireland. In *Progress in Experimental Tumor Research* (eds Dawe, C. J., Scarpelli, D. G., Wellings, S. R.), pp. 129–140. S. Karger, Basel, Switzerland.

Murad, A., Houston, A. H. (1988). Leucocytes and leucopoietic capacity in goldfish, *Carassius auratus* exposed to sublethal levels of cadium. *Aquat. Toxicol.* **5**, 141–155.

Nagae, M., Fuda, H., Hara, A., Kawamura, H., Yamauchi, K. (1993). Changes in serum immunoglobulin M (IgM) concentrations during early development of chum salmon (*Oncorhynchus keta*) as determined by sensitive ELISA technique. *Comp. Biochem. Physiol.* **106A**, 69–74.

Nakanishi, T. (1987a). Histocompatibility analysis in tetraploids induced from clonal triploid crucian carp and in gynogenetic diploid goldfish. *J. Fish Biol.* **31** (Suppl. A), 35–40.

Nakanishi, T. (1987b). Transferability of immune plasma and pronephric cells in isogeneic, allogeneic and xenogeneic transfer systems in crucian carp. *Dev. Comp. Immunol.* **11**, 521–528.

Nakanishi, T. (1991). Ontogeny of the immune system in *Sebastiscus marmoratus*: histogenesis of the lymphoid organs and effects of thymectomy. *Exp. Biol. Fishes* **30**, 135–145.

Nakanishi, Y., Kodama, H., Murai, T., Mikami, T., Izawa, H. (1991). Activation of rainbow trout complement by C-reactive protein. *Am. J. Vet. Res.* **52**, 397–401.

Nash, K. A., Fletcher, T. C., Thomson, A. W. (1986). Migration of fish leucocytes *in vitro*: the effect of factors which may be involved in mediating inflammation. *Vet. Immunol. Immunopathol.* **12**, 83–92.

Navarro, V., Quesada, J. A., Abad, M. E., Taverne, N., Rombout, J. H. W. M. (1993). Immuno(cyto)chemical characterisation of monoclonal antibodies to gilthead seabream (*Sparus aurata*) immunoglobulin. *Fish Shellfish Immunol.* **3**, 167–177.

Nelson, G. (1989). Phylogeny of major fish groups. In *The Hierarchy of Life. Molecules and Morphology in Phylogenetic Analysis* (eds Fernholm, B., Bremer, K., Jörnvall, H.), pp. 325–336. Elsevier Science Publishers B.V., Amsterdam, The Netherlands.

Nelson, J. S. (1994). *Fishes of the World*, 3rd edn. John Wiley and Sons, New York, USA.

Nigrelli, R. F. (1947). Spontaneous neoplasms in fishes. III. Lymphosarcoma in *Astyanax* and *Esox*. *Zoologica NY* **32**, 101–108.

Nonaka, M., Yamaguchi, N., Natsuume-Sakai, S., Morinobu, T. (1981a). The complement system of rainbow trout (*Salmo gairdneri*). I. Identification of the serum lytic system homologous to mammalian complement. *J. Immunol.* **126**, 1489–1494.

Nonaka, M., Natsuume-Sakai, S., Takahashi, M. (1981b). The complement system of rainbow trout (*Salmo gairdneri*). II. Purification and characterisation of the fifth component (C5). *J. Immunol.* **126**, 1495–1498.

Nonaka, M., Fujii, T., Kaidoh, T., Natsuume-Sakai, S., Yamaguchi, N., Takahashi, M. (1984). Purification of a lamprey complement protein homologous to the third component of mammalian complement system. *J. Immunol.* **133**, 3242–3249.

Nonaka, M., Irie, M., Tanabe, K., Kaidoh, T., Natsuume-Sakai, S., Takahashi, M. (1985). Identification and characterisation of a variant of the third component of complement (C3) in rainbow trout (*Salmo gairdneri*) serum. *J. Biol. Chem.* **260**, 809–815.

Nonaka, M., Takahashi, M. (1992). Complete complementary DNA sequence of the third component of complement of the lamprey: implications for the evolution of thioester containing proteins. *J. Immunol.* **148**, 3290–3295.

Nunomura, W. (1991). C-reactive protein in eel: purification and agglutinating activity. *Biochim. Biophys. Acta* **1076**, 191–196.

O'Neill, J. G. (1979). The immune response of the brown trout, *Salmo trutta*, L. to MS2 bacteriophage: immunogen concentration and adjuvants. *J. Fish Biol.* **15**, 237–248.

O'Neill, J. G. (1981). Effects of intraperitoneal lead and cadmium on the humoral response of *Salmo trutta*. *Bull. Environ. Contam. Toxicol.* **27**, 42–48.

Obenauf, S. D., Smith, S. H. (1985). Chemotaxis of nurse shark leukocytes. *Dev. Comp. Immunol.* **9**, 221–230.

Olesen, N. J., Vestergård Jørgensen, P. E. (1986). Quantification of serum immunoglobulin in rainbow trout *Salmo gairdneri* under various environmental conditions. *Dis. Aquat. Org.* **1**, 183–189.

Olivier, G., Evelyn, T. P. T., Lallier, R. (1985). Immunity to *Aeromonas salmonicida* in coho salmon (*Oncorhynchus kisutch*) induced by modified Freund's complete adjuvant: its non-specific nature and the probable role of macrophages in the phenomenon. *Dev. Comp. Immunol.* **9**, 419–432.

Ono, H., Klein, D., Vincek, V., Figueroea, F., O'hUigin, C., Tichy, H., Klein, J. (1992). Major histocompatibility complex class II genes in zebrafish. *Proc. Natl Acad. Sci. USA* **89**, 11886–11890.

Ono, H., Figueroea, F., O'hUigin, C., Klein, J. (1993a). Cloning of the β_2-microglobulin gene in the zebrafish. *Immunogenetics* **38**, 1–10.

Ono, H., O'hUigin, C., Vincek, V., Stet, R. J. M., Figueroea, F., Klein, J. (1993b). New β chain-encoding MHC class II genes in the carp. *Immunogenetics* **38**, 146–149.

Ono, H., O'hUigin, C., Vincek, V., Klein, J. (1993c). Exon-intron organization of fish major histocompatibility complex class II B genes. *Immunogenetics* **38**, 223–234.

Ortiz-Muniz, G., Sigel, M. M. (1971). Antibody synthesis in lymphoid organs of two marine teleosts. *J. Reticuloendothel. Soc.* **9**, 42–52.

Ourth, D. D. (1980). Secretory IgM, lysozyme and lymphocytes in the skin mucus of the channel catfish, *Ictalurus punctatus*. *Dev. Comp. Immunol.* **4**, 65–74.

Ourth, D. D. (1986). Purification and quantitation of channel catfish (*Ictalurus punctatus*) immunoglobulin M. *J. Appl. Ichthyol.* **3**, 140–143.

Ourth, D. D., Wilson, E. A. (1982a). Bactericidal serum response of the channel catfish against gram-negative bacteria. *Dev. Comp. Immunol.* **6**, 579–583.

Ourth, D. D., Wilson, E. A. (1982b). Alternate pathway of complement and bactericidal response of the channel catfish to *Salmonella paratyphi*. *Dev. Comp. Immunol.* **6**, 75–85.

Ozaki, H., Ohwaki, M., Fukada, T. (1983). Studies on lectins of amago (*Oncorhynchus rhodurus*) I. Amago lectin and its receptor on homologous macrophages. *Dev. Comp. Immunol.* 7, 77–87.

Palenzuela, O., Sitjà-Bobadilla, A., Alvarez-Pellitero, P. (1996). Isolation and partial characterization of serum immunoglobulins from sea bass (*Dicentrarchus labrax* L.) and gilthead sea bream (*Sparus aurata* L.). *Fish Shellfish Immunol.* **6**, 81–94.

Partula, S., Charlemagne, J. (1993). Characterization of serum immunoglobulins in a chondrostean fish, *Acipenser baeri*. *Dev. Comp. Immunol.* **17**, 515–524.

Partula, S., Fellah, J. S., de Guerra, A., Charlemagne, J. (1994). Identification of cDNA clones encoding the T-cell receptor β chain in the rainbow trout (*Oncorhynchus mykiss*). *C.R. Acad. Sci. D* **317**, 765–770.

Partula, S., de Guerra, A., Fellah, J. S., Charlemagne, J. (1995). Structure and diversity of the T cell antigen receptor β-chain in a teleost fish. *J. Immunol.* **155**, 699–706.

Partula, S., de Guerra, A., Fellah, J. S., Charlemagne, J. (1996a). Structure and diversity of the T cell receptor α-chain in a teleost fish. *J. Immunol.* **157**, 207–212.

Partula, S., Schwager, J., Timmusk, S., Pilström, L., Charlemagne, J. (1996b). A second immunoglobulin light chain isotype in rainbow trout. *Immunogenetics* **45**, 44–51.

Passer, B. J., Chen, C. H., Miller, N. W., Cooper, M. D. (1997). Identification of a T lineage antigen in the catfish. *Dev. Comp. Immunol.* **20**, 441–450.

Pedrera, I. M., Collazos, M. E., Ortega, E., Barriga, C. (1992). *In vitro* study of the phagocytic process in splenic granulocytes of the tench (*Tinca tinca* L.). *Dev. Comp. Immunol.* **16**, 431–439.

Peleteiro, M. C., Richards, R. H. (1985). Identification of lymphocytes in the epidermis of the rainbow trout, *Salmo gairdneri* Richardson. *J. Fish Dis.* **8**, 161–172.

Pettersen, E. F., Fyllingen, I., Kavlie, A., Maaseide, N. P., Glette, J., Endressen, C., Wergeland, H. I. (1995). Monoclonal antibodies reactive with serum IgM and leukocytes from Atlantic salmon (*Salmo salar* L.). *Fish Shellfish Immunol.* **5**, 275–287.

Pettitt, T. R., Rowley, A. F., Secombes, C. J. (1989). Lipoxins are major lipoxygenase products of rainbow trout macrophages. *FEBS Lett.* **259**, 168–170.

Pettitt, T. R., Rowley, A. F., Barrow, S. E., Mallet, A. I., Secombes, C. J. (1991). Synthesis of lipoxins and lipoxygenase products by macrophages from rainbow trout, *Oncorhynchus mykiss*. *J. Biol. Chem.* **266**, 8720–8726.

Phillips, J. O., Ourth, D. D. (1986). Isolation and molecular weight determination of two immunoglobulin heavy chains in the channel catfish, *Ictalurus punctatus*. *Comp. Biochem. Physiol.* **85B**, 49–54.

Piganelli, J. D., Zhang, J. A., Christensen, J. M., Kaattari, S. L. (1994). Enteric coated microspheres as an oral method for antigen delivery to salmonids. *Fish Shellfish Immunol.* **4**, 179–188.

Pilström, L., Bengten, E. (1996). Immunoglobulin in fish – genes, expression and structure. *Fish Shellfish Immunol.* **6**, 243–262.

Pilström, L., Petersson, A. (1991). Isolation and partial characterization of immunoglobulin from the cod (*Gadus morhua* L.). *Dev. Comp. Immunol.* **15**, 143–152.

Pinto, R. M., Jofre, J., Bosch, A. (1993). Interferon-like reactivity in sea bass affected by viral erythrocytic infection. *Fish Shellfish Immunol.* **3**, 89–96.

Pitombeira, M., Martins, J. M., Gomes, F. V. B. (1975). Leukemia in fish. *Medicinia* **35**, 251–256.

Plehn, M. (1924). *Praktikum der Fischkrankheiten*. E. Schweizerbart, Stuttgart, Germany.

Postlethwait, J. H., Johnson, S. L., Midson, C. N., Talbot, S. W., Gates, M., Ballinger, E. W., Africa, D., Andrews, R., Carl, T., Eisen, J. S., Horne, S., Kimmel, C. B., Hutchinson, M., Johnson, M., Rodriguez, A. (1994). A genetic linkage map for the Zebrafish. *Science* **264**, 699–703.

Press, C. M., Lillehaug, A. (1995). Vaccination in European salmonid aquaculture: A review of practices and prospects. *Br. Vet. J.* **151**, 45–69.

Press, C. M., Dannevig, B. H., Landsverk, T. (1994). Immune and enzyme histochemical phenotypes of lymphoid and nonlymphoid cells within the spleen and head kidney of Atlantic salmon (*Salmo salar* L.). *Fish Shellfish Immunol.* **4**, 79–93.

Pulsford, A., Tomlinson, M. G., Lemaire-Gony, S., Glynn, P. J. (1994). Development and immunocompetence of juvenile flounder, *Platichtys flesus*, L. *Fish Shellfish Immunol.* **4**, 63–78.

Rainger, G. E., Rowley, A. F., Pettitt, T. R. (1992). Effect of inhibitors of eicosanoid biosynthesis on the immune reactivity of the rainbow trout, *Oncorhynchus mykiss*. *Fish Shellfish Immunol.* **2**, 143–154.

Raison, R. L., Hull, C. J., Hildemann, W. H. (1978a). Characterization of immunoglobulin from the Pacific hagfish, a primitive vertebrate. *Proc. Natl Acad. Sci. USA* **75**, 5679–5682.

Raison, R. L., Hull, C. J., Hildemann, W. H. (1978b). Production and specificity of antibodies to streptococci in the Pacific hagfish, *Eptatretus stoutii*. *Dev. Comp. Immunol.* **2**, 253–262.

Rasquin, P., Hafner, E. (1951). Response of a fish lymphosarcoma to mammalian ACTH. *Zoologica NY* **36**, 163–169.

Rast, J. P., Litman, G. W. (1994). T-cell receptor gene homologs are present in the most primitive vertebrates. *Proc. Natl Acad. Sci. USA* **91**, 9248–9252.

Rast, J. P., Anderson, M. K., Ota, T., Litman, R. T., Margittai, M., Shamblott, M. J., Litman, G. W. (1994). Immunoglobulin light chain class multiplicity and alternative organizational forms in early vertebrate phylogeny. *Immunogenetics* **40**, 83–99.

Rast, J. P., Anderson, M. K., Litman, G. W. (1995a). The structure and organization of immunoglobulin genes in lower vertebrates. In *Immunoglobulin Genes*, 2nd edn (eds Honjo, T., Alt, F. W.), pp. 315–341. Academic Press Ltd., London, UK.

Rast, J. P., Haire, R. N., Litman, R. T., Pross, S., Litman, G. W. (1995b). Identification and characterization of T-cell antigen receptor-related genes in phylogenetically diverse vertebrate species. *Immunogenetics* **42**, 204–212.

Razquin, B. E., Castillo, A., Lopez-Fierro, P., Alvarez, F., Zapata, A., Villena, A. J. (1990). Ontogeny of IgM-producing cells in the lymphoid organs of rainbow trout, *Salmo gairdneri* Richardson: an immuno- and enzyme histochemical study. *J. Fish Biol.* **36**, 159–173.

Renault, T., Torchy, C., de Kinkelin, P. (1991). Spectrophotometric method for titration of trout interferon, and its application to rainbow trout fry experimentally infected with viral haemorrhagic septicaemia virus. *Dis. Aquat. Org.* **10**, 23–29.

Richter, R., Ambrosius, H. (1972). Beiträge zur immunbiologie poikilothermer wirbeltiere. VIII. Ein hochmolekulares immunglobulin des flussbarsches (*Perca fluviatilis* L.). *Acta Biol. Med. Germ.* **26**, 309–318.

Richter, R., Frenzel, E.-M., Hädge, D., Kopperschläger, G., Ambrosius, H. (1973). Strukturelle und immunchemische untersuchungen am immunoglobulin des karpfens (*Cyprinus carpio* L.). I. Analsyse am gesamtmolekül. *Acta Biol. Med. Germ.* **30**, 735–749.

Rijkers, G. T. (1982a). Non-lymphoid defense mechanisms in fish. *Dev. Comp. Immunol.* **6**, 1–13.

Rijkers, G. T. (1982b). Kinetics of humoral and cellular immune reactions in fish. *Dev. Comp. Immunol.* **2**, 93–100.

Rijkers, G. T., van Oosteron, R., van Muiswinkel, W. B. (1981). The immune system of cyprinid fish. Oxytetracycline and the regulation of the humoral immunity in carp. *Vet. Immunol. Immunopathol.* **2**, 281–290.

Roald, S. O., Håstein, T. (1979). Lymphosarcoma in an Atlantic salmon *Salmo salar*. *J. Fish Dis.* **2**, 249–251.

Roberts, R. J. (1975). Melanin-containing cells of teleost fish and their relation to disease. In *The Pathology of Fishes* (eds Ribelin, W. E., Migaki, G.), pp. 399–428. University of Wisconsin Press, Madison, WI, USA.

Roberts, R. J. (1989). *Fish Pathology*, 2nd edn. Bailliere Tindall, London, UK.

Robertsen, B., Engstad, R. E., Jørgensen, J. B. (1994). β-Glucans as immunostimulants in fish. *Modulators Fish Immune Responses* **1**, 83–99.

Robertsen, B., Trobridge, G., Leong, J. C. (1997). Molecular cloning of double-stranded RNA inducible Mx genes from Atlantic salmon (*Salmo salar* L.). *Dev. Comp. Immunol.* **21**, 397–412.

Robohm, R. A. (1986). Paradoxical effects of cadmium exposure on antibacterial antibody responses in two fish species: inhibition in cunners (*Tautogolabrus adspersus*) and enhancement in striped bass (*Morone saxitalis*). *Vet. Immunol. Immunopathol.* **12**, 251–262.

Rodrigues, P. N. S., Hermsen, T. T., Rombout, J. H. W. M., Egberts, E., Stet, R. J. M. (1995). Detection of MHC class II transcripts in lymphoid tissues of the common carp (*Cyprinus carpio*). *Dev. Comp. Immunol.* **19**, 483–496.

Rogel-Gaillard, C., Chilmonczyk, S., de Kinkelin, P. (1993). *In vitro* induction of interferon-like activity from rainbow trout leucocytes stimulated by Egtved virus. *Fish Shellfish Immunol.* **3**, 383–394.

Roman, T., Charlemagne, J. (1994). The immunoglobulin repertoire of the rainbow trout (*Oncorhynchus mykiss*): definition of nine Igh-V families. *Immunogenetics* **40**, 210–216.

Roman, T., De Guerra, A., Charlemagne, J. (1995). Evolution of specific antigen recognition: size reduction and restricted length distribution of the CDRH3 regions in the rainbow trout. *Eur. J. Immunol.* **25**, 269–273.

Roman, T., Andersson, E., Bengten, E., Hansen, J., Kaattari, S., Pilström, L., Charlemagne, J., Matsunaga, T. (1996). Unified nomenclature of Ig V_H genes in rainbow trout (*Oncorhynchus mykiss*): definition of eleven V_H families. *Immunogenetics* **43**, 325–326.

Rombout, J. H. W. M., van den Berg, A. A. (1989). Immunological importance of the second gut segment of carp. I. Uptake and processing of antigens by epithelial cells and macrophages. *J. Fish Biol.* **35**, 13–22.

Rombout, J. H. W. M., Lamers, C. H. J., Helfrich, M. H., Dekker, A., Taverne-Thiele, J. J. (1985). Uptake and transport of intact macromolecules in the intestinal epithelium of carp (*Cyprinus carpio* L.) and the possible immunological implications. *Cell Tissue Res.* **239**, 519–530.

Rombout, J. H. W. M., Blok, L. J., Lamers, C. H. J., Egberts, E. (1986). Immunization of carp (*Cyprinus carpio*) with a *Vibrio anguillarum* bacterin: indications for a common mucosal immune system. *Dev. Comp. Immunol.* **10**, 341–351.

Rombout, J. H. W. M., Bot, H. E., Taverne-Thiele, J. J. (1989a). Immunological importance of the second segment of carp. II. Characterization of mucosal leucocytes. *J. Fish Biol.* **35**, 167–178.

Rombout, J. H. W. M., van den Berg, A. A., van den Berg, C. T. G. A., Witte, P., Egberts, E. (1989b). Immunological importance of the second gut segment of carp. III. Systemic and/or mucosal immune response after immunization with soluble or particulate antigen. *J. Fish Biol.* **35**, 179–186.

Rombout, J. H. W. M., Taverne-Thiele, J. J., Villena, M. I. (1993a). The gut associated lymphoid tissue (GALT) of carp (*Cyprinus carpio* L.): an immunocytochemical analysis. *Dev. Comp. Immunol.* **15**, 55–66.

Rombout, J. H. W. M., Taverne, N., Van de Kamp, M., Taverne-Thiele, A. J. (1993b). Differences in mucus and serum immunoglobulin of carp (*Cyprinus carpio* L.). *Dev. Comp. Immunol.* **17**, 309–317.

Romestand, B., Breuil, G., Bourmaud, C. A. F., Coeurdacier, J. L., Bouix, G. (1995). Development and characterisation of monoclonal antibodies against sea bass immunoglobulin, *Dicentrarchus labrax* Linnaeus. *Fish Shellfish Immunol.* **5**, 347–357.

Rosenshein, I. L., Marchalonis, J. J. (1987). The immunoglobulins of carcharine sharks: a comparison of serological and biochemical properties. *Comp. Biochem. Physiol.* **86B**, 737–747.

Rosenshein, I. L., Schluter, S. F., Vasta, G. R., Marchalonis, J. J. (1985). Phylogenetic conservation of heavy chain determinants of vertebrates and protochordates. *Dev. Comp. Immunol.* **9**, 783–795.

Rowley, A. F., Hunt, T. C., Page, M., Mainwaring, G. (1988). Fish. In *Vertebrate Blood Cells* (eds Rowley, A. F., Ratcliffe, N. A.), pp. 19–127. Cambridge University Press, Cambridge, UK.

Rowley, A. F., Pettitt, T. R., Secombes, C. J., Sharp, G. J. E., Barrow, S. E., Mallet, A. I. (1991). Generation and biological activities of lipoxins in the rainbow trout – an overview. *Adv. Prostaglan. Thrombox. Leukotriene Res.* **218**, 557–560.

Ruben, L. N., Warr, G. W., Decker, J. M., Marchalonis, J. J. (1977). Phylogenetic origins of immune recognition: lymphocyte heterogeneity and hapten/carrier effect in the goldfish, *Carrassius auratus*. *Cell. Immunol.* **31**, 266–283.

Rudikoff, S., Voss, E. W., Sigel, M. M. (1970). Biological and chemical properties of natural antibodies in the nurse shark. *J. Immunol.* **105**, 1344–1352.

Russell, W. J., Voss, E. W., Sigel, M. M. (1970). Some characteristics of anti-dinitrophenyl antibody of the gray snapper. *J. Immunol.* **105**, 262–264.

Sailendri, K., Muthukkaruppan, V. R. (1975). Morphology of lymphoid organs in a cichlid teleost *Tilapia mossambica* (Peters). *J. Morphol.* **147**, 109–122.

Sakai, D. K. (1981). Heat inactivation of complements and immune hemolysis reaction in rainbow trout, masu salmon, coho salmon, goldfish and tilapia. *Bull. Jap. Soc. Sci. Fish.* **47**, 979–991.

Sakai, D. K. (1983). Lytic and bactericidal properties of salmonid sera. *J. Fish Biol.* **23**, 457–466.

Sakai, D. K. (1984). The non-specific activation of rainbow trout,

Salmo gairdneri Richardson, complement by *Aeromonas salmonicida* extracellular products and the correlation of complement activity with the inactivation of lethal toxicity products. *J. Fish Dis.* **23**, 457–466.

Sakai, D. K. (1992). Repertoire of complement in immunological defense mechanisms of fish. *Annu. Rev. Fish Dis.* **2**, 223–247.

Sakai, D. K., Okada, H., Koide, N., Yoshiharu, T. (1987). Blood type compatibility of lower vertebrates: phylogenetic diversity in blood transfusion between fish species. *Dev. Comp. Immunol.* **11**, 105–115.

Sanchez, C., Dominguez, J. (1991). Trout immunoglobulin populations differing in light chains revealed by monoclonal antibodies. *Mol. Immunol.* **28**, 1271–1277.

Sanchez, G. A., Gajardo, M. K., De Ioannes, A. E. (1980). Ig-M-like natural hemagglutinin from ratfish serum: isolation and physio-chemical characterization (*Callorhynchus callorhynchus*). *Dev. Comp. Immunol.* **4**, 667–678.

Sanchez, C., Dominguez, J., Coll, J. (1989). Immunoglobulin heterogeneity in the rainbow trout, *Salmo gairdneri* Richardson. *J. Fish Dis.* **12**, 459–465.

Sanchez, C., Lopez-Fierro, P., Zapata, A. G., Dominguez, J. (1993a). Characterization of monoclonal antibodies against heavy and light chains of trout immunoglobulin. *Fish Shellfish Immunol.* **3**, 237–251.

Sanchez, C., Babin, M., Tomillo, J., Ubeira, F. M., Dominguez, J. (1993b). Quantification of low levels of rainbow trout immunoglobulin by enzyme immunoassay using two monoclonal antibodies. *Vet. Immunol. Immunopathol.* **36**, 65–74.

Sato, A., Klein, D., Sultmann, H., Figueroa, F., O'hUigin, C., Klein, J. (1997). Class I MHC genes of cichlid fishes: identification, expression and polymorphism. *Immunogenetics* **46**, 63–72.

Scapigliati, G., Mazzini, M., Mastrolia, L., Romano, N., Abelli, L. (1995). Production and characterization of a monoclonal antibody against the thymocytes of the sea bass *Dicentrachus labrax* (L.) (Teleostea, Percicthydae). *Fish Shellfish Immunol.* **5**, 393–405.

Scapigliati, G., Romano, N., Picchietti, S., Mazzini, M., Mastrolia, L., Scalia, D., Abelli, L. (1996). Monoclonal antibodies against sea bass *Dicentrarchus labrax* (L.) immunoglobulins: immunolocalisation of immunoglobulin-bearing cells and applicability in immunoassays. *Fish Shellfish Immunol.* **6**, 383–401.

Schluter, S. F., Hohman, V. S., Edmundson, A. B., Marchalonis, J. J. (1989). Evolution of immunoglobulin light chains: cDNA clones specifying sandbar shark constant regions. *Proc. Natl Acad. Sci. USA* **86**, 9961–9965.

Schluter, S. F., Beichel, C. J., Martin, S. A., Marchalonis, J. J. (1990). Sequence analysis of homogeneous peptides of shark immunoglobulin light chain by tandem mass spectrometry: correlation with gene sequence and homologies among variable and constant region peptides of shark and mammals. *Mol. Immunol.* **27**, 17–23.

Secombes, C. J. (1985). The *in vitro* formation of teleost multinucleate giant cells. *J. Fish Dis.* **8**, 461–464.

Secombes, C. J. (1994a). The phylogeny of cytokines. In *The Cytokine Handbook* (ed. Thompson, A. W.), pp. 567–594. Academic Press, London, UK.

Secombes, C. J. (1994b). Macrophage activation in fish. *Modulators Fish Immune Response* **1**, 49–57.

Secombes, C. J., Fletcher, T. C. (1992). The role of phagocytes in the protective mechanisms of fish. *Annu. Rev. Fish Dis.* **2**, 53–71.

Secombes, C. J., Manning, M. J. (1980). Comparative studies on the immune system of fishes and amphibians. I. Antigen localization in the carp *Cyprinus carpio*. *J. Fish Dis.* **3**, 399–412.

Secombes, C. J., van Groningen, J. J. M., Egberts, E. (1983a). Separation of lymphocyte subpopulations in carp *Cyprinus carpio* L. by monoclonal antibodies: immunohistochemical studies. *Immunology* **48**, 165–175.

Secombes, C. J., van Groningen, J. J. M., van Muiswinkel, W. B., Egberts, E. (1983b). Ontogeny of the immune system in carp (*Cyprinus carpio* L.). The appearance of antigenic determinants on lymphoid cells detected by mouse anti-carp thymocyte monoclonal antibodies. *Dev. Comp. Immunol.* **7**, 455–464.

Secombes, C. J., White, A., Fletcher, T. C., Stagg, R., Houlihan, D. F. (1995). Immune parameters of plaice, *Pleuronectes platessa* L. along a sewage sludge gradient in the Firth of Clyde, Scotland. *Ecotoxicology* **4**, 239–340.

Secombes, C. J., Hardie, L. J., Daniels, G. (1996). Cytokines in fish: an update. *Fish Shellfish Immunol.* **6**, 291–304.

Sekellick, M. J., Ferrandino, A. F., Hopkins, D. A., Marcus, P. I. (1994). Chicken interferon gene: cloning, expression, and analysis. *J. Interferon Res.* **14**, 71–79.

Shalev, A., Greenberg, A. H., Lögdberg, L., Björck, L. (1981). β_2-microglobulin-like molecules in low vertebrates and invertebrates. *J. Immunol.* **127**, 1186–1191.

Shalev, A., Caspi, R. R., Lögdberg, L., Björck, L., Segal, S., Avtalion, R. R. (1984). Specific activity of carp antisera against β_2-microglobulin (β_2m) and evidence for a β_2m homologue in carp (*Cyprinus carpio*). *Dev. Comp. Immunol.* **8**, 639–648.

Shamblott, M. J., Litman, G. W. (1989a). Complete nucleotide sequence of primitive vertebrate immunoglobulin light chain genes. *Proc. Natl Acad. Sci. USA* **86**, 4684–4688.

Shamblott, M. J., Litman, G. W. (1989b). Genomic organization and sequences of immunoglobulin light chain genes in a primitive vertebrate suggest coevolution of immunoglobulin gene organization. *EMBO J.* **8**, 3733–3739.

Shankey, T. V., Clem, L. W. (1980a). Phylogeny of immunoglobulin structure and function. VIII. Intermolecular heterogeneity of shark 19S IgM antibodies to pneumoccal polysaccharide. *Mol. Immunol.* **17**, 365–375.

Shankey, T. V., Clem, L. W. (1980b). Phylogeny of immunoglobulin structure and function. IX. Intramolecular heterogeneity of shark 19S IgM antibodies to the dinitrophenyl hapten. *J. Immunol.* **125**, 2690–2698.

Sharp, G. J. E., Pettitt, T. R., Rowley, A. F., Secombes, C. J. (1992). Lipoxin-induced migration of fish leucocytes. *J. Leuk. Biol.* **51**, 140–145.

Sharp, G. J. E., Pike, A. E., Secombes, C. J. (1991). Leukocyte migration in rainbow trout (*Oncorhynchus mykiss* (Walbaum)): optimisation of migration conditions and responses to host-pathogen (*Diphyllobothrium dendriticum* (Nitzsch)) derived chemoattractants. *Dev. Comp. Immunol.* **15**, 295–305.

Shelton, E., Smith, M. (1970). The ultrastructure of carp (*Cyprinus carpio*) immunoglobulin: a tetrameric macroglobulin. *J. Mol. Biol.* **54**, 615–617.

Shephard, K. L. (1994). Functions of fish mucus. *Rev. Fish Biol. Fisheries* **4**, 401–429.

Shiomi, K., Utsumi, K., Tsuchiya, S., Shimakura, K., Nagashima, Y. (1994). Comparison of proteinaceous toxins in the skin mucus from three species of eels. *Comp. Biochem. Physiol.* **107**, 389–394.

Shum, B. P., Azumi, K., Zhang, S., Kehrer, S. R., Raison, R. L., Detrich, H. W., Parham, P. (1996). Unexpected β_2-microglobulin sequence diversity in individual rainbow trout. *Proc. Natl Acad. Sci. USA* **93**, 2779–2784.

Sigel, M. M., Voss, E. W., Rudikoff, S. (1972). Binding properties of shark immunoglobulins. *Comp. Biochem. Physiol.* **42A**, 249–259.

Sigel, M. M., Hamby, B. A., Huggins, E. M. (1986). Phylogenetic studies on lymphokines. Fish lymphocytes respond to human IL-1 and epithelial cells produce an IL-1 like factor. *Vet. Immunol. Immunopathol.* **12**, 47–58.

Sin, Y. M., Ling, K. H., Lam, T. T. (1994). Passive transfer of protective immunity against Ichthyophthiriasis from vaccinated mother to fry in tilapias, *Oreochromis aureus*. *Aquaculture* **120**, 229–237.

Siwicki, A. K., Anderson, D. P., Dixon, O. W. (1989). Comparisons of nonspecific and specific immunomodulation by oxolinic acid, oxytetracycline and levamisole in salmonids. *Vet. Immunol. Immunopathol.* **23**, 195–200.

Sizemore, R. C., Miller, N. W., Cuchens, M. A., Lobb, C. J., Clem, L. W. (1984). Phylogeny of lymphocyte heterogeneity: the cellular requirements for *in vitro* mitogenic response of channel catfish leukocytes. *J. Immunol.* **133**, 2920–2924.

Skarmeta, A. M., Bandin, I., Santos, Y., Toranzo, A. E. (1995). *In vitro* killing of *Pasteurella piscicida* by fish macrophages. *Dis. Aquat. Org.* **23**, 51–57.

Slierendrecht, W. J., Olesen, N. J., Lorenzen, N., Jørgensen, P. E. V., Gottschau, A., Koch, C. (1996). Genetic alloforms of rainbow trout (*Oncorhynchus mykiss*) complement component C3 and resistance to viral haemorrhagic septicaemia under experimental conditions. *Fish Shellfish Immunol.* **6**, 235–237.

Small, P. A., Klapper, D. G., Clem, L. W. (1970). Half-lives, body distribution and lack of interconversion of serum 19S and 7S IgM of sharks. *J. Immunol.* **105**, 29–37.

Smith, A. M., Wivel, N. A., Potter, M. (1970). Plasmacytosis in the pronephros of the carp *Cyprinus carpio*. *Anat. Rec.* **167**, 351–370.

Smith, G. M., Coates, C. W., Strong, L. C. (1936). Neoplastic diseases in small tropical fishes. *Zoologica NY* **21**, 219–224.

Smith, P. D., Braun-Nesje, R. (1982). Cell mediated immunity in the salmon: lymphocyte and macrophage stimulation, lymphocyte/macrophage interactions and the production of lymphokine-like factors by stimulated lymphocytes. *Dev. Comp. Immunol.* **6** (Suppl. 2), 233–238.

Sonstegard, R. A. (1975). Lymphosarcoma in muskellunge (Esox masquinongy). In *The Pathology of Fishes* (eds Ribelin, W. E., Migaki, G.), pp. 907–924. University of Wisconsin Press, Madison, WI, USA.

Sonstegard, R. A. (1976). Studies on the etiology and epizootiology of lymphosarcoma in *Esox* (*Esox lucius* L. and *Esox masquinongy*). In *Progress in Experimental Tumor Research* (eds Dawe, C. J., Scarpelli, D. G., Wellings, S. R.), pp. 141–155. S. Karger, Basel, Switzerland.

St. Louis-Cormier, E. A., Osterland, C. K., Anderson, P. D. (1984). Evidence for a cutaneous secretory immune system in rainbow trout (*Salmo gairdneri*). *Dev. Comp. Immunol.* **8**, 71–80.

Stanley, K. K., Herz, J., Dickinson, J. (1987). Topological mapping of complement component C9 by recombinant DNA techniques suggests a novel mechanism for its insertion into target membranes. *EMBO J.* **6**, 1951–1957.

Stave, J. W., Roberson, B. S. (1985). Hydrocortisone suppresses the chemiluminescent response of striped bass phagocytes. *Dev. Comp. Immunol.* **9**, 77–84.

Steffensen, J. F., Lomholt, J. P. (1992). The secondary vascular system. In *The Cardiovascular System* (eds Hoar, W. S., Randall, D. J., Farrell, A. P.), pp. 185–217. Academic Press, London, UK.

Stet, R. J. M., Egberts, E. (1991). The histocompatibility system in teleostean fishes: from multiple histocompatibility loci to a major histocompatibility complex. *Fish Shellfish Immunol.* **1**, 1–16.

Stet, R. J., van Erp, S. H. M., Hermsen, T., Sultmann, H., Egberts, E. (1993). Polymorphism and estimation of the number of MHC-*Cyca* class I and class II genes in laboratory strains of the common carp (*Cyprinus carpio* L.). *Dev. Comp. Immunol.* **17**, 141–156.

Stolen, J. S., Gahn, T., Kasper, V., Nagle, J. J. (1984). The effect of environmental temperature on the immune response of a marine teleost (*Paralichthys dentatus*). *Dev. Comp. Immunol.* **8**, 89–98.

Stuge, T. B., Miller, N. W., Clem, L. W. (1995). Channel catfish cytotoxic effectors from peripheral blood and pronephron are different. *Fish Shellfish Immunol.* **5**, 469–471.

Sultmann, H., Mayer, W. E., Figueroea, F., O'hUigin, C., Klein, J. (1993). Zebrafish MHC class II α chain-encoding genes: polymorphism, expression and function. *Immunogenetics* **38**, 408–420.

Suzuki, Y. (1986). Neutrophil chemotactic factor in eel blood plasma. *Bull. Jap. Soc. Sci. Fish.* **52**, 811–816.

Suzuki, Y., Iida, T. (1992). Fish granulocytes in the process of inflammation. *Annu. Rev. Fish Dis.* **2**, 149–160.

Suzumura, E., Kurata, O., Oramoto, N., Ikeda, Y. (1994). Characteristics of natural killer-like cells in carp. *Fish Pathol.* **29**, 199–203.

Tachibana, K. (1988). Chemical defense in fishes. *Bioorg. Mar. Chem.* **2**, 117–138.

Tahir, A., Secombes, C. J. (1996). Modulation of dab (*Limanda limanda* L.) macrophage respiratory burst activity. *Fish Shellfish Immunol.* **6**, 135–146.

Takahashi, Y., Kawahara, E. (1987). Maternal immunity in newborn fry of the ovoviviparous guppy. *Nipp. Suis. Gakk.* **53**, 721–735.

Takemura, A. (1993). Changes in an immunoglobulin (IgM) like protein during larval stages in tilapia, *Oreochromis mossambicus*. *Aquaculture* **115**, 233–241.

Takeuchi, H., Figueroa, F., O'hUigin, C., Klein, J. (1995). Cloning and characterization of class I MHC genes of the zebrafish, *Brachydanio rerio*. *Immunogenetics* **42**, 77–84.

Tamai, T., Sato, N., Kimura, S., Shirata, S., Murakami, H. (1992). Cloning and expression of flatfish interleukin 2 gene. In *Animal Cell Technology: Basic & Applied Aspects* (ed. Murakami, H.), pp. 509–514. Kluwer Academic Publishers, Dordrecht, The Netherlands.

Tamai, T., Shirahata, S., Noguchi, T., Sato, N., Kimura, S., Murakami, H. (1993a). Cloning and expression of flatfish (*Paralichthys olivaceus*) interferon cDNA. *Biochem. Biophys. Acta* **1174**, 182–186.

Tamai, T., Shirahata, S., Sato, N., Kimura, S., Nonaka, M.,

Murakami, H. (1993b). Purification and characterization of interferon-like antiviral protein derived from flatfish (*Paralichthys olivaceus*) lymphocytes immortalized by oncogenes. *Cytotechnology* **11**, 121–131.

Tatner, M. F. (1985). The migration of labelled thymocytes to the peripheral lymphoid organs in the rainbow trout, *Salmo gairdneri* Richardson. *Dev. Comp. Immunol.* **9**, 85–91.

Tatner, M. F., Horne, M. T. (1983). Susceptibility and immunity to *Vibrio anguillarum* in post-hatching rainbow trout fry, *Salmo gairdneri* Richardson 1836. *Dev. Comp. Immunol.* **7**, 465–472.

Tatner, M. F., Manning, M. J. (1983). The ontogeny of cellular immunity in the rainbow trout, *Salmo gairdneri* Richardson, in relation to the stage of development of the lymphoid organs. *Dev. Comp. Immunol.* **7**, 69–75.

Tatner, M. F., Manning, M. J. (1985). The ontogenic development of the reticuloendothelial system in the rainbow trout, *Salmo gairdneri* Richardson. *J. Fish Dis.* **8**, 35–41.

Temkin, R. J., McMillan, D. B. (1986). Gut-associated lymphoid tissue (GALT) of the goldfish, *Carassius auratus*. *J. Morphol.* **190**, 9–26.

Thompson, I., Fletcher, T. C., Houlihan, D. F., Secombes, C. J. (1994). The effect of dietary vitamin A on the immunocompetence of Atlantic salmon (*Salmo salar* L.). *Fish Physiol. Biochem.* **12**, 513–523.

Thompson, J. S. (1982). An epizootic on lymphoma in northern pike, *Esox lucius* L., from the Aland islands of Finland. *J. Fish Dis.* **5**, 1–11.

Thorburn, M. A., Jansson, E. K. (1988). Frequency distribution in rainbow trout populations of absorbance values from an ELISA for *Vibrio anguillarum* antibodies. *Dis. Aquat. Org.* **5**, 171–177.

Thuvander, A. (1989). Cadmium exposure of rainbow trout, *Salmo gairdneri* Richardson: effects on immune functions. *J. Fish Biol.* **35**, 521–529.

Thuvander, A., Hongslo, T., Jansson, E., Sundquist, B. (1987). Duration of protective immunity and antibody titres measured by ELISA after vaccination of rainbow trout, *Salmo gairdneri* Richardson, against vibriosis. *J. Fish Dis.* **10**, 479–486.

Thuvander, A., Fossum, C., Lorenzen, N. (1990). Monoclonal antibodies to salmonid immunoglobulin: characterization and applicability in immunoassays. *Dev. Comp. Immunol.* **14**, 415–423.

Tomlinson, S., Stanley, K. K., Esser, A. F. (1993). Domain structure, functional activity, and polymerisation of trout complement protein C9. *Dev. Comp. Immunol.* **17**, 67–76.

Tomonaga, S., Kobayashi, K. (1985). A second class of immunoglobulin in the cartilaginous fishes. *Dev. Comp. Immunol.* **9**, 797–802.

Tremmink, J. H. M., Bayne, C. J. (1987). Ultrastructural characterization of leukocytes in the pronephros of carp (*Cyprinus carpio* L.). *Dev. Comp. Immunol.* **11**, 125–137.

Trobridge, G. D., Leong, J. C. (1995). Characterization of a rainbow trout Mx gene. *J. Interferon Cytokine Res.* **15**, 691–702.

Trobridge, G. D., Chiou, P. P., Leong, J. C. (1997). Cloning of the rainbow trout (*Onchorynchus mykiss*) Mx2 and Mx3 cDNAs and characterization of trout Mx protein in salmon cells. *J. Virol.* **71**, 5304–5311.

Trump, G. N. (1970). Goldfish immunoglobulins and antibodies to bovine serum albumin. *J. Immunol.* **104**, 1267–1275.

Tutter, A., Riblet, R. (1989). Conservation of an immunoglobulin variable-region family indicates a specific, noncoding function. *Proc. Natl Acad. Sci. USA* **86**, 7460–7464.

Tutter, A., Brodeur, P., Shlomchik, M., Riblet, R. (1991). Structure, map position and evolution of two newly diverged mouse Ig V_H gene families. *J. Immunol.* **147**, 3215–3223.

Uhr, J. W., Finkelstein, M. S., Franklin, E. C. (1962). Antibody response to bacteriophage FX 174 in non-mammalian vertebrates. *Proc. Soc. Exp. Biol. Med.* **111**, 13–15.

Vallejo, A. N., Miller, N. W., Jørgensen, T., Clem, L. W. (1990). Phylogeny of immune recognition: antigen processing/presentation in channel catfish immune responses to hemocyanins. *Cell. Immunol.* **130**, 364–377.

Vallejo, A. N., Miller, N. W., Clem, L. W. (1991a). Phylogeny of immune recognition: processing and presentation of structurally-defined antigens. *Dev. Immunol.* **1**, 137–148.

Vallejo, A. N., Ellsaesser, C. F., Miller, N. W., Clem, L. W. (1991b). Spontaneous development of functionally active long-term monocyte-like cell lines from channel catfish. *In Vitro Cell. Dev. Biol.* **27A**, 279–286.

Vallejo, A. N., Miller, N. W., Clem, L. W. (1992). Antigen processing and presentation in teleost immune responses. *Annu. Rev. Fish Dis.* **2**, 73–89.

van der Heijden, M. H. T., Rooijakkers, J. B. M. A., Booms, G. H. R., Rombout, J. H. W. M., Boon, J. H. (1995). Production, characterisation and applicability of monoclonal antibodies to european eel (*Anguilla anguilla* L.) immunoglobulin. *Vet. Immunol. Immunopathol.* **45**, 151–164.

van Dijk, K. W., Mortari, F., Kirkham, P. M., Schroeder Jr, H. W., Milner, E. C. (1993). The human immunoglobulin V_H7 gene family consists of a small polymorphic group of six to eight gene segments dispersed throughout the V_H locus. *Eur. J. Immunol.* **23**, 832–839.

van Erp, S. H. M., Dixon, B., Figueroa, F., Egberts, E., Stet, R. J. M. (1996a). Identification and characterization of a new major histocompatibility complex class I gene in carp (*Cyprinus carpio* L.). *Immunogenetics* **44**, 49–61.

van Erp, S. H. M., Egberts, E., Stet, R. J. M. (1996b). Characterization of class IIA and B genes in a gynogenetic carp clone. *Immunogenetics* **44**, 192–202.

van Loon, J. J. A., van Oosterom, R., van Muiswinkel, W. B. (1981). Development of the immune system in carp, *Cyprinus carpio* L. In *Aspects of Developmental and Comparative Immunology* (ed. Solomon, J. B.), pp. 469–470. Pergamon Press, Oxford, UK.

van Muiswinkel, W. B. (1995). The piscine immune system: innate and acquired immunity. In *Fish Diseases and Disorders* (ed. Woo, P. T. K.), pp. 729–750. CAB International, Cambridge, UK.

van Muiswinkel, W. B., Anderson, D. P., Lamars, C. H. J., Egberts, E., van Loon, J. J. A., Ijssel, J. P. (1985). Fish immunology and fish health. In *Fish Immunology* (eds Manning, M. J., Tatner, M. F.), pp. 1–8. Academic Press, London, UK.

van Muiswinkel, W. B., Lamers, C. H. J., Rombout, J. H. W. M. (1991). Structural and functional aspects of the spleen in bony fish. *Res. Immunol.* **142**, 362–366.

Varner, J., Neame, P., Litman, G. W. (1991). A serum heterodimer from hagfish (*Eptatretus stoutii*) exhibits structural similarity and partial sequence identity with immunoglobulin. *Proc. Natl Acad. Sci. USA* **88**, 1746–1750.

Vazquez, M., Mizuki, N., Flajnik, M. F., McKinney, E. C.,

Kasahara, M. (1992). Nucleotide sequence of a nurse shark immunoglobulin heavy chain cDNA clone. *Mol. Immunol.* **29**, 1157–1158.

Vazquez-Moreno, L., Porath, J., Schluter, S. F., Marchalonis, J. J. (1992). Purification of a novel heterodimer from shark (*Carcharhinus plumbeus*) serum by gel-immobilized metal chromatography. *Comp. Biochem. Physiol.* **103B**, 563–568.

Ventura-Hollman, T., Jones, C. J., Ghaffari, S. H., Lobb, C. J. (1994). Structure and genomic organisation of V_H gene segments in the channel catfish: members of different V_H gene families are interspersed and closely linked. *Mol. Immunol.* **31**, 823–832.

Ventura-Hollman, T., Ghaffari, S. H., Lobb, C. J. (1996). Characterization of a seventh family of immunoglobulin heavy chain V_H gene segments in the channel catfish, *Ictalurus punctatus*. *Eur. J. Immunogenet.* **23**, 7–14.

Verburg-van Kemenade, L. B. M., Weyts, F. A. A., Debets, R., Flik, G. (1995). Carp macrophages and neutrophilic granulocytes secrete an interleukin-1-like factor. *Dev. Comp. Immunol.* **19**, 59–70.

Viale, G., Calamari, D. (1984). Immune response in rainbow trout *Salmo gairdneri* after long-term treatment with low levels of Cr, Cd, and Cu. *Environ. Poll.* **35**, 247–257.

Vilain, C., Wetzel, M.-C., Du Pasquier, L., Charlemagne, J. (1984). Structural and functional analysis of spontaneous anti-nitrophenyl antibodies in three cyprinid fish species: carp (*Cyprinus carpio*), goldfish (*Carassius auratus*), and tench (*Tinca tinca*). *Dev. Comp. Immunol.* **8**, 611–622.

Vogel, W. O. P., Claviez, M. (1981). Vascular specialization in fish, but no evidence for lymphatics. *Z. Naturforsch. Sect. C Bio.* **36**, 490–492.

Voss, E. W., Sigel, M. M. (1972). Valence and temporal changes in affinity of purified 7S and 18S nurse shark anti-2,4 dinitrophenyl antibodies. *J. Immunol.* **109**, 665–673.

Voss, E. W., Fryer, J. L., Banowetz, G. M. (1978a). Isolation, purification, and partial characterization of a lectin from chinook salmon ova. *Arch. Biochem. Biophys.* **186**, 25–34.

Voss, E. W., Groberg, W. J., Fryer, J. L. (1978b). Binding affinity of tetrameric coho salmon Ig anti-hapten antibodies. *Immunochem.* **15**, 459–464.

Walker, R. A., McConnell, T. J. (1994). Variability in an MHC class II β chain-encoding gene in striped bass (*Morone saxitilis*). *Dev. Comp. Immunol.* **18**, 325–342.

Walter, R. B., Morizot, D. C. (1996). Conservation of genome and gene structure from fishes to mammals. *Adv. Struct. Biol.* **4**, 1–24.

Wang, R., Neumann, N. F., Shen, Q., Belosevic, M. (1995). Establishment of a macrophage cell line from the gold fish. *Fish Shellfish Immunol.* **5**, 329–346.

Wardle, C. S. (1971). New observations on the lymph system of the plaice (*Pleuronectes platessa*) and other teleosts. *J. Mar. Biol. Ass. UK* **51**, 977–990.

Warr, G. W. (1982). Behaviour of unreduced polymeric and monomeric immunoglobulins in sodium dodecyl sulfate-polyacrylamide gel electrophoresis. *Mol. Immunol.* **19**, 75–81.

Warr, G. W. (1983). Immunoglobulin of the toadfish, *Spheroides glaber*. *Comp. Biochem. Physiol.* **76B**, 507–514.

Warr, G. W. (1995). The immunoglobulin genes in fish. *Dev. Comp. Immunol.* **19**, 1–12.

Warr, G. W., Marchalonis, J. J. (1977). Lymphocyte surface immunoglobulin of the goldfish differs from its serum counterpart. *Dev. Comp. Immunol.* **1**, 15–22.

Warr, G. W., Griffin, B. R., Anderson, D. P., McAllister, P. E., Lidgerding, B., Smith, C. E. (1984). A lymphosarcoma of thymic origin in the rainbow trout, *Salmo gairdneri* Richardson. *J. Fish Dis.* **7**, 73–82.

Warr, G. W., Middleton, D. L., Miller, N. W., Clem, L. W., Wilson, M. R. (1991). An additional family of VH sequences in the channel catfish. *Eur. J. Immunogenet.* **18**, 393–397.

Warr, G. W., Miller, N. W., Clem, L. W., Wilson, M. R. (1992). Alternate splicing pathways of the immunoglobulin heavy chain transcript of a teleost fish, *Ictalurus punctatus*. *Immunogenetics* **35**, 253–256.

Waterstrat, P. R., Ainsworth, A. J., Capley, G. (1988). Use of discontinuous gradient technique for the separation of channel catfish, *Ictalurus punctatus* (Rafinesque), peripheral blood leukocytes. *J. Fish Dis.* **11**, 289–294.

Watkins, W. M. (1987). Biochemical genetics of blood group antigens: retrospect and prospect. *Biochem. Soc. Transact.* **15**, 620–624.

Weeks, B. A., Warinner, J. E. (1986). Functional evaluation of macrophages in fish from a polluted estuary. *Vet. Immunol. Immunopathol.* **12**, 313–320.

Weeks, B. A., Sommer, S. R., Dalton, H. P. (1988). Chemotactic response of fish macrophages to *Legionella pneumophila*: correlation with pathogenicity. *Dis. Aquat. Org.* **5**, 35–38.

Weeks-Perkins, B. A., Ellis, A. E. (1995). Chemotactic responses of Atlantic salmon (*Salmo salar*) macrophages to virulent and attenuated strains of *Aeromonas salmonicida*. *Fish Shellfish Immunol.* **5**, 313–323.

Weinheimer, P. F., Mestecky, J., Acton, R. T. (1971). Species distribution of J chain. *J. Immunol.* **107**, 1211–1212.

Wester, P. W., Vethaak, A. D., van Muiswinkel, W. B. (1994). Fish as biomarkers in immunotoxicology. *Toxicology* **86**, 213–232.

Wetzel, M. C., Charlemagne, J. (1985). Antibody diversity in fish. Isoelectrofocalisation study of individually-purified specific antibodies in three teleost fish species: tench, carp and goldfish. *Dev. Comp. Immunol.* **9**, 261–270.

Whittington, R. J. (1993). Purification and partial characterisation of serum immunoglobulin of the European perch (*Perca fluviatilis* L.). *Fish Shellfish Immunol.* **3**, 331–343.

Whyte, S. K., Chappell, L. H., Secombes, C. J. (1989). Cytotoxic reactions of rainbow trout, *Salmo gairdneri* Richardson, macrophage for larvae for the eye fluke *Diplostomum spathaceum* (Digenea). *J. Fish Biol.* **35**, 333–345.

Wiegertjes, G. F., Stet, R. J. M., van Muiswinkel, W. B. (1994). Divergent selection for antibody production in common carp (*Cyprinus carpio*) using gynogenesis. *Anim. Genet.* **25**, 251–257.

Wiegertjes, G. F., Stet, R. J. M., van Muiswinkel, W. B. (1995). Investigations into the ubiquitous nature of high and low immune responsiveness after divergent selection for antibody production in common carp (*Cyprinus carpio* L.). *Vet. Immunol. Immunopathol.* **48**, 355–366.

Williams, G. (1931). On various fish tumors. *Proc. Trans. Lpool Biol. Soc.* **45**, 98–109.

Wilson, M. R., Warr, G. W. (1992). Fish immunoglobulins and the genes that encode them. *Annu. Rev. Fish Dis.* **2**, 201–221.

Wilson, M. R., Wang, A.-C., Fish, W. W., Warr, G. W. (1985). Anomalous behaviour of goldfish IgM heavy chains in sodium dodecylsulphate polyacrylamide gel electrophoresis. *Comp. Biochem. Physiol.* **82B**, 41–49.

Wilson, M. R., Marcus, A., van Ginkel, F., Miller, N. W., Clem, L. W., Middleton, D., Warr, G. W. (1990). The immunoglobulin M heavy chain constant region gene of the channel catfish, *Ictalurus punctatus*: an unusual mRNA splice pattern produces the membrane form of the molecule. *Nucl. Acids Res.* **18**, 5227–5233.

Wilson, M. R., Middleton, D., Warr, G. W. (1991). Immunoglobulin VH genes of the goldfish, *Carassius auratus*: a reexamination. *Mol. Immunol.* **28**, 449–457.

Wilson, M. R., van Ravenstein, E., Miller, N. W., Clem, L. W., Middleton, D. L., Warr, G. W. (1995a). cDNA sequences and organisation of IgM heavy chain genes in two holostean fish. *Dev. Comp. Immunol.* **19**, 153–164.

Wilson, M. R., Ross, D. A., Miller, N. W., Clem, L. W., Middleton, D. L., Warr, G. W. (1995b). Alternate pre-mRNA processing pathways in the production of membrane IgM heavy chains in holostean fish. *Dev. Comp. Immunol.* **19**, 165–177.

Wilson, M. R., Zhou, H., Bengtén, E., Clem, L. W., Stuge, T. S., Warr, G. W., Miller, N. W. (1998). T cell receptors in channel catfish: structure and expression of TCR α and β genes. *Mol. Immunol.*, in press.

Wilson, V., Jeffreys, A. J., Barrie, P. A., Boseley, P. G., Slocombe, P. M., Easton, A., Burke, D. C. (1983). A comparison of vertebrate interferon gene families detected by hybridization with human interferon DNA. *J. Mol. Biol.* **166**, 457–475.

Winkelhake, J. L., Chang, R. J. (1982). Acute phase (C-reactive) protein-like macromolecules from rainbow trout (*Salmo gairdneri*). *Dev. Comp. Immunol.* **6**, 481–489.

Wishkovsky, A., Avtalion, R. R. (1982). Induction of helper and suppressor functions in carp (*Cyprinus carpio*) and their possible implication in seasonal disease in fish. *Dev. Comp. Immunol.* **2**, 28–91.

Wittbrodt, J., Adam, D., Malitschek, B., Maueler, W., Raulf, F., Telling, A., Robertson, S., Schartl, M. (1989). Novel putative receptor tyrosine kinase encoded by the melanoma-inducing *Tu* locus in *Xiphophorus*. *Nature* **341**, 415–421.

Wolke, R. E. (1992). Piscine macrophage aggregates: a review. *Annu. Rev. Fish Dis.* **2**, 91–108.

Wolke, R. E., Wyand, D. S. (1969). Ocular lymphosarcoma of an Atlantic cod. *Bull. Wildl. Dis. Ass.* **5**, 401–403.

Wongtavatchai, J., Conrad, P. A., Hedrick, R. P. (1995). Effect of microsporidian *Enterocytozoon salmonis* on the immune response of chinook salmon. *Vet. Immunol. Immunopathol.* **48**, 367–374.

Yamaga, K. M., Kubo, R. T., Etlinger, H. M. (1978). Studies on the question of conventional immunoglobulin on thymocytes from primitive vertebrates. II. Delineation between Ig-specific and cross-reactive membrane components. *J. Immunol.* **120**, 2074–2079.

Yamamoto, F.-I., Clausen, H., White, T., Marken, J., Hakomori, S.-I. (1990). Molecular genetic basis of the histo-blood group ABO system. *Nature* **345**, 229–233.

Yano, T., Nakao, M. (1994a). Isolation of a carp complement protein homologous to mammalian factor D. *Mol. Immunol.* **31**, 337–342.

Yano, T., Nakao, M. (1994b). The complement system of vertebrates. *Dev. Comp. Immunol.* **18**, S76.

Yano, T., Matsuyama, H., Nakao, M. (1988). Isolation of the first component of complement (C1) from carp serum. *Nipp. Suis. Gakk.* **54**, 851–859.

Yoshida, S. H., Stuge, T. B., Miller, N. W., Clem, L. W. (1995). Phylogeny of lymphocyte heterogeneity: cytotoxic activity of channel catfish peripheral blood leukocytes directed against allogeneic targets. *Dev. Comp. Immunol.* **19**, 71–77.

Zapata, A. (1979). Ultrastructural study of the teleost fish kidney. *Dev. Comp. Immunol.* **3**, 55–65.

Zapata, A. (1980). Ultrastructure of elasmobranch lymphoid tissue. 1. Thymus and spleen. *Dev. Comp. Immunol.* **4**, 459–472.

Zapata, A. (1981a). Ultrastructure of elasmobranch lymphoid tissue. 2. Leydigs and epigonal organs. *Dev. Comp. Immunol.* **5**, 43–52.

Zapata, A. (1981b). Lymphoid organs of teleost fish. II. Ultrastructure of renal lymphoid tissue of *Rutilus rutilus* and *Gobio gobio*. *Dev. Comp. Immunol.* **5**, 685–690.

Zapata, A., Cooper, E. L. (1990). *The Immune System: Comparative Histophysiology*. John Wiley & Sons, Chichester, UK.

Zelikoff, J. T. (1993). Metal pollution-induced immunomodulation in fish. *Annu. Rev. Fish Dis.* **3**, 305–325.

Zelikoff, J. T., Enane, N. A., Bowser, D., Squibb, K. S., Frenkel, K. (1991). Development of fish peritoneal macrophages as a model for higher vertebrates in immuno-toxicological studies. I. Characterisation of trout macrophage morphological, functional and biochemical properties. *Fund. Appl. Toxicol.* **16**, 576–589.

Zhou, H., Bengtén, E., Miller, N. W., Warr, G. W., Clem, L. W., Wilson, M. R. (1997). T cell receptor sequences in the channel catfish. *Dev. Comp. Immunol.* **21**, 238.

Zou, J., Cunningham, C., Secombes, C. J. (1997). Rainbow trout recombinant interleukin 1β: expression, renaturation and determination of the biological activities. *Dev. Comp. Immunol.* **21**, 192.

III Immunology of Amphibians

1. Introduction

Living amphibians are the descendants of the first tetrapod vertebrates that diverged from their lungfish-like ancestors at the Devonian period, about 450 million years ago. *Urodela* (newts, salamanders), *Anura* (toads, frogs) and *Gymnophiona* (limbless coecilians) are the three Orders of the *Lissamphibia* Infraclass which is considered to be monophyletic. They actually represent about 4000 species. The oldest known frog fossil is from the Early Triassic. Frogs and toads have an highly modified hind-limb and pelvis for their jumping mode of locomotion, and a short and flat head specialized for processing living prey. Newts and salamanders are much less specialized: the oldest well-preserved fossil salamander is from the Late Jurassic and modern urodeles do not appear very different, at a first examination, from *Ichthyostega* (Devonian), who may be the first sister-group of all tetrapods. All modern amphibian groups probably arose in the Triassic and expanded in the Cenozoic.

From their key position in vertebrate evolution, amphibians represent important models for the developmental and comparative analysis of the immune system. Two species are currently studied: the New Word urodele *Ambystoma mexicanum* (axolotl, a neotenic species) from the *Ambystomatidae* family, and the African anuran *Xenopus laevis* (South African toad, or clawed toad), from the *Pipidae* family. Both species are worldwide laboratory species, they breed in captivity and have, for several decades, been models of choice for many investigations, including development, regeneration and immunity. More limited results have been obtained from several other urodele species from Europe, North Africa (*Triturus alpestris*, *Pleurodeles waltl*), or North America (*Triturus viridescens*) and from two New World anuran species, *Rana pipiens* (leopard frog) and *Rana catesbeiana* (bull-frog).

2. Lymphoid Organs and Lymphocytes

Thymus

The amphibian thymus anlagen arise from the dorsal epithelium of the second (anuran), or third, fourth and fifth (urodeles) pharyngeal pouches. In *Xenopus*, the thymus buds (one pair) differentiate 3 days after fertilization, are colonized during the following days by mesenchyme-derived haemopoietic precursors, and by days 6–8, the cortex–medulla architecture becomes visible. The larval thymus reach a peak size (1–2 $\times$ 10^6 thymocytes) just before metamorphosis, but transitionally involutes during metamorphosis, where most thymocytes are lost and eliminated by macrophages. An efficient regeneration then occurs in the metamorphosed froglet (4–5 $\times$ 10^7 thymocytes at 2–3 months post metamorphosis). Thymus then regresses at the time of sexual maturity, during the second year of life. The *Xenopus* thymus contains most of the cell types that constitute the stroma of mammalian thymus, like macrophages, nurse-like cells, cortical large dendritic cells and different types of epithelial cells. Unlike in mammals, a distinct physical barrier, rich in blood vessels and IgM-producing B cells separates the cortex and the medulla areas (reviewed by Du Pasquier *et al.*, 1989).

Thymus development is much slower in urodeles where the thymic anlagen appear at the time of hatching, 3 weeks after fertilization. In the axolotl, the three pairs of thymus epithelial buds are invaded by mesenchyme-derived precursors over a period of 10–15 days starting 12 days after hatching, and develop slowly to reach a total of about 0.5 $\times$ 10^5 cells 2.5 months after fertilization, and a peak size (about 5 $\times$ 10^6 cells) at the time that precedes sexual maturity (about 12 months) (Tournefier *et al.*, 1990). In metamorphosing urodele species (*Triturus*, *Pleurodeles*), only one pair of thymus develops from the fifth pharyngeal pouches. Thymus then develops (about 10^6 cells) until premetamorphosis (2–5 months after fertilization), and involutes at the time of sexual maturity. Interestingly, the urodele thymus have no cortex–medulla differentiation. However, the presence of three different types of epithelial cells, reticular cells, macrophages and epithelial cysts clearly confirm the complexity of the stromal architecture (Tournefier *et al.*, 1990).

Spleen

The spleen appears about 12–14 days after fertilization in *Xenopus*, thus about 10 days after thymus. The mature spleen (about 4 $\times$ 10^7 lymphocytes in adults) is a spherical

Handbook of Vertebrate Immunology
ISBN 0-12-546401-0

hematopoietic organ with well-defined red (mainly erythropoietic) and white (mainly lymphopoietic) pulp, but no structure reminiscent of the avian and mammalian germinal centres. Lymphocytes are organized in a follicular-like B-cell rich structure centred around a central arteriole, which is lined by a boundary layer of cells and surrounded by scattered nodular thymus-derived cells (Du Pasquier *et al.*, 1989). This peripheral area is a site of antigen retention (Collie, 1974). Surface immunoglobulin positive (sIg^+) B-cells first appear in the *Xenopus* spleen about 15 days post-fertilization, and the spleen is the main source of IgM antibody forming cells in adults.

The axolotl spleen anlagen appears at the time of hatching (about 3 weeks after fertilization), and develops slowly to reach about 10^5 cells 2.5 months after fertilization, half of them being B Cells. Although some red/white pulp-like organization can be seen macroscopically, there is no true follicular-like structure and there are no germinal centres in the spleen, even following several rounds of immunization with the same antigen. At 3.5 months and later, B cells appear scattered in all the organ, with no structural organization (Fellah *et al.*, 1989). Erythropoiesis occurs in the red pulp throughout life, and thrombocytes, macrophages and granulocytes are scattered into the white and red pulps. An axolotl adult spleen may contain up to 5×10^7 leukocytes.

Other sites of lymphocytopoiesis

The lymphoid nodules (cavity bodies) which are associated with the *Xenopus* larvae pharynx contain thymus-derived cells and disappear at metamorphosis (Du Pasquier *et al.*, 1989). B-cell rich lymphoid nodules then arise post-metamorphosis in the intestine mucosa (*lamina propria*). Although amphibian gills, pharynx and mesenteries frequently contain more or less structured lymphoid nodules, these structures, which are not sensitive to antigenic stimulation, cannot be considered as lymph node equivalents.

In urodeles and anurans, the hematopoietic peripheral layer of the liver supports B-cell lymphopoiesis which appears to persist throughout life in *Xenopus* (but not in other anuran species), but disappears at the time of sexual maturity (12 months in axolotl). This bone marrow-like tissue also supports permanent granulocytopoiesis.

Rare IgM-producing cells can be seen along the digestive tract in axolotl and an important accumulation of secretory IgY molecules accumulates in the epithelial digestive cells of young axolotls and disappears at 7–8 months, in correlation with the diminution of the production of a secretory component-like molecule. It is not known if these secretory IgY molecules are produced locally or trapped from blood or lymph by the intestinal cells (Fellah *et al.*, 1992a). In *Xenopus*, IgM-, IgX- but no IgY-producing cells can be found in the digestive tract. The anuran, but not the urodele kidneys contain lymphocytes, mainly B cells, and can retain some antigens in the intertubular tissue. Amphibians posses B cells and T-like cells in the blood and it has been demonstrated that these cells are fully mature and can efficiently collaborate *in vitro* in an MHC-restricted manner for the production of antibodies to T-cell dependent antigens (Blomberg *et al.*, 1990).

Surface lymphocyte markers

B cells can be easily recognized in axolotl and *Xenopus* by using monoclonal antibodies (mAb) specific for the different isotypes of the heavy (H) and the light (L) chains of the Ig molecule (Hsu and Du Pasquier, 1984; Tournefier *et al.*, 1988a). More difficult is the production of mAbs specific for T cells, or T-cell subpopulations. In the axolotl, mAb 34.38.6 recognizes 65–72 kDa surface polypeptides and labels, in immunofluorescence, nearly all thymocytes, 60–63% sIg^- splenic lymphocytes of normal animals, but only 9% splenic lymphocytes in thymectomized animals. mAb 34.38.6 does not recognize sIg^+ (B) cells, but labels haemopoietic stem cells, granulocytes and macrophages (Tournefier *et al.*, 1988b). A polyclonal antibody (L12) was raised against a 38 kDa axolotl thymocyte membrane polypeptide and stained 80–86% of thymocytes and 40–46% of sIg^- lymphoid cells in the spleen. L12 co-immunoprecipitates several (38, 43 and 22 kDa) covalently linked molecules that form a multimeric complex on the T-cell surface (Kerfourn *et al.*, 1992).

A membrane glycoprotein of 120 kDa had been identified on *Xenopus* T-cells with mAb XT-1 (Nagata, 1985). This marker is restricted to the *Xenopus* J strain and was very useful in tracing thymus-derived cells in XT-1 negative frogs (Horton *et al.*, 1992). It stained 92–98% thymocytes, 22–37% sIg^- splenic lymphocytes, and 20–30% peripheral lymphocytes. Some bone marrow cells are also labelled. mAb AM22 recognize *Xenopus* cortical thymocytes and a subpopulation of peripheral T-cells and may bind to a CD8 equivalent since it immunoprecipitated 35 kDa surface polypeptides which form heterodimers (Du Pasquier and Flajnik, 1990). Another mAb, AM15, recognizes a 18 kDa protein of a subpopulation of *Xenopus* T cells (Flajnik *et al.*, 1990a).

Two mAbs were derived from mice immunized with *Xenopus* thymocytes (mAb 1S9-2) or with the T-cell like B3B7 cell line (mAb X71). These mAbs stained a subpopulation (65–80%) of thymocytes, but no cells in peripheral tissues. The stained cells represent a subpopulation of cortical thymocytes which is also labelled by mAb AM22 (anti-CD8-like), and thus could be the *Xenopus* equivalent of the avian and mammalian cortical double positive (DP) thymocytes. A single 55–60 kDa glycoprotein (CTX) was immunoprecipitated from the *Xenopus* cell line ff-2. The CTX gene was cloned and the deduced protein was shown to be composed of two V and C2 Ig-like domains followed by a large transmembrane domain (Chrétien *et al.*, 1996).

A mAb specific for thymocytes has also been described in the bullfrog (Sugiyama *et al.*, 1990).

In vitro stimulation

In tadpoles and adult *Xenopus*, the use of classical mitogenic agents such as LPS (lipopolysaccharide), PHA (phytohaemagglutinin), Con A (concanavalin A) and PPD (purified protein derivative of tuberculin), stimulate *in vitro* B cells and T cells in the same patterns as in mammals (Williams and Horton, 1980). The PHA and mixed leukocyte (MLR) responses are sensitive to thymectomy (Du Pasquier and Horton, 1976). Furthermore, a T-cell growth factor (TCGF) is produced by stimulated T lymphocytes and is biochemically similar to interleukin 2 (IL-2) (Turner *et al.*, 1991; Haynes and Cohen, 1993a,b).

The use of different mitogenic agents in urodele was unsuccessful until optimum culture conditions were established (Koniski and Cohen, 1992, 1994). In these amphibians the use of fetal calf serum strongly inhibits *in vitro* immune functions and different protein-free medium must be used with each mitogen.

The axolotl has a population of B lymphocytes that proliferate specifically with a high stimulation index to LPS. This proliferative capacity is observed throughout ontogenesis without significant changes (Salvadori and Tournefier, 1996). The T cell subpopulation proliferates significantly in response to PHA and Con A, although splenic T lymphocytes of young axolotl (less than 10 months) do not have this functional ability. Axolotl lymphocytes are able to proliferate *in vitro* with a significant stimulation index to staphylococcal enterotoxins A and B (SEA and SEB) (Salvadori and Tournefier, 1996). The fact that these superantigens can activate lymphocytes from a primitive vertebrate suggests a striking conservation of molecular structures implied in superantigen presentation and recognition.

3. Lymphocyte Antigen-specific Receptors

T-cell receptors (TCR)

Despite many efforts, no antibodies have yet been produced against TCR-equivalent molecules in any cold-blooded vertebrate species and all our present knowledge on these molecules comes from recombinant DNA studies. Genes encoding the α and β chains of the axolotl TCR have been recently cloned (Fellah *et al.*, 1993a, 1996). Five V_α segments were identified, each of them belonging to a separate family. The best identity scores for these axolotl V_αs were provided by sequences belonging to the human $V_\alpha1$ family and the mouse $V_\alpha3$ and $V_\alpha8$ families. A total of 14 different J_α segments were identified from 44 V_α–J_α regions sequenced, suggesting a large repertoire of J_α segments. The structure of the axolotl α-chain CDRα3 loop is in good agreement with that of mammals, including a majority of small hydrophobic residues at position 92 and of charged, hydrophilic or polar residues at positions 93 and 94, which are highly variable and correspond to the V_α–J_α junction. This suggests that some positions of the axolotl CDRα3 loop are positively selected during T-cell differentiation, particularly around residue 93 which could be selected for its ability to make contacts with MHC-associated antigenic peptides, as in mammals. The axolotl C_α domain had the typical structure of mammalian and avian C_αs, including the charged residues in the TM segment that are thought to interact with other proteins in the membrane, as well as most of the residues forming the conserved antigen receptor transmembrane (CART) motif.

Fourteen V_β families were identified in the axolotl (Fellah *et al.*, 1994; J. S. Fellah, unpublished results). There appears to be a greater genetic distance between the axolotl V_β families than between the different V_β families of any mammalian species examined to date: most of the axolotl V_βs are less than 35% identical (nucleotides) and the less related families ($V_\beta4$ and $V_\beta8$) have no more than 23.2% identity (13.5% at the amino acid level). Despite their great mutual divergence, several axolotl V_β are sequence-related to some mammalian V_β genes, such as the human $V_\beta13$ and $V_\beta20$ segments and their murine $V_\beta8$ and $V_\beta14$ homologues. However, the axolotl $V_\beta8$ and $V_\beta9$ families are not significantly related to any other V_β sequence at the nucleotide level and show limited amino acid similarity to mammalian V_α, V_κIII or V_H sequences.

Four C_β isotypes have been detected in the axolotl (Fellah *et al.*, 1993a; J. S. Fellah, unpublished results). The extracellular domains of the $C_\beta1$, $C_\beta2$ and $C_\beta3$ isotypes show an impressively high degree of identity (83–85%), suggesting that a very efficient mechanism of gene correction has been in operation to preserve this structure. However, the $C_\beta4$ isotype is clearly divergent (43–48% identity with $C_\beta1$–3). The transmembrane axolotl C_β domains have been less well conserved when compared with the mammalian C_β but they do maintain the lysine residue which is thought to be involved in the charged interaction between the TCR$\alpha\beta$ heterodimer and the CD3 complex.

A large number of independent rearrangements were sequenced, in which a defined V_β ($V_\beta7.1$) segment is associated with either $C_\beta1$, $C_\beta2$, $C_\beta3$ or $C_\beta4$. Three $J_\beta1$ segments were associated with the V_β–$C_\beta1$, six with V-$C_\beta2$, three with V-$C_\beta3$ and four with V-$C_\beta4$ rearrangements. Three different D_β-like sequences were identified ($D_\beta1$ and $D_\beta4$ are identical); these can be productively read in their three putative reading frames. This suggests that the axolotl TCRβ locus is organized into, at least, four independent (D_β–J_β–C_β) clusters that use the same collection of V_β segments. About 40% of the β-chain VDJ junctions in 2.5-month-old axolotl nonimmunocompetent larvae had N nucleotides, compared with 73% in 10–25-month-old animals. The β-chain VDJ junctions had about 30% of defective rearrangements at any stage of development, which could be due to the slow rate of cell division in the axolotl lymphoid organs, correlated to the large genome in this urodele. A significant proportion of the axolotl CDRβ3 sequences deduced for in frame VDJ rearrangements are the same in animals of different origins. The molecular mechanism of this redundancy is unknown, but it could represent a selective advantage in the young axolotl for the recognition of some pathogens in their environment (Kerfourn *et al.*, 1996).

Nine V_β families were detected in *Xenopus*, which are rather divergent from each other: the best similarity (48%) is found between $V_\beta4$ and $V_\beta5$ (Chrétien *et al.*, 1997). Two members were detected in the $V_\beta1$ and $V_\beta8$ families and Southern experiments confirmed that the *Xenopus* V_β segments are organized in families that contain one, or only few members. The analysis of a large number of VDJ rearrangements allowed the detection of eight J_β elements with the canonical structure of all J segments. Two D_β core segments were also detected: GGGAGAGGGGGC, which is identical to trout D_β and mouse, rat and human $D_\beta1$ segments, and GGGACTGGGGGGGC, which is similar to axolotl, rat, mouse and rabbit $D_\beta2$ segments. Despite these large numbers of different J_β segments, and of the possible equivalent of $D_\beta1$ and $D_\beta2$ segments, a single C_β was found in *X. laevis*, and Southern blot experiment data are consistent with the presence of a single C_β locus and two alleles in all *Xenopus* species, except in *Xenopus ruwenzoriensis* (the *X. laevis* C_β probe does not hybridize with *Xenopus tropicalis* DNA).

B-cell receptors (BCR)

Two high molecular weight (IgM and IgX) and one low molecular weight (IgY) Ig classes have been characterized in *Xenopus*. The anti-IgM mAbs stained 15–25% of splenocytes, 1–2% of thymocytes and some lymphocytes from the intestinal mucosa (IEL). The anti-IgY mAbs stained only 1–2% splenocytes and less than 1% of thymocytes and IEL. Almost all surface IgY positive ($sIgY^+$) B cells also express surface IgM molecules ($sIgM^+$). Serum IgX was detected by immunoprecipitation using mAbs directed against Ig light (L) chains or V_H determinants, and can also be detected in the supernatants of spleen cells and IEL cultures. The μ (72 kDa) and χ (80 kDa) heavy (H) chains are highly glycosylated, but the υ chain is present in two molecular forms of 69.1 and 66.4 kDa, the lighter one being devoid of carbohydrates. Up to 60% of the intestinal B cells are IgX positive, whereas in spleen and liver these cells were hardly detectable. The three H isotypes can associate to 25, 27 and 29 kDa independent L chains. However, υ chains preferentially associate with the slower-migrating L chains and almost no υ chain is associated to the 27 kDa L chains (Hsu and Du Pasquier, 1984; Hsu *et al.*, 1985).

The genes encoding *Xenopus* H_μ, H_υ and H_χ chains, and the three L chains (ρ, σ and type III-λ-like) isotypes have been recently cloned. The three H chain isotypes have four constant domains. The *Xenopus* C_μ region is well related to other vertebrate C_μ, the $C_\mu3$ and $C_\mu4$ domains are the most conserved (47% and 42% identities with mammalian IgM) whereas $C_\mu1$ and $C_\mu2$ show little homology. The C_υ region of the IgY molecule, which is considered in *Xenopus* as the physiological equivalent of the mammalian IgG, is best structurally related to avian υ and mammalian ε chains. The C_χ region is not clearly phylogenetically related to any other vertebrate isotype. Putative switch (S) regions were detected 5′ to the C_υ and C_χ genes and it was demonstrated that a DNA recombination event similar to that occurring during H chain class switch recombination in mammals leads to the expression of IgX in *Xenopus* (Schwager *et al.*, 1988; Amemiya *et al.*, 1989; Haire *et al.*, 1989). A comparative analysis of the structure of the transmembrane (TM) exons of *Xenopus* μ, υ and χ genes confirmed the strong homology of the amphibian μ gene to the μ genes of other vertebrates. The TM exon of the υ and χ genes have few conserved residues with avian and mammalian γ and ε TM exons, suggesting that these two mammalian isotypes might share a common ancestor with amphibian υ genes and that the υ, γ and ε modern isotypes probably diverged early in evolution, at or before primitive amphibians (Mußmann *et al.*, 1996).

Eleven V_H families, about 15 D_H segments, and nine J_H segments have been detected into the *Xenopus* IgH locus. The V_H families are interspersed in the V_H locus and are as structurally different as in mammals, although the CDR1 regions from a given V_H family appear to be more conserved in *Xenopus*. The V, D and J elements can rearrange randomly during B-cell differentiation, thus providing an important potential of combinatory diversity. Junctional diversity also arises at the V–D and D–J junctions, with the possible occurrence of random nucleotide deletion (nibbling), and template (P), or nontemplate (N) nucleotide addition. Initially, the genetic potentialities for the diversification of the H chain repertoire do not appear to be fundamentally different in *Xenopus*, compared with mammals (Schwager *et al.*, 1989; Haire *et al.*, 1990, 1991).

The *Xenopus* L_ρ chains are homologous to the mammalian L_κ chains. One C_ρ and five J_ρ elements were detected, that rearrange randomly with several members of a single V_ρ family. This combinatory potential, and efficient variability at the V_ρ–J_ρ junction allow a substantial diversity of the L_ρ repertoire. The L_λ (also called Type III) chains are built from at least six V_λ families, two J_λ and two C_λ segments: this is the most diverse *Xenopus* L chain isotype that compares in complexity to both mammalian κ and λ chains. The L_σ chains consist of two ($C_\sigma1$ and $C_\sigma2$) isotypes and two ($V_\sigma1$ and $V_\sigma2$) families that rearrange with their own set of $J_\sigma1$ and $J_\sigma2$ elements. Many $V_\sigma1$, but few $V_\sigma2$ genes were detected and a relatively poor diversity of the expressed L_σ chains was found, compared with L_ρ and L_λ (Schwager *et al.*, 1991; Zezza *et al.*, 1991, 1992; Stewart *et al.*, 1993; Haire *et al.*, 1996).

High molecular weight (IgM) and low molecular weight (IgY, 172 kDa) Ig classes were characterized in the axolotl. The anti-IgM mAbs stained about 40% splenocytes and <1% thymocytes; the anti-IgY mAbs stained 15–20% splenocytes and <1% thymocytes. Bμ and Bυ cells are stained by mAb 33.101.2, specific for axolotl L chains. No B cells double stained by the anti-μ and anti-υ mAbs were found, suggesting that Bμ and Bυ cells might represent independent B-cell lineages. The H_μ chains (76 kDa) are highly glycosylated and, as in *Xenopus*, the H_υ chain is present in two antigenically similar forms of 68 and 66 kDa of different carbohydrate composition (Fellah and Charlemagne, 1988; Tournefier *et al.*, 1988a).

The genes encoding axolotl H_μ and H_υ chains were cloned (Fellah *et al.*, 1992b, 1993b). The axolotl C_μ is divided into four typical domains with an overall identity (nucleotides) of 56% with *Xenopus* C_μ (74% identity between the two $C_\mu4$ domains). The axolotl C_υ is most closely related to *Xenopus* C_υ (40% overall identity at the amino acid level). Additional cysteines in $C_\upsilon1$ and $C_\upsilon2$ domains is consistent with an additional intra-domain S-S bond similar to the avian C_υ and the human C_ε. This confirms that an ancestral relationship might have occurred between amphibian, avian υ chains and mammalian ε chains. However, the corresponding Ig classes molecules have different biological properties in these three groups of vertebrates (see later).

Eleven V_H families were detected in the axolotl, some of which are clearly related to *Xenopus* V_H, but others cannot be compared with any described V_H. Six JH segments were identified and DH segments were used by a majority of the V–D–J junctions. The same collection of V_H segments appear to be used by IgM and IgY molecules, however it is not known whether, as in *Xenopus*, an IgM > IgY switch may have occurred in the individual B cells (Golub and Charlemagne, 1998).

A cDNA segment encoding a C_L segment similar to C_λ was recently cloned in the axolotl (S. André, J. S. Fellah and J. Charlemagne, in preparation).

4. The Major Histocompatibility Complex (MHC)

The major histocompatibility (MHC) class I and class II proteins initiate immune responses by presenting self and foreign peptides to T cells. The MHC genomic organization has been extensively studied in mammals and also in lower vertebrates.

MHC genes have been characterized in *X. laevis* and in the axolotl. Most genes were identified in *Xenopus* by the immunoscreening of a cDNA expression library using alloantibodies (Flajnik *et al.*, 1986), or by PCR in the axolotl, using degenerated oligonucleotides, which allowed the cloning of genes with low similarity to mammalian homologues (Sammut *et al.*, 1996).

In *Xenopus*, the MHC is a single genetic region encoding classical class I (class Ia) and class II molecules (Shum *et al.*, 1993). Heat-shock protein (HSP) 70 and complement genes are also linked to this region. A distinct chromosome harbours a large family of nonclassical class I genes (class Ib) composed of nine subfamilies (Flajnik *et al.*, 1993). Only a single complete classical class Ia locus is present in *Xenopus* MHC, in contrast to all other species of vertebrates examined, and only one polymorphic class Ia molecule is expressed per MHC haplotype although *X. laevis* is a tetraploid species (Shum *et al.*, 1993). Regardless of the number of chromosomes in other *Xenopus* species, only two MHC haplotypes (i.e. a diploid set of genes) are expressed. It thus appears that polyploid species of *Xenopus* have developed a mechanism by which to silence all but one diploid set of MHC genes.

The MHC of *X. laevis* contains class II β genes. Class II β chains are polymorphic and have only one N-linked glycan in the β_1 domain. The pattern of nucleotide substitutions in the β_1 distal domain is essentially similar to that of functional mammalian MHC class II genes, which is a good argument for *Xenopus* class II antigen presentation ability. The known CD4 binding site, located in mammals between strands 3 and 4 of the membrane proximal domain, is not well conserved in *Xenopus* class II β chain sequences (Sato *et al.*, 1993).

In the axolotl, classical class Ia genes have been recently characterized (Sammut *et al.*, 1996) and show a strong homology with *Xenopus* class Ia genes. As opposed to *Xenopus*, the axolotl displays 10–12 class Ia genes detected by Southern blot analysis. The class II genes have recently been cloned and shown to be genetically linked to class I genes (B. Sammut, personal communication).

In *Xenopus* and axolotl, MHC class I molecules are highly polymorphic, and display the classic features of mammalian class I molecules. The α chain is about 350 amino acids long and consist of three extra cellular domains α_1, α_2 and α_3, a transmembrane region, and a cytoplasmic tail. The α_1 and α_2 domains form an intramolecular dimer comprising the peptide-binding groove, which is bordered by two α-helices and a floor of β-pleated sheets. There is a strong conservation of amino acids in the peptide-binding region that have been shown in mammals to anchor peptides at their *N*- and *C*-termini. These anchoring positions are conserved throughout evolution in class Ia molecules but can be variable in class Ib proteins of *Xenopus*. Amino acids presumed to interact with β_2-microglobulin in the loop between the fourth and fifth β strands of the α domain are also well conserved, although the β_2-microglobulin gene and molecule have not yet been characterized in amphibians. All these observations suggest that MHC molecules of amphibians present peptides to T cells in the same way as described in mammals (Flajnik and Du Pasquier, 1990; Kaufman *et al.*, 1990; Shum *et al.*, 1993; Kasahara *et al.*, 1995; Sammut *et al.*, 1996). Conversely, amino acids of the class Ia α chain which interact with the CD8 molecule in mammals are not conserved in amphibians or in fish.

The tissue distribution of MHC class I and class II molecules in adult amphibian are essentially the same as in mammals, as assessed by northern blotting, immunofluorescence and immunoprecipitation experiments. Class Ia are ubiquitously expressed in all tissues except brain and gonads, class II are expressed by B lymphocytes and by T lymphocytes at the same level of expression. Class II molecules are 'differentiation antigens' in *Xenopus* since adult thymocytes in the thymic medulla express more class II than immature thymocytes (Du Pasquier and Flajnik, 1990; Flajnik *et al.*, 1990b). Most notably, MHC class I molecules are not expressed in *Xenopus* tadpoles and appear first after metamorphosis (Flajnik *et al.*, 1986). Class II molecules are expressed mainly on B lymphocytes. The shift to expression by all lymphocytes appears to be dependent on the cell changes that occur at metamorphosis.

Class I molecules are detectable at hatching (Sammut *et al.*, 1996) in the axolotl.

The *Xenopus* MHC genetic region encoding class I and class II molecules directs rapid graft rejection, MLR and T–B collaboration (Flajnik and Du Pasquier, 1990), although MHC-restriction does not appear as exclusive as in mammals after restoration of the humoral response in thymectomized animals implanted with MHC-incompatible thymus (Du Pasquier and Horton, 1982; Gearing *et al.*, 1984). Nevertheless, *Xenopus* T cells interact with antigen-presenting cells in a manner similar to mammals, and MHC class II restriction of secondary proliferative responses *in vitro* is present in this anuran amphibian (Harding *et al.*, 1993).

5. Immunobiology of T and B Cells

Immunological reactivity to allografts occurred about 12 days after fertilization in *Xenopus* (stage 49) but graft rejection is impaired in a period ranging from 15 days before to 30 days after metamorphosis, which correlates with transitory thymus involution. Adult frogs reject allografts about 20 days after a first transplantation and about 9 days following a second-set graft. Thymectomy in *Xenopus* either decreases or abolishes allograft rejection, *in vitro* MLR and PHA responsiveness, IgY antibody synthesis and antibody responses to thymus-dependent antigens. It does not abolish *in vivo* or *in vitro* responses to thymus-independent antigens and B-cell polyclonal activation (LPS). *Xenopus* T-cells efficiently responds *in vitro* to polyclonal activators (PHA, Con A) and cytotoxic lymphocytes can be generated in *in vitro* MLR experiments using cells from *in vivo* allostimulated donors (Du Pasquier *et al.*, 1989).

Thymectomy between 4 and 7 days after fertilization delays or suppresses allograft rejection and IgY responses but affects only the antibody response when performed later (up to 40 days after fertilization). This suggests that T cytotoxic and T helper cells mature at different times in the thymus and that full T-cell help occurs later in differentiation. *In vitro* assays using purified T and B-cells from carrier- or hapten-primed *Xenopus* of various MHC types allowed the conclusion that T-cell help is involved in the differentiation of thymus-dependent, antigen-primed B cells and that this collaboration is MHC-restricted. Several sets of experiments (reviewed by Du Pasquier *et al.*, 1989) demonstrated that, although it was not absolute, there was an MHC-dependent selection of the T-cell repertoire in *Xenopus* and that T-cell clones may be positively selected in the thymus.

The antibody response is slower in *Xenopus* than in mammals. IgM is first produced, then it is produced in conjunction with IgY. The maximum response to DNP is reached 2 weeks after first immunization. A second injection generates a significant stronger ($\times$ 10–100), mainly IgY, immediate secondary response, but the peak of this response occurs at the same time as the primary one. There is a poor enhancement of the antibody affinity during the secondary response and it was suggested that, although somatic mutations occurs in *Xenopus* V_H segments at almost the same rate as in mammals, these mutations may not be properly selected, perhaps because of the absence of germinal centres in the lymphoid organs (Du Pasquier *et al.*, 1989).

Adult urodeles reject allografts in a chronical manner, 15–50 days after a first transplantation, and about 11 days following a second-set graft. Larval thymectomy suppresses allograft rejection in subsequent adult urodeles, but does not abolish *in vivo* responses to particular thymus-dependent antigens, such as sheep or horse erythrocytes (Charlemagne and Houillon, 1968; Charlemagne and Tournefier, 1977; Charlemagne, 1979). In fact, larval and adult thymectomy, and also low dose (0.05–1.5 Gy) irradiation and hydrocortisone (HC) treatments significantly enhanced specific antibody synthesis against erythrocyte antigens (Charlemagne, 1981; Tournefier, 1982). These results could indicate that T-cell help is impaired in urodeles, but that some type of T-cell dependent suppression acts on antibody production. Urodeles cannot be immunized against soluble antigens and, in normal and thymectomized axolotls, IgM is the single antigen-sensitive Ig class; specific IgY are not produced, even after several rounds of immunization with the same particular antigen. It must also be noted that thymectomy has no effect on the IgY serum level in the axolotl. The kinetics of the antibody response is slow – maximum anti-erythrocytes or anti-DNP titres arise 60–100 days after primary immunization – and no typical secondary response occurs following hyperimmunization. Thus, although the *Xenopus* and axolotl IgY molecules are clearly homologous at the molecular level, their respective physiological functions are radically different: they can be considered as IgG-like (thymus-dependent; sensitive to thymectomy) in anurans, but IgA-like in urodeles, at least in the first 7–8 months of development where most IgY are found associated with the digestive epithelium and can be secreted in the gut lumen following transepithelial transport in association with secretory component-like molecules (Fellah *et al.*, 1992a).

6. Ontogeny of the Immune Response

Only 50–70% of *Xenopus* tadpoles splenic lymphocytes, including B cells are MHC class II-positive, whereas all adult splenic lymphocytes are positive. Furthermore, classical polymorphic class I MHC molecules are not expressed before metamorphosis. Thus, the *Xenopus* alloimmune capacity is not fully developed before metamorphosis and, in the absence of class I molecules, the capacity of tadpoles to differentiate classical cytotoxic T-cells is questionable. However, *Xenopus* tadpoles develop allorecognition (graft rejection, MLR) around stage 49 (about 12 days after fertilization) but the capacity to reject or to tolerate grafts depends on genetic conditions (minor non-MHC, or MHC differences), and on the balance of several complex parameters such as the size of the grafts, age of tadpoles and breeding conditions. Metamorphosis represents a crucial period when a high frequency of allogenic grafts can be tolerated. This appears to be an active phenomenon, because thymectomy before metamorphosis decreases the number of tolerance cases, and also because immunocompetent cells from metamorphosing animals can inhibit graft rejection when injected to isogenic adults. MLR and the splenic response to T-cell mitogens are depressed at metamorphosis and MLR does not totally recover until 2 months later in the young froglet. This might be related to the transient decline of the number of lymphocytes at the time of metamorphosis (90–95% in the thymus, 50% in the spleen), and could be indicative of the loss of some immunocompetent cells and transient impairment of immunity. This transient phase of immunodeficience could be a means for the metamorphosing tadpole to remain unreactive to numerous new self antigens that appear during the metamorphosis climax, and an opportunity for the young froglet to build a new repertoire of self tolerant T-cells (Du Pasquier *et al.*, 1986, 1989; Flajnik *et al.*, 1987).

Different sets of antibodies are produced by *Xenopus* tadpoles and adults. The anti-(2,4) dinitrophenol (DNP) antibodies isoelectric focusing patterns (spectrotypes) are different and less heterogeneous in tadpoles compared with adults, even when animals from the same isogenic clone were compared, suggesting that pre- and post-metamorphosed *Xenopus* express different, or partially different, antibody repertoires. The first pre-B and B cells can be detected in the liver at stage 45 (3–4 days), IgM and IgY are detected in serum at 12 and 15 days, respectively. A second wave of pre-B cell production possibly arises after metamorphosis, and this could account for the detection of two successive antibody repertoires. When tadpoles are blocked in their metamorphosis by inhibitors of the thyroid function, the larval B-cell repertoire does not switch to production of the adult repertoire, despite other changes in gene expression such as the synthesis of adult haemoglobin and the appearance of MHC class I antigens on leukocytes (Du Pasquier *et al.*, 1986, 1989; Flajnik *et al.*, 1987). Sequential rearrangement of V_H genes occurs in the developing tadpoles: members of the V_H1 family are first used from day 5 to 8, then V_H3, 6, 9 and 10 rearrange from days 9–10, and at day 13, all V_H families can be used. All the J_H and D_H elements are expressed randomly at early larval stages, but at later stages (56–58, about 40 days), J_H5, 7 and 3 and D_H1, 2 and 10 are overexpressed. There is virtually no N-diversity in tadpoles. The ability of *Xenopus* tadpoles to produce specific antibodies arises around stage 51–52 (about 10 days). Most of these antibodies are IgM; switching to IgY appears not to be efficient before metamorphosis, although Igs of the three classes are present in tadpole serum. The tadpole and adult B (and T) cell repertoires are not completely independent, since memory for antigenic stimulations, such as the injection of soluble antigens or allostimulation appears, at least to some extent, to be transferred through metamorphosis.

As previously noted, urodele development is much slower than anuran and lymphoid organs begin to develop at the time of hatching, 2–3 weeks after fertilization. *Ambystoma tigrinum* (tiger salamander) larvae begin to reject allografts at about 40 days post fertilization and appear to be fully competent for rejection at 100 days. Thymectomy efficiently suppresses allograft rejection when performed at 50–60 days and still has some effect on transplantation immunity at 80 days (Cohen *et al.*, 1969). Thymic anlagen appear in *Pleurodeles* at the time of hatching, about 15 days after fertilization and are invaded in the following days by a limited number of stem cells. Small lymphocytes are first seen at 55 days and intense proliferation begins by about 70 days. Skin grafts are tolerated in almost all *Pleurodeles* larvae thymectomized before 70 days. Stage 52 (75–80 days), that represents the mid-larval stage, is the end point for complete efficiency of thymectomy and thymectomy at 90–100 days is always ineffective (Fache and Charlemagne, 1975). In another urodele amphibian, *Triturus alpestris*, the thymus and spleen anlagen appear only 35 days after fertilization and small lymphocytes are first seen at 70 days, 7–15 days before metamorphosis. Thymectomy still prevents allograft rejection when performed up to 1 week before metamorphosis (Tournefier, 1973). Clusters of IgM and

IgY-producing B-cells are first seen in spleen sections of 70-day-old axolotls, about 30 days after differentiation of the spleen anlage. At about 100 days, the relative proportion of IgM- and IgY-producing cells in the spleen is the same as that in adult animals. IgM are first detected in the serum of 90-day-old axolotls (4% of the adult value) and their concentration linearly increases until 13 months. Interestingly, low molecular weight IgY is first detected in the serum at 7 months (5% of the adult value) and the concentration increases rapidly until 11 months. However, abundant secretory IgY molecules are present in the stomach and intestinal mucosa of young axolotls from day 50, and until 7 months. Thereafter, IgY progressively disappears from the gut and becomes detectable in the serum. Intraepithelial axolotl IgY is closely associated with secretory component-like molecules whose expression down-regulates at the same time IgY disappears from the gut (Fellah *et al.*, 1989).

7. Tumors of the Immune System

Although several tumors of lymphoid origin have been described in the last 30 years in several anuran and urodele species, none were fully characterized and no *in vitro* cell lines were derived from them. In recent years, several spontaneous thymus lymphoid tumors have been detected in the *Xenopus* colony at the Basel Institute for Immunology (Du Pasquier and Robert, 1992; Robert *et al.*, 1994; Du Pasquier *et al.*, 1995). The first one (MAR1) was from a 4-year-old *Xenopus* of the *ff* strain and appeared to be a thymus tumor which was transplantable in isogenic *ff* tadpoles and froglets, but not in adult animals. Tumor fragments grew locally and metastases occured mainly in thymus, but also in spleen, liver and muscles around the injection site. *In vitro* cell lines were derived from MAR1 (e.g. B3B7). Three other tumors were then discovered, one from the *ff* strain (ff-2), and two from the LG15 strain (LG15/0 and LG15/40). All these tumors are sIg^- and classical MHC class I molecules are not expressed on the surface of B3B7, 15/0 and 15/40, but are present (with β_2-microglobulin) on the surface of ff-2 cells. All the lines were stained by mAbs F1F6, RC47 or AM14 (general lymphocyte specificities), mAb X21.2 (thymocytes) and mAb AM22 (CD8 equivalent); lines B3B7, ff-2 and 15/40 express large quantities of the CTX antigen (mAbs 1S9-2 and X71) which is thought to be a marker of the cortical DP lymphocytes. The 15/0 and 15/40 lines express μ chain mRNA, but no Ig proteins. The B3B7 line is aneuploid (41 chromosomes instead of $2n = 36$ in *Xenopus laevis*). The two alleles of the IgH locus are rearranged in frame (V_H3-D_H10-J_H7/V_H4-D_H10-J_H3) but are not transcribed in the B3B7 line. The two alleles of the IgLρ locus appears to be deleted and the two alleles of the IgLσ locus are rearranged, one is out of frame, and the second uses a V_σ pseudogene. These *Xenopus* lymphoid cell lines resemble some rare types of mammalian leukaemias and represent very useful tools for the analysis of lymphoid cell differentiation and Ig and TcR gene expression.

8. Conclusion

Through the development of the recombinant DNA techniques, considerable progress has been made during these last 10 years, in analysis and comprehension of the amphibian immune system. Many antibody and molecular probes are now available to analyse further the cellular and molecular aspects of the amphibian immune response. However, some are still unavailable, such as the probes that would be useful for the cellular analysis of T-cells, like anti-TcR antibodies. It is now possible to analyse precisely the strategies that have been developed in these nonamniotic tetrapods to efficiently protect against the multiple potential bioagressors that they encounter from their early stages of development.

9. References

Amemiya, C. T., Haire, R. N., Litman, G. W. (1989). Nucleotide sequence of a cDNA encoding a third distinct *Xenopus* immunoglobulins heavy chain isotype. *Nucl. Acids Res.* **17**, 5368.

Blomberg, B., Bernard, C. C. A., Du Pasquier, L. (1980). *In vitro* evidence for T–B lymphocyte collaboration in the clawed toad, *Xenopus*. *Eur. J. Immunol.* **10**, 869–876.

Charlemagne, J. (1979). Thymus-independent response to erythrocytes and suppressor T cells in the Mexican axolotl *(Amphibia, Urodela, Ambystoma mexicanum)*. *Immunology* **36**, 643–647.

Charlemagne, J. (1981). Suppressor T cells and antibody synthesis in the X-irradiated Mexican axolotl. *Eur. J. Immunol.* **11**, 717–721.

Charlemagne, J., Houillon, C. (1968). Effets de la thymectomie larvaire chez l'Amphibien Urodèle *Pleurodeles waltlii* Michah. Production à l'état adulte d'une tolérance aux homogreffes cutanées. *C.R. Acad. Sci.* Sér. D, **267**, 253–256.

Charlemagne, J., Tournefier, A. (1977). Anti-horse red blood cells antibody synthesis in the Mexican axolotl (*Ambystoma mexicanum*). Effects of thymectomy. In *Developmental Immunobiology* (eds Solomon, J. B., Horton, J. P.), pp. 267–273. Elsevier/North-Holland Biomedical Press, Amsterdam, The Netherlands.

Chrétien, I., Robert, J., Marcuz, A., Garcia-Sanz, J. A., Courtet, M., Du Pasquier, L. (1996). CTX, a novel molecule specifically expressed on the surface of cortical thymocytes in *Xenopus*. *Eur. J. Immunol.* **26**, 780–791.

Chrétien, I., Marcuz, A., Fellah, J. S., Charlemagne, J., Du Pasquier, L. (1997). The T-cell receptor beta genes of *Xenopus*. *Eur. J. Immunol.* **27**, 763–771.

Cohen, N. (1969). Immunogenetic and developmental aspects of tissue transplantation immunity in Urodele Amphibians. In *Biology of Amphibian Tumors* (ed. Mizell, M.), pp. 153–168. Springer-Verlag, New York, USA.

Collie, M. H. (1974). The location of soluble antigen in the spleen of *Xenopus laevis*. *Experientia* **30**, 1205–1207.

Du Pasquier, L., Flajnik, M. F. (1990). Expression of MHC class II antigens during *Xenopus* development. *Devel. Immunol.* **1**, 75–95.

Du Pasquier, L., Horton, J. D. (1976). The effect of thymectomy on the mixed leukocyte reaction and phytohemagglutinin responsiveness in the clawed toad, *Xenopus laevis*. *Immunogenetics* **3**, 105–112.

Du Pasquier, L., Robert, J. (1992). *In vitro* growth of thymic tumor cell lines from *Xenopus*. *Devel. Immunol.* **2**, 295–307.

Du Pasquier, L., Flajnik, M. F., Hsu, E., Kaufman, J. F. (1986). Ontogeny of the immune system in anuran amphibians. In *Progress in Immunology VI* (eds Cinader, B., Miller, R. G.), pp. 1079–1088.

Du Pasquier, L., Schwager, J., Flajnik, M. F. (1989). The immune system of *Xenopus*. *Annu. Rev. Immunol.* **7**, 251–275.

Du Pasquier, L., Courtet, M., Robert, J. (1995). A *Xenopus* lymphoid tumor cell line with complete Ig genes rearrangements and T-cell characteristics. *Mol. Immunol.* **32**, 523–593.

Fache, B., Charlemagne, J. (1975). Influence on allograft rejection of thymectomy at different stages of larval development in Urodele Amphibian *Pleurodeles waltlii* Michah. *(Salamandridae)*. *Eur. J. Immunol.* **5**, 155–157.

Fellah, J. S., Charlemagne, J. (1988). Characterization of an IgY-like low molecular weight immunoglobulin class in the Mexican axolotl. *Mol. Immunol.* **25**, 1377–1386.

Fellah, J. S., Vaulot, D., Tournefier, A., Charlemagne, J. (1989). Ontogeny of immunoglobulin expression in the Mexican axolotl. *Development* **107**, 253–263.

Fellah, J. S., Iscaki, S., Vaerman, J. P., Charlemagne, J. (1992a). Transient developmental expression of IgY and secretory component-like protein in the gut of the axolotl *(Ambystoma mexicanum)*. *Devel. Immunol.* **2**, 181–190.

Fellah, J. S., Wiles, M. V., Charlemagne, J., Schwager, J. (1992b). Evolution of vertebrate IgM: complete amino-acid sequence of the constant region of *Ambystoma mexicanum* μ-chain deduced from cDNA sequence. *Eur. J. Immunol.* **22**, 2595–2601.

Fellah, J. S., Kerfourn, F., Guillet, F., Charlemagne, J. (1993a). Conserved structure of amphibian T-cell antigen receptor β chain. *Proc. Natl Acad. Sci. USA* **90**, 6811–6814.

Fellah, J. S., Wiles, M. V., Charlemagne, J., Schwager, J. (1993b). Phylogeny of immunoglobulin heavy chain isotypes: structure of the constant region of *Ambystoma mexicanum* υ chain deduced from cDNA sequence. *Immunogenetics* **38**, 311–317.

Fellah, J. S., Kerfourn, F., Charlemagne, J. (1994). Evolution of T cell receptor genes. Extensive diversity of Vβ families in the Mexican axolotl. *J. Immunol.* **153**, 4539–4542.

Fellah, J. S., Kerfourn, F., Dumay, A-M., Aubet, G., Charlemagne, J. (1996). Structure and diversity of the T-cell receptor α-chain in the Mexican axolotl. *Immunogenetics* **45**, 235–241.

Flajnik, M. F., Du Pasquier, L. (1990). The major histocompatibility complex of frogs. *Immunol. Res.* **113**, 47–63.

Flajnik, M. F., Kaufman, J. F., Riegert, P., Du Pasquier, L. (1984). Identification of class I major histocompatibility complex encoded molecules in the amphibian *Xenopus*. *Immunogenetics* **20**, 433–442.

Flajnik, M. F., Kaufman, J. F., Hsu, E., Manes, M., Parisot, R., Du Pasquier, L. (1986). Major histocompatibility complex-encoded class I molecules are absent in immunologically competent *Xenopus* before metamorphosis. *J. Immunol.* **197**, 3891–3899.

Flajnik, M. F., Hsu, E., Kaufman, J. F., Du Pasquier, L. (1987). Changes in the immune system during metamorphosis of *Xenopus*. *Immunol. Today* **8**, 58–64.

Flajnik, M. F., Camel, C., Kramer, J., Kasahara, M. (1990a). Evolution of the major histocompatibility complex: molecular cloning of MHC class I from the amphibian *Xenopus*. *Proc. Natl Acad. Sci. USA* **78**, 537–541.

Flajnik, M. F., Ferrone, S., Cohen, N., Du Pasquier, L. (1990b). Evolution of the MHC: antigenicity and unusual tissue distribution of *Xenopus* (frog) class II molecules. *Mol. Immunol.* **27**, 451–462.

Flajnik, M. F., Kasahara, M., Shum, B. P., Salter-Lid, L., Taylor, E., Du Pasquier, L. (1993). A novel type of class I gene organization in vertebrates: a large family of non-MHC-linked class I genes is expressed at the RNA level in the amphibian *Xenopus*. *EMBO J.* **12**, 4385–4396.

Gearing, A. J. H., Cribbin, F. A., Horton, J. D. (1984). Restoration of the antibody response to sheep erythrocyte in thymectomized *Xenopus* implanted with MHC-compatible or MHC-incompatible thymus. *J. Embryol. Exp. Morph.* **84**, 287–302.

Golub, R., Charlemagne, J. (1998). Structure and repertoire of VH families in the Mexican axolotl. *J. Immunol.* **160**.

Haire, R. N., Shamblott, M. J., Amemiya, C. T., Litman, G. W. (1989). A second *Xenopus* immunoglobulin constant region isotype gene. *Nucl. Acids Res.* **17**, 1776.

Haire, R. N., Amemiya, C. T., Suzuki, D., Litman, G. W. (1990). Eleven distinct VH gene family and additional patterns of sequence variation suggest a high degree of immunoglobulin gene complexity in a lower vertebrate, *Xenopus laevis*. *J. Exp. Med.* **171**, 1721–1737.

Haire, R. N., Ohta, Y., Litman, R. T., Amemiya, C. T., Litman, G. W. (1991). The genomic organization of immunoglobulin VH genes in *Xenopus laevis* shows evidence for interspersion of families. *Nucl. Acids Res.* **19**, 3061–3066.

Haire, R. N., Ohta, Y., Rast, J. P., Litman, R. T., Chan, F. Y., Zon, L. I., Litman, G. W. (1996). A third Ig light chain gene isotype in *Xenopus laevis* consists of six distinct VL families and is related to mammalian λ genes. *J. Immunol.* **157**, 1544–1550.

Harding, F. A., Flajnik, M. F., Cohen, N. (1993). MHC restriction of T-cell proliferative responses in *Xenopus*. *Devel. Comp. Immunol.* **17**, 425–437.

Haynes, L., Cohen, N. (1993a). Transforming Growth Factor Beta (TGFβ) is produced by and influence the proliferative response of *Xenopus laevis* lymphocytes. *Devel. Immunol.* **3**, 223–230.

Haynes, L., Cohen, N. (1993b). Futher characterization of an Interleukin-2-cytokine produced by *Xenopus laevis* T lymphocytes. *Devel. Immunol.* **3**, 331–338.

Horton, J. D., Horton, T. L., Ritchie, P., Varley, C. A. (1992). Skin xenograft rejection in *Xenopus* immunohistology and effect of thymectomy. *Transplantation* **53**, 473–476.

Hsu, E., Du Pasquier, L. (1984). Studies on *Xenopus* immunoglobulins using monoclonal antibodies. *Mol. Immunol.* **21**, 257–270.

Hsu, E., Flajnik, M. F., Du Pasquier, L. (1985). A third immunoglobulin class in amphibians. *J. Immunol.* **135**, 1998–2004.

Kasahara, M., Flajnik, M. F., Ishibashi, T., Natori, T. (1995). Evolution of the major histocompatibility complex: a current overview. *Transplant. Immunol.* **3**, 1–20.

Kaufman, J. F., Skoedt, K., Salomonsen, J. (1990). The MHC molecules of non-mammalian vertebrates. *Immunol. Rev.* **113**, 83–117.

Kerfourn, F., Charlemagne, J., Fellah, J. S. (1996). The structure, rearrangement, and ontogenic expression of *DB* and *JB* gene segments of the Mexican axolotl T-cell antigen receptor beta chain *(T C R B)*. *Immunogenetics* **44**, 275–285.

Koninsky, A., Cohen, N. (1992). Reproductible proliferative response of salamander (*Ambystoma mexicanum*) lymphocytes cultures with mitogens in serum free medium. *Devel. Comp. Immunol.* **16**, 441–451.

Koninsky, A., Cohen, N. (1994). Mitogen activated axolotl (*Ambystoma mexicanum*) splenocytes produce a cytokine that promotes growth of homologous lymphoblasts. *Devel. Comp. Immunol.* **18**, 239–250.

Mußmann, R., Wilson, M., Marcuz, A., Courtet, M., Du Pasquier, L. (1996). Membrane exon sequences of the three *Xenopus* Ig classes explain the evolutionary origin of mammalian isotypes. *Eur. J. Immunol.* **26**, 409–414.

Nagata, S. (1985). A cell surface marker of thymus-dependent lymphocytes in *Xenopus laevis* is identifiable by mouse monoclonal antibody. *Eur. J. Immunol.* **15**, 837–841.

Robert, J., Guiet, C., Du Pasquier, L. (1994). Lymphoid tumors of *Xenopus laevis* with different capacities for growth in larvae and adults. *Devel. Immunol.* **3**, 297–307.

Salvadori, F., Tournefier, A. (1996). Activation by mitogens and superantigens of axolotl lymphocytes; functional characteristics and ontogenic study. *Immunology* **88**, 586–592.

Sammut, B., Laurens, V., Tournefier, A. (1997). Isolation of classical class I cDNAs from the axolotl, *Ambystoma mexicanum*. *Immunogenetics* **45**, 285–294.

Sato, K., Flajnik, M. F., Du Pasquier, L., Katagiri, M., Kasahara, M. (1993). Evolution of the MHC: isolation of class II β-chain cDNA clones from the amphibian *Xenopus laevis*. *J. Immunol.* **150**, 2831–2843.

Schwager, J., Mikoryack, C. A., Steiner, L. A. (1988). Amino acid sequence of heavy chain from *Xenopus laevis* IgM deduced from cDNA sequence: implication for evolution of immunoglobulin domains. *Proc. Natl Acad. Sci. USA* **85**, 2245–2249.

Schwager, J., Bürckert, N., Courtet, M., Du Pasquier, L. (1989). Genetic basis of the antibody repertoire in *Xenopus*: analysis of the VH diversity. *EMBO J.* **8**, 2989–3001.

Schwager, J., Bürckert, N., Schwager, M., Wilson, M. (1991). Evolution of immunoglobulin light chain genes: analysis of *Xenopus* IgL isotypes and their contribution to antibody diversity. *EMBO J.* **10**, 505–511.

Shum, B. P., Avila, D., Du Pasquier, L., Kasahara, M., Flajnik, M. F. (1993). Isolation of a classical MHC class I cDNA from an Amphibian. Evidence for only one class I locus in the *Xenopus* MHC. *J. Immunol.* **151**, 5376–5386.

Stewart, S. E., Du Pasquier, L., Steiner, L. A. (1993). Diversity of expressed V and J regions of immunoglobulin light chains in *Xenopus laevis*. *Eur. J. Immunol.* **23**, 1980–1986.

Sugiyama, K., Amenomori, A., Hatakeyama, K. (1990). Leukocyte surface markers in *Rana catesbiana*, identified using mouse monoclonal antibodies. *Cell Diff. Devel.* **29**, 105–112.

Tournefier, A. (1973). Développement des organes lymphoides chez l'Amphibien Urodèle *Triturus alpestris* Laur.; tolérance des allogreffes après la thymectomie larvaire. *J. Embryol. Exp. Morph.* **29**, 383–396.

Tournefier, A. (1982). Corticosteroid action on lymphocyte subpopulations and humoral immune response of axolotl (Urodele Amphibian). *Immunology* **46**, 155–162.

Tournefier, A., Fellah, J. S., Charlemagne, J. (1988a). Monoclonal antibodies to axolotl immunoglobulins specific for different heavy chains isotypes expressed by independent lymphocyte subpopulations. *Immunol. Letters* **18**, 145–148.

Tournefier, A., Guillet, F., Ardavin, C., Charlemagne, J. (1988b). Surface markers of axolotl lymphocytes as defined by monoclonal antibodies. *Immunology* **63**, 269–276.

Tournefier, A., Lesourd, M., Gounon, P. (1990). The axolotl thymus. Cell types of the microenvironment, a scanning and electron microscopic study. *Cell Tissue Res.* **262**, 387–389.

Turner, S. L., Horton, T. L., Ritchie, P., Horton, J. D. (1991). Splenocyte response to T-cell derived cytokines in thymectomized *Xenopus*. *Devel. Comp. Immunol.* **15**, 319–328.

Williams, N. H., Horton, J. D. (1980). Ontogeny of mitogen responsiveness in thymus and spleen of *Xenopus laevis*. In *Developmental Aspects of Developmental and Comparative Immunology* (ed. Solomon, J. B.), pp. 493–494. Pergamon Press, Oxford and New York.

Zezza, D. J., Mikoryak, C. A., Schwager, J., Steiner, L. A. (1991). Sequence of C region of L chains from *Xenopus laevis* Ig. *J. Immunol.* **146**, 441–4047.

Zezza, D. J., Stewart, S. E., Steiner, L. A. (1992). Genes encoding *Xenopus laevis* Ig L chains. Implications for the evolution of κ and λ chains. *J. Immunol.* **149**, 3968–3977.

IV AVIAN IMMUNOLOGY

1. Introduction

Studies on the ontogeny of the immune system of the chicken have contributed significantly to the advances in the knowledge of the immune system of mammals, including man. The bursa of Fabricius and the thymus, two major primary lymphoid organs that regulate the development of humoral and cellular compartments of the immune system, respectively, are located at anatomically diverse locations in the chicken. The thymus is a multilobular structure that spans the cervical area whereas the bursa is an extension of the hind gut and is located near cloaca. It is possible to selectively remove the bursa or the thymus and induce selective immune deficiency. Studies on experimentally induced selective immune deficiencies in the chicken pioneered the concept of the dichotomy of the immune system into B- and T-cell compartments and ushered an era of intense research activity in this area. Immunology is now one of the fastest growing disciplines in biology. As a result, the immunologic concepts are constantly being redefined.

Although the knowledge of the mechanisms of the immune system has accelerated at a much faster pace in mammals than in the avian species, avian immunology has made great strides forward in recent years. With the increasing availability of reagents that specifically recognize avian immune cells or cell products, new information is being generated at a rapid rate. Data indicate that the overall organization and functions of the avian immune system are similar to those of the mammalian system. There are also some interesting and important differences. We have made efforts to highlight these differences in various chapters of this section.

Among the avian species, the immune system of the chicken has been studied most extensively. Thus, much of the information presented is based on studies in chickens. It is likely that the broad principles of immunology identified in chickens will apply to other members of the avian species.

All contributing authors are recognized avian immunologists currently active in their field of research. Gathering information from diverse sources and compiling it into short summaries to conform to the format of this volume is a time-consuming task and I am greatly indebted to each of the authors for making a contribution to this section.

2. Lymphoid Organs and Their Anatomical Distribution

Introduction

Descriptions and illustrations of the gross anatomy and histology of the lymphoid system of the chicken are available (Payne, 1971; Hodges, 1974; Rose, 1981; Riddell, 1987). The microanatomy of lymphoid tissue in the digestive system has been described by Calhoun (1954).

General description

Thymus and bursa of Fabricius (BF) are the central (primary) lymphoid organs (Table IV.2.1). Peripheral (secondary) lymphoid organs (Table IV.2.1) include spleen, Harderian (paraocular, nictitating membrane) glands, bone marrow, conjunctival-associated lymphoid tissue (CALT), bronchial-associated lymphoid tissue (BALT), and gut-associated lymphoid tissue (GALT), of which the cecal tonsils are most prominent. Chickens do not have lymph nodes but do have lymphoid nodules associated with lymphatics.

Diffuse lymphoid tissue is scattered in the parenchyma of many organs including liver (see Figure IV.2.3c), pancreas, and kidney. It is irregularly shaped, unencapsulated, and characterized by sheets of medium to small lymphocytes, and may contain one or more germinal centers. Germinal centers (nodular lymphoid tissue, lymphoid nodules) are circumscribed, round to oval, and composed of a mixed population of large (immature), medium, and small (mature) lymphocytes. Immature cells are pale because they stain lightly with hematoxylin and eosin in contrast to more intensely stained mature lymphocytes. Germinal centers (see Figure IV.2.2c) are separated from surrounding tissue by a fine connective tissue capsule. Cells of germinal centers and plasma cells are derived from the BF and are referred to as bursa-dependent tissue; other scattered collections of lymphocytes are thymus dependent (Payne, 1971). Lymphoid tissue is a 'three-dimensional meshwork' consisting of reticular fibers and fixed cells

Handbook of Vertebrate Immunology
ISBN 0-12-546401-0

Table IV.2.1 Lymphoid organs of the chicken

Organ	*Characteristics*		*Selected references*
Thymus	Location	Seven lobes on each side of the neck in close association with jugular veins and vagus nerves; extends from anterior cervical regions into the thorax and portions may be embedded in the thyroid glands; in close association with parathyroid and ultimobranchial glands; innervated during embryogenesis through a topographical distribution of fibers from the central nervous system.	Payne (1971), Hodges (1974), Rose (1981), White (1981), Bulloch (1988)
	Ontogeny	Arises from 3rd and 4th branchial pouches; visible by embryo day 5; stem cells of yolk-sac origin begin populating by embryo day 6.5; first $CD3^+$ cells appear on embryo day 9; $TCR1^+$ and $TCR2^+$ cells appear on embryo day 12; cortical and medullary zones present by embryo day 13; $TCR1^+$ cells in cortex and medulla on embryo day 13; number of cells with T-cell surface markers increases to include nearly 100% of cells by hatching; grows to 12–16 weeks after hatch then regresses with onset of sexual maturity.	Toivanen *et al.* (1981), Bucy *et al.* (1990), Chen *et al.* (1990)
	Function	Primary (central) lymphoid organ for development of helper and suppressor cells that modulate antibody production; delayed hypersensitivity reactions; graft versus host reactions; macrophage activation; and cytotoxic responses.	Warner *et al.* (1962), Cooper *et al.* (1966, 1969), Chi *et al.* (1981)
	Comments	Also functions as a secondary lymphoid organ; after hatching contains 5–15% B cells; has endocrine functions.	Ivanyi (1981), Eerola *et al.* (1987), Brewer *et al.* (1989), Marsh and Scanes (1994)
Bursa of Fabricius	Location	Dorsal to the cloaca in the caudal body cavity.	Payne (1971), Rose (1981)
	Ontogeny	First primordium detectable on embryo day 4; stem cells of yolk-sac origin begin populating on embryo day 7.5; cells bearing IgM appear on embryo day 10 and are present in significant numbers from embryo day 12; IgG-positive cells by embryo day 14; IgA-positive cells by embryo day 16; attains maximum weight by 3–10 weeks post-hatching; regresses with onset of sexual maturity and is fibrotic by 23.5 weeks post-hatching; some redevelopment occurs with molting.	Moore and Owen (1966), Glick (1978, 1995), Toivanen *et al.* (1981), Romppanen (1982), Boyd and Ward (1984), Naukkarinen and Sorvari (1984), Pink and Lassila (1987), Masteller and Thompson (1994), Masteller *et al.* (1995)
	Function	Primary (central) lymphoid organ for development (diversification and amplification) of antibody producing cells; by 2–3 weeks after hatching, about 5% of bursal cells leave daily, going first into the blood and then into B-cell areas of the spleen, thymus, and cecal tonsils; about 1% of the blood B-cell pool is replaced each hour by cell migration from the BF.	Glick *et al.* (1956), Warner *et al.* (1962), Cooper *et al.* (1966, 1969), Weill and Reynaud (1987), Weill *et al.* (1987), Lassila (1989), Lassila *et al.* (1989), Salant *et al.* (1989), Mansikka *et al.* (1990), Reynaud *et al.* (1992), Paramithiotis and Ratcliffe (1993, 1994)
	Comments	Also functions as a secondary lymphoid organ as part of the GALT; uptake of colloidal carbon demonstrated; stalk contains a diffuse lymphoid cell infiltration of T- and B-cells; has endocrine functions. Fibronectin is a component of the extracellular matrix.	Odend'hal and Breazile (1980), Naukkarinen (1982), Naukkarinen and Sorvari (1982), Naukkarinen and Syrjanen (1984), Eerola *et al.* (1987), Hippelainen *et al.* (1987), Naukkarinen and Hippelainen (1989), Palojoki *et al.* (1993)

(continued)

Table IV.2.1 *Continued*

Organ	*Characteristics*		*Selected references*
Spleen	Location	Lies dorsal and to the right of the proventriculus; the largest secondary lymphoid organ; accessory spleens may be present.	White (1981)
	Ontogeny	First detectable on embryo day 5; red pulp develops by embryo day 8; white pulp emerges after embryo day 12; cells with lymphoid cell markers appear by embryo days 10 or 11; TCR1$^+$ cells found on embryo day 15; TCR2$^+$ cells appear on embryo day 19; ellipsoid development completed by embryo day 18; germinal centers by day 10 post-hatching – earlier if chicks are antigenically stimulated; as chicks age, up to 65% of cells are T-cells; approximately a fivefold weight increase during the first six weeks after hatching.	Chi *et al.* (1981), Toivanen *et al.* (1981), Olah *et al.* (1985), Bucy *et al.* (1990), Kasai *et al.* (1995)
	Function	A major site for hemopoietic activity in the developing embryo; important for antigen processing and antibody production after hatching.	Payne (1971), Hodges (1974), Rose (1981), White (1981), Del-Cacho *et al.* (1995)
	Comments	Phagocytosis of effete erythrocytes after hatching.	Hodges (1974)
Cecal tonsils	Location	Inner facing wall of ceca at ileo-cecal junctions.	Payne (1971), Rose (1981)
	Ontogeny	Not present at hatching; develop shortly afterwards; readily identified by 10 days old; size increases up to about 12 weeks old, depending on degree of antigenic stimulation.	Befus *et al.* (1980), Toivanen *et al.* (1981)
	Function	Largest collection of gut-associated lymphoid tissue (GALT); contains both T (35%) and B cells (45–55%); involved in antibody production and cell-mediated immune functions.	Ivanji (1981), Chi *et al.* (1981), Gallego *et al.* (1995)
	Comment	Peyer's patches and lymphoid tissue at esophageal–proventricular junction, distal cecal lymphoid nodules, Meckel's diverticulum, and bursa of Fabricius, in addition to other collections of lymphoid tissue scattered from the pharynx to the cloaca have similar structure and function; may be alternative site for B-cell differentiation.	Payne (1971), Hoshi and Mori (1973), Befus *et al.* (1980), Rose (1981), Toivanen *et al.* (1981), Del-Cacho *et al.* (1993)
Harderian gland	Location	Ventral and posteriomedial to the eyeball within the orbit; secretory duct opens on the surface of the nictitating membrane; rich supply of autonomic innervation.	Bang and Bang (1968), Payne (1994)
	Ontogeny	Progressive increase in number of plasma cells by 2 weeks old; very few IgA-positive cells at 1.5 weeks post-hatching, but 46% IgA-positive by 3.5 weeks old.	Bang and Bang (1968), Wight *et al.* (1971), Tsuji *et al.* (1993)
	Function	Contains numerous plasma cells which produce and secrete primarily IgA and other immunoglobulins; major secondary (peripheral) lymphoid organ of the head-associated lymphoid tissue (HALT); B cells comprise 80–90% of lymphoid cell population.	Survashe and Aitkin (1977), Ivanji (1981), Baba *et al.* (1988, 1990), Gallego *et al.* (1992a, b), Maslak and Reynolds (1995)

(continued)

Table IV.2.1 *Continued*

Organ	*Characteristics*		*Selected references*
Harderian gland	Comments	Other lymphoid tissue in this region with similar functions include Harderian gland ducts, paranasal glands, lateral nasal ducts, and conjunctival-associated lymphoid tissue (CALT); Harderian gland may be relatively bursa-independent; intraepithelial lymphocytes speculated to be involved in antigen transport.	Kittner and Olah (1980), Fix and Arp (1991), Del-Cacho *et al.* (1992), Scott *et al.* (1993)
Bronchial-associated lymphoid tissue (BALT)	Location	Diffuse and nodular lymphoid tissue in the lamina propria of the respiratory epithelium from the nares to bronchi and at the openings of the three most caudal secondary bronchi.	Riddell (1987), Fagerland and Arp (1993b)
	Ontogeny	Lymphoid deposits are present from 1 day old and increase in size and number to 8 weeks old; germinal centers present by 2 weeks old.	Fagerland and Arp (1993a)
	Function	B- and T-cell functions.	Payne (1971), Rose (1981)
	Comments	Lymphoepithelium associated with the lymphoid follicles is described; microfold (M) cells are not present.	Bienenstock *et al.* (1973), Fagerland and Arp (1993a, b)
Other lymphoid tissue	Location	About 80% of lymphocytes in peripheral blood are T cells; bone marrow contains B cells (about 15–25% of mononuclear cells) and T cells (about 80%); mural nodules are located along lymphatic vessels; most prominent in the pelvic region of the chicken; lymphoid tissue is located in mucosa of the gall bladder and in visceral organs including liver, pancreas, kidney, oviduct, and pineal gland; chickens do not have lymph nodes.	Biggs (1957), Payne (1971), Hodges (1974), Chi *et al.* (1981), Ivanji (1981), Rose (1981), Olah and Glick (1984), Sugimura *et al.* (1987), Kimijima *et al.* (1990), Paramithiotis and Ratcliffe (1993), Glick (1995)
	Ontogeny	Lymphoid cells appear at hatch and increase in number and size with age and antigenic stimulation. No lymphoid tissue found in mucosa of gall bladder at day of age, some found at 8 days old, large numbers of lymphoid cells found at 49 days old.	Leslie *et al.* (1976), Toivanen *et al.* (1981)
	Functions	B- and T-cell functions; immunoglobulins can be demonstrated in bile; IgA-positive cells located in mucosa of gall bladder.	Katz *et al.* (1974), Leslie *et al.* (1976), Glick and Rosse (1981), Jeurissen *et al.* (1988)
	Comments	Views that lymphoid tissue in visceral tissues was abnormal have been replaced by concept that these collections are part of the lymphoid system and increase with age and antigenic stimulation; lymphoid tissue in the wall of the oviduct (middle infundibulum and in regions from isthmus to vagina) has little relation to the secretion of antibodies into the yolk by epithelial glandular cells of the magnum.	Bang and Bang (1968), Payne (1971), Hodges (1974), Rose (1981), Kimijima *et al.* (1990)

(macrophages and reticular cells) within which are lymphocytes, plasma cells, and granulocytes (Rose, 1981). The supporting reticular network is not readily apparent without special stains.

Monoclonal antibodies can be used to identify specific cell types within the lymphoid system that express specific surface markers (Table IV.2.2).

Central (primary) organs of the chicken lymphoid system

Thymus

Thymic lobes (Figure IV.2.1a) are divided into lobules (Figure IV.2.1b) by fine, connective tissue. Lobules have a

Table IV.2.2 Summary of selected reports on use of monoclonal antibodies to define the anatomy of the chicken lymphoid system

Components identified	*References*
Secretory bursal epithelial cells	Blauer and Tuohimaa (1995)
T-cell subset distribution in spleen, bursa of Fabricius, and intestine	Bucy *et al.* (1990)
Distribution of thymocytes in blood and spleen	Chan *et al.* (1988)
Distribution of thymocytes, based on presence of TCR1, TCR2, and TCR3 markers, in blood, thymus, spleen, and intestinal epithelium	Chen *et al.* (1994)
Presence of cells bearing a common leukocyte antigen in spleen, thymus, bursa of Fabricius, cecal tonsil, and peripheral blood	Chung *et al.* (1991)
T-cell distribution in diffusely infiltrated area of bursa of Fabricius	Cortes *et al.* (1995)
Colonization of spleen and intestinal epithelium by γ/δ T-cells	Dunon *et al.* (1993)
B- and T-cell subsets in GALT	Fagerland and Arp (1993b)
Location of dendritic cells in medulla of bursal follicle, T-dependent areas of spleen, Peyer's patches, cecal tonsil, and Harderian gland	Gallego *et al.* (1992c)
T-cell subset in spleen, blood, bone marrow, and bursa of Fabricius	Houssaint *et al.* (1985)
B-cell regions of spleen, thymus, and germinal centers	Huffnagle *et al.* (1989)
Cell components of spleen	Jeurissen (1993)
B- and T-cells in bone marrow, thymus, bursa of Fabricius, and spleen	Jeurissen *et al.* (1988)
Nonlymphoid cells in spleen	Jeurissen *et al.* (1992)
Structure of ellipsoids of spleen	Kasai *et al.* (1995)
Immunoglobulin-containing cells in oviduct	Kimijima *et al.* (1990)
T-cells in thymus, blood and spleen	Knabel and Loesch (1993)
T-lymphocytes, splenic lymphocytes, blood leukocytes	Kondo *et al.* (1990)
Distribution of T-cell subsets in thymus, cecal tonsil, spleen, bone marrow, and bursa of Fabricius	Kon-Ogura *et al.* (1993)
T-cell subsets in jejunum	Lillehoj and Chung (1992)
T-cell subsets in thymus, spleen, and blood	Lillehoj *et al.* (1988a)
Distribution of cells expressing class II antigen in bursa of Fabricius, thymus, spleen	Lillehoj *et al.* (1988b)
Appearance of follicle-associated epithelial cells and reticuloepithelial cells in bursa of Fabricius	Lupetti *et al.* (1990)
Bursal and peripheral B-cells	Mansikka *et al.* (1989)
IgA positive cells at corticomedullary border of thymus	Olah and Glick (1992)
Location of vimentin positive cells in bursa of Fabricius and cecal tonsil	Olah and Glick (1995)
B-lymphocyte subsets in yolk sac and embryonic bursa of Fabricius	Olson and Ewert (1990)
Multilineage cell distribution patterns in yolk sac, thymus, bursa of Fabricius, bone marrow, spleen, cecal tonsil, and Harderian gland	Olson and Ewert (1994)
T- and B-cell subsets in small intestine	Rothwell *et al.* (1995)
Sites of fibronectin in bursa of Fabricius	Palojoki *et al.* (1993)
Identification of $CD8^+$ cells in intestinal epithelium	Tregaskes *et al.* (1995)

distinct outer cortex composed of densely packed small lymphocytes and an inner medulla containing less densely packed lymphocytes, reticular cells, and islands of epithelial cells (Figure IV.2.1c) called Hassall's corpuscles (Payne, 1971; Hodges, 1974; Rose, 1981), variable numbers of granulocytes are commonly found in these corpuscles.

Bursa of Fabricius

The BF mucosa has 11–13 longitudinal folds (Figure IV.2.1f) covered by specialized follicular epithelium, which forms the raised follicular pad, and columnar or pseudostratified interfollicular epithelium. The underlying connective tissue contains 8000–12 000 lymphoid (bursal)

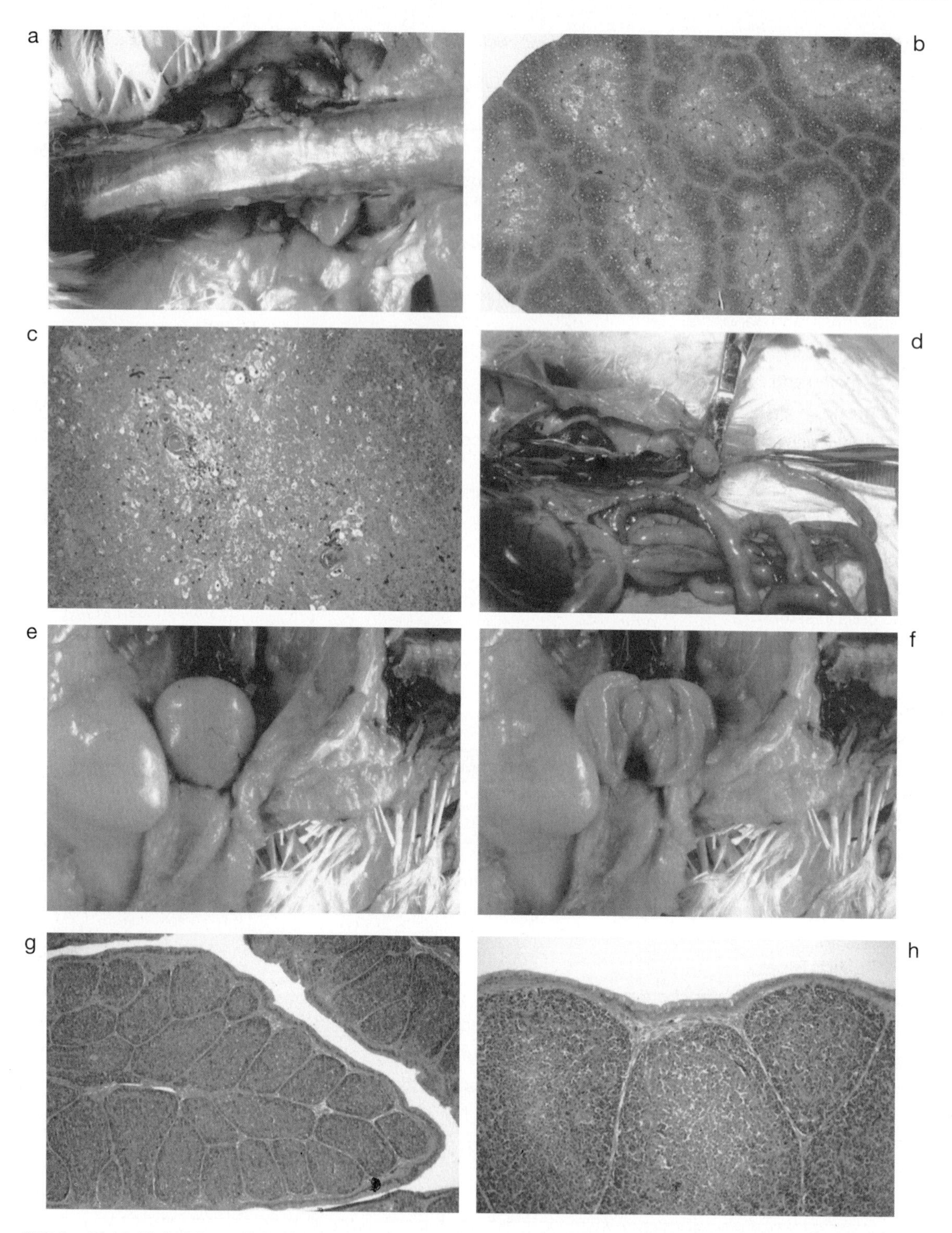

Figure IV.2.1 (a) Multiple lobes of the thymus lie on each side of the trachea. (b) The multilobular histologic structure of the thymus is evident; each lobule is composed of the darker-staining cortex and the pale medulla. (c) Several Hassall's corpuscles are in this section of the thymus. (d) The bursa of Fabricius is at the tip of the forceps. (e) The outer (serosal) surface of the bursa of Fabricius and the relationship of the bursa to the cloaca. (f) The bursa is opened to show the inner folds or plicae. (g) Bursal lymphoid follicles are covered by the epithelium of the plicae. (h) Bursal follicles are separated by thin connective tissue septae; note the cortex and medulla as well as the relationship of the overlying plical epithelium.

follicles (Figure IV.2.1g) separated from each other by delicate connective tissue (Olah and Glick, 1978). Each bursal follicle has an outer cortex containing densely packed lymphocytes and an inner medulla (Figure IV.2.1h), which contains loosely packed lymphocytes and reticular cells. The cortex is separated from the medulla by a single layer of cuboidal epithelial cells resting on a basement lamina, which is continuous with the basal cell layer of the interfollicular epithelium. Small blood vessels are present in the cortex but not medulla. A diffuse collection of lymphocytes just dorsal to the opening of the bursal duct contains numerous thymus-dependent (T) cells, indicating the BF also functions as a secondary lymphoid organ (Odend'hal and Breazile, 1980). Active uptake of particulate matter (Naukkarinen, 1982) and bursal duct ligation experiments (Dolfi *et al.*, 1989) provide further evidence of its secondary role as part of the gut-associated lymphoid tissue (GALT).

Peripheral (secondary) lymphoid organs of the chicken

Spleen

White and red pulp comprise about 80% of splenic tissue (Hodges, 1974). They are not sharply distinct from each other in the chicken spleen. White pulp consists of periarterial sheaths (periarterial lymphoid sheaths, PALS) surrounding medium and small branches of central splenic arteries (Figure IV.2.2b) that contain small, T-dependent lymphocytes. Germinal centers (B-dependent tissue) are often located adjacent to central arteries within these T-dependent sheaths (Figure IV.2.2c). Penicillar arterioles at the periphery of the white pulp give rise to capillaries, which become sheathed with reticular cells forming ellipsoids (Payne, 1971). These vessels have high endothelial cells, thick basement laminae, and intimate association with reticular cells. Ellipsoidal cells, periellipsoid B-cell sheaths, and surrounding macrophages form a complex considered to be the functional equivalent of the marginal zone in the mammalian spleen (Jeurissen *et al.*, 1992).

Red pulp is a loose spongy tissue with chords of reticular cells located between venous sinuses that contains lymphocytes, macrophages, granulocytes, and plasma cells. The relationship of T- and B-dependent areas to blood vessels in the chicken spleen (Cheville and Beard, 1972), and blood flow from the central artery through the periarterial lymphoid sheath, the periarteriolar reticular sheath, and red pulp into the venous sinus of the turkey, which is identical to that in the chicken, have been described elsewhere (Cheville and Sato, 1977).

Gut-associated lymphoid tissue (GALT)

Cecal tonsils (Figure IV.2.2d) contain dense masses of small lymphocytes and large numbers of immature and mature plasma cells. Lymphoid tissue with a similar histological structure to cecal tonsils is also found in the distal region of each cecum about 3 cm from the ileo-cecal junction (Del-Cacho *et al.*, 1993). Peyer's patches, located in the small intestinal mucosa, are structurally similar to cecal tonsils. Epithelium covering Peyer's patches contains numerous lymphocytes, few, if any, goblet cells, and lacks a continuous basal lamina. Subjacent to the epithelium is a heavy B-dependent lymphocytic infiltration. A dense core of T-dependent lymphoid tissue containing B-dependent lymphoid follicles lies deeper in the lamina propria (Hoshi and Mori, 1973; Befus *et al.*, 1980). Peyer's patches in chickens share several characteristics with mammalian Peyer's patches including a specialized lymphoepithelium, presence of microfold (M) cells, follicular structure, active particle uptake, ontogenic development, and age-associated involution. The majority of intraepithelial lymphocytes in the intestine are T cells (Lawn *et al.*, 1988). Lymphoid aggregates in the urodeum and proctodeum are also part of the GALT.

Head-associated lymphoid tissue (HALT)

Head-associated lymphoid tissue is found in the Harderian (paraocular) and paranasal glands, lachrymal and lateral nasal ducts, and conjunctival lymphoid tissue (CALT) (Figure IV.2.2e) (Bang and Bang, 1968; Fix and Arp, 1991). The Harderian gland (HG) has large numbers of plasma cells in subepithelial connective tissue (Figure IV.2.2f–g). Testosterone treatment does not inhibit HG development, which suggests that this lymphoid organ is relatively BF independent (Kittner and Olah, 1980). Stromal elements of the HG may produce secretions that influence proliferation and differentiation of plasma cells (Scott *et al.*, 1993).

Bronchial-associated lymphoid tissue (BALT)

Bronchial epithelium overlying lymphoid tissue is primarily squamous and nonciliated at day 1 and week 1, becoming progressively more columnar and ciliated with age. It does not contain M cells (Fagerland and Arp, 1993a). Occasional lymphoid nodules can be found in the lung as isolated foci not associated with primary bronchi (Figure IV.2.3).

Mural nodules

Mural lymphoid nodules are closely associated with lymph vessels. They are circular, elongated, or oval, nonencapsulated, and contain diffuse lymphoid tissue within which are usually found three or four germinal centers (Biggs, 1957; Payne, 1971; Rose, 1981).

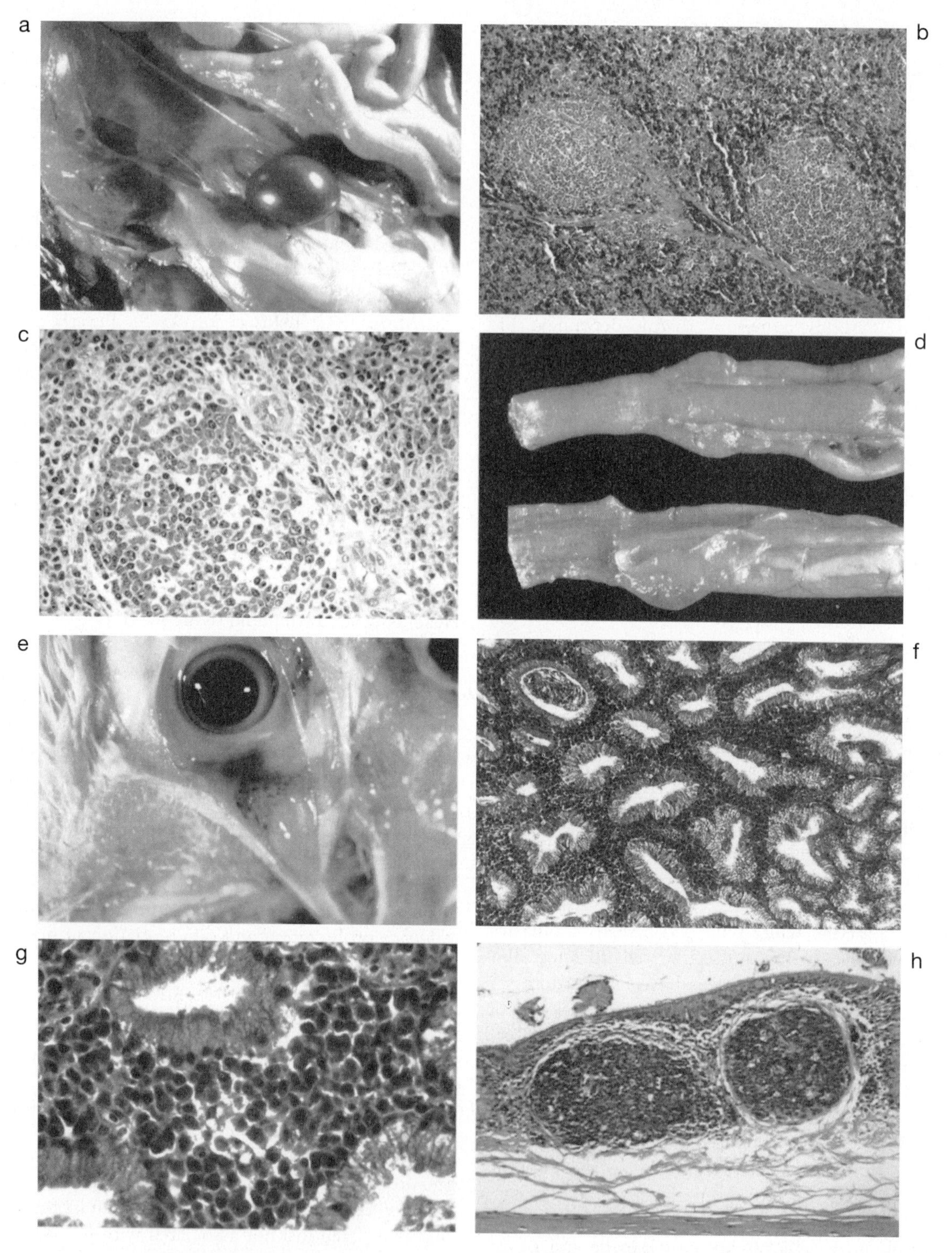

Figure IV.2.2 (a) The spleen is the the oval organ in the central area of the photograph. (b) Periarterial lymphoid sheaths are located in the white pulp of the spleen. (c) A bursa-dependent lymphoid follicle is located adjacent to a small artery and surrounded by thymus dependent lymphoid cells. (d) Cecal tonsils unopened (top) and opened (bottom). (e) Small nodules in the conjunctiva are the conjunctival-associated lymphoid tissue (CALT). (f) The Harderian gland contains lymphoid cells in the connective tissue between the glands. (g) Plasma cells are the predominant cell population in the Harderian gland. (h) Nodular deposits of lymphoid tissue are located in the mucosa of the trachea.

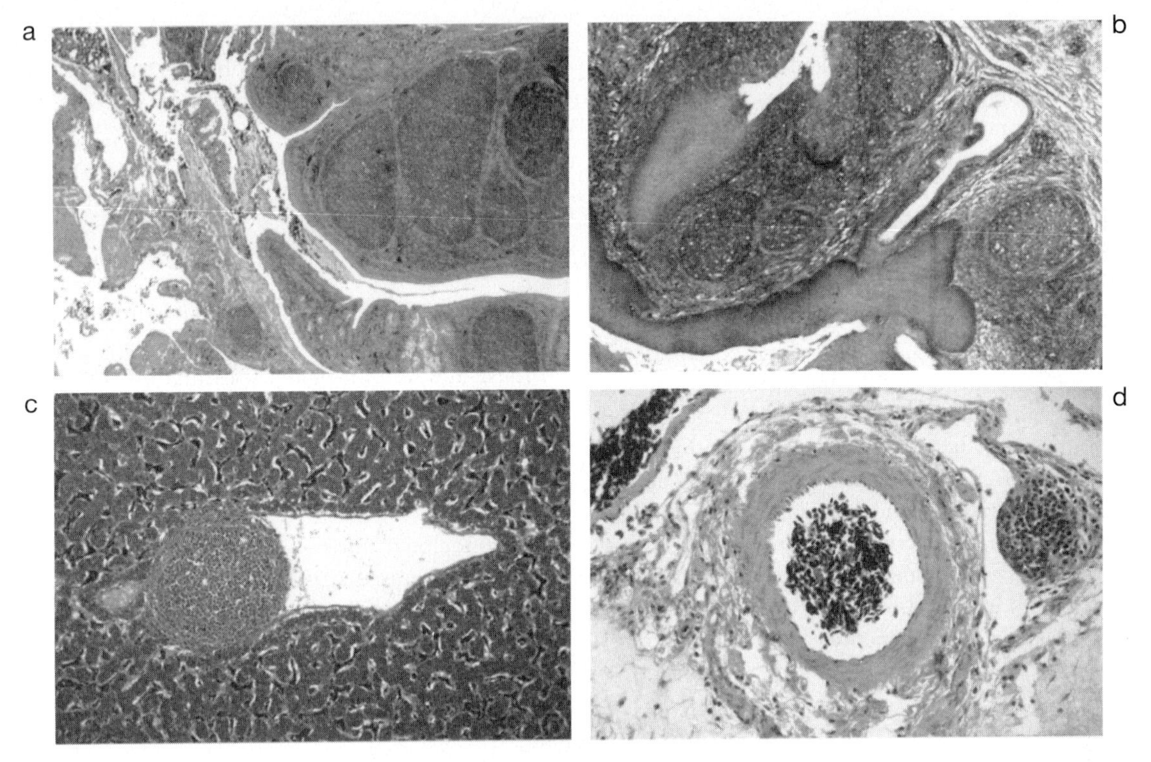

Figure IV.2.3 (a) Nodular and diffuse collections of lymphoid tissue (BALT) are located beneath the bronchial epithelium. (b) Nodular and diffuse collections of lymphoid tissue are located beneath the epithelium at the junction of the esophagus and proventriculus. (c) Lymphoid nodule located adjacent to the central vein of the liver. (d) Lymphoid nodule located in the wall of a lymphatic vessel.

3. Leukocyte Markers in the Chicken

Introduction

To discriminate between the various populations of leukocytes and their function in the immune system, monoclonal antibodies are the main tool. Since the publication of the monoclonal antibody technique by Köhler and Milstein in 1975, several thousands of monoclonal antibodies have been developed for leukocyte phenotyping. In order to obtain some clarity and uniformity in the abundance of monoclonal antibodies, an international nomenclature has been created, first for human leukocyte markers but soon thereafter also for other species. In this nomenclature monoclonal antibodies are divided in antibody-clusters, based on their reactivity with the same cellular marker. These clusters have their own unique code, the CD (cluster of differentiation or cluster designation) code. In principle a CD code is only given when at least three monoclonal antibodies of different laboratories recognize the same antigen with a certain molecular weight and distribution. In some clusters, however, fewer antibodies are available or information is missing on the antigen; in that case the letter 'w' (workshop) is added to the CD code, indicating the provisional clustering. Within a CD cluster subdivisions can be made with letter codes to indicate alternative splicing products of one gene complex (e.g. CD45 cluster) or indicate similar but not identical proteins of a multigene family (e.g. CD11).

The avian CD nomenclature committee was established in 1991 under the auspices of the Veterinary Immunology Committee of the International Union of Immunology Societies. Since then, avian CD workshops have been held in Montreal, Canada (1991), Budapest, Hungary (1992), Tours, France (1992), and Reading, UK (1994). In 1991, it was decided to establish an avian nomenclature system based as much as possible on the mammalian CD system, but with some exceptions (Ratcliffe *et al.*, 1993). Obviously, owing to the limited number of laboratories and the lack of cloned genes, the formal requirements for clustering were incompletely met. Therefore chicken antigens were designated ch, quail antigens q, and turkey antigens t. For antigens predominantly expressed on T lymphocytes T was used, for B-cell antigens B, and for antigens present on more than one lineage of leukocytes, L. Numbers were assigned sequentially on the basis of when application was made to the committee. An exception was made for antigens with an expected potential homology with a previously defined mammalian CD antigen; in those cases then the same number was used. Since 1991, however, an increasing number of antigens have been cloned showing significant homology to mammalian counterparts. These antigens are designated with the mammalian CD numbers.

In addition to the CD antigens, several other groups of antigens occur on the cell surface of leukocytes which can be used for phenotyping. These antigens comprise the immunoglobulins on B lymphocytes and plasma cells, T-cell receptors on T lymphocytes, and MHC class II molecules on cells involved in antigen presentation. Monoclonal antibodies for these types of antigen are included here only when they exhibit a nonallotypic recognition pattern. Monoclonal antibodies specific for leukocytes (Table IV.3.1), for T lymphocytes (Table IV.3.2), and for B lymphocytes (Table IV.3.3) are listed according the recognized antigen, including the molecular mass, the homology with the human gene, and the cellular and tissue distribution. In the text a short description of the (putative) function of each antigen is given as well as any remarkable chicken-specific information. With respect to mononuclear phagocytes, the number of available monoclonal antibodies is insufficient to perform clustering. In addition, the techniques used for detection and characterization of the recognized antigen differ from those used for lymphocyte antigens. Therefore, monoclonal antibodies specific for the mononuclear phagocyte lineage are listed according their cellular distribution pattern in Table IV.3.4.

Leukocytes

ChL2, ChL3, ChL4, ChL5, ChL7, and ChL10

Of these molecules no more information is available than given in Table IV.3.1.

ChL6

The monoclonal antibody A2B5 specific for ChL6 was originally generated against embryonic retinal cells (Eisenbarth *et al.*, 1979). The antigen in chick retina is a sphingoglycolipid that is highly resistent against proteases.

Table IV.3.1 Monoclonal antibodies for chicken leukocyte antigens

Antigen	*Monoclonal antibodies*	*Molecular mass (kDa)*	*Homology (%)*	*Distribution*	*References*
ChL2	CLA-2	180 and 100		All leukocytes	Chen *et al.* (1991)
ChL3	CLA-3	90–100		All leukocytes	Chen *et al.* (1991)
ChL4	86B5	162 and 142		Thymocytes, T and B lymphocytes, some macrophages	Miller and Pink (1986)
ChLw5	Hy5M19	2 × 128		Most bone marrow leukocytes, T and some B lymphocytes, mucosal plasma cells	
ChL6	A2B5			Bursacytes, some B and T lymphocytes	Eisenbarth *et al.* (1979)
ChL7	L43	23		Most leukocytes	Pink and Rijnbeek (1983)
ChL9	maEE1	48	?	All leukocytes, embryonic erythrocytes	Miller and Pink (1986)
ChL10	K11	92, 42, 41		All leukocytes	Chung *et al.* (1991)
ChL12	11A9	38–40		Embryonic hemopoietic cells, thymocytes, T lymphocytes, some bursacytes and B cells	Houssaint *et al.* (1991)
ChL13	BEN	95–110	?	Immature thymocytes and myeloid cells, activated T lymphocytes	Corbel *et al.* (1992), Pourquié *et al.* (1992)
CDw45	LT40, HIS-C7, CL-1, L17	180–215		All leukocytes	Houssaint *et al.* (1985), Jeurissen *et al.* (1988), Paramithios *et al.* (1991)
CD45R1	CLA-1	180 and 195		Thymocytes, T lymphocytes, embryonic bursacytes, some macrophages	Chen and Cooper (1987)
CD57	HNK-1	110		Most B and T lymphocytes	Péault *et al.* (1987)
HEMCAM	c264	98	?	Embryonic hemopoietic progenitor cells, thymocytes, capillary endothelial cells	Vainio *et al.* (1996)

Table IV.3.2 Monoclonal antibodies for chicken T lymphocyte antigens

Antigen	*Monoclonal antibodies*	*Molecular mass (kDa)*	*Homology (%)*	*Distribution*	*References*
ChT1	CT1, CT1a, $T_{10}A_6$, RR5–89, MUI83	63 and 45 and dimers	0	Thymocytes, some T lymphocytes	Chen *et al.* (1984), Houssaint *et al.* (1985), Boyd *et al.* (1992)
CD3	CT3	20,19,17,16	36–40	All T lymphocytes	Chen *et al.* (1986), Bernot and Auffray (1991)
CD4	CT4, 2-6, 2-35	64	23	Subpopulation of $\alpha\beta$ T lymphocytes and thymocytes	Chan *et al.* (1988), Luthala *et al.* (1993)
CD5	2-191, 3-58	64	38	T and B lymphocytes	R. Koskinen and O. Vainio, unpublished data
CDw6	S3	110		Splenic $\gamma\delta$ and most $\alpha\beta$ T lymphocytes, some thymocytes	
ChT6	INN-CH-16	50		Activated T lymphocytes	Schauenstein *et al.* (1988)
ChT7		110		Activated T lymphocytes	Lee and Tempelis (1992)
CD8α	CT8, EP72, 11-39, 3-298, AV12, AV13, AV14, CVI-ChT-74.1	34	37	Subpopulations of $\alpha\beta$, $\gamma\delta$, and NK-like T lymphocytes and thymocytes	Chan *et al.* (1988), Noteburn *et al.* (1991), Luhtala *et al.* (1993), Tregaskes *et al.* (1995)
CD8β	EP42	34	34	Subpopulation of $\alpha\beta$ T lymphocytes	
ChT11	A19	120, 90, 28		Intestinal and activated splenic T lymphocytes	Haury *et al.* (1993)
CD28	2-4, 2-102, AV7	40	50	$\alpha\beta$ T lymphocytes	Vainio *et al.* (1991), Young *et al.* (1994)
$\gamma\delta$ TCR	TCR1	50, 40	30–33	$\gamma\delta$ T lymphocytes	Sowder *et al.* (1988)
$\alpha\beta_1$ TCR	TCR2	50, 40	26–35	Subpopulation of $\alpha\beta$ T lymphocytes	Chen *et al.* (1988), Cihak *et al.* (1988)
$\alpha\beta_2$ TCR	TCR3	48, 40	26–35	Subpopulation of $\alpha\beta$ T lymphocytes	Chen *et al.* (1989), Char *et al.* (1990)

In addition this molecule is expressed on bursacytes and some peripheral B and T cells (Ratcliffe, 1989).

ChL9

ChL9 is a highly glycosylated molecule expressed on embryonic erythrocytes and all leukocytes.

ChL12

ChL12 is an allotypic molecule of 40 kDa found on most embryonic hematopoietic cells (Houssaint *et al.*, 1991). After hatching, however, ChL12 is found predominantly on T cells and most peripheral B cells, including those that have recently migrated from the bursa of Fabricius. It is also expressed on a small number of bursal cells. The molecule is also known as the Ov antigen.

ChL13

ChL13 has been cloned and appears to be the same as DM-GRASP and Sc-1 (Pourquié *et al.*, 1992). The molecule is a surface glycoprotein of the immunoglobulin superfamily. It is expressed on subpopulations of myeloid and lymphoid cells (among others activated T-cells) in both chicken and quail, on neural cells and bursal epithelium. Recently, the human homologue has been cloned and the molecule demonstrated to be the ligand for CD6, and thus an activated leukocyte-cell adhesion molecule (Bowen *et al.*, 1995).

CD45

The CD45 molecule is characterized by its intrinsic phosphotyrosine phophatase activity. Using the monoclonal

Table IV.3.3 Monoclonal antibodies for chicken B lymphocyte antigens

Antigen	*Monoclonal antibodies*	*Molecular mass (kDa)*	*Homology (%)*	*Distribution*	*References*
ChB1	CB1	2 × 55	0	Bursacytes, B lymphocytes[dull]	Chen and Cooper (1987)
ChB2	CB2	80–125		Bursacytes	Chen and Cooper (1987)
ChB3	CB3	50		Bursacytes, B lymphocytes	Pickel *et al.* (1990)
ChB4	CB4	107, (53 and 39)		Bursacytes, B lymphocytes	Chen and Cooper (1987)
ChB5	CB5	167		Bursacytes, B lymphocytes	Chen and Cooper (1987)
ChB6	AV20, HIS-C1	2 × 70–75	0	Bursacytes, B lymphocytes	Jeurissen *et al.* (1988), Tregaskes *et al.* (1996)
ChB6.1	L22, 21-1A4				Pink and Rijbeek (1983), Veromaa *et al.* (1988)
ChB6.2	5-11G2				Veromaa *et al.* (1988)
ChB7	CB7			Bursacytes	Olsen and Ewert (1990)
ChB8	CB8			Bursacytes, some B lymphocytes	Olsen and Ewert (1990)
ChB9	CB9	2 × 65		Bursacytes, some B lymphocytes	Olsen and Ewert (1990)
ChB10	CB10	220		Bursacytes, some B lymphocytes	Olsen and Ewert (1990)
ChB11	CB11	37		Some bursacytes, B lymphocytes	Olsen and Ewert (1990)
ChB12	HY30	2 × 67		Bursacytes, B lymphocytes	Huffnagle *et al.* (1989)
IgM	M-1, HIS-C12, CVI-ChIgM-59.7	2 × 79, 2 × 25		Most bursacytes, some B lymphocytes and some plasma cells	Chen *et al.* (1982), Jeurissen *et al.* (1988), Koch and Jongenelen (1988)
IgG	G-1, CVI-ChIgG-47.3	2 × 70, 2 × 25		Few bursacytes, some B lymphocytes and some plasma cells	Chen *et al.* (1982), Jeurissen *et al.* (1988), Koch and Jongenelen (1988)
IgA	A-1, CVI-ChIgA-46.5	2 × 76, 2 × 25		Few bursacytes, some B lymphocytes and some plasma cells	Chen *et al.* (1982), Jeurissen *et al.* (1988), Koch and Jongenelen (1988)
L chain	L-1, CVI-ChIgL-47.5	25		Most bursacytes, some B lymphocytes and all plasma cells	Chen *et al.* (1982), Koch and Jongenelen (1988)

antibodies listed in Table IV.3.1 immunoprecipitates were obtained from leukocyte cell surfaces with molecular masses in the range of 180–215 kDa that expresses phophotyrosine phophatase activity (Paramithiotis *et al.*, 1991). Interestingly, recognition by the monoclonal antibody CLA-1 is restricted to low molecular weight isoforms of CD45 which exhibit a restricted tissue distribution.

CD57

This molecule was designated CD57 because it was characterized with a monoclonal antibody that was raised against human natural killer cells and showed similar characteristics in the chicken. CD57 appears to be a carbohydrate epitope since some ChL13 molecules express the CD57 epitope and some do not.

HEMCAM

HEMCAM belongs to the immunoglobulin superfamily and consists of five domains, V–V–C2–C2–C2 (Vainio *et al.*, 1996). It is very similar to gicerin, a molecule involved in neurite outgrowth in the chicken and significantly homologous with human MUC18, a molecule that mediates melanoma progression and metastasis. HEMCAM-positive, *c*-kit-positive embryonic bone marrow cells

Table IV.3.4 Monoclonal antibodies for chicken mononuclear phagocyte antigens

Monoclonal antibodies	*Molecular mass (kDa)*	*Distribution*	*References*
Mononuclear phagocyte lineage markers			
47/83		Monocytes, interdigitating cells, macrophages	Kornfeld *et al.* (1983)
MYL51/2	170	Monocytes, interdigitating cells, macrophages, granulocytes	Kornfeld *et al.* (1983)
CVI-ChNL-68.1	87	Monocytes, interdigitating cells, macrophages, B $\text{lymphocytes}_{\text{dull}}$	Jeurissen *et al.* (1988)
Mononuclear phagocyte subset markers			
HUM1		Monocytes	Matsuda *et al.* (1990)
MUI66		Macrophages	Boyd *et al.* (1990, 1992)
MUI79		Monocytes, macrophages	Boyd *et al.* (1990, 1992)
MEP17	125/150	Monocytes, macrophages, (pro)eosinophils, lymphocytes	McNagny *et al.* (1992)
CVI-ChNL-74.2		Macrophages	Jeurissen *et al.* (1992)
K1	61–68/135	Macrophages, thrombocytes	Kaspers *et al.* (1993), Lillehoj *et al.* (1993)
KUL1		Monocytes, macrophages	Mast and Cooper (1995)
Mononuclear phagocyte activation markers			
CMTD-1		Activated (peritoneal) macrophages	Trembicki *et al.* (1986)
CMTD-2		Carbohydrate-activated (peritoneal) macrophages	Trembicki *et al.* (1986)

contain an extremely high frequence of T-cell precursors, as well as progenitors for other hemopoietic cell lineages. HEMCAM has three different RNA splice variants. It mediates cell–cell adhesion and cell spreading of pro-T lymphocytes.

T Lymphocytes

ChT1

ChT1 has recently been cloned, but no human homologue is found (K. Katevuo and O. Vainio, unpublished results). The molecule is present on all cortical thymocytes, to a lesser extent on medullary thymocytes, and on few splenic T cells (Chen *et al.*, 1984). ChT1 therefore seems to be involved in T-cell differentiation and maturation in the thymus.

CD3–TCR complex

Like mammalian T lymphocytes, chicken T lymphocytes are characterized by their T-cell receptor complex, which consists of two antigen-binding receptors, TCR$\alpha\beta$ or TCR$\gamma\delta$, in association with two signalling molecules CD3. The diversification of TCR molecules, necessary for antigen recognition, occurs by rearrangement of a single V, D and J segment derived from the multiple polymorphic copies in the genome. The chicken TCRβ locus contains two V_β families, $V_\beta1$ with 17 and $V_\beta2$ with five individual segments. Within the V_β families the homology is 90%, between the families less than 30% (Lahti *et al.*, 1991; Chen *et al.*, 1994). Based on these homologies, monoclonal antibodies have been developed against the $V_\beta1$ family members (TCR2) and against the $V_\beta2$ family members (TCR3; Chen *et al.*, 1990). CD3 is the only molecule present on the surface of all peripheral T lymphocytes. In addition, cytoplasmic CD3 molecules can be detected in NK cells (Göbel *et al.*, 1994).

CD4 and CD8

Both CD4 (R. Koskinen and O. Vainio, unpublished results) and CD8 (Tregaskes *et al.*, 1995) have been cloned. CD4 is a monomeric glycoprotein of 64 kDa. CD8 is a dimer of either two CD8α chains of 34 kDa or of a CD8α and a CD8β chain, each of 34 kDa. Generally, T lymphocytes bearing CD8$\alpha\beta$ heterodimers are found in systemic organs, whereas those cells bearing CD8$\alpha\alpha$ homodimers are predominantly found in the intestines. Both CD4 and CD8 function as adhesion molecules, strengthening the binding of TCR to its ligand and are involved in signaling by means of their association with tkl, the avian homologue of mammalian p56lck.

CD5 and CDw6

CD5 and CDw6 both belong to the highly conserved family of multiple repeats of the cysteine-rich macrophage scavenger receptor type I motif (Göbel *et al.*, 1996). They are monomeric, highly N-glycosylated proteins that are constitutively phosphorylated. Both molecules are supposed to play a role in signal transduction. CD5, which has been recently cloned (R. Koskinen and O. Vainio, unpublished data), is present on peripheral $\alpha\beta$TCR and $\gamma\delta$TCR T cells. In addition, it is found at a low level on most B cells. CDw6 is found on systemic but not intestinal $\alpha\beta$TCR T cells and on some splenic $\gamma\delta$TCR T cells.

ChT6

ChT6 most likely resembles the α-chain of the IL-2 receptor and is thus present on mitogen-activated T cells (Schauenstein *et al.*, 1988).

ChT11

The ChT11 molecule has properties of an integrin and is thought to represent the $\alpha^E\beta_7$ molecule.

CD28

CD28 is a glycoprotein, that in contrast to mammals does not form disulfide linked homodimers because the chicken sequence lacks cysteine (Young *et al.*, 1994). It is present on all peripheral $\alpha\beta$TCR T cells and on a subpopulation of $\gamma\delta$TCR T cells. CD28 acts as a costimulatory molecule in addition to the primary signal from the CD3–TCR complex (Arstila *et al.*, 1994).

B lymphocytes

ChB1

ChB1 has recently been cloned, but it shows no significant homology with any human molecule cloned so far (R. Goitsuka and C.-L. Chen, unpublished data). The molecule is structurally related to the C-type lectin superfamily with less than 30% homology to CD72 and CD23 identities.

ChB3

ChB3 is a molecule on the surface of bursacytes and B cells that is associated with β_2-microglobulin (Pickel *et al.*, 1990).

ChB2, ChB4 and ChB5

No more information is available for these molecules than given in Table IV.3.3.

ChB6

The ChB6 molecule, which is better known as Bu-1, has also recently been cloned but no mammalian homologue for this highly glycosylated protein has been found (Tregaskes *et al.*, 1996). It is an allotypic antigen and the allotypes ChB6.1 and ChB6.2 can be differentiated by monoclonal antibodies. AV20 is an antibody that recognizes both allotypes. In addition, HIS-C1 appears to recognize an intracellular part of the ChB6 molecule (J. Young, personal communication). ChB6 is expressed on the earliest B cells, continues to be expressed on B cells throughout ontogeny, but is lost when B cells develop into plasma cells. Because ChB6 expression is restricted to B cells yet is ubiquitously expressed on B cells, it appears to be an important molecule in B cell development and maturation, although the precise function is unknown. ChB6 appears to be a prerequisite for bursal stem cells to be able to enter the bursa during ontogeny (Houssaint *et al.*, 1991).

ChB7, ChB8, ChB9, ChB10, and ChB11

No more information is available for these molecules than that given in Table IV.3.3.

ChB12

ChB12 is better known as Bu-2. Like ChB6, this molecule is expressed selectively on cells committed to the B-cell lineage. ChB12 and ChB6 are difficult to separate by biochemical properties or cell distribution. Typically, however, expression of ChB12 remains high on REV-T transformed B-cell lines, whereas expression of ChB6 is rapidly downregulated (Huffnagle *et al.*, 1989).

Immunoglobulins

In the chicken, three immunoglobulin isotypes exist: IgM, IgG (sometimes called IgY), and IgA (see also Section 5, this chapter). Although the structure of these molecules is comparable to those in mammals, the mechanisms used to obtain diversity in the variable region differs markedly. Both heavy and light chain are based on a single V gene and diversification is by gene conversion using homologous, non-functional pseudogenes as donor elements (Reynaud *et al.*, 1994).

Mononuclear phagocytes

In the group of monoclonal antibodies specific for mononuclear phagocytes, insufficient antibodies with similar characteristics are present to perform clustering of antigens. Therefore, the monoclonal antibodies have been listed according a more general division into those recognizing the entire lineage from bone marrow monocytes to mature tissue macrophages, those recognizing a subset of mononuclear phagocytes, and those recognizing an activation marker. In contrast to cell determinants on lymphocytes, which are mostly found on the cell surface, cell determinants specific for mononuclear phagocytes are often located intracellularly. This phenomenon hampers the isolation and characterization of the recognized molecules, therefore, only few molecules have been investigated in detail.

4. T-Cell Receptors

T-cell receptor repertoire ontogeny

Chicken T-cells have been well characterized using polyclonal and monoclonal antibody raised against functionally important T-cell molecules (Table IV.4.1), such as T-cell receptors. Three major lineages of chicken T cells have been defined, TCR1, TCR2 and TCR3, named for their ontogenetic order (Cooper *et al.*, 1991).

Ontogeny of T-cell subsets

The developing chicken thymus exhibits three periods of receptivity to thymocyte precursor influx, followed by three waves of differentiation of all three T-cell lineages. Chicken $\gamma\delta$ T cells ($TCR1^+$) are detected at day 11 of embryogenesis (E11), and peak in relative cell numbers (30%) at E15. $\alpha\beta$ T cells expressing $V_\beta1$ ($TCR2^+$) appear at E15, and predominate in the thymus by E17–18, whereas $\alpha\beta$ T cells expressing $V_\beta2$ ($TCR3^+$) develop around E18 and are maintained as a smaller subset (Chen *et al.*, 1989; Char *et al.*, 1990; Lahti *et al.*, 1991).

Based on developmental parallels with mammals, chicken $\alpha\beta$ T-cell maturation involves the same positive and negative selection mechanisms as in mammals (Cooper *et al.*, 1991). Chicken $\alpha\beta$ T cells pass slowly through the thymic cortex, gradually increase surface TCR expression, and exhibit the same pattern of CD4/CD8 expression during thymic maturation as for mammalian $\alpha\beta$ T cells. In contrast, chicken $\gamma\delta$ T cells do not appear to undergo intrathymic selection, as they migrate quickly from the thymus without clonal expansion, immediately express high levels of surface TCR and are resistant to receptor modulation.

In contrast to reports that some mammalian T cells may be of extrathymic origin, most early chicken T-cell development appears to be thymus-dependent (Chen *et al.*, 1996; Dunon and Imhof, 1996). Experimental evidence includes data from chick–quail chimeras (Coltey *et al.*, 1989; Bucy *et al.*, 1989), and from chimeras of congenic chicken strains differing in the ov alloantigen T-cell marker (Dunon *et al.*, 1993a, b).

The waves of $V_\beta1$ T cells vary in their homing preferences to peripheral organs: the first wave preferentially migrates to the spleen, the second wave to the spleen and intestine, and the third wave and later T cells to the spleen (Dunon *et al.*, 1993b, 1994; Dunon and Imhof, 1996). Homing of $V_\beta1$ and $V_\beta2$ T cell subsets to the intestine differs markedly given the relative absence of $V_\beta2$ T cells in the intestine (Char *et al.*, 1990; Dunon *et al.*, 1994).

Table IV.4.1 TCR-associated proteins and accessory molecules involved in TCR signalling

Molecule	*mAb*	*References*	*Notes*
CD2	OC2	Knabel *et al.* (1993)	
CD3 complex	CT3	Chen *et al.* (1986)	16, 17, 19, 20 kDa chains
		Bernot and Auffray (1991)	19 kDa chain (CD3 γ/δ), gene cloned
CD4	CT4	Chan *et al.* (1988)	64 kDa
	EP96	Marmor *et al.* (1993)	
	2–6	Luhtala *et al.* (1993)	
CD5	OC5	Knabel *et al.* (1993)	
	2–191	Vainio and Imhof (1996)	64 kDa
CD6 (candidate)	S3	Göbel *et al.* (1996b)	110 kDa
CD8 α-chain	CT8	Chan *et al.* (1988)	
	AV12	Tregaskes *et al.* (1995)	24 kDa core protein
	74.1	Noteborn *et al.* (1991)	
	EP72	Tregaskes *et al.* (1995)	
	11–39	Luhtala *et al.* (1995)	
CD8 β-chain	EP42	Tregaskes *et al.* (1995)	21 kDa core protein
CD25	INN-CH-16	Schauenstein *et al.* (1988)	50 kDa, IL-2 receptor α-chain
CD28	AV7	Young *et al.* (1994)	40 kDa, gene cloned
	2–4	Vainio *et al.* (1991)	Originally reported as anti-CD2 candidate
CD45	HIS-C7	Jeurissen *et al.* (1988a)	180 kDa
	CL-1	Houssaint *et al.* (1987)	
	LT40	Paramithiotis *et al.* (1991)	
	CLA-1	Chen and Cooper (1987)	Restricted isoform, 195 and 180 kDa
lck	—	Chow *et al.* (1992)	$pp56^{lck}$, 56 kDa, *tkl* oncogene
TCR $\gamma\delta$	TCR1	Sowder *et al.* (1988)	All $\gamma\delta$ T cells
TCR $\alpha\beta$ ($V_\beta1$)	TCR2	Cihak *et al.* (1988)	$V_\beta1$-expressing $\alpha\beta$ T cells
TCR $\alpha\beta$ ($V_\beta2$)	TCR3	Char *et al.* (1990)	$V_\beta2$-expressing $\alpha\beta$ T cells

Ontogeny of TCR gene repertoires

Despite low homologies to their mammalian homologues, genes for all four chicken TCR loci have been cloned (Table IV.4.2). The genomic organizations of all four loci are not completely known; however, conserved features of mammalian and avian TCR genes include consensus amino acids for structural requirements, overall genomic organization (although chicken loci encode fewer gene segments), and location of TCR δ genes within the TCR α locus (Kubota *et al.*, 1995; Chen *et al.*, 1996). Somatic diversity of rearranged TCR genes appears to be generated through combinatorial and junctional mechanisms, as in mammalian TCR genes, rather than by gene conversion as in avian immunoglobulin genes (McCormack *et al.*, 1991a, b; Cooper *et al.*, 1991).

The embryonic thymus $V_\beta 1$ repertoire of all three waves of T-cell development shows no evidence for preferential $V_\beta 1$ or J_β usage or selection for CDR3 lengths, and therefore no preselection of $V_\beta 1$ T cells for colonization into spleen or intestine (Dunon *et al.*, 1994; Dunon and Imhof, 1996).

Distribution of $\alpha\beta$ and $\gamma\delta$ TCR populations

Migration of T cell subsets from the thymus to the periphery occurs in the same order as their development, i.e. $\gamma\delta$ T cells appear in spleen around E15–17, $V_\beta 1^+$ $\alpha\beta$ T cells appear around E19, and $V_\beta 2^+$ $\alpha\beta$ T cells appear at 3 days posthatching (Chen *et al.*, 1988; Coltey *et al.*, 1989; Char *et al.*, 1990). Dunon and coworkers (1993a, b; 1994) have shown that emigration of $\gamma\delta$ and $V_\beta 1$ T-cell subsets into spleen and intestine occurs in waves, following each wave of thymic T-cell development.

After migration from the thymus, most $V_\beta 1^+$ $\alpha\beta$ T cells express CD4 and are localized in the splenic periarteriolar sheath and intestinal lamina propria. Chicken $\alpha\beta$ T cells are responsive to T-cell mitogens, are capable of graft-versus-host (GVH) alloreactivity, and $CD4^+$ cells secrete lymphokines (Cooper *et al.*, 1991). Thymectomy at hatch results in only a small decrease in the numbers of $\alpha\beta$ T cells in the periphery, owing to T-cell migration from the thymus by the time of hatching.

Splenic $\gamma\delta$ T cells in chickens are localized to the sinusoids of the red pulp and remain dispersed (Bucy *et al.*, 1988). $\gamma\delta$ T cells predominate in the intestinal epithelium, but are minor populations or absent in lamina propria, Peyer's patches, cecal tonsils, and normal skin (Bucy *et al.*, 1988). Thymic $\gamma\delta$ T cells enter the intestinal epithelium at all levels of the villi just after hatching, and persist longer than in the spleen due to an apparent higher capacity for self renewal (Dunon *et al.*, 1993a, b). Although $\gamma\delta$ T cells in the thymus and blood are $CD4^-CD8^-$, about two-thirds of the $\gamma\delta$ T cells in the spleen and intestine express CD8. Intestinal $CD8^+$ cells are further divided into $CD8\alpha\alpha$ and $CD8\alpha\beta$ subsets (Tregaskes *et al.*, 1995).

Avian $\alpha\beta$ T cells respond to *in vitro* stimulation via mitogens and TCR-crosslinking, but $\gamma\delta$ T cells ($CD8^+$) respond only in the presence of $CD4^+$ T cells or soluble factors derived from them (Arstila *et al.*, 1993; Kasahara *et al.*, 1993), and have an activated phenotype based on size and MHC class II expression (Ewert *et al.*, 1984). Only $CD4^+$ cells ($V_\beta 1$ and $V_\beta 2$) are capable of inducing GVH lesions in chick embryos, and $\gamma\delta$ T cells are recruited by the alloreactive $\alpha\beta$ T cells (Tsuji *et al.*, 1995; Chen *et al.*, 1996). These data suggest mutual regulatory functions between $\alpha\beta$ and $\gamma\delta$ T cells in avian immune responses (Cooper and Chen, 1993; Kasahara *et al.*, 1993).

Table IV.4.2 Chicken T-cell receptor genes

Locus	*Gene segments*	*References*	*Notes*
TCR β	6–8 $V_\beta 1$	Tjoelker *et al.* (1990)	22% amino acid identity to mammalian $V_\beta I$
	3–5 $V_\beta 2$	Tjoelker *et al.* (1990)	46% amino acid identity to mammalian $V_\beta II$
	1 D_β	McCormack *et al.* (1991b)	
	4 J_β	Cooper *et al.* (1991)	
	1 C_β	Chen *et al.* (1996)	Locus mapped to chromosome 1
TCR α	~25 $V_\alpha 1$	Göbel *et al.* (1994b)	
	~10 $V_\alpha 2$	Kubota *et al.* (1995)	24% amino acid identity with $V_\alpha 1$
	~25 J_α and 1 C_α	Göbel *et al.* (1994b)	
TCR γ	$V_\gamma 1$ family	Rast and Litman (1994)	
	$V_\gamma 2$ family	Six *et al.* (1995)	33% amino acid identity with $V_\gamma 1$
	$V_\gamma 3$ family	Six *et al.* (1995)	25–29% amino acid identity with $V_\gamma 1$ and $V_\gamma 2$
	3 J_γ	Six *et al.* (1995)	60–65% amino acid identity
	1 C_γ	Chen *et al.* (1996)	30% amino acid identity to mammalian C_γ
TCR δ	$V_\delta 1$ and $V_\delta 2$ family	Chen *et al.* (1996)	
	2 D_δ and 1 J_δ	C. H. Chen (personal communication)	
	1 C_δ	Kubota *et al.* (1995)	33% amino acid identity to mammalian C_δ

~, approximately.

Species-specific aspects of avian T-cells

TCR genes

Unique features of avian TCR genes include a single exon encoding the leader and variable region in TCR α and γ loci (Göbel *et al.*, 1994; Six *et al.*, 1995), and an alternatively spliced TCR β transcript encoding an invariant truncated β chain of unknown function (Dunon *et al.*, 1995). Although mammalian V_β genes are divided into 20–30 V_β subfamilies, the families of mammalian V_β genes grouped by structural features, V_βI and V_βII, correspond to the only two chicken V_β families, $V_\beta1$ and $V_\beta2$ (Tjoelker *et al.*, 1990; Cooper *et al.*, 1991; Lahti *et al.*, 1991).

Segregation of $\alpha\beta$ and $\gamma\delta$ T-cell lineages by differential TCR gene expression during avian T-cell development may differ from that in mammals. Although the deletion of TCR δ genes occurs in chicken $\alpha\beta$ T cells, as it does in mammals (Kubota *et al.*, 1995), transcriptional silencing of TCR γ genes does not appear to occur in all chicken $\alpha\beta$ T cells (Chen *et al.*, 1996).

Function of T-cell subsets

Chicken $V_\beta1$ and $V_\beta2$ T-cell subpopulations appear to have some distinct differences in ontogeny and tissue distribution (discussed above), and in function. Monoclonal antibody depletion experiments revealed that $V_\beta1^+$ T cells are essential for IgA production and normal mucosal antibody responses (Cihak *et al.*, 1991). Functional differences between the $V_\beta1$ and $V_\beta2$ TCR repertoires are also revealed by GVH reactions using various strains differing in MHC class II alleles (Chen *et al.*, 1996).

The avian immune system is characterized by a higher frequency of $\gamma\delta$ T cells than mouse and human, and may reach 30–50% of peripheral blood lymphocytes (Sowder *et al.*, 1988). Peripheral blood $\gamma\delta$ T cells also respond to androgens, resulting in an expanded population in males (Arstila and Lassila, 1993).

5. Cell Surface and Secreted Immunoglobulins in B Cell Development

Chicken immunoglobulin loci

The chicken Immunoglobulin (Ig) light (L) and heavy (H) chain loci are depicted in Figure IV.5.1. At the heavy-chain locus, the designation of three constant region genes is based on their cloning from cDNA and does not exclude the possible existence of other isotypes. Unlike mammalian immunoglobulins, there is a single light-chain locus. It is unclear at present to which chromosomes the heavy- and light-chain loci map.

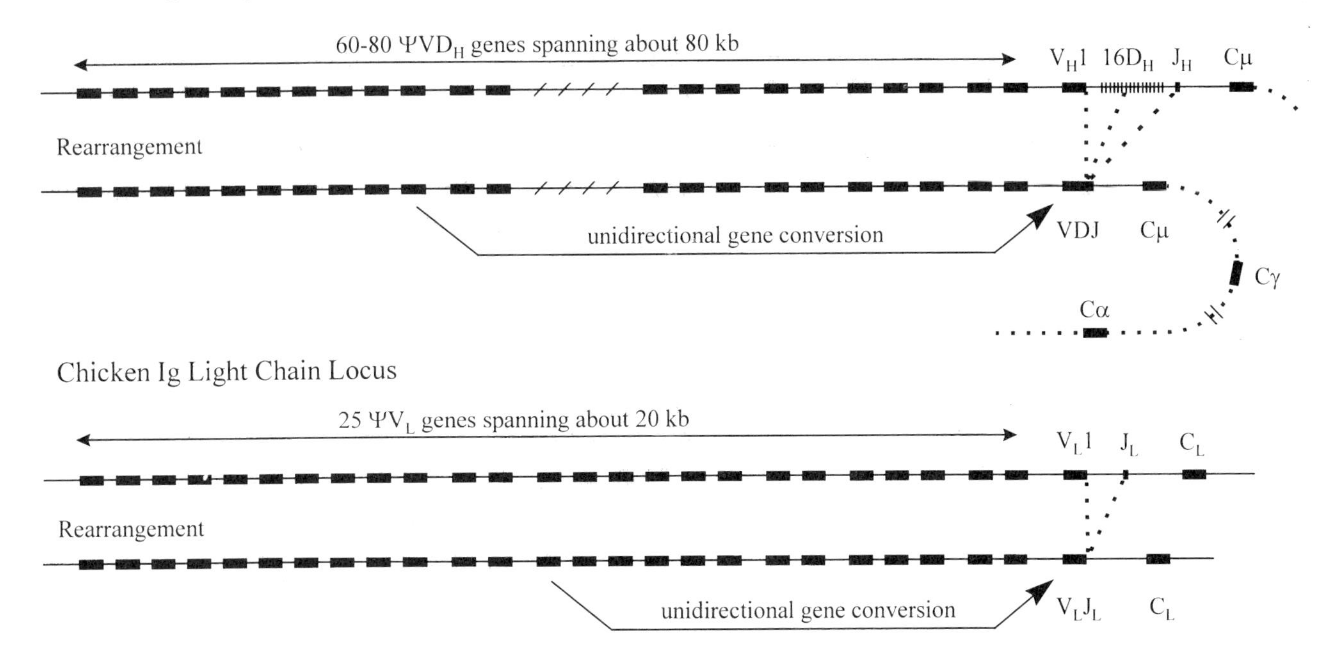

Figure IV.5.1 The chicken Ig loci. In the heavy chain locus, the 16 D_H segments are arrayed in a region of about 15 kb between V_H1 and J_H, which in turn lies about 22 kb 5′ of the C_μ gene. The positions of C_γ and C_α relative to each other and distances from C_μ are currently unknown. At the light chain locus, V_L1 lies 1.8 kb 5′ of J_L, which in turn lies 1.6 kb 5′ of C_L.

Rearrangement of chicken Ig genes

To be expressed, all vertebrate Ig genes must undergo rearrangement from a germline configuration to a more condensed, transcribeable form. At the molecular level, rearrangement of the chicken Ig heavy and light chain genes is indistinguishable from that seen in the well-studied murine and human models (as reviewed by Ratcliffe and Jacobsen, 1994). The RAG-1 and RAG-2 gene products, required for murine Ig gene rearrangement (Shinkai *et al.*, 1992; Mombaerts *et al.*, 1992) have also been observed in chicken embryos at the appropriate time (Carlson *et al.*, 1991; Reynaud *et al.*, 1992). Furthermore, the highly conserved recombination signal sequences (RSSs) which target the recombinase complex have been identified in the chicken Ig light and heavy chain loci (Reynaud *et al.*, 1985, 1989, 1991).

However, there are several striking differences between chicken and mammalian Ig gene rearrangement and repertoire generation. In mice and humans, variable, diversity (in heavy chain), and joining regions have evolved into multimembered families of functional gene segments, capable of generating considerable recombinatorial diversity (Tonegawa, 1983). In the chicken, however, unique variable regions (V_H1 or V_L1) and junctional segments (J_H or J_L) participate in heavy- and light-chain gene rearrangement (Reynaud *et al.*, 1985, 1989) (Figure IV.5.1). Thus, during light-chain rearrangement, in all B cells, V_L1 rearranges to the J_L segment located 1.8 kb downstream. Similarly, rearrangement at the heavy-chain locus commences with the rearrangement of one of the 16 possible D_H segments to the unique J_H segment. Heavy chain rearrangement is completed upon joining of the unique V_H1 to the resulting D_H–J_H complex (Reynaud *et al.*, 1991). This process generates minimal recombinatorial diversity since there is little sequence variation among the 16 D elements. The chicken might represent an extreme example of the generalized avian Ig loci since Muscovy ducks contain several functional Ig V_L genes. Junctional imprecision generated by the resolution of hairpin intermediate structures can result in short sequence palindromes (P nucleotides) at V(D)J junctions which have been clearly identified in chicken Ig gene rearrangement (McCormack *et al.*, 1989). In contrast, the addition of random nucleotides (N nucleotides), catalysed by terminal deoxyribonucleotidyl transferase (Tdt) has not been observed in chicken Ig rearrangement.

Diversification of the primary immune repertoire

In chickens, the rearranged light and heavy chain V(D)J complexes serve as substrates for a process of gene conversion. Clusters of highly homologous pseudogenes, ΨV_L and ΨVD_H, are located upstream of the functional V_L1 and V_H1 genes respectively and serve as donors in the intrachromosomal transfer of sequence to the rearranged genes (Figure IV.5.1). About 25 ΨV_L and 80 ΨVD_H have been identified, in regions spanning approximately 20 kb and 80 kb respectively (Reynaud *et al.*, 1987, 1989). The ΨV_L are highly homologous to the V_L1 gene, with differences found mainly in and around the hypervariable or CDR regions. The ΨVD_H are homologous to fusions between a V_H gene and a D_H segment and, again, sequence variation occurs primarily in the CDR regions. Neither ΨV_L or ΨVD_H have functional RSS sequences. In addition, both ΨV_L and ΨVD_H lack leader introns. Hence, neither pseudogene family members can undergo rearrangement with the downstream J segment and neither is transcribed.

While the molecular mechanism of gene conversion is currently unknown, preferred donor pseudogenes are those which are most proximal to V_L1, have the highest degree of homology to the target sequence and are in the opposing orientation to the functional V gene (McCormack and Thompson, 1990; McCormack *et al.*, 1991).

Immunoglobulin isotype switching

Following antigen exposure and T cell help, mammalian B cells are induced to undergo isotype switching, a process in which intervening constant region genes are deleted so that the functional VDJ sequence ends up 5′ to an alternative constant region gene. Analysis of chicken cDNA sequences encoding γ and α Ig heavy chains is consistent with an equivalent deletion occurring at the chicken IgH locus.

The bursa of Fabricius: organ of B cell expansion and immunoglobulin diversity

Prebursal

Cells committed to the B lymphocyte lineage, as determined by the presence of surface B cell markers, Ig gene rearrangements, and sIg expression have been identified in extrabursal compartments of the developing embryo (Benatar *et al.*, 1991; Reynaud *et al.*, 1992) demonstrating that the bursa is not required for Ig gene rearrangement, although there is the likelihood that the bursal microenvironment is required for high rate V gene conversion.

In striking contrast to most mammalian models of B-cell development, the rearrangement of chicken Ig genes is restricted to a short window of time during embryonic life (reviewed by Ratcliffe, 1989). In addition, following DJ rearrangement, either H or L chain loci complete rearrangement in a random order (Benatar *et al.*, 1992), in contrast to mammalian V genes, where H chain rearrangement typically precedes that of L chain.

In mammals, an appreciable proportion of mature B cells have both Ig loci rearranged; only one locus being productively rearranged, resulting in a monospecific B-cell.

In chicken, few mature B-cells contain nonproductive V gene rearrangements, suggesting a distinct mechanism for allelic exclusion (see Ratcliffe and Jacobsen, 1994).

Bursal

The bursa of Fabricius plays a key role in avian B-cell development and antibody diversification (for a recent review, see Ratcliffe *et al.*, 1996). Following colonization by a small number of B-cell precursors during embryonic life, cells expressing surface immunoglobulin undergo rapid proliferation, such that by about 2 months of age there are approximately 10 000 follicles in the bursa, with each containing about 10^5 B lymphocytes (Olah and Glick, 1978). Seeding B lymphocytes from the bursa into the periphery begins at about the time of hatch, and continues until the bird has reached approximately 4–6 months old, at which time the bursa begins to atrophy.

Postbursal

B cells which have migrated from the bursa to the periphery include those cells which have the potential to respond to antigen and subsequently go on to secrete Ig. In addition, the postbursal B-cell compartment includes the capacity for self-renewal since the bursa undergoes functional involution by about 6 months of age. Recent evidence has demonstrated that functional B-cell heterogeneity established in the bursa is reflected in discrete populations of B cells in the periphery, although the physiological basis for this heterogeneity remains speculative (Paramithiotis and Ratcliffe, 1993, 1996).

Avian immunoglobulin isotypes and their roles in the periphery

Early attempts to characterize chicken immunoglobulin identified three Ig classes, each of which has subsequently been cloned at the cDNA level. (Dahan *et al.*, 1983; Parvari *et al.*, 1988; Mansikka, 1992). Some of the basic properties of chicken immunoglobulins are described in Table IV.5.1.

IgM: Immunoglobulin of primary humoral responses

Chicken IgM is the first antibody observed after primary immunization of chickens and the high molecular weight form of serum IgM (~900 K) can be reduced to heavy chains (70 K) and light chains (22 K) predicting a pentameric structure similar to mammalian IgM. This notion is supported by the amino acid sequence of the μ-chain which maintains key amino acids required for pentamer assembly and binding to J-chain, despite overall homology to the mammalian μ of 28–36%.

IgM is found on the surface of most chicken B cells (Kincade and Cooper, 1971) and can transduce signals to the B-cell cytoplasm (Ratcliffe and Tkalec, 1990). Mammalian sIgM is part of a signaling complex which includes the associated Igα and Igβ chains, collectively referred to as the B-cell antigen receptor. Signals are transduced downstream of Igα/β by a cascade of protein tyrosine kinases in which $p72^{syk}$ is critical and which may also include lyn, fyn and blk (reviewed by Pleiman *et al.*, 1994). At this time, there is only circumstantial evidence to support the existence of the chicken homologues of the Igα and Igβ coreceptor molecules (O. Vainio, personal communication). However, recent cloning of the transmembrane and cytoplasmic domains of chicken sIgM (Figure IV.5.2a) has revealed extensive (78%) homology to their mammalian equivalents, and chicken homologues to syk, lyn, fyn and

Table IV.5.1 Properties of chicken immunoglobulin isotypes. Data taken from Ratcliffe (1996) and references therein

Isotype	*Heavy chain (kDa)*	*# H chain Ig domains*	*Homology to mammalian*	*Serum concentration*	*Sources*	*Structure and comments*
IgM	70 kDa	5	About 30%	1–2 mg/ml	Serum	900 kDa, consistent with heavily glycosylated $(\mu_2L_2)_5$ plus a J chain
			78% for TM[a]		Cell surface	μ_2L_2 monomer of membrane IgM, no J chain
IgG	67 kDa	4	30–35%	5–10 mg/ml	Serum	175 kDa, γ_2L_2 monomeric form
					Egg yolk	γ_2L_2, high concentrations (10 mg/ml) of IgG are found in egg yolk (low concentrations in egg white)
IgA	65 kDa	4	32–41%	≈3 mg/ml	Serum	170 kDa, α_2L_2-momeric form without J-chain
					Bile	350 kDa, consistent with $(\alpha_2L_2)_2$ plus a J-chain
					Mucosa (tears, saliva)	600–700 kDa, consistent with $(\alpha_2L_2)_4$ plus a J-chain

[a]TM refers to the transmembrane and cytoplasmic domains of chicken sIgM described in Figure IV.5.2.

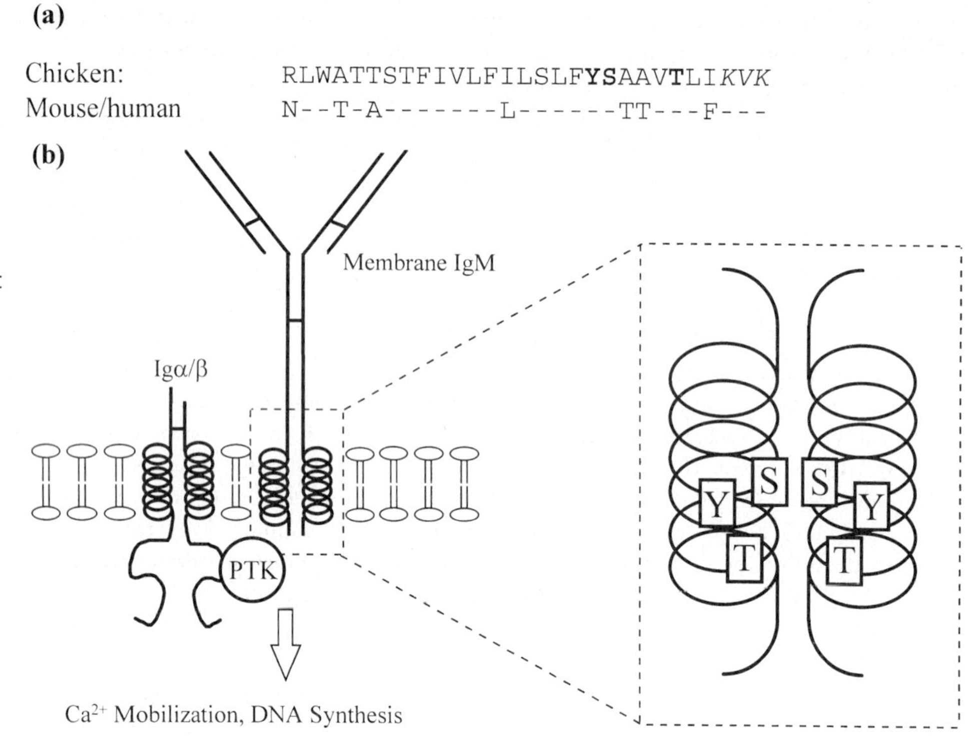

Figure IV.5.2 Signal transduction through chicken sIgM. (a) Comparison of the amino acid sequence of the recently cloned transmembrane and cytoplasmic domains of chicken μ chains with their mammalian counterparts. Amino acids in bold represent those demonstrated to be required for association of the mammalian μ chains with Igα/β. Amino acids in italics (*KVK*) represent the intracytoplasmic tail of the sIgM molecule. (b) sIgM associating with Igα/β and downstream protein tyrosine kinases of which p72^{syk} has been shown to be critical.

blk appear to have equivalent signal-transducing functions (Takata *et al.*, 1994). These considerations support the existence of a chicken signaling complex (Figure IV.5.2b) equivalent to mammalian sIgM.

IgG: A mammalian homologue or unique isotype?

Chicken IgG is functionally homologous to mammalian IgG in that it participates in the recall response to antigen. However, analysis of structure and sequence of chicken IgG has demonstrated that, evolutionarily, it is as similar to mammalian IgE as it is to IgG. This has led to the suggestion that the chicken molecule is the evolutionary ancestor to both IgE and IgG in mammals (Parvari *et al.*, 1988).

IgA: major immunoglobulin of the mucosal immune system

In mammals, IgA is the primary isotype produced in the mucosal immune system. In external secretions, IgA exists in a dimeric or tetrameric form of IgA monomers joined by a J chain, whereas serum IgA is monomeric. Cloning of the cDNA of C_α from a chicken Harderian cDNA library demonstrated that the C_α chain is divided into four Ig domains, three of which have 32–41% homology to human C_α (Mansikka, 1992). In mammalian species α heavy chains have three C_α Iγ domains and a hinge region between $C_\alpha1$ and $C_\alpha2$. This hinge region may have resulted from deletions during evolution from a Cα2 Ig domain in the primordial C_α gene, which has been more conserved in chickens.

6. Major Histocompatibility Complex (MHC) Antigens

Introduction

Genetic and functional studies on products of the 'B-complex' (chicken MHC), which was originally described as a blood group locus (Briles, 1962), revealed that it consists of at least three loci: F (class I antigens), L (class II antigens) and G locus (class IV antigens) (Pink *et al.*, 1977). The first crossing over in MHC was described between F/L and G loci (Hala, 1977). The following methods have been used for studies of MHC antigens: hemagglutination, lymphoagglutination, skin transplantation, graft-versus-host reaction (Hala, 1977), mixed lymphocyte reaction, biochemical analysis (Pink *et al.*, 1977), flow cytometry (Kaufman *et al.*, 1995b) and molecular biology (Guillemot and Auffray, 1989).

The regions, genes and their functions or antigens of the MHC are presented in Table IV.6.1.

Tissue distribution, structure and function of proteins, and minor antigens

The tissue distributions, structures and functions of the B-F and B-L molecules encoded by the B complex appear to be very similar to human classical class I and class II molecules.

B-F molecules are expressed in a wide variety of cells, including erythrocytes, as occurs in all tetrapod verte-

Table IV.6.1 Genes on the MHC microchromosome and related regions

Region	*Gene*	*Antigen/function*	*References*
B-F/B-L	(cosmid cluster I of roughly 130 kB)		
	12.3	G protein	Guillemot *et al.* (1988, 1989)
	8.5	B-G chain	Guillemot *et al.* (1988), Kaufman *et al.* (1991)
	B-LBI	Class II β chain	Guillemot *et al.* (1988), Zoorob *et al.* (1993)
	8.4	Cell surface molecule	Guillemot *et al.* (1988)
	B-LBII	Class II β chain	Guillemot *et al.* (1988), Zoorob *et al.* (1990, 1993)
	RINGIII	Putative transcription factor	Thorpe *et al.* (1996)
	21.6	Complement component C4	Guillemot *et al.* (1988), Beck and J. Kaufman, unpublished data
	B-FI	Class I α chain	Guillemot *et al.* (1988), Kaufman *et al.* (1995b)
	21.7	Transporter for antigen presentation	Guillemot *et al.* (1988), Kaufman *et al.* (1995b)
	B-FIV	Class I α chain	Guillemot *et al.* (1988), Kroemer *et al.* (1990), Kaufman *et al.* (1995b)
B-G	(cosmid cluster V of 60 kb) 0.05 cM from B-F/B-L region		Hala *et al.* (1981),
	43KCN, 43B8, 43F, 43Y, 43A	B-G chains	Kaufman *et al.* (1991), Doehring *et al.* (1993)
	(cosmid cluster VI of 75kb)		
	FB1, FB3, 22E, 13B, F8, F11	B-G chains	Kaufman *et al.* (1991), Doehring *et al.* (1993)
Rfp-Y	(cosmid cluster II/IV of 100 kb, unlinked to other clusters)		
	Y-FV	Class I α chain	Guillemot *et al.* (1988), Miller *et al.* (1994)
	17.8	C-type lectin	Guillemot *et al.* (1988), Bernot *et al.* (1994)
	Y-FVI	Class I α chain	Guillemot *et al.* (1988), Miller *et al.* (1994)
	Y-LBIII	Class II β chain	Guillemot *et al.* (1988), Zoorob *et al.* (1990, 1993), Miller *et al.* (1994)
	Y-LBIV	Class II β chain	Guillemot *et al.* (1988), Zoorob *et al.* (1993), Miller *et al.* (1994)
	(cosmid cluster III of 100 kb, unlinked to other clusters)		
	Y-LBV	Class II β chain	Guillemot *et al.* (1988), Zoorob *et al.* (1993)
	13.1	Unknown	
	rRNA	5S and 12S rRNA	Bloom and Bacon (1985)
Other genes			
	BLA	Class II α (5 cM from cluster I)	Kaufman *et al.* (1995a)
	47c	B-G chain (unknown location)	Kaufman *et al.* (1991)
	Y-LBVI	Class II β chain (unknown location)	Zoorob *et al.* (1993)
	B2M	β_2-microglobulin (chromosome 9 or 10)	Riegert *et al.* (1996)
	fB	F_2/factor B (non-MHC)	Koch (1986c)

brates except mammals. Expression of B-L molecules is much more limited, they are present on B cells, monocyte-macrophages and activated T cells (Pink *et al.*, 1977).

B-F molecules are composed of a polymorphic glycosylated transmembrane α-chain noncovalently associated with the invariant β_2-microglobulin subunit. B-L molecules are composed of a nonpolymorphic glycosylated transmembrane α-chain noncovalently associated with a polymorphic glycosylated transmembrane β-chain, much like human HLA-DR and H-2E molecules. Models of B-F and B-L molecules based on human structures and chicken sequences indicate that certain residues which are important for the function of mammalian homologs, including some involved in binding the antigenic peptide and others involved in binding coreceptors (CD8 for class I and CD4 for class II), are highly conserved. In addition, most of the residues responsible for polymorphism are located in the peptide-binding cleft, as in mammals (Pink *et al.*, 1977; Guillemot *et al.*, 1988; Kaufman *et al.*, 1992).

All evidence to date suggests that the polymorphic B-F and B-L molecules have exactly the same function as the mammalian homology – they bind antigenic peptides and present them to T lymphocytes (Vainio *et al.*, 1984, 1988; Maccubbin and Schierman, 1986; Fulton *et al.*, 1995). The cell biology of antigen presentation is apparently also similar, as indicated by the existence of TAP genes and invariant chain molecules (Guillemot *et al.*, 1986; Kaufman *et al.*, 1995a). However, certain MHC haplotypes express far fewer class I molecules on the surface than others, which may have an important but as yet unknown effect on function (Kaufman *et al.*, 1995b).

In contrast to B-F and B-L molecules, the class I-like and class II-like molecules from the RFP-Y locus, Y-F and Y-L, are not highly polymorphic, diverse in sequence, or expressed at high levels (Zoorob *et al.*, 1993; J. Kaufman, unpublished data). In addition, a chicken B-cell specific class I-like molecule has been described (Pickel *et al.*, 1990).

Of the multigene family of B-G genes, many are abundantly expressed on erythrocytes and thrombocytes. Others are expressed at a low level on leukocytes, while others are expressed on intestinal epithelial cells. B-G molecules are generally disulfide-linked dimers of two nonglycosylated transmembrane chains, which each consist of an extracellular immunoglobulin V-like gene and a cytoplasmic region of heptad repeats, indicating coiled coils. The function of B-G molecules is unknown (Kaufman *et al.*, 1991).

Table IV.6.2 Association between MHC and other traits[a]

Trait	*References*
Immune response	
Dinitrophenol (B9-low, B12-high)	Balcarova *et al.* (1973)
(T-G) – A–L	Günther *et al.* (1974)
GAT	Benedict *et al.* (1975)
GAT (gene complementation)	Steadham and Lamont (1993)
Autoimmune disease	
Thyroiditis (severity)	Bacon *et al.* (1974)
Thyroiditis (onset)	Neu *et al.* (1986)
Infection	
Marek's disease	
B19-low, B21-high	Hansen *et al.* (1967)
B2/B21-highest resistance	Briles *et al.* (1980)
Transient paralysis	Schierman and Fletcher (1980)
Sarcoma tumors	
B2-high, B5-low	Collins *et al.* (1977)
B6-high, B13-low	Schierman *et al.* (1977)
B12-high, B4-low	Plachy *et al.* (1979)
Resistance to	
Fowl cholera (B1-high, B19-low)	Lamont *et al.* (1987)
Coccidiosis (B5, B15-high, B2, B12, B19-low)	Ruff and Bacon (1984)
Productivity	
Hatchability (heterozygous matings best)	Briles and Kruger (1955)
Egg production (B15 favored over B19)	Simonsen *et al.* (1982)

[a] Hala *et al.* (1981), Bacon (1987), Gravora (1990), Plachy *et al.* (1992).

Regulation of immune response

The MHC of the chicken regulates immune responsiveness to a variety of antigens, disease resistance, response to autoantigens and productivity (Table IV.6.2; Hala *et al.*, 1981; Bacon, 1987; Plachy *et al.*, 1992; Kaufman *et al.*, 1995b). The importance of other loci outside the MHC for the regulation of all these traits is well documented (Hala, 1977; Palladino *et al.*, 1977; Gavora, 1990).

Genetic studies in different chicken models in many laboratories have established that MHC genotype shows by far the largest influence on the regression of Rous sarcomas. An example of the hierarchy of MHC (B) haplotypes (Plachy, 1988) governing genetic resistance to RSV challenge in defined animal models is given in Figure IV.6.1. To explain these observations, a quantitative difference in the presence of the B-F/L molecules on immune cells has been proposed (Plachy *et al.*, 1992). Some of the data can be explained also in terms for peptide-binding specificity of dominantly-expressed class I molecules (Kaufman *et al.*, 1995b).

That non-MHC genes might be associated with further variation in regression can be inferred from the considerable residual variation in tumor-growth parameters of MHC-genotypic regressor groups in F_2 crosses between two MHC nonidentical lines as well as from differences in regression between inbred lines thought to be MHC identical. The decisive role of T-cell mediated immunity in Rous sarcoma regression has been confirmed. It appears that different antigens (viral envelope antigens, viral group-specific antigens, tumor-specific antigens, embryonic antigens, or other nondefined antigens) might stimulate an immune response to Rous sarcomas in different experimental situations.

Specificity of the response (regression or progression) of particular inbred lines to different strains of RSV has also been acknowledged. RSV tumorigenesis is mediated by the oncogene v-src. It has been demonstrated that v-src DNA alone can induce sarcomas *in vivo*. Good correlation between the growth of RSV-induced and v-src-DNA-induced tumors has been observed in independent chicken genetic models of regressor and progressor lines, and src-specific immunity has also been demonstrated.

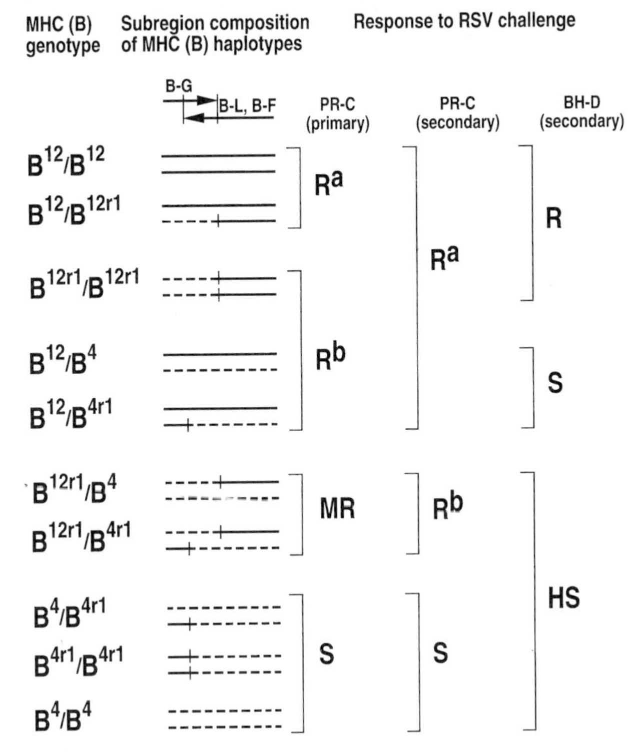

Figure IV.6.1 A hierarchy of the B haplotypes in response to Rous sarcoma virus (RSV) challenge. Chickens from Prague recombinant congenic lines CB(B^{12}/B^{12}), CC(B^4/B^4), CB.R1(B^{12r1}/B^{12r1}), CC.R1(B^{4r1}/B^{4r2}), and their F_1 hybrids and a number of backcross matings of these lines were used. PR-C = Prague strain of Rous sarcoma virus subgroup C; BH-D = Bryan high-titer pseudotype of Rous sarcoma virus subgroup D. R = resistant (regression of tumors $\geqslant$ 95%); MR = moderately resistant (regression of tumors by approximately 50%); S = susceptible (progression of tumors $\geqslant$ 95%); HS = highly susceptible (progression of tumors 100%, rapid growth). a,b = significant ($P < 0.05$) differences in the regression time. Chickens of all genotypes are highly susceptible to primary challenge with BH-D. Data according to Plachy (1988) and Plachy *et al.* (1992).

Species-specific aspects of MHC genetics or protein structure and function

The MHC of the chicken exhibits some significant departures in organization, appears to be simpler and smaller than the MHC of mammals, and it may contain only the 'minimal essential' number of genes that is required for the function of an MHC (Kaufman *et al.*, 1995b).

Chicken MHC (B) harbours a large family of highly polymorphic B-G genes with a unique structure comprising an immunoglobulin-like domain of the variable region type (Ig V-like), a transmembrane domain, and intracytoplasmic heptad domains. These genes are only distantly related to butyrophilin and myelin–oligodendrocyte glycoprotein that are encoded by genes at the periphery of the human and mouse MHC (Kaufman *et al.*, 1991).

All of the class II β, class I α and β_2m, that have been sequenced are very small, with many introns around 100 bp. Some portions of these genes are extremely G + C rich. In most MHC haplotypes, unlike in mammals, only one class I molecule is highly expressed at the protein level in blood and spleen cells. Further difference is in the distribution of antigens – both class I and IV are expressed on erythrocytes.

A second MHC gene cluster (Rfp-Y) with two class I α and two class II β genes was described. Rfp-Y segregates independently of the MHC (B) but could be located on the same microchromosome. A 'hot spot' of recombination between these two complexes has been suggested, perhaps being the very large and repetitive, and therefore recombinogenic, NOR. Nonclassical class I genes have been found outside the MHC in mammals, but the Rfp-Y complex represents a novel type of gene organization with a combination of both class I and class II genes in an independent gene cluster (Miller *et al.*, 1994).

An interesting difference in evolution of class II beta genes of birds and mammals has been described recently (Edwards *et al.*, 1995). The evolutionary tree of avian class II β genes reveals that orthologous relationships have not been retained as in placental mammals and that genes in songbirds and chickens have had very recent common ancestors within their respective groups.

Acknowledgment

This work was supported in part by the grants from the Austrian Ministry of Science and Research (Austrian–Czech Cooperation Program) and from the Agency of the Czech Republic (Project No. 523/96/0670).

7. Cytokines

Cytokines are soluble proteins that are involved in diverse physiological and pathological processes and in regulating immunity. In mammals, a large number of cytokines with well-defined biological functions have been identified and characterized. The genes coding for many of these cytokines have been cloned and expressed. Although many biological equivalents of mammalian cytokines have been identified in birds, only a few avian cytokines have been purified and characterized. In recent years, genes coding for some of the avian cytokines and their receptors have been cloned. These include chicken types I and II IFN, TGF-β, cMGF, 9E3/CEF4 (chicken IL-8), IL-2, IL-1 receptor and the α-chain of the IL-2 receptor. Recently, the gene for turkey type I IFN was also cloned. With the exception of TGF-β, there is little sequence homology between the genes of mammalian and avian cytokines. In general, mammalian and avian cytokines show little cross-species biological activity.

In this section, we have tabulated (see Tables IV.7.1–

Table IV.7.1 Avian cytokines

ATH (avian thymic hormone)	
Homology	80% with human α-parvalbumin (Brewer *et al.*, 1989, 1991) at the amino acid level
Source	Chicken thymus (Barger *et al.*, 1991), chicken muscle (Brewer *et al.*, 1991)
Target	Chicken bone marrow cells (Murthy *et al.*, 1984)
Activity	Promotes maturation (Murthy *et al.*, 1984)
Bursin (bursapoietin)	
Source	Chicken B cells (Guellati *et al.*, 1991; Jankovic, 1987)
Target	Chicken hypothalamo-hypophysial-adrenocortical axis (Guellati *et al.*, 1991)
Activity	Restoration of immune abilities (Audhya *et al.*, 1986) and adrenocorticotropic abilities (Guellati *et al.*, 1991)
FGF-2 (fibroblast growth factor):	
Source	Human recombinant (Hossain *et al.*, 1996)
Target	Chicken sensory neurons, cochleovestibular ganglion of chicken embryos (Hossain *et al.*, 1996)
Activity	Increased explant growth, increased neuroblast migration, increased neurite outgrowth (Hossain *et al.*, 1996)
G-CSF (granulocyte stimulating factor)	
Source	Chicken macrophage cell line (HD11) (Leutz *et al.*, 1988, 1989), chicken serum (Byrnes *et al.*, 1993b)
Target	Chicken myeloblasts (Leutz *et al.*, 1988, 1989), chicken bone marrow cells (Byrnes *et al.*, 1993b)
Activity	Increased colony formation (Byrnes *et al.*, 1993b; Leutz *et al.*, 1988, 1989)
IL-1 (interleukin-1)	
Source	Chicken macrophages (Hayari *et al.*, 1982; Klasing and Peng, 1990; Bombara and Taylor, 1991; Byrnes *et al.*, 1993a), chicken splenocytes (Brezinschek *et al.*, 1990)
Target	Chicken thymocytes (Hayari *et al.*, 1982; Klasing and Peng, 1990; Bombara and Taylor, 1991; Byrnes *et al.*, 1993a)
Activity	Co-mitogen (Hayari *et al.*, 1982; Klasing and Peng, 1990; Bombara and Taylor, 1991; Byrnes *et al.*, 1993a) corticosterone upregulation in chickens *in vivo* (Brezinschek *et al.*, 1990)
IL-2 (interleukin-2, T-cell growth factor)	
Homology	23.8% identical and 46.2% similar to bovine IL-15; 24.5% identical and 44.1% similar to bovine IL-2; 25% identity and 44% similarity to the genetic sequence of bovine IL-15 (Sundick and Gill-Dixon, 1997)
Source	Chicken peripheral blood lymphocytes, spleen cells (Vaino *et al.*, 1986; Fredericksen and Sharma, 1987; Myers *et al.*, 1992)
Target	Chicken lymphocytes (Vaino *et al.*, 1986; Fredericksen and Sharma, 1987; Myers *et al.*, 1992) and T cell blasts (Sundick and Gill-Dixon, 1997)
Activity	Mitogenic (Vaino *et al.*, 1986; Fredericksen and Sharma, 1987; Myers *et al.*, 1992; Sundick and Gill-Dixon, 1997)
IL-3 (interleukin-3)	
Source	Human recombinant (Chang *et al.*, 1990)
Target	Chicken intestine (Chang *et al.*, 1990)
Activity	Anion secretion (Chang *et al.*, 1990)
IL-6 (interleukin-6)	
Source	Duck monocytes (Higgins *et al.*, 1993), chicken ascites (Rath *et al.*, 1995)
Target	Murine cell line (7TD) (Higgins *et al.*, 1993) and B9-hybridoma (Rath *et al.*, 1995)
Activity	Induction of cell proliferation (Higgins *et al.*, 1993; Rath *et al.*, 1995)
IL-8 (interleukin-8, 9E3/CEF4)	
Homology	51% to mammalian IL-8 at the amino acid level, 45% to melanoma growth-stimulatory activity (Stoeckle and Barker, 1990)
Source	Chicken blood lymphocytes (Barker *et al.*, 1993), chicken fibroblasts (Rot, 1991; Gonneville *et al.*, 1991; Barker *et al.*, 1993)
Target	Chicken heterophils (Barker *et al.*, 1993), chicken monocytes (Gonneville *et al.*, 1991)
Activity	Chemotaxis (Rot, 1991; Barker *et al.*, 1993)
Type I IFN (type I interferon)	
Homology	Chicken type I IFN has at the amino acid/nucleic acid level homology with mammalian IFN: α (24/23%), β (20/24%), ω (23/42%), τ (20/43%), γ (3/31%) and flatfish IFN (16/35%) (Sekellick *et al.*, 1994)
Source	Chicken spleen cells (Fredericksen and Sharma, 1987; Dijkmans *et al.*, 1990), recombinant chicken IFN (Sekellick *et al.*, 1994), recombinant turkey IFN (Suresh *et al.*, 1995)
Target	Chicken spleen cells (Dijkmans *et al.*, 1990; Sekellick *et al.*, 1994), chicken embryo fibroblasts (Fredericksen and Sharma, 1987), turkey embryo fibroblasts (Suresh *et al.*, 1995)
Activity	Antiviral (Fredericksen and Sharma, 1987; Dijkmans *et al.*, 1990; Sekellick *et al.*, 1994; Suresh *et al.*, 1995)

(continued)

Table IV.7.1 *Continued*

IFN-γ (interferon-γ, macrophage activating factor)	
Homology	35% with equine and 32% with human IFN-γ; 15% with type I chicken IFN (Digby and Lowenthal, 1995) at the protein level
Source	Chicken T-cell line (Lowenthal *et al.*, 1995), recombinant chicken IFN-γ (Digby and Lowenthal, 1995)
Target	Chicken macrophages (Kaspers *et al.*, 1994; Digby and Lowenthal, 1995; Lowenthal *et al.*, 1995)
Activity	Induction of nitric oxide production (Digby and Lowenthal, 1995; Lowenthal *et al.*, 1995); upregulation of MHC class II expression (Kaspers *et al.*, 1994)
cMGF (chicken myelomonocytic growth factor)	
Homology	41% to human IL-6 (Leutz *et al.*, 1988, 1989), 56% to human G-CSF (Sterneck *et al.*, 1992) at the amino acid level
Source	Chicken macrophages (Amrani *et al.*, 1986), chicken fibroblasts (Samad *et al.*, 1993), human recombinant (Amrani, 1990), chicken recombinant (Leutz *et al.*, 1988, 1989)
Target	Chicken hepatocytes (Amrani *et al.*, 1986; Amrani, 1990; Samad *et al.*, 1993)
Activity	Increased fibrinogen synthesis (Amrani *et al.*, 1986; Amrani, 1990), increased fibronectin synthesis (Samad *et al.*, 1993)
MIF (migration inhibitory factor)	
Homology	Similar in physicochemical properties to human MIF (Joshi and Glick, 1990)
Source	Chicken lymphocytes (Joshi and Glick, 1990), embryonic chicken lens (Wistow *et al.*, 1993)
Target	Chicken peripheral blood lymphocytes (Joshi and Glick, 1990)
Activity	Decreased macrophage migration (Joshi and Glick, 1990)
PDGF (platelet-derived growth factor)	
Source	Recombinant human PDGF (Sieweke *et al.*, 1990)
Target	Chicken fibroblasts (Sieweke *et al.*, 1990)
Activity	Mitogenic (Sieweke *et al.*, 1990)
TGF-β (transforming growth factor-β)	
Activity	Regulation of proliferation and differentiation of a variety of cells (Roberts and Sporn, 1990), mitogenic (Sieweke *et al.*, 1990)
TGF-β1	
Homology	Identical to human TGF-β1, 71% identity to amino acid sequences of processed human TGF-β2, 78% and 80% to chicken TGF-βs 3 and 4, respectively (Jakowlew *et al.*, 1988a)
Source	Chicken chondrocytes (Jakowlew *et al.*, 1988b)
Target	Chicken fibroblasts (Jakowlew *et al.*, 1988b)
Activity	Mitogenic (Jakowlew *et al.*, 1988b)
TGF-β2	
Source	Chicken embryonic fibroblasts (Burt and Paton, 1991)
TGF-β3	
Homology	76% with human TGF-β1, 79% with human TGF-β2 (Jakowlew *et al.*, 1988b; Sieweke *et al.*, 1990) at the amino acid level
Source	Chicken chondrocytes, chicken embryo fibroblasts, Rous sarcoma virus-transformed chicken embryonic fibroblasts (Jakowlew *et al.*, 1988b), chicken ascites (Rath *et al.*, 1995)
TGF-β4	
Homology	81%, 64% and 71% to human TGF-β1, human TGF-β2 and chicken TGF-β3, respectively (Jakowlew *et al.*, 1988c) at the amino acid level
Source	Chicken embryo chondrocytes (Jakowlew *et al.*, 1988c), chicken embryo fibroblasts (Burt and Paton, 1991)
TNF-α (tumor necrosis factor)	
Source	Chicken macrophages (Klasing and Peng, 1990; Qureshi *et al.*, 1990; Byrnes *et al.*, 1993a; Zhang *et al.*, 1995), chicken neurons (Gendron *et al.*, 1991)
Target	Chicken and mouse cell lines (Klasing and Peng, 1990; Qureshi *et al.*, 1990; Byrnes *et al.*, 1993a; Zhang *et al.*, 1995)
Activity	Cytolysis (Klasing and Peng, 1990; Qureshi *et al.*, 1990; Byrnes *et al.*, 1993a; Zhang *et al.*, 1995)
Thymulin	
Homology	To synthetic mammalian thymulin (Chang and Marsh, 1993), thymulin-like activity of avian sera (Marsh *et al.*, 1991)
Source	Chicken thymus (Marsh and Scanes, 1993)
Target	Chicken T cells (Marsh and Scanes, 1993)
Activity	Induction of maturation (Marsh and Scanes, 1993)
Thrombocyte colony stimulatory factor	
Source	Spleen-conditioned medium (Nicolas-Bolnet *et al.*, 1995)
Target	Late thrombocyte progenitors (Nicolas-Bolnet *et al.*, 1995)
Activity	Colony formation (Nicolas-Bolnet *et al.*, 1995)

Table IV.7.2 Avian cytokine genes

Cytokine	*Protein size*	*Cross-species activity*	*Gene structure cloned*	*Primer sequence used*	*Reference*
9E3/CEF4	11 kDa	Turkey	cDNA	Not published	Sugano *et al.* (1987), Suresh *et al.* (1995)
cIFN type I	193 aa	Turkey	cDNA	Sense: 5′-TTGGCCATCTATGAGA TGCTCCAGMANATHTT-3′ Antisense: 5′-CGGACCACTGTCCA NGCRCA-3′	Sekellick *et al.* (1994)
tIFN type I	162 aa	Chicken	cDNA	Sense: 5′-ATGGCTGTGCCTGC AAGCCCA-3′ Antisense: 5′-AGTGCGCGTGTTGC CTGTGA-3′	Suresh *et al.* (1995)
IFN-γ	145 aa		cDNA	Not published	Digby and Lowenthal (1995)
IL-2	121 aa (mature protein)		cDNA	Not published	Sundick and Gill-Dixon (1997)
cMGF	24 kDa		cDNA	Not published	Leutz *et al.* (1989)
MIF	115 aa 10 kDa		cDNA	5′-CAGGATCCCGATGTT CA(TC)C(GA)TA(AC) ACACCAA-3′ (5064) 5′-TAGTCGACGGT (GATC)GA(GA)TT(GA)TT CCA(GC)CC-3′ (5065)	Wistow *et al.* (1993)
TGF-β1	25 kDa dimer		cDNA	Human cDNA-probe	Jakowlew *et al.* (1988a)
TGF-β2	112 aa monomer		Genomic DNA	Simian TGF-β 2 cDNA-primer: TCCATCTACAA CAGCACCAGGGACTT GCTCCAGGAGAAG Chicken cDNA-TGF-β2-primer: CGAGACCC GCGCCTTTGAGTT	Burt and Paton (1991)
TGF-β3	112 aa monomer		cDNA	Human cDNA-probe	Jakowlew *et al.* (1988a, c)
TGF-β4	114 aa monomer		cDNA	Human cDNA-probe	Jakowlew *et al.* (1988b)

Table IV.7.3 Avian cytokine receptors characterized

Receptor	*Expressing cells*	*Immunological reagent*	*Reference*
CD25 (α-chain of IL-2)	Thymocytes, peripheral T cells	Monoclonal antibody INN-CH-16	Hála *et al.* (1986); Schauenstein *et al.* (1988); Fedecka-Bruner *et al.* (1991)
IL-1 (type I)	Chicken embryo fibroblasts		Guida *et al.* (1992)

Table IV.7.4 Methods used to detect avian cytokines

Cytokine	*Bioassay*	*Immunoassay*	*Susceptible cells/commercially available immunoreagents*	*Reference*
TNF-α	Cytotoxicity assay		L929 mouse cells	Byrnes *et al.* (1993a, b)
			CU-25	Qureshi and Miller (1991)
			MDTC-RP-9	van Ostade *et al.* (1991)
			CHCC OU-2	Zhang *et al.* (1995)
		Immunocytochemistry	Polyclonal rabbit-anti mouse TNF-α (Genzyme)	Wride and Sanders (1993)
		Western blot	Monoclonal rat-anti mouse TNF-α (Endogen Inc./ UBI Inc.)	
IL-1	IL-1 proliferation assay		Thymocytes	Weiler and von Bulow (1987)
			Human fibroblasts	Loppnow *et al.* (1989)
Type I IFN	Antiviral IFN-assay		Chicken embryo-fibroblasts	Sekellick *et al.* (1994)
IFN-γ	Nitrite assay		Chicken macrophages (HD11)	Lowenthal *et al.* (1995)
Thymulin	Thymulin assay			Coudert *et al.* (1979)
		Immunoassay		Marsh *et al.* (1991)
IL-2	T cell proliferation assay		T cell blasts	Vaino *et al.* (1986), Kaplan *et al.* (1993)
	Spleen blast cell proliferation assay			Kaplan *et al.* (1992), Sundick and Gill-Dixon (1997)
TGF-β	Radioreceptor assay on A549 cells			Wakefield (1987)
	Mink lung epithelial cell inhibition assay	Neutralizing ab, also against TGF-β		Danielpour *et al.* (1989)
IL-6		hIL-6 ELISA (Intertest 6 × human IL-6 Elisa Kit)		
	B9 hybridoma proliferation assay			Rath *et al.* (1995)
	7TD-1 hybridoma proliferation assay		7TD-1 murine cell line	Higgins *et al.* (1993)
cMGF	Colony formation	Polyclonal antiserum	E26-transformed myeloblasts, bone marrow granulocytes	Leutz *et al.* (1984)

IV.7.4) the data on avian cytokines and cytokine receptors. Excellent summaries of the data on avian cytokines were published by Klasing (1994) and Lillehoj (1993). There is considerable current interest in avian cytokines and new findings are imminent.

8. Complement

Introduction

Lytic complement (C) activity in avian sera was first reported in the early part of the century (Muir, 1911). More recently it has been demonstrated that the sera of avian species contain macromolecules that structurally and/or functionally resemble mammalian C components. Because both birds and mammals encounter a battery of comparable pathogens, these observations would appear to suggest that these groups have evolved C systems that are similar: indeed, in some respects they are. However, although the avian C system is not as well understood as that of mammals, there are indications that the C systems of species in these groups might possess fundamental differences. This section presents information on individual components, activation pathways, and the ontogeny of the avian C system, and also provides comparisons between the C systems of birds and mammals.

Avian complement components

The molecular properties of avian C components which have been characterized are listed in Table IV.8.1. Chicken

Table IV.8.1 Molecular properties of avian complement components

Component	Species	*Whole molecule (kDa)*	*Separate chains (kDa)*	*Approximate serum concentration (μg/ml)*	*Reference*
C3	Chicken	185	118 68		Laursen and Koch (1989)
		180	116 67	500	Mavroidis *et al.* (1995)
	Quail	183	110 73		Kai *et al.* (1983)
Factor B	Chicken	95		50–100[a]	Koch (1986b)
Clq	Chicken	504	6 × 25.9 6 × 24.8 6 × 24.8	50–70	Yonemasu and Sasaki (1986)

[a] Estimated from data in Koch (1986b).

and quail C3 are similar in size to their counterparts in other vertebrates; they also possess a reactive thiolester bond, and upon activation are cleaved into fragments that resemble mammalian C3a and C3b (Kai *et al.*, 1983; Laursen and Koch, 1989). The cDNA sequence of chicken C3 has recently been reported (Mavroidis *et al.*, 1995) and the primary structure deduced from it shows identity with human C3 and those of other species in several functionally important regions.

When chicken factor B is activated it is cleaved at sites identical to those in mammalian factor B, producing 37 kDa and 60 kDa fragments (Koch, 1986b). This component also exhibits genetic polymorphism, but does not appear to be linked to the MHC, as is the case in mammals (Koch, 1986c). Chicken and human Clq are also similar. Both are large macromolecules containing six subunits with each subunit composed of three distinct polypeptide chains (Yonemasu and Sasaki, 1986). These Clq molecules also resemble one another in carbohydrate, hydroxyproline, and hydroxylysine content, as well as in their susceptibility to degradation by collagenase.

C4 activity has been detected in chicken serum using human and guinea pig reagents (Barta and Hubbert, 1978) and in pigeon serum using C4-deficient guinea pig serum (Koppenheffer and Russell, 1986). Because the activity measured in both instances was very low, it is uncertain whether avian C4 is only minimally compatible with mammalian C reagents or whether these avian sera contain low amounts of this component. It is also possible that the putative C4 activity might be a result of interactions between C components other than avian C4. Nevertheless, Wathen *et al.* (1987) found that chicken plasma contains a protein that cross-reacts with an antihuman C4 antiserum. After electrophoresis, plasma of adult females exhibited three bands while that of immature birds and adult males exhibited two bands. This, together with the low levels of C4 activity detected in chicken and pigeon sera, suggests the possibility that avian species possess a functional C4 component.

Immunoelectrophoretic and hemolytic tests were used to detect C6 activity in a seabird, the manx shearwater (Whitehouse, 1988). Interestingly, this component was found to be highly polymorphic with seven distinct allelic variants. Additional studies have demonstrated that enzymes present in chicken plasma possess cleavage activity equivalent to that of mammalian factor I and its cofactor (factor H) (Kaidoh and Gigli, 1987). This is not surprising in view of the studies on quail (Kai *et al.*, 1983) and chickens (Laursen and Koch, 1989) which showed that C3 of these species is degraded by serum enzymes, and the finding that factor I cleavage sites are conserved in chicken C3 (Mavroidis *et al.*, 1995). The presence in chicken serum of the factor B-activating enzyme, the serine esterase factor D, has been inferred by the loss of C3-activating capacity which occurs in serum depleted of low molecular weight substances by gel filtration chromatography (Jensen and Koch, 1991).

Complement activation

Table IV.8.2 lists a number of properties exhibited by avian C. As in other vertebrate species, avian antibody-independent alternative complement pathway (ACP) activity has been demonstrated with the use of heterolo-

Table IV.8.2 Characteristics of avian complement

- Antibody-independent ACP activity is demonstrable *in vitro*
- Microbial parasites activate the ACP *in vivo*
- The ACP is activated by avian antibodies
- Hemagglutinating levels of Ab produce maximum lysis
- CCP activity is difficult to demonstrate
- Both the ACP and CCP might utilize factor B
- Cobra venom factor does not uniformly activate avian C
- The level of hemolytic C in chickens is MHC-linked

gous erythrocytes (Ohta *et al.*, 1984). It is responsible for the bactericidal activity of turkey and chicken plasma (Snipes and Hirsh, 1986; Wooley *et al.*, 1992), and has also been shown to provide protection in both embryos and chicks against infection by a strain of fowl pox virus (Ohta *et al.*, 1984), as well as natural resistance against infection by a protozoan parasite in chickens (Kierszenbaum *et al.*, 1976).

An interesting feature of avian Abs is their ability to activate homologous C by a pathway which exhibits characteristics of the ACP (Koch, 1986a; Koppenheffer and Russell, 1986). It is important to note that in several studies which employed hemolytic tests, dilutions of avian antisera or antibodies that produced maximal lysis caused marked agglutination of target cells (Gewurz *et al.*, 1966; Gabrielsen *et al.*, 1973a; Koppenheffer and Russell, 1986). Whether similar levels of target cell sensitization occur *in vivo* is questionable, thus raising the possibility that activation of the ACP by avian antibodies may have little physiological significance. This is puzzling in light of the fact that activation of an avian classical complement pathway (CCP) by homologous antibodies has not been demonstrated, although mammalian antibodies have been shown to activate chicken Clq (Yonemasu and Sasaki, 1986) as well as a lytic pathway that requires both Ca^{2+} and Mg^{2+} (Barta and Barta, 1975; Koch, 1986a) – a characteristic of the vertebrate CCP. Moreover, in the latter studies by Koch (1986a) depletion of factor B abolished lytic activity, leading to the suggestion that chicken factor B participates in both the CCP and ACP. This idea is consistent with the recent finding that the chicken factor-B cleavage fragment, Ba, has a similar degree of amino acid sequence identity with Ba and the low molecular weight C2 fragment in the human and mouse (Kjalke *et al.*, 1993). However, it cannot explain the observation that cobra venom factor (CVF) depletes ACP activity in this species, while the activity of the CCP, as measured by the lysis of RBC sensitized with rabbit antibodies, is unaffected (Ohta *et al.*, 1984). The latter finding suggests that a component unique to the chicken ACP is inactivated by CVF, and thus the existence of two distinct C pathways cannot be ruled out.

MHC and ontogeny

While no regions of the avian MHC have been identifed that are homologous to those which encode mammalian C components, linkage has been demonstrated between hemolytic C levels and the chicken MHC (Chanh *et al.*, 1976; Shen *et al.*, 1984). A dominant gene appears to control high serum C levels, but neither the locus nor gene product responsible for producing this effect has yet been identified.

C activity has been detected early in avian development. In Japanese quail, hemolytic C activity was demonstrated in 10-day embryos and C3 was found to be present in 7-day embryos (Kai, 1985). In chickens, Gabrielsen *et al.* (1973b) showed that sera of 13-day embryos possessed C activity. In both species, C activity increased rapidly during the first week after hatching, and following a plateau, it increased steadily for several weeks until adult levels were attained.

Summary and conclusions

Several key avian C components have been characterized and shown to resemble their mammalian counterparts. While the avian ACP appears to function in a manner similar to that of other vertebrates, the nature of the C pathways activated by avian antibodies remains uncertain. The gene(s) encoding chicken factor B is not linked to the MHC, although linkage has been demonstrated between hemolytic C levels and the MHC in this species. Birds show evidence of C activity during embryonic development, and a similar ontogenic pattern of hemolytic C activity occurs in chickens and quail. Cellular C receptors have been given little attention since the work of Thunold (1981) and there is no information available concerning the biological activities of the small C-cleavage fragments. Moreover, useful information about the avian C system has been obtained from a limited number of species. It is evident that much work remains to be done in order to better understand this important immune effector system in birds.

9. Ontogeny of the Immune System

Introduction

Chicken has long been a favorite model of both developmental biologists and immunologists. It is therefore not surprising that many of the key insights into the ontogeny of immune system have come from studies on the chicken. One obvious reason is the accessibility of the avian embryo to manipulation and study at any stage of development. Another advantage is the clear dichotomy of lymphoid development in the chicken. Both T and B cells develop in distinct organs, thymus and the bursa of Fabricius, respectively (Table IV.9.1). This contrasts with mammals, in which B cell development is spread to fetal liver and multiple sites of bone marrow. Combined, these two features have guaranteed the appeal of chicken as an experimental animal. In recent years many of the molecules important in the immune system have been characterized on both genetic and protein level, producing tools with which to probe the development of the immune system with increasing accuracy. These studies have shown that while many of the developmental pathways are similar to mammalian species, some aspects are unique to the chicken. In this section we review the development

Table IV.9.1 Ontogeny of the main cell lineages of the immune system

Cell type	*Main origin*	*First observed*	*Notes*
B cells	Bursa of Fabricius	Lymphoid follicles in bursa develop from ED8	Rearrangement of Ig genes occurs prebursally. Diversification takes place within bursa by somatic gene conversion
T cells	Thymus	$\gamma\delta$ thymocytes, ED12 $\alpha/V_{\beta}1$ thymocytes, ED15 $\alpha/V_{\beta}2$ thymocytes, ED18	Thymus is seeded in three waves. Unlike the Ig genes, diversification of TCR genes results from recombination and junctional variation
Phagocytes	Bone marrow	Phagocytic activity in yolk sac on ED3, in liver on ED4	Main phagocytic cell types are monocytes/macrophages and heterophils
NK cells	Bone marrow	In spleen, ED8	NK cells are thymus-independent, characterized by cytoplasmic CD3

of avian immune system – both its common and unique features.

Early precursors

Early hematopoiesis in the chicken begins in yolk sac on embryonic day (ED) 1–2 and probably plays an important role in embryonic erythropoiesis (Martin *et al.*, 1978). It is unlikely that cells originating in the yolk sac participate in the generation of lymphoid precursors. Intra-embryonic hematopoiesis begins on ED4. Hematopoietic stem cells in the early embryo localize first to intra-aortic cell clusters and at a later stage to para-aortic mesenchyme, ventrally of the aorta (Lassila *et al.*, 1978; Dieterlen-Lievre and Martin, 1981; Cormier *et al.*, 1986; Toivanen and Toivanen, 1987; Cormier and Dieterlen-Lievre, 1988). At present, the stem cells can only be identified functionally. When transferred into irradiated hosts, cells from para-aortic mesenchyme are able to generate both B and T cells of the donor type. During normal development these stem cells seed the primary lymphoid organs and thus generate the various lymphocyte populations. Bone marrow is also seeded by the stem cells and, after hatching, becomes a major site of hematopoiesis. Its role in lymphopoiesis, however, is less clear in the chicken than in mammals, especially during embryonic period (Toivanen and Toivanen, 1987).

B-cell development

The bursa of Fabricius develops as an outgrowth of cloacal epithelium and is seeded between ED8 and ED14 by stem cells originating in the para-aortic area (Toivanen and Toivanen, 1973; Houssaint *et al.*, 1976). In irradiated hosts, these stem cells are capable of reconstituting the entire B-cell lineage. These cells can first be found in embryonic spleen, from which they migrate to bursa and give rise to lymphoid follicles. The rearrangement of immunoglobulin genes occurs prior to the entry into the bursa, e.g., in yolk sac, spleen, blood and bone marrow (Ratcliffe *et al.*, 1986; Weill *et al.*, 1986; Mansikka *et al.*, 1990a). Unlike mammalian species, however, rearrangement in chickens does not generate significant diversity in the immunoglobulin genes. Because the chicken only has one functional V and J gene segment in both the heavy and light chain locus, each of the B-cell precursors expresses practically identical immunoglobulins (Reynaud *et al.*, 1985). It has been suggested that when expressed on cell surface, this prototype immunoglobulin molecule can bind to an as yet unknown self ligand, triggering proliferation and further differentiation (Masteller and Thompson, 1994).

The bursa has an essential role in B-cell development because it is the site of immunoglobulin gene diversification and, in bursectomized animals, only oligoclonal antibodies are observed (Weill *et al.*, 1986; Reynaud *et al.*, 1987; Mansikka *et al.*, 1990b). The precursors entering bursa give rise to lymphoid follicles that start with only a few precursors but after proliferation contain approximately 100 000 cells each (Pink *et al.*, 1985). Within these follicles the developing B cells undergo gene conversion, a process in which parts of nonfunctional pseudogenes are copied into the rearranged immunoglobulin gene (Weill *et al.*, 1986; Reynaud *et al.*, 1987). The heterogeneity of the developing B cells within the follicles increases from ED 15 onwards, and almost all of the immunoglobulin gene diversity in the chicken is due to gene conversion. Although rearrangement is clearly independent of primary lymphoid organs, gene conversion only takes place in the bursa.

Similar to mammals, the primary lymphoid organs of chicken are sites of extensive cell death (Motyka and Reynolds, 1991; Neiman *et al.*, 1991). In the bursa, it has been estimated that only 5 % of the total cell numbers survive to form the mature B-cell population (Lassila, 1989). It has been reported that bursal cells undergoing apoptotic cell death down-modulate the expression of surface immunoglobulin (Paramithiotis *et al.*, 1995). It is thus probable that one reason leading to cell death is inability to express a functional immunoglobulin. Other possible reasons may include expression of autoreactive antigen receptor, but details of the B-cell repertoire selection are poorly understood. The minority of cells which survive start to emigrate out of bursa around hatching.

T-cell development

It is currently held that practically all T-cell development in the chicken is thymus-dependent (Dunon *et al.*, 1993a, b). Thus, unlike in the mouse, in which extrathymic T-cell populations have been described, in chicken all T cells, including the intestinal intraepithelial T cells, originate in thymus from which they emigrate to form the peripheral cell pool. Thymus is seeded in three waves of blood-borne precursors (Jotereau *et al.*, 1980). The first wave starts to arrive in thymus on ED6, the second on ED12 and the third on ED18. Each of these waves gives rise to all T-cell subsets. The first T-cell receptor-expressing thymocytes are $\gamma\delta$ TCR^+, first observed on ED12. $\alpha\beta$ TCR^+ thymocytes expressing $V_\beta1$ family TCR genes develop next, on ED15, and $V_\beta2$–TCR^+ thymocytes on ED18 (Bucy *et al.*, 1990b; Cooper *et al.*, 1991). The mature T cells probably emigrate to periphery in distinct waves, as well. The first peripheral TCR^+ cells can be observed in spleen, on ED15. These cells express the $\gamma\delta$ TCR, $\alpha\beta$ T cells appear 3 days later.

The genes encoding the TCR chains in the chicken have been described, and their main characteristics resemble the mammalian genes (Tjoelker *et al.*, 1990; Göbel *et al.*, 1994b; Kubota *et al.*, 1995; Six *et al.*, 1995; Chen *et al.*, 1996). Thus, they consist of multiple gene segments which can, on the basis of sequence comparison, be divided into families. As in mammals, and unlike in the avian immunoglobulin genes, diversity of the TCR genes results mainly from rearrangement and junctional variation. The mechanism determining the segregation of thymocytes to either $\gamma\delta$ or $\alpha\beta$ T-cell lineage is unclear. It has been suggested that $\alpha\beta$ thymocytes may represent those cells which fail to express $\gamma\delta$ TCR (de Villartay *et al.*, 1988). In these cells the TCR δ locus, situated within the α locus, is deleted during rearrangement. Support for this scenario is provided by studies of chicken $\alpha\beta$ T cell lines, in which δ locus was shown to be deleted from both alleles (Kubota *et al.*, 1995). In contrast, the TCR γ genes have been reported to be frequently rearranged and transcribed in $\alpha\beta$ T-cell lines (Chen *et al.*, 1996), which argues against the model of transcriptional silencing of γ genes being the determinative factor.

Studies in mouse have shown that during the maturational process within the thymus T cells are subjected to negative and positive selection (Nossal, 1994; von Boehmer, 1994). These events result in the death of most of the thymocytes and modify the repertoire of the surviving cells to imprint MHC restriction and self tolerance. Data obtained in the chicken suggest that $\alpha\beta$ T cells develop through a similar pathway (George and Cooper, 1990; Cooper *et al.*, 1991). They pass from CD4/CD8 double-negative stage to a double-positive stage in the thymic cortex, and thereafter one of the coreceptors is down-modulated. The $\alpha\beta$ TCR is first expressed at low levels, and at a certain stage the cells are susceptible to TCR down-modulation triggered by receptor cross-linking with mAb. It is less clear what kind of selection, if any, the $\gamma\delta$ TCR^+ thymocytes are subject to (George and Cooper, 1990; Cooper *et al.*, 1991). They express high levels of TCR immediately and appear to pass quickly through the thymus. Cross-linking of $\gamma\delta$ TCR has no effect on its expression, and cyclosporin A treatment, which arrests the development of $\alpha\beta$ T cells, does not affect $\gamma\delta$ T-cell development. After thymic maturation both $\alpha\beta$ and $\gamma\delta$ T cells emigrate to all lymphoid tissues in the periphery. An interesting exception is the homing pattern of $\alpha\beta$ T cells expressing TCR $V_\beta2$-gene family (Lahti *et al.*, 1991). These cells are very rare or absent in the intestinal epithelium and perhaps because of this cannot provide help for antibody responses of the IgA class (Cihak *et al.*, 1991). The reasons for this difference between the two $\alpha\beta$ T cell lineages are not known.

Other cell types

Nonlymphoid cells participating in immune function are also similar to those described in mammals (Powell, 1987b). The main phagocytic cells are monocytes/macrophages and, in analogy to mammalian neutrophils, heterophils. These cells are bone marrow derived and are continuously released to circulation. Such macrophage-like cells as Kupffer cells of liver and osteoclasts, although originally from bone marrow, may also divide locally. Natural killer cells are a distinct lymphoid lineage, characterized in the chicken as cells expressing at least some components of CD3 in cytoplasm but not on the cell surface (Göbel *et al.*, 1996a). First observed in spleen on ED8, the NK cells probably share a common precursor with T and B cells, but their development is thymus-independent.

Concluding remarks

Although a significant part of what is known about the ontogeny of immune system has been obtained from studies in the chicken, many aspects remain unclear. Lack of suitable markers for the hematopoietic stem cells is a major obstacle to understanding the early events of lymphopoiesis. B-cell precursors entering the bursa have already rearranged their immunoglobulin genes, but for T-cell precursors the degree of lineage commitment prior to reaching thymus remains a matter of speculation. The developmental pathways within the primary lymphoid organs and the formation of mature lymphocyte repertoire remain areas for further study. The development of $\gamma\delta$ T cells is of considerable interest because the chicken is a species with a very high frequency of these cells. The recent characterization of genes encoding the TCR chains and the growing list of molecules defined by monoclonal antibodies will undoubtedly bring new insights. In this way, the chicken continues to contribute to the understanding of developmental immunology in general.

10. Immunocompetence of the Embryo and the Newly-Hatched Chick

Physiology

The microenvironment of the bursa of Fabricius is required for the generation of antibody diversity (gene conversion) and thus for the development of the humoral immune response (Huang and Dreyer, 1978; Weill and Reynaud, 1987; Ratcliffe, 1989; Benatar *et al.*, 1992; Vainio and Imhof, 1995). Gene conversion starts between embryonation days (ED) 15 and 17 and continues until approximately 6 weeks old. Mature lymphocytes from the bursa then seed secondary lymphoid organs just prior to and post hatching. If the bursa is destroyed before the majority of gene conversion events have occurred, serum antibody levels may approach normal levels but the antibodies will be essentially non-functional (Granfors *et al.*, 1982). In the chicken, the bursa loses it's critical role around 3 weeks, although involution does not occur until sexual maturity.

The thymus is the primary organ for production and maturation of T cells. By the time of hatch most peripheralization of T cells has been accomplished and thymectomy post hatching does not destroy cell-mediated immunity (Cooper *et al.*, 1991). Thymopoetin is a polypeptide that is secreted by thymic epithelial cells, probably thymic nurse cells, and it selectively induces the differentiation of committed precursor cells to early T cells (Tempelis, 1988). This action is mediated by an intracellular cAMP signal. Thymic nurse cells employ positive or negative (deletion or inactivation of self-reactive T cells) selection of immature T-cells by engulfing the thymocytes in vacuoles lined by epithelial cell membranes (Rieker *et al.*, 1995). The number of thymocytes entering a single nurse cell appears to be species specific and is limited to four in the chicken. Normal embryos also have suppressor T-cells that suppress both humoral and cell mediated immunity (Tempelis, 1988). These cells are lost shortly after hatch.

Spleen germinal center formation is age dependent and reaches a maximum number at 4–5 weeks old. Lymphoid aggregates in the intestine (Peyer's patches) are absent at hatch, become visible at days 1–2 post hatching (PH), reach maximal numbers at around 16 weeks old and almost completely disappear by 52 weeks old (Lillehoj and Trout, 1996). In addition, the distribution of various T-cell subpopulations in the intestine do not reach adult proportions until 4–5 weeks PH (Lillehoj, 1991; Lillehoj and Trout, 1996). It is unclear how the differing distribution of T-cells in the early post hatching period affects the capability of the chick to mount an effective cellular immune response. Although early reports suggested that chicks in the first 2 weeks post hatching lack adequate immunologic responsiveness (Droege, 1971, 1976; Seto, 1981), the recent success of embryo vaccination indicates that the immunological immaturity of the late embryo may not have a major practical implication.

Factors that can impact the developing bursa and thymus include genetics, infectious agents, nutrition, environment and chemical and/or hormonal interventions. The influence of the MHC on disease resistance has been well documented in the chicken (Dietert *et al.*, 1991). The MHC may also exert an influence on the developing immune organs. Both the size of the bursa and the thymus and the number of mature lymphocytes produced by these organs are influenced by the number of MHC copies found in each cell. In research using MHC dosage mutants (disomic, trisomic and tetrasomic birds), Hemendinger *et al.* (1992), found that the greater the number of MHC copies per cell, the greater the cellular expression of MHC class II antigens and the more severe the reduction in bursal and thymic organ weights and T- and B-lymphocyte numbers.

Infectious agents that can impact the developing immune system are numerous and include infectious bursal disease virus (IBDV) (Saif, 1991), chicken infectious anemia virus (Pope, 1991), Marek's disease virus (MDV) (Rivas and Fabricant, 1988), hemorrhagic enteritis (HE) virus (Sharma, 1991) and mycotoxins (Corrier, 1991). Many of these agents are acquired vertically or horizontally within the first week of life, often causing permanent immunosuppression. Early lymphoid organ development can be impacted by nutritional deficiencies. Marsh *et al.* (1986) demonstrated that experimentally induced dietary deficiencies of vitamin E and selenium can impair bursal development. Environmental factors that affect immunocompetence have recently been reviewed by Dietert and Golemboski (1994). Of particular importance to the young chick is the amount of time spent in the hatcher. Wyatt *et al.* (1986) reported that holding chicks in the hatcher for 30 h post hatching had significant detrimental effects on both bursa and spleen weights and increased the severity of vaccine reactions. Finally, experimentally induced immunosuppression by cyclophosphamide (Misra and Bloom, 1991) and testosterone (Olah *et al.*, 1986) have been well documented.

Nonspecific immune defenses

Although no specific markers for NK cells have been identified in the chicken, numerous reports state that cells possessing NK-like activity do exist (Fleischer, 1980; Leibold *et al.*, 1980; Sharma and Okazaki, 1981; Chai and Lillehoj, 1988). These cells have been isolated from the intestine, bursa, spleen, thymus and peripheral blood. NK-like activity increases with age and does not reach adult levels until approximately 6 weeks post hatching, depending upon the genetic lineage (Lillehoj and Chai, 1988). Yamada and Hayami (1983) reported that α-fetoprotein in quail or chicken amniotic fluid stimulated suppressor cells which then reduced NK activity. In another report, injection of thymulin caused a reduction in NK activity in chickens that were infected with MDV (Quere *et al.*, 1989). NK cell activity may be important in the resistance to MDV (Sharma, 1981), a disease commonly acquired in the early post hatching period.

Cells of the monocyte/macrophage lineage form early in the development of the embryo, around day 3 (Comeir *et al.*, 1986; Dieterlen-Lievre, 1989) and exhibit enough function to respond to some bacterial pathogens during the second week of incubation (Klasing, 1991). The availability of specific antibody and/or complement can be a limiting factor in early embryonic macrophage responsiveness, and the rapid immunologic response to certain pathogens immediately post hatching has been associated with an increase in complement availability (Powell, 1987; Klasing, 1991).

Embryo vaccination

Chicken embryo vaccination is unique as it is the first widespread commercial use in any species of prenatal vaccination. The concept was initially devised by Sharma and Burmester (1982) to protect chicks from virulent MDV exposure that occurred too early for adequate protection by conventional at-hatch vaccination. Vaccination of 18-day-old embryos with turkey herpesvirus (HVT) protected 80–90% of chicks from challenge-exposure to virulent MDV at 3 days post hatching compared with 16–22% of chicks vaccinated at hatch with HVT. No deleterious effect on hatching was observed in these trials. Timing of vaccination was critical because embryos inoculated with HVT prior to ED 16 sustained extensive embryonic and extraembryonic tissue damage (Longenecker *et al.*, 1975; Sharma *et al.*, 1976). Embryo vaccination also has the additive benefits of ensuring the precise delivery of vaccines to each individual and of labor savings via automation of the delivery system. Experimental vaccination in chickens has been successful for infectious bronchitis virus (Wakenell and Sharma, 1986) and IBDV (Sharma, 1985) alone or in combination with HVT, and HE and Newcastle disease virus (NDV) in turkeys (Ahmad and Sharma, 1993). The commercial use of embryo vaccination for protection against MDV (Sharma *et al.*, 1995) is widespread, with 75–80% of all commercial broilers being vaccinated embryonically.

Maternal antibodies

IgM and IgA are located in the amniotic fluid, thus swallowing by the embryo corresponds to colostrum ingestion in mammals (Rose *et al.*, 1974), although minimal transfer occurs. IgG is found in the yolk and begins to be absorbed in the late stages of embryonic development until shortly after hatch (Kowalczyk *et al.*, 1985; Powell, 1987). Failure of absorption can affect transfer of maternal immunity and results in an immunocompromised chick. Chick IgG half-life is approximately two times that of the adult bird in order to compensate for the time it takes to fully absorb the yolk. Serum IgA appears at approximately 10 days old and IgM at 4 days old. The amount of antibody transferred from hen to chick can vary with the age of the hen and the point of time in lay, and also with the titer level in the hen's serum. Increasing a hen's serum titer will not necessarily stimulate a corresponding degree of increase in titer in the embryo (P. S. Wakenell, unpublished data). Although maternal antibodies provide variable degrees of protection against pathologic organisms (Powell, 1987), interference with certain embryonic or at-hatch vaccines can be substantial. Of particular importance to the commercial industry is IBDV infection where vaccination of hens results in transfer of high levels of maternal antibodies to their progeny (Wyeth and Cullen, 1976; Wood *et al.*, 1981; Naqi *et al.*, 1983). Although these antibodies are fairly effective in protecting the chick until approximately 21 days post hatching, interference with initial vaccination will often completely prevent development of active immunity; therefore predicting the timing of IBDV vaccination can be difficult (Solano *et al.*, 1986).

11. Mucosal Gut Immunity

Introduction

In chickens, as in mammalian systems, there exists a separate mucosal immune system that exhibits a number of unique features, including antigen-presenting cells, immunoregulatory cells, and effector cell types, that are distinct from their counterparts in the systemic immune system (reviewed in Lillehoj, 1996). It is now widely accepted that the common mucosal immune system consists of two separate but interconnected compartments: mucosal inductive sites, which include the nasal-associated and gut-associated lymphoid tissues strategically located where they encounter environmental antigens, and mucosal effector sites, which include the lamina propria (LP) of the intestine and the upper respiratory tract (reviewed by McGhee *et al.*, 1989).

The gut-associated immune system

Intraepithelial and lamina propria lymphocytes

The gut-associated lymphoid tissue (GALT) in chickens include organized lymphoid structures such as the bursa of Fabricius, cecal tonsils (CT), Peyer's patches (PP), Meckel's diverticulum and lymphocyte aggregates scattered along the intraepithelium and LP of the gastrointestinal tract. The bursa of Fabricius, a hollow oval sac located dorsally to the cloaca, is the central lymphoid organ for B-cell lymphopoiesis and lymphocyte maturation, where generation of antibody diversity occurs (reviewed by Ratcliffe, 1989). The CT are discrete lymphoid nodules located at the proximal ends of the ceca near the ileocolonic junction (Befus *et al.*, 1980). The CT contain central crypts, diffuse lymphoid tissues, and germinal centers (Glick *et al.*, 1981). Both T and B cells are present in the germinal centers; plasma cells expressing surface IgM, IgG and IgA are also located in the CT (Jeurissen *et al.*, 1989). The function of CT is unknown, but active uptake of orally administered carbon particles has been shown, suggesting a role in antigen sampling (Befus *et al.*, 1980).

The PP are lymphoid aggregates in the intestine which possess a morphologically distinct lymphoepithelium with microfold (M) cells, follicles and a B-cell-dependent subepithelial zone and a T-cell-dependent central zone; they possess no goblet cells (Owens and Jones, 1974; Befus *et al.*, 1980). The PP represent the major inductive site for IgA responses to pathogenic microorganisms and ingested antigens in the gastrointestinal tract. The PP and CT of chickens are easily identified at 10 days post hatching (Befus *et al.*, 1980). As the birds age, the intestinal lymphoid aggregates undergo involution such that by 20 weeks, the lymphoid follicles become less distinct and fewer in number and there appear to be a relative depopulation of the subepithelial zone in both the CT and PP. The PP can be identified in the intestine by day 1 or 2 and increase to a maximum at 5–16 weeks old. Their number then decreases through morphological involution and at 52 and 58 weeks old, only a single PP is evident in the ileum anterior to the ileocecal junction. The Meckel's diverticulum is a remnant of the yolk on the small intestine, and usually persists as a discrete structure for the lifetime of the chickens. The exact function of this lympho-epithelial structure is not known but it contains germinal centers with B cells and macrophages (Befus *et al.*, 1980).

Within the gastrointestinal mucosa, intestinal lymphocytes are present in two anatomic compartments: the epithelium (intra-epithelial lymphocytes, IEL) and the lamina propria (lamina propria lymphocytes, LPL). The composition of IEL and CT is shown in Tables IV.11.1 and IV.11.2. The predominant subset of IEL T lymphocytes expresses the CD3 polypeptides (γ, δ, ε, and ζ) noncovalently associated with the $\gamma\delta$ chain receptor heterodimer of the antigen specific T cell receptor (TCR) (Bucy *et al.*, 1988). Another subset of T lymphocytes expresses the CD3 polypeptide chains in association with the $\alpha\beta$ chain receptor. Among the 85% of blood T cells that are CD3$^+$, 16% are TCR$\gamma\delta^+$ (Sowder *et al.*, 1988). TCR$\gamma\delta^+$ cells in the blood and the thymus lack both CD4 and CD8 molecules and most of the TCR$\gamma\delta^+$ cells, which are localized in the splenic sinusoids and in the intestinal epithelium, express the CD8 homologue (Bucy *et al.*, 1988). The changes in T-lymphocyte subpopulations in the intestine reflect age-related maturation of the GALT (Lillehoj and Chung, 1990). The ratios of TCR$\gamma\delta$ to TCR$\alpha\beta$ cells in intraepithelium and the lamina propria in SC chickens were 0.96 and 1.23 respectively at 8 weeks and 4.29 and 2.15 respectively at 12 weeks. Jejunum CD8$^+$ IEL increase gradually until 4–6 weeks old and subsequently decline as chickens age. CD4$^+$ cells represented a minor subpopulation among the IEL. The percentages of TCR$\gamma\delta^+$ to TCR$\alpha\beta^+$ T cells in

Table IV.11.1 Percentage composition of intestinal intraepithelial and caecal tonsil lymphocytes of 6- to 8-week-old White Leghorn chickens

Cell[a] type	*Intraepithelial lymphocytes[b] (percentage)*		
	Duodenum	*Jejunum*	*Caeca*
$\alpha\beta$TCR	33	33	37
$\gamma\delta$TCR	32	33	19
CD3	72	69	62
CD8	37	25	28
CD4	ND	7	ND

[a] Percentage refers to positive cells compared with total number of viable lymphocytes.
[b] Cells were isolated by treating intestinal tissue segments with dithiothreitol followed by EDTA in Ca^{2+} and Mg^{2+} free Hanks balanced salt solution.

Table IV.11.2 Percentage of lymphocyte subpopulation in caecal tonsil

Cell[a] type	*Caecal tonsil[b] (percentage)*	
	3 weeks	*7 weeks*
$\gamma\delta$TCR	11–13	13–14
$\alpha\beta$TCR	37–50	48–52
CD3	51–54	47–60
CD8	19–29	24–26
CD4	23–29	24–31

[a] Percentage refers to positive cells compared with total number of viable lymphocytes.
[b] Cells were prepared by gentle homogenization through a stainless steel mesh.

newborn and adult White leghorn chickens showed that the numbers of these cells in IEL do not reach adult levels until 4–5 weeks after hatching (Lillehoj and Chung, 1990). The composition of various T-cell subpopulations in the intestine depended upon host age, the regions of the gut examined and host genetic background.

Germinal centers seen in the cecal tonsil contain scattered TCR$\alpha\beta^+$, CD4$^+$ cells, but no TCR$\gamma\delta^+$ cells (Bucy *et al.*, 1988). Neither subset is present in the intestinal mucosa at hatching, and only occasional TCR$\gamma\delta^+$ or TCR$\alpha\beta^+$ cells are seen in the intestine of 3-day-old chicks. By 6 days post hatching, both subsets are present and the number of these cells reach adult levels by 1 month old. A third type of cell mediating intestinal immunity is the NK cell. NK cells constitute a population of non-T, non-B, nonmacrophage mononuclear cells with characteristic morphology that are capable of spontaneous cytotoxicity against a wide variety of syngeneic, allogenic, and xenogeneic target cells (Herberman, 1978). NK cell activity has been demonstrated in the spleen and peripheral blood, thymus, bursa and intestine (Chai and Lillehoj, 1988; Lillehoj and Chai, 1988). Great variability in cytotoxic potential has been observed among NK cells from different lymphoid organs. Furthermore, a substantial strain variation in NK cell activity has also been demonstrated (Chai and Lillehoj, 1988). NK cell activity increases with their age and their cytotoxic potential is not fully developed until 6 weeks after hatching (Lillehoj and Chai, 1988). There is much confusion about the phenotypic characterization of chicken leukocytes mediating natural cytotoxicity. A unique IEL subpopulation, termed TCR0 cells, showing cytoplasmic CD3 and lacking surface TCR/CD3 complex, which was detected mainly in the intra-epithelium where most express CD8 antigen (Bucy *et al.*, 1990), show cytotoxicity against NK susceptible target cells in chickens (reviewed by Kasahara *et al.*, 1994).

Cytokines in the intestine

Growth stimulatory and growth inhibitory autocrine growth factors are potential modulators of intestinal epithelial cell growth. Transforming growth factor-beta (TGF-β) has been shown to potently inhibit normal epithelial cell growth and modulate differentiation in several cell types, including bronchial epithelial cells, osteoblasts, adrenocortical cells and myoblasts, among others (reviewed by Roberts and Sporn, 1990; Massague, 1990; Sporn and Roberts, 1990). Three different isoforms of TGF-β including TGF-β2, 3 and 4 have been identified in the chicken (Jakowlew *et al.*, 1988a, 1988b, 1990). Expression of the mRNAs for TGF-β2, 3 and 4 can be detected in the intestine by day 8 of embryogenesis, and expression of TGF-β4 mRNA, in particular, appears to increase steadily with development of this tissue. Because expression of TGF-β4 protein detected by immunohistochemical staining is prominent in the tips of intestinal villus by day 19 of embryogenesis, it is possible TGF-β4 may play a role in modulating growth of the intestinal villus. Further studies will be necessary to investigate biological relevance of the TGF-β mRNAS in the embryonic development of intestine in the chicken.

Effector functions associated with GALT

The roles of various components of GALT in host defense against microbial infections has been extensively studied (Brandzaeg *et al.*, 1987). Three general functions of gut-associated immune system include: (1) processing and presentation of antigen, (2) production of local antibodies, and (3) activation of cell-mediated immunity.

In chickens, IgA and IgM are the predominant immunoglobulins in the external intestinal secretions. Although IgG is found in the gut, it is believed to be derived from the circulation or leaked from the lymphatics following permeability changes which occur during infection. Secretory IgM, which is pentameric, is effective in elimination of microbes. However, several distinctive features are important for IgA to function as a secretory antibody. One is the ability of IgA monomer to polymerize. Other properties of secretory IgA are its ability to associate with a 15 kDa peptide-joining (J) chain and a 70 kDa protein, the secretory component (SC) produced by epithelial cells. The IgA–SC complex is internalized in endocytic vesicles, transported across the cytoplasm and exocytozed onto the external surface of the epithelium. A functional homologue of mammalian SC has been described in chickens (Parrard *et al.*, 1983). A minor source of IgA in secretions is derived from blood via the hepatobiliary IgA transport system. In contrast to the transepithelial IgA pathway, hepatocytes express a specific receptor for blood IgA (Allen *et al.*, 1987). Polymeric IgA injected into the blood was cleared into the bile in less than 3 h (Sanders and Case, 1977).

The major functions of secretory IgA (sIgA) include prevention of environmental antigen influx into internal body compartments, neutralization of viruses and microbial toxins and prevention of adherence to, and colonization of mucosal surfaces by microbial pathogens. Secretory antibodies may bind to the pathogen's surface and prevent binding to the epithelium by direct blocking, by steric hindrance, by induction of conformational changes or by reduction of motility. In this manner, microorganisms would be susceptible to the natural cleaning functions of the mucosae. Animals infected with *Eimeria* spp. produce parasite-specific antibodies in both the circulation and mucosal secretions (Trees *et al.*, 1985; Lillehoj, 1987). Circulating antibodies consist of IgM, IgG, and IgA (Trees *et al.*, 1985; Lillehoj and Ruff, 1986) whereas secretory IgA has been detected in bile and gut washings of infected animals (Lillehoj and Ruff, 1986). Challenge infection with *Eimeria tenella* or *Eimeria acervulina* did not elicit an anamnestic SIgA response (Lillehoj, 1987). Although antibodies are produced, *in vivo* studies using hormonal and chemical bursectomy (Giambrone *et al.*, 1981; Lillehoj, 1987) clearly indicated that antibodies play a minor role in protection against coccidiosis (Lillehoj, 1987) and leukocytozoan parasites (Isobe *et al.*, 1989).

Cell-mediated immune responses include both antigen-specific and antigen nonspecific activation of various cell populations, including T lymphocytes, NK cells and macrophages. Although CTL activity has been demonstrated in the intestine of mammals, MHC-restricted IEL CTL activity has yet to be shown in chickens. The importance of T cells in immune responses to coccidia has been well documented (reviewed by Lillehoj and Trout, 1993, 1994). The number of duodenal IELs expressing the CD8 antigen increases in SC and TK chickens following primary infection with *E. acervulina* (Lillehoj, 1994). Following secondary infection, a significantly greater number of $CD8^+$ IEL were observed in the SC chickens which showed a lower level of oocyst production compared with TK chickens. SC chickens showed increased $TCR\alpha\beta^+CD8^+$ cells following shortly after challenge infection with *E. acervulina* (Trout and Lillehoj, 1995). Furthermore, depletion of either $CD8^+$ cells or $TCR\alpha\beta^+$ cells resulted in substantial increases in oocyst production following challenge *E. acervulina* infection in chickens (Trout and Lillehoj, 1996). These results suggest that a significant increase in $TCR\alpha\beta^+CD8^+$ IEL in SC chickens may reflect enhanced acquired immune status in these chickens compared with TK chickens. The observation that chicken intestinal IEL contain NK cells that mediate spontaneous cytotoxicity (Chai and Lillehoj, 1988) suggests that NK cells may play an important role in local defense. Positive correlations between NK cell activity and genetically determined disease resistance to coccidia (Lillehoj, 1989) have been noted.

TGF-β1 has been shown to inhibit the proliferation of T-lymphocytes (Kehrl *et al.*, 1986). After infection with the coccidian parasite, expression of TGF-β4 mRNAs increased in chicken intestinal intra-epithelial lymphocytes (Jakowlew *et al.*, 1997). This increase in expression of TGF-β4 in response to infection with the coccidian parasite is similar to the induction of TGF-β1 in the mouse following infection with the protozoan parasites *Leishmania braziliensis* and *Trypanosoma cruzi*. In these parasitic infections, TGF-β1 has been shown to influence the replication of these parasites in macrophages and acts to increase the intracellular replication of these parasites, thus providing a mechanism for these parasites to escape the immune protection system (Silva *et al.*, 1991; Barral-Netto and Barral, 1994; Barral *et al.*, 1994). Whether TGF-β plays a similar role in coccidiosis may be determined in further studies.

Cytokines and lymphokines have been shown to influence coccidial infections. Cell culture supernatant from Con A-stimulated lymphocytes was capable of inhibiting the replication of *Eimeria* parasites in MDBK cell cultures and reduced oocyst production following *E. acervulina* and *E. tenella* infections in chickens (Lillehoj *et al.*, 1989). Sporozoites and merozoites of *E. tenella* induced *in vitro* TNF-like factor production by normal peripheral blood derived macrophages (Zhang *et al.*, 1995). Treatment of chickens with anti-TNF antibody resulted in a partial abrogation of *E. tenella*-induced body weight-loss in SC chickens but not in TK chickens (Zhang *et al.* 1995). Lymphocytes from *Eimeria*-infected chickens produced higher level of γ-IFN when induced with Con A compared with lymphocytes from uninfected chickens (Martin *et al.*, 1994). Strain differences in *Eimeria*-induced γ-IFN production was observed (Martin *et al.*, 1994). Although the exact role of these cytokines in coccidiosis needs to be further elucidated, this result may explain the severity of clinical signs associated with coccidial infection.

Conclusions

Intestinal immune responses to enteric pathogens that lead to protective immunity involve the complex interplay of soluble factors, leukocytes, epithelial cells, endothelial cells, and other physiological factors of the GALT. Different effector mechanisms may be involved depending on the type of enteric pathogens and on whether a primary or secondary response occurs. It is likely that these complex interactions have contributed to the difficulties in developing an effective vaccine against enteric pathogens. In contrast to mammals, limited information is available about the intestinal immune system of chickens. However, recent technical advances in molecular and cellular immunology will facilitate our understanding of the ontogeny, structure, and function of the GALT and may lead to new approaches to vaccination against enteric pathogens.

12. Tumors of the Immune System

Introduction

Tumors involving lymphocytes and other cells of the avian immune system are frequently caused by viral agents. Five groups of viruses are associated with these tumors in domesticated birds (Table IV.12.1). Marek's disease (MD), caused by Marek's disease herpesvirus (MDV), and lymphoid leukosis (LL), caused by avian leukosis virus (ALV), are diseases of great economic importance for the poultry industry. Detailed information on the pathology, pathogenesis and virology of MD and LL is available in recent review articles by Calnek and Witter (1991, 1997), Payne and Purchase (1991) and Payne and Fadly (1997). Acute defective leukemia viruses (DLV) are of minor economic importance, but are of interest because they contain different *onc* genes (Enrietto and Hayman, 1987). Reticuloendotheliosis virus (REV) has recently

Table IV.12.1 Viral etiology of tumors of the immune system in domesticated birds

Virus family	*Virus*	*Serotypes/subgroups*	*References*
Herpesviridae	Marek's disease virus	Serotype 1 (MDV-1)	Calnek and Witter (1991)
	Marek's disease virus	Serotype 2 (MDV-2)[a]	Calnek and Witter (1991)
	Herpesvirus of turkeys	Serotype 3 (HVT)[a]	Calnek and Witter (1991)
Retroviridae	Avian leukosis virus (ALV)	(A) Exogenous, subgroups A–D, J	Payne and Purchase (1991) Payne *et al.* (1991a, b)
		(B) Endogenous, subgroup E[a]	Payne and Purchase (1991)
	Acute defective leukemia virus (DLV)	Several subgroups	Payne and Purchase (1991) Enrietto and Hayman (1987)
	Lymphoproliferative disease retrovirus (LPDV)	Unknown	Biggs (1991)
	Reticuloendotheliosis virus (REV)	(A) Nondefective	Witter (1991)
		(B) Defective	Witter (1991)

[a]These viruses are nononcogenic.

become an economically important pathogen in chickens; REV-induced diseases have recently been reviewed by Witter (1991, 1997). Lymphoproliferative disease virus (LPDV) is the cause of a lymphoproliferative disorder in turkeys (Biggs, 1991).

Natural and experimental hosts

Tables IV.12.2 and IV.12.3 list the natural and experimental hosts in which virally induced tumors of the immune system have been described.

Table IV.12.2 Phenotypic characterization of Marek's disease virus-induced lymphoid tumors in avian species

Avian species	*Natural/experimental infection*	*Target cells for transformation*	*References*
Chicken	Natural	$CD4^+CD8^-TCR\alpha\beta^+$ and some other T cell subsets[a]	Calnek and Witter (1991) Schat *et al.* (1991)
Japanese quail	Natural	Lymphocytes (T cells?)	Pradhan *et al.* (1987) Imai *et al.* (1991)
Turkey	Experimental	$CD4^+$ T cells, B cells	Nazerian *et al.* (1982) Powell *et al.* (1984) Schat *et al.* (1988)

[a]See text for details.

Table IV.12.3 Mechanism of transformation and phenotypic characterization of immune cells transformed by avian retroviruses

Virus	*Host*	*Target cells*	*References*	*Transformation mechanism (Ref)*
ALV	Chickens	Immature, μ^+ B cells	Payne and Purchase (1991)	*c-myc* activation (Kung and Maihle, 1987)
		Myelocytes	Payne *et al.* (1991b)	*c-onc* activation
DLV	Chickens	Many cell types	Enrietto and Hayman (1987)	*v-onc* genes
LPDV	Turkeys	Pleomorphic cells	Biggs (1991)	Unknown (Sarid *et al.*, 1994)
	Chickens	Pleomorphic cells[a]	Ianconescu *et al.* (1983)	Unknown
REV	Chickens	Many cell types	See text	Defective REV: *v-rel*
	Japanese quail	Pleomorphic cells	Schat *et al.* (1976)	Nondefective REV: *c-myc*
	Turkeys	Lymphoblastoid cells	McDougall *et al.* (1978)	activation (Swift *et al.*, 1985)
	Duck species	Lymphoblastoid cells	Li *et al.* (1983)	
	Pheasants	Lymphoblastoid cells	Drén *et al.* (1983)	
	Geese	B cells (?)	Drén *et al.* (1988)	

[a]Only after experimental infection.

Target cells for transformation and pathogenesis

Target cells for transformation by MDV

All MDV-transformed lymphocytes are T cells expressing class I and class II major histocompatibility complex (MHC) antigens, CD3, and TCR$\alpha\beta$1 or TCR$\alpha\beta$2. MD cell lines have also been established from $CD4^-CD8^-$ or CD^-CD8^+ T cells expressing $CD3^+$TCR$\alpha\beta$1 or TCR$\alpha\beta$2 (see below).

Pathogenesis of MD

The pathogenesis of MD is summarized in Figure IV.12.1 (Calnek, 1985, 1986; Schat, 1987). The following sequential events lead to tumor formation in genetically susceptible strains of chickens: (1) cytolytic infection of B cells provokes inflammation and immune responses, including activation of T cells which then also become infected; (2) latency develops in both B and T cells coincident with early immune responses, perhaps due to the production of cytokines (Buscaglia and Calnek, 1988); (3) proliferation of latently infected T cells leads to an undefined change in the virus–host cell relationship which causes neoplastic transformation of the cell; and (4) permanent immunosuppression coincides with the second phase of cytolytic infection, impairing surveillance mechanisms and allowing tumor development. Activation of T cells by inoculation with MDV-infected, allogenic chick kidney cells results in transformation of additional subpopulations of T cells (Calnek *et al.*, 1989). The MDV genome has not yet been completely sequenced and, as a consequence, the molecular basis for transformation by MDV has not yet been elucidated.

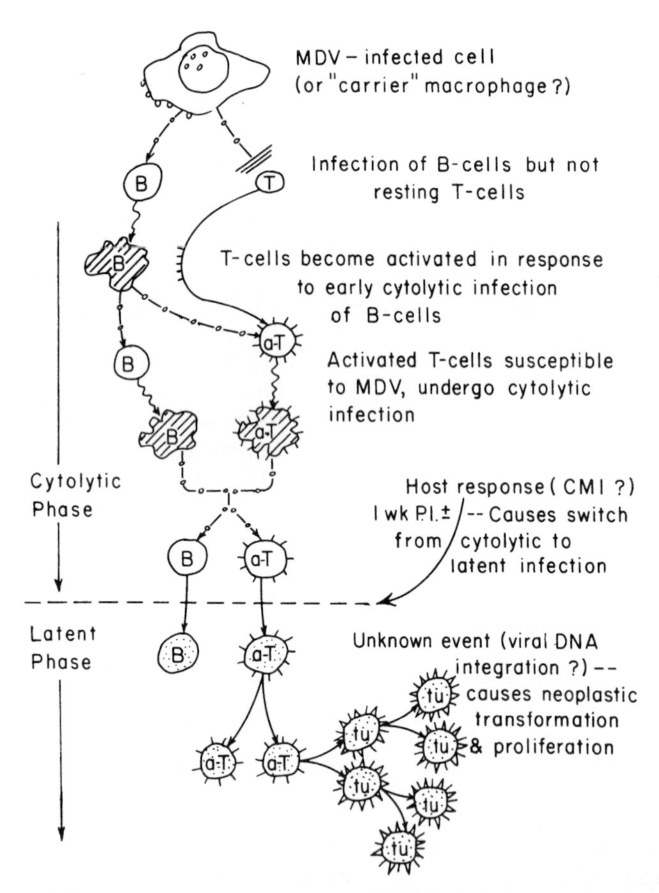

Figure IV.12.1 Schematic representation of sequential events in lymphocyte infections with MDV leading to the formation of tumors. From Calnek, B. W. (1985). Pathogenesis of Marek's disease – a review. In *International Symposium on Marek's Disease* (eds Calnek, B. W., Spencer, J. L.), pp. 374–390. AAAP, Kennett Square, PA, USA. Reproduced with permission of the American Association of Avian Pathologists.

Target cells for transformation by avian retroviruses

In contrast to ALV, nondefective REV can transform both IgM-bearing B cells (Witter *et al.*, 1981) and T cells (Witter *et al.*, 1986) in chickens. The target cells for transformation by REV are poorly defined in species other than chickens, mainly because monoclonal antibodies (mAb) are not available for differentiation antigens on lymphoid cells in these species. Defective REV can transform many subsets of B and T cells and also cells from the myeloid series (Beug *et al.*, 1981; Lewis *et al.*, 1981; Chen *et al.*, 1988; Barth *et al.*, 1990; Schat *et al.*, 1992; Marmor *et al.*, 1993). DLV can also transform many different cells of the immune system, depending on the *onc* gene that is present in a specific DLV strain (Enrietto and Hayman, 1987).

Pathogenesis of retrovirus-induced tumors

Two mechanisms of transformation are used by avian retroviruses to transform cells. Defective viruses carry an *onc* gene, which replaces one or more of the genes required for viral replication. As a consequence these viruses require a helper virus for their replication. The *v-onc* gene is inserted into the genomic DNA under control of the viral long-terminal repeat (LTR), and the *v-onc* protein is directly involved in the transformation of the cell. Tumors usually develop after a short incubation period, especially after inoculation with the defective REV-T strain. The *v-rel* oncogene of REV-T is a repressor of the NF-κB transcription factor complex and probably functions by a direct nuclear blockade of the κB enhancer (Walker *et al.*, 1992).

The nondefective ALV and REV transform lymphoid cells by insertion of the viral LTR in the *c-myc* region (Kung *et al.*, 1992). The 3′ LTR is responsible for the transcription of *c-myc* in B-cell lymphomas (Boerkool and Kung, 1992). Retrovirus LTRs contain many enhancer and promoter elements, some of which can be influenced by infection with MDV-2 (Table IV.12.1), leading to

Table IV.12.4 Effects of viral replication (R) or tumor cells (T) on immune functions in chickens infected with MDV, ALV, or REV[a]

Virus	*R or T*	*T cells*	*B cells, and antibodies (Ab)*	*Macrophages*	*NK cells*
MDV	R	Thymic atrophy	Bursa atrophy	Suppressor macrophages ⇑	Lysis of LSCC-RP9 ⇑
		Mitogen responses (MR): transient ⇓	Ab production ⇓		
	T	MR: permanent ⇓	See above	–	Lysis of LSCC-RP9 ⇓[b]
ALV	R/T	Tolerance to ALV proteins[c]	Tolerance to ALV proteins[c]	–	–
REV	R/T	MR: ⇓	Ab production ⇓	Suppressor macrophages ⇑ (?)	–
			Tolerance to REV proteins[c]		

[a] For references see Schat (1996).
[b] Probably caused by α-fetal protein or chicken fetal antigens.
[c] After *in ovo* infection or infection of immunoincompetent chicks.
⇑, increased activity; ⇓, decreased activity.

enhanced transcription (reviewed by Ruddell, 1995). This enhanced transcription could explain the augmentation of LL in chickens infected with the MDV-2.

The pathogenesis of tumor development has been reviewed by Neiman (1994). Briefly, 4–8 weeks post infection bursal follicles containing preneoplastic stem cells can be observed. These cells have an increased expression of *myc* and will give rise to bursal lymphomas, which will eventually migrate from the bursa causing metastatic tumors.

Effects of tumors on the immune function

The effect of tumors on the immune functions have been examined mostly in chickens (reviewed by Schat, 1996). Table IV.12.4 summarizes the immunosuppressive effects of infection and tumor development on immune functions. Unfortunately, there is a paucity of detailed information on the functional consequences of subsets of lymphoid cells affected by the tumors. There is also a lack of knowledge on the mechanism(s) causing immunosuppression at the molecular level.

13. Autoimmunity in Avian Model Systems

Autoimmunity arises when the immune system inappropriately begins to react to self. In chickens, as in humans, many independent genetic and environmental factors must be present for severe disease to develop. Primarily, the immune system must be capable of reacting against an autoantigen. Tolerance is maintained through several mechanisms including clonal deletion in the thymus and clonal inactivation both in the thymus and in the periphery. Thus, MHC haplotype and any deviation in the regulation of lymphoid development can contribute to the activation of autoreactive cells. In the case of organ specific autoimmunity, there are very often defects in the target tissue which facilitate immune infiltration. Three spontaneous models of autoimmune disease in chickens exemplify this paradigm (Kaplan *et al.*, 1991; Table IV.13.1). The Obese strain (OS) chicken develops an autoimmune thyroiditis which in many ways is similar to Hashimoto's disease in humans. UCD line 200 chickens develop a condition which resembles scleroderma. Finally, the Smyth–DAM (delayed amelanosis) line chicken has a progressive loss of retinal and feather pigmentation similar to vitiligo. These three strains have been studied exten-

Table IV.13.1 Defects associated with avian autoimmune diseases

Strain	*Immune defect*	*Target organ/pathology associated defect*
Obese strain	T-cell hyperactivity Anti-thyroglobulin antibodies	Sensitivity to iodine-mediated thyroid infiltration
UCD line 200	T-cell hypoactivity Deficiency in thymic architecture Antibodies to Ig, ssDNA, type II collagen	Hyperactive fibroblast
Smyth–DAM line	B-cell hyperactivity Antimelanocyte antibodies	Hyperactive melanocyte

sively for their utility in understanding the genesis of autoimmune attack, something that cannot be done using induced models.

The OS chicken thyroiditis is characterized by extensive mononuclear cell infiltration of the thyroid three to five weeks after hatching (Wick *et al.*, 1970a, b). By 7 weeks the thyroid is completely destroyed. Immune attack is accompanied by high titers of antithyroglobulin antibodies (Polley *et al.*, 1981; Bagchi *et al.*, 1985). While transfer of these antibodies can affect disease severity (Jaroszewski *et al.*, 1978), infiltration is mediated by T cells (Pontes de Carvalho *et al.*, 1981). The infiltrate is composed of activated $\alpha\beta$ TCR-positive T cells, plasma cells and macrophages (Krömer *et al.*, 1985; Cihak *et al.*, 1995) and have a higher percentage of CD8-positive cells than is seen in the periphery (Bagchi *et al.*, 1995). Several genetic lesions which affect immune responses have been found in OS chickens. OS lymphocytes are hyperactive in response to the T-cell mitogen concanavalin A, and demonstrate increased T-cell growth factor secretion and surface expression of activated T-cell markers (Livezey *et al.*, 1981; Schauenstein *et al.*, 1985). OS birds also have increased levels of corticosterone-binding globulin, which results in decreased levels of circulating corticosterone and could contribute towards increased immune responses *in vivo* (Fässler *et al.*, 1988). Furthermore, OS chickens fail to show a transient increase in corticosterone following immunization – a defect which maps to the hypothalamo-hypophyseal axis (Schauenstein *et al.*, 1987). However, none of these defects strictly correlate with disease occurence in breeding analysis (Neu *et al.*, 1985, 1986; Krömer *et al.*, 1989). There has also been some suggestion of a generalized defect in thymic development in the OS chicken, although this has not been studied extensively (Welch *et al.*, 1973; Jakobisiak *et al.*, 1976).

Analysis of OS thyroid function showed a decreased susceptibility to thyroxine-mediated suppression of radioiodide uptake and organification (Sundick *et al.*, 1979). This thyroid-stimulating hormone (TSH) autonomy is shared with the Cornell strain (CS) chicken, which develops mild thyroiditis at a low frequency compared with the OS. TSH autonomy is not due to thyroid-stimulating antibodies (Sundick and Wick, 1974). Additional studies of embryonic thyroid cells demonstrated that OS cells had decreased proliferation and function in culture, which likely result from a lack of autocrine growth factor secretion (Truden *et al.*, 1983). It is not clear, however, that either of these phenotypes has a relationship to disease. In contrast, there is a definitive role for iodide in disease progression. Large population studies of CS chickens correlated increased dietary iodide with increased disease incidence, severity and autoantibody production (Bagchi *et al.*, 1985). Thyroidal iodine is also crucial for the development of OS thyroiditis. A treatment regimen consisting of potassium perchlorate (to inhibit active transport of iodine) and mononitrotyrosine (to promote the loss of iodotyrosines) starting *in ovo* and continued past hatching, decreased thyroid infiltration to 2% of control levels and decreased antithyroglobulin antibody titers for as long as 9 weeks (Brown *et al.*, 1991). Other antithyroid drugs such as propylthiouracil and aminotriazole also decreased disease severity. This suggests that iodine is important for the immunogenicity of thyroglobulin (Tg), which is the target of autoimmune attack. Indeed, highly iodinated Tg is a better immunogen (Sundick *et al.*, 1987). However, OS chickens make Tg with a slightly lower than normal iodide content and their autoantibodies to Tg react equally well with high or low iodide Tg (Sundick *et al.*, 1987). Since iodination of other proteins in the OS thyroid cannot be detected (Sundick *et al.*, 1991), it is likely that iodine plays an additional role distinct from the generation of an autoimmune epitope. In support of this, studies have shown that treatment of OS birds with antioxidants decreases disease severity and can inhibit initial immune infiltration (Bagchi *et al.*, 1990). The potential role of iodide or oxygen-free radicals which specifically lead to disease in the OS are currently being studied.

The University of California, Davis (UCD) line 200 chicken develops a scleroderma like syndrome with symptoms very similar to those seen in human scleroderma (Gershwin *et al.*, 1981). Between 3 and 5 weeks after hatching birds develop comb lesions, polyarthritis and dermal lesions (Gershwin *et al.*, 1981). There is also a high rate of mortality in this line. Dermal pathology shows an intense mononuclear infiltrate composed largely of activated T cells (van de Water *et al.*, 1989) which are skewed towards CD8 cells. During infiltration, small vessels proliferate and fibroblasts migrate into the dermis leading to extensive collagen deposition, which in turn causes thickening and occlusion of blood vessels (Gershwin *et al.*, 1981). A large percentage of the flock characteristically has rheumatoid factor and autoantibodies to type II collagen, ssDNA and additional nuclear and cytoplasmic antigens by six months of age (Gershwin *et al.*, 1981; Haynes and Gershwin, 1984).

In contrast to the OS chicken, the UCD line 200 birds have a hyporesponsive immune system. Lymphocytes show a decreased response to T-cell growth factor, concanavalin A and anti-CD3 (van de Water *et al.*, 1990; Wilson *et al.*, 1992). Within the thymus there is evidence of a developmental defect in subcapsular thymic epthelium and increased MHC class II expression, particularily in the thymic cortex (Boyd *et al.*, 1991). Identifying target organ defects is difficult in this system because many of the autoantigens are widely distributed in the organism. Furthermore, the skin is a diverse organ and many defects could contribute towards autoimmune attack. It has been shown that fibroblasts from UCD line 200 birds are activated and secrete increased amounts of collagen and glucosaminoglycan (Duncan *et al.*, 1992). This phenotype is dependent on factors secreted by T cells in the dermal infiltrate and is probably responsible for a considerable portion of disease pathology (Duncan *et al.*, 1995).

The Smyth–DAM line chicken has a high incidence of feather and retinal pigmentation loss which is phenotypically similar to the human disease vitiligo (Smyth *et al.*, 1981). DAM autoimmunity is strongly correlated with both T- and B-cell activity. B-cell hyperactivity has been documented following immunization (Lamont *et al.*, 1982). Furthermore, there is a correlation between levels of antimelanocyte antibody and disease incidence (Lamont and Smyth, 1981; Austin *et al.*, 1992). T cells with a low CD4:CD8 ratio are found infiltrating the feather pulp (Erf *et al.*, 1995). Treatment of DAM chickens with immunosuppresive agents such as cyclosporin A decreases amelanosis (Pardue *et al.*, 1987), although the condition was shown to worsen following discontinuation of the treatment. Pathogenesis in this model appears to be mediated by specific destruction of melanocytes in the feathers and retinas of the birds (Austin *et al.*, 1992). Indeed, a melanocyte-specific defect has been identified. DAM melanocytes display hypermelanogenesis and have increased tyrosinase activity (Boissy *et al.*, 1986). This is of specific interest because DAM autoantibodies detect a tyrosinase related protein (Austin and Boissy, 1995). DAM sera also detect both cell surface and cytoplasmic antigens from isolated melanocytes (Austin *et al.*, 1992; Searle *et al.*, 1993). However, it has not been demonstrated directly that these antibodies are the cause of the disease.

These systems create a novel view of autoimmunity. One trend that appears is the role for a specific T-cell subset in disease pathology. Studies of all three strains show increased percentages of CD8 cells at the sites of infiltration compared with peripheral blood and that CD4:CD8 ratios skew even further towards CD8 as disease progresses. Furthermore, these T cells are $\alpha\beta$ TCR-positive, despite a large percentage of $\gamma\delta$ T cells in chicken peripheral blood. Whether these cells are specific for autoantigens and how they play a role in disease pathogenesis remains to be determined. Despite evidence of immune dysregulation in all three strains, autoimmunity is limited to specific target sites. Attempts to induce autoimmunity in organs other than the normal site of pathology have shown no significant differences in tissue infiltration between normal and autoimmune-prone chickens. Thus, the presence of immune dysregulation itself is not sufficient for the development of a tissue-specific autoimmune condition. These systems establish the requirement for multiple genetic defects in the development of autoimmune disease in chickens.

14. Effect of Viruses on Immune Functions

The intensive rearing conditions of commercial poultry operations provides an environment conducive to the selection and spread of infectious agents. Numerous viruses, many of which are highly virulent, are endemic in most of the world's poultry-producing areas. In the face of these virulent viruses, routinely used vaccines are often ineffective in providing an acceptable level of protection against clinical disease, death, and virus shedding. In addition to these direct effects of virus infection, a number of avian viruses are capable of causing immunosuppression. Immunosuppression has great economic significance for the commercial poultry industry because affected flocks are susceptible to opportunistic infections, respond poorly to vaccines, and generally perform less well than unaffected flocks. This section attempts to summarize information on some common immunosuppressive viruses in poultry and it is not intended to be a comprehensive summary of the literature.

Effect of viruses on immune functions

See Tables IV.14.1 and IV.14.2.

15. Concluding Remarks

Availability of reagents particularly monoclonal antibodies that can identify antigens in avian cells and cell products have facilitated significant advances in the knowledge of avian immunology. The avian immune system conforms to the basic immunologic mechanisms identified in mammals. As in mammals, chicken T cells have two antigen-binding receptors, i.e., TCR$\alpha\beta$ and TCR$\gamma\delta$. There is a higher frequency of $\gamma\delta$T cells in birds than in mammals. Further, in the chicken, the TCRβ locus has two V_β families recognized by two different monoclonal antibodies. The TCR2 antibody recognizes $V_\beta 1$ family and TCR3 antibody recognizes $V_\beta 2$ family. Chicken B cells produce IgM, IgG (or IgY) and IgA. The mechanism of attaining diversity in the variable region of Ig is different from that in mammals. In the chicken, rearrangement does not generate adequate diversity because both heavy and light chains of Ig have only one functional V and J gene segment. Much of the diversity is attained by somatic gene conversion by incorporating copies of regions from nonfunctional pseudogenes into rearranged Ig genes. The molecular mechanism(s) of gene conversion needs to be elucidated. The process of gene conversion and hence acquisition of Ig diversity occurs in the bursa. Thus, the bursa, an organ that is unique to birds, is essential for normal ontogeny of the B-cell system. Studies on the characteristics and functions of the cells of the monocyte/macrophage lineage have lagged behind those conducted on lymphocytes and await the development of appropriate reagents.

The MHC in chickens is designated as B complex, which consists of three major loci, i.e., B-F (equivalent of class I MHC), B-L (equivalent of class II MHC) and B-G

Table IV.14.1 Immunosuppressive viruses

	Marek's disease virus (DNA)	*Hemorrhagic enteritis virus (DNA)*
Lymphoid organs		
Bursa	Atrophy (Adldinger and Calnek, 1973; Payne and Rennie, 1973; Frazier, 1974)	
Thymus	Atrophy (Adldinger and Calnek, 1973; Payne and Rennie, 1973; Frazier, 1974)	
Spleen	Splenomegaly (Adldinger and Calnek, 1973; Payne and Rennie, 1973; Frazier, 1974)	Splenomegaly (Gross and Dommermuth, 1975)
B-cell functions	Cytolytic and latent infection of B cells (Shek *et al.*, 1983; Calnek *et al.*, 1984)	
Number	Reduced (Zhonggui *et al.*, 1996)	Reduced due to lytic infection (Suresh and Sharma, 1995, 1996)
Antibody production	Reduced (Purchase *et al.*, 1968; Burg *et al.*, 1971; Evans *et al.*, 1971; Jakowski, 1973; Payne *et al.*, 1976; Yamamoto *et al.* 1995; Cui and Qin, 1996)	Reduced (Nagaraja *et al.*, 1982b)
T-cell functions		
Number	Decreased (Zhonggui *et al.*, 1996)	Unchanged (Suresh and Sharma, 1995)
CD4:CD8 ratio	Reduced (Yamamoto *et al.* 1995; Morimura *et al.*, 1996); Increased (Lessard *et al.*, 1996)	Increased in acute infection, decreased in chronic infection (Suresh and Sharma, 1995)
In vitro proliferation	Reduced (Burg *et al.*, 1971; Alm *et al.*, 1972; Lu and Lapen, 1974; Theis *et al.*, 1975; Lee *et al.*, 1978a; Schat *et al.*, 1978; Powell, 1980; Theis, 1981; Liu and Lee, 1983; Quere, 1992; Lessard *et al.*, 1996; Morimura *et al.*, 1996; Zhonggui *et al.*, 1996)	Reduced in acute infection (Nagaraja *et al.*, 1982a, 1985)
Cutaneous hypersensitivity	Reduced (Rusov *et al.*, 1996)	
Graft rejection	Delayed (Purchase *et al.*, 1968)	
Tumor rejection	Delayed (Calnek *et al.*, 1975, 1978)	
Natural killer cell functions		
Activity	Reduced by the presence of tumor cells (Heller and Schat, 1985) Enhanced (Lessard *et al.*, 1996) Enhanced in resistant chickens and reduced in susceptible chickens (Sharma, 1981) Enhanced at the site of the tumor (Sharma, 1983)	
Soluble mediator production		
Interferon	Induced (Sharma, 1989)	
Interleukin-2	Reduced (Zhonggui *et al.*, 1996)	
Suppressor factors	Present (Bumstead *et al.*, 1985; Bumstead and Payne, 1987)	
Suppressor cell induction	Reduced embryonal suppressor cells (Sharma, 1988) Macrophages (Lee *et al.*, 1978a, b; Theis, 1981) Suppressor or transformed T cells (Theis *et al.*, 1975; Theis, 1977, 1981; Quere, 1992)	
Response to extraneous vaccines	Enhanced early/reduced late (Lessard *et al.*, 1996)	Reduced NDV efficacy (Nagaraja *et al.*, 1985)
Secondary infection susceptibility	Increased (Biggs *et al.*, 1968)	Increased incidence of colibacillosis (Domermuth and Larsen, 1984; Sponenberg *et al.*, 1985; van den Hurk *et al.*, 1994)

Table IV.14.2 Immosuppressive viruses

	Chicken anemia (DNA)	*Infectious bursal disease (RNA)*	*Reovirus (RNA)*
Lymphoid organs			
Bursa	Atrophy (Goyro *et al.*, 1989; Lucio *et al.*, 1990; Cloud *et al.*, 1992a)	Atrophy and lymphocyte depletion (Helmboldt and Garner, 1964; Müller *et al.*, 1979; Ley *et al.*, 1983; Ezeokoli *et al.*, 1990; Ramm *et al.*, 1991)	Transient atrophy (Montgomery *et al.*, 1986)
Thymus	Atrophy (Goyro *et al.*, 1989; Lucio *et al.*, 1990; Cloud *et al.*, 1992a; Jeurissen *et al.*, 1992a, b)	Atrophy (Cheville, 1967; Mazariegos *et al.*, 1990; Sharma *et al.*, 1993; Inoue *et al.*, 1994)	Transient atrophy (Montgomery *et al.*, 1986)
Spleen			Splenomegaly (Kerr and Olson, 1969; Tang *et al.*, 1987a, b)
B-cell functions			
Number	Unchanged (Cloud *et al.*, 1992a)	Reduced due to lytic infection (Hudson *et al.*, 1975; Hirai *et al.*, 1979; Hirai and Calnek, 1979; Nakai and Hirai, 1981; Sivanandan and Maheswaran, 1981; Müller, 1986; Becht and Müller, 1991; Rodenberg *et al.*, 1994)	
Antibody production	Unchanged (Goodwin *et al.*, 1992)	Reduced (Ivanji, 1975; Ivanji and Morris, 1976; Hirai *et al.*, 1979; Hopkins *et al.*, 1979; Sharma *et al.*, 1989; Craft *et al.*, 1990)	Reduced (Rinehart and Rosenberger, 1983)
In vitro proliferation	Increased in spleen, decreased in peripheral blood (Cloud *et al.*, 1992b)	Reduced (Sivanandan and Maheswaran, 1980b)	
T-cell functions			
Number	Decreased cytotoxic T cell number in spleen and thymus (Cloud *et al.*, 1992a; Bounous *et al.*, 1995)	Unchanged (Hirai *et al.*, 1979; Cloud *et al.*, 1992a; Rodenberg *et al.*, 1994). Reduced if birds infected prior to 3 weeks old, enhanced if infected after 3 weeks old (Sivanandan and Maheswaran, 1980a)	Unchanged in spleen/reduced in peripheral blood in acute infection (Pertile, 1995)
CD4:CD8 ratio	Increased in spleen and thymus (Cloud *et al.*, 1992a)	Unchanged (Rodenberg *et al.*, 1994)	Unchanged (Pertile, 1995)
In vitro proliferation	Reduced in acute infection (Adair *et al.*, 1991; Cloud *et al.*, 1992b; Bounous *et al.*, 1995)	Reduced in acute infection (Sivanandan and Maheswaran, 1980b; Confer *et al.*, 1981; Confer and MacWilliams, 1982; Sharma and Lee, 1983; Montgomery *et al.*, 1986; Sharma and Fredrickson, 1987; Nusbaum *et al.*, 1988)	Reduced in acute infection (Rinehart and Rosenberger, 1983; Montgomery *et al.*, 1986; Sharma *et al.*, 1994; Pertile *et al.*, 1995, 1996)
Graft rejection		Increased time (Panigrahy *et al.*, 1977) or unchanged (Hudson *et al.*, 1975; Giambrone *et al.*, 1977)	

(continued)

Table IV.14.2 *Continued*

	Chicken anemia	*Infectious bursal disease*	*Reovirus*
Natural killer cell functions			
Number	Reduced in acute infection (Bounous *et al.*, 1995)		
Activity		Unchanged (Sharma and Lee, 1983)	
Macrophage functions			
Number			Increased in spleen (Kerr and Olson, 1969; Tang *et al.*, 1987a, b; Pertile *et al.*, 1995)
Nitric oxide production			Increased (Sharma *et al.*, 1994; Pertile *et al.*, 1995, 1996)
Phagocytosis	Reduced (McConnell *et al.*, 1993a, b)	Reduced (Santivatr *et al.*, 1981)	Unchanged (Pertile *et al.*, 1995)
Bactericidal activity	Reduced (McConnell *et al.*, 1993a, b)		
Fc expression	Reduced (McConnell *et al.*, 1993a, b)		
Soluble mediator production			
Interferon	Elevated early/decreased late (Adair *et al.*, 1991; McConnell *et al.*, 1993a, b)	Induced (Gelb *et al.*, 1979)	Enhanced production by attenuated virus/no change with pathogenic virus (Ellis *et al.*, 1983a, b) Induces *in vitro* production by chicken embryo fibroblasts (Winship and Marcus, 1980) Reduced production by mitogen stimulated spleen cells/ normal following macrophage removal (Pertile *et al.*, 1996)
Interleukin-1	Reduced (McConnell *et al.*, 1993a, b)		
Interleukin-2	Reduced in acute infection (Adair *et al.*, 1991; McConnell *et al.*, 1993b)		Reduced, but normal following macrophage removal (Pertile *et al.*, 1996)
Suppressor factor(s)		Produced (Sharma and Fredrickson, 1987)	Produced (Pertile, 1995)
Suppressor cell induction		Suppressor macrophages (Sharma and Lee, 1983)	Suppressor macrophages (Pertile *et al.*, 1996)
Response to vaccines	Reduced protection by MDV, NDV and ILT vaccines (Box *et al.*, 1988; Otaki *et al.*, 1989; Cloud *et al.*, 1992a)	Reduced vaccine efficacy (Allan *et al.*, 1972; Faragher *et al.*, 1974; Hirai *et al.*, 1974; Pejkovski *et al.*, 1979; Sharma, 1984; Ezeokoli *et al.*, 1990; Mazeriegos *et al.*, 1990; Higashihara *et al.*, 1991; Cloud *et al.*, 1992b)	Reduced MDV vaccine efficacy (Rinehart and Rosenberger, 1983)
Secondary infection susceptibility		Increased (Cho, 1970; Wyeth, 1975; Fadley *et al.*, 1976; Anderson *et al.*, 1977; Rosenberger and Gelb, 1978; Pejokovski *et al.*, 1979; Yuasa *et al.*, 1980; Santivatr *et al.*, 1981; Sharma, 1984; Moradin *et al.*, 1990)	

which does not have a mammalian equivalent. The function of the B-G locus in the chicken is not known. The chicken B complex participates in immune functions much like mammalian MHC and has been shown to influence disease resistance, tumor regression and immune responsiveness against certain antigens. Advances are being made in defining the molecular structure of the B complex.

There is much current interest in avian cytokines. Although the genes for only a few avian cytokines have been cloned and expressed, a number of cytokines have been identified by their biological activity. The avian cytokines appear to be biological equivalents of mammalian cytokines and engage in similar immunologic and physiologic functions. The interest in avian cytokines results from their possible prophylactic and therapeutic use in poultry. Because commercial poultry populations receive multiple vaccinations against infectious disease, it is hoped that immunomodulating cytokines may enhance vaccinal protection and reduce the negative impact of immunosuppression. Cytokine genes vectored into fowl pox or herpes viruses are being tested for beneficial immunomodulatory effects.

Vaccine delivery in commercial chicken populations is undergoing a dramatic change. Conventionally, the vaccines are administered in hatched chickens but now certain vaccines are being injected into embryonated eggs several days before hatch. Laboratory and field data have shown that the vaccine-exposed embryos develop protective immunity before they hatch. These observations indicate that embryos are more immunologically competent than previously estimated. *In ovo* vaccination in poultry may lead the way for prenatal immunization in mammals.

Immunosuppression and accompanying poor performance of birds is a major problem in the poultry industry. In addition to environmental stresses associated with intensive rearing conditions, a number of immunosuppressive agents, particularly viruses have been identified. Many of these viruses are endemic in poultry-producing areas. The mechanism by which the viruses cause immunosuppression has not been well defined and is probably unique for each immunosuppressive virus. It is hoped that improved understanding of the avian immune system will unravel strategies for improved management of viral immunosuppression.

16. References

Adair, B. M., McNeilly, F., McConnell, C. D., Todd, D., Nelson, R. T., McNulty, M. S. (1991). Effects of chicken anemia agent on lymphokine production and lymphocyte transformation in experimentally infected chickens. *Avian Dis.* **35**, 783–792.

Adldinger, H. K., Calnek, B. W. (1973). Pathogenesis of Marek's disease: early distribution of virus and viral antigens in infected chickens. *J. Natl Cancer Inst.* **50**, 1287–1298.

Ahmad, J., Sharma, J. M. (1993). Protection against hemorrhagic enteritis and Newcastle disease in turkeys by embryo vaccination with monovalent and bivalent vaccines. *Avian Dis.* **37**, 485–491.

Allan, W. H., Faragher, J. T., Cullen, G. A. (1972). Immunosuppression by the infectious bursal agent in chickens immunized against Newcastle disease. *Vet. Rec.* **90**, 511–512.

Allen, P. M., Matsueda, G. R., Evans, R. J., Dunbar Jr, J. B., Marshall, G., Unanue, E. R. (1987). Identification of the T-cell and Ia contact residues of a T-cell antigenic peptide. *Nature* **327**, 713–715.

Alm, G. V., Siccardi, F. J., Peterson, R. D. (1972). Impairment of the lymphocyte response to phytohemagglutinin in chickens with Marek's disease. *Acta Pathol. Microbiol. Scand.* [A] **80**, 109–114.

Amrani, D. L. (1990). Regulation of fibrinogen biosynthesis: glucocorticoid and interleukin-6 control. *Blood Coagulation Fibrin.* **1**, 443–446.

Amrani, D. L., Mauzy-Melitz, D., Mosesson, M. W. (1986). Effect of hepatocyte-stimulating factor and glucocorticoids on plasma fibronectin levels. *Biochem. J.* **238**, 365–371.

Anderson, W. I., Reid, W. M., Lukert, P. D., Fletcher, O. J. (1977). Influence of infectious bursal disease on the development of immunity to *Eimeria tenella*. *Avian Dis.* **21**, 637–641.

Arstila, T. P., Lassila, O. (1993). Androgen-induced expansion of the peripheral blood $\gamma\delta$ T cell population in the chicken. *J. Immunol.* **151**, 6627–6633.

Arstila, T. P., Toivanen, P., Lassila, O. (1993). Helper activity of $CD4^+$ $\alpha\beta$ T cells is required for the avian $\gamma\delta$ T cell response. *Eur. J. Immunol.* **23**, 2034–2037.

Arstila, T. P., Vainio, O., Lassila, O. (1994). Evolutionarily conserved function of CD28 in $\alpha\beta$ T cell activation. *Scand. J. Immunol.* **40**, 368–371.

Audhya, T., Kroon, D., Heavner, G., Viamontes, G., Goldstein, G. (1986). Tripeptide structure of bursin, a selective B cell differentiating hormone of the bursa of Fabricius. *Science* **231**, 997–999.

Austin, L. M., Boissy, R. E. (1995). Mammalian tyrosinase-related protein-1 is recognized by autoantibodies from vitiliginous Smyth chickens. An avian model for human vitiligo. *Am. J. Pathol.* **146**, 1529–1541.

Austin, L. M., Boissy, R. E., Jacobson, B. S., Smyth Jr, J. R. (1992). The detection of melanocyte autoantibodies in the Smyth chicken model for vitiligo. *Clin. Immunol. Immunopathol.* **64**, 112–120.

Baba, T., Masumoto, K., Nishida, S., Kajikawa, T., Mitsui, M. (1988). Harderian gland dependency of immunoglobulin A production in the lacrimal fluid of chicken. *Immunology* **65**, 67–72.

Baba, T., Kawata, T., Masumoto, K., Kajikawa, T. (1990). Role of the Harderian gland in immunoglobulin A production in chicken lacrimal fluid. *Res. Vet. Sci.* **49**, 20–24.

Bacon, L. D. (1987). Influence of the major histocompatibility complex on disease resistance and productivity. *Poult. Sci.* **66**, 802–811.

Bacon, L. D., Kite Jr, J. H., Rose, N. R. (1974). Relation between the major histocompatibility (B) locus and autoimmune thyroiditis in obese chickens. *Science* **186**, 274–275.

Bacon, L. D., Witter, R. L., Fadly, A. M. (1989). Augmentation of retrovirus-induced lymphoid leukosis by Marek's disease herpesviruses in white leghorn chickens. *J. Virol.* **63**, 504–512.

Bagchi, N., Brown, T. R., Urdanivia, E., Sundick, R. S. (1985).

Induction of autoimmune thyroiditis in chickens by dietary iodine. *Science* **230**, 325–327.

Bagchi, N., Brown, T. R., Herdegen, D. M., Dhar, A., Sundick, R. S. (1990) Antioxidants delay the onset of thyroiditis in obese strain chickens. *Endocrinology* **127**, 1590–1595.

Bagchi, N., Brown, T. R., Sundick, R. S. (1995). Thyroid cell injury is an initial event in the induction of autoimmune thyroiditis by iodine in obese strain chickens. *Endocrinology* **136**, 5054–5060.

Balcarova, J., Hala, K., Hraba, T. (1973). Differences in antibody formation to the dinitrophenol group in inbred lines of chickens. *Folia Biol.* (Prague) **19**, 19–24.

Bang, B. G., Bang, F. B. (1968). Localized lymphoid tissues and plasma cells in paraocular and paranasal organ systems in chickens. *Am. J. Path.* **53**, 735–751.

Barger, B., Pace, J. L., Ragland, W. L. (1991). Purification and partial characterization of an avian thymic hormone. *Thymus* **17**, 181–197.

Barker, K. A., Hampe, A., Stoeckle, M. Y., Hanafusa, H. (1993). Tranformation-associated cytokine 9E3/CEF4 is chemotactic for chicken peripheral blood mononuclear cells. *J. Virol.* **67**, 3528–3533.

Barral, A., Barral-Netto, M., Yong, E. C., Brownell, C. E., Twardzik, D. R., Reed, S. G. (1994). Transforming growth factor-β as a virulence mechanism for *Leishmania Braziliensis*. *Proc. Natl Acad. Sci. USA* **90**, 3442–3446.

Barral-Netto, M., Barral, A. (1994). Transforming growth factor-β in tegumentary leishmaniasis. *Brazilian J. Med. Biol. Res.* **27**, 1–9.

Barta, D., Barta, V. (1975). Chicken (*Gallus gallus*) hemolytic complement: optimal conditions for its titration. *Immunol. Communication* **4**, 337–351.

Barta, D., Hubbert, N. L. (1978). Testing of hemolytic complement components in domestic animals. *Am.J. Vet. Res.* **39**, 1303–1308.

Barth, C. F., Ewert, D. L., Olson, W. C., Humphries, E. H. (1990). Reticuloendotheliosis virus REV-T (REV-A)-induced neoplasia: development of tumors within the T-lymphoid and myeloid lineages. *J. Virol.* **64**, 6054–6062.

Becht, H., Muller, H. (1991). Infectious bursal disease-B cell dependent immunodeficiency syndrome in chickens. *Behring Inst. Mitt.* **89**, 217–225.

Befus, A. D., Johnston, N. Leslie, G. A., Bienenstock, J. (1980). Gut-associated lymphoid tissue in the chicken. I. Morphology, ontogeny, and some fundamental characteristics of Peyer's patches. *J. Immunol.* **125**, 2626–2632.

Benatar, T., Iacampo, S., Tkalec, L., Ratcliffe, M. J. H. (1991). Expression of immunoglobulin genes in the avian embryo bone marrow revealed by retroviral transformation. *Eur. J. Immunol.* **21**, 2529–2536.

Benatar, T., Tkalec, L., Ratcliffe, M. J. H. (1992). Stochastic rearrangement of chicken immunoglobulin variable region genes in chicken B-cell development. *Proc. Natl Acad. Sci. USA* **89**, 7615–7619.

Benedict, A. A., Pollard, L. W., Morrow, P. R., Abplanalp, H. A., Maurer, P. H., Briles, W. E. (1975). Genetic control of immune responses in chickens. I. Responses to terpolymer of poly ($Glu^{60}Ala^{30}Tyr^{10}$) associated with major histocompatibility complex. *Immunogenetics* **2**, 313–324.

Bernot, A., Auffray, C. (1991). Primary structure and ontogeny of an avian CD3 transcript. *Proc. Natl Acad. Sci. USA* **88**, 2550–2554.

Bernot, A., Zoorob, R., Auffray, C. (1994). Linkage of a new member of the lectin supergene family to chicken MHC genes. *Immunogenetics* **39**, 221–229.

Beug, H., Müller, H., Grieser, S., Doederlein, G., Graf, T. (1981). Hematopoietic cells transformed *in vitro* by REV_T avian reticuloendotheliosis virus express characteristics of very immature lymphoid cells. *Virology* **115**, 295–309.

Bienenstock, J., Johnson, N., Perey, D. Y. E. (1973). Bronchial lymphoid tissue I. Morphologic characteristics. *Lab. Invest.* **28**, 686–692.

Biggs, P. M. (1957). The association of lymphoid tissue with the lymph vessels in the domestic chicken (*Gallus domesticus*). *Acta Anat.* **29**, 36–47.

Biggs, P. M. (1991). Lymphoproliferative disease of turkeys. In *Diseases of Poultry*, 9th edn (eds Calnek, B. W., Barnes, H. J., Beard, C. W., Reid, W. M., Yoder Jr, H. W.), pp. 456–459. Iowa State University Press, Ames, IA, USA.

Biggs, P. M., Long, P. L., Kenzy, S. G., Rootes, D. G. (1968). Relationship between Marek's disease and coccidiosis. II. The effect of Marek's disease on the susceptibility of chickens to coccidial infection. *Vet. Rec.* **83**, 284–289.

Blauer, M., Tuohimaa, P. (1995). Activin beta-A and beta-B subunit expression in the developing chicken bursa of Fabricius. *Endocrinology* **136**, 1482–1487.

Bloom, S., Bacon, L. (1985). Linkage of the major histocompatibility (B) complex and the nucleolar organizer region in the chicken. *J. Heredity* **76**, 146–154.

Boerkoel, C. F., Kung, H.-J. (1992). Transcriptional interaction between retroviral long terminal repeats (LTRs): mechanism of 5′ LTR suppression and 3′ LTR promoter activation of *c-myc* in avian B-cell lymphomas. *J. Virol.* **66**, 4814–4823.

Boissy, R. E., Moellmann, G., Trainer, A. T., Smyth Jr, J. R., Lerner, A. B. (1986). Delayed amelanotic (DAM or Smyth) chicken: melanocyte dysfunction *in vivo* and *in vitro*. *J. Invest. Dermatol.* **86**, 149–156.

Bombara, C. J., Taylor, R. L. (1991). Signal transduction events in chicken interleukin-1 production. *Poult. Sci.* **70**, 1372–1380.

Bounous, D. I., Goodwin, M. A., Brooks, R. L., Lamichhane, C. M., Campagnoli, R. P., Brown, J., Snyder, D. B. (1995). Immunosuppression and intracellular calcium signaling in splenocytes from chicks infected with chicken anemia virus, CL-1 isolate. *Avian Dis.* **39**, 135–140.

Bowen, M. A., Patel, D. D., Li., X., Modrell, B., Malacko, A. R., Wang, W. C., Marquardt, H., Neubauer, M., Pesando, J. M., Francke, U., Haynes, B. F., Aruffo, A. (1995). Cloning, mapping, and characterization of activated leucocyte-cell adhesion molecule (ALCAM), a CD6 ligand. *J. Exp. Med.* **181**, 2213–2220.

Box, P. G., Holmes, H. C., Bushell, A. C., Finney, P. M. (1988). Impaired response to killed Newcastle disease vaccine in chickens possessing circulating antibody to chicken anemia agent. *Avian Pathol.* **17**, 713–723.

Boyd, R. L., Ward, H. A. (1984). Lymphoid antigenic determinants of the chicken: ontogeny of bursa-dependent lymphoid tissue. *Immunol.* **8**, 149–167.

Boyd, R. L., Wilson, T. J., Ward, H. A., Mitrangas, K. (1990). Phenotypic characterization of chicken bursal stromal elements. *Dev. Immunol.* **1**, 41–51.

Boyd, R. L., Wilson, T. J., van de Water, J., Haapanen, L. A., Gershwin, M. E. (1991). Selective abnormalities in the thymic microenvironment associated with avian scleroderma, an inherited fibrotic disease of L200 chickens. *J. Autoimmun.* **4**, 369–380.

Boyd, R. L., Wilson, T. J., Bean, A. G., Ward, H. A., Gershwin, M. E. (1992). Phenotypic characterization of chicken thymic stromal elements. *Dev. Immunol.* **2**, 51–66.

Brandtzaeg, P., Baklien, K., Bjerke, K., Rognum, T. O., Scott, H., Valnes, K (1987). Nature and properties of the human gastro-intestinal immune system. In *Immunology of the Gastro-intestinal Tract* (eds Miller, K., Nicklin, S.), pp. 1–85. CRC Press, Boca Raton, FL, USA.

Brewer, J. M., Wunderlich, J. K., Kim, D. H., Carr, M. Y., Beach, G. G., Ragland, W. L. (1989). Avian thymic hormone (ATH) is a parvalbumin. *Biochem. Biophys. Res. Commun.* **160**, 1155–1161.

Brewer, J. M., Arnold, J., Beach, G. G., Ragland, W. L., Wunderlich, J. K. (1991). Comparison of the amino acid sequences of tissue-specific parvalbumins from chicken muscle and thymus and possible evolutionary significance. *Biochem. Biophys. Res. Commun.* **181**, 226–231.

Brezinschek, H. P., Faessler, R., Klocker, H., Kroemer, G., Sgonc, R., Dietrich, H., Jakober, R., Wick, G. (1990). Analysis of the immune-endocrine feedback loop in the avian system and its alteration in chickens with spontaneous auto-immune thyroiditis. *Eur. J. Immunol.* **20**, 2155–2159.

Briles, W. E. (1962). Additional blood group systems in the chicken. *Ann. NY Acad. Sci.* **97**, 173–183.

Briles, W. E., Krueger, W. F. (1955). The effect of parental B blood group genotypes on hatchability and livability in Leghorn inbred lines. *Poult. Sci.* **34**, 1182 (abstract).

Briles, W. E., Briles, D. L., Pollock, D. L., Pattison, M. (1980). Marek's disease resistance in chickens affected by complementation of B alloalleles in a cross of commercial parent stocks. *Anim. Blood Groups Biochem. Genet.* **2** (Suppl. 1), 27–28.

Brown, T. R., Sundick, R. S., Dhar, A., Sheth, D., Bagchi, N. (1991). Uptake and metabolism of iodine is crucial for the development of thyroiditis in obese strain chickens. *J. Clin. Invest.* **88**, 106–111.

Bucy, R. P., Chen, C. H., Cihak, J., Lösch, U., Cooper, M. D. (1988). Avian T cells expressing $\gamma\delta$ receptors localize in the splenic sinusoids and the intestinal epithelium. *J. Immunol.* **141**, 2200–2205.

Bucy, R. P., Coltey, M., Chen, C. H., Char, D., LeDouarin, N. M., Cooper, M. D. (1989). Cytoplasmic $CD3^+$ surface $CD8^+$ lymphocytes develop as a thymus-independent lineage in chick-quail chimeras. *Eur. J. Immunol.* **19**, 1449–1455.

Bucy, R. P., Chen, C. H., Cooper, M. D. (1990a). Development of cytoplasmic $CD3^+$/T cell receptor-negative cells in the peripheral lymphoid tissues of chickens. *Eur. J. Immunol.* **20**, 1345–1350.

Bucy, R. P., Chen, C. H., Cooper, M. D. (1990b). Ontogeny of T cell receptors in the chicken thymus. *J. Immunol.* **144**, 1161–1168.

Bulloch, K. (1988). A comparative study of the autonomic nervous system innervation of the thymus in the mouse and chicken. *Int. J. Neurosci.* **40**, 129–140.

Bumstead, J. M., Payne, L. N. (1987). Production of an immune suppressor factor by Marek's disease lymphoblastoid cell lines. *Vet. Immunol. Immunopathol.* **16**, 47–66.

Bumstead, J. M., Flack, I. H., Payne, L. N. (1985). Studies on the role of prostaglandin E2 in Marek's disease. In *International Symposium on Marek's Disease* (eds Calnek, B. W., Spencer, J. L.), pp. 268–285. American Association of Avian Pathologists, Kennett Square, PA, USA.

Burg, R. W., Feldbush, T., Morris, C. A., Maag, T. A. (1971). Depression of thymus and bursa dependent immune systems of chicks with Marek's disease. *Avian Dis.* **15**, 662–671.

Burt, D. W., Paton, I. R. (1991). Molecular cloning and primary structure of the chicken transforming growth factor-β2 gene. *DNA Cell Biol.* **10**, 723–734.

Buscaglia, C., Calnek, B. W. (1988). Maintenance of Marek's disease herpesvirus latency *in vitro* by a factor found in conditioned medium. *J. Gen. Virol.* **69**, 2809–2818.

Byrnes, S., Eaton, R., Kogut, M. (1993a). *In vitro* interleukin-1 and tumor necrosis factor-alpha production by macrophages from chickens infected with either *Eimeria maxima* or *Eimeria tenella*. *Int. J. Parasitol.* **23**, 639–645.

Byrnes, S., Emerson, K., Kogut, M. (1993b). Dynamics of cytokine production during coccidial infections in chickens: colony-stimulating factors and inferferon. *FEMS Microbiol. Immunol.* **6**, 45–52.

Calhoun, M. L. (1954). *Microscopic Anatomy of the Digestive System of the Chicken.* Iowa State University Press, Ames, IA, USA.

Calnek, B. W. (1985). Pathogenesis of Marek's disease – a review. In *International Symposium on Marek's Disease* (eds Calnek, B W., Spencer, J. L.), pp. 374–390. American Association of Avian Pathologists, Kennett Square, PA, USA.

Calnek, B. W. (1986). Marek's disease – a model for herpesvirus oncology. *CRC Crit. Rev. Microbiol.* **12**, 293–320.

Calnek, B. W., Witter, R. L. (1991). Marek's disease. In *Diseases of Poultry*, 9th edn (eds Calnek, B. W., Barnes, H. J., Beard, C. W., Reid, W. M., Yoder Jr, H. W.), pp. 342–385. Iowa State University Press, Ames, IA, USA.

Calnek, B. W., Witter, R. L. (1997). Marek's disease. In *Diseases of Poultry*, 10th edn (eds Calnek, B. W., Barnes, H. J., Beard, C. W., McDougald, L. R., Saif, Y. M.), pp. 369–413. Iowa State University Press, Ames, IA, USA.

Calnek, B. W., Higgins, D. A., Fabricant, J. (1975). Rous sarcoma regression in chickens resistant or susceptible to Marek's disease. *Avian Dis.* **19**, 473–481.

Calnek, B. W., Fabricant, J., Schat, K. A., Murthy, K. K. (1978). Rejection of a transplantable Marek's disease lymphoma in normal vs. immunologically deficient chickens. *J. Natl Cancer Inst.* **60**, 623–630.

Calnek, B. W., Schat, K. A., Ross, L. J. N., Shek, W. R., Chen, C.-L. H. (1984). Further characterization of Marek's disease virus-infected lymphocytes. I. *In vivo* infection. *Int. J. Cancer* **33**, 289–298.

Calnek, B. W., Lucio, B., Schat, K. A., Lillehoj, H. S. (1989). Pathogenesis of Marek's disease virus-induced local lesions. 1. Lesion characterization and cell line establishment. *Avian Dis.* **33**, 291–302.

Carlson, L. M., Oettinger, M. A., Schatz, D. G., Masteller, E. L., Hurley, E. A., McCormack, W. T., Baltimore, D., Thompson, C. B. (1991). Selective expression of RAG-2 in chicken B cells undergoing immunoglobulin gene conversion. *Cell* **64**, 201–208.

Chai, J. Y., Lillehoj, H. S. (1988). Isolation and functional characterization of chicken intestinal intra-epithelial lymphocytes showing natural killer cell activity against tumour target cells. *Immunology* **63**, 111–117.

Chan, M. M., Chen, C.-L. H., Ager, L. L., Cooper, M. D. (1988).

Identification of the avian homologues of mammalian CD4 and CD8 antigens. *J. Immunol.* **140**, 2133–2138.

Chang, E. B., Musch, M. W., Mayer, L. (1990). Interleukins 1 and 3 stimulate anion secretion in chicken intestine. *Gastroenterology* **98**, 1518–1524.

Chang, W. P., Marsh, J. A. (1993). The effect of synthetic thymulin on cell surface marker expression by avian T cell precursors. *Dev. Comp. Immunol.* **17**, 85–96.

Chanh, T. C., Benedict, A. A., Abplanap, H. (1976). Association of serum hemolytic complement levels with the major histocompatibility complex in chickens. *J. Exp. Med.* **144**, 555–561.

Char, D., Sanchez, P., Chen, C. H., Bucy, R. P., Cooper, M. D. (1990). A third sublineage of avian T cells can be identified with a T cell receptor-3-specific antibody. *J. Immunol.* **145**, 3547–3555.

Chen, C.-L. H., Cooper, M. D. (1987). Identification of cells surface molecules on chicken lymphocytes with monoclonal antibodies. In *Avian Immunology: Basis and Practice*, Vol. I (eds Toivanen, A., Toivanen, P.), pp. 137–154, CRC Press, Boca Raton, FL, USA.

Chen, C. H., Lehmeyer, J. E., Cooper, M. D. (1982). Evidence for an IgD homologue on chicken lymphocytes. *J. Immunol.* **129**, 2580.

Chen, C.-L. H., Chanh, T. C., Cooper, M. D. (1984). Chicken thymocyte-specific antigen identified by monoclonal antibodies: ontogeny, tissue distribution and biochemical characterization. *Eur. J. Immunol.* **14**, 385–391.

Chen, C. H., Ager, L. L., Gartland, G. L., Cooper, M. D. (1986). Identification of a T3/T cell receptor complex in chickens. *J. Exp. Med.* **164**, 375–380.

Chen, C. H., Cihak, J., Lösch, U., Cooper, M. D. (1988a). Differential expression of two T cell receptors, TCR1 and TCR2, on chicken lymphocytes. *Eur. J. Immunol.* **18**, 539–543.

Chen, L., Lim, M. Y., Bose Jr, H., Bishop, J. M. (1988b). Rearrangements of chicken immunoglobulin genes in lymphoid cells transformed by the avian retroviral oncogene *v-rel*. *Proc. Natl Acad. Sci. USA* **85**, 549–553.

Chen, C. H., Sowder, J. T., Lahti, J. M., Cihak, J., Lösch, U., Cooper, M. D. (1989). TCR3: a third T cell receptor in the chicken. *Proc. Natl Acad. Sci.* **86**, 2351–2355.

Chen, C. L. H., Bucy, R. P., Cooper, M. D. (1990). T cell differentiation in birds. *Semin. Immunol.* **2**, 79–86.

Chen, C. H., Pickel, J. M., Lahti, J. M., Cooper, M. D. (1991). Surface markers on avian immune cells. In *Avian Cellular Immunology* (ed. Sharma, J. M.), pp. 1–22. CRC Press, Boca Raton, FL, USA.

Chen, C. H., Göbel, T. W. F., Kubota, T., Cooper, M. D. (1994). T cell development in the chicken. *Poult. Sci.* **73**, 1012–1018.

Chen, C. H., Six, A., Kubota, T., Tsuji, S., Kong, F.-K., Göbel, T. W. F., Cooper, M. D. (1996). T cell receptors and T cell development. *Curr. Topics Microbiol. Immunol.* **212**, 37–53.

Cheville, N. F. (1967). Studies on the pathogenesis of Gumboro disease in the bursa of Fabricius, spleen, and thymus of the chicken. *Am. J. Pathol.* **51**, 527–551.

Cheville, N. F., Beard, C. W. (1972). Cytopathology of Newcastle disease. The influence of bursal and thymic lymphoid systems in the chicken. *Lab. Invest.* **27**, 129–143.

Cheville, N., Sato, S. (1977). Pathology of adenoviral infection in turkeys (*Meleagris gallopavo*) with respiratory disease and colisepticemia. *Vet. Pathol.* **14**, 567–581.

Chi, D. S., Galton, J. E., Thorbecke, J. (1981). Role of T cells in immune responses of the chicken. In *Avian Immunology* (eds Rose, M. E., Payne, L. N., Freeman, B. M.), pp. 102–134. Clark Constable Limited, Edinburgh, UK.

Cho, B. R. (1970). Experimental dual infections of chickens with infectious bursal and Marek's disease agents. I. Preliminary observation on the effect of infectious bursal agent on Marek's disease. *Avian Dis.* **14**, 665–675.

Chow, L. M., Ratcliffe, M. J., Veillette, A. (1992). *tkl* is the avian homolog of the mammalian *lck* tyrosine protein kinase gene. *Mol. Cell. Biol.* **12**, 1226–1233.

Chung, K. S., Lillehoj, H. S., Jenkins, M. C. (1991). Avian leukocyte common antigens: molecular weight determination and flow cytometric analysis using new monoclonal antibodies. *Vet. Immunol. Immunopathol.* **28**, 259–273.

Cihak, J., Ziegler-Heitbrock, H. W., Trainer, H., Schranner, I., Merkenschlager, M., Lösch, U. (1988). Characterization and functional properties of a novel monoclonal antibody which identifies a T cell receptor in chickens. *Eur. J. Immunol.* **18**, 533–538.

Cihak, J., Hoffmann-Fezer, G., Ziegler-Heibrock, H. W. L., Stein, H., Kaspers, B., Chen, C. H., Cooper, M. D., Lösch, U. (1991). T cells expressing the Vβ1 T-cell receptor are required for IgA production in the chicken. *Proc. Natl Acad. Sci. USA* **88**, 10951–10955.

Cihak, J., Hoffmann-Fezer, G., Koller, A., Kaspers, B., Merkle, H., Halá, K., Wick, G., Lösch, U. (1995). Preferential TCR V beta 1 gene usage by autoreactive T cells in spontaneous autoimmune thyroiditis of the obese strain chicken. *J. Autoimmun.* **8**, 507–520.

Cloud, S. S., Rosenberger, J. K., Lillehoj, H. S. (1992a). Immune dysfunction following infection with chicken anemia agent and infectious bursal disease virus. I. Kinetic alterations of avian lymphocyte subpopulations. *Vet. Immunol. Immunopathol.* **34**, 337–352.

Cloud, S. S., Rosenberger, J. K., Lillehoj, H. S. (1992b). Immune dysfunction following infection with chicken anemia agent and infectious bursal disease virus. II. Alterations of *in vitro* lymphoproliferation and *in vivo* immune responses. *Vet. Immunol. Immunopathol.* **34**, 353–366.

Collins, W. M., Briles, W. E., Zsigray, R. M., Dunlop, W. R., Corbet, A. C., Clark, K. K., Marks, J. L., McGrail, T. P. (1977). The B locus (MHC) in the chicken: association with the fate of RSV-induced tumors. *Immunogenetics* **5**, 333–343.

Coltey, M., Bucy, R. P., Chen, C. H., Cihak, J., Lösch, U., Char, D., LeDouarin, N. M., Cooper, M. D. (1989). Analysis of the first two waves of thymus homing stem cells and their T cell progeny in chick quail chimeras. *J. Exp. Med.* **170**, 543–557.

Confer, A. W., MacWilliams, P. S. (1982). Correlation of hematological changes and serum monocyte inhibition with the early suppression of phytohemagglutinin stimulation of lymphocytes in experimental infectious bursal disease. *Can. J. Comp. Med.* **46**, 169–175.

Confer, A. W., Springer, W. T., Shane, S. M., Conovan, J. F. (1981). Sequential mitogen stimulation of peripheral blood lymphocytes from chickens inoculated with infectious bursal disease virus. *Am.J. Vet. Res.* **42**, 2109–2113.

Cooper, M. D., Chen, C. H. (1993). Avian T cell development and the $\gamma\delta/\alpha\beta$ T cell connection. In *Proc. Xth World Vet. Poultry Assoc. Congress*, Sydney, pp. 27–31.

Cooper, M. D., Peterson, R. D. A., South, M. A., Good, R. A.

(1966). The functions of the thymus system and the bursa system in the chicken. *J. Exp. Med.* **123**, 75–102.

Cooper, M. D., Cain, W. A., Van Alten, P. J., Good, R. A. (1969). Development and function of the immunoglobulin producing system. I. Effect of bursectomy at different stages of development on germinal centers, plasma cells, immunoglobulin, and antibody production. *Int. Arch. Allergy* **35**, 242–252.

Cooper, M. D., Chen, C. H., Bucy, R. P., Thompson, C. B. (1991). Avian T cell ontogeny. *Adv. Immunol.* **30**, 87–117.

Corbel, C., Bluestein, H. G., Pourquie, O., Vaigot, P., Le Douarin, N. (1992). An antigen expressed by avian neuronal cells is also expressed by activated T lymphocytes. *Cell. Immunol.* **141**, 99–110.

Cormier, F., Dieterlen-Lievre, F. (1988). The wall of the chick embryo aorta harbors M-CFC, G-CFC and BFU-E. *Development* **102**, 279–285.

Cormier, F., de Paz, P., Dieterlen-Lievre, F. (1986). *In vitro* detection of cells with monocytic potentiality in the wall of the chick embryo aorta. *Dev. Biol.* **118**, 167–175.

Corrier, D. E. (1991). Mycotoxicosis: mechanisms of immunosuppression. *Vet. Immunol. Immunopathol.* **30**, 73–87.

Cortes, A., Fonfria, J., Vicente, A., Varas, A., Moreno, J., Zapata, A. G. (1995). T-dependent areas in the chicken bursa of Fabricius: an immunological study. *Anat. Rec.* **242**, 91–95.

Coudert, F., Cauchy, L., Dambrine, G. (1979). Evidence for a thymic factor in sera of chickens. *Folia Biol.* (Prague) **25**, 317–318.

Craft, D. W., Brown, J., Lukert, P. D. (1990). Effects of standard and variant strains of infectious bursal disease virus on infections of chickens. *Am. J. Vet. Res.* **51**, 1192–1197.

Cui, Z., Qin, A. (1996). Immunodepressive effects of the recombinant 38KD phosphorylated protein of Marek's disease virus on chicks. *Proc. 4th Int. Symp. Marek's Disease*, p. 35.

Dahan, A., Reynaud, C. A., Weill, J. C. (1983). Nucleotide sequence of a chicken μ heavy chain mRNA. *Nucl. Acids Res.* **11**, 5381–5389.

Danielpour, D., Kim, K. Y., Dart, L. L., Watanabe, S., Roberts, A. B., Sporn, M. B. (1989). Sandwich enzyme-linked immunosorbent assays (SELISAs) quantitate and distinguish two forms of transforming growth factor-beta (TGF-beta 1 and TGF-beta 2) in complex biological fluids. *Growth Fact.* **2**, 61–71.

de Villartay, J.-P., Hockett, R. D., Coran, D., Korsmeyer, S. J., Cohen, D. I. (1988). Deletion of the human T cell receptor α-gene by a site-specific recombination. *Nature* **335**, 170–174.

Del-Cacho, E., Gallego, M., Bascuas, J. A. (1992). Ultrastructural localization of a soluble antigen in the chicken Harderian gland. *Immunol.* **16**, 209–219.

Del-Cacho, E., Gallego, M., Sanz, A., Zapata, A. (1993). Characterization of distal lymphoid nodules in the chicken caecum. *Anat. Rec.* **237**, 512–517.

Del-Cacho, E., Gallego, M., Arnal, C., Bascuas, J. A. (1995). Localization of splenic cells with antigen-transporting capability in the chicken. *Anat. Rec.* **241**, 105–112.

Dieterlen-Lièvre, F. (1989). Development of the compartments of the immune system in the avian embryo. *Dev. Comp. Immunol.* **13**, 303–311.

Dieterlen-Lièvre, F., Martin, C. (1981). Diffuse intraembryonic hemopoiesis in normal and chimeric avian development. *Dev. Biol.* **85**, 180–191.

Dietert, R. R., Golemboski, K. A. (1994). Environment-immune interactions. *Poult. Sci.* **73**, 1062–1076.

Dietert, R. R., Taylor, R. L., Dietert, M. F. (1991). Biological function of the chicken major histocompatibility complex. *CRC Crit. Rev. Poult. Biol.* **3**, 111–129.

Digby, M. R., Lowenthal, J. W. (1995). Cloning and expression of the chicken interferon-γ-gene. *J. Interferon Cytok. Res.* **15**, 933–938.

Dijkmans, R., Creemers, J., Billiau, A. (1990). Chicken macrophage activation by interferon: do birds lack the molecular homologue of mammalian interferon-γ? *Vet. Immunol. Immunopathol.* **26**, 319–332.

Doehring, C., Riegert, P., Salomonsen, J., Skjoedt, K., Kaufman, J. (1993). The extracellular Ig V-like regions of the polymorphic B-G antigens of the chicken MHC lack structural features expected for antibody variable regions. In *Avian Immunology in Progress* (ed. Coudert, F.), pp 145–152. Inst. Nat. de la Rech. Agronomique, les Colloques no. 62, Paris, France.

Dolfi, A., Bianchi, F., Lupetti, M., Michelucci, S. (1989). The significance of the intestinal flow in the maturing of B lymphocytes and the chicken antibody response. *J. Anat.* **166**, 233–242.

Domermuth, C. H., Larsen, C. T. (1984). Vaccination against hemorrhagic enteritis of turkeys: an update. *J. Am. Vet. Med. Assoc.* **185**, 337.

Drén, C. N., Sághy, E., Glávits, R., Rátz, F., Ping, J., Sztojkov, V. (1983). Lymphoreticular tumour in pen-raised pheasants associated with a reticuloendotheliosis-like virus infection. *Avian Pathol.* **12**, 55–71.

Drén, C. N., Németh, I., Sári, I., Rátz, F., Glávits, R., Somogyi, P. (1988). Isolation of a reticuloendotheliosis-like virus from naturally occurring lymphoreticular tumours of domestic goose. *Avian Pathol.* **17**, 259–277.

Droege, W. (1971). Amplifying and suppressive effect of thymus cells. *Nature* **234**, 549–551.

Droege, W. (1976). The antigen-inexperienced thymic suppressor cells: a class of lymphocytes in the young chicken thymus that inhibits antibody production and cell-mediated immune responses. *Eur. J. Immunol.* **6**, 279–287.

Duncan, M. R., Wilson, T. J., van de Water, J., Berman, B., Boyd, R., Wick, G., Gershwin, M. E. (1992). Cultured fibroblasts in avian scleroderma, an autoimmune fibrotic disease, display an activated phenotype. *J. Autoimmun.* **5**, 603–615.

Duncan, M. R., Berman, B., van de Water, J., Boyd, R. L., Wick, G., Gershwin, M. E. (1995). Mononuclear cells from fibrotic skin lesion in avian scleroderma constitutively produce fibroblast-activating cytokines and immunoglobulin M. *Int. Arch. Aller. Immunol.* **107**, 519–526.

Dunon, D., Imhof, B. A. (1996). T cell migration during ontogeny and T cell repertoire generation. *Curr. Topics Microbiol. Immunol.* **212**, 79–93.

Dunon, D., Cooper, M. D., Imhof, B. A. (1993a). Thymic origin of embryonic intestinal $\gamma\delta$ T cells. *J. Exp. Med.* **177**, 257–263.

Dunon, D., Cooper, M. D., Imhof, B. A. (1993b). Migration patterns of thymus-derived $\gamma\delta$ T cells during chicken development. *Eur. J. Immunol.* **23**, 2545–2550.

Dunon, D., Schwager, J., Dangy, J. P., Cooper, M. D., Imhof, B. A. (1994). T cell migration during development: homing is not related to TCR Vβ1 repertoire selection. *EMBO J.* **13**, 808–815.

Dunon, D., Schwager, J., Dangy, J. P., Imhof, B. A. (1995).

Ontogeny of TCR Vβ1 expression revealed novel invariant alternative transcripts. *J. Immunol.* **154**, 1256–1264.

Edwards, S. V., Wakeland, E. K., Potts, W. K. (1995). Contrasting histories of avian and mammalian MHC genes revealed by class II beta sequences from songbirds. *Proc. Natl Acad. Sci. USA* **92**, 12200–12204.

Eerola, E., Veromaa, T., Toivanen, P. (1987). Special features in the structural organization of the Avian lymphoid system. In *Avian Immunology: Basis and Practice* (eds Toivanen, A., Toivanen, P.), pp. 9–21. CRC Press Boca Raton, FL, USA.

Eisenbarth, G. S., Walsh, F. S., Nirenberg, M. (1979). Monoclonal antibody to a plasma membrane antigen of neurons. *Proc. Natl Acad. Sci. USA* **76**, 4913–4917.

Ellis, M. N., Eidson, C. S., Brown, J., Kleven, S. H. (1983a). Studies on interferon induction and interferon sensitivity of avian reoviruses. *Avian Dis.* **27**, 927–936.

Ellis, M. N., Eidson, C. S., Fletcher, O. J., Kleven, S. H. (1983b). Viral tissue tropisms and interferon production in White Leghorn chickens infected with two reovirus strains. *Avian Dis.* **27**, 644–651.

Enrietto, P. J., Hayman, M. J. (1987). Structure and virus-associated oncogenes of avian sarcoma and leukemia viruses. In *Avian Leukosis* (ed. de Boer, G. F.), pp. 29–46. Martinus Nijhoff Publishing, Dordrecht, The Netherlands.

Erf, G. F., Trejo-Skalli, A. V., Smyth Jr, J. R. (1995). T cells in regenerating feathers of Smyth line chickens with vitiligo. *Clin. Immunol. Immunopathol.* **76**, 120–126.

Evans, D. L., Beasley, J. N., Patterson, L. T. (1971). Correlation of immunological competence with lesions in selected lymphoid tissues from chickens with Marek's disease. *Avian Dis.* **15**, 680–687.

Ewert, D. L., Munchus, M. S., Chen, C. H., Cooper, M. D. (1984). Analysis of structural properties and cellular distribution of avian Ia antigen by using monoclonal antibodies to monomorphic determinants. *J. Immunol.* **132**, 2524–2530.

Ezeokoli, C. D., Ityondo, E. A., Nwannenna, A. I., Umoh, J. U. (1990). Immunosuppression and histopathological changes in the bursa of Fabricius associated with infectious bursal disease vaccination in chicken. *Comp. Immunol. Microbiol. Infect. Dis.* **13**, 181–188.

Fadley, A. M., Winterfield, R. W., Olander, H. J. (1976). Role of the bursa of Fabricius in the pathogenicity of inclusion body hepatitis and infectious bursal disease virus. *Avian Dis.* **20**, 467–477.

Fagerland, J. A., Arp, L. H. (1993a). Structure and development of bronchus-associated lymphoid tissue in conventionally reared broiler chickens. *Avian Dis.* **37**, 10–18.

Fagerland, J. A., Arp, L. H. (1993b). Distribution and quantitation of plasma cells, T cells, lymphocyte subsets, and B lymphocytes in bronchus-associated lymphoid tissue of chickens. *Region. Immunol.* **5**, 28–36.

Faragher, J. T., Allan, W. H., Wyeth, C. J. (1974). Immunosuppressive effect of infectious bursal agent on vaccination against Newcastle disease. *Vet. Rec.* **95**, 385–388.

Fässler, R., Dietrich, H., Krömer, G., Schwarz, S., Brezinschek, H. P., Wick, G. (1988). Diminished glucocorticoid tonus in Obese strain (OS) chickens with spontaneous autoimmune thyroiditis: increased plasma levels of a physicochemically unaltered corticosteroid binding globulin but normal total corticosterone plasma concentration and normal glucocorticoid receptor contents in lymphoid tissue. *J. Steroid Biochem.* **30**, 375–379.

Fedecka-Bruner, B., Penningers, J. Vaigot, P., Lehmann, A., Martínez A. C., Kroemer, G. (1991). Developmental expression of IL-2-receptor light chain (CD25) in the chicken embryo. *Dev. Immunol.* **1**, 237–242.

Fix, A. S., Arp, L. H. (1991). Morphologic characterization of conjunctiva-associated lymphoid tissue in chickens. *Am. J. Vet. Res.* **52**, 1852–1859.

Fleischer, B. (1980). Effector cells in avian spontaneous and antibody-dependent cell-mediated cytotoxicity. *J. Immunol.* **125**, 1161–1166.

Frazier, J. A. (1974). Ultrastructure of lymphoid tissue from chicks infected with Marek's disease virus. *J. Natl Cancer Inst.* **52**, 829–837.

Fredericksen, T. L., Sharma, J. M. (1987). Purification of avian T cell growth factor and immune interferon using gel filtration high resolution chromatography. In *Avian Immunology* (eds Weber, W. T., Ewert, D. L., Liss, A. R.), p. 145. Academic Press, New York, USA.

Fulton, J. E., Thacker, E. L., Bacon, L. D., Hunt, H. D. (1995). Functional analysis of avian class I (BFIV) glycoproteins by epitope tagging and mutagenesis *in vitro*. *Eur. J. Immunol.* **25**, 2069–2076.

Gabrielsen, A. E., Pickering, R. J., Good, R. A. (1973a). Haemolysis in chicken serum I. The ionic environment. *Immunology* **25**, 167–177.

Gabrielsen, A. E., Pickering, R. J., Linna, T. J., Good, R. A. (1973b). Haemolysis in chicken serum. II. Ontogenetic development. *Immunology* **25**, 179–184.

Gallego, M., Del-Cacho, E., Arnal, C., Bascuas, J. A. (1992a). Local immune response in the chicken Harderian gland to antigen given by different ocular routes. *Res. Vet. Sci.* **52**, 38–43.

Gallego, M., Del-Cacho, E., Felices, C., Bascuas, J. A. (1992b). Immunoglobulin classes synthesized by the chicken Harderian gland after local immunization. *Res. Vet. Sci.* **52**, 44–47.

Gallego, M., Del-Cacho, E., Arnal, C., Felices, C., Lloret, E., Bascuas, J. A. (1992c). Immunocytochemical detection of dendritic cells by S-100 protein in the chicken. *Eur. J. Histochem.* **36**, 205–213.

Gallego, M., Del-Cacho, E., Bascuas, J. A. (1995). Antigen-binding cells in the cecal tonsil and Peyer's patches of the chicken after bovine serum albumin administration. *Poult. Sci.* **74**, 472–479.

Gavora, J. S. (1990). Disease genetics. In *Poultry Breeding and Genetics* (ed. Crawford, R. D.), pp 805–846. Elsevier Science Publ., B.V., Amsterdam, The Netherlands.

Gelb, J., Eidson, C. S., Fletcher, O. J., Kleven, S. J. (1979). Studies on interferon induction by infectious bursal disease virus (IBDV). II. Interferon production in White Leghorn chickens infected with an attenuated or pathogenic isolate of IBDV. *Avian Dis.* **23**, 634–645.

Gendron, R. L., Nestel, F. P., Lapp, W. S. (1991). Expression of tumor necrosis factor alpha in the developing nervous system. *Int. J. Neurosci.* **60**, 129–136.

George, J. F., Cooper, M. D. (1990). γ/δ T cells and α/β T cells differ in their developmental patterns of receptor expression and modulation requirements. *Eur. J. Immunol.* **20**, 2177–2181.

Gershwin, M. E., Abplanalp, H., Castles, J. J., Ikeda, R. M., van de Water, J., Eklund, J., Haynes, D. (1981). Characterization of a spontaneous disease of White Leghorn chickens resembling progressive systemic sclerosis (scleroderma). *J. Exp. Med.* **153**, 1640–1659.

Gewurz, H., South, M. A., Good, R. A. (1966). The ontogeny of complement activity. Complement titers in the developing chick embryo during graft-vs.-host reactions. *Proc. Soc. Exp. Biol. Med*: **23**, 718–721.

Giambrone, J. J., Donahoe, J. P., Dawe, D. L., Eidson, C. S. (1977). Specific suppression of the bursa-dependent immune system of chicks with infectious bursal disease virus. *Am. J. Vet. Res.* **38**, 581–583.

Giambrone, J. J., Klesius, P. H., Eckamn, M. K., Edgar, S. A. (1981). Influence of hormonal and chemical bursectomy on the development of acquired immunity to coccidia in broiler chickens. *Poult. Sci.* **60**, 2612–2618.

Glick, B. (1978). The immune response in the chicken: lymphoid development of the bursa of Fabricius and thymus and an immune response role for the gland of Harder. *Poult. Sci.* **57**, 1441–1444.

Glick, B. (1995). Embryogenesis of the bursa of Fabricius: stem cell, microenvironment, and receptor–paracrine pathways. *Poult. Sci.* **74**, 419–426.

Glick, B., Rosse, C. (1981). Cellular composition of the bone marrow in the chicken: II. The effect of age and the influence of the bursa of Fabricius on the size of cellular compartments. *Anat. Rec.* **200**, 471–479.

Glick, B., Chang, T. S., Japp, R. G. (1956). The bursa of Fabricius and antibody production. *Poult. Sci.* **35**, 224–225.

Glick, B., Holbrook, K. A., Olah, I., Perkins, W. D., Stinson, R. (1981). An electron and light microscopy study of caecal tonsil. *Dev. Comp. Immunol.* **5**, 95–104.

Göbel, T. W. F., Chen, C. H., Shrimpf, J., Grossi, C. E., Bernot, A., Bucy, R. P., Auffray, C., Cooper, M. D. (1994a). Characterization of avian natural killer cells and their intracellular CD3 protein complex. *Eur. J. Immunol.* **24**, 1685–1691.

Göbel, T. W. F., Chen, C. H., Lahti, J. M., Kubota, T., Kuo, C. L., Aebersold, R., Hood, L., Cooper, M. D. (1994b). Identification of T cell receptor α-chain genes in the chicken. *Proc. Natl Acad. Sci. USA* **91**, 1094–1098.

Göbel, T. W. F., Chen, C. H., Cooper, M. D. (1996a). Avian natural killer cells. *Curr. Topics Microbiol. Immunol.* **212**, 107–117.

Göbel, T. W. F., Chen, C. H., Cooper, M. D. (1996b). Expression of an avian CD6 candidate is restricted to $\alpha\beta$ T cells, splenic $CD8^+$ $\gamma\delta$ T cells and embryonic NK cells. *Eur. J. Immunol.* **26**, 1743–1747.

Gonneville, L, Martins, T. J., Bédard, P.-A. (1991). Complex expression pattern of the CEF-4 cytokine in transformed and mitogenically stimulated cells. *Oncogene* **6**, 1825–1833.

Goodwin, M. A., Brown, J., Smeltzer, M. A., Girshick, T., Miller, S. L., Dickson, T. G. (1992). Relationship of common avian pathogen antibody titers in so-called chicken anemia agent (CAA)-antibody-positive chicks to titers in CAA-antibody negative chicks. *Avian Dis.* **36**, 356–358.

Goryo, M., Suwa, T., Umemura, T., Hakura, C., Yamashiro, S. (1989). Histopathology of chicks inoculated with chicken anemia agent (MSB1-TK 5803 strain). *Avian Pathol.* **18**, 73–89.

Granfors, K., Martin, C., Lassila, O., Suvitaival, R., Toivanen, A., Toivanen, P. (1982). Immune capacity of the chicken bursectomized at 60 h of incubation: production of immunoglobulins and specific antibodies. *Clin. Immunol. Immunopathol.* **23**, 549–569.

Gross, W. B., Domermuth, C. H. (1975). Spleen lesions of hemorrhagic enteritis of turkeys. *Avian Dis.* **20**, 455–466.

Guellati, M., Ramade, F., Le Nguyen, D., Ibos, F., Bayle, J. D. (1991). Effects of early embryonic bursectomy and opotherapic substitution on the functional development of the adrenocorticotropic axis. *J. Dev. Physiol.* **15**, 357–363.

Guida, S., Heguy, A., Melli, M. (1992). The chicken IL-1 receptor: differential evolution of the cytoplasmic and extracellular domains. *Gene* **111**, 239–243.

Guillemot, F., Auffray, C. (1989). The molecular biology of the chicken major histocompatibility complex. *CRC Critical Rev. Poult. Biol.* **2**, 255–275.

Guillemot, F., Turmel, P., Charron, D., Le Douarin, N., Auffray, C. (1986). Structure, polymorphism and biosynthesis of chicken MHC class II (B-L) antigens and associated molecules. *J. Immunol.* **137**, 1251–1257.

Guillemot, F., Billault, A., Pourquie, O., Behar, G., Chausse, A.-M., Zoorob, R., Kreiblich, G., Auffray, C. (1988). A molecular map of the chicken major histocompatibility complex: the class II beta genes are closely-linked to the class I genes and the nucleolar organizer. *EMBO J.* **7**, 2775–2785.

Guillemot, F., Billault, A., Auffrey, C. (1989). Physical linkage of a guanine nucleotide-binding protein related gene to the chicken major histocompatibility complex. *Proc. Natl Acad. Sci. USA* **86**, 4594–4598.

Günther, E., Balcarova, J., Hala, K., Ruede, E., Hraba, T. (1974). Evidence for an association between immune responsiveness of chickens to (T,G)-A-L and the major histocompatibility system. *Eur. J. Immunol.* **4**, 548–553.

Hála, K. (1977). The major histocompatibility system of the chicken. In *The Major Histocompatibility System in Man and Animals* (ed. Goetze, G.), pp. 291–312. Springer-Verlag, Berlin, Heidelberg, New York.

Hála, K., Boyd, R., Doick, G. (1981). Chicken major histocompatibility complex and disease. *Scand. J. Immunol.* **14**, 607–616.

Hála, K., Schauenstein, K., Neu, N., Kroemer, G., Wolf, H., Boeck, G., Wick, G. (1986). A monoclonal antibody reacting with a membrane determinant expressed on activated chicken T lymphocytes. *Eur. J. Immunol.* **16**, 1331–1336.

Hansen, M. P., Law, G. R. J., van Zandt, J. N. (1967). Differences in susceptibility to Marek's disease in chickens carrying two different B locus blood group alleles. *Poult. Sci.* **46**, 1268 (abstract).

Haury, M., Kasahara, Y., Schaal, S., Bucy, R. P., Cooper, M. D. (1993). Intestinal T lymphocytes in the chicken express an integrin-like antigen. *Eur. J. Immunol.* **23**, 313–319.

Hayari, Y., Schauenstein, K., Globerson, A. (1982). Avian lymphokines, IL: interleukin-1 activity in supernatants of stimulated adherent splenocytes of chickens. *Dev. Comp. Immunol.* **6**, 785–788.

Haynes, D. C., Gershwin, M. E. (1984). Diversity of autoantibodies in avian scleroderma: an inherited fibrotic disease of White Leghorn chickens. *J. Clin. Invest.* **73**, 1557–1568.

Heller, E. D., Schat, K. A. (1985). Inhibition of natural killer activity in chickens by Marek's disease transformed cell lines. In *International Symposium on Marek's Disease* (eds Calnek, B. W., Spencer, J. L.), pp. 286–294. American Association of Avian Pathologists, Kennett Square, PA, USA.

Helmboldt, C. F., Garner, E. (1964). Experimentally induced Gumboro disease (IBA). *Avian Dis.* **8**, 561–575.

Hemendinger, R. A., Putnam, J. R., Bloom, S. E. (1992). MHC dosage effects on primary immune organ development in the chicken. *Dev. Comp. Immunol.* **16**, 175–186.

Herberman, R. B., Holden, H. T. (1978). Natural cell-mediated immunity. *Adv. Can. Res.* **27**, 305–377.

Higashihara, M., Saijo, K., Fujisaki, Y., Matumoto, M. (1991). Immunosuppressive effect of infectious bursal disease virus strains of variable virulence for chickens. *Vet. Microbiol.* **26**, 241–248.

Higgins, D. A., Cromie, R. L., Srivastava, G., Herzbeck, H., Schlüter, C., Gerdes, J. Diamantstein, T., Flad, H.-D. (1993). An examination of the immune system of the duck (*Anas platyrhynchos*) for factors resembling some defined mammalian cytokines. *Dev. Comp. Immunol.* **17**, 341–355.

Hippelainen, M., Naukkarinen, A., Alhava, E., Sorvari, T. E. (1987). Immunization of chickens via the bursa of Fabricius isolated from the rest of the gut-associated lymphoid tissue using four surgical techniques. *Poult. Sci.* **66**, 514–520.

Hirai, K., Calnek, B. W. (1979). *In vitro* replication of infectious bursal disease virus in established lymphoid cells lines and chicken B lymphocytes. *Infect. Immun.* **25**, 964–970.

Hirai, K., Shimakura, S., Kawamoto, E., Taguchi, F., Kim, S. T., Change, C. N., Iritani, Y. (1974). The immunodepressive effect of infectious bursal disease virus in chickens. *Avian Dis.* **18**, 50–57.

Hirai, K., Kunihiro, K., Shimakura, S. (1979). Characterization of immunosuppression in chickens by infectious bursal disease virus. *Avian Dis.* **23**, 950–965.

Hodges, R. D. (1974). *The Histology of the Fowl.* Academic Press, NY, USA.

Hopkins, I. G., Edwards, K. R., Thornton, D. H. (1979). Measurement of immunosuppression in chickens caused by infectious bursal disease vaccines using *Brucella abortus* strain 19. *Res. Vet. Sci.* **27**, 260–261.

Hoshi, H., Mori, T. (1973). Identification of the bursa-dependent and thymus-dependent areas in the *Tonsilla cecalis* of chickens. *Japan. J. Exp. Med.* **111**, 309–322.

Hossain, W. A., Zhou, X., Rutledge, A., Baier, C., Morest, D. K. (1996). Basic fibroblast growth factor affects neuronal migration and differentiation in normotypic cell cultures from the cochleovestibular ganglion of the chick embryo. *Exp. Neurol.* **138**, 121–143.

Houssaint, E., Belo, M., Le Douarin, N. M. (1976). Investigations on cell lineage and tissue interactions in the developing bursa of Fabricius through interspecific chimeras. *Dev. Biol.* **53**, 250–264.

Houssaint, E., Diez, E., Jotereau, F. V. (1985). Tissue distribution and ontogenic appearance of a chicken T lymphocyte differentiation marker. *Eur. J. Immunol.* **15**, 305–308.

Houssaint, E., Tobin, S., Cihak, J., Lösch, U. (1987). A chicken leukocyte common antigen: biochemical characterization and ontogenetic study. *Eur. J. Immunol.* **17**, 287–290.

Houssaint, E., Mansikka, A., Vainio, O. (1991). Early separation of B and T lymphocyte precursors in chick embryo. *J. Exp. Med.* **174**, 397–406.

Huang, H. V., Dreyer, W. M. (1978). Bursectomy *in ovo* blocks the generation of immunoglobulin diversity. *J. Immunol.* **121**, 1738–1747.

Hudson, L., Pattison, H., Thantrey, N. (1975). Specific B lymphocyte suppression by infectious bursal agent (Gumboro disease virus) in chickens. *Eur. J. Immunol.* **5**, 675–679.

Huffnagle, G. B., Ratcliffe, M. J. H., Humphries, E. H. (1989). Bu-2, a novel avian cell surface antigen on B cells and a population of non-lymphoid cells, is expressed homogeneously in germinal centers. *Hybridoma* **8**, 589–604.

Ianconescu, M., Yaniv, A., Gazit, A., Perk, K., Zimber, A. (1983). Susceptibility of domestic birds to lymphoproliferative disease virus (LPDV) of turkeys. *Avian Pathol.* **12**, 291–302.

Imai, K., Yuasa, N., Furuta, K., Narita, M., Banba, H., Kobayasi, S., Horiuchi, T. (1991). Comparative studies on pathogenical, virological and serological properties of Marek's disease virus isolated from Japanese quail and chicken. *Avian Pathol.* **20**, 57–65.

Inoue, M., Fukuda, M., Miyano, K. (1994). Thymic lesions in chicken infected with infectious bursal disease virus. *Avian Dis.* **38**, 839–846.

Isobe, T., Kohno, M., Suzuki, K., Yoshihara, S. (1989). *Leucocytozoon caulleryi* infection in bursectomized chickens. In *Coccidia and Intestinal Coccidiomorphs* (ed. INRA), pp. 79–85. INRA, Tours, France.

Ivanyi, J. (1975). Immunodeficiency in the chicken. II. Production of monomeric IgM following testosterone treatment of infection with Gumboro disease. *Immunology* **28**, 1015–1021.

Ivanyi, J. (1981). Functions of the B-lymphoid system in chickens. In *Avian Immunology* (eds Rose, M. E, Payne, L. N., Freeman, B. M.), pp. 63–101. Clark Constable Limited, Edinburgh, UK.

Ivanyi, J., Morris, R. (1976). Immunodeficiency in the chicken. IV. An immunologic study of infectious bursal disease. *Clin. Exp. Immunol.* **23**, 154–165.

Jakobisiak, M., Sundick, R. S., Bacon, L. D., Rose, N. R. (1976). Abnormal response to minor histocompatability antigens in Obese strain chickens. *Proc. Natl Acad. Sci. USA* **73**, 2877–2880.

Jakowlew, S. B., Dillard, P. J., Sporn, M. B, Roberts, A. B. (1988a). Nucleotide sequence of chicken transforming growth factor-beta 1 (TGF-beta 1). *Nucl. Acids Res.* **16**, 8730.

Jakowlew, S. B., Dillard, P. J., Sporn, M. B., Roberts, A. B. (1988b). Complementary deoxyribonucleic acid cloning of a messenger ribonucleic acid encoding transforming growth factor 4 from chicken embryo chondrocytes. *Mol. Endocrinol.* **2**, 1186–1195.

Jakowlew, S. B., Dillard, P. J., Kondaiah, P., Sporn, M. B, and Roberts, A. B. (1988c). Complementary deoxyribonucleic acid cloning of a novel transforming growth factor-beta messenger ribonucleic acid from chick embryo chondrocytes. *Mol. Endocrinol.* **2**, 747–755.

Jakowlew, S. B., Dillard, P. J., Sporn, M. B., Roberts, A. B. (1990). Complementary deoxyribonucleic acid cloning of an mRNA encoding transforming growth factor-β2 from chicken embryo chondrocytes. *Growth Factors* **2**, 123–133.

Jakowlew, S. B., Mathias, A., Lillehoj, H.S. (1997). Transforming growth factor-β isoforms in the developing chicken intestine and spleen: increase in transforming growth factor-β4 with coccidia infection. *Vet. Immunol. Immunopathol.* **55**, 321–339.

Jakowski, R. M., Fredrickson, T. N., Luginbuhl, R. E. (1973). Immunoglobulin response in experimental infection with cell-free and cell-associated Marek's disease virus. *J. Immunol.* **111**, 238–248.

Jankovic, B. D. (1987). Neuroimmune interactions: experimental and clinical strategies. *Immunol. Lett.* **16**, 341–354.

Jaroszewski, M., Sundick, R. S., Rose, N. R. (1978). Effects of antiserum containing thyroglobulin antibody on the chicken thyroid gland. *Clin. Immunol. Immunopathol.* **10**, 95–103.

Jensen, L. B., Koch, C. (1991). An assay for complement factor B in species at different levels of evolution. *Develop. Comp. Immunol.* **15**, 173–179.

Jeurissen, S. H. M. (1993). The role of various compartments in the chicken spleen during an antigen-specific humoral response. *Immunology* **80**, 29–33.

Jeurissen, S. H., Janse, E. M., Ekino, S., Nieuwenhuis, P., Koch, G., de Boer, G. F. (1988a). Monoclonal antibodies as probes for defining cellular subsets in the bone marrow, thymus, bursa of Fabricius, and spleen of the chicken. *Vet. Immunol. Immunopathol.* **19**, 225–238.

Jeurissen, S. H. M., Janse, E. M., Koch, G., de Boer, G. F. (1988b). The monoclonal antibody CVI–ChNl–68.1 recognizes cells of the monocyte-macrophage lineage in chickens. *Dev. Comp. Immunol.* **12**, 855–864.

Jeurissen, S. H. M., Janse, E. M., Koch, G., de Boer, G. F. (1989). Postnatal development of mucosa-associated lymphoid tissue in chickens. *Cell Tissue Res.* **258**, 119–124.

Jeurissen, S. H., Claassen, E., Janse, E. M. (1992). Histological and functional differentiation of non-lymphoid cells in the chicken spleen. *Immunology* **77**, 75–80.

Jeurissen, S. H., Janse, M. E., Van Roozelaar, D. J., Koch, G., de Boer, G. F. (1992a). Susceptibility of thymocytes for infection by chicken anemia virus is related to pre- and posthatching development. *Dev. Immunol.* **2**, 123–129.

Jeurissen, S. H., Wagenaar, F., Pol, J. M., van der Eb, A. J., Noteborn, M. H. (1992b). Chicken anemia virus causes apoptosis of thymocytes after *in vivo* infection and of cell lines after *in vitro* infection. *J. Virol.* **66**, 7383–7388.

Joshi, P., Glick, B. (1990). Lymphocyte inhibitory and chemotactic factors produced by bursal and thymic lymphocytes. *Poult. Sci.* **69**, 249–258.

Jotereau, F. V., Houssaint, E., Le Douarin, N. M. (1980). Lymphoid stem cell homing to the early thymic primordium of the avian embryo. *Eur. J. Immunol.* **10**, 620–627.

Kai, C., Yoshikawa, Y., Yamanouchi, K., Okada, H. (1983). Isolation and identification of the third component of complement of Japanese quails. *J. Immunol.* **130**, 2814–2820.

Kai, C., Yoshikawa, Y., Yamanouchi, K., Okada, H., Morikwawa, S. (1985). Ontogeny of the third component of complement of Japanese quails. *Immunology* **54**, 463–470.

Kaidoh, T., Gigli, I. (1987). Phylogeny of C4b-C3b cleaving activity: similar fragmentation patterns of human C4b and C3b produced by lower animals. *J. Immunol.* **139**, 194–201.

Kaplan, M. H., Sundick, R. S., Rose, N. R. (1991). Autoimmune diseases. In *Avian Cellular Immunology* (ed. Sharma, J. M.), pp. 183–197. CRC Press, Boca Raton, FL, USA.

Kaplan, M. H., Dhar, A., Brown, T. R., Sundick, R. S. (1992). Marek's disease virus-transformed chicken T-cell lines respond to lymphokines. *Vet. Immunol. Immunopathol.* **34**, 63–79.

Kaplan, M. H., Smith, D. I., Sundick, R. S. (1993). Identification of G protein coupled receptor induced in activated T cells. *J. Immunol.* **151**, 628–636.

Kasahara, Y., Chen, C. H., Cooper, M. D. (1993). Growth requirements for avian $\gamma\delta$ T cells include exogenous cytokines, receptor ligation and *in vivo* priming. *Eur. J. Immunol.* **23**, 2230–2236.

Kasahara, Y., Chen, C. H., Gobel, T. W. F., Bucy, R. P., Cooper, M. D. (1994). Intraepithelial lymphocytes in birds. In *Advances in Host Defense Mechanisms* (eds Gallin, J. I., Fauci, A. S.), pp. 163–174. Raven Press, New York, USA.

Kasai, K., Nakayama, A., Ohbayashi, M., Nakagawa, A., Ito, M., Saga, S., Asai, J. (1995). Immunohistochemical characteristics of chicken spleen ellipsoids using newly established monoclonal antibodies. *Cell Tissue Res.* **281**, 135–141.

Kaspers, B., Lillehoj, H. S., Lillehoj, E. P. (1993). Chicken macrophages share a common cell surface antigen defined by a monoclonal antibody. *Vet. Immunol. Immunopathol.* **36**, 333–346.

Kaspers, B., Lillehoj, H.S., Jenkins, M.C., Pharr, G.T. (1994). Chicken interferon-mediated induction of major histocompatibility complex class II antigens on peripheral blood monocytes. *Vet. Immunol. Immunopathol.* **44**, 71–84.

Katz, D., Kohn, A., Arnon, R. (1974). Immunoglobulins in the airway washings and bile secretions of chickens. *Eur. J. Immunol.* **4**, 494–499.

Kaufman, J., Skjoedt, K., Salomonsen, J. (1991). The B-G multigene family of the chicken major histocompatibility complex. *Crit. Rev. Immunol.* **11**, 113–143.

Kaufman, J., Andersen, R., Avila, D., Engberg, J., Lambris, J., Salomonsen, J., Welinder, K., Skjoedt, K. (1992). Different features of the MHC class I heterodimer have evolved at different rates: chicken B-F and beta2-microglobulin sequences reveal invariant surface residues. *J. Immunol.* **148**, 1532–1546.

Kaufman, J., Bumstead, N., Miller, M., Riegert, P., Salomonsen, J. (1995a). The chicken class II alpha gene is located outside of the B complex. In *Advances in Avian Immunology Research* (eds Davison, T. F., Bumstead, N., Kaiser, P.), pp. 119–127. Carfax, Abingdon, UK.

Kaufman, J., Voelk, H., Walny, H.-J. (1995b). A 'minimal essential MHC' and an 'Unrecognized MHC': two extremes in selection for polymorphism. *Immunol. Rev.* **143**, 63–88.

Kehrl, J. H., Wakefield, L. M., Roberts, A. B., Jakowlew, S. B., Alvarez-Mon, M., Derynck, R., Sporn, M. B., Fauci, A. S. (1986). Production of transforming growth factor β by human T lymphocytes and its potential role in the regulation of T cell growth. *J. Exp. Med.* **163**, 1037–1050.

Kerr, K. M., Olson, N. O. (1969). Pathology of chickens experimentally inoculated or contact-infected with an arthritis-producing virus. *Avian Dis.* **13**, 729–745.

Kierszenbaum, F., Ivanyi, J., Budzko, D. (1976). Mechanisms of natural resistance to trypanosomal infection. *Immunology* **30**, 1–6.

Kimijima, T., Hashimoto, Y., Kitagawa, H., Kon, Y., Sugimura, M. (1990). Localization of immunoglobulins in the chicken oviduct. *Jap. J. Vet. Sci.* **52**, 299–306.

Kincade, P. W., Cooper, M. D. (1971). Development and distribution of immunoglobulin-containing cells in the chicken: an immunofluorescent analysis using purified antibodies μ, γ, and light chains. *J. Immunol.* **106**, 371–382.

Kittner, Z., Olah, I. (1980). Contribution of chicken's central lymphoid organs to the cellular composition of the gland of Harder. *Acta Biol. Acad. Sci. Hung.* **31**, 177–185.

Kjalke, M., Welinder, K. G., Koch, C. (1993). Structural analysis of chicken factor B-like protease and comparison with mammalian complement proteins factor B and C. *J. Immunol.* **151**, 4147–4152.

Klasing, K. C. (1991). Avian inflammatory response: mediation by macrophages. *Poult. Sci.* **70**, 1176–1186.

Klasing, K. C. (1994). Avian leukocytic cytokines. *Poult. Sci.* **73**, 1035–1043.

Klasing, K. C., Peng, R. K. (1990). Monokine-like activities released from a chicken macrophage line. *Anim. Biotech.* **1**, 107–120.

Knabel, M., Loesch, U. (1993). Characterization of new monoclonal antibodies identifying avian T lymphocyte antigens. *Immunology* **188**, 415–429.

Knabel, M., Cihak, J., Lösch, U. (1993). Characterization of new monoclonal antibodies identifying avian T lymphocyte antigens. *Immunobiology* **188**, 415–429.

Koch, C. (1986a). Complement system in avian species. In *Avian Immunology: Basis and Practice* (eds Toivanen, A., Toivanen, P.), pp. 43–55. CRC Press, Boca Raton, FL, USA.

Koch, C. (1986b). The alternative complement pathway in chickens. Purification of factor B and production of a monospecific antibody against it. *Acta Path. Microbiol. Immunol. Scand., Se. C* **94**, 253–259.

Koch, C. (1986c). A genetic polymorphism of the complement component factor B in chickens not linked to the major histocompatibility complex (MHC). *Immunogenetics* **23**, 364–367.

Koch, G., Jongenelen, I. M. C. A. (1988). Quantification and class distribution of immunoglobulin-secreting cells in the mucosal tissues of the chicken. In *Histophysiology of the Immune System* (eds Fossum, S., Rolstad, B.), pp. 633–639. Plenum Publishing, New York, USA.

Köhler, G., Milstein, C. (1975). Continuous cultures of fused cells secreting antibody of defined specificity. *Nature* **256**, 495–497.

Kondo, T., Hattori, M., Kodama, H., Onuma, M., Mikami, T. (1990). Characterization of two monoclonal antibodies which recognize different subpopulations of chicken T lymphocytes. *Japan. J. Vet. Res.* **38**, 11–18.

Kon-Ogura, T., Kon, Y., Onuma, M., Kondo, T., Hashimoto, Y., Sugimura, M. (1993). Distribution of T cell subsets in chicken lymphoid tissues. *J. Vet. Med. Sci.* **55**, 59–66.

Koppenheffer, T. L., Russell, B. (1986). Unusual complement activation properties of serum immunoglobulins of the pigeon *Columba livia*. *Immunology* **57**, 473–478.

Kornfeld, S., Beug, H., Doderlein, G., Graf, T. (1983). Detection of avian hemopoietic surface antigens with monoclonal antibodies to myeloid cells: their distribution on normal and leukemic cells of various lineages. *Exp. Cell Res.* **143**, 383–394.

Kowalczyk, K., Daiss, J., Halpern, J., Roth, T. F. (1985). Quantitation of maternal-fetal IgG transport in the chicken. *Immunology* **54**, 755–762.

Kroemer, G., Zoorob, R., Auffray, C. (1990). Structure and expression of a chicken MHC class I gene. *Immunogenetics* **31**, 405–409.

Krömer, G., Sundick, R. S., Schauenstein, K., Halá, K., Wick, G. (1985). Analysis of lymphocytes infiltrating the thyroid gland of Obese strain chickens. *J. Immunol.* **135**, 2452–2457.

Krömer, G., Neu, N., Kuehr, T., Dietrich, H., Fässler, R., Halá, K., Wick, G. (1989). Immunogenetic analysis of spontaneous autoimmune thyroiditis of Obese strain chickens. *Clin. Immunol. Immunopathol.* **52**, 202–213.

Kubota, T., Chen, C. H., Hockett, R., Göbel, T. W. F., Cooper, M. D. (1995). Conservation of the T cell receptor α/δ locus in chickens. *FASEB J.* **9**, 4736.

Kung, H.-J., Maihle, N. J. (1987). Molecular basis of oncogenesis by non-acute avian retroviruses. In *Avian Leukosis* (ed. de Boer, G. F.), pp. 77–99. Martinus Nijhoff Publishing, Dordrecht, The Netherlands.

Kung, H.-J., Boerkoel, C., Carter, T. H. (1992). Retroviral mutagenesis of cellular oncogenes: a review with insights into the mechanisms of insertional activation. *Curr. Topics Microbiol. Immunol.* **171**, 1–26.

Lahti, J. M., Chen, C. H., Tjoelker, L. W., Pickel, J. M., Schat, K. A., Calnek, B. W., Thompson, C. B., Cooper, M. D. (1991). Two distinct $\alpha\beta$ T-cell lineages can be distinguished by the differential usage of T-cell receptor Vβ gene segments. *Proc. Natl Acad. Sci. USA* **88**, 10956–10960.

Lamont, S. J., Smyth Jr, J. R. (1981). Effect of bursectomy on development of a spontaneous postnatal amelanosis. *Clin. Immunol. Immunopathol.* **21**, 407–411.

Lamont, S. J., Boissy, R. E., Smyth Jr, J. R. (1982). Humoral immune response and expression of spontaneous postnatal amelanosis in DAM line chickens. *Immunol. Commun.* **11**, 121–127.

Lamont, S. J., Bohn, C., Cheville, N. (1987). Genetic resistance to fowl cholera is linked to the major histocompatibility complex. *Theor. Appl. Genet.* **25**, 284–289.

Lassila, O. (1989). Emigration of B cells from chicken bursa of Fabricius. *Eur. J. Immunol.* **19**, 955–958.

Lassila, O., Eskola, J., Toivanen, P. (1978). The origin of lymphoid stem cells studied in chicken yolk-sac-embryo chimaeras. *Nature* **272**, 353–354.

Lassila, O., Lefkovits, I., Alanen, A. (1989). Immunoglobulin diversification in bursal duct-ligated chickens. *Eur. J. Immunol.* **19**, 1343–1345.

Laursen, I., Koch, C. (1989). Purification of chicken C3 and a structural and functional characterization. *Scand. J. Immunol.* **30**, 529–538.

Lawn, A. M., Rose, M. E., Bradley, J. W., Rennie, M. C. (1988). Lymphocytes of the intestinal mucosa of chickens. *Cell. Tissue Res.* **252**, 189–195.

Lee, L. F., Sharma, J. M., Nazerian, K., Witter, R. L. (1978a). Suppression and enhancement of mitogen response in chickens infected with Marek's disease virus and the herpesvirus of turkeys. *Infect. Immun.* **21**, 474–479.

Lee, L. F., Sharma, J. M., Nazerian, K., Witter, R. L. (1978b). Suppression of mitogen-induced proliferation of normal spleen cells by macrophages from chickens inoculated with Marek's disease virus. *J. Immunol.* **120**, 1554–1565.

Lee, T. H., Tempelis, C. H. (1992). A possible 110-kDa receptor for interleukin-2 in the chicken. *Dev. Comp. Immunol.* **16**, 463–472.

Leibold, W., Janotte, G., Peter, H. H. (1980). Spontaneous cell mediated cytotoxicity (SCMC) in various mammalian species and chickens: selective reaction pattern and different mechanisms. *Scand. J. Immunol.* **11**, 203–222.

Leslie, G. A., Stankus, R. P., Martin, L. N. (1976). Secretory immunological system of fowl. V. The gallbladder: an integral part of the secretory immunological system of fowl. *Int. Arch. Allergy Appl. Immunol.* **51**, 175–185.

Lessard, M., Hutchings, D. L., Spencer, J. L., Lillehoj, H. S., Gavora, J. S. (1996). Influence of Marek's disease virus strain AC-1 on cellular immunity in birds carrying endogenous viral genes. *Avian Dis.* **40**, 645–653.

Leutz, A., Beug, H., Graf, T. (1984). Purification and characterization of cMGF, a novel chicken myelomonocytic growth factor. *EMBO J.* **3**, 3191–3197.

Leutz, A., Beug, H., Walter, C., Graf, T. (1988). Hematopoietic growth factor glycosylation: multiple forms of chicken myelomonocytic growth factor. *J. Biol. Chem.* **263**, 3905–3911.

Leutz, A., Damm, K., Sterneck, E., Kowenz, E., Ness, S., Frank, R., Gausepohl., H., Pan, Y.-C., Smart, J., Hayman, M., Graf, T. (1989). Molecular cloning of chicken myelomonocytic growth factor (cMGF) reveals relationship to interleukin 6 and granulocyte colony stimulating factor. *EMBO J.* **8**, 175–181.

Lewis, R. B., McClure, J., Rup, B., Niesel, D. W., Garry, R. F., Hoelzer, J. D., Nazerian, K., Bose Jr, H. R. (1981). Avian reticuloendotheliosis virus: identification of the hematopoietic target cell for transformation. *Cell* **25**, 421–431.

Ley, D. H., Yamamotao, R., Bickford, A. A. (1983). The pathogenesis of infectious bursal disease: serologic, histopathologic, and clinical chemical observations. *Avian Dis.* **27**, 1060–1085.

Li, J., Calnek, B. W., Schat, K. A., Graham, D. L. (1983). Pathogenesis of reticuloendotheliosis virus infection in ducks. *Avian Dis.* **27**, 1090–1105.

Lillehoj, H. S. (1987a). Effects of immunosuppression on avian coccidiosis: cyclosporin A but not hormonal bursectomy abrogates host protective immunity. *Infec. Immun.* **55**, 1616–1621.

Lillehoj, H. S. (1987b). Secretory IgA response in SC and FP chickens experimentally inoculated with *Eimeria tenella* and *acervulina*. In *Recent Advances in Mucosal Immunology* (eds McGhee, J. R., Bienenstock, J., Orga, P. L.), pp. 977–980. Plenum Publishing Corp., New York, USA.

Lillehoj, H. S. (1989). Intestinal intraepithelial and splenic natural killer cell responses to Eimerian infections in inbred chickens. *Infec. Immun.* **57**, 1879–1884.

Lillehoj, H. S. (1991a). Cell-mediated immunity in parasitic and bacterial diseases. In *Avian Cellular Immunology* (ed. Sharma, J. M.), pp. 155–181. CRC Press, Boca Raton, FL, USA.

Lillehoj, H. S. (1991b). Lymphocytes involved in cell-mediated immune responses and methods to assess cell-mediated immunity. *Poult. Sci.* **70**, 1154–1164.

Lillehoj, H. S. (1993). Avian interleukin-2 and interferon. *Avian Immunology in Progress*, Tours (France), August 31–September 2, pp. 105–111.

Lillehoj, H. S. (1994). Analysis of *Eimeria acervulina*-induced changes in intestinal T lymphocyte subpopulations in two inbred chickens showing different levels of disease susceptibility to coccidia. *Res. Vet. Sci.* **56**, 1–7.

Lillehoj, H. S., Chai, J. Y. (1988). Comparative natural killer cell activities of thymic, bursal, splenic and intestinal intraepithelial lymphocytes. *Dev. Comp. Immunol.* **12**, 629–643.

Lillehoj, H. S., Cheung, K. S. (1992). Postnatal development of T lymphocyte subpopulations in the intestinal intraepithelium and lamina propria in chickens. *Vet. Immunol. Immunopathol.* **31**, 347–360.

Lillehoj, H. S., Ruff, M. D. (1986). Comparison of disease susceptibility and subclass specific antibody response in SC and FP chickens experimentally inoculated with *E. tenella*, *E. acervulina* or *E. maxima*. *Avian Dis.* **31**, 112–119.

Lillehoj, H. S., Trout, J. M. (1993). Coccidia: a review of recent advances in immunity and vaccine development. *Avian Pathol.* **22**, 3–21.

Lillehoj, H. S., Trout, J. M. (1994). $CD8^+$ T cell coccidia interactions. *Parasit. Today* **10**, 10–14.

Lillehoj, H. S., Trout, J. M. (1996). Avian gut-associated lymphoid tissues and intestinal immune responses to *Eimeria* parasites. *Clin. Microbiol. Rev.* **9**, 349–360.

Lillehoj, H. S., Lillehoj, E. P., Weinstock, D., Schat, K. A. (1988a). Functional and biochemical characterization of Avian T lymphocyte antigens identified by monoclonal antibodies. *Eur. J. Immunol.* **18**, 2059–2066.

Lillehoj, H. S., Kim, S., Lillehij, E. P., Bacon, L. D. (1988b). Quantitative differences in Ia antigen expression in the spleens of 15I-5-B congenic and inbred chickens as defined by a new monoclonal antibody. *Poult. Sci.* **67**, 1525–1535.

Lillehoj, H. S., Kang, S. Y. Keller, L, Sevoian, M. (1989). *Eimeria tenella* and *E. acervulina*: lymphokines secreted by an avian T cell lymphoma or by sporozoite-stimulated lymphocytes protect chickens against avian coccidiosis. *Exper. Parasitol.* **69**, 54–64.

Lillehoj, H. S., Isobe, T., Weinstock, D. (1993). Tissue distribution and cross-species reactivity of new monoclonal antibodies detecting chicken T lymphocytes and macrophages. In *Avian Immunology in Progress* (ed. Coudert, F.), pp. 37–42. INRA editions, Paris, France.

Liu, X., Lee, L. F. (1983). Kinetics of phytohemagglutinin response in chickens infected with various strains of Marek's disease virus. *Avian Dis.* **27**, 660–666.

Livezey, M. D., Sundick, R. S., Rose, N. R. (1981). Spontaneous autoimmune thyroiditis in chickens: II. Evidence for autoresponsive thymocytes. *J. Immunol.* **127**, 1469–1472.

Longenecker, B. M., Pazderka, F., Stone, H. S., Gavora, J. S., Ruth, R. F. (1975). *In ovo* assay for Marek's disease virus and turkey herpesvirus. *Infec. Immun.* **11**, 922–931.

Loppnow, H., Flad, H.-D., Dürrbaum, I., Musehold, G., Fetting, R., Ulmer, A. J., Herzbeck, H., Brandt, E. (1989). Detection of interleukin 1 with human dermal fibroblasts. *Immunobiology* **179**, 283–291.

Lowenthal, J. W., Digby, M. R., York, J. J. (1995). Production of interferon-γ by chicken T cells. *J. Interferon Cytok. Res.* **15**, 933–938.

Lu, Y.-S., Lapen, R. L. (1974). Splenic cell mitogenic response in Marek's disease: comparison between noninfected tumor-bearing and nontumor-bearing infected chickens. *Am. J. Vet. Res.* **35**, 977–980.

Lucio, B., Schat, K. A., Shivaprasad, H. L. (1990). Identification of the chicken anemia agent, reproduction of the disease, and serological survey in the United States. *Avian Dis.* **34**, 146–153.

Luhtala, M., Salomonsen, J., Hirota, Y., Onodera, T., Toivanen, P., Vainio, O. (1993). Analysis of chicken CD4 by monoclonal antibodies indicates evolutionary conservation between avian and mammalian species. *Hybridoma* **12**, 633–646.

Luhtala, M., Koskinen, R., Toivanen, P., Vainio, O. (1995). Characterization of chicken CD8-specific monoclonal antibodies recognizing novel epitopes. *Scand. J. Immunol.* **42**, 171–174.

Lupetti, M., Dolfi, A., Giannessi, F., Bianchi, F., Michelucci, S. (1990). Reappraisal of histogenesis in the bursal lymphoid follicle of the chicken. *Am. J. Anat.* **187**, 287–302.

Maccubin, D. L., Schierman, L. W. (1986). MHC restricted cytotoxic response of chicken T cells: expression, augmentation and clonal characterisation. *J. Immunol.* **136**, 12–16.

Mansikka, A. (1992). Chicken IgA H chains. Implications concerning the evolution of H chain genes. *J. Immunol.* **149**, 855–861.

Mansikka, A., Veromaa, T., Vainio, O., Toivanen, P. (1989). B-cell differentiation in the chicken: expression of immunoglobulin genes in the bursal peripheral lymphocytes. *Scand. J. Immunol.* **29**, 325–332.

Mansikka, A., Sandberg, M., Lassila, O., Toivanen, P. (1990a). Rearrangement of immunoglobulin light chain genes in the chicken occurs prior to colonization of the embryonic bursa of Fabricius. *Proc. Natl Acad. Sci. USA* **87**, 9416–9420.

Mansikka, A., Jalkanen, S., Sandberg, M., Granfors, K., Lassila, O., Toivanen, P. (1990b). Bursectomy of chicken embryos at 60 h of incubation leads to an oligoclonal B cell compartment and restricted Ig diversity. *J. Immunol.* **145**, 3601–3609.

Marmor, M. D., Benatar, T., Ratcliffe, M. J. H. (1993). Retroviral transformation *in vitro* of chicken T cells expressing either α/β or γ/δ T cell receptors by reticuloentheliosis virus strain T. *J. Exp. Med.* **177**, 647–656.

Marsh, J. A. (1993). The humoral activity of the avian thymic microenvironment. *Poult. Sci.* **72**, 1294–1300.

Marsh, J. A., Scanes, C. G. (1994). Neuroendocrine-immune interactions. *Poult. Sci.* **73**, 1049–1061.

Marsh, J. A., Combs Jr, G. F., Whitacre, M. E., Dietert, R. R. (1986). Effect of selenium and vitamin E dietary deficiencies on chick lymphoid organ development. *Proc. Soc. Exp. Biol. Med.* **182**, 425–436.

Marsh, J. A., Johnson, E., Safieh, B., Kendall, M., Dardenne, M., Scanes, C. G. (1991). Assessment of thymulin activity in Cornell K and SLD chickens. *Poult. Sci.* **70** (Suppl. 1), 24. (abstract).

Martin, A., Lillehoj, H. S., Kasper, B., Bacon, L. D. (1994). Mitogen-induced lymphocyte proliferation and interferon production induced by coccidia infection. *Avian Dis.* **38**, 262–268.

Martin, C., Beaupain, D., Dieterlen-Lievre, F. (1978). Developmental relationships between vitelline and intraembryonic haematopoiesis studied in avian yolk sac chimeras. *Cell Differ.* **7**, 115–130.

Maslak, D. M., Reynolds, D. L. (1995). B cells and T lymphocyte subsets of the head-associated lymphoid tissues of the chicken. *Avian Dis.* **39**, 736–742.

Massague, J. (1990). The transforming growth factor-β family. *Annu. Rev. Cell Biol.* **6**, 597–641.

Mast, J., Cooper, M. D. (1995). Monoclonal antibodies reactive with the chicken monocyte/macrophage lineage. In *Advances in Avian Immunology Research* (eds Davison, T. F., Bumstead, N., Kaiser, P.), pp. 39–48. Abingdon, Oxfordshire, UK.

Masteller, E. L., Thompson, C. B. (1994). B cell development in the chicken. *Poult. Sci.* **73**, 998–1011.

Masteller, E. L., Larsen, R. D., Carlson, L. M., Pickel, J. M., Nickoloff, B., Lowe, J., Thompson, C. B., Lee, K. P. (1995). Chicken B cells undergo discrete development changes in surface carbohydrate structure that appear to play a role in directing lymphocyte migration during embryogenesis. *Development* **121**, 1657–1667.

Matsuda, H., Inoue, M., Negoro, Murata, M. (1990). A monoclonal antibody against chicken monocytic leukemia cell line. *Jap. J. Vet. Sci.* **52**, 1285–1288.

Mavroidis, M., Sunyer, J., Lambris, J. D. (1995). Isolation, primary structure, and evolution of the third component of chicken complement and evidence for a new member of the α_2-macroglobulin family. *J. Immunol.* **154**, 2164–2174.

Mazariegos, L. A., Lukert, P. D., Brown, J. (1990). Pathogenicity and immunosuppressive properties of infectious bursal disease 'intermediate' strains. *Avian Dis.* **34**, 203–208.

McConnell, C. D., Adair, B. M., McNulty, M. S. (1993a). Effects of chicken anemia virus on cell-mediated immune function in chickens exposed to the virus by a natural route. *Avian Dis.* **37**, 366–374.

McConnell, C. D., Adair, B. M., McNulty, M. S. (1993b). Effects of chicken anemia virus on macrophage function in chickens. *Avian Dis.* **37**, 358–365.

McCormack, W. T., Thompson, C. B. (1990). Chicken IgL variable region gene conversions display pseudogene donor preference and 5′ to 3′ polarity. *Genes Dev.* **4**, 548–558.

McCormack, W. T., Tjoelker, L. W., Carlson, L. M., Petryniak, B., Barth, C. (1989). Chicken Ig_L gene rearrangement involves deletion of a circular episome and addition of single nonrandom nucleotides to both coding segments. *Cell* **56**, 785–791.

McCormack, W. T., Tjoelker, L. W., Thompson, C. B. (1991a). Avian B-cell development: generation of an immunoglobulin repertoire by gene conversion. *Annu. Rev. Immunol.* **9**, 219–241.

McCormack, W. T., Tjoelker, L. W., Stella, G., Postema, C., Thompson, C. B. (1991b). Chicken T-cell receptor β-chain diversity: an evolutionarily conserved $D\beta$-encoded glycine turn within the hypervariable CDR3 domain. *Proc. Natl Acad. Sci. USA* **88**, 7699–7703.

McDougall, J. S., Biggs, P. M., Shilleto, R. W. (1978). A leukosis in turkeys associated with infection with reticuloendotheliosis virus. *Avian Pathol.* **7**, 557–568.

McGhee, J. R., Mestecky, J., Elson, C. O., Kiyono, H. (1989). Regulation of IgA synthesis and immune response by T cells and interleukins. *J. Clin. Immunol.* **9**, 175–199.

McNagny, K. M., Lim, F., Grieser, S., Graf, T. (1992). Cell surface proteins of chicken hematopoietic progenitors, thrombocytes and eosinophils detected by novel monoclonal antibodies. *Leukemia* **6**, 975–984.

Miller, M. M., Pink, J. R. L. (1986). Chick embryonic erythrocyte antigen-tissues sharing expression. *Exp. Cell Res.* **167**, 295–310.

Miller, M. M., Goto, R., Bernot, A., Zoorob, R., Auffray, C., Bumstead, N., Briles, W. E. (1994). Two MHC class I and two MHC class II genes map to the chicken Rfp-Y system outside the B complex. *Proc. Natl Acad. Sci. USA* **91**, 4397–4401.

Misra, R. R., Bloom, S. E. (1991). Roles of dosage, pharmacokinetics, and cellular selectivity to damage in selective toxicity of cyclophosphamide towards B and T cells in development. *Toxicology* **66**, 239–256.

Mombaerts, P., Iacomini, J., Johnson, R. S., Herrup, K., Tonegawa, S., Papaioannou, V. E. (1992). RAG-1 deficient mice have no mature T and B lymphocytes. *Cell* **68**, 869–877.

Montgomery, R. D., Villegas, P., Dowe, D. L., Brown, J. (1986). A comparison between the effect of avian reoviruses and infectious bursal disease virus on selected aspects of the immune system of the chicken. *Avian Dis.* **30**, 298–308.

Moore, M. A. S., Owen, J. J. T. (1966). Experimental studies on the development of the bursa of Fabricius. *Dev. Biol.* **14**, 40–51.

Moradian, A., Thorsen, J., Julian, R. J. (1990). Single and combined infections of specific-pathogen-free chickens with infectious bursal disease virus and an intestinal isolate of reovirus. *Avian Dis.* **34**, 63–72.

Morimura, T., Ohashi, K., Hattori, M., Sugimoto, C., Onuma, M. (1996). Apoptosis and nonresponsiveness of T cells in chickens infected with Marek's disease virus. *Proc. 4th Int. Symp. Marek's Disease*, p. 55.

Motyka, B., Reynolds, J. D. (1991). Apoptosis is associated with extensive B cell death in the sheep ileal Peyer's patch and the chicken bursa of Fabricius: a possible role in B cell selection. *Eur. J. Immunol.* **21**, 1951–1958.

Muir, R. (1911). Relationships between the complements and immune bodies of different animals: the mode of action of immune body. *J. Path. Bact.* **16**, 523–524.

Müller, H. (1986). Replication of infectious bursal disease virus in lymphoid cell. *Arch. Virol.* **87**, 191–203.

Müller, R., Kaufer, I., Reinacher, M., Weiss, E. (1979). Immunofluorescent studies of early virus propagation after oral infection with infectious bursal disease virus (IBDV). *Zentralbl. Veterinaermed. Reihe B* **26**, 345–352.

Murthy, K. K., Beach, F. G., Ragland, W. L. (1984). Expression of T cell markers on chicken bone marrow precursor cells incubated with an avian thymic hormone. In *Thymic Hormones and Lymphokines* (ed. Goldstein, A. L.), pp. 375–82. Plenum Press, New York, USA.

Myers, T. J., Lillehoj, H. S., Fetterer, R. H. (1992). Partial purification and characterization of chicken interleukin-2. *Vet. Immunol. Immunopathol.* **34**, 97–114.

Nagaraja, K. V., Emery, D. A., Patel, B. S., Pomeroy, B. S., Newman, J. A. (1982a). *In vitro* evaluation of B-lymphocyte function in turkeys infected with hemorrhagic enteritis virus. *Am. J. Vet. Res.* **43**, 502–504.

Nagaraja, K. V., Patel, B. L., Emery, D. A., Pomeroy, B. A., Newman, J. A. (1982b). *In vitro* depression of the mitogenic response of lymphocytes from turkeys infected with hemorrhagic enteritis virus. *Am. J. Vet. Res.* **43**, 134–136.

Nagaraja, K. V., Kang, S. Y., Newman, J. A. (1985). Immunosuppressive effect of virulent strain of hemorrhagic enteritis virus in turkeys vaccinated against Newcastle disease. *Poult. Sci.* **64**, 588–590.

Nakai, T., Hirai, K. (1981). *In vitro* infection of fractionated chicken lymphocytes by infectious bursal disease virus. *Avian Dis.* **25**, 831–838.

Naqi, S. A., Marquez, B., Sahin, N. (1983). Maternal antibody and its effect on infectious bursal disease immunization. *Avian Dis.* **27**, 623–631.

Naukkarinen, A. (1982). Transport of colloidal carbon from the lymphoid follicles to other tissue compartments in the chicken bursa of Fabricius. *J. Reticuloendothel. Soc.* **31**, 423–432.

Naukkarinen, A., Hippelainen, M. (1989). Development of the peripheral immune function in the chicken. A study on the bursa of Fabricius isolated from the rest of the gut-associated lymphoid tissue (GALT). *APMIS* **97**, 787–792.

Naukkarinen, A., Sorvari, T. E. (1982). Morphological and histochemical characterization of the medullary cells in the bursal follicles of the chicken. *Acta Pathol. Microbiol. Immunol. Scand.* **90**, 193–199.

Naukkarinen, A., Sorvari, T. E. (1984). Involution of the chicken bursa of Fabricius: a light microscopic study with special reference to transport of colloidal carbon in the involuting bursa. *J. Leukoc. Biol.* **35**, 281–290.

Naukkarinen, A., Syrjanen, K. J. (1984). Effects of anti-T lymphocyte serum on immunological reactivity and on the T cell area of the cloacal bursa in the chicken. *Acta Pathol. Microbiol. Immunol. Scand.* **92**, 145–152.

Nazerian, K., Elmubarak, A., Sharma, J. M. (1982). Establishment of B-lymphoblastoid cell lines from Marek's disease virus-induced tumors in turkeys. *Intl. J. Cancer* **29**, 63–68.

Neiman, P. E. (1994). Retrovirus-induced B cell neoplasia in the bursa of Fabricius. *Adv. Immunol.* **56**, 467–484.

Neiman, P. E., Thomas, S. J., Loring, G. (1991). Induction of apoptosis during normal and neoplastic B cell development in the bursa of Fabricius. *Proc. Natl Acad. Sci. USA* **88**, 5857–5861.

Neu, N., Halá, K., Dietrich, H., Wick, G. (1985). Spontaneous autoimmune thyroiditis in Obese strain chickens: a genetic analysis of target organ abnormalities. *Clin. Immunol. Immunopathol.* **37**, 397–405.

Neu, N., Halá, K., Dietrich, H., Wick, G. (1986). Genetic background of spontaneous autoimmune thyroiditis in the Obese strain of chickens studied in hybrids with an inbred line. *Int. Arch. Allergy Appl. Immun.* **80**, 168–173.

Nicolas-Bolnet, C., Johnston, P. A., Kemper, A. E., Ricks, C., Petitte, J. N. (1995). Synergistic action of two sources of avian growth factors on proliferative differentiation of chick embryonic hematopoietic cells. *Poult. Sci.* **74**, 1102–1116.

Nossal, G. J. V. (1994). Negative selection of lymphocytes. *Cell* **76**, 229–239.

Noteborn, M. H. M., de Boer, G. F., van Roozelaar, D. J., Karreman, C., Kranenburg, O., Vos, J. G., Jeurissen, S. H., Hoeben, R. C., Zantema, A., Koch, G. (1991). Characterization of cloned chicken anemia virus DNA that contains all elements for the infectious replication cycle. *J. Virol.* **65**, 3131–3139.

Nusbaum, K. E., Lukert, P. D., Fletcher, O. J. (1988). Experimental infection of one-day-old poults with turkey isolates of infectious bursal disease virus. *Avian Pathol.* **17**, 51–62.

Odend'hal, S., Breazile, J. E. (1980). An area of T cell localization in the cloacal bursa of White Leghorn chickens. *Am. J. Vet. Res.* **41**, 255–258.

Ohta, H., Yoshikawa, Y., Kai, C., Yamanouchi, K., Okada, H. (1984). Lysis of horse red blood cells mediated by antibody-independent activation of the alternative pathway of chicken complement. *Immunology* **52**, 437–443.

Olah, I., Glick, B. (1978). The number and size of the follicular epithelium and follicles in the bursa of Fabricius. *Poult. Sci.* **57**, 1445–1450.

Olah, I., Glick, B. (1984). Lymphopineal tissue in the chicken. *Dev. Comp. Immunol.* **8**, 855–862.

Olah, I., Glick, B. (1992). Non-lymphoid cells produce IgA-like substance in the corticomedullary zone of the chicken thymus. *Thymus* **19**, 105–110.

Olah, I., Glick, B. (1995). Dendritic cells in the bursal follicles and germinal centers of the chicken's caecal tonsil express vimentin but not desmin. *Anat. Rec.* **243**, 384–389.

Olah, I., Glick, B., Taylor Jr, R.L. (1985). Effect of surgical bursectomy on the ellipsoid, ellipsoid-associated cells, and periellipsoid region of the chicken's spleen. *J. Leukoc. Biol.* **38**, 459–469.

Olah, I., Glick, B., Toro, I. (1986). Bursal development in normal and testosterone-treated chick embryos. *Poult. Sci.* **65**, 574–588.

Olson, W. C., Ewert, D. L. (1990). Markers of B lymphocyte differentiation in the chicken. *Hybridoma* **9**, 331–350.

Olson, W. C., Ewert, D. L. (1994). A novel multilineage cell-surface antigen expressed on terminally differentiated chicken B cells in mucosal tissues. *Cell. Immunol.* **154**, 328–341.

Otaki, Y., Nunoya, T., Tajima, M., Kato, A., Nomura, Y. (1988). Depression of vaccinal immunity to Marek's disease by infection with chicken anemia agent. *Avian Pathol.* **17**, 333–347.

Owen, R. L., Jones, A. L. (1974). Epithelial cell specialization within human Peyer's patches: an ultrastructural study of intestinal lymphoid follicles. *Gastroenterology* **66**, 189–203.

Palladino, M. A., Gilmour, D. G., Scafuri, A. R., Stone, H. A., Thorbecke, G. J. (1977). Immune response differences between two inbred chicken lines identical at the major histocompatibility complex. *Immunogenetics* **5**, 253–259.

Palojoki, E., Toivanen, P., Jalkanen, S. (1993). Chicken B cells adhere to the CS-1 site of fibronectin throughout their bursal and postbursal development. *Eur. J. Immunol.* **23**, 721–726.

Panigrahy, B., Misra, L. K., Naqi, S. A., Hall, C. F. (1977). Prolongation of skin allograft survival in chickens with infectious bursal disease virus. *Poult. Sci.* **56**, 1745.

Paramithiotis, E., Ratcliffe, M. J. H. (1993). Bursa-dependent subpopulations of peripheral B lymphocytes in chicken blood. *Eur. J. Immunol.* **23**, 96–102.

Paramithiotis, E., Ratcliffe, M. J. H. (1994). Survivors of bursal B cell production and emigration. *Poult. Sci.* **73**, 991–997.

Paramithiotis, E., Ratcliffe, M. J. H. (1996). Evidence for phenotypic heterogeneity among B cells emigrating from the bursa of Fabricius: a reflection of functional diversity? *Curr. Top. Microbiol. Immunol.* **212**, 29–36.

Paramithiotis, E., Tkalec, L., Ratcliffe, M. J. H. (1991). High levels of CD45 are expressed coordinately with CD4 and CD8 on avian thymocytes. *J. Immunol.* **147**, 3710–3717.

Paramithiotis, E., Jacobsen, K. A., Ratcliffe, M. J. H. (1995). Loss of cell surface immunoglobulin expression precedes B cell death by apoptosis in the bursa of Fabricius. *J. Exp. Med.* **181**, 105–113.

Pardue, S. L., Fite, K. V., Bengston, L., Lamont, S. J., Boyle III, M. L., Smyth Jr, J. R. (1987). Enhanced integumental and ocular amelanosis following the termination of cyclosporine administration. *J. Invest. Dermatol.* **88**, 758–761.

Parrard, J. V., Rose, M. E., Hesketh, P. (1983). A functional homologue of mammalian SC exists in chickens. *Eur. J. Immunol.* **13**, 566–570.

Parvari, R., Avivi, A., Lentner, F., Ziv, E., Tel-Or, S., Burnstein, I., Schecter, I. (1988). Chicken immunoglobulin γ-heavy chains: limited VH gene repertoire, combinatorial diversification by D segments and evolution of the heavy chain locus. *EMBO J.* **7**, 739–744.

Payne, A. P. (1994). The Harderian gland: a tercentennial review. *J. Anat.* **185**, 1–49.

Payne, L. N. (1971). The lymphoid system. In *Physiology and Biochemistry of the Domestic Fowl* (eds Bell, D J., Freeman, B. M.), pp. 985–1038. Academic Press, NY, USA.

Payne, L. N., Fadly, A. M. (1997). Leukosis/sarcoma group. In *Diseases of Poultry*, 10th edn (eds Calnek, B. W., Barnes, H. J., Beard, C. W., McDougald, L. R., Saif, Y. M.), pp. 414–466. Iowa State University Press, Ames, IA, USA.

Payne, L. N., Purchase, H. G. (1991). Leukosis/sarcoma group. In *Diseases of Poultry*, 9th edn. (eds Calnek, B. W., Barnes, H. J., Beard, C. W., Reid, W. M., Yoder Jr, H. W.), pp. 386–439. Iowa State University Press, Ames, IA, USA.

Payne, L. N., Rennie, M. (1973). Pathogenesis of Marek's disease in chicks with and without maternal antibody. *J. Natl Canc. Inst.* **51**, 1559–1573.

Payne, L. N., Frazier, J. A., Powell, P. C. (1976). Pathogenesis of Marek's disease in chicks with and without maternal antibody. *Int. Rev. Exp. Pathol.* **16**, 59–154.

Payne, L. N., Brown, S. R., Bumstead, N., Howes, K., Frazier, J. A., Thouless, M. E. (1991a). A novel subgroup of exogenous avian leukosis virus in chickens. *J. Gen. Virol.* **72**, 801–807.

Payne, L. N., Gillespie, A. M., Howes, K. (1991b). Induction of myeloid leukosis and other tumours with the HPRS-103 strain of ALV. *Vet. Rec.* **129**, 447–448.

Péault, B., Chen, C. H., Cooper, M. D., Barbu, M., Lipinski, M., LeDouarin, N. M. (1987). Phylogenetically conserved antigen on nerve cells and lymphocytes resembles myelin-associated glycoprotein. *Proc. Natl Acad. Sci. USA* **84**, 814–818.

Pejkovski, C., Davelaar, F. G., Kouwenhoven, B. (1979). Immunosuppressive effect of infectious bursal disease virus on vaccination against infectious bronchitis. *Avian Pathol.* **8**, 95–106.

Pertile, T. L. (1995). Cellular immune functions in chickens infected with avian reovirus. PhD thesis, University of Minnesota, Minneapolis, MN, USA.

Pertile, T. L., Sharma, J. M., Walser, M. M. (1995). Reovirus infection in chickens primes splenic adherent macrophages to produce nitric oxide in response to T cell-produced factors. *Cell. Immunol.* **164**, 207–216.

Pertile, T. L., Karaca, K., Walser, M. M., Sharma, J. M. (1996). Suppressor macrophages mediate depressed lymphoproliferation in chickens infected with avian reovirus. *Vet. Immunol. Immunopathol.* **53**, 129–145.

Pickel, J. M., Chen, C. H., Cooper, M. D. (1990). An avian B-lymphocyte protein associated with beta 2-microglobulin. *Immunogen.* **32**, 1–7.

Pink, J. R. L., Lassila, O. (1987). B cell commitment and diversification in the bursa of Fabricius. In *Current Topics in Microbiology and Immunology*, Vol. 135 (eds Paige, C. J., Gisler, R. H.), pp. 57–64. Springer-Verlag, Heidelberg, Germany.

Pink, J. R., Rijnbeek, A.-M. (1983). Monoclonal antibodies against chicken lymphocyte surface antigens. *Hybridoma* **2**, 287–296.

Pink, J. R. L., Droege, W., Hala, K., Miggiano, V. C., Ziegler, A. (1977). A three-locus model for chicken major histocompatibility complex. *Immunogenetics* **5**, 203–216.

Pink, J. R., Vainio, O., Rijnbeek, A. (1985). Clones of B lymphocytes in individual follicles of the bursa of Fabricius. *Eur. J. Immunol.* **15**, 83–87.

Plachy, J. (1988). An analysis of the response of recombinant congenic lines of chickens to RSV challenge provides evidence for further complexity of the genetic structure of the chicken MHC(B). *Folia Biol.* (Prague) **34**, 170–181.

Plachy, J., Hala, K., Benda, V. (1979). Regression of tumours induced by RSV in different inbred lines of chickens. *Folia Biol.* (Prague) **25**, 335–336.

Plachy, J., Pink, J. R. L., Hala, K. (1992). Biology of the chicken MHC (B complex). *Crit. Rev. Immunol.* **12**, 47–79.

Pleiman, C. M., D'Ambrosio, D., Cambier, J. C. (1994). The B-cell antigen receptor complex: structure and signal transduction. *Immunol. Today* **15**, 393–399.

Polley, C. R., Bacon, L. D., Rose, N. R. (1981). Spontaneous autoimmune thyroiditis in chickens: I. The effect of bursal reconstitution. *J. Immunol.* **127**, 1465–1468.

Pontes de Carvalho, L. C., Wick, G., Roitt, I. M. (1981). Requirement of T cells for the development of spontaneous autoimmune thyroiditis in the Obese strain (OS) chickens. *J. Immunol.* **126**, 750–753.

Pope, C. R. (1991). Chicken anemia agent. *Vet. Immunol. Immunopath.* **30**, 51–65.

Pourquié, O., Corbel, C., le Caer, J.-P., Rossier, J., le Douarin, N. M. (1992). BEN, a surface glycoprotein of the immunoglobulin superfamily, is expressed in a variety of developing systems. *Proc. Natl Acad. Sci. USA* **89**, 5261–5265.

Powell, P. C. (1980). *In vitro* stimulation of blood lymphocytes by phytohemagglutinin during the development of Marek's disease. *Avian Pathol.* **9**, 471–475.

Powell, P. C. (1987a). Immune mechanisms in infections of poultry. *Vet. Immunol. Immunopathol.* **15**, 87–113.

Powell, P. C. (1987b). Macrophages and other nonlymphoid cells contributing to immunity. In *Avian Immunology: Basis and Practice* (eds Toivanen, A., Toivanen, P.), pp. 195–212. CRC Press, Boca Raton, FL, USA.

Powell, P. C., Howes, K., Lawn, A. M., Mustill, B. M., Payne, L. N., Rennie, M., Thompson, M. A. (1984). Marek's disease in turkeys: the induction of lesions and the establishment of lymphoid cell lines. *Avian Pathol.* **13**, 201–214.

Pradhan, H. K., Mohanty, G. C., Mukit, A., Pattnaik, B. (1987). Experimental studies on Marek's disease in Japanese quail (*Coturnix coturnix japonica*). *Avian Dis.* **31**, 225–233.

Purchase, H. G., Chubb, R. C., Biggs, P. M. (1968). Effect of lymphoid leukosis and Marek's disease on the immunological responsiveness of the chickens. *J. Natl Cancer Inst.* **40**, 583–592.

Quere, P. (1992). Suppression mediated *in vitro* by Marek's disease virus-transformed T-lymphoblastoid cell lines: effect on lymphoproliferation. *Vet. Immunol. Immunopathol.* **32**, 149–164.

Quere, P., Dambrine, G., Bach, M. A. (1989). Influence of thymic hormone on cell-mediated and humoral immune responses in Marek's disease. *Vet. Microbiol.* **19**, 53–59.

Qureshi, M. A., Miller, L. (1991). Signal requirements for the acquisition of tumoricidal competence by chicken peritoneal macrophages. *Poult. Sci.* **70**, 530–538.

Qureshi, M. A., Miller, L., Lillehoj, H. S., Ficken, M. D. (1990). Establishment and characterization of a chicken mononuclear cell line. *Vet. Immunol. Immunopathol.* **26**, 237–250.

Ramm, H. C., Wilson, T. J., Boyd, R. L., Ward, H. A., Mitrangas, K., Fahey, K. J. (1991). The effect of infectious bursal disease virus on B lymphocytes and bursal stromal components in specific pathogen-free (SPF) White Leghorn chickens. *Dev. Comp. Immunol.* **15**, 369–381.

Rast, J. P., Litman, G. W. (1994). T cell receptor gene homologs are present in the most primitive jawed vertebrates. *Proc. Natl Acad. Sci. USA* **91**, 9248–9252.

Ratcliffe, M. J. H. (1989). Development of the avian B cell lymphocyte lineage. *CRC Crit. Rev. Poultry Biol.* **2**, 207–234.

Ratcliffe, M. J. H. (1996). Chicken immunoglobulin isotypes and allotypes. In *Handbook of Experimental Immunology*, 5th edn. (eds Weir, D. W., Herzenberg, L. A.), pp. 24.1–24.15. Blackwell Scientific Publications, Oxford, UK.

Ratcliffe, M. J. H., Jacobsen, K.A. (1994). Rearrangement of immunoglobulin genes in chicken B cell development. *Semin. Immunol.* **6**, 175–184.

Ratcliffe, M. J. H., Tkalec, L. (1990). Cross-linking of the surface immunoglobulin on lymphocytes from the bursa of Fabricius results in second messenger generation. *Eur. J. Immunol.* **20**, 1073–1078.

Ratcliffe, M. J. H., Lassila, O., Pink, J. R. L., Vainio, O. (1986). Avian B cell precursors: surface immunoglobulin expression is an early, possibly bursa-independent event. *Eur. J. Immunol.* **16**, 129–133.

Ratcliffe, M. J. H., Boyd, R., Chen, C., Vainio, O. (1993). Avian CD nomenclature workshops: Montreal, June 1992, Budapest, August 1992 and Tours, September 1992. *Vet. Immunol. Immunopathol.* **38**, 375–386.

Ratcliffe, M. J. H., Paramithiotis, E., Coumidis, A., Sayegh, C., Demaries, S. L., Martinez, O., Jacobsen, K. A. (1996). The bursa of Fabricius and its role in avian B lymphocyte development. In *Poultry Immunology* (eds Davison, T. F., Morris, T. R., Payne, L. N.), pp. 11–30. Carfax Publishing Co., Abingdon, UK.

Rath, N. C., Huff, W. E., Bayyari, G. R., Balog, J. M. (1995). Identification of transforming growth factor-beta and interleukin-6 in chicken ascites fluid. *Avian Dis.* **39**, 382–389.

Reynaud, C. A., Anquez, V., Dahan, A., Weill, J. C. (1985). A single rearrangement event generates most of the chicken immunoglobulin light chain diversity. *Cell* **40**, 283–291.

Reynaud, C. A., Anquez, V., Grimal, H., Weill, J. C. (1987). A hyperconversion mechanism generates the chicken light chain pre-immune repertoire. *Cell* **48**, 379–388.

Reynaud, C. A., Dahan, A., Anquez, V., Weill, J.-C. (1989). Somatic hyperconversion diversifies the single V_H gene of the chicken with a high incidence in the D region. *Cell* **59**, 171–183.

Reynaud, C. A., Anquez, V., Weill, J.-C. (1991). The chicken D locus and its contribution to the immunoglobulin heavy chain repertoire. *Eur. J. Immunol.* **21**, 2661–2670.

Reynaud, C. A., Imhof, B. A., Anquez, V., Weill, J. C. (1992). Emergence of committed B lymphoid progenitors in the developing chicken embryo. *EMBO. J.* **11**, 4349–4358.

Reynaud, C. A., Bertocci, B., Dahan, A., Weill, J.-C. (1994). Formation of the chicken B-cell repertoire: ontogenesis, regulation of Ig gene rearrangement, and diversification by gene conversion. *Adv. Immunol.* **57**, 353–378.

Riddell, C. (1987). *Avian Histopathology* 1st edn. American Association of Avian Pathologists, Kennett Square.

Riegert, P., Andersen, R., Bumstead, N., Doehring, C., Dominguez-Steglich, M., Enberg, J., Salomonsen, J., Schmidt, M., Skjoedt, K., Schwager, J., Kaufman, J. (1996). The chicken beta2-microglobulin gene is located on a non-MHC microchromosome: a small, G + C rich gene with X and Y boxes in the promoter. *Proc. Nat. Acad. Sci. USA* **93**, 1243–1248.

Rieker, T., Penninger, J., Romani, N., Wick, G. (1995). Chicken thymic nurse cells: an overview. *Dev. Comp. Immunol.* **19**, 281–289.

Rinehart, C. L., Rosenberger, J. K. (1983). Effects of avian reovirus on the immune responses of chickens. *Poult. Sci.* **62**, 1488–1489.

Rivas, A. L., Fabricant, J. (1988). Indications of immunodepression in chickens infected with various strains of Marek's disease virus. *Avian Dis.* **32**, 1–8.

Roberts, A. B. (1988). Complementary deoxyribonucleic acid cloning of a novel transforming growth factor-β messenger ribonucleic acid from chick embryo chondrocytes. *Mol. Endocrinol.* **2**, 747–755.

Roberts, A. B., Sporn, M. B. (1990). The transforming growth factor-βs. In *Handbook of Experimental Pharmacology. Peptide Growth Factors and Their Receptors*, Vol. 95/I (eds Sporn, M. B., Roberts, A. B.), pp. 419–472. Springer-Verlag, Heidelberg, Germany.

Rodenberg, J., Sharma, J. M., Belzer, S. W., Nordgren, R. M., Naqi, S. (1994). Flow cytometric analysis of B cell and T cell subpopulations in specific-pathogen-free chickens infected with infectious bursal disease virus. *Avian Dis.* **38**, 16–21.

Romppanen, T. (1982). Postembryonic development of the chicken bursa of Fabricius: a light microscopic histoquantitative study. *Poult. Sci.* **61**, 2261–2270.

Rose, M. E. (1981). Lymphatic system. In *Form and Function in Birds*, Vol. 2 (eds King, A. S., McLelland, J.), pp. 341–384. Academic Press, NY, USA.

Rose, M. E., Orlans, E., Buttress, N. (1974). Immunoglobulin classes in the hen's egg: their segregation in yolk and white. *Eur. J. Immunol.* **4**, 521–523.

Rosenberger, J. K., Gelb, J. (1978). Response to several avian respiratory viruses as affected by infectious bursal disease virus. *Avian Dis.* **22**, 95–105.

Rot, A. (1991). Chemotactic potency of recombinant human neutrophil attractant/activation protein-1 (interleukin-8) for polymorphonuclear leukocytes of different species. *Cytokine* **3**, 21–27.

Rothwell, L., Gramzinski, R. A., Rose, M. E., Kaiser, P. (1995). Avian coccidiosis: changes in intestinal lymphocyte populations associated with the development of immunity to Eimeria maxima. *Parasite Immunol.* **17**, 525–533.

Ruddell, A. (1995). Transcription regulatory elements of the avian retroviral long terminal repeat. *Virology* **206**, 1–7.

Ruff, M. D., Bacon, L. D. (1984). Coccidiosis in 15.B congenic chicks. *Poult. Sci.* **63** (Suppl. 1), 172–173.

Rusov, C., Miljkovic, B., Jojic-Malicevic, L., Zivkovic, R. (1996). Immunosuppressive effect of oncogenic Marek's disease virus on cutaneous hypersensitivity to phytohemagglutinin in young chickens. *Proc. 4th Int. Symp. Marek's Disease*, p. 56.

Saif, Y. M. (1991). Immunosuppression induced by infectious bursal disease virus. *Vet. Immunol. Immunopathol.* **30**, 45–50.

Salant, E. P., Pink, R. L., Steinberg, C. M. (1989). A model of bursal colonization. *J. Theor. Biol.* **139**, 1–16.

Samad, F., Bergtrom, G., Eissa, H., Amrani, D. L. (1993). Stimulation of chick hepatocyte fibronectin production by fibroblast-conditioned medium is due to interleukin 6. *Biochim. Biophys. Acta* **1181**, 207–213.

Sanders, B. G., Case, W. L. (1977). Chicken secretory immunoglobulin: chemical and immunological characterization of chicken IgA. *Comp. Biochem. Physiol.* **56B**, 273–278.

Santivatr, D. S., Maheswaran, S. K., Newman, J. A., Pomeroy, B. S. (1981). Effect of infectious bursal disease virus infection on the phagocytosis of *Staphylococcus aureus* by mononuclear phagocytic cells of susceptible and resistant strains of chickens. *Avian Dis.* **25**, 303–311.

Sarid, R., Chajut, A., Gak, E., Kim, Y., Hixson, C. V., Oroszlan, S., Tronick, S. R., Gazit, A., Yaniv, A. (1994). Genome organization of a biologically active molecular clone of the lymphoproliferative disease virus of turkeys. *Virology* **204**, 680–691.

Sarma, G., Greer, W., Gildersleeve, R. P., Murray, D. L., Miles, A. M. (1995). Field safety and efficacy of *in ovo* administration of HVT + SB-1 bivalent Marek's disease vaccine in commercial broilers. *Avian Dis.* **39**, 211–217.

Schat, K. A. (1987). Marek's disease: a model for protection against herpesvirus induced tumours. *Cancer Surveys* **6**, 1–37.

Schat, K. A. (1996). Immunity to Marek's disease, lymphoid leukosis and reticulodenotheliosis. In *Poultry Immunology*, Poultry Symposium Series Vol. 24 (eds Davison, F., Morris, T. R., Payne, L. N.), pp. 209–234. Carfax, Abingdon, UK.

Schat, K. A., Gonzalez, J., Solorzano, A., Avila, E., Witter, R. L. (1976). A lymphoproliferative disease in Japanese Quail. *Avian Dis.* **20**, 153–161.

Schat, K. A., Schultz, R. D., Calnek, B. W. (1978). Marek's disease: effect of virus pathogenicity and genetic susceptibility on responses of peripheral blood lymphocytes to concanavalin A. In *Advances in Comparative Leukemia Research* (eds Bentvelzen, P., Hilgers, J., Yohn, D. S.), pp. 183–186. Elsevier/North-Holland, Amsterdam, The Netherlands.

Schat, K. A., Chen, C.-L., Lillehoj, H., Calnek, B. W., Weinstock, D. (1988). Characterization of Marek's disease cell lines with monoclonal antibodies specific for cytotoxic and helper T cells. In *Advances in Marek's Disease Research*, Proc. 3rd Int. Symp. on Marek's Disease (eds Kato, S., Horiuchi, T., Mikami, T., Hirai, K.), pp. 220–226. Osaka, Japan.

Schat, K. A., Chen, C.-L. H., Calnek, B. W., Char, D. (1991). Transformation of T-lymphocyte subsets by Marek's disease herpesvirus. *J. Virol.* **65**, 1408–1413.

Schat, K. A., Pratt, W. D., Morgan, R., Weinstock, D., Calnek, B. W. (1992). Stable transfection of reticuloendotheliosis virus-transformed lymphoblastoid cell lines. *Avian Dis.* **36**, 432–439.

Schauenstein, K., Krömer, G., Sundick, R. S., Wick, G. (1985). Enhanced response to Con A and production of TCGF by lymphocytes of Obese strain (OS) chickens with spontaneous autoimmune thyroiditis. *J. Immunol.* **134**, 872–879.

Schauenstein, K., Fässler, R., Dietrich, H., Schwarz, S., Krömer, G., Wick, G. (1987). Disturbed immune-endocrine communication in autoimmune disease: lack of corticosterone response to immune signals in Obese strain chickens with spontaneous autoimmune thyroiditis. *J. Immunol.* **139**, 1830–1833.

Schauenstein, K., Krömer, G., Hála, K., Böck, G., Wick, G. (1988). Chicken-activated-T-lymphocyte-antigen (CATLA) recognized by monoclonal antibody INN-CH-16 represents the IL-2 receptor. *Dev. Comp. Immunol.* **12**, 823–831.

Schierman, L. W., Fletcher, O. J. (1980). Genetic control of Marek's disease virus-induced transient paralysis: association with the major histocompatibiliuty complex. In *Resistance and Immunity to Marek's Disease* (ed. Biggs, P. M.), pp 429–440. ECSC-EEC-EAEC, Brussels, Belgium.

Schierman, L. W., Watanabe, D. H., McBride, R. A. (1977). Genetic control of Rous sarcoma regression in chickens: linkage with the major histocompatibility complex. *Immunoglobulins* **5**, 325–332.

Scott, T. R., Savage, M. L., Olah, I. (1993). Plasma cells of the chicken Harderian gland. *Poult. Sci.* **72**, 1273–1279.

Searle, E. A., Austin, L. M., Boissy, Y. L., Zhao, H., Nordlund, J. J., Boissy, R. E. (1993). Smyth chicken melanocyte autoantibodies: cross-species recognition, *in vivo* binding and plasma membrane reactivity of the antiserum. *Pigment Cell Res.* **6**, 145–157.

Sekellick, M. J., Ferrandino, A. F., Hopkins, D. A., Marcus, P. I. (1994). Chicken interferon gene: cloning, expression and analysis. *J. Interferon Res.* **14**, 71–79.

Seto, F. (1981). Early development of the avian immune system. *Poult. Sci.* **60**, 1981–1995.

Sharma, J. M. (1981). Natural killer cell activity in chickens exposed to Marek's disease virus: inhibition of activity in susceptible chickens and enhancement of activity in resistant and vaccinated chickens. *Avian Dis.* **25**, 882–893.

Sharma, J. M. (1983). Presence of adherent cytotoxic cells and nonadherent natural killer cells in progressive and regressive Marek's disease tumors. *Vet. Immunol. Immunopathol.* **5**, 125–140.

Sharma, J. M. (1984). Effect of infectious bursal disease virus on protection against Marek's disease by turkey herpesvirus vaccine. *Avian Dis.* **28**, 629–640.

Sharma, J. M. (1985). Embryo vaccination with infectious bursal disease virus alone or in combination with Marek's disease vaccine. *Avian Dis.* **29**, 1155–1169.

Sharma, J. M. (1988). Presence of natural suppressor cells in the chicken embryo spleen and the effect of virus infection of the embryo on suppressor cell activity. *Vet. Immunol. Immunopathol.* **19**, 51–66.

Sharma, J. M. (1989). *In situ* production of interferon in tissues of chickens exposed as embryos to turkey herepesvirus and Marek's disease virus. *Am. J. Vet. Res.* **50**, 882–886.

Sharma, J. M. (1991). Hemorrhagic enteritis of turkeys. *Vet. Immunol. Immunopathol.* **30**, 67–71.

Sharma, J. M., Burmester, B. R. (1982). Resistance to Marek's disease at hatching in chickens vaccinated as embryos with the turkey herpesvirus. *Avian Dis.* **26**, 134–149.

Sharma, J. M., Fredrickson, T. L. (1987). Mechanism of T cell immunodepression by infectious bursal disease virus of chickens. In *Avian Immunology*, pp. 283–294. Alan R. Liss, New York, USA.

Sharma, J. M., Lee, L. F. (1983). Effect of infectious bursal disease on natural killer cell activity and mitogenic response of chickens lymphoid cells: role of adherent cells in cellular immune suppression. *Infect. Immun.* **42**, 747–754.

Sharma, J. M., Okazaki, M. (1981). Natural killer cell activity in specific pathogen-free chickens. *J. Natl Cancer Inst.* **63**, 527–531.

Sharma, J. M., Dohms, J. E., Metz, A. L. (1989). Comparative pathogenesis of serotype 1 and variant serotype 1 isolates of infectious bursal disease virus and the effect of those viruses on humoral and cellular immune competence of specific pathogen free chickens. *Avian Dis.* **33**, 112–124.

Sharma, J. M., Dohms, J., Walser, M., Snyder, D. B. (1993). Presence of lesions without virus replication in the thymus of chickens exposed to infectious bursal disease virus. *Avian Dis.* **37**, 741–748.

Sharma, J. M., Karaca, K., Pertile, T. (1994). Virus-induced immunosuppression in chickens. *Poult. Sci.* **73**, 1082–1086.

Shek, W. R., Calnek, B. W., Schat, K. A., Chen, C.-L. H. (1983). Characterization of Marek's disease virus-infected lymphocytes: discrimination between cytolytically and latently infected cells. *J. Natl Cancer Inst.* **70**, 485–491.

Shen, P. F., Smith, E. J., Bacon, L. D. (1984). The ontogeny of blood cells, complement, and immunoglobulins in 3- to 12-week old $15I_5$-B congenic White Leghorn chickens. *Poult. Sci.* **63**, 1083–1093.

Shinkai, Y., Rathburn, G., Lam, K. P., Oltz, E. M., Stewart, V., Medelsohn, M., Charron, J., Datta, M., Young, F., Stall, A. M., Alt, F. W. (1992). RAG-2 deficient mice lack mature lymphocytes owing to inability to initiate V(D)J rearrangement. *Cell* **68**, 855–867.

Sieweke, M. H., Thompson, N. L., Sporn, M. B., Bissell, M. J. (1990). Mediation of wound-related Rous sarcome virus tumorigenesis by TGF-β. *Science* **248**, 1656–1660.

Silva, J. S., Twardzik, D. R., Reed, S. G. (1991). Regulation of trypanosoma cruzi infections *in vitro* and *in vivo* by transforming growth factor β (TGF-β). *J. Exp. Med.* **174**, 539–545.

Simonsen, M., Kolstad, N., Edfors-lilja, I., Liledahl, L. E., Sorensen, P. (1982). Major histocompatibility genes in egg-laying hens. *Am. J. Reprod. Immunol.* **2**, 148–152.

Sivanandan, V., Maheswaran, S. K. (1980a). Immune profile of infectious bursal disease: I. Effect of infectious bursal disease virus on peripheral blood T and B lymphocytes of chickens. *Avian Dis.* **24**, 715–725.

Sivanandan, V., Maheswaran, S. K. (1980b). Immune profile of infectious bursal disease: II. Effect of IBD virus on pokeweed-mitogen stimulated peripheral blood lymphocytes. *Avian Dis.* **24**, 734–742.

Sivanandan, V., Maheswaran, S. K. (1981). Immune profile of infectious bursal disease: III. Effect of infectious bursal disease virus on the lymphocyte responses to phytomitogens and on mixed lymphocyte reaction of chickens. *Avian Dis.* **25**, 112–120.

Six, A., Rast, J. P., Li, Y., Cooper, M. D., Chen, C. H., Litman, G. W. (1995). Identification of the chicken T cell receptor $\gamma\delta$ genes. *FASEB J.* **9**, 4737.

Smith, E. J. (1987). Endogenous avian leukemia viruses. In *Avian Leukosis* (ed. de Boer, G. F.), pp. 101–120. Martinus Nijhoff, Dordrecht, The Netherlands.

Smyth Jr, J. R., Boissy, R. E., Fite, K. V. (1981). The DAM chicken: a spontaneous postnatal cutaneous and ocular amelanosis. *J. Hered.* **72**, 150–156.

Snipes, K. P., Hirsh, D. C. (1986). Association of complement sensitivity with virulence of *Pasteurella multocida* isolated from turkeys. *Avian Dis.* **30**, 500–504.

Solano, W., Giambrone, J. J., Williams, J. C., Lauerman, L. H., Panangala, V. S., Garces, C. (1986). Effect of maternal antibody on timing of initial vaccination of young White Leghorn chickens against infectious bursal disease virus. *Avian Dis.* **30**, 648–652.

Sowder, J. T., Chen, C. H., Ager, L. L., Chan, M. M., Cooper, M. D. (1988). A large subpopulation of avian T cells express a homologue of the mammalian T $\gamma\delta$ receptor. *J. Exp. Med.* **167**, 315–322.

Sponenberg, D. P., Domermuth, C. H., Larsen, C. T. (1985). Field outbreaks of colibacillosis of turkeys associated with hemorrhagic enteritis virus. *Avian Dis.* **29**, 838–842.

Sporn, M. B., Roberts, A. B. (1990). The multifunctional nature of peptide growth factors. In *Handbook of Experimental Pharmacology. Peptide Growth Factors and Their Receptors*, Vol. 95/I (eds Sporn, M. B., Roberts, A. B.), pp. 3–15. Springer-Verlag, Heidelberg, Germany.

Steadham, E. M., Lamont, S. J. (1993). Gene complementation in biological crosses for humoral immune response to glutamic acid-alanine-tyrosine. *Poult. Sci.* **72**, 76–81.

Sterneck, E., Blattner, C., Graf, T., Leutz, A. (1992). Structure of the chicken myelomonocytic growth factor gene and specific activation of its promotor in avian myelomonocytic cells by protein kinases. *Mol. Cell. Biol.* **12**, 1728–1735.

Stoeckle, M. Y., Barker, K. A. (1990). Two burgeoning families of platelet factor 4-related proteins: mediators of inflammatory responses. *New Biologist* 2, 313–323.

Sugano, S., Stoeckle, M. Y., Hanafusa, H. (1987). Transformation by Rous sarcoma virus induces a novel gene with homology to a mitogenic platelet protein. *Cell* 49, 321–328.

Sugimura, M., Suzuki, Y., Atojo, Y., Hashimoto, Y. (1987). Accumulation of Kupffer cells within lymphocyte aggregates in the livers of chickens injected with colloidal carbon and erythrocytes. *Jap. J. Vet. Sci.* 49, 771–778.

Sundick, R. S., Gill-Dixon, C. (1997). A cloned chicken lymphocyte homologous to both mammalian IL-2 and IL-15. *J. Immunol.* 159, 720–725.

Sundick, R. S., Wick, G. (1974). Increased 131-I uptake by the thyroid glands of Obese strain (OS) chickens derived from non-protomone supplemented hens. *Clin. Exp. Immunol.* 18, 127–139.

Sundick, R. S., Bagchi, N., Livezey, M. D., Brown, T. R., Mack, R. E. (1979). Abnormal thyroid regulation in chickens with autoimmune thyroiditis. *Endocrinol.* 105, 493–498.

Sundick, R. S., Herdegen, D., Brown, T. R., Bagchi, N. (1987). The incorporation of dietary iodine into thyroglobulin increases its immunogenicity. *Endocrinology* 120, 2078–2084.

Sundick, R. S., Herdegen, D., Brown, T. R., Dhar, A., Bagchi, N. (1991). Thyroidal iodine metabolism in obese strain chickens before immune-mediated damage. *J. Endocrinol.* 128, 239–244.

Suresh, M., Sharma, J. M. (1995). Hemorrhagic enteritis virus induced changes in the lymphocyte subpopulations in turkeys and the effect of experimental immunodeficiency on viral pathogenesis. *Vet. Immunol. Immunopathol.* 45, 139–150.

Suresh, M., Sharma, J. M. (1996). Pathogenesis of type II avian adenovirus infection in turkeys: *in vivo* immune cell tropism and tissue distribution of the virus. *J. Virol.* 70, 30–36.

Suresh, M., Karaca, K., Foster, D., Sharma, J. M. (1995). Molecular and functional characterization of turkey interferon. *J. Virol.* 69, 8159–8163.

Survashe, B. D., Aitkin, I. D. (1977). Removal of the lacrimal gland and ligation of the Harderian gland duct in the fowl (*Gallus domesticus*): procedures and sequelae. *Res. Vet. Sci.* 22, 113–119.

Swift, R. A., Shaller, E., Witter, R. L., Kung, H.-J. (1985). Insertional activation of *c-myc* by reticuloendotheliosis virus in chicken B lymphoma: nonrandom distribution and orientation of the provirus. *J. Virol.*, 54, 869–872.

Takata, M., Sabe, H., Hata, A., Inazu, T., Homma, Y., Nukada, T., Yamamura, H., Kurosaki, T. (1994). Tyrosine kinases Lyn and Syk regulate B cell receptor-coupled Ca^{2+} mobilization through distinct pathways. *EMBO J.* 13, 1341–1349.

Tang, K. N., Fletcher, O. J., Villegas, P. (1987a). Comparative study of the pathogenicity of avian reoviruses. *Avian Dis.* 31, 577–583.

Tang, K. N., Fletcher, O. J., Villegas, P. (1987b). The effect on newborn chicks of oral inoculation of reovirus isolated from chickens with tenosynovitis. *Avian Dis.* 31, 584–590.

Tempelis, C. H. (1988). Ontogeny and regulation of the immune response in the chicken. *Prog. Vet. Microbiol. Immun.* 4, 1–20.

Theis, G. A. (1977). Effects of lymphocytes from Marek's disease infected chickens on mitogen response of syngeneic normal chicken spleen cells. *J. Immunol.* 118, 887–894.

Theis, G. A. (1981). Subpopulations of suppressor cells in chickens infected with cells of a transplantable lymphoblastic leukemia. *Infec. Immun.* 34, 526–534.

Theis, G. A., McBride, R. A., Scierman, L. W. (1975). Depression of *in vitro* responsiveness to phytohemagglutinin in spleen cells cultured from chickens with Marek's disease. *J. Immunol.* 115, 848–853.

Thompson, C. B., Neiman, P. E. (1987). Somatic diversification of the chicken immunoglobulin light chain gene is limited to the rearranged variable gene segment. *Cell* 48, 369–378.

Thorpe, K. L., Abdulla, S., Kaufman, J., Trowsdale, J., Beck, S. (1996). Phylogeny and structure of the RING3 gene. *Immunogenetics* 44, 391–396.

Thunold, S., Schauenstein, K., Wolf, H., Thunold, K. S., Wick, G. (1981). Localization of IgG Fc and complement receptors in chicken lymphoid tissue. *Scand. J. Immunol.* 14, 145–152.

Tjoelker, L. W., Carlson, L. M., Lee, K., Lahti, J., McCormack, W. T., Leiden, J. M., Chen, C. H., Cooper, M. D., Thompson, C. B. (1990). Evolutionary conservation of antigen recognition: the chicken T cell receptor β chain. *Proc. Natl Acad. Sci. USA* 87, 7856–7860.

Toivanen, A., Toivanen, P. (1987). Stem cells of the lymphoid system. In *Avian Immunology: Basis and Practice* (eds Toivanen, A., Toivanen, P.), pp. 23–37. CRC Press, Boca Raton, FL, USA.

Toivanen, A., Toivanen, P., Eskola, J., Lassila, O. (1981). Ontogeny of the chicken lymphoid system. In *Avian Immunology* (eds Rose, M. E., Payne, L. N., Freeman, B. M.), pp. 45–62. Clark Constable Limited, Edinburgh, UK.

Toivanen, P., Toivanen, A. (1973). Bursal and post-bursal stem cells in the chicken. Functional characteristics. *Eur. J. Immunol.* 3, 585–595.

Tonegawa, S. (1983). Somatic generation of antibody diversity. *Nature* 302, 575–581.

Trees, A. J., Crozier, S. J., McKellar, S. B., Wachira, T. M. (1985). Class-specific circulating antibodies in infections with *Eimeria tenella*. *Vet. Parasitol.* 18, 349–357.

Tregaskes, C. A., Kong, F., Paramithiotis, E., Chen, C.-L. H., Ratcliffe, M. J. H., Davison, T. F., Young, J. R. (1995). Identification and analysis of the expression of CD8$\alpha\alpha$ and CD8$\alpha\beta$ isoforms in chickens reveals a major TCR-$\gamma\delta$ CD8$\alpha\beta$ subset of intestinal intraepithelial lymphocytes. *J. Immunol.* 154, 4485–4494.

Tregaskes, C. A., Bumstead, N., Davison, T. F, Young, J. (1996). Chicken B-cell marker ChB6 (Bu-1) is a highly glycosylated protein of unique structure. *Immunogenetics* 44, 212–217.

Trembicki, K. A., Qureshi, M. A., Dietert, R. R. (1986). Monoclonal antibodies reactive with chicken peritoneal macrophages: identification of macrophage heterogeneity. *Proc. Soc. Exp. Biol. Med.* 183, 28–41.

Trout, J. M., Lillehoj, H. S. (1995). *Eimeria acervulina*: evidence for the involvement of $CD8^+$ T lymphocytes in sporozoite transport and host protection. *Poult. Sci.* 74, 1117–1125.

Trout, J. M., Lillehoj, H. S. (1996). T lymphocyte roles during *Eimeria acervulina* and *E. tenella* infections. *Vet. Immunol. Immunopathol.* 53, 163–172.

Truden, J. L., Sundick, R. S., Levine, S., Rose, N. R. (1983). The decreased growth rate of Obese strain (OS) chicken thyroid cells provides *in vitro* evidence for a primary organ abnormality in chickens susceptible to autoimmune thyroiditis. *Clin. Immunol. Immunopathol.* 29, 294–305.

Tsuji, S., Baba, T., Kawata, T., Kajikawa, T. (1993). Role of

Harderian gland on differentiation and proliferation of immunoglobulin A-bearing lymphocytes in chicken. *Vet Immunol. Immunopathol.* **37**, 271–283.

Tsuji, S., Illges, H., Kubota, T., Char, D., Bucy, P., Simonsen, M., Chen, C. H., Cooper, M. D. (1995). Donor $\alpha\beta$ T cell clones recruit host $\gamma\delta$ T cells in acute GVH reactions. *9th Intl. Congress Immunol.*, p. 251, abstract 1486.

Vainio, O., Imhof, B. A. (1995). The immunology and developmental biology of the chicken. *Immunol. Today* **16**, 365–370.

Vainio, O., Imhof, B.A. (1996). Immunology and developmental biology of the chicken. *Curr. Topics Immunol. Microbiol.* **212**, 263–274.

Vainio, O., Koch, C., Toivanen, A. (1984). B-L antigens (class II) of the chicken major histocompatibility complex control T-B cell interaction. *Immunogenetics* **19**, 131–140.

Vainio, O., Ratcliffe, M. J. H., Leanderson, T. (1986). Chicken T-cell growth factor: use in the generation of a long-term cultured T-cell line and biochemical characterization. *Scand. J. Immunol.* **23**, 135–142.

Vainio, O., Veromaa, T., Eerola, E., Toivanen, P., Ratcliffe, M. J. H. (1988). Antigen-presenting cell–T cell interaction in the chicken is MHC class II antigen restricted. *J. Immunol.* **140**, 2864–2868.

Vaino, O., Riwar, B., Brown, M. H., Lassila, O. (1991). Characterization of the putative avian CD2 homologue. *J. Immunol.* **147**, 1593–1599.

Vainio, O., Dunon, D., Aïssi, F., Dangy, J. P., McNagny, K. M., Imhof, B. A. (1996). HEMCAM, an adhesion molecule expressed by c-kit$^+$ hemopoietic progenitors. *J. Cell Biol.* **135**, 1655–1668.

van den Hurk, J., Allan, B. J., Riddell, C., Watts, T., Potter, A. A. (1994). Effect of infection with hemorrhagic enteritis virus on susceptibility of turkeys to *Escherichia coli*. *Avian Dis.* **38**, 708–716.

van de Water, J., Haapanen, L., Boyd, R., Abplanalp, H., Gershwin, M. E. (1989). Identification of T cells in early dermal infiltrates in avian scleroderma. *Arthritis Rheum.* **32**, 1031–1040.

van de Water, J., Wilson, T. J., Haapanen, L. A., Boyd, R. L., Abplanalp, H., Gershwin, M. E. (1990). Ontogeny of T cell development in avian scleroderma. *Clin. Immunol. Immunopathol.* **56**, 169–184.

van Ostade, X., Tavernier, J., Prange, T., Fiers, W. (1991). Localization of the active site of human tumour necrosis factor (hTNF) by mutational analysis. *EMBO J.* **10**, 827–836.

Veromaa, T., Vainio, O., Eerola, E., Toivanen, P. (1988). Monoclonal antibodies against chicken Bu-1a and Bu-1b alloantigens. *Hybridoma* **7**, 41–48.

von Boehmer, H. (1994). Positive selection of lymphocytes. *Cell* **76**, 219–228.

Wakefield, L. M. (1987). An assay for type-beta transforming growth factor receptor. *Methods Enzymol.* **146**, 167–173.

Wakenell, P. S., Sharma, J. M. (1986). Chicken embryonal vaccination with avian infectious bronchitis virus. *Am. J. Vet. Res.* **47**, 933–938.

Walker, W. H., Stein, B., Ganchi, P. A., Hoffman, J. A., Kaufman, P. A., Ballard, D. W., Hannink, M., Greene, W. C. (1992). The *v-rel* oncogene: insights into the mechanism of transcriptional activation, repression, and transformation. *J. Virol.* **66**, 5018–5029.

Warner, N. L., Szenberg, A., Burnet, F. M. (1962). The immunological role of different lymphoid organs in the chicken. I. Dissociation of immunological responsiveness. *Aust. J. Exp. Biol. Med.* **40**, 373–387.

Wathen, L. K., Leblanc, D., Warner, C. M., Lamont, S. J., Noroskog, A. W. (1987). A chicken sex-limited protein that crossreacts with the fourth component of complement. *Poult. Sci.* **66**, 162–165.

Weiler, H., von Bulow, V. (1987). Development of optimal conditions for lymphokine production by chicken lymphocytes. *Vet. Immunol. Immunopathol.* **14**, 257–262.

Weill, J. C., Reynaud, C. A. (1987). The chicken B cell compartment. *Science* **238**, 1094–1098.

Weill, J. C., Reynaud, C. A., Lassila, O., Pink, J. R. L. (1986). Rearrangement of chicken immunoglobulin genes is not an ongoing process in the embryonic bursa of Fabricius. *Proc. Natl Acad. Sci. USA* **83**, 3336–3340.

Weill, J. C., Leibowitch, M., Reynaud, C. A. (1987). Questioning the role of the embryonic bursa in the molecular differentiation of B lymphocytes. In *Current Topics in Microbiology and Immunology*, Vol. 135 (eds Paige, C. J., Gisler, R. H.), pp. 111–124. Springer-Verlag, Heidelberg, Germany.

Welch, P., Rose, N. R., Kite Jr, J. H. (1973). Neonatal thymectomy increases spontaneous autoimmune thyroiditis. *J. Immunol.* **110**, 575–577.

White, R. G. (1981). The structural organization of Avian lymphoid tissues. In *Avian Immunology* (eds Rose, M. E., Payne, L. N., Freeman, B. M.), pp. 21–44. Clark Constable Limited, Edinburgh, UK.

Whitehouse, D. B. (1988). Genetic polymorphisms of animal complement components. *Exper. Clin. Immunogenet.* **5**, 143–164.

Wick, G., Kite Jr, J. H., Cole, R. K., Witebsky, E. (1970a). Spontaneous thyroiditis in the Obese strain of chicken: III. The effect of bursectomy on the development of disease. *J. Immunol.* **104**, 45–53.

Wick, G., Kite Jr, J. H., Witebsky, E. (1970b). Spontaneous thyroiditis in the Obese strain of chicken: IV. The effect of thymectomy and thymo-bursectomy on the development of disease. *J. Immunol.* **104**, 54–62.

Wight, P. A. L., Burns, R. B., Rothwell, B., MacKenzie, G. M. (1971). The Harderian gland of the domestic fowl. I. Histology, with reference to the genesis of plasma cells and Russell bodies. *J. Anat.* **110**, 307–315.

Wilson, T. J., van de Water, J., Mohr, F. C., Boyd, R. L., Ansari, A., Wick, G., Gershwin, M. E. (1992). Avian scleroderma: evidence for qualitative and quantitative T cell defects. *J. Autoimmun.* **5**, 261–276.

Winship, T. R., Marcus, P. I. (1980). Interferon induction by viruses. VI. Reovirus: virion genome dsRNA as the interferon inducer in aged chick embryo cells. *J. Interferon Res.* **1**, 155–167.

Wistow, G. J., Shaughnessy, M. P., Lee, D. C., Hodin, J., Zelenka, P. S. (1993). A macrophage migration inhibitory factor is expressed in the differentiating cells of the eye lens. *Proc. Natl Acad. Sci. USA* **90**, 1272–1275.

Witter, R. L. (1991). Reticuloendotheliosis. In *Diseases of Poultry*, 9th edn (eds Calnek, B. W., Barnes, H. J., Beard, C. W., Reid, W. M., Yoder Jr, H. W.), pp. 439–456. Iowa State University Press, Ames, IA, USA.

Witter, R. L. (1997). Reticuloendotheliosis. In *Diseases of Poultry*, 10th edn (eds Calnek, B. W., Barnes, H. J., Beard, C. W., McDougald, L. R., Saif, Y. M.), pp. 467–484. Iowa State University Press, Ames, IA, USA.

Witter, R. L., Smith, E. J., Crittenden, L. B. (1981). Tolerance, viral shedding, and neoplasia in chickens infected with non-defective reticuloendotheliosis viruses. *Avian Dis.* **25**, 374–394.

Witter, R. L., Sharma, J. M., Fadly, A. M. (1986). Nonbursal lymphomas induced by nondefective reticuloendotheliosis virus. *Avian Pathol.*, **15**, 467–486.

Wood, G. W., Muskett, J. C., Thornton, D. H. (1981). The interaction of live vaccine and maternal antibody in protection against infectious bursal disease. *Avian Path.* **10**, 365–373.

Wooley, R. E., Spears, K. R., Brown, J., Nolan, L. K., Fletcher, O. J. (1992). Relationship of complement resistance and selected virulence factors in pathogenic avian *Escherichia coli*. *Avian Dis.* **36**, 679–684.

Wride, M. A., Sanders, E. J. (1993). Expression of tumor necrosis factor-α (TNF-α)-cross-reactive proteins during early chick embryo development. *Dev. Dynamics* **198**, 225–239.

Wyatt, C. L., Weaver Jr, W. D., Beane, W. L., Denbow, D. M., Gross, W. B. (1986). Influence of hatcher holding times on several physiologic parameters associated with the immune system of chickens. *Poult. Sci.* **65**, 2156–2164.

Wyeth, P. J. (1975). Effect of infectious bursal disease on the response of chickens to *S. typhimurium* and *E. coli* infections. *Vet. Rec.* **96**, 238–243.

Wyeth, P. J., Cullen, G. A. (1976). Maternally derived antibody-effect on susceptibility of chicks to infectious bursal disease. *Avian Path.* **5**, 253–260.

Yamada, A., Hayami, M. (1983). Suppression of natural killer cell activity by chicken alpha-fetoprotein in Japanese quails. *J. Natl Cancer Inst.* **70**, 735–738.

Yamamoto, H., Hattori, M., Ohashi, K., Sugimoto, C., Onuma, M. (1995). Kinetic analysis of T cells and antibody production in chickens infected with Marek's disease virus. *J. Vet. Med. Sci.* **57**, 945–946.

Yonemasu, K., Sasaki, T. (1986). Purification, identification and characterization of chicken Clq, a subcomponent of the first component of complement. *J. Immunol. Meth.* **88**, 245–253.

Young, J. R., Davison, T. F., Tregaskes, C. A., Rennie, M. C., Vainio, O. (1994). Monomeric homologue of mammalian CD28 is expressed on chicken T cells. *J. Immunol.* **152**, 3848–3851.

Yuasa, N., Taniguchi, T., Noguchi, T., Yoshida, I. (1980). Effect of infectious bursal disease virus infection on incidence of anemia by chicken anemia agent. *Avian Dis.* **24**, 202–209.

Zhang, S., Lillehoj, H. S., Ruff, M. D. (1995). Chicken tumor necrosis-like factor. I. *In vitro* production by macrophages stimulated with *Eimeria tenella* or bacterial lipopolysaccharide. *Poult. Sci.* **74**, 1304–1310.

Zhonggui, L., Qingzhang, L., Rong, G., Yun, L., Shimin, Z., Zhiyong, Z. (1996). Immunosuppressive mechanism of chicken Marek's disease. *Proc. 4th Int. Symp. Marek's Disease*, p. 45.

Zoorob, R., Behar, G., Kroemer, G., Auffray, C. (1990). Organization of a functional chicken class II B gene. *Immunogenetics* **31**, 179–187.

Zoorob, R., Bernot, G., Renoir, D. M., Choukri, F., Auffray, C. (1993). Chicken major histocompatibility complex class II B genes: analysis of interallelic and interlocus sequence variance. *Eur. J. Immunol.* **23**, 1139–1145.

V IMMUNOLOGY OF THE RAT

1. Introduction

The rat is undoubtedly one of the most commonly used animals in experimental biological and medical research. There are two species of rats, the brown rat, *Rattus norvegicus*, and the black rat, *Rattus rattus*, which both originate from South East Asia (India and China). Together with animals such as the mouse and guinea pig, rats belong to the largest order of mammals, the Rodentia. The laboratory rat has been developed from *R. norvegicus*, at the beginning of the twentieth century and most laboratory rat strains originate from animals held at the Wistar Institute of Anatomy and Biology at Philadelphia, PA, USA. A large number of genetically well-defined inbred laboratory rat strains are now available for experimental research (Festing, 1979). Rats are very strong animals that do not have many requirements with respect to their food and housing; they are easy to handle and to maintain, breed well in captivity, have a limited lifespan and are economic in their usage as experimental animals. Importantly, although rats are small animals they are big enough to carry out a wide variety of experimental manipulations relatively easily (for an overview refer to e.g. van Dongen *et al.*, 1990; Waynforth and Flecknell, 1992). Over the years much became known about their physiology, anatomy, histology, genetics and, traditionally, also about their behaviour. Many disease models have been developed in this species, further establishing their firm position as experimental animals in both fundamental and applied medicine.

Likewise, rats are also useful in immunological research and are the second most frequently used animal (after the mouse) in this field. Although rats contribute significantly to all aspects of immunology, their larger size offer unique possibilities for the application of microsurgical techniques (van Dongen *et al.*, 1990; Waynforth and Flecknell, 1992). Both well-established microsurgical techniques such as thoracic duct cannulation and novel techniques such as vascular thymus transplantation (Kampinga *et al.*, 1990, 1997) are easier in rats than in mice, and have proven their value to address fundamental immunological issues. For this reason, these animals are often the first choice for transplantation studies and it is therefore not surprising that recent technology to induce indefinite allogenic graft survival by intrathymic injection of allo-antigen was first developed in rats (Klatter *et al.*, 1995; Posselt *et al.*, 1990). A variety of congenic, mutant and recombinant rat strains of immunological interest have been developed during recent years, further expanding the possibilities of experimentation. Furthermore, excellent autoimmune disease models are available for rats. These diseases develop either spontaneously as in the case of, for example, the diabetes-prone BB rat strain (Chappel and Chappel, 1983) or can be induced by autoantigen administration, resulting in autoimmune diseases varying from the widely used model of experimental autoimmune encephalomyelitis (EAE) (Swanborg, 1995) to the more recently developed model of antimyeloperoxidase-associated proliferative glomerulonephritis (Brouwer *et al.*, 1993). In this chapter an overview is given of our expanding knowledge of the immune system of the laboratory rat, *R. norvegicus*. Information about the immune system of second rat species, the black rat, *R. rattus*, is very scarce and is beyond the scope of this chapter.

2. Lymphoid Organs: Their Anatomical Distribution and Cell Composition

The rat lymphoid system

As in other mammals, the rat lymphoid system can be distinguished in two parts: (1) a *central* and (2) a *peripheral* part (Figure V.2.1).

Central part

Essentially, the major role of the central part is to provide the peripheral part – via the circulation – with appropriate immunologically competent cells (ICC), i.e. T and B cells. To this end, B cells and precursor T cells are continuously generated, derived from hemopoietic stem cells in the bone marrow. B-cell production and maturation takes place mainly inside the bone marrow. Precursor T cells (prothymocytes), once released into the circulation, gain access to the thymic parenchyma, where, through processes of proliferation, maturation and selection, newly formed T cells are eventually generated, which, as recent thymic emigrants, enter into the circulation, where further maturation and differentiation may take place.

Handbook of Vertebrate Immunology
ISBN 0-12-546401-0

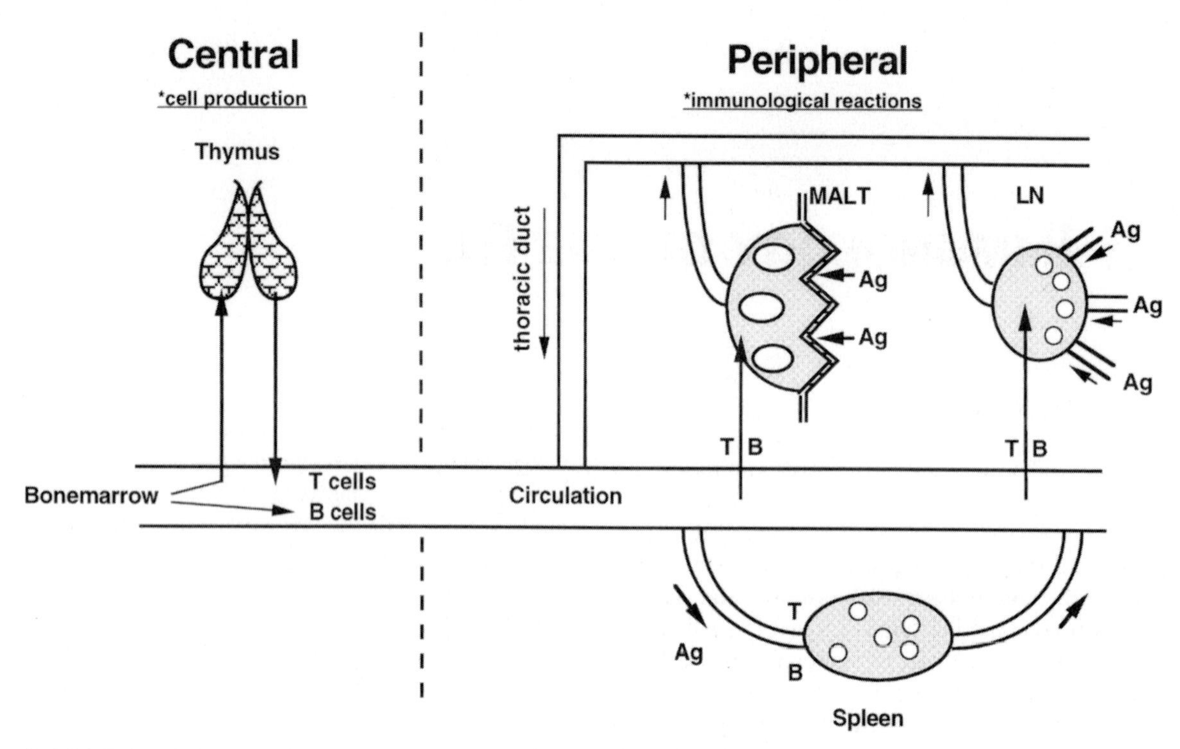

Figure V.2.1 Schematic representation of central and peripheral components of the rat lymphoid system. In the *central* part, T and B cells are formed, which, upon entering the circulation can enter *peripheral* lymphoid organs like spleen, lymph nodes and MALT. From here T and B cells, after a short interval of hours or days, will re-enter the circulation, either directly (spleen) or indirectly, through efferent lymphatics (lymph nodes, MALT). LN, lymph node; MALT, mucosa associated lymphoid tissue; Ag, antigen; T, T cells; B, B cells.

Peripheral part

In the peripheral part, the T and B cells home to the respective lymphoid organs, from where, after a short period of time (hours/days) they re-enter the circulation, either directly (spleen) or indirectly through efferent lymphatics (e.g. lymph nodes). In these lymphoid organs highly specialized microenvironments exist that allow the interaction between antigen (brought there by means of antigen presenting cells, APC) and these recirculating cells so as to perform their respective immunological functions (cellular immunity, humoral immunity, immunological memory) in an antigen-specific way. Through this recirculation process naive cells, effector cells and memory cells are continuously redistributed all over the body (surveillance function). In the peripheral part most lymphoid structures are lymph-associated (lymphatic system) apart from the spleen, which is essentially blood-associated.

The Lymphatic System

A major purpose of the lymphatic system is to collect surplus tissue fluid (resulting from the capillary filtration process and incomplete protein reabsorption at the venous side) and to transport this back to the circulation so as to avoid stagnation of tissue fluid (seen in pathological conditions as edema) (Yoffey and Courtice, 1956). In addition, these channels are used to transport antigens (either as such or processed by APCs) that have penetrated either the skin or the mucosal lining of the respective tracts (digestive, respiratory, urogenital) to the local draining lymph nodes (the ‘meeting point’ of antigen/APC and ICC). Focusing on this lymphatic system and the source of the lymph contained within, one can make the following distinctions: (Figure V.2.2)

Skin-draining lymph nodes: (e.g. axillary, popliteal and inguinal lymph nodes). Here, primary lymph is collected from dermal and subcutaneous connective tissue, which under normal circumstances will be virtually antigen free. Accordingly, immune reactivity in these nodes is usually low, except for the normal (physiological) recirculation of T and B cells through these nodes. These nodes usually have multiple afferent lymphatics, but only one efferent lymphatic through which the collected lymph is transported to a subsequent station (secondary lymph node or major lymph vessel).

Mucosa-draining lymph nodes (e.g. mesenteric lymph nodes, hilar lymph nodes, cervical lymph nodes). Because the mucosal lining of the respective tracts is under constant threat of penetration by all types of bacteria and viruses (and, for example, in the small intestine translocation of bacteria is a normal daily phenomenon) the draining lymph nodes usually show a high degree of immune reactivity, among others witnessed by the occurrence of numerous germinal centres (GC) inside B-lym-

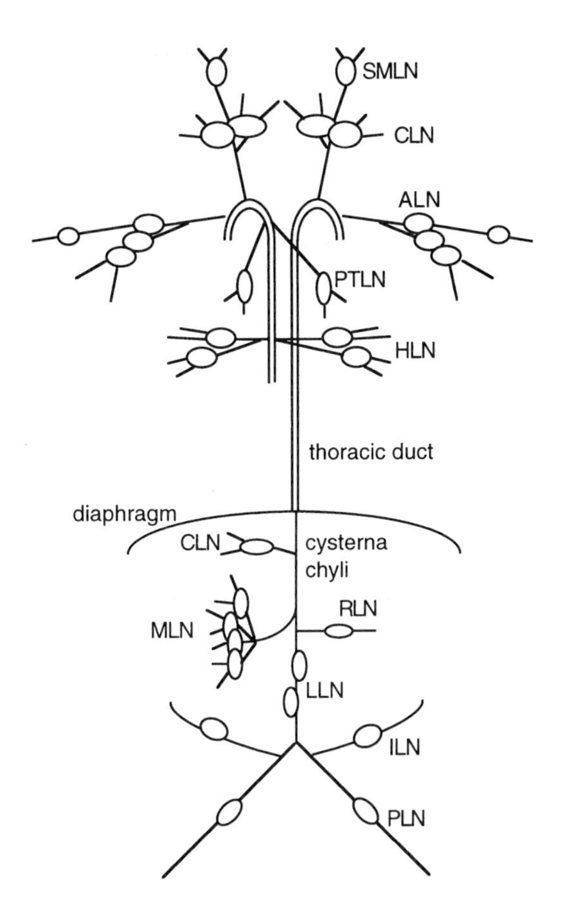

Figure V.2.2 Lymphatic system of the rat. ALN, axillar lymph nodes; CLN, cervical lymph nodes; CLN, celiac lymph node; HLN, hilar lymph nodes; ILN, inguinal lymph node; LLN, lumbar lymph nodes; MLN, mesenteric lymph nodes; PLN, popliteal lymph node; PTLN, parathymic lymph node; RLN, renal lymph node; SMLN, submandibular lymph node. After Greene (1935), *Anatomy of the Rat*, in *Trans. Am. Phil. Soc.*, New Series, vol. XXVII, p. 330.

phocyte follicles. Like skin draining lymph nodes, they have multiple afferent and only a single efferent lymphatic.

In addition to these mucosa-draining lymph nodes the mucosa itself may contain organized lymphoid structures that are either genetically predetermined (like Peyer's patches) or reactive by nature (solitary nodules). Together, these structures are referred to as MALT (mucosa associated lymphoid tissue). Examples of MALT are:

- for the digestive tract: NALT (nasal associated lymphoid tissue) (Spit *et al.*, 1989; Kuper *et al.*, 1990), Peyer's patches and PCLT (proximal colonic lymphoid tissue) (Crouse *et al.*, 1989; de Boer *et al.*, 1992); the last two collectively referred to as GALT (gut associated lymphoid tissue).
- for the respiratory tract: BALT (bronchus associated lymphoid tissue) (Plesch *et al.*, 1982, 1983).

No predetermined structures have been identified for the urogenital tract.

MALT can be considered as a lymph node situated immediately underneath the mucosal lining without, however, afferent lymphatics. Antigen transport across the mucosal lining is mediated by antigen sampling membranous epithelial cells (M-cells) (Owen *et al.*, 1986). Thus the antigen content of the associated lumen is constantly sampled and, accordingly, these structures usually show a high degree of immune reactivity, again indicated, for example, by the occurrence of numerous GC in these structures (absent in germfree animals).

Organ/tissue-draining lymph nodes. In addition to the above-mentioned lymphoid structures, which are all associated with the body's outer epithelial lining, some organs have their own associated lymph nodes as in the CLN (celiac lymph nodes) for the liver, RLN (renal lymph node, kidney) and PTLN (parathymic lymph node, thymus). In general, all the lymph produced in any organ, gland, muscular tissue, connective tissue, bone marrow, etc., usually passes through one (or more) lymph nodes before entering into the two major lymph-draining vessels (right and left thoracic duct) before gaining access to the circulation again. Because the lymph derived from these organs/tissues generally will be antigen free, immune reactivity in the draining lymph nodes is accordingly rather low.

Morphological and functional aspects of respective lymphoid organs

Central lymphoid organs

The bone marrow

Rat bone marrow has a fairly high content of lymphoid cells, especially B-cell progenitors in various stages of differentiation. From 12.5×10^7 nucleated cells isolated from one femur about 25% belong to the B-cell lineage, as determined by the monoclonal antibody HIS24 (Deenen *et al.*, 1987). During B-cell development in the bone marrow, both cell proliferation and cell loss (through apoptosis?) occur in the various compartments of differentiating cells in such a way that from the daily production of 650×10^6 cells (total bone marrow compartment) only 165×10^6/day turn into newly formed B cells (NFB) of which only 16×10^6 cells/day gain access to the pool of mature B cells. As the total B cell pool in the periphery has been estimated to be approximately 1×10^9 cells, daily renewal rate (turnover) has been estimated to be approximately 1–2% (Deenen and Kroese, 1993). B-cell development in rat bone marrow will be described in more detail in section 8.

B-cell differentiation in the rat bone marrow occurs in a centripetal way, the more immature forms being detected in the periphery of the marrow cavity immediately adjacent to the endosteum. Moreover, early B-cell formation appears to occur as a clonal event, presumably associated with specialized local microenvironments (Hermans and Opstelten, 1991).

No quantitative data are available on the production of T cell precursors. However, when young adult rats were treated with a single dose of Adriamycin, resulting in total depletion of all nucleated cells in the bone marrow (except for the stem cells) and thymus, it took from 15 days (Albino Oxford (AO) rats) to 22 days (Brown Norway (BN) rats) before early thymocyte progenitors could be detected again in the thymus (P. Nieuwenhuis, unpublished data). Using an *in vitro* culture system Prakapas *et al.* (1993) have identified a 10% lymphoid subset in the bone marrow with thymus repopulating activity which did not express B-lymphocyte associated antigens.

The thymus

The rat thymus is situated in the upper part of the mediastinum anterius just behind the sternum, with two small tips extending upwards lying along the carotid arteries. The thymus has no afferent lymphatics. Lymph formed in its capsule (and trabeculae) and in the medulla drain through an efferent lymphatic to reach the parathymic lymph nodes.

The thymus parenchyma is made up of a reticulum of epithelial cells among which T cells in various stages of differentiation are situated. Epithelial cells can be recognized by various monoclonal antibodies and can thus be distinguished in various subtypes, a major distinction being between cortical and medullary (Kampinga *et al.*, 1989).

Both cortical and medullary epithelial cells express MHC class II determinants. In addition, the medulla contains MHC class II positive dendritic cells, randomly dispersed between the epithelial cells with a tendency to concentrate at the corticomedullary junction.

In the experiments using Adriamycin (see the Bone Marrow, above) thymus regeneration (starting at days 15 and 22, respectively) took approximately 1 and 2 weeks, respectively, for a qualitatively and quantitatively normal medullary T-cell population to reappear.

From experiments using vascular thymus transplantation turnover of the total T cell population in the thymus has been estimated to take ± 28 days (Kampinga *et al.*, 1990a). From X-ray irradiation and bone marrow reconstitution experiments turnover time for dendritic cells was found to be in the same order of magnitude. Cortical macrophages, however, turn over rather slowly (Kampinga *et al.*, 1990b).

Studies on the blood–thymus barrier in the rat indicate that, although cortical capillaries are indeed impenetrable for Ig molecules, these molecules may leave the capsular capillaries (having a fenestrated endothelium) to reach the outer cortical zone (subcapsular zone) from where, along with a flux of tissue fluid, these molecules are transported in a centripetal way through the meshes of the cortical parenchyma towards the corticomedullary junction, from where they are drained into the efferent lymph vessels. Thus the cortical parenchyma, despite the presence of a blood–thymus barrier at the level of the cortical capillaries, is accessible for a substantial proportion of serum proteins, up to a certain molecular weight (Nieuwenhuis *et al.*, 1988; Nieuwenhuis, 1990).

For details of the innervation of the rat thymus see Kendall and Al-Shawaf (1991).

Peripheral lymphoid organs

The spleen

The rat spleen is composed of two compartments designated white pulp and red pulp. The white pulp contains all the lymphoid elements and is essentially associated with the branching arterial/arteriolar system. Small arterioles are surrounded by the PALS (peri-arteriolar lymphocyte sheath), mainly containing T cells (and interdigitating dendritic cells, IDCs), onto which numerous B cells containing follicular structures (also containing follicular dendritic cells, FDCs) are attached (van Ewijk and Nieuwenhuis, 1985). These follicles may or may not show GC activity.

A major feature of these follicular structures in the rat is the presence of a well-developed marginal zone containing a peculiar subset of B cells not to be found in other lymphoid tissues (lymph nodes, MALT): MZ B cells are predominantly $\mu^{hi}\delta^{lo}$ cells in contrast to recirculating follicular B cells which are $\mu^{lo}\delta^{hi}$ (Kroese *et al.*, 1995). Presumably, these cells play a role in the T-cell independent humoral immunity against bacterial capsular antigens (polysaccharides). However, very little is known about the actual function of this cell type and its origin. MZ B cells do not recirculate although they may develop from thoracic duct B lymphocytes. Presumably owing to the presence of this population the T/B cell ratio in the spleen is significantly lower than in the blood or in lymph nodes (Table V.2.1).

T cell areas consist of a mixture of $CD4^+$ and $CD8^+$ T cells (average ratio $\pm 2{:}1$). In GCs a particular subset of T cells can be found which are $CD4^+/ER3^+$ and presumably play a role in regulating memory B cell production (Kroese *et al.*, 1985).

The red pulp consists of cords of Billroth interspersed with venous sinuses draining the blood that enters these spaces from the perifollicular marginal sinus and the terminal arteriolar branches (penicillary arterioles). Up to the age of 12–20 weeks areas of extensive erythropoiesis as well as numerous megakaryocytes may be found in the red pulp. In contrast to mice no cells from the myeloid series (like PMNs) are formed in the rat spleen, apart from under pathological conditions. Stem cell content of the rat spleen, also in contrast to mice, is usually negligible.

Lymph nodes

As outlined above, three types of lymph node can be distinguished. However, in cell composition, one type of lymph node stands out against the other two with an on average lower T/B cell ratio (see Table V.2.1, cervical and mesenteric lymph nodes). These lymph nodes are all of the

Table V.2.1 Absolute numbers of total lymphoid cells and percentages of lymphoid subsets in various lymphoid organs of the adult male Lewis rat[a]

Organ	*Absolute number* $\times 10^{-6}$	*%B*	*%T*	*%CD4*	*%CD8*	*T/B ratio*
Bone marrow	9950	15.1	2.1	4.6	1.9	0.14
Thymus	1450	0.14	94.6	94.3	94.4	831.0
Blood	140	19.3	67.7	57.9	22.6	3.6
Spleen	310	42.9	44.1	39.0	19.8	1.0
Peripheral lymph node	130	20.0	73.1	58.4	21.5	3.8
Cervical lymph node	100	28.5	67.4	51.1	19.9	2.5
Mesenteric lymph node	230	30.5	61.6	53.1	17.4	2.0
Peyer's Patches	100	64.4	15.5	14.3	4.5	0.3

[a] After Westermann *et al.* (1989).

mucosa-associated type and, as a result of continuous antigen exposure, these lymph nodes invariably show numerous GC in their cortical area, thereby increasing their content of B cells and thus lowering the T/B cell ratio.

The peripheral lymph nodes from Table V.2.1 (representing mostly skin-associated lymph nodes but the same also holds for the other organ-associated type) have virtually the same T/B cell ratio as that in the blood, suggesting that entry into and transit times for both T and B cells in these nodes are within the same order of magnitude and constitute a stochastic process. However, for thoracic duct lymph, draining both skin-associated and mucosa-associated lymph nodes below the level of the diaphragm, faster transit times for T cells compared with B cells have been found (12–18 h versus 24–36 h, respectively).

In the paracortex (T-cell area) high endothelial venules (HEVs) are a common feature through which both T and B cells have been shown to enter into these nodes (Nieuwenhuis and Ford, 1976).

In the paracortex a special type of antigen-presenting cell (interdigitating or dendritic cells) has been described by Veldman *et al.* (1970). Later, these cells were found to be part of the ubiquitous dendritic cell system (Nieuwenhuis, 1996), containing all dendritic cells, wherever located, with the sole exception of the follicular dendritic cells.

Mucosa-associated lymphoid tissues

Essentially, all the various components such as NALT, Peyer's patches, PCLT and BALT have the same general structure, i.e. rows of B-cell follicular structures invariably showing high GC activity, interspersed by interfollicular areas, populated by T cells, where HEVs may be found. The domes of the follicular structures are situated immediately underneath the epithelial lining containing M cells. The T/B cell ratio is rather low as a result of GC being a dominant feature of these structures.

From the present data there is no indication that any of these lymphoid structures has a central lymphoid function comparable to the bursa of Fabricius in the chicken. In contrast to man the rat has no tonsils, although Waldeyer's ring-equivalent lymphoid tissue resembling the pharyngeal tonsil of man has been described (Koornstra *et al.*, 1991). In contrast to man the BALT is relatively well developed.

Blood

In Table V.2.1 the T/B cell ratio for blood (in Lewis rats) is given as 3.6. Within the T cell compartment $CD4^+$ T cells outnumber $CD8^+$ T cells by a factor of 2.6. Lewis rats are generally considered to be typical T_H1 responders, which is reflected in their susceptibility to the induction of EAE (a T_H1-mediated autoimmune disease) but not to the induction of $HgCl_2$-induced immune complex glomerulonephritis (ICGN) (a T_H2-mediated autoimmune disease). In contrast, BN rats are highly susceptible for the induction of ICGN but not for EAE (Table V.2.2).

When we analysed BN blood for the respective T-cell subsets it appeared not only that they had a quite different ratio of proposed T_H2 over proposed T_H1 cells (28) (see also Section 6) but also a quite different CD4/CD8 ratio (5.7 – almost double the value for Lewis rats; in our stock 2.9). Consequently we set out to determine these ratios for various other rat strains and tried to correlate these data with the (known) susceptibility for either type of autoimmune disease. Data are shown in Table V.2.2 (Groen *et al.*, 1993).

The most striking observations were (1) that the CD4/CD8 ratio can differ over a wide range (from 19 to 1.1) and (2) that the T_H2/T_H1 ratio was also found to vary between 28 and 0.5, apparently quite independently from the CD4/CD8 ratio. There appeared to be no correlation with MHC haplotype and available data do not allow any conclusion as to a correlation between inducibility of respective types of autoimmune disease and MHC haplotype, CD4/CD8 ratio or T_H2/T_H1 ratio. These data thus are difficult to reconcile with the notion of T_H2 versus T_H1 respondership and susceptibility to respective types of autoimmune disease. Clearly, other as yet unknown aspects will have to be taken into account to establish what makes a certain strain susceptible or not to the induction of a particular form of autoimmune disease.

Table V.2.2 Variability in CD4/CD8⁻ and T_H2/T_H1-ratios in blood of various rat strains

Strain	*MHC*	*CD4*	*CD8*	*CD4/CD8*	T_H2/T_H1	*ICGN*	*EAE*
DPBB	u	95	5	19	0.1	?	(diab.)
MAXX	n	88	9	9.7	3	+++	?
BN	n	85	15	5.7	28	+++	– –
LewxBN	l/n	82	17	4.8	0.5?	+++	+++
PVG	c	80	20	4	8	++	–
U	u	79	19	4.2	3.4	?	++
DA	a	76	23	3.3	1.5	?	(RA)
Lew	l	72	25	2.9	0.9	– –	+++
DZB	u	70	28	2.5	3.1	+++	?
AO	u	63	36	1.8	3.1	+	– –
Buff	b	62	34	1.8	2	?	+++
WF	u	58	39	1.5	3.7	?	?
LOU	u	52	45	1.2	4.9	?	?
AO1p	ullu	51	45	1.1	2.8	– –	?

ICGN, Immune Complex Glomerulo-Nephritis; EAE, Experimental Allergic Encephalomyelitis; (diab.), spontaneous autoimmune diabetes; (RA), rheumatoid arthritis.
After H. Groen *et al.* (1993).

3. Rat Cluster of Differentiation (CD) Antigens

For over two decades many monoclonal antibodies to rat antigens have been developed. Originally, these antigens were described independently by each individual investigator. The First International Workshop on Human Leukocyte Differentiation Antigens (HLDA) in 1982 designed a unifying Cluster of Differentiation (CD) nomenclature for humans. This proved very useful and was extended at the 6th HLDA workshop in November 1996 in Kobe, Japan to bring the number of CD antigens to 166 (Kishimoto *et al.*, 1997). The database of the 6th HLDA workshop is accessible at http://mol.genes.nig.ac.jpn./hlda/.

Molecules of significant function are relatively conserved and many CD homologues of rat origin have been reported. The advancement of new molecular biological techniques has rapidly expanded the number of rat CD homologues. Claims of equivalence are supported by many parameters of homology such as cDNA sequence, protein homology, antigen distribution and function.

While many rat molecules exhibit great similarity with their human homologues, some rat antigens differ in protein sequence, antigen distribution and perhaps function. For example, the restricted expression of alternative splicing variants of the CD45 molecule is distributed differently in rat T-cell subsets compared with human T cells. Another example is the expression of CD5 where both in humans and in mice a distinct subset of $CD5^+$ B cells (with possibly unique functions) can be found, while in rats such a subset cannot be discriminated on the basis of CD5 expression (Vermeer *et al.*, 1994).

In both humans and rats many new antigens are being discovered and this has encouraged to find their respective homologues in the other species. For example, biologically important antigens such as CD40 and CD95 (Fas, APO-1) were first discovered in humans and now rat homologues are being cloned on the basis of sequence homology (P. M. Dammers and F. G. M. Kroese, unpublished data). On the other hand the OX-40 antigen (belonging to the TNFR/NGR family) was first characterized in rats and has now been officially assigned as CD134 at the sixth HLDA workshop. Another future candidate for the CD cluster nomenclature is the rat RT6 molecule which is present on subsets of T cells which might be involved in the regulation of tolerance (Greiner *et al.*, 1988). The human homologue gene has been cloned, but was shown to be a pseudogene. Whether other human molecules have taken over the function of RT6 is presently unknown.

Table V.3.1 is an extension of a previously published chart (Lai *et al.*, 1997) and has been updated (to March 1997). We realize that an overview of this still rapidly expanding field will always be incomplete but, hopefully, it will provide a basis for further exploration of biologically important molecules in the rat immune system. This rat database will also be made accessible on the World Wide Web.

4. Leukocyte Traffic and Associated Molecules

Introduction

Lymphocytes are unique among leukocytes with respect to their continuous traffic through the body, using the blood and lymph as routes to connect the lymphoid and non-

Table V.3.1 Rat cluster of differentiation (CD) homologues. This chart was originally published in Lai *et al.* (1997) and is reproduced here in modified and extended form with the permission of the publisher (Blackwell Science Inc.)

Cluster of differentiation (CD)	*Alternative names*	*Reduced molecular mass (kDa); molecular structure*	*Reported cellular distribution*	*Known or proposed function*	*Name of clone*	*References*
CD1d		45 kDa type I transmembrane associated with β_2 microglobulin	Liver, intestine, kidney, heart, thymus, lymph node, spleen	Antigen-presentation; FcR-binding	Gene cloned, polyclonal Ab; crossreactive mAb 1H1, 3C11	Burke *et al.* (1994), Ichimiya *et al.* (1994)
CD2	LFA-2; sheep red blood cell receptor	50 kDa type I transmembrane	T cells, thymocytes, NK cells, macrophages	T-cell activation	OX-34, OX-53, OX-54, OX-55	Jefferies *et al.* (1985), Williams *et al.* (1987), Clark *et al.* (1988)
CD3		Multichain complex associated with TCR; eta (22 kDa) and zeta (16 kDa) gene cloned; epsilon (24 kDa)	T cells	Signal transduction; T-cell receptor complex assembly	G4.18; 1F4	Tanaka *et al.* (1989), Itoh *et al.* (1993), Nicolls *et al.* (1993)
CD4		53 kDa	Thymocytes, T-cell subsets, monocytes, subpopulation macrophages	MHC class II co-receptor	W3/25, OX-35, OX-36, OX-37, OX-38, ER2, RIB 5/2	Williams *et al.* (1977), Webb *et al.* (1979), Jefferies *et al.* (1985), Joling *et al.* (1985), Lehmann *et al.* (1992)
CD5	Ly-1; Lyt-1	67–69 kDa type I transmembrane	Thymocytes, T cells; no distinct CD5 expressing B cell subset found	Regulation of cell activation	OX-19, HIS47, R1–3B3	Dallman *et al.* (1984), Matsuura *et al.* (1984), Vermeer *et al.* (1994)
CD8a		32–34 kDa homodimer (CD8a/CD8a) or heterodimer (CD8a/CD8b)	Most thymocytes, T-cell subset, NK cells; intraepithelial lymphocytes	MHC class I co-receptor	OX-8, G28, G41, G162, N42	Brideau *et al.* (1980), Thomas and Green (1983), Torres-Nagel *et al.* (1992)
CD8b		32–34 kDa heterodimer (CD8a/CD8b)	Most thymocytes, T-cell subset, intraepithelial lymphocytes; not on fresh NK cells	MHC class I co-receptor	341	Torres-Nagel *et al.* (1992)
CD9		26 kDa	Neural system; epithelia, haematopoeitic cells	Intercellular signalling	ROCA2	Tole and Patterson (1993)
CD10	CALLA; neutral endopeptidase (NEP)	94–100 kDa	Epithelial cells	Neutral endopeptidase	Gene cloned, mAb not found; some antihuman mAb crossreact	Malfroy *et al.* (1987), Mahendran *et al.* (1989), Helene *et al.* (1992)
CD11a	LFA-1 α chain	160–170 kDa; noncovalently pairs with CD18 to form LFA-1 integrin	All leukocytes, except for peritoneal macrophages	Adhesion to ICAM-1	TA-3, WT.1	Tamantani *et al.* (1991), Issekutz and Issekutz (1992)
CD11b	Mac-1 α chain; CR3 α chain	140–160 kDa noncovalently pairs with CD18 to form Mac-1 integrin	Myeloid progenitors, monocytes, macrophages, granulocytes	Adhesion to ICAM-1	OX-42, WT.5, ED7, ED8, 1B6C	Robinson *et al.* (1986), Damoiseaux *et al.* (1989), Tamantani *et al.* (1991), Huitinga *et al.* (1993), Mulligan *et al.* (1993)
CD11c	C3bi-receptor	120–130 kDa; noncovalently pairs with CD18 to form p150,95 integrin	Monocytes, granulocytes, macrophages	Adhesion to iC3b	OX-42 recognizes a shared epitope on CD11b and CD11c	Robinson *et al.* (1986), Tamantani *et al.* (1991)
CD13	Aminopeptidase N	140 kDa transmembrane glycoprotein	Kidney, lung	Zinc metalloproteinase	Gene cloned, mAb not found	Watt and Yip (1989)
CD14	LPS receptor	58 kDa transmembrane protein	Macrophages, granulocytes, monocytes, glial cells	LPS receptor	ED9	Damoiseaux *et al.* (1989), Tracy and Fox (1995), Galea *et al.* (1996)
CD15			Brain		MMA	Reifenberger *et al.* (1987)
CD16	FcγRIII	Multiple isoforms, PI-linked transmembrane protein	NK, monocytes, neutrophils, B cell lymphoma	Low-affinity IgG receptor	Genes cloned, mAb not found	Farber and Sears (1991), Farber *et al.* (1993)
CD18	β_2 integrin, β chain of LFA-1 family	90–100 kDa; pairs with CD11a, b, c	Leukocytes: see alpha partner for distribution	Adhesion; pairs with CD11a, b, c	NG2B12, WT.3, 1F12, CL-26	Tamantani *et al.* (1991), Bautista *et al.* (1993), Mulligan *et al.* (1993), Pavlovic *et al.* (1994)
CD23	FcεRII	45 kDa type II transmembrane glycoprotein	Alveolar macrophages, B cells	Low-affinity IgE receptor	Polyclonal Rb5; crossreactive anti-human mAb BB10	Mencia-Huerta *et al.* (1991), Flores-Romo *et al.* (1993)
CD24	Heat-stable antigen	PI-anchored transmembrane glycoprotein	B cells and B-cell progenitors		HIS50	Kroese *et al.* (1991), Tong *et al.* (1993), Hermans *et al.* (1997)
CD25	IL-2R α chain, TAC	55 kDa type I integral membrane	Broad: activated T cells, B cells, NK cells, and macrophages; upregulated upon activation	Receptor for IL-2, T cell activation and proliferation	ART-18, ART-65, ART-35, ART-38, ART-75, ART-94, OX-39, NDS61, NDS63, NDS66	Osawa and Diamantstein (1983), Mouzaki and Diamantstein (1987), Mouzaki *et al.* (1987), Paterson *et al.* (1987), Tellides *et al.* (1989), Page and Dallman (1991), Wood *et al.* (1992)

(*continued*)

Table V.3.1 *Continued*

Cluster of differentiation (CD)	*Alternative names*	*Reduced molecular mass (kDa); molecular structure*	*Reported cellular distribution*	*Known or proposed function*	*Name of clone*	*References*
CD26	Dipeptidyl peptidase IV	220 kDa homodimer	Thymocytes, B and T cells, epithelial cells and certain endothelial cells	Ectopeptidase; cell activation	OX-61, F2/8A18	Ronco *et al.* (1984), McCaughan *et al.* (1990), Gorrell *et al.* (1991)
CD28	Tp44; Receptor for B7 family	45 kDa disulfide-linked homodimer	All α,β T cells, most γ,δ T cells, half of NK cells, immature T cells	Co-stimulation; ligand for B-7	JJ319, JJ316	Clarke and Dallman (1992), Mitnacht *et al.* (1995), Tacke *et al.* (1995)
CD29	Integrin β1 chain	130 kDa; forms heterodimer with integrin α subunit (CD49)	Broad, including lymphocytes, endothelia, smooth muscle cells, epithelia	Adhesion	Ha2/5, Ha2/11	Mendrick and Kelly (1993)
CD31	PECAM-1	130 kDa type I membrane	Endothelia; weak on lymphocytes, neutrophils	Adhesion and migration	Polyclonal antibody	Vaporciyan *et al.* (1993), Wakelin *et al.* (1996)
CD32	FcγRII	Type I transmembrane protein	Granulocytes, mast cells	Low-affinity Fc receptor for aggregated immunoglobulin, immune complexes	Gene cloned, mAb not found	Bocek and Pecht (1993), Isashi *et al.* (1995)
CD35	CR1; C3b receptor	200 kDa type I transmembrane	B cells, monocytes, granulocytes, macrophages	Phagocytosis	Polyclonal crossreactive	Quigg *et al.* (1993)
CD36	Platelet GPIV	85 kDa palmitoylated transmembrane protein	Adipocytes	Broad	mAb not found	Jochen and Hays (1993)
CD37		Transmembrane 4 superfamily			Gene cloned, mAb not found	Tomlinson and Wright (1996)
CD38	T10	34 kDa type II transmembrane protein	Pancreas, brain, duodenum, heart (RNA analysis)	ADP-ribosyl cyclase; hydrolase; cell activation	Gene cloned, mAb not found	Koguma *et al.* (1994)
CD40			B cells	Co-stimulation in B cell activation	Gene cloned, mAb not found	Dammers *et al.* (1998)
CD43	Leukosialin; sialophorin	100 kDa highly sialylated transmembrane membrane protein	Broad and varied: stem cells, hematopoietic precursor cells; thymocytes, T cells, plasma cells; monocytes	Adhesion	W3/13, HIS17, ER1, OX-56, OX-57, OX-58, OX-74, OX-75, 5H4, 8B8, 5G7	Williams *et al.* (1977), Joling *et al.* (1985), Kroese *et al.* (1985), Killeen *et al.* (1987), Cyster *et al.* (1991), Cyster and Williams (1992), Howell *et al.* (1994)
CD44	Pgp-1; H-CAM		Varied and broad; all leukocytes, increased after activation; connective tissue	Adhesion; homing	OX-49, OX-50	Paterson *et al.* (1987), Wirth *et al.* (1993), Westermann *et al.* (1994)
CD44r	CD44v	120–200 kDa; splice variants of standard CD44	Epithelia	Adhesion, metastasis	1.1ASML	Günthert *et al.* (1991), Seiter *et al.* (1993), Wirth *et al.* (1993)
CD45	Leukocyte common antigen	Multiple isoforms: 180, 190, 200, 220 240 kDa	Broad and varied expression on leukocytes; not on mature erythrocytes	Cell differentiation/activation; tyrosine phosphatase	OX-1, OX-28, OX-29, OX-30	Sunderland *et al.* (1979), Thomas *et al.* (1985), Woollett *et al.* (1985)
CD45.1	RT7.1	190 kDa on thymocytes; expressed on rat strains ACI, AO, BN, DA, DZB, LEW, MAXX, PVG, SHR, WAG	Hematopoietic precursor cells; all leukocytes		NDS58, BC84	Newton *et al.* (1986), Mojcik *et al.* (1987), Kampinga *et al.* (1990)
CD45.2	RT7.2	Expressed on rat strains BUF, LOU and WF	Hematopoietic precursor cells; all leukocytes		HIS41, 8G6.1	Mojcik *et al.* (1987), Kampinga *et al.* (1990)
CD45R	Restricted LCA		Subset of CD8$^+$ T cells	Inhibition of antibody production	RTS-1	Nagoya *et al.* (1991)
CD45R	B220; restricted LCA	205 kDa	Pro- and pre-B cells, B cells, weak on marginal zone B cells, small subset of T cells		HIS24	Kroese *et al.* (1987, 1990)
CD45RA/B	Restricted LCA; exon A or A/B dependent	240 kDa	B cells	Signal transduction	OX-33	Woollett *et al.* (1985)
CD45RC	Restricted LCA; exon C dependent	190, 200, 220 kDa	Subset of thymocytes, CD8+ T cells, subset of CD4+ T cells, B cells, NK cells	Signal transduction	OX-22, OX-31, OX-32, HIS25	Spickett *et al.* (1983), Kroese *et al.* (1985), Woollett *et al.* (1985), Law *et al.* (1989), Powrie and Mason (1989), McCall *et al.* (1992), Nagai *et al.* (1993)
CD46	Membrane cofactor protein (MCP)		Neonatal astrocytes		Polyclonal antihuman crossreactive	Yang *et al.* (1993)
CD48	BCM-1; BLAST-1	43–45 kDa member GPI-linked Ig superfamily	Mature hematopoietic cells and their bone marrow precursors, vascular endothelium and some connective tissue	Adhesion; costimulation; counter receptor of CD2 (rats)	OX-45, OX-46	Arvieux *et al.* (1986), Van der Merwe *et al.* (1993, 1994)

(*continued*)

Table V.3.1 *Continued*

Cluster of differentiation (CD)	*Alternative names*	*Reduced molecular mass (kDa); molecular structure*	*Reported cellular distribution*	*Known or proposed function*	*Name of clone*	*References*
CD49a	Integrin α1 chain; α1 chain of VLA-1	180–200 kDa transmembrane glycoprotein; non-covalently pairs with CD29 to form VLA-1	Broad	Laminin/collagen receptor	3A3, 1B1, Ha31/8	Turner *et al.* (1989), Mendrick *et al.* (1995)
CD49b	Integrin α2 chain; α2 chain of VLA-2	160 kDa transmembrane glycoprotein; non-covalently pairs with CD29 to form VLA-2	Activated lymphocytes, epithelial cells, platelets	Laminin/collagen receptor	Ha1/29	Mendrick and Kelly (1993), Mendrick *et al.* (1995)
CD49d	Integrin α4 chain; α4 chain of VLA-4	150 kDa transmembrane glycoprotein; associates non-covalently with integrin β1 subunit (CD29) to form VLA-4 (or LPAM-2) or with integrin β7 subunit to form LPAM-1	Broad, B and T cells, monocytes, thymocytes, mast cells	Adhesion; VLA-4 is ligand for VCAM-1 (CD106) and LPAM-1 is ligand for MacCam-1	TA-2, MRα4–1; crossreactive anti human mAb HP2/1	Issekutz and Wykretowicz (1991), Molina *et al.* (1994), Yasuda *et al.* (1995)
CD49e	Integrin α5 chain; α5 chain of VLA-5	Non-covalently pairs with integrin β1 subunit (CD29) to form VLA-5	Peripheral B cells, peritoneal mast cells, endothelium	Fibronectin receptor	HMα5–1	Yasuda *et al.* (1995)
CD51	α chain of vitronectin receptor; osteoclast functional antigen (OFA)	Forms complex with β3 (CD61) chain to form vitronectin receptor	Osteoclasts		Crossreactive antihuman mAb 23C6	Horton *et al.* (1991)
CD53	OX-44 antigen	43 kDa, transmembrane 4 superfamily	All myeloid cells and peripheral lymphoid cells and their precursors; only a small subset of thymocytes	Signal transduction	OX-44, 2D1, 6E2, 7D2	Paterson *et al.* (1987), Bellacosa *et al.* (1991), Bell *et al.* (1992), Tomlinson *et al.* (1993)
CD54	ICAM-1	85 kDa type I glycoprotein; Ig superfamily	Broad; endothelial cells, including high endothelial venules; lymphocytes, epithelial cells, subpopulation of macrophages	Cell adhesion mediated by LFA-1 (CD11/CD18)	1A29, 6A22, 10A25, 10A56	Tamatani and Miyasaka (1990), Colic and Drabek (1991), Kanagawa *et al.* (1991), Tamantani *et al.* (1991), Kita *et al.* (1992), Christensen *et al.* (1993)
CD55	Decay-accelerating factor (DAF)		Astrocytes		Polyclonal antibody	Yang *et al.* (1993)
CD59	Rat inhibitory protein (RIP)	21 kDa PI-linked glycoprotein	Broad; erythrocytes, widely expressed in kidney	Inhibition of membrane attack complex (MAC) of complement; interaction with C-8 and C-9	6D1, TH9	Hughes *et al.* (1992, 1993), Rushmere *et al.* (1994)
CD61	β3 chain of αIIb (CD41) and αv (CD51)	100 kDa; non-covalently pairs with CD51 to form the vitronectin receptor and CD41	Platelets, megakaryocytes and osteoclasts, mast cells	Extracellular adhesion	F4, F11, HMβ3-1	Helfrich *et al.* (1992), Yasuda *et al.* (1995)
CD62E	E-selectin; ELAM	60 kDa	Endothelium	Endothelial-leukocyte interaction	ARE-5; cross-reactive antihuman mAbs; CL-3 and CL-37	Mulligan *et al.* (1991), Fries *et al.* (1993), Billups *et al.* (1995), Misugi *et al.* (1995)
CD62L	L-selectin; LECAM-1; A.11	65 kDa on lymphocytes; 62 kDa on neutrophils	Most peripheral lymphocytes and neutrophils; small subpopulation of thymocytes	Endothelial-leukocyte interaction	HRL1, HRL2, HRL3, HRL4, OX-85	Saoudi *et al.* (1986), Watanabe *et al.* (1992), Tamatani *et al.* (1993a, b), Westermann *et al.* (1994), Sackstein *et al.* (1995)
CD62P	P-selectin; GMP-140; PADGEM	140 kDa type I transmembrane glycoprotein	Platelets, endothelial cells	Endothelial-leukocyte interaction; interaction of platelets and neutrophils	Crossreactive antihuman mAbs PB1.3, (CY1747), PNB1.6 and LYP20	Mulligam *et al.* (1992), Winocour *et al.* (1992), Auchampach *et al.* (1994), Chignier *et al.* (1994), Zimmerman *et al.* (1994), Papayianni *et al.* (1995)
CD63	AD1 antigen; ME491	50–60 kDa; transmembrane 4 superfamily glycoprotein	Mast cells, bone marrow, platelets	Secretory process/signal transduction	AD1	Kitani *et al.* (1991), Nishikata *et al.* (1992)
CD66a	BGP.1; ecto-ATPase	Multiple isoforms 105 and 110 kDa	Liver cells, epithelia in gastrointestinal tract, liver, some secretory glands, vagina, kidney and lung, granulocytes	L-form (105 kDa) involved in adhesion	362.50	Hixson *et al.* (1985), Odin and Obrink (1987), Lin *et al.* (1991), Culic *et al.* (1992), Cheung *et al.* (1993)
CD66e	CEA	350 kDa	Carcinoma cell-line RCA-1	Adhesion	5B1	Kim and Abeyounis (1988, 1990)
CD68		90–110 kDa, intracellular protein	Monocytes, macrophages		ED1	Dijkstra *et al.* (1985), Damoiseaux *et al.* (1994)

(continued)

Table V.3.1 *Continued*

Cluster of differentiation (CD)	*Alternative names*	*Reduced molecular mass (kDa); molecular structure*	*Reported cellular distribution*	*Known or proposed function*	*Name of clone*	*References*
CD71	Transferrin receptor	95 kDa homodimeric type II membrane protein	Proliferating cells, activated T and B cells, endothelium, neuronal cells, epithelial cells	Activation, iron metabolism	OX-26	Jefferies *et al.* (1984, 1985), Giometto *et al.* (1990)
CD73	Ecto-5′-nucleotidase	67–68 kDa	Broad, including kidney; B and T cell subsets	Ecto-5′-nucleotidase activity	5NE5, 5NH3, 5N4–2, B5	Siddle *et al.* (1981), Bailyes *et al.* (1984), Carvalho *et al.* (1987), Dawson *et al.* (1989), Gandhi *et al.* (1990)
CD74	Ii, MHC class II invariant (γ) chain	Gene cloned	MHC class II positive cells	Intracellular sorting of class II molecules	Gene cloned, mAb not found	McKnight *et al.* (1989), Neiss and Reske (1994)
CD75	Alpha 2,6-sialyltransferase (SiaT-1)	Gene cloned	Liver, kidney	Alpha 2,6-sialyltransferase activity	Gene cloned, mAb not found	O'Hanlon *et al.* (1989)
CD77	Globotriaosyl-ceramide		Intestinal epithelium and capillary endothelium		BGR23	Kotani *et al.* (1994)
CD79a	Ig alpha; mb-1	38 kDa type I glycoprotein, subunit of B-cell antigen receptor complex	B cells	Signal transduction	Crossreactive mAb HM57	Jones *et al.* (1993)
CD79b	Ig beta; B29	36 kDa type I glycoprotein, subunit of B-cell antigen receptor complex	B cells	Signal transduction	Crossreactive mAb B29/123	Jones *et al.* (1993)
CD80	B7/BB1; B7-1	80–90 kDa type I transmembrane protein; Ig superfamily	Activated B cells, T cells, dendritic cells, macrophages	Counter-receptor of CD28/ CD152 (CTLA-4); cell activation; co-stimulation	3H5	Judge *et al.* (1995), Maeda *et al.* (1997)
CD81	Target of antiproliferative antibody (TAPA-1)	Tetramembrane spanning protein	Astrocytes	Regulation of mitotic activity	AMP1	Geisert *et al.* (1996)
CD86	B7-2	90–100 kDa type I transmembrane protein; Ig superfamily	T cells, macrophages	Counter-receptor of CD28/ CD152 (CTLA-4); cell activation; co-stimulation	24F	Maeda *et al.* (1997)
CD90	Thy-1	18 kDa PI-linked Ig superfamily	Broad and varied; subset T cells; precursor and immature B cells; stem cells; brain	T cell activation, signal transduction; apoptosis	OX-7, HIS51, ER4	Mason and Williams (1980), Campbell *et al.* (1981), Bukovsky *et al.* (1983), Joling *et al.* (1985), Hermans and Opstelten (1991), Kroese *et al.* (1995)
CD91	α2 macroglobulin receptor (A2MR); lipoprotein receptor-related protein (LRP)	600 kDa membrane protein	Monocytes/macrophages, liver, adipocytes	Multifunctional: lipoprotein metabolism, hemostasis, proteinase inhibition	Phage antibody scFv7; polyclonal Ab	Kowal *et al.* (1989), Meilinger *et al.* (1995), Hodits *et al.* (1995)
CD95	Fas; APO-1	Gene cloned	Heart myocytes; variant form in liver; B cells	Apoptosis	Gene cloned, mAb not found; polyclonal available	Kimura *et al.* (1994), Tanaka *et al.* (1994), Dammers *et al.* (1998)
CD95L	Fas ligand	Type II membrane protein	Activated lymphocytes, CTLs	Apoptosis	Gene cloned, mAb not found; polyclonal available	Mita *et al.* (1994)
CD103	HML-1	Heterodimer of 100 and 120 kDa	Intraepithelial lymphocytes		RGL-1	Cerf-Bensussan *et al.* (1986)
CD106	VCAM-1	Gene cloned	High endothelium, inflamed blood vessels	Adhesion molecule, ligand for VLA-4 (CD49d/CD29)	5F10	Williams *et al.* (1992), May *et al.* (1993)
CD107a	LAMP-1, lgp120	107 kDa lysosomal membrane sialoglycoprotein	Liver		Gene cloned, mAb not found; polyclonal Ab	Howe *et al.* (1988), Himeno *et al.* (1989), Akasaki *et al.* (1993)
CD107b	LAMP-2	96 kDa lysosomal membrane sialoglycoprotein	Liver		Gene cloned, mAb not found; polyclonal Ab	Noguchi *et al.* (1989), Akasaki *et al.* (1993)
CD115	M-CSFR; CSF-1R; c-fms	Gene cloned	Macrophages, myoblasts	Signal transduction; cell differentiation	Gene cloned, mAb not found; polyclonal Ab	Raivich *et al.* (1991), Borycki *et al.* (1992)
CD117	*c*-kit; stem cell factor receptor	Gene cloned	Broad (brain, liver, testis, pancreas)	Tyrosine kinase	Gene cloned, mAb not found; polyclonal available	Tsujimura *et al.* (1991), Fujio *et al.* (1996), Oberg-Welsh and Welsh (1996)
CD120a	TNF R, type I	48 kDa		Binds TNF-α and TNF-β	Gene cloned, mAb not found	Himmler *et al.* (1990)
CD121a	Il-IR, type I	80 kDa	Broad	Binds IL-1	Gene cloned, crossreactive antihuman mAb	Hart *et al.* (1993), Sutherland *et al.* (1994), Mugridge *et al.* (1995), Scherzer *et al.* (1996)

(continued)

Table V.3.1 *Continued*

Cluster of differentiation (CD)	*Alternative names*	*Reduced molecular mass (kDa); molecular structure*	*Reported cellular distribution*	*Known or proposed function*	*Name of clone*	*References*
CD121b	IL-1R, type II; p68	60 kDa Ig superfamily glycoprotein	Broad	Binds IL-1; ? not involved in signaling	Gene cloned, crossreactive antihuman mAb; ALVA 42	Luheshi *et al.* (1993), Bristulf *et al.* (1994), Mugridge *et al.* (1995), Scherzer *et al.* (1996)
CD122	IL-2R β chain; p75	85–100 kDa hematopoietin receptor family	NK cells and T cells; upregulated upon activation	Signalling; high IL-2 binding; pairs with CD25 and γc chains to form high affinity receptor	L316	Page and Dallman (1991), Park *et al.* (1996)
CD124	IL-4R	Cytokine receptor superfamily	T cells	IL-4 receptor	Gene cloned, mAb not found	Richter *et al.* (1995)
CDw130	IL-6R	Gene cloned	Broad	IL-6 receptor	Gene cloned, mAb not found; crossreactive antimouse mAb	Baumann *et al.* (1990), Schobitz *et al.* (1993), Greenfield *et al.* (1995)
CDw131	Common β chain	Pairs with α chain of IL-3R, IL-5R and GM-CSFR	Microglia	Signal transducing subunit of receptor for IL-3, IL-5 and GM-CSF	Gene cloned, mAb not found	Appel *et al.* (1995)
CD134	OX-40	50 kDa, transmembrane protein, member of TNF receptor/nerve growth receptor superfamily	Activated $CD4^+$ T cell blasts		OX-40	Paterson *et al.* (1987), Mallett *et al.* (1990)
CD138	Syndecam-1	Transmembrane proteoglycan	Broad		Gene cloned, mAb not found; polyclonal Ab	Cizmeci-Smith *et al.* (1992), Carey *et al.* (1994), Kovalsky *et al.* (1994)
CD140a	Platelet-derived growth factor receptor (PDGFR) alpha	Gene cloned, member of the receptor tyrosine kinase family	Endothelial cells; brain		Gene cloned, mAb not found; polyclonal Ab	Lee *et al.* (1990), Lindner and Reidy (1995)
CD140b	Platelet-derived growth factor receptor (PDGFR) beta		Endothelial cells, lipocytes, lung		Gene cloned, mAb not found; crossreactive antihuman mAb; polyclonal Ab	Sarzani *et al.* (1992), Wong *et al.* (1994), Liu *et al.* (1995)
CD141	Thrombomodulin		Endothelial cells		mAb not found; polyclonal Ab	Sabolic *et al.* (1992), Arai *et al.* (1995)
CD143	Angiotensin-converting enzyme (ACE)	160 kDa	Endothelial cells		α-ACE3.1.1, A10-E3, F9-F9, A24; crossreactive antihuman mAb 9B9	Auerbach *et al.* (1982), Moore *et al.* (1984), Strittmatter and Snyder (1984), Danilov *et al.* (1989)
CD147	Neurothelin, basignin, OX-47, CE-9	40–68 kDa type I membrane glycoprotein	Broad, including spleen and thymus; levels increase on lymphocytes on activation		OX-47	Paterson *et al.* (1987), Fossum *et al.* (1991), Nehme *et al.* (1993, 1995)
CD148	HPTP-eta	180–220 kDa transmembrane protein	Endothelial cells, megakaryocytes, smooth muscle cells, platelets	Adhesion; protein tyrosine phosphatase activity	Gene cloned, mAb not found	Borges *et al.* (1996)
CD152	CTLA-4	Gene cloned	T cells	Co-stimulation	Gene cloned, mAb not found	Oaks *et al.* (1996)
CD157	BST-1	Gene cloned	Islets of Langerhans		Gene cloned, mAb not found	Furuya *et al.* (1995)

lymphoid organs of the body. Granulocytes, in contrast, enter a tissue, die, and never return to the blood. Very little is known about the kinetics of monocytes. This is probably due to the problem of labeling monocytes for trafficking experiments without activating them, which would affect their migratory routes. Data on normal rather than the pathological conditions is discussed in this section.

The recruitment of lymphocytes to one specific organ or its compartments depends on the interaction of adhesion molecules on leukocytes and endothelial cells (entry into the tissue), the migration through the parenchyma of an organ (transit), and mechanisms to exit the organ (exit), e.g. via efferent lymphatics in lymph nodes or via the venous blood in the spleen. In the transit phase lymphocytes can be activated and they start to proliferate or they die, e.g. by apoptosis. Thus, recruitment is the net effect of all these aspects. Quantitatively, many more lymphocytes migrate daily to organs without high endothelial venules (HEVs), such as the spleen, lung and liver, than to those with HEVs, e.g. lymph nodes and Peyer's patches (Pabst and Westermann, 1997). Most studies have focused on the entry phase in organs with high endothelial venules (HEV), that is the interaction between endothelial cells and lymphocytes mediated by different adhesion molecules. There is a certain preferential migration of lymphocytes to peripheral lymph nodes rather than to Peyer's patches but not an exclusive migration to one organ only. The preferential accumulation of lymphoblasts in the rat, e.g. gut-derived blasts migrating to the lamina propria of the small intestine and Peyer's patches as well as other mucosal organs such as the lung, is summarized by Smith and Ford (1983).

Species-specific aspects in leukocyte traffic

The rat was the experimental animal used in most of the early experiments documenting the recirculation of lymphocytes which were performed by Gowans and others in the early 1960s (for review see Ford, 1980). The rat has several advantages as an experimental model for lymphocyte migration studies compared with other small laboratory animals such as mouse, hamster and guinea pig, or larger animals such as rabbit, pig, sheep and monkey.

1. As in mice, there are a series of inbred strains that enable leukocyte transfers, nude rats lacking T lymphocytes, and congenic strains in which lymphocyte transfers without any *in vitro* labeling are possible when an antibody against the other strain is used, e.g. leukocyte common antigen (RT7.1 and RT7.2).
2. Physiologically migrating lymphocytes can be obtained much more easily in the rat than in the mouse, e.g. by taking the blood or lymph from the thoracic duct. Mesenteric lymph nodes can be removed in the rat and later lymphocytes derived directly from the gut wall can be obtained and used for lymphocyte traffic experiments. In the mouse, mainly cell suspensions from the spleen or lymph nodes are used but these contain many nonmigratory lymphocytes which might also be damaged by the separation procedure. Repeated blood samples can be obtained from individual rats and long term intravenous (i.v.) infusions and thoracic duct cannulations for kinetic experiments can be carried out for up to 7 days (Westermann *et al.*, 1994a).
3. Organs, e.g. the lung, can be transplanted and used as models for transplantation immunology and also for study of the basic mechanisms of lymphocyte traffic (Westermann *et al.*, 1996a) or the effects of blocking antibodies against adhesion molecules in preventing reperfusion injuries.
4. Intravital videomicroscopy has been used to study the interaction of lymphocytes with the vessel wall in rat Peyer's patches (Miura *et al.*, 1995), which would not be possible in larger animals owing to the size of the organs.
5. Some basic molecular mechanisms of the role of adhesion molecules on leukocytes and endothelial cells can best be studied in *in vitro* tests. The adhesion of lymphocytes to HEV on frozen sections was first established on rat tissues (Stamper and Woodruff, 1976, 1977) and later improved in this species (Willführ *et al.*, 1990a). Endothelial cells from HEV of rat lymph nodes have successfully been cultured (reviewed in Ager, 1994).
6. Tamatani and Miyasaka (1990) and Tamatani *et al.* (1991, 1993) described monoclonal antibodies (mAbs) against rat ICAM-1, CD11/CD18 and L-selectin and Issekutz and Wykretowicz (1991) described mAbs against VLA-4. Details of the expression of adhesion molecules on lymphocyte subsets and different vessels in the rat are compiled in Tables V.4.1 and V.4.2. All these reagents enable detailed studies on lymphocyte traffic in rats. In general, the functions of the adhesion molecules are similar to those described for mice and humans.
7. In rats a large number of models for human disease have been established. They have been used to interfere with lymphocyte traffic by applying antibodies against adhesion molecules, e.g. in models for inflamed skin (Issekutz *et al.*, 1991) or rheumatoid arthritis (Issekutz *et al.*, 1996).

Table V.4.1. Constitutive expression of adhesion molecules involved in the traffic of blood leukocyte subsets[a]

		Adhesion molecules[b]					
Subsets		*L-selectin (CD62L)*	*α4-integrins (CD49d)*	*LFA-1 (CD11a/18)*	*ICAM-1 (CD54)*	*LFA-2 (CD2)*	*Pgp-1 (CD44)*
B lymphocytes	(15%)[c]	85%	40%	65%	95%	5%	45%
T lymphocytes	(65%)	95%	30%	75%	15%	100%	100%
CD8	(15%)	75%	15%	60%	30%	95%	100%
CD4	(50%)	90%	20%	60%	5%	100%	100%
NK cells[d]	(5%)	70%	10%	100%	80%	75%	100%
Monocytes[e]	(5%)	60%	95%	100%	95%	25%	100%
Granulocytes[f]	(10%)	15%	15%	95%	25%	35%	100%

[a] The references are summarized in Westermann *et al.* (1994b, 1996b) and Klonz *et al.* (1996).
[b] Percentage of adhesion molecule positive cells in each subset.
[c] The data in brackets indicate the frequency of the subset among all blood leukocytes.
[d] Identified with the antibody 3.2.3.
[e] Identified by low CD4 expression and high forward-scatter characteristics.
[f] Identified by forward and side-scatter characteristics.

Table V.4.2. Expression of adhesion molecules on various types of endothelial cells

	Adhesion molecules[a]					
Endothelial cells	*ICAM-1 (CD54)*	*VCAM-1 (CD106)*	*E-selectin (CD62E)*	*P-selectin (CD62P)*	*PECAM-1 (CD31)*	*GlyCAM-1*[b]
Arterial	(+)	–	–	nd	+	nd
Capillary	+	nd	↑	↑	nd	nd
Postcapillary						
flat	+	↑	nd	↑	nd	nd
high lymph node	+	+	nd	nd	nd	+
Peyer's patches	+	nd	nd	nd	nd	–
Venous	+	–	↑	nd	+	nd

+, constitutively expressed; ↑, expressed on stimulation; (+), weakly expressed; –, not expressed; nd, not determined.
[a]No antibodies for rat MadCAM-1 are available. The references for the adhesion molecules described are summarized in Mulligan *et al.* (1991), May *et al.* (1993), Seekamp *et al.* (1993), Tamatani *et al.* (1993), Vaporciyan *et al.* (1993), Yamazaki *et al.* (1993), Matsuo *et al.* (1994), van Oosten *et al.* (1995), Tipping *et al.* (1996).
[b]Identified by the ability to bind L-selectin (CD62L).

The combination of these advantages makes the rat a valuable model for many basic and applied experiments on lymphocyte migration. The large litter size, easy breeding, low costs and genetic characterizations are all advantages of using the rat as an experimental animal rather than larger animals such as the rabbit, pig or sheep.

In addition to adhesion molecules, an additional class of molecules, the chemokines (formerly called chemoattractant cytokines) and their receptors, are of increasing interest because of the effects on lymphocyte migration but also on T-cell activation and hematopoiesis (Prieschl *et al.*, 1995; Mackay, 1996). In contrast to the mouse and humans, little is known about these molecules in the rat. Examples of these chemokines are the macrophage inflammatory protein-2 (MIP-2) (Wu *et al.*, 1995; Schmal *et al.*, 1996), the monocyte chemoattractant protein-1 (MCP-1) (Berman *et al.*, 1996) and the cytokine-induced neutrophil chemoattractant (CINC) (Wu *et al.*, 1995). The role of chemokines in the rat has been studied in respect to the recruitment of neutrophils. There are no data about effects of chemokines on leukocyte migration kinetics or potential differential effects on lymphocyte subsets. It is probable that many more functions of chemokines in rat leukocyte traffic will be published.

Based on the advantages of the rat for studies on lymphocyte traffic some concepts of lymphocyte migration, which had been generalized as valid for all species, have been questioned or modified by data obtained in the rat. Some examples of lymphocyte traffic experiments only performed in the rat so far are also given in the following paragraphs. Several years ago it was proposed that B lymphocytes migrate prefentially to Peyer's patches and T lymphocytes to peripheral lymph nodes. When physiologically migrating lymphocytes were used no such preference could be found *in vitro* or *in vivo* (Westermann *et al.*, 1992; Walter *et al.*, 1995). Mackay (reviewed in 1993) proposed the concept that memory T-lymphocytes migrate to tissues such as the skin and arrive at the regional lymph nodes via afferent lymphatics and 'virgin' lymphocytes enter lymph nodes via HEVs when the expression of the CD45R isoforms is taken as a marker for memory and naive lymphocytes. In the rat, however, neither *in vitro* nor *in vivo* did traffic experiments support this concept (reviewed in Westermann and Pabst, 1996). The effects of different cytokines on the expression of adhesion molecules of endothelial cells have been studied in detail *in vitro*, but much less is known about the effects *in vivo*. In the rat, the effect of interferon γ on lymphocyte kinetics has been studied. While the number of lymphocytes in the blood were unaffected by continuous IFN-γ infusions, the recirculation into the thoracic duct was greatly reduced with different effects on lymphocyte subsets (Westermann *et al.*, 1993, 1994a). Recently, Anderson and Shaw (1993) summarized data from rats to formulate a new concept: cytokines, which reach a lymph node by the afferent lymphatics, are transported by a fibroblastic reticular cell conduit system to the HEV, the entry site for lymphocytes into a lymph node. Not only the interaction of lymphocytes with endothelial cells regulates lymphocyte traffic but the transit through the organs might be of even greater importance. The route of B and T lymphocytes through the different compartments of the spleen and lymph node have been studied in great detail only in the rat. Different labeling techniques have been combined with morphometry and immunohistology to document the route of lymphocyte subsets and the total numbers per compartment (Willführ *et al.*, 1990b; Blaschke *et al.*, 1995). Combining the use of purified T-lymphocyte subsets from the thoracic duct with surface markers such as CD45RC and blocking antibodies against the α_4 integrin resulted in surprising results, e.g. mature recirculating CD45RC$^-$CD4$^+$ T lymphocytes enter the thymus by α_4 integrin–VCAM-1 interaction (Bell *et al.*, 1995). Thus, the rat is an important species in challenging those hypotheses of lymphocyte migration which generalize a phenomenon described only in the mouse as valid for all other species.

5. Cytokines and their Receptors

Molecular cloning techniques have revolutionized the large scale production of a wide variety of prokaryotic and eukaryotic proteins including cytokines. These latter (glyco)proteins have received much attention in recent years mainly because of their potential clinical value. The term 'cytokine' denotes to a diverse group of glycoproteins which affect various cell functions and have a myriad of biological activities. Originally, cytokines were named according to cell source or earliest identified biological activity. However, designations as lymphokines, monokines, interferons and growth factors have lost much of their meaning because subsequent studies have shown that these glycoproteins could be produced by numerous cell types and exert overlapping and often similar regulatory actions. At present, more than 50 genes encoding for different cytokine/receptor combinations have been identified, cloned and thoroughly characterized; new cytokines are still discovered with such frequency that it is difficult to maintain an overall picture. A major problem in cytokine research is the species specificity, which means that human cytokines show little if any biological activity in animal systems. Even between closely related animal species, such as mice and rats, interchangeable use is sometimes not possible.

In this section, we have summarized available data from literature describing the isolation, characterization and expression of chromosomal and cDNA genes encoding rat cytokines and their specific cell-surface receptors. The data are presented in tabular form (Tables V.5.1 and V.5.2) and provide the reader with a brief overview of the present state of affairs of rat cytokines and their receptors at both the nucleotide and protein level. We have chosen to review the most important chemical properties and biological activities of the different recombinant proteins. Detailed information on gene structure, amino acid sequences and other biophysical and biological aspects can be found in the references provided.

It is possible that we may have missed a number of papers with crucial data on this subject and apologize for any unintentional omission of important contributions. In general, information on rat cytokines is fragmented and incomplete. This is quite different from mouse and human cytokines, which have always been the cytokines of choice for many research groups all over the world. Numerous human and mouse cytokine genes have been thoroughly characterized and most of them have been expressed in high yield in either prokaryotic or eukaryotic cells. That rat cytokines have always been a 'changeling' in the field of cytokine research is not surprising. High cross-species reactivity of mouse cytokines in the rat system has always been an argument against the need of cloning and expressing rat cytokines. However, it is realized more and more that for *in vivo* studies, homologous cytokines are preferable. This is because of the possible antigenicity of heterologous cytokines upon long-term *in vivo* administration. Although a relatively low number of rat cytokine genes have been isolated and characterized to date, there is a clear growing interest in rat cytokines, which can be expected to increase in the coming years. At present, a relatively low number of cytokines are expressed in appropriate host/vector systems for large-scale production. Consequently, the biological activities in the tables have not all been established with the recombinant proteins themselves but adopted from bioactivities exerted by mouse and/or human cytokine analogues. We do not think that this adaptation depreciate the information provided because of the high degree of functional conservation between rat, mouse and human cytokines. There are also a number of gaps in the table which may indicate that this information is not available, or resides in laboratory journals somewhere, but has not appeared in scientific papers, or the experiments have simply not been performed. Originally, we also planned to include immunoassays and bioassays in this survey. However, it was discovered that, with some notable exceptions, bioassays specific for mouse cytokines can often be used to measure the biological activity of rat cytokines and these assays are described elsewhere in this book. The list of immunoassays (particularly ELISAs) specific for rat cytokines is growing rapidly and can be found in the catalogues of numerous biotechnological firms. For that reason, we have omitted this information from the tables.

6. T-Cell Development

Intra-thymic T-cell development

In terms of TCR, CD2, CD4, and CD8 expression, thymocyte differentiation in the rat largely resembles that in the mouse (Aspinall *et al.*, 1991). In addition, some markers more or less unique for the rat have been used in thymocyte differentiation studies, MRC OX-44 (Paterson and Williams, 1987; Paterson *et al.*, 1987a,c), MRC OX-22 (anti-CD45RC) (Law *et al.*, 1989), ER3 (Joling *et al.*, 1985), HIS44 (Aspinall *et al.*, 1991) and HIS45 (Kampinga *et al.*, 1990). OX-44 was shown to label virtually all $CD4^{-}8^{-}$ double negative (DN) cortical and mature CD4 or CD8 single positive (SP) medullary thymocytes. Thymopoietic potency, as demonstrated by intrathymic injection and proliferation upon allogenic or mitogenic stimulation are accounted for by $OX\text{-}44^{+}$ DN thymocytes (Paterson *et al.*, 1987a). Regeneration of the thymic cortex after irradiation, however, occurs mainly by proliferation of $CD4^{-}8^{+}/OX\text{-}44^{-}$ thymocytes (Paterson and Williams, 1987). CD45RC is expressed by ≈60% of DN thymocytes, and these thymocytes have thymus regenerative capacity, whereas $CD45RC^{-}$ DN thymocytes have not (Law *et al.*, 1989). Proliferation, as measured by BrdU incorporation is identical among $CD45RC^{+}$ and $CD45RC^{-}$ DN thymocytes (Law *et al.*, 1989). The ER3 determinant was shown to label virtually all cortical thymocytes and CD8 SP medullary thymocytes (Joling *et al.*, 1985). The mAbs HIS44 and HIS45 recognize cortical and medullary thymocytes respectively (Kampinga and Aspinall, 1990b; Aspinall *et al.*, 1991).

A discrepancy with mouse data is the fact that rat $TCR^-/CD4^-8^-$ thymocytes do not appear to express IL-2Rα (CD25) or other IL-2-binding proteins (Paterson and Williams, 1987; Takacs *et al.*, 1988). This has raised some debate on the issue of whether IL-2 is an essential growth factor for thymocytes in the DN preselectional stage of differentiation (Aspinall *et al.*, 1991; Kroemer and Martinez, 1991; Kampinga, 1991). Another feature in which rat intrathymic T-cell development appears to differ from that in mouse and humans is that most immature $TCR^{low}/CD4^+8^+$ thymocytes in the rat express little or no CD28, whereas CD28 expression on $TCR^{intermediate}$ and TCR^{high} thymocytes is high (Tacke *et al.*, 1995). CD28 and other cell interaction molecules (CD2, CD5, CD53 and, to a lesser extent, CD11a and CD44) on selectable rat thymocytes were found to be rapidly upregulated upon TCR ligation *in vitro*. Their kinetics and cell distribution were elegantly demonstrated by Mitnacht *et al.* (1995).

An outline of fetal rat thymocyte development is given in Figure V.6.1.

Post-thymic T-cell development

Much of the study on phenotypic peripheral T-cell development in the rat involves the markers CD4, CD8, Thy-1, RT6, and CD45RC. As in mice and humans, except for a small (3–5%) subset, most rat peripheral T cells show mutually exclusive expression of CD4 and CD8, identifying two functionally different T-cell subsets (Brideau *et al.*, 1980). The presence or absence of the remaining three markers has been found to represent differences in maturational stage, but also in function of peripheral T cells in the rat.

In adult rats, Thy-1 is expressed by a subset of bone marrow cells, all thymocytes and a small population of peripheral leukocytes, including a small subset of T lymphocytes (Ritter *et al.*, 1978; Crawford and Goldschneider, 1980). In neonatal rats Thy-1 is expressed by the vast majority of peripheral T cells (Thiele *et al.*, 1987), and, after adult thymectomy the percentage of $Thy\text{-}1^+$ T cells in the periphery rapidly declines (Hosseinzadeh and Goldschneider, 1993; Groen *et al.*, 1995). From these and other findings it was concluded that Thy-1 expression is lost during the early stages of post-thymic T-cell development. In terms of T-cell function, it was found that $Thy\text{-}1^+$ T cells in the rat exhibit delayed allograft rejection (Yang and Bell, 1992).

The alloantigen RT6 is expressed on 50–85% (depending on age and strain) of peripheral $CD4^+$ and $CD8^+$ T cells in the rat (Greiner *et al.*, 1982; Mojcik, 1988, 1991), on a subset of intraepithelial lymphocytes (Fangmann *et al.*, 1990), a subset of NK cells (Wonigeit, 1996), but not on thymocytes and bone marrow cells (Mojcik, 1988). $RT6^+$ T cells have been shown to play an important regulatory role in the prevention of autoimmunity (see Section 17). These cells suppress mixed lymphocyte reactivity, and cytotoxic reactions have been documented to be mediated by their ($CD8^+$) $RT6^-$ counterparts (Greiner *et al.*, 1988). Acquisition of RT6 expression on T cells occurs almost simultaneously with the loss of Thy-1 (Thiele *et al.*, 1987; Groen *et al.*, 1996b).

In the rat, CD45RC is expressed by all B cells, two-third of $CD4^+$ and ≈90% of $CD8^+$ T cells (Spickett *et al.*, 1983). $CD4^+/CD45RC^+$ T cells provide primary B-cell help *in vivo*, proliferate vigorously in MLR, produce high amounts of IL-2 and IFN-γ and are associated with cellular autoimmunity (Powrie and Mason, 1990b; reviewed by Fowell *et al.*, 1991) and the prevention of humoral autoimmunity (Mathieson *et al.*, 1993). The expression and loss of CD45RC on T cells has been shown to be cyclic and to reflect changes in the state of activation of T cells (Bell and Sparshott, 1990; Sparshott and Bell, 1994).

With respect to phenotypic development of rat T cells, two studies examining the combined expression of Thy-1, RT6 and CD45RC are of particular interest. Hosseinzadeh and Goldschneider (1993) used the intra-thymic FITC injection technique, and Kampinga *et al.* (1992, 1997) used the technique of vascular thymus transplantation in RT7 congeneic PVG rats to study phenotypic changes in developing rat peripheral T cells. These studies demonstrated that recent thymic migrants/emigrants (RTM or RTE, respectively) in the rat can unequivocally be identified by the expression of Thy-1 and the absence of both RT6 and CD45RC. In diabetes-prone BB (DPBB) rats, both $RT6^+$ and $CD45RC^+$ T cells are severely under-represented (Greiner *et al.*, 1986; Groen *et al.*, 1989), whereas percentages of $Thy\text{-}1^+$ T cells are over-represented (Groen *et al.*, 1995). A relatively high proportion of T cells in these rats is of the $Thy\text{-}1^+/RT6^-/CD45RC^-$ phenotype (Groen *et al.*, 1996b). A high frequency of these immature peripheral T cells is in agreement with a short life span observed for a majority of T cells in these rats (Groen *et al.*, 1996a). Upon the loss of Thy-1 expression during peripheral maturation in normal rats, RT6 expression is acquired and, slightly later, CD45RC expression also. The phenotype thus acquired ($Thy\text{-}1^-/RT6^+/CD45RC^+$) represents the bulk of peripheral T cells in the rat and these cells were therefore termed common peripheral T cells (CPT). In addition, Kampinga *et al.* (1996) showed that, secondarily, both RT6 and CD45RC are lost from a small subset of formerly $Thy\text{-}1^-/RT6^+/CD45RC^+$ T cells, leaving them $Thy\text{-}1^-/RT6^-/CD45RC^-$, $Thy\text{-}1^-/RT6^-/CD45RC^+$ or $Thy\text{-}1^-/RT6^+/CD45RC^-$. Since RT6 and CD45RC have separately been shown to be lost upon activation (Hunt and Lubaroff, 1987; Powrie and Mason, 1990a), and since the

Table V.5.1. Recombinant rat cytokines: cloning, expression and characteristics

Abbreviations	*Cytokine*	*Alternative name*	*Expression system*	*Chromosomal gene structure*	*Amino acid content (mature protein)*	*Potential glycosylation sites (mature protein)*	*Cysteine residues (mature protein)*	*Molecular mass (kDa) (mature protein)*
IFN-α	Interferon alpha	Type I/leukocyte IFN	*Escherichia coli* Baculovirus	Multiple copies, no introns	169	1	5	18–19
IFN-β	Interferon beta	Type I/fibroblast IFN	CHO cells	Single copy, no introns	163	4	1	16–40 (heavily glycosylated)
IFN-γ	Interferon gamma	Type II/immune IFN	CHO cells	Single copy, three introns	137	2	3	14–25 glycosylated
IL-1α	Interleukin-1 alpha	Catabolin, tumor-inhibitory factor-2, lymphocyte activating factor	COS-1 *E. coli*	Not reported	156	Not reported	0	$\approx$17 (nonglycosylated)
IL-2	Interleukin-2	T-cell growth factor	CHO	Not reported	135	One putative *O*-glycosylation site	Not reported	$\approx$17
IL-3	Interleukin-3	Multi-colony stimulating factor (multi-CSF)	COS-1	Four introns	140	2	4	22.5–26
IL-4	Interleukin-4	B-cell growth factor-I, B-cell stimulatory factor-1 (BSF-1)	CHO	Single copy, three introns	123	4	7	17–20
IL-5	Interleukin-5	B cell growth factor-2, T cell replacement factor, eosinophil differentiation factor (EDF)	Retroviral expression in T88M cells	Three introns, single copy gene	132	Not reported	4	$\approx$14.5 (predicted)
IL-6	Interleukin-6	IFN-β2, B cell stimulatory factor-2, B cell differentiation factor	Transection of murine L cells and human HeLa cells with rat IL-6 cDNA; *E. coli* (produced as a fusion protein)	Four introns, single copy gene	187	None	5	$\approx$22
IL-10	Interleukin-10	Cytokine synthesis inhibitory factor (CSIF), B-cell derived T-cell growth factor	COS-7 cells, *E. coli* (fusion protein)	Not reported	160	2	5	$\geqslant$18.6 (predicted)

Abbreviations	Amino acid sequence homology with cytokine counterparts of other species (mature protein)	Nucleotide sequence data (GenBank/ EMBL Accession numbers)	Cross-species activity	Major cell source or location	Main biological activities	Other characteristics	References
IFN-α	Mouse: 81%	Not reported	No activity on human cells	Various cell types, most prominently macrophages	Exerts antiviral activity; has antiproliferative properties; induces MHC class I expression; influences lymphocyte traffic; augments NK activity	Acid stable	Dijkema *et al.* (1984), van der Meide *et al.* (1986a), Spanjaard *et al.* (1989), P. H. van der Meide *et al.*, unpublished data
IFN-β	Mouse: 76% Human: 46%	Not reported	Approximately 10% antiviral activity on human cells	Fibroblasts, macrophages, epithelial cells	Similar to IFN-α	Acid stable, heat labile	Ruuls *et al.* (1996), P. H. van der Meide *et al.*, unpublished data
IFN-γ	Mouse: 87% Human: 39%	M29315-17	Full activity on mouse cells; no activity on human cells	T cells, NK cells	Activates monocytes/ macrophages; induces MHC class I and II molecules on various cell types; exerts antiviral activity	Acid labile	Dijkema *et al.* (1985), van der Meide *et al.* (1986b)
IL-1α	Mouse: 83% Human: 65% (including precursor region)	Not reported	Active on human cells (GIF activity)	Monocyte/ macrophages endothelial cells, glial cells, T and B cells, NK cells, fibroblasts	Stimulates synthesis of acute phase proteins; activates resting T cells; stimulates growth fibroblasts and astrocytes; stimulates CRF release; augments PGE and collagenase synthesis		Nishida *et al.* (1989)
IL-2	Mouse: 78% Human: 66%	M 22899	High crossreactivity on mouse cells	T cells	Growth factor for activated T cells; activates cytotoxic lymphocytes; stimulates production of secondary cytokines; stimulates proliferation of B lymphocytes		McKnight *et al.* (1989), McKnight and Classon (1992)
IL-3	Mouse: 54%	Not reported	No or poor cross-species activity on human and mouse cells, respectively	T cells, mast cells, macrophages, microglial cells	Supports growth of pluripotent bone marrow cells; growth factor for mast cells; synergistic activity with other hematopoietic growth factors; survival factor for cholinergic neurons		Cohen *et al.* (1986), Kamegai *et al.* (1990), Gebicke-Haerter *et al.* (1994)
IL-4	Mouse: 57–61% Human: 42%	X 16058	No biological activity on mouse and human cells; mouse IL-4 also fails to act on rat cells	T cells, mast cells, B cells, basophils	Growth factor for activated B and resting T cells; promotes T_H2 development; induces MHC class II expression on B cells; promotes IgE production from B cells; upregulates the expression of CD23	Acts as a monomer	Leitenberg and Feldbush (1988), Richter *et al.* (1990), McKnight *et al.* (1991), McKnight and Classon (1992)
IL-5	Mouse: 92% Human: 68%	X54419	Strong biological activity on mouse cells	T cells (T_H2-type), mast cells, eosinophils	Stimulates growth and differentiation of eosinophils; stimulates B cell differentiation		Überla *et al.* (1991)
IL-6	Mouse: 93% Human: 58%	M 26744 and M 26745	Full activity on mouse cells	T cells, monocytes/ macrophages, fibroblasts, endothelial cells, B cells, synovial cells, keratinocytes, astrocytes	Induces terminal differentiation of activated B cells into Ig-secreting plasma cells; stimulates production of acute phase proteins by hepatocytes; acts as a hybridoma/plasmocytoma growth factor; acts in synergy with IL-3 to support proliferation of cultured mutlipotential hemopoietic progenitor cells	A unique property of the rat IL-6 gene is the presence of two different mRNA species differing by 1.2 Kb in their 3′-nontranslated regions	Fey *et al.* (1989), Northemann *et al.* (1989), Frorath *et al.* (1992)
IL-10	Mouse: 83% Human: 75% Rabbit: 71%	X 60675	Active on mouse cells	T cells, B cells, macrophages	Inhibition of cytokines released by T_H1 cells; autocrine inhibition of antigen-presenting functions by monocytes and macrophages	Striking homology with cDNA and protein sequence of BCRF1, a gene encoded by the Epstein–Barr virus genome	Goodman *et al.* (1992a), Feng *et al.* (1993)

(continued)

Table V.5.1. *Continued*

Abbreviations	*Cytokine*	*Alternative name*	*Expression system*	*Chromosomal gene structure*	*Amino acid content (mature protein)*	*Potential glycosylation sites (mature protein)*	*Cysteine residues (mature protein)*	*Molecular mass (kDa) (mature protein)*
IL-13	Interleukin-13	P600 (mouse)	Baculovirus and *E. coli*	Not reported	111	4	4	12.1[a] (predicted from cDNA sequence)
TNF-α	Tumor necrosis factor alpha	Cachectin	CHO	Not reported	156	Not reported	2	17–22
SCF	Stem cell factor	*Steel* factor, mast cell growth factor, kit ligand	COS-1 *E. coli*	At least five introns	164–165	Five *N*-linked and numerous *O*-linked carbohydrate attachment sites	Not reported	18.5 (deglycosylated CSF)
CNTF	Ciliary neurotrophic factor	Not reported	Not applicable	One intron	200	Not reported	Not reported	22
TGF-β_3	Transforming growth factor beta-3	None	Not applicable	Not reported	410	Not reported	14	≈45 (predicted)
TGF-α	Transforming growth factor alpha	No	Not applicable	Five introns	50	Not reported	6	5.5 (predicted)
EGF	Epidermal growth factor	No	Not applicable	More than one copy	53	Not reported	6	5.8 (predicted)
KGF	Keratinocyte growth factor	Heparin-binding growth factor type 7	Not applicable	Not reported	194	Not reported	7	21.3 (predicted)
HB-EGF	Heparin-binding epidermal growth factor-like growth factor	No	Not applicable	Not reported	86	One *O*-linked glycosylation site	6	9.5 (predicted)

Abbreviations	*Amino acid sequence homology with cytokine counterparts of other species (mature protein)*	*Nucleotide sequence data (GenBank/ EMBL Accession numbers)*	*Cross-species activity*	*Major cell source or location*	*Main biological activities*	*Other characteristics*	*References*
IL-13	Mouse: 79% Human: 63%	Not reported		T cells (T_H2 type), mast cells	Downregulates cytotoxic and inflammatory functions of monocytes and macrophages; induces IL-4-independent IgE synthesis, shares many of its biological actions with IL-4 and IL-10	Does not share any significant amino acid homology with rat IL-4	Lakkis and Cruet (1993), F. G. Lakkis, unpublished data
TNF-α	Mouse: 91% (P. H. van der Meide, unpublished data)	M 98820	Full biological activity on mouse cells	T cells, monocytes/ macrophages, fibroblasts astrocytes, microglial cells	Inductor acute phase responses; selective cytotoxic activity; various effects on different cell types by modulating gene expression for other cytokines, acute phase proteins, etc.	Contiguous genetic arrangement of TNF-α and TNF-β	Chung and Benveniste (1990), Rothe *et al.* (1992), Appel *et al.* (1995a), P. H. van der Meide *et al.*, unpublished data
SCF	Not reported	Not reported	Strong cross-species activity on human and mouse cells	Bone marrow stromal cells, endothelial cells	Augments proliferation of both myeloid, erythroid and lymphoid hematopoietic progenitors in bone marrow cultures. SCF acts in synergy with other factors such as IL-7, GM-CSF, C-CSF, IL-3, IL-6 and erythropoietin; this suggests that SCF stimulate stem cells in combination with a variety of other cytokines	Natural SCF is heavily glycosylated; in addition the molecule is very acidic and acts as a dimer under nondenaturating conditions. SCF is analogous to PDGF and CSF-1	Martin *et al.* (1990), Zsebo *et al.* (1990)
CNTF	Not reported	Not reported	Not reported	Schwann cells, subpopulation of type 1 astrocytes	Promotes survival and maturation of cultured oligodendrocytes; induces cholinergic properties in sympathetic neurons; induces acute-phase protein expression in hepatocytes	The expression of CNTF is restricted to the postnatal period	Stöckli *et al.* (1991), Carroll *et al.* (1993)
TGF-β_3	Mouse: 99%	Not reported	Not reported	Fetal lung fibroblasts	Suppresses both basal and estradiol-induced prolactine release		Wang *et al.* (1995)
TGF-α	Human: 92%	M 31076	Not reported	Tumor cells and various transformed cells	Autocrine growth regulator (EGF-related cytokine)	Synthesized as a transmembrane glycoprotein precursor of 159 or 160 amino acids	Blasband *et al.* (1990)
EGF	Human: 68%	M 63585	Not reported	Kidney, small intestine, mammary gland	EGF stimulates proliferation and differentiation of cells of ectodermal and mesodermal origin and may play a crucial role in early tissue development	Synthesized as a precursor (ppEGF) of 1133 amino acids	Saggi *et al.* (1992), Abraham *et al.* (1993)
KGF	Human: 92%	X 56551	Not reported	Candidate for a stromal-derived growth factor that acts in strictly a paracine mode of action	Supports the compensatory growth of epithelial cells in a variety of tissues	Belongs to the heparin-binding (fibroblast) growth factor (HBGF/FGF) family	Yan *et al.* (1991)
HB-EGF	Mouse: 87% Human: 76% Monkey: 76%	L 05489	Not reported	Transcript expression in multiple tissues, particularly lung, skeletal muscle, brain and heart	Mitogen for smooth muscle cells, fibroblasts and keratinocytes	Belongs to the EGF protein family. The secreted HB-EGF is derived from a 208-residue precursor that includes a 23 amino acid signal peptide, a 86 amino acid mature HB-EGF polypeptide and a transmembrane domain of 24 amino acids	Abraham *et al.* (1993)

(*continued*)

Table V.5.1. *Continued*

Abbreviations	*Cytokine*	*Alternative name*	*Expression system*	*Chromosomal gene structure*	*Amino acid content (mature protein)*	*Potential glycosylation sites (mature protein)*	*Cysteine residues (mature protein)*	*Molecular mass (kDa) (mature protein)*
bFGF	Basic fibroblast growth factor	No	Not applicable	One putative intron	145	Not reported	4	16 (predicted)
CINC-1	Cytokine-induced neutrophil chemo-attractant-1	Rat GRO Rat KC	*E. coli*	Single-copy gene, three introns	72	0	4	8
CINC-2α	Cytokine-induced neutrophil chemo-attractant-2α	No	*E. coli*	Single copy gene, three introns	69	0	4	8
CINC-2β	Cytokine-induced neutrophil chemo-attractant-2β	No	*E. coli*	Single copy, four introns	68	0	4	8
CINC-3	Cytokine-induced neutrophil chemo-attractant-3	Macrophage inflammatory protein-2 (MIP-2)	*E. coli*		69	0	4	8
IL-12 (p40)	Interleukin-12 p40 chain	No	Not applicable	Probably identical to mouse	Not reported	Not reported	Not reported	Not reported
vgr	Vegetal related cytokine	TGF-β related cytokine	Not applicable	Not reported	133	Not reported	7	Not reported
IGF-II	Insulin-like growth factor II	Somatomedin	Not applicable	Single copy gene. Transcription is initiated from two alternative promotors, which produce at least four different mRNA species ranging in size from 1 to 4.6 Kb. There are three protein-coding exons	67	Not reported	6	7.5

Abbreviations	Amino acid sequence homology with cytokine counterparts of other species (mature protein)	Nucleotide sequence data (GenBank/ EMBL Accession numbers)	Cross-species activity	Major cell source or location	Main biological activities	Other characteristics	References
bFGF	Human: 97%	Not reported	Probably high; there are only five conservative amino acid substitutions and one amino acid deletion between rat and human sequences	Brain cortex, hypothalamus	A potent mitogen for various cell types of mesodermal or neuroectodermal origin	bFGF may be stored in a biononavailable form	Kurokawa *et al.* (1988), Shimasaki *et al.* (1988)
CINC-1	Human IL-8: 47% Human GRO-α: 68% Mouse KC: 92% Mouse MIP-2: 71%	M 86536	Not reported	Macrophages, fibroblasts and probably various other cell types	Plays a decisive role in the activation and the infiltration of neutrophils into inflammatory sites; the neutrophil chemotactic activities of all CINCs *in vitro* are quite similar	Member of the pro-inflammatory 'chemokine' superfamily of chemotactic cytokines; all CINCs are counterparts of human GRO. CINC-2α and CINC-2β are isoforms; the former is produced as a major chemokine by rat macrophages and in rat inflammation *in vivo*. CINC-2β differ from CINC-2α in the terminal three amino acids	Watanabe *et al.* (1989), Huang *et al.* (1992), Konishi *et al.* (1993), Zagorski and DeLarco (1993), Nakagawa *et al.* (1994)
CINC-2α	Human IL-8: 44% Human GRO-α: 57% Mouse KC: 62% Mouse MIP-2: 78%	Not reported	Not reported				Nakagawa *et al.* (1994, 1996a, b), Shibata *et al.* (1995, 1996)
CINC-2β	Human IL-8: 43% Human GRO-α: 59% Mouse KC: 65% Mouse MIP-2: 81%	Not reported	Not reported				Nakagawa *et al.* (1994, 1996b)
CINC-3	Human IL-8: 49% Human GRO-α: 64% Mouse KC: 65% Mouse MIP-2: 90%	RNIMIP-2, X 65647	Not reported	Macrophages, fibroblasts and probably various other cell types	Plays a decisive role in the activation and infiltration of neutrophils into inflammatory sites	As described for CINC-1, -2α and -2β	Nakagawa *et al.* (1994, 1996b), Shibata *et al.* (1995)
IL-12 (p40)	94% identity with predicted amino acid sequence of mouse IL-12 p40	M 86771 (IL-12p35: M 86672)	Not known	Monocytes/ macrophages	Functions as a heterodimeric (p35 and p40 chains) cytokine; promotes the development of T_H1-type of immune responses	A partial cDNA clone of the p40 subunit has been analyzed	Mathieson and Gillespie (1996)
vgr	Human: 98% Mouse: 98%	X 58830	High	Constitutively expressed in the CNS	Suggested to have trophic effects on cells of the nervous system and to play a role in inflammatory and degenerative diseases of the CNS	This cytokine belongs to a subfamily of TGF-β related cytokines that are believed to play a regulatory role in embryonic development and organogenesis	Sauermann *et al.* (1992)
IGF-II	Not reported	J 02637	Not reported	Fetal or neonatal tissues in all of the developmental stages	IGF-II is a mitogenic polypeptide that has structural similarity to proinsulin. It plays an important role in fetal development and in the function of the CNS. Serum levels are increased at birth and decrease upon aging	The chromosomal gene encodes a precursor protein (pre-pro-IGF-II) that comprises 179 amino acids consisting of a signal peptide (23 amino acids), the mature IGF-II (67 amino acids) and a trailer polypeptide of 89 residues. The gene is located on chromosome I and linked to the insulin gene	Frunzio *et al.* (1986), Soares *et al.* (1986), Chiariotti *et al.* (1988)

Table V.5.2. Recombinant rat cytokine receptors and binding proteins: cloning, expression and characterization

Cytokine receptor abbreviation	*Cytokine receptor name*	*Chromosomal gene structure*	*Amino acid content (mature protein)*	*Potential glycosylation sites (mature protein)*	*Molecular mass (kDa) (mature protein)*	*Amino acid sequence homology with receptors of other species*	*Ligand(s)*	*Expression in vivo*	*Nucleotide sequence data (GenBank/EMBL Accession numbers)*	*Molecular structure*	*Other characteristics*	*References*
FGFR4	Fibroblast growth factor receptor subtype 4	Not reported	650	Not reported	72 (predicted)	Human: 92%	aFGF bFGF	Lung and kidney	M 91599	Two potential immunoglobulin-like domains, 21 hydrophobic amino acids encoding a potential transmembrane domain and a split tyrosine kinase motif	FGFR4 belongs to the Ig-like superfamilies of hormone receptors. The cloned cDNA lacks a signal sequence and acidic box and may represent an intracellular form of the FGFR4 receptor	Horlick *et al.* (1992)
TGF-βIIR	Transforming growth factor-β type II receptor	Not reported	544	2		Human: 92% Mink: 91%	TGF-β	Ovary, lung and kidney	L 09653	A cysteine-rich extracellular domain (136 amino acids), a transmembrane domain of 30 amino acids and an intracellular protein kinase domain of 378 amino acids	12 cysteine residues in mature protein	Tsuchida *et al.* (1993)
KGFR	Keratinocyte growth factor receptor	Not reported	822 (isoform b)	10 (isoform a and b)	130–150	Human: 97%	aFGF KGF	Parathyroid cells	Z 35138 and Z 35139	Two isoforms: a receptor molecule with 2 Ig-like domains (a) and one with 3 Ig-like domains (b). Both isoforms contain a transmembrane domain and two tyrosine kinase domains	Isoform b has three possible glycosaminoglycan-attachment sites; there are 19 cysteine residues in the mature isoform b	Takagi *et al.* (1994)
IL-4R	Interleukin-4 receptor	Not reported	775	5	Not reported	Mouse: 78% Human: 52%	IL-4	Various cell types	X 69903	The receptor consists of an extracellular domain of 207 amino acids, a transmembrane region of 24 amino acids and a cytoplasmic domain of 544 amino acids	26 cysteine residues; for signalling, the IL-4R interact with the γ chain initially identified for the IL-2 receptor	Richter *et al.* (1995)
IL-2Rα	Interleukin-2 receptor alpha chain	Not reported	267	2	Not reported	Human: 60% Mouse: 82% Bovine: 52%	IL-2	Activated T, B and NK cells	M 55049	Extracellular domain of 214 amino acids, a transmembrane region of 19 amino acids and a cytoplasmic tail of 13 amino acids	12 cysteine residues out of 14 are conserved between human, mouse, rat and bovine species; redundant in IL-2-mediated signaling	Page and Dallman (1991)
IL-2Rβ	Interleukin-2 receptor beta chain	Not reported	511	4	≈ 80	Human: 60% Mouse: 83%	IL-2	T and NK cells	M 55050	Extracellular domain of 213 amino acids, a transmembrane region of 28 amino acids and a cytoplasmic domain of 270 amino acids	Essential for signal transduction; less than 10% of IL-2Rβ+ cells co-express the IL-2Rα chain; a strong TCR-mediated stimulus leads to induction of IL-2Rβ chain which combines with the γ chain to form a functional IL-2R in the absence of IL-2Rα	Page and Dallman (1991), Park *et al.* (1996)

IL-3Rβ	Interleukin-3 receptor beta chain (βc)	Single gene	874	4	Not reported	77.7% (AIC2A) and 79.5% (AIC2B) with mouse βc (two IL-3Rβ subunits in the mouse)	The α-subunit binds IL-3; association between the α and β subunit results in a high-affinity receptor for IL-3; βc functions as signal transducer. GM-CSF and IL-5 may also use βc as transducer	Rapid induction of βc mRNA in microglia cells located in various rat brain regions after systemic administration of LPS	Not reported	A transmembrane region of 26 amino acids	Transducing subunit of the IL-3 high-affinity receptor; this subunit is most likely the rat equivalent of the human IL-3Rβ-subunit; the signal peptide of the gene product consists of 22 amino acids; the external domain contains eight conserved cysteine residues; no kinase domain was found in the cytoplasmic region. IL-3R mRNA α- and β-subunits detected in isolated microglial cells	Appel *et al.* (1995b)
GPR14	G protein coupled receptor related to the human somastatin receptor (SSTR)	No introns	386	Not reported	Not reported	Overall amino acid identity is approximately 27% with human SSTR4	Most likely a peptidergic ligand	Not known	U32673	SSTR-like	Shares identity with the somatostatin and opioid receptors; signaling function for an as yet unidentified peptidergic ligand	Marchese *et al.* (1995)
TGF-β-MP$_{ls}$ or TGF-β_1-BP	Transforming growth factor type β masking protein (large subunit) or Transforming growth factor type β_1 binding protein	Not known	841	3	110–120 (M_r) 91.6 (calculated M_r)	Human: 90%	TGF-β_1 and TGF-β_2	Various tissues	M 55431	The large latent TGF-β_1 complex (isolated from platelets) is composed of the TGF-β_1 molecule noncovalently associated with a masking protein; the masking protein consists of a disulfide-bonded complex of a dimer of the *N*-terminal propeptide of the TGF-β_1 precursor and a third component denoted TGF-β_1-BP (identical to TGF-β-MP$_{ls}$). TGF-β_1-BP does not bind directly to active TGF-β_1	TGF-β_1-BP may play a role in the activation mechanism of the latent TGF-β_1 rather than in masking the activity of TGF-β_1; the full-length cDNA encodes a pre-pro-precursor of 1712 amino acids, a pro-precursor of 1692 amino acids and a mature protein of 841 amino acids; the mature protein is acid-stable, endoglycosydase H-sensitive and contains 13 EGF-like domains and two cysteine-rich internal repeats	Tsuji *et al.* (1990)
IL-1RI	Interleukin-1 receptor type I	Not reported	Not reported	Not reported	80.5	High degree of homology with both the human and mouse type I IL-1R	IL-1α,β	Various cell types. Rarely co-expressed with IL-1RII	Not reported	A cytoplasmic domain of approximately 215 amino acids	Dissociation constant (K_d) of 1.2×10^{-10} M on rat epithelial cells. Partial cDNA sequence analysed; high degree of homology with both mouse and human type I IL-1R; >92% with mouse IL-1R and >82% with human IL-1R	Hart *et al.* (1993), Sutherland *et al.* (1994)
IL-1RII	Interleukin-1 receptor type II		403		≈68	Human: 62% Mouse: 90%	IL-1α, β	Various cell types including insulinoma cells	Z22812	An extracellular domain of 342 amino acids, a transmembrane of region of 26 amino acids and a cytoplasmic tail of 35 amino acids	Eight conserved cysteine residues; an endogenously produced soluble form of IL-1RII may act as a physiological IL-1 antagonist glucocorticoids upregulate IL-1RII on polymorpho-nuclear cells	Bristulf *et al.* (1994)
IL-1ra	Interleukin-1 receptor antagonist	Three introns	152	One	16.7 (predicted)	IL-1 receptor	Monocytes/ macrophages	M63099–M63101	Not reported	Mouse: 90% Human: 75%	Blocks the binding of both IL-1α and IL-1β to the IL-1 receptor without inducing a signal of its own	Eisenberg *et al.* (1991)

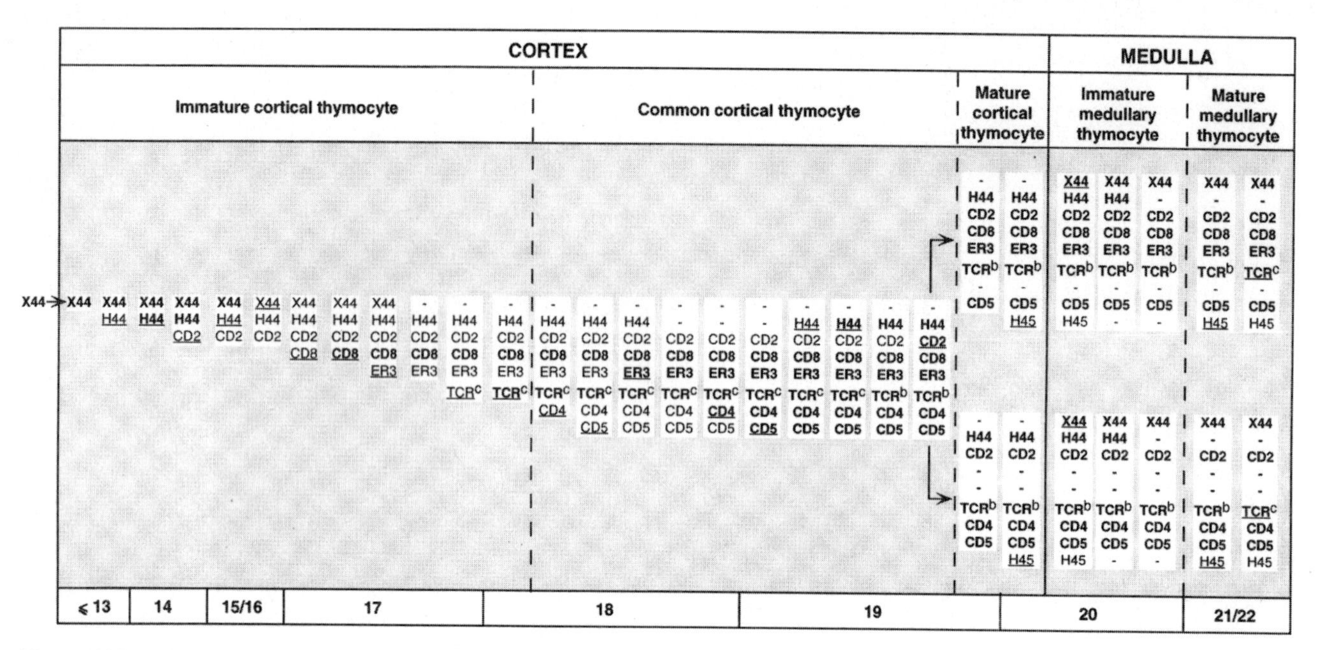

Figure V.6.1 This scheme of thymocyte differentiation in the fetal rat is based on immunohistochemical analysis of frozen sections of rat thymus taken at different stages of gestation. Changes in phenotype between two successive stages are underlined and high expression of a molecule is marked in bold. The antibody against the TCR-$\alpha\beta$ detects a determinant expressed first in the cytoplasm (TCR^c), then both on the surface and in the cytoplasm (TCR^b) and later only on the surface (TCR^s). The days of gestation are shown at the bottom of the diagram and an attempt to fit the stages of differentiation into a possible scheme of events in the adult thymus is shown. The terms 'mature' and 'immature' refer to phenotype and not to function. The different compartments are not to scale: in the adult thymus $\approx$80% of thymocytes is found in the population of 'common' cortical thymocytes, with the immature and mature populations of cortical thymocytes each containing <5% of the cortical population. The mature and immature medullary thymocyte populations contain $\approx$5% and $\approx$8%, respectively, of the total number of thymocytes (Groen *et al.*, 1996a). H and X represent the antibody nomenclature prefixes HIS and MRC-OX respectively. This figure was reproduced with permission of Kampinga and Aspinall (1990a).

Thy-1$^-$/RT6$^-$/CD45RC$^-$ subset was substantially increased in an antigen-rich environment, such as Peyer's patches, it was suggested that activation renders T cells negative for both RT6 and CD45RC. Recent experiments, however, show that loss of RT6 and CD45RC does not occur within 48 h after *in vitro* mitogenic stimulation for a majority of T cells, which is when all T cells in culture have upregulated CD25 expression and most have downregulated quiescent cell antigen-1 (QCA-1, see below) expression (H. Groen, unpublished data). The remaining Thy-1$^-$/RT6$^-$/CD45RC$^+$ and Thy-1$^-$/RT6$^+$/CD45RC$^-$ T-cell subsets were, based on their relative increase after adult thymectomy, marked as post-CPT stages of development, and are likely to be resting memory T cells. Based on the properties and cytokine profiles of these two subsets (Fowell *et al.*, 1991; Fowell and Mason, 1993; Beijleveld *et al.*, 1996), and the association between inducibility of cellular versus humoral autoimmune diseases on one side, and balances between the two subsets mentioned on the other (Groen *et al.*, 1993), Thy-1$^-$/RT6$^-$/CD45RC$^+$ and Thy-1$^-$/RT6$^+$/CD45RC$^-$ CD4$^+$ T cells were suggested to be T$_H$1 and T$_H$2 resting memory subsets. This suggestion was supported by the fact that cellular autoimmunity is associated with CD4$^+$/Thy-1$^-$/RT6$^-$/CD45RC$^+$ T cells (Beijleveld *et al.*, 1996) and can be prevented with CD4$^+$/Thy-1$^-$/RT6$^+$/CD45RC$^-$ T cells (Fowell and Mason, 1993). Conversely, the BN rat, having high basic levels of IgE and a high percentage of IL-4-producing T cells, and being susceptible to autoantibody-mediated mercury-chloride-induced glomerulonephritis (Aten *et al.*, 1991; van der Meide *et al.*, 1995), has high percentages of CD4$^+$/Thy-1$^-$/RT6$^+$/CD45RC$^-$ T cells (Groen *et al.*, 1993). Mercury-induced autoimmunity was found to be suppressed by CD4$^+$/CD45RC$^+$ T lymphocytes (Mathieson *et al.*, 1993). The phenotypic changes mentioned above are depicted in Figure V.6.2 for both CD4$^+$ and CD8$^+$ T cells. It can be seen in the figure that the proposed routes for phenotypic development of CD4$^+$ and CD8$^+$ T cells are virtually identical. Percentages of peripheral blood T cells expressing Thy-1, RT6 or CD45RC of a number of rat strains are listed in Table V.6.1. Percentages of CD4$^+$ and CD8$^+$ T cells and the balances between the proposed T$_H$1 and T$_H$2 resting memory subsets are listed in Section 2.

Other markers of interest for T cell developmental and functional studies in the rat are the ones recognized by the MRC OX-39, OX-40 and HIS45 monoclonal antibodies. MRC OX-39 recognizes CD25, the IL-2-receptor (Paterson *et al.*, 1987b). The OX-40 marker (CD134) is expressed only by CD4$^+$ T blasts (Paterson *et al.*,

Table V.6.1 Percentages of Thy-1⁺, RT6⁺ and CD45RC⁺ T cells in blood of 20-week-old rats of different strains

Strain	*MHC*	*Thy-1*	*RT6.1*	*RT6.2*	*CD45RC*
AO	u	14.7 ± 0.5	—	79.3 ± 1.7	67.4 ± 1.3
AO1p	ullu	21.2 ± 1.3	—	78.3 ± 0.6	68.2 ± 3.0
DPBB	u	38.3 ± 1.8	9.6 ± 0.3	—	16.0 ± 4.7
BN	n	14.4 ± 0.7	—	83.2 ± 3.3	41.0 ± 2.1
BUF	b	12.7 ± 1.6	76.1 ± 1.2	—	67.9 ± 0.3
DZB	u	16.3 ± 0.9	—	85.9 ± 1.5	75.6 ± 0.7
LEW	l	13.1 ± 1.0	71.9 ± 0.8	—	70.6 ± 2.6
LOU	u	14.4 ± 1.9	—	79.3 ± 1.7	67.4 ± 1.3
MAXX	n	14.3 ± 1.4	—	86.8 ± 1.9	68.5 ± 2.8
PVG	c	6.0 ± 0.3	75.8 ± 1.5	—	65.4 ± 1.1
WF	u	14.8 ± 2.1	—	75.7 ± 1.5	64.0 ± 1.5

1987b), whereas the HIS45 mAb (where T cells are concerned) recognizes a 43–47 kDa protein expressed by resting T cells only (Kampinga *et al.*, 1990). Mitogenic stimulation of T cells causes the loss of the HIS45 epitope, which is why it was called quiescent cell antigen-1 (QCA-1). QCA-1 is expressed by medullary thymocytes and also by the vast majority (>95%) of Thy-1⁺ RTMs. Peripheral T cells not expressing QCA-1 are mostly found among the mature Thy-1⁻ phenotype (H. Groen, unpublished data).

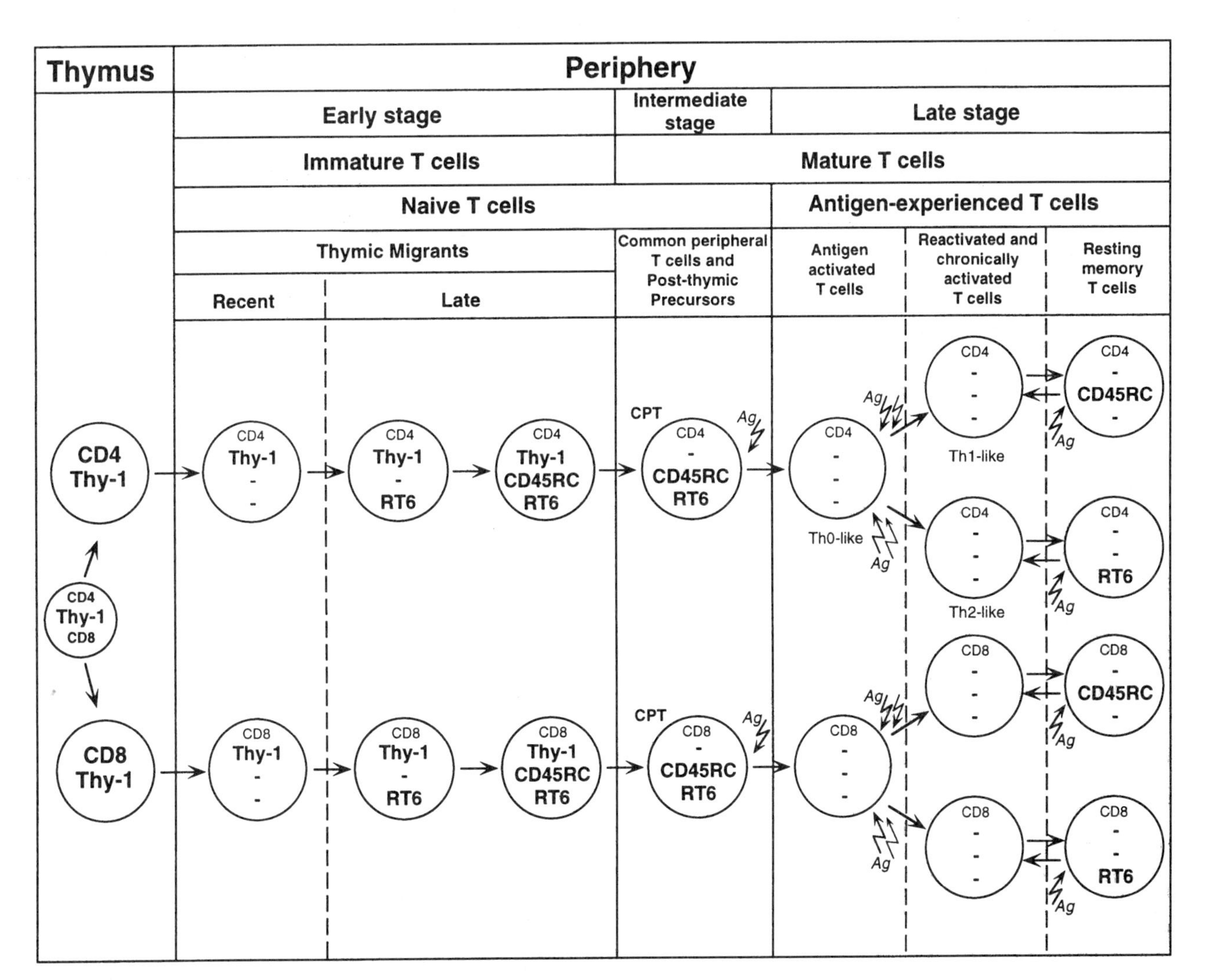

Figure V.6.2 A concept of development of peripheral CD4⁺ and CD8⁺ T cells in the rat. The sequence of events depicted in the diagram was deduced from the vascular thymus transplantation studies described by Kampinga *et al.* (1996).

7. The T-cell Receptor

T-cell receptor sequences

Available rat TCR sequences reveal a high degree of similarity to the homologous mouse gene segments. In Tables V.7.1 and V.7.2, V, (D) and J sequences of the *Tcra* and *Tcrb* loci are compiled and ordered according to their similarity to mouse sequences. With the exception of the few genomic sequences available (see table footnotes), cDNA sequences are given. Note that the homologous mouse sequences are given in the new official *TCRV* nomenclature (Arden *et al.*, 1995). *JA* and *JB* nomenclatures follow those developed by Koop *et al.* (1994) and by Gascoigne (1995), repectively.

To date, a total of 63 *VA*, 46 *JA*, 25 *VB* and 12 *JB* functional rat TCR segments have been identified, suggesting a degree of complexity similar to that in mice. As in that species, there is one *TCRAC* and two *TCRBC* loci. Genomic deletions of TCR segments, as observed in some mouse strains, have not been described in the rat.

With few exceptions, all known *TCRA* sequences were derived from the LEW strain (Table V.7.1). Although not yet shown by sequence, the rat *Tcra* locus is at least dimorphic, as indicated by the lack of reactivity of certain TCRVα-specific monoclonal antibodies (mAbs) with T-cells and thymocytes of the LEW but not of the DA background (N. Torres-Nagel *et al.*, unpublished data).

There are two major *Tcrb* haplotypes. Table V.7.2 lists all currently available sequences, including different

Table V.7.1 Sequences of LEW[a] T-cell receptor α (*Tcra*) segments

Most homologous mouse segment[b]	*Rat clone name or designation*	*Accession No. (EMBL/GenBank)*	*Homology with mouse at nucleotide (amino acid) level*	*Source*	*Current numbering (family.member)*
AC[c]	TRA29[a]	M18853	84	1	—
AV1S2	TRA29[a]	M18853	ND	1	1
	T48a2	X62325	82	3	1
AV1S3	rVA14	L37959	86.6	4	1
	rVA48	L11034	86.1	4	1
	rVA32	L11029	88.6	4	1
	rVA16	L37961	88.6	4	1
	rVA46	—	85.2	4	1
AV1S5	rVA23	L11025	87.7	4	1
AV1S8	rVA33	L37963	87.0	4	1
	rVA56	L37966	90.9	4	1
	rVA83	L37973	84	4	1
AV2S1	510	X14318	(77)	2	2
AV2S3	rVA37	L11030	88.9	4	2
AV3S2	rVA41	L11033	90.8	4	3
AV3S5	rVA68	L37970	88.8	4	3
	rVA04	L37957	82.9	4	3
AV3?	rVA22	L11024	88.7	4	3
AV4S1	rVA27	L11021	86.2	4	4
	4.12[d]	X81302	ND	5	4
	4.13[d]	X81303	ND	5	4
	4.14[d]	X81304	ND	5	4
AV4S2	rVA52	L11036	89.8	5	4
	4.22[d]	X81305	ND	5	4
	4.23[d]	X81306	ND	5	4
	4.25[d]	X81308	ND	5	4
	4.26[d]	X81309	ND	5	4
	4.27[d]	X81310	ND	5	4
AV4S3 or *AV4S10*	P2.3	X75647	90.1 90.1	5	4
	4.32[d]	X81311	ND	5	4
	4.33[d]	X81312	ND	5	4
	4.34[d]	X81313	ND	5	4
	4.35[d]	X81314	ND	5	4
	4.36[d]	X81315	ND	5	4

(continued)

Table V.7.1 *Continued*

Most homologous mouse segment[b]	*Rat clone name or designation*	*Accession No. (EMBL/GenBank)*	*Homology with mouse at nucleotide (amino acid) level*	*Source*	*Current numbering (family.member)*
AV4S11	4.41[d]	X81316	89.2	5	4
AV5S1	rVA24	L11026	82.5	4	5
	rVA30	L11027	86.0	4	5
AV6S1	Q4a12	X62324	92	3	6
ADV7S1	rVA03	L37956	82.1	4	7
AV8S2	8.4[d]	X80511	(88.6)	6	8.4
AV8S4	8.2[d]	X80509	(92.0)	6	8.2
	DP15	Y09187	(90.9)	7	—
AV8S10	8.5[d]	X80512	(86.0)	6	8.5
	8.6[d]	X80513	(87.2)	6	8.6
	8.7[d]	X80514	(82.5)	6	8.7
AV8S11	DP10	Y09185	(88.6)	7	—
	DP118	Y09186	(89.8)	7	—
	8.3[d]	X80510	(88.6)	6	8.3
AV10S1	rVA15	L37960	88.6	4	10
	T48a3	X62328	90.5	3	10.1
	T48a5	X62330	87.1	3	10.2
AV10S2	rVA60	L37968	87.1	4	10
AV11S2	Q4a13	X62329	92	3	—
AV11S3	rVA69	L37971	85.1	4	11
AV12S1	rVA26	L37962	90.0	4	12
	rVA38	L11031	84.5	4	12
AV15S1	rVA12	—	89.8	4	13
AV17S1	rVA10	L11019	91.9	4	16
	rVA31	L11028	91.0	4	16
	rVA66	L37969	90.3	4	16
AV?	rVA47	L37965	88.3	4	15
AV20S1	rVA39	L11032	85.9	4	23
	rVA51	L11035	84.7	4	23
	Q4a23	X62327	—	3	—
AV21	rVA59	L37967	59	4	24
AJ2	VA8s2F10	Y09175	92.4	7	—
AJ4	rVA27	L11021	93.2	4	4
AJ5	rVA68	L37970	87.1	4	5
AJ6	VA8s2F15	Y09176	84.9	7	—
AJ7	VA8S2F23	Y09177	78.3	7	—
AJ9	rVA59	L37967	82.1	4	8
AJ11	rVA48	L11034	93.1	4	9
AJ12	VA8s2A2	Y09174	86.8	7	—
AJ13	rVA83	L37973	85.7	4	11
AJ15	rVA12	—	78.3	4	12
AJ16	rVA47	L37965	87.7	4	13
AJ17	rVA44	L37964	79.0	4	14
AJ18	rVA64	—	90.6	4	15
AJ21	rVA37	L11030	85.7	4	16
AJ22	T48a5	X62330	ND	3	—
AJ23	rVA04	L37957	96.5	4	18
AJ24	VA8s2F39	Y09178	90.6	7	—
AJ26	rVA25	L11020	87.9	4	20
AJ27	TRA29[a]	M18853	ND	1	—
AJ28	rVA60	L37968	87.9	4	22
AJ30	VA8s2A16	Y09170	84.0	7	—
AJ31	rVA33	L37963	83.9	4	24

(continued)

Table V.7.1 *Continued*

Most homologous mouse segment[b]	*Rat clone name or designation*	*Accession No. (EMBL/GenBank)*	*Homology with mouse at nucleotide (amino acid) level*	*Source*	*Current numbering (family.member)*
AJ32	VA8s2F5	Y09179	89.6	7	—
AJ33	rVA14	L37959	100	4	26
AJ34	VA8s2A41	Y09171	74.5	7	—
AJ37	VA8s2F22	Y09181	89.6	7	—
AJ38	rVA30	L11027	95.1	4	31
AJ39	T48a2	—	ND	3	—
AJ40	rVA74	L37972	93.3	4	33
	Q4a21	—	ND	3	—
AJ42	VA8s2F57	Y09182	86.8	7	—
AJ43	510	X14318	ND	2	—
AJ44	VA8s2L73	Y09168	81.1	7	—
AJ45	VA8s2F13	Y09183	87.7	7	—
	rVA51	L11035	94.5	4	37
AJ47	rVA16	L37961	88.9	4	39
AJ48	rVA19	L11022	90.4	4	40
AJ49	rVA03	L37956	87.9	4	41
AJ50	VA8s2L90	Y09169	86.8	7	—
AJ52	VA8s2DP47	Y09172	80.2	7	—
AJ53	VA8s2F32	Y09184	86.8	7	—
AJ56	rVA52	L11036	83.9	7	47
AJ57	rVA39	L11032	89.2	4	48
AJ58	rVA66	L37969	89.1	4	49
AJ62	VA8s2F6	Y09180		7	—
AJ63	VA8s2L15	Y09173		7	—

Sources: 1, Morris *et al.* (1988); 2, Burns *et al.* (1989); 3, Hinkkanen *et al.* (1993); 4, Shirwan *et al.* (1995); 5, Stangel *et al.* (1994); 6, Giegerich *et al.* (1995); 7, N. Torres-Nagel *et al.*, unpublished data.
[a] All but the three sequences labelled are derived from LEW. [a] labelled sequences are from AO strain.
[b] WHO-IUIS nomenclature (Arden *et al.*, 1995).
[c] *AC* LEW sequence has been published by Burns *et al.* (1989).
[d] Genomic sequences.

alleles, together with the strains of origin. Very little sequence information is available about *Tcrg* and *Tcrd* loci. Table V.7.3 lists the four cDNA sequences currently available, along with the gene segments from which they are presumed to be derived (mouse nomenclature according to Garman *et al.* (1986)).

mAbs to rat TCR

mAbs identifying rat TCR are listed in Table V.7.4. They cover about 30% of the V_β repertoire but only about 5% of that of the V_α. One subset (V_γ?) specific mAb to $\gamma\delta$ TCR as well as mAbs to constant determinants of the $\alpha\beta$- and $\gamma\delta$TCR are available.

TCR associated proteins and accessory molecules involved in TCR signalling

Available evidence indicates that the components of the TCR signal transduction complex, which has been studied in greater detail in mice and humans (see the corresponding chapters) are also present in the rat. Table V.7.5 lists only those components and their interactions that have been identified in the rat.

Ontogeny of the rat TCR repertoire

In fetal life, $\gamma/\delta TCR^+$ cells are first detected in the thymus at day 16 of gestation (Kühnlein *et al.*, 1995), followed by the appearance of $\alpha/\beta TCR^{low}$ cells around day 20 and of $\alpha/\beta TCR^{high}$ cells at birth (about day 23 of gestation, with strain variations) (Lawetzky *et al.*, 1991). The early

Table V.7.2 Sequences of T-cell receptor β-chain (*Tcrb*) segments. Genomic location: chromosome 4 (Dissen *et al.*, 1993)

Most homologous mouse segment[a]	*Rat clone name or designation/rat strain*	*Accession No. (EMBL/GenBank)*	*Homology with mouse at nucleotide (amino acid) level*	*Source*	*Current numbering (family.member)*
BC1	(C_β1)[b]/LEW	X14319	—	1	
		M65136	(89.6)	2	
	(C_β1)[b]/DA[c]	M63793	(87.4)	2	
	(C_β1)/LER[c]		—	3	
BC2	TRB4/AO	M18854	92 (88)	4	
	(C_β2)[b]/DA[c]	M63794	—	5	
BV1SA1	V_β1/LEW	—	86 (78)	6	1
BV2S1	V_β2/LEW	—	94 (92)	6	2
BV3S1A1	V_β3.1/LEW		91 (84)	6	3.1
BV4S1	V_β4/LEW	—	88 (85)	6	4
	TRB70/AO	M18843	88 (85)	4	
BV5S1	V_β5.1/LEW	—	88 (85)	6	5.1
BV5S2A1	V_β5.2/LEW	—	92 (91)	6	5.2
BV6S1	V_β6/LEW	X80525	90 (84)	4, 6, 18	6
	TRB4/A0	M18845	90 (84)		
BV7S1	V_β7/LEW	—	88 (84)	6	7
BV8S1	V_β8.1/LEW	—	88 (80.6)	6	8.1 (8.6)
	V_β8.6/LOU/M	M58628	90 (84)	7	
	V_β8.2/LEW	X14973	87.2 (80.6)	8	8.2
	Tcrb-V8.2^{a}/DA[c]	X98251	—	9	
	V_β8.2/LER[c]	—	—	10	
BV8S2A1	Tcrb-V8.4^{1}/LEW (pseudogen)	X97672	89 (83)	9	8.4
	Tcrb-V8.2^{F344}/ F344/DA[c] (functional)	X77995	89 (83)	11	
	V_β8.4/LER[c]	—	—	10	
	510E/LEW (variant of Tcrb-V8.4^{1}?)	U06104	—	12	—
	510C (Pseudogen)	U06102	86 (88)	13	—
BV8S3	V_β8.3/LEW	M58627	—	6	8.3 (8.5)
	510D/LEW	U06103	87 (85)	13	
	V_β8.5(LOU/M)	—	87 (85)	14	
	V_β8.5/DA[c]	—	—	15	
	V_β8.3/LER[c]	—	—	10	
BV9S1	V_β9/LEW	—	88 (78)	6	9
	CTRB39/DA	M23887	88 (78)	16	
BV10S1	V_β10/LEW	—	91 (84)	6	10
	TRB12/AO	M18839	91 (84)	4	
BV11S1	V_β11/LEW	—	91 (88)	6	11
BV12S1	V_β12/LEW	X80522	92 (87)	6, 18	12
	TRB100/AO	M18841	92 (87)		
BV13S1	V_β13/LEW		87 (72)	6	13
BV14S1	V_β14/LEW	X80523	86 (81)	6, 18	14
	CTRB188/DA	M23889	86 (81)	16	
	TRB15/AO	M18842	86 (81)	4	
BV15S1	V_β15/LEW	—	88 (79)	6	15
	CTRB29/DA[c]	M23885	—	16	
BV16S1	V_β16/LEW	—	88 (87)	6	16
	(allele in BN)	—	—	17	
BV17S1A1	V_β17/LEW	—	88 (77)	6	17
	CTRB3/DA	M23882	88 (77)	16	

(continued)

Table V.7.2 *Continued*

Most homologous mouse segment[a]	*Rat clone name or designation/rat strain*	*Accession No. (EMBL/GenBank)*	*Homology with mouse at nucleotide (amino acid) level*	*Source*	*Current numbering (family.member)*
BV18S1	V_β18/LEW	–	92 (89)	6	18
BV19S1	V_β19/LEW	–	88 (79)	6	19
	CTRB4/DA[c]	M23883	–	16	
BV20S1	V_β20/LEW	–	88 (92)	6	20
	CTRB9/DA[c]	–	–	16	
BV22P	V_β3.3/LEW	X80524	91 (75)	6, 18	3.3
BJ1S1	J_β1.1/LEW	X14973	93 (100)	6, 8	1.1
	Jβ1.1[b]/DA[c]	M63793	–	16	
	TR64/A0[c]	M18846	–	3	
BJ1S2	J_β1.2/LEW	–	93 (87)	6	1.2
	J_β1.2[c]/DA	M63793	–	5	
BJ1S3	J_β1.3/LEW	–	88 (88)	5	1.3
	J_β1.3[b]/DA	M63793	–	4	
	TRB91/AO	M18848	–	3	
BJ1S4	J_β1.4/LEW	–	90 (94)	6	1.4
	J_β1.4[b]/DA[c]	M63793	–	5	
BJ1S5	J_β1.5/LEW	X80525	78 (73)	6, 18	1.5
	J_β1.5[b]/DA[c]	M63793	–	5	
	TRB77/AO	M18847	–	4	
BJ1S6	J_β1.6/LEW	–	98 (100)	6	1.6
	J_β1.6[b]/DA[c]	M63793	–	5	
BJ2S1	J_β2.1/LEW	M74472	92 (88)	6, 7	2.1
	J_β2.1[b]/DA	M63794	92 (88)	5	
	TRB67/AO[c]	M18849	–	4	
BJ2S2	J_β2.2/LEW	X97672	98 (100)	6, 9	2.2
	J_β2.2[b]/DA	M63794	98 (100)	5	
	TRB89	M18850	98 (100)	4	
BJ2S3	J_β2.3/LEW	M74473	85 (69)	6, 7	2.3
	J_β2.3[b]/DA	M63794	85 (69)	5	
BJ2S4	J_β2.4/LEW	–	92 (94)	6	2.4
	J_β2.4/DA[b]	M63974	92 (94)	5	
BJ2S5	J_β2.5/LEW	X80523	92 (94)	6, 18	2.5
	J_β2.5[b]/DA[c] (Pseudogene)	M63974	–	5	
	TRB100/AO	M18841	92 (94)	4	
BJ2S6	J_β2.6/LEW	X80522	92 (94)	6, 7	2.6 (2.7)
	J_β2.6[b]/DA	M63974	92 (94)	5	
	TRB9 (J_β2.7)/AO	M18851	92 (94)	4	
BD1	D_β1[b]/DA	M63793	100 (100) 1bp shorter	5	
BD2	D_β2[b]/DA	M63794	100 (100) 1bp longer	5	

Sources: 1, Burns *et al.* (1989); 2, Blankenhorn *et al.* (1992); 3, Blankenhorn *et al.* (1991); 4, Morris *et al.* (1988); 5, Williams *et al.* (1991); 6, Smith *et al.* (1991); 7, Offner *et al.* (1991); 8, Chluba *et al.* (1989); 9, Asmuß *et al.* (1996); 10, Blankenhorn *et al.* (1995); 11, Herrmann *et al.* (1994); 12, Zhang *et al.* (1994); 13, Zhang and Heber-Katz (1992); 14, Hashim *et al.* (1991); 15, Gold *et al.* (1994); 16, Williams and Gutman (1989); 17, Smith *et al.* (1992); 18, Gold *et al.* (1995).

[a] WHO-IUIS nomenclature for BV segments (Arden *et al.*, 1995), nomenclature for BJ segments according to Gascoigne (1995).

[b] Genomic sequence.

[c] Protein difference from LEW sequence.

Table V.7.3 Sequences of cDNA clones encoding γ and δ chains. Putative genomic location of *Tcrg* on chromosome 17 (Yasue *et al.*, 1992)

Rat clone name	*Rat clone name (designation)[a]/ rat strain*	*Accession No. (EMBL/GenBank)*	*Homology to mouse at nucleotide (amino acid) level*	*Reference*
RG4	$V_\gamma 2/J_\gamma 1/C_\gamma 1$/F344	S75435	(87)/(79)/(91)	Kinebuchi *et al.* (1994)
RG7	$V_\gamma 1.1/J_\gamma/C_\gamma 4$/F344	S75437	(78)/(84)/(76)	Kinebuchi *et al.* (1994)
$DETC_\gamma$	$V_\gamma 3/J_\gamma 1$/LEW	—	88 (95)	Kühnlein *et al.* (1996)
$DETC_\delta$	$V_\delta 1/J_\delta 2/D_\delta 2$/LEW	—	88 (92)	Kühnlein *et al.* (1996)

[a] Nomenclature according to Garman *et al.* (1986).

Table V.7.4 Monoclonal antibodies to rat T-cell receptors

mAb	*Specificity[a]*	*References*
G99	At least two $V_\alpha 4$ family members (clones P2.3 and 4.36)	Stangel *et al.* (1994), Torres-Nagel *et al.* (1994a)
G177	$V_\alpha 8.2$	Torres-Nagel *et al.* (1994a)
R73	C_β	Hünig *et al.* (1989)
R78	$V_\beta 8.2^{1b}$ and $V_\beta 8.4^{ab}$	Torres-Nagel *et al.* (1993), Herrmann *et al.* (1994), Asmuß *et al.* (1996)
B73	$V_\beta 8.3$ ($V_\beta 8.5$)	Torres-Nagel *et al.* (1993)
G101	$V_\beta 10$	Torres-Nagel *et al.* (1993)
HIS42	$V_\beta 16$	Kampinga *et al.* (1989), Torres-Nagel *et al.* (1993)
V65	Pan $\gamma\delta$ TCR	Kühnlein *et al.* (1994)
V45	Subset of γ/δ TCR	Kühnlein *et al.* (1996)

[a] According to current numbering (see Tables V.7.1 and V.7.2).
[b] Detects the $V_\beta 8.2$ allele of LEW but not of DA and the $V_\beta 8.4$ allele of DA. $V_\beta 8.4$ LEW is a pseudogene.

Table V.7.5 Molecular interactions identified in the rat TCR complex

Molecule	*Associations/comments*	Reference
CD2	Coprecipitated in mild detergents, mechanism of association unknown	Beyers *et al.* (1992)
CD3	ε and ζ components identified. Non-covalently associated with TCR	Tanaka *et al.* (1989), Lawetzky *et al.* (1990), Beyers *et al.* (1992)
CD4	Via associated $p56^{lck}$	Beyers *et al.* (1992)
CD8	Via associated $p56^{lck}$	Beyers *et al.* (1992)
CD5	Coprecipitated in mild detergents, mechanism of association unknown	Beyers *et al.* (1992)
$p56^{lck}$	Interaction with CD4, CD8 and TCR/CD3 complex shown.	Beyers *et al.* (1992)
$p59^{fyn}$	Coprecipitated in mild detergents.	Beyers *et al.* (1992)

'waves' of γ/δTCR with restricted diversity have not yet been identified in rats, but the presence of thymus-dependent dendritic epidermal T cells with a canonical γ/δ TCR highly homologous to that of mice suggests similarities in γ/δ T-cell development (Kühnlein *et al.*, 1996).

Both in ontogeny and during adult T-cell development, $CD3^-4^-8^-$ thymocytes progress through an immature $CD4^-8^+$ stage which contains the direct precursors of $CD4^+8^+$ thymocytes (Paterson and Williams, 1987; Lawetzky *et al.*, 1991) from which the mature $CD4^+8^-$ and $CD4^-8^+$ α/β T-cell subsets are selected. The branchpoint for the α/β versus γ/δ lineage decision has not been exactly defined, but the absence of *TCRGC* mRNA from $CD4^+8^+$ thymocytes and its presence in their direct precursors, the immature $CD4^-8^+$ thymocytes suggests that it can occur up to that stage (H. J. Park and T. Hünig, unpublished data). Immature $CD4^-8^+$ thymocytes also contain *TCRBC* but not *TCRAC* mRNA (H. J. Park and T. Hünig, unpublished data), and exhibit weak cell surface reactivity with the TCRβ-specific mAbs R73 (Hünig, 1988). Presumably, the TCR detected on these cells is the pre-TCR consisting of a mature β and a pre-Tα chain(s). When $CD4^-8^+$ immature thymocytes progress to the $CD4^+8^+$ stage *in vitro*, RAG-1 expression is reinitiated and *TCRAC* mRNA appears (H. J. Park and T. Hünig, unpublished data). $CD4^+8^+$ cells express $\alpha\beta$TCR at their surface at approximately 1/5 of the density of mature T cells (Hünig *et al.*, 1989). Thymic selection of the TCR repertoire has been shown in terms of both MHC class-specific and allele-specific 'overselection' of V_β and V_α segments (Torres-Nagel *et al.*, 1994b). No evidence for thymic negative selection by endogenous superantigens has been obtained, although the specificity of rat V_β segments for mouse *mtv*-encoded and bacterial superantigens has been defined (Herrmann *et al.*, 1994; Surh *et al.*, 1994). Athymic rats contain reduced numbers of $\alpha\beta$ T-cells which increase with age. Similar, to the situation in mice, the interindividual variation of V-gene usage indicates that the extrathymic TCR repertoire is oligoclonal. This also holds true for $\alpha\beta TCR^+$ intestinal intraepithelial lymphocytes (iIEL), although the extrathymic origin of these cells has not been formally proven (N. Torres-Nagel *et al.*, unpublished data).

Distribution and phenotype of $\alpha\beta$ and $\gamma\delta$ T-cell populations

$\alpha\beta$ T cells

$\alpha\beta$ T cells are present in peripheral lymphoid organs and blood in frequencies and CD4/8 ratios comparable to those found in mice and humans (Hünig *et al.*, 1989). In contrast to mice but as in humans, activated CD4 T-cells can express CD8$\alpha\alpha$ (Torres-Nagel *et al.*, 1992) and MHC II (Broeren *et al.*, 1995; Seddon and Mason, 1996). A small subset coexpresses the NK cell activation receptor NKR-P1 (Brisette-Storkus *et al.*, 1994). $\alpha\beta$ T cells predominate among iIEL where most of them lack CD2 (Fangman *et al.*, 1991) and CD8β (Torres-Nagel *et al.*, 1992), and $CD4^+CD8\alpha\alpha$ cells are a major subset (Torres-Nagel *et al.*, 1992).

$\gamma\delta$ T cells

$\gamma\delta$ T cells are a small minority in thymus, peripheral lymphoid organs, blood and (in contrast to mice, where they are much more frequent) also among iIEL (Lawetzky *et al.*, 1990; Kühnlein *et al.*, 1994, 1995). Whereas in lymph nodes and spleen, 90% are $CD2^+CD4^-CD8\alpha\beta^+CD5^+$, those of the gut epithelium are $CD2^-CD4^-CD8\alpha^+\beta^-$ $CD5^-$ (Kühnlein *et al.*, 1994). In the skin, thymus-dependent dendritic epidermal T cells (DETC) express a canonical receptor highly homologous to that of mice (Kühnlein *et al.*, 1996). This cell type is not found in humans (Elbe *et al.*, 1996).

8. B-Cell Development

B-cell generation in adult bone marrow

As in other mammalian species, most if not all generation of new, virgin (or naive) B cells in the rat takes place in the bone marrow (BM). At this location surface IgM^+ cells are formed in extremely large numbers from pluripotent hemopoietic stem cells, through a series of steps that include proliferation and differentiation of precursor cells. During these steps immunoglobulin heavy and light chain genes are rearranged and expressed followed by selection of functional B cells. Rat pluripotent hemopoietic stem cells express high levels of Thy-1 antigen (CD90) and CD43 (leukosialin, sialophorin) only weakly; they lack the leukocyte common antigen (L-CA)/CD45R molecules detected by mAbs OX-22 and HIS24 (Goldschneider *et al.*, 1978; Hunt, 1979; McCarthy *et al.*, 1987). Various stages of B-cell development can be distinguished phenotypically in rat BM (Hunt *et al.*, 1977; Opstelten *et al.*, 1986; Kroese *et al.*, 1990, 1995; McKenna and Goldschneider, 1993; McKenna *et al.*, 1994; Hermans *et al.*, 1997). A model of virgin B-cell development is shown in Figure V.8.1. All B-lymphoid precursor cells express CD45R/HIS24 and Thy-1. The heat-stable antigen (HSA)/CD24, recognized by mAb HIS50 (Hermans *et al.*, 1997) is expressed on all B-lineage cells in the BM, except for the earliest B-cell precursors, the pre-pro-B cells. Pre-pro-B cells thus are $Thy\text{-}1^+CD45R^+HSA^-$ cells and represent less than 2% of total nucleated BM cells. These cells are thought to give rise to $Thy\text{-}1^+CD45R^+HSA^+$ cells (pro-B cells; 5% of all nucleated BM cells) which still have no detectable levels of cytoplasmic or surface Ig μ chains. Approximately half of the $Thy\text{-}1^+CD45R^+HSA^+$ (but μ-$chain^-$) cells in the BM expresses detectable levels of the nuclear enzyme terminal deoxynucleotidyl transferase (TdT) which plays a role during immunoglobulin gene rearrangements and contributes to the diversity of Ig V regions. Conversely, more than 95% of the TdT-expressing cells have this phenotype ($Thy\text{-}1^+CD45R^+HSA^+\mu^-$). The next stage of B lymphopoiesis is represented by the pre-B cells (20% of all nucleated BM cells). These cells are also $Thy\text{-}1^+CD45R^+HSA^+$ but are characterized by the presence of free μ-chains in their cytoplasm only. Finally (small) pre-B cells differentiate to newly formed (immature) B cells (NF-B cells), which express high levels of surface IgM, but little or no surface IgD (sIgD) (and still are $Thy\text{-}1^+CD45R^+HSA^+$) (Crawford and Goldschneider, 1980; Kroese *et al.*, 1990, 1995; Deenen and Kroese, 1993; Soares *et al.*, 1996; Hermans *et*

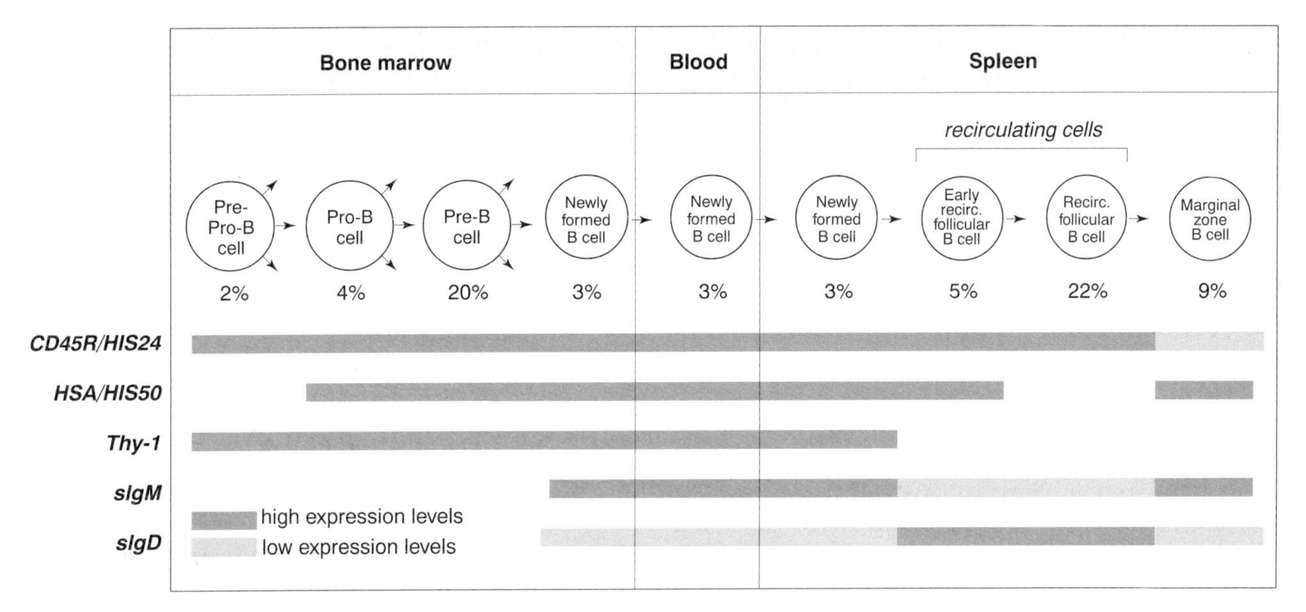

Figure V.8.1 A model of virgin B-cell development in the rat. For explanation see text. Numbers represent proportion of cells relative to total nucleated cells (bone marrow) or lymphoid cells (blood, spleen) in PVG rats.

al., 1997). These cells leave the BM and enter the blood circulation.

The total number of cells produced by the precursor B cell compartment (CD45R$^+$sIgM$^-$) is approximately 650 million cells per day (Deenen *et al.*, 1987, 1990). Most cells are produced in the compartment of (large) pre-B cells (approximately 320 million cells per day). The sum of all peripheral B cells per rat is in the order of 10^9 cells (Gray, 1988). Together with the notion that mature B cells turn over slowly (1–2% per day) (Gray, 1988; Chan and MacLennan, 1993; Deenen and Kroese, 1993; Kroese *et al.*, 1995), the observed production rate of B cells in the BM is far beyond the need to supply the animal with new mature cells. Indeed, the entire peripheral B-cell pool is reconstituted in a few days in B-cell depleted (anti-IgM treated) rats (Bazin *et al.*, 1985). In physiological situations most B cells produced in the BM never reach their final stage and die owing to nonproductive rearrangements or selection mechanisms leading to apoptosis. Information about the molecular events that take place during B-cell generation in rat BM is extremely limited. Some evidence suggests that rat pre-B cells (and part of the TdT$^+$ cells) may have already rearranged both their heavy chains and κ light chains, albeit that κ chains are not yet expressed as proteins (Opstelten *et al.*, 1985; Hunt *et al.*, 1988). Recent studies have shown that the homolog of the murine $\lambda 5$ gene also exists in the rat (Dammers *et al.*, unpublished data). This gene encodes for part of the surrogate light chain and plays a decisive role in B cell development (Melchers *et al.*, 1993). Similar to mice, the rat $\lambda 5$ equivalent is expressed (mRNA) in pre- and pro-B cells but not in mature B cells.

B-cell genesis takes place within the meshes of the three-dimensional network of stromal elements of the BM. Immunohistological analysis of rat BM reveals that the subendosteal area is enriched in Thy-1$^+$ cells, pre-B cells and TdT$^+$ cells (Hermans *et al.*, 1989; Hermans and Opstelten, 1991), suggesting that most B-cell genesis occurs in the peripheral parts of the BM, giving rise to a progeny that differentiates and migrates to more centrally located areas of the BM to exit via the endothelium of the sinus walls. Studies in chimeric rats have shown that once formed, B cells rapidly disperse through the BM (Hermans *et al.*, 1992).

B-cell generation in fetal liver

During fetal and neonatal life B cells are also generated in the liver (Veldhuis and Opstelten, 1988). Pre-B cells appear by day 17 of gestation followed by IgM$^+$ B cells which are first detected 2 days later. B-cell genesis in fetal and neonatal rat liver proceeds, similar to mice, without expression of TdT. Fetal liver B-cell lymphopoiesis may represent a developmental pathway, distinct from adult B-cell development in the BM (Lam and Stall, 1994).

Peripheral virgin B-cell differentiation

The majority of mature, resting B cells are recirculating follicular B cells (RF-B) which express high levels sIgD and low levels sIgM, recirculate through the body and are located in follicles of lymphoid organs (Bazin *et al.*, 1982; Gray *et al.*, 1982; MacLennan *et al.*, 1982; Kroese *et al.*, 1990). In rats, mature B cells lack Thy-1/CD90 expression, which distinguishes them from their Thy-1$^+$ immature predecessors (Crawford and Goldschneider, 1980; Deenen and Kroese, 1993; Soares *et al.*, 1996). Two subsets of immature, Thy-1$^+$ B cells can be distinguished in the periphery on the basis of relative levels of sIgM and sIgD: newly formed B cells (NF-B cells) (Thy-1^{+}IgM^{bright}-IgD^{dull}) and early recirculating B cells (ERF-B cells) (Thy-1^{+}IgM^{dull}IgD^{bright}). These subsets probably represent two sequential stages of B-cell differentiation towards mature (Thy-1^{-}) B cells (Kroese *et al.*, 1995) (see Figure V.8.1). NF-B cells are probably not yet fully immunocompetent (Whalen and Goldschneider, 1993) and have only a limited capability to recirculate, as illustrated by their absence from thoracic duct lymph and lymph nodes. Recent BM emigrants have restricted capacity to enter follicles of secondary lymphoid organs (Lortan *et al.*, 1987). Most likely, NF-B cells differentiate subsequently to a second immature Thy-1^{+} B-cell subset (ERF-B cells), which start to recirculate (as witnessed by their presence in all lymphoid organs, blood and thoracic duct lymph) and become IgM^{dull}IgD^{bright}, similar to mature, RF-B cells, giving them their name. Finally, ERF-B cells are thought to mature further, downregulate Thy-1 expression, and become mature RF-B cells (Thy-1^{-}IgM^{dull}IgD^{bright}).

Differentiation from pre-B cells to mature RF-B cells proceeds without cell divisions (Kroese *et al.*, 1995). In terms of absolute numbers of cells significant cell loss is observed at the transition of the NF-B cell stage in the BM and those in the periphery, probably reflecting antibody repertoire selection mechanisms (see Figure V.8.2). At later stages of their differentiation towards mature RF-B cells relatively few cells die. At the level of V_H gene usage there is no evidence for extensive selection between ERF-B cells and RF-B cells (Vermeer, 1995). Once cells reach the RF-B cell stage they become relatively long-lived with a lifespan in the order of months and a turnover rate of only 1–2% per day (Chan and MacLennan, 1993; Deenen and Kroese, 1993; Kroese *et al.*, 1995).

Marginal zone B cells

In addition to follicles, a significant proportion of splenic B cells is located in 'marginal zones' (MZs), which surround lymphoid follicles and T-cell areas in spleen. MZs are absent in peripheral lymph nodes and Peyer's patches. In rats in particular, MZ can be distinguished easily and are prominently present in their spleens (MacLennan *et al.*, 1982). Rat MZ-B cells are slightly larger cells, are IgM^{bright}IgD^{dull} and Thy-1^{-} and characteristically express low levels of the CD45R determinant recognized by monoclonal antibody HIS24 (Bazin *et al.*, 1982; Gray *et al.*, 1982; Kroese *et al.*, 1985, 1990, 1995) (see Figure V.8.1). They also express CD21 (CR2, C3d receptor) and CD35 (CR1, C3b receptor) (Gray *et al.*, 1984). A newly developed mAb, HIS57, reacts strongly with MZ-B cells and only weakly with a minor proportion of RF-B cells, but not with other B-lineage cells (Deenen *et al.*, 1997). MZ-B cells are sessile cells and some (circumstantial) evidence indicates that they may play a role in antibody responses to TI-2 antigens (MacLennan *et al.*, 1982; Lane *et al.*, 1986). In addition to virgin (i.e. antigen-inexperienced) MZ-B cells, memory cells also appear to be present in splenic MZ (Liu *et al.*, 1988, 1991). The origin of (virgin) MZ-B cells is incompletely known. It has been speculated that they may belong to a distinct lineage of B cells (Bazin *et al.*, 1982; MacLennan *et al.*, 1982). However, there is evidence indicating that (virgin) MZ-B cells are directly derived from RF-B cells (Kumamaratne and MacLennan, 1981; Kroese *et al.*, 1986; Lane *et al.*, 1986; MacLennan and Gray, 1986; De Boer, 1994).

B-1 cells in the rat?

In the mouse a minor, but important, subpopulation of B cells is constituted by the B-1 cells (also called Ly-1 B cells or CD5^{+} B cells). These cells differ from the main population of BM-derived conventional B cells (or B-2 cells) in many respects (for review see Kantor and Herzenberg, 1993). B-1 cells of the mouse are strongly enriched in peritoneal and pleural cavities and have a distinct phenotype, being IgM^{bright}IgD^{dull}CD43^{+}CD11b^{+}. Part of them expresses CD5 (B-1a cells) and another part bears no detectable CD5 (B-1b or 'sister' cells). B-1 cells might well be a distinct B-cell lineage: they arise during ontogeny from progenitor cells located in fetal omentum and fetal liver, while in the adult animal they are largely self-replenishing and not significantly produced by the BM. Antibodies produced by B-1 cells tend to have a low

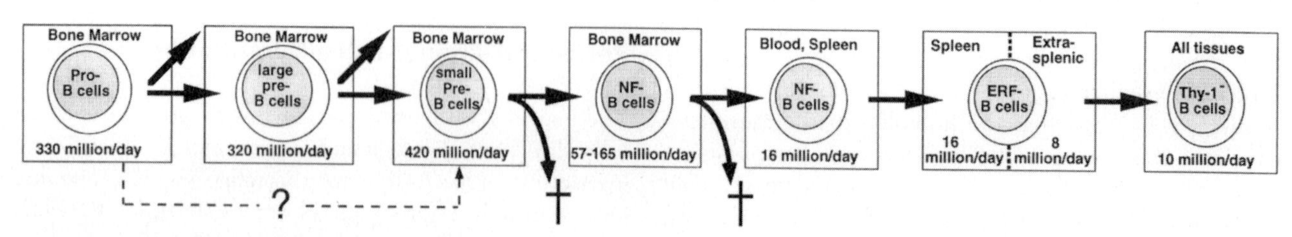

Figure V.8.2 Model of B (precursor) cell dynamics in bone marrow and periphery in the rat. B (precursor) cells are defined by HIS24 (B220/CD45R), TdT, cytoplasmic μ chains, sIgM, sIgD and Thy-1 (see text for explanation of subsets and differentiation pathway). Pro-B cells and large pre-B cells are cycling cells, whereas the other B (precursor) cells are non-dividing. The model shows that most of the generated B cells die at the transition between the stage of the small pre-B cells and the stage of the Newly Formed (NF) B cells. Data are obtained from Deenen *et al.* (1987, 1990), Deenan and Kroese (1993) and Kroese *et al.* (1995), and represent production rate in the cycling compartments and renewal rates in the non-dividing compartments.

affinity, are multireactive and are frequently directed to autoantigens and bacteria-related determinants. To date, no B-1 cells have been unequivocally detected in rats (Kroese *et al.*, 1990; De Boer *et al.*, 1994; Vermeer *et al.*, 1994). In adult rats a distinct subpopulation of CD5 expressing ($sIgM^+$) cells appears to be absent, while in neonatal rat spleen all B cells express low amounts of CD5, which decrease gradually with age. In marked contrast with mice, only ≈1% of cells recovered from the rat peritoneal cavity are B cells, which hampers detection of B-1 lineage cells in the rat. This relative absence of B cells in the peritoneal cavity of the rat is not due to intrinsic properties of rat lymphoid cells since transfer of rat fetal liver cells into SCID mice (Surh and Sprent, 1991) results in high numbers of rat B cells in the peritoneal cavity of the chimera (Deenen *et al.*, 1997). A significant proportion of peritoneal B cells in these xenogeneic chimeras have a unique phenotype not detected elsewhere. These cells resemble MZ-B cells being $IgM^{high}IgD^{low}CD45R/HIS24^{low}$ $HIS57^{high}$, but differ from them in their high expression levels of Thy-1 (Deenen *et al.*, 1997). This newly identified B cell subset might be a candidate for the homologue of B-1 cells in the mouse.

In vivo humoral immune response

In vivo antigenic stimulation of mammals leads to two types of humoral immune responses in their secondary lymphoid organs: (1) the generation of antibody-secreting plasma cells (i.e. plasmacellular reaction), and (2) the generation of germinal centers for memory B-cell production (i.e. germinal center reaction). In rats, antigen (hapten)-specific antibody-forming cells have been detected in lymphoid tissues by a few days after immunization through the appropriate route both with T-cell dependent and T-cell independent antigens (van der Brugge-Gamelkoorn *et al.*, 1986; Claassen *et al.*, 1986; Liu *et al.*, 1991; van der Dobbelsteen *et al.*, 1992). Initially, these antibody-forming cells are located in the interdigitating cell-rich areas of the T-cell zones, and later they are predominantly found in red pulp of the spleen or medulla of the lymph nodes. Rat plasma cells lack CD45R/HIS24 but are $CD43^+$ positive (Kroese *et al.*, 1987a) and these cells may be either short-lived or long-lived (Ho *et al.*, 1986). As in other species, rat germinal centers arise in B-cell follicles as a cluster of proliferating (antigen-specific) B blasts between the meshes of the long and slender processes of immune complex trapping follicular dendritic cells (Liu *et al.*, 1991). Studies in rats were the first to demonstrate that *de novo* germinal centers develop oligoclonally from one to three precursor cells (Kroese *et al.*, 1987b). Rat germinal center B-cells have a distinct phenotype and are characterized by the absence of sIgD and CD45R/HIS22 and their expression of CD45R/HIS24 and HSA/HIS50 (Bazin *et al.*, 1982; Kroese *et al.*, 1985; Hermans *et al.*, 1997). There is evidence that the germinal center precursor cells may be found both among $sIgD^+$ (Seijen *et al.*, 1988) and $sIgD^-$ (Vonderheide and Hunt, 1991) B-cell populations. Virtually nothing is known about the molecular events such as somatic hypermutation, isotype switching, selection and apoptosis that are thought to take place during the germinal center reaction and the formation of memory B cells in rats.

9. Rat Immunoglobulins

Nomenclature

Grabar and Courcon (1958) and Escribano and Grabar (1962), using the newly discovered immunoelectrophoretic analysis described the first data on the rat serum proteins. Two lines of precipitation were identified in the gamma-globulin area and were supposed to be rat immunoglobulins. Arnason *et al.* (1963, 1964) later identified three proteins with antibody activities and called them IgM, IgX and IgG. The IgM protein was characterized by its susceptibility to cysteine and its molecular weight. In these publications (Arnason *et al.*, 1963, 1964), two other antibody isotypes were shown, IgX and IgG, which were later called IgA and IgG (Arnason *et al.*, 1964). As in the mouse species, the present IgG1 isotype was misnamed and called IgA. Attention must be paid to publications of that time, in which the IgG1 isotype has the wrong appellation of IgA (see Table V.9.1). Finally, the various IgG subisotypes were correctly identified, IgG1 (Jones, 1969), IgG2a and

Table V.9.1 Historical nomenclature of the rat immunoglobulin (sub)isotypes

Reference	*IgM*[a] *(γM,γ₁M,β₂M)*	*IgA* *(β₂A,γ₁A,γA)*	*IgD* *(γD)*	*IgE* *(γE)*	*IgG1*	*IgG2a*	*IgG2b* *(7Sγ,γss, or γG)*	*IgG2c*
Arnason *et al.* (1963)	IgM				IgX	IgG		
Arnason *et al.* (1964)	IgM				IgA	IgG		
Binaghi and Sarando de Merlo (1966)						IgGa	IgGb	
Jones (1969)	γM	γA			7Sγ1	7Sγ2		
Bazin *et al.* (1973)	IgM	IgA		IgE	IgG1	IgG2a	IgG2b	IgG2c
Bazin *et al.* (1978)			IgD					

[a] Synonyms (old usage) are given in brackets.

IgG2b and IgG2c (Bazin *et al.*, 1974a). IgA was recognized by a cross reaction between rat serum and a strong polyclonal antimouse IgA (Nash *et al.*, 1969) and then shown to have the main physicochemical and biological properties of human or mouse IgA. IgE was distinguished by the reaginic properties of immunized rats which were similar to those of man. Stechschulte *et al.* (1970) and Jones and Edwards (1971) published a rat immunoglobulin equivalent to human IgE. This result was confirmed by Bazin *et al.* (1974b) with the help of IgE myeloma protein from the LOU rat strain. The first indication of a rat IgD isotype was given by the binding between a chicken serum antihuman IgD and the membrane of rat lymphocytes (Ruddick and Leslie, 1977). Rat IgD was fully characterized with the help of LOU rat IgD monoclonal immunoglobulins (Bazin *et al.*, 1978).

According to the recommendation of the committee of immunoglobulin nomenclature of the World Health Organization (1973), the two initial letters 'Ig' must be adopted for all immunoglobulin isotypes and all species. All the letters must also be written on the same line, thus IgG1 and not IgG_1. The division of IgG isotype into subisotypes (or subclasses) based on their electrophoretic mobility, can be maintained even at the present stage of knowledge.

The rat IgM, IgD, IgG, IgA and IgE isotypes are the equivalent of their human or mouse homologues, at least, most of their physicochemical and biological properties are similar. Rat light chains were characterized by Querinjean *et al.* (1972, 1975) who established the presence of κ and λ chains as described for other species (Hood *et al.*, 1967).

However, if most of the physicochemical and biological properties of the five immunoglobulin isotypes are common to all mammal species in which they have been studied (Bazin *et al.*, 1990), it is difficult to give an exact correlation between the IgG subclasses from different species (Figure V.9.1). From work by Der Balian *et al.* (1980) and Nahm *et al.* (1980) on biological properties, by Brüggemann *et al.* (1988) on sequence homology, and Bazin *et al.* (unpublished data) on antigenic cross-reactivity, rat IgG2c and mouse IgG3 immunoglobulins are clearly equivalent. Other IgG subisotypes are more difficult to correlate. Bazin (1967a, b) showed that when thymodependent antigens such as human serum albumin or horse ferritin are given to mice by the intraperitoneal route, antibodies of the IgG1 isotype appear first followed by IgG2. Similarly, thymodependent antigen (mouse monoclonal antibody antirat IgD) injected to rats gives predominantly antibodies of the IgG1 isotype (Soares *et al.*, 1996), which indicates that T-dependent immune response, both in mice and rats, produces mostly IgG1 antibodies.

Molecules of the same immunoglobulin isotype (or subisotype) may have different amino acid sequences and may express different antigenic determinants. They can present two or more antigenic forms which are called allotypes in the Oudin terminology (1956), and individuals of the same species may be divided into as many allotypic groups as antigenic forms. Each individual expresses one allotype. Allotypic specificities correspond to structural differences determined by genes which are inherited as Mendelian characters. Immunoglobulin genes are codominant and, as such, always expressed in a heterozygote; however every cell expresses only one of the allelic genes. Four allotypes have been identified in the rat species.

The first rat allotype discovered was reported by Barabas and Kelus (1967) immunizing, 'black and white hooded rats' with Wistar immunoglobulins. Later, Wistar (1969) showed the presence of an allotypic difference in rat light chain immunoglobulins. This difference was confirmed by others (Armeding, 1971; Gutman and Weisman, 1971; Humphrey and Santos, 1971). Rocklin *et al.* (1971) suggested that this marker was located on the κ light chains, which was confirmed by Beckers *et al.* (1974) and Nezlin *et al.* (1974) using monoclonal and polyclonal purified κ light chains, respectively. A common nomenclature adopted by the scientists working in the field (Gutman *et al.*, 1983) defines an allotype called IgK-1 and two alleles, IgK-1a (reference strain, LOU/C) and IgK-1b (reference strain, DA). The differences between the two alleles were studied by Gutman *et al.* (1975) who described one sequence gap and many amino acid substitutions

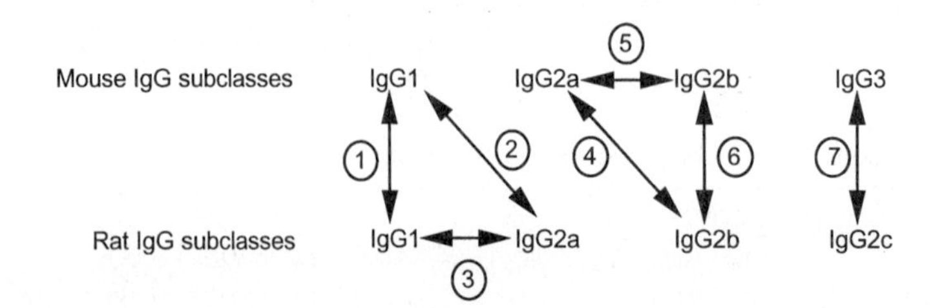

Figure V.9.1 Correlation between mouse and rat IgG subclasses (Bazin *et al.*, 1990). The various homologies are respectively supported: (1) antigenic cross-reactions (Carter and Bazin, 1980) and sequence homology (Brüggemann *et al.*, 1986; Brüggemann, 1988); (2) homocytotropic antibodies and sequence homology (Brüggemann *et al.*, 1986; Brüggemann, 1988); (3) antigenic crossreaction (H. Bazin, unpublished data) and sequence homology (Brüggemann, 1988); (4) sequence homology (Brüggemann *et al.*, 1986; Brüggemann, 1988); (5) sequence homology (Ollo *et al.*, 1981; Brüggeman, 1988); (6) sequence homology (Brüggeman, 1988); (7) homologies based on physicochemical and biological properties (Der Balian *et al.*, 1980; Nahm *et al.*, 1980; Brüggeman, 1988; Brüggemann *et al.*, 1988).

Table V.9.2 Allotypic markers in normal and congenic rat strains

Rat strains	Genetic background	Number of backcrosses	Allotypic markers		
			IgK-1	*IgH-1*	*IgH-2*
LOU/C	LOU/C	—	a	a	a
LOU/C[a].IgK-1b(OKA)	LOU/C	12	b	a	a
LOU/C.IgH-2b(OKA)	LOU/C	12	a	b	b
OKA	OKA	—	b	b	b
PVG-1a[b]	PVG/c	—	a		
PVG-1[b]	PVG/c	15	b		
DA	DA	—	b		

[a]From H. Bazin and A. Beckers (1976) *Molecular and Biological Aspects of the Acute Allergic Reaction*, p. 125. With permission.
[b]Hunt and Gowans, personal communication.

between the κ chains of the Lewis and DA rat strains. Bazin *et al.* (1974c) described another allotype located on the rat α heavy-chain of immunoglobulins. This allotype is now called IgH-1 with two alleles: IgH-1a (reference strain, LOU/C) and IgH-non1a (reference strain, OKA). The IgH-1a determinant is located on 100% of the α chains from LOU/C rats. Beckers and Bazin (1975) described an antigenic difference for the rat γ2b heavy chains (IgH-2) with two alleles IgH-2a (reference strain, LOU/C) and IgH-2b (reference strain, OKA), according to the commonly adopted nomenclature (Gutman *et al.*, 1983). A third allotype of rat immunoglobulin heavy chains has been associated with the rat IgG2c subisotype (Leslie, 1984) and designated IgH-3 with two alleles – IgH-3a on the heavy chain of COP, F344, BN, DA and IgH-3b on Wistar/Fu, LOU and SHR rat strains. IgH-1 and IgH-2 loci are linked to each other (Beckers and Bazin, 1975) but not to the IgK-1 locus (Beckers *et al.*, 1974). The heavy chain genes are not linked to the gene(s) coding for sex, coat pattern and color and eye color (Bazin *et al.*, 1974c). The IgK-1 locus is not linked to the major histocompatibility complex and three coat-color genes (Gutman and Weissman, 1971). Congenic strain rats for kappa Ig allotypes have been developed as shown in Table V.9.2 and can be used as markers for B cells.

10. Rat Ig Variable Region Genes

Generation of immunoglobulin diversity

As in other mammalian species, in rats the diversity of the B-cell receptor is generated by recombination of many different genetic elements before the complete antibody is expressed. Diversity is generated by the combination of heavy and light chains, each consisting of a variable and a constant region. The variable region of a heavy chain is constituted from a variable (V_H), diversity (D) and joining (J_H) gene segment, while the variable region of a light chain is a combination of only a variable (V_L) and joining (J_L) gene segment. The constant regions encode the different isotypes. In rats the isotypes IgM, IgD, IgG1, IgG2a, IgG2b, IgG2c, IgA and IgE are identified (Bazin *et al.*, 1974). The constant regions encoding the IgM, IgG1, IgG2a, IgG2b, IgA and IgE molecules have been (partly) sequenced (Hellman *et al.*, 1982a, b; Brüggemann *et al.*, 1986; Brüggemann, 1988), and the constant genes for the kappa and lambda light chains are also known (Querinjean *et al.*, 1975; Sheppard and Gutmann, 1981, 1982). Genomic sequence data have shown that there are at least three rat λ-like constant region genes, of which only Igλ-C2 seems to be functional (Steen *et al.*, 1987; Frank and Gutman, 1988). During early B-cell development the heavy-chain genes are first rearranged, first by D–J joining, after which the V–DJ joining is completed. The different recombination signals (RSS) necessary in this process are heptamer and nonamer sequences which are either separated by 12 or 23 base pair spacers. In the rat, we recently confirmed the presence of such RSS signals at the 5′ end of nonfunctionally rearranged D–J genes (Vermeer, 1995). The RSS sequences at both sides of the J_H gene segment and of the most 3′ DQ52 element have also been completely described at the genomic level (Brüggemann *et al.*, 1986; Lang and Mocikat, 1991).

After complete assembly of the heavy chain, it is first combined with a surrogate light chain protein encoded by a combination of λ5 and V-preB genes. Recently, the rat λ5 gene was cloned and sequenced (Dammers *et al.*, 1998a). After successful expression of these proteins, the light chain genes are rearranged and either lambda or kappa light chains are combined with the heavy chain to complete the expression of the B cell receptor. The accessory α and β (CD79a and b) proteins that accompany the membrane Ig molecule in the mouse and in humans have not yet been described in the rat.

Ig heavy-chain variable region sequences

Available rat heavy chain variable region (V_H) sequences reveal a high degree of similarity to the homologous mouse

Table V.10.1 Rat variable heavy chain genes

V_H gene family	*Clone name or designation*	*Strain*	*Configuration*	*Isotype*	*Accession number*	*Source*
3609	ERF2.21	PVG/E	R	IgM	Z93360	Vermeer (1995)
	ERF2.67	PVG/E	R	IgM	Z93368	Vermeer (1995)
J558	11-160	CBH/Cbi	R	IgG2b	X91994	Léger *et al.* (1995)
	53R-1	—	R	IgG2a	L07396	Karp (1993)
	56R-3	—	R	IgG2a	L07398	Karp (1993)
	53R-4	—	R	IgG2a	L07397	Karp (1993)
	56R-9	—	R	IgG1	L07400	Karp (1993)
	ALN/9/94	CBH/Cbi	R	IgG2b	X91996	Léger *et al.* (1995)
	ERF1.25	PVG/E	R	IgM	Z93350	Vermeer (1995)
	ERF1.72	PVG/E	R	IgM	Z93355	Vermeer (1995)
	ERF2.13	PVG/E	R	IgM	Z93359	Vermeer (1995)
	ERF2.37	PVG/E	R	IgM	Z93363	Vermeer (1995)
	GK1.5	Lewis	R	—	M84149	A. Rashid *et al.* (unpublished data)
	GLVH6	PVG/E	G	—	X86661	Vermeer (1995)
	GLVH7	PVG/E	G	—	X86667	Vermeer (1995)
	GLVH9	PVG/E	G	—	X86665	Vermeer (1995)
	GLVH17	PVG/E	G	—	X86678	Vermeer (1995)
	GLVH18	PVG/E	G	—	X86663	Vermeer (1995)
	GLVH19	PVG/E	G	—	X86664	Vermeer (1995)
	GLVH28	PVG/E	G	—	X86668	Vermeer (1995)
	MEC6	DZB	R	IgM	X78893	Aten *et al.* (1995), Dammers *et al.* (1998c)
	MEC7	DZB	R	IgM	X68782	Aten *et al.* (1995), Dammers *et al.* (1998c)
	RNIGVHL	BN	G	—	L07403	Karp (1993)
	RNIGVIL	BN	G	—	L07404	Karp (1993)
	TDL31	PVG/E	R	IgM	[a]	Vermeer (1995)
	YFC51.1.1	DA	R	—	M87787	Sims *et al.* (1991)
J606	ERF3.23	PVG/E	R	IgM	Z93370	Vermeer (1995)
PC7183	ERF1.1	PVG/E	R	IgM	Z93349	Vermeer (1995)
	ERF1.82	PVG/E	R	IgM	Z93356	Vermeer (1995)
	ERF1.83	PVG/E	R	IgM	Z93357	Vermeer (1995)
	ERF2.4	PVG/E	R	IgM	Z93364	Vermeer (1995)
	ERF2.12	PVG/E	R	IgM	Z93358	Vermeer (1995)
	ERF2.29	PVG/E	R	IgM	Z93362	Vermeer (1995)
	ERF2.52	PVG/E	R	IgM	Z93366	Vermeer (1995)
	ERF2.74	PVG/E	R	IgM	Z93369	Vermeer (1995)
	ERF3.8	PVG/E	R	IgM	Z93371	Vermeer (1995)
	GLVH4	PVG/E	G	—	X86672	Vermeer (1995)
	GLVH10	PVG/E	G	—	X86680	Vermeer (1995)
	GLVH11	PVG/E	G	—	X86670	Vermeer (1995)
	GLVH12	PVG/E	G	—	X86674	Vermeer (1995)
	GLVH15	PVG/E	G	—	X86673	Vermeer (1995)
	GLVH16	PVG/E	G	—	X86671	Vermeer (1995)
	GLVH20	PVG/E	G	—	X86658	Vermeer (1995)
	GLVH21	PVG/E	G	—	X86662	Vermeer (1995)
	GLVH24	PVG/E	G	—	X86669	Vermeer (1995)
	GLVH25	PVG/E	G	—	X86676	Vermeer (1995)
	GLVH26	PVG/E	G	—	X86677	Vermeer (1995)
	GLVH27	PVG/E	G	—	X86675	Vermeer (1995)
	HC4A	Lewis	R	IgG2a	L22652	M. A. Agius and S. Bharati (unpublished data)
	Hg12	BN	R	IgE	Z75897	Hirsch *et al.* (1986), Varga *et al.* (1995), Dammers *et al.* (1998c)

(*continued*)

Table V.10.1 *Continued*

V_H gene family	Clone name or designation	Strain	Configuration	Isotype	Accession number	Source
PC7183	Hg15	BN	R	IgG1	Z75898	Hirsch *et al.* (1984), Guéry *et al.* (1990), Dammers *et al.* (1998c)
	Hg16	BN	R	IgG1	Z75899	Guéry *et al.* (1990), Dammers *et al.* (1998c)
	Hg17	BN	R	IgG1	Z75900	Guéry *et al.* (1990), Dammers *et al.* (1998c)
	Hg33	BN	R	IgE	Z75903	Hirsch *et al.* (1986), Varga *et al.* (1995), Dammers *et al.* (1998c)
	MEC1	DZB	R	IgG2a	Z75904	Aten *et al.* (1995), Dammers *et al.* (1998c)
	MEC2	DZB	R	IgG1	Z75905	Aten *et al.* (1995), Dammers *et al.* (1998c)
	MEC3	DZB	R	IgM	X78897	Aten *et al.* (1995), Dammers *et al.* (1998c)
	MEC8	DZB	R	IgM	X78895	Aten *et al.* (1995), Dammers *et al.* (1998c)
	NC1/34HL	—	R	IgG	M62827	Piccioli *et al.* (1991)
	PC3	PVG/E	G	—	A002461	Dammers *et al.* (1998b)
	PC4	PVG/E	G	—	A002462	Dammers *et al.* (1998b)
	PC6	PVG/E	G	—	A002463	Dammers *et al.* (1998b)
	PC9	PVG/E	G	—	A002464	Dammers *et al.* (1998b)
	PC10	PVG/E	G	—	A002465	Dammers *et al.* (1998b)
	PC11	PVG/E	G	—	A002466	Dammers *et al.* (1998b)
	PC12	PVG/E	G	—	[a]	Dammers *et al.* (1998b)
	PC13	PVG/E	G	—	A002467	Dammers *et al.* (1998b)
	PC17	PVG/E	G	—	A002468	Dammers *et al.* (1998b)
	PC18	PVG/E	G	—	A002471	Dammers *et al.* (1998b)
	PC19	PVG/E	G	—	A002469	Dammers *et al.* (1998b)
	PC20	PVG/E	G	—	A002472	Dammers *et al.* (1998b)
	TDL5	PVG/E	R	IgM	[a]	Vermeer (1995)
	TDL14	PVG/E	R	IgM	[a]	Vermeer (1995)
	TDL15	PVG/E	R	IgM	[a]	Vermeer (1995)
	TDL21	PVG/E	R	IgM	[a]	Vermeer (1995)
	TDL27	PVG/E	R	IgM	[a]	Vermeer (1995)
	TDL28	PVG/E	R	IgM	[a]	Vermeer (1995)
	TDL30	PVG/E	R	IgM	[a]	Vermeer (1995)
	VHHAR1/VH1.1	Lewis	R	IgM	S81282	Vermeer (1995)
	Y13-259	—	R	IgG	X55179	Furth *et al.* (1982), Werge *et al.* (1990)
	YNB46.1.8 SG2B1.19	PVG	R	IgG2a	M61885	Mathieson *et al.* (1990), Gorman *et al.* (1991)
	YTH906.9.21	—	R	—	M87783	A. P. Lewis (unpublished data)
Q52	57R-1	—	R	IgG2a	L07401	Karp (1993)
	57R-2	—	R	IgG1	L07402	Karp (1993)
	Alpha D11	Sprague–Dawley	R	—	L17077	Cattaneo *et al.* (1988), Ruberti *et al.* (1994)
	Alpha D11	Sprague–Dawley	R	—	L17079	Cattaneo *et al.* (1988), Ruberti *et al.* (1994)
	ERF1.5	PVG/E	R	IgM	Z93354	Vermeer (1995)
	ERF1.30	PVG/E	R	IgM	Z93351	Vermeer (1995)
	ERF1.49	PVG/E	R	IgM	Z93353	Vermeer (1995)
	ERF2.27	PVG/E	R	IgM	Z93361	Vermeer (1995)
	ERF2.49	PVG/E	R	IgM	Z93365	Vermeer (1995)
	ERF2.59	PVG/E	R	IgM	Z93367	Vermeer (1995)

(*continued*)

Table V.10.1 *Continued*

V_H gene family	Clone name or designation	Strain	Configuration	Isotype	Accession number	Source
	GLVH5	PVG/E	G	—	X86660	Vermeer (1995)
	GLVH13	PVG/E	G	—	X86659	Vermeer (1995)
	Hg18	BN	R	IgM	Z75901	Guéry *et al.* (1990), Dammers *et al.* (1998c)
	RNIGVJL	BN	G	—	L07405	Karp (1993)
	TDL24	PVG/E	R	IgM	[a]	Vermeer (1995)
	TDL34	PVG/E	R	IgM	[a]	Vermeer (1995)
S107	132A	Lewis	R	IgG2a	L22654	M. A. Agius and S. Bharati (unpublished data)
	YTH 34.5HL	DA	R	IgG2a	X60290; X07387	Reichmann *et al.* (1988), Foote and Winter (1992)
VGAM3.8	ERF1.34	PVG/E	R	IgM	Z93352	Vermeer (1995)
VH10	56R-7	—	R	IgG1	L07399	Karp (1993)
VH11	MEC4	DZB	R	IgG2a	X78894	Aten *et al.* (1995), Dammers *et al.* (1998c)
	MEC5	DZB	R	IgG1	X78896	Aten *et al.* (1995), Dammers *et al.* (1998c)
	YTH655 (5) 6	—	R	—	M87785	J. Shearin (unpublished data)
X24	Hg32	BN	R	IgE	Z75902	Hirsch *et al.* (1986), Varga *et al.* (1995), Dammers *et al.* (1998c)

[a] Accession numbers will be available soon.

gene segments. In Table V.10.1 V_H sequences are compiled and classified according to the homologous mouse V_H gene family names, together with their references. Of the 105 sequences known at present, 80 were recently derived at our laboratories; 36 are germline derived, while from the remainder the cDNA sequence of the rearranged VDJ is given. Pseudogenes were excluded from the list.

The D sequences are not very well described at the genomic level, except for the most 5′ DQ52 (Brüggemann *et al.*, 1986), but some can be deduced from both functional and nonfunctional D–J joinings sequenced. The rat J_H locus is completely cloned and sequenced and confirms the existence of four functional J_H sequences (J_H1–4) and one pseudo J-gene located upstream of J_H1 (Lang and Mocikat, 1991).

Complexity of the Ig V_H locus

On the basis of sequence homology all the known rat V_H genes can be divided into V_H gene families, as in the mouse. The existence of rat V_H gene families, named after the mouse homologue has been confirmed for the J558, PC7183, X24, J606, Q52, X24, S107, V_H10, V_H11, VGAM3.8 and 3609 families (Table V.10.1). Members of the V_H12, V_H13 and V_H14 are not (yet) described.

Southern blotting analysis of *Eco*RI digested genomic DNA suggest the complexity of the V_H gene locus is of the same size as in the mouse (Table V.10.2), although the hybridization with the probes specific for the mouse V_H gene families VGAM3.8 and V_H12 was very weak (unpub-

Table V.10.2 Complexity of rat V_H genes

	Complexity[a]	
V_H gene family	Number of hybridized EcoRI RF	% V_H genes
J558	16	11
Q52	14	10
VGAM3.8	5	4
PC7183	18	13
3609	10	7
J606	13	9
3660	6	4
S107	9	6
X24	18	13
V_H10	11	8
V_H11	11	8
V_H12	4	3
V_H13	4	3
V_H14	1	1
Total	140	100

[a] Complexity is defined as the maximal number of different restriction fragments that can be found after hybridization of *Eco*rI-digested rat liver DNA (AO, LOU, PVG/E, BN strains) to mouse V_H gene family-specific probes (Vermeer, 1995).

lished data). The number of V_H genes within the different families is however different between the rat and the mouse. While in the mouse the J558 V_H gene family is by far the largest family, containing approximately 50% of all the mouse V_H genes, in the rat this V_H gene family contains about 20% of the V_H genes. Conversely, the small mouse X24 family (two members) shows at least 18 bands with Southern blotting. The PC7183 family also appears larger in the rat, which has been confirmed at the genomic sequence level, because, currently, 55 different V_H genes are described at the sequence level of which at least 24 are germline derived, while the mouse PC7183 family contains only 18 members (Carlsson *et al.*, 1992).

In mice, the V_H gene locus show many differences in numbers and sizes of hybridizing fragments between different mouse strains. When we analyzed the rat strains AO, Louvain, PVG and BN, little polymorphism was detected. The only strain in which some differences could be detected was the BN strain (Vermeer, 1995). Because fewer bands were observed in the V_H gene families PC7183, J606, S207 and X24 in BN rats compared with the other strains, it could mean that the BN rat has a partial deletion of the V_H gene locus. Whether this observation has any relevance to the observed sensitivity to develop antibody-mediated autoimmune diseases, e.g. $HgCl_2$-induced immune complex glomerulonephritis in this rat strain, is presently unknown (Aten *et al.*, 1988).

Ig light-chain sequences

There are only a limited number of rat light chain variable region sequences available, both for the κ and for the λ locus. The few available are compiled in Table V.10.3. Most of them are all cDNA derived sequences. The λ locus consists of at least 10–15 gene segments, representing at least four distantly related subfamilies (Aguilar and Gutman, 1992). All V_λ gene segments analyzed use only $J_\lambda 2$ and $C_\lambda 2$, which appears to be the only functional constant lambda gene (Steen *et al.*, 1987). In total 18 V_κ genes have been described (Table V.10.3), of which two are germline derived. The V_κ genes are not defined as subgroups as described in the mouse. The J_κ locus has been well described and consist of six functional J_κ genes and one pseudogene (Sheppard and Gutmann, 1982).

Development of Ig repertoire

Whenever a complete Ig is expressed on newly formed (NF) B cells, these cells are subject to both positive and negative selection. Kinetic studies have shown that many cells are lost between the NF-B cells in the bone marrow and those in the spleen (Deenen *et al.*, 1990). Early recirculating B cells (ERF-B) that just have left the bone marrow can be recognized due to the presence of Thy-1, in contrast to the normal recirculating, follicular B cells (RF-B) (Deenen and Kroese, 1993; Kroese *et al.*, 1995). Analysis of usage of V_H genes in spleen-derived ERF-B and RF-B cells from the ductus thoracticus suggests that both populations have undergone a selection process, since the usage of V_H gene families is nonrandom according to the complexity of the V_H gene families. Members from the PC7183 V_H gene family are more frequently used than would be expected, while members from the V_H gene families S104, X24, V_H10 and V_H11 are underrepresented (Vermeer, 1995). This nonrandom usage is stronger in RF-B than in ERF-B cells, suggesting an ongoing process of selection of B cells on the basis of the repertoire utilized.

11. B Lymphoid Tumors or Immunocytomas

Major progress in the studies of immunoglobulins has been made by using monoclonal immunoglobulins or myeloma proteins (or paraproteins) synthesized by B lymphoid tumor cells. The most studied were those of men (McIntyre, 1950) and mice (Dunn, 1975). These spontaneous cell growths were rather rare and were called myeloma tumors, plasmacytomas or immunocytomas according to their origins and their main characteristics. The first model of immunoglobulin secreting tumors was developed by Potter *et al.* (1957), in BALB/c mice.

Lymphoid tumors arising in the ileocaecal area of rats have been reported in the scientific literature since the 1930s (Curtis and Dunning, 1940; Roussy and Guérin, 1942; Dunning and Curtis, 1946; Guérin, 1954). Some or all these growths were immunocytomas, some of them probably secreting monoclonal immunoglobulins. However, none of these observations indicated an incidence greater than 3%. In the late 1950s, Maisin and his colleagues, working in the Faculty of Medicine of the University of Louvain, Belgium, described a special type of lymphoid tumors appearing in the ileocaecal area of their rats (Maisin *et al.*, 1957; Maldague *et al.*, 1958). Some of these tumors secreted monoclonal immunoglobulins (Maisin *et al.*, 1964). In 1970, Bazin and Beckers started breeding rats from different rat nuclei at the University of Louvain. They bred 28 different and distinct lines. The line presenting the highest incidence of malignant immunocytoma was chosen as histocompatibility reference and called LOU/C. Immunocytomas appeared spontaneously in the ileocaecal lymph node(s) (Moriamé *et al.*, 1977) of 8-month-old LOU/C rats or older. Their incidence was up to 34% in males and 17% in females (Bazin *et al.*, 1972).

These tumors were rapid growing and killed their hosts within the month following their detection by abdomen palpation. Ascitic fluids were sometimes present. Metastases were often found in the mediastinal lymph nodes or in the pleural cavity. The histological aspect of these tumors has been described by Maldague *et al.* (1958). The

Table V.10.3 Rat variable light chain genes

Light chain isotype	*Clone name or designation*	*Strain*	*Vλ gene subfamily*	*Configuration*	*Accession number*	*Source*
λ	RLV007	DA	V1	R	M77356	Aguilar and Guttman (1992)
	RLV010	DA	V2	R	M77357	Aguilar and Guttman (1992)
	RLV031	DA	V2	R	M77359	Aguilar and Guttman (1992)
	RLV052	DA	V1	R	M77360	Aguilar and Guttman (1992)
	RLV070	DA	V2	R	M77361	Aguilar and Guttman (1992)
	RLV086	DA	V2	R	M77362	Aguilar and Guttman (1992)
	RLV100	DA	V2	R	M77363	Aguilar and Guttman (1992)
	RLV118	DA	V2	R	M77365	Aguilar and Guttman (1992)
	RLV124	DA	V2	R	M77366	Aguilar and Guttman (1992)
	RLV130	DA	V2	R	M77367	Aguilar and Guttman (1992)
	RLV144	DA	V1	R	M77368	Aguilar and Guttman (1992)
	RLV154	DA	V4	R	M77370	Aguilar and Guttman (1992)
	RLV155	DA	V2	R	M77371	Aguilar and Guttman (1992)
	RLV163	DA	V2	R	M77372	Aguilar and Guttman (1992)
	RLV191	DA	V2	R	M77373	Aguilar and Guttman (1992)
κ	11/160	CBH/Cbi	—	R	X91995	Léger *et al.* (1995)
	21/61	Louvain	—	R	X75533	Kütemeier *et al.* (1994)
	132A	Lewis	—	R	L22655	M. A. Agius and S. Bharati (unpublished data)
	53R-1	—	—	R	L07406	Karp (1993)
	53R-4	—	—	R	L07407	Karp (1993)
	56R-3	—	—	R	L07408	Karp (1993)
	56R-7	—	—	R	L07409	Karp (1993)
	57R-1	—	—	R	L07410	Karp (1993)
	Alpha D11	Sprague–Dawley	—	R	L17078	Ruberti *et al.* (1994)
	GK1.5	Lewis	—	R	M84148	Rashid *et al.* (unpublished data)
	HC4A	Lewis	—	R	L22653	M. A. Agius and S. Bharati (unpublished data)
	IR2	LOUC/Wsl	—	R	M14434	Hellman *et al.* (1985)
	IR162	LOUC/Wsl	—	R	M15402	Hellman *et al.* 1985)
	RNIGKVFL	—	—	G	L07411	Karp (1993)
	RNIGKVGL	—	—	G	L07412	Karp (1993)
	Y3-Ag 1.2.3.	Louvain	—	R	X16129	Crowe *et al.* (1989)
	Y13-259	—	—	R	X55180	Werge *et al.* (1990)
	YNB46.1.8SG2 B1.19	PVG	—	R	M61884	Gorman *et al.* (1991)
	YTH 34.5HL	DA	—	R	X60291 X07383	Reichmann *et al.* (1988), Foote and Winter (1992)

secreting tumor cells could not be described as plasma cells but exhibited a marked uniformity in size, a granular nucleus with relatively small nucleoli, and a rim of deeply basophilic and pyroniphilic cytoplasm (Bazin *et al.*, 1972, 1973). Around 80–90% of these tumors secreted monoclonal immunoglobulins with or without Bence–Jones proteins. Most, if not all tumors were transplantable in histocompatible rats and some have been transplanted from animal to animal for years without any change in their immunoglobulin production.

The class distribution of the various proteins is given in Table V.11.1. The LOU rats gave a very high percentage of IgE secreting tumors. This model provided the first monoclonal immunoglobulins of the IgE (Bazin *et al.*, 1974a) and of the IgD (Bazin *et al.*, 1978) isotypes in other species than humans. The model also permitted a correct nomenclature of the rat immunoglobulins and the discovery of the IgG2c subisotype in the rat species (Bazin *et al.*, 1974b). More details of the model can be found in Bazin (1990).

Mouse plasmacytomas and also Burkitt's lymphoma and rat immunocytomas carry an Ig/myc translocation which appears to play a decisive role in the genesis of these tumors, acting through the consecutive activation of myc which makes it refractory to regulation (Pear *et al.*, 1986, 1988).

Table V.11.1 Class distribution of complete monoclonal immunoglobulins synthesized by immunocytomas appearing in LOU rats or their F_1 hybrids

	Immunocytomas	*Number of Ig-secreting immunocytomas*	*IgM*	*IgD*	*IgG1*	*IgG2a*	*IgG2b*	*IgG2c*	*IgE*	*IgA*	*ND*[a]
Number	691	440	13	4	167	27	3	20	190	13	3
Percentage	—	100	2.95	0.91	37.90	6.13	0.68	4.54	43.18	2.95	0.68

[a] Not determined.

12. The Major Histocompatibility Complex

History of the rat major histocompatibility complex (MHC)

In 1938, Lumsden made the observation that the serum of rats that rejected a transplanted sarcoma agglutinated the erythrocytes of other rats (Lumsden, 1938). Although not recognized as such, it is quite likely that this was the first indication ever for the presence of anti-MHC antibodies in these sera. It was not until 1960 that two groups of investigators, Stark and colleagues in Prague, Czechoslovakia, and Bogden and Aptekman in New York, USA, independently reported that inbred rats after tissue transplantation produced hemagglutinating antibodies specific for the major histocompatibility complex of that species. Stark and colleagues called the complex *RtH-1* (Frenzl *et al.*, 1960), whereas the New York group designated it as *Ag-B* (Bogden and Aptekman, 1960). By that time several blood group antigens (*Ag*) had been described in the rat by a number of investigators and designated A, B, C, etc; blood group antigen B proved to be identical to the MHC.

At the Second International Workshop on Alloantigenic Systems in the Rat in 1979 it was recommended that the rat MHC be designated as *RT1*.

The RT1 complex

RT1 is located on chromosome 20 (Locker *et al.*, 1990). The various MHC and MHC-linked loci that have been identified to date are arranged in five clusters according to the structure of the molecules they encode for: class I genes, class II genes, class III genes, loci of the *grc*, and genes coding for various enzymes (summarized in Table V.12.1). Class I molecules consist of two polypeptide chains. One of these, the heavy chain, which has covalently attached to it one or two carbohydrate moieties, has a comparable molecular weight in several species: 45 000 in the rat (Blankenhorn, 1978). This heavy chain consists of an intracellular, a transmembrane and an extracellular segment. This last part consists of two disulfide loops and an amino terminal segment. A light chain, β_2-microglobulin (β_2M), is noncovalently attached to the heavy chain. β_2M is a polypeptide, with a molecular weight of 12 000, which is folded in a single disulfide loop and has no direct attachment to the membrane. Class I antigens are found on all nucleated cells and on red blood cells. Class II molecules consist of two noncovalently bound polypeptides: an α-chain with a molecular weight of 31 000 to 34 000 and a β-chain with a molecular weight of 26 000–29 000 (Fukumoto *et al.*, 1982). Class II antigens are normally found only on certain cells of the immune system (B lymphocytes, macrophages, dendritic cells, and antigen-presenting cells (APCs) in general), but can also be expressed or upregulated on other tissues during activation (Mason, 1985), especially under the influence of certain cytokines (e.g. IFN-γ) (Markmann *et al.*, 1987; Rayner *et al.*, 1987; Steiniger *et al.*, 1989). Class II molecules are encoded for by the RT1.B, RT1.D and RT1.H loci. Blankenhorn *et al.* (1983) demonstrated that the RT1.B and RT1.D loci code for I-A and I-E (mouse) like class II molecules. They are associated with a strong mixed lymphocyte reaction (MLR), i.e. the proliferation of cells that is found, when lymphocytes of two genetically different individuals are co-cultured. The control of capability to respond to small polypeptide antigens is also located in this region (Lobel and Cramer, 1981). Class I molecules provide the major stimulus for cytotoxic T-cell activity (Liebert *et al.*, 1982, 1983).

The relative chromosomal localization of the various loci is shown in Figure V.12.1 (after Gill *et al.*, 1995). The MHC map of the rat shows a high degree of similarity with that of the mouse (Gill *et al.*, 1995).

Table V.12.1 Rat major histocompatibility complex loci

Class/region	*Loci*
I	*A; K (Pa); F, E; N_1, N_2, N_3; O; G; C; 11/35R; LW2; L; R**
II	*H; B; D*
III	*TAP1, TAP2; TNFα; C2; 21-OH; C4; Bf; Hsp70*
grc	*grc; rcc; ft; dw-3*
Other	*Glo-1; Acry-1; Neu-1*

* Formerly *M*. The *M* designation is now the homologue of the *H-2M* locus of the mouse (K. Wonigeit *et al.*, unpublished data). More information about rat MHC loci can be found in Gill *et al.* (1995).

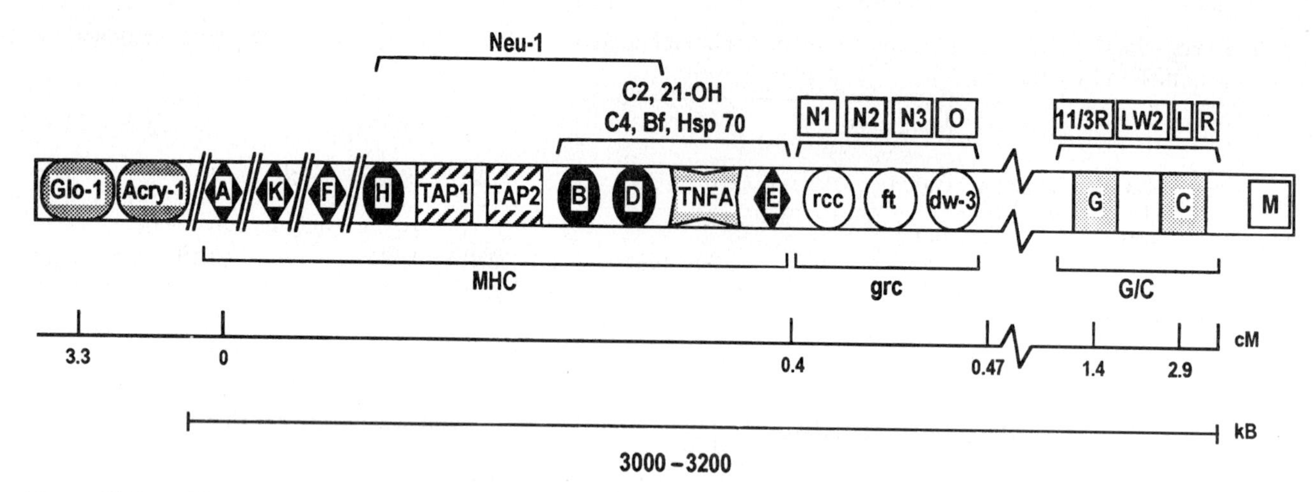

Figure V.12.1 Map of MHC and MHC-linked loci in the rat: ♦, classical class I (class Ia) antigens; ▒, non-classical class I (class Ib) antigens; ●, class II antigens; ▨, transporter proteins for antigenic peptides; ▩, tumor necrosis factor α; ○, loci of the *grc* (*rcc*, resistance to chemical carcinogenesis; *ft*, fertility; *dw-3*, body size); and ⬭, various enzymes. Loci mapped only to a region are given above the brackets: *Neu-1*, neuraminidase-1; C2, C4 and Bf, complement components; 21-OH, 21-hydroxylase; Hsp-70, heat shock protein 70; N1, N2 and N3, TL-like loci; and O, Q-like locus. Adapted from Gill *et al.* (1995).

Although most of the original work on MHC typing was done using serological (hemagglutination for class I antigens) and cellular (MLR typing for class II antigens) techniques, major progress in the study of rat MHC has been made over recent years using various molecular approaches (reviewed by Gill *et al.*, 1995). An extensive list of monoclonal antibodies against rat alloantigens, including antibodies defining both rat MHC class I and class II antigens, was published by Butcher (1987).

Rat strains

RT1 haplotypes have been allocated based primarily on serological and cellular typing results. Representative strains for the various haplotypes are given in Table V.12.2 (Gill *et al.*, 1995). Several of these haplotypes share characteristics with other haplotypes, i.e. the RP-strain is serologically (defining class I MHC antigens) identical to other $RT1^{u}$ rat strains (WAG, WF), and in MLR-typing (defining class II MHC antigens) identical to other $RT1^{l}$ rat strains (LEW) (Vaessen *et al.*, 1979). Because the origin of this mixture of RT1 characteristics cannot be deduced, these strains are described as 'natural' RT1 recombinant rat strains and are assigned a unique RT1 haplotype (see Table V.12.3). This contrasts with the laboratory derived RT1 recombinant rat strains (examples given in Table V.12.4). These strains and there haplotypes are derived from crosses between known parental strains, therefore the exact nature of the recombination within the MHC locus is known and can be used for comparisons with the original parental strains in studies in which the association of certain functional characteristics with certain loci of the RT1 complex is investigated.

Table V.12.2 RT1 haplotypes and representative strains

RT1 haplotype	*Strains*	*RT1 haplotype*	*Strains*
a	AVN	*k*	WKA/Hok, WKAH, KYN, OKA, SHR
av1	DA, ACI, ACP, COP	*l*	LEW
b	BUF, ALB	*lvl*	F344
c	AUG, PVG	*m*	MNR
d	BDV	*n*	BN, MAXX
dv1	BDIX, TAL	*o*	MR
e	BDVII	*p*	RP
f	AS2	*q*	NIG III
g	KGH	*sa*	NSD
h	HW	*sb*	WIN
i	BI	*u*	WF, TO, WAG, SDJ, YO, BB/W
j	LEJ		

More information about inbred strains can be found in *Rat News Letter*, No. 26, January 1992, and in *Genetic Monitoring of Inbred Strains of Rats*, edited by H. J. Hedrich, Gustav Fisher Verlag, New York, 1990.

Table V.12.3 'Natural' RT1 recombinant rat strains

Strain	*Haplotype*	*MHC Typing**
BDVII	*e*	A^aL^c
BI	*i*	A^nL^a
KGH	*g*	A^gL^l
LEJ	*j*	A^uL^b
MNR	*m*	A^aL^c
MR	*o*	A^dL^a
RP	*p*	A^uL^l

* MHC Typing was done with serological (*A*gglutination) and lymphocyte (Mixed *L*ymphocyte Reactivity) typing. More information about 'natural' RT1 recombinant strains can be found in Section 103 of *Handbook of Experimental Immunology*, 4th edition, edited by D. M. Weir, Blackwell Scientific Publications, Oxford, UK, 1986.

Several groups have put much effort in developing RT1 congenic rat strains, in which the MHC locus of one strain of rats (for example the $RT1^l$ haplotype of the LEW rat strain) is placed on the background of a different strain (for instance the BN rat strain) by multiple backcrossing and selection for the donor RT1 haplotype (resulting in our example in the BN.1L congenic rat strain). Several series of RT1 congenic rat strains have been developed (see Table V.12.5) and have been and still are extremely useful for studying the role of MHC-encoded proteins in a variety of physiological processes. The list given in Table V.12.5 is not complete and one should also be aware of the fact that some of these congenic strains have become extinct over the years. For further details and current availability please refer to the contacts given in the list.

Associations of RT1 with disease and/or immunological (dys)function

Although originally recognized as a gene complex encoding for tranplantation (histocompatibility) antigens, the MHC has over the years been shown to be involved in a variety of biological functions. The most important is probably MHC-restriction as the basis for self/nonself discrimination of the immune system (Zinkernagel, 1997). It is therefore not surprising that the MHC and MHC-encoded proteins exert an especially strong influence on immunologically mediated diseases and disorders. Various autoimmune models in rats display a virtually absolute

Table V.12.4 Laboratory-derived RT1 recombinant rat strains

Haplotype name Recombinant	*RT1 regions*							*Strains of origin*	*Carrier of recombinant haplotype*
	A		*B*		*D*		*C*		
r1	*a*	//	*c*		*c*		*c*	DA; PVG	PVG.R1
r2	*a*	//	*u*		*u*		*u*	LEW.1A(AVN); LEW.1U(WP)	LEW.1AR1
r3	*a*		*a*		*a*	//	*u*	LEW.1A(AVN); LEW.1U(WP)	LEW.1AR2
r4	*u*		*u*		*u*	//	*a*	LEW.1U(WP); LEW.1A(AVN)	LEW.1WR1
*r5**	*n*		*n*		*n*	//	*l*	BN; LEW	LEW.1N(2R)*
r6	*u*	//	*a*		*a*		*a*	LEW.1WR1; LEW.1A(AVN)	LEW.1WR2

* The *r5* haplotype is now extinct. More information about RT1 recombinant rat strains (R1 to R33) can be found in Section 103 of *Handbook of Experimental Immunology*, 4th edition, edited by D. M. Weir, Blackwell Scientific Publications, Oxford, 1986, and in Kunz *et al.* (1989).

Table V.12.5 RT1 congenic rat strains

Name	*Differential RT1 haplotype*	*Donor strain*	*Developed and maintained by*
The BN series			Initiated by J. Palm *et al.*, Philadelphia, PA, USA. Continued and extended by others, particularly H. Kunz and T. Gill, Pittsburgh, PA, USA
BN.1A	*av1*	DA	
BN.1B	*b*	BUF	
BN.1C	*c*	AUG	
BN.1E	*e*	BDVII	
BN.1G	*g*	KGH	
BN.1H	*h*	HW	
BN.1I	*i*	BI	
BN.1K	*k*	WKA	
BN.1L	*l*	LEW	
BN.1L	*lv1*	F344	
BN.1U	*u*	WF	
BN.1U	*u*	YOS	

(continued)

Table V.12.5 *Continued*

Name	*Differential RT1 haplotype*	*Donor strain*	*Developed and maintained by*
The LEW series			Initiated by P. Ivanyi, O. Stark and V. Kren in Prague, Czech Republic. Maintained and extended by O. Stark and V. Kren in Prague, Czech Republic and H. Hedrich, Hannover, Germany
LEW.1A	*a*	AVN	
LEW.1A	*av1*	DA	
LEW.1B	*b*	BP	
LEW.1C	*c*	AUG	
LEW.1C	*c*	PVG	
LEW.1D	*d*	BDV	
LEW.1D	*dv1*	BDIX	
LEW.1F	*f*	AS2	
LEW.1K	*k*	SHR	
LEW.1N	*n*	BN	
LEW.1W	*u*	WP	
The PVG series			Developed in part and maintained by G. Butcher, Cambridge, UK
PVG.*RT1*^{a}	*av1*	DA	
PVG.*RT1*^{l}	*lv1*	F344	
PVG.*RT1*^{l}	*l*	LEW	
PVG.*RT1*^{n}	*n*	BN	
PVG.*RT1*^{o}	*o*	MR	
PVG.*RT1*^{u}	*u*	AO	
The DA series			H. Kunz and T. Gill, Pittsburgh, PA, USA
DA.1D	*d*	BDV	
DA.1F	*f*	AS2	
DA.1I	*i*	BI	
DA.1M	*m*	MNR	
DA.1N	*n*	BN	
DA.1O	*o*	MR	
The WKA series			T. Natori, Sapporo, Japan
WKA.1A	*av1*	ACI	
WKA.1A	*av1*	DA	
WKA.1B	*b*	ALB	
WKA.1B	*b*	BUF	
WKA.1C	*c*	PVG	
WKA.1D	*dv1*	BDIX	
WKA.1J	*j*	LEJ	
WKA.1L	*l*	LEW	
WKA.1N	*n*	MAXX	
WKA.1U	*u*	WF	
Others			
AO.1P	*p*	RP	P. Nieuwenhuis and J. Rozing, Groningen, The Netherlands
AUG.1N	*n*	BN	H. Kunz and T. Gill, Pittsburgh, PA, USA
LOU.RT1^{k}	*k*	OKA	H. Bazin, Brussels, Belgium
W Hok.1L	*lv1*	F344	T. Natori, Sapporo, Japan
W Hok.1T	*t*	TO	
W Hok.1U	*u*	SDJ	

Contacts for further information on the above-mentioned congenic rat strains:
H. Bazin, Experimental Immunology Unit, Faculty of Medicine, University of Louvain, Clos Chapelle aux Champs 30, 1200 Brussels, Belgium;
G. Butcher, ARC Institute of Animal Physiology, Babraham, Cambridge, CB2 4AT, UK;
H. Hedrich, Zentralinstitut fur VersuchtierKunde, Medische Hochschule Hannover, Postfach 610180, 30625-Hannover, Germany;
V. Kren, Dept. Biology, 1st Medical Faculty, Charles University, Albertov 4, Praha 2, 12800, Czech Republic;
H. Kunz and T. J. Gill III, Dept. Pathology, University of Pittsburgh, School of Medicine, Pittsburgh, PA 15261, USA;
T. Natori, Dept. Pathology, Hokkaido University, School of Medicine, 060 Sapporo, Japan;
P. Nieuwenhuis and J. Rozing, Dept. Histology and Cell Biology, University of Groningen, School of Medicine, Oostersingel 69/1, 9713 EZ Groningen, The Netherlands.
More information on RT1 congenic rat strains can be found in Section 103 of *Handbook of Experimental Immunology*, 4th edition edited by D. M. Weir (Blackwell Scientific Publications, Oxford, UK), 1986.

Table V.12.6 Associations of RT1 with disease and/or immunological (dys)function

Disease/(dys)function	*Strain*	*RT1*	*References*
Spontaneous IDDM	DPBB	*u*	Awata *et al.* (1995), Ellerman and Like (1995)
Thymectomy + repetitive low dose irradiation-induced IDDM	PVG.RT1^u	*u*	Fowell *et al.* (1991)
Repetitive low dose streptozotocin-induced IDDM	Lew, DA	*l, av1*	Lukic *et al.* (1990)
RT6-depletion-induced IDDM	DRBB	*u*	Greiner *et al.* (1987)
Adjuvant-induced arthritis	Lew, DA	*l, av1*	Vingsbo *et al.* (1995)
Collagen-induced arthritis	WF, DA	*u, av1*	Griffiths *et al.* (1992)
Acute EAE	Lew, DA	*l, av1*	Mustafa *et al.* (1993)
Relapsing EAE	BUF, WAG	*b, u*	van Gelder *et al.* (1996)
Spontaneous thyroiditis	DPBB	*u*	Awata *et al.* (1995)
$HgCl_2$-induced glomerulonephritis	BN	*n*	Aten *et al.* (1988, 1991)
Spontaneous hypertension	Lew.1WR2	*WR2*	Kunes *et al.* (1996)
HY-(*un*)responsiveness	—	*a, n, (u)*	Geginat *et al.* (1993)
Growth, reproduction, cancer resistance	—	*grc*-complex	Gill (1996)

IDDM, insulin dependent diabetes mellitus.
EAE, experimental autoimmune encephalomyelitis.

association with certain RT1 haplotypes (see Table V.12.6): The RT1^u haplotype is associated with collagen-induced arthritis (Griffiths *et al.*, 1992), relapsing experimental autoimmune encephalomyelitis (EAE) (van Gelder *et al.*, 1996), and spontaneous thyroiditis (Awata *et al.*, 1995). Furthermore, both the spontaneous form of insulin dependent diabetes mellitus (IDDM) (Awata *et al.*, 1995; Ellerman and Like, 1995) and also the inducible models for IDDM using thymectomy and repetitive low-dose irradiation (Fowell *et al.*, 1991) or RT6 depletion (Greiner *et al.*, 1987) are uniquely associated with the RT1^u haplotype. Conversely, the repetitive low-dose streptozotocin-induced form of IDDM, is associated with the RT1^l haplotype (Lukic *et al.*, 1990), as are adjuvant-induced arthritis (Vingsbo *et al.*, 1995) and the acute model for EAE (Mustafa *et al.*, 1993). These last autoimmune models are not only associated with the RT1^l haplotype, but also with the RT1^{av1} haplotype. Severe $HgCl_2$-induced glomerulonephritis with proteinuria and linear glomerular autoantibody deposition appears to have a unique association with the RT1^n haplotype (Aten *et al.*, 1988, 1991). However, one has to consider the fact that the RT1^n haplotype is always derived from the BN rat, implying that BN non-RT1 antigens may contribute to the susceptibility or severity of disease. The latter is suggested by the fact that the mild nonproteinuric form of glomerulonephritis observed in AO rats (RT1^u) is worsened when BN non-RT1 antigens are also present ((AOxBN)F1 and MAXX) (Aten *et al.*, 1991).

Rats with an RT1^n haplotype display strong HY-responsiveness, as do rats with the RT1^a haplotype, whereas RT1^u rats are HY-unresponsive (Geginat and Günther, 1993). Finally, genes in the MHC, especially at the *grc*-locus, play an important role in the control of reproduction, growth and development, and cancer resistance (Gill, 1996).

Acknowledgments

The authors express their gratitude to various authors that have gathered much of the information on RT1 used for this section in earlier publications and to Thomas J. Gill III in particular for his long-lasting contribution to this field and the provision of regular updates on the current knowledge on RT1.

13. Nonspecific Immunity

Introduction

Nonspecific immunity or innate immunity comprises defense mechanisms against pathogens that are not raised nor directed against a particular pathogen, but that exist before exposure to pathogens has occurred and that protect against a broad range of pathogens. Nonspecific immunity is not accompanied by the induction of immunological memory. Mechanical, biochemical and cellular barriers are encountered before pathogens invade a host and give rise to an immune response.

Skin and mucosa

The intact epidermal and mucosal epithelia provide an important impediment for microbes to invade a potential host. An overview of the mechanical and biochemical barriers that microbes encounter are given in Table V.13.1.

These epithelial cells and the glands associated with the epithelia produce secretions that contribute to this first line of defense. For example, bacteria bind to mucus in the respiratory tract and are thus removed by cilia of the

Table V.13.1 Natural barriers and mechanisms of defense

Physical	Epithelial cells of skin and mucosal surfaces Adherence of microbes to mucus, ciliar transport of mucus
Chemical	pH Enzymes: lysozyme, pepsin Antibacterial peptides: defensins, calprotectin complement
Microbiological	Population of nonpathogenic, commensal flora

respiratory epithelium (Plotkowski *et al.*, 1993). In the rat intestine the mucosal gel layer appeared to be protective against translocation of bacteria after an ischemia/reoxygenation model (Maxson *et al.*, 1994). The low pH of the stomach kills a wide variety of pathogens. In the upper respiratory tract and lung antimicrobial proteins are present, produced by epithelial cells and/or granulocytes and monocytes (van Iwaarden, 1992; van Golde, 1995).

The colonization of pathogenic microbes is also impeded by the overgrowth of skin and epithelia by commensal, nonpathogenic microorganisms (Salminen *et al.*, 1995).

However, on certain sites of mucosal surfaces (e.g. where it is covered by M-cells) living microbes can easily enter the submucosal tissues. Bacterial products, such as endotoxin, can cause increased permeability to bacteria across ileal mucosal segments in rats (Go *et al.*, 1995). Furthermore the integrity of skin and mucosa may be disrupted and wounds occur, allowing pathogens to enter the subepithelial tissues. In the subepithelial tissues, the pathogens encounter a second line of defense, the tissue macrophages. Inflammatory cells (PMNs, monocytes) are recruited upon epithelial damage by the production of chemokines (e.g. IL-8) by epithelial cells (Jung *et al.*, 1995). In addition to these mechanisms to remove the microbes, processes are initiated to promote, in case of disrupted integrity of the epithelial layer, the repair of the epithelial integrity: wound healing. In the process of wound healing macrophages and cytokines (e.g. TGF-β; Shah *et al.*, 1992) play an important role. Local GM-CSF application promotes wound healing in rats (Jyung *et al.*, 1994). Once the pathogens have entered disrupted epithelia, adaptive immune responses are initiated by dendritic cells carrying immunogenic epitopes of the pathogens via the afferent lymph to the draining lymph node.

Macrophages and granulocytes

Different cell types of the immune system contribute to the innate immunity. In addition, soluble factors produced by cells of the immune system are important for the destruction of pathogens (e.g. complement). Macrophages and granulocytes are major cell types contributing to nonspecific immunity. Eosinophils, mast cells and NK cells are beyond the scope of this section.

Monocytes and granulocytes (neutrophils) originate from a common myelomonocytic precursor in the bone marrow under the influence of growth factors such as GM-CSF, G-CSF and M-CSF. In addition to the myelomonocytic precursors (myeloblast, monoblast, promonocyte) mature granulocytes and macrophages are also present in the bone marrow cavity. Two types of resident macrophages are typical for the bone marrow: the osteoclasts along the bony surfaces (Sminia *et al.*, 1986) and macrophages in the center of erythroblastic islets (Barbé *et al.*, 1996). Rat bone marrow can easily be obtained from the femoral medulla for histology since there are no cancellous spicules and thus there is no need for decalcification (Barbé *et al.*, 1997). Cell suspensions are prepared by flushing out the marrow from the femoral medulla after cutting of both ends with a pair of scissors.

After maturation monocytes and granulocytes leave the bone marrow to enter the peripheral blood stream. In rats, monocytes comprise about 10% of the white blood cell count and neutrophils 20–25% (Valli *et al.*, 1983). Scriba *et al.* (1996) have described a useful method to obtain 90% pure monocytes from rats, and their yield is very high: 30–40 × 10^6 per adult (350 g) Lewis rat.

Monocytes can leave the blood stream and enter the tissues and organs of the body to become tissue macrophages. Different organs and tissues have their own characteristic type(s) of macrophages (listed in Table V.13.2) (Dijkstra and Damoiseaux, 1993). Tissue macrophages exert different functions depending on their site of residence.

Local inflammation

When an infectious organism or inflammatory agent crosses the epithelial barrier and enters the tissue it is initially recognized and phagocytosed by resident macrophages. As a consequence of phagocytosis, the macrophage secretes a number of inflammatory mediators that initiate and regulate changes in vascular permeability, production of other inflammatory mediators and cytokines, increased expression of vascular adhesion molecules and, consequently, mediate the recruitment of granulocytes and monocytes from the circulation. In concert, these processes contribute to an effective elimination of the infectious agent. For the rat a variety of models have been used to study local inflammation *in vivo* and components of this process *in vitro*. The peritoneal cavity is often used in the rat as a site for administration of an inflammatory stimulus (e.g. thioglycolate). Although perhaps not representative for a local infection in the connective tissue of e.g. the skin, this method is very convenient for harvesting and analysis of infiltrated cells.

Table V.13.2 Phenotype of monocyte/macrophage populations with tissue-specific functions

	ED1	*ED2*	*ED3*	*CR3*	*MHCII*	*Reference*
Bone marrow						
Monoblasts/promonocytes	+					Damoiseaux *et al.* (1989a), Keller *et al.* (1989)
Nurse cells	+	+				Barbé *et al.* (1996)
Osteoclasts	+					Smina *et al.* (1986a)
Blood						
Monocytes	+					Damoiseaux *et al.* (1989a)
Lung						
Alveolar macrophages	+					Dijkstra *et al.* (1985)
Interstitial macrophages	+	+				Dijkstra *et al.* (1985)
Peribroncheal DC	+	+			+	Holt *et al.* (1987)
Liver						
Kupffer cells	+					Dijkstra *et al.* (1985)
Peritoneal cavity						
Resident macrophages	+	+/−				Beelen *et al.* (1987)
Exudate macrophages	+					Beelen *et al.* (1987)
Spleen						
Red pulp macrophages	+	+				Dijkstra *et al.* (1985)
White pulp macrophages	+	+				Dijkstra *et al.* (1985)
Marginal metallophils	+		+			Dijkstra *et al.* (1985)
Marginal zone macrophages	+		+			Dijkstra *et al.* (1985)
Interdigitating cells	+			+	+	Dijkstra *et al.* (1985), Damoiseaux *et al.* (1989b)
Lymph node						
Subsinusoidal macrophages	+		+	+		Dijkstra *et al.* (1985), Damoiseaux *et al.* (1989b)
Medullary macrophages	+	+/−	+	+		Dijkstra *et al.* (1985), Damoiseaux *et al.* (1989b)
Connective tissue						
Resident macrophages (histiocytes)	+	+				Dijkstra *et al.* (1985)
Joints						
Synovial lining cells	+	+				Dijkstra *et al.* (1987)
Brain						
Microglial cells			+			Robinson *et al.* (1986)
Meningeal macrophages	+	+				Hickey and Kimura (1988)
Perivascular macrophages	+	+				Hickey and Kimura (1988)
Thymus						
Cortical macrophages	+	+				Sminia *et al.* (1986b)
Medullary macrophages	+					Sminia *et al.* (1986b)

Recognition of microorganisms and their components

In order to recognize microorganisms or their components, macrophages express a number of different receptors that can mediate binding and phagocytosis. These include the mannose receptor, scavenger receptors, receptors for lipopolysaccharide (CD14), Fc receptors, and complement receptors.

The mannose receptor is a member of the C-type lectin family that recognizes mannose, a carbohydrate that is highly exposed on the surface of many microorganisms (reviewed by Drickamer *et al.*, 1993). For the rat, the mannose receptor can be demonstrated using mannose conjugates (Chroneos *et al.*, 1995). It is expressed selectively by subpopulations of macrophages and is not present on monocytes. A closely related molecule, the mannose-binding protein (MBP) in plasma has been studied in detail in the rat (Weis *et al.*, 1992; Ng *et al.*, 1996). A different class of receptors involved in phagocytosis by macrophages are the scavenger receptors. The molecular properties of these receptors, which recognize a relatively broad range of substances, have been characterized in detail in the mouse and human systems (reviewed by Krieger *et al.*, 1994). The expression of scavenger receptors on rat macrophages has been demonstrated using standard ligands (e.g. acetylated/oxidized LDL) (Miyazaki *et al.*, 1993). CD14 is a phosphoinositol-linked molecule expressed by monocytes, macrophages and granulocytes (reviewed by Ulevitch *et al.*, 1995). CD14 functions as a receptor for the complex of lipopolysaccharide (LPS) and the plasma protein LPS-binding protein (LBP). Although other macrophage receptors have also been shown to bind LPS, CD14 is considered to be primarily responsible for the LPS-induced signalling that leads to the production of nitric oxide, TNF-α, IL-1 and IL-6. Although most information on the structure and function of CD14 comes from studies on the human molecule, a cDNA clone encoding rat CD14 has been reported (Galea *et al.*, 1996), and rat macrophages respond to LPS in a way that is suggestive of CD14-mediated signalling (Stadler *et al.*, 1993; Broug-Holub *et al.*, 1995). Another class of receptors involved in phagocytosis are the Fc receptors. Fc receptors probably only play a significant role in phagocytosis by macrophages during the later stages of inflammation, when specific antibodies have been formed. A number of different Fc receptors with different binding affinities for different immunoglobulin classes and isotypes have been described in detail in humans (van de Winkel *et al.*, 1993). Rat macrophages have also been demonstrated to express Fc receptors, but these have not been characterized as extensively as in human or mouse (Chao *et al.*, 1989; Prokhorova *et al.*, 1994). Members of the β_2-integrins, CR3 (also called MAC-1 or $\alpha_M\beta_2$) and p150/95 ($\alpha_X\beta_2$), function as receptors for the C3bi-fragment of complement on monocytes, macrophages and granulocytes. These receptors also interact with several other ligands, including the cellular adhesion receptor ICAM-1, which is expressed by e.g. inflammatory endothelial cells. In the rat CR3 (recognized by the mAb ED7, ED8 and OX42), and probably also p150/95, mediate the binding and ingestion of C3bi-coated particles by macrophages (Robinson *et al.*, 1986). Anti-CR3 administration *in vivo* has also been shown to prevent the inflammation and clinical signs in autoimmune experimental allergic encephalomyelitis (EAE) in rats (Huitinga *et al.*, 1993). In addition to integrins, rat Kupffer cells express a nonintegrin that appears to function as a complement receptor (Maruiwa *et al.*, 1993).

Production of inflammatory mediators and cytokines

In response to an inflammatory stimulus a number of factors can be secreted by local macrophages and other cells, that mediate the inflammatory response and its local symptoms (e.g. redness, swelling, pain, heat) as well as the systemic consequences. These factors include small inflammatory mediators, cytokines and chemokines. As for most (if not all) species, small inflammatory mediators in the rat include eicosanoids, vasoactive peptides, reactive oxygen species (ROS) and nitric oxide (NO). Eicosanoids, which can be released by macrophages and (in the case of trauma) by platelets, cause vasodilatation and increased vascular permeability, and function as chemoattractants for phagocytic cells. Important vasoactive peptides include serotonin, histamin (both released e.g. by mast cells) and the potent platelet-activating factor (PAF) produced by macrophages. ROS are produced by macrophages and granulocytes and have important antimicrobial and tissue-destructive activity. Their action *in vivo* can be inhibited by injection of scavenger-enzymes (Ruuls *et al.*, 1995). Macrophages can also produce NO radicals, which, in addition to their cytotoxic effects, have strong immunomodulatory effects; NO synthetase inhibitors have been used to study this in the rat *in vivo* (Ruuls *et al.*, 1996).

Activated tissue macrophages and (after infiltration into the inflamed site) monocytes are the most prominent source of the cytokines TNF-α, IL-1 and IL-6. From these TNF-α and IL-1 are released soon after stimulation and act on many other cells (both locally and systemically) by binding to specific surface receptors. Their effects include the stimulation of eicosanoid release by various cells, increased expression of adhesion receptors on endothelial cells, and the secretion of a special set of cytokines, called the chemokines, which orchestrate the infiltration of granulocytes and monocytes into the inflamed tissue (Schall, 1991). IL-6 is produced relatively late after macrophage activation and has profound systemic effects, which include the induction of fever and the acute phase response in the liver (see below). The production of TNF-α, IL-1, IL-6, and NO in response to stimuli such as LPS has been demonstrated for different rat macrophage populations (Stadler *et al.*, 1993; Broug-Holub *et al.*, 1995; van der Laan *et al.*, 1996). Many of these cytokines (including TNF-α, IL-1, IL-6; Nishida *et al.*, 1989; Northemann *et al.*, 1989) and their receptors (Baumann *et al.*, 1990; Himmler *et al.*, 1990) have now been cloned and recombinant proteins are available. A number of rat chemokines have also now been identified and their chemotactic potential been investigated (Nakagawa *et al.*, 1996; Schmal *et al.*, 1996).

Infiltration of inflammatory cells

The recruitment of leukocytes (predominantly neutrophils and monocytes) from the circulation is a crucial step in the host defence against infection. In general, the accumulation of granulocytes into inflamed sites is most notable early (1–6 h), whereas mononuclear cell infiltration is typically evident in a delayed time frame (24–48 h). Such kinetics are also seen in the rat peritoneal cavity after injection of an inflammatory stimulus (Beelen *et al.*, 1983). The immigration of cells involves attachment to the endothelium of the vessel wall and subsequent transmigration. The initial adhesion of leukocytes to the endothelium, a process known as 'rolling', is mediated by adhesion receptors of the selectin-family, whereas the firm attachment to the vessel wall is mediated by molecules such as intercellular adhesion molecule-1 (ICAM-1) and vascular cell adhesion molecule-1 (VCAM-1) on endothelial cells and their integrin ligands on leukocytes. In the mesenteric venules of rat peritoneum the processes of rolling and firm attachment can be followed (Norman *et al.*, 1995), and evidence for the role of many of these adhesion molecules has been obtained now (Misugi *et al.*, 1995; Issekutz *et al.*, 1996). Local cytokines (TNF-α, IL-1) and LPS contribute to the efficiency of the process by inducing/enhancing the expression of adhesion receptors (E-selectin, ICAM-1, VCAM-1) on the endothelium, while chemokines, eicosanoids, vasoactive peptides and chemotactic complement fragments (e.g. C3a) can further contribute to the recruitment of cells to the inflammatory site.

Systemic effects of inflammation

In addition to their local affects the cytokines TNF-α, IL-1 and IL-6 released from the site of inflammation can also have dramatic systemic effects. They act on the hypothalamus to modulate body temperature and cause fever (Dinarello, 1988; Navarra *et al.*, 1992). IL-1 and IL-6 also act on the hypothalamus to increase corticotropin releasing hormone levels (Berkenbosch *et al.*, 1987; Navarra *et al.*, 1991), and they subsequently affect the pituitary gland to induce ACTH (Naitoh *et al.*, 1988), which in turn increases glucocorticoid (in the rat, mainly corticosterone) production by the adrenal gland (Tilders *et al.*, 1994). This hypothalamic–pituitary–adrenal (HPA) axis has been well characterized in the rat and its activity varies considerably among different rat strains (Sternberg *et al.*, 1989, 1990). Another major target of these cytokines is the liver, where they modulate the production of a set of plasma proteins, called acute-phase proteins (APPs), by hepatocytes. APPs whose production is down-regulated include albumin and transferrin. The particular set of APPs that is upregulated varies among species. In the rat, where assays for analyzing many APPs are available (Ikawa *et al.*, 1990), a typical acute-phase response can be induced by systemic administration of LPS, or local administration of Freund's complete adjuvant or turpentine (Geisterfer *et al.*, 1993). The effect can be mimicked by cytokine administration *in vivo* (Kampschmidt *et al.*, 1986; Geiger *et al.*, 1988) and an optimal response requires glucocorticoids (Baumann *et al.*, 1990). Various rat hepatocyte culture systems have been developed to study the regulation of APP and these have confirmed that some APPs (type 1) can be induced by either IL-6 or IL-1/TNF-α, whereas others (type 2) are regulated exclusively by IL-6 (and 'IL-6-like' cytokines oncostatin M, leukemia-inhibitory factor, and IL-11) (Schreiber *et al.*, 1989; Baumann *et al.*, 1991, 1993). The typical APP in the rat are listed in Table V.13.3. Particular features of the rat acute phase response include the relatively high (10–100-fold) induction of α_2-macroglobulin and cysteine protease inhibitor (Schreiber *et al.*, 1989).

Table V.13.3 Rat acute-phase proteins

TYPE 1 (IL-6 or IL-1/TNF-α)	*TYPE 2 (IL-6)*
α_1-Acid glycoprotein	α_1-Antichymotrypsin
C-reactive protein	α_1-Proteinase inhibitor
Complement C3	α_2-Macroglobulin
Complement factor B	Ceruloplasmin
Haptoglobin	Cysteine proteinase inhibitor
Hemopexin	Fibrinogen
Serum amyloid A protein	
Serum amyloid P	

14. Components of the Mucosal Immune System

Introduction

At several sites in the mucosal lining organized accumulations of lymphocytes and macrophages are present directly beneath the epithelium. Together with solitary lymphoid nodules, these accumulations are the components of the mucosa-associated lymphoid tissue (MALT). The structure and development of lymphoid tissue at three separate mucosal sites have been studied in rodents: (1) the gut-associated lymphoid tissue (GALT); (2) the bronchus-associated lymphoid tissue (BALT); and (3) the nasal-associated lymphoid tissue (NALT) (Table V.14.1).

The NALT is the component of MALT in the upper respiratory tract. Organized lymphoid tissue in the mucosa of the nasal cavity is present in rats on both sides of the entrance of the pharyngeal duct, closely associated with the respiratory epithelium (Spit *et al.*, 1989). NALT in rats

Table V.14.1 Characteristics of PP (GALT), NALT and BALT in Wistar rats (modified from Kuper *et al.*, 1992)

	Peyers patches (GALT)	*NALT*	*BALT*
Ontogeny			
Presence of tissue	Before birth	At birth	Day 4
First lymphocytes	T cells	T cells	B cells
Distinct T and B areas	At day 10	At day 10	>day 21
Adult tissue			
Germinal centers	++	±	±
Follicular dendritic cells	+	+	–
Intraepithelial lymphocytes	++	++	±
T/B ratio	0.2	0.9	0.7
CD4/CD8 ratio	5.0	2.4	2.6
Surface IgA^+ B cells	++	±	+

is composed of a loose reticular network with lymphocytes and macrophages, covered by epithelium. The epithelium is infiltrated with B cells, $CD4^+$ and $CD8^+$ lymphocytes, and $ED1^+$ macrophages. The B cells are IgM^+ or IgG^+, surface IgA or IgE is rare in the rat NALT (Kuper *et al.*, 1989). NALT is present at birth as a small accumulation of mainly T lymphocytes in a meshwork of reticulum cells; distinct areas of T and B cells appear at 10 days after birth (Hameleers *et al.*, 1989; van Poppel *et al.*, 1993). Although it has been suggested that the NALT plays a role in the uptake and response to inhaled antigens, because it contains M cells (Spit *et al.*, 1989), the response to intranasally applied nonviable, nonreplicating artificial antigens (i.e., TNP-KLH, TNP-LPS) is restricted to the lymph nodes of the upper respiratory tract (Hameleers *et al.*, 1991). Conversely, inoculation with live bacteria may induce specific antibody production in the rat NALT.

The lower respiratory tract in rat contains distinct units of mucosa-associated lymphoid tissue, defined as BALT (Bienenstock *et al.*, 1973; Sminia *et al.*, 1989). BALT is present primarily at sites between an artery and bronchus epithelium. The appearance of BALT is antigen-driven, as BALT in rat is detectable fom day 4 after birth (Plesch *et al.*, 1983). Rats that are kept under routine laboratory conditions and rats that are kept under specific-pathogen-free (SPF) circumstances have BALT. However, the presence of BALT also varies with species, as mice that are kept conventionally usually have BALT, whereas SPF mice often have none (van Rees *et al.*, 1996). However, the microbial status does not alter or increase the expression of class II MHC molecules or the distribution of macrophages in rat respiratory organs (Steiniger and Sickel, 1992).

In rat BALT, T and B cell areas are present from 4 weeks after birth (Plesch *et al.*, 1983), but do not have a distinct location (van der Brugge Gamelkoorn and Sminia, 1985). Among the epithelial cells covering BALT, cells bearing microfolds are present, which resemble the microfold cells (M cells) in the Peyer's patch (PP) epithelium. These cells play a pivotal role in the uptake of antigens. Intratracheal administration of antigen leads to a BALT-confined plasmacellular reaction (van der Brugge-Gamelkoorn *et al.*, 1986).

The GALT is composed of PPs, solitary follicles, lymphocytes in the lamina propria and epithelium, proximal colonic tissue (PCLT) and appendix. NALT and mesenteric lymph nodes are often also included in GALT. PPs are located in the small intestine. Each PP can be divided into three regions on both anatomical and functional bases. First, the dome that has a unique lymphoepithelium in which M-cells occur; second, the follicles or B-cell zones with prominent germinal centers; and third, the parafollicular region or T-cell dependent area. The germinal centers of PPs contain many B cells with surface IgA. PPs are important IgA-inductive sites. During ontogeny, on day 20 of gestation in rats, the gut wall has distinct spots in which villi are absent and a dome-shaped epithelium with cuboidal epithelial cells becomes prominent, which is indicative of a developing PP (Wilders *et al.*, 1983; van Rees *et al.*, 1988).

Lymphoid cell aggregates are also found along the large intestine. An organized lymphoepithelial structure located at 25% of the distance from the caecum to the rectum has been termed PCLT in rat and mouse colon (Perry and Sharp, 1988; Crouse *et al.*, 1989; De Boer *et al.*, 1992). The general structure of these organized lymphoid tissues is comparable with PPs, and no evident phenotypical differences were found between the lymphocyte populations of PCLT and either jejunal or ileal PP (De Boer *et al.*, 1992). Although some differences in functional aspects have been described (Perry and Sharp, 1991), PCLT in the rat appears to be a PP-like structure with some unusual features, rather than a bursa equivalent (De Boer *et al.*, 1992).

Transport of antigens

M cells play a pivotal role in transport of antigens over the epithelial layer. Although they represent a very small minority in the epithelium, M cells play a primary role in transport of antigens to the inductive arm of the mucosal immune system. Their functional significance is a consequence of their position – within the epithelium covering the PP and therefore in close vicinity of lymphocytes and antigen-presenting cells. Intact liposomes are endocytosed by M cells (Childers *et al.*, 1990). It is not known yet if specific receptors exist on the surface of M cells to facilitate the transport of macromolecules across the epithelium.

It is generally believed that M cells do not modify the antigens they transport from the intestinal lumen. However, M cells from rat jejunal PP express MHC Class II determinants (Jarry *et al.*, 1989; Allan *et al.*, 1993) indicating that they have some capacity to present endocytosed antigens directly to lymphocytes (Allan *et al.*, 1993).

Epithelial cells that are not M cells also play a role in the uptake of antigen in the rat gut. Soluble antigen is taken up primarily through the villi outside PPs and can be traced into the lamina propria whereas particles are taken up by M cells and can be found in PPs (Sminia *et al.*, 1991). Epithelial cells of the normal small intestine express MHC class II molecules. In adult rat jejunum class II MHC molecules have been detected in association with intracellular organelles (Mayrhofer and Spargo, 1990). Isolated enterocytes from rat small intestine can present antigen to primed T cells (Bland and Warren, 1986). However, the rate of antigen uptake through M cells varies depending on the species. Uptake via M cells in rabbits appears to be more efficient than in rat or mice (Sminia *et al.*, 1991).

It has been suggested that uptake of antigens via enterocytes is more likely to induce tolerance whereas uptake via M cells may induce specific immune responses (Biewenga *et al.*, 1993). In support of this hypothesis, antigen presentation by enterocytes *in vitro* results in

CD8 stimulation rather than $CD4^+$ proliferation (Mayer and Shlien, 1987).

Production and transport of mucosal Igs

Essential components of a secretory immune system are a secretory immunoglobulin isotype (IgA or IgM) and a receptor–transporter system to facilitate transport of the IgA across the epithelium into the gut lumen or into secretions such as milk. Dimeric IgA with J chain binds to the polymeric Ig receptor (pIgR) on the internal surface of the epithelial cells and is released at the luminal surface, together with secretory component (SC), derived by cleavage of the pIgR. SC has been identified in rat (Vaerman and Lemaitre-Coelho, 1979; Acosta-Altamirano *et al.*, 1980).

Secretory (sIgA) antibodies inhibit microbial adherence and prevent absorption of antigens from mucosal surfaces (McGhee and Mestecky, 1990). Other functions of IgA, also aimed at exclusion of pathogenic antigens, are to neutralize intracellular microbial pathogens directly within the epithelial cells, and to bind antigens in the lamina propria and excrete them through the adjacent epithelium into the lumen of the gut (Mazanec *et al.*, 1993).

After being taken up by the gut, antigen is being processed by macrophages (predominantly in the PPs) and presented to T-helper cells, leading to the activation of B cells in the PPs. The first step in IgA B-cell differentiation is an IgA-directed switch of $sIgM^+$ B cells, derived from the bone marrow. The committed B cells subsequently migrate to the mesenteric lymph nodes, and enter the blood circulation via the thoracic duct. In the spleen the B cells may differentiate further, and migrate into the gut lamina propria to develop into mature IgA-producing cells.

Cytokines play a central role in the generation of IgA responses. First, cytokines influence the isotype to switch to IgA, and second cytokines induce $sIgA^+$ B cells to differentiate into IgA plasma cells. TGF-β, IL-4 and IL-5 have a significant influence in this process (reviewed by Strober and Ehrhardt, 1994).

In mice, B-1 cells have been described that may play an important role in mucosal IgA responses (Kroese *et al.*, 1994). However, it is not clear whether a population with similar characteristics is present in the rat (Vermeer *et al.*, 1994).

Passive immunity

Passive transfer of immunity from mother to offspring occurs prenatally through placenta or yolk sac, and postnatally via milk. Species differ in the contribution each route makes to the transfer of immunity (Waldman and Strober, 1969). Three groups can be determined based on these differences (using either the prenatal or postnatal transfer route only, or using both routes; Renegar and Small, 1994). In rats, prenatal transmission occurs via the yolk sac/placenta and the fetal gut (Waldman and Strober, 1969). However, most transport of antibody occurs postnatally via colostrum and milk (Heiman and Weisman, 1989).

There is evidence that milk-borne T cells can protect the rat neonate against infectious disease. T cells ($OX\text{-}19^+$) make up 45% of rat milk lymphocyte population. Within T cells, the ratio $CD4^+/CD8^+$ is 1.03, whereas in lactating dams the corresponding ratio in peripheral blood is 2.8 (Na *et al.*, 1992). Because the percentage of NK cells in milk from infected dams (65%) is higher than that from control milk (21%), it has been suggested that NK cells are selectively passaged into rat milk (Na *et al.*, 1992). Baby rats, suckling on a mother immune to *Trichinella spiralis* are protected against challenge infection (Kumar *et al.*, 1989).

Weaning in rats take place at days 21 and 22 after birth, inducing activation of the mucosal immune system of the pup. Weaning leads to intestinal maturation, as assessed by villus area, crypt length, crypt-cell production rate and disaccharidase activity (Cummins *et al.*, 1991).

Structure of the placenta

In the rat, the hemochorial uteroplacental unit is comparable to that in mice. The following description of the general histology of the term placenta starts at the maternal side of the uteroplacentar unit. The uterus is composed of myometrium (smooth muscle), endometrium and a pregnancy-dependent zone, the metrial gland. This zone develops in the rodent myometrium during pregnancy. The decidua basalis consists of endometrium infiltrated by fetal trophoblast; it is drained exclusively by a maternal endothelial-lined venous system. In the fetal part of the placenta, trophoblast-lined maternal blood spaces (lacunae) are present. Adjacent to the decidua basalis, in the spongiotrophoblast, only maternal blood circulates through fetal trophoblast tissue. The part which is also perfused by fetal blood has been termed the labyrinth. The exchange of fetal and maternal products is controlled at this level. Finally, the chorioallantoic plate consists of nontrophoblastic fetal connective tissue transmitting fetal blood vessels into the labyrinth.

To identify the different compartments in the rat placenta, demonstration of alkaline phosphatase activity by enzyme histochemistry is useful. Alkaline phosphatase activity is abundant in the labyrinth, is present to a lesser extent in the spongiotrophoblast and absent in other fetal layers and maternal tissue (van Oostveen *et al.*, 1992).

Gestation in the rat lasts for 21 days. Throughout gestation, maternal macrophages are present in large numbers in the myometrium, the endometrium and the metrial gland. In the labyrinth, macrophages are present at early stages and are probably of fetal origin. In the spongiotrophoblast and the decidua basalis, layers of the placenta containing both maternal and fetal cells, only few macrophages are present. Throughout gestation very few B and T lymphocytes are present in the placenta and are restricted mostly to the maternal compartments (van Oostveen *et al.*, 1992).

The metrial gland contains a unique cell type, the granulated metrial gland cell (GMG) in large numbers (Bulmer, 1968; Peel *et al.*, 1983). Because these cells are bone-marrow derived (Mitchell, 1983) and since more GMG cells are present in allogeneic than in syngeneic pregnancies (Kanbour-Shakir *et al.*, 1990), an immunological function for GMG cells has been suggested (Kanbour-Shakir *et al.*, 1990).

The monoclonal antibody ED11, raised against rat reticulum cells (van den Berg, 1989) stained fibers lining the sinuses in the spongiotrophoblast. As in this compartment of the placenta maternal blood circulates through fetal tissue, and ED11 plays a role in trapping of immune complexes (van den Berg *et al.*, 1991), the spongiotrophoblast may be important in protecting the fetus from circulating immune complexes (van Oostveen *et al.*, 1992).

15. The Complement System

Introduction

The complement system, together with antibodies and phagocytes, plays an essential role in our humoral immune defence. By binding to specific complement receptors, complement is also able to activate various cell types such as leukocytes, endothelial cells, fibroblasts, renal cells and astrocytes. The mechanisms of activation of the complement system have been elucidated in recent years; almost all molecules playing a role in the complement cascade have been cloned and a start has been made in the generation of mice with specific deletions in genes for encoding complement molecules. In this section we describe the mechanisms of complement activation followed by a section about the complement system of the rat.

Mechanisms of complement activation

During the past 40 years, extensive research has led to detailed insight in the mechanisms of complement activation. Most complement components are present in the circulation or in other body fluids in a proenzymatic, nonactive form. Evidence, obtained mainly in the early seventies, suggests that the main organ responsible for the maintenance of circulating complement levels is the liver, but recent studies in rats, employing liver transplantation models, indicate that extrahepatic synthesis of complement may be responsible for about 25% of the levels of C6 and C2 (Brauer *et al.*, 1994; Timmerman, 1996).

To appreciate the full potential of biological activities of complement the system requires activation. There are two pathways which lead to activation of the central

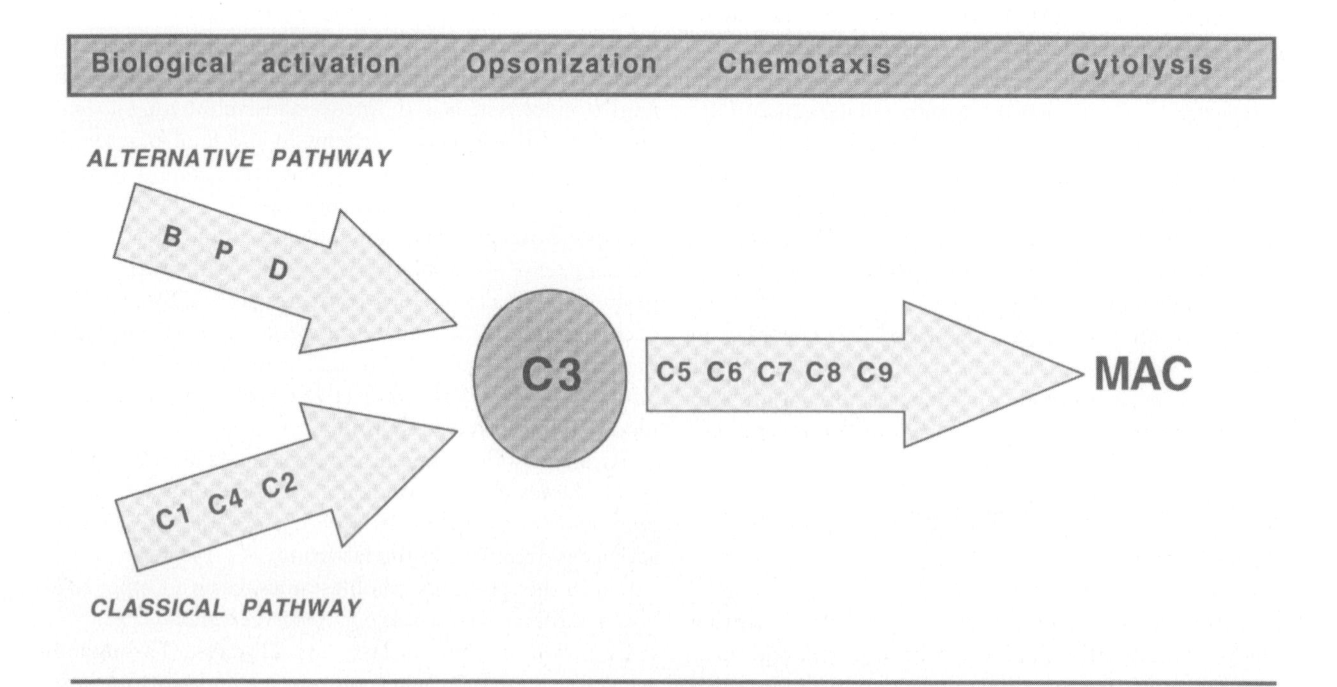

Figure V.15.1 Activation of C3, the third component of complement, may occur either via the classical or the alternative pathway, leading to the generation of activated C3 and further recruitment of the terminal sequence of complement C5–C9 and assembly of the membrane attack complex (MAC). During activation of the different pathways, different biological functions of complement are expressed, such as opsonization, chemotaxis and cytolysis.

component of complement C3 (Figure V.15.1). The classical pathway, initiated not only by immune complexes composed of IgM and IgG antibodies but also by substances such as lipopolysaccharides and DNA, leads to binding and activation of C1, the first component of complement, and to activation of its natural substrates C4 and C2. The complex of activated C4 (C4b) and C2 (C2b) has enzymatic activity and is able to convert C3 into C3b and C3a. Like C4b, C3b possesses a labile thioester which is able to covalently attach C4b and C3b to activating surfaces or to bystander molecules, tissue and cells. Generation of C3b is the first step leading to amplification of C3 cleavage, and finally results in deposition of multiple C3b molecules on bacteria and viruses, and enhanced recognition of substances by phagocytic cells via C3 receptors. The presence of multiple C3b molecules on the activator results in recruitment of the terminal complement components and activation of complement up to C9, resulting in the generation of the membrane attack complex (MAC).

Initial cleavage of C3 to C3b may also occur independently of C1, C4 and C2 via the alternative pathway. It has been shown that Gram-negative bacteria, meningococci and polymeric IgA are especially efficient activators of the alternative pathway. Recently, a third pathway of complement activation was proposed (Lu *et al.*, 1990), initiated via mannose-binding protein (MBP). It is thought that MBP functions like C1q, and is associated with uncleaved C1r and C1s in a complex, simular to C1q in C1. The MBP pathway merges into the classical pathway at the C4 and C2 level. The generation of MAC, with its strong cytolytic potential is perhaps the best known activity of complement, however, current opinion is that opsonization of foreign pathogens and the generation of anaphylactic and chemotactic fragments may be more relevant. For example, after the recently described chemotactic chemokines, such as monocyte chemoattractant protein-1 (MCP-1) and IL-8, C5a is known as one of the most potent chemotactic agents for neutrophils and monocytes. Moreover, fragments of C4, C3 and C5 are also directly involved in enhancement of local vascular permeability.

Because complement activation is enzymatic in nature, regulation of this system is essential to prevent potential depletion of complement levels. Efficient control of classical pathway activation is achieved by C1-esterase inhibitor (C1-Inh) and C4-binding protein (C4bp). C1-Inh inactivates activated C1 and prevents further activation of C4 and C2 and, thereby, the formation of the classical C3 convertase. C4bp binds to C4b and prevents the formation of the C4b2a enzyme.

C4bp also functions as a cofactor for the serine protease factor I in the further degradation of exposed C4b. Factor I not only degrades C4b in the classical pathway C3 convertase, but also inactivates C3b, together with factor H, preventing further amplification of C3 cleavage and generation of the MAC. Fluid-phase inhibitors of the MAC are vitronectin and clusterin (SP40,40). Vitronectin, previously known as S-protein, binds to C5b-C9 and prevents its insertion in biomembranes. Clusterin influences MAC function in a similar way.

As mentioned above, the complement components C4 and C3 possess a thioester which, upon activation, allows these components to bind covalently to neighbouring molecules, cells and tissue. Potentially, the activated C4b and C3b fragments function as a focus for further complement activation, which might lead to damage of host cells. To prevent this, a number of membrane-bound regulators of complement activation are present on homologous cells and tissue. Decay-accelerating factor (DAF, CD55), membrane cofactor protein (MCP, CD46) and complement receptor type I (CR1) share properties with the fluid-phase regulators C4bp, factor H and factor I. The encoding genes are all located in the gene cluster for 'regulators of complement activation (RCA)' on chromosome 1 q (Seya, 1995). These membrane RCA are of great importance since it has been shown that deficiencies are associated with various types of illnesses such as paroxysmal nocturnal hemoglobulinuria (PNH) and vasculitic syndromes. In this sense the membrane-bound regulator of complement, homologous restriction factor (HRF, MIRL, CD59) is also of great importance because it controls the generation of MAC on homologous tissue.

Soluble forms of most of these membrane regulators of complement have also been shown to inhibit complement in the fluid phase, making them prime candidates for exogenous administration in order to downregulate complement activation during disease. Soluble CRI (sCRI) has been shown to efficiently downregulate complement activation at the C3 level under various conditions in experimental models.

The complement system of the rat

While detailed procedures for the isolation, purification and measurement of most human complement components have been reported, relatively few reports are available on the isolation of rat complement proteins. Complement components can be either measured with sensitive hemolytic assays or, when antibodies are available, with various immunological techniques such as ELISA, radioimmunoassay and immunodiffusion. Using sera deficient in a single complement component it is relatively easy to assess almost every complement component of the rat, except C2 (Leenaerts *et al.*, 1994). Complement deficient sera in single components may be obtained from patients with specific complement deficiencies or from guinea pigs (C4, C2, C3), rabbits (C6) or rats (C6). It is also possible to prepare complement-deficient reagents using immunoabsorption with specific antibodies. The most difficult rat complement component to assess is C2, which is measured using intermediates of sheep erythrocytes sensitized with antibody and human C4 (Brandt *et al.*, 1996).

Methods for the purification of a number of rat complement components such as C3, C4, B, H, C9 and C1q are available (Daha and van Es, 1979, 1980, 1982; Daha *et al.*, 1979; Wing *et al.*, 1993; Liu *et al.*, 1995). Furthermore the availability of antibodies against these components provides us with the possibility of measuring both classical and alternative pathway activation *in vivo* at both the circulation and at the tissue level (Boyers *et al.*, 1993). However, attention must be turned to purification of missing components. The availability of purified membrane regulators of rat complement is limited. CRYY/P65 and CD59, which regulate the classical and terminal sequence respectively, have been shown to be involved in complement activation at the tissue level (Quigg *et al.*, 1995). In contrast to, for example, guinea pigs, complement deficiencies in rats are relatively rare. In 1974, a genetic deficiency of C4 was reported in Wistar rats, but since that time no further details have been published (Arroyave *et al.*, 1979). Recently, we described a strain of PVG/c rats with a complete deficiency of C6 and a partial deficiency of C2 (Leenaerts *et al.*, 1994). This rat strain now provides the opportunity to analyze the role of C5–C9 in a number of experimental diseases. The appropriate size of rats for microsurgery provides us with a powerful tool for further research in fields such as xenotransplantation. To be able to generate rats with specific complement deficiencies in the near future more effort is required in the purification and cloning of the remaining key components of the rat complement system.

16. Models of Immunodeficiency in the Rat

Introduction

This section reviews a variety of natural immunodeficient rat strains (Table V.16.1), using a rather broad definition of the term 'immunodeficient'. It does not deal with the various inducible models in which by manipulation of the immune system either by depletion with antibodies, by treatment with immunosuppressive agents or irradiation, or through (a)specific tolerance induction a state of 'immunodeficiency' can be achieved in rats. The main focus will be on rat strains exhibiting a variety of levels of immunodeficiency, ranging from complete absence of certain immune functions and/or subsets to situations in which strong skewing of the regulatory branch of the immune system results in a certain degree of immunodeficiency, as evidenced by an increased risk of these animals to develop autoimmunity. The present state of the field of genetic manipulation in rats for the generation of new models of immunodeficiency is reviewed.

Nude rats

The nude mutation first appeared in 1953 in a rat colony at the Rowett Research Institute, Aberdeen, Scotland. The mutation was lost, but reappeared in the same colony in the 1970s (Festing, 1978). The autosomal recessive muta-

Table V.16.1 Natural immunodeficient rat strains

Strain	*Mutation*	*Defect*	*Consequence (phenotype)*	*References*
Nude rat	*rnu*	Thymic aplasia	Loss of T-cell dependent immunity	Festing (1978), Hougen (1991), Schuurman (1992, 1995)
Nude rat	*rnu^{nz}*	Thymic aplasia	Loss of T-cell dependent immunity	Berridge (1979), Douglas-Jones (1981), Marshall and Miller (1981), Schuurman (1992)
DP-BB	*lyp*; ??	Virtual absence of RT6 positive T cells	Lack of immune regulatory functions: high incidence of autoimmunity (diabetes, thyroiditis)	Chappel (1983), Mordes (1987), Crisa (1992)
LEC	*thid*	Block in differentiation from $CD4^+CD8^+$ (DP) to $CD4^+CD8^-$ (SP) thymocytes	Virtual absence of peripheral $CD4^+$ T cells and associated functions (note: not true for IEL)	Agui (1990, 1991), Yamada (1991)
BN	??	TH_2-skewed	High sensitivity for (induced) Ab-mediated autoimmunity (e.g. $HgCl_2$-induced glomerulonephritis)	Rozing (1979), Aten (1988)
Lewis	??	TH_1-skewed	High sensitivity for cell mediated immunopathology, incl. (inducible) autoimmunity (e.g. EAE)	Pat Happ (1988), Mustafa (1994)

tion is designated *rnu* and is linked with the myeloperoxidase locus in the rat (Murakumo, 1995). Approximately at the same time, another nude mutation was observed in a rat colony maintained at the Victoria University, Wellington, New Zealand, designated *nznu* or rnu^{nz} (Berridge, 1979; Douglas-Jones, 1981; Schuurman, 1992). The most obvious phenotypical characteristic of the nude rat is its hairlessness. The main immunological feature of the nude rat is the congenital aplasia of the thymus. Owing to the absence of this primary lymphoid organ nude rats are deficient for a wide variety of thymus-dependent cell mediated immune functions measured both *in vitro* and *in vivo* (reviewed by Hougen, 1991; Schuurman, 1992, 1995). Nude rats have been widely used for a variety of experiments, including studies on human tumor growth and treatment, xenotransplantation, extrathymic T-cell differentiation, thymus or thymus-fragment implantation, infection, toxicology, and functional studies of transferred cells (reviewed by Hougen, 1991; Schuurman, 1992, 1995). The original *rnu* mutation has been backcrossed to a number of strains and is commercially available from a number of breeders.

DP-BB rat

The BB rat was discovered in 1974 in a commercial colony of Wistar-derived rats at the Bio-Breeding Laboratories, Ottawa, Canada (Chappel, 1983). The DP-BB (Diabetes Prone) substrain develops spontaneously diabetes at high frequency between 60 and 120 days old (reviewed by Mordes, 1987; Crisa, 1992). All BB rats are descendants of the original Ottawa litters (Prins, 1991), but BB rats in various colonies vary with respect to frequency and severity of diabetes (Mordes, 1987). BB rats develop also thyroiditis spontaneously at a high frequency (Mordes, 1987; Crisa, 1992). DP-BB rats are severely lymphopenic (Crisa, 1992). They lack $RT6^+$ T cells (Crisa, 1992). RT6 is expressed on 60–80% of peripheral T cells, both on $CD4^+$ and $CD8^+$ T cells (Greiner, 1987). $RT6^+$ cells have been shown to have immunoregulatory properties (Greiner, 1987). Elimination of $RT6^+$ cells using monoclonal antibody treatment in young nonlymphopenic nondiabetic DR-BB (Diabetes Resistant) rats (Greiner, 1987) turns such rats into diabetic animals, thereby highlighting the protective immunoregulatory role of the $RT6^+$ T cells. The linkage between lymphopenia and the development of diabetes in DP-BB rats is nevertheless not an absolute one, since Like *et al.* (1986) have developed a line of BB rats that are nonlymphopenic, have normal numbers of T cells, but still are diabetic – the 'nonlymphopenic diabetic' (NLD) Wor-BB rats. The principal international resource colony of BB rats is presently kept at the University of Massachusetts Medical Center, Worcester, USA, under the aegis of the National Institutes of Health (NIH).

LEC rat

The LEC mutant rat displays a maturational arrest in the development of $CD4^+CD8^+$ double positive (DP) into $CD4^+$ single positive (SP), but not into $CD8^+$ SP thymocytes (Agui, 1990, 1991). This maturation arrest is caused by a single resessive gene system defined as *thid* (T-helper immunodeficiency) (Yamada, 1991). This mutation is present in bone marrow-derived cells, but not in thymic stromal cells in the LEC rat (Agui, 1991). The maturational defect in the thymus results in a significant reduction of the number of peripheral $CD4^+$ T cells. $CD4^+$ T cells, however, are not completely absent, but show functional abnormalities (Sakai, 1995): they are incapable of IL-2 production upon Con A stimulation. The intrathymic maturational defect in LEC rats appears not to influence the development of $CD4^+$ intra-epithelial lymphocytes (IEL), since normal numbers of both $CD4^+CD8^+$ and $CD4^+CD8^-$ IEL can be found in LEC rats (Sakai, 1994). The LEC mutant strain is a Long Evans substrain and is kept at the Institute for Animal Experimentation, University of Tokushima, Japan.

BN rat

Brown Norway (BN) rats have generally been considered as poor immunological responders in various *in vivo* and *in vitro* assays (Rozing, 1979). Moreover, BN rats are highly susceptible for the induction of autoimmune glomerulonephritis with a short course of $HgCl_2$ (Aten, 1988). Using phenotypical analysis with antibodies allowing a TH1/TH2-like distinction (Groen, 1993) in combination with cytokine profiling, it was recently shown that the $CD4^+$ T-cell system of BN rats is skewed towards the (predominantly IL-4, IL-6 and IL-10 mediated) T_H2-type (Mathieson, 1992). This T_H2-type responsiveness might well explain the low responsiveness of BN rats in certain (IL-2 dependent) assay systems (Rozing, 1979), and the high sensitivity for antibody-mediated forms of autoimmunity (Aten, 1988). The 'immunodeficiency' of the BN rat is thereby merely a reflection of an unbalanced T helper system. BN rats are available from a number of commercial breeders.

Lewis rat

In contrast to BN rats Lewis rats have strong cellular immune responses (Rozing, 1979). Lewis rats, however, are skewed to T_H1-type cellular responses (predominantly IL-2 and IFN-γ mediated responses) (Groen, 1993; Beijleveld, 1996). This preferential T_H1-response profile Lewis rats might well be involved in the high susceptibility of Lewis rats to cell-mediated immunopathology and T_H1-mediated inducible autoimmunity such as experimental autoimmune encephalomyelitis (EAE) (Pat Happ, 1988;

Mustafa, 1994). As for the BN rat, the 'immunodeficiency' of the Lewis rat is then a reflection of an unbalanced T helper system. Lewis rats are available from a number of commercial breeders.

SCID rat

During backcrossing of the RP MHC locus on the Albino Oxford (AO) background at the University of Groningen, The Netherlands, a spontaneous mutation occurred with great resemblance to the SCID phenotype found in mice (defective in both the B- and T-cell system). During subsequent mating this mutation was lost. No other SCID-like rats have been reported to date.

Virus-induced/acquired immunodeficiency in the rat

Although a broad spectrum of viral infections has been described in the rat, with both suppressive and stimulatory effects on the immune system of the infected animal, no virus with properties comparable with human immunodeficiency virus (HIV), resulting in a severe acquired immunodeficiency syndrome, has been described in rats to date.

Gene-targeted genetic manipulation in rats

Transgenesis in the rat is now a well-established procedure for a number of specialized labs (Mullins, 1993). This technology obviously will also enable the generation of new immunodeficient rat models by silencing genes, through homologous recombination, that code for immunologically important proteins, as has been successfully applied in a wide variety of mouse genes in knock-out (KO) mice. In order to do so, however, *in vitro*-growing embryonic stem (ES) cells are essential for this procedure; at present rat ES-cells are not available. At the 11th International Workshop on Alloantigenic Systems in the Rat (Toulouse, France, 1996) preliminary data on the pluripotency of isolated rat ES cells were reported by P. M. Iannaccone and colleagues (Northwestern University Medical School, Chicago, USA). The development of a new generation of immunodeficient KO rats can therefore be expected in the nearby future. Another genetic approach for creating functional knock-out animals using antisense RNA transgenesis technology has already been applied succesfully in the rat (Matsumoto, 1995).

17. Autoimmunity

Overview

The rat has been used historically as an animal model for studies of drug toxicity, transplantation, immune reactivity, tumor biology, genetic analysis and autoimmunity. The many forms of autoimmunity that have been studied in rats can be divided into three general categories: (1) spontaneous, (2) induced, and (3) transgenic. The induced models of autoimmunity have been developed using a variety of methods, including immune manipulation, immunization (Table V.17.1), chemical administration, diet manipulation or infection. The most intensively studied autoimmune disorder in the rat is the sponta-

Table V.17.1 Immunization-induced autoimmune diseases in the rat

Autoimmune syndrome	*Abbreviation*	*Immunizing antigen*	*Model of human disease*
Adjuvant arthritis	AA	Incomplete Freund's adjuvant + *Mycobacterium tuberculosis* $H37_{Ra}$	Rheumatoid arthritis
Collagen-induced arthritis	CIA	Collagen in CFA	Rheumatoid arthritis
Dementia	EAD	High molecular weight neurofilament protein	Dementia
Encephalomyelitis	EAE	Myelin basic protein, proteolipid protein	Multiple sclerosis
Glomerulonephritis	GN	Glomerular basement membrane	Glomerulonephritis
Hepatitis	EAH	Liver proteins	Autoimmune hepatitis
Myasthenia gravis	EAMG	Torpedo acetylcholine receptor, AChR	Myasthenia gravis
Myocarditis	EAM	Myosin	Autoimmune myocarditis
Nephritis	HN	Renal tubular antigen	Heymann's nephritis
Neuritis	EAN	Peripheral nerve myelin	Guillain–Barré syndrome
Orchitis	EAO	Rat testicular antigen	Autoimmune orchitis
Pinealitis	EAP	Interphotoreceptor retinoid-binding protein, retinal S-antigen	Unknown
Sialadenitis	EAS	Salivary glands	Sjogren's syndrome
Thyroiditis	EAT	Thyroglobulin	Hashimoto's thyroiditis
Uveoretinitis	EAU	Interphotoreceptor retinoid-binding protein, retinal S-antigen	Autoimmune uveitis

neously diabetic BB rat model of human autoimmune insulin-dependent diabetes mellitus (IDDM). This model has been the subject for over 1400 scientific reports since its discovery in the mid 1970s (Crisá *et al.*, 1992; Rossini *et al.*, 1995; Mordes *et al.*, 1996b). Another widely studied rat model of autoimmunity is the induced model of experimental autoimmune encephalomyelitis (EAE). EAE is an animal model for human multiple sclerosis, and is the standard model for investigation of antigen-specific induced autoimmune syndromes (Swanborg, 1988; Swanborg and Stepaniak, 1996). Many of the findings obtained by investigating rat models of autoimmunity, particularly by the analysis of IDDM and EAE in rats, have general applicability, and have been translated into clinical therapeutic modalities. It is these 'general principals' that have been translated to the human autoimmune diseases that they model.

Spontaneous models of autoimmunity

The diabetes-prone bio-breeding (DP-BB) rat

The DP-BB rat spontaneously develops autoimmune insulin-dependent diabetes and thyroiditis (Crisá *et al.*, 1992; Rossini *et al.*, 1995; Mordes *et al.*, 1996b). The DP-BB rat was derived from a colony of outbred Wistar rats at the Bio-Breeding Laboratories (Ottawa, Canada) that were observed to spontaneously develop hyperglycemia and ketoacidosis. Affected animals at the sixth generation were used as founders of the inbred strains that are currently maintained in an international resource colony at the University of Massachusetts at Worcester which is sponsored by the National Institutes of Health (NIH). Greater than 90% of both sexes of DP-BB rats develop hyperglycemia between 50 and 90 days old which is associated with infiltration of the islets of Langerhans (insulitis) and selective destruction of the insulin-producing beta cells (Crisá *et al.*, 1992; Rossini *et al.*, 1995; Mordes *et al.*, 1996b). This is followed by the appearance of 'end stage' islets that are small, distorted, and lack β cells. Without exogenous insulin treatment, ketoacidosis ensues, followed rapidly by severe dehydration and death within a few days of the onset of hyperglycemia. Depending on the subline of DP-BB rats, the incidence of thyroiditis varies between 5 and 60% at 110 days old (Crisá *et al.*, 1992; Rossini *et al.*, 1995; Mordes *et al.*, 1996). The incidence of thyroiditis increases in older animals, but does not result in clinical symptoms of thyroid deficiency.

Several lines of evidence support the autoimmune nature of diabetes and thyroiditis disease in the DP-BB rat. First, there is a selective inflammatory infiltrate of the islets and thyroids. Second, autoantibodies to islet and thyroid antigens are detected in the circulation. Third, the disease can be adoptively transferred by bone marrow cells, T cells, and T cell lines. Fourth, both diabetes and thyroiditis can be prevented by immunosuppressive treatments such as neonatal thymectomy or treatment with immunosuppressive drugs. Finally, the disease can be prevented by adoptive transfer of T lymphocytes obtained from normal, disease-free, histocompatible strains of rats into DP-BB rats (Crisá *et al.*, 1992; Rossini *et al.*, 1995; Mordes *et al.*, 1996b).

A unique immunological feature of the DP-BB rat that is not observed in humans with IDDM is a severe deficiency of mature T cells, predominately $CD8^+$ and $RT6^+$ T lymphocytes. $CD8^+$ is expressed on a subset of T cells that mediate cytotoxic effector T cell function. RT6 is expressed on $\approx 50\%$ of $CD4^+$ (helper/inducer) T cells and $\approx 80\%$ of $CD8^+$ T cells. The deficiency of $RT6^+$ T cells in DP-BB has been found to be particularly important in the spontaneous development of the autoimmune disorders (Greiner *et al.*, 1987).

Studies of disease pathogenesis in the autoimmune DP-BB strain of rats has been complemented by studies of its induction in the congenic diabetes-resistant (DR-BB) strain. At the sixth generation, progeny resistant to diabetes were also selected for breeding, and the resistant lines are currently housed in the NIH-sponsored colony and are completely free of spontaneous disease (Crisá *et al.*, 1992; Rossini *et al.*, 1995; Mordes *et al.*, 1996). However, the diabetes and thyroiditis can be induced by a variety of interventions, including infection with Kilham rat virus (KRV), low-dose irradiation or cyclosphosamide, and *in vivo* depletion of $RT6^+$ T cells with an anti-RT6.1 cytotoxic monoclonal antibody (see below). The ability to induce disease in resistant DR-BB rats, and to prevent disease in susceptible DP-BB rats, led to the formulation of a 'balance hypothesis' of autoimmune IDDM (Mordes *et al.*, 1987, 1996a, b), a paradigm that has since been extended to many autoimmune diseases (Mordes *et al.*, 1996a, b). This general principal postulates that the expression of autoimmunity is a function of the relative balance between autoreactive $RT6^-$ effector T cells and $RT6^+$ regulatory T cells. The availability of the RT6 alloantigen expressed on the 'regulatory' T cell population and its absence of expression on the 'effector' T cell population has allowed a clear distinction between these two interactive T cell populations and intense investigation of their respective roles in the expression of autoimmunity.

BBZ/Wor rat model of obese autoimmune non-insulin-dependent diabetes

The spontaneously diabetic DP-BB rat was crossed with the obese Zucker rat, and numerous lines derived from these crosses have been characterized (Guberski *et al.*, 1988). Of particular interest are the obese diabetic line of BBZ/Wor rats. These rats develop a non-insulin-dependent diabetes associated with insulitis and substantial β cell loss, but can survive for months without the need for administration of exogenous insulin. The presence of insulitis suggests an autoimmune origin to this disorder. It was also discovered that the diabetes in the obese line of rats developed earlier than in the line of lean diabetic rats, suggesting that increased β-cell activity may allow enhanced targeting/susceptibility of the β cells to autoimmune destruction (Guberski *et al.*, 1993). Further characterization of this obese model of autoimmune diabetes may allow a clearer understanding of the relationship between obesity, β-cell metabolism, autoimmune insulitis and the genetic factors that are important in autoimmunity.

The female BUF rat model of spontaneous autoimmune thyroiditis

The appearance of thyroiditis in older female BUF rats led to the hypothesis that this was a spontaneous model of autoimmune thyroiditis (Rose *et al.*, 1976). This model was investigated intensely in the early 1970s, and the presence of inflammatory thyroid infiltrates, autoantibodies to thyroglobulin, and the provocation by neonatal thymectomy or treatment with methylcholanthrene, supported an autoimmune nature of its pathogenesis. However, upon rederivation of the BUF colony into viral antibody-free (VAF) status, where potentially immunomodulatory viruses and bacteria were eliminated from the colony, the spontaneous incidence of thyroiditis has decreased dramatically, and is now extremely rare in older, unmanipulated VAF female BUF rats. BUF rats have retained their susceptibility to autoimmunity as demonstrated by the facility of simple manipulation of the immune system to increase the frequency of thyroiditis dramatically (see below).

Autoimmunity induced by immune system modulation

Immune system modulation has been found to result in the expression of autoimmunity in 'normal' animals that never spontaneously develop autoimmunity. Much of this work can be interpreted in light of the balance hypothesis of autoimmunity that was developed in the BB rat model of IDDM. Observations across autoimmune disorders and in both normal and autoimmune-prone rat strains suggest that this paradigm may be an important component of the development and expression of autoimmunity not only in rats, but in other species as well.

Induced models of IDDM

As noted above, a subpopulation of nondiabetic BB rats at the sixth generation of inbreeding were selected and bred for resistance to IDDM. These rats are now designated diabetes-resistant (DR-BB) rats, which do not spontaneously develop IDDM. DR-BB rats display normal levels of all T-cell populations, but retain their susceptibility to IDDM induction. This is shown most dramatically by simple elimination of a subset of peripheral T cells that express the RT6 alloantigen. In conjunction with a low dose of an immune system activator, poly I:C, the incidence of diabetes approaches 100% in RT6-depleted DR-BB rats. IDDM can also be induced in DR-BB rats by low-dose irradiation or cyclosphosphamide, or by high doses of poly I:C (Mordes *et al.*, 1987, 1996a, b).

Diabetes can also be induced in DR-BB rats by an infectious agent, suggesting environmental control of autoimmune expression. Upon injection of high doses of KRV, up to 30% of DR-BB rats develop diabetes (Guberski *et al.*, 1991). When combined with either low dose poly I:C or depletion of $RT6^+$ T cells, the incidence of diabetes in KRV-infected DR-BB rats approaches 100%. These results support the 'balance hypothesis' of autoimmunity by demonstrating that alteration of this balance by reduction of a $RT6^+$ regulatory T cells, or by infection with an immunomodulatory virus, such as KRV, tips the balance in favor of the effector phase of autoimmunity. This paradigm has recently been extended to other rat strains that fail spontaneously to develop autoimmunity. Greater than 90% of PVG.$RT1^u$ or LEW.WR1 (RT1 $A^uB/D^u C^a$) congenic rats infected with KRV and injected with poly I:C (Ellerman *et al.*, 1996), and >75% of PVG.$RT1^u$ rats that are treated with anti-RT6-antibody plus poly I:C, express autoimmune IDDM (Whalen *et al.*, 1997).

Autoimmune diabetes can also be induced in normal PVG rats by thymectomy and irradiation (Stumbles and Penhale, 1993). Up to 34% of such manipulated animals will develop diabetes within 10 weeks of treatment. The disease in these animals appears to be mediated by a population of $CD4^+CD45^{hi}$-expressing T cells, and can be prevented by a population of $CD4^+RT6^+CD45^{lo}$-expressing T cells (Fowell and Mason, 1993; Fowell *et al.*, 1993). The protective population of $CD4^+$ T cells expresses IL-4 and is T_H2-like, suggesting that the protective regulatory T cells may mediate their immunoregulatory effects upon effector T cells through cytokine secretion.

Syngeneic graft-vs-host (GvH) reactivity

A GvH like syndrome can be induced in LEW rats by irradiation, syngeneic bone marrow reconstitution, and a short-term treatment with cyclosporin A followed by withdrawal of the drug treatment (Fischer *et al.*, 1989). Within 2 months of drug withdrawal, an autoimmune disease syndrome resembling syngeneic GvH is observed. As in $HgCl_2$-induced autoimmune glomerulonephritis (see below), LEW rats are susceptible to induction of autoimmunity using this protocol while BN rats are resistant (Ohajekwe *et al.*, 1995). This model has also been suggested to correlate with autoimmune scleroderma in humans (Bos *et al.*, 1989). The amount of irradiation administered alters the severity of the induced syngeneic GvH, and the autoimmune nature of the disease has been demonstrated by adoptive transfer, which can be inhibited by co-administration of normal T cells (Fischer *et al.*, 1989). Again, the data suggest that a balance exists between regulatory and effector cells in normal rats, and that alteration of this balance can lead to the expression of autoimmunity.

Induced BUF rat model of thyroiditis

Female BUF rats maintained under VAF conditions fail spontaneously to develop thyroiditis with aging. However, thyroiditis in these animals can be induced by neonatal thymectomy, and the incidence can be increased by administration of exogenous iodine in their food or drinking water (Allen and Braverman, 1990; Cohen *et al.*, 1990). These observations support the concept that both immune manipulation and environmental influences can modulate the expression of autoimmunity in susceptible strains of rats.

Autoimmunity induced by immunization

Multiple autoimmune diseases have been experimentally induced in rats by immunization with syngeneic tissues in complete Freund's adjuvant. The female LEW rat is the most commonly used strain to model immunization-induced autoimmunity, and the most studied disease syndrome is experimental autoimmune encephalomyelitis (EAE). Many common principles have been elucidated using this experimental system. The role of effector cells, regulatory cells, and tolerization protocols, including orally induced tolerance, in autoimmunity have been investigated using this model system. In addition, the role of specific immunomodulatory T-cell subsets, cytokines, and autoantigen recognition have been defined in the LEW rat EAE model (Swanborg, 1988; Swanborg and Stepaniak, 1996). This model is easily studied owing to (1) its relative ease of inducibility, (2) its reproducibility, and (3) its high incidence of severe disease. It has also been used to demonstrate that two separate autoantigens, myelin basic protein (MBP) and proteolipid protein (PLP) can induce similar disease syndromes, and that the immune response to these defined autoantigens are MHC-restricted.

This model is a classic example of the balance hypothesis of autoimmunity. MBP-autoreactive T cells have been demonstrated in the circulation of both susceptible and resistant strains of rats (Swanborg, 1995). This observation engendered analysis of circulating lymphocyte populations in humans, where MBP autoreactive T cells can also be demonstrated (Swanborg, 1995). The presence of these autoreactive cell populations in the absence of pathology or disease has been hypothesized to be due to the presence of immunoregulatory elements in the normal immune system that keep the autoreactive T cells in check. The mechanism by which this occurs is not known, but these observations lend support to the balance hypothesis of autoimmunity formulated in the BB rat model of autoimmunity.

Another widely used rat model of autoimmunity induced by immunization is adjuvant arthritis (van Eden *et al.*, 1996). This disease syndrome is easily induced in rats by immunization with mycobacteria suspended in oil, while mice are generally not susceptible. Adjuvant arthritis in the rat is used to model human rheumatoid arthritis because of the lack of well-defined autoantigen targets and the generalized inflammatory nature of the disease (van Eden *et al.*, 1996). This latter characteristic has led to extensive use of this model for testing anti-inflammatory drugs.

The general observation in immunization-induced autoimmunity in rats is that almost any autoantigen emulsified in complete Freund's adjuvant can lead to induction of autoimmunity. The autoantigens encompass a wide array of cell types and tissues, and the induced autoimmune disorders have been postulated to represent many autoimmune human disorders with similar pathology (see Table V.17.1).

Chemically induced autoimmunity

The rat has been used to study the immunotoxic effects of drugs, and in particular the effect of chemically-induced autoimmunity by exposure to mercury (Bigazzi, 1994; Druet, 1995). Three rat strains are susceptible to the autoimmune effects of mercury, the BN, MAXX, and DZB inbred strains, while all other strains tested to date are resistant. Following multiple low doses of $HgCl_2$, a self-limiting course of autoimmune glomerulonephritis develops, peaking at around 2 weeks, and regressing spontaneously by around 4 weeks. Continuous treatment with $HgCl_2$ fails to alter the monophasic course of disease and, following recovery, the animals are resistant to further treatment with mercury. The glomerulonephritis is characterized by proteinuria caused by the deposition of autoantibodies to epitopes of the renal glomerular basement membrane, including laminin. The disease is autoimmune in that autoantibodies are produced, and the disease can be transferred by lymphocytes to otherwise resistant strains. In addition, it has been postulated that the expression of autoimmunity following exposure to mercury is the result of a direct effect on regulatory cell activity (Kosuda *et al.*, 1991; Druet, 1994, 1995). The level of $RT6^+$ T cells, a population known to prevent autoimmunity in the BB strain, decreases concomitantly with the induction of glomerulonephritis in susceptible strains (Kosuda *et al.*, 1991). Mercury exposure is postulated to induce suppressor cell activity in resistant strains such as the LEW rat. The ability of mercury to modulate autoreactive effector and regulatory cell function, and either induce or prevent autoimmunity in rats, may lead to our understanding of the divergent outcomes that are observed following exposure of humans to xenobiotics.

In addition to mercury, cadmium, gold salts and D-penicillamine are also able to induce autoimmunity in BN rats (Bigazzi, 1994; Druet, 1994, 1995).

Transgenic rat models of autoimmunity

Although many transgenic mouse models of autoimmunity have been developed, there are relatively few transgenic rat models available, and the only transgenic rat model of autoimmunity developed to date is the HLA-B27 transgenic rat. This transgenic animal develops a spontaneous multisystem inflammatory disease that resembles human B27-associated disease (Breban *et al.*, 1996). The pathology includes skin lesions with marked psoriasiform dermatitis and progressive alopecia, chronic colitis resembling ulcerative colitis, joint inflammation, and male genital inflammatory lesions. The development of these systemic inflammatory lesions does not occur when the transgenic rats are maintained in a germfree state (Taurog *et al.*, 1994). The autoimmune disease can be adoptively transferred by bone marrow or fetal liver cells that express the B27 transgene, and is T cell dependent (Breban *et al.*, 1996).

Other transgenic rats have been developed, including rat transgenic models for Charcot–Marie–Tooth disease (Sereda *et al.*, 1996), hypertension (Wagner *et al.*, 1996), obese non-insulin responsive glucose intolerance (Rosella *et al.*, 1995), and many other syndromes. However, none of the other transgenic rats display diseases of an autoimmune pathogenesis.

Additional rat models of autoimmunity

Additional rat models of autoimmunity include dietary induction of thyroiditis by high levels of iodine intake, (Mooij *et al.*, 1994), and autoimmunity induced by infectious agents. These latter models include measles virus-induced autoimmune reactions against brain antigen in rats (ter Meulen and Liebert, 1993), and *Mycoplasma arthritidis* infection, which causes severe polyarthritis and resembles human rheumatoid arthritis (Kirchhoff *et al.*, 1989). These models demonstrate environmental influences on the induction of autoimmunity in otherwise normal strains of rats.

Conclusion

The rat has been used for the study of many spontaneous and induced autoimmune syndromes. The observations obtained from these investigations have provided insights into the function of the normal immune system, and the regulatory imbalances that lead to the expression of autoimmunity. The utility of the rat models has been extensive, and has led to numerous clinical trials and the development of many drugs for treatment of human diseases. It should be cautioned, however, that no rodent model of human autoimmunity completely parallels the human syndrome in pathogenesis, etiology or clinical outcome. However, general principals revealed in the rat models, such as the balance hypothesis of regulatory and effector T cell modulation of the expression of autoimmunity, appear to be translated across many autoimmune disorders and across many species. It is with this understanding that the rat models of autoimmunity will lead to rational and therapeutic application to human autoimmune diseases.

18. References

Abraham, J. A., Damm, D., Bajardi, A., Miller, J., Klagsbrun, M., Ezekowitz, R. A. B. (1993). Heparin-binding EGF-like growth factor: characterization of rat and mouse cDNA clones, protein domain conservation across species, and transcript expression in tissues. *Biochem. Biophys. Res. Commun.* **190**, 125–133.

Acosta-Altamirano, G., Burranco-Acosta, C., van Roost, E., Vaerman, J. P. (1980). Isolation and characterization of secretory IgA (SIgA) and free secretory component (FSC) from rat bile. *Mol. Immunol.* **17**, 1525–1537.

Ager, A. (1994). Lymphocyte recirculation and homing: Roles of adhesion molecules and chemoattractants. *Trends Cell Biol.* **4**, 326–333.

Agui, T., Oka, M., Yamada, T., Sakai, T., Izumi, K., Ishida, Y., Himeno, K., Matsumoto, K. (1990). Maturational arrest from $CD4^+8^+$ to $CD4^+8^-$ thymocytes in a mutant strain (LEC) of rats. *J. Exp. Med.* **172**, 1615–1624.

Agui, T., Sakai, T., Himeno, K., Matsumoto, K. (1991a). Bone marrow-derived progenitor T cells convey the origin of maturational arrest from $CD4^+CD8^+$ to $CD4^+CD8^-$ thymocytes in LEC mutant rats. *Eur. J. Immunol.* **21**, 2277–2280.

Agui, T., Sakai, T., Matsumoto, K. (1991b). Ontogeny of T-cell maturation in LEC mutant rats which bear a congenital arrest of maturation from $CD4^+CD8^+$ to $CD4^+8^-$thymocytes. *Eur. J. Immunol.* **21**, 2537–2541.

Aguilar, B. A., Gutman, G. A. (1992). Transcription and diversity of immunoglobulin lambda chain variable genes in the rat. *Immunogenetics* **37**, 39–48.

Akasaki, K., Fukuzawa, M., Kinoshita, H., Furuno, K., Tsuji, H. (1993). Cyclin of two endogenous lysosomal membrane proteins, lamp-2 and acid phosphatase, between the cell surface and lysosomes in cultured rat hepatocytes. *J. Biochem. Tokyo* **114**, 598–604.

Allan, C. H., Mendrick, D. L., Trier, J. S. (1993). Rat intestinal M cells contain acidic endosomal-lysosomal compartments and express class II major histocompatibility complex determinants. *Gastroenterology* **104**, 698–708.

Allen, E. M., Braverman, L. E. (1990). The effect of iodine on lymphocytic thyroiditis in the thymectomized buffalo rat. *Endocrinology* **127**, 1613–1616.

Anderson, A. O., Shaw, S. (1993). T-cell adhesion to endothelium: the FRC conduit system and other anatomic and molecular features which facilitate the adhesion cascade in lymph nodes. *Sem. Immunol.* 5, 271–282.

Appel, K., Honegger, P., Gebicke-Haerter, P. J. (1995a). Expression of interleukin-3 and tumor necrosis factor-β mRNAs in cultured microglia. *J. Neuroimmunol.* **60**, 83–91.

Appel, K., Buttini, M., Sauter, A., Gebicke-Haerter, P. J. (1995b). Cloning of rat interleukin-3 receptor beta-subunit from cultured microglia and its mRNA expression *in vivo*. *J. Neurosci.* **15**, 5800–5809.

Arai, M., Mochida, S., Ohno, A., Ogata, A., Obama, H., Maruyama, I., Fujiwara, K. (1995). Blood coagulation equilibrium in rat liver microcirculation as evaluated by endothelial cell thrombomodulin and macrophage tissue factor. *Thromb. Res.* **15**, 113–123.

Arden, B., Clark, S. P., Kabelitz, D., Mak, T. W. (1995). Mouse T-cell receptor variable gene segment families. *Immunogenetics* **42**, 501–530.

Armeding, D. (1971). Two allotypic specificities of rat immunoglobulins. *Eur. J. Immunol.* **1**, 39–45.

Arnason, B. G., de Vaux St. Cyr, C., Grabar, P. (1963). Immunoglobulin abnormalities of the thymectomized rats. *Nature* **199**, 1199–1200.

Arnason, B. G., de Vaux St. Cyr., C., Relyveld, E. H. (1964). Role of the thymus in immune reactions in rats. IV. Immunoglobulins and antibody formation. *Int. Arch. Allergy Appl. Immunol.* **25**, 206–224.

Arroyave, C. M., Levy, R. N, Johnson, J. S. (1979). Genetic deficiency of the fourth component (C4) in wistar rats. *Immunology* **33**, 453–459.

Arvieux, J., Willis, A. C., Williams, A. F. (1986). MRC OX-45: a leukocyte/endothelium rat membrane glycoprotein of 45,000 molecular weight. *Mol. Immunol.* **23**, 983–990.

Asmuß, A., Hofmann, K., Hochgrebe, T., Giegerich, G., Hünig, T., Herrmann, T. (1996). Alleles of highly homologous rat T-cell receptor β-chain variable segments 8.2 and 8.4. *J. Immunol.* **157**, 4436–4441.

Aspinall, R., Kampinga, J., van Den Bogaerde, J. (1991). T-cell development in the fetus and the invariant series hypothesis. *Immunol. Today* **12**, 7–10.

Aten, J., Bosman, C. B., Rozing, J., Stijnen, T., Hoedemaeker, P. J., Weening, J. J. (1988). Mercuric chloride-induced autoimmunity in the Brown Norway rat. Cellular kinetics and major histocompatibility complex antigen expression. *Am. J. Pathol.* **133**, 127–138.

Aten, J., Veninga, A., de Heer, E., Rozing, J., Nieuwenhuis, P., Hoedemaeker, P. J., Weening, J. J. (1991). Susceptibility to the induction of either autoimmune reactivity or non-specific immunosuppression by mercuric chloride is related to the MHC class II haplotype. *Eur. J. Immunol.* **21**, 611–616.

Aten, J., Veninga, A., Coers, W., Sonnenberg, A., Timpl, R., Claessen, N., van Eendenburg, J. D. H., De Heer, E., Weening, J. J. (1995). Autoantibodies to the laminin P1-fragment in $HgCl_2$-induced membranous glomerulopathy. *Am. J. Pathol.* **146**, 1467–1480.

Auchampach, J. A., Oliver, M. G., Anderson, D. C., Wanning, A. M. (1994). Cloning, sequence comparison and *in vivo* expression of the gene encoding rat P-selectin. *Gene* **145**, 251–255.

Auerbach, R., Alby, L., Grieves, J., Joseph, J., Lindgren, C., Morrissey, L. W., Sidky, Y. A., Tu, M., Watt, S. L. (1982). Monoclonal antibody against angiotensin-converting enzyme: its use as a marker for murine, bovine, and human endothelial cells. *Proc. Natl Acad. Sci. USA* **79**, 7891–7895.

Awata, T., Guberski, D. L., Like, A. A. (1995). Genetics of the BB rat: association of autoimmune disorders (diabetes, insulitis, and thyroiditis) with lymphopenia and major histocompatibility complex class II. *Endocrinology* **136**, 5731–5735.

Bailyes, E. M., Soos, M., Jackson, P., Newby, A. C., Siddle, K., Luzio, J. P. (1984). The existence and properties of two dimers of rat liver ecto-5′-nucleotidase. *Biochem. J.* **221**, 369–377.

Barabas, A. Z., Kelus, A. S. (1967). Allotypic specificity of serum protein in inbred strains of rats. *Nature* **215**, 155.

Barbé, E., Huitinga, I., Döpp, E. A., Bauer, J., Dijkstra, C. D. (1996). A novel bone marrow frozen section assay for studying hemopoietic interactions *in situ*: the role of stromal macrophages in erythroblast binding. *J. Cell Sci.* **109**, 2937–2945.

Baumann, H., Gauldie, J. (1990). Regulation of hepatic acute phase plasma protein genes by hepatocyte stimulating factors and other mediators of inflammation. *Mol. Biol. Med.* 7, 149–159.

Baumann, H., Schendel, P. (1991). Interleukin-11 regulates the hepatic expression of the same plasma protein genes as interleukin-6. *J. Biol. Chem.* **266**, 20424–20427.

Baumann, M., Baumann, H., Fey, G. H. (1990). Molecular cloning, characterization and functional expression of the rat liver interleukin 6 receptor. *J. Biol. Chem.* **15**, 19853–19862.

Baumann, H., Morella, K. K., Wong, G. H. (1993). TNF-alpha, IL-1 beta, and hepatocyte growth factor cooperate in stimulating specific acute phase protein genes in rat hepatoma cells. *J. Immunol.* **151**, 4248–4257.

Bautista, A. P., Spolarics, Z., Jaeschke, H., Smith, C. W., Spitzer, J. J. (1993). Monoclonal antibody against the CD18 adhesion molecule stimulates glucose uptake by the liver and hepatic nonparenchymal cells. *Hepatology* **17**, 924–931.

Bazin, H. (1967a). Les immunoglobulines de la souris. *Nouv. Rev. Fr. Hematol.* 7, 507–526.

Bazin, H. (1967b). Mouse immunoglobulins. II. Study of the antibodies synthesized in the various classes of immunoglobulins in response to the injection of 2 soluble protein antigens. *Ann. Inst. Pasteur Paris* **112**, 162–172.

Bazin H. (ed.) (1990). *Rat Hybridomas and Rat Monoclonal Antibodies*. CRC Press Inc., Boca Raton, FL, USA.

Bazin, H., Beckers, A. (1976). IgE myelomas in rats. Nobel Symposium No. 33. Molecular and biological aspects of the acute allergic reaction, Stockholm (eds Johansson, S. G. O., Stranberg, K., Uvnas, B.), p. 125–152. Plenum Company, New York, USA.

Bazin, H., Deckers, A., Beckers, A., Heremans, J. F. (1972). Transplantable immunoglobulin-secreting tumours in rats. I. General features of LOU/Wsl strain rat immunocytomas and their monoclonal proteins. *Int. J. Cancer* **10**, 568–580.

Bazin, H., Beckers, A., Deckers, C., Moriamé, M. (1973). Transplantable immunoglobulin-secreting tumours in rats. V. Monoclonal immunoglobulins secreted by 250 ileocecal immunocytomas of the LOU/Wsl rats. *J. Natl Cancer Inst.* **51**, 1359–1361.

Bazin, H., Beckers, A., Querinjean, P. (1974a). Three classes and four (sub)classes of rat immunoglobulins: IgM, IgA, IgE, and IgG1, IgG2a, IgG2b, IgG2c. *Eur. J. Immunol.* **4**, 44–48.

Bazin, H., Querinjean, P., Beckers, A., Heremans, J. F., Dessy, F. (1974b). Transplantable immunoglobulin-secreting tumours in rats. IV. Sixty-three IgE-secreting immunocytoma tumours. *Immunology* **26**, 713–723.

Bazin, H., Beckers, A., Vaerman, J. P., Heremans, J. F. (1974c). Allotypes of rat immunoglobulins. I. An allotype at the alpha-chain locus. *J. Immunol.* **112**, 1035–1041.

Bazin, H., Beckers, A., Urbain-Vansanten, G., Pauwels, R., Bruyns, C., Tilkin, A. F., Platteau, B., Urbain, J. (1978). Transplantable IgD immunoglobulin-secreting tumours in rats. *J. Immunol.* **121**, 2077–2082.

Bazin, H., Gray, D., Platteau, B., MacLennan, I. C. M. (1982). Distinct $\delta+$ and $\delta-$ B lymphocyte lineages in the rat. *Ann. NY Acad. Sci.* **399**, 157–174.

Bazin, H., Platteau, B., MacLennan, I. C. M., Johnson, G. D. (1985). B-cell production and differentiation in adult rats. *Immunology* **54**, 79–88.

Bazin, H., Rousseaux, J., Rousseaux-Prévost, R., Platteau, B., Querinjean, P., Malache, J. M., Delaunay, T. (1990). Rat immunoglobulins. In *Rat Hybridomas and Rat Monoclonal Antibodies* (ed. Bazin, H.), pp. 5–42. CRC Press Inc., Boca Raton, FL, USA.

Beckers, A., Bazin, H. (1975). Allotypes of rat immunoglobulins. III. An allotype of the gamma2b-chain locus. *Immunochemistry* **12**, 671–675.

Beckers, A., Querinjean, P., Bazin, H. (1974). Allotypes of rat immunoglobulins. II. Distribution of the allotypes of kappa and alpha chain loci in different inbred strains of rat. *Immunochemistry* **11**, 605–609.

Beelen, R. H. J., Walker, W. S. (1983). Dynamics of cytochemically distinct subpopulations of macrophages in elicited rat peritoneal macrophages. *Cell. Immunol.* **82**, 246.

Beelen, R. H. J., Eestermans, I. L., Döpp, E. A., Dijkstra, C. D. (1987). Monoclonal antibodies ED1, ED2 and ED3 against rat macrophages: expression of recognized antigens in different stages of differentiation. *Transpl. Proc. XIX* **3**, 3166–3170.

Beijleveld, L. J. J., Groen, H., Broeren, C. M., Klatter, F. A., Kampinga, J., Damoiseaux, J. G. M. C., van Breda Vriesman, P. J. C. (1996). Susceptibility to clinically manifest cyclosporin A (CsA)-induced autoimmune disease is associated with interferon-gamma producing $CD45RC^{+}RT6^{-}$ T helper cells. *Clin. Exp. Immunol.* **105**, 486–496.

Bell, E. B., Sparshott, S. M. (1990). Interconversion of CD45R subsets of CD4 T cells *in vivo*. *Nature* **348**, 163.

Bell, E. B., Sparshott, S. M., Ager, A. (1995). Migration pathways of CD4 T cell subsets *in vivo*: the $CD45RC^{-}$ subset enters the thymus via α4 integrin–VCAM-1 interaction. *Int. Immunol.* **7**, 1861–1871.

Bell, G. M., Seaman, W. E., Niemi, E. C., Imboden, J. B. (1992). The OX-44 molecule couples to signalling pathways and is associated with CD2 on rat T lymphocytes and a natural killer cell line. *J. Exp. Med.* **175**, 527–536.

Bellacosa, A., Lazo, P. A., Bear, S. E., Tsichlis, P. N. (1991). The rat leukocyte antigen MRC OX-44 is a member of a new family of cell surface proteins which appears to be involved in growth regulation. *Mol. Cell. Biol.* **11**, 2864–2872.

Berkenbosch, F., van Oers, J., Del Rey, A., Tilders, F., Besedovsky, H. (1987). Corticotropin-releasing factor-producing neurons in the rat activated by interleukin-1. *Science* **238**, 524–526.

Berman, J. W., Guida, M. P., Warren, J., Amat, J., Brosnan, C. F. (1996). Localization of monocyte chemoattractant peptide-1 expression in the central nervous system in experimental autoimmune encephalomyelitis and trauma in the rat. *J. Immunol.* **156**, 3017–3023.

Berridge, M. V., O'Kech, N., McNeilage, L. J., Heslop, B. F., Moore, R. (1979). Rat mutant (NZNU) showing 'nude' characteristics. *Transplantation* **6**, 410–413.

Beyers, A. D., Spruyt, L. L., Williams, A. F. (1992). Molecular associations between the T-lymphocyte antigen receptor complex and the surface antigens CD2, CD4, or CD8 and CD5. *Proc. Natl Acad. Sci. USA* **89**, 2945–2949.

Bienenstock, J., Johnston, N., Perey, D. Y. E. (1973). Bronchial lymphoid tissue. I. Morphologic characteristics. *Lab. Invest.* **28**, 686–692.

Biewenga, J., van Rees, E. P., Sminia, T. (1993). Induction and regulation of IgA responses in the microenvironment of the gut. *Clin. Immunol. Immunopathol.* **67**, 1–13.

Billups, K. L., Sherley, J. L., Palladino, M. A., Tindall, J. W., Roberts, K. P. (1995). Evidence for E-selectin complement regulatory domain mRNA splice variants in the rat. *J. Lab. Clin. Med.* **126**, 580–587.

Bigazzi, P. E. (1994). Autoimmunity and heavy metals. *Lupus* **3**, 449–453.

Binaghi, R., Sarando de Merlo, E. (1966). Characterization of rat IgA and its non-identity with the anaphylactic antibody. *Int. Arch. Allergy Appl. Immunol.* **30**, 589–596.

Bland, P. W., Warren, L. G. (1986). Antigen presentation by epithelial cells of the rat small intestine. I. Kinetics, antigen specificity and blocking by anti-Ia sera. *Immunology* **58**, 1–8.

Blankenhorn, E. P., Cecka, J. M., Götze, D., Hood, L. (1978). Partial *N*-terminal amino acid sequence of rat transplantation antigens. *Nature*, **274**, 90–92.

Blankenhorn, E. P., Symington, F. W., Cramer, D. V. (1983). Biochemical characterization of Ia antigens encoded by the RT1.B and RT1.D loci in the rat MHC. *Immunogenetics* **17**, 475–484.

Blankenhorn, E. P., Stranford, S., Smith, P. D., Hickey, W. F. (1991). Genetic differences in the T cell receptor alleles of LEW rats and the encephalomyelitis resistant derivative, LER; and their impact on the inheritance of EAE resistance. *Eur. J. Immunol.* **21**, 2033–2041.

Blankenhorn, E., Smith, P. D., Williams, C. B., Gutman, G. A. (1992). Alleles of the rat T-cell receptor β chain complex. *Immunogenetics* **35**, 324–331.

Blankenhorn, E. P., Stranford, S. A., Martin, A.-M., Hickey, W. F. (1995). Cloning of myelin basic-protein reactive T-cells from the experimental allergic encephalomyelitis-resistant rat strain, LER. *J. Neuroimmunol.* **59**, 173–183.

Blasband, A. J., Rogers, K. T., Chen, X., Azizkhan, J. C., Lee, D. C. (1990). Characterization of the rat transforming growth factor alpha gene and identification of promoter sequences. *Mol. Cell. Biol.* **10**, 2111–2121.

Blaschke, V., Micheel, B., Pabst, R., Westermann, J. (1995). Lymphocyte traffic through lymph nodes and Peyer's patches of the rat: B- and T-cell-specific migration patterns within the tissue, and their dependence on splenic tissue. *Cell Tissue Res.* **282**, 377–386.

Bocek, P., Pecht, I. (1993). Cloning and sequence of the cDNA coding for rat type II Fc gamma receptor of mast cells. *FEBS Lett.* **331**, 86–90.

Bogden, A. E., Aptekman, P. M. (1960). The 'R-1' factor: a histocompatibility antigen in the rat. *Cancer Res.* **20**, 1372.

Bogers, W. M. J. M., van Rooijen, N., Janssen, D. J., van Es,

L. A., Daha, M. R. (1993). Complement enhances the elimination of soluble aggregates of IgG by rat liver endothelial cells *in vivo*. *Eur. J. Immunol.* 23, 433–438.

Borges, L. G., Seifert, R. A., Grant, F. J., Hart, C. E., Disteche, C. M., Edelhoff, S., Solca, F. F., Liberman, M. A., Lindner, V., Fischer, E. H., Lok, S., Bowen-Pope, D. F. (1996). Cloning and characterization of rat density-enhanced phosphatase-1, a proteintyrosine kinase phosphatase expressed by vascular cells. *Circ. Res.* 79, 570–580.

Borycki, A. G., Guillier, M., Leibovitch, M. P., Leibovitch, S. A. (1992). Molecular cloning of CSF-1 receptor from rat myoblasts. Sequence analysis and regulation during myogenesis. *Growth Factors* 6, 209–218.

Bos, G. M., Majoor, G. D., Willighagen, R. G., van Breda Vriesman, P. J. (1989). Chronic cyclosporine-induced autoimmune disease in the rat: a new experimental model for scleroderma. *J. Invest. Dermatol.* 93, 610–615.

Brandt, J., Pippin, J., Schulze, M., Hänsch, G. M., Alpers, C. E., Johnson, R. J., Gordon, K., Couser, W. G. (1996). Role of the complement membrane attack complex (C5–C9) in mediating experimental menangio proliferative glomerulonephritis. *Kidney Int.* 49, 335–343.

Brauer, R. B., Baldwin III, W. M., Wang, D., Hirwitz, L. R., Hess, A. D., Klein, A. S., Sanfilipo, S. (1994). Hepatic and extrahepatic biosynthesis of complement factor C6 in the rat. *J. Immunol.* 153, 3168–3176.

Breban, M., Fernandez-Sueiro, J. L., Richardson, J. A., Hadavand, R. R., Maika, S. D., Hammer, R. E., Taurog, J. D. (1996). T cells, but not thymic exposure to HLA-B27, are required for the inflammatory disease of HLA-B27 transgenic rats. *J. Immunol.* 156, 794–803.

Brideau, R. J., Carter, P. B., McMaster, W. R., Masson, D. W., Williams, A. F. (1980). Two subsets of rat T lymphocytes defined with monoclonal antibodies. *Eur. J. Immunol.* 10, 609–615.

Brisette-Storkus, C., Kaufmann, C. L., Pasewitz, L., Worsey, H. M., Lakomy, R., Ildstad, S. T., Chambers, W. H. (1994). Characterization and function of the NKR-P1dim/T cell receptor-$\alpha\beta+$ subset of rat T cells. *J. Immunol.* 152, 388–396.

Bristulf, J., Gatti, S., Malinowsky, D., Bjork, L., Sundgren, A. K., Bartfai, T. (1994). Interleukin-1 stimulates the expression of type I and type II interleukin-1 receptors in the rat insulinoma cell line Rinm5F: sequencing a rat type II interleukin-1 receptor cDNA. *Eur. Cytokine Netw.* 5, 319–330.

Broeren, C. P., Wauben, M. H., Lucassen, M. A., van Meurs, M., van Kooten, P. J., Boog, C. J., Claassen, E., van Eden, W. (1995). Activated rat T-cells synthesize and express functional major histocompatibility class II antigens. *Immunology* 84, 193–201.

Broug-Holub, E., Persoons, J. H., Schornagel, K., Kraal, G. (1995). Changes in cytokine and nitric oxide secretion by rat alveolar macrophages after oral administration of bacterial extracts, *Clin. Exp. Immunol.* 101, 302–307.

Brouwer, E., Huitema, M. G., Klok, P. A., De Weerd, H., Tervaert, J. W., Weening, J. J., Kallenberg, C. G. (1993). Antimyeloperoxidase-associated proliferative glomerulonephritis: an animal model. *J. Exp. Med.* 177, 905–914.

Brüggemann, M. (1988). Evolution of a rat immunoglobulin gamma heavy-chain gene family. *Gene* 74, 473–482.

Brüggemann, M., Free, J., Diamon, A., Hobard, J., Cobbold, S., Waldmann, H. (1986). Immunoglobulin heavy chain locus of the rat: stricking homology to mouse antibody genes. *Proc. Natl Acad. Sci. USA* 83, 6075–6079.

Brüggemann, M., Delmastro-Galgre, P., Waldmann, H., Calabi, F. (1988). Sequence of a rat immunoglobulin gamma 2c heavy chain constant region cDNA: extensive homology to mouse gamma 3. *Eur. J. Immunol.* 18, 317.

Bukovsky, A., Presl, J., Zidovsky, J., Mancal, P. (1983). The localization of Thy-1.1, MRC OX 2 and Ia antigens in the rat ovary and fallopian tube. *Immunology* 48, 587–596.

Bulmer, D. (1968). Further studies on the granulated metrial gland cells of the pregnant rat. *J. Anat.* 103(3), 479–489.

Burke, S., Landau, S., Green, R., Tseng, C. C., Nattakom, T., Canchis, W., Yang, L., Kaiserlai, D., Gespach, C., Balk, S., Blumberg, R. (1994). Rat cluster of differentiation molecule 1: expression on the surface of intestinal epithelial cells and hepatocytes. *Gastroenterology* 106, 1143–1149.

Burns, F. R., Li, X., Shen, N., Offner, H., Chou, Y. K., Vandenbark, A. A., Heber-Katz, E. (1989). Both rat and mouse T cell receptors specific for the encephalitogenic determinant of myelin basic protein use similar $V\alpha$ and $V\beta$ chain genes even though the major histocompatibility complex and encephalitogenic determinants being recognized are different. *J. Exp. Med.* 169, 27–39.

Butcher, G. W. (1987). A list of monoclonal antibodies specific for alloantigens of the rat. *Immunogenetics* 14, 163–176.

Campbell, D. G., Gagnon, J., Reid, K. B., Williams, A. F. (1981). Rat brain Thy-1 glycoprotein. The amino acid sequence, disulphide bonds and an unusual hydrophobic region. *Biochem J.* 195, 15–30.

Carey, D. J., Stahl, R. C., Cizmecci-Smith, G., Asundi, V. K. (1994). Syndecam-1 expressed in Schwann cells cause morphological transformation and cytoskeletal reorganization and associates with actin during cell spreading. *J. Cell Biol.* 124, 161–170.

Carlsson, L., Övermo, C., Holmberg, D. (1992). Developmentally controlled selection of antibody genes: characterization of individual V_H7183 genes and evidence for stage-specific somatic diversification. *Eur. J. Immunol.* 22, 71–78.

Carroll, P., Sendtner, M., Meyer, M., Thoenen, H. (1993). Rat ciliary neurotrophic factor (CNTF): gene structure and regulation of mRNA levels in glial cell cultures. *Glia* 9, 176–187.

Carter, P., Bazin, H. (1980). Immunology. In *The Laboratory Rat*, Vol. 2, pp. 181–212. Academic Press, New York, USA.

Carvalho, G. S., Luzio, J. P., Siddle, K., Coombs, R. R. (1987). Detection of ecto-5′-nucleotidase on rat B and T lymphocyte subpopulations using a monoclonal antibody in rosetting reactions. *J. Immunol. Methods* 102, 119–126.

Cattaneo, A., Rapposelli, B., Calissano, P. (1988). Three distinct types of monoclonal antibodies after long-term immunization of rats with mouse nerve growth factor. *J. Neurochem.* 50, 1003–1010.

Cerf-Bensussan, N., Guy-Grand, D., Lisowska-Grospierre, B., Griscelli, C., Bhan, A. K. (1986). A monoclonal antibody specific for rat intestinal lymphocytes. *J. Immunol.* 136, 76–82.

Chan, E. Y. T., MacLennan, I. C. M. (1993). Only a small proportion of splenic B cells in adults are short-lived cells. *Eur. J. Immunol.* 20, 557–363.

Chao, D., MacPherson, G. G. (1989). Lymph node macrophage heterogeneity: the phenotypic and functional characterization of two distinct populations of macrophages from rat lymph node. *Eur. J. Immunol.* 19, 1273–1261.

Chappel, C. I., Chappel, W. R. (1983). The discovery and development of the BB rat colony: an animal model of spontaneous diabetes mellitus. *Metabolism* **32**, 8–10.

Cheung, P. H., Thompson, N. L., Earley, K., Culic, O., Hixson, D., Lin, S. H. (1993). Cell-CAM105 isoforms with different adhesion functions are coexpressed in adult rat tissues and during liver development. *J. Biol. Chem.* **268**, 6139–6146.

Chiariotti, L., Brown, A. L., Frunzio, R., Clemmons, D. R., Rechler, M. M., Bruni, C. B. (1988). Structure of the rat insulin-like growth factor II transcriptional unit: heterogeneous transcripts are generated from two promoters by use of multiple polyadenylation sites and differential ribonucleic acid splicing. *Mol. Endocrinol.* **2**, 1115–1126.

Chignier, E., Sparagano, M. H., McGregor, L., Thillier, A., Pellecchia, D., McGregor, J. L. (1994). Two sites (23–30, 76–90) on rat P-selectin mediate thrombin activated platelet-neutrophil interactions. *Comp. Biochem. Physiol. A. Physiol.* **109**, 881–886.

Childers, N. K., Denys, F. R., McGee, N. F., Michalek, S. M. (1990). Ultrastructural study of liposome uptake by M cells of rat Peyer's patch: an oral vaccine system for delivery of purified antigen. *Region. Immunol.* **3**, 8–16.

Chluba, J., Steeg, C., Becker, A., Wekerle, H., Epplen, J. T. (1989). T cell receptor β chain usage in myelin basic protein specific rat T lymphocytes. *Eur. J. Immunol.* **19**, 279–284.

Christensen, P. J., Kim, S., Simon, R. H., G.B., T., Paine, R. (1993). Differentation-related expression of ICAM-1 by rat alveolar epithelial cells. *Am. J. Respir. Cell Mol. Biol.* **8**, 9–15.

Chroneos, Z., Shepert, V. L. (1995). Differential regulation of the mannose receptor and SP-A receptors on macrophages. *Am. J. Physiol.* **269**, 721–726.

Chung, I. Y., Benveniste, E. N. (1990). Tumor necrosis factor-α production by rat astrocytes. Induction by lipopolysaccharide, IFN-γ and IL-1β. *J. Immunol.* **144**, 2999–3007.

Cizmeci-Smith, G., Asundi, V., Stahl, R. C., Teichman, L. J., Chernousov, M., Cowan, K., Carey, D. J. (1992). Regulated expression of syndecan in vascular smooth muscle cells and cloning of rat syndecan core protein cDNA. *J. Biol. Chem.* **267**, 15729–15736.

Claassen, E., Kors, N., Dijkstra, C. D., van Rooijen, N. (1986). Marginal zone of the spleen and the development and localization of specific antibody forming cells against thymus-dependent and thymus-independent type 2 antigens. *Immunology* **57**, 399–403.

Clarke, G. J., Dallman, M. J. (1992). Identification of a cDNA encoding the rat CD28 homologue. *Immunogenetics* **35**, 54–57.

Clark, S. J., Law, D. A., Paterson, D. J., Puklavec, M., Williams, A. F. (1988). Activation of rat T lymphocytes by anti-CD2 monoclonal antibodies. *J. Exp. Med.* **167**, 1861–1872.

Cohen, D. R., Hapel, A. J., Young, I. G. (1986). Cloning and expression of the rat interleukin-3 gene. *Nucl. Acids Res.* **14**, 3641–3658.

Cohen, S. B., Diamantstein, T., Weetman, A. P. (1990). The effect of T cell subset depletion on autoimmune thyroiditis in the Buffalo strain rat. *Immunol. Lett.* **23**, 263–268.

Colic, M., Drabek, D. (1991). Expression and function of intercellular adhesion molecule 1 (ICAM-1) on rat thymic macrophages in culture. *Immunol. Lett.* **28**, 251–257.

Crawford, J. M., Goldschneider, I. (1980). Thy-1 antigen, and B-lymphocyte differentiation in the rat. *J. Immunol.* **124**, 969–976.

Crisá, L., Mordes, J. P., Rossini, A. A. (1992). Autoimmune diabetes mellitus in the BB rat. *Diab./Metab. Rev.* **8**, 9–37.

Crouse, D. A., Perry, G. A., Murphy, B. O., Sharp, J. G. (1989). Characteristics of submucosal lymphoid tissue located in the proximal colon of the rat. *J. Anat.* **162**, 53–65.

Crowe, J. S., Smith, M. A., Cooper, H. J. (1989). Nucleotide sequence of Y3-Ag 1.2.3. rat myeloma immunoglobulin kappa chain cDNA. *Nucl. Acids Res.* **17**, 7992.

Cuida, M., Brun, J. G., Tynning, T., Jonsson, R. (1995). Calprotectin levels in oral fluids: the importance of collection site. *Eur. J. Oral Sci.* **103**, 8–10.

Culic, O., Huang, Q. H., Flanagan, D., Hixson, D., Lin, S. H. (1992). Molecular cloning and expression of a new rat liver cell-CAM105 isoform. Differential phosphorylation of isoforms. *Biochem. J.* **285**, 47–53.

Cummins, A. G., Thompson, F. M., Mayrhofer, G. (1991). Mucosal immune activation and maturation of the small intestine at weaning in the hypothymic (nude) rat. *J. Pediatr. Gastr. Nutr.* **12**, 361–368.

Curtis, M. R., Dunning, W. F. (1940). Transplantable lymphosarcomata of the mesenteric lymph nodes of rats. *Am. J. Cancer* **40**, 299–309.

Cyster, J., Williams, A. (1992). The importance of cross-linking in the homotypic aggregation of lymphocytes induced by anti-leukosyalin (CD43) antibodies. *Eur. J. Immunol.* **22**, 2565–2572.

Cyster, J., Shotton, D., Williams, A. (1991). The dimension of the T lymphocyte glycoprotein leukosialin and identification of linear protein epitopes that can be modified by glycosylation. *EMBO J.* **10**, 893–902.

Daha, M. R., van Es, L. A. (1979). Isolation of the fourth component (C4) of rat complement. *J. Immunol.* **123**, 2261–2264.

Daha, M. R., van Es, L. A. (1980). Isolation and characterization of rat complement factor B and its interaction with cell-bound human C3. *Immunology* **41**, 849–855.

Daha, M. R., van Es, L. A. (1982). Isolation, characterization, and mechanism of action of rat β1H. *J. Immunol.* **128**, 1839–1843.

Daha, M. R., Stuffers-Heiman, M., Kijlstra, A., van Es, L. A. (1979). Isolation and characterization of the third component of rat complement. *Immunology* **36**, 63–70.

Dallman, D. J., Thomas, M. L., Green, J. R. (1984). MRC OX19: a monoclonal antibody that labels rat T lymphocytes and augments *in vitro* proliverative reponses. *Eur. J. Immunol.* **14**, 260–267.

Dammers, P., De Boer, T., Deenen, G. J., Kroese, F. G. M. (1998a). Molecular cloning of rat λ5 cDNA. (Manuscript in preparation).

Dammers, P., Vermeer, L. A., De Vries, A. F., Bakker, R., Boonstra, A., Bos, N. A. (1998b). Molecular cloning and sequence analysis of rat germline VH genes. (Manuscript submitted).

Dammers, P. M., Bun, J. C. A. M., Bellon, B., Kroese, F. G. M., Aten, J., Bos, N. A. (1998c). Isotype switched B cells in HgCl2-induced autoimmune glomerulopathy utilize predominantly germline V_H genes. (Manuscript submitted).

Damoiseaux, J. G. M. C., Döpp, E. A., Beelen, R. H. J., Dijkstra, C. D. (1989a). Rat bone marrow and monocyte cultures: influence of culture time and lymphokines on the expression of macrophage differentiation antigens. *J. Leukoc. Biol.* **46**, 246–253.

Damoiseaux, J. G. M. C., Döpp, E. A., Neefjes, J. J., Beelen, R. H. J., Dijkstra, C. D. (1989b). Heterogeneity of macrophages in the rat evidenced by variability in determinants: two new anti-rat macrophage antibodies against a heterodimer of 160 and 95 kDa (CD11b/CD18). *J. Leukoc. Biol.* **46**, 556–564.

Damoiseaux, J. G., Döpp, E. A., Calame, W., Chao, D., MacPherson, G. G., Dijkstra, C. D. (1994). Rat macrophage lysosomal membrane antigen recognized by monoclonal antibody ED1. *Immunology* **83**, 140–147.

Danilov, S., Sakharov, I., Martynov, A., Faerman, A., Muzykantov, V., Klibanov, A., Trakht, I. (1989). Monoclonal antibodies to angiotensin-converting enzyme: a powerful tool for lung and vessel studies. *J. Mol. Cell. Cardiol.* **21** (Suppl. 1), 165–170.

Dawson, T. P., Gandhi, R., Le Hir, M., Kaissling, B. (1989). Ecto-5′-nucleotidase: localization in rat kidney by light microscopic histochemical and immunohistochemical methods. *J. Histochem. Cytochem.* **37**, 39–47.

de Boer, N. K. (1994). B cell lineages in the rat. Thesis, University of Groningen, The Netherlands.

de Boer, N. K., Kroese, F. G. M., Sharp, J. G., Perry, G. A. (1992). Immunohistological characterization of proximal colonic lymphoid tissue in the rat. *Anat. Rec.* **233**, 569–576.

de Boer, N. K., Meedendorp, B., Ammerlaan, W. A. M., Nieuwenhuis, P., Kroese, F. G. M. (1994). B cells specific for bromelain treated erythrocytes are not derived from adult rat bone marrow. *Immunobiology* **190**, 105–115.

Deenen, G. J. D., Kroese, F. G. M. (1993). Kinetics of B cell subpopulations in peripheral lymphoid tissues: evidence for the presence of phenotypically distinct short-lived and long-lived B cell subsets. *Int. Immunol.* **5**, 735–741.

Deenen, G. J., Hunt, S. V., Opstelten, D. (1987). A stathmokinetic study of B lymphocytopoiesis in rat bone marrow: proliferation of cells containing cytoplasmic μ-chains, terminal deoxynucleotidyl transferase and carrying HIS 24 antigen. *J. Immunol.* **139**, 702–710.

Deenen, G. J., van Balen, I., Opstelten, D. (1990). In rat B lymphocyte genesis sixty percent is lost from the bone marrow at the transition of nondividing pre-B cell to $sIgM^+$ B lymphocyte, the stage of Ig light chain gene expression. *Eur. J. Immunol.* **20**, 557–564.

Deenen, G. J., Dammers, P. M., De Boer, T., Kroese, F. G. M. (1997). Identification of a novel rat B cell subset in the peritoneal cavity of xenogeneic rat to mouse scid chimeras. *Transpl. Proc.* **29**, 1752–1753.

Der Balian, G., Slack, J., Clevinger, B., Bazin, H., Davie, J. M. (1980). Antigenic similarities of rat and mouse IgG subclasses associated with anti-carbohydrate specificities. *Immunogenetics* **152**, 209–218.

Dijkema, R., Pouwels, P., de Reus, A., Schellekens, H. (1984). Structure and expression in *Escherichia coli* of a cloned rat interferon-α gene. *Nucl. Acids Res.* **12**, 1227–1242.

Dijkema, R., van der Meide, P. H., Pouwels, P. H., Caspers, M., Dubbeld, M., Schellekens, H. (1985). Cloning and expression of the chromosomal immune interferon gene of the rat. *EMBO J.* **4**, 761–767.

Dijkstra, C. D., Damoiseaux, J. G. M. C. (1993). Macrophage heterogeneity established by immunocytochemistry. *Progr. Histochem. Cytochem.* **27**, 1–65.

Dijkstra, C. D., Döpp, E. A., Joling, P., Kraal, G. (1985). The heterogeneity of mononuclear phagocytes in lymphoid organs: distinct macrophage subpopulations in the rat recognized by monoclonal antibodies ED1, ED2 and ED3. *Immunology* **54**, 589–599.

Dijkstra, C. D., Döpp, E. A., Vogels, I. C. M., van Noorden, C. J. F. (1987). Macrophages and dendritic cells in antigen induced arthritis: an immunohistochemical study using cryostat sections of the whole knee joint of rat. *Scand. J. Immunol.* **26**, 513–523.

Dinarello, C. A. (1988). Biology of interleukin-1. *FASEB J.* **2**, 108–115.

Dissen, E., Hunt, S. V., Rolstad, B., Fossum, S. (1993). Localization of the rat T-cell receptor β-chain and carboxypeptidase A1 loci to chromosome 4. *Immunogenetics* **37**, 153–156.

Douglas-Jones, A., Nelson, J., Jansen, V., Miller, T. (1981). Characterization of the nude rat (rnunz): morphological characteristics of the lymphoid system. *Austr. J. Exp. Biol. Med. Sci.* **59**, 277–286.

Drickamer, K., Taylor, M. E. (1993). Biology of animal lectins. *Annu. Rev. Cell. Biol.* **9**, 237–264.

Druet, P. (1994). Metal-induced autoimmunity. *Arch. Toxicol. Suppl.* **16**, 185–191.

Druet, P. (1995). Metal-induced autoimmunity. *Human Exp. Tox.* **14**, 120–121.

Dunn, T. B. (1975). Plasma-cell neoplasms beginning in the ileocaecal area in strain C3H mice. *J. Natl Cancer Inst.* **19**, 371–391.

Dunning, W. F., Curtis, M. R. (1946). The respective roles of longevity and genetic specificity in the occurrence of spontaneous tumors in the hybrids between two inbred lines of rats. *Cancer* **6**, 61–81.

Eisenberg, S. P., Brewer, M. T., Verderber, E., Heimdal, P., Brandhuber, B. J., Thompson, R. C. (1991). Interleukin-1 receptor antagonist is a member of the interleukin 1 gene family: evolution of a cytokine control mechanism. *Proc. Natl Acad. Sci. USA* **88**, 5232–5236.

Elbe, A., Foster, C., Stingl, G. (1996). T cell receptor $\alpha\beta$ and $\gamma\delta$ T cells in rat and human skin – are they equivalent? *Semin. Immunol.* **8**, 341–349.

Ellerman, K. E., Like, A. A. (1995). A major histocompatibility complex class II restriction for BioBreeding/Worcester diabetes-inducing T cells. *J. Exp. Med.* **182**, 923–930.

Ellerman, K. E., Richards, C. A., Guberski, D. L., Shek, W. R., Like, A. A. (1996). Kilham rat virus triggers T-cell-dependent autoimmune diabetes in multiple strains of rat. *Diabetes* **45**, 557–562.

Escribano, M. J., Grabar, P. (1962). L'analyse immunoélectrophorétique du sérum de rat normal. *C.R. Acad. Sci.* **255**, 206–208.

Fangmann, J., Schwinzer, R., Winkler, M., Wonigeit, K. (1990). Expression of RT6 alloantigens and the T-cell receptor on intestinal intraepithelial lymphocytes of the rat. *Transpl. Proc.* **22**, 2543–2544.

Fangmann, J., Schwinzer, R., Wonigeit, K. (1991). Unusual phenotype of intestinal intraepithelial lymphocytes in the rat: predominance of T cell receptor $\alpha/\beta^+/CD2^-$ cells and high expression of the RT6 alloantigen. *Eur. J. Immunol.* **21**, 753–760.

Farber, D. L., Sears, D. W. (1991). Rat CD16 is defined by a family of class III Fc gamma receptors requiring co-expression of heteroprotein subunits. *J. Immunol.* **146**, 4352–4361.

Farber, D. L., Giorda, R., Nettleton, M. Y., Trucco, M., Kochan,

J. P., Sears, D. W. (1993). Rat class III Fc gamma receptor isoforms differ in IgG subclass binding specificity and fail to associate productively with rat CD3 zeta. *J. Immunol.* **150**, 4364–4375.

Feng, L., Tang, W. L., Chang, J. C. C., Wilson, C. B. (1993). Molecular cloning of rat cytokine synthesis inhibitory factor (IL-10) cDNA and expression in spleen and macrophages. *Biochem. Biophys. Res. Commun.* **192**, 452–458.

Festing, M. F. W. (1979). *Inbred Strains in Biomedical Research.* Macmillan, Basingstoke, UK.

Festing, M. F. W., May, D., Connors, T. A., Lovell, D., Sparrow, S. (1978). An athymic nude mutation in the rat. *Nature* **274**, 365–366.

Fey, G. H., Hattori, M., Northemann, W., Abraham, L. J., Baumann, M., Braciak, T. A., Fletcher, R. C., Gauldie, J., Lee, F., Reymond, M. F. (1989). Regulation of rat liver acute phase genes by interleukin-6 and production of hepatocyte stimulating factors by rat hepatoma cells. *Ann. NY Acad. Sci.* **557**, 317–331.

Fischer, A. C., Beschorner, W. E., Hess, A. D. (1989). Requirements for the induction and adoptive transfer of cyclosporine-induced syngeneic graft-versus-host disease. *J. Exp. Med.* **169**, 1031–1041.

Flores-Romo, L., Shields, J., Humbert, Y., Graber, P., Aubry, J. P., Gauchat, J. F., Ayala, G., Allet, B., Chavez, M., Bazin, H. (1993). Inhibition of an *in vivo* antigen-specific IgE response by antibodies to CD23. *Science* **261**, 1038–1041.

Foote, J., Winter, G. (1992). Antibody framework residues affecting the conformation of the hypervariable loops. *J. Mol. Biol.* **224**, 487–499.

Ford, W. L. (1980). The lymphocyte – its transformation from a frustrating enigma to a model of cellular function. In *Blood, Pure and Eloquent: a Story of the Discovery of People and Ideas* (ed. Wintrobe, M. M.), pp. 457–508. McGraw Hill, New York, USA.

Fossum, S., Mallett, S., Barclay, A. N. (1991). The MRC OX-47 antigen is a member of the immunoglobulin superfamily with an unusual transmembrane sequence. *Eur. J. Immunol.* **21**, 671–679.

Fowell, D., Mason, D. (1993). Evidence that the T cell repertoire of normal rats contains cells with the potential to cause diabetes. Characterization of the $CD4^+$ T cell subset that inhibits this autoimmune potential. *J. Exp. Med.* **177**, 627–636.

Fowell, D., McKnight, A. J., Powrie, F., Dyke, R., Mason, D. (1991). Subsets of $CD4^+$ T cells and their roles in the induction and prevention of autoimmunity. *Immunol. Rev.* **123**, 37–64.

Frank, M. B., Gutman, G. A. (1988). Two pseudogenes among three rat immunoglobulin lambda chain genes. *Molec. Immunol.* **25**, 953–960.

Frenzl, B., Kren, V., Stark, O. (1960). Attempt to determine blood groups rats. *Folia Biologica Prag.* **6**, 121–126

Fries, J. W., Williams, A. J., Atkins, R. C., Newman, W., Lipscomb, M. F., Collins, T. (1993). Expression of VCAM-1 and E-selectin in an *in vivo* model of endothelial activation. *Am. J. Pathol.* **143**, 725–737.

Frorath, B., Abney, C. C., Berthold, H., Scanarini, M., Northemann, W. (1992). Production of recombinant rat interleukin-6 in *Escherichia coli* using a novel highly efficient expression vector pGEX-3T. *BioTechniques*, **12**, 558–563.

Frunzio, R., Chiariotti, L., Brown, A. L., Graham, D. E., Rechler, M. M., Bruni, C. B. (1986). Structure and expression of the rat insulin-like growth factor II (rIGF-II) gene. rIGF-II RNAs are transcribed from two promoters. *J. Biol. Chem.* **261**, 17138–17149.

Fujio, K., Hu, Z., Evarts, R. P., Marsden, E. R., Niu, C. H., Thorgeirsson, S. S. (1996). Co-expression of stem cell factor and c-kit in embryonic and adult liver. *Exp. Cell. Res.* **224**, 243–250.

Fukumoto, T., McMaster, W. R., Williams, A. F. (1982). Mouse monoclonals against rat major histocompatibility antigens. Two Ia antigens and expression of Ia and class I antigens in rat thymus. *Eur. J. Immunol.* **12**, 237–243.

Furth, M., Davis, L., Fleurdelys, B., Scolnick, E. (1982). Monoclonal antibodies to the p21 products of the transforming gene of Harvey Murine Sarcoma virus and of the cellular ras gene family. *J. Virology* **43**, 294–304.

Furuya, Y., Takasawa, S., Yonekura, H., Tanaka, T., Takahara, J., Okamoto, H. (1995). Cloning of a cDNA encoding rat bone marrow stromal cell antigen 1 (BST-1) from islets of Langerhans. *Gene* **165**, 329–330.

Galea, E., Reis, D. J., Fox, E. S., Xu, H., Feinstein, D. L. (1996). CD14 mediate endotoxin induction of nitric oxide synthase in cultured brain glial cells. *J. Neuroimmunol.* **64**, 19–28.

Gandhi, R., Le Hir, M., Kaiserling, B. (1990) Immunolocalization of ecto-5′-nucleotidase in the kidney by a monoclonal antibody. *Histochemistry* **95**, 165–174.

Garman, R. D., Doherty, P., Raulet, D. H. (1986). Diversity, rearrangement and expression of murine T-cell gamma genes. *Cell* **45**, 733–742.

Gascoigne, N. R. J. (1995). Genomic organisation of T-cell receptor genes in the mouse. In *T-cell Receptors* (eds Bell, J. I., Owen, M. J., Simpson, E.), pp. 288–300. Oxford University Press, Oxford, UK.

Gebicke-Haerter, P. J., Appel, K., Taylor, G. D., Schobert, A., Rich, I. N., Northoff, H., Berger, M. (1994). Rat microglial interleukin-3. *J. Neuroimmunol.* **50**, 203–214.

Geginat, G., Günther, E. (1993). Genetic control of the cellular *in vitro* response to the H-Y antigen in the rat. *Transplantation* **56**, 448–452.

Geiger, T., Andus, T., Klapproth, J., Hirano, T., Kishimoto, T., Heinrich, P. C. (1988). Induction of rat acute-phase proteins by interleukin-6 *in vivo*. *Eur. J. Immunol.* **18**, 717–721.

Geisert, E. E. J., Yang, L., Irwin, M. H. (1996). Astrocyte growth, reactivity, and the target of the antiproliferative antibody TAPA. *J. Neurosci.* **16**, 5478–5487.

Geisterfer, M., Richards, C. D., Baumann, M., Fey, G., Gwynne, D., Gauldie, J. (1993). Regulation of IL-6 and hepatic IL-6 receptor in acute inflammation *in vivo*. *Cytokine* **5**, 1–7.

Giegerich, G., Stangel, M., Torres-Nagel, N. E., Hünig, T., Toyka, K. (1995). Sequence and diversity of rat T-cell receptor *Tcra V8* gene segments. *Immunogenetics* **41**, 125–130.

Gill III, T. J. (1996). Role of the Major Histocompatibility Complex region in reproduction, cancer, and autoimmunity. *Am. J. Reprod. Immunol.* **35**, 211–215.

Gill III, T. J., Natori, T., Salgar, S. K., Kunz, H. W. (1995). Current status of the major histocompatibility complex in the rat. *Transpl. Proc.* **27**, 1495–1500.

Giometto, B., Bozza, F., Argentiero, V., Gallo, P., Pagni, S., Piccinno, M. G., Tavolato, B. (1990). Transferrin receptors in rat central nervous system. An immunocytochemical study. *J. Neurol. Sci.* **98**, 81–90.

Go, L. L., Healey, P. J., Watkins, S. C., Simmons, R. L., Rowe,

M. I. (1995). Effect of endotoxin on intestinal mucosal permeability to bacteria *in vitro*. *Arch. Surg.* **130**, 53–58.

Gold, D. P., Surh, C. D., Sellins, K. S., Schroder, K., Sprent, J., Wilson, D. P. (1994). Rat T-cell responses to superantigens. II. Allelic differences in Vβ8.2 and Vβ8.5 β chains determine responsiveness to staphylococcal enterotoxin B and mouse mammary tumor virus-encoded products. *J. Exp. Med.* **179**, 63–69.

Gold, R., Giegerich, G., Hartung, H. P., Toyka, K. (1995). T-cell receptor (TCR) usage in Lewis rat experimental autoimmune encephalomyelitis: TCR beta-chain-variable-region V beta 8.2-positive T cells are not essential for induction and course of disease. *Proc. Natl Acad. Sci. USA* **82**, 5850–5854.

Goldschneider, I., Gordon, L. K., Morris, R. J. (1978). Demonstration of Thy-1 antigen on pluripotent hemopoietic stem cells in the rat. *J. Exp. Med.* **148**, 1351–1366.

Goodman, R. E., Oblak, J., Bell, R. G. (1992). Synthesis and characterization of rat interleukin-10 (IL-10) cDNA clones from the RNA of cultured OX8$^-$ OX22$^-$ thoracic duct T cells. *Biochem. Biophys. Res. Commun.* **189**, 1–7.

Gorman, S., Clark, M., Routledge, E., Cobbold, S. (1991). Reshaping a therapeutic CD4 antibody. *Proc. Natl Acad. Sci. USA* **88**, 4181–4185.

Gorrell, M. D., Wickson, J., McCaughan, G. W. (1991). Expression of the rat CD26 antigen (dipeptidyl peptidase IV) on subpopulations of rat lymphocytes. *Cell. Immunol.* **134**, 205–215.

Grabar, P., Courcon, J. (1958). Etude des sérums de cheval, lapin, rat et souris par l'analyse immunoélectrophorétique. *Bull. Soc. Chim. Biol.* **40**, 1993–2003.

Gray, D. (1988). Population kinetics of rat peripheral B cells. *J. Exp. Med.* **167**, 805–816.

Gray, D., MacLennan, I. C. M., Bazin, H., Khan, M. (1982). Migrant $\mu^+\delta^+$ and static $\mu^+\delta^-$ B lymphocyte subsets. *Eur. J. Immunol.* **12**, 564–569.

Gray, D., McConell, I., Kumararatne, D. S., Humphrey, J. H., Bazin, H. (1984). Marginal zone cells express CR1 and CR2 receptor. *Eur. J. Immunol.* **14**, 47–52.

Greene, E. C. (1935). Anatomy of the rat. *Trans. Am. Phil. Soc., New Ser.*, **XXVII**, 1–370.

Greenfield, E. M., Shaw, S. M., Gornik, S. A., Banks, M. A. (1995). Adenyl cyclase and interleukin-6 are downstream effectors of parathyroid hormone resulting in stimulation of bone resorption. *J. Clin. Invest.* **96**, 1238–1244.

Greiner, D. L., Reynolds, C., Lubaroff, D. M. (1982). Maturation of functional T-lymphocyte subpopulations in the rat. *Thymus* **4**, 77–90.

Greiner, D. L., Handler, E. S., Nakano, K., Mordes, J. P., Rossini, A. A. (1986). Absence of the RT6 T cell subset in diabetes-prone BB/W rats. *J. Immunol.* **136**, 148–151.

Greiner, D. L., Mordes, J. P., Handler, E. S., Angelillo, M., Nakamura, N., Rossini, A. A. (1987). Depletion of RT6.1$^+$ T lymphocytes induces diabetes in resistant BioBreeding/Worcester (BB/W) rats. *J. Exp. Med.* **166**, 461–475.

Greiner, D. L., Mordes, J. P., Angelillo, M., Handler, E. S., Mojcik, C. F., Nakamura, N., Rossini, A. A. (1988). Role of regulatory RT6$^+$ T-cells in the pathogenesis of diabetes mellitus in BB/Wor rats. In *Frontiers in Diabetes Research: Lessons from Animal Diabetes II* (eds Shafrir, E., Renold, A. E.), pp. 58–67. John Libbey, London, UK.

Griffiths, M. M., Cremer, M. A., Harper, D. S., McCall, S., Cannon, G. W. (1992). Immunogenetics of collagen-induced arthritis in rats. Both MHC and non-MHC gene products determine the epitope specificity of immune response to bovine and chick type II collagens. *J. Immunol.* **149**, 309–316.

Groen, H., van der Berk, J. M. M. M., Nieuwenhuis, P., Kampinga, J. (1989). Peripheral T cells in diabetes prone (DP) BB rats are CD45R-negative. *Thymus*, **14**, 145–150.

Groen, H., Klatter, F. A., van Petersen, A. S., Pater, J. M., Nieuwenhuis, P., Kampinga, J. (1993). Composition of rat CD4$^+$ resting memory T-cell pool is influenced by major histocompatibility complex. *Transpl. Proc.* **25**, 2782–2783.

Groen, H., Klatter, F. A., Brons, N. H. C., Wubbena, A. S., Nieuwenhuis, P., Kampinga, J. (1995). High-frequency, but reduced absolute numbers of recent thymic migrants among peripheral blood T lymphocytes in diabetes-prone BB rats. *Cell. Immunol.* **163**, 113–119.

Groen, H., Klatter, F. A., Brons, N. H. C., Mesander, G., Nieuwenhuis, P., Kampinga, J. (1996a). Abnormal thymocyte subset distribution and differential reduction of CD4$^+$ and CD8$^+$ T cell subsets during peripheral maturation in diabetes-prone BB rats. *J. Immunol.* **156**, 1269–1275.

Groen, H., Pater, J. M., Klatter, F. A., Nieuwenhuis, P., Rozing, J. (1996b). Expression of RT6, but not CD45RC is disturbed on immature peripheral T cells in the BB rat. *Adv. Exp. Med. Biol.* **419**, 253–256.

Guberski, D. L., Butler, L., Like, A. A. (1988). The BBZ/Wor rat: an obese animal with autoimmune diabetes. In *Frontiers in Diabetes Research: Lessons from Animal Diabetes II* (eds Renold, A., Shafrir, E.), pp. 182–185. John Libbey, London, UK.

Guberski, D. L., Thomas, V. A., Shek, W. R., Like, A. A., Handler, E. S., Rossini, A. A., Wallace, J. E., Welsh, R. M. (1991). Induction of type 1 diabetes by Kilham's rat virus in diabetes resistant BB/Wor rats. *Science* **254**, 1010–1013.

Guberski, D. L., Butler, L., Manzi, S. M., Stubbs, M., Like, A. A. (1993). The BBZ/Wor rat: clinical characteristics of the diabetic syndrome. *Diabetologia* **36**, 912–919.

Guérin, M. (1954). *Tumeurs Spontanées des Animaux de Laboratoire*. A. Legrand et Cie, Paris, France.

Guéry, J.-C., Druet, E., Glotz, D., Hirsch, F., Mandet, C., De Heer, E., Druet, P. (1990). Specificity and cross-reactive idiotypes of anti-glomerular basement membrane autoantibodies in $HgCl_2$-induced autoimmune glomerulonephritis. *J. Immunol.* **20**, 93–100.

Günthert, U., Hofmann, M., Rudy, W., Reber, S., Zöller, M., Haussmann, I., Matzku, S., Wenzel, A., Ponta, H., Herrlich, P. (1991). A new variant of glycoprotein CD44 confers metastatic potential to rat carcinoma cells. *Cell* **65**, 13–24.

Gutman, G. A., Weissman, I. L. (1971). Inheritance and strain distribution of a rat immunoglobulin allotype. *J. Immunol.* **107**, 1390–1393.

Gutman, G. A., Loh, E., Hood, L. (1975). Structure and regulation of immunoglobulins: kappa allotypes in the rat have multiple amino acid differences in the constant region. *Proc. Natl Acad. Sci. USA* **72**, 5046–5050.

Gutman, G. A., Bazin, H., Rocklin, C. V., Nezlin, R. S. (1983). A standard nomenclature for rat immunoglobulin allotypes. *Transplant. Proc.* **15**, 1685–1686.

Hameleers, D. M. H., van der Ende, M., Biewenga, J., Sminia, T. (1989). An immunohistochemical study on the postnatal development of rat nasal-associated lymphoid tissue (NALT). *Cell Tissue Res.* **256**, 431–438.

Hameleers, D. M. H., van der Ven, I., Biewenga, J., Sminia, T.

(1991). Mucosal and systemic antibody formation in the rat after intranasal administration of three different antigens. *Immunol. Cell Biol.* **69**, 119–125.

Hart, R. P., Liu, C., Shadiack, A. M., McCormack, R. J., Jonakait, G. M. (1993). An mRNA homologous to interleukin-1 receptor type I is expressed in cultured rat sympathetic ganglia. *J. Neuroimmunol.* **44**, 49–56.

Hashim, G., Vandenbark, A. A., Gold, D. P., Diamanduros, T., Offner, H. (1991). T cell lines specific for an immunodominant epitope of human basic protein define an encephalotigenic determinant for experimental autoimmune encephalomyelitis resistant LOU/M rats. *J. Immunol.* **146**, 515–520.

Heiman, H. S., Weisman, L. E. (1989). Transplacental or enteral transfer of maternal immunization-induced antibody protects suckling rats from type III group B streptococcal infections. *Pediatr. Res.* **26**, 629–632.

Helene, A., Milhiet, P. E., Haouas, H., Boucheix, C., Beaumint, A., Roques, B. P. (1992). Effects of monoclonal antibodies against the common acute lymhoblastic leukemic antigen on endopeptidase-24.11 activity. *Biochem. Pharmacol.* **43**, 809–814.

Helfrich, M. H., Nesbitt, S. A., Horton, M. A. (1992). Integrins on rat osteoclasts: characterization of two monoclonal antibodies (F4 and F11) to rat beta 3. *J. Bone Miner. Res.* **7**, 345–351.

Hellman, L., Petterson, U., Engström, Å., Karlsson, T., Bennich, H. (1982a). Structure and evolution of the heavy chain from rat immunoglobulin E. *Nucl. Acids Res.* **10**, 6041–6049.

Hellman, L., Pettersson, U., Bennich, H. (1982b). Characterization and molecular cloning of the mRNA for the heavy (ε) chain of rat immunoglobulin E. *Immunology* **79**, 1264–1268.

Hellman, L., Engström, Å., Bennich, H., Pettersson, U. (1985). Structure and expression of kappa-chain genes in two IgE-producing rat immunocytomas. *Gene* **40**, 107–114.

Hermans, M. H. A., Opstelten, D. (1991). *In situ* visualization of hemopoietic cell subsets and stromal elements in rat and mouse bone marrow by immunostaining of frozen sections. *J. Histochem. Cytochem.* **39**, 1627–1634.

Hermans, M. H. A., Hartsuiker, H., Opstelten, D. (1989). An *in situ* study of B lymphopoiesis in rat bone marrow: topographical arrangement of terminal deoxynucleotidyl transferase positive cells and pre-B cells. *J. Immunol.* **142**, 67–73.

Hermans, M. H., Wubbena, A. S., Kroese, F. G. M., Hunt, S. V., Cowan, R., Opstelten, D. (1992). The extent of clonal structure in different lymphoid organs. *J. Exp. Med.* **175**, 1255–1269.

Hermans, M., Deenen, G. J., De Boer, N., Bo, W., Kroese, F. G. M., Opstelten, D. (1997). Expression of HIS50 Ag: a rat homologue of mouse heat-stable antigen and human CD24 on B lymphoid cells in the rat. *Immunology* **90**, 14–22.

Herrmann, T., Hochgrebe, T., Torres-Nagel, N., Huber, B., Hünig, T. (1994). Control of the rat T cell response to retroviral and bacterial superantigens by class II MHC products and *Tcrb-V8.2* alleles. *J. Immunol.* **152**, 4300–4309.

Hickey, W. F., Kimura, H. (1988). Perivascular microglial cells of the CNS are bone-marrow derived and present antigen *in vivo*. *Science* **239**, 290–292.

Himeno, M., Noguchi, Y., Sasaki, H., Tanaka, Y., Furuno, K., Kono, A., Sasaki, Y., Kato, K. (1989). Isolation and sequencing of a cDNA clone encoding 107 kDa sialoglycoprotein in rat liver lysosomal membranes. *FEBS Lett.* **244**, 351–356.

Himmler, A., Maurer-Fogy, I., Kronke, M., Scheurich, P., Pfizenmaier, K., Lantz, M., Olsson, I., Hauptmann, R., Stratowa, C., Adolf, G. R. (1990). Molecular cloning and expression of human and rat tumor necrosis factor receptor chain (p60) and its soluble derivative, tumor necrosis factor-binding protein. *DNA Cell Biol.* **9**, 705–715.

Hinkkanen, A. E., Määttä, J., Qin, Y.-F., Linington, C., Salmi, A., Wekerle, H. (1993). Novel Tcra-V and -J transcripts expressed in rat myelin-specific T-cell lines. *Immunogenetics* **37**, 235–238.

Hirsch, F., Druet, E., Vendeville, B., Cormont, F., Bazin, H., Druet, P. (1984). Production of monoclonal anti-glomerular basement membrane antibodies during autoimmune glomerulonephritis. *Clin. Immunol. Immunopathol.* **33**, 425–430.

Hirsch, F., Kuhn, J., Ventura, M., Vial, M.-C., Fournie, G., Druet, P. (1986). Autoimmunity induced by $HgCl_2$ in Brown-Norway rats. *J. Immunol.* **136**, 3272–3276.

Hixson, D., McEntire, K., Obrink, B. (1985). Alterations in the expression of a hepatocyte cell adhesion molecule by transplantable rat hepatocellular carcinomas. *Cancer Res.* **45**, 3742–3749.

Ho, F., Lortan, J. E., MacLennan, I. C. M., Khan, M. (1986). Distinct short-lived and long-lived antibody-producing cell populations. *Eur. J. Immunol.* **16**, 1297–1301.

Hodits, R. A., Nimpf, J., Pfistermueller, D. M., Hiesberger, T., Schneider, W. J., Vaughan, T. J., Johnson, K. S., Haumer, M., Kuechler, E., Winter, G., Blaas, D. (1995). An antibody fragment from a phage display library competes for ligand binding to the low density lipoprotein receptor family and inhibits rhinovirus infection. *J. Biol. Chem.* **270**, 24078–24085.

Holt, P. G., Schon-Hergard, M. A. (1987). Localization of T cells, macrophages and dendritic cells in rat respiratory tract tissue: implications for immune function studies. *Immunology* **62**, 349–356.

Hood, L., Gray, W. R., Sanders, B. G., Dreyer, W. J. (1967). Light chain evolution. *Cold Spring Harbor Symp. Quant. Biol.* **22**, 133.

Horlick, R. A., Stack, S. L., Cooke, G. M. (1992). Cloning, expression and tissue distribution of the gene encoding rat fibroblast growth factor receptor subtype 4. *Gene* **120**, 291–295.

Horton, M. A., Taylor, M. L., Arnett, T. R., Helfrich, M. H. (1991). Arg–Gly–Asp (RGD) peptides and the anti-vitronectin receptor antibody 23C6 inhibit dentine resorption and cell spreading by osteoclasts. *Exp. Cell Res.* **195**, 368–375.

Hosseinzadeh, H., Goldschneider, I. (1993). Recent thymic emigrants in the rat express a unique phenotype and undergo post-thymic maturation in peripheral tissues. *J. Immunol.* **150**, 1670–1679.

Hougen, H. P. (1991). The athymic nude rat. *Act. Pathol. Microbiol. Immunol. Scand.* **99**, 9–39.

Howe, C. L., Granger, B. L., Hull, M., Green, S. A., Gabel, C. A., Helenius, A., Mellman, I. (1988). Derived protein sequence, oligosaccharides, and membrane insertion of the 120 kDa lysosomal membrane glycoprotein (lgp120): identification of a highly conserved family of lysosomal membrane glycoproteins. *Proc. Natl Acad. Sci. USA* **85**, 7577–7581.

Howell, D. N., Anuja, V., Jones, L., Blow, O., Saufilippo, F. P. (1994). Differential expression of CD43 (leukosialin, sialophorin) by mononuclear phagocyte populations. *J. Leukoc. Biol.* **55**, 536–544.

Huang, S., Paulauskis, J. D., Kobzik, L. (1992). Rat KC cDNA cloning and mRNA expression in lung macrophages and fibroblasts. *Biochem. Biophys. Res. Commun.* **184**, 922–929.

Hughes, T. R., Piddlesden, S. J., Williams, J. D., Harrison, R. A., Morgan, B. P. (1992). Isolation and characterization of a membrane protein from rat erythrocytes which inhibits lysis by the membrane attack complex of rat complement. *Biochem. J.* **284**, 169–176.

Hughes, T. R., Meri, S., Davies, M., Williams, J. D., Morgan, B. P. (1993). Immunolocalization and characterization of the rat analogue of human CD59 in kidney and glomerular cells. *Immunology* **80**, 439–444.

Huitinga, I., Damoiseaux, J. G. M. C., Döpp, E. A., Dijkstra, C. D. (1993). Treatment with anti-CR3 antibodies suppresses experimental allergy in Lewis rats. *Eur. J. Immunol.* **23**, 709–715.

Humphrey, R. L., Santos, G. S. (1971). Serum protein allotype markers in certain inbred rat strains. *Fed. Proc. Fed. Am. Soc. Exp. Biol.* **30**, 248.

Hünig, T. (1988). Crosslinking of the T-cell antigen receptor interferes with the generation of $CD4^+8^+$ thymocytes from their immediate $CD4^-8^+$ precursors. *Eur. J. Immunol.* **18**, 2089–2092.

Hünig, T., Wallny, H.-J., Hartley, J. K., Lawetzky, A., Tiefenthaler, G. (1989). A monoclonal antibody to a constant determinant of the rat T cell antigen receptor that induces T cell activation. *J. Exp. Med.* **169**, 73–86.

Hunt, H. D., Lubaroff, D. M. (1987). Changes in membrane antigen phenotype of T cells during lectin activation. *Transpl. Proc.* **19**, 3179–3180.

Hunt, S. V. (1979). The presence of Thy-1 on the surface of rat lymphoid stem cells and colony forming units. *Eur. J. Immunol.* **9**, 853–859.

Hunt, S. V., Mason, D. W., Williams, A. F. (1977). In rat bone marrow Thy-1 antigen is expressed on cells with membrane immunoglobulin and on precursors of peripheral B lymphocytes. *Eur. J. Immunol.* **7**, 817–823.

Hunt, S. V., Medlock, E. S., Greiner, D. L., Goldschneider, I., Opstelten, D. (1988). Rat immunoglobulin genes have comparable patterns of JH rearrangement in normal peripheral and in pre-B and cultured TdT^+ cells from bone marrow. *Adv. Exp. Med. Biol*, **237**, 63–68.

Ichimiya, S., Kikuchi, K., Matsuura, A. (1994). Structural analysis of the rat homologue of CD1. Evidence for evolutionary conservation if the CD1D class and widespread transcription by rat cells. *J. Immunol.* **153**, 1112–1123.

Ikawa, M., Shozen, Y. (1990). Quantification of acute phase proteins in rat serum and in the supernatants of a cultured rat hepatoma cell line and cultured primary hepatocytes by an enzyme-linked immunosorbent assay. *J. Immunol. Meth.* **134**, 101–106.

Isashi, Y., Tamakoshi, M., Nagai, Y., Sudo, T., Murakami, M., Uede, T. (1995). The rat neutrophil low-affinity receptor for IgG: molecular cloning and functional characterization. *Immunol. Lett.* **46**, 157–163.

Issekutz, A. C., Issekutz, T. B. (1992). The contribution of LFA-1 (CD11a/CD18) and MAC-1 (CD11b/CD18) to the *in vivo* migration of polymorphonuclear leucocytes to inflammatory reactions in the rat. *Immunology* **76**, 655–661.

Issekutz, A. C., Ayer, L., Miyasaka, M., Issekutz, T. B. (1996). Treatment of established adjuvant arthritis in rats with monoclonal antibody to CD18 and very late activation antigen-4 integrins suppresses neutrophil and T-lymphocyte migration to the joints and improves clinical disease. *Immunology* **88**, 569–576.

Issekutz, T. B. (1991). Inhibition of *in vivo* lymphocyte migration to inflammation and homing to lymphoid tissues by the TA-2 monoclonal antibody. A likely role of VLA-4 *in vivo*. *J. Immunol.* **147**, 4178–4184.

Issekutz, T. B., Wykretowicz, A. (1991). Effect of a new monoclonal antibody, TA-2, that inhibits lymphocyte adherence to cytokine stimulated endothelium in the rat. *J. Immunol.* **147**, 109–116.

Issekutz, T. B., Miyasaka, M., Issekutz, A. C. (1996). Rat blood neutrophils express very late antigen 4 and it mediates migration to arthritic joint and dermal inflammation. *J. Exp. Med.* **183**, 2175–2184.

Itoh, Y., Matsuura, A., Kinebuchi, M., Honda, R., Takayama, S., Ichimya, S., Kon, S., Kikuchi, K. (1993). Structural analysis of the CD3 zeta/eta locus of the rat. Expression of zeta but not eta transcripts by rat T cells. *J. Immunol.* **151**, 4705–4717.

Jarry, A., Robaszkiewicz, M., Brousse, N., Potet, F. (1989). Immune cells associated with M cells in the follicle-associated epithelium of Peyer's patches in the rat. An electron- and immuno-electron microscopic study. *Cell Tissue Res.* **255**, 293–298.

Jefferies, W. A., Brandon, M. R., Hunt, S. V. (1984). Transferrin receptor on endothelium of brain capillaries. *Nature* **312**, 162–163.

Jefferies, W. A., Brandon, M. R., Williams, A. F., Hunt, S. V. (1985a). Analysis of lymphopoietic stem cells with a monoclonal antibody to the rat transferrin receptor. *Immunology* **54**, 333–341.

Jefferies, W. A., Green, J. R., Williams, A. F. (1985b). Authentic T helper CD4 (W3/25) antigen on rat peritoneal macrophages. *J. Exp. Med.* **162**, 117–127.

Jochen, A., Hays, J. (1993). Purification of the major substrate for palmitoylation in rat adipocytes: N terminal homology with CD36 and evidence for cell surface acylation. *J. Lipid Res.* **34**, 1783–1792.

Joling, P., Tielen, F. J., Vaessen, L. M. B., Hesse, C. J., Rozing, J. R. (1985a). Intrathymic differentiation in the rat. *Adv. Exp. Med. Biol.* **186**, 235–244.

Joling, P., Tielen, F. J., Vaessen, L. M. B., Huijbregts, J. M. A., Rozing, J. (1985b). New markers on T cell subpopulations defined by monoclonal antibodies. *Transpl. Proc.* **17**, 1857–1860.

Jones, M., Cordell, J. L., Beyers, A. D., Tse, A. G., Mason, D. Y. (1993). Detection of T and B cells in many animal species using cross-reactive anti-peptide antibodies. *J. Immunol.* **150**, 5429–5435.

Jones, V. E. (1969). Rat 7S immunoglobulins: characterization of gamma2- and gamma1-anti-hapten antibodies. *Immunology* **16**, 589–599.

Jones, V. E., Edwards, A. J. (1971). Preparation of an antiserum specific for rat reagin (rat gammaE?). *Immunology* **21**, 383–385.

Judge, T. A., Liu, M., Christensen, P. J., Fak, J. J. (1995). Cloning the rat homolog of the CD28/CTLA-4 ligand B7–1; structural and functional analysis. *Int. Immunol.* **7**, 171–178.

Jung, H. C., Ecjmann, L., Yang, S. K., Panja, A., Fierer, J., Morzycka-Wroblewska, E., Kagnoff, M. F. (1995). A distinct array of proinflammatory cytokines is expressed in human colon epithelial cells in response to bacterial invasion. *J. Clin. Invest.* **95**, 55–65.

Jyung, R. W., Wu, L., Pierce, G. F., Mustoe, T. A. (1994). Granulocyte-macrophage colony-stimulating factor and granulocyte colony-stimulating factor: differential action on incisional wound healing. *Surgery* **115**, 325–334.

Kamegai, M., Niijima, K., Kunishita, T., Nishizawa, M., Ogawa, M., Araki, M., Ueki, A., Konishi, Y., Tabira, T. (1990). Interleukin-3 as a trophic factor for central cholinergic neurons *in vitro* and *in vivo*. *Neuron* **2**, 429–436.

Kampinga, J. (1991). Invariant involvement of IL-2 in thymocyte differentiation (Reply). *Immunol. Today* **12**, 246–247.

Kampinga, J., Aspinall, R. (1990). Thymocyte differentiation and thymic micro-environment development in the fetal rat thymus: an immunohistological approach. In *Thymus Update 3* (eds Kendall, M. D., Ritter, M. A.), pp. 149–186. Harwood Academic Publishers, London, UK.

Kampinga, J., Berges, S., Boyd, R. L., Brekelmans, P., Colic, M., van Ewijk, W., Kendall, M., Ladyman, H., Nieuwenhuis, P., Ritter, M. A., Schuurman, H. J., Tournefier, A. (1989a). Thymic epithelial antibodies: immunohistological analysis and introduction of nomenclature. Summary of the Epithelium Workshop held at the 2nd Workshop on the Thymus: Histophysiology and Dynamics in the Immune System. *Thymus* **13**, 165–173.

Kampinga, J., Kroese, F. G. M., Pol, G. H., Niewenhuis, P., Haag, F., Singh, P. B., Roser, B., Aspinall, R. (1989b). A monoclonal antibody to a determinant of the rat T cell antigen receptor expressed by a minor subset of T cells. *Int. Immunol.* **1**, 289–295.

Kampinga, J., Schuurman, H. J., Pol, G. H., Bartels, H., Vaessen, L. M. B., Tielen, F. J., Rozing, J., Blaauw, E. H., Roser, B., Aspinall, R., Nieuwenhuis, P. (1990a). Vascular thymus transplantation in rats. Technique, morphology and function. *Transplantation* **50**, 669–678.

Kampinga, J., Nieuwenhuis, P., Roser, B., Aspinall, R. (1990b). Differences in turnover between thymic medullary dendritic cells and a subset of cortical macrophages. *J. Immunol.* **145**, 1659–1663.

Kampinga, J., Kroese, F. G. M., Pol, G. H., Opstelten, D., Seijen, H. G., Boot, J. H., Roser, B., Nieuwenhuis, P., Aspinall, R. (1990c). RT7-defined alloantigens in rats are part of the leucocyte common antigen family. *Scan. J. Immunol.* **31**, 699–710.

Kampinga, J., Kroese, F. G. M., Pol, G. H., Meedendorp, B., van Eendenburg, J., Groen, H., van Den Bogaerde, J., Nieuwenhuis, P., Roser, B., Aspinall, R. (1990d). Inhibition of T cell responses *in vitro* by an antibody against a novel lymphocyte surface molecule (QCA-1). *Int. Immunol.* **2**, 915–920.

Kampinga, J., Groen, H., Klatter, F. A., Meedendorp, B., Aspinall, R., Roser, B., Nieuwenhuis, P. (1992). Postthymic T cell development in rats: an update. *Biochem. Soc. Trans.* **20**, 191–197.

Kampinga, J., Groen, H., Klatter, F. A., Pater, J. M., van Petersen, A. S., Roser, B., Nieuwenhuis, P., Aspinall, R. (1997). Postthymic T cell development in the rat. *Thymus*, **24**, 173–200.

Kampschmidt, R. F., Mesecher, M. (1985). Interleukin-1 from P388D: effects upon neutrophils plasma iron, and fibrinogen in rats, mice, and rabbits. *Proc. Soc. Exp. Biol. Med.* **179**, 197–200.

Kanagawa, K., Ishikura, H., Takahashi, C., Tamatani, T., Myisaka, M., Koyanagi, T., Yoshiki, T. (1991). Identification of ICAM-1 positive cells in the nongrafted and transplanted rat kidney – an immunohistochemical and ultrastructural study. *Transplantation* **52**, 1052–1062.

Kanbour-Shakir, A., Kunz, H. W., Gill, T. J., Armstrong, D. T., MacPherson, T. A. (1990). Morphologic changes in the rat uterus following natural mating and embryo transfer. *Am. J. Reprod. Immunol.* **23**, 78–83.

Kantor, A. B., Herzenberg, L. A. (1993). Origin of murine B cell lineages. *Annu. Rev. Immunol.* **11**, 501–538.

Karp, S. L., Kieber-Emmons, T., Sun, M.-J., Wolf, G., Neilson, E. G. (1993). Molecular structure of a cross-reactive idiotype on autoantibodies recognizing parenchymal self. *J. Immunol.* **150**, 867–879.

Keller, R., Joller, P. W., Keist, R. (1989). Surface phenotype of rat bone marrow derived mononuclear phagocytes. *Cell. Immunol.* **120**, 277–285.

Kendall, M. D., Al-Shawaf, A. (1991). The innervation of the rat thymus. *Brain Behav. Immun.* **5**, 9–28.

Killeen, N., Barclay, A. N., Willis, A. C., Williams, A. F. (1987). The sequence of rat leukosialin (W3/13 antigen) reveals a molecule with O-linked glycosylation of one third of its extracellular amino acids. *EMBO J.* **6**, 4029–4034.

Kim, J. G., Abcyounis, C. J. (1988). Monoclonal rat antibodies to rat carcinoembryonic antigen. *Immunol. Invest.* **17**, 41–48.

Kim, J. G., Abeyounis, C. J. (1990). Isolation and characterization of rat carcinoembryonic antigen. *Int. Arch. Allergy Appl. Immunol.* **92**, 43–49.

Kimura, K., Wakatsuki, T., Yamamoto, M. (1994). A variant mRNA species encoding a truncated form of Fas antigen in the rat liver. *Biochem. Biophys. Res. Commun.* **198**, 666–674.

Kinebuchi, M., Kikuchi, K., Matsuura, A. (1994). Rat T-cell receptor γ chain sequences – identification of conserved gene segments and a unique chimeric constant region gene, C4L. *Immunogenetics* **40**, 449–455.

Kirchhoff, H., Binder, A., Runge, M., Meier, B., Jacobs, R., Busche, K. (1989). Pathogenetic mechanisms in the Mycoplasma arthritidis polyarthritis of rats. *Rheumatol. Int.* **9**, 193–196.

Kishimoto, T., Goyert, S., Kikutani, H., Mason, D., Miyasaka, M., Moretta, L., Ohno, T., Okumura, K., Shaw, S., Springer, T. A., Sugamura, K., Sugawara, H., Von dem Borne, A. E. G. K., Zola, H., eds. (1997). Leucocyte typing VI: White Cell Differentiation Antigens, Garland Publishers (in press).

Kita, Y., Takashi, T., Tamatani, T., Miyasaka, M., Horiuchi, T. (1992). Sequence and expression of rat ICAM-1. *Biochim. Biophys. Acta* **1131**, 108–110.

Kitani, S., Berenstein, E., Mergenhagen, S., Tempst, P., Siraganian, R. P. (1991). A cell surface glycoprotein of rat basophilic leukemia cells close to the high affinity IgE receptor (FcεRI). *J. Biol. Chem.* **266**, 1903.

Klatter, F. A., Raué, H. P., Bartels, H. L., Pater, J. M., Groen, H. G., Nieuwenhuis, P., Kampinga, J. (1995). Simultaneous transplantation and intrathymic tolerance induction. A method with clinical potential. *Transplantation* **60**, 1208–1210.

Klonz, A., Wonigeit, K., Pabst, R., Westermann, J. (1996). The marginal blood pool of the rat contains not only granulocytes but also lymphocytes, NK-cells and monocytes: a second intravascular compartment, its cellular composition, adhesion molecule expression and interaction with the peripheral blood pool. *Scand. J. Immunol.* **44**, 461–469.

Koguma, T., Takawasa, S., Togho, A., Karawasa, T., Furuya, Y., Yonekura, H., Okamoto, H. (1994). Cloning and characterization of cDNA encoding rat ADP-ribosyl cyclase/cyclic ADP-ribose hydrolase (homologue to human CD38) from islets of Langerhans. *Biochim. Biophys. Acta* **1223**, 160–162.

Konishi, K., Takata, Y., Watanabe, K., Date, K., Yamamoto, M., Murase, A., Yoshida, H., Suzuki, T., Tsurufuji, S., Fujioka, M. (1993). Recombinant expression of rat and human GRO proteins in *Escherichia coli. Cytokine* **5**, 506–511.

Koop, B. F., Rowen, L., Wang, K., Kuo, C. L., Seto, D., Lenstra, J. A., Howard, S., Shan, W., Deshpande, P., Hood, L. (1994). The human T-cell receptor TCRAC/TCRDC ($C\alpha/C\delta$) region: Organization, sequence, and evolution of 97.6 Kb of DNA. *Genomics* **19**, 478–493.

Koornstra, P. J., de Jong, F. I., Vlek, L. F., Marres, E. H., van Breda-Vriesman, P. J. (1991). The Waldeyer ring equivalent in the rat. A model for analysis of oronasopharyngeal immune responses. *Actas Otolaryngol. Stockholm* **111**, 591–599.

Kosuda, L. L., Wayne, A., Nahounou, M., Greiner, D. L., Bigazzi, P. E. (1991). Reduction of the RT6.2+ subset of T lymphocytes in Brown Norway rats with mercury-induced renal autoimmunity. *Cell Immunol.* **135**, 154–167.

Kotani, M., Kawashima, I., Ozawa, K., Ogura, K., Ariga, T., Tai, T. (1994). Generation of one set of murine monoclonal antibodies specific for globo-series glycolipids: evidence for differential distribution of the glycolopids in rat small intestine. *Arch. Biochem. Biophys.* **310**, 89–96.

Kovalsky, H., Gallai, M., Armbrust, T., Ramadori, G. (1994). Syndecan-1 gene expression in isolated rat liver cells (hepatocytes, Kupffer cells, endothelial cells and Ito cells). *Biochem. Biophys. Res. Comm.* **204**, 944–949.

Kowal, R. C., Herz, J., Goldstein, J. L., Esser, V., Brown, M. S. (1989). Low density lipoprotein receptor-related protein mediates uptake of cholesteryl esters derived from apoprotein E-enriched lipoproteins. *Proc. Natl Acad. Sci. USA* **86**, 5810–5814.

Krieger, M., Herz, J. (1994). Structures and functions of multiligand lipoprotein receptors: macrophage scavenger receptors and LDL receptor-related protein. *Annu. Rev. Biochem.* **63**, 601–637.

Kroemer, G., Martinez, A. C. (1991). Invariant involvement of IL-2 in thymocyte differentiation (Letter). *Immunol. Today* **12**, 246.

Kroese, F. G. M., Opstelten, D., Wubbena, A. S., Deenen, G. J., Aten, J., Schwander, E. H., De Leij, L., Nieuwenhuis, P. (1985a). Monoclonal antibodies to rat B lymphocyte (sub-) population. *Adv. Exp. Med. Biol.* **186**, 81–89.

Kroese, F. G. M, Wubbena, A. S., Joling, P., Nieuwenhuis, P. (1985b). T-lymphocytes in rat lymphoid follicles are a subset of T helper cells. In *Advances in Experimental Medicine and Biology* (ed. Klaus, G. G. B.), 186, 443. Plenum Press, New York, USA.

Kroese, F. G. M., Wubbena, A. S., Nieuwenhuis, P. (1986). Germinal centre formation and follicular antigen trapping in the spleen of lethally X-irradiated and reconstituted rats. *Immunology* **57**, 99–104.

Kroese, F. G. M., Wubbena, A. S., Opstelten, D., Deenen, G. J., Schwander, E. H., De Leij, L., Vos, H., Poppema, S., Volberda, J., Nieuwenhuis, P. (1987a). B lymphocyte differentiation in the rat: production and characterization of monoclonal antibodies to B lineage-associated antigens. *Eur. J. Immunol.* **17**, 921–928.

Kroese, F. G. M., Wubbena, A. S., Seijen, H. G., Nieuwenhuis, P. (1987b). Germinal centers develop oligoclonally. *Eur. J. Immunol.* **17**, 1069–1072.

Kroese, F. G. M., Butcher, E. C., Lalor, P. A., Stall, A. M., Herzenberg, L. A. (1990). The rat B cell system: the anatomical localization of flow cytometry-defined B cell subpopulations. *Eur. J. Immunol.* **20**, 1527–1534.

Kroese, F. G. M., Hermans, M. H., De Boer, N. K., Lalor, P. A., Stall, A. M., Butcher, E. C., Herzenberg, L. A. (1991). Rat B cell subsets identified by flow cytometry and their anatomical counterparts. In *Lymphatic Tissues and* in vivo *Immune Responses* (eds Ezine, S., Berrih-Aknin, S., Imhof, B.), pp. 397–402. Marcel Dekker, New York, USA.

Kroese, F. G. M., Kantor, A. B., Herzenberg, L. A. (1994). The role of B-1 cells in mucosal immune responses. In *Handbook of Mucosal Immunology* (ed. Ogra, P. L.), pp. 217–224. Academic Press, Inc., New York, USA.

Kroese, F. G. M., de Boer, N. K., de Boer, T., Nieuwenhuis, P., Kantor, A. B., Deenen, G. J. (1995). Identification and kinetics of two recently bone marrow-derived B cell populations in peripheral lymphoid tissues. *Cell. Immunol.* **162**, 185–193.

Kühnlein, P., Park, H.-J., Herrmann, T., Elbe, A., Hünig, T. (1994). Identification and characterization of rat γ/δ T-lymphocytes in peripheral lymphoid organs, small intestine, and skin with a monoclonal antibody to a constant determinant of the γ/δ T-cell receptor. *J. Immunol.* **153**, 979.

Kühnlein, P., Vicente, A., Varas, A., Hünig, T., Zapata, A. (1995). γ/δ T-cells in fetal, neonatal and adult rat lymphoid organs. *Developmental Immunology* **4**, 181–188.

Kühnlein, P., Mitnacht, R., Torres-Nagel, N. E., Herrmann, T., Elbe, A., Hünig, T. (1996). The canonical T-cell receptor of dendritic epidermal γ/δ T-cells is highly conserved between rats and mice. *Eur. J. Immunol.* **26**, 3092–3097.

Kumamaratne, D. S., MacLennan, I. C. M. (1981). Cells of the marginal zone of the spleen are lymphocytes derived from recirculating precursors. *Eur. J. Immunol.* **11**, 865–869.

Kumar, S. N., Stewart, G. L., Steven, W. M., Seelig, L. L. (1989). Maternal to neonatal transmission of T-cell mediated immunity to *Trichinella spiralis* during lactation. *Immunology* **68**, 87–95.

Kunes, J., Kohoutova, M., Zicha, J. (1996). Major histocompatibility complex in the rat and blood pressure regulation. *Am. J. Hypertens.* **9**, 675–680.

Kunz, H. W., Cortese Hassett, A. L., Inomata, T., Misra, D. N., Gill, T. J. (1989). The RT1.G locus in the rat encodes a Qa/TL-like antigen. *Immunogenetics* **30**, 181–187.

Kuper, C. F., Hemleers, D. M., Bruijntjes, J. P., van der Ven, I., Biewenga, J., Sminia, T. (1990). Lymphoid and non-lymphoid cells in nasal-associated lymphoid tissue (NALT) in the rat. An immuno- and enzyme-histochemical study. *Cell Tissue Res.* **259**, 371–377.

Kuper, C. F., Koornstra, P. J., Hameleers, D. M. H., Biewenga, J., Spit, B. J., Duijvestijn, A. M., van Breda Vriesman, P. J. C., Sminia, T. (1992). The role of nasopharyngeal lymphoid tissue. *Immunol. Today* **13**, 219–224.

Kurokawa, T., Seno, M., Igarashi, K. (1988). Nucleotide sequence of rat basic fibroblast growth factor cDNA. *Nucl. Acids Res.* **16**, 5201–5202.

Kütemeier, G., Hoehne, W., Werner, T., Shuh, R., Mozikat, R. (1994). Assembly of humanized antibody genes from synthetic oligonucleotides using a single-round PCR. *Biotechniques* **17**, 242–246.

Lai, L., Alaverdi, N., Chen, Z., Kroese, F. G. M., Bos, N. A.,

Huang, E. C. M. (1997). Monoclonal antibodies to human, mouse, and rat cluster of differentiation antigens. In *The Handbook of Experimental Immunology*, 5th edn (eds Weir, L. A. H. D., Herzenberg, L. A.), pp. 61.1–61.37. Blackwell Science Inc., Cambridge, MA, USA.

Lakkis, F. G., Cruet, E. N. (1993). Cloning of rat interleukin-13 (IL-13) cDNA and analysis of IL-13 gene expression in experimental glomerulonephritis. *Biochem. Biophys. Res. Commun.* **197**, 612–618.

Lam, K. P., Stall, A. M. (1994). Major histocompatibility complex class II expression distinguishes two distinct B cell developmental pathways during ontogeny. *J. Exp. Med.* **180**, 507–516.

Lane, P. J. L., Gray, D., Oldfields, S., MacLennan, I. C. M. (1986). Differences of recruitment of virgin B cells into antibody responses to thymus-dependent and thymus-independent type-2 antigens. *Eur. J. Immunol.* **16**, 719–726.

Lang, P., Mocikat, R. (1991). Immunoglobulin heavy chain joining genes in the rat: comparison with mouse and human. *Gene* **102**, 261–264.

Law, D. A., Spruyt, L. L., Paterson, D. J., Williams, A. F. (1989). Subsets of thymopoietic rat thymocytes defined by expression of the CD2 antigen and the MRC OX-22 determinant of the leukocyte-common antigen CD45. *Eur. J. Immunol.* **19**, 2289–2295.

Lawetzky, A., Tiefenthaler, G., Kubo, R., Hünig, T. (1990). Identification and characterization of rat T-cell subpopulations expressing α/β and γ/δ T-cell receptors. *Eur. J. Immunol.* **20**, 343–349.

Lawetzky, A., Kubbies, M., Hünig, T. (1991). Rat 'first-wave' mature thymocytes: cycling lymphoblasts that are sensitive to activation-induced cell death but rescued by interleukin 2. *Eur. J. Immunol.* **21**, 2599–2604.

Lee, K. H., Bowen-Pope, D. F., Reed, R. R. (1990). Isolation and characterization of the alpha platelet-derived growth factor from rat olfactory epithelium. *Mol. Cell Biol.* **10**, 2237–2246.

Leenaerts, P. L., Stadt, R. K., Hall, B. M., van Damme, B. J., Vanrenterghem, Y., Daha, M. R. (1994). Hereditary C6 deficiency in a strain of PVG/c rats. *Clin. Exp. Immunol.* **97**, 1–5.

Léger, O., Jackson, E., Dean, C. (1995). Primary structure of the variable regions encoding antibody to NG2, a tumour-specific antigen on the rat chondrosarcoma HSN. Correlation of idiotypic specificities with amino acid sequences. *Mol. Immunol.* **32**, 697–709.

Lehmann, M., Sternkopf, F., Metz, F., Brock, J., Docke, W. D., Plantikow, A., Kuttler, B., Hahn, H. J., Ringel, B., Volk, H. D. (1992). Induction of long-term survival of rat skin allografts by a novel, highly efficient anti-CD4 monoclonal antibody. *Transplantation* **54**, 959–962.

Leitenberg, D., Feldbush, T. L. (1988). Lymphokine regulation of surface Ia expression on rat B cells. *Cell. Immunol.* **111**, 451–460.

Leslie, G. A. (1984). Allotypic determinants (Igh-3) associated with the IgG2c subclass of rat immunoglobulins. *Mol. Immunol.* **21**, 577–580.

Liebert, M., Kunz, H. W., Gill III, T. J., Cramer, D. V. (1982). CML characterization of a product of a second class I locus in the rat MHC. *Immunogenetics* **16**, 143–155.

Liebert, M., Kunz, H. W., Gill III, T. J., Cramer, D. V. (1983). Comparison of antigens in the rat MHC that act as CML determinants. *Int. Arch. Allergy Appl. Immunol.* **72**, 279–283.

Like, A. A., Guberski, D. L., Butler, L. (1986). Diabetic BB/Wor rats need not to be lymphopenic. *J. Immunol.* **136**, 3254–3258.

Lin, S. H., Culic, O., Flanagan, D., Hixson, D. C. (1991). Immunochemical characterization of two isoforms of rat liver ecto-ATPase that show an immunological and structural identity with a glycoprotein cell-adhesion molecule with Mw 105,000. *Biochem. J.* **278**, 155–161.

Lindner, V., Reidy, M. A. (1995). Platelet-derived growth factor ligand and receptor expression by large vessel endothelium *in vivo*. *Am. J. Pathol.* **146**, 1488–1487.

Liu, Y. J., Oldfield, S., MacLennan, I. C. M. (1988). Memory B cells in T cell-dependent antibody responses colonize the splenic marginal zones. *Eur. J. Immunol.* **18**, 355.

Liu, Y. J., Zhang, J., Lane, P. J. L., Chan, E. T. Y., MacLennan, I. C. M. (1991). Sites of specific B cell activation in primary and secondary responses to T cell-dependent and T cell-independent antigens. *Eur. J. Immunol.* **21**, 2951–2962.

Liu, M., Liu, J., Buch, S., Tanswell, A. K., Post, M. (1995). Antisense oligonucleotides for PDGF-B and its receptor inhibit mechanical strain-induced fetal lung cell growth. *Am. J. Physiol.* **269**, L178-L184.

Liu, L., Tornquist, E., Mattson, P., Eriksson, N. P., Persson, J. K., Morgan, B. P., Aldskogius, H., Svensson, M. (1995). Complement and clustrin in the spinal cord dorsal horn and gracile nucleus following sciatic nerve injury in the adult rat. *Neuroscience* **68**, 167–169.

Lobel, S., Cramer, D. V. (1981). Demonstration of a new genetic locus in the major histocompatibility system of the rat. *Immunogenetics* **13**, 465–473.

Locker, J., Gill III, T. J., Kraus, J. P., Ohura, T., Swarop, M., Riviere, M., Islam, M. Q., Levan, G., Szpirer, J., Szpirer, C. (1990). The rat MHC and cystathione beta-synthase gene are syntenic on chromosome 20. *Immunogenetics* **31**, 271–274.

Lortan, J. E., Roobottom, C. A., Oldfield, S., MacLennan, I. C. M. (1987). Newly produced virgin B cells migrate to secondary lymphoid organs but their capacity to enter follicles is restricted. *Eur. J. Immunol.* **17**, 1311–1316.

Lu, J. H., Thiel, S., Wiedeman, A., Timpl, R., Reid, K. B. (1990). Binding of the pentamer/hexamer forms of mannan-binding protein to zymosan activates the proenzyme C1r2Cs2 complex of the classical pathway of complement without involvement of C1q. *J. Immunol.* **144**, 2287–2294.

Luheshi, G., Hopkins, S. J., Lefeuvre, R. A., Dascombe, M. J., Ghiara, P., Rothwell, N. J. (1993). Importance of brain IL-1 type II receptors in fever and thermogenesis in the rat. *Am. J. Physiol.* **265**, E585–E591.

Lukic, M. L., Mostarica, M., Ejdus, L., Bonaci, B. (1990). Strain differences in alloreactivity in rats: the roles of T-cell subsets. *Transpl. Proc.* **22**, 2549–2550.

Lumsden, T. (1938). Agglutination tests in the study of tumour immunity, natural and acquired. *Am. J. Cancer* **32**, 395–417.

Mackay, C. R. (1993). Homing of naive, memory and effector lymphocytes. *Curr. Opin. Immunol.* **5**, 423–431.

Mackay, C. R. (1996). Chemokine receptors and T cell chemotaxis. *J. Exp. Med.* **184**, 1–4.

MacLennan, I. C. M., Gray, D. (1986). Antigen-driven selection of virgin and memory B cells. *Immunol. Rev.* **91**, 61–85.

MacLennan, I. C. M., Gray, D., Kumararatne, D. S., Bazin, H. (1982). The lymphocytes of splenic marginal zones: a distinct B cell lineage. *Immunol. Today* **3**, 305–307.

Maeda, K., Sato, T., Azuma, M., Yagita, H., Okumura, K.

(1997). Characterization of rat CD80 and CD86 by molecular cloning and mAb. *Int. Immunol.* **9**, 993–1000.

Mahendran, R. S., O'Hare, M. J., Ormerod, M. G., Edwards, P. A., McIlhinney, R. A., Gusterson, B. A. (1989). A new monoclonal antibody to a cell surface antigen that distinguishes luminal epithelial cells in the mammary gland. *J. Cell. Sci.* **94**, 545–552.

Maisin, J., Maldague, P., Dunjic, A., Maisin, H. (1957). Syndromes mortels et effets tardifs des irradiations totales et subtotales chez le rat. *J. Belge Radiol.* **40**, 346–398.

Maisin, J., Maldague, P., Deckers, C., Gond-Que, P. (1964). Le leucosarcome du rat. *Symp. Lymph. Tumours Afr.* **1963**, 341–354.

Maldague, P., Maisin, J., Dunjic, A., Pham-Hong-Que, Maisin, H. (1958). Le leucosarcome chez le rat. L'incidence de ce néoplasme après irradiation totale, subtotale et locale. *Sang* **24**, 751.

Malfroy, B., Shofield, P., Kuang, W. J., Seeburg, P. H., Mason, A. J., Henzel, W. J. (1987). Molecular cloning and amino acid sequence of rat enkephalinase. *Biochem. Biophys. Res. Commun.* **144**, 59–66.

Mallett, S., Fossum, S., Barclay, A. N. (1990). Characterization of the MRC OX-40 antigen of activated CD4 positive T lymphocytes: a molecule related to nerve growth factor receptor. *EMBO J.* **9**, 1063–1068.

Marchese, A., Heiber, M., Nguyen, T., Heng, H. H. Q., Saldivia, V. R., Cheng, R., Murphy, P. M., Tsui, L.-C., Shi, X., Gregor, P., George, S. R., O'Dowd, B. F., Docherty, J. M. (1995). Cloning and chromosomal mapping of three novel genes, GPR9, GPR10, and GPR14, encoding receptors related to interleukin-8, neuropeptide Y, and Somatostatin receptors. *Genomics* **29**, 335–344.

Markmann, J. F., Hickey, W. F., Kimura, H., Woehrle, M., Barker, C. F., Naji, A. (1987). Gamma interferon induces novel expression of Ia antigens by rat pançreatic islet endocrine cells. *Pancreas* **2**, 258–261.

Marshall, E., Miller, T. (1981). Characterization of the nude rat (rnunz). Functional characteristics. *Aust. J. Exp. Biol. Med. Sci.* **59**, 287–296.

Martin, F. H., Suggs, S. V., Langley, K. E., Lu, H. S., Ting, J., Okino, K. H., Morris, C. F., McNiece, I. K., Jacobson, F. W., Mendiaz, E. A., Birkett, N. C. *et al.* (1990). Primary structure and functional expression of rat and human stem cell factor DNAs. *Cell* **63**, 203–211.

Maruiwa, M., Mizoguchi, A., Russel, G. J., Narula, N., Stronska, M., Mizoguchi, E., Rabb, H., Arnaout, M. A., Bhan, A. K. (1993). Anti-KCA-3 a monoclonal antibody reactive with a rat C3 receptor distinguishes Kupffer cells from other macrophages. *J. Immunol.* **150**, 4019–4030.

Mason, D. W. (1985). The possible role of class II major histocompatibility antigens in self tolerance. *Scand. J. Immunol.* **21**, 397–400.

Mason, D. W., Williams, A. F. (1980). The kinetics of antibody binding to membrane antigens in solution at the cell surface. *Biochem. J.* **187**, 1–20.

Mathieson, P. W. (1992). Mercuric chloride induced autoimmunity. *Autoimmunity* **13**, 243–247.

Mathieson, P. W., Gillespie, K. M. (1996). Cloning of a partial cDNA for rat interleukin-12 (IL-12) and analysis of IL-12 expression *in vivo*. *Scand. J. Immunol.* **44**, 11–14.

Mathieson, P., Cobbold, S., Hale, G., Clark, M., Oliveira, D., Lockwood, C., Waldmann, H. (1990). Monoclonal antibody therapy in systemic vasculitis. *New. Eng. J. Med.* **323**, 250–254.

Mathieson, P. W., Thiru, S., Oliveira, D. B. G. (1993). Regulatory role of OX22^{high} T cells in mercury-induced autoimmunity in the Brown Norway rat. *J. Exp. Med.* **177**, 1309–1316.

Matsumoto, K., Kakidani, H., Anzai, M., Nagata, N., Takahashi, A., Takahashi, Y., Miyata, K. (1995). Evaluation of an antisense RNA transgene for inhibiting growth hormone gene expression in transgenic rats. *Dev. Genet.* **16**, 273–277.

Matsuo, Y., Onoder, H., Shiga, Y., Shozuhara, H., Ninomiya, M., Kihara, T., Tamatani, T., Miyasaka, M., Kogure, K. (1994). Role of cell adhesion molecules in brain injury after transient middle cerebral artery occlusion in the rat. *Brain Res.* **656**, 344–352.

Matsuura, A., Ishii, Y., Yuasa, H., Narita, H., Kon, S., Takami, T., Kikuchi, K. (1984). Rat T lymphocyte antigens comparable with mouse Lyt-1 and Lyt2,3 antigenic systems: characterization by monoclonal antibodies. *J. Immunol.* **132**, 316–322.

Maxson, R. T., Dunlap, J. P., Tryka, F., Jackson, R. J., Smith, S. D. (1994). The role of the mucus gel layer in intestinal bacterial translocation. *J. Surg. Res.* **57**, 682–686.

May, M. J., Entwistle, G., Humphries, M. J., Ager, A. (1993). VCAM-1 is a CS1 peptide-inhibitable adhesion molecule expressed by lymph node high endothelium. *J. Cell Sci.* **106**, 109–119.

Mayer, L., Shlien, R. (1987). Evidence for function of Ia molecules on gut epithelial cells in man. *J. Exp. Med.* **166**, 1471–1483.

Mayrhofer, G., Spargo, L. D. J. (1990). Distribution of class II major histocompatibility antigens in enterocytes of the rat jejunum and their association with organelles of the endocytic pathway. *Immunology* **70**, 11–19.

Mazanec, M. B., Nedrud, J. G., Kaetzel, C. S., Lamm, M. E. (1993). A three-tiered view of the role of IgA in mucosal defense. *Immunol. Today* **14**, 430–434.

McCall, M. N., Shotton, D. M., Barclay, A. N. (1992). Expression of soluble isoforms of rat CD45. Analysis by electron-microscopy and use in epitope mapping of anti-CD45R monoclonal antibodies. *Immunology* **76**, 310–317.

McCarthy, K. F., Hale, M. L., Fehnel, P. L. (1987). Purification of rat hematopoietic stem cells by flow cytometry. *Cytometry* **8**, 269–305.

McCaughan, G. W., Wickson, J. E., Creswick, P. F., Gorell, M. D. (1990). Identification of the bile canalicular cell surface molecule GP110 as the ectopeptidase dipeptidyl peptidase IV: an analysis by tissue distribution, purification and N-terminal amino acid sequence. *Hepatology* **11**, 534–544.

McGhee, J. R., Mestecky, J. (1990). In defense of mucosal surfaces. Development of novel vaccines for IgA responses protective at the portals of entry of microbial pathogens. *Infect. Dis. Clin. N. Am.* **4**, 315–320.

McIntyre, W. (1950). Case of mollities and fragilitas ossuim accompanied with urine strongly charged with animal matter. *Med. Chir. Soc. Trans.* **33**, 211.

McKenna, S. D., Goldschneider, I. (1993). A selective culture system for generating terminal deoxynucleotidyl transferase-positive lymphoid precursor cells *in vitro*. V. Detection of stage-specific pro-B-cell stimulating activity in medium conditioned by mouse bone marrow stromal cells. *Dev. Immunol.* **3**, 181–195.

McKenna, S. D., Medlock, E. S., Greiner, D. L., Goldschneider, I.

(1994). A selective culture system for generating terminal deoxynucleotidyl transferase-positive lymphoid precursor cells *in vitro*. IV. Properties and developmental relationships of the lymphoid cells in the adherent and nonadherent compartments of the culture. *Exp. Hematol.* **22**, 1164–1170.

McKnight, A. J., Classon, B. J. (1992). Biochemical and immunological properties of rat recombinant interleukin-2 and interleukin-4. *Immunol.* **75**, 286–292.

McKnight, A. J., Mason, D. W., Barclay, A. N. (1989). Sequence of a rat MHC class II-associated invariant chain cDNA clone containing a 64 amino acid thyroglobulin-like domain. *Nucl. Acids Res.* **17**, 3983–3984.

McKnight, A. J., Barclay, A. N., Mason, D. W. (1991). Molecular cloning of rat interleukin-4 cDNA and analysis of the cytokine repertoire of subsets of $CD4^+$ T cells. *Eur. J. Immunol.* **21**, 1187–1194.

Meilinger, M., Haumer, M., Szakmary, K. A., Steinbock, F., Scheiber, B., Goldenberg, H., Huettinger, M. (1995). Removal of lactoferrin from plasma is mediated by binding to low density lipoprotein receptor-related protein/alpha 2-macroglobulin receptor and transport to endosomes. *FEBS Lett.* **360**, 70–74.

Melchers, F., Karasuyama, H., Haasner, D., Bauer, S., Kudo, A., Sakaguchi, N., Jameson, B., Rolink, A. (1993). The surrogate light chain in B-cell development. *Immunol. Today* **14**, 60–68.

Mencia-Huerta, J. M., Dugas, B., Boichot, E., Petit-Frere, C., Paul-Eugene, N., Lagente, V., Capron, M., Liu, F. T., Braquet, P. (1991). Pharmacological modulation of the antigen-induced expression of the low-affinity IgE receptor (FC epsilon receptor RII/CD23) on rat alveolar macrophages. *Int. Arch. Allergy Appl. Immunol.* **94**, 295–298.

Mendrick, D. L., Kelly, D. M. (1993). Temporal expression of VLA-2 and modulation of its ligand specificity by rat glomerular epithelial cells *in vitro*. *Lab. Invest.* **72**, 367–702.

Mendrick, D. L., Kelly, D. M., DuMont, S. S., Sandstrom, D. J. (1995). Glomerular epithelial and mesangial cells differentially modulate the binding of specificities of VLA-1 and VLA-2. *Lab. Invest.* **72**, 367–375.

Misugi, E., Kawamura, N., Imanishi, N., Tojo, S. J., Morooka, S. (1995). Sialyl Lewis X moiety on rat polymorphonuclear leukocytes responsible for binding to rat E-selectin. *Biochem. Biophys. Res. Commun.* **215**, 547–554.

Mita, E., Hayashi, N., Iio, S., Takehara, T., Hijioka, T., Kasahara, A., Fusamoto, H., Kamada, T. (1994). Role of Fas ligand in apoptosis induced by hepatitis C virus infection. *Biochem. Biophys. Res. Commun.* **204**, 468–474.

Mitchell, B. (1983). The distribution of immune cell markers in the rat metrial gland. *J. Anat.* **137**, 799.

Mitnacht, R., Tacke, M., Hunig, T. (1995). Expression of cell interaction molecules by immature rat thymocytes during passage through the $CD4^+8^+$ compartment: developmental regulation and induction by T cell receptor engagement of CD2, CD5, CD28, CD11a, CD44 and CD53. *Eur. J. Immunol.* **25**, 328–332.

Miura, S., Tsuzuki, Y., Fukumura, D., Serizawa, H., Suematsu, M., Kurose, I., Imaeda, H., Kimura, H., Nagata, H., Tsuchiya, M., Ishii, H. (1995). Intravital demonstration of sequential migration process of lymphocyte subpopulations in rat Peyer's patches. *Gastroenterology* **109**, 1113–1123.

Miyazaki, A., Sakai, M., Yamaguchi, E., Sakamoto, Y., Shichiri, M., Horiuchi, S. (1993). Two independent macrophage receptors for acetylated high-density lipoprotein. *Biochim. Biophys. Acta* **1170**, 143–150.

Mojcik, C. F., Greiner, D. L., Goldschneider, I., Lubaroff, D. M. (1987). Monoclonal antibodies to RT7 and LCA antigens in the rat: cell distribution and segregation analysis. *Hybridoma* **6**, 531–543.

Mojcik, C. F., Greiner, D. L., Medlock, E., S., Komschlies, K. L., Goldschneider, I. (1988). Characterization of RT6 bearing rat lymphocytes. I. Ontogeny of the $RT6^+$ subset. *Cell. Immunol.* **114**, 336–346.

Mojcik, C. F., Greiner, D. L., Goldschneider, I. (1991). Characterization of RT6-bearing rat lymphocytes. II. Developmental relationships of $RT6^-$ and $RT6^+$ T cells. *Dev. Immunol.* **1**, 191–201.

Molina, A., Sanchez-Madrid, F., Bricio, T., Martin, A., Barat, A., Alvarez, V., Mampaso, F. (1994). Prevention of mercuric chloride-induced nephritis in the Brown Norway rat by treatment with antibodies against the alpha 4 integrin. *J. Immunol.* **153**, 2313–2320.

Mooij, P., de Wit, H. J., Drexhage, H. A. (1994). A high iodine intake in Wistar rats results in the development of a thyroid-associated ectopic thymic tissue and is accompanied by a low thyroid autoimmune reactivity. *Immunology* **81**, 309–316.

Moore, M. G., Chrzanowski, R. R., McCormick, J. R., Cieplinski, W., Schwink, A. (1984). Production of monoclonal antibodies to rat lung angiotensin-converting enzyme. *Clin. Immunol. Immunopathol.* **33**, 301–312.

Mordes, J. P., Desemone, J., Rossini, A. A. (1987). The BB rat. *Diab./Metab. Rev.* **3**, 725–750.

Mordes, J., Bortell, R., Doukas, J., Rigby, M. R., Whalen, B. J., Zipris, D., Greiner, D. L., Rossini, A. A. (1996a). The BB/Wor rat and the balance hypothesis of autoimmunity. *Diab./Metab. Rev.* **2**, 103–109.

Mordes, J. P., Greiner, D. L., Rossini, A. A. (1996b). Animal models of autoimmune diabetes mellitus. In *Diabetes Mellitus. A Fundamental and Clinical Text* (eds LeRoith, D., Taylor, S. I., Olefsky, J. M.), pp. 349–360. Lippincott-Raven, Philadelphia, USA.

Moriamé, M., Beckers, A., Bazin, H. (1977). Decrease in the incidence of malignant ileo-caecal immunocytoma in LOU/C rats after surgical removal of the ileo-caecal lymph nodes. *Cancer Lett.* **3**, 139–143.

Morris, M., Barclay, A. N., Williams, A. F. (1988). Analysis of T-cell receptor β chains in rat thymus, and rat $C\alpha$ and $C\beta$ sequences. *Immunogenetics* **27**, 174–179.

Mouzaki, A., Diamantstein, T. (1987). Four epitopes on the rat 55-kDa subunit of the interleukin-2 receptor as defined by newly developed mouse anti-rat interleukin-2 receptor monoclonal antibodies. *Eur. J. Immunol.* **17**, 1661–1664.

Mouzaki, A., Volk, H. D., Osawa, H., Diamantstein, T. (1987). Blocking of interleukin 2 (IL 2) binding to the IL 2 receptor is not required for the *in vivo* action of anti-IL 2 receptor monoclonal antibody (mAb). I. The production, characterization and *in vivo* properties of a new mouse anti-rat IL-2 receptor mAb that reacts with an epitope different to the one that binds to IL 2 and the mAb ART-18. *Eur. J. Immunol.* **17**, 335–341.

Mugridge, K. G., Perretti, M., Ghiara, P., Galeotti, C. L., Melli, M., Parenti, L. (1995). Gastric antisecretory and anti-ulcer actions of rat IL-1 in rat involve different IL-1 receptor types. *Am. J. Physiol.* **259**, G763–769.

Mulligan, M. S., Varani, J., Dame, M. K., Lane, C. L., Smith, C. W., Anderson, D. C., Ward, P. A. (1991). Role of endothelial-leukocyte adhesion molecule 1 (ELAM-1) in neutrophil-mediated lung injury in rats. *J. Clin. Invest.* **88**, 1396–1406.

Mulligan, M. S., Polley, M. J., Bayer, R. J., Nunn, M. F., Paulson, J. C., Ward, P. A. (1992). Neutrophil-dependent acute lung injury. Requirement for P-selectin (GMP-140). *J. Clin. Invest.* **90**, 1600–1607.

Mulligan, M. S., Johnson, K. J., Todd, R. F., Issekutz, T. B., Miyasaka, M., Tamatani, T., Smith, C. W., Anderson, D. C., Ward, P. A. (1993). Requirements for leucocyte adhesion molecules in nephrotoxic nephritis. *J. Clin. Invest.* **91**, 577–587.

Mullins, J. J., Mullins, L. J. (1993). Trangenesis in nonmurine species. *Hypertension* **22**, 630–633.

Murakumo, Y., Takahashi, M., Hayashi, N., Taguchi, M., Arakawa, A., Sharma, N., Sakata, K., Saito, M., Amo, H., Katoh, H. *et al.* (1995). Linkage of the athymic nude locus with the myeloperoxidase locus in the rat. *Pathol. Int.* **45**, 261–265.

Mustafa, M., Vingsbo, C., Olsson, T., Ljungdahl, A., Hojeberg, B., Holmdahl, R. (1993). The major histocompatibility complex influences myelin basic protein 63–88-induced T cell cytokine profile and experimental autoimmune encephalomyelitis. *Eur. J. Immunol.* **23**, 3089–3095.

Mustafa, M., Vingsbo, C., Olsson, T., Issazadeh, S., Ljungdahl, A., Holmdahl, R. (1994). Protective influences on experimental autoimmune encephalomyelitis by MHC class I and class II alleles. *J. Immunol.* **153**, 3337–3344.

Na, H. R., Hiserodt, J. C., Seelig, L. L. Jr (1992). Distribution of lymphocyte subsets in rat milk from normal and *Trichinella spiralis*-infected rats. *J. Reproduct. Imm.* **22**, 269–279.

Nagai, Y., Inobe, M., Kikuchi, K., Uede, T. (1993). Functional and phenotypical analysis of subsets of rat $CD4^+$ T cells. *Microbiol. Immunol.* **37**, 623–632.

Nagoya, S., Kikuchi, K., Uede, T. (1991). Monoclonal antibody to a structure expressed on a subpopulation of rat CD8 T cell subsets. *Microbiol. Immunol.* **35**, 895–911.

Nahm, N., Der Balian, G., Venturi, D., Bazin, H., Davie, J. M. (1980). Antigenic-similarities of rat and mouse IgG subclasses associated with anti-carbohydrate specificities. *Immunogenetics* **11**, 199–203.

Naitoh, Y., Fukata, J., Tominaga, T. (1988). Interleukin-6 stimulates the secretion of adrenocorticotrophic hormone in conscious, freely moving rats. *Biochem. Biophys. Res. Commun.* **155**, 1459–1463.

Nakagawa, H., Komorita, N., Shibata, F., Ikesue, A., Konishi, K., Fujioka, M., Kato, H. (1994). Identification of cytokine-induced neutrophil chemoattractants (CINC), rat GRO/CINC-2α and CINC-2β, produced by granulation tissue in culture: purification, complete amino acid sequences and characterization. *Biochem. J.* **301**, 545–550.

Nakagawa, H., Shiota, S., Takano, K., Shibata, F., Komorita, N., Kato, H. (1996a). Cytokine-induced neutrophil chemoattractant (CINC)-2α, a novel member of rat GRO/CINCs, is a predominant cytokine produced by activated macrophages and granulation tissue in culture. *Eur. Cytokine Netw.* 7, 513 (abstract).

Nakagawa, H., Shiota, S., Takano, K., Shibata, F., Kato, H. (1996b). Cytokine-induced neutrophil chemoattractant (CINC)-2 alpha, a novel member of rat GRO/CINCs, is a predominant chemokine produced by lipopolysaccharide-stimulated rat macrophages in culture. *Biochem. Biophys. Res. Commun.* **220**, 945–948.

Nash, D. R., Vaerman, J. P., Bazin, H., Heremans, J. F. (1969). Identification of IgA in rat serum and secretions. *J. Immunol.* **103**, 145–148.

Navarra, P., Tsagarakis, S., Faria, M. S., Rees, L. H., Besser, Grossman, A. B. (1991). Interleukins-1 and -6 stimulate the release of corticotropin-releasing hormone-41 from rat hypothalamus *in vitro* via the eicosanoid cyclooxygenase pathway. *Endocrinology* **128**, 37–44.

Navarra, P., Pozzoli, G., Brunetti, L., Ragazzoni, E., Besser, M., Grossman, A. (1992). Interleukin-1beta and interleukin-6 specifically increase the release of prostaglandin E_2 from rat hypothalamic explants *in vitro*. *Neuroendocrinology* **56**, 61–68.

Nehme, C. L., Cesario, M. M., D.G., M., Koppel, D. E., Bartles, J. R. (1993). Breaching the diffusion barries that compartmentalizes the transmembrane glycoprotein CE9 to the posterior-tail plasma membrane domain of the rat spermatozoon. *J. Cell Biol.* **120**, 687–694.

Nehme, C. L., Fayos, B. E., Bartles, J. R. (1995). Distribution of the integral membrane glycoprotein CE9 (MRC OX-47) among rat tissues and its induction by diverse stimuli of metabolic activation. *Biochem. J.* **310**, 693–698.

Neiss, U., Reske, K. (1994). Non-coordinate synthesis of MHC class II proteins and invariant chains by epidermal Langerhans cells derived from short-term *in vitro* cultures. *Int. Immunol.* **6**, 61–71.

Newton, M. R., Wood, K. J., Fabre, J. W. (1986). A new monoclonal alloantibody detecting a polymorphism of the rat leukocyte common (LC) antigen. *J. Immunogenet.* **13**, 41–50.

Nezlin, R. S., Vengerova, T. I., Rockhlin, O. V., Machulla, H. K. G. (1974). Localization of allotypic markers of kappa light chains of rat immunoglobulins in the constant part of the chain. *Fed. Eur. Biochem. Soc. Meet. (Proc.)* **36**, 93.

Ng, K. K., Drickamer, K., Weis, W. I. (1996). Structural analysis of monosaccharide recognition by rat liver mannose-binding protein. *J. Biol. Chem.* **271**, 663–674.

Nicolls, M. R., Aversa, C. G., Pearce, N. W., Spinelli, A., Berger, M. F., Gurley, K. E., Hall, B. M. (1993). Induction of long-term specific tolerance in rats by therapy with an anti-CD3-like monoclonal antibody. *Transplantation* **55**, 459–468.

Nieuwenhuis, P. (1990). Self-tolerance induction and the blood-thymus barrier. In *Thymus Update III* (eds Kendall, M. D., Ritter, M. A.), pp. 31–51, Harwood Academic Publ., London, UK.

Nieuwenhuis, P. (1996). Histophysiology of the lymphoid system: The Thymus and T-cells, In *The Physiology of Immunity* (eds Marsh, J. A., Kendall, M.), 3–32, CRC Press, Boca Raton, Florida, USA.

Nieuwenhuis, P., Ford, W. L. (1976). Comparative migration of B- and T-lymphocytes in the rat spleen and lymph nodes. *Cell. Immunol.* **23**, 254.

Nieuwenhuis, P., Stet, R. J. M., Wagenaar, J. P. A., Wubbena, A. S., Kampinga, J., Karrenbeld, A. (1988). The transcapsular route: a new way for (self) antigens to by-pass the blood-thymus barrier? *Immunol. Today* **9**, 372–375.

Nishida, T., Nishino, N., Takano, M., Sekiguchi, Y., Kawai, K., Mizuno, K., Nakai, S., Masai, Y., Hirai, Y. (1989). Molecular cloning and expression of rat interleukin-1 alpha cDNA. *J. Biochem.* **105**, 351–357.

Nishikata, H., Oliver, C., Mergenhagen, S. E., Siraganian, R. P. (1992). The rat mast cell antigen AD1 (homologue to human CD63 or melanoma antigen ME491) is expressed in other cells in culture. *J. Immunol.* **149**, 862–870.

Noguchi, Y., Himeno, M., Sasaki, H., Tanaka, Y., Kono, A., Ssaki, Y., Kato, K. (1989). Isolation and sequencing of a cDNA clone encoding rat liver lysosomal cathepsin D and the structure of three forms of mature enzymes. *Biochem. Biophys. Res. Comm.* **164**, 1113–1120.

Norman, K. E., Moore, K. L., McEver, R. P., Ley, K. (1995). Leukocyte rolling *in vivo* is mediated by P-selectin glycoprotein ligand-1. *Blood* **86**, 4417–4421.

Northemann, W., Braciak, T. A., Hattori, M., Lee, F., Fey, G. H. (1989). Structure of the rat interleukin-6 gene and its expression in macrophage-derived cells. *J. Biol. Chem.* **264**, 16072–16082.

Oaks, M. K., Penwell, R. T., Tector, A. J. (1996). Nucleotide sequence of ACI rat CTLA-4 molecule. *Immunogenetics* **43**, 173–174.

Oberg-Welsh, C., Welsh, M. (1996). Effects of certain growth factors on *in vitro* maturation of rat fetal islet-like structures. *Pancreas* **12**, 334–339.

Odin, P., Obrink, B. (1987). Quantitative determination of the organ distribution of the cell adhesion molecule cell-CAM 105 by radioimmunoassay. *Exp. Cell Res.* **171**, 1–15.

Offner, H., Vainiene, M., Gold, D. P., Morisson, W. J., Wang, R.-Y., Hashim, G. A., Vandenbark, A. A. (1991). Protection against experimental encephalomyelitis. Idiotypic autoregulation induced by a nonencephalitogenic T-cell clone expressing a cross-reactive T cell receptor V gene. *J. Immunol.* **146**, 4165–4172.

Ohajekwe, O. A., James, T., Hardy, M. A., Oluwole, S. F. (1995). Prevention of cyclosporine-induced syngeneic graft-versus-host disease in bone marrow transplantation by UV-B irradiated bone marrow cells. *Bone Marrow Transpl.* **15**, 627–632.

O'Hanlon, T. P., Lau, K. M., Wang, X. C., Lau, J. T. (1989). Tissue-specific expression of beta galactoside alpha-2,6-sialyltransferase. Transcript heterogeneity predicts a divergent polypeptide. *J. Biol. Chem.* **264**, 17389–17394.

Ollo, R., Auffray, C., Morchamps, C., Rougeon, F. (1981). Comparison of mouse immunoglobulin gamma 2a and gamma 2b chain genes suggests that exons can be exchanged between genes in a multigenic family. *Proc. Natl Acad. Sci. USA* **78**, 2442.

Opstelten, D., Deenen, G. J., De Jong, B., Idenburg, V. J., Hunt, S. V. (1985). Ig light chain rearrangement and chromosomal abnormality in the LAMA, early B-lineage tumour in the rat. *Adv. Exp. Med. Biol.* **186**, 27–33.

Opstelten, D., Deenen, G. J., Rozing, J., Hunt, S. V. (1986). B lymphocyte associated antigens on terminal deoxynucleotidyl transferase positive-cells and pre-B cells in bone marrow of the rat. *J. Immunol.* **137**, 76–84.

Osawa, H., Diamantstein, T. (1983). The characterization of a monoclonal antibody that binds specifically to rat lymphoblasts and inhibits IL2 receptor functions. *J. Immunol.* **130**, 51–55.

Oudin, J. (1956). L'allotypie de certains antigènes protéidiques du sérum. *C.R. Acad. Sci.* **242**, 2606–2608.

Owen, R. L., Apple, R. T., Bhalla, D. K. (1986). Morphometrical and cytochemical analysis of lysosomes in rat Peyer's patch follicle epithelium: their reduction in volume fraction and acid phosphatase content in M cells compared to adjacent enterocytes. *Anat. Rec.* **216**, 521–527.

Pabst, R., Westermann, J. (1997). Lymphocyte traffic to lymphoid and non-lymphoid organs in different species is regulated by several mechanisms. In *Adhesion Molecules and Lymphocyte Trafficking* (ed. Hamann, A.), pp. 21–37. Harwood Academic Publishers, Chur, Switzerland.

Page, T. H., Dallman, M. J. (1991). Molecular cloning of cDNAs for the rat interleukin 2 receptor alpha and beta chain genes: differentially regulated gene activity in response to mitogenic stimulation. *Eur. J. Immunol.* **21**, 2133–2138.

Papayianni, A., Serhan, C. N., Phillips, M. L., Rennke, H. G., Brady, H. R. (1995). Transcellular biosynthesis of lipoxin A4 during adhesion of platelets and neutrophils in experimental immune complex glomerulonephritis. *Kidney Int.* **47**, 1295–1302.

Park, J. H., Hanke, T., Hünig, T. (1996). Identification and cellular distribution of the rat interleukin-2 receptor β chain: induction of the IL-2R$\alpha^{-}\beta^{+}$ phenotype by major histocompatibility complex class I recognition during T cell development *in vivo* and by T cell receptor stimulation of CD4^{+}8^{+} immature thymocytes *in vitro*. *Eur. J. Immunol.* **26**, 2371–2375.

Paterson, D. J., Williams, A. F. (1987). An intermediate cell in thymocyte differentiation that expresses CD8 but not CD4 antigen. *J. Exp. Med.* **166**, 1603–1608.

Paterson, D. J., Green, J. R., Jefferies, W. A., Puklavec, M., Williams, A. F. (1987a). The MRC OX-44 antigen marks a functionally relevant subset among rat thymocytes. *J. Exp. Med.* **165**, 1–13.

Paterson, D. J., Jefferies, W. A., Green, J. R., Brandon, M. R., Corthesy, P., Puklavec, M., Williams, A. F. (1987b). Antigens of activated rat T lymphocytes including a molecule of 50,000 Mw detected only on CD4 positive T blasts. *Mol. Immunol.* **24**, 1281–1290.

Pat Happ, M., Wettstein, P., Dietzschold, B., Heber-Katz, E. (1988). Genetic control of the development of Experimental Allergic Encephalomyelitis in rats. *J. Immunol.* **141**, 1489–1494.

Pavlovic, M. D., Colic, M., Pejnovic, N., Tamatani, T., Miyasaka, M., Dujic, A. (1994). A novel anti-rat CD18 monoclonal antibody triggers lymphocyte homotypic aggregation and granulocyte adhesion to plastic: different intracellular pathways in resting versus activated thymocytes. *Eur. J. Immunol.* **24**, 1640–1648.

Pear, W. S., Ingvarsson, S., Steffen, D., Munke, M., Francke, U., Bazin, H., Klein, G., Sümegi, J. (1986). Multiple chromosomal rearrangements in a spontaneously arising t(6;7) rat immunocytoma juxtapose c-myc and immunoglobulin heavy chain sequences. *Proc. Natl Acad. Sci. USA* **83**, 7376–7380.

Pear, W. S., Nelson, S. F., Axelson, H., Wahlström, G., Bazin, H., Klein, G., Sümegi, J. (1988). Aberrant class switching juxtaposes c-myc with a middle repetitive element (LINE) and an IgH intron in two spontaneously arising rat immunocytomas. *Oncogene* **2**, 499–507.

Peel, S., Stewart, I. J., Bulmer, D. (1983). Experimental evidence for the bone marrow origin of granulated metrial gland cells of the mouse uterus. *Cell Tissue Res.* **233**, 647–656.

Perry, G. A., Sharp, J. G. (1988). Characterization of proximal colonic lymphoid tissue in the mouse. *Anat. Rec.* **220**, 305–312.

Perry, G. A., Sharp, J. G. (1991). Functional and morphological aspects of proximal colonic lymphoid tissue. In *Lymphatic Tissues and* in vivo *Immune Responses* (eds Imhof, B. A., Berrih-Aknim, S., Ezine, S.), pp. 487–492. Marcel Dekker, Inc., New York, USA.

Piccioli, P., Ruberti, F., Biocca, S., Di Luzio, A., Werge, T. M., Bradbury, A., Cattaneo, A. (1991). Neuroantibodies: molecular cloning of a monoclonal antibody against substance P for expression in the central nervous system. *Proc. Natl Acad. Sci. USA* **88**, 5611–5615.

Plesch, B. E. C. (1982). Histology and immunohistochemistry of bronchus associated lymphoid tissue (BALT) in the rat. In In vivo *Immunology. Histophysiology of the Lymphoid System* (eds Nieuwenhuis, P., van den Broek, A. A., Hanna, M. G.), pp. 491–497. Plenum Press, New York/London.

Plesch, B. E. C., Gamelkoorn, G. J., Vande Ende, M. (1983). Development of bronchus-associated lymphoid tissue (BALT) in the rat, with special reference to T- and B-cells. *Dev. Comp. Immunol.* **7**, 179–188.

Plotkowski, M. C., Bakolet-Laudinat, O., Puchelle, E. (1993). Cellular and molecular mechanisms of bacterial adhesion to respiratory mucosa. *Eur. Resp. J.* **6**, 903–916.

Posselt, A. M., Barker, C. F., Tomaszewenski, J. E., Markmann, J. F., Choti, M. A., Naji, A. (1990). Introduction of donor specific unresponsiveness by intrathymic islet transplantation. *Science* **249**, 1293–1295.

Potter, M., Fahey, J. L., Pilgrim, H. J. (1957). Abnormal serum protein and bone destruction in transmissible mouse plasma-cell neoplasms (multiple myeloma). *Proc. Soc. Exp. Biol. Med.* **94**, 327–333.

Powrie, F., Mason, D. (1989). The MRC OX-22$^-$CD4$^+$ T cells that help B cells in secondary immune responses derive from naive precursors with the MRC OX-22$^+$CD4$^+$ phenotype. *J. Exp. Med.* **169**, 653–662.

Powrie, F., Mason, D. (1990a). Subsets of rat CD4$^+$ T cells defined by their differential expression of variants of the CD45 antigen: developmental relationships and *in vitro* and *in vivo* functions. *Curr. Top. Microbiol. Immunol.* **159**, 79–96.

Powrie, F., Mason, D. (1990b). OX-22^{high} CD4$^+$ T cells induce wasting disease with multiple organ pathology: prevention by the OX-22^{low} subset. *J. Exp. Med.* **172**, 1701–1708.

Prakapas, Z., Denoyelle, M., Dargemont, C., Kroese, F. G. M., Thiery, J. P., Deugnier, M.-A. (1993). Enrichment and characterization of thymus repopulating cells in stroma-dependent cultures of rat bone marrow. *J. Cell Science*, **104**, 1039–1048.

Prieschl, E. E., Kulmburg, P. A., Baumruker, T. (1995). The nomenclature of chemokines. *Int. Arch. Allergy Immunol.* **107**, 475–483.

Prins, J., Herberg, L., den Bieman, M., van Zutphen, B. F. M. (1991). Genetic characterization and interrelationship of inbred lines of diabetes-prone and non diabetes-prone BB rats. In *Frontiers in Diabetes Research. Lessons from Animal Diabetes III* (ed. Shafrir, E.), pp. 19–24. Smith-Gordon, London, UK.

Prokhorova, S., Lavnikova, N., Laskin, D. L. (1994). Functional characterization of interstitial macrophages and subpopulations of alveolar macrophages from rat lung. *J. Leukoc. Biol.* **55**, 141–146.

Querinjean, P., Bazin, H., Beckers, A., Deckers, C., Heremans, J. F., Milstein, C. (1972). Transplantable immunoglobulin-secreting tumors in rats. Purification of chemical characterization of four kappa chains from LOU/Wsl rats. *Eur. J. Biochem.* **31**, 354–359.

Querinjean, P., Bazin, H., Kehoe, J. M., Capra, J. D. (1975). Transplantable immunoglobulin-secreting tumours in rats. VI. N-terminal sequence variability in LOU/Wsl rat monoclonal heavy chains. *J. Immunol.* **114**, 1375–1378.

Quigg, R. J., Galishoff, M. L., Sneed, A. E., Kim, D. (1993). Isolation and characterization of complement receptor type 1 from rat glomerular epithelial cells. *Kidney Int.* **43**, 730–736.

Quigg, R. J., Morgan, B. P., Holers, V. M., Adler, S., Sheed, A. E., Lo, C. F. (1995). Complement regulation in the rat glomerulus: CRRY and CD59 regulate complement in glomerular mesangial and endothelial cells. *Kidney Int.* **48**, 412–421.

Raivich, G., Gehrmann, J., Kreutzberg, G. W. (1991). Increase of macrophage colony-stimulating factor and granulocyte-macrophage colony-stimulating factor receptors in the regenerating rat facial nucleus. *J. Neurosci. Res.* **30**, 682–686.

Rayner, D. C., Lydyard, P. M., de Assis Paiva, H. J., Bidey, S., van der Meide, P. H., Varey, A. M., Cooke, A. (1987). Interferon-mediated enhancement of thyroid major histocompatibility complex antigen expression. A flow cytometric analysis. *Scand. J. Immunol.* **25**, 621–628.

Reichmann, L., Clark, M., Waldmann, H., Winter, G. (1988). Reshaping human antibodies for therapy. *Nature* **332**, 323–327.

Reifenberger, G., Mai, J. K., Krajewski, S., Wechsler, W. (1987). Distribution of anti-leu-7, anti-leu-11a, and anti-leu-M1 immunoreactivity in the brain. *Cell Tissue Res.* **248**, 305–313.

Renegar, K. B., Small, P. A. (1994). Passive immunization: systemic and mucosal. In *Handbook of Mucosal Immunology* (ed. Ogra, P. L.), pp. 347–356. Academic Press, Inc., New York, USA.

Richter, G., Blankenstein, T., Diamantstein, T. (1990). Evolutionary aspects, structure, and expression of the rat interleukin 4 gene. *Cytokine* **2**, 221–228.

Richter, G., Hein, G., Blankenstein, T., Diamantstein, T. (1995). The rat interleukin 4 receptor: co-evolution of ligand and receptor. *Cytokine* **7**, 237–241.

Ritter, M. A., Gordon, L. K., Goldschneider, I. (1978). Distribution and identity of Thy-1-bearing cells during ontogeny in rat hemopoietic and lymphoid tissues. *J. Immunol.* **121**, 2463–2471.

Robinson, A. P., White, T. M., Mason, D. W. (1986). Macrophage heterogeneity in the rat as delineated by two monoclonal antibodies MRC OX-41 and MRC OX-42, the latter recognizing complement receptor 3. *Immunology* **57**, 239–247.

Rockhlin, O. V., Vengerova, T. L., Nezlin, R. S. (1971). Allotypes of light chains of rat immunoglobulins. *Immunochemistry* **8**, 525–538.

Ronco, P., Melcion, C., Geniteau, M., Ronco, E., Reininger, L., Galceran, M., Verroust, P. (1984). Production and characterization of monoclonal antibodies against rat brush border antigens of the proximal convoluted tubule. *Immunology* **53**, 87–95.

Rose, N. R., Bigazzi, P. E., Noble, B. (1976). Spontaneous autoimmune thyroiditis in the BUF rat. *Adv. Exp. Med. Biol.* **73** Pt B, 209–216.

Rosella, G., Zajac, J. D., Baker, L., Kaczmarczyk, S. J., Andrikopoulos, S., Adams, T. E., Proietto, J. (1995). Impaired glucose tolerance and increased weight gain in transgenic rats overexpressing a non-insulin-responsive phosphoenolpyruvate carboxykinase gene. *Molec. Endocrinol.* **9**, 1396–1404.

Rossini, A. A., Mordes, J. P., Handler, E. S., Greiner, D. L. (1995). Human autoimmune diabetes mellitus: lessons from BB rats and NOD mice–*Caveat emptor. Clin. Immunol. Immunopathol.* **74**, 2–9.

Rothe, H., Schuller, I., Richter, G., Jongeneel, C. V., Kiesel, U., Diamantstein, T., Blankenstein, T., Kolb, H. (1992). Abnormal TNF production in prediabetic BB rats is linked to defective CD45R expression. *Immunology* **77**, 1–6.

Roussy, G., Guérin, P. (1942). Les lymphosarcomes du rat. *Bull. Assoc. fr. Etude Cancer* **30**, 17–28.

Rozing, J., Vaessen, L. M. B. (1979). Mitogen responsiveness in the rat. *Transpl. Proc.* **11**, 1657–1659.

Ruberti, F., Cattaneo, A., Bradbury, A. (1994). The use of the RACE method to clone hybridoma cDNA when V region primers fail. *J. Immunol. Methods* **173**, 33–39.

Ruddick, J. H., Leslie, G. A. (1977). Structure and biological functions of human IgD. XI. Identification and ontogeny of rat lymphocyte immunoglobulin having antigenic cross-reactivity with human IgD. *J. Immunol.* **118**, 1025–1031.

Rushmere, N. K., Harrison, R. A., van den Berg, C. W., Morgan, B. P. (1994). Molecular cloning of the rat analogue of human CD59: structural comparison with human CD59 and identification of a putative active site. *Biochem. J.* **304**, 595–601.

Ruuls, S. R., Bauer, J., Sontrop, K., Huitinga, I., 'T Hart, B., Dijkstra, C. D. (1995). Reactive oxygen species are involved in the pathogenesis of experimental allergic encephalomyelitis. *J. Neuroimmunol.* **56**, 207–217.

Ruuls, S. R., De Labie, M. C. D. C., Weber, K. S., Botman, C. A. D., Groenestein, R. J., Dijkstra, C. D., Olsson, T., van der Meide, P. H. (1996). The length of treatment determines whether IFN-β prevents or aggravates experimental autoimmune encephalomyelitis in Lewis rats. *J. Immunol.* **157**, 5721–5731.

Sabolic, I., Culic, O., Lin, S. H., Brown, D. (1992). Localization of ecto-ATPase in rat kidney and isolated renal cortical membrane vesicles. *Am. J. Physiol.* **262**, F217-F218.

Sackstein, R., Meng, L., Xu, X. M., Chin, Y. H. (1995). Evidence of post-transcriptional regulation of L-selectin gene expression in rat lymphoid cells. *Immunology* **85**, 198–204.

Saggi, S. J., Safirstein, R., Price, P. M. (1992). Cloning and sequencing of the rat preproepidermal growth factor cDNA: comparison with mouse and human sequences. *DNA and Cell Biol.* **11**, 481–487.

Sakai, T., Agui, T., Matsumoto, K. (1994). Intestinal intraepithelial lymphocytes in LEC mutant rats. *Immunol. Lett.* **41**, 185–189.

Sakai, T., Agui, T., Matsumoto, K. (1995). Abnormal CD45RC expression and elevated CD45 protein tyrosine phosphatase activity in LEC rat peripheral $CD4^+$ T cells. *Eur. J. Immunol.* **25**, 1399–1404.

Salminen, S., Isolauri E., Onnela, T. (1995). Gut flora in normal and disordered states (review). *Chemotherapy* **41** (Suppl. 1), 5–15.

Saoudi, A., Seddon, B., Nicholson, M., Mason, D. (1986). $CD4^+CD8^+$ thymocytes that express L-selectin mediate both normal T cell function and protection of rats from diabetes upon adoptive transfer. *Abstr. XIth Int. Workshop Alloantigenic Systems in the Rat*, p. 21.

Sarzani, R., Arnaldi, G., De Pirro, R., Moretti, P., Schiaffino, S., Rapelli, A. (1992). A novel endothelial tyrosine kinase cDNA homologous to platelet derived growth factor receptor cDNA. *Biochem. Biophys. Res. Commun.* **186**, 706–714.

Sauermann, U., Meyermann, R., Schluesener, H. J. (1992). Cloning of a novel TGF-β related cytokine, the vgr, from rat brain: cloning of and comparison to homologous human cytokines. *J. Neuroscience Res.* **33**, 142–147.

Schall, T. J. (1991). Biology of the RANTES/SIS cytokine family. *Cytokine* **3**, 1–18.

Scherzer, W. J., Ruutiainen-Altman, K. S., Putowski, L. T., Kol, S., Adashi, E. Y., Rohan, R. M. (1996). Detection and *in vivo* hormonal regulation of ovarian type I and type II interleukin-1 receptor mRNAs: increased expression during periovulatory period. *J. Soc. Gynecol. Investig.* **3**, 131–139.

Schmal, H., Shanley, T. P., Jones, M. L. Friedl, H. P., Ward, P. A. (1996). Role for macrophage inflammatory protein-2 in lipopolysaccarode-induced lung injury in rats. *J. Immunol.* **156**, 1963–1972.

Schobitz, B., De Kloet, E. R., Sutanto, W., Holsboer, F. (1993). Cellular localization of interleukin 6 mRNA and interleukin 6 receptor mRNA in rat brain. *Eur. J. Neurosci.* **5**, 1426–1435.

Schreiber, G., Tsykin, A., Aldred, A. R., Thomas, T., Fung, W. P., Dickson, P. W., Cole, T., Birch, H., De Jong, F. A., Milland, J. (1989). The acute phase response in the rodent. *Ann. NY Acad. Sci.* **557**, 61–85.

Schuurman, H. (1995). The nude rat. *Hum. Exp. Toxicol.* **14**, 122–125.

Schuurman, H., Hougen, H. P., van Loveren, H. (1992). The rnu (Rowett Nude) and rnunz (New Zealand Nude) rat: an update. *ILAR News* **34**, 3–12.

Scriba, A., Luciano, L., Steinger, B. (1996). High yield purification of rat monocytes by combined density gradient and immunomagnetic separation. *J. Immunol. Meth.* **189**, 203–216.

Seddon, B. S., Mason, D. (1996). Effects of cytokines and glucocorticoids on endogenous class II MHC antigen expression by activated rat $CD4^+$ T cells. Mature $CD4^+8^-$ thymocytes are phenotypically heterogeneous on activation. *Int. Immunol.* **8**, 1185–1193.

Seekamp, A., Warren, J. S., Remick, D. G., Till, G. O., Ward, P. A. (1993). Requirements for tumor necrosis factor-α and interleukin-1 in limb ischemia/reperfusion injury and associated lung injury. *Am. J. Pathol.* **143**, 453–463.

Seijen, H. G., Bun, J. C. A. M., Wubbena, A. S., Löhlefink, K. G. (1988). The germinal center precursor cell is mu and delta positive. *Adv. Exp. Med. Biol.* **237**, 233–237.

Seiter, S., Arch, R., Reber, S., Komitowski, D., Hofmann, M., Ponta, H., herrlich, P., Matzku, S., Zoller, M. (1993). Prevention of tumor metastasis formation by anti-variant CD44. *J. Exp. med.* **177**, 443–455.

Sereda, M., Griffiths, I., Puhlhofer, A., Stewart, H., Rossner, M. J., Zimmerman, F., Magyar, J. P., Schneider, A., Hund, E., Meinck, H. M., Suter, U., Nave, K. A. (1996). A transgenic rat model of Charcot–Marie–Tooth disease. *Neuron* **16**, 1049–1060.

Seya, T. (1995). Human regulator of complement activation (RCA) gene family proteins and their relationship to microbiol infection. *Microbiol. Immunol.* **39**, 295–305.

Shah, M., Foreman, D. M., Ferguson, M. W. (1992). Control of scarring in adult wounds by neutralising antibody to transforming growth factor beta. *Lancet* **339**, 213–214.

Sheppard, H. W., Gutmann, G. A. (1981). Allelic forms of rat kappa chain genes: evidence for strong selection at the level of nucleotide sequence. *Proc. Natl Acad. Sci. USA* **78**, 7064–7068.

Sheppard, H. W., Gutmann, G. A. (1982). Rat kappa chain J-segment genes: two recent gene duplications separate rat and mouse. *Cell* **29**, 121–127.

Shibata, F., Konishi, K., Kato, H., Komorita, N., Al-Mokdad, M., Fujioka, M., Nakagawa, H. (1995). Recombinant production and biological properties of rat cytokine-induced neutrophil chemoattractants, GRO/CINC-2a, CINC-2b and CINC-3. *Eur. J. Biochem.* **231**, 306–311.

Shibata, F., Kato, H., Konishi, K., Okumura, A., Ochiai, H., Nakajima, K., Al-Mokdad, M., Nakagawa, H. (1996). Differential changes in the concentrations of cytokine-induced neutrophil chemoattractant (CINC)-1 and CINC-2 in exudate during rat lipopolysaccharide-induced inflammation. *Cytokine* **8**, 222–226.

Shimasaki, S., Emoto, N., Koba, A., Mercado, M., Shibata, F., Cooksey, K., Baird, A., Ling, N. (1988). Complementary DNA cloning and sequencing of rat ovarian basic fibroblast growth factor and tissue distribution study of its mRNA. *Biochem. Biophys. Res. Commun.* **157**, 256–263.

Shirwan, H., Barwari, L., Fuss, I., Makowka, L., Cramer, D. V. (1995). Structure and repertoire usage of rat TCR α-chain genes in T-cells infiltrating heart allografts. *J. Immunol.* **154**, 1964–1972.

Siddle, K., Bailyes, E., Luzio, J. (1981). A monoclonal antibody inhibiting rat liver 5′-nucleotidase. *FEBS Lett.* **128**, 103–107.

Sims, M., Hassal, D., Brett, S., Rowan, W., Lockyer, M., Angel, A., Lewis, A., Hale, G., Waldmann, H., Crowe, J. (1991). A humanized CD18 antibody can block function without cell destruction. *J. Immunol.* **151**, 2296–2308.

Sminia, T., Dijkstra C. D. (1986). The origin of osteoclasts: an immunohistochemical study in embryonic rat bone. *Calcif. Tiss.* **39**, 263–266.

Sminia, T., van Asselt, A. A., van de Ende, M. B., Dijkstra, C. D. (1986). Rat thymus macrophages: an immunohistochemical study on fetal, neonatal and adult thymus. *Thymus* **8**, 141–150.

Sminia, T., van der Brugge-Gamelkoorn, G. J., Jeurissen, S. H. M. (1989). Structure and function of bronchus-associated lymphoid tissue (BALT). *Crit. Rev. Immunol.* **9**, 119–150.

Sminia, T., Twisk, K., van der Ven, I., Soesatyo, M. (1991). Uptake of soluble and particulate antigen by the gut. In *Lymphatic Tissues and Immune Responses* (eds Imhof, B. *et al.*), pp. 505–508. Marcel Dekker Inc., New York, USA.

Smith, L. R., Kono, D. H., Theofilopolous, A. N. (1991). Complexity and sequence identification of 24 rat V_{β} genes. *J. Immunol.* **147**, 375–379.

Smith, L. R., Kono, D. H., Kammuller, M. E., Balderas, R. S., Theofilopoulos, A. N. (1992). V_{β} repertoire in rats and implications for endogenous superantigens. *Eur. J. Immunol.* **22**, 641–645.

Smith, M. E., Ford, W. L. (1983). The recirculating lymphocyte pool of the rat: a systematic description of the migratory behaviour of recirculating lymphocytes. *Immunology* **49**, 83–94.

Soares, M. B., Turken, A., Ishii, D., Mills, L., Episkopou, V., Cotter, S., Zeitlin, S., Efstratiadis, A. (1986). Rat insulin-like growth factor II gene. A single gene with two promoters expressing a multitranscript family. *J. Mol. Biol.* **192**, 737–752.

Soares, M., Havaux, X., Nisol, F., Bazin, H., Latinne, D. (1996). Modulation of rat B cell differentiation *in vivo* by the administration of an anti-mu monoclonal antibody. *J. Immunol.* **156**, 108–118.

Spanjaard, R. A., van Dijk, M. C. M., Turion, A. J., van Duin, J. (1989). Expression of the rat interferon-α1 gene in *Escherichia coli* controlled by the secondary structure of the translation initiation region. *Gene* **80**, 345–351.

Sparshott, S. M., Bell, E. B. (1994). Membrane CD45R isoform exchange on CD4 T cells is rapid, frequent and dynamic *in vivo*. *Eur. J. Immunol.* **24**, 2573–2578.

Spickett, G. P., Brandon, M. R., Mason, D. W., Williams, A. F., Woollett, G. R. (1983). MRC OX-22, a monoclonal antibody that labels a new subset of T lymphocytes and reacts with the high molecular weight form of the leukocyte-common antigen. *J. Exp. Med.* **158**, 795–810.

Spit, B. J., Hendriksen, E. G., Bruijntjes, J. P., Kuper, C. F. (1989). Nasal lymphoid tissue in the rat. *Cell Tissue Res.* **255**, 193–198.

Stadler, J., Harbrecht, B. G., Di Silvio, M., Curran, R. D., Jordan, M. L., Simmons, R. L., Billiar, T. R. (1993). Endogenous nitric oxide inhibits the synthesis of cyclooxygenase products and interleukin-6 by rat Kupffer cells. *J. Leukoc. Biol.* **53**, 165–172.

Stamper, H. B., Woodruff, J. J. (1976). Lymphocyte homing into lymph nodes: *in vitro* demonstration of the selective affinity of recirculating lymphocytes for high-endothelial venules. *J. Exp. Med.* **144**, 828–833.

Stamper, H. B., Woodruff, J. J. (1977). An *in vitro* model of lymphocyte homing. I. Characterization of the interaction between thoracic duct lymphocytes and specialized high-endothelial venules of lymph nodes. *J. Immunol.* **119**, 722–780.

Stangel, M., Giegerich, G., Torres-Nagel, N. E., Hünig, T., Hartung, H.-P. (1994). Structural analysis of the rat T-cell receptor TCRA V4 gene family. *Immunogenetics* **41**, 125–130.

Stechschulte, D. J., Orange, R. P., Austen, K. F. (1970). Immunochemical and biologic properties of rat IgE. I. Immunochemical identification of rat IgE. *J. Immunol.* **104**, 1082–1086.

Steen, M. L., Hellman, L., Petterson, U. (1987). The immunoglobulin lambda locus in rat consists of two $C\lambda$ genes and a single $V\lambda$ gene. *Gene* **55**, 75–84.

Steiniger, B., Sickel, E. (1992). Class II MHC molecules and monocytes/macrophages in the respiratory system of conventional, germ-free and interferon-gamma-treated rats. *Immunobiology* **184**, 295–310.

Steiniger, B., Falk, P., Lohmuller, M., van der Meide, P. H. (1989). Class II MHC antigens in the rat digestive system. Normal distribution and induced expression after interferon-gamma treatment *in vivo*. *Immunology* **68**, 507.

Sternberg, E. M., Hill, J. M., Chrousos, G. P., Kamilaris, T., Listwak, S. J., Gold, P. W., Wilder, R. L. (1989). Inflammatory mediator-induced hypothalamic–pituitary–adrenal axis activation is defective in streptococcal cells wall arthritis-susceptible Lewis rats. *PNAS* **86**, 2374–2378.

Sternberg, E. M., Wilder, R. L., Gold, P. W., Chrousos, G. P. (1990). A defect in the central component of the immune system–hypothalamic–pituitary–adrenal axis feedback loop is associated with susceptibility to experimental allergic arthritis and other inflammatory diseases. *Ann. NY Acad. Sci.* **594**, 289–292.

Stöckli, K. A., Lillien, L. E., Näher-Noe, M., Breitfeld, G., Hughes, R. A., Raff, M. C., Thoenen, H., Sendtner, M. (1991). Regional distribution, developmental changes and cellular localization of CNTF-mRNA and protein in the rat brain. *J. Cell. Biol.* **115**, 447–459.

Strittmatter, S. M., Snyder, S. H. (1984). Angiotensin-converting enzyme in the male rat reproductive system: autoradiographic visualization with [^{3}H]captopril. *Endocrinology* **115**, 2332–2341.

Strober, W., Ehrhardt, R. O. (1994). Regulation of IgA B cell development (1994). In *Handbook of Mucosal Immunology* (ed. Ogra, P. L.), pp. 159–176. Academic Press, New York, USA.

Stumbles, P. A., Penhale, W. J. (1993). IDDM in rats induced by thymectomy and irradiation. *Diabetes* **42**, 571–578.

Sunderland, C. A., McMaster, W. R., Williams, A. F. (1979). Purification with monoclonal antibody of a predominant leukocyte-common antigen and glycoprotein from rat thymocytes. *Eur. J. Immunol.* **9**, 155–159.

Surh, C. D., Sprent, J. (1991). Long-term xenogeneic chimeras. Full differentiation of rat T and B cells in SCID mice. *J. Immunol.* **147**, 2148–2154.

Surh, C. D., Gold, D. P., Wiley, S., Wilson, D. B., Sprent, J. (1994). Rat T-cell response to superantigens. I. Vβ-restricted clonal deletion of rat T-cells differentiating in rat–mouse chimeras. *J. Exp. Med.* **179**, 57–62.

Sutherland, D. B., Varilek, G. W., Neil, G. A. (1994). Identification and characterization of the rat intestinal epithelial cell (IEC-18) interleukin-1. *Am. J. Physiol.* **266**, C1198–C1203.

Swanborg, R. H. (1988). Experimental allergic encephalomyelitis. *Methods Enzymol.* **162**, 413–421.

Swanborg, R. H. (1995). Experimental autoimmune encephalomyelitis in rodents as a model for human demyelinating disease. *Clin. Immunol. Immunopath.* **77**, 4–13.

Swanborg, R. H., Stepaniak, J. A. (1996). Experimental autoimmune encephalomyelitis in the rat. In *Current Protocols in Immunology* (eds Coligan, J. E., Kruisbeek, A. M., Margulies, D. H., Shevach, E. M., Strober, W.), pp. 15.2.1–15.2.14. John Wiley & Sons, New York, USA.

Tacke, M., Clark, G. J., Dallman, M. J., Hunig, T. (1995). Cellular distribution and costimulatory function of rat CD28. Regulated expression during thymocyte maturation and induction of cyclosporin A sensitivity of costimulated T cell responses by phorbol ester. *J. Immunol.* **154**, 5121–5127.

Takacs, L., Ruscetti, F. W., Kovacs, E. J., Rocha, B., Brocke, S., Diamantstein, T., Mathieson, B. J. (1988). Immature, double negative ($CD4^-$,$CD8^-$) rat thymocytes do not express IL-2 receptors. *J. Immunol.* **141**, 3810–3818.

Takagi, Y., Shrivastav, S., Miki, T., Sakaguchi, K. (1994). Molecular cloning and expression of the acidic fibroblast growth factor receptors in a rat parathyroid cell line (PT-r). *J. Biol. Chem.* **269**, 23743–23749.

Tamatani, T., Miyasaka, M. (1990). Identification of monoclonal antibodies reactive with the rat homolog of ICAM-1, and evidence for a differential involvement of ICAM-1 in the adherence of resting versus activated lymphocytes to high endothelial cells. *Int. Immunol.* **2**, 165–171.

Tamatani, T., Kotani, M., Miyasaka, M. (1991). Characterization of the rat leukocyte integrin CD11/CD18, by the use of LFA-1 subunit-specific monoclonal antibodies. *Eur. J. Immunol.* **21**, 627–633.

Tamatani, T., Kitamura, F., Kuida, K., Shirao, M., Mochizuki, M., Suematsu, M., Schmid-Schonbein, G. W., Watanabe, K., Tsurufuji, S., Miyasaka, M. (1993a). Characterization of rat LECAM-1 (L-selectin) by the use of monoclonal antibodies and evidence for the presence of soluble LECAM-1 in rat sera. *Eur. J. Immunol.* **23**, 2181–2188.

Tamatani, T., Kuida, K., Watanabe, T., Koike, S., Miyasaka, M. (1993b). Molecular mechanisms underlying lymphocyte recirculation. III. Characterization of the LECAM-1 (L-selectin)-dependent adhesion pathway in rats. *J. Immunol.* **150**, 1735–1745.

Tanaka, M., Ito, H., Adachi, S., Akimoto, H., Nishikawa, T., Kasajima, T., Marumo, F., Hiroe, M. (1994). Hypoxia induces apoptosis with enhanced expression of Fas antigen messenger RNA in cultured neonatal rat cardiomyocytes. *Circ. Res.* **75**, 426–433.

Tanaka, T., Masuko, T., Yagita, H., Tamura, T., Hashimoto, Y. (1989). Characterization of a CD3-like rat T cell surface antigen recognized by a monoclonal antibody. *J. Immunol.* **142**, 2791–2795.

Taurog, J. D., Richardson, J. A., Croft, J. T., Simmons, W. A., Zhou, M., Fernandez-Sueiro, J. L., Balish, E., Hammer, R. E. (1994). The germfree state prevents development of gut and joint inflammatory disease in HLA-B27 transgenic rats. *J. Exp. Med.* **180**, 2359–2364.

Tellides, G., Dallman, M. J., Morris, P. J. (1989). Mechanism of action of interleukin-2 receptor (IL2R) monoclonal antibody (mAb) therapy: target cell depletion or inhibition of function? *Transpl. Proc.* **21**, 997–998.

ter Meulen, V., Liebert, U. G. (1993). Measles virus-induced autoimmune reactions against brain antigen. *Intervirology* **35**, 86–94.

Thiele, H.-G., Koch, F., Kashan, A. (1987). Postnatal distribution profiles of Thy-1^+ and $RT6^+$ cells in peripheral lymph nodes of DA rats. *Transpl. Proc.* **19**, 3157–3160.

Thomas, M. L., Green, J. R. (1983). Molecular nature of the W3/25 and OX-8 marker antigens for rat T lymphocytes, comparisons with mouse and human antigens. *Eur. J. Immunol.* **13**, 855–858.

Thomas, M. L., Barclay, A. N., Gagnon, J., Williams, A. F. (1985). Evidence from cDNA clones that the rat leucocyte common antigen (T200) spans the lipid bilayer and contains a cytoplasmatic domain of 80,000 Mr. *Cell* **41**, 83–93.

Tilders, F. J. H., DeRijk, R. H., van Dam, A. M., Vincent, V. A. M., Schotanus, K., Persoons, J. H. A. (1994). Activation of the hypothalamus–pituitary–adrenal axis by bacterial endotoxins: routes and intermediate signals. *Psychoneuroendocrinology* **19**, 209–232.

Timmerman, J. J. (1996). Extrahepatic synthesis of complement component C6 is not upregulated during an LPS-induced acute phase response. Thesis, University of Leiden, The Netherlands.

Tipping, P. G., Huang, X. R., Berndt, M. C., Holdsworth, S. R. (1996). P-selectin directs T lymphocyte-mediated injury in delayed-type hypersensitivity responses: studies in glomerulonephritis and cutaneous delayed-type hypersensitivity. *Eur. J. Immunol.* **26**, 454–460.

Tole, S., Patterson, P. H. (1993). Distribution of CD9 in the developing and mature rat nervous system. *Dev. Dyn.* **197**, 94–106.

Tomlinson, M. G., Wright, M. D. (1996). Characterisation of mouse CD37: cDNA and genomic cloning. *Mol. Immunol.* **33**, 867–872.

Tomlinson, M. G., Williams, A. F., Wright, M. D. (1993). Epitope mapping of anti-rat CD53 monoclonal antibodies. Implications for the membrane orientation of the Transmembrane 4 Superfamily. *Eur. J. Immunol.* **23**, 136–140.

Tong, T., Hunt, S., Opstelten, D. (1993). HIS50, a rat homolog to mouse J11d/heat stable antigen and human CD24: cDNA isolation and sequence. *J. Cell Biochem.* **17** (Suppl.), 242.

Torres-Nagel, N., Kraus, E., Brown, M. H., Tiefenthaler, G., Mitnacht, R., Williams, A. F., Hünig, T. (1992). Differential thymus dependence of rat CD8 isoform expression. *Eur. J. Immunol.* **22**, 2841–2848.

Torres-Nagel, N. E., Gold, D. P., Hünig, T. (1993). Identification of rat TCR Vβ 8.2, 8.5, and 10 gene products by monoclonal antibodies. *Immunogenetics* **37**, 304–308.

Torres-Nagel, N., Giegerich, G., Gold, D. P., Hünig, T. (1994a). Identification of rat TCRA V 4 and 8 gene products by monoclonal antibodies and cDNA sequence. *Immunogenetics* **39**, 367–370.

Torres-Nagel, N., Herrmann, T., Giegerich, G., Wonigeit, K., Hünig, T. (1994b). Preferential TCR V usage in positive repertoire selection and alloreactivity of rat T-lymphocytes. *Int. Immunol.* **6**, 1367–1373.

Tracy, T. F. J., Fox, E. S. (1995). CD14-lipopolysaccharide receptor activity in hepatic macrophages after cholestic liver injury. *Surgery* **118**, 371–377.

Tsuchida, K., Lewis, K. A., Mathews, L. S., Vale, W. W. (1993). Molecular characterization of rat transforming growth factor-β type II receptor. *Biochem. Biophys. Res. Commun.* **191**, 790–795.

Tsuji, T., Okada, F., Yamaguchi, K., Nakamura, T. (1990). Molecular cloning of the large subunit of transforming growth factor type B masking protein and expression of the mRNA in various rat tissues. *Proc. Natl Acad. Sci. USA* **87**, 8835–8839.

Tsujimura, T., Hirota, S., Nomura, S., Niwa, Y., Yamazaki, M., Tono, T., Morii, E., Kim, H. M., Kondo, K., Nishimune, Y. E. A. (1991). Characterization of Ws mutant allele of rats: a 12-base deletion in tyrosine kinase domain of *c*-kit gene. *Blood* **78**, 1942–1946.

Turner, D. C., Flier, L. A., Carbonetto, S. (1989). Identification of a cell-surface protein involved in PC12 cell substratum adhesive and neurite outgrowth on laminin and collagen. *J. Neurosci.* **9**, 3287–4690.

Überla, K., Li, W., Qin, Z., Richter, G., Raabe, T., Diamantstein, T., Blankenstein, T. (1991). The rat interleukin-5 gene: characterization and expression by retroviral gene transfer and polymerase chain reaction. *Cytokine* **3**, 72–81.

Ulevitch, R. J., Tobias, P. S. (1995). Receptor-dependent mechanisms for stimulation by bacterial endotoxin. *Annu. Rev. Immunol.* **13**, 437–457.

Vaerman, J. P., Lemaitre-Coelho, I. (1979). Transfer of circulating human IgA across the rat liver into bile. In *Transmission of Proteins Through Living Membranes* (ed. Hemmings, W. A.), pp. 383–398. Elsevier, Amsterdam, The Netherlands.

Vaessen, L. M., Lameijer, L., D., Carpenter, C. B., Cramer, D. V., Rozing, J. (1979). The R1-rat: a new RT1 haplotype. *Transpl. Proc.* **11**, 1565–1567.

Valli, V. E., Villeneuve, D. C., Reed, B., Barsoum, N., Smith, G. (1983). Evaluation of blood and bone marrow, rat. In *Hemopoietic System* (eds Jones, T. C., Ward, J. M., Hunt, R. D.), pp. 9–26. Springer Verlag, Berlin, Germany.

van de Winkel, J. G., Capel, P. J. (1993). Human IgG Fc receptor heterogeneity: molecular aspects and clinical implications. *Immunol. Today* **14**, 215–221.

van den Berg, T. K., Dopp, E. A., Kraal, G., Dijkstra, C. D. (1989). The heterogeneity of rat peripheral lymphoid organs identified by monoclonal antibodies. *Eur. J. Immun.* **19**, 1747–1756.

van den Berg, T. K., Dopp, E. A., Kraal, G., Dijkstra, C. D. (1991). Selective inhibition of immunecomplex trapping by FDC with monoclonal antibody against rat C3. *Eur. J. Immunol.* **22**, 957–962.

van den Dobbelsteen, G. P., Brunekreef, K., Sminia, T., van Rees, E. P. (1992). Effect of mucosal and systemic immunization with pneumococcal polysaccharide type 3,4 and 14 in the rat. *Scand. J. Immunol.* **36**, 661–669.

van der Brugge-Gamelkoorn, G. J., Sminia, T. (1985). T cells and T cell subsets in rat bronchus associated lymphoid tissue (BALT) *in situ* and in suspension. In *Microenvironments in the Lymphoid System*, p. 323. Plenum Publishing Corporation.

van der Brugge-Gamelkoorn, G. J., Claassen, E., Sminia, T. (1986). Anti-TNP-forming cells in bronchus associated lymphoid tissue (BALT) and paratracheal lymph node (PTLN) of the rat after intratracheal priming and boosting with TNP-KLH. *Immunology* **57**, 405–409.

van der Laan, L. J. W., Ruuls, S. R., Weber, K. S., Lodder, I. J., Döpp, E. A., Dijkstra, C. D. (1996). Macrophage phagocytosis of myelin *in vitro* determined by flow cytometry: phagocytosis is mediated by CR3 and induces production of TNF-α and nitric oxide. *J. Neuroimmunol.* **70**, 145–152.

van der Meide, P. H., Dijkema, R., Caspers, M., Vijverberg, K., Schellekens, H. (1986a). Cloning, expression and purification of rat IFN-α1. *Meth. Enzymology* **119**, 441–453.

van der Meide, P. H., Dubbeld, M., Vijverberg, K., Kos, T., Schellekens, H. (1986b). The purification and characterization of rat gamma interferon by use of two monoclonal antibodies. *J. Gen. Virol.*, **67**, 1059–1071.

van der Meide, P. H., Groenestein, R. J., De Labie, M. C. D. C., Aten, J., Weening, J. J. (1995). Susceptibility to mercuric chloride-induced glomerulonephritis is age-dependent: study of the role of IFN-gamma. *Cell. Immunol.* **162**, 131–137.

van der Merwe, P. A., McPherson, D. C., Brown, M. H., Barclay, A. N., Cyster, J. G., Williams, A. F., Davis, S. J. (1993). The NH_2-terminal domain of rat CD2 binds rat CD48 with a low affinity and binding does not require glycosylation of CD2. *Eur. J. Immunol.* **23**, 1373–1377.

van der Merwe, P. A., Barclay, A. N., Mason, D. W., Davies, E. A., Morgan, B. P., Tone, M., Krishnam, A. K., Ianelli, C., Davis, S. J. (1994). Human cell-adhesion molecule CD2 binds CD58 (LFA-3) with a very low affinity and an extremely fast dissociation rate but does not bind CD48 or CD59. *Biochemistry* **33**, 10149–10160.

van Dongen, J. J., Remie, R., Rensema, J. W., van Wunnik, G. H. J. (1990). *Manual of Microsurgery on the Laboratory Rat. Techniques in the Behavioral and Neural Sciences*, Vol. 4 (ed. Huston, J. P.). Elsevier Science Publishers B.V., Amsterdam, The Netherlands.

van Eden, W., Wagenaar-Hilbers, J. P. A., Wauben, M. H. M. (1996). Adjuvant arthritis in the rat. In *Current Protocols in Immunology* (eds Coligan, J. E., Kruisbeek, A. M., Margulies,

D. H., Shevach, E. M., Strober, W.), pp. 15.4.1–15.4.8. John Wiley & Sons, New York, USA.

van Ewijk, W., Nieuwenhuis, P. (1985). Compartments, domains and migration pathways of lymphoid cells in the splenic pulp. *Experientia* **41**, 199–209.

van Gelder, M., Kinwell-Bohre, E. P. M., Mulder, A. H., van Bekkum, D. W. (1996). Both bone marrow- and non-bone marrow-associated factors determine susceptibility to experimental autoimmune encephalomyelitis of BUF and WAG rats. *Cell. Immunol.* **168**, 39–48.

van Golde L. M. G. (1995). Potential role of surfactant proteins A and D in innate lung defense against pathogens. *Biol. Neonate* **67** (Suppl.), 2–17.

van Iwaarden, J. J. (1992). Surfactant and the pulmonary defense system. In *Pulmonary Surfactant: from Molecular Biology to Clinical Practice* (eds Robertson, B., van Golde, L. M. G., Batenburg, J. J.), pp. 215–227. Elsevier Science Publishers, Amsterdam, The Netherlands.

van Oostveen, D. C., van den Berg, T. K., Damoiseaux, J. G. M. C., van Rees, E. P. (1992). Macrophage subpopulations and reticulum cells in rat placenta. *Cell Tissue Res.* **268**, 513–519.

van Oosten, M., van de Bilt, E., de Vries, H. E., van Berkel, T. J. C., Kuiper, J. (1995). Vascular adhesion molecule-1 and intercellular adhesion molecule-1 expression on rat liver cells after lipopolysaccharide administration *in vivo*. *Hepatology* **22**, 1538–1546.

van Poppel, M. N., van den Berg, T. K., van Rees, E. P., Sminia, T., Biewenga, J. (1993). Reticulum cells in the ontogeny of nasal-associated lymphoid tissue (NALT) in the rat. *Cell Tissue Res.* **273**, 577–581.

van Rees, E. P., Dijkstra, C. D., van der Ende, M. B., *et al.* (1988). The ontogenetic development of macrophage subpopulations and Ia-positive non-lymphoid cells in gut-associated lymphoid tissue of the rat. *Immunology* **63**, 79–85.

van Rees, E. P., Dijkstra, C. D., Sminia, T. (1996). The structure and development of the lymphoid organs. In *Pathobiology of the Aging Mouse* (ed. Mohr, U.), pp. 173–187. ILSI Press, Washington, USA.

Vaporciyan, A. A., DeLisser, H. M., Yan, H. C., Mendiguren, I. I., Thom, S. R., Jones, M. L., Ward, P. A., Albeda, S. M. (1993). Involvement of platelet-endothelial cell adhesion molecule-1 in neutrophyl recruitement *in vivo*. *Science* **262**, 1580–1582.

Varga, J. M., Kalchschmid, G., Bellon, B., Kuhn, J., Druet, P., Fritsch, P. (1995). Mechanism of allergic cross-reactions. *Int. Arch. Allergy Immunol.* **108**, 196–199.

Veldhuis, G. J., Opstelten, D. (1988). B cell precursor populations in fetal and neonatal rat liver: frequency, topography and antigenic phenotype. *Adv. Exp. Med. Biol.* **237**, 57–62.

Veldman, J. E. (1970). Histophysiology and electron microscopy of the immune response. Acad. Thesis University of Groningen, Drukkerij van Denderen, Groningen, The Netherlands.

Vermeer, L. (1995). Molecular analysis of rat B cell differentiation. Thesis, University of Groningen, Groningen, The Netherlands.

Vermeer, L. A., De Boer, N. K., Bucci, C., Bos, N. A., Kroese, F. G. M., Alberti, S. (1994). MRC OX 19 recognizes the rat CD5 surface glycoprotein, but does not provide evidence for a population of $CD5^{bright}$ B cells. *Eur. J. Immunol.* **24**, 585–592.

Vingsbo, C., Jonsson, R., Holmdahl, R. (1995). Avridine-induced arthritis in rats: a T cell-dependent chronic disease influenced both by MHC genes and by non-MHC genes. *Clin. Exp. Immunol.* **99**, 359–363.

Vonderheide, R. H., Hunt, S. V. (1991). Comparison of IgD+ and IgD− thoracic duct B lymphocytes as germinal center precursor cells in the rat. *Int. Immunol.* **3**, 1273–1281.

Wagner, J., Thiele, F., Ganten, D. (1996). The renin-angiotensin system in transgenic rats. [Review]. *Pediat. Nephrol.* **10**, 108–112.

Wakelin, M. W., Sanz, M. J., Dewar, A., Albelda, S. M., Larkin, S. W., Boughton-Smith, N., Williams, T. J., Nourshargh, S. (1996). An anti-platelet-endothelial cell adhesion molecule-1 antibody inhibits leukocyte extravasation from mesenteric microvessels *in vivo* by blocking the passage through the basement membrane. *J. Exp. Med.* **184**, 229–239.

Waldman, T. A., Strober, W. (1969). Metabolism of immunoglobulins. *Progr. Allergy* **13**, 1–110.

Walter, S., Micheel, B., Pabst, R., Westermann, J. (1995). Interaction of B and T lymphocyte subsets with high endothelial venules in the rat: binding *in vitro* does not reflect homing *in vivo*. *Eur. J. Immunol.* **25**, 1199–1205.

Wang, J., Kuliszewski, M., Yee, W., Sedlackova, L., Xu, J., Tseu, I., Post, M. (1995). Cloning and expression of glucocorticoid-induced genes in fetal rat lung fibroblasts. *J. Biol. Chem.* **270**, 2722–2728.

Watanabe, K., Konishi, K., Fujioka, M., Kinoshita, S., Nakagawa, H. (1989). The neutrophil chemoattractant produced by the rat kidney epithelioid cell line NRK-52E is a protein related to the KC/gro protein. *J. Biol. Chem.* **264**, 19559–19563.

Watanabe, T., Song, Y., Hirayama, Y., Tamatani, T., Kuida, K., Miyasaka, M. (1992). Sequence and expression of a rat cDNA for LECAM-1. *Biochim. Biophys Acta* **1131**, 321–324.

Watt, V. M., Yip, C. C. (1989). Amino acid sequence deduced from a rat kidney cDNA suggests it encodes the Zn-peptidase aminopeptidase N. *J. Biol. Chem.* **5**, 5480–5487.

Waynforth, H. B., Flecknell, P. A. (1992). *Experimental and Surgical Technique in the Rat*, 2nd edn. Academic Press, London, UK.

Webb, M., Mason, D. W., Williams, A. F. (1979). Inhibition of the mixed lymphocyte response with a monoclonal antibody specific for a rat T lymphocyte subset. *Nature* **282**, 841–843.

Weis, W. I., Drickamer, K., Hendrickson, W. A. (1992). Structure of a C-type mannose binding protein complexed with an oligosaccharide. *Nature* **360**, 127–134.

Werge, T., Biocca, S., Cattaneo, A. (1990). Intracellular immunization. Cloning and intracellular expression of a monoclonal antibody to the p21 ras protein. *FEBS Lett.* **274**, 193–198.

Westermann, J., Pabst, R. (1996). How organ-specific is the migration of naive and memory T cells? *Immunol. Today* **17**, 278–282.

Westermann, J., Ronneberg, S., Fritz, F. J., Pabst, R. (1989). Proliferation of lymphocyte subsets in the adult rat: a comparison of different lymphoid organs. *Eur. J. Immunol.* **19**, 1087–1093.

Westermann, J., Blaschke, V., Zimmermann, G., Hirschfeld, U., Pabst, R. (1992). Random entry of circulating lymphocyte subsets into peripheral lymph nodes and Peyer's patches: *in vivo* no evidence of a tissue specific migration of B and T lymphocytes at HEV level. *Eur. J. Immunol.* **22**, 2219–2223.

Westermann, J., Persin, S., Matyas, J., van der Meide, P., Pabst, R. (1993). IFN-gamma influences the migration of thoracic duct B and T lymphocyte subsets *in vivo*. Random increase in disappearance from the blood and differential decrease in reappearance in the lymph. *J. Immunol.* **150**, 3843–3852.

Westermann, J., Matyas, J., Persin, S., van der Meide, P., Heerwagen, C., Pabst, R. (1994a). B- and T-lymphocyte subset numbers in the migrating lymphocyte pool of the rat: the influence of IFN-gamma on its mobilization monitored through blood and lymph. *Scand. J. Immunol.* **39**, 395–402.

Westermann, J., Nagahori, Y., Walter, S., Heerwagen, C., Miyasaka, M., Pabst, R. (1994b). B and T lymphocyte subsets enter peripheral lymph nodes and Peyer's patches without preference *in vivo*: no correlation occurs between their localization in different types of high endothelial venules and the expression of CD44, VLA-4, LFA-1, ICAM-1, CD2 or L-selectin. *Eur. J. Immunol.* **24**, 2312–2316.

Westermann, J., Smith, T., Peters, U., Tschernig, T., Pabst, R., Steinhoff, G., Sparshott, S. M., Bell, E. B. (1996a). Both activated and nonactivated leukocytes from the periphery continuously enter the thymic medulla of adult rats: phenotypes, sources and magnitude of traffic. *Eur. J. Immunol.* **26**, 1866–1874.

Westermann, J., Walter, S., Nagahori, Y., Heerwagen, C., Miyasaka, M., Pabst, R. (1996b). Blood leucocyte subsets of the rat: expression of adhesion molecules and localization within high endothelial venules. *Scand. J. Immunol.* **43**, 297–303.

Whalen, B. J., Goldschneider, I. (1993). Identification and characterization of B cell precursors in rat lymphoid tissues. I. Adoptive transfer assays for precursors of TI-1, TI-2, and TD antigen-reactive B cells. *Cell. Immunol.* **151**, 168–186.

Whalen, B. J., Doukas, J., Mordes, J. P., Rossini, A. A., Greiner, D. L. (1997). Induction of insulin-dependent diabetes mellitus (IDDM) in PVG.RT1^u rats. *Transplant. Proc.* **29**, 1684–1685.

Wilders, M. M., Sminia, T., Janse, E. M. (1983). Ontogeny of non-lymphoid cells in the rat gut with special reference to large mononuclear Ia-positive dendritic cells. *Immunology* **50**, 303–314.

Willführ, K. U., Hirschfeld, U., Westermann, J., Pabst, R. (1990a). The *in vitro* lymphocyte/endothelium binding assay. An improved method employing light microscopy. *J. Immunol. Methods* **130**, 201–207.

Willführ, K. U., Westermann, J., Pabst, R. (1990b). Absolute numbers of lymphocyte subsets migrating through the compartments of the normal and transplanted spleen. *Eur. J. Immunol.* **20**, 903–911.

Williams, A. F., Galfre, G., Milstein, C. (1977). Analysis of cell surfaces by xenogeneic myeloma-hybrid antibodies: differentiation antigens of rat lymphocytes. *Cell* **12**, 663–673.

Williams, A. F., Barclay, A. N., Clark, S. J., Paterson, D. J., Willis, A. C. (1987). Similarities in sequence and cellular expression between rat CD2 and CD4 antigens. *J. Exp. Med.* **165**, 368–380.

Williams, A. F., Atkins, R. C., Fries, J. W., Gimbrone, M. A., Cybulski, M. I., Collins, T. (1992). Nucleotide sequence of rat vascular cell adhesion molecule-1 cDNA. *Biochim. Biophys. Acta* **1131**, 214–216.

Williams, C. B., Gutman, G. A. (1989). T cell receptor β-chain genes in the rat. *J. Immunol.* **142**, 1027–1035.

Williams, C. B., Blankenhorn, E. P., Byrd, K. E., Levinson, G., Gutman, G. A. (1991). Organisation and nucleotide sequence of the rat T cell receptor β-chain complex. *J. Immunol.* **146**, 4406–4413.

Wing, M. G., Seilly, D. J., Bridgman, D. J., Harrison, R. A. (1993). Rapid isolation and biochemical characterization of rat C1 and C1q. *Molle. Immunol.* **30**, 433–440.

Winocour, P. D., Chignier, E., Parmentier, S., McGregor, J. L. (1992). A member of the selectin family (GMP-140/PADGEM) is expressed on thrombin-stimulated rat platelets *in vitro*. *Comp. Biochem. Physiol. A.* **102**, 265–271.

Wirth, K., Arch, R., Somasundaram, C., Hofmann, M., Weber, B., Herrlich, P., Matzku, S., Zöller, M. (1993). Expression of CD44 isoforms carrying metastasis-associated sequences in newborn and adult rats. *Eur. J. Cancer* **29a**, 1172–1177.

Wistar, R., Jr (1969). Immunoglobulin allotype in the rat. Localization of the specificity to the light chain. *Immunology* **17**, 23.

Wong, L., Yamasaki, G., Johnson, R. J., Friedman, S. L. (1994). Induction of beta-platelet-derived growth factor in rat hepatic lipocytes during cellular activation *in vivo* and in culture. *J. Clin. Invest.* **94**, 1563–1569.

Wonigeit, K., Dinkel, A., Fangmann, J., Thude, H. (1996). Expression of the ectoenzyme RT6 is not restricted to resting peripheral T cells and is differently regulated in normal peripheral T cells, intestinal IEL, and NK cells. *Adv. Exp. Med. Biol.* **419**, 257–264.

Wood, M. J., Sloan, D. J., Dallman, M. J., Charlton, H. M. (1992). A monoclonal antibody to the interleukin-2 receptor enhances the survival of neural allografts: a time-course study. *Neuroscience* **49**, 409–418.

Woollett, G. R., Barclay, A. N., Puklavec, M., Williams, A. F. (1985). Molecular and antigenic heterogeneity of the rat leukocyte-common antigen from thymocytes and T and B lymphocytes. *Eur. J. Immunol.* **15**, 168–173.

World Health Organization (1973). Nomenclature of human immunoglobulins. *Bull. WHO* **48**, 373.

Wu, X., Dolecki, G. J., Lefkowith, J. B. (1995). GRO chemokines: a transduction, integration, and amplification mechanism in acute renal inflammation. *Am. J. Physiol.* **269**, F248–F256.

Yamada, T., Natori, T., Izumi, K., Sakai, T., Agui, T., Matsumoto, K. (1991). Inheritance of T helper immunodeficiency (thid) in LEC mutant rats. *Immunogenetics* **33**, 216–219.

Yamazaki, T., Seko, Y., Tamatani, T., Miyasaka, M., Yagita, H., Okumura, K., Nagai, R., Yazaki, Y. (1993). Expression of intercellular adhesion molecule-1 in rat heart with ischemia/reperfusion and limitation of infarct size by treatment with antibodies against cell adhesion molecules. *Am. J. Pathol.* **143**, 410–418.

Yan, G., Nikolaropoulos, S., Wang, F., McKeehan, W. L. (1991). Sequence of rat keratinocyte growth factor (heparin-binding growth factor type 7). *In vitro Cell. Dev. Biol.* **27A**, 437–438.

Yang, C., Jones, J. L., Barnum, S. R. (1993). Expression of decay-accelerating factor (CD55), membrane cofactor protein (CD46) and CD59 in the human astroglioma cell line, D54-MG, and primary rat astrocytes. *J. Neuroimmunol.* **47**, 123–132.

Yang, C. P., Bell, E. B. (1992). Functional maturation of recent thymic emigrants in the periphery; development of alloreactivity correlates with the cyclic expression of CD45RC isoforms. *Eur. J. Immunol.* **22**, 2261–2269.

Yasuda, M., Hasunuma, Y., Adachi, H., Sekine, C., Sakanishi, T., Hashimoto, H., Ra, C., Yagita, H., Okumura, K. (1995). Expression and function of fibronectin binding integrins on rat mast cells. *Int. Immunol.* **7**, 251–258.

Yasue, M., Serikawa, T., Kuramoto, T., Mori, M., Higashigushi, T., Ishizaki, K., Yamada, J. (1992). Chromosomal assignment of 17 structural genes and 11 related DNA fragments in rats (*Rattus norvegicus*) by Southern blot analysis of rat x mouse somatic cell hybrids. *Genomics* **12**, 659–664.

Yoffey, M. J., Courtice, F. C. (1956). *Lymphatics, Lymph and Lymphoid Tissue*, 2nd edn. Edward Arnold (Publishers) Ltd, London, UK.

Zagorski, J., DeLarco, J. E. (1993). Rat CINC (cytokine-induced neutrophil chemoattractant) is the homolog of the human GRO proteins but is encoded by a single gene. *Biochem. Biophys. Res. Commun.* **190**, 104–110.

Zhang, X.-M., Heber-Katz, E. (1992). T cell receptor sequences from encephalitogenic T cells in adult Lewis rats suggests an early ontogenic origin. *J. Immunol.* **148**, 746–752.

Zhang, X.-M., Esch, T. R., Clark, L., Gregorian, S., Rostami, A., Otvos, L., Heber-Katz, E. (1994). Neuritogenic Lewis rat T-cells use Tcrb chains that include a new Tcrb-V8 family member. *Immunogenetics* **40**, 266–270.

Zimmerman, B. J., Paulson, J. C., Arrhenius, T. S., Gaeta, F. C., Granger, D. N. (1994). Thrombin receptor peptide-mediated rolling in rat mesenteric venules: roles of P-selectin and sialyl Lewis X. *Am. J. Physiol.* **267**, H1049–H1053.

Zinkernagel, R. M. (1997). The discovery of MHC restriction. *Immunol. Today* **18**, 14–17.

Zsebo, K. M., Wypych, J., McNiece, I. K., Lu, H. S., Smith, K. A., Karkare, S. B., Sachdev, R. K., Yuschenkoff, V. N., Birkett, N. C., Williams, L. R., Satyagal, V. N., Tung, W., Bosselman, R. A., Mendiaz, E. A., Langley, K. E. (1990). Identification, purification and biological characterization of hematopoietic stem cell factor from Buffalo rat liver conditioned medium. *Cell* **63**, 195–201.

VI IMMUNOLOGY OF LAGOMORPHS

1. Introduction

The rabbit is still the animal of choice for production of many polyclonal antibodies and most commercial sources of antibodies still feature reagents produced in the rabbit. Although the mouse is used in many studies as a model for human diseases because of its smaller size and the availability of inbred, transgenic and 'knock-out' strains, it is noteworthy that studies of DNA and protein sequences suggest that lagomorphs are more closely related to primates than are rodents (Li *et al.*, 1990). Analyses of protein sequences also showed that the order Lagomorpha is more closely related to primates than to rodents (Graur *et al.*, 1996; Novacek, 1996). A rabbit model of human hemolytic disease of the newborn is described below in Section 9. Complement deficiencies are described in Section 12 and autoimmune diseases in Section 16. The rabbit is also useful for studies of various infectious diseases including syphilis (Sell and Hsu, 1993), tuberculosis (reviewed by Dannenberg, 1991), virus-induced papilloma, myxoma virus, HTLV1 and HIV (see also Sections 14 and 15). A useful volume on the rabbit in contemporary immunological research appeared in 1987 (Dubiski, 1987).

The rabbit has been the model animal with which a number of major immunological concepts were delineated. A few of these are described below.

Major contributions of the rabbit to current immunological concepts

Allelic exclusion

The expression of only one of two parental chromosomal products in individual B lymphocytes was documented in early studies using the rabbit model (Davie *et al.*, 1971; Hayward *et al.*, 1978; Loor and Kelus, 1978; Gathings *et al.*, 1981, 1982). The presence of independent genes contributing to the single heavy polypeptide chain was documented genetically when recombination between V_H and C_H genes was observed in laboratory rabbits (Mage *et al.*, 1971, 1982, 1992; Kelus and Steinberg, 1991; Newman *et al.*, 1991).

Allotype suppression

If developing heterozygous rabbits are exposed to antibodies to paternal allotypes (see Section 7) during fetal and/or neonatal life, expression of that type is markedly depressed for the entire life of the animal (chronic allotype-suppression, reviewed by Horng *et al.*, 1980; Mage, 1975).

V_H gene diversification mechanisms

Although the role of gut associated lymphoid tissues (GALT) in development of the immune system was suggested many years ago (Cooper *et al.*, 1968), a resurgence of interest in the role of GALT has come from recent studies that suggest that primary repertoire development occurs when rabbit B lymphocytes reside in specialized tissues of the GALT (Weinstein *et al.*, 1994a, b). This is discussed further in Sections 2, 7 and 13.

Idiotypes and network interactions in immune regulation (latent allotypes)

The original descriptions of idiotypes and the production of a series of interacting idiotypes and anti-idiotypes (idiotype networks) used the rabbit model. Some observations of 'latent allotypes' may have been due to idiotypic mimicry. Excellent reviews of the field of latent allotypes have been prepared by Roux (1983) and McCartney-Francis (1987).

2. Lymphoid Organs, Their Anatomical Distribution and Ontogeny of the Immune System

Table VI.2.1 shows the percentage of B and T lymphocytes within rabbit lymphoid tissues. Figure IV.2.1 is a diagrammatic representation that identifies the novel aspects of the gut-associated lymphoid tissues (GALT) of the rabbit, locating three different lymphoid organs present at the distal end of the ileum and the proximal end of the caecum. The overall organization of the rabbit lymphoid

Handbook of Vertebrate Immunology
ISBN 0-12-546401-0

Table VI.2.1 Percentage of B and T lymphocytes within rabbit lymphoid tissues according to Fujiwara *et al.* (1974) (A) and Rudzik *et al.* (1975) (B)

	T Cells		*B Cells*	
Cell source	*A*	*B*	*A*	*B*
Appendix	23.3	ND	53.0	ND
BALT	ND	18.4	ND	41.5
Bone marrow	15.5	ND	37.2	ND
Lamina propria	ND	11.1	ND	ND
Lymph node	57.0	ND	37.3	ND
PBL	40.1	43.9	42.4	ND
Peyer's patch	ND	16.6	ND	78.7
Spleen	37.8	19.7	47.4	ND
Thymus	94.5	94.0	5.0	ND

ND, not determined; PBL, peripheral blood lymphocytes; BALT, bronchus-associated lymphatic tissue.

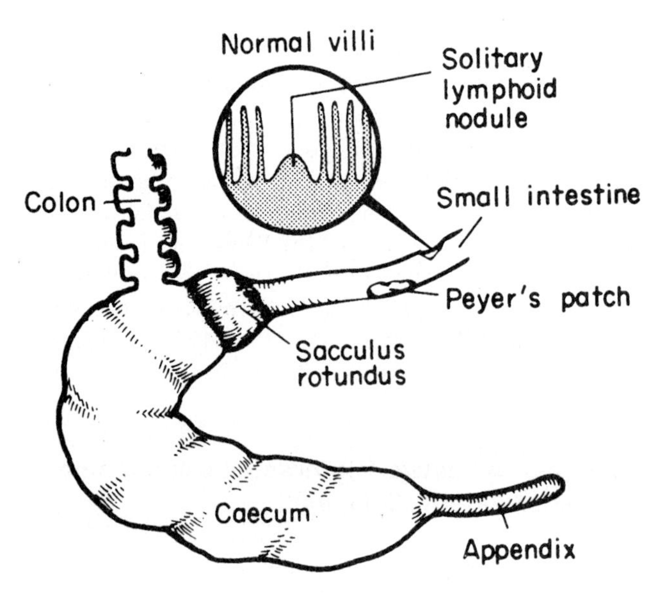

Figure VI.2.1 Diagrammatic representation identifying the novel aspects of the gut-associated lymphoid tissues of the rabbit. From Befus and Bienenstock (1982), with permission.

system is similar to that of other mammalian species, with the exception of the appendix, which is present at the caudal end of the caecum, and the sacculus rotundus located at the ileo-caecal junction. These two GALT have functions that have been identified only in rabbit tissue. The third GALT identified in the diagram is the Peyer's patch, which is located along the small intestine. In the rabbit, there are between two and 10 Peyer's patches along the small intestine (Befus and Bienenstock, 1982).

The rabbit spleen (Figure VI.2.2A) is composed of red and white pulp compartments. The red pulp comprises venous sinuses supported by a characteristic reticular meshwork (not visible at this magnification). Within the red pulp the interstitium contains clusters of mononuclear phagocytes, hematopoietic colonies and differentiated plasma cells. The white pulp is organized centripetally around central arterioles. The periarteriolar lymphatic sheath (PALS) contains a preponderance of small T cells but other recirculating cells and antigen-presenting cells resembling interdigitating dendritic cells are also present. In the periphery of the PALS is the marginal zone and B-cell follicles. The marginal zone is separated from the PALS by a line of antigen-presenting cells. Enzymatic activities and phenotypic markers suggest that these cells differ from antigen-presenting cells in the marginal zone, and from those in PALS and germinal centers. The organization of the B-cell follicles also follows a centripetal relationship to the central arteriole. The mantle zone containing small IgM^+B cells is furthest away from the arteriole. The germinal center (GC) itself is composed of two distinct regions, the light zone and the dark zone (defined by characteristic hematoxylin and eosin staining). The germinal center dark zone (containing centroblasts) rests against the border between the follicle and the PALS, and the light zone (containing centrocytes) fills in the remaining space between the mantle zone and the dark zone. Within these two zones B cells can undergo developmental changes, including DNA base pair changes as part of affinity maturation. The marginal zone, containing intermediate sized B-cells, reticulum and antigen-presenting cells forms a belt separating the white pulp from the red pulp. This reticulum is interrupted by numerous venus sinuses (marginal sinuses) supplied by vessels originating from the central arterioles. These marginal sinuses are where recirculating lymphocytes in the blood attach before migrating into the splenic white pulp. Recently, the MAdCAM-1 vascular addressin (mucosal addressin which is a ligand for $\alpha 4\ \beta 7$ integrin on mouse lymphocytes) has been identified on the endothelial lining of marginal sinuses (Kraal *et al.*, 1995). The spleen is the only peripheral lymphoid organ that does not have high endothelial venules (HEVs) supporting lymphocyte recirculation. Lyons and Parish (1995) have suggested that marginal-zone macrophages in the mouse are the splenic analog of HEV, forming the port of entry of lymphocytes into the white pulp of the spleen.

The rabbit lymph node (Figure VI.2.2B) is organized into compartments similar to those of most mammalian species. There is a cortex comprised of superficial cortex below the subcapsular lymphatic sinus where B-cell-follicles are present and a deep cortex where recirculating cells ($T \gg B$) emigrate from the blood. In the superficial cortex lymphoid follicles are organized into at least three compartments similar to those described in lymph nodes of other species such as mouse. It is interesting to note that the mantle zone is closest to the areas of antigen uptake and the GC dark zone is closest to deep cortex, where recirculating cells enter via HEVs. Below the deep cortex are areas containing B-cells in various states of maturation from lymphoblasts to plasma cells. These areas become the medullary plasma cell cords and have been termed (for the rabbit) 'follicular funnels' by Kelly *et al.* (1972) because

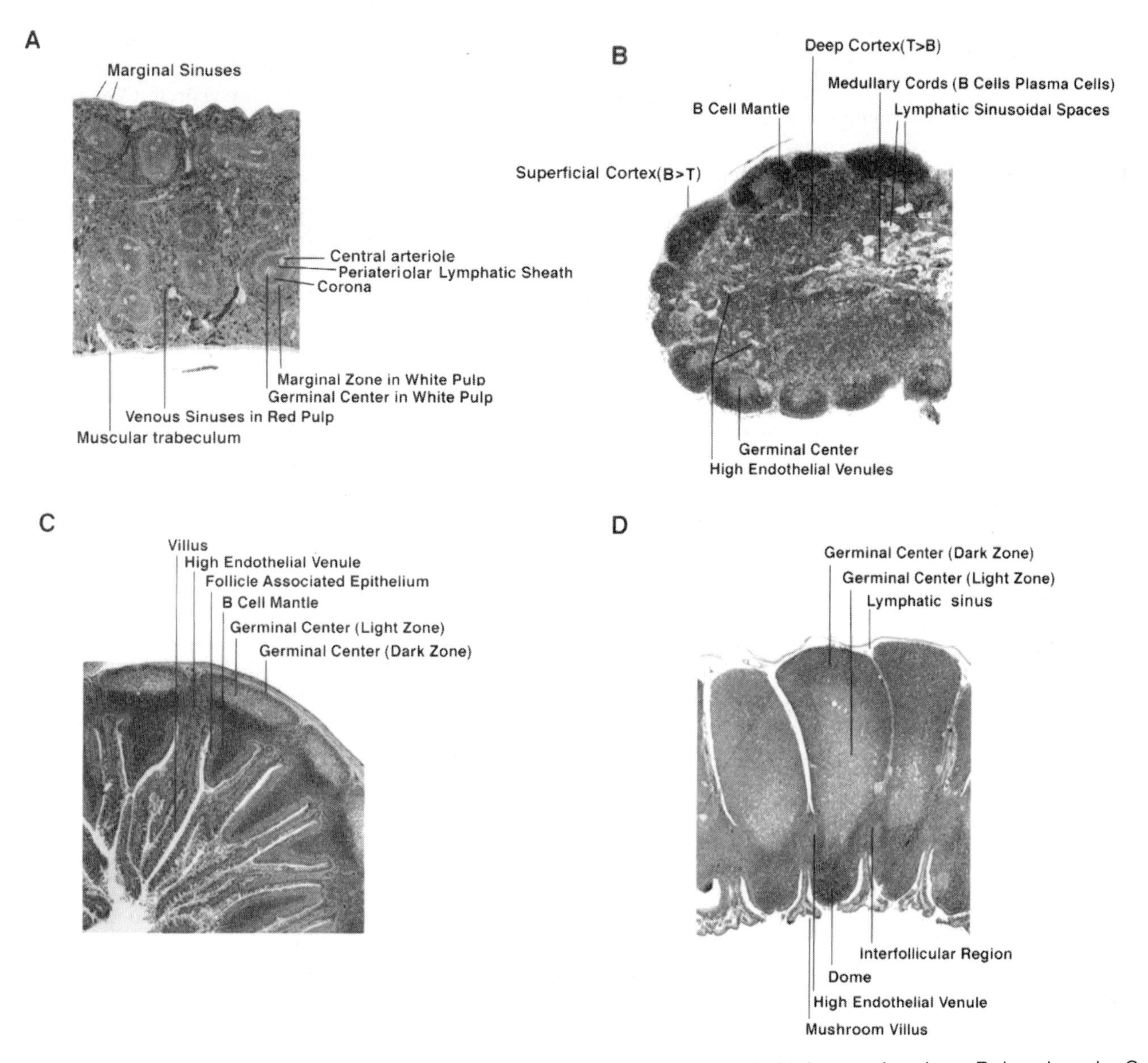

Figure VI.2.2 Haematoxylin and eosin stained sections of various rabbit lymphoid tissues. A: spleen; B: lymph node; C: Peyer's patch; D: appendix. Labels identify important features.

they appear to be organized across columnar areas of cortex below individual B-cell follicles. Separating plasma cell cords are medullary lymphatic sinuses which contain lymphoid cells enroute out into efferent lymphs. Phagocytic cells anchored to reticular fibers that cross intranodal lymphatic sinuses contribute to the removal of particulates from lymph.

Craig and Cebra in 1971 used the rabbit as an animal model to identify the Peyer's patch (PP) as a site for commitment of B cells to IgA production (Figure VI.2.2C). They found that B cells isolated from PP subsequently made IgA in the spleen after reinfusion into recipient rabbits. After expansion in the spleen, these B cells could circulate to a site along the intestinal tract, where they lodged and secreted antibody which was carried across epithelial cells into the gut lumen in endosomes. The organization of the rabbit PP is similar to that of other mammalian species, with a dome region extending into the lumen of the gut and the follicle germinal center abutting the muscular wall. The dome is covered by a special epithelium containing M or microfold cells which transport antigen into the PP in endosomes. Beneath the dome is the GC. It has an easily discernible dark zone surrounding a light zone. Unlike the spleen and lymph node, the mantle zone of the PP and other GALT structures does not surround the germinal center but rests above it, filling the space defined by the dome epithelium and connecting with the interfollicular areas laterally (this is most pronounced in the appendix and sacculus rotundus and least evident in the PP). Between the dome–GC region are T-cell rich areas known as the interfollicular regions. It is within the interfollicular regions that recirculating cells enter PP via HEV.

The rabbit appendix is composed of several-hundred lymphoid follicles. As shown in Figure VI.2.2D, the organization of these follicles is similar to the PP, except that the young rabbit has GCs that are larger than those found in the corresponding PP. By adulthood, the GC of the appendix is similar in size and cellular composition to that of the PP. In addition, the villus of the appendix has a characteristic shape that is unique to the appendix and sacculus rotundus. The villi in these two GALT have been given the name mushroom villi, based on their identifiable shape.

Species-specific aspects of organ development, structure and function

Most lymphoid organs of the rabbit bear striking similarity in terms of organ development, structure and function to their counterparts in other mammalian species. One lymphoid tissue that does not share these characteristics with other mammalian species is the rabbit appendix, a caecal diverticulum.

The rabbit appendix is about 2.5 cm long at birth and contains no organized B- or T-cell follicular regions. However, IgM, but not IgA B cells can be detected. By 6 weeks old the appendix reaches its largest size, measuring about 9 cm long and is composed of several hundred individual follicles, each characterized by several distinct morphological regions (Figure VI.2.3). B cells are located in the germinal center and dome region, the latter lying directly under the follicle associated epithelium. T cells are not found in either of these regions at 6 weeks old, but can be detected in the interfollicular region found between follicles (Table VI.2.2). Starting at 9 weeks after birth and ending at adulthood the rabbit appendix undergoes changes in gross morphology and B and T cell distribution that ultimately result in a lymphoid organ in the adult rabbit that bears little resemblance to that of the young rabbit appendix. In fact, the adult rabbit appendix appears very similar in structure and cell distribution to the Peyer's patches (Table VI.2.2 and Figure VI.2.3). The adult rabbit appendix contains germinal centers that are smaller than those found at 6 weeks after birth. In addition, T cells can now be found dispersed throughout the formerly B-cell only zones. Most of these T cells express CD4 and are distributed throughout both the dome region and the germinal centers. CD8-expressing T cells can also be detected in the adult appendix, with most located in the interfollicular regions and just under the follicle associated epithelium (Table VI.2.2 and Figure VI.2.3).

The changes seen in the appendix from the young to the adult rabbit (Figures VI.2.3 and VI.2.4) may correspond to changes in function (Crabb and Kelsall, 1940; Weinstein *et al.*, 1994a, and unpublished results). The young rabbit appendix is involved in diversification of the B-cell antibody repertoire (Weinstein *et al.*, 1994b).

The sacculus rotundus of the 6-week-old rabbit, another GALT, which is located at the ileo-cecal junction, bears a striking resemblance to the appendix of a similar aged rabbit. The few differences between these tissues are smaller germinal centers and fewer follicles. Like the appendix, the sacculus rotundus also changes as the rabbit gets older, with the result that by adulthood it also bears a strong resemblance to a Peyer's patch. The function of the sacculus rotundus may be the same as the appendix at equivalent stages of development.

3. Rabbit Leukocyte Markers

More than 30 surface markers have been identified for rabbit leukocytes. At present, approximately 20 of these markers can be identified with monoclonal antibodies (mAbs) and 15 of them with rabbit-specific DNA probes. Rabbit leukocyte antigens for which either mAb or DNA probes are available are listed in Table VI.3.1, along with their cellular distribution. The GenBank accession numbers are provided for all DNA probes. For some molecules, including CD4, CD5, MHC class I, and MHC class II, both mAb and DNA probes are available. Several

Table VI.2.2 Summary of IgM, IgA and CD4 staining levels in different lymphoid regions of rabbit appendix during development

	IgM					*IgA*					*CD4*				
	FAE	*D*	*LZ*	*DZ*	*IF*	*FAE*	*D*	*LZ*	*DZ*	*IF*	*FAE*	*D*	*LZ*	*DZ*	*IF*
1 day	–[a]	+	–	–	–	–	–	–	–	–	–	–	–	–	–
2 week	++	++	++	+	–	++	++	++	–	–	–	–	–	–	+
6 week	++	++	++	+	–	++	++	++	–	–	–	–	–	–	+
9 week	++	++	++	+	–	++	++	++	–	–	–	±	–	–	+
5 month	++	++	++	+	–	++	++	++	–	–	–	+	+	–	+
9 month	++	++	++	+	–	++	++	++	–	–	–	+	+	–	+
1.5 year	+	++	+	–	–	++	++	++	+	–	–	+	+	+	+
4 year	+	++	+	–	–	++	++	++	+	–	–	+	+	+	+
Jejunal Peyer's patch	+	++	+	–	–	++	++	++	+	–	–	+	+	+	+

[a]++ = stains darkly; + = stains moderately; ± = stains faintly; – = no staining seen.
FAE, follicle associated epithelium; D, dome; LZ, light zone; DZ, dark zone, IF, interfollicular.

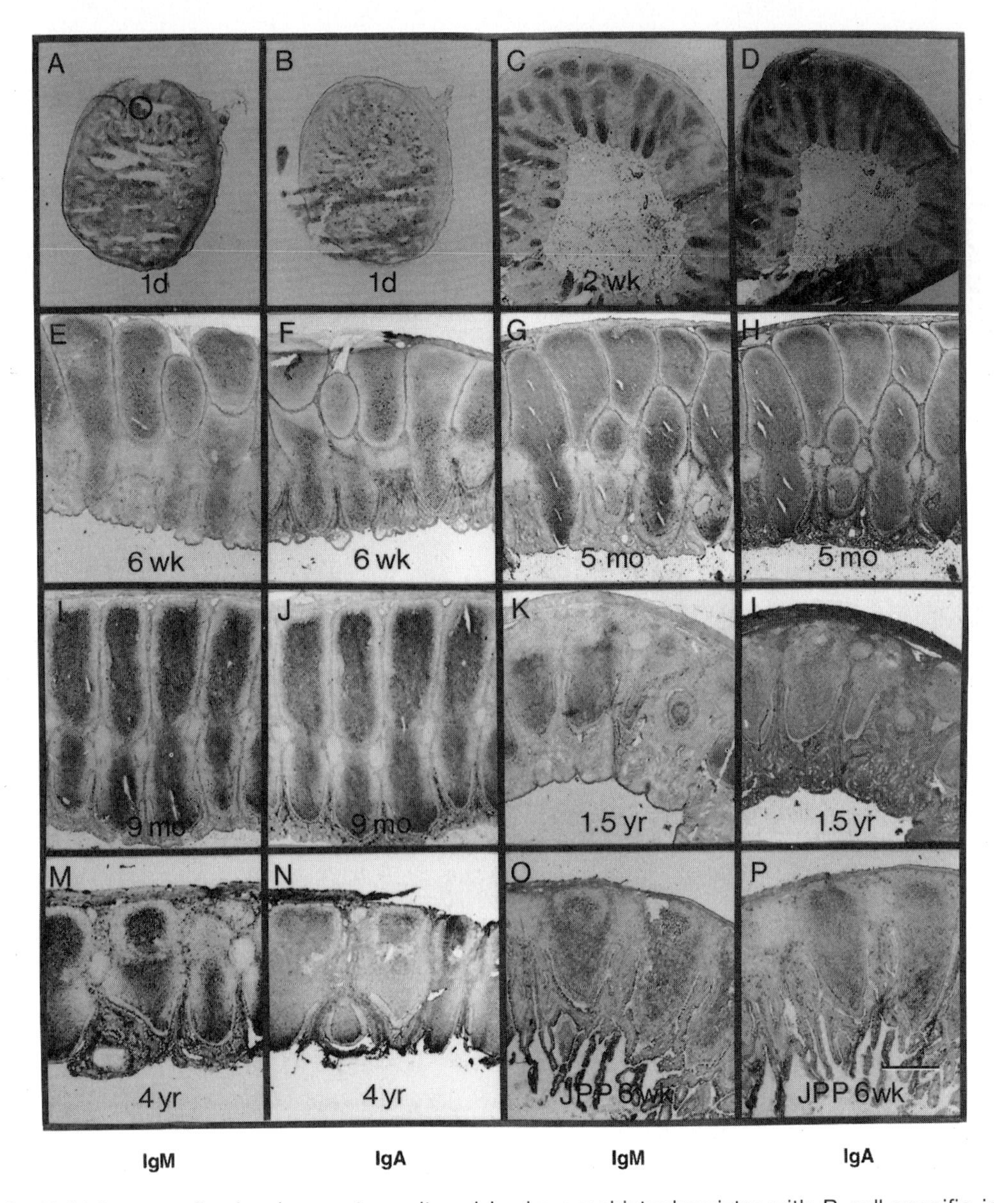

Figure VI.2.3 Rabbit appendix development monitored by immunohistochemistry with B cell-specific immunoglobulin markers. Rabbit appendix tissue was taken 1 day (A, B), 2 weeks (C, D), 6 weeks (E, F), 5 months (G, H), 9 months (I, J), 1.5 years (K, L), and 4 years (M, N) after birth. Rabbit jejunal Peyer's patch (JPP) tissue was taken 6 weeks (O, P) after birth. Semi-thin (7-μm) sections of rabbit appendix and JPP were stained with either a mouse anti-rabbit IgM (A, C, E, G, I, K, M, O) or a mouse anti-rabbit IgA (B, D, F, H, J, L, N, P). Note change in size and shape of the rabbit appendix from 6 weeks to 1.5 years. The rabbit appendix at 1.5 and 4 years (K–N) bears a strong resemblance to a JPP (O–P) in shape, immunoglobulin (Ig) staining patterns, and size. Most GC staining for IgM and IgA is in the LZ with DZ B cells expressing little surface Ig. Scale bar = 5 mm.

other rabbit leukocyte cell surface molecules have been identified by mAb (McNicholas *et al.*, 1981; De Smet *et al.*, 1983; Loar *et al.*, 1986), however since the molecular specificities of these antibodies are not known, we did not include them in Table VI.3.1.

Specificity of reagents

Most rabbit leukocyte mAb are produced by immunizing mice with rabbit cells, rather than with purified molecules. The target of the selected mAb is usually identified by comparing the cellular distribution and the molecular weight of the immunoprecipitated molecule with known mouse and human CD molecules. We have concerns about the accuracy of this approach to determining antibody specificity. This concern is illustrated by the following example. Kotani *et al.* (1993a) developed the antirabbit CD5 mAb KEN-5 by immunizing mice with rabbit T cells. This mAb reacted with all rabbit T cells and with <1% B cells. It immunoprecipitated a molecule of 67 kDa. Raman and Knight (1992) also developed

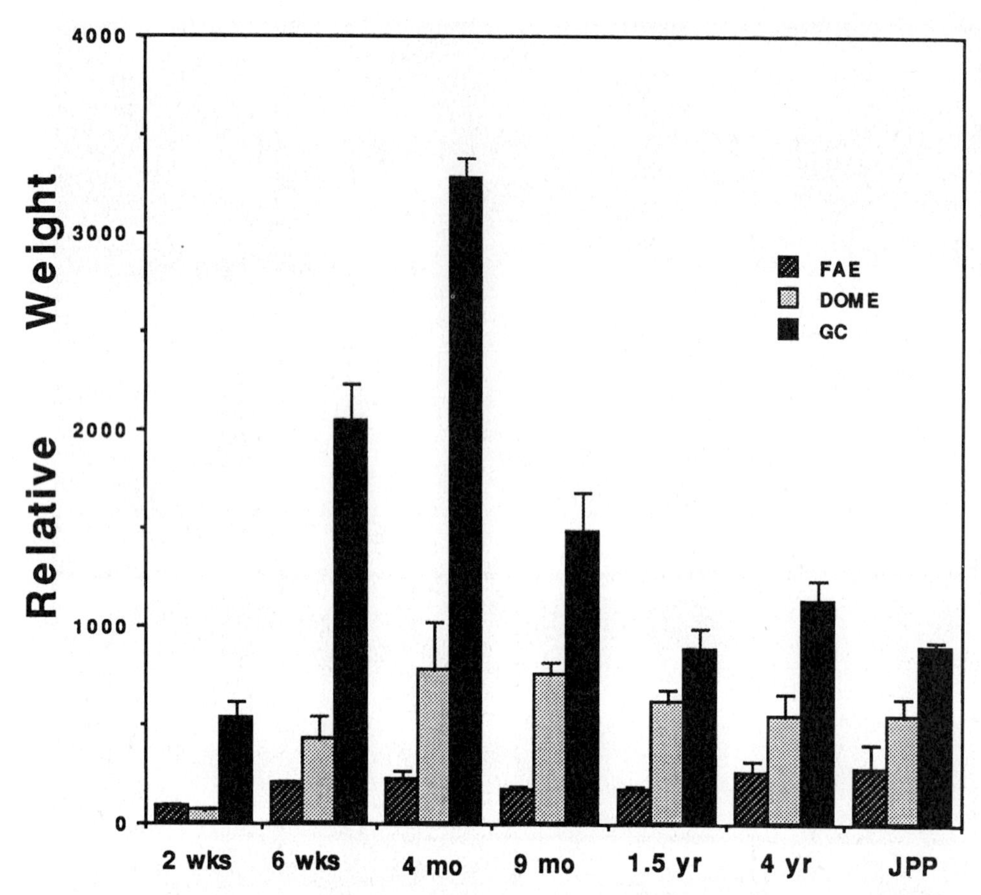

Figure VI.2.4 Changes in size of different rabbit appendix B-cell follicular regions. Mean weights of follicle-associated epithelium (FAE), dome region and germinal centers (GC) on photographs were measured. Note dramatic rise then fall in GC size during rabbit appendix development. By 1.5 years, the size of appendix GC were similar to those in JPP.

an antirabbit CD5 mAb; however, they cloned the rabbit CD5 gene and immunized mice with murine T cells that had been transfected with this gene. Unlike the KEN-5 mAb, the Raman and Knight anti-CD5 mAb (R-CD5) reacted with essentially all rabbit B cells. The KEN-5 mAb did not react with murine cells transfected with the rabbit CD5 gene. We conclude that the KEN-5 mAb is not directed against rabbit CD5 but instead is directed against a T-cell molecule of unknown identity.. We suggest that before a mAb can be identified as reacting with the rabbit homologue of a mouse or human molecule, the specificity must be characterized by reactivity with molecules known, either by amino acid or by nucleotide sequence analysis, to be homologous to the respective mouse or human molecule.

Several of the mAb used to identify rabbit cell surface molecules are cross-reacting antibodies that were developed against mouse or human CD immunogens. Although in our experience most such mAb raised against mouse or human CD molecules do not cross-react with rabbit leukocytes, we encourage investigators to test antimouse and antihuman mAb for cross-reactivity with rabbit leukocytes before developing rabbit-specific reagents. It is more likely that antihuman mAb developed in mouse, a species phylogenetically distant from human, will cross-react with rabbit leukocytes than will the antimouse mAb developed in other rodents. We have included cross-reacting antihuman mAb in Table VI.3.1 for those rabbit leukocyte cell-surface molecules for which no rabbit-specific reagents are available.

Lymphocyte markers

B cells can be easily identified by readily available polyclonal anti-Ig reagents, and as a result, relatively few antirabbit Ig mAbs have been developed. Although B cells can also be identified by anti-MHC class II mAb, these mAb are less reliable because they react with monocytes, macrophages, and activated T cells, as well as with B cells. No pan-B lineage-specific mAbs for molecules such as rabbit B220, CD19, or CD20 have been reported. Similarly, although several T-cell-specific mAb have been developed, notably anti-CD4 and anti-CD8, no reagents directed specifically against the rabbit T-cell specific antigen, CD3, are currently available. Two commonly used antirabbit T cell mAbs, anti-CD43 and 9AE10, are not optimal reagents because, in addition to reacting with

Table VI.3.1 Rabbit leukocyte antigens

Molecule	*Cell distribution*	*Monoclonal antibodies*	*Probe*	*Reference*
CD1	Thymocytes, dendritic cells		M26249	Calabi *et al.* (1989)
$CD3_{\varepsilon}$	T cells	PC3/188A[a]		Jones *et al.* (1993)
CD4	T cell subset	KEN-4	S44055 M92840	Kotani *et al.* (1993a) Hague *et al.* (1992)
CD5	T and B cells	KEN-5[b] R-CD5	 L03204	Kotani *et al.* (1993a) Raman and Knight (1992)
$CD8_{\beta}$	T cell subset	12.C7	 L22293	De Smet *et al.* (1983) Sawasdikosol *et al.* (1993)
CD9	Pan-leukocyte, platelets, fibroblasts	MM2/57[a]		Hornby *et al.* (1991), Wilkinson *et al.* (1992b)
CD11a (LFA-1 α chain)	Pan-leukocyte	NR185 KEN-11		Blackford and Wilkinson (1993) Kotani *et al.* (1993a)
CD11b (MAC-1; CR3α chain)	Neutrophils, macrophages, NK cells	198		Smet *et al.* (1986), Galea-Lauri *et al.* (1993b)
CD11c (P150,95; CR4α chain)	Neutrophils, macrophages, NK cells	3/22		Blackford *et al.* (1996)
CD18 (LFA-1 β_2 chain)	Pan-leukocyte	L13/64 RCN1/21		Jackson *et al.* (1983) Wilkinson *et al.* (1984), Galea-Lauri *et al.* (1993a)
CD25 ($IL2R_{\alpha}$)	Activated T cells	KEI-α1		Kotani *et al.* (1993b), Sawasdikosol *et al.* (1993)
CD28	T cells (most $CD4^+$, some $CD8^+$ cells)		D49841	Isono and Seto (1995)
CD43 (leukosialin)	T cells, thymocytes	L11/135		Jackson *et al.* (1983), Wilkinson *et al.* (1992a)
CD44	Pan-leukocyte	W4/86 RPN 3/24		Jackson *et al.* (1983) Wilkinson *et al.* (1984), Galea-Lauri *et al.* (1993b)
CD45	Pan-leukocyte	1.24 L12/201 L12/27		De Smet *et al.* (1983) Jackson *et al.* (1983) Wilkinson *et al.* (1993)
CD49d (VLA-4 α_4 subunit)	Monocytes, lymphocytes	HP1/2[a]		Kling *et al.* (1995)
CD54 (ICAM-1)	Widely distributed; many activated cells	Rb2/3		Richardson *et al.* (1994)
CD58 (LFA-3)	Pan-leukocyte, red blood cells, platelets	VC21		Wilkinson *et al.* (1992b)
CD62E (E-selectin)	Endothelial cells	9H9 14G2		Olofsson *et al.* (1994)
CD62L (L-selectin)	White blood cells	Dreg-200[a] LAM1-3[a]		Garcia *et al.* (1995) Ley *et al.* (1993)
CD80 (B7-1)	Activated B cells, macrophages, dendritic cells		D49843	Isono and Seto (1995)
CD86 (B7-2)	Activated B cells, macrophages, dendritic cells		D49842	Isono and Seto (1995)
CD106 (VCAM-1)	Endothelium (inflammatory settings); nonvascular cells	Rb1/9		Richardson *et al.* (1994)

(continued)

Table VI.3.1 *Continued*

Molecule	*Cell distribution*	*Monoclonal antibodies*	*Probe*	*Reference*
CTLA-4	Activated T cells		D49844	Isono and Seto (1995)
MHC-I	Nucleated cells	61	K02819	Marche *et al.* (1985), LeGuern *et al.* (1987)
MHC-II $DP_{\alpha/\beta}$	B cells, macrophages		M22640 (α chain) M21465-8 (β chain)	Sittisombut and Knight (1986) LeGuern *et al.* (1985)
MHC II $DQ_{\alpha/\beta}$	B cells, macrophages	2C4	M15557 (α chain)	Lobel and Knight (1984), LeGuern *et al.* (1986), Sittisombut and Knight (1986)
MHC II $DR_{\alpha/\beta}$	B cells, macrophages	RDR34	M28161 (α chain)	Sittisombut and Knight (1986), Laverrier *et al.* (1989), Spieker-Polet *et al.* (1990)
MHC II Do_{β}	Epithelium of thymic medulla, B cells		M96942	Chouchane *et al.* (1993)
TCRα/β	T cells		M12885 (α chain) M14576-7, D17416-26 (β chain)	Marche and Kindt (1986a, b) Angiolillo *et al.* (1985), Isono *et al.* (1994)
TCRγ/δ	T cells	CγM1[a]	L22290, D38134-44, D42090 (γ chain) L22291, D26555, D38118-21 (δ chain)	Sawasdikosol *et al.* (1993) Isono *et al.* (1995) Kim *et al.* (1995)

[a] mAb directed against immunogen of human origin.
[b] KEN-5 mAb was made against rabbit thymocytes as immunogen (Kotani *et al.*, 1993a); R-CD5 was made against the product of the cloned rabbit CD5 gene (Raman and Knight, 1992).

T cells, anti-CD43 may react with progenitor B cells (Hardy *et al.*, 1991) and 9AE10 reacts with neutrophils (McNicholas *et al.*, 1981; Chen *et al.*, 1984). These reagents can be used to distinguish T cells from B cells, but they must be used cautiously if other leukocytes are present in the cells being examined. Analysis of rabbit T cells will be simplified when a reagent specific for rabbit CD3 becomes available.

Summary

The number of rabbit-specific leukocyte markers is considerably less than the number of markers available for mouse and human leukocytes. With the introduction of gene cloning methods, it is now a straightforward experiment, albeit time consuming, to clone the rabbit homologue of a mouse or human gene of interest and develop mAb against the product of the expressed gene. We anticipate that as the methods for immunization with DNA are refined, mAbs of desired specificity will be easily obtained and the number of mAbs specific for rabbit leukocytes will increase significantly.

4. Leukocyte Traffic and Associated Molecules

Evidence for organ-specific leukocyte recruitment

Rabbits have been used extensively for *in vivo* studies of leukocyte recruitment as a result of the species' size, availability, the availability of immunological reagents, and their apparent similarity to humans in both leukocyte–endothelial adhesion pathways and leukocyte recruitment mechanisms. In the systemic microcirculation, intravital microscopy of small venules in both the abdominal mesentery and the tenuissimus muscle (Lindbom *et al.*, 1990; Granert *et al.*, 1994; Olofsson *et al.*, 1994) demonstrate that leukocyte–endothelial interactions utilize a multistep adhesion cascade similar to that proposed for other species (Springer, 1995; Butcher and Picker, 1996). This cascade (see 'Molecules Involved in Leukocyte–Endothelial Interactions' and 'Molecules Involved in Leukocyte Migration' later in this section) consists of five sequential events:

(1) inflammatory activation of the endothelial cell creating a proadhesive condition;

(2) random contact of circulating leukocytes with the vessel wall that precedes low affinity, adhesion-dependent leukocyte rolling (selectin-mediated);
(3) local activation of the proximate leukocyte resulting in firm adhesion to endothelium (β_1 and β_2 integrin-mediated);
(4) leukocyte diapedesis between endothelial cells into extravascular tissue;
(5) subendothelial migration of leukocytes along chemotactic gradients.

Leukocyte recruitment in the heart, skin, skeletal muscle, intestine, liver, and central nervous system has been shown to be attenuated by adhesion molecule antagonists (e.g., monoclonal antibodies, soluble carbohydrates). These studies are consistent with the sequential activation of first selectin, followed by β_2 integrin (CD11/CD18) adherence mechanisms (reviewed by Harlan *et al.*, 1992; Talbott *et al.*, 1994), as described above. However, there are organ-specific and stimulus-dependent exceptions to this standard neutrophil adhesion cascade paradigm. In the systemic circulation, intraperitoneal *S. pneumonia* normally elicits neutrophil emigration by a CD18-dependent mechanism that can be converted to a CD18-independent pathway by addition of macrophages to the peritoneum (Mileski *et al.*, 1990). Similarly, early (4 h) neutrophil emigration elicited by intraperitoneal protease peptone is CD18-dependent, whereas emigration at later time points (24 h) is CD18-independent (Winn and Harlan, 1993). This alternative pathway appears to require cytokine or chemokine signaling initiated by either resident or elicited peritoneal macrophages. Neutrophil emigration in the pulmonary circulation of rabbits also utilizes alternative adhesion/emigration pathways (e.g., selectin- and/or CD18-independent) (reviewed by Hogg and Doershuck, 1995). In contrast to the systemic circulation where neutrophil–endothelial adherence and emigration occur in the relatively large-diameter post-capillary venules, in the pulmonary vasculature these processes occur in the capillaries themselves, where their small diameter induces neutrophil deformation and neutrophil sequestration, and may eliminate the requirement for initial neutrophil 'rolling' (low-affinity endothelial adhesion) prior to emigration (Wiggs *et al.*, 1994; Hogg and Doershuck, 1995).

Tissue-specific recruitment of leukocyte subpopulations

Cultured rabbit vascular endothelial cells have been used for the study of leukocyte–endothelial adhesion and emigration (Kume *et al.*, 1992; Kim *et al.*, 1994), but to a limited extent owing to technical challenges in maintaining these cell lines. *In vivo* studies in rabbits emphasize neutrophil (see previous subsection) and monocyte recruitment mechanisms, with scant information pertaining to lymphocytes. Neutrophil emigration is dependent on β_2 integrin-mediated adhesion in tissues not possessing resident macrophages, whereas mononuclear cell emigration utilizes both β_1 and β_2 integrin-mediated pathways (Winn and Harlan, 1993; Kling *et al.*, 1995).

Molecules involved in leukocyte-endothelial interactions

The molecular components of the leukocyte-endothelial adhesion cascade have been described in detail (Carlos and Harlan, 1994). Table VI.4.1 is generated from a variety of references and describes those adhesion molecules (left

Table VI.4.1 Leukocyte and endothelial adhesion molecules described in rabbits

Molecule	*Distribution*	*Ligands*	*Regulation of expression*
Selectins			
L-selectin (CD62L)	WBC	Sle^x, MAdCAM-1, CD34, GlyCAM-1	CE; ↓ Upon inflam. activation[a]
E-selectin (CD62E)	EC	Sle^x, PSGL-1, ESL-1	↑ Upon inflam. activation[a]
P-selectin (CD62P)	P, EC	Sle^x, PSGL-1	↑ Upon inflam. activation[b]
β_1 Integrins			
VLA-4 ($\alpha_4\beta_1$, CD49d/CD29)	L, M, neural crest cells	VCAM-1, FN	CE; not inducible
β_2 Integrins			
LFA-1 (CD11a/CD18)	WBC	ICAM-1, ICAM-2	CE; ↑ Upon inflam. activation[a]
Mac-1 (CD11b/CD18)	PMN, M, ±L	ICAM-1, other?	CE; ↑ Upon inflam. activation[a]
Immunoglobulin superfamily			
ICAM-1 (CD54)	EC, M, L, EpC	LFA-1, Mac-1	CE; ↑ Upon inflam. activation[a]
VCAM-1 (CD106)	EC	VLA-4	↑ Upon inflam. activation[a]
PECAM-1 (CD31)	EC, WBC, P	PECAM-1, other?	CE; not inducible

CE = constituitively expressed; WBC = all leukocytes; EC = vascular endothelium; PMN = neutrophils; L = lymphocytes; M = monocytes; P = platelets; FN = fibronectin; EpC = epithelial cells.
[a]TNF-α, IL-1α, lipopolysaccharide. [b]Histamine, oxidants.
↑ = increased; inflamm. = inflammatory.

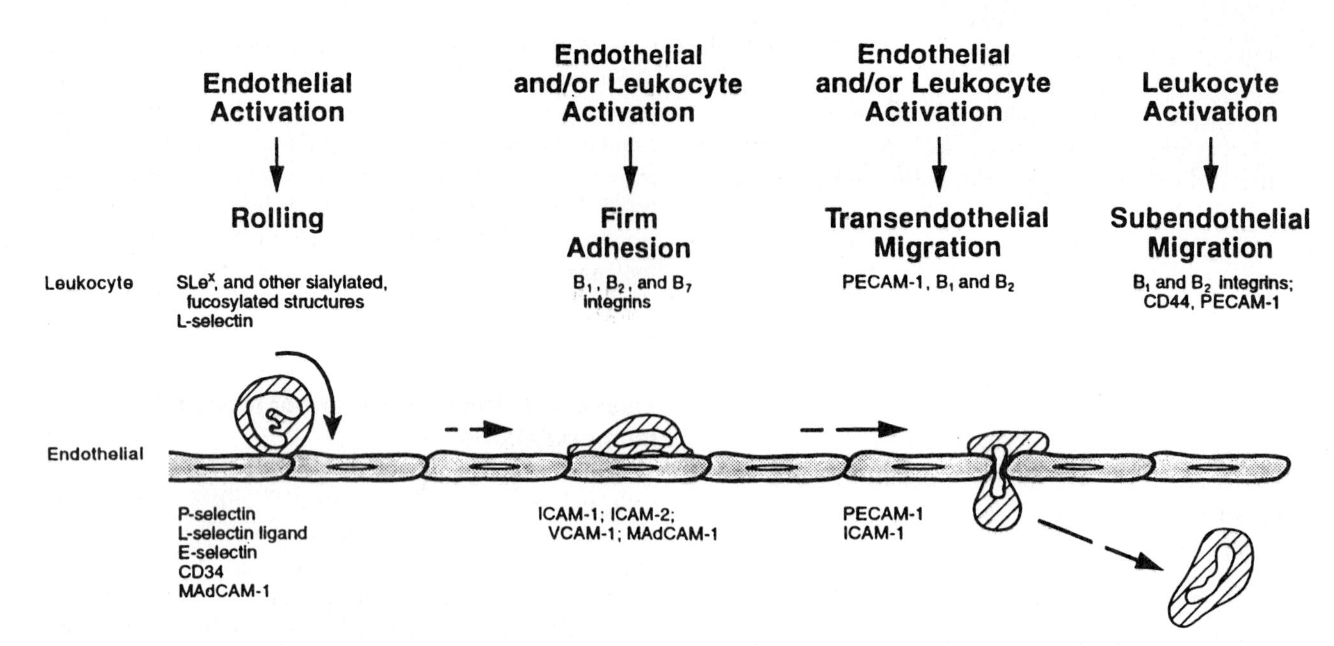

Figure VI.4.1 The multistep adhesion cascade for leukocyte–endothelial adhesion and transendothelial migration (modified from S. R. Sharar, R. K. Winn and J. M. Harlan (1995). The adhesion cascade and anti-adhesion therapy: an overview. *Springer Semin. Immunopathol.* **16**, 359, and reproduced with permission of Springer-Verlag GmbH & Co. KG).

column) demonstrated (by flow cytometry, immunocytochemistry, or *in vivo* inhibition studies) to play a role in leukocyte recruitment in rabbits. Of these various adhesion molecules, rabbit-specific DNA cloning has been reported only for E-selectin (Larigan *et al.*, 1992). The counter-receptor ligands listed in Table VI.4.1 include receptors identified in humans that presumably play a role in leukocyte-endothelial adhesion and recruitment in rabbits.

Molecules involved in leukocyte migration

The molecular components of leukocyte migration have been described in detail (Carlos and Harlan, 1994; Butcher and Acker, 1996). Figure VI.4.1 describes the generalized leukocyte adhesion/emigration cascade; please refer to the previous subsection for those components of the cascade that have been identified in rabbits.

Table VI.4.2 Chemotactic molecules described in rabbits

Chemotactic molecule	*Leukocytes recruited*	*Organ studied*	*Reference*
C5a	G	Skin, lung, mesentery	Arfors *et al.* (1987), Argenbright *et al.* (1991), van Osselaer *et al.* (1993), Hellewell *et al.* (1994)
fMLP	G	Skin, peritoneum	Drake (1993), van Osselaer *et al.* (1993)
LTB_4	G	Skin	Arfors *et al.* (1987)
Histamine	G	Skin	Arfors *et al.* (1987)
IL-1α	G	Skin, lung, eye	Rosenbaum and Boney (1993), Hellewell *et al.* (1994)
IL-1β	P, M, L	Articular joint	McDonnell (1992)
IL-2	L, ?G	Skin, lung	Ohkubo (1991), Wiebke *et al.* (1995)
IL-8	G	Mesentery, lung, skin, articular joint	Hechtman *et al.* (1991), Drake and Issekutz (1993), Ley *et al.* (1993), Folkesson *et al.* (1995), Matsukawa *et al.* (1995)
GRO	G	Lung	Johnson *et al.* 1996
LPS	G, M	Skin, lung, eye, peritoneum	Doerschuk *et al.* (1990), Drake and Issekutz (1993), Rosenbaum and Boney (1993)
TNF-α	G	Skin	Drake and Issekutz (1993)

G = granulocytes; L = lymphocytes; LPS = lipopolysaccharide; M = monocytes.

Chemotactic molecules and populations recruited

Limited information is available describing specific chemotactic or other inflammatory molecules and their function in rabbits. A partial list of these molecules implicated by a variety of *in vivo* studies in rabbit species is shown in Table VI.4.2. Of these various chemotactic molecules, rabbit-specific DNA cloning has been reported for IL-8 (Yoshimura and Yuhki, 1991), MCP-1 (Yoshimura and Yuhki, 1991; Kajikawa *et al.*, 1993), and GRO (Johnson *et al.*, 1996). *In vivo* experiments in these studies suggest that the biological effects of rabbit-specific chemokines are different from those seen with human-derived chemokines when both are administered to rabbits.

Species-specific aspects of leukocyte traffic

To date, no leukocyte adhesion or emigration pathways have been described that are unique to rabbit species. The overwhelming majority of *in vivo* and *in vitro* experiments in rabbits have utilized New Zealand White rabbits or their cultured vascular endothelium (see sub-section on tissue-specific recruitment of leukocytes). Watanabe rabbits with heritable hyperlipidemia (Watanabe, 1980) have also been studied to determine the role of leukocyte adhesion and emigration in the development of atherosclerosis and related lesions (Calderon *et al.*, 1994).

5. Cytokines

The data on cytokines in the rabbit are still incomplete and scattered. The cDNA sequences of rabbit IL-1 (GenBank accession number M26295; Cannon *et al.*, 1989) and a partial sequence of IL-2 (GenBank accession number Z36904; A. Schock and C. McInnes, unpublished data) are in the GenBank database. A number of papers report measurements of rabbit cytokines and their roles in various disease models. Immunoassays for detection of IL-1α and IL-1β have been described (Clark *et al.*, 1991) and kits based on these assays are commercially available. Although a crystal structure of rabbit IFN-γ was reported in 1991 (Samudzi *et al.*, 1991) the full sequence is not yet available in the GenBank database.

6. T-cell Receptor (TCR)

TCR genes, their chromosome location and homology with other species

The organization of the entire TCR gene loci is much the same as that of human and mouse. There are a single gene

Table VI.6.1 Rabbit TCR gene segments and their similarity to human and mouse homologues

Gene segment	Percentage nucleotide (amino acid) identity[a] to: Human	Mouse	Reference
α Chain			
C_α	75.3 (68.6)	71.5 (65)	Marche and Kindt (1986a)
V_α	59 (44)	75 (70)	Marche and Kindt (1986a)
β Chain			
$C_\beta1$	83.9 (76.3)	80.3 (76.6)	Angiolillo *et al.* (1985)
$C_\beta2$	83 (76)	81 (78)	Marche and Kindt (1986b)
Chimeric C_β			Komatsu *et al.* (1987), Harindranath *et al.* (1989b)
$D_\beta2$–$J_\beta2$	74		Harindranath *et al.* (1991)
$V_\beta1$	78.7	73.7	Lamoyi *et al.* (1986)
$V_\beta2$	74.5	64.8	Marche and Kindt (1986b)
$V_\beta3$	83.6	80.1	Lamoyi and Mage (1987)
$V_\beta4$	67.0	65.2	Lamoyi and Mage (1987)
$V_\beta5$	79.7	76.1	Isono *et al.* (1994)
$V_\beta6$	82.8	74.1	Isono *et al.* (1994)
$V_\beta7$ (7.1–7.3)	74.5–76.2	66.4–68.6	Isono *et al.* (1994)
$V_\beta8$	80.4	78.7	Isono *et al.* (1994)
$V_\beta9$	81.8	70.4	Isono *et al.* (1994)
$V_\beta10$	80.7	74.3	Isono *et al.* (1994)
$V_\beta11$	82.7	65.2	Isono *et al.* (1994)
γ Chain			
C_γ	82.3	80.3	Sawasdikosol *et al.* (1993), Isono *et al.* (1995)
$J_\gamma1$	83	74	Isono *et al.* (1995)
$J_\gamma2$	82	54	Isono *et al.* (1995)
$V_\gamma1$(1.1–1.5T)	65–72	66–69	Isono *et al.* (1995)
$V_\gamma2$	78	61	Isono *et al.* (1995)
δ Chain			
C_δ	80.6	77.7	Sawasdikosol *et al.* (1993), Kim *et al.* (1994)
$V_\delta1$	77.0	66.2	Kim *et al.* (1994)
$V_\delta2$T	65.9	64.7	Kim *et al.* (1995)
$V_\delta3$T	81.2	65.4	Kim *et al.* (1995)
$V_\delta4$	74.9	73.6	Kim *et al.* (1995)
$V_\delta5$	86.7	77.1	Kim *et al.* (1995)

[a] Per cent identity was shown for the closest counterpart.

segment each of C_α, C_γ and C_δ, two tandemly arranged C_β gene segments, and clusters of V, D (for β and δ chains) and J gene segments. The structural map of these segments has not been completely determined and their chromosome location is not known. Allelic variation has been identified for C_β and C_γ segments (Komatsu *et al.*, 1987; Harindranath *et al.*, 1989a; Isono *et al.*, 1995), and some rabbits have a third copy of C_β gene or the chimeric gene that may have arisen by an unequal crossing-over event (Komatsu *et al.*, 1987; Harindranath *et al.*, 1989b). Table VI.6.1 lists TCR gene segments of published nucleotide sequences and their similarity to human and mouse homologues.

TCR-associated proteins and accessory molecules involved in TCR signaling

Table VI.6.2 lists TCR-associated proteins and accessory molecules examined with monoclonal antibody and/or whose nucleotide sequence has been determined.

TCR repertoire ontogeny: thymic dependent and thymic independent

No data relating to the TCR repertoire ontogeny are available.

Distribution of $\alpha\beta$ and $\gamma\delta$ TCR populations

Antibodies against rabbit $\alpha\beta$ and $\gamma\delta$ TCR are not yet available and the tissue distribution of T cells carrying these TCRs therefore remains mostly unknown. Cross-reactive antibodies to homologous molecules of other species and/or nucleotide probes for reverse-transcription polymerase chain reaction (RT-PCR) provides only limited information. The $\gamma\delta$T-cell population in the peripheral blood, at 23% (Sawasdikosol *et al.*, 1993), is relatively high compared with human and mouse; $V_\gamma 2$ and $V_\delta 1$ genes are dominantly expressed in these cells (Kim *et al.*, 1995).

Species-specific aspects of the structure, function or development of the TCR and associated proteins

Comparative analysis of nucleotide sequences of TCR gene segments revealed a greater similarity of rabbit to human than to mouse. A high percentage nucleotide identity is observed in almost all gene segments and, in addition, the similarity in C_γ chain gene is remarkable in that the gene of both these species, and not others thus far reported, contain repetitive exons that have lost the cysteine residue involved in the interchain disulfide bond (Isono *et al.*, 1995). Genetic similarity between human and rabbit is also reported for the absence of linkage of TCRβ chain genes to immunoglobulin C_κ chain genes; these genes are linked in mouse (Hole *et al.*, 1988). T-cell costimulatory molecules also displayed greater amino acid identity to human than mouse homologues (Isono and Seto, 1995).

Table VI.6.2 Rabbit TCR-associated proteins and accessory molecules involved in TCR signaling

Molecule (synonym)	*Molecular mass (kDa)*	*Monoclonal antibody[a] and nucleotide sequence database accession no.*	*Reference*
CD3ε	—[c]	PC3/188A[b]	Jones *et al.* (1993)
CD4	50	KEN-4	Kotani *et al.* (1993)
		M92840	Hague *et al.* (1992)
CD8β	28	12.C7	De Smet *et al.* (1983)
		L22293	Sawasdikosol *et al.* (1993)
CD11a(LFA-1α)	150	KEN11	Kotani *et al.* (1993)
CD18(LFA-1β)	90	L13/64	Jackson *et al.* (1983)
		RCN1/21	Galea-Lauri *et al.* (1993)
CD28	—	D49841	Isono and Seto (1995)
CD45	200	I.24	De Smet *et al.* (1983)
		L12/201	Jackson *et al.* (1983)
CD54(ICAM-1)	—	R6.5[b]	Argenbright *et al.* (1991)
CD58(LFA-3)	42	VC21	Wilkinson *et al.* (1992)
CD80(B7)	—	D49843	Isono and Seto (1995)
CD86(B7-2)	—	D49842	Isono and Seto (1995)
CTLA-4	—	D49844	Isono and Seto (1995)

[a] Representative antibodies alone are cited.
[b] Cross-reactive antibody produced against the molecule from other species.
[c] Not available.

Table VI.7.1 Rabbit immunoglobulin classes

Rabbit Ig classes	*MW × 10^{-3}*	*Concentration in serum (mg/ml)*	*Distribution*	*Function*
IgM	800–950	~1[a]	Serum Lymph	Primary Immune responses Complement Fixation
IgG	149	(≈5–20)[b]	Serum Lymph	Secondary Immune responses
IgA	158–162	(≈3–4)	Mucosal secretions[c] and serum	Local and systemic protection
IgE	185–190[d]	[e]	Serum Lymph Fixed on basophils/mast cells	Histamine release from mast cells, IgE-mediated Anaphylaxis release of platelet Activating factor from basophils (Betz *et al.*, 1980)

[a]The concentration of IgM in rabbit serum may be elevated during infections. *Trypanosoma equiperdum* has been used to experimentally elevate IgM levels prior to isolation of IgM (*Van Tol et al.*, 1978; Gilman-Sachs and Dray, 1985).
[b]The concentration of IgG in rabbit sera varies with age, tending to be elevated in older animals. Hyperimmunization with streptococcal, pneumococcal or micrococcal vaccine can cause markedly elevated IgG concentrations greater than 50 mg/ml (Krause, 1970; Haber, 1970; Mage *et al.*, 1977).
[c]Spieker-Polet *et al.* (1993) found that there are at least 10 different IgA isotypes that are expressed (see also Table VI.7.2). mRNA corresponding to the isotype encoded by the gene that maps most 5′ ($C_\alpha 4$), was found expressed in a variety of tissues including small intestine, appendix, mesenteric lymph node, mammary tissue, salivary gland, lung, tonsil and Peyer's patches. This was the only detectable isotype in lung and tonsil. In other tissues the additional isotypes were found to be expressed at various levels.
[d]Lindqvist (1968).
[e]The absolute amount of IgE in rabbit sera probably varies as it does in man depending upon genetic and environmental factors. It also becomes elevated upon hyperimmunization (Kindt and Todd, 1970).

7. Immunoglobulins

Function

The rabbit's immunoglobulin classes and some functions are summarized in Table VI.7.1. There is no definitive evidence for IgD in the rabbit. Although early studies suggested there might be another immunoglobulin on the surface of IgM^+ B cells that was not IgA or IgG (Wilder *et al.*, 1979; Eskinazi *et al.*, 1979; Sire *et al.*, 1979), no gene encoding the δ heavy chain has been identified in the region downstream of the C_μ gene.

Ig gene loci

An extensive review of rabbit immunoglobulin genes that were originally detected serologically as allelic forms (allotypes), including descriptions of methods and reagents can be found in Roux and Mage (1996). Additional information about rabbit V_H, IgM and IgA can be found in: Gilman-Sachs *et al.* (1969); Knight and Hanly (1975); Currier *et al.* (1985); Gallarda *et al.* (1985); Allegrucci *et al.* (1990) and Becker and Knight (1990).

The Ig heavy-chain gene loci with some known allelic forms, heavy chain locus haplotypes, and light chain loci with some known allelic forms are summarized in Tables VI.7.2, VI.7.3 and VI.7.4 respectively. Some GenBank locus names and accession numbers are also listed. Only germline V_H that are known to be expressed are included. There are more than 100 rabbit germline V_H gene (and pseudogene) or cDNA sequences containing rearranged VDJ in the GenBank database. Similarly, only some light chain germline V region sequences are included. Many of the C_L cDNA sequences listed also have associated rearranged V_L sequences. In addition to the GenBank, the '*Kabat Database of Sequences of Proteins of Immunological Interest*' (http://immuno.bme.nwu.edu/) is another valuable resource available on the world wide web. The rabbit Ig gene sequences are tabulated and updated frequently within this compendium.

Accessory molecule function

Information about accessory molecules on rabbit B lymphocytes is limited. Homologues of the Igα and Igβ components of the B-cell receptor complex have been discovered (Fitts *et al.*, 1995). CD5 is expressed on the majority of rabbit B lymphocytes (Raman and Knight, 1992). CD5 appears to bind to the V_H region of certain immunoglobulins with framework region specificity (Pospisil *et al.*, 1996). It may thus affect selective expansion of certain subsets of B cells (Pospisil *et al.*, 1995, 1996). Its role as a coreceptor has been suggested.

Table VI.7.2 Rabbit immunoglobulin genes: heavy chains

		GenBank locus name	*Accession number*	*Reference*
Heavy chain variable region (V_H)				
V_H1a1[a]	multiple amino	RABIGHVXA	M93171, J04864	Knight and Becker (1990)
V_H1a2	acid differences	RABIGHVXB	M93172, J04865	Knight and Becker (1990)
V_H1a3	in framework regions 1 and 3	RABIGHVXC	M93173, J04866	Knight and Becker (1990)
y33- x32-[b] z[b]		RABIGHVDBQ	M77083	Short *et al.* (1991)
Heavy chain constant regions (C_H)				
IgM[c]				
The cDNA sequences of one secreted μ and		RABIGHAB	J00666	Bernstein *et al.* (1984)
membrane *C*-terminus are available		RABIGHAC	K01357	Bernstein *et al.* (1984)
IgG				
'Hinge region' between C_H1 and C_H2 position 225				
d11 Met, d12 Thr				
C_H2 domain of IgG position 309				
e14 Thr, e15 Ala				
Haplotypes	d11e15	RABIGCMB	M16426	Martens *et al.* (1982)
	d12e14	RABIGHAD	K00752, M12187	Bernstein *et al.* (1983a)
	d12e15	RABIGCA	L29172, N00008	Martens *et al.* (1984)
IgA[c]				
One cDNA of g75 type		OCIG02	X00353	Knight *et al.* (1984)
13 germline genes (α13 is g75) α5 and α6 are f72 all others f71		OCCALPHA1–OCCALPHA13	X51647, X82108–X82119	Burnett *et al.* (1989)

[a] Most allelic forms were originally defined serologically. The a-locus alleles found in domestic and laboratory rabbits correspond to the genes mapping most 3′ (closest to the D_H and J_H regions) (Knight and Becker, 1990). In most rabbit B lymphocytes this first V_H1 gene is found rearranged and expressed (Becker *et al.*, 1990; Allegrucci *et al.*, 1991). The locus contains 100–200 genes. Although some other V_H genes rearrange, they may primarily function as donors for 'gene conversion' of the rearranged V_H1 sequence (Becker and Knight, 1990; Weinstein *et al.*, 1994). In normal rabbits 'a-negative' Igs occur in 10 to 30% of the total IgG or Ig-bearing B cells. Their expression is elevated in mutant Alicia rabbits (Kelus and Weiss, 1986) and in allotype-suppressed rabbits (Short *et al.*, 1991) (see section 16).
[b] The germline x32 and z gene sequences are not known but several recurring cDNA sequences are probably close to the germline sequence.
[c] Serologically defined types and haplotypes are shown in Table VI.7.3. At least 12 of the 13 C_α genes are expressed *in vitro* (Schneiderman *et al.*, 1989) and 10 *in vivo* (Spieker-Polet *et al.*, 1993).

Ontogeny of the Ig repertoire in rabbits

The newborn rabbit is born with passively acquired maternal protective immunity. Using immunoglobulin allotypic markers, it has been possible to track the disappearance of the maternal immunoglobulins and also to detect the appearance of immunoglobulins synthesized by the young rabbit that could only have been inherited from their sires (reviewed by Mage, 1967).

Repertoire generation

A current working model of V_H repertoire development in the rabbit is that in the fetal and neonatal period, $V_HD_HJ_H$ and V_LJ_L rearrangements occur in bone marrow, fetal liver, and perhaps omentum, but there is limited receptor diversity because of rearrangement and utilization of mainly a single V_H gene – V_H1 that carries the V_Ha allotypic sequences (see Table VI.7.2). Cells from these sources seed the gut-associated lymphoid tissues. In particular, appendix follicles form, driven by gut antigens and perhaps superantigens; there is B-cell expansion. By 6 weeks of age the appendix is highly cellular and there is development of the primary high copy number repertoire, V_H gene diversification by gene conversion and somatic mutation. Positive and negative selective events occur and cells that survive the selection exit to the periphery to participate in further encounters with foreign antigens.

Species specific aspects of Ig structure

The rabbit (and probably other Lagomorphs) have a duplication of the entire C_κ locus (Benammar and Caze-

Table VI.7.3 Rabbit immunoglobulin genes: heavy chain haplotypes

	V_H			μ		γ		α	
Designation	*a*	*x*	*y*	*ms*[b]	*n*[b]	*d*	*e*	*f*	*g*
A	1	—	—	16	(81)	12	15	73	74
B	1	—	33, 30	17	(80)	12	15	71	75
C	1	—	33, 30	17	(80)	11	15	72	74
I	1	—	33, 30	17	(80)	12	14	69	77
J	1	—	—	16	(81)	12	15	70	76
R2M [I(F-C)][a]	1	—	33, 30	17	(80)	11	15	72	74
R6K[a1F-(F-I)][a]	1			17	(80)	12	14	69	77
a1FK	1			17	(80)	12	15	71	75
R3K F-C[a]	2	32	33	17	(80)	11	15	72	74
E	2	32	33, —	17	(80)	12	15	71	75
F	2	32	33, —	17	(80)	12	15	69	77
R1M F-I[a]	2	32	33, —	17	(80)	12	14	69	77
Ali F-I	(2)	32	33, —	17	(80)	12	14	69	77
M	2	32	33, —	16	(81)	12	15	73	74
G	3	32	—	17	(80)	12	15	71	75
H	3	32	—	16	(81)	11	15	72	74
R4K H-I[a]	3	32	—	17	(80)	12	14	69	77
R7K [H(*ali* F-I)][a]	3	32	—	16	(81)	12	14	69	77
R8K [a3F-(F-I)][a]	3			17	(80)	12	14	69	77
a3FK	3			17	(80)	12	15	71	75

[a] These haplotypes were derived from recombinations that occurred during breeding of laboratory rabbits (Mage *et al.*, 1982, 1992; Kelus and Steinberg, 1991). The parental haplotypes from which the recombinants were derived are shown in brackets. The a1FK and a3FK are the parental haplotypes from which R6K and R8K were derived, respectively.
[b] There are two parallel nomenclatures for μ alleles, ms16/17 and m80/81.

Table VI.7.4 Rabbit immunoglobulin genes: light chains

Comments	*Allele*	*GenBank cDNA*	*Genomic DNA*	*References*
Light chain variable regions				
V_κ				
Only a few germline V_κ and V_λ sequences are available				
V_κ 18a			OCIG06 (X00977)	Heidmann and Rougeon (1984)
V_κ 18b			OCIG07 (X02336)	Heidmann and Rougeon (1984)
V_κ 19a			OCIG08 (X02337)	Heidmann and Rougeon (1984)
V_κ 19b			OCIG09 (X02338)	Heidmann and Rougeon (1984)
V_κ 20			RABIGKVA (K02131)	Leiberman *et al.* (1984)
V_λ			RABIGLFUNA (M27840)	Hayzer and Jaton (1989)
			RABIGLFUNB (M27841)	Hayzer and Jaton (1989)
C_κ1 Four alleles in domestic and laboratory rabbits				
Originally defined serologically, the alleles have multiple amino acid sequence differences. b[bas] has a mutation of K1b9	b4	RABIGKAA (K10358, J00667)	RABIGKCA (K01360) RABIGKCC2 (K01362)	Heidmann *et al.* (1981), Dreher *et al.* (1983), Emorine and Max (1983), Emorine *et al.* (1983), Heidmann and Rougeon (1983)
	b5	RABIGKAB (K00751)	RABIGKCD (K01363, M29144)	Bernstein *et al.* (1983b), Pavirani *et al.* (1983), Emorine *et al.* (1984), Esworthy and Max (1986)
	b6	OCIGK1B6 (M37809)	OCIGK1B6 (M37809, M29583)	Dreher *et al.* (1990)
	b9	RABIGKAC (K01359)	OCIGK (X00674)	Akimeno *et al.* (1984), McCartney-Francis *et al.* (1984)
	b[bas]		OCIGKM (X03050)	Lamoyi and Mage (1985)

(*continued*)

Table VI.7.4 *Continued*

Comments	*Allele*	*cDNA*	*Genomic DNA*	*References*
$C_\kappa 2$, two alleles known				
In mutant Basilea and wild-type b9k rabbits-Leu is at C_κ position 204	bas1	RABIGKAD (K01361)		Bernstein *et al.* (1984)
Found in most laboratory strains – Pro at position 204	bas2		OCIG01 (V00885)	Heidmann and Rougeon (1983)
			OCIGK2B6 (X05800)	Mariame *et al.* (1987)
			OCIGK2B9 (X05801)	Mariame *et al.* (1987)
$J_\kappa 1$				
There are different J_κ genes associated with each b allotype	b4		RABIGKCC1 (K01361)	Heidmann and Rougeon (1983)
	b5		RABIGKCD (K01363, M29144)	Esworthy and Max (1986)
A distinct set of J_κ genes are associated with the $C_\kappa 2$ alleles	b6		RABIGK1B6 (M37809, M29583)	Dreher *et al.* (1990)
	b9		RABIGKC01 (M14602–M14067)	Akimeno *et al.* (1986)
$J_\kappa 2$				
A distinct set of J_κ genes are associated with the $C_\kappa 2$ gene.			OCIG05 (X00232)	Emorine and Max (1983)
C_λ				
Several isotypic forms – gene organization may differ in different rabbit strains.	(c7)	OCIGLC7 (X57729, X57843)		Hayzer *et al.* (1990)
The serologically detectable types c7 and c21 appear to be products of linked C_λ genes rather than truly allelic forms	(c21)	RAGIGLAAA (D00091, M15807, N00091)	RABIGLCA (M12388)	Duvoisin *et al.* (1986), Hayzer and Jaton (1987)
		RABIGVLCLF (M17645, M25622)	RABIGLCB (M12761)	Duvoisin *et al.* (1984), Duvoisin *et al.* (1986), Hayzer *et al.* (1987)
			RABIGLCC (M12762)	Duvoisin *et al.* (1986)
			RABIGLCD (M12763)	Duvoisin *et al.* (1986)

nave, 1982; Heidmann and Rougeon, 1983b). The second C_κ gene, $\kappa 2$ maps about 1 Mb from $C_\kappa 1$ (Hole *et al.*, 1991) and is normally expressed at very low levels. However, in a mutant rabbit strain, Basilea (Kelus and Weiss, 1977), there is a defect in the mRNA splice acceptor site required for splicing of J_κ to C_κ mRNA to form mature light-chain message (Lamoyi and Mage, 1985). In these mutants there is elevated expression of the $\kappa 2$ light chains in addition to elevated production of λ-type light chains (see also Section 16).

An unusual feature of the $\kappa 1$ light chains of the rabbit (and probably other lagomorphs), is an extra disulphide bond that is not found in $\kappa 2$ or in κ light chains of other species such as mouse, rat and man. The disulphide bond forms between the V_L and C_L domains of the light chains and may stabilize their structure. Interestingly, one of the four major allotypes, b9, lacks the Cys found at position 80 in V_L sequences of the other allotypes and has instead a Cys in the J_H region leading to another form of disulphide bonding (McCartney-Francis *et al.*, 1983). The evolutionary and functional significance of these structural differences has been discussed (McCartney-Francis *et al.*, 1983; Mage *et al.*, 1987).

8. MHC Antigens

The multiplicity of histocompatibility genes in rabbit, as examined through skin graft exchanges among F_2 and F_3 generations of the cross between the B and Y inbred rabbit lines was estimated at 19 unlinked loci (Chai, 1974). Among these, the major histocompatibility complex system, RLA, has been characterized at the level of proteins and genes (Table VI.8.1); the minor loci remain undetermined.

Serologically defined RLA-A antigens were originally identified by determining the effect of matching of leukocyte alloantigens on graft survival (Tissot and Cohen, 1972). A series of leukocytotoxic antisera obtained from rabbits receiving skin grafts from donors within or between different inbreeding lines defined up to 13 distinct RLA-A alleles (Table VI.8.2). However, as the antisera are no longer available, typing of the MHC class I region in rabbit is now superseded by restriction fragment length polymorphism (RFLP) analysis (Table VI.8.2 and VI.8.3).

The RLA-D locus was first defined as a genetic locus controlling *in vitro* mixed lymphocyte reactivity (MLR) (Tissot and Cohen, 1974), it also affected the fate of

Table VI.8.1 Classification of RLA genes

A. Class I RLA genes	8–12 heavy chain genes classified into 4 subfamilies. Clone 19–1 and the corresponding pR9 cDNA clone represent the first subfamily which is most abundantly expressed.[a] The cDNA clones, pR11 and pR27, are derived from the second and third subfamilies, respectively.[b]
B. Class II RLA genes	RLA-DP subregion: 2α, 1β: DPα2-DPα1-DPβ[c]
	RLA-DQ subregion: 1α, 2β: DOβ-(DQα-DQβ)[d]
	RLA-DR subregion: 1α, 5β: DRα-DRβ1, DRβ2-DRβ3-DRβ4,[e] DRβ5
Unclassified	2α: DFα,[f] HLA-DNA homolog[g]

[a] Marche *et al.* (1985).
[b] Rebierre *et al.* (1987b).
[c] There is evidence for an additional RLA-DPβ-like gene located outside the RLA-DP cluster but it has not been cloned (Sittisombut *et al.*, 1989).
[d] The RLA-DQα and RLA-DQβ genes are closely linked (LeGuern *et al.*, 1985, 1987; Sittisombut and Knight, 1986). The RLA-DOβ gene is located on the same 150 kb NotI-MluI fragment as the RLA-DQα gene in the RLA-D1 haplotype; the distance has not yet been determined (Chouchane and Kindt, 1992; Chouchane *et al.*, 1993).
[e] The RLA-DRβ3 and RLA-DRβ4 genes are more homologous to each other than to the rest of RLA-DRβ genes (Sittisombut *et al.*, 1989).
[f] The RLA-DFα gene is a pseudogene containing only exon 4 coding sequence but has been detected in all rabbits tested and in hare and cottontail (Marche *et al.*, 1991).
[g] Previously designated as the RLA-DPα2 gene but is unlinked to other RLA-DP genes (LeGuern *et al.*, 1985); may represent HLA-DNA homolog (T. J. Kindt, personal communication).

allogeneic skin grafts because compatibility at both of the RLA-A and RLA-D loci resulted in the longest graft survival time (Cohen and Tissot, 1974). Six distinct RLA-D alleles have been identified by one-way MLR assay but the assay is also replaced by RFLP analysis (Tables VI.8.2 and VI.8.4). The RLA-A and RLA-D loci are closely linked to the He blood group locus in rabbit linkage group VII (Tissot and Cohen, 1974).

Antirabbit β_2-microglobulin has been used to purify and characterize β_2-microglobulin-associated molecules from solubilized membrane glycoproteins of RL-5, a T-lymphocyte cell line derived from an inbred B/J rabbit (Kimball *et al.*, 1979). Following dissociation of β_2-microglobulin with acid, analysis of the 43 kDa protein peak revealed a major protein with high level of sequence identity with HLA-B7 and H-2K^b molecules. Corresponding mRNAs and their relative expression in RL-5 cells were derived by screening a cDNA library with the H-2L^d probe (Tykocinski *et al.*, 1984). Restriction enzyme analysis of randomly picked clones revealed a major group of abundantly expressed

Table VI.8.2 Assignment of rabbit MHC haplotypes

	RLA-A haplotype		*RLA-D haplotype*		
RLA haplotype	*SD*[a]	*RFLP*[b]	*MLC*[c]	*RFLP*[b]	*Rabbit strain*[d]
1	A1	A1	D1	D1	III/Dw, A
2	A2	A2	D2	D2	B
3	A3	A3	D3c	D3c	C
4	A4	nd	D2	nd	na
5	A5	A5	D4	D4	E
6	A6	A6	D3f	D3a	F
7	A7	A7	D4	D7	na
8	A8	A8	D3	D6	na
9	A9	nd	D3	nd	na
10	A10	A7	D2	D2	na
11	A11	A11	D3l	D3a	B/J
12	A12	nd	D5	D5	na
13	A13	A13	D4	D8	na
14	nd	nd	D10	D10	III/J
15	nd	A15	nd	D15	na
16	nd	A16	nd	D2	na
17	nd	A1	nd	D17	na

[a] Alleles defined by serological typing.
[b] Haplotypes defined by RFLP analysis.
[c] Alleles defined by mixed lymphocyte culture test.
[d] III/Dw, B/J and III/J are inbred strains, the others (A, B, C, E and F) are RLA homozygous rabbits maintained by selective breeding; nd, not done; na, not available. Reprinted from Sittisombut (1996), with permission. Copyright CRC Press, Boca Raton, Florida.

Table VI.8.3 List of RLA-A haplotypes as defined by RFLP analysis[a]

	Probe (restriction enzyme)[b]							
RLA-A haplotype	*RLA-A exon 4 (19-1-3′) (BamHI)*							*pR27-3′ (BglII)*
A1	18[c]	11	8.5	3.7				5.6
A2	18	6.1	4.7	3.7				5.6
A3	18	11	6.1	3.7				7.2
A5	18	11	8.5[d]	6.7	6.1	4.7[d]	3.7	7.2
A6	18	11	10	6.7	6.1	4.7	3.7	5.6
A7	18	14	6.7	3.7				5.6
A8	25	18	11	6.1	3.7			7.2
A11	18	11	10	6.7	6.1	4.7	3.7	3
A13	18	11	8.5	4.7	3.7			4.4
A15	25	18	11	6.1	4.7			10
(A16)	18	14	11[e]	6.7	6.1[e]	3.7		5.6

[a] Adapted from Marche *et al.* (1989) with permission.
[b] Probes are listed in Table VI.8.5.
[c] Size of restriction fragment in kb.
[d] May be doublets.
[e] The status of haplotype 16 with respect to these fragments is not yet clear.
Reprinted from Sittisombut (1996), with permission. Copyright CRC Press, Boca Raton, Florida.

cDNA clones and three other groups of rare and structurally atypical clones. These mRNA species are products of approximately 8–12 class I genes in the rabbit genome (Kindt and Singer, 1987). Differential hybridization with multiple cDNA and oligonucleotide probes allows subdivision of these class I RLA genes into four subfamilies (Kindt and Singer, 1987; Rebiere *et al.*, 1987b). Allotypic variation of rabbit class I genes and corresponding proteins is not yet known. However, RFLP patterns of all class I genes in individual rabbits correlate well with serologically defined RLA-A alleles. With *Bam*HI digestion, distinct patterns of bands hybridized to a class I exon 4 probe correspond to distinguishable RLA-A alleles in nine out of ten haplotypes tested (Tables VI.8.2 and VI.8.3) (Marche *et al.*, 1989).

The heterodimeric nature of RLA-D antigens and the

Table VI.8.4 List of RLA-D haplotypes as defined by RFLP analysis

	Locus (restriction enzyme)						
RLA-D haplotype	*DPα1 (BamHI)*	*DPα2 (HindIII)*	*DPβ (EcoRI)*	*DQα (PvuII/EcoRI)*	*DQβ (HindIII)*	*DRα (EcoRI)*	*DRβ[a] (BamHI)*
D1	14[b]	8.3	[3.8, 8.5][cd]	8/2.8	3.3	6	2.4, 5.2, 7.4, 9.2, 27.5
D2	14	8.3	[3.8, 8.5]	1.8/6.3	[5.5, 6.1]	5.4	2.4, 5.2, 7.2, 10.5, 12, 27.5
D3a	14	8.3	[2.7, 8.5]	5.9/3.7	[2.8, 8.4]	4.8	2.4, 4.5, 7.2, 10.5, 12, 27.5
D3c	24[e]	7.6	[3.8, 8.5]	5.9/3.7	[2.8, 8.4]	4.8	2.4, 4.5, 7.2, 10.5, 12, 27.5
D4	14	nd	nd	3/9.6	nd	6	nd
D5	24[e]	8.3	[3.8, 8.5]	nd/7.8	16.5	5.6	2.4, 5.9, 7.2, 12, 27.5
D6	14	nd	nd	6.3/2.8	nd	4.8	nd
D7	24	nd	nd	4.7/6.3	nd	5.4	nd
D8	14	nd	nd	6.3/6.3	nd	6	nd
D10	14	8.3	[3.8, 8.5]	nd/3.7	[2.8, 8.4][f]	4.8	2.4, 4.5, 7.3, 12, 18.5
D15	4	nd	nd	3/3.3	nd	6	nd
D17	14	nd	nd	3.7/3.7	nd	6	nd

[a] Pattern represents all five RLA-DRβ genes.
[b] Size of DNA fragment in kb.
[c] [ ] Represents cleavage within an allelic form.
[d] Additional hybridizing band due to another DPβ-like gene is observed (Sittisombut *et al.*, 1989).
[e] Reported as the >27.5 kb band by Sittisombut *et al.* (1989).
[f] Reported as the [2.9, 8.4] kb fragments by Sittisombut *et al.* (1989), but are likely to be identical to the [2.8, 8.4] fragments of D3 haplotypes.
Reprinted from Sittisombut (1996), with permission. Copyright CRC Press, Boca Raton, Florida.

linkage of corresponding genes with the RLA-D locus defined previously by MLR and classical genetic analysis were established by using polyclonal antisera directed against nonimmunoglobulin, polymorphic surface antigens of rabbit lymphoid cells (Knight *et al.*, 1980). A monoclonal antibody (2C4), which strongly inhibited the MLR and proliferative response to soluble antigens (Lobel and Knight, 1984), precipitated from metabolically labeled spleen cell lysate multiple α-chain and β-chain spots in two-dimensional nonequilibrium pH gradient electrophoretic gel (Knight *et al.*, 1987). Most spots were homologous to HLA-DQα or -DQβ chains at the *N*-termini whereas an α-chain spot resembled the HLA-DRα chain (Knight *et al.*, 1987). Following the cloning of the RLA-DRα gene and several DRβ genes (LeGuern *et al.*, 1985; Sittisombut and Knight, 1986; Laverriere *et al.*, 1989; Sittisombut *et al.*, 1989), an additional monoclone (RDR34) was raised by immunization with transfected cells expressing products of the RLA-DRα and RLA-DRβ2 genes (Spieker-Polet *et al.*, 1990).

Cross hybridization with human and murine class II probes identified homologues of the HLA-DQ, -DP, -DR, -DO, and -DF genes in the rabbit genome (Table VI.8.1). As shown by pulsed-field gel analysis, these genes are located within approximately 630 and 1200 kb of DNA, respectively, in the RLA-1 and -2 haplotypes in the following order: DP-DOβ-DQ-DR (Chouchane and Kindt, 1992). The RLA-A and RLA-D regions are closely linked in a few rabbits examined (Chouchane and Kindt, 1992). Comparatively higher level of allotypic variation of RFLP pattern was detected in the RLA-DQα gene (Table VI.8.4). As a group, the RLA-DRβ genes are quite polymorphic; distinct RFLP patterns of these genes also correlate with the RLA-D typing data in a limited set of animals tested (Table VI.8.4).

Based on the studies of inbred rabbits and rabbits carrying known serologically defined and MLR-defined haplotypes, RFLP analysis is sufficiently sensitive to distinguish 11 RLA-A alleles and 12 RLA-D alleles with available DNA probes (Table VI.8.5) (Marche *et al.*, 1989; Sittisombut *et al.*, 1989). Additional RLA class I haplotypes are likely to exist in outbred colonies (Boyer *et al.*, 1995) and

Table VI.8.5 List of probes and alleles/patterns detected by RFLP analysis

				RFLP alleles/patterns	
Locus	*Probe*[a]	*Region*	*Enzyme*	*Number of alleles/ haplotypes tested*	*Alleles/patterns (kb)*
Class I					
R9	19-1-3′ (g)	Exons 6-8 (0.5 kb)[b]	*Bam*HI	10/10	10 patterns of multiple fragments (see Table VI.8.3)
R27	R27-3′ (c)	3′ UT (0.35 kb)	*Bgl*II	5/10	3, 4.4, 5.6, 7.2, 10
Class II					
DPα1	DPα,DPα1 (g)	Exons 2, 3[c] (0.6 kb)	*Bam*HI	3/12	4, 14, 24[d]
DPα2	DPα2 (g)	Exons 3, 4 (1.45 kb)	*Hind*III	2/6	7.6, 8.3
DPβ	DPβ (g)	Exon 3 (2 kb)	*Eco*RI	2/6	[2.7, 8.5][e], [3.8, 8.5][e]
DQα	DQα (g)	Exons 3, 4[c] (0.5 kb)	*Pvu*II	6/10	1.8, 3, 4.7, 5.9, 6.3, 8
			*Eco*RI	6/12	2.8, 3.3, 3.7, 6.3, 7.8, 9.6
DQβ	DQβ (g)	Exon 3 (2 kb)	*Hind*III	4/6	[2.8, 8.4][e], 3.3, [5.5, 6.1][e], 16.5
DRα	DRα (g)	Exons 3, 4[c] (0.8 kb)	*Eco*RI	4/12	2.4[f], 4.8, 5.4, 5.6, 6
DRβ	7.1-3′ (c)	Exon 3 and other 3′ sequences (0.7 kb)	*Bam*HI	5/6	Five patterns of multiple fragments (see Table VI.8.4)
DFα	DN,RDF (g)	Intron 4 and exon 4 (0.5 kb)	*Bgl*II	4/10	2.7, 4.7, 8, 10

[a]Probes were derived from either genomic clone (g) or cDNA clone (c).
[b]Size of probe in kb.
[c]Multiple similar probes have been employed but only the smallest probe is shown.
[d]Reported as the $>$27.5 kb band by Sittisombut *et al.* (1989).
[e]Represents cleavage within an allelic form.
[f]Present in an outbred rabbit of unknown haplotype.

wild rabbits. However, many of the putatively novel RLA class I and class II RFLP alleles/haplotypes detected in New Zealand White rabbits (Boyer *et al.*, 1995) most likely reflect heterozygosity among outbred animals.

Expression and tissue distribution

In the virally transformed RL-5 T cell line, at least four class I genes are transcriptionally active. Various mRNA species, 1.4, 1.9, 2.8 and 3.9 kb in size, are detected with a conserved exon 4 probe (Tykocinski *et al.*, 1984; Rebierre *et al.*, 1987a). However, protein products of only one gene, 19-1, are demonstrated by precipitation with anti-β_2-microglobulin and two anti-class I monoclones (Kimball *et al.*, 1979; Wilkinson *et al.*, 1982).

By employing the same class I exon 4 probe, the 1.4 kb and 2.8 kb mRNA species are detected in thymus whereas only the 1.4 kb form is present in lymph node and liver (Rebierre *et al.*, 1987a). Transcripts of the unusual gene corresponding to pR27 are found in thymus but not in brain, testis, lymph node, liver or muscle (Rebierre *et al.*, 1987a).

Relatively high levels of the mRNA transcripts of RLA-DQα, RLA-DQβ, RLA-DRα and RLA-DRβ genes are present in rabbit spleen; the RLA-DPα1, RLA-DPα2 and RLA-DPβ mRNA are much less abundant (Kulaga *et al.*, 1987; Spieker-Polet *et al.*, 1990). These class II mRNA are also detected in bone marrow, appendix, lymph nodes and thymus (Kulaga *et al.*, 1987). The RLA-DOβ mRNA are found at low levels in appendix, spleen and lymph nodes (Kulaga *et al.*, 1987).

Among the class II RLA proteins, the RLA-DQα and RLA-DQβ chains are readily identifiable from rabbit spleen by 2C4 monoclone (Knight *et al.*, 1987). They are also induced in RL-5 cells productively infected with HTLV-1 (Zhao *et al.*, 1995). Expressibility of the RLA-DR and RLA-DP genes has been examined by transfecting corresponding genes into the A20 B-lymphoma cell line. With the RLA-DRα and RLA-DRβ2 genes, stably transfected cells are recognized by the monoclonal antibody RDR34, which stained about 50% of rabbit spleen cells similar to 2C4 (Spieker-Polet *et al.*, 1990). Although other RLA-DRβ genes have not yet been tested by transfection, transcripts of the RLA-DRβ1 and RLA-DRβ3 or RLA-DRβ4 genes can be found in rabbit spleen (Spieker-Polet *et al.*, 1990) and cDNA clone derived from the -DRβ4 gene is isolated (Sittisombut, 1987). Cell surface expression of the RLA-DP proteins has not yet been observed. In the RLA-D10 haplotype, this may be due to unfavorable nucleotides surrounding the first AUG codon of the RLA-DPβ mRNA (Sittisombut *et al.*, 1988).

Disease association

Regression and malignant conversion of virus-induced papilloma are associated with specific alleles of RLA genes (Han *et al.*, 1992). Shope cottontail rabbit papillomavirus infection in rabbits results in skin warts which may regress spontaneously or progress into malignant lesion. Regression occurs in 10–40% of papilloma-bearing domestic rabbits whereas epidermoid carcinoma develops in as high as 73% (Syverton, 1952; Kreider and Bartlett, 1981). These changes are associated with contrasting histological features and systemic immunologic responses (Okabayashi *et al.*, 1991, 1993; Lin *et al.*, 1993; Selvakumar *et al.*, 1994; Hagari *et al.*, 1995) which may reflect differential expression of viral genes (Zeltner *et al.*, 1994). In New Zealand White rabbits, significant association between early papilloma regression and the 5.6 kb (*Eco*RI) RLA-DRα allele and the 3.0 kb (*Pvu*II) RLA-DQα allele is evident (Han *et al.*, 1992). In contrast, malignant conversion is associated with the 1.8 kb (*Pvu*II) RLA-DQα allele. The association of RLA system with papilloma progression is parallel to the linkage between HLA-DQw3 and carcinoma of the cervix in human (Wank and Thomssen, 1991; Helland *et al.*, 1992; Apple *et al.*, 1994; Gregoire *et al.*, 1994).

Species-specific aspect

The second exon of the RLD-DPβ gene in the RLA-D10 haplotype contains a complex deletion/insertion mutation (Sittisombut *et al.*, 1988) which is absent from known HLA-DPβ alleles (Marsh and Bodmer, 1995) but resembles the ones found at similar position of the H-2Aβ1 gene of five mouse strains (Estess *et al.*, 1986; Acha-Orbea and McDewitt, 1987). This mutation is assumed to occur by a gene conversion-like event after speciation (Sittisombut *et al.*, 1988), but it is possible, according to the trans-species hypothesis of MHC evolution (Klein *et al.*, 1993), that such mutations occur in a member of the pool of ancestral class II β genes which existed even before mammalian diversification. The RLA-DPβ allele and the H-2Aβ1 alleles bearing a similar mutation descend from the mutant allele whereas the HLA-DPβ alleles derive from another β allele in the same pool.

9. Rabbit Blood Groups

The rabbit blood group system was first described by Levine and Landsteiner (1929) and the genetics of the system were first examined by Castle and Keeler (1933). From the mid-1950s to the late 1970s, numerous papers describing the serology, biochemistry and genetics of rabbit blood groups appeared authored by Carl Cohen and Robert G. Tissot. Drawing from this large body of work, Cohen (1982) summarized the cellular antigen systems, including the red cell antigens, of the rabbit.

The rabbit blood group system is summarized in Table VI.9.1. To date, all markers have been identified serologically in direct hemagglutination (HA) assays using trypsin-treated target red cells.

Table VI.9.1 Rabbit blood groups

Locus[a]	Alleles	Number of phenotypes	Detection method
Hg	Hg^A, Hg^D, Hg^F	6	Direct HA
Hb	Hb^B, Hb^M	3	Direct HA
Hc	Hc^C, Hc^L	3	Direct HA
He	*He, he*	2	Direct HA
Hh	*Hh, hh*	2	Direct HA
Hq	Hq^Q, Hq^{QS}, Hq^S	3	Direct HA
Hu	Hu^U, Hu^Y	3	Direct HA

[a] Modified from Cohen (1982).

The Hq blood group system was originally thought to consist of two alleles, Hq^Q and Hq^S, but later studies (R. G. Tissot, unpublished data) showed that Q and S can also be found in the same haplotype. No rabbits have been found lacking both Q and S.

The Hg blood group system in the rabbit was the most thoroughly studied of the seven rabbit blood groups. The Hg system has strong biochemical and genetic resemblances to the human Rh system. Twelve antisera were produced defining red cell antigens coded at the *Hg* locus. Five antisera were produced which recognized epitopes (I, J, T, V, and W), which presumably resulted from interactions between products of the Hg^A, Hg^D and/or Hg^F alleles. Three antisera recognized epitopes (K, P, and R) shared between products of the Hg^A, Hg^D and/or Hg^F alleles. The Hg blood group system is summarized in Table VI.9.2. One antiserum, defining blood group N, was produced with specificity for antigens on the red blood cells of one litter of inbred rabbits which should have been blood group Hg^D, based on their parentage, but whose red cells failed to react with standard anti-Hg^D typing serum. Unfortunately, this litter of rabbits failed to reproduce and the unique N blood group was never identified again. Thus, the unique twelfth Hg blood group N is not included in Table VI.9.2.

Table VI.9.3 summarizes the known combinations of epitopes recognized by the five antisera which detect the ‘associative’ or ‘interactive’ epitopes which can be detected only when the appropriate combinations of epitopes are present on the red cell.

In summary, seven rabbit blood groups are known. The most widely studied of these is coded at the *Hg* locus and consists of three codominant autosomal allelic genes, Hg^A, Hg^D and Hg^F. These code for six major phenotypes (A/A, A/D, A/F, D/F, D/D, and F/F). The antigens described by Cohen (1982) using the anti-I, -J, -T, -V, and -W antisera appear to be epitopes formed by associations of the products of the Hg^A, Hg^D and/or Hg^F alleles. Epitopes defined by the anti-K, -P, and -R antisera are shared between products of the Hg^A, Hg^D and/or Hg^F alleles.

Table VI.9.2 The rabbit Hg blood group

Genotype	Reactions with typing sera: anti- A	D	F	K	P	R
Hg^A/Hg^A	+	–	–	–	+	+
Hg^D/Hg^D	–	+	–	+	–	+
Hg^F/Hg^F	–	–	+	+	+	–
Hg^A/Hg^F	+	–	+	+	+	+
Hg^D/Hg^F	–	+	+	+	+	+
Hg^A/Hg^D	+	+	–	+	+	+

Table VI.9.3 Associative epitope summary

Antiserum	Epitope combinations detected
Anti-I	A/D
Anti-J	K/R
Anti-T	A/K
Anti-V	P/D
Anti-W	A/A

Rabbit blood groups remain of interest for the same reason that they originally were studied: the blood groups serve as the best and most convenient small-animal model for human blood group studies, particularly for the human Rh blood group with which the rabbit Hg blood group appears to be phylogenetically analogous. Recent studies to recreate and to refine the rabbit model of hemolytic disease of the fetus/newborn based on the Hg blood groups have been successful (Moise *et al.*, 1992, 1994, 1995a, b) and homozygous breeding stock have been reestablished. The rabbit Hg blood group recently provided the model leading to successful isolation of human Rh D antigen in a soluble, serologically active form (Yared *et al.*, in press).

10. Passive Transfer of Immunity

An excellent review of the anatomy and development of fetal membranes, uterine development and implantation structure of the placenta, yolk sac, chorion, amnion and allantois, and transmission to fetal circulation can be found in Brambell (1970).

11. Nonspecific Immunity

Defensins, antimicrobial and cytotoxic peptides produced by mammalian cells have been studied in rabbit models of several infectious diseases (reviewed by Leherer *et al.*, 1993).

12. Complement

The information available on rabbit complement is summarized in Tables VI.12.1 and VI.12.2.

Table VI.12.1 Complement components in the rabbit

Pathway				
Symbol	*Name*	*Function/characteristics*	*Molecular mass (kDa)*	*Genes*
Classical pathway				
C4	C4	A component of the C3 and C5 convertase of the classical pathway. Combines with C2/Thiolester-containing structural protein. Single binding specificity (human C4B-like). Incompatibility between rabbit C4 and human C2/guinea pig C2 (von Zabern, 1988)	α92, β78, γ27 (Dodds and Law, 1990); α90, β75, γ32 (Takata *et al.*, 1985)	
C2	C2	Cleaves C3 and C5/serine protease. Incompatibility between rabbit C2 and human C4/guinea pig C4 (von Zabern, 1988)		Upstream sequence element (USE) for the poly A site (PCR amplification) (Moreira *et al.*, 1995)
Alternative pathway				
C3	C3	Attaches covalently after proteolytic activation generates metastable C3b. C3b fragment is part of C3/C5 convertase. Combines with C5/ Thiolester-containing structural protein. Half life is 29 h. Functional catabolic rate is 4.3%/h (Peake *et al.*, 1991) Functional catabolic rate is 2.7 $\pm$ 0.3%/h. Synthesis rate is 1.0 $\pm$ 0.2 mg C3/kg/h (Manthei *et al.*, 1984). C3α contains the binding sites for factor H, properdin, CR2, CR3 and the factor I cleavage sites. Rabbit C3 binds human factor H, CR1, CR2, and MCP (Becherer *et al.*, 1990). Rabbit C3 contains Con A-binding carbohydrates in both α- and β-chains. Rabbit C3 binds human factor H, CR1, and CR2 (Alsenz *et al.*, 1992). Total carbohydrate content is half that of human H. Compatibility between rabbit C3b and human B (Horstman and Müller-Eberhard, 1985).	171 (α123, β70) (Giclas *et al.*, 1981); 195 (Horstman and Müller-Eberhard, 1985); α116, β72 (Peake *et al.*, 1991); α115, β72 (Komatsu *et al.*, 1988); α119, β72 (Alsenz *et al.*, 1992) Serum C3 concentration: 880 μg/ml (range 610–1120) (Horstman and Müller-Eberhard, 1985) Plasma C3 concentration 1.23 $\pm$ 0.3 mg/ml (Manthei *et al.*, 1984)	C3α chain cDNA 2182 bp; 726 amino acid residues. This region shares 78% similarity with the human and mouse sequences with conservation of the cysteinyl residues (Kusano *et al.*, 1986)
B	Factor B	Binds to C3b forming the precursor of the C3/C5 convertase (C3b,Bb)/Bb subunit of this complex is a serine protease. Rabbit C3b,Bb enzyme resembles the human analogue with respect to half-life, control by Factor H, and stabilization by nickel ions. Compatibility between rabbit B and human D (Horstman and Müller-Eberhard, 1985)	85; Serum concentration is 89 μg/ml (range 68–108) (Horstman and Müller-Eberhard, 1985)	
H	Factor H	Inactivates the C3/C5 convertase by dissociating its subunits; also a cofactor for factor I/Serine protease. Total carbohydrate content is half that of the human H (Horstman and Müller-Eberhard, 1985)	155; Serum concentration is 128 μg/ml (range 83–185) (Horstman and Müller-Eberhard, 1985)	

(*continued*)

Table VI.12.1 *Continued*

Pathway				
Symbol	*Name*	*Function/characteristics*	*Molecular mass (kDa)*	*Genes*
Alternative pathway				
P	Properdin	Regulator (positive) that enhances activation by binding to and stabilizing the C3/C5 convertase. Structural protein. The amino terminal 36 residues have 78% homology to the equivalent region of human P (Nakano *et al.*, 1986)	58 (Nakano *et al.*, 1986)	
Effector pathway				
C5	C5	Anchoring molecule for C6/Structural protein, member of the thiolester family but lacking the thiolester.	171 (α 129, β 88) (Giclas *et al.*, 1981)	
C6	C6	Anchoring molecule for C7/Structural protein.	95 (Arroyave and Müller-Eberhard, 1971)	
C8	C8	Anchoring molecule for C9/Structural protein	175 (α–γ 114, β 61, γ 26) (Komatsu *et al.*, 1985) α 64 (555 amino acid residues) β 61 (536 amino acid residues) γ 26 (182 amino acid residues) All three subunits are strikingly similar to human with regard to length, molecular weight, *N*-, and *C*-terminal residues, conserved cysteines, and overall sequence (White *et al.*, 1994)	C8 α cDNA 2070 bp C8 β cDNA 2052 bp C8 γ cDNA 751 bp (White *et al.*, 1995)
C9	C9	Forms pores in the membrane/ Structural proteins	70 (Ganzler and Hansch, 1984)	C9 cDNA 2019 bp; 597 amino acid residues; 72% identity to human C9 (Hüsler *et al.*, 1995)
C4BP	C4-binding protein	Masking of C2-binding site on C4/ Structural proteins. No complex form (unlike human C4BP)		C4BP α cDNA 1794 bp; 597 amino acid residues; 63% identity to human C4BP; 50% identity to mouse C4BP (de Frutos and Dahlback, 1995)

13. Mucosal Immunity

Transport of antigens and the function of antigen-presenting cells in mucosal tissues

Transcytosis of antigen appears to be a prerequisite for mounting an effective mucosal immune response. In rabbit gut-associated lymphoid tissue (GALT), antigen is transported from the lumen of the intestine into organized lymphoid tissue by specialized follicle associated epithelium (FAE) cells known as M (membranous or microfold) cells. M cells are characterized by several features including: a pocket on their basolateral surface within which reside lymphocytes and accessory cells, lack of poly-Ig receptors on basolateral plasma membranes, and a lack of the highly organized apical brush borders and stalked, membrane anchored, hydrolases found in absorptive enterocytes (Neutra and Kraehenbuhl, 1992). The rate of

Table VI.12.2 Genetic complement deficiency states and induced complement deficiency by CVF[a] in rabbits

	C6 deficiency	*C8 α–γ deficiency*	*C3 hypo-complementemia*	*Induced complement deficiency by CVF*
Hemolytic activity (CH50)	Absent	Absent	27–37% of normal	12–15% of normal
Bactericidal activity	Absent	Diminished	Diminished	—
Serum C3 level	Normal	Normal	10% of normal	Diminished
Other components	Normal	Normal	Normal	C6 1/3–1/2 of normal
C3 catabolism *in vivo*	—	Normal	Normal	—
C3 inhibitor	—	Absent	Absent	—
Antibody production	Normal	Normal (against BSA)	Normal (against BSA)	—
BCG skin reaction (Delayed-type hypersensitivity)	Diminished	Enhanced	Diminished	—
Clotting time	Diminished (prolonged)	—	—	Normal
Phagocytosis	Normal	—	—	Normal
Immuno adherence	Normal	—	—	1% of normal
Production of antinuclear antibody	—	Absent	Present	—
Other characteristics				
Immuno clearance	Normal	—	—	—
Active Arthus reaction	Normal	—	—	—
Passive Arthus reaction	Diminished	—	—	—
Serum sickness nephritis	Diminished	—	—	—
Membranous nephropathy	Diminished	—	—	—
Local Shwartzman reaction	Diminished	—	—	—
Endotoxin tolerance	Diminished	—	—	—
Skin allograft rejection	Diminished	—	—	—
Growth rate	Normal	Dwarf	Normal	—
Reproductive activity	Normal	Diminished	Normal	—
Survival rate	Normal	Diminished	Diminished	—
Molecular basis	—	Abnormal cotranslational processing of C8α. (A mutation in an exon–intron junction of C8α)	Pretranslational defect resulting from mutations within the C3 gene	
References	Hammer *et al.* (1981), Rother (1986)	Komatsu *et al.* (1985, 1990, 1991), Kaufman *et al.* (1991)	Komatsu *et al.* (1988, 1992)	Cochrane *et al.* (1970)

[a] CVF = cobra venom factor.

antigen uptake by M cells is partly dependent on the ability of antigens to adhere to their luminal surfaces (Pappo and Ermak, 1989). M cells express surface receptors with lectin-binding activity and for binding bacterial polysaccharides, in addition to receptors with specificity for secretory IgA (Pappo, 1989). Particulate antigens in the lumen that localize to M cell apical surfaces, are engulfed by coated and uncoated pits and are shuttled in vesicles to underlying antigen-presenting cells. Antigen is not processed while it transits across FAE because of the lack of lysosomes in M cells. M cells are responsible for the transport of more than 99% of the antigens that find their way into GALT follicles (Neutra and Kraehenbuhl, 1992) with transit rates averaging 2 μm/min. In studies using fluorescent-labeled polystyrene microspheres, uptake by rabbit Peyer's patch (PP) M cells was at least one order of magnitude greater than that observed for murine PP, with internalization of particles within 10 min of *in vivo* administration. The kinetics of antigen uptake by rabbit PP M cells suggests that antigen is taken up in a synchronous wave from the lumen of the intestinal tract (Pappo and Ermak, 1989; Pappo *et al.*, 1991). The M cell content of the FAE of rabbit GALT is approximately 50%, compared with the 10% detected in murine FAE. Rabbit M cells identified by staining with either vimentin or peanut agglutinin show localization to the periphery of the FAE, with few found at the apex of the dome (Jepson *et al.*, 1993).

Within M-cell pockets antigen-processing cells, such as monocytes or immature dendritic cells, ingest and degrade antigens transcytosed for use by Ia antigen-expressing cells. Processed antigen is then disseminated throughout the follicular and interfollicular regions by these cells for stimulation of the mucosal B- and T-cell immune responses.

Production and transport of mucosal Igs

Following stimulation of B cells in an organized mucosal lymphoid tissue with antigen taken in from the gut lumen, the expressed heavy-chain isotype changes from IgM or IgG to IgA (Weinstein and Cebra, 1991). Most of these mucosally derived IgAs have been shown to be directed against microbial surface components, which is consistent with the role of secretory IgA in recognition and aggregation of intact microorganisms on mucosal surfaces. Following the heavy-chain isotype switch and other maturation events, B cells leave and migrate to lamina propria anywhere in the body or back to an organized GALT and secrete IgA antibody. This ability of GALT B lymphoblasts to populate mucosal sites in the conjunctiva, upper respiratory tract, bronchi, mammary glands and gastrointestinal tract is regarded as 'the common mucosal immune system'. IgA antibody is the primary isotype produced at mucosal surfaces for secretion into the gastric mucosa. B cells secreting IgA are located in the Peyer's patch and laminal propria, but may also reside within the rabbit appendix and sacculus rotundus. IgA antibody produced by mucosal plasma cells is arranged tail to tail by oligomerization via the J chain. The IgA dimer then binds to the poly-Ig receptor-containing secretory component (SC) located on the basal membrane of epithelial cells lining the gut and is transported through the cell and secreted into the mucosa. In rabbits, unlike other mammalian species, the poly-Ig SC receptor may lack domains II and III, which prevents the covalent disulfide linkage formed between SC and IgA found in other mammalian species. SC of the rabbit binds the Fc portion of the IgA dimer. By binding to the IgA dimer, SC stabilizes the antibody molecules and masks the sites for protease cleavage located in the hinge region of IgA. The effect is to make the IgA dimer particularly well suited for survival in the protease-rich mucosal environment (Kraehenbuhl and Neutra, 1992).

Species-specific aspects of the structure or function of the mucosal immune system

The mucosal immune system of the lagomorph serves several different functions. As in other mammalian species, the mucosal immune system is responsible for the immune response to bacteria, viruses, protozoa, or other traumatizing agents that infect the mammal by way of the gut, respiratory, or genital tract mucosa. In both the young and adult rabbit these immune responses are mediated primarily by B and T cells that reside in organized lymphoid tissues at mucosal sites such as the Peyer's patch. However, the mucosal immune system of rabbits is not limited to protective functions; it also serves as a site for the generation of the rabbit's diverse antibody repertoire.

B-cell development in lagomorphs is distinct from that of other mammalian species. In species such as humans, mice and rats, combinatorial assortment of B cell antibody genes in the bone marrow contributes to antibody diversity. This random combinational assortment by joining of one of many distinct germline heavy-chain variable region (V_H), diversity (D_H), and joining (J_H) gene segments for the heavy chain; V_L and J_L for the light chain has been described as the generation of a high copy-number primary antibody repertoire. While B cells in the rabbit appear to rearrange their V_H and V_L genes in the bone marrow, the rearrangement process uses a limited number of V_H genes, with one V_H gene, V_H1, utilized by 80–95% of B cells with productive rearrangements (Allegrucci *et al.*, 1990; Becker and Knight, 1990; Becker *et al.*, 1990). This limited gene usage restricts the development of a diverse antibody repertoire with specificities for the limitless number of novel antigens found in the environment.

To cope with this predicament, the rabbit has developed a system similar to that found for chicken B-cell development. In chickens, with only one rearranging V_H and V_L gene, upstream V_H and V_L pseudogenes are used as DNA donors to diversify the rearranged variable region by gene conversion (Reynaud *et al.*, 1987, 1989, 1991). Some junctional diversity and somatic hypermutation also occur. This occurs early in life within a GALT found only in avian species known as the bursa of Fabricius. In lagomorphs, a process similar to that of the chicken begins by week 6 after birth and occurs within a GALT, the appendix. The young rabbit appendix bears a striking similarity in both gross morphological and lymphoid cellular distribution to the bursa of the chicken, while sharing few of the features associated with more conventional GALT such as PPs. Rearranged V_H genes from B cells of young rabbits undergo a somatic diversification process that includes both gene conversion-like and somatic hypermutation mechanisms (Weinstein *et al.*, 1994). It is not known whether the young rabbit appendix also plays a role in protective functions.

In the chicken, diversification in the bursa is a short-term affair and is complete within 12–15 weeks after birth, at which time the bursa begins to involute and no longer appears as an organized lymphoid tissue by adulthood (Reynolds, 1983). It has not been clearly defined what the timeline of diversification is in the rabbit. Unlike the chicken bursa, the rabbit appendix does not involute, but instead goes through morphological and cellular distribution changes that result in a GALT in the adult which looks more like a PP, while bearing little resemblance to the appendix of a young rabbit (Figure VI.2.3; P. D. Weinstein, unpublished data). The change in structure of the appendix may or may not be an indicator of changes in rabbit appendix function. These new functions may include antigen-specific B-cell responses within the germinal centers, leading to production of secretory IgA B cells. Further investigation will be needed to determine whether the adult rabbit appendix continues to be involved in V_H diversification.

Another rabbit GALT, the sacculus rotundus appears to have some of the same functions as the rabbit appendix. Included in these functions may be the diversification of the V_H gene repertoire. However, at present there is little data to support this assumption. Other GALT, including the PP are similar in structure and function to those of other species such as mice. The only differences appear to be the number of follicles found in each Peyer's patch and the number of patches found along the intestinal tract.

14. Immunodeficiencies

Mutant strains Basilea (*bas*) and Alicia (*ali*) were both discovered by Kelus and Weiss (1977, 1986).

ali-Mutant rabbits with the ali trait have a deletion encompassing the V_Ha allotype-encoding V_H1 gene (see Section 7). In heterozygotes most B cells rearrange and express the normal allele, be it a1, a2 or a3. In homozygous *ali/ali* rabbits, B cells rearrange upstream genes, including V_H4, *y33*, *x*, *z* and perhaps others. Some of these rearranged genes undergo further changes in sequence presumed to be via somatic gene conversion (Chen *et al.*, 1993, 1995). In the appendix of developing mutant rabbits, selection appears to favor growth and expansion of cells bearing V_H genes with sequences more similar to the deleted V_H1a2 gene (Pospisil *et al.*, 1995).

bas-Mutant rabbits with the bas trait have a defective $C_\kappa1$ gene derived from the b9 allotype. The acceptor site for J_κ to C_κ mRNA splicing is mutated (Lamoyi and Mage, 1985). Because of this splice site defect, $C_\kappa1$-type light chains are not produced. The homozygous mutants produce elevated levels of the $C_\kappa2$ gene product as well as elevated levels of λ type light chains. In heterozygotes the product of the normal allele is expressed in the majority of B lymphocytes.

Rabbits infected with human immunodeficiency virus (HIV-1) show no signs of disease that mimics the acquired immune deficiency (AIDS) disease in humans. While HIV-1 infection of rabbits has been demonstrated by the detection of viral nucleic acids and proteins from primary tissues of animals injected with the virus, there have been few reports of disease associated with infection (Kulaga *et al.*, 1989; Reina *et al.*, 1993). However, one study found that rabbits transfused with blood from HIV-1-infected rabbits developed a $CD4^+$ lymphocytopenia and two rabbits in this study died as a result of lymphocyte depletion in lymph nodes and thymus (Simpson *et al.*, 1995).

The myxoma virus of the poxvirus family has a profound effect on the immune system of infected rabbits. The production of virus-encoded proteins subvert cytokine networks by mimicking various cytokine receptors and this results in immunosuppression leading to the death of infected animals (McFadden and Graham, 1994; Mossman *et al.*, 1996). In addition, downregulation of CD4 has been demonstrated following the *in vitro* infection of a rabbit $CD4^+$ cell line with myxoma virus (Barry *et al.*, 1995).

15. Tumors of the Immune System

Leukemias and lymphomas are the predominant immune system tumors found in rabbits. These tumors have been found in rabbits experimentally infected with viruses or in rabbits expressing a c-myc transgene. Except for a lymphosarcoma found in the WH strain of rabbits, spontaneous tumors of the immune system have not been reported. Table VI.15.1 outlines the occurrence of immune system tumors found in rabbits, as well as the etiology and pathology of those tumors.

16. Autoimmune Diseases

Table VI.16.1 lists some models of autoimmune diseases in the rabbit. In the following discussion we will briefly annotate one of them.

In the 1950s, using rabbits, Rose and Witebsky (Rose and Witebsky, 1956; Witebsky and Rose, 1956) succeeded in producing autoantibodies to thyroid antigens and in inducing autoimmune thyroiditis. This now classical work established convincingly that Ehrlich and Morgenroth's theoretical concept of 'autoantibodies' was a realistic one, and that the histopathology of Hashimoto's thyroiditis could be reproduced rather well in an animal model by autoimmunization.

T cells in autoimmune thyroiditis have been explored in both obese strain (OS) chickens and in murine autoimmune thyroiditis. A central role for T-helper cells in autoimmunization is well established, and a role for both $CD4^+$ and $CD8^+$ T cells in the effector limb leading to tissue injury appears likely (Charreire, 1989; Wick *et al.*, 1989).

However, over the last two decades, humoral immunity in autoimmune thyroiditis has received rather little attention. Doubts were cast early on about a primary role of antibodies in tissue damage of autoimmune thyroiditis by the discordance of antibody titers and disease severity and by the difficulties encountered in attempts to transfer the disease by antibodies into normal recipients. Nevertheless, a few reports appeared to support a major role for autoantibodies in the pathogenesis of autoimmune thyroiditis; immune complex deposits were reported in human (Kalderon and Bogaars, 1977; Weetman *et al.*, 1989), murine (Clagett *et al.*, 1974; Tomazic and Rose, 1975;

Table VI.15.1 Tumors of the immune system

Tumor	*Etiology*	*Target population*	*Pathology and effects on immune function*
Leukemia	Experimental infection with HTLV-1. Usual route is by the IV injection of HTLV-1-infected cell lines. Transmission can also occur from mother to offspring.	T cells	Infection usually results in seroconversion but the rabbits remain healthy. In certain inbred strains of rabbit, disease with an ATL like pathology is observed, with cellular infiltration of organs and elevated leukocyte counts (Seto *et al.*, 1987). Infection with one cell line has been found to result in acute lymphoproliferative disease (Simpson *et al.*, 1996)
Lymphoma	Experimental infection with herpesvirus *Macaca arctoides* (HVMA). The route of infection was via intramscular injection of cell-free virus.	ND	Fifty per cent of the infected animals reveal a lymphoproliferative disease classified as malignant lymphoma, lymphoblastic lymphoma or lymphoid hyperplasia. In infected animals the spleen, lung, liver, kidney and suprarenal glands are infiltrated by lymphoid cells (Wutzler *et al.*, 1995)
Leukemia	Transgenic rabbits expressing the c-*myc* gene with the immunoglobulin μ heavy chain enhancer.	Pre-B cells	Early development of leukemia (17–20 days old). Large atypical lymphocytes of oligoclonal origin with infiltration of spleen, liver, bone marrow and kidney with peripheral blood counts of 100 000–500 000/mm^3 (Knight *et al.*, 1988)
Lymphoma	Trangenic rabbits expressing the c-*myc* gene with the immunoglobulin κ light-chain enhancer.	B cells	Tumors developed at 5–11 months old. Lymph node structure consists of tumor cell aggregates. The kidneys, gastro-intestinal tract, ovaries and tissues adjacent to lymph nodes are infiltrated by lymphoid cells (Sethupathi *et al.*, 1994)
Lymphosarcoma	Occurred in WH strain of rabbit and exhibited an autosomal recessive pattern of inheritance. Type C retrovirus induced in lymphosarcomatous tissue (Bedigian *et al.*, 1978).	ND	The lymphosarcoma involves visceral lymph nodes, kidneys, spleen, liver and adrenal glands. Infiltrates of lymphoid cells are found in many organs of affected rabbits (Fox *et al.*, 1970)
Malignant lymphoma	Induced in several strains of rabbits following innoculation of rabbits with herpesvirus saimiri via several different routes (Daniel *et al.*, 1974).	T cells	The type of lymphoma found varied from poorly differentiated to well differentiated. Although the spleen, lymph nodes, liver and lungs showed marked infiltration, a diffuse infiltrate was found in most organs examined (Ablashi *et al.*, 1980; Faggioni *et al.*, 1982)

Vladutiu, 1990) and avian (Wick and Graf, 1972; Katz *et al.*, 1981; Kofler *et al.*, 1983) thyroid lesions, and potentiating (Jaroszewski *et al.*, 1978) and precipitating (Katz *et al.*, 1986) effects were ascribed to autoantibodies in OS chickens, but the pathogenicity of humoral antibody-mediated mechanisms remained uncertain.

Recently, additional evidence for a significant role of autoantibodies in the pathogenesis of autoimmune thyroiditis was obtained in rabbits. In the first group of animals, the superior thyroid artery was cannulated, allowing for administration of serum containing autoantibodies directly to the target organ over several days (Inoue *et al.*, 1993a). Autoimmune thyroiditis ensued in most perfused rabbits. A broad range of histopathological lesions was observed, and many animals had rather extensive cellular infiltrates, as well as IgG and C3 deposits in the follicular laminae. These results suggest that the earlier failures of transfer experiments may be explained by dilution and loss of antibodies because of a systemic administration.

A second series of experiments explored the role of the late complement components in the pathogenesis of auto-

Table VI.16.1 Selected rabbit models of autoimmune diseases

Animal model	*Reference*
Experimental autoimmune thyroiditis	Rose and Witebsky (1956)
Antiglomerular basement membrane nephritis/membranous glomerulonephritis	Lerner and Dixon (1966)
Membranous glomerulonephritis	Shibata *et al.* (1972)
Interstitial nephritis	Klassen *et al.* (1971)
Mercuric chloride-induced nephritis	Roman-Franco *et al.* (1976, 1978)
Mercuric chloride-induced systemic immune complex disease	Albini and Andres (1983)
Experimental autoallergic encephalitis	Waksman (1959)
Galactocerebroside-induced neuritis	Saida (1981)
Autoimmune hepatitis	Dienes (1995)
Postvasectomy autoimmune orchitis	Bigazzi *et al.* (1976)

immune thyroiditis (Inoue *et al.*, 1993b). Very little was known about complement in autoimmune thyroiditis; most intriguing was a report on decreases in serum complement concentrations in Lewis rats immunized with thyroid antigens. Thus, it appeared useful to explore the ability of complement C6-deficient rabbits to develop experimental autoimmune thyroiditis. Rabbits were immunized with a crude thyroid extract of rabbit thyroids to encompass as many autoantigens as possible. Normocomplementemic and C6-deficient rabbits had the same titers of thyroid antibodies and immune deposits in the follicular laminae, but the former had much more extensive cellular infiltrates of thyroids than the latter.

Intraperitoneal administration of serum from normocomplementemic rabbits into C6-deficient animals restored substantially their ability to develop extensive cellular infiltrates in their thyroids. The terminal complement components appear to influence significantly the severity of autoimmune thyroiditis in rabbits. Conversely, a 'residual' thyroiditis occurred in C6-deficient animals, suggesting the cooperation of at least two distinct effector mechanisms in tissue lesions of experimental autoimmune thyroiditis. Indirectly, the requirement for terminal complement components demonstrates also the requirement for antibodies in the development of full-blown autoimmune thyroiditis of the rabbit.

Ultimately, studies on autoimmune diseases initiated by Rose and Witebsky led to the formulation of the Witebsky postulates, which still are the 'golden rule' in research on autoimmunity.

17. Conclusions

Although the rabbit is used less frequently in studies of cellular immunology than in the past, there continues to be active research into the organization, regulated expression and diversification of the immune repertoire in the rabbit. Recent studies have renewed interest in the important role that gut-associated lymphoid tissue (GALT) plays in development of the immune system and antibody repertoire in rabbit (Weinstein *et al.*, 1994a, b). The mechanism of gene conversion-like alterations of V_H gene sequences has yet to be described in detail. Such studies of the rabbit have also renewed interest in the extent to which the GALT plays a role in early development of the human immune system.

Although the rabbit has been used extensively in studies of leukocyte traffic (see Section 4), relatively little is known about lymphocytes compared with neutrophils. The circulation and recirculation patterns of rabbit B and T lymphocytes have not been delineated in any detail. Recent evidence supports the hypothesis that self-renewal of B lymphocytes makes an important contribution to the B-lymphocyte compartment of adult rabbits as there is very little continuous B-lymphopoiesis (Crane *et al.*, 1996). Because of the demonstrated role of GALT in repertoire development, it will be important in the future to identify the rabbit equivalent of 'post bursal stem cells'.

Very little specific information and few clones of the rabbit cytokines are currently available. Neonatal immune responses have not been examined in light of modern understanding of B and T lymphocyte development. There is also relatively little specific information about nonspecific immunity and NK cells of the rabbit.

Model for disease-related research

There are numerous rabbit models for diseases. Among those with immunological relevance are: various infectious diseases including syphilis (Sell and Hsu, 1993), tuberculosis (Dannenberg, 1991), virus-induced papilloma, myxoma virus, HTLV1 and HIV (see sections 14 and 15). NZW rabbits infected with rabbit papilloma virus are models for a virus–MHC connection in malignant conversion (Han *et al.*, 1992). A rabbit model of human hemolytic disease of the newborn was described in section 9. Complement deficiencies are described in section 12 and autoimmune diseases in Section 16.

18. References

Ablashi, D., Sundar, K., Armstrong, G., Golway, P., Valerio, M., Bengali, Z., Lemp, J., Fox, R. (1980). Herpesvirus saimiri-induced malignant lymphoma in inbred strain III/J rabbits (*Oryctolagus cuniculus*). *J. Cancer Res. Clin. Oncol.* **98**, 165–172.

Acha-Orbea, H., McDevitt, H. O. (1987). The first external domain of the nonobese diabetic mouse class II I-Aβ chain is unique. *Proc. Natl Acad. Sci. USA* **84**, 2435–2439.

Akimenko, M. A., Heidmann, O., Rougeon, F. (1984). Complex allotypes of the rabbit immunoglobulin kappa light chains are encoded by structural alleles. *Nucl. Acids Res.* **12**, 4691–4701.

Akimenko, M. A., Mariame, B., Rougeon, F. (1986). Evolution of the immunoglobulin kappa light chain locus in the rabbit: evidence for differential gene conversion events. *Proc. Natl Acad. Sci. USA* **83**, 5180–5183.

Albini, B., Andres, G. (1983). Autoimmune disease induced in rabbits by administration of mercuric chloride. In *Immune Mechanisms in Renal Disease* (eds Cummings, N. B., Michael, A. F., Wilson, C. B.), pp. 249–260. Plenum, New York.

Allegrucci, M., Newman, B. A., Young-Cooper, G. O., Alexander, C. B., Meier, D., Kelus, A. S., Mage, R. G. (1990). Altered phenotypic expression of immunoglobulin heavy-chain variable-region (V_H) genes in Alicia rabbits probably reflects a small deletion in the V_H genes closest to the joining region. *Proc. Natl Acad. Sci. USA* **87**, 5444–5448.

Allegrucci, M., Young-Cooper, G. O., Alexander, C. B., Newman, B. A., Mage, R. G. (1991). Preferential rearrangement in normal rabbits of the 3′ V_Ha allotype gene that is deleted in Alicia mutants: somatic hypermutation/conversion may play a major role in generating the heterogeneity of rabbit heavy chain variable region sequences. *Eur. J. Immunol.* **21**, 411–417.

Alsenz, J., Avila, D., Huemer, H. P., Esparza, I., Becherer, J. D., Kinoshita, T., Wang, Y., Oppermann, S., Lambris, J. D. (1992). Phylogeny of the third component of complement, C3: analysis of the conservation of human CR1, CR2, H, and B binding sites, concanavalin A binding sites, and thiolester bond in the C3 from different species. *Dev. Comp. Immunol.* **16**, 63–76.

Angiolillo, A. L., Lamoyi, E., Bernstein, K. E., Mage, R. G. (1985). Identification of genes for the constant region of rabbit T-cell receptor β chains. *Proc. Natl Acad. Sci. USA* **82**, 4498–4502.

Apple, R. J., Erlich, H. A., Klitz, W., Manos, M. M., Becker, T. M., Wheeler, C. M. (1994). HLA DR-DQ Associations with cervical carcinoma show papillomavirus-type specificity. *Nature Genet.* **6**, 157–162.

Arfors, K. E., Lundberg, C., Lindbom, L., Lundberg, K., Beatty, P. G., Harlan, J. M. (1987). A monoclonal antibody to the membrane glycoprotein complex CD18 inhibits polymorphonuclear leukocyte accumulation and plasma leakage *in vivo*. *Blood* **69**, 338–340.

Argenbright, L. W., Letts, L. G., Rothlein, R. (1991). Monoclonal antibodies to the leukocyte membrane CD18 glycoprotein complex and to intercellular adhesion molecule-1 inhibit leukocyte–endothelial adhesion in rabbits. *J. Leukocyte Biol.* **49**, 253–257.

Arroyave, C. M., Müller-Eberhard, H. J. (1971). Isolation of the sixth component of complement from human serum. *Immunochemistry* **8**, 995–1006.

Barry, M., Lee, S., Boshkov, L., McFadden, G. (1995). Myxoma virus induces extensive CD4 downregulation and dissociation of $p56^{lck}$ in infected rabbit $CD4^+$ T lymphocytes. *J. Virol.* **69**, 5243–5251.

Becherer, J. D., Alsenz, J., Lambris, J. D. (1990). Molecular aspect of C3 interactions and structural/functional analysis of C3 from different species. *Curr. Top. Microbiol. Immunol.* **153**, 45–72.

Becker, R. S., Knight, K. L. (1990). Somatic diversification of immunoglobulin heavy chain VDJ genes: evidence for somatic gene conversion in rabbit. *Cell* **63**, 987–997.

Becker, R. S., Suter, M., Knight, K. L. (1990). Restricted utilization of V_H and C_H genes in leukemic rabbit B cells. *Eur. J. Immunol.* **20**, 397–402.

Bedigian, H., Fox, R., Meier, H. (1978). Induction of type C RNA virus from cultured rabbit lymphosarcoma cells. *J. Virol.* **27**, 313–319.

Befus, A. D., Bienenstock, J. (1982). The mucosa-associated immune system of the rabbit. In *Animal Models of Immunological Processes*, pp. 167–220. Academic Press, New York, USA.

Benammar, A., Cazenave, P. A. (1982). A second kappa isotype. *J. Exp. Med.* **156**, 585–589.

Bernstein, K. E., Alexander, C. B., Mage, R. G. (1983a). Nucleotide sequence of a rabbit IgG heavy chain from the recombinant F-I haplotype. *Immunogenetics* **18**, 387–397.

Bernstein, K. E., Skurla, R. M., Mage, R. G. (1983b). The sequences of rabbit κ light chains of b4 and b5 allotypes differ more in their constant regions than in their 3′ untranslated regions. *Nucl. Acids Res.* **11**, 7205–7214.

Bernstein, K. E., Alexander, C. B., Reddy, E. P., Mage, R. G. (1984a). Complete sequence of a cloned cDNA encoding rabbit secreted μ chain of V_Ha2 allotype: comparisons with V_Ha1 and membrane μ sequences. *J. Immunol.* **132**, 490–495.

Bernstein, K. E., Lamoyi, E., McCartney-Francis, N., Mage, R. G. (1984b). Sequence of a cDNA encoding Basilea kappa light chains (K2 isotype) suggests a possible relationship of protein structure to limited expression. *J. Exp. Med.* **159**, 635–640.

Betz, S. J., Lotner, Z., Henson, P. M. (1980). Generation and release of platelet-activating factor (PAF) from enriched preparations of rabbit basophils; failure of human basophils to release PAF. *J. Immunol.* **125**, 2749–2755.

Bigazzi, P. E., Kosuda, L. L., Hsu, K. C., Andres, G. A. (1976). Immune complex orchitis in vasectomized rabbits. *J. Exp. Med.* **143**, 382–404.

Blackford, J., Wilkinson, J. M. (1993). A monoclonal antibody which recognises rabbit CD11a and which inhibits homotypic T cell aggregation. *Biochem. Soc. Trans.* **21**, 198S.

Blackford, J., Reid, H. W., Pappin, D. J. C., Bowers, F. S., Wilkinson, J. M. (1996). A monoclonal antibody, 3/22, to rabbit CD11c which induces homotypic T cell aggregation: evidence that ICAM-1 is a ligand for CD11c/CD18. *Eur. J. Immunol.* **26**, 525–531.

Boyer, M. I., Bowen, C. V. A., Danska, J. S. (1995). Restriction fragment-length polymorphism analysis of the major histocompatibility complex in New Zealand White rabbits. *Transplantation* **59**, 1043–1046.

Brambell, F. W. R. (1970). The transmission of passive immunity from mother to young. In *Frontiers of Biology*, Vol. 18 (eds

Neuberger, A., Tatum, E. L., Holborow, E. J.), pp. 42–79. North Holland Publishing Co., London, UK.

Burnett, R. C., Hanly, W. C., Zhai, S. K., Knight, K. L. (1989). The IgA heavy-chain gene family in rabbit: cloning and sequence analysis of 13 Cα genes. *EMBO J.* **8**, 4041–4047.

Butcher, E. C., Picker, L. J. (1996). Lymphocyte homing and homeostasis. *Science* **272**, 60–66.

Calabi, F., Belt, K. T., Yu, C. Y., Bradbury, A., Mandy, W. J., Milstein, C. (1989). The rabbit CD1 and the evolutionary conservation of the CD1 gene family. *Immunogenetics* **30**, 370–377.

Calderon, T. M., Factor, S. M., Hatcher, V. B., Berliner, J. A., Berman, J. W. (1994). An endothelial cell adhesion protein for monocytes recognized by monoclonal antibody IG9: expression of *in vivo* inflamed human vessels and atherosclerotic human and Watanabe rabbit vessels. *Lab. Invest.* **70**, 836–849.

Cannon, J. G., Clark, B. D., Wingfield, P., Schmeissner, U., Losberger, C., Dinarello, C. A., Shaw, A. R. (1989). Rabbit IL-1: cloning, expression, biologic properties, and transcription during endotoxemia. *J. Immunol.* **142**, 2299–2306.

Carlos, T. M., Harlan, J. M. (1994). Leukocyte-endothelial adhesion molecules. *Blood* **84**, 2068–2101.

Castle, W. E., Keeler, C. D. (1933). Blood group inheritance in the rabbit. *Proc. Natl Acad. Sci. USA* **19**, 92–98.

Chai, C. K. (1974). Genetics studies of histocompatibility in rabbits: identification of major and minor genes. *Immunogenetics* **1**, 126–132.

Charreire, J. (1989). Immune mechanisms in autoimmune thyroiditis. *Adv. Immunol.* **46**, 263–334.

Chen, H., Alexander, C., Young-Cooper, G., Mage, R. (1993). VH gene expression and regulation in the mutant Alicia rabbit. Rescue of VHa2 allotype expression. *J. Immunol.* **150**, 2783–2793.

Chen, H., Alexander, C., Mage, R. (1995). Characterization of a rabbit germline VH gene that is a candidate donor for VH gene conversion in mutant Alicia rabbits. *J. Immunol.* **154**, 6365–6371.

Chen, Z., Metzger, D. W., Adler, F. L. (1984). Characterization of a monoclonal antibody reactive with rabbit T lymphocytes and neutrophils. *Cell. Immunol.* **85**, 297–308.

Chouchane, L., Kindt, T. J. (1992). Mapping of the rabbit MHC reveals that class I genes are adjacent to the DR subregion and defines an insertion/deletion-related polymorphism in the class II region. *J. Immunol.* **149**, 1216–1222.

Chouchane, L., Brown, T. J., Kindt, T. J. (1993). Structure and expression of a nonpolymorphic rabbit class II gene with homology to HLA-DOB. *Immunogenetics* **38**, 64–66.

Clagett, J. A., Wilson, C. B., Weigle, W. O. (1974). Interstitial immune complex thyroiditis in mice: the role of autoantibody to thyroglobulin. *J. Exp. Med.* **140**, 1349–1456.

Clark, B. D., Bedrosian, I., Schindler, R., Cominelli, F., Cannon, J. G., Shaw, A. R., Dinarello, C. A. (1991). Detection of interleukin 1α and 1β in rabbit tissues during endotoxemia using sensitive radioimmunoassays. *J. Appl. Physiol.* **71**, 2412–2418.

Cochrane, C. G., Müller-Eberhard, H. J., Aikin, B. S. (1970). Depletion of plasma complement *in vivo* by a protein of cobra venom: its effect on various immunologic reactions. *J. Immunol.* **105**, 55–69.

Cohen, C. (1982). The immunogenetics of cellular antigen systems of the rabbit. In *Oral Immunogenetics and Tissue Transplantation* (eds Riviere, G. R., Hildemann, W. H.), pp. 181–197. Elsevier North Holland, Inc., Amsterdam, The Netherlands.

Cohen, C., Tissot, R. G. (1974). The effect of the RL-A locus and the MLC locus on graft survival in the rabbit. *Transplantation*, **18**, 150–154.

Cooper, M. D., Perey, D. Y., Gabrielsen, A. E., Sutherland, D. E. R., McKneally, M. F., Good, R. A. (1968). Production of an antibody deficiency syndrome in rabbits by neonatal removal of organized intestinal lymphoid tissues. *Int. Arch. Allergy* **33**, 65–88.

Crabb, E. D., Kelsall, M. A. (1940). Organization of the mucosa and lymphatic structures in the rabbit appendix. *J. Morphology* **67**, 351–367.

Craig, S. W., Cebra, J. J. (1971). Peyer's patches: an enriched source of precursors for IgA producing immunocytes in the rabbit. *J. Exp. Med.* **134**, 188–200.

Crane, M. A., Kingzette, M., Knight, K. L. (1996). Evidence for limited B-lymphopoiesis in adult rabbits. *J. Exp. Med.* **183**, 2119–2127.

Currier, S. J., Gallarda, J. L., Knight, K. L. (1988). Partial molecular genetic map of the rabbit V_H chromosomal region. *J. Immunol.* **140**, 1651–1659.

Daniel, M., Melendez, L., Hunt, R., King, N., Anver, M., Barahona, H., Baggs, R. (1974). Herpesvirus saimiri: VII. Induction of malignant lymphoma in New Zealand white rabbits. *J. Natl Cancer Inst.* **53**, 1803–1807.

Dannenberg, A. M., Jr (1991). Delayed-type hypersensitivity and cell-mediated immunity in the pathogenesis of tuberculosis. *Immunol. Today* **12**, 228–233.

Davie, J. M., Paul, W. E., Mage, R. G., Goldman, M. B. (1971). Membrane-associated immunoglobulin of rabbit peripheral blood lymphocytes: allelic exclusion at the *b* locus. *Proc. Natl Acad. Sci.* **68**, 430–434.

de Frutos, P. G., Dahlbaeck, B. (1995). cDNA structure of rabbit C4b-binding protein alpha-chain. Preserved sequence motif in complement regulatory protein modules which bind C4b. *Biochem. Biophys. Acta* **1261**, 285–289.

De Smet, W., Vaeck, M., Smet, E., Brys, L., Hamers, R. (1983). Rabbit leukocyte surface antigens defined by monoclonal antibodies. *Eur. J. Immunol.* **13**, 919–928.

Dienes, H. P. (1995). Autoimmune hepatitis. *Verh. Deutsch GesPathol.* **79**, 177–185.

Dodds, A. W., Law, S. K. (1990). The complement component C4 of mammals. *Biochemical J.* **265**, 495–502.

Doerschuk, C. M., Winn, R. K., Coxson, H. O., Harlan, J. M. (1990). CD18-dependent and -independent mechanisms of neutrophil emigration in the pulmonary and systemic microcirculation of rabbits. *J. Immunol.* **144**, 2327–2333.

Drake, W. T., Issekutz, A. C. (1993). Transforming growth factor-beta enhances polymorphonuclear leucocyte accumulation in dermal inflammation and transendothelial migration by a priming action. *Immunology* **78**, 197–204.

Dreher, K. L., Emorine, L., Kindt, T. J., Max, E. E. (1983). cDNA clone encoding a complete rabbit immunoglobulin kappa light chain of b4 allotype. *Proc. Natl Acad. Sci. USA* **80**, 4489–4493.

Dreher, K. L., Asundi, V., Wolf, B., Bruggeman, L. (1990). Rabbit Ig kappa 1b6 gene structure. *J. Immunol.* **145**, 325–330.

Dubiski, S. (ed.). (1987). *The Rabbit in Contemporary Immunological Research*. Longman Scientific and Technical, Harlow, England.

Duvoisin, R. M., Kocher, H. P., Garcia, I., Rougeon, F., Jaton,

J. C. (1984). Nucleotide sequence of a cDNA encoding the constant region of a rabbit immunoglobulin light chain of the lambda type. *Eur. J. Immunol.* **14**, 379–382.

Duvoisin, R. M., Heidmann, O., Jaton, J. C. (1986). Characterization of four constant region genes of rabbit immunoglobulin-lambda chains. *J. Immunol.* **136**, 4297–4302.

Emorine, L., Max, E. E. (1983). Structural analysis of a rabbit kappa 2 J-C locus reveals multiple deletions. *Nucl. Acids Res.* **11**, 8877–8890.

Emorine, L., Dreher, K., Kindt, T. J., Max, E. E. (1983). Rabbit immunoglobulin kappa genes: structure of a germline b4 allotype J-C locus and evidence for several b4-related sequences in the rabbit genome. *Proc. Natl Acad. Sci. USA* **80**, 5709-5713.

Emorine, L., Sogn, J. A.,Trinh, D., Kindt, T. J., Max, E. E. (1984). A genomic gene encoding the b5 rabbit immunoglobulin kappa constant region: implications for latent allotype phenomenon. *Proc. Natl Acad. Sci. USA* **81**, 1789-1793.

Eskinazi, D. P., Bessinger, B. A., McNicholas, J. M., Leary, A. L., Knight, K. L. (1979). Expression of an unidentified immunoglobulin isotype on rabbit Ig-bearing lymphocytes. *J. Immunol.* **122**, 469–474.

Estess, P., Begovich, A. B., Koo, M., Jones, P., McDevitt, H. O. (1986). Sequence analysis and structure-function correlations of murine q, k, u, s, and f haplotype I-Aβ cDNA clones. *Proc. Natl Acad. Sci. USA* **83**, 3594–3598.

Esworthy, S., Max, E. E. (1986). The rabbit kappa-1b5 immunoglobulin gene: another J region gene cluster with only one functional J gene segment? *J. Immunol.* **136**, 1107–1111.

Faggioni, A., Ablashi, D., Armstrong, G., Sundar, S., Loeb, W., Martin, D., Valerio, M., Parker, G., Fox, R. (1982). Herpesvirus saimiri induced malignant lymphoma of the poorly and well differentiated types in three inbred strains of New Zealand white rabbits. In: *Advances in Comparative Leukemia Research 1981* (eds Yohn, D. S., Blakeslee, J. R.). Elsevier, North Holland.

Fitts, M. G., Metzger, D. W., Hendershot, L. M., Mage, R. G. (1995). The rabbit B cell receptor is noncovalently associated with unique heterotrimeric protein complexes: possible insights into the membrane IgM/IgD coexpression paradox. *Mol. Immunol.* **32**, 753–759.

Folkesson, H. G., Matthay, M. A., Hebert, C. A., Broaddus, V. C. (1995). Acid aspiration-induced lung injury in rabbits is mediated by interleukin-8-dependent mechanisms. *J. Clin. Invest.* **96**, 107–116.

Fox, R., Meier, H., Crary, D., Myers, D. O., Norberg, R., Laird, C. (1970). Lymphosarcoma in the rabbit: genetics and pathology. *J. Natl Cancer Inst.* **45**, 719–729.

Fujiwara, S., Armstrong, R. M., Cinader, B. (1974). Age and organ dependent variations in the number of thymus derived RTLA bearing cells. *Immun. Commun.* **3**, 275–284.

Galea-Lauri, J., Blackford, J., Wilkinson, J. M. (1993a). The expression of CD11/CD18 molecules on rabbit leucocytes: identification of monoclonal antibodies to CD18 and their effect on cellular adhesion processes. *Molec. Immunol.* **30**, 529–537.

Galea-Lauri, J., Wilkinson, J. M., Evans, C. H. (1993b). Characterization of monoclonal antibodies against rabbit CD44: evidence of a role for CD44 in modulating synoviocyte metabolism. *Molec. Immunol.* **30**, 1383–1392.

Gallarda, J. L., Gleason, K. S., Knight, K. L. (1985). Organization of rabbit immunoglobulin genes. I. Structure and multiplicity of germ-line V_H genes. *J. Immunol.* **135**, 4222–4228.

Ganzler, F., Hansch, G. M. (1984). Isolation of functionally active complement components from rabbit serum: purification and characterization of rabbit C9. *Immunobiol.* **168**, 99.

Garcia, N., Mileski, W. J., Lipsky, P. (1995). Differential effects of monoclonal antibody blockade of adhesion molecules on *in vivo* susceptibility to soft tissue infection. *Infect. Immun.* **63**, 3816–3819.

Gathings, W. E., Mage, R. G., Cooper, M. D., Lawton, A. R., Young-Cooper, G. O. (1981). Immunofluorescent studies on the expression of V_Ha allotypes by pre-B and B cells of homozygous and heterozygous rabbits. *Eur. J. Immunol.* **11**, 200–206.

Gathings, W. E., Mage, R. G., Cooper, M. D., Young-Cooper, G. O. (1982). A subpopulation of small pre-B cells in rabbit bone marrow expresses light chains and exhibits allelic exclusion of *b* locus allotypes. *Eur. J. Immunol.* **12**, 76–81.

Giclas, P. C., Keeling, P. J., Henson, P. M. (1981). Isolation and characterization of the third and fifth components of rabbit complement. *Mol. Immunol.* **18**, 113–123.

Gilman-Sachs, A., Dray, S. (1985). Allotypic and isotypic specificities of rabbit IgM; localization to Fabμ or Fcμ fragments. *Molec. Immunol.* **22**, 57–65.

Gilman-Sachs, A., Mage, R. G., Young, G. O., Alexander, C., Dray, S. (1969). Identification and genetic control of two rabbit immunoglobulin allotypes at a second light chain locus, the *c* locus. *J. Immunol.* **103**, 1159–1167.

Granert, C., Raud, J., Xie, X., Lindquist, L., Lindbom, L. (1994). Inhibition of leukocyte rolling with polysaccharide fucoidin prevents pleocytosis in experimental meningitis in the rabbit. *J. Clin. Invest.* **93**, 929–936.

Graur, D., Duret, L., Gouy, M. (1996). Phylogenetic position of the order Lagomorpha (rabbits, hares and allies). *Nature* **379**, 333–335.

Gregoire, L., Lawrence, W. D., Kukuruga, D., Eisenbrey, A. B., Lancaster, W. D. (1994). Association between HLA-DQB1 alleles and risk for cervical cancer in African-American women. *Int. J. Cancer* **15**, 504–507.

Haber, E. (1970). Antibodies of restricted heterogeneity for structural study. *Fed. Proc.* **29**, 66–71.

Hagari, Y., Budgeon, L. R., Pickel, M. D., Kreider, J. W. (1995). Association of tumor necrosis-alpha gene expression and apoptotic cell death with regression of Shope papillomas. *J. Invest. Dermatol.* **104**, 526–529.

Hague, B. F., Sawasdikosol, S., Brown, T. J., Lee, K., Recker, D. P., Kindt, T. J. (1992). CD4 and its role in infection of rabbit cell lines by human immunodeficiency virus type 1. *Proc. Natl Acad. Sci. USA* **89**, 7963–7967.

Hammer, C. H., Gaither, T., Frank, M. M. (1981). Complement deficiencies of laboratory animals. In *Immunologic Defects in Laboratory Animals* 2 (eds Gershwin, M. E., Merchant, B.), pp. 207–240. Plenum Press, New York.

Han, R., Breitburd, B., Marche, P. N., Orth, G. (1992). Linkage of regression and malignant conversion of rabbit viral papillomas to MHC class II genes. *Nature* **356**, 66–68.

Hardy, R. R., Carmack, C. E., Shinton, S. A., Kemp, J. D., Hayaka, K. (1991). Resolution and characterization of pro-B and pre-pro-B cell stages in normal mouse bone marrow. *J. Exp. Med.* **173**, 1213–1225.

Harindranath, N., Lamoyi, E., Mage, R. G. (1989a). Allotypes of the constant region of the rabbit T cell receptor β-2 chain. *J. Immunol.* **142**, 3292–3297.

Harindranath, N., Lamoyi, E., Mage, R. G. (1989b). Genomic sequence and organization of rabbit Tcrb constant region genes. *Immunogenetics* **30**, 465–474.

Harindranath, N., Alexander, C. B., Mage, R. G. (1991). Evolutionarily conserved organization and sequences of germline diversity and joining regions of the rabbit T-cell receptor β2 chain. *Molec. Immunol.* **28**, 881–888.

Harlan, J. M., Winn, R. K., Vedder, N. B., Doerschuk, C. M., Rice, C. L. (1992). *In vivo* models of leukocyte adherence to endothelium. In *Adhesion: Its Role in Inflammatory Disease* (eds Harlan, J. M., Liu, D. Y.), pp. 117–150. Freeman, New York, USA.

Hayward, A. R., Simons, M. A., Lawton, A. R., Mage, R. G., Cooper, M. D. (1978). Pre-B and B cells in rabbits: ontogeny and allelic exclusion of kappa light chain genes. *J. Exp. Med.* **148**, 1367–1377.

Hayzer, D. J., Jaton, J. C. (1987). Nucleotide sequence of a cDNA clone encoding a rabbit immunoglobulin lambda light chain: the V lambda region differs markedly from that of other species. *J. Immunol.* **138**, 2316–2322.

Hayzer, D. J., Jaton, J. C. (1989). Cloning and sequencing of two functional rabbit germline immunoglobulin V-lambda genes. *Gene* **80**, 185–191.

Hayzer, D. J., Duvoisin, R. M., Jaton, J. C. (1987). cDNA clones encoding rabbit immunoglobulin lambda chains. Evidence for length variation of the third hypervariable region and for a novel constant region. *Biochem. J.* **245**, 691–697.

Hayzer, D. J., Young-Cooper, G. O., Mage, R. G., Jaton, J. C. (1990). cDNA clones encoding immunoglobulin lambda chains from rabbit expressing the phenotype c7. *Eur. J. Immunol.* **20**, 2707–2712.

Hechtman, D. H., Cybulsky, M. I., Fuchs, H. J., Baker, J. B., Gimbrone, M. A. (1991). Intravascular IL-8: inhibitor of polymorphonuclear leukocyte accumulation at sites of acute inflammation. *J. Immunol.* **147**, 883–892.

Heidmann, O., Rougeon, F. (1983a). Diversity in the rabbit kappa variable regions is amplified by nucleotide deletions and insertions at the V-J junction. *Cell* **34**, 767–777.

Heidmann, O., Rougeon, F. (1983b). Multiplicity of constant kappa light chain genes in the rabbit genome: a b4b4 homozygous rabbit contains a kappa-bas gene. *EMBO J.* **2**, 437–441.

Heidmann, O., Rougeon, F. (1984). Immunoglobulin kappa light-chain diversity in rabbit is based on the 3′ length heterogeneity of germ-line variable genes. *Nature* **311**, 74–76.

Heidmann, O., Auffray, C., Cazenave, P.-A., Rougeon, F. (1981). Nucleotide sequence of constant and 3′ untranslated regions of a kappa immunoglobulin light chain mRNA of a homozygous b4 rabbit. *Proc. Natl Acad. Sci. USA* **78**, 5802–5806.

Helland, A., Borresen, A. L., Kaern, J., Ronningen, K. S., Thorsby, E. (1992). HLA antigens and cervical carcinoma (correspondence). *Nature* **356**, 23.

Hellewell, P. G., Young, S. K., Henson, P. M., Worthen, G. S. (1994). Disparate role of the b2-integrin CD18 in the local accumulation of neutrophils in pulmonary and cutaneous inflammation in the rabbit. *Am. J. Respir. Cell Mol. Biol.* **10**, 391–398.

Hogg, J. C., Doershuck, C. M. (1995). Leukocyte traffic in the lung. *Annu. Rev. Physiol.* **57**, 97–114.

Hole, N. J. K., Lamoyi, E., Komatsu, M., Harindranath, N., Young-Cooper, G. O., Mage, R. G. (1988). Linked genetic markers of the rabbit κ light chain are not linked to the Tcrβ chain genes. *Immunogenetics* **28**, 99–107.

Hole, N. J., Young-Cooper, G. O., Mage, R. G. (1991). Mapping of the duplicated rabbit immunoglobulin kappa light chain locus. *Eur. J. Immunol.* **21**, 403–409.

Hornby, E. J., Brown, S., Wilkinson, J. M., Mattock, C., Authi, K. S. (1991). Activation of human platelets by exposure to a monoclonal antibody. PM6/248, to glycoprotein IIb-IIIa. *Br. J. Haematol.* **79**, 277–285.

Horng, W. J., Gilman-Sachs, A., Dray, S. (1980). *In vivo* immunoglobulin allotype suppression in the rabbit. In *Regulation of Function of Lymphocytes by Antibodies* (eds Bona, C., Cazenave, P. A.), pp. 139–155. Wiley, New York, USA.

Horstmann, R. D., Müller-Eberhard, H. J. (1985). Isolation of rabbit C3, factor B, and factor H and comparison of their properties with those of the human analog. *J. Immunol.* **134**, 1094–1100.

Hüsler, T., Lockert, D. H., Kaufman, K. M., Sodetz, J. M., Sims, P. J. (1995). Chimeras of human complement C9 reveal the site recognized by complement regulatory protein CD59. *J. Biol. Chem.* **270**, 3483–3486.

Inoue, K., Nielsen, N., Milgrom, F., Albini, B. (1993a). Transfer of experimental autoimmune thyroiditis by *in situ* perfusion of thyroids with immune sera. *Clin. Immunol. Immunopathol.* **66**, 11–17.

Inoue, K., Nielsen, N., Biesecker, G., Milgrom, F., Albini, B. (1993b). Role of late complement components in experimental autoimmune thyroiditis. *Clin. Immunol. Immunopathol.* **66**, 1–10.

Isono, T., Seto, A. (1995). Cloning and sequencing of the rabbit gene encoding T-cell costimulatory molecules. *Immunogenetics* **42**, 217–220.

Isono, T., Isegawa, Y., Seto, A. (1994). Sequence and diversity of variable gene segments coding for rabbit T-cell receptor beta chains. *Immunogenetics* **39**, 243–248.

Isono, T., Kim, C. J., Seto, A. (1995). Sequence and diversity of rabbit T-cell gamma chain genes. *Immunogenetics* **41**, 295–300.

Jackson, S., Chused, T. M., Wilkinson, J. M., Leiserson, W. M., Kindt, T. J. (1983). Differentiation antigens identify subpopulations of rabbit T and B lymphocytes: definition by flow cytometry. *J. Exp. Med.* **157**, 34–46.

Jaroszewski, J., Sundick, R. S., Rose, N. R. (1978). Effects of antiserum containing thyroglobulin antibody on the chicken thyroid gland. *Clin. Immunol. Immunopathol.* **10**, 95–103.

Jepson, M. A., Simmons, N. L., Hirst, G. L., Hirst, B. H. (1993). Identification of M cells and their distribution in rabbit intestinal Peyer's patches and appendix. *Cell Tissue Res.* **273**, 127–136.

Johnson, M. C., Kajikawa, O., Goodman, R. B., Wong, V. A., Mongovin, S. M., Wong, W. B., Fox-Dewhurst, R., Martin, T. R. (1996). Molecular expression of the α-chemokine rabbit GRO in *Escherichia coli* and characterization of its production by lung cells *in vitro* and *in vivo*. *J. Biol. Chem.* **271**, 10853–10858.

Jones, M., Cordell, J. L., Beyers, A. D., Tse, A. G. D., Mason, D. Y. (1993). Detection of T and B cells in many animal species using cross-reactive anti-peptide antibodies. *J. Immunol.* **150**, 5429–5435.

Kajikawa, O., Goodman, R. B., Mongovin, S. M., Wong, V. A., Martin, T. R. (1993). *Am. Rev. Respir. Dis.* **147**, A751.

Kalderon, A. E., Bogaars, H. A. (1977). Immune complex deposits in Graves' disease and Hashimoto's thyroiditis. *Am. J. Med.* **63**, 729–734.

Katz, D. V., Kite, J. H. Jr, Albini, B. (1981). Immune complexes in tissues of Obese strain (OS) chickens. *J. Immunol.* **126**, 2296–2301.

Katz, D. V., Albini, B., Kite, J. H. Jr (1986). Materno-embryonally transferred antibodies precipitate autoimmune thyroiditis in Obese strain (OS) chickens. *J. Immunol.* **137**, 542–545.

Kaufman, K. M., Letson, C. S., Komatsu, M., Sodetz, J. M. (1991). Molecular basis of C8α–γ deficiency in the rabbit: identification of an abnormal leader sequence in the C8α subunit. *Complement and Inflamm.* **8**, 171–172.

Kelly, R. H., Wolstencroft, R. A., Dumonde, D. C., Balfour, B. M. (1972). Role of lymphocyte activation products (LAP) in cell-mediated immunity. II. Effects of lymphocyte activation products on lymph node architecture and evidence for peripheral release of LAP following antigenic stimulation. *Clin. Exp. Immunol.* **10**, 49–65.

Kelus, A. S., Steinberg, C. M. (1991). Is there a high rate of mitotic recombination between the loci encoding immunoglobulin V_H and C_H regions in gonial cells? *Immunogenetics* **33**, 255–259.

Kelus, A. S., Weiss, S. (1977). Variant strain of rabbits lacking immunoglobulin κ polypeptide chain. *Nature* **265**, 156–158.

Kelus, A. S., Weiss, S. (1986). Mutation affecting the expression of immunoglobulin variable regions in the rabbit. *Proc. Natl Acad. Sci. USA* **83**, 4883–4886.

Kim, C. J., Isono, T., Tomoyoshi, T., Seto, A. (1994). Expression of T-cell receptor gamma/delta chain genes in a rabbit killer T-cell line. *Immunogenetics* **39**, 418–422.

Kim, C. J., Isono, T., Tomoyoshi, T., Seto, A. (1995). Variable-region sequences for T-cell receptor-γ and -δ chains of rabbit killer cell lines against Shope carcinoma cells. *Cancer Lett.* **89**, 37–44.

Kim, J. A., Territo, M. C., Wayner, E., Carlos, T. M., Parhami, F., Smith, C. W., Haberland, M. E., Fogelman, A. M., Berliner, J. A. (1994). Partial characterization of leukocyte binding molecules on endothelial cells induced by minimally oxidized LDL. *Arterioscler. Thromb.* **14**, 427–433.

Kimball, E. S., Coligan, J. E., Kindt, T. J. (1979). Structural characterization of antigens encoded by rabbit RLA-11 histocompatibility genes. *Immunogenetics* **8**, 201–211.

Kindt, T. J., Todd, C. W. (1970). Homocytotropic antibody in primary responder and hyperimmune rabbits. *J. Immunol.* **104**, 1491–1496.

Kindt, T. J., Singer, D. S. (1987). Class I major histocompatibility complex genes in vertebrate species: what is the common denominator? *Immunol. Res.* **6**, 57–66.

Klassen, J., McCluskey, R. T., Milgrom, F. (1971). Nonglomerular renal disease produced in rabbits by immunization with homologous kidney. *Am. J. Pathol.* **63**, 333–358.

Klein, J., Satta, Y., O'hUigin, C., Takahata, N. (1993). The molecular descent of the major histocompatibility complex. *Annu. Rev. Immunol.* **11**, 269–295.

Kling, D., Fingerle, J., Harlan, J. M., Lobb, R. R., Lang, F. (1995). Mononuclear leukocytes invade rabbit arterial intima during thickening formation via CD18- and VLA-4-dependent mechanisms and stimulate smooth muscle migration. *Circ. Res.* **77**, 1121–1128.

Knight, K. L., Becker, R. S. (1990). Molecular basis of the allelic inheritance of rabbit immunoglobulin V_H allotypes: implications for the generation of antibody diversity. *Cell* **60**, 963–970.

Knight, K. L., Hanly, W. C. (1975). Genetic control of α chains of rabbit IgA: allotypic specificities on the variable and the constant regions. In *Contemporary Topics in Molecular Immunology* (eds Inman, F. P., Mandy, W. J.), pp. 55–88. Plenum Press, New York, USA.

Knight, K. L., Leary, A. L., Tissot, R. G. (1980). Identification of two Ia-like alloantigens on rabbit B lymphocytes. *Immunogenetics* **10**, 443–453.

Knight, K. L., Martens, C. L., Stoklosa, C. M., Schneiderman, R. D. (1984). Genes encoding alpha-heavy chains of rabbit IgA: characterization of cDNA encoding IgA-g subclass alpha-chains. *Nucl. Acids Res.* **12**, 1657–1670.

Knight, K. L., Johnson, A., Coligan, J. E., Kindt, T. J. (1987). Partial amino acid sequence analysis of rabbit MHC class II molecules isolated from two-dimensional polyacrylamide gels. *Mol. Immunol.* **24**, 449–454.

Knight, K. L., Spieker-Polet, H., Kazdin, D., Oi, V. (1988). Transgenic rabbits with lymphocytic leukemia induced by the c-*myc* oncogene fused with the immunoglobulin heavy chain enhancer. *Proc. Natl Acad. Sci. USA* **85**, 3130–3134.

Kofler, H., Kofler, R., Wolf, H., Wick, G. (1983). Immunofluorescence studies on the codistribution of immune deposits and complement in the thyroid glands of Obese strain (OS) chickens. *Immunobiology* **4**, 390–401.

Komatsu, M. (1992). Molecular biology for genetic deficiency complement components in rabbits: C8α–γ deficiency and C3 hypocomplemententemia. *JARQ* **26**, 48–54.

Komatsu, M., Yamamoto, K., Kawashima, T., Migita, S. (1985). Genetic deficiency of the α–γ-subunit of the eight complement component in the rabbit. *J. Immunol.* **134**, 2607–2609.

Komatsu, M., Lamoyi, E., Mage, R. G. (1987). Genomic DNA encoding rabbit T cell receptor β-chains: isotypes and allotypes of Cβ. *J. Immunol.* **138**, 1621–1626.

Komatsu, M., Yamamoto, K., Nakano, Y., Nakazawa, M., Ozawa, A., Mikami, H., Tomita, M., Migita, S. (1988). Hereditary C3 hypocomplementemia in the rabbit. *Immunology* **64**, 363–368.

Komatsu, M., Imaoka, K., Satoh, M., Mikami, H. (1990). Hereditary C8α–γ deficiency associated with dwarfism in the rabbit. *J. Heredity* **81**, 413–417.

Komatsu, M., Yamamoto, K., Mikami, H., Sodetz, J. M. (1991). Genetic deficiency of complement component C8 in the rabbit: evidence of a translational defect in expression of the α–γ subunit. *Biochem. Genet.* **29**, 271–274.

Kotani, M., Yamamura, Y., Tamatani, T., Kitamura, F., Miyasaka, M. (1993a). Generation and characterization of monoclonal antibodies against rabbit CD4, CD5 and CD11a antigens. *J. Immunol. Methods* **157**, 241–252.

Kotani, M., Yamamura, Y., Tsudo, M., Tamatani, T., Kitamura, F., Miyasaka, M. (1993b). Generation of monoclonal antibodies to the rabbit interleukin-2 receptor α chain (CD25) and its distribution in HTLV-1-transformed rabbit T cells. *Jpn. J. Cancer Res.* **84**, 770–775.

Kraal, G., Schornagel, K., Streeter, P. R., Holzmann, N., Butcher, E. C. (1995). Expression of the mucosal vascular addressin MadCAM-1 on sinus-lining cells in the spleen. *Am. J. Pathol.* **147**, 763–771.

Krause, R. M. (1970). Factors controlling the occurrence of antibodies of uniform properties. *Fed. Proc.* **29**, 59–65.

Kraehenbuhl, J.-P., Neutra, M. R. (1992). Transepithelial transport and mucosal defence II: secretion of IgA. *Trends Cell Biol.* **2**, 170–174.

Kreider, J. W., Bartlett, G. L. (1981). The Shope papilloma-carcinoma complex of rabbits: a model system for neoplastic progression and spontaneous regression. *Adv. Cancer Res.* **35**, 81–110.

Kulaga, H., Sogn, J. A., Weissman, J. D., Marche, P. N., LeGuern, C., Long, E. O., Kindt, T. J. (1987). Expression patterns of MHC class II genes in rabbit tissues indicate close homology to human counterparts. *J. Immunol.* **139**, 587–592.

Kulaga, H., Folks, T., Rutledge, R., Truckenmiller, M., Gugel, E., Kindt, T. (1989). Infection of rabbits with human immunodeficiency virus 1. A small animal model for acquired immunodeficiency syndrome. *J. Exp. Med.* **169**, 321–326.

Kume, N., Cybulsky, M. I., Gimbrone, M. A. (1992). Lysophosphatidylcholine, a component of atherogenic lipoproteins, induces mononuclear leukocyte adhesion molecules in cultured human and rabbit arterial endothelial cells. *J. Clin. Invest.* **90**, 1138–1144.

Kusano, M., Choi, N.-H., Tomita, M., Yamamoto, K., Migita, S., Sekiya, T., Nishimura, S. (1986). Nucleotide sequence of cDNA and derived amino acid sequence of rabbit complement component C3 α-chain. *Immunol. Invest.* **15**, 365–378.

Lamoyi, E., Mage, R. G. (1985). Lack of K1b9 light chains in basilea rabbits is probably due to a mutation in an acceptor site for mRNA splicing. *J. Exp. Med.* **162**, 1149–1160.

Lamoyi, E., Mage, R. (1987). A cluster of rabbit T-cell β-chain variable region genes. *Immunogenetics* **25**, 55–62.

Lamoyi, E., Angiolillo, A. L., Mage, R. G. (1986). An evolutionarily conserved rabbit T-cell receptor β-chain variable region. *Immunogenetics* **23**, 266–270.

Larigan, J. D., Tsang, T. C., Rumberger, J. M., Burns, D. K. (1992). Characterization of cDNA and genomic sequences encoding rabbit ELAM-1: conservation of structure and functional interactions with leukocytes. *DNA Cell. Biol.* **11**, 149–162.

Laverriere, A., Kulaga, H., Kindt, T. J., LeGuern, C., Marche, P. N. (1989). A rabbit class II MHC gene with strong similarities to HLA-DRalpha. *Immunogenetics* **30**, 137–140.

LeGuern, C., Marche, P. N., Kindt, T. J. (1985). Molecular evidence of five distinct MHC class II α genes in the rabbit. *Immunogenetics* **22**, 141–148.

LeGuern, C., Weissman, J. D., Marche, P. N., Jouvin-Marche, E., Laverriere, A., Bagnato, M. R., Kindt, T. J. (1987). Sequence determination of a transcribed rabbit class II gene with homology to HLA-DQα. *Immunogenetics* **25**, 104–109.

LeGuern, A., Wetterskog, D., Marche, P. N., Kindt, T. J. (1987). A monoclonal antibody directed against a synthetic peptide reacts with a cell surface rabbit class I MHC molecule. *Molec. Immunol.* **24**, 455–461.

Lehrer, R. I., Lichtenstein, A. K., Ganz, T. (1993). Defensins: antimicrobial and cytotoxic peptides of mammalian cells. *Annu. Rev. Immunol.* **11**, 105–128.

Lerner, R. A., Dixon, F. J. (1966). Transfer of ovine experimental allergic glomerulonephritis (EAG) with serum. *J. Exp. Med.* **124**, 431–442.

Levine, P., Landsteiner, K. (1929). On immune isoagglutinins in rabbits. *J. Immunol.* **17**, 559–565.

Ley, K., Baker, J. B., Cybulsky, M. I., Gimbrone, M. A. Jr, Luscinskas, F. W. (1993). Intravenous interleukin-8 inhibits granulocyte emigration from rabbit mesenteric venules without altering L-selectin expression or leukocyte rolling. *J. Immunol.* **151**, 6347–6357.

Li, W. H., Gouy, M., Sharp, P. M., O'hUigin, C., Yang, Y. W. (1990). Molecular phylogeny of Rodentia, Lagomorpha, Primates, Artiodactyla, and Carnivora and molecular clocks. *Proc. Natl Acad. Sci. USA* **87**, 6703–6707.

Lieberman, R., Emorine, L., Max, E. E. (1984). Structure of a germline rabbit immunoglobulin V kappa-region gene: implications for rabbit V kappa–J kappa recombination. *J. Immunol.* **133**(5), 2753–2756.

Lin, Y.-L., Borenstein, L. A., Selvakumar, R., Ahmed, R., Wettstein, F. O. (1993). Progression from papilloma to carcinoma is accompanied by changes in antibody response to papillomavirus proteins. *J. Virol.* **67**, 382–389.

Lindbom, L., Lundberg, C., Prieto, J., Raud, J., Nortamo, P., Gahmberg, C. G., Patarroyo, M. (1990). Rabbit leukocyte adhesion molecules CD11/CD18 and their participation in acute and delayed inflammatory responses and leukocyte distribution *in vivo*. *Clin. Immunol. Immunopathol.* **57**, 105–119.

Lindqvist, K. J. (1968). A unique class of rabbit immunoglobulins eliciting passive cutaneous anaphylaxis in homologous skin. *Immunochemistry* **5**, 525–542.

Loar, L. S., Dennison, D., Sell, S. (1986). Production and characterization of monoclonal antibodies to rabbit lymphocyte subpopulations. I. Tissue immunofluorescence and flow cytometric analysis. *J. Immunol.* **137**, 2784–2790.

Lobel, S. A., Knight, K. L. (1984). The role of rabbit Ia molecules in immune functions as determined with the use of an anti-Ia monoclonal antibody. *Immunology* **51**, 35–43.

Loor, F., Kelus, A. S. (1978). Allelic exclusion in the B lineage cells of the rabbit. *Eur. J. Immunol.* **8**, 315–324.

Lyons, A. B., Parish, C. R. (1995). Are murine marginal-zone macrophages the splenic white pulp analog of high endothelial venules? *Eur. J. Immunol.* **25**, 3165–3172.

McCartney-Francis, N. L. (1987). Latent allotype expression in the rabbit. In *The Rabbit in Contemporary Immunological Research* (ed. Dubiski, S.), pp. 134–147. Longman Scientific & Technical, Harlow, England.

McCartney-Francis, N., Skurla, R., Mage, R. G., Bernstein, K. E. (1984). Kappa chain allotypes and isotypes in the rabbit: cDNA sequence of clones encoding b9 suggest an evolutionary pathway and possible role of the interdomain disulphide bond in quantitative allotype expression. *Proc. Natl Acad. Sci. USA* **81**, 1794–1798.

McDonnell, J., Hoerrner, L. A., Lark, M. W., Harper, C., Dey, T., Lobner, J., Eiermann, G., Kazazis, D., Siner, I. I., Moore, V. L. (1992). Recombinant human interleukin-1 beta-induced increase in levels of proteoglycans, stromelysin, and leukocytes in rabbit synovial fluid. *Arthritis Rheum.* **35**, 799–805.

McFadden, G., Graham, K. (1994). Modulation of cytokine networks by poxvirus: the myxoma virus model. *Semin. Virol.* **5**, 421–429.

McNicholas, J. M., Raffeld, M., Loken, M. R., Reiter, H., Knight, K. L. (1981). Monoclonal antibodies to rabbit lymphoid cells: preparation and characterization of a T-cell-specific antibody. *Molec. Immunol.* **18**, 815–822.

Mage, R. G. (1967). Quantitative studies on the regulation of expression of genes for immunoglobulin allotypes in heterozygous rabbits. *Cold Spring Harbor Symp. Quant. Biol.* **32**, 203–210.

Mage, R. G. (1975). Allotype suppression in rabbits: effects of anti-allotype antisera upon expression of immunoglobulin genes. *Transplant. Rev.* **26**, 84–99.

Mage, R. G., Young-Cooper, G. O., Alexander, C. (1971). Genetic control of variable and constant regions of immunoglobulin H chains. *Nature* **230**, 63–64.

Mage, R. G., Rejnek, J., Young-Cooper, G. O., Alexander, C. (1977). Are 'hidden' genes for immunoglobulin allotypes of rabbit heavy and light chains expressed upon hyperimmunization? *Eur. J. Immunol.* **7**, 460–468.

Mage, R., Dray, S., Gilman-Sachs, A., Hamers-Casterman, C., Hamers, R., Hanly, W., Kindt, T., Knight, K., Mandy, W., Naessens, J. (1982). Rabbit heavy chain haplotypes – allotypic determinants expressed by V_H–C_H recombinants. *Immunogenetics* **15**, 287–297.

Mage, R. G., McCartney-Francis, N. L., Komatsu, M., Lamoyi, E. (1987). Evolution of genes for allelic and isotypic forms of immunoglobulin kappa chains and of the genes for T-cell receptor beta chains in rabbits. *J. Molec. Evol.* **25**, 292–299.

Mage, R. G., Young-Cooper, G. O., Alexander, C. B., Hanly, W. C., Newman, B. A. (1992). A new V_H–C_H recombinant in the rabbit. *Immunogenetics* **35**, 131–135.

Manthei, U., Strunk, R. C., Giclas, P. C. (1984). Acute local inflammation alters synthesis, distribution, and catabolism of the third component of complement (C3) in rabbits. *J. Clin. Invest.* **74**, 424–433.

Marche, P. N., Kindt, T. J. (1986a). Two distinct T-cell receptor α-chain transcripts in a rabbit T-cell line: implications for allelic exclusion in T cells. *Proc. Natl Acad. Sci. USA* **83**, 2190–2194.

Marche, P. N., Kindt, T. J. (1986b). A variable region gene subfamily encoding T cell receptor β-chains is selectively conserved among mammals. *J. Immunol.* **137**, 1729–1734.

Marche, P. N., Tykocinski, M. L., Max, E. E., Kindt, T. J. (1985). Structure of a functional rabbit class I MHC gene: similarity to human class I genes. *Immunogenetics* **21**, 71–82.

Marche, P. N., Rebiere, M. C., Laverriere, A., English, D. W., LeGuern, C., Kindt, T. J. (1989). Definition of rabbit class I and class II MHC gene haplotypes using molecular typing procedures. *Immunogenetics* **29**, 273–276.

Marche, P. N., Laverriere, A., LeGuern, C., Kindt, T. J. (1991). A gene remnant in the rabbit MHC related to HLA-DRα. *Res. Immunol.* **142**, 525–532.

Mariame, B., Akimenko, M. A., Rougeon, F. (1987). Interallelic and intergenic conversion events could induce differential evolution of the two rabbit immunoglobulin kappa light chain genes. *Nucleic Acids Res.* **15**(15), 6171–6179.

Marsh, S. G. E., Bodmer, J. G. (1995). HLA class II region nucleotide sequences, 1995. *Tissue Antigens* **45**, 258–280.

Martens, C. L., Moore, K. W., Steinmetz, M., Hood, L. E., Knight, K. L. (1982). Heavy chain genes of rabbit IgG: isolation of a cDNA encoding gamma heavy chain and identification of two genomic C-gamma genes. *Proc. Natl Acad. Sci. USA* **79**, 6018–6022.

Martens, C. L., Currier, S. J., Knight, K. L. (1984). Molecular genetic analysis of genes encoding the heavy chains of rabbit IgG. *J. Immunol.* **133**, 1022–1027.

Matsukawa, A., Yoshimura, T., Maeda, T., Ohkuwara, S., Takagi, K., Yoshinaga, M. (1995). Neutrophil accumulation and activation by homologous IL-8 in rabbits: IL-8 induces destruction of cartilage and production of IL-1 and IL-1 receptor antagonist *in vivo*. *J. Immunol.* **154**, 5418–5425.

Mileski, W., Harlan, J., Rice, C., Winn, R. (1990). *Streptococcus pneumoniae*-stimulated macrophages induce neutrophils to emigrate by a CD18-independent mechanism of adherence. *Circ. Shock* **31**, 259–267.

Moise, K. J., Hesketh, D. E., Belfort, M. M., Saade, G., van den Veyver, I. B., Hudson, K. M., Rodkey, L. S. (1992). Ultrasound-guided blood sampling of rabbit fetuses. *Lab. An. Sci.* **42**, 398–401.

Moise, K. J., Saade, G., Knudsen, L., Valdez-Torres, A., Belfort, M. A., Hsu, H., Harvey, S. C., Hudson, K. M., Rodkey, L. S. (1994). Ultrasound-guided cardiac blood sampling of the rabbit fetus. *Fetal Diagn. Ther.* **9**, 331–336.

Moise, K. J., Rodkey, L. S., Saade, G., Gei, A., Dure, M., Graham, A., Creech, C. (1995a). An animal model for hemolytic disease of the fetus and newborn. I. Alloimmunization techniques. *Am. J. Obstet. Gynecol.* **173**, 51–55.

Moise, K. J., Rodkey, L. S., Saade, G. R., Dure, M., Dorman, K., Mayes, M., Graham, A. (1995b). An animal model for hemolytic disease of the fetus and newborn. II. Fetal effects in New Zealand rabbits. *Am. J. Obstet. Gynecol.* **173**, 747–753.

Moreira, A., Wollerton, M., Monks, J., Proudfoot, N. J. (1995). Upstream sequence elements enhance poly(A) site efficiency of the C2 complement gene and are phylogenetically conserved. *EMBO J.* **14**, 3809–3819.

Mossman, K., Nation, P., Macen, J., Garbutt, M., Lucas, A., McFadden, G. (1996). Myxoma virus M-T7, a secreted homolog of the interferon-γ, is a critical virulence factor for the development of myxomatosis in European rabbits. *Virology* **215**, 17–30.

Nakano, Y., Masuda, T., Sakamoto, T., Tomita, M. (1986). Isolation and characterization of rabbit properdin of the alternative complement pathway. *J. Immunol. Methods* **90**, 77–83.

Neutra, M. R., Kraehenbuhl, J.-P. (1992). Transepithelial transport and mucosal defence I: the role of M cells. *Trends Cell Biol.* **2**, 134–138.

Newman, B. A., Young-Cooper, G. O., Alexander, C. B., Becker, R. S., Knight, K. L., Kelus, A. S., Meier, D., Mage, R. G. (1991). Molecular analysis of recombination sites within the immunoglobulin heavy chain locus of the rabbit. *Immunogenetics* **34**, 101–109.

Novacek, M. (1996). Where do rabbits and kin fit in? *Nature* **379**, 299–300.

Ohkubo, C., Bigos, D., Jain, R. K. (1991). Interleukin 2 induced leukocyte adhesion to the normal and tumor microvascular endothelium *in vivo* and its inhibition by dextran sulfate: implications for vascular leak syndrome. *Cancer Res.* **51**, 1561–1563.

Okabayashi, M., Angell, M. G., Christensen, N. D., Kreider, J. W. (1991). Morphometric analysis and identification of infiltrating leukocytes in regressing and progressing Shope rabbit papillomas. *Int. J. Cancer* **49**, 919–923.

Okabayashi, M., Angell, M. G., Christensen, N. D., Kreider, J. W. (1993). Shope papilloma cell and leukocyte proliferation in regressing and progressing lesions. *Am. J. Pathol.* **142**, 489–496.

Olofsson, A. M., Arfors, K.-E., Ramezani, L., Wolitsky, B. A., Butcher, E. C., von Andrian, U. H. (1994). E-selectin mediates leukocyte rolling in interleukin-1-treated rabbit mesentery venules. *Blood* **84**, 2749–2758.

Pappo, J. (1989). Generation and characterization of monoclonal antibodies recognizing follicle epithelial M cells in rabbit gut-associated lymphoid tissue. *Cell. Immunol.* **120**, 31–41.

Pappo, J., Ermak, T. H. (1989). Uptake and translocation of fluorescent latex particles by rabbit Peyer's patch follicle epithelium: a quantitative model for M cell uptake. *Clin. Exp. Immunol.* **76**, 144–148.

Pappo, J., Ermak, T. H., Steger, H. J. (1991). Monoclonal antibody-directed targeting of fluorescent polystyrene microspheres to Peyer's patch M cells. *Immunology* **73**, 277–280.

Pavirani, A., McCartney-Francis, N., Jacobsen, F., Mage, R. G., Reddy, P. E., Fitzmaurice, L. C. (1983). Analyses of the splenic mRNA expressed by rabbits of different immunoglobulin kappa-light chain allotypes: conserved sequences in the 3′ untranslated region and allotype-specific probes. *J. Immunol.* **131**, 1000–1006.

Peak, P. W., Charlesworth, J. A., Pussell, B. A. (1991). Activation of rabbit C3: studies of the generation of cleavage products *in vitro* and of their metabolism *in vivo*. *Complement Inflamm.* **8**, 261–270.

Pospisil, R., Young-Cooper, G. O., Mage, R. G. (1995). Preferential expansion and survival of B lymphocytes based on V_H framework 1 and framework 3 expression: 'Positive' selection in appendix of normal and V_H-mutant rabbits. *Proc. Natl Acad. Sci. USA* **92**, 6961–6965.

Pospisil, R., Fitts, M. G., Mage, R. G. (1996). CD5 is a potential selecting ligand for B-cell surface immunoglobulin framework region sequences. *J. Exp. Med.* **184**, 1279–1284.

Raman, C., Knight, K. L. (1992). $CD5^+$ B cells predominate in peripheral tissues of rabbit. *J. Immunology* **149**, 3858–3864.

Rebierre, M. C., Marche, P. N., Kindt, T. J. (1987a). A rabbit class I major histocompatibility complex gene with a T cell-specific expression pattern. *J. Immunol.* **139**, 2066–2074.

Rebierre, M. C., Marche, P. N., LeGuern, C., Kindt, T. J. (1987b). Molecular studies of the rabbit major histocompatibility complex. In *The Rabbit in Contemporary Immunological Research* (ed. Dubisky, S.), pp. 42–62. Longman Scientific, Harlow, UK.

Reina, S., Markham, P., Gard, E., Rayed, F., Reitz, M., Gallo, R., Varnier, O. (1993). Serological, biological and molecular characterization of New Zealand White rabbits infected by intraperitoneal inoculation with cell-free human immunodeficiency virus. *J. Virol.* **67**, 5367–5374.

Reynaud, C. A., Anquez, V., Grimal, H., Weill, J. C. (1987). A hyperconversion mechanism generates the chicken light chain preimmune repertoire. *Cell* **48**, 379–388.

Reynaud, C. A., Dahan, A., Anquez, V., Weill, J. C. (1989). Somatic hyperconversion diversifies the single VH gene of the chicken with a high incidence in the D region. *Cell* **59**, 171–183.

Reynaud, C. A., Mackay, C. R., Muller, R. G., Weill, J. C. (1991). Somatic generation of diversity in a mammalian primary lymphoid organ: the sheep ileal Peyer's patch. *Cell* **64**, 995–1005.

Reynolds, J. D., Morris, B. (1983). The evolution and involution of Peyer's patches in fetal and postnatal sheep. *Eur. J. Immunol.* **13**, 627–635.

Richardson, M., Kurowska, E. M., Carroll, K. K. (1994). Early lesion development in the aortas of rabbits fed low-fat, cholesterol-free, semipurified casein diet. *Atherosclerosis* **107**, 165–178.

Roman-Franco, A. A., Turiello, M., Albini, B., Ossi, E., Milgrom, F., Andres, G. A. (1976). Anti-basement membrane antibody (A-BM Ab) and immune complexes (IC) in rabbits injected with mercuric chloride ($HgCl_2$). *Kidney Int.* **10**, 549 (abstract).

Roman-Franco, A. A., Turiello, M., Albini, B., Ossi, E., Milgrom, F., Andres, G. A. (1978). Anti-basement membrane antibodies and antigen–antibody complexes in rabbits injected with mercuric chloride. *Clin. Immunol. Immunopathol.* **9**, 464–481.

Rose, N. R., Witebsky, E. (1956). Studies on organ specificity. V. Changes in the thyroid glands of rabbits following active immunization with rabbit thyroid extracts. *J. Immunol.* **76**, 417–427.

Rosenbaum, J. T., Boney, R. S. (1993). Efficacy of antibodies to adhesion molecules CD11a or CD18 in rabbit models of uveitis. *Curr. Eye Res.* **12**, 827–831.

Rother, K. (1986). Rabbits deficient in C6. In *Hereditary and Acquired Complement Deficiencies in Animals and Man* (eds Rother, K., Rother, U.), Progress in Allergy, Vol. 39, pp. 192–201. Karger, Basel.

Roux, K. H. (1983). Rabbit latent allotypes: current status and future directions. *Surv. Immunol. Res.* **2**, 342–350.

Roux, K. H., Mage, R. G. (1996). Rabbit immunoglobulin allotypes. In *Handbook of Experimental Immunology*, 5th ed. (eds Weir, D. M., Herzenberg, L. A., Blackwell, C., Herzenberg, L. A.), Vol. I, Chapter 26, pp. 26.1–26.17. Blackwell Sciences Inc., Cambridge, MA, USA.

Rudzik, O., Perey, D. Y. E., Bienenstock, J. (1975). Differential IgA repopulation after transfer of autologous and allogeneic rabbit Peyer's patch cells. *J. Immunol.* **114**, 40–44.

Saida, T., Saida, K., Silberberg, D. H., Brown, M. J. (1981). Experimental allergic neuritis induced by galactocerebroside. *Ann. Neurol.* **9** (Suppl.), 87–101.

Samudzi, C. T., Burton, L. E., Rubin, J. R. (1991). Crystal structure of recombinant rabbit interferon-gamma at 2.7 Å resolution. *J. Biol. Chem.* **266**, 21791–21797.

Sawasdikosol, S., Hague, B. F., Zhao, T. M., Bowers, F. S., Simpson, R. M., Robinson, M., Kindt, T. J. (1993). Selection of rabbit $CD4^-CD8^-TCR\gamma\delta$ cells by *in vitro* transformation with HTLV-1. *J. Exp. Med.* **178**, 1337–1345.

Schneiderman, R. D., Hanly, W. C., Knight, K. L. (1989). Expression of 12 rabbit IgA $C\alpha$ genes as chimeric rabbit-mouse IgA antibodies. *Proc. Natl Acad. Sci. USA* **86**, 7561–7565.

Sell, S., Hsu, P-L. (1993). Delayed hypersensitivity, immune deviation, antigen processing and T-cell subset selection in syphilis pathogenesis and vaccine development. *Immunol. Today* **14**, 577–582.

Selvakumar, R., Borenstein, L. A., Lin, Y.-L., Ahmed, R., Wettstein, F. O. (1994). T cell response to Cottontail rabbit papillomavirus structural proteins in infected rabbits. *J. Virol.* **68**, 4043–4048.

Seto, A., Kawanishi, M., Matsuda, S., Ogawa, K., Miyoshi, I. (1987). Induction of preleukemic stage of adult T cell leukemia-like disease in rabbits. *Jpn. J Cancer Res.* **78**, 1150–1155.

Setupathi, P., Spieker-Polet, H., Polet, H., Yam, P., Tunyaplin, C., Knight, K. (1994). Lymphoid and non-lymphoid tumors in E_k-*myc* transgenic rabbits. *Leukemia* **8**, 2144–2155.

Shibata, S., Sakaguchi, H., Nagasawa, T., Naruse, T. (1972).

Nephritogenic glycoprotein. II. Experimental production of membranous glomerulonephritis in rats by a single injection of homologous renal glycopeptide. *Lab. Invest.* **27**, 457–465.

Short, J. A., Sethupathi, P., Zhai, S. K., Knight, K. L. (1991). VDJ genes in V_Ha2 allotype-suppressed rabbits. Limited germline V_H gene usage and accumulation of somatic mutations in D regions. *J. Immunol.* **147**, 4014–4188.

Simpson, R., Hubbard, B., Alling, D., Teller, R., Fain, M., Bowers, F., Kindt, T. (1995). Rabbits transfused with human immunodeficiency virus Type 1-infected blood develop immune deficiency with $CD4^+$ lymphocytopenia in the absence of clear evidence for HIV type 1 infection. *AIDS Res. Hum. Retrovirus.* **11**, 297–305.

Simpson, R., Zhao, T., Hubbard, B., Sawasdikosol, S., Kindt, T. (1996). Experimental acute T cell leukemia-lymphoma is associated with thymic atrophy in human T cell leukemia virus type I infection. *Lab. Invest.* **74**, 696–710.

Sire, J., Colle, A., Burgois, A. (1979). Identification of an IgD-like surface immunoglobulin on rabbit lymphocytes. *Eur. J. Immunol.* **9**, 13–16.

Sittisombut, N. (1987). Structure and organization of the rabbit major histocompatibility complex class II genes. Ph.D. Thesis, University of Illinois at Chicago, Chicago.

Sittisombut, N. (1996). The major histocompatibility complex of the rabbit. In *The Major Histocompatibility Complex Region of Domestic Animal Species* (eds Shook, L. B., Lamont, S. J.), pp. 269–292. CRC Press, Boca Raton, FL, USA.

Sittisombut, N., Knight, K. L. (1986). Rabbit major histocompatibility complex. I. Isolation and characterization of three subregions of class II genes. *J. Immunol.* **136**, 1871–1875.

Sittisombut, N., Mordacq, J., Knight, K. L. (1988). Rabbit MHC II. Sequence analysis of the R-DPα- and β-genes. *J. Immunol.* **140**, 3237–3243.

Sittisombut, N., Tissot, R., Knight, K. L. (1989). Rabbit major histocompatibility complex. III. Multiple class II DRβ genes and restriction fragment length polymorphism of the Class II α and β genes. *J. Immunogenet.* **16**, 63–75.

Smet, E. G., De Smet, W., Byrs, L., De Baetselier, P. C. (1986). Mab.198: a monoclonal antibody recognizing the complement type 3 receptor (CR3) in the rabbit. *Immunology* **59**, 419–425.

Spieker-Polet, H., Sittisombut, N., Yam, P. C., Knight, K. L. (1990). Rabbit major histocompatibility complex. IV. Expression of major histocompatibility complex class II genes. *J. Immunogenet.* **17**, 123–132.

Spieker-Polet, H., Yam, P.-C., Knight, K. L. (1993). Differential expression of 13 IgA-heavy chain genes in rabbit lymphoid tissues. *J. Immunol.* **150**, 5457–5464.

Springer, T. A. (1995). Traffic signals on endothelium for lymphocyte recirculation and leukocyte emigration. *Annu. Rev. Physiol.* **57**, 827–872.

Syverton, J. T. (1952). The pathogenesis of the rabbit papilloma-to-carcinoma sequence. *Ann. NY Acad. Sci.* **54**, 1126–1140.

Talbott, G. A., Sharar, S. R., Harlan, J. M., Winn, R. K. (1994). Leukocyte-endothelial interactions and organ injury: the role of adhesion molecules. *New Horizons* **2**, 545–554.

Takata, Y., Hozono, H., Takeda, J., Tanaka, E., Hong, K., Kinoshita, T., Inoue, K. (1985). Two types of rabbit C4, one of which consists of two peptide chains and the other of three chains. *Complement* **2**, 77.

Tissot, R. G., Cohen, C. (1972). Histocompatibility in the rabbit. Identification of a major locus. *Tissue Antigens* **2**, 267–279.

Tissot, R. G., Cohen, C. (1974). Histocompatibility in the rabbit. Linkage between RL-A, MLC, and the He blood group loci. *Transplantation* **18**, 142–149.

Tomazic, V., Rose, N. R. (1975). Autoimmune murine thyroiditis. VII. Induction of the thyroid lesions by passive transfer of immune serum. *Clin. Immunol. Immunopathol.* **4**, 511–518.

Tykocinski, M. L., Marche, P. N., Max, E. E., Kindt, T. J. (1984). Rabbit class I MHC genes: cDNA clones define full-length transcripts of an expressed gene and a putative pseudogene. *J. Immunol.* **133**, 2261–2269.

van Osselaer, N., Herman, A. G., Rampart, M. (1993). Dextran sulphate inhibits neutrophil emigration and neutrophil-dependent plasma leakage in rabbit skin. *Agents Actions* **38**, C51–53.

van Tol, M., Veenhoff, E., Seijen, H. (1978). Rabbit IgM: its isolation in high yields by a convenient procedure using serum from trypanosome-infected animals. *J. Immun. Methods* **21**, 125–131.

Vladutiu, A. O. (1990). Thyroglobulin antibodies in experimental autoimmune thyroiditis. In *Organ-Specific Autoimmunity* (eds Bigazzi, P. E., Wick, G., Wicher, K.), pp. 241–266. Marcel Dekker, New York, USA.

von Zabern, I. (1988). Species-dependent incompatibilities. In *The Complement System* (eds Rother, K., Till, G. O.), pp. 196–202. Springer, Berlin Heidelberg New York.

Waksman, B. H. (1959). Experimental encephalomyelitis and the autoallergic disease. *Int. Arch. Allergy* **14** (Suppl.), 1–87.

Wank, R., Thomssen, C. (1991). High risk of squamous cell carcinoma of the cervix for women with HLA-DQw3. *Nature* **352**, 723–725.

Watanabe, Y. (1980). Serial inbreeding of rabbits with hereditary hyperlipidemia (WHHL-rabbit): incidence and development of atherosclerosis and xanthoma. *Atherosclerosis* **36**, 261–267.

Weetman, A. P., Cohen, S. B., Oleesky, D. A., Morgan, B. P. (1989). Terminal complement complexes and C1/C1 inhibitor complexes in autoimmune thyroid disease. *Clin. Exp. Immunol.* **77**, 25–30.

Weinstein, P. D., Cebra, J. J. (1991). The preference for switching to IgA expression by Peyer's patch germinal center B cells is likely due to the intrinsic influence of their microenvironment. *J. Immunol.* **147**, 4126–4135.

Weinstein, P. D., Mage, R. G., Anderson, A. O. (1994a). The appendix functions as a mammalian bursal equivalent in the developing rabbit. In *Proceedings of the 11th International Conference on Lymphoid Tissues and Germinal Centers* (ed. Heinen, E.), pp. 249–253. Plenum Press, New York, USA.

Weinstein, P. D., Anderson, A. O., Mage, R. G. (1994b). Rabbit IgH sequences in appendix germinal centers: V_H diversification by gene conversion-like and hypermutation mechanisms. *Immunity* **1**, 647–659.

White, R. V., Kaufman, K. M., Letson, C. S., Platteborze, P. L., Sodetz, J. M. (1994). Characterization of rabbit complement component C8: functional evidence for the species-selective recognition of C8 alpha by homologous restriction factor (CD59). *J. Immunol.* **152**, 2501–2508.

Wick, G., Graf, J. (1972). Electron microscopic studies in chickens of the Obese strain (OS) with spontaneous hereditary autoimmune thyroiditis. *Lab. Invest.* **27**, 400–411.

Wick, G., Breznischek, H. P., Hala, K., Dietrich, H., Wolf, H., Kroemer, G. (1989). The Obese strain of chickens. An animal model with spontaneous autoimmune thyroiditis. *Adv. Immunol.* **47**, 433–450.

Wiebke, J. L., Quinlan, W. M., Doyle, N. A., Sligh, J. E., Smith, C. W., Doerschuk, C. M. (1995). Lymphocyte accumulation during *Pseudomonas aeruginosa*-induced pneumonia in rodents does not require CD11a and intercellular adhesion molecule-1. *Am. J. Respir. Cell Mol. Biol.* **12**, 513–519.

Wiggs, B. R., English, D., Quinlan, W. M., Doyle, N. A., Hogg, J. C., Doerschuk, C. M. (1994). Contributions of capillary pathway size and neutrophil deformability to neutrophil transit through rabbit lungs. *J. Appl. Physiol.* 77, 463–470.

Wilder, R. L., Yuen, C. C., Coyle, S. A., Mage, R. G. (1979). Demonstration of a rabbit cell surface Ig that bears light chain and V_H, but lacks μ-, α-, and γ-allotypes – rabbit IgD? *J. Immunol.* **122**, 464–468.

Wilkinson, J. M., Tykocinski, M. L., Coligan, J. E., Kimball, E. S., Kindt, T. J. (1982). Rabbit MHC antigens: occurrence of non-β2-microglobulin-associated class I molecules. *Mol. Immunol.* **19**, 1441–1451.

Wilkinson, J. M., Wetterskog, D. L., Sogn, J. A., Kindt, T. J. (1984). Cell surface glycoproteins of rabbit lymphocytes: characterization with monoclonal antibodies. *Molec. Immunol.* **21**, 95–103.

Wilkinson, J. M., Galea-Lauri, J., Reid, H. W. (1992a). A cytotoxic rabbit T-cell line infected with a γ-herpes virus which expresses CD8 and class II antigens. *Immunology* 77, 106–108.

Wilkinson, J. M., Galea-Lauri, J., Sellars, R. A., Boniface, C. (1992b). Identification and tissue distribution of rabbit leucocyte antigens recognized by monoclonal antibodies. *Immunology* **76**, 625–630.

Wilkinson, J. M., McDonald, G., Smith, S., Galea-Lauri, J., Lewthwaite, J., Henderson, B., Revelle, P. A. (1993). Immunohistochemical identification of leucocyte populations in normal tissue and inflamed synovium of the rabbit. *J. Pathol.* **170**, 315–320.

Winn, R. K., Harlan, J. M. (1993). CD18-independent neutrophil and mononuclear leukocyte emigration into the peritoneum of rabbits. *J. Clin. Invest.* **92**, 1168–1173.

Witebsky, E., Rose, N. R. (1956). Studies on organ specificities. IV. Production of rabbit thyroid antibodies in the rabbit. *J. Immunol.* **76**, 408–416.

Wutzler, P., Meerbach, A., Farber, I., Wolf, H., Scheibner, K. (1995). Malignant lymphomas induced by an Epstein–Barr virus-related herpesvirus from *Macaca arctoides* – a rabbit model. *Arch. Virol.* **140**, 1979–1995.

Yared, M. A., Moise, K. J., Rodkey, L. S. (1998). Stable solid-phase Rh antigen. *Transfusion Medicine* (in press).

Yoshimura, T., Yuhki, N. (1991). Neutrophil attractant/activation protein-1 and monocyte chemoattractant protein-1 in rabbit: cDNA cloning and their expression in spleen cells. *J. Immunol.* **146**, 3483–3488.

Zeltner, R., Borenstein, L. A., Wettstein, F. O., Iftner, T. (1994). Changes in RNA expression pattern during the malignant progression of cottontail rabbit papillomavirus-induced tumors in rabbits. *J. Virol.* **68**, 3620–3630.

Zhao, T. M., Robinson, M. A., Bowers, F. S., Kindt, T. J. (1995). Characterization of an infectious molecular clone of human T-cell leukemia virus type I. *J. Virol.* **69**, 2024–2030.

VII IMMUNOLOGY OF THE DOG

1. Introduction

The dog has long been an important research model in two major areas. Dogs play an important role in the investigation of new drugs since they are one of the major models used in toxicity trials, including the effects of investigational drugs on the immune system. Historically, the dog has been a valuable model for bone marrow transplantation, with many of the advances made in the dog being directly transferrable to human clinical bone marrow transplantation protocols. More recently, the dog has become an important model in the study of primary immunodeficiency disease. For example, the determination of the immunologic defect in X-linked severe combined immunodeficient (XSCID) dogs led to the discovery of the gene responsible for XSCID.

Because dogs develop many of the same immunologic diseases as humans, they represent an ideal large animal model in which to study the immunology and pathogenesis of these diseases in a compressed period of time. Previously, the major limitation of the use of the dog as an experimental model in immunological research has been the paucity of immunological reagents available to dissect the canine immune system. Over the past few years, great strides forward have been made in the development of these reagents.

2. Lymphoid Organs and their Anatomical Distribution

Thymus

The thymus is believed to be the main site of T cell development in the dog. It consists of two main lobes and is situated in the cranial mediastinum. A thymic anlage is observed in 25-day-old embryos (Kelly, 1963). Lymphocytes are present in the thymus of 35-day-old fetuses and Hassall's corpuscles appear at 45 days (Kelly, 1963; Snyder *et al.*, 1993). The thymus undergoes rapid postnatal growth and reaches maximum size at 1–2 months of age as percentage of body weight, and at 6 months of age in absolute terms (Yan and Gawlak, 1989). The thymus involutes in older dogs, but remnants of thymic tissue can usually be identified even in old animals.

The thymus of 4–8-week-old dogs contains approximately 15% double negative ($CD4^-CD8^-$) cells, 70% double-positive ($CD4^+CD8^+$) cells, 12% $CD4^+CD8^-$ cells, and 3% $CD4^-CD8^+$ cells (Somberg *et al.*, 1994).

The thymus receives its arterial blood supply from the internal thoracic arteries, the brachiocephalic artery and the left subclavian artery (Bezuidenhout, 1993). Efferent lymph vessels drain into the cranial mediastinal and sternal lymph nodes. The thymus contains sympathetic and parasympathetic nerves.

Spleen

The spleen is a boot-shaped organ in the abdominal cavity. Its exact location depends on the fullness of the stomach. It has a fibroelastic capsule that allows great distention. Trabeculae extend from the capsule into the parenchyma. A rudimentary spleen is present in 27–28-day-old embryos, but lymphocytic infiltration does not occur until gestation day 52 (Bryant and Shifrine, 1972). Germinal centers and plasma cells are not present until after birth. The white pulp consists of lymphoid follicles and periarteriolar lymphoid sheaths (PALS) surrounded by a marginal zone. The PALS contains predominantly $CD5^+CD4^+$ T cells and fewer $CD5^+CD8^+$ T cells (Rabanal *et al.*, 1995). The red pulp consists of a network of reticular cells and macrophages, and venous sinuses.

The splenic artery is a branch of the celiac or cranial mesenteric artery which forms 25 branches that enter the spleen via the hilus. The arteries pass through the trabeculae, forming smaller branches that enter the white pulp. These branches end in the marginal zone or form connections with the venous sinuses of the red pulp. The spleen is mainly innervated by sympathetic nerves from the celiac plexus (Bezuidenhout, 1993).

The main immune function of the spleen is the response to blood-borne pathogens.

Lymph nodes

The lymph nodes of the dog with their afferent and efferent lymphatics are listed in Table VII.2.1, and their distribution is illustrated in Figure VII.2.1.

The lymph node can be divided into a cortex with

Handbook of Vertebrate Immunology
ISBN 0-12-546401-0

Table VII.2.1 Lymph nodes (ln) in the dog and their afferent and efferent lymphatics

Lymph node	*Draining area*	*Efferent lymphatics*
Parotid	Superficial head regions caudal of eyes, external ear, eyelid, masticatory muscles	Medial retropharyngeal ln
Mandibular	Head, salivary glands	Medial retropharyngeal ln
Lateral pharyngeal	Adjacent structures	Medial retropharyngeal ln
Medial retropharyngeal	Parotid and mandibular lymph nodes, tongue, larynx, esophagus	Tracheal trunks
Superficial cervical	Superficial tissues of neck, head, thoracic limb and wall	Right tracheal trunk, thoracic duct, external jugular veins
Deep cervical	Esophagus, trachea, last 5–6 cervical vertebrae	Tracheal trunk, thoracic duct or mediastinal ln
Axillary	Superficial tissues of thoracic and ventral abdomen, mammary glands	Tracheal trunk, thoracic duct, external jugular veins
Accessory axillary	Superficial tissues of thoracic and ventral abdomen, mammary glands	Tracheal trunk, thoracic duct, external jugular veins
Sternal	Thoracic wall, ventral abdominal wall, diaphragm, mediastinum, thymus	Right lymphatic duct, thoracic duct
Intercostal	Ribs, pleura, intercostal tissues, vertebrae	Cranial mediastinal ln
Cranial mediastinal	Thymus, trachea, esophagus, heart, pleura, pulmonary ln, intercostal ln, tracheobronchial ln	Tracheal trunks, right lymphatic duct, thoracic duct
Tracheobronchial	Lung, thoracic viscera	Cranial mediastinal ln
Pulmonary	Lung	Tracheobronchial ln
Lumbar	Lumbar vertebrae, aorta, adrenal glands, kidneys, diaphragm, ovaries, uterus	Lumbar trunk, cisterna chyli
Medial iliac	Pelvis, urogenital organs, deep and superficial inguinal ln, left colic, sacral and hypogastric ln	Lumbar trunk, lumbar ln
Hypogastric iliac	Hindlimb, urogenital organs	Medial iliac, lumbar trunk
Sacral	Rectum, genital organs	Medial and hypogastric iliac
Hepatic	Stomach, duodenum, liver, pancreas	Intestinal trunk
Splenic	Esophagus, stomach, pancreas, spleen, liver, omentum, diaphragm	Intestinal trunk
Gastric	Esophagus, stomach, diaphragm, peritoneum	Intestinal trunk
Pancreatico-duodenal	Duodenum, pancreas, omentum	Intestinal trunk
Jejunal	Jejunum, ileum, pancreas	Intestinal trunk
Colic	Ileum, cecum, colon	Intestinal trunk, medial iliac ln, lumbar ln
Popliteal	Pelvic limb distal to node	Medial iliac ln
Iliofemoral	Medial surface of pelvic limb, knee and hock joint	Medial iliac ln
Superficial inguinal	Abdominal wall, caudal mammae, penis, scrotum	Medial iliac ln

follicles, a paracortex, and a medulla. The paracortical cells are mainly $CD5^+$ T cells, 70–80% of which express CD4 and 20–30% express CD8 (Rabanal *et al.*, 1995). The follicles are populated by B cells and a few $CD4^+$ T cells. Medullary cords contain mostly plasma cells and macrophages.

The arterial blood supply to the lymph node is variable among dogs and even among lymph nodes within an animal (Salvador *et al.*, 1992; Belz and Heath, 1995a). In some dogs, the main artery enters the node via the hilus, as is usually the case in other species. However, in many nodes, arteries form a network that surrounds the lymph node, and from which branches penetrate the capsule. These branches may provide the bulk of the arterial blood supply to the lymph node. Venous blood leaves the lymph node via the hilus. High endothelial venules are present in the deep cortex (Belz and Heath, 1995a). Afferent lymph vessels divide into terminal lymphatics that deliver lymph into the subcapsular or trabecular sinuses (Belz and Heath, 1995b). Lymph flows through a continuous network of the subcapsular, trabecular, cortical and medullary sinuses and leaves the lymph node via the hilus. The sympathetic innervation of the mesenteric lymph node of the dog has been studied in some detail (Popper *et al.*, 1988). The sympathetic nerves enter the lymph node via the hilus. They are present in medullary trabeculae and are associated with blood vessels in the medulla, paracortex and capsule (Popper *et al.*, 1988). This suggests a possible role

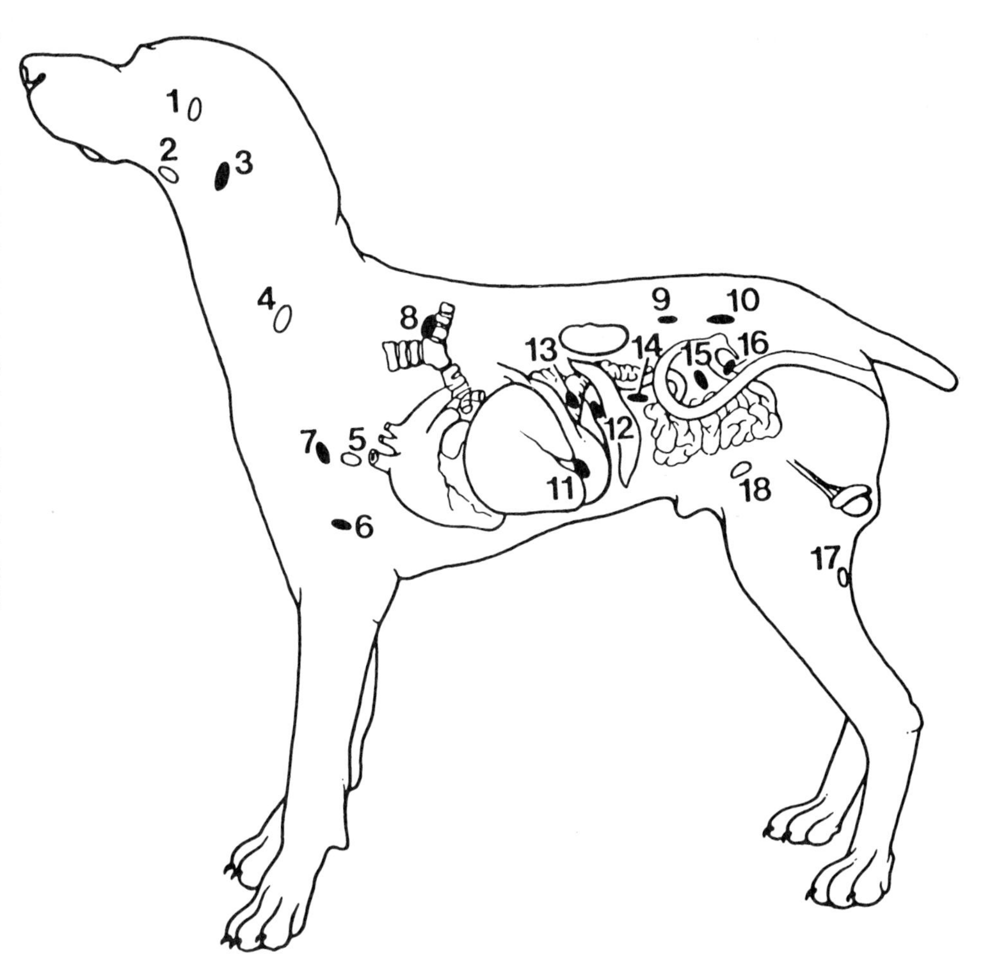

Figure VII.2.1 Distribution of lymph nodes present in the dog. The open symbols indicate superficially located lymph nodes (ln) and the filled symbols internal lymph nodes. Adapted from *Anatomy of the Dog. An Illustrated Text*, Budras, K. D., Fricke, W., McCarthy, P. H. (1994). 1, parotid; 2, mandibular; 3, medial retropharyngeal; 4, superficial cervical; 5, axillary; 6, sternal; 7, cranial mediastinal; 8, tracheobronchial; 9, lumbar; 10, medial iliac; 11, hepatic; 12, splenic; 13, gastric; 14, pancreaticoduodenal; 15, jejunal; 16, colic; 17, popliteal; 18, superficial inguinal.

of sympathetic nerves in the regulation of the blood flow through the lymph node.

Tonsils

The palatine tonsils are paired lymphoid organs, located in the lateral wall of the pharynx. Smaller and less well-defined lymphoid tissues of the oropharyngeal cavity are the lingual and pharyngeal tonsils. The palatine tonsils consist of follicles populated by B cells and a few $CD4^+$ T cells, and an interfollicular area with T cells. Squamous cell epithelium overlies the tonsils and forms crypts that extend deeply into the lymphoid tissue. The arterial blood supply comes from the tonsilar artery, a branch of the lingual artery. Numerous efferent lymphatics merge into two or three lymph vessels that drain into the medial retropharyngeal lymph node (Belz and Heath, 1995b).

Peyer's patches

The dog has two types of Peyer's patches (HogenEsch *et al.*, 1987a; HogenEsch and Felsburg, 1992a). The duodenal and jejunal Peyer's patches are secondary lymphoid tissues presumably of importance to the mucosal immune response. The single ileal Peyer's patch has several features of a primary lymphoid organ, including predominant IgM production and early involution (HogenEsch and Felsburg, 1992b). The follicles of the ileal Peyer's patch of young dogs are large and populated by IgM B cells (HogenEsch and Felsburg, 1992a). The interfollicular areas are small. The duodenal and jejunal Peyer's patches have relatively smaller follicles and large domes and interfollicular areas. The interfollicular areas contain predominantly T cells. The lymph vessels from the Peyer's patches form a plexus beneath the follicles from which efferent lymph vessels drain to the jejunal lymph nodes and to the hepatic lymph nodes (duodenal Peyer's patches).

3. Leukocytes and their Antigens

In 1993, the First International Canine Leukocyte Antigen Workshop (CLAW) was held to identify monoclonal antibodies (mAb) to canine leukocyte antigens and canine equivalents of documented CD antigens (Cobbold

Table VII.3.1 Monoclonal antibodies specific for canine leukocyte antigens

Antigen	*Other name(s)*	*MW*	*Cellular distribution*
CD1a		49	Thymocytes, Langerhans and dendritic cells
CD1c		43	Thymocytes, Langerhans and dendritic cells
CD4	MHC class II receptor	58	Thymocytes, T cell subset, neutrophils
CD5		67	Thymocytes, T cells, B cell subset
CD8α	MHC class I receptor	35	Thymocytes, T cell subset
CD11a	LFA-1	180	Leukocytes
CD11b	Mac-1, Mo-1, CR3	175	Granulocytes, monocytes
CD11c		150	Granulocytes, monocytes
CD18	β_2 integrin	95	Leukocytes
CD21	p150/95	150	B cells, follicular dendritic cells
CDw41	GPIIb/IIIa		Platelets, megakaryocytes
CD44	Ppg-1	90	Leukocytes, activated T cells
CD45	LCA	180–220	Pan leukocytes
CD45RA		205, 220	B cells, T cell subset
CD49d	VLA-4	130, 150	Thymocytes, T cells, myeloid cells
CD54	ICAM-1	95	Broad
CD62	P-selectin, GMP-140	140	Platelets, megakaryocytes, endothelial cells
CD90	Thy-1	29	Thymocytes, T cells
Miscellaneous			
MHC class II		28, 34	B cells, T cells, monocytes, dendritic cells
TCR$\alpha\beta$			Thymocytes, T cells
TCR$\gamma\delta$			Thymocytes, T cells
αd	CD11d?	155	Macrophage subset
T cells			Thymocytes, pan T cells
B cells			Mature B cells
Granulocytes			Granulocytes
Monocytes			Monocytes, macrophages

and Metcalfe, 1994). Nineteen groups submitted 127 mAb to the workshop; 58 of the 127 mAb were clustered within several CD designations. Table VII.3.1 lists the defined mAb in the dog (Moore *et al.*, 1990, 1992, 1994, 1996; Sandmaier *et al.*, 1990; Smith *et al.*, 1991; Danilenko *et al.*, 1992, 1995; Gebhard and Carter, 1992; Dore *et al.*, 1993; Moore and Rossitto, 1993; Cobbold and Metcalfe, 1994).

Canine monoclonal antibodies are only now becoming commercially available. Table VII.3.2 lists the canine specific monoclonal antibodies that have either been documented at the 1993 CLAW meeting or characterized in other publications. For those mAb not commercially available, the source of the antibody is listed with the reference for the antibody.

Chabanne *et al.* (1994) recently screened a large number of human mAb for their cross-reactivity against canine leukocytes. Table VII.3.3 lists those human mAb that were shown to be cross-reactive.

The cDNAs for six canine leukocyte antigens have been cloned including CD3ε (Nash *et al.*, 1991a), CD4 and CD8α (Gorman *et al.*, 1994), CD28 (Pastori *et al.*, 1994), CD34 (McSweeney *et al.*, 1996), and CD38 (Uribe *et al.*, 1997). Table VII.3.4 compares the canine cDNAs with those of other species.

4. Leukocyte Traffic and Associated Molecules

No detailed studies of organ- and tissue-specific recruitment of the various leukocyte populations in the dog have been reported. Much of what is known about leukocyte trafficking in the dog is limited to the interaction of the leukocytes with the vascular endothelium (Smith *et al.*, 1991; Spertini *et al.*, 1991; Youker *et al.*, 1992; Dore *et al.*, 1993, 1995; Lefer *et al.*, 1994; Jerome *et al.*, 1994). Table VII.4.1 lists the canine leukocyte adhesion molecules that have been identified with their corresponding endothelial ligand. The major endothelial cell ligand, ICAM-1, has recently been cloned and shows 61% identity with human ICAM-1 (Manning *et al.*, 1995). Endothelial cell activation by LPS, IFN-γ, IL-1β, and TNF-α results in the up-regulation of ICAM-1. Several endothelial ligands that have been identified in other species such as ICAM-2, E-selectin (ELAM-1), and VCAM-1 have yet to be identified in the dog.

In vitro and *in vivo* studies have identified chemotactic factors that are responsible for the recruitment of leukocyte populations in the dog and are listed in Table VII.4.2 (Thomsen and Strom, 1989; Strom and Thomsen, 1990; Thomsen and Thomsen, 1990; Thomsen *et al.*, 1991;

Table VII.3.2 Available canine monoclonal antibodies

Antigen	*mAb*	*Isotype*	*Source*
CD1a	CA 9.AG5	IgG1	P. Moore (CLAW)
CD1c	CA 13.9H11	IgG1	P. Moore (CLAW)
CD4	YKIX 302.9.3.7	Rat IgG2a	Biosource
	LSM 12.125	IgG1	Custom Monoclonals, Sacramento, CA, USA
	LSM 8.53	IgM	D. Gebhard (CLAW)
	DH 29A	IgM	VMRD, Pullman, WA, USA
	CA 13.1E4	IgG1	P. Moore (CLAW)
CD5	YKIX 322.3.2	Rat IgG2a	S. Cobbold
	DOG 17-4-8	Rat IgG2a	Connex, Martinsed, Germany
	DH 3B	IgG1	VMRD, Pullman, WA, USA
	DH 13A	IgM	VMRD, Pullman, WA, USA
	DH 14A	IgM	VMRD, Pullman, WA, USA
	DH 39C	Ig?	W. Davis (CLAW)
	DH 60A	IgM	W. Davis (CLAW)
CD8α	YCATE 55.9.1	Rat IgG1	Biosource
	YCATE 60.3.9	Rat IgG1	S. Cobbold (CLAW)
	DOG 10-1-1	IgG1	Connex, Martinsed, Germany
	LSM 1.140	IgG1	Custom Monoclonals, Sacramento, CA, USA
	LSM 4.78	IgM	Custom Monoclonals, Sacramento, CA, USA
	CA 9.JD3	IgG2a	P. Moore (CLAW)
CD11a	CA 11.4D3	IgG1	P. Moore (CLAW)
CD11b	CA 16.3E10	IgG1	P. Moore (CLAW)
CD11c	CA 11.6A1	IgG1	P. Moore (CLAW)
CD18	CA 1.4E9	IgG1	P. Moore (CLAW)
	CA 16.3C10	IgG1	P. Moore (CLAW)
	H20A	IgG1	VMRD, Pullman, WA, USA
	BA030A	IgG1	VMRD, Pullman, WA, USA
CD21	CA 2.1D6	IgG1	P. Moore (CLAW)
CDw41	2F9	Ig?	S. Burnstein (CLAW)
	DOG 20-4	IgG2a	Connex, Martinsed, Germany
	DOG 21-5	Rat IgG2a	Connex, Martinsed, Germany
CD44	S5	IgG1	B. Sandmaier (CLAW)
	YKIX 337.8.7	Rat IgG2a	Biosource
CD45	YKIX 716.13.2	Rat IgG2b	Biosource
	DOG 23-4	Rat IgG2a	Connex, Martinsed, Germany
	DOG 32-1	Rat IgG2a	C. Voss (CLAW)
	CA 12.10C12	IgG1	P. Moore (CLAW)
	7B6	IgG1	R. Alejandro (CLAW)
CD45RA	CA 4.1D3	IgG1	P. Moore (CLAW)
CD49d	CA 4.5B3	IgG1	P. Moore (CLAW)
CD54	CL 18.1D8	IgG1	C. Smith (Smith *et al.*, 1991)
CD62	MD3	IgG1	C. Smith (Dore *et al.*, 1993)
CD90	F3-20-7	Ig?	Biosource, Serotec
	DOG 13-1	Rat IgG2a	Connex, Martinsed, Germany
	DOG 14-2	Rat IgG2a	Connex, Martinsed, Germany
	DH 2A	IgM	VMRD, Pullman, WA, USA
	DH 24H	Ig?	VMRD, Pullman, WA, USA
	YKIX 337.217.7	Rat IgG2a	Biosource
	CA 1.4G8	IgG1	P. Moore (CLAW)
	5G2	IgG2a	R. Alejandro (CLAW)
Miscellaneous			
MHC class II	YKIX 334.2.1	Rat IgG2a	Biosource
	DOG 26-1	Rat IgG2a	Connex, Martinsed, Germany
	TGD 12	IgG2a	H. Grosse-Wilde (CLAW)
	CA 2.1C2	IgG1	P. Moore (CLAW)
	B1F6	IgG2a	R. Alejandro (CLAW)
TCR$\alpha\beta$	CA 15.8G7	IgG1	P. Moore (Moore and Rossitto, 1993)
TCR$\gamma\delta$	CA 20.8H1	IgG2a	P. Moore (Moore *et al.*, 1994)
αd	CA 11.8H2	IgG1	P. Moore (Danilenko *et al.*, 1995)
T cells (pan)	LSM 8.358	IgM	Custom Monoclonals, Sacramento, CA, USA
B cells (mature)	LSM 11.425	IgG1	Custom Monoclonals, Sacremento, CA, USA
Granulocytes	DOG 15-7	Rat IgG2a	Connex, Martinsed, Germany
Monocytes	DH 59B	IgG1	VMRD, Pullman, WA, USA

Table VII.3.3 Human monoclonal antibodies that cross-react with canine leukocytes

Antigen	*mAb*	*Isotype*	*Dog reactivity*	*Source*
CD6	MT 606	IgG2b	Thymocytes, T cells	Boehringer-Mannheim
CD8	MT 811	IgG1	Thymocytes, T cell subset	Boehringer-Mannheim
CD18	MHM 23	IgG1	Leukocytes	Dako
	MCA 503	IgG2b	Leukocytes	Serotec
	IOT 18	IgG1	Leukocytes	Immunotech
	AM 18	IgG1	Leukocytes	Biosys
CD21	IOB 1a	IgG1	B cells	Immunotech
	S-B2	IgG1	B cells	Biosys
CDw29	4B4	IgG1	Activated T cells	Coulter

Table VII.3.4 Cloned cDNAs for canine leukocyte antigens

		Amino acid homology (%)		
Antigen	*Predicted protein (aa)*	*Human*	*Mouse*	*Rat*
CD3ε	202	58	58	
CD4	439	61		
CD8α	218	69		
CD28	173	85	78	75
CD34	389	69	62	
CD38	308	56	56	58

Table VII.4.1 Leukocyte–endothelial cell interactions in the dog

Leukocyte determinants		*Endothelial ligands*	
Molecule	*Cell distribution*	*Molecule*	*Cell distribution*
CD11b/CD18 (Mac-1)	Neutrophils, monocytes, and eosinophils	ICAM-1	Unstimulated and activated endothelial cells
CD11a/CD18 (LFA-1)	Leukocytes	ICAM-1	Activated endothelial cells
L-selectin (LAM-1)	Leukocytes	?	Activated endothelial cells
Sialyl Lewis X	Neutrophils and monocytes	P-selectin	Activated endothelial cells
CD44 (HCAM, Pgp-1)	Leukocytes	?	?

Table VII.4.2 Chemotactic molecules in the dog

Molecule	*Responding cells*
IL-8	Neutrophils
RANTES	Monocytes, eosinophils
Macrophage chemoattractant protein (MCP-1)	Monocytes
Leukotriene B_4	Neutrophils
Platelet activating factor (PAF)	Neutrophils
C5a	Neutrophils
IL-1α	Neutrophils

Dreyer *et al.*, 1992; Meurer *et al.*, 1993; van Riper *et al.*, 1993; Zwahlen *et al.*, 1994).

5. Cytokines

Cytokines

There are few reagents commercially available for the analysis of cytokines in canine immune responses. Several human and murine cytokines have an effect on canine cells and are presumably cross-reactive, although it is generally unknown if the canine receptor interacts with the same affinity with the heterologous cytokine as with the homologous cytokine. These include IL-1β (Tamaoki *et al.*, 1994), IL-2 (Helfand *et al.*, 1992), IL-6 (Peng *et al.*, 1994), IL-7 (P. J. Felsburg *et al.*, unpublished data), IL-8 (Meurer *et al.*, 1993; Zwahlen *et al.*, 1994), IL-10 (Galkowska *et al.*, 1995), IL-11 (Nash *et al.*, 1995), and RANTES (Meurer *et al.*, 1993).

The cDNAs for several canine cytokines have been cloned, including IL-2 (Dunham *et al.*, 1995; Knapp *et al.*, 1995), IL-8 (Ishikawa *et al.*, 1993; Matsumoto *et al.*, 1994), IFN-γ (Devos *et al.*, 1992; Zucker *et al.*, 1992), GM-SCF (Nash *et al.*, 1991b), TNF-α (Zucker *et al.*, 1994), and stem cell factor (Shull *et al.*, 1992). Canine specific polyclonal antibody has been generated against canine recombinant IL-8 (Massion *et al.*, 1995) and canine specific monoclonal antibodies have been produced against canine recombinant IFN-γ (Fuller *et al.*, 1994) and stem cell factor (Huss *et al.*, 1995).

Measurement of cytokines can be performed at the mRNA and protein level. These are complementary approaches since there is not necessarily a good correlation between mRNA and protein expression. Reverse transcription-polymerase chain reaction (RT-PCR) is a useful procedure for the detection of cytokine mRNA. In addition to the use of canine specific primers for those cytokines that have been cloned, generic RT-PCR primers have been designed that amplify a portion of IL-2, IL-4, IL-6, IL-10, IL-12, IFN-γ and TNF-α of various domestic species including the dog (Rottman *et al.*, 1996).

Protein-based assays include bioassays, ELISAs, and immunohistochemistry. Bioassays are available for IL-1β (Klaich and Hauter, 1994; Yamashita *et al.*, 1994a; Bravo *et al.*, 1996), IL-2 (Helfand *et al.*, 1992), IL-6 (Rivas *et al.*, 1992; Yamashita *et al.*, 1994a; HogenEsch *et al.*, 1995), IFN-γ (Zucker *et al.*, 1993), GM-CSF (Nash *et al.*, 1991b), and TNF-α (Yamashita *et al.*, 1994a). These assays appear to be relatively specific for their respective cytokines. However, it is often necessary to subject samples, especially serum samples, to treatment that removes or inactivates inhibitors (HogenEsch *et al.*, 1995; Bravo *et al.*, 1996).

ELISAs for human or mouse cytokines generally do not detect canine cytokines with the possible exception of IL-8 and other chemokines. A dog-specific ELISA has been developed for IL-8 (Massion *et al.*, 1995) and IFN-γ (Fuller *et al.*, 1994). Polyclonal antisera against human cytokines have been used for *in situ* detection of IL-1β, IL-6 and TNF-α (Day, 1996) and TGF-β (Vilafranca *et al.*, 1995).

Table VII.5.1 summarizes the currently available methods for evaluating canine cytokines.

Table VII.5.1 Currently available methods for canine cytokine analysis

Cytokine	*Cloning*	*Detection method*	*Dog cross-reactivity*
IL-1β		Bioassay, immunohistochemistry	Human
IL-2	Yes	Bioassay, RT-PCR[a]	Human, mouse
IL-4		RT-PCR	
IL-6		Bioassay, RT-PCR, immunohistochemistry	Human, mouse
IL-7			Human, mouse
IL-8	Yes	ELISA[b], RT-PCR	Human
IL-10		RT-PCR	Human
IL-11			Human
IL-12		RT-PCR	
IL-15			Human
IFN-γ	Yes	Bioassay, RT-PCR, ELISA	
GM-CSF	Yes	Bioassay, RT-PCR	Human
TNF-α	Yes	Bioassay, RT-PCR, immunohistochemistry	
TGF-β		Immunohistochemistry	
Stem cell factor (SCF)	Yes	Immunohistochemistry, RT-PCR	
RANTES			Human

[a] RT-PCR, reverse transcription polymerase chain reaction.
[b] Enzyme-linked immunosorbent assay.

Cytokine receptors

The high-affinity canine IL-2 receptor binds human IL-2. This feature has been used to detect canine IL-2 receptors on activated T cells and leukemic cells employing radioactively labeled (Helfand *et al.*, 1995) or phycoerythrin-labeled (Somberg *et al.*, 1992) human recombinant IL-2. The high-affinity IL-2 receptor consists of three chains, α, β, and γ. The γ chain of the IL-2 receptor has been cloned (Henthorn *et al.*, 1994).

6. T-cell Receptor

Knowledge of the T cell receptor (TCR) in the dog is limited to the recent cloning of the constant regions of the canine TCRα and TCRβ (Ito *et al.*, 1993; Takano *et al.*, 1994). Table VII.6.1 illustrates the identity of the canine TCRα–C and TCRβ–C amino acid sequences with those of other species.

Table VII.6.1 Cloned canine T-cell receptor cDNAs

	Amino acid homology (%)	
Compared with	*TCRα-C*	*TCRβ-C*
Human	46	84
Mouse	48	80
Rat	47	79
Bovine	53	81
Sheep	46	
Rabbit	44	77
Chicken		40

7. Immunoglobulins

The basic characteristics of canine immunoglobulins and their concentrations in various body fluids is listed in Table VII.7.1 (Reynolds and Johnson, 1970; Heddle and Rowley, 1975; Halliwell and Gorman, 1989; Mazza and Whiting, 1994; Yang *et al.*, 1995; Tizard, 1996).

The molecular characterization of the canine immunoglobulin genes lags far behind those of other species. Only two of the canine immunoglobulin genes have been cloned – IgA and IgE (Patel *et al.*, 1995). Table VII.7.2 summarizes the homology of the alpha and epsilon constant region genes with the human and murine.

The mechanism and accessory molecules involved in B cell signaling and activation have not been studied in the dog.

The pattern of immunoglobulin isotype development in the dog is similar to that of other mammalian species, including humans, as illustrated in Table VII.7.3.

8. Major Histocompatibility Complex (MHC) Antigens

Historically, the dog has been a useful model for solid organ and hematopoietic stem cell transplantation (Deeg and Storb, 1994). While an understanding of the MHC (termed DLA for dog leukocyte antigen) is important in all of these applications, molecular analysis of the DLA has lagged behind that of the mouse and human as well as several agricultural animals.

Early understanding of canine MHC involved primarily cellular, serological, and immunochemical analyses. The

Table VII.7.1 Properties of canine immunoglobulins

Property	*IgG*	*IgM*	*IgA*	*IgE*	*IgD*
Size (kDa)	180 000	900 000	360 000	200 000	180 000
Concentration (mg/dl)					
Serum	700–2000 IgG1: 300–1400 IgG2: 300–1400 IgG3: < 1–200 IgG4: < 4–200	70–270	20–150	0.7–72	?
Saliva	0.5–5	0.5–7	17–125		
Colostrum	500–2200	70–370	150–340		
Milk	10–30	10–54	110–620		
Fecal extract	90–1070	50–160	80–540		
Placental transfer	Yes	No	No	No	No
Complement fixation	Yes	Yes	No	?	?
Opsonization	Yes	Yes	No	?	?
Neutralization	Yes	Yes	Yes	?	?
Agglutination	Yes	Yes	Yes	?	?
Reaginic activity	No	No	No	Yes	No

Table VII.7.2 Canine immunoglobulin constant region genes

		Amino acid homology	
Gene	*Chromosome*	*Human*	*Mouse*
IgA	?		
CH1		57%	52%
CH2		70%	67%
CH3		82%	73%
Total		70%	65%
IgE	?		
CH1		59%	42%
CH2		53%	42%
CH3		62%	55%
CH4		55%	56%
Total		57%	49%

Table VII.7.3 Age-related development of canine immunoglobulins

	Serum concentrations (mg/dl)		
Age (weeks)	*IgG*	*IgM*	*IgA*
<1	1110 ± 334	24 ± 11	5 ± 2
2	264 ± 70	64 ± 28	3 ± 1
4	152 ± 24	103 ± 34	4 ± 1
8	492 ± 218	170 ± 44	7 ± 2
12	636 ± 250	226 ± 40	25 ± 8
16	790 ± 263	218 ± 42	26 ± 9

DLA is divided into three serologically defined antigens: DLA-A with five specificities, DLA-B with four specificities, and DLA-C with three specificities (Bull *et al.*, 1987). A fourth antigen, DLA-D with 10 specificities, is defined by mixed leukocyte culture (Deeg *et al.*, 1986). The DLA-A antigens are characterized as class I molecules by their association with β_2-microglobulin (Krumbacher *et al.*, 1986). An immunochemical analysis of glycosylated and nonglycosylated DLA molecules suggests that the products of one predominant class I locus are present on the surface of peripheral blood leukocytes and DLA-C antigens are thought to be weakly expressed class I antigens (van der Feltz and Ploegh, 1984; Doxiadis *et al.*, 1986). Conversely, DLA-B antigens when studied by two-dimensional gel electrophoresis and lysostrip experiments exhibit typical class II properties with a high level of serological polymorphism in the β chain and no serological polymorphism in the α chain (Doxiadis *et al.*, 1989). Interestingly, in contrast to mice and humans, class II gene products were absent on almost all lymphocytes. DLA-A and DLA-B gene products have been defined by one-dimensional isoelectric focusing and immunoblotting and there is a high degree of correlation between the biochemically defined antigens and the serological specificities (Kubens *et al.*, 1995).

Beginning in the late 1980s, a molecular analysis of the canine MHC began. Using an HLA-B7 cDNA probe and studying the patterns on Southern analysis from 40 dogs, Sarmiento and Storb concluded that there are at least eight class I genes (Sarmiento and Storb, 1989). The same group used a similar approach with various class II probes to study the number of class IIA and IIB genes (Sarmiento and Storb, 1988a,b). Sarmiento and Storb also probed a canine cDNA library with human probes and isolated a DRB cDNA clone called DRB5 in addition to a class I cDNA clone called I16 (Sarmiento and Storb, 1990a,b). Using reverse transcription-polymerase chain reaction (RT-PCR), Sarmiento *et al.* (1990, 1992, 1993) found that at least three class II loci, DRB, DQA, and DQB are polymorphic. The DRA locus appears to be monomorphic (Wagner *et al.*, 1995).

Recently, a better understanding of the canine MHC loci has begun to emerge. In the cases of both class I and class II loci, the genomic clones were isolated using canine cDNA probes that previously had been isolated using HLA probes. Using I16 as a probe, seven distinct class I loci have been isolated (Burnett and Geraghty, 1995; R. Burnett, unpublished data). One locus, designated DLA-79, has shown limited polymorphism and relatively low mRNA expression in a wide variety of tissues and has been designated a class Ib gene (Burnett and Geraghty, 1995).

Three other class I loci, termed DLA-88, DLA-12, and DLA-64, appear to be complete genes by sequence analysis, and all three are transcribed in canine peripheral blood leukocytes (R. Burnett, unpublished data). DLA-88 and DLA-64 appear to be more polymorphic than DLA-12 and DLA-79 (J. Wagner, unpublished data). Two other genes, DLA-53 and DLA-12a, are truncated class I pseudogenes. C1pg-26 is a processed gene located outside the DLA (R. Burnett, unpublished data). Neither the tissue expression nor the function of any of the class I gene are known at present although one could infer by analogy from other species that DLA class I genes could serve as cytotoxic T lymphocyte targets.

Class II genes, in contrast to class I genes, have strong homology between species. There is one highly polymorphic DRB gene designated DRB-B1 and many dogs have a DRB pseudogene termed DRB-B2 (Wagner *et al.*, 1996a,b). There is one DQA gene with a limited amount of polymorphism, one polymorphic DQB gene, and one DQB pseudogene (Wagner *et al.*, 1996c; Wagner, unpublished data). No DP-like genes have been identified to date in the dog. A summary of known DLA loci is shown in Table VII.8.1.

Two polymorphic satellite markers, one located in the class I region and one located in the class II region, have been recently described and shown to be useful for intrafamilial typing (Wagner *et al.*, 1996d).

Table VII.8.1 Documented DLA loci

Locus name	*Gene type*	*Class*	*Number of alleles*
DRA	Complete	II	1
DRB-1	Complete	II	18
DRB-B2	Pseudo	II	
DQA	Complete	II	9
DQB-B1	Complete	II	7
DQB-B2	Pseudo	II	
DLA-88	Complete	I?a	22
DLA-79	Complete	Ib	3
DLA-12	Complete	I?a or b	5
DLA-53	Pseudo	I	
DLA-12a	Pseudo	I	

9. Red Blood Cell Antigens

More than a dozen blood group or blood type systems have been described in dogs. The first four canine blood groups were recognized in 1910 based upon naturally occurring agglutinins (Von Dungern and Hirschfield, 1910). Subsequently, alloimmunizations by repeated transfusions and transplantations were more rewarding and led to the description of up to 13 blood types in various studies (Suzuki *et al.*, 1975; Colling and Saison, 1980a; Bell, 1983). However, owing to the unavailability of serological reagents for comparisons, the exact total number of different blood groups identified remains unknown (Bell, 1983; Symons and Bell, 1992).

The blood types recognized as international standards are listed in Table VII.9.1 (Bull *et al.*, 1987). The original blood type designation by capital letters of the alphabet is presently used along with the DEA (dog erythrocyte antigen) numbers. Based upon limited family studies, the mode of inheritance is autosomal dominant (Colling and Saison, 1980a). For all blood group systems other than the DEA 1 system, a red blood cell can be positive or negative for that blood type, e.g., a dog's red blood cells can be DEA 4 positive or negative. The DEA 1 system, however, has two well-recognized blood types: DEA 1.1 also known as A_1, and DEA 1.2 or A_2. In this three-allele system, a dog erythrocyte can be positive for any one of these two types or negative. Furthermore, a probable third blood type (DEA 1.3, A_3) has recently been discovered in this system (Symons and Bell, 1992). Finally, the DEA 7 system may also have an additional subtype described with the letter O (Bell, 1983).

Table VII.9.1 Dog erythrocyte antigens (DEA) based on international standards

Blood groups		*Frequency*	*Antigenicity in transfusion reactions*
DEA 1.1	(A_1)	33–45%	+++
DEA 1.2	(A_2)	4–22%	+
DEA 3	(B)	5–24%	+
DEA 4	(C)	56–98%	ND
DEA 5	(D)	8–25%	+
DEA 6	(F)	60–99%	ND
DEA 7	(Tr)	8–45%	+
DEA 8	(He)	17–40%	ND

The frequencies of these blood types in dogs on limited surveys over the past 35 years are listed in Table VII.9.1. There are likely large differences between geographic locations and breeds. The structure and function of the canine blood types remain unknown. The DEA 7 system appears serologically related to the human A, pig A, cattle J, and sheep R antigen (Colling and Saison, 1980b). DEA 7 is not a typical red blood cell antigen as it is produced by tissues and absorbed onto red blood cells.

Serological reagents for canine blood typing have not been standardized and most of them are not generally available; naturally occurring antibodies are not useful. Because of the variation in antigenicity, dogs have to be repeatedly alloimmunized to produce adequate antisera. Although attempts to produce antisera in rabbits failed, a murine monoclonal antibody against DEA 1.1 (Andrews *et al.*, 1992) has been generated and forms the basis for a commercially available simple card test (Rapid Vet TMH canine 1.1; DMS Laboratories, Inc., 2 Darts Mill Rd, Flemington, NJ 08822, USA, 1-800-567-4367). Polyclonal reagents for DEA 1.1/1.2, 3, 4, 5, and 7 are presently available from Dr Robert Bull (R228 Life Sciences Building, Michigan State University, East Lansing, MI 48824, USA, 1-517-355-4616). Reagents for DEA 1.1 and 1.2 are also available from other institutions. All these reagents act as saline agglutinins at 4°C. DEA 1.1 red blood cells agglutinate better in the presence of fresh autogenous serum and are rapidly hemolyzed in the presence of complement. The agglutination reactions may be variable and enhanced by canine antiglobulins.

Red blood cell antigens are responsible for blood incompatibility reaction in dogs (Young *et al.*, 1949, 1952). Because of a lack of clinically significant naturally occurring alloantibodies, blood incompatibilities have only been described in sensitized dogs, i.e. dogs that have previously received a transfusion. Based upon experimental and clinical data, DEA 1.1 appears to be the most antigenic blood type and has caused severe acute hemolytic transfusion reactions (Giger *et al.*, 1995). An acute hemolytic transfusion reaction against an unidentified common erythrocyte antigen has also been described in a previously sensitized dog (Callan *et al.*, 1995). However, no clinical cases of hemolysis of the newborn have been documented in dogs.

Table VII.10.1 Histological development of key organs of immunity in canine fetuses

Day of gestation	*Lymphoid organ development*
27–28	Primordia of the spleen and thymus is evident.
35	Thymic primordium descends from the cervical region into the anterior thoracic cavity. It is composed of epithelial lobules and mesenchymal stroma only.
35–40	The thymus becomes actively lymphopoietic and begins to show corticomedullary demarcation. Hassall's corpuscles begin to appear between days 38 and 40. The spleen and lymph nodes are devoid of lymphopoiesis.
45	The thymic microenvironment has assumed its normal postnatal histologic appearance.
45–52	Lymphocytic infiltration of lymph nodes and spleen with evidence of T cell dependent zones. Bone marrow matrix becomes heavily cellular and contains abundant hematopoietic stem cells.
60–63	Prominent postcapillary venule development in peripheral lymphoid tissues.

Table VII.10.2 Functional development of the immune system in fetal dogs

Day of gestation	*Immunological function*
38–43	Appearance of CFU-g/m[a] in fetal liver.
40	Respond to primary immunization with bacteriophage ϕX174 as determined by antibody titers 30 days post immunization.
43	100% susceptible to leukemia allotransplantation.
45	Lymphocytes from spleen and lymph node respond to PHA.
48	Slight antibody response to immunization with ovine red blood cells as determined by presence of antibody on day 56.
50	50% susceptible to leukemia allotransplantation. Fetal thymocytes become responsive to PHA. Respond to immunization with live *Brucella canis*, as determined by presence of antibody on day 59.
Birth	Antibody response to KLH.

[a] CFU-g/m = Colony forming units - granulocytes/macrophage.

10. Ontogeny of the Immune System

The dog is a multiparous animal with a gestation period of 60–63 days. Table VII.10.1 describes the fetal development of lymphoid organs in the dog (Kelly, 1963; Bryant and Shifrine, 1972; Miller and Benjamin, 1985; Snyder *et al.*, 1993).

Our knowledge of the ontogeny of immune responses in the dog is extremely limited and is summarized in Table VII.10.2 (Jacoby *et al.*, 1969; Shifrine *et al.*, 1971; Bryant *et al.*, 1973; Klein *et al.*, 1983; Prummer *et al.*, 1985). The ability to respond to immunization includes only those antigens that have been used in both fetal and neonatal studies. The fetal development of B and T lymphocytes has not been described.

11. Passive Transfer of Immunity

The placentation in the dog differs from the hemochorial placenta of humans in which the blood of the mother is in direct contact with the trophoblast, permitting direct entry of maternal IgG into the fetal bloodstream. Dogs have an endotheliochorial placenta in which four structures separate the maternal and fetal blood – the endothelium of the uterine vessels and the chorion, mesenchyme (connective tissue), and the endothelium of the fetal tissues. These four layers of tissue between the maternal and fetal circulation in the dog limit the *in utero* transfer of maternal IgG to the fetus. Thus, only 5–10% of maternal antibody in the dog is obtained *in utero* through the placenta.

Because newborn dogs are essentially devoid of maternal antibody when they are born, mammary secretions, colostrum and milk from the mother are critical in terms of immune protection of the newborn puppy. The immunoglobulin composition of canine colostrum consists of: IgG: 500–2200 mg/dl; IgM: 70–370 mg/dl; and IgA: 150–340 mg/dl (Reynolds and Johnson, 1970). When a newborn puppy suckles colostrum within the first few hours after birth, the proteins contained in the colostrum, including the immunoglobulins, are transported across the small intestine in the ileum by pinocytosis and very quickly reach the bloodstream, resulting in a massive transfusion of maternal antibody. The levels of serum IgG in newborn puppies that receive colostrum approach those levels found in adults. As in other species, there is a very limited period of time in which the newborn can absorb intact IgG from the intestinal tract, usually 12–36 h. After that time, the newborn becomes incapable of absorbing intact proteins, a phenomenon referred to as closure. There are two main physiological reasons for the transient permeability of the neonatal intestine. First, the presence of trypsin inhibitors in the colostrum and the neutral pH of the digestive tract during the first few days of life prevents the proteolytic digestion of ingested proteins. As the pH of the stomach assumes its normal acidic pH, proteolytic enzymes are activated. The second reason is the rapid turnover of the immature epithelial cells lining the intestinal tract that absorb immunoglobulins with more mature epithelial cells.

Following closure, and until weaning, secretory IgA from the milk, protected from proteolytic digestion by its secretory component, is continuously present in the intestine of the neonate and is the most important factor protecting them from enteric infections. The immunoglobulin concentration of canine milk consists of: IgG, 10–30 mg/dl; IgM, 10–54 mg/dl; and IgA, 110–620 mg/dl (Tizard, 1996). Other proteins such as lactoferrin and lactoperoxidase in canine colostrum and milk also contribute to the protection of the neonate. The composition and function of the cellular elements in canine colostrum has not been evaluated.

Failure to receive colostrum is not recognized as common a clinical problem in the dog as it is in other species of domestic animals, particularly the horse. However, puppies that fail to receive colostrum are severely hypogammaglobulinemic during the first weeks of life until their own immune system has had a chance to mount a primary and secondary humoral immune response to the antigens in its environment. There are several reasons why individual puppies may not receive adequate colostrum. These include large litters, in which the puppies are competing for the finite amount of colostrum available, poor mothering, especially in young mothers, and puppies, for whatever reason, not suckling within the first 24 h.

The duration of passive immunity afforded by maternal antibody in the dog is derived from studies evaluating antibody titers to common canine viruses in the dam and in their puppies (Gillespie *et al.*, 1958; Baker *et al.*, 1959; Carmichael *et al.*, 1962; Winters, 1981; Pollock and Carmichael, 1982). The titer of maternal antibody in the serum of the neonate depends upon the amount of antibody received *in utero* and following nursing and the titer of the dam. For those puppies that do not receive colostrum, the amount of maternal antibody received *in utero* through placental transfer averages 3–5.5% of the mother's antibody titer. Conversely, the amount of maternal antibody in puppies that have received and absorbed colostrum is approximately 77–92% of the mother's antibody titer. Since the half-life of maternal antibody in the dog is approximately 8.4 days, the average protection from maternal antibody in the neonate is between 8–16 weeks.

12. Neonatal Immune Responses

Unlike many other species, little is known about the neonatal immune response in the dog. This is especially true for the role the physiology of the neonate has on immune function and the function of the innate immune system in the neonate.

With regard to specific immune defenses in the neonatal dog, much of what we know is related to the ability of the neonate to respond to specific vaccination. In the 1950s, when vaccines were being introduced to veterinary medicine, it was thought that neonatal dogs were not immunologically competent since there was a high rate of vaccination failures in these dogs. It soon became evident that the reason for the high vaccination failure rate in neonatal dogs was due to the presence of maternal antibody. The fact that neonatal dogs possess a functional humoral immune system was demonstrated by Jacoby *et al.* (1969). In this study, colostrum-deprived, gnotobiotic puppies were vaccinated within the first 24 h of birth with the T cell-dependent antigen, bacteriophage ϕX174. All the neonatal puppies developed a primary and secondary specific antibody response following immunization. The only difference between the neonatal group and an adult group of dogs was the magnitude of the response. These studies documented that neonatal dogs possess a functional B cell and T cell system at birth. Dogs immunized intranasally with a modified live vaccine within the first week of life also develop a protective immune response, even in the presence of maternal antibody.

The age-related development of serum immunoglobulins is described in Section 7. Serum IgM concentrations reach normal adult levels by 4–8 weeks old. Following the decline in maternal IgG, the neonate produces its own IgG with normal adult levels being approached by 16 weeks of age. As in other species, the synthesis of serum IgA lags behind the other isotypes and does not reach adult levels until approximately 1 year of age (P. Felsburg *et al.*, unpublished data). This may result in a transient IgA deficiency in the dog, as seen in children (P. J. Felsburg, 1994).

The phenotype of the lymphocyte subpopulations in neonatal dogs differs significantly from that of adult dogs, as shown in Table VII.12.1. By 16 weeks of age, the

Table VII.12.1 Phenotype of neonatal canine peripheral blood lymphocytes

Age (weeks)	Phenotype (%)			
	B cells	CD3	CD4	CD8
<1	33.4±8.1	45.0±10.4	38.1±6.4	7.5±5.9
2	14.3±5.9	58.2±4.6	51.0±4.2	8.0±2.4
4	17.5±6.7	59.4±5.8	50.4±5.6	8.8±3.3
8	12.9±3.5	66.1±5.8	54.8±4.9	10.5±4.1
16	5.6±2.4	90.5±6.5	77.8±6.2	13.7±3.6
24	5.1±3.1	88.9±5.6	72.1±4.6	16.5±6.6

Table VII.12.2 Blastogenic response of neonatal lymphocytes to PHA

Age (weeks)	*Blastogenic response (cpm)*
<1	19312±5766
2	25893±4390
4	26574±2816
8	34361±5914
16	30409±6124
24	35024±6826

phenotype of the peripheral lymphocytes have attained a normal adult phenotype (Somberg *et al.*, 1994, 1996; P. Felsburg *et al.*, unpublished data).

Although the proportion of peripheral T cells in the neonatal dog is significantly lower than that in the adult, they are functionally competent as shown in Table VII.12.2 by their ability to proliferate in response to mitogenic stimulation (Felsburg *et al.*, 1992; Somberg *et al.*, 1994; P. Felsburg *et al.*, unpublished data).

Thus, unlike rodents, dogs appear to be immunologically mature at birth by the criteria examined.

13. Nonspecific Immunity

Physical and anatomical barriers, especially the skin and mucosal surfaces, are the dog's first line of defense against infection. The physiological barriers include temperature, pH, oxygen tension and various soluble chemical factors. The skin and the surface of mucous membranes provide an effective barrier to the entry of most microorganisms. In order for bacteria to gain entry, most pathogens need to colonize and penetrate the mucous membrane barrier, but such colonization and penetration is inhibited by the various innate defenses listed in Table VII.13.1 (Halliwell and Gorman, 1989; Felsburg, 1994; Tizard, 1996).

The proteins that have been documented to be involved in the canine acute phase response are listed in Table VII.13.2 (Eckersall and Conner, 1988; Stellar *et al.*, 1991;

Table VII.13.1 Natural barriers in the dog and their mechanism of defense

Barrier	*Mechanism of defense*
Skin	Physical barrier, desquamation, dessication, acidic pH owing to fatty acids, and competition by normal bacterial flora.
Gastrointestinal tract	Hydrolytic enzymes in the saliva, acidic pH of the stomach, proteolytic enzymes in the small intestine, motility of the gastrointestinal tract, shedding of intestinal crypt epithelial cells, and competition by normal bacterial flora.
Respiratory tract	Turbulence, mucus, ciliary action, and cough reflex.
Urinary tract	Low pH, and flushing action.
Mammary gland	Nonlactating: keratin plugging of teats. Lactating: flushing action, lactoferrin and lactoperoxidase.

Table VII.13.2 Canine acute phase response (APR) proteins

APR protein	*Function*
Major APR Proteins	
C reactive protein (CRP)	Acts as an opsonin and can activate the classical complement pathway, enhances chemotaxis and phagocytosis of neutrophils and macrophages, and binds to nuclear chromatin of damaged cells and may be involved in the degradation of nuclear components of damaged cells.
Serum amyloid A (SAA)	Inhibits IL-1β and TNF-α induced fever, inhibits platelet aggregation, and inhibits neutrophil oxidative burst.
Other APR Proteins	
α_1 acid glycoprotein	Suppression of lymphocyte blastogenic response and antibody production.
α_2-macroglobulin	Protease inhibitor in the dog that functions by neutralizing lysosomal hydrolases released by activated neutrophils and macrophages. Unlike other species, α_1-antitrypsin and α_1-antichymotrypsin are not APR proteins in the dog.
Ceruloplasmin	Scavenges superoxide released by neutrophils.
Complement components	Enhances the local accumulation of neutrophils, macrophages, and plasma proteins.
Fibrinogen	Promotes wound healing.
Haptoglobin	Inhibits bacterial growth by binding iron and preventing its uptake by bacteria.

Table VII.13.3 Polymorphonuclear leukocyte populations in the dog

Cells	*Function*
Basophils	Although basophils play an important role in Type I hypersensitivity reactions in various species, their function in the dog is unclear.
Eosinophils	The major biological function of eosinophils is the destruction of parasites, particularly helminthic parasites. Phagocytosis of IgE opsonized antigens is promoted by the presence of FcεRII and CR3 receptors on the eosinophil.
Mast cells	Although mast cells are not a polymorphonuclear leukocyte, they play a predominant role in IgE-mediated inflammation. They are the tissue equivalent of the circulating basophil. They possess the high-affinity cell surface IgE receptor (FcεRI). Mast cells are the principal mediator cell in Type I hypersensitivity reactions in the dog.
Neutrophils	Neutrophils are the major cellular elements in acute inflammation. They also play a primary role in maintaining normal host defense against microorganisms through phagocytosis and destruction of foreign antigens. Phagocytosis is aided by the presence of FcγRII and CR1 receptors on their cell surface. Destruction of the foreign antigen is mediated through neutrophil-generated toxic oxygen radicals and granule-derived cytotoxic proteins.

Gruys *et al.*, 1994; Yamashita *et al.*, 1994b). Serum amyloid A (SAA), one of the two major acute phase response proteins in the dog, has been cloned and the predictive protein shows 80% homology to human and murine serum amyloid A (Stellar *et al.*, 1991).

Inflammation is an important component of the nonspecific (innate) immune system. The polymorphonuclear leukocytes are the cellular elements involved in acute inflammatory responses in response to infection with microorganisms or to tissue injury. Table VII.13.3 describes the polymorphonuclear leukocyte populations in the dog.

The major population of cells involved in chronic inflammatory reactions are the circulating monocytes and the various tissue-resident macrophages (Table VII.13.4). Documented functions of monocytes/macrophages in the dog include secretion of inflammatory cytokines (IL-1, IL-6 and TNF), secretion of inhibitory cytokines (IL-10), phagocytosis and destruction of bacteria, destruction of tumor cells, repair of tissue damage, and antigen presentation (Tizard, 1996). The binding of opsonized antigen is mediated through their surface FcγRI and CR1 receptors.

The clearance of bacteria and other particles from the bloodstream in the dog is similar to that observed in humans and predominantly through macrophages in the liver and spleen (approximately 80%). This is in contrast to the clearance of bacteria in other species of domestic animals in which approximately 90% of the particles in the bloodstream are cleared through pulmonary intravascular macrophages in the lung.

Table VII.13.4 Monocyte/macrophage populations in the dog

Cells	*Tissue distribution*
Alveolar macrophage	Alveoli of the lung
Histiocyte	Connective tissue, skin
Kupffer cell	Liver
Macrophage	Spleen, lymph node
Microglial cells	Brain
Monocyte	Blood, lymph node
Osteoclast	Bone
Pulmonary intravascular macrophage	Capillaries of the lung

14. Complement System

The complement system is composed of a series of plasma proteins and cellular receptors which play important roles in host defense and inflammation (Winkelstein *et al.*, 1995). Individual components of the complement system possess opsonic (C3b), anaphylatoxic (C3a, C5a), and chemotactic (C5a) activities. In addition, the assembly of the terminal components (C5, C6, C7, C8 and C9) into the membrane attack complex (C5b-C9) generates cytolytic activities for certain mammalian cells, such as erythrocytes, and for Gram-negative bacteria. Finally, the complement system also plays an important role in processing and clearance of immune complexes and is critical for the generation of primary antibody responses and for the generation and maintenance of B cell memory.

Most human complement components have been characterized in detail at both the level of the protein and the gene. Although components of the complement system have been identified in all nonhuman mammalian species that have been examined to date, they have been characterized in most species in much less detail than their human counterparts.

The canine complement system

All of the individual components of the classical pathway (C1 through C9) and C1 esterase inhibitor have been detected in canine sera using functional assays (Sargent and Austen, 1970; Barta and Hubbert, 1978; Sargent and Johnson, 1980). These functional assays assess the ability of the test serum to lyse antibody-sensitized sheep erythrocytes in the presence of hemolytic reagents that contain all the components of the classical pathway in excess, but lack the specific component in question. In general, functional titers of a given canine component are highest when tested in hemolytic assays that utilize developing reagents that are composed of canine components rather than the more commonly available guinea pig or human components (Sargent and Austen, 1970; Sargent and Johnson, 1980).

The canine complement system is capable of generating most of the biologic activities of the complement system (Winkelstein *et al.*, 1982). Total hemolytic activity, which is dependent on the activation of the complete cascade from C1 through C9, is present in dog serum at levels comparable to that of other mammalian species. In addition, the canine complement system is capable of generating C3-dependent opsonic activity and C5a-dependent serum chemotactic activity (Winkelstein *et al.*, 1982).

Although all of the components of the classical pathway have been purified from canine serum as distinct functional activities (Sargent *et al.*, 1976; Tamura and Nelson, 1968), only C1q (Wu *et al.*, 1988) and C3 (Johnson *et al.*, 1985) have been purified to biochemical homogeneity. The first component of complement (C1) is composed of three biochemically and genetically distinct subcomponents, C1q, C1r and C1s. Canine C1q has been purified from serum using affinity chromatography with columns containing IgG and Protein A and has three chains of 27 kDa, 26 kDa and 23 kDa (Wu *et al.*, 1988). As with C1q from other mammalian species, it is able to bind to immune complexes and thus has been used to detect circulating immune complexes in dogs with systemic lupus erythematosus and rheumatoid arthritis (Wu *et al.*, 1988). Canine C3 has also been purified to biochemical homogeneity using sequential column chromatography with DEAE-Sephacel and CM-Sepharose (Johnson *et al.*, 1985). It has a molecular mass of 179 kDa and like its orthologs in other species is composed of two disulfide linked chains of 114 kDa (α) and 65 kDa (β) respectively.

Compared with man and some other mammalian species such as the mouse and guinea pig, relatively little is known about the molecular biology and genetics of the complement system in the dog. In man, two loci located within the class III region of the MHC encode for C4 (O'Neill *et al.*, 1978). The gene products of these two loci (termed C4A and C4B) share most of their structural and functional characteristics but differ in minor antigenic determinants and binding specificities. Other primates, as well as certain ungulates, also appear to have two different isotypes of C4, while the dog appears to possess only one isotype (Kay and Dawkins, 1984; Dodds and Law, 1990). As in other species, the gene for canine C4 is closely linked to the MHC and is in linkage disequilibrium with specific DLA-B haplotypes (Grosse-Wilde *et al.*, 1983). Canine C4, like its human counterpart, is highly polymorphic (Grosse-Wilde, 1983; Kay and Dawkins, 1984; Kay *et al.*, 1984; O'Neill *et al.*, 1984) with two common allotypes and at least seven less common ones (Doxiadis *et al.*, 1987). Canine C3 (Gorman *et al.*, 1981), C6 (Anderson *et al.*, 1983; Eldridge *et al.*, 1983; Shibata *et al.*, 1995), and C7 (Johnson *et al.*, 1986) have also been found to be polymorphic with polymorphisms inherited as autosomal codominant traits. In fact, the inheritance patterns of C7 polymorphisms are consistent with two closely linked loci for this component, which in turn are linked to C6 (Eldridge *et al.*, 1983).

15. Mucosal Immunity

The mucosal immune system encompasses the immune system of the gastrointestinal tract, the respiratory tract, the genital tract, the conjunctiva, and the mammary glands. It can be divided into inductive sites and effector sites. The former are the organized lymphoid tissues in which the immune response is initiated resulting in activated B and T lymphocytes. The inductive sites include Peyer's patches (PPs) in the small intestine, the solitary lymphoid nodules in the stomach, small and large intestine, the tonsils, lymphoid nodules in the third eyelid, and the lymph nodes draining the mucosal tissues. Activated B and T cells preferentially migrate to the effector sites of the same organ as the inductive sites, but migration to other mucosal tissues may occur (common mucosal immune system). The effector sites are the epithelium and the lamina propria (propria mucosae) of the mucosal tissues. They are populated by plasma cells, cytotoxic and/or cytokine-secreting effector T cells, macrophages, mast cells, and granulocytes. In addition, these tissues contain dendritic cells which transport antigens to the draining lymph nodes for the generation of an immune response.

Antigen transport

Little is known about antigen transport in the mucosal immune system of the dog. The follicle-associated epithelium of the duodenal and jejunal Peyer's patches contain M-cells interspersed among enterocytes (HogenEsch and Felsburg, 1990). These cells have a poorly developed brush border with weak alkaline phosphatase activity, short microvilli and invaginations of the basal cell membrane, which form pockets that contain lymphocytes (HogenEsch and Felsburg, 1990). The ileal Peyer's patch predominantly consists of epithelial cells with ultrastructural characteristics of M-cells (HogenEsch and Felsburg, 1990). There appear to be no functional studies on the uptake of antigen by M-cells in the dog. Microparticles (about 1 μm diameter) injected into loops of the small intestine were observed several days later in phagocytic cells in the jejunal lymph nodes (Wells *et al.*, 1988). This transport was mediated by macrophages and apparently independent of the presence of Peyer's patches (Wells *et al.*, 1988). Pulmonary macrophages play an important role in the uptake and transport of microparticles instilled in the lung to the tracheobronchial lymph nodes (Harmsen *et al.*, 1985).

Secretory immunoglobulins

Immunoglobulin A (IgA) is the major immunoglobulin of most external secretions in the dog (Vaerman and Heremans, 1969a; Heddle and Rowley, 1975). Serum IgA and secretory IgA exist predominantly as dimers with a J-chain. Serum IgA may be largely derived from plasma cells in the intestinal lamina propria (Vaerman and Heremans, 1970), although significant IgA secretion by spleen and bone marrow cells has been observed *in vitro* (HogenEsch and Felsburg, 1992a). IgA is the predominant immunoglobulin isotype produced by plasma cells in the intestinal lamina propria (Vaerman and Heremans, 1969b; Hart, 1979; Willard and Leid, 1981; HogenEsch and Felsburg, 1992b). The number of IgA-secreting plasma cells decreases from the duodenum to the ileum (Hart, 1979; Willard and Leid, 1981; HogenEsch and Felsburg, 1992b). Transport of IgA across the epithelium into the lumen occurs via the polymeric immunoglobulin (poly-Ig) receptor that is expressed on the basolateral surface of the epithelial cells. The poly-Ig receptor binds the dimeric IgA molecule and translocates to the apical side of the cell. Proteolytic cleavage of the poly-Ig receptor generates a 70–75 kDa fragment that remains bound to the IgA molecule, the secretory component. Both IgA and IgM in mucosal secretions of the dog contain the secretory component (Ricks *et al.*, 1970; Thompson *et al.*, 1975; Thompson and Reynolds, 1977), implying that the transport of IgA and IgM across the epithelia into mucosal secretions occurs via the poly-Ig receptor. The bile fluid is not a major contributor to the intestinal pool of IgA in the dog. In contrast to species in which bile is a major source of intestinal IgA, poly-Ig receptors are expressed on bile epithelial cells, but not on hepatocytes (Delacroix *et al.*, 1983, 1984). Intestinal fluids also contain a considerable amount of IgG. Relatively few IgG plasma cells are present in the lamina propria, therefore, serum is probably a significant source of intestinal IgG (Vaerman and Heremans, 1969b). The domes of Peyer's patches in dogs contain many IgG plasma cells and may account for much of the local IgG production (HogenEsch and Felsburg, 1992b).

Intraepithelial lymphocytes

Intraepithelial lymphocytes (IEL) form a phenotypically heterogeneous population of cells in humans and mice. Phenotypic studies have not been performed in dogs. The number of IELs averages 10–15 per 100 villus enterocytes (Thomas and Anderson, 1982).

Peyer's patches (PPs) and solitary lymphoid nodules

The dog has 26–29 PPs (HogenEsch *et al.*, 1987). The PPs in the duodenum and jejunum appear similar in morphology and function (Table VII.15.1). The dome epithelium contains M-cells interspersed among enterocytes, and very few goblet cells compared with intestinal villi (HogenEsch and Felsburg, 1990). These PPs have large domes and follicles separated by well-developed interfollicular T cell areas. The domes contain IgG and IgA plasma cells, as well as B and T cells and macrophages (HogenEsch and Felsburg, 1992a). Cells from the duodenal and jejunal PPs can be induced to secrete a large amount of IgA and IgG *in vitro*, consistent with a role as a secondary lymphoid organ (HogenEsch and Felsburg, 1992b). The ileal PP is 26–30 cm long in 4–6-month-old dogs. It forms a complete ring around the distal 6–10 cm of the small intestine and tapers to a 1 cm wide band on the antimesenteric side of the intestine proximally (HogenEsch *et al.*, 1987). Its surface area accounts for approximately 80% of that of all PPs combined. In young dogs, the ileal PP contains large follicles, sparsely populated interfollicular areas and small domes. The dome epithelium consists mostly of M-cells (HogenEsch and Felsburg, 1990). The domes contain IgA and IgG plasma cells, as well as lymphocytes and macrophages, consistent with a role as a secondary lymphoid organ. However, *in vitro* stimulated ileal PP cells secrete predominantly IgM and a small amount of IgA and IgG (HogenEsch and Felsburg, 1992b). In contrast to the duodenal and jejunal PPs, ileal PP undergoes involution

Table VII.15.1 Distribution, life history, and function of Peyer's patches in the dog

Peyer's patch	*Number*	*Length (mm)*	*Ontogeny*	*Early involution*	*Function*
Duodenum	4–5	10–15	45 days (fetal)	No	Secondary lymphoid organ
Jejunum	20–23	10–36	?	No	Secondary lymphoid organ
Ileum	1	300	55 days (fetal)	Yes	Primary and secondary lymphoid organ

after the dog reaches sexual maturity (HogenEsch and Felsburg, 1992b). These features suggest a dual role of the ileal PP in the dog, i.e. as a primary and secondary lymphoid organ.

In addition to their presence in Peyer's patches, solitary lymphoid nodules occur in the stomach, small intestine, cecum, colon, and rectum. Lymphoid nodules are present in the lamina propria of the stomach with a mean density of 0.8–15.6/cm^2, depending on the region (Kolbjornsen *et al.*, 1994). They consist of a B cell follicle with an adjacent mixed B and T cell area, but lack a dome or a follicle-associated epithelium. The cecum and colon of the dog contain submucosal lymphoid nodules, lymphoglandular complexes, that are invaginated by the intestinal mucosa (Atkins and Schofield, 1972). There are approximately three nodules per cm^2 in the cecum, and smaller numbers in the proximal colon (Atkins and Schofield, 1972). Hebel identified similar structures throughout the large intestine, including the rectum (Hebel, 1960). The invaginated epithelium of the lymphoid nodules in the cecum and colon contains few goblet cells and resembles the follicle-associated epithelium of Peyer's patches. These lymphoid nodules are not present in fetuses, and may be first identified in the cecum of 1-week-old pups (Atkins and Schofield, 1972). It is likely that these structures are secondary lymphoid organs and play the same role in the large intestine as the duodenal and jejunal PPs in the small intestine.

16. Immunodeficiencies

Although there are over 50 primary (genetic) immunodeficiency diseases documented in humans, the study of these diseases in the dog is still in its infancy. The primary immunodeficiencies described in Table VII.16.1 are the genetic disorders of the specific and nonspecific components of the immune system that have been well-documented in the dog and that are associated with increased susceptibility to infections. These include X-linked severe combined immunodeficiency (Felsburg *et al.*, 1992; Henthorn *et al.*, 1994; Somberg *et al.*, 1994, 1995), selective IgA deficiency (Whitbread *et al.*, 1984; Felsburg *et al.*, 1985; Moroff *et al.*, 1986), leukocyte adhesion deficiency (Giger *et al.*, 1987; Trowalk-Wigh *et al.*, 1992), C3 deficiency

Table VII.16.1 Genetic immunodeficiencies

Disease	*Inheritance*	*Defective protein*	*Immune dysfunction*
C3 deficiency	Autosomal recessive	C3	Defective bacterial opsonization and phagocytosis.
Cyclic hematopoiesis	Autosomal recessive	?	Cyclic neutropenia with metabolic defect in neutrophils including myeloperoxidase deficiency and defective iodination resulting in impaired bactericidal activity.
Leukocyte adherence deficiency	Autosomal recessive	β integrin CD18	Impaired leukocyte adherence due to defective CD11a/CD18, CD11b/CD18, and CD11c/CD18 integrins.
Selective IgA deficiency	Autosomal recessive	Secretory IgA	Failure of terminal differentiation in IgA$^+$B cells resulting in the lack of secretory IgA production.
Transient hypogammaglobulinemia of infancy	?	?	Delayed immunoglobulin production.
X-linked severe combined immunodeficiency	X-linked	Common gamma (γc) chain	Defective receptors for IL-2, IL-4, IL-7, IL-9, and IL-15 resulting in a profound defect in T cell maturation and function and the inability of B cells to undergo isotype class-switching.

Table VII.16.2 Primary immunodeficiencies with probable genetic basis

Disease	*Inheritance*	*Defective protein*	*Immune dysfunction*
Immunodeficiency syndrome in Shar-Peis	?	?	Variable immunoglobulin deficiency, decreased blastogenic response to stimulation, and decreased *in vitro* synthesis of IL-6.
Neutrophil deficiency in Doberman Pinschers	?	?	Impaired bactericidal activity of neutrophils.
Neutrophil deficiency in Weimaraners	?	?	Neutrophil oxidative metabolic defect.
Systemic avian mycobacteriosis in Basset Hounds	?	?	An interleukin or RAMP protein deficiency is suspected.

Table VII.16.3 Acquired immunodeficiencies

Disease	*Immunological abnormality*
Failure to receive colostrum	95–99% of maternal antibody (IgG) is transferred to the neonate through the colostrum. Failure to receive colostrum results in a hypogammaglobulinemic state until the neonate's own immune system produces antibody.
Metabolic diseases	
Growth hormone deficiency	T cell deficiency.
Zinc deficiency	T cell deficiency.
Viral infections	
Canine distemper virus	Decreased immunoglobulin production and lymphocyte blastogenic response to mitogenic stimulation due to direct lytic effect, decreased IL-1 production, and increased production of prostaglandin by monocytes.
Canine parvovirus	T cell deficiency due to direct lytic effect of the virus.

(Winkelstein *et al.*, 1981, 1995; Blum *et al.*, 1985), cyclic hematopoiesis (Jones *et al.*, 1975; Lothrop *et al.*, 1988), and transient hypogammaglobulinemia of infancy (Felsburg, 1994).

There are several primary immunodeficiencies in the dog that most likely have a genetic basis to them since they have been documented in related dogs (Table VII.16.2). These include immunodeficiency syndrome in Shar-Peis (Rivas *et al.*, 1995), neutrophil deficiencies in Doberman Pinschers and Weimaraners (Studdert *et al.*, 1984; Breitschwerdt *et al.*, 1987; Couto *et al.*, 1989) and an immunodeficiency in Bassett hounds that results in an increased susceptibility to systemic avian tuberculosis (Carpenter *et al.*, 1988).

Acquired immunodeficiencies that are secondary to some other event are the most commonly encountered and documented immunodeficiencies in the dog. Table VII.16.3 lists the major acquired immunodeficiencies in the dog (Roth *et al.*, 1980, 1984; Jezyk *et al.*, 1986; Krakowka, 1992; Felsburg, 1994).

17. Tumors of the Immune System

Table VII.17.1 describes the characteristics of the tumors of the canine immune system. These include lymphomas and lymphoid leukemias (Grindem and Buoen, 1986; Nolte *et al.*, 1993; Standstrom and Rimaila-Parmaen, 1979; Tomley *et al.*, 1983; Onions, 1984; Momoi *et al.*, 1993; Teske *et al.*, 1994; Fisher *et al.*, 1995; Helfand *et al.*, 1995), plasma cell tumors (Thrall, 1981), malignant histiocytosis (Walton *et al.*, 1996; Wellman *et al.*, 1985, 1988), and myeloid leukemias (Kawakami *et al.*, 1989; Perk *et al.*, 1992; Sykes *et al.*, 1985; Felsburg *et al.*, 1994; Couto, 1985; Facklam and Kociba, 1985; Grinden *et al.*, 1986).

The etiology of most tumors of the canine immune system is unknown. The presence of cytogenetic abnormalities and apparent breed predilections suggest a genetic origin for some tumors, and the presence of reverse transcriptase activity or retroviral-like particles have suggested a possible viral etiology for canine lymphoma and AML.

The lineage of the tumor cell population can be determined by cytochemistry, immunophenotyping, or examining Ig and TCR gene rearrangements.

Little information is available on the effects that tumors of the canine immune system have on immune function.

18. Autoimmunity

The following diseases represent those with a documented or suspected autoimmune origin, and does not include the many other immune-mediated diseases that have been described in the dog. For most of these diseases in the dog, the immunoregulatory abnormalities and mechanisms of immune dysfunction have not been studied. The clinical and immunological characteristics of the following autoimmune diseases have been the subject of recent reviews (Halliwell and Gorman, 1989; Bistner, 1994; Greco and Harpold, 1994; Kemppainen and Clark, 1994; Lewis, 1994a,b; Happ, 1995; Hoenig, 1995; Tizard, 1996).

Systemic lupus erythematosus (SLE)

Canine SLE is a generalized immune complex disease that has clinical features similar to human SLE, including fever, polyarthritis, glomerulonephritis, and mucocutaneous lesions. As in humans, canine SLE is a chronic disease with alternating subacute periods and relapses. The pathogenesis of canine SLE is typical of a systemic immune complex disease with circulating immune complexes being deposited in the synovia of joints, producing a polyarthritis, in the glomeruli of the kidney, producing a glomerulonephritis, and at the basement membrane of the skin, producing a dermatitis. Circulating immune complexes

Table VII.17.1 Tumors of the immune system

Tumor type	*Etiology*	*Target population*	*Pathogenesis*
Lymphoma and lymphoid leukemia	Genetic (?)	Lymphoid precursors: B cells, T cells, and NK cells	Clonal expansion and selection of transformed lymphocytes that may be related to aberrant expression of oncogenes or deletion of tumor suppressor genes.
	Viral (?)		Unknown. Possible insertion of retrovirus near oncogene sequence or disruption of tumor suppressor gene expression.
Malignant histiocytosis	Unknown Genetic (?)	Monocyte and macrophage	Clonal expansion of transformed monocyte. Tissues involved are generally those where there are resident macrophages: lung, skin, lymph node, spleen, CNS are common sites.
Myeloid leukemia	Radiation	Monocyte (AMoL)	Clonal expression with development of leukemia with cytogenetic abnormalities.
	Viral (?)	Myeloid precursors (AML)	Unknown. Possible insertion of retrovirus near oncogene sequence or disruption of tumor suppressor gene expression.
	Genetic (?)	Myeloid precursors (AML, AMoL, CML)	Clonal expansion and selection of transformed monocytic or myeloid precursor that may be related to aberrant expression of oncogenes or deletion of tumor suppressor genes.
	Unknown	Granulocytic precursors (CML)	Expression of transformed cell committed to the granulocytic lineage with a large population of its progeny undergoing complete differentiation.
Plasma cell tumors	Unknown	B cells	Transformed clone is committed to undergo plasma cell differentiation.

can also attach to erythrocytes and platelets resulting in immune-mediated hemolytic anemia and thrombocytopenia. The most common clinical presentation in canine SLE is a nonerosive polyarthritis.

Although the principal autoantibody in canine SLE is directed against the cell nucleus (antinuclear antibodies) detectable by immunofluorescence, the distribution of the autoantibodies are more restricted in the dog than they are in human SLE in that they are primarily IgG antibodies directed against histones, particularly H1, H2A, H3, and H4. Another difference is the pattern of reactivity to individual histones. In human SLE, the anti-histone antibodies are directed against the trypsin-sensitive regions of the histones, whereas in canine SLE the anti-histone antibodies are reactive with the trypsin-resistant regions.

A possible role of a genetic factor(s) in canine SLE is suggested by the predilection of SLE in collies, Shetland sheepdogs, beagles, German shepherds, Irish setters, and poodles.

Rheumatoid arthritis (RA)

Canine RA is a progressive, bilaterally symmetrical, erosive polyarthritis. The disease affects young to middle-aged dogs and primarily occurs in small and toy breeds. The joint changes observed in canine RA are very similar to those seen in human RA. As in human RA, the autoantibody, rheumatoid factor (RF), involved in canine RA is directed against autologous IgG. The pathogenesis of canine RA has not been studied in detail, but it is most likely similar to that of human RA. The histopathological findings in canine RA include the presence of synovial hyperplasia, fibrinous deposits, subchondral bone erosions, and a marked mononuclear cell infiltrate. In many dogs, the synovial membranes exhibit a diffuse T-cell infiltrate adjacent to focal aggregates of B cells.

Although there are many clinical and histopathological similarities between human and canine RA, there are several differences. Canine RF is usually an IgG antibody rather than an IgM antibody; features of systemic RA that often accompany joint disease in humans are not seen in the dog, and rheumatoid nodules which occur subcutaneously in people with RA have not been reported in the dog.

Although the stimulus that initiates RA in susceptible animals has not been determined, most theories implicate an infectious agent. Canine distemper virus has been suggested as playing a role in the pathogenesis of canine RA. Dogs with RA have been shown to have antibodies to canine distemper virus in their synovial fluids and these antibodies are not found in dogs with osteoarthritis. Immune complexes precipitated from the synovial fluid of dogs with RA have been shown to contain canine distemper virus by Western blotting.

Sjögren's syndrome

Sjögren's syndrome represents a triad of keratoconjunctivitis sicca, xerostomia, and the presence of connective tissue disease such as rheumatoid arthritis or systemic lupus erythematosus. Sjögren's syndrome was originally identified in a colony of dogs with SLE, and subsequently in the general dog population. Affected dogs develop antibodies to the epithelial cells of the nictitating membrane, and to the lacrimal and salivary glands which leads to conjunctival dryness (keratoconjunctivitis sicca) and mouth dryness (xerostomia). In a recent study of 50 affected dogs, 40% had evidence of rheumatoid factor and 42% had antinuclear antibodies. Histological examination of affected lacrimal and salivery glands showed extensive infiltration with lymphocytes and monocytes.

Autoimmune skin disease

The pemphigus group of autoimmune skin diseases result from formation of autoantibodies against the intercellular cement of the skin and consists of four diseases that have been documented in both humans and dogs. The pemphigus diseases are characterized by the formation of bullae (vesicles) within the epidermis. Direct immunofluorescence of the skin lesions reveals linear deposits of immunoglobulin surrounding the epidermal cells giving a honeycomb appearance. Pemphigus vulgaris is the most severe of the pemphigus diseases in the dog and has a predilection for the mucocutaneous junctions and oral mucosa. Bullae formation in pemphigus vulgaris results from autoantibody production against desmoglein-3, a protein in desmosomes which is responsible for squamous cell adhesion. The bullae in pemphigus vulgaris are localized to the suprabasilar region of the lower epidermis. Pemphigus vegetans is a rare, mild variant of pemphigus vulgaris. Pemphigus foliaceous is the most common autoimmune skin disease in the dog and is characterized by superficial subcorneal bullae. The autoantibody in pemphigus foliaceous is directed against desmoglein-1, a protein found in squamous cell desmosomes. Unlike pemphigus vulgaris, pemphigus foliaceous does not have a predilection for mucocutaneous junctions in the dog. It usually involves the head and nose, but may become generalized. Pemphigus erythematosus is a hybrid between pemphigus foliaceous and systemic lupus erythematosus with immunological features of both diseases. It is confined primarily to the face and ears. Direct immunofluorescence of skin lesions shows intercellular immunofluorescence typical of the pemphigus disease and also concomitant immune complex deposition at the dermo-epidermal junctions. Some dogs may have antinuclear antibodies.

Bullous pemphigoid is a bullous autoimmune skin disease that clinically resembles pemphigus vulgaris. Although the disease has a predilection for mucocutaneous junctions, the groin, and axilla, it differs from pemphigus vulgaris in that the bullae are located in the subepidermis. The autoantibody in bullous pemphigoid are directed against desmoplakins – proteins in hemidesmosomes that attach cells to the basement membrane of the skin. Direct immunofluorescence reveals linear deposits of immunoglobulin and/or complement at the dermoepidermal junction. A breed predilection for collies, Shetland sheepdogs, and Doberman pinschers has been reported.

Discoid lupus erythematosus (DLE) clinically resembles the cutaneous manifestation of SLE in dogs, but there is no internal organ involvement, and antinuclear antibody tests are negative. The lesions are primarily isolated to the face and are exacerbated by sunlight. Direct immunofluorescence reveals immune complex deposition at the dermo-epidermal junction that are usually restricted to involved skin only which is in contrast to SLE in which the immune complex deposition extends to adjacent normal skin. A breed predilection for DLE has been observed for collies, Shetland sheepdogs, German shepherds, and Siberian huskies.

Autoimmune hemolytic anemia (AIHA)

AIHA in the dog is caused by the production of autoantibodies against antigens on the red cell membrane. The pathogenesis of AIHA in the dog is similar to that in other species with the destruction of the red cells either through intravascular hemolysis mediated by complement or, more commonly, through phagocytosis of the red cell by the macrophages of the spleen or liver. The mechanism of destruction is dependent, in part, on the isotype of the autoantibody involved, IgG or IgM. Cold agglutinin disease is a variant of AIHA that is primarily manifested clinically by necrosis of the extremities rather than anemia. The autoantibody involved is an IgM antibody. A genetic predisposition is suggested by the apparent breed predilection in German shepherds, Irish setters, Old English sheepdogs, miniature dachshunds, Scottish terriers, vizslas, and cocker spaniels.

Autoimmune thrombocytopenia (AITP)

Autoantibodies to platelets results in a shortened life span of the platelets owing to extravascular destruction in the spleen that results in the inability to repair damage to capillaries, resulting in petechial hemorrhage. The autoantibody in the dog recognizes an antigen present on both platelets and megakaryocytes. A breed predilection for AITP has been observed in Old English sheepdogs, cocker spaniels, and poodles.

Addison's disease

Several reports of Addison's disease have been reported in the dog based upon the clinical findings of hypoadrenocorticism and the presence of autoantibody activity against adrenal tissue by indirect immunofluorescence. Histologically, the disease is characterized by a lymphocytic infiltrate and atrophy of the gland. One case also had a concomitant hypothyroidism with autoantibodies against thyroid tissue resembling immunoendocrinopathy syndrome or Schmidt's syndrome in humans.

Insulin-dependent diabetes mellitus (IDDM)

IDDM, or type 1 diabetes mellitus, in humans is an autoimmune disease that is characterized by a genetic susceptibility and immunological destruction of the islet β cells. Patients show evidence of autoantibodies to multiple components of the islet β cells, and lymphocytic infiltration of the islets of Langerhans. Although IDDM occurs spontaneously in the dog, an autoimmune basis for the disease remains to be determined. Approximately 50% of diabetic dogs have evidence of β-cell antibodies, however infiltration of islets with lymphocytes is rarely seen. This may be due to the fact that diabetes is usually not diagnosed until dogs are extremely hyperglycemic and the pancreas is atrophied or fibrosed.

Lymphocytic thyroiditis

Canine lymphocytic thyroiditis closely resembles Hashimoto's thyroiditis (lymphocytic thyroiditis) in humans. Histologically, canine lymphocytic thyroiditis is characterized by infiltration of the thyroid gland by lymphocytes, plasma cells, and macrophages often forming germinal centers (lymphoid nodules). Affected dogs possess autoantibodies to various thyroid antigens. As in humans, approximately 60% of the dogs have antithyroglobulin antibodies. One difference between canine and human patients is that autoantibodies to microsomal antigens is present in $< 30\%$ of dogs compared to $> 85\%$ of humans. Another difference between canine and human patients is that dogs appear predisposed to producing autoantibodies to triiodothyronine (T_3) rather than thyroxine (T_4), whereas human patients show an equal distribution between the two.

A genetic predisposition to the disease is suggested by the increased incidence in beagles, golden retrievers, Doberman pinschers, Shetland sheepdogs, great Danes, Irish setters and Old English sheepdogs. Additional evidence for a genetic component to the disease in dogs is that relatives of affected animals often have antithyroid antibodies.

Vogt–Koyanagi–Harada syndrome

Vogt–Koyanagi–Harada syndrome is an autoimmune disease of humans and dogs which consists of a uveitis of one or both eyes characterized by acute iridocyclitis, choroiditis, and retinal detachment. The ocular lesions may be accompanied by depigmentation of the iris and retina. The disease is also characterized by depigmentation of the skin, primarily of the eyelids and nasal planum, and whitening of the hair. These features of the disease usually occur after the ocular lesions. Histologically, the disease in both humans and dogs is characterized by diffuse infiltration of the uvea with lymphocytes, plasma cells, and macrophages including epithelioid cells and giant cells. In the skin, there is an infiltration of lymphocytes and macrophages at the dermo-epidermal junction.

The disease was originally described in Akitas, but has subsequently been documented in Samoyeds, Irish setters, golden retrievers, Saint Bernards, Australian sheepdogs, and Shetland sheepdogs.

Myasthenia gravis

Myasthenia gravis is an autoimmune disease of humans and dogs that is characterized by muscle weakness due to a disorder of neuromuscular transmission. The pathogenesis of the disease in dogs is similar to that in humans with over 90% of affected dogs exhibiting antibodies to acetylcholine receptors. These autoantibodies, which bind at the postsynaptic membrane of the neuromuscular junction, interrupt transmission by producing a functional blockade to the binding of acetylcholine, increasing the endocytosis of acetylcholine receptors, and form immune complexes that bind complement, causing further destruction of the postsynaptic membrane.

Polymyositis

Canine polymyositis is a generalized myositis that affects primarily large breed dogs. An autoimmune basis for the disease is suspected based upon the finding that approximately 50% of affected dogs have antinuclear antibodies and/or antisarcolemmal antibody. Histologically, there is a lymphocytic and plasma cell infiltration of involved muscle.

Autoimmune masticatory myopathy

A myositis that is restricted to the muscles of mastication (masseter, temporalis, and pterygoid muscles) has been described in the dog. These muscles are derived from the mesoderm rather than myotomes from which other skeletal muscle is derived, and contain two distinct types of myofibers (1 and 2M). Affected dogs have circulating antibodies to myofiber type 2M, and immunoglobulin deposition can be demonstrated in biopsies of affected muscles. Histologically, there is myofiber degeneration with atrophy and fibrosis and an infiltration of the affected area with lymphocytes and plasma cells.

19. Conclusions

Although considerable advances have been made in recent years in the development of immunological and molecular reagents for studying the canine immune system, the availability of these reagents still lags behind those available for other species. These endeavors must remain a priority in order to be able to promote the dog as an experimental model.

Perhaps the major value of the dog in immunological research is its great potential as a naturally occurring animal model for human disease. Not only will these models provide insight into the immunopathogenesis of the various disease homologues, they will also serve as valuable preclinical models in which to develop and evaluate immunological and genetic therapeutic intervention protocols for these diseases.

20. References

Anderson, J. E., Ladiges, W. C., Giblett, E. R., Weiden, P., Storb, R. (1983). Polymorphism of the sixth component of complement (C6) in dogs. *Animal Genet.* **21**, 155–160.

Andrews, G. A., Chavey, P. S., Smith, J. E. (1992). Production, characterization, and applications of a murine monoclonal antibody to dog erythrocyte antigen 1.1. *J. Am. Vet. Med. Assoc.* **201**, 1549–1552.

Atkins, A. M., Schofield, G. C. (1972). Lymphoglandular complexes in the large intestine of the dog. *J. Anat.* **113**, 169–178.

Baker, J. A., Robson, D. S., Gillespie, J. H., Burgher, J., Doughty, M. F. (1959). A normograph that predicts the age to vaccinate puppies against distemper. *Cornell Vet.* **49**, 158–167.

Barta, O., Hubbert, N. L. (1978). Testing of hemolytic complement components in domestic animals. *Am. J. Vet. Res.* **39**, 1303–1308.

Bell, K. (1983). The blood groups of domestic animals. In *Red Blood Cells of Domestic Animals* (eds Agar, A. S., Board, P. G.), pp. 133–164. Elsevier Science Publishers, Amsterdam, The Netherlands.

Belz, G. T., Heath, T. J. (1995a). Pathways of blood flow to and through superficial lymph nodes in the dog. *J. Anat.* **187**, 413–427.

Belz, G. T., Heath, T. J. (1995b). Lymph pathways of the medial retropharyngeal lymph node in dogs. *J. Anat.* **186**, 517–526.

Bezuidenhout, A. J. (1993). The lymphatic system. In *Miller's Anatomy of the Dog*, 3rd edn. (ed. Evans, H. E.), pp. 717–757. W.B. Saunders, Philadelphia, PA, USA.

Bistner, S. (1994). Allergic- and immunologic-mediated diseases of the eye and anexae. *Vet. Clinics NA* **24**, 711–734.

Blum, J. R., Cork, L. C., Morris, J. M., Olson, J. L., Winkelstein, J. A. (1985). The clinical manifestations of a genetically-determined deficiency in the third component of complement in the dog. *Clin. Immunol. Immunopathol.* **24**, 304–315.

Bravo, L., Legendre, A. M., Hahn, K. A., Rohrbach, B. W., Braha, T., Lothrop, C. D. (1996). Serum granulocyte colony-stimulating factor (G-CSF) and interleukin-1 (IL-1) concentrations after chemotherapy-induced neutropenia in normal and tumor-bearing dogs. *Exp. Hematol.* **24**, 11–17.

Breitschwerdt, E. B., Brown, T. T., DeBuyssher, E. V., Anderson, B. R., Thrall, D. E., Hager, E., Ananaba, G., Degen, M. A., Ward, M. D. (1987). Rhinitis, pneumonia and defective neutrophil function in the Doberman Pinscher. *Am. J. Vet. Res.* **48**, 1054–1062.

Bryant, B. J., Shifrine, M. (1972). Histiogenesis of lymph nodes during development of the dog. *J. Reticulo. Soc.* **12**, 96–107.

Bryant, B. J., Shifrine, M., McNeil, C. (1973). Cell-mediated immune response in the developing dog. *Int. Arch. Allergy Appl. Immunol.* **45**, 937–942.

Bull, R. W., Vriesendorp, H. M., Cech, R., Grosse-Wilde, H., Bijma, A. M., Ladiges, W. L., Krumbacher, K., Doxiadis, I., Ejima, H., Templeton, J., Albert, E. D., Storb, R., Deeg, H. J. (1987). Joint report of the third international workshop on canine immunogenetics. II. Analysis of the serological typing of cells. *Transplantation* **43**, 154–161.

Burnett, R. C., Geraghty, D. E. (1995). Structure and expression of a divergent canine class I gene. *J. Immunol.* **155**, 4278–4285.

Callan, M. B., Jones, L. T., Giger, U. (1995). Hemolytic transfusion reactions in a dog with an alloantibody to a common antigen. *J. Vet. Intern. Med.* **9**, 277–279.

Carpenter, J. L., Myers, A. M., Conner, M. W., Schelling, S. H., Kennedy, F. A., Reimann, K. A. (1988). Tuberculosis in five Bassett hounds. *J. Am. Vet. Med. Assoc.* **192**, 1563–1568.

Carmichael, L. E., Robson, D. S., Barnes, F. D. (1962). Transfer and decline of maternal infectious canine hepatitis antibody in puppies. *Proc. Soc. Exp. Biol. Med.* **109**, 677–681.

Chabanne, L., Marchal, T., Kaplanski, C., Fournel, C., Magnol, J. P., Monier, J. C., Rigal, D. (1994). Screening of 78 monoclonal antibodies directed against human leukocyte antigens for cross-reactivity with surface markers on canine lymphocytes. *Tissue Antigens* **43**, 202–205.

Cobbold, S., Metcalfe, S. (1994). Monoclonal antibodies that define canine homologues of human CD antigens: Summary of the First International Canine Leukocyte Antigen Workshop (CLAW). *Tiss. Antigens* **43**, 137–154.

Colling, D. T., Saison, R. (1980a). Canine blood groups. I. Description of new erythrocyte specificities. *Anim. Genet.* **11**, 1–12.

Colling, D. T., Saison, R. (1980b). Canine blood groups. II. Description of a new allele in the Tr blood group system. *Anim. Genet.* **11**, 13–20.

Couto, C. G. (1985). Clinicopathologic aspects of acute leukemias in the dog. *J. Am. Vet. Med. Assoc.* **186**, 681–685.

Couto, C. G., Krakowka, S., Johnson, G., Ciekot, P., Hill, R., Lafrado, L., Kociba, G. (1989). *In vivo* immunologic features of Weimaraner dogs with neutrophil abnormalities and recurrent infections. *Vet. Immunol. Immunopathol.* **23**, 103–112.

Danilenko, D. M., Moore, P. F., Rossitto, P. V. (1992). Canine leukocyte cell adhesion molecules (LeuCAMs): characterization of the CD1/CD18 family. *Tissue Antigens* **40**, 13–21.

Danilenko, D. M., Rossitto, P. V., der Vieren, M. V., Trong,

H. L., McDonough, S. P., Affolter, V. K., Moore, P. F. (1995). A novel canine leukointegrin, $\alpha d\beta 2$, is expressed by specific macrophage subpopulations in tissue and a minor $CD8^+$ lymphocyte subpopulation in peripheral blood. *J. Immunol.* **155**, 35–44.

Day, M. J. (1996). Expression of interleukin-1β, interleukin-6 and tumor necrosis factor-α by macrophages in canine lymph nodes with mineral-associated lymphadenopathy, granulomatous lymphadenitis or reactive hyperplasia. *J. Comp. Pathol.* **114**, 31–42.

Deeg, H. J., Storb, R. (1994). Canine bone marrow transplantation models. *Curr. Topics Vet. Res.* **1**, 103–114.

Deeg, H. J., Raff, R. F., Grosse-Wilde, H., Bijma, A. M., Buurman, W. A., Doxiadis, I., Kolb, H. J., Drumbacher, K., Ladiges, W. L., Losstein, K. L., Schoch, G., Westbrock, D. L., Bull, R. W., Storb, R. (1986). Joint report of the third international workshop on canine immunogenetics. I. Analysis of homozygous typing cells. *Transplantation* **41**, 111–117.

Delacroix, D. L., Furtado-Barreira, G., de Hemptinne, B., Goudswaard, J., Dive, C., Vaerman, J. P. (1983). The liver in the IgA secretory immune system. Dogs, but not rats and rabbits, are suitable models for human studies. *Hepatology* **3**, 980–988.

Delacroix, D. L., Furtado-Barreira, G., Rahier, J., Dive, C., Vaerman, J. P. (1984). Immunohistochemical localization of secretory component in the liver of guinea pigs and dogs versus rats, rabbits, and mice. *Scand. J. Immunol.* **19**, 425–434.

Devos, K. F., Buerinck, F., Van Audenhove, K., Fiers, W. (1992). Cloning and expression of the canine interferon-gamma gene. *J. Interferon Res.* **12**, 95–102.

Dodds, A. W., Law, S. K. A. (1990). The complement component C4 of mammals. *J. Biochem.* **265**, 495–502.

Dore, M., Hawkins, H. K., Entman, M. L., Smith, C. W. (1993a). Production of a monoclonal antibody against GMP-140 (P-selectin) and studies of its vascular distribution in canine tissues. *Vet. Pathol.* **30**, 213–222.

Dore, M., Korthuis, R. J., Granger, D. N., Entman, M. L., Smith, C. W. (1993b). P-selectin mediates spontaneous leukocyte rolling *in vivo*. *Blood* **82**, 1308–1316.

Dore, M., Simon, S. I., Hughers, B. J., Entman, M. L., Smith, C. W. (1995). P-selectin and CD18 mediated recruitment of canine neutrophils under conditions of shear stress. *Vet. Pathol.* **32**, 258–268.

Doxiadis, G., Schoen, W., Doxiadis, I., Deeg, H. J., O'Neill, G. J., Grosse-Wilde, H. (1987). Statement on the nomenclature of dog C4 allotypes. *Immunogenetics* **25**, 167–170.

Doxiadis, I., Krumbacher, K., Neefhes, J. J., Ploegh, H. L., Grosse-Wilde, H. (1986). Canine MHC biochemical definition of class I, class II, and class III determinants: similarities and differences to the human and murine systems. *Immunobiology* **173**, 264–265.

Doxiadis, I., Krumbacher, K., Neefjes, J. J., Ploegh, H. L., Grosse-Wilde, H. (1989). Biochemical evidence that the DLA-B locus codes for a class II determinant expressed on all canine peripheral blood lymphocytes. *Exp. Clin. Immunogenet.* **6**, 219–224.

Dreyer, W. J., Michael, L. H., Nguyen, T., Smith, C. W., Anderson, D. C., Entman, M. L., Rossen, R. D. (1992). Kinetics of C5a release in cardiac lymph of dogs experiencing coronary artery ischemia-reperfusion injury. *Circ. Res.* **71**, 1518–1524.

Dunham, S. P., Argyle, D. J., Onions, D. E. (1995). The isolation and sequence of canine interleukin-2. *DNA Seq.* **5**, 177–180.

Eckersall, P. D., Conner, J. G. (1988). Bovine and canine acute phase proteins. *Vet. Res. Comm.* **12**, 169–178.

Eijima, H., Kurokawa, K., Ikemoto, S. (1986). Phenotype and gene frequency of red blood cell groups in dogs of various breeds reared in Japan. *Jpn. J. Vet. Sci.* **48**, 363–368.

Eldridge, P. R., Hobart, M. J., Lachmann, P. J. (1983). The genetics of the sixth and seventh components of complement in the dog: polymorphism, linkage, locus duplication and silent alleles. *Biochem. Genet.* **21**, 81–91.

Facklam, N. R., Kociba, G. J. (1985). Cytochemical characterization of leukemic cells in 20 dogs. *Vet. Pathol.* **22**, 363–369.

Felsburg, P. J. (1994). Overview of the immune system and immunodeficiency diseases. *Vet. Clinics NA* **24**, 629–653.

Felsburg, P. J., Glickman, L. T., Jezyk, P. F. (1985). Selective IgA deficiency in the dog. *Clin. Immunol. Immunopathol.* **36**, 297–305.

Felsburg, P. J., Somberg, R. L., Perryman, L. E. (1992). Domestic animal models of severe combined immunodeficiency: canine X-linked severe combined immunodeficiency and severe combined immunodeficiency in horses. *Immunodeficiency Rev.* **3**, 277–303.

Felsburg, P. J., Somberg, R. L., Krakowka, G. S. (1994). Acute monocytic leukemia in a dog with X-linked severe combined immunodeficiency. *Clin. Diag. Lab. Immunol.* **1**, 379–384.

Fisher, D. J., Naydan, N., Werner, L. L., Moore, P. F. (1995). Immunophenotyping lymphomas in dogs: a comparison of results from fine needle aspirate and needle biopsy samples. *Vet. Clin. Pathol.* **24**, 118–123.

Fuller, L., Carreno, M., Esquenazi, V., Zucker, K., Zheng, S., Roth, D., Burke, G., Nery, J., Asthana, D., Olson, L., Miller, J. (1994). Characterization of anti-canine cytokine monoclonal antibodies specific for IFN-γ: effect of anti-IFN-γ on renal transplant rejection. *Tissue Ag.* **43**, 163–169.

Galkowska, H., Wojewodzka, U., Olszewski, W. L. (1995). Cytokines and adherence molecules involved in spontaneous dendritic cell-lymphocyte clustering in skin afferent lymph. *Scand. J. Immunol.* **42**, 324–330.

Gebhard, D. H., Carter, P. B. (1992). Identification of canine T lymphocyte subsets with monoclonal antibodies. *Vet. Immunol. Immunopathol.* **32**, 187–199.

Giger, U., Boxer, L. A., Simpson, P. J., Lucchesi, B. R., Todd, R. F. (1987). Deficiency of leukocyte surface glycoproteins Mo1, LFA-1 and Leu M5 in a dog with recurrent bacterial infections: an animal model. *Blood* **69**, 1622–1630.

Giger, U., Gelens, C. J., Callan, M. B., Oakley, D. A. (1995). An acute hemolytic transfusion reaction caused by dog erythrocyte antigen 1.1 incompatibility in a previously sensitized dog. *J. Am. Vet. Med. Assoc.* **206**, 9–14.

Gillespie, J. H., Baker, J. A., Burgher, J., Robson, D., Gilman, B. (1958). The immune response of dogs to distemper virus. *Cornell Vet.* 103–126.

Gorman, N. T., McConnell, I., Lachman, P. N. (1981). Characterization of the third component of canine and feline complement. *Vet. Immunol. Immunopathol.* **2**, 309–316.

Gorman, S. D., Frewin, M. R., Cobbold, S. P., Waldmann, H. (1994). Isolation and expression of cDNA encoding the canine CD4 and CD8α antigens. *Tissue Antigens* **43**, 184–188.

Greco, D. S., Harpold, L. M. (1994). Immunity and the endocrine system. *Vet. Clinics NA* **24**, 765–782.

Grinden, C. B., Buoen, I. C. (1986). Cytogenetic analysis of leukemic cells in the dog. *J. Comp. Pathol.* **196**, 623–635.

Grinden, C. B., Stevens, J. B., Perman, V. (1986). Cytochemical reactions in cells from leukemic dogs. *Vet. Pathol.* **23**, 103–109.

Grosse-Wilde, H., Doxiadis, G., Krumbacher, K., Dekkers-Bijma, A., Kolb, H. J. (1983). Polymorphism of the fourth complement component in the dog and linkage to the DLA system. *Immunogenetics* **18**, 537–540.

Gruys, E., Obwolo, M. J., Toussaint, M. J. M. (1994). Diagnostic significance of the major acute phase proteins in veterinary clinical chemistry. *Vet. Bulletin* **64**, 1009–1018.

Halliwell, R. E. W., Gorman, N. T. (1989). *Veterinary Clinical Immunology*. W.B. Saunders, Philadelphia, PA, USA.

Happ, G. M. (1995). Thyroiditis – A model canine autoimmune disease. *Adv. Vet. Sci. Comp. Med.* **39**, 97–129.

Harmsen, A. G., Muggenburg, B. A., Snipes, M. B., Bice, D. E. (1985). The role of macrophages in particle translocation from lungs to lymph nodes. *Science* **230**, 1277–1280.

Hart, I. R. (1979). The distribution of immunoglobulin-containing cells in canine small intestine. *Res. Vet. Sci.* **27**, 269–274.

Hebel, R. (1960). Untersuchungen uber das vorkommen von lymphotischen darmkrypten in der tunica submucosa des darmes von schwein, rind, schaf, hund und katze. *Anat. Anz.* **109**, 7–27.

Heddle, R. J., Rowley, D. (1975). Dog immunoglobulins. I. Immunochemical characterization of dog serum, parotid saliva, colostrum, milk and small bowel fluid. *Immunology* **29**, 185–195.

Helfand, S. C., Modiano, J. F., Nowell, P. C. (1992). Immunophysiological studies of interleukin-2 and canine lymphocytes. *Vet. Immunol. Immunopathol.* **33**, 1–16.

Helfand, S. C., Modiano, J. F., Moore, P. F., Soergel, S. A., MacWilliams, P. S., Dubielzig, R. D., Hank, J. A., Gelfand, E. W., Sondel, P. M. (1995). Functional interleukin-2 receptors are expressed on natural killer-like leukemic cells from a dog with cutaneous lymphoma. *Blood* **86**, 636–645.

Henthorn, P. S., Somberg, R. L., Fimiani, V. M., Puck, J. M., Patterson, D. F., Felsburg, P. J. (1994). IL-2Rγ gene microdeletion demonstrates that canine X-linked severe combined immunodeficiency is a homologue of the human disease. *Genomics* **23**, 69–74.

Hoenig, M. (1995). Pathophysiology of canine diabetes. *Vet. Clinics NA* **25**, 553–561.

HogenEsch, H. (1989). The morphology and function of Peyer's patches in the dog. PhD. Thesis, University of Illinois, IL, USA.

HogenEsch, H., Felsburg, P. J. (1990). Ultrastructure and alkaline phosphatase activity of the dome epithelium of canine Peyer's patches. *Vet. Immunol. Immunopathol.* **24**, 177–186.

HogenEsch, H., Felsburg, P. J. (1992a). Immunohistology of Peyer's patches in the dog. *Vet. Immunol. Immunopathol.* **30**, 147–160.

HogenEsch, H., Felsburg, P. J. (1992b). Isolation and phenotypic and functional characterization of cells from Peyer's patches in the dog. *Vet. Immunl. Immunopathol.* **31**, 1–10.

HogenEsch, H., Housman, J. M., Felsburg, P. J. (1987). Canine Peyer's patches: macroscopic, light microscopic, scanning electron microscopic and immunohistochemical investigations. *Adv. Exp. Med. Biol.* **216A**, 249–256.

HogenEsch, H., Snyder, P. W., Scott-Moncrieff, J. C., Glickman, L. T., Felsburg, P. J. (1995). Interleukin-6 activity in dogs with juvenile polyarteritis syndrome: effect of corticosteroids. *Clin. Immunol. Immunopathol.* 77, 107–110.

Huss, R., Hong, D. S., Beckham, C., Kimball, L., Myerson, D. H., Storb, R., Deeg, H. J. (1995). Ultrastructural localization of stem cell factor in canine marrow-derived stromal cells. *Exp. Hematol.* **23**, 33–40.

Ishikawa, J., Suzuki, S., Hotta, K., Hirota, Y., Mizuno, S., Suzuki, K. (1993). Cloning of a canine gene homologous to the human interleukin-8-encoding gene. *Gene* **131**, 305–306.

Ito, K., Tsunoda, M., Watanabe, K., Ito, K., Kashiwagi, N., Obata, F. (1993). Isolation and sequence analysis of cDNA for the dog T cell receptor Tcrα and Tcrβ chains. *Immunogenetics* **38**, 60–63.

Jacoby, R. O., Dennis, R. A., Griesemer, R. A. (1969). Development of immunity in fetal dogs: humoral responses. *Am. J. Vet. Res.* **30**, 1503–1510.

Jerome, S. N., Dore, M., Paulson, J. C., Smith, C. W., Korthuis, R. J. (1994). P-selectin and ICAM-1 dependent adherence reactions: role in the genesis of postischemic no-reflow. *Am. J. Physiol.* **266**, H1316–H1321.

Jezyk, P. F., Haskins, M. E., MacKay-Smith, W. E., Patterson, D. F. (1986). Lethal acrodermatitis in bull terriers. *J. Am. Vet. Med. Assoc.* **88**, 833–839.

Johnson, J. P., Hammer, C., Winkelstein, J. A. (1985). Purification of the third component of canine complement. *Vet. Immunol. Immunopathol.* **8**, 377–389.

Johnson, J. P., McLean, R. H., Cork, L. C., Winkelstein, J. A. (1986). Genetic analysis of an inherited deficiency of the third component of complement in the dog. *Am. J. Med. Genet.* **25**, 557–562.

Jones, J. B., Lange, R. D., Jones, E. S. (1975). Cyclic hematopoiesis in a colony of dogs. *J. Am. Vet. Med. Assoc.* **166**, 365–367.

Kawakami, T., Cain, G., Taylor, N. (1989). Establishment and partial characterization of a radiation-induced canine monocytic leukemic cell line (RK9ML-1). *Leukemia Res.* **13**, 709–714.

Kay, P. H., Dawkins, R. L. (1984). Genetic polymorphism of complement Cr in the dog. *Tissue Antigens* **23**, 151–155.

Kay, P. H., Dawkins, R. L., Penhale, J. W. (1985). The molecular structure of different polymorphic forms of C3 and C4. *Immunogenetics* **21**, 313–319.

Kelly, W. D. (1963). The thymus and lymphoid morphogenesis in the dog. *Fed. Proc.* **20**, 600.

Kemppainen R. J., Clark, T. P. (1994). Etiopathogenesis of canine hypothyroidism. *Vet. Clincs NA* **24**, 467–476.

Klaich, G. M., Hauter, P. M. (1994). Induction of dog IL-1 by free and liposomal encapsulated doxorubicin. *Anticancer Drugs* **5**, 355–360.

Klein, A. K., Dyck, J. A., Stitzel, K. A., Shimizu, M. J., Fox, L. A., Taylor, N. (1983). Characterization of canine fetal lymphohematopoiesis. *Exp. Hematol.* **11**, 263–274.

Knapp, D. W., Williams, J. S., Andrisani, O. M. (1995). Cloning of the canine interleukin-2-encoding cDNA. *Gene* **159**, 281–282.

Kolbjornsen, O., Press, C. M., Moore, P. F., Landsverk, T. (1994). Lymphoid follicles in the gastric mucosa of dogs. Distribution and lymphocyte phenotypes. *Vet. Immunol. Immunopathol.* **40**, 299–312.

Krakowka, S. (1992). Acquired immunodeficiency diseases. In *Current Veterinary Therapy, XI: Small Animal Practice* (eds

Kirk, R. W., Bonagura, J. D.), pp. 453–457. W.B. Saunders, Philadelphia, PA, USA.

Krumbacher, K., van der Feltz, M. J. M., Happel, M., Gerlach, C., Losslein, L. K., Grosse-Wilde, H. (1986). Revised classification of the DLA loci by serological studies. *Tissue Antigens* 27, 262–268.

Kubens, B. S., Krumbacher, K., Grosse-Wilde, H. (1995). Biochemical definition of DLA-A and DLA-B gene products by one-dimensional isoelectric focusing and immunoblotting. *Eur. J. Immunogenet.* **22**, 199–207.

Lefer, D. J., Flynn, D. M., Phillips, M. L., Ratcliffe, M., Buda, A. J. (1994). A novel sialyl Lewis analog attenuates neutrophil accumulation and myocardial necrosis after ischemia and reperfusion. *Circulation* **90**, 2390–2401.

Lewis, R. M. (1994a). Rheumatoid arthritis. *Vet. Clinics NA* **24**, 697–701.

Lewis, R. M. (1994b). Immune-mediated muscle disease. *Vet. Clinics NA* **24**, 703–710.

Lothrop, C. D. Jr, Warren, D. J. (1988). Correction of canine cyclic hematopoiesis with recombinant human granulocyte colony-stimulating factor. *Blood* **72**, 1324–1328.

Manning, A. M., Lu, H. F., Kukielka, G. L., Olivr, M. G., Ty, T., Toman, C. A., Drong, R. F., Slightom, J. L., Ballantyne, C. M., Entman, M. L., Smith, C. W., Anderson, D. C. (1995). Cloning and comparative sequence analysis of the gene encoding canine intercellular adhesion molecule-1 (ICAM-1). *Gene* **156**, 291–295.

Massion, P. P., Hebert, C. A., Leong, S., Chan, B., Inoue, H., Gratton, K., Sheppard, D., Nadel, J. A. (1995). *Staphylococcus aureus* stimulates neutrophil recruitment by stimulating interleukin-8 production in dog trachea. *Am. J. Physiol.* **268**, L85–94.

Matsumoto, Y., Mohamed, A., Onodera, T., Kato, H., Ohashi, T., Goitsuka, R., Tsujimoto, H., Hasegawa, A., Furusawa, S., Yoshihara, K. (1994). Molecular cloning and expression of canine interleukin 8 cDNA. *Cytokine* **6**, 455–461.

Mazza, G., Whiting, A. H. (1994). Development of an enzyme-linked immunosorbent assay for the detection of IgG subclasses in the serum of normal and diseased dogs. *Res. Vet. Sci.* **57**, 133–139.

McSweeney, P. A., Rouleau, K. A., Storb, R., Bolles, L., Wallace, P. M., Beauchamp, M., Krizanac-Bengez, L., Moore, P., Sale, G., Sandmaier, B., de Revel, T., Appelbaum, F. R., Nash, R. M. (1996). Canine CD34: cloning of the cDNA and evaluation of an antiserum to recombinant protein. *Blood* **68**, 1992–2003.

Meurer, R., Van Riper, G., Feeney, W., Cunningham, P., Hora, D., Springer, M. S., MacIntyre, D. E., Rosen, H. (1993). Formation of eosinophilic and monocytic intradermal inflammatory sites in the dog by injection of human RANTES but not human monocyte chemoattractant protein 1, human macrophage inflammatory protein 1α, or human interleukin 8. *J. Exp. Med.* **178**, 1913–1921.

Miller, G. K., Benjamin, S. A. (1985). Radiation-induced quantitative alterations in prenatal thymic development in the beagle dog. *Lab. Invest.* **52**, 224–231.

Momoi, Y., Nagase, M., Okamoto, Y., Okuda, M., Sasaki, N., Watari, T., Goitsuka, R., Tsujimoto, H., Hasegawa, A. (1993). Rearrangements of immunoglobulin and T-cell receptor genes in canine lymphoma/leukemia cells. *J. Vet. Med. Sci.* **55**, 775–780.

Moore, P. F., Rosin, A. (1986). Malignant histiocytosis of Bermese mountain dogs. *Vet. Pathol.* **23**, 1–10.

Moore, P. F., Rossitto, P. V. (1993). Development of monoclonal antibodies to canine T cell receptor complex (TCR/CD3) and their utilization in the diagnosis of T cell neoplasia. *Vet. Pathol.* **30**, 457.

Moore, P. F., Rossitto, D. M., Danilenko, D. M. (1990). Canine leukocyte integrins: characterization of a CD18 homologue. *Tissue Antigens* **36**, 211–220.

Moore, P. F., Rossitto, P. V., Danilenko, D. M., Wielenga, J. J., Raff, R. F., Severns, E. (1992). Monoclonal antibodies specific for canine CD4 and CD8 define functional T lymphocyte subsets and high density expression of CD4 by canine neutrophils. *Tissue Antigens* **40**, 75–85.

Moore, P. F., Rossitto, P. V., Olivry, T. (1994). Development of monoclonal antibodies to canine T cell receptor $\gamma\delta$ and their utilization in the diagnosis of epidermotropic cutaneous T cell lymphoma. *Vet. Pathol.* **31**, 597.

Moore, P. F., Schrenzel, M. D., Affolter, V. K., Olivry, T., Naydan, D. (1996). Canine cutaneous histiocytoma is an epidermotropic Langerhans cell histocytosis that expresses CD1 and specific β_2 molecules. *Am. J. Pathol.* **148**, 1699–1708.

Moroff, S. D., Hurvitz, A. I., Peterson, M. E., Saunders, L., Noone, K. E. (1986). IgA deficiency in Shar-Pei dogs. *Vet. Immunol. Immunopathol.* **13**, 181–188.

Nash, R. A., Scherf, U., Storb, R. (1991a). Molecular cloning of the CD3ε subunit of the T cell receptor/CD3 complex in dog. *Immunogenetics* **33**, 396–398.

Nash, R. A., Schuening, F. G., Appelbaum, F. R., Hammond, W. P., Boone, T., Morris, C. F., Slichter, S. J., Storb, R. (1991b). Molecular cloning and *in vivo* evaluation of canine granulocyte-macrophage colony-stimulating factor. *Blood* **78**, 930–937.

Nash, R. A., Seidal, K., Storb, R., Slichter, S., Schuening, F. G., Appelbaum, F. R., Becker, A. B., Bolles, L., Deeg, H. J., Graham, T., Hackman, R. C., Burstein, S. A. (1995). Effects of rhIL-11 on normal dogs and after sublethal radiation. *Exp. Hematol.* **23**, 389–396.

Nolte, M., Werner, M., Noke, J., Georgii, A. (1993). Different cytogenetic findings in two clinically similar leukemic dogs. *J. Comp. Pathol.* **108**, 337–342.

O'Neill, G. J., Yand, S. Y., DuPont, B. (1978). Two HLA linked loci controlling the fourth component of human complement. *Proc. Natl Acad. Sci.* **75**, 5165–5169.

O'Neill, G. J., Lang, M., Neri, C., Deeg, H. J. (1984). C4 polymorphism in the dog: molecular heterogeneity of the C4 alpha and C4 gamma subunit chains. *Immunogenetics* **20**, 649–654.

Onions, D. E. (1984). A prospective survey of familial canine lymphosarcoma. *J. Nat. Cancer Inst.* **72**, 909–912.

Pastori, R. L., Milde, K. F., Alejandro, R. (1994). Molecular cloning of the dog homologue of the lymphocyte antigen CD28. *Immunogenetics* **39**, 373.

Patel, M., Selinger, D., Mark, G. E., Hickey, G. J., Hollis, G. F. (1995). Sequence of the dog immunoglobulin alpha and epsilon constant region genes. *Immunogenetics* **41**, 282–286.

Peng, J., Frises, P., George, J. N., Dale, G. L., Burstein, S. A. (1994). Alteration of platelet function in dogs mediated by interleukin-6. *Blood* **83**, 398–403.

Perk, K., Safran, N., Dahlberg, J. E. (1992). Propagation and characterization of novel canine lentivirus isolated from a dog. *Leukemia* **6**, 1558–1578.

Pollock, R. V. H., Carmichael, L. E. (1982). Maternally derived immunity to canine parvovirus infection: transfer, decline, and interference with vaccination. *J. Am. Vet. Med. Assoc.* **180**, 37–42.

Popper, P., Mantyh, C. R., Vigna, S. R., Maggio, J. E., Mantyh, P. W. (1988). The localization of sensory nerve fibers and receptor binding sites for sensory neuropeptides in canine mesenteric lymph nodes. *Peptides* **9**, 257–267.

Prummer, O., Raghavachar, A., Werner, C., Calvo, W., Carbovonell, F., Steinbach, I., Fliedner, T. M. (1985). Fetal liver transplantation in the dog. *Transplantation* **39**, 349–355.

Rabanal, R. M., Ferrer, L., Else, R. W. (1995). Immunohistochemical detection of canine leukocyte antigens by specific monoclonal antibodies in canine normal tissues. *Vet. Immunol. Immunopathol.* **47**, 13–23.

Reynolds, H. Y., Johnson, J. S. (1970). Quantitation of canine immunoglobulins. *J. Immunol.* **105**, 698–703.

Ricks, J., Roberts, M., Patterson, R. (1970). Canine secretory immunoglobulins: identification of secretory component. *J. Immunol.* **105**, 1327–1333.

Rivas, A. L., Tintle, L., Kimball, E. D., Scarlett, J., Quimby, F. W. (1992). A canine febrile disorder associated with elevated interleukin-6. *Clin. Immunol. Immunopathol.* **64**, 36–45.

Rivas, A. L., Tintle, L., Argentieri, D., Kimball, E. S., Goodman, M. G., Anderson, D. W., Capetola, R. J., Quimby, F. W. (1995). A primary immunodeficiency syndrome in Shar-Pei dogs. *Clin. Immunol. Immunopathol.* **74**, 243–251.

Roth, J. A., Lamox, L. G., Altszuler, N., Hampshire, J., Kaeberle, M. L., Shelton, M., Draper, D. D., Ledet, A. E. (1980). Thymic abnormalities and growth hormone deficiency in dogs. *Am. J. Vet. Res.* **41**, 1256–1262.

Roth, J. A., Kaeberle, M. L., Grier, R. L., Hopper, J. G., Spiegel, H. E., McAllister, H. A. (1984). Improvement in clinical condition and thymus morphologic features associated with growth hormone treatment of immunodeficient dwarf dogs. *Am. J. Vet. Res.* **45**, 1151–1155.

Rottman, J. B., Tompkins, W. A. F., Tompkins, M. B. (1996). A reverse transcription-quantitative competitive polymerase chain reaction (RT-qcPCR) technique to measure cytokine gene expression in domestic mammals. *Vet. Pathol.* **33**, 242–248.

Salvador, A. C., Pereira, A. S., de Sa, C. M., Grande, N. R. (1992). Blood vasculature of the lymph node in the dog: anatomical evidence for participation of extrahilar arterial vessels in the blood supply of the cortex. *Acta Anat.* **143**, 41–47.

Sandmaier, B. M., Storb, R., Appelbaum, F. R., Gallatin, W. M. (1990). An antibody that facilitates hematopoietic engraftment recognizes CD44. *Blood* **76**, 630–636.

Sargent, A. U., Austen, K. F. (1970). The effective molecular titration of the early components of dog complement. *Proc. Soc. Exp. Biol. Med.* **133**, 1117–1122.

Sargent, A. U., Johnson, S. B. (1980). The hemolytic equivalence of human, guinea pig, and canine complement proteins. *Immunol. Comm.* **9**, 453–463.

Sargent, A. U., Johnson, S. B., Richardson, A. K. (1976). The isolation and functional characterization of the first seven components of canine hemolytic complement. *Immunochemistry* **13**, 823–829.

Sarmiento, U. M., Storb, R. (1988a). Characterization of class II alpha genes and DLA-D region allelic associations in the dog. *Tissue Antigens* **32**, 224–234.

Sarmiento, U. M., Storb, R. (1988b). Restriction fragment length polymorphism of the major histocompatibility complex of the dog. *Immunogenetics* **28**, 117–124.

Sarmiento, U. M., Storb, R. (1989). RFLP analysis of DLA class I genes in the dog. *Tissue Antigens* **34**, 158–163.

Sarmiento, U. M., Storb, R. (1990a). Nucleotide sequence of a dog DRB cDNA clone. *Immunogenetics* **31**, 396–399.

Sarmiento, U. M., Storb, R. (1990b). Nucleotide sequence of a dog class I cDNA clone. *Immunogenetics* **31**, 400–404.

Sarmiento, U. M., Sarmiento, J. I., Storb, R. (1990). Allelic variation in the DR subregion of the canine major histocompatibility complex. *Immunogenetics* **32**, 13–19.

Sarmiento, U. M., DeRose, S., Sarmiento, J. I., Storb, R. (1992). Allelic variation in the DQ subregion of the canine major histocompatibility complex. I. DQA. *Immunogenetics* **35**, 416–420.

Sarmiento, U. M., DeRose, S., Sarmiento, J. I., Storb, R. (1993). Allelic variation in the DQ subregion of the canine major histocompatibility complex. II. DRQ. *Immunogenetics* **37**, 148–152.

Shibata, T., Abe, T., Tanabe, Y. (1995). Genetic polymorphism of the sixth component of complement (C6) in dogs. *Animal Genet.* **26**, 105–106.

Shifrine, M. N., Smith, J. B., Bulgin, M. S., Bryant, J. B., Zec, Y. C., Osburn, B. I. (1971). Response of canine fetuses and neonates to antigenic stimulation. *J. Immunol.* **107**, 965–970.

Shull, R. M., Suggs, S. V., Langley, K. E., Okino, K. H., Jacobsen, F. W., Martin, F. H. (1992). Canine stem cell factor (c-kit ligand) supports the survival of hematopoietic progenitors in long-term canine marrow culture. *Exp. Hematol.* **20**, 1118–1124.

Smith, C. W., Entman, M. L., Lane, C. L., Beaudet, A. L., Ty, T. I., Youker, K., Hawkins, H. K., Anderson, D. C. (1991). Adherence of neutrophils to canine cardiac myocytes *in vitro* is dependent on intercellular adhesion molecule-1. *J. Clin. Invest.* **88**, 1216–1223.

Snyder, P. W., Kazacos, E. A., Felsburg, P. J. (1993). Histologic characterization of the thymus in canine X-linked severe combined immunodeficiency. *Clin. Immunol. Immunopathol.* **67**, 55–67.

Somberg, R. L., Robinson, J. P., Felsburg, P. J. (1992). Detection of canine interleukin-2 receptors by flow cytometry. *Vet. Immunol. Immunopathol.* **33**, 17–24.

Somberg, R. L., Robinson, J. P., Felsburg, P. J. (1994). T lymphocyte development and function in dogs with X-linked severe combined immunodeficiency. *J. Immunol.* **153**, 4006–4015.

Somberg, R. L., Pullen, R. P., Casal, M. L., Patterson, D. F., Felsburg, P. J., Henthorn, P. S. (1995). A single nucleotide insertion in the canine interleukin-2 receptor gamma chain results in X-linked severe combined immunodeficiency disease. *Vet. Immunol. Immunopathol.* **47**, 203–213.

Somberg, R. L., Tipold, A., Hartnett, B. J., Moore, P. F., Henthorn, P. S., Felsburg, P. J. (1996). Postnatal development of T cells in dogs with X-linked severe combined immunodeficiency. *J. Immunol.* **156**, 1431–1435.

Spertini, O., Kansas, G. S., Reimann, K. A., Mackay, C. R., Tedder, T. F. (1991). Function and evolutionary conservation of distinct epitopes on the leukocyte adhesion molecule-1 (TQ-1, Leu-8) that regulate leukocyte migration. *J. Immunol.* **147**, 942–949.

Stellar, G. C., DeBeer, M. C., Lelias, J. M., Snyder, P. W.,

Glickman, L. T., Felsburg, P. J., Whitehead, A. S. (1991). Dog serum amyloid A protein. Identification of multiple isoforms defined by cDNA and protein analysis. *J. Biol. Chem.* **266**, 3505–3510.

Strandstrom, H. V., Rimaila-Pananen, E. (1979). Canine atypical malignant lymphomas. *Am. J. Vet. Res.* **40**, 1033–1034.

Strom, H., Thomsen, M. K. (1990). Effects of proinflammatory mediators on canine neutrophil chemotaxis and aggregation. *Vet. Immunol. Immunopathol.* **25**, 209–217.

Studdert, V. P., Phillips, W. A., Studdert, M. J., Hosking, C. S. (1984). Recurrent and persistent infections in related Weimaraner dogs. *Aust. Vet. J.* **61**, 261–263.

Suzuki, Y., Stormont, C., Morris, B. G., Shifrine, M., Dobrucki, R. (1975). New antibodies in dog blood groups. *Transplant. Proc.* **7**, 365–367.

Swisher, S. N., Young, L. E. (1961). The blood group system of dogs. *Physiol. Rev.* **41**, 495–520.

Swisher, S. N., Bull, R., Bowdler, J. (1973). Canine erythrocyte antigens. *Tissue Antigens* **3**, 164–165.

Sykes, G. P., King, J. M., Cooper, B. C. (1985). Retrovirus-like particles associated with myeloproliferative disease in the dog. *J. Comp. Pathol.* **95**, 559–564.

Symons, M., Bell, K. (1992). Canine blood groups: description of 20 specificities. *Anim. Genet.* **23**, 509–515.

Takano, M., Hayashi, N., Goitsuka, R., Okuda, M., Momoi, Y., Youn, H. Y., Watari, T., Tsujimoto, H., Hasegawa, A. (1994). Identification of dog T cell receptor β chain genes. *Immunogenetics* **40**, 246.

Tamaoki, J., Yamawakai, I., Takeyama, K., Chiyotani, A., Yamauchi, F., Konno, K. (1994). Interleukin-1 beta inhibits airway smooth muscle contraction via epithelium-dependent mechanism. *Am. J. Resp. Crit. Care Med.* **149**, 134–137.

Tamura, N., Nelson, R. A. (1968). The purification and reactivity of the first component from guinea pig, human, and canine sera. *J. Immunol.* **101**, 1333–1345.

Teske, E., Wisman, P., Moore, P. F., van Heerde, P. (1994). Histologic classification and immunophenotyping of canine non-Hodgkin's lymphoma: unexpected high frequency of T cell lymphomas with B cell morphology. *Exp. Hematol.* **22**, 1179–1187.

Thomas, J., Anderson, N. V. (1982). Interepithelial lymphocytes in the small intestinal mucosa of conventionally reared dogs. *Am. J. Vet. Res.* **43**, 200–203.

Thompson, R. E., Reynolds, H. Y., Waxdal, M. J. (1975). Structural composition of canine secretory component and immunoglobulin A. *Biochemistry* **14**, 2852–2860.

Thompson, R. E., Reynolds, H. Y. (1977). Isolation and characterization of canine secretory immunoglobulin M. *J. Immunol.* **118**, 323–329.

Thomsen, M. K., Strom, H. (1989). Biological variation in random and leukotriene B4-directed migration of canine neutrophils. *Vet. Immunol. Immunopathol.* **21**, 219–224.

Thomsen, M. K., Thomsen, H. K. (1990). Effects of interleukin-1α on migration of canine neutrophils *in vitro* and *in vivo*. *Vet. Immunol. Immunopathol.* **26**, 385–393.

Thomsen, M. K., Larsen, C. G., Thomsen, H. K., Kirstein, D., Skak-Nielsen, T., Ahnfelt-Ronne, I., Thestrup-Pedersen, K. (1991). Recombinant human interleukin-8 is a potent activator of canine neutrophil aggregation, migration, and leukotriene B4 biosynthesis. *Lab. Invest.* **96**, 260–266.

Thrall, M. A. (1981). Lymphoproliferative disorders: lymphocytic leukemia and plasma cell myelomas. *Vet. Clin. NA* **11**, 321–347.

Tizard, I. R. (1996). *Veterinary Immunology*, 5th edn. W.B. Saunders, Philadephia, PA, USA.

Tomley, F. M., Armstrong, S. J., Mahy, B. W., Owen, L. N. (1983). Reverse transcriptase activity and particles of retroviral density in cultured canine lymphosarcoma cells. *Br. J. Cancer* **47**, 277–284.

Trowalk-Wigh, G., Hakansson, L., Johannisson, A., Nborrgren, L., Segerstad, C. H. (1992). Leukocyte adhesion protein deficiency in Irish setter dogs. *Vet. Immunol. Immunopathol.* **32**, 261–280.

Uribe, L. P., Henthorn, P. S., Grimaldi, J. C., Somberg, R., Suter, S., Moore, P., Felsburg, P. J., Weinberg, K. (1997). Cloning, sequencing and expression of the cDNA encoding the canine hematopoietic antigen, CD38. *Blood*, in press.

Vaerman, J. P., Heremans, J. F. (1969a). The immunoglobulins of the dog. II. The immunoglobulins of canine secretions. *Immunochemistry* **6**, 779–786.

Vaerman, J. P., Heremans, J. F. (1969b). Distribution of various immunoglobulin containing cells in canine lymphoid tissue. *Immunology* **17**, 627–633.

Vaerman, J. P., Heremans, J. F. (1970). Origin and size of immunoglobulin A in the mesenteric lymph of the dog. *Immunology* **18**, 27–38.

van der Feltz, M. J. M., Ploegh, H. L. (1984). Immunochemical analysis of glycosylated and nonglycosylated DLA class I antigens. *Immunogenetics* **19**, 95–108.

van Riper, G., Siciliano, S., Fischer, P. A., Meurer, R., Springer, M. S., Rosen, H. (1993). Characterization and species distribution of high affinity GTP-coupled receptors for human Rantes and monocyte chemoattractant protein 1. *J. Exp. Med.* **177**, 851–856.

Vilafranca, M., Wohlsein, P., Borras, D., Pumarola, M., Domingo, M. (1995). Muscle fibre expression of transforming growth factor-β1 and latent transforming growth factor-β binding protein in canine masticatory myositis. *J. Comp. Pathol.* **112**, 299–306.

Von Dungern, E., Hirschfeld, L. (1910). Uber Nachweis der Vererbung biochemischer Strukturen. *Z. Immunitatsforsch.* **4**, 531–540.

Wagner, J. L., DeRose, S., Burnett, R. C., Storb, R. (1995). Nucleotide sequence and polymorphism analysis of canine DRA cDNA clones. *Tissue Antigens* **45**, 284–287.

Wagner, J. L., Burnett, R. C., DeRose, S. A., Francisco, L. V., Storb, R., Ostrander, E. A. (1996a). Molecular analysis of DLA-DRBB1 polymorphism. *Tissue Antigens* **48**, 554–561.

Wagner, J. L., Burnett, R. C., Storb, R. (1996b). Molecular analysis of the DLA-DR subregion. *Tissue Antigens* **48**, 549–553.

Wagner, J. L., Burnett, R. C., DeRose, S. A., Francisco, L. V., Storb, R., Ostrander, E. A. (1996c). Molecular analysis and polymorphism of the DLA-DQA gene. *Tissue Antigens* **48**, 199–204.

Wagner, J. L., Burnett, R. C., DeRose, S. A., Francisco, L. V., Storb, R., Strander, E. A. (1996d). Histocompatibility testing of dog families with polymorphic microsatellite markers. *Transplantation* **62**, 876–877.

Walton, R., Modiano, J. F., Thrall, M. A., Wheeler, S. (1996). Bone marrow cytological findings in 4 dogs and 1 cat with hemophagocytic syndrome. *J. Vet. Inter. Med.* **10**, 7–14.

Wellman, M. L., Davenport, D. I., Morton, D., Jacobs, R. M.

(1985). Malignant histiocytosis in four dogs. *J. Am. Vet. Med. Assoc.* **187**, 919–921.

Wellman, M. L., Krakowka, S., Jacobs, R. M., Kociba, G. J. (1988). A macrophage-monocyte cell line from a dog with malignant histiocytosis. *In Vitro Cell. Dev. Biol.* **24**, 223–229.

Wells, C. L., Maddaus, M. A., Eriandsen, S. L., Simmons, R. L. (1988). Evidence for the phagocytic transport of intestinal particles in dogs and rats. *Infect. Immun.* **56**, 278–282.

Whitbread, T. J., Batt, R. M., Garthwaite, G. (1984). Relative deficiency of serum IgA in the German shepherd dog: a breed abnormality. *Res. Vet. Sci.* **37**, 350–352.

Willard, M. D., Leid, R. W. (1981). Nonuniform horizontal and vertical distributions of immunoglobulin A cells in canine intestines. *Am. J. Vet. Res.* **42**, 1573–1580.

Winkelstein, J. A., Cork, L. C., Griffin, D. E., Griffin, J. W., Adams, R. J., Price, D. L. (1981). Genetically determined deficiency of the third component of complement in the dog. *Science* **212**, 1169–1170.

Winkelstein, J. A., Johnson, J. P., Swift, A. J., Ferry, F., Yolken, R., Cork, L. C. (1982). Genetically determined deficiency of the third component of complement in the dogs: *In vitro* studies of the complement system and complement-mediated serum activities. *J. Immunol.* **129**, 2598–2602.

Winkelstein, J. A., Sullivan, K. E., Colten, H. R. (1995). Genetically determined deficiencies of complement. In *Metabolic Basis of Inherited Disease* (eds Scriver, C. R., Beaudet, A. L., Sly, W. S., Valle, D. L.). McGraw-Hill Book Co., New York, USA.

Winters, W. D. (1981). Time dependent decreases of maternal canine virus antibodies in newborn puppies. *Vet. Rec.* **107**, 295–299.

Wu, C. C., Bey, R. F., Loken, K. I. (1988). Purification and characterization of the C1q subcomponent of canine complement and its use in the ^{125}I-C1q binding assay for detection of immune complexes. *Am. J. Vet. Res.* **49**, 865–869.

Yamashita, K., Fujinaga, T., Hagio, M., Miyamoto, T., Izumisawa, Y., Kotani, T. (1994a). Bioassay for interleukin-1, interleukin-6, and tumor necrosis factor-like activities in canine sera. *J. Vet. Med. Sci.* **56**, 103–107.

Yamashita, K., Fujinaga, T., Miyamoto, T., Hagio, M., Izumisawa, Y., Kotani, T. (1994b). Canine acute phase response: relationship between serum cytokine activity and acute phase protein in dogs. *J. Vet. Med. Sci.* **56**, 487–492.

Yang, M., Becker, A. B., Simons, F. E. R., Peng, Z. (1995). Identification of a dog IgD-like molecule by a monoclonal antibody. *Vet. Immunol. Immunopathol.* **47**, 215–224.

Yang, T. J., Gawlak, S. L. (1989). Lymphoid organ weights and organ:body weight ratios of growing beagles. *Lab. Anim.* **23**, 143–146.

Youker, K., Smith, C. W., Anderson, D. C., Miller, D., Michael, L. H., Rossen, R. D., Entman, M. L. (1992). Neutrophil adherence to isolated adult cardiac myocytes. *J. Clin. Invest.* **89**, 602–609.

Young, L. E., Ervin, D. M., Yuile, C. L. (1949). Hemolytic reactions produced in dogs by transfusion of incompatible dog blood and plasma. *Blood* **4**, 1218–1231.

Young, L. E., O'Brien, W. A., Swisher, S. N. (1952). Blood groups in dogs – their significance to the veterinarian. *Am. J. Vet. Res.* **13**, 207–213.

Zucker, K., Lu, P., Esquenazi, V., Miller, J. (1992). Cloning of the cDNA for canine interferon-gamma. *J. Interferon Res.* **12**, 191–194.

Zucker, K., Lu, P., Asthana, D., Carreno, M., Yang, W. C., Esquenazi, V., Fuller, L., Miller, J. (1993). Production and characterization of recombinant canine interferon-gamma from *E. coli*. *J. Interferon Res.* **13**, 91–97.

Zucker, K., Lu, P., Fuller, L., Asthana, D., Esquenazi, V., Miller, J. (1994). Cloning and expression of the cDNA for canine tumor necrosis factor-alpha in *E. coli*. *Lymphokine Cytokine Res.* **13**, 191–196.

Zwahlen, R. D., Spreng, D., Wyder-Walther, M. (1994). *In vitro* and *in vivo* activity of human interleukin-8 in dogs. *Vet. Pathol.* **31**, 61–66.

VIII Immunology of the Cat

1. Introduction

During the past 20 to 30 years, there has been a growing interest in the cat as a companion animal. In many countries, the number of cats has exceeded that of dogs. For instance Switzerland with 1.3 million cats has about three times the number of dogs. As a consequence, veterinary immunologists in industry and academia have devoted increasing efforts in the study of the feline immune system, pursuing both applied and basic research. The applied research is devoted mainly to improving the cat's health by studying pathogenesis, immune reactions and diagnostic possibilities of various diseases and to the development and improvement of vaccines. Indeed, the numbers of vaccines for use in cats has grown tremendously during the last 10 years. In addition to the applied research from which cats profit directly, the cat has attracted much interest from basic research as a model for various human diseases. There were two events which triggered increased study of the feline immunology: the first was the discovery of the feline leukemia virus (FeLV) in Scotland by Jarrett's group (Jarrett *et al.*, 1964) and the second was the isolation of the feline immunodeficiency virus (FIV) in California by Pedersen and coworkers (Pedersen *et al.*, 1987).

The discovery that FeLV caused transmissible lymphosarcomas in an outbred species was considered important for tumor genesis also in man. The event triggered a great number of studies relating to mechanisms of tumor development and epidemiology of FeLV. As a consequence of the hypothesis that the FeLV-induced feline oncornavirus associated cell membrane antigen (FOCMA) supposedly a nonvirus coded tumor specific antigen, (Essex *et al.*, 1975) might protect cats against FeLV-induced tumors, many studies were initiated relating to the immune mechanisms that could be involved in protection against these tumors. When it became clear that cats are capable of spontaneously overcoming FeLV viremia, many groups started on the development of a vaccine against FeLV infection. Several FeLV vaccines, including a highly efficacious recombinant one, are now available to veterinarians.

The discovery of FIV attracted much interest from basic researchers as FIV was immediately recognized as an important model for the study of HIV and AIDS (Gardner *et al.*, 1991, Bendinelli *et al.*, 1991). Many research groups switched their interest to the study of FIV in cats. In the early days of FIV research, few reagents existed to investigate the feline immune system during FIV infection. This soon changed in Europe with the support from the European Concerted Action on FIV Vaccination and after the first description of antifeline CD4$^+$ antibodies (Ackley *et al.*, 1990), a number of different antifeline leukocyte differentiation antigens are now available and the methods for their use have been established. To date, few assays exist that allow the quantification of feline cytokines on the protein level. However, sequences of most cytokines described in mice and other species have also been obtained in the feline system.

It is the goal of this chapter to provide an overview on current state of knowledge of the cat's immune system and on the source of reagents available. The topics of the chapter on the immunology of the cat are listed in analogy to those of the other chapters in this handbook. Because little information is available on the T-cell receptor and on leukocyte traffic, no specific sections can be listed on these topics. Diseases of the immune system are important in the cat; therefore, a separate section on immunological diseases is included. The editor and authors of this chapter hope that the information may help researchers with previous experience or new interest in the immunology of the cat to gain a rapid overview on what has been done and can be done in feline immunology.

2. Lymphoid Organs and their Anatomical Position

The lymphoid organs can be classified as either central or peripheral. The central lymphoid organs of the cat are the thymus and the bone marrow which are identified as areas of lymphocyte development and differentiation.

Lymph nodes, tonsils, spleen and Peyer's patches of the gut are classified as peripheral lymphoid organs (Vollmerhaus *et al.*, 1981, 1994; Gershwin *et al.*, 1995). They are used as a platform for lymphoid cells which have achieved immune capability to come into contact with antigens. Thus, the peripheral lymphoid organs serve as sites of development for specific features of immunity.

As far as known, the lymphoid organs of the cat show characteristics in development, structure and function

Handbook of Vertebrate Immunology
ISBN 0-12-546401-0

similar to that of other mammals. Knowledge of topography and function of lymphatic tissue has become important in connection with feline retrovirus infections as model for HIV and AIDS in humans. Therefore, the information presented in this section should help identify lymphatic tissue which can be surgically removed without sacrificing the animal.

Central lymphoid organs

Thymus

The feline thymus is an elongated multilobed structure located in the thoracic mediastinum. Each thymic lobule is clearly divided into an outer cortex region and an inner medulla. The cortex is composed of dense aggregates (no follicles) of small lymphocytes (thymocytes) which are surrounded by reticular connective tissue. It can be seen by FACS analysis that 93% of thymic cells are stained by a feline pan-T cell marker (Tompkins *et al.*, 1990b). The medulla shows a relatively low cellular density and consists of epitheloid reticulum cells, sparse lymphocyte populations and Hassall's corpuscles. The thymus weighs about 0.32–0.39 g (0.4% of bodyweight) in neonates; it increases in weight and size, and undergoes progressive involution in which the thymic tissue is successively replaced by connective and adipose tissue, starting in the fourth month post partum. Only efferent lymphatics are present in the thymus, which lead to the nodi lymphatici (nll.) mediastinales. The thymus is innervated by the vagus nerve and the truncus sympathicus (Vollmerhaus *et al.*, 1981, 1994; König, 1992; Gershwin *et al.*, 1995).

Bone marrow

The bone marrow in the central cavities of developing long bones is the source of hematopoietic cells-including immature lymphocytes (which are derived mainly from mitotic divisions of lymphoblasts or prolymphocytes) (Gershwin *et al.*, 1995). Islands of hematopoietic cells and their stem cell precursors are surrounded by a supporting network of primitive reticulin-producing cells, fibroblasts, collagen and adipose tissues.

Lymphocytes leaving the bone marrow by efferent lymphatic vessels either migrate to the spleen and lymph nodes directly if they participate in antibody production, or they pass through the thymus, and then to the lymph nodes if they are destined for cellular immunity. This cellular migration mostly occurs before birth, indicating that lymphoid tissues of the neonate are well developed (Friess *et al.*, 1990; Gershwin *et al.*, 1995).

Peripheral lymphoid organs

Lymph node

The main function of the lymph node is to filtrate the lymph and to sift out foreign antigenic material.

The node consists of a capsule, a cortex region and a medulla. The cortex contains a dense mass of lymphocytes in which germinal centers are embedded. In these germinal centers bone marrow-derived B lymphocytes undergo differentiation into antibody-producing plasma cells. Mature B cells then move to the peripheral area (mantle) of the germinal center and then to the sinus areas in the medulla. The medulla contains large blood vessels that branch and form lymph cords. These are supported by a loose network of reticular tissue. Between these cords there is a loose network of lymphatic sinuses which contain macrophages, lymphocytes, and plasma cells. The paracortical, or T-cell-dependent area located between the medulla and the cortex area is involved in cellular immunity (Friess *et al.*, 1990; König, 1992; Gershwin *et al.*, 1995). FACS-analysis of single cell suspensions obtained from feline popliteal lymph nodes reveals that up to 49% of the cells belong to the T-cell population (Tompkins *et al.*, 1990).

Size, cellular density and composition of the lymph node change during the lifespan. This is mostly due to antigen-stimulation and consecutive enhanced 'trapping' of circulating lymphocytes followed by a transient shutdown in the exit rate of lymphocytes. Later, most of the cells in the node are derived from clones of these antigen-stimulated lymphocytes (Gershwin *et al.*, 1995).

The lymph enters the node via the afferent route at the capsular surface and exits via the efferent route at the hilus neighbored by venoles and arterioles.

Spleen

The main function of the spleen is vascular filtration of foreign material and removal of damaged or aged erythrocytes. In addition, the feline spleen is able to hold up to 16% of the total blood in order to support the general circulation. This can be actively done by contraction of smooth muscles infiltrating the capsule and the trabeculae of the organ. The feline spleen is classified as a reticular spleen as opposed to the sinusoid spleen of the dog (Friess *et al.*, 1990; Vollmerhaus *et al.*, 1994).

The spleen is a large vascular organ located in the gastrosplenic ligament along the greater curvature of the stomach, 114–185 × 14–31 mm in size, in the adult cat, depending on the amount of blood contained (König, 1992; Vollmerhaus *et al.*, 1994). The organ is surrounded by a capsule and divided by a trabecular framework into lobules. Central arterioles entering at the hilus continue in the neighborhood of the trabeculae as smaller straight arterioles and end as capillaries. These capillaries are embedded in a reticular network; this is the red pulp of

Table VIII.2.1 Palpable lymph nodes of the cat[a]

Lymph node	*Topography*	*Number[b]/Size*	*Lymphatic drainage*
nll. Mandibulares	Postero-lateral to the proc. angularis of the mandibula, medial and lateral to the vena linguofacialis	2/12–15 × 8–10 × 6–8 mm	Lips, chin region, mouth cavity, cheek glands, eyelids
nll. Retro-pharyngei laterales	Behind the ear and caudal of the parotid gland along the vena auricularis caudalis	1–7/0.5–15 × 1–3 × 0.5–3 mm	Region of the ear, eye, forehead, neck and parotid gland
nll. Cervicales superficiales dorsales[c]	In front of and below the cervical part of the musculus trapezius	1–3/8–20 × 5–6 × 3–4 mm	Dorsal area of the neck, shoulder and the forelimb
nl. Axillaris propius	In the fork between vena thoracica lateralis and vena axillaris	1–2/3–6 × 4–5 × 2–4 mm	Skin and subcutaneous region of the medial aspect of the upper and lower arm, lateral thoracic wall
nll. Axillares accessorii[c]	Along the vena thoracica lateralis from the level of the 3rd to 6th intercostal spaces	2–5/5–10 × 2–2 × 2–3 mm	Skin of the inside of the upper and lower forelimbs, lumbal region, lateral and dorsal chest wall
nl. Inguinalis superficialis[c]	Embedded in adipose tissue between arteria and vena pudenda externa in the femoral canal	1/5–15 × 3–5 × 2–3 mm	Inguinal and gluteal regions, in female cats the posterior half of the mamma
nll. Epigastrici caudales[c]	Embedded in adipose tissue along arteria and vena epigastrica caudalis superficialis	1–5/3–8 × 2–5 × 1–3 mm	Caudal part of the ventral abdominal wall and subcutis of the thigh
nl. Popliteus superficialis	Subcutaneously in the flexor region of the knee beneath vena saphena lateralis	1/6–7 × 4–5 × 4 mm	Cutis, subcutis and muscles of the shank and hind foot

nl., nodus lymphaticus; nll., nodi lymphatici.
[a]References: Vollmerhaus *et al.*, 1981, 1994; Meier, 1989.
[b]On each side of the cat.
[c]Palpable only in young and lean cats.

the spleen which has no afferent lymphatics. The white pulp is also embedded in a reticular network and is the actual lymphatic tissue of the spleen. It is composed of Malpighian bodies, representing the sites of B-lymphocyte differentiation which are present at or near the divisions of central arterioles. The areas surrounding the central arterioles consist of T lymphocytes and also belong to the white pulp as well. In the feline spleen, up to 65% of the cells are T-cells (Tompkins *et al.*, 1990).

Tonsils

The tonsils located in the pharynx consist both of T lymphocytes and lymph follicles containing B lymphocytes, but lack afferent lymphatics and unlike the lymph node, a definite capsule (Friess *et al.*, 1990; Vollmerhaus *et al.*, 1994; Gershwin *et al.*, 1995).

Peyer's patches

The Peyer's patches are situated in the gut. With respect to function and morphology, they are analogous to the tonsils. However, there are, in addition, numerous solitary lymphoid nodules spread throughout the lamina of the gut. Their main function is the production of immunoglobulin A (IgA) along with other types of immunoglobulins. The epithelial cells covering areas of Peyer's patches can be identified as 'M' cells with micropinocytic properties, allowing the cells to sample antigens. Beneath this area of specialized epithelium there are aggregates of T cells, B cells and plasma cells. Lymphocytes that home to Peyer's patches show an adhesion molecule called VLA-4 (very late antigen) (Gershwin *et al.*, 1995).

Lymph nodes and lymph collecting ducts of the cat

The anatomically most accessible lymphatic tissue are the lymph nodes. Therefore, many experimental studies involving the cat as an animal model for various diseases focus mainly on the immunological and pathological changes in lymph nodes.

The lymphatic system of the cat was first described in particular by Sugimura *et al.* (1955, 1956, 1958, 1959). These first descriptions of size, number and weight of feline lymph nodes were followed by adaptations and variations such as number of nodes and nomenclature (Vollmerhaus *et al.*, 1981, 1994; Meier, 1989).

The size of a lymph node depends on age and weight of the animal and presence or absence of pathological conditions. The largest sizes are seen under pathological conditions (Meier, 1989).

A vital condition for the immune functions of the lymphoid organs is the anatomical connection of the circulating blood system and the lymph vessels. The peripheral lymph nodes are connected with the blood stream via postcapillary venules in the cortex of the node. Lymphocytes exit into the lymph node by receptor-ligand-binding between their cell adhesion molecules, called integrins and selectins and high endothelial cell venules. L-selectin was found to be such a homing receptor on lymphocytes for peripheral nodes (Gershwin *et al.*, 1995).

While T cells circulate through the paracortical area, B cells circulate through the germinal centers of the cortex. They leave the lymph node through efferent lymph vessels leading to the next lymph node or eventually into lymph collecting ducts. These ducts finally empty into the venous blood system to complete the lymphocyte circulation (Vollmerhaus *et al.*, 1981, 1994; Gershwin *et al.*, 1995).

Palpable lymph nodes of the cat

The feline lymph nodes which are palpable and constant in occurrence under physiologic conditions are listed in Table VIII.2.1. In addition, lymph nodes are indicated which are palpable only in young and lean cats.

The diagrammatic outline of the lymph nodes listed in Table VIII.2.1 are indicated in Figure VIII.2.1.

Physiologically nonpalpable lymph nodes of the cat

The various lymph centers, their most important lymph nodes and the corresponding drainage areas of the non-palpable feline lymph nodes are listed in Table VIII.2.2.

Figure VIII.2.2 gives a schematic survey of most of the lymph nodes listed in Tables VIII.2.1 and VIII.2.2.

Lymph collecting ducts of the cat

The lymph collection ducts carry the lymph from the superficial and visceral lymph centers to the venous blood stream.

Figure VIII.2.1 Diagram of the palpable lymph nodes of the cat (1) nodi lymphatici (nll.) mandibulares; (2) nll. retropharyngei laterales; (3) nll. cervicales superficiales dorsales[a]; (4) nodus lymphaticus (nl.) axillaris propius; (5) nll. axillares accessorii[a]; (6) nl. inguinalis superficialis[a]; (7) nll. epigastrici caudales[a]; (8) nl. popliteus superficialis. [a]Palpable only in young and lean cats (reproduced from J. Frewein and Vollmerhaus, 1994, with kind permission of Prof. Frewein and the publisher).

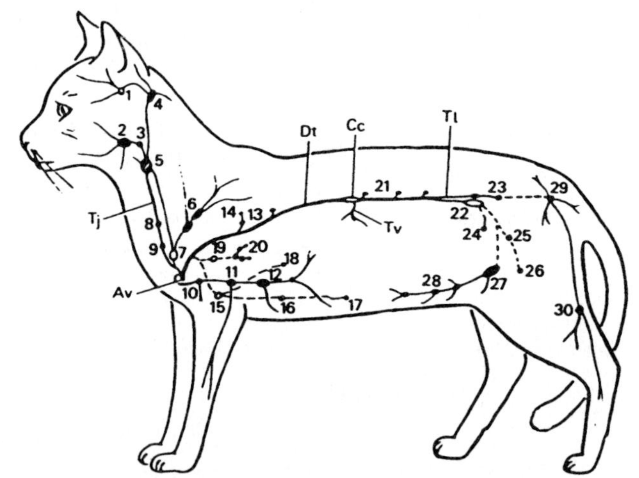

Figure VIII.2.2 Diagram of the lymph nodes and lymph collecting ducts of the cat without the intestinale nodes (Av) angulus venosus; (Cc) cysterna chyli; (Dt) ductus thoracicus; (Tj) Truncus jugularis; (Tl) Truncus lumbalis; (Tv) truncus visceralis. (1) nodus lymphaticus (nl.) parotideus[c]; (2) nodi lymphatici (nll.) mandibulares[e]; (3) nll. mandibulares accessorii[b]; (4) nll. retropharyngei laterales[e]; (5) nl. retropharyngeus medialis; (6) nll. cervicales superficiales dorsales[d]; (7) nl. cervicales superficialis ventralis[b,c]; (8) nl. cervicalis profundus medius[a]; (9) nl. cervicalis profundus caudalis[b]; (10) nl. axillaris primae costae[a]; (11) nl. axillaris proprius[e]; (12); nll. axillares accessorii[d]; (13) nll. thoracici aortici[a]; nl. intercostalis[a]; (15) nl. sternalis craniales; (16) nl. sternalis caudalis[a]; (17) nl. epigastricus cranialis[a]; (18) nl. phrenicus[a]; (19) nll. mediastinales craniales; (20) nll. bifurcationis and pulmonales[a]; (21) nll. lumbales aortici; (22) nll. iliaci mediales; (23) nll. sacrales; (24) nl. subiliacus[a]; (25) nl. iliofemoralis[a]; (26) nl. femoralis[a]; (27) nl. inguinalis superficialis[d]; (28) nll. epigastrici caudales[d]; (29) nl. ischiadicus[c]; (30) nl. popliteus superficialis[e]. [a]Inconsistent in occurrence; [b]Almost constant in occurrence; [c]Palpable when pathologically enlarged; [d]Palpable only in young and lean cats; [e]Palpable under physiologic conditions (reproduced from J. Frewein and Vollmerhaus, 1994, with kind permission of Prof. Frewein and the publisher).

Table VIII.2.2 Main lymph centers of the cat and their lymph nodes[a]

Lymphocentrum	*Lymph node*	*Number*	*Lymphatic drainage*
Lc. Parotideum	nl. Parotideus[d]	1–2	Upper eyelid, parotid gland, parts of the upper half of the head
Lc. Mandibulare	nll. Mandibulares	2	Upper and lower lips, chin region, mouth cavity, cheek glands, eyelids
	nll. Mandibulares access.[b]	1–4	
Lc. Retropharyngeum	nll. Retropharyngei laterales [f]	1–7	See Table VIII.2.1; oral cavity and tongue, cervical section of the oesophagus and trachea, thyroid and mandibular gland, parotid gland
	nl. Retropharyngeus medialis	1	
Lc. Cervicale superficiale	nll. Cervicales superficiales dorsales[e]	1–3	See Table VIII.2.1
	nl. Cervicales superficialis ventralis[c,d]	1–2	Ventral part of the neck, thoracic inlet
Lc. Cervicale profundum	nl. Cervicalis profundus medius[b]	1 (unilateral)	Thyroid gland, cervical parts of trachea and oesophagus
	nl. Cervicalis profundus caudalis[c]	1–6	Trachea, oesophagus, thyroid gland
Lc. Axillare	nl. Axillaris proprius[f]	1–2	See Table VIII.2.1
	nl. Axillaris primae costae[b]	1	Lateral wall of the thorax
	nll. Axillares accessorii[e]	3–7	See Table VIII.2.1
Lc. Thoracicum dorsale	nll. Thoracici aortici[b]	1–5	Peritoneum
	nl. Intercostalis[b]	1–2	Pleura
Lc. Thoracicum ventrale	nl. Sternalis craniales	1–5	Ventral thoracic and abdominal wall, diaphragm, pericardium
	nl. Sternalis caudalis[b]	1	
	nl. Epigastricus cranialis[b]	1	
Lc. Mediastinale	nl. Phrenicus[b]	1	Diaphragm; heart, trachea, thymus, oesophagus, pleura, pericardium
	nll. Mediastinales craniales	2–8	
Lc. Bronchale	nl. Bifurcationis seu tracheobronchialis dexter	1–2	Mainly the lung, but also heart, pericardium, mediastinum and diaphragm
	nl. Bifurcationis seu tracheobronchialis sinister	1–2	
	nl. Bifurcationis seu tracheobronchialis medius	1–2	Lungs
	nl. Pulmonalis[b]	1	
Lc. Lumbale	nll. Lumbales aortici	2–7	Diaphragm, kidneys, adrenals, dorsal abd. wall, ovary, oviduct, uterus, testes
Lc. Coeliacum	nll. Lienales	1–3	Spleen, greater curvature of the stomach, left lobe of the pancreas; stomach, liver oesophagus, diaphragm, duodenum, pancreas
	nll. Gastrici	1–4	
	nll. Hepatici	2–4	
	nl. Pancreaticoduodenalis	1–2	
Lc. Mesentericum craniale	nll. Jejunales	2–20	Small intestine and body of pancreas
	nll. Caecales	1–3	Caecum, ileum
	nll. Colici	3–9	Ileum, ascending and transverse colon, descending colon and caecum
Lc. Mesentericum caudale	nll. Mesenterici caudales	1–3	Descending colon and rectum
Lc. Iliosacrale	nll. Iliaci mediales	2–4	Pelvic wall, pelvic limb, uterus, urinary bladder, sometimes oviduct or testis
	nll. Sacrales	1–6	Rectum, uterus, vagina, urinary bladder, ureter, wall of the pelvis, pelvic outlet, tail and hind limb
Lc. Inguinale profundum	nl. Iliofemoralis[b]	1	Parts of the neighboring ventral abdominal wall, gluteal region, thigh
	nl. Femoralis[b]	1	
Lc. Inguinale superficiale	nl. Inguinalis superficialis[e]	1	Inguinal and gluteal regions, posterior half of the mamma; caudal part of the ventral abd. wall and subcutis of the thigh
	nll. Epigastrici caudales	1–5	
	nl. Subiliacus[b]	1	
Lc. Ischiadicum	nl. Ischiadicus[d]	1	Skin, subcutis, fasciae of the thigh, anal region, hind limb
Lc. Popliteum	nl. Popliteus superficialis[f]	1	See Table VIII.2.1.

[a]References: Vollmerhaus *et al.* (1981, 1994); Meier (1989).
[b]Inconsistent in occurrence.
[c]Almost constant in occurrence.
[d]Palpable when pathologically enlarged.
[e]Palpable only in young and lean cats.
[f]Palpable under physiologic conditions.
Lc., lymphocentrum; nl., nodus lymphaticus; nll., nodi lymphatici.

The truncus jugularis collects the lymph from the lymph nodes of the head and throat and joins the venous angle. The truncus viscerales is formed by the efferent vessels of the lymph nodes of the coeliac and cranial mesenteric lymph centers. The truncus lumbalis collects the lymph from the pelvis, the nll. sacrales and the nll. iliaci mediales. The trunci viscerales and lumbales form the cysterna chyli which is 7–30 mm long and lies between the renal vessels and the crura of the diaphragm. The ductus thoracicus developed from the cysterna chyli is formed like a rope-ladder and end as a single trunk into the venous angle. The diagrammatic outline of these ducts are seen in Figure VIII.2.2.

3. Leukocyte Differentiation Antigens

Introduction

The development of antibodies recognizing feline leukocyte differentiation antigens is central to the study of the feline immune system in health and disease. However, the production of monoclonal antibodies can be a costly and time-consuming process and researchers studying the feline immune system seldom undertake monoclonal antibody production unless there is a pressing need for a reagent with a defined specificity. Thus, the identification of antibodies that recognize feline leukocyte differentiation antigens has tended to reflect areas of most active investigation, for example the discovery of feline immunodeficiency virus (FIV) fuelled the production of antibodies against the feline homologues of CD4 and CD8 in order that T lymphocyte subsets could be monitored in FIV-infected cats, and that the question of whether feline CD4 acts as a receptor for FIV could be addressed.

The expression of leukocyte differentiation antigens in the feline immune system has been reviewed by Willet and Callanan (1995). This section will focus on recent progress in the study of feline leukocyte differentiation antigens, highlighting areas of similarity and difference from the human immune system.

T-cell antigens

The feline homologue of CD4 is perhaps the best studied of the feline leukocyte differentiation antigens, primarily because CD4 was identified as the primary binding receptor for human immunodeficiency virus and researchers sought to define the role of the feline homologue of CD4 in FIV infection. Expression of CD4 in the feline immune system resembles that of murine CD4 in that expression is restricted solely to helper T lymphocytes and their thymic precursors and is notably absent from monocytes and macrophages (Ackley and Cooper, 1992). In contrast, CD4 expression in the human and rat immune systems extends to monocytes and macrophages. Indeed, CD4 expression in the human immune system has been studied extensively in relation to the cell tropism of human immunodeficiency virus (HIV) and it would appear to be widespread, with levels of CD4 being detected on eosinophils, follicular dendritic cells, megakaryocytes and microglia. The sole example of a non-T-lineage cell that expresses CD4 in the cat is the Langerhans cell (Tompkins *et al.*, 1990b) although only a single anti-CD4 antibody, vpg39, has been demonstrated to recognize these cells. cDNAs encoding the feline homologue of CD4 have been cloned (Dumont-Drieux *et al.*, 1992; Norimine *et al.*, 1992) and have revealed that the feline CD4 molecule contains a 17-amino-acid insertion in D2. Structural studies predict that this insertion may affect the flexibility of the hinge region between D2 and D3. Feline CD4 is also unusual in that the first cysteine involved in intrasheet disulphide bridge in D2 of human CD4 (residues 130 and 159) has been replaced by a tryptophan residue, ruling out the possibility of a similar disulphide bridge forming in feline CD4. Several antibodies have been described which recognize feline CD4, including Fel7, vpg30–39 and cat 30A. The Fel7 antibody and the antibodies of the vpg30–39 series have been epitope mapped using soluble feline CD4 produced in CHO cells and recognize five distinct epitopes on the molecules. Whether the epitopes can be related to functional properties of CD4 such as MHC class II binding or a putative interaction with interleukin 16 has yet to be established, although the epitope recognized by vpg39 is polymorphic in nature and absent from PBMC of some cats (Willet *et al.*, 1994a).

Several antibodies have been described which recognize the feline homologue of CD8 (Klotz and Cooper, 1986; Tompkins *et al.*, 1990b). As in other species, feline CD8 appears to exist as a heterodimer of α *and* β chains; immunoprecipitations with the FT2 and 3.357 antibodies yielded two distinct protein species and northern blotting analysis of mRNA from feline PBMC with a murine Ly-3 probe suggested the existence of CD8β in PBMC (Pecoraro *et al.*, 1994b). A cDNA has been cloned encoding the feline CD8α chain (Pecoraro *et al.*, 1994a) and the expression product is recognized by the cross-species reactive antihuman CD8-antibody OKT8. Interestingly, the feline CD8 specific antibodies FT2, 3.357 and vpg9 do not recognize the α-chain cDNA clone when expressed on either CrFK or CHO cells suggesting that either they recognize the α/β heterodimer or are β-chain specific. Molecular cloning of the feline CD8β-chain should confirm the reactivity of antifeline CD8 antibodies and establish whether feline CD8 exists solely as an α/β heterodimer or whether α/α homodimers/multimers exist.

CD4 and CD8 positive lymphocytes together comprise the majority of T lymphocytes in the peripheral circulation. All CD4 and CD8 positive lymphocytes are recognized by the 43-Pan T antibody which recognizes a CD5-

like molecule (Ackley and Cooper, 1992). The 43-Pan T precipitates a single 72 kDa glycoprotein which is phosphorylated in both resting and activated T cells and is upregulated by stimulation of T cells with phorbol ester or phytohaemagglutinin. Moreover, addition of the 43 Pan-T antibody to cultures of Con A stimulated PBMC augments the proliferative response. The 43-Pan T antibody therefore displays many of the characteristics of the feline homologue of CD5. Although other Pan T antibodies have been described, these antibodies have still to be characterized further. In combination with the anti-CD4 and CD8 antibodies the 43-Pan T antibody can be utilized to enumerate T lymphocyte numbers in the peripheral blood or to define the T-cell compartment in lymphoid tissues. There is no evidence to suggest that the 43-Pan T ligand is expressed on a minority of B cells, as is observed with human CD5. Thymocytes express the ligand recognized by the Fel 5F4 antibody (P. F. Moore, unpublished data), a prospective antifeline CD1a antibody. As with human CD1a, the Fel 5F4 antibody also stains Langerhans cells, intra-epithelial dendritic cells and interdigitating dendritic cells in the lymph node (Tompkins *et al.*, 1990b).

Immunohistochemical studies of feline T lymphocyte subpopulations can be achieved using antipeptide reagents such as anti-CD3 (Dako), an example of a polyclonal serum raised against a conserved intercellular domain of a cell surface molecule that shows broad cross-species reactivity.

B-cell antigens

The enumeration of feline B cells in peripheral blood has previously only been possible using reagents which recognize surface immunoglobulins. While cross-species reactive antibodies such as BE-5 (CD21), RFB4 (CD22) and B-B20 (CD40) can be utilized for the enumeration of feline B cells the reliability of these antibodies is somewhat erratic. While true Pan-B-cell reagents for use in the cat have yet to be identified there are now good cross-species reactive antibodies in RA3.6B2 (CD45R) and CA2.1D6 (CD21). The anti-CD45R antibody is a rat antibody raised against murine B cells but cross-reacts with both human and feline cells. Although the principal reactivity of this antibody is with B cells, the ligand for the antibody (the high molecular weight form of leukocyte common antigen) is also present on murine NK cells and non-MHC restricted CTLs. The antibody has been used immunohistochemically to enumerate B cells in both frozen and paraffin-embedded lymph node sections, and by flow cytometry to enumerate B cells in peripheral blood where less than 1.0% of RA3.6B2-positive cells were co-stained with a Pan-T antibody CF255 (Monteith *et al.*, 1996). The data suggest that greater than 99% of the cells recognized in peripheral blood by RA3.6B2 were B lymphocytes. Whether this antibody will prove useful for the enumeration of B cells in diseased cats where subpopulations of leukocytes may have undergone expansion remains to be seen. However, it has been used very successfully to immunophenotype feline lymphosarcomas (Jackson *et al.*, 1996).

The anticanine CD21 antibody CA2.1D6 cross-reacts well with feline B lymphocytes. B cells are also the principal CD21-expressing cell type, although follicular dendritic cells and some epithelial cells have been demonstrated to be CD21-positive in humans. The RFB9 antibody, an antihuman CD21 antibody, can be used to stain the dendritic cell network in the germinal centres of the feline lymph node; however, this antibody fails to recognize B cells in peripheral blood. Thus, the CA2.1D6 appears to be a true anti-B cell antibody and for flow cytometric analyses would be the reagent of choice. The utility of this antibody in the study of the feline immune system was illustrated by the studies of Dean *et al.* (1996b) where CD21 expression was shown to be mutually exclusive of expression of CD4 and CD8, and in those of Quackenbush *et al.* (1996) where CA2.1D6 was used to investigate the replication of feline leukaemia virus in B cells.

Adhesion molecules

There has been relatively little research on adhesion molecules in the feline immune system, however some notable advances have been made. Cross-species reactive monoclonal antibodies recognizing antigens on the surface of endothelial cells have been described (Weyrich *et al.*, 1995; Buerke *et al.*, 1996). The antibodies recognizing P-selectin (PB1.3), ICAM-1 (RR1/1), E-selectin (Cy1787) will be of value in the study of neutrophil trafficking and adherence. The cross-species reactivity of several antibodies recognizing the β1-integrins has been evaluated and antibodies recognizing the feline homologues of VLA2 (CD49b), VLA4 (CD49d) and VLA6 (CD49f) identified (B. Willett, unpublished observations). These reagents will be of value in the study of lymphocyte migration and adhesion.

Nonlineage

The search for the cellular receptor for FIV identified a monoclonal antibody, vpg15, which recognized the feline homologue of CD9 (Hosie *et al.*, 1993; Willett *et al.*, 1994a). Subsequently a cDNA encoding feline CD9 was cloned and characterized (Willett and Neil, 1995). Feline CD9 resembles CD9 from other species except for a unique sequence at the crown of the first predicted extracellular loop. The first extracellular loop of human, bovine and murine CD9 contains a putative site for *N*-linked glycosylation; this is absent from feline CD9 (Willett and Neil, 1995). Despite this divergence of amino acid sequence, the feline CD9 clone proved to be functional in assays of B-cell migration (Shaw *et al.*, 1995). CD9 is expressed on activated but not resting T cells and thus may be used as a marker for activation. T cell activation can also be evaluated using the 9F23 antibody which recognizes the interleukin-2 receptor alpha (IL2-Rα) chain. This antibody recognizes a binding site on the IL-2Rα distinct from the IL-2-binding site and has been used to evaluate the responsiveness of feline PBMC to mitogens in FIV-infected cats. The authors concluded that induction of IL2-Rα expression on Con A-stimulated PBMC was significantly depressed in FIV-infected cats (Ohno *et al.*, 1992a). A cDNA encoding feline IL-2Rα has been cloned and expressed in CrFK cells. The expression product binds recombinant human IL-2 and is recognized by the 9F23 antibody (Goitsuka and Hasegawa, 1995).

Monoclonal antibodies have been generated which define feline bone marrow erythroid (FeEr1), myeloid (FeMy) and lymphoid (FeLy) cell lineages (Groshek *et al.*, 1994). Although the identity of the antigens recognized by these antibodies remains to be established, they should prove useful reagents in the immunophenotyping of feline haemopoietic neoplasia.

NK cells

There has been a single report of the use of an antibody specific for feline NK cells. The study by Zhao *et al.* (1995) utilized a cross-species reactive antihuman CD57 monoclonal antibody to evaluate LAK cells in FIV-infected cats. CD57 is expressed by approximately 50% of resting NK cells in man, but is rapidly lost after activation. It is also expressed on a subset of T cells and it is this CD57$^+$ T cell subset that appears to be elevated in HIV-infected individuals. A list of some of the feline leukocyte differentiation antigens identified to date is shown in Table VIII.3.1.

Conclusions

The identification of antibodies that can be used for the enumeration of feline lymphocytes has provided researchers in feline immunology with the opportunity to study the major lymphocyte subsets in the feline immune system in health and disease. However, there is clearly a pressing need for reagents specific for NK cells and monocytes. So far, screening of antibodies against leukocyte differentiation antigens of other species has failed to identify cross-species reactive antibodies. Molecular cloning of feline leukocyte differentiation antigens may provide a source of pure antigen for immunization; specificities that have proven intractable may then be obtained by either expressing the cDNA in murine cells and using the transfected cells to immunize mice, or mice may be inoculated directly with the eukaryotic expression vector DNA itself, i.e. DNA immunization. Previous studies have illustrated the similarities and differences between the immune systems of the cat and man, the species-specific patterns of CD4 expression should remind

Table VIII.3.1 Feline leukocyte differentiation antigens for which specific monoclonal antibodies have been identified

Specificity	*Antibody*	*Accession number*	*References*
CD4	Fel7, vpg30–39, cat 30A		Ackley *et al.* (1990); Tompkins *et al.* (1990b); Dumont-Drieux *et al.* (1992); Norime *et al.* (1992); Willett *et al.* (1994a)
CD8	FT2, vpg9, 3.357, OKT8	CD8a-D16536	Klotz and Cooper (1986); Tompkins *et al.* (1990b); Willett *et al.* (1993); Pecoraro *et al.* (1994a)
CD5	43 Pan-T		Ackley and Cooper (1992)
CD9	vpg15, FMC56	L35275, D30786	Hosie *et al.* (1993); Willett *et al.* (1994a)
CD10	J5		Horton *et al.* (1988)
CD18	MHM23		Dakopatts Ltd[a]
CD21	CA2.1D6		Dean *et al.* (1996b); Quackenbush *et al.* (1996)
CD25	9F23	D16143	Ohno *et al.* (1992b)
CD29	4B4	U27351	Coulter Immunology Ltd[b]
CD35	To5		Aasted *et al.* (1988)
CD49b	16B4		Serotec Ltd[c]
CD49d	44H6		Serotec Ltd
CD49f	4F10		Serotec Ltd
CD45R	RA3.6B2		Pharmingen[d]
CD54	RR1/1		Weyrich *et al.* (1995)
CD62E	Cy1787		Weyrich *et al.* (1995)
CD62L	huDREG-200		Buerke *et al.* (1996)
CD62P	PB1.3		Weyrich *et al.* (1995)

[a]Dako Ltd, 16 Manor Courtyard, Hughenden Avenue, High Wycombe, Bucks HP13 5RE, UK.
[b]Coulter Electronics Ltd, Northwell Drive, Luton, Beds LU3 3RH, UK.
[c]Serotec, 22 Bankside, Station Approach, Kidlington, Oxford OX5 1JE, UK.
[d]Pharmingen Deutschland GmbH, Flughafenstrasse 54, Haus A, 22335 Hamburg, Germany.

researchers in immunology that we cannot simply extrapolate from the human or murine immune systems and expect the same rules to apply in other species. Even when the feline homologue of a human leukocyte differentiation antigen has been cloned and sequenced, functional studies should follow to confirm that the molecule has the same function in the cat as in man. The knowledge gained by such studies will benefit both the study of feline immunology, and the understanding of the function of the immune system in general.

4. Cytokines

Cytokines are molecules which are produced and secreted by leukocytes and several other cells. They act as intercellular mediators similar to hormones. The many lymphokines, monokines, interleukins, interkrines, chemokines and hormones, activating or inhibiting factors, or even growth factors are now generally and collectively known and described as cytokines. These factors are important for the differentiation and division of hematopoietic stem cells, for the activation or inhibition of lymphocyte functions and lymphocyte proliferation, and for the activation of phagocytes. In addition, they can function as chemokines (chemoattractants) and as cytotoxins. They act mostly in paracrine action on cells other than those from which they originate. However, they may also regulate the cell that produces them (autocrine action). Communication between immune regulator and effector cells, between antigen-presenting cells, lymphocytes and other cells occurs not only through direct cell to cell contact but also by secreting a plethora of soluble factors, leading to an intricate network between these cells.

During the last 10 years it has become clear that the activation of the cellular (T helper 1 pathway TH1) and humoral (T helper 2 pathway TH2) arms of the immune system are driven by different sets of cytokines (Mosmann and Coffman, 1989). It is mainly in this context that interest in feline cytokines, especially in connection with retrovirus and coronavirus infections, has arisen during the early 1990s.

Several feline cytokines have been cloned and sequenced and some have also been expressed. However, for most feline cytokines no specific antibodies are available that can be used for the detection of the cytokines in organ sections by immunological methods or for their quantification in cell culture supernatants or in body fluids. In some cases, antibodies specific for cytokines of other species show immunological cross-reactivity that allows application in the feline system. In this section, the known feline cytokines are listed together with information on their sequence, cross-reactivity with those of other species, cell source, function and methods of detection and quantification.

IL-1

Goitsuka *et al.* (1987) were the first to describe IL-1 at the protein level in cats. They showed that IL-1 is released by LPS-stimulated alveolar macrophages and by peritoneal exudate cells collected from cats with feline infectious peritonitis (FIP). Three isoforms were described, with isoelectric points of 4.1, 4.8 and 5.3, respectively. It is unclear which of these isoforms correspond to the human IL-1α and IL-1β. Hasegawa and Hasegawa (1991) demonstrated by *in situ* hybridization IL-1α mRNA to be produced by macrophages present in the inflammatory lesions in cats with FIP. Induction of IL-1 by LPS-stimulated macrophages was also confirmed by Daniel *et al.* (1993). The sequence of feline IL-1 has not been published.

IL-2

IL-2 which was originally designated T-cell growth factor (TCGF) is produced by lectin- or antigen-stimulated mature T-lymphocytes; the gene coding for IL-2 is located on the cat's chromosome B 1 (Seigel *et al.*, 1984). Tompkins *et al.* (1987) described a cell culture assay that allowed quantification of feline IL-2. Recombinant human IL-2 is highly capable of stimulating feline T-lymphocytes, an observation made by many research groups but never specifically investigated and published. After systemic application, human IL-2 induced eosinophilia in cats (Tompkins *et al.*, 1990a). Feline IL-2 has been sequenced and expressed in COS cells; it consists of 154 amino acids, including a putative signal sequence, and has 81%, 69%, 60% and 64% identity to human, bovine, murine and rat IL-2, respectively (Cozzi *et al.*, 1993; for Accession number see Table VIII.4.1). Feline IL-2 shows a similarity of 90% with canine IL-2 (Dunham *et al.*, 1995). Interestingly, while recombinant human IL-2 promotes proliferation of both, human and feline leukocytes, recombinant feline IL-2 only promotes proliferation of feline cells, but not human cells (Cozzi *et al.*, 1995). Around 20% of feline lymphocytes in the peripheral blood stream are positive for the IL-2 receptor (Iwamato *et al.*, 1989); in that study, the IL2 receptor was detected by a monoclonal antibody to human IL-2 receptor cross-reacting with the feline counterpart. In feline immunodeficiency virus (FIV) infection, expression of IL-2 is markedly inhibited (Tompkins *et al.*, 1989a; Bishop *et al.*, 1992; Lawrence *et al.*, 1992, 1995). In conjunction with nucleoside analogues and other cytokines, IL-2 has been used with some success to treat feline leukemia virus (FeLV) infections (Zeidner *et al.*, 1990, 1993).

IL-4

Feline interleukin-4 has been cloned and sequenced by Schijns *et al.* (1995a; for EMBL Accession number see

Table VIII.4.1 EMBL GenBank Accession numbers, characteristics and references of characterized feline cytokines

Gene product	*Accession no.*	*Length (bp)*	*Last update (m/y)*	*Reference*
IL-1β	M92060	804	9/93	Daniel *et al.* (unpublished data)
IL-2	L25408	462 (partial)	11/94	
	L19402	779	10/93	Cozzi *et al.* (1993)
IL-4	U39634	561	2/97	Lerner and Elder (unpublished data)
	U82193	284 (partial)	1/97	
	X87408	521	10/95	Schijns *et al.* (1995a)
IL-6	D 13227			Ohashi *et al.* (1989)
	L16914	724	3/94	Bradley *et al.* (1993)
IL-10	U39569	737	11/95	Scott and O'Reilly (unpublished data)
IL-12 (p35)	U83184		12/96	Fehr *et al.* (1997)
	Y07761			Schijns *et al.* (1997)
IL-12 (p40)	U83185		12/96	Fehr *et al.* (1997)
	Y07762			Schijns *et al.* (1997)
IL-16	AF003701	390		Leutenegger *et al.* (1998)
IFN-β	U81267	561	2/97	Lyons *et al.* (1997)
IFN-γ	X86972	543	3/96	Schijns *et al.* (1995b)
	D30619	567	6/96	Argyle *et al.* (1995)
TNF-α	M92061	705	9/93	Daniel *et al.* (1993)
	U82193	284 (partial)	1/97	Lyons *et al.* (1997)
	X5400	1722	3/93	McGraw *et al.* (1990)
TNF-R (p80)	U51429	247 (partial)	6/96	Duthie *et al.* (1996)
TNF-R (p60)	U72344	542 (partial)	10/96	Duthie *et al.* (1996)
SCF (stem cell factor)	D50833	953	2/97	Dunham and Onions (1996b)
MGF (mast cell growth factor)	U82188	430	1/97	Lyons *et al.* (1997)
NGF-β (nerve growth factor-β)	U82190	136	1/97	Lyons *et al.* (1997)

Table VIII.4.1). The IL-4 nucleotide sequence was found to be 83%, 82%, 80%, 79%, and 61% homologous to that of IL-4 from killer whale, pig, human, manatee, and rat, respectively. Dean *et al.* (1996a) measured the level of mRNA expression in peripheral and intestinal lymph nodes during the early phase of FIV infection and found that increased levels of IL-4 mRNA levels were paralleled by increased IL-10 mRNA expression.

IL-6

Feline IL-6, which was originally found to have a molecular weight of 30 000–40 000, has biological properties similar to IL-6 of human and murine origin but is not neutralized by antihuman IL-6 antiserum (Ohashi *et al.*, 1989; Goitsuka *et al.*, 1990).

Feline IL-6 was sequenced (Bradley *et al.*, 1993; Ohashi *et al.*, 1993; for the EMBL Accession number see Table VIII.4.1) and was found to be 752 bp long. It shows homology of 81%, 76%, 63%, and 61% with pig, human, rat, and mouse IL-6, respectively. The predicted amino acid sequence exhibits 66%, 53%, 37%, and 30% homology with pig, human, rat, and mouse IL-6, respectively. After transfection with IL-6 cDNA, Crandell feline kidney cells (CFK) produced biologically active proteins that showed hybridoma growth-promoting activity (Ohashi *et al.*, 1993). It is not clear how the difference between the apparent molecular weight of 30 000–40 000 described (Ohashi *et al.*, 1989; Goitsuka *et al.*, 1990) and the predicted molecular weight is explained. In humans, IL-6 is known to be important in the differentiation of activated B cells into Ig-secreting cells (Chen-Kiang, 1995). FIP and FIV infections are diseases in which B cells are greatly stimulated. The observation by Goitsuka *et al.* (1990) and Ohashi *et al.* (1992) that in cats with FCoV and FIV infection plasma IL-6 activity was significantly higher than in healthy controls, suggests that in the cat also, IL-6 plays an important role in B-cell activation. IL-6 was produced by peritoneal exudate cells collected from cats with FIP (Goitsuka *et al.*, 1990) which suggests that macrophages might be an important origin of IL-6.

IL-10

Feline IL-10 was recently cloned and sequenced by Scott and O'Reilly (1993; for EMBL Accession number, see Table VIII.4.1) and expressed in *Escherichia coli* (Leutenegger *et al.*, 1998b). In mice, IL-10 is known to inhibit antigen-specific T-cell activation by suppressing IL-12 synthesis (D'Andrea *et al.*, 1993). This leads to down-regulation of antigen presentation and accessory cell functions of monocytes, macrophages, Langerhans cells and

Table VIII.4.2 Homologies between known IL-12 sequences of different species: percentages of nucleotide (NA) and amino acids (AA) identity between the feline, canine, human, bovine, porcine and murine IL-12-p35 (top) and p40 (bottom) sequences

	NA (%)					
AA (%)	*Feline IL-12*	*Canine IL-12*	*Human IL-12*	*Bovine IL-12*	*Porcine IL-12*	*Murine IL-12*
p35						
Feline IL-12		93.3	89.9	87.4	85.8	66.8
Canine IL-12	91.5		88.0	nd	87.4	69.4
Human IL-12	87.8	85.0		86.9	85.8	73.1
Bovine IL-12	82.0	81.1	83.1		nd	nd
Porcine IL-12	85.3	83.4	84.5	89.2		67.5
Murine IL-12	57.9	55.6	60.2	59.3	59.7	
p40						
Feline IL-12		93.3	89.9	87.4	85.8	66.8
Canine IL-12	91.5		88.0	nd	87.4	69.4
Human IL-12	87.8	85.0		86.9	85.8	73.1
Bovine IL-12	82.0	81.1	83.1		nd	nd
Porcine IL-12	85.3	83.4	84.5	89.2		67.5
Murine IL-12	57.9	55.6	60.2	59.3	59.7	

dendritic cells. IL-10 is considered to be a potent immunosuppressant, both *in vitro* and *in vivo* and it attracts much interest as a potentially important immunoregulatory protein in the control of inflammatory, autoimmune and other immune-mediated diseases (de Vries, 1995). No *in vitro* assays are available to quantify feline IL-10. When IL-10 mRNA was determined by reverse transcription-polymerase chain reaction (RT-PCR) in LPS stimulated alveolar macrophages, it was reported to be significantly increased in FIV infected cats (Ritchey and Tompkins, 1996). In FIV infection, production of IL-10 mRNA in different lymphnodes was found to be increased after onset of viremia (Dean *et al.*, 1996a).

IL-12

As with other species, feline IL-12 is a heterodimeric protein consisting of 2 chains (p35 and p40) which are linked through disulfide bonds. While p35 was sequenced by Bush *et al.* (1994), both chains were sequenced and the sequences deposited (Fehr *et al.*, 1998; Schijns *et al.*, 1996; for EMBL Accession number see Table VIII.4.1). IL-12, originally known as natural killer cell stimulatory factor (NKCSF, Kobayashi *et al.*, 1989), is a key cytokine of the T_H1 pathway inducing IFN-γ (Trinchieri, 1995). Feline IL-12 is closely related to the IL-12 of other species (Table VIII.4.2). Using RT-PCR, Rottmann et al. (1996) measured IL-12 and IL-10 expression in bronchial lymph nodes of FIV positive cats with experimental *Toxoplasma gondii* infection. After *T. gondii* infection, levels of IL-12 and IL-10 mRNA were decreased in FIV-negative cats while FIV infected cats did not show this reduction. These results suggest that in early asymptomatic FIV infection cytokine regulation is impaired.

IL-16

IL-16, formerly known as lymphocyte chemoattractant factor (LCF) is a chemoattractant factor produced by $CD8^+$ T cells with predominant chemotactic effects for $CD4^+$ T cells (Cruikshank and Center, 1982). Although LCF does not induce T cell proliferation in lymphocytes, the factor induces IL-2 receptor and MHC II (HLA-DR) upregulation. *In vitro*, IL-16 is secreted from lymphocytes after Con A, histamine or serotonin stimulation (Center *et al.*, 1983; Laberge *et al.*, 1995, 1996). Although mRNA for IL-16 is also detectable in $CD4^+$ cells, the protein is translated and stored in biologically active form mostly in $CD8^+$ cells (Theodore *et al.*, 1986). After stimulation by histamine, IL-16 is secreted within 4 h.

The T-cell specific chemoattractant factor was described as a 14 kDa protein which appears to be linked noncovalently to form a tetrameric bioactive molecule with a mass of 56 kDa (Cruikshank and Center, 1982). At that time, only IL-1 was known to exist in multimeric forms with various degrees of bioactivity (Furutani *et al.*, 1986). The sequencing of the feline IL-16 cDNA revealed that codon 26 which is deleted in some of the human and also in some of the African green monkey genes (Bayer *et al.*, 1995) appears to be deleted also in the feline gene (Table VIII.4.3; Leutenegger *et al.*, 1998a).

Table VIII.4.3 Homologies of nucleotide (N) and amino acid (AA) sequences of recombinant IL-16 (Needleman and Wunsch Algorithm)

	N (%)		
AA (%)	*Human*	*AGM*[1]	Feline
Human	—	96.7%	85.6%
Africa green monkey (AGM)	95.4%	—	88.9%
Feline	84.6%	86.9%	—

[1]African green monkey.

IL-18

A novel cytokine designated interferon-γ-inducing factor (IGIF) was described in the human and murine system (Okamura *et al.*, 1995; Micallef *et al.*, 1996; Ushio *et al.*, 1996). This cytokine, which has also been designated IL-18, has been cloned, sequenced and expressed from a human cDNA library (Ushio *et al.*, 1996). In synergy with IL-12, IL-18 was found to induce IFN-γ and to decrease IL-10 expression. The feline counterpart of IL-18 was amplified from alveolar macrophages by RT-PCR (Argyle *et al.*, 1996).

IFN-α

IFN-α has not yet been sequenced. However, based on its cross-species activity, recombinant human IFN-α has been reported to be successful for the treatment of FeLV infection (Weiss *et al.*, 1991).

IFN-β

The feline analogue to IFN-β was sequenced by Lyons *et al.* (1993; for EMBL Accession number see Table VIII.4.1). In contrast to IFN-α, IFN-β is strictly species specific. No information is available on the effect of feline IFN-β.

IFN-γ

Feline IFN-γ has recently been sequenced after RT-PCR amplification (Argyle *et al.*, 1995; Schijns *et al.*, 1995b; for EMBL Accession number see Table VIII.4.1). At the nucleotide level, it shares 78% and 63% homology with the cDNA of human and murine IFN-γ. At the amino acid level, the feline IFN-γ shares 63% and 43% homology with human and murine homologs, respectively. For many years, IFN-γ has been known to have profound effects on the different functions of the immune system, especially on the T_H1 pathway (Young and Hardy, 1995). In cats, IFN-γ activity has been studied only in FIV infection. During the acute phase of FIV infection, IFN-γ and IL-12 mRNAs rise transiently to very high levels; thereafter these cytokines return to baseline concentrations (Dean *et al.*, 1996a). In contrast, while the mRNA levels of IFN-γ and IL-12 return to the original concentrations, IL-4 and IL-10 mRNAs remain elevated. These results suggest that during early FIV infection a T_H1 immune response is triggered which later switches to a T_H2 type of immune response. It has to be stressed that studies dealing with expression of IL-12 and IFN-γ are based on quantification of mRNA expression: it is important that these findings are confirmed by quantification of the respective proteins.

TNF-α

TNF-α has been sequenced (McGraw *et al.*, 1990; Daniel *et al.*, 1993; Lyons *et al.*, 1993; for EMBL Accession number see Table VIII.4.1) and expressed in *E. coli* as functional protein (Rimstad *et al.*, 1995). The biologically active protein had a molecular mass of 17 kDa. Feline TNF-α immunologically cross-reacts with antibodies specific for human TNF-α (Lehmann *et al.*, 1992; Rimstad *et al.*, 1995). TNF-α production is increased in FIV infected cats (Lehmann *et al.*, 1992) probably reflecting increased production by macrophages infected by FIV (Lin and Bowman, 1993). FeLV infection of macrophages also has been reported to trigger increased TNF-α production (Khan *et al.*, 1993). In a L929 cytotoxic assay, TNF-α was found to display CD50 activity at 15 ng/ml (Rimstad *et al.*, 1995). Cats given recombinant feline (rf)TNF-α intravenously manifested the typical biological effects of TNF-α, including fever, depression, and piloerection. The rfTNF-α upregulated IL-2 receptor and MHC-II antigen expression on peripheral blood mononuclear cells stimulated *in vitro*, but had no effect on TNF-α receptor and MHC I antigen expression (Rimstad *et al.*, 1995). When BAL cells collected from cats with long-term FIV infection were stimulated by LPS or Con A and phorbol myristate acetate, less TNF-α was produced than by cells from cats not infected by FIV (Ma *et al.*, 1995). However, in the acute phase of FIV infection, significantly more TNF-α was produced than in noninfected cats. According to Kraus *et al.* (1996) expression of FIV p24 during development of viremia during the acute phase of FIV infection is closely associated with expression of TNF-α, which suggests a close interrelationship between FIV and TNF-α expression. TNF-α expression was detected in feline hearts shortly after endotoxin application (Kapadia *et al.*, 1995). Ohno *et al.* (1993) demonstrated that TNF-α induces apoptosis in CmFK cells chronically infected by FIV while apoptosis was not found in noninfected cells. These findings may be important in the understanding of the pathogenesis of FIV induced immunosuppression. The feline TNF-receptor was recently sequenced (Duthie *et al.*, 1996; for EMBL Accession number see Table VIII.4.1)

Stem cell factor

Stem cell factor (SCF) has been found to be an essential hematopoietic cytokine that interacts with other cytokines to facilitate survival of hematopoietic stem and progenitor cells, to influence their entry into the cell cycle and to stimulate their proliferation and differentiation (Hassan and Zander, 1996). Two isoforms of feline stem cell factor (fSCF) have been sequenced by RT-PCR and expressed in *E. coli* (Dunham and Onions, 1996a,b; for EMBL Accession number see Table VIII.4.1), The two isoforms had 274 amino acids and 246 amino acids. Feline SCF shows a high degree of homology to the SCFs of other species at both the nucleic acid and protein level. *In vitro*, the recombinant protein was found to induce cell proliferation in different cell lines (EC50 4–40 ng/ml). When injected into specific pathogen-free (SPF) cats, SCF was found to induce neutrophilia and to increase the number of peripheral blood colony forming units (Dunham and Onions, 1996b). Feline SCF may prove to be a useful cytokine for therapy of retrovirus induced anemias and neutropenias.

Mast cell growth factor

The mast cell growth factor (MGF) stimulates growth of mast cells. The sequence of feline MGF (Lyons *et al.*, 1993; for EMBL Accession number see Table VIII.4.1) shows a 37% similarity with that of feline SCM sequenced by Dunham and Onions (1996b; for EMBL Accession number see Table VIII.4.1).

Nerve growth factor

The human nerve growth factor (NGF) consists of a family of factors responsible for the survival, differentiation and functional activities of sensory and sympathetic neurons in the peripheral nervous system (Barde, 1990). It also supports the development and functional activities of cholinergic neurons in the central nervous system. The 7S form of NGF is a complex of three proteins (α, β and γ), The 26 kD β subunit is a homodimer of two disulfide-bonded proteins with a length of 118 amino acids and displays the biological activity of NGF. Feline NGF-β has been sequenced (Lyons *et al.*, 1993; for EMBL Accession number see Table VIII.4.1). However, the biological effects of the feline NGF-β have not been investigated thoroughly.

Acknowledgements

Some of the authors own work was supported by Swiss National Science Foundation grant no. 31-42486.94. The authors are indebted to the Union Bank of Switzerland on behalf of a customer and the Winn Foundation, New Jersey, USA for financial support.

5. The Major Histocompatibility Complex

Introduction

Early studies on the feline major histocompatibility complex (MHC) by Pollack *et al.* (1982) using *in vitro* lymphocytotoxicity assays suggested that cats fail to develop lymphocytotoxic antibodies in response to pregnancy or transfusion. Moreover, while immunization with foreign cells induced lymphocytotoxic antibodies, the response was not allospecific, as observed in other species. Similarly, only weak mixed lymphocyte reaction (MLR) responses were detected between unrelated cats of different breeds. The authors concluded that there may be limited polymorphism in the feline MHC. Subsequent studies by Winkler *et al.* (1989) investigated the development of cytotoxic antibodies in the cat following skin grafting between unrelated cats. The alloantisera generated displayed lymphocytotoxicity and permitted the identification of six clusters of overlapping feline MHC (termed FLA) specificities. Using these sera the authors were able to define FLA haplotypes and show that the specificities segregated as a single Mendelian complex.

Genetic characterization of the feline major histocompatibility complex

The genetic characterization of the feline major histocompatibility complex MHC has been reviewed in detail by Yuhki (1995). The feline MHC class I and II (FLA I and FLA II) loci map to the centromeric region of chromosome B2. The FLA I molecule is encoded from a single open reading frame and gives rise to a 363 amino acid (a.a.) protein similar in structure to human HLA I. Following a 24 a.a. leader sequence there are three extracellular domains of 90 a.a. (a1), 92 a.a. (a2) and 92 a.a. (a3), a 31 a.a. transmembrane domain and a 34 a.a. cytoplasmic domain. Cysteine residues at positions 101, 164, 203 and 259, and the predicted site for *N*-linked glycosylation (amino acids 86–88) are conserved between feline, human and murine class I molecules (Yuhki and O'Brien, 1988). Interestingly, Winkler *et al.* (1989) were able to identify only one transcriptionally active MHC class I locus (with two alleles), perhaps explaining previous reports suggesting limited diversity in the feline MHC. However, abundant polymorphisms were observed at both the MHC class I and class II loci.

Screening of feline cDNA libraries using human MHC class II specific (DRA and DRB) probes identified three DRA and DRB equivalents and suggested the existence of at least two DRA and two DRB loci. Screening of the libraries with human MHC class II specific (DQA and DQB) probes failed to identify cross-hybridizing clones, suggesting divergence at this locus. Feline MHC class II DRA molecule consists of a 25 a.a. leader sequence followed by an 84 a.a. a1 domain, a 107 a.a. a2 domain, a 23 a.a. transmembrane domain and a 15 a.a. cytoplasmic domain. The DRB molecule consists of a 29 a.a. leader sequence followed by a 95 a.a. b1 domain, 104 a.a. b2 domain, 23 a.a. transmembrane and 16 a.a. cytoplasmic domain. Interestingly, although polymorphic residues were observed between feline MHC DRA chains, none of the mutations resided in either of the a1 domains, the domains of the molecule containing the antigen recognition site. In contrast, polymorphisms were observed in the b1 domain of feline MHC DRB, in the region comprising the antigen recognition site (Yuhki, 1995).

Biochemical characterization of feline MHC antigens

Early studies by Neefjes *et al.* (1986) analysed the biochemical characteristics of feline MHC molecules. Detergent extraction followed by immunoprecipitation demonstrated abundant MHC class I molecules, and low levels of MHC class II antigens in unstimulated human PBLs. In contrast, similar analyses of feline PBLs demonstrated high levels of both MHC class I and MHC class II on unstimulated PBLs. In the murine and human immune systems MHC class II expression is restricted, with predominant expression on B cells, macrophages, dendritic cells and activated T cells. In contrast, both resting and activated T cells express high levels of MHC class II molecules in the cat (see below). The α-chain of feline MHC class II has an acidic pI and apparent M_r of 35×10^3 whereas the β-chain has a basic pI and an apparent M_r of 30×10^3 (Neefjes *et al.*, 1986). The feline MHC class I α-chain has an apparent M_r of approximately 45×10^3. Interestingly immunoprecipitation of feline MHC class I from cultured feline lymphocytes with the cross-species reactive antibody W6/32 did not yield β_2-microglobulin (Neefjes *et al.*, 1986). While this may reflect exchange of β_2-microglobulin with bovine β_2-microglobulin from the culture medium, it is possible that W6/32 recognizes feline MHC class I in the absence of β_2-microglobulin, as has been reported with other species.

Expression of MHC molecules in the feline immune system

Several reagents have been identified with which the expression of MHC antigens in the feline immune system can be studied. Early studies on the feline MHC utilized cross-species-reactive antibodies such as W6/32 (anti-human MHC class I) and ISCR3 (antimurine IE). Recently, reagents specific for feline MHC class II have been developed including 42-3H2 (Rideout *et al.*, 1990) and vpg3 (Willett *et al.*, 1995). These antibodies were generated as a fortuitous byproduct of feline immunodeficiency virus research, both antibodies resulting from fusions of splenocytes from mice immunized with purified FIV. It has been shown that as lentiviruses bud from cells they incorporate a range of cellular proteins into the viral envelope (Arthur *et al.*, 1992). FIV appears to incorporate a significant quantity of MHC class II molecules and several groups have reported anti-MHC class II antibodies predominating during immunization of mice with FIV.

Flow cytometric analysis of feline peripheral blood mononuclear cells using the 42-3H2 antibody revealed that MHC class II molecules are expressed on the majority of T and B lymphocytes (Rideout *et al.*, 1990). Importantly, both resting and activated T cells express MHC class II molecules, giving rise to a characteristic bimodal histogram when MHC class II expression is analysed on lymphocytes. The high-intensity peak consists predominately of B lymphocytes whereas the low-intensity peak consists predominately of T lymphocytes (Rideout *et al.*, 1990; Willett and Callanan, 1995). In the FIV-infected cat higher levels of MHC class II expression are detected on a subset of $CD8^+$ lymphocytes which usually express reduced levels of MHC class II ($CD8^{low}$), suggesting that this subpopulation may represent activated $CD8^+$ lymphocytes (Willett and Callanan, 1995). While Ohno *et al.* (1992a) reported that higher levels of MHC class II could be detected on PBMC from FIV-infected cats compared with specific pathogen-free (SPF) cats, the authors did not establish whether this was due to an increase in the number of B lymphocytes or upregulation of T cell MHC class II expression. In contrast, Rideout *et al.* (1992) demonstrated that there was a persistent elevation in the percentage of $CD4^+$ and $CD8^+$ T lymphocytes expressing MHC class II antigens shortly after FIV infection but that a similar elevation was present in cats chronically infected with FeLV. The data suggest that despite a basal level of MHC class II expression on feline T cells, expression can be upregulated following T cell activation.

An early study of the feline MHC by Pollack *et al.* (1982) using a cross-reactive antimurine IE antibody suggested that activated and resting feline T lymphocytes could be differentiated on the basis of expression of an IE-like molecule. Therefore, it would appear that while some feline MHC class II molecules are expressed on all T cells (42-3H2 and vpg3 reactive), additional MHC class II molecules are expressed only upon activation (IE-like). The expression of MHC class II molecules on resting T cells is not unique to the cat, similar phenomena have been described in the dog, pig and horse (Thistlewaite *et al.*, 1983; Doveren *et al.*, 1985; Crepaldi *et al.*, 1986). The differentiation of activated T lymphocytes on the basis of MHC class II expression has been observed in the horse where IA-like molecules are constitutively expressed on all T cells while IE-like molecules are restricted to activated T cells.

Table VIII.6.1 The feline AB blood group system

Type	*Disialoganglioside* $(NeuNAc)_2GD_3$	*Intermediate x*	*Intermediate y*	$(NeuNGc)_2GD_3$	*Alloantibodies*	*Frequency*
A	+	+	+	+++	+	Most common blood type
B	++++	—	—	—	+++	Common in certain breeds and geographic locations
AB	++	++	++	++	—	Extremely rare

6. Red Blood Cell Antigens

Although naturally occurring alloantibodies were recognized in cats in 1915, it was not until the second half of the twentieth century that two major feline blood types, known as type A and B, were discovered. In 1981, Auer and Bell (1981) also identified an extremely rare blood type AB that reacted with anti-A and anti-B reagents. These three blood types form the only known blood group system in cats designated as the AB system, although it is not related to the human ABO system (Bell, 1983) (Table VIII.6.1).

In contrast to most blood types in other species, type-A and type-B antigens are not codominantly inherited (Giger *et al.*, 1991a). The A allele is dominant over the B allele. Thus, all type-B cats are homozygous for the B allele (genotype B/B), whereas type-A cats can be homozygous or heterozygous for the A allele (genotype A/A or A/B). The AB allele is recessive to the A allele, but dominant over the B allele. There may be an additional genetic mechanism responsible for the inheritance of the AB blood type in cats, but they are not generally produced by breeding type-A cats to type-B cats (Griot-Wenk *et al.*, 1996).

Specific neuraminic acids on gangliosides, containing ceramide dihexoside (Galb1-4Glc-cer) as a backbone, correlate with the feline AB blood group antigens (Andrews *et al.*, 1992; Griot-Wenk *et al.*, 1993). Although disialogangliosides predominated, mono- and tri-sialogangliosides were also isolated. Type B cats express only *N*-acetyl-neuraminic acid on these gangliosides. A-type red cells express predominantly *N*-glycol-neuraminic acid containing gangliosides, but also some *N*-acetyl neuraminic acids. Equal amounts of the above two neuraminic acids containing disialogangliosides and two intermediary forms were found on type AB erythrocytes. In addition to these glycolipid patterns, differences in the glycoprotein patterns were also identified.

Naturally-occurring alloantibodies have been used for blood typing cats (Bücheler and Giger, 1993). All type-B cats have very strong anti-A alloantibodies which act as strong ($>1:64$) agglutinins and hemolysins. Type-A cats have generally weak anti-B alloantibodies (1:2). Therefore, the anti-B serum has been replaced with *Triticum vulgaris* lectin which strongly agglutinates type B cells (Butler *et al.*, 1991). A simple card test is now available to differentiate between type A, B, and AB cats (DMS Laboratories, 2 Darts Mill Rd, Flemington, NJ 08822; 1-800-567-4367).

The frequency of blood type A and B among domestic shorthair and longhair cats differs markedly between parts of the United States and other parts of the world (Giger *et al.*, 1991a,b, 1992). The distribution of type A and B among purebred cats varies even more, although no geographic variation has been noted. Knowledge of the blood type frequency in each breed is important when estimating the risk of AB incompatibility reactions. The AB blood type has been recognized in domestic shorthair and longhair cats and purebred cats with blood type B, but occurs extremely rarely (Griot-Wenk *et al.*, 1996). With less than 1%, type AB is not listed in Tables VIII.6.2 and VIII.6.3.

Although the function of these red blood cell antigens remains unknown, they are responsible for serious blood incompatibility reactions. Owing to the fact that cats have naturally occurring alloantibodies, even a first mismatched

Table VIII.6.2 Frequency of blood type A in domestic shorthair and longhair cats

	Type A (%)
Asia, Tokyo, Japan	90.0
Australia, Brisbane	73.7
Europe	
Austria	97.0
England	97.1
Finland	100
France	85.1
Germany	94.0
Italy	88.8
Netherlands	96.1
Scotland	97.1
Switzerland	99.6
North America, United States	
Northeast	99.7
North Central/Rocky Mountains	99.6
Southeast	98.5
Southwest	97.5
West Coast	95.3
South America, Argentina	97.0

Type B = 100-type A; type AB is less than 1%.

Table VIII.6.3 Frequency of blood type A in purebred cats in the United States

	Type A (%)
Abyssinian	86
Birman	84
British shorthair	60
Burmese	100
Cornish rex	66
Devon rex	59
Exotic shorthair	75
Himalayan	93
Japanese Bobtail	84
Maine Coon	98
Norwegian Forest	93
Persian	84
Russian blue	100
Scottish fold	82
Siamese	100
Somali	83
Sphinx	82
Tonkinese	100

Type B = 100-type A; type AB frequency is less than 1%.

blood transfusion can result in life-threatening hemolytic transfusion reactions particularly when a type-B cat receives type A blood (Giger and Bücheler, 1991). Furthermore, type A and AB kittens born to queens, even when primiparous, with blood type B may develop neonatal isoerythrolysis when receiving colostrum during the first day of life (Giger, 1991; Casal *et al.*, 1996). Kittens at risk may be blood typed at birth with cord blood and type A and AB kittens born to type B queens may be successfully foster-nursed for 24 h.

7. Immunoglobulins

Introduction

As with any mammalian species analyzed to date, cats have different immunoglobulin (Ig) isotypes designated IgA, IgG and IgM (Aitken, *et al.*, 1967; Okoshi *et al.*, 1968; Klotz *et al.*, 1985; Grant, 1995; Paul, 1995). However, the number of subclasses identified for different Ig isotypes varies considerably from species to species. The mouse has one IgA class, human has two but the rabbit has as many as 13 different subclasses of IgA (Spieker-Polet *et al.*, 1993). In contrast, rabbits have only one class of IgG, mice and humans have four but pigs have at least five different subclasses of IgG (Butler and Brown, 1994). The analysis of the Ig isotypes summarized above was made by analyzing the genes encoding the different Ig molecules. This has provided conclusive evidence for the structure of Igs which was not possible to elucidate otherwise because only a few amino acid differences may represent different isotypes, allotypes or Ig subclasses.

To date, no data have been published analyzing genes encoding cat Ig. Therefore, the information available has been obtained by using monoclonal antibodies (mAb) specific to cat Ig and biochemical methods to analyze Ig. This information forms the basis for the discussion of cat Ig.

Cat Ig light chains

Cats have two different classes of Ig light chains: κ and λ each with a molecular mass of 25 kDa. The ratio between κ and λ light chains expressed on Ig in various tissues and fluid has been estimated to be 1 : 3 (Hood *et al.*, 1967; Klotz *et al.*, 1985). mAb specific for κ and λ light chains have been described (Klotz *et al.*, 1985; Grant, 1995).

Cat Ig heavy chains

IgM

Cat IgM was first analyzed by Aitken *et al.* (1967) by agar gel diffusion techniques. As in other mammals, there appears to be only one IgM class. Some cat IgM binds to *Staphylococcus aureus* Protein A (SpA; Goudswaard *et al.*, 1978; Lindmark *et al.*, 1983). However, it cannot be excluded that variable regions representing V_HIII like families (Paul, 1995) from cat bind to SpA rather than the Ig heavy chains from IgM (Sasso *et al.*, 1991). There are mAbs specific to cat IgM (Klotz *et al.*, 1985; Grant, 1995). The serum concentration of IgM in specific pathogen-free cats is 0.32 mg/ml ($\pm$SD 0.27; n = 22; Grant, 1995).

IgA

At least one IgA isotype is present in cats but some workers have found evidence for two subclasses of IgA based on differential binding to SpA or binding to a mAb (Grant, 1995). However, different glycosylation of IgA, dimeric IgA or allotypes of IgA (Brown *et al.*, 1995) cannot be excluded. The serum concentration of IgA in specific pathogen-free cats is 0.31 mg/ml ($\pm$SD 0.24; n = 22; Grant, 1995).

IgG

There is evidence for at least two (Schultz *et al.*, 1974) and possibly up to four IgG subclasses in cat named IgG1 to IgG4 based on anionic exchange chromatography (Grant, 1995), the analysis of differential binding of cat IgG to SpA using the methods of Seppala *et al.* (1981) and mAb to cat IgG (Klotz *et al.*, 1985; Grant, 1995). The heavy chain of IgG has been estimated to be 50–55 kDa. There is no mAb which exclusively recognizes one of the postulated four

IgG subclasses (Klotz *et al.*, 1985; Grant, 1995). mAb to cat Ig, which have been described in some detail by Grant (1995) have been evaluated for cross-reactivty to human, cow, dog, goat and horse Ig. Some cross-reactivity was found among the same isotypes present in different species.

IgE

Clinical evidence of type I allergies in cats associated with IgE implies that IgE is present in cats. However, although all mammalian species analyzed in any detail have IgE, no cat IgE has yet been described.

Conclusion

mAb against cat κ and λ light chains and mAb against IgM, IgA and IgG exist (Klotz *et al.*, 1985; Grant, 1995). There is no specific mAb against the putative subclasses of cat IgA or IgG. Ongoing analysis of Ig genes in some laboratories will improve understanding of the structure of Ig in cats.

Acknowledgements

I would like to thank Dres. V. Rutten, University of Utrecht and Ch. Grant, Custom Monoclonals Sacramento Ca USA for their willingness to discuss data on cat IgG and to make mAbs available for testing. Mathias Ackerman made valuable suggestions to improve the manuscript.

8. Passive Transfer of Maternal Immunity

Introduction

At birth, kittens are highly susceptible to microbial infections because of their immature immune system. During the course of maturation, kittens are protected from microbial infection mostly by humoral immunity transferred from their mothers.

Transplacental immunity

The structure of the feline placenta is different from those of most domestic animals. The fetal chorionic epithelium in feline placenta is in close contact with the endothelium of maternal capillaries, and this type of placenta is called endotheliochorial placenta (Leiser and Kaufmann, 1994). This type of placentation allows a small amount (5–10%) of maternal immunoglobulins (primarily IgG) to transfer to the fetus (Scott *et al.*, 1970; Schultz *et al.*, 1974). The amount of immunoglobulin transferred to the fetus is very difficult to determine by standard quantitative assays, such as radial immunodiffusion and immunoelectrophoresis (Schulz *et al.*, 1974). Consequently, detection of transplacental immunoglobulins in the fetus has been demonstrated more readily by assays which detect antigen-specific antibodies in the serum of presuckling kittens, such as neutralizing antibody assay, enzyme-linked immunosorbent assay, and immunoblot analysis (Harding *et al.*, 1961; Scott *et al.*, 1970; Pu *et al.*, 1995). There is no information available about transplacental cell-mediated immunity in cats.

Immunity transferred by colostrum and milk

Newborn kittens receive a majority of maternal antibodies via ingestion of colostrum. During late pregnancy and under the influence of hormones, the mammary glands of queens secrete milk into mammary alveoli. The milk preformed before parturition is called colostrum. Colostrum is rich in immunoglobulins and the immunoglobulin fraction consists predominantly of IgG; IgA and IgM are present in lower concentration (Table VIII.8.1). The concentration of IgG in colostrum is two- to four-fold higher than that found in serum. A majority of IgG and a considerable portion of IgA in the colostrum are actively transported from the circulation into mammary alveoli by specific receptors on the acinar epithelial cells of the mammary glands (Gorman and Halliwell, 1989). Colostrum contains trypsin inhibitors, as well as antimicrobial factors such as lysozyme, lactoferin, and lactoperoxidase (Stabenfeldt, 1992). In addition, colostrum has high concentrations of lipids and lipid-soluble vitamins (particularly vitamin A), proteins (such as caseins and albumin), and minerals, and also low levels of carbohydrates (Reece, 1991; Stabenfeldt, 1992). All of these components are nutritionally important to newborn kittens.

Milk is produced by queens immediately after the consumption of colostrum by kittens. The composition of milk is considerably different from that of colostrum. Milk has low levels of IgG and IgA and lacks IgM (Table VIII.8.1). IgG is the predominant immunoglobulin class in the milk of cats, unlike other nonruminant animals such as dogs, pigs and horses, in which IgA is the predominant class (Tizard, 1996). A majority of IgG and IgA is synthesized locally in the mammary glands, unlike colostrum which consists of immunoglobulins transported from the serum. Milk also differs from colostrum in that it contains less lipids, proteins, and minerals, and has more carbohydrates (Reece, 1991; Stabenfeldt, 1992).

Within the first 24–36 h after birth, colostral immunoglobulins (IgG, IgA, and IgM) are absorbed by the intestinal epithelial cells of newborn kittens (Figures VIII.8.1 and VIII.8.2). The trypsin inhibitors present in colostrum decrease the proteolytic activity in the gastrointestinal tract of newborn kittens, thereby enabling the immunoglobulins to retain their structure and function

Table VIII.8.1 Immunoglobulin levels (mg/dl) in serum and milk of nursing queens[a]

		Average immunoglobulin concentration (range) at postpartum (weeks)						
Sample	*Ig Isotypes*[b]	*0*	*1*	*2*	*3*	*4*	*6*	*References*
Serum	IgG	764 (420–1550)	594 (350–840)	611 (340–840)	434 (385–495)	906 (660–1450)	742 (480–920)	Pu *et al.* (unpublished data)
	IgA	138 (66–310)	77 (26–175)	66 (0–210)	88 (41–135)	175 (26–500)	179 (41–260)	
	IgM	444 (230–700)	402 (245–560)	420 (0–800)	405 (200–690)	576 (208–800)	525 (265–750)	
Milk	IgG	1572 (685–3150)	189 (60–400)	99 (0–230)	90 (80–120)	110 (60–220)	117 (60–200)	
	IgA	57 (26–127)	14 (0–35)	7 (0–23)	15 (0–26)	5 (0–21)	7 (0–21)	
	IgM	47 (0–280)	0	0	0	0	0	
Serum	IgG	1375						Pedersen (1987)
	IgA	215						
	IgM	116						
Milk	IgG	4400	440	360	300	315	100	
	IgA	340	44	20	20	24	24	
	IgM	58	2	0	0	0	0	
Serum	IgG	1894 (1171–2258)						Gorman and Halliwell (1989)
	IgA	285 (102–582)						
	IgM	247 (60–390)						
Milk	IgG	3570 (2750–4674)			189 (94–255)			
	IgA	254 (50–488)			13 (9–20)			
	IgM	110 (31–300)			20 (10–40)			

[a]Specific pathogen-free cats were used in all studies except for the study by Gorman and Halliwell (1989) which did not specify the cat source.
[b]Immunoglobulin (Ig) isotypes were quantified using radial immunodiffusion assay plates (Bethyl Laboratories, Inc., Montgomery, Texas, USA) by Pu *et al.*, and unspecified methods by Pedersen (1987) and Gorman and Halliwell (1989).

(Tizard, 1996). The ingested immunoglobulins bind first to specific Fc receptors on the surface of intestinal epithelial cells. Subsequently, immunoglobulins are taken up by the epithelial cells via pinocytosis and transferred into the lacteals and then the circulation (Tizard, 1996). Once maternal immunoglobulins are absorbed into the circulation, a small portion is released back onto the mucosal surface, thereby providing the local immunity. However, a majority of maternal immunoglobulins remain in the circulation and confer systemic immunity (Tizard, 1996). In contrast, immunoglobulins present in milk remain mainly in the intestine, and consequently, provide local immunity for the gastrointestinal tract (Tizard, 1996).

Upon passive transfer, the levels of serum immunoglobulins in newborn kittens approach those found in adults (Scott *et al.*, 1970) (Figure VIII.8.1). The transferred immunoglobulins are important in protecting newborn kittens against infections, such as feline leukemia virus (FeLV) (Jarrett *et al.*, 1977), feline panleukopenia virus (FPV) (Scott *et al.*, 1970) and feline immunodeficiency virus (FIV) (Pu *et al.*, 1995). The lack of protection in some kittens may result from failure of queens to produce or transfer sufficient amounts of protective antibodies specific for pathogens encountered during this critical period (Tizard, 1996).

It is still unclear whether maternal immune cells that are present in the colostrum and milk can be transferred from queens to kittens. However, studies with other domestic animals indicate that maternal immune cells can be transferred via colostrum and milk to the newborns and provide selected cell-mediated immunity (Tizard, 1996).

Vaccination of kittens

Vaccination of kittens should be scheduled according to the level and nature of the passive immunity received

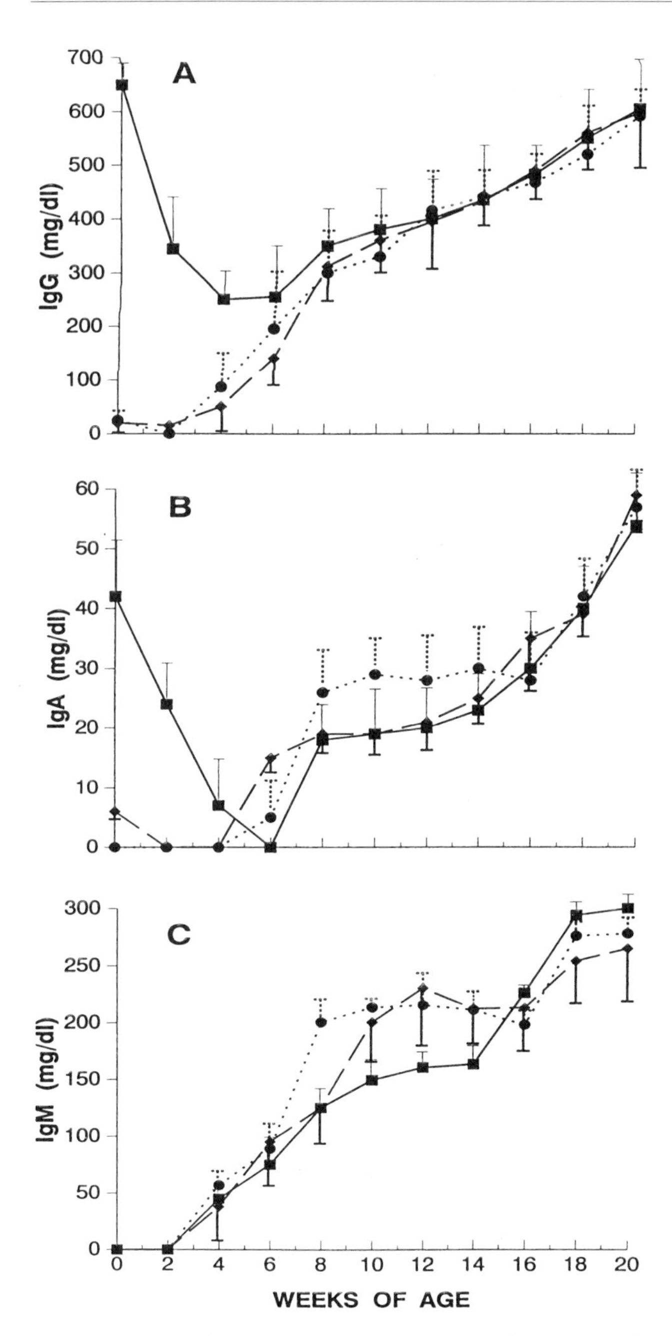

Figure VIII.8.1 Serum levels of IgG (A), IgA (B), and IgM (C) in kittens ($n = 6$) receiving either colostrum plus milk (——), milk alone (– – –), or milk replacement (.....). The immunoglobulin levels were determined by radial immunodifusion assays (Bethyl Laboratories, Inc., Montgomery, Texas, USA).

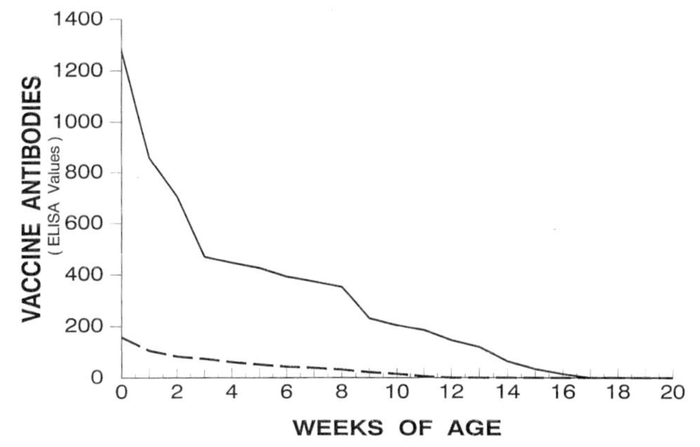

Figure VIII.8.2 FIV-specific antibodies detected in serum samples of kittens born to FIV vaccinated queens. Immediately after birth, kittens ($n = 4$) were either nursed by their mothers (——) or FIV antibody-free surrogate queens (– – –). FIV-specific antibodies were detected by FIV transmembrane peptide ELISA.

by the kittens. This is based on the concept that maternal antibodies reactive to vaccine antigens can interfere with the development of the immunity elicited by the vaccination (Scott *et al.*, 1970; Scott, 1971). Vaccine interference can result in delay and/or failure in the development of protective levels of vaccine-induced immunity. In general, the level of vaccine interference correlates directly with the titer of specific antibodies present in the serum of kittens. Thus, the vaccination schedule of kittens should be designed in a manner which avoids or overcomes potential vaccine interference. Ideally, kittens should receive the first vaccination at a time when pathogen-specific maternal antibodies are depleted. It is possible to predict this time-point based on the titer of specific antibodies in queens and the half-life of specific maternal antibodies in kittens. The half-life of maternally derived antibodies varies with their isotypes; for example 15 days for FeLV antibodies (Jarrett *et al.*, 1977), 9.5 days for FPV antibodies (Scott *et al.*, 1970), 7 days for feline enteric coronavirus antibodies (Pedersen *et al.*, 1981), and 18.5 days for feline rhinotracheitis virus antibodies (Gaskell, 1975). The half-life of antibodies may also vary between kittens from different litters (Scott *et al.*, 1970). Generally, specific antibody titers in kittens and queens are not usually determined before vaccination, thus, vaccination schedules are designed according to the known half-life of specific antibodies. For example, if the half-life of antibodies for a specific pathogen is 15 days, the maternal antibodies remaining in the serum of kittens will drop to insignificant levels (approximately 1.5%) by 3 months old. Because the half-life of antibodies against a majority of pathogens is equal to or less than 15 days, the primary vaccination should be administered to kittens at about 3 months old. However, it is important to note that low titers of residual maternal antibodies to FPV (Scott *et al.*, 1970) and FIV (Figure VIII.8.2) have been detected in the serum of kittens at 4 months old. Thus, to overcome the potential vaccine interference, vaccine boost at 4 months old is important in ensuring successful vaccination, as has been suggested by others (Scott *et al.*, 1970; Scott, 1971). This view is further supported by results from experimental vaccine studies which demonstrate the ability of vaccine boosts to overcome maternal antibody interference (Scott, 1971) (Figure VIII.8.3).

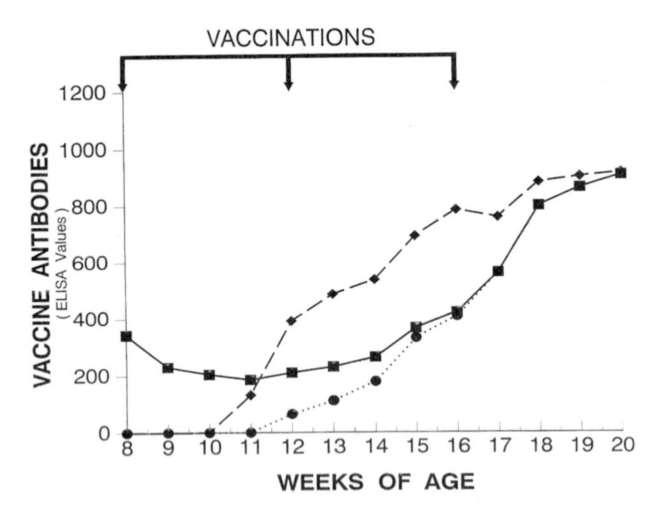

Figure VIII.8.3 Vaccine antibody titers of kittens ($n = 6$) that received colostrum and milk from either vaccinated (——) or unvaccinated (– – –) mothers prior to vaccinations at 8, 12, and 16 weeks old. The actual antibody titers produced by kittens (.....) after vaccination were calculated using the following formula: antibody titer in vaccinated kittens subtracted by antibody titer in unvaccinated littermate.

Failure of passive transfer of immunity

Compared with large newborn domestic animals (i.e. calves, foals, and lambs), failure of passive transfer of maternal immunity to kittens may not play a key role in neonatal mortality. Newborn kittens can survive without ingestion of colostrum providing that they receive an appropriate milk replacement containing the adequate nutritional content (Figure VIII.8.1). In addition, such newborn kittens require special care, such as maintenance of proper hygiene and keeping them in sanitary containers under proper temperature and humidity conditions.

9. Neonatal Immune Response

Although little is known about the development of the immune system in cats, it most likely follows a pattern similar to other mammals in that the thymus is the first lymphoid organ to develop, followed by the secondary lymphoid tissue. The feline thymus first appears during the fifth week of gestation and occupies the cranioventral mediastinal space. Lymphocytes from the bone marrow seed the thymus by 40 days of gestation. The secondary lymphoid tissue such as spleen and lymph nodes are then seeded with mature T cells (from the thymus) and B cells (from the bone marrow). In the cat, population of the secondary lymphoid tissue is largely complete at the time of birth.

Thymectomy performed at 5 weeks old had no direct effect on numbers of circulating B and T lymphocytes or on the responsiveness of peripheral lymphocytes to concanavalin A or pokeweed mitogen. Survival of skin allografts performed 3 weeks after thymectomy was slightly prolonged compared with nonthymectomized littermates. Kittens appear to be similar to the cow, sheep, and pig in that neonatal thymectomy does not have important immunologic or physiologic consequences (Hoover *et al.*, 1978). This contrasts with mice and puppies which develop a severe wasting and immunodeficiency syndrome following early thymectomy (Roth, 1967).

Thymic aplasia and hairlessness has been reported in Birman kittens. Histological examination of 2-day-old kittens revealed reduced germinal centers and paracortical depletion of lymphocytes in the lymph nodes, spleen, and Peyer's patches. Thymic tissue could not be identified. Pedigree analysis of five litters containing both normal and affected kittens suggested an autosomal recessive mode of inheritance. Although the clinical and histologic appearance was similar to congenital hypotrichosis with thymic aplasia in other species, further immunological studies have not been performed (Casal *et al.*, 1993).

In contrast to surgical thymectomy, neonatal infections with lymphotropic viruses can lead to premature thymic atrophy and a wasting syndrome associated with immune deficiency. The immune deficiency occurs because lymphocytes in the secondary lymphoid tissue, as well as those in the thymus, are depleted. The most common cause of thymic atrophy in kittens is infection with feline leukemia virus (FeLV), and affected kittens have increased survival of skin allografts, indicating defective T-cell immunity (Perryman *et al.*, 1972). Feline panleukopenia virus (FPV) also induces severe thymic atrophy along with lymphocyte depletion of the spleen, lymph nodes, and Peyer's patches, and destruction of bone marrow myeloid cells, leading to marked panleukopenia (Rohovsky and Griesemer, 1967; Larson *et al.*, 1976). FPV causes a lytic infection of rapidly dividing cells. Thus, infection of feline fetuses with FPV at 35 days gestation, during early thymic development, results in immune dysfunction, as demonstrated by delayed skin allograft rejection, while infection at 45 or 55 days of gestation, after the thymus is well developed, has no effect. Postnatally infected cats of various ages had no alteration in graft rejection times or humoral immune responses and only transient depression in lymphocyte responses to the T-cell mitogens Con A and PHA. Thus, the immunosuppression of FPV is due to transient panleukopenia and depressed T-cell responses, and is not as profound as that induced by FeLV infection (Schultz *et al.*, 1976).

Although found in secondary lymphoid tissue prior to birth, there is evidence that lymphocytes in the neonatal cat continue to mature both phenotypically and functionally for a period of time after birth. Analysis of lymphocyte subsets in perinatal kittens demonstrated a steady decrease in null lymphocytes, with a concomitant increase in T and B lymphocytes from 56 days of gestation through to 8 weeks old (Sellon *et al.*, 1996). In addition, the $CD4^+:CD8^+$ ratios of kittens are much higher than in adult cats due to high numbers of $CD4^+$ cells and very low numbers of $CD8^+$ cells. As kittens mature from birth to nearly 1 year old their $CD4^+$:$CD8^+$ ratios slowly decline as $CD8^+$ cells increase (English *et al.*, 1994; Sellon *et al.*, 1996). This gradual increase in $CD8^+$ cell numbers is most likely a reflection of a $CD8^+$ response to increasing antigen exposure.

It is generally believed that most mammals are able to mount an immune response at the time of birth but that the magnitude of the response is much less than that of an adult. In support of this, IL-2 production in response to Con A stimulation is much lower in kittens less than 10 weeks of age compared to adults (M. B. Tompkins, unpublished data). This suggests that, although neonatal kittens have high numbers of $CD4^+$ cells, these cells are not able to function at full capacity. Resistance to FeLV infection is also a reflection of functional maturation of the immune response. One-hundred per cent of neonates and 85% of weanling kittens become persistently infected with FeLV upon challenge, while only 15% of cats older than 4 months develop a persistent infection (Hoover *et al.*, 1976). This age-related resistance has been shown to be associated with maturation of macrophage function. Macrophages from neonatal kittens are highly permissive for FeLV infection and replication, while macrophages from adult cats are not (Hoover *et al.*, 1981).

Immunological tolerance may be induced in neonatal kittens up to 25 days old. Unresponsiveness may be induced following exposure to antigens administered either orally or parenterally. As in other species, tolerance is more likely to develop to simple soluble antigens than to complex antigens (Gorham *et al.*, 1971)

The neonatal kitten depends primarily upon maternal antibody for protection from infectious diseases. Newborn kittens obtain nearly all of their maternal antibody from colostrum, although a very small amount of IgG also may be transferred across the placenta. The colostral phase of lactation lasts approximately 5 days, during which time the concentration of IgG and IgA in milk is several-fold greater than that in the dam's serum (Yamada *et al.*, 1991) The neonate is able to absorb the antibodies most efficiently during the first 16 h of life (Casal *et al.*, 1994), although this period may be extended somewhat if feeding is delayed. The transfer of colostral antibodies, especially IgG, is so efficient that immunoglobulin levels in the neonate may exceed those of the dam shortly after birth (Yamada *et al.*, 1991).

The half-life of maternally derived immunoglobulin is variable and depends, in part, on its isotype. As in other species, the half-life of transferred IgG (4.14 ± 1.29 days), IgA (2.03 ± 0.33 days), and IgM (2.2 ± 1.2 days) is shorter than that of endogenously synthesized immunoglobulin. Transferred maternal antibody wanes to a nadir at 4–5 weeks old, leading to a period of increased susceptibility to pathogens before endogenous antibody production reaches adult levels at about 12 weeks old (Yamada *et al.*, 1991). While passive immunity is vital to protection from infection of the newborn kitten, immune responses to both live and inactivated antigens are suppressed by high levels of maternal antibodies. Interference with oronasal (herpes virus, calicivirus) and parenteral (herpesvirus, panleukopenia virus) vaccination is a well-documented phenomenon of cats that nurse immune queens (Pedersen, 1987).

Failure of passive transfer occurs in kittens that fail to nurse vigorously in the first 24 h after birth and often leads to death from bacterial sepsis during the first and second weeks of life. Hypoimmunoglobulinemia in these kittens may be corrected by the parenteral administration of 3–5 cm^3 of serum from a healthy, immunized cat, preferably one from the same environment in which the kitten is reared.

In addition to systemic passive transfer, maternal antibodies in the milk (lactogenic immunity) provide nursing kittens with continuous local protection against pathogens in the oral cavity and gastrointestinal tract during the first weeks of life prior to the development of local mucosal immunity (Pedersen, 1987).

10. Nonspecific Immunity in the Cat

Introduction

In mammalian species, the immune system can be categorized into two distinct but overlapping components. The 'innate' or nonspecific immune system consists of natural anatomical barriers, macrophages, natural killer cells, neutrophils, and eosinophils. The nonspecific immune system is the first to make contact with pathogens and is characterized by an immediate response, lack of antigen specificity, and lack of immunological memory. The cells of the innate immune system interact directly with pathogens and subsequently secrete cytokines which enhance their microbicidal activity and serve as cofactors for initiating the second component of the immune response, the acquired antigen-specific, T-cell immune response.

The specific interactions which provide innate immunity in the cat have not been studied extensively. Innate immunity in the cat is presumed to be similar to that in other species. This brief review focuses on the individual components of the feline innate immune system.

Natural barriers

The natural barriers, such as the skin or mucosal surfaces which line the respiratory or gastrointestinal tract, form the first line of defense against invading pathogens. Specific information which evaluates nonspecific immunity in the context of natural barriers in the cat is currently not available. It is assumed that the components of these are similar to other mammals. The anatomical location of mucosa-associated lymphoid tissue (tonsils, Peyer's patches) in the cat has been described previously (Pedersen, 1987; Tompkins, 1993).

Neutrophils

The role of the feline neutrophil in nonspecific immunity is the phagocytosis and killing of microbes. In other species, the neutrophils have been shown to secrete cytokines (Lloyd and Oppenheim, 1992), but this has not been investigated in the cat. Feline peripheral blood polymorphonuclear cells can be separated from mononuclear cells by utilizing a double-density gradient technique (Toth *et al.*, 1992). Feline neutrophils are morphologically similar to those in other species. The morphometry of granule genesis which accompanies feline neutrophil maturation has been studied (Fittschen, 1988a). The complex carbohydrate staining of feline neutrophil primary and secondary granules resembles that in humans and rabbits; however, cats lack tertiary granules analogous to those found in other species (Fittschen, 1988b). Cat neutrophils also have a third (late-forming) type of granule which has not been described in other species (Fittschen *et al.*, 1988b). Recombinant canine granulocyte colony-stimulating factor (rcG-CSF) and glucocorticoids have been shown to cause significant elevation in peripheral neutrophil counts in cats (Obradovich *et al.*, 1993; Duncan *et al.*, 1994).

Studies of feline neutrophil function have evaluated chemotaxis, phagocytosis, and oxidative burst activity during disease states. Feline neutrophils, similar to equine, porcine, bovine, and canine neutrophils, do not respond chemotactically to *N*-formyl-methionylleucyl-phenylalanine (FMLP) (Gray *et al.*, 1986). Neutrophil and endothelial cell interactions involving integrins and selectins have been well studied in the cat and serve as a model for neutrophil-induced endothelial damage in myocardial ischemia (Ma *et al.*, 1991; Lefer and Ma, 1994; Murohara *et al.*, 1994). Supernatant from cultured peritoneal exudate cells (PEC) from cats with effusive feline infectious peritonitis (FIP) was chemotactic for peripheral blood neutrophils from healthy cats (Tsuji *et al.*, 1989). Peripheral blood neutrophils collected from FIP-infected cats also showed reduced chemotaxis to zymosan-activated serum while showing similar chemotactic responses to control neutrophils when exposed to PEC supernatants (Tsuji *et al.*, 1989). Neutrophils from FeLV-viremic clinically affected cats had significantly lower chemotactic responses than did those from subclinically affected FeLV-viremic cats (Kiehl *et al.*, 1987). The chemiluminescence response as an indicator of oxidative burst activity in feline neutrophils has been found deficient in cats infected with feline leukemia virus (FeLV) and feline immunodeficiency virus (Lafrado *et al.*, 1987; Hanlon *et al.*, 1993).

Eosinophils

The role of the eosinophil in the cat is thought to be anthelmintic immunity and allergy. Specific information on characterization of feline eosinophil granule components, surface receptor expression, chemotactic stimuli, and effector functions is not available.

A technique for isolation of feline eosinophils via peritoneal lavage has been previously described (Moriello *et al.*, 1993). In the lung, eosinophils have been found to be a significant component of the cell population obtained by bronchoalveolar lavage in specific pathogen free and conventional cats (Hawkins and DeNicola, 1989; McCarthy and Quin, 1989). Human recombinant interleukin-2 (rHuIL-2) induces a peripheral eosinophilia in cats secondary to an enhanced maturation response in bone marrow precursor cells (Tompkins *et al.*, 1990a). In addition to this maturation signal, rHuIL-2 induces a potent activation signal for eosinophils, as measured by a decrease in density and an increase in longevity in culture (Tompkins *et al.*, 1990).

Globule leukocyte

Although not a polymorphonuclear cell, the globule leukocyte is mentioned here because of its morphologic similarity to the eosinophil. The globule leukocyte is slightly larger than an eosinophil and has eosinophilic cytoplasmic granules on hematoxylin and eosin staining. Unlike eosinophils, the globule leukocyte has a single, excentric nucleus and not a bilobate nucleus as seen in feline eosinophils. Globule leukocytes are found in the mucosa of the intestinal, respiratory, and urogenital systems of many species including the cat. Although their function is poorly understood, their mucosal location indicates a possible role in mucosal innate immunity. The origin of the globule leukocyte is also a matter of debate, with current opinions focusing on mast cell or lymphocytic lineage. A neoplasm of globule leukocytes in the cat has been reported (Honor *et al.*, 1986).

Natural killer cells

Classically, the NK cell's role in nonspecific immunity is the non-MHC restricted recognition and removal of viral-infected and neoplastic transformed cells. These functions have been described in the cat. Morphologically, natural killer cells, along with lymphokine-activated killer (LAK) cells and cytotoxic T lymphocytes (CTL), have large, azurophilic cytoplasmic granules and are collectively known as 'large granular lymphocytes'. Phenotypically, NK cells express CD16, CD56, and CD57 but do not express CD3, CD4, or the T-cell receptor. A unique subpopulation of NK cells that express CD8 and exhibit non-MHC-restricted cytotoxic activity has been reported in cats (Zhao *et al.*, 1995). A method of induction of cytotoxic large granular lymphocytes from peripheral blood mononuclear cells has been described in cats by Tompkins *et al.* (1989b).

Most studies on NK cell function in cats have concentrated on NK cell cytotoxicity to sensitive tumor cell lines or virally infected targets. For example, a reduction of NK cell-mediated cytotoxicity against baby hamster kidney cells was demonstrated in cats infected with feline immunodeficiency virus (FIV) (Hanlon *et al.*, 1993). Other studies have shown that NK cells from FIV-infected cats are still able to bind to target cells but have reduced ability to kill them, and the defect could not be overcome by *in vitro* treatment of effector cells with IL-2 (Zaccaro *et al.*, 1995). A population of $CD57^+CD8^+$ LAK cells obtained from FIV-infected cats had cytotoxic activity against FIV-infected target cells which exceeded the cytotoxic activity of LAK cells derived from non FIV-infected cats (Zhao, 1995). The same population of cells also exhibited cytotoxic activity against FeLV-infected cells, indicating the cytotoxic activity was not antigen specific.

The mononuclear phagocytic system

The mononuclear phagocytic system of cats is similar to other species and consists of blood monocytes and tissue or resident macrophages. The resident macrophages are the first to contact pathogens and, through phagocytosis and cytokine production, regulate activation of NK cells (IL-12) and recruitment of neutrophils (IL-1, IL-8). Activation of NK cells results in NK cell-derived IFN-γ, which activates the macrophage, resulting in augmented effector functions. In secondary lymphoid organs, the macrophage, via upregulation of MHC II molecules and elaboration of cytokines, presents antigen to T cells, thus completing the bridge between nonspecific and antigen-specific immunity.

Methods have been published for the isolation of feline macrophages from the lung (BAL), peritoneum, and bone marrow (Stoddart and Scott, 1988; Hawkins and DeNicola, 1989; Daniel *et al.*, 1993). Pulmonary intravascular macrophages have been demonstrated in cats and removal of blood-borne pathogens/particulates is performed by pulmonary intravascular macrophages and not by spleen or liver macrophages as in humans, mice, and dogs (Winkler, 1988).

Feline macrophages have also been observed as secretory cells. Bioactive TNF has been detected from feline alveolar macrophages stimulated with LPS (J. W. Ritchey, unpublished data) and thioglycolate-elicited peritoneal macrophages (Lin and Bowman, 1993). TNF mRNA and IL-1 mRNA and protein have been demonstrated in bone marrow-derived macrophages stimulated with LPS (Daniel *et al.*, 1993). Constitutive cultures of feline alveolar macrophages express mRNA for TNF, IL-6, IL-10, and IL-12, but do not have detectable levels of IL6 or TNF as measured by bioassay (J. W. Ritchey, unpublished data). In addition to TNF, feline alveolar macrophages stimulated with LPS produce bioactive IL-6 (J. W. Ritchey, unpublished data). Macrophage function has also been studied in disease states. Alveolar macrophages from FIV-infected cats constitutively produce TNF and IL-6 as early as 4 weeks post infection (J. W. Ritchey, unpublished data). Peritoneal macrophages from cats infected with FIV had decreased IL-1 secretion, although the macrophages had enhanced antimicrobial activity compared to controls (Lin and Bowman, 1992). This may be related to macrophage activation by *in vivo* conditioning from IFNγ, which our laboratory has demonstrated to be upregulated in the lymph nodes of FIV-infected cats (J. Levie, unpublished data). Differences also exist between different populations of resident macrophages because feline peritoneal macrophages had higher microbicidal activity and released more IL-1 in response to LPS stimulation than feline alveolar macrophages (Lin and Bowman, 1991). Lastly, the intrinsic resistance to infection and replication of coronaviruses in feline peritoneal macrophages correlates with *in vivo* virulence and the development of feline infectious peritonitis (Stoddart and Scott, 1989).

Conclusions

Although much remains to be examined of the feline nonspecific immune system, what is known suggests that this system in the cat is very similar to other mammalian species. With the continued development of reagents for use in feline immunological research and the continued application of the cat as models for disease (AIDS, myocardial ischemia), the future holds promise for further, indepth examination of feline nonspecific defense mechanisms.

11. The Complement System

The first documented quantitation of cat complement activity appeared in 1938 (Dingle *et al.*, 1938). Apparently typical classical and alternative complement pathways have been identified in cat serum (Grant, 1977; Grant *et al.*, 1979; Day *et al.*, 1980; Kobilinsky *et al.*, 1980; Grant and Michalek, 1981; Goddard *et al.*, 1987; Hosoi *et al.*, 1990; Fevereiro *et al.*, 1993; Buerke *et al.*, 1995). Cat C1 has been partially purified (Olsen *et al.*, 1974), and cat C3 was purified and identified as a dimer of two polypeptide chains of 128 and 71 kDa (Jacobse-Geels *et al.*, 1980). Cat complement (C′) components appear to be interchangeable with those of other species, e.g. cat C1 will initiate the classical pathway when subsequent components are derived from guinea-pig (Olsen *et al.*, 1974); guinea-pig C2 will activate purified cat C3, and the resulting C5 convertase will initiate formation of the membrane attack complex with purified human C′ components C5 through

C9 (J. M. McCarty and C. K. Grant, unpublished data). As in other species, spontaneous activation of feline C1 is prevented by C1 esterase inhibitor C1 INH (Buerke *et al.*, 1995), and cat C3 is inhibited by cobra venom factor (CVF). Like humans, mice, rabbits and cows, the cat C2 gene is very closely spaced to the related Factor B gene, and the upstream sequence element of all five mammalian species is highly conserved (Moreira *et al.*, 1995).

Human serum C′ has been proven to inactivate some retroviruses by an antibody independent mechanism whereby C1 binds directly to a receptor on the transmembrane region, and subsequent virolysis occurs via the classical pathway (Bartholomew *et al.*, 1978). Cat serum possesses the same property and will effectively inactivate human immunodeficiency virus type-1 (HIV-1) (Hosoi *et al.*, 1990). Cat retroviruses – feline leukemia virus (FeLV) or feline immunodeficiency virus (FIV) – do not appear to be efficiently inactivated by cat C′ alone. In the presence of cat antibodies to retrovirus envelope proteins, however, virolysis is mediated by cat C′ via the classical activation pathway.

Cat C′ appears to be relatively weak when compared with other species in hemolytic or C′ fixation tests employing rabbit antisheep red blood cell (RBC) serum and sheep RBC targets. If bovine or human RBC are substituted (with the appropriate rabbit antibovine RBC or rabbit antihuman RBC immune serum), then cat serum exhibits average or strong hemolytic functions respectively (Dingle *et al.*, 1938; Grant, 1977).

To summarize, Cat C′ factors appear to be similar to the C′ components of other mammalian species, and *in vitro* interchangeability with human or guinea-pig components has been demonstrated. As such, valid inferences can probably be drawn from studies of C′ metabolism in health or disease of other species; this is important because a paucity of information exists on the interactions of feline C′ and pathogenesis. Complement level fluctuations, and in particular hypocomplementemia, have been regularly observed in FeLV infections, FeLV-related tumors and nonregenerative anemias, but C′ levels and C′ consumption are factors in the pathogenesis of other viral infections such as feline infectious peritonitis (FIP), feline herpesvirus type-1 (FHV), and probably FIV and parasitic infections. Complement-dependent antibodies lyse viruses and other pathogenic organisms, but they are less effective if C′ levels are severely reduced or not available, hence further studies of cat C′ levels during the courses of pathogenesis are required.

Table VIII.11.1 Roles for cat complement in feline immune responses to infectious disease agents: summary of citations

Infectious agent	*C′-related observations*	*References*
FeLV	(a) C′ depletion by cobra venom factor did not promote viremia in early stage experimental infection, but the CVF was administered before cytotoxic anti-FeLV gp70 antibodies might appear.	Kraut *et al.* (1985); Johnson *et al.* (1988)
	(b) Classical pathway of the cat C′ system is activated *in vitro* by incubation of FeLV with normal cat serum.	Kraut *et al.* (1987)
	(c) C′ levels in naturally infected, healthy, cats are relatively stable. Hypocomplementemia observed in 55% cats with leukemias or lymphomas and 45% of cats with nonregenerative anemias.	Grant *et al.* (1979)
	(d) Hypocomplementemia, and increased circulating immune complexes (CIC), noted in infected cats, and cats bearing virus-related tumors.	Day *et al.* (1980)
	(e) Virus infected erythroid, and granulocyte-macrophage bone-marrow progenitors, lysed by anti-gp70 antibodies and C′.	Kobilinsky *et al.* (1980)
	(f) Virus-producing cat leukemias and lymphomas are lysed by cat anti-gp70 antibodies and cat C′.	Grant *et al.* (1977, 1979); Grant and Michalek (1981)
	(g) Deposition of IgM and C3 detected in the kidneys of the majority of naturally infected cats. Kidney functions were severely disrupted in 40% of these cats.	Olsen *et al.* (1974)
FIV	No evidence found for Ig and C′ mediated lysis of bone marrow white cell progenitors in FIV infections (cf. Kobilinsky *et al.*, 1980)	Linenberger and Abkowitz (1992)
FIP	(a) Hypocomplementemia detected in 45% of natural infected symptomatic cases.	Grant *et al.* (1979)
	(b) Experimental infection resulted in CIC formation, initial C′ concentration increases, then hypocomplementemia and death.	Horimoto *et al.* (1989); Hosoi *et al.* (1990)
FHV	(a) Cat anti-FHV antibodies plus C′ lyse FHV-infected cells via the alternative pathway of C′ activation.	Goddard *et al.* (1987)
	(b) Neutralization of FHV increased by addition of C′.	Jacobse-Geels *et al.* (1982)
Brugia pahangia	(a) C′ mediates cat granulocytes to adhere to and then kill microfilariae.	Jacobse-Geels *et al.* (1980)

12. Ontogeny of the Immune System

The cat has an endotheliochorial type of placenta and receives its passively acquired immunoglobulin both pre- and post-natally. The transmission of immunoglobulin across the small intestine is quantitatively the most important (Harding *et al.*, 1961), and whilst in pre-suckled kitten serum IgG can be quantified, little or no IgA or IgM can be detected (Schultz *et al.*, 1974). To date, there have been relatively few age-related studies to the development of cat serum immunoglobulins (Barlough *et al.*, 1981). Early studies would indicate that by 50 days of gestation there are significant serum immunoglobulin levels (Okoshi *et al.*, 1968), which are comparable to that found in presuckled neonates. Following birth serum IgG levels fall over the first month of life, presumably reflecting the half-life of the passively transferred immunoglobulin. Subsequently, they rise but adult levels are not achieved until more than 1 year old (Yamada *et al.*, 1991).

B lymphocyte system

To date only one detailed study of the ontogeny of the feline B lymphocyte system has been reported (Klotz *et al.*, 1985). Lymphocytes first appear in fetal circulation at about 25 days of gestation. Studies in other species have shown that pre-B cells (cytoplasmic μ^+, surface Ig^-) are first detectable in fetal liver and later in bone marrow, where they are generated throughout life. Pre-B cells have been detected in the fetal liver of cats of 42 days of domestic gestation, but earlier fetuses have not been examined. Surface IgM^+ B cells were also detectable in 42-day-old fetal liver, and it would appear that, as in other species, feline liver and bone marrow are the sites of origin and differentiation of cells of the B-lineage. Following birth the frequency of splenic B cells increases rapidly, with no further increase observed between 12 weeks old and adulthood. The majority of splenic, blood and bone marrow B-cells express surface IgM, with a smaller number expressing surface IgG. In cats, plasma cells secreting IgM are most frequent in the bone marrow from 1-week-old animals, but as animals mature the frequency of IgG plasma cells increases, reflecting isotype switching and clonal expansion in response to environmental antigens. In contrast to humans, rodents and rabbits, feline B cells preferentially express λ over κ light chains ($\approx 4:1$), but this would appear to be independent of age and environmental exposure (Klotz *et al.*, 1985).

T lymphocyte system

Following birth the absolute numbers of lymphocytes increase until adult numbers are achieved by 13–15 weeks old. A recently published study (Sellon *et al.*, 1996) has extended these findings and shown that during the last few weeks of gestation there were significant changes in the numbers and proportion of lymphocyte subsets. The most dramatic was a threefold increase in the proportion of cells staining for a pan-T cell marker, and it has been speculated that this may reflect hormonal influences associated with parturition. During the first 4 weeks after birth there is a dramatic increase in the number of Ig^+ B cells, probably reflecting a response to environmental antigens. Interestingly, at this time there is also an increase in the absolute number and proportion of null cells (sIg^-, pan T^-). Kittens are born with greater CD4 : CD8 ratios ($\approx 3.5:1$) than adults ($\approx 1.5:1$). Changes in the ratio towards an adult phenotype would appear to be primarily due to an increase in $CD8^+$ T cells, but this may not be reached until about 1 year old, and this would appear to be sensitive to antigenic exposure.

As a consequence of the paucity of data upon the feline immune system it is not beneficial to compare and contrast features of the ontogeny of its immune system with that found in other species. It is sufficient to say that it would not appear to be significantly different from other species with an endotheliochorial placenta.

13. Mucosal Immunity

Introduction

Compared with rodents and man and even some other domestic species such as the pig and cow, the feline mucosal immune system has received scant attention. The current interest in the cat as an experimental model for *Helicobacter pylori* infection, feline immunodeficiency virus (FIV) mucosal vaccine studies and small bowel transplantation, would suggest that this neglect may be rapidly rectified.

The gastrointestinal immune system

The gastrointestinal tract is one of the largest immunological organs of the body, containing more lymphocytes and plasma cells than the spleen, bone marrow and lymph nodes combined. The gut-associated lymphoid tissue comprises cells organized within the lymphoid follicles of the Peyer's patches as well as those distributed throughout the lamina propria and intestinal epithelium.

Secretory immunoglobulins

As with other mammals, IgA in the cat is the predominant immunoglobulin in mucosal secretions (Vaerman *et al.*, 1969). It is found in large amounts in saliva, tears, respiratory and intestinal secretions, milk and bile

(Schultz *et al.*, 1974). In serum cat IgA exists mainly as a dimer with only trace amounts in the monomeric form. In the small intestinal lamina propria IgA-producing cells predominate, accounting for between 40% and 80% of the total plasma cell pool. In contrast, IgG-positive cells (51%) are more frequent in colonic tissues, with smaller numbers of IgA- and IgM-producing cells (Klotz *et al.*, 1985).

The cells of the gastrointestinal immune system

The general anatomy and morphology of the feline intestinal tract have not been comprehensively described. The intestinal lamina propria is heavily populated with 'immune cells' and, in the colon, it has been shown that greater than 90% of the cells stain with an anti-fCD45 monoclonal antibody (Sturgess, 1997). In the majority of species studied to date there is a preponderance of $CD4^+$ over $CD8^+$ T cells in the lamina propria. In the cat small intestine this distribution is less marked and approximately equal numbers of fCD4 and fCD8 cells have been enumerated by both immunohistology on tissue sections and flow cytometry of cells isolated by collagenase digestion (Durgut, 1996). In the feline colonic lamina propria the fCD4:fCD8 ratio is always less than 1 (Sturgess, unpublished data). Furthermore it may be of significance that whereas in the colon the sum of the $fCD4^+$ and $fCD8^+$ cells approximate to the number of cells expressing the fCD5 marker, in the small intestine there is a greater number of $fCD4^+$ and $fCD8^+$ cells.

Intraepithelial cells (IELs) may be enumerated on a morphological basis by identifying those cells which are located towards the lumen from the basement membrane. As in other species IELs are more numerous in the feline small intestine ($\approx$23 IELs/100 epithelial cells, EC) than colon ($\approx$4 IELs/100 EC). At both sites the majority of feline IELs are $fCD8^+$, $fCD5^+$. The intestinal epithelium of rodents, man and guinea pigs express MHC class II antigens, whereas that of pigs, sheep and cattle do not. In species expressing epithelial class II it is restricted to the small intestine, the large bowel epithelium being negative or only very weakly positive in healthy animals (Bland, 1988). The tissue distribution is restricted to the fully differentiated absorptive enterocytes on the upper two-thirds of the villus. In the cat, the colonic epithelial cells are uniformly negative for MHC class II antigens, while a small proportion of the small intestinal epithelial cells contain MHC class II antigens intracellularly, within vesicular like structures in the supranuclear region. The biological role of epithelial class II antigens is unclear but it has been shown in species with constitutive expression of class II that it can be further enhanced in response to intestinal inflammation. Under these conditions not only is there increased expression upon mature small intestinal enterocytes, but expression may also be induced upon immature crypt cells and, in inflammatory bowel disease, upon colonic epithelium. No results of similar studies in the cat have been reported.

Functional studies of the gut immune system

It is generally accepted that the Peyer's patches are the major site of induction of mucosal responses. In contrast, it is suggested that the lamina propria is essentially an effector organ involved in surveillance and in the provision of help during the rapid responses to recall antigens. Such mechanisms may include both active protective responses against potential pathogens as well as the prevention of damaging allergic responses to dietary and environmental antigens. Difficulties associated with the isolation of viable cells from the lamina propria and Peyer's patches have delayed functional analysis of the cellular and molecular basis of these mechanisms. However, it has been shown that whereas Peyer's patch cells produce relatively more IL-2 and IFN-γ, lamina propria cells synthesized relatively more IL-4, IL-5 and IL-6. Recent studies in cats have focused upon reverse transcription-polymerase chain reaction (RT-PCR) analysis of cells isolated from the colon and have detected mRNA encoding for a number of cytokines. While the detailed analysis of local cytokine production and receptor expression in this species has yet to be completed, it is noteworthy that in asymptomatic feline immunodeficiency virus (FIV)-infected cats there was enhanced expression of IL-2, IL-6 and IL-10 (Sturgess, unpublished observations).

Infection via the intestinal tract

Helicobacter pylori infection has recently been described in domestic cats and in view of the possible zoonotic implications this finding has promoted considerable attention. *Helicobacter pylori* infection is associated with a lymphofollicular gastritis, consisting of IgM^+ B cells assembled into multiple lymphoid follicles surrounded by clusters of $fCD4^+$ and $fCD8^+$ T cells (Fox *et al.*, 1996).

Although biting probably provides the major route of transmission of FIV it has been shown that it is possible to infect cats across the intact vaginal and rectal surfaces. Moreover, following infection it has been possible to detect proviral DNA. Replicating virus and viral proteins can be detected in the colonic follicle associated epithelial cells and occasional cells in the lamina propria (Bishop *et al.*, 1996). Cats infected with FIV via the rectal route remain asymptomatic for more than 1 year and this is associated with an increase in the number of colonic lamina propria $fCD8^+$ T cells and increased expression of IL-2, IL-6 and IL-10 mRNA. These results might suggest that lamina propria cytotoxic cells (CTLs) may play a role in the control of infection during the pre-AIDS period but this remains to be determined.

14. Immunological Diseases

Introduction

Diseases of an immunological nature can be classified into six categories: (1) disorders of immediate hypersensitivity or allergies, (2) diseases resulting from the reaction of autoantibodies and alloantibodies, (3) conditions resulting from the deposition of immune complexes, (4) diseases mediated by cellular immunity, (5) gammopathies, and (6) acquired and congenital immunodeficiencies. Each of these categories of immune disease is a result of normal immune reactions that have been subverted or perturbed in some manner. This section deals with the first five categories of immune diseases. Immunodeficiencies, because of their great current interest, will be covered in a separate chapter.

The types of immune diseases that occur in cats parallel those of dogs, humans, and other species of animals. As in humans and dogs, a common group of cofactors link all immunological diseases of cats, regardless of category. These cofactors include: (1) genetic predisposition, (2) drug therapy for other disorders, (3) infectious diseases, particularly those of a persistent nature, (4) cancer, (5) diet, (6) age, particularly the very young and old, and (7) gender, with sexually intact females suffering more than intact males and with male/female castrates having similar and intermediate disease incidence.

Mechanisms of allergy

Allergy is defined as a disease or reaction caused by an immune response to one or more environmental antigens, resulting in tissue inflammation and organ dysfunction. Allergic diseases are mediated by the IgE system. This system is an integral part of the skin and mucosal defenses to parasite invasion and migration. Parasites shed surface proteins into the surrounding tissues during invasion and migration, especially at key stages in their growth, such as molting. These antigenic substances stimulate the production of specific IgE antibodies by plasma cells in local diffuse lymphoid aggregates. Once produced, IgE antibodies circulate only briefly before becoming firmly bound to the surface of tissue mast cells and basophils. Basophils and mast cells that are coated with specific IgE antibodies are said to be 'sensitized'. Sensitized basophils and mast cells are potent deterrents to further parasite migration or invasion. Antigens released by the parasites during invasion or migration will bind to specifically sensitized mast cells in the immediate area and, if sufficient amounts of specific IgE are complexed, the mast cells (or basophils in the case of blood-borne parasites) will release their granules into the surrounding milieu. These granules contain factors such as prostaglandins, leukotrienes, histamine, eosinophil chemotactic factor, slow-reacting substance of anaphylaxis, and platelet-activating factor. The resultant reaction is characterized by increased vascular permeability at the site, local edema, fibrin meshing, smooth muscle contraction and the influx of large numbers of eosinophils. This inflammatory reaction slows the migration of the parasites and provides a milieu in which the immune attack can be launched. The primary effector cell is the eosinophil, which, when activated, has the ability to contact directly and kill parasites through potent oxidants and cationic substances present in its granules.

Allergic diseases involve the same mechanisms as parasite immunity, except that IgE antibodies are inappropriate, i.e. usually nonparasitic, antigens. These inappropriate antigens are referred to as 'allergens'. Allergens are most likely to be taken into the body by the same routes as parasites – through oral and respiratory mucous membranes and the skin.

Specific allergic diseases

Systemic allergies

Systemic allergies result from the entry of allergens into the bloodstream, either via injection or absorption across the mucous membranes. The most frequent cause of systemic allergic reactions in cats are vaccinations, followed by injectable medications. Oral medications, certain ingested foods and insect bites are less common offenders. The mildest form of systemic allergy is urticaria. Urticaria, or hives, are small, raised, circular areas of edema, hyperemia and pruritus that appear on the skin within minutes of systemic allergenic exposure. Angioneurotic edema (facial-conjunctival edema) is the next most severe form of systemic allergy. This reaction is characterized by the rapid appearance of edema around the eyes, face and lips, and mild signs of anaphylactic shock. Anaphylactic shock is the most severe form of systemic allergy. The target organ for anaphylactic shock in the cat is the portal vasculature; affected cats rapidly exhibit nausea, incoordination, pallor and discoloration of the mucous membranes of the mouth, rapid thready pulse, vomiting or diarrhea and, in severe cases, death. The cause of death is shock brought about by the rapid pooling of venous blood in the intestines, spleen and liver.

Allergies of the skin

Nonseasonal allergic reactions of the skin of cats are usually associated with dietary allergens. Two types of lesions are seen. The first consists of highly pruritic plaque-like lesions about the head and neck, which are frequently subject to self-mutilation by scratching and further complicated by secondary bacterial infection. Miliary dermatitis is the second and more common manifestation of skin allergies of dietary origin. It is characterized by numerous small scabs, usually along the dorsum, that are sloughed off with small tufts of hair.

Seasonal skin allergies, most often from pollens, are uncommon in cats. They are usually characterized by pruritus, erythema and hair loss on the ears and about the eyes and face. Allergies to topical ocular and otic medications are common in young cats being treated for infections of the ears or eyes such as ear mites or chlamydiosis. This phenomenon is akin to 'allergic breakthrough' in humans, which is often associated with common infections. The infections appear to induce a dysregulation of normal T-cell regulation and a heightened IgE response to potential allergens. The resulting allergic otitis or conjunctivitis can closely resemble the infectious diseases that initiated the treatment, making it difficult to determine when the infection ends and the allergy starts. Once medication is stopped, however, the otitis or conjunctivitis rapidly resolve.

Flea allergic dermatitis is much less common in cats than in dogs, probably because the cat is the natural host for the common flea. The cat, therefore, has evolved a more benign host–parasite relationship than the dog. Flea allergic dermatitis, when it occurs in cats, is often a mixture of immediate (IgE-mediated), intermediate (immune complex or Arthus-mediated) and delayed (T-cell mediated) hypersensitivities.

Eosinophilic granuloma complex

Eosinophilic granulomas occur as ulcerative lesions on the margins of the upper lips (rodent ulcers), as tumor-like proliferations on the dorsum of the tongue or on the hard palate, or linear encrustations on the backs of the legs and paws. The central lesion is a peculiar necrosis of underlying collagen; bundles of necrotic collagen are surrounded by dense infiltrates of closely adherent eosinophils, reminiscent of the IgE-mediated attack on parasites. Although eosinophils are prominent in the lesions, the role of allergy, if any, is uncertain. The author has found no response to strict dietary control and to confinement away from common biting insects. The disease has a strong genetic component and, like many other immune diseases, intact females have the most severe lesions and intact males the least severe. Ovariohysterectomy and castration tends to equalize the incidence and severity at an intermediate level. It also responds to glucocorticoid therapy.

Allergies of the respiratory tract

Allergic conjunctivitis, associated with mild to moderate reddening of the conjunctiva and excessive tearing, has been observed as a seasonal condition in cats. The probable cause is pollen. A chronic mild to moderately severe conjunctivitis has been linked to continuous household exposure to cosmetics and cigarette smoke. A chronic hyperplastic conjunctivitis, characterized by intense inflammation and granular hyperplasia of the conjunctival membranes, is an uncommon condition of cats. Allergic rhinitis, a common condition in humans, is uncommon in cats but has been seen under similar conditions as allergic conjunctivitis. Seasonal allergies of the nasal passages and conjunctiva require only temporary treatment, while chronic allergies are best treated by changes in environment.

Allergic bronchitis, a common disease of dogs, is rarely seen in cats. Allergic bronchiolitis, however, is frequent in cats but uncommon in dogs. The disease is triggered by environmental allergens that are minute enough in size to be inhaled into the smaller airways. Allergic bronchiolitis is characterized by intermittent coughing or retching, mild to moderately severe diffuse peribronchiolar infiltrates on chest radiographs, and eosinophilia in tracheal wash or bronchiolar brushes. Severe cases may be accompanied by weight loss and fatigue. An important differential diagnosis is chronic lung-worm infestation.

Cats are the only animal species that suffer from a true allergic asthma. However, true asthma in cats is often grouped together with the aforementioned chronic allergic bronchiolitis under the term 'feline asthma'. True asthma in cats, as in man, is characterized by the sudden and transient attacks of severe bronchiolar constriction, the production of a thick tenacious mucus, and expiratory dyspnea. Asthmatic attacks in cats occur more often in summer, especially after they have been put outdoors. Status asthmaticus is not unusual in asthmatic cats and death can ensue if untreated. For this reason, owners of asthmatic cats will often keep injectable epinephrine (adrenaline) at hand.

Allergic bowel disease

Allergic bowel disease is extremely common in cats, probably because modern commercial cat foods are alien to cats' ancestral diets. Wild cats are entirely carnivorous, while modern domestic cats are usually fed foods rich in cereals, egg and milk by-products, and soy protein. Meats such as beef and fish, common in commercial cat foods, are also unnatural in wild cat's diets. Allergies of the stomach and upper small intestine are very common in cats and manifested by vomiting of food within 30 min or less of eating. Allergies centered in the jejunum and ileum produce loose stools of normal frequency and volume. Because a cat's stool is often buried in the litter pan, owners will frequently not notice that it is abnormal. The stool may be extremely odorous, however, in which case the owner will be alerted to a problem every time the cat has a bowel movement. Allergic colitis is most frequently accompanied by fresh red blood in the stool and mucus. Frequent, high volume, loose stools are seen in only severe cases of colitis. Eosinophilia is often mild to nonexistent in allergic enteritis and colitis, although eosinophilic infiltrates are common in biopsies of inflamed tissues. In addition to changes in the consistency or odor of the stool, many cats with allergic bowel disease will have hair coats that are dry, thin from excessive shedding, and lackluster. Plaque-like pruritic lesions about the head and neck, or miliary dermatitis, are accompanying features in some cats.

Eosinophilic enteritis is a term used for allergic bowel disease that is accompanied by a significant blood eosinophilia. Eosinophilic enteritis is much more severe than allergic enteritis and is frequently accompanied by significant diarrhea and weight loss. Blood and mucus is observed in the stool when the colon is also involved.

Bowel allergies, regardless of their site, are diagnosed and treated in the same manner, by feeding a hypoallergenic diet composed of a single animal meat, such as lamb, rabbit, or turkey. Severe cases, especially when eosinophilia is present, benefit from added corticosteroid treatment. When the bowel disease has subsided, a satisfactory long-term diet is found by trial and error with introduction of one new food at a time.

Allergic breakthrough leading to allergic enteritis is also frequent in young cats undergoing common intestinal infections. Typically, a group of cats develop an infectious type of enteritis, which clears up in most but becomes chronic in a small proportion. The tendency is to continue to search for an infectious cause in these latter animals, while the true cause is food allergy. The cats will respond well to hypoallergenic diet and after several months can be reintroduced to normal food.

Autoantibody and alloantibody diseases

Diseases involving alloantibodies

Only two major blood types have been recognized in cats, type A and type B (rarely AB); type A is dominant to B. Up to 98% of random outbred cats are blood type A, depending on the area of the world (highest in USA and Europe, and somewhat lower in Australia). Cats with blood type A often have preformed alloantibodies to type-B red cells, as do type-B cats to type-A cells. Alloantibodies against type-A cells are usually at higher titer and much more lytic than antibodies to type-B red cells. As a result, the most severe alloantibody reactions occur when type-A red cells (type-A donor) are exposed to type-A antibodies (type-B recipient), a situation occurring in neonatal isoerythrolysis and mismatched blood transfusions.

Neonatal isoerythrolysis is a common problem in certain pedigree breeds of cats that have inadvertently accumulated a high incidence of the type-B gene. Queens homozygous for type-B red cells, when bred to a homozygous blood type-A tom, will produce mostly type-A kittens. Following nursing, highly lytic type-A alloantibodies are passed in the colostrum, triggering an acute hemolytic episode in the newborns following nursing. Kittens usually die within the first day or two of life, and show signs of depression, hemoglobinuria and pallor.

Blood transfusion reactions are, fortunately, rare in outbred cats, mainly because of the extremely high incidence of blood type A, making it unlikely for randomly selected donors and recipients to have different blood types. The situation may be quite different among certain pedigree breeds.

Diseases involving autoantibodies

Autoantibodies are produced in a number of different situations. Some cats, especially pedigreed animals, have a genetic predisposition towards autoantibody formation. This predisposition, acting with unknown environmental triggers, leads to a breakdown in normal self/nonself recognition. Certain chronic infectious diseases, such as feline immunodeficiency virus (FIV), feline leukemia virus (FeLV) and haemobartonellosis are frequently associated with autoantibody diseases. Drugs administered chronically for other disease conditions, such as propylthiouracil for hyperthyroidism, can cause autoantibodies to be produced. Certain types of neoplasms, especially lymphoid and myeloid tumors, have also been associated with autoantibodies.

Autoimmune hemolytic anemia (AIHA) in cats, which is similar to its human counterpart but unlike the canine disease, is more likely to be secondary than primary (idiopathic). The most common secondary causes are FeLV infection and haemobartonellosis. Lymphoid and myeloid neoplasia, cancer of other types, systemic lupus erythematosus (SLE) and drugs (e.g. propylthiouracil) are less common predisposing conditions.

AIHA in cats is more likely to be chronic than peracute or acute in nature. Except for cats with underlying infections such as *Haemobartonella* or FeLV, most AIHA cases in cats are Coombs' antibody and in-saline agglutination negative. A chronic, nonresponsive, Coombs' antibody-negative anemia suggests that the immune attack is directed against immature stages of red blood cells in the marrow and not against mature cells in the circulation. Coombs'-positive AIHA is more likely to be acute, responsive, and associated with autoantibodies against mature circulating red blood cells. The frequent secondary nature of AIHA in cats makes treatment more difficult and the prognosis worse. Secondary conditions must be searched for and eliminated if possible. Treatment, when indicated, usually involves corticosteroids or a combination of corticosteroids and other immunosuppressive drugs.

Autoimmune thrombocytopenia (AITP) is uncommon in cats, and like AIHA, it is often secondary. The most common predisposing disorders are FeLV or FIV infections. Propylthiouracil, used to treat hyperthyroidism, has also been linked to both AITP and AIHA. Cats are much more resistant to clinical disease when thrombocytopenic than dogs, so many cases may go undiagnosed or are picked up coincidentally during diagnostic work-ups for other complaints. Echymotic and petechial hemorrhages of the skin, epistaxis, melaena, and hematuria have all been associated with AITP in cats. Prognosis and treatment are the same as for AIHA.

Diseases caused by autoantibodies to structures other than blood cell membranes have been described in cats, but are much less common in this species than in dogs. Pemphigus foliaceous is a chronic exfoliating dermatitis that commonly affects the ears, nasal planum, orbital ridges and feet. The condition is caused by autoantibodies against the intracellular cement substance that binds the uppermost layer of noncornified epidermal cells to each other and to the overlying cornified cells. It is usually idiopathic in nature, although antibiotic treatment has apparently triggered one case and purebreds are more likely to be affected than outbred cats. The resultant autoantibody binding causes a subcorneal blister that rapidly ruptures. Pemphigus erythematosus is a more virulent version of pemphigus foliaceous. Lesions are more widespread and severe, with a tendency to involve mucocutaneous junctions. Cats with pemphigus erythematosus are positive for antinuclear antibodies as well as having typical pemphigus-type bullous disease. Pemphigus vulgaris and bullous pemphigoid, two common disorders of dogs, are extremely rare in cats. The former is caused by autoantibodies to the intracellular cement binding the basal cells to their overlying epidermal cells, while the latter is caused by antibodies to the basal lamina binding the entire epidermis to the subcutis.

Myasthenia gravis is caused by autoantibodies against the acetylcholine receptors of skeletal muscle. The condition is very rare in cats, although common in dogs. In one feline case, the disease was associated with cysts in the thymus. Myasthenia gravis in people is often associated with thymoma. The disease is associated with rapid muscle weakness upon exercise with improvement of muscle strength following rest.

Immune complex diseases

Immune complex diseases are caused by the chronic deposition of antigen–antibody–complement complexes within the basement membranes of blood vessels. The requirements for immune complex disease are a chronic source of antigen, a concomitant antibody response to the antigen, complement binding, and intravascular or intrabasement membrane deposition of the antigen–antibody–complement complex. The hallmark lesion of immune complex disease is vasculitis of either acute or chronic nature.

Like other immune diseases, immune complex disorders are either idiopathic or secondary. The most common secondary causes are chronic infectious diseases such as FIV or FeLV infections.

Systemic lupus erythematosus (SLE)

SLE has been referred to as the penultimate immunological disease, both for the high incidence in man and animals and the wide range of immunologic and immunopathologic manifestations. SLE has three immunological components: (1) immune complex disease (vasculitis) involving one or more organ systems, (2) heightened B-cell responses, and (3) autoantibody production, especially to various nuclear proteins (i.e., antinuclear antibody or ANA) and to cell surface antigens. Like its counterparts in dogs and humans, SLE in cats has a strong genetic component, and as with dogs, it is more likely to be seen in pedigreed than outbred animals (in particular Persians, Siamese and related breeds).

Cats with SLE are usually from $1\frac{1}{2}$ to 6 years old at presentation. The most frequent presenting signs are bizarre psychotic behavior (hiding, fear, apprehension), twitching of facial muscles, and intermittant fever/weight loss. Although most cats with SLE have significant polyarthritis, lameness is only observed in one-third of animals. A chronic crusty skin disease, especially on the face and head, is the third most common presenting complaint, and renal disease the fourth. AIHA and AITP are less frequent presenting disorders. Physical and laboratory findings include fever, generalized lymphadenopathy, leukopenia and suppurative synovial fluid taps from tarsal and carpal joints. Conjunctivitis and palatine/glossal ulcers are occasionally observed. The ANA test is usually positive at titers ranging from 1:10 to 1:400, with either a speckled or homogenous pattern. Care must be taken not to over-interpret positive ANAs in normal cats or cats with disease signs incompatible with SLE. Many normal cats have low titers of ANA (less than 1:10) and ANAs at even high titer develop in a number of infectious conditions such as acute FeLV and FIV.

A significant proportion of cats presenting with signs of SLE go on to develop progressive glomerulonephritis and renal failure. Indeed, about 5% of younger pedigree cats that present with signs of chronic renal failure are probably suffering from SLE. This aspect of SLE of cats more closely resembles lupus in humans than in dogs. Cats with SLE should be treated chronically with a combination of corticosteroids and chlorambucil or cyclophosphamide. Remission is achieved once the ANA titer declines to zero. Kidney function should be periodically monitored and the disease should be considered lifelong in most animals.

Discoid lupus erythematosus

This is a common disorder of dogs, but is rare in cats. The lesion is usually localized to the nasal planum and is worsened by exposure to sunlight. Histopathological changes in the affected skin, characterized by liquefaction of the basal cell layer, subdermal inflammation and immune complex deposition in the basal lamina are identical to those seen in SLE. Affected cats are usually ANA-negative, however, and lesions are not present in other organs.

Chronic progressive polyarthritis

Two forms of the disease are recognized, a periarticular osteoproliferative arthritis similar to human Reiter's disease and a chronic erosive polyarthritis similar to human rheumatoid arthritis. The Reiter's form of the disease is the more common and occurs exclusively in younger male cats from $1\frac{1}{2}$ to 4 years old. Affected cats present with acute polyarthritis, high fever, lymphadenopathy, pronounced reluctance to move and pain on joint palpation. Affected joints, especially distal limb joints, are often swollen and reddened. The high fever lessens somewhat after several weeks and the condition becomes lower grade with progressive periarticular new bone formation, weight loss and malaise. Synovial fluid taps demonstrate high numbers of nondegenerative neutrophils.

The rheumatoid form of chronic progressive polyarthritis tends to affect older male cats and is much more chronic and insidious in its course, and fever and other constitutional signs are absent or mild. Instead of periarticular new bone formation, there is a destruction of the joint surfaces of more distal joints with associated subluxations, luxations and deformities. Affected cats are often not febrile and, except for the progressive joint disease, may even ambulate without much apparent pain.

Chronic progressive polyarthritis appears to be caused by immune reactions against a common retrovirus of cats, feline syncytium-forming virus. Because up to 60% of outdoor cats in the United States will ultimately acquire this lifelong infection without obvious clinical signs, cats which develop chronic progressive polyarthritis must be predisposed by both gender and genetics. In these respects, chronic progressive polyarthritis of cats resembles Reiter's disease of humans.

Treatment with combination immunosuppressive drug therapy (glucocorticoids and cyclophosphamide) has been only partly rewarding. Remission is difficult to achieve and, even if obtained, it is hard to sustain.

Arthritis associated with FIV infection

Lameness associated with one or several limb joints has been observed in cats with FIV infection. Synovial fluid demonstrates an increased number of nondegenerative neutrophils. The condition responds to corticosteroid treatment. The condition is either of immune complex origin, or results from inflammation evoked by the presence of virus infected macrophages in the synovium.

Idiopathic polyarthritis

A nonerosive polyarthritis has been observed on occasion, mainly in younger female cats. Because of cat's propensity to mask clinical signs of joint pain, many more cases probably go unrecognized. Affected cats show minimal signs of lameness and stiffness and fever, when present, is often mild. Lymphadenopathy is seen in a proportion of cases. The diagnosis is by joint taps of both tarsal and carpal joints and the demonstration of large numbers of nondegenerative neutrophils. Affected cats should be tested for predisposing conditions that may present in a similar manner, such as FeLV, FIV or SLE. The condition is often self-limiting after several weeks or months, but can be hastened to resolution by corticosteroid treatment.

Idiopathic systemic vasculitis

Cats occasionally present with severe vasculitis of unknown origin. In one form, usually seen in young cats, ulcers appear on the hard palate, tongue, cornea, foot pads, and skin. Ulcers appear in waves, heal slowly, then reappear. Deep biopsies will usually demonstrate a leukocytoclastic vasculitis. A second form of vasculitis is associated with severe edema and subcutaneous bruising that is widespread on the trunk or, in rare cases, within the abdomen. These cats are more likely to have associated clotting disorders, probably owing to concomitant disseminated intravascular coagulopathy. Affected cats are prone to develop thrombosis of pulmonary or coronary arteries and sudden death. The former condition responds to a combination of corticosteroids and cyclophosphamide or chlorambucil, but treatment is often indefinite. The latter condition responds poorly to treatment and has a much more grave prognosis.

A condition analagous to polyarteritis nodosa of humans has been described in cats, but is now considered to be a manifestation of feline infectious peritonitis (FIP). Indeed, the classical pyogranulomatous peri-venular lesions of effusive FIP are typical of an Arthus-type vasculitis.

Toxic epidermal necrolysis (TEN) is an acute arteritis characterized by diamond-shaped full-thickness necrosis of segments of skin; necrotic areas correspond to the area perfused by the effected vessels. The condition in cats is usually associated with injections of steroids and/or antibiotics several days earlier and, in rare circumstances, it is either idiopathic or associated with infections such as FIP. Areas of affected skin are pruritic at the onset, due to hypoxia of the nerves. This pain disappears within 24–48 h as the affected skin and its nerves die. The affected skin then becomes discolored and undergoes dry gangrene. If enough of the skin is affected, the condition can be fatal. However, this is seldom the case and the condition is managed by excising dead skin and allowing the wounds to heal by secondary intention or with grafting.

Glomerulonephritis

Glomerulonephritis is also either idiopathic or secondary. Cats with idiopathic disease are more apt to present mainly with signs of renal disease, whereas cats with secondary disease usually present for the underlying disease and glomerular disease is detected only during the clinical work-up. Common secondary causes include SLE, FIV and FeLV infections.

Glomerulonephritis in cats differs greatly from the canine disease in one important aspect; proteinuria is much less severe than in dogs, even given similar degrees of histopathology. Cats with glomerulonephritis are therefore much less likely to present with classical nephrotic syndrome (hyperproteinuria, hypoproteinemia, edema, hypercholesterolemia, hyperfibrinogenemia and hypercoagulability) than dogs. Rather, glomerulonephritis in cats is more likely to end in renal insufficiency with elevated blood urea nitrogen and creatinine and reduced urine concentrating ability. Because of the tendency of cats to be less proteinuric, the diagnosis of glomerulonephritis often rests on kidney biopsy or post-mortem examination.

Idiopathic glomerulonephritis is very uncommon in cats compared with dogs. As discussed above, affected cats usually present with renal insufficiency and, infrequently, with nephrotic syndrome.

Immunological diseases caused by cellular immunity

Cell-mediated immunity is the major host defense mechanism against altered host or foreign (allogenic or xenogenic) cells. Alterations to host cells can occur through intracellular microbial infection, chemical alterations of cell surface proteins, or malignant transformation. Cellular immunity can be innate (natural killer or NK cells, lymphokine-activated killer (LAK) cells) or specific (cytotoxic T cells, CTLs). Specific T cell-mediated killing can be further augmented by the activation of macrophages. The hallmark lesion of cell mediated immune injury is the infiltration of lymphocytes, plasma cells and macrophages.

Delayed-type hypersensitivity disease of the skin

Cats can develop chronic delayed-type hypersensitivity disease of the skin underlying plastic flea collars and in the areas of flea bites. These reactions are often very pruritic and persist for many days after the initiating stimuli are eliminated.

Chronic plasmacytic/lymphocytic enteritis

Cats commonly develop a chronic inflammation of the small bowel characterized by villus atrophy, thickening of the bowel wall, and moderate to dense plasmacytic/lymphocytic cell infiltrates. The condition manifests as loose stools and weight loss in the face of a good or even excessive appetite. This type of bowel disease is different in immunopathogenesis from allergic and eosinophilic enteritis, which involve IgE and immediate hypersensitivity. However, both allergic enteritis and chronic plasmacytic/lymphocytic enteritis are caused by antigenic material in the diet. The treatment of chronic plasmacytic/lymphocytic enteritis also involves the feeding of a hypoallergenic diet and corticosteroid treatment.

Cell-mediated diseases of unknown (autoimmune?) etiology

Over 50% of cats more than 10 years old suffer from a characteristic and progressive chronic interstitial nephritis. The disease is characterized by intratubular infiltrates of lymphocytes and plasma cells, wedge-shaped infarcts extending from the pelvis to cortex that result in scarring, decrease in kidney mass, and irregularity in shape and feel. Affected cats slowly lose renal function over a period of several years, eventually becoming polyuric, thin, hypertensive and anemic. This disorder is a major cause of death among older cats, along with cancer. There is no known treatment to halt the relentless destruction of the kidney.

'Big pad' disease is another disorder peculiar to cats and of unknown etiology. The pads of the feet, in particular the metacarpal and metatarsal pads become grossly enlarged and pillow-like to palpation. As they become larger, the pads often crack, causing the cat to become sore-footed. Biopsies show a dense plasmacytic/lymphocytic infiltrate. Corticosteroid treatment is used in cats showing clinical signs.

Polychondritis of the cartilage of the ears has been observed in cats. The ear pinnae become progressively thickened and contorted, much like the proverbial 'boxer's ear'. Biopsies show islands of ear cartilage that are isolated and surrounded by a predominantly lymphocytic infiltrate. The condition will only rarely involve other cartilage in the body and is self-limiting as the ear cartilage is destroyed.

Cholangiohepatitis is the feline equivalent of chronic active hepatitis in dogs and humans. The disease is characterized by a lymphocytic/plasmacytic infiltration around cholangioles, cholangiolar proliferation, biliary stasis, fibroplasia and parenchymal scarring. Cats are markedly icteric, but disproportionately healthy. Many of the cats with this disease have high titers of ANA, but it is uncertain whether this is part of the etiology of the disease (i.e., a form of lupoid hepatitis) or whether the ANA is merely a byproduct of slow hepatocyte death, nucleic acid release, and enhanced state of immune reactivity. Treatment is with glucocorticoids or combination immunosuppressive drug therapy.

Granulomatous diseases of infectious etiology

Granulomatous diseases result from inadequate cellular immunity, usually against intracellular microbial pathogens. The production of strong cellular immunity will lead to rapid containment of the invading microbes and mild or negligible disease signs. At the opposite extreme, a complete lack of cellular immunity will lead to overwhelming systemic disease. Individuals that mount partial cellular immunity, however, develop granulomatous inflammation. The granuloma is a partially successful attempt to limit the spread of an intracellular pathogen. The offending microbes are found in the center of the granuloma within macrophages, while radiating outward are areas of neutrophilic infiltrate, lymphocytic/plasmacytic inflammation and, in chronic granulomas, fibrous tissue. Because the microbes are only weakly contained, living organisms frequently break out of the lesion, often within macrophages, and initiate new granulomas in adjacent or distant sites. The course of disease, i.e. rapid recovery, disseminated illness or granulomatous disease, is a function of the immune system and modulating genetic and environmental factors.

Two classic granulomatous diseases in cats include the dry or noneffusive form of FIP and chronic ulceroproliferative faucitis/periodontitis associated with chronic feline calicivirus infection. The former condition is untreatable and inevitably fatal, while the latter may be slowed with corticosteroid treatment and judicious removal of teeth as periodontal tissue becomes involved. Granulomatous inflammation is also associated with a number of superficial skin infections of cats, including focal mycobacteriosis (*Mycobacterium fortuitum*, *M. cheloni*, *M. smegmatis*, *M. lepraemurium*), acintomycosis/nocaridosis, bacterial L-forms, sporothricosis, miscellaneous saprophytic soil fungi, protothecosis, and leishmaniasis, to name but a few.

Gammopathies (dysproteinemias or paraproteinemias)

The term gammopathy refers to excessive levels of antibody globulins in the blood. Two general types of gammopathies are recognized: (1) polyclonal, and (2) monoclonal. Polyclonal gammopathies involve increases in all major immunoglobulin classes (IgG, IgA and IgM), while monoclonal gammopathies involve only a single immunoglobulin class.

Polyclonal gammopathy

A polyclonal gammopathy is seen in a number of chronic infectious diseases of cats, but the most noteworthy is FIP. Most cats with FIP, especially those with the dry form, will demonstrate progressive rises in all immunoglobulin classes. The antibodies produced are not just against viral antigens, because actual FIP virus antibody titers often bear no direct relationship with the actual level of immunoglobulins in the blood.

A benign polyclonal gammopathy is seen in many aged cats. This is probably associated with a asynchronous aging of the T-cell compared with the B-cell system. Although aged cats with a polyclonal gammopathy should be checked for underlying diseases, care must be taken not to over-react to this finding in otherwise healthy animals.

Monoclonal gammopathy

Monoclonal gammopathies are usually caused by plasma cell tumors, i.e., multiple myeloma. Less commonly, they are associated with chronic lymphocytic leukemias, infectious diseases, or benign (idiopathic) causes. The term multiple myeloma comes from human medicine, where this particular tumor has a predilection to spread widely in the marrow cavities of flat bones. This is a misnomer for the cat, however, because most plasma cell tumors of cats arise from the viscera (liver, spleen, intestine, chest cavity), and a smaller proportion infiltrate bone. The most common immunoglobulin type seen in feline plasma cell tumors is IgG, followed distantly by IgM and then IgA.

Plasma cell tumors can present in a number of manners. If the tumor infiltrates and destroys tissues locally, signs will be referable to the organ involved just as any other type of tumors. Plasma cell tumors can also cause disease through the proteins that they produce. High levels of monoclonal antibody globulin, especially IgM or IgA, can cause the blood to become highly viscous, leading to heart disease and hemorrhage in organs such as the eyes and brain. Some myeloma proteins can agglutinate or congeal in the cold (cryoglobulins), leading to thrombosis of small vessels in the cooler extremities such as the ears. Pathological accumulations of a monoclonal antibody are almost always associated with a significant decrease in normal immunoglobulins (probably by a negative-feedback mechanism). If normal antibody levels are decreased too much, immunodeficiency can occur.

15. Immunodeficiency Diseases

Introduction

Immunodeficiency diseases were at one time considered rare and personified by the famous 'boy in the glass bubble', who suffered from a severe combined immunodeficiency disorder of genetic origin and lived his short life in strict isolation. With the world-wide human immunodeficiency virus (HIV) pandemic, however, immunodeficiency has become the byword of the age. It is clear, however, that immunodeficiency results from a myriad of causes, ranging from subtle to catastrophic in its clinical appearance. Animals are also not unique in suffering from various immunodeficiencies and all forms of the disorder described in humans have been reported in one or another species of animals, including the cat.

The hallmark of immunodeficiency is inappropriate infections: infections caused by microbes that are normally not pathogenic, infections that are unusually severe or persistent, multiple infections in the same individual, the recurrence of infections that occurred in a subclinical form earlier in life, and infections that are unexplainably difficult to treat.

Innate versus adaptive immunity

Immunodeficiency results from inadequacies in either innate (natural) or adaptive (acquired) immunity. Adaptive immunity can be subdivided into two major processes, i.e., humoral immunity and cell-mediated immunity. Humoral immunity involves the production of antigen-specific antibodies, while cell-mediated immunity is mediated by antigen specific cytotoxic T lymphocytes (CTLs). A second characteristic feature, in addition to specificity, of adaptive immunity is immunological memory. Immunological memory is the ability to recognize specific foreign substances much more rapidly and vigorously on subsequent compared to primary exposure. Innate immunity involves all host defense mechanisms that are nonspecific in nature and do not engender an immunologic memory. Innate immunity is usually present in more or less the same state before, during, and after an infection.

Innate immunity

Although many people think of innate immunity mainly in terms of natural killer (NK) cells, lymphokine-activated killer (LAK) cells, the properdin system for alternative complement pathway activation, defensins, mannose-binding proteins, collectins, interferons and numerous other soluble inhibitors, natural defense mechanisms are far broader in scope. Skin and mucous membranes are the most important components of innate immunity. Skin has an outer cornified layer that resists microbial growth and invasion. Glandular secretions onto the skin and mucous membranes also contain nonspecific antimicrobial substances. The cilia of certain mucous membranes, such as those of the upper respiratory and lower urogenital tracts, act as brooms to trap and sweep out foreign material. The ciliary layer also greatly increases the surface area and acts as a trap to hold mucus secretions, the latter being rich in specific and nonspecific immune substances and phagocytic cells. The urine of cats is usually of very high specific gravity, which inhibits bacterial growth.

These innate mucous membrane and skin defenses are highly efficient. Bacteria are present in the mouth, oropharynx and upper esophagus, but have all but disappeared by the stomach. A tremendous bacterial flora exists in the rectum and colon, but is absent from the lower small bowel. Bacteria are present in the oropharynx and the upper trachea, but are absent at all places distal to these. Organisms can also be cultured from the prepuce, vagina and distal urethra, but are absent from the proximal urethra, bladder, ureters and kidneys. Bacteria are present on the surface of the skin but are absent from layers below the stratum corneum. Invasion of microbes from their normal sites to deeper tissues, whether it is gut, skin, respiratory tract or urinary system, can only occur if this normal barrier is in some manner breached or bypassed.

Adaptive immunity

Cell-mediated immunity

Cellular immunity occurred earliest in evolution and is absolutely required for host survival. Specific cell-mediated immunity involves CTLs bearing the CD8 cell surface marker. Nonactivated CTLs possess surface immunoglobulin-like receptors that act to recognize foreign antigens; each cell bearing receptors for a single antigen. If a host cell becomes infected or in some other manner antigenically altered, some of the foreign antigens will be arrayed on its surface in conjunction with the proteins of MHC I. The CTLs will then come in intimate contact with the abnormal cell by receptor/ligand interactions. Antigen recognition will cause this specific subset of cells to clonally expand, yielding a larger population of specifically activated CTLs and memory cells. Specifically activated CTLs will then come into intimate contact with the antigenically altered cells through the same recognition mechanism and secrete substances that will cause their death. CTLs also send chemical signals to macrophages, causing them to activate and become cell killers or to destroy any microbes that they may have internalized. CTLs are helped in their activity by $CD4^+$ T cells (T-helper cells). Therefore, deficiencies in specific cellular immunity can arise from deficiencies in number and/or function of CTLs, T-helper cells or macrophages. This system also has inhibitory controls mediated by a subset of $CD8^+$ T-cells known as T-suppressor cells.

Humoral immunity

Humoral immunity involves the production of antigen specific antibodies of IgM, IgG, IgA, and IgE subclasses. IgM antibodies are very large, consisting of five basic immunoglobulin molecules connected by a joining protein. IgM antibodies are produced in tonsils, diffuse lymphoid aggregates, spleen and lymph nodes and circulate freely in the bloodstream; they are also secreted into mucous membranes. IgG antibodies are monomeric and are also produced in the same tissues and are found mainly in the bloodstream. IgA antibodies are produced mainly by tonsils and diffuse lymphoid aggregates underlying mucosal surfaces. IgA in the cat is dimeric in the bloodstream; a small secretory protein is added by the epithelial cells as it is transported from the blood side to luminal side of the mucosa. IgA is the principal secretory protein found in mucus. IgE is a monomer and is rapidly bound to basophils and mast cells.

The production of specific immunoglobulin is triggered when antigen-presenting cells (APCs) take up the antigen and array it on their surfaces in association with MHC II. Antibody production is carried out specifically by a subset of lymphocytes called B cells. Most B cells are also under both positive and negative regulatory control by T-helper cells ($CD4^+$ T cells) and T-suppressor cells ($CD8^+$ T cells), respectively. The activity of antibodies are also modulated by other proteins, in particular proteins of the complement cascade. IgM and certain subtypes of IgG are good complement binders, while IgA and IgE are poor. Although it is antibody that targets specific foreign proteins, it is complement factors that modulate the function of the bound antibody, e.g., lysis, opsonization, neutrophil attraction, etc. Because of the close interaction between antibodies and complement proteins, deficiencies in complement may cause similar symptoms as deficiencies in antibodies and vice versa.

Maternal immunity

The newborn cat is born with negligible levels of immunoglobulin in its blood and with only its IgM system functional. In order to protect the newborn against infections until its immune system is mature, the queen transfers antibodies to the kitten in the colostrum. Colostrum is the first milk; it is rich in IgG and IgA, but contains little IgM. There are two possible explanations for the absence of IgM in colostrum : (1) it is not required because the IgM system is fully functional at birth, and (2) if present, it would inhibit normal IgM production in the same manner as maternal IgG inhibits host IgG. The early maturity of the IgM system is an example of ontogeny recapitulating phylogeny; IgM appears early in chordate evolution, followed by IgG and then IgA. The ability of the kitten to produce IgM antibodies from birth onwards is an important backup to maternal immunity. It is noteworthy that IgM is both systemic and secretory, combining the activities of IgG and IgA.

The intestinal epithelium of the newborn kitten is open for the absorption of IgG and IgA for the first 24 h, after which time no further immunoglobulin is absorbed from the gut lumen to the bloodstream. Therefore, the higher the levels of antibodies in the colostrum, the more likely the kitten will receive adequate levels of antibodies from its mother. This first pulse of maternally derived antibody provides 'passive systemic immunity'. IgG remains in the circulation with a half-life of around 7 days. In a similar manner, the passively acquired IgA will be slowly secreted from the blood into the mucous membranes. Active production of IgM is detectable from birth onward, IgG from about 4–6 weeks of age onward, and IgA from 6–8 weeks onward (following weaning).

In addition to passive systemic immunity, the queen continues to provide considerable amounts of IgG and IgA in the milk. Although the IgG is largely degraded in the stomach, it plays an important role in preventing infection in the oropharynx. Almost all pathogens enter by way of the mouth and the oropharyx and tonsils are often the first tissues to be infected. IgA antibodies consumed in the milk enter the intestine in an intact form and become part of the mucous film of the intestine. Antibodies that are provided in milk from 12–24 h of age to weaning at 6–8 weeks provide what is called 'passive local immunity'.

Specific immunodeficiencies

Disorders associated with innate immunity

Deficiencies in classical innate immune defenses (NK cells, LAK cells, properdin, defensins, etc.) have not been described in the cat. There are, however, several anatomical defects in innate defenses that lead to immunodeficiency states. Kittens suffering from severe herpesvirus rhinitis are often left with badly scarred and atrophied turbinates. If the damage is severe enough, the local mucosal barrier is rendered inoperative and these animals suffer for the rest of their life from recurrent bacterial rhinitis/sinusitis. Some kittens are born with small congenital oronasal fistulas; these are often on the midline just behind the upper incisors and are easily overlooked. Similar fistulae can develop following recovery from palate and skull fractures. In addition to fistulae in the hard palate, some kittens are born with a cleft in their soft palate, reminiscent of the normal anatomical appearance of the throat of birds. Food can pass from the oral to the nasal cavity during eating and cause a chronic rhinitis.

Tom cats with urethral obstructions from feline urologic syndrome (FUS) were formerly treated with a perineal urethrostomy. About one-fourth or more of these animals suffered later from chronic urinary tract infections because of the removal of the distal defense mechanisms in the penile urethra. This, along with the development of preventative diets reducing the incidence of FUS, is why perineal urethrostomy is no longer used as a routine treatment for urethral obstruction in cats. Cats that are being treated with fluid and electrolyte solutions while urinary catheters are in place are very prone to urinary tract infections. Normal cat urine, which is of high specific gravity, is inhibitory to bacteria. By lowering the specific gravity of the urine, while bypassing the distal urethral defenses with a catheter, bacteria can both invade into the bladder and replicate. A similar situation occurs in aged cats with chronic kidney disease, polyuria/polydypsia and low specific gravity urine; such cats sometimes develop severe bacterial infections (usually hemolytic strains of *Escherichia coli*).

Secondary bacterial infections of the skin often accompany allergic dermatitis; pruritus induces excessive scratching and abrasion of the skin, thus allowing local bacteria to breach the corneum.

Cats do not cough nearly as well as humans or dogs, and when they develop pneumonia, it is very difficult for them to clear exudates. This is one reason why pneumonias in cats are more likely to be severe or fatal than in species exhibiting a strong cough reflex.

Innate immune defenses are often overwhelmed in young kittens that are hand-reared and fed high-energy supplements. The overfeeding of such foods, often in bolus form, leads to the dumping of large amounts of undigested nutrients into the lower bowel. This causes an overgrowth of bacteria, especially of nonresident types, in the colon and retrograde bacterial colonization of the small bowel. The result is an intractable enteritis and, if associated with deficiencies in passive maternal immunity (see preceding section), to bacterial sepsis.

Deficiencies in adaptive immunity

Deficiencies in adaptive immunity are either congenital or acquired and involve cellular or humoral arms, or both. Congenital deficiencies of specific immunoglobulin subclasses, complement components, or severe combined immunodeficiency have not been recognized in cats, but undoubtedly exist. Several acquired immunodeficiencies occur in cats, however. These include undefined immunodeficiencies to specific diseases, age-associated immunodeficiency, stress-induced immunodeficiencies, failure of maternal immunity, and retrovirus-associated immunodeficiencies.

Persian cats as a breed are more susceptible to dermatomycosis (ringworm); they get more lesions, are more difficult to treat, carry and shed the fungi for much longer periods of time, and are much more susceptible to invasive forms of the infection (mycetomas). This immunodeficiency is quite specific because the breed is not more susceptible to other types of infections. Siamese and related breeds of cats are more susceptible to severe feline herpesvirus rhinitis and, as such, they are much more likely to develop chronic rhinitis and sinusitis as a consequence (see preceding discussion on innate immunity).

Cats as a species have very little acquired resistance to the feline infectious peritonitis virus (FIPV) both experimentally and in the field. Over 90% of experimentally and naturally infected cats will die if they are infected with this virus. FIPV is a simple mutant of the much more ubiquitous feline enteric coronavirus (FECV), and in highly FECV endemic environments (pedigreed catteries, shelters and large multiple cat households), about 5% of FECV-infected cats will succumb to FIP. The reason for this susceptibility is probably linked to a lack of extended evolutionary adaptation. FIP was not seen in cats prior to the 1950s, either because the parent FECV virus had not yet evolved as a feline pathogen or because drastic changes in feline husbandry associated with urbanization, pedigreed cat breeding, cat shelters, and the keeping of multiple cats indoors as pets, have greatly enhanced FECV infection.

Age-related immunodeficiency is a common factor in disease of kittens. Kittens infected with feline leukemia virus (FeLV) at 6 weeks old will almost always become persistently viremic, while a majority of kittens exposed to the same dose of FeLV at 16 weeks old will recover. Six-week-old kittens infected with feline enteric coronavirus (FECV) will shed virus at high levels in their feces for many months, whereas cats infected as adolescents will shed much lower levels of virus for only a few weeks. The much higher rate and longer duration of FECV replication in the 6-week-old kittens leads to a much higher incidence of feline infectious peritonitis (FIP). The FIP virus is a mutant of FECV and the greater and longer virus replication, the greater chance that the mutation will occur. Age resistance can also be shown for virtually all common infections of cats, including feline herpesvirus, feline calicivirus, *Chlamydia psittaci*, and *Bordetella bronchiseptica*. Neonatal streptococcal infections also have an interesting age-related pattern in cats. Young primiparous queens have much higher levels of β-hemolytic streptococci (*Streptococcus canis)* in their genital tract than older queens, and whereas the numbers of streptocci decrease during pregnancy in older multiparous queens, the numbers increase in primiparous queens. As a result, young queens often infect their young during parturition, with considerable neonatal mortality.

Stress-related immunodeficiency occurs in any situation where large numbers of cats are kept indoors, closely confined, and in intimate contact with each other. It is particularly intense in multi-cat environments where kittens are also reared. The mechanism of this immunodeficiency, while assumed to be stress-related, is not precisely known. Cats, especially kittens, raised in such environments handle a number of common infections far worse than well-managed specific pathogen free laboratory cats. Feline herpesvirus (FHV), feline calicivirus (FCV), FECV, feline leukemia virus (FeLV), *C. psittaci*, *Bartonella henselae* (the cat scratch agent), *Microsporum canis*, *Giardia*, *Cyptosporidia*, and *Coccidia* are all infectious agents that are far more severe in catteries, shelters and large multiple cat households than among free-roaming cats or in experimentally infected laboratory cats. Primary infections are clinically more severe, a larger proportion of cats will carry the organisms following recovery and for longer periods of time, and complications are far more common. This factor is the primary reason why most laboratory cats are now maintained in a specific pathogen free state; prior to the use of SPF cats, infectious disease problems among non-SPF laboratory cats precluded their use from many long-term experiments.

Failure of maternal immunity has two components: (1) failure to receive or absorb colostrum, thus causing a deficiency of passive systemic immunity, and (2) a failure to receive mother's milk causing a deficiency in passive local immunity. Kittens orphaned or taken from their mother prior to nursing and fed artificially will be deficient in both passive systemic and passive local immunity and are extremely difficult to raise to maturity. Despite artificial nursing and nurturing, many will develop fatal bacterial sepsis within the first 2 weeks of life, and those that do not frequently suffer from severe and chronic enteritis.

Kittens that receive colostrum, but are weaned shortly thereafter, are much easier to rear artificially but still suffer inordinately from enteric infections. A failure to absorb colostral antibodies has also been described in a proportion of kittens not taken from their mother. It is unclear whether the intestinal tract of these kittens closed prematurely to the absorption of immunoglobulin or if they failed to nurse in a timely manner. Such kittens will still receive passive local immunity from their mothers but are, nonetheless, much more prone to sepsis and early death. Kittens that are known to be deficient in passive systemic immunity should be given injections of serum from older cats. Small amounts of serum can also be given with the artificial milk in orphaned kittens to help provide passive local immunity. The last two steps have been used effectively to rear valuable orphaned and endangered species of wild felids.

Acquired immunodeficiencies are associated with both FeLV and feline immunodeficiency virus (FIV) infections. The half-life of cats with persistent FeLV infection is around 1 year; thus most chronic FeLV carriers will be dead after 3–4 years. The greatest causes of death in FeLV carriers are T-cell lymphomas (mainly thymic, ocular, neurologic or generalized), myeloproliferative diseases, and aplastic anemias.

Immunodeficiency is probably associated with about 1 in 5 or so of FeLV-infected cats. The most noteworthy immunodeficiency is to FIPV infection. When FeLV was still a common infection in cats (in the 1960s and 1970s), about one-third or more of cats with FIP were FeLV positive. It has since been shown that FeLV infection has a highly specific inhibitory effect on FIPV immunity. Cats that are subclinically infected with FIPV will become clinically ill with FIP within several months of FeLV infection. Other types of FeLV immunodeficiency tend to be associated with the profound neutropenias that are often seen in cats with aplastic anemia or myeloproliferative disease. *Haemobartonellosis felis* is commonly seen as an underlying or overlying infection in such cats. Acute necrotizing ulcerative gingivitis is seen almost exclusively in severely neutropenic cats, as is a characteristic multifocal suppurative and necrotizing pneumonia caused by the EF4 bacterium. Other infections seen in leukopenic FeLV-infected cats include tooth root infections, suppurative otitis, and bacterial peritonitis.

The acquired immunodeficiency of FIV infection is highly specific compared with the immunodeficiencies caused by FeLV. One-half or more of FIV-infected cats develop a progressive deficiency in $CD4^+$ T cells, defective macrophage function, follicular atrophy, and loss of follicular dendritic cells. The loss of $CD4^+$ cells appears to be the central lesion in the overall deficiency and tends to impact on cellular immunity more than humoral. Once the $CD4^+$ cell numbers fall below 100–200 cells/μl blood, affected cats often develop chronic infections of the skin, respiratory tract, and intestines.

16. Tumors of the Immune System

Tumors of the immune system of the cat are the most common neoplasms of cats. They are of great scientific importance since for the majority of them a known etiology exists which allows the formulation of hypotheses about the pathogenesis of the tumors. Furthermore, these neoplasms are of high practical importance in veterinary medicine and effective methods for prophylaxis are available.

One of the difficulties of reviewing the tumors of the immune system is that there is no clear-cut border between tumors of immune cells, tumors in primary and secondary immune organs, and tumors of the hematopoietic system in general. These three approaches will therefore be dealt with. The main topic, however, is the discussion of tumors of lymphatic cells whereas tumors of the erythrocytic and megakaryocytic series are mentioned only for completeness in systematic tables. Metastatic tumors in immune organs and tissues as well as tumors in these organs and tissues arising from cells not specific for the immune system (e.g., vascular cells, fibroblasts, epithelium in tonsils) are not dealt with.

As in every tumor classification, a differentiation of tumors is preferable which is based on the origin of the tumor cells. These are the cell lines in bone marrow (granulocytic – neutrophil, eosinophil, basophil – monocytic, megakaryocytic, erythrocytic) and lymphatic organs and tissues (B and T lymphocytes, plasma cell, dendritic cells, macrophages) and distributed cells of the immune system (e.g., Langerhans cells, mast cells).

There are three general methods of classifying the tumors of the immune system in cats. These are: (1) by histological and cytological criteria including, histochemical and immunohistochemical reactions, (2) by surface molecules expressed on these cells (clusters of differentiation, CD), and (3) for lymphomas, also by macroscopical findings. All three approaches have their advantages and their limitations.

The classifications based on the classical pathological differentiation of the hematopoietic tumors of the cat are systematics of human pathology which have been adopted directly or been slightly modified to better fit to the situation in the cat. These classifications are numerous and no one method is the accepted gold standard. The most important systems include the Kiel classification, the French–American–British cooperative group systematic, the National Cancer Institute working formulation, and the Animal Leukemia Study Group classification. These classifications or combinations of them are used mostly by pathologists.

The classification by surface molecules specific for a certain cell type is still limited by several factors. First, not enough monoclonal antibodies against feline immune cell surface molecules have been produced and characterized. Thus, only some of the many CDs known in other animal species and man are in use in the feline system. Second, at least in some tumors the neoplastic cells express unusual combinations of CDs which do not occur on normal cells. Third, the knowledge of the biological importance of the findings obtained in such studies is still not very broad and is equivocal. These classifications are mostly used by immunologists and pathologists with a special interest in this scientific field.

The classification according to the macroscopical appearance of lymphomas in the cat takes account of the presence or absence of a main tumor mass and its location. In leukemias, the tumor cells primarily infiltrate bone marrow, red pulp of the spleen, and medulla of the lymph nodes without forming a solid tumor mass; in lymphomas, such solid tumors do exist. They can occur in a wide variety of organs and their localization bears some implications concerning biological behavior, etiology and pathogenesis of the neoplasms. This classification is used mostly by clinicians and in everyday diagnostic pathology.

An attempt to show the most important and most often used criteria for the classification of tumors of the immune system of the cat is presented in Table VIII.16.1. Tumors of the hemopoietic system account for about 30–50% of the tumors of the cat. About 10–15% of them are myeloproliferative diseases and about another 10–15% are lymphoid leukemias. The remaining 70–80% are lymphomas.

Myeloproliferative disease is usually preceded by a prodromal myelodysplastic phase which is readily recognized in experimental induction of the condition. The most important etiology of myelo- and lympho-proliferative diseases in the cat is the feline leukemia virus (FeLV), a retrovirus of the genus mammalian type C retrovirus group which multiplies in immune cells. About two-thirds of all cats with myelo- or lympho-proliferative diseases are persistently viremic with FeLV. Provirus DNA could be demonstrated in lymphomas in about 20% more cases than virus production. This is interpreted as evidence to suggest FeLV involvement in at least a portion of lymphomas in FeLV-negative cats.

Between the different forms of lymphomas there is a clear difference in the association with FeLV. Mediastinal lymphomas occur in younger animals, are in 90% of the cases T-cell lymphomas (predominantly $CD4^{+}/CD8^{+}$ or $CD4^{-}/CD8^{+}$) and never B-cell lymphomas, and are in 80–90% of the cases FeLV-producers. The other extreme in conjunction with FeLV infection is the intestinal lymphoma, which occurs predominantly in older animals, has the highest percentage of B-cell lymphomas and only in less than 20% of the cases produces FeLV. FeLV infection was not demonstrated in association with plasma cell tumors, mast cell tumors, and thymomas. The relative risk of developing a tumor is in FeLV-positive cats 29 times higher than in FeLV-negative cats for mediastinal lymphoma, nine times higher for multicentric lymphoma, and six times higher for the miscellaneous lymphoma group. The overall risk of developing the myelo- or lympho-proliferative disease is for a FeLV-positive cat seven times higher than for a FeLV-negative cat.

FeLV does not carry an *onc* gene. Cellular *onc* genes and other cell development regulating genes, however, can be picked up and transduced after infection of a cell and thus foster tumor formation. A transforming effect is also

Table VIII.16.1 Classification of tumors of the feline immune system

1. Myeloproliferative disease
   - 1.1. Acute leukemia
     - 1.1.1. Myeloblastic/myelocytic leukemia
     - 1.1.2. Monocytic leukemia
     - 1.1.3. Malignant histiocytosis
     - 1.1.4. Erythroleukemia
     - 1.1.5. Megakaryoblastic leukemia
   - 1.2. Chronic leukemia
     - 1.2.1. Chronic granulocytic leukemia
     - 1.2.2. Mast cell leukemia
     - 1.2.3. Megakaryocytic myelosis
     - 1.2.4. Polycythemia vera
2. Myelodysplastic syndromes
   - 2.1. Refractory anemia with excess blast cells
   - 2.2. Chronic myelomonocytic leukemia
   - 2.3. Myeloid metaplasia with myelofibrosis
3. Lymphoproliferative disease
   - 3.1. Lymphoid leukemia
     - 3.1.1. Acute lymphoid leukemia
     - 3.1.2. Chronic lymphoid leukemia
   - 3.2. Lymphoma
     - 3.2.1. Multicentric lymphoma
     - 3.2.2. Mediastinal lymphoma
     - 3.2.3. Intestinal lymphoma
     - 3.2.4. Miscellaneous lymphomas
       - 3.2.4.1. Renal lymphoma
       - 3.2.4.2. Cutaneous lymphoma
         - 3.2.4.2.1. Classical cutaneous lymphoma
         - 3.2.4.2.2. Mycosis fungoides (epitheliotropic T-cell lymphoma)
       - 3.2.4.3. Nasal lymphoma
       - 3.2.4.4. Single lymph node lymphoma
       - 3.2.4.5. Neural lymphoma
       - 3.2.4.6. Ocular lymphoma
       - 3.2.4.7. Other extranodal lymphomas (gingiva, larynx, trachea, muscle)
4. Plasma cell tumors
   - 4.1. Medullary plasmacytoma
   - 4.2. Extramedullary plasmacytoma
5. Peripheral mast cell tumors
   - 5.1. Cutaneous mast cell tumor
   - 5.2. Visceral (multiple) mast cell tumor
6. Thymoma
   - 6.1. Epithelial thymoma
   - 6.2. Lymphoid thymoma

possible via sequences of provirus DNA integrated into the cellular genome which may influence the expression of cellular genes.

The second agent known to enhance the risk for the development of lymphomas is the feline immunodeficiency virus (FIV), a retrovirus from the genus lentivirus. A FIV-positive cat has a five times greater chance of developing lymphomas than a FIV-negative cat. These lymphomas are mostly high-grade B-cell lymphomas of the centroblastic or immunoblastic subtypes as in human and simian immunodeficiency virus infection. Their pathogenesis is thought to be associated with the polyclonal B-cell stimulation occurring after FIV infection. A direct transforming effect of FIV is not to be expected because FIV provirus is not demonstrable in the lymphoid tumor cells of FIV-positive cats.

Most of the tumors mentioned in this chapter lead sooner (within months; e.g., acute leukemia, high-grade lymphomas) or later (within years; e.g., chronic leukemia, low-grade lymphomas) to the death of the cat. The exceptions include local mast cell tumors, cutaneous plasmacytomas, and Mycosis fungoides which can be completely cured by surgery.

17. Conclusions

In recent years, much progress has been made in feline immunology because natural retrovirus infections exist in the cat allowing its use as model for retrovirus infections in man. With the increasing availability of reagents and techniques, research will without doubt continue to progress at a rapid pace. Novel approaches in vaccinology based on new adjuvant preparations including cytokines such as IL-12, IFN-γ and IL-18, newly produced antigens for mucosal immunization, DNA plasmid injections and others may help to further improve the cat's health. The likelihood that an efficacious feline immunodeficiency virus (FIV) vaccine will be introduced to the field looks promising as many groups could repeat the first successful results published by Yamamoto (1991, 1993). The detection and quantification of cytokines will help characterize the pathogenesis of diseases and assays to measure cytokines may soon become important not only for research but also for veterinary practice. In addition, use of cytokines may eventually become important for immunotherapy of diseases such as FIV infection and feline infectious peritonitis, an almost always fatal disease associated with feline coronavirus infection which affects between 5 and 12% of all cats under 1 year old.

To speed progress, it will be important not only to have additional techniques and reagents available, but also to have exchange between laboratories. The European Concerted Action on FIV Vaccination which was initiated in 1990 and has obtained funding until 1999 has proven ideal for the exchange of knowledge, reagents and personnel between groups at relatively little cost. In addition, it will be important to further improve immunology education not only for researchers but also for veterinary practitioners.

For obvious reasons, the major deficit that currently exists in the feline system, is the difficulty with which cytotoxic T cell studies can be performed. It would be highly desirable if uncomplicated standard reagents could be developed that allow labelling of autologous cells with the target antigen to be studied. To further study the T_H1/T_H2 pathways and the cytokines involved, it would be highly desirable if well characterized T_H1 and T_H2 cell lines were available, as they are in the murine system.

18. References

Aasted, B., Blixenkrone-Moller, M., Larsen, E. B., Ohmann, H. B., Simensen, R. B., Uttenthal, A. (1988). Reactivity of 11 anti-human leucocyte monoclonal antibodies with lymphocytes from several domestic animals. *Vet. Immunol. Immunopathol.* **19**, 31–38.

Ackley, C. D., Cooper, M. D. (1992). Characterization of a feline T-cell-specific monoclonal antibody reactive with a CD5-like molecule. *Am. J. Vet. Res.* **53**, 466–471.

Ackley, C. D., Hoover, E. A., Cooper, M. D. (1990). Identification of CD4 homologue in the cat. *Tissue Antigens* **35**, 92–98.

Aitken, I. D., Olafsdottir, E., McCusker, H. B. (1967). Immunological studies in the cat 1. The serological response to some foreign proteins. *Res. Vet. Sci.* **8**, 234–241.

Andrews, G. A., Chavey, P. S., Smith, J. E., Rich, L. (1992). *N*-glycolylneuraminic acid, and *N*-acetylneuraminic acid define feline blood group A, and B antigens. *Blood* **79**, 2485–2491.

Argyle, D. J., Smith, K., McBride, K., Fulton, R., Onions, D. E. (1995). Nucleotide, and predicted peptide sequence of feline interferon-gamma (IFN-γ). *DNA Seq.* **5**, 169–171.

Argyle, D. J., Hanlon, L., Onions, D. E. (1996). The cloning of feline specific TH1 cytokines, and their potential in therapy, and vaccination. In *Workshop on Cytokines in FIV and HIV Infection, Pisa, Italy, Nov. 15–16.* p. 14.

Arthur, L. O., Bess, J. W. J., Sowder, R. C., Benveniste, R. E., Mann, D. L., Chermann, J. C., Henderson, L. E. (1992). Cellular proteins bound to immunodeficiency viruses: implications for pathogenesis and vaccines. *Science* **258**, 1935–1938.

Auer, L., Bell, K. (1981). The A. B. blood group system of cats. *Anim. Blood Groups Biochem. Genet.* **12**, 287–297.

Baier, M., Werner, A., Bannert, N., Metzner, K., Kurth, R. (1995). HIV suppression by interleukin-16. *Nature* **378**, 563.

Barde, Y. A. (1990). The nerve growth factor family. *PGFR* **2**, 237–248.

Barlough, J. E., Jacobson, R. H., Scott, F. W. (1981). The immunoglobulins of the cat. *Cornell Vet.* **71**, 397–407.

Bartholomew, R. M., Esser, A. F., Muller-Eberhard, H. J. (1978). Lysis of oncornaviruses by human serum. Isolation of the viral complement (C1) receptor, and identification as p15e. *J. Exp. Med.* **147**, 844–853.

Bell, K. (1983). Blood groups of domestic animals. In *Red Blood*

Cells of Domestic Mammals (eds Agar, N. S., Board, D. G.), pp. 137–164. Elsevier, Amsterdam, The Netherlands.

Bendinelli, M., Pistello, M., Lombardi, S., Poli, A., Garzelli, C., Matteucci, D., Ceccherini-Nelli, L., Malvaldi, G., Tozzini, F. (1995). Feline immunodeficiency virus: an interesting model for A.I.D.S studies and an important cat pathogen. *Clin. Microbiol. Rev.* **8**, 87–112.

Bishop, S. A., Williams, N. A., Gruffydd-Jones, T. J., Harbour, D. A., Stokes, C. R. (1992). Impaired T-cell priming, and proliferation in cats infected with feline immunodeficiency virus. *AIDS* **6**(3), 287–293.

Bishop, S. A., Stokes, C. R., Gruffydd-Jones, T. J., Whiting, C. V., Harbour, D. A. (1996). Vaginal and rectal infection of cats with feline immunodeficiency virus. *Vet. Microbiol.* **51**, 217–227.

Bland, P. W. (1988). MHC class II expression by the gut epithelium. *Immunol. Today* **9**, 174–178.

Bradley, W. G., Gibbs, C., Kraus, C., Good, R. A., Day, N. K. (1993). Molecular cloning and characterization of a cDNA encoding feline interleukin-6. *Proc. Soc. Exp. Biol. Med.* **204**, 301–305.

Brown, W. R., Kacskovics, I., Amendt, B. A., Blackmore, N. B., Rothschild, M., Shinde, R., Butler, J. E. (1995). The hinge deletion allelic variant of porcine IgA results from a mutation at the splice acceptor site in the first C alpha intron. *J. Immunol.* **154**, 3836–3842.

Bücheler, J., Giger, U. (1993). Alloantibodies against A and B blood types in cats. *Vet. Immunol. Immunopathol.* **38**, 283–295.

Buerke, M., Murohara, T., Lefer, A. M. (1995). Cardioprotective effects of a C1 esterase inhibitor in myocardial ischemia and reperfusion. *Circulation* **91**, 393–402.

Buerke, M., Weyrich, A. S., Murohara, T., Queen, C., Klingbeil, C. K., Co, M. S., Lefer, A. M. (1996). Humanized monoclonal antibody DREG-200 directed against L-selectin protects in feline myocardial reperfusion injury. *J. Pharmacol. Exp. Therap.* **271**, 134–142.

Bush, K., Day, N. K., Kraus, L. A., Good, R. A., Bradley, W. G. (1994). Molecular cloning of feline interleukin 12 p.35 reveals the conservation of leucine-zipper motifs present in human, and murine IL-12 p.35. *Mol. Immunol.* **31**(17), 1373–1374.

Butler, J. E., Brown., W. R. (1994). The immunoglobulins and immunoglobulin genes of swine. *Vet. Immunol. Immunopathol.* **43**, 5–12.

Butler, M., Andews, G. A., Smith, J. E. (1991). Reactivity of lectins with feline erythrocytes. *Comp. Haematol. Int.* **1**, 217–219.

Casal, M. L., Straumann, U., Sigg, C., Arnold, S., Rüsch, P. (1993). Congenital hypotrichosis with thymic aplasia in nine Birman kittens (abstract). *Proc. Am. Coll. Vet. Intern. Med. Forum* **11**, 941.

Casal, M. L., Giger, U., Jerzyk, P. F. (1994). Transfer of colostral antibodies from the queen to the kitten (abstract). *Proc. Am. Coll. Vet. Intern. Med. Forum* **12**, 1002.

Casal, M. L., Jezyk, P. F., Giger, U. (1996). Transfer of colostral andibodies from queens to neonatal kittens. *Am J. Vet. Res.* **57**, 1653–1658.

Center, D. M., Cruikshank, W. W., Berman, J. S., Beer, D. J. (1983). Functional characteristics of histamine receptor-bearing mononuclear cells. I. Selective production of lymphocyte chemoattractant lymphokines with histamine used as a ligand. *J. Immunol.* **131**(4), 1854–1859.

Chen-Kiang, S. (1995). Regulation of terminal differentiation of human B-cells by IL–6. *Curr. Top. Microbiol. Immunol.* **194**, 189–198.

Cozzi, P. J., Padrid, P. A., Takeda, J., Alegre, M. A., Yuhki, N. (1993). Sequence and functional characterization of feline interleukin 2. *Biochem. Biophys. Res. Commun.* **194**, 1038–1043.

Cozzi, P. J., Padrid, P., Tompkins, M. B., Alegre, M. L., Takeda, J., Leff, A. R. (1995). Bioactivity of recombinant feline interleukin-2 on human, and feline leukocytes. *Vet. Immunol. Immunopathol.* **48**(1–2), 27–33.

Crepaldi, T., Crump, A., Newman, M., Ferrone, S., Antczak, D. F. (1986). Equine T lymphocyte express MHC II antigens. *J. Immunogenet.* **13**, 349–360.

Cruikshank, W., Center, D. M. (1982). Modulation of lymphocyte migration by human lymphokines. I. Purification of a lymphotactic factor (LCF). *J. Immunol.* **128**(6), 2569–2574.

D'Andrea, A., Aste-Amezaga, M., Valiante, N. M., Ma, X. Kubin, M., Trinchieri, G. (1993). Interleukin 10 (IL-10) inhibits human lymphocyte interferon gamma-production by suppressing natural killer cell stimulatory factor/IL-12 synthesis in accessory cells. *J. Exp. Med.* **178**(3), 1041–1048.

Daniel, S. L., Legendre, A. M., Moore, R. N., Rouse, B. T. (1993). Isolation and functional studies on feline bone marrow derived macrophages. *Vet. Immunol. Immunopathol.* **36**, 107–122.

Day, N. K., O'Reilly Felice, C., Hardy, Jr., W. D., Good, R. A., Witkin S. S. (1980). Circulating immune complexes associated with naturally occurring lymphosarcoma in pet cats. *J. Immunol.* **125**, 2363–2366.

de Vries, J. E. (1995). Immunosuppressive, and anti-inflammatory properties of interleukin 10. *Ann. Med.* **27**(5), 537–541.

Dean, G. A., Lavoy A., Pedersen, N. C. (1996a). FIV-associated changes in cytokine mRNA *in vivo*. In *3rd Int. Feline Retrovirus Research Symposium, March 3–9, 1996*, Fort Collins Colorado, USA.

Dean, G. A., Reubel, G. H., Moore, P. F., Pedersen, N. C. (1996b). Proviral burden and infection kinetics of feline immunodeficiency virus in lymphocyte subsets of blood and lymph node. *J. Virol.* **70**, 5165–5169.

Dingle, J. H., Fothergill, L. D., Chandler, C. A. (1938). Studies on *Haemophilus influenzae*. III. The failure of complement of some animal species, notably the guinea-pig, to activate the bactericidal function of sera of certain other species. *J. Immunol.* **34**, 357–391.

Doveren, R. F. C., Buurman, W. A., Schutte, B., Groenewegen, G., van der Linden, C. J. (1985). Class II antigens on canine lymphocytes. *Tissue Antigens* **25**, 255–265.

Dumont-Drieux, A. M., de Parseval, A., Heiber, M., Salmon, P., Pancino, G., Sonigo, P. (1992). Unusual amino acid sequence of the second Ig-like domain of the feline CD4 protein. *Aids Res. Hum. Retroviruses* **8**, 1581–1591.

Duncan, J. R., Prasse, K. W., Mahaffey, E. A. (1994). *Veterinary Laboratory Medicine,* 3rd edn, pp. 51–52. Iowa State University Press, Ames, IA, USA.

Dunham, S. P., Onions, D. E. (1996a). Feline recombinant stem cell factor expression, purification, and bioactivity *in vitro*, and *in vivo*. In *Workshop on Cytokines in FIV and HIV Infection, Pisa, Italy, Nov. 15–16*. p. 13.

Dunham, S. P., Onions, D. E. (1996b). The cloning and sequencing of cDNAs encoding two isoforms of feline stem cell factor. *DNA Seq.* **6**(4), 233–237.

Dunham, S. P., Argyle, D. J., Onions, D. E. (1995). The isolation, and sequence of canine interleukin–2. *DNA Seq.* 5(3), 177–180.

Durgut, R. (1996). A study of the pathogenesis of inflammatory bowel disease in the cat. MSc thesis, University of Bristol, UK.

Duthie, S., Nasir, L., Eckersall, P. D. (1996). *Felis catus* tumor necrosis factor receptor p60 (TNFR-1). In *Workshop on Cytokines in FIV and HIV Infection, Pisa, Italy, Nov. 15–16.* p. 12.

English, R. V., Nelson, P., Johnson, C. M., Nassise, M., Tompkins, W. A., Tompkins, M. B. (1994). Development of clinical disease in cats experimentally infected with feline immunodeficiency virus. *J. Inf. Dis.* **170**, 543–552.

Essex, M., Grant, C. K., Cotter, S. M., Sliski, A. H., Hardy, Jr., W. D. (1979). Leukemia specific antigens: FOCMA and immune surveillance. [Review]. *Hämatologie Bluttransfusion* **23**, 453–486.

Fehr, D., Dean, G. A., Huder, J., Zhanyun, F., Huettner, S., Higgins, J. W., Pedersen N. C., Lutz, H. (1998). Nucleotide, and predicted peptide sequence of feline interleukin-12 *DNA Seq.*

Fevereiro, M., Roneker, C., DeNoronha, F. (1993). Enhanced neutralization of feline immunodeficiency virus by complement viral lysis. *Vet. Immunol. Immunopathol.* **36**, 191–206.

Fittschen, C., Parmley, R. T., Bishop, S. P., Williams, J. C. (1988a). Morphometry of feline neutrophil granule genesis. *Am. J. Anat.* **181**, 195–202.

Fittschen, C., Parmley, R. T., Austin, R. L. (1988b). Ultrastructural cytochemistry of complex carbohydrates in developing feline neutrophils. *Am. J. Anat.* **181**, 149–162.

Fox, J. G., Perkins, S., Yan, L., Shen, Z., Attardo, L., Pappo, J. (1996). Local immune response in *Helicobacter pylori* infected cats and identification of *H. pylori* in saliva, gastric fluid and faeces. *Immunology* **88**, 400–406.

Frewein, J., Vollmerhaus, B. (eds) (1997). *Anatomie von Hund und Katze.* Blackwell, Berlin.

Friess, A. E., Schlüns, J. (1990). Das Immunsystem und die Organe der Abwehr. In *Zytologie, Histologie und Mikroskopische Anatomie der Haussäugetiere* (eds Mosimann, W., Kohler, T.), pp. 114–144. Verlag Paul Parey, Berlin, Germany.

Furutani, Y., Notake, M., Fukui, T., Ohne, M., Nomura, H., Yamada, M., Nakamura, S. (1986). Complete nucleotide sequence of the gene for human interleukin-1 alpha. *Nucl. Acids Res.* **143**, 3167–3179.

Gardner, M. B. (1991). Simian and feline immunodeficiency viruses: animal lentivirus models for evaluation of AIDS vaccines and antiviral agents. *Antiviral Res.* **15**(4), 267–286.

Gaskell, R. M. (1975). Studies on feline viral rhinotracheitis with particular reference to the carrier state. Ph.D. diss., Department of Veterinary Medicine, University of Bristol, UK.

Gershwin, L. J., Krakowa, S., Olsen, R. G. (1995). The lymphoid system, cells of the lymphoid system, In *Immunology and Immunopathology of Domestic Animals*, 2nd edn (eds Gershwin, L. J., Krakowa, S., Olsen, R. G.), pp. 13–29. Mosby, New York, USA.

Giger, U. (1991). Feline neonatal isoerythrolysis: a major cause of the fading kitten syndrome. In: *Proc Am Coll Vet Intern Med*, pp. 347–350.

Giger, U., Bücheler, J. (1991). Transfusion of type-A, and type-B blood to cats. *J. Am Vet Med Assoc*, **198**, 411–418.

Giger, U., Bücheler, J., Patterson, D. F. (1991a). Frequency, and inheritance of A and B blood types in feline breeds of the United States. *J. Hered.* **82**, 15–20.

Giger, U., Griot-Wenk, M. E., Bücheler, J. (1991b). Geographical variation of the feline blood type frequencies in the United States. *Feline Pract.* **19**, 21–27.

Goddard, L. E., Wardley, R. C., Gaskell, R. M., Gaskell, C. J. (1987). Antibody, complement mediated lysis of felid herpesvirus 1 infected cells *in vitro*. *Res. Vet. Sci.* , **42**, 307–312.

Goitsuka, R., Hasegawa, A. (1995). Detection of feline cytokine and cytokine receptor mRNA transcripts. In *Feline Immunology and Immunodeficiency* (eds Willett, B. J., Jarrett, O.), pp. 84–94. Oxford University Press, Oxford, UK.

Goitsuka, R., Hirota, Y., Hasegawa, A., Tomoda, I. (1987). Feline interleukin 1 derived from alveolar macrophages stimulated with lipopolysaccharide. *Jap. J. Vet. Sci.* **49**(4), 631–636.

Goitsuka, R., Onda, C., Hirota, Y., Hasegawa, A., Tomoda, I. (1988). Feline interleukin 1 production induced by feline infectious peritonitis virus. *J. Vet. Sci.* **50**(1), 209–214.

Goitsuka, R., Ohashi, T., Ono, K., Yasukawa, K., Koishibara, Y., Fukui, H., Ohsugi, Y., Hasegawa, A. (1990). IL-6 activity in feline infectious peritonitis. *J. Immunol.* **144**(7), 2599–2603.

Goitsuka, R., Ohno, K., Matsumoto, Y., Hayashi, N., Momoi, Y., Okamoto, Y., Watari, T., Tsujimoto, H., Hasegawa A. (1993). Establishment, and characterization of a feline large granular lymphoma cell line expressing interleukin 2 receptor alpha-chain. *J. Vet. Med. Sci.* **55**(5), 863–865.

Gorham, J. R., Henson, J. B., Dodgen, C. J. (1971). Basic principles of immunity in cats. *J. Am. Vet. Med. Assoc.* **158**, 846–853.

Gorman, N. T., Halliwell., R. E. W. (1989). Immunoglobulin quantitation and clinical interpretation. In *Veterinary Clinical Immunology* (eds Gorman, N. T., Halliwell, R. E. W.), pp. 55–73. W.B. Saunders Co., Philadelphia, PA, USA.

Goudswaard, J., van der Donk, J. A., Noordszij, A., van Dam, R. H., Vaerman., J. P. (1978). Protein A reactivity of various mammalian immunoglobulins. *Scand. J. Immunol.* **8**, 21–28.

Grant, C. K. (1977). Complement 'specificity', interchangeability: measurement of hemolytic complement levels, and use of the complement-fixation test with sera from common domesticated animals. *Am. J. Vet. Res.* **38**, 1611–1617.

Grant, C. K. (1995). Purification and characterization of feline IgM, IgA isotypes and three subclasses of IgG. In *Feline Immunology and Immunodeficiency* (eds Willett, B. J., Jarret, O.), Chapter 7, pp. 95–107. Oxford Science Publications, Oxford University Press, Oxford, UK.

Grant, C. K., DeBoer, D. J., Essex, M., Worley, M. B., Higgins, J. (1977). Antibodies from healthy cats exposed to feline leukemia virus lyse feline lymphoma cells slowly with cat complement. *J. Immunol.* **119**, 401–406.

Grant, C. K., Pickard, D. K., Ramaika, C., Madewel, B. R., Essex, M. (1979). Complement, tumor antibody levels in cats, and changes associated with natural feline leukemia virus infection, and malignant disease. *Cancer Res.* **39**, 75–81.

Grant, C. K., Michalek, M. T. (1981). Feline leukemia – unique, cross-reacting antigens on individual virus-producing tumors identified by complement dependent antibody. *Int. J. Cancer* **28**, 209–217.

Gray, G. D., Ohlman, G. M., Morton, D. R., Schaub, R. G. (1986). Feline polymorphonuclear leukocytes respond chemotactically to leukotriene B4 and activated serum but not to F-met-leu-phe. *Agents Actions* **18**, 401–406.

Griot-Wenk, M. E., Pahlsson, P., Chisholm-Chait, A., Spitalnik, P. F., Spitalnik, S. L., Giger, U. (1993). Biochemical characterization of the feline AB blood group system. *Anim Genet.* **24**, 401–407.

Griot-Wenk, M. E., Callan, M. B., Casal, M. L., Chisolm-Chait, A., Spitalnik, S. L., Patterson, D. F., Giger, U. (1996). Blood type AB in the feline AB blood group system. *Am J. Vet Res.* 57, 1438–1442.

Groshek, P. M., Dean, G. A., Hoover, E. A. (1994). Monoclonal antibodies identifying feline haematopoietic cell lineages. *Comp. Haematol. Int.* **4**, 181–191.

Hanlon, M. A., Marr, J. M., Hayes, K. A., Mathes, L. E., Stromberg, P. C., Ringler, S., Krakowka, S., Lafrado L. J. (1993). Loss of neutrophil and natural killer cell function following feline immunodeficiency virus infection. *Viral Immunol.* **6**, 119–124.

Harding, S. K., Brunner, D. W., Bryant, I. W. (1961). The transfer of antibody from mother cat to her kittens. *Cornell Vet.* **51**, 531–539.

Hasegawa, T., Hasegawa, A. (1991). Interleukin 1 alpha mRNA-expressing cells in the local inflammatory response in feline infectious peritonitis. *J. Vet. Med. Sci.* , **53**(6), 995–999.

Hassan, H. T., Zander, A. (1996). Stem cell factor as a survival, and growth factor in human normal, and malignant hematopoiesis. *Acta Haematol.* **95**(3–4), 257–262.

Hawkins, E. C., DeNicola, D. B. (1989). Collection of bronchoalveolar lavage fluid in cats using an endotracheal tube. *Am. J. Vet. Res.* **50**, 855–860.

Hiraga, C., Kanki, T., Ichikawa, Y. (1981). Immunobiological characteristics of germfree and specific pathogen-free cats. *Lab. Anim. Sci.* **31**, 391–396.

Honor, D. J., DeNicola, D. B., Turek, T. J., Brown, P. J. (1986). A neoplasm of globule leukocytes in a cat. *Vet. Pathol.* **23**, 287–292.

Hood, L., Gray, W. R., Sanders, B. G., Dreyer, W. J. (1967). Light chain evolution. *Cold Spring Harbour Symp. Quant. Biol.* **32**, 133–146.

Hoover, E. A., Olsen, R. G., Hardy, W. D., Schaller, J. P., Mathes, L. E. (1976). Feline leukemia virus infection: age-related variation in response of cats to experimental infection. *J. Natl Cancer Inst.* **57**, 365–369.

Hoover, E. A., Krakowka, S., Cockerell, G. L., Olsen, R. G. (1978). Thymectomy in preweanling kittens: technique and immunologic consequences. *Am. J. Vet. Res.* **38**, 99–103.

Hoover, E. A., Rojko, J. L., Wilson, P. L., Olsen, R. G. (1981). Determinants of susceptibility and resistance to feline leukemia virus. I. Role of macrophages. *J. Natl Cancer Inst.* **67**, 889–898.

Horimoto, T., Limcumpao, J. A., Tohya, Y., Takahashi, E., Mikami, T. (1989). Enhancement of neutralizing activity of anti-feline herpes type-1 sera by complement supplementation. *Nippon Juigaku Zasshi* **51**, 1025–1027.

Horton, M. A., Fowler, P., Simpson, A., Onions, D. (1988). Monoclonal antibodies to human antigens recognize feline myeloid cells. *Vet. Immunol. Immunopathol.* **18**, 213–217.

Hosie, M. J., Willett, B. J., Dunsford, T. H., Jarrett, O., Neil, J. C. (1993). A monoclonal antibody which blocks infection with feline immunodeficiency virus identifies a possible non-CD4 receptor. *J. Virol.* **67**, 1667–1671.

Hosoi, S., Borsos, T., Dunlop, N., Nara, P. L. (1990). Heat-labile, complement-like factor(s) of animal sera prevent (s) HIV-1 infectivity *in vitro*. *J. Acquir. Immune Defic. Syndr.* 3, 366–371.

Iwamoto, K., Takeishi, M., Takagi, K., Yukawa, M., Kuyama, T., Ishida, M., Kuramochi, T. (1989). Expression of interleukin-2 receptor (IL-2R) in feline peripheral blood lymphocytes. *Cell. Molec. Biol.* **35**(3), 279–284.

Jackson, M. L., Wood, S. L., Misra, V., Haines, D. M. (1996). Immunohistochemical identification of B and T lymphocytes in formalin-fixed, paraffin-embedded feline lymphosarcomas: relation to feline leukemia virus status, tumor site, and patient age. *Can. J. Vet. Res.* **60**, 199–204.

Jacobse-Geels, H. E. L., Daha, M. R., Horzinek, M. C. (1980). Isolation, characterization of feline C3, and evidence for the immune complex pathogenisis of feline infectious peritonitis. *J. Immunol.* **125**, 1606–1610.

Jacobse-Geels, H. E. L., Daha, M. R., Horzinek, M. C. (1982). Antibody, immune complexes, and complement activity fluctuations in kittens with experimentally induced feline infectious peritonitis. *Am. J. Vet. Res.* **43**, 666–670.

Jarrett, O., Russell, P. H., Stewart, M. F. (1977). Protection of kittens from feline leukaemia virus infection by maternally derived antibody. *Vet. Rec.* **101**, 304–305.

Jarrett, W. F. H., Crawford, E. M., Martin, W. B., Davie, F. (1964). A virus-like particle associated with leukemia (lymphosarcoma). *Nature* **202**, 567–568.

Johnson, P., Mackenzie, C. D., Denham, D. A., Suswillo, R. R. (1988). The effect of diethylcarbamazine on the *in vitro* serum-mediated adherence of feline granulocytes to microfilariae of *Brugia pahangi*. *Trop. Med. Parasitol.* **39**, 291–294.

Kapadia, S., Lee, J., Torre-Amione, G., Birdsall, H. H., Ma, T. S., Mann D. L. (1995). Tumor necrosis factor-alpha gene, and protein expression in adult feline myocardium after endotoxin administration. *J. Clin. Invest.* **96**(2), 1042–1052.

Khan, K. N., Kociba, G. J., Wellman, M. L. (1993). Macrophage tropism of feline leukemia virus (FeLV) of subgroup-C, and increased production of tumor necrosis factor-alpha by FeLV-infected macrophages. *Blood* **81**(10), 2585–2590.

Kiehl, A. R., Fettman, M. J., Quackenbush, S. L., Hoover, E. A. (1987). Effects of feline leukemia virus infection on neutrophil chemotaxis *in vitro*. *Am. J. Vet. Res.* **48**, 76–80.

Klotz, F. W., Cooper, M. D. (1986). A feline thymocyte antigen defined by a monoclonal antibody (FT2) identifies a subpopulation of non-helper cells capable of specific cytotoxicity. *J. Immunol.* **136**, 2510–2514.

Klotz, F. W., Gathings, W. E., Cooper, M. D. (1985). Development and distribution of B-lineage cells in the domestic cat: analysis with monoclonal antibodies to cat μ-, γ-, κ- and λ-chains and heterologous anti-α antibodies. *J. Immunol.* **134**, 95–100.

Kobayashi, M., Fitz, L., Ryan, M., Hewick, R. M., Clark, S. C., Chan, S., Loudon, R., Sherman, F., Perussia, B., Trinchieri, G. (1989). Identification, and purification of natural killer cell stimulatory factor (NKSF), a cytokine with multiple biologic effects on human lymphocytes. *J. Exp. Med.* **170**(3), 827–845.

Kobilinsky, L., Hardy, Jr., W. D., Ellis, R., Witkin, S. S., Day, N. K. (1980). *In vitro* activation of feline complement by feline leukemia virus. *Infect. Immun.* **29**, 165–170.

König, H. E. (1992). Immunsystem und lymphatische Organe. In *Anatomie der Katze* (ed. König, H.), pp. 124–131. Gustav Fischer Verlag, Stuttgart, Germany.

Kraus, L. A., Bradley, W. G., Engelman, R. W., Brown, K. M., Good, R. A., Day, N. K. (1996). Relationship between tumor necrosis factor alpha, and feline immunodeficiency virus expressions. *J. Virol.* 70(1), 566–569.

Kraut, E. H., Rojko, J. L., Olsen, R. G., Tuomari, D. L. (1985). Effects of cobra venom factor treatment on latent feline leukemia virus infection. *J. Virol.* **54**, 873–875.

Kraut, E. H., Rojko, J. L., Olsen, R. G., Tuomari, D. L. (1987). Effects of treatment with cobra venom factor on experimentally induced feline leukemia. *Am J. Vet. Res.* **48**, 1063–1066.

Laberge, S., Cruikshank, W., Kornfeld, H., Center, D. M. (1995). Histamine-induced secretion of lymphocyte chemoattractant factor from CD8$^+$ T cells is independent of transcription and translation. *J. Immunol.* **155**, 2902–2910.

Laberge, S., Cruikshank, W. W., Beer, D. J., Center, D. M. (1996). Secretion of IL-16 (lymphocyte chemoattractant factor) from serotonin-stimulated CD8$^+$ T cells *in vitro*. *J. Immunol.* **156**(1), 310–315.

Lafrado, L. J., Lewis, M. G., Mathes, L. E., Olsen, R. G. (1987). Suppression of *in vitro* neutrophil function by feline leukaemia virus (FeLV) and purified FeLV-p15E. *J. Gen. Virol.* **68** (Pt2), 507–513.

Larson, S., Flagstad, A., Aalbæk, B. (1976). Experimental feline panleukopenia in the conventional cat. *Vet. Pathol.* **13**, 216–240.

Lawrence, C. E., Callanan, J. J., Jarrett, O. (1992). Decreased mitogen responsiveness, and elevated tumor necrosis factor production in cats shortly after feline immunodeficiency virus infection. *Vet. Immunol. Immunopathol.* **35** (1–2), 51–59.

Lawrence, C. E., Callanan, J. J., Willett, B. J., Jarrett, O. (1995). Cytokine production by cats infected with feline immunodeficiency virus: a longitudinal study. *Immunology* **85**(4), 568–574.

Lefer, A. M., Ma, X. L. (1994). PMN adherence to cat ischemic-reperfused mesenteric vascular endothelium under flow: role of P selectins. *J. Appl. Phys.* **76**, 33–38.

Lehmann, R., Joller, H., Haagmans, B. L., Lutz, H. (1992). Tumor necrosis factor alpha levels in cats experimentally infected with feline immunodeficiency virus: effects of immunization, and feline leukemia virus infection. *Vet. Immunol. Immunopathol.* 35(1–2), 61–69.

Leiser, R., Kaufmann, P. (1994). Placental structure: in a comparative aspect. *Exp. Clin. Endocrinol.* **102**, 122–134.

Leutenegger, C. M., Huder, J., Lutz, H. (1998a). Cloning, molecular characterization and expression of feline interleukin-16. *DNA Seq.*, in press.

Leutenegger, C. M., Huder, J., Lutz, H. (1998b). Cloning, molecular characterization and expression of feline IL-10, in preparation.

Lin, D.-S. (1992). Feline Immune System. *Comp. Immun. Microbiol. Infect. Dis.* **15**, 1–11.

Lin, D.-S., Bowman, D. D. (1991). Cellular responses of cats with primary toxoplasmosis. *J. Parasitol.* 77, 272–279.

Lin, D.-S., Bowman, D. D. (1992). Macrophage functions in cats experimentally infected with feline immunodeficiency virus and *Toxoplasma gondii*. *Vet. Immunol. Immunopathol.* **33**, 69–78.

Lin, D.-S., Bowman, D. D. (1993). Detection of tumor necrosis factor activity in the supernatants of feline macrophages and lymphocytes. *Comp. Immunol. Microbiol. Infect. Dis.* **16**, 91–94.

Lindmark, R., Thoren-Tolling, K., Sjoquist, J. (1983). Binding of immunoglobulins to protein A and immunoglobulin levels in mammalian sera. *J. Immunol.* **62**, 1–13.

Linenberger, M. L., Abkowitz, J. L. (1992). *In vivo* infection of marrow stromal fibroblasts by feline leukemia virus. *Exp. Hematol.* **20**, 1022–1027.

Linenberger, M. L., Shelton, G. H., Persik, M. T., Abkowitz, J. L. (1991). Hematopoiesis in asymptomatic cats infected with feline immunodeficiency virus. *Blood* 78: 1963–1968.

Lloyd, A. R., Oppenheim, J. J. (1992). Poly's lament: the neglected role of the polymorphonuclear neutrophil in the afferent limb of the immune response. *Immunol. Today* **13**, 169–172.

Lyons, L. A., Langhlin, T. F., Copeland, N. G., Jenkins, N. A., Womack, J. E., O'Brien, S. J. (1997). Comparative anchor tagged sequences (CATS) for integrative mapping of mammalian genomes. *Nature Genetics* **15**(1), 47–56.

Ma, J., Kennedy-Stoskopf, S., Sellon, R., Tonkonogy, S., Hawkins, E. C., Tompkins, M. B., Tompkins, W. A. (1995). Tumor necrosis factor-alpha responses are depressed, and interleukin-6 responses unaltered in feline immunodeficiency virus infected cats. *Vet. Immunol. Immunopathol.* **46**(1–2), 35–50.

Ma, X. L., Tsao, P. S., Lefer, A. M. (1991). Antibody to CD-18 exerts endothelial and cardiac protective effects in myocardial ischemia and reperfusion. *J. Clin. Invest.* **88**, 1237–1243.

McCarthy, G. M., Quinn, P. J. (1989). Bronchoalveolar lavage in the cat: cytologic findings. *Can. J. Vet. Res.* **53**, 259–263.

McGraw, R. A., Coffee, B. W., Otto, C. M., Drews, R. T., Rawlings, C. A. (1990). Gene sequence of feline tumor necrosis factor alpha. *Nucl. Acids Res.* **18**(18), 5563.

Meier, A. (1989). Zur deskriptiven Anatomie der tastbaren Lymphknoten der Hauskatze. Diss. Med. Vet. München, Germany.

Micallef, M. J., Ohtsuki, T., Kohno, K., Tanabe, F., Ushio, S., Namba, M., Tanimoto, T., Torigoe, K., Fujii, M., Ikeda, M., Fukuda, S., Kurimoto, M. (1996). Interferon-gamma-inducing factor enhances T helper 1 cytokine production by stimulated human T cells: synergism with interleukin-12 for interferon-gamma production. *Eur. J. Immunol.* **26**(7), 1647–1651.

Monteith, C. E., Chelack, B. J., Davis, W. C., Haines, D. M. (1996). Identification of monoclonal antibodies for immunohistochemical staining of feline B lymphocytes in frozen and formalin-fixed paraffin-embedded tissues. *Can. J. Vet. Res.* **60**, 193–198.

Moreira, M., Wollerton, M., Monks, J., Proudfoot, N. J. (1995). Upstream sequence elements enhance poly(A) site efficiency of the C.2 complement gene, and are phylogenetically conserved. *EMBO J.* **14**, 3809–3819.

Moriello, K. A., Young, K. M., Cooley, A. J. (1993). Isolation of feline eosinophils via peritoneal lavage. *Am. J. Vet. Res.* **54**, 223–227.

Mosmann, T. R., Coffman, R. L. (1989). Th1 and Th2 cells: different patterns of lymphokine secretion lead to different functional properties. *Annu. Rev. Immunology* 7, 145–173.

Murohara, T., Buerke, M., Lefer, A. M. (1994). Polymorphonuclear leukocyte-induced vasocontraction and endothelial dysfunction. Role of selectins. *Arterio Thrombosis* **14**, 1509–1519.

Neefjes, J. J., Hensen, E. J., de Kroon, T. I. P., Ploegh., H. L. (1986). A biochemical characterization of feline MHC products: unusually high expression of class II antigens on peripheral blood lymphocytes. *Immunogenetics* **23**, 341–347.

Norimine, J., Miyazawa, T., Kawaguchi, Y., Tomonaga, K., Shin, Y.-S., Toyosaki, T., Mikami, T. (1992). A cDNA encoding feline CD4 has a unique repeat sequence downstream of the V-like region. *Immunology* **76**, 74–79.

Oberlin, E., Amara, A., Bachelerie, F., Bessia, C., Virelizier, J.-L., Arenzana-Seisdedos, F., Schwartz, O., Heard, J.-M., Clark-Lewis, I., Legler, D. F., Loetscher, M., Baggiolini, M., Moser, B. (1996). The CXC chemokine receptor SDF-1 is the ligand for LESTR/fusin and prevents infection by T-cell-line adapted HIV-1. *Nature* **382**, 833–835.

Obradovich, J. E., Ogilvie, G. K., Stadler-Morris, S., Schmidt, Br, Cooper, M. F., Boone, T. C. (1993). Effect of recombinant canine granulocyte colony-stimulating factor on peripheral blood neutrophil counts in normal cats. *J. Vet. Int. Med.* **7**, 65–67.

Ohashi, T., Goitsuka, R., Ono, K., Hasegawa, A. (1989). Feline hybridoma growth factor/interleukin-6 activity. *J. Leukocyte Biol.* **46**(6), 501–507.

Ohashi, T., Goitsuka, R., Watari, T., Tsujimoto, H., Hasegawa, A. (1992). Elevation of feline interleukin 6-like activity in feline immunodeficiency virus infection. *Clin. Immunol. Immunopathol.* **65**(3), 207–211.

Ohashi, T., Matsumoto, Y., Watari, T., Goitsuka, R., Tsujimoto, H., Hasegawa, A. (1993). Molecular cloning of feline interleukin-6 cDNA. *J. Vet. Med. Sci.* **55**(6), 941–944.

Ohno, K., Watari, T., Goitsuka, R., Tsujimoto, H., Hasegawa, A. (1992a). Altered surface antigen expression on peripheral blood mononuclear cells in cats infected with feline immunodeficiency virus. *J. Vet. Med. Sci.* **54**, 517–522.

Ohno, K., Goitsuka, R., Kitamura, K., Hasegawa, A., Tokunaga, T., Honda., M. (1992b). Production of a monoclonal antibody that defines the alpha-subunit of the feline IL-2 receptor. *Hybridoma* **11**, 595–605.

Ohno, K., Nakano, T., Matsumoto, Y., Watari, T., Goitsuka, R., Nakayama, H., Tsujimoto, H., Hasegawa A. (1993). Apoptosis induced by tumor necrosis factor in cells chronically infected with feline immunodeficiency virus. *J. Virol.* **67**(5), 2429–2433.

Okamura, H., Tsutsi, H., Komatsu, T., Yutsudo, M., Hakura, A., Tanimoto, T., Torigoe, K., Okura, T., Nukada, Y., Hattori, K. *et al.* (1995). Cloning of a new cytokine that induces IFN-gamma production by T cells. *Nature* **378**(6552), 88–91.

Okoshi, S., Tomoda, I., Makimura, S. (1968). Analysis of normal cat serum by immunoelectrophoresis. *Jap. J. Vet. Sci.* **29**, 337–345.

Olsen, R. G., Krakowka, S. G., Mathes, L. E., Yohn, D. S. (1974). Composite complement: a mixture of feline C1, guinea-pig complement as a reagent compatible with feline antibodies. *Am. J. Vet. Res.* **35**, 1389–1392.

Paul, W. E. (1995). *Fundamental Immunology*, 3rd edn. Raven Press, New York, USA.

Pecoraro, M. R., Kawaguchi, Y., Miyazawa, T., Norimine, J., Maeda, K., Toyosaki, T., Tohya, Y., Kai, C., Mikami, T. (1994a). Isolation, sequence and expression of a cDNA encoding the α-chain of the feline CD8. *Immunology* **81**, 127–131.

Pecoraro, M. R., Kawaguchi, Y., Okita, M., Inoshima, Y., Tohya, Y., Kai, C., Mikami, T. (1994b). Stable expression of the cDNA encoding the feline CD8-alpha gene. *J. Vet. Med. Sci.* **56**, 1001–1003.

Pedersen, N. C. (1987). Basic and clinical immunology. In *Diseases of the Cat: Medicine and Surgery* Vol. 1. (ed. Holzworth, J.), pp. 146–181. W.B. Saunders, Philadelphia, PA, USA.

Pedersen, N. C., Boyle, J. F., Floyd, K., Fudge, A., Barker, J. (1981). An enteric coronavirus infection of cats and its relationship to feline infectious peritonitis. *Am. J. Vet. Res.* **42**, 368–377.

Pedersen, N. C., Ho, E. W., Brown, M. L., Yamamoto, J. K. (1987). Isolation of a T-lymphotropic virus from domestic cats with an immunodeficiency-like syndrome. *Science* **235**, 790–793.

Perryman, L. E., Hoover, E. A., Yohn, D. S. (1972). Immunologic reactivity of the cat: immunosuppression in experimental feline leukemia. *J. Natl Cancer Inst.* **49**, 1357–1365.

Poli, A., Abramo, Taccini, F. C., Guidi, G., Barsotti, P., Bendinelli, M., Malvaldi, G. (1993). Renal involvement in feline immunodeficiency virus infection: a clinicopathological study. *Nephron* **64**, 282–288.

Pollack, M. S., Mastrota, F., Chin-Louie, J., Mooney, S., Hayes, A. (1982). Preliminary studies of the feline histocompatibility system. *Immunogenetics* **16**, 339–347.

Pu, R., Okada, S., Little, E. R., Xu, B., Stoffs, W., Yamamoto, J. K. (1995). Protection of neonatal kittens against feline immunodeficiency virus infection with passive maternal antiviral antibodies. *AIDS* **9**, 235–242.

Quackenbush, S. L., Mullins, J. I., Hoover, E. A. (1996). Replication kinetics and cell tropism of an immunosuppresive feline leukaemia virus. *J. Gen. Virol.* **77**, 1411–1420.

Reece, W. O. (1991). *Physiology of Domestic Animals*, pp. 317–329. Lea and Febiger, Philadelphia, PA, USA.

Rideout, B. A., Moore, P. F., Pedersen, N. C. (1990). Distribution of MHC class II antigens in feline tissues and peripheral blood. *Tissue Antigens* **36**, 221–227.

Rideout, B. A., Moore, P. F., Pedersen, N. C. (1992). Persistent up-regulation of MHC class II antigen on lymphocytes-T from cats experimentally infected with feline immunodeficiency virus. *Vet. Immunol. Immunopathol.* **35**, 71–81.

Rimstad, E., Reubel, G. H., Dean, G. A., Higgins, J., Pedersen, N. C. (1995). Cloning, expression, and characterization of biologically active feline tumour necrosis factor-alpha. *Vet. Immunol. Immunopathol.* **45**(3–4), 297–310.

Ritchey, J., Tompkins, M. (1996). Alveolar macrophage function during early post-FIV infection In *3rd. Int. Feline Retrovirus Research Symposium, March 3–9, 1996*, Fort Collins Colorado, USA.

Rohovsky, M. W., Griesemer, R. A. (1967). Experimental feline infectious enteritis in the germfree cat. *Path. Vet.* **4**, 391–410.

Roth, J. A. (1967). Possible association of thymus dysfunction with fading syndromes in puppies and kittens. *Vet. Clin. North Am.* **17**, 601–616.

Rottman, J. B., Freeman, E. B., Tonkonogy, S., Tompkins, M. B. (1996). A reverse transcription-polymerase chain reaction technique to detect feline cytokine genes. *Vet. Immunol. Immunopathol.* **45**(1–2): 1–18.

Sasso, E. H., Silverman, G. J., Mannik, M. (1991). Human IgA and IgG (Fab')2 that bind to Staphylococcal protein A belong to the VHIII subgroup. *J. Immunol.* **147**, 1877–1883.

Schijns, V. E., Wierda, C. M., van Dam, E. J., Vahlenkamp, T. W., Horzinek, M. C. (1995a). Molecular cloning of cat interleukin-4. *Immunogenetics.* **42**(5), 434–435.

Schijns, V. E., Wierda, C. M., Vahlenkamp, T., Horzinek, M. C., deGroot, R. J. (1995b). Molecular cloning, and expression of cat interferon-γ. *Immunogenetics* **42**, 440–441.

Schijns, V. E., Wierda, C. M. H., Vahlenkamp, T. W., Horzinek, M. C. (1997). Molecular cloning of cat interleukin-12. *Immunogenetics* **45**(6), 462–463.

Schultz, R. D., Scott, F. W., Duncan, J. R., Gillespie, J. H. (1974). Feline immunoglobulins. *Infect. Immun.* **9**, 391–393.

Schultz, R. D., Mendel, H., Scott, F. W. (1976). Effect of feline panleukopenia virus infection on development of humoral and cellular immunity. *Cornell Vet.* **66**, 324–332.

Scott, F. W. (1971). Comments on feline panleukopenia biologics. *JAVMA* **158**, 910–915.

Scott, F. W., Csiza, C. K., Gillespie, J. H. (1970). Maternally derived immunity to feline panleukopenia. *JAVMA* **156**, 439–453.

Seigel, L. J., Harper, M. E., Wong-Staal, F., Gallo, R. C., Nash, W. G., O'Brien, S. J. (1984). Gene for T-cell growth factor: location on human chromosome 4q, and feline chromosome B1. *Science* **223**(4632), 175–178.

Sellon, R. K., Levy, J. K., Jordan, H. L., Gebhard, D. H., Tompkins, M. B., Tompkins, W. A. (1996). Changes in lymphocyte subsets with age in perinatal cats: late gestation through eight weeks. *Vet. Immunol. Immunopathol.* **53**, 105–113.

Seppala, I., Sarvas, H., Peterfy, F., Mekela, O. (1981). The four subclasses of IgG can be isolated from mouse serum by using Protein A-Sepharose. *Scand. J. Immunol.* **14**, 335–340.

Shaw, A. R. E., Domanska, A., Mak, A., Gilchrist, A., Dobler, K., Visser, L., Poppema, S., Fliegel, L., Letarte, M., Willett, B. J. (1995). Ectopic expression of human and feline CD9 in a human B cell line confers b1 integrin-dependent motility on fibronectin and laminin substrates, and enhanced tyrosine phosphorylation. *J. Biol. Chem.* **41**, 24092–24099.

Spieker-Polet, Yam, H. P. C., Knight, K. L. (1993). Differential expression of 13 IgA-heavy chain genes in rabbit lymphoid tissues. *J. Immunol.* **150**, 5457–5465.

Stabenfeldt, G. H. (1992). The mammary gland. In *Textbook of Veterinary Physiology* (ed. Cunningham, J. G.), pp. 467–484. W.B. Saunders Co., Philadelphia, PA, USA.

Stewart, S. J., Fujimoto, J., Levy, R. (1986). Human T lymphocytes and monocytes bear the same Leu 3 (T4) antigen. *J. Immunol.* **136**, 3773–3778.

Stoddart, C. A., Scott, F. W. (1988). Isolation and identification of feline peritoneal macrophages for in vitro studies of coronavirus–macrophage interactions. *J. Leukoc. Biol.* **44**, 319–328.

Stoddart, C. A., Scott, F. W. (1989). Intrinsic resistance of feline peritoneal macrophages to coronavirus infection correlates with *in vivo* virulence. *J. Virol.* **63**, 436–440.

Sturgess, C. P. (1997). Studies on mucosal effector mechanisms in Feline Immunodeficiency Virus (FIV) infection in cats. PhD thesis, University of Bristol.

Sugimura, M., Kudo, N., Takahata, K. (1955). Studies on the lymphonodi of cats I. Macroscopical observations on the lymphonodi of heads and necks. *Jap. J. Vet. Res.* **3**, 90–104

Sugimura, M., Kudo, N., Takahata, K. (1956). Studies on the lymphonodi of cats II. Macroscopical observations on the lymphonodi of the body surfaces, thoracic and pelvic limbs. *Jap. J. Vet. Res.* **4**, 101–112.

Sugimura, M., Kudo, N., Takahata, K. (1958). Studies on the lymphonodi of cats III. Macroscopical observations on the lymphonodi in the abdominal and pelvic cavities. *Jap. J. Vet. Res.* **6**, 69–88.

Sugimura, M., Kudo, N., Takahata, K. (1959). Studies on the lymphonodi of cats IV. Macroscopical observations on the lymphonodi in the thoracic cavity, supplemental observations on those in the head and neck. *Jap. J. Vet. Res.* **7**, 27–52.

Theodore, A. C., Center, D. M., Cruikshank, W. W., Beer, D. J. (1986). A human T-cell hybridoma-derived lymphocyte chemoattractant factor. *Cell Immunol.* **98**(2), 411–421.

Thistlewaite, J. R. J., Pennington, L. R., Lunney, J. K., Sachs, D. H. (1983). Immunological characterization of MHC recombinant swine. *Transplantation* **35**, 394–400.

Tizard, I. R. (1996). *Veterinary Immunology an Introduction*, 5th edn, pp. 237–325. W.B. Saunders Co., Philadelphia, PA, USA.

Tompkins, M. B. (1993). Lymphoid system. In *Atlas of Feline Anatomy for Veterinarians*,. 1st edn (eds Hudson, L. C., Hamilton W. P.), pp. 113–126. W.B. Saunders, Philadelphia, PA, USA.

Tompkins, M. B., Ogilvie, G. K., Franklin, R. A., Kelley, K. W., Tompkins, W. A. (1987). Induction of IL-2, and lymphokine activated killer cells in the cat. *Vet. Immunol. Immunopathol.* **16**(1–2), 1–10,

Tompkins, M. B., Ogilvie, G. K., Gast, A. M., Franklin, R., Weigel, R., Tompkins, W. A. (1989a). Interleukin-2 suppression in cats naturally infected with feline leukemia virus. *J. Biol. Response Mod.* **8**(1), 86–96.

Tompkins, M. B., Pang, V. F., Michaely, P. A., Feinmehl, R. I., Basgall, E. J., Baszler, T. V., Zachary, J. F., Tompkins, W. A. (1989b). Feline cytotoxic large granular lymphocytes induced by recombinant human IL-2. *J. Immunol.* **143**, 749–754.

Tompkins, M. B., Novotney, C., Grindem, C. B., Page, R., English, R., Nelson, P., Tompkins, W. A. (1990a). Human recombinant interleukin-2 induces maturation and activation signals for feline eosinophils *in vivo*. *J. Leukoc. Biol.* **48**, 531–540.

Tompkins, M. B., Gebhard, D. H., Bingham, H. R., Hamilton, M. J., Davis, W. C., Tompkins, W. A. F. (1990b). Characterization of monoclonal antibodies to feline T lymphocytes and their use in the analysis of lymphocyte tissue distribution in the cat. *Vet. Immunol. Immunopathol.* **26**, 305–317.

Toth, T. E., Smith, B., Pyle, H. (1992). Simultaneous separation and purification of mononuclear and polymorphonuclear cells from the peripheral blood of cats. *J. Virol. Methods* **36**, 185–195.

Trinchieri, G. (1995). Interleukin-12: a proinflammatory cytokine with immunoregulatory functions that bridge innate resistance, and antigen-specific adaptive immunity. *Annu. Rev. Immunol.* **13**, 251–276.

Tsuji, M., Goitsuka, R., Hirota, Y., Hasegawa, A. (1989). Chemotactic responses of neutrophils in cats with spontaneous feline infectious peritonitis. *Nippon Juigaku Zasshi* **51**, 917–923.

Ushio, S., Namba, M., Okura, T., Hattori, K., Nukada, Y., Akita, K., Tanabe, F., Konishi, K., Micallef, M., Fujii, M., Torigoe, K., Tanimoto, T., Fukuda, S., Ikeda, M., Okamura, H., Kurimoto, M. (1996). Cloning of the cDNA for human IFN-gamma-inducing factor, expression in *Escherichia coli*, and studies on the biologic activities of the protein. *J. Immunol.* **156**(11), 4274–4279.

Vaage, J. (1994). Microvascular injury induced by intravascular platelet aggregation. An experimental study. *Scand. J. Thorac. Cardiovasc. Surg.* **28**, 127–133.

Vaerman, J. P., Heremans, J. F., van Kerckhoven, G. (1969). Identification of IgA in several mammalian species. *J. Immunol.* **103**, 1421–1423.

Vollmerhaus, B. (1981). Lymphatic system. In *The Anatomy of the Domestic Animals*, Vol. 3 (eds Nickel, R., Schummer, A., Seiferle, E.), pp. 269–440. Verlag Paul Parey, Berlin, Germany.

Vollmerhaus, B., Friess, A., Waibl, H. (1994). Immunorgane und Lymphgefässe. In *Anatomie von Hund und Katze* (eds Frewein, J., Vollmerhaus, B.), pp. 298–315. Blackwell Wissenschafts Verlag, Berlin, Germany.

Weiss, R. C., Cummins, J. M., Richards, A. B. (1991). Low-dose orally administered alpha interferon treatment for feline leukemia virus infection. *J. Am. Vet. Med. Assoc.* **199**(10), 1477–1481,

Weyrich, A. S., Buerke, M., Albertine, K. H., Lefer, A. M. (1995). Time course of coronary vascular endothelial adhesion molecule expression during reperfusion of the ischemic feline myocardium. *J. Leukocyte Biol.* 57, 45–55.

Willett, B. J., Callanan, J. J. (1995). The expression of leucocyte differentiation antigens in the feline immune system. In *Feline Immunology and Immunodeficiency* (eds Willett, B. J., Jarrett, O. J.), pp. 3–15. Oxford University Press, Oxford, UK.

Willett, B. J., Neil, J. C. (1995). cDNA cloning and eukaryotic expression of feline CD9. *Mol. Immunol.* **32**, 417–423.

Willett, B. J., Hosie, M. J., Callanan, J. J., Neil, J. C., Jarrett, O. (1993). Infection with feline immunodeficiency virus is followed by the rapid expansion of a $CD8^+$ lymphocyte subset. *Immunology* 78, 1–6.

Willett, B. J., de Parseval, A., Peri, E., Rocchi, M., Hosie, M. J., Randall, R., Klatzmann, D., Neil, J. C., Jarrett, O. (1994a). The generation of monoclonal antibodies recognising novel epitopes by immunisation with solid matrix antigen-antibody complexes reveals a polymorphic determinant on feline CD. *J. Immunol. Methods* **176**, 213–220.

Willett, B. J., Hosie, M. J., Jarrett, O., Neil, J. C. (1994b). Identification of a putative cellular receptor for feline immunodeficiency virus as the feline homologue of CD9. *Immunology* **81**, 228–233.

Winkler, G. C. (1988). Pulmonary intravascular macrophages in domestic animal species: review of structural and functional properties. *Am. J. Anat.* **181**, 217–234.

Winkler, C., Schultz, A., Cevario, S., O'Brien, S. J. (1989). Genetic characterization of FLA, the cat major histocompatibility complex. *Proc. Natl Acad. Sci. USA* **86**, 943–947.

Yamada, T., Nagai, Y., Matsuda, M. (1991). Changes in serum immunoglobulin values in kittens after ingestion of colostrum. *Am. J. Vet. Res.* **52**, 393–396.

Yamamoto, J. K., Okuda, T., Ackley, C. D., Louie, H., Pembroke, E., Zochlinski, H., Munn, R. J., Gardner, M. B. (1991). Experimental vaccine protection against feline immunodeficiency virus. *AIDS Res. Hum. Retroviruses* 7(11), 911–922.

Yamamoto, J. K., Hohdatsu, T., Olmsted, R. A., Pu, R., Louie, H., Zochlinski, H. A., Acevedo, V., Johnson, H. M., Soulds, G. A., Gardner, M. B. (1993). Experimental vaccine protection against homologous, and heterologous strains of feline immunodeficiency virus. *J. Virol.* **67**(1), 601–605.

Young, H. A., Hardy, K. J. (1995). Role of interferon-γ in immune cell regulation. *J. Leukocyte Biol.* **58**(4), 373–381.

Yuhki, N. (1995). The feline major histocompatibility complex. In *Feline Immunology and Immunodeficiency*, pp. 45–64. Oxford University Press, Oxford, UK.

Yuhki, N., O'Brien, S. J. (1988). Molecular characterization and genetic mapping of class I and class II MHC genes of the domestic cat. *Immunogenetics* **16**, 414–425.

Zaccaro, L., Falcone, M. L., Silva, S., Bigalli, L., Cecchettini, A., Malvaldi, G., Bendinelli, M. (1995). Defective natural killer cell cytotoxic activity in feline immunodeficiency virus-infected cats. *AIDS Res. Hum. Retroviruses* **11**, 747–752.

Zeidner, N. S., Rose, L. M., Mathiason-DuBard, C. K., Myles, M. H., Hill, D. L., Mullins, J. I., Hoover, E. A. (1990). Zidovudine in combination with alpha interferon, and interleukin-2 as prophylactic therapy for FeLV-induced immunodeficiency syndrome (FeLV-FAIDS). *J. AIDS* 3(8), 787–796

Zeidner, N. S., Mathiason-DuBard, C. K., Hoover, E. A. (1993). Reversal of feline leukemia virus infection by adoptive transfer of lectin/interleukin-2-activated lymphocytes, interferon-alpha, and zidovudine. *J. Immunotherapy* **14**(1), 22–32.

Zhao, Y., Gebhard, D., English, R., Sellon, R., Tompkins, M., Tompkins, W. (1995). Enhanced expression of novel $CD57^+CD8^+$ LAK cells from cats infected with FIV. *J. Leukoc. Biol.* **58**, 423–431.

Suggested reading for section 14

Auer, L., Bell, K. (1980). The AB blood group system in the domestic cat. *Anim. Blood Groups Biochem. Genet.* **11**, 63–64.

Auer, L., Bell, K., Coates, S. (1982). Blood transfusion reactions in the cat. *J. Am. Vet. Med. Assoc.* **180**, 729–730.

Day, M. J. (1996a). Diagnostic assessment of the feline immune system, part 1. *Feline Pract.* **24**(2), 24–27.

Day, M. J. (1996b). Diagnostic asssesment of the feline immune system, part 2. *Feline Pract.* **24**(3), 14–25.

Day, M. J. (1996c). Diagnostic assessment of the feline immune system, part 3. *Feline Pract.* **24**(4), 7–12.

Day, M. J., Hanlon, L., Powell, L. M. (1993). Immune mediated skin diseases in the dog and cat. *J. Comp. Pathol.* **109**, 395–407.

Giger, U., Griot-Wenk, M., Bucheler, J., Smith, S., Diserens, D., Hale, A. (1991). Geographic variation of the feline blood type frequencies in the United States. *Feline Pract.* **19**(6), 21–26.

Griot-Wenk, M., Pahisson, P., Chisholm-Chait, A., Spitalnik, B. E., Spitalnik, S. L., Giger, U. (1993). Biochemical characterization of the feline AB blood group system. *Anim. Genet.* **24**, 401–407.

Hawkins, E. C., Feldman, B. F., Blanchard, P. C. (1986). Immunoglobulin A myeloma in a cat with pleural effusion and serum hyperviscosity. *J. Am. Vet. Med. Assoc.* **188**, 876–878.

Kalaher, K. M., Scott, D. M. (1991). Discoid lupus erythematosus in a cat. *Feline Pract.* **19**(1), 7–11.

O'Dair, H. A., Holt, P. E., Pearson, G. R., Gruffyd-Jones, T. J. (1991). Acquired immune mediated myasthenia gravis in a cat associated with a cystic thymus. *J. Sm. Anim. Pract.* **32**, 198–202.

Pedersen, N. C. (1987). Basic and clinical immunology. In *Diseases of the Cat* (ed. Holzworth, J.), pp. 146–181. W.B. Saunders, Philadelphia, PA, USA.

Pedersen, N. C. (1988). *Feline Infectious Diseases*. American Veterinary Publications, Goleta, CA, USA.

Pedersen, N. C. (1991). Common infectious diseases of multiple-cat environments. In *Feline Husbandry* (ed. Pedersen, N. C.), pp. 163–288. American Veterinary Publications, Goleta, CA, USA.

Pedersen, N. C., Barlough, J. E. (1991). Systemic lupus erythematosus in the cat. *Feline Pract.* **19**(3), 5–13.

Pedersen, N. C., Pool, R. R., O'Brien, T. (1980. Feline chronic progressive polyarthritis. *Am. J. Vet. Res.* **41**, 522–535.

Scott, D. W., Miller, W. H., Griffith, C. E. (1995). *Small Animal Dermatology*, 5th edn. W.B. Saunders, Philadelphia, PA, USA.

Warner, L. L., Gorman, N. T. (1984). Immune-mediated disorders of cats. *Vet. Clin. North Am.* **14**, 1039–1064.

Suggested reading for section 15

Day, M. J. (1996a). Diagnostic assessment of the feline immune system, part 1. *Feline Pract.* **24**(2), 24–27.

Day, M. J. (1996b). Diagnostic assessment of the feline immune system. Part 2. *Feline Pract.* **24**(3), 14–25.

Day, M. J. (1996c). Diagnostic assessment of the feline immune system, part 3. *Feline Pract.* **24**(4), 7–12.

Pedersen, N. C. (1987). Basic and clinical immunology. In *Diseases of the Cat* (ed. Holzworth, J.), pp. 146–181. W.B. Saunders, Philadelphia, PA, USA.

Pedersen, N. C. (1988). Feline leukemia virus. In *Feline Infectious Diseases*, pp. 83–106. American Veterinary Publications, Goleta, CA, USA.

Pedersen, N. C. (1991). Common infectious diseases of multiple-cat environments. In *Feline Husbandry* (ed. Pedersen, N. C.), pp. 163–288. American Veterinary Publications, Goleta, CA, USA.

Pedersen, N. C. (1993). The feline immunodeficiency virus. In: *The Retroviridae*, Vol. 2 (ed. Levy, J. A.), pp. 181–228. Plenum Press, New York, USA.

Willett, B. J., Jarrett, O. (1995). *Feline Immunology and Immunodeficiency*. Oxford Science Publications, Oxford, UK.

Suggested reading for section 16

Callanan, J. J., Jones, B. A., Irvine, J., Willett, B. J., McCandlish, I. A. P., Jarrett, O. (1996). Histologic classification, and immunophenotype of lymphosarcomas in cats with naturally, and experimentally acquired feline immunodeficiency virus infections. *Vet. Pathol.* **33**, 264–272.

Evans, R. J., Gorman, N. T. (1987). Myeloproliferative disease in the dog and cat: definition, aetiology, and classification. *Vet. Rec.* **121**, 437–443.

Grindem, C. B., Perman, V., Stevens, J. B. (1985). Morphological classification, and clinical, and pathological characteristics of spontaneous leukemia in 10 cats. *J. Am. Anim. Hosp. Assoc.* **21**, 227–236.

Jackson, M. L., Haines, D. M., Meric, S. M., Misra, V. (1993). Feline leukemia virus detection by immunohistochemistry, and polymerase chain reaction in formalin-fixed, paraffin-embedded tumor tissue from cats with lymphosarcoma. *Can. J. Vet. Res.* **57**, 269–276.

Jackson, M. L., Wood, S. L., Misra, V., Haines, D. M. (1996). Immunohistochemical identification of B and T lymphocytes in formalin-fixed, paraffin-embedded feline lymphosarcomas: relation to feline leukemia virus status, tumor site, and patient age. *Can. J. Vet. Res.* **60**, 199–204.

Jain, N. C. (1993). Classification of myeloproliferative disorders in cats using criteria proposed by the animal leukaemia study group: a retrospective study of 181 cases (1969–1992). *Comp. Haematol. Int.* **3**, 125–134.

Reinacher, M., Theilen, G. H. (1987). The frequency, and significance of FeLV infection in necropsied cats. *Am. J. Vet. Res.* **48**, 939–945.

Reinacher, M., Wittmer, G., Koberstein, H., Failing, K. (1995). Untersuchungen zur Bedeutung der FeLV-Infektion fuer Erkrankungen bei Sektionskatzen. *Berl. Muench. Tieraerztl. Wschr.* **108**, 58–60.

Rezanka, L. J., Rojko, J. L., Neil, J. C. (1992). Feline leukemia virus: pathogenesis of neoplastic disease. *Cancer Invest.* **10**, 371–389.

Shelton, G. H., Grant, C. K., Cotter, S. M., Gardner, M. B., Hardy, W. D., Jr., DiGiacomo, R. F. (1990). Feline immunodeficiency virus, and feline leukemia virus infections, and their relationships to lymphoid malignancies in cats: a retrospective study (1968–1988). *J. Acq. Immunodef. Syndr.* **3**, 623–630.

Valli, V. E. O., Parry, B. W. (1992a). The hematopoietic system: The leukon: IV, Myeloproliferative disease; V, Myelodysplastic syndromes; VI, Lymphoproliferative disease. In *Pathology of Domestic Animals*, Vol. 3, 4th edn (eds Jubb, K. V. F., Kennedy, P. C., Palmer, N.), pp. 114–157. Academic Press, San Diego, CA, USA.

Valli, V. E. O., Parry, B. W. (1992b). Lymphoreticular tissues: I, Thymus; E, Hyperplastic and neoplastic diseases of the thymus. In *Pathology of Domestic Animals*, Vol. 3, 4th edn (eds Jubb, K. V. F., Kennedy, P. C., Palmer, N.), pp. 218–221. Academic Press, San Diego, CA, USA.

IX IMMUNOLOGY OF MUSTELIDAE

1. Introduction

This review includes the following species: Eurasian badger (*Meles meles*), domestic ferret (*Mustela putorius furo*), marten (*Martes martes/foina/americana*), mink (*Mustela vison*), river otter (*Lutra canadensis*), sea otter (*Enhydra lutris*), striped skunk (*Mephitis mephitis*) and weasel (*Mustela erminia/frenata/rixosa/sibirica*). Little information on immunological parameters is available for most of these species. No relevant immunological data are presently available for marten and weasel.

2. Badger

In an assessment for a potential vaccine, a vaccinia recombinant virus expressing the rabies glycoprotein was tested in several wild animal species. The immunization of badgers against rabies with the use of this recombinant virus was also investigated. All badgers showed seroconversion, however, only 50% of the badgers orally administered this vaccine were protected against rabies (Brochier *et al.*, 1989).

Badgers appear to be highly susceptible to tuberculosis and are an epidemiological reservoir for *Mycobacterium bovis* in England (Little *et al.*, 1982). Enzyme-linked immunosorbent assays (ELISA) for the detection of antibodies to *M. bovis* in infected badgers have been developed (Mahmood *et al.*, 1987a; Goodger *et al.*, 1994a,b; Clifton-Haley *et al.*, 1995) and a monoclonal mouse anti-badger IgG has been produced (Goodger *et al.*, 1994a). Infected wild badgers may have serum antibodies to *M. bovis* and/or may show delayed type hypersensitivity, as demonstrated by positive skin test (Higgins, 1985; Mahmood *et al.*, 1987b). In experimental *M. bovis* infections, badgers have a positive skin test to tuberculin and positive lymphocyte transformation to whole BCG *in vitro* during the early stage of infection. With progression of the disease the lymphocyte responses decrease whereas antibody titers increase markedly (Mahmood *et al.*, 1987a). Thus, the spread of the infection in the body correlates with waning of the cell-mediated immunity.

3. Ferret

Although the ferret is a popular research animal (Frederick and Babish, 1985), there is very little information on the ferret's immune system. Infections of ferrets have been used as animal models for infections with human pathogens, such as infection with *Strongyloides stercoralis* (Davidson, 1988), *Pneumocystis carinii* (Stokes *et al.*, 1987; Gigliotti and Hughes, 1988), *Helicobacter pylori/H. mustelae* (Lee, 1995a,b), influenza virus (Small *et al.*, 1976; Barber and Small, 1978; Bird *et al.*, 1983) and measles virus (Brown *et al.*, 1985). Measles virus-specific IgG has been found in the brains of patients with subacute sclerosing panencephalitis (SSPE), a slowly progressing central nervous system disease affecting children. Immunoglobulin could also be demonstrated in the brain of ferrets infected with measles virus. Evidence was provided that the immunoglobulin found in the SSPE ferret brains was actively synthesized in the central nervous system and that some of the immunoglobulin was present in the form of immune complexes (Brown *et al.*, 1985).

Ferrets immunized against the canine heartworm *Dirofilaria immitis* by infection and subsequent ivermectin treatment were partially resistant to challenge infections (Blair and Campbell, 1981). Rabies virus vaccines and distemper vaccines have been evaluated in ferrets by protection tests or measurements of antibody responses (Appel and Harris, 1988; Hoover *et al.*, 1989; Rupprecht *et al.*, 1990b). Depending on the dose, killed organisms of *Mycobacterium bovis* either suppressed or enhanced transformation of ferret peripheral blood mononuclear cells (Thorns and Morris, 1984).

Ferret sera evaluated for complement activity were found to have strong hemolytic activity for sheep and rabbit erythrocytes (Higgins and Langley, 1985; Ish *et al.*, 1993). There is no information available on individual complement components.

4. River Otter

Significant increases in serum virus-neutralizing antibodies were found in river otters after their vaccination with canine adenovirus type 2 and feline calici virus (Hoover *et al.*, 1985).

Handbook of Vertebrate Immunology
ISBN 0-12-546401-0

5. Sea Otter

Interleukin-6 (IL-6) cDNA fragments of the sea otter have been cloned and sequenced. IL-6 of the sea otter shares 40 invariant amino acids with 11 other mammalian species (King *et al.*, 1996).

6. Skunk

Skunks can chronically harbor leptospires in the tubular lumina of their kidneys (Tabel and Karstad, 1967). Local production of antibodies to *Leptospira pomona* has been demonstrated to occur in kidneys of chronically infected skunks (Tabel, 1970).

Skunks are a major reservoir for rabies virus in North America. Oral administration of modified live rabies virus vaccines did not provide protection against challenge (Tolson *et al.*, 1988; Rupprecht *et al.*, 1990a). Intranasal instillation of modified live rabies vaccines caused disease (Tolson *et al.*, 1988; Rupprecht *et al.*, 1990a). Intramuscular injection of modified live virus vaccine (ERA/BHK-21) resulted in protection against challenge (Tolson *et al.*, 1988). It has been reported that intramuscular injections of modified live rabies vaccine into skunks that were trapped and released successfully immunized 54–72% of the skunk population in an area of Toronto, Canada (Rosatte *et al.*, 1993).

7. Mink

Introduction

Aleutian disease of mink, also called viral plasmacytosis, is caused by Aleutian disease virus (ADV) (reviewed by Porter *et al.*, 1980; Porter, 1986; Bloom *et al.*, 1994). ADV is a parvovirus which causes persistent infection. It plays havoc with the mink's immune system. Infected mink develop a marked hypergammaglobulinemia (Tabel and Ingram, 1970). Serum IgG concentrations may rise up to 75 mg/ml and as much as 80% of the IgG has been found to be antibody to ADV (Porter *et al.*, 1980; Porter, 1986). Lethal outcome of the disease results from continued formation of immune complexes, leading to severe glomerulonephritis (Porter *et al.*, 1980). The mechanism of the exessive humoral immune response is not known. There is evidence that macrophages are the major target cell of the virus (Kanno *et al.*, 1993a,b) and that infection of macrophages leads to enhanced production of interleukin 6 (IL-6) (Bloom *et al.*, 1994). Table IX.7.1 presents some references on aspects of this viral plasmacytosis.

Table IX.7.1 Aleutian disease of mink (viral plasmacytosis)

Reviews	Porter *et al.*, 1980; Porter, 1986; Bloom *et al.*, 1994
Target cell of ADV: macrophages, follicular dendritic cells	Mori *et al.*, 1991; Kanno *et al.*, 1993a,b; Bloom *et al.*, 1994
Measurements of antiviral antibodies	Cho and Greenfield, 1978; Porter *et al.*, 1980; Aasted and Cohn, 1982
Effect of antiviral antibodies on virus multiplication; antibody-dependent enhancement of infection	Porter, 1986; Alexandersen *et al.*, 1989, 1994; Bloom *et al.*, 1994
T cell responses	An and Wilkie, 1981; Aasted, 1989
Hypergammaglobulinemia	Tabel and Ingram, 1970; Bazeley, 1976; Porter *et al.*, 1980
Pathogenesis	Cheema *et al.*, 1972; Portis and Coe, 1979; Porter *et al.*, 1980; Porter, 1986; Lodmell *et al.*, 1990; Mori *et al.*, 1991; Kanno *et al.*, 1993a,b; Bloom *et al.*, 1994

Leukocytes and their markers

Aasted *et al.* (1988) tested nine commercially available monoclonal antibodies (mAbs) specific for human leukocyte antigens on mink lymphocytes. A mAb anti-CD4 and an anti-CD8 reacted with mink lymphocytes; a mAb anti-HLA-DR also reacted with mink lymphocytes. During the development of Aleutian disease, $CD8^+$ T cells were found to double in numbers (Aasted, 1989). Jacobson *et al.* (1993) tested 20 mAbs specific for human leukocyte antigens on mink leukocytes. An anti-CD14 mAb reacted with mink monocytes and an anti-CD18 mAb reacted with mink lymphocytes, monocytes and granulocytes. Miyazawa *et al.* (1994) generated several hybridoma clones which produced mAbs reacting with subpopulations of mink lymphocytes. mAbs MTS-4.3 and MTS-9.3 reacted with mink T-cells whereas MTS-5.6 reacted with dendritic cells.

Cytokines

Nucleotide sequences of mink IL-6 cDNA have close identity with those of sea otter and dog (King *et al.*, 1996). There is enhanced IL-6 production by macrophages after infection with ADV (Bloom *et al.*, 1994). Computer analysis of the ADV genome sequence revealed three copies of CTGGGA, a sequence identical to the consensus for an IL-6-responsive enhancer element (Bloom *et al.*, 1994). A mink lung epithelial cell line, Mv1Lu (CCL64) has been extensively used for a bioassay of TGF-β (Garrigue-Antar *et al.*, 1995).

Immunoglobulins (Table IX.7.2)

Immunoelectrophoretic analysis provided evidence that mink have four subclasses of IgG: IgG1, IgG2a, IgG2b and IgG2c (Tabel and Ingram, 1972). The mink is a unique species with respect to expressing allotypic polymorphism of Ig lambda chains (Fomicheva, 1991). The gene for mink lambda light chain constant polypeptide was assigned to mink chromosome 4 (Khlebodarova *et al.*, 1992). The frequencies of IgG heavy chain allotypes H3 and H4 were significantly higher in ADV-infected mink than in normal mink from the same population (Fomicheva, 1991).

Table IX.7.2 Immunoglobulins of mink

Immune responses to	
Bacille Calmette-Guerin	Smits *et al.*, 1996
DNP-KLH	Smits and Godson, 1996
Distemper virus	Blixenkrone-Moller *et al.*, 1991
Classes, subclasses	
IgM	Coe and Hadlow, 1972; Coe and Race, 1978; Blixenkrone-Moller *et al.*, 1991
IgG	Coe and Hadlow, 1972; Tabel and Ingram, 1972; Coe and Race, 1978; Galakhar *et al.*, 1988; Ufimtseva and Galakhar, 1991; Peremislov *et al.*, 1992
IgA	Coe and Hadlow, 1972; Coe and Race, 1978; Portis and Coe, 1979
Allotypes of	
IgG, heavy chain	Baranov *et al.*, 1981; Taranin *et al.*, 1987; Fomicheva and Volkova, 1990; Fomicheva *et al.*, 1990, 1991a,b; Fomicheva, 1991; Mechetina *et al.*, 1992a,b
Light chain	Volkova *et al.*, 1987; Najakshin *et al.*, 1990, 1991, 1993; Khlebodarova *et al.*, 1992; Bovkun *et al.*, 1993

Major histocompatibility complex (MHC) antigens

As already stated a mAb anti-HLA-DR reacted with mink lymphocytes (Aasted *et al.*, 1989). A study of mink MHC I molecules indicated a restricted polymorphism of the MHC I molecules within the population of mink investigated (Wienberg and Aasted, 1991). One group of 110 mink were all infected with ADV and when the profiles of these mink were analysed for progressive versus nonprogressive disease status, progressive Aleutian disease was associated almost exclusively with only two MHC profiles.

Ontogeny and passive transfer of immunoglobulins

No IgM and IgA was found in serum of newborn mink kits who had not yet suckled (Coe and Race, 1978). An average of 0.56 mg IgG/ml was found in serum of kit which had not suckled. Serum IgG levels of suckling kits steadily increased from the time of birth. Surprisingly, they reached a peak (7–10 mg/ml) as late as 8 days postpartum. IgG serum levels decreased, before weaning, during week 5 of life to a mean of 2.9 mg/ml. IgM and IgG synthesis was found in cultures of neonatal spleen cells collected 1–2 days postpartum (Coe and Race, 1978).

Passive transfer of antiviral antibodies restricts replication of ADV in infected kits (Alexandersen *et al.*, 1989).

Nonspecific immunity

ADV appears to replicate in macrophages of infected mink (Mori *et al.*, 1991) and was found to impair phagocytosis by the monophagocytic system of infected mink (Lodmell *et al.*, 1990).

Complement

Complement component C3 products have been demonstrated by immunofluorescence in glomeruli of mink affected by progressive Aleutian disease (Cheema *et al.*, 1972; Porter *et al.*, 1980).

Immunodeficiency

Certain mink breeds have been found to carry hereditary abnormal granules in their leukocytes (Leader *et al.*, 1963; Padgett *et al.*, 1963). This genetic condition called Chediak–Higashi syndrome is an autosomal recessive disease which has been described in humans, mink, cats, cattle, mice, killer whales, blue foxes, and silver foxes (Penner and Prieur, 1987). Complementation analysis indicates that the same or homologous genes are defective in mice, mink and humans (Penner and Prieur, 1987; Perou and Kaplan, 1993; Perou *et al.*, 1996). Mononuclear leukocytes derived from mink afflicted with the Chediak–Higashi syndrome were found to be defective in chemotaxis (Gallin *et al.*, 1975). Mink carrying the Chediak–Higashi trait are highly susceptible to progressive Aleutian disease (Porter *et al.*, 1980).

Autoimmunity

Mink affected by Aleutian disease have been found to have higher than normal levels of autoantibodies to erythrocytes, IgG and DNA (Saison and Karstad, 1968; Hahn and Kenyon, 1980; Mouritsen *et al.*, 1989). There is, however, no conclusive evidence that these antibodies are a major contributing factor to the pathology of Aleutian disease.

8. Conclusions

This chapter has shown that there is scant information on the immune system of most of the mustelid species. The economic importance of mink for the fur industry stimulated a modest interest into research on mink. Aleutian disease of mink is an intriguing affliction problem. Research into the mechanism of the excessive humoral immune response to AVD might provide some clues to the pathogenesis of Castleman's disease in humans (Bloom *et al.*, 1994).

9. References

Aasted, B. (1989). Mink infected with Aleutian disease virus have an elevated level of CD8-positive T-lymphocytes. *Vet Immunol. Immunopathol.* **203**, 375–385.

Aasted, B., Cohn, A. (1982). Inhibition of precipitation in counter current electrophoresis. A sensitive method for detection of mink antibodies to Aleutian disease virus. *Acta Pathol. Microbiol. Immunol. Scand. C.* **90**, 15–19.

Aasted, B., Blixenkrone-Moller, M., Larsen, E. B., Bielefeldt-Ohmann, H., Simesen, R. B., Uttenthal, A. (1988). Reactivity of eleven anti-human leucocyte monoclonal antibodies with lymphocytes from several domestic animals. *Vet Immunol. Immunopathol.* **19**, 31–38.

Alexandersen, S., Larsen, S., Cohn, A., Uttenthal, A., Race, R. E., Aasted, B., Hansen, M., Bloom, M. E. (1989). Passive transfer of antiviral antibodies restricts replication of Aleutian mink disease parvovirus *in vivo*. *J. Virol.* *63*, 9–17.

Alexandersen, S., Storgaard, T., Kamstrup, N., Aasted, B., Porter, D. D. (1994). Pathogenesis of Aleutian mink disease parvovirus infection: effects of suppression of antibody response on viral mRNA levels and on development of acute disease. *J. Virol.* **68**, 738–749.

An, S. H., Wilkie, B. N. (1981). Mitogen- and viral antigen-induced transformation of lymphocytes from normal mink and from mink with progressive or nonprogressive Aleutian disease. *Infect. Immun.* **34**, 111–114.

Appel, M. J., Harris, W. V. (1988). Antibody titers in domestic ferret jills and their kits to canine distemper virus vaccine. *Am. J. Vet. Med. Assoc.* **193**, 332–333.

Baranov, O. K., Fomicheva, I. I., Ternovsky, D. V., Ternovskaya, J. G. (1981). Interspecific distribution of allotypic mink (*Mustela vison*) IgG antigens. *J. Immunogenet.* **8**, 249–256.

Barber, W. H, Small, Jr, P. A. (1978). Local and systemic immunity to influenza virus infections in ferrets. *Infect. Immun.* **21**, 221–228.

Bazeley, P. L. (1976). The nature of Aleutian disease in mink. I. Two forms of hypergammaglobulinemia as related to method of disease transmission and type of lesion. *J. Inf. Dis.* **134**, 252–257.

Bird, R. A., Sweet, C., Husseini, R. H., Smith, H. (1983). The similar interaction of ferret alveolar macrophages with influenza virus strains of differing virulence at normal and pyrexial temperatures. *J. Gen. Virol.* **64**, 1807–1810.

Blair, L. S., Campbell, W. C. (1981). Immunizations of ferrets against *Dirofilaria immitis* by means of chemically abbreviated infections. *Parasite. Immunol.* **3**, 143–147.

Blixenkrone-Moller, M., Pedersen, I. R, Appel, M. J., Griot, C. (1991). Detection of IgM antibodies against canine distemper virus in dog and mink sera employing enzyme-linked immunosorbent assay (ELISA). *J. Vet. Diagn. Invest.* **3**, 3–9.

Bloom, M. E., Kanno, H., Mori, S., Wolfinbarger, J. B. (1994). Aleutian mink disease: puzzles and paradigms. *Infec. Agent. Dis.* **3**, 179–301.

Bovkun, L. A., Peremislov, V. V., Nayakshin, A. M., Belousov, E. S., Mechetina, L. V., Aasted, B., Taranin, A. V. (1993). Expression of immunoglobulin kappa and lambda chains in mink. *Eur. J. Immunol.* **23**, 1929–1934.

Brochier, B., Blancou, J., Thomas, I., Languet, B., Artois, M., Kieny, M. P., Lecocq, J. P., Costy, F., Desmettre, P., Chappuis, G., Pastoret, P. P. (1989). Use of recombinant vaccinia-rabies glycoprotein virus for oral vaccination of wildlife against rabies: innocuity to several non-target bait consuming species. *J. Wildl. Dis.* **25**, 540–547.

Brown, H. R., Pessolo, T. L., Nostro, A. F., Thormar, H. (1985). Demonstration of immunoglobulin in brains of ferrets inoculated with an SSPE strain of the measles virus: use of protein A conjugated to horseradish peroxidase. *Acta Neuropathol. Berl.* **65**, 195–201.

Cheema, A., Henson, J. B., Gorham, J. R. (1972). Aleutian disease of mink. Prevention of lesions by immunosuppression. *Am. J. Pathol.* **66**, 543–556.

Cho, H. J., Greenfield, J. (1978). Eradication of Aleutian disease of mink by eliminating positive counterimmunoelectrophoresis reactors. *J. Clin. Microbiol.* **7**, 18–22.

Clifton-Haley, R. S., Sayer, A. R., Stock, M. P. (1995). Evaluation of an ELISA for *Mycobacterium bovis* infection in badgers (*Meles meles*). *Vet. Rec.* **137**, 555–558.

Coe, J. E., Hadlow, W. J. (1972). Studies on immunoglobulins of mink: definition of IgG, IgA and IgM. *J. Immunol.* **108**, 530–537.

Coe, J. E., Race, R. E. (1978). Ontogeny of mink IgG, IgA and IgM. *Proc. Soc. Exp. Biol. Med.* **157**, 289–292.

Davidson, R. A. (1988). *Strongyloides stercoralis* infection in the ferret. *J. Parasitol.* **74**, 177–179.

Fomicheva, I. I. (1991). IgG allotypes of the domestic mink:

genetics, expression and evolution. *Exp. Clin. Immunogenet.* **8**, 185–218.

Fomicheva, I. I., Volkova, O. Yu. (1990). Genetic polymorphism of immunoglobulin G in the mink. VI. A regulatory gene controlling the expression of the gamma-chain constant region allotype of mink immunoglobulin. *Exp. Clin. Immunogenet.* 7, 213–220.

Fomicheva, I. I., Popova, N. A., Tservadze, D. K., Volkova, O. Yu., Kochlashvili, T. I., Baranov, O. K. (1990). Mink IgG-allotypes and Aleutian disease. *Genetika* **26**, 109–113.

Fomicheva, I. I., Popova, N. A., Tsertsvadze, D. K. (1991a). Genetic polymorphism of IgG in the mink. 7. Expression of the c gamma-allotypes in domestic mink infected with the Aleutian disease virus. *Exp. Clin. Immunogenet.* **8**, 107–114.

Fomicheva, I. I., Tsertsvadze, D. K., Volkova, O. Yu., Popova, N. A., Smirnykh, S. I., Kisteneva, N. A., Kuznetsov, K. N., Kudashev, V. F., Kaveshnikov, Yu. D. (1991b). Activation of expression of two immunoglobulin SN-genes in the American mink with Aleutian disease. *Genetika* **27**, 895–902.

Frederick, K. A., Babish, J. G. (1985). Compendium on the recent literature on the ferret. *Lab. Anim. Sci.* **35**, 298–318.

Galakhar, N. L., Djatchenko, S. N., Fomicheva, I. I., Mechetina, L. V., Taranin, A. V., Belousov, E. S., Nayakshin, A. M., Baranov, O. K. (1988). Mink–mouse hybridomas that secrete mink immunoglobulin G. *J. Immunol. Methods* **115**, 39–43.

Gallin, J. I., Klimerman, J. A., Padgett, G. A., Wolff, S. M. (1975). Defective mononuclear leukocyte chemotaxis in the Chediak–Higashi syndrome of humans, mink, and cattle. *Blood* **45**, 863–870.

Garrigue-Antar, L., Barbieux, I., Lieubeau, B., Boisteau, O., Gregoire, M. (1995). Optimisation of CCL64-based bioassay for TGF-β. *J. Immunol. Methods* **1862**, 267–274.

Gigliotti, F., Hughes, W. T. (1988). Passive immunoprophylaxis with specific monoclonal antibody confers partial protection against *Pneumocystis carinii* pneumonitis in animal models. *J. Clin. Invest.* **81**, 1666–1668.

Goodger, J., Russel, W. P., Nolan, A., Newell, D. G. (1994a). Production and characterization of a monoclonal badger anti-immunoglobulin G and its use in defining the specificity of *Mycobacterium bovis* infection in badgers by western blot. *Vet. Immunol. Immunopathol.* **40**, 243–252.

Goodger, J., Nolan, A., Russel, W. P., Dalley, D. J., Thorns, C. J., Stuart, F. A., Croston, P., Newell, D. G. (1994b). Serodiagnosis of *Mycobacterium bovis* infection in badgers: development of an indirect ELISA using a 25 kDa antigen. *Vet. Rec.* **135**, 82–85.

Hahn, E. C., Kenyon, A. J. (1980). Anti-deoxyribonucleic acid antibody associated with persistent infection of mink with Aleutian disease virus. *Infec. Immun.* **29**, 452–458.

Higgins, D. A. (1985). The skin inflammatory response of the badger (*Meles meles*). *Brit. J. Exp. Pathol.* **66**, 643–653.

Higgins, D. A., Langley, D. J. (1985). A comparative study of complement activation. *Vet. Immunol. Immunopathol.* **9**, 1983–1989.

Hoover, J. P., Baldwin, C. A., Rupprecht, C. E. (1989). Serological response of domestic ferrets (*Mustela putorius furo*) to canine distemper and rabies virus vaccine. *J. Am. Vet. Med. Assoc.* **194**, 234–238.

Hoover, J. P., Castro, A. E., Nieves, M. A. (1985). Serologic evaluation of vaccinated river otters. *J. Am. Vet. Med. Assoc.* **187**, 1162–1165.

Ish, C., Ong, G. L., Desai, N., Mattes, M. J. (1993). The specificity of alternative complement pathway-mediated lysis of erythrocytes: a survey of complement and target cells from 25 species. *Scand. J. Immunol.* **38**, 113–122.

Jacobsen, C. N., Aasted, B., Broe, M. K., Petersen, J. L. (1993). Reactivities of 20 anti-human monoclonal antibodies with leucocytes from ten different animal species. *Vet. Immunol. Immunopathol.* **39**, 461–466.

Kanno, H., Wolfinbarger, J. B., Bloom, M. E. (1993a). Aleutian mink disease parvovirus infection of mink peritoneal macrophages and human macrophage cell lines. *J. Virol.* **67**, 2075–2082.

Kanno, H., Wolfinbarger, J. B., Bloom, M. E. (1993b). Aleutian mink disease parvovirus infection of mink macrophages and human macrophage cell line U937: demonstration of antibody-dependent enhancement of infection. *J. Virol.* **67**, 7017–7024.

Khlebodarova, T. M., Matveeva, N. M., Serov, O. L., Najakshin, A. M., Belousov, E. S., Bogachev, S. V., Baranov, O. K. (1992). The mink gene for the lambda light immunoglobulin chain: characterization of cDNA and chromosomal localization. *Mamm. Genome* **2**, 96–99.

King, D. P., Schrenzel, M. D., McKnight, M. L., Reidarson, T. H., Hanni, K. D., Stott, J. L., Ferrick, D. A. (1996). Molecular cloning and sequencing of interleukin 6 cDNA fragments from the harbor seal (*Phoca vitulina*), killer whale (*Orcinus orca*), and southern sea otter (*Enhydra lutris nereis*). *Immunogenetics* **43**, 190–195.

Leader, R. W., Padgett, G. A., Gorham, J. R. (1963). Studies of abnormal leucocyte bodies in mink. *Blood* **22**, 477–484.

Lee, A. (1995a). Animal models and vaccine development. *Ballieres Clin. Gastroenterol.* **9**, 615–632.

Lee, A. (1995b). *Helicobacter* infections in laboratory animals: a model for gastric neoplasia? *Ann. Med.* **27**, 575–582.

Little, T. W., Naylor, P. F., Wilesmith, J. W. (1982). Laboratory study of *Mycobacterium bovis* infection in badgers and calves. *Vet. Rec.* **111**, 550–557.

Lodmell, D. L., Bergman, R. K., Bloom, M. E., Ewalt, L. C., Hadlow, W. J., Race, R. E. (1990). Impaired phagocytosis by the mononuclear phagocytic system in sapphire mink affected with Aleutian disease. *Proc. Soc. Exp. Biol. Med.* **195**, 75–78.

Mahmood, K. H., Rook, G. A., Stanford, J. L., Stuart, F. A., Pritchard, D. G. (1987a). The immunological consequences of challenge with bovine tubercle bacilli in badgers (*Meles meles*). *Epidemiol. Infect.* **98**, 155–163.

Mahmood, K. H., Stanford, J. L., Rook, G. A., Stuart, F. A., Pritchard, D. G., Brewer, J. I. (1987b). The immune response in two populations of wild badgers naturally infected with bovine tubercle bacilli. *Tubercle* **68**, 119–125.

Mechetina, L. V., Fomicheva, I. I., Taranin, A. V. (1992a). Genetic polymorphism of IgG in the mink. VIII. A quantitative study of the expression of C gamma-allotypes (H3, H4, H6, H8) in sera. *Exp. Clin. Immunogenet.* **9**, 24–32.

Mechetina, L. V., Olimova, D. C., Taranin, A. V. (1992b). Genetic polymorphism of IgG in the mink. IX. High proportion of allotype-producing lymphocytes in individuals with minor level of allotypes H3 and H4 in serum. *Exp. Clin. Immunogenet.* **9**, 141–148.

Miyazawa, M., Mori, S., Spangrude, G. J., Wolfinbarger, J. B., Bloom, M. E. (1994). Production and characterization of new monoclonal antibodies that distinguish subsets of mink lymphoid cells. *Hybridoma* **13**, 107–114.

Mori, S., Wolfinbarger, J. B., Miyazawa, M., Bloom, M. E.

(1991) Replication of Aleutian *mink* disease parvovirus in lymphoid tissues of adult mink: involvement of follicular dendritic cells and macrophages. *J. Virol.* **65**, 952–956.

Mouritsen, S., Aasted, B., Hoier-Madsen, M. (1989). Mink with Aleutian disease have autoantibodies to some autoantigens. *Vet. Immunol. Immunopathol.* **23**, 179–186.

Najakshin, A. M., Belousov, E. S., Taranin, A. V., Bogachev, S. S., Rogozin, I. B., Baranov, O. K. (1990). Cloning and sequencing of immunoglobulin lambda-chain cDNA in American mink (*Mustela vison*). *Genetika* **26**, 1527–1531.

Najakshin, A. M., Belousov, E. S., Aliabyev, B. Yu., Taranin, A. V. (1991). Genes of lambda-chain immunoglobulins from mink (*Mustela vison*). *Dokl. Akad. Nauk. SSSR* **319**, 1477–1479.

Najakshin, A. M., Belousov, E. S., Alabyev, B. Yu., Bogachev, S. S., Taranin, A. V. (1993). cDNA clones encoding mink immunoglobulin lambda chains. *Mol. Immunol.* **30**, 1205–1212.

Padgett, G. A., Leader, R. W., Gorham, J. B. (1963). Hereditary abnormal leukocyte granules in mink. *Fedn. Proc., Fedn. Am. Socs. Exp. Biol.* **22**, 428.

Penner, J. D., Prieur, D. J. (1987). Interspecific genetic complementation analysis with fibroblasts from humans and four species of animals with Chediak–Higashi syndrome. *Am. J. Med. Genet.* **28**, 455–470.

Peremislov, V. V., Mechetina, L. V., Taranin, A. V. (1992). Monoclonal antibodies against heavy and light chains of domestic mink IgG. *Hybridoma* **11**, 629–638.

Perou, C. M., Justice, M. J., Pryor, R. J., Kaplan, J. (1996). Complementataion of the beige mutation in cultured cells by episomally replicating murine yeast artificial chromosomes. *Proc. Natl Acad. Sci. USA* **93**, 5905–5909.

Perou, C. M., Kaplan, J. (1993). Complementation analysis of Chediak-Higashi syndrome: the same gene may be responsible for the defect in all patients and species. *Somat. Cell. Mol. Genet.* **19**, 459–468.

Porter, D. D. (1986). Aleutian disease: a persistent parvovirus infection of mink with a maximal but ineffective host humoral response. *Progr. Med. Virol.* **33**, 42–60.

Porter, D. D., Larsen, A. E., Porter, H. G. (1980). Aleutian disease of mink. *Adv. Immunol.* **29**, 261–286.

Portis, J. L., Coe, J. E. (1979). Deposition of IgA in renal glomeruli of mink affected with Aleutian disease. *Am. J. Pathol.* **96**, 227–236.

Rosatte, R. C., MacInnes, C. D., Power, M. J., Johnston, D. H., Bachmann, P., Nunan, C. P., Wannop, C., Pedde, M., Calder, L. (1993). Tactics for the control of wildlife rabies in Ontario (Canada). *Rev. Sci. Tech.* **12**, 95–98.

Rupprecht, C. E., Charlton, K. M., Artois, M., Casay, G. A., Webster, W. A., Campbell, J. B., Lawson, K. F., Schneider, L. G. (1990a). Ineffectiveness and comparative pathogenicity of attenuated rabies virus vaccines for the striped skunk (*Mephitis mephitis*). *J. Wildl. Dis.* **26**, 99–102.

Rupprecht, C. E., Gilbert, J., Pitts, R., Marshall, K. R., Koprowski, H. (1990b). Evaluation of an inactivated rabies virus vaccine in domestic ferrets. *J. Am. Vet. Med Assoc.* **196**, 1614–1616.

Saison, R., Karstad, L. (1968). Evidence of an auto-immune reaction in viral plasmacytosis (Aleutian disease) of mink as demonstrated by the Coombs test. *Bibl. Haematol.* **29**, 486–493.

Small, P. A., Jr, Waldman, R. H., Bruno, J. C., Gifford, G. E. (1976). Influenza infections in ferrets: role of serum antibody in protection and recovery. *Infec. Immun.* **13**, 417–424.

Smits, J. E., Godson, D. L. (1996). Assessment of humoral immune response in mink (*Mustela vison*): antibody production and detection. *J. Wildl. Dis.* **32**, 358–361.

Smits, J. E., Blakley, B. R., Wobeser, G. A. (1996). Immunotoxicity studies in mink (*Mustela vison*) chronically exposed to dietary bleached kraft pulp mill effluent. *J. Wildl. Dis.* **32**, 199–208.

Stokes, D. C., Gigliotti, F., Regh, J. E., Snellgrove, R. L., Hughes, W. T. (1987). Experimental *Pneumocystis carinii* pneumonia in the ferret. *Brit. J. Exp. Pathol.* **68**, 267–276.

Tabel, H. (1970). Local production of antibodies to *Leptospira pomona* in kidneys of chronically infected skunks (*Mephitis mephitis*). *J. Wildl. Dis.* **6**, 299–304.

Tabel, H., Ingram, D. G. (1970). The immunoglobulins in Aleutian Disease (viral plasmacytosis) of mink. Different types of hypergammaglobulinemias. *Can. J. Comp. Med.* **34**, 329–332.

Tabel, H., Ingram, D. G. (1972). Immunoglobulins of mink. Evidence for five immunoglobulin classes of 7S type. *Immunology* **22**, 933–942.

Tabel, H., Karstad, L. (1967). The renal carrier state of experimental *Leptospira pomona* infections in skunks (*Mephitis mephitis*). *Am. J. Epidemiol.* **85**, 9–16.

Taranin, A. V., Mechetina, L. V., Volkova, O. Yu., Fomicheva, I. I., Baranov, O. K., Belyaev, D. K. (1987). Genetic polymorphism of IgG in the mink. III. Instability of expression and the problem of the genetic control of C gamma-allotypes. *Exp. Clin. Immunogenet.* **4**, 73–80.

Thorns, C. J., Morris, J. A. (1984). Suppression and enhancement of transformation of ferret peripheral blood mononuclear cells by mycobacteria. *Res. Vet. Sci.* **36**, 345–347.

Tolson, N. D., Charlton, K. M., Lawson, K. F., Campbell, J. B., Stewart, R. B. (1988). Studies of ERA/BHK-21 rabies vaccine in skunks and mice. *Can. J. Vet. Res.* **52**, 58–62.

Ufimtseva, E. G., Galakhar, N. L. (1991) Interspecies mouse–mink hybridomas as producers of mink immunoglobulin. *Tsitol. Genet.* **25**, 35–40.

Volkova, O. Yu, Fomicheva, I. I., Taranin, A. V., Baranov, O. K., Belyaev, D. K. (1987). Genetic polymorphism of IgG in the mink. IV. Identification and genetic control of L3 allotype of the light chains. *Exp. Clin. Immunogenet.* **4**, 81–88.

Wienberg, L., Aasted, B. (1991). Investigation of mink MHC (MhcMuvi) class I molecules by isoelectric focusing (IEF). *Eur. J. Immunogen.* **18**(3), 165–173.

X Immunology of Horses and Donkeys

1. Introduction

The horse and donkey play diverse and changing roles in society around the world. While they are increasingly viewed as companion animals in western society their role as agricultural animals remains important in all cultures, and in many societies they are still an important beast of burden. While the health and well-being of horses are therefore important concerns for mankind, they remain subject to several diseases whose control requires an improved understanding of equine immunological defenses. For example, current vaccinations against equine viral respiratory infections are of limited efficacy (Hannant, 1991; Crabb and Studdert, 1990). However, our ability to design better vaccines has been limited in the past by our knowledge of equine immunobiology (Morrison, 1991).

While the number of scientists investigating equine immunology remains small, in recent years many new tools have become available for studying leukocyte antigens, immunoglobulins, and cytokines. Often these developments have resulted from major collaborative efforts, such as in the case of the international workshops that have characterized equine leukocyte antigens (Kydd *et al.*, 1994; Lunn *et al.*, 1998). Nevertheless, there are many gaps in the arsenal of the equine immunologist, and it is no time for investigators to relax their efforts. This chapter attempts to provide a current review of our understanding of the components and peculiarities of the equine immune system.

2. Leukocytes and their Antigens

Both innate and adaptive immunological responses are mediated by leukocytes, and among these cells the lymphocytes are the pivotal population. Lymphocytes have many diverse roles in immune responses, but apart from some differences in size and granularity there are no morphological features that indicate their different functions. This problem has been addressed by studies of cell surface glycoprotein molecules using monoclonal antibodies (mAbs), which has resulted in enormous advances in cellular immunology in the past 15 years. The application of this technology to human leukocyte antigens resulted in a series of workshops, the most recent of which was the Fifth International Workshop and Conference held in Boston in 1993. This workshop evaluated 1450 mAbs submitted by 500 laboratories (Schlossman *et al.*, 1995). Similar workshops have been held to address similar goals in the domestic species including the horse. The scale of the equine workshops is necessarily much smaller. The First International Workshop on Equine Leukocyte Antigens was held in July, 1991, in Cambridge, UK (Kydd *et al.*, 1994). That first workshop evaluated 86 mAbs from five laboratories. In addition to performing a detailed analysis of the reactivity of mAbs the first workshop also established criteria for giving equine cluster designations (EqCDs) to equine leukocyte antigens. Currently, antibodies directed against equine leukocyte antigens are characterized by a combination of the tissue distribution of the antigen together with functional data in some instances, and either the molecular weight of the antigenic molecule or the genetic sequence of the molecule. Studies of the homology of equine leukocyte antigens with the antigens of other species would be furthered by the availability of information regarding the genetic sequence of equine antigens, but currently this information is available only for the EqCD2 and EqCD44 molecules (Tavernor *et al.*, 1993b, 1994). As a result of this limitation the current criteria for establishing an equine cluster of determination homologous to a human CD do not always require genetic sequence information.

The Second Equine Leukocyte Antigen Workshop (ELAW II) Meeting was held in July, 1995, in Squaw Valley, California. This basically continued the work of the first meeting, and evaluated a total of 113 mAbs submitted by 10 laboratories (Lunn *et al.*, 1998). In addition to analyzing leukocyte antigens, ELAW II also undertook analysis of mAbs recognizing equine immunoglobulins, and the results of this study are described in Section 3 of this chapter. Details of the EqCD's established by ELAW II are presented in Table X.2.1. In addition, several important anti-equine leukocyte antigen reagents have been described which have not been studied by the workshops, and these are listed separately in Table X.2.2. A discussion of the equine leukocyte antigens identified by these various antibodies follows.

Handbook of Vertebrate Immunology
ISBN 0-12-546401-0

Table X.2.1 EqCD antigens identified at international workshops (Kydd *et al.*., 1991; Lunn *et al.*, 1997): typical ranges of expression in leucocyte populations, and molecular weight of antigen in SDS–PAGE analysis in reducing conditions

	Percentage of cells expressing antigen by FACS						
Antigen	*Lymphocytes*	*T-cells*	*B-cells*	*Granulocytes*	*Thymocytes*	*kDa*	*Additional references*
EqCD2	65–80	90–95	0–7	0–10	53–72	58	Tavernor *et al.* (1994); Tumas *et al.* (1994a)
EqCD3	65–90	95–99	0–5	0–5	60–95	20	Blanchard-Channell *et al.* (1994)
EqCD4	55–70	80–90	0–3	0–5	55–80	58	Lunn *et al.* (1991a)
EqCD5	60–90	90–95	0–10	0–10	90–99	69	Lunn *et al.* (1991a)
EqCD8	10–20	15–25	0–2	0–5	70–90	32, 39	Lunn *et al.* (1991a)
EqCD11a/18	60–99	70–99	40–80	90–99	95–99	100, 180	
EqCD13	0–8	0	0	75–95	0–35	140–150	Schram *et al.* (1996)
EqCD44	80–99	80–99	55–80	95–99	95–99	76	Tavernor *et al.* (1993b)
EqMHC I	95–100	95–99	95–99	90–99	95–99	45, 12	Hesford *et al.* (1989); Kydd *et al.* (1994)
EqMHC II	70–85	80–99	65–90	0–15	55–75	34	Crepaldi *et al.* (1986); Hesford *et al.* (1989); Monos *et al.* (1989); Lunn *et al.* (1993)
EqWC1	45–70	65–75	10–35	35–75	40–55	33	Lunn *et al.* (1994a)
EqWC2	60–90	95–99	0–20	75–99	75–99	145–180	
EqWC4 (EqCD28)	5–15	5–15	0–3	0	Not done	46	Byrne *et al.* (1996)
B cells	10–20	0–5	50–95	0	0–15	Various	Zhang *et al.* (1994)
Macrophages	0–20	0–3	0–3	0–4	0–15	Unknown	

Table X.2.2 Summary of mAbs not examined in workshops that recognize equine leukocyte antigens

Specificity	*Clone*	*Notes*	*References*
CD3	PC3/188A	Cross-reactive antibody raised against synthetic peptide sequences from cytoplasmic region of antigen.	Jones *et al.* (1993)
CD18	MHM23	An anti-human CD18 mAb that recognizes all equine leucocytes on FACS analysis.	Jacobsen *et al.* (1993)
CD18	CA16.2G1	An anti-canine CD18 mAb that recognizes all equine leucocytes on FACS analysis.	P. F. Moore[1], Davis, CA.
CD41/61	Co.35E4 and Co.2oA1	Characterized by cellular distribution and immunoprecipitation as recognizing equine glycoprotein IIb/IIIa (the homologous human antigen is designated CD41/61). This $\beta 3$ integrin adhesion receptor is expressed on platelets only.	Pintado *et al.* (1995)
CD79a	HM57; M7051, Dako	Cross-reactive antibody raised against synthetic peptide sequences from cytoplasmic region of CD79a, a B-cell marker also known as MB-1 which is part of the B cell antigen receptor complex. CD79a is expressed at the pre-B cell stage and continues to be present on B-cells throughout their differentiation.	Jones *et al.* (1993)
MHC II	7 reagents	A series of rat and murine mAbs recognizing HLA-DR, DQ and DP antigens were shown to recognize class II MHC antigens on equine lymphocytes on FACS analysis.	Monos *et al.* (1989)
B cells	B29A	Partially characterized in the First Equine Leucocyte Workshop as mAB WS68, this reagent was subsequently described in a second publication as recognizing a complex B cell surface antigen complex not described in other species.	Kydd *et al.* (1994); Tumas *et al.* (1994a)
NK cells	5C6	A mAb that recognizes a function-associated molecule on fish NK cells that is evolutionarily conserved in man, sheep, cattle, and horses.	Harris *et al.* (1993)
Macrophages	1.646	Recognizes a cytoplasmic antigen of equine mononuclear phagocytes consisting of two proteins (150 and 30 kDa). This product works on deparaffinized formalin fixed tissues for immunohistological staining.	Sellon *et al.* (1993)
Granulocytes and macrophages	DH59B	Partially characterized in the First Equine Leucocyte Workshop as mAB WS25, this reagent was subsequently described in a second publication as recognizing a pan-granulocyte/monocyte 96 kDa antigen.	Kydd *et al.* (1994); Tumas *et al.* (1994a)

[1]personal communication

Specific equine leukocyte antigens

EqCD2 (EqWC3)

This equine antigen was designated a 'workshop cluster' antigen by the first workshop, but it has recently been shown to be an orthologue of the human CD2 T cell marker on the basis of expression cloning and gene sequencing (Tavernor *et al.*, 1994). The EqCD2 antigen has a molecular weight of 58 kDa (Tumas *et al.*, 1994a), consistent with the characteristics of human CD2 (Barclay *et al.*, 1993). In immunohistochemistry, EqCD2 antibodies recognize all cells in T-dependent regions of lymph nodes and the majority of mature medullary thymocytes together with many cortical thymocytes (Kydd *et al.*, 1994). FACS analysis of peripheral blood leukocytes also confirms that EqCD2 is restricted to T lymphocytes (Lunn *et al.*, 1998). The distribution of EqCD2 is similar to human CD2, in contrast to the mouse where CD2 is also present on B cells (Barclay *et al.*, 1993). The human CD2 T cell antigen binds its ligand, CD58, on cytotoxic targets, antigen presenting cells, or memory T cells and binding of CD2 can mediate T-cell activation via its cytoplasmic domain (Barclay *et al.*, 1993). In the horse, Kydd and Hannant (1997) demonstrated that anti-EqCD2 mAbs partially inhibit PHA-induced lymphoproliferation.

EqCD3

One mAb has been described that recognizes an equine homologue of the CD3 antigen (Blanchard-Channell *et al.*, 1994; Lunn *et al.*, 1998). This antigen is exclusively restricted to equine T lymphocytes both in FACS analysis of peripheral blood and in immunohistochemical examination of lymphoid tissues. It is not known which component of the EqCD3 complex is recognized by the available mAb; however, this antibody can induce interleukin-2 receptor expression on T lymphocytes (Blanchard-Channel *et al.*, 1994), which is a feature shared with antibodies specific for the epsilon chain of human and murine CD3. An additional cross-reactive mAb has been described that recognizes CD3 orthologues in several species including the horse (Jones *et al.*, 1993). This reagent was raised against synthetic peptide sequences from cytoplasmic regions of the antigen and therefore may be most effective in immunohistological studies.

EqCD4

The EqCD4 antigen is the orthologue of the human CD4 antigen (Kydd *et al.*, 1994; Lunn *et al.*, 1991a, 1998). Its expression on equine T cells is mutually exclusive with EqCD8 in all extrathymic locations, as is the case in humans and other domestic species but in contrast to the pig (Pescovitz *et al.*, 1985; Parnes, 1989). Kydd and Hannant (1997) demonstrated that anti-EqCD4 mAbs partially inhibit PHA-induced lymphoproliferation, as has been reported in other species when a soluble anti-CD4 reagent is used (Parnes, 1989; Band and Chess, 1985).

EqCD5

Reagents recognizing putative equine orthologues of CD5 were among the first well characterized equine leukocyte antigen reagents (Crump *et al.*, 1988; Wyatt *et al.*, 1988). Several reagents have been identified which recognize EqCD5 (Kydd *et al.*, 1994; Lunn *et al.*, 1991a, 1998; Blanchard-Channell *et al.*, 1994), which has generally been described as a T-cell restricted antigen in the horse. While there is some evidence for low level EqCD5 expression on a subset of equine B lymphocytes (Crump *et al.*, 1988) this phenomenon has not been investigated in detail. Anti-EqCD5 mAbs enhance PHA-induced lymphoproliferation consistent with a role for the EqCD5 antigen in T cell activation (Kydd and Hannant, 1997). There is one report of the use of anti-EqCD5 mAbs to deplete equine T lymphocytes *in vivo* (Tumas *et al.*, 1994b).

EqCD8

Several mAbs have been identified which recognize EqCD8 (Lunn *et al.*, 1991a, 1998; Kydd *et al.*, 1994), which has been characterized as a covalently linked heterodimeric structure expressed on a subset of T lymphocytes in peripheral blood and lymphoid organs. EqCD8 expression among T lymphocytes is mutually exclusive with EqCD4 expression (Lunn *et al.*, 1991a). There is some evidence that the available mAbs may be able to distinguish between the α and β chains of the molecule. In functional analysis equine MHC I-restricted cytoxicity has been shown to be restricted to EqCD8$^+$ T lymphocytes (O'Brien *et al.*, 1991; McGuire *et al.*, 1994; Allen *et al.*, 1995). Blocking of the EqCD8 molecule by labeling with mAbs has been shown to have a number of biological effects. When used as a soluble antibody, anti-EqCD8 mAbs can inhibit PHA-induced lymphoproliferation (Kydd and Hannant, 1997), or abrogate CTL activity (O'Brien *et al.*, 1991). In contrast when the anti-EqCD8 mAbs are immobilized in agarose beads, binding of the EqCD8 molecule can serve as an activation signal (Lunn *et al.*, 1996). These functional effects of binding the EqCD8 receptor with soluble antibody, or cross-linking it with immobilized antibody are very probably a result of blocking or activating the cytoplasmic protein-tyrosine kinase p56^{lck}, which is complexed to the α-chain of the CD8 molecule in humans (Rudd, 1990).

While generally regarded as being a T lymphocyte restricted antigen, the CD8 molecule is also expressed on natural killer (NK) cells in some species, such as the rat (Trinchieri, 1989), while in humans CD8 expression has been reported on 30–50% of NK cells (Perussia *et al.*, 1983). There is evidence that equine NK cells may express the EqCD8 antigen (Lunn *et al.*, 1995b), and this is discussed in Section 6.

EqCD11a/18

The EqCD11a/18 molecule consists of a noncovalently linked heterodimer which is expressed on all cells of hemopoietic origin (Kydd *et al.*, 1994; Lunn *et al.*, 1998). Human CD11a/18 (also known as LFA-1) is a β_2 integrin which functions as a cellular adhesion molecule (Kishimoto *et al.*, 1989). Functional studies of EqCD11a/18 have been limited. In one study no clear effect of anti-EqCD11a/18 mAbs on PHA-induced lymphoproliferation could be demonstrated (Kydd and Hannant, 1997), while in a study of equine LAK cells it was shown that an anti-EqCD11a/18 mAb was capable of blocking LAK activity, consistent with the critical role of this antigen in intercellular adhesion events (Lunn *et al.*, 1994b). Two-color FACS analysis demonstrates that EqCD11a/18 expression is higher on a subpopulation of T cells (Lunn *et al.*, 1998). This range of T cell expression may be explained by the higher levels of expression on memory compared with naive T cells (Wallace and Beverley, 1990; Dustin and Springer, 1991).

EqCD13

Only one mAb has been described which appears to recognize an orthologue of the human CD13 antigen (Kydd *et al.*, 1994; Lunn *et al.*, 1998). The human CD13 antigen was among the first myeloid antigens to be identified (Look *et al.*, 1989a), and is currently widely used for the identification of cells of the myeloid lineage (Landay and Muirhead, 1989; Kidd and Vogt, 1989). The CD13 antigen is a 150–170 kDa structure present on all blood neutrophils, basophils, eosinophils, monocytes, myelomonocyte precursors, and myelogenous leukaemias, but not on T or B cells. On non-hematopoietic cells, CD13 is found on fibroblasts, osteoclasts, small intestinal epithelial cells, proximal renal tubule cells, and synaptic membranes of the central nervous system (Look *et al.*, 1989a). Analysis of cDNA sequences demonstrates that CD13 is identical to aminopeptidase N (Look *et al.*, 1989b), and expression of CD13 correlates with cell surface aminopeptidase activity (Ashmun and Look, 1990; Ashmun *et al.*, 1992). This membrane-bound zinc-binding metalloprotease is thought to be involved in the metabolism of regulatory peptides by many diverse cell types (Look *et al.*, 1989a).

The equine orthologue of CD13 has broadly similar characteristics, being expressed on cells of the myeloid lineage, and also on enterocytes (Kydd *et al.*, 1994; Schram *et al.*, 1996). The affinity purified EqCD13 molecule has aminopeptidase N activity, and a molecular mass of 150 kDa. The EqCD13 molecule differs from its human counterpart in that there is no evidence of expression on renal proximal tubule cells. The anti-EqCD13 antibody may prove valuable in differentiating cells of the myeloid and lymphoid lineage, particularly in poorly differentiated tumors of the hematopoietic system (Schram *et al.*, 1996).

EqCD44

The EqCD44 molecule is a heavily glycosylated monomeric antigen present on all leukocytes but with a higher density on T lymphocytes (Kydd *et al.*, 1994; Lunn *et al.*, 1998). The EqCD44 molecule has been confirmed to be an orthologue of the human CD44 antigen on the basis of expression cloning and gene sequencing (Tavernor *et al.*, 1993b).

EqMHC I and EqMHC II

A detailed description of equine MHC molecules is presented in Section 5.

EqWC1

This 'workshop cluster' antigen was characterized at the first equine leukocyte antigen workshop (Kydd *et al.*, 1994) and investigated in detail in an associated research paper (Lunn *et al.*, 1994a). Several allotypes of this antigen can be differentiated with mAbs. The EqWC1 antigen is expressed on a major T lymphocyte subset, all medullary thymocytes, and granulocytes, and has no obvious orthologue in other species. EqWC1 antibodies have no consistent effect on PHA-induced lymphoproliferation (Kydd and Hannant, 1997) and are incapable of blocking LAK activity (Lunn *et al.*, 1994b).

EqWC2

This 'workshop cluster' antigen was characterized at the first equine leukocyte antigen workshop (Kydd *et al.*, 1994), and is expressed on all T lymphocytes and granulocytes. EqWC2 antibodies have no consistent effect on PHA-induced lymphoproliferation (Kydd and Hannant, 1997), but are capable of blocking LAK activity (Lunn *et al.*, 1994b), consistent with a role in intercellular interactions. The identity of the EqWC2 antigen remains unknown, although its biochemical characteristics are consistent with the low molecular weight isoform of CD45 (CD45RO), or a member of the CD49 integrin family (Barclay *et al.*, 1993). However, in the absence of sequence information the homology of this antigen remains unknown.

EqWC4 (EqCD28)

This 'workshop cluster' antigen was characterized at the first equine leukocyte antigen workshop (Kydd *et al.*, 1994) and is defined by two mAbs. A significant body of evidence exists to suggest that it represents an orthologue of the human CD28 antigen (Lunn *et al.*, 1996a; Byrne *et al.*, 1996). Expression is restricted to a small subset of

$EqCD8^+$ and $EqCD4^+$ T lymphocytes in both peripheral blood and lymphoid organs. The EqWC4 molecule has a molecular mass of 46 kDa, but can exist as covalently linked homodimer. The proportion and phenotype of cells expressing EqWC4 (Byrne *et al.*, 1996) is somewhat different from the characterized populations of human and mouse lymphocytes expressing CD28 (Barclay *et al.*, 1993). Therefore, for the present, it is not possible to definitively identify the EqWC4 molecule as an orthologue of CD28.

B cell markers

A series of mAbs have been described which identify nonimmunoglobulin antigens which are restricted to B lymphocytes (Kydd *et al.*, 1994; Zhang *et al.*, 1994; Lunn *et al.*, 1998). Various identities have been proposed for these reagents, including an equine homologue of the CD19 antigen and the B cell isoform of the CD45 molecule (Barclay *et al.*, 1993). With the information available it has not proven possible to assign specificities to these reagents beyond their reactivity with B lymphocytes.

Macrophage markers

Two mAbs have been described which recognize macrophages in peripheral blood and are not expressed on either peripheral blood lymphocytes or granulocytes (Kydd *et al.*, 1994; Lunn *et al.*, 1998). No further characterization of these reagents has been performed, however, in FACS analysis both reagents represent useful macrophage markers. An additional macrophage restricted marker has been described in detail by Sellon *et al.* (1993). This mAb recognizes a cytoplasmic antigen consisting of two proteins of 150 and 30 kDa.

The equine T-cell receptor

No mAbs have been described which recognize the equine TCR. While the varying roles and distributions of $\alpha\beta$ and $\gamma\delta$ T cells have been documented in humans and domestic ruminants, they remain largely uninvestigated in equids. In two recent papers Schrenzel *et al.* (1994, 1995) described the cloning and sequencing of both the $\alpha\beta$ and $\gamma\delta$ TCR genes. Both $\alpha\beta$ and $\gamma\delta$ T cells were found to be widely distributed throughout the body of adult horses, with the exception of the bone marrow where only $\alpha\beta$ T cells were found (Schrenzel and Ferrick, 1995). These pioneering studies should facilitate future investigations of the equine TCR in normal immune responses and immune-related diseases.

Advances in equine immunology dependent on leukocyte markers

The development of a comprehensive panel of mAbs recognizing differentiation markers of equine leukocytes has been a relatively recent development. Nevertheless, it has already precipitated advances in several areas of equine immunological study. In the field of infectious disease the role of CTLs in defense against equine herpes virus (Allen *et al.*, 1995; Ellis *et al.*, 1995; Edens *et al.*, 1996) and equine infectious anemia virus (McGuire *et al.*, 1994) has been defined. The dynamic changes in lymphocytes subpopulations in peripheral blood (Lunn *et al.*, 1991b; Allen *et al.*, 1995; Hines *et al.*, 1996a), the small airways (Hines *et al.*, 1996a; Kydd *et al.*, 1996), the eye (Kalsow *et al.*, 1992), and the uterus (Watson and Dixon, 1993) in response to infectious or inflammatory disease have been examined. In addition lymphocyte subpopulation changes in both primary and secondary immunodeficiency diseases have been investigated (Weldon *et al.*, 1992; Boy *et al.*, 1992; Dascanio *et al.*, 1992; Lunn *et al.*, 1995b).

3. Immunoglobulins

The immunoglobulin classes of the horse are complex, and include at least four IgG subisotypes (IgGa, b, c, IgG(T)), IgA, IgM, and IgE (Rockey, 1967; Zolla and Goodman, 1968; McGuire and Crawford, 1972; McGuire *et al.*, 1973; Roberts, 1975; Suter and Fey, 1983); there are no reports of the identification of equine IgD. Normal concentrations of these immunoglobulins in serum and other fluids has been reviewed elsewhere (Gorman and Halliwell, 1989). The distribution of equine immunoglobulins as determined by immunohistochemistry has also been reported for lymphatic tissue (Khaleel *et al.*, 1975), reproductive and intestinal tracts (Widders *et al.*, 1984), the respiratory tract (Mair *et al.*, 1987, 1988), the aqueous humor (Matthews *et al.*, 1983), and the anterior uvea (Matthews, 1989).

The study of equine immunoglobulins has a long history owing to the early use of hyperimmune equine antisera in the therapy and prophylaxis of human disease. As reviewed by Roberts (1975) equine IgG and IgM were first distinguished in 1939, and a year later an additional 'T-protein' was identified as resulting from diptheria or tetanus toxoid immunization. Recent investigations of equine immunoglobulins have typically focused on the increased morbidity and mortality associated with failure of passive transfer of immunity to the equine neonate (McGuire *et al.*, 1977), and on specific immunodeficiencies such as selective IgM deficiency (McClure *et al.*, 1996). Initial investigations of the functional characteristics and classification of equine immunoglobulin isotypes and subisotypes were conducted more than 20 years ago (McGuire *et al.*, 1973; Montgomery, 1973), but have subsequently been limited by a lack of well-characterized reagents. However, in recent years several papers have described the development of equine immunoglobulin-secreting heterohybridomas (Appleton *et al.*, 1989; Richards *et al.*, 1992; Wagner *et al.*, 1995), the cloning of specific equine immunoglobulin genes (Home *et al.*, 1992a; Ford *et al.*, 1994; Marti *et al.*, 1995; Navarro *et al.*, 1995), and finally the development of a panel of monoclonal antibodies capable of distinguishing equine immunoglobulin isotypes and sub-isotypes (McGuire *et al.*, 1983; Lunn *et al.*, 1995a, 1998; Sheoran *et al.*, 1998; Sugiura *et al.*, 1998).

Immunoglobulin G subisotypes

Four IgG subisotypes have been well described, three are designated as IgGa, IgGb and IgGc on the basis of their increasing anodal mobility in immuno-electrophoresis (Rockey *et al.*, 1964; Klinman *et al.*, 1965; Rockey, 1967). However, while the separation of IgGa from IgGb is possible in the donkey (Allen and Dalton, 1975), it has proven difficult in the horse due to the similarity in charge of these two molecules and studies have often been restricted to partial purification resulting in IgGab and IgGc (Rockey, 1967; McGuire *et al.*, 1973). Recently, the separation of IgGa and IgGb in the horse has been described using protein A and protein G affinity chromatography (Sheoran and Holmes, 1996), or a combination of ion-exchange chromatography and affinity chromatography (Sugiura *et al.*, 1998). Some confusion exists in early reports about the status of IgG(T) as a subisotype of IgG because some authors proposed that IgG(T) be classified as a separate Ig isotype, IgT (Dorrington and Rockey, 1968; Montgomery *et al.*, 1969; Montgomery, 1973). Widders *et al.* (1986) purified IgG(T) free from contamination with IgG and conducted investigations of the antigenic relationship between IgG and IgG(T) using immunoelectrophoresis and double immunodiffusion. These experiments unequivocally classified IgG(T) as a subisotype of IgG. Details of amino acid composition and terminal peptide sequences, cyanogen bromide fragments, and tryptic peptide mapping of the IgG subisotypes has also been reported (Montgomery, 1973).

Limited studies of the functional characteristics of equine IgG subisotypes have been performed. A consistent finding is that IgG(T) differs from the other equine immunoglobulins in its behavior in precipitation reactions (McGuire *et al.*, 1979; McGuire *et al.*, 1973). Two types of precipitin curves can be produced with equine antisera, one is similar to that seen with antisera from other species, while the other is characterized by a narrow zone of precipitation and a zone of soluble complexes in antibody excess described as a flocculation reaction. There is good evidence that IgG(T) is responsible for the flocculation reaction, which may result from IgG(T) preferentially binding to single antigen molecules with both antigen-binding sites for steric reasons, i.e. favoring intramolecular as opposed to intermolecular binding (Banks and McGuire, 1975). Additional features of IgG(T) include its failure to fix complement (a characteristic shared with IgGc) and its ability to inhibit complement fixation by IgGa and IgGb (McGuire *et al.*, 1973). It has also been shown that equine monocytes and neutrophils do not appear to have Fc receptor sites that will bind IgG(T), whereas they can bind other IgG subisotypes and IgM (Banks and McGuire, 1975). These features suggest that IgG(T) is likely to be best adapted to toxin neutralization, and less efficient in complement fixation, opsonization, and antibody-dependent cellular cytotoxicity.

Several investigators have recently described mAbs recognizing IgGa, IgGb, IgGc, and IgG(T) (Lunn *et al.*, 1995a; Sheoran *et al.*, 1998; Sugiura *et al.*, 1998), and the specificity of these reagents was confirmed by the ELAW II workshop (Lunn *et al.*, 1997). Using these antibodies in affinity chromatography purification of equine immunoglobulins has allowed the accurate determination of the molecular weights of the heavy and light chains (Table X.3.1) (Lunn *et al.*, 1997). Several heterohybridomas secreting equine IgG subisotypes have been described (Appleton *et al.*, 1989; Richards *et al.*, 1992; Wagner *et al.*, 1995), and some of these have been classified as being of the IgGa, IgGb or IgG(T) subisotypes on the basis of reactivity with anti-equine immunoglobulin mAbs (Richards *et al.*, 1992; Lunn *et al.*, 1995a, 1998; A. Sheoran, personal communication).

An anti-IgG subisotype mAb has been reported to be a

Table X.3.1 Molecular weights of equine immunoglobulins

Immunoglobulin	*Heavy chain*	*Light chain*	*Whole molecule*
IgGa	50 000	27 000	178 000
IgGb	53 000	27 000	160 000
IgGc	52 000	27 000	169 000
IgG(T)	58 000	27 000	188 000
IgA	61 000	27 000	150 000–700 000
IgM	89 000	–	900 000

This chart is derived from information presented in several publications (McGuire *et al.*, 1973; Hirano *et al.*, 1990; Lunn *et al.*, 1995a, 1996; Sheoran *et al.*, 1998).

marker for B lymphocytes (Lunn *et al.*, 1995a). One possible explanation for this unusual phenomenon was that the mAb was recognized IgG molecules attached to B lymphocytes by Fc receptors. However incubation at 37°C for 3–4 h did not decrease labeling and therefore this explanation seemed improbable. In a recent study this observation was extended and it was found that many anti-equine Ig mAbs recognized B lymphocytes, however prolonged incubation (periods of 8 h or more) at 37°C decreased this labeling except in the case of anti-IgM or anti-light chain mAbs (Lunn *et al.*, 1998). The presumed explanation for these observations is that equine B lymphocytes are particularly efficient at binding serum immunoglobulins, and require extensive incubation in serum-free medium for capping and internalization of these Ig molecules to occur.

One additional form of equine Ig has been reported which may represent an IgG subisotype. This Ig with $\gamma 1$ mobility exists as a salt-dissociable noncovalently linked aggregate, was first isolated from hyperimmune equine antipneumococcal serum (Sandor *et al.*, 1964) and subsequently designated as aggregating immunoglobulin (AI) (Zolla and Goodman, 1968). Studies have shown that AI is closely related to IgG but there has been no agreement on whether it represents a subisotype of IgG or is a distinct isotype. It has been designated by various investigators as $\gamma 1$ component (Helm and Allen, 1970), as an Ig isotype IgB (Montgomery, 1973), and as a subisotype of IgG termed IgG(B) (Allen and Johnson, 1972).

Immunoglobulin A

The existence of equine IgA was first discovered when an immunological cross-reaction was observed between an immunoglobulin in equine serum and milk with anti-human alpha-chain antiserum (Vaerman *et al.*, 1971). Later, secretory IgA and free secretory component were identified and isolated in equine milk and other secretions (McGuire and Crawford, 1972; Pahud and Mach, 1972). Antigenic determinants specific for secretory IgA associated with secretory component were demonstrated by McGuire and Crawford (1972). Serum IgA exists predominantly as dimers while monomers, trimers and tetramers also occur. The molecular mass of serum IgA extends from 150 kDa to about 700 kDa with the majority being about 350 kDa, suggesting a predominant dimeric form in serum similar to the cow and the pig, but distinct from human serum IgA which exists largely as monomers (Roberts, 1975). The molecular mass range of secretory IgA is similar to that of serum IgA but the majority is larger than 350 kDa, the difference being assumed to be due to the presence of a secretory component of 80 kDa (McGuire *et al.*, 1973). Monoclonal antibodies recognizing equine IgA have been described (Sugiura *et al.*, 1997; Lunn *et al.*, 1997) and the molecular weights of heavy and light chains are given in Table X.3.1.

Immunoglobulin E

Matthews *et al.* (1983) and Suter and Fey (1983) provided preliminary evidence of the presence of IgE and described its partial purification and the preparation of antisera. The existence of IgE in the horse has now been firmly established with the recent characterization of its cDNA and deduced amino acid sequence (Navarro *et al.*, 1995; Marti *et al.*, 1995). The C_ε heavy chain had 54% homology with human C_ε and 52% with ovine. No mAbs specific for equine IgE are currently available, although polyclonal antisera have been described (Halliwell *et al.*, 1993). Given the importance of hypersensitivity diseases in horses, and development an anti-IgE mAbs for the horse remains a major goal for equine immunologists.

Immunoglobulin M

IgM exists in serum as a single subisotype in a pentameric structure with a molecular mass of 990 kDa (Kabat, 1939; McGuire *et al.*, 1973) which after reduction and alkylation separates into five monomeric subunits each capable of antigen binding, but with no precipitating and agglutinating properties (Hill and Cebra, 1965). The structure of IgM appears to be similar to that of man with a single J chain of molecular weight of 15 kDa associated covalently with each pentamer (Mitchell *et al.*, 1977). The molecular weights of the heavy and light chains are given in Table X.3.1. To date no subisotypes or allotypes of IgM have been demonstrated. There is a single report of the production of heterohybridomas producing equine IgM (Lunn *et al.*, 1998). A number of mAbs are available with specificity for equine IgM (McGuire *et al.*, 1983; Lunn *et al.*, 1998; Sugiura *et al.*, 1998), including some reagents that recognize feline IgM and cross-react with equine IgM (Lunn *et al.*, 1998). Anti-IgM mAbs have been shown to be effective B-cell markers in both fluorescence-activated cell sorting (FACS) and immunohistochemistry (Lunn *et al.*, 1998; McGuire *et al.*, 1983).

Light chains

Equine immunoglobulins possess two types of L chains, κ and λ (Allen *et al.*, 1968). Equine Ig λ genes have been sequenced, revealing that over 90% of the L chains in horse serum immunoglobulins are λ (Home *et al.*, 1992b). The reason for this is uncertain. The horse does possess a functional κ locus with associated variable region gene segments at least as extensive as those for the λ locus (Home *et al.*, 1992b; Ford *et al.*, 1994). At least one mAb is available which recognizes an equine light chain (Sheoran *et al.*, 1998; Lunn *et al.*, 1998).

Table X.4.1 Cloned equine cytokines

Cytokine	% similarity to Human	Murine	Bovine	Porcine	Reference
Interleukin-1α	72	60	—	—	Kato *et al.* (1995)
Interleukin-1β	67	62	—	—	Kato *et al.* (1995)
Interleukin-2	72	54	62	70	Tavernor *et al.* (1993); Vandergrifft and Horohov (1993)
Interleukin-4	62	40	—	—	Vandergrifft and Horohov (1994)
Interleukin-6	62	—	—	71	C. E. Swiderski *et al.* (personal communication)
Interleukin-10	—	—	—	—	Swiderski and Horohov (1995)
Interferon-α	71–77	—	—	—	Himmler *et al.* (1986)
Interferon-β	59	—	—	—	Himmler *et al.* (1986)
Interferon-ω	—	—	—	—	Himmler *et al.* (1986)
Interferon-γ	67	—	78	—	Curran *et al.* (1994); Grunig *et al.* (1994)
Tumor necrosis factor-α	84	—	—	—	Su *et al.* (1992)

4. Cytokines

As the field of immunology has focused more recently on the role of cytokines in regulating immune responses, so too have equine immunologists. While relatively few groups have been involved in such efforts, the tools of modern biotechnology have helped to accelerate the rate of advancement. Thus, the cloning, sequencing and expression of a number of equine cytokines has been accomplished (Table X.4.1). The availability of these reagents will help lead to a better understanding of the equine immune system and, perhaps, to the development of novel therapeutic strategies for treating diseases of the horse.

Characterized equine cytokines

IL-1

Equine IL-1 exists in two forms, IL-1α and IL-1β, which may be identified by chromatographic separation and isoelectric focusing (May *et al.*, 1990). Equine IL-1 drives IL-2 production by the murine EL-4 cell line and is cytocidal for D10.G4.1 cells. Equine articular cells respond to recombinant human IL-1 (May *et al.*, 1992). The cDNA for equine IL-1α and IL-1β were recently cloned and sequenced (Kato *et al.*, 1995). The deduced amino acid sequence of equine IL-1α exhibits 72% and 60% similarity with that of human and murine IL-1α, respectively; the amino acid sequence of equine IL-1β exhibits 67% and 62% similarity with that of human and murine IL-1β, respectively.

IL-2

This was one of the first equine cytokines to be cloned and sequenced (Vandergrifft and Horohov, 1993; Tavernor *et al.*, 1993a) yielding a protein core of predicted relative mass of 14 891 Da. Equine IL-2 shares 72% amino acid similarity with the human sequence, 70% with porcine, 62% with bovine and ovine, 56% with rat and 54% amino acid sequence similarity with the mouse. Recombinant equine IL-2 expressed in COS and CHO cells induces proliferation and lymphokine activated killer (LAK) cell activity in equine peripheral blood mononuclear cells, but has no effect on murine CTLL-2 cells (probably because of unique amino acid substitutions in regions relevant to receptor binding) (Vandergrifft and Horohov, 1993; Horohov *et al.*, 1996b). Human IL-2 induces proliferation and LAK cell activity in equine PBMC, although it has lower specific activity (Fenwick *et al.*, 1988; Hormanski *et al.*, 1992).

IL-4

Equine IL-4 cDNA was cloned and sequenced from mitogen-stimulated equine PBMC and shown to share homology with IL-4 sequences from other species (Vandergrifft and Horohov, 1994). The equine IL-4 amino acid sequence shares 62% similarity with the human IL-4 and 40% sequence similarity with the mouse and rat IL-4 amino acid sequences. The precursor equine IL-4 protein has a predicted relative mass of 15 283 Da. Equine IL-4 expressed in COS cells augments the proliferative response of B cells to mitogens and inhibits the generation of lymphokine-activated killer cells (Horohov *et al.*, 1996b).

IL-6

Equine IL-6 activity can be detected using the murine hybridoma cell line B 13.29 clone B.9 (Morris *et al.*, 1992). Equine IL-6 cross-reacts with antibodies to human IL-6 in neutralization and slot-blot radioimmunoassays (Billinghurst *et al.*, 1995). The cDNA for equine IL-6 has recently been cloned and sequenced (C. E. Swiderski *et al.*, personal communication). The deduced amino acid sequence yields a 20 556 Da protein which exhibits 71% homology to porcine IL-6 and 62% amino acid homology with human IL-6.

IL-10

Equine IL-10 has recently been cloned and sequenced (Swiderski and Horohov, 1995). Regions of equine IL-10 exhibit homology to the IL-10-like sequence of equine herpesvirus type 2 (EHV-2) (Rode *et al.*, 1993). Both genes share homology with IL-10 of humans and the Epstein–Barr virus protein BCRF1.

TNF-α

Equine TNF-α activity can be detected with an *in vitro* cytotoxicity assay using the murine fibrosarcoma cell line WEHI 164 clone 13 (Morris *et al.*, 1990) or L929 cells (MacKay *et al.*, 1991). The molecular cloning and nucleotide sequence of the 2610-bp genomic sequence encoding equine TNF-α has been reported (Su *et al.*, 1992). The deduced amino acid sequence of equine TNF-α exhibits 84% homology with its human counterpart. Monoclonal antibodies and antisera to recombinant equine TNF-α have been produced and a capture ELISA has been reported (Su *et al.*, 1992).

Interferons

Type I (α, β and ω) (Himmler *et al.*, 1986) and Type II (γ) (Grunig *et al.*, 1994) interferons of the horse have been cloned and expressed as recombinant proteins and preliminary investigations into their clinical application begun. Natural human interferon-α has already been evaluated as a treatment for inflammatory airway disease in racehorses (Moore *et al.*, 1995). The cytokine was administered orally in a double-blind, randomized trial, and led to a decrease in the number of inflammatory cells in the lower airways. A review of the potential use of interferons in large domestic species was recently published (Moore, 1996).

The role of cytokines in equine disease

It has become clear that the induction of protective resistance or the exacerbation of disease is dependent on the pattern of cytokine gene expression during an immune response (Liles and Van Voorhis, 1995). The two major areas of emphasis for identifying the role of cytokines in disease processes are inflammatory responses where certain cytokines, notably IL-1, IL-6 and TNF-α, play a central role in the disease process and T-cell-mediated immune responses where the cytokines produced by specific T helper cell subsets may either prevent or exacerbate disease.

Cytokine production during septicemia

Septicemia is a common problem in young foals, occurring when intact bacteria or their toxins gain access to the systemic circulation. Of the potentially toxic bacterial components, cytokine responses to the lipopolysaccharide (LPS) portion of the Gram-negative bacterial cell wall have been evaluated extensively in a number of species, including the horse (MacKay *et al.*, 1991; Morris and Moore, 1991; MacKay and Lester, 1992). Upon entering the circulation, LPS initiates a cascade of inflammatory mediators that target multiple organs resulting in the syndrome of clinical signs termed endotoxemic shock. The horse, by virtue of its profound physiological responses to relatively low doses of endotoxin, is categorized as an endotoxin-sensitive species (Morris, 1991; Breuhaus and DeGraves, 1993). A growing body of evidence indicates that TNF-α, IL-1, and IL-6 act as early mediators of equine endotoxemic shock (MacKay *et al.*, 1991; Morris and Moore, 1991; MacKay and Lester, 1992).

TNF-α plays a pivotal role in initiating systemic toxemic reactions during Gram-negative bacterial and endotoxemic sepsis (Beutler *et al.*, 1985). TNF-α is released rapidly by LPS-stimulated macrophages in many species including the horse and TNF-α infusion mimics many of the signs of endotoxemic shock (Beutler *et al.*, 1985; Han *et al.*, 1990; MacKay *et al.*, 1991; Allen *et al.*, 1993). TNF-α causes fever, progressive acidosis, hemoconcentration, hypotension, and coagulation abnormalities that are characteristic of the toxemic syndrome (Grunfeld and Palladino, 1990). Neutralization of TNF-α activity by infusion of anti-TNF-α has been shown to improve the clinical course of endotoxemic and Gram-negative septic shock in a variety of species, including horses (Beutler *et al.*, 1985; Cargile *et al.*, 1995). TNF-α activity has been detected in adult horses and foals after experimental induction of endotoxemia, in horses with colic attributable to gastrointestinal tract disease, and in foals with presumed sepsis (Morris *et al.*, 1990; Morris and Moore, 1991; Morris *et al.*, 1991). Significantly, as demonstrated in other species, high TNF-α activity has been shown to correlate with disease severity and negative outcomes in foals with presumed sepsis and adult horses with gastrointestinal disease (Morris and Moore, 1991; Morris *et al.*, 1991). Although TNF-α appears to be the most proximal cytokine mediator of endotoxemic and Gram-negative bacterial septic shock, TNF-α induces in a cascade manner both IL-1 and IL-6 which have demonstrated roles in the pathogenesis of toxemic shock (Dinarello *et al.*, 1986; Shalaby *et al.*, 1989).

Equine leukocytes have been shown *in vitro* to release sequentially TNF-α followed by IL-1 in response to stimulation by endotoxin (Seethanathan *et al.*, 1990). TNF-α induces IL-1 production by monocytes (Dinarello *et al.*, 1986). Both TNF-α and IL-1 induce the production of IL-6 which appears to provide negative feedback for TNF-α and IL-1 production (Aderka *et al.*, 1989; Shalaby *et al.*, 1989). The biological activities of IL-1 make it significant in the pathogenesis of fever, depression, hypoglycemia, negative protein balance, acidosis, edema, and coagulation defects demonstrated in septicemic shock (Dinarello, 1988). Although *in vivo* production of IL-1 in equine septic shock has yet to be documented, *in vitro* evidence supports its production *in vivo*.

IL-6, which is produced primarily by mononuclear phagocytes, is an extremely pleotrophic cytokine whose biological effects have been extensively reviewed (Hirano *et al.*, 1990). Gram-negative endotoxin is a potent stimulus for IL-6 production both directly and indirectly via TNF-α and IL-1 induction (Dinarello *et al.*, 1986; Shalaby *et al.*, 1989; Morris *et al.*, 1992). Increases in IL-6 activity have been documented following endotoxin infusion in horses and foals (MacKay and Lester, 1992; Robinson *et al.*, 1993; Durando *et al.*, 1994). The role of IL-6 endotoxemia has not been fully defined, but potentially beneficial effects of IL-6 production during endotoxemia have been described. These include decreased production of TNF-α and IL-1 by monocytes (Aderka *et al.*, 1989; Shalaby *et al.*, 1989), and induction of the full range of acute-phase proteins that not only augment nonspecific and specific immunological defenses against microorganisms but also mediate protection and repair from tissue damage caused by inflammatory mediators such as TNF-α and IL-1 (Hirano *et al.*, 1990; Almawi *et al.*, 1991; Lue *et al.*, 1991). Other studies have indicated that IL-6 may contribute to the pathogenic effects of Gram-negative infections or endotoxemia and that administration of IL-6 antagonists may actually be beneficial in these disease states (Tiao *et al.*, 1994; Meyer *et al.*, 1995).

In summary, the contribution of inflammatory cytokines in the systemic toxemic reactions to endotoxin in the horse parallels that seen in humans and experimental models. The continued development of equine-specific cytokine reagents can be expected to lead to better characterization, diagnosis and treatment of Gram-negative infections or endotoxemia in the horse.

Possible equine models for T_H1 and T_H2 responses

The existence of two defined $CD4^+$ T helper cell subsets based on distinct patterns of cytokine gene expression was initially recognized in the mouse (Mosmann *et al.*, 1986). These subsets have been designated either T_H1 or T_H2 where T_H1 cells produce IL-2, IFN-γ, and TNFα and T_H2 cells produce IL-4, IL-5, and IL-10 (Mosmann, 1991). Although both subsets of cells may be induced during an immune response, it has been shown that one or the other population often dominates (Mosmann and Coffman, 1989). Protection from intracellular parasites generally stimulates a T_H1 response whereas helminth parasites evoke a T_H2 response. Deviation from this pattern, as has been demonstrated using different strains of mice, is associated with susceptibility to the invading organism.

Less information is available regarding T_H1 and T_H2 responses in other species. While cytokine responses characteristic of T_H1 and T_H2 cells have been identified in the bovine (Brown *et al.*, 1994), the correlation of a defined cytokine pattern with immune responses associated with either protective resistance or disease of non-murine species remains unproven. The existence of T_H1 and T_H2 responses in the horse has yet to be demonstrated. Because T_H2 subsets have been associated with immune responses to metazoan parasites in other species (Urban *et al.*, 1992), a similar condition in the horse may provide evidence for this response. Acquired resistance to *Strongylus vulgaris* in the horse, as demonstrated by multiple or single experimental infections, is associated with an eosinophilia and other signs of immediate-type hypersensitivity (Dennis *et al.*, 1993). *In vitro* stimulation of peripheral blood mononuclear cells from infected ponies yielded supernatants containing an eosinophil chemotactic factor. While it is not known whether this chemotactic factor is IL-5, this type of response is consistent with the induction of the T_H2 subset. Comparisons of ponies vaccinated with either parenteral inoculations of *S. vulgaris* antigen, or oral administration of radiation-attenuated larvae showed markedly different outcomes following challenge infections, and responses characteristic of either a T_H1 or a T_H2 cytokine response (Monahan *et al.*, 1994). Those ponies receiving the parenteral administration of antigen showed exacerbated signs of disease, without an eosinophilia following challenge infestations, while those receiving the oral larval vaccine were protected from infection and exhibited an eosinophilia characteristic of a T_H2 response.

Atopic individuals exhibit characteristic Type I hypersensitivity responses with elevated levels of IL-4 leading to production of IgE antibodies (Kapsenberg *et al.*, 1992). Chronic obstructive pulmonary disease (COPD) is a respiratory disease of horses which has long been suspected to be immunological. The demonstration of fungal antigen-specific degranulation in basophils from COPD horses (Dirscherl *et al.*, 1993) and the presence of allergen-specific IgE in the bronchoalveolar lavage fluids of affected horses (Halliwell *et al.*, 1993) supports the contention that some forms of the disease are immune mediated. Because IgE production is under the control of IL-4, it is possible that affected horses exhibit increased T_H2 activity. The finding that PBMC from affected horses exhibited no blastogenic activity to the allergens is consistent with the absence of a delayed-type hypersensitivity (Type IV) response (Dirscherl *et al.*, 1993).

Studies of host defenses in other species have demonstrated the importance of both local humoral responses in preventing infection and cell-mediated immunity for viral clearance and recovery from infection (Bender and Small, 1992). Infection with influenza viruses leads to the induction of a T_H1 response in humans (Romagnani, 1991). This is not surprising considering the well-recognized role cell-mediated immune responses play in the recovery from influenza virus infections. It is likely that the protective response to equine influenza virus would also be a T_H1

response since immunized horses exhibit cell-mediated immune responses to viral antigens *in vitro* (Hannant and Mumford, 1989). However, at mucosal surfaces a T_H2 response may be necessary in order to induce IgA production (Kiyono *et al.*, 1990; McGhee *et al.*, 1991). Differences in the antibody responses generated by vaccination versus infection have also been demonstrated in both humans and horses (Bender and Small, 1992; Ben Ahmeida *et al.*, 1994; Nelson *et al.*, 1995). Thus, the immune response to a primary infection with equine influenza virus is characterized by serum and nasal IgGa, IgGb, and IgA antibody responses while vaccination induces only IgGc and IgG(T) responses (Nelson *et al.*, 1998). Since isotype switching is under the control of T-cell cytokines (Schultz and Coffman, 1991), it is likely that these differences in isotype production between infected and vaccinated horses are the result of differential expression of those cytokines.

A somewhat different situation may exist in ponies infected with equine infectious anemia virus (EIAV). Peripheral blood mononuclear cells from EIAV-infected ponies exhibit marked *in vitro* lymphoproliferative responses to viral antigens early in the infection, however, the intensity of this response decreases with time (Issel *et al.*, 1992). While cell-mediated immune response can be significantly depressed with disease progression, the infected ponies continue to mount a strong antibody response to the virus (Newman *et al.*, 1991). These results mimic the T_H1 to T_H2 switch which is thought to occur in AIDS patients (Clerici and Shearer, 1994; Mosmann, 1994). Whether this is characteristic of lentivirus infections has not been determined.

5. Major Histocompatibility Complex (MHC) Antigens

Genetic, biochemical and functional characteristics of the equine MHC

The equine MHC extends over 4000 kb on chromosome 20 (Antczak, 1989), and a genetic map has been produced (Weitkamp and Sandberg, 1990). Both EqMHC I and EqMHC II genes have been cloned (Barbis *et al.*, 1994b; Szalai *et al.*, 1994a, b). At least two polymorphic class I loci have been identified: ELA-A and ELA-B (Bernoco *et al.*, 1987a; Lazary *et al.*, 1988; Antczak, 1989). In addition soluble class I molecules have been detected in equine serum (Lew *et al.*, 1986a, b). The biochemical characteristics of the equine MHC I and II antigens have proven similar to those of other species, with the heavy chain of MHC I precipitated by alloantisera or mAbs of 44 kDa (Hesford *et al.*, 1989; Kydd *et al.*, 1994), and MHC II consisting of a noncovalently linked dimer of a 30–33 kDa α chain, and a 28–30 kDa β chain (Crepaldi *et al.*, 1986; Hesford *et al.*, 1989; Monos *et al.*, 1989; Lunn *et al.*, 1993). Monoclonal antibodies specifically produced against equine MHC I (Donaldson *et al.*, 1988; Kydd *et al.*, 1991, 1994; Lunn *et al.*, 1997) and MHC II have been described (Crepaldi *et al.*, 1986; Monos *et al.*, 1989; Lunn *et al.*, 1993, 1998; Kydd *et al.*, 1994). However, it is possible to study equine MHC antigens using cross-reactive mAbs produced against the MHC of other species (Crepaldi *et al.*, 1986; Monos *et al.*, 1989). This results from the extensive conservation of the structure of MHC antigens between species. Some anti-EqMHC I mAbs have been shown to block MHC I-restricted CTLs *in vitro* (O'Brien *et al.*, 1991), although they have no effect on PHA-induced lymphoproliferation (Kydd and Hannant, 1996).

Equine MHC polymorphism

Extensive investigations of equine MHC (or ELA) polymorphism have taken place for over 15 years (Antczak, 1992). In the horse, ELA class I antigens have been identified and their polymorphic determinants studied with alloantisera (Bright *et al.*, 1978; Lazary *et al.*, 1980a, b, 1986), resulting in a series of workshops which have identified at least 13 of these highly variable ELA class I gene products (Lazary *et al.*, 1988). The alloantisera used for these studies were derived from pregnant mares, which, in contrast to other species, normally become sensitized to paternal MHC antigens (Bright *et al.*, 1978).

ELA class II polymorphic antigens have also been identified (Hesford *et al.*, 1989). Studies of ELA polymorphism have also been completed using biochemical methods such as isoelectric focusing (Schuberth *et al.*, 1992), and RFLP analysis using human cDNA probes (Hanni *et al.*, 1988). A major reason for the study of these polymorphic MHC antigens is the determination of any disease associations (McClure, 1988), and conditions in which positive associations have been made include equine sarcoid (Lazary *et al.*, 1985, 1994; Angelos *et al.*, 1988; Marti *et al.*, 1993), and dermal insect bite hypersensitivity (Weitkamp and Sandberg, 1990; Lazary *et al.*, 1994). There is limited evidence for associations with arytenoid chondritis and cryptorchidism (McClure *et al.*, 1988), while results of studies of associations between ELA and fertility were equivocal (MacCluer *et al.*, 1988; Park *et al.*, 1989).

Regulation of equine MHC expression

One of the principle focuses of study of equine MHC expression has been in the area of reproductive immunology, and specifically the endometrial cup (Antczak and Allen, 1989). This area is discussed in detail in Section 10. A second area of study has been the distribution of equine MHC II products on PBLs. An unusual characteristic of equine MHC II antigens is that they are present on a large subpopulation of T lymphocytes in addition to all B lymphocytes (Crepaldi *et al.*, 1986; Lunn *et al.*, 1993), this is also a feature of MHC II in dogs and swine (Crepaldi *et al.*, 1986; Monos *et al.*, 1989; Holmes and Lunn, 1994). In contrast, MHC expression in man and rodents is restricted to antigen-presenting cells (Raulet and Eisen, 1990). Recently, it has been shown that this equine T lymphocyte MHC II expression occurs in both the EqCD4 and EqCD8 subsets, is virtually absent at birth, but it subsequently increases during the first 6 months of life (Lunn *et al.*, 1993). Both $CD4^+$ and $CD8^+$ T lymphocytes can effectively process and present antigens to MHC II-restricted T lymphocytes, provided that they can initially capture the antigen (Lanzavecchia *et al.*, 1988). Therefore, while T lymphocytes appear to be able to present antigens in association with MHC II, this process may be limited due to an inability to capture antigens. This form of antigen presentation may be of importance when antigens are captured after binding to T lymphocyte surface molecules, or when the antigen is produced endogenously by the T lymphocyte itself (Lanzavecchia, 1990).

While MHC II expression by resting T lymphocytes does not occur in most species, MHC II has been described as an activation marker on mature memory human and ovine T lymphocytes (Hopkins *et al.*, 1989; Wallace and Beverley, 1990). Therefore, it is possible that constitutive expression of MHC II on T lymphocytes in the horse may identify memory T-cells (Lanzavecchia, 1990). Testing this hypothesis will require the differentiation of equine naive and memory T-lymphocytes, either on the basis of cell surface phenotype or the ability to respond to recall antigens. The lack of MHC II expression on T lymphocytes of 1-day-old foals may indicate the absence of 'memory' T-lymphocytes in the peripheral circulation at this age, consequent to the lack of previous exposure to foreign antigens (Lunn *et al.*, 1993). This is comparable to man where $CD45RO^+$ memory T-cells are almost absent at birth, but increase in numbers throughout the first 20 years of life (Hayward *et al.*, 1989).

The level of equine MHC II expression on lymphocytes is also determined by haplotype. It has been demonstrated that low expression of MHC II antigens is linked to the D3 haplotype (Barbis *et al.*, 1994a). The functional significance of this phenomenon is unknown, but it is interesting to note that the D3 allele has been associated with increased susceptibility to equine sarcoid (Lazary *et al.*, 1985).

Non-MHC lymphocyte alloantigens

A variety of equine lymphoid antigens exhibit polymorphisms that are not controlled by the MHC, including CD11a/18 (Kydd *et al.*, 1994) and EqWC1 (Kydd *et al.*, 1994; Lunn *et al.*, 1994a). However, the ELY system is the best characterized, and is encoded by the ELY-1 and ELY-2 loci, with two alleles at each locus (Antczak, 1992). Alloantibodies to the ELY antigens naturally occur in pregnant mares and can complicate analysis of alloantisera. The identity of the ELY antigen is unknown, but it has some of the characteristics of the human CD45 antigen (Byrns *et al.*, 1987).

6. Innate Immunity

Natural killer and lymphokine activated killer cells

Non-MHC restricted cytotoxic lymphocytes represent an important component of mammalian nonadaptive innate immunity and are a critical defense mechanism in viral infectious disease and neoplasia. This natural cytotoxicity is predominantly mediated by natural killer (NK) cells, which are identified by a $CD3^-CD16^+CD56^+$ phenotype, and to a lesser extent by non-MHC restricted cytotoxic T cells (Lanier *et al.*, 1986; Fitzgerald-Bocarsly *et al.*, 1988). It is possible to demonstrate a further form of non-MHC restricted cytotoxicity by culturing peripheral blood lymphocytes in the presence of high concentrations of interleukin-2 (IL-2) to give rise to lymphokine-activated killer (LAK) cells (Lydyard and Grossi, 1993b). These LAK cells exhibit a broad range of non-MHC restricted cytolytic activity against tumors and some normal tissues, and even against previously NK-resistant targets (Trinchieri, 1989). The majority of human LAK cells are derived from activated NK cells (Phillips and Lanier, 1986; Roberts *et al.*, 1987), although a component of this LAK activity can be mediated by cells of the T-cell lineage (Ortaldo *et al.*, 1986).

In the horse, one of the first investigations of natural cytotoxic cells was made in foals suffering from severe combined immunodeficiency (SCID), a primary immunodeficiency of horses (discussed in detail in Section 8). SCID is a fatal, inherited disease of Arabian foals first discovered in 1973 (McGuire and Poppie, 1973). The disease is an autosomal recessive condition that results in the absence of both B and T lymphocytes, but a NK lymphocyte population persists. In 1987 Magnuson *et al.* demonstrated that PBMCs isolated from SCID foals had the morphological characteristics of large granular lymphocytes and, when incubated for 24 h with 100 units/ml of recombinant human IL-2, demonstrated cytotoxic activity against YAC-1 lymphoma or K562 erythroleukemia NK target cell lines. These results indicated that cytotoxic cells with morphological and functional characteristics of natural killer cells are produced by horses with SCID. When PBMCs from normal horses were used in the same assays they were unable to mediate the same cytotoxic activity, although they could mediate antibody-dependent cytotoxicity (ADCC) against the same targets. The reason for this failure is presumably the low precursor frequency of NK cells in the peripheral blood of normal horses.

The next substantial investigation of natural cytotoxicity in horses was published by Hormanski *et al.* (1992) who described the generation of equine LAK cells by incubation of PBMCs from normal horses with high concentrations IL-2 (up to 500 units/ml for 3 days). The investigators used a novel equine lymphoma target cell line, EqT8888, which has subsequently proven extremely useful in a number of equine immunological studies. The LAK cells generated in this study failed to lyse xenogeneic targets cells (Daudi and EL4) owing to a failure to bind the target cells. The identity of the cells mediating this LAK activity was uncertain, although it was proposed that they were of the T-cell lineage (Hormanski *et al.*, 1992).

An investigation of SCID foals using mAbs demonstrated that a large percentage of the residual lymphocyte population expressed the EqCD8 antigen, while failing to express the EqCD3 antigen (Lunn *et al.*, 1995b). The implication of this finding was that equine NK cells, which should represent the major lymphoid population in SCID foals, expressed the EqCD8 antigen. Expression of CD8 or its homologue on NK cells has been recorded in other species such as the rat (Trinchieri, 1989), while in humans CD8 expression has been reported on 30–50% of NK cells (Perussia *et al.*, 1983). A further investigation was performed to determine whether the EqCD8 molecule could play any role in equine non-MHC restricted killing. The molecules involved in non-MHC restricted killing are poorly defined in most species, and a number of theoretical interactions between these cytotoxic cells and their targets have been proposed (Trinchieri, 1989; Karre, 1995). It is likely that NK cells possess more than one recognition mechanism in order to account for their ability to kill targets either on the basis of failure to express MHC I, or in other instances independent of their MHC I status (Lanier and Phillips, 1992). In addition, the molecular interactions involved in target recognition may involve NK-specific receptors for NK target cell molecules, or multiple adhesion molecule interactions (Garni-Wagner *et al.*, 1993). A series of experiments were conducted that demonstrated that positive selection of LAK precursors based on the expression of the EqCD8 antigen resulted in significant increases in LAK activity (Lunn *et al.*, 1997). In addition, blocking specific EqCD8 epitopes with mAbs significantly decreased LAK activity. This result indicated that the EqCD8 molecule could play a substantial role as an activation molecule in LAK; however, the study failed to demonstrate whether the LAK precursor cells prepared by EqCD8 selection were NK cells or T lymphocytes.

It is not yet possible to define equine NK cells in terms of a unique cell surface phenotype, although there is evidence that they may be $EqCD8^{+}/EqCD3^{-}$. In addition, NK activity is difficult to demonstrate in horses without IL-2 augmentation of cytotoxic activity to generate LAK cells. However, LAK activity is a functionally defined phenomenon that may depend on both T lymphocytes and NK cells (Ortaldo *et al.*, 1986; Lanier *et al.*, 1986), and therefore does not represent an effector function of a single cell lineage. A mAb that reacts with an evolutionarily conserved NK molecule that cross-reacts with a number of species, including the horse, has been described by Harris *et al.* (1993) (see Table X.2.2). Further evaluation of this reagent may prove it to be a valuable tool for studying NK activity in horses.

Neutrophil and macrophage function

There have been a limited number of studies of neutrophil function (Hietala and Wolf, 1987; Morris *et al.*, 1987; Slauson *et al.*, 1987; Foerster and Wolf, 1990; Grunig *et al.*, 1991; Wichtel *et al.*, 1991; Wong *et al.*, 1992b), and in one instance of the antibacterial activity of neutrophil lysozyme (Pellegrini *et al.*, 1991). Many of these studies have focused on the developmental aspects of neutrophil function, and these are discussed in Section 7. One interesting study demonstrated that exercise may transiently impair neutrophil antimicrobial functions (Wong *et al.*, 1992b). Studies of equine macrophage function have principally been conducted in the context of studies of *Rhodococcus equi* infection of foals and are discussed in Section 8.

Complement

As discussed in Section 3, the various equine IgG subisotypes have different capacities to fix complement (McGuire *et al.*, 1973; Montgomery, 1973). The purification and characterization of equine complement has been described (Boschwitz and Timoney, 1993), together with a hemolytic assay system measuring its concentration (Reis, 1989).

7. Development of Equine Immunity

Ontogeny of the equine immune system

There have been few studies of the prenatal development of the equine immune system. As in other species, the thymus is the first lymphoid organ to develop and mitogen-responsive cells can be identified there from day 80 of the 340-day gestational period of the horse (Perryman *et al.*, 1980). Subsequently, these cells appear in peripheral blood at 120 days, lymph nodes at 160 days, and the spleen at 200 days. Cells responsive in mixed lymphocyte reactions are detectable in the thymus from 100 days, and the spleen at 200 days. Immunoglobulin production is detectable prior to 200 days, and newborn foals typically have IgM concentrations in their serum of approximately 165 μg/ml. It appears that functional T lymphocytes are present by day 100 and B lymphocytes by day 200 of gestation. When immunological competence of the equine fetus is assessed in terms of specific antibody responses, *in utero* immunization of foals in late gestation with keyhole limpet hemocyanin in an alum adjuvant results in detectable specific antibody production and T-cell responsiveness at the time of birth (Hannant *et al.*, 1991). In addition, the equine fetus can respond to coliphage T2 at 200 days, and Venezuelan equine encephalitis virus at 230 days (Martin and Larson, 19173; Morgan *et al.*, 1975).

Detailed studies of the appearance of lymphocyte subpopulations defined by mAbs have not been performed in the equine fetus. However, some information regarding the maturation of thymocytes in young horses is available. During thymic maturation of T cells, stem cells migrate into the thymus, and mature into T cells under the influence of the epithelial microenvironment (Boyd *et al.*, 1993; Lydyard and Grossi, 1993a). In this process different patterns of cell surface differentiation antigen expression distinguish successive stages of thymocyte maturation. In humans, the earliest thymic precursor cells express low levels of CD4 (Godfrey and Zlotnik, 1993). This CD4 expression is lost as early thymocytes become double negative $CD4^-CD8^-$ cells and then demonstrate their T-cell commitment by TCRβ gene rearrangement, which is an essential trigger for subsequent events and leads to low levels of expression of a cell surface TCRβ-CD3 complex (Palmer *et al.*, 1993). Intermediate thymocytes are $CD4^{lo}CD8^{lo}$, but after TCRα gene rearrangement and expression of cell surface TCR$\alpha\beta$ they rapidly become $CD4^{hi}CD8^{hiTCR-}CD3^{hi}$ (Godfrey and Zlotnik, 1993). Subsequently, thymocytes selected on the basis of productive TCR gene rearrangement and lack of self reactivity become mature T cells expressing either CD4 or CD8 (single positive) in combination with high levels of TCR^+CD3. Using two color fluorescence-activated cell staining analysis, it is possible to demonstrate similar patterns of EqCD3, EqCD4 and EqCD8 antigen expression in the equine thymus (Lunn *et al.*, 1991a; Blanchard-Channell *et al.*, 1994).

Immunocompetence in foals

Infectious disease in neonatal foals is associated with high morbidity and mortality. While failure of passive transfer is a major cause of this problem, as discussed in Section 8, immaturity of the immune system has also been considered a potential contributing factor. As a result, a number of studies of neonatal immunocompetence have been completed. Neutrophils are generally found to be fully functional from birth (Morris *et al.*, 1987; Wichtel *et al.*, 1991), however, their function is significantly impaired prior to absorption of colostral antibodies which are required for opsonization (Bernoco *et al.*, 1987b). Alveolar macrophages recovered from bronchoalveolar lavages may be low in number up to 2 weeks of age and have impaired chemotactic function (Liu *et al.*, 1987). Lymphocyte blastogenesis responses may be depressed on the day of birth, but they subsequently rapidly rise to adult levels (Sanada *et al.*, 1992). There are no markers available for the development of memory lymphocytes in horses, although the pattern of T lymphocyte MHC II expression in foals may be relevant to this issue (see Section 5). The foal can respond to foreign antigens from the day of birth, but a factor that significantly affects *de novo* immune response in foals is the suppressive effect of passively transferred maternal antibodies. The rate of decline of these antibodies varies for both individuals and different infectious agents. For many important pathogens, the concentration of maternal antibodies in foals falls to nonprotective levels by 2–3 months old (Galan *et al.*, 1986; Gibbs *et al.*, 1988; Robinson *et al.*, 1993). However, the remaining antibody can still render the foal unresponsive to vaccination for weeks or even months. In the case of equine influenza virus infection, maternal antibodies can persist until 6 months of age and prevent immune responses in foals vaccinated prior to reaching that age (van Maanen *et al.*, 1992).

8. Disorders of the Immune System

The horse suffers from a similar spectrum of immune-mediated diseases as do other species, and descriptions are presented in standard equine internal medicine texts. Conditions of principal interest to the comparative immunologist are the focus of this section, and these include the various forms of equine immunodeficiency, and neonatal isoerythrolysis. Hypersensitivity (allergic) diseases also represent an important form of equine immune-mediated disease that provide many opportunities for comparative study. The immunopathogenesis of obstructive bronchitis in particular is a critical area of equine immunological research (Halliwell *et al.*, 1979, 1993; Lawson *et al.*, 1979; Mair *et al.*, 1988; Dirscherl *et al.*, 1993; McGorum *et al.*, 1993a, b). However, progress in this area is currently limited by the lack of reagents for studying equine IgE and the biology of equine mast cells.

Primary immunodeficiencies

Several primary immunodeficiency conditions of horses have been identified (Perryman *et al.*, 1987; McClure *et al.*, 1996). The most common and best characterized is severe combined immunodeficiency, which represents one of the most important areas of comparative immunological study in the horse.

Severe combined immunodeficiency

Severe combined immunodeficiency (SCID) of Arabian foals is a fatal, autosomal recessive disease first discovered in 1973 (McGuire and Poppie, 1973), and subsequently extensively investigated (Buening *et al.*, 1978; Perryman and McGuire, 1978; Lew *et al.*, 1980; Yilma *et al.*, 1982; Magnuson *et al.*, 1985; Wyatt *et al.*, 1987, 1988; Perryman *et al.*, 1988b). Most SCID foals become ill by 1 month old, and the majority die by 5 months. It is estimated that approximately 2–3% of Arabian foals have SCID which would suggest that 25% of Arabian horses carry the recessive gene (Poppie and McGuire, 1977). There is no test to detect carriers other than costly and impractical multiple breedings with known carriers. Severe combined immunodeficiency foals lack functional B and T lymphocytes, while neutrophils, macrophages, natural killer (NK) cells, and the complement system function normally. Prothymocytes are present in the thymus of SCID foals, but do not mature into functional T lymphocytes (Wyatt *et al.*, 1987; Perryman *et al.*, 1988b). The rare lymphocytes that are present in the circulation of SCID foals have the functional characteristics of NK cells (Magnuson *et al.*, 1987). The successful immunological reconstitution of one SCID foal with a histocompatible bone marrow transplant indicates that the architecture of the thymus and other lymphoid organs is intact (Perryman *et al.*, 1988a). This result supports the hypothesis that a biochemical defect in prolymphocytes results in their inability to mature into T and B lymphocytes (Perryman *et al.*, 1988b).

In an immunohistological study of SCID foals, it was demonstrated that the majority of the cells in the thymus and lymph nodes of SCID foals were of the lymphoid lineage, based on their patterns of expression of the EqCD11a/18, EqCD44, and EqCD13 antigens (Lunn *et al.*, 1995b). While approximately 10–20% of these cells expressed the early T-lymphocyte marker EqCD5, as previously described by Wyatt *et al.* (1987), there were few cells expressing the mature T-lymphocyte marker $EqCD3^+$ and no detectable expression of EqCD4. However, the majority of these lymphoid cells expressed the EqCD8 marker. Given the previous finding that the rare lymphocytes of SCID foals are of the NK lineage (Magnuson *et al.*, 1987), this led to the proposal that $EqCD3^-EqCD4^-EqCD8^+$ cells may represent a normal equine NK cell population: a $CD3^-CD4^-CD8^+$ NK cell surface antigen phenotype is seen in other species. In humans NK cells do not express CD3 or CD4, but 30–50% do express low levels of CD8 (Perussia *et al.*, 1983; Trinchieri, 1989), and in the rat all NK cells have a $CD3^-CD4^-CD8^+$ phenotype (Barclay *et al.*, 1993). If the SCID $EqCD3^-EqCD4^-EqCD8^+$ cells are NK cells, then the presence of large numbers of these cells in the thymus of these foals is analogous to the situation in SCID mice. In normal mice, as in humans, it is very difficult to find any evidence for the presence of NK cells in the thymus. However, in SCID mice Garni-Wagner *et al.* (1990) demonstrated that approximately 20% of thymic cells were mature NK cells with normal functional and phenotypic characteristics. This NK cell population would be difficult to identify in the thymus of normal mice as it would constitute a very small percentage of the total cell population due to the presence of normal T cells and their precursors. The role of NK cells in the thymus is uncertain, as NK cell precursors appear to be bone marrow derived, and NK cell differentiation does not require the thymus (Trinchieri, 1989; Garni-Wagner *et al.*, 1990). However, it is possible that a mature thymic NK cell population could have a regulatory role in thymic maturation and intrathymic selection (Garni-Wagner *et al.*, 1990).

Severe combined immunodeficiency has also been described in mice, dogs and humans. In mice, SCID results from defective rearrangement of T-cell receptor (TCR) and immunoglobulin genes, consequent to an autosomal recessive defect in the topisomerase enzyme (Bosma *et al.*, 1983; Fulop and Phillips, 1990). The lymphocytes of SCID mice do not express CD3, CD4 or CD8, and show no rearrangements of the TCRβ chain, although they do maintain a normal NK cell population (Habu *et al.*, 1987; Garni-Wagner *et al.*, 1990; Bosma and Carroll, 1993). In humans there are a variety of causes, but among the most common are defects in the purine metabolism pathway that result in failure of lymphocyte maturation (Rosen *et al.*, 1984). The cause of equine SCID has been a subject of intense investigation and speculation for many years. In equine SCID there is no evidence of decreased adenosine deaminase (ADA) or purine nucleoside phosphorylase (PNP) activity typical of human SCID (Perryman *et al.*, 1987; Rosen, 1993), and therefore the characteristics of equine SCID appear to have more in common with the murine form of the disease. Strong evidence for this proposal has recently been presented by Wiler *et al.* (1995), who demonstrated a defect in V(D)J recombination in SCID foals which was most probably due to a defect in expression of DNA-dependent protein kinase activity. The same group of researchers led by Dr Katheryn Meek went on to identify a frame-shift mutation, specifically a five nucleotide deletion, in the DNA-dependent protein kinase catalytic subunit (DNA-PKCS) resulting in the complete absence of full length transcripts (Shin *et al.*, 1997). The equine SCID defect results in a complete absence of DNA-PKCS activity and failure of both coding or signal joint ligation. In contrast, coding ligation can be demonstrated in SCID mice (leaky phenotype). This difference may be because the murine defect occurs in a different region of the DNA-PKCS gene (Blunt *et al.*, 1996; Danska *et al.*, 1996) and consequently some kinase activity may remain and be sufficient to support coding ligation (Shin *et al.*, 1997).

Selective IgM deficiency

Selective IgM deficiency was first described by Perryman *et al.* (1977), and is now an established clinical entity (McClure *et al.*, 1996). The condition is most common in Arabians and Quarter Horses, although the heritability is unknown. Three forms are recognized. The most common presentation affects foals, and is associated with severe infections invariably leading to death before 10 months of age. A second form affects foals, leading to recurrent unresponsive infections and poor development, although these animals may survive 1–2 years. The third form is seen in horses of 2–5 years of age, which are not necessarily presented for recurrent infections, but about half of which will develop lymphosarcoma. A recent study of an affected foal demonstrated normal responses to T-cell mitogens, but a failure of response to the B-cell mitogen lipopolysaccharide (Weldon *et al.*, 1992).

Other primary immunodeficiencies of horses

Equine medical texts typically list transient hypogammaglobulinemia and agammaglobulinemia as established syndromes in horses (McClure *et al.*, 1996). However, these conditions have been infrequently reported (McGuire *et al.*, 1975; Banks *et al.*, 1976; Deem *et al.*, 1979) and it is uncertain whether they represent true primary immunodeficiencies. Transient hypogammaglobulinemia is a delayed onset of autologous production of immunoglobulins in foals leading to low Ig levels between 2–4 months and resulting in recurrent bacterial and viral infections. Therapy is aimed at supporting the foal until autologous immunoglobulin production starts. Agammaglobulinemia results from complete B-cell dysfunction that may be X-linked (it has only been seen in males). Affected animals can survive up to 1–2 years, despite recurrent bacterial infections.

A further form of immunodeficiency affecting humoral immunity was described by Boy *et al.* (1992) in a 10-month-old Arabian colt. The animal exhibited an absence of serum IgM, IgA, and IgG(T), and a normal concentration of IgG. *In vitro* testing of PBMCs with T-cell mitogens elicited normal responses, while responses to B-cell mitogens were weak. On post mortem examination there was generalized lymphocyte depletion of lymphoid organs.

Secondary immunodeficiencies

Failure of passive transfer of immunity

There is no transplacental transfer of immunoglobulins across the diffuse epitheliochorial placenta of the foal. Consequently foals are born essentially agammaglobulinemic, and must absorb passively transferred maternal immunoglobulins from colostrum if they are to survive (McClure *et al.*, 1996). While foals are capable of *de novo* immunoglobulin synthesis at birth, endogenously produced immunoglobulins will not reach adequate levels until 2 months. The importance of failure of passive transfer of immunoglobulins in causing morbidity and mortality in foals has been extensively documented (Jeffcott, 1974; Clabough *et al.*, 1991; Stoneham *et al.*, 1991), and represents a major problem in the horse industry. Equine colostral antibodies include all the isotypes present in serum, and in similar proportions and concentrations (Gorman and Halliwell, 1989). It is possible to substitute bovine colostrum and achieve a level of immune protection in foals, although the resulting immune protection may be suboptimal (Holmes and Lunn, 1991).

Other secondary immunodeficiencies of horses

Secondary immunodeficiency in the horse can be associated with malnutrition, *in utero* equine herpes virus-1 infection, lymphosarcoma, and pregnancy (Yednock and Rosen, 1989; McClure, 1995). In addition Prescott (1993) proposed immunodeficiency as an explanation for the high frequency of respiratory tract infections affecting foals in the period from 1 to 6 months of age (Prescott *et al.*, 1991; Hoffman *et al.*, 1993a, b). One particularly important form of respiratory disease in horses of this age-group is *Rhodococcus equi* infection (Prescott, 1991). This pathogen can cause epidemic disease in foals, and currently the only effective prophylactic measure for foals is the administration of plasma from hyperimmunized donor horses prior to exposure (Martens *et al.*, 1987, 1989). This intracellular pathogen parasitizes macrophages, and appears able to resist the defense mechanisms in the age groups of foals in which it causes disease (Zink *et al.*, 1985; Martens *et al.*, 1988).

While there is considerable circumstantial evidence for decreased resistance to infectious disease in foals, no specific immunodeficiency syndrome has been defined which explains this phenomenon. In certain instances infections with pathogens such as *Pneumocystis carinii* (Ainsworth *et al.*, 1993) or *Candida albicans* (McClure *et al.*, 1985) in young horses are consistent with a diagnosis of immunodeficiency. However, given the current limitations of objectively measuring immunocompetence in horses in field conditions (Lunn and McClure, 1993), it is difficult to determine whether the high rate of respiratory infections in young horses represents an immunodeficiency state or a normal age-dependent susceptibility to respiratory infections, which is exacerbated by current husbandry practices.

Allogeneic incompatibilities – neonatal isoerythrolysis

Of the domestic species, the horse is the only animal in which neonatal isoerythrolysis occurs naturally (McClure *et al.*, 1996). The condition results in the destruction of the foal's erythrocytes after absorption of maternal alloantibodies from colostrum. The alloantibodies are produced in response to exposure to incompatible blood groups antigens, typically as a result of pregnancy or transfusion. The incidence of these antibodies is high, typically 1–2% in the general mare population, but up to 10% in mule pregnancies (McClure *et al.*, 1996). While there are seven blood group systems in horses, each with numerous individual factors, it is the Aa and Qa antigens that are responsible for the majority of neonatal isoerythrolysis cases. Therefore, Aa- or Qa-negative mares are at risk of developing alloantibodies against their foal's paternal antigens. In mule pregnancies (donkey sire, horse dam), a unique donkey antigen is responsible for development of alloantibodies (McClure *et al.*, 1994; Traub-Dargatz *et al.*, 1995). Antibodies can also form to the Ca antigen, although this can occur in the absence of exposure to Ca-positive red cells ('natural' antibody formation). Rather than causing disease, these anti-Ca antibodies may actually be protective, preventing development of alloantibodies against the Aa and Qa antigens through antibody-mediated immunosuppression of the immune response (Bailey *et al.*, 1988).

9. Tumors of the Immune System

Lymphosarcoma

Equine lymphosarcoma has been extensively reviewed (Collatos, 1992; Carlson, 1995, 1996). Lymphosarcoma is the most common hematopoietic tumor of the horse, comprising 1–3% of all equine tumors and frequently proving fatal (Collatos, 1992). No viral etiologic agent has been associated with equine lymphosarcoma. There is no sex or breed predisposition, and lymphosarcoma typically affects horses from 5–10 years old, although it can occur in almost all age groups. Four forms of the condition are recognized (Collatos, 1992, Carlson, 1996). In the *generalized* or *multicentric* form, tissues affected are widespread and include most lymph nodes, liver, spleen and other internal organs. Lymphadenopathy is common, but often internal. In the *alimentary* or *intestinal* form, affected horses are frequently less than 5 years old and suffer from malabsorption. Intestinal lesions predominantly involve the small bowel, together with intestinal, splenic, and hepatic lymph nodes. Bowel infiltrates can be diffuse or discrete. The *mediastinal* or *thymic* form of the disease is seen in adult horses and is associated with respiratory signs of disease. Finally the *cutaneous* form of the disease causes multiple subcutaneous nodules which can be localized or generalized. While local lymph nodes can be involved, generalized lymphadenopathy and internal organ involvement are not common. This form of the disease can be tolerated for many years and is the most susceptible to therapy. Leukemia is rare in lymphosarcoma (Lester *et al.*, 1993; Turrel, 1995; Carlson, 1996) but sporadic cases do occur and recently both a T- and B-cell form of lymphocytic leukemia were reported (Stewart, 1992).

Various immunological disorders are associated with lymphosarcoma, and IgM deficiency is the most common of these (Carlson, 1996). Immunodeficiency and gammopathies have been reported in several cases of equine lymphosarcoma (Collatos, 1992; Furr *et al.*, 1992; Ansar Ahmed *et al.*, 1993). Despite the availability of appropriate mAbs (Aida *et al.*, 1992; Schram *et al.*, 1996; Lunn *et al.*, 1997), few investigations of the lymphoid phenotype of equine lymphosarcoma have been reported. Investigations of the EqT8888 equine lymphoma cell line showed it to be poorly differentiated, expressing MHC I and MHC II antigens, and EqCD44 (Hormanski *et al.*, 1992; Lunn *et al.*, 1997). There was no evidence of surface immunoglobulin or T lymphocytes marker expression. In a study of a generalized case of lymphosarcoma, Asahina *et al.* (1994) suggested that the cells were of the B-cell lineage. Future investigations of equine lymphosarcoma can now benefit from the extensive array of newly developed equine immunological reagents, and a better understanding of this important form of equine neoplasia is likely.

Myeloproliferative disease

Myeloproliferative disease occurs rarely in horses, but can affect any of the stem cell lines in the bone marrow (Carlson, 1996). Affected horses are often less than 5 years old. The majority of signs relate to the failure of normal production of marrow cells, resulting in anemia, thrombocytopenia, and neutropenia. Atypical cells can appear in the periphery and several forms of leukemia have been reported including myelomonocytic (nonlymphoid), monocytic, chronic lymphocytic, granulocytic, eosinophilic, and erythrocytic (Turrel, 1995; Carlson, 1996). In a recent report of a subleukemic acute myelomonocytic leukemia, the phenotype of the cells were identified with the available mAbs (Buechner-Maxwell *et al.*, 1994).

Plasma cell myeloma

Plasma cell myelomas are rare in horses (Edwards *et al.*, 1993; Turrel, 1995; Carlson, 1996), occurring in all ages and typically resulting in anemia with circulating plasma cells present in less than half of the reported cases. Monoclonal proteins have been reported in the $\alpha 2$, β, or γ region by serum electrophoresis, and have been characterized as subclasses of IgG.

10. Reproductive Immunology

Of all the areas of comparative immunological study in the horse, the field of reproductive immunology has proven one of the most fertile and exciting. Maternal immunological interactions with the fetus involve a unique set of events which must prevent maternal rejection of trophoblast tissue invading the uterus, and at the same time control this invasion to regulate growth and prevent damage to maternal tissues (Daya and Clark, 1990). Tolerance of the fetus therefore represents successful transplantation of an allograft (i.e. tissue from another member of the same species). Unique features of placentation in the horse make it exceptionally well-suited to studying these interactions.

The mare has an epitheliochorial placentation which represents the least invasive of all placentation. However, a subpopulation of highly invasive trophoblast cells of the equine conceptus differentiates between days 25 and 36 (day 0 = day of ovulation) to form the chorionic girdle (Allen, 1982). Distinct from noninvasive trophoblasts of the equine placenta, chorionic girdle cells express MHC I and begin to attach and to invade the uterine epithelium by day 35 (Enders and Liu, 1991). Within a 48-h period, girdle cells aggressively migrate through the uterine epithelium into the endometrial stroma where they form distinct nodules, called endometrial cups. These endometrial cups are, in effect, individual fetal allografts and they secrete equine chorionic gonadotrophin (eCG). The role of eCG in the horse is unknown, although it has profound gonadotrophic effects when administered to other species (Murphy and Martinuk, 1991). At the time of invasion, the uterine epithelium is rich in large granular lymphocytes (Enders *et al.*, 1991) and invasion further stimulates small lymphocyte migration into the stroma (Enders and Liu, 1991). Thus, the girdle cells migrate into a stroma rich in lymphocytes, macrophages and plasma cells. Immunologically, chorionic girdle cells are distinguished by briefly expressing high levels of MHC I during the period of uterine invasion, and subsequently losing this expression by day 45 as these cells differentiate and form endometrial cups (Donaldson *et al*, 1990, 1992; Kydd *et al.*, 1991). This regulation of MHC I expression occurs at the transcriptional level (Maher *et al.*, 1996). Maternal leukocytic infiltration of the endometrial cups is coincident with these changes in MHC I expression. In addition, there is a rise in the antipaternal antibody in the mare's blood 2–4 weeks after girdle cell invasion (Crump *et al.*, 1987) – a time consistent with a primary immune response. It appears, therefore, that the mare is capable of mounting an immune response to the fetal-derived girdle cells. The formation of endometrial cups, a trophoblast-derived tissue which detaches from the conceptus to form discrete islands of tissue within the uterine stroma, and the corresponding local immune response is unique (Allen, 1982). It has been proposed that the profound local maternal cellular response associated with the formation of endometrial cups may play a role in the ultimate degeneration and desquamation of the endometrial cups at approximately 120 days of pregnancy (Allen, 1982). The principal focus of reproductive immunological research in the horse is the interaction between the placental tissues in endometrial cups and the maternal immune system.

During fetal development in all species there is a bidirectional interaction between the maternal immune system, and the feto-placental unit (Wegmann *et al.*, 1993). Maternal tolerance of the fetus may result from lack of expression of MHC class I and II antigens by the trophoblast, local immunosuppressive factors, or suppressor cell activity (Antczak, 1989; Daya and Clark, 1990; Roth *et al.*, 1990; Starkey, 1992). The maternal uterine lymphocyte population associated with fetal tissue in other species generally lack CD4, CD8 or T-cell receptor antigens on their surface, yet express natural killer or lymphokine-activated killer (LAK) cell markers (Croy *et al.*, 1988; Lee *et al.*, 1988; Starkey *et al.*, 1988; Head, 1989; Parr *et al.*, 1990; King and Loke, 1991; Starkey, 1992). The majority of decidua cells express CD56 strongly but are negative for CD16; a minor population of these cells is negative or dim for CD56 but positive for CD16 (King and Loke, 1991; Starkey, 1992). These cells may mediate the maternal immune regulation of fetal development and appear to play a role in early placentation. Cytotoxicity of these large granular lymphocytes against placental tissues may be controlled by their interaction with the unusual monomorphic MHC molecule (HLA-G in humans) expressed only by the extravillous trophoblast (Kovats *et al.*, 1990). Through this mechanism the large granular lymphocytes may recognize the placenta as 'self', and tolerate trophoblast invasion of the uterine wall and the establishment of placentation (King and Loke, 1991). However, activation of these uterine large granular lymphocytes by lymphokines can generate LAK cells which can lyse extravillous trophoblast. A balance between these factors inducing large granular lymphocyte tolerance and cytotoxicity may explain maternal tolerance of the fetus, and control of placental growth.

Additional evidence for the immunological interaction between the fetus and dam comes from studies of maternal cytokine production during pregnancy (Wegmann *et al.*, 1993). Much of this cytokine production can be attributed to the maternal lymphocyte T_H1 and T_H2 subsets. The T_H1 cytokines are generally harmful to the maintenance of pregnancy, and when T_H1 cytokines predominate there is excessive induction of natural killer cells by IL-2 and IFN-γ, and an associated increase in fetal resorption. It has been proposed that pregnancy causes the maternal immune system to preferentially mount T_H2 responses (Wegmann *et al.*, 1993). This maternal trend towards T_H2 responses appears to be brought about by the feto-placental unit, as only fetal tissues and not maternal uterine lymphoid cells are found to produce T_H2 cytokines. Furthermore, it appears that some of these cytokines, such as IL-10, are produced by nonlymphoid cells in the feto-placental unit.

The regulation of MHC expression by the chorionic girdle tissues appears to play an important role in the maintenance of pregnancy. The girdle cells express MHC I from the time of invasion until day 45, as described above (Donaldson *et al.*, 1992). To date, no evidence has been found for expression of an equine equivalent of the HLA-G molecule described in humans (Antczak *et al.*, 1996). The importance of this MHC I expression is demonstrated by the failure of the majority of donkey-in-horse extra-specific pregnancies, which typically do not form endometrial cups, while other hybrid conceptuses which do form cups are successful (Allen, 1982). These donkey-in-horse pregnancies can be rescued by immunization of the mare with paternal donkey lymphocytes, which maintains these pregnancies despite the absence of endometrial cup tissues. This suggests that exposure to paternal MHC I antigens is a key factor in the maintenance of these pregnancies, and that this is an important function of invading girdle cells during cup formation. This explanation is consistent with the hypothesis that successful pregnancy depends on promotion of maternal T_H2-mediated humoral immune responses (Wegmann *et al.*, 1993). The brief expression of MHC I antigens by the invading girdle cells appears to stimulate such a T_H2 humoral immune response, and may in this way help maintain pregnancy by establishing a suitable maternal immune environment.

As described above murine and human decidua contain numerous leukocytes characteristic of natural killer or large granular lymphocytes, yet only a small percentage (8%) of T cells (Starkey, 1992). In contrast the maternal leukocytic response to endometrial cups includes large numbers of T lymphocytes with a normal $EqCD4^+$ and $EqCD8^+$ phenotype (Grunig *et al.*, 1995). The role of these cells is uncertain, although a number of possibilities have been proposed. In the first instance it is possible that these maternal cells contain an NK population, or at least a population capable of mounting a LAK response. This is based on the proposal that equine NK cells may have an EqCD8 phenotype (Lunn *et al.*, 1995b), and $EqCD8^+$ cells can mediate LAK cytotoxicity (Lunn *et al.*, 1997); recently, it was demonstrated that $EqCD8^+$ equine LAK cells can lyse cultured chorionic girdle cells (Vagnoni *et al.*, 1996). Alternatively, it is possible that MHC-restricted CTLs may play a role in regulating development of endometrial cups (Baker *et al.*, 1996), although this could presumably only be of importance in interactions during the early phase of cup development prior to day 45 when MHC I expression is down-regulated (Donaldson *et al.*, 1992).

These studies suggest that cytotoxicity may play a role in the life cycle of the endometrial cups and their ultimate elimination, and therefore regulation of maternal restricted cytotoxic cells may be of critical importance. Factors which regulate cell-mediated immunity could originate either from the maternal or placental tissue, both of which have been reported, in other species, to produce immunosuppressive factors (Clark *et al.*, 1988; Daya and Clark, 1990; Feinberg *et al.*, 1992; Suzuki *et al.*, 1995). Immunosuppressive activity of the placenta has also been demonstrated in the horse (Roth *et al.*, 1990, 1992; Lea and Bolton, 1991; Watson and Zanecosky, 1991). Lea and Bolton (1991) found that day 80 endometrial cup tissue extracts suppressed mixed lymphocyte responses and blastogenesis, and demonstrated that this suppression was not due to eCG. Roth *et al.* (1990) described a >30 000 kDa substance produced in culture by day 9–20 horse embryos which was shown to suppress lymphocyte proliferation in several mitogen-stimulated blastogenesis assays, and appeared to be acting by inhibiting responses to IL-2 (Baxevanis *et al.*, 1993). Grunig and Antczak (1995) recently demonstrated that endometrium and endometrial cup tissue expressed TNF-α, IFN-γ, IL-2 and IL-4; however, cultured chorionic girdle cells expressed only TNF-α, suggesting the other factors may be of maternal origin.

These data cannot yet explain the regulation of endometrial cup lifespan. However, studies of equine endometrial cups have generated considerable advances in our understanding of reproductive immunology, and have also provided the impetus for many advances in equine immunology. Research in this area promises to continue to provide new insights into a unique immunological interaction.

11. Conclusions

It is apparent that there are many areas of equine immunology where there is a pressing need for further research. Investigators urgently need reagents for equine IgE, information about immunoglobulin heavy chain gene sequences, more cytokine reagents, and information about naive and memory T lymphocytes. Nevertheless numerous important discoveries have been made in recent years and our understanding of immunity to infectious disease in the horse is increasing at a rapid pace. In addition, it is apparent that comparative immunology studies in the horse have enormous scope. In the case of reproductive immunology in particular, the horse offers a unique model for examining immunological regulatory processes. One extremely promising area for equine investigation is the effect of exercise stress on immune function. In this area the equine athlete offers unique opportunities for advances (Wong *et al.*, 1992a, b; Hines *et al.*, 1996b, Horohov *et al.*, 1996a). Given the important place of the horse in our society, and the unique features of this species, studies of equine immunology are likely to remain a key area of veterinary and comparative investigation.

12. References

Aderka, D., Le, J. M., Vilcek, J. (1989). IL-6 inhibits lipopolysaccharide-induced tumor necrosis factor production in cultured human monocytes, U937 cells, and in mice. *J. Immunol.* **143**, 3517–3523.

Aida, Y., Okada, K., Kageyama, R., Amanuma, H. (1992). Cross-reactivity between a monoclonal antibody that recognizes a tumor-associated antigen on bovine lymphosarcoma cells and blood lymphocytes from various mammalian species. *Am. J. Vet. Res.* **53**, 1988–1991.

Ainsworth, D. M., Weldon, A. D., Beck, K. A., Rowland, P. H. (1993). Recognition of *Pneumocystis carinii* in foals with respiratory disease. *Equine Vet. J.* **25**, 103–109.

Allen, G. P., Yeargan, M., Costa, L. R. R., Cross, R. (1995). Major histocompatibility complex class I-restricted cytotoxic

T-lymphocyte responses in horses infected with equine herpesvirus I. *J. Virol.* **69**, 606–612.

Allen, G. K., Green, E. M., Robinson, J. A., Garner, H. E., Loch, W. E., Walsh, D. M. (1993). Serum tumor necrosis factor alpha concentrations and clinical abnormalities in colostrum-fed and colostrum-deprived neonatal foals given endotoxin. *Am. J. Vet. Res.* **54**, 1404–1410.

Allen, P. Z., Dalton, E. J. (1975). Studies on equine immunoglobulins IV. Immunoglobulins of the donkey. *Immunology*, **28**, 187–197.

Allen, P. Z., Johnson, J. S. (1972). Studies on equine immunoglobulins. 3. Antigenic interrelationships among horse and dog G globulins. *Comp. Biochem. Physiol. B. Comp. Biochem.* **41**, 371–383.

Allen, P. Z., Berger, B. M., Helms, C. M. (1968). An examination of the cross reactivity of human and equine gammaG globulins and their fragments by immunodiffusion. *J. Immunol.* **101**, 1023–1035.

Allen, W. R. (1982). Immunological aspects of the endometrial cup reaction and the effect of xenogeneic pregnancy in horses and donkeys. *J. Reprod. Fert. Suppl.* **31**, 57–94.

Almawi, W. Y., Lipman, M. L., Stevens, A. C., Zanker, B., Hadro, E. T., Strom, T. B. (1991). Abrogation of glucocorticoid-mediated inhibition of T cell proliferation by the synergistic action of IL-1, IL-6, and IFN-gamma. *J. Immunol.* **146**, 3523–3527.

Angelos, J., Oppenheim, Y., Rebhun, W., Mohammed, H., Antczak, D. F. (1988). Evaluation of breed as a risk factor for sarcoid and uveitis in horses. *Anim. Genet.* **19**, 417–425.

Ansar Ahmed, S., Furr, M., Chickering, W. R., Sriranganathan, N., Sponenberg, D. P. (1993). Immunologic studies of a horse with lymphosarcoma. *Vet. Immunol. Immunopathol.* **38**, 229–239.

Antczak, D. F. (1989). Biology of the major histocompatibility complex. In *Veterinary Clinical Immunology* (eds, Halliwell, R. E. W., Gorman, N. T.), pp. 473–492. W.B. Saunders Company, Philadelphia, PA, USA.

Antczak, D. F. (1992). The major histocompatibility complex of the horse. In *Equine infectious diseases IV: Proceedings of the Sixth International Conference, Cambridge 1991* (eds Plowright, W., Rossdale, P. D., Wade, J. F.), pp. 99–112. R & W Publications, Newmarket, UK.

Antczak, D. F., Allen, W. R. (1989). Maternal immunological recognition of pregnancy in equids. *J. Reprod. Fert. Suppl.* **37**, 69–78.

Antczak, D. F., Davies, C. J., Ellis, S. (1996). Expression of major histocompatibility complex (MHC) class I genes in invasive horse trophoblast. Presented at the 13th Rochester Trophoblast Conference in Banff, 1996.

Appleton, J. A., Gagliardo, L. F., Antczak, D. F., Poleman, J. C. (1989). Production of an equine monoclonal antibody specific for the H7 hemagglutinin of equine influenza virus. *Vet. Immunol. Immunopathol.* **23**, 257–266.

Asahina, M., Murakami, K., Ajito, T., Goryo, M., Okada, K. (1994). An immunohistochemical study of an equine B-cell lymphoma. *Comp. Pathol.* **111**, 445–451.

Ashmun, R. A., Look, A. T. (1990). Metalloprotease activity of CD13/aminopeptidase N on the surface of human myeloid cells. *Blood* **75**, 462–469.

Ashmun, R. A., Shapiro, L. H., Look, A. T. (1992). Deletion of the zinc-binding motif of CD13/aminopeptidase N molecules results in loss of epitopes that mediate binding of inhibitory antibodies. *Blood* **79**, 3344–3349.

Bailey, E., Albright, D. G., Henney, P. J. (1988). Equine neonatal isoerythrolysis: evidence for prevention by maternal antibodies to the Ca blood group antigen. *Am. J. Vet. Res.* **49**, 1218–1222.

Baker, J. M., Bamford, A., Antczak, D. F. (1996). Anti-paternal cytotoxic lymphocyte (CTL) activity during allogeneic or xenogeneic pregnancy in equids. *Placenta* **17** (abstract).

Band, I., Chess, L. (1985). Perturbation of the T4 molecule transmits a negative signal to T cells. *J. Exp. Med.* **162**, 1294–1303.

Banks, K. L., McGuire, T. C. (1975). Surface receptors on neutrophils and monocytes from immunodeficient and normal horses. *Immunology* **28**, 581–588.

Banks, K. L., McGuire, T. C., Jerrells, R. (1976). Absence of B lymphocytes in a horse with primary agammaglobulinaemia. *Clin. Immunol. Immunopathol.* **5**, 282–290.

Barbis, D. P., Bainbridge, D., Crump, A. L., Zhang, C. H., Antczak, D. F. (1994a). Variation in expression of MHC class II antigens on horse lymphocytes determined by MHC haplotype. *Vet. Immunol. Immunopathol.* **42**, 103–114.

Barbis, D. P., Maher, J. K., Stanek, J., Klaunberg, B. A., Antczak, D. F. (1994b). Horse cDNA clones encoding two MHC class I genes. *Immunogenetics* **40**, 163.

Barclay, A. N., Birkeland, M. L., Brown, M. H., Beyers, A. D., Davis, S. J., Somoza, C., Williams, A. F. (1993). *The Leucocyte Antigen FactsBook*. Academic Press, London, UK.

Baxevanis, C. N., Reclos, G. J., Gritzapis, A. D., Dedousis, G. V. Z., Missitzis, I., Papamichail, M. (1993). Elevated prostaglandin E2 production by monocytes is responsible for the depressed levels of natural killer and lymphokine activated killer cell function in patients with breast cancer. *Cancer* **12**, 491–501.

Ben Ahmeida, E. T., Potter, C. W., Gregoriadis, G., Adithan, C., Jennings, R. (1994). IgG subclass response and protection against challenge following immunisation of mice with various influenza A vaccines. *J. Med. Microbiol.* **40**, 261–269.

Bender, B. S., Small, P. A. J. (1992). Influenza: pathogenesis and host defense. *Semin. Respir. Infect.* **7**, 38–45.

Bernoco, D., Byrns, G., Bailey, E., Lew, A. M. (1987a). Evidence of a second polymorphic ELA class I (ELA-B) locus and gene order for three loci of the equine major histocompatibility complex. *Anim. Genet.* **18**, 103–118.

Bernoco, M., Liu, I. K. M., West-Ehlert, C. J., Miller, M. E., Bowers, J. (1987b). Chemotactic and phagocytic function of peripheral blood polymorphonuclear leucocytes in newborn foals. *J. Reprod. Fert. Suppl.* **35**, 599–605.

Beutler, B., Milsark, I. W., Cerami, A. C. (1985). Passive immunization against cachectin/tumor necrosis protects mice from lethal effect of endotoxin. *Science* **229**, 869–871.

Billinghurst, R. C., Fretz, P. B., Gordon, J. R. (1995). Induction of intra-articular tumour necrosis factor during acute inflammatory responses in equine arthritis. *Equine Vet. J.* **27**, 208–216.

Blanchard-Channell, M., Moore, P. F., Stott, J. L. (1994). Characterization of monoclonal antibodies specific for equine homologues of CD3 and CD5. *Immunology* **82**, 548–554.

Blunt, T., Gell, D., Fox, M., Taccioli, G. E., Lehmann, A. R., Jackson, S. P., Jeggo, P. A. (1996). Identification of a nonsense mutation in the carboxyl-terminal region of DNA-dependent protein kinase catalytic subunit in the scid mouse. *Proc. Natl. Acad. Sci. USA* **93**, 10285–10290.

Boschwitz, J. S., Timoney, J. F. (1993). Purification and characterization of equine complement factor C3. *Vet. Immunol. Immunopathol.* **38**, 139–153.

Bosma, G. C., Custer, R. P., Bosma, M. J. (1983). A severe combined immunodeficiency mutation in the mouse. *Nature* **301**, 527–530.

Bosma, M. J., Carroll, A. M. (1993). The SCID mouse mutant: definition, characterization, and potential uses. *Ann. Rev. Immunol.* **9**, 323–349.

Boy, M. G., Zhang, C., Antczak, D. F., Hamir, A. N., Whitlock, R. H. (1992). Unusual selective immunoglobulin deficiency in an arabian foal. *J. Vet. Intern. Med.* **6**, 201–205.

Boyd, R. L., Tucek, C. L., Godfrey, D. I., Izon, D. J., Wilson, T. J., Davidson, N. J., Bean, A. G. D., Ladyman, H. M., Ritter, M. A., Hugo, P. (1993). The thymic microenvironment. *Immunol. Today* **14**, 445–459.

Breuhaus, B. A., DeGraves, F. J. (1993). Plasma endotoxin concentration in clinically normal and potentially septic equine neonates. *J. Vet. Int. Med.* **7**, 296–302.

Bright, S., Antczak, D. F., Ricketts, S. (1978). Studies on equine leukocyte antigens. In *Proceedings of the 4th International Conference on Equine Infectious Diseases* (eds. Bryans, J. T., Gerber, H.), pp. 229–236. Veterinary Publications, Princeton, USA.

Brown, W. C., Woods, V. M., Chitko-McKown, C. G., Hash, S. M., Rice-Ficht, A. C. (1994). Interleukin-10 is expressed by bovine type 1 helper, type 2 helper, and unrestricted parasite-specific T-cell clones and inhibits proliferation of all three subsets in an accessory-cell-dependent manner. *Infec. Immun.* **62**, 4697–4708.

Buechner-Maxwell, V., Zhang, C., Robertson, J., Jain, N. C., Antczak, D. F., Feldman, B. F., Murray, M. J. (1994). Intravascular leukostasis and systemic aspergillosis in a horse with subleukemic acute myelomonocytic leukemia. *J. Vet. Int. Med.* **8**, 258–263.

Buening, G. M., Perryman, L. E., McGuire, T. C. (1978). Immunoglobulins and secretory component in the external secretions of foals with combined immunodeficiency. *Infec. Immun.* **19**, 695–698.

Byrne, K. M., Davis, W. C., Holmes, M. A., Brassfield, A. L., McGuire, T. C. (1997). Cytokine RNA expression in an equine CD4+ subset differentiated by expression of a novel 46 kDa surface protein. *Vet. Immunol. Immunopathol.* **56**, 191–204.

Byrns, G., Crump, A. L., LaLonde, G., Bernoco, D., Antczak, D. F. (1987). The ELY-1 locus controls a di-allelic alloantigenic system on equine lymphocytes. *J. Immunogen.* **14**, 59–71.

Cargile, J. L., MacKay, R. J., Dankert, J. R., Skelley, L. (1995). Effect of treatment with a monoclonal antibody against equine tumor necrosis factor (TNF) on clinical, hematologic, and circulating TNF responses of miniature horse given endotoxin. *Am. J. Vet. Res.* **56**, 1451–1459.

Carlson, G. P. (1995). Lymphosarcoma in horses. *Leukemia* **9**(Suppl 1), S101.

Carlson, G. P. (1996). Diseases of the hematopoietic and hemolymphatic systems. In *Large Animal Internal Medicine* 2nd edn (ed. Smith, B. P.), pp. 1189–1257. C.V. Mosby Company, St. Louis.

Clabough, D. L., Levine, J. F., Grant, G. L., Conboy, H. S. (1991). Factors associated with failure of passive transfer of colostral antibodies in Standardbred foals. *J. Vet. Intern. Med.* **5**, 335–340.

Clark, D. A., Falbo, M., Rowley, R., Banwatt, D., Stedronska-Clark, J. (1988). Active suppression of host-versus-graft reaction in pregnant mice. IX. Soluble suppressor activity obtained from allopregnant mouse decidua that blocks cytolytic effector response to interleukin 2 is related to TGF-β. *J. Immunol.* **14**, 3833–3840.

Clerici, M., Shearer, G. M. (1994). The Th1–Th2 hypothesis of HIV infection: new insights. (Review). *Immunol. Today* **15**, 575–581.

Collatos, C. (1992). Lymphoproliferative and myeloproliferative disorders. In *Current Therapy in Equine Medicine – 3*, 3rd edn (ed. Robinson, N. E.), pp. 513–516. W.B. Saunders Company, Philadelphia.

Crabb, B. S., Studdert, M. J. (1990). Comparative studies of the proteins of equine herpesviruses 4 and 1 and asinine herpesvirus 3: antibody response of the natural hosts. *J. Gen. Virol.* **71**, 2033–2041.

Crepaldi, T., Crump, A., Newman, M., Ferrone, S., Antczak, D. F. (1986). Equine T lymphocytes express MHC Class II antigens. *J. Immunogen.* **13**, 349–360.

Croy, B. A., Waterfield, A., Wood, A., King, G. J. (1988). Normal murine and porcine embryos recruit NK cells to the uterus. *Cell. Immunol.* **115**, 471–480.

Crump, A., Donaldson, W. L., Miller, J., Kydd, J. H., Allen, W. R., Antczak, D. F. (1987). Expression of major histocompatibility complex (MHC) antigens on horse trophoblast. *J. Reprod. Fert. Suppl.* **35**, 379–388.

Crump, A. L., Davis, W. C., Antczak, D. F. (1988). A monoclonal antibody identifying a T-cell marker in the horse. *Anim. Genet.* **19**, 349–357.

Curran, J. A., Argyle, D. J., Cox, P., Onions, D. E., Nicolson, L. (1994). Nucleotide sequence of the equine interferon gamma cDNA. *DNA Sequence* **4**, 405–407.

Danska, J. S., Holland, D. P., Mariathasan, S., Williams, K. M., Guidos, C. J. (1996). Biochemical and genetic defects in the DNA-dependent protein kinase in murine scid lymphocytes. *Mol. Cell. Biol.* **16**, 5507–5517.

Dascanio, J. J., Zhang, C. H., Antczak, D. F., Blue, J. T., Simmons, T. R. (1992). Differentiation of chronic lymphocytic leukemia in the horse. *J. Vet. Intern. Med.* **6**, 225–229.

Daya, S., Clark, D. A. (1990). Immunoregulation of the materno-fetal interface. *Immunol. Allergy Clin. N. Am.* **10**, 49–64.

Deem, D. A., Traver, D. S., Thacker, H. L., Perryman, L. E. (1979). Agammaglobulinaemia in a horse. *J. Am. Vet. Med. Assoc.* **175**, 469–472.

Dennis, V. A., Klei, T. R., Chapman, M. R. (1993). Generation and partial characterization of an eosinophil chemotactic cytokine produced by sensitized equine mononuclear cells stimulated with *Strongylus vulgaris* antigen. *Vet. Immunol. Immunopathol.* **37**, 135–149.

Dinarello, C. A., Cannon, J. G., Wolff, S. M., Bernheim, H. A., Beutler, B., Cerami, A., Figari, I. S., Palladino Jr, M. A., O'Connor, J. V. (1986). Tumor necrosis factor (cachectin) is an endogenous pyrogen and induces production of interleukin 1. *J. Exp. Med.* **163**, 1433–1450.

Dinarello, C. A. (1988). Biology of interleukin 1. *FASEB J.* **2**, 108–115.

Dirscherl, P., Grabner, A., Buschmann, H. (1993). Responsiveness of basophil granulocytes of horses suffering from chronic obstructive pulmonary disease to various allergens. *Vet. Immunol. Immunopathol.* **38**, 217–227.

Donaldson, W. L., Crump, A. L., Zhang, C. H., Kornbluth, J., Kamoun, M., Davis, W. C., Antczak, D. F. (1988). At least two loci encode polymorphic class I MHC antigens in the horse. *Anim. Genet.* **19**, 379–390.

Donaldson, W. L., Zhang, C. H., Oriol, J. G. Antczak, D. F. (1990). Invasive equine trophoblast expresses conventional class I major histocompatibility complex antigens. *Development* **110**, 63–71.

Donaldson, W. L., Oriol, J. G., Plavin, A., Antczak, D. F. (1992). Developmental regulation of class I histocompatibility complex antigen expression by equine trophoblastic cells. *Differentiation* **52**, 69–78.

Dorrington, K. J., Rockey, J. H. (1968). Studies on the conformation of purified human and canine gamma-A-globulins and equine gamma-T-globulin by optical rotatory dispersion. *J. Biol. Chem.* **243**, 6511–6519.

Durando, M. M., MacKay, R. J., Stephen, L., Skelley, L. A. (1994). Effects of polymyxin B and *Salmonella typhimurium* antiserum on horses given endotoxins intravenously. *Am. J. Vet. Res.* **55**, 921–927.

Dustin, M. L., Springer, T. A. (1991). Role of lymphocyte adhesion receptors in transient interactions and cell locomotion. *Annu. Rev. Immunol.* **9**, 27–66.

Edens, L. M., Crisman, M. V., Toth, T. E., Ahmed, S. A., Murray, M. J. (1996). *In vitro* cytotoxic activity of equine lymphocytes on equine herpesvirus-1 infected allogenic fibroblasts. *Vet. Immunol. Immunopathol.* **52**, 175–189.

Edwards, D. F., Parker, J. W., Wilkinson, J. E., Helman, R. G. (1993). Plasma cell myeloma in the horse. A case report and literature review (Review). *J. Vet. Int. Med.* **7**, 169–176.

Ellis, J. A., Bogdan, J. R., Kanara, E. W., Morley, P. S., Haines, D. M. (1995). Cellular and antibody responses to equine herpesvirus 1 and 4 following vaccination of horses with modified-live and inactivated viruses. *J. Am. Vet. Med. Assoc.* **206**, 823–832.

Enders, A. C., Liu, I. K. M. (1991). Trophoblast-uterine interaction during equine chorionic girdle cell maturation, migration, and transformation. *Am. J. Anat.* **192**, 366–381.

Enders, A. C., Liu, I. K. M., Allen, W. R. (1991). Endometrial lymphocytes and lymphatic vessels at the time of girdle cell migration in the horse. *Am. J. Anat.* **192**, 366–381.

Feinberg, B. B., Tans, N. S., Walsh, S. W., Brath, P. C., Gonik, B. (1992). Progesterone and estradiol suppress human mononuclear cell cytotoxicity. *J. Reprod. Immunol.* **21**, 139–148.

Fenwick, B. W., Schore, C. E., Osburn, B. I. (1988). Human recombinant interleukin-2(125) induced *in vitro* proliferation of equine, caprine, ovine, canine and feline peripheral blood lymphocytes. *Comp. Immunol. Microbiol. Infect. Dis.* **11**, 51–60.

Fitzgerald-Bocarsly, P., Herberman, R., Hercend, T., Hiserodt, J., Kumar, V., Lanier, L. L., Ortaldo, J. R., Pross, H., Reynolds, C., Welsh, R., Wigzell, H. (1988). A definition of natural killer cells. *Immunol. Today* **9**, 292.

Foerster, R. J., Wolf, G. (1990). Phagocytosis of opsonized fluorescent microspheres by equine polymorphonuclear leukocytes. *Zentralblatt. Fur. Veterinarmedizin – Reihe. B.* **37**, 481–490.

Ford, J. E., Home, W. A., Gibson, D. M. (1994). Light chain isotype regulation in the horse. Characterization of Ig kappa genes. *J. Immunol.* **153**, 1099–1111.

Fulop, G. M., Phillips, R. A. (1990). The scid mutation in mice causes a general defect in DNA repair. *Nature* **347**, 479–482.

Furr, M. O., Crisman, M. V., Robertson, J., Barta, O., Swecker Jr, W. S. (1992). Immunodeficiency associated with lymphosarcoma in a horse. *J. Am. Vet. Med. Assoc.* **201**, 307–309.

Galan, J. E., Timoney, J. F., Lengemann, F. W. (1986). Passive transfer of mucosal antibody to *Streptococcus equi* in the foal. *Infec. Immun.* **54**, 202–206.

Garni-Wagner, B. A., Purohit, A., Mathew, P. A., Bennett, M., Kumar, V. (1993). A novel function-associated molecule related to non-MHC-restricted cytotoxicity mediated by activated natural killer cells and T cells. *J. Immunol.* **151**, 60–70.

Garni-Wagner, B. A., Witte, P. L., Tutt, M. M., Kuziel, W. A., Tucker, P. W., Bennett, M., Kumar, V. (1990). Natural killer cells in the thymus. Studies in mice with severe combined immune deficiency. *J. Immunol.* **144**, 796–803.

Gibbs, E. P., Wilson, J. H., All, B. P. (1988). Studies on passive immunity and the vaccination of foals against eastern equine encephalitis in Florida. *Equine Infect. Dis. V: Proc. 5th Int. Conf.* **5**, 201–205 (Abstract).

Godfrey, D. I., Zlotnik, A. (1993). Control points in early T-cell development. *Immunol. Today* **14**, 547–553.

Gorman, N. T., Halliwell, E. W. (1989). Immunoglobulin quantitation and clinical interpretation. In *Veterinary Clinical Immunology* (eds Halliwell, E. W., Gorman, N. T.), pp. 55–74. W.B. Saunders Company, Philadelphia.

Grunfeld, C., Palladino Jr, M. A. (1990). Tumor necrosis factor: immunologic, antitumor, metabolic, and cardiovascular activities (Review). *Adv. Int. Med.* **35**, 45–71.

Grunig, G., Antczak, D. F. (1995). Horse trophoblast produces tumor necrosis factor α but not interleukin 2, interleukin 4, or interferon γ. *Biol. Repro.* **52**, 531–539.

Grunig, G., Witschi, U., Winder, C., Hermann, M., von Fellenberg, R. (1991). Neutrophil migration induced by equine respiratory secretions, bronchoalveolar lavage fluids and culture supernatants of pulmonary lavage cells. *Vet. Immunol. Immunopathol.* **29**, 313–328.

Grunig, G., Himmler, A., Antczak, D. F. (1994). Cloning and sequencing of horse interferon-gamma cDNA. *Immunogenetics* **39**, 448–449.

Grunig, G., Triplett, L., Canady, L. K., Allen, W. R., Antczak, D. F. (1995). The maternal leucocyte response to the endometrial cups in horses is correlated with the developmental stages of the invasive trophoblast cells. *Placenta* **16**, 539–559.

Habu, S., Kimura, M., Katsuki, M., Hioki, K., Nomura, T. (1987). Correlation of T cell receptor gene rearrangements to T cell surface antigen expression and to serum immunoglobulin level in scid mice. *Eur. J. Immunol.* **17**, 1467–1471.

Halliwell, R. E., Fleischman, J. B., Mackay Smith, M., Beech, J., Gunson, D. E. (1979). The role of allergy in chronic pulmonary disease of horses. *J. Am. Vet. Med. Assoc.* **174**, 277–281.

Halliwell, R. E., McGorum, B. C., Irving, P., Dixon, P. M. (1993). Local and systemic antibody production in horses affected with chronic obstructive pulmonary disease. *Vet. Immunol. Immunopathol.* **38**, 201–215.

Han, J., Brown, T., Beutler, B. (1990). Endotoxin-responsive sequences control cachectin/tumor necrosis factor biosynthesis at the translational level. *J. Exp. Med.* **171**, 465–475.

Hannant, D. (1991). Immune responses to common respiratory pathogens: problems and perspectives in equine immunology. *Equine Vet. J.* **12** Suppl., 10–19.

Hannant, D., Mumford, J. A. (1989). Cell mediated immune

responses in ponies following infection with equine influenza virus (H3N8): the influence of induction culture conditions on the properties of cytotoxic effector cells. *Vet. Immunol. Immunopathol.* **21**, 327–337.

Hannant, D., Rossdale, P. D., McGladdery, A. J., O'Neill, T., Ousey, J. C. (1991). Immune response of the equine foetus to protein antigens. *Proc. Sixth Int. Con. Equine Inf. Dis., Cambridge, UK*, p. 86. (abstract).

Hanni, K., Hesford, F., Lazary, S., Gerber, H. (1988). Restriction fragment length polymorphisms of horse class II MHC genes observed using various human alpha- and beta-chain cDNA probes. *Anim. Genet.* **19**, 395–408.

Harris, D. T., Camenisch, T. D., Jaso-Friedmann, L., Evans, D. L. (1993). Expression of an evolutionarily conserved function associated molecule on sheep, horse and cattle natural killer cells. *Vet. Immunol. Immunopathol.* **38**, 273–282.

Hayward, A. R., Lee, J., Beverley, P. C. L. (1989). Ontogeny of expression of UCHL1 antigen on Tcr-1+ (CD4/8) and TcRδ+ T cells. *Eur. J. Immunol.* **19**, 771–773.

Head, J. R. (1989). Can trophoblast be killed by cytotoxic cells? *In vitro* evidence and *in vivo* possibilities. *Am. J. Reprod. Immunol.* **20**, 100–105.

Helm, C. M., Allen, P. Z. (1970). Studies on equine immunoglobulins II. Antigenic interrelationships among horse IgG IgG(T) and antipneumoccal γ1-component. *J. Immunol.* **105**, 1253–1263.

Hesford, F., Lazary, S., Curty-Hanni, K., Gerber, H. (1989). Biochemical evidence that leukocyte antigens W13, W22 and W23 are present on horse major histocompatibility complex class II molecules. *Anim. Genet.* **20**, 415–420.

Hietala, S. K., Ardans, A. A. (1987). Neutrophil phagocytic and serum opsonic response of the foal to *Corynebacterium equi*. *Vet. Immunol. Immunopathol.* **14**, 279–294.

Hill, W. C., Cebra, J. J. (1965). Horse anti-SI immunoglobulins. I. Properties of gamma-M-antibody. *Biochemistry* **4**, 2575–2584.

Himmler, A., Hauptmann, R., Adolf, G. R., Swetly, P. (1986). Molecular cloning and expression in *Escherichia coli* of equine type I interferons. *DNA* **5**, 345–356.

Hines, M. T., Palmer, G. H., Byrne, K. M., Brassfield, A. L., McGuire, T. C. (1996a). Quantitative characterization of lymphocyte populations in bronchoalveolar lavage fluid and peripheral blood of normal adult Arabian horses. *Vet. Immunol. Immunopathol.* **51**, 29–37.

Hines, M. T., Schott, H. C., Bayly, W. M., Leroux, A. J. (1996b). Exercise and immunity: a review with emphasis on the horse. *J. Vet. Intern. Med.* **10**, 280–289.

Hirano, T., Akira, S., Taga, T., Kishimoto, T. (1990). Biological and clinical aspects of interleukin 6 (Review). *Immunol. Today* **11**, 443–449.

Hoffman, A. M., Viel, L., Juniper, E., Prescott, J. F. (1993a). Clinical and endoscopic study to estimate the incidence of distal respiratory tract infection in thoroughbred foals on Ontario breeding farms. *Am. J. Vet. Res.* **54**, 1602–1607.

Hoffman, A. M., Viel, L., Prescott, J. F., Rosendal, S., Thorsen, J. (1993b). Association of microbiologic flora with clinical, endoscopic, and pulmonary cytologic findings in foals with distal respiratory tract infection. *Am. J. Vet. Res.* **54**, 1615–1622.

Holmes, M. A., Lunn, D. P. (1991). A study of bovine and equine immunoglobulin levels in pony foals fed bovine colostrum. *Equine Vet. J.* **23**, 116–118.

Holmes, M. A., Lunn, D. P. (1994). Variation of MHC II expression on canine lymphocytes with age. *Tissue Antigens* **43**, 179–183.

Home, W. A., Ford, J. E., Gibson, D. M. (1992). L chain isotype regulation in horse. I. Characterization of Ig lambda genes. *J. Immunol*, **149**, 3927-3936.

Hopkins, H., Dutia, B. M., Budjdoso, R., McConnell, I. (1989). *In vivo* modulation of CD1 and MHC class II expression by sheep afferent lymph dendritic cells. *J. Exp. Med.* **170**, 1301–1319.

Hormanski, C. E., Truax, R., Pourciau, S. S., Folsom, R. W., Horohov, D. W. (1992). Induction of lymphokine-activated killer cells of equine origin: specificity for equine target cells. *Vet. Immunol. Immunopathol.* **32**, 25–36.

Horohov, D. W., Keadle, P. L., Pourciau, S. S., Littlefield-Chabaud, M. A., Kamerling, S. G., Keowen, M. L., French, D. D., Melrose, P. A. (1996). Mechanism of exercise induced augmentation of lymphokine-activated killer (LAK) cell activity in horses. *Vet. Immunol. Immunopathol.* **53**, 221–233.

Issel, C. J., Horohov, D. W., Lea, D. F., Adams Jr, W. V., Hagius, S. D., McManus, J. M., Allison, A. C., Montelaro, R. C. (1992). Efficacy of inactivated whole-virus and subunit vaccines in preventing infection and disease caused by equine infectious anemia virus. *J. Virol.* **66**, 3398–3408.

Jacobsen, C. N., Aasted, B., Broe, M. K., Petersen, J. L. (1993). Reactivities of 20 anti-human monoclonal antibodies with leucocytes from ten different animal species. *Vet. Immunol. Immunopathol.* **39**, 461–466.

Jeffcott, L. B. (1974). Studies on passive immunity in the foal. *J. Comp. Path.* **84**, 93–101.

Jones, M., Cordell, J. L., Beyers, A. D., Tse, A. G., Mason, D. Y. (1993). Detection of T and B cells in many animal species using cross-reactive anti-peptide antibodies. *J. Immunol.* **150**, 5429–5435.

Kabat, F. A. (1939). The molecular weight of antibodies. *J. Exp. Med.* **69**, 103–119.

Kalsow, C. M., Dwyer, A. E., Smith, A. W., Nifong, T. P. (1992). Pinealitis coincident with recurrent uveitis: immunohistochemical studies. *Curr. Eye Res.* **11** (Suppl), 147–151.

Kapsenberg, M. L., Jansen, H. M., Bos, J. D., Wierenga, E. A. (1992). Role of type 1 and type 2 T helper cells in allergic diseases (Review). *Curr. Opin. Immunol.* **4**, 788–793.

Karre, K. (1995). Express yourself or die: peptides, MHC molecules, and NK cells. *Science* **267**, 978–979.

Kato, H., Ohashi, T., Nakamura, N., Nishimura, Y., Watari, T., Goitsuka, R., Tsujimoto, H., Hasegawa, A. (1995). Molecular cloning of equine interleukin-1 alpha and beta cDNAs. *Vet. Immunol. Immunopathol.* **48**, 221–231.

Khaleel, S. A., Kenney, R. M., Allen, P. Z. (1975). Distribution of immunoglobulins in equine tissues by indirect immunofluorescence. *J. Comp. Path.* **85**, 611–622.

Kidd, P. G., Vogt, R. F. (1989). Report of the workshop on the evaluation of T-cell subsets during HIV infection and AIDS. *Clin. Immunol. Immunopathol.* **52**, 3–9.

King, A., Loke, Y. W. (1991). On the nature and function of human uterine granular lymphocytes. *Immunol. Today* **12**, 432–435.

Kishimoto, T. K., Larson, R. S., Corbi, A. L., Dustin, M. L., Staunton, D. E., Springer, T. A. (1989). The leucocyte integrins. *Adv. Immunol.* **46**, 149–182.

Kiyono, H., Taguchi, T., Aicher, W. K., Beagley, K. W., Fujihashi, K., Eldridge, J. H., McGhee, J. R. (1990). Immunoregulatory confluence: T cells, Fc receptors and cytokines for IgA immune responses (Review). *Int. Rev. Immunol.* **6**, 263–273.

Klinman, N. R., Rockey, J. H., Karush, F. (1965). Equine anti-hapten antibody II. The γG (7Sγ) components and their specific interaction. *Immunochemistry* **2**, 51–60.

Kovats, S., Main, E. K., Librach, C., Stubblebine, M., Fisher, S. J., DeMars, R. (1990). A class I antigen, HLA-G, expressed in human trophoblast. *Science* **248**, 220–223.

Kydd, J. H., Hannant, D. (1997). Evaluation of phytohaemagglutinin-induced lymphoproliferation as an aid to clustering ELAW II monoclonal antibodies. *Vet. Immunol. Immunopathol.* (in press).

Kydd, J. H., Butcher, G. W., Antczak, D. F., Allen W. R. (1991). Expression of major histocompatibility complex (MHC) class 1 molecules on early trophoblast. *J. Reprod. Fert. Suppl.* **44**, 463–477.

Kydd, J. H., Antczak, D. F., Allen, W. R., Barbis, D., Butcher, G., Davis, W., Duffus, W. P. H., Edington, N., Grunig, G., Holmes, M. A., Lunn, D. P., McCullock, J., O'Brien, M. A., Perryman, L. E., Tavernor, A. S., Williamson, S., Zhang, C. (1994). Report of the First International Workshop on Equine Leucocyte Antigens, Cambridge, U.K., July 1991. *Vet. Immunol. Immunopathol.* **42**, 1–60.

Kydd, J. H., Hannant, D., Mumford, J. A. (1996). Residence and recruitment of leucocytes to the equine lung after EHV-1 infection. *Vet. Immunol. Immunopathol.* **52**, 15–26.

Landay, A. L., Muirhead, K. A. (1989). Procedural guidelines for performing immunophenotyping by flow cytometry. *Clin. Immunol. Immunopathol.* **52**, 48–60.

Lanier, L. L., Phillips, J. H. (1992). Natural killer cells. *Curr. Opin. Immunol.* **4**, 38–42.

Lanier, L. L., Le, A. M., Cwirla, S., Federspiel, N., Phillips, J. H. (1986). Antigenic, functional, and molecular genetic studies of human natural killer cells and cytotoxic T lymphocytes not restricted by the major histocompatibility complex. *Fed. Proc.* **45**, 2823–2828.

Lanzavecchia, A. (1990). Receptor-mediated antigen uptake and its effect on antigen presentation to class II-restricted T lymphocytes. *Annu. Rev. Immunol.* **8**, 773–793.

Lanzavecchia, A., Roosnek, E., Gregory, T., Berman, P., Abrignani, S. (1988). T cells can present antigens such as HIV gp120 targeted to their own surface molecules. *Nature* **334**, 530–532.

Lawson, G. H., McPherson, E. A., Murphy, J. R., Nicholson, J. M., Wooding, P., Breeze, R. G., Pirie, H. M. (1979). The presence of precipitating antibodies in the sera of horses with chronic obstructive pulmonary disease (COPD). *Equine Vet. J.* **11**, 172–176.

Lazary, S., Bullen, S., Muller, J., Kovacs, G., Bodo, I., Hockenjos, P., De Weck, A. L. (1980a). Equine leukocyte antigen system II. Serological and mixed lymphocyte reactivity studies in families. *Transplantation* **30**, 210–215.

Lazary, S., De Weck, A. L., Bullen, S., Straub, R., Gerber, H. (1980b). Equine leukocyte antigen system I. Serological studies. *Transplantation* **30**, 203–209.

Lazary, S., Gerber, H., Glatt, P. A., Straub, R. (1985). Equine leucocyte antigens in sarcoid-affected horses. *Equine Vet. J.* **17**, 283–286.

Lazary, S., Dubath, M. L., Luder, C., Gerber, H. (1986). Equine leucocyte antigen system. IV. Recombination within the major histocompatibility complex (MHC). *J. Immunogenet.* **13**, 315–325.

Lazary, S., Antczak, D. F., Bailey, E., Bell, T. K., Bernoco, D., Byrns, G., McClure, J. J. (1988). Joint report of the Fifth International Workshop on Lymphocyte Alloantigens of the Horse, Baton Rouge, Louisiana, 31 October–1 November 1987. *Anim. Genet.* **19**, 447–456.

Lazary, S., Marti, E., Szalai, G., Gaillard, C., Gerber, H. (1994). Studies on the frequency and associations of equine leucocyte antigens in sarcoid and summer dermatitis. *Anim. Genet.* **25** (Suppl 1), 75–80.

Lea, R. G., Bolton, A. E. (1991)l. The effect of horse placental tissue extract and equine chorionic gonadotrophin in the proliferation of horse lymphocytes stimulated *in vitro*. *J. Reprod. Immunol.* **19**, 13–23.

Lee, C. S., Gogolin-Ewens, K. J., Brandon, M. R. (1988). Identification of a unique lymphocyte subpopulation in the sheep uterus. *Immunology* **63**, 157–164.

Lester, G. D., Alleman, A. R., Raskin, R. E., Meyer, J. C. (1993). Pancytopenia secondary to lymphoid leukemia in three horses. *J. Vet. Int. Med.* **7**, 360–363.

Lew, A. M., Hosking, C. S., Studdert, M. J. (1980). Immunologic aspects of combined immunodeficiency disease in Arabian foals. *Am. J. Vet. Res.* **41**, 1161–1166.

Lew, A. M., Bailey, E., Valas, R. B., Coligan, J. E. (1986a). The gene encoding the equine soluble class I molecule is linked to the horse MHC. *Immunogenetics* **24**, 128–130.

Lew, A. M., Valas, R. B., Maloy, W. L., Coligan, J. E. (1986b). A soluble class I molecule analogous to mouse Q10 in the horse and related species. *Immunogenetics* **23**, 277–283.

Liles, W. C., van Voorhis, W. C. (1995). Review: nomenclature and biologic significance of cytokines involved in inflammation and the host immune response. [Review]. *J. Inf. Dis.* **172**, 1573–1580.

Liu, I. K. M., Walsh, E. M., Bernoco, M., Cheung, A. T. W. (1987). Bronchalveolar lavage in the newborn foal. *J. Reprod. Fert. Suppl.* **35**, 587–592.

Look, A. T., Ashmun, R. A., Shapiro, L. H., O'Connell, P. J., Gerkis, V., D'Apice, A. J., Sagawa, K., Peiper, S. C. (1989a). Report on the CD13 (amnopeptidase N) cluster workshop. In *Leucocyte Typing IV: White Cell Differentiation Antigens.* (eds Knapp, W., Dorken, B., Rieber, E. P., Stein, H., Gilks, W. R., Schmidt, R. E., von dem Borne, A. E. G. Kr.), pp. 784–787. Oxford University Press, Oxford.

Look, A. T., Ashmun, R. A., Shapiro, L. H., Peiper, S. C. N. (1989b). Human myeloid plasma membrane glycoprotein CD13 (gp 150) is identical to aminopeptidase N. *J. Clin. Invest.* **83**, 1299.

Lue, C., Kiyono, H., McGhee, J. R., Fujihashi, K., Kishimoto, T., Hirano, T., Mestecky, J. (1991). Recombinant human interleukin 6 (rhIL-6) promotes the terminal differentiation of *in vivo*-activated human B cells into antibody-secreting cells. *Cell. Immunol.* **132**, 423–432.

Lunn, D. P., McClure, J. T. (1993). Clinico-pathological diagnosis of immunodeficiency. *Equine Vet. Educ.* **5**, 30–32.

Lunn, D. P., Holmes, M. A., Duffus, W. P. H. (1991a). Three monoclonal antibodies identifying antigens in all equine T-lymphocytes, and two mutually exclusive T-lymphocyte subsets. *Immunology* **74**, 251–257.

Lunn, D. P., Holmes, M. A., Gibson, J., Field, H. J., Kydd, J. H., Duffus, W. P. H. (1991b). Haematological changes and equine lymphocyte subpopulation kinetics during primary infection and attempted re-infection of specific pathogen free foals with EHV-1. *Equine Vet. J.* **12** (Suppl.), 35–41.

Lunn, D. P., Holmes, M. A., Duffus, W. P. H. (1993). Equine T lymphocytes MHC II expression: variation with age and subset. *Vet. Immunol. Immunopathol.* **35**, 225–238.

Lunn, D. P., Holmes, M. A., Duffus, W. P. H. (1994a). Polymorphic expression of an equine T lymphocyte and neutrophil subset marker. *Vet. Immunol. Immunopathol.* **42**, 83–89.

Lunn, D. P., Schram, B. R., Vagnoni, K. E., Schobert, C. S., Truax, R., Horohov, D. W. (1994b). Receptor interactions in equine lymphokine-activated killing. In *Equine Infectious Diseases VII. Proceedings of the Seventh International Conference, Tokyo 1994* (Eds Nakajima, H., Plowright, W.), pp. 133–138. R & W Publications, Newmarket, UK.

Lunn, D. P., Holmes, M. A., Schram, B. R., Duffus, W. P. H. (1995a). Monoclonal antibodies specific for equine IgG subisotypes including an antibody which recognizes B lymphocytes. *Vet. Immunol. Immunopathol.* **47**, 239–251.

Lunn, D. P., McClure, J. T., Schobert, C. S., Holmes, M. A. (1995b). Abnormal patterns of equine leucocyte differentiation antigen expression in SCID foals suggests the phenotype of normal equine NK cells. *Immunology*, **84**, 495–499.

Lunn, D. P., Schram, B. R., Vagnoni, K. E., Schobert, C. S., Horohov, D. W., Ginther, O. J. (1996). Positive selection of EqCD8+ precursors increases equine lymphokine-activated killing. *Vet. Immunol. Immunopathol.* **53**, 1–13.

Lunn, D. P., Holmes, M. A., Antczak, D. F., Baker, J. M., Bendali-Ahcene, S., Blanchard-Channell, M., Byrne, K. M., Cannizzo, K., Davis, W., Hamilton, M. J., Hannant, D., Kondo, T., Kydd, J. H., Monier, M. C., Moore, P. F., Neeraj, N., Schram, B. R., Sheoran, A. S., Stott, J. L., Sugiura, T., Vagnoni, K. E. (1998). Report of the Second Equine Leucocyte Antigen Workshop. *Vet. Immunol. Immunopathol.* (in press).

Lydyard, P., Grossi, C. (1993a). Development of the immune system. In *Immunology* 3rd edn (eds Roitt, I. M., Brostoff, J., Male, D. K.), pp. 11.1–11.16. Mosby, St. Louis, USA.

Lydyard, P., Grossi, C. (1993b). Cells involved in immune responses. In *Immunology* 3rd edn (eds Roitt, I. M., Brostoff, J., Male, D. K.), pp. 2.2–2.20. Mosby, St. Louis, USA.

MacCluer, J. W., Bailey, E., Weitkamp, L. R., Blangero, J. (1988). ELA and fertility in American Standard bred horses. *Anim. Genet.* **19**, 359–372.

MacKay, R. J., Lester, G. D. (1992). Induction of the acute-phase cytokine, hepatocyte-stimulating factor/interleukin 6, in the circulation of horses treated with endotoxin. *Am. J. Vet. Res.* **53**, 1285–1289.

MacKay, R. J., Merritt, A. M., Zertuche, J. M., Whittington, M., Skelley, L. A. (1991). Tumor necrosis factor activity in the circulation of horses given endotoxin. *Am. J. Vet. Res.* **52**, 533–538.

Magnuson, N. S., Perryman, L. E., Suttle, D. P., Robinson, J. L., Mason, P. H., Marta, K. M. (1985). Metabolic investigations of fibroblasts from horses, *Equus caballus*, with hereditary severe combined immunodeficiency. *Comp. Biochem. Physiol. B* **81**, 781–786.

Magnuson, N. S., Perryman, L. E., Wyatt, C. R., Mason, P. H., Talmadge, J. E. (1987). Large granular lymphocytes from SCID horses develop potent cytotoxic activity after treatment with human recombinant interleukin 2. *J. Immunol.* **139**, 61–67.

Maher, J. K., Tresnan, D. B., Deacon, S., Hannah, L., Antczak, D. F. (1996). Analysis of MHC Class I expression in equine trophoblast cells using in situ hybridization. *Placenta* **17**, 351–359.

Mair, T. S., Stokes, C. R., Bourne, F. J. (1987). Quantification of immunoglobulins in respiratory tract secretions of the horse. *Vet. Immunol. Immunopathol.* **14**, 197–203.

Mair, T. S., Stokes, C. R., Bourne, F. J. (1988). Increased IgA production in chronic obstructive pulmonary disease. *Equine Vet. J.* **20**, 214–216.

Martens, J. G., Martens, R. J., Renshaw, H. W. (1988). *Rhodococcus* (*Corynebacterium*) *equi*: bacterial capacity of neutrophils from neonatal and adult horses. *Am. J. Vet. Res.* **49**, 295–299.

Martens, R. J., Martens, J. G., Renshaw, H. W., Hietala, S. K. (1987). *Rhodococcus equi*: equine neutrophil chemiluminescent and bactericidal responses to opsonizing antibody. *Vet. Microbiol.* **14**, 277–286.

Martens, R. J., Martens, J. G., Fiske, R. A., Hietala, S. K. (1989). *Rhodococcus equi* foal pneumonia: protective effects of immune plasma in experimentally infected foals. *Equine Vet. J.* **21**, 249–255.

Marti, E., Lazary, S., Antczak, D. F., Gerber, H. (1993). Report of the first international workshop on equine sarcoid (Review). *Equine Vet. J.* **25**, 397–407.

Marti, E., Szalai, G., Bucher, K., Dobbelaere, D., Gerber, H., Lazary, S. (1995). Partial sequence of the equine immunoglobulin epsilon heavy chain cDNA. *Vet. Immunol. Immunopathol.* **47**, 363–367.

Martin, B. R., Larson, K. A. (1973). Immune response of the equine fetus to coliphage T2. *Am. J. Vet. Res.* **34**, 1363–1364.

Matthews, A. G. (1989). Immunohistochemical investigation of the distribution of immunoglobulins G, A and M within the anterior uvea of the normal equine eye. *Equine Vet. J.* **21**, 438–441.

Matthews, A. G., Imlah, P., McPherson, E. A. (1983). A reagin-like antibody in horse serum. 1. Occurrence and some biological properties. *Vet. Res. Commun.* **6**, 13–23.

May, S. A., Hooke, R. E., Lees, P. (1990). The characterisation of equine interleukin-1. *Vet. Immunol. Immunopathol.* **24**, 169–175.

May, S. A., Hooke, R. E., Lees, P. (1992). Interleukin-1 stimulation of equine articular cells. *Res. Vet. Sci.* **52**, 342–348.

McClure, J. J. (1988). Equine disease association studies: a clinician's perspective. *Anim. Genet.* **19**, 409–415.

McClure, J. J. (1995). Diseases of the immune system. In *The Horse: Diseases and Clinical Management* (eds Kobluk, C. N., Ames, T. R., Geor, R. J.), pp. 1051–1063. W.B. Saunders Company, Philadelphia, USA.

McClure, J. J., Addison, J. D., Miller, R. I. (1985). Immunodeficiency manifested by oral candidiasis and bacterial septicemia in foals. *J. Am. Vet. Med. Assoc.* **186**, 1195–1197.

McClure, J. J., Koch, C., Powell, M., McClure, J. R. (1988). Association of arytenoid chondritis with equine lymphocyte antigens but no association with laryngeal hemiplegia, umbilical hernias and cryptorchidism. *Anim. Genet.* **19**, 427–433.

McClure, J. J., Koch, C., Traub-Dargatz, J. (1994). Characterization of a red blood cell antigen in donkeys and mules associated with neonatal isoerythrolysis. *Anim. Genet.* **25**, 119–120.

McClure, J. J., Parish, S. M., Hines, M. T. (1996). In *Large Animal Internal Medicine* (ed. Smith, B. P.), pp. 1844–1873. C. V. Mosby Company, St. Louis, MO, USA.

McGhee, J. R., Fujihashi, K., Beagley, K. W., Kiyono, H. (1991). Role of interleukin-6 in human and mouse mucosal IgA plasma cell responses (Review). *Immunologic. Res.* **10**, 418–422.

McGorum, B. C., Dixon, P. M., Halliwell, R. E. (1993a). Phenotypic analysis of peripheral blood and bronchoalveolar lavage fluid lymphocytes in control and chronic obstructive pulmonary disease affected horses, before and after 'natural (hay and straw) challenges'. *Vet. Immunol. Immunopathol.* **36**, 207–222.

McGorum, B. C., Dixon, P. M., Halliwell, R. E. W., Irving, P. (1993b). Comparison of cellular and molecular components of bronchoalveolar lavage fluid harvested from different segments of the equine lung. *Res. Vet. Sci.*, **55**, 57–59.

McGuire, T. C., Crawford, T.B. (1972). Identification and quantitation of equine serum and secretory immunoglobulin A. *Infec. Immun.* **6**, 610–615.

McGuire, T. C., Poppie, M. J. (1973). Hypogammaglobulinemia and thymic hypoplasia in horses: a primary combined immunodeficiency disorder. *Infec. Immun.* **8**, 272–277.

McGuire, T. C., Crawford, T. B., Henson, J. B. (1973). The isolation, characterisation and functional properties of equine immunoglobulin classes and subclasses. In *Proceedings of the 3rd International Conference on Equine Infectious Diseases, Paris 1972* (eds, Bryans, J. T., Gerber, H.), pp. 364–381. Karger, Basel, Switzerland.

McGuire, T. C., Poppie, M. J., Banks, K. L. (1975). Hypogammaglobulinaemia predisposing to infection in foals. *J. Am. Vet. Med. Assoc.* **166**, 71–75.

McGuire, T. C., Crawford, T. B., Hallowell, A. L., Macomber, L. E. (1977). Failure of colostral immunoglobulin transfer as an explanation for most infections and deaths of neonatal foals. *J. Am. Vet. Med. Assoc.* **170**, 1302–1304.

McGuire, T. C., Archer, B. G., Crawford, T. B. (1979). Equine IgG and IgG(T) antibodies: dependence of precipitability on both antigen and antibody structure. *Mol. Immunol.* **16**, 787–790.

McGuire, T. C., Perryman, L. E., Davis, W. C. (1983). Analysis of serum and lymphocyte surface IgM of healthy and immunodeficient horses with monoclonal antibodies. *Am. J. Vet. Res.* **44**, 1284–1288.

McGuire, T. C., Tumas, D. B., Byrne, K. M., Hines, M. T., Leib, S. R., Brassfield, A. L., O'Rourke, K. I., Perryman, L. E. (1994). Major histocompatibility complex-restricted CD8+ cytotoxic T lymphocytes from horses with equine infectious anemia virus recognize Env and Gag/PR proteins. *J. Virol.* **68**, 1459–1467.

Meyer, T. A., Wang, J., Tiao, G. M., Ogle, C. K., Fischer, J. E., Hasselgren, P. O. (1995). Sepsis and endotoxemia stimulate intestinal interleukin-6 production. *Surgery* **118**, 336–342.

Mitchell, K. F., Karush, F., Morgan, D. O. (1977). IgM antibody – II The isolation and characterization of equine J chain. *Immunochemistry* **14**, 233–236.

Monahan, C. M., Taylor, H. W., Chapman, M. R., Klei, T. R. (1994). Experimental immunization of ponies with *Strongylus vulgaris* radiation-attenuated larvae or crude soluble somatic extracts from larval or adult stages. *J. Parasit.* **80**, 911–923.

Monos, D. S., Wolf, B., Radka, S. F., Rifat, S., Donawick, W. J., Soma, L. R., Zmijewski, C. M., Kamoun, M. (1989). Equine class II MHC antigens: identification of two sets of epitopes using anti-human monoclonal antibodies. *Tissue Antigens* **34**, 111–120.

Montgomery, P. C., Dorrington, K. J., Rockey, J. H. (1969). Equine antihapten antibody. The molecular weights of the subunits of equine immunoglobulins. *Biochemistry* **8**, 1247–1258.

Montgomery, P. C. (1973). Molecular aspects of equine antibodies. In *Proceedings of the 3rd International Conference on Equine Infectious Diseases, Paris 1972* (eds Bryans, J. T., Gerber, H.), pp. 341–363. Karger, Basel, Switzerland.

Moore, B. R. (1996). Clinical application of interferons in large animal medicine (Review). *J. Am. Vet. Med. Assoc.* **208**, 1711–1715.

Moore, B. R., Krakowka, S., Robertson, J. T., Cummins, J. M. (1995). Cytologic evaluation of bronchoalveolar lavage fluid obtained from standardbred racehorses with inflammatory airway disease. *Am. J. Vet. Res.* **56**, 562–567.

Morgan, D. O., Bryans, J. T., Mock, R. E. (1975). Immunoglobulins produced by the antigenised equine foetus. *J. Reprod. Fert. Suppl.* **23**, 735–738.

Morris, D. D. (1991). Endotoxemia in horses. *J. Vet. Int. Med.* **5**, 167–181.

Morris, D. D., Moore, J. N. (1991). Tumor necrosis factor activity in serum from neonatal foals with presumed septicemia. *J. Am. Vet. Med. Assoc.* **199**, 1584–1589.

Morris, D. D., Gaulin, G., Strzemienski, P. J., Spencer, P. (1987). Assessment of neutrophil migration, phagocytosis and bactericidal capacity in neonatal foals. *Vet. Immunol. Immunopathol.* **16**, 173–184.

Morris, D. D., Crowe, N., Moore, J. N. (1990). Correlation of clinical and laboratory data with serum tumor necrosis factor activity in horses with experimentally induced endotoxemia. *Am. J. Vet. Res.* **51**, 1935–1940.

Morris, D. D., Moore, J. N., Crowe, N. (1991). Serum tumor necrosis factor activity in horses with colic attributable to gastrointestinal tract disease. *Am. J. Vet. Res.* **52**, 1565–1569.

Morris, D. D., Crowe, N., Moore, J. N., Moldawer, L. L. (1992). Endotoxin-induced production of interleukin 6 by equine peritoneal macrophages *in vitro*. *Am. J. Vet. Res.* **53**, 1298–1301.

Morrison, W. I. (1991). Towards an understanding of the immune responses that control infectious diseases. *Equine Vet. J.* **12** (Suppl.), 6–9.

Mosmann, T. R. (1991). Regulation of immune responses by T cells with different cytokine secretion phenotypes: role of a new cytokine, cytokine synthesis inhibitory factor (IL10) (Review). *Int. Arch. Allergy Appl. Immunol.* **94**, 110–115.

Mosmann, T. R. (1994). Cytokine patterns during the progression to AIDS. *Science* **265**, 193–194.

Mosmann, T. R., Coffman, R. L. (1989). TH1 and TH2 cells: different patterns of lymphokine secretion lead to different functional properties (Review). *Annu. Rev. Immunol.* **7**, 145–173.

Mosmann, T. R., Cherwinski, H., Bond, M. W., Giedlin, M. A., Coffman, R. L. (1986). Two types of murine helper T cell clone. I. Definition according to profiles of lymphokine activities and secreted proteins. *J. Immunol.* **136**, 2348–2357.

Murphy, B. D., Martinuk, S. D. (1991). Equine chorionic gonadotrophin. *Endoc. Rev.* **12**, 27–44.

Navarro, P., Barbis, D. P., Antczak, D., Butler, J. E. (1995). The complete cDNA and deduced amino acid sequence of equine IgE. *Mol. Immunol.* **32**, 1–8.

Nelson, K. M., Schram, B. R., McGregor, M. W., Sheoran, A. S., Olsen, C. W., Lunn, D. P. (1997). Local and systemic isotype-specific antibody responses to equine influenza virus infection versus conventional vaccination. *Vaccine* (in press).

Newman, M. J., Issel, C. J., Truax, R. E., Powell, M. D., Horohov, D. W., Montelaro, R. C. (1991). Transient suppression of equine immune responses by equine infectious anemia virus (EIAV). *Virology* **184**, 55–66.

O'Brien, M. A., Holmes, M. A., Lunn, D. P., Duffus, W. P. H. (1991). Evidence for MHC class-I restricted cytotoxicity in the one-way, primary mixed lymphocyte reaction. *Equine Vet. J.* **12** (Suppl.), 30–35.

Ortaldo, J. R., Mason, A., Overton, R. (1986). Lymphokine-activated killer cells, analysis of progenitors and effectors. *J. Exp. Med.* **164**, 1193–1205.

Pahud, J. J., Mach, J. P. (1972). Equine secretory IgA and secretory component. *Int. Arch. Allergy* **42**, 175–186.

Palmer, D. B., Hayday, A., Owen, M. J. (1993). Is TCR β expression an essential event in early thymocyte development? *Immunol. Today* **14**, 460–462.

Park, C. A., Hines, H. C., Threlfall, W. R. (1989). Equine lymphocyte antigens and reproduction in the Standardbred mare. *Anim. Genet.* **20**, 99–104.

Parnes, J. R. (1989). Molecular biology and function of CD4 and CD8. *Adv. Immunol.* **44**, 265–312.

Parr, E. L., Young, L. H. Y., Parr, M. B., Young, J. D.-E. (1990). Granulated metrial gland cells of pregnant mouse uterus are natural killer-like cells that contain perforin and serine esterase. *J. Immunol.* **145**, 2365–2372.

Pellegrini, A., Waiblinger, S., von Fellenberg, R. (1991). Purification of equine neutrophil lysozyme and its antibacterial activity against Gram-positive and Gram-negative bacteria. *Vet. Res. Com.* **15**, 427–435.

Perryman, L. E., McGuire, T. C. (1978). Mixed lymphocyte culture responses in combined immunodeficiency of horses. *Transplantation* **25**, 50–52.

Perryman, L. E., McGuire, T. C., Hilbert, B. J. (1977). Selective immunoglobulin M deficiency in foals. *J. Am. Vet. Med. Assoc.* **170**, 212–215.

Perryman, L. E., McGuire, T. C., Torbeck, R. L. (1980). Ontogeny of lymphocyte function in the equine fetus. *Am. J. Vet. Res.* **41**, 1197–1200.

Perryman, L. E., Magnuson, N. S., Bue, C. M., Wyatt, C. R., Riggs, M. W. (1987). Selective and combined immunodeficiencies in horses. In *Immune-deficient Animals in Biomedical Research. 5th International Workshop on Immune-deficient Animals 1985* (eds Rygaard, J., Spang-Thomsen, M.), pp. 34–46. Karger, Basel, Switzerland.

Perryman, L. E., Bue, C. M., Magnuson, N. S., Mottironi, V. D., Ochs, H. S., Wyatt, C. R. (1988a). Immunological reconstitution of foals with combined immunodeficiency. *Vet. Immunol. Immunopathol.* **17**, 495–508.

Perryman, L. E., Wyatt, C. R., Magnuson, N. S., Mason, P. H. (1988b). T lymphocyte development and maturation in horses. *Anim. Genet.* **19**, 343–348.

Perussia, B., Fanning, V., Trinchieri, G. (1983). A human NK and K cell subset shares with cytotoxic T cell expression of the antigen recognized by antibody OKT8. *J. Immunol.* **131**, 223–231.

Pescovitz, M. D., Lunney, J. K., Sachs, D. H. (1985). Murine anti-swine T4 and T8 monoclonal antibodies: distribution and effects on proliferative and cytotoxic T cells. *J. Immunol.* **134**, 37–44.

Phillips, J. H., Lanier, L. L. (1986). Dissection of the lymphokine-activated killer phenomenon. Relative contribution of peripheral blood natural killer cells and T lymphocytes to cytolysis. *J. Exp. Med.* **164**, 814–825.

Pintado, C. O., Friend, M., Llanes, D. (1995). Characterisation of a membrane receptor on ruminants and equine platelets and peripheral blood leukocytes similar to the human integrin receptor glycoprotein IIb/IIIa (CD41/61). *Vet. Immunol. Immunopathol.* **44**, 359–368.

Poppie, M. J., McGuire, T. C. (1977). Combined immunodeficiency in foals of Arabian breeding; evaluation of mode of inheritance and estimation of prevalence of affected foals and carrier mares and stallions. *J. Am. Vet. Med. Assoc.* **170**, 31–33.

Prescott, J. F. (1991). *Rhodococcus equi*: an animal and human pathogen. *Clin. Micro. Rev.* **4**, 20–34.

Prescott, J. F. (1993). Immunodeficiency and serious pneumonia in foals: the plot thickens (editorial comment). *Equine Vet. J.* **25**, 88–89.

Prescott, J. F., Wilcock, B. P., Carman, P. S., Hoffman, A. M. (1991). Sporadic, severe bronchointerstitial pneumonia of foals. *Can. Vet. J.* **32**, 421–425.

Raulet, D. H., Eisen, H. N. (1990). Cellular basis for immune responses. In *General Immunology*, 2nd edn (ed. Eisen, H. N.), pp. 81–124. J.B. Lippincott Company, Philadelphia, USA.

Reis, K. J. (1989). A hemolytic assay for the measurement of equine complement. *Vet. Immunol. Immunopathol.* **23**, 129–137.

Richards, C. M., Aucken, H. A., Tucker, E. M., Hannant, D., Mumford, J. A., Powell, J. R. (1992). The production of equine monoclonal immunoglobulins by horse–mouse heterohybridomas. *Vet. Immunol. Immunopathol.* **33**, 129–143.

Roberts, K., Lotze, M. T., Rosenberg, S. A. (1987). Separation and functional studies of the human Lymphokine-activated Killer Cell. *Cancer Res.* **47**, 4366–4371.

Roberts, M. C. (1975). Equine immunoglobulins and the equine immune system. In *The Veterinary Annual* 15th edn (eds Grunsell, C. S. G., Hill, F. W. G.), pp. 192–203. Wright-Scientifica, Bristol, UK.

Robinson, J. A., Allen, G. K., Green, E. M., Garner, H. E., Loch, W. E., Walsh, D. M. (1993). Serum interleukin-6 concentrations in endotoxin-infused neonatal foals. *Am. J. Vet. Res.* **54**, 1411–1414.

Rockey, J. H. (1967). Equine antihapten antibody. *J. Exp. Med.* **125**, 249–275.

Rockey, J. H., Klinman, N. R., Karush, F. (1964). Equine antihapten antibody I. 7S β2A and 10S γ1-globulin components of purified anti-β-lactoside antibody. *J. Exp. Med.* **120**, 589–609.

Rode, H. J., Janssen, W., Rosen-Wolff, A., Bugert, J. J., Thein, P., Becker, Y., Darai, G. (1993). The genome of equine herpesvirus type 2 harbors an interleukin 10 (IL10)-like gene. *Virus Genes* **7**, 111–116.

Romagnani, S. (1991). Human T_H1 and T_H2 subsets: doubt no more. *Immunol. Today* **12**, 256–257.

Rosen, F. S., Cooper, M. D., Wedgwood, R. J. P. (1984). The primary immunodeficiencies. *N. Engl. J. Med.* **311**, 235–242.

Rosen, F. S. (1993). The primary specific immunodeficiencies. In *Clinical Apects of Imunology* 5th edn (eds Lachmann, P. J., Peters, D. K., Rosen, F. S., Walport, M. J.), pp. 1271–1284. Blackwell Scientific Publications, Oxford, UK.

Roth, T. L., White, K. L., Thompson, D. L. J., Barry, B. E.,

Capehart, J. S., Colborn, D. R., Rabb, M. H. (1990). Suppression of lymphocyte proliferation by a greater than 30,000 molecular weight factor in horse conceptus-conditioned medium. *Biol. Reprod.* **43**, 298–304.

Roth, T. L., White, K. L., Thompson, D. L., Rahmanian, S., Horohov, D. W. (1992). Involvement of interleukin 2 receptors in conceptus-derived suppression of T and B cell proliferation in horses. *J. Reprod. Fert.* **96**, 309–322.

Rudd, C. E. (1990). CD4, CD8 and the TCR-CD3 complex: a novel class of protein-tyrosine kinase receptor. *Immunol. Today* **11**, 400–406.

Sanada, Y., Noda, H., Nagahata, H. (1992). Development of lymphocyte blastogenic response in the neonatal period of foals. *Zentralblatt. Fur. Veterinarmedizin – Reihe. A* **39**, 69–75.

Sandor, G., Korach, S., Mattern, P. (1964). 7S globulin, immunologically identical to 19S gamma-1(β)-M-globulin, a new protein of horse serum. *Nature* **204**, 795–796.

Schlossman, S. F., Boumsell, L., Gilks, W., Harlan, J. M., Kishimoto, T., Morimoto, C., Ritz, J., Shaw, S., Silverstein, R., Springer, T., Tedder, T. F., Todd, R. F. (1995). *Leucocyte Typing V. White Cell Differentiation Antigens: Proceedings of the Fifth International Workshop and Conference, Boston, 1993*, 2044 pp. Oxford University Press, Oxford, UK.

Schram, B. R., Holmes, M. A., Lunn, D. P. (1996). Characterization of an equine homologue of the myeloid differentiation antigen CD13. *Conference of Research Workers in Animal Disease, Chicago, November 1996*. Abstract 128.

Schrenzel, M. D., Watson, J. L., Ferrick, D. A. (1994). Characterization of horse (*Equus caballus*) T-cell receptor beta chain genes. *Immunogenetics* **40**, 135–144.

Schrenzel, M. D., Ferrick D. A. (1995). Horse (*Equus caballus*) T-cell receptor alpha, gamma, and delta chain genes: nucleotide sequences and tissue-specific gene expression. *Immunogenetics* **42**, 112–122.

Schuberth, H. J., Anders, I., Pape, U., Leibold, W. (1992). One-dimensional isoelectric focusing and immunoblotting of equine major histocompatibility complex class I antigens. *Anim. Genet.* **23**, 87–95.

Schultz, C. L., Coffman, R. L. (1991). Control of isotype switching by T cells and cytokines (Review). *Curr. Opin. Immunol.* **3**, 350–354.

Seethanathan, P., Bottoms, G. D., Schafer, K. (1990). Characterization of release of tumor necrosis factor, interleukin-1, and superoxide anion from equine white blood cells in response to endotoxin. *Am. J. Vet. Res.* **51**, 1221–1225.

Sellon, D. C., Cullen, J. M., Whetter, L. E., Gebhard, D. H., Coggins, L., Fuller, F. J. (1993). Production and characterization of a monoclonal antibody recognizing a cytoplasmic antigen of equine mononuclear phagocytes. *Vet. Immunol. Immunopathol.* **36**, 303–318.

Shalaby, M. R., Waage, A., Aarden, L., Espevik, T. (1989). Endotoxin, tumor necrosis factor-alpha and interleukin 1 induce interleukin 6 production *in vivo*. *Clin. Immunol. Immunopathol.* **53**, 488–498.

Sheoran, A. S., Holmes, M. A. (1996). Separation of equine IgG subclasses (IgGa, IgGb and IgG(T)) using their differential binding characteristics for staphylococcal protein A and streptococcal proteins. *Vet. Immunol. Immunopathol.* **55**, 33–430.

Sheoran, A. S., Lunn, D. P., Holmes, M. A. (1998). Monoclonal antibodies to subclass specific antigenic determinants on equine immunoglobulin gamma chains and their characterizations. *Vet. Immunol. Immunopathol.*, in press.

Shin, E. K., Perryman, L. E., Meek, K. (1997). A kinase-negative mutation of DNA-PKCS in equine SCID results in defective coding and signal joint formation. *J. Immunol.* **158**, 3565–3569.

Slauson, D. O., Skrabalak, D. S., Neilsen, N. R., Zwahlen, R. D. (1987). Complement-induced equine neutrophil adhesiveness and aggregation. *Vet. Pathol.* **24**, 239–249.

Starkey, P. M. (1992). Natural killer cells/large granular lymphocytes in pregnancy. In *The Natural Killer Cell* (eds Lewis, C. E., McGee, J. O'D.), pp. 206–240. IRL Press, New York, USA.

Starkey, P. M., Sargent, I. L., Redman, W. G. (1988). Cell populations in human early pregnancy decidua: characterization and isolation of large granular lymphocytes by flow cytometry. *Immunology* **65**, 129–134.

Stewart, C. (1992). Clinical applications of flow cytometry. *Cancer* **69**, 1543–1552.

Stoneham, S. J., Wingfield Digby, N. J., Ricketts, S. W. (1991). Failure of passive transfer of colostral immunity in the foal: incidence, and the effect of stud management and plasma transfusions. *Vet. Rec.* **128**, 416–419.

Su, X., Morris, D. D., Crowe, N. A., Moore, J. N., Fischer, K. J., McGraw, R. A. (1992). Equine tumor necrosis factor alpha: cloning and expression in *Escherichia coli*, generation of monoclonal antibodies, and development of a sensitive enzyme-linked immunosorbent assay. *Hybridoma* **11**, 715–727.

Sugiura, T., Kondo, T., Imagawa, H., Kamada, M. (1998). Production of monoclonal antibodies to six isotypes of horse immunoglobulin. *Vet. Immunol. Immunopathol.*, in press.

Suter, M., Fey, H. (1983). Further purification and characterization of horse IgE. *Vet. Immunol. Immunopathol.* **4**, 545–553.

Suzuki, T., Hiromatsu, K., Ando, Y., Okamoto, T., Tomoda, Y., Yoshikai, Y. (1995). Regulatory Role of $\gamma\delta$ T cells in uterine intraepithelial lymphocytes in maternal antifetal immune response. *J. Immunol.* **154**, 4476–4484.

Swiderski, C. E., Horohov, D. W. (1995). Molecular cloning and sequencing of equine interleukin 10 (IL-10) and beta-actin. *Proceedings of the 70th Annual Meeting of the Conference of Research Workers in Animal Diseases, Chicago, IL, Nov. 13–14*. Abstract 125.

Szalai, G., Antczak, D. F., Gerber, H., Lazary, S. (1994a). Molecular cloning and characterization of horse DQB cDNA. *Immunogenetics* **40**, 458.

Szalai, G., Antczak, D. F., Gerber, H., Lazary, S. (1994b). Molecular cloning and characterization of horse DQA cDNA. *Immunogenetics* **40**, 457.

Tavernor, A. S., Allen, W. R., Butcher, G. W. (1993a). cDNA cloning of equine interleukin-2 by polymerase chain reaction. *Equine Vet. J.* **25**, 242–243.

Tavernor, A. S., Deverson, E. V., Coadwell, W. J., Lunn, D. P., Zhang, C., Davis, W., Butcher, G. W. (1993b). Molecular cloning of equine CD44 by a COS cell expression system. *Immunogenetics* **37**, 474–477.

Tavernor, A. S., Kydd, J. H., Bodian, D. L., Jones, E. Y., Stuart, D. I., Davis, S. J., Butcher, G. W. (1994). Expression cloning of an equine T-lymphocyte glycoprotein CD2 cDNA, structure-based analysis of conserved sequence elements. *Eur. J. Biochem.* **219**, 969–976.

Tiao, G., Rafferty, J., Ogle, C., Fischer, J. E., Hasselgren, P. O. (1994). Detrimental effect of nitric oxide synthase inhibition during endotoxemia may be caused by high levels of tumor necrosis factor and interleukin-6. *Surgery*, **116**, 332–337.

Traub-Dargatz, J. L., McClure, J. J., Koch, C., Schlipf Jr, J. W. (1995). Neonatal isoerythrolysis in mule foals. *J. Am. Vet. Med. Assoc.* **206**, 67–70.

Trinchieri, G. (1989). Biology of natural killer cells. *Adv. Immunol.* **47**, 187–375.

Tumas, D. B., Brassfield, A. L., Travenor, A. S., Hines, M. T., Davis, W. C., McGuire, T. C. (1994a). Monoclonal antibodies to the equine CD2 T lymphocyte marker, to a pan-granulocyte/monocyte marker and to a unique pan-B lymphocyte marker. *Immunobiology* **192**, 48–64.

Tumas, D. B., Hines, M. T., Perryman, L. E., Davis, W. C., McGuire, T. C. (1994b). Corticosteroid immunosuppression and monoclonal antibody-mediated CD5+ T lymphocyte depletion in normal and equine infectious anaemia virus-carrier horses. *J. Gen. Virol.* **75**, 959–968.

Turrel, J. M. (1995). Oncology. In *The Horse: Diseases and Clinical Management* (eds Kobluk, C. N., Ames, T. R., Geor, R. J.), pp. 1111–1136. W.B. Saunders Company, Philadelphia, PA, USA.

Urban Jr, J. F., Madden, K. B., Svetic, A., Cheever, A., Trotta, P. P., Gause, W. C., Katona, I. M., Finkelman, F. D. (1992). The importance of Th2 cytokines in protective immunity to nematodes (Review). *Immunol. Rev.* **127**, 205–220.

Vaerman, J. P., Querinjean, P., Heremans, J. F. (1971). Studies on the IgA system of the horse. *Immunology* **21**, 443–454.

Vagnoni, K. E., Schram, B. R., Ginther, O. J., Lunn, D. P. (1996). Susceptibility of equine chorionic girdle cells to lymphokine-activated killer cell activity. *Am. J. Reprod. Immunol.* **36**, 184–190.

van Maanen, C., Bruin, G., de Boer Luijtze, E., Smolders, G., de Boer, G. F. (1992). Interference of maternal antibodies with the immune response of foals after vaccination against equine influenza. *Vet. Q.* **14**, 13–17.

Vandergrifft, E. V., Horohov, D. W. (1993). Molecular cloning and expression of equine interleukin 2. *Vet. Immunol. Immunopathol.* **39**, 395–406.

Vandergrifft, E. V., Horohov, D. W. (1994). Molecular cloning and sequencing of equine interleukin 4. *Vet. Immunol. Immunopathol.* **40**, 379–384.

Wagner, B., Radbruch, A., Richards, C., Leibold, W. (1995). Monoclonal equine IgM and IgG immunoglobulins. *Vet. Immunol. Immunopathol.* **47**, 1–12.

Wallace, D. L., Beverley, P. C. L. (1990). Phenotypic changes associated with activation of CD45RA+ and CD45RO+ T cells. *Immunology* **69**, 460–467.

Watson, E. D., Dixon, C. E. (1993). An immunohistological study of MHC class II expression and T lymphocytes in the endometrium of the mare. *Equine Vet. J.* **25**, 120–124.

Watson, E. D, Zanecosky, H. G. (1991). Regulation of mitogen and TCGF-induced lymphocyte blastogenesis by prostaglandins and supernatants from equine embryos and endometrium. *Res. Vet. Sci.* **51**, 61–65.

Wegmann, T. G., Lin, H., Guilbert, L., Mosmann, T. R. (1993). Bidirectional cytokine interactions in the maternal-fetal relationship: is successful pregnancy a TH2 phenomenon? *Immunol. Today* **14**, 353–356.

Weitkamp, L. R., Sandberg, K. (1990). Horse (*Equus caballus*). In *Genetic Maps* 5th edn (ed. O'Brien, S. J.), pp. 4.107–4.109. Cold Spring Harbor Laboratory Press, Maine, USA.

Weldon, A. D., Zhang, C., Antczak, D. F., Rebhun, W. C. (1992). Selective IgM deficiency and abnormal B-cell response in a foal. *J. Am. Vet. Med. Assoc.* **201**, 1396–1398.

Wichtel, M. G., Anderson, K. L., Johnson, T. V., Nathan, U., Smith, L. (1991). Influence of age on neutrophil function in foals. *Equine Vet. J.* **23**, 466–469.

Widders, P. R., Stokes, C. R., David, J. S. E., Bourne, F. J. (1984). Quantitation of the immunoglobulins in reproductive tract secretions of the mare. *Res. Vet. Sci.* **37**, 324–330.

Widders, P. R., Stokes, C. R., Bourne, F. J. (1986). Investigation of the antigenic relationship between equine IgG and IgGT. *Vet. Immunol. Immunopathol.* **13**, 255–259.

Wiler, R., Leber, R., Moore, B. B., VanDyk, L. F., Perryman, L. E., Meek, K. (1995). Equine severe combined immunodeficiency: a defect in V(D)J recombination and DNA-dependent protein kinase activity. *Proc. Natl. Acad. Sci. USA* **92**, 11485–11489.

Wong, C. W., Smith, S. E., Thong, Y. H., Opdebeeck, J. P., Thornton, J. R. (1992). Effects of exercise stress on various immune functions in horses. *Am. J. Vet. Res.* **53**, 1414–1417.

Wyatt, C. R., Magnuson, N. S., Perryman, L. E. (1987). Defective thymocyte maturation in horses with severe combined immunodeficiency. *J. Immunol.* **139**, 4072–4076.

Wyatt, C. R., Davis, W. C., McGuire, T. C., Perryman, L. E. (1988). T lymphocyte development in horses I. Characterization of monoclonal antibodies identifying three stages of T lymphocyte differentiation. *Vet. Immunol. Immunopathol.* **18**, 3–18.

Yednock, T. A., Rosen, S. D. (1989). Lymphocyte homing. *Adv. Immunol.* **44**, 313–378.

Yilma, T., Perryman, L. E., McGuire, T. C. (1982). Deficiency of interferon-gamma but not interferon-beta in Arabian foals with severe combined immunodeficiency. *J. Immunol.* **129**, 931–933.

Zhang, C. H., Donaldson, W. L., Antczak, D. F. (1994). An equine B cell surface antigen defined by a monoclonal antibody. *Vet. Immunol. Immunopathol.* **42**, 91–102.

Zink, M. C., Yager, J. A., Prescott, J. F., Wilkie, B. N. (1985). *In vitro* phagocytosis and killing of *Corynebacterium equi* by alveolar macrophages of foals. *Am. J. Vet. Res.* **46**, 2171–2174.

Zolla, S., Goodman, J. W. (1968). An aggregating immunoglobulin in hyperimmune equine anti-pneumococcal sera. *J. Immunol.* **100**, 880–897.

XI Immunology of the Pig

1. Introduction

For a species to be a relevant model of human disease, it must emulate human conditions in both disease and health. Over the years, swine have been one of the most frequent and useful such models (see Tumbleson *et al.*, 1996, for review). Swine resemble humans in almost all aspects analyzed. They are similar in size, feeding patterns, immune system, skin structure, renal, cardiac, and pulmonary physiology and anatomy to man. The similarity has been so great that pig skin is used as a temporary covering in severe burns, heart valves are used to replace human valves damaged by disease and, more recently, it has resulted in the pig being the most widely considered xenogeneic organ donor to man. Swine have been used as models of 'alcoholism, allotransplantation, atherosclerosis, congenital abnormalities, dermal healing, diabetes ... hypertension ... kidney disease, melanoma ... septic shock...' to name a few (Tumbleson *et al.*, 1996). Although nonhuman primates are more closely related to humans than are pigs, their scarcity (resulting from the breeding characteristics), expense, and ethical concerns limit their use. Although much attention has been paid to swine as a model of human disease, clearly the species has also great economic value. Analysis of the immune system in particular, has been conducted to identify the response of swine to infectious agents, and endogenous antigens. These studies could ultimately improve the health of the species with resultant economic benefit.

Although the pig is an excellent model for human disease, the two species are clearly not identical. Many aspects of the immune system differ, reflecting evolutionary diversion. These differences, which are discussed below in greater detail, are anatomical (the unusual structure of porcine lymph nodes), physiological (the large number of γ/δ T-cells) and genetic (the extreme diversity within the T-cell receptor gene complex). The analyses of these differences will provide a fruitful area for future research.

2. Lymphoid Organs and their Anatomical Distribution

In the pig, as in other species, the size of lymphoid organs and their different compartments is largely dependent upon the age and environmental status of the individual animal (Rothkötter *et al.*, 1991; Joling *et al.*, 1994), for example small lymph nodes without germinal centers are present in germ-free animals, whereas well-developed large lymph nodes are present in conventional pigs. Therefore, the breed, age, and antigenic stimulation (specific pathogen free) must be stated in all studies. In addition, not only the relative, but also the absolute number, of lymphocytes per gram of tissue and per whole lymphoid organ must be given in any experiment dealing with alterations in the immune system. Because many reports did not address these issues, it is impossible to give a list of absolute or even relative numbers of lymphocyte subsets in different organs for the pig in general.

The structure and functional anatomy of lymphoid organs and the number of lymphoid cells in non-lymphoid organs of the pig has been extensively analyzed (Binns, 1982; Binns *et al.*, 1986; Binns and Pabst, 1988, 1994). In addition to the classical lymphoid organs, nonlymphoid organs, such as the different compartments of the pig lung (Pabst and Binns, 1994) and the skin (Fritz *et al.*, 1990), contain many lymphocytes.

Species-specific aspects of the development, structure and function of lymphoid organs in the pig

Owing to the multilayered placenta, there is no transfer of maternal immunoglobulins into the pig fetal circulation. Therefore, fetal pigs and colostrum-deprived piglets reared in a germ-free environment are excellent models in which to differentiate between innate immune reactions and those resulting from external antigenic stimuli (Tlaskalowa-Hogenova *et al.*, 1994). Several obvious advantages of the pig as an experimental model for such immune reactions include the large litter size, the availability of germ-free, specific-pathogen-free and conventional animals, and the range of different sized animals, e.g. minipigs and micropigs. Strains inbred for alloantigens allow cell transfer experiments without rejection. The recently produced CD45 congenic strain is an excellent model for lymphocyte migration studies since lymphocytes can be transferred without any *in vitro* labeling procedure and identified in the recipient by using the respective antibody (Binns *et al.*, 1995). With regard to these

Handbook of Vertebrate Immunology
ISBN 0-12-546401-0

aspects, other large animals, e.g. sheep, cannot compare. The size of the pig enables surgical procedures, such as selective perfusion-labeling of lymph nodes, spleen, thymus and bone marrow, that permit the study of lymphocyte production and migration without disruption of the structure of these lymphoid organs (for review see Pabst and Binns, 1986).

Similar to ruminants, the pig has two types of Peyer's patches (PP) in the small intestine. There are about 20 discrete individual PP in the upper small intestine and one continuous (>1 m long) PP in the terminal ileum. These two types of PP differ in their development and regression during life, and lymphocyte subset composition and lymphocyte entry in migration experiments (see Binns and Pabst, 1988; Rothkötter and Pabst, 1989). M cells in the covering epithelium of the dome of PP of pigs can be identified by staining with antibodies against cytokeratin 18 (Gebert *et al.*, 1994). The arrangement of the lymphatics in the continuous PP differs from that in sheep and rabbits (Lowden and Heath, 1994). In the large intestine, PP-like structures called lymphoglandular complexes are formed by diverticula of the epithelium surrounded by organized lymphoid tissue (Morfitt and Pohlenz, 1989).

The inverted structure of the pig lymph node is peculiar but not restricted to pigs; this anatomical structure is also found in elephant, rhinoceros, dolphin, hippopotamus and warthog lymph nodes (for review see Binns, 1982). Most larger lymph nodes are formed by the aggregation of several embryological nodelets, each of which results in a system that is inverted: the 'cortex' is in the central parts, the 'paracortex' is found at the periphery. The afferent lymph flows through paratrabecular sinuses to the central cortex and then percolates peripherally through the paracortex to the medulla and finally the efferent lymphatics. The medulla is much denser than in other species, and is filled with macrophages, eosinophils and some plasma cells. Despite this inverted structure most lymph node functions of pigs are similar to those of other species (Binns, 1982). Lymphocyte traffic through the lymph nodes, however, is completely different (see Section 4). In light microscopy the high endothelial venules (HEV) of pig lymph nodes are similar to those of rats and humans. There are unique cells bridging the adluminal wall, as documented by electron microscopy (Sasaki *et al.*, 1994).

A further interesting aspect of the pig immune system is the large number of intravascular macrophages in the lung, resulting in a high clearance capacity of antigenic particulate material (Winkler, 1988).

The size of the pig enables long-term studies with repeated blood sampling in individual animals, e.g. after neonatal thymectomy to document the lack of γ/δ null T cells and an increased number of $CD4^+CD8^+$ double-positive cells (Licence and Binns, 1995). Prolonged drainage of lymph makes restraining cages necessary in rats and, to a lesser degree, in sheep. In pigs, however, the intestinal lymph has been continuously collected for up to 7 days without restraining the animals (Rothkötter *et al.*, 1995).

The peculiar structure of some lymphoid organs and the effect of age and breeding conditions on lymphocyte numbers must be taken into consideration in all quantitative studies on the immune system in pigs. The various porcine lymphoid organs and their makeup is summarized in Table XI.2.1.

3. Leukocytes and their Markers

The pig initially lagged behind other species in the availability of leukocyte markers. For many years, porcine cells were primarily identifiable by nylon–wool adherence (B cells) and nylon–wool nonadherence (T cells which were also sheep red blood cell rosette positive, and Null cells which were not) (Binns, 1982). Phagocytic cells could be identified by the ability to ingest particles such as latex beads (Pescovitz *et al.*, 1984). There were several early attempts to prepare polyclonal antisera against porcine T-cells of which the preparation by Johnson *et al.* (1980) was perhaps the best characterized. The first monoclonal antibodies against porcine leukocyte markers became available in 1984 (Pescovitz *et al.*, 1984). With the increase in reagents, the First International Swine Cluster of Differentiation (CD) Workshop was held in Budapest, Hungary in 1992 (Lunney *et al.*, 1994). This was followed 3 years later by the Second Workshop held in Davis, California in 1995 (Saalmüller, 1996). As a result of these two workshops, CD markers were established for swine. Most of these are analogous to well-defined markers already identified in other species, whereas others are, to date, unique to the pig. The various markers for which monoclonal antibodies, as defined by the Workshops, are available are listed (Table XI.3.1). In addition, putative cross-reactive monoclonal antibodies, not subject to rigorous analysis as part of workshops, are available to CD18 (Lunney *et al.*, 1994) and CD2, CD11b, CD21, CD47, CD49e (VLA), CD49f (VLA), CD79b, CD80 (B7), CD90 (Thy-1) (Pharmagen, San Diego, CA, USA, and Batten *et al.*, 1996) and CD86 (Davis *et al.*, 1996). The list of antibodies reactive with these markers is constantly changing, but the most complete listing compiling both the first and second workshop results has been published by Saalmüller (1996). The antibodies are available from several sources, including The American Type Tissue Collection, individual investigators, and various companies.

For several of the identified CD markers, sufficient data is available detailing specific aspects of distribution and function. These are summarized below.

CD2

The large population of nylon wool nonadherent cells were initially characterized as Null cells by the absence of

Table XI.2.1 Porcine lymphoid organs and their cellular makeup

Organ	*Lymphocyte numbers*[a]	*Lymphocyte compartments*	*Lymphocyte subsets*[b,c,d]	*Proliferation*[g]	*Percentage of newly formed lymphocytes leaving*[h]
Blood	3%		T > B[b] CD4 > CD8		
Thymus	44%	Cortex	T ≫ B[b]	++	<3%
		Medulla	T, few B	+	
Bone marrow	5%		B > T		
Spleen	9%	Follicles	B > T[b]	+++	
		PALS	T > B	++	≈17%
		Marginal zone	T	+	
		Red Pulp	B > T	+	
Lymph nodes	12%	Cortex	B > T	++	
		Paracortex	T > B	+	≈40%
		Medulla	T ~ B	+	
Tonsils	2%	Follicles	B ≫ T	++	
		Interfollicular area	T > B	+	
Jejunal PP		Follicles	B ≫ T[e]	++	
		Interfollicular area	T > B	+	
Ileal PP		Dome	B ≫ T	+	
Small intestine	5%				
Lamina propria			CD4 > CD8[f]	+[c]	
Intraepithelial lymphocytes			CD8 > CD4	−	
Lung	0.2%	Intravascular pool	Not determined	Not determined	
		Interstitial pool			
		Bronchoalveolar space	T ≫ B		

[a] Pabst and Trepel (1975) (26 kg body weight, conventional pigs, total number of lymphocytes *c.* 320 × 10^9); for details of lymphocyte subsets in the lymphoid organs see [b] Joling *et al.* (1994), [c] Rothkötter *et al.* (1991), [d] Bianchi *et al.* (1992), [e] Rothkötter and Pabst (1989), [f] Rothkötter *et al.* (1994), Vega-López *et al.* (1993), [g] Pabst and Fritz (1986), [h] summarized in Pabst and Binns (1986).

surface Ig and their inability to rosette with sheep red blood cells (Binns *et al.*, 1977; Binns, 1982). The CD2 marker is the rosette receptor, and in pigs, antibodies to CD2 have been shown to block rosette formation (Hammerberg *et al.*, 1985). The use of such anti-CD2 mAb along with other markers of T cells, such as those to the γ/δ receptor, have confirmed that most Null cells are in fact T cells that fail to express CD2 (Licence *et al.*, 1995). Cloning of the CD2 gene has not yet been reported in pigs. Elimination of the CD2 population from *in vitro* cell populations or its blockade eliminates proliferative responses. Despite the availability of a large panel of anti-CD2 monoclonal antibodies, only one epitope has been identified (Saalmüller, 1996).

CD3

One of the most important markers to identify T cells, monoclonal antibodies to CD3 have only recently become available (Kirkham *et al.*, 1996; Yang *et al.*, 1996; Pescovitz *et al.*, 1998). Antibodies against CD3 have been hard to identify and those generated by Yang *et al.* were made against affinity purified porcine CD3, although several others have been identified by screening of antiporcine lymphocyte antibodies (Pescovitz *et al.*, 1998). All of the anti-CD3 antibodies generated to date react with the ε chain. There is biochemical evidence of the existence of multiple CD3 chains consistent with the γ, δ, and ζ chains (Kirkham *et al.*, 1996). The reason for this lack of monoclonal antibodies reactive with the other chains may reflect conservation of sequence between pig and mouse (the parent hybridoma species) or result from the screening process itself (Yang *et al.*, 1996). As with other species, monoclonal antibodies against CD3 can stimulate Ca^{2+} flux, *in vitro* proliferation, and retargeting of cytotoxic T-cell killing. CD3 clearly identifies the majority of the Null cells as T cells that predominately express γ/δ (Yang *et al.*, 1996). These γ/δ cells express CD3 at a higher level than the α/β cells. This has been hypothesized to correct for the absence of other co-receptors such as CD2, CD4 and CD8 (Yang *et al.*, 1996).

CD4

A monoclonal antibody to CD4 was among the first of the antiporcine reagents identified (Pescovitz *et al.*, 1984, 1985). Like other species, porcine CD4 marks cells of the helper phenotype that tend to respond to class II MHC

Table XI.3.1 Porcine leukocyte markers confirmed in the First or Second International Swine Cluster of Differentiation Workshop

Cell subset	*Antigen*	*Molecular weight*	*Gene homology*
T cell	CD2	50–55	
	CD3ε	23	65%
	CD4	55	72%
	CD5	63	84%
	wCD6	110	
	wCD8	35/35	
	SWC1	19	
	SWC2	49–51	
Activation	wCD25	65–70	
B-cells	wCD1	40/12	
	wCD21	150	
	SWC7	90	
Macrophages	SWC3	97/68	
	CD14	43	
	SWC9	>205/130	
Null cells	SWC4	Unknown	
	SWC5	Unknown	
	SWC6	Unknown	
Lymphoid cells	CD16		
	CD18	166/155/95	
	wCD29	16/32/>132	
	SWC8	32	
	wCD44	80	
	CD45	226/21/190	Variable

differences (Pescovitz *et al.*, 1985). Of the several monoclonal antibodies reactive with porcine CD4, all react with the same or a closely related epitope (Pescovitz *et al.*, 1994). Despite this, there is an interesting polymorphism reported in porcine CD4 (Sundt *et al.*, 1992; Gustafsson *et al.*, 1993). The functional significance of this substantial polymorphism that completely eliminates reactivity with any of the reported anti-CD4 reagents is unknown. CD4 mRNA is present but the epitope with which the CD4 monoclonal antibodies react has been destroyed. There does not appear to be any detectable effect on various immune parameters.

CD5

The porcine CD5 gene has been partially sequenced and shown to have 96% DNA homology to murine and 84% to human CD5 (Appleyard *et al.*, 1998). There are eight monoclonal antibodies which are all reactive with the same epitope of porcine CD5 (Saalmüller, 1996). Within the thymus, the more mature cells express higher levels of CD5 (Saalmüller *et al.*, 1994b). Within the peripheral cell population, which is represented by three different levels of CD5 expression, the $CD5^-$ cells contain all of the NK activity, the $CD5^{dim}$ cells contain γ/δ cells, and the $CD5^{bright}$ cells contain the rest (Saalmüller *et al.*, 1994b,c). Some B cells also express CD5 (Saalmüller *et al.*, 1994; Appleyard *et al.*, 1998b). As opposed to other species, however, fetal IgM^+ liver B-cells, and newborn and adult nonactivated splenic and peripheral blood IgM^+ B cells did not express CD5. The relevance of this to the porcine immune response is unknown (Cukrowska *et al.*, 1996). As with other species, porcine CD5 appears to be able to transmit an activation signal (Appleyard *et al.*, 1998a).

CD6

Three monoclonal antibodies have been identified as reactive with porcine CD6 (Saalmüller, 1996). CD6 is not present on B cells or monocytes. It is present on 95% of thymocytes but is only expressed by 76% of peripheral T cells (Saalmüller *et al.*, 1994a). All of the $CD4^+$ peripheral T cells express CD6, whereas only a portion of the $CD8^+$ cells express CD6. Those $CD8^+$ cells that fail to express CD6, express CD8 at a reduced level (Saalmüller *et al.*, 1994a). The γ/δ T-cell receptor subset also does not express CD6. Functional analysis has now demonstrated that all of the NK activity is within the $CD6^-$ population (Pauly *et al.*, 1996). Putting this data together with that from CD5, it is apparent that the phenotype of porcine NK cells is $CD5^-CD6^-CD8^{dim}$.

CD8

The Second International CD Workshop greatly expanded the knowledge of porcine CD8 through the development of several new reagents (Zuckermann *et al.*, 1998b). One of these reacts only with the $CD8^{bright}$ population of cells, i.e. the $CD4^-/CD8^+$ cells. This indicates that the CD8 expressed on these cells is qualitatively and also quantitatively different from that expressed on the CD4/CD8 dual expressing cells. It has been suggested that the dual expressing cells have only $CD8\alpha/\alpha$ homodimers on the surface whereas the $CD4^-/CD8^+$ cells express the $CD8\alpha/\beta$ heterodimer. This will be confirmed only when the porcine CD8 genes are cloned and sequenced.

CD44

CD44 is an adhesion molecule that is involved in lymphocyte homing, T-cell activation, and intercellular interactions. Porcine CD44 was first identified in a soluble form in intestinal efferent lymph using the anti-human CD44 monoclonal antibody, Hermes-1 (Yang *et al.*, 1993). It was characterized as 48 000 to 70 000 kDa soluble and 90 000 kDa membrane bound molecule. Porcine CD44 was subsequently purified and both polyclonal and monoclonal antibodies were generated (Yang *et al.*, 1993). Both types of antibodies were widely species cross-reactive. The monoclonal antibodies failed to stain porcine erythrocytes but did stain all mononuclear cells compared with the situation in humans, where red cells do express CD44. It is possible that porcine red cells also express CD44 but a distinct isoform of it (Yang *et al.*, 1993). The level of CD44 appeared to correlate with homing capacity of the particular lymphocyte populations (Yang *et al.*, 1993).

CD45

The CD45 marker is characterized by extensive recombination events producing different expressed proteins. These different isoforms have been associated with different states of activation and antigen exposure, and are differentially expressed among the various cell lineages. Within the pig, eight isoforms are possible, but only three have been detected at the molecular level and four at the protein level (Schnitzlein *et al.*, 1996). The genes encoding the various isoforms have been expressed in CHO cells thereby allowing careful mapping of the various anti-CD45 monoclonal antibodies to the various isoforms (Schnitzlein *et al.*, 1998; Zuckermann *et al.*, 1998a).

CD86

One of the current paradigms of immune activation is the requirement for a second signal in addition to that delivered through the T-cell receptor. One of these second signals is the CD28 pathway through its interaction with the B7-1 and B7-2 proteins. Because of the great interest in swine as a possible xenogeneic donor, attention has been focused on interactions between human and porcine cells, such as the vascular endothelium. It has been shown that porcine vascular endothelium cells express B7-2 (CD86), a 79 kDa protein, and that this can function as a stimulatory molecule for xenogeneic proliferation (Davis *et al.*, 1996).

Unique features of leukocyte marker expression

Although the pig generally has and expresses the same markers as found in other species, there are two features that are relatively unique to pigs. The first of these is the very high frequency of cells that were initially identified as Null cells. These have now been shown to be γ/δ T-cell receptor expressing cells. This type of cell is discussed elsewhere in this section. The other feature is the wide variety of T cells, as distinguished by the presence of CD4 and CD8. When monoclonal antibodies to CD4 and CD8 were first analyzed in the pig, a difference from other species was immediately apparent. Unlike humans and mice, where expression of CD4 and CD8 were essentially mutually exclusive on the surface of peripheral T cells, in pigs a large number, in fact sometimes the majority, of such cells express both CD4 and CD8 (Pescovitz *et al.*, 1985). This dual expressing phenotype was similar to thymic T-cells. The early hypothesis that the dual expressing cells were simply premature thymic emigrants was discounted by the presence of CD1 on thymic but not on peripheral dual-expressing cells (Pescovitz *et al.*, 1990). More recent analysis suggests that the dual-expressing population represents a memory type of cell. This is consistent with the increased percentage of these cells present in older animals (Zuckermann and Husmann, 1996). However, there has been a suggestion that some CD4 cells may derive from the dual-expressing population by loss of CD8 (Licence *et al.*, 1995).

4. Leukocyte Traffic and Associated Molecules

Lymphoid cells do not migrate randomly but accumulate preferentially in the different compartments of various tissues. For example, there is a predominance of B and T cells in the follicles and the paracortex in lymph nodes, respectively. The migration of plasma cell precursors into the lamina propria of the gut wall is another well-known example of preferential lymphocyte migration into nonlymphoid tissue. In pigs, T, B and γ/δ T-cells contribute to the circulating pool of lymphoid cells. T and B cells migrate to all lymphoid and many other tissues, while Null cells are present in small numbers in lymphoid organs but are often found in nonlymphoid tissues (Pabst and Binns, 1989; Rothkötter *et al.*, 1990, 1993; Barman *et al.*, 1996). The accumulation of lymphoid subsets in various tissues is a result of several mechanisms: first, the 'entry' into an organ, which is regulated by the interaction of adhesion molecules with endothelial cells and lymphocytes; second the 'transit' of lymphocytes through the parenchyma of the organ; and finally the 'exit' either directly into the blood or into lymph vessels. During the transit phase, lymphocytes can be activated to start proliferation or cells can die by apoptosis. Thus, the number of lymphocytes found at a given time in one compartment of an organ is the net effect of all these parameters and not just the result of preferential 'entry'. Cell proliferation has been observed in all lymphoid organs of the pig (see Section 2). Newly formed cells have been detected even in the epithelium and lamina propria of the intestine (Rothkötter *et al.*, 1994). Apoptosis of lymphocytes has not been studied in detail in the pig.

Lymphoid cells enter lymphoid organs such as lymph nodes, Peyer's patches and tonsils via high endothelial venules (HEV). Although many more lymphocytes recirculate through 'non-HEV' organs, (e.g. the spleen and bone marrow; Pabst and Binns, 1989), lymphocyte migration studies have often been restricted to the regulation of lymphocyte entry via HEV into lymphoid organs. In mice and humans, many adhesion molecules and other factors

Table XI.4.1 Adhesion molecules in the pig

Adhesion molecule	*Site of expression*	*Available data*
Integrins		
CD18	Lymphocytes, natural killer cells, polymorphonuclear leukocytes	Antibody against porcine CD18a (Kim *et al.*, 1994; Saalmüller, 1996) Cross-reacting antibodies (Walsh *et al.*, 1991; Windsor *et al.*, 1993)
CD11/CD18 (LFA-1)		Antibodies against porcine CD11/18 (Hildreth *et al.*, 1989)
β_1 comparable subunit	Skin basal cells	Molecule (King *et al.*, 1991)
Immunoglobulin superfamily		
VCAM-1	Endothelial cells	Cloned molecule (Tsang *et al.*, 1994) Cross-reacting Ab (Tsang *et al.*, 1994; Batten *et al.*, 1996)
Selectins		
E-selectin	Endothelial cells	Cloned molecule (Tsang *et al.*, 1995) Cross-reacting Ab (Keelan *et al.*, 1994: Tsang *et al.*, 1995)
Ligand of L-selectin	Endothelial cells	Cross-reacting antibody (Whyte *et al.*, 1994, 1995)
CD44	Lymphocytes	Porcine Ab (Yang and Binns, 1993a, b; Saalmüller, 1996)

have been described that regulate the three steps of cell adhesion at these venules (for review see Springer, 1994). To date, only a few porcine homologues of adhesion molecules known from other species have been described on lymphoid cells and on the endothelium (reviewed by Binns and Pabst, 1994). However, interest is now focusing on the expression of adhesion molecules on other leukocytes (e.g. polymorphonuclear leukocytes) and on endothelium of larger vessels in the pig, because it serves as an animal model for many studies related to clinical problems in human medicine (Binns and Pabst, 1996). The adhesion molecules currently known in the pig are summarized in Table XI.4.1.

Aspects of adhesion molecule detection

Only a limited number of adhesion molecules have been detected in the pig using cross-reacting antibodies against adhesion molecules of other species. One reason may be the differences in the distribution of saccharides on the HEV of the pig compared with other species (Whyte *et al.*, 1993). Furthermore, the specificity of these cross-reactions is open to debate (e.g. the reaction of anti L-selectin or LAM-1 antibodies to pig HEV is controversial; Spertini *et al.*, 1991; Whyte *et al.*, 1994, 1995). So far, receptor-ligand pairs cannot be described in the pig in detail because of the lack of experimental data. Antibodies have been developed that may detect adhesion molecules, but the ligands detected have not yet been characterized (Haverson *et al.*, 1994). Despite these limitations, current knowledge provides evidence that the basic mechanisms of lymphocyte-HEV interactions in pigs are comparable to those in other species.

Adhesion molecules and experimental models

The expression of adhesion molecules has often been examined with respect to experimental models rather than basic immunology. E-selectin and vascular cell adhesion molecule-1 (VCAM-1) have been induced by human cytokines on porcine endothelial cells (Batten *et al.*, 1996), and also by inflammatory agents (Whyte *et al.*, 1994; Woolley *et al.*, 1995; Binns *et al.*, 1996). The characterization of E-selectin expression *in vivo* has also been performed (Keelan *et al.*, 1994). Anti-CD18 treatment was used in experimental models for heart surgery (Aoki *et al.*, 1995), in a peritonitis model (Wollert *et al.*, 1993), and in experimental sepsis (Walsh *et al.*, 1991; Windsor *et al.*, 1993). CD44 is involved in lymphocyte migration via HEV but the mechanism has not been resolved. In pig models, CD44 was indirectly involved in lymphocytes binding to cultured endothelial cells (Yang and Binns, 1993a). Until recently, the development of HEV in nonlymphoid organs has been described only in chronic inflammation. Binns *et al.* (1990, 1992b) have now documented the appearance of HEV in the skin of pigs within a day after injection of PHA, TNF-α, and IL-2. Thus, these vessels specialized for lymphocyte entry into tissues, are much more dynamic than previously thought.

Differences in lymphocyte migration in the pig

There is a major difference in lymphocyte migration in the pig compared with other species. As first described 30 years ago, efferent lymph draining from pig lymph nodes contains very few lymphoid cells compared with other species (Binns and Hall, 1966). This was explained by the structure of the lymph node, which provided the functional basis for the re-entry of lymphocytes into the blood via HEV within the lymph node parenchyma (Pabst and Binns, 1989; Whyte *et al.*, 1993; Sasaki *et al.*, 1994) and not the expression of adhesion molecules on the lymphocyte surface. This has been documented in lymphocyte tracing studies in pigs using *in vitro* labeled lymphocytes of sheep. The xenogeneic cells showed a migration pattern comparable to that of autologous lymphocytes (Binns and Licence, 1990). It has not been clarified whether there are distinct HEV for entry and exit or whether the HEV work as a revolving door.

Lymphoblasts

It has been shown in the pig that only a certain number of lymphoblasts from different tissues accumulate in their tissue of origin, while large numbers were found in nonlymphoid organs, e.g. lung, liver and muscles (Binns *et al.*, 1992a). It is not known whether the recirculation of these lymphoblasts is regulated by a special set of adhesion molecules. The higher expression of porcine CD44 on lymphoblasts than other lymphocytes might be such a mechanism involved in lymphoblast migration (Yang and Binns, 1993b).

Leukocyte migration in the pig

Although few adhesion molecules have been described in the pig, much is known about the basic mechanisms of porcine lymphocyte migration. Porcine lymphoid cells have been labeled using radioactivity or fluorochromes. Furthermore, it is possible to use a polymorphic genetic marker in MHC-homogeneous pig strains, thus the cells of a marker-positive donor can be detected in the host using immunofluorescence or immunohistochemistry (Binns *et al.*, 1995). There are very few species in which lymphocyte migration to so many organs has been studied. In an extensive series of experiments, lymphocytes from the peripheral blood were labeled with ^{51}Cr and their traffic studied in different age groups from fetal to adult pigs. The different entry of migrating lymphocytes into various lymph node groups, different types of tonsils and, in particular, into the two types of Peyer's patches in the small intestine, the bone marrow and the thymus has been described (Binns and Licence, 1985; Binns and Pabst, 1988, 1996). In addition, cell labeling within the organs and surgery to collect lymph from various sites are techniques that make the pig useful for *in vivo* migration studies (Pabst *et al.*, 1993).

5. Cytokines and Interferons

Cytokines play a central role in modulation of immunological and physiological processes under both homeostatic and disturbed conditions. The rapidly expanding status of cytokine reagent development in swine reflects the current interest in porcine immunology and disease pathogenesis, the potential of pigs as organ donors for xenotransplantation in humans, and the use of swine as biomedical research models. The number of cytokines cloned or described in pigs has increased dramatically since 11 were listed by Murtaugh in 1994 (Table XI.5.1). Public websites are available at the URLs `http://www.public. iastate.edu/~pigmap` for genetic information and `http://kbot.mig.missouri.edu:443/cytokines/ explorer.html` for cytokine reagent availability. Cytokine regulation and function are largely similar in swine and other mammalian species, as exemplified by mouse and man. The following information emphasizes the aspects of cytokine biology and measurement that are relevant or unique to swine. Information on porcine cytokine assay methods is provided in Table XI.5.2 and nucleotide sequence information for polymerase chain reaction (PCR) detection and molecular cloning is provided in Table XI.5.3.

Inflammatory cytokines and chemokines

The inflammatory cytokines and chemokines IL-1α, IL-1β, IL-6, IL-8, TNF-α, TNF-β, MCP-1, and MCP-2 are described in pigs (Table XI.5.1). Inflammatory cytokine and chemokine expression is induced principally by cell wall products of Gram-negative bacteria and activators of the transcription factor NF-κB. However, the endotoxin of *Serpulina hyodysenteriae*, a Gram-negative swine enteric pathogen, is a weak or ineffective inducer of inflammatory cytokine expression (Greer and Wannemuehler, 1989; Sacco *et al.*, 1996). Although TNF-α appears to play a central role in the pathophysiology of septic lung injury, which resembles acute porcine pleuropneumonia (Olson, 1988; Kiorpes *et al.*, 1990), TNF levels are not increased in the lung in response to infection with *Actinobacillus pleuropneumoniae*. IL-1 and IL-6, however, are present at high levels (Baarsch *et al.*, 1995).

IL-1 appears to be an important inflammatory mediator involved in the destruction of cartilage and bone that is a feature of atrophic rhinitis, erysipelas and arthritis. IL-1 localizes to inflammatory cells of affected joints (Davies *et al.*, 1992), and it activates chondrocytes (Dingle *et al.*, 1990), increases collagenase activity (Richards *et al.*, 1991), and suppresses collagen synthesis (Tyler and Benton, 1988). IL-1 is assumed to play an important role in other inflammatory diseases of swine, but definitive studies require the generation of purified proteins and specific antibodies.

IL-6 is induced in lymphoid cells by mitogenic stimulation, in myeloid cells by lipopolysaccharide, and in fibroblasts by inflammatory cytokines and viral RNA mimics (Scamurra *et al.*, 1996). In fibroblasts, IL-6 but not IL-1 or TNF secretion is induced by *Pasteurella multocida* toxin,

Table XI.5.1 Characteristics and properties of porcine cytokines

Cytokine	*Alternative names*	*Species crossreactivity*	*Principal cell source*	*Regulation or site of expression*	*Known and presumed (?) activities in pigs*	*Chromosome location*
Interleukin-1α	IL-1α	H → P[a]	Macrophages	Sites of bacterial infection, stress	Inflammation	3q1.2–q1.3
Interleukin-1β	IL-1β	H → P, B → P	Macrophages	Sites of bacterial infection, stress	Inflammation	3q1.1–q1.4
Interleukin-2	IL-2	P ↔ H, P → B, P → M	T cells	Activated T cells	T cell proliferation and activation	8
Interleukin-3	IL-3				Hematopoietic stem cell differentiation	
Interleukin-4	IL-4	H → P	T cells		Macrophage suppression, T cell growth and activation (?)	
Interleukin-6	IL-6	H → P	Macrophages, T cells	Sites of bacterial infection, stress	Acute phase response, T cell and B cell stimulus	
Interleukin-8	IL-8, AMCF-I	H → P	Macrophages	Sites of bacterial infection	Neutrophil chemokine	8q
Interleukin-10	IL-10	H → P	T cells, macrophages	Activation of T cells and macrophages	Macrophage suppression	
Interleukin-12	IL-12	H → P	Macrophages, dendritic cells	Activation by LPS or bacteria	IFN-γ production, NK cell activation, and Th-1 induction (?)	
Interleukin-15	IL-15, γ/δ T cell growth factor		T cells	Activated T cells	γ/δT cell proliferation and activation (?)	
Interleukin-18	IL-18, IFN-γ inducing factor, IL-1γ		Macrophages	Activation by LPS or bacteria	IFNγ induction (?)	
Tumor necrosis factor-α	TNF-α	H → P	Macrophages	LPS and Gram-positive bacterial toxins	Inflammation	7p1.1–q1.2
Tumor necrosis factor-β	TNF-β, lymphotoxin		T cells		Inflammation	7p1.1–q1.2
Macrophage chemotactic protein I	MCP I					
Macrophage chemotactic protein II	MCP II					12q
Alveolar macrophage chemotactic factor II	AMCFII					
Transforming growth factor β_1	TGF-β_1	P ↔ H	Lymphocytes	Constitutive in T cells		6cen–q2.1
Transforming growth factor β_2	TGF-β_2	P ↔ H	Lymphocytes	Constitutive in T cells		10p
Transforming growth factor β_3	TGF-β_3	P ↔ H	Lymphocytes	Not active		7q
Erythropoietin	EPO			Bone marrow	Erythrocyte differentiation	
Macrophage-colony stimulating factor	M-CSF, CSF-1				Myeloid differentiation and activation	
Granulocyte macrophage-colony stimulating factor	GM-CSF				Granulocyte and macrophage differentiation and activation	
Stem cell factor	SCF, c-kit ligand, MCGF				Stem cell proliferation and differentiation	
Thrombopoietin	TPO			Hematopoietic cell precursors	Megakaryocytopoiesis	
Interferon γ	IFN-γ		T cells, NK cells	IL-12 on macrophages, NK cells	Macrophage activation	5p1.2–q1.1
Interferon α	IFN-α		Various	Viral infection	Antiviral activity	1q2.4–q2.6
Interferon β	IFN-β		Various	Viral infection	Antiviral activity	1q2.3–q2.6
Interferon ω	IFN-ω		Various		Antiviral activity, reproduction	
Short porcine type I interferon	spI IFN		Trophoblasts		Antiviral activity	
Interleukin-1 receptor antagonist	IL-1ra	H → P	Macrophages	Inflammatory stimuli	Inhibitor of IL-1 activity	

[a] H (human), P (pig), B (bovine), arrow indicates cytokine direction of action.

Table XI.5.2 Known methods of cytokine detection

Cytokine	*Bioassay*	*Reference*	*Immunoassay*	*Reference or source*	*Genetic*[a]	*Reference*
IL-1α, IL-1β	D10.G4.1 cell proliferation	Hopkins and Humphreys (1989); Winstanley and Eckersall (1992)			*In situ* hybridization of mRNA	Baarsch *et al.* (1995)
IL-2	CTLL-2 cell proliferation	Iwata *et al.* (1993)				
IL-6	B9 cell proliferation	Scamurra *et al.* (1996)			*In situ* hybridization of mRNA	Baarsch *et al.* (1995)
IL-8	Neutrophil chemotaxis	Lin *et al.* (1994)			*In situ* hybridization of mRNA	Baarsch *et al.* (1995)
IL-10	Inhibition of IFN-γ activity	Blancho *et al.* (1995)				
IL-12	Lymphoblast proliferation	Gately *et al.* (1992)				
TNF-α	L929 cytotoxicity	Baarsch *et al.* (1991)	ELISA	Endogen, Inc.	*In situ* hybridization of mRNA	Baarsch *et al.* (1995)
TNF-β	PK15 cytotoxicity	Pauli *et al.* (1994)				
M-CSF (CSF-1)			Antihuman CSF-1 monoclonal antibody	Tuo *et al.* (1995)		
IFN-γ	MHCII induction	Le Moal *et al.* (1989)	RIA, western blots Single cell detection	Gonzalez Juarrero *et al.* (1994); Trebichavsky *et al.* (1993)		
IFN-α	Antiviral activity	L'Haridon *et al.* (1991)	Anti pIFN-α monoclonal antibody ELISPOT assay	L'Haridon *et al.* (1991),	Nowacki *et al.* (1993)	
IFN-β	Antiviral activity					
spI IFN	Antiviral activity	Niu *et al.* (1995)	anti-spI serum	Niu *et al.* (1995)		

[a] PCR and Northern blot sources are not included since the information is available in Table XI.4.1.

Table XI.5.3 Oligonucleotide primer sequences for cytokine detection using PCR

Cytokine	*Genbank number or reference*	*Forward and reverse primers*	*Source*
IL-1α	M86730		
IL-1β	M86725, X74568		
IL-2	X58428, X56750, S37892	F AACCTCAACTCCTGCCAC R TCCTTGATATTTGCTGAGTCA	Y. Zhou, F. A. Zuckermann and M. P. Murtaugh, unpublished
IL-3	Yang *et al.*, 1996		
IL-4	L12991, X68330	F CACAAGTGCGACATCAC R TCAACACTTTGAGTATTTC	Y. Zhou and M. P. Murtaugh, unpublished
IL-6	M86722, M80258, M80255		
IL-8	M86923, X61151, M99367		
IL-10	L20001	F GCGACTTGTTGCTGACCGG R GAACCTTGGAGCAGATTTTG	M. P. Murtaugh *et al.*, unpublished
IL-12	L35765 (35 kDa)	p35 F CTCCCAAAATCTGCTGAAGG R CATTCTGTCGATGGTCACCG	Foss and Murtaugh (1997)
	U08317 (40 kDa)	p40 F GATGCTGGCCAGTACACCTG(TCG) R (TC)CCTGATGAAGAAGCTGCT(GG)	
IL-15	U58142		
IL-18	U68701	F GACAATTGCATCAACTTTGTGG R GGTCTCTCTCTTTTTCACAAGC	Foss and Murtaugh (1997)
TNF-α	X57321, M29079, X54001, X54859		
TNF-β	X54859		
MCP I	Z48479		
MCP II	Z48480		
RANTES	F14636		
AMCFII	M99368		
TGF-β_1	M23703	F GCCCTGGATACCAACTACTG R TCAGCTGCACTTGCAGGAAC	Y. Zhou, unpublished
TGF-β_2	L08375, X70142, S48994	F CGGAAGAAGCGTGCTTTGGATGC R GCTGCATTTGCAAGACTTTAC	Y. Zhou, unpublished
TGF-β_3	X14150		
EPO	L10607		
M-CSF	Tuo *et al.* (1995)	F ACTGTAGCCACATGATTGGGA R GCCTCTCCAGAAGCTTCTTCT	Tuo *et al.* (1995)
GM-CSF	U61139, D21074	F CTGGCAGCATGTGGATGC R CTTTTGAAGGTGATAGACTGGG	Foss and Murtaugh (1997)
SCF	L07786		
TPO	Gurney *et al.* (1995)		
IFN-γ	S63967	F GCAGAAGAAAGGTCAGC R AGCTACCTTTAGGAACCT	Y. Zhou, F. A. Zuckerman and M. P. Murtaugh, unpublished
IFN-α	M28623, X57191		
IFN-β	M86726		
IFN-ω	X57194, X57195, X57196		
spl IFN	Z22707		
IL-1ra	L38849		
HPRT[a]	U32316	F TGAACGTCTTGCTCGAGATG R TCAAATCCAACAAAGTCTGGC	Foss and Murtaugh (1997)

[a] Hypoxanthine phosphoristosyl transferase (HPRT) is used as a control for mRNA equivalency.

suggesting a role for IL-6 in atrophic rhinitis (Rosendal *et al.*, 1995). Administration of LPS or *Pneumococcus in vivo* increases blood levels of IL-6 (Klosterhalfen *et al.*, 1991; Ziegler-Heitbrock *et al.*, 1992). The presence of IL-6 mRNA in the preimplantation conceptus also indicates a role in reproduction (Mathialagan *et al.*, 1992).

IL-8 is homologous to alveolar macrophage chemotactic factor-I (AMCF-I) (Goodman *et al.*, 1992). In macrophages the gene is exquisitely sensitive to LPS, indicating that the recruitment of neutrophils to the lung in response to bacterial infection may be due to IL-8 expression (Lin *et al.*, 1994).

Immune response cytokines

The elegant T_H1–T_H2 paradigm of immune responsiveness to infection based on murine models has strongly influenced the design and interpretation of disease resistance and vaccinology studies in swine even though its essential features remain unproven in nonmurine species. In swine, it is known that IL-2 shares a high degree of sequence conservation with other mammalian IL-2 molecules (Bazan, 1992). It is secreted by mitogen-activated lymphocytes and supports the growth of lymphocytes in culture and increases natural killer activity in adult, but not newborn, animals (Charley *et al.*, 1985; Charley and Fradelizi, 1987). Infection by viruses, including African swine fever virus, poxviruses and paramyxoviruses, induces IL-2 and killer cell activity in porcine peripheral blood mononuclear cells (Scholl *et al.*, 1989; Steinmassl and Wolf, 1990). Various immunomodulating substances, including ACTH and imuthiol (sodium diethyldithiocarbamate) are reported to suppress IL-2 production (Flaming *et al.*, 1988; Klemcke *et al.*, 1990).

IL-4 is active on porcine macrophages and NK cells (Knoblock and Canning, 1992; Zhou *et al.*, 1994) but its effects on T cell proliferation and activation are not known in swine. IL-10 suppresses macrophage functions and is readily expressed in pigs (Blancho *et al.*, 1995). IL-4 expression is exceedingly low based on reverse transcription-polymerase chain reaction (RT-PCR) assays (Y. Zhou and M. P. Murtaugh, unpublished data), and TGF-β expression is primarily constitutive (Zhou *et al.*, 1992), thus IL-10 may be the principal macrophage suppressor and initiator of T_H2-type responses in the pig. TGF-β has a variety of physiological effects but immunological functions have not been described in pigs (reviewed by Murtaugh, 1994). IL-12 was recently cloned but scant information is available on its role in immune responsiveness. Similarly, IL-15 and IL-18 have been cloned but no biological studies have been reported (Table XI.5.1).

Hematopoietic growth factors

A variety of hematopoietic growth and differentiation factors from pigs have been cloned in *E. coli* as indicated in Tables XI.5.1 and XI.5.3. In so far as is known, they appear to have the same activities as their human and murine counterparts. Human and murine IL-6, IL-11, G-CSF, GM-CSF, stem cell factor (SCF) and erythropoietin (Epo) induced swine progenitor colony formation to varying degrees, but IL-3 showed no effect (Emery *et al.*, 1996). Authentic porcine IL-3 promoted pig hematopoiesis in SCID mouse recipients and the effect was enhanced by the presence of porcine GM-CSF (Yang *et al.*, 1996).

Interferons

Four subfamilies of antiviral interferons (IFN-α, -β, -ω and short porcine type I (spI) and one immune interferon (IFN-γ) are described in pigs. IFN-ω and spI IFN, in addition to possessing the classical antiviral activity of type I interferons, may be involved in maternal recognition during pregnancy since they are expressed in the conceptus and trophoblast, respectively (Mege *et al.*, 1991; Lefevre and Boulay, 1993).

Future prospects

Only a few cytokine receptor molecules have been cloned from the pig or are identified by monoclonal antibodies. They include the cloned receptors for TGF-β type 3 (Genbank no. L07595) and TNF p55 (Genbank no. U19994), and the IL-2 receptor (Bailey *et al.*, 1992), respectively. The high level of nucleotide sequence similarity between pigs and humans will facilitate PCR cloning of additional cytokines and their receptors; it will also provide the necessary information for expression of recombinant proteins, production of monoclonal antibodies and development of immunoassays. It is clear from examination of Tables XI.5.2 and XI.5.3 that the detection and localization of cytokines and their receptors is less advanced than that of the cognate mRNA molecules and is a principal obstacle to the characterization of cytokine functions in swine. Of equal importance is the need for long-term culture of swine immune cells, including myeloid cells and antigen-specific T cells and B cells, in order to elucidate cytokine regulation and function in cognate immune responses.

6. T-cell Receptor

The study of the porcine T-cell receptor is still in its infancy with the first sequence data reported in 1993 (Thome *et al.*, 1993). As seen with other species, there are two types of receptors composed of α/β and γ/β heterodimers. Similar to immunoglobulins, the T-cell receptors are made of both variable and constant regions. The different identified porcine T-cell receptor genes and their amino acid homology to other species are shown (Table XI.6.1). Independent sequencing of the δ chain by Grimm and Misfeldt (1994) found a slightly lower degree of homology. There are no data on the chromosomal localization of the genes. There is little data on the molecules associated with T-cell receptor. When CD3 was immunoprecipitated with antiporcine CD3 monoclonal antibody,

Table XI.6.1 Porcine T-cell receptor genes and deduced amino acid sequence homology to various species

Gene	Gene number	Molecular weight	Per cent species homology			
			Human[a]	Mouse	Sheep	Cattle
α	1	46	63.8	63.6	60.1	59.4
β	1	47	80.3	77.0	82.0	84.0
γ	3 (?4)[b]	31, 35, 46	56.9	52.9	69.2	67.6
δ	1	40	73.9	69.7	77.0	74.5

[a] Data presented are from Thome *et al.* (1993). The molecular mass is in kDa.
[b] The γ chain has three identified genes of 31, 35, and 35 (deduced) kDa respectively. There is a 46 kDa precipitable protein for which a gene has not been identified.

55 kDa and 43 kDa dimers were coprecipitated. These bands are consistent with the porcine α/β and γ/δ T-cell receptor proteins, respectively (Hirt *et al.*, 1990; Yang *et al.*, 1996; Pescovitz *et al.*, 1998).

The ontogeny and thymic dependency of particular T-cell receptor-subset positive cells is best defined for the γ/δ cells. Binns first showed that neonatal thymectomy led to subsequent loss of the Null cell (γ/δ) population (Binns *et al.*, 1977). With use of monoclonal antibodies specific for the γ/δ cells they recently confirmed these early findings of the Null population and demonstrated the strong thymus dependence of these cells (Licence *et al.*, 1995). They also found a small, apparently thymic independent, population of γ/δ cells in animals followed for almost 2 years after thymectomy. The thymic nature of the γ/δ is also supported by the ability to detect these cells within the thymus (Hirt *et al.*, 1990). The thymus dependence/independence of the α/β cells is still not clear. Although overall, $CD2^+$ (i.e. α/β cells) did not change after thymectomy, the phenotype of the cells with regard to CD4 and CD8 was thymic dependent. The dual expressing CD4/CD8 cells were much more common after thymectomy. These cells, which are α/β positive, might have developed extrathymically, while the CD4 and CD8 single expressing cells developed intrathymically (Licence *et al.*, 1995). The α/β cells are present in low numbers in young pigs and increase with age (Yang *et al.*, 1996).

With the availability of antibodies to the relevant T-cell markers, it is finally possible to analyze the distribution of the various T-cell receptor populations. Within the peripheral blood, the α/β cells can be divided into four populations ($CD4^+/CD8^-$, $CD4^+/CD8^{low}$, $CD4^-/CD8^{low}$, $CD4^-/CD8^{high}$) while the γ/δ cells are divided into three populations ($CD2^-/CD4^-/CD8^-$, $CD2^+/CD4^-/CD8^{low}$, $CD2^+/CD4^-/CD8^-$). These same populations occurred in the peripheral lymphoid tissue although the γ/δ cells were lower and most were either $CD2^+$ or $CD8^+$ (Yang *et al.*, 1996).

In addition to the expression of multiple T-cell populations, there are also several other unique characteristics of the porcine T-cell receptor (Thome *et al.*, 1993). As opposed to human, mice, and cattle who have 2 C_β genes, the pig appears to only have one. The pig has three identified C_γ genes and four C_γ proteins that are more diverse in the hinge region than humans. Within the V_δ, diversity results from joining steps between four V_δ, three D_δ and one J_δ. This V_δ diversity is the greatest of all animal T-cell receptors studied (Yang *et al.*, 1995). The V_δ region also appears to have an Ig-like CDR3 region thereby suggesting that it may recognize conformational epitopes in the context of MHC.

7. Immunoglobulins

Table XI.7.1 summarizes the immunoglobulin (Ig) classes of the swine including what is known about their concentrations in serum, colostrum and milk whey. Because antibodies that recognize the many IgG subclasses are unavailable, their distribution is unknown. However, based on transcript frequency, IgG1 appears to be the most abundant (Kacskovics *et al.*, 1994). For the same reason, nothing is known about the biological functions of the various subclasses. The roles of IgG, IgA and IgM appear to be the same as that described for their counterparts in other well-studied species.

Because body fluids other than lacteal secretions are heavily influenced by transudation, secretion rate and sampling procedures, only relative rather than absolute values are meaningful (Butler and Hamilton, 1991). In this regard, IgA (primarily as SIgA) is the major Ig in all exocrine body fluids except colostrum, alveolar fluids and some urogenital secretions. In parotid saliva, nasal and upper bronchial fluids, 80–90% of the Ig is IgA (Butler and Brown, 1994). In serum, IgA occurs as a dimer whereas colostrum contains IgA ranging in size from 6.4S to 15S with the majority being 11S SIgA (Porter, 1973).

Allotypic variants of IgA and several of the IgG subclasses are known. Best described are the IgA^a and IgA^b Mendelian variants of IgA (Brown *et al.*, 1995). The latter lacks most of its structural hinge owing to a splice site mutation in the first intron. Older serological data and recent molecular studies indicate that allotypic variants also occur among the various IgG subclasses, although the relationship between the older serological data and recent sequence data is unclear.

The Ig genes of swine are the best studied of those among common farm animals. Information about them is

Table XI.7.1 The swine Igs and their distribution in serum and lacteal secretions

Ig	*Mol. wt* $\times 10^{-3}$	*H-chain mol. wt* $\times 10^{-3}$ *PAGE*[a]	*AA Comp*[b]	*Concentration (mg/ml)*[c] *Serum*	*Colostrum whey*	*Milk whey*
IgM	950	72	72.4	1.1 (4.5)	9.1 (7.2)	1.4 (18)
dIgA[d]	320	58	51.4	1.8 (7.3)	?	?
SIgA	400	58	51.4	ND[e]	21.2 (16.8)	5.6 (72)
IgG (Total)	150	52	49.5	21.5 (88.0)	95.6 (75.9)	0.8 (10)
IgG1	150	52	49.5	?[e]	?	?
IgG2a	150	52	49.5	?	?	?
IgG2b	150	52	49.5	?	?	?
IgG3	150	52	49.5	?	?	?
IgG4	150	52	49.5	?	?	?
IgE	200	?	60.7	?	?	?

[a] Estimated from polyacrylamide gel electrophoresis (PAGE).
[b] Based on amino acid composition. Does not include carbohydrate. Length of H-chain will vary depending on the degree of substitution in CDR3 of the V_H region.
[c] Value in parenthesis is the percentage of total Ig in that body fluid. From Butler (1995).
[d] dIgA = dimeric IgA; SIgA = secretory IgA; concentration of IgA does not reflect differences in allotype distribution.
[e] ND = not detectable.
? = no data available.

summarized in Table XI.7.2 although the most significant features of these Ig genes do not lend themselves to tabular description.

Swine C_μ is highly conserved as it is in all mammals that have been studied. Like all swine Ig genes, it is highly homologous to the human C_μ (80% in $C_\mu 4$; Sun and Butler, 1997). The greatest homology is with that of sheep. In fact, the amino acid sequence of membrane IgM (μM) is identical between sheep and swine. Soluble IgM (S_μ) is also most homologous to human S_μ although sequence data for S_μ are only available for comparison from three species – human, mouse and swine.

The various swine IgG subclass genes (C_γ) show the greatest intersubclass homology of any species so far studied (Kacskovics *et al.*, 1994). This fact, combined with the virtual absence of plasmacytomas in swine (see Section 17), no doubt contributed to the failure of investigators to separate the subclass proteins using biochemical separation procedures. The nomenclature used is based on order of discovery and frequency of occurrence. The designations have no homologous relationship to IgG subclasses in other species since subclass diversification occurred after mammalian speciation. The exception to this rule is closely-related species such as rats and mice and sheep and cattle. Like mice, the major sequence differences among the subclasses are found in the hinge and C-terminal portion of $C_\gamma 3$ domain (Kacskovics *et al.*, 1994). Whether these differences translate into functional differences among swine IgG subclasses, has not been determined.

The C_α genes of swine are most similar to human $C_\alpha 2$ (Brown and Butler, 1994) and bovine C_α (Brown *et al.*, 1997), since human $C_\alpha 1$ is unusual among all species so far studied in having a 39 nucleotide (nt) hinge insertion. This insertion encodes the hinge sequence which is attacked by IgA proteases made by various human bacterial pathogens.

Species homology among genes for C_μ and C_α is highest at the 3′ end and lowest at the 5′ end (Brown and Butler, 1994; Sun and Butler, 1997). The single exception is the

Table XI.7.2 The Ig genes of swine

Ig gene	*Linkage group*	*Homology with other species*
C_μ	7	68–75% with all mammals studied, μM excluded
S_μ	7	Higher with human S_μ than mouse S_μ
$C_\gamma 1$, $C_\gamma 2a$, $C_\gamma 2b$ $C_\gamma 3$, $C_\gamma 4$	7	Highest overall homology to human C_γ Inter-subclass variation most similar to mouse C_γ
$C_\alpha{}^a$, $C_\alpha{}^b$	7	Highest homology to human $C_\alpha 2$ and bovine $C\alpha$
C_ε	7	High homology shared with human, horse and bovine
C_λ		Highest homology to human
C_κ		Highest homology to human
V_H1, V_H2, V_Hn	7	Most homologous to V_H III of human; single V_H family
D_H	7	
J_H	7	Single J_H genes most homology to human J_H6 and mouse J_H4

$C_\mu 2$ exon, which is considered to be the ancestral sequence for the hinge of IgG and IgA. $C_\mu 2$ is variable among species as are the sequences encoding the hinges of IgG and IgA within and between species.

The exact number of C_γ genes remains unknown, although at least eight are detectable in genomic Southern blots (Kacskovics *et al.*, 1994). Whether these are all expressible has not been determined. Allotype variants of both $C_\gamma 1$ and $C_\gamma 2a$ are known from sequence analysis.

There are single C_ε, C_μ and C_α genes in swine, although the two discrete allelic variants of C_α discussed above occur in most breeds studied.

The variable region of the swine heavy chain locus is highly characteristic and perhaps even unique among farm animals. There are approximately 20 V_H genes, all of which belong to a highly homologous (> 90%) family with homology to human V_HIII (Sun *et al.*, 1994). In addition, the swine has only a single J_H segment. The number of D_H segments is still unknown although newborn piglets predominately use only two, D_H1 and D_H2 (Sun and Butler, 1996). Whether multiple C_λ and C_κ genes occur in swine and just how many V_κ, V_λ, J_κ and J_λ segments occur, remains unknown.

The ontogeny of the Ig repertoire

Early studies indicated that Ig (IgM) could be detected in 74-day fetuses (Schultz *et al.*, 1971) and that immunizations of fetuses *in utero* via laparotomy, resulted in immune responses as early as 55 days (Tlaskalova-Hogenova *et al.*, 1994). Potent responses by fetuses immunized by *in utero* catheterization have been observed (Butler *et al.*, 1986). Recently, Sun *et al.* (1998) showed that VDJ rearrangement is detectable by day 30 *in utero* and primarily four V_H genes are used throughout gestation. Although V_HB is the most 3′ functional gene in the heavy chain locus of swine (Sun and Butler, 1996) it is not preferentially used during fetal life (Sun *et al.*, 1998).

Studies in newborn piglets reveal that swine express a very limited number of V_H genes and D_H segments and that these show very little somatic (point) mutation (Butler *et al.*, 1996; Sun and Butler, 1996). During fetal life Sun *et al.* (1997) observed no somatic point mutation. However, newborn piglets do express hybrid V_H genes which can be interpreted to mean that swine utilize somatic gene conversion in the development of their antibody repertoire. Analyses of V_H sequences from adult swine reveal somatic hypermutation of the extent that germline V_H and D_H segment are no longer recognizable. Thus, affinity maturation and somatic mutation most likely proceed in lymph node and spleen germinal centers, as it does in mice and humans and this process begins after birth.

Antibody repertoire development in swine must be viewed in relation to other species. Swine appear to follow the pattern of repertoire development that was earlier observed in chickens and rabbits and more recently in ruminants. These species have been designated the C–L–A (chicken–lagomorph–artiodactyl) group (Butler, 1997). It appears that these species differ significantly from the primate–rodent (P–R) group in Ig gene organization, lymphoid anatomy and repertoire development. The latter utilize large numbers of V_H genes from 7 to 14 V_H families to generate a potentially large repertoire by combinatorial joining, somatic (point) mutation and life-long generation of B-cells. In contrast, swine and other C–L–A group species utilize hindgut lymphoid follicles to generate diversity from a single V_H gene family by somatic gene conversion (chicken) or somatic gene conversion and somatic point mutation (sheep, cattle, swine, rabbit). In some species of the C–L–A group (cattle) extensive length polymorphism of CDR3, also appears to be important.

The concept reviewed above and published elsewhere (Butler, 1997), still requires empirical support. Because ileal Peyer's patches (IPP) are important in sheep and they involute early in life, Butler suggested that the extensive development of IPP in fetal and neonatal piglets, and their subsequent involution, parallels that in sheep and most likely serves the same function. However, Reynaud *et al.* (1991) did not observe gene conversion but did observe point mutation in sheep. Perhaps gene conversion would be undetectable in animals of the age they studied, since somatic point mutation would probably have masked the evidence. Since gene conversion has been shown in cattle (Jackson *et al.*, 1996) it is likely to be a common feature of all Artiodactyla.

There are currently no data on Igα and Igβ in swine. The only Ig-associated molecules that have been studied are secretory component and J-chain. The latter is found in both polymeric IgM and polymeric IgA as in other species. Porcine J-chain has a molecular weight similar to human J-chain (15 600; Zikan, 1973). Porcine SC is apparent after reduction and alkylation of SIgA and migrates in PAGE with a molecular mass of about 68 kDa which is smaller than that seen for human and bovine SC. The poly Ig receptor of swine has not been cloned and sequenced.

There are no unusual data on natural antibodies in swine. Swine no doubt have low affinity, IgM-associated antibodies to most non-self antigens as do all other species when sensitive methods of detection are employed. Gnotobiotic animals have low levels of these antibodies presumably owing to lack of stimulation of the immune system by microbial antigens, B cell superantigens, mitogens or maternal regulatory factors obtained from colostrum.

Species-specific aspects of swine Igs

Information to date indicates that two features of swine Igs and Ig genes are unique. However, there are a number of features of the swine Igs that are highly characteristic, albeit shared with other members of the C–L–A group

(Butler, 1997). Both the unique and the characteristic aspects are worthy of discussion.

The swine is unique among mammals examined to date in that they have only a single J_H (Butler *et al.*, 1996); this is a feature they share with the chicken. The swine is also unique because of the occurrence of two allotypes of IgA that differ at a splice site mutation which virtually eliminates the hinge of the IgA^b variant.

Compared with species such as mice and humans, the swine Igs are highly characteristic and quite different from these two well-studied species. Swine lack a gene for IgD (Butler *et al.*, 1996), have but a single family of V_H genes (V_H3), probably utilize the ileal Peyer's patch in repertoire development and both somatic gene conversion and point mutation. These are features which the swine share with ruminants, the rabbit and the chicken. However, they are 'species-specific' if the frame of reference is humans and mice. Swine have the largest number of IgG subclasses of species examined to date. Since most species have been poorly studied, this feature may be more characteristic than unique. Finally, all swine IgGs share with bovine IgG2, a four-amino acid deletion in the lower hinge. In mice and humans, this is believed important for recognition by the high affinity Fcγ receptor. This deletion is not shared by human or mouse IgGs or by bovine IgG1. The functional significance of these more compact IgGs remains to be shown.

8. Major Histocompatibility Complex (MHC) Antigens

The swine major histocompatibility complex (swine leukocyte antigen, SLA) located on chromosome 7 and overlapping the centromere (Smith *et al.*, 1995), has been the subject of increased attention over the past several years. This has been both because analysis of the MHC may allow the breeding of healthy or more disease-resistant animals and because a detailed understanding of the MHC will make the species more useful as a potential xenogeneic organ donor. The current knowledge of the SLA has recently been extensively reviewed (Schook *et al.*, 1996; Lunney and Butler, 1997). Therefore, this section will summarize the more salient features.

Methods to characterize MHC antigens

Class I antigens

Serology with alloantisera has been and continues to be the most reliable method for assigning class I alleles to individuals from uncharacterized families. Panels of alloantisera have been characterized most extensively by Vaiman in France and Kristensen in Denmark (see Schook *et al.*, 1996; Lunney and Butler, 1997) for a complete list of references). The last international test of alloantisera was reported by Renard *et al.* (1988). Monoclonal antibodies (mAb) have been produced against class I antigens but are mostly reactive against monomorphic determinants. Lunney (1994) summarized the reactivity of known mAb with cells from inbred pigs and with expressed cloned class I gene products. A few mAb reactive with specific class I gene products have been produced (Ivanoska *et al.*, 1991) and some of these show clear reactivity with only one SLA specificity (Kristensen *et al.*, 1992).

The class I antigens have been cloned and sequenced (Singer *et al.*, 1982, 1988) and used as probes for class I polymorphism. The high degree of polymorphism of class I genes and the limited number of expressed class I gene products makes restriction fragment length polymorphism (RFLP) analyses difficult to interpret (Singer *et al.*, 1988; Jung *et al.*, 1989; Shia *et al.*, 1991; Smith *et al.*, 1995).

Class II antigens

Serological methods have produced anti-class II alloantisera, but usually in the presence of class I antibodies. Using SLA-defined recombinant pigs, specific class II alloantisera can be produced and used to type these antigens (Lunney *et al.*, 1978; Thistlethwaite *et al.*, 1983). The mAb generated to date have mostly shown reactivity to monomorphic determinants of class II antigens (Lunney, 1994). Because of molecular weight differences, SLA-DR can be differentiated from SLA-DQ on nonreducing gels (Schook *et al.*, 1996). The class II genes of pigs have been defined (Sachs *et al.*, 1988) and most genes have been cloned (Hirsch *et al.*, 1990, 1992; Pratt *et al.*, 1990). Using both of these porcine genes as probes, and specific human probes, enabled many labs to perform RFLP analyses of porcine class II genes (Sachs *et al.*, 1988; Shia *et al.*, 1991; Smith *et al.*, 1995). More recently defined methods for PCR RFLP, have enabled scientists to clearly type SLA-DRB and SLA-DQB alleles quickly (Shia *et al.*, 1995; Brunsberg *et al.*, 1996; Komatsu *et al.*, 1996). This is currently the best method for SLA typing outbred pigs.

Class III genes

Early data used serologic criteria to evaluate C2, C4 and Bf levels (Lie *et al.*, 1987; Chardon *et al.*, 1988; Geffrotin *et al.*, 1990). Most of the genes in the region have been defined by cloning (Peelman *et al.*, 1995, 1996; Chardon *et al.*, 1996).

Distribution, structure, and function of MHC antigens

The class I antigens are widely expressed in porcine tissues. Assays have shown expression in some embryonic tissue but it is restricted. The class II antigens show definite restricted expression. As expected, they are expressed in B cells and activated macrophages, but they are also

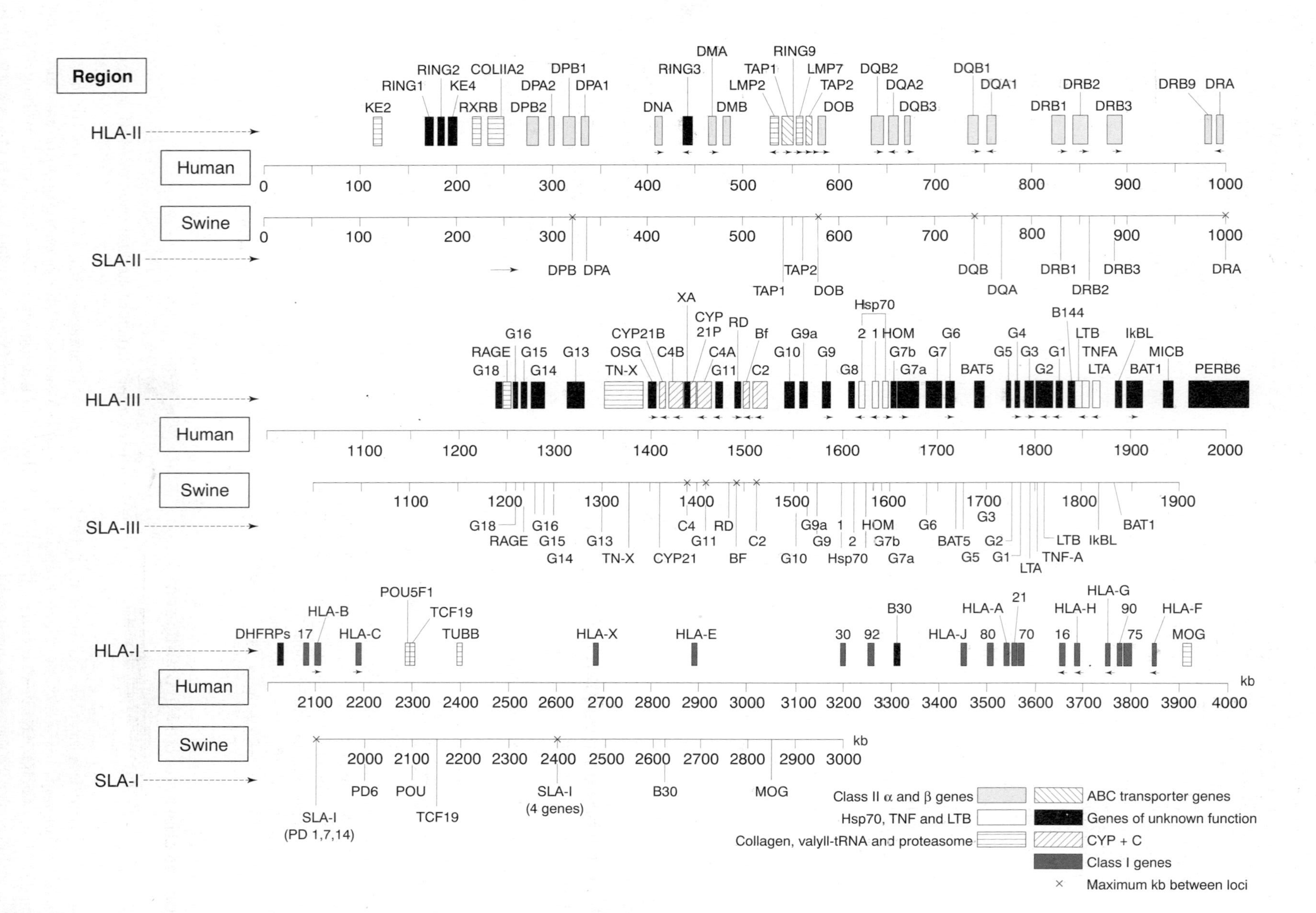

Figure XI.8.1 Comparative map of the major histocompatibility complex of humans and swine.

expressed on a subset of $CD8^+$ T cells (Lunney *et al.*, 1987) and vascular endothelium (Pescovitz *et al.*, 1983).

The structure of the expressed class I and class II genes have been described in detail (Lunney *et al.*, 1986; Chardon *et al.*, 1988). This has been confirmed in studies of the expressed cloned class I.

The class I genes serve to restrict T cell activity, particularly antiviral responses (Schook *et al.*, 1996). The class II genes are associated with controlling immune responses, as shown by production of antibody against defined antigens (Lunney *et al.*, 1986; Schook *et al.*, 1996). Mallard and colleagues have tried to define a series of responses associated with SLA class I and II genes (Mallard *et al.*, 1989a, b, c). The association of SLA genes with production traits have recently been reviewed (Schook *et al.*, 1996).

Species specific aspects of the MHC

The SLA complex has been proven to map across the centromere of chromosome 7; no other species has shown such mapping (Figure XI.8.1) (Smith *et al.*, 1995). Singer *et al.* (1982) first showed that swine had four class I genes. This has been substantiated by later studies with classical and nonclassical class I probes (Singer *et al.*, 1988). More recently, Chardon *et al.* (1996) cloned the swine class I region and proved that there is less than 1000 kb between BAT1 and MOG, proving that the swine class I region is among the smallest of any known MHC region. This also corresponds with the data that indicates that swine have only 7–8 class I genes compared with 50–80 for mice and humans. There is no expressed SLA-DP class II gene yet known. Finally, for class III genes, unlike other higher mammals, swine have been found to have only one C2 and one C3 gene.

9. Red Blood Cell Antigens

The discovery of the human ABO blood group system by Landsteiner (1900) with naturally occurring anti-A and anti-B and the demonstration by Dungern and Hirzfeld (1910) that the ABO antigens followed a Mendelian pattern of inheritance was the impetus for subsequent research into human and animal blood groups. All species of domestic animals were examined for the existence of systems similar to the human ABO. In many species, the search was successful with the discovery of the J, R and A–O systems of cattle, sheep and pigs, respectively. Two different approaches were used to identify blood groups in pigs other than the A–O groups. One was examination of immune antisera from sows that had been vaccinated with crystal violet swine fever vaccine whose piglets were subsequently born with hemolytic disease (Saison, 1958). This approach was taken because of the obvious possibility of detecting red cell isoantibodies responsible for the hemolytic disease. The other approach was to examine immune antisera from sows injected repeatedly with nonpooled donor blood (Andresen, 1962). The blood groups of the pig were subject of intensive studies during the 1960s and 1970s and now rival the bovine blood groups in complexity.

Blood groups, as any other hereditary factor, are the phenotypic expression of genes present on chromosomes. All porcine blood group substances studied to date are mucopolysaccharides and glycolipids (Thiele and Hojny, 1971). Thus, these substances cannot be the primary products of genetic activity. Recently, Gonzalez *et al.* (1995) stated that the B system is placed on the glycophorin A protein, similar to human blood group M–N. Nevertheless, blood groups are inherited, with only a few exceptions, as codominant Mendelian traits. This implies that red blood cell antigens present in a descendant must also be present at least in one of its parents. No detailed chemical studies, aimed at determining the specificities of the various porcine blood group antigens have been undertaken. Most of the studies of porcine blood groups have been concerned with genetic and serological aspects and, therefore, most of the reported studies in this section are of this nature.

Terminology

Early in the development of the porcine blood groups, a nomenclature was adapted by Andresen (1962) that is still in use today. Andersson *et al.* (1993) changed the loci designation encoding the erythrocyte antigens to the form EAX (EA = erythrocyte antigen).

- *Blood group systems* have upper case letters (examples are: EAA, EAB and EAC).
- *Blood factors* are designated with four letters, three upper case letters, corresponding to the blood group system, followed by a lower case letter (except the factors A and O of EAA) indicating the blood factor (examples are: EAEa, EAEb and EAEd).
- *Phenogroups* have upper case letters corresponding to the blood group system, followed by lower case letters corresponding to the blood factors which characterize the phenogroup. In some cases a phenogroup is characterized by one blood factor only and then the factor symbol is the same as the phenogroup symbol (examples are: EADa, EAEbdgkmps and EAEdeghkmnps).
- *Allele (gene) symbols* are derived directly from the corresponding phenogroup symbols simply by converting the lower case letters to superscripts and italicizing the term (examples are: EAD^{a}, $EAE^{bdgkmps}$ and $EAE^{deghkmnps}$). Whether the phenogroups within such complex systems as EAE and EAM of pigs are coded for by single series of allelic genes or by clusters of closely linked genes has been debated. Operationally, however, the phenogroups are distributed within breeds as if

coded for by a single allelic series of genes. The superscript minus indicates that the system is still open, with the product of at least one allele not yet identified by a blood factor, as distinguished from a closed system (examples are: EAC^{-} and EAK^{-}).

- *Genotypes* are symbolized by pairing the appropriate allele symbols as in the examples: $EAE^{bdgkmps}$ $EAE^{deghkmnps}$ ($EAE^{bdgkmps/deghkmnps}$) and EAH^{-} EAH^{-} ($EAH^{-/-}$).

Andresen (1962) described 10 genetic systems, EAA, EAB, EAE, EAF, EAG, EAH, EAI, EAJ, EAK and EAL. Later, a further six systems, EAC, EAD, EAM, EAN, EAO and EAP were discovered (Nielsen, 1961; Andresen and Baker, 1964; Hála and Hojny, 1964; Hojny and Hála, 1965; Saison *et al.*, 1967; Hojny *et al.*, 1994) and many of the original systems expanded with the discovery of new factors.

The current status of porcine blood groups is outlined in Table XI.9.1.

Serology

Reagents

Porcine blood typing reagents are prepared mainly from alloimmune sera of pigs produced by immunizing a recipient pig intramuscularly or subcutaneously with 50% saline suspensions of red blood cells of a single donor pig (Andresen, 1963). Vögeli (1990) used complete adjuvants in his immunizations for increased responses.

Serological test

Four types of serological tests have been outlined by Andresen (1963): direct agglutination test, dextran test, hemolytic test and indirect antiglobulin (Coombs) test. Vögeli (1990) listed the serological characteristics and type of test of nearly all reagents. For the detection of the activity of hemolytic antibodies, rabbit complement has to be added to the antigen–antibody complex. However, it is first necessary to absorb the rabbit serum once at 4°C with porcine red blood cells (1:4) to remove the natural heterolysins and heteroagglutinins. The preparation of rabbit antiswine globulin and the mechanics of the antiglobulin test have been described by Vögeli (1990). The antiglobulin serum requires careful absorption with swine red blood cells in order to prevent false positive or negative reactions in the antiglobulin test. Because of the time-consuming nature of the antiglobulin test, not all porcine blood-typing laboratories routinely type for blood factors detected by this test. Andresen (1963) proposed the use of the modified dextran test, pointing out that this test can often be an alternative for the antiglobulin test and can increase the sensitivity of the agglutination reaction. In fact, the dextran test is important for the elucidation of the EAL and EAO systems.

Historical development

The early history of pig blood groups has been reviewed by Andresen (1962). From a simple beginning of two blood types, A and −, detected by naturally occurring anti-A, the porcine blood groups now are divided into 16 genetic systems consisting of 78 blood factors and 82 alleles (Table XI.9.1).

EAA system

The EAA system is peculiar both from a serological and genetic point of view and because of its similarity to certain blood group systems in other mammalian species, notably the J-Oc system in cattle and the R-O system in sheep (Sprague, 1958). Each of the three species has a substance, A, J and R, respectively, that cross-reacts with the bovine anti-J and, likewise, each of them has another substance, O, Oc and O, respectively, that cross-reacts with the bovine anti-O. Genetic studies have shown that the antigenic expression of the phenotypes depends first on the EAA system, with the two antigenic factors A and O (factor A being dominant over O), and on a second epistatic S system with the two alleles *S* and *s* (Rasmusen, 1964). In adults the four phenotypes are regulated by the following genotypes: 'A' ($EAA^{a/a}$ or $EAA^{a/o}$) + ($S^{S/S}$ or $S^{S/s}$); 'O' ($EAA^{o/o}$) + ($S^{S/S}$ or $S^{S/s}$); 'A_{weak}' ($EAA^{a/a}$ or $EAA^{a/o}$) + ($S^{s/s}$); and 'dash' ($EAA^{o/o}$) + ($S^{s/s}$) (Vögeli *et al.*, 1983). The porcine A and O antigens are not true red cell antigens. They are serum and tissue antigens that are secondarily attached to red cells when the serum concentrations are sufficiently high (Hojny and Hradecky, 1971). The epistatic S system is reported to be undetectable in pigs under 5 weeks old. Such animals always show type *s/s*, implying that the phenotypic expression of factors A and O is suppressed in early life. Soluble A substance is present in the sera and saliva of all A-positive pigs. O substance is present in the sera of O-positive pigs but is absent from the sera of A-positive pigs. In the sera of '−' pigs, O substance occurs in less than half of these pigs and the levels are low (Camacho and Hojny, 1979). The distributions of A and O substances in milk and saliva have been determined in the same report and in the report of Hojny and Glasnák (1970).

EAB, EAD, EAG, EAI and EAO systems

All these systems are closed and two blood factors and two alleles are known (Andresen, 1957, 1962, 1964a; Andresen and Wroblewski, 1961; Hojny and Hála, 1965; Hojny *et al.*, 1967; Saison *et al.*, 1967; Hradecky and Linhart, 1970).

Table XI.9.1 Current status of porcine blood groups

Systems	*Blood factors*	*Alleles*[a]	*Test method*[b]
EAA	A (A,Aw), O	EAA^{a}, EAA^{o}	1, 4
EAB	a, b	EAB^{a}, EAB^{b}	1
EAC	a	EAC^{a}, EAC^{-}	4
EAD	a, b	EAD^{a}, EAD^{b}	1
EAE	a, b, d, e, f, g, h, i, j, k, l, m, n, o, p, r, s, t	$EAE^{bdgkmps}$, $EAE^{deghkmnps}$, EAE^{aeglns}, $EAE^{defhkmnps}$, $EAE^{bdfkmps}$, EAE^{aeflns}, $EAE^{degklnps}$, EAE^{aegils}, $EAE^{deghjmnt}$, $EAE^{abgklps}$, $EAE^{abgkmps}$, $EAE^{aegmnops}$, $EAE^{bdgklps}$, $EAE^{deghjmnr}$, $EAE^{abgkmops}$, EAE^{bdgjmt}, EAE^{bdgjmr}	1
EAF	a, b, c, d	EAF^{ac}, EAF^{ad}, EAF^{bc}, EAF^{bd}	1
EAG	a, b	EAG^{a}, EAG^{b}	1
EAH	a, b, c, d, e	EAH^{a}, EAH^{b}, EAH^{ab}, EAH^{cd}, EAH^{bd}, EAH^{be}, EAH^{-}	4
EAI	a, b	EAI^{a}, EAI^{b}	2
EAJ	a, b	EAJ^{a}, EAJ^{b}, EAJ^{-}	2
EAK	a, b, c, d, e, f, g	EAK^{acf}, EAK^{bf}, EAK^{acef}, EAK^{ade}, EAK^{adeg}, EAK^{-}	4
EAL	a, b, c, d, f, g, h, i, j, k, l, m	EAL^{adhi}, EAL^{bcgi}, EAL^{bdfi}, EAL^{agim}, EAL^{adhjk}, EAL^{adhjl}	2,3
EAM	a, b, c, d, e, f, g, h, i, j, k, m	$EAM^{ab(e)}$, EAM^{aem}, EAM^{aejm}, $EAM^{ade(m)}$, EAM^{b}, EAM^{bcd}, EAM^{bcdi}, EAM^{bd}, EAM^{bdg}, $EAM^{be(f)m}$, EAM^{cd}, EAM^{cdi}, (EAM^{cdk}), EAM^{d}, EAM^{djk}, EAM^{dk}, EAM^{ef}, EAM^{efm}, EAM^{h}, EAM^{-}	2,4
EAN	a, b, c	EAN^{a}, EAN^{b}, EAN^{bc}	2
EAO	a, b	EAO^{a}, EAO^{b}	3
EAP	a	EAP^{a}, EAP^{-}	2

[a] The genes for which phenogroups have not yet been detected (open systems) are symbolized by adding a dash (–) as a superscript to the symbol for the blood group system (Examples C^{-} and H^{-}).
[b] 1, direct agglutination test; 2, antiglobulin (Coombs) test; 3, dextran test; 4, haemolytic test.
Data obtained in part from: Bell (1983), Vogeli (1990), Fries *et al.* (1990), Hojny and Nielsen (1992), Gonzalez *et al.* (1992) and Andersson *et al.* (1993).

EAC, EAJ and EAP systems

The EAC and EAP systems are one-factor, two-allelic open systems (Andresen and Baker, 1964; Hojny *et al.*, 1994). After screening three generations of families for the EAM blood group system, it was shown that EAMl is controlled by an allele from another new blood group system. Hojny *et al.* (1994) proposed to designate this system EAP and changed the factor designation from EAMl to EAPa. The EAJ system is open and consists of two factors and three alleles (EAJ^{a}, EAJ^{b} and EAJ^{-}).

EAE system

Eighteen antibodies have been discovered to detect blood factors of this system and the genetic control is by 17 alleles (Hojny and Nielsen, 1992). This system has gradually developed from being a simple blood group system to become one of the most elaborate swine blood group systems. Five closed subsystems have been found in EAE (EAEa–EAEd, EAEb–EAEe, EAEf–EAEg, EAEl–EAEm and EAEj–EAEs).

EAF system

This is recognized as a four allelic (EAF^{ac}, EAF^{ad}, EAF^{bc} and EAF^{bd}) closed system (Hojny *et al.*, 1984).

EAH system

Andresen (1964b) described three factors (EAHa, EAHb, and EAHc) in this system forming five different alleles. Later, factors EAHd and EAHe were detected extending the number of alleles to seven (EAH^{a}, EAH^{b}, EAH^{ab}, EAH^{cd}, EAH^{bd}, EAH^{be} and EAH^{-} (Hojny, 1973).

EAK system

Initially, Brucks (1966) used five reagents (EAKa, EAKb, EAKc, EAKd and EAKe) to determine genetic relations between the corresponding factors in Landrace pigs; subsequently, two other factors, EAKf and EAKg were discovered. The present status of EAK is a seven-factor and six-allelic open system (Nielsen and Vögeli, 1982).

EAL system

Twelve factors have been detected in this closed system controlling six complex alleles (EAL^{adhi}, EAL^{adhjk}, EAL^{adhjl}, EAL^{agim}, EAL^{bcgi} and EAL^{bdfi}) (Linhart, 1971). Factors EALa-EALb, EALd-EALg and EALi-EALj form three closed subsystems and factors EALk and EALl are a subgroup of EALj.

EAM system

EAM is the most complex blood group system in the pig. The number of alleles exceeds even that of EAE and is still not definitive. It consists of 12 blood factors and 20 alleles and is classified as an open system (Hojny *et al.*, 1979; Hojny and van Zeveren, 1985). Difficulties in the serological determination of the EAM blood factors limit the practical use of the EAM system.

EAN system

The EAN system is recognized as a three-allelic (EAN^{a}, EAN^{b} and EAN^{bc}) closed system (Saison, 1967). Soluble Na and Nb substances are present in the sera of all animals that have the corresponding antigens on the red cells. These substances also occur in the milk (Hojny and Glasnák, 1970).

Table XI.9.2 Chromosomal location of porcine blood group loci

Chromosome	*Blood group locus*	*Reference*
1	*EAA*	Ellegren *et al.* (1994)
4	*EAL*	Archibald *et al.* (1995)
6	*EAH, EAO, S*	Juneja *et al.* (1983), Vögeli *et al.* (1988), Archibald *et al.* (1995), Cepica *et al.* (1996)
7	*EAL, EAJ, EAC*	Andresen and Baker (1964), Archibald *et al.* (1995)
8	*EAF*	Röhrer *et al.* (1996)
9	*EAE, EAK, EAN*	Ellegren *et al.* (1993), Röhrer *et al.* (1996)
11	*EAM*	Archibald *et al.* (1995)
12	*EAD*	Cepica *et al.* (1996)
15	*EAG*	Fries *et al.* (1983)
18	*EAI*	Röhrer *et al.* (1996)

Significance and application of porcine blood group systems and red cell antigens

Parentage testing

Blood typing of pigs for parentage control is regularly used in most countries with intensive pig breeding. Both the blood groups and protein and enzyme polymorphisms are used. Vögeli (1990) showed that wrongly assigned parentages can be as high as 5%, even in herds where active marker typing is practiced. In Switzerland and Germany, the blood typing scheme consists of 9–10 blood group systems and 4–10 protein systems. This allows an exclusion probability of about 95–97%.

Mapping of blood group systems

The polymorphic blood group systems were typed in reference families of the ongoing mapping projects. This has resulted in mapping of 14 of the 16 blood group loci to different chromosomes (Table XI.9.2). Porcine blood group loci are quite evenly distributed over the genome.

Heterozygosity and genetic distances

In pigs, it is of interest to know the degree of heterozygosity within populations and the relationship between different breeds. The average heterozygosity estimates based on quite a large number of polymorphic blood marker loci have been reported in a few studies (van Zeveren *et al.*, 1990; Cepica *et al.*, 1995). In these studies, the average degree of heterozygosity ranged from 30 to 40% in the different Euro-American breeds. The breeds with a larger population size (e.g., Landrace and Large White) showed a higher degree of heterozygosity than those breeds with smaller population size (e.g. Pietrain). Because heterosis is an effect of increased heterozygosity, pig breeds with a greater degree of heterozygous marker genes appear healthier than breeds with a lower degree (Dinklage and Gruhn, 1969). The number of pigs born and reared was found to be greater when the matings resulted in heterozygous offspring with regard to blood groups (Dinklage and Hohenbrink, 1970).

The relationships among the four major pig breeds, namely, the two European ones (Landrace and Large White) and the two American ones (Duroc and Hampshire), were given in some reports. The most distant of the four breeds was found to be Duroc by some (Oishi *et al.*, 1990; Vögeli, 1990), and Hampshire by others (Cepica *et al.*, 1995). Major (1968) and Vögeli (1990), on the basis of blood group data, classified the European Landrace into two groups – the one consisting of the Landraces of Germany, Holland, Hungary and Switzerland and the other of Denmark, Sweden, Czechoslovakia and Russia.

Relationship to production performance and diseases

Several reasons for the selective advantage of these immunogenetic and biochemical polymorphisms have been proposed. One suggestion is that they are, or have been associated with, health of the animals. Therefore, many attempts have been made to relate the polymorphic blood group traits in pigs to quantitative traits of economic importance. However, when different traits are considered in various populations, several inconsistencies become apparent. Most interesting seems to be the EAH blood group system. Jensen *et al.* (1968) found a major and consistent effect of alleles at the EAH system on reproductive performance. Differences in fitness among genotypes for *EAH* alleles were later observed by Rasmusen and Hagen (1973). Alleles of *EAH* and *S* are also found to be associated with the porcine stress syndrome (Rasmusen and Christian, 1976; Vögeli *et al.*, 1994). Five blood genetic markers were shown to be closely linked to the halothane (HAL) locus. In Sweden, Switzerland and some other countries, the typing of these five blood groups and protein loci in conjunction with the halothane testing of the offspring was used to deduce HAL-marker loci haplotypes of each member of a given family (Gahne and Juneja, 1985; Vögeli *et al.*, 1988).

Blood group incompatibility causes erythroblastosis fetalis in man and a similar hemolytic disease of the newborn piglet. Much research has been done to identify the antigens responsible for hemolytic disease. Cases have been reported on the interaction of antigens, EABa, EABb, EAFa, EAFb, EAEa, EAEb, EAHa, EAKa and EAKb with their corresponding antibodies (Andresen and Baker, 1963; Vögeli, 1990).

The blood group antigens are complicated biological structures of the cell membrane and may serve not only as receptors for physiological substances but also for disease-producing agents (i.e., toxins). Resistance to porcine edema disease and post-weaning diarrhea is associated with the absence of colonization of the intestine with toxigenic *Escherichia coli* strains belonging to certain serotypes. Colonization depends on specific binding between adhesive bacterial fimbriae (i.e., fimbriae F18) (Bertschiner *et al.*, 1990) and the brush border enterocytes. The specific binding is mediated by receptors expressed on the surface of the brush borders. Adhesion of fimbriae F18 has been shown to be genetically controlled by the host (Bertschinger *et al.*, 1993). Adhesion in edema or diarrhea susceptible animals is dominant over absence of adhesion in resistant animals. The corresponding genetic locus (*ECF18R*) was mapped to porcine chromosome 6, based on its close linkage to the epistatic *S* locus controlling the expression of EAA blood group antigens (Vögeli *et al.*, 1996). The explanation for the association between the blood group phenotype and edema disease and post-weaning diarrhea may relate to the expression of fucosylated blood group antigens in the intestinal surface mucous cells.

Xenotransplantation

Red cell surface antigens other than blood groups have received special attention of late (Table XI.9.3). Some of these antigens, e.g. CD59 or CD46, are involved in complement action and are crucial in pigs studied as xenogeneic organ donors. The porcine analogue of human CD59 was effective in the protection of pig

Table XI.9.3 Human red blood cell surface proteins in humans (from Anstee, 1990) and their counterparts in swine

Protein	*Copies per cell in human RBC ($\times 10^3$)*	*Data in swine RBC*	*Reference*
Anion transport protein	1000		
Glycophorin A	1000	mAb and protein	Llanes *et al.* (1992)
Glucose transport protein	500	mAb and protein only in neonatal pig	Craik *et al.* (1988)
Glycophorin B	250		
Rh polypeptides 30K	100–200		
Rh polypeptides	100–200		
Glycophorin C/D	100		
CD47	10–50		
CD59	20–40	Protein	van den Berg *et al.* (1995)
Kell	4–18		
Fy	10–12		
CD55	6–14		
CD44	5–10	mAb	Saalmüller *et al.* (1998)
Nucleoside transporter	10	mAb and protein	Good *et al.* (1987)
CD58	3–8		
LW	3–5		
Lu	1.5–4		
Acetylcholinesterase	3		
CR1	1		
12E7	1		
HLA Class I	0.3–0.7	mAb, protein and gene	Andersson *et al.* (1993)
CD46 (MCP)	Not detected	mAb, protein and gene	van den Berg *et al.* (1997)

erythrocytes against lysis by serum from a variety of species including man. This finding will be of particular relevance to current attempts to render swine organs resistant to human complement (van den Berg *et al.*, 1995).

Species-specific aspect of RBC antigens

Erythrocytes from adult pigs lack glucose transport activity whereas erythrocytes from neonatal pigs are able to transport glucose. Neonatal pig red cell membranes contain an identifiable polypeptide band which also mediated nucleoside transport in humans (Good *et al.*, 1987; Craik *et al.*, 1988).

Recently, a protein with cofactor activity for factor I of the human and pig complement 3b (C3b) have been described on swine erythrocytes. This is the pig analog of human MCP or CD46 (van den Berg *et al.*, 1997). This molecule was not present on human erythrocytes but it is present on erythrocytes of other primates such as orangutan; it is used by measles virus to attach to human cells. There is no data identifying clinical consequences of the differential expression between human and swine CD46 (van den Berg *et al.*, 1997).

10. Ontogeny of the Immune System

The sequence of immune organ development and appearance of lymphoid populations is given in Table XI.10.1. Similarly, the gestational time line of immune responsiveness is shown in Table XI.10.2.

Table XI.10.1 Gestational sequence of organ development and appearance of lymphoid populations

Day of gestation	*Developmental event*
16	Blood islands in the yolk sac, vitelline veins
17	Liver rudiment
20	Liver erythropoiesis, platelets
21	Thymus rudiment
22	Spleen rudiment
24	End of yolk sac hemopoiesis Macrophage precursors in the yolk sac
25	Liver macrophages
28	Lymphoid cells and myeloblasts in the liver Lymphoid cells in thymic area
32	Spleen erythropoiesis, lymphatic vessels
40	γ/δ T-cells and pre-B-cells in the liver, lymphocytes in omentum
44	CD4$^+$ and CD8$^+$ T cells in the thymus, sIgM$^+$ B-cells in the liver lymphocytes in the spleen and lymph nodes
50	Spontaneous secretion of IgM by liver B-cells T-cells in the spleen and skin, thymus cortex and medulla lymphocytes in bone marrow and intestines
60	FcR on splenocytes
64	Hassall's bodies in the thymus
67	Thymic B cells secrete minute amounts of IgM, IgG, IgA
77	Peyer's patches, follicular structures formed
1 day after birth	Lymphocytes and macrophages appear in alveolar lavages
Postnatal	Germinal centers appear in conventional animals

Development of T lymphocytes

The pig embryonic liver and thymus each contain about 10 000 lymphoid cells at 28 days of gestation. Coincident with this, the number of stem cells in the circulation decreases. The fetal liver and thymus contain the majority of all embryonic lymphocytes through day 38. By day 50, however, the majority of lymphocytes are within the thymus.

CD2$^+$ cells can first be detected by magnetic sorting on day 30 with the monoclonal antibody MSA4. Intracytoplasmic CD3 immunoreactivity was found in hepatic lymphocytes also by day 30. Both Thy-1 mRNA and protein were found in the liver on day 35 using Northern blot.

γ/δ T-cells were found first extrathymically (in the liver) by day 40. They appear in the spleen by day 50 but only later in the thymus. By day 73, near the end of gestation, they form 3% of blood leukocytes but are frequent in the gut-associated lymphocytes (GALT) (Trebichavsky, 1994).

CD4 and CD8 are expressed by thymocytes by day 44 but not before day 40 of gestation. They are found in the spleen and liver by day 50 (F. Kováru *et al.*, 1994). Their localization in the skin correlates with the beginning of cellular immune response against PPD (Trebichavsky *et al.*, 1992). CD44 is strongly expressed in the thymus by day 40 of gestation.

CD4$^+$CD8^{dull} (double-positive) lymphocytes, that are hypothesized to be preactivated (memory) CD4$^+$ helper cells in pigs, appear in pig fetuses at the end of gestation; their number slowly increases with the age. Because no exogenous antigens are present in pig fetuses, these cells could be memory T cells with receptors specific for autoantigens (Zuckermann and Husmann, 1996).

Development of B lymphocytes

Because pig fetuses develop in the absence of maternal immunoglobulins, precolosttral fetal serum contains the products of differentiated B cells before antigenic stimulation. In addition, the development of B cells is not influenced by maternal regulatory idiotypic-antiidiotypic network.

Table XI.10.2 Gestational sequence of ontogeny of immune responses

Day of gestation	*Developmental event*
25	Virus activated liver cells produce IFN-α
34	*In vitro* activated lymphocytes produce TNF-α
44	Liver B cells stimulated *in vitro* synthesize Ig
50	PHA, PWM, and ConA elicit lymphocyte proliferation *in vitro*
55	Antibody response against flagellin administered to fetuses *in vivo*
	Antibodies against parvovirus crossing the placental barrier
	Cellular response against PPD
60	Immune reaction against a leptospiral antigen
80	Plasmocytes in immunized fetuses
	Immune reaction against allogenic lymphocytes
Hours after birth	Colostrum absorbed by the intestinal wall
1 day	Rise of serum Igs in suckling animals, gut closure
1 week	Macrophages exert full activity
3–4 weeks	Production of autologous antibodies

The first cells containing cytoplasmic μ-chain can be detected in the liver around 40 days of gestation by immunohistological methods. Four days later, the first B cells expressing surface IgM (sIgM) are found. The sIgM$^+$ cells appear in the spleen on day 50, and in the bone marrow on day 60 of gestation (Jarosková, 1982). Only later do they appear in the omentum (Trebichavsky *et al.*, 1981). Fetal IgM$^+$ B-cells, in contrast to human and murine B-1 cells, do not express the CD5 molecule (Cukrowska, 1996a).

Spontaneous production of minute amounts of IgM, IgG and IgA has been detected using autoradiography in the first half of gestation (Prokesová, 1981). Using the ELISPOT method, cells spontaneously secreting IgM were first found on gestational day 50 in the liver and later in the spleen. In contrast to fetal liver and splenic cells secreting exclusively IgM, fetal thymic B cells were found to undergo spontaneous isotype switching to IgG and IgA (Cukrowska, 1996b). Peyer's patches were detectable between 50 and 70 days, with follicle size increasing throughout gestation (Chapman *et al.*, 1974).

The functional maturation of the fetal B-cell compartment was studied using different B-cell mitogens. *In vitro* responses of fetal liver, splenic and thymic cells to bacterial mitogens resulted in pronounced Ig synthesis. In addition, intrauterine injection of pig fetuses with these antigens resulted in a rise of serum IgM, IgG and IgA (Tlaskalová-Hogenová, 1994).

Preimmune natural antibodies produced in the fetal period spontaneously, or after polyclonal stimulation, were found to react with phylogenetically conserved molecules, autoantigens (myosin, ssDNA, TRG, FSH, thymocytes) and bacterial components (*Escherichia coli*, tetanic anatoxin) (Tlaskalová-Hogenová, 1994; Cukrowska, 1996a).

Specific antibody responses of fetuses were analyzed after direct intrauterine immunization; sheep red blood cells, phages, flagellin and other proteins were administered as antigens (Tlaskalová, 1970; Sterzl, 1977). Reactivity against DNP-flagellin was detectable by the plaque method on day seven after immunization of 55-day-old fetuses. Antiflagellin antibodies were found in these immunized fetuses. Further, stimulation of MHC II antigen expression was found in the spleen and liver (Trebichavsky *et al.*, 1988). Recently, analysis of the pig Ig diversification has been described, indicating gene conversion as the main mechanism of B-cell repertoire development (Sun, 1994).

11. Passive Transfer of Immunity

Placentation

The pig has an epitheliochorial placenta and thus the thickness of the partition between the maternal and fetal blood consists of a maximum of six layers. The surface of contact between maternal and fetal tissue is increased by the development of outgrowths on the surface of the chorion called villi. In the pig, the villi are scattered all over the surface of the chorion (known as the diffuse type of placenta) and they interdigitate with the corresponding depressions in the uterine wall. Given the thickness of the pig placenta, it is not surprising that transfer of passive immunity is entirely postnatal.

Functions of the mammary gland

In addition to their well established nutritional role, mammary gland secretions provide the suckling piglet with a protective cover from infection while its own immune system is developing. The vast majority of accumulated data in this area relates to the transfer of passive immunity; however, there is an accumulating body of evidence showing that mammary secretions may also provide an educational role (Newby and Stokes, 1982).

Origins and composition of colostrum and milk

The two mammary secretions, colostrum and milk differ markedly in their composition and origin (Table XI.11.1). In pig colostrum, IgG is the major immunoglobulin isotype with smaller amounts of IgA and IgM. Significant proportions of all three isotypes are derived from serum but the

Table XI.11.1 Concentration of immunoglobulin isotypes (mg/ml; percentage derived by transudation from serum)

	IgG	*IgA*	*IgM*
Serum	24.3	2.4	2.9
Colostrum	61.8 (100%)	9.7 (40%)	3.2 (85%)
Milk	1.4 (10%)	3.0 (10%)	0.9 (30%)

Data prepared after Curtis and Bourne (1971) and Bourne (1973).

colostral concentrations exceed those found in serum. The mechanisms by which IgG is concentrated into colostrum is not fully understood, but data derived from cattle (Hammer and Mossman, 1978) indicate that this may be receptor mediated. In contrast, IgA predominates in milk, the vast majority of which is locally synthesized in the mammary gland (Bourne and Curtis, 1973; Brown *et al.*, 1975).

Intestinal absorption of macromolecules in the neonate

The small intestine of the newborn pig is able to absorb macromolecules nonselectively over its entire length, but this is maximal in the mid-portion with a sharp 'tailing off' at either end (Pierce and Smith, 1967; Smith *et al.*, 1979). Absorption of macromolecules is associated with the formation of characteristic vacuoles in the villus epithelial cells (Clarke and Hardy, 1971; Smith *et al.*, 1979). Small intestinal enterocytes from suckling newborn pigs have a characteristic apical tubulovesicular membrane system (ATVMS). More recent studies have begun to determine the ATVMS role in the uptake of IgG. At least four steps in this process have been identified: the passage of immunoglobulin between microvilli; binding to apical areas of the plasma membrane; accumulation in invaginations of the plasma membrane and noncoated vesicles; and storage in granules in the apical cytoplasm (Komuves and Heath, 1992). There is also a single large subnuclear granule that contains a concentration of IgG at least twice that found in the apical vesicles (Komuves *et al.*, 1993). The intracellular mechanisms by which IgG is successively concentrated in these vesicles has not been determined.

In addition to the well-recognized effect of colostrum in providing the young piglet with circulating antibody (Curtis and Bourne, 1971), absorbed IgA molecules may also be further redistributed to mucosal secretions including the respiratory tract (Morgan and Bourne, 1981). Whether this is purely the 11S secretory IgA molecule, is not known, but based upon the serum half-lives of the 11S and 9S molecules (Bourne *et al.*, 1978), it is most likely the 11S form.

Closure

Various studies have suggested that the formation of vacuoles in epithelial cells and the transport of immunoglobulin into plasma is largely dependent upon the concentration of protein presented to the intestinal mucosa (for review see Baintner, 1986). During the first few hours of life, the piglet's ability to absorb immunoglobulin falls rapidly and 'closure' is generally complete by 24–48 h. The mechanisms responsible for closure have not been fully elucidated, but it has been shown that its onset may be influenced by a variety of colostral factors including lactose (Lecce *et al.*, 1964) and colostral trypsin inhibitor (Baintner, 1979).

The transition from colostrum to milk

During the postpartum period, the composition of colostrum changes rapidly to that of mature milk, with a progressive change in the IgA:IgG ratio from 0.16 at birth, to 0.32 at 24 h, 0.33 at 48 h, 1.78 at day 5 and 2.22 from day 8. These changes in immunoglobulin isotype coincide with closure and a shift in the major site of activity of the passively transferred antibodies from the systemic circulation to locally within the gut lumen. The role of milk antibody in protecting the suckling offspring from enteric infection is well documented. In the pig it is perhaps best illustrated with *Escherichia coli* (Nagy *et al.*, 1976) and transmissible gastroenteritis virus (TGEV) (Bohl and Saif, 1975). In this context, it is noteworthy that studies in pigs infected with TGEV, where the first evidence of simultaneous appearance of IgA antibody in gut secretions and milk was seen, demonstrated the immunological link between gut and mammary gland that was subsequently seen in other species (Bohl *et al.*, 1972).

The role of cells in mammary secretions

Leukocytes are also a normal component of mammary secretions, reaching concentrations of greater than 10^7 cells/ml in colostrum, but falling to less than one-tenth of that number in milk. Polymorphonuclear leukocytes are the predominant cell type, but there are also significant populations of lymphocytes, macrophages, and epithelial cells (Evans *et al.*, 1982). The majority of lymphocytes in colostrum are T cells, with $CD8^+$ cells predominating over $CD4^+$ cells (Le Jan, 1994). Colostral leukocytes are absorbed from the intestinal lumen by intercellular migration and rapidly appear in blood within 2 h and in liver, spleen, lung and lymph nodes within 24 h (Tuboly *et al.*, 1988). The precise biological role of these cells is not known but several studies have indicated that they may have an immunomodulatory role (Williams *et al.*, 1993).

The 'educational role' of mammary secretions

There is a strong relationship between colostral intake during the first 6 h of life and the ability of piglets to survive and gain weight through the first 3 weeks of life. Furthermore, it has been shown that mammary secretions are also able to affect the capacity of piglets to respond immunologically. This effect lasts long beyond the disappearance of maternally derived antibody. Evidence for an 'educational role' was first provided by Hoerlien (1957). He showed that, while the immune responsiveness to a whole range of antigens was diminished if colostrum was withheld, feeding colostrum from sows immunized with *Brucella abortus* suppressed the piglets primary immune response to that antigen. In contrast, other studies showed that mixing bacterial toxoids with highly dilute immune sera could enhance the specific immune response (Segre and Kaeberle, 1962). Such studies serve to illustrate the complexity of the immunological messages passing from the dam to her suckling offspring. A number of factors contained in colostrum and milk have all been shown to influence the outcome of this interaction. These include antibody isotype and concentration, cells and cellularly derived soluble factors, as well as the age when the offspring is first challenged. With such a complex array of effects, it is not surprising that, to date, it has not been possible to manipulate reliably the offsprings' response via this route. The possible outcome of transfer is further complicated by the presence of antigen. For example, following the introduction of ovalbumin (OVA) into the diet of sows during gestation and or lactation, OVA could be detected in colostrum and milk (Telemo *et al.*, 1991). In piglets born to sows fed OVA during gestation and lactation, there was significant transfer of both antigen and specific antibody. When the piglets were then weaned onto OVA containing diets at 3 weeks old the outcome differed, with those piglets born to sows fed OVA during lactation, alone, developing a strong serum IgG to OVA and clinical diarrhea. A partial explanation of these results was provided in a related study in which piglets were fed by gastric tube with a specific antigen at birth and then challenged with the same antigen in the weaning diet at 3 weeks old. The results showed that feeding antigen at birth resulted in a dose-dependent suppression of the response to antigen fed at weaning (Bailey *et al.*, 1994).

During the first few weeks of life the protective cover provided by passively transferred maternal immunity wanes and the piglets become dependent upon their own capacity to make appropriate immune responses for survival. To date, there have been no definitive functional studies to determine the age at which the porcine immune system is fully mature. Based upon anatomic studies and phenotypic analysis, it has been suggested that all the key elements of the systemic immune system are present soon after birth (see Section 10). Similar studies of the gut mucosal immune system would suggest that maturation of this compartment of the immunological armoury may be somewhat delayed (see Section 15). In all these studies, it is not possible to separate what is the consequence of antigen exposure (or rather the lack of it) from an inherent capacity to respond. One approach used to begin to address this question is to attempt to relate age of weaning to the ability to develop an 'appropriate response' to antigens in the postweaning diet. Using a family pen system in which piglets are left with their mothers and gradually wean themselves over a 12-week period it was concluded that oral tolerance to dietary antigens was not fully established until 12 weeks (Miller *et al.*, 1994). Although such studies cannot be used to determine the age at which the piglet's gastrointestinal immune system is first able to recognize dietary antigen, they clearly suggest that the age at which the piglet may cease to benefit from the consumption of maternal milk may be considerably later than would be suggested by current husbandry practices.

12. Neonatal Immune Responses

Physiology of the neonate which may influence the immune system

The swine epitheliochorial placenta prevents transfer of immunoglobulins (Sterzl, 1966) (see also Section 11). Consequently, maternal colostrum represents a vital source of antibodies for neonates. The piglet exhibits no hydrochloric acid and pepsin activity in stomach during the first 3 weeks after birth and milk proteins including immunoglobulins are merely coagulated by chymosin. A trypsin inhibitor present in colostrum partially blocks degradation of proteins in the upper part of the intestine. The ability to absorb Igs decreases rapidly during the first hours of feeding. Between 2 h (duodenum) and 48 h (ileum), the lining of the gut changes (gut closure) and further absorption of intact colostral Igs is prevented (Murata, 1977). Even after this closure of the intestinal wall, colostrum protects against perorally introduced pathogenic microorganisms. Enterocytes of the small intestine of newborn piglets contain great vacuoles that disappear after gut closure. Similar cells are found in germ-free animals during the first 5–6 weeks of postnatal life (Mandel, 1987).

Nonspecific immune defenses in neonates

In newborn piglets, the total peripheral white blood cell count (WBC) of 4.2–9 $\times$ 10^6/ml is made up of about 60% neutrophils and 38% lymphocytes. This ratio 'flips' by day 10 after birth. In contradiction, the proportion of lymphocytes decreases in germ-free piglets after birth. The absolute number of leukocytes decreases to 3 $\times$ 10^6/ml in germ free state, i.e. one-quarter of that of conventional piglets at 1 week after birth (Mandel, 1982).

Peripheral blood monocytes of newborn pigs are immature with *in vitro* phagocytic activity and metabolic burst similar to those of fetal macrophages. This low activity persists during the first week of life after which it increases to the normal state. The same conspicuous increase is seen in monocytes of germ-free pigs after their monocontamination with microorganisms of endogenous microflora such as *Enterococcus faecalis*.

Natural killer (NK) cytotoxicity is rather low in newborn pigs, whereas antibody-dependent cell-mediated cytotoxicity (ADCC) is comparable with that of older piglets (Kim, 1981). NK cytotoxicity believed until recently to develop only postnatally, was confirmed to develop before birth as in other animals (Kovárù, H. *et al.*, 1994). The ability to produce IFN-α is not reduced in newborn pigs (Artursson, 1992).

Specific immune defenses in neonates

Neonatal pigs, born into an environment rich in multiple potentially harmful agents, are immediately protected after birth by a supply of nonspecific factors and specific antibodies present in maternal colostrum (Klobasa, 1981). To analyze the specific antibody response in newborn piglets lacking transferred maternal antibodies, colostrum-deprived piglets reared in germ-free conditions were studied (Mandel, 1987). Germ-free piglets were colonized with a defined bacterial strain of *Escherichia coli* or were orally immunized with sheep erythrocytes and the antibody response was studied on both cellular and serological levels. Specific IgM antibodies were detected during the first several days. Later, a pronounced IgA response prevailed both in sera and in intestinal contents (Tlaskalová-Hogenová, 1981). In experiments analyzing the role of antibodies on the immune response of germ-free piglets, the importance of quantitative relations between transferred or preformed antibodies and antigen dose (antigen–antibody complexes) was demonstrated (Sterzl, 1977).

In addition to antibody-specific responses, polyclonal B-cell activation occurred in germ-free piglets, especially when microbial components were used. Recently, the effect of active molecules derived from *Rhodococcus opacus* on mucosal immunity and resistance to infection has been studied (Tlaskalová-Hogenová, 1994) as has the participation of cytokines in specific antibody responses in newborn piglets (Bailey, 1992).

Species-specific aspects of immune responses in neonates

Immunodeficiency in pigs extends from birth up to 4 weeks. This is observed in serum antibody concentrations, in response to antigenic challenge exposure, and in leukocyte proliferative responses (Metzger, 1978).

Colostrum intake in the first 3–6 h of life may be of paramount importance. The last piglets born in a large litter do not achieve as high a level of Ig as their siblings. IgG constitutes over 80% of the total colostrum immunoglobulin, whereas IgA represents only a minor component (Porter, 1969; Bourne, 1973). Colostrum-deprived piglets fail to produce antibodies following antigenic stimulation that induces antibodies in colostrum-fed littermates (Segre, 1966).

The immune system of newborn pigs has been extensively studied. Localization of immune cells in lymphatic organs has been recently described by Bianchi (1992). The $sIgM^+$ B cells are frequent in newborn pigs and are localized mainly in lymphatic follicles. Lymphocytes with cytoplasmic IgM, IgG or IgA are rarely detected in the spleen, lymph nodes or GALT, but they are relatively frequent in the thymus. Neither germinal centers nor plasma cells are present in lymphatic organs of newborn pigs. Conversely, the T-cell compartments appear to be fully developed in lymphoid organs before birth. Only the small intestine seems to have a major influx of T cells, which begins postnatally. $CD2^+/CD8^+$ cells are present in epithelia, whereas $CD4^+$ cells only appear in intestinal villi of older animals (beginning by 14 days old). The SLA-DR molecules are well expressed in lymphatic organs of newborn pigs.

13. Nonspecific Immunity

The observed hypersusceptibility of animals with various forms of immunodeficiency to overwhelming microbial infection does not in any way negate the importance of nonspecific defense mechanisms to 'maintenance of health' in individuals. Nowhere is this more important that at the mucosal surfaces, where a wide variety of nonspecific defense mechanisms act in concert with elements of the specific acquired response. There are a large number of mechanisms that may be included under the headings of nonspecific defense. Broadly, these may be classified under four headings: natural/physical barriers (e.g. skin, gut pH, peristalsis, cervix); acute phase and antimicrobial proteins (e.g. IL-1, IL-6, IFN-α, lactoferrin, lactoperoxidase); granulocytic leukocytes; and the entire mononuclear phagocyte system. Many of these functions are common to all species and thus the aim of this section is to highlight examples in which studies in pigs have greatly enhanced the understanding of natural defense.

Natural and physical barriers

Respiratory diseases are a major cause of morbidity and mortality in young pigs, and studies of respiratory defense mechanisms in this species are of considerable practical significance. The anatomical organization of the respiratory tract provides an important barrier against infection and disease. In the upper respiratory tract, bony, scroll-like turbinates divide the nasal cavity into conducting channels that induce turbulence in the inspired air resulting in impaction of particles larger than 10 μm. The deposited particles are then removed from the respiratory tract by the ciliated mucus membrane. At the bronchial level, further turbulence filters out particles of 3–10 μm, while smaller particles may enter the alveoli. The mucociliary apparatus extends from the nostrils to the terminal bronchioles and consists of cilia that beat in the inner sol layer of the mucus. Particles may be trapped in the outer gel layer. These particles are carried to the pharynx or the bronchus-associated lymphoid tissue by the action of cilia on the inner sol layer. Such clearance can be augmented by the cough reflex. The two mechanisms combined have been estimated to clear up to 90% of material deposited in the tracheobronchial tree. The tracheobronchial tree can be physically cleared within 1 hour.

Soluble antimicrobial factors

Soluble factors also play an important role in nonspecific defense against bacteria and viruses. For example, within the gastrointestinal tract lactoperoxidase has been shown to protect neonatal colostrum-deprived piglets from experimental *Escherichia coli* infection. Systemically, microbial infection may trigger the activation of complement by the alternative pathway, resulting in phagocytosis and killing of the organism. In comparison with opsonization mediated by specific antibody and activation of the classical pathway of complement, this process is relatively slow and inefficient, but it still provides an important first line of defense. Following phagocytosis, macrophages may be activated by bacterial LPS to produce monokines. This structurally diverse group of molecules includes IL-1, IL-6, IL-8, IL-12 and TNF-α. Many of these have been cloned for the pig (see Section 5) and appear to have properties similar to those reported for other species. Porcine IFN-α and IFN-β have also been analyzed (for review see LaBonnardiere *et al.*, 1994). In pigs, interferon production has been detected following infection with a range of viruses including rotavirus, transmissible gastroenteritis virus (TGEV) and pseudorabies.

Granulocytes

Circulating pig neutrophils express cell-surface antigens SWC1, SWC3, SWC8, and react with the unclustered antibodies MIL2 and MIL4 (for references to these antigens and monoclonal antibodies see Section 3). Expression of the MIL4 epitope is bimodal (Haverson *et al.*, 1994). These antibodies clearly recognize a series of myeloid-specific cell-surface molecules that currently have no obvious human homologues. *In vivo* pig models have demonstrated rapid extravasation of neutrophils following intravenous challenge with live bacterial or local intradermal challenge with lectin. Sepsis-induced tissue neutrophils upregulate surface CD18 and Fc receptors. *In vivo* this can be induced by a number of mediators including leukotriene B_4, and TNF-α and can be blocked by receptor antagonists (Windsor *et al.*, 1994; Vandermeer *et al.*, 1995). Localization in tissues is dependent on expression of E-selectin by endothelial cells (Ridings *et al.*, 1995; Binns *et al.*, 1996). Neutrophils isolated by bronchoalveolar lavage are uniformly MIL4$^+$, indicating selective transendothelial migration of progressive differentiation.

A number of peptides with antimicrobial activity, originally identified from intestine or bone marrow, have now been characterized from pig neutrophils. These peptides, termed cathelicidins (or bactenecins), possess a conserved, cathelin *N*-terminal domain and divergent, *C*-terminal protegrin domain and have probably arisen as a result of extensive gene duplication (Zanetti *et al.*, 1995). Genetic analysis has also demonstrated 5′ binding sites for IL-6 and NF-κB, suggesting that transcription in neutrophils may be upregulated as part of the acute-phase response (Zhao *et al.*, 1995a, b). Synthetic protegrin domains have been demonstrated to promote phagocytosis of *Salmonella* by neutrophils (Shi *et al.*, 1996) and to kill Gram-positive and Gram-negative bacteria, probably by membrane permeabilization (Storici *et al.*, 1994; Tossi *et al.*, 1995). *In vivo*, the LPS-binding peptide CAP18, blocked LPS-induced increases in serum TNFα in pigs and attenuated concurrent shock (cardiac output, arterial pressure and $P\text{O}_2$) (Vandermeer *et al.*, 1995).

Despite their potential for bacterial phagocytosis and bacterial killing, the involvement of extravasated neutrophils in local tissue damage has been demonstrated in septic shock and ischemic reperfusion models. The mechanism of damage is still unclear but is likely to be associated, in part, with release of oxygen radicals, although the ability of alveolar PMN to generate O_2^- was decreased compared with circulating neutrophils in sepsis (Windsor *et al.*, 1994). In the lung, lipid peroxidation resulted in decreased surfactant activity (Gilliard *et al.*, 1994). Release of platelet activating factor (PAF) by activated neutrophils may also contribute to tissue injury, although depletion studies have indicated that platelet deposition on endothelium can occur independently of neutrophils (Merhi *et al.*, 1994) and that PAF may be required for their extravasation.

Eosinophils are present in relatively low numbers in the blood of pigs but accumulate in the intestinal lamina propria. Both peripheral blood and intestinal eosinophils express the SWC3 and SWC8 but not the SWC1 or MIL2 antigens. Expression of LFA-1 is lost between blood and tissue (Magyar *et al.*, 1995). All tissue eosinophils express the MIL4 antigen. Morphologically, intestinally derived eosinophils are larger, lighter, and contain fewer granules, comparable to the activated eosinophils described in allergic and parasitized humans (Haverson *et al.*, 1994). Phagocytosis has not been described by pig eosinophils but the initial isolation of antimicrobial peptides from the intestinal mucosa, a site containing few neutrophils but large numbers of eosinophils, suggests that they may be capable of killing bacteria.

Comparison of blood and intestinal mast cells has demonstrated low levels of the SWC3, SWC8, and MIL4 antigens (Haverson *et al.*, 1994). The demonstration of acute, anaphylactic shock in sensitized pigs injected with *Ascaris* antigen implies activation by ligation of surface IgE as in other species. It is also likely that pig mast cells associate with, and can be activated by, local innervation.

14. Complement System

Components of the porcine complement system

The complement system of the pig has received surprisingly little attention despite the economic importance of this species and the potential importance of complement in diseases of the pig. The porcine complement system is similar to the human complement system, with functional classical and alternative pathways in serum. Furthermore, pig serum can restore lytic activity to human serum depleted of individual components, making it likely that the pig complement system closely resembles that in man. Most complement components have been specifically detected (Table XI.14.1), either functionally or antigenically (Day *et al.*, 1969; Geiger *et al.*, 1972; Barta and Hubbert, 1978; Nakashima *et al.*, 1992; van den Berg and Morgan, 1994; Davies *et al.*, 1994; Høgåsen *et al.*, 1995). The sequence of factor B, clusterin and MCP have been determined by cDNA cloning (Table XI.14.1). Factor D, properdin and CR1 have not been specifically detected, but are most likely present. Close association of C2, Bf and C4 within the swine major histocompatibility complex (SLA) locus has been reported (Lie *et al.*, 1987). This association is analogous to the known linkage of C2, Bf and C4 of human and mouse complement to their respective histocompatibility loci. This linkage to the SLA locus is reflected in the markedly different levels of haemolytic complement activity found in pigs with different SLA haplotypes (Vaiman *et al.*, 1978).

In man, complement plays an important role in the clearance of immune complexes, tagging them (through CR1) to erythrocytes and aiding their elimination in liver and spleen. Immune complex clearance in pigs is also complement dependent but, in contrast to the situation in man, there is little or no binding of complexes to erythrocytes and the lung is the major site of elimination (Davies *et al.*, 1995).

Only low levels of complement are present in newborn pigs, but the serum hemolytic activity approaches adult levels by 1 week after birth. The serum complement level increases much more rapidly in nursing piglets compared with artificially fed piglets, suggesting colostral uptake of complement (Rice and L'Ecuyer, 1963).

Species restriction of the alternative pathway has been thoroughly studied (Ish *et al.*, 1993). Porcine serum lysed erythrocytes from (listed in decreasing titer) guinea pig, mouse, duck, turkey, rat, goose, burro, human, chicken, rabbit and horse; no lysis of erythrocytes from the following species was observed: ferret, ox, dog, cat, goat or

Table XI.14.1 Known components of the porcine complement system

Component	*Chromosome*	*Homology with Man*	*Reference for sequence*
C1			
C2	7		
C3		C3a 69%	Corbin and Hugli (1976)
C4	7		
C5		C5a 68%	Gerard and Hugli (1980)
C6			
C7			
C8			
C9			
Factor B	7	87%	Peelman *et al.* (1991)
Factor I			
Factor H		*N*-terminal 58%	Høgåsen *et al.* (1995)
C1 inhibitor			
C4BP			
Vitronectin (S protein)		*N*-terminal 67%	Nakashima *et al.* (1992)
Clusterin		72%	Diemer *et al.* (1992)
DAF			
MCP		42%	Toyomura *et al.* (1996)
CD59		*N*-terminal 46%	van den Berg *et al.* (1995)

sheep. Porcine erythrocytes were lysed by serum from (listed in decreasing titer) ferret, mink, fox, dog, armadillo, raccoon, opossum, mouse and rat but not by serum from human, cat, ox, goat, sheep, monkey, guinea pig, hamster, rabbit or horse.

Complement deficiency states

C6 deficiency was detected in two out of 241 pigs examined in a Japanese study of C6 polymorphism (Shibata *et al.*, 1993). The pigs had no classical pathway hemolytic activity. There was no specific clinical data, but since the animals were from ordinary breeding stocks, they were probably healthy. Inheritance was not investigated (Table XI.14.2).

Hereditary factor H deficiency has been reported in the Norwegian Yorkshire breed (Høgåsen *et al.*, 1995). The mode of transmission is autosomal recessive with full penetrance (Jansen *et al.*, 1995). The gene frequency in Norwegian Yorkshire was 0.07. The condition is especially interesting since all deficient animals developed lethal membranoproliferative glomerulonephritis (MPGN) type II (Jansen, 1993; Jansen *et al.*, 1993, 1995). Neither factor H deficiency nor MPGN type II has previously been described in species other than in humans. Factor H-deficient pigs appear healthy at birth, but develop rapidly progressive renal failure within a few weeks. The median survival among 25 animals was only 37 days, and maximum survival was 72 days (Jansen, 1993). Partial factor H replacement by regular transfusions of normal porcine plasma significantly increased median survival to 82 days (Høgåsen *et al.*, 1995). Human factor H deficiency is also associated with MPGN type II, but much less closely, since some had other types of nephritis or no (clinical) nephritis at all. Thus, factor H deficiency is obviously much more dangerous for pigs than for humans (Table XI.14.2).

Regulators of complement activation

In all mammals, complement is tightly regulated by a battery of fluid-phase and cell membrane proteins (Morgan and Meri, 1994). Little is known about the presence and distribution of complement regulatory molecules in the pig. However, the recent interest in using pig organs for xenotransplantation into humans has prompted an examination of complement regulation in the pig. In order to circumvent complement-mediated hyperacute rejection, an inevitable consequence of pig–human transplants, pigs are now being bred that express human complement regulators on endothelium (Morgan, 1995). The contribution of the endogenous pig inhibitors to complement regulation remains unassessed. Of the pig membrane inhibitors, only CD59 and membrane cofactor protein (MCP, CD46) have been purified and characterized (van den Berg *et al.*, 1994; van den Berg *et al.*, 1997). These regulatory molecules function in a manner indistinguishable from their human analogues and function across species – pig MCP can function as a cofactor for the degradation of human C3b by human factor I and pig CD59 is an efficient inhibitor of human C8 and C9.

Table XI.14.2 Complement deficiency states

	C6 deficiency	*Factor H deficiency*
Inheritance	?	Autosomal recessive
Frequency	1% (2/241)	0.5% (13.5% heterozygotes)
Country	Japan	Norway
Breed	Large White and Berkshire	Yorkshire
Clinical features	Healthy (?)	MPGN type II[a]

[a] Membranoproliferative glomerulonephritis type II.

Infectious disease

Only a few studies in the pig have examined the involvement of complement in disease. In humans, complement plays important roles in the innate resistance against viral and bacterial infection and in immune-complex handling. A role for complement has been suggested in two important viral diseases of the pig. Aujeszky's disease is caused by pseudorabies virus; in infected pigs, serum complement levels are markedly reduced (Silmanowicz *et al.*, 1978), suggesting that the virus activates complement. *In vitro*, it has been shown that viral surface glycoproteins bind porcine C3 but not human C3 (Huemer *et al.*, 1992). Whether complement activation has a protective effect in this disease has not been shown. Evidence for a protective role of complement has been provided in a second viral disease of pigs, African swine fever. Pigs infected with African swine fever virus have low complement levels and immune complexes containing viral antigens, antibody and C3 are found in kidney, liver and lungs (Slauson and Sanchez-Vizcaino, 1981; Fernandez *et al.*, 1992). Cells infected with African swine fever virus are readily lysed by the classical pathway of pig complement *in vitro*, suggesting that complement may be protective (Norley and Wardley, 1982; Wardley and Wilkinson, 1985).

Animal models of human disease

Pigs have been used to generate models of many human diseases that are known to be dependent on complement activation, including adult respiratory distress syndrome (ARDS; Borg *et al.*, 1984), sepsis models and cardiopulmonary bypass. For example, when endotoxin was used to induce ARDS in pigs, the highest responders were found to have the highest initial complement levels (Modig and Borg, 1986). Complement depletion using cobra venom factor (CVF) is protective in this model and also reduced the effects of injection of *Pseudomonas aeruginosa* in a pig model of severe septic acute respiratory failure (Derhing *et al.*, 1987). In a model of pulmonary hypertension induced by the systemic activation of complement using different forms of CVF, all forms of CVF (C3-activating) caused pulmonary hypertension but only forms that also activated C5 (generating C5a) induced pulmonary hypertension (Cheung *et al.*, 1989).

Xenotransplantation

Owing to the increasing shortage of organs for human transplantation, the pig is under evaluation as a potential future xenotransplant donor. The violent hyperacute rejection that appears when porcine organs are perfused with human blood is caused by high titers of natural human antibodies reactive with carbohydrate structures present on porcine vascular endothelial cells; these activate complement which damages the cells. Because membrane-bound complement inhibitors show more or less species restriction, vigorous efforts have been made to make strains of transgenic pigs that express a high density of one or more of the human regulators DAF, MCP and CD59 on the vasculature. This strategy has so far been promising (McCurry *et al.*, 1995).

15. Mucosal Immunity

Oropharynx

The major immunological component of the oropharynx is the palatine tonsil. This consists of organized lymphoid tissue covered by stratified squamous epithelium but penetrated by branching crypts covered with nonkeratinized epithelium. The organized tissue contains B cell follicles and T cells. The crypt epithelium is a lymphoepithelium containing goblet cells, microfold cells (M cells) and intraepithelial lymphoid cells (Belz and Heath, 1996). Plasma cells appear rapidly, within a few days after birth (Brown and Bourne, 1976b). The majority of plasma cells are IgG^+, although some IgM^+ and IgA^+ cells are also present (Bradley *et al.*, 1976a). The lamina propria of the oropharynx contains relatively few bone-marrow derived cells.

Small intestine

Peyer's patch

Approximately 11–26 discrete patches have been reported scattered through the jejunum, the number and position of which probably remains constant throughout life (Pabst *et al.*, 1988). Each patch contains multiple B cell follicles separated by interfollicular areas dominated by T cells. Plasma cells containing IgM, IgG and IgA are present in the dome and between the bases of the follicles (Brown and Bourne, 1976a). Microfold cells have been described in the overlying lymphoepithelium, and are demonstrable by staining for cytokeratin-18 (Gebert *et al.*, 1994). Between the epithelium and the follicle is a dome region containing discrete cells expressing high levels of MHC class II antigens and with morphologic characteristics of dendritic cells (Wilders *et al.*, 1983).

Peyer's patches are not fully formed at birth. Accumulations of leukocytes are visible but extend and organize rapidly in the first few days of life. Increases in size and organization are at least partially dependent on antigen (Rothkötter and Pabst, 1989). A single, large Peyer's patch develops in the ileum and is heavily populated with B cells in young animals. With age, this patch regresses, becoming a series of isolated patches and its content of B and T cells becomes comparable to that of the discrete patches.

Discrete, ileal Peyer's patches probably serve as inductive sites for T and B cell responses. Recirculation studies have demonstrated low levels of T cell traffic through the ileal patch (Binns and Licence, 1985), suggestive of a primary B cell organ comparable to that extensively studied in the sheep.

Mesenteric lymph node

This node has a similar architecture to that described for systemic lymph nodes. The number of plasma cell numbers is low in newborn animals but increases to adult levels by 2-weeks-old (Brown and Bourne, 1976b). Memory B cells are present following oral immunization with transmissible gastroenteritis (TGE) (Berthon *et al.*, 1990).

Lamina propria

The intestinal lamina propria is heavily populated with leukocytes in mature animals. Plasma cells and B cells predominate around the crypts and T cells in the villi. Plasma cells are predominantly IgA^+ and IgM^+ but some IgG^+ cells are also present (Brown and Bourne, 1976a). In the T-cell-dominated villi there is clear spatial separation of $CD8^+$ cells in and under the epithelium, with $CD4^+$ cells in the lamina propria deep in the capillary plexus (Vega-Lopez *et al.*, 1993). There is extensive expression of MHC class II antigens in the villi and crypts, although a proportion of this is associated with capillary endothelium (Wilson *et al.*, 1996). Large numbers of eosinophils and cells of the mast cell/basophil series are also present (Pabst and Beil, 1989; Haverson *et al.*, 1994). Resident mast cells and eosinophils are larger than elsewhere, resembling the hypodense, 'activated', cells described in parasitic and allergic disease. Mast cells from the mucosa are biochemically distinguishable from those in other sites and may be comparable to mucosal mast cells described in other species (Xu *et al.*, 1993).

Young piglets are born with virtually no T cells or plasma cells in the villus lamina propria. Development appears to be antigen-driven and occurs in phases. Plasma cells accumulate in the first 4 weeks of life. The majority of such cells are IgM^+ initially but IgA^+ cells rapidly predominate (Brown and Bourne, 1976b). γ/δ T cells are present in the fetus (Trebichavsky *et al.*, 1995), although in low numbers. During the first week, $CD2^+4^-8^-$ cells enter the intestine; $CD2^+4^+$ cells appear at around 3 weeks, $CD2^+8^+$ at around 7 weeks (Rothkötter *et al.*, 1991; Bianchi *et al.*, 1992; Vega-Lopez *et al.*, 1995). The number of cells expressing myeloid-associated antigen SWC1 also increases with age, probably reflecting an increase in granulocyte (eosinophil, mast cell) numbers.

IgA is the major immunoglobulin in intestinal secretions and the majority is probably synthesized locally in the lamina propria. Crypt epithelial cells contain IgA and IgM but not IgG, indicating selective transport into the lumen within the crypts (Brown and Bourne, 1976a). Luminal IgA is associated with secretory component, consistent with epithelial transport using secretory component as a receptor (Bourne *et al.*, 1971). IgA is also selectively transported into bile (Sheldrake *et al.*, 1991). Immunoglobulin G is present in the intracellular spaces and is likely to be important during local inflammation. Within the T cell population, antigen-specific responses to pathogens have been difficult to demonstrate, even following enteric infections (Brim *et al.*, 1995). Activation of isolated lamina propria T cells *in vitro* has demonstrated transcription of IL-4 but failure to transcribe or secrete IL-2, despite expression of high levels of wCD25 (Bailey *et al.*, 1994b). These patterns of cytokine secretion suggest highly differentiated cells susceptible to apoptosis, perhaps involved in immunoregulation rather than expression of memory. Macrophages isolated from the lamina propria can trigger T cells in primary mixed-lymphocyte reactions, but poorly by comparison with systemic macrophages (Kambarage *et al.*, 1994).

Epithelium

The majority of lymphocytes in the intestinal epithelium (IEL) express CD2. In mature animals a high proportion also express $CD8^+$. Available antibodies have not allowed identification of $CD8^+\alpha\alpha$ homodimer IEL described in humans.

Development of the epithelial compartment is comparable to that of the lamina propria. In young pigs, IEL are mostly $CD2^-CD4^-CD8^-$. During the first few weeks of life, $CD2^+CD4^-CD8^-$ appear but $CD8^+$ IEL do not appear until 7 weeks onwards (Whary *et al.*, 1995). The ability of IEL from young piglets to respond to mitogens is poor and develops with time (Wilson *et al.*, 1986b), although this can be delayed by early weaning. Failure to proliferate may reflect their inability to produce IL-2 at this age since the cells are capable of responding to IL-2 *in vitro* (Whary *et al.*, 1995). Intraepithelial cells from mature pigs can produce IL-2 and interferon following activation *in vitro* and also engage in limited cytotoxicity (Wilson *et al.*, 1986a).

Respiratory tract

The architecture of bronchial lymph nodes is similar to that described for systemic LN. In mature animals, IgG^+ plasma cells predominate over IgA^+ and IgM^+ cells. Newborn piglets have few plasma cells but the number increases slowly up to 5 weeks old (Brown and Bourne, 1976b).

Clusters of leukocytes present in the walls of bronchi and bronchioli form the bronchial-associated lymphoid tissue (BALT). Both T cells and Ig-positive cells are present. These, however, are not organized into follicles comparable with lymph nodes or with the intestinal Peyer's patches (Delventhal *et al.*, 1992a). Plasma cells are predominantly IgA^+ with some IgG^+ and IgM^+ (Bradley *et al.*, 1976b). The development of BALT is antigen-dependent.

IgA is the major immunoglobulin in tracheal secretions and the majority is produced locally (Morgan *et al.*, 1980). In the bronchi, IgG is the predominant immunoglobulin in secretions and a significant proportion is also synthesized locally. There is evidence for localization of primed T and B cells following intestinal exposure to pathogenic microorganisms and following intraperitoneal immunization (Sheldrake, 1990; Hensel *et al.*, 1995).

Intraalveolar leukocytes

In normal animals, the majority of bronchoalveolar cells are macrophage-like (Haverson *et al.*, 1994). Alveolar macrophages can phagocytose bacteria and produce IL-1 and TNF-α but may be relatively poor at intracellular killing (Chitko-McKown *et al.*, 1991). Following challenge with bacteria the numbers of T-cells, plasma cells and granulocytes increases (Delventhal *et al.*, 1992b). There is evidence for recirculation of lymphocytes, particularly of the Null-cell subset, from the bronchoalveolar spaces back into regional draining lymph nodes (Pabst and Binns, 1995).

Intravascular leukocytes

Macrophages are normally apposed to the lung capillary endothelium. Intravascular macrophages are capable of phagocytosis, intracellular killing and production of IL-1 and TNF-α (Chitko-McKown *et al.*, 1991). Their ability to process and present antigen to T cells is unclear. A large population of T cells is present within the vasculature of the pig lung, apparently retained by interaction with endothelial cells.

Reproductive tract

Plasma cells are present throughout the lamina propria. In the endometrium, IgG^+ cells predominate while in the cervix and vagina IgA^+ cells are more common (Hussein *et al.*, 1983a). The total numbers of plasma cells increase from the Fallopian tube to the vagina and during estrus. Few leukocytes are present in the endometrium of non-pregnant animals but the number increases during early pregnancy, following implantation (White and Binns, 1994). Infiltrating cells are primarily of the $SWC1^+$, myeloid series: γ/δ T cells are also present. Immunoglobulins are present in uterine fluids. Despite the presence of plasma cells, the majority of IgG is serum-derived rather than locally synthesized. Similarly, more than half of the secreted IgA is also serum-derived (Hussein *et al.*, 1983b).

Mammary glands

Tissue leukocytes are rare in the mammary gland of normal, nonlactating sows. After farrowing, the number of plasma cells increases rapidly (Brown *et al.*, 1975). Cells containing IgM, IgG and IgA are present but the numbers of IgA^+ cells is greatest at all stages of lactation. This is reflected in the composition of milk, which contains more IgA than IgG or IgM. During lactation, leukocytes appear in milk; these are predominantly neutrophils capable of limited phagocytosis (Evans *et al.*, 1982), but also some lymphocytes. Antigen-specific T cells have been detected in milk following intramammary immunization.

Functional organization

The immune system must be capable of induction and expression of protective responses against pathogenic microorganisms infecting across mucosal surfaces. However, in pigs as in other species, significant quantities of fed protein are absorbed immunologically intact across the intestinal mucosa (Wilson *et al.*, 1989; Telemo *et al.*, 1991). Immune responses to harmless dietary components must be regulated to prevent tissue damage and impaired function of mucosal surfaces (absorption of macromolecules, gaseous exchange), and systemic tolerance to fed proteins (oral tolerance) has been demonstrated in the pig (Newby *et al.*, 1979; Bailey *et al.*, 1993). Thus, the immunological structures associated with mucosal surfaces have evolved to generate different types of response to pathogens and to harmless environmental antigens. The sites at which such value judgments are made are unclear in any species, but models of the function of mucosal-associated lymphoid tissues must account for both of these outcomes.

The mucosal-associated lymphoid tissue can be divided into two major compartments: that consisting of the organized lymphoid structures (tonsils, Peyer's patches, mesenteric and bronchial lymph nodes, etc.) and that occurring in tissues specialized for other functions (the intestinal lamina propria, pulmonary intravascular leukocytes, etc.). The lymphoid tissue comprising the mucosal immune system of the pig has received a significant amount of attention. In particular, considerations of scale have made studies of the diffuse lymphoid components of the lung, lamina propria and the mucosal vasculature possible. In the conventional model, the organized tissues are inductive sites populated by naive cells: following priming they recirculate as memory cells through the diffuse effector sites, such as the intestinal lamina propria. However, recent evidence suggests that the conventional model of recirculation of naive and memory T cells may be less well understood than had been thought (Westermann and Pabst, 1996; Meeusen *et al.*, 1996). The unusual migration pathway of T-cells within the lymph nodes of the pig has made interpretation of experiments in this species difficult (Binns *et al.*, 1985). Both T and B cells do appear in afferent lymph draining from the intestine and there is evidence for local proliferation, particularly of T cells (Rothkötter *et al.*, 1995). Retransfused cells from afferent lymph return selectively to the intestine, although whether these are from Peyer's patch or lamina propria has not been determined, nor whether this is the sole route of exit (Rothkötter *et al.*, 1993). Following oral or intraperitoneal immunization, increases in T and B cells and in antigen-specific immunoglobulins have been seen in bronchoalveolar lavage fluid (Sheldrake, 1990; Hensel *et al.*, 1995). Thus, there is evidence for spread of effector cells from one mucosal surface to another – the common mucosal system.

Although the organized lymphoid tissues of the newborn piglet rapidly develop and express immunological function, the immunological architecture of diffuse effector sites such as the intestinal lamina propria can take up to 7–9 weeks to develop, depending on antigenic challenge. While this is consistent with a need to produce effector cells, the time course is delayed, despite continuous environmental challenge. At 3 weeks old the young piglet is capable of active immune responses to live virus and to dietary components with a magnitude comparable to that with systemic antigenic challenge (Welch *et al.*, 1988; Derbyshire and Lesnick, 1990; Bailey *et al.*, 1994). However, tolerance to continuously fed proteins is not fully achieved until after 8 weeks of age: similarly, the magnitude of primary responses to novel dietary components is reduced with age (Wilson *et al.*, 1989; Miller *et al.*, 1994). This early failure to regulate responses to harmless dietary proteins or commensal bacteria has been proposed as contributory to post-weaning diarrhea in early-weaned piglets (Miller *et al.*, 1984; Li *et al.*, 1991). The ability to regulate such responses develops in parallel with the immunological architecture of the lamina propria. Together with recent studies on the nature of local T cells (Bailey *et al.*, 1994), it appears likely that, in the pig, the diffuse mucosal immune system is involved in immunoregulation in addition to expression of effector function.

16. Immunodeficiencies

There are no reports of primary immunodeficiency in swine. Because the care and evaluation of an animal with immunodeficiency would be expensive, such animals that would probably present with infections would be killed. As noted above there is a report of C6 deficiency in swine with a resultant absence of the classical complement pathway. The animals were apparently asymptomatic. Acquired immunodeficiencies with relevant references are listed (Table XI.16.1).

17. Tumors of the Immune System

Spontaneous tumors of pigs are quite rare. It has been hypothesized that this rarity is due to majority of pigs being slaughtered before they reach an age at which cancer frequency would increase (Fisher *et al.*, 1978). The majority of reports of cancers are those seen at slaughter houses. In these series of observations, the most common tumor is one of the immune system, the lymphosarcoma, which has been reported to occur at 15–60 cases per million slaughtered animals (Busse *et al.*, 1978; Marcato, 1987). Other immune system cancers of the pig, have included thymoma, plasmacytoma, myeloid leukemia, and mast cell leukemia (Kadota *et al.*, 1986; Marcato, 1987).

The lymphosarcomas are a heterogeneous group of tumors that have been characterized as both of B-cell and T-cell origin. Recently non-T, non-B lymphosarcomas have been described that neither rosette with sheep red cells nor express surface immunoglobulin (Kadota *et al.*, 1990). A major deficiency of all of the reports on lymphosarcomas in the pig to date is the primitive phenotyping of the cells since the cases were analyzed prior to the availability of anti-T-cell and B-cell monoclonal antibodies.

The etiology of porcine lymphosarcoma falls into two groups: one hereditary and the other spontaneous. The hereditary form occurs in Large White pigs and presents before 6 months of age (McTaggart *et al.*, 1971, 1979; Head *et al.*, 1974). The animals present with stunted growth and lymph node enlargement. In the peripheral circulation, lymphocytes predominate. All animals die before 15 months of age and never reach sexual maturity. The lymph nodes draining the gut and lung are primarily involved with liver and spleen in late stages (Head *et al.*, 1974). Serum γ-globulin levels are elevated and IgG has been found in the urine suggesting a B cell phenotype (Imlah *et al.*, 1979). The disease has been shown to be autosomal recessive in that 25% of progeny of carriers developed lymphoma (McTaggart *et al.*, 1971, 1979). In one study, C-type viral particles have been seen in one case, raising the issue of a viral etiology on a permissive background (Campbell, 1977). The spontaneous type can occur in older animals (Bostock and Owen, 1973) and presents in several forms such as multicentric or alimentary (Bostock and Owen, 1973; Ito and Fujita, 1977). The majority of these have been reported to have B-cell characteristics of Ig production (Kadota *et al.*, 1986; Hayashi *et al.*, 1988; Kadota *et al.*, 1988; Tanimoto *et al.*, 1994; Nakajima *et al.*, 1989). A C-type RNA tumor virus has been isolated from a spontaneous porcine lymphosarcoma (Busse *et al.*, 1978; Suzuka *et al.*, 1985). Antibodies against it have been rare in screened normal animals as well as those with lymphosarcoma. The virus may be an endogenous virus and its causative association with lymphosarcoma is still not clear (Busse *et al.*, 1978).

Thymomas occur only rarely with one series finding none in almost 4 million slaughtered pigs (Sandison *et al.*, 1969). Three cases have been identified (Kadota *et al.*, 1986, 1990). All three cases were identified as T cell in origin by the ability of the malignant cells to rosette with sheep red blood cells and the absence of surface Ig. No clinical description is provided about the health of the

Table XI.16.1 Acquired immunodeficiencies

Type	*Age of animals*	*Effect*	*Reference*
Surgical			
Thymectomy	Neonate	Decreased Null cell	Binns *et al.* (1977)
Splenectomy	3 month	Minimal effect	Izbicki *et al.* (1989)
Radiation	Neonatal	Dose dependent	Case *et al.* (1972)
	3 months	Lymphopenia	Pennington *et al.* (1988)
Stress			
Cold	1–2 months	Increased antibody response	Blecha *et al.* (1981)
Crowding	1–2 months	Decreased *in vivo* PHA response	Yen *et al.* (1987)
Nutrition			
Vitamin E	Neonate	Increased *in vitro* PHA response	Larsen *et al.* (1981)
Selenium	Neonate	Increased *in vitro* PHA response	Larsen *et al.* (1981)
Colostrum deprived	Neonate	Increased infections	Lecce *et al.* (1991)

animals other than to report that one was normal. The etiology is unknown.

In contrast to the usual age of animals affected with lymphosarcoma, myeloid leukemia is a disease of older animals (Kadota *et al.*, 1987). Various levels of maturation have been described. Undifferentiated tumors may be difficult to distinguish from lymphosarcoma; the greenish color of the neoplastic tissue from myeloperoxidase can be helpful (Kashima *et al.*, 1982). C-type viral particles have been found associated with these tumors (Kadota *et al.*, 1984).

Mast cell leukemia with systemic mastocytosis has been reported in a 9-year-old pig (Bean-Knudsen *et al.*, 1989). Mast cell tumors are the more common presentation and are themselves quite rare. The etiology is unknown.

18. Autoimmunity

Two forms of glomerulonephritis occurring spontaneously in pigs have been reported. The first, the inherited Factor H deficiency in Norwegian Yorkshire breed is described above (Section 14). An identical form of membranoproliferative glomerulonephritis has been found in man (Levy *et al.*, 1986). The second form of glomerulonephritis is structurally related to human IgA nephropathy. The disease has been reported in asymptomatic Japanese swine slaughtered at about 6 months old at very high frequency (26%, Shirota *et al.*, 1986). On closer examination, animals with disease can have mild proteinuria and microhematuria, in keeping with the human form (Yoshie, 1991). Histologically, the lesions in the kidney are found within the glomerulus resulting in mesangial enlargement. On immunofluorescence, granular deposits of IgA, IgG, IgM, C3 are found in the mesangial area (Shirota *et al*, 1986, 1988; Yoshie, 1991). Ultrastructure examination disclosed spherical microparticles in the mesangium, intramembranous, subendothelial, and subepithelial areas that were strikingly similar to lesions found in human IgA disease (Shirota *et al.*, 1988; Yoshie, 1991). Circulating immune complexes of IgA and *Mycoplasma hyorhinis* antigen have been identified, suggesting a possible etiological agent (Yoshie, 1991). Other chronic infections such as African Swine fever virus have also been associated with immune complex glomerulonephritis.

Autoimmune thrombocytopenia and hemolytic disease in the neonatal period have been described (Nordstoga, 1965; Lie, 1968; Linklater *et al.*, 1973; Dimmock *et al.*, 1982). Similar conditions, particularly hemolytic, have been noted in neonatal humans and constitute, in the extreme case, the syndrome of erythroblastosis fetalis. The human disease is caused by the passive transfer of maternal antibodies against paternal antigens, particularly Rh, to the fetus *in utero*. The different structure of the porcine placenta (reviewed above) that prohibits passive transfer of maternal antibodies, protects the pig fetus. However, after birth, the large amount of maternal Ig absorbed from the colostrum results in the hemolytic and thrombocytopenic state that develops in the first several days. An interesting additional and not completely explained feature, is the recurrence of thrombocytopenia that can develop at about day 11 (Linklater *et al.*, 1973). This state is marked by the absence of megakaryocytes, and possibly associated with pancytopenia including neutrophils, basophils, eosinophils, and monocytes. It is possible that this later stage is the result of a cellular graft-versus-host disease caused by the transfer of maternal lymphocytes contained within colostrum (see above). These cells would need to expand in the presence of paternal antigens in the neonate to a number sufficient to cause disease. Similar cases of delayed antibody mediated graft-versus-host disease, have been reported in humans after organ transplant and probably result from the transfer of immunocompetent cells.

Rheumatoid arthritis, a crippling disease common to man, has a related disease in pigs that has been studied as such for over 25 years (Drew, 1972). The disease is common in pigs kept under field conditions. Human rheumatoid arthritis is characterized histologically by villous hypertrophy, proliferation of superficial synovial cells, a marked infiltrate of lymphocytes with $CD4^+$ cells more common than $CD8^+$, and increased expression of human leukocyte antigen (HLA)-DR; there is an associated autoantibody to Ig called rheumatoid factor, and rheumatoid (subcutaneous) nodules. The porcine form is histologically very similar (Sikes *et al.*, 1969; Drew, 1972). The infiltrating cells are predominantly $CD4^+$ and MHC class II expression is increased (as reviewed in Franz *et al.*, 1995). The detection of rheumatoid factor and nodules has been less consistent (Drew, 1972). The etiology of the human disease is unknown. The porcine form has been shown to be both spontaneously associated with, and experimentally to be induced by, bacterial infections. Two different bacterial organisms have been isolated from involved joints: *Erysipelothrix rhusiopathiae* (Drew, 1972; Franz *et al.*, 1995) and *Mycoplasma hyorhinis* (Barden *et al.*, 1971, 1973). Recent data suggests that *E. rhusiopathiae* can survive within and be isolated by lysis of chondrocytes. This may serve as a reservoir of antigen that may then be presented using the MHC class II antigen expressed on these cells (Franz *et al.*, 1995). Of the two agents responsible for porcine arthritis, *M. hyorhinis* is also associated with renal disease as noted above, and *E. rhusiopathiae* is also a pathogen in man. There is no evidence that *E. rhusiopathiae* is a causative agent of human rheumatoid arthritis.

19. Conclusions

Although the pig immune system has been under study for many years, knowledge is still in its infancy. The pig is now benefiting from the intense interest in its potential use as xenogeneic organ donor to man. This has resulted in an infusion of research support both from government and industry. Areas that need to be further studied include all aspects of the immune system.

Although rather few of the components and regulators of the pig complement system have been studied in isolation, it is clear that pigs possess a fully functional complement system that closely resembles that in man. The likely importance of complement in protection against infections and in the generation of immune complex diseases should encourage further investigation of pig complement. Given the current enthusiasm for the use of pig organs for transplantation in humans, there is a particular need for regulators of complement in the pig to be identified and characterized. The major source of complement in all species is the liver; if pig to human liver transplants are to be countenanced then a more thorough understanding of the mechanisms of pig complement, which will be present in the serum of the human recipient, becomes essential.

The interaction between the antigen-presenting cells and those of humans need to be analyzed to identify routes of tolerance induction, as perhaps through CD28 pathway blockade. The role of virus in the albeit rare lymphoid tumors of pigs needs to be clarified to prevent the transmission to the recipient.

The pig, however, deserves analysis for its own value, as a commercially important source of food. Analysis of inherited traits that either weaken or strengthen the response to specific pathogens will benefit pig farmers worldwide. In conclusion, the study of the immune system of the pig should be a fruitful area of research for many more years.

20. References

Andersson, L., Archibald, A. L., Gellin, J., Schook, L. B. (1993). First Pig gene mapping Workshop (PGM1). *Animal Genet.* **24**, 205–216.

Andresen, E. (1957). Investigations on blood groups of the pig. *Nord. Vet. Med.* **9**, 274–284.

Andresen, E. (1962). Blood groups in pigs. *Ann NY Acad. Sci.* **97**, 205–225.

Andresen, E. (1963). A study of blood groups of the pig. Thesis, Veterinary and Agricultural University, Copenhagen, Munksgaard, Copenhagen, Denmark.

Andresen, E. (1964a). The inheritance of the blood factors Ia and Ib in pig of the Duroc and Hampshire breeds. *Vox Sang.* **9**, 617–621.

Andresen, E. (1964b). Further studies on the H blood group system in pigs, with special reference to a new red-cell antigen Hc. *Acta Genet.* **14**, 319–326.

Andresen, E., Baker, L. N. (1963). Hemolytic disease in pigs caused by anti-Ba. *J. Anim. Sci.* **22**, 720–725.

Andresen, E., Baker, L. N. (1964). The C blood-group system in pigs and the detection and estimation of linkage between the C and J systems. *Genetics* **49**, 379–386.

Andresen, E., Wroblewski, A. (1961). The G and H blood group systems of the pig. *Acta Vet. Scand.* **2**, 267–280.

Anstee, D. J. (1990). The nature and abundance of human red cell surface glycoproteins. *J. Immunogen.* **17**, 219–225.

Aoki, M., Jonas, R. A., Nomura, F., Kawata, H., Hickey, P. R. (1995). Anti-CD18 Attenuates Deleterious Effects of Cardiopulmonary Bypass and Hypothermic Circulatory Arrest in Piglets. *J. Card. Surg.* **10**, 407–417.

Appleyard, G. D., Wilke, B. N. (1998). Porcine CD5 gene and gene product identified on the basis of inter-species conserved cytoplasmic domain sequences. *Vet. Immunol. Immunopathol.* (in press).

Appleyard, G. D., Wilke, B. N. (1998a). Lymphocyte activation events involving porcine CD5. *Clin. Exp. Immunol.* (in press).

Appleyard, G. D., Wilke, B. N. (1998b). Characterization of porcine CD5 and $CD5^+$ B cells. *Clin. Exp. Immunol.* **111**, 225–230.

Archibald, A. L., Haley, C. S., Brown, J. F. *et al.* (1995). The PigMaP consortium linkage map of the pig (Sus scrofa). *Mamm. Genome* **6**, 157–175.

Artursson, K., Lindersson, M., Wallgren, P. (1992). A study of the interferon-production in newborn piglets. *Vet. Immunol. Immunopathol.* **35**, 114–115.

Baarsch, M. J., Scamurra, R. W., Burger, K., Foss, D. L., Maheswaran, S. K., Murtaugh, M. P. (1995). Inflammatory cytokine expression in swine experimentally infected with *Actinobacillus pleuropneumoniae. Infect. Immun.* **63**, 3587–3594.

Baarsch, M. J., Wannemuehler, M. J., Molitor, T. W., Murtaugh, M. P. (1991). Detection of tumor necrosis factor in porcine alveolar macrophages using an L929 fibroblast bioassay. *J. Immunol. Methods* **140**, 15–22.

Bailey, M., Clarke, C. J., Wilson, A. D., Williams, N. A., Stokes, C. R. (1992). Depressed potential for interleukin-2 production following early weaning of piglets. *Vet. Immunol. Immunopathol.* **34**, 197–207.

Bailey, M., Stevens, K., Bland, P. W., Stokes, C. R. (1992). A monoclonal antibody recognising an epitope associated with pig interleukin-2 receptors. *J. Immunol. Methods* **153**, 85–91.

Bailey, M., Miller, B. G., Telemo, E., Stokes, C. R., Bourne, F. J. (1993). Specific immunological unresponsiveness following active primary responses to proteins in the weaning diet of pigs. *Int. Arch. Allergy Immunol.* **101**, 266–271.

Bailey, M., Miller, B. G., Telemo, E., Stokes, C. R., Bourne, F. J. (1994a). Altered immune responses to proteins fed after neonatal exposure of piglets to antigen. *Int. Arch. Allergy Immunol.* **103**, 183–187.

Bailey, M., Hall, L., Bland, P. W., Stokes, C. R. (1994b). Production of cytokines by lymphocytes from spleen, mesenteric lymph node and intestinal lamina propria of pigs. *Immunology* **82**, 577–583.

Baintner, K. (1979). Biochemical markers of selective and non-selective protein absorption, with special reference to intestinal proteolysis. In *Protein Transmission Through Living Membranes* (ed. Hemmings, W. A.), pp. 213–224. Elsevier/North Holland Amsterdam, The Netherlands.

Baintner, K. (1986). *Intestinal Absorption of Macromolecules and Immune Transmission from Mother to Young.* CRC Press, Boca Raton, FL, USA.

Barden, J., Decker, J. L, Dalgard, D. W., *et al.* (1973). Mycoplasma hyorhinis swine arthritis. III. Modified disease in Piney Woods Swine. *Infect. Immun.* **8**, 887–890.

Barden, J. A., Decker, J. L. (1971). *Mycoplasma hyorhinis* swine arthritis I. clinical and microbiologic features. *Arthritis Rheum.* **14**, 193–201.

Barman, N. N., Bianchi, A. T. J., Zwart, R. J., Pabst, R., Rothkötter, H. J. (1997). Jejunal and ileal Peyer's patches in pigs differ in their postnatal development. *Anat. Embryol.* **195**, 41–50.

Barta, O., Hubbert, N. L. (1978). Testing of hemolytic complement components in domestic animals. *Am. J. Vet. Res.* **39**, 1303–1308.

Batten, P., Yacoub, M. H., Rose, M. L. (1996). Effect of human cytokines (IFN-γ, TNF-α, IL-1β, IL-4) on porcine endothelial cells: induction of MHC and adhesion molecules and functional significance of these changes. *Immunology* **87**, 127–133.

Bazan, J. F. (1992). Unraveling the structure of IL-2 (letter). *Science* **257**, 410–413.

Bean-Knudsen, D. E., Caldwell, C. W., Wagner, J. E., *et al.* (1989). Porcine mast cell leukemia with systemic mastocytosis. *Vet Pathol.* **26**, 90–92.

Bell, K. (1983). The blood groups of domestic mammals. In *Red Blood Cell of Domestic Mammals* (eds Agar, N.S., Board, P. G.), pp. 143–146. Elsevier, Amsterdam, The Netherlands.

Belz, G. T., Heath, T. J. (1996). Tonsils of the soft palate of young pigs – crypt structure and lymphoepithelium. *Anatomical Record* **245**, 102–113.

Berthon, P., Bernard, S., Salmon, H., Binns, R. M. (1990). Kinetics of the *in vitro* antibody response to transmissible gastroenteritis (TGE) virus from pig mesenteric lymph node cells, using the Elispot and ELISA tests. *J. Immunol. Meth.* **131**, 173–182.

Bertschinger, H. U., Bachmann, M., Mettler, C., Pospischil, A., Schraner, E. M., Stamm, M., Sydler, T., Wild, P. (1990). Adhesive fimbriae produced *in vivo* by *Escherichia coli* O139:K12(B):H1 associated with enterotoxaemia in pigs. *Vet. Microbiol.* **25**, 267–281.

Bertschinger, H. U., Stamm, M., Vögeli, P. (1993). Inheritance of resistance of oedema disease in the pig: experiments with an *Escherichia coli* strain expressing fimbriae 107. *Vet. Microbiol.* **35**, 79–89.

Bianchi, A. T. J., Zwart, R. J., Jeurissen, S. H. M., Moonen-Leusen, H. W. M. (1992). Development of the B- and T-cell compartments in porcine lymphoid organs from birth to adult life: an immunohistological approach. *Vet. Immunol. Immunopathol.* **33**, 201–221.

Binns, R. M. (1982). Organisation of the lymphoreticular system and lymphocyte markers in the pig. *Vet. Immunol. Immunopath.* **3**, 95–146.

Binns, R. M., Hall, J. G. (1966). The paucity of lymphocytes in the lymph of unanaesthetised pigs. *Br. J. Exp. Pathol.* **47**, 275–280.

Binns, R. M., Licence, S. T. (1985). Patterns of migration of labelled blood lymphocyte populations: Evidence for two types of Peyer's patches in the young pig. *Adv. Exp. Med. Biol.* **168**, 661–668.

Binns, R. M., Licence, S. T. (1990). Exit of recirculating lymphocytes from lymph nodes is directed by specific exit signals. *Eur. J. Immunol.* **20**, 449–452.

Binns, R. M., Pabst, R. (1988). Lymphoid cell migration and homing in the young pig: alternative immune mechanisms in action. In *Migration and Homing of Lymphoid Cells*, Vol. II (ed. Husband, A. J.), pp. 137–174. CRC Press, Boca Raton, FL, USA.

Binns, R. M., Pabst, R. (1994). Lymphoid tissue structure and lymphocyte trafficking in the pig. *Vet. Immunol. Immunopathol.* **43**, 79–87.

Binns, R. M., Pabst, R. (1996). The functional structure of the pig's immune system, resting and activated. In *Advances in Swine in Biomedical Research* (eds Tumbleson, M., Schook, L.), pp. 253–265. Plenum Press, New York, USA.

Binns, R. M., Pallares, V., Symons, D. B. A., *et al.* (1977). Effect of thymectomy on lymphocyte subpopulations in the pig. *Int. Archs. Allergy Appl. Immun.* **55**, 96–101.

Binns, R. M., Pabst, R., Licence, S. T. (1985). Lymphocyte emigration from lymph nodes by blood in the pig and efferent lymph in the sheep. *Immunology* **54**, 105–111.

Binns, R., Pabst, R., Licence, S. T. (1986). The behavior of pig lymphocyte populations in vivo. In *Swine in Biomedical Research* (ed. Tumbleson, M. E.), Vol. 3, pp. 1837–1852. Plenum Press, New York, USA.

Binns, R. M., Licence, S. T., Wooding, F. B. P. (1990). PHA induces major short-term protease-sensitive lymphocyte traffic involving HEV-like blood vessels in acute DTH-like reactions in skin and tissues. *Eur. J. Immunol.* **20**, 1067–1071.

Binns, R. M., Licence, S. T., Pabst, R. (1992a). Homing of blood, splenic, and lung emigrant lymphoblasts: comparison with the behaviour of lymphocytes from these sources. *Int. Immunol.* **4**, 1011–1019.

Binns, R. M., Licence, S. T., Wooding, F. B., Duffus, W. P. H. (1992b). Active lymphocyte traffic induced in the periphery by cytokines and phytohemagglutinin: three different mechanisms? *Eur. J. Immunol.* **22**, 2195–2203.

Binns, R. M., Licence, S. T., Whyte, A., Wilby, M., Rothkötter, H. J., Bacon, M. (1995). Genetically determined CD45 variant of value in leukocyte tracing *in vivo* in the pig. *Immunology* **86**, 25–33.

Binns, R. M., Licence, S. T., Harrison, A. A., Keelan, E. T. D., Robinson, M. K., Haskard, D. O. (1996). *In vivo* E-selectin up-regulation correlates early with infiltration of PMN, later with PBL entry -mabs block both. *Am. J. Physiol-Heart Cir. Physiol.* **39**, H183.

Blancho, G., Gianello, P., Germana, S., Baetscher, M., Sachs, D. H., LeGuern, C. (1995). Molecular identification of porcine IL-10: regulation of expression in a kidney allograft model. *Proc. Natl. Acad. Sci. USA* **92**, 2800–2804.

Blecha, F., Kelley, K. W. (1981). Effects of cold and weaning stressors on the antibody-mediated immune response of pigs. *J. Anim. Sci.* **553**, 439–447.

Bohl, E. H., Gupta, R. K. P., Olquin, M. V. F., Saif, L. J. (1972). Antibody responses in serum, colostrum and milk of swine after infection or vaccination with TGE virus. *Infect. Immunity* **6**, 289–301.

Bohl, E. H., Saif, L. J. (1975). Passive immunity in transmissible gastroenteritis of swine: Immunoglobulin characteristics of antibodies in milk after inoculating virus by different routes. *Infect. Immunity* **11**, 23–32.

Borg, T., Gerdin, B., Hallgren, R., *et al.* (1984). Complement activation and its relationship to adult respiratory distress syndrome. An experimental study in pigs. *Acta Anaesthesiolog. Scand.* **28**, 158–165.

Bostock, D. E., Owen, L. N. (1973). Porcine and ovine lymphosarcoma: A review. *J. Natl. Cancer Inst.* **50**, 9–939.

Bourne, F. J. (1973). The immunoglobulin system of the suckling pig. *Proc. Nutr. Soc.* **32**, 205–215.

Bourne, F. J., Curtis, J. (1973). The transfer of immunoglobulin

IgG, IgA and IgM from serum to colostrum and milk in the sow. *Immunology* **24** 157–162.

Bourne, F. F., Pickup, J., Honour, J. W. (1971). Intestinal immunoglobulins in the pig. *Biochi. Biophys. Acta* **229**, 18–25.

Bourne, F. J., Curtis, J., Johnson, R. H. (1974). Antibody formation in porcine fetuses. *Res. Vet. Sci.* **16**, 223–227.

Bourne, F. J., Newby, T. J., Evans, P., Morgan, K. L. (1978). The immune requirements of the newborn pig and calf. *Ann. Rech. Vet.* **9**, 239–244.

Bradley, P. A., Bourne, F. J., Brown, P. J. (1976a). The respiratory immune system in the pig. II. Associated lymphoid structures. *Vet. Pathol.* **13**, 90–97.

Bradley, P. A., Bourne, F. J., Brown, P. J. (1976b). The respiratory tract immune system in the pig. I. Distribution of immunoglobulin-containing cells in the respiratory tract mucosa. *Vet. Pathol.* **13**, 81–89.

Brim, T. A., Vancott, J. L., Lunney, J. K., Aif, L. J. (1995). Cellular immune responses of pigs after primary inoculation with porcine respiratory coronavirus or transmissible gastroenteritis virus and challenge with transmissible gastroenteritis virus. *Vet. Immunol. Immunopathol.* **48**, 35–54.

Brown, P. J., Bourne, F. J. (1976a). Distribution of immunoglobulin staining cells in alimentary tract, spleen and mesenteric lymph node of the pig. *Am. J. Vet. Res.* **37**, 9.

Brown, P. J., Bourne, F. J. (1976b). Development of immunoglobulin-containing cell populations in intestine, spleen and mesenteric lymph node of the young pig as demonstrated by peroxidase-conjugated antiserums. *Am. J. Vet. Res.* **37**, 1309–1314.

Brown, P. J., Bourne, F. J., Denny, H. R. (1975). Immunoglobulin-containing cells in pig mammary gland. *J. Anatomy* **120**, 329–335.

Brown, W. R., Butler, J. E. (1994). Characterization of the single Cα gene of swine. *Mol. Immunol.* **31**, 633–642.

Brown, W. R., Kacskovics, I., Amendt, B., Shinde, R., Blackmore, N., Rothschild, M., Butler, J. E. (1995). The hinge deletion variant of porcine IgA results from a mutation at the splice acceptor site in the first Cα intron. *J. Immunol.* **154**, 3836–3842.

Brown, W. R., Rabbani, H., Butler, J. E., Hammarstrom, L. (1997). Characterization of the bovine Cα gene. *Immunology* **91**, 1–6.

Brucks, R. (1966). A study of the K blood group system of swine. In *Polymorphismes Biochimiques des Animaux*, pp. 167–170. INRA, Paris.

Brunsberg, U., Edfors-Lilj, I., Andersson, L., Gustafsson, K. (1996). Structure and organization of pig MHC class II DRB genes: evidence for genetic exchange between loci. *Immunogenetics* **44**, 1–8.

Busse, C., Marschall, H. J., Moennig, V. (1978). Further investigations on the porcine lymphoma C-type particle (PLCP) and the possible biological significance of the virus in pigs. *Ann. Rech. Vet.* **9**, 651–658.

Butler, J. E. (1995). Antigen receptors, their immunomodulation and the immunoglobulin genes of cattle and swine. *Livestock Production Science* **42**, 105–121.

Butler, J. E. (1997). Immunoglobulin gene organization and the mechanism of repertoire development. *Scand. J. Immunol.* **45**, 455–462.

Butler, J. E., Brown, W. R. (1994). The immunoglobulins and immunoglobulin genes of swine. In *Porcine Immunity* (ed. Lunney, J. K.), pp. 5–12. *Vet. Immunol. Immunopath.*, Special Edition.

Butler, J. E., Hamilton, R. G. (1991). Quantitation of specific antibodies: Methods of expression, standards, solid-phase considerations and specific applications. In *Immunochemistry of Solid-phase Immunoassay* (ed. Butler, J. E.), pp. 173–198. CRC Press, Boca Raton, FL.

Butler, J. E., Klobasa, F., Werhahn, E., Cambier, J. C. (1986). Swine as a model for the study of maternal-neonatal immunoregulation. In *The Swine in Biomedical Research* (ed. Tumbleson, M. E.) pp. 1883–1899. Plenum Press, New York.

Butler, J. E., Sun, J., Navarro, P. (1996). The swine immunoglobulin heavy chain has a single J_H and no IgD. *International Immunology* **8**, 1897–1904.

Camacho, A., Hojny, J. (1979). Soluble substance 0 (zero) in sera and saliva of the pig. *Proc. 16th Int. Conf. Anim. Blood Groups Biochem. Polymorphism, Leningrad, 1978, Vol. III*, pp. 125–129.

Campbell, J. G. (1977). The ultrastructure of a porcine hereditary lymphoma with some observations on cell cultures and enzyme cytochemistry. *J. Path.* **122**, 191–200.

Case, M. T., Simon, J. (1972). Whole-body gamma irradiation of newborn pigs: hematologic changes. *Am. J. Vet. Res.* **333**, 1217–1223.

Cepica, S., Wolf, J., Hojny, J., Vacková, I., Schröffel, Jr, J. (1995). Relations between genetic distance of parental pig breeds and heterozygosity of their F1 crosses measured by genetic markers. *Anim. Genet.* **26**, 135–140.

Cepica, S., Moser, G., Schröffel Jr, J., Knorr, C., Geldermann, H., Stratil, A., Hojny, J. (1996). Chromosomal assignment of porcine EAD, EAO, LPR and P3 genes by linkage analysis. *Anim. Genet.* **27**, 109–111.

Chapman, H. A., Johnson, J. S., Cooper, M. D. (1974). Ontogeny of Peyer's patches and immunoglobulin-containing cells in pigs. *J. Immunol.* **112**, 555–563.

Chardon, P., Nunes, M., Dezeure, F., *et al.* (1988). Genetic organization of the SLA complex. In *The Molecular Biology of the Major Histocompatibility Complex of Domestic Animal Species* (eds Warner, C., Rothschild, M., Lamont, S.), pp. 63–78. ISU Press, Ames, IA.

Chardon, P., Rogel-Gaillard, C., Save, J. C., *et al.* (1996). Establishment of the physical continuity between the pig SLA class I and class III regions. *Anim. Genet.* **27** (Suppl. 2), 76.

Charley, B., Fradelizi, D. (1987). Differential effects of human and porcine interleukin 2 on natural killing (NK) activity of newborn piglets and pdult pigs lymphocytes. *Ann. Rec. Vet.* **18**, 227–232.

Charley, B., Petit, E., Leclerc, C., Stefanos, S. (1985). Production of porcine interleukin-2 and its biological and antigenic relationships with human interleukin-2. *Immunol. Lett.* **10**, 410–413.

Cheung, A. K., Porter, C. J., Wilcox, L. (1989). Effects of two types of cobra venom factor on porcine activation and pulmonary artery pressure. *Clin. Exp. Immunol.* **78**, 299–306.

Chitko-McKown, C. G., Chapes, S. K., Brown, R. E., Phillips, R. M., McKown, R. D., Blecha, F. (1991). Porcine alveolar and pulmonary intravascular macrophages: comparison of immune functions. *J. Leukocyte Bio.* **50**, 364–372.

Clarke, R. M., Hardy, R. N. (1971). Histological changes in the small intestine of the young pig and their relation to macromolecular uptake. *J. Anatomy* **108** 63–77.

Corbin, N. C., Hugli, T. E. (1976). The primary structure of porcine C3a anaphylatoxin. *J. Immunol.* **117**, 990–995.

Craik, J. D., Good, A. H., Gottschalk, R., Jarvis, S. M., Paterson, F. A. P., Cass, E. C. (1988). Identification of glucose and nucleoside transport proteins in neonatal pig erythrocytes using monoclonal antibodies against band 4.5 polypeptides of adult human and pig erythrocytes. *Biochemical Cell Biology* **66**, 839–852.

Cukrowska, B., Sinkora, J, Rehakova, Z., *et al.* (1996a). Isotype and antibody specificity of spontaneously formed immunoglobulins in pig fetuses and germ-free piglets: production by $CD5^-$ B cells. *Immunology* **88**, 611–617.

Cukrowska, B., Sinkora, J., Mandel, L., Splíchal, I., Bianchi, A. T. J., Kováru, F., Tlaskalová-Hogenová, H. (1996b). B cells of the pig fetuses and germ-free pigs spontaneously produce IgM, IgG and IgA: detection by Elispot method. *Immunology* **87**, 487–492.

Curtis, J., Bourne, F. J. (1971). Immunoglobulin quantitation in sow serum, colostrum and milk and in the serum of young pigs. *Biochim. Biophys. Acta* **236**, 319–332.

Davies, K. A., Chapman, P. T., Norsworthy, P. J., Jamar, F., Athanassiou, P., Keelan, E. T. M., Harrison, R. A., Binns, R. M., Haskard, D. O., Walport, M. J. (1995). Clearance pathways of soluble immune complexes in the pig: insights into the adaptive nature of antigen clearance in humans. *J. Immunol.* **155**, 5760–5768.

Davies, M. E., Horner, A., Franz, B., Schuberth, H. J. (1992). Detection of cytokine activated chondrocytes in arthritic joints from pigs infected with *Erysioelothrix rhusiopathiae*. *Ann. Rheum. Dis.* **51**, 978–982.

Davies, M. E., Horner, A., Loveland, B. E., McKenzie, I. F. (1994). Upregulation of complement regulators MCP (CD46), DAF (CD55) and protectin (CD59) in arthritic joint disease. *Scand. J. Rheumatol.* **23**, 316–321.

Davis, T. A., Craighead, N., Williams, A. J., *et al.* (1996). Primary porcine endothelial cells express membrane-bound B7-2 (CD86) and a soluble factor that co-stimulate cyclosporin A-resistant and CD28-dependent human T cell proliferation. *Int. Immunol.* **8**, 1099–1111.

Day, N. K. B., Pickering, R. J., Gewurz, H., Good, R. A. (1969). Ontogenetic development of the complement system. *Immunology*, **16**, 319–326.

Dehring, D. J., Steinberg, S. M., Wismar, B. L. (1987). Complement depletion in a porcine model of septic acute respiratory disease. *J. Trauma* **27**, 615–625.

Delventhal, S., Hensel, A., Petzoldt, K., Pabst, R. (1992a). Effects of microbial stimulation on the number, size and activity of bronchus-associated lymphoid tissue (BALT) structures in the pig. *Int. J. Exp. Pathol.* **73**, 351–357.

Delventhal, S., Hensel, A., Petzoldt, K., Pabst, R. (1992b). Cellular changes in the bronchoalveolar lavage (BAL) of pigs, following immunization by the enteral or respiratory route. *Clin. Exp. Immunol.* **90**, 223–227.

Derbyshire, J. B., Lesnick, C. E. (1990). The effect of interferon induction in newborn piglets on the humoral immune response to oral vaccination with transmissible gastroenteritis virus. *Vet. Immunol. Immunopath.* **24**, 227–234.

Diemer, V., Hoyle, M., Baglioni, C., Millis, A. J. (1992). Expression of porcine complement cytolysis inhibitor mRNA in cultured aortic smooth muscle cells. Changes during differentiation *in vitro*. *J. Biol. Chem.* **267**, 5257–5264.

Dimmock, C. K., Webster, W. R., Shiels, I. A., *et al.* (1982). Isoimmune thrombocytopenic purpura in piglets. *Aust. Vet. J.* **59**, 157–158.

Dingle, J. T., Davies, M. E., Mativi, B. Y., Middleton, H. F. (1990). Immunological identification of interleukin-1 activated chondrocytes. *Ann. Rheum. Dis.* **49**, 889–892.

Dinklage, H., Gruhn, R. (1969). Blutgruppen- und Serumproteinpolymorphismus bei verschiedenen in Deutschland vorhandenen Schweinerassen. *Z. Tierzücht. Züchtungsbiol.* **86**, 136–146.

Dinklage, H., Hohenbrink, R. (1970). Untersuchung über den Einfluss heterozygoter Blutgruppengenorte auf Merkmale der Zuchtleistung bei der Deutschen Landrasse. *Züchtungskunde* **4**, 284–293.

Dungern, E. v., Hirzfeld, L. H. (1910). Ueber Nachweis und Vererbung biochemischer Strukturen. *Zeitschr. Immunol. Forsch.* **4**, 531.

Drew, R. A. (1972). Erysipelothrix arthritis in pigs as a comparative model for rheumatoid arthritis. *Proc. Roy. Soc. Med.* **65**, 42–46.

Ellegren, H., Johansson, M., Chowdhary, B. P., Marklund, S., Ryter, D., Marklund, L., Nielsen, P. B., Edors-Lilja, I., Gustavsson, I., Juneja, R. K., Andersson, L. (1993). Assignment of 20 microsatellite markers to the porcine linkage map. *Genomics* **16**, 431–439.

Ellegren, H., Chowdhary, B. P., Fredholm, M., Hoyheim, B., Johansson, M., Nielsen, P. B., Thomsen, P. D., Andersson, L. (1994). A physically anchored linkage map of pig chromosome 1 uncovers sex- and position-specific recombination rates. *Genomics* **24**, 342–350.

Emery, D. W., Sachs, D. H., Leguern, C. (1996). Culture and characterization of hematopoietic progenitor cells from miniature swine. *Exper. Hematol.* **24**, 927–935.

Evans, P. A., Newby, T. J., Stokes, C. R., Bourne, F. J. (1982). A study of cells in the mammary secretions of sows. *Vet. Immunol. Immunopathol.* **3**, 515–527.

Fernandez, A., Perez, J., de las Mulas, J. M., Carrasco, L., Dominguez, J., Sierra, M. A. (1992). Localization of African swine fever viral antigen, swine IgM, IgG and Clq in lung and liver tissues of experimentally infected pigs. *J. Comp. Pathol.* **107**, 81–90.

Flaming, K. P., Thaler, R. C., Blecha, F., Nelssen, J. L. (1988). Influence of sodium diethyldithiocarbamate (Imuthiol) on lymphocyte function and growth in weanling pigs. *Comp. Immunol. Microbial. Inf. Dis.* **11**, 181–187.

Fisher, L. F., Olander, H. J. (1978). Spontaneous neoplasms of pigs – A study of 331 cases. *J. Comp. Path.* **88**, 505–517.

Foss, D. L., Murtaugh, M. P. (1997). Molecular cloning and mRNA expression of porcine interleukin-12. *Vet. Immunol. Immunolpath.* **57**, 121–134.

Franz, B., Davies, E. M., Horner, A. (1995). Localization of viable bacteria and bacterial antigens in arthritic joints of *Erysipelothrix rhusiopathiae*-infected pigs. *FEMS Immunol. Med. Micro.* **12**, 137–142.

Fries, R., Stranzinger, G., Vögeli, P. (1983). Provisional assignment of the G blood group locus to chromosome 15 in swine. *J. Hered.* **74**, 426–430.

Fries, R., Vögeli, P., Stranzinger, G. (1990). Gene mapping in the pig. *Advances in Veterinary Science and Comparative Medicine*. **34**, 273–303.

Fritz, F. J., Pabst, R., Binns, R. M. (1990). Lymphocyte subsets and their proliferation in a model for a delayed-type hypersensitivity reaction in the skin. *Immunology* **71**, 508–516.

Gahne, B., Juneja, R. K. (1985). Prediction of the halothane (Hal) genotypes of pigs by deducing Hal, Phi, Po2, Pgd haplotypes of parents and offspring: results from a large-scale practice in Swedish breeds. *Anim. Blood Groups Biochem. Genet.* **16**, 265–283.

Gately, M. K., Chizzonite, R. (1992). Measurement of human and mouse interleukin 12. In *Current Protocols in Immunology* 6 (eds Ausubel, F. M., *et al.*), 16.1–16.8.

Geffrotin, C., Chardon, P., de Andres-Cara, D. F., *et al.* (1990). The swine steroid 21-hydroxylase gene (CYP21): cloning and mapping within the swine leucocyte antigen complex. *Anim. Genet.* **21**, 1–13.

Gebert, A., Rothkötter, H. J., Pabst, R. (1994). Cytokeratin 18 is an M-cell marker in porcine Peyer's patches. *Cell Tissue Res.* **276**, 213–221.

Gilliard, N., Heldt, G. P., Loredo, J., Gasser, H., Redl, H., Merritt, T. A., Spragg, R. G. (1994). Exposure of the hydrophobic components of porcine lung surfactant to oxidant stress alters surface-tension properties. *J. Clin. Invest.* **93**, 2608.

Gonzalez Juarrero, M., Mebus, C., Garmendia, A. E. (1994). Porcine lymphocyte gamma interferon responses to mitogenic stimuli monitored by a direct immunoassay. *Vet. Immunol. Immunopathol.* **40**, 201–212.

Geiger, H., Day, N., Good, R. A. (1972). The ontogenetic development of the later complement components in fetal piglets. *J. Immunol.* **108**, 1098–1104.

Gerard, C., Hugli, T. E. (1980). Amino acid sequence of the anaphylatoxin from the fifth component of porcine complement. *J. Biol. Chem.* **255**, 4710–4715.

Gonzalez, A., Friend, M., Morera, L., Vögeli, P., Llanes, D. (1992). Mouse monoclonal antibodies to swine blood group antigens. *Animal Genetics* **23**, 469–473.

Gonzalez, A., Friend, M., Pintado, C. O., Vögeli, P., Llanes, D. (1995). A monoclonal antibody to swine erythrocytes recognizes the blood group on the major glycophorin. *Anim. Genet.* **26**, 351–354.

Good, A. H., Craik, J. D., Jarvis, S. M., Kwong, F. Y. P., Young, J. D., Paterson, A. R. P., Cass, C. E. (1987). Characterization of monoclonal antibodies that recognize band 4.5 polypeptides associated with nucleoside transport in pig erythrocytes. *Biochem. J.* **244**, 749–755.

Goodman, R. B., Foster, D. C., Mathewes, S. L., Osborn, S. G., Kuijper, J. L., Forstrom, J. W., Martin, T. R. (1992). Molecular cloning of porcine alveolar macrophage-derived neutrophil chemotactic factors I and II; identification of porcine IL-8 and another intercrine-alpha protein. *Biochem.* **31**, 10483–10490.

Greer, J. M., Wannemuehler, M. J. (1989). Pathogenesis of *Treponema hyodysenteriae*: Induction of interleukin-1 and tumor necrosis factor by a treponemal butanol/water extract (Endotoxin). *Microb. Pathog.* 7, 279–288.

Grimm, D. R., Misfeldt, M. L. (1994). Partial cloning and sequencing of the gene encoding the porcine T-cell receptor δ-chain constant region. *Gene* **144**, 271–275.

Gurney, A. L., Kuang, W. J., Xie, M. H., Malloy, B. E., Eaton, D. L., de Sauvage, F. J. (1995). Genomic structure, chromosomal localization, and conserved alternative splice forms of thrombopoietin. *Blood* **85**, 981–988.

Gustafsson, K., Germana, S., Sundt III, T. M., *et al.* (1993). Extensive allelic polymorphism in the CDR2-like region of the miniature swine CD4 molecule. *J. Immunol.* **151**, 1365–1370.

Hála, K, Hojny, J. (1964). Blood group of the N system in pigs. *Fol. Biol. Praha* **10**, 239–244.

Hammer, D. K., Mossman, H. (1978). The importance of membrane receptors in the transfer of immunoglobulins from plasma to colostrum. *Ann. Rech. Vet.* **9**, 229–234.

Hammerberg, C., Schurig, G. G. (1986). Characterization of monoclonal antibodies directed against swine leukocytes. *Vet. Immunol. Immunopathol.* **11**, 107–121.

Haverson, K., Bailey, M., Higgins, V. R., Bland, P. W., Stokes, C. R. (1994). Characterisation of monoclonal antibodies specific for porcine monocytes, macrophages and granulocytes from peripheral blood and mucosal tissues. *J. Immunol. Meth.* **170**, 233–245.

Hayashi, M., Tsuda, H., Okumura, M., *et al.* (1988). Histopathological classification of malignant lymphomas in slaughtered swine. *J. Comp. Path.* **98**, 11–21.

Head, K. W., Campbell, J. G., Imlah, P., *et al.* (1974). Hereditary lymphosarcoma in a herd of pigs. *Vet. Rec.* **95**, 523–527.

Hensel, A., Stockhifezurwieden, N., Petzold, K., Lubitz, W. (1995). Oral immunisation of pigs with viable or inactivated *Actinobacillus pleuropneumoniae* serotype 9 induces pulmonary and systemic antibodies and protects against homologous aerosol challenge. *Infect. Immun.* **63**, 3048–3053.

Hildreth, J. E., Holt, V., August, J. T., Pescovitz, M. D. (1989). Monoclonal antibodies against porcine LFA-1: species cross-reactivity and functional effects of β-subunit-specific antibodies. *Mol. Immunol.* **26**, 883–895.

Hirsch, F., Sachs, D. H., Gustafsson, K., *et al.* (1990). Class II genes of miniature swine. III. Characterization of an expressed pig class II gene homologous to HLA-DQA. *Immunogenetics* **31**, 52–56.

Hirsch, F., Germana, S., Gustafsson, K., *et al.* (1992). Structure and expression of class II alpha genes in miniature swine. *J. Immunol.* **149**, 841–846.

Hirt, W., Saalmüller, A., Reddehase, M. J. (1990). Distinct γ/δ T cell receptors define two subsets of circulating porcine $CD2^-CD4^-CD8^-$ T lymphocytes. *Eur. J. Immunol.* **20**, 265–269.

Hoerlein A. (1957). The influence of colostrum on antibody response in baby pigs. *J. Immunol.* **78**, 112–117.

Høgåsen, K., Jansen, J. H., Mollnes, T. E., Hovdenes, J., Harboe, M. (1995). Hereditary porcine membranoproliferative glomerulonephritis type II is caused by factor H deficiency. *J. Clin. Invest.* **95**, 1054–1061.

Hojny, J. (1973). Further contribution to the H blood group system in pigs. *Anim. Blood Groups Biochem. Genet.* **4**, 161–168.

Hojny, J., Glasnák, V. (1970). A comparison of A, Na and Nb substances in the blood, milk and saliva of sows. *Anim. Blood Groups Biochem. Genet.* **1**, 47–51.

Hojny, J., Hála, K. (1965) Blood group system O in pig. *Proc. 9th Eur. Conf. Anim. Blood Groups* (Prague, 1964) (ed. Matousek, J.), pp. 163–168. Pub. House Czech. Acad. Sci., Prague.

Hojny, J., Hradecky, J. (1971). The time of appearance of blood group antigens in miniature pigs. *Anim. Blood Groups Bioch. Genet.* **2**, 105–112.

Hojny, J., Nielsen, P. B. (1992). Allele E^{bdgjmr} (E17) in the pig E blood group system. *Anim. Genet.* **23**, 523–524.

Hojny, J., van Zeveren, A. (1985). M1, a new factor in the porcine M blood group system. *Anim. Blood Groups Biochem. Genet.* **16**, 69–72.

Hojny, J., Gavalier, M., Hradecky, J., Linhart, J. (1967). New blood factors in pigs. *Proc. 10th Eur. Conf. Anim. Blood Groups Biochem. Polymorphism, Paris 1966*, pp. 151–158.

Hojny, J., Hradecky, J., Camacho, A. (1979). Further factors and alleles of the M blood group system in pigs. *Proc. 16th Int. Conf. Anim. Blood Groups Biochem. Polymorphism, Leningrad, 1978, Vol. III*, pp. 114–120.

Hojny, J., Hradecky, J., Linhart, J. (1984). New blood group allele (F^{ad}) in the pig. *Anim. Blood Groups Biochem. Genet.* **15**, 227–228.

Hojny, J., Schröffel Jr, J., Geldermann, H., Cepica, S. (1994). The porcine M blood group system: evidence to suggest assignment of its M1 factor to a new system (P). *Anim. Genet.* **25**(Suppl. 1), 99–101.

Hopkins, L. J., Humphreys, M. (1989). Simple, sensitive and specific bioassay of interleukin-1. *J. Immunol. Methods* **120**, 271–276.

Hradecky, J., Linhart, J. (1970). Db-next blood group factor of the D system in pigs. *Anim. Blood Groups Biochem. Genet.* **1**, 65–66.

Huemer, H. P., Larcher, C., Coe, N. E. (1992). Pseudorabies virus glycoprotein III derived from virions and infected cells bind to the third component of complement. *Virus Res.* **23**, 271–280.

Hussein, A. M., Newby, T. J., Bourne, F. J. (1983a). Immunohistochemical studies of the local immune system in the reproductive tract of the sow. *J. Reprod. Immunol.* **5**, 1–15.

Hussein, A. M., Newby, T. J., Stokes, C. R., Bourne, F. J. (1983b). Quantitation and origin of immunoglobulins A, G, and M in the secretions and fluids of the reproductive tract of the sow. *J. Reprod. Immunol.* **5**, 17–26.

Imlah, P., Brownlie, S. E., Head, K. W., *et al.* (1979). Serum gamma globulin levels and the detection of IgG heavy chain and light chain in the serum and urine of cases of pig hereditary lymphosarcoma. *Eur. J. Cancer* **15**, 1337–1349.

Ish, C., Ong, G. L., Desai, N., Mattes, M. J. (1993). The specificity of alternative complement pathway-mediated lysis of erythrocytes: a survey of complement and target cells from 25 species. *Scand. J. Immunol.* **38**, 113–122.

Ito, T., Fujita, N. (1977). Pathological studies on lymphosarcoma in swine. *Jap. J. Vet. Sci.* **39**, 599–608.

Ivanoska, D., Sun, D. C., K, L. J. (1991). Production of monoclonal antibodies reactive with polymorphic and monomorphic determinants of SLA class II gene products. *Immunogenetics* **33**, 220–223.

Iwata, H., Ueda, T., Takayanagi, K., Wada, M., Inoue, T. (1993). Swine interleukin 2 activity produced by mesenteric lymph node cells. *J. Vet. Med. Sci.* **55**, 729–734.

Izbicki, J. R., Ziegler-Heitbrock, H. W. L., Meier, M., *et al.* (1989). The impact of splenectomy on antibody response in the porcine model. *J. Clin. Lab. Immunol.* **30**, 13–19.

Jackson, S. M., Osborne, B. A., Goldsby, R. A. (1996). Diversification of immunoglobulin V_H genes in individual follicles of bovine IPP. *FASEB J.* **10**, A1031, Abstr. #181.

Jansen, J. H. (1993). Porcine membranoproliferative glomerulonephritis with intramembranous dense deposits (porcine dense deposit disease). *APMIS* **101**, 281–289.

Jansen, J. H., Høgåsen, K., Molines, T. E. (1993). Extensive complement activation in hereditary porcine membranoproliferative glomerulonephritis type-II (porcine dense deposit disease). *Am. J. Pathol.* **143**, 1356–1365.

Jansen, J. H., Høgåsen, K., Grøndahl, A. M. (1995). Porcine membranoproliferative glomerulonephritis type II: an autosomal recessive deficiency of factor H. *Vet. Rec.* **137**, 240–244.

Jarosková, L., Kovàru, F., Tlaskalová, H., Trebichavsky, I., Fornùsek, L., Holub, M. (1982). The development of B lymphocytes and their reactivity in pig fetuses. *Adv. Exp. Med. Biol.* **19**, 25–30.

Jensen, E. L., Smith, C., Baker, L. N., Cox, D. F. (1968). Quantitative studies on the blood group and serum protein systems in pigs. II. Effects on production and reproduction. *J. Anim. Sci.* **27**, 856–862.

Johnson Jr, H., Lunney, J. K., Sachs, D. H., *et al.* (1980). Preparation and characterization of an antiserum specific for T cells of pigs. *Transplantation* **29**, 477–483.

Joling, P., Bianchi, A. T. J., Kappe, A. L., Zwart, R. J. (1994). Distribution of lymphocyte subpopulations in thymus, spleen, and peripheral blood of specific pathogen free pigs from 1 to 40 weeks of age. *Vet. Immunol. Immunopathol.* **40**, 105–117.

Juneja, R. K., Gahne, B., Edfors-Lilja, I., Andresen, E. (1983). Genetic variation at a pig serum protein locus, Po-2 and its assignment to the Phi, Hal, S, H, Pgd linkage group. *Anim. Blood Groups Biochem. Genet.* **14**, 27–36.

Jung, Y. C., Rothschild, M. F., Flanagan, M. P., *et al.* (1989). Restriction fragment length polymorphisms in the swine major histocompatibility complex. *Theor. Appl. Genet.* **77**, 271–274.

Kacskovics, I., Sun, J., Butler, J. E. (1994). Five IgG subclasses identified from cDNA sequences from a single swine. *J. Immunol.* **153**, 3565–3573.

Kadota, K., Nakajima, H. (1988). Histological progression of follicular centre cell lymphomas in immunoglobulin-producing tumours in two pigs. *J. Comp. Path.* **99**, 146–158.

Kadota, K., Niibori, S. (1985). A case of swine follicular lymphoma with intracytoplasmic immunoglobulin inclusions. *J. Comp. Path.* **95**, 599–608.

Kadota, K., Yamazaki, M., Ishino, S., *et al.* (1984). Ultrastructure and C-type particles in myeloid leukemia of a pig. *Vet. Pathol.* **21**, 263–265.

Kadota, K., Ishino, S., Nakajima, H. (1986a). Immunological and ultrastructural observations on swine thymic lymphoma. *J. Comp. Path.* **96**, 371–378.

Kadota, K., Nemoto, K., Mabara, S., *et al.* (1986b). Three types of swine immunoglobulin-producing tumours: lymphoplasmacytic lymphosarcoma, immunoblastic lymphosarcoma and plasmacytoma. *J. Comp. Path.* **96**, 541–550.

Kadota, K., Akutsu, H., Saito, M., *et al.* (1987) Ultrastructure of swine myelogenous leukaemic cells, with particular reference to intracytoplasmic granules. *J. Comp. Path.* **97**, 401–406.

Kadota, K., Ishino, S., Hashimoto, N., *et al.* (1990a). Malignant lymphomas of thymus origin in two sows. *J. Vet. Med.* **37**, 592–600.

Kadota, K., Nakajima, H., Hashimoto, N. (1990b). Non-T, Non-B lymphoblastic lymphoma in two pigs. *J. Vet. Med.* **37**, 104–112.

Kambarage, D., Bland, P., Stokes, C. (1994). The accessory cell activity of porcine intestinal macrophages in the induction of T-cell responses. *J. Vet. Med. Sci.* **56**, 1135–1138.

Kashima, T., Nomura, T. (1982). A case of myeloid leukaemia (eosinophilic) in swine. *Jpn. J. Vet. Sci.* **44**, 529–533.

Keelan, E. T., Licence, S. T., Peters, A. M., Binns, R. M., Haskard, D. O. (1994). Characterization of E-selectin expression *in vivo* with use of a radiolabeled monoclonal antibody. *Am. J. Physiol.* **266**, H279–290.

Kim, Y. B., Huh, N. D. (1981). Natural killer and killer (NK/K) cell system in gnotobiotic miniature swine. In *Recent Advances in Germfree Research* (eds Sasaki, S., Ozawa, A., Hashimoto, K.), pp. 585–593. Tokai University Press, Tokyo, Japan.

Kim, Y. B., Zhang, J., Davis, W. C., Lunney, J. K. (1994). CD11/CD18 Panel Report for Swine CD Workshop. *Vet. Immunol. Immunopathol.* **43**, 289–291.

King, I. A., Tabiowo, A., Fryer, P. R., Purkis, P. E., Leigh, I. (1991). Basal cell glycoprotein in pig epidermis closely resembles the β-1 subunit of the integrin family of cell adhesion molecules. *J. Invest. Dermatol.* **97**, 501–505.

Kiorpes, A. L., MacWilliams, P. S., Schenkman, D. I., Backstrom, L. R. (1990). Blood gas and hematological changes in experimental peracute porcine pleuropneumonia. *Can. J. Vet. Res.* **54**, 164–169.

Kirkham, P. A., Takamatsu, H., Yang, H., *et al.* (1996). Porcine CD3ε: its characterization, expression and involvement in activation of porcine T-lymphocytes. *Immunol.* **87**, 616–623.

Klemcke, H. G., Blecha, F., Nienaber, J. A. (1990). Pituitary-adrenocortical and lymphocyte responses to bromocriptine-induced hypoprolactinemia, andrenocorticotropic hormone, and restraint in swine. *Proc. Soc. Exp. Biol. Med.* **195**, 100–108.

Klobasa, F., Werhahn, E., Butler, J. E. (1981). Regulation of humoral immunity in the piglet by immunoglobulins of maternal origin. *Res. Vet. Sci.* **31**, 195–206.

Klosterhalfen, B., Horstmann-Jungemann, K., Vogel, P., Dufhues, G., Simon, B., Kalff, G., Kirkpatrick, C. J., Mittermayer, C., Heinrich, P. C. (1991). Hemodynamic variables and plasma levels of PG12, TXA2 and IL-6 in porcine model of recurrent endotoxemia. *Circ. Shock* **35**, 237–244.

Komatsu, M., Kawakami, K., Maruno, H., *et al.* (1996). RT-PCR-based genotyping for swine major histocompatibility complex (SLA) class II genes. *Anim. Technol. Japan.* **67**, 211–217.

Komuves, L. G., Heath, J. P. (1992). Uptake of maternal immunoglobulins in the enterocytes of suckling piglets: Improved detection with a streptavidin-biotin bridge gold technique. *J. Histochem. Cytochem.* **40**, 1637–1646.

Komuves, L. G., Nichols, B. L., Hutchens, W., Heath, J. P. (1993). Formation of crystalloid inclusions in the small intestine of neonatal pigs: an immunocytochemical study using colloidal gold. *Histochem. J.* **25**, 19–29.

Kováru, F., Kováru, H., Halouzka, R., Zendulka, J., Drábek, J. (1994a). Immunophenotypic characterization of pig lymphocytes in prenatal development. *Cell Biol. Internat.* **18**, 531.

Kováru, H., Kováru, F., Halouzka, R., Kozáková, H. (1994b). Pig development of natural killer cytotoxicity with modified C-6 glioma cell line as target. *Cell Biol. Internat.* **18**, 531.

Knoblock, K. F., Canning, P. C. (1992). Modulation of *in vitro* porcine natural killer cell activity by recombinant interleukin-1 alpha, interleukin-2 and interleukin-4. *Immunology* **76**, 299–304.

Kristensen, B., Renard, C., Ostergard, H., *et al.* (1992). Reactivity of murine monoclonal anti-SLA class I antibodies on cells from outbred swine. *Anim. Genetics* **23**, 41.

Landsteiner, K. (1900). Zur Kenntnis der antifermentativen, lytischen und agglutinierenden Wirkungen des Blutserums. *Zentbl. Bakt. Parasitkde.* **27**, 357–362.

La Bonnardiere, C., Lefevre, F., Charley, B. (1994). Interferon responses in pigs: molecular and biological aspects. *Vet. Immunol. Immunopathol.* **43**, 29–36.

Larsen, H. J., Tollersrud, S. (1981). Effect of dietary vitamin E and selenium on the phytohaemagglutinin response of pig lymphocytes. *Res. Vet. Sci.* **31**, 301–305.

Le Jan, C. (1994). A study by flow cytometry of lymphocytes in sow colostrum. *Res. Vet. Sci.* **57**, 300–304.

Lecce, J. G., Morgan, D. O., Matronr, G. (1964). Effect of feeding colostral and milk components on the cessation of intestinal absorption of large macromolecules (closure) in neonatal pigs. *J. Nutr.* **84**, 43–54.

Lecce, J. G., Leary, J., Lee, H., Clare, D. A., *et al.* (1991). Protection of agammaglobulinemic piglets from porcine rotavirus infection by antibody against simian rotavirus SA-11. *J. Clin. Microbiol.* **29**, 1382–1386.

Lefevre, F., Boulay, V. (1993). A novel and atypical type one interferon gene expressed by trophoblast during early pregnancy. *J. Biol. Chem.* **268**, 19760–19768.

Le Moal, M. A., Motta, I., Truffa-Bachi, P. (1989). Improvement of an ELISA bioassay for the routine titration of murine interferon-gamma. *Res. Immunol.* **140**, 613–624.

L'Haridon, R. M., Bourget, P., Lefevre, F., La Bonnardiere, C. (1991). Production of an hybridoma library to recombinant porcine alpha I interferon: a very sensitive assay (ISBBA) alows the detection of a large number of clones. *Hybridoma* **10**, 35–47.

Levy, M. L., Halbwachs-Mecarelli, M. C., Gubler, G., *et al.* (1986). H deficiency in two brothers with atypical dense intramembranous deposit disease. *Kidney Int.* **30**, 949–956.

Li, D. F., Nelssen, J. L., Reddy, P. G., Blecha, F., Klemm, R., Goodband, R. D. (1991). Interrelationship between hypersensitivity to soybean proteins and growth performance in early-weaned pigs. *J. Anim. Sci.* **69**, 4062–4069.

Licence, S. T., Binns, R. M. (1995). Major long-term changes in $\gamma\delta$ T-cell receptor-positive and $CD2^+$ T-cell subsets after neonatal thymectomy in the pig: a longitudinal study lasting nearly 2 years. *Immunol.* **85**, 276–284.

Lie, H. (1968). Thrombocytopenic purpura in baby pigs. Clinical studies. *Acta Vet. Scand.* **9**, 285–301.

Lie, W. R., Rothschild, M. F., Warner, C. M. (1987). Mapping of C2 Bf, and C4 genes to the swine major histocompatibility complex (swine leukocyte antigen). *J. Immunol.* **139**, 3388–3394.

Lin, G., Pearson, A. E., Scamurra, R. W., Zhou, Y., Baarsch, M. G., Weiss, D. J., Murtaugh, M. P. (1994). Regulation of interleukin-8 expression in porcine alveolar macrophages by bacterial lipopolysaccharide. *J. Biol. Chem.* **269**, 77–85.

Linhart, J. (1971). Lm, a new blood group factor of the L system in pigs. *Anim. Blood Groups Biochem. Genet.* **2**, 243–245.

Linklater, K. A., McTaggart, H. S., Imlah, P. (1973). Haemolytic disease of the newborn, thrombocytopenic purpura and neutropenia occurring concurrently in a litter of piglets. *Br. Vet. J.* **129**, 36–46.

Llanes, D., Marisa Nogal, Prados, F., Margarita del Val y Viuela, E. (1992). A erythroid species specific antigen of swine detected by a monoclonal antibody. *Hybridoma* **11:6**, 757–764.

Lowden, S., Heath, T. (1994). Ileal Peyer's patches in pigs: intercellular and lymphatic pathways. *Anat. Rec.* **239**, 297–305.

Lunney, J. K. (1994). The swine leukocyte antigen complex. *Vet. Immunol. Immunopathol.* **43**, 19–28.

Lunney, J. K., Butler, J. E. (1998). Immunogenetics. In *Genetics of the Pig* (eds Rothschild, M. F., Ruvinsky, A.) pp. 163–167. CAB International, Wallingford, UK.

Lunney, J. K., Sachs, D. H. (1978). Transplantation in miniature swine. IV. Chemical characterization of MSLA and Ia-like antigens. *J. Immunol.* **120**, 607–612.

Lunney, J. K., Pescovitz, M. D. (1987). Phenotypic and functional characterization of pig lymphocyte populations. *Vet. Immunol. Immunopathol.* **17**, 135–144.

Lunney, J. K., Pescovitz, M. D., Sachs, D. H. (1986). *The Swine Major Histocompatibility Complex: its Structure and Function.* In *Swine in Biomedical Research* (ed. M. E. Tumblesom) 3, 1821–1836. Plenum Press, New York, USA.

Lunney, J. K., Walker, K., Goldman, T. (1994). Analyses of anti-human CD monoclonal antibodies for cross reactions with swine cell antigens. *Vet Immunol. Immunopathol.* **43**, 207–210.

Lunney, J. K., Walker, K., Goldman, T., *et al.* (1994). Overview of the First International Workshop to define swine leukocyte cluster differentiation (CD) antigens. *Vet. Immunol. Immunopathol.* **43**, 193–206.

Magyar, A., Mihalik, R., Olah, I. (1995). The surface phenotype of swine blood and tissue eosinophil granulocytes. *Vet. Immunol. Immunopathol.* **47**, 273.

Major, F. (1968). Untersuchungen über die verwandtschaftlichen Beziehungen zwischen verschiedenen europäischen Landrasse-Populationen mit Hilfe von Blutgruppenfaktoren. Thesis, Univ. Göttingen (Germany).

Mallard, B. A., Wilkie, B. N, Kennedy, B. W. (1989a). Genetic and other effects on antibody and cell mediated immune response in swine leucocyte antigen (SLA) defined miniature pigs. *Anim. Genet.* **20**, 167–175.

Mallard, B. A., Wilkie, B. N., Kennedy, B. W. (1989b). Influence of major histocompatibility genes on serum hemolytic complement activity in miniature swine. *Am. J. Vet. Res.* **50**, 389–397.

Mallard, B. A., Wilkie, B. N., Kennedy, B. W. (1989c). The influence of the swine major histocompatibility genes (SLA) on variation in serum immunoglobulin (Ig) concentration. *Vet. Immunol. Immunopathol.* **21**, 139–151.

Mandel, L., Trávnícek, J. (1982). Hematology of conventional and germfree miniature Minnesota piglets. *Z. Versuchstierk.* **24**, 299–307.

Mandel, L., Trávnícek, J. (1987). The minipig as a model in gnotobiology. *Die Nahrung* **5-6**, 613–618.

Marcato, P. S. (1987). Swine lymphoid and myeloid neoplasms in Italy. *Vet. Research Com.* **11**, 325–337.

Mathialagan, N., Bixby, J. A., Roberts, R. M. (1992). Expression of interleukin-6 in porcine, ovine, and bovine preimplantation conceptuses. *Mol. Reprod. Dev.* **32**, 324–330.

McCurry, K. R., Kooyman, D. L., Alvarado, C. G., Cotterell, A. H., Martin, M. J., Logan, J. S., Platt, J. L. (1995). Human complement regulatory proteins protect swine-to-primate cardiac xenografts from humoral injury. *Nature. Med.* **1**, 423–427.

McTaggart, H. S., Head, K. W., Laing, A. H. (1971). Evidence for a genetic factor in the transmission of spontaneous lymphosarcoma (leukaemia) of young pigs. *Nature* **232**, 557–558.

McTaggart, H. S., Laing, A. H., Imlah, P., *et al.* (1979). The genetics of hereditary lymphosarcoma of pigs. *Vet. Rec.* **105**, 189.

Meeusen, E. N. T., Premier, R., Brandon, M. R. (1996). Tissue-specific migration of lymphocytes: a key role for Th1 and Th2 cells? *Immunol Today* **17**, 421–424.

Mege, D., Lefevre, F., La Bonnardiere, C. (1991). The porcine family of interferon-omega: cloning, structural analysis, and functional studies of five related genes. *J. Interferon Res.* **11**, 341–350.

Merhi, Y., Llacoste, L., Lam, J. Y. T. (1994). Neutrophil implications in platelet deposition and vasoconstriction after deep arterial injury by angioplasty in pigs. *Circulation* **90**, 997-1002.

Meri, S., Morgan, B. P. (1994). Membrane proteins that protect against complement lysis. *Springer Semin. Immunopathol.* **15**, 369–396.

Metzger, J. J., Ballet-Lapierre, C., Houdayer, M. (1978). Partial inhibition of the humoral immune response of pigs after early postnatal immunization. *Am. J. Vet. Res.* **39**, 627–631.

Miller, B. G., Newby, T. J., Stokes, C. R., Hampson, D. J., Brown, P. J., Bourne, P. J. (1984). The importance of dietary antigen in the cause of postweaning diarrhoea in pigs. *Am. J. Vet. Res.* **45**, 1730–1733.

Miller, B. G., Whittemore, C. T., Stokes, C. R., Telemo, T. (1994). The effect of delayed weaning on the development of oral tolerance to soya-bean protein in pigs. *Brit. J. Nutr.* **71**, 615–625.

Modig, J., Borg, T. (1986). Biochemical markers in a porcine model of adult respiratory distress syndrome induced by endotoxemia. *Resuscitation* **14**, 225–236.

Morfitt, D. C., Pohlenz, J. F. L. (1989). Porcine colonic lymphoglandular complex: distribution, structure, and epithelium. *Am. J. Anat.* **184**, 41–51.

Morgan, B. P. (1995). Complement regulatory molecules: application to therapy and transplantation. *Immunol. Today* **16**, 257–259.

Morgan, B. P., Meri, S. (1994). Membrane proteins that protect against complement lysis. *Springer Sem. Immunopathol.* **15**, 369–396.

Morgan, K. L., Bourne, F. J. (1981). Immunoglobulin content of the respiratory tract secretions of piglets from birth to 10 weeks old. *Res. Vet. Sci.* **31**, 40–42.

Morgan, K. L., Hussein, A. M., Newby, T. J., Bourne, F. J. (1980). Quantification and origin of the immunoglobulins in porcine respiratory tract secretions. *Immunology* **41**, 729–736.

Murata, H., Namioka, S. (1977). The duration of colostral immunoglobulin uptake by the epithelium of the small intestine of neonatal pigs. *J Comp Pathol.* **87**, 431–439.

Murtaugh, M. P. (1994). Porcine cytokines. *Vet. Immunol. Immunopathol.* **43**, 37–44.

Nagy, L. K., Mackenzie T., Bharucha Z. (1976). *In vitro* studies on the antimicrobial effects of colostrum and milk from vaccinated and unvaccinated pigs on *Escherichia coli*. *Res. Vet. Sci.* **21**, 132–140.

Nakajima, H., Mabara, S., Ishino, S., *et al.* (1989). Malignant lymphomas of follicular centre cell origin in 14 pigs. *J. Vet. Med.* **36**, 621–630.

Nakashima, N., Miyazaki, K., Ishikawa, M., Yatohgo, T., Ogawa, H., Uchibori, H., Matsumoto, I., Seno, N., Hayashi, M. (1992). Vitronectin diversity in evolution but uniformity in ligand binding and size of the core polypeptide. *Biochim. Biophys. Acta* **1120**, 1–10.

Newby, T. J., Stokes, C. R., Bourne, F. J. (1982). Immunological activities of milk. *Vet. Immunol. Immunopathol.* 3, 67–94.

Newby, T. J., Stokes, C. R., Huntley, J., Evans, P., Bourne, F. J. (1979). The immune response of the pig following oral immunisation with soluble protein. *Vet. Immunol. Immunopathol.* **1**, 37.

Nielsen, P. B. (1961). The M blood group system of the pig. *Acta Vet. Scand.* **2**, 246–253.

Nielsen, P. B., Vögeli, P. (1982). A new Kd subgroup designated Kg in the porcine K blood group system. *Anim. Blood Groups Biochem. Genet.* **13**, 65–66.

Niu, P. D., Lefevre, F., Mege, D., La Bonnardiere, C. (1995). Atypical porcine type I interferon – biochemical and biological characterization of the recombinant protein expressed in insect cells. *Euro. J. Biochem.* **230**, 200–206.

Norley, S. G., Wardley, R. C. (1982). Complement-mediated lysis of African swine fever virus-infected cells. *Immunology* **46**, 75–82.

Nordstoga, K. (1965). Thrombocytopenic purpura in baby pigs caused by maternal isoimmunization. *Path. Vet.* **2**, 601–610.

Nowacki, W., Cederblad, B., Renard, C., La Bonnardiere, C., Charley, B. (1993). Age-related increase of porcine natural interferon alpha producing cell frequency and of interferon yield per cell. *Vet. Immunol. Immunopathol.* **37**, 113–122.

Oishi, T., Tanaka, K., Tonita, T., *et al.* (1990). Genetic variations of blood groups and biochemical polymorphism in Jinhua pigs (in Japanese). *Jap. J. Swine Sci.* **27**, 202–208.

Olson, N. C. (1988). Biochemical, physiological, and clinical aspects of endotoxemia. *Mol. Aspects Med.* **10**, 511–629.

Pabst, R. Beil, W. (1989). Mast-cell heterogeneity in the small-intestine of normal, gnotobiotic and parasitized pigs. *Int. Arch. Aller. Appl. Immunol.* **88**, 363–366.

Pabst, R., Binns, R. M. (1986). Comparison of lymphocyte production and migration in pig lymph nodes, tonsils, spleen, bone marrow and thymus. In *Swine in Biomedical Research*, Vol. 3 (ed. Tumblesom, M. E.), pp. 1865–1871. Plenum Press, New York, USA.

Pabst, R., Binns, R. M. (1989). Heterogeneity of lymphocyte homing physiology: several mechanisms operate in the control of migration to lymphoid and non-lymphoid organs *in vivo*. *Immunol. Rev.* **108**, 83–109.

Pabst, R., Binns, R. M. (1994). The immune system of the respiratory tract in pigs. *Vet. Immunol. Immunopathol.* **43**, 151–156.

Pabst, R., Binns, R. M. (1995). Lymphocytes migrate from the bronchoalveolar space to regional bronchial lymph-nodes. *Am. J. Resp. Crit. Care Med.* **151**, 495–499.

Pabst, R., Fritz, F. J. (1986). Comparison of lymphocyte production in lymphoid organs and their compartments using the metaphase arrest technique. *Cell Tissue Res.* **245**, 423–430.

Pabst, R., Trepel, F. (1975) Quantitative evaluation of the total number and distribution of lymphocytes in young pigs. *Blut* **31**, 77–86.

Pabst, R., Geist, M., Rothkötter, H. J., Fritz, F. J. (1988). Postnatal development and lymphocyte production of jejunal and ileal Peyer's patches in normal and gnotobiotic pigs. *Immunology* **64** 539–544.

Pabst, R., Binns, R. M., Rothkötter, H. J., Westermann, J. (1993). Quantitative analysis of lymphocyte fluxes *in vivo*. *Curr. Top. Microbiol. Immunol.* **184**, 151–159.

Pauli, U., Bertoni, G., Duerr, M., Peterhans, E. (1994). A bioassay for the detection of tumor necrosis factor from eight different species: evaluation of neutralization rates of a monoclonal antibody against human TNF-alpha. *J. Immunol. Meth.* **171**, 263–265.

Pauly, T., Weiland, E., Hirt, W., *et al.* (1996). Differentiation between MHC-restricted and non-MHC-restricted porcine cytolytic T lymphocytes. *Immunology* **88**, 238–246.

Peelman, L. J., Chardon, P., Nunes, M., *et al.* (1995). The BAT1 gene in the MHC encodes an evolutionarily conserved putative nuclear RNA helicase of the DEAD family. *Genomics* **26**, 210–218.

Peelman, L. J., Chardon, P., Vaiman, M., *et al.* (1996). A detailed physical map of the porcine major histocompatibility complex (MHC) class III region: comparison with human and mouse MHC class III region. *Mamm. Genome.* **7**, 363–367.

Peelman, L. J., Van de Weghe, A. R., Coppieters, W. R., van Zeveren, A. J., Bouquet, Y. H. (1991). Cloning and sequencing of the porcine complement factor B. *Immunogenetics* **34**, 192–195.

Pennington, L. R., Sakamoto, K., Popitz-Bergez, F. A., *et al.* (1988). Bone marrow transplantation in miniature swine I. Development of the model. *Transplantation* **45**, 21–26.

Pescovitz, M. D., Sachs, D. H., Lunney, J. K., *et al.* (1983). Localization of class II MHC antigens on porcine renal vascular endothelium. *Transplantation* **37**, 627–630.

Pescovitz, M. D., Lunney, J. K., Sachs, D. H. (1984). Preparation and characterization of monoclonal antibodies reactive with porcine PBL. *J. Immunol.* **133**, 368–375.

Pescovitz, M. D., Lunney, J. K., Sachs, D. H. (1985). Murine anti-swine T4 and T8 monoclonal antibodies: Distribution and effects on proliferative and cytotoxic T cells. *J. Immunol.* **134**, 37–44.

Pescovitz, M. D., Hsu, S.-M., Katz, S. I., *et al.* (1990). Characterization of a porcine CD1-specific mAb that distinguishes CD4/CD8 double-positive thymic from peripheral T lymphocytes. *Tissue Antigens* **35**, 151–156.

Pescovitz, M. D., Aasted, B., Canals, A., *et al.* (1994). Analysis of monoclonal antibodies reactive with the porcine CD4 antigen. *Vet. Immunol. Immunolpathol.* **43**, 233–236.

Pescovitz, M. D., Book, B. K., Aasted, B., *et al.* (1998). Analyses of monoclonal antibodies reacting with porcine CD3: Results from the Second International Swine CD Workshop. *Vet. Immunol. Immunopathol.*, in press.

Pierce, A. E., Smith, M. W. (1967). The intestinal absorption of pig and bovine immune lactoglobulin and human serum albumin by the new-born pig. *J. Physiol. (Lond.)* **190**, 1–18.

Porter, P. (1969). Transfer of immunoglobulins IgG, IgA and IgM to lacteal secretions in the parturient sow and their absorption by the neonatal pig. *Biochem. Biophys. Acta* **181**, 381–392.

Porter, P. (1973). Studies of porcine secretory IgA and its component chains in relation to intestinal absorption of colostral immunoglobulins by the neonatal pig. *Immunology* **24**, 163–174.

Pratt, K., Sachs, D. H., Germana, S., *et al.* (1990). Class II genes of miniature swine. II. Molecular identification and characterization of DRB(b) genes from the SLA^c haplotype. *Immunogenetics* **31**, 1–13.

Prokesová, L., Trebichavsky I., Kováru, F., Kostka, J., Rejnek, J. (1981). Ontogeny of immunoglobulin synthesis. Production of IgM, IgG and IgA in pig foetuses. *Develop. Comp. Immunol.* **5**, 491–499.

Rasmusen, B. A. (1964). Gene interaction and the A-O blood group system in pigs. *Genetics* **50**, 191–198.

Rasmusen, B. A., Christian, L. L. (1976). H blood types in pigs as predictors of stress susceptibility. *Science* **191**, 947–948.

Rasmusen, B. A., Hagen, K. L. (1973). The H blood-group system and reproduction in pigs. *J. Anim. Sci.* **37**, 568–573.

Renard, C., Kristensen, B., Gautschi, C., *et al.* (1988). Joint report of the first international comparison test on swine leukocyte antigens. *Anim. Genet.* **19**, 63–72.

Reynaud, C.-A., McKay, C. R., Müller, R. G., Weill, J.-C. (1991). Somatic generation of diversity in a mammalian primary lymphoid organ: The sheep ileal Peyer's patches. *Cell* **64**, 995–1005.

Rice, C. E., L'Ecuyer, C. (1963). Complement titers of naturally and artificially raised piglets. I. In piglets of different birth weights. *Can. J. Comp. Med. Vet. Sci.* **27**, 157–161.

Richards, C. D., Rafferty, J. A., Reynolds, J. J., Saklatvala, J. (1991). Porcine collagenase from synovial fibroblasts: cDNA sequence and modulation of expression of RNA *In vitro* by various cytokines. *Matrix* **11**, 161–167.

Ridings, P. C., Bloomfield, G. L., Holloway, S., Windsor, A. C. J., Jutila, M. A., Fowler, A. A., Sugerman, H. J. (1995). Sepsis-induced acute lung injury is attenuated by selectin blockade following the onset of sepsis. *Arch. Surg.* **130**, 1199–1208.

Rohrer, G. A., Vögeli, P., Stranzinger, G., Alexander, L. J., Beattie, C. W. (1996). Mapping 28 erythrocyte antigen, plasma protein and enzyme polymorphisms using an efficient genomic scan of the porcine genome. *Anim. Genet.* **28**, 328–330.

Rosendal, S., Frandsen, P. L., Nielsen, J. P., Gallily, R. (1995). *Pasteurella multocida* toxin induces IL-6, but not IL-1α or TNF-α in fibroblasts. *Can. J. Vet. Res.* **59**, 154–156.

Rothkötter, H. J., Pabst, R. (1989). Lymphocyte subsets in jejunal and ileal Peyer's patches in normal and gnotobiotic minipigs. *Immunology* **67**, 103–108.

Rothkötter, H. J., Zimmermann, H. J., Pabst, R. (1990). Size of jejunal Peyer's patches and migration of lymphocyte subsets in pigs after resection or transposition of the continuous ileal Peyer's patch. *Scand. J. Immunol.* **31**, 191–197.

Rothkötter, H. J., Ulbrich, H., Pabst, R. (1991). The postnatal development of gut lamina propria lymphocytes: number, proliferation and T and B cell subsets in conventional and germfree pigs. *Pediatr. Res.* **29**, 237–242.

Rothkötter, H. J., Huber, T., Barman, N. N., Pabst, R. (1993). Lymphoid cells in afferent and efferent intestinal lymph: lymphocyte subpopulations and cell migration. *Clin. Exp. Immunol.* **92**, 317–322.

Rothkötter, H. J., Kirchhoff, T., Pabst, R. (1994). Lymphoid and non-lymphoid cells in the epithelium and lamina propria of intestinal mucosa of pigs. *Gut* **35**, 1582–1589.

Rothkötter, H. J., Hriesik, C., Pabst, R. (1995). More newly formed T than B lymphocytes leave the intestinal mucosa via lymphatics. *Eur. J. Immunol.* **25**, 866–869.

Saalmüller, A. (1996). Characterization of swine leukocyte differentiation antigens. *Immunol. Today* **17**, 352–354.

Saalmüller, A., Aasted, B., Canals, A., *et al.* (1994a). Analyses of monoclonal antibodies reactive with porcine CD6. *Vet. Immunol. Immunopathol.* **43**, 243–247.

Saalmüller, A., Aasted, B., Canals, A., *et al.* (1994b). Analyses of monoclonal antibodies reactive with porcine CD5. *Vet. Immunol. Immunopathol.* **43**, 237–242.

Saalmüller, A., Hirt, W., Maurer, S., *et al.* (1994c). Discrimination between two subsets of porcine $CD8^+$ cytolytic T lymphocytes by the expression of CD5 antigen. *Immunology* **81**, 578–583.

Saalmüller, A., Denham, S., Haverso, K., Davis, B., Dominguez, J., Pescovitz, M. D., Stokes, C., Zuckerman F., Lunney, J. (1998). The second international swine CD workshop. *Vet. Immunol. Immunopath.*, in press.

Sacco, R. E., Nibbelink, S. K., Baarsch, M. J., Murtaugh, M. P., Wannemuehler, M. J. (1996). Induction of Interleukin (IL)-1β and IL-8 mRNA Expression in Porcine Macrophages by Lipopolysaccharide from *Serpulina hyodysenteriae*. *Infect. Immun.* **64**, 4369–4372.

Sachs, D. H., Germana, S., El-Gamil, M., *et al.* (1988). Class II genes of miniature swine. *Immunogenetics* **28**, 22–32.

Saison, R. (1958). Report of a blood group system in swine. *J. Immunol.* **80**, 463–467.

Saison, R. (1967). Two new antibodies, anti-Nb and anti-Nc in the N blood-group system in pigs. *Vox Sang.* **12**, 215–220.

Saison, R., Rasmusen, B. A., Hradecky, J. (1967). Da, a factor in a new blood group system in pigs. *Canad. J. Genet. Cytol.* **9**, 794–798.

Sandison, A. T., Anderson, L. J. (1969). Tumors of the thymus in cattle, sheep and pigs. *Cancer Res.* **29**, 1146–1150.

Sasaki, K., Pabst, R., Rothkötter, H. J. (1994). The unique ultrastructure of high-endothelial venules in inguinal lymph nodes of the pig. *Cell Tissue Res.* **276**, 85–90.

Scamurra, R. W., Arriaga, C., Sprunger, L., Baarsch, M. J., Murtaugh, M. P. (1996). Regulation of interleukin-6 expression in porcine immune cells. *J. Interferon Cytok. Res.* **16**, 289–296.

Schnitzlein, W. M., Zuckermann, F. (1996). Defining the isoforms of porcine CD45. In *Advances in Swine in Biomedical Research* (eds Tumbleson, M., Schook, L.), pp. 345–357. Plenum Press, New York, USA.

Schnitzlein, W., Zuckermann, F. (1998). Determination of the specificity of CD45 and CD45R monoclonal antibodies through the use of transfected hamster cells producing individual porcine CD45 isoforms. *Vet. Immunol. Immunopath*, in press.

Scholl, T., Lunney, J. K., Mebus, C. A., Duffy, E., Martins, C. L. (1989). Virus-specific cellular blastogenesis and interleukin-2 production in swine after recovery from african swine fever. *Am. J. Vet. Res.* **50**, 1781–1786.

Schook, L. B., Rutherford, M. S., Lee, J.-K., *et al.* (1996). The swine major histocompatibility complex. *The Major Histo-Compatibility Complex Region of Domestic Animal Species* (ed. Schook, L. B., Lamont, S. J.), pp. 213–243. CRC Press, New York, NY.

Schultze, R. D., Wang, J. T., Dunne, H. W. (1971). Development of humoral immune response of pig. *Am J. Vet. Res.* **32**, 1331–1336.

Segre, D. (1966). Swine in immunologic research. In *Swine in Biomedical Research* (eds Bustad, L. K., McClellan, R. O., Burns, M. P.), pp. 263–272. Battelle Northwest, Richland.

Segre, D., Kaeberle, M. L. (1962). The immunologic behaviour of baby pigs. Production of antibody in three week old pigs. *J. Immunol.* **89**, 782–789.

Sheldrake, R. F. (1990). Memory associated with IgA in the porcine respiratory tract. *Res. Vet. Sci.* **48**, 47–52.

Sheldrake, R. F., Romalis, L. F., Saunders, M. M. (1991). Origin of antibody in porcine bile after intraperitoneal immunisation. *Res. Vet. Sci.* **50**, 242–244.

Shi, J. H., Ross, C. R., Chengappa, M. M., Sylte, M. J., McVey, D. S., Blecha, F. (1996). Antibacterial activity of a synthetic peptide (pr-26) derived from pr-39, a proline-arginine-rich neutrophil antimicrobial peptide. *Antimicrobial Agent Chemotherap.* **40**, 115–121.

Shia, Y.-C., Gautschi, C., Ling, M.-S., *et al.* (1991). RFLP analysis of SLA haplotypes in Swiss Large White and American Hampshire pigs using SLA class I and class II probes. *Anim. Biotech.* **2**, 75–91.

Shia, Y.-C., Bradshaw, M., Rutherford, M. S., *et al.* (1995). PCR-based genotyping for characterization of SLA-DQB and SLA-DRB alleles in domestic pigs. *Anim. Genet.* **26**, 91–99.

Shibata, T., Akita, T., Abe, T. (1993). Genetic polymorphism of the sixth component of complement (C6) in the pig. *Anim. Genet.* **24**, 97–100.

Shirota, K., Nomura, Y. (1988). Ultrastructure of glomerulopathy in swine. *Jpn. J. Vet. Sci.* **50**, 1–8.

Shirota, K., Koyama, R., Nomura, Y. (1986). Glomerulopathy in swine: Microscopic lesions and IgG or C3 deposition in 100 pigs. *Jpn. J. Vet. Sci.* **48**, 15–22.

Sikes, D., Crimmins, L. T., Fletcher, O. J. (1969). Rheumatoid arthritis of swine: A comparative pathologic study of clinical spontaneous remissions and exacerbations. *Am. J. Vet. Res.* **30**, 753–769.

Silmanowicz, P., Truchlinska, J., Lipinska, M. (1978). Immunological experimental arthritis in pigs II. The level of complement in the pigs immunized with the Erysipelothrix rhusiopathiae and virus of Aujeszky's disease. *Arch. Immunol. Ther. Exp.* **26**, 813–817.

Singer, D. S., Camerini-Otero, R. D., Satz, M. L., *et al.* (1982). Characterization of a porcine genomic clone encoding a major histocompatibility antigen: Expression in mouse L cells. *Proc. Natl Acad. Sci. USA* **79**, 1403–1407.

Singer, D. S., Ehrlich, R., Golding, H., *et al.* (1988). *The Molecular Biology of the MHC of Domestic Animal Species* (eds. Ames, W. C. R. M., Ames, L. S.), pp. 53–62. ISU Press: IA.

Slauson, D. O., Sanchez-Vizcaino, J. M. (1981). Leukocyte-dependent platelet vasoactive amine release and immune complex deposition in African swine fever. *Vet. Pathol.* **18**, 813–826.

Smith, M. W., Burton, K. A., Munn, E. A. (1979). Vacuolation and non-specific protein transport by the new-born pig intestine. In *Protein Transmission through Living Membranes* (ed. Hemmings, W. A.), pp. 197–212. Elsevier/North Holland Amsterdam, The Netherlands.

Smith, T. P. L., Roher, G. A., Alexander, L. J., *et al.* (1995). Direct integration of the physical and genetic linkage maps of swine chromosome 7 reveals that the SLA spans the centromere. *Genome Res.* **5**, 259–271.

Spertini, O., Kansas, G. S., Reimann, K. A., Mackay, C. R., Tedder, T. F. (1991). Function and evolutionary conservation of distinct epitopes on the leukocyte adhesion molecule-1 (TQ-1, Leu-8) that regulate leukocyte migration. *J. Immunol.* **147**, 942–949.

Sprague, L. M. (1958). On the distribution and inheritance of a natural antibody in cattle. *Genetics* **43**, 913–918.

Springer, T. A. (1994). Traffic signals for lymphocyte recirculation and leukocyte emigration: the multistep paradigm. *Cell* **76**, 301–314.

Steinmassl, M., Wolf, G. (1990). Formation of interleukin-2 and interferon alpha by mononuclear leukocytes of swine after *in vitro* stimulation with different virus preparations. *Zentralblatt Fur Veterinarmedizin – Reihe B.* **37**, 321–331.

Sterzl, J., Kováru, F. (1977). Development of lymphatic tissue and immunocompetency in pig fetuses and germ-free piglets. *Acta Vet. Brno* **46** (Suppl. 4), 13–53.

Sterzl, J., Rejnek, J., Trávnícek, J. (1966). Impermeability of pig placenta for antibodies. *Folia microbiologica* **11**, 7–10.

Storici, P., Scocchi, M., Tossi, A., Gennaro, R., Zanetti, M. (1994). Chemical synthesis and biological-activity of a novel antibacterial peptide deduced from a pig myeloid cDNA. *FEBS Letters* **337**, 303–307.

Sun, J., Butler, J. E. (1996). Molecular characteristics of VDJ transcripts from a newborn piglet. *Immunology* **88**, 331–339.

Sun, J., Butler, J. E. (1997). Sequence analysis of pig switch μ, $C\mu$ and $C\mu m$. *Immunogenetics* **46**, 452–460.

Sun, J., Kacskovics, I., Brown, W. R., Butler, J. E. (1994). Expressed swine V_H genes belong to a small V_H gene family homologous to human V_HIII. *J. Immunol.* **153**, 5618–5627.

Sun, J., Hayword, C., Shinde, R., Christensen, R., Butler, J. E. (1998). Antibody repertoire development in fetal and neonatal piglets. I. Four V_H genes account for 80% of V_H usage during 84 days of fetal life. *J. Immunol.*, in press.

Sundt III, T. M., LeGuern, C., Germana S., *et al.* (1992). Characterization of a polymorphism of CD4 in miniature swine. *J. Immunol.* **148**, 3195–3201.

Suzuka, I., Sekiguchi, K., Kodama, M. (1985). Some characteristics of a porcine retrovirus from a cell line derived from swine malignant lymphomas. *FEBS Lett.* **183**, 124–128.

Tanimoto, T., Minami, A., Yano, S., *et al.* (1994). Ileal lymphoma in swine. *Vet. Pathol.* **31**, 629–636.

Telemo, E., Bailey, M., Miller, B. G., Stokes, C. R., Bourne, F. J. (1991). Dietary antigen handling by mother and offspring. *Scand. J. Immunol.* **34**, 689–696.

Thiele, O. W., Hojny, J. (1971). The chemical nature of the pig blood-group substances dissolved in the serum. *Experienta* **27**, 447–448.

Thistlethwaite Jr, J. R., Pennington, L. R., Lunney, J. K., *et al.* (1983). Immunologic characterization of MHC recombinant swine. Production of SLA class specific antisera and detection of Ia antigens on both B and non-B PBL. *Transplantation* **35**, 394–400.

Thome, M., Saalmüller, A., Pfaff, E. (1993). Molecular cloning of porcine t cell receptor α, β, γ and δ chains using polymerase chain reaction fragments of the constant regions. *Eur. J. Immunol.* **23**, 1005–1010.

Tlaskalová, H., Sterzl, J., Hájek, P., Pospícil, M., Ríha, I., Marvanová, H., Kamarytová, V., Mandel, L., Kruml, J., Kováøù, F. (1970). The development of antibody formation during embryonic and postnatal periods. In *Developmental Aspects of Antibody Formation and Structure* (eds Sterzl, J., Ríha, I.), pp. 767–792. Academia, Prague.

Tlaskalová-Hogenová, H., Mandel, L., Trebichavsky, I., Kováru, F., Barot, R., Sterzl, J. (1994). Development of immune responses in early pig ontogeny. *Vet. Immunol. Immunopathol.* **43**, 135–142.

Tlaskalová-Hogenová, H., Cerná, J., Mandel, L. (1981). Peroral immunization of germfree piglets, appearance of antibody-forming cells and antibodies of different isotypes. *Scand. J. Immunol.* **13**, 467–472.

Tossi, A., Scocchi, M., Zanetti, M., Storici, P., Gennaro, R. (1995). Pmap-37, a novel antibacterial peptide from pig myeloid cells – cDNA cloning, chemical synthesis and activity. *Europ. J. Biochem.* **228**, 941–946.

Toyomura, K., Fujimura, T., Murakami, H., Inoue, N., Takeda, J., Kinoshita, T. (1996). Expression cloning and functional analysis of pig homologue of MCP (PMCP). *Mol. Immunol.* **33**, 10.

Trebichavsky, I., Holub, M., Jarosková, L., Mandel, L., Kováru, F. (1981). Ontogeny of lymphatic structures in the pig omentum. *Cell and Tissue Res.* **215**, 437–442.

Trebichavsky, I., Kováru, F., Nemec, M. (1988). Expression of MHC class II antigens and immunoglobulins in immunized pig fetuses. *Folia Biologica* **34**, 53–57.

Trebichavsky, I., Kováru, F., Reháková, Z., Mandel, L., Pospícil, R. (1992). Skin response to tuberculin in pig fetuses and germ-free piglets. *Immunol. Lett.* **33**, 271–276.

Trebichavsky, I., Barotciorbaru, R., Charley, B., Splichal, I. (1993). Induction of inflammatory cytokines by nocardia fractions. *Folia Biol.* **39**, 243–249.

Trebichavsky, I., Sinkora, J., Reháková, Z., Splíchal, I., Whyte, A., Binns, R., Pospícil, R., Tucková, L. (1995). Distribution of gamma/delta T cells in the pig fetus. *Folia Biologica* **41**, 227–237.

Tsang, Y. T., Haskard, D. O., Robinson, M. K. (1994). Cloning and expression kinetics of porcine vascular cell adhesion molecule. *Biochem. Biophys. Res. Commun.* **201**, 805–812.

Tsang, Y. T. M., Stephens, P. E., Licence, S. T., Haskard, D. O., Binns, R. M. (1995). Porcine E-selectin: cloning and functional characterization. *Immunology* **85**, 140–145.

Tuboly, S., Bernath, S., Glavits, R., Medveczky, I. (1988). Intestinal absorption of colostral lymphoid cells in newborn pigs. *Vet. Immunol. Immunopathol.* **20**, 75–85.

Tumbleson, M., Schook, L. B. (1996). Advances in swine in biomedical research. *Advances in Swine in Biomedical Research* (eds Tumbleson, M., Schook, L. B.), pp. 1–4. Plenum Press, New York, USA.

Tuo, W. B., Harney, J. P., Bazer, F. W. (1995). Colony-stimulating factor-1 in conceptus and uterine tissues in pigs. *Biol. Reprod.* **53**, 133–142.

Tyler, J. A., Benton, H. P. (1988). Synthesis of type II collagen is decreased in cartilage cultured with interleukin 1 while the rate of intracellular degradation remains unchanged. *Coll. Relat. Res.* **8**, 393–405.

Vaiman, M., Hauptmann, G., Mayer, S. (1978). Influence of the major histocompatibility complex in the pig (SLA) on serum haemolytic complement levels. *J. Immunogenet.* **5**, 59–65.

van den Berg, C. W., Morgan, B. P. (1994). Complement-inhibiting activities of human CD59 and analogues from rat, sheep, and pig are not homologously restricted. *J. Immunol.* **152**, 4095–4101.

van den Berg, C. W., Harrison, R. A., Morgan, B. P. (1995). A rapid method for the isolation of analogs of human CD59 by preparative SDS-PAGE: application to pig CD59. *J. Immunol. Meth.* **179**, 223–231.

van den Berg, C. W., De la Lastra, J. M., Llanes, D., Morgan, B. P. (1997). Purification and characterization of the pig analogue of human membrane cofactor protein (CD46/MCP). *J. Immunol.* **158**, 1703–1709.

Vandermeer, T. J., Menconi, M. J., Osullivan, B. P., Larkin, V. A., Wang, H. L., Sofia, M., Fink, M. P. (1995). Acute lung injury in endotoxemic pigs – role of leukotriene B-4. *J. Appl. Physiol.* **78**, 1121–1131.

van Zeveren, A., Bouquet, Y., Van de Weghe, A., Coppieters, W. (1990). A genetic blood marker study on 4 pig breeds. I. Estimation and comparison of within-breed variation. *J. Anim. Breed Genet.* **107**, 104–112.

Vega-López, M. A., Telemo, E., Bailey, M., Stevens, K., Stokes, C. R. (1993). Immune cell distribution in the small intestine of the pig: immunohistological evidence for an organized compartmentalization in the lamina propria. *Vet. Immunol. Immunopathol.* **37**, 49–60.

Vega-López, M. A., Bailey, M., Telemo, E., Stokes, C. R. (1995). Effect of early weaning on the development of immune cells in the pig intestine. *Vet. Immunol. Immunopathol.* **44**, 319–327.

Vögeli, P. (1990). Blood groups of pigs, serological and genetical studies (in German). PhD habilitation, Federal Institute of Technology (ETH), Zürich, Switzerland.

Vögeli, P., Gerwig, C., Schneebeli, H. (1983). The A-O and H blood group systems, some enzyme systems and halothane sensitivity of two divergent lines of Landrace pigs using index selection procedures. *Livest. Prod. Sci.* **10**, 159–169.

Vögeli, P., Kühne, R., Gerwig, C., Kaufmann, A., Wysshaar, M., Stranzinger, G. (1988). Prediction of the halothane genotypes with the aid of the S, PI II, HAL, H, PO2, PGD haplotypes of parents and offspring in Swiss Landrace pigs (in German). *Züchtungskunde* **60**(1), 24–37.

Vögeli, P., Bolt, R., Fries, R., Stranzinger, G. (1994). Co-segregation of the malignant hyperthermia and the Arg^{615}-Cys^{615} mutation in the skeletal muscle calcium release channel protein in five European Landrace and Pietrain pig breeds. *Anim. Genet.* **25**, Suppl. 1, 59–66.

Vögeli, P., Bertschinger, H. U., Stamm, M., Stricker, C., Hagger, C., Fries, R., Rapacz, J., Stranzinger, G. (1996). Genes specifying receptors for F18 fimbriated *Escherichia coli*, causing oedema disease and postweaning diarrhoea in pigs map to chromosome 6. *Anim. Genet.* **27**, 321–328.

Walsh, C. J., Carey, P. D., Cook, D. J., Bechard, D. E., Fowler, A. A., Sugerman, H. J. (1991). Anti-CD18 antibody attenuates neutropenia and alveolar capillary-membrane injury during Gram-negative sepsis. *Surgery* **110**, 205–211.

Wardley, R. C., Wilkinson, P. (1985). An immunological approach to vaccines against African swine fever virus. *Vaccine* **3**, 54–56.

Welch, S. K., Saif, L. J., Ram, S. (1988). Cell-mediated immune responses of suckling pigs inoculated with attenuated or virulent transmissible gastroenteritis virus. *Am. J. Vet. Res.* **49**, 1228–1234.

Westermann, J., Pabst, R. (1996). How organ-specific is the migration of 'naive' and 'memory' T cells? *Immunol. Today* **17**, 278–282.

Whary, M. T., Zarkower, A., Confer, F. L., Ferguson, F. G. (1995). Age-related differences in subset composition and activation responses of intestinal intraepithelial and mesenteric lymph-node lymphocytes from neonatal swine. *Cellular Immunol.* **163**, 215–221.

Whyte, A. Binns, R. M. (1994). Adhesion molecule expression and infiltrating maternal leucocyte phenotypes during blastocyst implantation in the pig. *Cell Biol. Int.* **18**, 759–766.

Whyte, A., Garratt, L., James, P. S., Binns, R. M. (1993). Distribution of saccarides in pig lymph-node high endothelial venules and associated lymphocytes visualized using fluorescent lectins and confocal microscopy. *Histochem. J.* **25**, 726–734.

Whyte, A., Haskard, D. O., Binns, R. M. (1994). Infiltrating $\gamma\delta$ T-cells and sSelectin eEndothelial ligands in the cutaneous phytohaemagglutinin-induced inflammatory reaction. *Vet. Immunol. Immunopathol.* **41**, 31–40.

Whyte, A., Wooding, P., Nayeem, N., Watson, S. R., Rosen, S. D., Binns, R. M. (1995). The L-selectin counter-receptor in porcine lymph nodes. *Biochem. Soc. Trans.* **23**, 159S.

Wilders, M. M., Drexhage, H. A., Weltevreden, E. F., Mullink, H., Duijvestijn, A., Meuwissen, S. G. M. (1983). Large mononuclear Ia-positive veiled cells in Peyer's patches. 1. Isolation and characterisation in rat, guinea-pig and pig. *Immunology* **48**, 453–460.

Williams, P. P. (1993). Immunomodulating effects of intestinal absorbed maternal colostral lymphocytes by neonatal pigs. *Can. J. Vet. Res.* **57**, 1–8.

Wilson, A. D., Stokes, C. R., Bourne, F. J. (1986a). Responses of intraepithelial lymphocytes to T-cell mitogens: a comparison between murine and porcine responses. *Immunology* **58**, 621–625.

Wilson, A. D., Stokes, C. R., Bourne, F. J. (1986b). Morphology and functional characteristics of isolated porcine intraepithelial lymphocytes. *Immunology* **59**, 109–113.

Wilson, A. D., Stokes, C. R., Bourne, F. J. (1989). Effect of age on absorbtion and immune responses to weaning or introduction of novel dietary antigens in pigs. *Res. Vet. Sci.* **46**, 180–186.

Wilson, A. D., Haverson, K., Bland, P. W., Stokes, C. R., Bailey, M. (1996). Expression of class II MHC antigens on normal pig intestinal endothelium. *Immunology* **88**, 98–103.

Windsor, A. C., Walsh, C. J., Mullen, P. G., Cook, D. J., Fisher, B. J., Blocher, C. R., Leeper-Woodford, S. K., Sugerman, H. J., Fowler, A. A. D. (1993). Tumor necrosis factor-α blockade prevents neutrophil CD18 receptor upregulation and attenuates acute lung injury in porcine sepsis without inhibition of neutrophil oxygen radical generation. *J. Clin. Invest.* **91**, 1459–1468.

Windsor, A. C. J., Carey, P. D., Sugerman, H. J., Mullen, P. G., Walsh, C. J., Fisher, B. J., Blocher, C. R., Fowler, A. A. (1994). Differential activation of alveolar, pulmonary arterial, and systemic arterial neutrophils demonstrates the existence of distinct neutrophil subpopulations in experimental sepsis. *Shock* **1**, 53–59.

Winkler, G. F. (1988). Pulmonary intravascular macrophages in domestic animal species. Review of structural and functional properties. *Am. J. Anat.* **181**, 217–234.

Wollert, S., Rasmussen, I., Lundberg, C., Gerdin, B., Arvidsson, D., Haglund, U. (1993). Inhibition of CD18-dependent Adherence of Polymorphonuclear Leukocytes does not Affect Liver Oxygen Consumption in Fecal Peritonitis in Pigs. *Circ. Shock* **41**, 230–238.

Winstanley, F. P., Eckersall, P. D. (1992). Bioassay of Bovine Interleukin-1-Like Activity. *Res. Vet. Sci.* **52**, 273–276.

Woolley, S. T., Whyte, A., Licence, S. T., Haskard, D. O., Wooding, F. B. P., Binns, R. M. (1995). Differences in E-selectin expression and leucocyte infiltration induced by inflammatory agents in a novel subcutaneous sponge matrix model. *Immunology*, **84**, 55–63.

Xu, L. R., Carr, M. M., Bland, A. P., Hall, G. A. (1993). Histochemistry and morphology of porcine mast cells. *Histochem. J.* **25**, 516–522.

Yang, H., Binns, R. M. (1993a). CD44 is not directly involved in the binding of lymphocytes to cultured high endothelial cells from peripheral lymph nodes. *Immunology* **79**, 418–424.

Yang, H., Binns, R. M. (1993b). Expression and regulation of the porcine CD44 molecule. *Cell. Immunol.* **149**, 117–129.

Yang, H., Binns, R. M. (1993c). Isolation and characterization of the soluble and membrane-bound porcine CD44 molecules. *Immunology* **78**, 547–554.

Yang, H, Parkhouse, R. M. (1996). Phenotypic classification of porcine lymphocyte subpopulations in blood and lymphoid tissues. *Immunology* **89**, 76–83.

Yang, H., Hutchings, A., Binns, R. M. (1993). Preparation and reactivities of anti-porcine CD44 monoclonal antibodies. *Scand. J. Immunol.* **37**, 490–498.

Yang, H., Oura, C. A., Kirkham, P. A., *et al.* (1996). Preparation of monoclonal anti-porcine CD3 antibodies and preliminary characterization of porcine T lymphocytes. *Immunol.* **88**, 577–585.

Yang, Y.-G., Ohta, S., Yamada, S., *et al.* (1995). Diversity of T cell receptor δ-chain cDNA in the thymus of a one-month-old pig. *J. Immunol.* **155**, 1981–1993.

Yang, Y. G., Sergio, J. J., Swinson, K., Glaser, R. M., Monroy, R., Sykes, M. (1996). Donor-specific growth factors promote swine hematopoiesis in severe combined immune deficient mice. *Xenotransplantation* **3**, 92–101.

Yen, J. R., Pond, W. G. (1987). Effect of dietary supplementation with vitamin C or carbadox on weaning pigs subjected to crowding stress. *J. Anim. Sci.* **64**, 1672–1681.

Yoshie, T. (1991). Study on the swine glomerulopathy resembling human IgA nephropathy. *Japan J. Nephrol.* **33**, 179–189.

Zanetti, M., Gennaro, R., Romeo, D. (1995). Cathelicidins – a novel protein family with a common proregion and a variable C-terminal antimicrobial domain. *FEBS Lett.* **374**, 1–5.

Zhao, C. Q., Ganz, T., Lehrer, R. I. (1995a). Structures of genes for 2 cathelin-associated antimicrobial peptides – prophenin-2 and pr-39. *FEBS Lett.* **376**, 130–134.

Zhao, C. Q., Ganz, T., Lehrer, R. I. (1995b). The structure of porcine protegrin genes. *FEBS Lett.* **368**, 197–202.

Zhou, Y., Murtaugh, M. P., Molitor, T. W. (1992). Characterization of transforming growth factor β expression in porcine immune cells. *Mol. Immunol.* **29**, 965–970.

Zhou, Y., Lin, G., Baarsch, M. J., Scamurra, R. W., Murtaugh, M. P. (1994). Interleukin-4 suppresses inflammatory cytokine gene transcription in porcine macrophages. *J. Leuk. Biol.* **56**, 507–513.

Ziegler-Heitbrock, H. W., Passlick, B., Kafferlein, E., Coulie, P. G., Izbicki, J. R. (1992). Protection against lethal pneumococcal septicemia in pigs is associated with decreased levels of interleukin-6 in blood. *Infect. Immun.* **60**, 1692–1694.

Zikan, J. (1973). J-chain in pig immunoglobulins. *Immunochemistry* **10**, 351–354.

Zuckermann, F., Peavey, C., Schnitzlein, W., *et al.* (1998a). Definition of the specificity of monoclonal antibodies against porcine CD45 and CD45R: report from the CD45/CD45R and CD44 subgroup of the Second International Swine CD Workshop. *Vet. Immunol. Immunopathol.*, in press.

Zuckermann, F., Pescovitz, M. D., Aasted, B., *et al.* (1998b). Report on the analyses of mAb reactive with porcine CD8 for the Second International Swine CD Workshop. *Vet. Immunol. Immunopathol.*, in press.

Zuckermann, F. A., Husmann, R. J. (1996). Functional and phenotypic analysis of porcine peripheral blood CD4/CD8 double positive T cells. *Immunology* **87**, 500–512.

XII Immunology of Camels and Llamas

1. Introduction

The *Camelidae* (camelids) is the only surviving family of the suborder of the Tylopoda, order of the Artiodactyla. Although they regurgitate their food like ruminants and have a similar digestive biochemistry, the anatomy of the three-compartment gastric system and the ultrastructure of the prostomach mucosa sets the Tylopodes apart from the Ruminentia. The camelids originated in North America during the Eocene era and the ancestral forms migrated via the Behring land bridge to the Eurasian continent and via the Caribbean land bridge into South America, the earliest camels appearing in Europe and Asia in the Upper Miocene (Pickford *et al.*, 1993). The present day camels of the old world and the llamas of the new world are characterized by anatomical and biochemical adaptations allowing them to survive in extremely harsh environments essentially hot, cold or altitude deserts. The dromedary camel (*Camelus dromedarius*) is slim legged and shorthaired and its habitats are warm arid and semiarid areas. In contrast, the Bactrian camel (*Camelus bactrianus*) is stocky and longhaired and its habitats are cold and mountainous deserts. Bactrian and Dromedary camels interbreed and are possibly divergent adaptations within a single species. Domestications occurred some 4000 years ago in the Arabian peninsula (Dromedary camel) and in Turkmenistan (Bactrian camel) (Zeuner, 1963; Wilson, 1984). At present, no original wild dromedary camels remain and the population of wild Bactrian camels in the Gobi desert is estimated at less than 1000. The global camel population is stagnating at approximately 20 million despite the growing interest in camel meat and milk products (FAO, 1990). Although, at the global level, camels are of limited economic significance, in arid or marginal areas they appear a most suitable and sustainable livestock (Farah, 1996). In recent years, a new and well-funded impetus for research in camel physiology and camel diseases came from the revival in the oil states of the age-old Bedouin tradition of camel racing, in addition to greater efforts in the Arabic countries (Wernery and Kaaden, 1995). One aspect in particular is directly of interest to immunology, namely the camel's resistance to numerous pathogenic microorganisms.

New World camelids are grouped in four species: the domesticated llama (*Lama glama*) and alpaca (*Lama pacos*) and the putative wild ancestors, the guanaco (*Lama guanicoe*) and the vicuña (*Lama vicugna*). The four 'species' interbreed and have fertile offspring. Mitochondrial DNA studies indicate that hybridization did occur during the 6000–7000 years of domestication (Stanley *et al.*, 1994; Kessler *et al.*, 1995).

In addition to the production of llama and alpacas for wool, the recent fashion of affluent countries of keeping llamas and alpacas as exotic pets has also contributed considerably to our recent knowledge of the biology and medicine of South American camelids (Fowler, 1989).

Although the immunology of the camelids is largely unknown, the findings of the last 5 years have completely upset the existing paradigms of antibody structure, antigen binding site and repertoire generation.

Three years ago, a stunning and completely fortuitous observation was published revealing that up to 75% of camelid antibodies belong to IgG subclasses devoid of light chains (Hamers-Casterman *et al.*, 1993).

At present, camelids are the only known vertebrates to possess naturally occurring heavy chain antibodies. That this discovery was only made recently can be attributed not only to the fact that camelids are largely unstudied at the biochemical level but probably also to a consensus attitude to science in which facts which do not fit current paradigms are perhaps given too little attention.

A few years ago, in a study on neonatal immunology, the consistent presence of an unknown 90 kDa molecule copurifying with IgG like material in camel serum was noted, but the observation was not followed up and the nature of the 90 kDa material retrospectively identified as heavy chain immunoglobulins was not established (Ungar-Waron *et al.*, 1987). The camelid heavy chain antibodies lack not only the light chain but also the first constant domain C_H1, bringing the V_H domain in a position adjacent to the hinge.

The existence of an extended antigen-binding repertoire within the heavy chain antibodies raises questions as to how diversity is achieved and how a single V domain scaffold can accommodate a variety of different types of antigen-binding sites. The findings postulate alternative pathways or modalities of B cell differentiation in which light chain rearrangement is bypassed. This chapter therefore differs from the other chapters by being oriented primarily towards the description of the immunoglobulin molecules, the genes underlying their assembly, the implications in phylogeny and ontogeny and the vast potential of application in medicine and biotechnology.

Handbook of Vertebrate Immunology
ISBN 0-12-546401-0

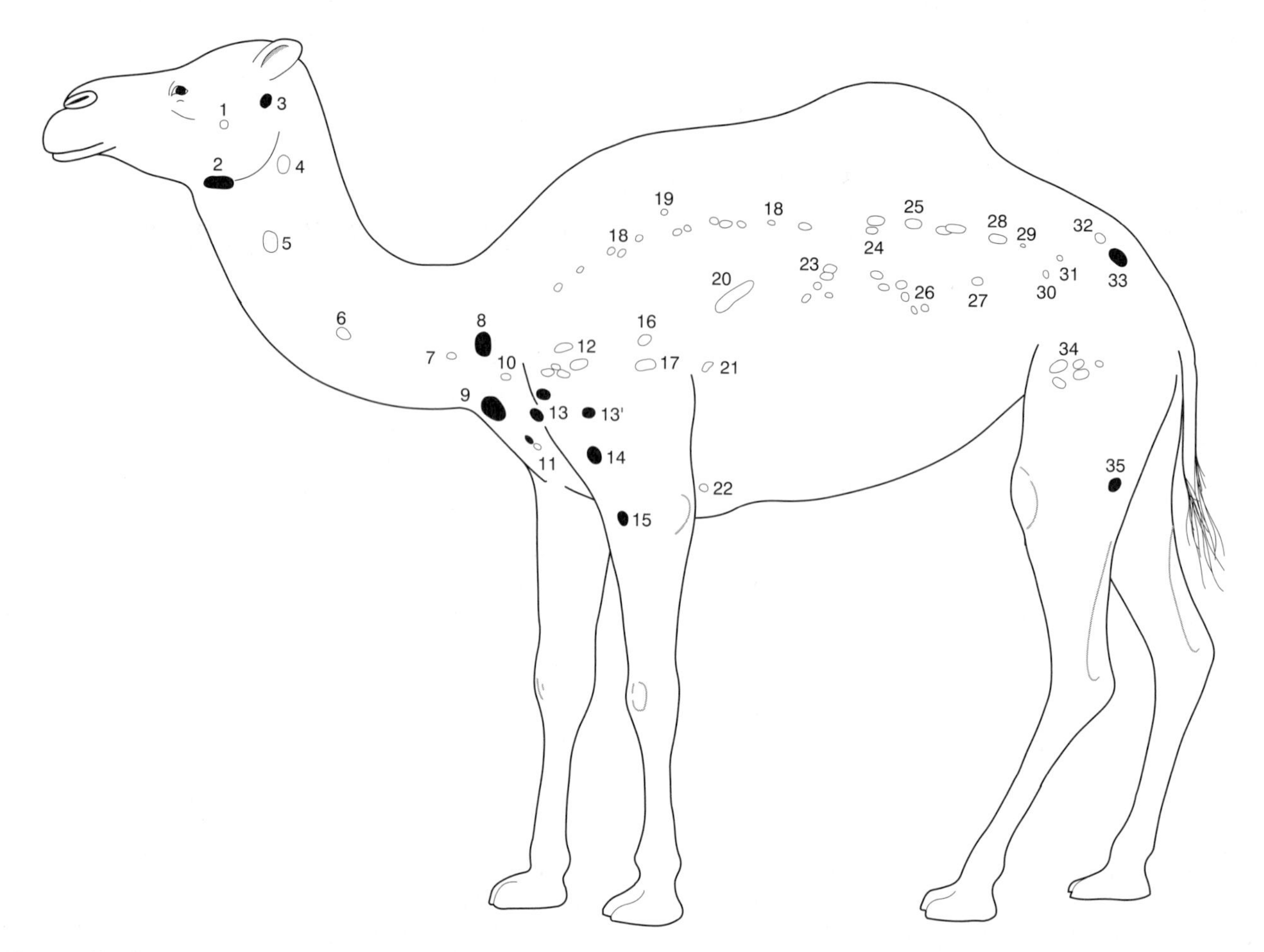

Figure XII.2.1 Schematic drawing of the lymph nodes (left view). Solid nodes are palpable. 1 Ln. pterygoideus; 2 Lnn. mandibulares; 3 Lnn. parotidei superficiales; 4 Lnn. retropharyngei mediales; 5 Lnn. cervicales profundi craniales; 6 Lnn. cervicales profundi medii; 7 Lnn. cervicales profundi caudales; 8 Lnn. cervicales superficiales dorsales; 9 Lnn. cervicales superficiales ventrales; 10 Lnn. mediastinales craniales; 11 Lnn. sternales craniales; 12 Lnn. mediastinales medii; 13 Lnn. axillares; 13′ Ln. axillaris accessorius; 14 Lnn. pectorales; 15 Lnn. cubitales; 16 Lnn. tracheobronchales medii; 17 Lnn. tracheobronchales sinistri; 18 Lnn. thoracici aortici; 19 Lnn. intercostales; 20 Lnn. mediastinales caudales; 21 Lnn. phrenici; 22 Lnn. sternales caudales; 23 Lymphocentrum celiacum; 24 Lnn. renales; 25 Lnn. lumbales aortici; 26 Lymphocentrum mesentericum craniale; 27 Lymphocentrum mesentericum caudale; 28 Lnn. iliaci mediales; 29 Lnn. sacrales; 30 Lnn. iliofemorales; 31 Lnn. hypogastrici; 32 Lnn. ischiadici; 33 Lnn. tuberales; 34 Lnn. inguinales superficiales (Lnn. scrotales mammarii); 35 Lnn. poplitei. Reproduced from Smuts and Bezuidenhout (1987) with permission.

2. Lymphoid Organs and their Anatomical Distribution

Lymph nodes in the camel are depicted in Figure XII.2.1 (Smuts and Bezuidenhout, 1987). Until quite recently, the only extensive description of the camel and llama lymph nodes was limited to difficult-to-access dissertations (Rogier, 1934; Cambirazio, 1967; Carrasco, 1968) or summarized in comprehensive treatises (Curasson, 1947; Grassé, 1972). Recently Agba *et al.* (1996a,b) published a detailed description of the location and drainage patterns of the lymph nodes of the camel. They note several discrepancies with the short description of Smuts and Bezuidenhout (1987). Small and multiple lymph nodes rather than single lymph nodes are frequent. A striking feature is the lack of lymphatic drainage of the large intestine apparently compensated by an extensive development of the Peyer's patches of which more than 700 have been enumerated.

Figure XII.2.2 shows enlarged cervical lymph nodes due to chronic lymphadenitis caused by *Corynebacterium pyogenes*, a disease well known among camel herders (Moustafa, 1994).

3. Leukocytes and their Markers

The proportion of different cell types in the camelids has been well established and camelids are characterized by high proportions of neutrophils and of lymphocytes (Tables XII.3.1. and XII.3.2). Purification of the polymor-

Figure XII.2.2 A camel with lymphadenitis. Enlargement of the inferior cervical lymph nodes at the base of the neck. Reproduced from Moustafa (1994) with permission.

Table XII.3.1 Normal blood parameters in dromedary camels

Erythrocytes	7.3–12.0 × $10^6/\mu l$
Leukocytes	6.0–13.5 × $10^3/\mu l$
Neutrophils	50–65%
Eosinophils	0–6%
Basophils	0–2%
Lymphocytes	30–45%
Monocytes	2–8%
Platelets	200–700 × $10^3/\mu l$

Adapted from Wernery (1995), Wernery and Kaaden (1995) and Gupta *et al.* (1979).
In *Bacillus aureus* infections the leukocyte count drops dramatically (1.2–5.1 × $10^3/mm^3$) and characterized by a considerable rise of the percentage of eosinophils (10–54%) and a drop of the percentage of neutrophils (4–44%) (Wernery *et al.*, 1992).

Table XII.3.2 Normal blood parameters in llamas

Erythrocytes	10.8–17.1 × $10^6/\mu l$
Leukocytes	7.5–23.8 × $10^3/\mu l$
Neutrophils	2.500–21.500/μl
Eosinophils	0–1.500/μl
Basophils	0–400/μl
Lymphocytes	1000–8000/μl
Monocytes	100–1500/μl
Platelets	200–600 × $10^3/\mu l$

Adapted from Fowler (1989).

phonuclear and the monocyte population obtained from camel blood by lysis of contaminating erythrocytes can be problematic as the oval shaped red blood cells, discovered more than 120 years ago (Gulliver, 1875), appear unusually

Table XII.3.3 Cross-reacting monoclonal antibodies against characterized antigens

Designation	*Reactions*	*Antigens*
PGBL6A	+	CD2a[a]
TUK4	++	CD14[a]
bIG-10	++	CD14[a]
bIG-13	++	CD14[a]
PNK-1	++	CD18[a]
FW4/101	Weak	CD29[a,b]
IL-A118	++	CD44[a]
IL-A148	++	CD44[a]
25-32		CD44[b]
B18		CD81[b]
FW3-181		MHC Class I[b]
bIG-6	++	β-Microglobulin[a]

[a]J. Naessens, ILRI, personal communication.
[b]3rd Ruminant Leukocyte workshop (Vilmos *et al.*, 1996)

resistant to osmotic lysis. Acridine orange, a nucleic acid dye, and fluorescence-labelled bacteria to follow phagocytosis, have been used to identify cells by cytofluorimetry. Three clusters can be separated: the eosinophil cluster, the neutrophile cluster and the mononuclear cells (Hageltorn and Saad, 1986). In contrast with bovids, the monocyte and lymphocyte populations could not be separated. However, the lymphocytes can be relatively well separated using Ficoll paque gradients (Zweygarth, 1984; Abdurahman and Saad, 1996).

Leukocyte markers were identified from cross-reactivity with several mouse monoclonal antibodies raised against ruminant (Vilmos *et al.*, 1996) or swine leukocytes (Saalmüller, 1996). Out of 159 antibodies reacting with bovine leukocytes, 34 (21%) reacted with camel cells. Of these, several have been characterized for the cell type and the antigen in cattle, swine and human and it is probable that they identify the same cells and the same antigen in camels, Table XII.3.3.

4. Immunoglobulins

IgM, IgG, IgA and even IgD have been detected in camel and llama sera on the basis of cross-reactivity with human immunoglobulins (Neoh *et al.*, 1973). However it has only recently been realized that camel and llama sera contain several IgG classes devoid of light chains (Figure XII.4.1) (Hamers-Casterman *et al.*, 1993). These heavy chain immunoglobulins also lack the first constant domain and hence have a six-domain structure (V, C_H2, $C_H3)_2$ and a molecular weight of *c.* 90 000. After reduction the molecular weight of these heavy chains drops to *c.* 45 000. Not only in early work but also more recently have these antibodies not been recognized as such and were either presented as a 40 kDa protein complexing IgG (Ungar-

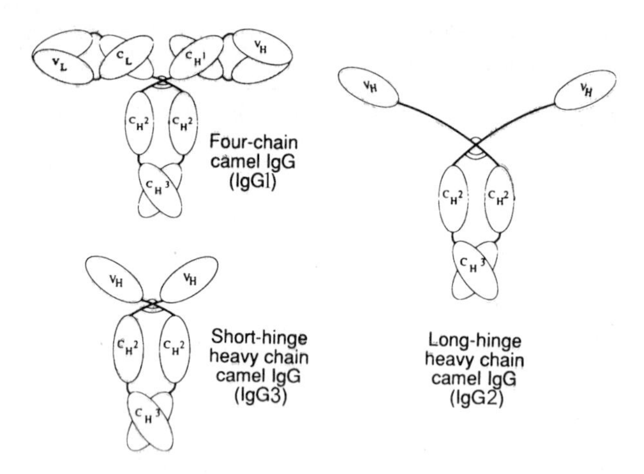

Figure XII.4.1 Schematic representation of the structural organization of the camel immunoglobulins. On the basis of size, the IgG1 fraction probably has the normal antibody assembly of two light and two heavy chains, IgG3 would have a hinge comparable in size to the human IgG1, IgG2 and IgG4. The two antigen-binding sites are much closer to each other, as camel IgG lacks the C_H1 domain. In camel IgG2, the long hinge, which is formed of Pro-X repeats (X = Glu, Gin or Lys), probably adopts a rigid structure. This long hinge could therefore substitute for the C_H1 domain and bring the two antigen-binding sites of IgG2 into normal positions. Reproduced from Hamers-Casterman *et al.* (1993) with permission.

Waron *et al.*, 1987; Azwai *et al.*, 1993) or went unnoticed (Elagamy *et al.*, 1992, 1996).

On a weight basis, the proportion of heavy chain antibodies in the dromedary and in the bactrian camel (50–75%) appear significantly higher than in the llama (30%). No estimates are available yet for the alpaca and the vicuña which also show the existence of heavy chain antibodies. On a molar basis, and this is more relevant to the antigen-binding capacity, the heavy chain antibodies with their six-domain structure represent a much higher proportion of the total serum immunoglobulins (camel 62–82%, llama 43%).

At present, no simple immunological methods exist to quantify the heavy chain antibodies.

Results obtained using conventional antibodies raised against camelid immunoglobulin (ELISA, radial immuno-diffusion) should be critically analysed as they will result from binding of both four-chain conventional Ig and heavy chain Ig. For the llama, discrepancies were found when different quantitative methods were used (Drew and Fowler, 1995; Hutchinson *et al.*, 1995a).

The absence of binding to protein G of some heavy chain immunoglobulin was instrumental in the discovery of the heavy chain immunoglobulins. After adsorption on protein G of camel serum, all the remaining and nonbinding IgGs are devoid of light chains and will bind to protein A. The combination of protein G and protein A chromatography allows a relatively specific isolation of camel and llama IgG subclasses valid with minor modifications for all camelid species (C. Hamers-Casterman, unpublished data).

In the camel and in the llama this purification has allowed the partial characterization of the IgG subclasses by protein sequence analysis, by sodium dodecyl sulphate (SDS)–PAGE and to a lesser extent by serology.

The major sequence data has, however, come from cDNA studies using PCR methods to recover expressed immunoglobulin gene sequence. The criteria used for attributing a sequence to a heavy chain immunoglobulin is the absence of the C_H1 domain and the usage of dedicated V_{HH} genes (see below).

The multiplicity of the heavy chain immunoglobulin subclasses and the absence of adequate typing reagents have hampered an adequate description of the Ig subclasses in camels and even more so in llamas and the correspondence between serological entities, cDNA sequences and isolated proteins is still fragmentary and much remains to be done.

The immunoglobulin G subclasses of the camelids

In the camel, three fractions of IgG can be purified. After adsorption of serum on protein G, differential acid elution will detach a first subclass of heavy chain immunoglobulin (IgG3) followed by conventional four-chain immunoglobulin (IgG1) usually displaying two components differing in SDS migration. The IgG being presumably more tightly bound by the contribution of the C_H1 binding (Derrick and Wigley, 1992).

The nonadsorbed fraction can then be adsorbed on a protein A column. The first acid-eluted fraction contains up to three different heavy chain IgG species differing in SDS–PAGE migration; two are usually present, with the third appearing during immunization (T. Serrao, unpublished data). A second fraction eluted at higher acidity contains essentially IgM, presumably molecules characterized by a subset of the V_HIII subgroup. These results would suggest the existence of six molecular species of IgG in the camels. It has not been determined whether all these molecular species correspond to subclasses or whether some are due to genetic polymorphisms (allotypes). To date, three molecular species have been identified from cDNA sequencing and one confirmed by protein sequencing. The major differences are located in the hinge region and are presented in Table XII.4.1.

In the llama, alpaca and vicuña, protein G and protein A adsorption also shows the presence of immunoglobulins devoid of light chains but their characterization at the protein level is practically nonexistent. Moreover, purification methods do not yield consistent results with sera from different individual llamas. This could be a consequence of different mixed ancestry (vicuña and/or guanaco) of the individual animals. At the cDNA level the description although limited to the llama is much more

Table XII.4.1 Hinge sequences

Ig	*Number of amino acids*		
Camel hinge sequences†‡*		*Two chain immunoglobulins*	
IgG1	12	EPHGGCPCPKCP	
IgG2	35	EPKIPQPQPKPQPQPQPQPKPQPKPEPECTCPKCP	cDNA PCR genomic DNA
IgG3	12	GTNEVCKCPKCP	cDNA PCR
		VCKCPKCP	protein sequence
Llama hinge sequences§		*Four chain immunoglobulins*	
IgG1a	19	ELKTPQPQSQPECRCKLCK	cDNA PCR
IgG1b	12	EPHGGCTCPQCP	cDNA PCR
		Two-chain immunoglobulins	
IgG2a	35	EPKIPQPQPKPQPQPQPQPKPQPKPEPECTCPKCP	cDNA PCR
IgG2b	29	EPKTPKPQPQPQPQPQPNPTTESKCPKCP	cDNA PCR
IgG3	12	ETNEVCKCPKCP	cDNA PCR

*C. Hamers-Casterman *et al.* (1993).
†Atarhouch *et al.* (1995); T. Atarhouch, unpublished.
‡Nguyen, unpublished.
§Vu *et al.* (1996a); Vu, in press.

complete. Two molecular species of heavy chains from the four-chain IgG1a, IgG1b and three molecular species of heavy chains from IgG2a, IgG2b and IgG3 have been completely sequenced. IgG2a and IgG3 are almost identical to dromedary camel IgG2 and IgG3; IgG2b appears to be slightly different (Vu *et al.*, 1996a). As is the case for the camel, it has not yet been determined whether all five molecular species of llama IgG belong to different subclasses or whether some represent allelic forms (allotypes) including those derived from mixed ancestry. In some species (e.g. rat and rabbit) allelic immunoglobulins can differ considerably in sequence (Sheppard and Gutman, 1981; van de Loo *et al.*, 1995).

Other immunoglobulin classes

Llama IgM has been highly purified by water precipitation followed by gel filtration and affinity chromatography on protein A (presumably a subpopulation of molecules containing the protein A-binding subset of the V_HIII subgroup). This immunoglobulin was used for producing specific anti-llama IgM and the resulting IgM was utilized to assay IgM in maternal and newborn serum and in colostrum (Garmendia and McGuire, 1987). Azwai *et al.* (1993) purified camel IgM from serum by a combination of ammonium sulphate precipitation and gel filtration. No evidence of IgM molecules devoid of light chain was found in camel IgM purified by water precipitation and gel filtration (N. Bendahman, unpublished data).

Azwai *et al.* were unable to detect IgA in camel serum but later (1996a) found evidence of IgA in camel colostrum using a monoclonal antibody raised against human IgA. It should be noted that Neoh *et al.* (1973) could detect IgG, IgA, IgM and even IgD with chicken sera raised against human immunoglobulins.

The light chains of conventional immunoglobulins

SDS–PAGE of reduced camel IgG reveals considerable size heterogeneity of these light chains. Three entities are consistently found of approximately 25 kDa, 29 kDa and 33 kDa. *N*-terminal sequencing suggests that the largest band is exclusively composed of λ chains whereas the 25 kDa and 29 kDa bands are a mixture of κ chains and λ chains (Legssyer *et al.*, 1995). The largest component appears to be *N* glycosylated. The physiological significance of any of this glycosylation is not known.

The unusual feature is that the glycosylation appears to be present in the constant C_λ domain and not as in the case in other species in the variable V_L domain where up to 20% (human or mouse) of the light chain have glycolysation sites and are effectively glycosylated. In the mouse $C_\lambda 1$ and $C_\lambda 2$ genes and in the rabbit the $C_\kappa 2$ gene of b9 allotype presents glycosylation sites but it is not known whether the light chain is effectively glycosylated (Figure XII.4.2).

In the camel, cDNA analysis has revealed the presence of the κ sequence and of two λ sequences differing by the presence or absence of a glycosylation site in the C_λ. We can therefore conclude that at least three light chain constant regions are consistently expressed in camels encoded by the C_κ, the $C_\lambda 1$ gene and the $C_\lambda 2$glyc gene. The λ_1 and λ_{2glyc} genes can be associated with a variety of V_λ domain. These V_λ sequences belong to the subgroups

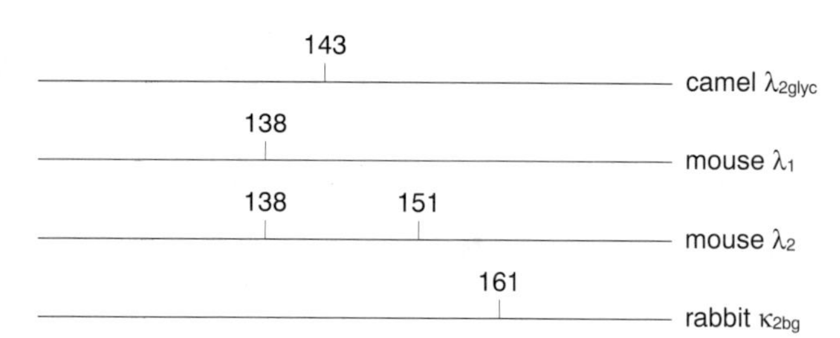

Figure XII.4.2 Position of glycosylation sites in camel, mouse and rabbit light chains.

$V_\lambda 2$ and $V_\lambda 3$ which are utilized in mammals which make extensive use of the V_λ repertoire. Little is known about the V_κ genes.

Characteristic features of the heavy chain immunoglobulins

The heavy chain antibodies lack the C_H1 domain. Hence the characteristic of heavy chain antibodies is the dimeric three-domain structure: the V_{HH} (V_H derived from heavy chain antibodies) linked via the hinge to the C_H2C_H3 domains.

In the human and mice isolated heavy chains never appear in the serum except in the so-called heavy chain disease in which they present an extensive deletion involving the C_H1 domain and part of the V_H (Seligman *et al.*, 1979). In four-chain immunoglobulin, a chaperon, the immunoglobulin heavy chain binding protein (BiP) inhibits heavy chain translocation from the endoplasmic reticulum to the golgi unless it has found a light chain partner (Hendershot, 1990). It appears to be the C_H1 and the V_H domains which are responsible for this BiP protein binding (Blond-Elguindi *et al.*, 1993; Knarr *et al.*, 1995). The absence of the C_H1 domain probably contributes to the fact that in camelids heavy chain antibodies can be secreted in absence of a light chain.

The absence of the C_H1 domain will normally result in a decrease in the span between binding sites and hence a decrease in the crossbinding capacity. Apparently, in the camelids, this is compensated by a diversity in heavy chain IgG subclasses presenting a variety of hinge sequences some of which are long enough to compensate completely the lack of the C_H1 domain (Table XII.4.1).

The long hinge of the camelid IgG2 has a structure that is very similar to the stem of the trypanosome procyclin (Roditi *et al.*, 1989) and to the *Escherichia coli* Ton B protein (Evans *et al.*, 1986) spanning the periplasmic space. The ProX repeat appears as a rigid structure capable of holding at a distance the V_{HH} domain. Analysis of the cDNA sequence of camel heavy chain antibody variable region V_{HH} reveals that these all belong to the V_HIII subgroup; at present the V_H of classical Ig is also known to contain members of the V_HII subgroup (Ghahroudi, 1996).

The absence of a light chain results in exposure to solvent of the hydrophobic face of the variable domain which would normally be in close contact with the variable domain of the light chain. This would lead to a lowered regional solubility of the variable heavy chain domain. In the camelids this is compensated by the replacement and reorientation of crucial amino acids with a concomitant increase in solubility. Figure XII.4.3 (see colour plate) shows the characteristic amino acid differences between the variable domain of four chain immunoglobulin and that of heavy chain immunoglobulins. The 'camelization' of human antibodies, a term coined by Davies and Riechman (1994) to describe the replacement of the amino acids G44, L45 and W47 in human cloned V_H domain by the camel specific amino acids (E44, R45 and G47) has resulted in better folding and solubility (numbering according to Kabat *et al.*, 1991). It is also these amino acids and in particular the Leu45 which are involved in the V_H contribution to BiP binding of conventional four-chain immunoglobulins. Heavy chain antibodies therefore appear to make use of a dedicated V_HIII subset justifying the designation of V_{HH}. This V_{HH} subset does not appear as a result of a developmental process but is already present in the germline (Nguyen *et al.*, 1997).

The features described for the dromedary camel are probably identical for the Bactrian camel in which the SDS–PAGE analysis of the serum immunoglobulin reveal an identical pattern (C. Hamers-Casterman, unpublished data).

An analysis of V_{HH} cDNA clones from camel (Muyldermans *et al.*, 1994) and llama (K. B. Vu, unpublished data), reveals additional important features. The CDR_1 and CDR_2 loops present greater size and structural variability than those of classical V_H domains. Often, they do not fit into the canonical structures described for these loops. Still more extraordinary is the length distribution of the CDR_3 regions which can be as long as 27 amino acids. In many cases an additional cysteine located in the CDR_2 or at position 45 in FRW_2 appears capable of making an additional stabilizing disulfide bridge with a cysteine in the CDR_3.

In summary, the camelid heavy chain antibodies possess remarkable structural peculiarities:

(1) Lack of C_H1 domain.
(2) Compensatory replacement of the missing CH_1 domain by stem hinges in IgG2 subclasses.
(3) Lack of light chain and compensatory replacements of key amino acids in V_{HH}.
(4) High regional solubility of V_{HH}.
(5) Absence of BiP binding (postulated).
(6) Possibility of extended CDR_3 region.
(7) Possibility of partial replacement of the light chain by long CDR_3 regions.
(8) Potential for novel binding site protruding configurations.
(9) Greater structural diversity of CDR_1 and CDR_2 regions.

Table XII.4.2 Incidence of antibodies to bacteria in camels

Bacteria	*Incidence of seropositivity*	*Disease susceptibility*
Pasteurella sp.	65–80%	Low, controversial
Leptospira		Low, subclinical
Brucella	0.3–24%	
Mycobacterium paratuberculosis	40% (Bactrian)	High
Tetanus	0–20% (Dromedary)	High
Chlamydia psittaci	+	Variable
Coxiella buneti	4–24%	Unknown
Rickettsia prowaseki	0.20%	No disease reported
Rickettsia mooseri	1.8–44.1%	No disease reported
Rickettsia rickettsii	1.6–26%	No disease reported
Rickettsia conorii	1.8–3.7%	No disease reported
Cowdria ruminantium	1%	No disease reported
Anaplasma	+	No disease reported
		Subclinical

Compiled from Wernery and Kaaden (1995).

Table XII.4.3 Incidence of antibodies to viruses in camels

Virus or disease	*Incidence (%)*	*Disease susceptibility*
Adenovirus	1.3	Unknown
African horse sickness	0.0–23	Not susceptible/reservoir?
Bluetongue	4.9–14.6	Not susceptible
Bovine herpesvirus 1	0.0–5.8	Not susceptible[a]
Bovine viral diarrhea/mucosal disease	0.0–15.5	?
Influenza	0.6–12.7	High
Parainfluenza	80–99	Low
Foot and mouth disease	0.0–2.6	Low (reservoir)
Rinderpest	0.0–97	Not susceptible
Respiratory syncytial disease	0.6	Unknown
Rift Valley fever	22–45	Low/unknown?
Rotavirus	50	Diarrhea
Rabies		Intermediate
Akabane virus	+	Unknown

Compiled from Wernery and Kaaden (1995).
[a]Llama and alpaca are susceptible.

Naturally occurring antibodies in infections

Antibodies linked to infections have been described in the camel and llama (reviewed in Wernery and Kaaden, 1995). Trypanosomiasis in camels was extensively studied and comparison made using ELISA and CATT (card agglutination tests) (Bajyana Songa and Hamers, 1988; Diall *et al.*, 1994). Antitrypanosome antibodies related to infection were detected in the IgM, IgG1 four-chain immunoglobulins and in the IgG2 and IgG3 heavy chain immunoglobulins (Hamers-Casterman *et al.*, 1993).

Prevalence of IgG and IgM antibody against the contagious echtyma (ort) virus and the camelpox virus, was conducted in Libya in healthy and diseased herds (Azwai *et al.*, 1995b, 1996b). These studies, carried out with purified camel pox antigen, confirm previous studies showing the widespread presence of camel pox antibodies in camels in nomadic herds. Mortality from camel pox appears to be higher in the first year of life and influenced by nutrition of the camel calves which compete with the camel owners for the available camel milk.

In most studies, however, there was no attempt to discriminate between IgG or IgM responses. Tables XII.4.2 and XII.4.3 present a compilation of naturally occurring immune responses against bacterial or viral antigens.

In the alpaca, an extensive New Zealand field study (Green *et al.*, 1996) involving nematode infections showed a negative correlation between serum antibody and faecal

Table XII.4.4 Experimental immunization of camels

Seroconversion by vaccination or experimental infection	
Clostridium perfringens	Alum and oil-based toxoid vaccine
Rickettsia mooseri	Infection
Rickettsia prowazeki	Infection
Rinderpest	Infection
Rabies	Alum-killed vaccine
Rift Valley fever	MVP22 attenuated
Influenza (bactrian)	Virus isolate/H_1N_1 subtype
Protection or cellular response induced by vaccination or experimental infection	
Camel pox	Resistance to camel pox
Parapox virus ovis	No protection
Parapox virus 'camellus'	Resistance to contagious ecthyma
Tuberculosis	Positive skin test
Papilloma	Receeding warts within 8–10 days

Compiled from Wernery and Kaaden (1995).

egg count (*Trichostrongylus colubriformis*). Antibodies against *Cooperia curticei, Ostertagia circumcincta* and *T. colubriformis* increased to more or less steady levels after 24–36 months.

Experimental immunizations

Most attempts to immunize camels were essentially linked to studies in pathology or to vaccination. Table XII.4.4 presents a compilation of these attempts.

Recently, experimental immunization was essentially directed towards analysing the contribution of the heavy chain antibodies to the immune response.

Experimental immunization of camels will nearly always result in antibodies appearing in the IgG1 four-chain immunoglobulins and in the IgG2 or IgG3 heavy chain immunoglobulin (Table XII.4.5).

Detection of the antibodies by ELISA indicates affinity of the order of 10^8 M. Hence, camels are capable of producing heavy chain antibodies of adequate affinity. So far, only *Salmonella* type XI polysaccharide antigen is unable to elicit heavy chain antibodies. Heavy chain antibodies could also be raised against haptens such as arsenylate, phenyloxazolone or phosphorylcholine. This is rather surprising as the antigen-binding site of antihapten antibodies is conventionally represented as being a cavity or groove between light and heavy chains. It is possible that in antihapten antibodies, the unusually large CDR_3 loop can contribute to the forming of a cavity. Camel antilysozyme, antitetanus and antiphenyloxazolone V_{HH} domains could successfully be cloned and expressed in *E. coli* (Desmyter *et al.*, 1996; Serrao *et al.*, 1996; Ghahroudi *et al.*, 1997) and llama antihuman chorionic gonadotrophin (HCG) was successfully expressed in yeast (Spinelli *et al.*, 1996). The antitetanus V_{HH} proved capable of protecting mice against a tenfold lethal dose of tetanus toxin (Figure XII.4.4) (Ghahroudi *et al.*, 1995b).

Recently, crystallographic studies have highlighted still further the uniqueness of camel antibodies and, in particular, the possible contribution of the large CDR_3 loop to novel antigen-binding sites. In a 2.5 Å resolution

Table XII.4.5 Experimental production of heavy chain antibodies in dromedary camels

	Response			
	Four-chain antibodies	*Two-chain antibodies*		
Antigen	*IgG1*	*IgG2*	*IgG3*	*Comments*
Cells, membranes, extracts				
Trypanosoma evansi	+	+	+	IgG3 agglutinates and fixes complement
Pseudomonas	+	–	+	
Salmonella	+	–	–	
Sheep red blood cells	+	+	+	
Proteins				
Tetanus toxoid	+	+	+	
Lysozyme	+	+	+	
Tobacco mosaic virus	+	+	+	
Haptens				
Dinitrophenol	–	–	–	
Arsenylate	+	+	+	
Phosphoryl choline	+	+	+	
Phenyl oxazolone	+	+	+	

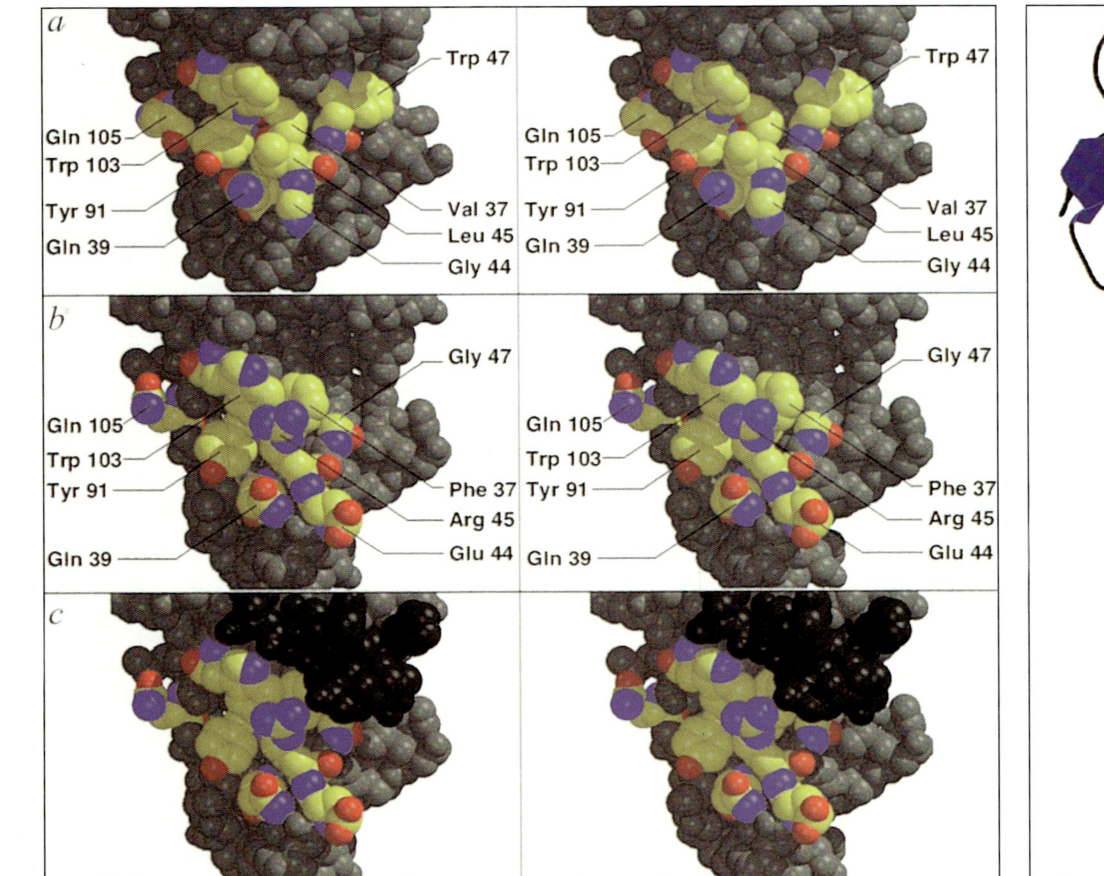

Figure XII.4.3. Stereo view of a space filling representation of the V_H, showing the side of the protein that interacts with the V_L in a conventional F_v. *a, The human Pot* V_H in which the amino acids conserved in human and mouse that are known to be important for the hydrophobic contact with the V_L domain, are coloured by atom type; all others are in grey. *b,* The camel cAb-Lys3 is shown from the same angle as in *(a)*. The amino acids of the CDR_3 part between Cys100e and Trp103 were removed. *c,* Same as *(b)* except that the removed CDR_3 part is included in black. Reproduced from Desmyter *et al.* (1996) with permission.

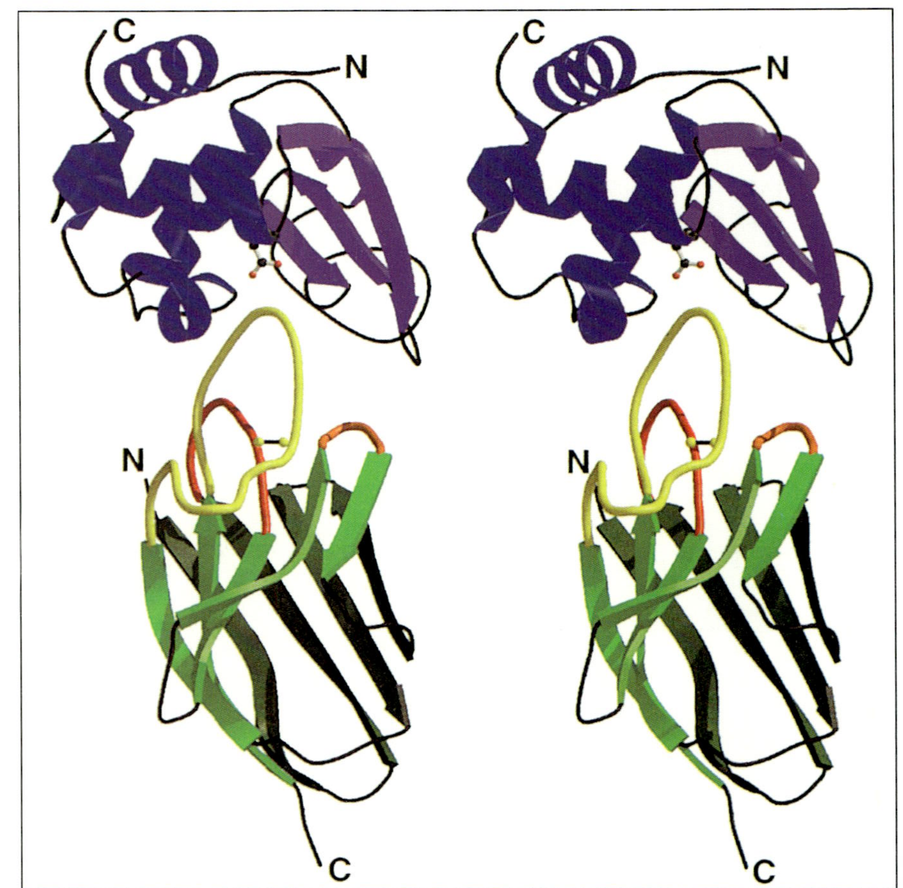

Figure XII.4.5. Stereo view of a ribbon representation of the X-ray structure of the cAb-Lys3 in complex with lysozyme. The α-helical domain of lysozyme is shown in blue, and the β-sheet domain in purple. The camel cAb-Lys3 framework β-strands are coloured dark green for the four-stranded β-sheet and lighter green for the five-standard β-sheet, which makes up the side of the protein that normally interacts with the V_L domain in a conventional F_v fragment. The CDR_1 region is shown in red, the CDR_2 loop in orange and the CDR_3 in yellow. The disulphide bond between CDR_1 and CDR_3, and the Glu 35 side-chain of lysozyme, are represented in ball-and-stick for reference. Reproduced from Desmyter *et al.* (1996) with permission.

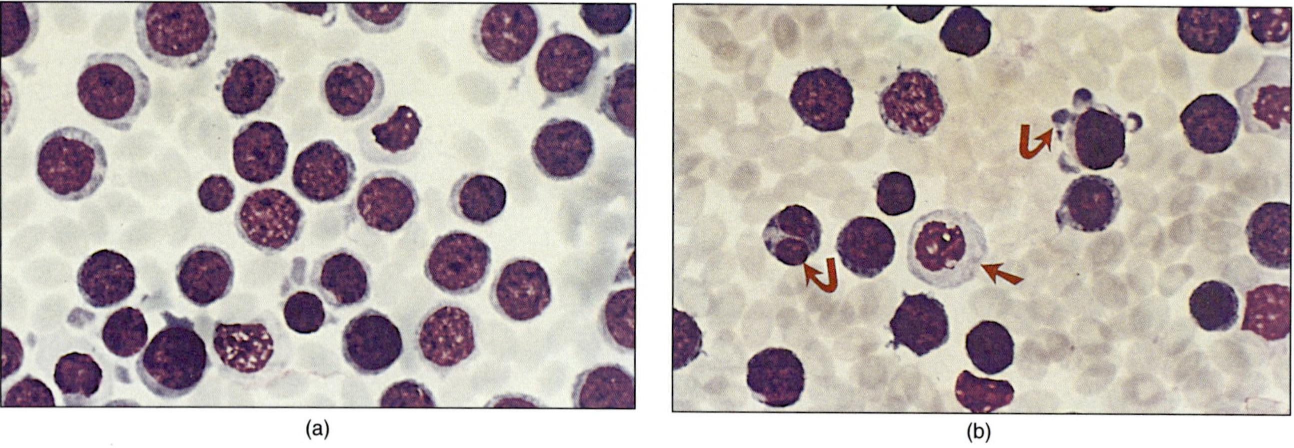

Figure XII. 12.1. (a). Peripheral blood film. Note marked variation in cell size of lymphocytes and lymphoblasts. (Giemsa's stain) (b). Peripheral blood film (Giemsa's stain). Lymphoblasts and atypical lymphocytes showing cytoplasmic blebs (curved arrow upper right), double nucleus (curved arriow left lower) and weak stainability of cytoplasm (short arrow centre). Reproduced from Tageldin *et al.* (1994).

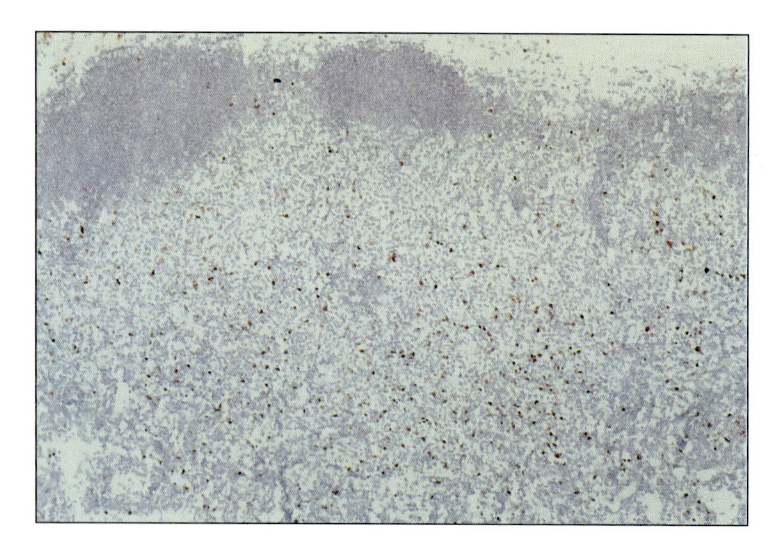

Figure XVI. 2.3. Partial view of a nu/nu mouse lymph node immunostained with anti-CD3. Primary follicles form the cortex. A few CD3-positive cells are scattered through the paracortex and medullary cords. Rare ones are seen in the follicles (Photomicrograph N. Antoine).

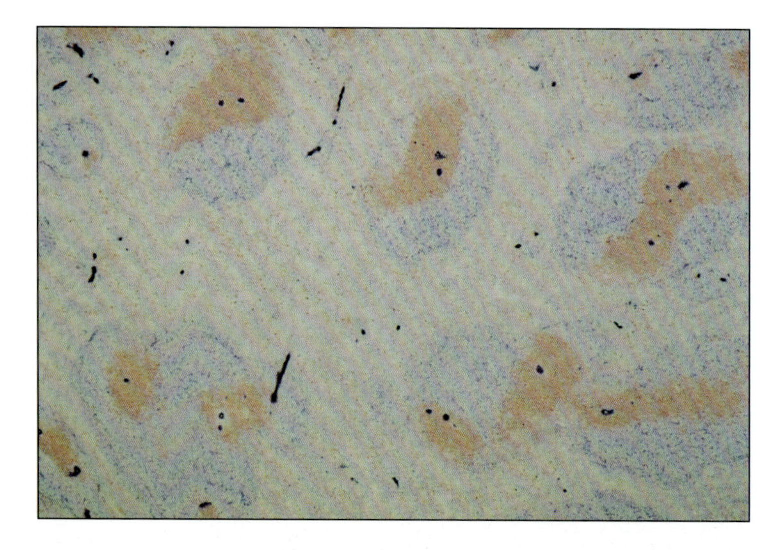

Figure XVI. 2.5. Low magnification of a mouse spleen after anti-B-cell (blue) and anti-T-cell (red) immunostaining. T cells occupy the periarteriolar sheath and B cells are gathered in the follicles and marginal zone. In the red pulp scattered B and T cells are found. Dark blue strands are alkaline phosphatase-positive blood vessels (Photograph I. Mancini).

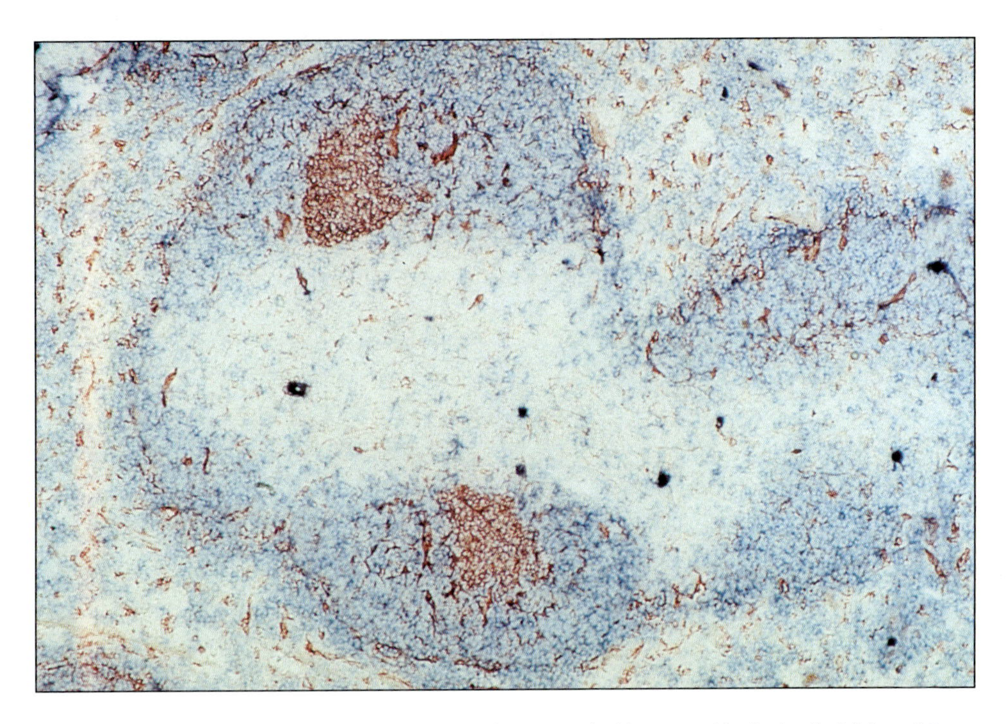

Figure XVI. 2.6. The white pulp of the mouse spleen revealed by peanut lectin (red) staining of the germinal centers and reticular fibers and by anti-B-cell immunolabelling (blue) of the follicular corona and marginal zones. The latter surround an unstained T-dependent periarteriolar sheath (Photomicrograph I. Mancini).

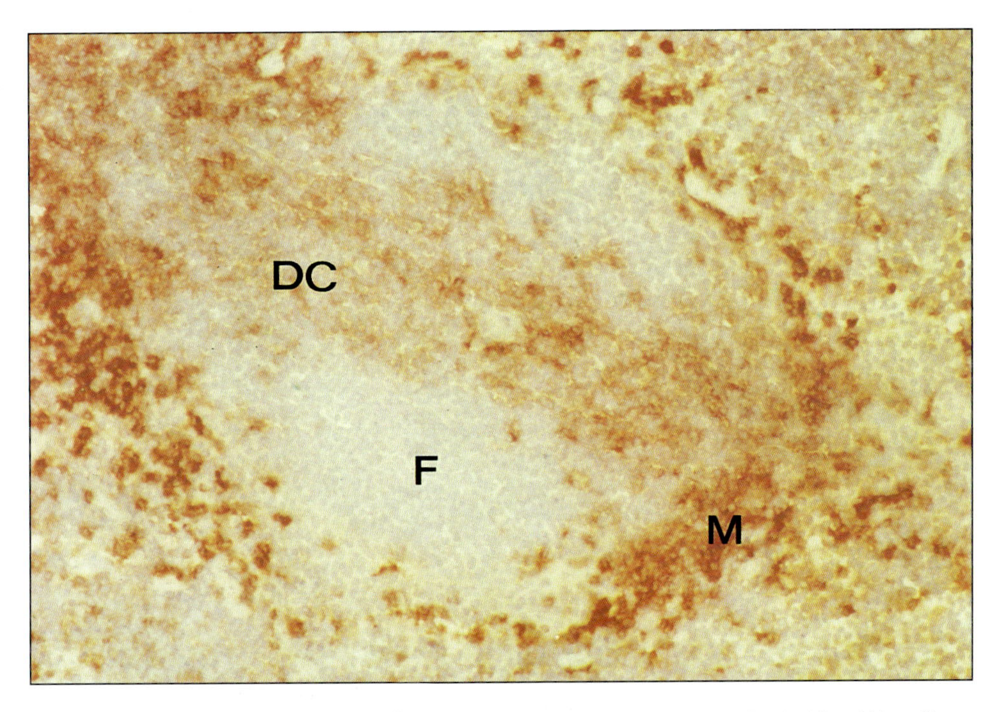

Figure XVI 2.7. Mouse spleen after immunolabeling with anti-N418 antigen. Typical dendritic cells (DC) arte labeled in the periarteriolar sheath. Marginal zone macrophages (M) react intensively. N418-positive cells are rare inside the follicles (F).

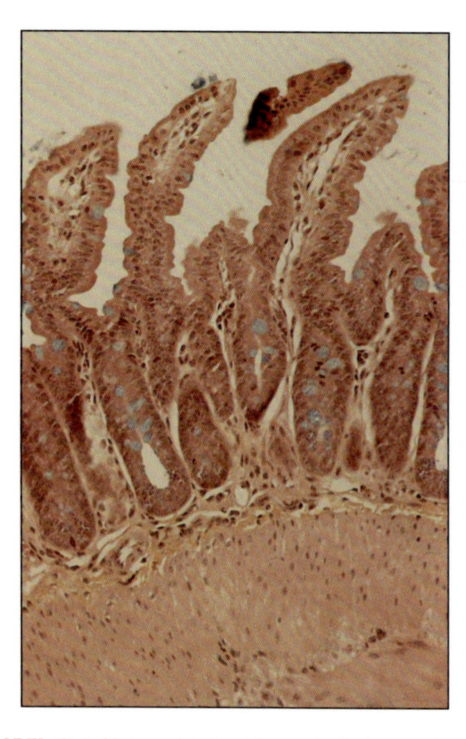

Figure XVII. 3.1. Human fetal gut transplanted subcutaneously into SCID mice.

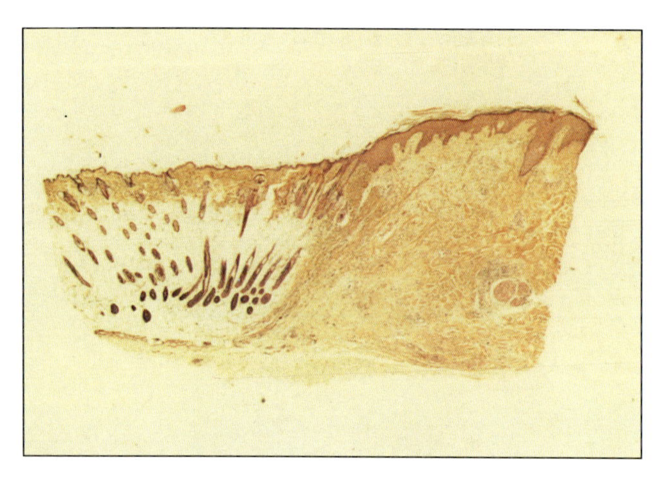

Figure XVII. 3.2. A hematoxilin and eosin micrograft of a junctional human (right)/mice (left) showed preservation of human epidermal and dermal structures.

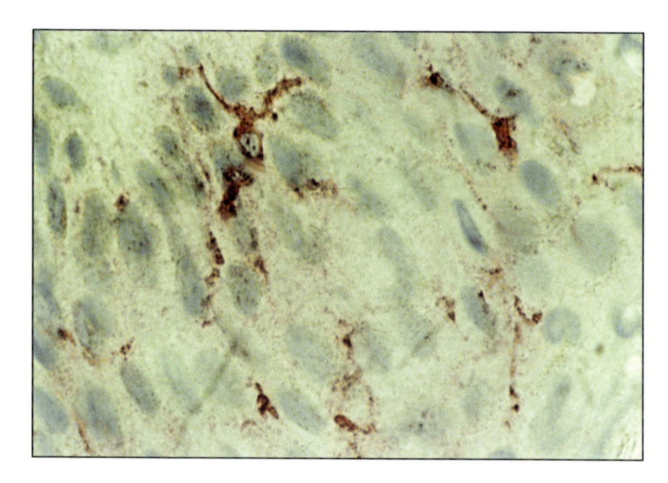

Figure XVII. 3.3. Human $CD1^{+}a$ cells can be observed in human skin engrafted onto Hu-PBL SCID mice, 1 week after intradermal immunization.

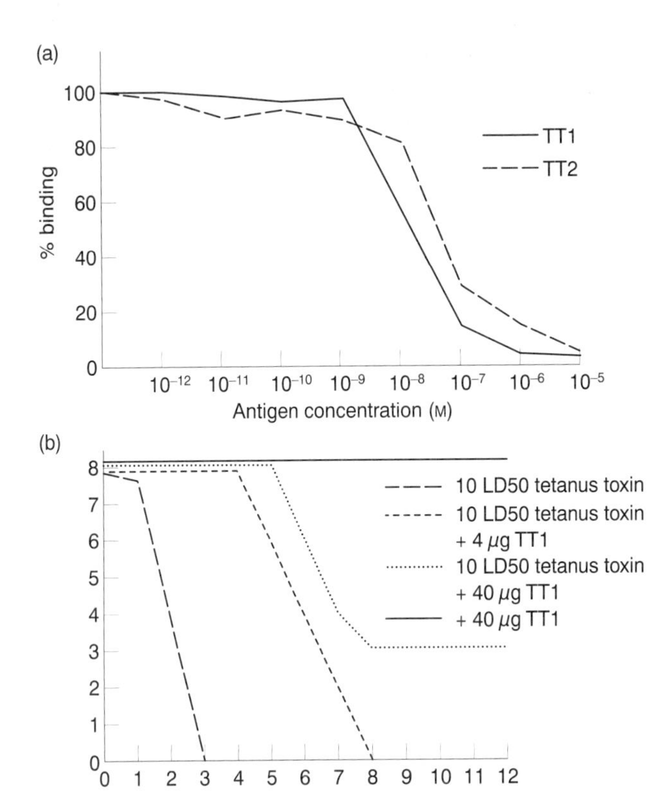

Figure XII.4.4 (a) Specificity of antigen binding of anti-tetanus toxin camel V_H, as measured by competitive ELISA. (b) Surviving mice after i.p. injection of tetanus toxin with/without camel V_H. Reproduced from Ghahroudi *et al.* (1995) with permission.

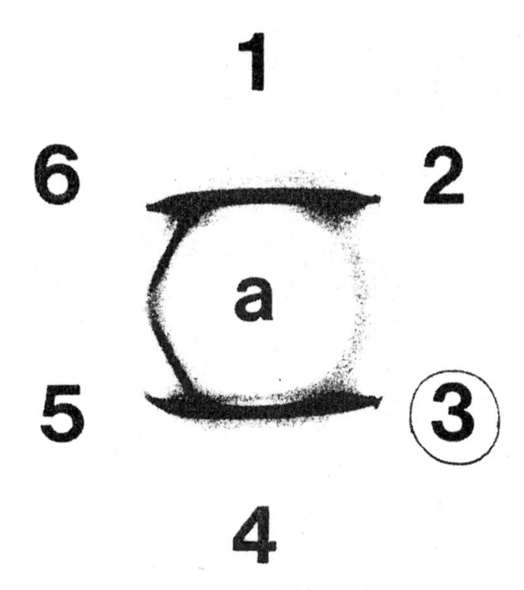

Figure XII.5.1 Immunodiffusion reaction of alpaca IgG with antiserum to human IgG. Human serum (1) and (4), serum collected after crias suckled (2, 5), purified alpaca IgG (3, 6), and antihuman IgG heavy chain specific antiserum (a). Reproduced from Garmendia and McGuire (1987) with permission.

of a cocrystal of camel α lysozyme V_{HH} and its antigen lysozyme, the unusually large CDR_3 partially replaces the missing light chain but, more surprisingly, protrudes as an arm which penetrates into the enzymatic site of the lysozyme. This is a completely novel type of antigen-binding site which might be immunologically important in penetrating viral, parasitic or bacterial structures (Figure XII.4.5 – see colour plate). The crystallographic structure of the camel antilysozyme and of the llama antiHCG also confirm the prediction that the CDR_1 loop does not confirm to the known canonical structures.

5. Phylogeny of the Camelid Immune System

From accepted classification, the Tylopoda, to which camels belong, are situated in the order of the Artiodactyls between the more primitive suiforms (pigs, hippos), and the ruminants.

However, comparative anatomy shows features more primitive than in the swine and paleontological records are compatible with an earlier emergence. The extensive structure differences of the stomach indicates that Tylopoda and Ruminentia evolved separately from a monogastric ancestor. Molecular data to confirm the conclusions based on comparative anatomy is barely available. In the various attempts to qualify or detect camel or llama immunoglobulins, sera raised against Ig of other species were used with somewhat unexpected results. In several instances, antihuman Ig reagents appear to give the best cross-reactions. This is the case for the mouse mAb anti-human IgA which apparently detects camel IgA better than anti-cat IgA or anti-horse IgA (Azwai *et al.*, 1996a) and a polyclonal anti-human IgG used to detect llama IgG (Garmendia and McGuire, 1987) (Figure XII.5.1). Other unexpected results are the cross-reaction obtained using anti-seal IgG by Azwai *et al.* (1993). Although comparative immunology is fraught with pitfalls, the results could confirm that the Tylopodes diverged very early in the Artiodactyl lineage (Romer, 1945), which may explain why their immunoglobulins display greater homology than expected with other mammalian groups such as the primates, and even the hoofed ungulates and the carnivores.

A feature of all the camelid immunoglobulin alignments is the close proximity to porcine sequences and, more unexpectedly, to human sequences (Muyldermans *et al.*, 1994; Atarhouch *et al.*, 1995; Legssyer *et al.*, 1995; Bang *et al.*, 1996b). This is not only to be found in the V_{HH}, but also in the C_H sequences and in the light chain sequences. Hence, the dedicated V_{HH} genes and the C_{HH} genes appear to originate within an anciently stabilized family of immunoglobulin genes. The generation of the V_{HH} repertoire characterized by specific key amino acids probably involved specific modes of selection which guaranteed their association with dedicated C_{HH} domains. A similar situation exists in the rabbit; lagomorphs have a dominant $\kappa 1$ light chain class which has, as particular feature, a disulphide bridge linking the $V_{\kappa}1$ to the $C_{\kappa}1$. Here also is raised the question of how a new V repertoire (in the case of the rabbit V_{κ} with a C80/108) was generated and how these V genes are organized in a way they will only be expressed with a dedicated C gene (in this case the $C_{\kappa}1$ with C171) (Mage, 1987; van der Loo and Verdoodt, 1992). A comparison of the V_{HH} loci of the camelids and the $V_{\kappa}1$ of the lagomorphs might reveal surviving features which were essential in the recent expansion of variable region repertoires (rabbit $V_{\kappa}1$ and camel V_{HH}).

6. Ontogeny of the Immune System

In the camel and llama, spleen and circulating lymphocytes have been used as a source of antibody-specific cDNA and therefore one can assume that blood and spleen contain immunocompetent cells. The generation of the repertoire of the heavy chain antibodies is of particular interest. The amino acid variation between the different V_{HH} analysed to date is much greater than within the members of the V_{HH}III subgroup of other species (mouse or man). This would suggest either a large and varied V_{HH} gene pool or alternatively a high level of maturation through gene conversion mutation or other genetic mechanisms. Ontogeny from a classical V_H gene repertoire is excluded by the discovery of V_{HH} genes in the unrearranged genome (Nguyen *et al.*, 1998). The ontogeny of the cells producing heavy chain immunoglobulins present an unusual dilemma. The accepted paradigm for the emergence of an immunocompetent B cell is that IgM (or IgG) appears on the cell surface as the result of a successful rearrangement of heavy and light chain immunoglobulin genes. The primary response will be the consequence of an antigen driven proliferation of those cells which have the right antigen-recognizing antibody on their surface. The heavy chain antibodies have a V_{HH} domain encoded by specific V_{HH} genes and the structure of the V_{HH} domain precludes close association with the light chains. To date, no evidence has been found of IgM lacking the light chains. Moreover, the sequences of the μ chain variable domains from camels obtained either by cDNA sequencing or by *N*-terminal amino acid sequencing bear the characteristics of V_H and not V_{HH} (R. Hamers, unpublished data; T. Atarhouch, unpublished data). These results, which have still to be confirmed, would imply that IgM is not involved in the primary response preceding secondary heavy chain IgG responses. One has therefore to postulate the existence of an alternative B cell lineage leading to cells producing heavy chain immunoglobulins (B_H cells). The way in which this putative B cell bypasses the classical IgM → IgG ontogeny necessitates further knowledge of the genomic organization of the V_{HH} and C_{HH} genes. At present, it is not known whether they are on the same chromosomes as the V_H and C_H genes let alone on which of the 37 chromosomes they reside (Hsu and Benischke, 1967–1974).

7. Passive Transfer of Immunity

The six-layer epitheliochorial placentation of the camelids prevents passage of the immunoglobulin to the fetus and all perinatal immunity is based on adequate absorption of colostrum and milk immunoglobulins, and on the eventual emergence of neonatal immunity. An extensive study carried out in the alpaca (*L. pacos*) shows that in the neonate presuckling cria's (newborn alpacas), the serum concentration of IgG and IgM are respectively of the order of 0.3 and 0.5 mg/ml. It remains to be confirmed whether these low levels correspond to neonatal synthesis of immunoglobulins. Maternal colostrum IgG which can

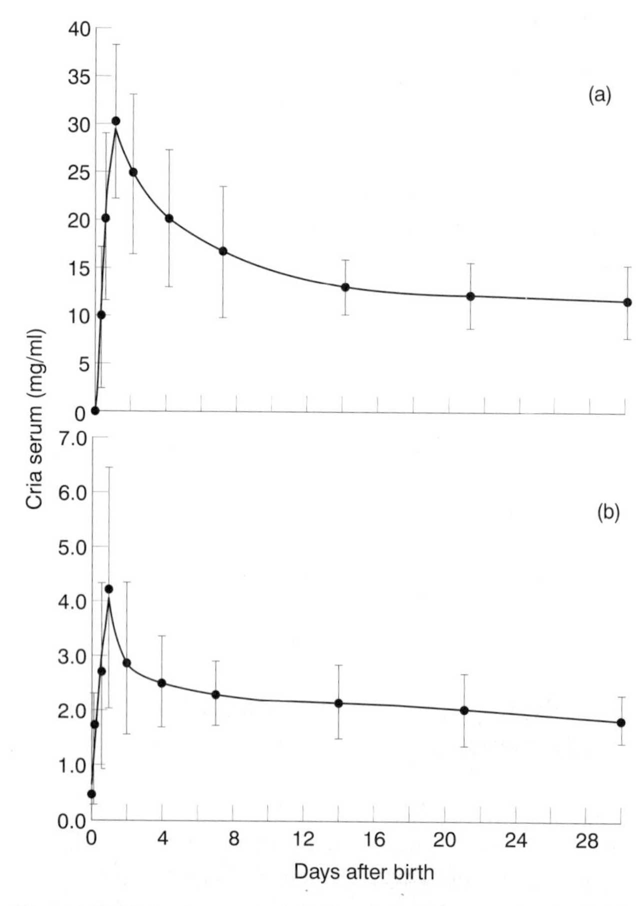

Figure XII.7.1 Immunoglobulin absorption from colostrum by the cria. Cria serum ($n = 10$) concentrations of IgG (a) and IgM (b) from birth to 30 days old. ● is the mean of serum IgG or IgM (mg/ml), and bar is 1 SD from the mean. Reproduced from Garmendia and McGuire (1987) with permission.

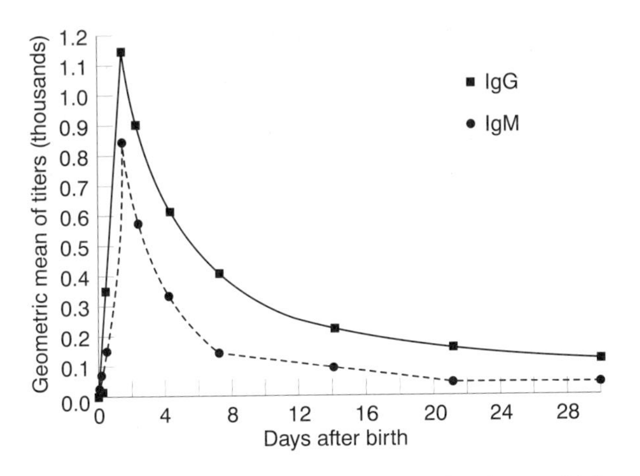

Figure XII.7.2 Absorption of specific antibodies to chicken RBC from colostrum by the cria. Geometric mean titres in serum from birth to 30 days old. Reproduced from Garmendia and McGuire (1987) with permission.

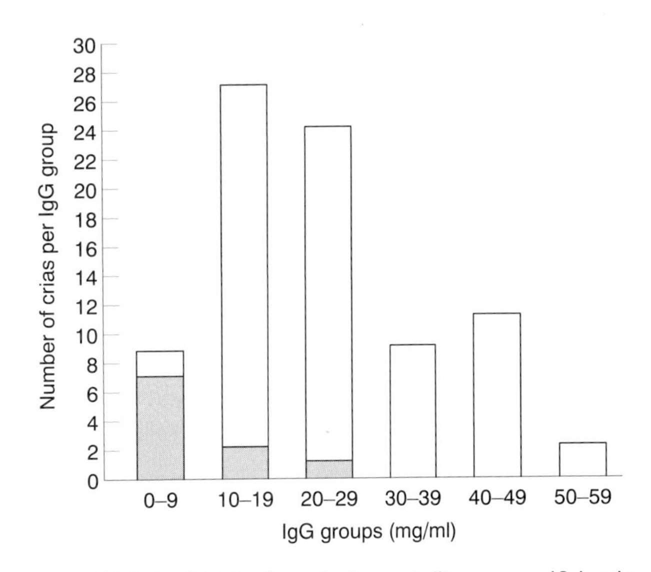

Figure XII.7.4 Distribution of cria mortality among 48-h cria serum IgG concentrations. ▒ = dead crias; □ = live crias. Reproduced from Garmendia *et al.* (1987) with permission.

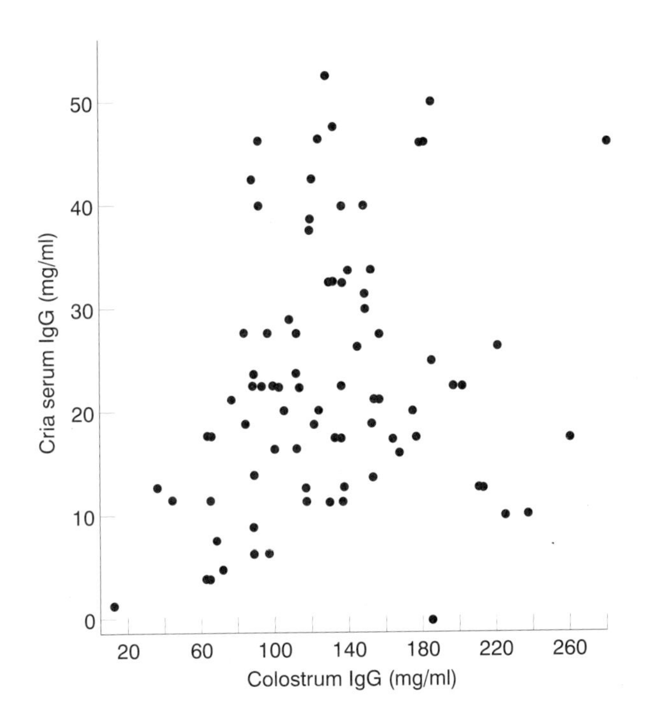

Figure XII.7.3 Relationship between colostrum collected before crias suckled and 48-h cria serum IgG concentrations (r = 0.23). Reproduced from Garmendia *et al.* (1987) with permission.

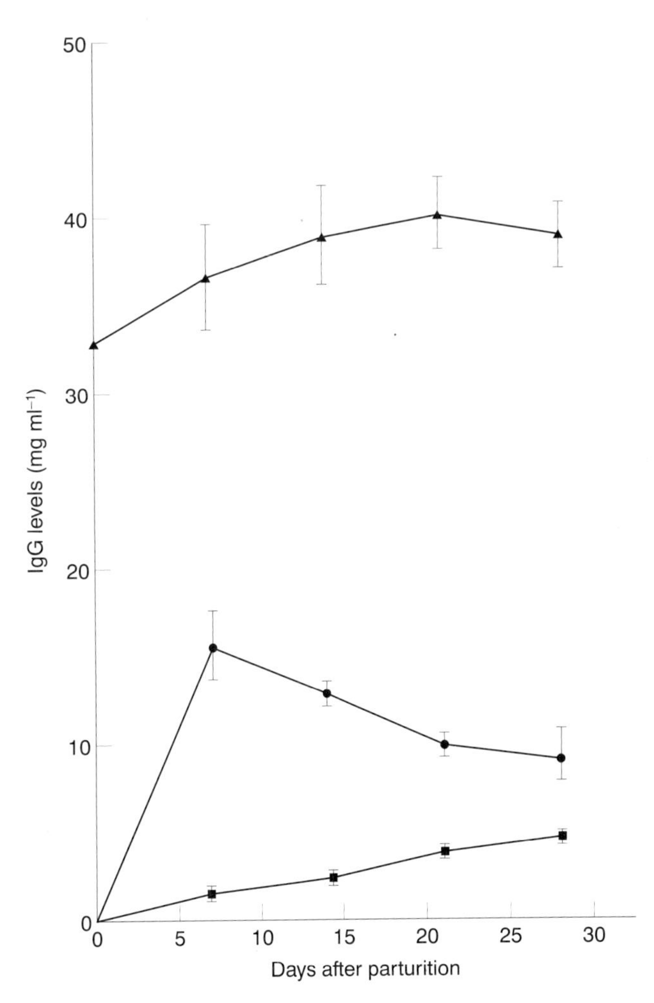

Figure XII.7.5 Camel serum immunoglobulin levels of ▲ – dams, ● – calves. Values are expressed as mean standard deviation. ■ – Immunoglobulin level in a calf which was refused by the dam for 4 days. Reproduced from Ungar-Waron *et al.* (1987) with permission.

reach the level of 250 mg/ml but presents extensive variations in concentration is the major source of perinatal immune protection (Garmendia and McGuire, 1987; Garmendia *et al.*, 1987).

The amount which is absorbed, measured by the serum Ig level in the suckling cria is weakly related to the presuckling colostrum variation (Figure XII.7.3). It will be important to determine to which extent the time of suckling affects the uptake. The serum IgG and IgM (which is also transferred) peak at 2 days after birth at an average level of respectively 30 mg/ml (IgG) and 4 mg/ml (IgM) after which they level off to respectively 10 mg/ml and 2 mg/ml (Figures XII.7.1 and XII.7.2).

Failure of passive Ig transfer appears to be a major determinant of mortality and can be only partially related to maternal insufficiency (Garmendia *et al.*, 1987) (Figures XII.7.3 and XII.7.4). In this study no distinction was made for four-chain and two-chain IgGs which had not yet been discovered.

In another study in the llama, neonatal transfer of antitetanus antibodies was firmly established (Murphy *et al.*, 1989).

In camel (*C. dromedarius*), colostrum has also been shown to contain not only IgG but also IgM. In addition, low amounts of IgA appear to be present and are detectable with antihuman IgA sera SDS–PAGE gels (Azwai *et al.*, 1996b). The figures published by Azwai clearly suggest the presence of all classes of IgG in the camel colostrum including the heavy chain immunoglobulins IgG2 and IgG3, which, predictably (Hamers-Casterman *et al.*, 1993), should be transported through the mammary gland (Jackson *et al.*, 1992). In the camel IgG uptake is similar to that in llama (Figure XII.7.5) (Ungar Waron *et al.*, 1987).

8. Blood Groups

Blood group polymorphisms are found in llamas but not in camels (Stormont, 1982). In llamas, five systems have been identified by cross immunization with erythrocytes (A–B, C^+C^-, D^+D^-, E^+E^- and F^+F^-) and are used for pedigree certification (Penedo *et al.*, 1988; Miller *et al.*, 1985).

9. Nonspecific Immunity

Nonimmunoglobulin antibacterial and antiviral activity was examined in camel milk. Babour *et al.* (1984) noted that while a low proportion of milk samples inhibited growth of *Clostridium perfringens* (7.5%), *Staphylococcus aureus* (4%), *Salmonella dysenteriae* (2%) and *Salmonella typhimurium* (1%), no inhibition was found for *Bacillus cereus* and *Escherichia coli*. Milk lysozyme appears to be active against *S. typhimurium* but contradictory results were obtained for *S. aureus*. Lactoferrin only affects *S. typhimurium* but this is probably related to iron metabolism. Lactoperoxidase appears to be active only against Gram-positive bacteria (*Lactococcus lactis*). Antirotavirus activity in milk is attributable to the milk IgG and IgA (Elagamy *et al.*, 1992, 1996).

10. Complement System

Very few studies have been made on camel complement (Bharnagar *et al.*, 1987; Olaho-Mukari *et al.*, 1995a,b) and these are limited to determining conditions of complement lysis. Evidence suggests that homologous erythrocytes and erythrocytes of goat or sheep are relatively refractory to complement lysis using the appropriate camel, goat or rabbit haemolysin. Fresh camel serum lyses unsensitized erythrocytes of chicken, rabbit and guinea pig and this activity has been attributed to activation of the alternative pathway.

11. Immunodeficiencies

In the llama, an immunodeficiency syndrome of unknown origin has been reported (Hutchinson *et al.*, 1995b). Juvenile immunodeficiency syndrome (JLIDS) affects young animals and is complex, affecting biochemical and cellular blood parameters. Mild normochromic, normocytic anaemia and hypoalbuminaemia characterize the disease in addition to low serum IgG concentration, poor vaccination response and areas of profound paracortical and cortical depletion in lymph node section. Mitogenic stimulation of lymphocytes by protein A-bearing bacteria or by pokeweed mitogen were also significantly reduced indicating severe B-cell function insufficiencies. In the camel, one of the major diseases is trypanosomiasis due to *Trypanosoma evansi*. Stress is a major factor exacerbating this disease. *T. evansi* is a salivarian trypanosome very closely related to TseTse fly-transmitted trypanosomes (*Trypanosoma brucei, Nanomonas congolense* and *Dutonella vivax* infecting cattle; *T. brucei gambiense* and *T. brucei rhodesiense* infecting man).

Trypanosoma evansi appears to be a variant of *T. brucei* which has lost its capacity to multiply in the TseTse fly vector and which has acquired an efficient capacity for being mechanically transmitted by biting flies. *T. brucei* infections in mice and cattle (Sileghem *et al.*, 1989; Sileghem and Flynn, 1992) and *T. evansi* infections in waterbuffalo (Minh Nguyen, unpublished data) have been shown to induce severe immunosuppression.

This effect can be mimicked *in vitro* and is due in part to macrophages which block CD4 differentiation, and downregulate IL-2 receptor (De Baetselier, 1996) (Figures XII.11.1 and XII.11.2) (Table XII.11.1). *In vivo*, this state of immunosuppression leads to opportunistic and lethal disease and is probably the primary cause of waterbuffalo mortality due to hemorraghic septicemia (*Pasteurella multicida* serotype E). This could also be the case in the camel and the potential role of trypanosome-induced immunosuppression should be taken into account in the epidemiology of disease and in the management of vaccination campaigns.

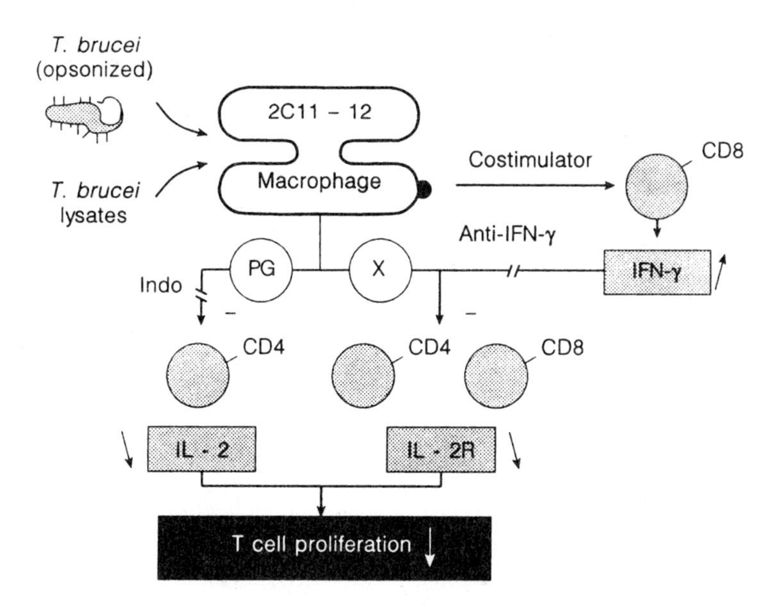

Figure XII.11.1 Schematic overview of the molecular and cellular interactions that may occur in the *in vitro* experimental model (2C11-12 macrophages) for *Trypanosoma brucei*-mediated immunosuppression. The following events may lead to suppression of T-cell proliferation. (1) Opsonized *T. brucei* or soluble *T. brucei* components will, upon interaction with 2C11-12 macrophages, trigger the expression of a suppressive phenotype. (2) Suppressive 2C11-12 cells will, upon cell–cell contact with $CD8^+$ T cells, sensitize these cells to produce IFN-γ. The requirement for cellular contact between 2C11-12P cells and $CD8^+$ T cells implies the existence of a membrane-bound costimulatory molecule (possibly TNF-α) on 2C11-12P cells. (3) 2C11-12P cells secrete soluble factor (X), which in conjunction with IFN-γ downregulate IL-2R on $CD4^+$ and $CD8^+$ T cells. Addition of anti-IFN-γ blocks this IFN-γ-mediated pathway of suppression. (4) 2C11-12P cells also secrete prostaglandin (PG) which blocks IL-2 secretion by $CD4^+$ T cells. This prostaglandin-mediated pathway of suppression is blocked by indomethacin (Indo). (5) Inhibition of IL-2 secretion and down-regulation of IL-2R expression results in reduced T-cell proliferation. From De Baetselier (1996). Reproduced with permission of S. Karger AG, Basel.

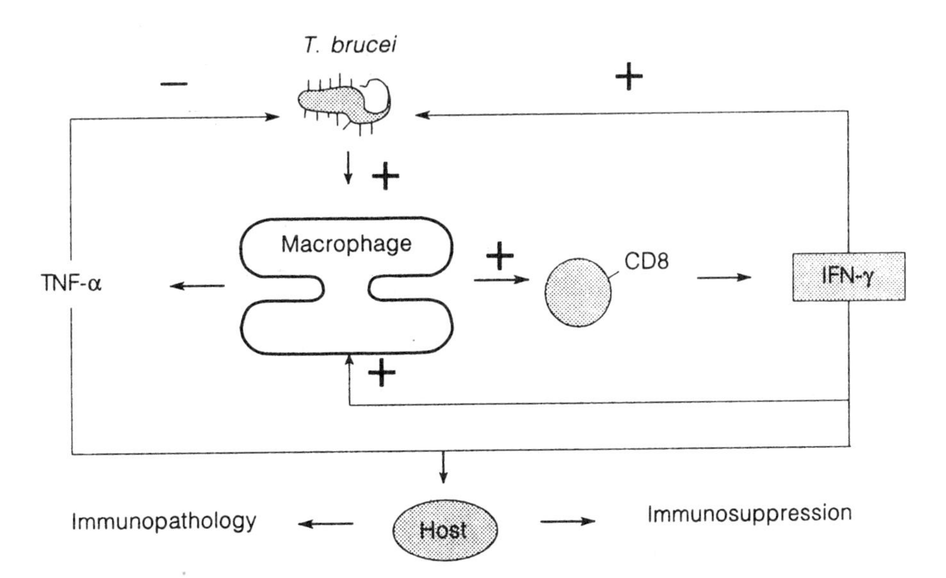

Figure XII.11.2 Schematic overview of the parasite–cytokine–host interactions that may occur during trypanosome infections. The following interactions are likely to occur. (1) Trypanosome components trigger macrophages to produce TNF-α and to signal $CD8^+$ T cells to produce IFN-γ. (2) IFN-γ and TNF-α exert, respectively, a growth-promoting and growth-inhibiting activity on trypanosomes. (3) TNF-α and IFN-γ exert noxious activities on the host by inducing a state of immunosuppression and by triggering immunopathological events such as inflammation of the central nervous system. From De Baetselier (1996). Reproduced with permission of S. Karger AG, Basel.

Table XII.11.1 *Trypanosoma brucei*-elicited immunosuppression: comparison of two lymphoid compartments

Response tests	*Lymphoid compartment analysed*	
	Lymph node	*Spleen*
T-cell proliferation	Suppressed (late)	Suppressed (early)
IL-2 production	Suppressed	Suppressed
IFN-γ production	Augmented	Suppressed
T-cell proliferation (treatment with anti-IFN-γ *in vivo*)	Normal	Suppressed
T-cell proliferation (treatment with anti-TNF-α *in vivo*)	Normal	Suppressed
Suppressive activity in cocultures	Yes	Yes

Reproduced from De Baetselier (1996).

12. Tumors of the Immune System

In recent years, four reports of acute lymphoblastic leukemia were published, all from the Arabian peninsula (Moustafa, 1994; Tageldin *et al.*, 1994; Afzal and Hussain, 1995; Wernery and Kaaden, 1995) (Figure XII.12.1 – see colour plate). Offspring of leukaemic dams did not develop the disease although followed for more than one year. Attempts to transfer the disease to other animals by blood injection proved unsuccessful. However, the geographical clustering would appear to indicate a retroviral infection. Lymphosarcoma has been reported in the llama (Fowler *et al.*, 1985; Underwood and Bell, 1993) and in the camel (Youssef *et al.*, 1987).

13. Conclusions

In addition to subjects which have already been highlighted as major foci for further research in camel immunology, such as an adequate understanding of the ontogeny of heavy chain antibodies and their particular involvement in disease control, other aspects at the fundamental immunological, veterinary and applied level should be emphasized. Unfortunately, the tools for studying the immune response at the molecular level are poorly developed, let alone at the cellular level.

Immunological reactions with antisera and in some instances, monoclonal antibodies raised against camel Ig or Igs from other species have been the only reagents available. Developed monoclonal antibodies cover only a limited range of IgG classes (Azwai *et al.*, 1995a). At present, the best tool remains immunoaffinity chromatography on protein G/protein A as a means of purifying and quantifying camel and llama IgGs. However, these methods do not permit the isolation in pure form of all the subclasses. Specific monoclonal antibodies, when developed, will probably be complementary. The IgG3 subclass of camel heavy chain antibodies is the only one which can be obtained at present in good yield and high purity from camel serum. The specific structure of the IgG3 hinge allows the enzymatic cleavage into useful fragments for binding or effector function studies. The fragments that can be obtained include the V_{HH}, the V_{HH} dimers encompassing the hinge, the Fc and the Fc dimer encompassing part of the hinge (Table XII.13.1) (Hamers-Casterman *et al.*, 1996).

The sequence C-terminal to the hinge disulphides has been designated in other species as being responsible for mammary transport (Jackson *et al.*, 1992). The availability of different fragments should allow confirmation of this role.

Native light chains prepared by chain separation under reducing conditions can be separated by lectin (SNA) chromatography into glycosylated λ and nonglycosylated (κ and λ) chains for studying the role or effect of glycosylation.

The unique structure of the camel heavy chain immunoglobulins and the novel antigen combining site overshadows all other aspects of camelid immunology by the scope of their biotechnological application. The high solubility of the V_{HH} allows specific antibody fragments to be made of such smaller size capable of reacting with a large diversity of antigens. The single-chain structure could allow economical development and production of

Table XII.13.1 Enzymic fragments obtained from camel IgG3

V_{HH}	Papain, <u>V_8 protease</u>
$(V_{HH})_2$	Papain pepsin <u>trypsin</u>
$(Fc)_2$	Trypsin <u>V_8 protease</u>
Fc	Trypsin
C_H3	Trypsin, <u>papain</u>
V_{HH}-Fc	<u>V_8 protease</u> – limited digest
$(V_{HH})_2$ ext.	<u>Pepsin</u>

C. Hamers-Casterman – unpublished data.
The underlined enzymes are ideal tools for producing heavy chain immunoglobulin fragments for experimental purposes insofar as they do not cleave within the extended CDR_3 loop.
V_8 protease cleaves almost exclusively the VE sequence of the hinge (see Table XII.4.1) yielding V_{HH} and disulfide linked Fc fragments and in the case of a single cleavage a V_{HH}-Fc fragment.
Trypsin cleaves both hinge KC sequences to give a good yield of dimeric $(V_{HH})_2$. Papain can be used to produce the relatively resistant CH_3 domain by cleavage at residue K360 (N terminal).
Pepsin is unique in cleaving within the CH_2 domain at residue V253/F254, to produce almost exclusively an extended $(V_{HH})_2$ dimer.

antibodies in bacteria (Bang *et al.*, 1995), yeasts, plants and animals which, in turn, should allow a more widespread use of engineered antibodies in everyday consumables and not only pharmaceuticals. The relatedness to human antibodies could lead to the development of camel antibody-based therapeutics and *in vivo* diagnostics; they also respond to the need for a good protein scaffold for engineering novel binding sites or synthetic enzymatic sites. The novel protruding binding site is potentially capable of entering not only enzymically active sites but also membrane receptors. Camel antibodies and the relevant CDR_3 loops might have a value as lead compounds in the development of new receptor-oriented pharmaceuticals. The camelids, which are among the least studied of the domestic animals could very well become the most important source of engineered proteins in applications ranging from plant improvements to human medicine.

Finally we would like to stress that currently very little is known about the immune system of the camel and of the llama. What is known, however, is completely upsetting paradigm and preconceived ideas, and is promoting biotechnology. Camel V_{HH}s represent the smallest antigen binding molecule yet found and as such will open new fields in protein engineering and in biotechnology (Sheriff and Constantine, 1996). It is hoped that these applied aspects will generate sufficient impetus to further also research towards understanding the ontogeny of a variant of the vertebrate humoral immune system which arose unexpectedly in the Tylopode lineage and which was able to dispense with the light chain.

14. Acknowledgements

The authors thank Professor M. E. Fowler for communicating material from the second edition of his book *Llama Medicine and Surgery*, thank their collaborators, students and old students, Dr J. Naessens, Dr T. Atarhouch, Mss I. Legssyer and T. Serrao, Messrs Vu, N. Bendahman and Nguyen for making available unpublished material and last but not least Ms L. Eeckhaudt for the infinite patience with which she typed the successive versions of this manuscript.

Much of the authors' work would not have been possible without the continuing support of the EU, the Fonds voor Geneeskundig Wetenschappelijk Onderzoek, the Vlaams Actieprogramma Biotechnology and the Vlaams Interuniversitair Instituut voor Biotechnology.

15. References

Abdurahman, O. A. S., Saad, A. M. (1996). Characterization of camel leukocytes by flow cytometry and microscopic evaluation of granulocyte phagocytosis of fluorescent bacteria. *Zbl. Vet. Med. A* 43(2), 111–117.

Afzal, M., Hussain, M. M. (1995). Acute prolymphocytic leukemia in the camel. *ACSAD Camel Newsl.* **11**, 22–24.

Agba, K. C., Gouro, S. A., Saley, M. (1996a). Les noeuds lymphatiques du dromadaire (*Camelus dromedarius*). 1. Noeuds lymphatiques du réseau vasculaire crânial (Tête, encolure, membre thoracique). *Rev. Méd. Vét.* **147**, 587–598.

Agba, K. C., Gouro, S. A., Saley, M. (1996b). Les noeuds lymphatiques du dromadaire (*Camelus dromedarius*). 2. Noeuds lymphatiques du membre pelvien, du bassin et de l'abdomen. *Rev. Méd. Vét.* **147**, 721–730.

Atarhouch, T., Hamers-Casterman, C., Muyldermans, S., Hamers, R. (1995). Complete sequence of the camel heavy chain IgG_3 isotype. *Arch. Physiol. Biochem.* **103**, B34.

Azwai, S.M., Carter, S.D., Woldehiwet, Z. (1993). The isolation and characterization of camel (*Camelus dromedarius*) immunoglobulin classes and subclasses. *J. Comp. Path.* **109**, 187–195.

Azwai, S.M., Carter, S.D., Woldehiwet, Z. (1995a). Monoclonal antibodies against camel (*Camelus dromedarius*) IgG, IgM and light chains. *Vet. Immunol. Immunopathol.* **45**, 175–184.

Azwai, S.M., Carter, S.D., Woldehiwet, Z. (1995b). Immune responses of the camel (*Camelus dromedarius*) to contagious ecthyma (Orf) virus infection. *Vet. Microbiol.* **47**, 119–131.

Azwai, S.M., Carter, S.D., Woldehiwet, Z. (1996a). Immunoglobulins of camel (*Camelus dromedarius*) colostrum. *J. Comp. Pathol.* **114**(3), 273–282.

Azwai, S.M., Carter, S.D., Woldehiwet, Z., Wernery, U. (1996b). Serology of *Orthopoxvirus cameli* infection in dromedary camels: analysis by Elisa and Western Blotting. *Comp. Immun. Microbiol. Infect. Dis.* **19**, 65–78.

Bajyana Songa, E., Hamers, R. (1988). A card agglutination test (CATT) for veterinary use based on an early VAT RoTat 1/2 of *T. evansi*. *Ann. Soc. Belge Med. Trop.* **68**, 233–240.

Vu, K. B., Ghahroudi, M. A., Atarhouch, T., Wyns, L., Hamers, R., Muyldermans, S. (1995). Expression of intact, functional heavy-chain immunoglobulins in *E. coli*. *Arch. Physiol. Biochem.* **103**, B68.

Vu, K. B., Muyldermans, S., Hamers, R., Wyns, L. (1996a). Identification of five llama IgG isotypes. *Arch. Physiol. Biochem.* **104**, B63.

Vu, K. B., Muyldermans, S., Wyns, L., Hamers, R. (1996b). Llama VH sequences of functional heavy-chain antibodies. *Arch. Physiol. Biochem.* **103**, B29.

Barbour, E. K., Nabbut, N. H., Frerichs, W. M., Al-Nakhli, H. M. (1984). Inhibition of pathogenic bacteria by camel's milk: relation to whey lysozyme and stage of lactation. *J. Food Protect.* **47**, 838–840.

Bhatnagar, R. N., Mittal, K. R., Jaiswal, T. N., Padamanaban, V. D. (1987). Levels of complement activities in sera of apparently healthy camels. *Indian Vet. J.* **64**, 192–195.

Blond-Elguindi, S., Cwirla, S. E., Dower, W. J., Lipshutz, R. J., Sprang, S. R., Sambrook, J. F., Gething, M-J. H. (1993). Affinity panning of a library of peptides displayed on bacteriophages reveals the binding specificity of BIP. *Cell* **75**, 717–728.

Carrasco, V. A. A. (1968). Vasos y centros linfaticos superficiales del miembro pelvico de la alpaca lama pacos (Vessels and superficial lymph nodes of the pelvic limb of the alpaca). Thesis, Fac. Med. Vet. Univ. Nac. Mayor San Marcos, Lima, Peru.

Curasson, C. (1947). *Le Chameau et ses Maladies*. Vigot Frères, Paris, VIe, France.

Davies, J., Riechmann, L. (1994). Camelising human antibody fragments: NMR studies on VH domains. *FEBS Lett.* **339**, 285–290.

De Baetselier, P. (1996). Mechanisms underlying trypanosome-induced T-cell immunosuppression. In *T-Cell Subsets and Cytokines Interplay in Infectious Diseases* (eds Mustafa, A. S., Al-attiyah, R. J., Nath, I., Chugh, T. D.), pp. 124–139. Karger, Basel, Switzerland.

Derrick, J. P., Wigley, D. B. (1992). Crystal structure of a streptococcal protein G domain bound to an Fab fragment. *Nature* **359**, 752–754.

Desmyter, A., Transue, T. R., Ghahroudi, M., Dao Thi, M-H., Poortmans, F., Hamers, R., Muyldermans, S., Wyns, L. (1996). Crystal structure of a camel single domain V_H antibody fragment in complex with lysozyme. *Nature Struct. Biol.* **3**, 803–811.

Diall, O., Bajyana Songa, E., Magnus, E., Kouyate, B., Diallo, B., Van Meirvenne, N., Hamers, R. (1994). Evaluation d'un test sérologique d'agglutination directe sur carte dans le diagnostic de la trypanosome caméline à *Trypanosoma evansi*. *Rev. Sci. Tech. Off. Int. Epiz.* **13**, 793–800.

Drew, M. L., Fowler, M. E. (1995). Comparison of methods for measuring immunoglobulin concentrations in neonatal llamas. *J. Am. Vet. Med. Assoc.* **206**, 1374–1380.

Elagamy, E. I., Ruppanner, R., Ismail, A., Champagne, C. P., Assaf, R. (1992). Antibacterial and antiviral activity of camel milk protective proteins. *J. Dairy Res.* **59**, 169–175.

Elagamy, E. I., Ruppanner, R., Ismail, A., Champagne, C. P., Assaf, R. (1996). Purification and characterization of lactoferrin, lactoperoxidase, lysozyme and immunoglobulins from camel's milk. *Int. Dairy J.* **6**, 129–145.

Evans, J. S., Levine, B. A., Trayer, I. P., Dorman, C. J., Higgins, C. F. (1986). Sequence imposed structural constraints in the Ton B protein of *E. coli*. *FEBS Lett.* **208**, 211–216.

FAO, OIE, WHO (1990). *Animal Health Yearbook*. Food and Agricultural Organisations of the United Nations, Rome.

Farah, Z. (1996). *Camel Milk Properties and Products*. SKAT, Swiss Centre for Development Cooperation in Technology and Management, St. Gallen, Switzerland.

Fowler, M. E. (1989). *Medicine and Surgery of South American Camelids*. Iowa State University Press, Ames, IA, USA.

Fowler, M. E., Gillespie, D., Harkema, J. (1985). Lymphosarcoma in a llama. *J. Am. Vet. Med. Assoc.* **187**, 1245–1246.

Gambirazio, C. (1967). Vasos linfaticos superficales y centrol linfacticos del miembro anterior de la alpaca, *Lama pacos* (Superficial lymphatic vessels and lymph nodes of the forelimb of the alpaca). Thesis, Fac. Med. Vet. Univ. Nac. Mayor San Marcos, Lima, Peru.

Garmendia, A. E., McGuire, T. C. (1987). Mechanism and isotypes involved in passive immunoglobulin transfer to the newborn alpaca (*Lama pacos*). *Am. J. Vet. Res.* **48**, 1465–1471.

Garmendia, A. E., Palmer, G. H., DeMartini, J. C., McGuire, T. C. (1987). Failure of passive immunoglobulin transfer: a major determinant of mortality in newborn alpacas (*Lama pacos*). *Am. J. Vet. Res.* **48**, 1472–1476.

Gharhoudi, M. A., Desmyter, A., Muyldermans, S., Hamers, R. (1995a). Identification of soluble, stable camel V_H antibody fragments expressed in *E. coli*, with specificity and neutralizing activity for tetanus toxoid. *Med. Fac. Landbouww. Univ. Gent* **60**, 2097–2100.

Ghahroudi, M. A., Muyldermans, S., Desmyter, A., Bendahman, N., Hamers, R. (1995b). Display of camel single domain antibodies on filamentous bacteriophages, and selection of two specific anti-tetanus toxin binders. *Arch. Physiol. Biochem.* **103**, B22.

Ghahroudi, M. A., Desmyter, A., Wyns, L., Hamers, R., Muyldermans, S. (1997). Selection and identification of single domain antibody fragments from camel heavy chain antibodies. *FEBS Lett.* **414**, 521–526.

Grassé, P. P. (1972). *Traité de Zoologie Tome XVI Mammifères, Fasc. IV*. Masson, Paris, France.

Green, R. S., Douch, P. G. C., Hill, F. I., Death, A. F., Wyeth, T. K., Donaghy, M. J. (1996). Antibody responses of grazing alpacas (*Lama pacos*) in New Zealand to intestinal nematodes. *Int. J. Parasitol.* **26**(4), 429–435.

Gulliver, G. (1875). Observations on the sizes and shapes of the red corpuscles of vertebrates. *Proc. Zool. Soc. Lond.* 474–495.

Gupta, G. C., Joshi, B. P., Rai, P. (1979). Observations on haematology of camel (*Camelus dromedarius* L). *Indian Vet. J.* **56**, 269–272.

Hageltom, M., Saad, A. M. (1986). Flow cytometric characterisation of bovine blood and milk leukocytes. *Am. J. Vet. Res.* **47**, 2012–2016.

Hamers-Casterman, C., Atarhouch, T., Muyldermans, S., Robinson, G., Hamers, C., Bajyana Songa, E., Bendahman, N., Hamers, R. (1993). Naturally occurring antibodies devoid of light chains. *Nature* **363**, 446–448.

Hamers-Casterman, C., Puype, M., Atarhouch, T., Robinson, G., Vandamme, J., Goethals, M., Vanderkerckhove, J., Hamers, R. (1998). Novel monovalent and divalent antigen binding fragments produced by enzyme digestion of the heavy chain camelid immunoglobulin IgG_3. Proceedings International Conference on Camelids: Science & Productivity. *J. Camel Pract. Res.*, in press.

Henderschot, L. M. (1990). Immunoglobulin heavy chain and binding protein complexes are dissociated *in vivo* by light chain addition. *J. Cell. Biol.* **111**, 829–837.

Hsu, T. C., Benischke, K. (1967–1974). *An Atlas of Mammalian Chromosomes*. Vol. 8, folio 389, Springer Verlag, Berlin, Germany.

Hutchinson, J. M., Salman, M. D., Garry, F. B., Johnson, L. W., Collins, J. K., Keefe, T. J. (1995a). Comparison of two commercially available single radial immunodiffusion kits for quantitation of llama immunoglobulin G. *J. Vet. Diagn. Invest.* 7(4), 515–519.

Hutchinson, J. M., Garry, F. B., Belknap, E. B., Getzy, D. M., Johnson, L. W., Ellis, R. P., Quackenbush, S. L., Rovnak, J., Hoover, E. A., Cockerell, G. L. (1995b). Prospective characterization of the clinicopathologic and immunologic features of an immunodeficiency syndrome affecting juvenile llamas. *Vet. Immunol. Immunopathol.* **49**, 209–227.

Jackson, T., Morris, B. A., Sanders, P. G. (1992). Nucleotide sequences and expression of cDNAs for a bovine anti-testosterone monoclonal IgG1 antibody. *Molec. Immunol.* **29**, 667–676.

Kabat, E. A., Wu, T. T., Perry, H. M., Gottesman, K. S., Foeler, C. (1991). *Sequences of Proteins of Immunological Interest*. US Public Health Services, NIH, Bethesda, MD, USA.

Kessler, M., Gauly, M., Frese, C., Hiendleder, S. (1995). DNA-studies on South American camelids. In *Proceedings 2nd European Symposium on South American Camelids*, Camerino, August 30–September 2 (eds Gerken, M., Renieri, C.). Universität Göttingen, Germany.

Knarr, G., Gething, M-J., Modrow, S., Buchner, J. (1995). BIP binding sequences in antibodies. *J. Biol. Chem.* **270**, 27589–27594.

Legssyer, I., Goethals, M., Puype, M., Van de Kerkhoven, J., Hamers-Casterman, C., Hamers, R. (1995). Characterization of light chains in camel immunoglobulins. *Arch. Physiol. Biochem.* **03**, B47.

Legssyer, I., Hamers, R. (1998). Glycosylation of light chains in camel IgG_1 and IgM immunoglobulins. Proceedings International Conference on Camelids: Science & Productivity. *J. Camel Pract. Res.*, in press.

Mage, R. G. (1987). Molecular biology of rabbit immunoglobulin and T-cell receptor genes. In *The Rabbit in Contemporary Immunological Research* (ed. Dubiski, S.), pp. 106–133. Wiley, New York, USA.

Miller, P. J., Hollander, P. J., Franklin, W. L. (1985). Blood typing South American camelids. *J. Hered.* **76**, 369–371.

Moustafa, A. M. (1994). First observation of camel (*Camelus dromedarius*) lymphadenitis in Libya. A case report. *Rev. Elev. Méd. Vét. Pays Trop.* **47**, 313–314.

Murphy, J. P., Gershwin, L. J., Tatcher, E. F., Fowler, M. E., Habig, W. H. (1989). Immune response of the llama (lama glama) to tetanus toxoid vaccination. *Am. J. Vet. Res.* **50**, 1279–1281.

Muyldermans, S., Atarhouch, T., Saldanha, J., Barbosa, J.A., Hamers, R. (1994). Sequence and structure of VH domain from naturally occurring camel heavy chain immunoglobulins lacking light chains. *Prot. Engng.* **7**, 1129–1135.

Neoh, S. H., Jahoda, D. M., Rowe, D. S. (1973). Immunoglobulin classes in mammalian species identified by cross-reactivity with antisera to human immunoglobulin. *Immunochemistry* **10**, 805–813.

Nguyen, V. K., Muyldermans, S., Hamers, R. (1997). The specific variable domain of camel heavy chain antibodies is encoded in the germline. *J. Mol. Biol.* **275**, 413–418.

Olaho-Mukani, W., Nyang'ao, J. N. M., Kimani, J. K., Omuse, J. K. (1995a). Studies on the haemolytic complement of the dromedary camel (*Camelus dromedarius*). I. Classical pathway haemolytic activity in serum. *Vet. Immunol. Immunopathol.* **46**, 337–347.

Olaho-Mukani, W., Nyang'ao, J. N. M., Kimani, J. K., Omuse, J. K. (1995b). Studies on the haemolytic complement of the dromedary camel (*Camelus dromedarius*). II. Alternate complement pathway haemolytic activity in serum. *Vet. Immunol. Immunopathol.* **48**, 169–176.

Penedo, M. C. T., Fowler, M. E., Bowling, A. T., Anderson, D. L., Gordon, L. (1988). Genetic variation in the blood of llamas. Lama glama and alpacas, lama pacos. *Animal Genet.* **19**, 267–276.

Pickford, M., Morales, J., Soria, D. (1993). First fossil camels from Europe. *Nature* **365**, 701.

Roditi, I., Schwarz, H., Pearson, T. W., Beecroft, R. P., Liu, M. K., Richardson, J. P., Bühring, H.-J., Pleiss, J., Bülow, R., Williams, R. O., Overath, P. (1989). Procyclin gene expression and loss of the variant surface glycoprotein during differentiation of *Trypanosoma brucei*. *J. Cell. Biol.* **108**, 737–746.

Rogier, F. (1934). Contribution à l'étude du système lymphatique du dromadaire. *Thèse Méd. Vét.* Toulouse.

Romer, A. S. (1945). *Vertebrate Paleontology*, pp. 453–460. University of Chicago Press, Chicago, IL, USA.

Saalmüller, A. (1996). Characterization of swine leukocyte differentiation antigens. *Immunol. Today* **17**, 352–354.

Seligmann, M., Mihaesco, E., Preud'homme, J. L., Danon, F., Brouet, J. C. (1979). Heavy chain diseases: current findings and concepts. *Immun. Rev.* **48**, 145–167.

Serrao, T., Muyldermans, S., Hamers, R., Wyns, L. (1996). Anti 2-phenyloxazolone γ-immunoglobulins from *Camelus dromedarius*. *Arch. Physiol. Biochem.* **104**, B62.

Sheppard, H. W., Gutman, G. A. (1981). Allelic forms of rat kappa-chain genes: evidence for strong selection at the level of the nucleotide sequence. *Proc. Natl Acad. Sci. USA* **78**, 7062–7068.

Sheriff, S., Constantine, K. L. (1996). Redefining the minimal antigen-binding fragment. *Nature Struct. Biol.* **3**, 733–736.

Silleghem, M., Darji, A., Hamers, R., Van de Winkel, M., De Baetselier, P. (1989). Dual role of macrophages in the suppression of interleukin-2 production and interleukin-2 receptor expression in *Trypanosoma*-infected mice. *Eur. J. Immunol.* **19**, 829–835.

Sileghem, M., Flynn, J. (1992). Suppression of interleukin 2 secretion and interleukin 2 receptor expression during tsetse-transmitted trypanosomiasis in cattle. *Eur. J. Immunol.* **22**, 767–773.

Smuts, M. M. S., Bezuidenhout, A. J. (1987). *Anatomy of the Dromedary*. Clarendon Press, Oxford, UK.

Spinelli, S., Frenken, L., Bourgeois, D., de Ron, L., Bos, W., Verrips, T., Anguille, C., Cambillau, C., Tegoni, M. (1996). The crystal structure of a llama heavy chain variable domain. *Nature Struct. Biol.* **3**, 752–757.

Stanley, H. F., Kadwell, M., Wheeler, J. C. (1994). Molecular evolution of the family Camelidae: a mitochondrial DNA study. *Proc. R. Soc. Lond.* **B256**, 1–6.

Stortmont, C. J. (1982). Blood groups in animals. *J. Am. Vet. Med. Assoc.* **181**, 1120–1123.

Tageldin, M. H., Al Sumry, H. S., Zakia, A. M., Fayza, A. O. (1994). Suspicion of a case of lymphocytic leukaemia in a camel (*Camelus dromedarius*) in Sultanate of Oman. *Rev. Elev. Méd. Vét. Pays Trop.* **47**, 157–158.

Underwood, W.J., Bell, T. H. (1993). Multicentric lymphosarcoma in a llama. *J. Vet. Diag. Investig.* **5**, 117–121.

Ungar-Waron, H., Elias, E., Gluckman, A., Trainin, Z. (1987). Dromedary IgG: purification, characterization and quantitation in sera of dams and newborns. *Isr. J. Vet. Med.* **43**, 198–203.

van der Loo, W., Bouton, C. E., Sanchez, M., Mougel, F., Castién, E., Hamers, R., Monnerot, M. (1995). Characterization and DNA sequence of the b6w2 allotype of the rabbit immunoglobulin kappa 1 light chain (*b* locus). *Immunogenetics* **42**, 333–341.

van der Loo, W., Verdoodt, B. (1992). Patterns of interallelic divergence at the rabbit b-locus of the immunoglobulin light chain constant region are in agreement with population genetical evidence for overdominant selection. *Genetics* **132**, 1105–1117.

Vilmos, P., Kurucz, E., Ocsovszki, I., Keresztes, G., Ando, I. (1996). Phylogenetically conserved epitopes of leukocyte antigens. *Vet. Immunol. Immunopathol.* **52**(4), 415–426.

Wernery, U. (1995). Blutparameter un Enzymwerte von gesunden und kranken Rennkamelen (*Camelus dromedarius*). *Tierärztl. Prax.* **23**, 187–191.

Wernery, U., Kaaden, O. R. (1995). *Infectious Diseases of Camelids*. Blackwell Wissenschafts-Verlag, Berlin, Germany.

Wernery, U., Schimmelpfenning, H. H., Seifert, H. S. H., Pohlenz, J. (1992). *Bacillus cereus* as a possible source of haemorhagic disease in dromedary camels (*Camelus dromedarius*). In *Proc. 1st. Int. Camel Conf.* (eds. Allen, W. R., Higgins, A. J., Mayhew, I. E., Snow, D. H., Wade, J. F.), pp. 51–58. Rand Publications, Newmarket, USA.

Wilson, R. T. (1984). *The Camel.* Longman, Harlow, UK.

Youssef, H. A., El Sebaie, A., Taha, M. M., Makady, F. (1987). Lymphosarcoma in a dromedary. *Vet. Med. Rev.* **1**, 68–71.

Zeuner, F. E. (1963). *A History of Domesticated Animals.* Hutchinson, London, UK.

Zweygarth, E. (1984). Isolation of mononuclear cells from the peripheral blood of camels (*Camelus dromedarius*). *Abl. Vet. Med.* **B31**, 786–789.

XIII IMMUNOLOGY OF CATTLE

1. Introduction

The world cattle population is approximately 1.2 billion, of which about 70 per cent reside in the developing countries. The beef industry is one of the most economical sources of meat in the world. This is due to the ruminant's ability to convert cellulose and hemicellulose into energy and produce proteins of high nutritional quality from inorganic nitrogen. This can occur even under harsh conditions on grasslands unsuitable for food crop production. Moreover, as cattle are important producers of milk for human consumption, efforts have been made to produce transgenic animals which secrete recombinant proteins in their milk.

Cattle must be protected from infectious diseases for optimal performance. There is a need to identify and characterize the protective immune mechanisms in cattle to identify the important immunoprotective antigens of pathogens for a rational production of efficient vaccines. For the last two decades there has been major progress at unravelling the bovine immune system and its functioning. Because of their placental structure, which is impermeable to immunoglobulins, cattle differ from man and other laboratory animals in their transfer of maternal immunity (see Section 9), immunoglobulin composition of colostrum and milk (see Section 9) and leukocyte distribution in the peripheral blood of the neonate ($\gamma\delta$ T cells in blood, see Section 3).

Since 1985, attention has been focused on the identification and functional characterization of bovine leukocyte populations, in particular the different populations of T cells and their receptors. Attention was also directed at the genomic organization, polymorphism and immunological role of the bovine major histocompatibility complex (MHC) with its bovine leukocyte antigens (BoLA). Since 1990, research extended to the bovine cytokines. Genes and proteins were sequenced and methods for their identification (monoclonal antibodies (mAb) and primers for polymerized chain reactions (PCR)) and quantification (bioassays as well as ELISA) were developed. Moreover, recombinant cytokines were produced to explore their immunomodulatory and therapeutic use in vaccine delivery and treatment.

In conclusion, bovine immunology is a rapidly evolving field of research in veterinary sciences. It is therefore most opportune to provide a review of current knowledge of the different immune components of the bovine immune system. It should also be mentioned that free-martinism in cattle played a crucial role in the elucidation of the mechanism of tolerance.

2. Lymphoid Organs

Thymus

The bovine thymus is a bilobed structure with thoracic and extrathoracic lobes situated in the anterior mediastinum and lower neck respectively. The lobes are made up of many lobules, each composed of an outer cortex and inner medulla; medullae of adjacent lobules are contiguous. The thymus is the principal site of T cell maturation and this process occurs sequentially in the cortex and medulla, with the more mature populations being found in the latter (Figure XIII.2.1). The major immune cell populations of the organ are T cells (*c.* 97%) and dendritic cells (*c.* 3%) and these are supported by a network of heterogeneous

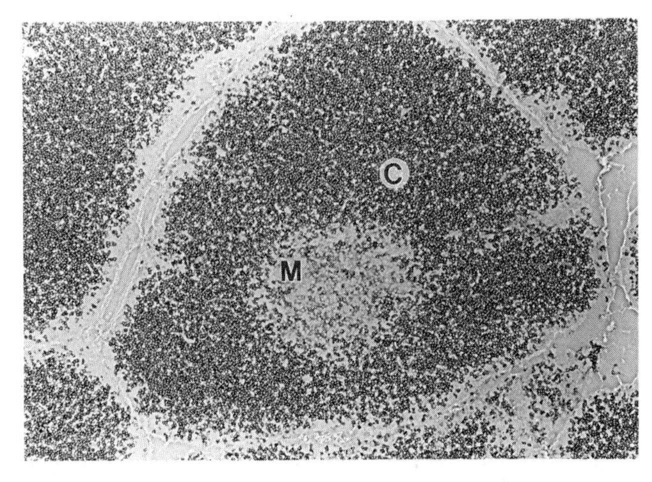

Figure XIII.2.1 Frozen section of bovine thymus stained with a mAb specific for CD1. The determinant is expressed in this tissue only by cortical thymocytes and therefore clearly discriminates the lymphocyte population of the cortex (C) from the more mature cells of the medulla (M). (×70)

Handbook of Vertebrate Immunology
ISBN 0-12-546401-0

Table XIII.2.1 Probable location of leukocyte populations in bovine thymus (modified from Morrison *et al.*, 1988)

Population	*Approximate %*	*Probable location*
$CD2^-CD4^-CD8^-$	10–20	Small numbers in outer cortex
$CD2^+CD4^+CD8^+$	50–60	Large numbers in cortex; absent in medulla
$CD2^+CD4^-CD8^+$	5–15	Moderate numbers in medulla; absent in cortex
$CD2^+CD4^+CD8^-$	5–15	Moderate numbers in medulla; absent in cortex
$\gamma\delta$ TCR^+	5–15	Moderate numbers in medulla; small numbers scattered in cortex
Interdigitating cells	ND	Scattered in medulla

ND, not done.

epithelial cells (Morrison *et al.*, 1986). The main T cell phenotypes found in the thymus and their approximate locations within the organ are summarized in Table XIII.2.1.

Lymph node

The bovine lymph node is similar in structural organization to that of the mouse and man. The node is an encapsulated organ composed of an outer cortex, an inner medulla and an intervening paracortical region. The cortex has a segmented appearance as a result of a number of connective tissue trabeculae that extend in from the capsule; the capsule and trabeculae are separated from the substance of the cortex by sinuses. Lymphoid cell populations within the node are supported by a framework of reticulin that is covered by reticulum cells. The substance of the node is interspersed by sinuses, which facilitate cell migration. The lymph node cortex is dominated by follicular areas, which lie adjacent to the sinuses adjoining the capsule and trabeculae. It is in the follicles that the greater proportion of the nodal B lymphocytes are found (Figure XIII.2.2). These areas also contain some $CD4^+$ T cells and small numbers of $CD8^+$ T cells. Up to 70% of follicles in lymph nodes of healthy 6–12-month-old animals contain germinal centres (Morrison *et al.*, 1986), which are composed largely of proliferating B lymphocytes and follicular dendritic cells (FDC). It is in germinal centres that maturation of the antibody response occurs, through the interaction of specific B cells with immune complexes trapped by FDC. The interfollicular cortex and paracortex are largely populated by $CD4^+$ and $CD8^+$ T lymphocytes interspersed by interdigitating dendritic cells (Figure XIII.2.3). Moderate numbers of $\gamma\delta$ T cells are found in the paracortex and this population also lines the subcapsular region of the cortex (Figure XIII.2.4). Relative proportions of T cells in these areas of the node are reflected in their phenotypic distribution in cell lymph node suspensions, as outlined in Table XIII.2.2. The lymph node medulla is composed of cord-like structures interspersed with sinuses that drain into the efferent lymphatic vessel. These sinuses contain populations of cells in transit from the node that presumably give rise to the cells of efferent lymph. Distribution of immune cell populations within the lymph node are summarized in Table XIII.2.3.

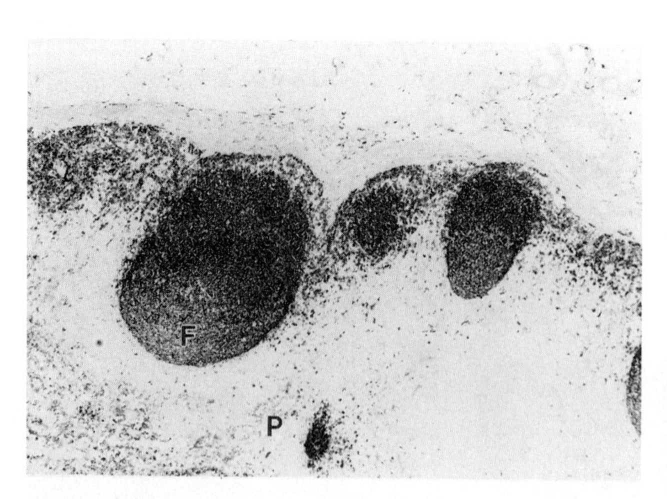

Figure XIII.2.2 Frozen section of bovine lymph node stained with a mAb specific for IgM. Staining is largely restricted to the follicular areas (F) of the cortex, although some positive cells are present in the paracortex (P). Dense aggregates of stain within the follicles are likely to represent immune complexes on follicular dendritic cells. (×76)

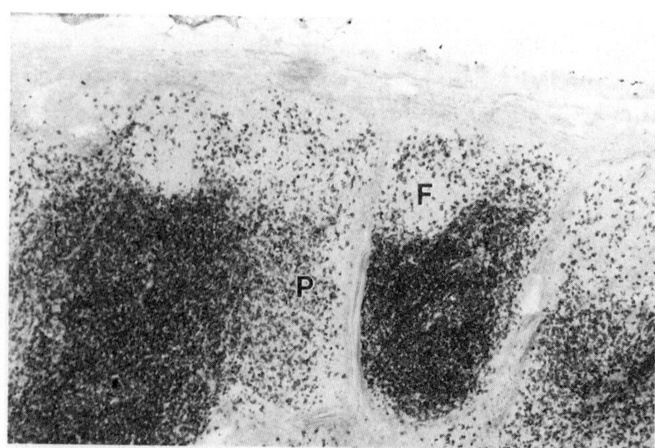

Figure XIII.2.3 Frozen section of bovine lymph node stained with a mAb specific for the bovine CD4 determinant. Staining is evident on the majority of paracortical lymphocytes (P) and is also sparsely distributed in the B cell follicles (F). (×68)

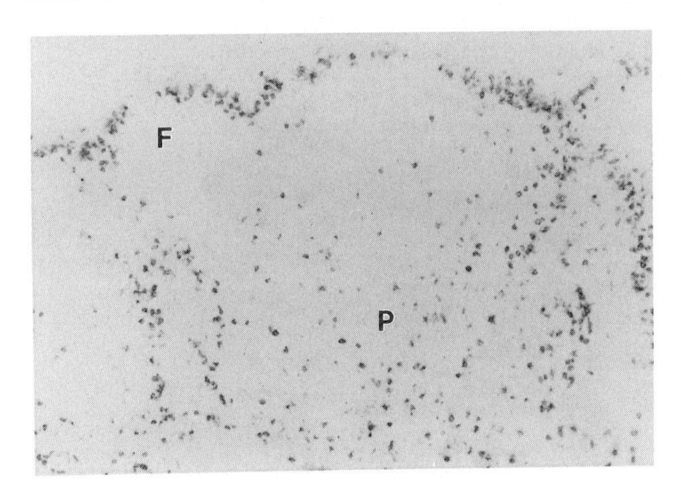

Figure XIII.2.4 Frozen section of bovine lymph node stained with a mAb specific for the WC1 determinant of bovine $\gamma\delta$ T cells. Positive cells are most evident in the subcapsular region of the cortex but are also seen throughout the paracortex (P). ($\times$ 60) (F = follicle).

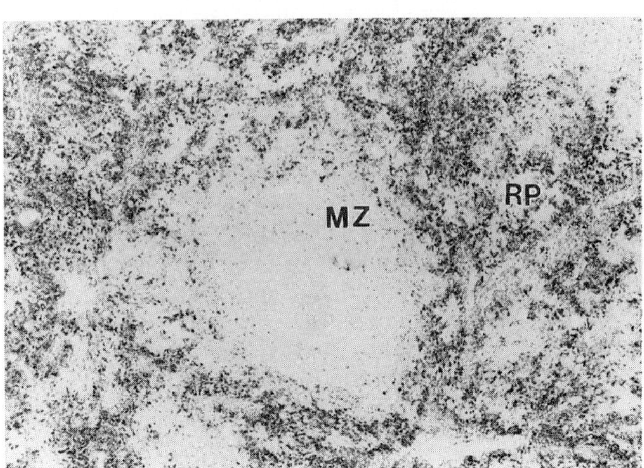

Figure XIII.2.5 Frozen section of bovine spleen stained with a mAb specific for the $\gamma\delta$ TCR. Positive cells are located predominantly in the marginal zones (MZ) and the red pulp (RP), but are also seen in the periarteriolar sheath. ($\times$ 60).

Table XIII.2.2 Distribution of the major leukocyte subpopulations in cell suspensions prepared from bovine lymphoid tissues[a]

Population	*Thymus*	*Lymph node*	*Spleen*
$CD3^+$ T cells	46%	25%	49%
$CD4^+$ T cells	59%	16%	6%
$CD8^+$ T cells	58%	9%	16%
mIg^+ B cells[b]	—	60%	32%
Macrophages	—	4%	13%
$\gamma\delta$ TCR^+ cells	15%	2%	44%
$WC1^+$ $\gamma\delta$ T cells	5%	3%	3%

[a]Data generated by FACS analysis of autopsy material taken from a single animal.
[b]B cells as identified by membrane-bound immunoglobulins.

Table XIII.2.3 Histological distribution of the major leukocyte subpopulations in bovine lymph node

Population	*Location*
$CD4^+$ T cells	Paracortex, scattered in follicles
$CD8^+$ T cells	Paracortex, sparse in follicles
$\gamma\delta$ T cells	Paracortex and cortex adjacent to sinuses; some present in medulla
mIg^+ B cells	Follicles, germinal centres
Interdigitating dendritic cells	Paracortex
Follicular dendritic cells	Germinal centres
Macrophages	Medulla, germinal centres, some in paracortex

The spleen

The bovine spleen is a large encapsulated organ situated on the dorsal surface of the rumen. Like that of other species, it is composed of red and white pulp areas. The red pulp constitutes the greater part of the organ and is largely composed of a system of venous sinusoids supported by a framework of reticular cells. The white pulp is associated with the arterial blood supply to the organ and forms sheaths around the arterioles (Figure XIII.2.5). $CD4^+$ and $CD8^+$ T cells and interdigitating dendritic cells are found in the inner or peri-arteriolar lymphoid sheath (PALS), while primary and germinal centre-containing B cell follicles locate to the outer sheath. The sheath is surrounded by a marginal zone, which is separated from it by the marginal sinus. The marginal zone is largely colonized by B cells, macrophages and $\gamma\delta$ T cells. As in the lymph node, FDC are found in the germinal centres and B-cell follicles contain some $CD4^+$ T cells and small numbers of $CD8^+$ T cells. The distribution of immune cell populations in the spleen of cattle is summarized in Table XIII.2.4. A significant feature of the bovine spleen is the large number of $\gamma\delta$ T cells that populate the red pulp (Figure XIII.2.5). The majority of these do not express the WC1 determinant found on most $\gamma\delta$ T cells of

Table XIII.2.4 Distribution of leukocyte populations in bovine spleen

Population	*Location*
$CD4^+$ T cells	PALS, follicles
$CD8^+$ T cells	PALS, sparse in follicles
$\gamma\delta$ T cells	Red pulp, marginal zones; small numbers in PALS
mIg^+ B cells	Follicles, germinal centres, marginal zones
Interdigitating cells	PALS
Follicular dendritic cells	Germinal centres
Macrophages	Marginal zones, germinal centres

Table XIII.2.5 Phenotypic distribution of bovine circulating leukocyte populations

Population	*PBMC*[a]	*Efferent lymph*	*Afferent lymph*	*Thoracic duct lymph*[b]
CD4$^+$ T cells	20–40%	50–60%	25–30%	56%
CD8$^+$ T cells	10–20%	20–30%	10–15%	10%
mIg$^+$ B cells	30–60%	10–20%	10–30%	25%
Macrophages	5–15%	—	—	—
$\gamma\delta$ TCR$^+$ cells	5–30%	5–15%	10–30%	8%
WC1$^+$ $\gamma\delta$ T cells	5–30%	5–15%	5–20%	ND
Dendritic cells	—	—	5–20%	—

ND, not done.
[a]Polymorphonuclear cells are removed during the preparation of PBMC.
[b]Data derived from a single animal.

blood, lymph and other lymphoid tissues (Tables XIII.2.2 and XIII.2.5).

Lymph and peripheral blood

A number of lymphatic vessels of cattle are readily accessible to surgical cannulation, including prescapular and prefemoral efferent lymphatics and the thoracic duct. Prior removal of the corresponding node allows collection of afferent lymph from the efferent vessel and lymph collected in this manner is usually referred to as pseudo-afferent.

Thoracic duct lymph

Thoracic duct lymph contains essentially the entire recirculating lymphocyte pool with the exception of that derived from the head, neck and forelimb on the right side, which drain into the right common jugular vein. In our experience, cell output from the bovine thoracic duct is of the order of 5×10^9/h. The phenotypic distribution of lymphocyte populations in thoracic duct lymph derived from a representative animal is provided in Table XIII.2.5.

Peripheral blood

In addition to those populations derived from thoracic duct lymph, the peripheral blood contains bone-marrow-derived cells of the granulocyte/macrophage lineages, including monocytes, neutrophils, eosinophils and basophils, as well as T- and B-cell precursors en route to the primary lymphoid tissues (Table XIII.2.5). It is also likely that the precursors of peripheral dendritic cells are present in bovine blood in small numbers, as has been demonstrated in man (van Voorhis *et al.*, 1982).

Efferent lymph

Studies in sheep have determined that cells of resting efferent lymph are largely (>90%) derived from blood lymphocytes that have circulated through the lymph node by crossing the high endothelial venules of the paracortex (Hall and Morris, 1965). The remainder originate from within the node and from afferent lymph cells that have percolated through its substance. Relative proportions of lymphocyte populations of efferent lymph are estimated in Table XIII.2.5, based on observations in a number of cattle. Cells of the granulocyte/macrophage lineage are not normally detected in efferent lymph. The flow rate of efferent lymph and the numbers and proportions of cells contained in it vary greatly with the state of activation of the node, but approximately 2–6 $\times$ 10^8 cells leave the resting node each hour.

Afferent lymph

Like that of other species, afferent lymph of cattle is characterized by a population of dendritic cells known as veiled or frilly cells. Derived from skin and associated tissues, these cells are highly efficient in antigen processing and presentation and there is strong evidence that they are the precursors of the interdigitating dendritic cells of the lymph node paracortex (McKeever *et al.*, 1992). At least a proportion are derived from Langerhans cells in the epidermis, as shown by the presence of cytoplasmic Birbeck granules (McKeever, 1994). Bovine veiled cells are morphologically and phenotypically heterogeneous and, at least for some antigens, this variation is reflected in substantial differences in antigen-presenting capacity (McKeever *et al.*, 1991). Lymphocyte populations of afferent lymph are almost invariably of resting morphology, with phenotypes distributed as outlined in Table XIII.2.5. Hourly cell output in afferent lymph varies greatly and in pseudo-afferent lymph is influenced by the degree to which efferent lymph from other nodes in the region contributes to the flow. In general, however, 1–3 $\times$ 10^7 cells are collected each hour.

Gut-associated lymphoid tissue (GALT)

The mucosa-associated lymphoid tissue is organized in aggregations of lymphoid structures in the mucosa and submucosa of the intestine, namely the Peyer's patches (PP), in isolated nodules in the lamina propria, and in isolated cells dispersed in the lamina propria and between the epithelial cells of the mucosa. The new-born calf has approximately 76 discrete PP (DPP) in the duodenum and jejunum, a single continuous PP (CPP) in the ileum, which extends proximally and terminally in the terminal jejunum and proximal cecum, respectively, and a single DPP in the proximal colon (Parsons *et al.*, 1989, 1991). At sexual maturity (18 months) the CPP is involuted and 18–40 DPP remain visible in adult cattle. As in sheep (Reynolds and Morris, 1983), the involuting CPP appears to be a primary lymphoid organ, namely a site of B-cell generation equivalent to the bursa of Fabricius of birds, where lymphopoiesis is not dependent upon antigen. In contrast, the DPP, cecal and colonic PP(CoPP) and propria nodules of the large intestine which have similar lymphocyte compositions, appear to be secondary lymphoid organs; they enlarge with age and develop germinal centres upon antigen stimulation.

The follicle-associated epithelium (FAE) and the membranous cell (M cell)

The epithelium covering the PP consists of the follicle-associated epithelium (FAE) and the interfollicular epithelium (IFE). The FAE, which is devoid of goblet cells, forms with the underlying mucosa a dome over the follicle and bulges into the lumen of the intestine (domed villus) (Parsons *et al.*, 1991). The FAE of the CPP consists of a homogeneous population of M cells with short, sparse microvilli and microfolds (Landverk, 1987). The FAE of the DPP also contains M cells, but these are scattered between absorptive enterocytes. The FAE of the proximal part of the CPP appears to be a transitional zone where the FAE of some domed villi are composed by pure M cells while others are composed of a mixture of M cells and absorptive enterocytes as in the DPP.

M cells are thought to be responsible for transport of macromolecules and processed proteins towards the follicles and thus to play a role in the uptake and presentation of antigen to the underlying tissue. Indeed, it has been shown that M cells endocytose bacteria (Ackermann *et al.*, 1988; Momotani *et al.*, 1988) and shed 50 nm carbonic anhydrase-positive membrane-bounded particles through the underlying intercellular spaces into the underlying follicles (Landverk, 1987; Parson *et al.*, 1991). Subepithelial and intraepithelial macrophages in the domed villi containing bacilli or bacterial debris have also been demonstrated and probably take part in antigen presentation.

The interfollicular areas are overlaid by the IFE which forms leaf-shaped absorptive villi, as opposed to the finger-like absorptive villi in other parts of the small intestine. The domed villi of the DPP are completely obscured by absorptive villi, while those of the CPP remain occasionally visible.

In the adult, isolated follicles remain in the ileum after involution of the CPP. The FAE of these isolated follicles contain clusters of M cells scattered among absorptive cells (Parsons *et al.*, 1991).

The lymphoid tissue of the large intestine comprises a single colonic DPP in the proximal colon, which extends into the cecum, and isolated propria nodules in the remaining colon, which decrease in number towards the rectum (Liebler *et al.*, 1988a; Parsons *et al.*, 1991). The colonic PP consists of lymphoglandular complexes with openings to deep invaginations into the submucosa and contains propria nodules with domed areas at the mucosal surface (Liebler *et al.*, 1988a,b; Parsons *et al.*, 1991); the FAE of the domed areas are of the CPP type while the FAE of the lymphoglandular complexes are of the DPP type.

Lymphocyte distribution

A detailed study of the lymphocyte distribution in the GALT of young and adult cattle was performed by Parsons *et al.* (1989). The DPP consists of round and pear-shaped (apex towards the lumen) B cell follicles along the intestinal mucosa, interspersed with T cell agglomerations (interfollicular T cell zones) and capped by domed villi. The CoPP have B follicles at the surface, sides and base of invaginations of the mucosa, with T cell zones between them. The jejunal and CoPP of calves are similar in B- and T-cell distribution (B/T ratio = 2:1), having characteristics of secondary lymphoid organs (Table XIII.2.6).

Conversely, the CPP is a B lymphoid structure with few T cells (B/T ratio = 20 : 1) indicating its possible primary lymphoid organ nature (Table XIII.2.6). The CPP is characterized by a sequential accumulation of pillar-like follicles (small diameter towards lumen) along the intestine; T cells form only occasionally small T-cell agglomerations between some follicles and are scanty in the apices of the domed villi and in absorptive villi (Parsons *et al.*, 1989).

After involution of the CPP, the ileum contains single B follicles with T cells in the domed and absorptive villi. In adult animals, T cell distributions of the jejunum (DPP), colon (CoPP) and ileum become similar with a B/T cell

Table XIII.2.6 Flowcytometric analysis of lymphocyte distribution (%)[a] in cell suspensions of the GALT (Parsons *et al.*, 1989)

	Jejunum/DPP	*Ileum/CPP*	*Colon/CoPP*
mIg$^+$ B cells	35 (36)$^{ns\,b}$	40 (24)^{ns}	51 (43)^{ns}
CD2$^+$ T cells	19 (39)$^{0.05}$	3 (29)$^{0.001}$	19 (33)^{ns}
CD4$^+$ T cells	11 (28)$^{0.05}$	2 (14)$^{0.05}$	9 (22)^{ns}
CD8$^+$ T cells	12 (10)^{ns}	3 (16)$^{0.01}$	11 (20)^{ns}
WC1$^+$ T cells	8 (6)^{ns}	2 (15)$^{0.05}$	9 (14)^{ns}

[a] Numbers without brackets are mean percentages in calves (<1 week; $n = 6$), numbers in brackets refer to adult cattle (2–6 years, $n = 6$).
[b] P between adult cattle and calves; [ns] not significant.

Table XIII.2.7 T lymphocyte subset distribution[a] by image analysis of cryostat sections of the discrete Peyer's patches (DPP, jejunum), continuous Peyer's patches (CPP, ileum) and CoPP (colon) (Parsons *et al.*, 1989)

	DPP	*CPP*	*CePP*	*CoPP*
CD4[+] T cells				
Dome	3 $(21)^{0.001b}$	2 (na)	2 $(4)^{0.05}$	2 $(19)^{0.01}$
Mucosa	1 $(10)^{0.001}$	<1 $(11)^{0.001}$	1 $(5)^{0.001}$	2 $(2)^{ns}$
CD8[+] T cells				
Dome	<1 $(8)^{0.001}$	<1 (na)	<1 $(2)^{0.05}$	<1 $(3)^{0.01}$
Mucosa	1 $(9)^{0.001}$	1 $(9)^{0.01}$	1 $(2)^{ns}$	2 $(8)^{0.01}$
WC1[+] T cells				
Dome	<1 $(<1)^{ns}$	<1 (na)	<1 $(<1)^{ns}$	<1 $(1)^{ns}$
Mucosa	<1 $(1)^{0.01}$	<1 $(<1)^{ns}$	<1 $(1)^{0.05}$	<1 $(<1)^{ns}$

[a]Mean of percentages of area stained in tissue sections by lymphocyte-specific monoclonal antibodies; numbers without brackets are values for calves (<1 week, $n = 4$), numbers in brackets refer to adult cattle (2–6 years, $n = 4$).
[b]*P* between adult cattle and calves; [ns] not significant.
na, not applicable.

ratio of 1:1, indicating an increase with age of T cells over B cells in secondary lymphoid organs (Table XIII.2.6). In adult cattle, CD4$^+$ T cells outnumber the CD8$^+$ T cells in the dome of the DPP and CoPP (CD4$^+$/CD8$^+$ = 3:1 to 6:1) while equal or higher numbers of CD8$^+$ T cells are observed in the mucosae (CD4$^+$/CD8$^+$ = 1:1 to 1:4) (Table XIII.2.7) (Parsons *et al.*, 1989).

3. The Leukocytes and their Markers

The investigation and characterization of cattle leukocyte differentiation antigens has resulted in the identification of a large number of surface molecules. Differential expression of these molecules has allowed the identification of functionally distinct subpopulations of cells, enabling *in vitro* and *in vivo* studies of their roles to be made in immune responses following infection or inoculation of antigens.

Three international workshops have been held to compare mAb to cattle leukocyte differentiation antigens (Howard *et al.*, 1991; Howard and Naessens, 1993; Naessens and Hopkins, 1996). Within these it has been agreed that since investigations in a number of mammalian species have established a considerable homology between the molecules expressed on the leukocyte surface, their function and the role of the cells that they identify, where possible, the cattle nomenclature should follow that adopted for human antigens. For an antigen to be given a bovine CD (boCD to distinguish the cattle molecule from the human one if necessary) number there should be very strong evidence that the molecule recognized is the homologue of a human CD antigen. This might include similarities in M_r, cell and tissue distribution, information on its function, sequence data, and cross-reactivity with human antigens. The mAbs that have been accepted as recognizing cattle CD antigens within the three workshops are listed in Table XIII.3.1 together with a summary of their cellular expression. The function of these molecules, in some cases, has been studied in cattle systems and shown to be the same as for their homologues in humans, in other cases it is assumed to be the same and immunological investigations with the mAb are made on this basis.

Within the workshops several clusters of mAb were identified that defined the same molecule but for which no human homologue was evident. These mAb are given workshop cluster (WC) numbers prefixed with ‘bo’ if necessary with the addendum that if the human CD homologue is established by subsequent studies a CD number, or other notation, should be substituted, as has occurred for WC2 (TCR-1), WC3 (CD21). Of these cattle molecules, the WC1 antigen has been the most studied. Sequence analysis of a cDNA clone showed that it belonged to an ancient family of cysteine-rich proteins while Southern blotting suggested that the bovine genome contains multiple WC1-like genes (Wijngaard *et al.*, 1992). The WC1 antigen is expressed by a majority of $\gamma\delta$ T cells in blood, where it comprises about 25% of peripheral blood mononuclear cells in young calves, and a minor subset of gut intra-epithelial lymphocytes (Clevers *et al.*, 1990). Subsequent studies (Wijngaard *et al.*, 1994) reported that the genes may be differentially expressed on subsets of WC1$^+$ cells while more recent investigations indicate a role in $\gamma\delta$ T cell activation (Hanby-Flarida *et al.*, 1996).

An alternative means of identifying cattle antigens has been to investigate mAbs to established human CD antigens for cross-reactivity on cattle cells. A survey of 772 mAbs submitted to the Fifth Human Leukocyte Antigen Workshop (Schlossman *et al.*, 1995) for ability to stain bovine leukocytes showed that 8.4% of mAbs stained cattle cells. A caveat to this approach is that specificity of staining must be confirmed. Thus, in this investigation of

Table XIII.3.1 Antibodies to bovine cell surface antigens

Antigen	Main cellular expression	Other names	Molecular mass (kDa)	mAbs
CD1	CT, IDC subset, ALVC subset		46	CC13[a], VPM5[d], TH97A[c]
CD1w1	CT, mono/macro, Lang, DDC, ALVC subset		46	SBU-20-27[g]
CD1w2	CT, IDC subset, ALVC subset	CD1b	46	CC14[a], CC20[a], CC40[a], CC90[a], CC122[a]
CD1w3	CT, B cells, mono/ macro, DDC, ALVC subset		44	CC43[a], CC118[a]
CD2	T, thy	SRBC, LFA-3 (CD58) receptor	58–62	CC42[a], IL-A26[b], IL-A42[b], IL-A43[b], IL-A45[b], 16-1E10[e], CH128A[c], CH132A[c], BAQ95A[c], BAT18A[c], BAT42A[c], BAT76A[c], CACT31A[c], MUC2A[c]
CD3	T	T cell receptor complex	12, 16, 22, 36 and 46	MM1A[c]
CD4	T subset (helper/ inducer)	T4, L3T4	50	ST-4A[f], CC8[a], CC26[a], CC30[a], IL-A11[b], IL-A12[b], GC50A, CACT83A[c], CACT87A[c], CACT138A[c], GC1A1[c]
CD5	T, B subset, thy	T1, Ly1	67	8C11[h], JP1D4[h], CC17[a], CC29[a], IL-A67[b], BLT-1[i], 79-5[g], SBU-25-91[g]
CD6	T subset, thy	T12	110	NAM3[h], CC38[a], IL-A27[b], IL-A28[b], IL-A57[b], BAQ82A[c], BAQ91A[c], CACT141A[c]
CD8	T subset (cytotoxic/ supressor)	T8, Lyt2, 3	34, 38	ST-8[f], 7C2[f], CC58[a], CC63[a], IL-A51[b], SBU-38-65[g], BAQ111A[c], BAT82A[c], CACT80[c], CACT88[c], CACT130[c]
CD11a	Leuk	LFA-1	180, 95	MD1H11[h], IVA35[j], IL-A99[b], CBU-72-87[g], BAQ30A[c], MUC76A[c]
CD11b	Mono/macro, B subset, gran	MAC-1	170, 95	CC94[a], CC104[a], CC125[a], CC126[a], IL-A15[b], IL-A130[b], MM10[c], MM13[c]
CD11c	Mono/macro, gran	CR4	160, 95	NAM4[h], IL-A16[b], C5-B6[e], BAQ153A[c]
CD14	Mono/macro	LPS receptor		CC-G33[a], VPM65[d], VPM66[d], VPM67[d]
CD18	Leuk	Integrin β_2 subunit	95	MF14B4[h]
CD21	B, ALVC subset, Alveolar macro	WC3, CR2	145	CC21[a], CC37[a], CC51[a], CC70[a], IL-A65[b], BAQ15A[c]
CD25	Act T	IL-2 R (p55), Tac	55	IL-A111[b], CACT108A[c], CACT109A[c], CACT116A[c]
CD41/CD61		$GPII_{\beta}$		IVA30[j], IVA31[j], IVA38[j], IVA125[j], IL-A164[b], IL-A166[b], CAPP2A[c]
CD44	Leuk, RBC, platelets	Pgp-1	95	Bufl[l], IL-A107[b], IL-A108[b], IL-A112[b], IL-A118[b], SBU-25-32[g]
CD45	Leuk	Leukocyte common antigen	180, 205, 220	151[f], 2E8[i], TD14[d], TD15[d], VPM18[d], SBU-1-28[g]
CD45R	T subset, B, thy subset		205, 220	IVA95[j], IVA103[j], IVA112[j], IVA313[j], IVA352[j], IVA373[j], CC31[a], CC76[a], CC77[a], CC99[a], CC103[a], Bo42[m], BAG36A[c], GC6A[c], GS5A[c], GX18A[c]
CD45RO	T subset mono/ macro, thy subset, gran		180	IL-A116[b]
CD49d	T, B, thy, mono	α_4 integrin, VLA-4		Clone 218[f]
CD62L	T subset, B subset, gran, mono/macro	L-selectin	90	Buf44[l], IVA94[j], CC32[a], BAQ92[c], Du-1-29[f]
CD71	Act T, Act B, macro	Transferrin receptor, T9		IL-A77[b], IL-A165[b]

(continued overleaf)

Table XIII.3.1 (*Continued*)

Antigen	*Main cellular expression*	*Other names*	*Molecular mass (kDa)*	*mAbs*
TCR1	T subset	WC2, $\gamma\delta$ TCR	37, 47	CACT16A[c], CACT17A[c], CACT18A[c], CACT61A[c], CACT71A[c], CACTB6A[c], CACTB12A[c], CACTB44A[c], CACTB81A[c], 86-D[f]
WC1	T subset	T19	215, 300	197[f], NAM2[h], CC15[a], CC39[a], CC101[a], CC115[a], CC117[a], IL-A29[b], BAG25A[c], BAQ4A[c], BAQ89A[c], BAQ90A[c], BAQ128A[c], BAQ159A[c], B7A1[c], CACTB1A[c], CACTB7A[c], CACTB15B[c], CACTB31A[c], CACTB32A[c]
WC4	B		90	CC55[a], CC57[a]
WC5	B, act T		46	IL-A54[b], IL-A55[b]
WC6	B, T subset, thy, ALVC		210–220	CC98[a], IL-A53[b], IL-A114[b]
WC7	B, T, thy		62, 69	TH1A[c], TH18A[c]
WC8	Act T		150	IL-A78[b], IL-A79[b], P13[b]
WC9	B, T subset, mono/ macro, gran		25	IVA31[j], IVA37[j], IVA50[j], IVA114[j], IL-A96[b], IL-A134[b], IL-A163[b], BAQ86A[c], MM41[c], RH1A[c], R18A[c], TH2A[c]
WC10	T subset, act T, B, thy		39, 115	CC28[a], CC62[a], CC69[a], IL-A56[b], CACT114A[c]
WC11	Leuk, platelets, RBC		52, (74) Gran 63, 85 PBM	IL-A117[b], IL-A136[b], CA26/1[m], CA17/1/6[m], 1C10[i], BAGB27A[c]
WC13				Co-3D1D4[k], Buf13[l]
WC14	Mono/macro, gran, T subset, B subset		150, 158	BT3/8.12[b], IL-A155[b]
WC15	RBC			IL-A135[b], IL-A137[b], IL-A138[b], IL-A160[b], ANA8[c]

Abbreviations: Act, activated; ALVC, afferent lymph veiled cells; B, B cells; CT, cortical thymocytes; DDC, dermal dendritic cells; FDC, follicular dendritic cells; Gran, granulocytes; IDC, interdigitating cells; Lang, Langerhans cells; Leuk, leukocytes; Macro, macrophages; Mega, megakaryocytes; Mono, monocytes; RBC, red blood cells; T, T cells; Thy, thymocytes.

[a]Institute for Animal Health, Compton Laboratory, Compton, Nr. Newbury, Berkshire, RG20 7NN, England. Contact: C. J. Howard. E mail: Chris.howard@afrc.ac.uk. Fax:++44-635-577237.*†

[b]International Lifestock Research Institute, P.O. Box 30709, Nairobi, Kenya. Contact: J. Naessens, E mail: j.naessens@cgnet.com. Fax:++254-2-631499.‡

[c]Department of Veterinary Microbiology & Pathology, Washington State University, Pullman, WA 99164-7040, USA. Contact: W. C. Davis. E mail: davisw@hawk.vetmed.wsu.edu. Fax: ++1-509-335-8328.§

[d]Department of Veterinary Pathology, University of Edinburgh, Summerhall, Edinburgh EH9 1QH, Scotland. Contact: J. Hopkins. E mail: JH@lab0.vet.edinburgh.ac.uk. Fax: ++44-31-650 6511.*

[e]University of Wisconsin, Department of Veterinary Science, Madison, WI 53706, USA. Contact: G. Splitter.

[f]Basel Institute for Immunology, Grenzacherstrasse 487, Postfach, CH–4005 Basel, Switzerland. Contact: P. Griebel. Fax: ++41-61-492380.†

[g]Centre for Animal Biotechnology, School of Veterinary Science, University of Melbourne, Parkville 3052, Victoria, Australia. Contact: A. Nash. E mail: U2605163@ucsvc.ucs.unimelb.edu.au. Fax: ++61-3-347 4083.

[h]Facultés Universitaires Notre Dame de la Paix, Immunology Unit, Rue de Bruxelles 61, B–5000, Namur, Belgium. Contact: J-J. Letesson.

[i]University of Connecticut, Department of Pathobiology, 61 N Eagleville Road, Storrs, CT 06269-3089, USA.

[j]Slovak Academy of Sciences, Institute of Animal Biochemistry and Genetics, 900 28 Ivanka pri Dunaji, Slovakia. Contact: R. Dusinsky. E mail: dusinsky@ubgz.savba.sk.

[k]Facultad de Veterinaria, Departemento de Genetica, Avda. de Medina Azahara, 9, 14005 - Cordoba, Spain. E mail: ge1llrud@lucano.uco.es.

[l]Biological Research Center, H.A.S., Institute of Genetics, POB 521, H–6701 Szeged, Hungary. Contact: I. Ando. Fax: ++36-62-433503.

[m]Immunology Unit, Veterinary School, Bischofsholer Damm 15, D–3000 Hannover 1, Germany. Contact: W. Leibold/H. J. Schuberth. E mail: jschub@immunologie.tiho-hannover.de. Fax:++49-511-856 7685.

Some of the mAbs listed from the laboratories marked *, †, ‡ or § (below) are available from the following depositories and commercial suppliers:

*Serotec Ltd., Bankside Station Approach, Kidlington, Oxon, OX5 1JE, UK. Fax: ++44-1865-379941 (USA 1-800-265-7376)

†European Collection of Animal Cell Cultures, Centre for Applied Microbiology & Research, Porton Down, Salisbury, SP4 0JG, UK. http://www.gdb.org/annex/ecacc/HTML/ecacc.html

‡American Type Culture Collection, 12301 Parklawn Drive, Rockville, MD 20852 USA. http://www.atcc.org

§VMRD, Inc., P.O. Box 502, Pullman, WA 99163, USA. Tel: ++1-509-334-5815. http://www.vmrd.com

Table XIII.3.2 Monoclonal antibodies to human leukocyte differentiation antigens that cross-react with bovine cells

Antigen	*Main cellular expression (in humans)*	*Other names*	*mAb*
CD11a	Leuk	LFA-1	CBR-LFA-1[a]
CD14	Mono/macro	LPS receptor	MY4, S39[a], M5E2, RPA.MI, biG10, biG12, biG13, biG14, TUK4
CD18	Leuk	Integrin β_2 subunit	6.7[a], CBR-LFA-1
CD21	B cells, FDC	C3d receptor, EBV receptor	WEHI-B2[a], HB5C, B2, KS-8, KS-9, BL13/10B1a, BU36, BU42
CD27	Thy, T cells		M-T271[a]
CD29	Leuk	Integrin β_1 subunit	5D9, K20, 4B4[a], TS2/10, JB1, mAb13
CD49a	Act T, mono	α_1 integrin, VLA-1	1B3.1[a], SR84
CD49b	B, mono, platelets	α_2 integrin, VLA-2	5E8, G1 14[a], Gi9
CD49d	B, thy	α_4 integrin, VLA-4	5D5, 9F10, L25[a]
CD49e	T subset, mono, platelets	α_5 integrin, VLA-5	X6, SAM-1[a]
CD51	Platelets, mega	αV integrin	13C2[a], 23C6
CD61a	Platelets, mega, macro	Integrin β_3 subunit	C5-1, LM609[a], AP6
CD62L	T subset, B subset, gran, mono/macro	L-selectin	LAMI-3[a], Dreg 56, FMC 46
CD62P	Platelets, mega, endothelium	P-selectin, PADGEM	sz-51[a]
CD63	Act platelets, mono/macro		46-4-5, 79-2-7, MOF11[a]
CDw78	B	Ba	FN4[a]
CD98	T, B, NK, gran, all human cell lines	4F2	2F3[a]
CD100	Broad expression on haemopoietic cells	GR3	BD16[a]

Abbreviations: act, activated; B, B cells; CT, cortical thymocytes; FDC, follicular dendritic cells; gran, granulocytes; IDC, interdigitating cells; Lang, Langerhans cells; leuk, leukocytes; macro, macrophages; mega, megakaryocytes; mono, monocytes; RBC, red blood cells; T, T cells; thy, thymocytes.
[a]Studied in two colour-staining experiments on bovine cells (Sopp and Howard, 1997). A full description of mAbs and a list of suppliers is included in *Leukocyte Typing V* (Schlossman *et al.*, 1995).

the CD groups that contained cross-reacting mAbs, 18 of the groups contained mAbs that showed similar expression of the target antigen on cattle and human cells (Table XIII.3.2) and these identified new cattle CD antigens not previously recognized within the three cattle workshops (Sopp and Howard, 1997). mAbs within nine CD groups stained cattle and human cells differently, indicating either a modified cell distribution for the antigen in cattle or cross-reaction with an epitope expressed on a different antigen. mAbs to defined antigens in other animal species that cross-react with cattle can be added to the list of reagents useful for bovine studies.

Expression of these leukocyte differentiation antigens has allowed the identification of the major populations of T lymphocytes and their subsets, which in some cases have distinct functional properties. Two major subpopulations comprise the $\alpha\beta$ TCR$^+$ T cells. These are the MHC class II restricted CD4$^+$ T cells that provide help for B-cell responses (Baldwin *et al.*, 1986; Howard *et al.*, 1989) and the MHC class I restricted CD8$^+$ T cells (Ellis *et al.*, 1986). Within the CD4$^+$ population subsets are evident and identified by expression of different isoforms of CD45. Cells responding in proliferation assays to soluble antigen are CD45RO$^+$ and IL-4 and IFN-γ production is evident within these cells (Bembridge *et al.*, 1995). This memory subset recirculates preferentially to the skin and mucosa as first described in sheep (Mackay *et al.*, 1990). Current investigations of a T_H1/T_H2 bias in cattle indicate that cells which differentially synthesize cytokines are present within the CD4$^+$ population (Estes *et al.*, 1995). The CD8$^+$ T cells have been shown to be cytolytic and to be central to immunity to certain virus infections (Goddeeris *et al.*, 1986; Taylor *et al.*, 1995).

Several subsets of $\gamma\delta$ TCR$^+$ T cells are evident in cattle. In the blood the WC1$^+$, CD2$^+$, CD8$^-$ phenotype predominates but other minor subsets that are WC1$^-$, CD2$^+$ and CD8$^+$ or CD8$^-$ are evident and these predominate in spleen and the gut mucosa (Clevers *et al.*, 1990; Wyatt *et al.*, 1994, 1996). The function of these cells has not been established but they appear to require exogenous IL-2 and a cellular signal to induce proliferation that is not via CD28 or MHC restricted (Collins *et al.*, 1996; Howard *et al.*, 1996; Okragly *et al.*, 1996).

4. Cytokines

Cytokines play a central role in immunological, physiological and pathological processes in animals. Our knowledge about bovine cytokines is increasing rapidly owing to the application of molecular cloning techniques. A better

Table XIII.4.1 Cloned bovine cytokines

Cytokine/receptor	*Database accession number*	*Tissue or cell source*	*Biological activity*	*Chromosomal location*	*References*
CD40L[c] (CD40 ligand)	Z48469	Lymphocytes			Mertens *et al.* (1995)
ENA[d] (epithelial-cell derived neutrophil attractant)	86149	Monocytes Macrophages Epithelial cells	Activates neutrophils Chemoattractant		Allmann-Iselin *et al.* (1994)
EPO (erythropoietin)	L41354	Liver Kidney Spleen	Stimulates proliferation and differentiation of erythroid progenitor cells	29	Agaba (1996) Suliman *et al.* (1996)
EPOR (EPO receptor)	U61398 U61399	Bone marrow			GenBank (1996)
FGFacidic[a]/basic[b] (fibroblast growth factor)	M35608 M97661 M13439 M13440	Retina Brain Pituitary Epithelial cells	Modulates the proliferation of several cell types Promotes cell survival Decreases inducible nitric oxide synthetase activity in endothelial cells	[a]7q31.3-31.2 [b]17	Abraham *et al.* (1986) Alterio *et al.* (1988) Goureau *et al.* (1995) Halley *et al.* (1988) Solinas-Toldo *et al.* (1995)
FGFR (FGF receptor)	Z68150	Oviduct Epithelial cells			GenBank (1995)
G-CSF (granulocyte colony stimulating factor)	P35833	Monocytes	Enhances the differentiation and activation of neutrophils		Kehrli *et al.* (1991) Lovejoy *et al.* (1993)
GM-CSF (granulocyte macrophage colony stimulating factor)	U22385	Monocytes	Growth-inducing factor for hematopoietic progenitor cells Enhances mature neutrophil functions	7q23-31	Daley *et al.* (1993) Maliszewski *et al.* (1988) Reddy *et al.* (1990)
GRO[d] (melanoma growth stimulatory activity)		Monocytes Epithelial cells Mesothelial cells	Chemotaxis on neutrophils		Rogivue *et al.* (1995)
ICAM (intracellular adhesion molecule)	U65789 L41844	Endothelial cells Mammary gland Lymph node			GenBank (1996)
IFN-α/β (interferon α and β)	>14 entries	Monocytes Macrophages B and T cells Endothelial cells Fibroblasts	Antiviral activity Immunomodulating effects Regulates functional activity of macrophages and neutrophils	8q15	Capon *et al.* (1985) Chaplin *et al.* (1996) Eggen and Fries (1995) Hansen *et al.* (1991) Leung *et al.* (1984) Tizard (1995) Velan *et al.* (1985)
IFN-γ (interferon γ)	M29867 Z54144	T cells NK cells	Stimulates macrophage activity Immunomodulating effects Increases surface expression of MHC class I and II on various cell types Stimulates cytotoxity	5q24.1	Cerretti *et al.* (1986) Tizard (1995) Solinas-Toldo *et al.* (1995)
IGF-1[a]/2[b] (insulin-like growth factor)	X15726 M60420 S76122 X53553	Liver Oviduct	Possible effects during ovulation Synergizes with EPO to enhance erythropoiesis	5q22-23[a] 29[b]	Brown *et al.* (1990) Easton *et al.* (1991) Fotsis *et al.* (1990) Li and Congote (1995) Schmidt *et al.* (1994) Schmutz *et al.* (1996)

(*continued*)

Table XIII.4.1 (*Continued*)

Cytokine/receptor	*Database accession number*	*Tissue or cell source*	*Biological activity*	*Chromosomal location*	*References*
IGF-1R[a]/2R[b] (IGF-1 and 2 receptors)				21[a] 9q25-27[b]	Moody *et al.* (1996) Solinas-Toldo *et al.* (1995)
LIF[e] (leukemia inhibitory factor)	D50337	Fibroblasts			GenBank (1995)
IL-1α/β (interleukin-1 α/β)	X12497 M37210 M35589 M37211	Monocytes Macrophages Neutrophils Endothelial cells	Pro-inflammatory cytokine Induces neutrophilia and monocytosis Activates osteoclasts Synergizes with other cytokines to promote B and T cell proliferation Immunoadjuvant Induces acute phase response		Collins and Oldham (1995) Godson *et al.* (1995) Lederer and Czuprynski (1995) Leong *et al.* (1988a,b) Maliszewski *et al.* (1988) Reddy *et al.* (1993)
IL-1R type I/II (IL-1 receptor)		Neutrophils Fibroblasts Leukocytes		11	Yoo *et al.* (1994)
IL-2 (interleukin 2)	M12791 M13204	T cells	Stimulates proliferation and differentiation of T lymphocytes Induces proliferation and immunoglobulin secretion by B lymphocytes Immunoadjuvant	17q26-27	Cerretti *et al.* (1986) Collins and Oldham (1995) Morsey *et al.* (1995) Reeves *et al.* (1986, 1987) Solinas-Toldo *et al.* (1995)
IL-2R α[a]/γ[b] (IL-2 receptor α/γ)	M20818 U24226	T and B cells Monocytes		13q13-q14[a] Xq23[b]	Weinberg *et al.* (1988) Yoo *et al.* (1995, 1996b)
IL-3	L31893	T cells Spleen	Stimulates early stages of hematopoietic cell differentiation Synergizes with other hematopoietic growth factors to promote proliferation of hematopoietic progenitor cells		Mwangi *et al.* (1995) Mertens *et al.* (1996)
IL-4	M77120 U14159 U14160	T and B cells Monocytes Macrophages	Upregulates various cell-surface markers on B cells and enhances the production of immunoglobulins Inhibits proliferation of Th1 and Th2-like clones	7q15-q21	Heussler *et al.* (1992) Estes *et al.* (1995) Buitkamp *et al.* (1995)
IL-5	Z67872	Lymphocytes			Mertens *et al.* (1996)
IL-6[e]	X57317 Z11749	T cells Monocytes Macrophages Endothelial cells	Stimulates the production of acute-phase proteins by hepatocytes Role in inflammatory reactions	4	Adams and Cruprynski (1995) Agaba (1996) Droogmans *et al.* (1992) Jian *et al.* (1995) Nakajima *et al.* (1993) Richards *et al.* (1995) Sterpetti *et al.* (1993)
IL-7	X64540			14	Agaba (1996) Cludts *et al.* (1992)
IL-8[d] (interleukin-8 like protein)	861479	Mononuclear cells Neutrophils	Activates neutrophils Chemoattractant for neutrophils		Hassfurther *et al.* (1994)

(*continued overleaf*)

Table XIII.4.1 (*Continued*)

Cytokine/receptor	*Database accession number*	*Tissue or cell source*	*Biological activity*	*Chromosomal location*	*References*
IL-8R (IL-8 receptor)	U19947			2	Li *et al.* (1996)
IL-10	U00799	T cells Monocytes B cells	Inhibits proliferation of all types of Th clones Downregulates IL2R expression and IFN-γ production Affects accessory cell function	1	Beever *et al.* (1996) Brown *et al.* (1994b) Chitko-McKown *et al.* (1995) Hash *et al.* (1994)
IL-12 p40/p35 (interleukin 12, subunits 40 and 35)	U11815 U14416	Lymphocytes Macrophages			Zarlenga *et al.* (1995a)
IL-15	U42433	Macrophages B cells Lymph node	Anabolic agent for muscle cells		GenBank (1996) Quin *et al.* (1996)
MCP-1/2 (monocyte chemotactic protein 1 and 2)	M84602 L32659 S67956	Monocytes Endothelial cells Seminal vesicles	Monocyte chemoattractant		Wempe *et al.* (1991, 1994) Kakizaki *et al.* (1995)
NGF[c] (nerve growth factor)	M26809		Neurotrophic effects Role in ontogenesis	3	Eggen and Fries (1995) Meier *et al.* (1986)
OSM[e] (oncostatin M)	S78434/5 S78487				Malik *et al.* (1995)
PECAM (platelet endothelial cell adhesion molecule)	U35433	Endothelial cells			Stewart *et al.* (1996)
PTN (pleiotrophin)	X52945	Uterus Cartilage Brain	Induces neurite outgrowth activity Mitogenic for endothelial cells		Delbe *et al.* (1995) Li *et al.* (1990)
SCF (stem cell factor)	D28934	Bone marrow Spleen Lymph node	Synergizes with other hematopoietic growth factors to stimulate myeloid and erythroid progenitor cells		Mertens *et al.* (1997) Zhou *et al.* (1994)
TGF-β^b (transforming growth factor β)	482694 M36371	Bone cells Macrophages Wart Mammary gland	Inhibits growth of several cell types Induces endothelin release Mediator of the inflammatory response		Kanse *et al.* (1991) Maier *et al.* (1991) Robey *et al.* (1987) van Obberghen-Schilling *et al.* (1987) Zurfluh *et al.* (1990)
TNF-$\alpha^{a,c}/\beta^c$ (tumor necrosis factor α/β)	Z14137 Z48808 Z14137	Macrophages Monocytes T and B cells Neutrophils Endothelial cells	Wide variety of effects on diverse cell types due to modulation of gene expression of growth factors and cytokines, inflammatory mediators and acute phase proteins	23[a]	Agaba *et al.* (1996) Cludts *et al.* (1993) Mertens *et al.* (1995) Myers and Murthaugh (1995)
VEGF (vascular endothelial growth factor)	M32976 M31836 M33750		Chemotaxin for endothelial cells Angiogenic mitogen		Koch *et al.* (1994) Leung *et al.* (1989) Tischer *et al.* (1989) Solinas-Toldo *et al.* (1995)
VEGFR (VEGF receptor)	X94263	Endothelial cells			Mandriota *et al.* (1996)

[ab] in column 1 refer to the chromosomal location of the genes in column 5.
[c] CD40L, NGF, TNF-α and TNF-β are members of the TNF gene family.
[d] ENA, GRO and IL-8 are members of the IL-8 gene family.
[e] LIF, IL-6 and OSM are members of the IL-6 gene family.

understanding of the role of cytokines in protective immune responses of cattle is of fundamental importance for improved health and vaccine development.

A list of cloned bovine cytokines and the corresponding GenBank accession numbers is presented in Table XIII.4.1. Information on the sources, the biological activities on bovine cells and the chromosomal location of the cytokine gene is included. Currently available methods for measuring cytokine levels involve the use of nucleic acid technology, bioassays and immunoassays. The investigation of cytokine gene expression by reverse transcription-polymerase chain reaction (RT-PCR) has become a standard procedure because of the high sensitivity of the technique and the lack of bovine specific assays for detection of the corresponding protein. Primer sequences specific for bovine cytokines are listed in Table XIII.4.2. Cytokine studies based on quantitative mRNA expression are readily performed but for each experimental system it must be shown that the level of cytokine mRNA is an accurate and reliable indicator of the cytokine protein levels. Reagents and assays available for bovine cytokines are given in Table XIII.4.3. In many cases, human cytokine reagents have been shown to cross-react with cattle and the degree of cross-reactivity is variable, depending on the cytokine. The use of the nonhomologous cytokine reagents may provide important insights into possible biological activities of the cytokine in cattle but potential differences in species specificity for the receptor and

Table XIII.4.2 Bovine cytokine-specific primer sequences for RT-PCR analyses

Cytokine	*Primers: FW (5′–3′)* *RV (5′–3′)*	*Product (bp)*	*References*
EPO	GCTGATGCTGTCCTTTCTGC GGGGAAAGCTGACTCTGTAC	520	Suliman *et al.* (1996)
FGF basic	TACAACTTCAAGCAGAAGAG CAGCTCTTAGCAGACATTGG	282	Watson *et al.* (1992)
GM-CSF	ATGTGGCTGCAGAACCTGCTTCTCC CTTCTGGGCTGGTTCCCAGCAGTCA	429	Ito and Kodama (1996)
IFN-γ	GGAGTATTTTAATGCAAGTAGCCC GCTCTCCGGCCTCGAAAGAGATT	387	Taylor *et al.* (1996)
IL-1α	CTCTCTCAATCAGAAGTCCTTCTATG CATGTCAAATTTCACTGCCTCCTCC	424	Ito and Kodama (1996)
IL-1β	AAACAGATGAAGAGCTGCATCCAA CAAAGCTCATGCAGAACACCACTT	394	Ito and Kodama (1996)
IL-2	ACGGGGAACACAATGAAAGGAAGT GGTAGGGCTTACAAAAAGAATCT	548	Covert and Splitter (1995)
IL-3	GTGACATGGAGGACTCCATA TCGCCCAAGTTCTTGTTCTC	280	Mertens *et al.* (1997)
IL-4	ATGGGTCTCACCTCCCAGCTGA GGTCTTGCTTGCCAAGCTGTTGA	312	Lutje *et al.* (1995)
IL-6	CCTTCACTCCATTCGCTGTC TTGCGTTCTTTACCCACTCG	534	Covert and Splitter (1995)
Il-7	GCCAGTAGCATCATCTGATTGTGA TAGTGGGTTGAGCTTCACTCAGGG	344	Mertens B. (unpublished data, 1996)
IL-10	CAGCAGCTGTATCCACTTGCCAAC CTCTCTTGGAGGTCACTGAAGACTC	375	Taylor *et al.* (1996)
IL-12 p40	TGGTATCCTGATGCTCCTGGAG TGCTCCAAGCTGACCTTCTCTG	444	Mertens B. (unpublished data, 1996)
SCF	CAACTGTCCTTGTAAGATTTGGTTG CAACTGTCAGTCAGCTTGACTG	663 or 579 (2 isoforms)	Mertens *et al.* (1997)
TGF-β_1	TGCTTCAGCTCCACAGAAAAGAAC CAGCTGCACTTGCAGGAGCGC	320	Mertens B. (unpublished data, 1995)
TNF-α	CTCGTATGCCAATGC AGGGCGATGATCCCAAAGTAGACC	378	Bienhoff and Allen (1995)
TNF-β	AGACCCCAGCACCCAGGACTCG GGAGATGCCATCTGTGTGAGTG	361	Covert and Splitter (1995)
Housekeeping gene: GAPDH	GATGCTGGTGCTGAGTATGTAGTG ATCCACAACAGACACGTTGGGAG	468	Taylor *et al.* (1996)
β-actin	ACCAACTGGGACGACATGGAGA AGCCATCTCCTGCTCGAAGTC	456	Lutje *et al.* (1995)

Note: Quantification of PCR products using competitive assays have been described (Bienhoff and Allen, 1994; Zarlenga *et al.*, 1995b; Taylor *et al.*, 1996) and also multiplex PCR for bovine cytokines (McKeever *et al.*, 1997).

Table XIII.4.3 Bovine cytokine reagents, cross-species reactivity and assays

Cytokine	*Bovine-specific reagents*			*Cross-species reactivity (human reagents on bovine cells)*[d]	*Assays*[e]	*References and sources*
	rec[a]	pAb[b]	mAb[c]			
ENA				anti-huENA Ab	Immunohistochemistry	Allmann-Iselin *et al.* (1994)[d,e]
EPO				huEPO		Adrianarivo *et al.* (1995)
FGF acidic				rhuFGFa anti-huFGFa Ab		Goureau *et al.* (1995)
FGF basic		x		rhuFGFb		Genzyme[b]; Goureau *et al.* (1995)[d]
G-CSF	x			rhuG-CSF		Amgen[a]; Kehrli *et al.* (1991), Kabbur *et al.* (1995)[d]
GM-CSF	x		x	rhuGM-CSF		Maliszewski *et al.* (1988)[a], VMRD[c], Kabbur *et al.* (1995)[d]
GRO				anti-huGRO Ab	Immunocytochemistry	Rovigue *et al.* (1995)[d,e]
IFN-α/β	x			huIFN-α	Antiviral neutralization assay with Vesicular stomatitis virus[1] and Madin–Darby bovine kidney cells or Semliki forest virus[2] and bovine BT-6 cells Immunosorbent binding assay[3] Induction of 2-5A synthetase activity[4]	Ciba-Geigy[a], Tizard (1995)[d], Brown *et al.* (1992)[e1], De Martini and Baldwin (1991)[e2], L'Haridon (1991)[e3], Totte *et al.* (1993)[e4]
IFN-γ	x		x	rhuIFN-γ	boIFN-γ ELISA-kit[1] Antiviral neutralization assay (see[e1] IFNα/β)[2] Induction of 2-5A synthetase activity (see [e4] IFN-α/β)[3]	Ciba-Geigy[a], Genentech, Wood *et al.* (1990)[c], Letesson J-J., Kabbur *et al.* (1995)[d], Tizard (1995), CSL[e1], Biosciences
IGF-2				rhuIGF-2		Li *et al.* (1995)
IL-1α	x			rhuIL-1α		Leong *et al.* (1988)[a], Blecha (1991)[d], Kabbur *et al.* (1995), Lederer and Czuprynski (1995)
IL-1β	x	x	x	rhuIL-1β	boIL-1β ELISA[1] Thymocyte costimulatory assay[2] Proliferation of IL-1-dependent cell line (LBRM-33)[3]	Collins and Oldham (1995)[a], Yoo *et al.* (1995)[b], Reddy *et al.* (1993)[c,d,e1,2], VMRD[c], Lederer and Czuprynski (1995)[d], Werling *et al.* (1995)[e1,3]
IL-1Ra antagonist				huIL-1Ra		Lederer and Czuprynski (1995)[d]
IL-2	x		x	rhuIL-2 anti-huIL-2 Ab	Proliferation assay of bovine IL-2 dependent T cell clone (300L1)[1] Proliferation assay on bovine T cell line (99.G1.G3)(IL-2/IL-4 assay)[2] Bioassay using IL2Rα Ab[3]	Collins and Oldham (1995)[a], Ciba-Geigy, VMRD[c], Collins *et al.* (1994)[d], Collins and Oldham (1993), Stevens and Olsen (1994)[e1], Brown *et al.* (1994a)[e2], Lutje *et al.* (1995)[e3]
IL-2Rα			x		Immunohistochemistry	Naessens *et al.* (1992), ILRI, VMRD
IL-3	x					ILRI

(*continued*)

Table XIII.4.3 (*Continued*)

Cytokine	*Bovine-specific reagents*			*Cross-species reactivity (human reagents on bovine cells)*[d]	*Assays*[e]	*References and sources*
	rec[a]	pAb[b]	mAb[c]			
IL-4	x		x	rhuIL-4	Proliferation assay on bovine T cell line (99.G1.G3)(IL2/IL4 assay)	Estes *et al.* (1995)[a], Letesson, J-J.[c], Belgium, Olsen and Stevens (1993)[d], Adler *et al.* (1995), Brown *et al.* (1994a)[e]
IL-6				rhuIL-6 Neutralizing antihuIL-6 Ab	huELISA[1] Proliferation of IL-6 sensitive cell line (7TDI)[2] or murine hybridoma cell line B9[3]	Jian *et al.* (1995)[d,e2], Sherpetti *et al.* (1993)[e1], Modat *et al.* (1990)[e3]
IL-7				rhuIL-7		Olsen and Stevens (1993)
IL-8				rhuIL-8	huELISA-kit Bovine neutrophil chemotactic assay	Hassfurther *et al.* (1994)[d,e]
IL-10				rhuIL-10 anti-huIL-10 Ab		Brown *et al.* (1994b)[d]
MCP				Neutralizing antihuMCP Ab	Immunocytochemistry	Kakizaki *et al.* (1995)[d,e]
OSM	x	x			Inhibition of proliferation of human A375 melanoma cells and mouse M1 myeloid leukemia cells	Malik *et al.* (1995)[a,b,d]
PTN				rhuPTN		Delbe *et al.* (1995)
TGF-α	x				Radioreceptor assay	Zurfluh *et al.* (1990)[a,e]
TGF-β				rhuTGF-β Neutralizing anti-huTGF-β Ab	huELISA-kit[1] Inhibition test on CCL64 cells[2] Inhibition of IL-5 induced proliferation of TF-1 cells[3] Luciferase transfected cells[4]	McCarthy and Bicknell (1993)[d], Kaji *et al.* (1994), Genzyme[e1], Adler *et al.* (1994)[e2], Randall *et al.* (1993)[e3], Kriegelstein and Unsicker (1995)[e4]
TNF-α	x	x	x	rhuTNF-α Neutralizing anti-huTNF-α Ab	Radioimmunoassay[1] ELISA[2] Cytotoxity on WEHI-164 cells Cytotoxity on murine L929[4] or swine PK15 cells[5]	Ciba-Geigy[a], Genentech, Ellis *et al.* (1993)[b,c,e3], Sileghem *et al.* (1992)[c], ILRI, Kabbur *et al.* (1995)[d], Pauli *et al.* (1994)[d,e5], Kenison *et al.* (1990)[e1], Werling *et al.* (1995)[e2,4], Adler *et al.* (1994)[e4,5]
TNF-β					Cytotoxity on WEHI-164 cells	Brown *et al.* (1993)

See Table XIII.4.1 for definitions.
[a]Recombinant protein.
[b]Polyclonal or [c] mAb to the bovine cytokine. [abc] in the references and sources refer to the corresponding bovine-specific reagents in column 2.
[d] Refers to cross-species reactivity in column 3.
[e1–e5] Refers to corresponding assays in column 4.
Sources: American Cyanamid and Immunex recently stopped the production of bovine specific cytokine reagents: VMRD Inc., Veterinary Medical Research and Development, Pullman, WA, USA (http:www.vmrd.com); CSL, Commonwealth Serum Laboratories, Parkville, Australia; ILRI, International Livestock Research Institute, Nairobi, Kenya (ilri@cgnet.com); Prof. J.-J. Letesson, Universite de Namur, Unite de Recherche Biologie Moleculaire, Namur, Belgium; Ciba-Geigy, Basel, Switzerland.
Updated databases on bovine cytokines are available through the Internet: Cytokine Explorer Database (http://kbot.mig.missouri.edu:443/cytokines) and the home pages of several companies (Amgen, Genzyme, R&D Systems, Promega) sometimes include information on cross-species reactivities.

immunological responses to nonhomologous proteins may occur. However, more bovine, sheep and goat cytokine and cytokine receptor reagents are becoming available. These reagents will allow the development of more accurate assays in the next few years and will consequently lead to a better understanding of immune and inflammatory processes in cattle.

5. The T-cell Receptor

General features of bovine TCR complexes

The TCR, which recognizes foreign antigens and MHC gene products, is assembled in the cytoplasm of T cells during their ontogeny. The TCR is composed of two disulphide-linked clonotypic heterodimer chains, $\alpha\beta$ and $\gamma\delta$. The TCR complexes (TCR–CD3 complexes) contain five noncovalently associated invariant components of CD3 complexes, CD3γ, CD3δ and CD3ε, and TCRζ and TCRη chains. The bovine TCR cDNA genes for α, β, γ and δ chains have been isolated by cDNA cloning and characterized during the last 7 years (Tanaka *et al.*, 1990; Takeuchi *et al.*, 1992; Ishiguro *et al.*, 1993). While the complete DNA sequences for CD3γ and δ and the TCRη chains have not been determined, the cDNA clones of bovine CD3ε and TCRζ components have been isolated from cDNA libraries by cross-hybridization with human and mouse counterpart probes (Clevers *et al.*, 1990; Hagens *et al.*, 1996). However, the chromosome locations in the bovine genome map and the genomic organization for bovine TCR, CD3 and TCR-associated genes have not been analysed in detail.

Molecular structure of bovine TCR genes

The primary structure of each TCR chain can be subdivided into specific regions: a hydrophobic leader (L) segment, a variable (V) segment, a diversity (D) segment in the case of β and δ chains, a joining (J) segment, and a constant (C) region. Table XIII.5.1 summarizes the lengths of amino acid sequences for bovine TCR α, β, γ and δ genes deduced from their DNA sequences. The structural sequences of bovine TCR genes show striking similarities to those of the TCR genes from other species, especially in the number and site locations of the cysteine in the disulfide bond for heterodimer formation, and the potential sites for *N*-linked glycosylation (Ishiguro and Hein, 1994). Table XIII.5.1 also shows the sequence similarities of each bovine TCR-C region to the corresponding regions of three different species (human, mouse and sheep) at both DNA and protein levels.

Bovine TCR $\alpha\beta$ chains

One C_α (144 amino acids) and two C_β ($C_\beta 1$ and $C_\beta 2$, 178 amino acids) genes were identified in cattle (Ishiguro *et al.*,

Table XIII.5.1 The bovine TCR genes

							Identity (%) between DNA/protein sequences for constant regions of three different species									
		Number of amino acids[c]					*Human*		*Mouse*			*Sheep*				
TCR chain	*mRNA*[a] *(kb)*	*L*	*V*	*D+J*	*C*		*C1*	*C2*	*C1*	*C2*	*C4*	*C1*	*C2*	*C3*	*C4*	*C5*
α	1.5	20–23	91–93 (0–2)[b]	20–24 (0–1)		140 (4)	73/60		68/55			90/81				
β	1.3	16–19	95–97 (0–2)	18–24 (0)	C1	178 (1)	84/81	83/82	76/75	78/75		96/94	95/94	95/94		
					C2	178 (2)	83/80	83/80	80/74	79/75		94/94	94/93	94/92		
γ	1.5	13–14	98–100 (0)	15–21 (0–1)	C1	222 (2)	63/48	66/47	71/64	70/64	62/42	76/69	91/87	68/58	68/58	64/54
					C2	211 (2)	68/51	64/51	65/65	65/65	62/45	81/72	91/86	70/61	76/62	65/54
					C3	194 (4)	70/53	70/65	56/53	63/51	65/51	73/64	69/58	94/87	71/60	66/53
					C4	210 (4)	72/56	67/54	65/51	66/67	64/48	97/95	81/71	73/64	79/67	65/59
δ	2.2	20	93–96 (0–1)	23–39 (0)		155 (2)	78/65		76/68			92/86				

[a]Size of mature transcript in kb.
[b]Number of *N*-linked glycosylation sites.
[c]Regions: L (leader), V (variable), D+J (diversity and joining) and C (constant).

Table XIII.5.2 EMBL Database accession numbers for bovine TCR–CD3 complexes

TCR or CD3 components	*Accession number*
TCR α chain	
V_{α}	D90010–D90029
C_{α}	D90030
TCR β chain	
V_{β}	D90121–D90133
$C_{\beta}1$	D90139
$C_{\beta}2$	D90140
TCR γ chain	
V_{γ}	D13648–D13661
$C_{\gamma}1$	D90413
$C_{\gamma}2$	D90415
$C_{\gamma}3$	D90414
$C_{\gamma}4$	X63680
TCR δ chain	
V_{δ}	D13655–D13661
C_{δ}	D90419
CD3 components	
CD3γ	X53268
CD3δ	X53269
CD3ε	U25687, X53270
CD3ζ (TCRζ)	U25688

1990; Tanaka *et al.*, 1990). The five bovine V_{α} families identified so far as full-length sequences show a 16–47% amino acid similarity between the families, while the amino acid similarity among nine bovine V_{β} families ranged from 18 to 53%, indicating that they are quite diverse. A bovine V_{β} family, termed the V_{β} 12-gene segment was isolated at a high frequency from a cDNA library, indicating that some V_{β} families may be used preferentially in T cell populations, as reported in human and mice. The EMBL database accession numbers for the TCR αβ gene segments are given in Table XIII.5.2. No mAb specifically recognizing the surface molecules of TCR α or β chain are available.

Bovine TCR γδ chains

Ruminant TCR γ chains, including those from sheep, have some unique features, especially in length and structure of the C_{γ} region. Four cDNA clones for the bovine C_{γ} gene have been identified (Ishiguro and Hein, 1994) (Table XIII.5.1). The difference in lengths for the bovine C_{γ} gene segment results from variability (about 27–54 amino acids) in the length of the connecting peptide or hinge segment between the extracellular domain and the transmembrane domain. The C_{γ} gene segment contains variable numbers of repeats of a five-amino-acid motif (consensus sequence TTEPP) which could be generated by duplication or triplication of the short exon. The characteristic consensus motif is repeated four times in the $C_{\gamma}1$ segment and once in the $C_{\gamma}2$, and slightly altered motifs (TAEPP or TTESP) are observed in $C_{\gamma}2$ and $C_{\gamma}4$ gene segments.

Three distinct V_{γ} families ($V_{\gamma}1$, $V_{\gamma}2$ and $V_{\gamma}5$) have been identified so far; the $V_{\gamma}5$ family is the predominant family of the V_{γ} genes expressed in peripheral blood lymphocytes (PBL). Comparison of protein sequences for the V_{γ} regions from three species showed that bovine V_{γ} genes are much more similar to sheep V_{γ} genes (22–89%) than those from human (18–53%) or mice (22–49%) genes. The DNA sequences of 12 functional V_{δ} genes which have been determined are highly similar to each other (75–97%) and are almost identical to the sheep $V_{\delta}1$ family (93%), compared with the human $V_{\delta}1$ family (62–69%) and the mouse $V_{\delta}6$ family (54–61%). The EMBL database accession numbers for the TCR γδ gene segments are given in Table XIII.5.2.

Invariant components of bovine TCR

The TCR chains are noncovalently associated with invariant components of CD3γ, CD3δ, CD3ε, TCRζ and TCRη chains. The CD3ε and TCRζ chains have an important role in signal transduction (Table XIII.5.3). Both TCR-associated molecules are the targets for tyrosine kinase, such as ZAP-70, lck and fyn, which stimulates the signal-

Table XIII.5.3 The bovine CD3 components and proportional identities with other species

Bovine CD3 chain	*Number of amino acids (partially determined)*	*Number of amino acids for CD3 components, and proportional identity (%) of sequences between bovine and the other species*		
		Human	*Mouse*	*Sheep*
CD3γ	(93)	NA[a]	NA	NA
CD3δ	(59)	NA	NA	NA
CD3ε	192	207 (73/62)[b]	189 (71/60)	192 (91/84)
TCRξ	166	168 (81/83)	164 (78/72)	167 (96/95)
TCRη	ND[c]	NA	NA	NA

[a]NA, not analyzed.
[b]Data shown as proportional identity (DNA/protein).
[c]ND, not detected.

ling pathways leading to T-cell proliferation and lymphokine secretion. The bovine CD3ε and TCRζ contain one and three typical ARAM (antigen recognition activation motif) sequences, respectively, in the cytoplasmic domain. Furthermore, the bovine TCRζ chain has a GDP/GTP-binding site with a C-terminal ARAM sequence. A mAb MM1A which recognizes native bovine CD3ε is available (Table XIII.3.1).

The EMBL database accession numbers for the CD3 genes are given in Table XIII.5.2.

Distribution of αβ and γδ TCR populations

Because the mAb which recognizes the TCR αβ chain has not been isolated, a detailed characterization of αβ T cell ontogeny and the distribution of αβ T cells in cattle have not been made. The number of γδ T cells in bovine PBL is usually very high, which is striking compared with the low proportion of γδ T cells found in humans and mice (Mackay and Hein, 1989; Hein and Mackay, 1991). It is generally accepted from experiments with anti-γδ mAb (for example: CACTB6A, CACTB81A and 86D) that positive proportions of γδ T cells in PBL are consistently higher in young calves (25–30%) than in older cows (3–10%), and that the population continues to decrease with advancing age. Why cattle have a high population of γδ T cells in PBL is unclear. The γδ T cells are usually localized with high frequency on the epithelial surface, especially within the skin, intestine, oesophagus and tongue, compared with their distribution in small numbers in the conventional T-cell domains of the spleen and lymph nodes. The BoWC1 molecule (WC1 family: 220 kDa) is the ruminant-specific differentiation antigen, which is expressed late in γδ T cell ontogeny.

6. Immunoglobulins

Ig isotypes and physicochemical properties

The nomenclature of bovine immunoglobulins follows the guidelines proposed by the nomenclature committee of IUIS (WHO, 1978). The different Ig isotypes identified in cattle are summarized in Table XIII.6.1 – four heavy chain classes (IgM, IgG, IgA and IgE), three IgG subclasses (IgG1, IgG2 and IgG3) and two light chain types (λ and κ). The production of mAb to different bovine isotypes (van Zaane and Ijzerman, 1984; Letesson *et al.*, 1985; Goldsby *et al.*, 1987; Naessens *et al.*, 1988; Tatcher and Gershwin, 1988; Estes *et al.*, 1990; Williams *et al.*, 1990) has allowed the establishment of isotype-specific assays. In cattle, no evidence has been found for the existence of IgD at the protein (Naessens and Williams, 1992; Naessens, 1997) or mRNA level (Butler *et al.*, 1996).

Bovine light chains are shared by all immunoglobulin classes. In contrast to primates, lagomorphs and rodents, the majority of light chains (over 90%) in cattle are of the λ type (Hood *et al.*, 1967). However, as no specific reagents to either λ or κ bovine isotypes exist, no further studies on λ : κ ratios have been made.

As in other mammals, bovine IgM exists as a pentameric form of the basic unit consisting of two μ heavy-chains and two light chains. It contains one covalently bound J-chain per molecule, which is necessary for the correct pentameric structure. The μ heavy-chain in cattle has a higher M_r than either the mouse or human μ chain, resulting in a larger IgM molecule. There is evidence for polymorphism on IgM as two mAbs detected an IgM-specific allotype and bound only half of the serum IgM and IgM-bearing cells in heterozygote animals (Naessens

Table XIII.6.1 Physicochemical properties of the different bovine heavy chain classes

	IgM	*IgG1*	*IgG2*[a]	*IgG3*[a]	*IgE*	*IgA*	*SC*
Heavy chain	μ	γ1	γ2	γ3	ε	α	
M_r Heavy chain, PAGE estimate (/1000)	80	58	55	58	68	60	
M_r Ig (1/1000)	1030	161–163	150–154			385–430	
Allotypes	B5.4	C1/C2	A1/A2				
	IL-A50	G1a,BA3,BA5					
Relative electrophoretic mobility	β2	β2	γ			β2	β2
S20,w	19.2–19.7	6.5–7.2	6.5–7.2			10.8–11	4.0–4.9
$S^{278}_{1\%,\ 1cm}$(UV-absorption)	11.8–12.6	12.1–13.7	12.3–13.5				
Carbohydrate (%)	10–12	2.8–3.1	2.6–3.0			6–10	5.9
Mannose (%)		0.53	0.57				1.92
Galactose (%)		0.37	0.45				0.19
Glucose (%)		0.62	0.45				0.45
Hexosamine (%)	2.8–3.9	0.9–1.72	0.8–1.6				2.5–3.1
Fucose (%)	1.25	0.10–0.26	0.16–0.26				0.24–0.80
Sialic acid (%)	1.39	0.28	0.15				0.4
Total SH/mol	43.6	37.5–43	32.5–35				30.0

[a]A newly proposed nomenclature replaces the terminology IgG2a and IgG2b by IgG2 and IgG3.
Based on Table I, Butler (1986) with permission from Karger, Basel.

et al., 1988; Williams *et al.*, 1990). No reagents for the allele are available.

Bovine IgA occurs predominantly in the dimeric form in serum, but higher polymers have been described. A J chain is associated with dimeric IgA. The genes for bovine C_μ and C_α have been isolated using a rabbit S_μ probe (Knight *et al.*, 1988). Southern blot analysis revealed one C_α gene and suggested two C_μ genes. However, the latter result was probably due to a restriction polymorphism.

sIgA, isolated from exocrine secretions, has a M_r of 420–550 × 10^3 because of an associated secretory component (SC). The SC can exist free or bound to polymeric immunoglobulin (pIg) and the bovine molecule has been well characterized biochemically (Butler, 1986). SC is the extracellular portion of the pIg receptor, which is responsible for binding and transcytosis of pIgA (and with lower efficiency, pIgM) through epithelial cells of mucosal and glandular tissues. At the apical surface, in contact with external secretions, the extracellular ligand-binding portion is proteolytically cleaved releasing the SC–Ig complex. An additional function of SC may be to render pIg more resistant to proteolytic cleavage in mucosal environments. The bovine cDNA sequence has been determined (Verbeet *et al.*, 1995) and a high degree of conservation among species (60–70%) observed.

A bovine IgE homologue, different from the other Ig classes, was identified in bioassays such as skin sensitization and mast cell binding (Nielsen, 1977), but the lack of an abundant source of IgE has prevented purification in sufficient quantities to acquire biochemical data. Specific antibodies have been raised (Nielsen, 1977; Gershwin and Dygert, 1983) and an ELISA test has been developed in which concentrations are measured in units, relative to a standard. One C_ε gene, 5′ of the C_α gene, was isolated and expressed as a heavy chain of 68 kDa in a chimeric bovine/murine Ig molecule (Knight *et al.*, 1988).

Two IgG subclasses, IgG1 and IgG2, were first identified as they eluted from an anion exchange column as two overlapping populations, with IgG1 being more tightly bound. They also had a different sensitivity for proteolysis, with IgG1 being more sensitive to pepsin, but more resistant to trypsin than IgG2. In a mixture of bovine IgG1 and IgG2, pepsin will cleave all IgG1 in $F(ab')_2$ and pFc′ fragments after 30 min, while leaving most of the IgG2. Comparisons of the $\gamma1$ and $\gamma2$ sequences show a major difference in the hinge-C_H2 region: the hinge regions are equal in length but show evolutionary divergence, and the $\gamma2$ chain has a 9 bp deletion at the start of the C_H2 domain (Symons *et al.*, 1989). This may explain the different capacity of the two subclasses for Fc receptor (FcR) binding on mammary cells and transport in colostrum (see below). A motif for the receptor site for C1q of complement, located in the C_H2 domain, is present in both subclasses (Mayans *et al.*, 1995). This is in agreement with the fact that both subclasses activate homologous complement, however, only IgG1 will fix heterologous complement *in vitro*.

While IgG1 antibodies behaved as a single subclass, heterogeneity was observed in the serological and electrophoretic behaviour of IgG2 antibodies (Butler, 1986). Subsequently, serological evidence for the existence of two isotypes among the IgG2 antibodies was demonstrated with polyclonal (Butler *et al.*, 1987) and mAb (Estes *et al.*, 1990). The two subclasses were tentatively called IgG2a and IgG2b. Further evidence for three subclasses came from studies in which three C_γ genes were isolated using a human C_γ probe (Knight *et al.*, 1988). The genes could be expressed as bovine/mouse IgG chimeras and were tested with isotype-specific antibodies. Two of the three genes corresponded to $C_\gamma1$ and $C_\gamma2$, while the third gene, $C_\gamma3$, probably codes for the 'IgG2b' heavy chain since its IgG product was not recognized by either anti-IgG1 or anti-IgG2. According to new nomenclature proposed by the Immunoglobulin Workshop of the Veterinary Immunology Committee, chaired by J. E. Butler, Iowa, USA, the three bovine IgG subclasses will be called IgG1, IgG2 (ex-IgG2a) and IgG3 (ex-IgG2b). Bovine C_γ pseudogenes also exist (Symons *et al.*, 1987; Knight *et al.*, 1988).

Allotypic determinants have been described on IgG1 and IgG2 subclasses (Table XIII.6.1). The best described allotypes are A1 and A2 on the IgG2 (ex-IgG2a) subclass. These will be renamed $IgG2^a$ and $IgG2^b$ according to the newly proposed nomenclature. Amino acid sequences of the two polymorphic heavy chains showed differences in four regions: (1) the region around the L-H bond, (2) the middle hinge, (3) a seven-amino-acid region at the beginning and (4) one Arg to Glu exchange at the end (position 419) of the C_H3 intradomain loop (Kacskovics and Butler, 1996). Analysis of pepsin cleavage fragments located the A1 allotype to the C_H3 domain (Heyermann and Butler, 1987), thereby restricting the position of A1 to the two sites in C_H3. The A2 allotype could not be located, and thus can be at any of the four sites.

A1 is an immunodominant epitope and a majority of heterologous antibodies to IgG2 contain specificity for the A1 allotype in addition to activity for the IgG2 isotype. A heterologous antiserum made by immunization with IgG2 of the A2 allotype contains mainly activity for the isotype, although some A2 allotype activity can be found. This bias in the reagents affects the correct estimation of IgG2 concentrations (Butler *et al.*, 1994).

It should be noted that the same nomenclature for subclasses (IgG1, IgG2, etc.) among different species often leads to the mistaken belief that these subclasses are homologous and have the same major functions. While this is generally true for the five known Ig classes (IgM, IgD, IgG, IgE and IgA), this is not the case for the subclasses. Most data are consistent with the hypothesis that speciation in mammals preceded subclass evolution (Kehoe and Capra, 1974).

Table XIII.6.2 Concentrations (mg/ml) of the Ig (sub)-classes in body fluids

	IgM	*IgG1*	*IgG2*	*sIgA*
Bile	–	0.10	0.09	–
Bronchoalveolar fluid	0.03	0.13	0.24	0.24
Colostrum	6.77	46.4	2.87	5.36
Intestinal fluid	Trace	0.25	0.06	0.24
Mature milk	0.086	0.58	0.005	0.081
Nasal secretions	0.04	1.56	Trace	2.81
Saliva	0.006	0.034	0.016	0.34
Seminal fluid	Trace	0.13	0.11	0.13
Serum	3.05	11.2	9.2	0.37
Synovial fluid	0.37	2.02	1.20	0.68
Tears	0.176	0.32	0.01	2.72
Urine	Trace	0.009	Trace	0.001

Reproduced from Table II, Butler 1986, with permission from Karger, Basel.

Ig isotypes: functional properties

Effector functions are associated with the presence or absence of receptor binding sites on the Fc part of the Ig molecule. The rearrangement of the V_H segment with different C_H regions (switch) thus allows recombination of an antigen-binding activity with different immunological functions (such as complement and Fc-receptor binding) and with dissemination in various body fluids (such as transfer through epithelial surfaces after binding to the pIg receptor).

Table XIII.6.2 summarizes concentrations of the different Ig classes in body fluids of normal cattle. Most of the data were collected before IgG3 was identified, so the column under IgG2 contains data for both IgG2 and IgG3.

A feature of the mucosal system in ruminants is the prominence of IgG1 relative to IgA, particularly in secretions from the mammary gland (Table XIII.6.2), where IgA is associated with milk fat globule membranes (Honkanen-Buzalski and Sandholm, 1981). This suggests the existence of an additional transport mechanism for IgG1 in ruminants. Because there is no placental transfer of maternal antibodies to the fetus in ruminants, transfer of passive immunity in cattle is ensured by the accumulation of extremely high concentrations of antibodies, particularly IgG1, in the colostrum and an efficient uptake of the intact proteins by the neonatal calf (see also Section 9, Passive Transfer of Immunity). IgA is more prominent in intestinal, nasal and lacrimal secretions. In the intestine IgA is produced locally, while most of the IgG is derived from plasma. In contrast, IgA in saliva is not produced locally but is selectively transferred from blood, presumably by virtue of its pIg receptor on the glandular epithelium (Lascelles *et al.*, 1986).

Homocytotropic antibodies, responsible for immediate hypersensitivity, are of the IgE class. They too are secreted in the colostrum and, like IgE in other species, they bind mast cells and play a role in allergies (de Weck, 1995) and also in resistance to parasites, particularly worm infections (Baker and Gershwin, 1992).

Other effector functions have been studied in detail for IgM, IgG1 and IgG2 only, mainly because no reagents or easy purification procedures for the others existed (data summarized in Table XIII.6.3). The two major IgG subclasses differ mainly in their ability to bind heterologous complement and to induce adherence to and phagocytosis by neutrophils and fresh monocytes. The observation that IgG2 does not fix heterologous complement, although it does fix bovine complement, may only be of relevance when testing antibodies by an *in vitro* complement assay. IgM is 10–20 times more effective at fixing complement than IgG, and this may explain its high concentration in serum of cattle compared with other species.

Binding of Ig to FcR on phagocytic cells and subsequent ingestion and killing by the cell may be a relevant parameter for immunity (Table XIII.6.3). IgG1 is the main antibody found in colostrum and milk, but does not have a specific receptor on resting neutrophils and monocytes. Although IgM did not reveal a specific FcR on phagocytic cells when tested for adherence and phagocytosis of erythrocytes, it was the only Ig class that induced uptake of *Staphylococcus aureus* and *Escherichia coli*. Using flow cytometry, both exogenous IgM and IgG2, but not IgG1, were shown to bind to neutrophils. Binding of IgG1 could be induced by treatment of neutrophils with IFN-γ (Worku *et al.*, 1994).

V genes

Three teams reported a number of independent V_L gene sequences from genomic and cDNA (Armour *et al.*, 1994; Parng *et al.*, 1995; Sinclair *et al.*, 1995). A striking feature is the extraordinarily high homology between cDNA V_L sequences and between germline V_L sequences. Only a small number of germline determinants were expressed and appear to be functional. The failure to find diversification of germline segments, even among the nonexpressed genes, led Sinclair *et al.* (1995) to conclude that light chains in bovine antibodies contribute little to recognition of antigen. Parng *et al.* (1995) suggested that V_L pseudogenes could be potential donors for gene conversion. They also observed point mutations which increased with the age of the animal. They proposed a model based on gene conversion, followed by somatic hypermutation, to account for diversity in the bovine V_L repertoire.

Variation among V_H sequences is also limited. Bovine V_H sequences (Jackson *et al.*, 1992; Armour *et al.*, 1994; Sinclair and Aitken, 1995; Saini *et al.*, 1997) show a very high nucleotide (85–90%) and amino acid (70–80%) homology, suggesting a single V_H gene family. This bovine V_H family has a high sequence homology with the single V_H family in sheep, with murine V_H Q-52 and with

Table XIII.6.3 Biological properties of bovine heavy chain classes

	IgM	*IgG1*	*IgG2*[a]	*IgG3*[a]	*IgE*	*IgA*
Binding to Protein A	–	–	weak		–	–
Binding to Protein G	–	++	++			–
Complement binding						
Bovine	+	+	+			
Guinea pig	+	+	–			
Rabbit	+	+	–			
Passive cutaneous anaphylaxis						
Homologous (bovine skin)		+	+			
Heterologous (rat skin)		+	–			
Binding to Fc-receptors						
Rosetting						
Neutrophils	–	–	+			
Fresh blood monocytes	–	–	+			
Cultured monocytes	+	+	+			
Alveolar macrophages		+	–			
Phagocytosis by neutrophils						
Erythrocytes	–	–	+			
Staphylococcus aureus/Escherichia coli	+	–	–			
Trypanosoma theileri	+	–	–			
Mycoplasma bovis	–	–	+			
FcR-binding of Ig (flowcytometry)						
Blood neutrophils	+	–	+			
Neutrophils + IFN-γ	+	+	++			

[a]A newly proposed nomenclature replaces the terminology IgG2a and IgG2b by IgG2 and IgG3. Most of the experiments with IgG2 antibodies were done before IgG3 was identified and probably included both subclasses.

human V_H4 (Sinclair *et al.*, 1995; Saini *et al.*, 1997). One unique characteristic of bovine V_H sequences is that most have a long CDR3 region (17 or more residues). Somatic hypermutations are an important mechanism for the generation of antibody diversity in cattle (Saini *et al.*, 1997).

Ig gene mapping

As in other species, the κ, λ and heavy chain loci have been mapped to three different chromosomes (Table XIII.6.4). Direct mapping of the heavy chain locus was obtained by *in situ* hybridization with a γ probe (Gu *et al.*, 1992), the other loci were indirectly mapped by linkage with known markers in synthenic groups. The locus of the pIg receptor was identified by *in situ* hybridization.

Table XIII.6.4 Bovine Ig genes

Locus	*Gene*	*Chromosome*	*References*
Lambda chain	IGL	17	Johnson *et al.* (1993)
Kappa chain	IGKC	11	Broad *et al.* (1995)
Heavy chain	IGHG	21q24	Gu *et al.* (1992)
	IGHM	21	Tobin-Janzen *et al.* (1992)
pIg-receptor	PIGR	16q13	Kulseth *et al.* (1994)

7. Major Histocompatibility Complex (MHC) Antigens

Methods to characterize MHC antigens

Class I antigens

The gene complex responsible for antigens encoded within the bovine MHC was named the bovine lymphocyte antigen (BoLA) system (Spooner *et al.*, 1979). Polymorphism of class I antigens has been studied using the classical immunogenetic approaches: transplantation, serology, allospecific T-cells, and one-dimensional isoelectric focusing (1D-IEF). International workshops have played a vital role in the standardization of serological reagents, typing methods and nomenclature of the BoLA system (Spooner *et al.*, 1979; Anonymous, 1982; Bull *et al.*, 1989; Bernoco *et al.*, 1991; Davies *et al.*, 1994b). More than 50 BoLA antigens have been recognized by the international community, nearly all of which behave as alleles of a single locus (BoLA-*A*). The BoLA serological specificities exhibit distinct differences in frequency in the different cattle breeds (Stear *et al.*, 1988b; Bull *et al.*, 1989).

One-dimensional isoelectric focusing has been successfully applied to problems of antigen and haplotype definition, and for investigation of the number of expressed class I products. This technique has been most useful for characterizing alleles not recognized by serological reagents and for defining new allelic subtypes of serologically defined antigens (Oliver *et al.*, 1989; Davies *et al.*, 1994b). Other class I typing techniques, such as cell-mediated lympholysis (Spooner and Morgan, 1981), immunoblotting (Viuff *et al.*, 1991) and Southern blotting (Lindberg and Andersson, 1988) have been used for demonstration of class I polymorphism. At the present time, serology remains the most universally accepted and easily performed method for class I typing.

Class II antigens

Cellular determinants detected in the mixed lymphocyte reaction (MLR) are encoded by BoLA class II antigens (Teale and Kemp, 1987). Because the MLR is difficult to replicate and standardize, this method has never become a widely-used typing tool, although it has been used for functional studies.

Serological reagents for detecting class II antigens have been characterized in independent comparison tests (Lewin *et al.*, 1991; Nilsson *et al.*, 1991) and in the fourth and fifth international BoLA workshops (Bernoco *et al.*, 1991; Davies *et al.*, 1994a). Five class II specificities (Dw designation) were accepted by the international community (Davies *et al.*, 1994a). Evidence for eight others was sufficient to warrant cluster designations (Dc series). Isoelectric focusing of immunoprecipitated class II antigens clearly demonstrated class II polymorphism at the protein level (Joosten *et al.*, 1989; Watkins *et al.*, 1989a). Discrimination of the products of the *DQ* and *DR* genes was achieved using locus-specific mAb (Bissumbhar *et al.*, 1994). A total of 12 IEF variants (DRBF designations) have been accepted by the international community (Davies *et al.*, 1994a).

Genetic organization of the bovine major histocompatibility complex

Using HLA genes as probes, nucleic acid hybridizations were used successfully to identify orthologous MHC genes, to reveal the extensive polymorphism of the class II genes and to delineate BoLA haplotypes (Andersson *et al.*, 1986a,b; Ennis *et al.*, 1988; Muggli-Cockett and Stone, 1988). Determination of genomic sequences permitted the application of PCR technology to the identification of sequence variation in coding exons of BoLA genes (Sigurdardóttir *et al.*, 1991; van Eijk *et al.*, 1992b). All these methods have been applied to research topics that range from MHC evolution (Andersson *et al.*, 1991) to disease association (Lewin and Bernoco, 1986).

The bovine MHC and at least 32 structural genes have been mapped to bovine autosome 23 (Figure XIII.7.1) (Fries *et al.*, 1993). Two class II subregions have been defined on the basis of genetic mapping (Andersson *et al.*, 1988; van Eijk *et al.*, 1993). These two subregions are separated by a relatively large distance, spanning at least 15 centimorgans (cM) from *DYA* to *DRB3*. Thus, while it may be convenient to refer to BoLA genes as a 'system' they are, in reality, independent clusters of genes. In addition to the class I and class II genes, the class III genes *CYP21* (Barendse *et al.*, 1994; Bishop *et al.*, 1994; Gwakisa *et al.*, 1994; van Eijk *et al.*, 1995), *BF* (Teutsch *et al.*, 1989), *HSP70* (Bishop *et al.*, 1994), *C4* (Andersson *et al.*, 1988) and *TNF* (Agaba *et al.*, 1996) have also been mapped to the bovine MHC. The overall size of the class I region is no less than 770 kb and probably no more than 1650 kb, similar to the organization of the HLA class I genes (Bensaid *et al.*, 1991). The *DQ* and *DR* genes are probably not more than 400 kb apart. In most haplotypes, the *DQA* and *DQB* genes have been duplicated (Andersson and Rask, 1988).

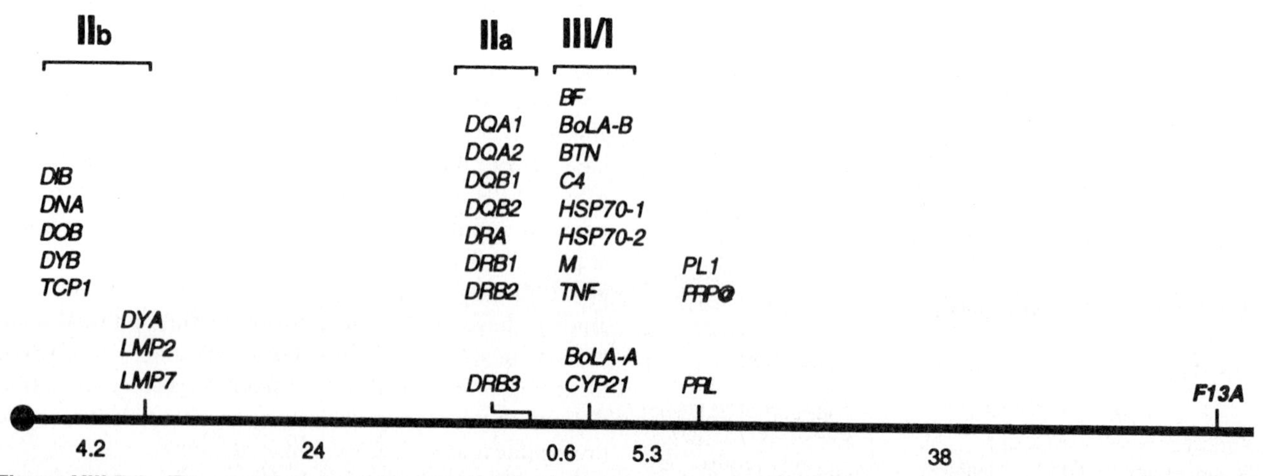

Figure XIII.7.1 The genetic map of structural genes located on bovine chromosome 23. Map distances are expressed in centimorgans and were estimated using the Illinois Reference/Resource families, a large collection of paternal half-sib families used for the construction of a bovine linkage map (Ma *et al.*, 1996). Map positions of loci located above and away from the indicate map marker were assigned based upon studies cited in the text.

An enormous number of BoLA haplotypes is theoretically possible given the extreme polymorphism of the BoLA-*A*, *DR* and *DQ* loci. Haplotype discrimination was enhanced by the use of serology in combination with 1D-IEF, restriction fragment length polymorphism (RFLP), polymerase chain reaction-RFLP (PCR-RFLP) and microsatellite typing techniques. At least 170 BoLA-*A*-*DRB3* haplotypes were identified in samples from most of the major dairy and beef breeds (Lewin, 1996). Among breeds, different haplotypic combinations are found, but certain haplotypes appear in several breeds (ancestral haplotypes). Haplotype data make it possible to elucidate the evolutionary relationships between cattle breeds and to relate the breed origins to human migration patterns (MacHugh *et al.*, 1994). Furthermore, haplotype discrimination has become increasingly important for dissecting MHC effects on diseases. Analysis of haplotypes in disease association studies can be helpful in mapping disease resistance and susceptibility to localized regions within the MHC (van Eijk *et al.*, 1992a; Xu *et al.*, 1993b). Haplotype diversity is

Table XIII.7.1 Disease associations

Disease	*Breed*	*BoLA*	*Effect*	*References*
Enzootic bovine leukosis				
Seroconversion	Holstein	A14	Late	Palmer *et al.* (1987), Lewin *et al.* (1988)
	Holstein	A15	Rapid	Palmer *et al.* (1987)
	Guernsey	A21	Late	Palmer *et al.* (1987)
	Guernsey	DA6.2, A12	Rapid	Palmer *et al.* (1987)
PL and B-cell numbers	Shorthorn	DA7	Resistance	Lewin and Bernoco (1986)
	Shorthorn	DA12.3	Susceptibility	Lewin and Bernoco (1986)
	I. Shorthorn	A6, EU28R	Susceptibility	Stear *et al.* (1988a)
	I. Shorthorn	A8	Resistance	Stear *et al.* (1988a)
	Holstein	A12 and A15	Susceptibility	Lewin *et al.* (1988)
	Holstein	A14 and A13	Resistance	Lewin *et al.* (1988)
	Holstein	DRB2*2A	Resistance	van Eijk *et al.* (1992a)
	Holstein	DRB2*1C	Susceptibility	van Eijk *et al.* (1992a)
	Holstein	DRB3(ER motif)	Resistance	Xu *et al.* (1993b)
	Holstein	DRB3(ER motif)	Resistance	Zanotti *et al.* (1996)
Mastitis				
Clinical mastitis	Norwegian Red	A2	Resistance	Solbu *et al.* (1982), Solbu and Lie (1990), Våge *et al.* (1992), Mejdell *et al.* (1994)
	Norwegian Red	A16	Susceptibility	Larsen *et al.* (1985)
	Norwegian Red	A11	Susceptibility	Solbu and Lie (1990)
	Swedish R & W	DQ1A	Susceptibility	Lundén *et al.* (1990)
	Holstein	A11	Resistance	Weigel *et al.* (1990)
	Holstein	CA42	Susceptibility	Mallard *et al.* (1995)
Subclinical mastitis	Icelandic	A19	Susceptibility	Oddgiersson *et al.* (1988)
(cell count)	Simmental (S) or S × Red Holstein	A15	High	Arriëns *et al.* (1994)
	Danish Black Pied	A11 and A30	Low	Aarestrup *et al.* (1995)
	Danish Black Pied	A21 and A26	High	Aarestrup *et al.* (1995)
California mastitis test	Holstein	A14	Low	Weigel *et al.* (1990)
Heminths				
Nematodes	Belmont Red	A7, CA36	Resistance	Stear *et al.* (1990)
	Africander × Hereford	A9	Susceptibility	Stear *et al.* (1988c)
Protozoa				
Theileria parva	*Bos indicus*	class I	Parasite entry	Shaw *et al.* (1995)
Ticks				
Boophilus microplus	Brahman × Shorthorn	A6, CA31	Susceptibility	Stear *et al.* (1989)
	Belmont Red	A19, A7	Resistance	Stear *et al.* (1990)
Posterior spinal paresis	Holstein	A8	Suspectibility	Park *et al.* (1993)
Ketosis	Norwegian Red	A2, A13	Resistance	Mejdell *et al.* (1994)
Retained placenta	Dutch Friesian	compatibility	Susceptibility	Joosten *et al.* (1991)

also a useful measure of genetic variability and can be used for genetic management of rare breeds and endangered species (Lewin *et al.*, 1993).

Tissue distribution, structure and function of BoLA molecules

Serological studies employing alloimmune reagents support the existence of at least one major expressed class I locus. One-dimensional IEF, using mAbs to phylogenetically conserved monomorphic class I determinants, revealed the presence of more than one class I gene product expressed by lymphocytes from homozygous individuals (Joosten *et al.*, 1988; Watkins *et al.*, 1989b). Mouse L cells transfected with genomic DNA from a homozygous cow produced by sire–daughter mating were shown to express two distinct class I products, as detected with allospecific cytotoxic T-lymphocytes and mAb to determinants on the *KN104* (African Boran class I) and *A10* molecules (Toye *et al.*, 1990).

Up to six distinct cDNA have been cloned from a single heterozygous bull indicating that there are three class I genes transcribed in bovine lymphocytes (Garber *et al.*, 1994). However, the functionality of these genes has not been clearly established.

Expression of classical class II heterodimers on cattle leukocytes was identified using cross-species reactive mAb (Lewin *et al.*, 1985a). While it is clear that distinct *DR* and *DQ* products are expressed on the cell surface, *DY*, *DI* or *DO* products have not been identified.

Cytoplasmic and cell-surface expression of class II antigens has been studied using immunofluorescence (Lewin *et al.*, 1985a), antibody-dependent complement-mediated cytotoxicity (Lewin *et al.*, 1985b), flow cytometry (Lewin *et al.*, 1985b; Davis *et al.*, 1987), immunoprecipitation (Hoang-Xuan *et al.*, 1982) and 1D-IEF (Joosten *et al.*, 1989; Watkins *et al.*, 1989a). Expression has been demonstrated on B cells (Lewin *et al.*, 1985a), activated T cells (Lalor *et al.*, 1986; Taylor *et al.*, 1993), cell lines (Ababou *et al.*, 1994), alveolar macrophages (Ohmann *et al.*, 1986), monocytes (Taylor *et al.*, 1993; Hughes *et al.*, 1994) and mammary (Fitzpatrick *et al.*, 1992) and bronchial (Spurzem *et al.*, 1992) epithelial cells.

Expression of the class II genes has been studied at the transcriptional level using Northern blot analysis (Burke *et al.*, 1991; Stone and Muggli-Cockett, 1993), cDNA cloning (Xu *et al.*, 1991, 1993a, 1994) and PCR–RFLP (Xu *et al.*, 1994). The class IIa region of cattle contains the known expressed class II genes: *DRA*, *DRB3*, *DQA*, *DQB1* and *DQB2*. In at least three haplotypes with duplicated *DQ* genes, both *DQB* genes are expressed (Xu *et al.*, 1994). Expression of other genes in the class IIa region, *DRB1* (a pseudogene), *DRB2*, *DQA2*, and the class IIb region genes *DYA*, *DYB*, *DIB*, and *DOB*, has not been demonstrated to date (Stone and Muggli-Cockett, 1993).

Disease associations

Research on the MHC of cattle is often justified in terms of using alleles or haplotypes for selection of disease-resistant animals. A synopsis of research on disease associations and immune responses can be found in Table XIII.7.1.

Unique properties of the bovine MHC

The *LMP2* proteosome subunit gene was mapped to the class IIb region (Shalhevet *et al.*, 1995) and provided further evidence of the distinctive organization of the BoLA system relative to HLA and H-2. A recent study (Park *et al.*, 1995) identified large variation in recombination rate in the class IIa–IIb interval using sperm typing: one bull had nearly double the recombination rate between *DYA* and the gene for prolactin (*PRL*), which is 2.5–4 cM telomeric to *DRB3*.

8. Red Blood Cell Antigens

The first papers on bovine red cell antigens appeared in the 1940s and the fascinating history of the early developments in this field has been traced by Stormont (1978). For more recent information see Bell (1983) and International Society for Animal Blood Group Research (1985). A blood group system is defined as antigenic specificities controlled by genes at a single locus or closely linked loci.

Within any one system, the antigenic determinants are often not inherited as separate entities but in groups, which behave as alleles ('multiple alleles'). The expression of the products of such multiple alleles is called a phenogroup and there has been much discussion on whether or not these are controlled by clusters of closely linked genes, since crossing over is sometimes observed. The nomenclature of cattle blood groups is confusing because research workers soon exhausted all the letters of the alphabet and had to resort to superscripts after the letters to name new factors, e.g. factors, K, K′ and K″ in the B system are different specificities. In addition, subtype relationships are often present where the same factor has different types of expression on the red cell membrane. These subtypes are indicated by a numerical subscript, e.g. F_1 and F_2. The serological test used to identify bovine blood group antigens is based on haemolysis in the presence of complement (usually rabbit). It is generally performed in 96-well microtitre plates.

Blood group systems

There are two types of systems in cattle depending on whether they are identified by naturally occurring antibodies or produced by deliberate immunization. Such immune (usually allo-) antibodies have to be extensively absorbed with different red cells to separate out the individual specificities which are needed to make nonspecific typing reagents. Monoclonal antibodies have been produced for blood typing purposes (Tucker *et al.*, 1986; Méténier *et al.*, 1991; Hønberg and Larsen, 1992) but have never replaced the normal panel of reagents used by service laboratories.

The natural blood group system

The J system is homologous to the R system of sheep, the J system of goats and the A system of pigs (Table XIII.8.1). Certain cattle have a 'natural' antibody, anti-J in their plasma and some very rare animals have an anti-O antibody. These two antisera are used to classify cattle within the J system. The J substance is not a true red cell antigen but is a soluble substance found in the plasma and other secretions, which only becomes attached to the red cells after birth. Some animals have J on their red cells and in their plasma (J^{cs}) while others do not have it on their red cells but only in the plasma and secretions (J^{s}). O substance has not been detected on red cells but is present in plasma and secretions of some individuals, sometimes together with J substance (Sprague, 1958). A fourth type of animal lacks both J and O substances and is therefore a 'nonsecretor'. The secretion of J and O substances is likely to be controlled, as in sheep (Rendel, 1957), by a secretor gene at a locus separate from that of J.

Red cell antigens identified by immune antisera

Ten systems have been identified in cattle and of these, nine have been assigned to particular chromosomes (Table XIII.8.1). With the exception of R′, and possibly F, all systems have a silent allele. There is no comprehensive list available of alleles in the complex B and C systems which have many antigen factors. The M locus is associated with the BoLA histocompatibility locus (Hines and Ross, 1987; Hønberg *et al.*, 1995).

Structure of the red cell antigens

Work on the chemical structure of the antigens has been sporadic and sparse. The F and V determinants of the F system appear to be surface sialoglycoproteins (Spooner and Maddy, 1971) and as such may be analogous to the MNSs blood group system of man. The J-system: J substance of the red cells is a lipid while that of the serum is composed of lipid and nonlipid fractions, the latter probably being a glycoprotein (Theile *et al.*, 1979). The M-system M′ antigen is an MHC class I-like molecule that is retained on the red cell during development (Hønberg *et al.*, 1995).

Table XIII.8.1 Bovine blood group systems

System (locus symbol)	*Chromosome assignment*	*Antigen factors*	*Minimum number of alleles*	*Alleles*
(a) J (*EAJ*)	Chr 11	J; O	4	*J; O* (not a simple inheritance)
(b) A (*EAA*)	Chr 15	A_1, A_2=A; D; H; Z′	11	A_1; *D*; *H*; A_1D; A_2D; A_1H; *DH*; A_1DH; AD_2Z'; A_2DH (*a* = silent allele)
B (*EAB*)	Chr 8	B_1; B_2=B; G_1; G_2=G; G_3; K; I_1; I_2; O_1; O_2; O_3; O_4; P_1=P; P_2; Q; T_1; T_2; Y_1; Y_2; A′; B′; D′; E'_1; E'_2; E'_3; E'_4; F'_1; F'_2=F′; G′; I'_1=I′; I'_2; J'_1=J′; J'_2; K′; O'_1; O'_2=O′; P'_1; P'_2=P′; Q′; Y′; A″; B″; D″; G″; I″; J″; F″; K″, O″	>600	Many
C (*EAC*)	Chr 18	C_1; C_2; E; R_1; R_2; W; X_0; X_1; X_2; C'_1; X′; L′; C''_1; C''_2=C″	77	Many
F (*EAF*)		F_1; F_2; V_1; V_2; (N′; V′)	5	F_1; F_2; V_1; V_2; F_2V_2
L (*EAL*)	Chr 3	L	2	*L*; *l* (*l* = silent allele)
M (*EAM*)	Chr 23	M_1; M_2=M; M′	3	M_1; M_2; *m* (*m* = silent allele)
R′ (*EAR′*)	Chr 16	R′; S′	2	*R′; S′*
S (*EAS*)	Chr 21	S; H′; U; U_2; U'_1=U′; U'_2; H″; S″; U″	15	*SH′; UH′; H′; U′; U′U″H′; U″H′; H′H″; SS″H; S″H′; U″*; U'_2U''; *UU″H′; H″U″H′; s* (*s* = silent allele)
Z (*EAZ*)	Chr 10	Z_1=Z; Z_2	3	*Z*; Z_2; *z* (*z* = silent allele)
T′ (*EAT′*)	Chr 19	T′	2	*T′*; *t* (*t* = silent allele)

Applications and significance of red cell antigen diversity

The primary reason for the early extensive research into bovine blood groups was to provide polymorphic genetic markers for parentage testing and identification of animals for pedigree registration. This is still the main objective of all the cattle blood typing service laboratories established throughout the world. With the discovery of highly polymorphic microsatellite markers, molecular biological techniques are beginning to replace those of serology and consequently blood typing research is on the decline. Very few new antigen factors have been discovered in recent years. However, in view of the exciting new findings at the molecular and biochemical level in the human blood group field, leading to discovery of structure–function relationships, it is anticipated that interest will be regenerated in farm animal species, including cattle. Soon, it may be possible to put forward explanations for the existence and maintenance of such great genetic variability within a species once the biochemical and physiological roles of the red cell antigens have been elucidated.

9. Passive Transfer of Immunity

Whereas in humans and some rodent species a haemochorial placentation facilitates molecular transport of immunoglobulins from maternal circulation to the fetus, ruminants have an epitheliochorial placentation which prevents this transport (see Table XIII.9.1). Thus, ruminants obtain no immunoglobulin prenatally from the maternal circulation and are born agammaglobulinaemic. However, large quantities of maternal IgG are selectively transported into the mammary gland just prior to parturition and these are ingested and absorbed intact into the circulation by the suckling neonate. The transport of maternal immunoglobulin into colostrum in ruminants is a highly-selective process in which immunoglobulin of the IgG1 isotype subclass is transported via a selective transport mechanism from maternal circulation into mammary secretion, again via IgG1 Fc receptor-mediated transport across the mammary epithelium, ensuring that IgG1 is the predominant isotype of immunoglobulin in ruminant colostrum (Lascelles, 1969). The intestine in newborn ruminants is consequently adapted for large-scale uptake of intact immunoglobulin across the gut into the neonatal circulation. Although it has been calculated that all immunoglobulin isotypes in colostrum are absorbed across the neonatal ruminant intestine with equal efficiency (Brandon and Lascelles, 1971) the predominance of IgG1 ensures that the serum immunoglobulin profile of post-suckled ruminants is similar to that of colostrum with a high predominance of immunoglobulin of the IgG1 isotype. This absorption, however, occurs only in the first 24–48 hours of life, during which time the large, palely staining vacuolated cells of the intestinal epithelium of the newborn ruminant are replaced by cells characteristic of the adult intestine (Simpson-Morgan and Smeaton, 1972) to achieve gut closure.

These adaptations to enable passive transfer of maternal antibody to the neonate are reflected in relative differences in the constitution of colostrum between species adapted to prenatal versus postnatal transfer. In humans, where IgG is transported placentally, there are very low levels of IgG in colostrum but relatively higher levels of IgA, whereas in ruminants the reverse applies with high levels of IgG in colostrum and relatively low levels of IgA, although a higher ratio of IgA/IgG occurs in later lactation and in involution secretion (Table XIII.9.2). It is noteworthy that while most of the IgA in milk is locally synthesized from plasma cells underlying the mammary epithelium, because serum IgA is dimeric in ruminants there is the potential for IgA produced at remote mucosal sites, but escaping secretion, to be transported from the blood circulation into milk after binding with the polymeric immunoglobulin (pIg) receptor on the mammary epithelial cells. Indeed, an inverse relationship exists between the proportion of serum-derived IgA in ruminant milk and the numbers of IgA plasma cells present in the gland, with relatively more serum-derived IgA in early

Table XIII.9.1 Correlation between placentation type and mode of passive antibody transfer from mother to young in various species (adapted from Brambell, 1958)

		Transmission of passive immunity				
		Prenatal		*Postnatal*		
Species	*Placentation*	*Proportion*	*Route*	*Proportion*	*Duration*	*Route*
Cattle, sheep, goat, horse, pig	Epitheliochorial	0	None	+++	36 h	Gut
Dog	Endotheliochorial	+	Transplacental	++	10 days	Gut
Mouse	Haemochorial	+	Transplacental	++	16 days	Gut
Rat	Haemochorial	+	Transplacental	++	20 days	Gut
Guinea pig	Haemochorial	+++	Transplacental	0	–	None
Rabbit	Haemochorial	+++	Transplacental	0	–	None
Man	Haemochorial	+++	Transplacental	0	–	None

Table XIII.9.2 Concentration of immunoglobulins (mg/ml) in colostrum and milk of various species

Species	*Secretion*	*Total IgG*		*IgA*	*IgM*	*References*
		IgG1	*IgG2*			
Cow	Colostrum	75.0	1.9	4.4	4.9	Mach and Pahud (1971)
	Milk	0.4	0.06	0.05	0.04	Mach and Pahud (1971)
Sheep	Colostrum	60.0	2.0	2.0	4.1	Pahud and Mach (1970)
	Milk (early lactaction)	NA	NA	2.3	NA	Sheldrake *et al.* (1984)
	Milk (mid-lactation)	NA	NA	0.7	NA	Sheldrake *et al.* (1984)
	Milk (late lactaction)	NA	NA	6.5	NA	Sheldrake *et al.* (1984)
	Involution	10.3	2.0	3.1	4.1	Watson *et al.* (1972)
Goat	Colostrum	58.0		1.7	3.8	Pahud and Mach (1970)
	Milk	0.3		0.06	0.03	Pahud and Mach (1970)
Horse	Colostrum	60.0		NA	NA	Rouse and Ingram (1970)
	Milk	0.4		0.8	0.07	McGuire and Crawford (1972)
Dog	Colostrum	14.5		3.1	2.2	Reynolds and Johnson (1970)
Rabbit	Colostrum	2.4		4.5	0.1	Eddie *et al.* (1971)
Pig	Colostrum	58.7		10.7	3.2	Porter (1969)
Human	Colostrum	0.2		17.9	0.8	Brandtzaeg *et al.* (1970)

NA, not assayed.

lactation than during late lactation and during involution when IgA plasma cell numbers in the gland increase dramatically (Sheldrake *et al.*, 1984). Although the reasons for the stage-dependent increase in IgA plasma cells in the gland are unknown, this phenomenon enables antibodies generated in response to antigens encountered at remote mucosal sites to be channelled into milk, especially during early lactation when local production of IgA is low (Sheldrake *et al.*, 1984). This same principle allows IgA antibodies absorbed from colostrum by the neonate to be secreted into mucosal secretions remote from the intestine by SC-dependent mechanisms (Scicchitano *et al.*, 1984, 1986). The increase in local production of IgA as lactation progresses parallels the decline in selective transport of serum-derived IgG1 into ruminant milk, which in turn appears to be inversely related to the synthetic activity of the mammary epithelium, although a role for hormonal factors in this phenomenon cannot be excluded (Lascelles and Lee, 1978).

While antibodies in colostrum are the predominant immune effectors transferred to the ruminant neonate, cells and other soluble factors in milk play an important role in terms of passive protection. There are normally large numbers of viable cells in bovine milk, varying from 5×10^4 to 2×10^6 cells/ml, although substantially higher numbers may occur if the gland becomes infected or inflamed. These cells are predominantly T cells of $\alpha\beta$ TCR^+, $CD8^+$ phenotype, with memory cell characteristics ($CD2^{high}$, $CD45R^{low}$) (Taylor *et al.*, 1994) and are predominantly derived from other mucosal sites (Manning and Parmely, 1980). To determine the extent to which these cells can be absorbed by the neonate Sheldrake and Husband (1985) administered radioactively labelled maternal lymphocytes intraduodenally to lambs within 14 h of birth. Transport of these cells from the duodenum via lacteal lymph and mesenteric lymph nodes was observed, indicating the potential for cells of maternal origin to have functional relevance in the passive protection of the neonate. Although the duration of their survival before elimination by the host immune response is not known, it is expected that this would be a transient event.

Soluble factors other than immunoglobulins in milk are also important effectors of passive immune protection, and part of the protective capacity of milk is attributable to the nonimmunoglobulin fraction. Antibacterial substances such as lactic acid, interferon, lysozyme, lactoferrin and the complement components C3 and C4 have been demonstrated in both human and ruminant milk and have been shown to affect the establishment of pathogenic bacteria either through inherent bactericidal activity or in conjunction with specific immune responses (Butler, 1979; Goldman *et al.*, 1982). The cytokine TGF-β has also been identified in bovine milk and has been shown to have a powerful modulatory effect on neonatal health and development (Tokuyama and Tokuyama, 1993).

Despite the beneficial effect of maternal antibody transfer, it is well-documented that maternal antibody, whether acquired prenatally by transplacental transfer or postnatally via colostrum, has a suppressive effect on the development of endogenous mucosal immune responses in the young animal. This can occur at two levels – systemically through circulating acquired antibody interfering with the development of gut responses, and locally by the presence of colostral and milk immunoglobulins in the lumen of the intestine. Indeed, feeding colostrum containing high titres of *Escherichia coli* antibodies to young calves inhibited the development of an active *E. coli* mucosal immune response in these animals in the perinatal period (Logan *et al.*, 1974). Conversely, calves deprived of colostrum have an earlier endogenous production of serum IgA (Husband and Lascelles, 1975).

The duration of protection provided by passively acquired maternal antibodies is short-lived. IgM and IgA antibodies are lost from the circulation more rapidly than IgG1 and IgG2 with estimates of the half-lives of the various passively acquired immunoglobulins in ruminants of 16–32 days for IgG1 and IgG2, 4 days for IgM and 2.5 days for IgA (Husband *et al.*, 1972). The extent to which the shorter half lives of IgA and IgM are due to their selective removal from the circulation by pIg receptors, enabling their secretion at mucosal surfaces, or is a true reflection of a shorter biological half life is unclear. Although the bulk of the protective effect of passively acquired immunoglobulins results from the systemic absorption of IgG in most species, there is an additional local role for colostral and milk antibody in the lumen of the intestine after its ingestion even in species that do not obtain passive antibody uptake via the gut, or in ruminants after gut closure has occurred. Antibodies of the IgA class increase in concentration as lactation progresses in ruminants and these are responsible for limiting intestinal colonization by enteric pathogens such as *E. coli* (Corley *et al.*, 1977).

The efficacy of passive transfer of antibody via colostrum in ruminants in protecting the neonate from specific diseases has been well documented. Calves that do not receive colostrum often succumb to disease at an early age, or, if they do survive, have restricted growth rates (Belknap *et al.*, 1991). Deliberate passive immunization against selected diseases has been highly effective and calves fed colostral antibodies from immunized cows have been demonstrated to be resistant to specific challenge with a range of selected pathogens including rotavirus (Saif *et al.*, 1983) and *E. coli* (Johnston *et al.*, 1977).

10. Nonspecific Immunity

The nonantigen-specific (or native) defence mechanisms tend to be highly conserved in vertebrates. Native defence mechanisms in cattle are basically similar to those in other species, but they have some unique aspects. Because of space limitations, the mechanism of action of native defence mechanisms will not be reviewed, instead, this chapter will emphasize those aspects of the native defence mechanisms that are unique in cattle.

The basic components of the nonspecific immune system are listed in Table XIII.10.1. These native defence mechanisms are very important for keeping animals healthy. If their function is disrupted for any reason, the animal becomes more susceptible to infection. In the presence of antibodies and/or a cell-mediated immune response some of the native defence mechanisms act more rapidly or more aggressively to help control infection. This is especially true for the complement system and the cellular components of nonspecific immunity listed in Table XIII.10.1.

Table XIII.10.1 Basic components of nonspecific immunity in cattle

Physical and chemical barriers to infection
Epithelial surfaces
Normal indigenous microflora
Normal excretory secretions and flow of body fluids
Mucus on mucosal surfaces
Rumen environment
Acid pH in abomasum
Anaerobic conditions in gastrointestinal tract
Bile
Molecular components of nonspecific immunity
Acute phase proteins (see Table XIII.10.2)
Complement system (see Section 11)
Conglutinin
Iron-binding proteins
Antibacterial cationic peptides (see Table XIII.10.3)
Cellular components of nonspecific immunity
Neutrophils
Mononuclear phagocytic system (macrophages)
Eosinophils
Mast cells
Natural killer cells

Physical and chemical barriers to infection

The major physical and chemical barriers to infection in cattle are basically similar to those in other species (Table XIII.10.1). The rumen environment provides a unique barrier to some infectious agents. The large volume, slow transit time, anaerobic environment, dense microbial flora, and high concentration of microbial digestive enzymes in the rumen inhibit the growth and survival of many pathogens.

Molecular components of nonspecific immunity

The acute phase response in cattle is triggered by proinflammatory cytokines IL-1, IL-6 and TNF, and involves many of the same components as in other species (Table XIII.10.2) (Godson *et al.*, 1995). One major difference is that C-reactive protein, a major component of the acute phase response in other species, does not increase during the acute phase response in cattle and is therefore not considered an acute-phase protein (Conner and Eckersall, 1988).

Conglutinin is a unique component of the nonspecific immune system in ruminants. Conglutinin is a large molecule found in the plasma which binds to the C3b component of complement on cell surfaces. The conglutinin molecule has multiple binding sites for C3b, therefore it clumps (conglutinates) particles coated with C3b. This is believed to facilitate the phagocytosis and removal of C3b-coated particles (see section 11).

Table XIII.10.2 Characteristics of the bovine acute phase response

(A) *Plasma proteins which increase during the acute phase response and their functions*	
Haptoglobin (Conner and Eckersall, 1986, 1988; Conner *et al.*, 1989; Hofner *et al.*, 1994; Godson *et al.*, 1995)	Binds free hemoglobin
Fibrinogen (Godson *et al.*, 1995)	Precursor of fibrin for coagulation
Serum amyloid A (Boosman *et al.*, 1989; Horadagoda and Eckersall, 1994)	Precursor of the AA class of amyloid fibril protein. Associates with high density lipoproteins in serum
Ceruloplasmin (Conner and Eckersall, 1986, 1988; Conner *et al.*, 1989)	Copper binding protein and oxygen radical scavenger
α_1 Antitrypsin (Conner and Eckersall, 1986, 1988; Conner *et al.*, 1989)	Protease inhibitor
α_1 Antichymotrypsin (Conner *et al.*, 1989)	Protease inhibitor
α_2 Macroglobulin (Conner *et al.*, 1989)	Protease inhibitor
Fetuin (Dziegielewska *et al.*, 1996)	Unknown function
Seromucoid (Conner and Eckersall, 1988; Conner *et al.*, 1989)	Unknown function
α_1 Acid glycoprotein (Godson *et al.*, 1995)	Transport protein
(B) *Other changes during the bovine acute phase response*	
Fever (Godson *et al.*, 1995)	
Decreased plasma albumin concentration (Conner and Eckersall, 1988)	
Decreased plasma Zinc and Iron concentration (Boosman *et al.*, 1989; Godson *et al.*, 1995)	
Leukocytosis (Godson *et al.*, 1995) or leukopenia (Boosman *et al.*, 1989), depending on the stimulus	

Cattle have several well-defined cationic antibacterial peptides (Table XIII.10.3). Most of these peptides are found in the tertiary granules of bovine neutrophils. In addition a unique cationic antibacterial peptide has been found in the tracheal mucosa of cattle (Diamond *et al.*, 1993). The cationic antibacterial peptides in the neutrophil granules are different from those described in other species. All of these cationic antibacterial peptides have been shown to be bactericidal for selected bacteria and are believed to be important components of native defence.

Table XIII.10.3 Cationic antibacterial peptides in cattle

Cationic antibacterial peptides	*Antimicrobial activity*
Neutrophil lysosomal cationic peptides	
Bactenecin (Romeo *et al.*, 1988)	*Staphylococcus aureus, Escherichia coli*
Bactenecin 5 (Gennaro *et al.*, 1989)	*Salmonella typhimurium, Klebsiella pneumoniae, E. coli, Leptospira interrogans*
Bactenecin 7 (Gennaro *et al.*, 1989)	*S. typhimurium, K. pneumoniae, E. coli, L. interrogans, Pseudomonas aeruginosa*
Indolicidin (Selsted *et al.*, 1992)	*S. aureus, E. coli*
β Defensins (13 related peptides) (Selsted *et al.*, 1993)	*S. aureus, E. coli*
Tracheal mucosa	
Tracheal antimicrobial peptide (Diamond *et al.*, 1991, 1993)	*S. aureus, K. pneumoniae, E. coli, P. aeruginosa, Candida albicans*

Cellular components of nonspecific immunity

Neutrophils and macrophages are the major phagocytic cell types in cattle and function similarly in cattle as in other species. They are an important first line of defence against infections, especially bacterial infections, and their activity is enhanced by the presence of antibody and T-cell derived cytokines; therefore, they also play a major role in acquired immunity.

Neutrophils

Cattle have a lower percentage of neutrophils in the peripheral blood than most species (reviewed by Roth, 1994). Healthy cattle have approximately half as many neutrophils as lymphocytes in the peripheral blood. A distinctive feature of ruminant neutrophils is the presence of a unique third granule type in the cytoplasm. These granules are larger than the azurophilic and specific granules and contain cationic antibacterial peptides (described above). The azurophilic and specific granules in bovine neutrophils contain many of the same enzymes found in neutrophils from humans and other species, but often in either higher or lower concentration. Notably, bovine neutrophils lack lysozyme, a major component of human neutrophil azurophilic granules.

Bovine neutrophils are chemotactically attracted to certain products of complement activation, and arachidonic acid metabolism. They are not chemotactically attracted to several factors that are known to attract human neutrophils, including formyl peptides, *Escherichia coli* culture filtrates, and platelet activating factor. Bovine neutrophils are unusual in that they have Fc receptors for IgM, and IgM can serve as an effective opsonin. C3b and

IgG2 (but not IgG1) also serve as opsonins. Some of the stimuli which induce the oxidative metabolism burst and release of primary granule contents also differ between bovine and human neutrophils.

Bovine neutrophil function has been shown to be suboptimal in young calves and in cows around the time of parturition. In addition, infectious agents, glucocorticoid therapy, stress, inadequate nutrition, and genetic defects have been associated with depressed neutrophil function and increase susceptibility to infection. Bovine neutrophil function has also been shown to be enhanced in the presence of several cytokines. In general, the cytokines are more effective at enhancing the function of neutrophils whose function was first depressed by one of the factors listed above.

Bovine neutrophils have been shown to be capable of mediating antibody-dependent cell-mediated cytotoxicity (ADCC) similar to neutrophils from other species. In addition, bovine neutrophils have been shown to have the unusual ability to mediate antibody-independent cell-mediated cytotoxicity (AICC) if they are activated by certain cytokines. TNF-α is one of the cytokines capable of inducing bovine neutrophil AICC. Because TNF-α can be released by macrophages during a nonantigen specific immune response, AICC by activated neutrophils may be a component of nonantigen specific host defence.

Macrophages

Macrophages are found in most tissues of cattle. Some of the more important tissue macrophages are: Kupffer cells in the liver sinusoids, microglial cells in the brain, alveolar macrophages, dendritic cells in the skin, macrophages in lymphoid tissue and mammary macrophages (Bielefeldt Ohmann and Babiuk, 1986; Bryan *et al.*, 1988). Cattle, like other ruminants and pigs, have high numbers of pulmonary intravascular macrophages (Winkler, 1988). These cells play a major role in clearing blood-borne bacteria and particulate material. In other species such as humans, dogs, and rats, which do not have high numbers of pulmonary intravascular macrophages, blood-borne pathogens are removed primarily in the spleen and bone marrow.

Bovine macrophages are important components of the native defence mechanisms. Resting macrophages are capable of phagocytosing and killing some infectious agents. They become much more aggressive and efficient at killing when activated by cytokines during a T-cell-mediated immune response. At least some of the populations of macrophages in cattle are capable of producing oxygen radicals and nitrite (NO_2^-) for killing. An important function of macrophages in native defence is the early release of the cytokines IL-1, IL-6, and TNF-α, in response to the presence of infectious agents. These cytokines initiate the inflammatory and acute phase responses.

Mast cells

Mast cells differ in a number of ways according to species and tissue sites, or even within the same tissue (Ruitenberg *et al.*, 1982; Goto *et al.*, 1984; Barret and Metcalfe, 1986; Gomez *et al.*, 1987; Shanahan *et al.*, 1987). Much of the knowledge on mast cells has come from studies comparing connective tissue mast cells and intestinal mucosal mast cells in rodents (Church, 1988).

There are three specific characteristics for bovine mast cells. First, mast cell density in the normal respiratory tract of cattle is far greater than in other species (Riley and West, 1953; Hebb *et al.*, 1968; Chen *et al.*, 1990). The second difference was first suspected when it was established that the concentration of dopamine in the lungs of ruminants is exceptionally high (Table XIII.10.4). This dopamine was later demonstrated to be concentrated in the mast cells (Bertler *et al.*, 1959; Coupland and Heath, 1961; Falck *et al.*, 1964) and to be liberated during experimentally induced anaphylaxis (Eyre and Deline, 1971). Finally, the control mechanisms of mediator release appear very different in the bovine species. In other mammals investigated to date, adrenergic control of the release of mediators of mast cells appears to have a consistent pattern, namely that β-agonists inhibit and that α-agonists and cholinergic agents enhance the release of mediators such as histamine and leukotrienes. In the bovine species it appears that excitation of either or both α- or β-adrenoreceptors causes inhibition of mediator release and that cholinomimetics and dopamine specifically induce enhancement (Burka and Eyre, 1976). This positive feedback may be augmenting the processes of inflammation and hypersensitivity by enhancing the liberation of other active substances.

Natural killer cells

Natural killer (NK) cells are lymphoid cells that mediate natural (spontaneous) non-MHC restricted cytotoxicity. In humans, most NK cells have the large granular lympho-

Table XIII.10.4 Dopamine content in the lungs of mammals

	Content (μg/g wet weight)	*References*
Cat	0.44	Aviado and Sadavongvivad (1970b)
Cow	11	Bertler and Rosengren (1969)
Dog	0.24	Aviado and Sadavongvivad (1970b)
Goat	6.4	Aviado and Sadavongvivad (1970b)
Guinea pig	0.22	Sadavongvivad (1970)
Man	0.59	Aviado and Sadavongvivad (1970b)
Mouse	0.15	Aviado and Sadavongvivad (1970a)
Rabbit	0.30	Sadavongvivad (1970)
Rat	0.50	Aviado and Sadavongvivad (1970b)
Sheep	4	Bertler and Rosengren (1969)

cyte (LGL) morphology. In cattle, the lymphoid cells with NK activity have been shown to be larger than average lymphocytes, but they have not been shown to have granules in their cytoplasm. Bovine NK cells are inefficient at lysing tumour target cells typically used to evaluate NK cell function in other species. They are efficient only at lysing these tumour cell targets if the NK cells are first activated by cytokines to become lymphokine activated killer (LAK) cells. In contrast nonactivated bovine NK cells have been shown to lyse efficiently various virus-infected cells, including cells infected with parainfluenza 3 virus, bovine leukaemia virus and bovine herpes virus 1. Bovine cells exhibiting NK activity have been shown to lack the CD3, CD4, CD5, CD6, and WC1 ($\gamma\delta$ T cell marker) molecules on their surfaces (reviewed by Roth, 1994).

11. Complement System

Relatively little is known about the bovine complement system compared with what is known about the human complement system. Historically, research on bovine complement started with the intent to use the gained knowledge for the development of complement-fixation tests designed for determining serum antibody levels (Rice and Crowson, 1950; Rice and Boulanger, 1952; Rice and Fuhamel, 1957; Fong *et al.*, 1971a,b). It is not the aim of this short review to describe the complex cascades of the complement system in detail. For this, the reader should consult more comprehensive reviews (Law and Reid, 1988; Holmskov *et al.*, 1994; Gewurz *et al.*, 1995).

Complement components

The components of the bovine complement system that have been investigated are listed in Table XIII.11.1. The relevant references are included.

Pathways of initiation

The classical pathway of complement activation is initiated when antibody is binding to an antigen. A concomitant change of conformation of the antibody leads to binding of C1q to the antibody. This pathway involves C1, C2, C4 and C3 and may lead to activation of the terminal pathway including C5, C6, C7, C8, C9. A complex of the last five components can form a pore in the membrane of a target cell and lead to cell lysis. The antibody-dependent classical pathway of complement is part of the adaptive immune system.

There are three known activation pathways of complement that can be ascribed as part of innate immunity. (1) Pentraxins bind, in a Ca^{2+}-dependent fashion, to surface carbohydrates of some microorganisms and, having done so, can bind C1q and thus activate the classical pathway (Gewurz *et al.*, 1995). (2) Mannose-binding protein (MBP), another bovine plasma protein, also can bind to some surface carbohydrates of certain microorganisms and can then, by bypassing C1q, activate the classical pathway (Holmskov *et al.*, 1995; Epstein *et al.*, 1996). (3) Finally, the alternative pathway of complement is activated in all mammals continuously, but at very low levels (Law and Reid, 1988). A certain number of molecules of complement compound C3 get hydrolysed to form C3(H_2O). C3(H_2O) assumes a conformation that allows the binding of factor B, which, when bound, is cleaved by factor D to Bb. Bb of this complex has enzymatic activity and the ability to convert C3 to C3b which, in turn, can bind more factor B and thus generate more C3-convertase C3bBb. This chain of events, which is a positive feedback mechanism, is kept in balance by control proteins H and I. When C3 is cleaved to C3b, an internal thiolester bond is exposed which has a short-lived ability to react randomly with a hydroxyl group on any foreign or autologous cell. Many invading micro-organisms will not allow the effective binding of control proteins H and I. Thus, the positive feedback will be amplified and lead to deposition of millions of C3b molecules on the microbe within minutes and make non-pathogenic microorganisms prone to either phagocytosis by phagocytic cells, that have receptors for C3b, or to lysis by the terminal complement pathway. The bovine alternative complement pathway discriminates between virulent and nonvirulent microbes (Howard, 1980; Eisenschenk *et al.*, 1995).

Control proteins

Some of the control proteins of the bovine complement system, i.e. factor H, factor I, properdin, C1-inhibitor, C4b-binding protein, membrane cofactor protein and S-protein, have been characterized to varying degrees. The counterparts of other control proteins of the human complement system, such as decay-accelerating factor, CD59 and homologous restriction factor, have not yet been described.

Cell receptors for cleavage products of complement components

The bovine receptors for C3b, iC3b, C3d and C5a have been investigated as indicated in Table XIII.11.1, but the bovine receptors for C3a and C4a have not.

Deficiencies

Deficiency of bovine CR3 (iC3b-receptor, Mac-1, CD18/CD11b) is the only reported deficiency of the bovine

Table XIII.11.1 Components of the bovine complement cascades

Complement components	*References*
C1, C1q, C1s	Barta (1976), Campbell *et al.* (1979a,b), Linscott and Triglia (1980), Triglia and Linscott (1980), Yonemasu *et al.* (1980), Sasaki and Yonemasu (1984), Yonemasu *et al.* (1980)
C2	Linscott and Triglia (1980), Triglia and Linscott (1980)
C3, C3b, iC3b	Linscott and Triglia (1980), Triglia and Linscott (1980), Menger and Aston (1985), Mhatre and Aston (1987b), Boulard (1989), Lu *et al.* (1993), Ogunremi and Tabel (1993a)
C4, C4a	Booth *et al.* (1979a,b), Linscott and Triglia (1980), Triglia and Linscott (1980), Smith *et al.* (1982), Groth *et al.* (1987), Dodds and Law (1990)
C5, C5a	Barta (1976), Linscott and Triglia (1980), Gennaro *et al.* (1986), Aston *et al.* (1990)
C6	Barta (1976), Linscott and Triglia (1980)
C7	Barta (1976), Linscott and Triglia (1980)
C8	Barta (1976), Linscott and Triglia (1980)
C9	Barta (1976), Linscott and Triglia (1980), Eisenschenk *et al.* (1992)
Factor B	Pang and Aston (1978), Tabel *et al.* (1984), Sethi and Tabel (1990a,b,c), Ogunremi and Tabel (1993a)
Factor D	Menger and Aston (1984)
Factor H	Mhatre and Aston (1987a), Tabel *et al.* (1990), Sakakibara *et al.* (1994), Soames *et al.* (1996)
Factor I (C3b-inactivator)	Barta (1976), Linscott and Triglia (1980), Triglia and Linscott (1980), Menger (1996), Soames *et al.* (1996)
Properdin	Nielsen *et al.* (1978)
C1-inhibitor	van Nostrand and Cunningham (1987)
C4-binding protein	Hillarp *et al.* (1994)
Decay-accelerating factor (CD55)	—[a]
Membrane cofactor protein (CD46)	Menger (1996)
CD59	—
Homologous restriction factor (C8-binding protein)	—
S protein (vitronectin)	Dahlbäck (1986), Mimuro and Loskutoff (1989), Filippsen *et al.* (1990), Hillarp *et al.* (1994)
C3b receptor (CR1; CD35)	Worku *et al.* (1995)
C3d receptor (CR2; CD21)	Naessens *et al.* (1990), Threadgill *et al.* (1994)
iC3b-receptor (CR3; CD11b/18; Mac-1)	Splitter and Morrison (1991), Gerardi (1996), Rutten *et al.* (1996)
C3a receptor	—
C4a receptor	—
C5a receptor	Perret *et al.* (1992)
Conglutinin	Lachmann (1967), Davis and Lachmann (1984), Strang *et al.* (1986), Loveless *et al.* (1989), Lee *et al.* (1991), Akiyama *et al.* (1992), Andersen *et al.* (1992), Hartshorn *et al.* (1993), Lim *et al.* (1993), Lu *et al.* (1993), Suzuki *et al.* (1993), Holmskov *et al.* (1994), Laursen *et al.* (1994), Reid and Turner (1994), Eda *et al.* (1996), Epstein *et al.* (1996)
Mannose-binding protein (MBP)	Kawasaki *et al.* (1985), Andersen *et al.* (1992), Holmskov *et al.* (1994), Epstein *et al.* (1996)
Pentraxins	Akiyama *et al.* (1992), Anderson *et al.* (1992), Sarikaputi *et al.* (1992), Gewurz *et al.* (1995)

[a]—Unknown.

complement system to date. This deficiency is associated with the bovine leukocyte adhesion deficiency (BLAD) (Shuster *et al.*, 1992; Gerardi, 1996; Rutten *et al.*, 1996; Cox *et al.*, 1997). BLAD results from a point mutation within the gene encoding bovine CD18 which is one chain of the heterodimeric β_2 integrin adhesion molecule (CD18/CD11a) on bovine leukocytes (Shuster *et al.*, 1992). Since CD18 is also part of the heterodimeric iC3b-receptor, which is present on neutrophils and mononuclear phagocytes, BLAD is associated with impaired microbicidal activities of phagocytes (Nagahata *et al.*, 1994, 1996).

Serum complement levels

Hemolytic serum complement levels initiated via the classical or alternative pathway and also C3 levels in calves during their first month of life are about one-third to one-half of the values measured in sera of their dams (Renshaw and Everson, 1979; Müller *et al.*, 1983). Bactericidal activities in calf sera correlate with complement activity (Barta *et al.*, 1972). Complement activity in bovine colostrum appears to be entirely mediated by the classical pathway (Brock *et al.*, 1975).

Differences between bovine complement and the complement system of other mammals

All Bovidae differ from other mammalian species in that their serum contains high levels of conglutinin (Lachmann, 1967; Ingram, 1982). The kinetics of the bovine alternative complement pathway appear to differ from the kinetics of the murine alternative complement pathway (Ogunremi and Tabel, 1993a,b).

Conglutinin belongs to the group of proteins called collectins, proteins that share structural features with collagen and functional properties with lectins (Reid and Turner, 1994; Epstein *et al.*, 1996). It binds to terminal *N*-acetyl glucosamine, mannose and fucose (Loveless *et al.*, 1989). It has the unique property of binding, in a Ca^{2+}-dependent fashion, to iC3b (Hirani *et al.*, 1985). There is circumstantial evidence that conglutinin may retard the dissociation of Bb from the C3-convertase C3bBb or from this complex after C3b has been degraded to iC3b (Tabel, 1996). Conglutinin that is bound to microorganisms, either directly via microbial carbohydrate or via iC3b, can mediate adherence to phagocytes via the C1q receptor of phagocytes (Reid and Turner, 1994). Conglutinin mediates a complement-dependent enhancement of the respiratory burst of phagocytes stimulated by *E. coli* (Friis *et al.*, 1991). It contributes to the bactericidal activity of bovine serum (Ingram, 1982; Friis-Christiansen *et al.*, 1990). Conglutinin is consumed during infections of cattle with *Babesia bovis* (Goodger *et al.*, 1981). Serum levels have been observed to be lowest at peaks of infection and to return to normal after infection (Ingram, 1982).

The kinetics of the bovine alternative complement pathway appear to proceed more slowly than the kinetics of the murine alternative complement pathway. By incubating zymosan with bovine or murine plasma, it was found that deposition of murine C3b onto zymosan particles peaked at 5 min (Ogunremi and Tabel, 1993b) whereas deposition of bovine C3b peaked at 20–30 min (Ogunremi and Tabel, 1993a).

12. References

Aarestrup, F. M., Jensen, N. E., Ostergård, H. (1995). Analysis of associations between major histocompatibility complex (BoLA) class I haplotypes and subclinical mastitis of dairy cows. *J. Dairy Sci.* **78**, 1684–1692.

Ababou, A., Goyeneche, J., Davis, W. C., Lévy, D. (1994). Evidence of the expression of three different BoLA-class II molecules on the bovine BL-3 cell line: determination of a non-DR non-DQ gene product. *J. Leukoc. Biol.* **56**, 182–186.

Abraham, J., Mergia, A., Whang, J., Tumolo, A., Friedman, J., Hjerrild, K., Gospodarowicz, D., Fiddes, J. (1986). Nucleotide sequence of a bovine clone encoding the angiogenic protein, basic fibroblast growth factor. *Science* **233**, 545–548.

Ackermann, M. R., Cheville, N. F., Deyoe, B. L. (1988). Bovine ileal dome lymphoepithelial cells: endocytosis and transport of *Brucella abortus* strain 19. *Vet. Pathol.* **25**, 28–35.

Adams, J., Czuprynski, C. (1995). *Ex vivo* induction of TNF-alpha and IL-6 mRNA in bovine whole blood by *Mycobacterium paratuberculosis* and mycobacterial cell wall components. *Microb. Pathog.* **19**, 19–29.

Adler, H., Frech, B., Thony, M., Pfister, H., Peterhans, E. and Jungi, T. (1995). Inducible nitric oxide synthetase in cattle. Differential cytokine regulation of nitric oxide synthetase in bovine and murine macrophages. *J. Immunol.* **154**, 4710–4718.

Agaba Kasigwa, M. (1996). Development of bovine type I markers and their application to investigating the trypanotolerance trait. Ph.D thesis, Department of Biology and Biochemistry, Brunel University, U.K.

Agaba, M., Kemp, S. J., Barendse, W., Teale, A. J. (1996). Polymorphism at the bovine tumor necrosis factor alpha locus and assignment to BTA 23. *Mamm. Genome* **7**, 186–187.

Akiyama, K., Sugii, S., Hirota, Y. (1992). Development of enzyme-linked immunosorbent assays for conglutinin, mannose-binding protein and serum amyloid-P component in bovine sera. *Am. J. Vet. Res.* **53**, 2102–2104.

Allmann-Iselin, I., Car, B., Zwahlen, R., Mueller-Schupbach, R., Wyder-Walther, M., Steckholzer, U., Walz, A. (1994). Bovine ENA, a new monocyte-macrophage derived cytokine of the interleukin-8 family. Structure, function, and expression in acute pulmonary inflammation. *Am. J. Pathol.* **145**, 1382–1389.

Alterio, J., Halley, C., Brou, C., Soussi, T., Courtois, Y., Laurent, M. (1988). Characterization of a bovine acidic FGF cDNA clone and its expression in brain and retina. *FEBS Lett.* **242**, 41–46.

Andersen, O., Friis, P., Holm-Nielsen, E., Vilsgaard, K., Leslie, R. G., Svehag, S. E. (1992). Purification, subunit characterization and ultrastructure of the three soluble bovine lectins: conglutinin, mannose-binding protein and the pentraxin serum amyloid-P-component. *Scand. J. Immunol.* **36**, 131–141.

Andersson, L., Rask, L. (1988). Characterization of the MHC class II region in cattle. The number of *DQ* genes varies between haplotypes. *Immunogenetics* **27**, 110–120.

Andersson, L., Böhme, J., Peterson, P. A., Rask, L. (1986a). Genomic hybridization of bovine class II major histocompatibility genes: 2. Polymorphism of DR genes and linkage disequilibrium in the DQ–DR region. *Anim. Genet.* **17**, 295–304.

Andersson, L., Böhme, J., Rask, L., Peterson, P. A. (1986b). Genomic hybridization of bovine class II major histocompatibility genes: 1. Extensive polymorphism of DQ and DQ genes. *Anim. Genet.* **17**, 95–112.

Andersson, L., Lundèn, A., Sigurdardottir, S., Davies, C. J., Rask, L. (1988). Linkage relationships in the bovine MHC region. High recombination frequency between class II subregions. *Immunogenetics* **27**, 273–280.

Andersson, L., Gustafsson, K., Jonsson, A.-K., Rask, L. (1991). Concerted evolution in a segment of the first domain exon of polymorphic MHC class II loci. *Immunogenetics* **33**, 235–242.

Adrianarivo, A., Muiya, P., Opollo, M., Logan-Henfry, L. (1995). *Trypanosoma congolense*: comparative effects of a primary infection on bone marrow progenitor cells from N'Dama and Boran cattle. *Exp. Parasitol.* **80**, 407–418.

Anonymous (1982). Proceedings of the Second International

Bovine Lymphocyte Antigen (BoLA) Workshop. *Anim. Blood Groups Biochem. Genet.* **13**, 33–53.

Armour, K. L., Tempest, P. R., Fawcett, P. H., Fernie, M. L., King, S. I., White, P., Taylor, G., Harris, W. J. (1994). Sequences of heavy and light chain variable regions from four bovine immunoglobulins. *Mol. Immunol.* **17**, 1369–1372.

Arriëns, M. A., Ruff, G., Schällibaum, M., Lazary, S. (1994). Possible association between a serologically detected haplotype of the bovine major histocompatibility complex and subclinical mastitis. *J. Anim. Breed. Genet.* **111**, 152–161.

Aston, W. P., Mhatre, A., Macrae, J. (1990). Isolation of the fifth component of the bovine complement system. *Vet. Immunol. Immunopathol.* **24**, 301–312.

Aviado, D. M., Sadavongvivad, C. (1970a). Pharmacological significance of biogenic amines in the lungs: histamine. *Br. J. Pharmacol.* **38**, 366–373.

Aviado, D. M., Sadavongvivad, C. (1970b). Pharmacological significance of biogenic amines in the lungs: noradrenaline and dopamine. *Br. J. Pharmacol.* **38**, 374–385.

Baker, D. G., Gershwin, L. J. (1992). Seasonal patterns of total and *Ostertagia*-specific IgE in grazing cattle. *Vet. Parasitol.* **44**, 211–221.

Baldwin, C. L., Teale, A. J., Naessens, J. G., Goddeeris, B. M., MacHugh, N. D., Morrison, W. I. (1986). Characterisation of a subset of bovine T lymphocytes that express BoT4 by monoclonal antibodies and function: similarity of lymphocytes defined by human T4 and murine L3T4. *J. Immunol.* **136**, 4385–4391.

Barendse, W., Armitage, S. M., Kossarek, L. M., Shalom, A., Kirkpatrick, B. W., Ryan, A. M., Clayton, D., Li, L., Neibergs, H. L., Zhang, N., Grosse, W. M., Weiss, J., Creighton, P., McCarthy, F., Ron, M., Teale, A. J., Fries, R., McGraw, R. A., Moore, S. S., Georges, M., Soller, M., Womack, J. E., Hetzel, D. J. S. (1994). A genetic linkage map of the bovine genome. *Nature Genet.* **6**, 227–235.

Barret, K. E., Metcalfe, D. D. (1986). Mast cell heterogeneity. Studies in non-human primates. In *Mast Cell Differentiation and Heterogeneity* (eds Befus, A. D., Bienenstock, J., Denburg, J. A.), pp. 231–238. Raven Press, New York, USA.

Barta, O. (1976). Separation of six bovine complement components and one inactivator. *Immunol. Commun.* **5**, 75–86.

Barta, O., Barta, V., Ingram, D. G. (1972). Postnatal development of bactericidal activity in serum from conventional and colostrum-deprived calves. *Am. J. Vet. Res.* **33**, 741–750.

Beever, J., Fisher, S., Lewin, H. (1996). Genetic mapping of 12 orthologs from human chromosomes 1 and 4 to bovine chromosomes 3, 6, 16 and 17. *Abstract XXVth International Conference on Animal Genetics, July 1996, Tours*, p. 92.

Belknap, E. B., Baker, J. C., Patterson, J. S., Walker, R. D., Haines, D. M., Clark, E. G. (1991). The role of passive immunity in bovine respiratory syncytial virus-infected calves. *J. Infect. Dis.* **163**, 470–476.

Bell, K. (1983). The blood groups of domestic mammals. In *Red Blood Cells of Domestic Animals* (eds Agar, N. S., Board, P. G.), pp. 133–164. Elsevier, Amsterdam, The Netherlands.

Bembridge, G. P., MacHugh, N. D., McKeever, D., Awino, E., Sopp, P., Collins, R. A., Gelder, K. I., Howard, C. J. (1995). CD45RO expression on bovine T cells: relation to biological function. *Immunology* **86**, 537–544.

Bensaid, A., Young, J. R., Kaushal, A., Teale, A. J. (1991). Pulsed-field gel electrophoresis and its application in the physical analysis of the bovine MHC. In *Gene Mapping Techniques and Applications* (eds Schook, L. B., Lewin, H. A., McLaren, D. G.), pp. 127–158. Marcel Dekker, Inc., New York, USA.

Bernoco, D., Lewin, H. A., Andersson, L., Arriens, M. A., Byrns, G., Cwik, S., Davies, C. J., Hines, H. C., Leibold, W., Lie, O., Meggiolaro, D., Oliver, R., Ostergård, H., Spooner, R. L., Stewart, J. A., Teale, A. J., Templeton, J. W., Zanotti, M. (1991). Joint report of the Fourth International Bovine Lymphocyte Antigen (BoLA) Workshop, East Lansing, Michigan, USA, August 25, 1990. *Anim. Genet.* **22**, 477–496.

Bertler, A., Rosengren, E. (1969). Occurrence and distribution of dopamine in brain and other tissues. *Experientia* **15**, 10–11.

Bertler, A., Falck, B., Hillarp, N. A., Rosengren, E., Torp, N. A. (1959). Dopamine and chromaffin cells. *Acta Physiol. Scand.* **47**, 251–258.

Bielefeldt Ohmann, H., Babiuk, L. A. (1986). Bovine alveolar macrophages: phenotypic and functional properties of subpopulations obtained by percoll density gradient centrifugation. *Eur. J. Biochem.* **218**, 669–677.

Bienhoff, S., Allen, G. (1995). Quantitation of bovine TNF-α mRNA by reverse transcription and competitive polymerase chain reaction amplification. *Vet. Immunol. Immunopathol.* **44**, 129–140.

Bishop, M. D., Kappes, S. M., Keele, J. W., Stone, R. T., Sunden, S. L. F., Hawkins, G. A., Toldo, S. S., Fries, R., Grosz, M. D., Yoo, J., Beattie, C. W. (1994). A genetic linkage map for cattle. *Genetics* **136**, 619–639.

Bissumbhar, B., Nilsson, P. R., Hensen, E. J., Davis, W. C., Joosten, I. (1994). Biochemical characterization of bovine MHC DQ allelic variants by one-dimensional isoelectric focusing. *Tissue Antigens* **44**, 100–109.

Blecha, F. (1991). Cytokines: applications in domestic food animals. *J. Dairy Sci.* **74**, 328–339.

Boosman, R., Niewold, T. A., Mutsaers, C., Gruys, E. (1989). Serum amyloid A concentrations in cows given endotoxin as an acute-phase stimulant. *Am. J. Vet. Res.* **50**, 1690–1694.

Booth, N. A., Campbell, R. D., Fothergill, J. E. (1979a). The purification and characterization of bovine C4, the fourth component of complement. *Biochem. J.* **177**, 959–965.

Booth, N. A., Campbell, R. D., Smith, M. A., Fothergill, J. E. (1979b). The isolation and characterization of bovine C4a, an activation fragment of the fourth component of complement. *Biochem. J.* **183**, 573–578.

Boulard, C. (1989). Degradation of bovine C3 by serine proteases from parasites *Hypoderma lineatum* (Diptera, Oestridae). *Vet. Immunol. Immunopathol.* **20**, 387–398.

Brambell, F. W. R. (1958). The passive immunity of the young mammal. *Biol. Rev.* **33**, 488.

Brandon, M. R., Lascelles, A. K. (1971). Relative efficiency of absorption of IgG1, IgG2, IgA and IgM in the newborn calf. *Aust. J. Exp. Biol. Med. Sci.* **49**, 629–633.

Brandtzaeg, P., Fjellanger, I., Gjeruldsen, S. T. (1970). Human secretory immunoglobulins. I. Salivary secretions from individuals with normal or low levels of serum immunoglobulins. *Scand. J. Haematol.* Suppl. **12**, 3–83.

Broad, T. E., Burkin, D. J., Cambridge, L. M., Jones, C., Lewis, P. E., Morse, H. G., Geyer, D., Pearce, P. D., Ansari, H. A., Maher, D. W. (1995). Six loci mapped on to human chromosome 2p are assigned to sheep chromosome 3p. *Anim. Genet.* **26**, 85–90.

Brock, J. H., Ortega, F., Pinerio, A. (1975). Bactericidal and hemolytic activity of complement in bovine colostrum and serum: effect of proteolytic enzymes and ethylene glycol tetraacetic acid (EFTA). *Ann. Immunol.* **126C**, 439–451.

Brown, W., Dziegielewska, K., Foreman, R., Saunders, N. (1990). The nucleotide and deduced amino acid sequences of insulin-like growth factor II cDNAs from adult bovine and fetal sheep liver. *Nucl. Acids Res.* **18**, 4614–4616.

Brown, W., Davis, W., Dobbelaere, D., Rice-Ficht, A. (1994a). $CD4^+$-cell clones obtained from cattle chronically infected with *Fasciola hepatica* and specific for adult worm antigen express both unrestricted and Th2 cytokine profiles. *Infect. Immun.* **62**, 818–827.

Brown, W., Woods, V., Chitko-McKown, C., Hash, S., Rice-Ficht, A. (1994b). Interleukin-10 is expressed by bovine type 1 helper, type 2 helper and unrestricted parasite-specific T-cell clones and inhibits proliferation of all three subsets in an accessory-cell-dependent manner. *Infect. Immun.* **62**, 4697–4708.

Brown, W., Zhao, S., Rice-Ficht, A., Logan, K., Woods, V. (1992). Bovine helper T-cell clones recognize five distinct epitopes on *Babesia bovis* merozoite antigens. *Infect. Immun.* **60**, 4364–4372.

Brown, W., Zhao, S., Woods, V., Tripp, C., Tetzlaff, C., Heussler, V., Dobbelaere, D., Rice-Ficht, A. (1993). Identification of two Th1 cell epitopes on the *Babesia bovis* – encoded 77-kilodalton merozoite protein (Bb-1) by use of truncated recombinant fusion proteins. *Infect. Immun.* **61**, 236–244.

Bryan, L. A., Griebel, P. J., Haines, D. M., Davis, W. C., Allen, J. R. (1988). Immunocytochemical identification of bovine Langerhans cells by use of a monoclonal antibody directed against class II MHC antigens. *J. Histochem. Cytochem.* **36**, 991–995.

Buitkamp, J., Schwaiger, F., Solinas-Toldo, S., Fries, R., Epplen, J. (1995). The bovine interleukin-4 gene: genomic organization, localization, and evolution. *Mamm. Genome* **6**, 350–356.

Bull, R. W., Lewin, H. A., Wu, M. C., Peterbaugh, K., Antczak, D., Bernoco, D., Cwik, S., Dam, L., Davies, C., Dawkins, R. L., Dufty, J. H., Gerlach, J., Hines, H. C., Lazary, S., Leibold, W., Leveziel, H., Lie, O., Lindberg, P. G., Meggiolaro, D., Meyer, E., Oliver, R., Ross, M., Simon, M., Spooner, R. L., Stear, M. J., Teale, A. J., Templeton, J. W. (1989). Joint report of the Third International Bovine Lymphocyte Antigen (BoLA) Workshop, Helsinki, Finland, 27 July 1986. *Anim. Genet.* **20**, 109–132.

Burka, J., Eyre, P. (1976). Modulation of the release of SRS-A from bovine lung *in vitro* by several autonomic and autocoid agents. *Int. Arch. Allergy Appl. Immunol.* **50**, 664–673.

Burke, M. G., Stone, R. T., Muggli-Cockett, N. E. (1991). Nucleotide sequences and Northern analysis of a bovine major histocompatibility class II DR-like cDNA. *Anim. Genet.* **22**, 343–352.

Butler, J. E. (1979). Immunologic aspects of breast feeding, antiinfectious activity of breast milk. *Sem. Perinatol.* **3**, 255–270.

Butler, J. E. (1983). Bovine immunoglobulins: an augmented review. *Vet. Immunol. Immunopathol.* **4**, 43–152.

Butler, J. E. (1986). Biochemistry and biology of ruminant immunoglobulins. *Prog. Vet. Microbiol. Immun.* **2**, 1–53.

Butler, J. E., Heyermann, H., Frenyo, L. V., Kieman, J. (1987). The heterogeneity of bovine IgG2. II. The identification of IgG2b. *Immunol. Letters* **16**, 31–46.

Butler, J. E., Navarro, P., Heyermann, H. (1994). Heterogeneity of bovine IgG2. VI. Comparative specificity of monoclonal and polyclonal capture antibodies for IgG2a(A1) and IgG2b(A2). *Vet. Immunol. Immunopathol.* **40**, 119–133.

Butler, J. E., Sun, J., Navarro, P. (1996). The swine Ig heavy chain locus has a single Jh and no identifiable IgD. *Int. Immunol.* **8**, 1897–1904.

Campbell, R. D., Booth, N. A., Fothergill, J. E. (1979a). Purification and characterization of subcomponent C1q of the first component of bovine complement. *Biochem. J.* **177**, 531–540.

Campbell, R. D., Booth, N. A., Fothergill, J. E. (1979b). The purification and characterization of subcomponent C1s of the first component of bovine complement. *Biochem. J.* **183**, 579–588.

Capon, D., Shepard, H., Goeddel, D. (1985). Two distinct families of human and bovine interferon-alpha genes are coordinately expressed and encode functional polypeptides. *Mol. Cell. Biol.* **5**, 768–779.

Cerretti, D., McKereghan, K., Larsen, A., Cantrell, M., Anderson, D., Gillis, S., Cosman, D., Baker, P. (1986a). Cloning, sequence and expression of bovine interleukin 2. *Proc. Natl Acad. Sci. USA* **83**, 3223–3227.

Cerretti, D., McKereghan, K., Larsen, A., Cosman, D., Gillis, S., Baker, P. (1986b). Cloning, sequence and expression of bovine interferon-gamma. *J. Immunol.* **136**, 4561–4564.

Chaplin, P., Parsons, K., Collins, R. (1996). The cloning of cattle interferon-A subtypes isolated from the gut epithelium of rotavirus-infected calves. *Immunogenetics* **44**, 143–145.

Chen, W., Alley, M. R., Manktelow, B. W., Slack, P. (1990). Mast cells in the bovine lower respiratory tract: morphology, density and distribution. *Br. Vet. J.* **146**, 425–436.

Chitko-McKown, C., Ruef, B., Rice-Ficht, A., Brown, W. (1995). Interleukin-10 downregulates proliferation and expression of interleukin-2 receptor p55 chain and interferon-gamma, but not interleukin-2 or interleukin-4, by parasite-specific helper T cell clones obtained from cattle chronically infected with *Babesia bovis* or *Fasciola hepatica*. *J. Interferon Cytokine Res.* **15**, 915–922.

Church, M. K. (1988). Mast cell heterogenieity. In *Mechanisms of Asthma, Pharmacology, Physiology, and Management* (eds Armour, C. L., Black, J. L.), pp. 15–23. Alan R. Liss, Inc., New York, USA.

Clevers, H., MacHugh, N. D., Bensaid, A., Dunlap, S., Baldwin, C. L., Kaushal, A., Iams, K., Howard, C. J., Morrison, W. I. (1990). Identification of a bovine surface antigen uniquely expressed on $CD4^-$ $CD8^-$ T cell receptor gamma/delta + T lymphocytes. *Eur. J. Immunol.* **20**, 809–817.

Cludts, I., Droogmans, L., Cleuter, Y., Kettman, R., Burny, A. (1992). Sequence of bovine interleukin 7. *DNA Seq.* **3**, 55–59.

Cludts, I., Cleuter, Y., Kettmann, R., Burny, A., Droogmans, L. (1993). Cloning and characterization of the tandemly arranged bovine lymphotoxin and tumour necrosis factor-alpha genes. *Cytokine* **5**, 336–341.

Collins, R., Oldham, G. (1993). Recombinant human interleukin 2 induces proliferation and immunoglobulin secretion by bovine B-cells: tissue differences and preferential enhancement of immunoglobulin A. *Vet. Immunol. Immunopathol.* **36**, 31–43.

Collins, R., Oldham, G. (1995). Effect of recombinant bovine IL-1 and IL-2 on B-cell proliferation and differentiation. *Vet. Immunol. Immunopathol.* **44**, 141–150.

Collins, R., Tayton, H., Gelder, K., Britton, P., Oldham, G. (1994). Cloning and expression of bovine and porcine interleukin-2 in baculovirus and analysis of species cross-reactivity. *Vet. Immunol. Immunopathol.* **40**, 313–324.

Collins, R. A., Sopp, R., Gelder, K. I., Morrison, W. I., Howard, C. J. (1996). Bovine γ/δ TCR + T lymphocytes are stimulated to proliferate by autologous *Theileria annulata*-infected cells in the presence of interleukin-2. *Scand. J. Immunol.* **44**, 444–452.

Conner, J. G., Eckersall, P. D. (1986). Acute phase response and mastitis in the cow. *Res. Vet. Sci.* **41**, 126–128.

Conner, J. G., Eckersall, P. D. (1988). Bovine acute phase response following turpentine injection. *Res. Vet. Sci.* **44**, 82–88.

Conner, J. G., Eckersall, P. D., Wiseman, A., Bain, R. K., Douglas, T. A. (1989). Acute phase response in calves following infection with *Pasturella haemolytica, Ostertagia ostertagi* and endotoxin administration. *Res. Vet. Sci.* **47**, 203–207.

Corley, L. D., Staley, T. E., Bush, L. J., Jones, E. W. (1977). Influence of colostrum on transepithelial movement of *Escherichia coli* 055. *J. Dairy Sci.* **60**, 1416–1421.

Coupland, R. E., Heath, I. D. (1961). Chromaffin cells, mast cells and melanin. II. The chromaffin cells of the liver capsule and gut in ungulates. *J. Endocrinol.* **22**, 71–76.

Covert, J., Splitter, G. (1995). Detection of cytokine transcriptional profiles from bovine peripheral blood mononuclear cells and $CD4^+$ lymphocytes by reverse transcriptase polymerase chain reaction. *Vet. Immunol. Immunopathol.* **49**, 39–50.

Cox, E., Mast, J., MacHugh, N., Swenger, B., Goddeeris, B. M. (1977). Expression of β_2 integrins on blood leukocytes of cows with and without bovine leukocyte adhesion deficiency. *Vet. Immunol. Immunopathol.* **58**, 249–263.

Dahlbäck, B. (1986). Inhibition of protein Ca cofactor function of human and bovine protein S by C4b-binding protein. *J. Biol. Chem.* **261**, 12022–12027.

Daley, M., Williams, T., Coyle, P., Furda, G., Dougherty, R., Hayes, P. (1993). Prevention and treatment of *Staphylococcus aureus* infections with recombinant cytokines. *Cytokine* **5**, 276–284.

Davies, C. J., Joosten, I., Andersson, L., Arriens, M. A., Bernoco, D., Bissumbhar, B., Byrns, G., van Eijk, M. J. T., Kristensen, B., Lewin, H. A., Mikko, S., Morgan, A. L. G., Muggli-Cockett, N. E., Nilsson, P. R., Oliver, R. A., Park, C. A., van der Poel, J. J., Polli, M., Spooner, R. L., Stewart, J. A. (1994a). Polymorphism of bovine MHC class II genes. Joint report of the Fifth International Bovine Lymphocyte Antigen (BoLA) Workshop, Interlaken, Switzerland, 1 August 1992. *Eur. J. Immunogenet.* **21**, 259–289.

Davies, C. J., Joosten, I., Bernoco, D., Arriens, M. A., Bester, J., Ceriotti, G., Ellis, S., Hensen, E. J., Hines, H. C., Horin, P., Kristensen, B., Lewin, H. A., Meggiolaro, D., Morgan, A. L. G., Morita, M., Nilsson, P. R., Oliver, R. A., Orlova, A., Ostergård, H., Park, C. A., Schuberth, H.-J., Simon, M., Spooner, R. L., Stewart, J. A. (1994b). Polymorphism of bovine MHC class I genes. Joint report of the Fifth International Bovine Lymphocyte Antigen (BoLA) Workshop, Interlaken, Switzerland, 1 August 1992. *Eur. J. Immunogenet.* **21**, 239–258.

Davis, A. E. 3rd, Lachmann, P. J. (1984). Bovine conglutinin is a collagen-like protein. *Biochemistry* **23**, 2139–2144.

Davis, W. D., Marusic, S., Lewin, H. A., Splitter, G. A., Perryman, L. E., McGuire, T. C., Gorham, J. R. (1987). The development and analysis of species specific and cross reactive monoclonal antibodies to leukocyte differentiation antigens and antigens of the major histocompatibility complex for use in the study of the immune system in cattle and other species. *Vet. Immunol. Immunopathol.* **15**, 337–376.

Delbe, J., Vacherot, F., Laaroubi, K., Barritault, D., Courty, J. (1995). Effect of heparin on bovine epithelial lens cell proliferation induced by heparin affin regulatory peptide. *J. Cell Physiol.* **164**, 47–54.

De Martini, J., Baldwin, C. (1991). Effects of gamma interferon, tumor necrosis factor alpha, and interleukin-2 on infection and proliferation of *Theileria parva*-infected bovine lymphoblasts and production interferon by parasitized cells. *Infect. Immun.* **59**, 4540–4546.

Diamond, G., Zasloff, M., Eck, H., Brasseur, M., Maloy, W. L., Bevins, C. L. (1991). Tracheal antimicrobial peptide, a cysteine-rich peptide from mammalian tracheal mucosa: peptide isolation and cloning of a cDNA. *Proc. Natl Acad. Sci. USA* **88**, 3952–3956.

Diamond, G., Jones, D. E., Bevins, C. L. (1993). Airway epithelial cells are the site of expression of a mammalian antimicrobial peptide gene. *Proc. Natl Acad. Sci. USA* **90**, 4596–4600.

Dodds, A. W., Law, S. K. A. (1990). The complement component of C4 of mammals. *Biochem. J.* **265**, 495–502.

Droogmans, L., Cludts, I., Cleuter, Y., Kettmann, R., Burny, A. (1992a). Nucleotide sequence of the bovine interleukin-6 gene promoter. *DNA Seq.* **3**, 115–117.

Droogmans, L., Cludts, I., Cleuter, Y., Kettmann, R., Burny, A. (1992b). Nucleotide sequence of the bovine interleukin-6 cDNA. *DNA Seq.* **2**, 411–413.

Dziegielewska, K. M., Brown, W. M., Gould, C. C., Matthews, N., Sedgwick, J. E. C., Saunders, N. R. (1996). Fetuin: an acute phase protein in cattle. *J. Comp. Physiol. B* **162**, 168–171.

Easton, A., Gierse, J., Seetharam, R., Klein, B., Kotts, C. (1991). Production of bovine insulin-like growth factor 2 (bIGF-2) in *Escherichia coli. Gene* **101**, 291–295.

Eda, S., Suzuki, Y., Kase, T., Kawai, T., Ohtani, K., Sakamoto, T., Kurimura, T., Wakamiya, N. (1996). Recombinant bovine conglutinin lacking the N-terminal and collagenous domains, has less conglutination activity but is able to inhibit haemagglutination by influenza A virus. *Biochem. J.* **316**, 43–48.

Eddie, D. S., Schulkind, M. L., Robbins, J. B. (1971). The isolation and biologic activities of purified secretory IgA and IgG anti-*Salmonella typhimurium* 'O' antibodies from rabbit intestinal fluid and colostrum. *J. Immunol.* **106**, 181–190.

Eggen, A., Fries, R. (1995). An integrated cytogenetic and meiotic map of the bovine genome. *Animal Genetics* **26**, 215–236.

Eisenschenk, F. C., Houle, J. J., Hoffmann, E. M. (1992). Purification of the ninth component of the bovine complement cascade. *Am. J. Vet. Res.* **53**, 435–439.

Eisenschenk, F. C., Houle, J. J., Hoffmann, E. M. (1995). Serum sensitivity of field isolates and laboratory strains of *Brucella abortus. Am. J. Vet. Res.* **56**, 1592–1598.

Ellis, J. A., Baldwin, C. L., MacHugh, N. D., Bensaid, A., Teale, A. J., Goddeeris, B. M., Morrison, W. I. (1986). Characterisation by a monoclonal antibody and functional analysis of a subset of bovine T lymphocytes that express BoT8, a molecule analogous to human CD8. *Immunology* **58**, 351–358.

Ellis, J., Godson, D., Campos, M., Sileghem, M., Babiuk, L. (1993). Capture immunoassay for ruminant tumor necrosis factor-α: comparison with bioassay. *Vet. Immunol. Immunopathol.* **35**, 289–300.

Ennis, P. D., Jackson, A. P., Parham, P. (1988). Molecular cloning of bovine class I MHC cDNA. *J. Immunol.* **141**, 642–651.

Epstein, J., Eichbaum, Q., Sheriff, S., Ezekowitz, R. A. B. (1996). The collectins in innate immunity. *Curr. Opin. Immunol.* **8**, 29–35.

Estes, D. M., Templeton, J. W., Hunter, D. M., Adams, L. G. (1990). Production and use of murine monoclonal antibodies reactive with bovine IgM isotype and IgG subisotypes (IgG1, IgG2a and IgG2b) in assessing immunoglobulin levels of serum of cattle. *Vet. Immunol. Immunopathol.* **25**, 61–72.

Estes, D., Hirano, A., Heussler, V., Dobbelaere, D. Brown, W. (1995). Expression and biological activities of bovine interleukin 4: effects of recombinant interleukin 4 on T cell proliferation and B cell differentiation and proliferation *in vitro*. *Cell Immunol.* **163**, 268–279.

Eyre, P., Deline, T. R. (1971). Release of dopamine from bovine lung by specific antigen and by compound 48/80. *Br. J. Pharmacol.* **42**, 423–427.

Falck, B., Nystedt, T., Rosengren, E., Stenflo, J. (1964). Dopamine and mast cells in ruminants. *Acta Pharmacol. Toxicol.* **21**, 51–58.

Filippsen, L. F., Valentin-Weigand, P., Blobel, H., Preisner, K. T., Chhatwal, G. S. (1990). Role of complement S protein (vitronectin) in adherence of *Streptococcus dysgalactiae* to bovine epithelial cells. *Am. J. Vet. Res.* **51**, 861–865.

Fitzpatrick, J. L., Cripps, P. J., Hill, A. W., Bland, P. W., Stokes, C. R. (1992). MHC class II expression in the bovine mammary gland. *Vet. Immunol. Immunopathol.* **32**, 13–23.

Fong, J. S. C., Muschel, L. H., Good, R. A. (1971a). Kinetics of bovine complement. I. Formation of a lytic intermediate. *J. Immunol.* **107**, 28–40.

Fong, J. S. C., Muschel, L. H., Good, R. A. (1971b). Kinetics of bovine complement. II. Properties of the lytic intermediate. *J. Immunol.* **107**, 34–40.

Fotsis, T., Murphy, C., Gannon, F. (1990). Nucleotide sequence of the bovine insulin-like growth factor 1 (IGF-1) and its IGF-1A precursor. *Nucl. Acids Res.* **18**, 676.

Fries, R., Eggen, A., Womack, J. E. (1993). The bovine genome map. *Mamm. Genome* **4**, 405–428.

Friis, P., Svehag, S. E., Andersen, O., Gahrn-Hansen, B., Leslie, R. G. (1991). Conglutinin exhibits a complement-dependent enhancement of the respiratory burst of phagocytes stimulated by *E. coli*. *Immunology* **74**, 680–684.

Friis-Christiansen, P., Thiel, S., Svehag, S. E., Dessau, R., Svendsen, P., Andersen, O., Laursen, S. B., Jensenius, J. C. (1990). *In vivo* and *in vitro* antibacterial activity of conglutinin, a mammalian plasma lectin. *Scand. J. Immunol.* **31**, 453–461.

Garber, T. L., Hughes, A. L., Watkins, D. I., Templeton, J. W. (1994). Evidence for at least three transcribed *BoLA* class I loci. *Immunogenetics* **39**, 257–265.

Gennaro, R., Simonic, T., Negri, A., Mottola, C., Secchi, C., Ronchi, S., Romeo, D. (1986). C5a fragment of bovine complement. Purification, bioassays, amino-acid sequence and other structural studies. *Eur. J. Biochem.* **155**, 77–86.

Gennaro, R., Skerlavaj, B., Romeo, D. (1989). Purification, composition, and activity of two bactenecins, antibacterial peptides of bovine neutrophils. *Infect. Immun.* **57**, 3142–3146.

Gerardi, A. S. (1996). Bovine leukocyte adhesion deficiency: a brief overview of a modern disease and its implications. *Acta Vet. Hungarica* **44**, 1–8.

Gershwin, L. J., Dygert, B. S. (1983). Development of a semi-automated microassay for bovine immunoglobulin E: definition and standardization. *Am. J. Vet. Res.* **44**, 891–895.

Gewurz, H., Zhang, X.-H., Lint, T. F. (1995). Structure and function of the pentraxins. *Curr. Opin. Immunol.* **7**, 54–64.

Goddeeris, B. M., Morrison, W. I., Teale, A. J. (1986). Generation of bovine cytotoxic cell lines, specific for cells infected with the protozoan parasite *Theileria parva* and restricted by products of the major histocompatibility complex. *Eur. J. Immunol.* **16**, 1243–1249.

Godson, D., Baca-Estrada, M., Hughes, H., Morsy, J., Van Donkersgoed, R., Harland, R., Shuster, D., Daley, M., Babiuk, L. (1995). Regulation of bovine acute phase responses by administration of interleukin-1. Abstract of Fourth International Veterinary Immunology Symposium, July 1995, Davis, p. 156.

Godson, D. L., Baca-Estrada, M. E., Van Kessel, A. G., Hughes, H. P. A., Morsy, M. A., Van Donkersgoed, J., Harland, R. J., Shuster, D. E., Daley, M. J., Babiuk, L. A. (1995). Regulation of bovine acute phase responses by recombinant interleukin-1β. *Can. J. Vet. Res.* **59**, 249–255.

Goldman, A. S., Garza, C., Nichols, B. L., Goldblum, R. M. (1982). Immunologic factors in human milk during the first year of lactation. *J. Pediatr.* **100**, 563–567.

Goldsby, R. A., Srikumaran, S., Arulanandam, B., Hague, B., Ponce de Leon, F. A., Sevoian, M., Guidry, A. J. (1987). The application of hybridoma technology to the study of bovine immunoglobulins. *Vet. Immunol. Immunopathol.* **17**, 25–35.

Gomez, E., Corrado, O. J., Davies, R. J. (1987). Histochemical and functional characteristics of the human nasal mast cell. *Int. Arch. Allergy Appl. Immunol.* **83**, 52–56.

Goodger, B. V., Wright, I. G., Mahoney, D. F. (1981). Changes in conglutinin, immunoconglutinin, complement C3 and fibronectin concentrations in cattle acutely infected with *Babesia bovis*. *Aust. J. Exp. Biol. Med. Sci.* **59**, 531–538.

Goto, T., Befus, D., Low, R., Bienenstock, J. (1984). Mast cell heterogeneity and hyperplasia in bleomycin-induced pulmonary fibrosis of rats. *Am. Rev. Respir. Dis.* **130**, 797–802.

Goureau, O., Faure, V., Courtois, Y. (1995). Fibroblast growth factors decrease inducible nitric oxide synthase mRNA accumulation in bovine retinal pigmented epithelial cells. *Eur. J. Biochem.* **230**, 1046–1052.

Groth, D. M., Wetherall, J. D., Umotong, B. A., Sparrow, P., Lee, I. R., Carrick, M. J. (1987). Purification and characterisation of the fourth component of bovine complement. *Complement* **4**, 1–11.

Gu, F., Chowdhary, B. P., Andersson, L., Harbitz, I., Gustavsson, I. (1992). Assignment of the bovine immunoglobulin gamma heavy chain (IGHG) gene to chromosome 21q24 by *in situ* hybridiziation. *Hereditas* **117**, 237–240.

Gwakisa, P., Mikko, S., Andersson, L. (1994). Close genetic linkage between *DRBP1* and *CYP21* in the MHC of cattle. *Mamm. Genome* **5**, 731–734.

Hagens, G., Galley, Y., Glaser, I., Davis, W. C., Baldwin, C. L., Clevers, H., Dobbelaere, D. A. E. (1996). Cloning, sequencing and expression of the bovine CD3ε and TCR-ζ chains, two invariant components of the T-cell receptor complex. *Gene* **169**, 165–171.

Hall, J. G., Morris, B. (1965). The origin of cells in the efferent lymph of a single node. *J. Exp. Med.* **121**, 901–910.

Halley, C., Courtois, Y., Laurent, M. (1988). Nucleotide sequence of a bovine acidic FGF cDNA. *Nucl. Acids Res.* **16**, 10913.

Hanby-Flarida, M. D., Trask, O. J., Yang, T. J., Baldwin, C. L. (1996). Modulation of WC1, a lineage-specific cell surface molecule of γ/δ T cells, augments cellular proliferation. *Immunology* **88**, 116–123.

Hansen, T., Leaman, D., Cross, J., Mathialagan, N., Bixby, J., Roberts, R. (1991). The genes for the trophoblast interferons and the related interferon-alpha-II possess distinct 5′-promoter and 3′-flanking sequences. *J. Biol. Chem.* **266**, 3060–3067.

Hartshorn, K. L., Sastry, K., Brown, K., White, M. R., Okarma, T. B., Lee, Y.-M., Tauber, A. I. (1993). Conglutinin acts as an opsonin for influenza A viruses. *J. Immunol.* **151**, 6265–6273.

Hash, S., Brown, W., Rich-Ficht, A. (1994). Characterization of a cDNA encoding bovine interleukin 10: kinetics of expression in bovine lymphocytes. *Gene* **139**, 257–261.

Hassfurther, R., Canning, P., Geib, R. (1994). Isolation and characterization of an interleukin-8-like peptide in the bovine species. *Vet. Immun. Immunopathol.* **42**, 117–126.

Hebb, C., Kasa, P., Mann, S. (1968). The relation between nerve fibres and dopamine cells of the ruminant lung. *Histochem. J.* **1**, 166–175.

Hein, W. R., Mackay, C. R. (1991). Prominence of $\gamma\delta$ T cells in the ruminant immune system. *Immunol. Today* **12**, 30–34.

Heussler, V., Eichhorn, M., Dobbelaere, D. (1992). Cloning of a full-length cDNA encoding bovine interleukin 4 by the polymerase chain reaction. *Gene* **144**, 273–278.

Heyermann, H., Butler, J. E. (1987). The heterogeneity of bovine IgG2. IV. Structural differences between IgG2a molecules of the A1 and A2 allotypes. *Mol. Immunol.* **12**, 1327–1334.

Hillarp, A., Thern, A., Dahlbäck, B. (1994). Bovine C4b binding protein. Molecular cloning of the α- and β-chains provides structural background for lack of complex formation with protein S. *J. Immunol.* **154**, 4190–4199.

Hines, H. C., Ross, M. J. (1987). Serological relationships among antigens of the BoLA and the bovine M blood group systems. *Anim. Genet.* **18**, 361–369.

Hirani, S., Lambris, J. D., Müller-Eberhard, H. J. (1985). Localization of the conglutinin binding site on the third component of human complement. *J. Immunol.* **134**, 1105–1109.

Hoang-Xuan, M., Charron, D., Zilber, M.-T., Levy, D. (1982). Biochemical characterization of class II bovine major histocompatibility complex antigens using cross-species reactive antibodies. *Immunogenetics* **15**, 621–624.

Hofner, M. C., Fosbery, M. W., Eckersall, P. D. (1994). Haptoglobin response to cattle infected with foot-and-mouth disease virus. *Res. Vet. Sci.* **57**, 125–128.

Holmskov, U., Malhotra, R., Sim, R. B., Jensenius, J. C. (1994). Collectins: collagenous C-type lectins of the innate immune defence system. *Immunol. Today* **15**, 67–74.

Holmskov, U. R., Laursen, S. B., Malhorta, R., Wiedemann, H., Timpl, R., Stuart, G. R., Tornoe, I., Madsen, P. S., Reid, K. B. M., Jensenius, J. C. (1995). Comparative study of the structural and functional properties of the bovine plasma C-type lectin, collectin-43 with other collectins. *Biochem. J.* **305**, 889–896.

Hood, L., Gray, W. R., Sanders, G., Dreyer, W. J. (1967). Light chain evolution. *Cold Spring Harbor Symp. Quant. Biol.* **32**, 133–145.

Hønberg, L. S., Larsen, B. (1992). Bovine monoclonal alloantibodies to blood group antigens prepared by murine × bovine or (murine × bovine) × bovine interspecies fusions. *Anim. Genet.* **23**, 497–508.

Hønberg, L. S., Larsen, B., Koch, C., Østergård, H., Skjødt, K. (1995). Biochemical identification of the bovine blood group M′ antigen as a major histocompatibility complex class I-like molecule. *Anim. Genet.* **26**, 307–313.

Honkanen-Buzalski, T., Sandholm, M. (1981). Association of bovine secretory immunoglobulins with milk fat globule membranes. *Comp. Immunol. Microbiol. Infect. Dis.* **4**, 329–342.

Horadagoda, A., Eckersall, P. D. (1994). Immediate responses in serum TNFα and acute phase protein concentrations to infection with *Pasteurella haemolytica* A1 in calves. *Res. Vet. Sci.* **57**, 129–132.

Howard, C. J. (1980). Variation in the susceptibility of bovine mycoplasmas to killing by the alternative complement pathway in bovine serum. *Immunology* **41**, 561–568.

Howard, C. J., Sopp, P., Parsons, K. R., Finch, J. (1989). *In vivo* depletion of BoT4 (CD4) and of non-T4/T8 lymphocyte subsets in cattle with monoclonal antibodies. *Eur. J. Immunol.* **19**, 757–764.

Howard, C. J., Morrison, W. I., Bensaid, A., Davis, W., Eskra, L., Gerdes, J., Hadam, M., Hurley, D., Leibold, W., Letesson, J. J. *et al.* (1991). Summary of workshop findings for leukocyte antigens of cattle. *Vet. Immunol. Immunopathol.* **27**, 21–27.

Howard, C. J., Naessens, J. (1993). Summary of workshop findings for cattle (tables 1 and 2). *Vet. Immunol. Immunopathol.* **39**, 25–47.

Howard, C. J., Sopp, P., Brownlie, J., Parsons, K. R., Kwong, L. S., Collins, R. A. (1996). Afferent lymph veiled cells stimulate proliferative responses in allogeneic CD4+ and CD8+ T cells but not $\gamma\delta$ TCR+ T cells. *Immunology* **88**, 558–564.

Hughes, H. P. A., Campos, M., McDougall, L., Beskorwayne, T. K., Potter, A. A., Babiuk, L. A. (1994). Regulation of major histocompatibility complex class II expression by *Pasteurella haemolytica* leukotoxin. *Infect. Immun.* **62**, 1609–1615.

Husband, A. J., Lascelles, A. K. (1975). Antibody responses to neonatal immunisation in calves. *Res. Vet. Sci.* **18**, 201–207.

Husband, A. J., Brandon, M. R., Lascelles, A. K. (1972). Absorption and endogenous production of immunoglobulins in calves. *Aust. J. Exp. Biol. Med. Sci.* **50**, 491–498.

Ingram, D. G. (1982). Comparative aspects of conglutinin and immunoconglutinins. In *Animal Models of Immunological Processes* (ed. Hay, J. B.), pp. 221–249. Academic Press, Toronto, Canada.

International Society for Animal Blood Group Research (1985). Notice from the standing committee on cattle blood groups and biochemical polymorphisms. *Anim. Blood Groups Biochem. Genet.* **16**, 249–252.

Ishiguro, N., Hein, W. R. (1994). The T cell receptor. In *Cell-mediated Immunity in Ruminants* (eds Goddeeris, B. M., Morrison, W. I.), pp. 59–73. CRC Press, Boca Raton, FL, USA.

Ishiguro, N., Tanaka, A., Shinagawa, M. (1990). Sequence analysis of bovine T-cell receptor α chain. *Immunogenetics* **31**, 57–60.

Ishiguro, N., Aida, Y., Shinagawa, T., Shinagawa, M. (1993). Molecular structures of cattle T-cell receptor gamma and delta chains predominantly expressed on peripheral blood lymphocytes. *Immunogenetics* **38**, 437–443.

Ito, T., Kodama, M. (1996). Demonstration by reverse transcription-polymerase chain reaction of multiple cytokine mRNA expression in bovine alveolar macrophages and peripheral blood mononuclear cells. *Res. Vet. Sci.* **60**, 94–96.

Jackson, T., Morris, B. A., Sanders, P. G. (1992). Nucleotide sequences and expression of cDNAs for a bovine anti-testosterone monoclonal IgG1 antibody. *Mol. Immunol.* **29**, 667–676.

Jian, Z., Yang, Z., Miller, M., Carter, C., Slauson, D., Bochsler,

P. (1995). Interleukin-6 secretion by bacterial lipopolysaccharide-stimulated bovine alveolar macrophages *in vitro*. *Vet. Immunol. Immunopathol.* **49**, 51–60.

Johnson, S. E., Barendse, W., Hetzel, D. J. (1993). The gamma fibrinogen gene (FGG) maps to chromosome 17 in both cattle and sheep. *Cytogen. Cell Genet.* **62**, 176–180.

Johnston, N. E., Estrella, R. A., Oxender, W. D. (1977). Resistance of neonatal calves given colostrum diet to oral challenge with a septicaemia-producing *Escherichia coli*. *Am. J. Vet. Res.* **38**, 1323–1326.

Joosten, I., Sanders, M. F., van der Poel, A., Williams, J. L., Hepkema, B. G., Hensen, E. J. (1989). Biochemically defined polymorphism of bovine MHC class II antigens. *Immunogenetics* **29**, 213–216.

Joosten, I., Oliver, R. A., Spooner, R. L., Williams, J. L., Hepkema, B. G., Sanders, M. F., Hensen, E. J. (1988). Characterization of class I bovine lymphocyte antigens (BoLA) by one-dimensional isoelectrofocusing. *Anim. Genet.* **19**, 103–113.

Joosten, I., Sanders, M. F., Hensen, E. J. (1991). Involvement of major histocompatibility complex class I compatibility between dam and calf in the aetiology of bovine retained placenta. *Anim. Genet.* **22**, 455–463.

Kabbur, M., Jain, N., Farvev, T. (1995). Modulation of phagocytic and oxidative burst activities of bovine neutrophils by human recombinant TNF-α, IL-lα, IFN-γ, G-CSF, GM-CSF. *Comp. Haematol. Int.* **5**, 47–55.

Kacskovics, I., Butler, J. E. (1996). The heterogeneity of bovine IgG2-VIII. The complete cDNA sequence of bovine IgG2a (A2) and an IgG1. *Mol. Immunol.* **33**, 189–195.

Kakizaki, Y., Waga, S., Sugimoto, K., Tanaka, H., Nukii, K., Takeya, M., Yoshimura, T., Yokoyama, M. (1995). Production of monocyte chemoattractant protein-1 by bovine glomeruler endothelial cells. *Kidney Int.* **48**, 1866–1874.

Kanse, T., Takahashi, K., Iam, I., Rees, A., Warren, J., Porta, M., Molinatti, P., Ghatei, M., Bloom, S. (1991). Cytokine stimulated endothelin release from endothelial cells. *Life Sci.* **48**, 1379–1384.

Kawasaki, N., Kawasaki, T., Yamashina, I. (1985). Mannan-binding protein and conglutinin in bovine serum. *J. Biochem.* **98**, 1309–1320.

Kehoe, J. M., Capra, J. D. (1974). Nature and significance of immunoglobulin subclasses. *NY St. J. Med.* **74**, 489–491.

Kehrli, M., Goff, J., Stevens, M., Boone, T. (1991). Effects of granulocyte colony stimulating factor administration to periparturient cows on neutrophils and bacterial shedding. *J. Dairy Sci.* **74**, 2448–2454.

Knight, K. L., Suter, M., Becker, R. S. (1988). Genetic engineering of bovine Ig. Construction and characterization of hapten-binding bovine/murine chimeric IgE, IgA, IgG1, IgG2 and IgG3 molecules. *J. Immunol.* **140**, 3654–3659.

Koch, A., Harlow, L., Haines, G., Amento, E., Unemori, E., Wong, W., Pope, R., Ferrara, N. (1994). Vascular endothelial growth factor. A cytokine modulating endothelial function in rheumatoid arthritis. *J. Immunol.* **152**, 4149–4156.

Kriegelstein, K., Unsicker, K. (1995). Bovine chromaffin cells release a transforming growth factor-beta-like molecule contained within chromaffin granules. *J. Neurochem.* **65**, 1423–1426.

Kulseth, M. A., Toldo, S. S., Fries, R., Womack, J., Lien, S., Rogne, S. (1994). Chromosomal localization and detection of DNA polymorphisms in the bovine polymeric immunoglobulin receptor gene. *Anim. Genet.* **25**, 113–117.

Lachmann, P. J. (1967). Conglutinin and immunoconglutinins. *Adv. Immunol.* **6**, 479–527.

Lalor, P. A., Morrison, W. I., Goddeeris, B. M., Jack, R. M., Black, S. J. (1986). Monoclonal antibodies identify phenotypically and functionally distinct cell types in the bovine lymphoid system. *Vet. Immunol. Immunopathol.* **13**, 121–140.

Landverk, T. (1987). The follicle-associated epithelium of the ileal Peyer's patch in ruminants is distinguished by its shedding of 50 nm particles. *Immunol. Cell Biol.* **65**, 251–261.

Larsen, B., Jensen, N. E., Madsen, P., Nielsen, S. M., Klastrup, O., Madsen, P. S. (1985). Association of the M blood group system with bovine mastitis. *Anim. Blood Groups Biochem. Genet.* **16**, 165–173.

Lascelles, A. K. (1969). Immunoglobulin secretion into ruminant colostrum. In *Lactogenesis* (eds Reynolds, M., Folley, S. J.), pp. 131–136. University of Pennsylvania Press, Philadelphia, PA, USA.

Lascelles, A. K., Lee, C. S. (1978). *Lactaction: A Comprehensive Treatise*, pp. 115–177. Academic Press, New York, USA.

Lascelles, A. K., Beh, K. J., Mukkur, T. K., Watson, D. L. (1986). The mucosal immune system with particular reference to ruminant animals. In *The Ruminant Immune System in Health and Disease* (ed Morrison, W. I.), pp. 429–457. University Press, Cambridge, UK.

Laursen, S. B., Thiel, S., Teisner, B., Holmskov, U., Wang, Y., Sim, R. B., Jensenius, J. C. (1994). Bovine conglutinin binds to an oligosaccharide determinant presented by iC3b, but not by C3, C3b or C3c. *Immunology* **81**, 648–654.

Law, S. K. A., Reid, K. B. M. (1988). *Complement*, IRL Press Ltd., Oxford, UK.

Lederer, J., Czuprynski, C. (1995). Interleukin-1. In *Cytokines in Animal Health and Disease* (eds Myers, M., Murtaugh, M.), pp. 59–88. Marcel Dekkers, Inc., New York, USA.

Lee, Y.-M., Leiby, K. R., Aller, J., Paris, K., Lerch, B., Okarma, T. B. (1991). Primary structure of bovine conglutinin, a member of the C-type animal lectin family. *J. Biol. Chem.* **266**, 2715–2723.

Leong, S., Flaggs, G., Lawman, M., Gray, P. (1988a). The nucleotide sequence for the cDNA of bovine interleukin-1 alpha. *Nucl. Acids Res.* **16**, 9053.

Leong, S., Flaggs, G., Lawman, M., Gray, P. (1988b). The nucleotide sequence for the cDNA of bovine interleukin-1 beta. *Nucl. Acids Res.* **16**, 9054.

Letesson, J. J., Lostrie-Trussart, N., Depelchin, A. (1985). Production d'anticorps monoclonaux specifiques d'isotypes d'immunoglobulines bovines. *Ann. Méd. Vét.* **129**, 131–137.

Leung, D., Capon, D., Goeddel, D. (1984). The structure and bacterial expression of three distinct bovine interferon-beta genes. *Biotechnology* **2**, 458–464.

Leung, D., Cachianes, G., Kuang, W., Goeddel, D., Ferrara, N. (1989). Vascular endothelial growth factor is a secreted angiogenic mitogen. *Science* **246**, 1306–1309.

Lewin, H. A. (1996). Genetic organization, polymorphism and function of the bovine major histocompatibility complex. In *The Major Histocompatibility Complex of Domestic Animal Species* (eds Schook, L. B., Lamont, S. J.), pp. 65–98. CRC Press, Inc., Boca Raton, FL, USA.

Lewin, H. A., Bernoco, D. (1986). Evidence for BoLA-linked resistance and susceptibility to subclinical progression of bovine leukaemia virus infection. *Anim. Genet.* **17**, 187–207.

Lewin, H. A., Calvert, C. C., Bernoco, D. (1985a). Cross reactivity of a monoclonal antibody (H4) with bovine, equine, ovine and porcine peripheral blood B lymphocytes. *Am. J. Vet. Res.* **46**, 785–788.

Lewin, H. A., Davis, W. C., Bernoco, D. (1985b). Monoclonal antibodies that distinguish bovine T and B lymphocytes. *Vet. Immunol. Immunopathol.* **9**, 87–102.

Lewin, H. A., Wu, M. C., Stewart, J. A., Nolan, T. J. (1988). Association between *BoLA* and subclinical bovine leukemia virus infection in a herd of Holstein-Friesian cows. *Immunogenetics* **27**, 338–344.

Lewin, H. A., Arnet, E. F., Lie, O., Bernoco, D., Davies, C. J., Nilsson, P. R., Ruff, G., Lazary, S. (1991). First international comparison test of serological reagents for typing class II products of the bovine major histocompatibility complex. *XXII International Conference on Animal Genetics. Anim. Genet.* **22** (Suppl. 1), 46–47.

Lewin, H. A., Spevak, E. M., Paige, K. N. (1993). Genetic management of herd health and productivity: molecular strategies. In *Proc. North American Public Bison Herds Symposium* (ed Walker, R. E.), pp. 38–52.

L'Haridon, R. M., Bourget, P., Lefevre, F., La Bonnardiere, C. (1991). Production of a hybridoma library to recombinant porcine alpha I interferon: a very sensitive assay (ISBBA) allows the detection of a large number of clones. *Hybridoma* **10**, 35–47.

Li, Q., Congote, L. (1995). Bovine fetal-liver stromal cells support erythroid colony formation: enhancement by insulin-like growth factor II. *Exp. Hematol.* **23**, 66–73.

Li, Y., Feng, J., Templeton, J. (1996). *Bos taurus* interleukin-8 receptor mRNA, complete cds. GenBANK accession U19947, unpublished.

Li, Y., Milner, P., Chauhan, A., Watson, M., Hoffman, R., Kodner, C., Milbrandt, J., Deuel, T. (1990). Cloning and expression of a developmentally regulated protein that induces mitogenic and neurite outgrowth activity. *J. Sci.* **250**, 1690–1694.

Liebler, E. M., Pohlenz, J. F., Woode, G. N. (1988a). Gut-associated lymphoid tissue in the large intestine of calves. I. Distribution and histology. *Vet. Pathol.* **25**, 503–508.

Liebler, E. M., Pohlenz, J. F., Cheville, N. F. (1988b). Gut-associated lymphoid tissue in the large intestine of calves. II. Electron microscopy. *Vet. Pathol.* **25**, 509–515.

Lim, B. L., Lu, J., Reid, K. B. M. (1993). Structural similarity between bovine conglutinin and bovine lung surfactant protein D and demonstration of liver as a site of synthesis of conglutinin. *Immunology* **78**, 159–165.

Lindberg, P.-G., Andersson, L. (1988). Close association between DNA polymorphism of bovine major histocompatibility complex class I genes and serological BoLA-A specificities. *Anim. Genet.* **19**, 245–255.

Linscott, W. D., Triglia, R. P. (1980). Methods for assaying nine bovine complement components and C3b-inactivator. *Mol. Immunol.* **17**, 729–740.

Logan, E. F., Stenhouse, A., Ormrod, D. J., Penhale, W. J. (1974). The role of colostral immunoglobulins in intestinal immunity to enteric colibacillosis in the calf. *Res. Vet. Sci.* **17**, 280–301.

Lovejoy, B., Cascio, D., Eisenberg, D. (1993). Crystal structure of canine and bovine granulocyte-colony stimulating factor (G-CSF). *J. Mol. Biol.* **234**, 640–653.

Loveless, R. W., Feizi, T., Childs, R. A., Mizuichi, T., Stoll, M. S., Oldroyd, R. G., Lachmann, P. J. (1989). Bovine serum conglutinin is a lectin which binds non-reducing terminal *N*-acetylglucosamine, mannose and fructose residues. *Biochem. J.* **258**, 109–113.

Lu, J., Laursen, S. B., Thiel, S., Jensenius, J. C., Reid, K. B. M. (1993). The cDNA cloning of conglutinin and identification of liver as a primary site of synthesis of conglutinin in members of the Bovidae. *Biochem. J.* **292**, 157–162.

Lundén, A., Sigurdardóttir, S., Edfors-Lilja, I., Danell, B., Rendel, J., Andersson, L. (1990). The relationship between bovine major histocompatibility complex class II polymorphism and disease studied by use of bull breeding values. *Anim. Genet.* **21**, 221–232.

Lutje, V., Mertens, B., Boulangé, A., Williams, D., Authié, (1995). *Trypanosoma congolense*: proliferative responses and interleukin production in lymph node cells of infected cattle. *Expt. Parasitol.* **81**, 154–164.

Ma, R. Z., Beever, J. E., Da, Y., Green, C. A., Russ, I., Park, C., Heyen, D. W., Everts, R. E., Overton, K. M., Fisher, S. R., Teale, A. J., Kemp, S. J., Hines, H. C., Guérin, G., Lewin, H. A. (1996). A male linkage map of the cattle (*Bos taurus*) genome. *J. Hered.* **87**, 261–271.

Mach, J. P., Pahud, J. J. (1971). Secretory IgA, a major immunoglobulin in most bovine external secretions. *J. Immunol.* **106**, 552–563.

Machugh, D. E., Loftus, R. T., Bradley, D. G., Sharp, P. M., Cunningham, P. (1994). Microsatellite DNA variation within and among European cattle breeds. *Proc. R. Soc. Lond.* B **256**, 25–31.

Mackay, C. R., Hein, W. R. (1989). A large proportion of bovine T cells express the $\gamma\delta$ T cell receptor and show a distinct tissue distribution and surface phenotype. *Int. Immunol.* **1**, 540–545.

Mackay, C. R., Marston, W. L., Dudler, L. (1990). Naive and memory T cells show distinct pathways of lymphocyte recirculation. *J. Exp. Med.* **171**, 801–817.

Maier, R., Schmid, P., Cox, D. (1991). Localization of transforming growth factor-beta 1, -beta 2 and -beta 3 gene expression in bovine mammary gland. *Mol. Cell Endocrinol.* **82**, 191–198.

Malik, N., Haugen, H., Modrell, B., Shoyab, M., Clegg, C. (1995). Developmental abnormalities in mice transgenic for bovine oncostatin M. *J. Mol. Cell Biol.* **15**, 2349–2358.

Maliszewski, C., Baker, P., Schoenborn, M., Davis, B., Cosman, D., Gillis, S., Cerretti, D. (1988). Cloning, sequence and expression of bovine interleukin 1-alpha and interleukin 1-beta complementary DNAs. *J. Mol. Immunol.* **25**, 429–437.

Maliszewski, C., Schoenborn, M., Cerretti, D., Wignall, J., Picha, K., Cosman, D., Tushinski, R., Gillis, S., Baker, P. (1988). Bovine GM-CSF: molecular cloning and biological activity of the recombinant protein. *Mol. Immunol.* **25**, 843–850.

Mallard, B. A., Leslie, K. E., Dekkers, J. C. M., Hedge, R., Bauman, M., Stear, M. J. (1995). Differences in bovine lymphocyte antigen associations between immune responsiveness and risk of disease following intramammary infection with *Staphylococcus aureus*. *J. Dairy Sci.* **78**, 1937–1944.

Mandriota, S., Menoud, P., Pepper, M. (1996). Transforming growth factor beta 1 downregulates VEGF-receptor 2/FLK-1 expression in vascular endothelial cells. *J. Biol. Chem.* **271**, 11500–11505.

Manning, L. S., Parmely, M. J. (1980). Cellular determinants of mammary cell-mediated immunity in the rat. I. The migration of radioisotopically labelled T lymphocytes. *J. Immunol.* **125**, 2508–2514.

Mayans, M. O., Coadwell, W. J., Beale, D., Symons, D. B. A., Perkins, S. (1995). Demonstration by pulsed neutron scattering that the arrangement of the Fab and Fc fragments in the overall structures of bovine IgG1 and IgG2 in solution is similar. *Biochem. J.* **311**, 283–291.

McCarthy, S., Bicknell, R. (1993). Inhibition of vascular endothelial cell growth by activin-A. *J. Biol. Chem.* **269**, 23066–23071.

McGuire, T. C., Crawford, T. B. (1972). Identification and quantitation of equine serum and secretory immunoglobulin A. *J. Am. Soc. Microbiol.* **6**, 610–615.

McKeever, D. J. (1994). Induction of T Cell immunity in ruminants. In *Cell-mediated Immunity in Ruminants* (eds Goddeeris, B. M., Morrison, W. I.), pp. 93–108. CRC Press, Boca Raton, FL, USA.

McKeever, D. J., MacHugh, N. D., Goddeeris, B. M., Awino, E., Morrison, W. I. (1991). Bovine afferent lymph veiled cells differ from blood monocytes in phenotype and are superior in accessory function. *J. Immunol.* **147**, 3703–3709.

McKeever, D. J., Awino, E., Morrison, W. I. (1992). Afferent lymph veiled cells prime $CD4^+$ T cell responses *in vivo*. *Eur. J. Immunol.* **22**, 3057–3061.

McKeever, D., Nyanjui, J., Ballingall, K. (1997). *In vitro* infection with *Theileria parva* is associated with IL10 expression in all bovine lymphocyte lineages. *Parasit. Immunol.* **19**, 319–324.

Meier, R., Becker-Andre, M., Goetz, R., Heumann, R., Shaw, A., Thoenen, H. (1986). Molecular cloning of bovine and chick nerve growth factor (NGF): delineation of conserved and unconserved domains and their relationship to the biological activity and antigenicity of NGF. *EMBO J.* **5**, 1489–1493.

Mejdell, C. M., Lie, O., Solbu, H., Arnet, E. F., Spooner, R. L. (1994). Association of major histocompatibility complex antigens (BoLA-A) with AI bull progeny test results for mastitis, ketosis and fertility in Norwegian Cattle. *Anim. Genet.* **25**, 99–104.

Menger, M. (1996). Characterization of the bovine alternative complement pathway C3-convertase. PhD thesis, Queen's University, Kingston, Ontario, Canada.

Menger, M., Aston, W. P. (1984). Factor D of the alternative pathway of bovine complement: isolation and characterization. *Vet. Immunol. Immunopathol.* **7**, 325–336.

Menger, M., Aston, W. P. (1985). Isolation and characterization of the third component of bovine complement. *Vet. Immunol. Immunopathol.* **10**, 317–331.

Mertens, B., Muriuki, C., Gaidulis, L. (1995). Cloning of two members of the TNF-superfamily in cattle: CD40 ligand and tumor necrosis factor alpha. *Immunogenetics* **42**, 430–431.

Mertens, B., Gorbright, E., Seow, H. (1996). Nucleotide sequence of the bovine interleukin-5 encoding cDNA. *Gene* **176**, 273–274.

Mertens, B., Muriuki, C., Muiya, P., Andrianarivo, A., Mwangi, S., Logan-Henfrey, L. (1997). Bovine stem cell factor: production of a biologically active protein and mRNA analysis in cattle infected with *Trypanosoma congolense*. *Vet. Immunol. Immunopathol.* **59**, 65–78.

Méténier, L., Nocart, M., Alaux, M. T. (1991). Mouse monoclonal antibodies to bovine blood groups antigens: additional results. *Anim. Genet.* **22**, 155–163.

Mhatre, A., Aston, W. P. (1987a). Isolation of bovine complement factor H. *Vet. Immunol. Immunopathol.* **14**, 357–375.

Mhatre, A., Aston, W. P. (1987b). A simple hemolytic assay for bovine complement component C3. *Vet. Immunol. Immunopathol.* **15**, 239–251.

Mimuro, J., Loskutoff, D. J. (1989). Purification of a protein from bovine plasma that binds to type-1 plasminogen-activator inhibitor and prevents its interaction with extracellular matrix. Evidence that the protein is vitronectin. *J. Biol. Chem.* **264**, 936–939.

Modat, G., Dornand, J., Junquero, D., Mary, A., Muller, A., Bonne, C. (1990). LPS-stimulated bovine aortic endothelial cells produce IL-1 and IL-6 like activities. *Agents Actions* **30**, 403–411.

Momotani, E., Whipple, D. L., Thiermann, B., Cheville, N. F. (1988). Role of M cells and macrophages in the entrance of *Mycobacterium paratuberculosis* into domes of ileal Peyer's patches in calves. *Vet. Pathol.* **25**, 131–137.

Moody, D., Pomp, P., Barendse, W. (1996). Linkage mapping of the bovine insulin-like growth factor-1 receptor gene. *Mamm. Genome* **7**, 168–169.

Morrison, W. I., Lalor, P. A., Christensen, A. K., Webster, P. (1986). Cellular constituents and structural organisation of the bovine thymus and lymph node. In *The Ruminant Immune System in Health and Disease* (ed. Morrison, W. I.), pp. 220–251. Cambridge University Press, Cambridge, UK.

Morrison, W. I., Baldwin, C. L., MacHugh, N. D., Teale, A. J., Goddeeris, B. M., Ellis, J. (1988). Phenotypic and functional characterisation of bovine lymphocytes. In *Progress in Veterinary Microbiology and Immunology* (ed. Padney, R.), pp. 134–164. S. Karger, Basel, Switzerland.

Morsey, M., Cox, G., Van Kessel, A., Campos, M., Babiuk, L. (1995). Interleukin-2. In *Cytokines in Animal Health and Disease* (eds Myers, M., Murtaugh, M.), pp. 89–119. Marcel Dekker, Inc., New York, USA.

Muggli-Cockett, N. E., Stone, R. T. (1988). Identification of genetic variation in the bovine major histocompatibility complex DR-like genes using sequenced bovine genomic probes. *Anim. Genet.* **19**, 213–225.

Müller, R., Boothby, J. T., Carroll, E. J., Panico, L. (1983). Changes of complement values in calves during the first month of life. *Am. J. Vet. Res.* **44**, 747–750.

Mwangi, S., Logan-Henfrey, L., McInnes, C., Mertens, B. (1995). Cloning of the bovine interleukin-3-encoding cDNA. *Gene* **162**, 309–312.

Myers, M., Murtaugh, M. (1995). Biology of tumor necrosis factor. In *Cytokines in Animal Health and Disease* (eds Myers, M., Murtaugh, M.), pp. 121–151. Marcel Dekker, Inc., New York, USA.

Naessens, J. (1997). Surface immunoglobulins on B lymphocytes from cattle and sheep. *Int. Immunol.* **9**, 349–354.

Naessens, J., Hopkins, J. (1996). Summary of third ruminant leukocyte workshop findings. *Vet. Immunol. Immunopathol.* **52**, 213–235.

Naessens, J., Williams, D. J. L. (1992). Characterization and measurement of $CD5^+$ B cells in normal and *Trypanosoma congolense*-infected cattle. *Eur. J. Immunol.* **22**, 1713–1718.

Naessens, J., Newson, J., Williams, D. J. L., Lutje, V. (1988). Identification of isotypes and allotypes of bovine immunoglobulin M with monoclonal antibodies. *Immunology* **63**, 569–574.

Naessens, J., Newson, J., McHugh, N., Howard, C. J., Parsons, K., Jones, B. (1990). Characterization of a bovine leucocyte differentiation antigen of 145,000 MW restricted to B lymphocytes. *Immunology* **69**, 525–530.

Naessens, J., Sileghem, M., MacHugh, N., Park, Y., Davis, W., Toye, P. (1992). Selection of BoCD25 monoclonal antibodies by screening mouse L-cells transfected with bovine p55-interleukin-2 (IL-2) receptor gene. *Immunology* **76**, 305–309.

Nagahata, H., Nochi, H., Sanada, Y., Tamoto, K., Noda, H., Kociba, G. J. (1994). Analysis of mononuclear cell functions in Holstein cattle with leukocyte adhesion deficiency. *Am. J. Vet. Res.* **55**, 1101–1106.

Nagahata, H., Higuchi, H., Noda, H., Tamoto, K., Kuwabara, M. (1996). Adhesiveness for extracellular matrixes and lysosomal enzyme release from normal and beta(2) integrin deficient bovine neutrophils. *Microbiol. Immunol.* **40**, 783–786.

Nakajima, Y., Momotani, E., Murakami, T., Ishikawa, Y., Morimatsu, M., Saito, M., Suzuki, H., Yasukawa, K. (1993). Induction of acute phase protein by recombinant human interleukin-6 (IL-6) in calves. *Vet. Immunol. Immunopathol.* **35**, 385–391.

Nielsen, K. H. (1977). Bovine reaginic antibody. III. Cross-reaction of human IgE and anti-bovine reaginic immunoglobulin antisera with sera from several species of mammals. *Can. J. Comp. Med.* **41**, 245–348.

Nielsen, K., Sheppard, J., Holmes, W., Tizard, I. (1978). Experimental bovine trypanosomiasis. Changes in serum immunoglobulins, complement and complement components in infected animals. *Immunology* **35**, 817–826.

Nilsson, P. H., van der Poel, J. J., van het Klooster, J. W., Davies, C. J., Joosten, I., Williams, J. L., Arriens, M. A. (1991). Comparison of BoLA class II serotyping and 1D-IEF in a Dutch herd of Holstein–Friesians. *Anim. Genet.* **22**, 47–48.

Oddgiersson, O., Simpson, S. P., Morgan, A. L. G., Ross, D. S., Spooner, R. L. (1988). Relationship between the bovine major histocompatibility complex (BoLA), erythrocyte markers and susceptibility to mastitis in Icelandic cattle. *Anim. Genet.* **19**, 11–16.

Ogunremi, O., Tabel, H. (1993a). A non-hemolytic assay for the activation of the alternative pathway of bovine complement. *Vet. Immunol. Immunopathol.* **38**, 155–167.

Ogunremi, O., Tabel, H. (1993b). Differences in the activity of the alternative pathway of complement in BALB/c and C57Bl/6 mice. *Exp. Clin. Immunogenet.* **10**, 31–37.

Ohmann, H. B., Davis, W. C., Babiuk, L. A. (1986). Surface antigen expression by bovine alveolar macrophages: functional correlation and influence of interferons *in vivo* and *in vitro*. *Immunobiology* **171**, 125–142.

Okragly, A. J., Hanby-Flarida, M., Mann, D., Baldwin, C. L. (1996). Bovine γ/δ T-cell proliferation is associated with self-derived molecules constitutively expressed *in vivo* on mononuclear phagocytes. *Immunology* **87**, 71–79.

Oliver, R. A., Brown, P., Spooner, R. L., Joosten, I., Williams, J. L. (1989). The analysis of antigen and DNA polymorphism within the bovine major histocompatibility complex: 1. The class I antigens. *Anim. Genet.* **20**, 31–41.

Olsen, S., Stevens, M. (1993). Effect of recombinant human cytokines on mitogen-induced bovine peripheral blood mononuclear cell proliferation. *Cytokine* **5**, 498–505.

Pahud, J. J., Mach, J. P. (1970). Identification of secretory IgA, free secretory piece and serum IgA in the ovine and caprine species. *Immunochemistry* **7**, 679–686.

Palmer, C., Thurmond, M., Picanso, J., Brewer, A. W., Bernoco, D. (1987). Susceptibility of cattle to bovine leukemia virus infection associated with BoLA type. Proc. of 91st Ann. Meeting of U.S. Anim. Health Assoc., pp. 218–228.

Pang, A. S. D., Aston, W. P. (1978). The alternative complement pathway in bovine serum: the isolation of a serum protein with factor B activity. *Immunochemistry* **15**, 529–534.

Park, C. A., Hines, H. C., Monke, D. R., Threlfall, W. T. (1993). Association between the bovine major histocompatibility complex and chronic posterior spinal paresis – a form of ankylosing spondylitis – in Holstein bulls. *Anim. Genet.* **24**, 53–58.

Park, C., Russ, I., Da, Y., Lewin, H. A. (1995). Genetic mapping of *F13SA* to BTA23 by sperm typing: difference in recombination rate between bulls in the *DYA-PRL* interval. *Genomics* **27**, 113–118.

Parng, C. L., Hansal, S., Goldsby, R. A., Osborne, B. A. (1995). Diversification of bovine λ-light chain genes. *Ann. NY Acad. Sci.* **764**, 155–157.

Parsons, K. R., Howard, C. J., Jones, B. V., Sopp, P. (1989). Investigation of bovine gut associated lymphoid tissue (GALT) using monoclonal antibodies against bovine lymphocytes. *Vet. Pathol.* **26**, 396–408.

Parsons, K. R., Bland, A. P., Hall, G. A. (1991). Follicle associated epithelium of the gut associated lymphoid tissue of cattle. *Vet. Pathol.* **28**, 22–29.

Pauli, U., Bertoni, G., Duerr, M., Peterhans, E. (1994). A bioassay for the detection of tumor necrosis factor from eight different species: evaluation of neutralization rates of a monoclonal antibody against human TNF-α. *J. Immunol. Methods* **171**, 263–265.

Perret, J. J., Raspe, E., Vassart, G., Parmentier, M. (1992). Cloning and functional expression of the canine anaphylatoxin C5a receptor. Evidence for high interspecies variability. *Biochem. J.* **288**, 911–917.

Porter, P. (1969). Transfer of immunoglobulins IgG, IgA and IgM to lacteal secretions in the parturient sow and their absorption by the neonatal piglet. *Biochim. Biophys. Acta* **181**, 381–392.

Randall, L. Wadhwa, M., Thorpe, R., Mire-Sluis, A. (1993). A novel sensitive bioassay for transforming growth factor beta. *J. Immunol. Methods* **164**, 61–67.

Reddy, D., Chitko-McKown, C., Reddy, P., Minocha, H., Blecha, F. (1993). Isolation and characterization of monoclonal antibodies to recombinant bovine interleukin-1 beta. *Vet. Immunol. Immunopathol.* **36**, 17–29.

Reddy, P., McVey, D., Chengappa, M., Blecha, F., Minocha, H., Baker, P. (1990). Bovine recombinant granulocyte-macrophage colony-stimulating factor enhancement of bovine neutrophil functions *in vitro*. *Am. J. Vet. Res.* **51**, 1395–1399.

Reeves, R., Spies, A., Nissen, M., Buck, C., Weinberg, A., Barr, P., Magnuson, N., Magnuson, J. (1986). Molecular cloning of a functional bovine interleukin 2 cDNA. *Proc. Natl Acad. Sci. USA* **83**, 3228–3232.

Reeves, R., Elton, T., Nissen, M., Lehn, D., Johnson, K. (1987). Posttranscriptional gene regulation and specific binding of the non histone protein HMG-1 by the 3′ untranslated region of bovine interleukin 2 cDNA. *Proc. Natl Acad. Sci. USA* **84**, 6531–6535.

Reid, K. B. M., Turner, M. W. (1994). Mammalian lectins in activation and clearance mechanisms involving the complement system. *Springer Semin. Immunopathol.* **15**, 307–325.

Rendel, J. (1957). Further studies on some antigenic characters of sheep blood determined by epistatic action of genes. *Acta Agric. Scand.* 224–259.

Renshaw, H. W., Everson, D. O. (1979). Classical and alternate complement pathway activities in paired dairy cow–newborn calf sera. *Comp. Immunol. Microbiol. Infect. Dis.* **1**, 259–267.

Renshaw, H. W., Eckblad, W. P., Tassinari, P. D., Everson, D. O. (1978). Levels of total haemolytic complement activity in paired cow-newborn calf sera. *Immunology* **34**, 801–805.

Reynolds, H. Y., Johnson, J. S. (1970). Quantitation of canine immunoglobulins. *J. Immunol.* **105**, 698–703.

Reynolds, J. D., Morris, B. D. (1983). The evolution and

involution of Peyer's patches in fetal and postnatal sheep. *Eur. J. Immunol.* **13**, 627–635.

Rice, C. E., Boulanger, P. (1952). The interchangeability of the complement components of different animal species. IV. In the haemolysis of rabbit erythrocytes sensitized with sheep antibody. *J. Immunol.* **68**, 197–205.

Rice, C. E., Crowson, C. N. (1950). The interchangeability of the complement components of different animal species. II. In the hemolysis of sheep erythrocytes sensitized with rabbit amboceptor. *J. Immunol.* **65**, 201–210.

Rice, C. E., Fuhamel, L. (1957). A comparison of the complement, conglutinin, and natural anti-sheep red cell antibody titers of the serum of newborn and older calves. *Can. J. Comp. Med.* **21**, 109–116.

Riley, J. F., West, G. B. (1953). The presence of histamine in tissue mast cells. *J. Physiol.* **120**, 528–537.

Richards, C., Scamurra, R., Murtaugh, M. (1995). Interleukin-6. In *Cytokines in Animal Health and Disease* (eds Myers, M., Murtaugh, M.), pp. 153–182. Marcel Dekker, Inc., New York, USA.

Robey, P., Young, M., Flanders, K., Roche, N., Kondaiah, P., Reddi, A., Termine, J., Sporn, M., Roberts, A. (1987). Osteoblasts synthesize and respond to transforming growth factor-type beta (TGF-beta) *in vitro*. *J. Cell Biol.* **105**, 457–463.

Rogivue, C., Car, B., Allmann-Iselin, R., Zwahlen, R., Walz, A. (1995). Bovine melanoma growth stimulatory activity: a new monocyte-macrophage-derived cytokine of the IL-8 family. Partial structure, function and expression in acute pulmonary inflammation. *Lab. Invest.* **72**, 689–695.

Romeo, D., Skerlavaj, B., Bolognesi, M., Gennaro, R. (1988). Structure and bactericidal activity of an antbiotic dodecapeptide purified from bovine neutrophils. *J. Biol. Chem.* **263**, 9573–9575.

Roth, J. A. (1994). Neutrophils and killer cells. In *Cell-mediated Immunity in Ruminants* (eds Goddeeris, B. M., Morrison, W. I.), pp. 127–142. CRC Press, Boca Raton, FL, USA.

Rouse, B. T., Ingram, D. G. (1970). The total protein and immunoglobulin profile of equine colostrum and milk. *Immunology* **19**, 901–907.

Ruitenberg, E. J., Gustowska, L., Elgersma, A., Ruitenberg, H. M. (1982). Effect of fixation on the light microscopical visualization of mast cells in the mucosa and connective tissue of the human duodenum. *Int. Arch. Allergy Appl. Immunol.* **67**, 233–238.

Rutten, V. P. M. G., Hoek, A., Muller, K. E. (1996). Identification of monoclonal antibodies with specificity to alpha or beta chains of beta(2) integrins using peripheral blood leucocytes of normal and bovine leucocyte adhesion deficient (BLAD) cattle. *Vet. Immunol. Immunopathol.* **52**, 341–345.

Sadavongvivad, C. (1970). Pharmacological significance of biogenic amines in the lungs: 5-hydroxytryptamine. *Br. J. Pharmac.* **38**, 353–365.

Saif, L. J., Redman, D. R., Smith, K. L., Theil, K. W. (1983). Passive immunity to bovine rotavirus in newborn calves fed colostrum supplement from immunized or non-immunized cows. *Infect. Immun.* **41**, 1118–1131.

Saini, S. S., Hein, W. R., Kaushik, A. (1997). A single predominantly expressed polymorphic Vh gene family, related to mammalian group I, class II, is identified in cattle. *Mol. Immunol.* **34**, 641–651.

Sakakibara, Y., Suiko, M., Fernando, P. H., Ohashi, T., Liu, M. C. (1994). Identification and characterization of a major bovine serum tyrosine-O-sulfate-binding protein as a complement factor H. *Cytotechnology* **14**, 97–107.

Sarikaputi, M., Morimatsu, M., Yamamoto, S., Syuto, B., Saito, M., Naiki, M. (1992). Latex agglutination test: a simple, rapid and practical method of bovine serum CRP determination. *Jap. J. Vet. Res.* **40**, 1–12.

Sasaki, T., Yonemasu, K. (1984). Comparative studies on biological activities of subcomponents C1q of the first component of human, bovine, mouse and guinea-pig complement. *Biochim. Biophys. Acta* **785**, 118–122.

Schlossman, S. F., Boumsell, L., Gilks, W. *et al.* (eds) (1995). *Leukocyte Typing V*. Oxford University Press, Oxford, UK.

Schmidt, A., Eispanier, R., Amselgruber, W., Sinowatz, F., Schams, D. (1994). Expression of insulin-like growth factor 1 (IGF-1) in the bovine oviduct during the oestrous cycle. *Exp. Clin. Endrocrinol.* **102**, 364–369.

Schmutz, S., Moker, J., Gallagher, D., Kappes, S., Womack, J. (1996). *In situ* hybridization mapping of LDHA and IGF2 to cattle chromosome 29. *Mamm. Genome* **7**, 473.

Scicchitano, R., Husband, A. J., Cripps, A. W. (1984). Immunoglobulin-containing cells and the origin of immunoglobulins in the respiratory tract of sheep. *Immunology* **52**, 529–537.

Scicchitano, R., Sheldrake, R. F., Husband, A. J. (1986). Origin of immunoglobulins in respiratory tract secretion and saliva of sheep. *Immunology* **58**, 315–321.

Selsted, M. E., Novotny, M. J., Morris, W. L., Tang, Y. Q., Smith, W., Cullor, J. S. (1992). Indolicidin, a novel bactericidal tridecapeptide amide from neutrophils. *J. Biol. Chem.* **267**, 4292–4295.

Selsted, M. E., Tang, Y., Morris, W. L., McGuire, P. A., Novotny, M. J., Smith, W., Henschen, A. H., Cullor, J. S. (1993). Purification, primary structures, and antibacterial activities of β-defensins, a new family of antimicrobial peptides from bovine neutrophils. *J. Biol. Chem.* **268**, 6641–6648.

Sethi, M., Tabel, H. (1990a). Activation of bovine monocytes and neutrophils by the Bb fragment of complement factor B: demonstration by the uptake of ^{3}H-deoxyglucose. *Can. J. Vet. Res.* **54**, 106–112.

Sethi, M., Tabel, H. (1990b). Fragment Bb of bovine complement factor B: stimulatory effect on the microbicidal activity of bovine monocytes. *Can. J. Vet. Res.* **54**, 405–409.

Sethi, M., Tabel, H. (1990c). Bb fragment of bovine complement factor B: stimulation of the oxidative burst in bovine monocytes. *Can. J. Vet. Res.* **54**, 410–414.

Shalhevet, D., Da, Y., Beever, J. E., van Eijk, M. J. T., Ma, R., Lewin, H. A., Gaskins, H. R. (1995). Genetic mapping of the *LMP2* proteasome subunit gene to the *BoLA* class IIb region. *Immunogenetics* **41**, 44–46.

Shanahan, F., Macniven, I., Dyck, N., Denburg, J. A., Bienenstock, J., Befus, A. D. (1987). Human lung mast cells: distribution and abundance of histochemically distinct subpopulations. *Int. Arch. Allergy Appl. Immunol.* **83**, 329–331.

Shaw, M. K., Tilney, L. G., Musoke, A. J., Teale, A. J. (1995). MHC class I molecules are an essential cell surface component involved in *Theileria parva* sporozoite binding to bovine lymphocytes. *J. Cell Sci.* **108**, 1587–1596.

Sheldrake, R. F., Husband, A. J. (1985). Intestinal uptake of intact maternal lymphocytes by neonatal rats and lambs. *Res. Vet. Sci.* **39**, 10–15.

Sheldrake, R. F., Husband, A. J., Watson, D. L., Cripps, A. W. (1984). Selective transport of serum-derived IgA into mucosal secretions. *J. Immunol.* **132**, 363–368.

Shuster, D. E., Kehrli, M. E. Jr, Ackerman, M. R., Gilbert, R. O.

(1992). Identification and prevalence of a genetic defect that causes leukocyte adhesion deficiency in Holstein cattle. *Proc. Natl Acad. Sci. USA* **89**, 9225–9229.

Sigurdardóttir, S., Borsch, C., Gustafsson, K., Andersson, L. (1991). Cloning and sequence analysis of 14 DRB alleles of the bovine major histocompatibility complex by using the polymerase chain reaction. *Anim. Genet.* **22**, 199–209.

Sileghem, M., Saya, R., Ellis, J., Flynn, J., Peel, J., Williams, D. (1992). Detection and neutralization of bovine tumor necrosis factor by a monoclonal antibody. *Hybridoma* **11**, 617–627.

Simpson-Morgan, M. W., Smeaton, T. C. (1972). The transfer of antibodies by neonates and adults. *Adv. Vet. Sci. Comp. Med.* **16**, 344–386.

Sinclair, M. C., Aitken, R. (1995). PCR strategies for isolation of the 5′ end of an immunoglobulin-encoding bovine cDNA. *Gene* 167, 285–289.

Sinclair, M. C., Gilchrist, J., Aitken, R. (1995). Molecular characterization of bovine V lambda regions. *J. Immunol.* **155**, 3068–3078.

Smith, M. A., Gerrie, L. M., Dunbar, B., Fothergill, J. E. (1982). Primary structure of bovine complement activation fragment C4a, the third anaphylatoxin. Purification and complete amino acid sequence. *Biochem. J.* **207**, 253–260.

Soames, C. J., Day, A. J., Sim, R. B. (1996). Predictions from sequence comparisons of residues of factor H in the interaction with complement component C3b. *Biochem. J.* **315** (Part 2), 523–531.

Solbu, H., Lie, O. (1990). Selection for disease resistance in dairy cattle. *Proceedings of the Fourth World Congress on Genetics Applied to Livestock Production*, **16**, 445–448.

Solbu, H., Spooner, R. L., Lie, O. (1982). A possible influence of the bovine major histocompatibility complex (BoLA) on mastitis. *Proceedings of the Second World Congress on Genetics Applied to Livestock Production*, **7**, 368–371.

Solinas-Toldo, S., Lengauer, C., Fries, R. (1995). Comparative genome map of human and cattle. *Genomics* **27**, 489–496.

Sopp, P., Howard, C. J. (1997). Cross reactivity of monoclonal antibodies to defined human leukocyte differentiation antigens with bovine cells. *Vet. Immunol. Immunopathol.* **56**, 11–25.

Splitter, G., Morrison, W. I. (1991). Individual antigens of cattle. Antigens expressed predominantly on monocytes and granulocytes: identification of bovine CD11b and CD11c. *Vet. Immunol. Immunopathol.* **27**, 87–90.

Spooner, R. L., Maddy, A. H. (1971). The isolation of ox red cell membrane antigens: antigens associated with sialoprotein. *Immunology* **21**, 809–816.

Spooner, R. L., Morgan, A. L. G. (1981). Analysis of BoLA w6. Evidence for multiple subgroups. *Tissue Antigens* **17**, 178–188.

Spooner, R. L., Oliver, R. A., Sales, D. I., McCoubrey, C. M., Millar, P., Morgan, A. G., Amorena, B., Bailey, E., Bernoco, D., Brandon, M., Bull, R. W., Caldwell, J., Cwik, S., van Dam, R. H., Dodd, J., Gahne, B., Grosclaude, F., Hall, J. G., Hines, H., Leveziel, H., Newman, M. J., Stear, M. J., Stone, W. H., Vaiman, M. (1979). Analysis of alloantisera against bovine lymphocytes. Joint report of the 1st International Bovine Lymphocyte Antigen (BoLA) Workshop. *Anim. Blood Groups Biochem. Genet.* **15**, 63–86.

Sprague, L. M. (1958). On the recognition and inheritance of the soluble blood group property 'Oc' of cattle. *Genetics* **43**, 906–912.

Spurzem, J. R., Sacco, O., Rossi, G. A., Beckmann, J. D., Rennard, S. I. (1992). Regulation of major histocompatibility complex class II gene expression on bovine bronchial epithelial cells. *J. Lab. Clin. Med.* **120**, 94–102.

Stear, M. J., Dimmock, C. K., Newman, M. J., Nicholas, F. W. (1988a). BoLA antigens are associated with increased frequency of persistent lymphocytosis in bovine leukaemia virus infected cattle and with increased incidence of antibodies to bovine leukemia virus. *Anim. Genet.* **19**, 151–158.

Stear, M. J., Pokorny, T. S., Muggli, N. E., Stone, R. T. (1988b). Breed differences in the distribution of BoLA-A locus antigens in American cattle. *Anim. Genet.* **19**, 171–176.

Stear, M. J., Tierney, T. J., Baldock, F. C., Brown, S. C., Nicholas, F. W., Rudder, T. H. (1988c). Class I antigens of the bovine major histocompatibility system are weakly associated with variation in faecal worm egg counts in naturally infected cattle. *Anim. Genet.* **19**, 115–122.

Stear, M. J., Nicholas, F. W., Brown, S. C., Holroyd, R. G. (1989). Class I antigens of the bovine major histocompatibility system and resistance to the cattle tick (*Boophilus microplus*) assessed in three different seasons. *Vet. Parasitol.* **31**, 305–315.

Stear, M. J., Hetzel, D. J. S., Brown, S. C., Gershwin, L. J., Mackinnon, M. J., Nicholas, F. W. (1990). The relationships among ecto- and endoparasite levels, class I antigens of the bovine major histocompatibility system, immunoglobulin E levels and weight gain. *Vet. Parasitol.* **34**, 303–321.

Sterpetti, A., Cucina, A., Morena, A., Di-Donna, S., D'Angelo, L., Cavallarro, A., Stipa, S. (1993). Shear stress increases the release of interleukin-1 and interleukin-6 by aortic endothelial cells. *Surgery* **114**, 911–914.

Stevens, M., Olsen, S. (1994). *In vitro* effects of live and killed *Brucella abortus* on bovine cytokine and prostaglandin E2 production. *Vet. Immunol. Immunopathol.* **40**, 149–161.

Stewart, R., Kashxur, T., Marsden, P. (1996). Vascular endothelial platelet endothelial adhesion molecule-1 (PECAM-1) expression is decreased by TNF-alpha and IFN-gamma. *J. Immunol.* **156**, 1221–1228.

Stone, R. T., Muggli-Cockett, N. E. (1993). *BoLA-DIB*: species distribution, linkage with *DOB*, and Northern analysis. *Anim. Genet.* **24**, 41–45.

Stormont, C. (1978). The early history of cattle blood groups. *Immunogenetics* **6**, 1–15.

Strang, C. J., Slayter, H. S., Lachmann, P. J., Davis, A. E. 3rd (1986). Ultrastructure and composition of bovine conglutinin. *Biochem. J.* **234**, 381–389.

Suliman, H., Majiwa, P., Feldman, B., Logan-Henfrey, L. (1996). Molecular cloning of a cDNA encoding bovine erythropoietin and analysis of its transcription in selected bovine tissues. *Gene* **171**, 275–280.

Suzuki, Y., Yin, Y., Makino, M., Kurimura, T., Wakamiya, N. (1993). Cloning and sequencing of a cDNA coding for bovine conglutinin. *Biochem. Biophys. Res. Commun.* **191**, 335–342.

Symons, D. B. A., Clarkson, C. A., Milstein, C. P., Brown, N. R., Beale, D. (1987). DNA sequence analysis of two bovine immunoglobulin C_H gamma pseudogenes. *J. Immunogenet.* **14**, 273–283.

Symons, D. B. A., Clarkson, C. A., Beale, D. (1989). Structure of bovine immunoglobulin constant region heavy chain gamma 1 and gamma 2 genes. *Mol. Immunol.* **26**, 841–850.

Tabel, H. (1996). Alternative pathway of complement in ruminants: role in infection. *Vet. Immunol. Immunopathol.* **54**, 117–121.

Tabel, H., Menger, M., Aston, W. P., Cochran, M. (1984). Alternative pathway of bovine complement: Concentration of factor B, hemolytic activity and heritability. *Vet. Immunol. Immunopathol.* **5**, 389–398.

Tabel, H., Groat, H. J., Kraay, G. J., Aston, W. P. (1990). Genetic polymorphism of complement factor H in cattle. *Anim. Genet.* **21**, 123–128.

Takeuchi, N., Ishiguro, N., Shinagawa, M. (1992). Molecular cloning and sequence analysis of bovine T-cell receptor γ and δ chain genes. *Immunogenetics* 35, 89–96.

Tanaka, A., Ishiguro, N., Shinagawa, M. (1990). Sequence and diversity of bovine T-cell receptor b-chain genes. *Immunogenetics* **32**, 263–271.

Tatcher, E. F., Gerschwin, L. J. (1988). Generation and characterization of murine monoclonal antibodies specific for bovine immunoglobulin E. *Vet. Immunol. Immunopathol.* **18**, 53–66.

Taylor, B. C., Choi, K. Y., Scibienski, R. J., Moore, P. F., Stott, J. L. (1993). Differential expression of bovine MHC class II antigens identified by monoclonal antibodies. *J. Leukoc. Biol.* 53, 479–489.

Taylor, B. C., Dellinger, J. D., Cullor, J. S., Stott, J. L. (1994). Bovine milk lymphocytes display the phenotype of memory T cells and are predominantly $CD8^+$. *Cell Immunol.* **156**, 245–253.

Taylor, G., Thomas, L. H., Wyld, S. G., Furze, J., Sopp, P., Howard, C. J. (1995). Role of lymphocyte subsets in recovery from respiratory syncytial virus infection of calves. *J. Virol.* **69**, 6658–6664.

Taylor, K., Lutje, V., Mertens, B. (1996). Nitric oxide synthesis is depressed in *Bos indicus* cattle infected with *Trypanosoma congolense* and *Trypanosoma vivax* and does not mediate Y-cell suppression. *Infect. Immun.* **64**, 4115–4122.

Teale, A. J., Kemp, S. J. (1987). A study of BoLA class II antigens with $BoT4^+$ T lymphocyte clones. *Anim. Genet.* 18, 17–28.

Teutsch, M. R., Beever, J. E., Stewart, J. A., Schook, L. B., Lewin, H. A. (1989). Linkage of complement factor B gene to the bovine major histocompatibility complex. *Anim. Genet.* **20**, 427.

Theile, O. W., Oulevey, J., Hennemuth, K., Koch, J. (1979). Studies on the chemical nature of the lipidic J blood-group substance of cattle. *Anim. Blood Groups Biochem. Genet.* **10**, 1–9.

Threadgill, D. S., Threadgill, D. W., Moll, Y. D., Weiss, J. A., Zhang, N., Davey, H. W., Wildeman, A. G., Womack, J. E. (1994). Syntenic assignment of human chromosome 1 homologous loci in the bovine. *Genomics* **22**, 626–630.

Tischer, E., Gospodarowicz, D., Mitchell, R., Silva, M., Schilling, J., Lau, K., Crisp, T., Fiddes, J., Abraham, J. (1989). Vascular endothelial growth factor: a new member of the platelet-derived growth factor gene family. *Biochem. Biophys. Res. Commun.* **165**, 1198–1206.

Tizard, I. (1995). Interferons. In *Cytokines in Animal Health and Disease* (eds Myers, M., Murtaugh, M.), pp. 1–57. Marcel Dekker, Inc., New York, USA.

Tobin-Janzen, T. C., Womack, J. E. (1992). Comparative mapping of IGHG1, IGHM, FES, and FOS in domestic cattle. *Immunogenetics* **36**, 157–165.

Tokuyama, Y., Tokuyama, H. (1993). Purification and identification of TGF-beta 2-related growth factor from bovine colostrum. *J. Dairy Res.* **60**, 99–109.

Totte, P., De Gree, A., Werenne, J. (1993). Role of interferons in infectious diseases in the bovine species: effect on viruses and rickettsias. *Revue Elev. Med. Vet. Pays. Trop.* **46**, 83–86.

Toye, P. G., MacHugh, N. D., Bensaid, A. M., Alberti, S., Teale, A. J., Morrison, W. I. (1990). Transfection into mouse L cells of genes encoding two serologically and functionally distinct bovine class I MHC molecules from a MHC-homozygous animal: evidence for a second class I locus in cattle. *Immunology* **70**, 20–26.

Triglia, R. P., Linscott, W. D. (1980). Titers of nine complement components, conglutinin and C3b-inactivator in adult and fetal bovine sera. *Mol. Immunol.* **17**, 741–748.

Tucker, E. M., Méténier, L., Grosclaude, L., Clarke, S. W., Kilgour, L. (1986). Monoclonal antibodies to bovine blood group antigens. *Anim. Genet.* **17**, 3–13.

Våge, D. I., Lingaas, F., Spooner, R. L., Arnet, E. F., Lie, O. (1992). A study on association between mastitis and serologically defined class I bovine lymphocyte antigens (BoLA-A) in Norwegian cows. *Anim. Genet.* **23**, 533–536.

van Eijk, M. J. T., Stewart-Haynes, J. A., Beever, J. E., Fernando, R. L., Lewin, H. A. (1992a). Development of persistent lymphocytosis in cattle is closely associated with *DRB2*. *Immunogenetics* **37**, 64–68.

van Eijk, M. J. T., Stewart-Haynes, J. A., Lewin, H. A. (1992b). Extensive polymorphism of the BoLA-DRB3 gene distinguished by PCR-RFLP. *Anim. Genet.* **23**, 483–496.

van Eijk, M. J. T., Russ, I., Lewin, H. A. (1993). Order of bovine *DRB3*, *DYA*, and *PRL* determined by sperm typing. *Mamm. Genome* **4**, 113–118.

van Eijk, M. J. T., Beever, J. E., Da, Y., Stewart, J. A., Nicholaides, G. E., Green, C. A., Lewin, H. A. (1995). Genetic mapping *BoLA-A*, *CYP21*, *DRB3*, *DYA* and *PRL* on BTA23. *Mamm. Genome* **6**, 151–152.

van Nostrand, W. E., Cunningham, D. D. (1987). Purification of a proteinase inhibitor from bovine serum with C1-inhibitor activity. *Biochim. Biophys. Acta* **923**, 167–175.

van Obberghen-Schilling, E., Kondaiah, P., Ludwig, R., Sporn, M., Baker, C. (1987). Complementary deoxyribonucleic acid cloning of bovine transforming growth factor-beta-1. *Mol. Endocrinol.* **1**, 693–698.

van Voorhis, W. C., Hair, L. S., Steinman, R. M., Kaplan, G. (1982). Human dendritic cells: enrichment and characterisation from peripheral blood. *J. Exp. Med.* **155**, 1172–1187.

van Zaane, D., Ijzerman, J. (1984). Monoclonal antibodies against bovine immunoglobulin and their use in isotype specific ELISAs for rotavirus antibody. *J. Immunol. Methods* **72**, 427–441.

Velan, B., Cohen, S., Grosfeld, H., Leitner, M., Shafferman, A. (1985). Bovine interferon alpha genes. *J. Biol. Chem.* **260**, 5498–5504.

Verbeet, M. P., Vermeer, H., Warmerdam, G. C. M., de Boer, H. A., Lee, S. H. (1995). Cloning and characterization of the bovine polymeric immunoglobulin receptor-encoding cDNA. *Gene* **164**, 329–333.

Viuff, B., Ostergård, H., Aasted, B., Kristensen, B. (1991). One-dimensional isoelectric focusing and immunoblotting of bovine major histocompatibility complex (BoLA) class I molecules and correlation with class I serology. *Anim. Genet.* **22**, 147–154.

Watkins, D. I., Shadduck, J. A., Rudd, C. E., Stone, M. E., Lewin, H. A., Letvin, N. L. (1989a). Isoelectric focusing of bovine major histocompatibility complex class II molecules. *Eur. J. Immunol.* **19**, 567–570.

Watkins, D. I., Shadduck, J. A., Stone, M. E., Lewin, H. A., Letvin, N. L. (1989b). Isoelectric focusing of bovine major histocompatibility complex class I molecules. *J. Immunogenet.* **16**, 233–245.

Watson, A., Hogan, T., Hahnel, A., Wiemer, K., Schultz, G. (1992). Expression of growth factor ligand and receptor genes in the preimplantation bovine embryo. *Mol. Reprod. Dev.* **31**, 87–95.

Watson, D. L., Brandon, M. R., Lascelles, A. K. (1972). Concentrations of immunoglobulins in mammary secretions of ruminants during involution with particular reference to selective transfer of IgG1. *Aust. J. Exp. Biol. Med. Sci.* **50**, 535–539.

de Weck, A. L. (1995). What can we learn from the allergic zoo? *Int. Arch. Allergy Immunol.* **107**, 13–18.

Weigel, K. A., Freeman, A. E., Kehrli, M. E. Jr, Stear, M. J., Kelley, D. H. (1990). Association of class I bovine lymphocyte antigen complex alleles with health and production traits in dairy cattle. *J. Dairy Sci.* **72**, 2538–2546.

Weinberg, A., Shaw, J., Paetkau, V., Bleackley, R., Magnuson, N., Reeves, R., Magnuson, J. (1988). Cloning of cDNA for the bovine IL-2 receptor (bovine Tac antigen). *Immunology* **63**, 603–610.

Wempe, F., Henschen, A., Scheit, K. (1991). Gene expression and cDNA cloning identified a major basic protein constituent of bovine seminal plasma as bovine monocyte-chemoattractant protein-1. (MCP-1). *DNA Cell Biol.* **10**, 671–679.

Wempe, F., Kuhlmann, J., Scheit, K. (1994a). Characterization of the bovine monocyte chemoattractant protein-1 gene. *Biochem. Biophys. Res. Commun.* **202**, 1272–1279.

Wempe, F., Hanes, J., Scheit, K. (1994b). Cloning of the gene for bovine monocyte chemoattractant protein-2. *DNA Cell Biol.* **13**, 1–8.

Werling, D., Sileghem, M., Lutz, H., Langhans, W. (1995). Effect of bovine leukemia virus infection on bovine peripheral blood monocyte responsiveness to lipopolysaccharide stimulation *in vitro*. *Vet. Immunol. Immunopathol.* **48**, 77–88.

WHO (1978). Proposed rules for the designation of immunoglobulins of animal origin. *WHO Bull.* **56**, 815–817.

Wijngaard, P. L., Metzelaar, M. J., MacHugh, N. D., Morrison, W. I., Clevers, H. C. (1992). Molecular characterization of the WC1 antigen expressed specifically on bovine CD4-CD8-gamma delta T lymphocytes. *J. Immunol.* **149**, 3273–3277.

Wijngaard, P. L., MacHugh, N. D., Metzelaar, M. J., Romberg, S., Bensaid, A., Pepin, L., Davis, W. C., Clevers, H. C. (1994). Members of the novel WC1 gene family are differentially expressed on subsets of bovine CD4-CD8-gamma delta T lymphocytes. *J. Immunol.* **152**, 3476–3482.

Williams, D. J. L., Newson, J., Naessens, J. (1990). Quantitation of bovine immunoglobulin isotypes and allotypes using monoclonal antibodies. *Vet. Immunol. Immunopathol.* **24**, 267–283.

Winkler, G. C. (1988). Pulmonary intravascular macrophages in domestic animal species: review of structural and functional properties. *Am. J. Anat.* **181**, 217–234.

Wood, P., Rothel, J., McWaters, P., Jones, S. (1990). Production and characterization of monoclonal antibodies specific for bovine gamma-interferon. *Vet. Immunol. Immunopathol.* **25**, 37–46.

Worku, M., Paape, M. J., Marquadt, W. W. (1994). Modulation of Fc receptors for IgG on bovine polymorphonuclear neutrophils by interferon-γ through de novo RNA transcription and protein synthesis. *Am. J. Vet. Res.* **55**, 234–238.

Worku, M., Paape, M. J., Di Carlo, A., Kehrli, M. E. Jr, Marquardt, W. W. (1995). Complement component C3b and immunoglobulin Fc receptors on neutrophils from calves with leucocyte adhesion deficiency. *Am. J. Vet. Res.* **56**, 435–439.

Wyatt, C. R., Madruga, C., Cluff, C., Parish, S., Hamilton, M. J., Goff, W., Davis, W. C. (1994). Differential distribution of gamma delta T-cell receptor lymphocyte subpopulations in blood and spleen of young and adult cattle. *Vet. Immunol. Immunopathol.* **40**, 187–199.

Wyatt, C. R., Brackett, E. J., Perryman, L. E., Davis, W. C. (1996). Identification of $\gamma\delta$ T lymphocyte subsets that populate calf ileal mucosa after birth. *Vet. Immunol. Immunopathol.* **52**, 91–103.

Yonemasu, K., Sasaki, T., Shinkai, H. (1980). Purification and characterization of subcomponent C1q of the first component of bovine complement. *J. Biochem.* **88**, 1545–1554.

Yoo, J., Stone, R., Kappes, S., Beattie, C. (1994). Linkage analysis of bovine interleukin 1 receptor types I and II (IL-1R I, II). *Mamm. Genome* **5**, 820–821.

Yoo, J., Ponce de Leon, F., Stone, R., Beattie, C. (1995). Cloning and chromosomal assignment of the bovine interleukin-2 receptor alpha (IL-2R alpha) gene. *Mamm. Genome* **6**, 751–753.

Yoo, J., Stone, R., Beattie, C. (1996a). Cloning and characterization of the bovine Fas. *DNA Cell Biol.* **15**, 227–234.

Yoo, J., Stone, R., Solinas-Toldo, S., Fries, R., Beattie, C. (1996b). Cloning and chromosomal mapping of bovine interleukin-2 receptor gamma gene. *DNA Cell Biol.* **15**, 453–459.

Xu, A., Clarke, T. J., Teutsch, M. R., Schook, L. B., Lewin, H. A. (1991). Sequencing and genetic analysis of a bovine DQB cDNA clone. *Anim. Genet.* **22**, 381–389.

Xu, A., McKenna, K., Lewin, H. A. (1993a). Sequencing and genetic analysis of a bovine *DQA* cDNA clone. *Immunogenetics* **37**, 231–234.

Xu, A., van Eijk, M. J. T., Park, C., Lewin, H. A. (1993b). Polymorphism in *BoLA-DRB3* exon 2 correlates with resistance to persistent lymphocytosis caused by bovine leukemia virus. *J. Immunol.* **151**, 6977–6985.

Xu, A., Park, C., Lewin, H. A. (1994). Both *DQB* genes are expressed in *BoLA* haplotypes carrying a duplicated *DQ* region. *Immunogenetics* **39**, 316–321.

Zanotti, M., Poli, G., Ponti, W., Polli, M., Rocchi, M., Bolzani, E., Longeri, M., Russo, S., Lewin, H. A., van Eijk, M. J. T. (1996). Association of BoLA class II haplotypes with subclinical progression of bovine leukaemia virus infection in Holstein-Friesian cattle. *Anim. Genet.* **27**, 337–341.

Zarlenga, D., Canals, A., Aschenbrenner, R., Gasbarre, L. (1995a). Enzymatic amplification and molecular cloning of cDNA encoding the small and large subunit of bovine interleukin 12. *Biochim. Biophys. Acta* **1270**, 215–217.

Zarlenga, D., Canals, A., Gasbarre, L. (1995b). Method for constructing internal standards for use in competitive PCR. *BioTechniques* **19**, 324–326.

Zhou, J., Hikono, H., Ohtaki, M., Kubota, T., Sakurai, M. (1994). Cloning and characterization of cDNAs encoding two normal isoforms of bovine stem cell factor. *Biochim. Biophys. Acta* **1223**, 148–150.

Zurfluh, L., Bolten, S., Byatt, J., McGrath, M., Tou, J., Zupec, M., Krivi, G. (1990). Isolation of genomic sequence encoding a biologically active bovine TGF-alpha protein. *Growth Factors* **3**, 257–266.

XIV SHEEP IMMUNOLOGY AND GOAT PECULIARITIES

1. Introduction

Small ruminants, which include sheep and goats, are important agricultural species with an estimated global population of 1.7 billion (Morris, 1995). Sheep have often been used as animal models for investigating the physiology of reproduction, endocrinology, cardiovascular function, pulmonary function, and the immune system. An understanding of host functions, relative to disease, and extensive knowledge of the physiology of this small ruminant contributes much to our ability to work effectively with this species for agrarian or scientific purposes. In fact, the sheep model has proven extremely useful for immunological research when addressing questions regarding the physiology of the immune system (reviewed by Hein, 1995).

Several sections in the present chapter exemplify the contributions made by the sheep model to the basic concepts of immunology. The sheep model has contributed greatly to our understanding of lymphocyte trafficking and the physiology of cell movement through lymphoid and nonlymphoid tissues (see Section 4). The surgical accessibility of the fetal lamb and its isolation from the maternal environment has offered an excellent opportunity to investigate the development of immunocompetence. This has offered rare insights into the ontogeny of the immune system (Section 10) and into the events that accompany the development of the T-cell receptor (Section 6), the Ig repertoire (Section 7), and self tolerance (Section 18).

Investigations of ovine immunology have also provided valuable insights into the evolution of the immune system. Investigations of the $\gamma\delta$ T-cell population in sheep have revealed marked structural and functional differences between ruminants and rodents (see Sections 5 and 18). Further investigations of these differences should enhance our understanding of the role that this enigmatic T-cell population plays in the physiology of the immune system and their contribution to host defense mechanisms. The sheep model also has a long history in contributing to our understanding of Peyer's patch structure and function. Investigations of B-cell development in this gut-associated lymphoid tissue have revealed that many mammalian species may utilize strategies for Ig repertoire diversification that are distinctly different from the model based on studies in rodents (reviewed by Griebel and Hein, 1996). In fact, rodents may prove to be the exception rather than the norm when studying mechanisms of mammalian B-cell development.

Extensive conservation of the structure and function of the mammalian immune system is also apparent in sheep. This is apparent in the conservation of surface molecules expressed on leukocytes (Section 3), the structure and function of numerous cytokines (Section 5), the organization of the major histocompatibility complex (Section 8), and the complement cascade (Section 14). This conservation of structure and function confirms that sheep provide a valuable model for investigating host responses to a wide variety of pathogens. In fact, as illustrated by studies of Maedi-Visna virus (Section 16) and bovine leukemia virus (Section 17) the sheep offers unique advantages when investigating the interaction between a pathogen and the immune system.

There are unique aspects to the development of the ruminant immune system that must be considered when using sheep as an immunological model. The passive transfer of immunity in the neonate (Section 11) has a major impact on the immune competence of the neonate (Section 12). Gut-associated lymphoid tissues also play a critical role in B-cell development (Section 2). Thus, any event that alters ileal Peyer's patch function may alter development of the Ig receptor repertoire.

In conclusion, the sheep model provides many unique advantages in the exploration of the immune system. An understanding of immune system structure and function in a variety of species will bring a much broader understanding of the physiological constraints that have influenced the evolution of this complex system.

2. Lymphoid Organs and their Anatomical Distribution

Introduction

A distinction should be made between the organized and diffuse lymphoid tissue, the latter found scattered throughout the loose connective tissues of the body. The

Handbook of Vertebrate Immunology
ISBN 0-12-546401-0

quantitative and functional contribution of the diffuse lymphoid tissue to the immune system tissue is considerable, exemplified by the diffuse lymphoid tissue of the gut, the respiratory tract and the skin. The leukocyte populations found in the epithelia of these organs likewise represent a very significant component. In this section, however, only lymphoid tissues fashioned into organized structures are dealt with.

Acquisition of immunological competence involves the ordered emergence of lymphoid organs which in sheep, as in other ruminants, develop in the fetus. The syndesmochorial placentation in sheep provides a protected environment for the fetus, allowing a development of lymphoid tissues in a milieu devoid of external antigens including antibodies from the ewe. The shaping of the early immune system in the sheep is thus determined by factors in the fetus itself. The extent to which the various components of the immune system develop during gestation in sheep presents a telling story about the endogenous potential.

Fetal haemopoiesis

The yolk sac, liver, thymus, spleen, bone marrow, lymph nodes and gut-associated lymphoid organs all contribute components to the haemopoietic tissues (Figure XIV.2.1). The primitive haemopoiesis in the yolk sac between 17 and 27 days gestation is essentially an erythropoiesis but may involve cells of the myeloid series. Thus, leukocytes reactive for CD45 can be found (Landsverk *et al.*, 1992) and leukocytic phagocytosis is an early capability. In embryos as young as 24 days gestation there is a population of free-floating macrophage cells which can engulf bacteria and other foreign material (Morris, 1984).

Whereas the emergence of truly lymphopoietic tissues with recognizable T and B lineages is a later event (35–40 days gestation), the interim period is dominated by intense haemopoiesis in the liver from about 22 days onwards. The haemopoiesis in the liver of sheep seems to be restricted to erythropoiesis, myelopoiesis and megakaryopoiesis (Miyasaka and Morris, 1988) even in later fetal age, although scattered cells reactive for IgM or CD5 can be found beyond 60 days.

Thymus

The thymus of sheep arises mainly from the ventral diverticulum of the paired third and fourth pharyngeal pouch. The endoderm gives rise to the epithelial reticulum that distinguishes the thymus from other lymphoid organs. Sheep retain the cervical part of the thymus in addition to its thoracic portion as the organ develops. The thymus becomes lymphoid about 35 days gestation when cells with the $CD45^+$ and $CD5^+$ phenotype appear and the particular association between lymphocytes and epithelial cells is established (Mackay *et al.*, 1986a; Maddox *et al.*, 1987b). A subdivision into cortex and medulla (Mackay *et al.*, 1986a) and primitive Hassal's corpuscles (Maddox *et al.*, 1987a) become apparent at 50 days. The particular association between $\gamma\delta$ T cells and Hassal's corpuscles apparent early in ontogeny has been noted (McClure *et al.*, 1989; Hein and Mackay, 1991).

Spleen

The spleen emerges at about 40 days gestation as a pale organ which is difficult to detect in its position in the dorsal mesentery. It takes 10 days before an increasing erythropoiesis gives the organ its typical dark red colour. The early spleen is remarkably diverse in the way it provides for the various haemopoietic cell lines supporting erythropoiesis, myelopoiesis and lymphopoiesis (Al Salami, 1995). The first IgM^+ cells have been reported at 48 days (Press *et al.*, 1993) and the spleen represents the first organ to provide an expansion of the B cell line. There is an early expression of the heavy chain of IgM at the mRNA level (C. M. Press and T. Landsverk, unpublished data) but at the present it is not known whether rearrangement of immunoglobulin genes takes place in the spleen. B-cell expansion is very significant in the early spleen and prior to 77 days of gestation IgM^+ cells account for over 80% of the emergent while pulp area (Press *et al.*, 1993).

The possibility that this early B-cell population might in fact be B-1 cells has attracted interest, however, FACS analysis has indicated just 1–2% $CD5^+$ B cells at 81 days (Press *et al.*, 1993). It is possible that sheep possess predominantly the B-1b ($CD5^-$) phenotype. Doubts exist, however, about the relevance of CD5 as a marker for the B-1 cells in sheep (Griebel and Ferrari, 1995). The role of the spleen in B-cell ontogeny is being addressed. Elimination of B cells by injection of anti-IgM antibody at 63 days results in a persistent deficiency of such cells (Press *et al.*, 1996), although it is uncertain whether the effect is due purely to an effect on the spleen populations. It is possible that the spleen represents a 'dead end' for colonizing cells as in the chicken (Reynaud *et al.*, 1992).

It is noteworthy that B-cell expansion at 50–70 days precedes, by a month, the emergence of the ileal Peyer's patch, which contributes the bulk of B cells to the postnatal immune system (reviewed in Section 10). It is of further interest that the B-cell expansion in the early spleen ($>$ 60 days) occurs before the major seeding of lymph nodes and gut with lymphocytes. It is not known whether the migrants to these organs actually originate in the spleen, but the metastatic behaviour of lymphocytes appears to be intrinsic and established early, although lymphocyte recirculation in its proper sense first commences at about 80 days (Pearson *et al.*, 1976).

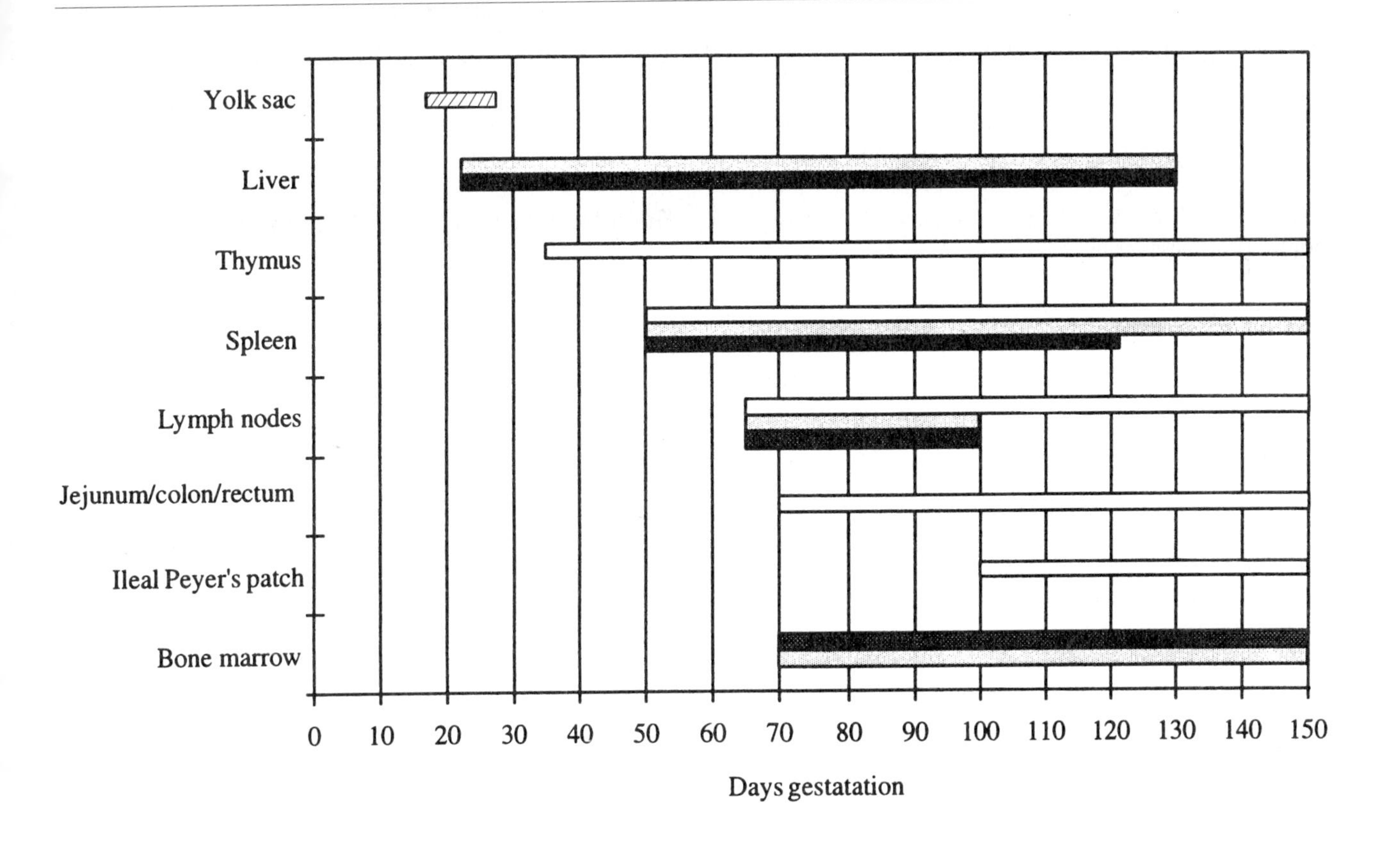

Fig. XIV.2.1 Haematopoietic tissues in fetal sheep.

Lymph nodes

At such an early stage (50–55 days gestation) no functional lymphatic system exists and the first lymph nodes to appear, the cervical superficial lymph nodes can first be discerned at 55–60 days gestation (Cole and Morris, 1973). The mediastinal lymph nodes are also early and in the next few days nodes such as the portal, subiliaci and jejunal develop (Morris and Al Salami, 1987). A connection of lymph nodes to peripheral and central lymphatics is established at 65–70 days gestation. This is when a functional blood and lymph network begins to function.

The first lymphocytes to appear in lymph nodes are mainly T cells; B cells appear to be somewhat later immigrants, with aggregates of B cells first appearing at about 80 days gestation, just preceded by the subdivision of the lymph node into cortical and medullary areas at 75–80 days (Al Salami, 1985; Al Salami *et al.*, 1985). These early B cells associate with specializations of stromal elements in the form of 5′-nucleotidase reactive cells, precursors of follicular dendritic cells (Halleraker *et al.*, 1994). About 10 days later such follicles appear also in the spleen. The emergence of primary lymphoid follicles in lymphoid organs such as the lymph nodes and spleen is a particularly noteworthy feature (Halleraker *et al.*, 1994).

Bone marrow

The bone marrow of sheep is involved in lymphopoiesis only to a minor extent. It assumes significance as a haemopoietic organ from about 70 days gestation. During fetal life the bone marrow rarely contains more than 5% cells with an identifiable lymphoid phenotype (Miyasaka and Morris, 1988). The extent of bone marrow erythropoiesis, granulopoiesis and megakaryopoiesis increases throughout fetal life so that from the second half of gestation it is the most important haemopoietic organ. Although lymphocytes and large blast resembling lymphoid precursor cells have been identified in the bone marrow, it is likely that they were migrant cells (Al Salami, 1985).

Gut-associated lymphoid tissue

The origin of the early migrants to the intestines is not known but they constitute both IgM- and CD5-reactive cells. Early seeding appears to be confined to intestinal segments such as sites close to the ileocaecal valve, with populations appearing both proximal and distal to the valve and to the segment in the rectum close to the rectoanal junction (Aleksandersen *et al.*, 1991). Whereas lymphocyte accumulations close to the ileacaecal valve appear around 67 days gestation (M. Aleksandersen and T. Landsverk, unpublished data) collections in the rectum are found at about 70 days (Aleksandersen *et al.*, 1991). Immigration to the jejunum also occurs at about this time (Reynolds and Morris, 1983). Although it appears that these early lymphocyte accumulations contain proliferating cells of both B and T lineages, no stromal specialization has been reported before the emergence of true follicles from about 100 days.

The distinction between the lymphoid tissues arising close to the ileo-caecal opening and the ileal Peyer's patch proper is an important one because the former precedes the latter by more than a month. The ontogeny and function of the ileal Peyer's patch is dealt with elsewhere (Immunoglobulins) but its rudiment can be identified at 100 days gestation by histochemical attributes which include carbonic anhydrase activity in the overlying epithelium (Landsverk *et al.*, 1987). This enzyme reactivity marks the commencement of transcytosis and the shedding of 50 nm membrane-bound particles to the lymphoid tissues (Landsverk *et al.*, 1990; Nicander *et al.*, 1991). This coincides with the first appearance of clusters of IgM^+ cells beneath the epithelium (Press *et al.*, 1992). This feature distinguishes the ileal Peyer's patch from the jejunal Peyer's patches (Landsverk, 1987). A full complement of stromal and accessory cells develop in all the intestinal sites (Halleraker *et al.*, 1990; Aleksandersen *et al.*, 1991; Nicander *et al.*, 1991) and the magnitude of the lymphopoiesis, in addition to other features, set the intestinal follicles apart from those of lymph nodes and spleen (Reynolds, 1985). It is thus apparent that the intestinal habitat contains unique components that mature lymphocyte proliferation prior to exposure to the postnatal antigen-rich milieu.

The intestinal mucosa is not the only mucosa to be equipped with follicles. The palatine tonsils develop from the second pharyngeal pouch and are significant, although often ignored lymphoid organs that develop in the fetus (T. Landsverk, unpublished data). Characteristic for this organ is a particular lympho-epithelial relationship with similarities to that of M cells (Oláh *et al.*, 1988). We do not know to what extent there is development of lymphoid follicles in the respiratory tract, the abomasum, the urogenital tract and around the eyes in the fetus, but they are present in the postnatal animal (Nicander *et al.*, 1993). The term mucosa-associated lymphoid tissue (MALT) is therefore appropriate. The constitutive character of the follicles in the respiratory tract may be questioned, and their emergence following infection by mycoplasma has been reported (Jones *et al.*, 1982).

Hemal nodes and mesenteric 'milk spots'

The enigmatic lymphoid organs in sheep, hemal nodes and mesenteric 'milk spots', also develop in the fetus. The haemal nodes are prevalent in the sublumbar area along the vena cava and abdominal aorta (Salazar, 1984). They develop from lymph node primordia that lose their lymph vessels (Nicander *et al.*, 1993) but contain sinuses filled with red blood cells (Macmillan, 1928). Lymphocyte accumulations may be found in the reticular parenchyma separating the sinuses and follicles may be present (Erencin, 1984). The milk spots emerge at 72 days gestation as small aggregations of cells along omental arteries, accumulating lymphocytes and macrophages as they develop (Shimotsuma *et al.*, 1994). The significance of these organs is unknown.

Stromal cells and accessory cells

In early lymphoid development the apparent lack of specialization in terms of stromal and accessory cells is a noteworthy feature and represents a possible explanation for an inability to mount antibody production against such antigens as *Salmonella* somatic antigens. It is therefore remarkable that *Salmonella* flagella antigens and a range of other antigens do result in antibody production upon challenge *in utero* at 65–80 days (Silverstein *et al.*, 1963; Fahey and Morris, 1978). Nevertheless it appears that the full complement of the immune system's repertoire of responsiveness requires some form of conditioning process which depends on specific structural relationships. Among such decisive components the development of lymphoid follicles may be important.

In sheep, the enzyme 5′-nucleotidase represents a marker of follicle stroma cells (Halleraker *et al.*, 1990). The first 5′-nucleotidase reaction appears in the cervical superficial lymph node about 80 days gestation and from 90 days gestation in the spleen. A similar reactivity is associated with the emerging follicles in the intestines. As the follicles in the spleen and lymph node develop, they acquire a full complement of stromal specializations, as well as lymphocyte phenotypes including the occurrence of scattered CD4 cells (Halleraker *et al.*, 1994). Mitoses occur among the lymphocytes and tingible body macrophages can be found. Such follicles are prevalent in late gestation fetuses and no longer correspond to the classical concept of primary follicles as 'rounded aggregations of small lymphocytes and reticular cells' (Nossal *et al.*, 1968) challenging prevailing ideas about the antigen dependency of the various stages through which primary follicles develop into secondary lymphoid follicles or germinal centres. Admittedly, such primary follicles in late gestation lack the marked zonation typical for germinal centres and the follicular dendritic cells lack the elaborate surface extensions seen in the light zone of germinal centres. All these features develop as the animal is exposed to external antigen after birth.

There is extensive development of intestinal lymphoid follicles *in utero* but the stromal cells of ileal Peyer's patch follicles never attain the elaborate surface extensions characteristic of germinal centre follicular dendritic cells (Nicander *et al.*, 1991). These mesenchyme-derived cells (Müller-Hermelink *et al.*, 1981) belong to a family of specialized cells that characterize the subdivisions of the follicle as they develop. Starting as primitive cells they develop into reticular fibroblasts in the dome area and the capsule. In the follicle proper, the fibroblasts develop an intimate relationship with lymphocytes which includes typical desmosome-like connections (Nicander *et al.*, 1991). In this respect, the fibroblasts resemble thymic epithelial cells. A further parallel exists between the thymus and the ileal Peyer's patch. There is an apparent endodermal lymphocyte interaction in the ileal Peyer's patch owing to the seeding of 50 nm membrane-bound epithelium-derived particles (CAPs) into the follicles and these particles appear to enter the lymphocytes (Landsverk *et al.*, 1990). In the jejunal Peyer's patch such a feature is notably represented by the enfolding of clusters of lymphocytes by the M cells (Landsverk, 1987). The implications of such interrelationships remain to be defined.

Lymphoid follicles are designed to provide microenvironments where processes of proliferation, expansion of receptor repertoire by somatic mutation, scrutiny and selection take place. All lymphoid organs, with the possible exception of the thymus, appear to be equipped with structures that ensure a confrontation with both endogenous and exogenous antigens. In fact, the follicle-associated epithelium, although incompletely mapped in sheep, is probably typical of all MALT. The epithelium has a remarkable capacity for uptake of material from the lumen; this is particularly prominent in the ileal Peyer's patch (Landsverk, 1988) and may be considered a counterpart to afferent lymphatics of lymph nodes and the blood capillaries supplying the marginal zone of the spleen. These sites are also equipped with dendritic cells thought to play a role in antigen presentation. In Peyer's patches of lambs dendritic cells can be identified by their MgATPase reactivity (Halleraker *et al.*, 1990). The descendants of the dendritic cells, the interdigitating cells, constitute an integral population of T-cell areas in both Peyer's patches and lymph nodes. Postcapillary venules provide the entry route to these tissues for both circulating lymphocytes and myeloid cells. In sheep, the postcapillary venules are usually lined with a conventional flattened endothelium (Schoefl, 1981), except during ontogeny when a high endothelium may be seen (Morris and Al Salami, 1987). Exit routes are provided by the terminal lymphatics which, both in the GALT and in lymph nodes, are conveniently interposed between the follicles and their adjoining T cell areas (Nicander *et al.*, 1991; Lowden and Heath, 1992). These lymphatics end blindly and contain lymph rich in proteins. Upon activation of lymph nodes such lymphatics tend to be crowded with lymphocytes (Nicander *et al.*, 1991).

Conclusions

The sheep model derives many of its attributes from the extent to which the organs discussed above may be studied as they develop in the fetus, shielded from external influence. The many issues that may be revisited include the pathways of B- and T-cell development, antigen dependent and independent components of lymphoid follicle development, recirculation patterns and forms of lymphocyte-stromal cell interactions. The primary immune responses of a naive immune system may likewise be explored.

3. Leukocytes and their Markers

Two reviews of sheep leukocyte antigens have been published (Mackay *et al.*, 1987; Mackay, 1988) and, more recently, two International Leukocyte Antigen Workshops, the first in 1991 (Hein *et al.*, 1991) and the second in 1993 (Hopkins *et al.*, 1993) were conducted to cluster monoclonal antibodies specific for sheep leukocyte antigens. Many of the workshop monoclonal antibodies (mAbs) bound to sheep leukocyte antigens that were identified as homologues of mouse or human leukocyte antigens. However, a few antigens could not be defined and some were unique to sheep. mAbs were clustered on the basis of immunohistochemistry, flow cytometry, and molecular weights for immunoprecipitated proteins. There was limited data on protein sequence, genetic structure, or the function of surface molecules. However, genes for three leukocyte antigens, the T19 molecule (Wijngaard *et al.*, 1992; Walker *et al.*, 1994), MHC class II (see Section 8) and the IL-2R-α chain (Verhagan *et al.*, 1992), have been cloned. The specificity of mAbs for individual MHC class II proteins (Ballingall *et al.*, 1995) and the IL-2R-α chain (Verhagaen *et al.*, 1993) have been confirmed through reactivity with proteins expressed in transfected cells. It was also apparent from the International Workshops that many mAbs cross-reacted extensively with leukocyte antigens of sheep, goat and cattle (Howard and Morrison, 1991).

Lymphocyte markers

T lymphocytes

Many of the major differentiation antigens defined for human and mouse T cells have been conserved on sheep T cells (Table XIV.3.1). However, there are some striking differences when sheep are compared with other species. First, the CD2 molecule is absent on ovine double-negative thymocytes and $\gamma\delta$ T cells (Mackay *et al.*, 1988c; Giegerich *et al.*, 1989). The low level expression of CD2 on ovine T cells may reflect the high level expression of LFA-1, a CD2 ligand, on sheep erythrocytes (Selvaraj *et al.*, 1987). Second, the complex T19 (WC1) molecule is expressed on most $\gamma\delta$ T cells of ruminants but is absent from mice and humans (Mackay *et al.*, 1986b, 1989; Wijngaard *et al.*, 1992). The CD5 molecule appears to be a pan-T cell antigen in sheep but there are $CD5^-CD8^+$ T cell populations in the mammary gland (Lee *et al.*, 1989) and the intestine (Gorrel *et al.*, 1988). As noted in Section 6, $\gamma\delta$ T cells are very numerous in young lambs. A monoclonal antibody has been raised to the $\gamma\delta$ T-cell receptor (TCR; Mackay *et al.*, 1989) but no mAbs have been produced for the $\alpha\beta$ TCR. Finally, expression of MHC class II molecules on T cells is usually restricted to activated T cells but in the blood of older sheep there is a marked increase in the frequency of MHC class II^+ T cells (Dutia *et al.*, 1993). Resting $CD4^+$ T cells in the blood of sheep are also characterized by the expression of the IL-2Rα chain on approximately 30% of the population (Verhagen *et al.*, 1993). Finally, an activation molecule, not associated with the TCR complex, has been characterized for sheep T cells but the identity of this molecule has not been established (Hein *et al.*, 1988).

B lymphocytes

A limited number of lineage specific antigens have been identified for sheep B cells (Table XIV.3.1). Monoclonal antibodies have been produced for all the major Ig isotypes: IgM (Beh, 1988), IgA (Beh, 1988), IgG1 (Beh, 1987), IgG2 (Beh, 1987) and IgE (Colditz *et al.*, 1994). Monoclonal antibodies have also been raised to Ig light chain (LC; Beh, 1988). The predominant form of Ig LC in sheep is λ LC. There is no evidence that IgD is expressed on sheep B cells but it appears that the Ig receptor associated molecule, mb-1 or Igα, has been conserved (Motyka and Reynolds, 1995).

The ileal Peyer's patch is the major site of B-cell production in young lambs. The majority of B cells in lymphoid follicles of the ileal Peyer's patch express surface IgM and are characterized by an absence of differentiation antigens found on B cells in blood and other lymphoid tissues (Hein *et al.*, 1988, 1989a; Griebel *et al.*, 1992). However, a mAb (SIC4.8R) has been produced that identifies a molecular complex unique to B cells present in the Peyer's patch of sheep and a similar molecular complex is also expressed on murine pre-B cells (Griebel *et al.*, 1996b). Thus, there may be a unique array of surface molecules that regulate the T cell-independent proliferation and differentiation of B cells in the ileal Peyer's patch (Griebel and Ferrari, 1994).

Myeloid markers

Very few lineage specific surface antigens have been well characterized for sheep myeloid cells. The percentage of cells expressing CD11a, CD11b, and CD11c varies in different populations of granulocytes, monocytes, macrophages and dendritic cells (Gupta *et al.*, 1993). This suggests that heterogeneous populations exist for each cell lineage. A careful examination of Fc receptor expression on afferent lymph dendritic cells confirmed the existence of at least four distinct subpopulations (Harkiss *et al.*, 1991). No lineage-specific mAbs have been identified for mast cells which express CD45 but not the CD44 isoform detected by the 25–32 mAb (Haig *et al.*, 1991).

Endothelial adhesion molecules

Monoclonal antibodies have been generated for a variety of endothelial adhesion molecules which may play an important role in the interaction between lymphocytes and endothelial cells. The specificity of these monoclonal antibodies for sheep endothelium was established on the basis of interspecies cross-reactivity of previously characterized mAbs or the molecular weights of immunoprecipitated proteins. These mAbs have been summarized (Mackay *et al.*, 1992a) and include: 218 (IgG2b isotype; α4), GoH3 (rat IgG; α6), CLB10G11 (IgG1; α2), 47 (IgG1; β1), Du1-29 (IgG1; L-selectin), HAE2-1 (IgG1; VCAM-1), MECA79 (rat IgG; MECA-79). The role of these molecules in sheep lymphocyte traffic is reviewed in Section 4.

Monoclonal antibody reagents

Monoclonal antibodies listed in Table XIV.3.2 are the primary reagent used to define a sheep CD antigen. The International Leukocyte Antigen Workshops have subsequently clustered numerous monoclonal antibodies with similar patterns of reactivity. Many of these mAbs and their laboratory of origin are listed in the Workshop summaries. There are few commercial sources for mAbs specific for sheep leukocyte antigens but many hybridomas and/or mAbs may be obtained from individual laboratories.

Table XIV.3.1 Leukocyte markers

CD	*Alternative names*	*Cell/tissue distribution*	*Protein structure*	*Genetics*	*References*
CD1b	T6	Dendritic cells in dermis, afferent lymph, lymph node, spleen, and thymus; cortical thymocytes	12 and 48 kDa (reduced and nonreduced)	2–7 genes Homologous to human CD1b	Bujdoso *et al.* (1989); Ferguson *et al.* (1994)
CD1c	T6	Dendritic cells in dermis, afferent lymph, lymph node, spleen, and thymus; cortical thymocytes; B cells	12 and 46 kDa (reduced and nonreduced)	ND	Mackay *et al.* (1985)
CD2	E-rosette receptor	CD4 and CD8 T cells; 60–70% thymocytes (DP); Dendritic cells and macrophages. Absent on B cells, $\gamma\delta$ T cells and DN thymocytes ($CD4^-$ $CD8^-$)	51–55 kDa (reduced and nonreduced)	ND	Mackay *et al.* (1988c); Giegerich *et al.* (1989)
CD3	TCR complex	All α/β and γ/δ TCR expressing cells	21 and 23 kDa (reduced)	γ, δ, ε, ζ are conserved γ and δ linked	Hein *et al.* (1989b); Hein and Tunnacliffe (1990, 1993)
CD4	ST-4 SBU-T4	Subset of α/β TCR expressing cells	56 kDa (reduced and nonreduced)	ND	Maddox *et al.* (1985a); Mackay *et al.*, (1986b)
CD5	ST-1 SBU-T1	Pan-T cell; Subset of B cells	67 kDa (reduced and nonreduced)	ND	Mackay *et al.* (1985); Beya *et al.* (1986); Griebel and Ferrari, (1995)
CD6	T12	Thymocytes and majority of T cells	100 kDa (nonreduced)	ND	Hopkins *et al.* (1993)
CD8	ST-8 SBU-T8	Subset of α/β TCR expressing cells	33 and 35 kDa (reduced)	ND	Maddox *et al.* (1985); Ezaki *et al.* (1987)
CD11a	LFA-1 α_L heavy chain associated with β_2 integrin	Lymphocytes, granulocytes, monocytes, macrophages	180 kDa (reduced and nonreduced)	ND	Mackay (1990)
CD11b	Mac-1; CR3 α_M chain associated with β_2 integrin	Expressed on alveolar macrophages, blood mononuclear cells, and granulocytes	170 kDa (reduced and nonreduced)	ND	Gupta *et al.* (1993)
CD11c	p150, 95 α_X chain associated with β_2 integrin	Alveolar macrophages, afferent lymph dendritic cells, and eosinophils	150 kDa (reduced and nonreduced)	ND	Gupta *et al.* (1993)
CD18	Integrin β_2	Lymphocytes, granulocytes, monocytes, macrophages	95 kDa (reduced and nonreduced)	ND	Gupta *et al.* (1993)
CD25	IL-2 receptor α chain	Activated T cells and 30–40% $CD4^+$ T cells in blood	47 kDa (reduced)	Chromosome 13q12–15; 71% homology with human	Verhagen *et al.* (1992, 1993)
CD40		B cells	ND	ND	Griebel and Ferrari (1995)
CD44	Pgp-1	Most leukocytes, thymocyte subpopulation, and many other tissues	94 kDa (reduced)	ND	Mackay *et al.* (1988a)
CD45	LCA	All lymphocytes, macrophages, and granulocytes	190, 210, 225 kDa (reduced)	ND	Maddox *et al.* (1985b)
CD45R	p220	Primarily B cells and some naive T cells	220 kDa (reduced)	ND	Mackay *et al.* (1987)
CD58	LFA-3	Mature and immature haematopoietic cells. Erythrocytes, vascular endothelium, smooth muscle	42 kDa	Ligand for CD2	Hunig *et al.* (1986)
CD59	Protectin	Expressed on erythrocytes and lymphocytes	19 kDa (nonreduced)	Restricts the activity of homologous complement	van den Burg *et al.* (1993)

(*continued*)

Table XIV.3.1 (*Continued*)

CD	*Alternative names*	*Cell/tissue distribution*	*Protein structure*	*Genetics*	*References*
WC1	T19	Expression restricted to $\gamma\delta$ T cells	220 kDa (nonreduced)	Multigene-family. Homology with SRCR	Mackay *et al.* (1986b); Wijngaard *et al.* (1992); Walker *et al.* (1994)
WC2 $\gamma\delta$ TCR	Null cell	Majority of intraepithelial T cells in the gut, skin and other mucosal surfaces. Majority of T cells in the blood of lambs	41–44 kDa and 36 kDa (reduced)	See Section 6	Mackay *et al.* (1989); Hein *et al.* (1990)
MHC class I	OLA I	All somatic cells except neurones	12 and 44 kDa (reduced and nonreduced)	See Section 8	Gogolin-Ewens *et al.* (1985)
MHC class II	OLA II	Most B cells, monocytes, macrophages, dendritic cells, thymic epithelial cells, and activated T cells	α-chain: 30–32 kDa; β-chain: 24–26 kDa	See Section 8	Puri *et al.* (1987)
mb-1	Igα protein of IgM receptor	B cells	42 kDa (reduced)	ND	Motyka and Reynolds (1995)

ND, not demonstrated in sheep; SRCR, scavenger receptor cysteine-rich family.

Table XIV.3.2 Monoclonal antibodies specific for leukocyte molecules

		Activity			
Molecule	*Clone(s)*	*FCM*	*IHC*	*Isotype*	*References*
CD1b	VPM5	+	+	IgM	Bujdoso *et al.* (1989)
CD1c	20.27	+	+	IgG1	Mackay *et al.* (1985)
CD2	36F	+	+	IgG2a	Mackay *et al.* (1988)
CD2	I35/A	+	+	IgG1	Giegerich *et al.* (1989)
CD3	Rabbit A/s	—	+	—	Ramos-Vara (1994)
CD4	17D-13	+	+	IgG1	Maddox *et al.* (1985)
CD5	ST-1a	+	+	IgG2a	Beya *et al.* (1986)
CD6	BAQ91A	+	+	IgG1	Hopkins *et al.* (1993)
CD8	E-95	+	+	IgM	Ezaki *et al.* (1987)
CD11a	F10–150	+	+	IgG1	Mackay *et al.* (1989)
CD11b	CC125	+	+	IgG1	Dutia *et al.* (1993)
CD11c	OM1	+	+	IgG1	Pepin *et al.* (1992)
CD18	MF13F5	+	+	IgG1	Gupta *et al.* (1993)
CD25	9–14	+	NR	IgG1	Verhagen *et al.* (1993)
CD44	25–32	+	+	IgG1	Mackay *et al.* (1988)
CD45	1.11.32	+	+	IgG1	Maddox *et al.* (1985)
CD45R	20–96	+	+	IgG1	Mackay *et al.* (1987)
CD58	L180/1	+	+	IgG1	Hunig *et al.* (1986)
WC1	197	+	+	IgG1	Mackay *et al.* (1986)
WC2 $\gamma\delta$ TCR	86-D	+	+	IgG1	Mackay *et al.* (1989)
T-cell activation	B5-5	+	+	IgG3	Hein *et al.* (1988)
mb-1	HM57	NR	NR	IgG1	Motyka and Reynolds (1995)
Ig LC	McM6	+	+	IgG3	Beh (1988)
IgM	McM9	+	+	IgG3	Beh (1988)
IgG1	McM1	+	+	IgG2a	Beh (1987)
IgG2	McM3	+	+	IgG3	Beh (1987)
IgA	McM10	+	+	IgG2a	Beh (1988)
IgE	Y41	NR	+	NR	Colditz *et al.* (1994)
MHC class I	41–19	+	+	IgG1	Gogolon-Ewens *et al.* (1985)
MHC class II	28.1	+	+	IgG1	Puri *et al.* (1985)

FCM, flow cytometry; IHC, immunohistochemistry; NR, not reported.

4. Leukocyte Migration

General considerations

The immune system consists of both fixed and migratory cell populations (Gowans, 1959; Reynolds *et al.*, 1982). Migratory cell populations carry out immune surveillance of tissues and disseminate immunological memory (Smith *et al.*, 1970a). Leukocytes gain access to the tissues under normal and pathological conditions by migrating out of the peripheral blood between endothelial cells lining the vasculature (Girard and Springer, 1995).

The distribution of leukocytes in blood and lymph is shown in Table XIV.4.1. All leukocytes appear to have the capability to migrate out of the peripheral blood and into the tissues under both normal and pathological conditions. Some of these cells can then be seen in the lymph draining those tissues (Hall and Morris, 1962; Smith *et al.*, 1970c). Sampling this lymph allows analysis of the cell types migrating through tissues. Only in a large animal is it possible to cannulate directly individual lymphatic vessels draining specific tissues; for this reason the sheep has proven to be a valuable model for the study of leukocyte migration (Young *et al.*, 1993; Hein 1995). Although all leukocytes can be found in the prenodal (peripheral) lymph, only lymphocytes are capable of continually recirculating between the blood and the tissues via postnodal (central) lymph (Hall and Morris, 1962; Smith *et al.*, 1970b).

Although fewer experiments have been performed to investigate the molecular mechanisms of leukocyte migration in sheep than in rodents and humans, considerably more is known about the physiology of leukocyte migration in the sheep than in any other species (Abernethy and Hay, 1992; Young *et al.*, 1993; Hein, 1995). Examination of the lymph draining normal and inflamed tissues allows analysis of the cellular events occurring within that tissue (Smith *et al.*, 1970c; Chin and Cahill, 1984). As a result, a considerable amount of information has become available regarding the cells trafficking throughout the body, and the mediators involved in these processes. Although the remainder of this section will concentrate on lymphocyte migration, information regarding the migration of other immature and mature leukocytes will also be presented when available.

Leukocyte traffic through unstimulated tissues

In mammals, lymphoid tissues constitute about 3% of total body weight (Trepel, 1974). For a 30 kg adult sheep, this would total 1 kg of lymphoid tissue, or about 1×10^{12} lymphocytes (Chin *et al.*, 1985). Based on a blood volume of 70 ml/kg and a peripheral blood lymphocyte count of 5×10^6/ml, only 1% of all lymphocytes (1×10^{10}) are in the peripheral blood at any time (Blunt, 1975). The total recirculating lymphocyte pool is $1–2 \times 10^{11}$ lymphocytes, or 10–20 times the peripheral blood pool (Schnappauf and Schnappauf, 1968). The remaining cells are resident within fixed lymphoid tissue, and do not recirculate in the same manner as those obtained from lymph (Reynolds *et al.*, 1982).

Extensive analysis has been made regarding the proportion of lymphocyte subsets in blood, peripheral and central lymph. A summary of this data is presented in Table XIV.4.2. These differences are taken to reflect preferential migration of lymphocyte subsets (Mackay *et al.*, 1988b).

Table XIV.4.1 Leukocyte values in blood and lymph

	Peripheral blood	*Peripheral lymph*	*Central lymph*
Lymphocytes	40–75%	70–95%	95–100%
Neutrophils	10–50%		
Eosinophils	0–15%	0–10%[a]	<1%[a]
Basophils	0–3%		
Macrophages, monocytes, dendritic cells	1–6%	5–20%	<1%
Cells/ml	$4–12 \times 10^6$	$1–20 \times 10^6$	$3–12 \times 10^7$

Data from Blunt (1975); Smith *et al.* (1970b).
[a]Pooled values including all granulocytes.

Table XIV.4.2 Lymphocyte subsets in blood and lymph

	Central lymph	*Peripheral lymph*	*Peripheral blood*
CD4 T cells	38.8 ± 7.7%	38.5 ± 3.7%	14.3 ± 3.0%
CD8 T cells	14.0 ± 1.8%	12.8 ± 1.1%	8.5 ± 0.5%
$\gamma\delta$ T cells[a]	10.0 ± 3.1%	27.8 ± 2.2%	11.8 ± 2.1%
B cells[b]	32.0 ± 7.0%	7.3 ± 0.3%	17.5 ± 5.3%

From Mackay *et al.* (1988b).
[a]SBU-T19-positive cells
[b]Surface Ig-positive cells

The typical output of cells in peripheral lymph is about 1×10^6 cells/h (Smith *et al.*, 1970b). This is considerably lower than the output of lymphocytes in the central lymph draining a lymph node (Hall and Morris, 1962). About 3×10^7 lymphocytes/h exit per gram of subcutaneous lymph node (Hay and Hobbs, 1977). In contrast to peripheral lymph which can contain all leukocyte subsets, only lymphocytes are found in central lymph (Smith *et al.*, 1970b). Studies of cell proliferation in unstimulated lymph nodes suggested that only about 5–10% of the lymphocytes found in central lymph are derived from cell division within the node (Hall and Morris, 1965). The remainder are derived from direct migration from the blood into the central lymph. Approximately one in every four lymphocytes delivered to the node vasculature exit the blood and passes into the node (Hay and Hobbs, 1977). In the rodent lymph node, the majority of cell traffic occurs between the 'high endothelial cells' (HECs) lining the post-capillary venules. Although sheep do not appear to possess these 'high endothelial venules' (HEVs), lymphocytes traffic through the post capillary venules in lymph nodes at similar high-levels (Harp *et al.*, 1990). Granulocytes are not usually prominent in either peripheral or central lymph, but their numbers may increase significantly after antigen stimulation (Smith *et al.*, 1970b).

When recirculating lymphocytes were labeled *in vitro*, reinjected intravenously, and their behaviour tracked *in vivo*, it was found that the average lymphocyte recirculates about once per day (Frost *et al.*, 1975; Cahill *et al.*, 1977; Chin and Hay, 1980). The time taken for a lymphocyte to extravasate and enter the lymph was independent of both the tissue examined and the presence of inflammation within the tissue (Cahill *et al.*, 1977; Chin and Hay, 1980).

Leukocyte traffic through stimulated tissues

Leukocyte traffic increases dramatically during an immune response (Smith *et al.*, 1970). Within lymph nodes, most of the increase can be accounted for by increased blood flow (Hay and Hobbs, 1977; Hay *et al.*, 1980). Mediators that increase blood flow to nonlymphoid tissues do not necessarily cause increased lymphocyte infiltration (A. N. Kalaaji and J. B. Hay, unpublished data). Many changes occurring in tissues during inflammation are mirrored in the composition of the draining lymph. For example, the output of cells in the peripheral lymph draining BCG-induced granulomas and allografted kidneys increases between 10 and 200 times baseline levels, and the cellular makeup is similar to those infiltrating the tissues (Smith *et al.*, 1970c; Pedersen *et al.*, 1975). Depending on the stimulus, the phenotypic makeup of the cells in the draining lymph has been found to vary considerably (Frost, 1978; Kimpton *et al.*, 1990; McClure *et al.*, 1991; Meeusen *et al.* 1991; Hare *et al.*, 1995). In many cases, antigens have been delivered directly to lymph nodes by infusing them into an afferent lymphatic vessel, and then monitoring the changes in the central lymph (Hall *et al.*, 1967; Smith and Morris, 1970; Smith *et al.*, 1970a). This results in a characteristic change in the traffic of lymphocytes through stimulated lymph nodes (Cahill *et al.*, 1974; Hay *et al.*, 1974). Stage 1 is characterized by a marked decrease in the output of lymphocytes in efferent lymph, and varies in both intensity and duration depending on the antigen used (Cahill *et al.*, 1974). The central lymph draining virally stimulated lymph nodes may become virtually acellular during this stage (Cahill *et al.*, 1976). This period of shutdown is unique to lymph nodes, and does not occur during inflammation in non-lymphoid tissues (Cahill *et al.*, 1976). This decreased cell output is due to arrest of recirculating cells within the node, rather than decreased migration from the blood (Cahill *et al.*, 1976; Hay *et al.*, 1980; Issekutz *et al.*, 1981). Stage 2 is characterized by a period of elevated output of lymphocytes, up to 10 times prestimulation levels. This is due to an increased recruitment of lymphocytes from the blood (Cahill *et al.*, 1974; Hay *et al.*, 1974). Antigen-specific cells functionally disappear from the central lymph during this time interval, and are likely retained within the node. This increased traffic of lymphocytes through stimulated lymph nodes likely maximizes the probability of antigen-specific cells encountering antigen (Cahill *et al.*, 1974). During stage 3, blast cells appear in the draining lymph, concomitant with the appearance of sensitized cells. This stage can last 2–3 days, after which the lymphocyte output returns to prestimulation levels. During a secondary response, the kinetics of this process are much more rapid. The phenotype of lymphocytes migrating out of lymph nodes stimulated with a variety of antigens has been studied, and appears to vary widely depending upon the antigen used (Frost, 1978; Kimpton *et al.*, 1990; McClure *et al.*, 1991; Meeusen *et al.*, 1991; Hare *et al.*, 1995).

Mediators of leukocyte migration

The sheep has proven to be a valuable model for investigating the roles of various mediators in leukocyte migration. Although *in vitro* models have been useful for screening potential mediators, fewer molecules have been found to be potent *in vivo* (Mulder and Colditz, 1993). Factors have been identified in the lymph draining antigen-stimulated tissues. For example, IL-2 and MIF activities have been identified in the lymph draining inflamed tissue (Hay *et al.*, 1973; Lowe and Lachman, 1974; Bujdoso *et al.*, 1990). In addition, factors effective at increasing the adhesion of lymphocytes to cultured endothelial cells have been identified in the lymph draining PPD-induced delayed type hypersensitivity (DTH) sites (Borron, 1991). An *in vivo* assay system has been developed in the sheep, which involves quantifying the migration of radiolabeled lymphocytes in response to subcutaneously injected mediators (Borgs and Hay, 1986). The migratory effect of

Table XIV.4.3 Mediators of leukocyte migration in sheep

Mediator	*Comments*	*References*
IL-1α	huIL-1α moderately effective at recruiting lymphocytes, causes accumulation of large numbers of neutrophils	Colditz and Watson (1992)
IL-2	huIL-2 has no effect on leukocyte migration; IL-2-like activity detected in lymph following antigen stimulation of lymph nodes	Colditz and Watson (1992); Bujdoso *et al.* (1990)
IL-8	huIL-8 causes slight lymphocyte accumulation, large numbers of neutrophils; ovIL-8 causes intense accumulation of neutrophils, slight accumulation of eosinophils	Colditz and Watson (1992); Mulder and Colditz (1993); Seow *et al.* (1994)
IFNα	huIFNα has no effect on the recruitment of leukocytes into skin, but promotes the migration of lymphocytes into but not out of lymph nodes	Kalaaji *et al.* (1988)
IFNγ	boIFNγ recruits large numbers of lymphocytes, no effect on neutrophils	Colditz and Watson (1992)
TNFα	huTNFα recruits large numbers of lymphocytes and neutrophils	Colditz and Watson (1992); Kalaaji *et al.* (1989)
ZAP	Causes slight lymphocyte accumulation, recruits large numbers of neutrophils	Colditz and Watson (1992); Mulder and Colditz (1993)
$PGE_{2\alpha}$	Appears to be involved in lymph node shutdown during antigen response. No effect on lymphocyte migration when injected subcutaneously	Hopkins *et al.* (1981); Kalaaji and Hay, unpublished
PAF	No effect on lymphocyte or neutrophil recruitment; recruits eosinophils	Colditz and Watson (1992); Mulder and Colditz (1993); Topper *et al.* (1992)
Bradykinin	Increases output of lymphocytes from lymph nodes	Moore (1984b)
LTB_4	No effect on leukocyte recruitment	Colditz and Watson (1992); Mulder and Colditz (1993)

Abbreviations: all cytokines used are recombinant; hu = human, bo = bovine, ov = ovine.

numerous mediators examined using this method are presented in Table XIV.4.3. Although recombinant human cytokines have been used in most of these studies, a number of sheep cytokines have been cloned in recent years, and are now available for study (McInnes, 1993). Neurologic and haemodynamic factors have also been found to play a role in lymphocyte migration through lymph nodes (Quin and Shannon, 1975; Moore, 1984a,b; Moore *et al.*, 1984).

Subset-specific and tissue-specific migration patterns of lymphocytes

A large amount of work has focused on linking the expression of specific molecules on the surface of lymphocytes with their *in vivo* homing patterns. Sheep contain at least three distinct pools of small lymphocytes which recirculate specifically through different tissues (Cahill *et al.*, 1977; Chin and Hay, 1980; Issekutz *et al.*, 1980). Pools of lymphocytes have been identified which recirculate specifically through mesenteric tissues, skin or subcutaneous lymph nodes. This is principally a characteristic of αβ T cells, whereas γδ T cells migrate randomly (Washington *et al.*, 1994). No such tissue specificity has been demonstrated for B cells. The molecules thought to be involved in regulating this process are discussed in the next section.

Lymphocytes have been collected from different sources and labeled *in vitro* with radioactive or fluorescent compounds. These cells were then reinjected into the same animal, and their distributions analyzed. Regardless of the label used, it was found that small lymphocytes labeled in this way reappear at peak concentrations in central lymph approximately 24 h following injection (Young *et al.*, 1993). It is interesting to note, however, that labeled cells can be detected in lymph as soon as 2–3 h following reinjection (Chin and Hay, 1984; Borgs and Hay, 1986). The differences in transit time likely reflect differences in the route of migration of individual lymphocytes through lymph nodes. There are also subset-specific differences in the migration of lymphocytes. $CD4^+$ and γδ TCR-positive lymphocytes are extracted from the blood by lymph nodes more efficiently than $CD8^+$ T cells, independent of tissue-specific migration patterns (Witherden *et al.*, 1990; Abernethy *et al.*, 1990, 1991). Although lymphoblasts also have tissue-specific homing characteristics, these cells do not recirculate (Hall *et al.*, 1970, 1977). Free-floating cells obtained from lymph appear to be unique in their capacity to migrate preferentially through specific tissues, since cells prepared from lymph nodes migrate randomly in similar assays (Reynolds *et al.*, 1982).

Molecules involved in leukocyte migration

In recent years, a model of leukocyte transendothelial migration has been developed, involving the sequential interaction of specific molecules on the surface of leukocytes with the endothelial cells lining blood vessels

Table XIV.4.4 Cell surface molecules involved in migration of sheep leukocytes

Molecule	*Comments*	*References*
L-selectin	Highly expressed on $\gamma\delta$ TCR$^+$ lymphocytes, 70% of central subcutaneous and intestinal lymph cells express L-selectin. Highest expression on CD4$^+$/CD45RA$^-$ T cells in central subcutaneous lymph. Shown to be involved in tethering and rolling on cultured lung endothelial cells	Mackay *et al.* (1992a); Abitorabi *et al.* (1996); Li *et al.* (1996)
MadCAM	Expressed on endothelium of mesenteric lymph nodes, Peyer's patches	C. Mackay, personal communication
β_7 Integrin	Correlates with gut-migrating T cells. Highly expressed on CD4$^+$/CD45R$^-$ T cells in central mesenteric lymph, moderate expression in lung central lymph, low expression in subcutaneous peripheral lymph	Abitorabi *et al.* (1996)
β_1 Integrin	Highly expressed on CD4$^+$/CD45R$^-$ T cells in subcutaneous peripheral lymph. Lower expression in mesenteric peripheral lymph	Mackay *et al.* (1992a); Abitorabi *et al.* (1996)
VCAM-1	Expressed on endothelium in stimulated subcutaneous lymph nodes	Mackay *et al.* (1992b)
CD44	Expressed on mature lymphocytes, medullary thymocytes and stromal cells, myeloid lineage stem cells. Absent from erythroid lineage stem cells. Expression on lymphocytes increased after mitogen stimulation	Mackay *et al.* (1988a); Haig *et al.* (1992)
CD11a	Expressed on recirculating lymphocytes at high levels, expressed on myeloid stem cells. Absent from erythroid stem cells. Shown to be involved in tethering and rolling on cultured lung endothelial cells	Mackay *et al.* (1990); Haig *et al.* (1992); Li *et al.* (1996)

(Springer, 1994). In other species, *in vitro* assays have been used to dissect the process of leukocyte migration through endothelium. Although similar studies have not been done widely in the sheep, a number of surface molecules believed to be involved in migration have been found on the surface of sheep leukocytes, and the process of leukocyte emigration is probably similar (Li *et al.*, 1996). The molecules identified on sheep leukocytes and their potential biological significance are summarized in Table XIV.4.4. The sheep model provides the added possibility of linking the expression of these molecules to the *in vivo* migration characteristics of lymphocytes (Mackay *et al.*, 1992a,b; Abitorabi *et al.*, 1996). For example, it has been proposed that memory T-cells migrate preferentially through nonlymphoid tissues such as skin and gut, whereas naive T cells migrate preferentially through lymph nodes (Mackay *et al.*, 1990). The β_1 integrin is highly expressed on the surface of memory-phenotype CD4$^+$ T cells in skin lymph, whereas the β_7 integrin is highly expressed on the surface of memory-phenotype CD4$^+$ T cells in gut lymph (Mackay *et al.*, 1992a; Abitorabi *et al.*, 1996).

Lymphocyte migration/recirculation in the sheep fetus

It is possible to study lymphocyte recirculation in the developing ruminant using techniques similar to those described for the adult. The sheep fetus provides a particularly interesting model because it develops in the absence of extrinsic antigen, thereby allowing study of lymphocyte migration in immunologically naive animals (Kimpton *et al.*, 1994). Tissue-specific migration pathways of lymphocytes present in the adult are not identifiable in the sheep fetus (Cahill *et al.*, 1979). When fetal lymphocytes were labeled and injected into mothers, they did not demonstrate tissue-specific recirculation patterns (Kimpton and Cahill, 1985). In the reciprocal experiment, maternal lymphocytes demonstrated tissue-specificity in the fetus. This appears to support the current model of tissue specific migration of memory lymphocytes. However, many newly formed lymphocytes lacking the CD45RA epitope migrate through fetal skin (Witherden *et al.*, 1994; Kimpton *et al.*, 1995). This demonstrates that naive lymphocytes can migrate efficiently through non-lymphoid tissues *in vivo*, although they may not demonstrate the tissue-specific recirculation patterns evident in adults.

Acknowledgements

The author thanks Wayne Hein and John Hansen for critical reading of the manuscript.

5. Sheep Cytokines

The characterization of sheep cytokines has progressed rapidly over the last 6–7 years. The impetus for this rapid progress has been the significance of sheep as a livestock species in many countries and, in addition, the preeminence of sheep as a large animal model for biomedical research. Unlike the cytokines of rodents and humans, molecular characterization of sheep cytokine cDNA has preceded functional characterization or characterization at the protein level. To date, in excess of 18 sheep cytokines have been characterized at the DNA level (Table XIV.5.1). Without exception this molecular characterization has been based on, and was possible as a result of, genetic similarities (primarily nucleotide sequence, but also gene linkage) between the cytokines of sheep and homologous molecules in rodents and humans. These similarities have facilitated the application of relatively simple cloning strategies such as PCR using primers based on regions of homology between rodent and human cytokine nucleotide sequences or alternatively, the low-stringency screening of sheep cDNA or genomic libraries with homologous human or rodent cDNA probes.

A more restricted number of sheep cytokines have been characterized at the protein and/or functional level. For the most part this has involved analysis of the activity of recombinant molecules expressed using either bacterial, yeast or mammalian expression systems (Table XIV.5.2). This characterization has frequently made use of the cross-species functional activity of many cytokines (Table XIV.5.3) and the availability of simple rodent cell based *in vitro* bioassays. Thus, it has been possible to assay recom-

Table XIV.5.1 Characteristics of cloned sheep cytokines and cytokine receptors

Cytokine	*Cell source/ expression*[a]	*Cloning source*	*Homologies*[b] *DNA*	*Protein*	*Chromosome location*	*Polymorphisms/ microsatellites*	*Accession number*	*References*
IL-1α	B	Stimulated macrophages	82%	73%	3p	Yes	X56754	Andrews *et al.* (1991); Hawken *et al.* (1996a)
IL-1β	B	Stimulated macrophages	75%	60%		Yes	X56755	Andrews *et al.* (1991)
IL-2	R (T cells)	Stimulated T lymphocytes	80%	65%	17		X53934	Goodall *et al.* (1990); Johnson *et al.* (1993)
IL-3	R (T cells)	Stimulated lymph node cells	58%	36%	5q13–15	Yes	Z18291	McInnes *et al.* (1994a,b); Hawken *et al.* (1996a,b)
IL-4	R (T cells, mast cells)	Stimulated lymph node cells	78%	57%	5q13–15	Yes	M96845	Seow *et al.* (1993); Engwerde *et al.* (1996); Hawken *et al.* (1996a,b)
IL-5	R (T cells, eosinophils)	Stimulated T cells	79%	65%	5q13–15	Yes	U17052, U17053	Hawken *et al.* (1996a,b)
IL-6	B	Stimulated macrophages	74%	53%	4	Yes	X62501	Andrews *et al.* (1993); Hawken *et al.* (1996a)
IL-7	B	Lymph node cells	84%	75%	9q		OAU10089	Barcham *et al.* (1995); Broad *et al.* (1995)
IL-8	B	Spleen	83%	78%			X7830	Seow *et al.* (1994a)
IL-10	B	Stimulated macrophages	85%	79%			U11421	Martin *et al.* (1995)
G-CSF	B	Stimulated macrophages	86%	82%			L07939	O'Brien *et al.* (1994)
GM-CSF	B	Stimulated macrophages	86%	85%	5q13–15	Yes	X53561	O'Brien *et al.* (1991); Hawken *et al.* (1996b)
TNF-α	B	Stimulated macrophages	85%	78%			X56756	Nash *et al.* (1991)
IFN-αII	B	Liver					X59067/8	Whaley *et al.* (1991)
IFN-γ	R (T cells, macrophages)	Stimulated lymph node cells	80%	76%		Yes	X52640	McInnes *et al.* (1991); Engwerde *et al.* (1996)
EPO	R (renal and hepatic cells)	Kidney		82%			Z24681	Fu *et al.* (1993)
TGF-β_1	B	Spleen	90%	96%		Yes	X76916	Robinson *et al.* (1994); Woodall *et al.* (1994)
IL-2Rα	R (T cells)	Stimulated T lymphocytes	71%	55%	13q12–15	Yes	Z11560	Verhagen *et al.* (1992); Mathews *et al.* (1994); Ansari *et al.* (1995)

[a]B = broad range of lymphoid and nonlymphoid cell types; R = restricted expression.
[b]Homology to human equivalent, coding region only for DNA.

Table XIV.5.2 Expression and activity of recombinant sheep cytokines and cytokine receptors

Cytokine	*Recombinant protein expression*	*Convenient bioassay*	*Specific activity*	*Other biological activities characterized*	*References*
IL-1α and IL-1β	*Escherichia coli* Yeast Mammalian	NOB-1/CTLL assay Ovine thymocyte costimulation (PHA)	1×10^7 U/mg	*In vivo* adjuvant activity (antibody, DTH) Pyrogenic Cartilage degradation	Andrews *et al.* (1991, 1994); Fiskerstrand *et al.* (1992); Nash *et al.* (1993)
IL-2	*Escherichia coli* Yeast	Proliferation of Con A activated IL-2 dependent sheep lymph node cells	1×10^7 U/mg	*In vivo* adjuvant activity (antibody) Generation of virus specific cytotoxic T cells	Nash *et al.* (1993); Bujdoso *et al.* (1995)
IL-3	Mammalian	Hematopoietic cell differentiation from bone marrow			McInnes *et al.* (1994a)
IL-6	Yeast	Mouse plasmacytoma/ hybridoma proliferation (B9 and 7TD1, etc.)	1×10^5 U/mg	Ig secretion by PMA-stimulated PBMC	Ebrahimi *et al.* (1995)
IL-7	*Escherichia coli*	Ovine thymocyte costimulation (con A, IL-2, IL-10)			Barcham *et al.* (1995); Martin *et al.* (1996)
IL-8	*Escherichia coli*	Chemotaxis of neutrophils		*In vivo* chemotactic activity for neutrophils, and to a lesser extent eosinophils and some T cells	Seow *et al.* (1994a); Haig *et al.* (1996)
IL-10	*Escherichia coli* Mammalian	Murine thymocyte costimulation (con A, IL-2, IL-7)		Inhibition of inflammatory cytokine synthesis (IL-1 and TNF-α) by macrophages	Martin *et al.* (1995, 1996)
GM-CSF	*Escherichia coli* Mammalian	Hematopoietic cell differentiation from bone marrow			McInnes *et al.* (1994a); O'Brien *et al.* (1995); Haig *et al.* (1996)
IFN-α	*Escherichia coli* Yeast Mammalian	Cytotoxic activity against actinomycin D-treated WEHI-164 cells Ovine thymocyte costimulation (PHA)	1×10^5 U/mg (cytotoxic assays)	Cartilage degradation	Nash *et al.* (1991, 1993); Green *et al.* (1993)
IFN-γ	Mammalian	Antiviral activity (MDBK bovine kidney cells and Semliki-Forrest virus)		Upregulation of MHC II expression on macrophages	Martin *et al.* (1996)
IL-2Rα	Mammalian	Binding of ^{125}I-IL-2	9 nm (affinity)		Verhagen *et al.* (1992)

binant sheep IL-1, IL-6, IL-7, IL-10, and TNF-α biological activity using either mouse cell lines (for example the various plasmacytomas and hybridomas that depend on IL-6 for growth) or dispersed mouse thymocytes in mitogen costimulation assays. In addition to the mouse cell based bioassays, these particular cytokines can also be assayed using sheep cell based methods including sheep thymocyte costimulation. In contrast, sheep cytokines such as IL-2, IL-3, IL-4, IL-8, GM-CSF and IFNγ have a more restricted species specificity and their functional characterization has

Table XIV.5.3 Some cross-species reactivities of ovine and human cytokines

	Cross-species activity[a]						
	Human to ovine			*Ovine to mouse*			
Cytokine	In vitro	In vivo	*Specific activity*	In vitro	In vivo	*Specific activity*	*References*
IL-1α	Yes		Reduced	Yes			Andrews *et al.* (1991)
IL-1β	Yes	Yes	Reduced	Yes	Yes	Retained	Andrews *et al.* (1991); Seow *et al.* (1994b)
IL-2	Yes		Retained	No			Verhagen *et al.* (1993)
IL-6				Yes			Andrews *et al.* (1993); Ebrahimi *et al.* (1995)
IL-7	Yes		Retained	Yes			Barcham *et al.* (1995)
IL-8	Yes	Yes	Retained				Seow *et al.* (1994a)
IL-10	Yes		Retained	Yes		Retained	Martin *et al.* (1995, 1996)
GM-CSF	No						O'Brien *et al.* (1991)
M-CSF	Yes		Retained				McInnes *et al.* (1994a)
TNFα	No	Yes		Yes		Reduced	Green *et al.* (1993)

[a]Unknown if not indicated.

been based on the use of sheep or bovine cell based assays. Sheep IL-2 and IL-4 can be assayed through their ability to support the proliferation of mitogen activated lymph node T cells, while IL-3 and GM-CSF both support the development of differentiated colonies in soft agar assays of sheep bone marrow cells. Sheep IFNγ activity is best assayed via its ability to protect a bovine cell line from the cytolytic activity of Semliki-Forest virus. An important point to note is that the species specificity of sheep cytokines is entirely consistent with the species specificity of their rodent and human homologues, although there are clearly some exceptions to this rule (for example, the broad cross-species reactivity of human IL-2 compared with the restricted activity of both sheep and mouse IL-2).

An important development ensuing from the expression of recombinant sheep cytokines, particularly in large quantities from bacteria and yeast, has been the production of cytokine-specific monoclonal and polyclonal antibodies. These reagents can be used in a variety of immunochemical based assay systems to quantify soluble cytokine levels or demonstrate individual cytokine secreting cells in normal and diseased tissue (Table XIV.5.4). While these immunochemical approaches to cytokine detection offer a number of advantages over bioassay, including both sensitivity and fidelity, there remain a number of limitations. Firstly, the number of sheep cytokines for which they are available is comparatively small, and secondly, their sensitivity, particularly with respect to

Table XIV.5.4 Ovine cytokine and cytokine receptor specific monoclonal antibodies

	Functional characteristics of ovine cytokine specific monoclonal antibodies[a]								
	Neutralization								
Ovine cytokine	In vivo	In vitro	*Western Blot*	*Immuno-histology*	*FACS*	*Immuno-precipitation*	*Immuno-assay*	*Sensitivity (pg/ml)*	*References*
IL-1α		Yes	Yes	Yes	Yes	Yes	Yes	5	Egan *et al.* (1994a)
IL-1β	Yes	Yes	Yes	Yes	Yes		Yes	5	Egan *et al.* (1994a)
IL-2	No	Yes	Yes			Yes	Yes	5000	Martin *et al.* (1996)
TNF-α		Yes	Yes	Yes	Yes	Yes	Yes	250	Egan *et al.* (1994b)
IFN-γ	Yes	Yes	Yes	Yes		Yes	Yes	5	Rothel *et al.* (1990)
GM-CSF		Yes			Yes		Yes	50	Entrican (1995)
IL-4		Yes							Rothel and Seow (1995)
IL-8		Yes		Yes				5	Rothel and Seow (1995)
IL-10			Yes			Yes			Martin *et al.* (1996)
IL-2Rα		Yes	Yes	Yes	Yes	Yes	Yes	?	Verhagen *et al.* (1993); Verhagen *et al.* (1994)

[a]Unknown if not indicated.

Table XIV.5.5 Ovine cytokine and cytokine receptor PCR oligonucleotide sequences

Cytokine			*Fragment size*	*References*
IL-1α	Forward	5′-GCT TCA AGG AGA ATG TGG-3′	338 bp	Andrews *et al.* (1991)
	Reverse	5′-GAG AAT CCT CTT CTG ATA C-3′		
IL-1β	Forward	5′-TAC AGT GAT GAG AAT GAG-3′	335 bp	Andrews *et al.* (1991)
	Reverse	5′-TCT CTG TCC TGG AGT TTG-3′		
IL-2	Forward	5′-AAC TCT TGT CTT GCA TTG-3′	436 bp	Seow *et al.* (1990)
	Reverse	5′-GAT GCT TTG ACA AAA GGT-3′		
IL-3	Forward	5′-ATG AGC AGC CTC TCT ATC TTG-3′	438 bp	McInnes *et al.* (1994a)
	Reverse	5′-ATA GTC TCT GCT GCT GCT AAG-3′		
IL-4	Forward	5′-GC CAC TTC GTC CAT GGA CAC-3′	302 bp	Seow *et al.* (1993)
	Reverse	5′-TTC CAA GAG GTC TCT CAG CG-3′		
IL-5	Forward	5′-ATT GAA GAA GTC TTT CAG GG-3′	184 bp	Genbank U17052/3
	Reverse	5′-ACA CCA AGG AAA ACT TGC AGG TA-3′		
IL-6	Forward	5′-GCT TCC AAT CTG GGT TCA-3′	347 bp	Andrews *et al.* (1993)
	Reverse	5′-CCA CAA TCA TGG GAG CCG-3′		
IL-7	Forward	5′-CCC GCC TCC CGC AGA CCA-3′	347 bp	Barcham *et al.* (1995)
	Reverse	5′-TGT GCC CTG TGA AAC TGT-3′		
IL-8	Forward	5′-ATG AGT ACA GAA CTT CGA-3′	222 bp	Seow *et al.* (1994a)
	Reverse	5′-TCA TGG ATC TTG CTT CTC-3′		
IL-10	Forward	5′-AGC TGT ACC CAC TTC CCA-3′	305 bp	Martin *et al.* (1995)
	Reverse	5′-GAA AAC GAT GAC AGC GCC-3′		
TNFα	Forward	5′-GGC TCT CCT GTC TCC CGT-3′	335 bp	Nash *et al.* (1991)
	Reverse	5′-GTT GGC TAC AAC GTG GGC-3′		
IFNγ	Forward	5′-ATG GCC AGG GCC CAT TTT-3′	338 bp	McInnes *et al.* (1991)
	Reverse	5′-ATT GAT GGC TTT GCG CTG-3′		
GM-CSF	Forward	5′-ATG TGG CTG CAG AAC CTG CTT CTC-3′	438 bp	O'Brien *et al.* (1991)
	Reverse	5′-CCT CTG GGC TGG TTC CCA GCA GTC-3′		
G-CSF	Forward	5′-ACC CCC CTT GGC CCT GCC-3′	522 bp	O'Brien *et al.* (1994)
	Reverse	5′-TCA GGG CTC AGC AAG GTA-3′		
TGF-β1	Forward	5′-ATG CCG CCT TCG GGG C-3′	1061 bp	Woodall *et al.* (1994)
	Reverse	5′-TCA GCT GCA CTT GCA GG-3′		
IL-2Rα	Forward	5′-ACT AGT CGA CCA ACA AGA GGC TG-3′	661 bp	Verhagen *et al.* (1992)
	Reverse	5′-CCG CGG ATC CTG AGC TGG GGC TG-3′		

the characterization of individual cytokine-secreting cells in tissue or dispersed cell populations, remains questionable. Alternative approaches available to researchers are based on detection of cytokine specific mRNA. These approaches include *in situ* hybridization and Northern blot analysis; however, the most sensitive and widely used technique is PCR analysis. The rapid progress in the molecular characterization of sheep cytokines means that appropriate primer sets are currently available for a wide range of sheep cytokines (Table XIV.5.5).

6. T-Cell Receptors

T-cell receptor genes

All known heterodimers and invariant proteins contributing to the sheep T-cell receptor complex have been cloned and are summarized in Tables XIV.6.1–XIV.6.3.

Surface expression of CD3/TCR proteins

Two CD3 chains of 21 kDa and 23 kDa that are noncovalently associated with the TCR heterodimers have been precipitated from sheep T cells using cross-reactive polyclonal antisera, although the precise identity of the CD3 components was not established (Hein *et al.*, 1989). Sheep $CD4^+$ and $CD8^+$ T cells express a surface TCRαβ heterodimer which migrates at 85 kDa under nonreducing conditions and resolves into 40 kDa and 50 kDa subunits after reduction (Hein *et al.*, 1989b). The sheep TCRγδ heterodimer migrates at 70–75 kDa without reduction (Hein *et al.*, 1989b), and, when precipitated by a TCR-specific mAb, resolves into 41–44 kDa and 36 kDa subunits after reduction (Mackay *et al.*, 1989). When anti-CD3 antibodies are used, a presumptive additional TCRγ chain of 55 kDa can also be precipitated (Mackay *et al.*, 1989). The heterogeneous sizes of sheep TCRγ proteins may reflect the diverse primary structure of the constant region genes (see below) or result from differences in post-translational glycosylation.

Table XIV.6.1 Heterodimer constant regions

Locus	*Number constant regions*	*Homology to human/mouse*[a]	*Chromosomal location*	*References*
TCRA	1	62/56	Unknown	Hein *et al.* (1991)
TCRB	2	81/78	4q32–qter	Grossberger *et al.* (1993); Pearce *et al.* (1995)
TCRG	5	61/60	Unknown	Hein *et al.* (1990); Hein and Dudler (1993a)
TCRD	1	68/67	Unknown	Hein *et al.* (1990)

[a]Homologies are shown as per cent protein identity.

Table XIV.6.2 Heterodimer variable regions

Locus	*D regions*	*J regions*	*V regions*	*References*
TCRA	—	Multiple	multiple	Hein *et al.* (1991)
TCRB	Several	Multiple	multiple	Grossberger *et al.* (1993)
TCRG	—	6	≈15	Hein and Dudler (1993a)
TCRD	Several	3	≈40	Hein and Dudler (1993)

Table XIV.6.3 Invariant components

Locus	*Homology to human/mouse*[a]	*Chromosomal location*	*References*
CD3G	72/65	Unknown	Hein and Tunnacliffe (1990)
CD3D	65/60	Unknown	Hein and Tunnacliffe (1990)
CD3E	54/61	Unknown	Hein and Tunnacliffe (1993)
TCRZ	82/72	1p14–p11	Hein and Tunnacliffe (1993); Ansari *et al.* (1994)

[a]Homologies are shown as per cent protein identity.

Table XIV.6.4 Frequency of $\alpha\beta$ and $\gamma\delta$ T cells in different tissues

Tissue	*Per cent $\alpha\beta$ T cells*[a]	*Per cent $\gamma\delta$ T cells*[a]
Thymus	96–98	2–4
Lymph nodes[b]	47–58	4–14
Spleen	55–75	5–8
Ileal Peyer's patch[c]	<1	—
Jejunal Peyer's patch[c]	10–15	1–2
PBMC	20–45	15–60
Efferent lymph	60–70	5–15
Afferent lymph	65–80	12–25

[a]Values shown are the percentage of each T cell type among the mononuclear leukocytes comprising each tissue.
[b]Peripheral lymph nodes (prescapular, prefemoral, popliteal) tend to contain relatively more T cells and are at the upper end of the ranges shown whereas gut-associated lymph nodes (mesenteric) are at the lower end of the range.
[c]The data for the Peyer's patches refers to the intrafollicular lymphocyte populations and does not include extrafollicular T cells which are prominent in the jejunal Peyer's patches but not the ileal Peyer's patches.

Prevalence of $\alpha\beta$ and $\gamma\delta$ T cells in different tissues

The frequency of T cells expressing the two types of TCR varies widely in sheep depending on the type of tissue and the age of the animal. Values typical of 3–4-month-old lambs are summarized in Table XIV.6.4.

Specific features of the sheep $\alpha\beta$ TCR

The sheep $\alpha\beta$TCR has not been extensively characterized but all indications show that its structure and expression resembles closely the patterns found in other mammals.

Specific features of the sheep $\gamma\delta$ TCR

There are a number of features about the sheep $\gamma\delta$TCR, and the cells expressing it, that differ from patterns seen in other mammals.

Number, diversity and expression of C_γ segments

Sheep express five C_γ segments that vary in length between 168 and 220 amino acids, making them the longest and most diverse of all known TCR heterodimer chains (Hein *et al.*, 1990; Hein and Dudler, 1993d). The additional C_γ sequence occurs in the connecting peptide region, between the transmembrane and immunoglobulin-like domains and results from a duplication or triplication of exon 2. In some cases, there are additional repeats of another 5-amino-acid motif located immediately 3′ of the immunoglobulin-like domain and either one or two additional cysteine residues are encoded in the extended segments. The expression of the five C_γ chains appears to be developmentally regulated; some are expressed only during fetal life while others are expressed after birth (Hein and Dudler, 1993a). Therefore, when the different C_γ genes are used in combination with the single C_δ gene, sheep preferentially express different isotypes of the $\gamma\delta$ TCR at different times of ontogeny.

Number and diversity of variable region genes

Sheep express a larger repertoire of V_γ and V_δ genes than other mammals. This is especially obvious in the case of V_δ genes, where there has been extensive duplication of the $V_\delta 1$ family (Hein and Dudler, 1993a). For the C_γ segments, the variable region repertoire is also developmentally regulated and fetal-specific and adult-specific expression of particular V-genes occurs. Sheep V_δ genes contain hypervariable regions similar to the CDRs in the Ig V-genes, suggesting that they have been selected on an evolutionary time-scale. There is to date no evidence that sheep V_δ genes can be expressed as part of the $\alpha\beta$TCR, as occurs in humans and mice. Sheep $\gamma\delta$ T cells migrating to tissues such as skin and gut also express a diverse V-gene repertoire comparable to that of blood-borne cells (Hein and Dudler, 1996b). To date, there is no evidence pointing to either tissue-specific repertoires or selective migration of sheep $\gamma\delta$ T cells.

Prevalence of sheep $\gamma\delta$ T cells

Sheep contain an unusually large number of $\gamma\delta$ T cells, particularly in blood (Hein and Mackay, 1991). Their prevalence is strongly age-dependent, being highest in perinatal and young animals where they may account for up to 60% of blood-borne mononuclear cells. TCR$\gamma\delta^+$ cells are also prominent, but do not form a majority, in other recirculating lymphocyte compartments such as those in efferent and afferent lymph. They are less numerous in solid peripheral lymphoid organs, where they comprise 1–14% of resident leukocytes (see Table XIV.3.1). In the regions of the jejunum that are free of Peyer's patches, $\gamma\delta$ T cells comprise around 18% of intraepithelial lymphocytes (Gyorffy *et al.*, 1992).

Recent reviews

More extensive reviews of the molecular genetics of sheep T-cell receptors have been published recently (Ishiguro and Hein, 1994; Hein, 1996). Another relevant review deals with the ontogeny of sheep T cells, including the development of the TCR repertoire (Hein, 1994).

Caprine T-cell receptor variable β chain (TCRVβ) repertoire

The TCR V_β repertoire was analyzed using blood mononuclear cells (PBMC) and synovial fluid cells collected from a 7-year-old Saanen goat, naturally infected with caprine arthritis encephalitis virus. Anchor-PCR was performed as described by Loh *et al.* (1989) using primers specific for sheep TCRVβ constant regions which were kindly donated by Dr. W. Hein, Basel Institute for Immunology, Switzerland (Grossberger *et al.*, 1993).

Full-length sequences of 55 PBMC-derived clones were obtained and sequenced and 45 PBMC-derived sequences were functionally rearranged and further analyzed. PBMC derived sequences were classified into 16 different V_β families (Table XIV.6.5) with nine V_β families ($V_\beta 1$; $V_\beta 2$; $V_\beta 3$; $V_\beta 4$; $V_\beta 6$; $V_\beta 7$; $V_\beta 15$; $V_\beta 17$; $V_\beta 90$) corresponding to bovine V_β families described in the literature (Tanaka *et al.*, 1990). A counterpart for the bovine $V_\beta 90$ family was not found in our original V_β fragment library, but was subsequently detected by reverse transcription-polymerase chain reaction (RT-PCR) using a goat V_β constant region primer and a bovine $V_\beta 90$ specific primer.

Compared with clones derived from PBMC an even larger number of clones (17 out of 28) from synovial fluid cells were nonfunctionally rearranged. All the productively rearranged V_β clones derived from synovial fluid cells belonged to families ($V_\beta 1$, $V_\beta 9$, $V_\beta 10$ and $V_\beta 25$) present in the V_β repertoire of PBMC (Table XIV.1.1). In conclusion, synovial fluid cells presented a completely different V_β repertoire when compared with blood. This restricted T-cell clonal expansion may have been dependent upon viral antigen stimulation.

Table XIV.6.5 Goat TCR V_β families in blood and synovial fluid

	Clone number	
TCR Vβ family[a]	*PBMC*[b] *(n = 45)*	*SFC*[c] *(n = 11)*
$V_\beta 1$	1,3,5,8,9,10,16,30,35,39,52,72	3,27
$V_\beta 2$	19,28,44	
$V_\beta 3$	69	
$V_\beta 4$	56	
$V_\beta 6$	14,53,55	
$V_\beta 7$	21,38,49	
$V_\beta 8$	2,4,13,25,29,68	
$V_\beta 9$	54	12
$V_\beta 10$	22	8
$V_\beta 13$	64,67	
$V_\beta 15$	42,66	
$V_\beta 17$	*	
$V_\beta 20$	17,33,48	
$V_\beta 24$	37,51,60,65	
$V_\beta 25$	27	9,11,14,15,19,20,25
$V_\beta 30$	58	

[a]Aligned DNA sequences showing over 75% homology were considered to belong to the same family. Goat Vβ families were named and numbered based on homology with human counterparts.
[b]Clones were generated using mRNA extracted from freshly isolated blood mononuclear cells of a 7-year-old Saanan goat, naturally infected with caprine arthritis encephalitis virus.
[c]Clones were generated using mRNA extracted from freshly isolated synovial fluid cells of a 7-year-old Saanan goat, naturally infected with caprine arthritis encephalitis virus.

7. Immunoglobulins

Immunoglobulin genes

The sheep heavy chain constant regions IgM, IgG1, IgG2 and IgE and light chain constant regions Igλ and Igκ have been cloned as cDNA. A number of V-genes at the *IgH*, *IgL* and *IgK* loci have been cloned, either as cDNA or as genomic DNA. Collectively, the available data give a good indication of the composition of the sheep immunoglobulin gene pool and are summarized in Tables XIV.7.1 and XIV.7.2. There is no evidence that sheep express IgD and, like chickens, rabbits, pigs and cattle, they probably lack the *IGDC* gene, although this has not been confirmed at the DNA level. Sheep express IgA but the *IGAC* gene has not been isolated and sequenced.

Expression and function

Immunoglobulins are expressed either as a membrane-anchored B cell antigen receptor or as secreted antibodies. Some physical and functional properties of sheep Igs are shown in Table XIV.7.3.

Concentration of Ig in biological fluids

The concentrations of sheep immunoglobulins found in several body fluids are given in Table XIV.7.4. The values are representative of levels reported in the literature (Curtain, 1975; Quin and Shannon, 1977; Gorin *et al.*, 1979; Lascelles *et al.*, 1981; Cripps *et al.*, 1985) and are shown in mg/ml unless indicated otherwise.

Ig repertoire ontogeny

The site of first development of B cells and the sequence of Ig gene rearrangements have not been defined in fetal lambs. Surface IgM^+ B cells circulate in the blood from around day 45 of ontogeny, although they remain a minor population and never account for more than 3–5% of PBMC during fetal life. There is a developmentally regulated usage of V_λ and V_κ genes during fetal development and a change in the relative usage of J_κ genes. As in adult sheep, fetal B cells express members of a single V_H gene family. Throughout fetal life among peripheral circulating B cells, there is little if any variability from germline sequence in the CDR1 and CDR2 regions of V genes. Some junctional or N-region diversity occurs at the V–J and V–D–J junctions contributing to CDR3 of the light and heavy chains respectively. In the case of light chains, this mostly occurs at V_κ–J_κ junctions whereas the V_λ–J_λ joins are almost invariant (Hein and Dudler, 1996).

The ileal Peyer's patch follicles become colonized with B cells at around 105 days of gestation and are a prominent site for B-cell expansion from day 120 onwards. As the B cells proliferate, their Ig genes are diversified by somatic hypermutation. Developing B cells accumulate mutations

Table XIV.7.1 Constant regions

Locus	*Constant regions*	*Homology to human/mouse*[a]	*Chromosomal location*	*References*
IGH	*IGMC*[b]	62/57	Unknown	Patri and Nau (1992); Hein and Dudler (1993b)
	IGGC1	67/62		Foley and Beh (1989)
	IGGC2	66/61		Clarkson *et al.* (1993)
	IGEC	47/46		Engewerda *et al.* (1992)
IGL	*IGLC* (2–3)[c]	74/74	Unknown	Foley and Beh (1989)
IGK	*IGKC* (1)[c,d]	58/60	3p22–p17	Foley and Beh (1992); Broad *et al.* (1995)

[a]Homologies shown as percent protein identity.
[b]Two allelic variants of sheep *IGMC* are known (Hein and Dudler, 1993).
[c]Numbers in brackets indicate the likely or known number of light chain constant region genes.
[d]A *Taq*I polymorphism has been identified at the *IGKC* gene (Parsons *et al.*, 1994).

Table XIV.7.2 Variable regions

Locus	*D regions*	*J regions*	*V regions*	*References*
IGH	Several	Several	10 (1 family)	Dufour *et al.* (1996)
IGL	–	1	90–100 (60–70 functional)	Reynaud *et al.* (1991, 1995)
IGK	–	3 (2 Functional, 1 Pseudogene)	10 (4 families)	Hein and Dudler (1996)

Table XIV.7.3 Physical and functional properties of sheep immunoglobulins

	B-cell antigen receptor[a]	*Secreted antibody*[b]	*Selective transport into secretions*	*Complement fixation*
IgM	75–80 kDa	19S	±	+
IgG1		7S	+	+
IgG2		7S	–	–
IgA		7S and 15S	+	–
IgE		8S	nd	nd
Igκ	25–30 kDa			
Igλ	25–30 kDa			

nd = No data reported.
[a]Molecular masses determined by immunoprecipitation using monoclonal antibodies specific for each Ig chain (Beh, 1987, 1988).
[b]Data from Curtain (1975).

Table XIV.7.4 Immunoglobulin concentration in body fluids

	IgM	*IgG1*	*IgG2*	*IgA*	*IgE*
Blood serum[a]	1.5–3	16–22	5–7	0.5–1	0.8–200[b]
Efferent lymph[c]	0.6–0.9	6–14	2.6–5.3	1–2	nd
Afferent lymph[c]	0.2–0.8	4–13	1.3–5.5	2.7[d]	nd
Mammary secretions:					
Colostrum	1.3	94.1	2.5	2.8	nd
Early lactation	0.4	5.0	0.1	2.3	nd
Mid lactation	0.1	1.4	0.2	0.7	nd
Involution	0.2	2.9	0.2	6.6	nd
Intestinal secretion	0.4–0.8	2.8–5.2	1.2–2.0	4.9–12.5	nd
Bile	0.2	0.05	0.04	1.6	nd

nd = No data reported.
[a]Values for blood serum are from mature postnatal animals. Fetal serum contains no measurable Igs up to around 120 days of gestational age. From 120 days until term, low levels of IgM ($<100\,\mu$g/ml) can sometimes be detected, and more rarely, trace amounts of IgG1. In heterozygotes, fetal antibodies always carry the paternal allotype and are therefore of fetal and not maternal origin (Curtain, 1975; Morris and Courtice, 1977).
[b]Concentrations of IgE is μg/ml.
[c]The concentration of Igs in lymph varies according to the body site from where the lymph originates and reflects the permeability of the vascular capillary bed. In general, the concentration in both afferent and efferent lymph increases in the order popliteal < renal < prescapular < hepatic (Quin and Shannon, 1977).
[d]Represents IgA level measured in intestinal afferent lymph (Cripps *et al.*, 1985). No data for afferent lymph from other sources.

specifically in the CDR regions. For the first few months after birth, the ileal Peyer's patch (IPP) is the major source of peripheral B cells, and the organ then begins to involute around 5–6 months old.

Special features of sheep immunoglobulins

Passive maternal transfer

The ovine placenta is impermeable to large molecules and Igs are not transferred prenatally from ewe to fetus. Antibodies are only transferred passively via colostrum and are absorbed nonspecifically across the gut-wall during the first 24 h after birth. The major secretory Ig in colostrum is IgG1, which occurs at levels much higher than in blood serum. IgG1 is concentrated in colostrum by selective transport across mammary epithelia although the receptors involved have not been identified. Structural features of immunoglobulin molecules that probably contribute to the preferential transport of IgG1 and not IgG2 have been elucidated (see below).

Structure of IgG

Sheep *IGEC1* has a structure that is comparable to other mammalian IgG molecules, but the hinge region of sheep *IGCC2* is shortened (Clarkson *et al.*, 1993). Modeling studies indicate that the close positioning of the Fab and Fc faces produces a compact molecule with little scope for angular rotational movements of Fab relative to Fc. The steric hindrance may limit the accessibility of the Ig molecule to Fc receptors presumptively expressed on other cells and could account for functional differences observed between sheep IgG isotypes. Sheep IgG1, but not IgG2, is selectively transported into colostrum (see above), milk and other exocrine secretions and sheep alveolar macrophages bind monomeric IgG1 but not IgG2 (see discussion in Clarkson *et al.*, 1993).

Repertoire diversification

Sheep Igs are diversified mainly by hypermutation of mature rearranged V-region segments. The rate and position of mutational change could not be influenced experimentally by regulating exposure to foreign antigens or T cells (Reynaud *et al.*, 1991, 1995). The hypermutation process therefore appears to occur as part of a developmental program that operates autonomously during B-cell ontogeny in the IPP. The pattern of nucleotide changes suggest that the mutation process operates mechanistically and is determined by the base pair composition of the V gene, perhaps implying that sheep Ig V-genes, or at least those used most commonly, have been specially selected during evolution.

Recent reviews

Two recent reviews have dealt with the IPP as a specialized site for B-cell development. One deals with more general aspects (Griebel and Hein, 1996) while the other looks at comparative molecular mechanism of Ig V-gene diversification (Weill and Reynaud, 1996).

8. Ovine Major Histocompatibility Complex Antigens (*Ovar*)

Methods used to characterize ovine MHC antigens

A number of different methods have been used to characterize MHC antigens of sheep (reviewed by Schwaiger *et al.*, 1996). The earliest characterization of ovine MHC antigens used sera from parous or immunized sheep in a standard histocompatibility microlymphocytotoxicity assay (Ford, 1975; Millot, 1979), and this method of ovine MHC typing is still used by some laboratories (Garrido *et al.*, 1995; Stear *et al.*, 1996). Using the microlymphocytotoxicity assay, evidence has been obtained for the existence of at least four genes with three showing some degree of linkage (Millot, 1979; Stear and Spooner, 1981; Cullen *et al.*, 1982). However, in general, serological work in sheep has been hampered owing to a lack of standard, shared reagents and serological typing methods have been largely replaced by methods detecting DNA polymorphisms. While two-dimensional electrophoretic techniques have been used to demonstrate protein polymorphisms in both class I and class II genes (Puri *et al.*, 1987), the logistical problems associated with this method have meant that no typing systems have been developed for sheep using this approach.

Early DNA polymorphism studies in sheep involved the detection of restriction fragment length polymorphisms (RFLPs) in Southern blotting studies using human probes (Chardon *et al.*, 1985; Blattman *et al.*, 1993; Hulme *et al.*, 1993). The low stringency conditions, necessitated by the use of heterologous probes, resulted in the detection of complex patterns containing large numbers of bands and it was difficult to assign bands to individual genes. The cloning of a number of sheep MHC genes enabled the human probes to be replaced by sheep probes making band scoring easier, and RFLP typing systems have been developed for a number of class II genes including *Ovar-DQA1*, *Ovar-DQA2*, *Ovar-DRA*, *Ovar-DRB*, and *Ovar-DQB* (Scott *et al.*, 1991a,b; Fabb *et al.*, 1993; Grain *et al.*, 1993; van Oorschot *et al.*, 1994). More recently, the many advantages of typing systems based on the use of the polymerase chain reaction (PCR) has resulted in the development of a range of PCR-based typing systems. These include systems based on microsatellites, single-strand conformational polymorphisms (SSCP), sequencing, and allele specific oligonucleotides. PCR-based systems are currently available for typing a large number of class I, class II and class III genes (Table XIV.8.1). It is likely that in the future the PCR-based typing systems will represent the methods of choice for typing the ovine MHC. To date, no attempt has been made to correlate the various serological, protein and DNA alleles.

Genetics of the ovine MHC

The MHC class I region has been mapped to chromosome 20q15–q23 by *in situ* hybridization using both human and porcine class I probes (Mahdy *et al.*, 1989; Hediger *et al.*, 1991). Linkage mapping has revealed that both the class II and class III regions are linked to the class I region on chromosome 20 with a genetic distance of 4–6 cM between class I and class II loci (Crawford *et al.*, 1995; Groth and Wetherall, 1995). While no detailed map of the overall MHC region exists for sheep, the available information indicates that the layout of the ovine MHC will be similar to that of cattle and humans.

Class I

A number of laboratories have attempted to examine the genetics of the ovine class I region using serological methods. At least two of the genes detected by serological methods, *OLA-A* (*Ovar-A*) and *OLA-B* (*Ovar-B*), are likely to be class I genes. *OLA-A* and *OLA-B* are separated by 0.6% recombination, and considerable linkage disequilibrium exists between them (Millot, 1979). A third serologically detected gene, *OL-X*, is separated by 26% recombination from the *OLA-A* and *OLA-B* genes; however, it is unknown whether *OL-X* represents a class I or class II locus. Both protein and gene cloning studies provide evidence for the existence of at least three expressed distinct polymorphic class I loci (Puri *et al.*, 1987; Grossberger *et al.*, 1990).

Class II

The majority of the work on the genetics of the ovine MHC has focused on the class II region. In 1991, Deverson and colleagues postulated that the ovine MHC class II region contained seven distinct class II α genes, 10 distinct β genes and 14 β-related sequences. However, subsequent work has revealed that this was an overestimate as it was not known at the time the extreme polymorphism of the ovine class II region. Five distinct α genes (*Ovar-DRA*, *Ovar-DQA1*, *Ovar-DQA2*, *Ovar-DYA*, *Ovar-DNA*) and seven distinct β genes (*Ovar-DRB1*, *Ovar-DRB2*, *Ovar-DQB1*, *Ovar-DQB2*, *Ovar-DYB*, *Ovar-DOB* and *Ovar-DMB*) have been cloned, and at least partially sequenced. Representatives from all class II gene families described for other species have been found in sheep, with the exception of *DP* genes. It is likely that an additional α gene (*Ovar-DMA*) will be found in sheep and possibly one or more extra β genes. Three genes from the class II region (*Ovar-DRB1*, *Ovar-DRB2* and *Ovar-DQA2*) have been mapped yielding recombination distances of 6 cM between the class I region and the *Ovar-DQA2* gene, and 4 cM between class I and the *Ovar-DRB* region (Schwaiger *et al.*, 1996; K. J. Snibson *et al.*, unpublished data).

Class III

Only a small number of genes from the class III region have been described for sheep. Haplotype variation has been reported for the number of C4 genes with sheep possessing either two or three copies of the C4 gene (Schwaiger *et al.*, 1996). No work has been reported on TAP, HSP70 or RING genes in sheep.

CD1

While several ovine CD1 genes have been cloned their chromosomal location has not been determined.

Tissue distribution, structure and function of ovine MHC antigens

A number of monoclonal antibodies have been made which bind to ovine MHC class I, class II and CD1 molecules (Gogolin-Ewens *et al.*, 1985; Mackay *et al.*, 1985; Puri *et al.*, 1985). The use of these antibodies has revealed that both the tissue distribution and structure of ovine MHC class I, class II and CD1 molecules are similar to their human homologues. Monoclonal antibodies to sheep MHC class I antigens precipitate a 44 kDa α-chain in association with a 12 kDa β_2-microglobulin chain (Gogolin-Ewens *et al.*, 1985). Four distinct polymorphic products, each consisting of an α (30–32 kDa) and a β chain (24–26 kDa), are recognized by monoclonal antibodies to ovine MHC class II molecules (Puri *et al.*, 1987). While some of these products have been identified as *Ovar-DR* and *Ovar-DQ* molecules, the identity of the other product(s) is unknown. Transfection studies have demonstrated that the *Ovar-DRA*/*Ovar-DRB1* and *Ovar-DQA1*/*Ovar-DQB1* gene products are capable of being expressed (Ballingall *et al.*, 1995). It is not known whether any products of the ovine *Ovar-DQA2*/*Ovar-DQB2*, *Ovar-DYA*/*Ovar-DYB*, *Ovar-DNA*/*Ovar-DOB* or *Ovar-DMA*/*Ovar-DMB* gene pairs are expressed, nor is it known whether any of the existing monoclonal antibodies to class II molecules could bind to these products. Monoclonal antibodies to the ovine CD1 molecule immunoprecipitate a 46 kDa α-chain in association with a 12 kDa β_2-microglobulin chain (Mackay *et al.*, 1985).

Ovine MHC class I molecules are expressed at varying levels on most nucleated cells (Gogolin-Ewens *et al.*, 1985) with the exception of neural cells. In contrast, the

Table XIV.8.1 Ovine MHC genes

Gene	*Comment*	*Mapping*	*Typing methods*[a]	*Polymorphism*	*References*
Class I	Evidence for at least three expressed class I loci and at least one pseudogene; at least one of the genes has an associated microsatellite	PL	Microlymphocytotoxicity **Microsatellite** SSCP	At least 17 different DNA (and peptide) sequences have been identified but these have not been assigned to genes	Puri *et al.* (1987); Grossberger *et al.* (1990); Garber *et al.* (1993); Groth and Wetherall (1994); Vicario *et al.* (1995); Novak, unpublished data
Class II					
Ovar-DRA	One expressed gene		RFLP sequencing	Limited polymorphism, one peptide allele and several RFLPs	Escayg *et al.* (1993); Fabb *et al.* (1993)
Ovar-DRB1	One expressed gene, contains a microsatellite in intron 2	L	**Microsatellite/ASO** Microsatellite SSCP - exon 2 RFLP - exon 2	At least 74 alleles (at least 56 peptide, at least 32 microsatellites)	Grain *et al.* (1993); Schwaiger *et al.* (1994); Schwaiger *et al.* (1996)
Ovar-DRB2	Pseudogene; contains a microsatellite in intron 5	L	**Microsatellite**	At least 13 alleles	Scott *et al.* (1991b); Blattman and Beh (1992)
	Preliminary evidence exists for the existence of two additional *Ovar-DRB* pseudogenes				Schwaiger *et al.* (1996)
Ovar-DQA1	One expressed gene; the gene is not found in all sheep	C	**SSCP** RFLP	At least seven peptide alleles, excluding the null	Maddox *et al.* (1994); Wright and Ballingall (1994)
Ovar-DQA2	One gene presumed to be expressed	LC	**SSCP** RFLP	At least 11 peptide alleles	Maddox *et al.* (1994); Wright and Ballingall (1994)
Ovar-DQB1, Ovar-DQB2	Two genes, both presumed to be expressed, can only be distinguished by their relative proximities to the *DQA* genes	C	**SSCP** RFLP	At least 14 peptide alleles	Grain *et al.* (1993); Wright and Ballingall (1994)
Ovar-DYA	Single gene, expression status unknown		**SSCP** -exon 2	At least two alleles	Wright *et al.* (1994); Davies and Maddox, unpublished data
Ovar-DYB	Single gene, expression status unknown, homologous to *Bota-DIB*		Not tested		Wright *et al.* (1994)
Ovar-DMA	Single gene, expression status unknown				Davies, K.P. and Maddox, J.F., unpublished data
Ovar-DMB	Single gene, expression status unknown		**SSCP**	At least three alleles	Davies, K.P., van Oorschot, R.A.H., Fabb, S.A. and Maddox, J.F., unpublished data
Ovar-DNA	Single gene		Not tested		Wright *et al.* (1995)
Ovar-DOB	Single gene		Not tested		Wright *et al.* (1996)
Ovar-DPA	No gene				
Ovar-DPB	No gene				

(continued)

Table XIV.8.1 (*Continued*)

Gene	*Comment*	*Mapping*	*Typing methods*[a]	*Polymorphism*	*References*
Class III					
Factor B	Single gene	L	**Microsatellite**	Seven alleles	Groth and Wetherall (1995)
TNFα	Single expressed gene		Sequence	At least two peptide alleles	Schwaiger *et al.* (1996)
C4	At least three genes represented by at least two functional isotypes; variation in number of genes per haplotype		RFLP	Two DNA alleles associated with one of the C4 genes	Groth *et al.* (1990); Ren *et al.* (1993); Schwaiger *et al.* (1996)
Steroid 21-hydroxylase	Two genes				Schwaiger *et al.* (1996)
Prolactin	Single gene		Sequence	At least two peptide alleles	Adams *et al.* (1989); Varma *et al.* (1989)
MHC-related genes					
CD1	2–7 Genes which are most homologous to human CD1B				Ferguson *et al.* (1994); Schwaiger *et al.* (1996)

ASO = allele specific oligonucleotide.
PL, physical assignment; L, linkage assignment; C, region spanned by overlapping cosmids.
[a]The most informative typing method used to date for each gene is in bold type.

distribution of MHC class II molecules is restricted to B cells, activated T-cells, monocytes, macrophages, dendritic cells and some endothelial (glomerular, kidney tubular, hepatic sinusoids) and epithelial (intestinal villi, cells lining endometrial glands) cells (Puri *et al.*, 1985). Ovine CD1 molecules have an even more limited distribution, and are found on cortical thymocytes, macrophages, peripheral blood monocytes, Langerhans cells, and at a low level on B lymphocytes (Mackay *et al.*, 1985). It is likely that there are at least two distinct CD1 molecules expressed in sheep. Separate molecules with distribution patterns analogous to that of human CD1b and CD1c molecules have been described, with the exception that sheep peripheral blood monocytes constitutively express CD1c (Dutia and Hopkins, 1991; Schwaiger *et al.*, 1996).

The construction of a panel of ovine class II *Ovar-DR* and *Ovar-DQ* transfectants has enabled the finer specificities of monoclonal antibodies to sheep class II molecules to be determined (Ballingall *et al.*, 1995). A small number of the antibodies tested bound to all the *Ovar-DQ* and *Ovar-DR* transfectants. Some antibodies were specific for *Ovar-DR* or an *Ovar-DR* subtype, others were specific for *Ovar-DQ* or an *Ovar-DQ* subtype, and the remainder recognized a mixture of *Ovar-DR* and *Ovar-DQ* types and subtypes.

The distribution of *Ovar-DR* and *Ovar-DQ* molecules on T lymphocytes has been shown to differ with immune status and age. *Ovar-DQ* is found on fewer T lymphocytes and its expression appears to be upregulated after recent activation. As sheep age, the level of MHC class II expression on T lymphocytes has been found to increase (Dutia *et al.*, 1993). Initially, T lymphocytes lack class II molecules, then the T lymphocyte progressively expresses *Ovar-DR* molecules, and finally both *Ovar-DR* and *Ovar-DQ* molecules.

Altered levels of expression of MHC molecules have been reported for a number of diseases. For example, the carpal joints of sheep suffering from visna viral inflammatory synovitis contain increased numbers of intensely class II-positive macrophages and dendritic cells, and the expression of MHC class II and CD1 molecules on chondrocytes is also altered (Harkiss *et al.*, 1991, 1995a). Similarly, the levels of both class II and CD1 molecules on dendritic cells have been shown to increase during secondary immune responses (Hopkins *et al.*, 1989). In addition, the presence of lower than normal levels of class I molecules on ovine squamous cell carcinomas is associated with invasiveness and their ability to evade immune surveillance (Townsend *et al.*, 1995).

Little work has been done on the interactions between ovine cytokines and MHC expression. However IFN-γ has been shown to upregulate the expression of both class I and class II molecules on alveolar macrophages (Nash *et al.*, 1992). Numerous studies have investigated a possible link between ovine MHC and disease or immune dysfunctions (Table XIV.8.2). Some associations between MHC haplotype and disease resistance or susceptibility have been made but these associations have not always been confirmed.

Table XIV.8.2 Associations between the ovine MHC and diseases or immune dysfunctions

Resistance to	*Breed*	*Typing system*	*Results*	*References*
Trichostrongylus colubriformis	Merino	Serological	SY1 phenotype present at a higher frequency in a line selected for high response to vaccination compared with one selected for low response	Outteridge *et al.* (1985, 1986, 1988)
	Merino	RFLP	No association with fecal egg count	Hulme *et al.* (1993)
Nematodes	Romney	Serological	SY1a + SY1b associated with low FEC, SY6 associated with high FEC – low numbers of sheep	Douch and Outteridge (1989)
Haemonchus contortus	Romanov	Serological	P14 phenotype associated with resistance	Luffau *et al.* (1990)
	Merino	Serological	No association found between MHC and FEC	Cooper *et al.* (1989)
		RFLP		Blattman *et al.* (1993)
Ostertagia circumcincta	Scottish Blackface	*DRB1* PCR typing	*Ovar-DRB1*0203L512* (G2) allele associated with low FEC	Buitkamp *et al.* (1994); Schwaiger *et al.* (1995, 1996)
		Serological		Stear *et al.* (1996)
Footrot	Merino	Serological	SY1b and SY6 associated with resistance	Outteridge *et al.* (1989)
		RFLP	Association between bands and susceptibility	Litchfield *et al.* (1992)
Corynebacterium pseudotuberculosis	Prealpe	Serological	Associations with time of onset of abscesses	Millot (1989)
Scrapie	Isle de France	Serological	A4, B6 haplotype associated with resistance to scrapie	Millot *et al.* (1988)
			More recent studies failed to confirm the association between the MHC and scrapie	Cullen *et al.* (1989)

Ovine-specific aspects of MHC genetics or protein structure and function

There appears to be little about the structure and organization of the ovine MHC that is specific to sheep. Both the types and genetics of ovine MHC molecules are very similar to those of cattle. The *Ovar-DQ1* and *Ovar-DQ2* gene duplication appears to reflect a *DQ* duplication in the artiodactyl lineage, and these genes both appear to be homologues of primate *DQ1* genes. No homologues of primate *DQ2* genes have been described for sheep or cattle. Similarly, *DY* genes appear to be an artiodactyl development as these genes have not been found in primates or rodents.

9. Sheep Red Blood Cell Antigens

Sheep red cell blood group antigens

Six genetic systems of red cell blood groups are currently recognized in sheep. The standard designations of these systems by capital letters A, B, C, D, M and R and the use of lower case letters to designate the well known antigenic specificities were adopted during the 1973 Workshop on Sheep Blood Groups held at Jouy-en-Josas (France) under the auspices of the International Society for Animal Blood Group Research (Nguyen, 1973).

In this standard system for nomenclature, the antigenic specificities, e.g. a and b detected in the A blood group system, are designated by the symbols Aa and Ab respectively. An exception is made for the R system (two antigens R and O), because this system differs in several aspects from the other blood group systems.

A list of 19 internationally accepted antigenic specificities belonging to six blood group systems of sheep is presented in Table XIV.9.1. The old symbols for specificities are shown in parentheses and the number of different alleles or phenogroups (i.e. haplotypes) observed in each system is indicated.

In sheep blood typing, rabbit complement is used in hemolytic test for detecting all blood group specificities, except for Da (D system) which is usually identified by agglutination test. All the sheep blood typing reagents are derived from immune-antisera, except for anti-R and anti-O which are natural antibodies found in the serum of certain sheep, goat and cattle.

The reactions of anti-R and anti-O against sheep red cells are mutually exclusive. Thus, sheep are classified, with respect to the R system, into three groups R, O and i, with i being the group of animals whose red cells are both

Table XIV.9.1 Sheep red cell blood groups

System symbol	*Internationally accepted specificities*[a]	*Number of recognized alleles or phenogroups*
A	Aa (A), Ab (K)	A^a, A^b, A^{ab}, A^-
B	Ba (P), Bb (B′), Bc (Y′), Bd (N′), Be (E′), Bf (E), Bg (O′), Bh (S), Bi (I)	About 100 phenogroups
C	Ca (C), Cb (C_x)	C^a, C^b, C^{ab}, C^-
D	Da (D)	D^a, D^-
M	Ma (M), Mb (L), Mc (M′)	M^a, M^b, M^{ac}, M^c
R	R and O antigens on red cells and in various body fluids and tissues	Interaction of genes at two separate loci R and I with two alleles at each locus: R^R, R^O; I^I, I^i

[a]All specificities are detected by hemolytic test, except for Da which is usually identified by agglutination test.

R and O negative. The R and O antigens are soluble substances which occur primarily in various body fluids and tissues. The antigens only appear on the red cells of lambs several days after birth. At least two separate loci (R and I) are involved in the expression of R and O blood group antigens (Rendel, 1957). From both serological and genetic points of view, the R system of sheep is homologous with certain blood group systems in other species, notably the J systems of cattle and goats and the A system of pigs (Stormont and Suzuki, 1961; Rasmusen, 1962, 1981).

The A system of sheep was originally described as a simple blood group system with one antigenic specificity Aa (A) and two phenotypes Aa and no-Aa (Rasmusen *et al.*, 1960). A second blood group specificity, Ab, detected later by several laboratories, permitted distinction of four phenotypes (Aa, Ab, AaAb and negative for both) and four phenogroups (A^a, A^b, A^{ab} and A^-).

The B system of sheep is, like its homologues in cattle and goats, very complex (Rasmusen, 1960; Nguyen and Ruffet, 1975; Nguyen, 1990). There is a large number of sheep B specificities, but only nine are internationally accepted specificities.

The C blood group system of sheep is also complex. The genes controlling sheep C specificities are located approximately nine map units from the I locus which influences the expression of R and O antigens (Nguyen, 1985). There is also a close linkage between the C system and the locus controlling amino-acid transport in sheep red cells (Tucker *et al.*, 1980).

Preliminary data indicate that the A, B, C and M blood group systems are mapped on sheep chromosomes 6, 10, 20 and 18, respectively (Crawford *et al.*, 1995).

The M system of sheep is a unique red cell blood group system which has a demonstrable physiological effect. The Mb (L) antigen acts as an inhibitor of active potassium transport in sheep red cells (Ellory and Tucker, 1970).

Blood transfusion and blood group incompatibility between mother and offspring have no practical importance in sheep breeding and there is no evidence for consistent associations of production or reproduction traits with sheep red cell blood groups. Therefore, the relevance of sheep blood polymorphisms is confined to the use of these genetic systems as markers in the detection of improperly recorded matings, in the investigations of genetic variation within and between sheep breeds and in the study of the evolutionary history of related species.

Other sheep red blood cell antigens

In addition to the red cell blood group allo-antigens whose chemical nature remains virtually unknown, sheep also carry other surface antigens on their erythrocytes.

Although sheep erythrocytes were often used as antigen in immunological works, e.g. the separation and functional assays of antigen specific mouse lymphocytes forming rosettes (Elliott, 1979) and the induction of a secondary antibody response *in vitro* with rabbit peripheral blood lymphocytes (Luzzati, 1979), the red cell surface molecules engaged in the antigenicity did not get much attention. Consequently, the most well known antigenic structure found on sheep red cells was the Forssman antigen, whose ubiquitous nature was intriguing and attractive.

Originally found in the alcoholic extract from guinea-pig kidney which inhibited the immune hemolysis of sheep red cells, Forssman antigen also occurred in various animal species such as horse, goat, dog, cat, mouse, chicken, etc. It was also recognized in certain bacteria and plants, but was not detectable in cattle, pig, rabbit and rat. In sheep, Forssman activity was observed in a glycolipid extracted from red cells (Diehl and Mallette, 1964). This ubiquitous antigen, whose chemical structure was identified as α-GaINAc Globoside, can also appear in certain human tumor cells from Forssman-negative patients (Hakamor, 1984).

10. Ontogeny of the Immune System

The average length of gestation in sheep is around 150 days, with minor breed-to-breed variation. A number of developmental changes that occur in the immune system of the fetal lamb over this interval have been well characterized.

Table XIV.10.1 Onset of haemopoietic activity in major lymphoid organs

Organ	*Onset of haemopoietic activity*	*Comments*[a]
Yolk sac	d16–17	Vitelline and embryonic circulations fuse on day 17–18 Yolk sac disappears around day 27
Liver	d21–22	Major haemopoietic organ until a few weeks before birth Site of generation for early lymphoid precursors. Myeloid and erythroid cells produced throughout fetal life
Thymus	d27–30	For first few days after colonization, myeloid and lymphoid cells occur in the thymus. Lymphocytes then become the predominant constituents
Spleen	d48–50	Supports production of all leukocyte lineages. May be an important site for early production of B cells. Distinct separation of red and white pulp evident by 80–85 days
Lymph nodes	d55–60	Lymph nodes in anterior part of body develop first especially parathymic and prescapular
Bone marrow	d70	Contains predominantly myeloid cells. Throughout fetal life, less than 5% of BM cells have a lymphoid phenotype
Peyer's patches		
Jejunal	d80–85	Follicles colonized specifically with B cells which then proliferate rapidly
Ileal	d105–110	Emigration of B cells to periphery may be limited during fetal life

[a]For further details see reviews by Al Salami *et al.* (1985) and Miyasaka and Morris (1988).

Development of lymphoid organs

The stage in fetal ontogeny when haemopoietic activity can first be detected in the major lymphoid organs is summarized in Table XIV.10.1. In most instances, the anlagae of the respective organs are visible as discrete aggregations of loose connective tissue one or two days before colonization with haemopoietic cells.

Development of T cells

The early fetal liver contains precursor lymphoid lineages but cells expressing mature T-cell markers are absent. The first $CD8^+$ T cells occur in the thymus at about d35 and the CD4 marker is first detectable a few days later (Mackay *et al.*, 1986a; Maddox *et al.*, 1987a). A few mature $CD8^+$ and $CD4^+$ T cells occur in the periphery for the first time about 1 week after the initial thymus colonization, at about d40 (Mackay *et al.*, 1986a). Using available reagents, $\gamma\delta$ T cells are first detected in thymus and periphery several days after the appearance of the $CD4^+$ and $CD8^+$ T cells (Maddox *et al.*, 1987a). In the thymus, the $\gamma\delta$ T cells constitute only 1–3% of thymocytes and localize prominently around Hassal's corpuscles (McClure *et al.*, 1989).

The number of T cells in the periphery increases steadily throughout fetal life and the solid lymphoid organs are progressively populated with them. The spleen, lymph nodes and jejunal Peyer's patches develop T-cell domains broadly similar to the adult-type pattern by the late stages of fetal ontogeny (Maddox *et al.*, 1987b,c). Both the $\alpha\beta$ and $\gamma\delta$ T cell lineages develop by a thymus-dependent pathway during fetal life (Hein *et al.*, 1990). However, the peripheral circulating $\gamma\delta$ subset is more severely abrogated after early fetal thymectomy and remains permanently depleted. Conversely, the peripheral $\alpha\beta$ T cell pool is able to regenerate itself after fetal thymectomy, either by the expansion of emigrants which left the thymus before its removal or by extra-thymic differentiation pathways (Hein, 1994). The rate and pattern of emigration of $\alpha\beta$ and $\gamma\delta$ T cells from the thymus changes during fetal and early postnatal life (Witherden *et al.*, 1994). After birth, the concentration of all T cell subsets in blood increases to reach a maximum between 6 and 8 months of age and then declines slowly. In the first few weeks to months after birth, $\gamma\delta$ T cells are unusually predominant and may comprise up to 50–60% of circulating T cells. A more extensive description of the ontogeny of sheep T cells is available in a recent review (Hein, 1994).

The usage of TCR C_γ, V_γ and V_δ genes is developmentally regulated and distinctive repertoires are expressed at different times of fetal and postnatal development (Hein and Dudler, 1993). This most likely reflects an ordered programme of rearrangement and splicing of gene segments during T cell development in the thymus, although differential selection of emigrants may also contribute. The usage of TCR V_β genes during fetal ontogeny is also developmentally regulated (Hein, unpublished results).

Development of B cells

The site where precursor fetal lamb B cells begin Ig gene rearrangement and then progress to form the first mature B cells remains unknown. Earlier reports described surface Ig^+ B cells in the blood of 45-day-old fetuses. Using more recently developed monoclonal antibodies, mature $sIgM^+$

B cells were detected for the first time in day 48 fetal spleen (Press *et al.*, 1993). B cells remain scarce in lymph nodes and most other tissues during fetal life and constitute only 3–5% of circulating cells. Lymph node primary follicles are first identifiable at day 80 (Halleraker *et al.*, 1994). B cells, or indeed any cells of lymphoid phenotype, form a minor population in the bone marrow. The follicles comprising the jejunal and ileal Peyer's patches (JPP and IPP, respectively) are a significant exception. Each of the many thousands of follicles becomes colonized with a small number of B-cell clones, at around day 80 and day 105 respectively (Reynolds and Morris, 1983), and the B cells then proliferate rapidly by a self-renewing mechanisms (Griebel and Ferrari, 1994) to form the predominant intrafollicular lymphoid population. B cells begin to mutate their Ig genes in these tissues, thereby diversifying the expressed antibody repertoire (Reynaud *et al.*, 1991). This process, which is antigen-independent, continues for the first few months after birth and the IPP is the major source of peripheral B cells at this time (Gerber *et al.*, 1986; Reynaud *et al.*, 1995). Before birth, the JPP has several functional similarities to the IPP but after birth it adopts a conventional role as an important site for the induction of antigen-induced mucosal immune responses (Griebel *et al.*, 1996a). Peripheral B cell numbers increase rapidly after birth and stabilize at about 6–8 months. This corresponds to the time at which the IPP undergoes programmed involution, a process which may be hastened by various types of stress. The precise contribution made by other organs to B lymphopoiesis after regression of the IPP has not been clarified. Germinal centres develop as prominent components of the spleen and lymph nodes, particularly those nodes exposed to chronic antigenic challenge from mucosal surfaces. Presumably, these act as sites for affinity maturation of antibodies and for the production of memory B cells.

Development of other cell lineages

Primitive myeloid cells (monocytes and macrophages, granulocytes, megakaryocytes) occur in the fetal yolk sac and liver at day 20. At this time, some of these cells are able to distinguish foreign material or cellular detritus and actively phagocytose particulate matter in the apparent absence of immunoglobulins, opsonins and complement. The frequency of myeloid cells decreases in the liver after day 75 and other tissues such as splenic red pulp and bone marrow become prominent sites for myelopoiesis. Early fetuses contain a distinctive free-floating population of macrophages in the circulation but these disappear by birth and are apparently replaced by conventional monocytes. As the lymph nodes develop from about day 60 onwards, the lymphatic sinuses become populated with phagocytic macrophages and polymorphonuclear cells (Miyasaka and Morris, 1988). The reticular cells contributing to the stroma of some lymphoid organs, such as the ileal Peyer's patches, undergo morphological and enzymatic changes during fetal ontogeny (Nicander *et al.*, 1991).

Development of lymphatic system and lymphocyte recirculation

Lymphatic vessels begin to develop as endothelial outgrowths at the base of the neck near the jugular venous confluence (jugular lymph sacs) and in the lumbar region near the bifurcation of the vena cava (ileal lymph sacs). Vessels then develop to give rise to the lymphatics of the head, neck and forelimbs and the hind-quarters and the cysterna chyli (Miyasaka and Morris, 1988). Lymph nodes develop at specific sites as a reticular framework around blood capillaries. They generate vascular sprouts that connect with the system of developing lymphatic vessels from around day 65 to establish the patency of the afferent and efferent drainage to and from the developing lymph node. Lymphocyte recirculation through tissues probably begins at a slow rate soon after this time, allowing for leukocyte entry into nodes chiefly via afferent lymphatics draining to the subcapsular sinus; this may play a significant role in establishing the cellular elements of lymph nodes (Morris and Al Salami, 1987). However, the parenchyma remains noticeably lymphopaenic until after day 70. By day 80, there is extensive blood–lymph recirculation of lymphocytes through specialized venules in lymph nodes and the cellularity of the parenchyma increases (Pearson *et al.*, 1976; Simpson-Morgan *et al.*, 1985). From this time, fetal lymph nodes produce a vigorous cellular and humoral response after antigenic challenge (Hugh *et al.*, 1985).

Ontogeny of immune responsiveness

In order to mount a specific immune response (antibody production and cell-mediated reactivity), the cellular elements that perform the recognitive and reactive tasks must be present and the lymphoid tissues must have a level of structural and functional organization that will allow the appropriate cellular interactions to proceed. The capacity to mount an immune response after deliberate antigenic challenge develops surprisingly early in fetal lambs but its qualitative nature changes during ontogeny. Also, the fetal lamb does not respond in a uniform way to all antigens. These differences in the phenotype of the expressed response, which are summarized in Table XIV.10.2, reflect the differential maturation of various structural elements and functional capabilities of the immune system that occur during ontogeny.

Table XIV.10.2 Development of fetal immune responsiveness

Class of response	*Stage of ontogeny*	*Comments*[a]
Tolerance to		
Allografts	d50–60	Allogenic adult, but not fetal, skin grafted before this time is accepted. Fetal skin may not possess dendritic cells capable of inducing tolerance
Border disease virus (BDV)		Infection of young fetuses with BDV induces tolerance and the animal harbours virus throughout life
Allograft rejection	d75	Fetal and adult skin allografts are invariably rejected when grafted onto fetal lambs from this age onwards
Antibody response to		
ϕx-174	d40	Fetal antibody response is primarily of IgM class and of low titre
Ferritin	d55	
Haemocyanin	d75	Secondary immunization does not induce class switching before d80–90
Akabane virus	d75	
Brucella spp.	d110	
Salmonella	First 1–2 months of postnatal life	Fetal lambs do not produce antibodies when immunized with these antigens. The capacity to mount a humoral response develops after birth
BCG		
Diphtheria		
Isotype switch IgM to IgG	d80–90	Capacity to produce IgG after secondary immunization first appears around this stage

[a]Further details may be found in Miyasaka and Morris (1988).

Recent reviews

For general reviews of the ontogeny of the immune system in the fetal lamb see Al Salami *et al.* (1985), Morris (1986), Miyasaka and Morris (1988) and Kimpton *et al.* (1994). For a specific review of T-cell development see Hein (1994). The role of the ileal Peyer's patch in B-cell development has been reviewed by Griebel and Hein (1996) and Reynolds (1997).

11. Transmission of Passive Immunity in the Sheep

There is effectively no prenatal transfer of immunoglobulin (or any other large protein) from the ewe to the fetal lamb. The placentation in sheep is syndesmochorial – the chorionic trophoblast is in direct contact with uterine subepithelial connective tissue, and the uterine epithelium is maintained throughout pregnancy in the intercaruncular areas but is eroded within the caruncles. Placental attachment occurs at the caruncles (specialized aglandular regions of the uterine epithelium). In the sheep there are usually 88–96 caruncles which develop an intimate physical association with the fetal cotyledons to form placentomes (Brambell, 1970). In most respects the placentation of sheep is essentially similar to that in the cow and goat.

The primary function of the mammary gland is to provide the infant with a rich supply of nutrients until it becomes more independent and self-sufficient. In the sheep (as for other ungulates) the mammary gland has assumed the additional role of being uniquely responsible for transferring passive immunity from mother to young. This contrasts with many other species of mammals in which immunoglobulin is transferred from maternal to fetal circulation via the placenta during the last trimester of pregnancy (e.g. human), or via the yolk sac (e.g. rabbit).

During the last 2–3 weeks of pregnancy in the ewe, the mammary gland accumulates immunoglobulin-rich colostrum, which is ingested by the neonatal lamb immediately after birth. The mechanism for transporting immunoglobulin from blood into the colostrum-forming mammary gland is both active and highly selective (Watson, 1973). Richards and Marrack (1963) were the first to observe that IgG1 (then known as 'fast' IgG) was selectively accumulated, relative to IgG2 or other macromolecules, in the colostrum of sheep. Subsequently, Mackenzie and Lascelles (1968) showed that selective transport of IgG1 (relative to IgG2) continued into lactation, although the magnitude was significantly less during lactation than during colostrogenesis.

The currently-accepted mechanism for selective transport of IgG1 into colostrum of ruminants has been developed from studies in both sheep (Watson, 1973) and cattle (Larson *et al.*, 1980), and is described in detail by Watson (1980). Briefly, IgG1 molecules in maternal blood attach to Fc-specific receptors on the basal or intercellular membranes of acinar epithelial cells. Pinocytosis of this membrane results in membrane-bound vesicles which contain more IgG1 than IgG2 or other macromolecules. The 'transport' vesicle then traverses the cytoplasm of the epithelial cell and discharges its contents, by reverse pinocytosis, into the lumen of the alveolus. This results in greater levels of IgG1 than IgG2 in colostrum, the concen-

Table XIV.11.1 Concentrations of immunoglobulins (mg/ml) in ovine blood serum, colostral whey and milk whey[a]

Body fluid	*IgG1*	*IgG2*	*IgM*	*IgA*
Blood serum (adult)	17.97	6.30	1.79	0.16
Colostral whey	80.60	2.47	4.09	3.62
Milk whey	0.70	0.08	0.03	0.08
Blood serum (neonate 24-h old)	8.95	0.35	1.34	1.15

[a]Data from Watson (1973) and Dawe *et al.* (1982).

tration ratio of the two isotypes being a function of the number and binding avidity of the Fc receptors which sequester IgG1 from blood. The magnitude of transport would be a function of frequency of vesicle formation, and transport activity across the cell. During lactation (compared with colostrogenesis) it is probable that there would be both fewer IgG1 receptors on the epithelial membrane and reduced transport activity. This explanation is supported by the observation that there is an inverse relationship between selective transport of IgG1 and *de novo* synthesis of milk-specific proteins and lactose by the mammary gland (Lascelles, 1969; Watson *et al.*, 1972).

The concentration of immunoglobulins in blood, colostrum and milk is shown in Table XIV.11.1.

Neonatal lambs are normally on their feet and suckling within 2 h after birth. Following ingestion of colostrum, clotted casein is retained in the stomach and IgG1-rich whey passes into the small intestine from where absorption of macromolecules into the circulation takes place (Simpson-Morgan and Smeaton, 1972). The absorption is nonselective and is mediated by highly-vacuolated, immature mucosal epithelial cells; it continues for 24–36 h after birth, at which time the immature cells are replaced by a mature epithelium that is incapable of absorbing macromolecules (Brandon and Lascelles, 1971): this process has become known as 'gut closure'. Although the mechanism of absorption of immunoglobulin from the intestinal lumen is nonselective, because of the preceding selectivity in favour of IgG1 in the colostrogenic mammary gland, the neonatal lamb's circulation carries much more maternal IgG1 than other immunoglobulin isotypes.

It is now generally accepted that, in contrast to the situation for nonruminant mammals, the IgA system in the mammary gland of the ewe is poorly developed (see Table XIV.11.1). Although a small amount of local IgA synthesis does occur in the gland, there is no selective transport mechanism to transfer IgA into mammary secretions. Studies in sheep have found little involvement of the mammary gland in a common mucosal immune system. In nonruminants IgA in milk is important in mediating immunological protection from gastro-enteric pathogens within the lumen of the alimentary tract. There is little evidence for an analogous role for milk-derived IgA in lambs; indeed, with IgG1 levels in milk almost 10 times higher than for IgA it is possible that IgG1 is more important than IgA in providing local protection in the gut during the nursing period (see Snodgrass and Wells, 1978).

The mammary tissues of ewes are heavily infiltrated with lymphocytes and macrophages during colostrogenesis (Lee and Lascelles, 1970), with most of the lymphocytes being $CD8^+$ $CD5^-$ (Lee *et al.*, 1989). However, the number of leukocytes in colostrum is quite variable with a predominance of neutrophils (41–84%), followed by macrophages (8–49%) and lymphocytes (6–11%) (Lee and Outteridge, 1981).

There is limited evidence, mainly from nonruminant species, that cell-mediated immunity may be transferred from mother to infant via immunocompetent lymphocytes in colostrum (Kumar *et al.*, 1989). Sheldrake and Husband (1985) found that mammary leukocytes could traverse the intestinal wall of the neonatal lamb, but the number of cells involved was extremely small and it is doubtful that any significant immune protection is conferred upon the lamb by this process.

Probably of much greater biological importance to the lamb than passive transfer of colostral leukocytes is the absorption of nonimmunoglobulin, immunomodulatory proteins or peptides from colostrum. It has been known for some time that the colostrum of various species contains a variety of proteins/peptides, usually in low concentration, that are involved in stimulation of cell growth and/or immunomodulation (reviewed in Watson, 1990). In this connection it has been found that bovine colostrum is a very rich source of insulin-like growth factors (IGFs) (Francis *et al.*, 1986), and an immunologically active nonapeptide has been isolated from ovine colostrum (Wieczorek *et al.*, 1979). Recent studies in our laboratory have confirmed that ovine colostral whey contains immunomodulatory proteins (Watson *et al.*, 1992; Wong and Watson, 1995). It is likely that such proteins, when absorbed from ingested colostrum into the circulation of the neonatal lamb, play a significant role in development of active immune responses in the infant.

In the 1970s reports in the literature estimated biological half-life of passively-acquired IgG in neonatal lambs in a range from 7 days (Cripps and Lascelles, 1974) to 25 days (Pearson and Brandon, 1976). The biological half-life of IgM was estimated to be 6 days and the biological half-life of IgA to be only 2 days (Pearson and Brandon, 1976). More recently, studies using antibody activity as a measure of the functional integrity of immunoglobulin, estimated the biological half-life in neonatal lambs to be 18–24 days (Watson, 1992). Of course, the duration of protective immunity conferred upon the lambs by colostral immunoglobulin is a function of the amount of specific antibody in colostrum and the volume of colostrum ingested as well as the biological half-life of the immunoglobulin in the lamb's circulation. It is therefore inevitable that the duration of the protection conferred by maternal immunoglobulin will be highly variable – passive immunity to gastrointestinal nematode parasites declines to negligible levels at about 10 weeks old (Watson and Gill, 1991). For viral (Cunliffe and Graves, 1970), bacterial (Sojka *et al.*, 1978) and unidentified microbial infections (Halliday, 1965) the duration of protection conferred to the lamb by colostral antibody is variable, but is generally considered to be at least 6 weeks and could be as long as 6 months (Halliday, 1978).

Failure of transfer of passive immunity in lambs has dire consequences. McGuire *et al.* (1983) reported that for lambs that failed to acquire maternal antibody in the first 36 h of life 45% died before the age of 3 weeks, whereas for lambs with adequate passive immunity the analogous mortality was 5%.

As for other ruminants (Husband and Lascelles, 1975), passive immunization of the neonatal lamb with colostral antibody may inhibit the development of active immune responses (Filmer and McClure, 1951), but the benefits of maternally-derived, passive immunity are overwhelming. Indeed, from the point of practical vaccination the following strategies are generally compatible and comprise an excellent disease-prevention protocol: immunization of the ewe in late pregnancy to maximize colostral antibody, and primary immunization of the lamb at 'marking' (castration and tail-docking at about 4 weeks old) followed by secondary immunization of the lamb at weaning (12–16 weeks old).

12. The Neonatal Immune System

Precocious development of the fetal immune system

T-cell development

The immune system of sheep, like that of humans, develops differently from that of mice. Whereas the peripheral immune system of mice undergoes very little expansion before birth, the peripheral lymphoid organs and thymus of humans and sheep develop substantially before birth. Immunological studies have shown essentially the same pattern of development for the thymus, spleen, lymph nodes and Peyer's patches in the human and ovine fetus, and the human infant and lamb reach a similar stage of immunological development at birth (Cahill and Trnka, 1980; Kimpton *et al.*, 1994, 1995). Gestation is relatively long in the sheep (150 days) and the placenta sequesters the developing fetal immune system from exogenous antigen and maternal Ig, ensuring that the developing fetus is truly immunologically naive, and the development of fetal immune competence occurs by a regulated program of development independent of antigen (Cahill and Trnka, 1980; Kimpton *et al.*, 1994).

T cells commence recirculating at least as early as day 75 of gestation and the size of the peripheral T-cell pool increases exponentially during intra-uterine life as new cells are added to the blood and lymphoid tissues (Pearson *et al.*, 1976; Cahill and Trnka, 1980, 1981; Kimpton *et al.*, 1994). The predominant lymphocyte recirculating in the fetus is the T cell and all the major T cell subsets including CD4, CD8, and $\gamma\delta$ T cells present in the adult are circulating in the fetus (Kimpton *et al.*, 1989, 1990, 1995). Early fetal thymectomy largely prevents the development of this peripheral T-cell pool (Pearson *et al.*, 1976; Hein *et al.*, 1990).

There is also an extensive circulation of naive T cells and dendritic cells through skin and extra-lymphoid tissues which is established as part of the normal development of the fetal immune system, indicating that developmentally programmed factors, as opposed to antigen, play a significant role in determining lymphocyte traffic patterns during T-cell ontogeny (Kimpton *et al.*, 1995). The development of naive T-cell traffic circuits through peripheral tissues during fetal life in the absence of foreign antigen is remarkable, and may be related to eliminating cells reactive to self components not expressed in the thymus.

B-cell development

The growth and development of, and lymphopoiesis in, fetal and postnatal Peyer's patches has been studied extensively by Reynolds who has established that the ileal Peyer's patch is a major site of lymphopoiesis and is the principle source of B cells in the sheep (Reynolds and Morris, 1983, 1984). Primordial Peyer's patches are first seen at 60 days gestation, lymphoid follicles first appear at 75 days and by 120 days Peyer's patches follicles are histologically mature. In the fetus there are fewer B cells than T cells. B cells make up 4–5% of cells circulating in blood and lymph except for lymph draining the ileal Peyer's patch which contains 15% B cells (Kimpton *et al.*, 1990). A high proportion of these cells are thought to be emigrating directly into lymph from their site of production in the ileal Peyer's patch.

The immune system at birth

The transition to postnatal life

At birth the neonatal lamb displays a highly developed immune system associated with a pool of more than 10^{10}

recirculating lymphocytes which has arisen as part of the normal growth and development of the fetus (Pearson *et al.*, 1976). Fetal recirculating lymphocytes are long-lived cells which continue to recirculate in the fetus from the time they first enter lymph, but within a few weeks after birth they have been largely replaced by a pool of short-lived lymphocytes (Cahill and Trnka, 1981; Cahill *et al.*, 1985).

The acquisition of immune competence in lambs occurs over a considerable period both *in utero* and during early postnatal life. During fetal development the lamb develops an extensive array of immune capabilities from quite early in gestation and is capable of generating antibody responses to a range of bacterial and viral antigens injected i.v. (Silverstein *et al.*, 1963; Fahey and Morris, 1978). Fetal lambs have also been able to respond to oral doses of ferritin (Husband and McDowell, 1975) or *Brucella abortus* (Richardson and Conner, 1972) and the cellular arm of the immune response is also well developed, with fetal lambs able to reject allogeneic skin grafts from around 70 days of gestation (Silverstein *et al.*, 1964). There are, however, certain antigens which stimulate no specific antibody response during fetal development, such as somatic antigens of *Salmonella*, which fail to stimulate a specific antibody response until some time after birth (Fahey, 1977).

At birth, a process is triggered whereby, over the next 10 days, 90% of these fetal lymphocytes are replaced by new lymphocytes which have been formed after birth (Cahill and Trnka, 1981). In this way the fetal immune system is largely replaced by a new neonatal immune system, which renews itself every week to 10 days. The relationship between the fetal immune system and the neonatal immune system, i.e., the relationship between the fetal lymphocytes and those new lymphocytes making up the immune system after birth is unknown.

The neonatal immune system during the first month

T-cell development

For the first few days after birth, the levels of white cells in the blood fall and then increase to reach a maximum around 6 months old. The rise in level of white cells in the blood is due essentially to an increase in the number of lymphocytes. The mean blood lymphocyte count in lambs at birth is 4×10^6/ml, and this rises to 9×10^6/ml in lambs 6 months old (Cole and Morris, 1973). The cell content of lymph increases manyfold after birth, particularly in the thoracic and intestinal ducts. From birth to 10 weeks old, the cell output in the intestinal lymph increases about sixfold from 2×10^8 cells per hour to around 12×10^8 cells per hour (Morris, 1986). A significant increase in the total cell output and in the number of blast cells in the intestinal and thoracic duct lymph occurs within 72 h of birth. No such changes occur in lymph from the lumbar trunk or in lymph from the popliteal node, and it appears likely that these alterations occur as the result of antigenic material penetrating the gut mucosa. The spleen and lymph nodes also increase substantially during this period, the spleen growing somewhat faster than the lymph nodes. The mesenteric lymph nodes and Peyer's patches have a more rapid growth rate than the popliteal, prefemoral, or lumbar nodes (Cole, 1969; Washington *et al.*, 1992).

The dramatic increase in the number of cells circulating through the gut after birth is associated with the appearance in the early weeks of life of a population of T cells which home specifically to the gut (Cahill *et al.*, 1977; Cahill and Trnka, 1981). The fact that gut-homing T cells only appear after birth and exposure to antigen when there is a rapid increase in the weight of the gut-associated lymphoid tissue and the circulation of cells through the gut, suggests a possible role for antigen in the generation of gut-homing. This is supported by experiments which show that gut homing properties do not have to be imprinted on T cells inside the thymus because late fetal thymectomy fails to prevent the development of this gut homing population after birth (Cahill *et al.*, 1996). It is also possible that there is a qualitative change in the export of T cells from the thymus after birth which results in the export of cells whose gut-homing properties are imprinted intrathymically.

In lambs during the first 1–2 weeks after birth, the number of $\alpha\beta$ T cells decrease slightly while the number of $\gamma\delta$ T cells expand rapidly (Hein and Mackay, 1991). Sheep, along with other ruminants, and chickens have a higher proportion of $\gamma\delta$ T cells in the peripheral T cell pool than do mice and rats. Although $\alpha\beta$ T cells constitute the majority of T cells in the periphery in sheep, both sheep and cattle show a high proportion of $\gamma\delta$ T cells in the blood which is age-related (Hein and Mackay, 1991; Washington *et al.*, 1992). The percentage of $\gamma\delta$ T cells in blood lymphocytes of sheep increases from around 15–20% at birth to 30–50% in the early months after birth, and then declines gradually with age to around 5–10% in animals of 5–8 years old (Hein and Mackay, 1991). This increase in blood $\gamma\delta$ T cells is abolished by fetal thymectomy (Hein *et al.*, 1990). Fetal thymectomized animals are born with only 10% of the normal number of circulating T cells. By 1 year old, however, CD4 and CD8 have reached about half the normal complement of $\alpha\beta$ T cells whereas the appearance of $\gamma\delta$ T cells in the periphery is virtually abolished (Hein and Mackay, 1991). In the absence of the thymus, therefore, $\gamma\delta$ T cells do not expand their numbers in the periphery and it appears that the increase in their numbers after birth is due mostly to a large increase in their rate of export from the thymus (Witherden *et al.*, 1994). In this context it has been suggested that peripheral $\gamma\delta$ T cells in young sheep are long-lived cells with a much lower proliferative rate than T cells of the $\alpha\beta$ lineage (McClure *et al.*, 1988a,b).

B-cell development

In addition to changes in the proportion of T cell subsets there is also a rapid expansion in the proportion of B cells circulating in neonatal lambs. During the first few weeks of postnatal life the proportion of B cells in blood and lymph increase from around 3–4% to 30–40% of the circulating lymphocyte population. Ileal Peyer's patches are a major site of B-cell production during development (Reynolds and Morris, 1983) and by 3–4 weeks after birth the Peyer's patches extend as a continuous band for about 1 m along the length of the small intestine back from the ileo-cecal junction (Morris, 1986). Removal of the terminal ileal Peyer's patch before birth prevents the tenfold increase in the percentage of B cells in blood and lymph which normally occurs in the first few weeks after birth (Gerber *et al.*, 1985; Morris, 1986). Although the large increase in the numbers of B cells in blood, lymph and lymphoid tissues probably reflects the rapid post-natal development of the ileal Peyer's patches, ileal Peyer's patches in isolated segments of bowel do not develop normally (Reynolds and Morris, 1984) and it is not known whether they need some form of nonspecific bacterial stimulation or whether some aspect of digestion is necessary for their normal development.

The B-cell repertoire is produced in ileal Peyer's patches by somatic hypermutation of light chain V_λ genes by mechanisms which target complementarily-determining regions and which are antigen independent. This generation of antigen receptor diversity among B cells is associated with massive cell death in ileal Peyer's patch, suggesting that a strong selective pressure operates on developing B cells (Reynaud *et al.*, 1995). Cell death in Peyer's patches is similar if not greater than that which occurs in the thymus. New B cells are formed in ileal Peyer's patches at 5–9 weeks old at 3×10^9 cells/h yet only 1.5×10^8 cells/h (or 5%) actually leave the Peyer's patch and enter the peripheral B-cell pool (Reynolds, 1986; Pabst and Reynolds, 1986).

In addition to the cells in the Peyer's patches there is a substantial population of small lymphocytes associated with the intestinal epithelium. In the fetus these numbers are sparse but shortly after birth the lamina propria increases in thickness and eventually contains large numbers of lymphocytes. Plasma cells are also present in the lamina propria by the end of the first two weeks after birth and the numbers of small lymphocytes within the intestinal epithelium increase rapidly. Reynolds (1976) calculated that at birth the gut had 3–4 lymphocytes per 100 epithelial cells, whereas 2 months after birth there were 30–35 lymphocytes per 100 epithelial cells. A substantial proportion of both intra-epithelial lymphocytes and lamina propria cells in the gut belong to the $\gamma\delta$ T cell lineage, which suggests that they are important in surveillance of intestinal pathogens and in early immune protection at mucosal surfaces.

Colostral immunoglobulins and immune defence mechanisms

Even with a range of immune capabilities at birth, the newborn lamb is relatively refractory to challenge with a range of antigens (Richardson *et al.*, 1971; Halliday, 1978). The impermeability of the ruminant placenta to maternal globulins results in the newborn lamb having extremely low levels of immunoglobulins (Pearson and Brandon, 1976; Campbell *et al.*, 1977). Unless specifically challenged with antigen, normal lambs deprived of colostrum show no increase in the amount of IgM until about the end of the second week after birth, with IgG appearing in significant amounts between 30 and 50 days after birth (Cole and Morris, 1973). However, specific antigen challenge at birth or soon after hastens the production of immunoglobulins in newborn lambs and stimulation of colostrum-deprived lambs with swine influenza virus can produce large quantities of immunoglobulins (Cole and Morris, 1973). Newborn lambs have been reported to be nonresponsive to Gram-negative bacterial antigens although this non-reactivity can be circumvented by prenatal immunization (Richardson *et al.*, 1968). Young lambs are generally more prone to infections than adults and give only a weak active immune response when vaccinated (Halliday, 1978). Newborn lambs are particularly vulnerable to infections because their immune systems are still developing and they usually have had no previous exposure to the antigens they are likely to encounter.

An important defence mechanism for neonatal lambs is the passive absorption of immunoglobulins in colostrum. The early ingestion of colostrum by the newborn lamb (the lambs small intestine becomes incapable of absorbing maternal immunoglobulins intact after 24 h) is of critical importance. After lambs ingest colostrum the concentrations of IgG1, IgM and IgA in serum increase sharply and reflect the quantities of these proteins in sheep colostrum, which is extremely rich in IgG1 (Pearson and Brandon, 1976; Campbell *et al.*, 1977). However, there is very little IgG2 in colostrum which is the only immunoglobulin class cytophilic for neutrophils (Watson, 1975), raising the possibility that neutrophil function may be less than optimal in neonatal lambs. A further study suggesting suboptimal neutrophil function found that *in vitro* Ig-mediated phagocytosis by neutrophils in 2-day-old lambs was only 30% of adult controls (Bernadina, 1991). It is, nonetheless, unknown whether neutrophil function in young lambs *in vivo* is sufficiently compromised to render them more susceptible to opportunistic infections. In 2-day-old lambs initially high concentrations of Ig were followed by decreasing values, with estimated half lives of maternal Ig being 24.3 days for IgG1, 6.4 for IgM and 1.7 days for IgA. From 16 to 32 days after lambs are born the concentrations of IgG1, IgG2 and IgM either remain stable or increase, indicating a balance between catabolism and endogenous production and all Ig classes increase by 64 days old (Pearson and Brandon, 1976).

It has also been demonstrated in neonatal lambs that lymphocytes of maternal origin present in colostrum are also able to enter the duodenum and be transported via lacteals to mesenteric lymph nodes during the period before gut closure; however, the significance of this finding to the neonates immune capabilities and development remains unclear (Sheldrake and Husband, 1985). In neonatal lambs the failure to ingest colostrum increases the lambs susceptibility to generalized bacterial infections, particularly of the coliform group (Campbell, 1974; Campbell *et al.*, 1977). It has also been shown that lambs ingesting colostrum will have better survival rates and faster weaning rates than those which do not (Treacher, 1973). In neonatal lambs, although absorption ends during the first 24 h with the establishment of gut closure, antibodies are still secreted in the milk and may help survival by combating pathogens such as rotaviruses in the pharynx and gut (Halliday, 1978).

Concluding remarks

The fetal lamb develops a large pool of peripheral T cells with extensive TCR diversity, and the ability to respond to a wide range of antigens. This confers a substantial capacity to mount immune responses to environmental pathogens in neonatal lambs, which, nonetheless, are severely compromised if they do not absorb colostral immunoglobulins during the first 24 h of life. Although extensive T-cell traffic through lymph nodes and extralymphoid tissues is established during fetal life, a gut-homing circuit develops only during the neonatal period, suggesting a key role for antigen in its generation.

13. Innate Immune Mechanisms in the Sheep

The natural barriers and defence mechanisms of sheep are summarized in Table XIV.13.1. The known acute phase proteins of sheep are summarized in Table XIV.13.2. The known properties and effector mechanisms are summarized for polymorphonuclear leukocytes (Table XIV.13.3) and monocyte/macrophages, natural killer cells, and dendritic cells (Table XIV.13.4).

The reticuloendothelial system and immune defence in sheep

There are only a few reports in the literature describing the ovine reticuloendothelial system. Histological examination of tissues reveals a distribution of mononuclear phagocytic cells similar to that of other mammals (e.g. alveolar macrophages, liver Kupffer cells, lymph node medullary sinus macrophages, microglial cells, osteoclasts and histiocytes). Tissue inflammatory reactions to subcutaneous carageenan have been detected as early as 75 days of the 147 days of gestation in the sheep (Kumta *et al.*, 1994) and phagocytosis of particles has been detected in fetal macrophages as early as 24 days of gestation (Al

Table XIV.13.1 Natural barriers and defence mechanisms

Organ	*Effector*	*Function*	*References*
Skin	Wool, hair, lanolin, keratin	Protection against the weather; maintenance of body temperature; protection against pathogens Desquamation sheds microorganisms	Jenkinson (1990, 1992)
	Sebaceous glands; sweat glands	Sebum and sweat on the skin surface forms protective 'cement'. Acidic pH on skin surface may restrict colonization	Jenkinson (1990, 1992)
	Langerhans cells; dermal dentritic cells	Foreign protein capture. Wound repair. Forms a cell network (pathogen trapping?)	Bujdoso *et al.* (1989); Jenkinson *et al.* (1990); Lear *et al.* (1996)
Rumen	pH 6.8, anaerobic environment	Supports commensal microorganisms which can prevent some pathogens from establishing. In the first 24 h of life, the rumen is not functional and may allow pathogens to establish in the abomasum which is at neutral pH at this time	Phillipson (1997)
Lung	Epithelial cells	Cationic β-defensins	Ganz *et al.* (1990)
	Surfactant	Anionic antimicrobial peptide	Brogden *et al.* (1996)
	Lung fluid	Antibodies and phagocytic cells protect lung epithelium. TGF-β detected	Burrells (1985); Begara *et al.* (1995)
Reproductive	Trophectoderm IFN-τ	Immunosuppressive. Antiviral. Pregnancy recognition signal. Inhibits GVH disease in mice. Upregulated by GM-CSF	Bazer *et al.* (1994); Jarpe *et al.* (1994); Assalmeliani *et al.* (1995); McGuire *et al.* (1995)

Table XIV.13.2 Ovine acute-phase proteins

Acute phase	*Protein*	*Function*	*References*
Early (local) Activated (e.g. LPS stimulated) epithelial, endothelial and macrophage cells	The cytokines: IL-1β, TNF-α, IL-6.	Not tested. Possibly act on liver cells to stimulate late-phase reactants	Fiskestrand and Sargan (1990); Young *et al.* (1990); Andrews *et al.* (1991); Green and Sargan (1991); Nash *et al.* (1991); Seow *et al.* (1991); Ebrahimi *et al.* (1995)
Late (systemic)	Haptoglobin, ceruloplasmin and fibrinogen Albumin	Not tested. Elevated in blood plasma following endotoxic shock or pulmonary damage Decreased in plasma	Pfeffer and Rogers (1989); Levieux and Venien (1991); Scott *et al.* (1992); Pfeffer *et al.* (1993) Skinner and Roberts (1994)

Table XIV.13.3 Ovine polymorphonuclear leukocytes and mast cells

Cell-type	*Granule mediators*	*Properties*	*References*
Neutrophil	Azurophilic and specific granules. Elastase; acid phosphatase; myeloperoxidase; alkaline phosphatase; superoxide; lactoferrin; β-glucuronidase	Phagocytosis. Increase in bacterial infections. Susceptibility to *P. haemolytica* cytolysin. Lower production of superoxide compared with bovine or human neutrophils. Lack of lysozyme. Lack of large granules seen in bovine neutrophils. C5a and IL-8-mediated migration. LPS-induced expression of TNF-α. Recruited by IL-1β, IL-8, GM-CSF, TNF-α. G-CSF augments neutrophilia	Rausch and Moore (1975); Buchta (1990); Junger *et al.* (1992); Rosolia *et al.* (1992); Mulder and Colditz (1993); Silflow and Foreyt (1994) Koizumi *et al.* (1993); Wickersham *et al.* (1993); Seow *et al.* (1994); Cirelli *et al.* (1995); Haig *et al.* (1995c); Persson *et al.* (1996a,b)
Eosinophil	Major basic protein; eosinophil cationic protein; peroxidase; arylsulphatase	Increase in helminth parasite infections. High frequency of precursors and mature cells in the bone marrow compared with the human and rodents. Express α_4-integrins (VLA-4). Respond to IL-5, PAF, IL-3 and GM-CSF. Accumulate in nematode antigen-challenged mammary gland. Depleted in CD4$^+$ T-cell-depleted and parasitized sheep	Dawkins *et al.* (1989); Haig *et al.* (1990, 1995b); Adams and Colditz (1991); Gill (1991); Buddle *et al.* (1992); Stevenson and Jones (1992); Topper *et al.* (1992); Gill *et al.* (1993); Rothwell *et al.* (1993); Abraham *et al.* (1994); Stevenson *et al.* (1995); Wunderlin and Palmer (1995); Woolaston *et al.* (1996)
Basophil	No mediators described. Smaller than mast cells and eosinophils	Basophil responses to various stimuli poor compared with guinea pigs	Yamada and Sonoda (1972); Rothwell *et al.* (1994)
Mast cell (mucosal and connective tissue)	Sheep mast cell protease (SMCP); histamine; β-hexosaminidase; arylsulphatase; leukotrienes (LTC_4); FcεR1$^+$. Few surface antigens shared with other leukocytes	SMCP has tryptase and chymase activity. Respond to 48/80. Activity regulated by heparin. Correlation with protection to certain nematodes. Globule leukocytes (degranulated mast cells) predominate in parasitized gut. IL-3-dependent development from haemopoietic precursor cells and proliferation	Huntley *et al.* (1984, 1986, 1987, 1992); Miller (1984); Haig *et al.* (1988, 1991); Chen *et al.* (1990); Gill *et al.* (1993); McInnes *et al.* (1993); Sture *et al.* (1995)

Table XIV.13.4 Ovine monocyte/macrophages, NK cells and dendritic cells

Cell-type	*Properties*	*Function*	*References*
Monocytes and macrophages	Phenotypic and biochemical heterogeneity in the tissues. In general, stain for nonspecific esterase. Phenotype: MHC-class-I and II, CD45, CD11a, CD11b; CD11c, CD44	Phagocytosis. Develop from haemopoietic precursors in GM-CSF, M-CSF, IL-3 and serum factors. Alveolar macrophages respond to LPS or infection to produce: IL-1β, TNF-α, IL-6, GM-CSF, G-CSF and respond to mitogens to produce prostaglandins, thromboxanes, leukotrienes and hydroxy-eicosatetraenoic acids. Recruited *in vivo* by LPS, IL-1β, IL-8, GM-CSF and TNF-α. TNF-α and IL-1-β receptors. Virus-induced procoagulant activity via tissue factor. Transformed myelomonocytic line developed	Haig *et al.* (1991); Lucio *et al.* (1992); McInnes *et al.* (1993); Francey *et al.* (1992); Ellis *et al.* (1991); Nash *et al.* (1992); Cirelli *et al.* (1995); O'Brien *et al.* (1994); Silflow *et al.* (1991); Persson *et al.* (1996a,b); Winstanley (1992); Fiskestrand *et al.* (1994); Lena *et al.* (1994); John *et al.* (1994)
NK cells	Large granular lymphocytes	Express the conserved NK function-associated molecule. Activated by type 1 interferon. Inhibited by endometrial serpin-like proteins. Active during sheep pox infection	Harris *et al.* (1993); Tuo *et al.* (1993); Liu and Hansen (1993); Bach *et al.* (1995)
Dendritic cells (DC)	Four phenotypes in skin: Langerhans cells and dermal DC (DDC) including migrating afferent lymph DC (ALDC). All MHC class II$^+$: Differ in expression of acetyl cholinesterase (LC only, CD1, and Coagulation factor XIII	Wound repair? ALDC function in antigen capture and presentation. DDC accumulate in cutaneous orf virus infections. DDC recruited into skin by GM-CSF and TNF-α. ALDC grow/survive in GM-CSF and TNF-α, inhibited by IFN-γ. ALDC have receptors for IL-1β and bind IgM and IgG	Jenkinson (1990); Lear *et al.* (1996); Morris and Courtice (1977); Miller and Adams (1977); Hopkins *et al.* (1989); Bujdoso *et al.* (1989); Harkiss *et al.* (1990); Hein *et al.* (1987); Haig *et al.* (1995a, 1995c); Fiskestrand *et al.* (1994)

Salami *et al.*, 1985). In adult sheep, aspects of alveolar macrophage phenotype and function have been studied (see references in Tables XIV.13.1–XIV.13.4). The cells of the mononuclear phagocytic system are the mature tissue-resident progeny of haemopoietic precursors in adult bone marrow and spleen. Ovine haemopoiesis is similar to that of other mammals (reviewed by Al Salami *et al.*, 1985; Haig, 1992), originating in the yolk sac at around 16 days of gestation, then in fetal liver up to approximately 140 days, and finally in the spleen and bone marrow in neonates and adult animals. The cytokines involved in maintaining haemopoiesis in the sheep (e.g. the colony-stimulating factors IL-3, GM-CSF, G-CSF, kit-ligand and IL-5) are structurally and functionally similar to the equivalent rodent and human cytokines (reviewed by Haig *et al.*, 1994).

Ovine-specific aspects of the structure and function of nonspecific immune mechanisms

The ovine innate immune system has been only tentatively studied. The available data suggest that the ovine system is overtly similar to that of other mammals. Ovine-specific responses (e.g. arachidonic acid metabolism by alveolar macrophages, Silflow *et al.*, 1991) may reflect a combination of the environment inhabited by sheep and the effect of selective breeding by man. Differences in wool thickness and length may influence breed differences in resistance to ectoparasites and other skin pathogens. Sheep in different geographical locations will experience infection with different pathogens. There are differences in this respect also between upland and lowland breeds (Henderson, 1990; Martin and Aitken, 1991). Ovine neutrophils are particularly sensitive to *Pasteurella haemolytica* cytotoxin (Silflow and Foreyt, 1994). Gastro-intestinal nematode infections are a cause of serious economic loss in farmed sheep world-wide, and persistent challenge during the evolution of the species may explain why sheep have large numbers of globule leukocytes in gastrointestinal epithelium (Huntley *et al.*, 1984) as well as eosinophils and their precursor cells in haemopoietic tissue compared with other species including rodents and man (Haig *et al.*, 1990, 1995b). Interferon-τ is a unique interferon species discovered in sheep and involved with the regulation of pregnancy (Bazer *et al.*, 1994). The anionic microbicidal peptide isolated from ovine lung surfactant fluid by Brogden *et al.* (1996) has also only been described in sheep. This is a member of the defensin family of peptides and has microbicidal activity against a range of bacteria in the presence of antibodies. The 175 antigen on the surface of ovine myeloid and erythroid cells has amino acid sequence similarity to a family of antimicrobial serine proteases found in neutrophils and macrophages (Deane *et al.*, 1995). No rodent or human counterpart has been described. Further work is required to determine whether there are any significant differences in innate immune mechanisms in sheep compared with other species.

Inducible nitric oxide synthase expression by goat macrophages

The term 'activated macrophage' describes a functional stage in which macrophages display enhanced antimicrobial and antitumoral activity. A key effector pathway of activated macrophages is the conversion of arginine into citrulline and nitric oxide (Green *et al.*, 1991; Nathan, 1992). The latter is a highly reactive diffusible gas with many molecular targets. In the context of activated macrophages, NO reacts with iron of enzymes essential for cell division, energy flow and survival. This limits growth of intracellular pathogens and neoplastic cells (Hibbs *et al.*, 1988; Nathan and Hibbs, 1991). NO generation by macrophages has been shown to be essential for growth control and/or elimination of protozoan parasites (Green *et al.*, 1991), intracellular bacteria (Flesch and Kaufmann, 1991) and viruses (Karupiah *et al.*, 1993).

Sustained NO generation by macrophage depends on the expression of an inducible nitric oxide synthase (iNOS). iNOS is not expressed by resting macrophages, but is induced by bacterial constituents such as lipopolysaccharide (LPS), by cytokines such as IFN-γ, or by a combination of the two (Ding *et al.*, 1988). This notion is entirely based on work performed with rodent macrophages. The role of NO generation and iNOS in human macrophages is highly controversial, and NO generation appears to occur at a lower level and under highly restricted activation conditions (James and Nancy, 1993; Schneemann *et al.*, 1993; Albina, 1995). This calls for an extension of investigations on macrophage iNOS in other species.

The expression of nitric oxide synthase by caprine macrophages activated with bacterial stimuli with or without IFN-γ was determined. Cultured goat macrophages produced one order of magnitude less nitric oxide (NO) than cultured bovine macrophages (Table XIV.13.5). Both stimulated bovine and caprine macrophages generate low amounts of nitrate, since there is little difference between assays for nitrite and nitrate combined and nitrite alone. Stimulated caprine macrophages were activated because there was a steady level of procoagulant activity (Table XIV.2.1) and indoleamine 2,3-dioxygenase. The activity of iNOS depends on a variety of cofactors, such as FAD, FMN (Stuehr *et al.*, 1991), NADPH and tetrahydrobiopterin (4HB) (Kwon *et al.*, 1989). However, 4HB was detected in lysates of activated caprine macrophages, which suggested that 4HB was not limiting NO synthesis.

Immunohistochemistry revealed that when appropriately stimulated the majority of bovine macrophages strongly express iNOS but only a minority of caprine macrophages express iNOS. The amount of iNOS expressed by activated macrophages was also determined by immunoblotting cell lysates with a specific monoclonal iNOS antibody. Activated bovine macrophages produced a conspicuous band at a molecular mass of 130 kDa but only a faint band was observed with lysates of caprine macrophages (Adler *et al.*, 1996). Finally, iNOS mRNA was determined by reverse transcription–polymerase chain reaction, using primers conserved between rat, human and bovine iNOS. Activated caprine macrophages displayed a stronger iNOS mRNA signal than nonactivated counterparts, but this signal was weaker than that of resting bovine macrophages. Thus, the lower level of NO production by goat macrophages resulted from a lower expression of steady-state mRNA and protein for inducible nitric oxide synthase (iNOS).

A sequence comparison showed that for a fragment close to the 5′ end of the coding sequence and consisting of

Table XIV.13.5 Induction of procoagulant activity and generation of nitrite by activated bovine and caprine macrophages derived from blood monocytes

		Response	
Species	*Stimulus*	*Recalcification time (s)*[a]	*Nitrite production in 24 h (μM)*
Cattle	None	158 ± 12	<2
	S. dublin (200 μg/ml)	25 ± 0	10.9 ± 1.7
	LPS (1 μg/ml)	58 ± 8	4.2 ± 0.5
Goat	None	172 ± 12	<2
	S. dublin (200 mug/ml)	25 ± 0	<2
	LPS (1 μg/ml)	65 ± 0	<2

[a]A decrease in recalcification time reflects enhanced expression of procoagulant activity, as there is a log–log relationship between coagulation shortening, as determined by turbidimetry, and concentration of cell surface-expressed tissue factor.

332 base pairs, there is a similarity of 97% at the mRNA level, and of greater than 99% at the protein level, between cattle and goat iNOS. The sequence similarity of this fragment between caprine and human iNOS is 94% and 98% for mRNA and protein, respectively. This indicates a high degree of conservation of this enzyme between species, which contrasts with the highly variable regulation. The latter may be due, in part, to differences in the structure of the promoter. No information on the sequence or activity of the 5′ flanking region of caprine or bovine iNOS is yet available.

Given the marked difference in iNOS expression by caprine and bovine macrophages it was of interest to look for iNOS expression *in vivo*, in lesions of animals suffering from bacterial infections. Listeric (*Listeria monocytogenes*) encephalitis is a lethal natural infectious disease of both cattle and goats. Brains of animals that succumbed to listeric encephalitis were sectioned and analyzed for expression of iNOS and for presence of *L. monocytogenes*. Remarkably, iNOS was expressed in lesions of cattle as well as of goats but expression was generally lower in goats than in cattle and the proportion of iNOS-expressing cells was slightly lower in goats than in cattle. Thus, the marked species difference observed in *in vitro* macrophage studies was not reflected to the same extent in the infected brains. *In vitro* assays indicate that the goat belongs to the low-responder species, together with sheep (Jungi *et al.*, 1996), rabbit (Schneemann *et al.*, 1993), pig (Turek *et al.*, 1994), dog (Tipold and Jungi, in preparation) and man (Schneemann *et al.*, 1993). However, expression of iNOS *in vivo*, in foci of bacterial infection, shows a significantly lower degree of species variation.

14. Complement

Early investigations of sheep serum complement activity used hemolytic assays and confirmed the existence of both the classical and alternative pathways of complement activation (Rice and Boulanger, 1952; Jonas and Broad, 1972; Barta *et al.*, 1975; Stankiewicz and Jonas, 1981). The role of complement in the pathophysiology of disease has been investigated for a variety of conditions (Stankiewicz *et al.*, 1981; Malu and Tabel, 1986; Colditz *et al.*, 1994; Heath *et al.*, 1994). These studies of complement activity have supported the assumption that the mammalian complement cascade is functionally and structurally conserved in sheep.

Studies of the protein and genetic structure of individual components of the complement cascade have been limited. The protein structure of only a few complement components has been described. The structure and function of a sheep CD59 analogue has been characterized for a protein isolated from erythrocyte membranes (van den Berg *et al.*, 1993). Sheep red blood cells also have a membrane protein with factor I-dependent cofactor activity, supporting the conversion of C3b to iC3b, but no decay-accelerating activity (Ezzel and Parker, 1992). The functional activities of this protein indicate that it may be a CD46 analogue. However, the presence of such a protein on sheep red blood cells would differ markedly from the pattern of CD46 expression in humans. Genetic studies have confirmed the presence of the C1r gene (Broad *et al.*, 1993) with at least four alleles (Phua and Wood, 1995), C4 with at least four genes and two functional isotypes (Groth *et al.*, 1988; Dodds and Law, 1990; Ren *et al.*, 1993; Schwaiger *et al.*, 1996) and a single Factor B gene that is linked to the major histocompatibility Class I microsatellite (Groth and Wetherall, 1995). Table XIV.14.1 summarizes the components of the complement cascade and indicates which components have been identified in sheep.

Very little work has been done to identify cell receptors for cleavage products of the complement components. Monoclonal antibodies specific for CD11b indicate that the CR3 receptor (iC3b receptor) is expressed on macrophages, granulocytes, and most lymphocytes (Gupta *et al.*, 1993). Several studies have been made with zymosan-activated sheep plasma (ZAP) which is a rich source of complement anaphylatoxins (i.e. C3a, C5a). The infusion of ZAP has dramatic effects on neutrophil activation (Perkowski *et al.*, 1983; Albertine *et al.*, 1989), degranulation (Rosolia *et al.*, 1992), margination (Meyrick and Brigham, 1984), and bone marrow development (Rosolia *et al.*, 1992). These observations indicate that neutrophils express receptors for anaphylatoxins and the expression of these receptors changes during the process of neutrophil maturation.

The level of complement activity varies among body fluids and changes with age. A hemolytic assay confirmed that all components of the classical complement cascade are present in normal sheep and fetal lamb serum (Jonas and Broad, 1972). The same hemolytic assay was negative when using cerebrospinal fluid, lymph, lung washes, and colostrum. However, heat-labile complement components could be detected in most of these fluids with antisera specific for sheep complement components. Studies of the classical and alternative complement cascade demonstrated significantly lower activity in newborn lambs (Oswald *et al.*, 1990). Activity of the classical pathway decays for approximately 3 months after birth and then reaches adult levels at 1 year old. In contrast, activity of the alternative pathway is low during the first 6 weeks after birth and then increases to adult levels at approximately 3 months. The post natal decay in complement activity and a low level of serum complement activity in newborn lambs, prior to suckling, indicates that there is a colostral transfer of complement components (Oswald *et al.*, 1990).

Studies of complement activity and protein and genetic data for sheep complement components support the conclusion that the mammalian complement cascade is conserved in sheep. There are no known acquired or inherited

Table XIV.14.1 Components of the sheep complement system

Complement component	*Protein structure/genetic information*	*References*
C1, C1q, C1s, C1r	C1r-Chromosome 3q; polymorphism	Broad *et al.* (1993); Phua and Wood (1995)
C2		
C3, C3b, iC3b		Rice and Boulanger (1952)
C4, C4a	Two functional isotypes, C4A (M_r 108 000) α chain. CB4 (M_r 95 000) α chain. At least three genes	Groth *et al.* (1988); Ren *et al.* (1993); Schwaiger *et al.* (1996)
C5, C5a		
C6		
C7		
C8		
C9		
Factor B	Single gene. Linked to MHC I on chromosome 20q15–q23	Groth and Wetherall (1995)
Factor D		
Factor H		
Factor I		
Properdin		
C1-inhibitor		
C4-binding protein		
Membrane cofactor protein (CD46)	High level activity on sheep red blood cells	Ezzel and Parker (1992)
Decay-accelerating protein (CD55)		
CD59	M_r = 19 000. Expressed on red blood cells and lymphocytes	van den Berg *et al.* (1993)
C8-binding protein		
S-protein (vitronectin)		
C3b receptor (CR1; CD35)		
C3d receptor (CR2; CD21)		
iC3b receptor (CR3; Mac-1; CD11b/CD18)	CD11b, M_r 170 000. Expressed on alveolar macrophages, blood mononuclear cells and granulocytes	Gupta *et al.* (1993)
C3a receptor		
C4a receptor		

complement deficiencies in sheep to provide further insight into the function of this system. However, unique aspects of sheep complement activation or regulation may exist if the observations for complement control proteins CD46 and CD59 are substantiated (Ezzel and Parker, 1992; van den Berg *et al.*, 1993). Thus, further investigations of sheep complement may provide greater insight into the development and functions of this integral component of the immune system.

15. Mucosal Immunity

Components of the mucosal immune system

The respiratory and alimentary tracts are the major components of the mucosal immune system (Table XIV.15.1). The mammary gland and the reproductive tract are also mucosal tissues, but do not appear to be important constitutive immunological tissues (Lee *et al.*, 1988, 1989; Outteridge and Lee, 1988; Alders and Shelton, 1990). The uterus has a population of granular intra-epithelial lymphocytes which appear to play a role in pregnancy or parturition (Meeusen *et al.*, 1993).

The liver and its draining lymph nodes are not usually considered part of the mucosal immune system but they are the sites of first immunological exposure to the many gut-absorbed antigens which enter the portal vein. These tissues should therefore be included in studies of gut-associated immunity. Normal liver contains B cells, CD4, CD8 and TCR $\gamma\delta$ cells in the portal tract and parenchyma, while WC1/T19 cells are rare and there is a vigorous cellular response in liver and hepatic lymph nodes during the immune response to local infection (Meeusen *et al.*, 1988, 1990).

Early findings suggest that local nerves also contribute

Table XIV.15.1 Components of the mucosal immune system (alimentary and respiratory)

Structure		*Leukocyte populations*[a]		*Function*	*References*
Constitutive[b]					
Organized	Peyer's patch	Follicle:	B cells, FDC	B cell development	Reynolds (1987); Landsverk *et al.* (1991); McClure and Emery (1993)
		Interfollicular:	T cells, IDC, mast cells, eosinophils, B cells	Induction of local immune response, effector of local immune response	
	Draining lymph node	Follicle:	B cells, FDC, T cells	Induction of local immune response, expansion of effector cells	Burrells (1985); Miyasaka and Morris (1988); Kimpton *et al.* (1989); Watt *et al.* (1992)
		Paracortex:	T cells, IDC		
		Medulla:	All leukocytes		
Diffused	Lamina propria	Basal:	Predominantly B cells, FDC, polymorphs	Induction of local immune response, expansion and modulation of local immune response; effector of local immune response	Winkler (1988); McClure and Emery (1994)
		Apical:	Predominantly T cells, IDC, mast cells/globule leukocytes		
		Intravascular:	Predominantly lymphocytes and macrophages	Modulator of systemic immune response	
	Mucosal epithelium	T cells, globule leukocytes		Effector of local immune response, modulator of systemic immune response, non-specific local protection	Brogden (1992); Gyorffy *et al.* (1992)
Reactive[c]					
Organized	Lymphoid follicles in lamina propria	B cells		Expansion of local immune response	Gorrell (1988c); Chen (1989)
Diffused	Intraluminal cells	Lymphocytes, macrophages, globule leukocytes			Burrells and Sutherland (1994); Stankiewicz *et al.* (1995)

[a] FDC, follicular dendritic cells; IDC, interdigitating dendritic cells.
[b] Constitutive tissues (present in the absence of foreign antigen) can be modified by local immune response to antigen.
[c] The systemic immune system also contributes to local mucosal defences, e.g. leakage of plasma IgG into gut mucus during the response to intestinal nematode parasites (McClure *et al.*, 1992).

to the mucosal immune response in a 'memory' fashion, possibly through local axon reflex arcs or CNS-derived impulses, to increase peristaltis, trigger inflammatory cell degranulation, and regulate lymphocyte activity and recirculation (Keefer and Mong, 1990; Stewart *et al.*, 1994; McClure *et al.*, 1994).

The function of the mucosal immune system, particularly that of intestinal responses, is closely tied to the nutritional status of the animal resulting from both nutrient intake and the presence of local lesions (Suttle *et al.*, 1992; Coop *et al.*, 1995; van Houtert *et al.*, 1995; Israf *et al.*, 1996).

Antigen transport and function of antigen-presenting cells

Very little is known of these functions in the sheep mucosal immune system other than the fact that they occur, as shown by the specific local responses which occur following intraluminal deposition of antigen (Davies *et al.*, 1986; Gorrell *et al.*, 1988a,b; McClure *et al.*, 1992). Macromolecular antigen uptake by M cells of jejunal Peyer's patches of lambs has been demonstrated, as has uptake by the follicle associated epithelium of ileal Peyer's patches, the latter uptake being the greater (Landsverk, 1987). The classical presentation of antigen by mucosal epithelial cells thought to occur in other species may not occur in sheep, whose intestinal epithelium does not express MHC Class II (Gorrell *et al.*, 1988b; Press *et al.*, 1991; McClure *et al.*, 1992). Cells bearing CD1 are present in intestinal lamina propria of adults and of fetal lambs by 85 days of gestation (McClure *et al.*, 1992; S. J. McClure and P. McCullagh, unpublished data), but their function has not been investigated.

Production and transport of mucosal immunoglobulin

The mucosal immune system is functionally distinct, although not separate, from the systemic system, and the rules governing systemic priming cannot be extrapolated to ensure induction of mucosal immune responses. Higher mucosal IgG and IgA levels are achieved by local immunological activity resulting, in turn, from local administration of antigen; systemic Ig levels are not a good indication of a protective mucosal response.

Secretory immunity involves the active transcellular transport of secretory immunoglobulins from tissue fluids to 'external' surfaces by epithelial cells of mucosal tissues. Mucosal immunoglobulins may originate locally or systemically, and immunoglobulin of mucosal origin may contribute to both the local humoral defence and to systemic antibody (Hall, 1986). All isotypes are present in gastro-intestinal secretions. Antigen-specific IgG and IgA antibodies and antibody-containing cells are found in mucus, lamina propria and draining lymph following intraluminal exposure to protein antigens and infective organisms (Curtain and Anderson, 1971; Smith *et al.*, 1984; Davies *et al.*, 1986; Donachie *et al.*, 1986; Stear *et al.*, 1995). Immunoglobulin in mucus results from plasma leakage and from active transport. The large number of IgG1-containing plasma cells in intestinal lamina propria is thought to contribute to the circulating IgG1 pool rather than to selectively secreted IgG1 in the gut, while most IgG1 and all IgG2 in intestinal secretions are derived from blood. However, virtually all IgA in intestinal secretions is locally produced (Lascelles *et al.*, 1985). About 90% of serum IgA is gut derived; some is contributed by lungs and mammary gland. Both migration of plasma cell precursors and selective transport of IgA from plasma to mucosal secretions contribute to IgA responses at remote sites (Husband *et al.*, 1987; Sheldrake, 1989). Little has been done with sheep IgE. It is present in lamina propria, germinal centres and mucus, but its source has not been identified.

Intramammary infusion of antigen results in IgG1 and IgA in mammary secretions, and either IgA- or IgG1-containing plasma cells in mammary tissue depending possibly on the antigen used (Sheldrake *et al.*, 1985; Lee *et al.*, 1992). While IgG1 and IgG2, IgM and IgA are present in milk, IgG1 is the predominant isotype in sheep colostrum (Lascelles *et al.*, 1985). Transfer of maternal immunoglobulins to the fetus occurs only in association with placental lesions (Miller, 1966; Poitras *et al.*, 1986).

Little is known of local antibody in the male reproductive tract other than that IgA is transferred selectively into secretions following local bacterial infection (Foster *et al.*, 1988).

Intra-epithelial lymphocytes (IELs)

The frequency, phenotype and distribution of IELs are summarized in Table XIV.15.2. Data is presented for mucosal sufaces and age groups that have been investigated.

Peyer's patches

Peyer's patches, organized aggregates of lymphoid follicles, are found throughout the small and large intestine. The lymphocyte composition, life history, and primary immune function of the various Peyer's patches are summarized in Table XIV.15.3.

The role of mast cells and eosinophils in mucosal immunity and inflammation

Mucosal inflammation has been studied in two systems, namely, gastro-intestinal parasitism and pulmonary hypersensitivity. Mast cell hyperplasia occurs during infection with nematode parasites (O'Sullivan and Donald, 1973), as does the appearance of large numbers of intra-epithelial globule leukocytes, the degranulated state of mucosal mast cells (Huntley, 1992). Both appear at the time the sheep develops protective immunity (Sommerville, 1956). Mast cells contribute to the local immediate hypersensitivity reaction that is responsible for rapid rejection by immune sheep of incoming parasites (Miller, 1985; McClure *et al.*, 1992; Wagland *et al.*, 1996). The mechanism probably involves inflammatory mediators released by primed mast cells when activated by specific antigen. These mediators include leukotrienes, histamine, prostaglandins, platelet-activating factor and sheep mast cell proteinase, and they are presumed to act to alter the permeability of intestinal epithelium, increase mucus secretion and induce epithelial shedding; some also affect motility of larval parasites (Douch *et al.*, 1983; Huntley *et al.*, 1986, 1987; Jones and Emery, 1991; Jones *et al.*, 1994; Bendixsen *et al.*, 1995). Association of mast cells and enteric nerve fibres suggests a functional inter-regulation (Stewart *et al.*, 1995).

Accumulation of eosinophils during nematode infection is more variable than mastocytosis, and may reflect more closely the presence of parasites than a protective immune response to them (Dineen and Windon, 1980; Douch *et al.*,

Table XIV.15.2 Intra-epithelial lymphocytes

Site	*Distribution*	*Phenotype*	*Frequency*	*References*
Fetal jejunum	Absorptive epithelium	CD4 CD8 TCR $\gamma\delta$	0.01% of epithelial area 0.6% of epithelial area (~5 mm) 3% of epithelial area	Gorrell *et al.* (1988b); S.J. McClure and P. McCullagh, unpublished data; Press *et al.* (1991)
Fetal jejunum	Follicle-associated epithelium	CD8 TCR $\gamma\delta$ IgM	<0.4% of epithelial area <1% ibid 12.4% ibid	Press *et al.* (1991)
Fetal ileum	Follicle-associated epithelium	CD8 TCR $\gamma\delta$ IgM	<0.4% of epithelial area <0.4% ibid 1.8% ibid	Press *et al.* (1991)
Young lamb jejunum	Absorptive epithelium	CD4 CD8 TCR $\gamma\delta$	0.8% of epithelial area 1.4% ibid 5.6% ibid	Press *et al.* (1991)
Young lamb jejunum	Follicle-associated epithelium	CD4 CD8 TCR $\gamma\delta$ IgM	<0.8% of epithelial area 0.1% ibid 2% ibid 0.7% ibid	Press *et al.* (1991)
Young lamb ileum	Follicle-associated epithelium	CD4 CD8 TCR $\gamma\delta$ IgM	<0.8% of epithelial area 0.1% ibid 1% ibid 0.8% ibid	Press *et al.* (1991)
Adult jejunum	Absorptive epithelium	CD4	5% of IEL	Gorrell *et al.* (1988b)
		CD5	50% of IEL	
		CD8	55% of IEL, 4% of epithelial area	Press *et al.* (1991)
		TCR $\gamma\delta$	18% of IEL, 3% of epithelial area	Gyoroffy *et al.* (1992)
		WCI/T19 CD2	<18% of IEL 75% of IEL	Mackay *et al.* (1989)
		MHCII CD8 CD5	70% of IEL 35% of CD8	Aleksandersen *et al.* (1995)
		CD8 MHCII TCR $\gamma\delta$ CD8 TCR $\gamma\delta$ CD5 TCR $\gamma\delta$ MHCII	90% of CD8 24% of $\gamma\delta$ 54% of $\gamma\delta$ 90% of $\gamma\delta$	McClure *et al.* (1992)
Uterus	Uterine and endometrial glandular epithelium	$CD8^+$ $CD45R^-$ TCR $\gamma\delta^-$	≈50% of IEL	Meeusen *et al.* (1993)
		$CD8^+$ $CD45R^+$ TCR $\gamma\delta^-$	≈25% of IEL	
		$CD8^+$ $CD45R^+$ TCR $\gamma\delta^+$	≈25% of IEL	

Intestinal epithelium does not express MHC Class II.
Intra-epithelial TCR $\gamma\delta$ cells differ from those in blood and lamina propria in not expressing the surface antigen WCI/T19.
The numbers of CD8 and TCR $\gamma\delta$ intra-epithelial lymphocytes (IEL) can increase or decrease dramatically on immunological stimulation or local infection (McClure *et al.*, 1992; Aleksandersen *et al.*, 1995).
Immunity to intestinal nematode parasites is associated with large numbers of intra-epithelial globule leukocytes, which are degranulated mast cells (Sommerville, 1956).
The author is not aware of published information characterizing the function of sheep intra-epithelial cells.

1986; McClure *et al.*, 1992). Most of these eosinophils in lamina propria of immune sheep express mRNA for IL-5, suggesting that autocrine regulation of eosinophilia occurs in the gut. TCR $\gamma\delta$ cells in the gut also express IL-5, and are therefore implicated in eosinophil regulation (Bao *et al.*, 1996). Parasite antigen may have a direct effect on eosinophils, as infusion into the mammary glands of worm-free sheep elicits an eosinophil-rich exudate (Adams and Colditz, 1991). Eosinophils also infiltrate the liver after local *Taenia* infection, but their role is unknown (Meeusen *et al.*, 1988).

Mast cells are also active in pulmonary immune

Table XIV.15.3 Intestinal Peyer's patches

Site	*Distribution*	*No.*	*Lymphocyte composition*		*Life history*	*Primary immune function*	*References*
Jejunum	Discrete	30–40	Interfollicular (large)	T cells: CD4, CD8, TCR $\gamma\delta$ B cells (few)	Appears 70 d.o.g.[b] Proliferation at 75 d.o.g. and persists throughout life	Intestinal sensitization to extrinsic antigen, proliferation and emigration of specific lymphocytes	1,2,3,4,5,6,7,8
			Follicular	B cells (40% $sIgM^+$) T cells (CD4)		B cell development role in fetus	
			FAE[a]	Clusters of IgM cells			
			Combined	B cells 50–75% of lymphocytes T cells: 16–25% $CD4^+$ 9% $CD8^+$; 3% TCR $\gamma\delta$			
Ileum	Continuous (1–2 m long)	1	Interfollicular (small) Follicle	T cells B cells (few) B cells ($\sim$80% of cells IgM^+) T cells (very few)	Appears 100 d.o.g. Proliferation at 110 d.o.g. Involutes about puberty	B-cell development in absence of extrinsic antigen – diversification of Ig repertoire by somatic hypermutation, deletion of putatively self-reactive cells	1,2,3,4,5,8,9, 10,11,12,13, 14
			FAE	No M cells, but particles shed from epithelium to central follicle		Export to periphery (fetus and young lamb)	
			Combined	B cells 70–95% T cells 1%		Sensitization to extrinsic antigen	
Large intestine	Discrete – ileocaecal	1	Morphology and lymphocytes as for jejunal Peyer's patches, except for greater number of TCR $\gamma\delta$ cells in rectum		Appears 65 d.o.g. Proliferation begins 70 d.o.g. Partial involution in adults	Developmental role in fetus (B and/or T cells)?	5,8,15,16
	proximal colon	1	FAE as for ileal Peyer's patch				
	proximal spiral colon	1	Combined: B cells 50^c–74% T cells 25–55%[c] (mostly CD4)				
Pharynx	Discrete	1			Persists		17,18

1, Larsen (1986); 2, Reynolds (1987); 3, Hein *et al.* (1989a); 4, Halleraker *et al.* (1990); 5, Landsverk *et al.* (1991); 6, Gyorhoffy *et al.* (1992); 7, Zanin *et al.* (1994); 8, Griebel and Hein (1996); 9, Pabst and Reynolds (1986); 10, Reynolds *et al.* (1991); 11, Reynaud *et al.* (1995); 12, Motyka *et al.* (1995); 13, Griebel and Ferrari (1995); 14, Renstrom *et al.* (1996); 15, Aleksandersen *et al.* (1990); 16, Aleksandersen *et al.* (1991); 17, Chen *et al.* (1989); 18, Chen *et al.* (1991).

[a]FAE, follicle-associated epithelium.

[b]d.o.g., days of gestation.

[c]Value is for young lambs.

responses, particularly in allergic responses. Both connective tissue and mucosal mast cell populations are present in sheep lung (Chen *et al.*, 1990). Hypersensitive sheep have more secretory granules in the mast cells than nonreactive sheep, although there are no differences in the number of mast cells or eosinophils. Thus, only mast cells appear to be involved in the initial development of allergic airway hypersensitivity (Chen *et al.*, 1991a). However, eosinophils may play a role in chronic allergic airway disease in sheep, as they are present in bronchoalveolar lavage fluid of affected sheep in elevated numbers (Bosse *et al.*, 1987).

16. Immunodeficiencies

Maedi-Visna virus (MVV) is a retrovirus of the subfamily lentivirinae which includes the immunodeficiency viruses, caprine arthritis encephalomyelitis virus (CAEV) and equine infectious anemia virus (EIAV). All these viruses infect cells of the monocyte/macrophage lineage but there is no lymphocyte infection with MVV, CAEV or EIAV (Joag *et al.*, 1996). MVV establishes a persistent infection of sheep but does not cause any of the gross immunodeficiencies which are seen with, for example, human or simian immunodeficiency viruses (Georgsson *et al.*, 1990; Blacklaws *et al.*, 1994b). This sheep lentivirus is therefore a valuable model for studying mechanisms of lentivirus persistence and pathology caused by infection of macrophages. This review concentrates on the contribution of the immune system to both the above aspects of the infection. MVV is very similar to CAEV and evidence may be taken from both virus systems in some instances.

There is some natural resistance to MVV among different sheep breeds although the basis for this is unknown (Petursson *et al.*, 1976; Narayan *et al.*, 1983; Cutlip *et al.*, 1986). The acquired immune response is now well documented and, at first sight, apparently normal. Cellular and humoral responses are seen within the lymphatics draining the site of subcutaneous infection by 10 days to 2 weeks post infection (Blacklaws *et al.*, 1995a). These include a polyclonal $CD8^+$ lymphocyte response (Bird *et al.*, 1993) as well as MVV-specific responses; a proliferative and cytotoxic T lymphocyte response and the induction of antiviral antibody (Bird *et al.*, 1993; Blacklaws *et al.*, 1995b). In blood, proliferative lymphocyte responses are seen 1–2 weeks after intracerebral or intranasal infections (Griffin *et al.*, 1978; Sihvonen, 1981; Larsen *et al.*, 1982b), cytotoxic T-lymphocyte responses by 1–3 months after subcutaneous infection (Lichtensteiger *et al.*, 1993; Blacklaws *et al.*, 1994a) and antibody by 1 month after infection by a variety of routes (Petursson *et al.*, 1976; Griffin *et al.*, 1978; Sihvonen, 1981; Larsen *et al.*, 1982b). The rate at which the immune response develops is slow compared with lytic virus infections (Whitton and Oldstone, 1996) but could be explained by the long replication time of MVV. In persistently infected animals lymphocyte proliferative and cytotoxic responses are detectable in a variety of tissues (Larsen *et al.*, 1982a; Sihvonen, 1984; Reyburn *et al.*, 1992; Torsteinsdottir *et al.*, 1992; Blacklaws *et al.*, 1994a; Lee *et al.*, 1994). Although the immune response may help to control the number of virus-infected cells (Bird *et al.*, 1993; Blacklaws *et al.*, 1995a,b) it never clears the infection completely. This may be due to dysfunction in the immune response to the virus or evasion of the immune response by the virus.

As the virus persists and continues to be expressed in infected sheep, pathology does develop. The most common clinical syndromes seen are interstitial pneumonia (Maedi or ovine progressive pneumonia), ataxia, wasting (Visna), mastitis, arthritis and lymphadenopathy as well as other tissues being affected (Sigurdsson, 1954; Sigurdsson and Palsson, 1958; Georgsson and Palsson, 1971; Schreiber *et al.*, 1975; Petursson *et al.*, 1976; Cutlip *et al.*, 1979; Georgsson *et al.*, 1982; Ellis and Demartini, 1985a; Vandermolen *et al.*, 1985; Kennedy-Stoskopf, 1989; Palfi *et al.*, 1989; Tuboly *et al.*, 1991; Narayan *et al.*, 1992). The pathology seen in all the different tissues affected by MVV is caused by a chronic active inflammatory process where there is the accumulation of lymphocytes and macrophages and germinal centre formation (Georgsson and Palsson, 1971; Petursson *et al.*, 1976; Deng *et al.*, 1986; Anderson *et al.*, 1992, 1994). There are also organ specific changes which include smooth muscle hyperplasia (lung) (Georgsson and Palsson, 1971), fibrosis (mammary glands) (Deng *et al.*, 1986) and gliosis (CNS) (Petursson *et al.*, 1976). Much of the pathology may be immune mediated and immunosuppression can reduce the number of lesions seen in the brain (Nathanson *et al.*, 1976). Immune activation may be caused by responses to persistent viral antigen expression or the induction of autoimmunity. The amount of viral antigen expressed correlates with the degree of pathology seen (Nathanson *et al.*, 1981; Klevjer-Anderson *et al.*, 1984; McGuire *et al.*, 1986; Brodie *et al.*, 1992) although, overall, little antigen is detected (Georgsson *et al.*, 1989; Brodie *et al.*, 1992; Watt *et al.*, 1992). Persistent antigen expression may allow the formation of immune complexes which help in the development of arthritis (Bertoni *et al.*, 1994). However, increased levels of T-cell proliferative and antibody responses to heat-shock protein 65 have also been documented in sheep with arthritis and since the levels of heat shock proteins are increased in inflamed joints there may be an autoimmune component of the disease (Harkiss *et al.*, 1995).

MVV is not a lytic virus infection and so does not cause major necrotic changes to cells, although nonproductive infection of oligodendrocytes has been postulated to lead to demyelination in Visna (Georgsson *et al.*, 1989). However, viral proteins may be directly toxic to host cells and indeed tat peptides are neurotoxic (Hayman *et al.*, 1993). Tat has also been implicated in helping to stimulate the lymphoproliferative component of the disease (Vellutini *et al.*, 1994).

Looking at the cell types which accumulate in lesions, there is an overall increase in all lymphocyte populations but especially $CD8^+$ lymphocytes (Harkiss *et al.*, 1991; Cordier *et al.*, 1992; Torsteinsdottir *et al.*, 1992; Watt *et al.*, 1992; Lujan *et al.*, 1993, 1995; Anderson *et al.*, 1994; Begara *et al.*, 1995b; Wilkerson *et al.*, 1995a,b). These lymphocytes appear to have impaired expression of the IL-2 receptor (Begara *et al.*, 1995a,b) and so may be dysfunctional. The macrophages present in lesions appear to be activated by increased MHC class II expression and fibronectin release but are not activated for phagocytic functions (Cordier *et al.*, 1990; Lujan *et al.*, 1993; Wilkerson *et al.*, 1995b; Lee *et al.*, 1996). Cytokine release from a variety of cell types may be altered (Lairmore *et al.*, 1988; Yilma *et al.*, 1988; Cordier *et al.*, 1990; Ellis *et al.*, 1991; Mdurvwa *et al.*, 1994; Werling *et al.*, 1994; Begara *et al.*, 1995c). Indeed, a cytokine called lentivirus interferon is released upon the interaction of MVV-infected macrophages and lymphocytes and this may play a role in the upregulation of MHC class II which is seen (Kennedy *et al.*, 1985; Narayan *et al.*, 1985; Zink *et al.*, 1987; Lairmore *et al.*, 1988; Yilma *et al.*, 1988; Zink and Narayan, 1989). The development of pathology may be determined by the type of T-cell response which is induced at infection. There is evidence in CAEV that a T_H2 response, which leads to antibody production, skews goats towards arthritis while those with a T_H1 response do not develop pathology (Perry *et al.*, 1995).

Some minor abnormalities in the immune response to MVV have been detected immediately after infection – loss of T cell proliferative responses to MVV antigen in the draining lymph node (Bird *et al.*, 1993; Blacklaws *et al.*, 1995b), together with reduced T-cell proliferation to mitogen and cytokine stimulation in efferent lymph (Bird *et al.*, 1993). It is also apparent that MVV-specific lymphocytes do not leave the lymph node until virus has disseminated through the body (Blacklaws *et al.*, 1995a). Whether the virus can specifically induce these effects to help it in its immune evasion or whether this is part of the control mechanism to stop inflammation and immune activation is not known.

The most important immune defect observed to date in persistently infected animals is a specific defect in the IgG2 response to MVV (Mehta and Thormar, 1974; Petursson *et al.*, 1983; Bird *et al.*, 1995). In ruminants, IgG2 may be important for antibody-dependent cell-mediated cytotoxicity (ADCC) reactions with neutrophil and monocyte/macrophage effectors (Feinstein and Hobart, 1969; Watson, 1975; Micusan and Borduas, 1977; Grewal and Rouse, 1979; McGuire *et al.*, 1979; Howard *et al.*, 1980; Yasmeen, 1981). MVV-neutralizing antibody is induced after infection (Gudnadottir and Palsson, 1965; Petursson *et al.*, 1976) but the affinity of these antibodies is lower for the virus than the affinity of the virus for its cellular receptor (Kennedy-Stoskopf and Narayan, 1986), and so may have no role to play *in vivo*. As most virus is cell associated, *in vivo* ADCC may be an important mechanism in viral clearance and a defect in IgG2 may therefore allow immune escape of virus-infected cells. Furthermore antibody titres generated to MVV are weak, being 10–20-fold lower than those generated by the parapox virus ORF (Bird *et al.*, 1995).

Immunodeficiencies are not a major symptom of MVV disease. However, lentivirus infected sheep and goats may display altered lymphocyte responses (Ellis and Demartini, 1985b,c; Begara *et al.*, 1995a) altered DTH responses to PPD (Pyrah and Watt, 1996), an increase in secondary *Pasteurella* infections (Myer *et al.*, 1988) and decreased macrophage function (Lee *et al.*, 1996). These observations suggest that the immune responses are suboptimal. These immune defects are not due to MVV infection of lymphocytes as this does not occur, but may be induced by viral infection of the dendritic cell compartment (Gorrell *et al.*, 1992). Similarly macrophage dysfunction caused by MVV infection may well play a role in abnormal DTH responses and increased bacterial infections.

Immune evasion by MVV may take the form of reduced antigen expression, increased mobility of infected macrophages or antigenic variation. There have been three replication states postulated for MVV: latent, where only proviral DNA is present; restricted, where there are a few hundred copies of viral RNA present but no antigen; and productive, where there are thousands of copies of viral RNA and antigen is seen (Haase *et al.*, 1977; Brahic *et al.*, 1981; Peluso *et al.*, 1985; Staskus *et al.*, 1991a; Brodie *et al.*, 1995; Haase, 1986). Proviral DNA has now been found in a number of cell types (Staskus *et al.*, 1991a; Brodie *et al.*, 1995). In the monocyte/macrophage lineage, MVV replication is tightly linked to the maturation state of macrophages (Narayan *et al.*, 1982, 1983; Gendelman *et al.*, 1986) through the control of cellular transcription factors for the LTR promoter (Small *et al.*, 1989). Overall, there is little antigen expressed, and few antigen positive cells are present, even after large infectious doses or at sites of pathology (Petursson *et al.*, 1976; Brodie *et al.*, 1992; Watt *et al.*, 1992; Blacklaws *et al.*, 1995b). In addition, the virus-infected macrophage may be particularly mobile, leaving the site of infection and primary infected lymph node to disseminate the infection rapidly throughout the body (Bird *et al.*, 1993; Blacklaws *et al.*, 1995a,b). Once an infected cell has left the original site of inoculation it may readily infect other tissues. MVV-infected macrophage progenitors have been detected in bone marrow (Gendelman *et al.*, 1985), which may be a continuous source of infected monocytes which will not express virus antigen until they have migrated into tissues and matured (Trojan Horses) (Gendelman *et al.*, 1985; Peluso *et al.*, 1985; Haase, 1986).

As with all retroviruses, replication of the MVV genome has no proof-reading function and so mutations accumulate rapidly (Wain-Hobson, 1996). If they prove to be advantageous to the virus they may be fixed in the virus population. Most MVV variation occurs in the *env* and *rev* genes (Narayan *et al.*, 1978; Scott *et al.*, 1979; Clements *et al.*, 1980, 1982; Sonigo *et al.*, 1985; Braun *et al.*, 1987; Stanley *et al.*, 1987; Querat *et al.*, 1990; Sargan *et al.*, 1991; Staskus *et al.*, 1991b). This may reflect selective pressure by the immune system and indeed neutralization-resistant mutants have been documented although the original virus strain is still detected late in infection (Narayan *et al.*, 1977, 1981; Scott *et al.*, 1979; Lutley *et al.*, 1983; Thormar *et al.*, 1983; Stanley *et al.*, 1987; Cheevers, 1993). There is no evidence for selective pressure on the LTR (long terminal repeat) sequence (Sargan *et al.*, 1995).

17. Tumors of the Immune System

Studies describing spontaneous lymphoid tumors in sheep are rarely reported (Johnstone and Manktelow, 1978; Di Guardo *et al.*, 1992; Lozano Alarcon *et al.*, 1992). Bovine leukemia virus (BLV) is not only the etiological agent of bovine leukosis, which is the most common neoplastic disease of cattle, but also of experimentally induced ovine leukosis. Natural transmission between sheep does not appear to occur (Djilali and Parodi, 1989) and was never observed in an experimental herd (Willems *et al.*, 1993). BLV, human T-lymphotropic viruses types I and II (HTLV I and II), simian T-lymphotropic virus and primate T-lymphotropic virus constitute a unique subgroup within the retrovirus family, characterized by a distinct genetic content, genomic organization, and strategy for gene expression (Burny *et al.*, 1994; Kettmann *et al.*, 1994; Cann and Chen, 1990; Gallo and Wong-Staal, 1990). The BLV sheep model is not only rewarding within itself but it also provides a valuable animal model for understanding some aspects of human retrovirus-induced diseases.

Clinical observations

Sheep are highly susceptible to BLV infection and have been preferentially used for experimental infection via several routes: by injection (i.p., i.v., i.d. or s.c.) or by oral administration (Mammerickx *et al.*, 1976). Specific antibodies can be detected 1–2 months after infection in all inoculated sheep following administration of various BLV-infected materials, i.e. whole blood, blood leukocytes, fresh or cultured lymphocytes derived from BLV-infected donors (cattle, sheep, goats) (Mammerickx *et al.*, 1988) as well as through the intradermal inoculation of proviral DNA mixed with a cationic liposome solution (Willems *et al.*, 1992).

Some infected animals develop a true lymphoid leukemia that persists for a few weeks or several months, until death of the animal, with severe leukemic changes in peripheral blood. White blood cell (WBC) counts range from 40 000 to 350 000/mm^3 at this terminal stage with an inverted B/T cell ratio, but no tumoral lesions are observed at necropsy (Boyt *et al.*, 1976; Olson and Baumgartener, 1976; Djilali *et al.*, 1987; Ishino *et al.*, 1989). Some animals develop lymphosarcoma, with no previous hematological disorders (Djilali and Parodi, 1989; Ohshima *et al.*, 1991). Lymph nodes are frequently altered by the extensive proliferation of neoplastic cells and the normal structure is no longer recognized. Neoplastic infiltration is also observed in different tissues (kidney, pancreas, intramuscular connective tissue of the heart, uterus, liver, lung, skin and spleen). Finally, lymphoid leukemia and localized lymphosarcoma frequently occur (Gatei *et al.*, 1989; Ishino *et al.*, 1989; Ohshima *et al.*, 1991; Murakami *et al.*, 1994a).

Although less than 5% of BLV-infected cattle develop tumors, nearly all experimentally infected sheep progress to and die during the tumor phase of the disease, with a shorter latency period than cattle (from 6 months to 6 years, average 3–5 years) (Djilali *et al.*, 1987; Djilali and Parodi, 1989; Gatei *et al.*, 1989). BLV infection, once established, is life-long.

Histological, immunological and biochemical studies

As in cattle, B lymphocytes contribute to any increase in blood lymphocytes during BLV infection of sheep (Djilali *et al.*, 1987; Djilali and Parodi, 1989; Gatei *et al.*, 1989; Ishino *et al.*, 1989; Ohshima *et al.*, 1991; Murakami *et al.*, 1994a,b; Van den Broeke, unpublished results). In tumors, cells react with mAbs against B-cell markers such as MHC class-II antigens, B2 and sIgM. In contrast, only minor populations express T-cell markers, such as CD4, CD8 and $\gamma\delta$TCR were observed (Dimmock *et al.*, 1990; Murakami *et al.*, 1994b; Van den Broeke, unpublished results). The percentage of sIg positive cells is at least 90% in blood of leukemic animals and more than 98% in BLV-induced lymphosarcoma (Ishino *et al.*, 1989; Djilali *et al.*, 1989). In sheep, neoplastic cells appear to be a monoclonal population of sIg-expressing B cells.

Expression of CD5 on transformed ovine B-lymphocytes is controversial; CD5$^-$ and CD5$^+$ B cell tumors have been described in sheep (Letesson *et al.*, 1990; Birkebak *et al.*, 1994; Murikami *et al.*, 1994a,b; Van den Broeke, unpublished data), whereas the transformed phenotype of BLV-target B lymphocytes is always CD5$^+$ in cattle (Depelchin *et al.*, 1989). These data suggest that in sheep the virus infects B cells with no preference for the CD5$^+$ B-cell population (Schwartz *et al.*, 1994).

As in cattle, sheep tumors appear to be a monoclonal expansion of cells carrying proviral information (Kettmann *et al.*, 1980b, 1984). Most of the tumor cells tested contained one BLV proviral copy per genome. In contrast, blood lymphocytes from aleukemic sheep and animals with early lymphocytosis are characterized by polyclonally integrated provirus. The appearance of a clonal subpopulation of cells among cells with polyclonally integrated provirus indicates the onset of leukemia (Rovnak *et al.*, 1993). However, tumors from different animals harbor the provirus at different sites, suggesting that the mechanisms for tumor initiation are independent of the integration site (Kettmann *et al.*, 1983; Gregoire *et al.*, 1984).

No p53 mutations were found in BLV-induced sheep tumors whereas in cattle, five out of 10 tumors harbored p53 mutations (Dequiedt *et al.*, 1995). It appears that p53 genomic alterations are not frequently (if at all) involved in BLV-induced leukemogenesis in sheep.

Viral expression

In vivo studies

Neither viral particles nor viral proteins or RNA have been readily detected in freshly isolated PBL or tumor cells (Baliga and Ferrer, 1977; Kettmann *et al.*, 1980a). Even transcriptionally competent proviruses are silent in BLV-infected sheep tumor cells (Van den Broeke *et al.*, 1988). Using *in situ* hybridization on freshly isolated PBL from clinically normal BLV-infected sheep, expression has only been detected in 1/2000 to fewer than 1/500 000 cells (Lagarias and Radke, 1989). In about 35% of the tumors, an average of 1/5000 cells contained viral RNA (Van den Broeke, unpublished results). The minimal *in vivo* expression of structural genes be a unique strategy allowing the virus to evade immunosurveillance and maintain permanent infection.

Ex vivo studies

Viral expression is induced in short-term cultures of blood lymphocytes isolated from sheep with elevated blood leukocyte numbers and an inverted B/T lymphocyte ratio (Stock and Ferrer, 1972; Baliga and Ferrer, 1977; Djilali *et al.*, 1987; Cockerell and Rovnak, 1988; Cornil *et al.*, 1988; Lagarias and Radke, 1989). Transcription could be detected as early as 30 min after blood leukocytes were placed in culture (Powers and Radke, 1992).

Transformed cell lines were established from tumor lymphocytes of BLV-infected animals with leukemia/lymphosarcoma (Kettmann *et al.*, 1985; Van den Broeke *et al.*, 1988, 1996; Rovnak *et al.*, 1993). The successfully established cell lines displayed the same surface antigenic profile and the same proviral integration sites as the original B-lymphocyte population, proving that the cells proliferating in culture were indeed derived from the clone present *in vivo*. In established ovine tumor cell lines the proviral information is silent and the culture is composed exclusively of B cells. *Ex vivo*, the morphologic and phenotypic homogeneity and the lack of viral expression in the BLV-induced tumor lymphocytes is unique to the sheep. The diseases caused by BLV in cattle and sheep are very closely related. However, differences between sheep and cattle in BLV transformation suggest that the ovine and bovine species require distinct events to achieve full malignancy.

Acknowledgement

Anne Van den Broeke is supported by Fondation MEDIC.

18. Experimental Investigation of Autoimmunity in the Sheep

General features

Autoimmune diseases of sheep have not been recognized as common conditions in clinical veterinary practice, although there has been some speculation that autoimmunity may contribute to development of the pathological processes in some infections. Similarly, the sheep has been employed only infrequently in experimental models of autoimmunity, reflecting perhaps both the uncommon nature of spontaneous autoimmune disease and logistic difficulties associated with this species, such as those generated by the inclusion of large numbers of animals in experimental groups. Nevertheless, the sheep affords some notable opportunities for investigation of autoimmunity, not all of which have been utilized.

Investigation of the acquisition of self/nonself discrimination can be undertaken in 'real time' in fetal lambs rather than being at the mercy of analogues that are believed to recapitulate developmental processes. An example of these is the use of irradiated animals reconstituted with various cell types as models of the fetal immune system. While such systems can provide data, it appears likely that a comprehensive understanding of the aberrations responsible for autoimmune disease will not be forthcoming until the processes normally responsible for the establishment of self-tolerance have been explained.

Use of fetal lambs for examination of self-tolerance or autoimmune reactions offers a rare opportunity to examine immune responses progressing in isolation from any influences of the external environment. While the maintenance of self-tolerance, as with any other immune phenomenon, has necessarily to occur in animals that are exposed to external environmental influences, the existence and generally uncontrolled nature of these will complicate, if not confound, the interpretation of observations. The degree of immunological isolation imposed upon the developing immune system of the fetal lamb by its placenta offers an opportunity to study regulation of self recognition in a simpler system more amenable to interpretation.

One attribute of than fetal lamb that reflects its isolation from the ewe is a marked hypogammaglobulinaemia. The fetal lamb possesses only that gammaglobulin that it has itself produced: in the absence of deliberate antigenic stimulation, its serum does not contain specific antibodies. As the nature and extent of contribution by B lymphocytes to the development and maintenance of autoimmune pathological features remains a confused issue in a number of experimental models and clinical presentations of autoimmune disease, study of autoimmune processes in the fetal lamb could permit some simplifications in interpretation.

Experimental induction of autoimmunity has most commonly been undertaken by challenging an animal with an autoantigen plus adjuvant. Another technique that can lead to autoimmunity, presumably by disturbing the regulation of antiself reactivity, entails disturbing the normal balance between regulatory and cytotoxic T-cells by thymectomy, with or without irradiation. A third experimental approach to elicit autoimmunity is to deprive the immune system of exposure to one or more autoantigens so that the capacity to recognize them as self is never acquired, or is lost. This approach can be initiated in postnatal life only if expression of the relevant autoantigen commences at an advanced stage of postnatal life (as, for example, is the case with autoantigens expressed only by mature gonads). Use of autoantigen deprivation as a strategy to study development of autoimmunity to other, earlier expressed, autoantigens requires intervention in fetal life and the sheep is especially suited to this.

Two other opportunities offered for extending existing investigation of autoimmunity in sheep have yet to be utilized, and these are discussed below.

Specific modifications of resting patterns of lymphocyte migration are usually considered to be an essential component of most immune responses although the precise nature of the changes and the mechanisms by which their effect on immune responsiveness is mediated are not well understood. Nevertheless, it would be surprising if alterations in lymphocyte migration were not also central to autoimmune processes. Investigation of the migration of specifically autoreactive lymphocytes during autoimmune reactions and of the significance of any modifications of migratory patterns for the progress of those reactions could best be undertaken by monitoring cell traffic in the lymphatics originating from the target organ. While this does not appear to have been done, the facility with which lymphatics from organs such as the testis can be cannulated in the sheep offers the opportunity to obtain novel data.

There is no *a priori* reason why autoimmune reactions and their pathological consequences should develop at a more rapid tempo in short-lived species of small laboratory animals than in longer-lived, larger animals. It is possible that more intensive study of experimentally induced and spontaneous autoimmunity in a species such as the sheep with a life span approaching 15 years could provide information relating to the regulation of clinical autoimmunity in human patients that is not readily apparent when the more rapid autoimmune processes that occur in short-lived species are studied.

Specific observations

Experimental allergic encephalomyelitis has been induced in sheep following the injection of whole sheep brain and complete Freund's adjuvant (Panitch *et al.*, 1976). The rationale in that experiment for essentially duplicating a procedure used many times in laboratory rodents was to test the hypothesis that some of the neuropathological features of Visna infections represented autoimmunity against central nervous system tissue. In the event, approximately half of the challenged sheep developed a fulminating lethal form of encephalomyelitis but the features of this were considered to differ from those of Visna to an extent that rendered an autoimmune contribution to the development of the latter unlikely.

Experimental allergic thyroiditis has been investigated in fetal lambs, taking advantage of several of the features listed above. Specifically, the development of self tolerance of thyroid autoantigens during fetal life was prevented by surgical removal of the thyroid gland at a stage before the lamb had the capacity to discriminate between self and nonself (McCullagh, 1989). In this experiment, the result recalled that of a similar experiment in larval amphibians performed much earlier. Removal of both thyroid lobes led to autoimmunity whereas hemithyroidectomy did not block self-tolerance of thyroid tissues. In a subsequent experiment, it was found that the transplantation of a fetal thyroid allograft into a fetus that had been thyroidectomized did not restore the normal process of learning to discriminate self from nonself (McCullagh, 1991). Thyroid tissue-specific antigens presented in association with nonself histocompatibility antigens were not adequate to educate the developing immune system. This remained so even if the thymus of the thyroidectomized fetus was also replaced by thymic tissue from the thyroid allograft donor (McCullagh, 1993).

While the experiments summarized above illustrated that self recognition of thyroid autoantigens by the fetal lamb is an acquired capacity, dependent upon their presentation in an appropriate context, they also provided a means to examine the nature of normal self-tolerance of these autoantigens. It is a consequence of the manner in which autoimmunity has been recognized and defined as a variation from the normal state of self-tolerance that this normal state may be taken to be 'background' in which nothing is observable. Reflecting this, investigation of self-tolerance has often been undertaken in animals that are autoimmune so that the features of self-tolerance have to be inferred in its absence. One means to study the normal operation of self tolerance without interrupting it would be to test *in vitro* the capacities of lymphocytes from normal animals that have had no stimulation of their immune system. Information gained from the earlier experiments in which antithyroid autoreactivity developed in thyroid-deprived fetuses was utilized to do this. T cells from these fetuses are cytotoxic for monolayers of autologous thyrocytes. However, if identical twin fetal lambs are produced by embryo microsurgery and one fetus is thyroidectomized to induce antithyroid autoimmunity, T cells from the other, normally self-tolerant fetus could be shown to suppress autoimmune reactivity (Chen *et al.*, 1995).

The significance of the suppression of autoreactivity by T cells from the normal identical twin is that it demonstrates that a normal animal, the immune system of which has not been exposed to any extrinsic stimulus, has, in the course of developing self tolerance, acquired cells with the capacity to suppress autoimmunity against one set of tissue-specific antigens. This strongly suggests that these cells make some contribution to the normal state of self-tolerance of those autoantigens.

19. Conclusions

A remarkable breadth of immunological knowledge has been generated through investigations of the development and function of the ovine immune system. During the last 20 years there has been much progress in the areas of leukocyte antigens, antigen receptors, ontogeny of the immune system, and cytokines. The advent of monoclonal antibodies and molecular biology techniques have facilitated many of these investigations. This increased knowledge of the immune system has also led to a greater understanding of disease pathogenesis and host responses to infection.

The present chapter also reveals a number of areas in which there is a paucity of information. Little is known about early events of B lymphopoiesis or the site where Ig gene rearrangement occurs during fetal development. There is limited information on the structure and function of leukocyte antigens; however, the T19 (WC1) antigen and the SIC4.8R monoclonal antibody suggest that there is much that is unique about sheep leukocyte molecules. Molecular biology should provide a powerful tool in this area of investigation as it has proven useful in the investigation of cytokines and MHC. Mapping the sheep genome will provide a vast amount of genetic information of comparative value in studies of the evolution of the immune system.

The strength of the sheep model has often been that it provided insight into the physiology of the immune system. With the advent of molecular biology it has become even more important to understand the relevance or function of individual genes or proteins within the context of an intact immune system. The sheep model provides direct access to a number of lymphoid microenvironments, through lymphatic cannulation and fetal surgery, and it is possible to follow the development of immune responses in a naive system. Thus, the sheep model provides an invaluable resource for understanding the complex interactions required to integrate immune responses and regulate the homeostasis of normal lymphoid development.

20. References

Abernethy, N. J., Hay, J. B. (1992). The recirculation of lymphocytes from blood to lymph: physiological considerations and molecular mechanisms. *Lymphology* **25**, 1–30.

Abernethy, N. J., Hay, J. B., Kimpton, W. G., Washington, E. A., Cahill, R. N. P. (1990). Non-random recirculation of small, CD4+ and CD8+ T lymphocytes in sheep: evidence for lymphocyte subset-specific lymphocyte-endothelial cell recognition. *Int. Immunol.* 2(3), 231–238.

Abernethy, N. J., Hay, J. B., Kimpton, W. G., Washington, E., Cahill, R. N. (1991). Lymphocyte subset-specific and tissue-specific lymphocyte-endothelial cell recognition mechanisms independently direct the recirculation of lymphocytes from blood to lymph in sheep. *Immunology* 72(2), 239–245.

Abitorabi, M. A., Mackay, C. R., Jerome, E. H., Osorio, O., Butcher, E. C., Erle, D. J. (1996). Differential expression of homing molecules on recirculating lymphocytes from sheep gut, peripheral, and lung lymph. *J. Immunol.* **156**, 3111–3117.

Abraham, W. M., Sielczak, M. W., Ahmed, A., Cortes, A., Lauredo, I. T., Kim, J., Pepinsky, B., Benjamin, C. D., Leone, D. R., Lobb, R. R., Weller, P. F. (1994). Alpha (4)-integrins mediated antigen induced late bronchial responses and prolonged airway hyper-responsiveness in sheep. *J. Clin. Invest.* **93**, 776–787.

Adams, D. B., Colditz, I. G. (1991). Immunity to *Haemonchus contortus* and the cellular response to helminth antigens in the mammary gland of non-lactating sheep. *Int. J. Parasitol.* **21**, 631–639.

Adams, T. E., Baker, L., Brandon, M. R. (1989). Cloning and sequence of and ovine prolactin cDNA. *Nucl. Acids Res.* **17**, 440.

Adler, H., Adler, B., Peveri, P., Werner, E. R., Wachter, H., Peterhans, E., Jurgi, T. W. (1996). Differential regulation of inducible nitric oxide synthase production in bovine and caprine macrophages. *J. Infect. Dis.* **173**, 971–978.

Albina, J. C. (1995). On the expression of nitric oxide synthase by human macrophages. Why no NO? *J. Leukoc. Biol.* **58**, 643–649.

Al Salami, M., Simpson-Morgan, M. W., Morris, B. (1985). Haemopoiesis and the development of immunological reactivity in the sheep fetus. In *Immunology of the Sheep* (eds Morris, B., Miyasaka, M.), pp. 19–46. Editiones Roche, Basle, Switzerland.

Al Salami, M. T. H. (1985). The ontogeny of the haemopoietic system in foetal sheep. PhD thesis in *Department of Immunology, John Curtin School of Medical Research*, p. 148. Australian National University, Canberra, Australia.

Albertine, K. H., Cerasoli, F., Jr, Tahamont, M. V., Ishihara, Y., Flynn, J. T., Peters, S. P., Gee, M. H. (1989). Zymosan-activated plasma causes prolonged decreases in PMN superoxide release in sheep. *J. Appl. Physiol.* **67**, 2481–2490.

Alders, D. B., Shelton, J. N. (1990). Lymphocyte subpopulations in lymph draining from the uterus and ovary. *J. Reprod. Immunol.* **17**, 27–40.

Aleksandersen, M., Hein, W. R., Landsverk, T., McClure, S. (1990). Distribution of lymphocyte subsets in the large intestinal lymphoid follicles of lambs. *Immunology* **70**, 391–397.

Aleksandersen, M., Nicander, L., Landsverk, T. (1991). Ontogeny, distribution and structure of aggregated lymphoid follicles in the large intestine of sheep. *Dev. Comp. Immunol.* **15**, 413–422.

Aleksandersen, M., Landsverk, T., Gjrde, B., Helle, O. (1995). Scarcity of gamma-delta T cells in intestinal epithelia containing coccidia despite general increase of epithelial lymphocytes. *Vet. Pathol.* **32**, 504–512.

Anderson, A., Harkiss, G., Watt, N. (1992). Immunohistological investigation of maedi-visna virus-associated synovitis in sheep. *Brit. J. Rheumatol.* **31**, 5.

Anderson, A. A., Harkiss, G. D., Watt, N. J. (1994). Quantitative analysis of immunohistological changes in the synovial membrane of sheep infected with maedi-visna virus. *Clin. Exp. Immunol.* **72**, 21–29.

Andrews, A. E., Barcham, G. J., Brandon, M. R., Nash, A. D. (1991). Molecular cloning and characterization of ovine IL-1α and IL-1β. *Immunology* **74**, 453–459.

Andrews, A. E., Bardham, G. J., Ashman, K., Meeusen, E. N. T., Brandon, M. R., Nash, A. D. (1993). Molecular cloning and characterization of IL-6. *Immunol. Cell Biol.* **71**, 341–348.

Andrews, A. E., Lofthouse, S. A., Bowles, V. M., Brandon, M. R., Nash, A. D. (1994). Production and *in vivo* use of recombinant ovine IL-1β as an immunological adjuvant. *Vaccine* **12**, 14–22.

Ansari, H. A., Pearce, P. D., Maher, D. W., Broad, T. E. (1994). Regional assignment of conserved reference loci anchors unassigned linkage and syntenic groups to ovine chromosomes. *Genomics* **24**, 451–455.

Ansari, H. A., Pearce, P. D., Maher, D. W., Broad, T. E. (1995). Human chromosome 10 loci map to three sheep chromosomes. *Mammalian Genome* **6**, 46–48.

Assalmeliani, A., Kinsky, R., Martal, J., Chaouat, G. (1995). *In-vivo* immunosuppressive effects of recombinant ovine interferon-TAU (Trophoblastin)-ROTP (Roifn-TAU) inhibits local GVH reaction in mice (PLN Assay), prevents fetal resorptions and favours embryo survival and implantation in the CBA/JX DBA/2 mice combination. *Am. J. Reprod. Immunol.* **33**, 267–275.

Bach, A. S., Ram, G. C., Bansal, M. P. (1995). Killer cell and natural killer cell cytotoxicity in sheep pox. *Indian J. Animal Sci.* **65**, 727–731.

Baglia, V., Ferrer, J. (1977). Expression of the bovine leukemia virus and its internal antigen in blood lymphocytes. *Proc. Soc. Exp. Biol. Med.* **156**, 156–388.

Ballingall, K. T., Dutia, B. M., Hopkins, J., Wright, H. (1995). Analysis of the fine specificities of sheep major histocompatibility complex class II-specific monoclonal antibodies using mouse L-cell transfectants. *Animal Genet.* **26**, 79–84.

Bao, S., McClure, S. J., Emery, D. L., Husband, A. J. (1996). Interleukin-5 mRNA expressed by eosinophils and γ/δ T cells in parasite-immune sheep. *Eur. J. Immunol.* **26**, 552–556.

Barcham, G. J., Andrews, A. E., Nash, A. D. (1995). Molecular cloning and expression of an interleukin-7 cDNA. *Gene* **154**, 265–269.

Barta, O., Barta, V., Shirley, R. A., McMurry, J. D. (1975). Hemolytic assay of sheep serum complement. *Zentralbl. Veterinaermed.* **B22**, 254–262.

Bazer, F. W., Ott, T. L., Spencer, T. E. (1994). Pregnancy recognition in ruminants, pigs and horses – signals from the trophoblast. *Theriogenology* **41**, 79–94.

Begara, I., Lujan, L., Collie, D. D. S., Miller, H. R. P., Watt, N. J. (1995a). *In vitro* response of lymphocytes from bronchoalveolar lavage fluid and peripheral blood to mitogen stimulation during natural maedi-visna virus infection. *Vet. Immunol. Immunopathol.* **49**, 75–88.

Begara, I., Lujan, L., Hopkins, J., Collie, D. D. S., Miller, H. R. P., Watt, N. J. (1995b). A study on lymphocyte activation in maedi-visna virus induced pneumonia. *Vet. Immunol. Immunopathol.* **45**, 197–210.

Begara, I., Lujan, L., McLaren, L., Collie, D. D. S., Miller, H. R. P., Watt, N. J. (1995c). Quantitation of transforming growth factor beta in plasma and pulmonary epithelial lining fluid of sheep experimentally infected with maedi visna virus. *Vet. Immunol. Immunopathol.* **48**, 261–273.

Beh, K. J. (1987). Production and characterization of monoclonal antibodies specific for sheep IgG subclasses IgG1 or IgG2. *Vet. Immunol. Immunopathol.* **14**, 187–196.

Beh, K. J. (1988). Monoclonal antibodies against sheep immunoglobulin light chain, IgM and IgA. *Vet. Immunol. Immunopathol.* **18**, 19–27.

Bendixsen, T., Emery, D. L., Jones, W. O. (1995). The sensitization of mucosal mast cells during infections with *Trichostrongylus colubriformis* or *Haemonchus contortus* in sheep. *Int. J. Parasitol.* **25**, 741–748.

Bernadina, W. E. (1991). Serum opsonic activity and neutrophil phagocytic capacity of newborn lambs before and 24–36 h after colostrum uptake. *Vet. Immunol. Immunopathol.* **29**, 127–138.

Bertoni, G., Zahno, M. L., Zanoni, R., Vogt, H. R., Peterhans, E., Ruff, G., Cheevers, W. P., Sonigo, P., Pancino, G. (1994). Antibody reactivity to the immunodominant epitopes of the caprine arthritis-encephalitis virus gp38 transmembrane protein associates with the development of arthritis. *J. Virol.* **68**, 7139–7147.

Beya, M.-F., Miyasaki, M., Dudler, L., Ezaki, T., Trnka, Z. (1986). Studies on the differentiation of T lymphocytes in sheep. II. Two monoclonal antibodies that recognize all ovine T lymphocytes. *Immunology* **57**, 115–121.

Bird, P., Blacklaws, B., Reyburn, H. T., Allen, D., Hopkins, J., Sargan, D., McConnell, I. (1993). Early events in immune evasion by the lentivirus maedi-visna occurring within infected lymphoid tissue. *J. Virol.* **67**, 5187–5197.

Bird, P., Reyburn, H. T., Blacklaws, B. A., Allen, D., Nettleton, P., Yirrell, D. L., Watt, N., Sargan, D., McConnell, I. (1995). The restricted IgG1 antibody response to maedi visna virus is seen following infection but not following immunization with recombinant *gag* protein. *Clin. Exp. Immunol.* **102**, 274–280.

Birkebak, T. A., Palmer, G. H., Davis, W. C., Knowles, D. P., McElwain, T. F. (1994). Association of gp51 expression and persistent $CD5^+$ B-lymphocyte expansion with lymphomagenesis in bovine leukemia virus infected sheep. *Leukemia* **8**, 1890.

Blacklaws, B. A., Bird, P., Allen, D., McConnell, I. (1994a).

Circulating cytotoxic T lymphocyte precursors in maedi-visna virus-infected sheep. *J. Gen. Virol.* 75, 1589–1596.

Blacklaws, B. A., Bird, P., McConnell, I. (1994b). Pathogenesis and immunity in lentivirus infections of small ruminants. In *Cell-Mediated Immunity in Ruminants* (eds Goddeeris, B. M. L., Morrison, W. I.), pp. 199–212. CRC Press, Boca Raton, FL, USA.

Blacklaws, B., Bird, P., McConnell, I. (1995a). Early events in infection of lymphoid tissue by a lentivirus, maedi-visna. *Trends Microbiol.* **3**, 434–440.

Blacklaws, B. A., Bird, P., Allen, D., Roy, D. J., MacLennan, I. C. M., Hopkins, J., Sargan, D. R., McConnell, I. (1995b). Initial lentivirus–host interactions within lymph nodes: a study of maedi-visna virus infection in sheep. *J. Virol.* **69**, 1400–1407.

Blattman, A. N., Beh, K. J. (1992). Dinucleotide repeat polymorphism within the ovine major histocompatibility complex. *Animal Genet.* **23**, 392.

Blattman, A. N., Hulme, D. J., Kinghorn, B. P., Woolaston, R. R., Gray, G. D., Beh, K. J. (1993). A search for associations between major histocompatibility complex restriction fragment polymorphism bands and resistance to *Haemonchus contortus* infection in sheep. *Animal Genet.* **24**, 277–282.

Blunt, M. H. (1975). *The Blood of Sheep: Composition and Function*. Springer Verlag, New York, USA.

Borgs, P., Hay, J. B. (1986). A quantitative lymphocyte localization assay. *J. Leukoc. Biol.* **39**, 333–342.

Borron, P. (1991). The role of the lymphatic endothelium in the lymph node's response to a delayed hypersensitivity stimulus. *Department of Immunology*, University of Toronto, Toronto, Canada.

Bosse, J., Boileau, R., Begin, R. (1987). Chronic allergic airway disease in the sheep model: functional and lung-lavage features. *J. Allergy Clin. Immunol.* **79**, 339–344.

Boyt, W. P., McKenzie, P. K. I., Emslie, V. W. (1976). Enzootic leukosis in a flock of sheep in Rhodesia. *Vet. Rec.* **98**, 112.

Brahic, M., Stowring, L., Ventura, P., Haase, A. T. (1981). Gene expression in visna virus infection in sheep. *Nature* **292**, 240–242.

Brambell, F. W. R. (1970). The transfer of passive immunity from mother to young. In *Frontiers of Biology* (eds Neuberger, A., Tatum, E. L.), vol. 18, pp. 201–233. Northern Holland, Amsterdam.

Brandon, M. R., Lascelles, A. K. (1971). Relative efficiency of absorption of IgG_1 and IgG_2, IgA and IgM in the newborn calf. *Aust. J. Exp. Biol. Med. Sci.* **49**, 629–633.

Braun, M. J., Clements, J. E., Gonda, M. A. (1987). The visna virus genome – evidence for a hypervariable site in the env-gene and sequence homology among lentivirus envelope proteins. *J. Virol.* **61**, 4046–4054.

Brightman, A. H., Wachsstock, R. S. *et al.* (1991). Lysozyme concentrations in the tears of cattle, goats, and sheep. *Am. J. Vet. Res.* **52**(1), 9–11.

Broad, T. E., Burkin, D. J., Jones, C., Lewis, P. E., Ansari, H. A., Pearce, P. D. (1993). Mapping of MYF5, C1R, MYHL, TPI1, IAPP, A2MR and RNR onto sheep chromosome 3q. *Animal Genet.* **24**(6), 415–419.

Broad, T. E., Burkin, D. J., Jones, C., Lewis, P. E., Ansari, H. A., Pearce, P. D. (1994). Seven loci on human chromosome 4 map onto sheep chromosome 6: a proposal to restore the original nomenclature of this sheep chromosome. *Mammalian Genome* **5**(7), 429–433.

Broad, T. E., Burkin, D. J., Cambridge, L.M., Jones, C., Lewis, P. E., Morris, H. G., Geyer, D., Pearce, P. D., Ansari, H. A., Maher, D. W. (1995a). Six loci mapped onto human chromosome 2p are assigned to sheep chromosome 3p. *Animal Genet.* **26**, 85–90.

Broad, T. E., Burkin, D. J., Cambridge, L. M., Maher, D. W., Lewis, P. E., Ansari, H. A., Pearce, P. D., Jones, C. (1995b). Assignment of five loci from human chromosome 8q on to sheep chromosome 9. *Cytogen. Cell Genet.* **68**, 102–106.

Brodie, S. J., Marcom, K. A., Pearson, L. D., Anderson, B. C., de la Concha-Bermejillo, A., Ellis, J. A., Demartini, J. C. (1992). Effects of virus load in the pathogenesis of lentivirus-induced lymphoid interstitial pneumonia. *J. Inf. Dis.* **166**, 531–541.

Brodie, S. J., Pearson, L. D., Zink, M. C., Bickle, H. M., Anderson, B. C., Marcom, K. A., Demartini, J. C. (1995). Ovine lentivirus expression and disease. Virus replication, but not entry, is restricted to macrophages of specific tissues. *Am. J. Pathol.* **146**, 250–263.

Brogden, K. A. (1992). Ovine pulmonary surfactant induces killing of *Pasteurella haemolytica, Escherichia coli*, and *Klebsiella pneumoniae* by normal serum. *Infect. Immun.* **60**(12), 5182–5189.

Brogden, K. A., De Lucca, A., Bland, J., Elliot, S. (1996). Isolation of an ovine pulmonary surfactant-associated anionic peptide bactericidal for *Pasteurella haemolytica*. *Proc. Natl Acad. Sci. USA* **93**, 412–416.

Buchta, R. (1990). Functional and biochemical properties of ovine neutrophils. *Vet. Immunol. Immunopathol.* **24**, 97–112.

Buddle, B. M., Jowett, G., Green, R. S., Douche, P. G. C., Risdon, P. L. (1992). Association of blood eosinophilia with the expression of resistance in Romney lambs to nematodes. *Int. J. Parasitol.* **22**, 955–960.

Buitkamp, J. G. D., Schwaiger, F.-W., Stear, M. J., Epplen, J. T. (1994). Association between the ovine Major Histocompatability Complex DRBI gene and resistance to *Ostertagia circumcincta* infestation. *Animal Genet.* **25**, 59–60.

Bujdoso, R., Hopkins, J., Dutia, B., Young, P., McConnell, I. (1989). Characterization of sheep afferent lymph dendritic cells and their role in antigen carriage. *J. Exp. Med.* **170**, 1285–1302.

Bujdoso, R., Young, P., Hopkins, J., McConnell, I. (1990). IL-2-like activity in lymph fluid following *in vivo* antigen challenge. *Immunology* **69**(1), 45–51.

Bujdoso, R., Williamson, D. R. M., Hunt, P., Blacklaws, B., Sargan, D., McConnell, I. (1995). Molecular cloning and expression of DNA encoding ovine interleukin-2. *Cytokine* **7**, 223–231.

Burny, A., Willems, L., Callebaut, I., Adam, E., Cludts, I., Dequiedt, F., Droogmans, L., Grimonpont, C., Kerkhofs, P., Mammerickx, M., Portetelle, D., Van den Broeke, A., Kettmann, R. (1994). Bovine leukemia virus: biology and mode of transformation. In *Viruses and Cancer* (eds Minson, A. C., Neil, J. C., McRae, M. A.), pp. 213–234. Cambridge University Press, Cambridge, UK.

Burrells, C. (1985). Cellular and humoral elements of the lower respiratory tract of sheep. Immunological examination of cells and fluid obtained by bronchoalveolar lavage of normal lungs. *Vet. Immunol. Immunopathol.* **10**, 225–243.

Burrells, C., Sutherland, A. D. (1994). Phenotypic analysis of lymphocytes obtained by bronchoalveolar lavage of normal sheep. *Vet. Immunol. Immunopathol.* **40**, 85–90.

Cahill, R. N. P., Trnka, Z. (1980). Growth and development of recirculating lymphocytes in the sheep fetus. In *Essays on the Anatomy and Physiology of Lymphoid Tissues* (eds Trnka, Z., Cahill, R. N. P.), pp. 38–49. Basle, Switzerland.

Cahill, R. N. P., Trnka, Z. (1981). Recirculation and life span of lymphocytes in the sheep fetus and in post-natal lambs. In *The Immune System* (eds Steinberg, C. M., Lefkovits, I.), pp. 367–374. S. Karger, Basle, Switzerland.

Cahill, R., Hay, J. B., Frost, H., Trnka, Z. (1974). Changes in lymphocyte circulation after administration of antigen. *Haematologia* **8**(1–4), 321–334.

Cahill, R. N. P., Frost, H., Trnka, Z. (1976). The effects of antigen on the migration of recirculating lymphocytes through single lymph nodes. *J. Exp. Med.* **143**, 870–888.

Cahill, R. N. P., Poskitt, D. C., Frost, H., Trnka, Z. (1977). Two distinct pools of recirculating T lymphocytes: migratory characteristics of nodal and intestinal T lymphocytes. *J. Exp. Med.* **145**, 420–428.

Cahill, R. N. P., Poskitt, D. C., Heron, I., Trnka, Z. (1979). Collection of lymph from single lymph nodes and intestines of fetal lambs in utero. *Int. Arch. Allergy Applied Immunol.* **59**, 117–120.

Cahill, R. N. P., Kimpton, W. G., Dudler, L., Trnka, Z. (1985). Lymphopoiesis in foetal and perinatal sheep. In *Immunology of the Sheep* (eds Morris, B., Miyasaka, M.), pp. 46–67. Editiones Roche, Basle, Switzerland.

Cahill, R. N. P., Kimpton, W. G., Washington, E. A., Dudler, L., Trnka, Z. (1996). Late term fetal thymectomy does not prevent the development of gut-homing T cells after birth. *Immunology* **88**, 130–133.

Campbell, S. G. (1974). Experimental colostrum deprivation in lambs. *Br. Vet. Rec.* **130**, 538–543.

Campbell, S. G., Siegel, M. J., Knowlton, B. J. (1977). Sheep immunoglobulins and their transmission to the neonatal lamb. *N.Z. Vet. J.* **25**, 361–365.

Cann, A. J., Chen, I. S. (1990). *Virology* (eds Fields, B. N., Knipe, D. M.), pp. 1501–1527. Raven Press, New York, USA.

Chardon, P. K. M., Cullen, P. R., Geffrotin, C., Auffray, C., Stominger, J. L., Cohen, D, Vaiman, M. (1985). Analysis of the sheep MHC using HLA class I, II, and C4 cDNA probes. *Immunogenetics* **22**, 349–358.

Cheevers, W. P. (1993). Failure of neutralising antibody to regulate CAE lentivirus expression *in vivo*. *Virology* **196**, 835–839.

Chen, W., Alley, M. R., Manktelow, B. W. (1989). Respiratory tract-associated lymphoid tissue in conventionally raised sheep. *J. Comp. Pathol.* **101**, 327–340.

Chen, W., Alley, M., Manktelow, B., Davey, P. (1990). Mast cells in the ovine lower respiratory tract: heterogeneity, morphology and density. *Int. Arch. Allergy Appl. Immunol.* **93**, 99–106.

Chen, W., Alley, M. R., Manktelow, B. W. (1991a). Airway inflammation in sheep with acute airway hypersensitivity to inhaled *Ascaris suum*. *Int. Arch. Allergy Appl. Immunol.* **96** 218–223.

Chen, W., Alley, M. R., Manktelow, B. W., Hopcroft, D., Bennet, R. (1991b). The potential role of the ovine pharyngeal tonsil in respiratory tract immunity: a scanning and transmission electron microscopy study of its epithelium. *J. Comp. Path.* **104**, 47–56.

Chen, X., Shelton, P. M. J., McCullagh, P. (1995). Suppression of anti-thyrocyte autoreactivity by the lymphocytes of normal fetal lambs. *J. Autoimm.* **8**, 539–559.

Chin, G. W., Cahill, R. N. P. (1984). The appearance of fluorescein-labelled lymphocytes in lymph following *in vitro* or *in vivo* labelling: the route of lymphocyte recirculation through mesenteric lymph nodes. *Immunology* **52**(2), 341–347.

Chin, G. W., Hay, J. B. (1980). A comparison of lymphocyte migration through intestinal lymph nodes, subcutaneous lymph nodes, and chronic inflammatory sites of sheep. *Gastroenterology* **79**, 1231–1242.

Chin, G. W., Hay, J. B. (1984). Distribution of radiolabelled lymph cells in lymph nodes and the migratory properties of blood lymphocytes in sheep. *Int. Arch. Allergy Appl. Immunol.* **75**(1), 52–57.

Chin, G. W., Pearson, L. D., Hay, J. B. (1985). Cells in sheep lymph and their migratory characteristics. *Experimental Biology of the Lymphatic Circulation* (ed. Johnston, M. G.), pp. 141–164. Elsevier Science Publishers, Amsterdam, The Netherlands.

Cirelli, R. A., Carey, L. A., Fisher, J. K., Rosolia, D. L., Elasser, T. H., Caperna, T. J., Gee, M. H., Albertine, K. H. (1995). Endotoxin infusion in anesthetized sheep is associated with intrapulmonary sequestration of leukocytes that immunohistochemically express tumor-necrosis-factor-alpha. *J. Leukoc. Biol.* **57**, 820–826.

Clarkson, C. A., Beale, D., Coudwell, J. W., Symons, D. B. A. (1993). Sequence of ovine Igγ2 constant region heavy chain cDNA and molecular modelling of ruminant IgG isotopes. *Molec. Immunol.* **30**, 1195–1204.

Clements, J. E., Pedersen, F. S., Narayan, O., Haseltine, W. A. (1980). Genomic changes associated with antigenic variation of visna virus during persistent infection. *Proc. Natl Acad. Sci. USA* **77**, 4454–4458.

Clements, J. E., D'Antonio, N., Narayan, O. (1982). Genomic changes associated with antigenic variation of visna virus. II. Common nucleotide sequence changes detected in variants from independent isolations. *J. Mol. Biol.* **158**, 415–434.

Cockerell, G., Rovnak, J. (1988). The correlation between the direct and indirect detection of bovine leukemia virus infection in cattle. *Leuk. Res.* **2**, 465.

Colditz, I. G., Watson, D. L. (1992). The effect of cytokines and chemotactic agonists on the migration of T lymphocytes into skin. *Immunology* **76**(2), 272–278.

Colditz, I. G., Woolaston, R. R., Lax, J., Mortimer, S. I. (1992). Plasma leakage in skin of sheep selected for resistance or susceptibility to fleece rot and fly strike. *Parasite Immunol.* **14**(6), 587–594.

Colditz, I. G., Lax, J., Mortimer, S. I., Clarke, R. A., Beh, K. J. (1994). Cellular inflammatory responses in skin of sheep selected for resistance or susceptibility to fleece rot and fly strike. *Parasite Immunol.* **16**(6), 289–296.

Cole, G. J. (1969). *The Lymphatic System and the Immune Response in the Lamb*. ANU, Canberra, Australia.

Cole, G. J., Morris, B. (1973). The lymphoid apparatus of the sheep: its growth, development and significance in immunologic reactions. *Adv. Vet. Sci. Comp. Med.* **17**, 225–263.

Coop, R. L., Huntley, J. F., Smith, W. D. (1995). Effect of dietary protein supplementation on the development of immunity to *Ostertertagia circumcincta* in growing lambs. *Res. Vet. Sci.* **59**, 24–29.

Cooper, D. W., Van Oorschot, R. A. H., Piper, L. R., Le Jambre, L. F. (1989). No association between the ovine leucocyte antigen (OLA) system in the Australian Merino and susceptibility to *Haemonchus contortus* infestation. *Int. J. Parasitol.* **19**, 695–697.

Cordier, G., Cozon, G., Greenland, T., Rocher, F., Guiguen, F., Guerret, S., Brune, J., Mornex, J. F. (1990). *In vivo* activation of alveolar macrophages in ovine lentivirus infection. *Clin. Immunol. Immunopathol.* **55**, 355–367.

Cordier, G., Guiguen, F., Cadore, J. L., Cozon, G., Jacquier, M. F., Mornex, J. F. (1992). Characterization of the lymphocytic alveolitis in visna-maedi virus-induced interstitial lung-disease of sheep. *Clin. Exp. Immunol.* **90**, 18–24.

Cornil, I., Delon, P., Parodi, A., Levy, D. (1988). T–B cell cooperation for bovine leukemia virus expression in ovine lymphocytes. *Leukemia* **2**, 313.

Coughlan, S. N., Harkiss, G. D., Dickson, L., Hopkins, J. (1996). Fcγ receptor expression on sheep afferent lymph dendritic cells and rapid modulation of cell surface phenotype following Fcγ receptor engagement *in vitro* and *in vivo*. *Scand. J. Immunol.* **43**, 31–38.

Crawford, A. M., Dodds, K. G., Ede, A. J., Pierson, C. A., Montgomery, G. W., Garmonsway, H. G., Beattie, A. E., Davies, K., Maddox, J. F., Kappes, S. W., Stone, R. T., Nguyen, T. C., Penty, J. M., Lord, E. A., Broom, J. E., Buitkamp, J., Schwaiger, W., Epplen, J. T., Matthew, P., Mattews, M. E., Hul, D. J. (1995). An autosomal genetic linkage map of the sheep genome. *Genetics* **140**, 703–724.

Cripps, A. W., Lascelles, A. K. (1974). The biological half-lives of IgG_1 and IgG_2 in milk-fed lambs and in non-pregnant colostrum-forming sheep. *Aust. J. Exp. Biol. Med. Sci.* **52**, 717–719.

Cripps, A. W., Husband, A. J., Scicchitano, R., Sheldrake, R. F. (1985). Quantitation of sheep IgG1, IgG2, IgA, IgM and albumin by radioimmunoassay. *Vet. Immunol. Immunopathol.* **8**, 137–147.

Cullen, P. R. (1989). Scrapie and the sheep MHC: claims of linkage refuted. *Immunogenetics* **29**, 414–418.

Cullen, P. R., Bunch, C., Brownlie, K., Morris, P. J. (1982). Sheep lymphocyte antigens: a preliminary study. *Animal Blood Groups Biochem. Genet.* **13**, 149–159.

Cunliffe, H. R., Graves, J. H. (1970). Immunologic response of lambs to emulsified Foot-and-Mouth Disease Vaccine. *Archiv fur die gesamte Viruforschung* **32**, 261–268.

Curtain, C. C. (1975). The ovine immune system. In *The Blood of Sheep: Composition and Function*, pp. 185–195. Springer Verlag, Berlin–Heidelberg.

Curtain, C. C., Anderson, N. (1971). Immunocytochemical localization of the ovine immunoglobulins IgA, IgG1, $IgG1_A$ and IgG2: effect of gastro-intestinal parasitism in the sheep. *Clin. Exp. Immunol.* **8**, 151–162.

Cutlip, R. C., Jackson, T. A., Lehmkuhl, H. D. (1979). Lesions of ovine progressive pneumonia: interstitial pneumonitis and encephalitis. *Am. J. Vet. Res.* **40**, 1370–1374.

Cutlip, R. C., Lehmkuhl, H. D., Brogden, K. A., Sacks, J. M. (1986). Breed susceptibility to ovine progressive pneumonia (maedi visna) virus. *Vet. Microbiol.* **12**, 283–288.

Davies, D. H., Long, D. H., McCarthy, A. R., Herceg, M. (1986). The effect of parainfluenza virus type 3 on the phagocytic cell response of the ovine lung to *Pasteurella haemolytica*. *Vet. Microbiol.* **11**, 125–144.

Dawe, S. J., Husband, A. J., Langford, C. M. (1982). Effects of induction of parturition in ewes with dexamethasone or oestrogen on concentrations of immunoglobulins by lambs. *Aust. J. Biolog. Sci.* **35**, 223–229.

Dawkins, H. J. S., Windon, R. G., Eagleson, G. J. (1989). Eosinophil responses in sheep selected for high and low responsiveness to *Trichinella colubriformis*. *Int. J. Parasitol.* **19**, 199–205.

Deane, D. L., Inglis, L., Haig, D. M. (1995). The 175 antigen expressed on myeloid and erythroid cells during differentiation is associated with serine protease activity. *Blood* **85**, 1215–1219.

Deng, P., Cutlip, R. C., Lehmkuhl, H. D., Brogden, K. A. (1986). Ultrastructure and frequency of mastitis caused by ovine progressive pneumonia virus infection in sheep. *Vet. Pathol.* **23**, 184–189.

Depelchin, A., Letesson, J., Lostrie, N., Mammerickx, M., Portetelle, D., Burny, A. (1989). Bovine leukemia virus (BLV) infected B cells express a marker similar to the CD4 T-cell marker. *Immunol. Lett.* **20**, 69.

Dequiedt, F., Kettmann, R., Burny, A., Willems, L. (1995). Mutations in the p53 tumor-suppressor gene are frequently associated with Bovine Leukemia Virus-induced leukemogenesis in cattle but not in sheep. *Virology* **209**, 676.

Deverson, E. V., Wright, H., Watson, S., Ballingall, K., Huskisson, N., Diamond, A. G., Howard, J. C. (1991). Class II major histocompatability complex genes of the sheep. *Animal Genet.* **22**, 211–225.

Di Guardo, G., Condoleo, R., Autorino, G. (1992). Multicentric lymphosarcoma associated with pulmonary adenomatosis, pulmonary lymphoid hyperplasia, and lymphoid interstitial pneumonia in a ewe. *Vet. Pathol.* **29**, 262.

Diehl, J. E., Mallette, M. F. (1964). Nature of the Forssman hapten from sheep erythrocytes. *J. Immunol.* **93**, 965–968.

Dimmock, C. K., Ward, W. H., Trueman, K. F. (1990). Lymphocyte subpopulations in sheep with lymphosarcoma resulting from experimental infection with bovine leukemia virus. *Immunol. Cell Biol.* **68**, 45.

Dineen, J. K., Windon, R. J. (1980). The effect of sire selection on the response of lambs to vaccination with irradiated *Trichostrongylus colubriformis* larvae. *Int. J. Parasitol.* **10**, 189–196.

Ding, A. H., Nathan, C. F., Stuchr, D. J. (1988). Release of reactive nitrogen in intermediates and reactive oxygen intermediates from mouse peritoneal macrophages. Comparison of activating cytokines and evidence for independent production. *J. Immunol.* **141**, 2407–2412.

Djilali, S., Parodi, A. (1989). The BLV-induced leukemia-lymphosarcoma complex in sheep. *Vet. Immunol. Immunopathol.* **22**, 233.

Djilali, S., Parodi, A. L., Cockerell, G. L. (1987). Development of leukemia and lymphosarcoma induced by Bovine Leukemia Virus in sheep: a cytopathological study. *Leukemia* **1**, 777.

Dodds, A. W., Law, S. K. A. (1990). The complement component C4 of mammals. *Biochemical J.* **265**(2), 495–502.

Donachie, W., Burrells, C., Sutherland, A. D., Gilmour, J. S., Gilmour, N. J. L. (1986). Immunity of specific pathogen-free lambs to challenge with an aerosol of *Pasteurella haemolytica* biotype A serotype 2. Pulmonary antibody and cell responses to primary and secondary infections. *Vet. Immunol. Immunopathol.* **11**, 265–279.

Douch, P. G., Outteridge, P. M. (1989). The relationship between ovine lymphocyte antigens and parasitological and production parameters in Romney sheep. *Int. J. Parasitol.* **19**, 35–41.

Douch, P. G. C., Harrison, G. B. L., Buchanan, L. L., Greer, K. S. (1983). *In vitro* bioassay of sheep gastrointestinal mucus for nematode paralysing activity mediated by substances with some properties characteristic of SRS-SA. *Int. J. Parasitol.* **13**, 207–212.

Douch, P. G. C., Harrison, G. B. L., Elliot, D. C., Buchanan, L. L., Greer, K. S. (1986). Relationship of gastrointestinal histology and mucus antiparasitic activity with the development of resistance to *Trichostrongylus* infections in sheep. *Vet. Parasitol.* **20**, 315–331.

Dufour, V., Nau, F., Malinge, S. (1996). The sheep Ig variable region repertoire consists of a single VH family. *J. Immunol.* **156**, 2163–2170.

Dutia, B. M., Hopkins, J. (1991). Analysis of the CD1 cluster in sheep. *Vet. Immunol. Immunopathol.* **27**, 189–194.

Dutia, B. M., McConnell, I., Bird, K., Keating, P., Hopkins, J. (1993). Patterns of major histocompatability complex class II expression on T cell subsets in different immunological compartments. I. Expression on resting T cells. *Eur. J. Immunol.* **23**, 2882–2888.

Ebrahimi, B., Roy, D. J., Bird, P., Sargan, D. R. (1995). Cloning, sequencing and expression of the ovine interleukin 6 gene. *Cytokine* **7**, 232–236.

Egan, P. J., Andrews, A. E., Barcham, G. J., Brandon, M. R., Nash, A. D. (1994a). Production and application of monoclonal antibodies to ovine interleukin 1α and interleukin 1β. *Vet. Immunol. Immunopathol.* **41**, 241–257.

Egan, P. J., Rothel, J. S., Andrews, A. E., Seow, H.-F., Wood, P. R., Nash, A. D. (1994b). Characterization of monoclonal antibodies to ovine tumour necrosis factor-α and development of a sensitive immunoassay. *Vet. Immunol. Immunopathol.* **41**, 259–274.

Elliott, B. E. (1979). A sensitive method for the separation of rosette forming cells. In *Immunological Methods* (eds Lefkovits, I., Pernis, B.), pp. 241–259. Academic Press, London, UK.

Ellis, J. A., Demartini, J. C. (1985a). Immunomorphologic and morphometric changes in pulmonary lymph nodes of sheep with progressive pneumonia. *Vet. Pathol.* **22**, 32–41.

Ellis, J. A., Demartini, J. C. (1995b). Evidence of decreased concanavalin A induced suppressor cell activity in the peripheral blood and pulmonary lymph nodes of sheep with ovine progressive pneumonia. *Vet. Immunol. Immunopathol.* **8**, 93–106.

Ellis, J. A., Demartini, J. C. (1985c). Ovine interleukin-2: partial purification and assay in normal sheep and sheep with ovine progressive pneumonia. *Vet. Immunol. Immunopathol.* **8**, 15–25.

Ellis, J. A., Lairmore, M., O'Toole, D., Campos, M. (1991). Differential induction of tumor necrosis factor alpha in ovine pulmonary alveolar macrophages following infection with *Corynebacterium pseudotuberculosis, Pasteurella haemolytica*, or lentivirus. *Infect. Immun.* **59**, 3254–3260.

Ellory, J. C., Tucker, E. M. (1970). Active potassium transport and L and M antigens of sheep and goat red cells. *Biochem. Biophys. Acta* **219**, 160–168.

Engewerda, C. R., Sandeman, R. A., Smart, S. J., Sandeman, R. M. (1992). Isolation and sequence of sheep immunoglobulin E heavy-chain complementary DNA. *Vet. Immunol. Immunopathol.* **34**, 115–126.

Engewerda, C. R., Dale, C. J., Sandeman, R. M. (1996). IgE, TNF-α, IL-1β, IL-4 and IFN-γ gene polymorphism in sheep selected for fleece rot and flystrike. *Int. J. Parasitol.* **26**, 781–791.

Entrican, G. (1995). Moredun Research Institute, Edinburgh, UK. Personal communication.

Erencin, Z. (1984). Hemolymph nodes in small ruminants. *Am. J. Vet. Res.* **9**, 286–295.

Escayg, A. P., Montgomery, G. W., Hickford, J. G., Bullock, D. W. (1993). A Bg111 RFLP at the ovine MHC class II DRA locus. *Animal Genet.* **24**, 217.

Ezaki, I., Miyasaki, M., Beya, M.-F., Dudler, L., Trnka, Z. (1987). A murine anti-sheep T8 monoclonal antibody, ST-8, that defines the cytotoxic T lymphocyte population. *Int. Arch. Allergy Appl. Immunol.* **82**, 168–177.

Ezzell, J. L., Parker, C. J. (1992). Cell-surface regulation of the human alternative pathway of complement: sheep but not rabbit erythrocytes express factor l-dependent cofactor activity. *Scand. J. Immunol.* **36**(1), 79–87.

Fabb, S. A., Maddox, J. F., Goglin-Ewens, K. J., Baker, L., Wu, M.-J., Brandon, M. R. (1993). Isolation, characterization and evolution of ovine MHC class II DRA and DQA genes. *Animal Genet.* **24**, 249–255.

Fahey, K. J. (1977). The response of foetal sheep to the somatic and flagellar antigens of *Salmonella typhimurium*. *Aust. J. Exp. Biol. Med. Sci.* **55**, 524–537.

Fahey, K. J., Morris, B. (1978). Humoral immune response in fetal sheep. *Immunology* **35**, 651–661.

Feinstein, A., Hobart, M. J. (1969). Structural relationship and complement fixing activity of sheep and other ruminant immunoglobulin G subclasses. *Nature* **223**, 950–952.

Ferguson, E. D., Dutia, B. M., Hein, W., Hopkins, J. (1994). *GenBank accession numbers: Z36890–92, X90567.*

Ferreira, A. M., Wuerzner, R., Hobart, M. J., Laehmann, D. J. (1995). Study of the *in vitro* activation of the complement alternative pathway by *Echinococcus granulosus* hydatid cyst fluid. *Parasite Immunol.* **17**(5), 245–251.

Filmer, D. B., McClure, T. J. (1951). Absorption of anti-nematode antibodies from ewes colostrum by the newborn lamb. *Nature* **168**, 170.

Fiskerstrand, C., Sargan, D. (1990). Nucleotide sequence of ovine interleukin-1 beta. *Nucl. Acids Res.* **18**, 7165–7166.

Fiskerstrand, C. E., Roy, D. J., Green, I., Sargan, D. R. (1992). Cloning expression and characterization of ovine interleukins 1α and 1β. *Cytokine* **4**, 418–428.

Fiskerstrand, C. E., Hopkins, J., Sargan, D. R. (1994). Interleukin-1 receptor expression by ovine afferent lymph dendritic cells: response to secondary antigen challenge. *Eur. J. Immunol.* **24**, 2351–2356.

Flesch, I. E. A., Kaufmann, S. H. E., (1991). Mechanisms involved in mycobacterial growth inhibition by gamma interferon-activated bone marrow macrophages: role of reactive nitrogen intermediates. *Infect. Immun.* **59**, 3213–3218.

Foley, R. C., Beh, K. J. (1989). Isolation and sequence of sheep IgH and L chain cDNA. *J. Immunol.* **142**, 708–711.

Foley, R. C., Beh, K. J. (1992). Analysis of immunoglobulin light chain loci in sheep. *Animal Genet.* **23**, 31–42.

Ford, C. H. J. (1975). Genetic studies of sheep leucocyte antigens. *J. Immunogenet.* **2**, 31–40.

Foster, R. A., Ladds, P. W., Husband, A. J., Hoffman, D. (1988). Immunoglobulins and immunoglobulin-containing cells in the reproductive tracts of rams naturally infected with *Brucella ovis*. *Aust. Vet. J.* **65**, 37–40.

Francey, T., Jungi, T. W., Rey, O., Peterhans, E. (1992). Culture of ovine bone-marrow derived macrophages and evidence for serum factors distinct from M-CSF contributing to their propagation *in vitro*. *J. Leukoc. Biol.* **51**, 525–534.

Francis, G. L., Read, L. C., Ballard, I. J., Bagley, C. J., Upton, F. M., Gravestock, P. M., Wallace, J. C. (1986). Purification and partial sequence analysis of insulin-like growth factor-1 from bovine colostrum. *Biochemical J.* **233**, 207–213.

Frost, H. (1978). The effect of antigen on the output of recirculating T and B lymphocytes from single lymph nodes. *Cell. Immunol.* **37**, 390–396.

Frost, H., Cahill, R. N. P., Trnka, Z. (1975). The migration of recirculating autologous and alogeneic lymphocytes through single lymph nodes. *Eur. J. Immunol.* **5**(12), 839–843.

Fu, P., Evans, B., Lim, G. B., Moritz, K., Wintour, E. M. (1993). The sheep erythropoietin gene: molecular cloning and effect of hemorrage on plasma erythropoietin and renal/liver messenger RNA in adult sheep. *Mol. Cell. Endocrinol.* **93**, 107–116.

Gallichan, W. S., Rosenthal, K. L. (1995). Specific secretory immune responses in the female genital tract following intranasal immunization with a recombinant adenovirus expressing glycoprotein B of herpes simplex virus. *Vaccine* **13**(16), 1589–1595.

Gallo, R., Wong-Staal, S. (1990). *Retrovirus Biology and Human Disease*, pp. 409. Marcel Dekker, New York, USA.

Ganz, T., Selstead, M. E., Lehrer, R. I. (1990). Defensins. *Eur. J. Haematol.* **44**, 1–8.

Garber, T. L., Hughes, A. L., Watkins, D. I., Templeton, J. W. (1993). *GenBank accession numbers: U03092–U03094.*

Garrido, J. J., Deandres, D. F., Pintado, C. O., Llanes, D., Stear, M. J. (1995). Serologically defined lymphocyte alloantigens in Spanish sheep. *Exper. Clin. Immunogenet.* **12**, 268–271.

Gatei, M. H., Brandon, R., Naif, H. M., Lavin, M. F., Daniel, R. C. W. (1989). Lymphosarcoma development in sheep experimentally infected with bovine leukemia virus. *J. Vet. Med.* **36**, 424.

Gatei, M. H., Good, M. F., Daniel, R. C. W., Lavin, M. F. (1993). T-cell responses to highly conserved CD4 and CD8 epitopes on the outer membrane protein of bovine leukemia virus: relevance to vaccine development. *J. Virol.* **67**(4), 1796–1802.

Gendelman, H. E., Narayan, O., Molineaux, S., Clements, J. E., Ghotbi, Z. (1985). Slow, persistent replication of lentiviruses: role of tissue macrophages and macrophage precursors in bone marrow. *Proc. Natl Acad. Sci. USA* **82**, 7086–7090.

Gendelman, H. E., Narayan, O., Kennedy-Stoskopf, S., Kennedy, P. G. E., Ghotbi, Z., Clements, J. E., Stanley, J., Pezeshkpour, G. (1986). Tropism of sheep lentiviruses for monocytes: susceptibilty to infection and virus gene expression increase during maturation of monocytes to macrophages. *J. Virol.* **58**, 67–74.

Georgsson, G., Palsson, P. A. (1971). The histopathology of maedi. *Vet. Pathol.* **8**, 63–80.

Georgsson, G., Martin, J. R., Klein, J., Palsson, P. A., Nathanson, N., Petursson, G. (1982). Primary demyelination in visna – an ultrastructural-study of Icelandic sheep with clinical signs following experimental-infection. *Acta Neuropathologica* **57**, 171–178.

Georgsson, G., Houwers, D. J., Palsson, P. A., Petursson, G. (1989). Expression of viral-antigens in the central nervous-system of visna-infected sheep – an immunohistochemical study on experimental visna induced by virus-strains of increased neurovirulence. *Acta Neuropathologica* **77**, 299–306.

Georgsson, G., Palsson, P. A., Petursson, G. (1990). Some comparative aspects of Visna and AIDS. In *Modern Pathology of AIDS and Other Retroviral Infections* (eds P. Racz, A. T. Hanse, J. C. Gluckman), pp. 82–98. Karger, Basle, Switzerland.

Gerber, H. A., Morris, B., Trevella, W. (1985). Humoral immunity in B cell depleted sheep. In *Immunology of the Sheep* (eds Morris, B., Miyasaka, M.), pp. 187–215. Edition Roche, Basle, Switzerland.

Gerber, H. A., Morris, B., Trevella, W. (1986). The role of gut associated lymphoid tissues in the generation of immunoglobulin bearing lymphocytes in sheep. *Aust. J. Biol. Med. Sci.* **64**, 201–213.

Giegerich, G. W., Hein, W. R., Miyasaka, M., Tiefenthaler, G., Hunig, T. (1989). Restricted expression of CD2 among subsets of sheep thymocytes and T lymphocytes. *Immunology* **66**, 354–361.

Gill, H. S. (1991). Genetic control of acquired resistance to haemonchosis in Merino lambs. *Para. Immunol.* **13**, 617–628.

Gill, H. S., Watson, D. L., Brandon, M. R. (1993). Monoclonal antibody to CD4$^+$ T-cells abrogates genetic resistance to *Haemonchus contortus* in sheep. *Immunology* **78**, 43–49.

Girard, J. P., Springer, T. A. (1995). High endothelial venules (HEVs): specialized endothelium for lymphocyte migration. *Immunol. Today* **16**(9), 449–457.

Gogolin-Ewens, K. J., Mackay, C. R., Mercer, W. R., Brandon, M. R. (1985). Sheep lymphocyte antigens (OLA) I. Major histocompatibility complex class I molecules. *Immunology* **56**, 717–723.

Goodall, J. C., Emery, D. C., Perry, A. C. F., English, L. S., Hall, L. (1990). cDNA cloning of ovine interleukin 2 by PCR. *Nucl. Acids Res.* **18**, 5883.

Gorin, A. B., Stewart, P., Gould, J. (1979). Concentration of immunoglobulin classes in subcompartments of the sheep lung. *Res. Vet. Sci.* **26**, 126–128.

Gorrell, M. D., Willis, G., Brandon, M. R., Lascelles, A. K. (1988a). Subpopulations of lymphocytes in the mammary gland of sheep. *Immunology* **66**, 388–393.

Gorrell, M. D., Willis, G., Brandon, M. R., Lascelles, A. K. (1998b). Lymphocyte phenotypes in the intestinal mucosa of sheep infected with *Trichostrongylus colubriformis*. *Clin. Exp. Immunol.* **72**, 274–279.

Gorrell, M. D., Miller, H. R. P., Brandon, M. R. (1988c). Lymphocyte phenotypes in the abomasal mucosa of sheep infected with *Haemonchus contortus*. *Parasite Immunol.* **10**, 661–674.

Gorrell, M. D., Brandon, M. R., Sheffer, D., Adams, R. J., Narayan, O. (1992). Ovine lentivirus is macrophagetropic and does not replicate productively in T lymphocytes. *J. Virol.* **66**, 2679–2688.

Gowans, J. L. (1959). The recirculation of lymphocytes from blood to lymph in the rat. *J. Physiol.* **146**, 54–69.

Grain, F., Nain, M.-C., Labonne, M.-P., Lantier, F., Lepochier, P., Gebuhrer, L., Asso, J., Maddox, J., Betuel, H. (1993). Restriction fragment length polymorphism of *DQB* and *DRB* class II genes of the ovine major histcompatibility complex. *Animal Genet.* **24**, 377–384.

Green, I. R., Sargan, D. R. (1991). Sequence of the DNA encoding tumor necrosis factor α: problems with cloning by inverse PCR. *Gene* **109**, 203–205.

Green, S. J., Nact, C. A., Meltzer, M. S. (1991). Cytokine-induced synthesis of nitrogen oxides in macrophages: a protective host response to Leishmania and other intracellular pathogens. *J. Leukoc. Biol.* **50**, 93–103.

Green, I. R., Fiskerstrand, C., Bertoni, G., Roy, D. J., Peterhans, E., Sagan, D. R. (1993). Expression and characterization of bioactive recombinant ovine TNF-α: some species specifity in cytotoxic response to TNF. *Cytokine* **5**, 213–223.

Gregoire, D., Couez, D., Deschamps, J., Heurtz, S., Hors-Cayla, M.-C., Szpirer, J., Szpirer, C., Burny, A., Huez, G., Kettmann, R. (1984). Different bovine leukemia virus-induced tumors harbor the provirus in different chromosomes. *J. Virol.* **50**, 275.

Grewal, A. S., Rouse, B. T. (1979). Characterisation of bovine leukocytes involved in antibody dependent cell cytotoxicity. *Int. Arch. Allergy Appl. Immunol.* **60**, 169–177.

Griebel, P. J., Ferrari, G. (1994). Evidence for a stromal cell-dependent self-renewing B cell population in lymphoid follicles of the ileal Peyer's patch of sheep. *Eur. J. Immunol.* **24**, 401–409.

Griebel, P., Ferrari, G. (1995). CD40 signalling in ileal Peyer's patch B cells: implications for T cell-dependent antigen selection. *Int. Immunol.* **7**, 369–379.

Griebel, P. J., Hein, W. R. (1996). Expanding the role of Peyer's patches in B-cell ontogeny. *Immunol. Today* **17**, 30–39.

Griebel, P. J., Kennedy, L., Graham, T., Davis, W. C., Reynolds, J. D. (1992). Characterization of B cell phenotypic changes during ileal and jejunal Peyer's patch development in sheep. *Immunology* **77**, 564–570.

Griebel, P. J., Kugelberg, B., Ferrari, G. (1996a). Two distinct pathways of B-cell development in Peyer's patches. *Develop. Immunol.* **4**, 263–277.

Griebel, P. J., Ghia, P., Grawunder, U., Ferrari, G. (1996b). A novel molecular complex expressed on immature B cells: a possible role in T cell-independent B cell development. *Develop. Immunol.* **5**, 67–78.

Griffin, D. E., Narayan, O., Adams, R. J. (1978). Early immune responses in visna, a slow viral disease of sheep. *J. Inf. Dis.* **138**, 340–350.

Grossberger, D., Hein, W., Marcuz, A. (1990). Class I major histocompatibility complex, cDNA clones from sheep thymus: alternative splicing could make a long cytoplasmic tail. *Immunogenetics* **32**, 77–87.

Grossberger, D., Marcuz, A., Fichtel, A., Dudler, L., Hein, W. R. (1993). Sequence analysis of sheep T-cell receptor β chains. *Immunogenetics* **37**, 222–226.

Groth, D. M., Wetherall, J. D. (1994). Dinucleotide repeat polymorphism within the ovine major histocompatibility complex class I region. *Animal Genet.* **25**, 61.

Groth, D. M., Wetherall, J. D. (1995). Dinucleotide repeat polymorphism adjacent to sheep complement factor B. *Animal Genet.* **26**, 282–283.

Groth, D. M., Wetherall, J. D., Taylor, L., Sparrow, P. R., Lee, I. R. (1988). Purification and characterization of ovine C4: evidence for two molecular forms in ovine plasma. *Molec. Immunol.* **25**, 577–584.

Groth, D. M., Lintorn-Terry, E., Outteridge, P., Windon, R. G., Wetherall, J. D. (1990). *Taq*I RFLP of the ovine complement component C4 gene. *Nucl. Acids Res.* **18**, 3102.

Gudnadottir, M., Palsson, P. A. (1965). Host–virus interaction in visna infected sheep. *J. Immunol.* **95**, 1116–1120.

Gupta, U. K., McConnell, I., Hopkins, J. (1993). Reactivity of the CD11/CD18 workshop monoclonal antibodies in the sheep. *Vet. Immunol. Immunopathol.* **39**, 93–102.

Gyorffy, E. J., Glogauer, M., Kennedy, L., Reynolds, J. D. (1992). T-cell receptor-γδ association with lymphocyte populations in sheep intestinal mucosa. *Immunology* **77**, 25–30.

Haase, A. T. (1986). Pathogenesis of lentivirus infections. *Nature* **332**, 130–136.

Haase, A. T., Stowing, L., Narayan, O., Griffin, D. E., Price, D. (1977). Slow persistent infection caused by visna virus: role of host restriction. *Science* **195**, 175–177.

Haig, D. M. (1992). Haemopoietic stem cells and the development of the blood cell repetoire. *J. Comp. Path.* **106**, 121–136.

Haig, D. M., Blackie, W., Huntley, J., MacKellar, A., Smith, W. G. (1988). The generation of ovine bone marrow-derived mast cells in culture. *Immunology* **65**, 199–203.

Haig, D. M., Brown, D., MacKellar, A. (1990). Ovine haemopoiesis: the development of bone marrow-derived colony-forming cells *in vitro* in the presence of factors derived from lymphoid cells and helper T-cells. *Vet. Immunol. Immunopathol.* **25**, 125–137.

Haig, D. M., Thomson, J., Dawson, A. (1991). Reactivity of the workshop monoclonal antibodies with ovine bone marrow cells and bone marrow-derived monocyte/macrophage and mast cell lines. *Vet. Immunol. Immunopathol.* **27**, 135–145.

Haig, D. M., Thomson, J., Percival, A. (1992). Purification and adhesion receptor phenotype of ovine bone marrow-derived haemopoietic colony-forming cells. *Vet. Immunol. Immunopathol.* **33**, 223–226.

Haig, D. M., McInnes, C., Wood, P., Seow, H.-F. (1994). The cytokines: origin, structure and function. In *Cell-Mediated Immunity in Ruminants* (eds Goddeeris, B., Morrison, W. J.), pp. 75–93. CRC Press, Boca Raton, FL, USA.

Haig, D. M., Percival, A., Mitchell, J., Green, I., Sargan, D. (1995a). The survival and growth of ovine afferent lymph dendritic cells in culture depends on tumour necrosis factor-α and is enhanced by granulocytic macrophage colony-stimulating factor but inhibited by interferon-γ. *Vet. Immunol. Immunopathol.* **45**, 221–236.

Haig, D. M., Stevenson, L. M., Thomson, J., Percival, A., Smith, W. D. (1995b). Haemopoietic cell responses in the blood and bone-marrow of sheep infected with the abomasal nematode *Telodorsagia circumcincta*. *J. Comp. Path.* **112**, 151–164.

Haig, D. M., Hutchison, G., Green, I., Sargan, D., Reid, H. W. (1995c). The effect of intradermal injection of GM-CSF and TNF-α on the accumulation of dendritic cells in ovine skin. *Vet. Dermatol.* **6**, 211–220.

Haig, D., Deane, P., Percival, A., Myatt, N., Thomson, J., Inglis, L., Rothel, J., Seow, H.-F., Wood, P., Miller, H. R. P., Reid, H. W. (1996). The cytokine response of afferent lymph following orf virus reinfection of sheep. *Vet. Dermatol.* **7**, 11–20.

Hakamori, S. (1984). Tumor associated carbohydrate antigens. *Annu. Rev. Immunol.* **2**, 103–126.

Hall, J. G. (1986). The physiology of secretory immunity. In *The Ruminant Immune System in Health and Disease* (ed. Morrison, W. I.), pp. 409–428. Cambridge University Press, Cambridge, UK.

Hall, J. G., Morris, B. (1962). The output of cells in lymph from the popliteal node of sheep. *Quart. J. Exp. Physiol.* **47**, 360–369.

Hall, J. G., Morris, B. (1965). The origin of the cells in the efferent lymph from a single lymph node. *J. Exp. Med.* **121**, 901–910.

Hall, J. G., Smith, M. E. (1970). Homing of lymph-borne immunoblasts to the gut. *Nature* **226**, 262–263.

Hall, J. G., Morris, B., Moreno, G. D., Bessis, M. C. (1967). The ultrastructure and function of the cells in lymph following antigenic stimulation. *J. Exp. Med.* **125**(1), 91–110.

Hall, J. G., Hopkins, J., Orlans, E. (1977). Studies on the lymphocytes of sheep. III. Destination of lymph-borne immunoblasts in relation to their tissue of origin. *Eur. J. Immunol.* 7, 30–37.

Halleraker, M., Landsverk, T., Nicander, L. (1990). Organization of ruminant Peyer's patches as seen with enzyme histochemical markers of stromal and accessory cells. *Vet. Immunol. Immunopathol.* **26**, 93–104.

Halleraker, M., Press, C. McL., Landsverk, T. (1994). Development and cell phenotypes in primary follicles of foetal sheep lymph nodes. *Cell Tissue Res.* **275**, 51–62.

Halliday, R. (1965). Failure of some hill lambs to absorb maternal gamma-globulin. *Nature* **205**, 614.

Halliday, R. (1978). Immunity and health in young lambs. *Vet. Record* **103**, 489–492.

Hare, G. M. T., Evans, P. J., Mackinnon, S. E., Wade, J. A., Young, A. J., Hay, J. B. (1995). Phenotypic analysis of migrant, efferent lymphocytes after implantation of cold preserved, peripheral nerve allografts. *J. Neuroimmunol.* **56**, 9–16.

Harkiss, G. D., Hopkins, J., McConnell, I. (1990). Uptake of antigen by afferent lymph dendritic cells mediated by antibody. *Eur. J. Immunol.* **20**(11), 2367–2374.

Harkiss, G. D., Watt, N. J., King, T. J., Williams, J., Hopkins, J. (1991). Retroviral arthritis: phenotypic analysis of cells in the synovial fluid of sheep with inflammatory synovitis associated with visna virus infection. *Clin. Immunol. Immunopathol.* **60**, 106–117.

Harkiss, G. D., Green, C., Anderson, A., Watt, N. J. (1995a). Immunoglobulin deposits in synovial membrane and cartilage and phenotype analysis of chondrocyte antigens in sheep infected with the visna retrovirus. *Rheumatol. Int.* **15**, 15–22.

Harkiss, G. D., Cattermole, J., Peterhans, E., Anderson, A., Vogt, H., Dickson, L., Watt, N. (1995b). T-cell and B-cell responses to mycobacterial 65-KDa heat-shock protein in sheep infected with maedi-visna virus. *Clin. Immunol. Immunopathol.* **74**, 223–230.

Harp, J. A., Pesch, B. A., Runnels, P. L. (1990). Extravasation of lymphocytes via paracortical venules in sheep lymph nodes: visualization using an intracellular fluorescent label. *Vet. Immunol. Immunopathol.* **24**, 159–167.

Harris, D. T., Camenisch, T. D., Jasofriedmann, L., Evans, D. L. (1993). Expression of an evolutionary conserved function associated molecule on sheep, horse and cattle natural killer cells. *Vet. Immunol. Immunopathol.* **38**, 273–282.

Hawken, R. J., Davies, K. P., Maddox, J. F. (1994). A dinucleotide repeat in the ovine interleukin-3 gene of limited polymorphism. *Animal Genet.* **25**, 286.

Hawken, R. J., Broom, M. F., van Stijn, T. C., Lumsden, J. M., Broad, T. E., Maddox, J. F. (1998). Mapping the ovine genes encoding IL-3, IL-4, IL-5 and CSF2 to sheep chromosome 5q13–15 by FISH. *Mammalian Genome*, **7**, 858–859, in press.

Hay, J. B., Hobbs, B. B. (1977). The flow of blood to lymph nodes and its relation to lymphocyte traffic and the immune response. *J. Exp. Med.* **145**(1), 31–44.

Hay, J. B., Lachmann, P. J., Trnka, Z. (1973). The appearance of migration inhibition factor and a mitogen in lymph draining tuberculin reactions. *Eur. J. Immunol.* **3**, 127–131.

Hay, J. B., Johnston, M. G., Vadas, P., Chin, W., Issekutz, T., Movat, H. Z. (1980). Relationships between changes in blood flow and lymphocyte migration induced by antigen. *Mongr. Allergy* **16**, 112–125.

Hayman, M., Arbuthnott, G., Harkiss, G., Brace, H., Filippi, P., Philippon, V., Thomson, D., Vigne, R., Wright, A. (1993). Neurotoxicity of peptide analogues of the transactivating protein tat from maedi-visna virus and human immunodeficiency virus. *Neuroscience* **53**, 1–6.

Heath, D. D., Holcman, B., Shaw, R. J. (1994). *Echinococcus granulosus*: the mechanism of oncosphere lysis by sheep complement and antibody. *Int. J. Parasitol.* **24**(7), 929–935.

Hediger, R., Ansari, H. A., Stranzinger, G. F. (1991). Chromosome banding and gene localizations support extensive conservation of chromosome structure between cattle and sheep. *Cytogenet. Cell Genet.* **57**, 127–134.

Hein, W., McClure, S., Miyasaka, M. (1987). Cellular composition of peripheral lymph and skin of sheep defined by monoclonal antibodies. *Int. Arch. Allergy Appl. Immunol.* **84**, 241–246.

Hein, W. R. (1994). Ontogeny of T cells. In *Cell-Mediated Immunity in Ruminants* (eds Goddeeris, B., Morrison, W. J.), pp. 19–36. CRC Press, Boca Raton, FL, USA.

Hein, W. R. (1995). Sheep as experimental animals in immunological research. *Immunologist* **3**(1), 12–18.

Hein, W. R. (1998). Molecular genetics of immune molecules. In *Genetics of the Sheep*. CAB International, Cambridge, in press.

Hein, W. R., Dudler, L. (1993a) Divergent evolution of T cell repertoires: extensive diversity and developmentally regulated expression of the sheep $\gamma\delta$ T cell receptor. *EMBO J.* **12**, 715–724.

Hein, W. R., Dudler, L. (1993b). Nucleotide sequence of the membrane form of sheep IgM and identification of two Cμ allotypes. *Mol. Immunol.* 30, 783–784.

Hein, W. R., Dudler, L. (1996). TCR $\gamma\delta$ T cells are prominent in normal bovine and express a diverse repertoire of receptors. *Immunology* **91**, 58–64.

Hein, W. R., Dudler, L. (1998). Ontogeny of the expressed immunoglobulin variable region gene repertoire in sheep. *J. Immunol.*, in press.

Hein, W. R., Mackay, C. R. (1991). Prominence of $\gamma\delta$ T cells in the ruminant immune system. *Immunol. Today* **12**, 30–34.

Hein, W. R., Tunnacliffe, A. (1990). Characterization of the CD3 γ and δ invariant subunits of the sheep T cell antigen receptor. *Eur. J. Immunol.* **20**, 1505–1511.

Hein, W. R., Tunnacliffe, A. (1993). Invariant components of sheep T-cell antigen receptor: cloning of the CD3 ε and Tcr ζ chains. *Immunogenetics* **37**, 279–284.

Hein, W. R., McClure, S., Beya, M.-F., Dudler, L., Trnka, Z. (1988). A novel glycoprotein expressed on sheep T and B lymphocytes is involved in a T cell activation pathway. *J. Immunol.* **140**, 2869–2875.

Hein, W. R., Dudler, L., Mackay, C. R. (1989a). Surface expression of differentiation antigens on lymphocytes in the ileal and jejunal Peyer's patches of lambs. *Immunology* **68**, 365–370.

Hein, W. R., Dudler, L., Beya, M.-F., Marcuz, A., Grossberger,

D. (1989b). T cell receptor gene expression in sheep: differential usage of TcR1 in the periphery and thymus. *Eur. J. Immunol.* **19**, 2297–2301.

Hein, W. R., Dudler, L., Morris, B. (1990a). Differential peripheral expansion and *in vivo* antigen reactivity of $\alpha\beta$ and $\gamma\delta$ T cells emigrating from the early fetal lamb thymus. *Eur. J. Immunol.* **20**, 1805–1813.

Hein, W. R., Dudler, L., Marcuz, A., Grossberger, D. (1990b). Molecular cloning of sheep T cell receptor γ and δ chain constant regions: unusual primary structure of γ chain hinge segments. *Eur. J. Physiol.* **20**, 1795–1804.

Hein, W. R., Marcuz, A., Fichtel, A., Dudler, L., Grossberger, D. (1991a). Primary structure of the sheep T-cell receptor α chain. *Immunogenetics* **34**, 39–41.

Hein, W. R., Dudler, L. *et al.* (1991b). Summary of workshop findings for leukocyte antigens of sheep. *Vet. Immunol. Immunopathol.* **27**, 28–30.

Hibbs, J. B. J., Tainter, R. R., Vavrin, Z., Rachlin, E. M. (1988). Nitric oxide: a cytotoxic activated macrophage effector molecules. *Biochem. Biophys. Res. Commun.* **157**, 87–94

Hopkins, J., McConnell, I., Pearson, J. D. (1981). Lymphocyte traffic through antigen-stimulated lymph nodes II. Role of prostaglandin E_2 as a mediator of cell shutdown. *Immunology* **42**, 225–231.

Hopkins, J., Dutia, B., Bujdoso, R., McConnell, I. (1989). *In vivo* modulation of CD1 and MHC Class II expression by sheep afferent lymph dendritic cells. *J. Exp. Med.* **170**, 1303–1318.

Hopkins, J., Ross, A. Dutia, B. M. (1993). Summary of workshop findings of leukocyte antigens in sheep. *Vet. Immunol. Immunopathol.* **39**, 49–59.

Howard, C. F., Taylor, G., Brownlie, J. (1980). Surface receptors for immunoglobulins on bovine polymorphonuclear neutrophils and macrophages. *Res. Vet. Sci.* **29**, 128–135.

Howard, C. J., Morrison, W. I. (1991). Comparison of reactivity of monoclonal antibodies on bovine, ovine and caprine tissues and cells from other animal species. *Vet. Immunol. Immunopathol.* **27**, 32–36.

Hugh, A. R., Trevella, W., Simpson-Morgan, M. W., Morris, B. (1985). The lymph-borne response of foetal lamb lymph nodes to challenge with *Brucella abortus in utero*. *Aust. J. Exp. Biolog. Med. Sci.* **63**, 381–395.

Hulme, D. J., Nicholas, F. W., Windon, R. G., Brown, S. C., Beh, K. J. (1993). The MHC class II region and resistance to an intestinal parasite in sheep. *J. Anim. Breed. Genet.* **110**, 459–472.

Hunig. T., Mitnacht, R., Tietenthaler, G., Kohler, C., Miyasaka, M. (1986). T11TS, the cell surface molecule binding the 'erythrocyte receptor' of T lymphocytes: cellular distribution, purification to homogeneity and biochemical properties. *Eur. J. Immunol.* **16**, 1615–1621.

Huntley, J., Newlands, G., Miller, H. R. P. (1984). The isolation and characterization of globule leukocytes: their derivation from mucosal mast cells in parasitised sheep. *Parasite Immunol.* **6**, 371–390.

Huntley, J. F., Gibson, S., Knox, D., Miller, H. R. P. (1986). The isolation and purification of a proteinase with chymotrypsin-like properties from ovine mucosal mast cells. *Int. J. Biochem.* **18**, 673–682.

Huntley, J., Gibson, S., Brown, D., Smith, W. D., Jackson, F., Miller, H. R. P. (1987). Systemic release of a mast cell protease following nematode infections in sheep. *Parasite Immunol.* **9**, 603–614.

Huntley, J., Haig, D. D., Irvine, J., Inglis, L., MacDonald, A., Rance, A., Moqbel, R. (1992). Characterization of ovine mast cells derived from *in vitro* culture of haemopoietic tissue. *Vet. Immunol. Immunopathol.* **32**, 47–64.

Huntley, J. F. (1992). Mast cells and basophils: a review of their heterogeneity and function. *J. Comp. Path.* **107**, 349–372.

Husband, A. J., Lascelles, A. K. (1975). Antibody responses to neonatal immunization in calves. *Res. Vet. Sci.* **18**, 201–207.

Husband, A. J., McDowell, G. H. (1975). Local and systemic immune responses following oral immunization of foetal lambs. *Immunology* **29**, 1019–1028.

Husband, A. J., Scicchitano, R., Sheldrake, R. F. (1987). Origin of IgA at mucosal sites. In *Recent Advances in Mucosal Immunology* (eds McGhee, J. R. *et al.*), pp. 1157–1162. Plenum Press, New York, USA.

Ishiguro, N., Hein, W. R. (1994). The T cell receptor. in *Cell-mediated Immunity in Ruminants* (eds Goddeeris, B., Morrison, W. J.), pp. 59–73. CRC Press, Boca Raton, FL, USA.

Ishino, S., Matsuda, I., Yamamoto, H., Yoshino, T., Sentisui, H., Mizuno, Y., Kono, Y. (1989). Pathological finding of two types of lympoid malignancy in sheep inoculated with bovine leukemia virus. *Jap. J. Vet. Sci.* **4**, 749.

Israf, D. A., Coop, R. L., Stevenson, L. M., Jones, D. G., Jackson, F., Jackson, E., MacKellar, A., Huntley, J. F. (1996). Dietary protein influences upon immunity to *Nematodirus battus* infection in lambs. *Vet. Parasitol.* **61**, 273–286.

Issekutz, T., Chin, W., Hay, J. B. (1980). Lymphocyte traffic through granulomas: differences in the recovery of indium-111-labeled lymphocytes in afferent and efferent lymph. *Cell. Immunol.* **54**(1), 79–86.

Issekutz, T. B., Chin, G. W., Hay, J. B. (1981). Lymphocyte traffic through chronic inflammatory lesions: differential migration versus differential retention. *Clin. Exp. Immunol.* **45**, 604–614.

James, S. L., Nacy, C. (1993) Effector functions of activated macrophages against parasites. *Curr. Opin. Immunol.* **5**, 518–523.

Jarpe, M. A., Johnson, H. M., Bazer, F. W., Ott, T. L., Curto, E. V., Krishna, N. R., Pontzer, C. H. (1994). Predicted structural motif of INF-TAU. *Prot. Engin.* **7**, 863–867.

Jenkinson, D. M. (1990). Sweat and sebaceous glands and their function in domestic animals. *Adv. Vet. Dermatol.* **1**, 229–251.

Jenkinson, D. M. (1992). The basis of the skin surface ecosystem. In *The Skin Microflora and Microbial Skin Disease* (ed. Noble, W. C.), pp. 1–33. Cambridge University Press, Cambridge, UK.

Jenson, W. A., Sheehy, S. E., Fox, M. H., Davis, W. C., Cockerell, G. L. (1990). *In vitro* expression of bovine leukemia virus in isolated B-lymphocytes of cattle and sheep. *Vet. Immunol. Immunopathol.* **26**(4), 333–342.

Joag, S. V., Stephens, E. B., Narayan, O. (1996). Lentiviruses. In *Virology* (eds Fields, B. N., Knipe, D. M., Howley, P. M.), pp. 1977–1996. Lippincott-Raven Publishers, Philadelphia, USA.

John, H. A., Deane, D., Haig, D. (1994). Generation of an ovine bone-marrow-derived myelomonocyte-like cell-line by retroviral-mediated transformation-immunological characterization and the effect of cytokines and lipopolysaccharides. *J. Leukoc. Biol.* **55**, 758–792.

Johnson, S. E., Barendse, W., Hetzel, D. J. (1993). The gamma fibrinogen gene (FGG) maps to chromosome 17 in both cattle and sheep. *Cytogenet. Cell Genet.* **62**, 176–180.

Johnstone, A., Manktelow, B. (1978). The pathology of spontaneous occurring malignant lymphoma in sheep. *Vet. Pathol.* **15**, 301.

Jonas, W., Broad, S. (1972). Complement activity of sheep body fluids and human and guinea pig colostrum. *Res. Vet. Sci.* **13**, 154–159.

Jones, G. E., Gilmour, J. S., Rae, A. G. (1982). The effect of different strains of *Mycoplasma ovipneumoniae* on specific pathogen-free and conventially reared lambs. *J. Comp. Pathol.* **92**, 267–272.

Jones, W. O., Emery, D. L. (1991). Demonstration of a range of inflammatory mediators released in trichostronglylosis of sheep. *Int. J. Parasitol.* **21**, 361–363.

Jones, W. O., Emery, D. L., McClure, S. J., Wagland, B. M. (1994). Changes in inflammatory mediators and larval inhibitory activity in intestinal contents and mucus during primary and challenge infections of sheep with *Trichostrongylus colubriformis*. *Int. J. Parasitol.* **24**, 519–525.

Junger, W. G., Hallstrom, S., Redl, H., Schlag, G. (1992). Inhibition of human, ovine and baboon neutrophil elastase with eglin-C and secretory leukocyte proteinase-inhibitor. *Biolog. Chem.* **373**, 119–122.

Jungi, T. W., Adler, H., Adler, B., Thony, M., Krampe, M., Peterhans, E. (1996). Inducible nitric oxide synthetase of macrophages. Present knowledge and evidence for species-specific regulation. *Vet. Immunol. Immunopath.* **54**, 323–330.

Kalaaji, A. N., Abernethy, N. J., McCullough, K., Hay, J. B. (1988). Recombinant bovine interferon-α1 inhibits the migration of lymphocytes from but not into lymph nodes. *Region. Immunol.* **1**, 56–61.

Kalaaji, A. N., McCullough, K., Hay, J. B. (1989). The enhancement of lymphocyte localization in skin sites of sheep by tumor necrosis factor alpha. *Immunol. Lett.* **23**, 143–148.

Kapil, A., Sharma, S. (1994). Anti-complement activity of oleanolic acid: an inhibitor of C-3-convertase of the classical complement pathway. *J. Pharm. Pharmacol.* **46**(11), 922–923.

Karupiah, G., Xie, Q. W., Butler, R. M., Nathan, C., Duarte, C., MacMicking, J. D. (1993). Inhibition of viral replication by interferon-gamma-induced nitric oxide synthase. *Science* **261**, 1445–1448.

Keefer, J. F., Mong, J. S. (1990). Identification and characterization of the substance P receptor in sheep intestinal smooth muscle membranes. *J. Pharmacol. Exp. Ther.* **255**, 120–127.

Kelsall, B. L., Strober, W. (1996). Distinct populations of dendritic cells are present in the subepithelial dome and T cell regions of the murine Peyer's patches. *J. Exp. Med.* **183**, 237–247.

Kennedy, P. G. E., Narayan, O., Ghotbi, Z., Hopkins, J., Gendelman, H. E., Clements, J. E. (1985). Persistent expression of Ia antigen and viral genome in visna-maedi virus-induced inflammatory cells: possible role of lentivirus-induced interferon. *J. Exp. Med.* **162**, 1970–1982.

Kennedy-Stoskopf, S. (1989). Pathogenesis of lentivirus-induced arthritis. *Rheumatol. Int.* **9**, 129–136.

Kennedy-Stoskopf, S., Narayan, O. (1986). Neutralising antibodies to visna lentivirus: mechanism of action and possible role in virus persistence. *J. Virol.* **59**, 37–44.

Kettmann, R. M. G., Cleuter, Y., Portetelle, D., Mammerickx, M., Burny, A. (1980a). Genomic integration of bovine leukemia provirus and lack of viral RNA expression in the target cells of cattle with different response to BLV infection. *Leuk. Res.* **4**, 509.

Kettmann, R., Cleuter, Y., Mammerickx, M., Meunier-Rotival, M., Bernadi, G., Burny, A., Chantrenne, H. (1980b). Genomic integration of bovine leukemia provirus: comparison between persistent lymphocytosis and lymph node tumor form of enzootic bovine leukosis. *Proc. Natl Acad. Sci. USA* **77**, 2577.

Kettmann, R., Deschamps, J., Couze, D., Claustriaux, J.-J., Palm, R., Burny, A. (1983). Chromosome integration domain for bovine leukemia provirus in tumors. *J. Virol.* **47**, 146.

Kettmann, R., Mammerickx, M., Portetelle, D., Gregoire, D., Burny, A. (1984). Experimental infection of sheep and goat with Bovine Leukemia Virus: localization of proviral information in the target cells. *Leuk. Res.* **8**, 937.

Kettmann, R., Cleuter, Y., Gregoire, D., Burny, A. (1985). Role of the 3′ long open reading frame region of bovine leukemia virus in the maintenance of cell transformation. *J. Virol.* **54**, 899.

Kettmann, R., Burny, A., Callebaut, I., Droogmans, L., Mammerickx, M., Willems, L., Portetelle, D. (1994). Bovine Leukemia Virus. In *The Retroviridae* 3rd edn. (ed. Levy, J. A.), pp. 39–81. Plenum Press, New York, USA.

Kimpton, W. G., Cahill, R. N. P. (1985). Circulation of autologous and allogeneic lymphocytes in lambs before and after birth. In *Immunology of the Sheep* (eds Morris, B., Miyasaka, M.), pp. 306–326. Editiones Roche, Basle, Switzerland.

Kimpton, W. G., Washington, E. A., Cahill, R. N. P. (1989a). Recirculation of lymphocyte subsets ($CD5^+$, $CD4^+$, SBU-$T19^+$ and B cells) through gut and peripheral lymph nodes. *Immunology* **66**, 69–75.

Kimpton, W. G., Washington, E. A., Cahill, R. N. P. (1989b). Recirculation of lymphocyte subsets ($CD5^+$, $CD4^+$, $CD8^+$, $T19^+$, and B cells) through fetal lymph nodes. *Immunology* **68**, 575–579.

Kimpton, W. G., Washington, E. A., Cahill, R. N. P. (1990). Non-random migration of $CD4^+$, $CD8^+$, and $\gamma\delta^+T19^+$ lymphocyte subsets following *in vivo* stimulation with antigen. *Cell. Immunol.* **130**, 236–243.

Kimpton, W. G., Washington, E. A., Cahill, R. N. P. (1994). The development of the immune system in the foetus. In *Textbook of Foetal Physiology* (eds Thorburn, G. D., Harding, R.), Oxford University Press, Oxford, UK.

Kimpton, W. G., Washington, E. A., Cahill, R. N. P. (1995). Virgin $\alpha\beta$ and $\gamma\delta$ T cells recirculate extensively through peripheral tissues and skin during normal development of the fetal immune system. *Int. Immunol.* 7(10), 1567–1577.

Klevjer-Anderson, P., Adams, D. S., Anderson, L. W., Banks, K. L., Mcguire, T. C. (1984). A sequential study of virus expression in retrovirus-induced arthritis in goats. *J. Gen. Virol.* **65**, 1519–1525.

Koistinen, V. (1992). Limited tryptic cleavage of complement factor H abrogates recognition of sialic acid-containing surfaces by the alternative pathway of complement. *Biochem. J.* **282**(2), 317-319.

Koizumi, T., Kubo, K., Shinozaki, S., Koyama, S., Kobayashi, T., Sekiguchi, M. (1993). Granulocyte colony-stimulating factor does not exacerbate endotoxin-induced lung injury in sheep. *Am. Rev. Respir. Dis.* **148**, 132–137.

Kumar, S. N., Kumar, W. M. S., Stewart, G. L., Seelig, L. L.

(1989). Maternal to neonatal transmission of T cell-mediated immunity to *Trichinella spiralis* during lactation. *Immunology* **68**, 87–92.

Kumta, S., Ritz, M., Hurley, J., Crowe, D., Romeo, R., O'Brein, B. (1994). Acute inflammation in fetal and adult sheep – the response to subcutaneous injection of turpentine and carrageenan. *Br. J. Plastic Surg.* **47**, 360–368.

Kwon, N. S., Nathan, C. F., Stuehr, D. J. (1989). Reduced biopterin as a cofactor in the generation of nitrogen oxides by murine macrophages.*J. Biol. Chem.* **264**, 20 496–20 501.

Lagarias, D., Radke, K. (1989). Transcriptional activation of bovine leukemia virus in blood cells from experimentally infected asymptomatic sheep with latent infection. *J. Virol* **63**, 2099.

Lairmore, M. D., Butera, S. T., Callahan, G. N., Demartini, J. C. (1988). Spontaneous interferon production by pulmonary leukocytes is associated with lentivirus-induced lymphoid interstitial pneumonia *J. Immunol.* **140**, 779–785.

Landsverk, T. (1987). The follicle-associated epithelium of the ileal Peyer's patch in ruminants is distinguishable by its shedding of 50 nm particles. *Immunol. Cell Biol.* **65**, 251–261.

Landsverk, T. (1988). Phagocytosis and transcytosis by the follicle-associated epithelium of the ileal Peyer's patch in calves. *Immunol. Cell Biol.* **66**, 261–268.

Landsverk, T., Jansson, Å., Nicander, L., Plîen, L. (1987). Carbonic anhydrase – a marker for particles shed from the epithelium to the lymphoid follicles of the ileal Peyer's patch in goat kids and lambs. *Immunol. Cell. Biol.* **65**, 425–429.

Landsverk, T., Trevella, W., Nicander, L. (1990). Transfer of carbonic anhydrase-positive particles from the follicle-associated epithelium to lymphocytes of Peyer's patches in foetal sheep and lambs. *Cell Tissue Res.* **261**, 239–247.

Landsverk, T., Halleraker, M., Aleksandersen, M., McClure, S., Hein, W., Nicander, L. (1991). The intestinal habitat for organised lymphoid tissues in ruminants: comparative aspects of structure, function and development. *Vet. Immunol. Immunopathol.* **28**, 1–16.

Landsverk, T., Press, C. M., Aleksandersen, M., Hein, W., Simpson-Morgan, M. (1992). B cell ontogeny in the sheep foetus. *Scand. J. Immunol.* **36**, 633.

Larsen, H. J. (1986). Distribution of T and B lymphocytes in jejunal and ileocaecal Peyer's patches of lambs. *Res. Vet. Sci.* **40**, 105–111.

Larsen, H. J., Hyllseth, B., Krogsrud, J. (1982a). Experimental maedi virus infection in sheep: cellular and humoral immune response during three years following intranasal inoculation. *Am. J. Vet. Res.* **43**, 384–389.

Larsen, H. J., Hyllseth, B., Krogsrud, J. (1982b). Experimental maedi virus infection in sheep: early cellular and humoral immune response following parenteral inoculation. *Am. J. Vet. Res.* **43**, 379–383.

Larson, B. L., Harvey, H. L., Devery, J. E. (1980). Immunoglobulin production and transport by the mammary gland. *J. Dairy Sci.* **63**, 665–671.

Lascelles, A. K. (1969). Immunoglobulin secretion into ruminant colostrum. In *Lactogenesis: the Initiation of Milk Secretion at Parturition*, pp. 131. University of Pennsylvania Press, Philadelphia, USA.

Lascelles, A. K., Beh, K. J., Husband, A. J. (1981). Origin of antibody in mammary secretion with particular reference to the IgA system. *Ruminant Imm. Sys. Adv. Exper. Med. Biol.* **137**, 493–511.

Lascelles, A. K., Beh, K. J., Mukkur, T. K., Watson, D. L. (1985). The mucosal immune system with particular reference to ruminant animals. In *The Ruminant Immune System in Health and Disease* (ed. Morrison, W. I.), pp. 391–408. Cambridge University Press, Cambridge, UK.

Lear, A., Hutchison, G., Reid, H. W., Norval, M., Haig, D. M. (1996). Phenotypic characterization of the dendritic cells accumulating in ovine dermis following primary and secondary virus infections. *Eur. J. Dermatol.* **6**, 135–140.

Lee, C. S., Lascelles, A. K. (1970). Antibody producing cells in antigenically stimulated mammary glands and in the gastro-intestinal tract of sheep. *Aust. J. Exp. Biol. Med. Sci.* **48**, 525–535.

Lee, C. S., Outteridge, P. M. (1981). Leucocytes of sheep colostrum, milk and involution secretion, with particular reference to ultrastructure and lymphocyte subpopulations. *J. Dairy Res.* **48**, 225–237.

Lee, C. S., Gogolin-Ewens, K., Brandon, M. R. (1988). Identification of a unique lymphocyte population in the sheep uterus. *Immunology* **63**, 157–164.

Lee, C. S., Meeusen, E., Brandon, M. R. (1989). Subpopulations of lymphocytes in the mammary gland of sheep. *Immunology* **66**, 388–393.

Lee, C. S., Meeusen, E., Brandon, M. R. (1992). Local immunity in the mammary gland. *Vet. Immunol. Immunopathol.* **32**, 1–11.

Lee, W. C., Bird, P., McConnell, I., Watt, N. J., Blacklaws, B. A. (1966). The phenotype and phagocytic activity of macrophages during maedi-visna virus infection. *Vet. Immunol. Immunopathol.* **51**, 113–126.

Lee, W. C., McConnell, I., Blacklaws, B. A. (1994). Cytotoxic activity against maedi-visna virus-infected macrophages. *J. Virol.* **68**, 8331–8338.

Lena, P., Freyria, A. M., Lyon, M., Cadore, J. L., Guiguen, F., Greenland, T., Belleville, J., Cordier, G., Mornex, J. F. (1994). Increased expression of tissue factor messenger-RNA and procoagulant anticoagulant activity in ovine lentivirus-infected alveolar macrophages. *Res. Virol.* **145**, 111–115.

Letesson, J., Mager, A., Mammerickx, M., Burny, A., Depelchin, A. (1990). B cells from bovine leukemia virus (BLV)-infected sheep with hematological disorders express the CD5 T-cell marker. *Leukemia* **4**, 377.

Levieux, D., Venien, A. (1991). A rapid and simple method for the preparation of a monospecific antibody to ovine haptoglobin. *J. Immunol. Methods* **141**, 111–115.

Li, X., Abdi, K., Rawn, J., Mackay, C. R., Mentzer, S. J. (1996). LFA-1 and L-selectin regulation of recirculating lymphocyte tethering and roling on lung microvascular endothelium. *Am. J. Res. Cell Mol. Biol.* **14**, 398–406.

Lichtensteiger, C. A., Cheevers, W. P., Davis, W. C. (1993). CD8+ cytotoxic T lymphocytes against antigenic variants of caprine arthritis-encephalitis virus. *J. Gen. Virol.* **74**, 2111–2116.

Litchfield, A. M., Raadsms, H. W., Hulme, D. J., Brown, S. C., Nicholas, F. W., Egerton, J. R. (1992). Disease resistance in Merino sheep. II. RFLPs in class II MHC and their association with resistance to footrot. *J. Anim. Breed. Genet.* **110**, 321–334.

Liu, W. J., Hansen, P. J. (1993). Effect of the progesterone-induced serpin-like proteins of the sheep endometrium on natural killer cell activity in sheep and mice. *Biol. Repro.* **49**, 1008–1014.

Loh, E. Y., Elliott, J. F., Cuirla, S., Lonier, L. L., Davis, M. M. (1989). Polymerise chain reaction with single-sided specificity: Analysis of T cell receptor delta chain. *Science* **243**, 217–220.

Lowden, S., Heath, T. (1992). Lymph pathways associated with Peyer's patches in sheep. *J. Anat.* **181**, 209–217.

Lowe, D. M., Lachmann, P. J. (1974). The fractionation of antigen-dependent macrophage migration inhibition and macrophage activation factors from lymph draining a tuberculin reaction. *Scand. J. Immunol.* **3**, 423–432.

Lozano Alarcon, F., Bradley, G., Allen, T., Reggiardo, C. (1992). Lymphosarcoma in a desert bighorn sheep. *J. Vet. Diagn. Invest.* **4**, 492.

Lucio, J., D'Brot, J., Guo, C.-B., Abraham, W. M., Lichtenstein, L., Kagey-Sobotka, A., Ahmed, T. (1992). Immunologic mast cell-mediated responses and histamine release are attenuated by heparin. *J. Appl. Physiol.* **73**, 1093–1101.

Luffau, G., Vu Tien Kang, J., Bouix, J., Nguyen, T. C., Cullen, P., Ricordeau, G. (1990). Resistance to experimental infections with *Haemonchus contortus* in Romanov sheep. *Genet. Select. Evol.* **22**, 205–229.

Lujan, L., Begara, I., Collie, D. D. S., Watt, N. J. (1993). Phenotypic analysis of cells in bronchoalveolar lavage fluid and peripheral-blood of maedi visna-infected sheep. *Clin. Exp. Immunol.* **91**, 272–276.

Lujan, L., Begara, I., Collie, D. D. S., Watt, N. J. (1995). $CD8^+$ lymphocytes in bronchoalveolar lavage and blood: *in vivo* indicators of lung pathology caused by maedi-visna virus. *Vet. Immunol. Immunopathol.* **49**, 89–100.

Lutley, R., Petursson, G., Palsson, P. A., Georgsson, G., Klein, J., Nathanson, N. (1983). Antigenic drift in visna: virus variation during long term infection of Icelandic sheep. *J. Gen. Virol.* **64**, 1433–1440.

Luzzati, A. L. (1979). Induction of secondary antibody response *in vitro* with rabbit peripheral blood lymphocytes. In *Immunological Methods* (eds Lefkovits, I., Pernis, B.), pp. 335–343. Academic Press, London, UK.

Mackay, C. R. (1988). Sheep leukocyte molecules: a review of their distribution, structure and possible function. *Vet. Immunol. Immunopathol.* **19**, 1–20.

Mackay, C. R., Maddox, J. F., Gogolin-Ewens, K., Brandon, M. R. (1985). Characterization of two sheep lymphocyte differentiation antigens, SBU-T1 and SBU-T6. *Immunology* **557**, 729–737.

Mackay, C. R., Maddox, J. F., Brandon, M. R. (1986a). Thymocyte subpopulations during early fetal development in sheep. *J. Immunol.* **136**, 1592–1599.

Mackay, C. R., Maddox, J. F., Brandon, M. R. (1986b). Three distinct subpopulations of sheep T lymphocytes. *Eur. J. Immunol.* **16**, 19–25.

Mackay, C. R., Maddox, J. F., Brandon, M. R. (1987). A monoclonal antibody to the p220 component of sheep LCA identifies B cells and a unique lymphocyte subset. *Cell. Immunol.* **110**, 46–55.

Mackay, C. R., Maddox, J. F., Wijffels, G. L., Mackay, I. R., Walker, I. D. (1988a). Characterization of a 95,000 molecule on sheep leucocytes homologous to murine Pgp-1 and human CD44. *Immunology* **65**, 93–99.

Mackay, C. R., Kimpton, W. G., Brandon, M. R., Cahill, R. N. P. (1988b). Lymphocyte subsets show marked differences in their distribution between blood and the afferent and efferent lymph of peripheral lymph nodes. *J. Exp. Med.* **167**(6), 1744–1765.

Mackay, C. R., Hein, W. R., Brown, M. H., Matzinger, P. (1988c). Unusual expression of CD2 in sheep: implications for T cell interactions. *Eur. J. Immunol.* **18**, 1681–1688.

Mackay, C. R., Beya, M.-F., Matzinger, P. (1989). γ/δ T cells express a unique surface molecule appearing late during thymic development. *Eur. J. Immunol.* **19**, 1477–1483.

Mackay, C. R., Marston, W. L., Dudler, L. (1990). Naive and memory T cells show distinct pathways of lymphocyte recirculation. *J. Exp. Med.* **171**, 801–817.

Mackay, C. R., Marston, W. L., Dudler, L., Spertini, O., Tedder, T. F., Hein, W. R. (1992a). Tissue-specific migration pathways by phenotypically distinct subpopulations of memory T cells. *Eur. J. Immunol.* **22**(4), 887–895.

Mackay, C. R., Marston, W., Dudler, L. (1992b). Altered patterns of T cell migration through lymph nodes and skin following antigen challenge. *Eur. J. Immunol.* **22**, 2205–2210.

Mackenzie, D. D. S., Lascelles, A. K. (1968). The transfer of $[^{131}I]$-labelled immunoglobulins and serum albumin from blood into milk of lactating ewes. *Aust. J. Exp. Biol. Med. Sci.* **46**, 285–294.

Macmillan, R. E. (1928). The so-called hemal nodes of the white rat, guinea pig, and sheep: a study of their occurrence, structure and significance. *Anat. Rec.* **39**, 155–169.

Maddox, J. F., Mackay, C. R., Brandon, M. R. (1985a). Surface antigens SBU-T4 and SBU-T8 of sheep lymphocytes subsets defined by monoclonal antibodies. *Immunology* **55**, 739–749.

Maddox, J. F., Mackay, C. R., Brandon, M. R. (1985b). The sheep analogue of leucocyte common antigen (LCA). *Immunology* **55**, 347–353.

Maddox, J. F., Mackay, C. R., Brandon, M. R. (1987a). Ontogeny of ovine lymphocytes. I. An immunohistological study on the development of T lymphocytes in the sheep embryo and fetal thymus. *Immunology* **62**, 97–105.

Maddox, J. F., Brandon, M. R., Mackay, C. R. (1987b). Ontogeny of ovine lymphocytes. II. An immunohistological study on the development of T lymphocytes in the sheep fetal spleen. *Immunology* **62**, 107–112.

Maddox, J. F., Brandon, M. R., Mackay, C. R. (1987c). Ontogeny of ovine lymphocytes. III. An immunohistological study on the development of T lymphocytes in sheep fetal lymph nodes. *Immunology* **62**, 113–118.

Maddox, J. F., Snibson, K. J., Ballingall, K. T., Fabb, S. A. (1994). The ovine *DQA1* and *DQA2* genes are both highly polymorphic. *Animal Genet.* **25**(Suppl.), 2–20.

Mahdy, E. A., Makinen, A., Chowdhary, B. P., Andersson, L., Gustavsson, I. (1989). Chromosomal localization of the ovine major histocompatibility complex (OLA) by *in situ* hybridization. *Hereditas* **111**, 87–90.

Malu, M. N., Tabel, H. (1986). The alternative pathway of complement in sheep during the course of infection with *Trypanasoma congolense* and after Berenil treatment. *Parasite Immunol.* **8**, 217–229.

Mammerickx, M., Dekegel, D., Burny, A., Portetelle, D. (1976). Study on the oral transmission of bovine leukosis to sheep. *Vet. Microbiol.* **1**, 347.

Mammerickx, M., Palm, R., Portetelle, D., Burny, A. (1988). Experimental transmission of enzootic bovine leukosis in sheep: latency period of the tumoral disease. *Leukemia* **2**, 103.

Martin, H. M., Barcham, G. J., Egan, P. J., Andrews, A. E., Nash, A. D. (1986). Centre for Animal Biotechnology, University of Melbourne, Australia. (Unpublished data.)

Martin, H. M., Nash, A. D., Andrews, A. E. (1995). Cloning and characterisation of an ovine interleukin 10 cDNA. *Gene* **159**, 187–191.

Martin, W. B., Aitken, I. D. (1991). *Diseases of Sheep*. Blackwell Scientific Publications, Oxford, UK.

Matthews, P., Maddox, J. F. (1994). A polymorphic dinucleotide repeat microsatellite detected in the ovine interleukin-2 receptor alpha (IL2-Rα, CD25) gene. *Animal Genet*. **25**, 200.

McClure, S. J., Emery, D. L. (1993). Recent advances in veterinary immunology, particularly mucosal immunity. In *Vaccines for Veterinary Applications* (ed. Peters, A. R.), pp. 1–30. Butterworth Heinemann, Oxford, UK.

McClure, S. J., Emery, D. L. (1994). Cell-mediated responses against gastrointestinal nematode parasites of ruminants. In *Cell-Mediated Immunity in Ruminants* (eds Godeeris, B. M. L., Morrison, W. I.), pp. 213–227. CRC Press, Boca Raton, FL, USA.

McClure, S., Dudler, L., Thorpe, D., Hein, W. R. (1988a). Analysis of cell division among subpopulations of lymphoid cells in sheep. I. Thymocytes. *Immunology* **65**, 393–399.

McClure, S., Dudler, L., Thorpe, D., Hein, W. R. (1988b). Analysis of cell division among subpopulations of lymphoid cells in sheep. II. Peripheral lymphocytes. *Immunology* **65**, 401–404.

McClure, S. J., Hein, W. R., Yamaguchi, K., Dudler, L., Beya, M.-F., Miyasaka, M. (1989). Ontogeny, morphology and tissue distribution of a unique subset of $CD4^-$, $CD8^-$ sheep T lymphocytes. *Immunol. Cell Biol*. **67**, 215–221.

McClure, S. J., Wagland, B. M., Emery, D. L. (1991). Effects of Freund's adjuvants on local, draining, and circulating lymphocyte populations in sheep. *Immunol. Cell Biol*. **69**, 361–367.

McClure, S. J., Emery, D. L., Wagland, B. M., Jones, W. O. (1992). A serial study of rejection of *Trichostrongylus colubriformis* by immune sheep. *Int. J. Parasitol*. **22**, 227–234.

McClure, S. J., Emery, D. L., Husband, A. J. (1994). Effects of adjuvant and route of immunization on the intestinal immune response of sheep to ovalbumin. *Reg. Immunol*. **6**, 210–217.

McCullagh, P. (1989). Interception of the development of self tolerance in fetal lambs. *Eur. J. Immunol*. **19**, 1387–1392.

McCullagh, P. (1991). Failure of foetal thyroid allographs to induce self-tolerance in thyroidectomized foetal lambs. *Immunology* **72**, 405–410.

McCullagh, P. (1993). The failure of a combination of thyroid and thymus allographs to prevent development of autoimmunity in thyroidectomized foetal lambs. *J. Autoimm*. **6**, 51–62.

McGuire, T. C., Musoke, A. J., Kurtti, T. (1979). Functional properties of bovine IgG1 and IgG2: interaction with complement, macrophages, neutrophils and skin. *Immunology* **38**, 249–256.

McGuire, T. C., Regnier, J., Kellow, T., Gates, N. L. (1983). Failure in passive transfer of immunoglobulin G_1 to lambs: measurement of immunoglobulin G_1 in ewe colostrums. *Am. J. Vet. Res*. **44**, 1064–1067.

McGuire, T. C., Adams, D. S., Johnson, G. C., Klevjer-Anderson, P., Barbee, D. D., Gorham, J. R. (1986). Acute arthritis in caprine arthritis-encephalitis virus challenge exposure of vaccinated or persistently infected goats. *Am. J. Vet. Res*. **47**, 537–540.

McGuire, W. J., Imakawa, K., Christenson, R. K. (1995). Granulocytic-macrophage colony-stimulating factor (GM-CFS) increases messenger RNA encoding for ovine interferon-TAU (OIFN-TAU) through the protein-kinase-C (PKC) 2nd messenger system. *Biol. Reprod*. **52**, 146.

McInnes, C. J. (1993). Current research on ovine cytokines. *Br. Vet. J*. **149**, 371–386.

McInnes, C. J., Logan, M., Redmond, J., Entrican, G., Baird, G. D. (1991). The molecular cloning of the ovine gamma-interferon cDNA using the polymerase chain reaction. *Nucl. Acids Res*. **18**, 4012.

McInnes, C. J., Haig, D. M., Logan, M., Entrican, G. (1994a). The cloning and expression of the gene for ovine interleukin-3 (multi-CSF) and a comparison of the *in vitro* haematopoietic activity of ovine IL-3 with GM-CSF and human M-CSF. *Exp. Hematol*. **21**, 1528–1534.

McInnes, C. J., Logan, M., Haig, D., Wright, F. (1994b). Cloning of cDna encoding for ovine interleukin-3. *Gene* **139**, 289–290.

Mdurvwa, E. G., Ogunbiyi, P. O., Gakou, H. S., Reddy, P. G. (1994). Pathogenic mechanisms of caprine arthritis-encephalitis virus. *Vet. Res. Comm*. **18**, 483–490.

Meeusen, E., Gorrell, M. D., Brandon, M. R. (1988). Presence of a distinct $CD8^+$ and $CD5^-$ leucocyte subpopulation in the sheep liver. *Immunology* **64**, 615–619.

Meeusen, E., Gorrel, M. D., Rickard, M. D., Brandon, M. R. (1989). Lymphocyte subpopulations of sheep in protective immunity to *Taenia hydatigena*. *Parasite Immunol*. **11**, 169–181.

Meeusen, E., Barcham, G. J., Gorrell, M. D., Rickard, M. D., Brandon, M. R. (1990). Cysticercosis: cellular immune responses during primary and secondary infection. *Parasite Immunol*. **12**, 403–418.

Meeusen, E., Lee, C.-S., Brandon, M. (1991). Differential migration of T and B cells during an acute inflammatory response. *Eur. J. Immunol*. **21**, 2269–2272.

Meeusen, E., Fox, A., Brandon, M., Lee, C. S. (1993). Activation of uterine intraepithelial gamma delta T cell receptor-positive lymphocytes during pregnancy. *Eur. J. Immunol*. **23**, 1112–1117.

Mehta, P. D., Thormar, H. (1874). Neutralising activity in isolated serum antibody fractions from visna-infected sheep. *Infect. Immun*. **10**, 678–680.

Meyrick, B. O., Brigham, K. L. (1984). The effect of a single infusion of zymosan-activated plasma on the pulmonary microcirculation of sheep. Structure-function relationships. *Am. J. Pathol*. **114**, 32–45.

Micusan, V. V., Borduas, A. G. (1977). Biological properties of goat immunoglobulin G. *Immunology* **32**, 373–381.

Miller, H. R. P. (1984). The protective mucosal response against gastrointestinal nematodes in ruminants and laboratory animals. *Vet. Immunol. Immunopathol*. **6**, 167–180.

Miller, H. R. P. (1985). Mucosal mast cells, basophils, immediate hypersensitivity reactions and protection against gastrointestinal nematodes. In *The Ruminant Immune System in Disease and Health* (ed. Morrison, W. I.), pp. 496–524. Cambridge University Press, Cambridge, UK.

Miller, H. R. P., Adams, E. P. (1977). Reassortment of lymphocytes in lymph from normal and allografted sheep. *Am. J. Pathol*. **87**, 59–80.

Miller, J. F. A. P. (1966). Immunity in the foetus and the newborn. *Brit. Md. Bull*. **22**, 21–26.

Millot, P. (1979). Genetic control of lymphocyte antigens in sheep: the *OLA* complex and two minor loci. *Immunogenetics* **9**, 509–534.

Millot, P. (1989). Sheep major histocompatibility (OLA) complex: apparent involvement in a flock endemic infection by *Corynebacterium pseudotuberculosis*. *Exp. Clin. Immunogen.* **6**, 225–235.

Millot, P., Chatelain, J., Dautheville, C., Salmon, D., Cathala, F. (1988). Sheep major histocompatibility (*OLA*) complex: linkage between a scrapie susceptibility/resistance locus and the *OLA* complex in Ile-de-France progenies. *Immunogenetics* **27**, 1–11.

Miyasaka, M., Morris, B. (1988). The ontogeny of the lymphoid system and immune responsiveness in sheep. *Prog. Vet. Microbiol. Immunol.* **4**, 21–55.

Moore, T. C. (1984a). Modification of lymphocyte traffic by vasoactive neurotransmitter substances. *Immunology* **52**, 511–518.

Moore, T. C. (1984b). The modulation by prostaglandins of increases in lymphocyte traffic induced by bradykinin. *Immunology* **54**, 455–460.

Moore, T. C., Lippmann, M., Khan, F., Locke, R. L. (1984). Depression of lymphocyte traffic in sheep by induced systemic arterial hypotension and by acute arterial occlusion. *Immunology* **53**, 667–682.

Morris, B. (1984). Experimental studies in the acquisition of an immunological identity. *Frontiers in Physiological Research: 29th International Congress of Physiological Sciences*, pp. 69–78. Cambridge University Press, Sydney, Australia.

Morris, B. (1986). The ontogeny and comportment of lymphoid cells in fetal and neonatal sheep. *Immunol. Rev.* **91**, 219–233.

Morris, B., Courtice, F. C. (1977). Cells and immunoglobulins in lymph. *Lymphology* **10**, 62–70.

Morris, B., Al Salami, M. (1987). The blood and lymphatic capillaries of lymph nodes in the sheep fetus and their involvement in cell traffic. *Lymphology* **20**, 244–251.

Morris, D. W. (1995). Earth's peeling veneer of life. *Nature* **373**, 25.

Motyka, B., Reynolds, J. D. (1995). Rescue of ileal Peyer's patch B cells from apoptosis is associated with induction of Bcl-2 expression. *Immunology* **84**, 383–387.

Motyka, B., Bhogal, H. S., Reynolds, J. D. (1995). Apoptosis of ileal Peyer's patch B cells is increased by glucocorticoids or anti-immunoglobulin antibodies. *Eur. J. Immunol.* **25**, 1865–1871.

Mulder, K., Colditz, I. G. (1993). Migratory responses of ovine neutrophils to inflammatory mediators *in vitro* and *in vivo*. *J. Leukoc. Biol.* **53**, 273–278.

Murakami, K., Okada, K., Ikawa, Y., Aida, Y. (1994a). Bovine leukemia virus induces $CD5^-$ B cell lymphoma in sheep despite temporarily increasing $CD5^+$ B cells in asymptomatic stage. *Virology* **202**, 458.

Murakami, K. Aida, Y., Kageyama, R., Numakunai, S., Ohshima, K., Okada, K., Ikawa, Y. (1994b). Immunopathologic study and characterization of the phenotype of transformed cells in sheep with bovine leukemia virus-induced lymphosarcoma. *Am. J. Vet. Res.* **55**, 72.

Müller-Hermelink, H. K., Gaudecker, B. von, Drenckhahn, D., Jaworsky, K., Feldmann, C. (1981). Fibroblastic and dendritic reticulum cells of lymphoid tissue. Ultrastructural, histochemical and 3H-thymidin labeling studies. *J. Canc. Res. Clin. Oncol.* **101**, 149–164.

Myer, M. S., Huchzermeyer, H. F. A. K., York, D. F., Hunter, P., Verwoerd, D. W., Garnett, H. M. (1988). The possible involvement of immunosuppression caused by a lentivirus in the aetiology of jaagsiekte and pasteurellosis in sheep. *Onderstepoort J. Vet. Res.* **55**, 127–133.

Narayan, O., Griffin, D. E., Chase, J. (1977). Antigenic shift of visna virus in persistently infected sheep. *Science* **197**, 376–378.

Narayan, O., Griffin, D. E., Clements, J. E. (1978). Virus mutation during 'slow infection': temporal development and characterisation of mutants of visna virus recovered from sheep. *J. Gen. Virol.* **41**, 343–352.

Narayan, O., Clements, J. E., Griffin, D. E., Wolinsky, J. S. (1981). Neutralizing antibody spectrum determines the antigenic profiles of emerging mutants of visna virus. *Infect. Immun.* **32**, 1045–1050.

Narayan, O., Wolinsky, J. S., Clements, J. E., Strandberg, J. D., Griffin, D. E., Cork, L. C. (1982). Slow virus-replication: the role of macrophages in the persistence and expression of visna viruses of sheep and goats. *J. Gen. Virol.* **59**, 345–356.

Narayan, O., Kennedy-Stoskopf, S., Sheffer, D., Griffin, D. E., Clements, J. E. (1983). Activation of caprine arthritis encephalitis virus expression during maturation of monocytes to macrophages. *Infect. Immun.* **41**, 67–73.

Narayan, O., Sheffer, D., Clements, J. E., Tennekoon, G. (1985). Restricted replication of lentiviruses: visna viruses induce a unique interferon during interaction between lymphocytes and infected macrophages. *J. Exp. Med.* **162**, 1954–1969.

Narayan, O., Zink, M. C., Gorrell, M., Mcentee, M., Sharma, D., Adams, R. (1992). Lentivirus induced arthritis in animals. *J. Rheumatol.* **19**, 25–34.

Nash, A. D., Barcham, G. J., Brandon, M. R., Andrews, A. E. (1991). Molecular cloning, expression and characterization of ovine TNF-α. *Immunol. Cell Biol.* **69**, 273.

Nash, A. D., Barcham, G. J., Andrews, A. E., Brandon, M. R. (1992). Characterization of ovine alveolar macrophages – regulation of surface-antigen expression and cytokine production. *Immunol. Immunopathol.* **31**, 77–94.

Nash, A. D., Lofthouse, S. A., Barcham, G. J., Jacobs, H. J., Ashman, K., Meeusen, E. N. T., Brandon, M. R., Andrews, A. E. (1993). Recombinant cytokines as immunological adjuvants. *Immunol. Cell Biol.* **71**, 367–379.

Nathan, C. F., Hibbs Jr., J. B. (1991). Role of nitric oxide synthesis in macrophage antimicrobial activity. *Curr. Opin. Immunol.* **3**, 65–70.

Nathanson, N., Panitch, H., Palsson, P. A., Petursson, G., Georgsson, G. (1976). Pathogenesis of visna. II. Effect of immunosuppression upon early central nervous system lesions. *Lab. Invest.* **35**, 444–451.

Nathanson, N., Martin, J. R., Georgsson, G., Palsson, P. A., Lutley, R. E., Petursson, G. (1981). The effect of post-infection immunization on the severity of experimental visna. *J. Comp. Pathol.* **91**, 185–191.

Nguyen, T. C. (1973). Report on the sheep blood group workshop (Jouy-en-Josas 1973). International comparison test and general rules for sheep blood group nomenclature. *Anim. Blood Groups Biochem. Genet.* **4**, 241–243.

Nguyen, T. C. (1985). Evaluation of linkage between the R-O-i and C blood group systems in sheep. *Anim. Blood Groups Biochem. Genet.* **16**, 13–17.

Nguyen, T. C. (1990). Genetic systems of red cell blood groups in goats. *Animal Genet.* **21**, 233–245.

Nguyen, T. C., Ruffet, G. (1975). Blood groups in sheep. 2. New blood group factors in the A, B, C, M systems and investigations on the gene frequencies in Berrichon-du-Cher, Ile-de-France and French Texel breeds. *Ann. Genet. Sel. Anim.* **7**, 145–157.

Nicander, L., Halleraker, T. L. M. (1991). Ontogeny of reticular cells in the ileal Peyer's patch of sheep and goats. *Am. J. Anat.* **191**(237), 249.

Nicander, L., Brown, E. M., Dellman, H. D., Landsverk, T. (1993). Lymphatic organs. In *Textbook of Veterinary Histology* (ed. Dellman, H. D.), pp. 120–134. Lea & Febiger, Philadelphia, USA.

Nicander, N., Nafstad, P., Landsverk, T., Engebretsen, R. H. (1991). A study of modified lymphatics in the deep cortex of ruminant lymph nodes. *J. Anat.* **178**, 203–212.

Nossal, J. V., Abbot, A., Mitchell, J., Lummus, Z. (1968). Antigens in immunity. XV. Ultrastructural features of antigen capture in primary and secondary lymphoid follicles. *J. Exp. Med.* **127**, 277–289.

O'Brein, P. M., Rothel, J. S., Seow, H.-F., Wood, P. R. (1991). Cloning and sequencing of the cDNA for ovine granulocyte-macrophage colony-stimulating factor (GM-CSF). *Immunol. Cell Biol.* **69**, 51–55.

O'Brein, P. M., Seow, H. F., Rothel, J. S., Wood, P. R. (1994). Cloning and sequencing of an ovine granulocyte colony-stimulating factor cDNA. *DNA Sequence* **4**, 339–342.

O'Brein, P. M., Seow, H.-F., Eutrican, G., Coupar, B. E., Wood, P. R. (1995). Production and characterization of ovine GM-CSF expressed in mammalian and bacterial cells. *Vet. Immunol. Immunopathol.* **48**, 287–298.

Ogmundsdottir, H. M., Hardarson, B., Steinarsdottir, M., Asgeirsson, B. (1991). The characteristics of macrophage-like cell lines derived from normal sheep spleens. *FEMS Microbiol. Immunol.* **89**(1), 21–32.

Ohshima, T., Aida, Y., Kim, J.-C., Okada, K., Chiba, T., Murakami, K., Ikawa, Y. (1991). Histopathology and distribution of cells harboring bovine leukemia virus (BLV) proviral sequences in ovine lymphosarcoma induced by BLV inoculation. *Vet. Med. Sci.* **53**, 191.

Okerengwo, A. A. (1990). The pathogenesis of *Plasmodium falciparum* associated tissue lesions: what role for C3b receptors? *Afr. J. Med. Medic. Sci.* **19**(2), 77–82.

Oláh, I., Takács, L., Törö, I. (1988). Formation of lymphoepithelial tissue in the sheep's palatine tonsils. *Acta Otolaryngol. Suppl.* **454**, 7–17.

Olson, C., Baumgartener, D. E. (1976). Pathology of lymphosarcoma in sheep induced with bovine leukemia virus. *Cancer Res.* **36**, 2365.

O'Sullivan, B. M., Donald, A. D. (1973). Responses to infection with the *Haemonchus contortus* in ewes of different reproductive status. *Int. J. Parasitol.* **3**, 521–531.

Oswald, I. P., Lantier, F., Bourgy, G. (1990). Classical and alternative pathway hemolytic activities of ovine complement: variation with age and sex. *Vet. Immunol. Immunopathol.* **24**(3), 259–266.

Outteridge, P. M., Lee, C. S. (1988). The defence mechanisms of the mammary gland of domestic ruminants. *Prog. Vet. Microbiol. Immunol.* **4**, 165–195.

Outteridge, P. M., Windon, R. G., Dineen, J. K. (1985). An association between a lymphocyte antigen in sheep and the response to vaccination against the parasite *Trichostrongylus colubriformis*. *Int. J. Parasitol.* **15**, 121–127.

Outteridge, P. M., Windon, R. G., Dineen, J. K., Smith, E. F. (1986). The relationship between ovine lymphocyte antigens and faecal egg count of sheep selected for responsiveness to vaccination against *Trichostrongylus colubriformis*. *Int. J. Parasitol.* **16**, 369–374.

Outteridge, P. M., Windon, R. G., Dineen, J. K. (1988). An ovine lymphocyte antigen marker for acquired resistance to *Trichostrongylus colubriformis*. *Int. J. Parasitol.* **18**, 853–858.

Outteridge, P. M., Stewart, D. J., Skerman, T. M., Dufty, J. H., Egerton, J. R., Ferrier, G., Marshall, D. J. (1989). A positive association between resistance to ovine footrot and particular lymphocyte antigen types. *Aust. Vet. J.* **66**, 175–179.

Pabst, R., Reynolds, J. D. (1986). Evidence of extensive lymphocyte death in sheep Peyer's patches. II. The number and fate of newly-formed lymphocytes that emigrate from Peyer's patches. *J. Immunol.* **136**, 2011–2017.

Palfi, V., Glavits, R., Hajtos, I. (1989). Testicular lesions in rams infected by maedi/visna virus. *Acta Veterinaria Hungarica* **37**, 97–102.

Panitch, H., Petorsson, G. G. G., Palsson, P. A., Nathanson, N. (1976). Pathogenesis of visna. III. immune responses to central nervous system antigens in experimental allergic encephalomyelitis and visna. *Lab. Invest.* **35**, 452–460.

Parsons, Y. M., Cooper, D. W., Piper, L. R., Beh, K. J. (1994). A *Taq*I polymorphism at the immunoglobulin kappa light chain locus in Merino sheep. *Animal Genet.* **25**, 54.

Patri, S., Nau, F. (1992). Isolation and sequence of a cDNA coding for the immunoglobulin μ chain of the sheep. *Molec. Immunol.* **29**, 829–836.

Pearce, P. D., Ansari, H. A., Maher, D. W., Broad, T. E. (1995). Five regional localizations to the sheep genome: first assignments to chromosomes 5 and 12. *Animal Genet.* **26**, 171–176.

Pearson, L. D., Brandon, M. R. (1976). Effect of fetal thymectomy on IgG, IgM and IgA concentrations in sheep. *Am. J. Vet. Res.* **37**, 1139–1141.

Pearson, L. D., Simpson-Morgan, M., Morris, B. (1976). Lymphopoiesis and lymphocyte recirculation in the sheep fetus. *J. Exp. Med.* **143**, 167–186.

Pedersen, N. C., Adams, E. P., Morris, B. (1975). The response of the lymphoid system to renal allografts in sheep. *Transplantation* **19**(5), 400–409.

Peluso, R., Haase, A., Stowring, L., Edwards, M., Ventura, P. (1985). A Trojan Horse mechanism for the spread of visna virus in monocytes. *Virology* **147**, 231–236.

Pepin, M., Cannella, D., Fontaine, J. J., Pittet, J. C., Le Pape, A. (1992). Ovine mononuclear phagocytosis in situ: identification by monoclonal antibodies and involvement in pyogranulomas in lambs. *J. Leukoc. Biol.* **51**, 188–198.

Perkowski, S. Z., Havill, A. M., Flynn, J. T., Gee, M. H. (1983). Role of intrapulmonary release of eicosanoids and superoxide anion as mediators of pulmonary dysfunction and endothelial injury in sheep with intermittent complement activation. *Circ. Res.* **53**, 574–583.

Perry, L. L., Wilkerson, M. J., Hullinger, G. A., Cheevers, W. P. (1995). Depressed $CD4^+$ T lymphocyte proliferative response and enhanced antibody response to viral antigen in chronic lentivirus-induced arthritis. *J. Inf. Dis.* **171**, 328–334.

Persson, K., Colditz, I., Flapper, P., Seow, H.-F. (1996a). Cytokine-induced inflammation in the ovine teat and udder. *Vet. Immunol. Immunopathol.* **53**, 73–86.

Persson, K., Colditz, I., Flapper, P., Seow, H.-F. (1996b). Leukocyte and cytokine accumulation in the ovine teat and udder during endotoxin-induced inflammation. *Vet. Res. Comm.*, in press.

Petursson, G., Nathanson, N., Georgsson, G., Panitch, H., Palsson, P. A. (1976). Pathogenesis of visna. I. Sequential virologic, serologic, and pathologic studies. *Lab. Invest.* **35**, 402–412.

Petursson, G., Douglas, B. M., Lutley, R. (1983). Immunoglobulin subclass distribution and restriction of antibody response in visna. In *Slow Viruses in Sheep, Goats and Cattle* (eds Sharp, J. M., Hoff-Jorgensen, R.), pp. 211–216. Commission of the European Communities, Brussels.

Pfeffer, A., Rogers, K. M. (1989). Acute phase response of sheep: changes in the concentrations of ceruloplasm in fibrinogen, haptoglobin and the major blood cell types associated with pulmonary damage. *Res. Vet. Sci.* **46**, 118–124.

Pfeffer, A., Rogers, K. M., O'Keefe, L., Osborn, P. J. (1993). Acute-phase protein response, food-intake, liveweight change and lesions following intrathoracic injection of yeast in sheep. *Res. Vet. Sci.* **55**, 360–366.

Phillipson, A. T. (1977). Ruminant digestion. In *Duke's Physiology of Domestic Animals* (ed. Swenson, M. J.), pp. 250–286. Cornell University Press, Ithaca, USA.

Phua, S. H., Wood, N. J. (1995). Rapid communication: a polymorphism detected in sheep using a complement subcomponent C1r (C1R) probe. *J. Anim. Sci.* **73**(2), 629.

Poitras, B. J., Miller, R. B., Wilkie, B. N., Bosu, W. T. K. (1986). The maternal to fetal transfer of immunoglobulins associated with placental lesions in sheep. *Can. J. Vet. Res.* **50**, 68–73.

Powers, M., Radke, K. (1992). Activation of bovine leukemia virus transcription in lymphocytes from infected sheep: rapid transition through early to late gene expression. *J. Virol.* **66**, 4769.

Press, C., McClure, S., Landsverk, T. (1991). Computer-assisted morphometric analysis of absorptive and follicle-associated epithelia of Peyer's patches in sheep foetuses and lambs indicates the presence of distinct T- and B-cell components. *Immunology* **72**, 386–392.

Press, C. M., Halleraker, M., Landsverk, T. (1992). Ontogeny of leukocyte populations in the ileal Peyer's patch of sheep. *Dev. Comp. Immunol.* **16**, 229–241.

Press, C. M., Hein, W. R., Landsverk, T. (1993). Ontogeny of leucocyte populations in the spleen of fetal lambs with emphasis on the early prominence of B cells. *Immunology* **80**, 598–604.

Press, C. M., Reynolds, J. D., McClure, S. J., Simpson-Morgan, M. W., Landsverk, T. (1996). Foetal lambs are depleted of IgM+ cells following a single injection of an α-IgM antibody early in gestation. *Immunology* **88**, 28–34.

Puri, N. K., Mackay, C. R., Brandon, M. R. (1985). Sheep lymphocyte antigens (OLA) II. Major histocompatibility complex class II molecules. *Immunology* **56**, 725–733.

Puri, N. K., Gogolin-Ewens, K. J., Brandon, M. R. (1987). Monoclonal antibodies to sheep MHC class I and class II molecules: biochemical characterization of three class I gene products and four distinct subpopulations of class II molecules. *Vet. Immunol. Immunopathol.* **15**, 59–86.

Pyrah, I. T. G., Watt, N. J. (1996). Immunohistological study of the depressed cutaneous DTH response of sheep naturally infected with an ovine lentivirus (maedi-visna virus). *Clin. Exp. Immunol.* **104**, 32–36.

Querat, G., Audoly, G., Sonigo, P., Vigne, R. (1990). Nucleotide sequence analysis of SA-OMVV, a visna-related ovine lentivirus: phylogenetic history of lentiviruses. *Virology* **175**, 434–447.

Quin, J. W., Shannon, A. D. (1975). The effect of anaesthesia and surgery on lymph flow, protein and leucocyte concentration in lymph of the sheep. *Lymphology* **8**, 126–135.

Quin, J. W., Shannon, A. D. (1977). The influence of the lymph node on the protein concentration of efferent lymph leaving the node. *J. Physiol.* **264**, 307–321.

Ramos-Vara, J. A. (1994). Reactivity of polyclonal human CD3 anti-serum in lymphoid tissues of cattle, sheep, goats, rats, and mice. *Am. J. Vet. Res.* **55**, 63–66.

Rasmusen, B. A. (1960). Blood groups in sheep. II. The B system. *Genetics* **45**, 1405–1417.

Rasmusen, B. A. (1962). Blood groups in sheep. In *Blood Groups in Infrahuman Species. Annals of the New York Academy of Sciences* (ed. Cohen, C.), pp. 306–319. New York, USA.

Rasmusen, B. A. (1981). Linkage of genes for PHI, halothane sensitivity, A-O inhibition, H red cell antigens and 6-PGD variants in pigs. *Anim. Blood Groups Biochem. Genet.* **12**, 207–209.

Rasmusen, B. A., Stormont, Y. S. C. (1960). Blood groups in sheep. III. The A, C, D, and M systems. *Genetics* **45**, 1595–1603.

Rausch, P. G., Moore, T. G. (1975). Granule enzymes of polymorphonuclear leukocytes. A polygenic comparison. *Blood* **46**, 913–919.

Ren, X. D., Dodds, A. W., Law, S. K. A. (1993). The thioester and isotypic sites of complement component C4 in sheep and cattle. *Immunogenetics* **37**(2), 120–128.

Rendel, J. (1957). Further studies on some antigenic characters of sheep blood determined by epistatic action of genes. *Acta Agr. Scand.* **7**, 224–259.

Renstrom, L. H. M., Press, C. McL., Trevella, W., Landsverk, T. (1996). Response of leucocyte populations in the ileal Peyer's patch of fetal lambs treated with ferritin *per os*. *Develop. Immunol.* **4**, 289–298.

Reyburn, H., McConnell, I. (1993). Maedi-visna and other lentiviruses: virus-host interactions. In *Viruses and the Cellular Immune Response* (ed. Thomas, D. B.), pp. 201–235. Marcel Dekker, Inc., New York, USA.

Reyburn, H. T., Roy, D. J., Blacklaws, B. A., Sargan, D. R., Watt, N. J., McConnell, I. (1992). Characteristics of the T cell-mediated immune response to maedi-visna virus. *Virology* **191**, 1009–1012.

Reynaud, C.-A., Mackay, C. R., Muller, K. G., Weill, J.-C. (1991). Somatic generation of diversity in a mammalian primary lymphoid organ: the sheep ileal Peyer's patches. *Cell* **64**, 995–1005.

Reynaud, C.-A., Imhof, B. A., Anquez, V., Weill, J.-C. (1992). Emergence of committed B lymphoid progenitors in the developing chick embryo. *EMBO J.* **11**, 4349–4352.

Reynaud, C.-A., Garcia, C., Hein, W. R., Weill, J. C. (1995). Hypermutation generating the sheep immunoglobulin repertoire is an antigen-independent process. *Cell* **80**, 115–125.

Reynolds, J., Heron, I., Dudler, L., Trnka, Z. (1982). T-cell recirculation in the sheep: migratory properties of cells from lymph nodes. *Immunology* **47**, 415–421.

Reynolds, J. D. (1976). The development and physiology of the gut-associated lymphoid system in lambs. Ph.D. thesis, Australian National University, Canberra, Australia.

Reynolds, J. D. (1985). Evidence of differences between Peyer's patches and germinal centers. *Adv. Exp. Biol. Med.* **186**, 101–109.

Reynolds, J. D. (1986). Evidence of extensive lymphocyte death in sheep Peyer's patches. I. A comparison of lymphocyte production and export. *J. Immunol.* **136**, 2005–2010.

Reynolds, J. D. (1987). Peyer's patches and the early development of B lymphocytes. *Current Topics in Microbiol. Immunol.* **135**, 43–56.

Reynolds, J. D. (1997). The genesis, tutelage and exodus of B cells in the ileal Peyer's patch of sheep. *Int. Rev. Immunol.* **15**, 265–299.

Reynolds, J. D., Morris, B. (1983). The evolution and involution of Peyer's patches in fetal and postnatal sheep. *Eur. J. Immunol.* **13**, 627–635.

Reynolds, J. D., Morris, B. (1984). The effect of antigen on the development of Peyer's patches in sheep. *Eur. J. Immunol.* **14**, 1–6.

Reynolds, J. D., Kennedy, L., Peppard, J., Pabst, R. (1991). Ileal Peyer's patch emigrants are predominantly B cells and travel to all lymphoid tissues in sheep. *Eur. J. Immunol.* **21**, 283–289.

Rice, C. E., Boulanger, P. (1952). The interchangeability of the complement components of different animal species. IV. In the hemolysis of rabbit erythrocytes sensitized with sheep antibody. *J. Immunol.* **68**, 197–205.

Richards, C. B., Marrack, J. R. (1963). Sheep serum γ globin. *Proteins of the Biological Fluids* **10**, 154–156.

Richardson, M., Conner, G. H. (1972). Prenatal immunization by the oral route: stimulation of Brucella antibody in fetal lambs. *Infect. Immun.* **5**, 454–460.

Richardson, M., Beck, C. C., Clark, D. T. (1968). Prenatal immunization of the lamb to Brucella: dissociation of immunocompetence and reactivity. *J. Immunol.* **101**, 1363–1366.

Richardson, M., Conner, G. H., Beck, C. C., Clark, D. T. (1971). Prenatal immunization of the lamb to brucella: secondary antibody response *in utero* and at birth. *Immunology* **12**, 795–803.

Robinson, N. A., Hygate, L. C., Mathews, M. E. (1994). Investigation of the effect of growth factor and protooncogene variants of wool quality. *Proceedings of the 5th World Congress on Genetics Applied to Livestock Production* **21**, 90–93.

Rosolia, D., McKenna, P., Gee, M., Albertine, K. (1992). Infusion of zymosan-activated plasma affects neutrophils in peripheral blood and bone marrow in sheep. *J. Leukoc. Biol.* **52**, 501–515.

Rothel, J. S., Seow, H.-F. (1995). CSIRO Division of Animal Health, Melbourne. Unpublished observations.

Rothel, J. S., Jones, S. L., Corner, L. A., Cox, J. C., Wood, P. R. (1990). A sandwich immunoassay for bovine interferon gamma and its use for the detection of tuberculosis in cattle. *Aust. Vet. J.* **67**, 134–137.

Rothwell, T. L. W., Windon, R. G., Horsburgh, B. A., Anderson, B. H. (1993). Relationship between eosinophilia and responsiveness to infection with *Trichostrongylus colubriformis* in sheep. *Int. J. Parasitol.* **23**, 203–211.

Rothwell, T. L. W., Horsburgh, B. A., France, M. P. (1994). Basophil leucocytes in response to parasitic infection and some other stimuli in sheep. *Res. Vet. Sci.* **56**, 319–324.

Rovnak, J., Boyd, A. L., Casey, J. W., Gonda, M, A., Jensen, W. A., Cockerell, G. L. (1993). Pathogenicity of molecularly cloned bovine leukemia virus. *J. Virol* **67**, 7096.

Ruffing, N. A., Anderson, G. B., Bondurant, R. H., Pashew, R. L., Bernoco, D. (1993). Antibody response of ewes and does to chimeric sheep-goat pregnancy. *Biol. Reprod.* **49**(6), 1260–1269.

Salazar, I. (1984). The relation of the lymphatic system to hemolymph nodes in the sheep. *Lymphology* **17**, 46–49.

Sargan, D. R., Bennet, I. D., Cousens, C., Roy, D. J., Blacklaws, B. A., Dalziel, R. G., Watt, N. J., McConnell, I. (1991). Nucleotide sequence of EV1, a British isolate of maedi-visna virus. *J. Genet. Virol.* **72**, 1893–1903.

Sargan, D. R., Sutton, K. A., Bennet, I. D., McConnell, I., Harkiss, G. D. (1995). Sequence and repeat structure variants in the long terminal repeat of maedi-visna virus EV1. *Virology* **208**, 343–348.

Schifferle, R. E., Wilson, M. E., Levine, M. J., Genco, R. J. (1993). Activation of serum complement by polysaccharide-containing antigens of *Porphyromonas gingivalis*. *J. Periodont. Res.* **28**(4), 248–254.

Schlueter, A. J., Segre, M., Segre, D. (1989). Detection and enumeration of immunoglobulin secreting cells. *J. Immunol. Methods* **124**(1), 35–42.

Schnappauf, H., Schnappauf, U. (1968). Drainage of the thoracic duct and amount of the 'easily mobilized' lymphocytes in calves, sheep and dogs. *Blut.* **16**(4), 209–220.

Schneemann, M., Schoedon, G., Hofer, S., Blau, N., Guerrero, L., Schaffner, A. (1993). Nitric oxide synthase is not a constituent of the antimicrobial armature of human mononuclear phagocytes. *J. Inf. Dis.* **167**, 1358–1363.

Schoefl, G. I. (1981). Observations on blood vessels in sheep lymph nodes. In *Festschrift for F. C. Courtice* (ed. Garlick D.), pp. 159–167. Sydney University, Sydney, NSW, Australia.

Schreiber, R. D., Noble, R. W., Reichlin, M. (1975). Restriction of heterogeneity of goat antibodies specific for human hemoglobin. *J. Immunol.* **114**, 170–175.

Schwaiger, F.-W., Weyers, E., Buitkamp, J., Ede, A. J., Crawford, A., Epplen, J. T. (1994). Interdependent MHC-DRB exon-plus-intron evolution in artiodactyls. *Mol. Biol. Evol.* **11**, 239–249.

Schwaiger, F.-W., Gostomski, D., Stear, M. J., Duncan, J. L., McKellar, Q. A., Epplen, J. T., Buitkamp, J. (1995). An ovine major histocompatibility complex DRB1 allele is associated with low faecal egg counts following natural, predominantly *Ostertagia circumcincta* infection. *Int. J. Parasitol.* **25**, 815–822.

Schwaiger, F.-W., Maddox, J., Ballingall, K., Buitkamp, J., Crawford, A., Dutia, B. M., Epplen, J. T., Ferguson, E. D., Groth, D., Hopkins, J., Rhind, S. M., Sargan, D., Wetherall, J., Wright, H. (1996). The major histocomaptibility complex of domestic animal species. In *The Ovine Major Histocompatibility Complex* (eds Schook, L., Lamont, S.), pp. 121–176. CRC Press, Boca Raton, FL, USA.

Schwartz, I., Bensaid, A., Polack, B., Perrin, B., Berthelemy, M., Levy, D. (1994). *In vivo* leukocyte tropism of bovine leukemia virus in sheep and cattle. *J. Virol.* **68**, 4589.

Scott, J. V., Stowring, L., Haase, A. T., Narayan, O., Vigne, R. (1979). Antigenic variation in visna virus. *Cell* **18**, 321–327.

Scott, P. C., Gogolin-Ewens, K. J., Adams, T. E., Brandon, M. R. (1991a). Nucleotide sequence, polymorphism, and evolution of ovine MHC class II *DQA* genes. *Immunogenetics* **34**, 69–79.

Scott, P. C., Maddox, J. F., Gogolin-Ewens, K. J., Brandon, M. R.

(1991b). The nucleotide sequence and evolution of ovine MHC class II *B* genes: *DQB* and *DRB*. *Immunogenetics* **34**, 80–87.

Scott, P. R., Murray, L. D., Penny, C. D. (1992). A preliminary study of serum haptoglobin concentration as a prognostic indicator of ovine dystocia cases. *Br. Vet. J.* **148**, 351–355.

Selvarej, P., Dustin, N. L., Mituacht, R., Hunig, T., Springer, T. A., Plunkett, M. L. (1987). Rosetting of human T lymphocytes with sheep and human erythrocytes. Comparison of human and sheep ligand binding purified E receptor. *J. Immunol.* **138**, 2690–2695.

Seow, H.-F., Rothel, J. S., Radford, A. J., Wood, P. R. (1990). The molecular cloning of ovine interleukin 2 gene by the polymerase chain reaction. *Nucl. Acids Res.* **18**, 171–175.

Seow, H.-F., Rothel, J., David, M. J., Wood, P. R. (1991). Nucleotide sequence of ovine macrophage interleukin-1β cDNA, *DNA Seq.* **1**, 423.

Seow, H.-F., Rothel, J. S., Wood, P. R. (1993). Cloning and sequencing an ovine interleukin-4 encoding cDNA. *Gene* **124**, 291–293.

Seow, H.-F., Yoshimura, T., Wood, P. R., Colditz, I. G. (1994a). Cloning, sequencing, expression and inflammatory activity in skin of ovine interleukin-8. *Immunol. Cell Biol.* **72**, 398–405.

Seow, H.-F., Rothel, J. S., Wood, P. R. (1994b). Expression and purification of recombinant ovine interleukin-1β from *Escherichia coli*. *Vet. Immunol. Immunopathol.* **41**, 229–239.

Sheldrake, R. F. (1989). *Mucosal Immunology: Concepts and Applications*. Postgraduate Committee of Veterinary Science Proceedings, p. 259. University of Sydney, Sydney, Australia.

Sheldrake, R. F., Husband, A. J. (1985). Intestinal uptake of intact maternal lymphocytes by neonatal rats and lambs. *Res. Vet. Sci.* **39**, 10–15.

Sheldrake, R. F., Husband, A. J., Watson, D. K. L., Cripps, A. W. (1985). The effect of intraperitoneal and intramammary immunization of sheep on the numbers of antibody-containing cells in the mammary gland, and antibody titres in blood serum and mammary secretions. *Immunology* **56**, 605–614.

Shimotsuma, M., Simpson-Morgan, M. W., Takahashi, T., Hagiwara, A. (1994). Ontogeny of milk spots in the fetal omentum. *Arch. Histol. Cytol.* **57**, 291–299.

Siebers, A., Finlay, B. B. (1996). M cells and the pathogenesis of mucosal and systemic infections. *Trends Microbiol.* **4**(1), 22–26.

Sigurdsson, B. (1954). Maedi, a slow progressive pneumonia of sheep: an epizoological and pathological study. *Br. Vet. J.* **110**, 255–270.

Sigurdsson, B., Palsson, P. A. (1958). Visna of sheep. A slow, demyelinating infection. *Brit. J. Exp. Pathol.* **39**, 519–528.

Sihvonen, L. (1981). Early immune responses in experimental maedi. *Res. Vet. Sci.* **30**, 217–222.

Sihvonen, L. (1984). Late immune responses in experimental maedi. *Vet. Microbiol.* **9**, 205–213.

Silflow, R. M., Foreyt, W. J. (1994). Susceptibility of phagocytes from elk, deer, bighorn, sheep and domestic sheep to *Pasteurella haemolytica* cytotoxins. *J. Wild. Dis.* **30**, 529–535.

Silflow, R. M., Foreyt, W. J., Taylor, S. M., Leagreid, W. W., Liggitt, H. D., Leid, R. W. (1991). Comparison of arachidonate metabolism by alveolar macrophages from bighorn and domestic sheep. *Inflammation* **15**, 43–53.

Silverstein, A. M., Uhr, J. W., Kraner, K. L., Lukes, R. J. (1963). Fetal responses to antigenic stimulus. II. Antibody production by the fetal lamb. *J. Exp. Med.* **117**, 799–812.

Silverstein, A. M., Prendergast, R. A., Kraner, K. L. (1964). Fetal responses to antigenic stimulus. IV. Rejection of skin homografts by the fetal lamb. *J. Exp. Med.* **119**, 799–812.

Simpson-Morgan, M. W., Smeaton, T. C. (1972). The transfer of antibodies by neonates and adults. *Adv. Vet. Sci. Comp. Med.* **16**, 355–386.

Simpson-Morgan, M. W., Trevella, W., Hugh, A. R., McClure, S. R., Morris, B. (1985). The long-term collection of lymph from single lymph nodes of foetal lambs *in utero*. *Aust. J. Exp. Biol. Med. Sci.* **63**, 397–409.

Skinner, J. G., Roberts, L. (1994). Haptoglobin as an indicator of infection in sheep. *Vet. Rec.* **134**, 33–36.

Small, J. A., Bieberich, C., Ghotbi, Z., Hess, J., Scangos, G. A., Clements, J. F. (1989). The visna virus long terminal repeat directs expression of a reporter gene in activated macrophages, lymphocytes, and the central nervous systems of transgenic mice. *J. Virol.* **63**, 1891–1896.

Smith, J. B., Morris, B. (1970). The response of the popliteal lymph node of the sheep to swine influenza virus. *Aust. J. Exp. Biol. Med. Sci.* **48**(1), 33–46.

Smith, J. B., Cunningham, A. J., Lafferty, K. J., Morris, B. (1970a). The role of the lymphatic system and lymphoid cells in the establishment of immunological memory. *Aust. J. Exp. Biol. Med. Sci.* **48**(1), 57–70.

Smith, J. B., McIntosh, G. H., Morris, B. (1970b). The traffic of cells through tissues: a study of peripheral lymph in sheep. *J. Anat.* **107**(1), 87–100.

Smith, J. B., McIntosh, G. H., Morris, B. (1970c). The migration of cells through chronically inflamed tissues. *J. Pathol.* **100**, 21–29.

Smith, W. D., Jackson, F., Jackson, E., Williams, J., Miller, H. R. P. (1984). Manifestations of resistance to ovine ostertagiosis associated with immunological responses in the gastric lymph. *J. Comp. Pathol.* **94**, 591–602.

Snibson, K. J., Maddox, J. F., Fabb, S. A., Brandon, M. R. (1998). Allelic variation of ovine MHC class II *DQA1* and *DQA2* genes. *Animal Genet.* (Submitted.)

Snodgrass, D. R., Wells, P. W. (1978). The immunoprophylaxis of rotavirus infections in lambs. *Vet. Rec.* **102**, 146–148.

Sojka, W. J., Wray, C., Morris, J. A. (1978). Passive protection of lambs against experimental enteric colibacillosis by colostral transfer of antibodies from K99-vaccinated ewes. *J. Med. Microbiol.* **11**, 493–499.

Sommerville, R. I. (1956). The histology of the ovine abomasum and the relation of the leucocyte to nematode infestation. *Aust. Vet. J.* **32**, 237–246.

Sonigo, P., Alizon, M., Staskus, K., Klatzmann, D., Cole, S., Danos, O., Retzel, E., Tiollais, P., Haase, A., Wainhobson, S. (1985). Nucleotide-sequence of the visna lentivirus – relationship to the aids virus. *Cell* **42**, 369–382.

Springer, T. A. (1994). Traffic signals for lymphocyte recirculation and leukocyte emigration: the multistep paradigm. *Cell* **76**(2), 301–314.

Stankiewicz, M., Jonas, W. (1981). Hemolysis of human erythrocytes heavily sensitized with sheep amboreceptor by sheep serum chelated with EGTA or $Mg2^+$ EGTA. *Vet. Immunol. Immunopathol.* **2**, 253–264.

Stankiewicz, M., Jonas, W., Elliot, D. (1981). Alternative pathway activation of complement in fetal lamb serum by *Trichostrongylus vitrinus* larvae. *Parasite Immunol.* **3**, 309–318.

Stankiewicz, M., Shaw, R. J., Jonas, W. E., Cabaj, W., Grimmett, D. J., Douch, P. G. C. (1994). A technique for the isolation and purification of viable mucosal mast cells/globule leukocytes from the small intestine of parasitised sheep. *Int. J. Parasitol.* **24**(2), 307–309.

Stankiewicz, M., Pernthaner, A., Cabaj, W., Jonas, W. E., Douch, P. G. C., Bisset, S. A., Rabel, B., Pfeffer, A., Green, R. S. (1995). Immunization of sheep against parasitic nematodes leads to elevated levels of globule leukocytes in the small intestine lumen. *Int. J. Parasitol.* **25**, 389–394.

Stanley, J., Bhaduri, L. M., Narayan, O., Clements, J. E. (1987). Topographical rearrangements of visna virus envelope glycoprotein during antigenic drift. *J. Virol.* **61**, 1019–1028.

Staskus, K. A., Couch, L., Bitterman, P., Retzel, E. F., Zupancic, M., List, J., Haase, A. T. (1991a). *In situ* amplification of visna virus DNA in tissue sections reveals a reservoir of latently infected cells. *Microbial Pathogen.* **11**, 67–76.

Staskus, K. A., Retzel, E. F., Lewis, E. D., Silsby, J. L., Stcyr, S., Rank, J. M., Wietgrefe, S. W., Haase, A. T., Cook, R., Fast, D., Geiser, P. T., Harty, J. T., Kong, S. H., Lahti, C. J., Neufeld, T. P., Porter, T. E., Shoop, E., Zachow, K. R. (1991b). Isolation of replication-competent molecular clones of visna virus. *Virology* **181**, 228–240.

Stear, M. J., Spooner, R. L. (1981). Lymphocyte antigens on sheep. *Anim. Blood Groups Biochem. Genet.* **12**, 265–276.

Stear, M. J., Bishop, S. C., Doligalska, M., Duncan, J. L., Holmes, P. H., Irvine, J., McCririe, L., McKellar, Q. A., Sinski, E., Murray, M. (1995). Regulation of egg production, worm burden, worm length and worm fecundity by host responses in sheep infected with *Ostertertagia circumcincta. Parasite Immunol.* **17**, 643–652.

Stear, M. J., Bairden, K., Bishop, S. C., Buitkamp, J., Epplen, J. T., Gostomski, D., McKellar, Q. A., Schwaiger, F.-W., Wallace, D. S. (1996). An ovine lymphocyte antigen is associated with reduced faecal egg counts in four-month-old lambs following natural, predominantly *Ostertagia circumcincta* infection. *Int. J. Parasitol.* **26**, 423–428.

Stevenson, L. M., Jones, D. G. (1992). Use of enzyme microassay to detect eosinophil potentiating activity of cytokines in sheep. *FEMS Microbiol. Immunol.* **105**, 325–330.

Stevenson, L. M., Huntley, J., Smith, W. D., Jones, D. G. (1994). Local eosinophil and mast cell related responses in abomasal nematode infections of lambs. *FEMS Immunol. Med. Microbiol.* **8**, 167–174.

Stewart, M. J., McClure, S. J., Rothwell, T. L. W., Emery, D. L. (1994). Changes in the enteric innervation of the sheep during primary or secondary immune responses to the nematode *Trichostrongylus colubriformis. Reg. Immunol.* **6**, 460–470.

Stock, N., Ferrer, J. (1972). Replicating C-type virus in phytohemagglutinin-treated buffy-coat cultures of bovine origin. *J. Natl Cancer Inst.* **48**, 985.

Stormont, C., Suzuki, Y. (1961). Blood group comparisons of cattle, sheep and goats. *Immunol. Lett.* **2**, 48–49.

Stuehr, D. J., Cho, H. J., Kwon, N. S., Weise, M. F., Nathan, C. F. (1991). Purification and characterization of the cytokine-induced macrophage nitric oxide synthase: an FAD- and FMN-containing flavoprotein. *Proc. Natl. Acad. Sci. USA* **88**, 7773–7777.

Sture, G., Huntley, J., MacKellar, A., Miller, H. R. P. (1995). Ovine mast cell heterogeneity is defined by the distribution of sheep mast cell protease. *Vet. Immunol. Immunopathol.* **48**, 275–285.

Suttle, N. F., Knos, D. P., Angus, K. W., Jackson, F., Coop, R. L. (1992). Effects of dietary molybdenum of nematode and host during *Haemonchus contortus* infection in lambs. *Res. Vet. Sci.* **52**, 230–235.

Tanaka, A., Ishiguro, N., Shinagawa, M. (199). Sequence and diversity of bovine T-cell receptor beta-chain genes.*Immunogenetics* **32**, 263–271.

Thormar, H., Barshatzky, M. R., Arnesen, K., Kozlowski, P. B. (1983). The emergence of antigenic variants is a rare event in long-term visna virus-infection *in vivo. J. Gen. Virol.* **64**, 1427–1432.

Topper, E. K., Colditz, I. G., Windon, R. G. (1992). Induction of tissue eosinophilia by platelet-activating factor in Merino sheep. *Vet. Immunol. Immunopathol.* **32**, 65–75.

Torsteinsdottir, S., Georgsson, G., Gisladottir, E., Rafnar, B., Palsson, P. A., Petursson, G. (1992). Pathogenesis of central nervous system lesions in visna: cell-mediated immunity and lymphocyte subsets in blood, brain and cerebrospinal fluid. *J. Neuroimmunol.* **41**, 149–158.

Townsend, W. L., Gorrell, M. D., Ladds, P. W. (1995). Major histocompatibility complex antigens in normal, acanthotic and neoplastic ovine skin: an association between tumor invasiveness and low level MHC class I expression. *Vet. Immunol. Immunopathol.* **45**, 237–252.

Treacher, T. T. (1973). Artificial weaning of lambs: a review. *Vet. Rec.* **92**, 311–315.

Trepel, F. (1974). Number and distribution of lymphocytes in man. A critical analysis. *Klin. Wochenschr.* **52**(11), 511–515.

Tuboly, S., Glavits, R., Megyeri, Z., Palfi, V. (1991). Immune-reactions and immunopathologic changes induced in ovine fetuses and lambs by maedi-visna virus. *Acta Veterinaria Hungarica* **39**, 139–147.

Tucker, E. M., Evans, R. S., Kilgour, L. (1980). Close linkage between the C blood group locus and the locus controlling amino acid transport in sheep erythrocytes. *Anim. Blood Groups Biochem. Genet.* **11**, 119–125.

Tuo, W. B., Ott, T. L., Bazer, F. W. (1993). Natural killer cell activity of lymphocytes exposed to ovine type-1 trophoblast interferon. *Am. J. Reprod. Immunol.* **29**, 26–34.

Turek, J. J., Schoenlein, I. A., Clark, L. K., van Alstine, W. G. (1994). Dietary polyunsaturated fatty acid effects on immune cells of the porcine lung. *J. Leukoc. Biol.* **56**, 599–604.

van Den Berg, C. W., Morgan, B. P. (1994). Complement-inhibiting activities of human CD59 and analogues from rat, sheep, and pig are not homologously restricted. *J. Immunol.* **152**(8), 4095–4101.

van Den Berg, C. W., Harrison, R. A., Morgan, B. P. (1993). The sheep analogue of human CD59: purification and characterization of its complement inhibitory activity. *Immunology* **78**(3), 349–357.

van den Broeke, A., Cleuter, Y., Chen, G., Portetelle, D., Mammerickx, M., Zagury, D., Fouchard, M., Coulombel, L., Kettmann, R., Burny, A. (1988). Even transcriptionally competent proviruses are silent in Bovine Leukemia Virus-induced sheep tumor cells. *Proc. Natl Acad. Sci. USA* **85**, 9263.

van den Broeke, A., Cleuter, Y., Droogmans, L., Burny, A., Kettmann, R. (1997). The isolation and culture of B lymphoblastoid cell lines from Bovine Leukemia Virus-induced tumors in sheep. In *Immunology Methods Manual. The Comprehensive Sourcebooks of Techniques*, vol. 4 (ed. Lefkovits, I.), pp. 2127–2132. Academic Press, London, UK.

van Houtert, M. F. J., Barger, I. A., Steel, J. W. (1995). Dietary protein for young grazing sheep-interactions with gastrointestinal parasitism. *Vet. Parasitol.* **60**, 283–295.

van Oorschot, R. A. H., Maddox, J. F., Adams, L. J., Fabb, S. A. (1994). Characterisation and evolution of ovine MHC class II *DQB* sequence polymorphism. *Animal Genet.* **25**, 417–424.

van Snick, U., Cayphas, S., Vink, A., Uttenhove, C., Coulie, P. G., Rubria, M. R., Simpson, R. J. (1986). Purification and NH_2 amino acid sequence of a T-cell-derived lymphokine with growth factor activity for B-cell hybridomas. *Proc. Natl Acad. Sci. USA* **83**, 9679–9683.

Vandermolen, E. J., Vecht, U., Houwers, D. J. (1985). A chronic indurative mastitis in sheep, associated with maedi visna virus-infection. *Vet. Quart.* 7, 112–119.

Varma, S., Kwok, S., Ebner, K. E. (1989). Cloning and nucleotide sequence of ovine prolactin cDNA. *Gene* 77, 349–359.

Vellutini, C., Philippon, V., Gambarelli, D., Horschowski, N., Nave, K.-A., Navarro, J. M., Auphan, M., Courcoul, M.-A., Filippi, P. (1994). The maedi-visna virus tat protein induces multiorgan lymphoid hyperplasia in transgenic mice. *J. Virol.* **68**, 4955–4962.

Verhagen, A. M., Andrews, A. E., Brandon, M. R., Nash, A. D. (1992). Molecular cloning, expression and characterisation of the ovine IL-2Rα chain. *Immunology* 76, 1–9.

Verhagen, A. E., Brandon, M. R., Nash, A. D. (1993). Characterization of the ovine IL-2Rα chain: differential regulation on ovine $\alpha\beta$ and $\gamma\delta$ T cells. *Immunology* **79**, 471–478.

Verhagen, A. M., Kimpton, W. G., Nash, A. D. (1994), Development of a sandwich immunoassay for the detection of soluble ovine IL-2Rα chain. *Vet. Immunol. Immunopathol.* **42**, 287–300.

Vicario, A. J. B., Martinez, N., Santiago, A., Aguirre, A. I., Mazon, L. I., Estomba, A. (1995). *GenBank accession numbers U25998–U26010.*

Wagland, B. M., Emery, D. L., McClure, S. J. (1996). Studies on the host–parasite relationship between *Trichostrongylus colubriformis* and susceptible and resistant sheep. *Int. J. Parasitol.* **26**, 1279–1286.

Wain-Hobson, S. (1996). Running the gamut of retroviral variation. *Trends Microbiol.* **4**, 135–141.

Walker, I. D., Glew, M. D., O'Keefe, M. A., Metcalffe, S. A., Clevers, H. A., Wijngaard, P., Hein, W. R. (1994). A novel multi-gene family of sheep gamma-delta T cells. *Immunology* **83**, 517–523.

Washington, E. A., Kimpton, W. G., Cahill, R. N. P. (1992). Changes in the distribution of $\alpha\beta$ and $\gamma\delta$ T cells in blood and in lymph nodes from fetal and postnatal lambs. *Dev. Comp. Immunol.* **16**, 493–501.

Washington, E. A., Katerelos, M., Cahill, R. N. P., Kimpton, W. G. (1994). Differences in the tissue-specific homing of $\alpha\beta$ and $\gamma\delta$ T cells to gut and peripheral lymph nodes. *Int. Immunol.* **6**(12), 1891–1897.

Watson, D. L. (1973). Local production and selective transfer of immunoglobulin by secretory organs in ruminants. PhD thesis, University of Sydney, Australia.

Watson, D. L. (1975). Cytophilic attachment of ovine IgG2 to autologous polymorphonuclear leucocytes. *Aust. J. Exp. Biol. Med. Sci.* **53**, 527–529.

Watson, D. L. (1980). Immunological functions of the mammary gland and its secretion – a comparative review. *Aust. J. Biol. Sci.* **33**, 403–422.

Watson, D. L. (1990). The breast. In *Mucosal Immunology: Proceedings of a Satellite Meeting of the World Congress of Gastroenterology* (ed. Cripps, E. A. W.), pp. 139–142. A. W. Cripps, Sydney, Australia.

Watson, D. L. (1992). Biological half-life of ovine antibody in neonatal lambs and adult sheep following passive immunization. *Vet. Immunol. Immunopathol.* **30**, 221–232.

Watson, D. L., Gill, H. S. (1991). Effect of weaning on antibody responses and nematode parasitism in Merino lambs. *Res. Vet. Sci.* **51**, 128–132.

Watson, D. L., Brandon, M. R., Lascelles, L. A. K. (1972). Concentrations of immunoglobulin in mammary secretion of ruminants during involution with particular reference to selective transfer of IgG_1. *Aust. J. Exp. Biol. Med. Sci.* **50**, 535–539.

Watson, D. L., Francis, G. L., Ballard, F. J. (1992). Factors in ruminant colostrum that influence cell growth and murine IgE antibody responses. *J. Dairy Res.* **59**, 369–380.

Watt, N. J., Macintyre, N., Collie, D., Sargan, D., McConnell, I. (1992). Phenotypic analysis of lymphocyte populations in the lungs and regional lymphoid-tissue of sheep naturally infected with maedi-visna virus. *Clin. Exp. Immunol.* **90**, 204–208.

Weill, J.-C., Reynaud, C.-A. (1996). Rearrangement/hypermutation/gene conversion: when, where, why? *Immunol. Today* **17**, 92–97.

Werling, D., Langhans, W., Geary, N. (1994). Caprine arthritis-encephalitis virus-infection changes caprine blood monocyte responsiveness to lipopolysaccharide stimulation in-vitro. *Vet. Immunol. Immunopathol.* **43**, 401–411.

Westermann, J., Pabat, R. (1996). How organ-specific is the migration of 'naive' and 'memory' T cells? *Immunol. Today* **17**(6), 278–282.

Whaley, A. E., Carroll, R. S., Imakawa, K. (1991). Cloning and analysis of the gene encoding ovine interferon alpha II. *Gene* **106**, 281–282.

Whitton, J. L., Oldstone, M. B. A. (1996). Immune response to viruses. In *Virology* (eds Fields, B. N., Knipe, D. M., Howley, P. M.), pp. 345–374. Lippincott-Raven Publishers, Philadelphia, USA.

Wickersham, N., Lloyd, J., Johnson, J., McCain, R., Christman, J. (1993). Acute inflammation in a sheep model of unilateral lung ischemia: the role of IL-8 recruitment of polymorphonuclear leukocytes. *Am. J. Respir. Cell. Mol. Biol.* **9**, 199–204.

Wieczorek, Z., Zimecki, M., Janusz, M., Lisowski, J. (1979). Proline-rich polypeptide from ovine colostrum: its effects on skin permeability and on the immune response. *Immunology* **36**, 875–881.

Wijngaard, P. L., Metzelaar, M. J., MacHugh, N. D., Morrison, W. I., Clevers, H. C. (1992). Molecular characterization of the WC1 antigen expressed specifically on bovine $CD4^-CD8^-\gamma\delta$ T lymphocytes. *J. Immunol.* **149**, 3272–3277.

Wilkerson, M. J., Davis, W. C., Baszler, T. V., Cheevers, W. P. (1995a). Immunopathology of chronic lentivirus-induced arthritis. *Am. J. Pathol.* **146**, 1433–1443.

Wilkerson, M. J., Davis, W. C., Cheevers, W. P. (1995b). Peripheral blood and synovial fluid mononuclear cell phenotypes in lentivirus induced arthritis. *J. Rheumatol.* **22**, 8–15.

Willems, L., Portetelle, D., Kerkhofs, P., Chen, G., Burny, A., Mamerickx, M., Kettmann, R. (1992). *In vivo* transfection of Bovine Leukemia Provirus into sheep. *J. Virol.* **189**, 775.

Willems, L., Kettmann, R., Dequiedt, F., Portetelle, D., Voneche, V., Cornil, I., Kerkhofs, P., Burny, A., Mammerickx, M.

(1993). *In vivo* infection of sheep by Bovine Leukemia Virus mutants. *J. Virol.* **189**, 4078.

Winkler, G. (1988). Pulmonary intravascular macrophages in domestic animal species – a review of structural and functional properties. *Am. J. Anat.* **181**, 217–234.

Winstanley, F. P. (1992). Detection of TNF-α receptors on ovine leuckocytes by flow cytofluorimetry. *Res. Vet. Sci.* **53**, 129–134.

Witherden, D. A., Kimpton, W. G., Washington, E. A., Cahill, R. N. P. (1990). Non-random migration of CD4$^+$, CD8$^+$, and $\gamma\delta$+T19+ lymphocytes through peripheral lymph nodes. *Immunology* **70**, 235–240.

Witherden, D. A., Abernethy, N. J., Kimpton, W. G., Cahill, R. N. P. (1994a). CD45RA expression on $\gamma\delta$ and $\alpha\beta$ T cells emigrating from the fetal and postnatal thymus. *Eur. J. Immunol.* **24**, 186–190.

Witherden, D. A., Kimpton, W. G., Abernethy, N. J., Cahill, R. N. P. (1994b). Changes in thymic export of $\alpha\beta$ and $\gamma\delta$ T cells during fetal and postnatal development. *Eur. J. Immunol.* **24**, 2329–2336.

Wong, C. W., Watson, D. L. (1995). Immunomodulatory effects of dietary whey proteins in mice. *J. Dairy Res.* **62**, 359–368.

Woodall, C. J., McLaren, L. J., Watt, N. J. (1994). Sequencing and chromosomal location of the gene encoding ovine latent transforming growth factor beta 1. *Gene* **150**, 371–373.

Woolaston, R. R., Mamueli, P., Eady, S. J., Barger, I. A., Lejambre, L. F., Banks, D. J. D., Windon, R. G. (1996). The value of circulating eosinophil count as a selection criterion for resistance of sheep to Trichostrongyle parasites. *Int. J. Parasitol.* **26**, 123–126.

Wright, H., Ballingall, K. T. (1994). Mapping and characterization of the *DQ* subregion of the ovine MHC. *Animal Genet.* **25**, 243–249.

Wright, H., Ballingall, K. T., Redmond, J. (1994). The *DY* subregion of the sheep MHC contains an A/B gene pair. *Immunogenetics* **40**, 230–234.

Wright, H., Redmond, J., Wright, F., Ballingall, K. T. (1995). The nucleotide sequence of the sheep MHC class II *DNA* gene. *Immunogenetics* **41**, 131–133.

Wright, H., Redmond, J., Ballingall, K. T. (1996). The sheep orthologue of the *HLA-DOB* gene. *Immunogenetics* **43**, 76–79.

Wunderlin, E., Palmer, D. G. (1995). Measurement of an eosinophil differentiation factor in sheep before and after infection with *Trichostrongylus colubriformis*. *Vet. Immunol. Immunopathol.* **49**, 169–175.

Yamada, Y., Sonoda, M. (1972). Basophils in ovine blood in electron microscopy. *Jap. J. Vet. Sci.* **34**, 29–32.

Yasmeen, D. (1981). Antigen specific cytophilic activity of sheep IgG1 and IgG2 antibodies. *Aust. J. Exp. Biol. Med. Sci.* **59**, 279–302.

Yilma, T., Owens, S., Adams, S. D. (1988). High levels of interferon in synovial fluid of retrovirus-infected goats. *J. Interferon Res.* **8**, 45–50.

Young, A. J., Hay, J. B., Chan, J. Y. C. (1990). Primary structure of ovine tumor necrosis factor alpha complementary DNA. *Nucl. Acids Res.* **18**, 6723.

Young, A. J., Hay, J. B., Mackay, C. R. (1993). Lymphocyte migration *in vivo*. *Curr. Top. Microbiol. Immunol.* **184**, 161–173.

Zanin, C., Bene, M. C., Martin, F., Perruchet, A. M., Borelly, J., Faure, G. C. (1994). Compartmentalization of specific B-cells in sheep mucosae associated lymphoid organs. *Vet. Immunol. Immunopathol.* **42**, 349–356.

Zink, M. C., Narayan, O. (1989). Lentivirus-induced interferon inhibits maturation and proliferation of monocytes and restricts the replication of caprine arthritis-encephalitis virus. *J. Virol.* **63**, 2578–2584.

Zink, M. C., Narayan, O., Kennedy, P. G. E., Clements, J. E. (1987). Pathogenesis of visna/maedi and caprine arthritis-encephalitis: new leads on the mechanism of restricted virus replication and persistent inflammation. *Vet. Immunol. Immunopathol.* **15**, 167–180.

XV Immunology of the Buffalo

1. Introduction

The buffalo belongs to the Class Mammalia, subclass Prototheria, order Artiodactyla, suborder Ruminantia, family Bovidae. Common names are Indian buffalo, River buffalo, Swamp buffalo, Arna, Carabao-Bubalus bubalis, and major breeds are Murrah, Nili/Ravi, Jafarabadi, Surti, Mehsana and Nagpuri.

The water buffalo (*Bubalus bubalis*) is the only buffalo species which has been domesticated. The other buffalo species, such as the gaur (*Bibos gaurus*), the cape buffalo (*Syncerus caffer*) and the North American (*Bison bison*) and European (*Bison bonasus*) bison are essentially wild animals and have a limited role in agriculture, but have zoological importance. The water buffalo is valued in many parts of tropical Asia, Europe and South America as a draught animal and provider of milk, meat and hides. The world population is considered to be 180 million (11% of the world's cattle) but their health and welfare have not been seriously researched until relatively recently, primarily because they have had little importance in western agriculture and are restricted mainly to less developed countries.

Most studies of the water buffaloes immune system have shown similarities with the immune system of the related species, the domestic cow (*Bos taurus*). However, there are distinct differences and these may explain the differences in disease susceptibility of these two species, e.g. buffaloes are better adapted to tropical climes and appear to be able to resist ticks and other pests, such as the warble fly and screw worm, and diseases of the feet are rare. Conversely, buffaloes are particularly susceptible to *Pasteurella multocida* infection. Understanding buffalo immunity can help to explain how this species deals with potentially fatal infections.

Most of the information in this chapter is on the water buffalo, but there is some cell marker information on the African buffalo.

2. Lymphoid Organs

There is little information on the immunological functions of the spleen, thymus or bone marrow of the buffalo. The distribution of superficial lymph nodes (LN) is shown in Figure XV.2.1.

In essence, the buffaloes lymphatic drainage is similar to the bovine (Saar and Getty, 1964). The parotid LN is large, single and elongated in adults and smaller in calves. Its afferent vessels drain the head, eye and neck and the efferents drain into the atlantal LN. The mandibular LN is single, large and located on the medial side of the angle of mandible and ventral part of the mandibular salivary gland. It drains the lips, cheek and muzzle and its three efferent vessels drain into the atlantal LN. Atlantal LN is located in the atlantal fossa below the mandibular salivary gland; in addition to draining the parotid and mandibular LNs, it drains the anterior neck region and its efferent vessels become the tracheal lymph duct. The prescapular LN is situated above the shoulder joint and drains the neck skin, shoulder, ventral and lateral surfaces of thorax with its three efferent vessels draining into the tracheal lymph duct. The axillary LN is the final LN draining the fore part of the buffalo and drains the medial aspect of the forelimb. The prefemoral LN is above the stifle joint and drains the rear of the thorax, the abdomen and lateral aspects of the rear limb. The popliteal LN is on the gastrocnemius muscle and drains the lower part of the rear limb. The superficial inguinal lymph node is at the neck of the scrotum which it drains along with the rear limb (Bagi *et al.*, 1991). Structural analysis of individual lymph nodes showed that the relative contributions to lymphatic drainage are fairly constant, as indicated by the measured sizes of nodes and germinal centres (Bagi *et al.*, 1992). The tracheal duct is similar to the bovine except that the left tracheal duct opens into the thoracic duct before its termination; in the bovine it forms a common trunk with the thoracic duct or opened into the left common jugular vein or branched and opened into the thoracic duct and common jugular vein (Dhablania *et al.*, 1994).

The mucosal associated lymphoid tissue has not been comprehensively described, lamina propria have been reported as diffuse lymphoid tissues in the interglandular spaces of the omaso-abomasal junction whereas the cardiac glands showed lymphocyte aggregates, extending to the muscularis mucosa (Kalita *et al.*, 1993). Studies of the testis show that there is no lymphatic connection between the two testes (Barnwal *et al.*, 1984).

Handbook of Vertebrate Immunology
ISBN 0-12-546401-0

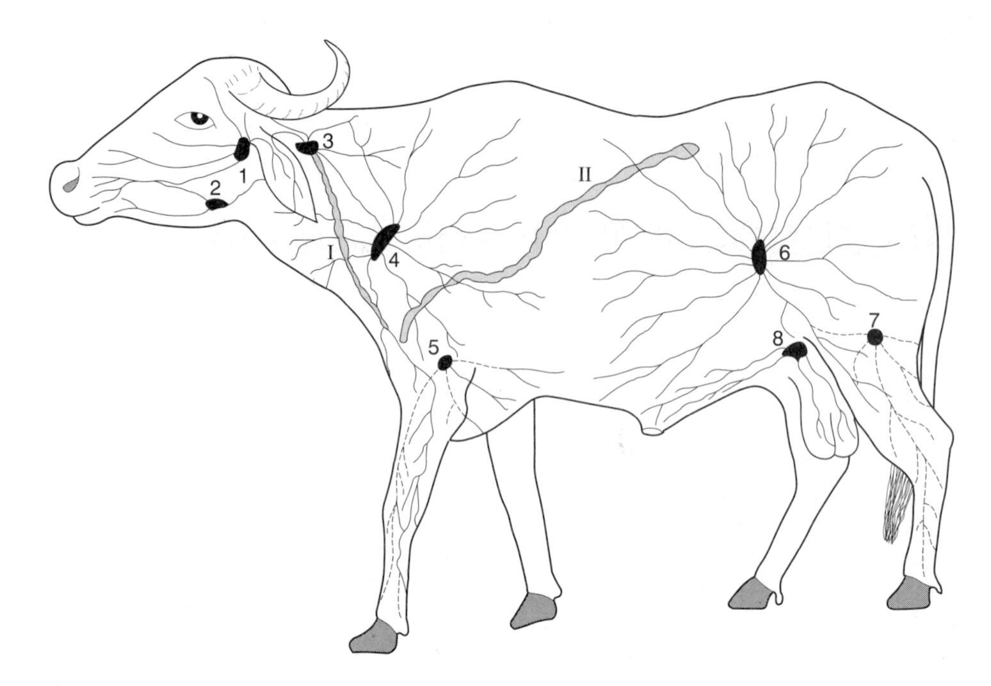

Figure XV.2.1 Schematic diagram of the Surti buffalo showing eight superficial lymph nodes and their respective draining areas. (1) Parotid lymph node (parotid lymphocentre); (2) mandibular lymph node (mandibular lymphocentre); (3) atlantal lymph node (atlantal lymphocentre); (4) prescapular lymph node (prescapular lymphocentre); (5) axillary lymph node (axillary lymphocentre); (6) prefemoral lymph node (prefemoral lymphocentre); (7) popliteal lymph node (popliteal lymphocentre); (8) superficial inguinal lymph node (superficial inguinal lymphocentre); I, Tracheal lymph duct; II, Thoracic duct. (After Bagi *et al.*, 1991).

3. Leukocytes and their Markers

Buffalo leukocytes have much in common with bovine cells and most current information is derived from studies analysing cross-reactions with bovine-specific reagents, primarily monoclonal antibodies.

Surface markers

In the First International Workshop on Leukocyte Antigens of cattle and sheep, Naessens (1991) analysed the cross-reactivity of monoclonal antibodies to bovine or ovine leukocyte antigens with cells of the African buffalo. This work was updated at the Second International Conference (Naessens *et al.*, 1993). These studies reported only positive or negative staining and cannot identify antigens bound, CD status of stained cells or their function. The monoclonal antibodies reacting with buffalo cells are shown in Table XV.3.1.

At the Third Workshop on Ruminant Leukocyte Antigens, monoclonal antibodies against bovine cells were checked against water buffalo peripheral blood leukocytes. Also, somatic cell hybrids between water buffalo and a chinese hamster cell line were used as targets for further analysis of antigen clusters (El Nahas *et al.*, 1996). A number of reagents cross-reacted with buffalo leukocytes but there was not enough data to conclusively ascribe CD nomenclature for the buffalo from this data. However, the cross-reacting reagents probably bind autologous molecules, but much more work will be needed to define water buffalo CD molecules and their functions. The results of this Workshop are summarized in Table XV.3.2.

E-rosetting cells form 36% of water buffalo peripheral blood lymphocytes, EAC rosettes are 17%, null cells forming the remaining 47% (Singh *et al.*, 1988). No further identification has been made of buffalo cell markers.

Mitogen stimulation experiments have shown that PHA, Con A and PWM can induce proliferation but not to very high levels (Bansal *et al.*, 1981). Responding cell types have not been identified. In one of the few immunological studies of the bison, it has been shown that lymphocytes can be easily separated from blood by Percoll density gradient centrifugation and that sheep E-rosettes can be used to detect (unidentified) subpopulations, as can fluorescein isothiocyanate (FITC)–peanut agglutinin (PNA) and FITC–Con A. Proliferation to LPS, Con A, PWM and PHA were similar to cattle and sheep and cultured lymphocytes could be induced to proliferate with rBoIL-2 (Nagi and Babiuk, 1989).

Buffalo also have differences to the bovine in susceptibility to specific infections. While it is possible to transform bovine lymphocytes *in vitro* with *Theileria annulata* and generate cell lines, this is not possible with buffalo

Table XV.3.1 Monoclonal antibodies to bovine CD markers: reagents cross-reacting with leukocytes of African buffalo (*Syncerus caffer*)

Putative antigen	*Workshop no.*	*mAb*	*Reactivity*
CD1	–	–	–
CD2	137	MUC2A	+
CD3	138	MM1A	+
CD4	133	GC150A1	Weak
CD5	–	–	–
CD6	132	BAQ91A	Weak
CD8	–	–	–
CD11a	44	72–87	+
	85	IL-A99	+
	99	MD1H11	+
	100	MD2B7	+
	178	BAQ30A	+
CD11b	11	IAH-CC126	+
	22	IAH-CC94	+
	23	IAH-CC125	+
	62	IL-A15	+
CD11c	180	BAQ153A	+
CD18	103	MF14B4	+
CD25	54	IL-A111	+
CD44	50	IL-A107	+
	51	IL-A108	+
	55	IL-A112	+
	61	IL-A118	+
	93	25–32	+
CD45R	28	IAH-CC77	+
	109	Bo42	+
MHC-I	63	IL-A19	+
MHC-II	10	IAH-CC85	+
	(+ Many more)		
Ig	69	IL-A58	+

Table XV.3.2 Monoclonal antibodies to bovine leukocyte antigens which bind water buffalo PBMC

	Workshop no.	*mAb*	*Reactivity*
Putative antigen			
CD5	406	IAH-CC17	+
CD8	309	IAH-CC58	+
CD11a	30	IVA35	+
CD11b	402	IAH-CC125	+
	354	IL-A130	+
	206	MM13	+
CD14	364	VPM66	+
	264	VPM67	+
CD41/CD61	232	CAPP2	+
CD44	258	Buff1	+
CD45R	225	IVA-112	+
CD62L	360	IAH-CC32	+
	452	Buff44	+
CD71	280	IL-A77	+
TCRgd	325	GB21A	+
WC1	336	BAQ89	+
	560	BAQ90	+
B cell antigens			
	20	DM7	+
	44	BAQ44	+
	56	IVA84	+
	242	IL-A65	+
	408	B18	+
	545	LCT27	+

lymphocytes, even though infection of lymphocytes could be detected, suggesting a mechanism for disease resistance in the buffalo (Chaudhri and Subramanian, 1992). There have been similar findings for the African buffalo, which would explain why this species also only has mild or asymptomatic infection (Steuber *et al.*, 1986). This difference may be due to a lack of a specific receptor on buffalo lymphocyte surfaces (Jura, 1986).

Buffalo skin is particularly sensitive to *Toxocara vitulorum* excretory–secretory antigens, as manifest by a hypersensitivity response with large dermal nodules caused by infiltration of lymphocytes, macrophages and eosinophils (Starke *et al.*, 1996).

4. Leukocyte Traffic

Adhesion molecules have not been recognized on the water buffalo but have been shown on the African buffalo (Naessens *et al.*, 1993). These have been identified by FACS staining with antibovine reagents but no function has been ascribed to bound ligands (Table XV.3.2).

5. Cytokines

Water buffalo IL-2 production by lymphoid cells can be induced by both Con-A and PHA stimulation, may be enriched from tissue culture medium by ammonium sulphate precipitation and will maintain homologous blasts from mitogen-stimulated PBMC. This buffalo IL-2 preparation also induces proliferation of Con-A blasts from cows, sheep and goats (Mathur *et al.*, 1992).

The water buffalo IFN-γ gene has been identified by *in situ* hybridization with a bovine cDNA probe (Hassanane *et al.*, 1994).

Naessens *et al.* (1993) showed that monoclonal antibody IL-A111 (from the Second International Workshop on Leukocyte antigens of cattle and sheep) directed against bovine IL-2R cross-reacted with African buffalo Con-A-activated lymphocytes.

6. T-Cell Receptor

One of the monoclonal antibodies to the bovine TCR$\gamma\delta$ (Workshop Ab325, clone GB21A) cross-reacts with water buffalo lymphocytes; however there is no data as to the nature of the cell surface ligand (Naessens *et al.*, 1993).

7. Major Histocompatibility Complex Antigens

Restriction fragment length polymorphism (RFLP) studies of buffalo DNA from blood leukocytes and using hybridization with a probe for human DQB gene product, after restriction enzyme digestion, have shown that water buffalo MHC is highly conserved, like cattle (Kumar *et al.*, 1992). Similar results were obtained with a range of restriction enzyme digests (Kumar *et al.*, 1993).

Naeseens *et al.* (1993) showed that monoclonal antibodies to bovine MHC-I cross-reacted with African buffalo as did many against MHC-II framework antigens (Table XV.7.1).

African buffalo MHC Class I maps to chromosome 2p (Iannuzzi *et al.*, 1993).

Table XV.7.1 Monoclonal antibovine MHC-I and MHC-II antibodies binding African buffalo lymphocytes

Antigen	*mAB*
MHC-I	IL-A19
	FW3-181
MHC-II	Th14B
	CC7
	CC85
	CC128
	IL-A113
	IL-A115
	IL-A6
	IL-77
	IL-80
	IL-81
	+ Many more

8. Immunoglobulins

IgG

In the water buffalo, there are two subclasses, IgG1 and IgG2, which may be separated by ion exchange chromatography or electrophoretic mobility as a result of differing charge. IgG2 is highly positively charged and thus elutes from an ion exchange (DEAE) column at low salt concentrations (Reddy and Giridhar, 1991) whereas IgG1 only elutes at high (>0.1 M tris/HCl) salt concentrations (Kakker and Goel, 1993a). IgG2 binds *Staphylococcus aureus* Protein A whereas IgG1 does not.

The African buffalo immunoglobulins are less well studied. The IgG subclasses are not described but reactions to Protein A have been demonstrated, as have reactions with a Protein A + Protein G construct, which are stronger (Kelly *et al.*, 1993).

In the water buffalo, unlike cattle, both subclasses have molecular masses of 162 kDa with heavy chains of 58 kDa and light chains of 24 kDa. Both subclasses are found in serum and colostrum and both fix complement, with IgG1 being the more effective of the two.

Antisera to the Ig subclasses have been raised in both rabbits and guinea pigs and their subclass specificity demonstrated after affinity chromatography removal of cross-reacting antibodies (Kakker and Goel, 1993b). Monoclonal antibodies to bovine IgG and IgM bind their buffalo equivalents; there is also a cross-reaction with monoclonal anti-bovine IgA (to molecules not identified as IgG1, IgG2 or IgM) but buffalo IgA has not been fully identified.

IgM

First described by Kulkarni *et al.* (1973), water buffalo IgM is similar in size to the bovine form; it is not a strong fixer of complement (Kakker and Goel, 1993b). Monoclonal antibodies to bovine IgM bind the buffalo equivalent.

Immunoglobulins in body fluids

A wide range of values have been given for adult water buffalo serum immunoglobulins; from 10 g/l (Verma and Joshi, 1994) to more than 30 g/l (Da Silva *et al.*, 1993). Serum gamma globulins are raised by levamisole treatment at vaccination (Verma and Joshi, 1994).

IgM is not detected in CSF, where mainly IgG is found with occasional IgA (Kulkarni, 1983). Nasal, tracheal and bronchopulmonary lavage fluids contain IgG, trace amounts of IgA, but no IgM (Kulkarni, 1982a).

Vaccine responses

Experiments with *Leptospira* serovars indicate that the water buffalo responds to the same antigen doses as cattle and develops better protection after boosting (Rao and Keshavamurthy, 1982). Oil–adjuvant vaccines have been used for *P. multocida* vaccination and all three vaccines tested in buffalo calves proved effective in generating sustained immunity beyond 270 days post-vaccination (Muneer *et al.*, 1994).

Serological testing

Polyclonal antisera have been used to determine the seroprevalence of antibodies to a number of infectious organisms and the degree of antibody production. Measurements of antibodies to *Toxocara vitulorum* in water buffalo calves showed an inverse correlation with faecal egg count (Rajapakse *et al.*, 1994). Serological detection of milk antibodies to *S. aureus* has been found to be more sensitive than direct bacteriological detection (D'Apice *et al.*, 1996). An ELISA developed to determine water buffalo antibodies to foot-and-mouth disease virus has shown itself faster and easier than conventional VN titration (Araujo *et al.*, 1996). Antibodies to hydatid antigens have been detected by ELISA but the assay has not been developed for diagnostic use (Deka and Gaur, 1993).

9. Passive Transfer of Immunity

The mammary gland is important for nutrition of the new born buffalo calf because of the high protein content, much of which (approximately 80%) is gamma globulin. As with cattle, colostral transfer of immunoglobulins is essential for calves which are born either hypogammaglobulinaemic or agammaglobulinaemic and can absorb antibodies from colostrum for several days. However, there appears to be a small transplacental transfer of proteins which include immunoglobulins. Water buffalo colostrum has a higher protein and immunoglobulin content than colostrum from dairy crossbred cattle (Singh *et al.*, 1993). The major class of Ig in colostrum is IgG and the vast majority is absorbed in the first feeding; 1 h later, intestinal absorptive capacity is severely reduced. However, calves can still absorb immunoglobulins from colostrum up to 96 h after birth. If colostrum is given immediately postpartum, globulins can be detected in calf serum 6 h after colostrum ingestion and successful transfer is indicated by calf serum gamma globulin levels that equal or exceed 30 g/l, equivalent to adult serum levels (Da Silva *et al.*, 1993). This drops to 20 g/l in 3-week-old calves (Kishtawaria *et al.*, 1983). The efficiency of immunoglobulin transfer is better in suckled calves than pail-fed calves (Joshi *et al.*, 1993). Colostral and neonate serum immunoglobulins are shown in Table XV.9.1 (Singh *et al.*, 1993; Joshi *et al.*, 1992).

Table XV.9.1 Passive transfer of immunoglobulins in the water buffalo

	Postpartum (h)						
	4	*16*	*28*	*40*	*52*	*64*	*70*
a. Colostral content of immunoglobulins							
Ig (%)	10.3	8.2	5.1	3.2	2.3	1.6	1.4
b. Proportions of IgM, IgA and IgG in colostrum							
	1	13	25	37	49	61	
IgM	0.2	0.2	0.4	0.3	1.5	1.1	
IgA	2.6	2.0	1.1	0.6	1.0	1.0	
IgG	89.2	95.7	96.5	97.3	94.1	95.2	

c. Calf serum immunoglobulins after multiple colostral intake (fed 1 h prior to each blood sampling)

	Postnatal (h)			
	8	*15*	*31*	*36*
Fed colostral Ig (g)	6.1	23.6	22.8	16.4
Absorption (g)	4.6	0.9	0.8	1.2

d. Serum immunoglobulins (mg/ml) in male and female calves

	Time (days) after colostral feeding								
	0	*0.5*	*1*	*2*	*3*	*7*	*14*	*21*	*28*
Male									
IgG	0.8	13	18	19	18	16	15	15	16
IgM	0	2	3	3	2	2	2	2	2
Female									
IgG	0.8	12	17	18	18	17	16	15	18
IgM	0	2	3	3	2	2	2	2	2

The success of passive transfer in buffalo has been shown by the strong correlation between colostrum antibody titre and the titre of passively acquired antibody in calf serum (Rajapkse *et al.*, 1994). Passively transferred specific antibodies decline over 3 weeks. The importance of passive transference has been highlighted by the inverse relationship between the 24-h serum level in calves and early mortality (Raghavan and Rai, 1987).

Antibodies to specific antigens, such as anti-*S. aureus* for mastitis testing, can be detected in milk, but these are probably related more to local production and will probably have no part in passive antibody transfer (D'Apice *et al.*, 1996).

10. Neonatal Immunity

Analysis of water buffalo fetal blood showed that 50% of samples contained small, unquantifiable amounts of IgG and IgM (Kulkarni, 1982b), but it is uncertain whether this is placentally derived or secreted by B cells of the fetus.

In young calves, enteric colibacillosis prevents efficient uptake of gammaglobulins, as well as other components, from colostrum. Normal serum immunoglobulin levels are 20 g/l in 3-week-old calves (Kishtawaria *et al.*, 1983). In 1–4-day-old calves the serum gamma globulin levels are equal to or more than 30 g/l – equivalent to adult serum levels (Da Silva *et al.*, 1993).

Neonatal immunity is not considered deficient since immunization is normal in young calves. Vaccination for foot-and-mouth virus is effective in 10-day-old calves, although boosters at weeks 4–5 are recommended (Christi and Mehta, 1993).

11. Nonspecific Immunity

Because *Pasteurella multocida* is a cause of high mortality in water buffaloes the oxidative microbicidal system of buffalo PMN was investigated and showed no apparent deficiency in that they were able to generate large amounts of H_2O_2 and NO^- and could be primed to secrete even more by infection or vaccination with *P. multocida* (Roy *et al.*, 1996). Buffalo PMN are also responsive, in terms of protein synthesis, thyroxine uptake and O^{2-} production, to digitonin, LPS and Con A. The anti-inflammatory drug, indomethacin, inhibited this activation (Sarmah and More, 1996). Antibacterial properties of cationic granule peptides have been identified in water buffalo PMN and shown to be active against both Gram-negative and Gram-positive bacteria (Sarmah *et al.*, 1993).

Table XV.12.1 Haemolytic complement in buffalo serum

Age (months)	*CH_{50}*	
	Range	*Mean*
3	337–596	401
4–6	418–640	486
7–9	664–881	730
10–12	697–823	777
13–15	832–874	847
16–18	848–1280	1106
19–21	1168–1531	1329
22–24	1331–2304	1599
25–36	2047–2725	2350
37–48	1741–2418	2156
48	1263–1883	1545

CH_{50}: units of serum lysing 50% of RBCs.

12. Complement

The complement system of the buffalo is similar to the bovine in many respects, including inefficient lysis of sensitized SRBCs and an inability to lyse homologous RBCs (Jain and Goel, 1989). Haemolytic ability is particularly enhanced at low (0.025 M) salt concentrations, more so than the complement-mediated haemolysis of other species. Complement is fully functional even in 1-month-old calves, but haemolytic levels increase with age to a peak between 2 and 3 years old. Levels decline from the age of 4 years (Table XV.12.1).

The alternative pathway of activation has been demonstrated in water buffaloes from 1 month to 12 years old and is Mg^{2+} dependent, inhibitable by zymosan, EGTA and heating at 50°C for 20 min (Arya and Goel, 1992). Studies of activation of the alternative pathway by different bacteria showed that while *Escherichia coli*, *Staphylococcus aureus* and *Mycobacterium bovis* were effective, that *Pasteurella multocida* failed to activate haemolytic complement and this could explain why, of all domestic animals, the water buffalo is particularly susceptible to haemorrhagic septicaemia (Singh *et al.*, 1995).

13. Immunomodulation

Levamisole *in vivo* induces increased IgG and IgM antibody responses to a number of vaccines, including rinderpest (Kumari *et al.*, 1987) and *Pasteurella multocida* (Verma and Joshi, 1994). Levamisole has also been used to reverse the immunosuppressive properties of prednisolone, including leucopaenia and reduced serum IgG (Keskar *et al.*, 1996).

14. Tumours of the Immune System

Lymphosarcomas are the only lymphoid associated tumour routinely reported in the water buffalo. The most commonly involved organ is the udder; the uterus is also commonly reported. Lesions are also seen in the testes, epididymis, seminal vesicles, prostate and urethra. Endocrine glands are also involved (Singh *et al.*, 1980). However, the only reported immunological effect of lymphosarcoma is of reduced proliferative responses of PBMC to mitogens (Con-A, PWM and PHA) (Bansal *et al.*, 1981).

15. References

Amerasinghe, P. H., Vasanthathilake, V. W. S. M., Lloyd, S., Fernando, S. T. (1994). Periparturient reduction in buffalo of mitogen-induced lymphocyte proliferation and antibody to toxocara-vitulorum. *Trop. Anim. Health Prod.* **26**, 109–116.

Araujo, J. P., Montassier, H. J., Pinto, A. A. (1996). Liquid phase blocking sandwich enzyme-linked immunosorbent assay for detection of antibodies against foot and mouth virus in water buffalo sera. *Am. J. Vet. Res.* **57**, 840–843.

Arya, A., Goel, M. C. (1992). Studies on activation and levels of haemolytic complement of buffalo (*Bubalus bubalis*). II. Alternate complement pathway activity in serum. *Vet. Immunol. Immunopathol.* **30**, 411–418.

Bagi, A. S., Vyas, K. N., Bhayani, D. M. (1991). Morphology, relationship and draining area of superficial lymph-nodes of the surti buffalo (*Bubalus-Bubalis*). *Indian Vet. J.* **68**, 1153–1156.

Bagi, A. S., Vyas, K. N., Panchal, K. M. (1992). Micrometry of several structures of superficial regional lymph-nodes in surti buffalo (*Bubalus-Bubalis*). *Indian Vet. J.* **69**, 45–48.

Bansal, M. P., Kumar, A., Sing, K. P. (1981). Effect of mitogens on DNA synthesis of buffalo lymphocytes in culture. *Indian Vet. J.* **58**, 434–437.

Barnwell, A. K., Roy, M. K., Singh, L. P. (1984). Lymphatic drainage of testis of buffalo. *Indian J. Anim. Sci.* **54**, 375–377.

Chaudhri, S. S., Subramanian, G. (1992). *In vitro* infectivity of sporozoites of *Theileria annulata* to viable lymphocytes of cattle and buffalo. *Indian J. Anim. Sci.*, **62**, 225–226.

Christi, K. S., Mehta, V. M. (1993). Effect of FMD vaccination on serum proteins in buffalo neonates. *Indian J. Anim. Sci.* **63**, 1109–1111.

D'Apice, L., Fenizia, D., Capparelli, R., Scala, F., Iannelli, D. (1996). Detection of antibodies to *Staphylococcus aureus* in water buffalo milk by flow cytometry. *Res. Vet. Sci.* **60**, 179–181.

Dhablania, D. C., Tyagi, R. P. S., Khatra, G. S. (1994). Morphological studies of the lymph nodes of head in buffalo calves. *Indian Vet. J.* **71**, 469–473.

Da Silva, M. C., Dequeiroz, W. T., Lau, H. D., Vale, W. G. (1993). Colostrum and serum protein levels in water buffalos. *Pesquisa Agropecuria Brasileira* **28**, 751–757.

Deka, D. K., Gaur, S. N. S. (1993). Elisa in the diagnosis of hydatidosis in buffalo. *Indian J. Anim. Sci.* **63**, 1254–1255.

El Nahas, S. M., Ramadam, H. A., Abou-Mosallem, A. A., Kurucz, E., Vilmos, P., Ando, I. (1996). Assignment of genes coding for leukocyte surface molecules to river buffalo chromosomes. *Vet. Immunol. Immunopathol.* **52**, 435–443.

Gandotra, V. K., Sharma, R. D. (1991). Immunological studies in normal and repeat breeder buffalos. *Indian J. Anim. Sci.* **61**(10), 1019–1023.

Hassanane, M. S., Chowhary, B. P., Gu, F., Andersson, L., Gustavsson, I. (1994). Mapping of the interferon-gamma (IFNG) gene in river and swamp buffalos by *in situ* hybridization. *Hereditas* **120**, 29–33.

Iannuzzi, L., Gallagher, D. S., Womack, J. E., Di Meo, G. P., Skow, L. C., Ferrara, L. (1993). Chromosomal localisation of the major histocompatibility complex in cattle and river buffalo by fluorescent *in situ* hybridisation. *Hereditas* **118**, 187–190.

Jain, A., Goel, M. C. (1989). Studies on activation and levels of haemolytic complement of buffalo (*Bubalus bubalis*) 1. Classical complement pathway. *Vet. Immunol. Immunopathol.* **23**, 267–277.

Joshi, V. B., Saini, S. S., Sodhi, S. S. (1992). Serum protein, immunoglobulin and haemolytic complement levels in the sera of buffalo neonates. *Indian J. Anim. Sci.* **62**, 728–731.

Joshi, V. B., Saini, S. S., Sodhi, S. S. (1993). Effect of management condition on serum proteins and immunoglobulin profile of buffalo calves. *Indian J. Anim. Sci.* **63**, 905–907.

Jura, W. G. Z. O. (1986). Invasion and intracellular development of *Theileria annulata* sporozoites in lymphoblastoid cell lines already transformed by *Theileria annulata* (Hissar), *Theileria annulata* (Ankara) and *Theileria annulata* (Mugaga). *Vet. Parasitol.* **22**, 203–214.

Kakker, N. K., Goel, M. C. (1993a). A methodological study on the preparation of subclass specific antisera to buffalo IgG. *Indian J. Anim. Sci.* **63**, 1112–1116.

Kakker, N. K., Goel, M. C. (1993b). Purification and characterisation of IgG1 and IgG2 from buffalo (*Bubalus bubalis*) serum and colostrum. *Vet. Immunol. Immunopathol.* **37**, 61–71.

Kalita, H. C., Prasad, R. V., Baishya, G. (1993). Occasional presence of parietal cells and distribution of lymphocytes in the cardiac glands of Indian buffalo (*Bubalus bubalis*). *Indian Vet. J.* **70**(9), 837.

Kelly, P. J., Tagwira, M., Matthewman, L., Mason, P. R., Wright, E. P. (1993). Reactions of sera from laboratory, domestic and wild animals in Africa with protein A and A recombinant chimeric protein-AG. *Comp. Immunol. Microbiol. Infect. Dis.* **16**, 299–305.

Keskar, D. V., Venkataraman, R., Srinivasan, S. R., Dhanapalan, P. (1996). Prednisolone induced immunodeficiency in buffalo calves and its correction with levamisole. *Indian Vet. J.* **73**, 102–103.

Kishtwaria, R. K., Misra, S. K., Choudhuri, P. C. (1983). Status of total protein, gamma-globulin, sodium and potassium in the serum of buffalo calves with enteric colibacillosis. *Indian J. Anim. Sci.* **53**, 558–560.

Kulkarni, B. A. (1982a). Immunoglobulins of the Indian buffalo-immunoglobulins in respiratory lavage fluids. *Indian Vet. J.* **59**, 103–106.

Kulkarni, B. A. (1982b). Note on the foetal immunoglobulins of the Indian buffalo. *Indian J. Anim. Sci.* **52**, 587–589.

Kulkarni, B. A. (1983). Immunoglobulins of the Indian buffalo-immunoglobulins in the CSF. *Indian Vet. J.* **60**, 607–609.

Kulkarni, B. A., Rao, S. S., Rindani, T. H. (1973).

Immunoglobulins of the Indian water buffalo. *Indian J. Biochem. Biophys.* **10**, 216–219.

Kumar, S., Bhat, P. N., Rasool, T. J., Bhat, P. P. (1992). Restriction fragment length polymorphism studies of the major histocompatibility locus of buffaloes-BuLA. *Indian J. Anim. Sci.* **62**, 448–451.

Kumar, S., Bhat, P. N., Rasool, T. J., Bhat, P. P. (1993). Evidence for the major histocompatibility complex in buffalo (BuLA) using restriction fragment length polymorphism technique. *Indian Vet. J.*, **70**, 183–184.

Kumari, K. N., Choudhuri, P. C., Krishnaswamy, S. (1987). Effect of levamisole on the immune response of buffalo calves to rinderpest tissue culture vaccine. *Indian Vet. J.* **64**, 984–985.

Mathur, B. B. L., Bansal, M. P., Ram, G. C., Subbarao, M. V. (1992). Buffalo interleukin-2 and its effects on proliferation of lymphocytes. *Indian J. Anim. Sci.* **62**, 291–293.

Muneer, R., Akhtar, S., Afzal, M. (1994). Evaluation of 3 oil adjuvant vaccines against *Pasteurella multocida* in buffalo calves. *Rev. Sci. Techn. l'Office Int. Epizooties* **13**, 837–843.

Nagi, A. M., Babiuk, L. A. (1989). Isolation, characterisation and *in vitro* mitogenic stimulation of peripheral blood lymphocytes from an American buffalo. *Can. J. Vet. Res.* **53**, 493–496.

Naessens, J. (1991). Characterisation of lymphocyte populations in African buffalo (*Syncerus caffer*) and waterbuck (*Kobus defassa*) with workshop monoclonal antibodies. *Vet. Immunol. Immunopathol.* **27**, 153–162.

Naessens, J., Olubayo, R. O., Davis, W. C., Hopkins, J. (1993). Cross-reactivity of workshop antibodies with cells from domestic and wild ruminants. *Vet. Immunol. Immunopathol.* **39**, 283–290.

Raghavan, K. C., Rai, A. V. (1987). Immunoglobulin levels and mortality rate in Surti buffalo calves. *Indian J. Anim. Sci.* **57**, 479–481.

Rajapkse, R. P. V. J., Lloyd, S., Fernando, S. T. (1994). *Toxocara vitulorum*: maternal transfer of antibodies from buffalo cows (*Bubalus bubalis*) to calves and levels of infection with *T. vitulorum* in the calves. *Res. Vet. Sci.* **57**, 81–87.

Rao, A. S., Keshavamurthy, B. S. (1982). Study of the immune response of buffalo calves to heat-killed pentavalent leptospiral antigens. *Indian Vet. J.* **62**, 357–361.

Reddy Y. K., Giridhar, P. (1991). Purification of buffalo IgG2 infraclass. *Indian J. Anim. Sci.* **61**, 1284–1286.

Roy, S. C., More, T., Pati, U. S., Srivastava, S. K. (1996). Effect of *Pasteurella multocida* vaccination on buffalo polymorphonuclear hydrogen peroxide and nitric acid production. *Vet. Immunol. Immunopathol.* **51**, 173–178.

Saar, L. I., Getty, R. (1964). Ruminant lymphatic system. In *Atlas of Applied Veterinary Anatomy*, 2nd edn (ed. Getty, R.), pp. 207–260. Iowa State University Press, Iowa, USA.

Sarmah, S., More, T. (1995). Extracellular release of antibacterial protein and peptide by buffalo polymorphonuclear cells in presence of *Escherichia coli*. *Indian Vet. J.* 72, 1265–1268.

Sarmah, S., More, T. (1996). Some biochemical responses of buffalo PMN cells to various stimuli. *Comp. Immunol. Microbiol. Inf. Dis.* **19**, 47–53.

Sarmah, S., Reddy, G. R., More, T. (1993). Cationic antibiotic granular peptides of buffalo polymorphonuclear cells. *Indian J. Anim. Sci.* **63**, 618–622.

Singh, A., Ahuja, S. P., Singh, B. (1993). Individual variation in the composition of colostrum and absorption of colostral antibodies by the precolostral buffalo calf. *J. Dairy Sci.* **76**, 1148–1156.

Singh, B., Singh, K. P., Parihar, N. S., Bansal, M. P., Singh, C. M. (1980). Lymphosarcomatous involvement of reproductive and endocrine organs in Indian buffalo. *Zbl. Vet. Med. A* **27**, 583–592.

Singh, D., Singh, N. P., Rae, R. B. (1982). Serum protein and IgM levels in buffalo calves experimentally infected with *Trypanosoma evansi*. *Phillipine J. Vet. Med.* **21**, 122–125.

Singh, S., Goel, M. C., Monga, D. P. (1993). Studies on activation and levels of haemolytic complement of buffalo (*Bubalus bubalis*). 3. C-3 haemolytic activity in health and chronic disease. *Vet. Immunol. Immunopathol.* **35**, 393–398.

Singh, S., Goel, M. C., Monga, D. P. (1995). Activation of alternative complement pathway in serum of buffalo by different bacteria. *Indian J. Anim. Sci.* **65**, 266–268.

Singh, V., Sharma, K. N., Raisinghani, P. M. (1988). Absolute leukocytes, E-rosettes and EAC rosette-forming lymphocytes of buffalo calves infected with *Trypanosoma evansi*. *Indian J. Anim. Sci.* **58**, 1288–1291.

Starke, W. A., Machado, R. Z., Bechara, G. H., Zocoller, M. C. (1996). Skin hypersensitivity test in buffaloes parasitized with *Toxocara vitulorum*. *Vet. Parasitol.* **63**, 283–290.

Steuber, S., Frevert, U., Ahmed, J. S., Hauschild, S., Schein, E. (1986). *In vitro* susceptibility of different mammalian lymphocytes to spororozoites of *Theileria annulata*. *Zeitschr. Parasit.* **72**, 831–834.

Third Workshop on Ruminant Leucocyte Antigen (1993). *Vet. Immunol. Immopathol.* **39**, pt 1–3.

Verma, R. S., Joshi, B. P. (1994). Observations on immunostimulation with levamisole in the Indian buffalo. *Indian Vet. J.* **71**, 740–742.

XVI The Mouse Model

1. Introduction

The text of this chapter is aimed at providing the reader with information about the main areas of mouse immunology that may aid understanding of the function of the immune system in other species less extensively studied. It is written with the goal of presenting recent developments.

Some aspects have been omitted – for example, the complement system is far better characterized in the guinea pig. The immune system is essentially entirely conserved throughout the mammalian species and the proteins are often interchangeable across species. In every mammalian species, the major histocompatibility complex appears to have overwhelming importance in determining the fate of allografts, in stimulating antibody production and cell-mediated lympholysis and in restricting the immune response. The mouse MHC was named H-2, whereas in most animal species, scientists have chosen to refer to the HLA nomenclature used in humans, which is particularly well characterized given the bridges provided by this system between basic sciences and practical clinical implications. The same is true for blood antigens, which are not described in this chapter.

Among the critical advances made in the last decade, are (1) a detailed understanding of the stages of differentiation of T cells, B cells and accessory cells and the surface antigens that distinguish them, (2) the identification of the critical molecules concerned with cell adhesion and with cell activation, (3) the discovery of increasing numbers of lymphokines with pleotropic activities, (4) the identification and cloning of the genes that code for the variable and constant segments of immunoglobulins that have demonstrated the contributions of genetic and somatic mechanisms for the origin of diversity, and, (5) the analysis of naturally occurring immune dysfunctions and the generation of transgenic models that have permitted the elucidation of the mechanisms of regulators.

The section, 'The Lymph Organs', deals with the functional anatomy and makes possible the understanding of TCR ontogeny and the functions of the mucosal immune system. The following section, 'Immunoglobulins', discusses the ontogeny and the genetics of immunoglobulins, as well as their role in the immune response and in the transfer of immunity. In the subsequent section, 'Cellular Immunology' we have listed the CD markers that help cell identification, whose state of activation can be determined or that play a role in cell traffic, and the main cytokines whose origin and role have been identified. The final section, 'Murine Models of Immunodeficiency', may be regarded as the review of the main genetic and acquired immunodeficiencies and of transgenic mouse models providing information enabling us to understand some mechanisms in the regulation of immune response.

Thus, rather than covering all aspects of mouse immunology, we have taken the position to centre around aspects which have been recently developed and that can further progress the understanding of immunology in other species.

2. The Lymphoid Organs

Introduction

In the mouse, as in other mammals, natural and specific immunity constitute the basis of immune defence. Specific immunity, also called adaptative immunity, results from specific recognition of antigen and appropriate interaction of lymphoid cells with accessory cells. In this way, the cell-mediated and antibody responses are efficiently carried out.

Lymphocytes are scattered throughout the body, with the exception of a few locations, but interact as if they belonged to a single organ. They recirculate via the blood and lymph and home to particular lymphoid organs. Stringent mechanisms control their behaviour. Two major evolutionary phases characterize the life of lymphocytes. The first, called **lymphopoiesis,** is antigen independent and takes place in the central lymph organs: the bone marrow for B cells and thymus for T cells. The second, called **immunopoiesis**, starts with antigen recognition and occurs in the peripheral lymphoid organs. The metabolically active lymphoid cells involved in lymphopoiesis are called B or T lymphoblasts; the cells taking part in immunopoiesis are called B or T immunoblasts; small, resting B and T cells are lymphocytes.

The central lymphoid organs

These organs, also called primary lymphoid organs, ensure production of mature B and T cells by governing

Handbook of Vertebrate Immunology
ISBN 0-12-546401-0

the progressive transformation of undifferentiated stem cells to clones expressing either the B-cell receptor (BCR) or the T cell receptor (TCR), in addition to other surface molecules and intracellular signalling systems enabling the lymphoid cells to recognize antigen and react adequately.

The bone marrow

In the mouse as in other mammals, the bone marrow produces precursors of B and T lymphoid cells over the animal's entire life. Located in cancellous bone, even that filling the long bones such as the femur, the bone marrow is a protected milieu sustained by a reticular tissue rich in blood vessels and adipose cells. The feeding arteries that cross the bone wall capillarize along it to form sinusoids sending confluent venules into the central part of the bone marrow; this creates different microenvironments through which the maturing cells progress. Among the nonhematopoietic cells are fibroblasts (reticular cells), macrophages, and adipocytes. Lymphopoiesis thus occurs in microenvironments conditioned by the blood circulation, by the progressing hematopoietic cells, and by stromal cells producing growth factors such as IL-3, IL-7, and several colony stimulating factors.

Precursors of B and T cells (pre-B, pre-T cells) proliferate at the periphery, then progressively move towards the centre to reach the bloodstream or, as happens to most of them, disappear by apoptosis. B cells evolve from pre-pre-B cells when D–J heavy chain rearrangements are detectable, to pre-B cells containing cytoplasmic μ chains, to immature IgM-expressing cells, and then to mature virgin B cells bearing MHC class II, Fc receptors, and other surface proteins. The last type of cells can migrate to the peripheral lymph organs. If immature B cells recognize an antigen, they undergo programmed cell death; there is thus clonal deletion, called central tolerance.

Pre-T cells do not rearrange their genes inside the bone marrow; they migrate to the thymic cortex to evolve there into mature T cells.

The thymus

The thymus is the primary lymphoid organ providing a microenvironment within which bone marrow-derived progenitors proliferate, mature, and undergo stringent selection procedures ('T cell education') to create a fully functional population of major histocompatibility complex (MHC)-restricted, self-tolerant T cells.

The thymus is a lobulated structure surrounded by a thin fibrous capsule; it is organized into three physically distinct areas, the outer subcapsular zone, the cortex and the inner medulla. The microenvironment is provided by an extracellular matrix and by stromal cells – epithelial cells, macrophages, dendritic cells, fibroblasts – which influence the developing thymocytes via cell-surface and secreted molecules.

Thymic fetal organ cultures, transgenic mice and chimeras have allowed us to propose the following scheme of T-cell differentiation. The most immature $CD4^-$ $CD8^-$ T-cell precursors are located mostly in the subcapsular zone, from which they migrate deeper in the thymus and enter the cortex, which is richly populated by a special type of epithelial cell that expresses high levels of class I and II MHC-encoded molecules. In the cortex, the T cell precursor starts to express both CD4 and CD8 and to rearrange their TCR α and β genes and as a consequence, each thymocyte express a different rearranged TCR-$\alpha\beta$ heterodimer. The specificity of the TCR expressed by a given $CD4^+$ $CD8^+$ thymocyte for a peptide–MHC complex present on another cell (stromal cell) will determine the further evolution of thymocytes. Evidence from fetal thymic organ cultures suggests that a $CD4^+$ $CD8^+$ thymocyte that expresses a TCR with low but measurable avidity for a self-peptide-MHC complex will be 'positively selected': it will be allowed to differentiate into a mature $CD4^+$ $CD8^-$ (if its TCR is specific for a self peptide-class II MHC complex) or a $CD4^-$ $CD8^+$ (if its TCR is specific for a self peptide-class I MHC complex) T cell able to exit the thymus and seed the secondary lymphoid tissues. In contrast, cortical $CD4^+$ $CR8^+$ thymocytes that fail to rearrange their TCR α and β genes or that express TCRs that have no avidity for self-peptide–MHC complexes do not survive and die by an apoptotic mechanism.

Cortical epithelial cells, in particular thymic nurse cells (TNC), are essential for the process of positive selection because they display the self peptide–MHC complexes that are recognized by $CD4^+$ $CD8^+$ thymocytes and also provide essential differentiation factors such as cytokines. Positive selected T cells migrate towards the central region of the thymus, encounter the cortico-medullary junction and finally enter the medulla. The medulla contains almost exclusively mature $CD4^+$ $CD8^-$ or $CD4^-$ $CD8^+$ T cells, that exit the thymus into the blood and seed the secondary lymphoid tissues.

The population of positively selected cortical T cells includes cells that have a strong avidity for self peptide–MHC complexes that are expressed on thymic dendritic cells. These cells are potentially autoreactive and must be eliminated. Data from many laboratories have shown that this occurs via physical deletion involving apoptosis. Experiments with TCR transgenic mice have shown that clonal deletion can occur early at the $CD4^+$ $CD8^+$ stage, either in the cortex or in the medulla. When T cells are eliminated in the cortex, the relevant self-peptide–MHC complex is presented by cortical epithelial cells that produce MHC molecules and peptides that appear to be involved not only in the education of T cells but also in their differentation. In the medulla, thymocytes are deleted by dendritic cells or interdigitating cells that either produce the relevant antigens or pick them up from the serum. This process occurs mainly at the corticomedullary junction where there is a high concentration of interdigitating cells.

The peripheral lymphoid organs

Peripheral, also called secondary, lymphoid organs are compound areas where various immune-response-related processes take place. All major peripheral lymphoid organs comprise:

- areas where antigen can either enter in its native form or be imported by antigen-presenting cells, mainly dendritic cells;
- areas into which lymphoid cells migrate, mostly through high endothelial venules (HEV); in the spleen, the open bloodstream provides access to lymphocytes;
- areas for lymphoids cell–accessory cell interactions for mutual activation;
- areas for T-cell proliferation and maturation;
- areas for B-cell proliferation and maturation;
- areas for effector cell differentiation: plasma cells or sensitized T cells.

Peripheral lymphoid organs are positioned at strategic places:

- the lymph nodes at lymphatic vessel confluence sites;
- the spleen in the bloodstream;
- the mucous associated lymphoid tissues (MALT) beneath epithelia through which antigen can pass: in the gut, respiratory airways, genito-urinary tract.

Effector cells can seed locally within special sites of the peripheral lymphoid organs but can also migrate to settle in the bone marrow, along secretory ducts, beneath epithelia, within synovial membranes, etc.

The lymph nodes

Mouse lymph nodes are numerous but usually small. In volume they reach a few mm^3 but lymph node volume may change according to the stimulation and differ according to the mouse strain. The lymph nodes of nude (nu/nu), SCID, and other immunocompromised mice are hardly detectable.

For experimental purposes certain lymph nodes tend to be preferred: the popliteal node draining the hind limb and posterior foot-pad; the inguinal node draining the skin of the posterior part of the body; the brachial and axillary nodes draining the front limb and anterior foot-pad; the mesenteric (intestinal) lymph-node chain draining the peritoneal cavity and a large section of the intestine; the aortic (lumbar and caudal) lymph nodes located at the fork of the abdominal aorta and draining the tail and genito-urinary area. Of these nodes the popliteal are the smallest, the mesenteric the largest. Dilute India ink can be injected to detect accurately the lymph nodes draining a given area; a grey-black stain appears within a few minutes of its administration and persists for many days. Lymph node dissection can be rendered difficult by surrounding adipose tissue that masks the nodes, which explains why cervical, mediastinic, and other mouse lymph nodes are less well studied. Their architecture includes four histological features:

- conjunctive tissue;
- lymphatic vessels;
- blood vessels;
- the presence of lymphoid and accessory cells.

A thin capsule envelops the nodes and, at the hilum, radiates into dense collagenous trabeculae. The latter envelop the blood vessels and anchor the reticular tissue.

The reticular tissue forms the internal stroma. Its structure and composition determine the organization of particular lymphoid areas, lymph flow, and migration or attachment of lymphoid and accessory cells. It is composed of type III collagen (reticular fibres) associated with proteoglycans and nonfilamentous proteins. Fibroblasts, the reticular cells, surround these fibres and participate in the construction of this meshwork. The reticular tissue not only provides a support but also stores growth factors which influence the defence cells.

Afferent lymphatic vessels pierce the capsule and open into the subcapsular sinus, a space limited by a continuous endothelium on the capsular side and a perforated internal endothelium. This sinus opens into intermediate sinuses which connect with the medullary sinuses. The latter form irregular spaces close to the hilum where they unite into one or a few efferent lymphatic vessels. The afferent lymph first encounters sinusal macrophages and then diffuses within the reticular tissue. The lymph flow is thus centripetal. It carries native antigen, antigen-presenting cells, and a few lymphoid cells. The efferent lymph contains sensitized or quiescent recirculating lymphocytes (rare accessory cells) and is enriched in antibodies.

The blood vessels enter through the hilum; the arteries subdivide to reach the cortical area where they produce capillaries which unite into postcapillary venules mainly located in the paracortex, also called the deep cortex. Postcapillary venules, also called high endothelial venules, are formed of cuboid or pyramidal endothelial cells; they express selectins promoting adherence and transepithelial passage of B, T and other cells (Figure XVI.2.1). This cell traffic depends on the arrival of antigen, the stimulation of lymphoid and accessory cells, and the capacity of endothelial cells to express certain selectins, also called addressins. The main venule leaves the node through the hilum.

The blood vessels thus contribute to creating compartments: the follicles develop around the cortical capillaries, the T-dependent zone around the paracortical HEV, and the medullary cords around larger vessels.

The lymphoid cells congregate in specialized areas: B cells gather in the follicles, T cells in the paracortex, and plasma cells in the medullary cords. Outside these areas of cell concentration, migrating B and T cells are encountered everywhere in the reticular tissue and sinuses.

Lymph nodes are classically subdivided into three areas: the cortex, the paracortex, and the medullary area. Functionally, however, one should consider the units composed

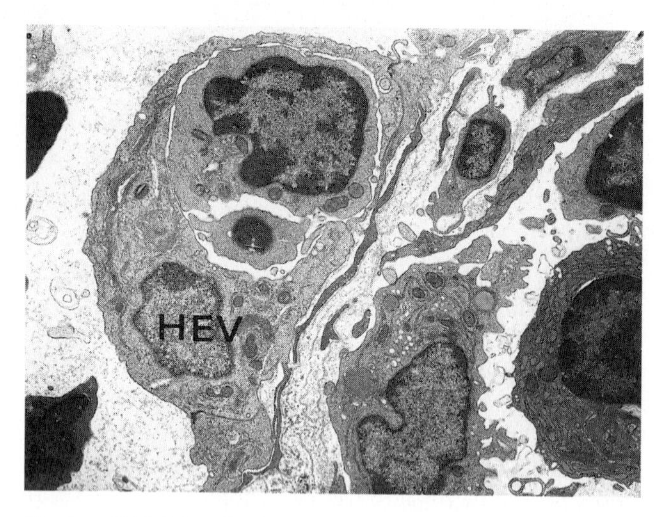

Figure XVI.2.1 Ultrastructural appearance of a high endothelial venule (HEV) with transiting lymphoid cells.

of a follicle and its adjacent T-dependent paracortical area, extending into the medullary cords (Figure XVI.2.2). These units are arranged in front of the openings of afferent lymphatics and thus subdivide the lymph node into radial compartments.

The T-dependent areas appear central to the development of both the cellular and antibody immune responses. Owing to the presence of HEV and their proximity to the sinuses, these areas are crossroads where lymphoid cells encounter native or presented antigen. For example, according to a widely accepted hypothesis, epidermal Langerhans cells loaded with antigen migrate to the

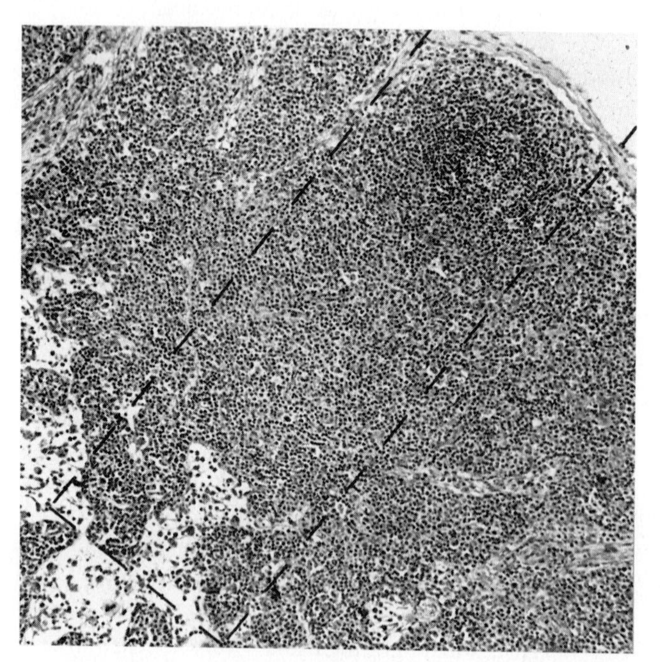

Figure XVI.2.2 Low magnification of a mouse lymph node, clearly showing a functional unit composed of a follicle and a paracortical T-zone. This radially positioned unit extends into medullary cords (Photomicrograph C. Kinet-Denoël).

dermis where they adopt the appearance of classical dendritic cells; they then enter the lymphatics to become veiled cells which pass into the afferent lymphatics, then into the sinuses, and finally into the paracortex to become interdigitating cells. Langerhans cells are skilled in the uptake and processing of antigen; interdigitating cells are 'professional' presenting cells capable of providing naive T cells with antigen and costimulating signals.

During this antigen presentation in the paracortex, T cells are activated. Both lymphoid and accessory cells release chemotactic factors and cytokines that can induce expression of selectins by HEV, thus promoting immigration of new cells. The activated T-cells proliferate or mature into effector or memory T-cells. Some migrate to the lymph follicles or medullary cords to influence B-cell activation, proliferation, maturation, or differentiation. Others leave the nodes via the efferent lymph.

B cells entering the nodes through HEV may recognize native antigen through their surface Ig and receive additional signals to become activated. Some migrate along blood vessels towards the medullary cords to accumulate there and become plasma cells, perhaps induced locally by T or other cells. Others migrate to the lymphoid follicles.

In nude (nu/nu) mice, few cells populate the paracortex, but among them are found CD3 T cells (Figure XVI.2.3 – see colour plate); their follicles are unstimulated, thus without a germinal centre.

Lymphoid follicles appear either as small, quiescent nodules called primary follicles, or as entities called secondary follicles which have enlarged owing to the development of a germinal centre. Primary follicles are composed of IgM^+, IgD^+ resting B cells, some T cells, and underdeveloped follicular dendritic cells (FDC). Activated B-cells can settle inside these primary follicles to proliferate and mature to memory cells or plasma-cell precursors. This induces profound modifications in the primary follicle, whose periphery is pushed into a crescent called corona or mantle zone. A germinal centre develops, composed at first only of centroblasts, then of two areas: a basal area in contact with the T-dependent zone and consisting of centroblasts, and an apical area beneath the mantle zone consisting of centrocytes. The basal part or dark zone and the apical area or light zone comprise the germinal centre which can be further subdivided on the basis of cell phenotypes. These changes undergone by a lymphoid follicle after antigenic stimulation can take 2–3 weeks and constitute the germinal centre reaction. B cells lose their ability to express IgD and acquire a fully activated status to become centroblasts which actively divide. Most enter apoptosis to end in huge local macrophages, the tingible body macrophages, filled with dark apoptotic lymphoid cells. During proliferation, B cells undergo somatic mutations in the IgV genes, which modify their affinity for antigen. After division they appear to migrate to the light zone where they decrease in size and exhibit a cleaved nucleus, features typical of centrocytes. Affinity selection for antigen apparently occurs in this area, as do the isotype switches to different Ig classes.

Errors during rapid cell division and Ig rearrangements or hypermutations followed by loss of antigen recognition explain the massive cell death observed inside the germinal centres. Affinity selection is made possibly by three concomitant phenomena: (1) somatic mutations; (2) loss of IgD expression, and (3) presentation of antigen by follicular dendritic cells (FDC). This cell type only populates lymphoid follicles and becomes fully differentiated inside the light zone (Figure XVI.2.4)

Derived from mesenchymal cells, the FDC acquire during the germinal center reaction a complex meshwork of cell extensions joined by desmosomes, maculae adherences, and gap-junctions. The FDC thus entirely envelop the germinal-centre lymphoid cells, creating special, protected microenvironments; by bearing many various adhesion molecules (ICAM-1, -2, V-CAM, integrins), they connect adjacent cells. Furthermore, the FDC bear abundant C3b receptors and, especially in mice, Fc receptors enabling them to bind to their surface, without endocytosis, huge amounts of immune complexes. To summarize the multiple roles of FDC, one can say that they improve the survival (i.e. the escape from apoptosis) of B and T lymphoid cells, allow affinity selection of B cells, and increase or decrease the proliferative activity of B cells according to the latter's activation status. FDC also appear to decrease Ig secretion and thus to favour maturation of centrocytes to resting memory B-cells. FDC are thus totally different from the bone marrow-derived dendritic cells specialized in antigen presentation to T cells.

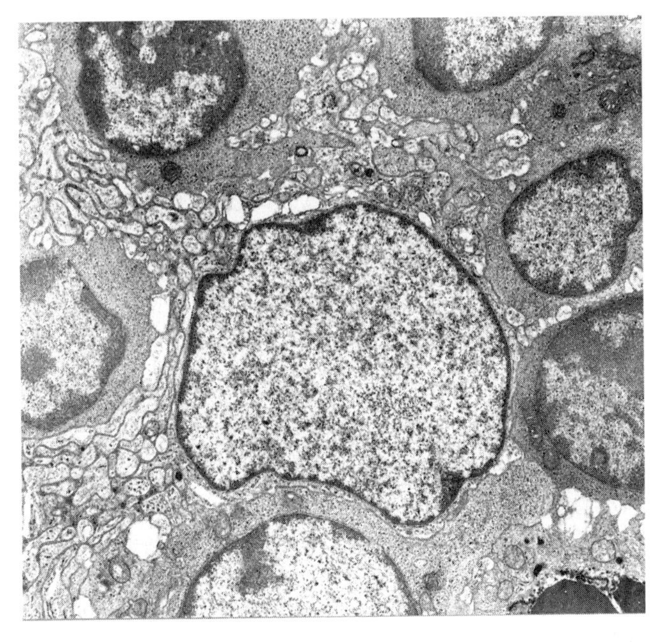

Figure XVI.2.4 Ultrastructural view of a follicular dendritic cell. Its nucleus contains euchromatin and a rim of dense chromatin along the envelope. The cytoplasm is subdivided into dendritic projections which envelop the adjacent lymphoid cells. The electron-dense extracellular material present in the dendritic network consists mainly of immune complexes and complement factors. An apoptotic cell is visible in the left lower corner.

According to Tew's and Szakal's group (1990), FDC differentiate from antigen-transporting cells of medullar origin which, inside the germinal centres, produce immune-complex-coated beds (iccosomes) that are taken up by B cells and presented to T cells. This attractive hypothesis is favoured by many groups. FDC are, however, long-lived sessile cells. During, or at the end of, the germinal centre reaction, they may release dendrites or small cytoplasmic fragments which can be taken up by adjacent cells, but this shedding phenomenon signals a renewal of the cell surface, not specific release of immune complexes.

T cells are present inside the follicles, especially in the germinal-centre light zone. They may account for 5–20% of the total cell population; observations on human lymph follicles suggest that they are of a special type. Most human follicular T cells are $CD4^+$ and express CD57, an NK-cell marker, but do not exhibit cytolytic activity. Their functions are unclear – they only weakly sustain B-cell proliferation or differentiation to Ig-producing cells and do not produce many cytokines. Follicular $CD4^+$ and $CD8^+$ T cells can divide and die locally. It is possible that T cells from the adjacent paracortex enter the follicles, but it also appears that B cells emigrating from the follicles can migrate to the T-dependent zone and there present to T cells the antigen collected from FDC, thus maintaining the memory T cell clones.

Cohorts of B cells escaping from the lymphoid follicles enter the subcapsular sinus to progress towards the efferent lymph. Others move along blood vessels to form medullary cords where B and T cells, plasma cells, macrophages, polymorphonuclears, and mast cells are found. After antigen stimulation, the cords become enlarged by the accumulation of plasma cells and protrude into the medullary sinuses, distending the sinusal endothelium.

A single lymph node may drain an area but usually several are interconnected in a network of superficial or deep nodes, filtering the lymph which collects in the abdominal cisterna chyli and then empties into the left jugulo-subclavian vein via the thoracic duct. Lymph can be obtained by tapping the cisterna chyli.

The spleen

This voluminous organ occupies the upper left area of the abdomen and is built on the bloodstream. In addition to its role in the immune responses, it ensures phagocytosis of foreign elements having entered the bloodstream, maturation of reticulocytes to erythrocytes, production and storage of platelets, and elimination of senescent or altered blood cells, notably erythrocytes. Despite these roles, the spleen is not indispensable. When it is surgically removed, most of its functions are taken over by other organs: lymph nodes, liver, and bone marrow.

General architecture
The capsule of the spleen is well developed and projects inside by trabeculae surrounding arteries and venules. These coarse structures sustain a reticular tissue which extends throughout the organ.

A freshly cut spleen displays two major areas distinguishable by the naked eye: the red pulp filled with erythrocytes and the white pulp rich in lymphoid cells. The red pulp accounts for about 80% of the spleen volume, the white for 10%, and the trabeculae for another 10%. These proportions may be profoundly modified upon antigen stimulation or other modulations of the immune response.

Vascularization
The general splenic anlage is constructed along the blood vessels. The hilar arteries pierce the capsule at different spots and enter the spleen enveloped by trabeculae. These trabecular arteries branch into smaller ones which leave the trabeculae to pass into the white pulp and form the central arteries. The latter leave the white pulp through the marginal zone. Upon entering the red pulp, the arteriolae subdivide into numerous small vessels, the penicillary arteriolae, which open directly into the reticular tissue or are connected to the sinusoids, which are large, irregular venous spaces bordered by a discontinuous endothelium. The sinusoids fuse to form the pulp venules which yield the trabecular veins sheathed in a thick conjunctive envelope.

Small efferent lymphatics are described. They start around the central arteriolae and extend to the trabeculae before leaving the spleen at the hilum.

The white pulp
The white pulp is organized along the central arteriolae and subdivided into T- and B-zones. A periarteriolar lymph sheath is present all around the central arteriolae and is composed mainly of $CD4^+$ and $CD8^+$ T cells. At the periphery, primary and secondary lymph follicles are inserted locally (Figure XVI.2.5 – see colour plate). In the secondary follicles, the germinal centres are always constructed with the dark zone adjacent to the periarteriolar T-cell sheath and the light zone oriented towards the marginal zone. The latter borders on the white pulp; it is a functionally important area (Figure XVI.2.6 – see colour plate). Composed of B and T cells, numerous macrophages, and dendritic cells (Figure XVI.2.7 – see colour plate), it ensures various functions:

- capture and presentation of antigens imported by the blood and released nearby by the open bloodstream;
- activation of B and T cells and their migration to the follicles (B cells) or periarteriolar sheath (T cells);
- accumulation of B memory cells;
- activation and proliferation of T-independent B cells in contact with special macrophages;
- migration of effector and memory cells into the red pulp;
- seeding of plasma cells which form foci or cords at certain stages of the immune response.

The red pulp
The red pulp comprises the sinusoids and the reticular tissue. The sinusoids are large, irregular venous spaces bordered by thin endothelial cells; because of the absence of a basal lamina, the endothelial cells are directly attached to reticular fibres and can thus be pulled apart to open intercellular gaps through which red and white cells can enter or exit.

The reticular tissue harbours many macrophages, lymphocytes, polymorphs, and platelets. Groups of plasma cells are found along the sinusoids, trabeculae, and marginal zones. The strands of reticular tissue between the sinusoids are also called Billroth cords.

The mucosa-associated lymph tissues (MALT)

The MALT system contains more lymphoid cells and produces more Ig than the lymph nodes or spleen. During ontogeny it develops before the other lymph organs and gives rise to the fetal liver and thymus. It thus occupies a special place in the defence system.

The acronym MALT designates all peripheral lymphoid tissues adjoining cavities connected to the external milieu; it thus comprises the tonsils, Peyer's patches, caecum, isolated lymphoid units formed from follicles and adjacent T-zones, and the dispersed effector cells located beneath and inside epithelia. Lymphatics drain these areas, thus uniting the MALT with lymph nodes and hence with the bloodstream.

In the MALT system compartments may be distinguished, such as the upper and lower respiratory tracts, the intestine, the genito-urinary system. The MALT system is thus heterogeneous, even though its different components share common features, such as production of IgA, special intracellular T cells, special mast cells, specific lymphocyte recirculation molecules, etc. Another common feature is that peripheral tolerance appears to be inducible at this level. Multifactorial mechanisms related to antigen presentation, suppressor cell induction, TFG-β secretion, and immune deviation interact to ensure antigen-specific tolerance induced by nasal, oral, or other administration.

Tonsils
Mice develop only small palatine tonsils. They possess, in the hard palate, a small lymphoid anlage formed of follicles and T-zones. The oral cavity and pharynx are drained by lymphatics converging into submaxillary and deep cervical lymph nodes.

Peyer's patches
In the ileal part of the intestine, mice have 9–14 Peyer's patches. These 3–4 mm wide patches may be detected by the naked eye owing to their whitish aspect and the bulge they produce in the intestinal wall at the opposite side of

the mesenteric insertion. Microscopically, they consist of numerous, large follicles extending from the lamina propria through the submucosa. These masses disrupt the mucosae muscularis and distend the muscle layer. Each Peyer's patch is composed of three to nine follicles, so the patches contain a total of 45–70 follicles.

Typical T-dependent zones separate the follicles. T, B, and accessory cells are grouped around high endothelial venules which express selectins for restricted mucosal recirculation of lymphoid cells. Numerous efferent lymphatics drain the patches and converge at the mesenteric lymph nodes. The follicles are usually hyperstimulated, perhaps because of the continuous penetration of protein antigens and organisms, some of which may release polyclonal mitogens (LPS) or superantigens (protein A). Centroblasts, centrocytes, T cells, follicular dendritic cells, and tingible body macrophages are found in the germinal centres. Immune complexes with antigens arising from the intestinal lumen also accumulate there. Most B cells formed in these follicles produce IgA.

Above each follicle, a 'suprafollicular dome' protrudes into the lumen. Owing to follicle aggregation, the villi are disorganized and pushed aside. The domes are covered by a modified intestinal epithelium, the follicle-associated-epithelium; the goblet cells have disappeared and are replaced by M cells interspersed among typical enterocytes (Figure XVI.2.8).

M cells are specialized cells found only along the mucosae above lymph organs such as the tonsils, Peyer's patches, and appendix. They derive from epithelial cells and do not form a classical brush border, but are covered with irregular microvilli, hence their name. M cells harbour clusters of lymphoid cells by emperipolesis, in deep cytoplasmic infoldings (Figure XVI.2.9). Their microvilli bear a cell coat rich in lectins capable of binding glycosylated proteins or microorganisms. These last are endocytosed and delivered by transcytosis, without fusion with lysosomes, to the basal side in the vicinity of lymphoid and accessory cells. The lymphocytes harboured by M cells are CD4 TCRα, β T cells associated with IgM, IgD B cells rich in MHC class II molecules. M cells do not express MHC class II molecules in the way that enterocytes can be induced to do and thus do not present antigens.

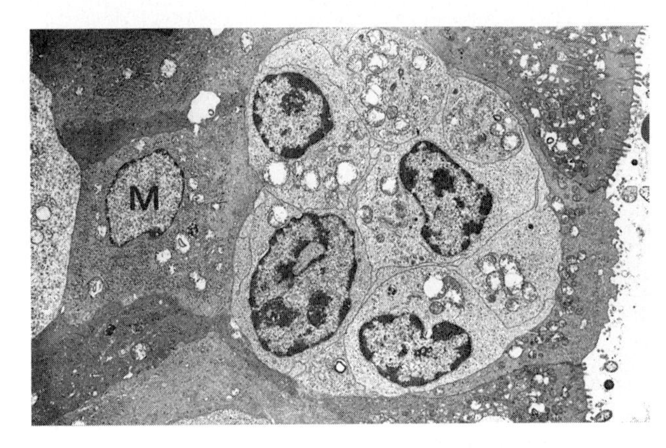

Figure XVI.2.9 Ultrastructural appearance of an M cell (M) harbouring, by emperipolesis, lymphoid cells in its apical cytoplasm.

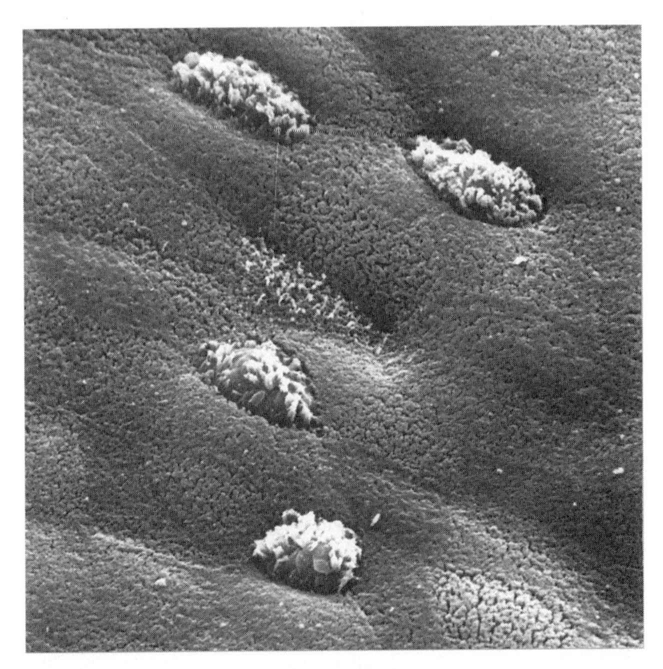

Figure XVI.2.8 M cells revealed by scanning electron microscopy. Their irregular apical microvilli are less well-ordered than those of enterocytes.

To terminally differentiate, M cells need signals from the environment; these signals are induced by antigen stimulation and lymphocyte activation.

Caecum

Mice do not develop an appendix, but at the blind end of the caecal diverticulum, follicles, adjoining T-zones typical follicle-associated-epithelia containing M cells are found. Here too, efferent lymphatics drain the lymphoid area.

The diffuse mucosal defence system

Along the mucosae, the secondary lymph organs occupy strategic positions, notably beneath antigen-penetration sites and afferent lymphatics. Most lymphoid cells, however, are scattered all along the mucosae either inside the epithelia or in the lamina propria. More than elsewhere, the lymphoid cells interact with elements previously viewed as acting only in the natural, nonspecific defence system.

Intra-epithelial cells comprise T cells and typical antigen-presenting dendritic cells. These contribute, along with the M cells, to antigen capture and import. Because they may be responsible for introducing virus particles, notably human immunodeficiency virus (HIV), through the rectal area, they deserve special attention. Intra-epithelial T cells are not, as proposed previously, dying cells desquamating with epithelial cells, but rather functionally active cells. They are mostly CD8$^+$ TCR$\alpha\beta$ or TCR$\gamma\delta$ cells. They appear to kill infected epithelial cells, thus blocking infectious agents at entrance. Intra-epithelial T cells influence MHC expression and cytokine production by enterocytes and stimulate mucus secretion. Because they are anchored to the epithelial cells, it would appear that they do not return to the lamina propria but terminate in the lumen. The lamina propria is rich in various mobile cells. Within the villi, CD4$^+$ T cells are seen in contact with dendritic cells, macrophages, mast cells, and polymorphonuclears.

Along the crypts and glands, especially the secretory ducts, T- and macrophage-associated plasma cells produce dimeric IgA. These molecules bind to specific receptors on the epithelial cells, are conveyed by transcytosis to the lumen, and released there bound to the secretory factor, an IgA receptor fraction protecting IgA from proteolysis. A large proportion of the IgA produced by bone-marrow plasma cells is transported by the blood and transferred through the epithelial cells of the bile, salivary, and pancreatic ducts.

In mice, mucosal plasma cells produce little IgM or IgG. This appears to reduce the risk of local inflammation by activation of the complement cascade. Other data indicate that mucosal T cells, macrophages, and mast cells also display less proinflammatory activity than do such cells elsewhere, but this depends on the location, since respiratory, alveolar, intestinal, and other mucosal areas exhibit their own particularities of defence.

3. Immunoglobulins

General structure of immunoglobulin

Physical and chemical aspects

Murine immunoglobulins (Igs) like Igs from other mammals are made up of four polypeptide chains: two identical heavy chains (H) of 55 kDa and around 450 amino acids and two identical light chains (L) of 25 kDa and around 220 amino acids. They are linked together by noncovalent interactions and stabilized by disulfide bonds. Allelic exclusion (see below) ensures that B cells express only one functionally rearranged heavy chain gene and one functional light chain gene (monospecificity of B lymphocyte).

Each chain contains both a constant and a variable region. The constant region is located at the C-terminal part of the chain (C_H and C_L). The mouse has five classes of Ig determined by the heavy constant region and termed isotypes, that is, IgM, IgD, IgG, IgE and IgA. The IgG class is further divided into four subclasses, IgG1, IgG2b, IgG2a and IgG3. The constant regions are constant only within isotypes, the different isotypes share only about 30% sequence identity. The H constant regions are constituted by three or four distinct domains termed C_H1 to C_H4 while the L constant region is formed by one domain. The H constant regions determine most of the metabolic and effector properties of the Ig. Physical and chemical properties of murine Ig are presented in Table XVI.3.1. Different Ig fragments can be obtained with proteolytic enzymes. Papain cleaves IgG above disulfide bond leading to two types of fragments, Fab (fraction antigen binding) with monovalent binding activity and Fc (fraction crystallizable) which is the H constant region without C_H1. A Fab is formed by the light chain and the V_H–C_H1 domain (Porter, 1958). Pepsine gives F(ab)′2 fragments with bivalent binding activity (Nisonoff *et al.*, 1960). These fragments can precipitate antigens. If F(ab)′2 fragments are reduced, two univalent Fab′ are obtained (Palmer and Nisonoff, 1964). The variable (V_H–V_L dimer) and the constant (C_H1–CL) parts of Fab are named Fv and Fb respectively. Fab and Fc are connected by a hinge region. This hinge region permits a flexible conformation for the C_H1 and C_H2 domains allowing the antibody to bind antigens at multiple sites to increase the strength of interaction. It contains the disulfide bonds that covalently link the two identical heavy chains.

The *N*-terminal region exhibits great differences between antibody molecules and is thus termed variable region (V_H and V_L). Assembly of H and L variable regions form a functional domain that provide ligand-binding specificity and idiotype. The variation of V_H and V_L is mainly concentrated in three particularly zones termed hypervariable regions or HVRs, while the remainder of the sequence shows the same level of variation as other proteins. The hypervariable regions of each H and L chains interact together to form the antigen-combining site (paratope) and are consequently named complementarity-determining regions, or CDRs. Regions surrounding the CDRs are termed framework regions (Frs) and present less variation than CDRs. Certain Fr residues are very conserved among variable regions. They appear crucial to folding, stabilization and dimerization of the domain. Detailed descriptions of the Ig structure are reviewed in Padlan (1996), McConnell and Martinez-Yamout (1996), Novotny and Bajorath (1996), Huston *et al.* (1996), Carayannopoulos and Capra (1993). Original references for the data can be found in those sources.

The variable regions are encoded by distinct gene segments, i.e., three gene segments namely V (variable), D (diversity) and J (joining) participate to the heavy variable region while only V and J gene segments encode for the light variable region. An elaborate genetic machinery recombines these genes. The CDR1 and CDR2 of H and L chains are encoded by the V_H and V_L genes. The CDR3 is mainly encoded by D and J gene segments.

Biological properties

During immune responses against different antigen challenges and infections, distinct isotypes dominate

Table XVI.3.1 Physical and chemical properties of different murine immunoglobulin classes

Property	*IgM*	*IgD*	*IgG*				*IgE*	*IgA*
Heavy chain	μ	δ	γ				ϵ	α
Usual molecular form	Pentamer	Monomer	Monomer				Monomer	Monomer–dimer
Other chains	J	None	None				None	J (dimer) Secretory
Light chain	95% of κ light chain, 5% of λ light chain							
Molecular mass (kDa)	900	180	150				187	160–400
Sedimentation constant (S)	19	7–8	6–7				8	7–13
Subclasses	None	None	G1	G2b	G2a	G3	None	None
Serum level (mg/ml)[a]	0.2	<0.001	0.5–2.5	0.2–0.6	0.4–1.5	0.1–0.2	0.001	0.4–1.3
Serum half life (days)[b]	2	Low level of circulating molecules	6–8	4–6	6–8	6–8	0.5	17–22 h
Classical complement activation	+	–	±	+	+	+	–	Alternative complement activation
Distribution in body fluids	Intravascular	Secreted at low level Receptor expressed with IgM on the surface of mature B lymphocytes	Intravascular, placental transfer				Mucosal surfaces in the gut and airways	External secretion Mucosal surfaces Colostrum, milk
Binding to								
A protein	± (Copurification with IgA)	–	± (Low affinity)[c]	++	++	++	–	±
G protein	–	–	+++	++	++	++	–	–

[a] These values are indicative only, they can vary from strain to strain and the status of immune system.
[b] Values based on the half lives of serum immunoglobulin in adult mice (Vieira and Rajewsky, 1988).
[c] Under alkaline high salt conditions.

because they have different effector functions (reviewed by Esser and Radbruch, 1990; Snapper and Finkelman, 1993 and references therein).

IgMs are secreted after a first antigenic contact. They generally express germline variable regions without somatic mutations and therefore they have a low affinity for antigen. Nevertheless, their pentameric form, leading to 10 antigen-binding sites can allow to bind with high avidity to an antigen presenting multiple similar epitopes. IgMs are very effective in complement activation and in enhancement of the opsonization of particulate antigen.

IgDs plasma cells are rare. Membrane IgD molecules are coexpressed with IgM in the surface of mature B cells. Activation of B cells by antigens leads to the down-regulation of the IgD expression. Despite the capacity of surface IgD to bind antigen, IgD function remains an enigma since various experiments report contradictory potential functions (reviewed by Colle *et al.*, 1990; Melchers *et al.*, 1995; Norvell and Monroe, 1996).

IgGs represent the most abundant antibodies in serum and their production is increased in B cells activated by antigens. In viral infection, IgG2a and to a lesser degree, IgG1 are dominant. IgG2a can efficiently activate the complement cascade and is a mediator of antibody-dependent cytotoxicity by binding to specific Fc receptors on macrophages. IgG1 is dominant in parasitic infections. It does not activate the complement very efficiently but can stimulate the phagocytosis. Moreover the ability of IgG1 to mediate mast cell degranulation may confer increased resistance to nematode parasites. Soluble protein antigens stimulate predominantly IgG1 responses. IgG2b and IgG3 are mainly induced by T-independent antigens such as carbohydrates. They activate the complement cascade. IgG3 plays an important role in antibacterial responses. IgG3s are very efficient in promoting phagocytosis and have the ability to self-aggregate after binding to antigen with multiple identical epitopes (Greenspan and Cooper, 1992). This property allows them, like pentameric IgM, to form strong avidity interactions with bacterial cell wall carbohydrate antigens because such antigens tend to induce low-affinity antibody responses. All IgGs can cross the placenta through binding to a specific receptor. They may play a protective role in neonates.

IgE production is linked to allergy. Mast cells and basophils which express high-affinity receptors for IgE (FcεRI) bind monomeric IgE in the absence of antigen. The presence of specific antigens (or allergens) induces the aggregation of IgE–FcεRI complexes leading to the secretion of histamine and other chemicals mediating the immediate hypersensitivity response (Snider *et al.*, 1994). Parasitic infection also induces IgEs. The binding of IgE via Fc-specific receptors on eosinophils and macrophages helps in the defence against parasites.

IgAs in the serum are in monomeric form, although secreted IgA are dimers linked together by a J chain. The association of IgA dimers with a molecule of secretory component highly glycosylated facilitate transport of this isotype into the gut lumen. Moreover IgA molecules have high resistance to digestion by proteolytic enzymes. IgAs play a key role in mucosal immunity (reviewed by Mazanec *et al.*, 1993; Kramer and Cebra, 1995). IgAs form an immune barrier by preventing the adherence and absorption of antigens, by neutralizing intracellular microbial pathogens directly within epithelial cells and by eliminating immune complexes locally by binding to the antigens and excreting them into the lumen. They are also found in extracorporeal fluids such as mucus and milk and thus, IgAs may play a role in neonatal immunity. They pass through epithelial cells and basal membrane via polymeric Ig receptors on the surface of mucosal epithelial cells. They lack the ability to fix complement by the classical pathway but they fix complement efficiently by the alternative pathway. IgA molecules are also able to induce eosinophil degranulation (role in antiparasite responses) and can serve as an opsonin for phagocytosis through a specific Fcα receptor. It is important to note that some functions ascribed to IgA molecules have been predicted from *in vitro* experiments.

Immunoglobulin genetics

During B-cell ontogeny, functional Ig genes are generated by somatic juxtaposition of gene segments which are separated in the germline (reviewed by Kofler *et al.*, 1992). H chains are encoded by gene segments termed variable (V), diversity (D), joining (J) and constant (C). The V_H locus that contains these gene segments is located on mouse chromosome 12. L chains result from the somatic recombination of V, J and C gene segments. There are two classes of L chains designated as kappa (κ) and lambda (λ). These gene families are located on separate chromosomes (6 (κ) and 16 (λ)). In mice, only 5% of antibodies possess a λ L-chain.

Heavy chains

V_H subgroup and family

Based on amino acid sequence similarity, Kabat *et al.* (1991) have divided murine V_H gene segments into major subgroups termed I, II, III and miscellaneous. These protein subgroups were further subdivided into 14 distinct V_H families based on nucleic acid sequence relatedness (more than 80% of similarity between members within a family and less than 70–75% similarity between families) (Brodeur and Riblet, 1984; Brodeur *et al.*, 1988; Kofler, 1988; Reininger *et al.*, 1988; Pennel *et al.*, 1989; Tutter *et al.*, 1991; Kofler *et al.*, 1992). Based on sequence similarity, the 14 V_H gene families can be assigned to three related groups suggesting that the V_H gene families evolved from three ancestral V genes 300 million years ago (Tutter and Riblet, 1988) (see Table XVI.3.2).

Table XVI.3.2 V_H classifications and V_H gene repertoire of mouse

	Protein subgroups		
Families[a]	*A*[b]	*B*[c]	*Complexity*[d]
Q52 (V_H2)	IB	I	15
3660 (V_H)3	IA		5–8
3609 (V_H8)	IB		7–10
CH27 (V_H12)	IA		1
J558 (V_H1)	IIA, IIB, VA, Miscellaneous	II	60–1000
VGAM 3.8 (V_H9)	IIA		5–7
SM7 (V_H14)	IIB, IIC		3–4
X-24 (V_H4)	IIB	III	2
7183 (V_H5)	IID, Miscellaneous		12
J606 (V_H6)	IIIC		10–12
S107 (V_H7)	IIIA		2–4
MRL-DNA4 (V_H10)	IIID		2–5
CP3 (V_H11)	Miscellaneous		1–6
3609N (V_H13)	IIIC		1

Table modified from Kofler *et al.* (1992).
[a] V_H gene families 1–7 (Brodeur and Riblet, 1984), 8 and 9 (Winter *et al.*, 1985), 12 (Pennell *et al.*, 1989), 13 and 14 (Tutter *et al.*, 1991), 11 (Reiniger *et al.*, 1988), 10 (Kofler, 1988).
[b] According to Kabat *et al.* (1991).
[c] According to Tutter and Riblet (1988).
[d] For references see footnote [a], Livant *et al.* (1986), Dzierzak *et al.* (1986), Perlmutter *et al.* (1984) and Siu *et al.* (1987).

V_H segment organization
Murine VH families are generally organized into clusters of related VH gene segments. Nevertheless some VH family members are interspersed, particularly J558 (V_H1) and 3609 (V_H8) at the 5′ end, and Q52 (V_H2) and 7183 (V_H5) at the 3′ end. Figure XVI.3.1 represents a schematic map of the mouse IgH locus.

V_H gene germline repertoire
The size of VH gene repertoire is relatively conserved between different inbred strains of mice. Nevertheless, certain families, particularly V_H1, present some important inter-strain differences (Meek *et al.*, 1990). Table XVI.3.2 illustrates the size of different V_H families. Only the size of the largest family, V_H1 is still controversial (estimates vary from 60 to 1000 gene segments) (Brodeur and Riblet, 1984; Livant *et al.*, 1986). Estimation of V_H1 usage in mitogen-stimulated splenocytes predict a V_H1 size closer to 60 than 1000. Using amplified cDNA and genomic libraries, Gu *et al.* (1991) estimate that the size of V_H1 is in the order of 100 genes. The relative under-representation of this V_H1 family in the expressed repertoire could be explained by the presence of multiple nonfunctional or identical V_H1 genes at the germline level (Blankenstein *et al.*, 1987).

D_H and J_H gene segments
V_H, D_H, J_H and C_H gene segments are arranged in contiguous but separated clusters on chromosome 12 (Sakano *et al.*, 1981; Kurosawa and Tonegawa, 1982; Wood and Tonegawa, 1983). Twelve D_H gene segments belong to three families (DQ52, DSP2 and DFL16) determined on the basis of coding and flanking region relatedness (Ichihara *et al.*, 1989). D_H gene segments other than DQ52 are clustered in the 60 kb region located between V_H and J_H gene segments. In BALB/c mice, DFL16.1 has been identified as the most upstream D_H gene segment characterized to date. A single DQ52 gene segment has been mapped to 750 bp 5′ of JH1. Recently, Feeney and Riblet (1993) identified a new functional D_H gene segment (DST4) unrelated to any of the known D_H families. DST4 gene segment has been located between 3′ end DSP2.8 and DQ52 gene segments. Four J_H gene segments termed J_H1–J_H4 are located approximately 7 kb upstream of C_H gene clusters.

C_H locus
Organization of C_H genes The general organization of the C_H gene locus is similar among laboratory strain mice. It consists of eight genes that cluster in a 200 kb region (Figure XVI.3.2a) (Shimizu *et al.*, 1982). Rats and mice present a striking homology in C_H gene organization. Murine $C_\gamma1$ gene is homologous to rat $C_\gamma2a$ and $C_\gamma1$ genes and murine $C_\gamma2b$ and $C_\gamma2a$ genes are close to rat $C_\gamma2b$ gene (see elsewhere in this book). Murine C_H locus differs from human C_H locus by the number of genes (8 and 9 respectively) and by the absence of pseudogenes (two in human). Rabbits have only one IgG isotype and 13 IgA isotypes. A gene encoding for rabbit C_δ has not been identified.

Structure of C_H genes. The heavy chain proteins can be expressed as secreted forms or membrane bound forms. Secretory forms of Ig are encoded by three exons (α) or four exons (μ, γ and ε) which correspond respectively to a functional and structural unit of the H chain, namely a domain or a hinge region (reviewed by Zhang *et al.*, 1995):

- C_δ is encoded by two exons separated by a hinge exon.
- all the C_γ genes contain three exons and a short hinge exon between C_H1 and C_H2. The greatest differences between the various C_γ genes are concentrated in the hinge exon.
- the C_α gene is formed by three exons with a hinge region encoded by the C_H2 exon.

The C_H exon nearest the 3′ end encodes the C-terminal part of the secreted form.

The Ig secreted form has a hydrophilic tail. Secreted IgM and IgA have additional sequences at the C-termini allowing intermolecular interactions and resulting in multimeric Ig molecules. The interactions are stabilized by a peptide called the J chain.

In the membrane-bound forms, one (α) or two (other isotypes) separate exons encode the hydrophobic transmembrane (26 amino acids) and the basic charged intracytoplasmic segments. These 'mini-exons' are located

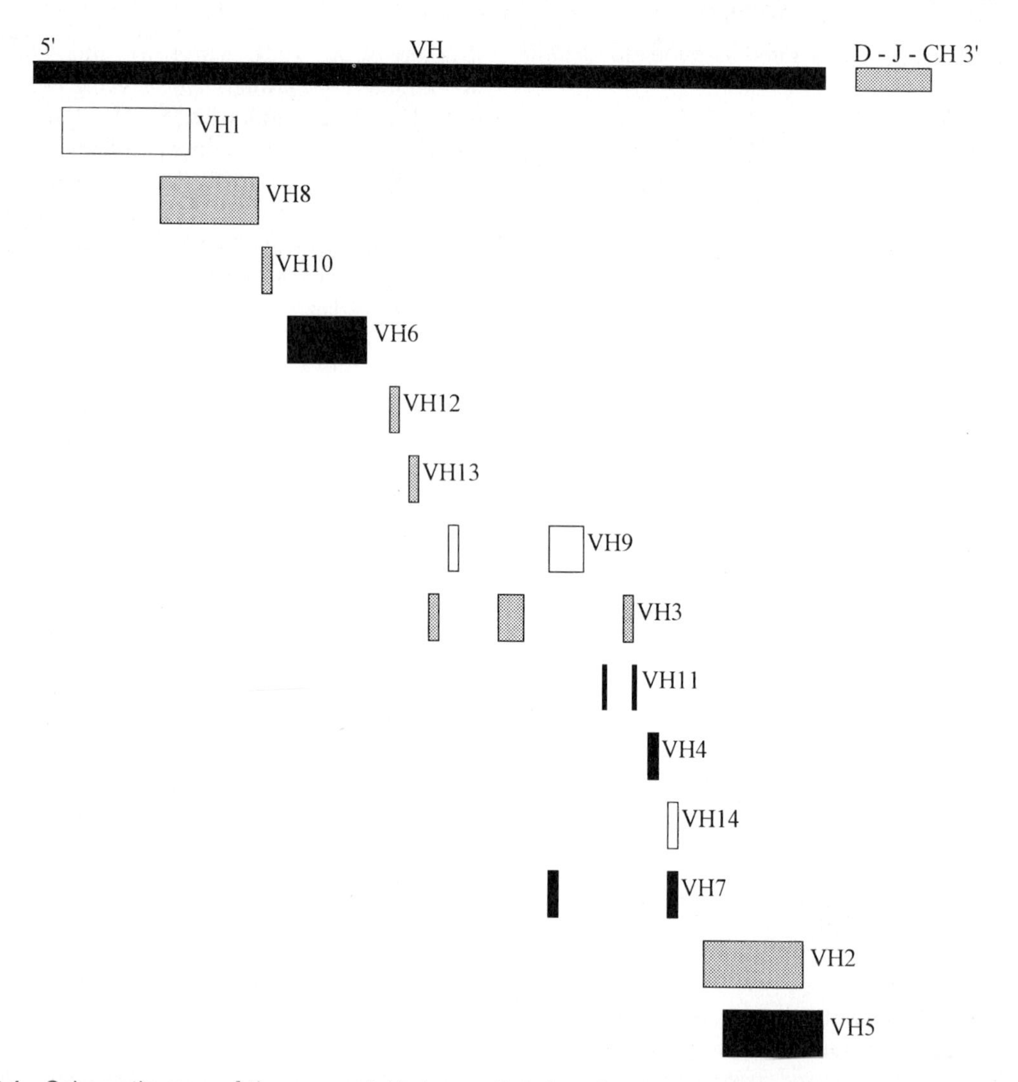

Figure XVI.3.1 Schematic map of the mouse IgV_H locus. Relative chromosomal position of the 14 mouse V_H families. Mapping information is from Meek *et al.* (1990), Pennel *et al.* (1989), Tutter *et al.* (1991) and Brodeur *et al.* (1988). Grey rectangles represent V_HI protein subgroups, open rectangles are V_HII protein subgroups, and filled rectangles are V_HIII protein subgroups. (Modified from Honjo and Matsuda, 1995.)

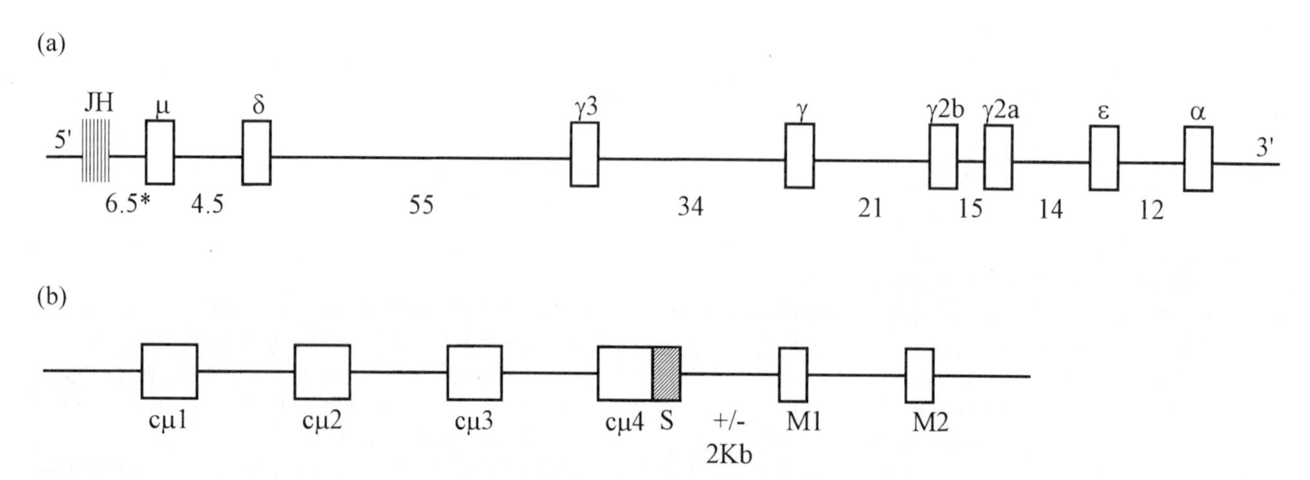

Figure XVI.3.2 (a) Heavy chain constant region locus of mouse (modified from Shimizu *et al.*, 1982); *approximate intergene distance (kb). (b) Germline configuration of C_μ gene: S, nucleotide sequence encoding the 20 *C*-terminal residues of the secretory form; M1 and M2, membrane exons encoding the transmembrane and intracytoplasmic segments.

downstream of the major exons (Figure XVI.3.2b). The intracytoplasmic segment vary from one isotype to another one: 27 residues for C_γ and C_ε chains, 14 residues for C_α chain and two residues for C_μ and C_δ chains. The shortness of the intracytoplasmic segments is incompatible with any enzymatic activity such as transduction of triggering signal of the antigen–antibody interaction. The transmembrane segments of all H chains share extensive homology and are involved in interacting with a heterodimer of membrane proteins namely Igα and Igβ. These two additional proteins possess long cytoplasmic regions (61 and 48 resides) and thus they play a major role in mediating signal transduction by the Ig receptors in B cells (see below).

Kappa light chains

V_κ subgroup and family

According to Strohal *et al.* (1989) and Kofler and Helmberg (1991), V_κ gene segments can be divided into 19 families which can be organized into seven phylogenetically related groups (Kroemer *et al.*, 1991) (see Table XVI.3.3).

V_κ classification has met with considerable difficulties since most V_κ gene families display higher similarities (75 to 79%) to other families. All these families form related groups (V_κ9A, 9B, 10, 11, 32, 33/34 and 38C). Few V_κ gene segments fall into typical families with > 80% similarity within and < 70–75% similarity between families (V_κ 4/5, 12/13, 20, 21, 23, RF) (reviewed by Kofler *et al.*, 1992).

Table XVI.3.3 V_κ classification and V_κ gene repertoire

V_κ gene family[a]	*V_κ protein group*[b]	*V_κ family complexity*[c]
9A	V	4–9
9B		2
10		2–3
11		4–6
12/13		2–8
32		4–8
33/34[d]		1–3
38C[e]		?
RF		0–1
21	III	6–13
23	V	2–4
8	I	5–16
19/28	V	4–6
22	I	1–2
4/5	IV, VI	25–50
1	II	4–6
2		1–6
24/25		6
20	VII	5–7

Table modified from Kofler *et al.* (1992).
[a] According to Strohal *et al.* (1989) and Kofler and Helmberg (1991).
[b] According to Kroemer *et al.* (1991).
[c] See footnote [a], Shefner *et al.* (1990) and Valiante and Caton (1990).
[d] Two groups (D'Hoostelaere and Klinman, 1990; Valiante and Caton, 1990) have independently described this family and termed it V_κ33 and V_κ34 respectively. Kofler and Helmberg (1991), have renamed V_κ33/34.
[e] This family has also been termed V_κ31 by Schlomchik *et al.* (1990).

V_κ segment organization. D'Hoostelaere and co-workers (1988) have suggested a V_κ gene order based on analyses of known Ig_κ recombinant mice and mouse strains with different Ig_κ haplotypes (see below and Figure XVI.3.3). Including more recent data (Kofler *et al.*, 1989; D'Hoostelaere and Klinman, 1990; Huppi and D'Hoostelaere, 1991) Kofler *et al.* propose another map based on the assumption that most V_κ gene families reside within individual clusters, since that has been confirmed for the V_κ21 family (reviewed by Kofler *et al.*, 1992).

V_κ gene germline repertoire. The entire V_κ repertoire in the germline is not precisely known. According to Kofler *et al.* (1989) using RFLP criteria, the entire V_κ germline repertoire would be formed by 160 genes.

J_κ and C_κ gene segments. The Ig_κ complex locus of inbred strains contains only one C_κ and five joining gene segments namely J_κ1 to J_κ5 (numbering from 5′ most J gene segment) located at 3.7 to about 2.5 kb 5′ of the C_κ region gene (Max *et al.*, 1979; Sakano *et al.*, 1979). J_κ3 encodes an amino acid sequence that has never been observed on a κ chain and it is assumed that J_κ3 represents a nonfunctional gene segment.

Lambda light chain

The low expression of Ig with the λ isotype in mice has

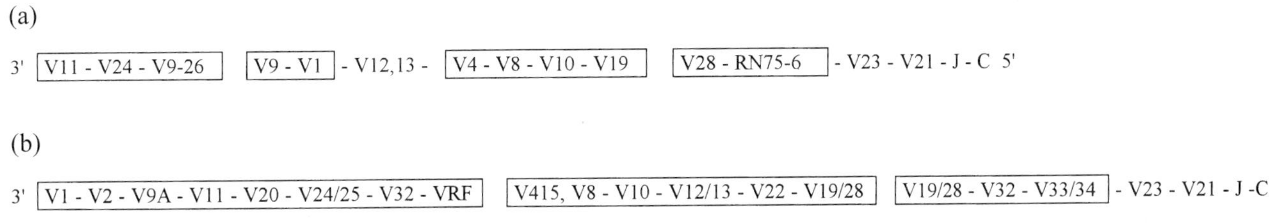

(a)

3′ [V11 - V24 - V9-26] [V9 - V1] - V12,13 - [V4 - V8 - V10 - V19] [V28 - RN75-6] - V23 - V21 - J - C 5′

(b)

3′ [V1 - V2 - V9A - V11 - V20 - V24/25 - V32 - VRF] [V415, V8 - V10 - V12/13 - V22 - V19/28] [V19/28 - V32 - V33/34] - V23 - V21 - J -C

Figure XVI.3.3 The mouse IgV_κ locus maps. Two alternative maps are represented according to D'Hoostelaere *et al.* (1988) (a) and Kofler *et al.* (1992) (b). The order of V_κ families in boxes is unknown.

complicated the analysis of λ chains. Thus, a major part of the knowledge of λ chains derives from spontaneous myelomas or hybridomas stimulated by antigens known to elicit a λ response. The λ gene family found in most laboratory mouse strains is one of the smallest immunoglobulin gene systems (chromosome 16). In contrast to the V_κ genes which are represented by several genes, only three V_λ functional genes have been found in BALB/c mice ($V_\lambda 1$, $V_\lambda 2$, $V_\lambda x$). Furthermore in contrast to the single mouse C_κ gene, four nonallelic mouse C_λ genes have been identified, $C_\lambda 1$, $C_\lambda 2$, $C_\lambda 3$ and $C_\lambda 4$ respectively. $C_\lambda 1$ is the most abundant among λ L-chains and $C_\lambda 3$ is the less utilized. The sequence of $C_\lambda 4$ does not contain the termination codons that would necessarily render it nonfunctional (reviewed by Eisen and Reilly, 1985).

V_λ gene germline repertoire

Each C_λ segment has an associated upstream J_λ segment (about 1.3 kb 5′ of the C). The J–$C_\lambda 3$ and J–$C_\lambda 1$ genes are organized in one cluster about 3 kb apart with the $V_\lambda 1$ gene lying about 19 kb upstream. A second C_λ cluster lying about 130 kb upstream of the $C_\lambda 3$–1 locus is constituted by J–$C_\lambda 2$ genes and the unexpressed genes J–$C_\lambda 4$. This cluster is flanked by two upstream V_λ genes, $V_\lambda 2$ and the rarely used $V_\lambda x$ (this last gene presents particular features – it encodes for a segment that has four additional amino acids in the third hypervariable region). The gene organization explains the common expression of $V_\lambda 2$ or $V_\lambda x$ with $C_\lambda 2$ and $V_\lambda 1$ with $C_\lambda 1$ or $C_\lambda 3$ (Figure XVI.3.4) (Storb *et al.*, 1989). The rearrangement of λ segments takes place within each V_λ J_λ C_λ cluster, thereby defining a unit of recombination (Sanchez *et al.*, 1991). These two clusters of J–C genes in mouse appear to arise by duplication of an ancestral V–J–C_λx–J–C_λy. It should be noted that some wild mice have larger numbers of λ genes.

$V_\lambda 5$ gene. A few years ago, a new murine gene, $\lambda 5$, was described (reviewed by Melchers *et al.*, 1993). The $\lambda 5$ gene has extensive homology to mouse λ genes and it is selectively expressed in pre-B-cell lines (see below).

Mice that express low levels of λ chains. Certain mouse strains, such as SJL, express lower serum levels of $\lambda 1$ chains than those found in most inbred strains like BALB/c or C57BL/6 (reviewed by Selsing and Daitch, 1995). The level of $\lambda 1$ molecules is about 50 times lower than the levels observed in BALB/c or C57BL/c. The SJL defect is $\lambda 1$-specific since K-Knock-out mice carrying the SJL locus have normal expression at the linked $\lambda 3$ gene. The $\lambda 1$ SJL molecules have a valine at amino acid position 155 instead of a glycine in $\lambda 1$ BALB/c molecules. This substitution results in a *Kpn*I cleavable site in the BALB/c $C\lambda 1$ gene which seems to be lost in SJL. Studies on recombinant-inbred and random-bred strains have indicated correlation of the $C\lambda 1$ *Kpn*I site polymorphism with low $\lambda 1$ serum levels, therefore the Gly–Val polymorphism appears to be directly involved in the low $\lambda 1$ phenotype.

The mechanism responsible for the low $\lambda 1$ expression in SJL mice are not completely understood. Because newborn SJL mice have approximately the same number of $\lambda 1$-bearing B cells as newborn BALB/c or C57BL/6 mice, the low $\lambda 1$ phenotype at the serum and B-cell levels emerge in the adult SJL mice. The low levels of $\lambda 1$-bearing IgG antibodies in SJL mice are not caused by a defect in the association of SJL $\lambda 1$ chains with H chains since hybridoma $\lambda 1$ chain combines with the $\gamma 2b$ heavy chain, suggesting that $\lambda 1$ chains in SJL mice can be associated with H chains other than μ.

Models proposed to date are based on an effect of the Gly–Val exchange on the ability of $\lambda+$ B cells to be stimulated by antigen binding or on regulation (perhaps by T cells) of λ-producing B cells in SJL and SJL like strains. Some mechanisms unrelated to the Gly–Val exchange may also regulate the low serum levels of λ chains expressed by some wild mice, since the mice exhibit $\lambda 1$ genes similar to those in BALB/c but do not show the *Kpn*I site polymorphism, suggesting a SJL-like allele.

Immunoglobulin gene polymorphisms

Although Ig loci have the same set of characteristics among inbred mouse strains, each strain has particular gene segments that encode Ig molecules with specific determinants recognized by specific antisera. These determinants have been termed either allotypes for constant regions or idiotypes roughly correlated with binding specificity. Historically, the study of idiotype expression has revealed the genetic polymorphism of immunoglobulin genes (see for example Tassignon *et al.*, 1993).

The entire complex of constant region and variable

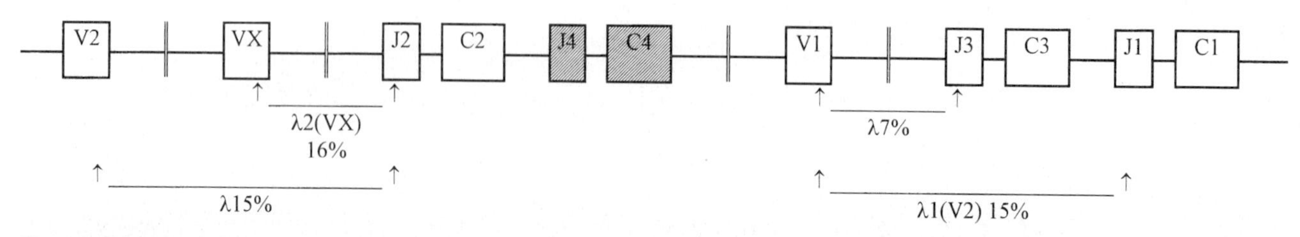

Figure XVI.3.4 Organization of the murine λ locus. Shaded boxes represent pseudogenes. All segments display the same transcriptional orientation. The levels of the most frequent combinations are indicated by lines under the gene segments (modified from Sanchez *et al.*, 1996).

Table XVI.3.4 Inbred mouse strains and their Igh-V, Igh-C and IgK haplotypes

				Igh-V[a]		
	Igh-V[c]	*Igh-C*[a]	*IgK*[b]	V_H*1-6-7-5*	V_H*3*	V_H*2*
BALB/c	a	a	c	a	a	a
SM	a	b	c	a	a	a
C58/J	a	a	d	a	a	a
DBA/2 DBA/1	c	c	c	c	c	c
RIII	a	g	c	a	a	a
PL	j°	j°	a	J	J	J
RF	c	c	a	c	c	c
NZB/J	d	n	b	J	J	d
AKR	d	d	a	J	J	d
CBA/C3H	J	J	c	j°	j°	j°
CE	f	f	?	f	f	f
BSV9	g	g	c	g	f	g
C57/BL6	b	b	c	b	b	b
SJL	b	b	e	b	b	b
A/J A/He	e	e	c	e	e	e
AL/N	e	o	c	e	e	e
SWR	g	p	c	g	f	g

[a] According to Brodeur and Riblet (1984).
[b] According to Kofler *et al.* (1989) and D'Hooselaere *et al.* (1988).
[c] According to Riblet *et al.* (1986) and Kofler *et al.* (1985).

region genes is called 'complex locus allele' or 'haplotype'. RFLP and nucleic acid sequence analyses have helped greatly in determining the polymorphism among genes coding for Igs in inbred mouse strains. Table XVI.3.4 illustrates major Igh-V, Igh-C and IgK haplotypes. Solin and Kaartinen (1992) have pointed out the existence of polymorphism at the Igh-J locus also among mice. Briefly, five haplotypes have been described:

- a for BALB/c, C58/J, RIII, DBA/2, RF, NZB/J, AKR
- j for CBA/C3H
- f for CE
- e for A/J
- b for C57BL, SJL

The Igh-D locus presents also a polymorphism between mouse strains (Brodeur and Riblet, 1984; Trepicchio and Barrett, 1985):

- a for BALB/c, RIII, C58/J, NZB/J, AKR
- c for DBA/2, RF
- e for A/J, AL, A/He
- g for BSVS, SWR, C3H, PL
- f for CE

B-cell receptor complex (BCR) structure and co-receptors

Mature B cells remain in a quiescent state until they encounter antigen in an appropriate context. Binding of antigen to Ig molecules on the cell surface of mature B lymphocytes leads to proliferation of clones of antigen-specific lymphocytes and to maturation into effector cells within these clones. It was originally supposed that this phenomenon was mediated only by the binding of antigen to membrane-bound Ig molecules (mIg). A great deal of research has revealed a more complex situation in which B cells can discriminate between simple ligands and those with a more complex nature and in which different surface molecules named co-receptors act negatively or positively on the ongoing B-cell proliferation and differentiation (reviewed by Cambier *et al.*, 1994).

BCR structure

Most of mature B cells co-express on their cell surface two classes of antigen receptors, mIgM and mIgD. During an immune response, class switching occurs and leads to the appearance of memory B-cells that express other classes of mIg (IgG, IgE and IgA). As mentioned above, the small cytoplasmic region of mIg molecules is unable to mediate transducing signals and therefore these mIg molecules constitute only a part of the antigen–receptor complex (BCR).

BCRs are noncovalently associated with disulfide-linked Igα (CD79a) and Igβ (CD79b) proteins (reviewed by Reth, 1992, 1994; Kim *et al.*, 1993; Pleiman *et al.*, 1994). The symmetry of Ig molecule leads one to assume that each mIg molecule is bound by two Igα/Igβ heterodimers. These heterodimers are also expressed with Ig-like complex on pre-B cells.

The Igα–Igβ proteins are structurally related and are encoded by the Ig superfamily genes mb-1 and B29 respectively. They are glycoproteins containing a single extracellular Ig-like domain, a single transmembrane domain and a cytoplasmic tail. Table XVI.3.5 depicts the

Table XVI.3.5 Characteristics of Igα and Igβ

				Amino acid of			
Protein	*kDa*	*Amino acids*	*N-Linked CHO*	*Extra cellular part*	*Transmembrane part*	*Cytoplasmic part*	*Isoelectric points*
Igα (CD79a)	34	220	2	109	22	61	4.7
Igβ (CD79b)	39	228	3	129	22	48	3.7

major characteristics of murine Igα–Igβ. The extracellular domains contain two (Igα) and three (Igβ) *N*-linked glycosylation sites and show all the hallmarks of proteins belonging to the Ig superfamily, such as two conserved cysteine residues forming the intradomain disulfide bond and the conserved tryptophan residue which is important in filling the hydrophobic space inside an Ig domain. The extracellular domains are followed by an extracellular spacer which contains a cysteine residue allowing the disulfide bond between the two proteins. Both transmembrane parts of Igα–Igβ show some similarity to each other (32% identity). The cytoplasmic parts of both proteins are highly conserved between mouse and human proteins (> 90% identity) and contain negatively charged amino acids and are thus acidic proteins. Both cytoplasmic sequences of Igα and Igβ possess four and two tyrosines respectively. Two tyrosines of Igα and both tyrosines of Igβ belong to a conserved motif found in other receptor complex such as TCR and Fc receptors and are named either ITAM (immunoreceptor tyrosine-based activation motif) or YxxL motif. Such motifs link the receptor to protein tyrosine kinases (PTKs) such as Syk and the Scr family members Fyn, Lyn, Blk and Btk. These PTKs act in concert with cytoplasmic and membrane bound phosphatases (reviewed by Goldman and De Franco, 1994; Pleiman *et al.*, 1994; Sefton and Taddie, 1994; Satterthwaite and Witte, 1996).

The transmembrane region of mIg molecules contains polar amino acids which play an important role in the binding of Igα–Igβ heterodimer (William *et al.*, 1990). The extracellular Ig-domain of Igα and Igβ may also play a role in binding the mIgM molecules (reviewed by Reth, 1992).

The BCR complex can therefore be compartmentalized into two distinct receptors subunits, the ligand-binding portion which is a tetrameric complex of Ig H chains and L chains and the surface Ig (sIg)-associated proteins which form the transducer unit of the BCR.

The mIg-associated proteins may play a role in transporting mIg, since most mIg molecules migrate to the cell surface when they are assembled with the Igα–Igβ heterodimer, although exceptions have been found to mIgD and mIgG (Williams *et al.*, 1993, 1994; Weiser *et al.*, 1994). The Igα/Igβ heterodimer also plays an important role in internalization of the BCR and transport to intracellular compartments where antigen processing takes place (Patel and Neuberger, 1993).

The Igα–Igβ heterodimer is also a critical signal mediator of the pre-B-cell receptor and its expression is required for allelic exclusion and during B-cell development to establish the peripheral pool of mature B-cells (Papavasiliou *et al.*, 1995a,b; Gong and Nussenzweig, 1996; Torres *et al.*, 1996).

Although Igα and Igβ are clearly the most readily detectable mIg-associated proteins, Cambier and coworkers (1994) identified an Igγ protein encoded by a second B29 gene. Igγ protein is identical in structure to Igβ but has a truncated cytoplasmic tail (by approximately 30 residues) leading to alteration of the ITAM motif (Friedrich *et al.*, 1993). It can be used instead of Igβ in low-density splenic B cells and bone marrow cells (Cambier *et al.*, 1994).

Recently, Reth's group described a new family of nonglycosylated proteins associated with murine BCR named BAPs (BCR associated proteins). Two of these proteins (BAP32 and BAP37) are specifically associated with IgM BCR (Terashima *et al.*, 1994), while two others (BAP29 and BAP31) bind preferentially to the IgD BCR (Kim *et al.*, 1994; Adachi *et al.*, 1996). BAP proteins bind specifically to mIg molecules in the transmembrane region which displays differences between mIg molecules. These BAP proteins may contribute to functional differences between IgM and IgD receptors and may represent the missing link between the BCR and the cytoskeleton (Kim and Reth, 1995a,b). Other potential functions of these BAP proteins are discussed by Adachi *et al.* (1996).

Co-receptors

As in T cells, the response of B cells to antigen can be modified by co-receptor molecules which can be divided into two subgroups. One includes co-receptors that participate nonspecifically in antigen recognition and act together with the BCR. At least three accessory membrane groups of proteins CD19–CD21–CD81–Leu13 complex, CD22 and FcγRIIb-1 have been described as modulating B-cell activation through mIg molecules (reviewed by Doody *et al.*, 1996). These molecules may recruit crucial signal transduction molecules into receptor complexes that may amplify the magnitude of the signal transduction cascade. They may also improve the overall avidity of antigen–BCR interaction and reduce the threshold of antigen required for signaling, which could be important under circumstances when antigen titers are limiting. The Fc receptors downregulate the proliferation and maturation of B cells. Other subgroup is referred to co-receptors that mediate the interaction between antigen-activated B cell and antigen specific helper T cell (CD40–CD40L, CD80/CD86–CD28/CTLA-4) (reviewed by Hathcock and Hodes, 1996; Clark *et al.*, 1996).

Co-receptors acting with BCR

CD19–CD21–CD81–Leu13 complex. CD19 is a surface glycoprotein of 95 kDa which belongs to the Ig superfamily and is expressed from the early stages of B cell development until the plasma cell stage. CD19 extracellular region contains two C2-type Ig-like domains separated by a smaller potentially disulfide-linked domain. CD19 cytoplasmic region has around 240 amino acids.

CD21 (CR2, complement receptor 2) is a 145 kDa glycoprotein member of the complement receptor family present on the B-cell lineage. CD21 is expressed on mature B cells and is rapidly lost following B-cell activation. CD21 is the B-cell receptor for the iC2b, C3dg and C3d fragments of the third component of complement. The CD21 extracellular region is formed by 15 or 16 repeating 60–70 amino acid structural elements named short consensus repeats (SCRs); the cytoplasmic domain has 34 amino acids.

CD19 and CD21 are noncovalently associated to form a multimolecular signal transduction complex on the B-cell surface. They are associated with other cell-surface proteins such as CD81 (TAPA-1) and Leu-13. CD81 is a 26 kDa protein expressed by a wide variety of cell types, including different lymphocyte lineages such as natural killer cells, thymocytes, eosinophils and B cells. The CD81 protein has a particular feature in the sense that *C*- and *N*-terminal ends are located within the cytoplasm, leading to a major portion of the protein exposed to the extracellular environment. Leu-13 is a 16 kDa cell surface protein expressed on leukocyte subsets.

Biological functions of this complex have been reviewed by Fearon and Carter (1995), Tedder *et al.* (1994) and Kehrl *et al.* (1994). Briefly, antigens with bound C3 fragments strongly enhance B-cell activation, particularly at lower antigen doses by the co-engagement of BCR and the CD19–CD21–CD81–Leu-13 complex (Dempsey *et al.*, 1996). This view is supported by two independent studies of CD19-deficient mice (Engel *et al.*, 1995; Rickert *et al.*, 1995).

CD22. CD22 is a 135 kDa phosphoglycoprotein belonging to the Ig gene superfamily and expressed on mature IgM^+/IgD^+ B cells (reviewed by Law *et al.*, 1994). During B-cell activation CD22 surface expression is increased but it is downregulated as B cells differentiate into plasma cells. Crosslinking of BCR induces a rapid phosphorylation of CD22 cytoplasmic tyrosines and thus CD22 can mediate signal transduction in B cells. Various *in vitro* experiments using antibodies to CD22 suggest two opposite roles of CD22. In some cases, the presence of CD22 increases signals delivered from BCR, in other cases CD22 inhibits these signals.

A recent study demonstrates that splenic B cells from mice with a disrupted CD22 gene are hyper-reactive to receptor signaling, which suggests that CD22 is a negative regulator of BCR signaling. CD22 on mature B cells may serve to raise the antigen concentration threshold required for B-cell triggering (O'Keefe *et al.*, 1996). CD22 is also an adhesion molecule through its lectin like-specificity which may restrict B-cell activation in secondary lymphoid organs to T-cell areas which provide necessary signals to B cells (reviewed by Doody *et al.*, 1996).

Fc receptors. In contrast to the positive effect on B-cell activation of antigen-bound C3 fragments, antigen bound to antibody induces negative signals to B cells. Antigen–antibody complexes can bring together the BCR and the FcγRIIb on B cells. FcγRIIb is a 40 kDa member of the Ig superfamily which is preferentially expressed on B cells. It is composed of two Ig-like domains, a single transmembrane region and a cytoplasmic tail (reviewed by Ravetch and Kinet, 1991; Fridman, 1993; Hulett and Hogarth, 1994). The inhibition effect mediated by the Fc receptor is dependent on a intracytoplasmic 13 amino acid motif containing a single tyrosine residue that mediates the negative modulation of mIg signaling by FcγRIIb (Muta *et al.*, 1994). This single tyrosine becomes phosphorylated upon co-crosslinking of mIg and FcγRIIb. This motif is called ITIM (immunoreceptor tyrosine-based inhibitory motif). Recent experiments using mice in which the gene for this Fc receptor has been disrupted suggest a role for FcγRIIB in regulating the serum level of antibody *in vivo* (Takai *et al.*, 1996).

Co-receptors implicated in B–T cell interaction

Antibody responses to antigens are heterogeneous with respect to signals required. Some responses are dependent on the presence of mature T helper cells whereas others are not. These distinct requirements depend on the properties of antigen such as physical structure and are referred to as T-dependent or T-independent antigens.

Antibody responses to soluble protein antigens require direct contact between antigen-specific B cells and T-helper cells that mediate crucial signals in both partners for the outcome of the response. Additional signals derived from T-helper cytokines are also necessary to achieve B-cell proliferation and differentiation. This phenomenon can be divided in two steps: first, signaling via BCR, and then signaling triggered by MHC plus peptide.

Surface BCRs with bound antigen are rapidly internalized and are delivered to processing compartments in which peptides are generated and are bound to MHC class II molecules (reviewed by Lanzavecchia, 1996 and references therein). MHC-bound peptides are expressed on the surface of B cells. Regulatory mechanisms involved in this process are still being investigated (Patel and Neuberger, 1993; Mitchell *et al.*, 1995). The recognition of the peptide bound to MHC by T-cell receptors allows the cognate interaction between T-helper cells and B lymphocytes. CD4 molecules, which can bind to monomorphic parts of MHC, act as coreceptors. This recognition leads to the upregulation of CD40 ligand (gp39) expression on T cells and its sequential interaction with CD40 on B cells. The CD40–CD40L interaction leads to the upregulation of the glycoproteins B7-1 (CD80) and B7-2 (CD86) on the B cells. T-cell activation is fully achieved by the binding of CD28/CTLA-4 to B7-1/B7-2 on B cells. Lymphokine production and secretion (IL-2, IL-4, IL-5, etc.) are induced in T cells and these lymphokines are bound to lymphokine receptors on B cells. Activated B cells can, therefore, proliferate and differentiate in effector cells such as antibody-producing cells. Generation of memory B-cells requires additional conditions, as reviewed by MacLennan (1994) and Gray *et al.* (1996).

Activation of B cells with T-independent antigens

T-independent antigens are subdivided into two categories based on their ability to elicit an antibody response in mice bearing the xid mutation (X-linked immunodeficiency). T-independent antigens that do or do not elicit antibody responses in xid mice was referred to as T-independent type I antigens or T-independent type II antigens, respectively (Mosier and Subbarao, 1982). Unresponsiveness of xid mice to T-independent type II antigens is linked to mutations of the BtK tyrosine kinase (Rawlings *et al.*, 1993; Kerner *et al.*, 1995; Khan *et al.*, 1995).

Most T-independent type I antigens are components of bacterial cell walls (LPS). They are directly mitogenic for murine B cells, irrespective of antigen specificity and induce polyclonal antibody production. Such antigens have the apparent capacity to replace T-helper cell requirements for B-cell activation. B cells may be activated by the crosslinking of putative LPS receptors (reviewed in Ulevitch and Tobias, 1995).

T-independent type II antigens are typically repeating polymers such as polysaccharide antigens (dextran, ficoll, etc.). The extensive surface Ig crosslinking by repetitive determinants of such antigens may trigger signal transduction in B cells (reviewed by Mond *et al.*, 1995). Recent data obtained by Torres *et al.* (1996) supports this view. Genetically altered mice in which the ITAM motifs in the Igα subunit of the BCR were inactivated were unable to respond to a T-independent polysaccharide antigen.

B-cell ontogeny and Ig repertoire expression

During the embryonic life, B cells are made in both liver and spleen. Shortly after birth, bone marrow becomes the primary site for *de novo* production of B cells throughout life. B-lymphocyte development from the earliest progenitors which have all Ig genes in germline configuration to the mature Ig-secreting plasma cells is characterized by various stages. The earliest stages are thought to be either antigen independent or dependent on some unknown self-antigens. The remaining stages in B-cell development (mature resting B cells and memory cells) are antigen sensitive. Antigen-dependent stages coincide with the occurrence of Ig receptors.

Establishment of V gene repertoire

Progenitor (pro) B cells which, on the basis of surface marker expression, clearly belong to the B lineage must have all Ig genes in germline configuration. Precursor (pre) B cells which have Ig gene loci at various stages of gene rearrangements can be divided into different subsets characterized by the expression of surface markers, intracytoplasmic markers and by growth properties as depicted in Table XVI.3.6.

The primary immunoglobulin repertoire results from somatic recombination of gene segments encoding variable region of H and L chains, the pairing of variable region genes and junctional diversity (addition or deletion of nucleotides during recombination process). After immunization, this repertoire is further expanded by somatic mutations taking place inside germinal centers (secondary antibody repertoire). Taking into account all variable gene segments encoding H and L chains, approximately 10^7 different immunoglobulins can be created. The junctional diversity increases this number to be greater than 10^9. Immunoglobulin gene rearrangement in B cells is highly regulated since various steps of rearrangement are required to have a functional gene encoding either heavy or light chains. As shown in Table XVI.3.6, the IgH locus rearrange first. H-chain gene assembly begins with the rearrangement of D gene segment to J gene segment (in both alleles), followed by the rearrangement of the V gene segment to the pre-assembled D–J segment (in only one allele). Transcripts issued from the V_H promoters that encode the V_H, D, J_H and C_H gene segments are differentially spliced and give rise to both the membrane bound and the secreted form of μ chain. The productive rearrangement of one allele prevents gene recombination on the second allele (allelic exclusion) and induces positive signal to trigger the rearrangement of Ig-L chains. The choice for a κ L-chain or λ L-chain is discussed below.

V(D)J recombination

All functional germline Ig-V region gene segments are flanked by conserved recombination signal sequences (RSS) which consist of a conserved heptamer and an A–T rich nonamer separated by either 12 or 23 bp spacer sequences (Tonegawa, 1983). The V(D)J recombination appears to be initiated by the recognition and the binding of RSSs by the enzymatic recombinase machinery (reviewed by Lewis, 1994). Recombination apparently occurs only between one sequence with a 12 bp spacer and another sequence with a 23 bp spacer, a requirement referred to as the 12 + 23 rule. This rule prevents recombinations between two V or two J gene segments. Both strands of the DNA segments are cleaved at the border of the two heptamers. These two are later joined together without modification. The two V gene segment ends that form the coding joints are released together to create hairpin loops.

The features of coding joints are complex. A variable number of bases are deleted by an exonuclease activity (to date no specific exonuclease has been definitively identified) from the ends of coding regions (opening of hairpin loops), nongermline nucleotides (*N* regions) are added in a coding joint by the terminal deoxynucleotide transferase (TdT) and, less frequently, extra bases interpreted as P nucleotides are added. P nucleotides are nucleotides joined to the end of an undeleted coding sequence to form a palindrome (hence P) with that sequence end. The length of P nucleotides is generally around 1 or 2 bp (Lafaille *et al.*, 1989). The resulting ends are ligated together by a DNA polymerase leading to the complete recombination event. TdT catalyzes the addition of nucleotides into the 3′ end of DNA strands and, particularly since no template specificity determines the nucleotides added, the enzyme

Table XVI.3.6 Phenotypic changes during B-cell development

Anatomical sites	*Bone marrow*				*Bone marrow periphery*	*Periphery*
A	Early Pro-B	Intermediate Pro-B	Late Pro-B/ large Pre-B	Small Pre-B	Immature B	Mature B
B	–	A PrePro B B Pro B C Pro B	C′	D Pre B	Immature B E	Mature B F
C	Pro-B	Pre-BI	Large Pre-BII	Small Pre-BII	Immature B	Mature B
Status of Ig gene rearrangements						
Heavy chain	germline	DJ	VDJ	VDJ	VDJ	VDJ
Light chain	germline	germline	germline	germline → VJ	VJ	VJ
Surface phenotype						
Ig like complexes:	surrogate H chain/ surrogate L chain (SL)	DJCμ protein (RfII)/SL	μH/SL	μH/SL μH/L	μH/L	μH/L δH/L
B220	–		Intermediate		High	
CD43	High		Low		–	
C-Kit		+			–	
CD25 (TAC)		–			+	
IL7-R		+			+	
Intracytoplasmic markers						
Tdt[a]		+			–	
RAG1/RAG2[b]		+	±	+		–
Growth properties on stromal cells and IL-7 *in vitro*		+			–	

Sites according to: A, Osmond (1986); B, Hardy *et al.* (1991b); C, Melchers *et al.* (1994).
[a]Terminal deoxynucleotidyl transferase.
[b]According to Grawunder *et al.* (1995).

adds dG residues preferentially (Desiderio *et al.*, 1984; Landau *et al.*, 1987; Lieber *et al.*, 1988). Heavy chains frequently contain more N regions than light chains, probably because the activity of TdT decrease as the B cells progress to the H and L chain recombination. The enzymatic recombinase machinery contributes greatly to generate diversity in the CDR3 of immunoglobulins since the CDR3 is generated by the joining of three gene segments (V–(D)–J) (reviewed by van Dyck and Meek 1992).

Proteins participating in V(D)J recombination. A few years ago, two genes, *RAG1* and *RAG2* (recombinase activating genes), were identified as key genes in the recombination process (Oettinger *et al.*, 1990; Mombaerts *et al.*, 1992; Shinkai *et al.*, 1992). They are not homologous to each other and are conserved in human, mouse and chicken. They are expressed together in pre-B and pre-T cells. Both genes are required for the recombination activity since mouse strains in which RAG genes have been turned off, fail to develop mature B and T cells (reviewed by Spanopoulou, 1996). These data support the view that RAG proteins are critical for V(D)J recombination.

Recently, two independent studies have described mechanisms that underlie DNA recognition by the RAG-1/RAG-2 proteins at the initial stages of V(D)J recombination (Difilippantonio *et al.*, 1996; Spanopoulou *et al.*, 1996). RAG-1 recognizes the nonamer and this interaction is stabilized by an additional interaction with the heptamer. Binding of RAG-2 is dependent upon the presence of RAG-1 and is more efficient with a 12-bp spacer than with a 23-bp spacer. The interaction between RAG-1 and RAG-2 (the nature of this interaction remains unknown) are required to form a cleavage-competent complex. The cleavage reaction depends on the presence of functional heptamer motif. Therefore one site of the RSS (nonamer) functions as high-affinity DNA-binding region that anchors the protein on the DNA, while the other (heptamer) is the site of cleavage.

Allelic exclusion and regulation of Ig gene rearrangement. Each B cell expresses only one V_H and one V_L region and thus a single antigen-binding site despite the two homologous chromosomal loci for H and L chain genes. This phenomenon is called allelic exclusion (reviewed by Rajwesky, 1996). Two general models have been proposed to explain allelic exclusion. The stochastic model is based on the observation of a high frequency of defective rearranged genes as an indication that functional rearrangements are rare. Consequently, the allelic exclusion would result from the low probability that two rare events take place in the same cell (Coleclough *et al.*, 1981). An alternative to the stochastic model is the 'regulated model' (Alt *et al.*, 1980). A functional rearrangement of one H- or L-chain gene in a B cell would inhibit further H- or L-chain gene rearrangement on the second allele in the same B cell. In the case of an unproductive rearrangement, no inhibitory effect could prevent B cell from continuing recombination until a productive rearrangement is performed.

A large range of distinct experiments supports the regulated model. We will mainly discuss novel investigations based on the introduction of transgenes into the genome of whole mice or mice in which Ig genes have been inactivated by targeted mutation (see also Karasuyama *et al.*, 1996). In transgenic mice carrying a transgenic functionally rearranged μ gene, the rearrangement of endogenous Ig genes is markedly suppressed (Weaver *et al.*, 1985). Endogenous IgH recombination is abolished with a transgene encoding the membrane form of μ rather than the secreted form (Nussenzweig *et al.*, 1987; Manz *et al.*, 1988; Picarella *et al.*, 1991). Experiments with mice presenting a disruption of the μ membrane exon supports the conclusion that a membrane form of μ chain is required to mediate allelic exclusion (Kitamura and Rajewsky, 1992). Interestingly, Iglesias *et al.* (1987) have shown that a δ transgene is able to inhibit endogenous V(D)J recombination.

Regulation of Ig gene rearrangements. The rearrangement of the H chain genes appears to be a prerequisite to the activation of the rearrangement of κ genes that occurs before the λ gene recombination (ordered model) (Reth *et al*, 1985). κ-Expressing cells without H chain gene recombination have never been observed. Reth *et al* (1987) have shown the activation of V_κ in pre-B cells transfected with a vector expressing C_μ. The membrane form of μ chain is the only effective mediator to activate the κ rearrangement. According to the regulated model, if the first κ rearrangement is unproductive, the second allele can proceed to another rearrangement. As soon as a productive κ chain appears, other κ rearrangements are suppressed. Allelic exclusion at the κ locus level is supported by experiments on transgenic mice carrying a functional rearrangement κ transgene. B cells that express the κ transgene with an endogenous H chain have their endogenous κ genes in germline configuration (Ritchie *et al.*, 1984). In the ordered model, the μ chain expression represents a key point in the development and triggering of rearrangements at IgH locus (allelic exclusion) and the initiation of rearrangements at the L-chain loci. The discovery that B-cell progenitors can express on their membrane a complex formed by a surrogate light chain with a μ chain (pBCR) has made sense of the ordered model (reviewed by Rajewsky, 1996; see Surrogate Light Chain and pBCR, below).

Isotype exclusion. Isotype exclusion phenomenon leads B cells to express only one L-chain allele despite the presence of two alleles at both κ and λ loci. Moreover, only one of the three λ isotypes is mainly expressed in λ-producing cells. For most B cells, the choice of L chain expressed is determined during pre-B stage development.

Two major contrasting models have been proposed to explain how differentiating B cells are committed to either κ or λ gene expression. The first is the stochastic model assuming that both κ and λ genes are activated for recombination concurrently during pre-B-cell development and that L-chain isotype expression is determined by the relative frequencies of forming either functional κ genes or λ genes. The second model is based on an ordered hierarchy of L-chain gene recombination in which κ genes are recombined first. In cells that fail to produce a functional κ gene, λ gene recombinations are activated.

The stochastic model is supported by the observed correlation between the proportions of κ and λ chains in the serum and the relative number of V_κ and V_λ genes present in the germline. This correlation appears compatible with the idea that the number of B cells committed to either κ or λ production reflects the relative number of genes that potentially can be recombined. The stochastic model would predict only a single L-chain gene recombination event either κ or λ genes. Although mouse κ-producing cells frequently exhibit one recombined allele and the others in germline configuration, λ-producing cells often have κ alleles that have recombined. These observations are consistent with the ordered model for L-chain gene recombination. Nevertheless, the stochastic model can be accepted by assuming that the rate of gene recombination differ for the κ and λ gene families (Ramsdem and Wu, 1991). Studies of A-MulV-transformed pre-B cells (Abelson murine leukemia virus) support the ordered model because many pre-B-cell lines recombine κ genes but not λ genes (Reth *et al.*, 1985). The mechanism of such apparent regulation of the λ gene rearrangement by the κ locus is unknown. Recently, data obtained from different types of κ Knock-out mice (deletion or disruption of either J_κ or C_κ or intronic κ enhancer region) suggest that even if the recombination of κ genes is not required to activate the λ locus, it could help the recombination of λ genes (reviewed by Sanchez *et al.*, 1996).

At this stage in our knowledge, the picture of L-chain gene recombination appears to represent a blend of the original stochastic and ordered models.

Factors participating in the development of B-cell lineage

Factors which contribute throughout B-cell differentiation process and signals that trigger genetic rearrangement of V genes are not fully understood. The bone marrow microenvironment, especially stromal cells, may provide differentiation signals for B cells. Some proliferation factors such as IL-7 produced by bone marrow stromal cells, IL-3, *c*-kit-ligand and flt3-ligand are known to be required for the proliferation of pro- and pre-B cells *in vitro* (Billips *et al.*, 1992; Galli *et al.*, 1994; Rosenberg and Kincade, 1994; Hirayama *et al.*, 1995; von Freeden-Jeffry *et al.*, 1995; Winkler *et al.*, 1995).

Fetal/neonatal V gene repertoires versus adult V gene repertoire

Fetal, neonatal and adult bone marrow V_H gene repertoires appear biased towards J_H proximal V_H gene families since an over-representation of V_H 7183 (V_H5) and V_H Q52 (V_H2) has been shown (Yancopoulos *et al.*, 1984; Perlmutter *et al.*, 1985; Jeong and Teale, 1988; Malynn *et al.*, 1990). The functional significance of nonrandom VH 7183 usage in mouse ontogeny remains enigmatic. Moreover Igs from fetal and newborn mice display very restricted junctional diversity. Some D segments (DFL16.2 and all DSP2 gene segments) are underexpressed and others are overexpressed (DFL16.1 and DQ52 gene segments). The germline-derived D segments are shorter and there is a lack of N regions related to lower TdT activity (reviewed by Feeney, 1992).

In peripheral B cells, the V_H gene families are expressed in more stochastic fashion and expressed V_H gene pattern roughly corresponds to the complexity of V_H families (Dildrop *et al.*, 1985; Wu and Paige, 1986; Schulze and Kelsoe, 1987; Yancopoulos *et al.*, 1988). Some authors have reported differences in V_H usage between murine strains (Jeong *et al.*, 1988; Yancopoulos *et al.*, 1988). Some data suggest that the usage of the V_H gene family in the adult reflects selection mechanisms (Gu *et al.*, 1991; Coutinho *et al.*, 1992; Grandien *et al.*, 1994). The exact rules of this selection are unknown but a functional BCR is required (Torres *et al.*, 1996). These rules could involve interactions with self-antigen and/or idiotype-antiidiotype interactions (reviewed by Urbain *et al.*, 1992; Kearney, 1993; Ryelandt *et al.*, 1995). As reviewed by Rajewsky (1996) most bone marrow B-cell precursors are destined to die. Only 3% of immature B-cells generated in the bone marrow appear to reach the mature long-lived peripheral B-cell pool. This observation could explain why the Ig repertoire expressed in the peripheral B cells differs from that expressed in the pre- and immature B cells in the bone marrow. Primary and secondary Ig repertoires are also shaped by tolerance phenomenon (reviewed by Klinman, 1996a). Ig-transgenic mice have been useful in the identification of major parameters of B-cell tolerance induction (reviewed by Goodnow, 1995; Carsetti *et al.*, 1995). In bone marrow, B cells expressing autoreactive antibodies are negatively selected. Depending on their affinity for antigen, autoreactive B cells may be rescued in bone marrow by further immunoglobulin rearrangement (receptor editing; reviewed by Melchers *et al.*, 1995; Nemazee, 1993; Radic and Weigert, 1994).

In contrast to V_H family expression in mouse ontogeny, expression of V_κ families does not reflect positional bias for J-proximal families. The relative few studies on V_κ gene usage are somewhat contradictory since one group report a preferential usage of V_κ families located in the middle of V_κ locus ($V_\kappa1$, $V_\kappa8$ and $V_\kappa9$) without a correlation with the complexity of the family (Kaushick *et al.*, 1989) and another one observes a bias for $V_\kappa4/5$ and $V_\kappa10$

and reduction of $V_\kappa1$ in functional early B cells (reviewed by Teale and Medina, 1992). In the adult functional repertoire, some groups report nonstoichiometric V_κ family usage (Kaushick *et al.*, 1989; Teale and Morris, 1989), while others favor a more randomized expression pattern (Kalled and Brodeur, 1991).

B-cell repertoire in species

Although in all vertebrates the genes encoding antibody variable regions are assembled during B-cell development through a process of site-specific recombination, species have selected different cellular and molecular strategies to create diversity in Ig repertoire. In the chicken, the unique functional V_λ and V_H are assembled with corresponding joining gene-segments in all B-cell progenitors and thus, all cells initially express the same or similar antibodies. Diversity is generated by a gene conversion process which occurs during intensive proliferation of B cells (reviewed by Reynaud *et al.*, 1994). Rabbits use the same strategy at least for the IgH locus (Knight and Crane, 1994). In the sheep, diverse V and J gene segments (less than in mice) are assembled but these rearrangements are further diversified by hypermutation mechanism without influence of external antigens (Reynaud *et al.*, 1995).

In mice and humans, B-cell generation begins in embryonic life in the para-aortic tissues, liver and spleen. After birth, it takes place in the bone marrow, where it continues throughout life. The situation is strikingly different in chicken and sheep. B-cell progenitors which arise early in ontogeny and have already undergone Ig gene rearrangements, colonize particular organs where the generation of diversity takes place. These organs are the bursa of Fabricius in the chicken and the ileal Peyer's patches (IPP) in the sheep. In both species, the building of B-cell repertoire starts during the embryonic development and continues for only a few weeks or months after birth (reviewed by Weill and Reynaud, 1995).

Surrogate light chain and pBCR

Gene and protein structure

Expression of Ig μ heavy chains on the surface of pre-B cell that do not make light chains has intrigued immunologists. Melchers and colleagues have identified on chromosome 16 a gene that has striking sequence similarity to the J and C regions of the λ locus (reviewed by Melchers, 1994; Rolink *et al.*, 1996; Karasuyama *et al.*, 1996). Because four murine C_λ genes were already known, they named it $\lambda5$. This gene is formed by three exons: the 3′ end of exon 1 presents sequence homologies to J_λ, the intron between exons 2 and 3, exon 3 itself and the 3′ untranslated region of the $\lambda5$ gene show high homology to $C\lambda$ sequences (Melchers *et al.*, 1993).

The use of a probe constructed with flanking regions of the genomic $\lambda5$ clone in pre-B cell mRNA has allowed identification of another transcribed segment located about 4.5 Kb 5′ of $\lambda5$. This latter region exhibited similarities to both V_λ and V_κ and it was consequently named Vpre-B. The mouse genome contains two nearly identical Vpre-B genes named Vpre-BI and Vpre-BII. The nucleotide sequences of Vpre-BII is 97% identical to Vpre-BI. The Vpre-B genes are pliced in two exons. The first exon and the intron preceding exon 2 show high sequence homologies to V_λ genes. The 5′ part of exon 2 shows weak homologies to V_κ V_λ and V_H. The Vpre-B gene encodes a polypeptide (ω) of 142 amino acids, including a signal peptide of 19 residues, while the $\lambda5$ gene codes for a polypeptide (ι) of 209 amino acids, including a signal peptide of 30 residues. The two Vpre-B proteins are different in four amino acids. Recently, an additional Vpre-B gene (called Vpre-B3 was isolated from a mouse pre-B cell line. The gene encodes a glycosylated protein of 123 amino acids.

Somatic recombination between $\lambda5$ and Vpre-B genes has not been found in B and pre-B cells, although these genes have no apparent defects that would prevent their expressions as proteins. They have typical consensus splice sites and initiation and termination codons. Therefore, it has been proposed that Vpre-B and $\lambda5$ proteins could associate with each other to form a light-chain-like structure now termed the surrogate light-chain (SL). Data from Tsubata and Reth (1990) based on gene transfection experiments have shown that surface μH chains are covalently linked to ω protein while ι protein is non-covalently associated. Each Vpre-B gene can participate to surrogate light chain in pre-B cell lines.

Expression of surrogate light chain in B cell development

According to Melchers and colleagues the surrogate light chain can be found covalently associated with the gp130/gp65 of the surrogate H chain complex on pro- and pre-BI cells, the $DHJHC_\mu$ proteins on pre-BI cells, which have DHJH rearrangements in reading frame II (RfII), and with μH chains expressed in pre-BII cells. Structure of these pre-B cells receptors is described by Melchers *et al.* (1993). Ig-like complexes are associated with Igα and Igβ. Table XVI.3.6 illustrates the different stages of B-cell development in murine bone marrow. Potential functions of μH chain–SL complex are well documented although they are still not fully understood (see below).

B-cell development in mice carrying a deficient $\lambda5$ gene

Studies of mice in which the $\lambda5$ gene has been inactivated by targeted mutation ($\lambda5$T) have been of great interest in understanding potential functions of SL chain in B-cell development (Kitamura *et al.*, 1992). The pro/pre-BI compartment has practically normal size in these mice. As in normal mice, this particular population of B cells decreases with age. *In vitro* these pro/pre-BI cells present the same properties as normal pro/pre-BI cells as reviewed by Rolink *et al.* (1996). These observations led the authors to propose that the SL was not essential for the formation of a pro/pre-BI compartment.

The most striking observation in these λ5T mice is the dramatic reduction of the pre-BII cell compartment with productive VHDJH rearrangements (20-fold), suggesting a major role of the SL in the selection and amplification of the pre-BII cells expressing membrane-bound μH chains. Consequent to this defect, the immature and mature B cell compartments are small in λ5T during the first 3 months of life. Later, mature B-cells gradually accumulate in the periphery. Despite such narrow B-cell compartments, λ5T mice are able to develop immune responses against T-dependent and T-independent antigens. SL–μH-chain complexes could signal pre-BII cells to proliferate in order to constitute rapidly the peripheral B-cell compartment.

In contrast to normal mice, λ5T mice have a high proportion of peripheral B cells that use rfII rearranged D–J alleles. This observation leads the authors to suggest that SL–DHJHC_μ protein complexes could induce deletion and/or could prevent the expansion of B cells using rfII rearrangement D–J alleles on pre-B cells.

Functions of μH chains–SL complex pre-BCR

Studies of λ5 knock-out mice suggest that the SL plays an important role in B-cell development. Karasuyama *et al.* (1996) have summarized data which define the possible functional roles ascribed to the SL in B-cell differentiation. We present here the major points:

- in bone marrow, pre-B cell II that have a productive rearrangement of μ heavy-chain are selectively expanded by signals delivered through pre-BCR formed by SL–μH chains and Igα–Igβ heterodimers;
- pre-BCR appears to participate in signaling allelic exclusion in the heavy-chain locus;
- the complex DHJHC_μ proteins with the SL appears to block the differentiation of pre-B cells which have rearranged the H-chain gene loci in RfII;
- The role of pre-BCR in promoting the light-chain gene rearrangement is still controversial since some experiments (Tsubata *et al.*, 1992) conclude to the necessity of a functional pre-BCR to have light-chain gene rearrangements and others (reviewed by Karasuyama *et al.*, 1996) demonstrate that neither μ heavy-chain nor SL is prerequisite for the rearrangement of light-chain loci. The authors propose that the membrane-bound μ heavy-chains, possibly in association with SL could only facilitate the rearrangement of light-chain genes;
- The possible functions of SL complexes with gp130/gp65 are still unknown.

B-cell subsets (and lineages) and natural antibodies

B lymphocytes can now be subdivided into subsets with distinct properties. A subpopulation of B cells which is numerically small in adults bears the pan-T cell glycoprotein Ly-1 (CD5). These cells are distinctive from the $IgD^{++}Ly1^-$ conventional B cells by surface phenotype ($IgM^{++}/IgD^{low}/B220^{low}$), anatomical localization (in spleen of neonates and in the peritoneal cavity in adult), early appearance in ontogeny, secretion of natural antibodies and certain autoreactive antibodies, extensive capacity for self-renewal and increased frequency in autoimmune mouse strains such as NZB and motheaten. These cells have been renamed B1 cells and conventional B cells, B2. Some B cells are similar in many respects to $CD5^+$B cells (phenotype and localization) but lack Ly-1/CD5 marker. They are named sister population or B-1b cells in contrast to B-1a cells which express the CD5 surface marker.

Many observations support the hypothesis that B1 cells derive from a separate lineage. Fewer B1-a cells are generated from adult bone marrow when compared with early B-cell progenitor sources such as fetal liver, bone marrow from young mice. The fetal omentum is a powerful source of B1-a and B1-b cells (reviewed by Hardy and Hayakawa, 1994; Solvason *et al.*, 1991; Herzenberg *et al.*, 1992; Kantor and Herzenberg, 1993; Marcos *et al.*, 1994).

CD5 B cells express an Ig repertoire distinct from that of conventional B cells (Tornberg and Holmberg, 1995) and a highly skewed V-gene repertoire toward the production of antibodies reactive with self antigen such as DNA, bromelian-treated mouse red blood cells (V_H11 and V_H12 families in context with $V_\kappa 9b$ and a $V_\kappa 20$ gene respectively) (Hardy, 1991).

The occurrence of different B-cell subsets is also of importance when considering the function and origin of normal immunoglobulins or natural antibodies (production of antibodies without intentional immunization or exposure to any environmental antigen). It is often assumed that natural antibodies represent the constitutive synthesis or leakage of small amounts of immunoglobulins by all B lymphocytes. This has never been proven but natural antibodies certainly contain IgG immunoglobulins secreted by memory B lymphocytes. Recent data indicate that a significant fraction of normal immunoglobulins (IgM and IgG) is synthesized by $CD5^+$ B lymphocytes. These antibodies were found to be polyreactive and sticky and can act as a first-line defense against bacteria, viruses and parasites. It has also been shown that B1 lymphocytes do not participate in most T-dependent immune response but can be induced by thymus-independent polymers, such as polysaccharides.

For the B-2 or conventional B-cell subset, it has not been established whether there is a single lineage giving rise to primary or secondary immune responses or whether there are separate precursors for primary and secondary repertoires already present prior to antigen arrival (reviewed by Klinman, 1996b).

Acknowledgements

The Laboratory of Animal Physiology is supported by grants from Belgian State and from Fonds National de la Recherche Scientifique. We are grateful to Dr A. van Acker for the critical reading of the manuscript and to M. Henneghien for invaluable help with computer problems.

4. Cellular Immunology

The CD markers

The CD nomenclature is an internationally recognized system for naming the cell surface molecules expressed on lymphocytes which are recognized by monoclonal antibodies; CD refers to 'cluster of differentiation'. Their nomenclature, the cells which express each antigen protein recognize and their function, when known, are listed in Table XVI.4.1.

Table XVI.4.1 Nomenclature of CD markers

CD	*Function*	*T*	*B*	*NK*	*Mo*	*Ma*	*G*	*Pl*	*Dc*	*Stc*
CD1a		Thc	S						X	
CD1b		Thc	S						CL	
CD1c		Thc	S						DC	
CD2	LFA-3 R	X		S						
CD2R		XX								
CD3	TCR sub-unit	X								
CD4	MHC II R	S			X					
CD5	CD72 R	X	S							
CD6		S	S							
CD7		S		X				X		
CD8	MHC I R	S		S						
CD9			pre-B		X			X		
CD10	CALLA		pre-B				X			
CD11a	LFA-1(α)	X	X	X	X	X	X			
CD11b	CR3(α)			X	X		X			
CD11c	CR4(α)		S	X	X		X			
CDW12					X		X			
CD13	Amino-peptidase				X		X			
CD14	LPS ligand				X		X		CL	
CD15	ELAM R				X		X			
CD16	FcγR III			X	X		X			
CDw17	Ceramide				X		X	X		
CD18	CD11β	X	X	X	X	X	X			
CD19			X							
CD20	Ion channel		X							
CD21	CR2		mat						CF D	
CD22			X							
CD23	FcεR II		mat XX		XX		E			
CD24			X				X			
CD25	IL-2 R	XX	XX		XX					
CD26		XX	X			X				
CD27		S								
CD28		S	XX							
CD29	VLAβ	X	X	X	X	X	X	X	X	
CD30		XX	XX							
CD31	PECAM		X		X		X	X		
CDw32	FcγR II		X			X	X			

(continued)

Table XVI.4.1 *(Continued)*

CD	*Function*	*T*	*B*	*NK*	*Mo*	*Ma*	*G*	*Pl*	*Dc*	*Stc*
CD33					X					X
CD34										X
CD35	CR1		X	S	X		X			
CD36			X		X			X		
CD37		X	mat			X				
CD38		Thc XX	prec							X
CD39			mat		X				Fdc	
CD40			x						Fdc	
CD41								X		
CD42a								X		
CD42b								X		
CD43	Sialine	X			X		X			
CD44		X	X	X	X	X	X			
CD45	Common leukocyte Ag(CLA)	X	X	X	X	X	X			
CD45RA	CLA isoform	S	X	X	X	X	X			
CD45RB	CLA isoform	X	X	X	X	X	X			
CD45RO	CLA isoform	S	X	X	X	X	X			
CD46	MCP	X	X	X	X	X	X	X		
CD47		X	X	X	X	X	X	X	X	X
CD48		X	X	X	X	X	X			
CDw49b	VLAα2	X	X	X	X	X	X	X		
CDw49d	VLAα4	X	X		X			X		
CD49f	VLAα6	X						X		
CDw50		X	X	X	X	X	X			
CD51	Vitronectin Rα	X	X	X	X	X	X	X		
CDw52		X	X	X	X	X	X			
CD53		X	X	X	X	X	X			
CD54	ICAM -1	X	X	X	X	X	X	X	X	X
CD55	DAF	X	X	X	X	X	X	X	X	X
CD56	NCAM	S		X						
CD57			S	X						
CD58	LFA3	X	X	X	X	X	X	X	X	X
CD59	TAP protectin	X	X	X	X	X	X	X	X	X
CDw60		S						X		
CD61	Vitronectin Rβ							X		
CD62	P-selectin							X		
CD63			X		X			X		
CD64	FcγR I				X					
CDw65					X		X			
CD66							X			
CD67							X			
CD68						X				
CD69		XX	XX	X		X				
CDw70		XX	XX							
CD71	Transferrin R	XX	XX	XX	XX	X				X
CD72	CD5 R		X							
CD73	Ecto-5′ nucleotidase	S								
CD74			X		X					
CDw75	α2,6 sialyltransferase	S	mat							
CD76		S	mat							
CD77			X							
CD7w78			X							

T = T lymphocytes; B = B lymphocytes; NK = NK cells; Mo = monocytes; Ma = macrophages; G = granulocytes; Pl = platelets; Dc = dendritic cells; Lc = Langerhans cells; Stc = stem cells.
X = present; XX = activated cells; S = subpopulation; Thc = thymocyte; mat = mature; Fdc = follicular dendritic cell; E = eosinophil.

Cytokines

Cytokines are a group of signalling molecules involved in the communication between cells including those of the immune system. Cytokine-mediated events occur during the initiation and effector stages of immune responses and the development of hematopoietic cells. An important aspect of cytokine activity is that they frequently work together to produce effects, rendering their activity difficult to describe. In this section, the emphasis is on the nomenclature, cellular source and cellular targets of the best characterized mouse cytokines which have been cloned and which influence cells of the immune systems. These properties are summarized in Table XVI.4.2.

Table XVI.4.2 Nomenclature, source and cellular effects of mouse cytokines

Cytokine	*Acronym*	*Alternative title*	*Cellular source*	*Cellular target*
Interferon-α	IFN-α		Leukocytes	Many cells
Interferon-β	IFN-β		Fibroblasts and epithelial cells	Many cells
Interferon-γ	IFN-γ		T cells, NK cells	Macrophages, T cells, B cells, NK cells
Interleukin 1α/ Interleukin 1β	IL-1α/ IL-1β	Lymphocyte-activating factor (LAF); B-cell activating factor (BAF); mononuclear cell factor (MCF)	Macrophages, endothelial cells, LGL, T and B cells, fibroblasts, epithelial cells, astrocytes, keratinocytes, osteoblasts	Thymocytes, neutrophils, hepatocytes, chondrocytes, muscle cells, endothelial cells, osteocytes, epidermal cells, macrophages, T cells, B cells, fibroblasts
Interleukin 2	IL-2	T-cell growth factor (TCGF)	T cells	T cells, B cells, macrophages
Interleukin 3	IL-3	Multi-potential colony stimulating factor (multi-CSF); mast-cell growth factor (MCGF); hematopoietic cell growth factor (HPGF)	T cells, mast cells	Multipotential stem cells, mast cells
Interleukin 4	IL-4	B-cell stimulating factor (BSF-1); T-cell growth factor II (TCGF-II); B-cell growth factor II (BCGF-II); mast-cell growth factor II (MCGF-II)	T cells, mast cells, bone marrow stromal cells	T cells, B cells, mast cells, macrophages, hematopoietic progenitors
Interleukin 5	IL-5	T-cell replacing factor (TRF); B-cell growth factor II (BCGF-II); eosinophil differentiation factor (EDF)	T cells	Eosinophils, B cells
Interleukin 6	IL-6	B-cell stimulation factor 2 (BSF-2); B-cell differentiation factor (BCDF); hepatocyte stimulating factor (HSF)	Fibroblasts, macrophages, endothelial cells	B cells, T cells, thymocytes, hepatocytes
Interleukin 7	IL-7	Pre-B-cell differentiation factor	Thymic stromal cells	Pro- and pre-B cells, thymocytes, activated mature T cells
Interleukin 8	IL-8	Monocyte-derived neutrophil chemotactic factor (MDNCF); neutrophil activating protein (NAP)	Monocytes, endothelial cells, epithelial cells, fibroblasts, chondrocytes, hepatocytes, keratinocytes, synoviocytes	Neutrophils, T cells, basophils
Interleukin 9	IL-9	Mast-cell growth factor (MCGF); T-cell growth factor (TGF)	T cells	$CD4^+$ T cells, mast cells
Interleukin 10	IL-10	B-cell derived thymocyte growth factor (B-TCGF); cytokine synthesis inhibiting factor (CSIF)	T cells (T_H2 and T_H0), macrophages, B cells	Macrophages, mast cells, NK cells

(continued)

Table XVI.4.2 (*Continued*)

Cytokine	*Acronym*	*Alternative title*	*Cellular source*	*Cellular target*
Interleukin 11	IL-11		Stromal cells	Hematopoietic progenitor cells
Interleukin 12	IL-12	Natural killer cell stimulating factor (NKSF); cytotoxic lymphocyte maturation factor (CLMF)	T cells, B lymphoblasts	T cells, NK cells, LAK cells
Granulocyte–macrophage colony stimulating factor	GM-CSF	Colony stimulating factor α, 2 (CSF-α, 2)	T cells, endothelial cells, fibroblasts, macrophages, mast cells	Multipotential stem cells, mature neutrophils and monocytes/macrophages
Macrophage colony stimulating factor	M-CSF	Colony stimulating factor 1 (CSF-1)	Fibroblasts, monocytes, endothelial cells	Multipotential stem cells, monocytes/macrophages
Granulocyte colony stimulating factor	G-CSF	Colony stimulating factor β (CSF-β)	Macrophages, fibroblasts, endothelial cells, mesothelial cells, T cells	Multipotential stem cells, granulocytes
Tumour necrosis factor	TNF	Cachectin, tumour necrosis factor α	Macrophages, T cells, thymocytes, B cells, NK cells	Tumour cells, tumour cell lines, neutrophils, osteoclasts, fibroblasts, macrophages, adipocytes, eosinophils, endothelial cells, chondrocytes, hepatocytes
Lymphotoxin	LT	Tumour necrosis factor β	T cells	Tumour cells, tumour cell lines, neutrophiols, osteoclasts
Transforming growth factor β	TGF-β		Platelets, activated macrophages, T cells, hepatocytes, thymocytes	Endothelial cells, T cells, B cells, keratinocytes, hepatocytes, fibroblasts, hematopoietic cells, osteoblasts

5. Murine Models of Immunodeficiency

This is a brief overview of murine models of immunodeficiency. Such models offer a double interest: first, some are very close or identical to immunodeficiency syndromes affecting humans or other animal species and can therefore be considered as disease models for the development of new treatments (including gene therapy); second, the phenotype of these immunodeficient mice often gives precious insight into more fundamental aspects of the immune system. This review addresses this double perspective.

Three types of models are considered here: (1) immunodeficiency caused by spontaneous mutations; (2) immunodeficiency caused by genetic manipulations; (3) acquired immunodeficiency due to nongenetic factors (Table XVI.5.1)

The search for the genes involved in naturally mutant strains such as nude or scid is fundamental because it often reveals unexpected interactions between nonimmune genes and various aspects of the immune responses. For example, the gene responsible for the nude phenotype was recently identified as a transcription factor controlling proliferation and differentiation of epithelial cells (Brissette *et al.*, 1996).

The analysis of genetically engineered animals often derives from a different intellectual process. On a theoretical basis, a candidate gene is proposed and the animal is genetically modified to confirm or refute the hypothesis.

Table XVI.5.1 Different types of immunodeficient mice

Spontaneously occurring mutant strains

Genetically modified strains
- Knock outs
- Transgenic
- Dominant negative mutations

Acquired immunodeficiency
- Thymectomy
- Retroviral infection
- Ionizing radiation
- Immunosuppressants
- Alkylating agents
- Alloimmune stimulation

Table XVI.5.2 Genetic mechanisms involved in immunodeficiency in murine inbred mice

Name, chromosome, transmission	*References*	*Gene involved and mechanism*	*Phenotype*
Nude (nu/nu), chromosome 11, AR	Brissette *et al.* (1996)	Winged-helix nude gene (whn) transcription factor expressed in hair, skin and thymus, controls balance between proliferation and differentiation.	Hairless, absence of thymus, T-cell lymphopenia, increased activity of NK cells and macrophages
Beige (bg/bg), chromosome 13, AR	Burkhardt *et al.* (1993), Fukai *et al.* (1996), Nagle *et al.* (1996)	Mutation specifically affects late endosomes and lysosomes; murine (bg) and human gene (CHS) show homology for yeast vacuolar sorting protein	Abnormal color of fur, impaired natural resistance to viral infections, defect of neutrophils and NK cells, proposed model for human Chediak–Higashi (CH) disease
SCID (scid/scid), chromosome 13, AR	Kirchgessner *et al.* (1995), Blunt *et al.* (1995), Wiler *et al.* (1995)	DNA-dependent kinase (p450), recruits RAG-1, RAG-2, and is involved in V(D)J recombination.	Severe combined immunodeficiency, T-cell and B-cell lymphopenia, hypogammaglobulinemia, proposed model for equine severe combined immunodeficiency
Xid, X chromosome, X-linked	Fukuda *et al.* (1996)	Bruton tyrosine kinase (btk) Reduced IP4-binding capacity	Lack of $CD5^+$ B cells, impaired immune responses to thymo-independent antigens (pneumococcal polysaccharides), proposed model for human X-linked agammaglobulinemia
Motheaten (me/me), chromosome 6, AR	Tsui *et al.* (1993), Lorenz *et al.* (1996)	Protein tyrosine phosphatase SHTP1. Hyperactivation of src-family tyrosine-kinases	Fatal immunodeficiency, thymic aplasia, autoimmunity (lupus-like syndrome)
Wasted (wst/wst), chromosome 2, AR	Abbot *et al.* (1994), Woloschak *et al.* (1996)	Gene unknown, increased frequency of apoptosis in thymus and spleen, faulty DNA repair, increased sensibility to ionizing radiation	Tremor, ataxia, thymic and splenic atrophy, death around fourth week of life
Alymphoplasia (aly/aly), chromosome 11, AR	Miyawaki *et al.* (1994), Koike *et al.* (1996)	Unknown	Deficiency in the systemic lymph nodes and Peyer's patches, absence of marginal zone, defective affinity maturation and Ig class switching, autoimmune exocrinopathy. Proposed model for Sjögren disease
Generalized lymphoproliferative disease (gld/gld)	Nagata and Suda (1995)	Fas ligand Impaired apoptosis	Lymphoproliferative disease, autoimmunity, immunodeficiency
Lymphoproliferation (lpr/lpr)	Nagata and Suda (1995)	Fas (CD95) Impaired apoptosis	Lymphoproliferative disease, autoimmunity, immunodeficiency

Three different types of genetically modified animals can be distinguished (Table XVI.5.1): (1) transgenic mice with an increased, constitutive expression of a given protein at a chosen site; (2) 'knockout' animals with a targeted disruption of a given gene; (3) animals with 'dominant negative mutations' which express abnormal, inhibitory proteins as transgenes.

Mice expressing a transgene coding for a normal murine protein do not usually develop immunodeficiency. In contrast, several autoimmune or inflammatory phenomena have been described in mice expressing increased amounts of a given cytokine in a target organ. For example, transgenic mice expressing interleukin-4 in the lung develop lymphocytic and eosinophilic inflammation of the trachea and bronchi (Rankin *et al.*, 1996) and mice constitutively expressing interleukin-2 in the Langerhans islets develop autoimmune diabetes (Elliot and Flavell, 1994).

The vast majority of genetically induced immunodeficiencies result from the targeted disruption of a gene of which product is involved at various levels of the immune responses. Caution is advised in the analysis of the phenotype of these animals. First, since the knockout is congenital, the development of the immune system is often profoundly modified, therefore it can be difficult to determine whether a given gene product is directly involved in an adult phenotype or if the abnormal adult phenotype is secondary to a more global immaturity of the immune system. Second, compensation mechanisms sometimes attenuate the impact of a given knockout of a phenotype. Although it may have an important role in physiological situations, a given gene may be knocked out without major impact on the immune responses because redundant genes can compensate and maintain an immunocompetent phenotype.

Using this principle, mice with dominant negative mutations have been designed (reviewed by Perlmutter and Alberola-Ila, 1996). These mice express transgenes which encode for abnormal, catalytically inactive proteins which inhibit a given pathway, even in the presence of multiple, redundant signals. Dominant negative mutations of certain genes have illustrated striking immune defects while the corresponding knockouts failed to induce immunodeficiency. For example, mice with a targeted disruption of N-Ras have a grossly normal phenotype while expression of dominant–negative Ras is associated with abnormal T-cell development and function (Swan *et al.*, 1995). Another related concept is the development of mice expressing a viral transgene, with potential immunosuppressive properties. For example, HIV-1 TAT transgenic mice have defective proliferative and cytotoxic response and an increased secretion of interleukin-10 (Garza *et al.*, 1996).

Spectacular progress has recently been made in the elucidation of the genetic mechanisms responsible for immunodeficiency in several murine inbred strains (Table XVI.5.2). In a few cases, similar mutations have been demonstrated to be responsible for immunodeficiency syndromes in humans or other animal species. The gene of which mutation is responsible for the beige phenotype is similar to the gene involved in human Chediak–Higashi disease (Nagle *et al.*, 1996). A deficiency of subunit p450 of a DNA-dependent kinase inhibits V(D)J recombination is responsible for a severe combined immunodeficiency in mice and in horses. The mechanism responsible for X-linked agammaglobulinemia involves Bruton tyrosine kinase in mice as in humans. In both cases, the mutation is associated with a loss of binding capacity for a metabolite of the inositol phosphate pathway (Table XVI.5.2).

More than 1000 articles have been published on targeted gene disruption and the potential impact on immune responses. It is therefore impossible to be exhaustive in the description of these knockout mice. Different categories of knockouts are presented together with a few examples inside each category (Tables XVI.5.3 and XVI.5.4). In view of the crucial role played by protein

Table XVI.5.3 Examples of knock-out mice with significant impact on immunocompetence

Cytokines

Interleukin 2 (IL-2): mild immunodeficiency, inflammatory bowel disease (Kündig *et al.*, 1993)

Interleukin 7 (IL-7): severe lymphopenia (B cells and T cells) (von Freeden-Jeffry *et al.*, 1995)

TNF-β (lymphotoxin): lack of lymph nodes, undifferentiated spleen, hypogammaglobulinaemia, defective Ig switch (De Togni *et al.*, 1994)

IFN-γ: increased proliferative responses to lectins or MLC, impaired expression of MHC class II products, reduced NK cell activity, increased secretion of T_H2 cytokines, impaired generation of reactive oxygen products, increased susceptibility to infection by intracellular bacteria (i.e. Mycobacterium) (Dalton *et al.*, 1993)

Cytokine receptors

Common cytokine (IL-2, -4, -7, -9, -15) receptor γ chain: severe combined immunodeficiency involving T cells, B cells and NK cells (Cao *et al.*, 1995; DiSanto *et al.*, 1995). Murine model for human (Noguchi, 1993) and canine (Henthorn *et al.*, 1994) X-linked severe combined immunodeficiency

Type I (IFN-α, -β, -ω) interferon receptor: increased susceptibility to viral infection (lymphochoriomeningitis virus, Semliki Forest virus, Theiler's virus) (reviewed in van den Broek, *et al.*, 1995)

Type II (IFN-γ) interferon receptor: phenotype grossly similar to IFN-$\gamma^{-/-}$ mice, increased susceptibility to infection by parasites (i.e. leishmania) and intracellular bacteria (i.e. mycobacterium) (van den Broek *et al.*, 1995).

(*continued*)

Table XVI.5.3 (*Continued*)

T-cell membrane coreceptors

CD4: mild immunodeficiency, normal development of $CD8^+$ T cells and cytotoxic responses, partial inhibition of TH responses (Rahemtulla *et al.*, 1991)

CD8 (α chain): absence of peripheral CD8+ T cells, inhibition of class I-restricted responses, normal CD4 responses, increased susceptibility to lymphochoriomeningitis virus infection (Fung-Leung *et al.*, 1991)

CD40 ligand: impaired T_H-dependent activation of B cells; normal activation of T cells (Oxenius *et al.*, 1996). Possible model for human X-linked immunodeficiency with IgM

CD28: normal thymocyte development, normal number of peripheral T cells, reduced secretion of IL-2, although initial response normal, failure to sustain a proliferation after polyclonal or antigen-specific stimulation (Shahinian *et al.*, 1993)

B-cell receptor (BCR)

J_H (segment of the heavy chain locus): absence of peripheral B cells, absence of sIg^+ B cells in the bone marrow, normal T cells and T-cell responses (Chen *et al.*, 1993)

B-cell coreceptors

CD40: impaired T_H-dependent activation of B cells; normal activation of T cells (Oxenius *et al.*, 1996)

CD43 (leukosialin, sialophorin): increased proliferative responses, increased homotypic adhesion, reduced viral clearance, possible model for Wiskott–Aldrich disease (Manjunath *et al.*, 1995)

Integrins

CD18 (β chain of the $\beta 2$ integrins LFA-1, Mac-1 and p150, 95): impaired inflammatory responses, allograft rejection, and leukocyte migration (Wilson *et al.*, 1993). Possible model for human leukocyte adhesion deficiency (LAD) type 1

ICAM-1: phenotype similar to $CD18^{-/-}$, in addition decreased contact hypersensitivity, increased resistance to septic shock induced by LPS (Sligh *et al.*, 1993; Xu *et al.*, 1994)

Major histocompatibility products, antigen processing and presentation

β_2-Microglobulin: phenotype similar to CD8α knock-out mice, absence of CTL responses, increased susceptibility to viral infection

I-Aβ (MHC class-II): phenotype grossly similar to $CD4^{-/-}$ mice; some $CD4^{low}$ T cells in LN; impaired T_H responses and decreased levels of IgG1 (Cosgrove, 1991). Possible model for human congenital immunodeficiency caused by defective MHC class II molecules

TAP-1: phenotype grossly similar to $CD8^{-/-}$ mice, impaired expression of class I molecules (van Kaer *et al.*, 1992)

CD3 chains

CD3 $\zeta\eta$: normal transition from DN to DP, reduced number of thymocytes, reduced expression of TCR, association of TCR with FcεRI γ chain on IEL

TCR chains

TCRα: normal expansion of DP cells in thymus, absence of SP thymocytes, absence of $\alpha\beta$ T cells, normal level of $\gamma\delta$ T cells, B cells and NK cells, impaired humoral responses to T cell-dependent antigens, absence of allograft rejection (Mombaerts *et al.*, 1992a)

TCRβ: small thymus, defect in the expansion of DP cells, absence of $\alpha\beta$ T cells, normal level of $\gamma\delta$ T cells, B cells and NK cells, impaired humoral responses to T cell-dependent antigens, absence of allograft rejection (Mombaerts *et al.*, 1992a)

TCRδ: normal thymocyte development, absence of $\gamma\delta$ T cells, normal humoral responses, normal level of $\alpha\beta$ T cells, B cells and NK cells, normal allograft rejection (Itohara *et al.*, 1993)

Transcription factors

IRF-1 (interferon regulatory factor-1): strongly decreased number of peripheral $CD8^+$ T cells, impaired IFN-α and IFN-β secretion after certain stimuli, increased susceptibility to viral infection (Matsuyama *et al.*, 1993)

NF-AT: lymphoproliferative disease, striking impairment of early IL-4 secretion, minor defect in IL-2 secretion (Hodge *et al.*, 1995). Human immunodeficiency due to mutated NF-AT gene is associated with strongly decreased IL-2 secretion

bcl-3 (member of the Iκ B family): normal development, severe defects in humoral response, susceptibility to infection with *Listeria monocytogenes* and *Streptococcus pneumoniae*, absence of germinal centers in the spleen (Schwartz *et al.*, 1997)

CREB: impaired thymocyte activation and IL-3 secretion (Barton *et al.*, 1996)

Cytotoxic process

Perforin: normal development of the immune system, normal activation of $CD8^+$ T cells; failure to eliminate viral infection (Walsh *et al.*, 1994)

Proteins involved in V(D)J recombination

RAG-1 and RAG-2: absent B cells and T cells (T^-B^-); thymocyte maturation arrested at early DN stage (Mombaerts *et al.*, 1992b; Shinkai *et al.*, 1992)

Ku86: combined immunodeficiency caused by defective V(D)J recombination (Zhu *et al.*, 1996)

Terminal deoxynucleotidyl transferase (TdT): persistence of a fetal antigen receptor repertoire

DNA synthesis

Adenosine deaminase: total deficiency is lethal, partial deficiency induces lymphopenia and mild immunodeficiency (Blackburn *et al.*, 1996) (proposed murine model for human ADA deficiency)

Table XVI.5.4 Genetically modified strains with abnormalities of TCR- and cytokine receptor-associated signaling pathways

Gene involved	*Ontogeny*	*T cells*	*B cells*
Antigen receptor signaling			
lck	Early block of thymocyte differentiation arrested transition from DN to DP cells (Molina *et al.*, 1992)	T-cell lymphopenia, slight inhibition of proliferative responses to TCR crosslinking, impaired cytotoxic activity	Normal responses to T-independent antigens
fyn	Grossly normal thymocyte differentiation, reduced proliferation of thymocytes (Appleby *et al.*, 1992; Stein *et al.*, 1992)	Normal proliferative responses	Normal responses to T-independent antigens
ZAP-70 (proposed model for human immunodeficiency due to ZAP-70 defect; Arpaia *et al.*, 1994)	Major defects in thymocyte differentiation, failure to generate SP $CD4^+$ and $CD8^+$ thymocytes. Altered negative selection (Chan *et al.*, 1994)	Refractory to mitogenic stimuli	Normal responses to T-independent antigens
Syk	Normal thymocyte differentiation	Normal proliferative responses	Profound abnormalities of B cell development (block at the pro-B stage) (Cheng *et al.*, 1995)
Itk (analog of Btk)	Decreased number of SP $CD4^+$ thymocytes	Reduced proliferative responses, increased by rIL-2	Normal responses to T-independent antigens
Cytokine receptor signaling			
JAK3 (proposed model for human autosomal severe combined immunodeficiency)	Small sized thymus but differentiation grossly normal. Increased CD4/CD8 ratio	T cell lymphopenia, absent IEL and NK cells. Increased CD4/CD8 ratio in the spleen with anergic CD4 T cells.	B cell lymphopenia, block at the pro-B stage
STAT-1		Absent lymph nodes increased susceptibility to viral infection, lack of response to IFN-α and IFN-γ, normal response to other cytokines (Durbin *et al.*, 1996; Meraz *et al.*, 1996)	Normal responses to T-independent antigens

kinases in regulating T-cell differentiation and activation processes, Table XVI.5.4 specifically overviews recent findings concerning the phenotype of mice with targeted disruption of protein tyrosine kinase genes.

Many of these knockouts are models for human or animal immunodeficiencies. For example, mice with targeted disruption of the gene encoding for the gamma chain of the common cytokine receptor present a severe combined immunodeficiency syndrome similar to human and canine SCID and mice with a knockout of JAK3 gene are similar to the autosomal form of human scid. Interestingly, despite the similarity of the gene involved, the phenotype is sometimes different in the murine model and in human disease. Humans with mutated ZAP-70 kinase present a severe T-cell defect with a lack of mature $CD8^+$ T cells, however $CD4^+$ T cells are present in the peripheral lymphoid, suggesting that another protein tyrosine kinase (i.e. syk) compensates the gene defect (Arpaia *et al.*, 1994). In constrast, mice with a targeted disruption of the ZAP-70 gene develop a defect involving $CD8^+$ and $CD4^+$ T-cells (Chan *et al.*, 1994).

Models of murine acquired immunodeficiency have become less relevant since the development of genetically engineered strains. Surgical (neonatal thymectomy), physical (ionizing radiation) or pharmacological suppression of the immune system (alkylating agents, cyclosporinA, corticosteroids) will not be discussed in this chapter. Infection of susceptible strains of mice with the Duplan strain of murine leukemia viruses (MuLV) induces a syndrome, termed murine acquired immune deficiency syndrome (AIDS) (MAIDS) which includes a striking lymphoproliferative disease, T-cell and B-cell immunodeficiency and the late development of B-cell lymphomas (Mistry and Duplan, 1973; Mosier *et al.*, 1985). A defective retrovirus is responsible for the disease and infects mostly B cells and not T cells (Huang *et al.*, 1991). Nevertheless, T cells present striking functional defects such as impaired calcium fluxes in response to mitogens or ionophores (Moutschen *et al.*, 1996) and are strictly required to sustain the lymphoproliferative process. MAIDS is clearly very different from human immunodeficiency virus (HIV) infection and should rather be considered as a model for ubiquitous immune defects as observed in lymphoproliferative or malignant states. SCID mice, grafted with human or feline (Johnson *et al.*, 1995) hematopoietic cells can develop an immune system which can be used as a model for species specific infectious diseases. The efficiency of the grafting process can be increased in SCID-transgenic mice expressing the genes for human cytokines such as interleukin-3 and granulocyte/macrophage-colony stimulating factor (GM-CSF) (Bock *et al.*, 1995). Such SCID-HU mice can be infected with the HIV virus (Mosier *et al.*, 1985) and provide useful information about the pathogenesis of HIV infection in a small animal model. Because of the recent identification of HIV coreceptors, a powerful approach could be the development of double transgenic mice expressing human CD4 and various chemokine receptors (i.e. CC-CR5).

Chronic alloimmune stimulation can also induce an immunodeficiency syndrome which bears some similarity to human AIDS. Allogeneic pregnancies followed by chronic immunization of the offspring with paternal lymphocytes induces lymphoproliferation, $CD4^+$ T cell lymphopenia, Kaposi-like sarcomas and B-cell lymphomas (Tergrigorov *et al.*, 1997).

6. References

Abott, C., Malas, S., Pilz, A., Pate, L., Ali, R., Peters, J. (1994). Linkage mapping around the ragged (Ra) and wasted (wst) loci on distal mouse chromosome 2. *Genomics* **20**, 94–98.

Adachi, T., Schamel, W. W. A., Kim, K. M., Watanabe, T., Becker, B., Nielsen, J., Reth, M. (1996). The specificity of association of the IgD molecule with the accessory proteins BAP31/BAP29 lies in the IgD transmembrane sequence. *EMBO J.* **15**, 1534–1541.

Alt, F. W., Anea, V., Bothwell, A. L., Baltimore, D. (1980). Activity of multiple light chain genes in murine myeloma cells producing a single, functional light chain. *Cell* **21**, 1–12.

Alt, F. W., Blackwell, T., Yancopoulous, G. (1987). Development of the primary antibody repertoire. *Science* **238**, 1079–1087.

Appleby, M. W., Gross, J. A., Cooke, M. P., Levin, S. D., Qian, X., Perlmutter, R. M. (1992). Defective T cell receptor signaling in mice lacking the thymic isoform of p59fyn. *Cell* **70**, 751–763.

Arpaia, E., Shahar, M., Dadi, H., Cohen, A., Roifman, C. M. (1994). Defective T cell receptor signaling and CD8+ thymic selection in humans lacking ZAP-70 kinase. *Cell* **76**, 947–958.

Barton, K., Muthusamy, N., Chanyangam, M., Fischer, C., Clendenin, C., Leiden, J. M. (1996). Defective thymocyte proliferation and IL-2 production in transgenic mice expressing a dominant-negative form of CREB. *Nature* **379**, 81–85.

Billips, L. G., Petitte, D., Dorshkind, K., Narayanan, R., Chia, C.-P., Landreth, K. S. (1992). Differential roles of stromal cells, interleukine-7 and kit-ligand in the regulation of B lymphopoiesis. *Blood* **79**, 1185–1192.

Blackburn, M. R., Datta, S. K., Wakamiya, M., Vartabedian, B. S., Kellems, R. E. (1996). Metabolic and immunologic consequences of limited adenosine deaminase expression in mice. *J. Biol. Chem.* **271**, 15203–15210.

Blankenstein, T., Bonhomme, E., Krawinkel, U. (1987). Evolution of pseudogenes in the immunoglobulin VH-gene family on the mouse. *Immunogenetics* **26**, 237–248.

Blunt, T., Finnie, N. J., Taccioli, G. E., Smith, G. C. M., Demengeot, J., Gottlieb, T. M., Mizuta, R., Varghese, A. J., Alt, F. W., Jeggo, P. A., Jackson, S. P. (1995). Defective DNA-dependent protein kinase activity is linked to V(D)J recombination and DNA repair defects associated with the murine acid mutation. *Cell* **80**, 813–823.

Bock, T. A., Orlic, D., Dunbar, C. E., Broxmeyer, H. E., Bodine, D. M. (1995). Improved engraftment of human hematopoietic cells in severe combined immunodeficient (SCID) mice carrying human cytokine transgenes. *J. Exp. Med.* **182**, 2037–2043.

Brissette, J. L., Li, J., Kamimura, J., Lee, D., Dotto, G. P. (1996).

The product of the mouse nude locus, Whn, regulates the balance between epithelial cell growth and differentiation. *Genes Dev.* **17**, 2212–2221.

Brodeur, P. H., Riblet, R. (1984). The immunoglobulin heavy chain variable region (Igh-V) locus in the mouse. I. One hundred Igh-V gene comprise seven families of homologous genes. *Eur. J. Immunol.* **14**, 922–930.

Brodeur, P. H., Osman, G. E., Mackle, J. J., Lalor, T. M. (1988). The organization of the mouse Igh-V locus. *J. Exp. Med.* **168**, 2261–2278.

Burkhardt, J. K., Wiebel, F. A., Hester, S., Argon, Y. (1993). The giant organelles in beige and Chediak–Higashi fibroblasts are derived from late endosomes and mature lysosomes. *J. Exp. Med.* **178**, 1845–1856.

Cambier, J. C., Pleiman, C. M., Clark, M. R. (1994). Signal transduction by the B cell antigen receptor and its coreceptors. *Annu. Rev. Immunol.* **12**, 457–486.

Cao, X., Shores, E. W., Hu-Li, J., Anver, M. R., Kelsall, B. L., Russel, S. M., Drago, J., Noguchi, M., Grinberg, A., Bloom, E. T. (1995). Defective lymphoid development in mice lacking expression of the common cytokine receptor γ chain. *Immunity* **2**, 223–228.

Carayannopoulos, L., Capra, D. (1993). Immunoglobulins, structure and function. In *Fundamental Immunology* (ed. Paul, W. E.), pp. 283–314. Raven Press Ltd, New York, USA.

Carsetti, R., Kohler, G., Lamers, M. C. (1995). Transitional B cells are the target of negative selection in the B cell compartment. *J. Exp. Med.* **181**, 2129–2140.

Chan, A. C., Van Oers, N. S. C., Tran, A., Turka, L., Law, C.-L., Ryan, J. C., Clark, E. A., Weiss, A. (1994). Differential expression of ZAP-70 and Syk protein tyrosine kinases, and the role of this family of protein tyrosine kinases in TCR signaling. *J. Immunol.* **152**, 4758–4766.

Chen, J., Trounstine, M., Alt, F. W., Young, F., Kurahara, C., Loring, J. F., Huszar, D. (1993). Immunoglobulin gene rearrangement in B cell deficient mice generated by targeted deletion of the J_H locus. *Int. Immunol.* **5**, 647–656.

Cheng, A. M., Rowley, B., Pao, W., Hayday, A., Bolen, J. B., Pawson, T. (1995). Syk tyrosine kinase required for mouse viability and B-cell development. *Nature* **378**, 303–306.

Clark, L. B., Foy, T. M., Noelle, R. J. (1996). CD40 and its ligand. *Adv. Immunol.* **63**, 43–78.

Coleclough, C., Perry, R. P., Karjalainen, K., Weigert, M. (1981). Aberrant rearrangements contribute significantly to allelic exclusion of immunoglobulin gene expression. *Nature* **290**, 372–378.

Colle, J. H., Le Moal, M. A., Truffa-Bachi, P. (1990). Immunological memory. *Critical Rev. Immunol.* **10**, 259–306.

Cosgrove, D., Gray, D., Dierich, A., Kaufman, J., LeMeur, M., Benoist, C., Mathis, D. (1991). Mice lacking MHC class II molecules. *Cell* **66**, 1051.

Coutinho, A., Freitas, A. A., Holmberg, D., Grandien, A. (1992). Expression and selection of murine antibody repertoires. *Int. Rev. Immunol.* **8**, 173–187.

Dalton, D. K., Pitts-Meek, S., Keshav, S., Figari, I. S., Bradley, A., Stewart, T. A. (1993). Multiple defects of immune cell function in mice with disrupted interferon-gamma genes. *Science* **259**, 1739.

D'Hoostelaere, L. A., Klinman, D. (1990). Characterization of a new mouse VK group. *J. Immunol.* **145**, 1706.

D'Hoostelaere, L. A., Huppi, K., Mock, B., Mallett, C., Potter, M. (1988). The Igk L chain allelic groups among the Igk haplotypes and Igk crossover populations suggest a gene order. *J. Immunol.* **141**, 652–661.

Dempsey, P. W., Allison, M. E. D., Akkaraju, S., Goodnow, C. C., Fearon, D. T. (1996). C3D of complement as a molecular adjuvant: bridging innate and acquired immunity. *Science* **271**, 348–350.

Desiderio, S. V., Yancopoulos, G. D., Paskind, M., Thomas, E., Boss, M. A., Landau, N., Alt, F. W., Baltimore, D. (1984). Insertion of N-regions into heavy chain genes is correlated with the expression of terminal deoxytransferase in B cells. *Nature* **311**, 752–755.

De Togni, P., Goellner, J., Ruddle, N. H., Streeter, P. R., Flick, A., Mariathasan, S., Smith, S. C., Carlson, R., Shornick, L. P., Strauss-Schoenberger, J., Russel, L. P., Karr, R., Chaplin, D. D. (1994). Abnormal development of peripheral lymphoid organs in mice deficient in lymphotoxin. *Science* **264**, 703–707.

Difilippantonio, M. J., McMahan, C. J., Eastman, Q. M., Spanopoulou, E., Schatz, D. G. (1996). RAG1 mediates signal sequence recognition and recruitment of RAG2 in V(D)J recombination. *Cell* **87**, 253–262.

Dildrop, R., Krawinkel, U., Winter, E., Rajewsky, K. (1985). VH gene expression in murine lipopolysaccharide blasts distributes over the nine known VH-gene groups and may be random. *Eur. J. Immunol.* **15**, 1154–1156.

DiSanto, J. P., Müller, W., Guy-Grand, D., Fischer, A., Rajewski, K. (1995). Lymphoid development in mice with a targeted deletion of the interleukin 2 receptor chain. *Proc. Natl Acad. Sci. USA* **92**, 377–381.

Doody, G. M., Dempsey, P. W., Fearon, D. T. (1996). Activation of B lymphocytes. Integration signals from CD19, CD22 and FCγRIIb1. *Cur. Opin. Immunol.* **8**, 378–382.

Durbin, J. E., Hackenmiller, R., Simon, M. C., Levy, D. E. (1996). Targeted disruption of the mouse Stat-1 gene results in compromised innate immunity to viral disease. *Cell* **84**, 443–450.

Dzierzak, E. A., Janeway, C. A. Jr, Richard, N., Bothwell, A. (1986). Molecular characterization of antibodies bearing Id-460. I. The structure of two highly homologous VH genes used to produce idiotype positive immunoglobulins. *J. Immunol.* **136**, 1864–1870.

Eisen, H. N., Reilly, E. B. (1985). Lambda chains and genes in inbred mice. *Annu. Rev. Immunol.* **3**, 337–365.

Elliot, E. A., Flavell, R. A. (1994). Transgenic mice expressing constitutive levels of IL-2 in islet beta cells develop diabetes. *Int. Immunol.* **6**, 1629–1637.

Engel, P., Zhou, L.-J., Ord, D. C., Sato, S., Koller, B., Leder, J. F. (1995). Abnormal B lymphocyte development, activation and differentiation in mice that lack or overexpress the CD19 signal transduction molecule. *Immunity* **3**, 39–50.

Esser, C., Radbruck, A. (1990). Immunoglobulin class switching: molecular and cellular analysis. *Annu. Rev. Immunol.* **8**, 717–735.

Fearon, D. T., Carter, R. H. (1995). The CD19/CR2/TAPA-1 complex of B lymphocytes: linking to acquired immunity. *Annu. Rev. Immunol.* **13**, 127–149.

Feeney, A. J. (1992). Comparison of junctional diversity in the neonatal and adult immunoglobulin repertoires. *Int. Rev. Immunol.* **8**, 113–122.

Feeney, A. J., Riblet, R. (1993). DST4: a new, and probably the last, functional DH gene in BALB/c mouse. *Immunogenetics* **37**, 217–221.

Fridman, W. H. (1993). Regulation of B-cell activation and

antigen presentation by Fc receptors. *Curr. Opin. Immunol.* 5, 355–360.

Friedrich, J., Campbell, K., Cambier, J. C. (1993). The gamma subunit of the B cell antigen receptor complex is a C-terminally truncated product of the B29 gene. *J. Immunol.* **150**, 2817–2822.

Fung-Leung, W. P., Shilham, M. W., Rahemtulla, A., Kündig, T. M., Vollenweider, M., Potter, J., van Ewijk, W., Mak, T. W. (1991). CD8 is needed for the development of cytotoxic T cells but not helper T cells. *Cell* **65**, 443.

Fukai, K., Oh, J., Karim, M. A., Moore, K. J., Kandil, H. H., Ito, H., Burger, J., Spritz, R. A. (1996). Homozygosity mapping of the gene for Chediak–Higashi syndrome to chromosome 1q42–q44 in a segment of conserved synteny that includes the mouse beige locus (bg). *Am. J. Hum. Genet.* **59**, 620–624.

Fukuda, M., Kojima, T., Kabayama, H., Mikoshiba, K. (1996). Mutation of the pleckstrin homology domain of Brutons tyrosine kinase in immunodeficiency impaired inositol 1,3,4,5-tetrakisphosphate binding capacity. *J. Biol. Chem.* **271**, 30303–30306.

Galli, S. J., Zsebo, K. M., Geissler, E. N. (1994). The kit ligand, stem cell factor. *Adv. Immunol.* **55**, 1–96.

Garza, H. H. Jr, Prakash, O., Carr, D. J. (1996). Aberrant regulation of cytokines in HIV-1 TAT72-transgenic mice. *J. Immunol.* **156**, 3631–3637.

Goldman, H. R., De Franco, A. L. (1994). Biochemistry of B lymphocyte activation. *Adv. Immunol.* **55**, 221–295.

Gong, S., Nussenzweig, M. C. (1996). Regulation of an early developmental checkpoint on the B cell pathway by Igβ. *Science* **272**, 411–414.

Goodnow, C. C., Cyster, J. G., Hartley, S. B. *et al.* (1995). Self-tolerance checkpoints in B lymphocyte development. *Adv. Immunol.* **59**, 279–368.

Grandien, A., Modigliani, Y., Freitas, A. A., Andersson, J., Coutinho, A. (1994). Positive and negative selection of antibody repertoires during B cell differentiation. *Immunol. Rev.* **137**, 53–89.

Grawunder, U., Leu, T. M. J., Schatz, D. G., Werner, A., Rolink, A. G., Melchers, F., Winkler, T. H. (1995). Down regulation of RAG1 and RAG2 gene expression in pre-B cells after functional immunoglobulin heavy chain rearrangement. *Immunity* **3**, 601–608.

Gray, D., Siepmann, K., Van Essen, D. *et al.* (1996). B–T lymphocyte interactions in the generation and survival of memory cells. *Immunol. Rev.* **150**, 45–61.

Greenspan, N. S., Cooper, L. J. N. (1992). Intermolecular cooperativity: active to why mice have IgG3? *Immunol. Today* **13**, 164–168.

Gu, H., Tarlinton, D., Müller, W., Rajeswky, K., Föster, I. (1991). Most peripheral B cells in mice are ligand selected. *J. Exp. Med.* **173**, 1357–1371.

Hardy, C. C. (1991). Variable gene usage, physiology and development of Ly-1+ (CD5) B cells. *Curr. Opin. Immunol.* **4**, 181–185.

Hardy, C. C., Carmack, C. E., Shinton, S. A., Kemp, J. D., Hayakawa, K. (1991). Resolution and characterization of pro-B and pre-pro-B cell stages in normal mouse bone marrow. *J. Exp. Med.* **173**, 1213–1225.

Hardy, R. R., Hayakawa, K. (1994). CD5 B cells, a fetal B cell lineage. *Adv. Immunol.* **55**, 297–339.

Hathcock, K. S., Hodes, R. J. (1996). Role of CD28-B7 costimulatory pathways in T cell-dependent B cell responses. *Adv. Immunol.* **62**, 131–166.

Henthorn, P. S., Somberg, R. L., Fimiani, V. M., Puck, J. M., Patterson, D. F., Felsburg, P. J. (1994). IL-2R gamma gene microdeletion demonstrates that canine X-linked severe combined immunodeficiency is a homologue of the human disease. *Genomics* **23**, 62–68.

Herzenberg, L. A., Haughton, G., Rajewsky, K., editors (1992). CD5 B cells in development and disease. *Ann. NY Acad. Sci.* **651**.

Hirayama, F., Lyman, S. D., Clark, S. C., Ogawa, M. (1995). The flt3 ligand supports proliferation of lymphohematopoietic progenitors and early B-lymphoid progenitors. *Blood* **85**, 1762–1769.

Hodge, M. R., Ranger, A. M., Charles de la Brousse, F., Hoey, T., Grusby, M. J., Glimcher, L. H. (1996). Hyperproliferation and dysregulation of IL-4 secretion in NF-ATP-deficient mice. *Immunity* **4**, 397–405.

Honjo, T., Matsuda, F. (1995). Immunoglobulin and heavy chain loci of mouse and human. In *Immunoglobulin Genes*, 2nd edn (eds Honjo, T., Alt, F. W.), pp. 145–172. Academic Press, London, UK.

Huang, M., Simard, C., Jolicoeur, P. (1991). The majority of cells infected with the defective murine AIDS virus belongs to the B-cell lineage. *J. Virol.* **65**, 6562.

Hulett, M. D., Hogarth, P. M. (1994). Molecular basis of Fc receptor function. *Adv. Immunol.* **54**, 1–127.

Huppi, K., D'Hoostelaere, L. A. (1991). Letter to the editor. *J. Immunol.* **146**, 4053.

Huston, J. S., Margolies, M. N., Haber, E. (1996). Antibody binding sites. *Adv. Protein Chem.* **49**, 330–450.

Ichihara, Y., Hayashida, H., Miyazawa, S., Kurosawa, Y. (1989). Only DF16.2, DSP2 and DQ52 gene families exist in mouse immunoglobulin heavy chain diversity gene loci, of which DF16.2 and DSP2 originate from the same primordial DH gene. *Eur. J. Immunol.* **19**, 1849–1854.

Iglesias, A., Lamers, M., Kohler, G. (1987). Expression of immunoglobulin γ chain causes allelic exclusion in transgenic mice. *Nature* **330**, 482–484.

Itohara, S., Mombaerts, P., Lafaille, J., Iacomini, J., Nelson, A., Clarke, A. R., Hooper, M. L., Farr, A., Tonegawa, S. (1993). T cell delta gene mutant mice: independent generation of alpha beta T cells and programmed rearrangement of gamma delta TCR genes. *Cell* **72**, 337.

Jeong, H. D., Teale, J. M. (1988). Comparison of the fetal and adult functional B cell repertoires by analysis of VH gene family expression. *J. Exp. Med.* **168**, 589–598.

Jeong, H. D., Komisar, J. L., Kraig, E., Teale, J. M. (1988). Strain-dependent expression of VH gene families. *J. Immunol.* **140**, 2436–2441.

Kabat, E. A., Wu, T. T., Perry, H., Gottesman, K. S., Foeller, C. (1991). *Sequences of Proteins of Immunological Interest.* US Department of Health and Human Services, MA, USA.

Kalled, S. L., Brodeur, P. H. (1991). Utilization of VK families and VK exons: implications for the available B cell repertoire. *J. Immunol.* **147**, 3194–3200.

Kantor, A. B., Herzenberg, L. A. (1993). Origin of murine B cell lineages. *Annu. Rev. Immunol.* **11**, 501–538.

Karasuyama, H., Rolink, A., Melchers, F. (1996). Surrogate light chain in B cell development. *Adv. Immunol.* **63**, 1–42.

Kaushik, A., Shulze, D. H., Bona, C. A., Kelsoe, G. (1989).

Murine VK gene expression violates the VH paradigm. *J. Exp. Med.* **169**, 1859–1864.

Kearney, J. F. (1993). Idiotypic network. In *Fundamental Immunology* (ed. Paul, W. E.), pp. 887–902. Raven Press Ltd, New York, USA.

Kehrl, J. H., Riva, A., Wilson, G. L., Thévenin, C. (1994). Molecular mechanisms regulating CD19, CD20 and CD22 gene expression. *Immunol. Today* **15**, 432–436.

Kerner, J. D., Appleby, M. W., Mohr, R. N. *et al.* (1995). Impaired expansion of mouse B cell progenitors lacking Btk. *Immunity* **3**, 301–312.

Khan, W. N., Alt, F. W., Gerstein, R. M. *et al.* (1995). Defective B cell development and function in Btk-deficient mice. *Immunity* **3**, 283–299.

Kim, K. M., Reth, M. (1995a). Function of B-cell antigen receptor of different classes. *Immunol. Lett.* **44**, 81–85.

Kim, K. M., Reth, M. (1995b). The B-cell antigen receptor of class IgD induces a stronger and more prolonged protein tyrosine phosphorylation than of class IgM. *J. Exp. Med.* **181**, 1005–1014.

Kim, K. M., Alber, G., Weiser, P., Reth, M. (1993). Signalling function of the B cell antigen receptors. *Immunol. Rev.* **132**, 125–146.

Kim, K. M., Adachi, T., Nielsen, P. J., Terashima, M., Lamers, M. C., Kohler, G., Reth, M. (1994). Two new proteins preferentially associated with membrane immunoglobulin D. *EMBO J.* **13**, 3793–3800.

Kirchgessner, C. U., Patil, C. K., Evans, J. W., Cuomo, C. A., Fried, L. M., Carter, T., Oettinger, M. A., Brown, J. M. (1995). DNA dependent kinase (p350) as a candidate gene for the murine SCID defect. *Science* **267**, 1178–1182.

Kitamura, D., Rajewsky, K. (1992). Targeted disruption of μ chain membrane exon causes loss of heavy-chain allelic exclusion. *Nature* **356**, 154–156.

Kitamura, D., Kudo, A., School, S., Muller, W., Melchers, F. (1992). A critical role of $\lambda 5$ protein in B cell development. *Cell* **69**, 823–831.

Klinman, N. R. (1996a). The clonal selection hypothesis and current concepts of B cell tolerance. *Immunity* **5**, 189–195.

Klinman, N. R. (1996b). *In vitro* analysis of the generation and propagation of memory B cells. *Immunol. Rev.* **150**, 91–111.

Knight, K. L., Crane, M. A. (1994). Generating the antibody repertoire in rabbit. *Adv. Immun.* **56**, 179–218.

Kofler, R. (1988). A new murine Ig VH gene family. *J. Immunol.* **140**, 4031–4034.

Kofler, R., Helmberg, A. (1991). Comment to the article 'A new Igk-V gene family in the mouse'. *Immunogenetics* **34**, 139–140.

Kofler, R., Perlmutter, R. M., Noonan, D. J., Dixon, F. J., Theofilopoulos, A. N. (1985). Ig heavy chain variable region gene complex of lupus mice exhibits normal restriction fragment length polymorphism. *J. Exp. Med.* **162**, 346–351.

Kofler, R., Duchosal, M. A., Dixon, F. J. (1989). Complexity, polymorphism and connectivity of mouse VK gene families. *Immunogenetics* **29**, 65–74.

Kofler, R., Geley, S., Kofler, H., Helmberg, A. (1992). Mouse variable-region gene families: complexity, polymorphism and use in non-autoimmune responses. *Immunol. Rev.* **128**, 5–21.

Koike, R., Nishimura T., Yasumizu, R., Tanaka, H., Hataba, Y., Hataba, Y., Watanabe, T., Miyawaki, S., Miyasaka, M. (1996). The splenic marginal zone is absent in alymphoplastic aly mutant mice. *Eur. J. Immunol.* **26**, 669–675.

Kramer, D. R., Cebra, J. J. (1995). Early appearance of 'natural' mucosal IgA responses and germinal centers in suckling mice developing in the absence of maternal antibodies. *J. Immunol.* **154**, 2051–2062.

Kroemer, G., Helmberg, A., Bernot, A., Auffray, C., Kofler, R. (1991). Evolutionary relationship between human and mouse immunoglobulin kappa light chain variable region genes. *Immunogenetics* **33**, 42–49.

Kündig, T. M., Schorle, H., Bachmann, M., Hengartner, H., Zinkernagel, R. M., Horak, I. (1993). Immune responses in interleukin-2 deficient mice. *Science* **262**, 1059–1061.

Kurosawa, Y., Tonegawa, S. (1982). Organization, structure and assembly of immunoglobulin heavy chain diversity DNA segments. *J. Exp. Med.* **155**, 201–218.

Lafaille, J. J., Decloux, A., Bonneville, M., Takagaki, Y., Tonegawa, S. (1989). Junctional sequences of T-cell receptor γ delta genes: implications for γ delta T cell lineages and for a novel intermediate of V-(D)-J joining. *Cell* **59**, 859–870.

Landau, N. R., Schatz, D. G., Rosa, M., Baltimore, D. (1987). Increased frequency of N-regional insertion in a murine pre-B-cell line infected with a terminal deoxynucleotidyl transferase retroviral expression vector. *Mol. Cell. Biol.* **7**, 3237–3243.

Lanzavecchia, A. (1996). Mechanisms of antigen uptake for presentation. *Curr. Opin. Immunol.* **8**, 348–354.

Law C.-L., Sidorenko, S. P., Clark, E. A. (1994). Regulation of lymphocyte activation by the cell-surface molecule CD22. *Immunol. Today* **15**, 442–448.

Lewis, S. M. (1994). The mechanism of V(D)J joining: lessons from molecular, immunological, and comparative analyses. *Adv. Immunol.* **56**, 27–150.

Lieber, M. R., Hesse, J. E., Mizuuchi, K., Geller, M. (1988). Lymphoid V(D)J recombination: nucleotide insertion at signal joints as well as coding joints. *Proc. Natl Acad. Sci. USA* **85**, 8588–8592.

Livant, D., Blatt, C., Hood, L. (1986). One heavy chain variable region gene segment subfamily in the BALB/c mouse contains 500–1000 or more members. *Cell* **47**, 461–470.

Lorenz, U., Ravichandran, K. S., Burakoff, S. J., Neel, B. G. (1996). Lack of SHTP1 results in src-family kinase hyperactivation and thymocyte hyperresponsiveness. *Proc. Natl Acad. Sci. USA* **93**, 9324–9629.

MacLennan, I. C. M. (1994). Germinal centers. *Annu. Rev. Immunol.* **12**, 117–139.

Malynn, B. A., Yancopoulos, G. D., Barth, J. E., Bona, C. A., Alt, F. W. (1990). Biased expression of JH-proximal VH genes occurs in the newly generated repertoire of neonatal and adult mice. *J. Exp. Med.* **171**, 843–859.

Manjunath, N., Correa, M., Ardman, M., Ardmann, B. (1995). Negative regulation of T-cell adhesion and activation by CD43. *Nature* **377**, 535.

Manz, J., Denis, K., Witte, O., Brenster, R., Storb, U. (1988). Feedback inhibition of immunoglobulin gene rearrangement by membrane μ, but not by secreted heavy chains. *J. Exp. Med.* **168**, 1363–1381.

Marcos, M. A., Godin, I., Cumano, A. *et al.* (1994). Developmental events from hemopoietic stem cells to B-cell populations and Ig repertoires. *Immunol. Rev.* **137**, 155–171.

Matsuyama, T., Kimura, T., Kitagawa, M., Pfeffer, K., Kawakami, T., Watanabe, N., Kündig, T. M., Amakawa, R., Kishihara, K., Wakeham, A., Potter, J., Furlonger, C. L., Narendran, A., Suzuki, H., Ohashi, P. S., Tanigushi, T., Mak, T. W. (1993). Targeted disruption of IRF-1 or IRF-2 results in abnormal type I IFN gene induction and aberrant lymphocyte development. *Cell* 75, 83.

Max, E. E., Seidman, J. G., Leder, P. (1979). Sequences of five potential recombination sites encoded close to an immunoglobulin K constant region gene. *Proc. Natl Acad. Sci. USA* **76**, 3450–3454.

Mazanec, M. B., Nedrud, J. G., Kaetzel, C. S., Lamm, M. E. (1993). A three-tiered view of the role of IgA in mucosal defense. *Immunol. Today* **14**, 430–495.

McConnell, H. M., Martinez-Yamout, M. (1996). Insight of antibody combining sites using nuclear magnetic resonance and spin labels haptens. *Adv. Protein Chem.* **49**, 135–149.

Meek, K., Rathbun, G., Reiniger, L., Jaton, J. C., Kofler, R., Tucker, P. W., Capra, J. D. (1990). Organization of murine immunoglobulin VH complex: placement of two new VH families (VH10 and VH11) and analysis of VH family clustering and interdigitation. *Mol. Immunol.* **27**, 1073–1081.

Melchers, F., Karasuyama, H., Haasner, D., Bauer, S., Kudo, A., Sakaguchi, N., Jameson, B., Rolink, A. (1993). The surrogate light chain in B-cell development. *Immunol. Today* **14**, 60–68.

Melchers, F., Haasner, D., Grawunder, U., Kalberer, C., Karasuyama, H., Winkler, T. H., Rolink, A. (1994). Role of IgH and L chains and surrogate H and L chains in the development of cells of the B lymphocytes lineage. *Annu. Rev. Immunol.* **12**, 209–225.

Melchers, F., Robert, A., Grawunder, U., Winkler, T. H., Karasuyama, H., Ghia, P., Andersson, J. (1995). Positively and negatively selecting events during B lymphopoiesis. *Curr. Opin. Immunol.* 7, 214–227.

Meraz, M. A., White, J. M., Sheehan, K. C., Bach, E. A., Rodig, S. J., Dighe, A. S., Kaplan, D. H., Riley, J. K., Greenlund, A. C., Campbell, D., Carver-Moore, K., DuBois, R. N., Clark, R., Aguet, M., Schreiber, R. D. (1996). Targeted disruption of the Stat-1 gene in mice reveals unexpected physiology in the JAK-STAT signaling pathway. *Cell* **84**, 431–432.

Mistry, P. B., Duplan, J. F. (1973). Propriétés biologiques d'un virus isolé d'une radioleucémie C57BL. I. Premiers passages du virus natif. *Bull. Cancer* **60**, 287.

Mitchell, R. N., Barnes, K. A., Grupp, S. A., Sanchez, M., Misulavin, Z., Nussenzweig, M. C., Abbas, A. K. (1995). Intracellular targeting of antigens internalized by membrane immunoglobulin in B lymphocytes. *J. Exp. Med.* **181**, 1705–1714.

Miyawaki, S., Nakamura, Y., Suzuka, H., Koba, M., Yasumizu, R., Ikehara, S., Shibata, Y. (1994). A new mutation, aly, that induces a generalized lack of lymph nodes accompanied by immunodeficiency in mice. *Eur. J. Immunol.* **24**, 429–434.

Molina, T. J., Kishihara, K., Siderovski, D. P., Ewik, W., Van Narendran, A., Timms, E., Wakeham, A., Paige, C. J., Hartmann, K.-U. U., Veilette, A. (1992). Profound block in thymocyte development in mice lacking p561ck. *Nature* **357**, 161–164.

Mombaerts, P., Clark, A. R., Rudnicki, M. A., Iacomini, J., Itohara, S., Lafaille, J. J., Wang, L., Ichikawa, Y., Jaenisch, R., Hooper, M. L., Tonegawa, S. (1992a). Mutations in the T-cell antigen receptor alpha and beta block thymocyte development at different stages. *Nature* **360**, 225.

Mombaerts, P., Iacomini, J., Johnson, R. S., Herrup, K., Tonegawa, S., Papaioannou, V. E. (1992b). RAG-1 deficient mice have no mature B and T lymphocytes. *Cell* **68**, 869.

Mond, J. J., Lees A., Snapper, C. (1995). T-cell independent antigens Type 2. *Annu. Rev. Immunol.* **13**, 655–692.

Mosier, D. E., Subbarao, B. (1982). Thymus-independent antigens: complexity of B lymphocyte activation revealed. *Immunol. Today* **3**, 217–222.

Mosier, D. E., Yetter, R. A., Morse, H. C. III (1985). Retroviral induction of acute lymphoproliferative disease and profound immunosuppression in adult C57BL/6 mice. *J. Exp. Med.* **161**, 766.

Moutschen, M., Trebak, M., Greimers, R., Colombi, S., Boniver, J. (1996). Subset-specific analysis of calcium fluxes in murine AIDS. *Int. Immunol.* **8**, 1715–1727.

Muta, T., Kurosaki, T., Misulavin, Z., Sanchez, M., Nussenzweig, M. C., Ravetch, J. V. (1994). A 13-amino-acid motif in the cytoplasmic domain of FCγRIIb modulates B-cell receptor signalling. *Nature* **368**, 70–73.

Nagata, S., Suda, T. (1995). Fas and Fas ligand: lpr and gld mutations. *Immunol. Today* **16**, 39–43.

Nagle, D. L., Karim, M. A., Woolf, E. A., Holmgren, L., Bork, P., Misumi, D. J., McGrail, S. H., Dussault, B. J. Jr, Perou, C. M., Boissy, R. E., Duyk, G. M., Spritz, R. A., Moore, K. J. (1996). Identification and mutation analysis of the complete gene for Chediak–Higashi syndrome. *Nat. Genet.* **14**, 307–311.

Nemazee, D. (1993). Promotion and prevention of autoimmunity by B lymphocytes. *Curr. Opin. Immunol.* **5**, 866–872.

Nisonoff, A., Wissler, F. C., Lipman, L. N. (1960). Properties of the major components of a peptic digest of rabbit antibody. *Science* **132**, 1770–1771.

Norvell, A., Monroe, J. G. (1996). Acquisition of surface IgD fails to protect from tolerance-induction: both surface IgM- and surface IgD-mediated signals induce apoptosis of immature murine B lymphocytes. *J. Immunol.* **156**, 1328–1332.

Noguchi, M., Yi, H., Rosenblatt, H. M., Filipovich, A. H., Adelstein, S., Modi, W. S., McBride, O. W., Leonard, W. J. (1993). Interleukin-2 receptor γ chain mutation results in X-linked severe combined immunodeficiency in humans. *Cell* **73**, 147–157.

Novotny, J., Bajorath, J. (1996). Computational biochemistry of antibodies and T-cell receptors. *Adv. Protein Chem.* **49**, 150–260.

Nussenzweig, M. C., Shaw, A. C., Sinn, E., Danner, D. B., Holmes, K. L., Morse, H., Leder, P. (1987). Allelic exclusion in transgenic mice that express the membrane form of immunoglobulin μ. *Science* **236**, 816–819.

O'Keefe, T. L., Williams, G. T., Davies, S. L., Clark, E. A. (1996). Hyperresponsive B cells in CD22-deficient mice. *Science* **274**, 798–801.

Oettinger, M. A., Schatz, D. G., Gorka, C., Baltimore, D. (1990). RAG-1 and RAG-2 adjacent genes that synergistically activate V(D)J recombination. *Science* **248**, 1517–1523.

Osmond, D. G. (1986). Population dynamics of bone marrow B lymphocytes. *Immunol. Rev.* **93**, 103–124.

Oxenius, A., Campbell, K. A., Maliszewski, C. R., Kishimoto, T., Kikutani, H., Hentgarnter, H., Zinkernagel, R. M., Bachmann, M. F. (1996). CD40–CD40 ligand interactions are critical in T–B cooperation but not for other anti-viral CD4+ T cell functions. *J. Exp. Med.* **3**, 2209–2218.

Padlan, E. P. (1996). X-ray crystallography of antibodies. *Adv. Protein Chem.* **49**, 57–134.

Palmer, J. L., Nisonoff, A. (1964). Dissociation of rabbit γ-globulin into half-molecules after reduction of one labile disulfide bond. *Biochemistry* **3**, 863–869.

Papavasiliou, F., Misulovin, S., Suh, H., Nussenzweig, M. C.

(1995a). The role of Igβ in precursor B cell transition and allelic exclusion. *Science* **268**, 408–411.

Papavasiliou, F., Misulovin, S., Suh, H., Nussenzweig, M. C. (1995b). The cytoplasmic domains of immunoglobulin Igα and Igβ can independently induce the precursor B cell transition and allelic inclusion. *J. Exp. Med.* **182**, 1389–1394.

Patel, K. J., Neuberger, M. S. (1993). Antigen presentation by the B cell antigen receptor is driven by the alpha/beta sheath and occurs independently of its cytoplasmic tyrosines. *Cell* **74**, 934–946.

Pennell, C. A., Sheehan, K. M., Brodeur, P. H., Clarke, S. H. (1989). Organization and expression of VH gene families preferentially expressed by Ly-1+ (CD5) B cells. *Eur. J. Immunol.* **19**, 2115–2121.

Perlmutter, R. M., Alberola-Ila, J. (1996). The use of dominant–negative mutations to elucidate signal transduction pathways in lymphocytes. *Curr. Opin. Immunol.* **8**, 285–290.

Perlmutter, R. M., Klotz, J. L., Bond, M. W., Nahm, M., Davie, J. M., Hood, L. (1984). Multiple VH gene segments encode murine antistreptococcal antibodies. *J. Exp. Med.* **159**, 179–192.

Perlmutter, R. M., Kearney, J. F., Chang, S. P., Hood, L. E. (1985). Developmentally controlled expression of immunoglobulin VH genes. *Science* **127**, 1597–1600.

Picarella, D., Serunian, L. A., Rosenberg, N. (1991). Allelic exclusion of membrane but not secreted immunoglobulin in a mature B cell line. *Eur. J. Immunol.* **21**, 55–62.

Pleiman, C. M., D'Ambrosio, D., Cambier, J. C. (1994). The B-cell antigen receptor complex: structure and signal transduction. *Immunol. Today* **15**, 393–398.

Porter, R. R. (1985). Separation and isolation of fractions of rabbit γ-globulin containing the antibody and antigenic combining sites. *Nature* **182**, 670–671.

Radic, M. Z., Weigert, M. (1994). Genetic and structural evidence for antigen selection of anti-DNA antibodies. *Annu. Rev. Immunol.* **12**, 487–520.

Rahemtulla, A., Fung-Leung, W. P., Schilham, M. W., Kündig, T. M., Sambhara, S. R., Narendran, A., Arabian, A., Wakeham, A., Paige, C. J., Zinkernagel, R. M. (1991). Normal development and function of CD8+ cells but markedly decreased helper cell activity in mice lacking CD4. *Nature* **353**, 180–184.

Rajewsky, K. (1996). Clonal selection and learning in the antibody system. *Nature* **381**, 751–758.

Ramsden, D. A., Wu, G. E. (1991). Mouse K light chain recombination signal sequences mediate recombination more frequently than do those of λ light chain. *Proc. Natl Acad. Sci. USA* **88**, 10721–10725.

Rankin, J. A., Picarella, D. E., Geba, G. P., Temann, U. A., Prasad, B., DiCosmo, B., Tarallo, A., Stripp, B., Whitsett, J., Flavell, R. A. (1996). Phenotypic and physiologic characterization of transgenic mice expressing interleukin 4 in the lung: lymphocytic and eosinophilic inflammation without airway hyperreactivity. *Proc. Natl Acad. Sci. USA* **93**, 7821–7825.

Ravetch, J. U., Kinet, J.-P. (1991). Fc receptors. *Annu. Rev. Immunol.* **9**, 457–492.

Rawlings, D. J., Saffran, D. C., Tsukoda, S. *et al.* (1993). Mutation of unique region of Bruton's tyrosine kinase in immunodeficient XID mice. *Science* **261**, 358–361.

Reiniger, L., Kaushik, A., Izui, S., Jaton, J.-C. (1988). A member of the new VH gene family encodes anti-bromelinized mouse red blood cell autoantibodies. *Eur. J. Immunol.* **18**, 1521–1533.

Reth, M. (1992). Antigen receptors on B lymphocytes. *Annu. Rev. Immunol.* **10**, 97–121.

Reth, M. (1994). B cell antigen receptors. *Curr. Opin. Immunol.* **6**, 3–8.

Reth, M. G., Ammirati, P., Jackson, S., Alt, F. W. (1985). Regulated progression of a cultured pre-B-cell line to the B-cell stage. *Nature* **317**, 353–355.

Reth, M., Petrac, E., Wiese, P., Label, L., Alt, F. W. (1987). Activation of VK gene rearrangement in pre-B-cells follows the expression of membrane-bound immunoglobulin heavy chain. *EMBO J.* **6**, 3299–3345.

Reynaud, C. A., Bertocci, B., Dahan, A., Weill, J. C. (1994). Formation of chicken B-cell repertoire: ontogenesis, regulation of Ig gene rearrangement and diversification by gene conversion. *Adv. Immun.* **57**, 353–378.

Reynaud, C. A., Garcia, C., Hein, W. R., Weill, J. C. (1995). Hypermutation generating the sheep immunoglobulin repertoire is an antigen-independent process. *Cell* **80**, 115–125.

Riblet, R., Tutter, A., Brodeur, P. (1986). Polymorphism and evolution of Igh-V gene families. *Curr. Top. Microbiol. Immunol.* **127**, 167–172.

Rickert, R. C., Rajewsky, K., Roes, J. (1995). Impairment of T-cell-dependent B-cell responses and B-1 cell development in CD19-deficient-mice. *Nature* **376**, 352–355.

Ritchie, K. A., Brinster, R. L., Storb, U. (1984). Allelic exclusion and control of endogenous immunoglobulin gene rearrangement in K transgenic mice. *Nature* **312**, 517–520.

Rolink, A., Haasner, D., Melchers, F., Andersson, J. (1996). The surrogate light chain in mouse B-cell development. *Intern. Rev. Immunol.* **13**, 341–356.

Rosenberg, N., Kincade, P. W. (1994). B-lineage differentiation in normal and transformed cells and the microenvironment that supports it. *Curr. Opin. Immunol.* **6**, 203–211.

Ryelandt, M., De Wit, D., Baz, A. *et al.* (1995). The perinatal presence of antigen (P-azophenylarsonate) or anti-μ antibodies lead to the loss of the recurrent idiotype (CRIA) in A/J mice. *Int. Immunol.* **7**, 645–652.

Sakano, H., Hüppi, K., Heinrich, G., Tonegawa, S. (1979). Sequences at the somatic recombination sites of immunoglobulin light chain genes. *Nature* **280**, 288–394.

Sakano, H., Kurosawa, Y., Weigert, M., Tonegawa, S. (1981). Identification and nucleotide sequence of a diversity DNA segment (D) of immunoglobulin heavy-chain genes. *Nature* **290**, 562–565.

Sanchez, P., Nadel, B., Cazenave, P.-A. (1991). Vλ–Jλ rearrangements are restricted within a V–J–C recombination unit in the mouse. *Eur. J. Immunol.* **21**, 907–911.

Sanchez, P., Rueff-Juy, D., Boudinot, P., Hachemi-Rachedi, S., Cazenave, P.-A. (1996). The λ B cell repertoire of κ-deficient mice. *Intern. Rev. Immunol.* **13**, 357–368.

Satterthwaite A., Witte, O. (1996). Genetic analysis of tyrosine kinase function in B cell development. *Annu. Rev. Immunol.* **14**, 131–154.

Schlomchik, M., Mascelli, M., Shan, H., Radic, M. Z., Pisetsky, D., Marshak-Rothstein, A., Weigert, M. (1990). Anti-DNA antibodies from autoimmune mice arise by clonal expansion and somatic mutation. *J. Exp. Med.* **171**, 265–297.

Schulze, D. H., Kelsoe, G. (1987). Genotypic analysis of B cell colonies by *in situ* hybridization. *J. Exp. Med.* **166**, 163–172.

Schwarz, E. M., Krimpenfort, P., Berns, A., Verma, I. M. (1997).

Immunological defects in mice with a targeted disruption in Bcl-3. *Genes Dev.* **11**, 187–197.

Shahinian, A., Pfeffer, K., Lee, K. P., Kündig, T. M., Kishihara, K., Wakeham, A., Kawai, K., Ohashi, P. S., Thompson, C. B., Mak, T. W. (1993). Differential T cell costimulation requirements in CD28-deficient mice. *Science* **261**, 609.

Shinkai, Y., Rathburn, G., Lam, K.-P., Oltz, E. M., Steward, V., Mendelsohn, M., Charron, J., Datta, M., Young, F., Stall, A. M., Alt, F. W. (1992). RAG-2-deficient mice lack mature lymphocytes owing to inability to initiate V(D)J rearrangement. *Cell* **68**, 855.

Sefton, B. M., Taddie, J. A. (1994). Role of tyrosine kinases in lymphocyte activation. *Curr. Opin. Immunol.* **6**, 372–379.

Selsing, E., Daitch, L. E. (1995). Immunoglobulin λ genes. In *Immunoglobulin Genes* (eds Honjo, T., Alt, F. W.), pp. 193–203. Academic Press, London, UK.

Shefner, R., Mayer, R., Kaushik, A., D'Eustachio, P., Bona, C., Diamond, B. (1990). Identification of a new VK gene family that is highly expressed in hybridomas from an autoimmune mouse strain. *J. Immunol.* **145**, 1609–1614.

Shimizu, A., Takahashi, N., Yaoita, Y., Honjo, T. (1982). Organization of the constant-region gene family of the mouse immunoglobulin heavy chain. *Cell* **28**, 499–506.

Shinkai, Y., Rathburn, J., Lam, K. P. *et al.* (1992). RAG-2 deficient mice lack mature lymphocytes owing to inability to initiate V(D)J rearrangement. *Cell* **68**, 855–867.

Siu, G., Springer, E. A., Huang, H. V., Hood, L. E., Crews, S. T. (1987). Structure of the T15 VH gene subfamily: identification of immunoglobulin gene promoter homologies. *J. Immunol.* **138**, 4466–4471.

Sligh, J. E., Ballantyne, C. M., Rich, S., Hawkins, H. K., Smith, C. W., Bradley, A., Beaudet, A. (1993). Inflammatory and immune responses are impaired in mice deficient in intercellular adhesion molecule 1. *Proc. Natl Acad. Sci. USA* **90**, 8529.

Snapper, C. M., Finkelman, F. D. (1993). Immunoglobulin class switching. In *Fundamental Immunology* (ed. Paul, W. E.), pp. 837–863. Raven Press Ltd, New York, USA.

Snider, D. P., Marshall, J. S., Perdue, M. H., Liang, H. (1994). Production of IgE antibody and allergic sensitization of intestinal and peripheral tissues after oral immunization with protein Ag and cholera toxin. *J. Immunol.* **153**, 647–657.

Solin, M.-L., Kaartinen, M. (1992). Allelic polymorphism of mouse Igh-J locus, which encodes immunoglobulin heavy chain joining (JH) segments. *Immunogenetics* **36**, 306–313.

Solvason, N., Lehuen, A., Kearney, J. F. (1991). An embryonic source of lyl but not conventional B cells. *Int. Immunol.* **3**, 543–550.

Spanopoulou, E., Zaitseva, F., Wang, F.-H., Santagata, S., Baltimore, D., Panayotou, G. (1996). The homeodomain region of Rag-1 reveals the parallel mechanisms of bacterial and V(D)J recombination. *Cell* **87**, 263–276.

Stein, P. L., Lee, H.-M., Rich, S., Soriano, P. (1992). p59fyn mutant mice display differential signaling in thymocytes and peripheral T cells. *Cell* **70**, 741–750.

Storb, U., Haasch, D., Arp, B., Sanchez, P., Cazenave, P.-A., Miller, J. (1989). Physical linkage of mouse λ genes by pulsed-field gel electrophoresis suggests that the rearrangement process favors proximate target sequence. *J. Mol. Cell. Biol.* **9**, 711–718.

Strohal, R., Helmberg, A., Keoemer, G., Kofler, R. (1989). Mouse VK gene classification by nucleic acid sequence similarity. *Immunogenetics* **30**, 475–493.

Takai, T., Ono, M., Hikida, M., Ohmori, H., Ravetch, J. V. (1996). Augmented humoral and anaphylactic responses in FCγRII-deficient mice. *Nature* **379**, 346–349.

Tassignon, J., Brait, M., Ismaili, J., Urbain, J., Gottlieb, P., Brown, A., Hasemann, C. A., Capra, J. D., Meek, K. (1993). Molecular characterization of monoclonal CRIA-positive anti-arsonate antibodies derived from idiotype-negative mice bearing a light chain polymorphism. *Proc. Natl Acad. Sci. USA* **90**, 9508–9512.

Teale, J. M., Morris, E. G. (1989). Comparison of VK gene family expression in adult and fetal B cells. *J. Immunol.* **143**, 2768–2772.

Teale, J. M., Medina, C. A. (1992). Comparative expression of adult and fetal V gene repertoire. *Intern. Rev. Immunol.* **8**, 95–111.

Tedder, T. F., Zhou, L.-J., Engel, P. (1994). The CD19/CD21 signal transduction complex of B lymphocytes. *Immunol. Today* **15**, 437–442.

Terashima, M., Kim, K. M., Adachi, T., Nielsen, P. J., Reth, M., Kohler, G., Lamers, M. C. (1994). *EMBO J.* **13**, 33782–33792.

Tergrigorov, V. S., Krifuks, O., Liubashevsky, E., Nyska, A., Trainin, Z., Toder, V. (1997). A new transmissible AIDS-like disease in mice induced by alloimmune stimuli. *Nature Medicine* **3**, 37–41.

Tonegawa, S. (1983). Somatic generation of antibody diversity. *Nature* **302**, 575–581.

Tornberg, U. C., Holmberg, D. (1995). B1-a, B1-b and B-2 B cells display unique VHDJH repertoires formed at different stages of ontogeny and under different selection pressures. *EMBO J.* **14**, 1680–1689.

Torres, R. M., Flaswinkel, H., Reth, M., Rajewsky, K. (1996). Aberrant B cell development and immune response in mice with a compromised BCR complex. *Science* **272**, 1804–1808.

Trepicchio W. Jr, Barret, K. J. (1985). The Igh-V locus of MRL mice: restriction fragment length polymorphism in eleven strains of mice as determined with VH and D gene probes. *J. Immunol.* **134**, 2734–2739.

Tsubata, T., Reth, M. (1990). The products of pre-B-cell specific (λ5 and VpreB) and the immunoglobulin μ chain form a complex that is transported onto the cell. *J. Exp. Med.* **172**, 973–976.

Tsubata, T., Tsubata, R., Reth, M. (1992). Crosslinking of the cell surface immunoglobulin chain (μ-surrogate light chains complex) on pre-B-cells induce activation of V gene rearrangements at the immunoglobulin K locus. *Int. Immunol.* **4**, 637–641.

Tsui, H. W., Siminovitch, K. A., de Souza, L., Tsui, F. W. (1993). Motheaten and viable motheaten mice have mutations in the haematopoietic cell phosphatase gene. *Nat. Gen.* **4**, 124–129.

Tutter, A., Riblet, R. (1988). Selective and neutral evolution in the murine Igh-V locus. *Curr. Top. Microbiol. Immunol.* **137**, 107–115.

Tutter, A., Brodeur, P. H., Shlomchik, M., Riblet, R. (1991). Structure, map position and evolution of two newly diverged mouse Ig VH gene families. *J. Immunol.* **147**, 3215–3223.

Ulevitch, R. J., Tobias, P. S. (1995). Receptor-dependent mechanisms of cell stimulation by bacterial endotoxin. *Annu. Rev. Immunol.* **13**, 437–457.

Urbain, J., Brait, M., De Wit, D. *et al.* (1992). B cell subsets, idiotype selection: positive selection for some B lymphocyte? *Int. Rev. Immunol.* **8**, 259–267.

Valiante, N. M., Caton, A. J. (1990). A new Igk-V gene family in the mouse. *Immunogenetics* **32**, 345–350.

van den Broek, M. F., Müller, U., Huang, S., Zinkernagel, R. M., Aguet, M. (1995). Immune defence in mice lacking type I and/ or type II interferon receptors. *Immunol. Rev.* **148**, 5–23.

van Dyk, L., Meek, K. (1992). Assembly of IgHCDR3: mechanism, regulation and influence on antibody diversity. *Int. Rev. Immunol.* **8**, 123–133.

van Kaer, L., Ashton-Rickardt, P. G., Ploegh, H. L., Tonegawa, S. (1992). TAP 1 mutant mice are deficient in antigen presentation, surface class I molecules, and CD4−CD8+ T cells. *Cell* **71**, 1205.

Vieira, P., Rajewsky, K. (1988). The half-lives of serum immunoglobulins in adult mice. *Eur. J. Immunol.* **18**, 313–316.

von Freeden-Jeffry, U., Vieira, P., Lucian, L. A., McNeil, T., Burdach, S. E. G., Murray, R. (1995). Lymphopenia in interleukin (IL)-7 gene-deleted mice identifies IL-7 as a nonredundant cytokine. *J. Exp. Med.* **181**, 1519–1526.

Walsh, C. M., Matloubian, M., Liu, C. C., Ueda, R., Kurahara, C. G., Christensen, J. L., Huang, M. T., Young, J. D., Ahmed, R., Clark W. R. (1994). Immune function in mice lacking the perforin gene. *Proc. Natl Acad. Sci. USA* **91**, 10854–10858.

Weaver, D., Costantini, F., Imanishi, K. T., Baltimore, D. (1985). A transgenic immunoglobulin μ gene prevents rearrangement of endogenous genes. *Cell* **42**, 117–127.

Weill, J. C., Reynaud, C. A. (1995). Generation of diversity by post-rearrangement diversification mechanisms: the chicken and the sheep antibody repertoires. In *Immunoglobulin Genes*, 2nd edn (eds Honjo, T., Alt, F. W.), pp. 267–288. Academic Press, London, UK.

Weiser P., Riesterer, C., Reth, M. (1994). The internalization of the IgG2a antigen receptor does not require the association with Ig-alpha and Ig-beta but the activation of protein kinases does. *Eur. J. Immunol.* **24**, 665–671.

Wiler, R., Leber, R., Moore, B. B., van Dyk, L. F., Perryman, L. E., Meek, K. (1995). Equine severe combined immunodeficiency: a defect in V(D)J recombination and DNA-dependent protein kinase activity. *Proc. Natl Acad. Sci. USA* **92**, 11485–11489.

Williams G. T., Venkitaranan, A. R., Gilmore, D. J., Neuberger, M. S. (1990). The sequence of the μ transmembrane segment determines the tissue specificity of the transport of immunoglobulin M to cell surface. *J. Exp. Med.* **171**, 947–952.

Williams, G. T., Dariavach, P., Venkitaraman, A. R., Gilmore, D. J., Neuberger, M. S. (1993). Membrane immunoglobulin without sheath or anchor. *Mol. Immunol.* **30**, 1427–1432.

Williams G. T., Peaker, C. J., Patel, K. J., Neuberger, M. S. (1994). The alpha/beta sheath and its cytoplasmic tyrosines are required for signaling by the B-cell antigen receptor but not for capping or for serine/threonine-kinase recruitment. *Proc. Natl Acad. Sci. USA* **91**, 474–478.

Wilson, R. W., Ballantyne, C. M., Smith, C. W., Montgomery, C., Bradley, A., O'Brien, W. E., Beaudet, A. L. (1993). Gene targeting yields a CD18-mutant mouse for study of inflammation. *J. Immunol.* **151**, 1571.

Winkler, T. H., Melchers, F., Rolink, A. G. (1995). Interleukine-3 and Interleukine-7 are alternative growth factors for the same B-cell precursors in the mouse. *Blood* **85**, 2045–2051.

Winter, E., Radbruch, A., Krawinkel, U. (1985). Members of novel VH gene families are found in VDJ regions of polyclonally activated B-lymphocytes. *EMBO J.* **4**, 2861–2867.

Woloschak, G. E., Chang-Liu, C. M., Chung, J., Libertin, C. R. (1996). Expression of enhanced spontaneous and gamma-ray-induced apoptosis by lymphocytes of the wasted mouse. *Int. J. Radiat. Biol.* **69**, 47–55.

Wood, C., Tonegawa, S. (1983). Diversity and joining segments of mouse immunoglobulin heavy chain genes are closely linked and in the same orientation: implications for the joining mechanism. *Proc. Natl Acad. Sci. USA* **80**, 3030–3034.

Wu, G. E., Paige, C. J. (1986). VH gene family utilization in colonies derived from B and pre-B cells detected by the RNA colony blot assay. *EMBO J.* **5**, 3475–3481.

Xu, H., Gonzalo, J. A., St. Pierre, Y., Williams, I. R., Kupper, T. S., Cotran, R. S., Springer, T. A., Guttierrez-Ramos, J.-C. (1994). Leukocytosis and resistance to septic shock in intercellular adhesion molecule 1-deficient mice. *J. Exp. Med.* **180**, 95.

Yancopoulos, G. D., Desiderio, D. V., Paskind, M., Kearney, J. F., Baltimore, D., Alt, F. W. (1984). Preferential utilization of the most JH proximal VH gene segments in pre-B cell lines. *Nature* **311**, 757–759.

Yancopoulos, G. D., Malynn, B. A., Alt, F. W. (1988). Developmentally regulated and strain-specific expression of murine VH gene families. *J. Exp. Med.* **168**, 417–435.

Zhang, J., Alt, F. W., Honjo, T. (1995). Regulation of class switch recombination of immunoglobulin heavy chain genes. In *Immunoglobulin Genes* (eds Honjo, T., Alt, F. W.), pp. 235–265. Academic Press, London, UK.

Zhu, C., Bogue, M. A., Lim, D. S., Hasty, P., Roth, D. B. (1996). Ku86-deficient mice exhibit severe combined immunodeficiency and defective processing of V(D)J recombination intermediates. *Cell* **86**, 379–389.

Further Reading for Section 2

Greene, E. C. (1968). *Anatomy of the Rat.* Hofner Publish. Comp., New York, USA.

Gründman, E., Wollmer, E. (1990). *Reaction Patterns of the Lymph Node.* Springer-Verlag, Berlin, Germany.

Heinen, E. (1995). *Follicular Dendritic Cells in Normal and Pathological Conditions.* Springer, New York, USA.

Imai, Y., Matsuda, M., Maeda, K., Yamakaw, M., Dobashi, M., Satoh, H., Terashima, K. (1991). In *Dendritic Cells in Lymphoid Tissues* (eds Imai, K. *et al.*), pp. 3–13. Elsevier, Amsterdam, The Netherlands.

Komazawa, S., Iki, H., Ohmura, M., Tsutui, S., Fujiwara, K. (1991). Comparative studies on distribution and fine morphology of the intestinal Peyer's patches in mongolian gerbils and mice. *J. Vet. Med. Sci.* **53**, 899–904.

Kosco-Vilbois, M. H. (1995). *An Antigen Depository of the Immune System: Follicular Dendritic Cells.* Springer-Verlag, Berlin, Germany.

Pastoret, P. P., Govaerts, A., Bazin, H. (1990). *Immunologie Animale.* Médecine-Sciences, Flammarion, Paris, France.

Sainte-Marie, G., Peng, F. S., Bélisle, C. (1982). Overall architecture and pattern of lymph flow in the rat lymph node. *Am. J. Anat.* **164**, 275–309.

Savidge, T. C. (1996). The life and times of an intestinal M cell. *Trends Microbiol.* **4**, 301–306.

Szakal, A. K., Kosco-Vilbois, M. H., Tews, J. G. (1989). *Annu. Rev. Immunol.* 7, 91–109.

Quimby, F. W. (1989). *Immunodeficient Rodents.* National Academic Press, Washington DC, USA.

Tew, J. G., Kosco, M. H., Burtong, F., Szakal, A. K. (1990). *Immunol. Rev.* **117**, 185–211.

XVII THE SCID MOUSE MUTANT: DEFINITION AND POTENTIAL USE AS A MODEL FOR IMMUNE DISORDERS

1. Introduction

Severe combined immunodeficiency (SCID) mice were described in 1983 by Bosma *et al.* (1983). For many years this model has been used to study V(D)J recombination, which is the only known form of site-specific DNA rearrangement in vertebrates. In 1988, two groups reported that the human immune system could be reconstituted in SCID mice (McCune *et al.*, 1988; Mosier *et al.*, 1988). Prior to this, several attempts were made instead to transfer critical elements of the human hematopoietic system into a mouse (Louwagie and Verwilghen, 1970; Kamelreid and Dick, 1988), although many of these experimental systems showed evidence of human hematopoiesis none presented the conditions necessary for long-term reconstitution with human cells of multiple lineages.

SCID mice with a fully functional human immune system would be extremely valuable in biomedical research, and such mouse models became very popular. In recent years, more than 1000 publications have described experiments related to the experimental use of these chimeric animals in studies including human immunodeficiency virus (HIV) infections, vaccine development, drug testing and autoimmune diseases. However, the rush to exploit the model and to publish results obtained from only a few animals have led to an overinterpretation of the relevance of these models. In addition this has detracted from the effort to define the nature of the chimerism and complicated the interpretation of already complex biological interpretations.

This review attempts to evaluate the advantages, the difficulties, and the limitations of each SCID model, in the light of our experience over the last five years (Mazingue *et al.*, 1991; Soulez *et al.*, 1991; Palluault *et al.*, 1992; Cesbron *et al.*, 1994; Lemaire *et al.*, 1994; Pestel *et al.*, 1994; Cadore *et al.*, 1995; Grandadam *et al.*, 1995; Tonnel *et al.*, 1995; Autran *et al.*, 1996; Lasnézas *et al.*, 1996; Delhem *et al.*, 1998). The reader is referred to an excellent review of Bosma and Carroll (1991) for comprehensive background information.

2. The SCID Mouse

The homozygous mutation of the *scid* gene which occurred spontaneously in the BALB/c C.B.-17 strain, results in a paucity of functional B and T cells (Bosma *et al.*, 1983). Tissues and cell lines derived from *scid* mice also exhibit a generalized hypersensitivity to γ irradiation and a defect in the repair of double-strand DNA breaks (Fulop and Phillips, 1990; Biedermann *et al.*, 1991; Hendrickson *et al.*, 1991). The SCID immune defect results from selective impairment of the joining of coding sequences during V(D)J recombination (Lieber *et al.*, 1988; Malynn *et al.*, 1988; Blackwell *et al.*, 1989). Correspondingly, broken molecules with coding ends are observed to accumulate in the *scid* mouse thymus (Roth *et al.*, 1992).

The mouse *scid* locus is located on chromosome 16, linked to the 15 and VpreB loci, at 1 cM centromeric from the IgL locus (Bosma *et al.*, 1989; Bosma and Carroll, 1991; Miller *et al.*, 1993). X-ray-hypersensitive Chinese hamster ovary cells carrying mutations in at least three different complementation groups exhibit impaired rearrangement of extrachromosomal V(D)J recombination substrates in cotransfection experiments with recombination-activating (RAG) genes (Pergola *et al.*, 1993; Taccioli *et al.*, 1994). One of these complementation groups, typified by the V3 mutation, is associated with preferential impairment of coding joint formation; this complementation group is represented in the mouse by the *scid* locus (Taccioli *et al.*, 1994). A second group of complementation which includes the CHO xrs-5 and xrs-6 mutations, encodes a subunit of the Ku antigen, a p86/70 heterodimer that binds DNA ends and associates with the catalytic subunit of a high molecular weight DNA-dependent protein kinase (DNA-PK_{CS}) (Getts and Stamato, 1994; Rathmell and Chu, 1994; Taccioli *et al.*, 1994). In humans, a gene that restores normal radio-resistance, double-strand-break repair and V(D)J recombination to *scid* cells has been mapped to human chromosome 8q11 (Itoh *et al.*, 1993; Kirchgessner *et al.*, 1993; Komatsu *et al.*, 1993; Banga *et al.*, 1994; Kurimasa *et al.*, 1994). No other synteny between mouse chromosome 16 and human chromosome 8 is known. The

Handbook of Vertebrate Immunology
ISBN 0-12-546401-0

gene encoding DNA-PK_{CS} maps to the same interval, and a yeast artificial chromosome clone spanning the DNA-PK_{cs} locus complements the V3 defect in radio-resistance and V(D)J recombination, suggesting that the *scid* mutation affects expression or activity of DNA-PK_{cs} (Blunt *et al.*, 1995; Kirchgessner *et al.*, 1995; Peterson *et al.*, 1995). No other human DNA repair or immune deficiency diseases that map to this region are known.

V(D)J recombination is the major source of B and T cell antigen receptor diversity. Consequently, the *scid* mutation leads to a failure of functional B and T cells. Natural killer (NK) cell function is unaffected (Dorshkind *et al.*, 1985) and apparently, myeloid cells function normally (Dorshkind *et al.*, 1984; Bancroft *et al.*, 1994). When housed in conventional facilities SCID mice die from opportunistic infections when they are 5–6 months old, whereas in a pathogen-free environment SCID mice survive normally for 1–2 years. However, 10% develop a T-cell lymphoma (Bosma *et al.*, 1983).

3. The SCID Model of Human Immune Responses

The SCID-hu mouse model

Description of the model

The SCID-hu mouse is created by surgical implantation of human fetal lymphoid organs into the SCID mouse (McCune *et al.*, 1988). Early constructs of SCID-hu mice were obtained with human fetal thymus harvested during the 18–24 weeks of gestation up to 18 h after interruption of pregnancy, and implanted under the kidney capsule of the mice. Postnatal human thymic tissues can be maintained in the SCID mouse but enlarge to a minimal extent, if at all (Barry *et al.*, 1991). The implanted human thymus quickly becomes vascularized and histological analysis shows that the development of the human tissue in SCID mice is almost indistinguishable from normal human ontogeny (McCune *et al.*, 1988). However, follow-up of the grafts shows a progressive decrease in the relative number of immature thymocytes, and finally a total thymic involution. Reasoning that the duration of T lymphopoiesis might be prolonged if human hematopoietic stem cells were provided to the thymus implant, fragments of fetal liver and thymus were co-implanted contiguous to one another. The liver implant did not persist, engrafted alone, or in the presence of thymus. For 6–11 months post-implantation, the thymus graft exhibited a full range of differentiating thymocytes, and its microanatomical structure was similar to that of a fresh, age-matched thymus (Namikawa *et al.*, 1990). It has been reported that the thymic interlobular septae contained hematopoietic foci, including blast cells, myelomonocytic elements, megararyocytes, clonogenic myeloerythroid progenitors (Namikawa *et al.*, 1990), or B cells (Vandekerckhove *et al.*, 1993). However, in our work, we have not observed such 'thymic islets' of human hematopoietic differentiation (Autran *et al.*, 1996).

Human T cells could be observed in the peripheral circulation for up to 10–12 months in 50% of chimeric animals (Namikawa *et al.*, 1990; Autran *et al.*, 1996). They are $CD3^+$, $\alpha\beta$ TCR^+, with a $CD4^+$ $CD8^+$ ratio of about 2:1. Approximately 75% are $CD45RA^+$, and thus naive T cells, whereas 18% express CD29, a marker of more mature or memory cells. T-lymphocytes from SCID-hu mice appeared functional, as suggested by the expression of the CD69 marker in response to a CD3 Sepharose or PHA stimulation, and by their proliferation after alloantigenic challenge in the presence of recombinant IL-2, or by their cytokine production (Krowka *et al.*, 1991; Vandekerckhove *et al.*, 1994; Vanhecke *et al.*, 1995). The TCR repertoire is polyclonal and very similar to that encountered in healthy humans (Vandekerckhove *et al.*, 1991, 1992a). These studies have also shown the lack of xenoreactive human cells in SCID-hu mice, whereas a graft-versus-host reaction (GvH) frequently occurs when adult human peripheral blood leukocytes (PBLs) are injected into SCID mice (see below). These, and additional data, have been reviewed previously in detail (McCune *et al.*, 1991).

The SCID-hu model provides a relatively large recipient for the human thymocytes (ranging between 10^7 and 10^8 cells. However, human T cells in the peripheral mouse circulation represented rarely more than 5% of the total leukocytes) (Kaneshima *et al.*, 1994; Autran *et al.*, 1996). No other human cell lineages are detectable in the periphery. No primary or secondary immune response is obtained in this model. Therefore, its primary application is in the analysis of human T-lymphoid differentiation, function, and pathology within the context of the organ itself, such as the induction of tolerance *in vivo* (Vandekerckhove *et al.*, 1992b; Baccala *et al.*, 1993; Kraft *et al.*, 1993). This approach should allow the development of quantitative and qualitative assays to aid the purification of immature human cells (Peault *et al.*, 1991; Baum *et al.*, 1992; Kraft *et al.*, 1993) and will be described in more detail below.

The SCID-hu mice model as an *in vivo* model for the physiopathological analysis of human viral diseases

Much of the work on this model has focused on HIV infection (Namikawa *et al.*, 1988). It has been shown that HIV infection is associated with suppression of human thymopoiesis (Aldrovandi *et al.*, 1993; Bonyhadi *et al.*, 1993; Autran *et al.*, 1996), while this is not the case when the human grafts are infected with human cytomegalovirus (Mocarski *et al.*, 1993; Brown *et al.*, 1995). HIV damage to thymic stromal cells might contribute to T-cell depletion and thymic dysfunction (Stanley *et al.*, 1993;

Autran *et al.*, 1996; Ogura *et al.*, 1996). Varicella zoster virus has also been demonstrated to replicate in T lymphocytes (Moffat *et al.*, 1995). Wild-type measles virus infection of the thymic stroma leads to induction of thymocyte apoptosis and may contribute to long-term alteration of immune responses (Auwaerter *et al.*, 1996). It is generally proposed that the extent of thymic disruption reflects the virulence of the virus studied, and therefore the SCID-hu mouse may serve as a model for the study of viral pathogenesis.

The SCID-hu chimera optimized by co-implanting various fetal human organs

The aim of these studies was to add either a microenvironment in which a pluripotent human hematopoietic stem cell could self-renew and differentiate, or a secondary functional compartment through which mature human progeny from the liver/thymus structure could migrate, and differentiate. Kyoizumi *et al.* (1992) implanted human fetal long bones into the subcutaneous space of the SCID mouse. While active multilineage hematopoiesis could be observed in the bones, no T lymphocytes or small CD19$^+$ and CD33$^+$ cells were found in the periphery (Kyoizumi *et al.*, 1992). More details are available in section 5 (The SCID Model of Normal and Neoplastic Human Hematopoiesis).

A methodology for implanting human lymph nodes has been developed. However, these organs disappeared progressively (McCune, 1992). Similarly, the implantation of human fetal lung into SCID mice resulted in the development of normal human alveolar and bronchiolar lung compartments as, indicated by the presence of mucus, cells with moving ciliae in the derived culture, and human macrophages (CD68$^+$) (Cesbron *et al.*, 1994). While this model provided the opportunity to study human viruses with lung tropism and for helping to define gene therapy protocols in human lung cells (Peault *et al.*, 1994), no improvement in the immune response was observed. Similarly, the SCID mouse model has been used to investigate human gastrointestinal ontogenesis. Human fetal gut is able to undergo region-specific morphogenesis and epithelial cytodifferentiation when transplanted subcutaneously into SCID mice (Savidge *et al.*, 1995). However, despite many efforts to develop this model, we were unable to engraft this tissue in a reproducible manner (Figure XVII. 3.1 – see colour plate).

Limitations of the SCID-hu model

Limitations of this system include a low level of migration of human immune cells through the murine tissues, the absence of an immune response and the dependence on fetal tissues which are not easy to procure (Cadore *et al.*, 1995).

The Hu-PBL-SCID mouse

Description of the model

An alternative approach toward introducing human immune cells into SCID mice involves intraperitoneal injection of adult human PBLs (Mosier *et al.*, 1988).

During the first 3 weeks after intraperitoneal injection of PBLs, most injected cells can be detected in the peritoneal cavity (Hoffmannfezer *et al.*, 1992), although their number progressively declines. Human cells are not detectable in the mouse organs that could potentially support the T-cell maturation (i.e. the thymus, the liver or the spleen). Injection of either cord blood, splenocytes or human bone marrow does not lead to any long-term engraftment, meaning that mice do not provide the appropriate environment for the establishment of human lymphopoiesis (Tary Lehmann and Saxon, 1992; Alegre *et al.*, 1994; Reinhardt *et al.*, 1994). However, after conditioning of SCID mice by irradiation, the additional injection of human lymphokines yields several human myeloid and lymphoid cell lineages (Lubin *et al.*, 1991; Lapidot *et al.*, 1992b; Chen *et al.*, 1995). This point will be developed in section 5.

After 1 month, human T and B leukocyte markers appear in the spleen, the liver, the intestines and the lung of the SCID mice, but not (or very few) in lymph nodes (Saxon *et al.*, 1991; Abedi *et al.*, 1992; Tary Lehmann and Saxon, 1992). Human T cells, which constitute the majority of these cells (>95%), appear in these organs in substantial numbers but are uniformly single-positive and express HLA-DR and CD45R0, which defines them as mature and activated/memory cells (Duchosal *et al.*, 1992b; Tary Lehmann and Saxon, 1992; Hoffmannfezer *et al.*, 1993). No human macrophage or other accessory cells are detectable. B cells are oligoclonal (Saxon *et al.*, 1991) and frequently cannot be detected (Carlsson *et al.*, 1992; Tary Lehmann and Saxon, 1992; Hoffmannfezer *et al.*, 1993). The peripheral distribution of T cells is characteristic of mature activated/memory T cells that have down-regulated their lymph node homing receptor; they are anergic and unresponsive to stimulation with anti-CD3 or mitogens (Tary Lehmann and Saxon, 1992). The human T cell repertoire is limited to xenoreactive clones against mouse antigens, which suggests a chronic GvH stimulation (Tary Lehmann *et al.*, 1994, 1995).

From these results, Tary Lehmann *et al.* (1995) have proposed that naive and memory T cells that do not encounter their cognate antigen during the first 2–3 weeks post-engraftment die, while the reactive clones against murine antigens are stimulated and clonally expand, becoming undetectable once more by month 5 post-engraftment. The compartmentalization of human accessory cells and human T cells within the peritoneal cavity appears to be critical: no recolonization could be observed when purified T cells were injected intraperitoneally or when PBLs were injected intravenously (Mosier *et al.*, 1988).

Significant levels of human immunoglobulins can be detected in the serum of the hu-PBL-SCID mice (Mosier *et al.*, 1988; Abedi *et al.*, 1992; Tary Lehmann and Saxon, 1992). During the first 3 weeks antigen-specific immune response can be induced in the chimeras. In agreement with the proposed model of hu-PBL-SCID chimerism (Tary Lehman *et al.*, 1995), an antibody response to recall antigen, against which the donor was sensitized, was often observed (Duchosal *et al.*, 1992a; Neil and Sammons, 1992; Aaberge *et al.*, 1996). Alternatively, the PBLs are challenged *in vitro* with the antigen before injection. Activated human B cells can subsequently be fused with myeloma cells (Carlsson *et al.*, 1992; Neil and Sammons, 1992). However, a primary immune response has been reported by various groups (Mazingue *et al.*, 1991; Aaberge *et al.*, 1992; Ifverson *et al.*, 1995; Bombil *et al.*, 1996).

Hu-PBLs SCID mice as a model for infectious diseases

While human engrafted lymphocytes are not functional after 3 weeks post-engraftment, they are still susceptible to lymphotropic viruses such as HIV (Mosier *et al.*, 1991), the Epstein–Barr (EB) virus (Rowe *et al.*, 1991) or prions (Lasmézas *et al.*, 1996). After the initial report (Mosier *et al.*, 1988) that the hu-PBL-SCID model was associated with a high frequency of outgrowth of B cells transformed with EB virus (Mosier *et al.*, 1992; Riddell and Greenberg, 1995), some reports claimed a strong correlation between EB virus seropositivity of donors and the efficiency of the immune reconstitution (Torbett *et al.*, 1991; Duchosal *et al.*, 1992b), although others did not report such a correlation (Aaberge *et al.*, 1992; Chiang *et al.*, 1995; Steinsvik *et al.*, 1995). Direct microorganism-mediated pathology can be studied in hu-PBL-SCID mice in the absence of pathology mediated by the immune response of the host. Effector functions of human T cells can also be tested *in vivo* by adoptive transfer into the chimeras (Riddell and Greenberg, 1995). SCID mice grafted with PBLs from gp160-vaccinated donors were shown to resist to HIV infection (Mosier *et al.*, 1993a). Adoptive transfer of the Nef-specific human CTL clone into SCID mice also protected them against an HIV challenge (van Kuyk *et al.*, 1994).

The ability to use a human–mouse chimeric model to generate CTLs *in vivo* would represent a further advance toward the establishment of a model for vaccine testing with naive lymphocytes. However, few investigators have succeeded in obtaining human T-cell responses in SCID mice by improving their engraftment. Segall *et al.* (1966) succeeded in obtaining an anti-Nef CTL response in SCID mice by immunization with a recombinant vaccinia–nef virus, within the first few weeks post-transplant through conditioning of recipient mice with sublethal irradiation and by using a large inoculum of human PBLs (80×10^6 per mouse). However, these authors did not derive a human T-cell line from the hu-PBL-SCID mice, rendering the demonstration of the MHC-restricted and Nef-specific response incompleted. In addition, the T cell response was only positive within the first weeks after the engraftment, before the antigenic repertoire of the human T cells deviates toward the xenoreactive clones (Tary Lehmann *et al.*, 1994). In another study, Malkowska *et al.* (1994) showed that V-g9/V-d2 cytotoxic T cells can be generated *in vivo* by immunization of hu-PBL-SCID mice with irradiated Daudi lymphoma, and that such mice are protected against subsequent IP challenge with the live lymphoma. However, these V-g9/V-d2 cytotoxic T cells are not conventional allogeneic CTLs because they recognize a homologue of the GroEL heat-shock protein family in an MHC-unrestricted manner, reminiscent of a superantigen response (Fisch *et al.*, 1990, 1992).

hu-PBLs SCID as a model for immune diseases

A detailed general review will not be undertaken, since this topic has been recently reviewed (Elkon *et al.*, 1993; Elkon and Ashany, 1994; Jorgensen *et al.*, 1995; Tonnel *et al.*, 1995).

Xenogeneic engraftment of SCID mice is not restricted to human tissues

Components of immune system from bovines (Boermans *et al.*, 1992; Balson *et al.*, 1993; Greenwood and Croy, 1993; Greenwood *et al.*, 1996), equines (Balson *et al.*, 1993), and felines (Meers *et al.*, 1993; Johnson *et al.*, 1994, 1995; Linenberger *et al.*, 1995) can be transferred to these immunodeficient mice. These studies are often performed in the context of infectious diseases such as theileriosis (Fell *et al.*, 1990; Tsuji *et al.*, 1992; Hagiwara *et al.*, 1993a, b, 1995; Nakamura *et al.*, 1995), babesiosis (Tsuji *et al.*, 1995) or the feline AIDS virus (Meers *et al.*, 1993; Johnson *et al.*, 1995; Linenberger *et al.*, 1995). Immune reconstitution with these species is limited by the same parameters that affect the human SCID model. It is also affected by the possible unavailability of recombinant growth factors or species-specific immunological reagents.

The SCID-Skin mouse model

Description of the model

Very few non-neoplastic tissues/cells from human adult, other than those from the immunological system, have been successfully engrafted into SCID mice. Levi and Bunge (1994). have analysed the capacity of human Schwann cells to form myelin around regenerating mouse axons. Valentine *et al.* (1994) showed that the SCID mouse allows the *in vivo* reconstitution of thyroid follicles from thyroid monolayer cells when transplanted subcutaneously within an extracellular basement membrane matrix. Normal human myoblasts transplanted in the anterior tibias of SCID mice were able to fuse and produce dystrophin in injected muscles of the mice (Huard *et al.*, 1994a, b).

The skin model has been more extensively developed because human skin is relatively easily obtained in collaboration with reconstitutive plastic surgeons. The demonstration of skin engraftment onto SCID mice was first described using grafts from recessive dystrophic epidermolysis bullosa patients (Kim *et al.*, 1992). Grafting of normal human skin onto SCID mice was also reported in the study of the regulation of human endothelial cell–leukocyte adhesion molecules (Yan *et al.*, 1993, 1994), the induction of human papillomavirus V-16 DNA replication (Brandsma *et al.*, 1995), or the role of UV light in carcinogenesis (Soballe *et al.*, 1996). Xenografted skin preserves a fully differentiated human epidermis and dermis, up to 12 months post-transplantation, with a success rate of >90% (Figure VXII. 3.2 – see colour plate). Kaufmann *et al.* (1993) have previously reported the presence of human skin immune cells (macrophages, lymphocytes and Langerhans cells) over 12 months of observation. However, in our hands, the Langerhans cells disappeared progressively after the third month post-engraftment (Delhem *et al.*, 1998).

In an effort to overcome the absence of consistent primary immune responses in the original SCID-hu mouse model, we, and others, have grafted human skin onto the backs of the mice and injected autologous PBLs. Studies of this combined human skin and PBL graft have shown the ability of human T cells to induce first-set rejection of allogeneic human skin (Alegre *et al.*, 1994; Murray *et al.*, 1994), and to induce a delayed-hypersensitivity reaction after intradermic injection of tetanus toxoid (Petzelbauer *et al.*, 1996). In this recent study (Petzelbauer *et al.*, 1996), the adoptive transfer was performed IP with human PBLs from donors immunized with tetanus toxoid. When PBLs from non-tetanus toxoid-vaccinated donors were used, no immune response was observed, indicating the absence of a primary immune response.

The obtention of a CTL response in the Hu-PBL-SCID skin model

Recently, using this human–mouse chimera, we evaluated the efficacy of raising a primary CTL response *in vivo* against HIV-LAI env gp160 protein by immunization with a recombinant vaccinia-Env virus (Delhem *et al.*, 1998). The injection of vCP-LAIgp160 into the skin graft induced a perivascular human $CD4^+$, $CD8^+$, CD3 or $CD45^+$ T cell infiltrate, and an epidermal recruitment of $CD1a^+$, $CD80^+$, $CD86^+$ Langerhans cells (LCs). In addition we were able to derive T cell lines from the human immunized engrafted skin, and observe *in vitro* that they exert an MHC-class II restricted cytolytic activity against target cells displaying the HIV-LAI env protein. The LCs play a pivotal role in the skin immune response (Steinman, 1991; Bos and Kapsenberg, 1993; Schmitt, 1995). The presence of activated LCs ($CD86^+$/$CD80^+$) may be crucial to the development of a specific T-cell response in the SCID mouse. The main difference between the experiment of Petzelbauer (1996), and ours was the supplementation with recombinant IL-2 (rIL-2). We clearly observed that in the absence of recombinant IL-2 supplementation, the immune response was weak or absent. In the SCID mouse model, an anti-CD3 activation of human PBLs and a subsequent administration of human IL-2 optimized human T-cell engraftment (Murphy *et al.*, 1993) and administration of rIL-2 improved the homing and engraftment of PBLs from rheumatoid arthritis patients (Kaul *et al.*, 1995). The improvement of engraftment by the administration of human interleukins will be further illustrated below.

In conclusion, the hu-PBL-SCID-Skin model represents an advance over *in vitro* and other *in vivo* SCID models. The ability to induce a CTL response against a viral antigen in naive lymphocytes, might provide a useful preclinical model for vaccine testing. Because many tumors or leukemias can also be transplanted into SCID mice, they could provide a system for generating anti-tumor T cells and for evaluating the efficiency of anti-tumor immunizations.

4. The SCID 'Leaky' Phenotype and New Immune-deficient Murine Strain

The 'leaky' phenotype

A few clones of antigen receptor positive B and T lymphocytes do appear in a variable proportion (2–25%) of young adult SCID mice and in virtually all old SCID mice (Bosma *et al.*, 1988; Carroll and Bosma, 1989; Mosier *et al.*, 1993b). Such mice are designated 'leaky' and have been described in detail previously (Bosma and Carroll, 1991). Two explanations have been advanced to account for these observations: either the SCID V(D)J recombinase is altered to permit normal recombination at a low frequency (Hendrickson *et al.*, 1990; Schuler *et al.*, 1990), or it may revert to a wild-type phenotype in some lymphocyte clones under selective pressure (Petrini *et al.*, 1990; Kotloff *et al.*, 1993). However, there is no conclusive evidence for either explanation. For a discussion see Schuler W. (1990), Bosma and Carroll (1991) and Weei-Chin and Desiderio (1995).

The residual immune system in leaky SCID mice which are capable of responding to mitogens and of producing immunoglobulins and cytokines (Bosma *et al.*, 1988; Carroll and Bosma, 1988; Carroll *et al.*, 1989), may develop reactions to allogeneic tissues/cells and subsequently affect the engrafted immune function. It has been suggested that this immune reaction might be the reason for the considerable variation in the human immune reconstitution among individual mice (Armstrong *et al.*, 1992; Tary Lehmann and Saxon. 1992; Bazin *et al.*, 1994; Steinsvik *et al.*, 1995). Because of this potential interference with graft acceptance, leaky mice are generally excluded from studies in which the transfer of human immune function is attempted (Torbett *et al.*, 1991). However, the number of leaky SCID mice, and the urgency of work using human tissue from clinical activities, make this precaution difficult in practice.

Other immune-deficient murine strains

In view of the limitations in the use of the SCID mice owing to the leaky phenotype, novel murine strains have been obtained to bypass this problem. The SCID mutation has been introduced into the CH3 strain by intensive backcrossing (Nonoyama *et al.*, 1993). Whereas 79% of 3-month-old SCID mice showed detectable immunoglobulins in the serum, only 15% of CH3-SCID mice have immunoglobulin and only at low levels (Nonoyama *et al.*, 1993). In our hands, however, the results of human immune reconstitution are not significantly affected, and these mice are more aggressive.

RAG knock-out mice

Two genes, RAG-1 and RAG-2 have been identified as being necessary for V(D)J recombination (Weei-Chin and Desiderio, 1995). These genes have been characterized and two strains of mice containing homozygous germline disruptions have been generated (Mombaerts *et al.*, 1992; Shinkai *et al.*, 1992). These animals, while fertile and apparently normal in all aspects, completely lack functional B and T cells. No leaky phenotype is observed. These strains are now available, but data supporting the notion that they will make superior recipients for immune system transfer are lacking. In contrast, Steinsvik *et al.* (1995) have shown that RAG mice support only limited survival of human transplanted PBLs compared to the SCID mice. This suggested, as discussed above, that a certain degree of xenoreactivity as found in SCID mice, improves the survival of human cells.

SCID and NK cell-deficient mice

Natural killer cells are present in SCID spleen and thymus (Dorshkind *et al.*, 1985; Garni *et al.*, 1990). NK activity is highly dependent on cytokines, and in particular interferon and IL-2, which are easily stimulated by bacterial or viral infections. Therefore, intercurrent infections of SCID mice lead to an elevation of NK levels and increased resistance to the engraftment of human PBLs (Mosier, 1990). The users of SCID models have to maintain SCID mice in very clean housing conditions. In addition, it has been suggested that following xenotransplantation, activated SCID NK cells could modulate the GvH reaction and might explain the variation in the efficiency in the immune reconstitution among individual mice (Armstrong *et al.*, 1992; Tary Lehman and Saxon, 1992; Bazin *et al.*, 1994; Steinsvik *et al.*, 1995). Thus, a few groups have obtained double mutant mice for the SCID mutation and NK cell deficiency (Mosier, 1990; Froidevaux and Loor, 1991; Mosier *et al.*, 1993b). The *beige* mutation causes a lysosomal storage defect that affects cytotoxic T cells and NK cells. Surprisingly, the introduction of the *beige* mutation into the SCID strain mutation depresses the leaky phenotype (Mosier *et al.*, 1993b). However, the use of SCID *beige* mice is not very popular, while nonobese diabetic/SCID (NOD/SCID) mice have become the recipient of choice for the study of human hematopoiesis.

5. SCID Model of Normal and Neoplastic Human Hematopoiesis

The absence of *in vivo* assays for human hematopoietic cells has been a major limitation in the characterization of the cellular and molecular mechanisms that regulate normal and pathological hematopoiesis, hence the interest in reconstituted SCID mice as a model system.

The SCID-Hu/bone model

Kyoizumi *et al.* (1992) implanted human fetal long bones into the subcutaneous space of the SCID-hu mouse. Mice were shown to sustain active human hematopoiesis *in vivo* for as long as 20 weeks after implantation and this human hematopoiesis was associated with multilineage differentiation in the engrafted bone. Thus, the bone marrow implants provided stem cells as well as the microenvironment requisite for their long-term maintenance and multilineage differentiation. This model was useful to assess the effect of various recombinant human hematopoietic growth factors, administered either alone, or in combination, on human hematopoiesis *in vivo* (Kyoizumi *et al.*, 1993). Furthermore, this model was applied to a better understanding of the differentiation of hematopoiesis, showing that a simple fractionation based on well-defined CD34 antigen levels could be used to reproducibly isolate cells highly enriched for *in vivo* long-term repopulating activity and for multipotent progenitors including T- and B-cell precursors ($CD34^{hi}$ population). Thus, the microenvironment of implanted fetal bone fragments has the ability to induce differentiation as well to maintain fetal and adult $CD34^+$ Lin^- selected hematopoietic progenitors (DiGiusto *et al.*, 1994). In addition, donor $CD34^+$ cells can repopulate secondary bone grafts, indicating that the fetal bone microenvironment in the SCID-hu mice is sufficient for maintenance and proliferation of early progenitors (Chen *et al.*, 1994). Microinjection of enriched human fetal liver and umbilical cord blood hematopoietic stem cell populations in irradiated SCID-hu transplanted with bone, thymus and spleen fragments resulted in donor-derived B cells, myeloid cells, immature and mature T cells (Fraser *et al.*, 1995).

The classical SCID-hu model is useful in the study of lymphoid reconstitution and differentiation. Peault *et al.* (1991) showed that low numbers of fetal $CD34^+$ progenitor cells can repopulate the lymphoid compartment in the human thymus. $CD34^+$ transduced with the neoR gene engrafted, reconstituted the lymphoid compartments of the human thymus in SCID-hu, potentially allowing testing of gene therapeutic reagents for both genetic and acquired diseases of the T-lymphoid cell lineage (Akkina *et al.*, 1991).

SCID bone marrow engrafting

In 1988, Kamelreid and Dick described a system closely modeled on conventional bone marrow transplantation assays employing intravenous injection of adult bone marrow into SCID mice conditioned by irradiation. Doses of 400–450 cGy of whole body irradiation are routinely used. Human macrophage progenitors migrate to the murine marrow, increase in number, and are maintained in this environment for several months; however, no mature cells are detected. The adjunction of cell growth factors (stem cell factor and PIXY 321) allowed the detection of immature stem cells, with differentiated human cells of multiple myeloid and lymphoid lineages (Lapidot *et al.*, 1992b). The transplantation of human cord blood cells resulted in high levels of multilineage engraftment, including myeloid and lymphoid lineages and the treatment with cytokines was not required, suggesting that neonatal cells provided their own cytokines in a paracrine fashion (Vormoor *et al.*, 1994b). Human peripheral blood stem cells (PBSCs) injected to SCID mice resulted in prolonged generation of physiological levels of human cytokines including IL-3, IL-6 and granulocyte macrophage colony-stimulating factor in the murine blood over a period of at least 4 months (Goan *et al.*, 1995). Similar results were observed in NOD/SCID with the engraftment of a small number of primitive cells that proliferate and differentiate in the murine microenvironment, producing large numbers of long-term culture-initiating cells (LTC-IC), *in vivo* colony-forming cells (CFC), immature $CD34^+CD38^-$ cells, and also mature myeloid, erythroid and lymphoid cells (Lapidot *et al.*, 1992b; Vormoor *et al.*, 1994a; Torbett *et al.*, 1995). The primitive cells that initiate the graft were operationally defined as SCID-repopulating cells (SRCs) (Vormoor *et al.*, 1995; Dick, 1996); the SRC, exclusively present in the $CD4^+CD8^-$ fraction, capable of multilineage repopulation of the bone marrow of NOD/SCID mice, are rarely transduced with retroviruses, distinguishing them from most CFCs and LTC-ICs (Larochelle *et al.*, 1996). This observation is consistent with the low level of gene marking seen in human gene therapy trials.

The limitations of both models are the low count of human cells harvested in the mouse, and variation in the reproducibility of this *in vivo* model for hematopoietic stem cell development and differentiation.

SCID mouse models of human hematological diseases

Many human solid tumors can be xenografted into immune deficient nude mice, but leukemias and lymphomas typically grow as an ascites or as localized subcutaneous tumors in these recipients, neither of which is analogous to the human disease.

Leukemia

The availability of SCID mice has allowed the examination of the *in vivo* homing, engraftment, and growth patterns of neoplastic human hematopoietic cells (Dick *et al.*, 1992). Primary leukemia cells or leukemia cell lines from patients with acute lymphoblastic leukemia (ALL), acute myeloblastic leukemia (AML), chronic myelocytic leukemia (CML), or chronic lymphocytic leukemia (CLL) have been found to cause overt leukemia in SCID mice, with a pattern reminiscent of human clinical disease (Uckun, 1996). Human ALL cells can cause overt leukemia in SCID mice, and SCID mice may provide an efficient and reproducible model to study the pathogenesis, and for evaluation of therapy (Cesano *et al.*, 1991; Gunther *et al.*, 1993; Uckun *et al.*, 1995b). Furthermore the ability of leukemic cells from newly diagnosed patients to cause leukemia are associated with poor event-free survival in patients from whom the cells were obtained (Cesano *et al.*, 1992; Uckun *et al.*, 1994). Taken together, these studies show that the SCID mouse model may provide important prognostic information for patients with ALL. The information obtained using these models may also be useful for planning the components of intensification and maintenance of therapy.

Despite the growth of cytokine-independent AML cell lines in SCID mice, the outcome of initial attempts to grow primary leukemic cells from adult patients with AML has been inconsistent (Cesano *et al.*, 1992; Sawyers *et al.*, 1992; De-Lord *et al.*, 1994). Contrary to the aforementioned studies, Lapidot *et al.* (1992) and Cesano *et al.* (1992) reported that primary leukemic cells from some patients with newly diagnosed or relapsed AML are able to cause disseminated leukemia in SCID mice. Lapidot *et al.* (1994) recently reported that a cytokine-dependent immature cell is responsible for initiating human AML after transplantation and leukemic cell proliferation was seen only with $CD34^+CD38^-$ immature leukemic cells. However, some studies have established the sublethally irradiated SCID mouse as an *in vivo* model system that could replicate human childhood AML without administration of cytokines (Chelstrom *et al.*, 1994).

However, cells obtained from patients in blast crisis show an invasive growth pattern, whereas cells from CML patients in the chronic phase show lower engraftment (Sirard *et al.*, 1996). Attempts to establish SCID mouse models of chronic lymphocytic leukemia (CCL) have been largely unsuccessful, with sequestration of leukemic cells in the spleen (Kobayashi *et al.*, 1992). Thus, SCID mice do not provide a suitable microenvironment for the growth of primary human CLL.

Epstein–Barr virus lymphomas

Burkitt's lymphoma, nasopharyngeal carcinoma and large cell lymphomas are associated with EB virus infection. *In vivo* studies of EB virus are hampered because the natural host range is restricted to humans. As pointed out in the previous section on hu-PBL-SCID, all mice reconstituted with adult PBL may develop human B cell malignancy, since the majority of adult humans are infected with this herpes virus. The appearance of tumors in hu-PBL-SCID mice depends on the number of PBLs injected and on the serological status of the donors. Typically, the lymphomas arising in SCID mice resemble large cell lymphomas (Okano *et al.*, 1990; Rowe *et al.*, 1990, 1991), and are characteristic of *in vitro* transformed lymphoblastoid cell lines. They have a diploid karyotype without rearrangement.

Burkitt-like tumor-related lymphoma (monoclonal and with genetic alterations) was developed successfully in hu-PBL-SCID mice infected with EB virus particles (Dosch *et al.*, 1991). Many aspects of EB virus lymphomagenesis, including the role of immune surveillance (Baiocchi *et al.*, 1995; Lacerda, 1996; Sutkowski *et al.*, 1996), might be studied in SCID mice given the remarkable frequency of tumor development in these animals.

The SCID mouse model is useful for testing innovative leukemia therapy programs

Uckun *et al.* (1995) have recently compared the antileukemic activity of 15 agents in the SCID mouse model of human ALL, and showed that inhibitors targeted to CD19 receptor associated tyrosine kinases may have therapeutic potential in the treatment of ALL. Cesano *et al.* (1994) used a SCID mouse model of human AML to determine the effectiveness of a novel adoptive transfer approach with a human killer T-cell clone. The cure of disseminated xenografted human Hodgkin's tumors by bispecific monoclonal antibodies and human cells has been shown in a SCID mouse model (Renner *et al.*, 1996), while Franken *et al.* (1996) showed that complete macroscopic regression of established B-cell lymphoma in mice was observed after the injection of an EBNA2-responsive EB virus promoter driving a suicide gene (Franken *et al.*, 1996).

SCID mice support the growth of human tumors of non-hematopoietic origin

Several reports suggest that SCID mice are good recipients for studying the growth of human tumor cell lines and primary tumor tissue (Mueller *et al.*, 1991; Nomura *et al.*, 1991; Yano *et al.*, 1996). Most important is the greater degree of metastatic spread observed in SCID mice compared with nude mice (Nomura *et al.*, 1991). More information is available in the review by Williams *et al.* (1993). In addition, canine (Sugimoto *et al.*, 1994; Anderson *et al.*, 1995) or feline (Maruo *et al.*, 1995; Shtivelman and Namikawa, 1995) tumors can also grow in SCID mice.

Conclusions

The SCID mouse model is useful in studying human leukemias, evaluating the effects of new drugs and patient prognosis. However, limitations of this model include the fact that the results of leukemia biology in a SCID mouse environment need to be interpreted with caution; the pharmacodynamic studies have to be clarified because therapeutic concentrations may not be achievable without toxicity in patients, and a drug specifically directed against human leukemia cells will not react with mouse tissues, leading to reduced systemic toxicity and greater antileukemic potency.

6. Use of the SCID Model Without Reconstitution

Non reconstituted SCID can be used to establish animal models for infectious diseases

While the SCID mice have been used extensively as xenografts recipients or in selective reconstitution experiments to elucidate the role of immune components in the physiopathology of the infectious diseases, non reconstituted SCID mice have also been used to establish animal models of opportunistic infection and of parasitic diseases such a filarialsis (Nelson *et al.*, 1991) or amebiasis (Cieslak *et al.*, 1992); see Seydel and Stanley (1996) for a review.

The role for the immune system in reproduction remains unclear

SCID mice may be used to study major questions in the field of reproductive immunology. For example, Croy (1993) showed that the transfer of *Mus caroli* embryos to the uteri of pseudopregnant *scid/scid* mice disproved the hypothesis that antigen-specific immune rejection of fetuses occurred in this model of midgestational pregnancy failure. Xenogeneic engraftment of embryonic and uterine tissues into *scid/scid* mice is also successful and has the potential for facilitating studies of the fetomaternal interface in domestic animal species, such as cattle and horses (Crepeau and Croy, 1988; Croy, 1993; Ossa *et al.*, 1994).

7. Acknowledgements

This work was supported by the Agence Nationale de Recherches sur le SIDA. Nadirah Delhem is a granted investigator of the SIDACTION association. We thank Ray Pierce for critical reading of the manuscript.

8. References

Aaberge, I. S., Michaelsen, T. E., Rolstad, A. K., Groeng, E. C., Solberg, P., Lovik, M. (1992). SCID-hu mice immunized with a pneumococcal vaccine produce specific human antibodies and show increased resistance to infection. *Infect. Immun.* **60**(10), 4146–4153.

Aaberge, I. S., Steinsvik, T. E., Groeng, E. C., Leikvold, R. B., Lovik, M. (1996). Human antibody response to a pneumococcal vaccine in SCID-PBL-hu mice and simultaneously vaccinated human cell donors. *Clin. Exp. Immunol.* **105**(1), 12–17.

Abedi, M. R., Christensson, B., Islam, K. B., Hammarstrom, L., Smith, C. I. E. (1992). Immunoglobulin production in severe combined immunodeficient (SCID) mice reconstituted with human peripheral blood mononuclear cells. *Eur. J. Immunol.* **22**(3), 823–828.

Akkina, R. K., Rosenblatt, J. D., Campbell, A. G., Chen, I. S., Zack, J. A. (1994). Modeling human lymphoid precursor cell gene therapy in the SCID-hu mouse. *Blood* **84**(5), 1393–1398.

Aldrovandi, G. M., Feuer, G., Gao, L. Y., Jamieson, B., Kristeva, M., Chem, I. S. Y., Zack, J. A. (1993). The SCID-hu mouse as a model for HIV-1 infection. *Nature* **363**(6431), 732–736.

Alegre, M. L., Peterson, L. J., Jeyarajah, D. R., Weiser, M., Bluestone, J. A., Thistlethwaite, J. R. (1994). Severe combined immunodeficient mice engrafted with human splenocytes have functional human T cells and reject human allografts. *J. Immunol.* **153**(6), 2738–2749.

Anderson, P. M., Meyers, D. E., Hasz, D. E., Covalcuic, K., Saltzman, D., Khanna, C., Uckun, F. M. (1995). *In vitro* and *in vivo* cytotoxicity of an anti-osteosarcoma immunotoxin containing pokeweed antiviral protein. *Cancer Res.* **55**(6), 1321–1327.

Armstrong, N., Cigel, F., Borcherding, W., Hong, R., Malkovska, V. (1992). *In vitro* preactivated human T cells engraft in SCID mice and migrate to murine lymphoid tissues. *Clin. Exp. Immunol.* **90**(3), 476–482.

Autran, B., Guiet, P., Raphael, M., Grandadam, M., Agut, H., Candotti, D., Grenot, P., Puech, F., Debre, P., Cesbron, J. Y. (1996). Thymocyte and thymic microenvironment alterations during a systemic HIV infection in a severe combined immunodeficient mouse model. *AIDS* **10**(7), 717–727.

Auwaerter, P. G., Kaneshima, H., McCune, J. M., Wiegand, G., Griffin, D. E. (1996). Measles virus infection of thymic epithelium in the SCID-hu mouse leads to thymocyte apoptosis. *J. Virol.* **70**(6), 3734–3740.

Baccala, R., Vandekerckhove, B. A. E., Jones, D., Kono, D. H., Roncarolo, M. G., Theofilopoulos, A. N. (1993). Bacterial superantigens mediate T-cell deletions in the mouse severe combined immunodeficiency-human liver/thymus model. *J. Exp. Med.* **177**(5), 1481–1485.

Baiocchi, R. A., Ross, M. E., Tan, J. C., Chou, C. C., Sullivan, L., Halder, S., Monne, M., Seiden, M. V., Narula, S. K., Sklar, J., Croce, C. M., Caligiuri, M. A. (1995). Lymphomagenesis in the SCID-hu mouse involves abundant production of human interleukin-10. *Blood* **85**(4), 1063–1074.

Balson, G. A., Croy, B. A., Ross, T. L., Yager, J. A. (1993). Demonstration of equine immunoglobulin in sera from severe combined immunodeficiency/beige mice inoculated with equine lymphocytes. *Vet. Immunol. Immunopathol.* **39**(4), 315–325.

Bancroft, G. J., Kelly, J. P., McDonald, V. (1994). Models of innate immunity and opportunistic infection in the SCID mouse. *Res. Immunol.* **145**(5), 344–347.

Banga, S. S., Hall, K. T., Sandhu, A. K., Weaver, D. T., Athwal, R. S. (1994). Complementation of V(D)J recombination defect and X-ray sensitivity of scid mouse cells by human chromosome 8. *Mutat. Res.-DNA Repair* **315**(3), 239–247.

Barry, T. S., Jones, D. M., Richter, C. B., Haynes, B. F. (1991). Successful engraftment of human postnatal thymus in severe combined immune deficient (SCID) mice: differential engraftment of thymic components with irradiation versus anti-asialo GM-1 immunosuppressive regimens. *J. Exp. Med.* **173**(1), 167–180.

Baum, C. M., Weissman, I. L., Tsukamoto, A. S., Buckle, A. M., Peault, B. (1992). Isolation of a candidate human hematopoietic stem-cell population. *Proc. Natl Acad. Sci. USA* **89**(7), 2804–2808.

Bazin, R., Boucher, G., Monier, G., Chevrier, M. C., Verrette, S., Broly, H., Lemieux, R. (1994). Use of hu-IgG-SCID mice to evaluate *in vivo* stability of human monoclonal IgG antibodies. *J. Immunol. Methods* **172**(2), 209–217.

Biedermann, K. A., Sun, J. R., Giaccia, A. J., Tosto, L. M., Brown, J. M. (1991). SCID mutation in mice confers hypersensitivity to ionizing radiation and a deficiency in DNA double-strand break repair. *Proc. Natl Acad. Sci. USA* **88**(4), 1394–1397.

Blackwell, T. K., Ferrier, P., Malynn, B. A., Pollock, R. R., Covey, L. R., Suh, H. Y., Heinke, L. B., Fulop, G. M., Phillips, R. A., Yancopoulos, G. D., *et al.* (1989). The effect of the scid mutation on mechanism and control of immunoglobulin heavy and light chain gene rearrangement. *Curr. Top. Microbiol. Immunol.* **152**(85), 85–94.

Blunt, T., Finnie, N. J., Taccioli, G. E., Smith, G. C. M., Demengeot, J., Gottlieb, T. M., Mizuta, R., Varghese, A. J., Alt, F. W., Jeggo, P. A., Jackson, S. P. (1995). Defective DNA-dependent protein kinase activity is linked to V(D)J recombination and DNA repair defects associated with the murine scid mutation. *Cell* **80**(5), 813–823.

Boermans, H. J., Percy, D. H., Stirtzinger, T., Croy, B. A. (1992). Engraftment of severe combined immune deficient beige mice with bovine foetal lymphoid tissues. *Vet. Immunol. Immunopathol.* **34**(3–4), 273–289.

Bombil, F., Kints, J. P., Scheiff, J. M., Bazin, H., Latinne, D. (1996). A promising model of primary human immunization in human-scid mouse. *Immunobiology* **195**(3), 360–375.

Bonyhadi, M. L., Rabin, L., Salimi, S., Brown, D. A., Kosek, J., McCune, J. M., Kaneshima, H. (1993). HIV induces thymus depletion *In vivo*. *Nature* **363**(6431), 728–732.

Bos, J. D., Kapsenberg, M. L. (1993). The skin immune system: progress in cutaneous biology. *Immunol. Today* **14**(2), 75–78.

Bosma, M. J., Carroll, A. M. (1991). The SCID mouse mutant: definition, characterization, and potential uses. *Annu. Rev. Immunol.* **9**(323), 323–350.

Bosma, G. C., Custer, R. P., Bosma, M. J. (1983). A severe combined immunodeficency mutation in the mouse. *Nature* **301**(5900), 527–530.

Bosma, G. C., Fried, M., Custer, R. P., Carroll, A., Gibson, D. M., Bosma, M. J. (1988). Evidence of functional lymphocytes in some (leaky) scid mice. *J. Exp. Med.* **167**(3), 1016–1033.

Bosma, G. C., Davisson, M. T., Reutsch, N. R., Sweet, H. O., Shultz, L. D., Bosma, M. J. (1989). The mouse mutation severe combined immune deficiency (scid) is on chromosome 16 [published erratum appears in *Immunogenetics* 1989; **29**(3), 224]. *Immunogenetics* **29**(1), 54–57.

Brandsma, J. L., Brownstein, D. G., Xiao, W., Longley, B. J. (1995). Papilloma formation in human foreskin xenografts after inoculation of human papillomavirus type 16 DNA. *J. Virol.* **69**(4), 2716–2721.

Brown, J. M., Kaneshima, H., Mocarski, E. S. (1995). Dramatic interstrain differences in the replication of human cytomegalovirus in SCID-hu mice. *J. Inf. Dis.* **171**(6), 1599–1603.

Cadore, B., Puech, F., Cesbron, J. Y. (1995). Ethics and fetal experimental grafts of tissue on scid/scid mice. *M. S.-Med. Sci.* **11**(5), 755–760.

Carlsson, R., Martensson, C., Kalliomaki, S., Ohlin, M., Borrebaeck, C. A. K. (1992). Human peripheral blood lymphocytes transplanted into SCID mice constitute an *in vivo* culture system exhibiting several parameters found in a normal humoral immune response and are a source of immunocytes for the production of human monoclonal antibodies. *J. Immunol.* **148**(4), 1065–1071.

Carroll, A. M., Bosma, M. J. (1988). Detection and characterization of functional T cells in mice with severe combined immune deficiency. *Eur. J. Immunol.* **18**(12), 1965–1971.

Carroll, A. M., Bosma, M. J. (1989). Rearrangement of T cell receptor delta genes in thymus of scid mice. *Curr. Top. Microbiol. Immunol.* **152**(63), 63–67.

Carroll, A. M., Hardy, R. R., Bosma, M. J. (1989). Occurrence of mature B (IgM^+, $B220^+$) and T ($CD3^+$) lymphocytes in scid mice. *J. Immunol.* **143**(4), 1087–1093.

Cesano, A., O'Connor, R., Lange, B., Finan, J., Rovera, G., Santoli, D. (1991). Homing and progression patterns of childhood acute lymphoblastic leukemias in severe combined immunodeficiency mice. *Blood* **77**(11), 2463–2474.

Cesano, A., Hoxie, J. A., Lange, B., Nowell, P. C., Bishop, J., Santoli, D. (1992). The severe combined immunodeficient (SCID) mouse as a model for human myeloid leukemias. *Oncogene* **7**(5), 827–836.

Cesano, A., Visonneau, S., Cioe, L., Clark, S. C., Rovera, G., Santoli, D. (1994). Reversal of acute myelogenous leukemia in humanized SCID mice using a novel adoptive transfer approach. *J. Clin. Invest.* **94**(3), 1076–1084.

Cesbron, J. Y., Agut, H., Gosselin B., Candotti, D., Raphael, M., Puech, F., Grandadam, M., Debre, P., Capron, A., Autran, B. (1994). SCID-hu mouse as a model for human lung HIV-1 infection. *C.R. Acad. Sci. III* **317**(7), 669–674.

Chelstrom, L. M., Gunther, R., Simon, J., Raimondi, S. C., Krance, R., Crist, W. M., Uckun, F. M. (1994). Childhood acute myeloid leukemia in mice with severe combined immunodeficiency. *Blood* **84**(1), 20–26.

Chen, B. P., Galy, A., Kyoizumi, S., Namikawa, R., Scarborough, J., Webb, S., Ford, B., Cen, D. Z., Chen, S. C. (1994). Engraftment of human hematopoietic precursor cells with secondary transfer potential in SCID-hu mice. *Blood* **84**(8), 2497–2505.

Chen, B. P., Fraser, C., Reading, C., Murray, L., Uchida, N., Galy, A., Sasaki, D., Tricot, G., Jagannath, S., Barlogie, B., *et al.* (1995). Cytokine-mobilized peripheral blood $CD34^+Thy^-1^+Lin^-$ human hematopoietic stem cells as target cells for transplantation-based gene therapy. *Leukemia* **9**(1), S17–25.

Chiang, B. L., Chou, C. C., Ding, H. J., Huang, M. S., Chen, J. M., Hsieh, K. H. (1995). Establishment of human IgE system in severe combined immunodeficient mice with peripheral blood mononuclear cells from asthmatic children. *J. Allergy Clin. Immunol.* **95**(1), 69–76.

Cieslak, P. R., Virgin, H. T., Stanley, S. J. (1992). A severe combined immunodeficient (SCID) mouse model for infection with *Entamoeba histolytica*. *J. Exp. Med.* **176**(6), 1605–1609.

Crepeau, M. A., Croy, B. A. (1988). Evidence that specific cellular immunity cannot account for death of *Mus caroli* embryos transferred to *Mus musculus* with severe combined immune deficiency diseases. *Transplantation* **45**(6), 1104–1110.

Croy, B. A. (1993). The application of scid mouse technology to questions in reproductive biology. *Lab. Anim. Sci.* **43**(2), 123–126.

Delhem, N., Hadida, F., Gorochov, G., Carpentier, F., de Cavel, J. P., Andréani, J. P., Autran, B., Cesbron, J. Y. (1998). Anti-HIVgp 160 cytotoxic T cells lines derived from Hu-PBL SCID coengrafted with human skin after primary immunization using recombinant canary pox. *J. Immunol.*, in press.

De-Lord, C., Clutterbuck, R., Powles, R., Morilla, R., Hanby, A., Titley, J., Min, T., Millar, J. (1994). Growth of primary human acute lymphoblastic and myeloblastic leukemia in SCID mice. *Leuk. Lymphoma* **16**(1–2), 157–165.

Dick, J. E. (1996). Normal and leukemic human stem cells assayed in SCID mice. *Semin. Immunol.* **8**, 197–206.

Dick, J. E., Sirard, C., Pflumio, F., Lapidot, T. (1992). Murine models of normal and neoplastic human haematopoiesis. *Cancer Surv.* **15**, 161–181.

DiGiusto, D., Chen, S., Combs, J., Webb, S., Namikawa, R., Tsukamoto, A., Chen, B. P., Galy, A. H. (1994). Human fetal bone marrow early progenitors for T, B, and myeloid cells are found exclusively in the population expressing high levels of CD34. *Blood* **84**(2), 421–432.

Dorshkind, K., Keller, G. M., Phillips, R. A., Miller, R. G., Bosma, G. C., O'Toole, M., Bosma, M. J. (1984). Functional status of cells from lymphoid and myeloid tissues in mice with severe combined immunodeficiency disease. *J. Immunol.* **132**(4), 1804–1808.

Dorshkind, K., Pollack, S. B., Bosma, M. J., Phillips, R. A. (1985). Natural killer (NK) cells are present in mice with severe combined immunodeficiency (scid). *J. Immunol.* **134**(6), 3798–3801.

Dosch, H. M., Cochrane, D. M., Cook, V. A., Leeder, J. S., Cheung, R. K. (1991). Exogenous but not endogenous EBV induces lymphomas in beige/nude/xid mice carrying human lymphoid xenografts. *Int. Immunol.* 3(7), 731–735.

Duchosal, M. A.., Eming, S. A., Fischer, P., Leturcq, D., Barbas, C. F., Mcconahey, P. J., Caothien, R. H., Thornton, G. B., Dixon, F. J., Burton, D. R. (1992a). Immunization of Hu-PBL-SCID mice and the rescue of human monoclonal Fab fragments through combinatorial libraries. *Nature* **355**(6357), 258–262.

Duchosal, M. A., Eming, S. A., Mcconahey, P. J., Dixon, F. J. (1992b). Characterization of hu-PBL-SCID mice with high human immunoglobulin serum levels and graft-versus-host disease. *Am. J. Pathol.* **141**(5), 1097–1113.

Elkon, K. B., Ashany, D. (1993). The SCID mouse as a vehicle to study autoimmunity. *Br. J. Rheumatol.* **32**(1), 4–12.

Elkon, K. B., Ashany, D. (1994). Autoimmunity versus allo- and xeno-reactivity in SCID mice. *Int. Rev. Immunol.* **11**(4), 283–293.

Fell, A. H., Preston, P. M., Ansell, J. D. (1990). Establishment of Theileria-infected bovine cell lines in scid mice. *Parasite Immunol.* **12**(3), 335–339.

Fisch, P., Malkovsky, M., Kovats, S., Sturm, E., Braakman, E., Klein, B. S., Voss, S. D., Morissey, L. W., DeMars, R., Welch, W. J., Bolhuis, R. L., Sondel, P. M. (1990). Recognition by human V gamma 9/V delta 2 T cells of a GroEL homolog on Daudi Burkitt's lymphoma cells. *Science* **250**, 1269–1273.

Fisch, P., Oettel, K., Fudim, N., Surfus, J. E., Malkovsky, M., Sondel, P. M. (1992). MHC-unrestricted cytotoxic and proliferative responses of two distinct human gamma/delta T cell subsets to Daudi cells. *J. Immunol.* **148**(8), 2315–2323.

Franken, M., Estabrooks, A., Cavacini, L., Scerbune, B., Wang, F., Scadden, D. (1996). Epstein–Barr virus driven gene therapy for EBV-related lymphomas. *Nature Med.* **2**, 1379–1381.

Fraser, C. C., Kaneshima, H., Hansteen, G., Kilpatrick, M., Hoffman, R., Chen, B. P. (1995). Human allogeneic stem cell maintenance and differentiation in a long-term multilineage SCID-hu graft. *Blood* **86**(5), 1680–1693.

Froidevaux, S., Loor, F. (1991). A quick procedure for identifying doubly homozygous immunodeficient scid beige mice. *J. Immunol. Methods* **137**(2), 275–279.

Fulop, G. M., Phillips, R. A. (1990). The scid mutation in mice causes a general defect in DNA repair. *Nature* **347**(6292), 479–482.

Garni, W. B., Witte, P. L., Tutt, M. M., Kuziel, W. A., Tucker, P. W., Bennett, M., Kumar, V. (1990). Natural killer cells in the thymus. Studies in mice with severe combined immune deficiency. *J. Immunol.* **144**(3), 796–803.

Getts, R. C., Stamato, T. D. (1994). Absence of a Ku-like DNA end binding activity in the xrs double-strand DNA repair-deficient mutant. *J. Biol. Chem.* **269**(23), 15981–15984.

Goan, S. R., Fichtner, I., Just, U., Karawajew, L., Schultze, W., Krause, K. P., von Harsdorf, S., von Schilling, C., Herrmann, F. (1995). The severe combined immunodeficient-human peripheral blood stem cell (SCID-huPBSC) mouse: a xenotransplant model for huPBSC-initiated hematopoiesis. *Blood* **86**(1), 89–100.

Grandadam, M., Cesbron, J. Y., Candotti, D., Vinatier, D., Pauchard, M., Capron, A., Debre, P., Huraux, J. M., Autran, B., Agut, H. (1995). Dose-dependent systemic human immunodeficiency virus infection of SCID-hu mice after intraperitoneal virus injection. *Res. Virol.* **146**(2), 101–112.

Greenwood, J. D., Croy, B. A. (1993). A study on the engraftment and trafficking of bovine peripheral blood leukocytes in severe combined immunodeficient mice. *Vet. Immunol. Immunopathol.* **38**(1–2), 21–44.

Greenwood, J. D., Bos, N. A., Croy, B. A. (1996). Offspring of xenogeneically-reconstituted scid scid mice are capable of a primary xenogeneic immune response to DNP-KLH. *Vet. Immunol. Immunopathol.* **50**(1–2), 145–155.

Gunther, R., Chelstrom, L. M., Finnegan, D., Tuelahlgren, L., Irvin, J. D., Myers, D. E., Uckun, F. M. (1993). *In vivo* anti-leukemic efficacy of anti-CD7-pokeweed antiviral protein immunotoxin against human T-lineage acute lymphoblastic leukemia/lymphoma in mice with severe combined immunodeficiency. *Leukemia* 7(2), 298–309.

Hagiwara, K., Tsuji, M., Ishihara, C., Tajima, M., Kurosawa, T., Iwai, H., Takahashi, K. (1993a). The Bo-RBC-SCID mouse model for evaluating the efficacy of anti-theilerial drugs. *Int. J. Parasitol.* **23**(1), 13–16.

Hagiwara, K., Tsuji, M., Ishihara, C., Tajima, M., Kurosawa, T., Iwai, H., Takahashi, K. (1993b). *Theileria sergenti* infection in the Bo-RBC-SCID mouse model. *Parasitol. Res.* **79**(6), 466–470.

Hagiwara, K., Tsuji, M., Ishihara, C., Tajima, M., Kurosawa, T., Takahashi, K. (1995). Serum from *Theileria sergenti*-infected cattle accelerates the clearance of bovine erythrocytes in SCID mice. *Parasitol Res.* **81**(6), 470–474.

Hendrickson, E. A., Qin, X. Q., Bump, E. A., Schatz, D. G., Oettinger, M., Weaver, D. T. (1991). A link between double-strand break-related repair and V(D)J recombination: the scid mutation. *Proc. Natl Acad. Sci. USA* **88**(10), 4061–4065.

Hendrickson, E. A., Schlissel, M. S., Weaver, D. T. (1990). Wild-type V(D)J recombination in scid pre-B cells. *Mol. Cell Biol.* **10**(10), 5397–5407.

Hoffmannfezer, G., Kranz, B., Gall, C., Thierfelder, S. (1992). Peritoneal sanctuary for human lymphopoiesis in SCID mice injected with human peripheral blood lymphocytes from Epstein–Barr virus-negative donors. *Eur. J. Immunol.* **22**(12), 3161–3166.

Hoffmannfezer, G., Gall, C., Zengerle, U., Kranz, B., Thierfelder, S. (1993). Immunohistology and immunocytology of human T-cell chimerism and graft-versus-host disease in SCID mice. *Blood* **81**(12), 3440–3448.

Huard, J., Roy, R., Guerette, B., Verreault, S., Tremblay, G., Tremblay, J. P. (1994a). Human myoblast transplantation in immunodeficient and immunosuppressed mice: evidence of rejection. *Muscle Nerve* **17**(2), 224–234.

Huard, J., Verreault, S., Roy, R., Tremblay, M., Tremblay, J. P. (1994b). High efficiency of muscle regeneration after human myoblast clone transplantation in SCID mice. *J. Clin. Invest.* **93**(2), 586–599.

Ifversen, P., Martensson, C., Danielsson, L., Ingvar, C., Carlsson, R., Borrebaeck, C. A. (1995). Induction of primary antigen-specific immune responses in SCID-hu-PBL by coupled T–B epitopes. *Immunology* **84**(1), 111–116.

Itoh, M., Hamatani, K., Komatsu, K., Araki, R., Takayama, K., Abe, M. (1993). Human chromosome-8 (p12–q22) complements radiosensitivity in the severe combined immune deficiency (SCID) mouse. *Radiat. Res.* **134**(3), 364–368.

Johnson, C. M., Selleseth, D. W., Ellis, M. N., Childers, T. A., Tompkins, M. B., Tompkins, W. A. (1994). Feline lymphoid tissues engrafted into scid mice maintain morphologic structure and produce feline immunoglobulin. *Lab. Anim. Sci.* **44**(4), 313–318.

Johnson, C. M., Selleseth, D. W., Ellis, M. N., Childers, T. A.,

Tompkins, M. B., Tompkins, W. A. (1995). Reduced provirus burden and enhanced humoral immune function in AZT-treated SCID-feline mice inoculated with feline immunodeficiency virus. *Vet. Immunol. Immunopathol.* **46**(1–2), 169–180.

Jorgensen, C., Bologna, C., Sany, J. (1995). Autoimmunity. Insights provided by the SCID mouse model. *Rev. Rhum. Engl. Ed.* **62**(7–8), 519–524.

Kamelreid, S., Dick, J. E. (1988). Engraftment of immune-deficient mice with human hematopoietic stem cells. *Science* **242**(4886), 1706–1709.

Kaneshima, H., Namikawa, R., McCune, J. M. (1994). Human hematolymphoid cells in SCID mice. *Curr. Opin. Immunol.* **6**(2), 327–333.

Kaufmann, R., Mielke, V., Reimann, J., Klein, C. E., Sterry, W. (1993). Cellular and molecular composition of human skin in long-term xenografts on SCID mice. *Exp. Dermatol* **2**(5), 209–216.

Kaul, R., Sharma, A., Lisse, J. R., Christadoss, P. (1995). Human recombinant IL-2 augments immunoglobulin and induces rheumatoid factor production by rheumatoid arthritis lymphocytes engrafted into severe combined immunodeficient mice. *Clin. Immunol. Immunopathol.* **74**(3), 271–282.

Kim, Y. H., Woodley, D. T., Wynn, K. C., Giomi, W., Bauer, E. A. (1992). Recessive dystrophic epidermolysis bullosa phenotype is preserved in xenografts using SCID mice – development of an experimental *in vivo* model. *J. Invest. Dermatol.* **98**(2), 191–197.

Kirchgessner, C. U., Patil, C. K., Evans, J. W., Cuomo, C. A., Fried, L. M., Carter, T., Oettinger, M. A., Brown, J. M. (1995). DNA-dependent kinase (p350) as a candidate gene for the murine SCID defect. *Science* **267**(5201), 1178–1183.

Kirchgessner, C. U., Tosto, L. M., Biedermann, K. A., Kovacs, M., Araujo, D., Stanbridge, E. J., Brown, J. M. (1993). Complementation of the radiosensitive phenotype in severe combined immunodeficient mice by human chromosome-8. *Cancer Res.* **53**(24), 6011–6016.

Kobayashi, R., Picchio, G., Kirven, M., Meisenholder, G., Baird, S., Carson, D. A., Mosier, D. E., Kipps, T. J. (1992). Transfer of human chronic lymphocytic leukemia to mice with severe combined immune deficiency. *Leuk. Res.* **16**(10), 1013–1023.

Komatsu, K., Ohta, T., Jinno, Y., Niikawa, N., Okumura, Y. (1993). Functional complementation in mouse–human radiation hybrids assigns the putative murine SCID gene to the pericentric region of human chromosome-8. *Hum. Mol. Genet.* **2**(7), 1031–1034.

Kotloff, D. B., Bosma, M. J., Ruetsch, N. R. (1993). Scid mouse pre-B cells with intracellular mu-chains – analysis of recombinase activity and IgH gene rearrangements. *Int. Immunol.* **5**(4), 383–391.

Kraft, D. L., Weissman, I. L., Waller, E. K. (1993). Differentiation of $CD3^-4^-8^-$ human fetal thymocytes *in vivo* – characterization of a $CD3^-4^+8^-$ intermediate. *J. Exp. Med.* **178**(1), 265–277.

Krowka, J. F., Sarin, S., Namikawa, R., McCune, J. M., Kaneshima, H. (1991). Human T cells in the SCID-hu mouse are phenotypically normal and functionally competent. *J. Immunol.* **146**(11), 3751–3756.

Kurimasa, A., Nagata, Y., Shimizu, M., Emi, M., Nakamura, Y., Oshimura, M. (1994). A human gene that restores the DNA-repair defect in SCID mice is located on 8P11.1–q11.1. *Hum. Genet.* **93**(1), 21–26.

Kyoizumi, S., Baum, C. M., Kaneshima, H., McCune, J. M., Yee, E. J., Namikawa, R. (1992). Implantation and maintenance of functional human bone marrow in SCID-hu mice. *Blood* **79**(7), 1704–1711.

Kyoizumi, S., Murray, L. J., Namikawa, R. (1993). Preclinical analysis of cytokine therapy in the SCID-hu mouse. Blood **81**(6), 1479–1488.

Lacerda, J. F. (1996). Human Epstein–Barr virus (EBV)-specific cytotoxic T lymphocytes home preferentially to and induce selective regressions of autologous EBV-induced B cell lymphoproliferations in xenografted CB-17 scid/scid mice. *J. Exp. Med.* **183**(3), 1215–1228.

Lapidot, T., Faktorowich, Y., Lubin, I., Reisner, Y. (1992a). Enhancement of T-cell-depleted bone marrow allografts in the absence of graft-versus-host disease is mediated by $CD8^+CD4^-$ and not by $CD8^-CD4^+$ thymocytes. *Blood* **80**(9), 2406–2411.

Lapidot, T., Pflumio, F., Doedens, M., Murdoch, B., Williams, D. E., Dick, J. E. (1992b). Cytokine stimulation of multilineage hematopoiesis from immature human cells engrafted in SCID mice. *Science* **255**(5048), 1137–1141.

Lapidot, T., Sirard, C., Vormoor, J., Murdoch, B., Hoang, T., Cacerescortes, J., Minden, M., Paterson, B., Caligiuri, M. A., Dick, J. E. (1994). A cell initiating human acute myeloid leukaemia after transplantation into SCID mice. *Nature* **367**(6464), 645–648.

Larochelle, A., Vormoor, J., Hanenberg, H., Wang, J. C. Y., Bhatia, M., Lapidot, T., Moritz, T., Murdoch, B., Xiao, X. L., Kato, I., Williams, D. A., Dick, J. E. (1996). Identification of primitive human hematopoietic cells capable of repopulating NOD/SCID mouse bone marrow: implications for gene therapy. *Nature Med.* **12**, 1329–1337.

Lasmézas, C. I., Cesbron, J. Y., Deslys, J. P., Demaimay, R., Adjou, K. T., Rioux, R., Lemaire, C., Locht, C., Dormont, D. (1996). Immune system-dependent and -independent replication of the scrapie agent. *J. Virol.* **70**(2), 1292–1295.

Lemaire, R., Flipo, R. M., Monte, D., Dupressoir, T., Duquesnoy, B., Cesbron, J. Y., Janin, A., Capron, A., Lafyatis, R. (1994). Synovial fibroblast-like cell transfection with the SV40 large T antigen induces a transformed phenotype and permits transient tumor formation in immunodeficient mice. *J. Rheumatol.* **21**(8), 1409–1419.

Levi, A. D. O., Bunge, R. P. (1994). Studies of myelin formation after transplantation of human Schwann cells into the severe combined immunodeficient mouse. *Exp. Neurol.* **130**(1), 41–52.

Lieber, M. R., Hesse, J. E., Lewis, S., Bosma, G. C., Rosenberg, N., Mizuuchi, K., Bosma, M. J., Gellert, M. (1988). The defect in murine severe combined immune deficiency: joining of signal sequences but not coding segments in V(D)J recombination. *Cell* **55**(1), 7–16.

Linenberger, M. L., Beebe, A. M., Pedersen, N. C., Abkowitz, J. L., Dandekar, S. (1995). Marrow accessory cell infection and alterations in hematopoiesis accompany severe neutropenia during experimental acute infection with feline immunodeficiency virus. *Blood* **85**(4), 941–951.

Louwagie, A. C., Verwilghen, R. L. (1970). Growth of haemopoietic spleen colonies after grafting of human bone marrow in mice. *Nature* **225**, 383.

Lubin, I., Faktorowich, Y., Lapidot, T., Gan, Y., Eshhar, Z., Gazit, E., Levite, M., Reisner, Y. (1991). Engraftment and development of human T and B cells in mice after bone marrow transplantation. *Science* **252**(5004), 427–431.

Malkovska, V., Cigel, F., Storer, B. E. (1994). Human T-cells in hu-PBL-SCID mice proliferate in response to Daudi lymphoma and confer anti-tumour immunity. *Clin. Exp. Immunol.* **96**(1), 158–165.

Malynn, B. A., Blackwell, T. K., Fulop, G. M., Rathbun, G. A., Furley, A. J., Ferrier, P., Heinke, L. B., Phillips, R. A., Yancopoulos, G. D., Alt, F. W. (1988). The scid defect affects the final step of the immunoglobulin VDJ recombinase mechanism. *Cell* **54**(4), 453–460.

Maruo, K., Sugimoto, T., Suzuki, K., Shirota, K., Ejima, H., Nomura, T. (1995). Xenotransplantation of high tumorigenicity of feline tumours in SCID mice. *J. Vet. Med. Sci.* **57**(5), 967–969.

Mazingue, C., Cottrez, F., Auriault, C., Cesbron, J. Y., Capron, A. (1991). Obtention of a human primary humoral response against schistosome protective antigens in severe combined immunodeficiency mice after the transfer of human peripheral blood mononuclear cells. *Eur. J. Immunol.* **21**(7), 1763–1766.

McCune, J., Kaneshima, H., Krowka, J., Namikawa, R., Outzen, H., Peault, B., Rabin, L., Shih, C. C., Yee, E., Lieberman, M., *et al.* (1991). The SCID-hu mouse: a small animal model for HIV infection and pathogenesis. *Annu. Rev. Immunol.* **9**(399), 399–429.

McCune, J. M. (1992). The SCID-hu mouse – a small animal model for the analysis of human hematolymphoid differentiation and function. *Bone Marrow Transpl.* **9**(S1), 74–76.

McCune, J. M., Namikawa, R., Kaneshima, H., Schultz, L. D., Lieberman, M., Weissman, I. L. (l988). The SCID-hu mouse: murine model for the analysis of human hematolymphoid differentiation and function. *Science* **241**(4873), 1632–1639.

Meers, J., Delfierro, G. M., Cope, R. B., Park, H. S., Greene, W. K., Robinson, W. F. (1993). Feline immunodeficiency virus infection – plasma, but not peripheral blood mononuclear cell virus titer is influenced by zidovudine and cyclosporine. *Arch. Virol.* **132**(1–2), 67–81.

Miller, R. D., Ozaki, J. H., Riblet, R. (1993). The mouse severe combined immune deficiency (scid) mutation is closely linked to the B-cell-specific developmental genes VpreB and lambda-5. *Genomics* **16**(3), 740–744.

Mocarski, E. S., Bonyhadi, M., Salimi, S., McCune, J. M., Kaneshima, H. (1993). Human cytomegalovirus in a SCID-hu mouse – thymic epithelial cells are prominent targets of viral replication. *Proc. Natl Acad. Sci. USA* **90**(1), 104–108.

Moffat, J. F., Stein, M. D., Kaneshima, H., Arvin, A. M. (1995). Tropism of varicella-zoster virus for human $CD4^+$ and $CD8^+$ T lymphocytes and epidermal cells in SCID-hu mice. *J. Virol.* **69**(9), 5236–5242.

Mombaerts, P., Iacomini, J., Johnson, R. S., Herrup, K., Tonegawa, S., Papaioannou, V. E. (1992). RAG-1-deficient mice have no mature lymphocytes-B and lymphocytes-T. *Cell* **68**(5), 869–877.

Mosier, D. E. (1990). Immunodeficient mice xenografted with human lymphoid cells: new models for *in vivo* studies of human immunobiology and infectious diseases. *J. Clin. Immunol.* **10**(4), 185–191.

Mosier, D. E., Gulizia, R. J., Baird, S. M., Wilson, D. B. (1988). Transfer of a functional human immune system to mice with severe combined immunodeficiency. *Nature* **335**(6187), 256–259.

Mosier, D. E., Gulizia, R. J., Baird, S. M., Wilson, D. B., Spector, D. H., Spector, S. A. (1991). Human immunodeficiency virus infection of human-PBL-SCID mice. *Science* **251**(4995), 791–794.

Mosier, D. E., Picchio, G. R., Kirven, M. B., Garnier, J. L., Torbett, B. E., Baird, S. M., Kobayashi, R., Kipps, T. J. (1992). EBV-induced human B-cell lymphomas in hu-PBL-SCID mice. *AIDS Res. Hum. Retroviruses* **8**(5), 735–740.

Mosier, D. E., Gulizia, R. J., Macisaac, P. D., Corey, L., Greenberg, P. D. (1993a). Resistance to human immunodeficiency virus-1 infection of SCID mice reconstituted with peripheral blood leukocytes from donors vaccinated with vaccinia gp160 and recombinant gp160. *Proc. Natl Acad. Sci. USA* **90**(6), 2443–2447.

Mosier, D. E., Stell, K. L., Gulizia, R. J., Torbett, B. E., Gilmore, G. L. (1993b). Homozygous scid/scid-beige/beige mice have low levels of spontaneous or neonatal T-cell-induced B-cell generation. *J. Exp. Med.* **177**(1), 191–194.

Mueller, B. M., Romerdahl, C. A., Trent, J. M., Reisfeld, R. A. (1991). Suppression of spontaneous melanoma metastasis in scid mice with an antibody to the epidermal growth factor receptor. *Cancer Res.* **51**(8), 2193–2198.

Murphy, W. J., Conlon, K. C., Sayers, T. J., Wiltrout, R. H., Back, T. C., Ortaldo, J. R., Longo, D. L. (1993). Engraftment and activity of anti-CD3-activated human peripheral blood lymphocytes tranferred into mice with severe combined immune deficiency. *J. Immunol.* **150**(8 Part 1), 3634–3642.

Murray, A. G., Petzelbauer, P., Hughes, C. C. W., Costa, J., Askenase, P., Pober, J. S. (1994). Human T-cell-mediated destruction of allogeneic dermal microvessels in a severe combined immunodeficient mouse. *Proc. Natl Acad. Sci. USA* **91**(19), 9146–9150.

Nakamura, Y., Tsuji, M., Arai, S., Ishihara, C. (1995). A method for rapid and complete substitution of the circulating erythrocytes in SCID mice with bovine erythrocytes and use of the substituted mice for bovine hemoprotozoa infections. *J. Immunol. Methods* **188**(2), 247–254.

Namikawa, R., Kaneshima, H., Lieberman, M., Weissman, I. L., McCune, J. M. (1988). Infection of the SCID-hu mouse by HIV-1. *Science* **242**(4886), 1684–1686.

Namikawa, R., Weilbaecher, K. N., Kaneshima, H., Yee, E. J., McCune, J. M. (1990). Long-term human hematopoiesis in the SCID-hu mouse. *J. Exp. Med.* **172**(4), 1055–1063.

Neil, G. A., Sammons, D. W. (1992). Immunization of SCID-hu mice and generation of anti-hepatitis B surface antigen-specific hybridomas by electrofusion. *Hum. Antibodies Hybridomas* **3**(4), 201–205.

Nelson, F. K., Greiner, D. L., Shultz, L. D., Rajan, T. V. (1991). The immunodeficient scid mouse as a model for human lymphatic filariasis. *J. Exp. Med*, **173**(3), 659–663.

Nomura, T., Takahama, Y., Hongyo, T., Takatera, H., Inohara, H., Fukushima, H., Ono, S., Hamaoka, T. (1991). Rapid growth and spontaneous metastasis of human germinal tumors ectopically transplanted into scid (severe combined immunodeficiency) and scid-nudestreaker mice. *Jpn. J. Cancer Res.* **82**(6), 701–709.

Nonoyama, S., Smith, F. O., Bernstein, I. D., Ochs, H. D. (1993). Strain-dependent leakiness of mice with severe combined immune deficiency. *J. Immunol.* **150**(9), 3817–3824.

Ogura, A., Noguchi, Y., Yamamoto, Y., Shibata, S., Asano, T., Okamoto, Y., Honda, M. (1996). Localization of HIV-1 in human thymic implant in SCID-hu mice after intravenous inoculation. *Int. J. Exp. Pathol.* **77**(5), 201–206.

Okano, M., Taguchi, Y., Nakamine, H., Pirruccello, S. J., Davis, J. R., Beisel, K. W., Kleveland, K. L., Sanger, W. G., Fordyce, R. R., Purtilo, D. T. (1990). Characterization of Epstein–Barr virus-induced lymphoproliferation derived from human peripheral blood mononuclear cells transferred to severe combined immunodeficient mice. *Am. J. Pathol.* **137**(3), 517–522.

Ossa, J. E., Cadavid, A. P., Maldonado, J. G. (1994). Is the immune system necessary for placental reproduction? A hypothesis of the mechanisms of alloimmunotherapy in recurrent spontaneous abortion. *Med. Hypotheses* **42**(3), 193–197.

Palluault, F., Soulez, B., Slomianny, C., Dei, C. E., Cesbron, J. Y., Camus, D. (1992). High osmotic pressure for *Pneumocystis carinnii* London Resin White embedding enables fine immunocytochemistry studies: I. Golgi complex and cell-wall synthesis. *Parasitol. Res.* 78(6), 482–488.

Peault, B., Weissman, I. L., Baum, C., McCune, J. M., Tsukamoto, A. (1991). Lymphoid reconstitution of the human fetal thymus in SCID mice with CD34+ precursor cells. *J. Exp. Med.* **174**(5), 1283–1286.

Peault, B., Tirouvanziam, R., Sombardier, M. N., Chen, S., Perricaudet, M., Gaillard, D. (1994). Gene transfer to human fetal pulmonary tissue developed in immunodeficient SCID mice. *Hum. Gene Ther.* 5(9), 1131–1137.

Pergola, F., Zdzienicka, M. Z., Lieber, M. R. (1993). V(D)J recombination in mammalian cell mutants defective in DNA double-strand break repair. *Mol. Cell Biol.* **113**(6), 3464–3471.

Pestel, J., Jeannin, P., Delneste, Y., Dessaint, J. P., Cesbron, J. Y., Capron, A., Tsicopoulos, A., Tonnel, A. B. (1994). Human IgE in SCID mice reconstituted with peripheral blood mononuclear cells from *Dermatophagoides pteronyssinus*-sensitive patients. *J. Imminol.* **153**(8), 3804–3810.

Peterson, S. R., Kurimasa, A., Oshimura, M., Dynan, W. S., Bradbury, E. M., Chen, D. J. (1995). Loss of the catalytic subunit of the DNA-dependent protein kinase in DNA double-strand-break-repair mutant mammalian cells. *Proc. Natl Acad. Sci. USA* **92**(8), 3171–3174.

Petrini, J. H., Carroll, A. M., Bosma, M. J. (1990). T-cell receptor gene rearrangements in functional T-cell clones from severe combined immune deficient (scid) mice: reversion of the scid phenotype in individual lymphocyte progenitors. *Proc. Natl Acad. Sci. USA* **87**(9), 3450–3453.

Petzelbauer, P., Groger, M., Kunstfeld, R., Petzelbauer, E., Wolff, K. (1996). Human delayed-type hypersensitivity reaction in a SCID mouse engrafted with human T cells and autologous skin. *J. Invest. Dermatol.* **107**(4), 576–581.

Rathmell, W. K., Chu, G. (1994). A DNA end-binding factor involved in double-strand break repair and V(D)J recombination. *Mol. Cell Biol.* **14**(7), 4741–4748.

Reinhardt, B., Torbett, B. E., Gulizia, R. J., Reinhardt, P. P., Spector, S. A., Mosier, D. E. (1994). Human immunodeficiency virus type 1 infection of neonatal severe combined immunodeficient mice xenografted with human cord blood cells. *Aids Res. Hum. Retroviruses* **10**(2), 131–141.

Renner, C., Bauer, S., Sahin, U., Jung, W. Van-Lier, R., Jacobs, G., Helde, G., Pfreundschuh, M. (1996). Cure of disseminated xenografted human Hodgkin's tumors by bi-specific monoclonal antibodies and human T cells: the role of human T-cells subsets in a preclinical model. *Blood* **87**(7), 2930–2937.

Riddell, S. R., Greenberg, P. D. (1995). Principles for adoptive T cell therapy of human viral diseases. *Annu. Rev. Immunol.* **13**, 545–586.

Roth, D. B., Menetski, J. P., Nakajima, P. B., Bosma, M. J., Gellert, M. (1992). V(D)J recombination – broken DNA molecules with covalently sealed (hairpin) coding ends in scid mouse thymocytes. *Cell* **70**(6), 983–991.

Rowe, M., Young, L. S., Rickenson, A. B. (1990). Analysis of Epstein–Barr virus gene expression in lymphomas derived from normal human B cells grafted into SCID mice. *Curr. Top. Microbiol. Immunol.* **166**(325), 325–331.

Rowe, M., Young, L. S., Crocker, J., Stokes, H., Henderson, S., Rickinson, A. B. (1991). Epstein–Barr virus (EBV)-associated lymphoproliferative disease in the SCID mouse model: implications for the pathogenesis of EBV-positive lymphomas in man. *J. Exp. Med.* **173**(1), 147–158.

Savidge, T. C., Morey, A. L., Ferguson, D. L., Fleming, K. A., Shmakov, A. N., Phillips, A. D. (1995). Human intestinal development in a severe-combined immunodeficient xenograft model. *Differentiation* **58**(5), 361–371.

Sawyers, C. L., Gishizky, M. L., Quan, S., Golde, D. W., Witte, O. N. (1992). Propagation of human blastic myeloid leukemias in the SCID mouse. *Blood* **79**(8), 2089–2098.

Saxon, A., Macy, E., Denis, K., Tary, L. M., Witte, O., Braun, J. (1991). Limited B cell repertoire in severe combined immunodeficient mice engrafted with peripheral blood mononuclear cells derived from immunodeficient or normal humans. *J. Clin. Invest.* **87**(2), 658–665.

Schmitt, D. (1995). Immune response in the skin. *Clin. Rev. Allergy Immunol.* **13**, 177–188.

Schuler, W. (1990). The scid mouse mutant: biology and nature of the defect. *Cytokines* **3**, 132–153.

Schuler, W., Schuler, A., Bosma, M. J. (1990). Defective V-to-J recombination of T cell receptor gamma chain genes in scid mice. *Eur. J. Immunol.* **20**(3), 545–550.

Segall, H., Lubin, I., Marcus, H., Canaan, A., Reisner, Y. (1996). Generation of primary antigen-specific human cytotoxic T lymphocytes in human/mouse radiation chimera. *Blood* **88**(2), 721–730.

Seydel, K. B., Stanley, S. L. (1996). SCID and the study of parasitic disease. *Clin. Microbiol. Rev.* **9**(2), 126.

Shinkai, Y., Rathbun, G., Lam, K. P., Oltz, E. M., Stewart, V., Mendelsohn, M., Charron, J., Datta, M., Young, F., Stall, A M., Alt, F. W. (1992). RAG-2-deficient mice lack mature lymphocytes owing to inability to initiate V(D)J rearrangement. *Cell* **68**(5), 855–867.

Shtivelman, E., Namikawa, R. (1995). Species-specific metastasis of human tumor cells in the severe combined immunodeficiency mouse engrafted with human tissue. *Proc. Natl Acad. Sci. USA* **92**(10), 4661–4665.

Sirard, C., Lapidot, T., Vormoor, J., Cashman, J. D., Doedens, M., Murdoch, B., Jamal, N., Messner, H., Addey, L., Minden, M., Laraya, P., Keating, A., Eaves, A., Lansdorp, P. M., Eaves, C. J., Dick, J. E. (1996). Normal and leukemic SCID-repopulating cells (SRC) coexist in the bone marrow and peripheral blood from CML patients in chronic phase, whereas leukemic SRC are detected in blast crisis. *Blood* **87**(4), 1539–1548.

Soballe, P. W., Montone, K. T., Satyamoorthy, K., Nesbit, M., Herlyn, M. (1996). Carcinogenesis in human skin grafted to SCID mice. *Cancer Res.* **56**(4), 757–764.

Soulez, B., Palluault, F., Cesbron, J. Y., Dei, C. E., Capron, A., Camus, D. (1991). Introduction of *Pneumocystis carinii* in a colony of SCID mice. *J. Protozool.* **38**(6).

Stanley, B. K., McCune, J., Kaneshima, H., Justement, J. S., Sullivan, M., Boone, E., Baseler, M., Adelsberger, J., Bonyhadi, M., Orenstein, J., Fox, C. H., Fauci, A. S. (1993). Human immunodeficiency virus infection of the human thymus and disruption of the thymic microenvironment in the SCID-hu mouse. *J. Exp. Med.* **178**, 1151–1163.

Steinman, R. M. (1991). The dendritic cell system and its role in immunogenicity. *Annu. Rev. Immunol.* **9**, 271–296.

Steinsvik, T. E., Gaarder, P. I., Aaberge, I. S., Lovik (1995). Engraftment and humoral immunity in SCID and RAG-2-deficient mice transplanted with human peripheral blood lymphocytes. *Scand. J. Immunol* **42**(6), 607–616.

Sugimoto, T., Maruo, K., Imaeda, Y., Suzuki, K., Shirota, K., Ejima, H., Endo, S., Nomura, T. (1994). Xenotransplantation of canine tumors into severe combined immunodeficient (SCID) mice. *J. Vet. Med. Sci.* **56**(6), 1087–1091.

Sutkowski, N., Palkama, T., Ciurli, C., Sekaly, R. P., Thorley-lawson, D. A., Huber, B. T. (1996). An Epstein–Barr virus-associated superantigen. *J. Exp. Med.* **184**(3), 971–980.

Taccioli, G. E., Gottlieb, T. M., Blunt, T., Priestley, A., Demengeot, J., Mizuta, R., Lehmann, A. R., Alt, F. W., Jackson, S. P., Jeggo, P. A. (1994). Ku80: product of the XRCC5 gene and its role in DNA repair and V(D)J recombination. *Science* **265**(5177), 1442–1445.

Tary Lehmann, M., Saxon, A. (1992). Human mature T cells that are anergic *in vivo* prevail in SCID mice reconstituted with human peripheral blood. *J. Exp. Med.* **175**(2), 503–516.

Tary Lehmann, M., Lehmann, P. V., Schols, D., Roncarolo, M. G., Saxon, A. (1994). Anti-SCID mouse reactivity shapes the human $CD4^+$ T cell repertoire in hu-PBL-SCID chimeras. *J. Exp. Med.* **180**(5), 1817–1827.

Tary Lehmann, M., Saxon, A., Lehmann, P. V. (1995). The human immune system in hu-PBL-SCID mice. *Immunol. Today* **16**(11), 529–533.

Tonnel, A. B., Pestel, J., Duez, C., Jeannin, P., Cesbron, J. Y., Capron, A. (1995). Human IgE in severe combined immunodeficiency mice reconstituted with peripheral blood mononuclear cells from *Dermatophagoides pteronyssinus*-sensitive patients. *Int. Arch. Allergy Immunol.* **107**(1–3), 223–225.

Torbett, B. E., Picchio, G., Mosier, D. E. (1991). hu-PBL-SCID mice: a model for human immune function, AIDS, and lymphomagenesis. *Immunol. Rev.* **124**(139), 139–164.

Torbett, B. E., Conners, K., Shao, L. E., Mosier, D. E., Yu, L. (1995). High level multilineage engraftment of human hematopoietic cells is maintained in NOD/SCID mice. *Blood* **86**, 436:a.

Tsuji, M., Hagiwara, K., Takahashi, K., Ishihara, C., Azuma, I., Siddiqui, W. A. (1992). *Theileria sergenti* proliferates in SCID mice with bovine erythrocyte transfusion. *J. Parasitol.* **78**(4), 750–752.

Tsuji, M., Terada, Y., Arai, S., Okada, H., Ishihara, C. (1995). Use of the Bo-RBC-SCID mouse model for isolation of a *Babesia* parasite from grazing calves in Japan. *Exp. Parasitol* **81**(4), 512–518.

Uckun, F. M. (1996). Severe combined immunodeficient mouse models of human leukemia. *Blood* **88**(4), 1135–1146.

Uckun, F. M., Downing, J. R., Chelstrom, L. M., Gunther, R., Ryan, M., Simon, J., Carroll, A. J., Tuelahlgren, L., Crist, W. M. (1994). Human t(4;11)(q21;q23) acute lymphoblastic leukemia in mice with severe combined immunodeficiency. *Blood* **84**(3), 859–865.

Uckun, F. M., Evans, W. E., Forsyth, C. J., Waddick, K. G., Ahlgren, L. T., Chelstrom, L. M., Burkhardt, A., Bolen, J., Myers, D. E. (1995a). Biotherapy of B-cell precursor leukemia by targeting genistein to CD19-associated tyrosine kinases. *Science* **267**(5199), 886–891.

Uckun, F. M., Sather, H., Reaman, G., Shuster, J., Land, V., Trigg, M., Gunther, R., Chelstrom, L., Bleyer, A., Gaynon, P., Crist, W. (1995b). Leukemic cell growth in SCID mice as a predictor of relapse in high-risk B-lineage acute lymphoblastic leukemia. *Blood* **85**(4), 873–878.

Valentine, M., Martin, A., Ungar, P., Katz, N., Schultz, L. D., Davies, T. F. (1994). Preservation of functioning human thyroid organoids in the severe combined immunodeficient mouse. 3. Thyrotropin independence of thyroid follicle formation. *Endocrinology* **134**(3), 1225–1230.

van Kuyk, R., Torbett, B. E., Gulizia, R. J., Leath, S., Mosier, D. E., Koenig, S. (1994). Cloned human $CD8^+$ cytotoxic T lymphocytes protect human peripheral blood leukocyte-severe combined immunodeficient mice from HIV-1 infection by an HLA-unrestricted mechanism. *J. Immunol.* **153**(10), 4826–4833.

Vandekerckhove, B. A., Krowka, J. F., McCune, J. M., de Vries, J. E., Spits, H., Roncarolo, M. G. (1991). Clonal analysis of the peripheral T cell compartment of the SCID-hu mouse. J. Immunol. **146**(12), 4173–4179.

Vandekerckhove, B. A., Baccala, R., Jones, D., Kono, D. H., Theofilopoulos, A. N., Roncarolo, M. G. (1992a). Thymic selection of the human T cell receptor V beta repertoire in SCID-hu mice. *J. Exp. Med.* **176**(6), 1619–1624.

Vandekerckhove, B. A., Namikawa, R., Bacchetta, R., Roncarolo, M. G. (1992b). Human hematopoietic cells and thymic epithelial cells induce tolerance via different mechanisms in the SCID-hu mouse thymus. *J. Exp. Med.* **175**(4), 1033–1043.

Vandekerckhove, B. A. E., Jones, D., Punnonen, J., Schols, D., Lin, H. C., Duncan, B., Bacchetta, R., Devries, J. E., Roncarolo, M. G. (1993). Human Ig production and isotype switching in severe combined immunodeficient-human mice. *J. Immunol.* **151**(1), 128–137.

Vandekerckhove, B. A. E., Barcena, A., Schols, D., Mohanpeterson, S., Spits, H., Roncarolo, M. G. (1994). *In vivo* cytokine expression in the thymus CD3(high) human thymocytes are activated and already functionally differentiated in helper and cytotoxic cells. *J. Immunol.* **152**(4), 1738–1743.

Vanhecke, D., Leclercq, G., Plum, J., Vandekerckhove, B. (1995). Characterization of distinct stages during the differentiation of human $CD69^+CD3^+$ thymocytes and identification of thymic emigrants. *J. Immunol.* **155**(4), 1862–1872.

Vormoor, J., Lapidot, T., Larochelle, A., Dick, J. E. (1995). Human hematopoiesis in SCID mice. In *Medical Intelligence Unit* (eds Roncarolo, M. G., Namikawa, R., Peault, B.), pp. 197–212. R. G. Landes Co., Austin, TX, USA.

Vormoor, J., Lapidot, T., Pflumio, F., Risdon, G., Patterson, B., Broxmeyer, H. E., Dick, J. E. (1994a). Immature human cord blood progenitors engraft and proliferate to high levels in severe combined immunodeficient mice. *Blood* **83**(9), 2489–2497.

Vormoor, J., Lapidot, T., Pflumio, F., Risdon, G., Patterson, B., Broxmeyer, H. E., Dick, J. E. (1994b). SCID mice as an *in vivo* model of human cord blood hematopoiesis. *Blood Cells* **20**(2–3), 316–322.

Weei-Chin, L., Desiderio, S. (1995). V(D)J recombination and the cell cycle. *Immunol. Today* **6**, 279–289.

Williams, S. S., Alosco, T. R., Croy, B. A., Bankert, R. B. (1993).

The study of human neoplastic disease in severe combined immunodeficient mice. *Lab. Anim. Sci.* **43**(2), 139–146.

Yan, H. C., Juhasz, I., Pilewski, J., Murphy, G. F., Herlyn, M., Albelda, S. M. (1993). Human/severe combined immunodeficient mouse chimeras. An experimental *in vivo* model system to study the regulation of human endothelial cell–leukocyte adhesion molecules. *J. Clin. Invest.* **91**(3), 986–996.

Yan, H. C., Delisser, H. M., Pilewski, J. M., Barone, K. M., Szklut, P. J., Chang, X. J., Ahern, T. J., Langersafer, P., Albelda, S. M. (1994). Leukocyte recruitment into human skin transplanted onto severe combined immunodeficient mice induced by TNF-alpha is dependent on E-selectin. *J. Immunol.* **152**(6), 3053–3063.

Yano, S., Nishioka, Y., Izumi, K., Tsuruo, T., Tanaka, T., Miyasaka, M., Sone, S. (1996). Novel metastasis model of human lung cancer in SCID mice depleted of NK cells. *Int. J. Cancer* **67**(2), 211–217.

XVIII XENOTRANSPLANTATION

1. Introduction

Organ allotransplantation represents an important medical advance of this century. The success of transplantation as the more appropriate treatment for an increasing number of acute and chronic pathologies, however, has engendered a major problem which is organ shortage (Cohen and Wight, 1993, Cooper *et al.*, 1994). In order to overcome the lack of organs, scientists recently reconsidered the possibility of using animals as potential donors for humans (Starzl *et al.*, 1994). Transplantation between species or **xenotransplantation** is not a new concept – in 1963, Reemtsma and colleagues carried out the first kidney xenotransplantation from chimpanzee to human (Reemtsma *et al.*, 1964) in New York. The kidney recipient lived for 9 months with a normal renal function but eventually died from intercurrent disease while the renal graft was normally functioning. This example certainly encouraged clinicians and scientists to continue to develop research in the xenotransplantation field.

The first species to be considered as a potential organ donor for human xenotransplantation is the nonhuman primate such as chimpanzee (Reemtsma *et al.*, 1964) or baboon (Starzl *et al.*, 1993). However, the possibility of virus, especially retrovirus, transmission from these primates to humans and the ethical and financial aspects rapidly negated this option. Most investigators therefore recently proposed swine as an alternative source of organs for clinical transplantation. Unlike nonhuman primates, pigs are easy to breed, have anatomical and physiological characteristics compatible with humans, and are well studied for several pathogens potentially transmissible to humans (Fishman, 1994; Sandrin and McKenzie, 1994). Moreover, genetic engineering is now easily achieved in pigs (Sachs and Bach, 1990). Pigs are phylogenetically distant from humans and the immune reaction across such a barrier is far more severe than with allografts. In fact, while a human allograft is usually rejected within 5–7 days, a primarily vascularized pig xenograft is rejected by primates or humans within minutes or hours owing to hyperacute rejection (HAR) (Perper and Najarian, 1966). The main obstacle to the achievement of discordant swine xenograft is the presence at the surface of the pig endothelium of glycosylated xenoantigens identified as Gal-α(1,3)-Gal epitope concomitantly with the presence in the human serum of preformed xenoantibodies (IgM, IgG) which recognize these carbohydrate residues. At unclamping of a vascularized pig organ into human, these natural antibodies would hence immediately recognize and bind the glycosylated determinants thereby activating the complement cascade and producing an HAR which is characterized by a rapid and complete destruction of the xenograft through interstitial hemorrhages, thromboses and infarct.

The observation that HAR takes place only when organ xenotransplantation is achieved across widely phylogenetically distant species led to a classification of xenografts into **concordant** versus **discordant**, based on the relative phylogenetic distance between the donor and recipient species and the presence or absence of circulating xenoreactive natural antibodies (XNA) in the serum of the recipient (Calne, 1970).

As with primarily vascularized discordant xenografts, most of ABO incompatible allografts are hyperacutely rejected by untreated human recipients and the pathogenesis of HAR has been shown to be mainly mediated by recipient circulating natural anti-A or -B isoagglutinins (Hammer *et al.*, 1973; Platt *et al.*, 1990b; Alexandre *et al.*, 1991a,b; Platt and Bach, 1991). In 1982, Alexandre and colleagues, demonstrated in ABO incompatible living-related kidney allografts that HAR can be overcome by depletion of recipient circulating natural antibodies (and other circulating proteins) using iterative plasmaphereses prior to transplantation (Alexandre *et al.*, 1991a,b). This was probably the turning point which clearly demonstrated the crucial role of natural antibodies in the pathogenesis of HAR and 'opened the door' to clinical organ allotransplantation across the ABO blood group barrier. Moreover, these pioneering studies suggested that HAR of primarily vascularized discordant xenografts might be achieved whether recipient circulating XNA could be removed (Sachs and Bach, 1990; Bach, 1991b). In the experimental setting, a significant prolongation of the survival time of pig to baboon kidney xenografts for up to 23 days (Alexandre, 1989; Gianello, 1995) was obtained by using a regimen consisting of iterative plasmaphereses prior to transplantation concomitantly to a quadruple drug therapy (antithymocyte globulins, cyclosporin A, azathioprine and prednisone). Although the graft survival was significantly extended, delayed rejection however ensued thereby suggesting the involvement of additional factors beside circulating XNA (Bijsterbosch *et al.*, 1985;

Handbook of Vertebrate Immunology
ISBN 0-12-546401-0

Platt *et al.*, 1990a; Bach, 1991a,b, 1994a,b; Platt and Bach, 1991a; Bach *et al.*, 1993; Grandien *et al.*, 1993; Platt, 1994; Platt and Holzknecht, 1994).

2. Hyperacute and Delayed Discordant Xenograft Rejection

Minutes or hours after unclamping a primarily vascularized discordant xenograft, HAR occurs and is characterized by a rapid destruction of the organ due to massive interstitial hemorrhages, edema, thromboses and marked endothelial cell (EC) destruction (Bach, 1991a; Platt and Bach, 1991a,b; Alexandre *et al.*, 1989). Although infiltration by neutrophils or macrophages might concomitantly occur, it remains focal and mild. During HAR, marked deposition of IgM and to a lesser extent IgG XNA is detected on the vascular endothelium along with complement C3, C4, C9 deposits (Platt *et al.*, 1990a; Platt, 1991a,b). This binding of circulating IgM XNA to endothelial cells (EC) also provokes an activation of the complement cascade, mainly through the classical pathway. The complement activation leads to EC activation and disruption and to the subsequent generation of intravascular pro-coagulant microenvironment (Bach, 1991a,b, 1994; Platt *et al.*, 1990a, 1991a; Bach *et al.*, 1993, 1994; Platt, 1994; Platt and Holzknecht, 1994). Thus, in presence of xenoreactive natural antibodies (XNA) and complement, the quiescent and anticoagulant vascular endothelium becomes activated and procoagulant initiates the destruction of the endothelial barrier (Bach, 1994). This first phase of events takes minutes or hours and transcription of EC genes or protein synthesis does not occur. If circulating XNA are, however, eliminated and/or the complement pathways inhibited, the survival of discordant xenografts might be significantly prolonged from few minutes or hours to several days or weeks (Alexandre *et al.*, 1989; Fischel *et al.*, 1990; Lesnikoski *et al.*, 1994; Leventhal *et al.*, 1994), but eventually acute rejection still occurs (Bach *et al.*, 1993, 1994). This second phenomenon called delayed xenograft rejection (DXR) includes gene transcription as well as protein synthesis and the pathophysiological basis underlying DXR is thought to be endothelial cell activation.

As mentioned above, during DXR, and contrary to hyperacute rejection (HAR), several genes are upregulated and protein synthesized in both EC and recipient circulating leukocytes (Bach *et al.*, 1993, 1994; Bach, 1994). Although the pathogenesis of DXR remains largely unknown, preliminary results demonstrate that during DXR, the xenograft is progressively infiltrated by activated macrophages, neutrophils and monocytes consequently producing a high level of TNF-α, IFN-γ and tissue factor (Lesnikoski *et al.*, 1994). Similar to HAR, there is no clear evidence for T cell activation or infiltration during DXR (Lesnikoski *et al.*, 1994). At the humoral level, massive EC vascular deposition of mainly IgM or IgG XNA is experienced, provoking processes of antibody-dependent cellular cytotoxicity (ADCC) mediated by activated macrophages, neutrophils and NK cells (Vercellotti *et al.*, 1991; Inverardi *et al.*, 1992, 1993; Galili, 1993a; Inverardi and Pardi, 1994; Schaapherder *et al.*, 1994). During both HAR and DXR, complement activation results in the generation of several membrane-bound complement proteolytic molecules, including iC3b, which are thought to contribute to neutrophil activation and EC injury upon the specific binding of neutrophils to iC3b-coated EC through the surface CD11b/CD18 receptor (Vercellotti *et al.*, 1991). Complement activation by XNA is also thought to contribute directly to the activation of EC (Bach *et al.*, 1994b; Platt, 1994a). Finally, xenogeneic T cell activation might also occur later during the process of DXR as suggested by the observation that human T cells can be activated by cultured EC (Rollins, 1994). The implication of unknown factors beyond antibodies and complement is however likely in the pathogenesis of both HAR and DXR (van Noesel *et al.*, 1993), (Figure XVIII.2.1)

3. Human Circulating Xenoreactive Natural Antibodies

The immune system of adult mammals, including humans, is characterized by the presence of peripheral self-reactive mature B cells which, upon activation produce polyreactive and monoreactive circulating natural IgM and IgG antibodies (Alt *et al.*, 1987; Swain and Reth, 1993). Natural antibodies recognize, in most cases, highly repeated polysaccharide antigens and are generated mainly by B cells from the B-1 lineage, probably in absence of T cell help (Alt *et al.*, 1987; Swain and Reth, 1993). Although natural antibodies represent a significant proportion of the repertoire of circulating antibodies, their role has not been established (Galili *et al.*, 1984, 1987; Kroese *et al.*, 1989; Avrameas, 1991). In humans, anti-blood-group (A or B) antibodies and also XNA which are responsible for the hyperacute rejection (HAR) of ABO incompatible allografts and discordant xenografts, respectively, belong to the repertoire of natural antibodies (Galili *et al.*, 1984, 1987; Kroese *et al.*, 1989; Avrameas, 1991).

Human natural antibodies directed against A and B tissue antigens recognize the GalNAc-α(1,3)(Fuc)-α(1,2)-Gal-β(1,3)-GlcNAc-R and the Gal-α(1,3)(Fuc)-α(1,2)-Gal-β(1,3)-GlcNAc-R oligosaccharide residues, respectively. In addition to these antiblood-group antibodies, humans and old-world primates also have in their serum circulating natural IgM and IgG antibodies which recognize the Gal-α(1,3)-Gal-β(β1,4)-GlcNAc-R oligosaccharide epitope, also termed α-galactosyl epitope (Galili *et al.*, 1987).

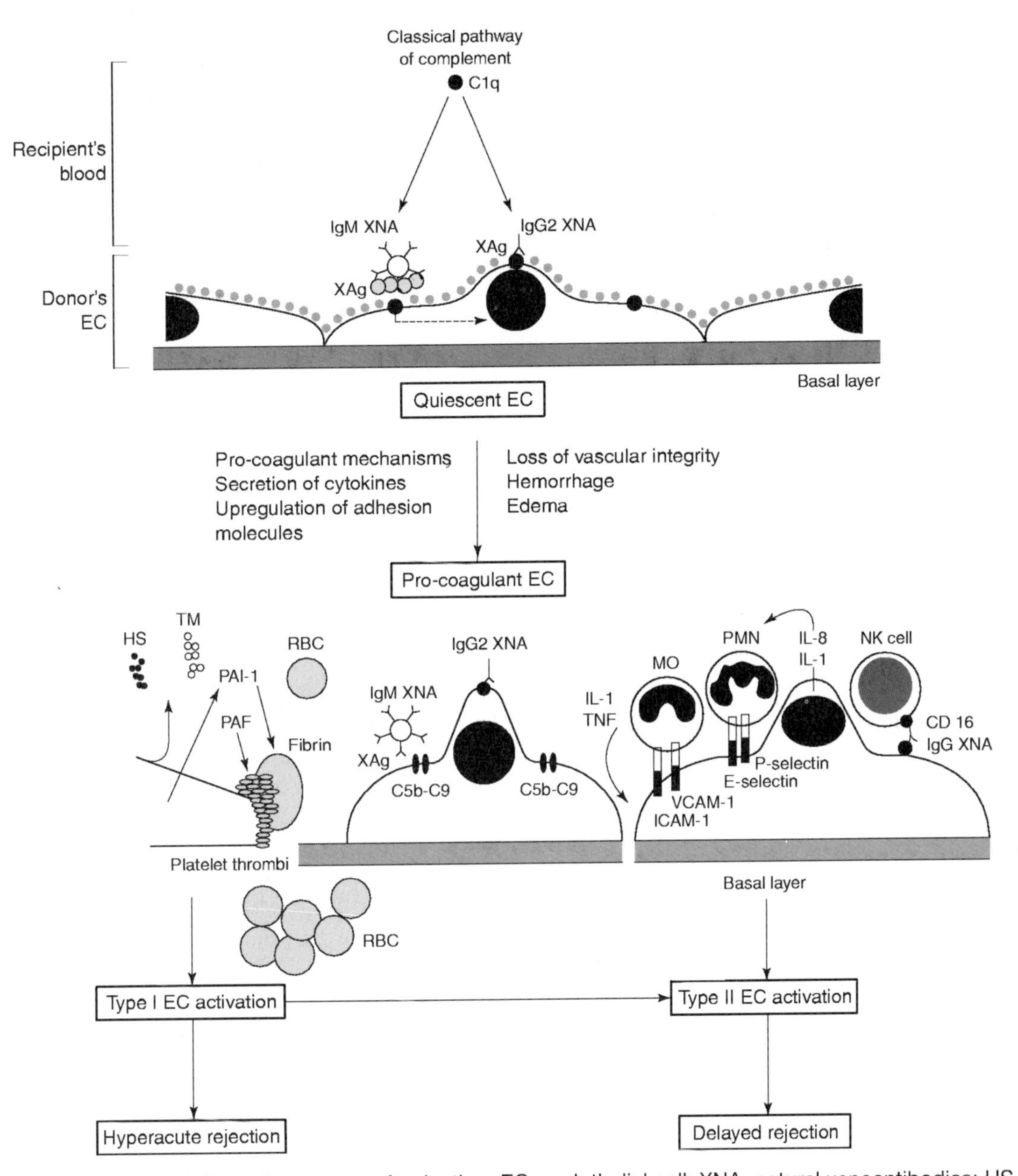

Figure XVIII.2.1. Model of discordant xenograft rejection: EC, endothelial cell; XNA, natural xenoantibodies; HS, heparan sulfate; TM, thrombomodulin; PAI-1, plasminogen activator inhibitor; PAF, platelet activating factor; Xag, xeno-antigen; MO, monocyte; PMN, polymorphonuclear cells; NK, natural killer cell; TNF, tumor necrosis factor; RBC, red blood cells. (See also Table XVIII.6.1.)

Recently, anti-α-galactosyl natural antibodies were found to be the main component among the repertoire of antiporcine circulating XNA in humans and Old-world primates (Good *et al.*, 1992; Galili *et al.*, 1993; Sandrin *et al.*, 1993; Sandrin and McKenzie, 1994). As for the A and B tissue antigens expressed on the surface of human EC, the α-galactosyl epitope is also present on several glycoproteins and glycolipids found on the surface of EC (Galili, 1988; Oriol, 1993). Upon recognition of these oligosaccharidic epitopes by circulating natural antibodies or xenoreactive natural antibodies (XNA), human or porcine EC is activated and HAR initiated (Cooper *et al.*, 1994; Sandrin and McKenzie, 1994).

The nature of human IgM and IgG XNA and their respective role in HAR of discordant xenografts remains controversial. Several studies tend to demonstrate that in human and nonhuman primate sera, exclusively polyreactive IgM XNA recognize EC *in vitro* (Platt *et al.*, 1990c, 1991a,b; Parker *et al.*, 1994) or *in vivo* (Platt *et al.*, 1991c; Geller *et al.*, 1993), whereas anti-α-galactosyl IgG antibodies are monoreactive (Alt *et al.*, 1987; Hamadeh *et al.*, 1992). However, other studies indicate that following *ex vivo* perfusion of porcine hearts with human blood, both IgM and IgG XNA are deposited on the vascular endothelium (Clark *et al.*, 1992; Tuso *et al.*, 1993) and once XNA are eluted from *ex vivo* perfused porcine xenografts, both

IgM and IgG antibodies are recovered (Ross *et al.*, 1993). As for the IgG anti-α-galactosyl antibodies, most of the XNA of the IgG isotype recovered in these experiments belong to the IgG2 subclass (Kirk *et al.*, 1993; Ross *et al.*, 1993). These data have recently been confirmed by the observation that in up to 100% human sera, circulating XNA which recognize cultured ECs belong mainly to the IgM and IgG2 isotypes (Bijsterbosch and Klaus, 1985; Gold *et al.*, 1990). In humans, up to 1% (50–100 μg/ml) of circulating IgG recognizes the α-galactosyl epitope and thus potentially recognizes ECs (Galili *et al.*, 1984, 1993), but the binding of such high concentration is not correlated with an intense staining upon immunohistology or ELISA. However, the low level of detection of circulating XNA of the IgG isotype may result from technical problems and in fact, most of the revealing antibodies used in these assays also bear the α-galactosyl epitope on their molecular structure thereby inducing their recognition by XNA which bind to the same epitope on the surface of EC (Fuller *et al.*, 1987; Thall and Galili, 1990; Borrebaeck *et al.*, 1993; Soares *et al.*, 1994a).

While in most mammals, B cells are probably rendered tolerant to the α-galactosyl 'self-antigen' during the early phase of B-cell development, in humans and anthropoid apes this oligosaccharide epitope is not recognized as part of a 'self-antigen' and mature anti-α-galactosyl B-cell clones can be generated in the peripheral primary B-cell repertoire (Goodnow, 1992; Galili *et al.*, 1984, 1987). However, in order to produce these anti-α-galactosyl antibodies, peripheral mature B cells must be activated by oligosaccharidic antigens. In humans, the sustained production of anti-α-galactosyl natural antibodies is thought to result from the continuous activation of specific mature peripheral B-cell clones by enteric bacteria presenting in their bacterial wall either the α-galactosyl molecule or a cross-reactive antigen (Brown *et al.*, 1993; Cambier *et al.*, 1994). As a result of this robust B-cell stimulation, up to 1% of circulating mature B cells in humans produce anti-α-galactosyl antibodies representing as much as 1% of the circulating IgG (Galili *et al.*, 1984, 1993). Interestingly, these monoreactive anti-α-galactosyl IgG XNA belong mainly to the IgG2 subclass which suggests their character to be T-independent (Hamadeh *et al.*, 1992; Galili, 1993; Galili *et al.*, 1993). In comparison, as much as 80% of human anti-porcine circulating XNA of the IgM isotype are polyreactive (Parker *et al.*, 1994) and represent only 0.1% of circulating IgM (Turman *et al.*, 1991; Vanhove and Bach, 1993; Parker *et al.*, 1994; Vanhove *et al.*, 1994).

Although anti-α-galactosyl antibodies of the IgM and IgG isotypes represent the majority of the anti-EC XNA repertoire, there is still no correlation between the binding of these anti-α-galactosyl antibodies to EC with the activation of these cells and consequently with the initiation of HAR. Using a porcine kidney cell line (PK15) expressing the α-galactosyl epitope, the detection of circulating anti-α-galactosyl XNA of both IgM and IgG isotypes is correlated with the cytotoxic effect of human and baboon sera (Kujundzic *et al.*, 1994). Conversely, the depletion of the anti-α-galactosyl antibodies inhibits the cytotoxic effect of these sera (Neethling *et al.*, 1994). Similar results have recently been reported using a α-(1,3)-galactosyl transferase transfected COS cell line (Vaughan *et al.*, 1994), but contrary to the PK15 or the transfected COS cell lines, the cytotoxic effect of human serum on cultured ECs is limited. This result indicates that the use of cell lines may not be representative of the type of EC injury caused by circulating XNA. There is no clear evidence that anti-α-galactosyl XNA of the IgG isotype may contribute *in vivo* to the HAR of pig to primate discordant xenografts. Conversely, circulating XNA of the IgG isotype have recently been suggested to inhibit EC activation *in vivo* and to protect primarily vascularized porcine xenografts from HAR (Magee *et al.*, 1994).

The role of XNA on the pathogenesis of pig to primate discordant primarily vascularized xenografts is actually difficult to assess because many methods used to eliminate circulating XNA *in vivo* also deplete additional factors which might be involved in HAR (Bach, 1994). The strongest evidence for the involvement of natural antibodies in the pathogenesis of HAR comes perhaps from studies showing that depletion of circulating natural antibodies (and other factors) allows to obtain long term survival of ABO incompatible allografts (Alexandre *et al.*, 1987, 1991a,b). As the oligosaccharidic epitopes expressed on EC and recognized by natural anti-A or -B antibodies are similar to those recognized by XNA, it is possible that circulating XNA may also be involved in the pathogenesis of HAR of pig to primate discordant xenografts. The central role of IgM XNA in the pathogenesis of HAR of discordant xenografts is supported by the following evidence: (1) anti-α-galactosyl XNA are absorbed during the *ex vivo* perfusion by porcine kidneys (Platt *et al.*, 1990d); (2) depletion of anti-α galactosyl circulating XNA is correlated with the prolongation of the survival time of pig to primate discordant xenografts (Platt *et al.*, 1990a,b,c,d); (3) The DXR observed upon depletion of circulating XNA, is correlated with the return of circulating XNA and their binding to α-galactosyl residues on EC (Soares, 1994b).

Finally, in order to better characterize the genes encoding for IgM natural antibodies in a hamster to rat xenograft model, Wu *et al.* (1997) examined the genetic structure of immunoglobulin V_H genes. Out of 18 individual cDNA libraries, they established that two germline genes encode the V_H genes of the preformed rat IgM antibodies. The comparison of the DNA sequences of the V_H regions revealed a high level of nucleic acids sequence identity (97–100%) between several antibodies, suggesting derivation from common family of germline V_H genes. The preferential use of a closely related group of germline genes is consistent with the hypothesis that the Ig genes encoding the xenoreactive antibodies in this model recognize a relative limited repertoire of target antigen epitopes. In addition the authors suggested that the humoral response between concordant or discordant species might use a common group of germline Ig V_H genes, originally selected as part of an innate immune response to infectious agents (Cramer *et al.*, 1996).

4. The Xenoantigens

On the surface of ECs, the xenoantigens recognized by circulating XNA from human and nonhuman primates are now becoming better understood. Using a sensitive cellular ELISA in which endothelial cells (EC)s are used as targets for human circulating xenoreactive natural antibodies (XNA), it has been demonstrated that IgM XNA and to less extent IgG XNA recognize a triad of glycoproteins that were termed following their molecular masses as gp 115, 125 and 135 (Platt *et al.*, 1990b,c,d). These gp 115/135 xenoantigens are shared by ECs and porcine platelets but not by porcine red blood cells or lymphocytes, which express xenoantigens of different molecular weights. Recently gp115, 125 and 135 xenoantigens have been demonstrated to be the porcine equivalents of human integrin β_3, αIIb, and α_2 glycoproteins (Platt and Holzknecht, 1994). Human circulating XNA also recognize on the surface of ECs a large glycoprotein (250 kDa) which is the porcine homolog of the human von Willebrand Factor (Platt, 1994; Platt and Holzknecht, 1994).

The epitope(s) recognized by circulating IgM XNA on the gp 115/135 xenoantigens are located on the *N*-linked biantennary oligosaccharide structure of these glycoproteins (Platt *et al.*, 1990b; Platt, 1994; Platt and Holzknecht, 1994). The specific elimination of the α-galactosyl epitope by digestion of the gp115/135 with α-galactosidase abrogates the binding of both human and baboon IgM XNA to the gp 115/135 xenoantigens thereby suggesting that the α-galactosyl residue is the main molecular epitope recognized on these *N*-linked-oligosaccharides (Collins *et al.*, 1994). However, the elimination of the α-galactosyl epitope from the surface of cultured endothelial cells (EC) decreases the binding of circulating XNA by only 70–80%, indicating that 20–30% of circulating XNA in human and Old-world primates recognize porcine xenoantigens other than the α-galactosyl epitope (Collins *et al.*, 1994). The recognition of other molecular structures by circulating XNA is further suggested by the observation that circulating XNA from New-world primates, which do not have circulating anti-α-galactosyl antibodies, recognize cultured ECs and circulating XNA from human or Old-world primates (Galili *et al.*, 1988; Collins *et al.*, 1994).

Lower molecular mass porcine xenoantigens of 62, 48, 42, 36, 34, 28, and 26 kDa have also been recently described as xenoantigens recognized by human circulating IgM XNA on the surface of EC (Tuso *et al.*, 1993a,b). Contrary to the gp 115/135 kDa xenoantigens, XNA would recognize the protein molecular structure of these lower molecular weight EC xenoantigens (Tuso *et al.*, 1993a,b). The binding of human IgM XNA to the majority of these low molecular weight EC xenoantigens is however not significantly depleted upon *ex vivo* perfusion of porcine organ or HAR of porcine xenografts, suggesting that these are not the main EC targets recognized *in vivo* by circulating XNA (Tuso *et al.*, 1993a,b; Platt, 1994; Platt and Holzknecht, 1994). More recently the number of glycoproteins expressing the α-galactosyl epitope on the surface of EC has been extended to at least 25, including the gp 115/135 (Platt *et al.*, 1990b), the lower molecular weight xenoantigens (Tuso *et al.*, 1993a,b) and several unidentified glycoproteins (Sandrin and McKenzie, 1994).

The α-galactosyl epitope recognized by circulating XNA has a molecular structure similar to the B tissue antigen epitope recognized by natural antibodies on the surface of human ECs. However, the L-fucose saccharide found on the B tissue antigen epitope is not present on the α-galactosyl residue (Galili *et al.*, 1987). In humans, up to 85% of circulating anti-B blood group natural antibodies are in fact XNA directed against the α-galactosyl epitope and do not recognize the L-fucose saccharide on the B tissue antigen (Galili *et al.*, 1987). The difference between the molecular structures of these two oligosaccharide epitopes results from their formation by two distinct enzymes, namely the α-(1,3)-D-galactosyl and α-(1,3)-galactosyl transferases, respectively (Thall and Galili, 1990; Galili *et al.*, 1993). Unlike the α-(1,3)-D-galactosyl transferase which is present in its active form exclusively in humans and certain nonhuman primates, the α-(1,3)-galactosyl transferase gene is detected in all mammals but has been functionally inactivated in anthropoid apes and humans (Galili and Swanson, 1991; Joziasse, 1992). Inactivation of the α-(1,3)-galactosyl transferase, by introduction of frameshift mutations generating stop codons in the coding gene, is thought to have occurred late during evolution (25–35 million years ago), as suggested by the observation that New-world primates produce this carbohydrate structure in most of their glycoproteins and glycolipids (Galili, 1991; Sandrin *et al.*, 1993). The selective pressure which may have favored the survival of anthropoid apes in which α-(1,3)-galactosyl transferase was inactivated is thought to be the presence of the *Plasmodium falciparum* parasite responsible for the malaria disease (Ramasamy, 1994).

5. Complement Activation

Upon binding of IgM xenoreactive natural antibody (XNA) to endothelial cells (ECs), activation of the classical pathway of complement generates several membrane-linked complement proteolytic products such as, C5b6 or C5b7, which initiate the morphological retraction of ECs and cause endothelial 'gaps' (Platt, 1994; Saadi *et al.*, 1994). The end-product of complement activation is C9; ECs are then perforated and lose their vascular barrier integrity. Although the classical pathway of complement certainly represents the major way of activation in mammals, the activation of the alternate pathway has also been demonstrated in the guinea pig to rat model. Together with circulating XNA, complement activation is certainly one of the main factors involved in the pathogenesis of both hyperacute rejection (HAR) and delayed xenograft rejection (DXR) of discordant xenografts (Platt *et al.*, 1990a, 1991a,b,c; Platt and Bach, 1991a,b; Deenen and Kroese, 1993; Platt, 1994; Platt and Holzknecht, 1994). Particularly in the porcine to primate xenograft model, activation of recipient complement is believed to act exclusively through the classical pathway, probably upon binding of circulating XNA of the IgM but not the IgG isotype to the gp115/135 kDa xenoantigens (Dalmasso *et al.*, 1992, 1993). The reasons why human complement is not directly activated through the alternative pathway by EC surface membrane, has not been established.

After binding of IgM XNA to the gp115/135 kDa xenoantigens, the complement is activated through the classical pathway generating iC3b-bound complement fragments which may probably also be found in the form of circulating gp115/135-bound immune complexes (Collins *et al.*, 1994). Generation of such immune complexes, has been recently shown to induce B-cell antigen processing and subsequent MHC class II restricted antigen presentation to specific T cells, resulting in the specific activation of these B cells and subsequent differentiation into antibody-secreting cells (Thornton *et al.*, 1994).

Complement activation is normally regulated by both soluble (C1 inhibitor, factor I, factor H, etc.) and membrane associated complement inhibitors (CD46, CD59 and CD55) which have the crucial role of continuously suppressing complement activation directed against 'self' cellular components in order to avoid a complement-mediated autoimmune disease (Mollnes and Lachmann, 1988; Lachmann, 1991). The importance of the complement inhibitors is well illustrated in a series of pathologies resulting from congenital deficiencies in genes encoding these regulatory proteins (Lachmann, 1991; Morgan and Wallport, 1991). Because the vascular endothelium is in constant contact with circulating complement proteins, EC express relatively high levels of surface membrane-associated complement inhibitors, including decay-accelerating factor (DAF, CD55), membrane cofactor protein (MCP, CD46) and membrane inhibitor of reactive lysis (CD59) (Brahim and Osmond, 1970; Gold *et al.*, 1990). However, these complement regulatory proteins are species-specific in order to exclusively allow complement activation against heterologous material such as xenogeneic cells or bacterial germs. While in physiological conditions membrane-bound complement inhibitor protects ECs from autologous complement-mediated injury, upon discordant xenografting these regulatory proteins are probably unable to suppress xenogeneic complement activation. EC activation observed during both HAR and delayed rejection in pig to primate xenografts may therefore result from the inability of EC membrane-bound regulatory proteins to inhibit the activation of primate complement; this situation would be similar in pig to human (Bach, 1991; Dalmasso *et al.*, 1991).

Several studies recently demonstrated the importance of these species-specific complement regulators on the protection of ECs from human complement mediated-injury including EC activation or lysis (Platt *et al.*, 1991a,b,c; Bach *et al.*, 1994; Platt and Holzknecht, 1994). The induction of human DAF (CD55) as well as MCP or CD59 expression on the surface of cultured ECs has been shown to protect these cells from human complement-mediated cell lysis (Dalmasso *et al.*, 1991). Similarly, soluble human C1 inhibitor also inhibits human complement-mediated lysis or EC activation (Dalmasso and Platt, 1993) and human sCR1 (soluble complement receptor 1) prolong the survival time of porcine cardiac xenografts perfused *ex vivo* with human blood (Yeh *et al.*, 1991; Pruitt *et al.*, 1994). These data strongly support the idea that upon discordant xenografting, complement activation, resulting from the inability of EC membrane-associated regulatory proteins to inhibit xenogeneic complement activation, may be one of the central factors involved in the rejection of discordant xenografts.

6. Endothelial Cell Activation

Normally, endothelial cells (ECs) provide a continuous but semipermeable barrier confining circulating molecules and cells inside the blood vessels, thus allowing the selective transit of some molecules or cells across this endothelial barrier. In order to maintain the vital exchange process between blood and neighboring tissues, the vascular endothelium insures the continuous flow of circulating soluble molecules and cells by generating an appropriate intraluminal anticoagulant microenvironment. At unclamping of discordant primarily vascularized xenografts, the maintenance of this anticoagulant environment is lost.

Recent discoveries demonstrate that the vascular endothelium represents an organ in which 'active' cells are able to respond to various stimuli by undergoing specific alterations in their cellular function, metabolism and structure now commonly referred as EC activation (Cotran, 1987). The process of EC activation is not immediate and there is at least two distinct phases that are characterized by the presence or absence of gene transcription and protein synthesis and are termed EC stimulation and EC activation, respectively. These two phases of EC activation have recently been named by Bach and colleagues as type I and type II EC activation (Bach, 1994; Bach *et al.*, 1994) (Table XVIII.6.1).

The main feature which characterizes the process of EC activation is the transition from a physiological quiescent and anti-coagulant EC phenotype into an activated and

Table XVIII.6.1 Endothelial cell (EC) activation

	EC phenotype
EC stimulation or EC type I activation	No detectable upregulation of gene transcription or protein synthesis. Morphological retraction leading to the formation of vascular 'gaps' probably responsible for edema and interstitial hemorrhage. Release of procoagulant factors such as heparan sulfate, thrombomodulin, PAF, PAI. Surface expression of adhesion molecules including P-selectin.
EC activation or EC type II activation	Detectable up-regulation of gene transcription or protein synthesis. Expression of several surface molecules including: tissue factor, MHC class II, E-selectin, ICAM-1, VCAM-1, P selectin. Generation and release of IL-8, PAI-1, IL-6, PAF, etc. Enhanced adhesion of circulating leukocytes including neutrophils, macrophages and NK cells.

PAF, platelet activating factor; ICAM, intercellular cell adhesion molecule; VCAM, vascular cell adhesion molecule; PAI-1, plasminogen activator inhibitor; IL, interleukin.

highly procoagulant one. Several stimuli including inflammatory cytokines such as IL-1, TNF-α and IFN-γ, and also histamine, thrombin, leukotrienes or bacterial LPS are well known to induce this EC phenotypic change (Nawroth and Stern, 1986; Pober *et al.*, 1986; Bevilacqua *et al.*, 1987; Schleef *et al.*, 1988; Brett *et al.*, 1989; Cavender *et al.*, 1989; Pober and Cotran, 1991). Moreover, the specific interactions between circulating leukocytes and ECs may also induce the expression of proinflammatory cytokines such as IL-1 and TNF-α initiating, therefore, activation of EC (Pober and Cotran, 1991; Kishihara *et al.*, 1993; Inverardi and Pardi, 1994). Leukocyte-mediated EC activation is able to activate circulating neutrophils, NK or T cells through activated ECs, thereby initiating the cellular rejection of primarily vascularized allografts and probably the delayed rejection of primarily vascularized discordant xenografts (Pardi *et al.*, 1987, Colson *et al.*, 1990; Pober and Cotran, 1991; Vercellotti *et al.*, 1991; Inverardi *et al.*, 1992, 1993; Kishihara *et al.*, 1993; Inverardi and Pardi, 1994). As for inflammatory cytokines, human circulating XNA and complement activation have recently been demonstrated to mediate type I and type II activation of EC (Bach *et al.*, 1993; Bach, 1994). In particular, human IgM xenoreactive natural antibodies (XNA) have recently been shown to activate cultured ECs, with or without the participation of complement activation (Bach, 1994; Bach *et al.*, 1994).

The type I activation of cultured EC by human or nonhuman primate sera is characterized by the rapid morphological retraction of EC and formation of 'endothelial holes' (Saadi *et al.*, 1994), the release of heparan sulfate (Platt *et al.*, 1990a,b,c,d, 1991a,b) and platelet-activating factor (PAF), as well as the surface expression of the P-selectin adhesion molecule (Coughlan *et al.*, 1993; Bach, 1994a,b). Type I activation of EC would be sufficient *per se* to induce the occurrence of edema, interstitial hemorrhage, mild neutrophil adhesion to EC, platelet aggregation and fibrin deposition which are the main pathological features of the HAR of discordant xenografts (Platt *et al.*, 1991a,b,c; Platt and Bach, 1991a,b).

As previously mentioned, in absence of circulating IgM XNA and/or in case of suppression of complement activation, hyperactive rejection (HAR) of discordant xenografts may be overcome but DXR still occurs (Bach *et al.*, 1993, 1994). Delayed xenograft rejection (DXR) is thought to be initiated mainly by type II or type I+II EC activation. Human serum mediates '*in vitro*' the upregulation of several EC genes including IL-8 and plasminogen activator inhibitor-1 as well as two new genes (ECI-6 and ECI-7) which are specifically expressed during type II EC activation (de Martin *et al.*, 1993; Bach, 1994; Bach *et al.*, 1994; Vanhove *et al.*, 1994). Furthermore, type II EC activation has also been demonstrated to occur '*in vivo*' during DXR (Lesnikoski *et al.*, 1994). For cultured ECs, *in vivo* activation of EC involves upregulation of ICAM-1 and VCAM-1, tissue factor and IL-8 expression (Lesnikoski *et al.*, 1994). The exact mechanism by which type II EC activation mediates DXR of discordant xenografts has not been demonstrated, but mechanisms occurring through NF-κB are now proposed (Cooper *et al.*, 1996).

Upon recognition of the α-galactosyl oligosaccharide residues (and probably other epitopes) on the gp115/135 kDa xenoantigens (and probably other xenoantigens), circulating XNA may initiate the pathogenesis of HAR by activating the EC lining the endothelium of primarily vascularized porcine xenografts. Furthermore, circulating XNA may either activate EC *per se* or through the activation of the classical pathway of complement, but these two hypotheses should not be considered as mutually exclusive (Bach, 1993, 1994; Bach *et al.*, 1994). Circulating XNA and particularly those of the IgM isotype and complement pathways contribute to EC activation (Bach *et al.*, 1994; Platt and Holzknecht, 1994). Cultured ECs release heparan sulfate following recognition of surface gp115/135 kDa xenoantigens by XNA of the IgM but probably not of the IgG isotype (Platt *et al.*, 1990a,b,c,d,e, 1991a). Heparan sulfate release is dependent on the generation of the C5a complement proteolytic product, thereby suggesting that XNA acts essentially as a 'bridge' to complement activation (Platt, 1991a,c,d; Dalmasso, 1992). Because the initiation of the activation of the complement cascade through the classical pathway by antibodies of the IgG2 isotype is extremely inefficient (Alt *et al.*, 1987; Grandien *et al.*, 1993), this may provide an explanation for the inability of XNA of the IgG isotype to activate cultured ECs (Platt *et al.*, 1990a, 1991b). Conversely, circulating IgM XNA have been shown to upregulate *per se* the expression of several genes involved in EC activation, indicating that even in the absence of complement activation, IgM XNA may initiate pathogenesis of DXR (Vanhove *et al.*, 1994).

7. Tolerance and Endothelial Cell Accommodation

As shown by Alexandre and colleagues, removal of recipient circulating xenoreactive natural antibodies (XNA) – and probably other factors – by iterative plasmaphereses is required for the acceptance of human ABO incompatible kidney allografts (Alexandre *et al.*, 1987, 1991a,b). Although anti-ABO natural antibodies regain normal serum levels in the few days following transplantation, hyperacute rejection (HAR) usually does not occur (Alexandre *et al.*, 1987; Latinne *et al.*, 1989). As first hypothesized by Bach (1991a,b), the acceptance of an ABO incompatible allograft could result from modification of endothelial cell (EC) physiological response characteristics leading to 'accommodation', i.e. allowing the survival of the transplant in presence of high levels of circulating IgM natural antibodies and complement, after a transient period of depletion (Bach, 1991a,b).

It is now well established that 'EC accommodation' may be linked to the inability of circulating newly formed antibodies to activate EC, once these EC have been allowed to 'heal-in' over a certain period of time in absence of circulating natural antibodies and complement activation (Bach, 1991a,b, 1993, 1994; Bach *et al.*, 1994). At least three hypotheses may contribute to explain the occurrence of EC accommodation: (1) after a period of depletion, circulating newly formed natural antibodies may have changed their specificity for the EC surface antigens and thus lose their ability to activate EC and initiate HAR; (2) during the period of antibody depletion some EC surface antigens (such as the A or B tissue antigens) may be actively modified and are no longer recognized by newly formed natural antibodies thereby losing their ability to activate EC and initiate HAR; (3) after a period of depletion, newly formed natural antibodies may still recognize the EC antigens which may, however, be expressed on the surface of 'accommodated' EC and thereby be unable to be activated at time of antibody recognition (Platt *et al.*, 1991a,b,c; Platt, 1994; Bach, 1994; Bach *et al.*, 1994).

In the pig to primate primarily vascularized discordant xenograft model, xenogeneic EC 'accommodation' appears to be more difficult to achieve compared with the ABO incompatible allograft condition. This is supported by the observation that 'accommodation' of pig to baboon kidney xenografts is not achieved using protocols similar to those demonstrated to induce the 'accommodation' of ABO incompatible kidney allografts (Alexandre *et al.*, 1987). However, several reports have demonstrated that following a brief period of depletion of circulating XNA, porcine xenografts sometimes survive from some days to several weeks in presence of high levels of circulating newly formed XNA, thus suggesting that xenogeneic EC 'accommodation' may also be possible in case of discordant xenograft (Bach, 1994; Bach *et al.*, 1994; Platt, 1994). Recent reports from Bach and colleagues (1997), demonstrate, however, that in a concordant rodent model (hamster to rat), accommodation may be obtained by a combined treatment including cobra venom factor and cyclosporine A. Tolerant rodents demonstrate several characteristics such as protective gene expression, T_H2 cytokines production and a high titer of IgG2c which does not fix the complement. These immune characteristics were, in contrast, absent in control animals, thereby probably explaining the accommodation phenomenon. Such tolerant rats also accepted a second hamster cardiac graft without immunosuppression (Hechenleitner, 1997).

EC accommodation has also recently been suggested as an active process involving the specific interaction between EC and serum-derived molecules such as heat-stable α-globulin which inhibits the upregulation of E-selectin, ICAM-1 and VCAM-1 gene expression upon LPS-driven type II EC activation (Stuhlmeier *et al.*, 1994). However, heat-stable α-globulin does not inhibit the expression of other type II EC genes such as IL-1, IL-8, or PAI-1, suggesting that other molecules may be involved in the induction of EC accommodation (Bach, 1994; Bach *et al.*, 1994; Stuhlmeier *et al.*, 1994). Therefore, the induction of EC accommodation in tolerant ABO incompatible living related kidney allografts may result from the active modification of EC metabolism which renders these cells 'resistant' to activation by natural antibodies or complement. Recently, Bach and colleagues communicated results showing that genes encoding for anti-apoptosis molecules such as A20 or Bcl-2, could be involved in the protection of EC activation. In fact, blocking A20 allows avoidance of EC activation and the mechanisms involved could depend on the inhibition of the phosphorylation of nuclear factors like NF-κB which cannot get into the nucleus to activate the transcription of several genes involved in EC activation (Cooper *et al.*, 1996).

8. Therapeutic Approaches: Future Trends

A great deal of effort has recently been directed into developing potential therapeutic approaches allowing long-term survival of pig to human discordant xenografts and opening the door to clinical xenografting. It has been postulated that if endothelial cell (EC) activation by human serum is suppressed for a sufficient period of time, the process of EC 'accommodation' would occur and hyperacute rejection (HAR) of porcine to human primarily vascularized xenografts could be avoided (Bach, 1993, 1994; Bach *et al.*, 1994). Given this assumption, three major lines of research are currently being attempted by most investigators in order to control the immunological mechanisms involved in the pathogenesis of HAR and delayed xenograft rejection (DXR) of porcine xenografts: (1) the depletion and the inhibition of production of circulating xenoreactive natural antibodies (XNA) (Alexandre *et al.*, 1989, 1991a,b; Soares *et al.*, 1993; Latinne *et al.*, 1994); (2) the inhibition of complement activation (Cairns *et al.*, 1993; Cooper *et al.*, 1993; Pruitt *et al.*, 1994); and (3) the development of transgenic pigs expressing on the surface of their vascular endothelium, human complement inhibitors such as DAF (CD55), MCP (CD46) or CD59 (Dalmasso *et al.*, 1991; Loveland *et al.*, 1993; Platt and Holzkenecht, 1994), as well as the generation of transgenic pigs lacking the expression of the α-galactosyl epitope on the EC vascular surface (Sandrin *et al.*, 1993; Sandrin and McKenzie, 1994). Finally, it has also been proposed that HAR of porcine to human xenografts potentially may be obviated if EC could be actively rendered 'accommodated' by suppressing the molecular mechanisms leading to the upregulation of certain genes responsible for EC activation while promoting the expression of other genes potentially responsible for EC 'accommodation' (Bach, 1994).

Many therapeutic approaches described to date focus on the depletion of recipient circulating IgM XNA. Following the demonstration that successful ABO incompatible kidney allograft could be achieved upon preoperative depletion of circulating natural antibodies (and other factors) by plasma exchange and splenectomy, similar protocols have been used to deplete circulating XNA (and other factors) and prolong xenograft survival time of discordant xenografts. Although splenectomy and plasma exchange have proven to be useful in depleting circulating XNA (80–85% of pretreatment levels) and prolonging the survival of discordant xenografts, HAR or DXR rejection always occurs upon return of circulating XNA. It has been found that following plasma exchanges, circulating XNA are significantly depleted for a short period (1–2 days), which does not allow EC accommodation (van de Stadt *et al.*, 1988; Alexandre *et al.*, 1989; Leventhal *et al.*, 1992; Figueroa *et al.*, 1993; Soares *et al.*, 1993; Latinne *et al.*, 1994).

Another approach which has been shown to be successful in achieving profound depletion of circulating IgM XNA in both humans and nonhuman primates consists of the extracorporeal perfusion of recipient blood through one or several porcine vascularized organs prior to transplantation (Fischel *et al.*, 1990; Tuso *et al.*, 1993a; Makowka *et al.*, 1993). In addition, extracorporeal perfusion of recipient serum using immunoabsorbent columns containing the oligosaccharidic epitopes recognized by circulating XNA (such as the α-galactosyl epitope) have also been reported to be useful in depleting these circulating XNA (Cairns *et al.*, 1993). In a pig to primate model, Kozlowski *et al.* (1997) compared three techniques of natural antibodies elimination: by laparotomy approach, the authors achieved in primates (1) either an *ex vivo* pig liver perfustion, or (2) a perfusion with a column bearing Gal α(1,3) Gal residues and (3) by internal jugular access, they tested a perfusion of plasma separated from cellular components by apheresis (CPA) using a similar affinity column. Each perfusion technique reduced the level of circulating natural antibodies by 80–90% and CPA was proposed as the first choice technique because it does not require repeated surgical procedures and appeared to be equally effective (Kozlowski *et al.*, 1997). These *ex vivo* perfusion techniques, however, have no effect on the level of XNA production and therefore will not avoid the XNA rebound. Continuous infusion of A or B tissue antigens bearing the α-galactosyl epitope have also been shown to be able to deplete significantly the level of circulating XNA (Cooper *et al.*, 1994). More recently the inhibition of natural IgM and IgG antibodies by using a synthetic octapeptide, Gal pep 1 (DAHWESWL) which mimics the carbohydrate epitope Gal α(1,3) has been proposed (Fryer, 1997). However, in order to achieve accommodation, total depletion of circulating XNA for a relatively long period may have to be achieved by removing not only the circulating pool of XNA but probably, more importantly, by inhibiting the production of XNA by B cells. The inhibition of B cells can be obtained by administration of high doses of an anti-μ mAb in adult rats allowing thus the total inhibition of B cell development at an early stage of maturation (Bazim *et al.*, 1978; Soares *et al.*, 1993).

Many scientists have concentrated their efforts on manipulating EC genes in order to eliminate the xenoantigen Gal α(1,3) Gal from the surface of the pig endothelium. Completely blocking the gene from the α1,3 GT is possible in mice (knock-out) but is currently not feasible in pigs because pig embryonic stem cells are not available. Therefore, Sandrin *et al.* (1993) examined the possibility of adding another carbohydrate determinant at the surface of the cell by transfecting the gene of the α1,2 fucosyltransferase (α1,2 FT) which uses the same substrate *N*-acetyl-galactosamine than α1,3 GT but leads to the expression at the cell surface of a **fucosyl epitope** which is the H substance of the human O blood group (McKenzie *et al.*, 1996). This H epitope is not recognized by human preformed antibodies and hyperacute rejection would therefore be avoided. Transgenic mice expressing α1,2 FT have a 90% reduction in the surface expression of Gal α(1,3) Gal epitope in all tissues examined and a 90% reduction in the binding of human preformed antibodies. The important reduction of Gal α(1,3) Gal expression in this case has been called 'transferase dominance', since the α1,2 FT uses the same substrate *N*-acetyl-galactosamine than the α1,3 GT, but the overexpression of the new gene produced a significant derivation to the production of H antigen instead of the Gal α(1,3) Gal epitope. In order to eliminate completely the expression at the cell surface of the Gal α(1,3) Gal epitope, the same authors examined the possibility to co-express into COS cells the galactosidase (Galdase) gene; this enzyme cleaves a linked-galactosyl residue to remove Gal α(1,3) Gal. The co-transfection into COS cells with α1,3 GT, α1,2 FT and Galdase cDNA showed no cell surface expression of Gal α(1,3) Gal or lysis in a cytotoxicity assay, thereby demonstrating the additive effect in their ability to reduce the expression of the galactosyl determinant. Following the same idea, Koike *et al.* (1997) microinjected pig embryos with a construct of human α1,2 FT cDNA and a transgenic pig which carried this human gene was established. Cytotoxicity of several cells derived from this transgenic pig was assessed, and showed a significant resistance to a challenge with human sera (Koike, 1997).

Another therapeutic approach is to inhibit the human complement activation by transfecting in pigs, human genes encoding for CD55, CD59 or CD46. In a pig to primate model, extended kidney graft survival from transgenic pigs which expressed the human DAF gene has recently been reported. Using a triple drug therapy including cyclosporin A, cyclophosphamide and methylprednisolone, the authors reported pig kidney graft survival from 6 days up to a maximum of 35 days (median = 13 days). Although these monkeys exhibited transient renal dysfunction during the follow-up, they all recovered normal creatinine levels and four animals were eventually sacrificed because of a very severe anemia. Upon histology, mild interstitial infiltrate and rejection was detected in three out of the four animals, whereas the remaining kidney had no evidence of rejection. These results show that despite the presence of preformed antibodies, the pig transgenic kidneys are able to resist hyperacute rejection, but that the recipient will eventually die from complications resulting from immunosuppression. The morphology of hDAF transgenic pig kidneys was compared with the histology of normal pigs following *ex vivo* xenoperfusion with human blood: the antibody deposition was similar in each group, whereas deposition of C4 and C9 on the glomerular and vascular endothelium was significantly reduced in transgenic kidneys. However, C4 deposits were similar between both groups. The expression of P selectin and the level of glomerula rupture and hemorrhage were significantly higher in kidneys from normal pigs than in kidneys from transgenic animals (Tolan, 1997). The same team also reported that using similar hDAF transgenic pig donors, they obtained prolonged cardiac graft survival in primates with a median survival of 40 days (range from 6 to 62 days). However, the authors demonstrated in this study that a high dose regimen of cyclophosphamide (CyP) is required in order to prolong the cardiac graft survival: in the first experimental group, the mean CyP daily dose was 21.8 mg/kg and the median graft survival 40 days, whereas in group 2, when the daily dose of CyP was reduced to 12.2 mg/kg, all five animals eventually rejected within a median of only 9 days. Interestingly, rejection of the xenograft was concomitant with a significant antibody rebound. In the first group, six animals out of 10 were sacrificed because of gastrointestinal problems and severe anemia but the cardiac graft was still beating. These important results show that severe immunosuppression significantly prolongs heart graft survival from discordant transgenic pigs expressing hDAF in primates, despite the presence of preformed antibodies. A less toxic anti-B cell agent will, however, be required in order to achieve such xenografts in humans. Finally, Norin *et al.* (1997) recently reported extended survival of orthotopic lung xenograft from transgenic pigs expressing CD59 (MAC) to baboons. In untreated recipients, a pig lung transplant is hyperacutely rejected within 3 h, whereas this graft survival was extended to 12 h when transgenic pigs expressing CD59 were used as donors. Using an immunosuppressive regimen including xenorgan perfusion to eliminate the preformed antibodies, CyA and total lymphoid irradiation, the authors reported discordant orthotopic lung graft survival up to 17 days (median = 7 days) (Norin *et al.*, 1997).

Several authors also recently reported that the co-expression of several genes encoding for inhibiting the expression of the galactosyl epitope or complement inhibitors have additive protective effects. Thus, α-1,3-galactosyltransferase knock-out mice express very little amount of Gal α(1,3) Gal residues and therefore are protected against human sera but the co-expression in these mice of the human DAF gene clearly demonstrated that dDAF/Gal Ko mice afforded a marked increase in protection from complement-mediated injury and lysis with human sera compared with cells from control group animals (van Denderen *et al.*, 1997). The same authors co-expressed both CD55 (DAF) and CD59 (MAC) cDNA in mice using the human ICAM-2 promotor to allow increased transgene expression and showed, with an *ex vivo* Langendorff system, that cotransfected CD55/CD59 mice had a graft survival significantly longer than 45 min in all the animals tested. Others demonstrated that CD46 (MCP) appears to provide better protection against complement activation than CD59 alone and they showed that transgenic mice expressing either human CD46 or CD59 provide a longer graft survival when perfused with human sera. Finally, they also demonstrated that the co-expression of CD59 in Gal-knockout mice provide an additional protective effect for the activation of the complement cascade by human sera.

In order to inhibit the complement cascade, other therapeutic tools have also been investigated and some authors developed complementary binding peptides to inhibit the initial step of complement activation from the classical pathway, the binding of C1q to xenoantibody. Using *in vitro* porcine endothelial culture, they showed that the lysis by human complement was significantly inhibited by one of the peptides which prevent C1q binding to xenoantibody bound to pig endothelial cells.

Others studied the effect of soluble complement receptor 1 (sCR1) in an experimental allograft rejection model and demonstrated that sCR1 is able to inhibit the complement activation and to decrease the B cell response in terms of antibody production. sCR1 is thus capable of blocking the classical complement pathway and a single dose of sCR1 is able to prolong the cardiac graft survival in sensitized rats and in discordant xenograft recipients (Pruitt *et al.*, 1994). Using a combination of CyA, CyP, steroids and sCR1, pig cardiac graft survivals were prolonged in primates up to 32 days.

An alternative to transgenic animals for avoiding hyperacute rejection of discordant xenografts, would be the induction of specific and permanent tolerance to xenoantigens. Although such protocols do not yet exist in human allotransplantation, scientists show that this goal appears possible in both concordant and discordant experimental models. Based upon the production of mixed lymphohematopoietic chimerism, Bartholomew *et al.* (1997) showed that tolerance to renal vascularized xenograft from baboon to cynomolgus primates is feasible. Using a nonlethal total body irradiation (250 cGy), a thymic irradiation (700 cGy) and 3 days of ATG prior to transplantation, long-term tolerance to kidney concordant xenografts is obtained for more than 6 months whether CyA (15 mg/kg.d) is given for 1 month and DSG (6 mg/kg) from operative day up to postoperative day 13. Thus, without chronic immunosuppression, a nonmyeloablative preconditioning protocol leads to long-term concordant xenograft survival but genuine tolerance is not induced since rejection of donor skin xenografts also provoked the rejection of the kidney xenograft (Bartholomew, 1997). Across a discordant species barrier, Zhao *et al.* (1997) demonstrated the possibility to induce specific skin graft tolerance in a pig to mouse model by grafting fetal or neonatal pig thymic tissue to thymectomized, T-cell depleted mice. In addition, they showed that normal maturation patterns are seen for mouse thymocytes developing in pig thymic grafts. Thymectomized pig THY-grafted B10 mice accepted the fetal pig father's skin for more than 200 days, while rejecting allogenic skin within 13 days. Swine leukocyte antigens histocompatibility system in pigs (SLA) mismatched third party skin grafts were rejected within 40 days. This result is the first demonstration that specific skin graft tolerance can be induced across a discordant species (Zhao, 1997).

Studying the hamster to rat model, Bach *et al.* (1997b) also demonstrated that accommodation of cardiac graft is obtained by using CyA (15 mg/kg.d) and cobra venom factor for 10 days. Dissecting the mechanisms of prolonged graft survival, the authors showed high titers of surface-bound rat IgM, IgG, C3, C6 and minor fibrin depositions. The mononuclear cell infiltration consisted mainly in macrophages, NK cells and activated T cells ($CD25^+$). The cytokine expression in the tolerated graft clearly was a T_H2 cytokine pattern (IL-4, IL-10, IL-13) whereas in rejected grafts, a T_H1 cytokine pattern (IL-2, IFN-γ) was indicated. In addition to these differences, the authors found a high expression of protective genes in tolerated grafts but no expression of such genes in rejected xenografts. One hypothesis being that T_H2 cytokines promote the induction of protective genes expression. Concomitantly to these findings, IgG2c was detected at high titre in tolerant animals whereas the titre was very low in rejecting rats. IgG2c does not fix the complement and thus would be consistent with a tolerant environment. Finally, in rejected xenografts, a high level of atherosclerosis was shown. Second cardiac grafts from hamster were tolerated by long-term tolerant rats of a first hamster graft (Bach *et al.*, 1997b).

9. Conclusions

Major progress has been made in the last few years in the understanding of the physiopathology of discordant xenografting. At the therapeutic level, more appropriate drugs for maintaining the XNA concentration at a low level and over a long period are required in addition to specific anti-B-cell drugs. On the donor side, transgenic pigs which will express human regulators of the complement as well as H antigen instead of the α-galactosyl epitope should be available within the next few years allowing combination of these 'universal donors' with an appropriate drug therapy in the recipient. The combination of both should open the door to clinical applications in the near future.

10. References

Alexandre, G. P. J., Squifflet, J. P., De Bruyère, M., Latinne, D., Reding, R., Gianello, P., Carlier, M., Pirson, Y. (1987). Present experiences in a series of 26 ABO-incompatible living donor renal allografts. *Transplant. Proc.* **19**, 4538–4542.

Alexandre, G. P. J., Gianello, P., Latinne, D., Carlier, M., Dewaele, A., van Obbergh, L., Moriau, M., Marbaix, E., Lambotte, J. L., Lambotte, L., Squifflet, J. P. (1989). Plasmapheresis and splenectomy in experimental renal xenotransplantation. *Xenograft* **25**, 259–266.

Alexandre, G. P. J., Latinne, D., Carlier, M., Moriau, M., Pirson, Y., Gianello, P., Squifflet, J. P. (1991a). ABO-incompatibility and organ transplantation. *Transplant Rev.* 5, 230–241.

Alexandre, G. P. J., Latinne, D., Gianello, P., Squifflet, J. P. (1991b). Preformed cytotoxic antibodies and ABO-incompatible grafts. *Clin. Transplantation* 5, 583–594.

Alt, F. W., Baltimore, D. (1982). Joining of immunoglobulin heavy chain gene segments: implications from a chromosome with evidence of three D–J_H fusion. *Proc. Natl Acad. Sci. USA* **79**, 4118–4122.

Alt, F. W., Blackwell, K., Yancopoulos, G. D. (1987). Development of the primary antibody repertoire. *Science* **238**, 1079–1087.

Avrameas, S. (1991). Natural autoantibodies: from 'horror autotoxicus' to 'gnothi seauton'. *Immunol. Today* **12**, 154–159.

Bach, F. H. (1991a). Revisiting a challenge in transplantation: discordant xenografting. *Human Immunol.* **30**, 262–269.

Bach, F. H. (1991b). Xenotransplantation: problems for consideration. *Clin. Transplantation* **5**, 595–599.

Bach. F. H. (1994). Discordant xenografts: problems beyond antibodies and complement. *Xeno* **2**, 57–59.

Bach, F. H., Turman, M. A., Vercellotti, G. M., Platt, J. L., Dalmasso, A. P. (1991). Accommodation: a working paradigm for progressing toward clinical discordant xenografting. *Transplant. Proc.* **23**, 205–207.

Bach, F. H., Blakely, M. L., van der Werf, W., Vanhove, B., Stuhlmeier, K. M., Hancock, W. W., de Martin, R., Winkler, H. (1993). Discordant xenografting: a working model of problems and issues. *Xeno* **1**, 8–16.

Bach, F. H., Robson, S. C., Ferran, C., Winkler, H., Millan, M. T., Stuhlmeier, K. M., Vanhove, B., Blakely, M. L., van der Werf, W. J., Hofer, E., de Martin, R., Hancock, W. W. (1994). Endothelial cell activation and thromboregulation during xenograft rejection. *Immunol. Rev.* **141**, 5–30.

Bach, F. H., Ferran, C., Caudinas, D., Miyatake, T., Koyamada, N., Mark, W., Hechenleitner, P., Hancock, W. W. (1997a). Accommodation of xenografts: expression of 'protective genes' in endothelial and smooth muscle cells. *Transplant Proc.* **29**, 56–58.

Bach, F. H., Ferran, C., Hechlenleitner, P., Marck, W., Kayamada, N., Miyatake, T., Winkler, H., Badrichani, A., Candinas, D., Hancock, W. W. (1997b). Accommodation of vascularized xenografts: expression of 'protective genes' by donor endothelial cells in a host Th_2 cytokine environment. *Nat. Med.* **3**, 196–204.

Bartholomew, A., Cosimi, A. B., Sachs, D. H., Bailin, M., Boskovic, S., Colvin, R., Hong, H., Johnson, M., Kimikawa, M., Meehan, S., Sablinski, T., Wee, S. L., Powelson, J. (1997). A study of tolerance in a concordant xenograft model. *Transplant Proc.* **29**, 923–924.

Bazin, H., Platteau, B., Beckers, A., Pauwels, R. (1978). Differential effect of neonatal injections of anti-μ or anti-δ antibodies on the synthesis of IgM, IgD, IgE, IgA, IgG1, IgG2a, IgG2b and IgG2c immunoglobulin classes. *J. Immunol.* **121**, 2083–2087.

Bazin, H., Gray, D., Platteau, B., MacLennan, I. C. M. (1982). Distinct δ^+ and δ^- B lymphocyte lineages in the rat. *Ann. NY Acad. Sci.* **399**, 157–174.

Bevilacqua, M., Pober, J. S., Mendrick, D. L., Cotran, R. S., Gimbrone, M. A. (1987). Identification of an inducible endothelial–leukocyte adhesion molecule. *Proc. Natl Acad. Sci. USA* **84**, 9238–9242.

Bevilacqua, M. P., Stengelin, S., Gimbrone, M. A., Seed, J. S. (1989). Endothelial leukocyte adhesion molecule 1: an inducible receptor for neutrophils related to complement regulatory proteins and lectins. *Science* **243**, 1160–1165.

Bijsterbosch, M. K., Klaus, G. G. B. (1985). Crosslinking of surface immunoglobulin and Fc receptors on B lymphocytes inhibits stimulation of inositol phospholipid breakdown via the antigen receptors. *J. Exp. Med.* **162**, 1825–1836.

Borrebaeck, C. A. K., Malmborg, A. C., Ohlin, M. (1993). Does endogenous glycosylation prevent the use of mouse monoclonal antibodies as cancer therapeutics? *Immunol. Today* **14**, 477–479.

Brahim, F., Osmond, D. G. (1970). Migration of bone marrow lymphocytes demonstrated by selective bone marrow labeling with thymidine-H^3. *Anat. Rec.* **168**, 139–160.

Brett, J., Gerlach, H., Nawroth, P., Steinberg, S., Godman, G., Stern, D. (1989). Tumor necrosis factor/cachectin increases permeability of endothelial cell monolayers by a mechanism involving regulatory G proteins. *J. Exp. Med.* **169**, 1977–1991.

Brown, J. H., Jardetzky, T. S., Gorga, J. C., Stern, L. J., Urban, R. G., Strominger, J. L., Wiley, D. C. (1993). Three-dimensional structure of the human class II histocompatibility antigen HLA-DR1. *Nature* **364**, 33–39.

Cairns, T., Karlsson, E., Holgersson, J., Taube, D., Welsh, K., Samuelsson, G. (1994). Confirmation of a major target epitope of human natural IgG and IgM anti-pig antibodies: terminal galactose alpha-1,3-galactose. *Transplant. Proc.* **26**, 1384.

Calne, R. T. (1970). Organ transplantation between widely discordant species. *Transplant Proc.* **2**, 550.

Cambier, J. C., Pleiman, C. M., Clark, M. R. (1994). Signal transduction by the B cell antigen receptor and its coreceptors. *Annu. Rev. Immunol.* **12**, 457–486.

Cavender, D. E., Edelbaum, D., Ziff, M. (1989). Endothelial cell activation induced by tumor necrosis factor and lymphotoxin. *Am. J. Pathol.* **134**, 551–559.

Clark, M. R., Cambell, K. S., Kazlauskas, A., Johnson, S. A., Hertz, M., Potter, T. A., Pleiman, C., Cambier, J. C. (1992). The B cell antigen receptor complex: association of Ig-α and Ig-β with distinct cytoplasmic effectors. *Science* **258**, 123–126.

Cohen, B., Wight, C. (1993). The shortage of donor organs: the European experience. *Xeno* **1**, 21–22.

Collins, B. H., Parker, W., Platt, J. L. (1994). Characterization of porcine endothelial cell determinants recognized by human natural antibodies. *Xenotransplantation* **1**, 36–46.

Colson, Y. L., Markus, B. H., Zeevi, A., Duquesnoy, R. J. (1990). Increased lymphocyte adherence to human arterial endothelial cell monolayers in the context of allorecognition. *J. Immunol.* **144**, 2975–2984.

Cooper, D. K. C., Ye, Y., Niekrasz, M., Kehoe, M., Martin, M., Neethling, F. A., Kosanke, S., DeBault, L. E., Worsley, G., Zuhdi, N., Oriol, R., Romano, E. (1993). Specific intravenous carbohydrate therapy. *Transplantation* **56**, 769–777.

Cooper, D. K. C., Koren, E., Oriol, R. (1994). Oligosaccharides and discordant xenotransplantation. *Immunol. Rev.* **141**, 31–58.

Cooper, J. T., Stroka, D. M., Brotstjan, C., Palmetshofer, A., Bach, F. H., Ferran, C. (1996). A20 inhibits endothelial cell activation by a mechanism involving NF-κB. *J. Biol. Chem.* **271**, 18068–18073.

Cotran, R. S. (1987). New roles for the endothelium in inflammation and immunity. *Am. J. Pathol.* **129**, 407–413.

Coughlan, A. F., Berndt, M. C., Dunlop, L. C., Hancock, W. W. (1993). *In vivo* studies of P-selectin and platelet-activating factor during endotoxemia, accelerated allograft rejection and discordant xenograft rejection. *Transplant Proc.* **25**, 2930.

Cramer, D. V., Wu, G., Borie, D. C., Shirwan, H., Miller, L., Makowka, L. (1996). Germline VH gene usage in xenoreactive monoclonal antibody. *Transplant Proc.* **28**, 541–542.

Dalmasso, A. P., Platt, J. L. (1993). Prevention of complement-mediated activation of xenogeneic endothelial cells in an *in vitro* model of xenograft hyperacute rejection by C1 inhibitor. *Transplantation* **56**, 1171–1176.

Dalmasso, A. P., Vercellotti, G. M., Platt, J. L., Bach, F. H. (1991). Inhibition of complement-mediated endothelial cell cytotoxicity by decay-accelerating factor. *Transplantation* **52**, 530–533.

Dalmasso, A. P., Vercellotti, G. M., Fischel, R. J., Bolman, R. M., Bach, F. H., Platt, J. L. (1992). Mechanism of complement activation in the hyperacute rejection of porcine organs transplanted into primate recipients. *Am. J. Pathol.* **140**, 1157.

de Martin, R., Vanhove, B., Cheng, Q., Hofer, E., Csizmadia, V., Winkler, H., Bach, F. H. (1993). Cytokine-inducible expression in endothelial cells of an I kappa Bα-like gene is regulated by NF-kappa B. *EMBO J.* **12**, 2773–2779.

Deenen, G. J., Kroese, F. G. M. (1993). Kinetics of B cell subpopulations in peripheral lymphoid tissues: evidence for the presence of phenotypically distinct short-lived and long-lived B cell subsets. *Int. Immunol.* **5**, 735–741.

Figueroa, J. Fuad, S. A., Kunjummen, B. D., Platt, J. L., Bach, F. H. (1993). Suppression of synthesis of natural antibodies by mycophenolate mofetil (RS-61443). Its potential use in discordant xenografting. *Transplantation* **55**, 1371.

Fischel, R. J., Bolman III, R. M., Platt, J. L., Najarian, J. S., Bach, F. H., Matas, A. J. (1990). Removal of IgM anti-endothelial antibodies results in prolonged cardiac xenograft survival. *Transplant. Proc.* **22**, 1077–1078.

Fishman, J. A. (1994). Miniature swine as organ donors for man: strategies for prevention of xenotransplant-associated infections. *Xenotransplantation* **1**, 47–57.

Fryer, J. P., Blondin, B., Stadler, C., Ivancic, D., Rattner, U., Kaplan, B., Kaufmann, D., Abecassis, M., Stuart, F., Anderson, B. (1997). Inhibition of porcine endothelial cell lysis by human serum using peptides which inhibit C19 *Transplant. Proc.* **29**, 883.

Fuller, L., Carreno, M., Esquenazi, V., Roth, D., Milgrom, M., Miller, J. (1987). Naturally occurring interspecies and intraspecies antiimmunoglobulin antibodies. *Transplantation* **44**, 712–715.

Galili, U. (1993). Interaction of the natural anti-Gal antibody with α-galactosyl epitopes: a major obstacle for xenotransplantation in humans. *Immunol. Today* **14**, 480–482.

Galili, U., Swanson, K. (1991). Gene sequences suggest inactivation of α-1,3-galactosyl-transferase in catarrhines after the divergence of apes from monkeys. *Proc. Natl Acad. Sci. USA* **88**, 7401.

Galili, U., Rachmilewitz, E. A., Peleg, A., Flechner, I. (1984). A unique natural human IgG antibody with anti-α-galactosyl specificity. *J. Exp. Med.* **160**, 1519–1531.

Galili, U., Buehler, J., Shohet, S. B., Macher, B. A. (1987). The human natural anti-Gal IgG. III. The subtlety of immune tolerance in man as demonstrated by crossreactivity between natural anti-Gal and anti-B antibodies. *J. Exp. Med.* **165**, 693–704.

Galili, U., Shohet, S. B., Kobrins, E., Stults, C. L. M., Macher, B. A. (1988). Man, apes, and old world monkeys differ from other mammals in the expression of α-galactosyl epitopes on nucleated cells. *J. Biol. Chem.* **263**, 17755–17762.

Galili, U., Anaraki, F., Thall, A., Hill-Black, C., Radic, M. (1993). One percent of human circulating B lymphocytes are capable of producing the natural anti-Gal antibody. *Blood* **82**, 2485–2493.

Geller, R. L., Bach, F. H., Turman, M. A., Casali, P., Platt, J. L. (1993). Evidence that polyreactive antibodies are deposited in rejected discordant xenografts. *Transplantation* **55**, 168.

Gianello, P. G., Latinne, D., Alexandre, G. P. G. (1995). Pig to baboon xenograft. *Xeno.* **3**, 26–30.

Gold, M. R., Law, D. A., DeFranco, A. L. (1990). Stimulation of protein tyrosine phosphorylation by the B-lymphocyte antigen receptor. *Nature* **345**, 810–813.

Good, A. H., Cooper, D. K. C., Malcolm, A. J., Ippolito, R. M., Koren, E., Neethling, F. A., Ye, Y., Zuhdi, N., Lamontagne, L. R. (1992). Identification of carbohydrate structures that bind human antiporcine antibodies: implications for discordant xenografting humans. *Transplant. Proc.* **24**, 559–562.

Goodnow, C. C. (1992). Transgenic mice and analysis of B-cell tolerance. *Annu. Rev. Immunol.* **10**, 489–518.

Grandien, A., Modigliani, Y., Coutinho, A., Andersson, J. (1993). Suppression of B cell differentiation by ligation of membrane-bound IgM. *Eur. J. Immunol.* **23**, 1561–1565.

Hamadeh, R. M., Jarvis, G. A., Galili, U., Mandrell, R. E., Zhou, P., McLeod Griffiss, J. (1992). Human natural anti-Gal IgG regulates alternative complement pathway activation on bacterial surfaces. *J. Clin. Invest.* **89**, 1223–1235.

Inverardi, L., Pardi, R. (1994). Early events in cell-mediated recognition of vascularized xenografts: cooperative interactions between selected lymphocyte subsets and natural antibodies. *Immunol. Rev.* **141**, 71–93.

Inverardi, L., Samaja, M., Motterlini, R., Mangili, F., Bender, J. R., Parid, R. (1992). Early recognition of a discordant xenogeneic organ by human circulating lymphocytes. *J. Immunol.* **149**, 1416.

Inverardi, L., Socci, C., Pardi, R. (1993). Leukocyte adhesion molecules in graft recognition and rejection. *Xeno* **1**, 35.

Joziasse, D. H., Sharper, N. L., Kim, D., Van den Eijnden, D. H., Sharper, J. H. (1992). Murine α1,3-galactosyltransferase a single gene locus specifies four isoforms of the enzyme by alternative splicing. *J. Biol. Chem.* **267**, 5534.

Kirk, A. D., Heinle, J. S., Mault, J. R., Sanfilippo, F. (1993). *Ex vivo* characterization of human anti-porcine hyperacute cardiac rejection. *Transplantation* **56**, 785–793.

Kishihara, J. F., Penninger, J., Wallace, V. A., Kündiz, T. M., Kazuhiro, K., Wekeham, A., Timms, E., Pfeffer, K., Ohashi, P. S., Thomas, M. L., Furlonger, C., Paige, C. J., Mak, T. W. (1993). Normal B lymphocyte development but impaired T cell maturation in CD45-exon6 protein tyrosine phosphatase-deficient mice. *Cell* **74**, 143–156.

Koike, C., Kannagi, R., Muramatsu, T., Akutsu, F., Yamakawa, H., Okada, H., Nakashima, I., Yokoyama, I., Tagaki, H. (1997). Introduction of human alpha 1-2 fucosyltransferase and its effect in transgenic pig. *Transplant. Proc.* **29**, 894.

Kozlowski, T., Fuchimoto, Y., Monroy, R., Martinez-Ruiz, R., Foley, A., Ritzenthaler, J., Xu, Y., Andrews, D., Awwad, M., Fishman, J., Sablinski, T., Sachs, D. H. (1996). Apheresis and column absorption for specific removal of Gal-alpha (1,3)-Gal natural antibody. *Transplant. Proc.* **29**, 961.

Kroese, F. G. M., Butcher, E. C., Stall, A. M., Lalor, P. A., Adams, S. (1989). Many of the IgA producing plasma cells in murine gut are derived from self-replenishing precursors in the peritoneal cavity. *Int. Immunol.* **1**, 75–84.

Kujundzic, M., Koren, E., Neethling, F. A., Milotic, F., Koscec, M., Kujundzic, T., Martin, M., Cooper, D. K. C. (1994). Variability of anti-αGal antibodies in human serum and their relation to serum cytotoxicity against pig cells. *Xenotransplantation* **1**, 58–65.

Lachmann, P. J. (1991). The control of homologous lysis. *Immunol. Today* **12**, 312–315.

Latinne, D., Soares, M., Havaux, X., Cormont, F., Lesnikoski,

B., Bach, F. H., Bazin, H. (1994). Depletion of IgM xenoreactive natural antibodies by injection of anti-μ monoclonal antibodies. *Immunol. Rev.* **141**, 94–125.

Latinne, D., Squifflet, J. P., De Bruyère, M., Pirson, Y., Gianello, P., Sokal, G., Alexandre, G. P. J. (1989). Subclasses of ABO isoagglutinins in ABO-incompatible kidney transplantation. *Transplant. Proc.* **21**, 641–642.

Lesnikoski, B. A., Shaffer, D. A., Van der Werf, W. J., Dalmasso, A. P., Soares, M., Latinne, D., Bazin, H., Hancock, W. W., Bach, F. H. (1995). Endothelial and host mononuclear cell activation and cytokine expression during rejection of pig-to-baboon discordant xenografts. *Transplant. Proc.* **27**, 290–291.

Leventhal, J. R., Flores, H. C., Gruber, S. A., Figueroa, J., Platt, J. L., Manivel, J. C., Bach, F. H., Matas, A. J., Bolman III, R. M. (1992). Evidence that 15-deoxyspergualin inhibits natural antibody production but fails to prevent hyperacute rejection in a discordant xenograft model. *Transplantation* **54**, 26–31.

Leventhal, J. R., Sakiyalak, P., Witson, J., Simone, P., Matas, A. J., Bolman, R. M., Dalmasso, A. P. (1994). The synergistic effect of combined antibody and complement depletion on discordant cardiac xenograft survival in nonhuman primates. *Transplantation* **57**, 974–978.

Loveland, B. E., Johnstone, R. W., Russell, S. M., Thorley, B. R., McKenzie, I. F. C. (1993). Different membrane cofactor protein (CD46) isoforms protect transfected cells against antibody and complement mediated lysis. *Transplant. Immunol.* **1**, 101–108.

Magee, J. C., Collins, B. H., Harland, R. C., Bollinger, R. R., Frank, M. M., Platt, J. L. (1995). Prevention of hyperacute xenograft rejection by intravenous immunoglobulin. *Transplant. Proc.* **27**, 271.

Makowka, L., Cramer, D. V., Hoffman, A., Sher, L., Podesta, L. (1993). Pig liver xenografts as a temporary bridge for human allografting. *Xeno* **1**, 27–29.

McKenzie, I. F. C., Osman, N., Cohney, S., Vaughan, H. A., Patton, K., Mouhtouris, E., Atkin, J. D., Elliot, E., Fodor, W. L., Squinto, S. P., Burton, D., Gallop, M. A., Oldenburg, K. R., Sandrin, M. S. (1996). Strategies to overcome the anti-Gal alpha (1,3) Gal reaction in xenotransplantation. *Transplant. Proc.* **28**, 537.

Mollnes, T. E., Lachmann, P. J. (1988). Regulation of complement. *Scand. J. Immunol.* **27**, 127–142.

Morgan, B. P. (1989). Complement membrane attack on nucleated cells: resistance, recovery and non-lethal effects. *Biochem. J.* **264**, 1–14.

Morgan, B. P., Walport, M. J. (1991). Complement deficiency and disease. *Immunol. Today* **12**, 301–306.

Nakamura, T., Sekar, M. C., Kubagawa, H., Cooper, M. D. (1993). Signal transduction in human B cells initiated via Igβ ligation. *Int. Immunol.* **5**, 1309–1315.

Nawroth, P. P., Stern, D. M. (1996). Modulation of endothelial cell hemostatic properies by tumor necrosis factor. *J. Exp. Med.* **163**, 740–745.

Neethling, F. A., Koren, E., Ye, Y., Richards, S. V., Kujundzic, M., Oriol, R., Cooper, D. K. C. (1994). Protection of pig kidney (PK15) cells from the cytotoxic effect of anti-pig antibodies by α-galactosyl oligosaccharides. *Transplantation* **57**, 959–963.

Norin, A. J., Brewer, R. J., Lawson, N., Grijalva, G. A., Vaynblat, M., Burack, J. H., Burton, W., Squinto, S. P., Kamholz, S. L., Fodor, W. L. (1997). Human CD59 transgene expression in swine: enhanced survival of orthotopic lung xenografts in baboons. *Transplant. Proc.* **29**, 884.

Oriol, R., Ye, Y., Koren, E., Cooper, D. K. C. (1993). Carbohydrate antigens of pig tissues reacting with human natural antibodies as potential targets for hyperacute vascular rejection in pig-to-man organ xenotransplantation. *Transplantation* **56**, 1433–1442.

Pardi, R., Bender, J. R., Engleman, E. G. (1987). Lymphocyte subsets differentially induce class II human leukocyte antigens on allogeneic microvascular endothelial cells. *J. Immunol.* **139**, 2585–2592.

Parker, W., Bruno, D., Holzknecht, Z. E., Platt, J. L. (1994). Characterization and affinity isolation of xenoreactive human natural antibodies. *J. Immunol.* **153**, 3791–3803.

Perper, R. J., Najarian, J. S. (1966). Experimental renal heterotransplantation. I. In widely divergent species. *Transplantation* **4**, 377.

Platt, J. L. (1994). A perspective on xenograft rejection and accommodation. *Immunol. Rev.* **141**, 127–149.

Platt, J. L., Bach, R. H. (1991a). Discordant xenografting: challenges and controversies. *Curr. Opin. Immunol.* **3**, 735–739.

Platt, J. L., Bach, F. H. (1991b). The barrier to xenotransplantation. *Transplantation* **52**, 937–947.

Platt, J. L., Holzknecht, Z. E. (1994). Porcine platelet antigens recognized by human xenoreactive natural antibodies. *Transplantation* **57**, 327–335.

Platt, J. L., Dalmasso, A. P., Vercellotti, G. M., Lindman, B. J., Turman, M. A., Bach, F. H. (1990a). Endothelial cell proteoglycans in xenotransplantation. *Transplant. Proc.* **22**, 1066.

Platt, J. L., Lindman, B. J., Chen, H., Spitalnik, S. L., Bach, F. H. (1990b). Endothelial cell antigens recognized by xenoreactive human natural antibodies. *Transplantation* **50**, 817.

Platt, J. L., Turman, M. A., Noreen, H. J., Fischel, R. J., Bolman III, R. M., Bach, F. H. (1990c). An ELISA assay for xenoreactive natural antibodies. *Transplantation* **49**, 1000–1001.

Platt, J. L., Vercellotti, G. M., Dalmasso, A. P., Matas, A. J., Bolman, R. M., Najarian, J. S., Bach, F. H. (1990d). Transplantation of discordant xenografts: a review of progress. *Immunol. Today* **11**, 450–457.

Platt, J. L., Vercellotti, G. M., Lindman, B. J., Oegema Jr, T. R., Bach, F. H., Dalmasso, A. P (1990e). Release of heparan sulfate from endothelial cells. Implications for pathogenesis of hyperacute rejection. *J. Exp. Med.* **171**, 1363–1368.

Platt, J. L., Dalmasso, A. P., Lindman, B. J., Ihrcke, N. A., Bach, F. H. (1991a). The role of C5a and antibody in the release of heparan sulfate from endothelial cells. *Eur. J. Immunol.* **21**, 2887.

Platt, J. L., Fishel, R. J., Matas, A. J., Reif, S. A., Bolman, R. M., Bach, F. H. (1991b). Immunopathology of hyperacute xenograft rejection in a swine-to-primate model. *Transplantation* **52**, 214–220.

Platt, J. L., Lindman, B. J., Geller, R. L., Noreen, H. J., Swanson, J. L., Dalmasso, A. P., Bach, F. H. (1991c). The role of natural antibodies in the activation of xenogenic endothelial cells. *Transplantation* **52**, 1037–1043.

Pober, J. S., Cotran, R. S. (1991). Immunologic interaction of T lymphocytes with vascular endothelium. *Adv. Immunol.* **50**, 261–301.

Pober, J. S., Gimbrone, M. A., Lapierre, L. A., Mendrick, D. L., Fiers, W., Rothlein, R., Springer, T. A. (1986). Overlapping patterns of activation of human endothelial cells by interleukin 1, tumor necrosis factor, and immune interferon. *J. Immunol.* **137**, 1893–1896.

Pruitt, S. K., Kirk, A. D., Bollinger, R. R., Marsh Jr, H. C., Collins, B. H., Levin, J. L., Mault, J. R., Heinle, J. S., Ibrahim, S., Rudolph, A. R., Baldwin III, W. M., Sanfilippo, F. (1994). The effect of soluble complement receptor type 1 on hyperacute rejection of porcine xenografts. *Transplantation* **57**, 363–370.

Ramasamy, R. (1994). Is malaria linked to the adherence of α-galactosyl epitopes in Old World primates? *Immunol. Today* **15**, 140.

Reemtsma, K., McCracken, B. H., Schlegel, J. U., Pearl, M. A. (1964). Heterotransplantation of the kidney: two clinical experiences. *Science* **143**, 700.

Rollin, S. A., Kennedy, S. P., Chodera, A. J., Elliott, E. A., Zavoico, G. B., Matis, L. A. (1994). Evidence that activation of human T cells by porcine endothelium involves direct recognition of porcine SLA and costimulation by porcine ligands for LFA-1 and CD2. *Transplantation* **57**, 1709–1716.

Ross, J. R., Kirk, A. D., Ibrahim, S. E., Howell, D. N., Baldwin III, W. M., Sanfilippo, F. P. (1993). Characterization of human anti-porcine 'natural antibodies' recovered from *ex vivo* perfused hearts – predominance of IgM and IgG2. *Transplantation* **55**, 1144–1150.

Saadi, S., Ihrcke, N. S., Platt, J. L. (1994). Endothelial cell shape and hyperacute rejection. *Transplant. Proc.* **26**, 1149.

Sachs, D. H., Bach, F. H. (1990). Immunology of xenograft rejection. *Human Immunol.* **28**, 245–251.

Sandrin, M. S., McKenzie, I. F. C. (1994). Galα(1,3)Gal, the major xenoantigen(s) recognized in pigs by human natural antibodies. *Immunol. Rev.* **141**, 169–190.

Sandrin, M. S., Vaughan, H. A., Dabkowski, P. L., McKenzie, I. F. C. (1993). Anti-pig IgM antibodies in human serum react predominantly with Gal(α1-3)Gal epitopes. *Immunology* **90**, 11391–11395.

Schaapherder, A. F. M., Daha, M. R., Te Bulte, M. J. W., van der Woude, F. J., Gooszen, H. G. (1994). Antibody-dependent cell-mediated cytotoxicity against porcine endothelium induced by a majority of human sera. *Transplantation* **57**, 1376–1382.

Schleef, R. R., Bevilacqua, M. P., Sawdey, M., Gimbrone, M. A., Loskutoff, D. J. (1988). Cytokine activation of vascular endothelium. Effects on tissue-type plasminogen activator and type 1 plasminogen activator inhibitor. *J. Biol. Chem.* **263**, 5797–5803.

Soares, M., Latinne, D., Elsen, M., Figueroa, J., Bach, F. H., Bazin, H. (1993). *In vivo* depletion of xenoreactive natural antibodies with an anti-μ monoclonal antibody. *Transplantation* **56**, 1427–1433.

Soares, M., Latinne, D., Bazin, H. (1994a). Can human anti-pig xenoreactive natural antibodies be detected accurately? *Xeno* **2**, 100–101.

Soares, M., Lu, X., Havaux, X., Baranski, A., Reding, R., Latinne, D., Daha, M., Lambotte, L., Bach, F. H., Bazin, H. (1994b). *In vivo* IgM depletion by anti-μ monoclonal antibody therapy: the role of IgM in hyperacute vascular rejection of discordant xenograft. *Transplantation* **57**, 1003–1009.

Starzl, T. E., Fung, J., Tzakis, A., Todo, S., Demetris, A. J., Marino, I. R., Doyle, H., Zeevi, A., Warty, V., Michaels, M., Kusne, S., Rudert, W. A., Trucco, M. (1993). Baboon-to-human liver transplantation. *Lancet* **341**, 65–71.

Starzl, T. E., Valdivia, L. A., Murase, N., Demetris, A. J., Fontes, P., Rao, A. S., Manez, R., Marino, I. R., Todo, S., Thomson, A. W., Fung, J. J. (1994). The biological basis of and strategies for clinical xenotransplantation. *Immunol. Rev.* **141**, 212–244.

Stuhlmeier, K. M., Csizmadia, V., Cheng, Q., Winkler, H., Bach, F. H. (1994). Selective inhibition of E-selectin, ICAM-1, and VCAM in endothelial cells. *Eur. J. Immunol.* **24**, 2186–2190.

Swain, S. L., Reth, M. (1993). Lymphocyte activation and effector functions. Editorial overview. Lymphocytes: the ultimate computers? *Curr. Opin. Immunol.* **6**, 355–358.

Thall, A., Galili, U. (1990). Distribution of Galα1 → 3Galβ1 → 4GlcNAc residues on secreted mammalian glycoproteins (Thyroglobulin, Fibrinogen, and immunoglobulin G) as measured by a sensitive solid-phase radioimmunoassay. *Biochemistry* **29**, 3959–3965.

Thornton, B. P., Vetvicka, V., Ross, G. D. (1994). Natural antibody and complement-mediated antigen processing and presentation by B lymphocytes. *J. Immunol.* **153**, 1728–1737.

Tolan, M. J., Friend, P., Lozzi, E., Waterworth, P., Langford, G. A., Dunning, J. (1997). Life-supporting transgenic kidney transplants in a pig to primate model. *Transplant. Proc.* **29**, 899.

Turman, M. A., Casali, P., Notkins, A. L., Bach, F. H., Platt, J. L. (1991). Polyreactivity and antigen specificity of human xenoreactive monoclonal and serum natural antibodies. *Transplantation* **52**, 710–717.

Tuso, P. J., Cramer, D. V., Yasunaga, C., Cosenza, C. A., Wu, G. D., Makowka, L. (1993a). Removal of natural human xenoantibodies to pig vascular endothelium by perfusion of blood through pig kidneys and livers. *Transplantation* **55**, 1375–1378.

Tuso, P. J., Cramer, D. V., Middleton, Y. D., Kearns-Jonker, M., Yasunaga, C., Cosenza, C. A., Davis, W. C., Wu, G. D., Makowka, L. (1993b). Pig aortic endothelial cell antigens recognized by human IgM natural antibodies. *Transplantation* **56**, 651–655.

Valdivia, L. A., Fung, J. J., Demetris, A. J., Celli, S., Pan, F., Tsugita, M., Starzl, T. W. (1994). Donor species complement after liver xenotransplantation. *Transplantation* **57**, 918–922.

van De Stadt, J., Vendeville, B., Weill, B., Crougneau, S., Michel, A., Filipponi, F., Icard, P., Renoux, M., Louvel, A., Houssin, D. (1988). Discordant heart xenografts in the rat. *Transplantation* **45**, 514–518.

van Denderen, B. J., Slavaris, E., Romanella, M., Aminian, A., Katerelos, M., Tange, M., Pearse, M., d'Apice, A. (1997). Combination of decay-accelerating factor expression and alpha-1,3-galactosyltransferase knockout affords added protection from human complement-mediated injury. *Transplantation* **64**, 882–888.

van Noesel, C. J. M., Lankester, A. C., van Lier, R. A. W. (1993). Dual antigen recognition by B cells. *Immunol. Today* **14**, 8–11.

Vanhove, B., Bach, F. H. (1993). Human xenoreactive natural antibodies. Avidity and targets on porcine endothelial cells. *Transplantation* **56**, 1251–1292.

Vanhove, B., de Martin, R., Lipp, J., Bach, F. H. (1994). Human xenoreactive natural antibodies of the IgM isotype activate pig endothelial cells. *Xenotransplantation* **1**, 17–23.

Vaughan, H. A., Loveland, B. E., Sandrin, M. S. (1994). GALα(1,3)GAL is the major xenoepitope expressed on pig endothelial cells recognized by naturally occurring cytotoxic human antibodies. *Transplantation* **58**, 879–882.

Vercellotti, G. M., Platt, J. L., Bach, F. H., Dalmasso, A. P.

(1991). Neutrophil adhesion to xenogeneic endothelium via iC3b. *J. Immunol.* **146**, 730–734.

Wu, G. D., Cramer, D. V., Shirwan, H., Borie, D., Chapman, F. A. (1997). Genetic evidence that the antibody response to xenografts is directed at a restricted number of target antigens. *Transplant. Proc.* **29**, 954.

Yeh, C. G., Marsh Jr, H. C., Carson, G. C., Berman, L., Concino, M. F., Scesney, S. M., Kuestner, R. E., Shibber, R., Donahue, K. A., Ip, S. H. (1991). Recombinant soluble human complement receptor type 1 inhibits inflammation in the reversed passive Arthus reaction in rat. *J. Immunol.* **146**, 250–256.

Zhao, Y., Khan, A., Sergio, J., Swenson, K., Oliveros, J., Arn, J. S., Barth, R. N., Sachs, D. H., Sykes, M. (1997). Specific tolerance induction across a discordant species barrier by grafting with fetal or neonatal pig thymus. *Transplant. Proc.* **29**, 1228–1229.

XIX Conclusion

As stated in the Introduction, species other than mouse and man have played a crucial role in illuminating our understanding of the structure and function of the immune system. Nevertheless, the mouse remains the primary immunology model owing to an extensive body of knowledge and immunological reagents, genetically defined mouse strains, and numerous disease models. Thus, for many aspects of immunology, the mouse model provides an invaluable research tool.

A comparison of the immune systems of different vertebrate species, from poikilothermic fish to homeothermic mammals, facilitates a more comprehensive understanding of how the immune system has evolved and integrated its many functions. With each species, novel responses and/or structures have developed in response to the changing physiology and environment of each animal. However, throughout evolution function has been conserved and the interspecies variation in structure informs us of the complexity, redundancy, and integration of the immune system.

Many of the primitive immunological responses first developed in fish or more primitive species have been retained by mammals, with further embellishments in structure or the regulation of function. Thus, any consideration of the vertebrate immune system cannot overlook the diverse collection of species that are fish. With over 20 000 members and an evolutionary span of 400–500 million years the immune system of fish provides an insight into the innovation, elaboration, and conservation of structure and function that has occurred in vertebrate immunology. Fish have developed a thymus and spleen, produce conventional immunoglobulin molecules, lymphatics are present, and a distinction can be made between mucosal and systemic immunity. Nevertheless, fish do not possess all features of the present mammalian immune system. They lack bone marrow and organized lymph nodes with germinal centres. However, the similarities between fish and mammals are more striking than the differences. Similarities among vertebrate immune systems are especially evident at the genetic level. This high degree of homology in vertebrate gene organization has facilitated numerous studies of fish T-cell receptor, Ig, and MHC genes.

Turning to birds, the immune system conforms to the basic immunological mechanisms described in mammals. Chicken T cells can express either TCR$\alpha\beta$ or TCR$\gamma\delta$ but there is, generally speaking, a higher frequency of TCR$\gamma\delta$ in birds than in mammals. Chickens produce the major Ig isotypes identified in mammals and there is maternal transfer of immunity through the yolk. However, the mechanism for generating IgV gene diversity is significantly different. In chickens, both the heavy and light chains of Ig have a single functional V and J gene segment. Therefore, Ig diversity is generated not by combinatorial rearrangements but rather through somatic gene conversion which incorporates segments of nonfunctional pseudogenes into the rearranged Ig gene. This process of somatic gene conversion occurs in the lymphoid follicles of the bursa of Fabricius. Thus, in birds a gut-associated lymphoid tissue and not the bone marrow functions as an essential organ for development of the B-cell system.

Among mammalian species there are several peculiarities worth noting. The camelid's unique Ig heavy chain structure and the novel antigen-combining site challenges established concepts of Ig structure and function. The fetal development of Peyer's patches in ruminants and numerous other domestic species also raises questions about the role of gut-associated lymphoid tissue in normal B-cell ontogeny. In a manner that is reminiscent of birds, studies in sheep indicate that for many mammalian species there are alternative mechanisms to combinatorial rearrangement for generating Ig diversity. Furthermore, the high frequency of TCR$\gamma\delta$ T cells in ruminants suggest further conservation of immune system structure and function between birds and mammals. Finally, the apparently inverted structure of the pig lymph node and the presence of double-positive ($CD4^+CD8^+$) T cells in the blood of pigs also challenges dogma regarding the structure and function of the mammalian immune system.

This book ends with two chapters on the mouse and one on xenografting. The chapters on mouse immunology acknowledge the significant role that mice continue to play in the development of immune concepts. A chapter on xenografting was considered important because mammalian species, such as the pig, are foreseen as potential graft donors for humans. A variety of forces drive immunology research and xenografting may become a strong motivation for further studies in comparative immunology. The editors of this book also hope that human curiosity continues to extend the pursuit of immunology into species that are of no immediate economic value. Each species is sure to offer fresh insights that challenge the present paradigms of the structure and function of the immune system. The voyage of discovery has just begun for immunologists.

Handbook of Vertebrate Immunology
ISBN 0-12-546401-0

Glossary

Introduction

This glossary is only intended to give definitions of most of the terms necessary for the understanding of this handbook, avoiding technical ones.

Many of the definitions are derived from *The Dictionary of Immunology*, Fourth edition (1995), W. John Herbert, Peter C. Wilkinson and David I. Stott, Academic Press, and we wish to extend our grateful thanks to the authors for permission to reproduce this material.

Abbreviations used in the entries:

cf. compare
e.g. for example
esp. especially
i.e. that is to say
q.v. which see (where used indicates important extension to the definition).

A

Accessory cell: cell that is essential for initiating T cell dependent immune responses usually by presenting antigen, bound to class II MHC antigen molecules, to $CD4^+$ T lymphocytes. Accessory cells include mainly class II MHC^+ mononuclear phagocytes, dendritic cells and B lymphocytes.

Addressin: term used to refer to the structures on a vascular endothelial cell that are recognized by homing receptors on leucocytes (particularly lymphocytes) thus allowing specific entry of leucocytes into a tissue in, for example, lymphocyte recirculation or inflammation.

Adenosine deaminase (ADA) deficiency: an enzyme defect inherited as an autosomal recessive trait. Presents as a form of severe combined immunodeficiency (SCID) in infants with deficiencies of both B lymphocyte precursors and T lymphocyte precursors. Adenosine deaminase is abundant in normal lymphocytes and in the thymus and its lack causes failure of adenosine metabolism and accumulation of toxic metabolites in lymphocyte precursors.

Adhesion molecules: extracellular matrix proteins that attract leucocytes from the circulation. For example, T and B lymphocytes possess lymph node homing receptors on their membranes which facilitate their passage through high endothelial venules.

Affinity maturation: increase in affinity of antibody occurring during the immune response. This is due to selection of high-affinity B cell clones under conditions of limiting antigen concentration. The high-affinity clones arise as a result of somatic hypermutation of V_H region and V_L region genes during proliferation of B cells in germinal centres.

Agammaglobulinaemia: absence of immunoglobulin.

Agranulocytosis: pathological fall in the level of circulating neutrophil leucocytes resulting from depression of myelopoiesis. Can develop without known cause or following administration of certain cytotoxic drugs and also as an idiosyncratic response to normally harmless doses of various chemicals or drugs.

Allelic exclusion: an exception to the rule that both genes (or alleles) are expressed at a particular locus by both of the chromosome pair that bear them. The heavy chain and light chain genes of immunoglobulins are examples of this exception in that only one allele is expressed by them. Thus, in animals heterozygous for immunoglobulin allotypes, individual B lymphocytes express only one of the allotypes, not both.

Allergic alveolitis: inflammation of the gas-exchanging part of the lungs consisting of an infiltrate mainly of lymphocytes. Caused by inhaling aerosols of organic particulates of about 1 μm diameter which are small enough to penetrate and sediment in the terminal airways. The main syndromes include bird fancier's lung (frequently pigeons or budgerigars), and farmer's lung.

Allergy: a synonym for hypersensitivity.

Allogeneic (allogenic): genetically dissimilar within the same species, usually used with respect to differences in MHC.

Allograft: graft exchanged between two genetically dissimilar individuals of the same species, e.g. members of an outbred population, or of two different inbred strains.

Allotope (allotypic determinant): the structural region of a protein that distinguishes it from the same protein in another individual of the same species.

Allotype: serologically identifiable difference between protein molecules that is inherited as an allele of a single genetic locus. Many allotypes have been correlated with amino acid substitutions in the heavy chains and light chains of immunoglobulins. The epitope formed by these amino acids is known as the allotope. Allotopes can be found in both the constant regions and variable regions of immunoglobulins.

Alternative pathway (alternate pathway): a pathway by which complement component C3 is cleaved and C5–C9 formed without a requirement for C1, C2 or C4. Can be activated by bacterial endotoxin, in the absence of specific antibody, and by polysaccharides from various fungal or bacterial sources.

Alveolar macrophage: a mononuclear phagocyte found loosely attached to the walls of pulmonary alveoli and derived from circulating monocytes which mature *in situ*.

Amyloidosis: disease characterized by deposition of insoluble protein fibrils in a variety of tissues, often leading to failure of affected organs.

Anaphylatoxins: a group of substances, mediators of inflammation, produced in serum during activation of the complement cascade. Anaphylatoxin activity is located in the low molecular weight fragments C3a and C5a (also C4a) that are formed and released after cleavage of C3, C5 and C4.

Anaphylaxis: a severe generalized form of immediate hypersensitivity due to the widespread effects of histamine and other vascular permeability factors. Mechanism identical to that of other type I hypersensitivity reactions, i.e. the result of reaction of antigen with mast cell-bound IgE antibody and subsequent release of vascular permeability factors and inflammatory substances. Symptoms vary in different species.

Anergy: failure of lymphocytes that have been primed by an antigen to respond on second contact with the antigen. Occurs in both T lymphocytes and B lymphocytes.

Antibody: protein with the molecular properties of an immunoglobulin q.v. and capable of specific combination with antigen. Carries antigen-binding sites that bind non-covalently with the corresponding epitope. Antibodies are produced in the body by the cells of the B lymphocyte series and are secreted by plasma cells in response to stimulation by antigen.

Antibody half-life: a measure of the average time of survival of any given antibody molecule after its synthesis. In practice, the time taken for the elimination of 50% of a measured dose of antibody from the body of the animal. The half-life of antibody will vary according to immunoglobulin class and animal species.

Antigen-presenting cell: cell that carries antigenic peptides bound to its own major histocompatibility complex (MHC) molecules in such a way that the peptide–MHC complex can be recognized by the T cell receptor (TCR).

Antigen processing: the intracellular mechanism by which protein antigens are broken down to form small peptides which, on binding to major histocompatibility complex (MHC) molecules, can be presented to the T cell receptor (TCR). The pathways of antigen processing are different for peptides that bind to class I MHC antigens or to class II MHC antigens.

Antiserum: serum from any animal, which contains antibodies against a stated antigen.

APC: see **antigen-presenting cell**.

Atopy: a constitutional or hereditary tendency to produce IgE antibody to common inhalant allergens, e.g. house dust mite (*Dermatophagoides pteronyssinus*) and grass pollen.

Autoantibody: antibody capable of specific reaction with an antigen that is a normal constituent of the body of the individual in whom that antibody was formed.

Autoantigen (self antigen): an antigen that is a normal constituent of the body and against which an immune response may be mounted by the same individual; this sometimes results in autoimmune disease.

Autograft (syngeneic graft): graft originated from, and applied to, the same individual, e.g. skin graft from back used for the repair of a facial burn.

Autoimmune disease: clinical disorder resulting from an immune response against autoantigen.

Autoimmunity: specific humoral immunity (autoantibody-mediated) or cell-mediated immunity to constituents of the body's own tissues (autoantigens). If reactions between autoantibody or T lymphocytes and autoantigen result in tissue damage they may be regarded as hypersensitivity reactions. When such damage is sufficient to cause any clinical abnormality, an autoimmune disease is present.

Autologous: derived from self; used in reference to grafts, antigens, etc.

Apoptosis: non-necrotic cell death in which cells shrink, show blebbing (zeiosis) with release of cell fragments, rounding-up of cell organelles and nucleus, and condensation of chromatin to give a sharp rim round the periphery of the nucleus. The DNA is cleaved by endonucleases. Apoptotic cells may not show impairment of membrane permeability, and death is not accompanied by release of cell contents as in necrosis. Macrophages are capable of removing large numbers of apoptotic cells without trace and this is probably an important disposal mechanism for senescent neutrophil leucocytes. Important mechanism for selection in maturation of both T cells and B cells. Of wide interest in development, both in immunology and elsewhere.

Frequently, but not invariably, the form taken by programmed cell death.

Arthus reaction: an inflammatory reaction, characterized by oedema, haemorrhage and necrosis, that follows the administration of antigen to an animal that already possesses precipitating antibody to that antigen.

Asthma: a common inflammatory lung disease characterized by general, but reversible, bronchial airway obstruction.

Axenic: adjective describing animals reared in isolation from all other organisms. The absence of bacteria and larger organisms is relatively easily achieved; freedom from viruses much more difficult, especially as the latter may be incorporated in the genome. Cf. **gnotobiotic, germ free.**

B

B cell: see **B lymphocyte**; the two names are interchangeable.

B lymphocyte (B cell): B lymphocytes are the mediators of humoral immunity and on stimulation by antigen they differentiate into antibody-forming plasma cells and B memory cells. In the case of thymus-dependent antigens this process requires cooperation with T lymphocytes. In birds, B lymphocyte maturation is determined by the bursa of Fabricius.

B lymphocyte receptor (B lymphocyte antigen receptor, B cell receptor): synonym for membrane immunoglobulin (mIg), the transmembrane antigen-recognizing unit of the B lymphocyte.

B lymphocyte repertoire: the number of different V_H region–V_L region combinations that the immune system is potentially capable of producing. This is considerably larger than the number of different B lymphocyte receptors present on the B cells of an individual at a given time.

BALT: Bronchus-associated lymphoid tissue.

Basophil leucocyte: a leucocyte derived from bone marrow and found in small numbers (less than 1%) in blood, which contains round granules of different sizes giving a basophilic reaction with normal stains. The granules contain heparin, also histamine and other vascular permeability factors that may be released at sites of inflammation or in immediate hypersensitivity reactions. Basophil leucocytes possess high affinity Fc receptors of IgE (FcεRI).

Bence–Jones protein: protein in urine of patients with myelomatosis. Consists of dimerized light chains of myeloma protein.

β_2 microglobulin: protein (12 kDa) structurally similar to a single immunoglobulin constant region which is found in free form in solution in biological fluids, whose major importance is that it is normally linked non-covalently to the class I MHC molecule and stabilizes that molecule in the correct conformation for antigen presentation.

Bird fancier's lung: restrictive lung disease (a syndrome of extrinsic allergic alveolitis, q.v.) caused by exposure to dust containing antigens derived from the blood plasma of birds, especially albumin and gammaglobulin. These are present in bird faeces and also in 'bloom' (dust) from the skin and feathers. Cross reactivity between avian species is enough to allow one antigen source, e.g. pigeon serum, to be used in tests for exposure to other birds.

Blast cell: a cell usually large (diameter $> 8\ \mu m$), with ill-differentiated cytoplasm rich in RNA and actively synthesizing DNA. The nuclear patterns of blast cells vary and help to determine morphologically the series to which it belongs, e.g. plasmablast, myeloblast (of myeloid cell series), etc.

Blood group: classification of isoantigens on the surfaces of erythrocytes. The most important blood groups in man are those of the ABO and Rhesus blood group systems.

Bone marrow: the soft tissue that fills the cavities of bones. Red marrow is actively haemopoietic (i.e. blood forming) and is found in developing bone, ribs, vertebrae and parts of long bones. It contains all the cells and corpuscles (with their precursors) of the circulating blood, and also megakaryocytes, reticulum cells, macrophages and plasma cells. It contains lymphocyte stem cells and is the principal site of formation of B lymphocytes in rodents and humans and pre-T lymphocytes (but not mature T lymphocytes) in the adult. In adult animals much of the red marrow is replaced by fatty tissue and becomes yellow marrow.

Bruton-type hypogammaglobulinaemia: see **X-linked agammaglobulinaemia.**

Buffy coat: the layer of white cells that forms between the red cell layer and the plasma when unclotted blood is centrifuged.

Bursa of Fabricius: a sac-like lympho-epithelial structure arising as a dorsal diverticulum from the cloaca of young birds. First described in 1621 by Hieronymus Fabricius, an Italian anatomist, it is composed entirely of plicae containing numerous lymphoid follicles. B cell lymphopoiesis takes place within these and continues until the structure involutes at about the time of sexual maturity. The bursa is associated with humoral immunity. Bursectomized chickens fail to make antibodies to a variety of antigens, and plasma cells and germinal centres are reduced or absent in their lymphoid tissues.

C

C reactive protein: serum protein of the pentraxin family normally present in serum but increased in concentration in many inflammatory processes.

Caecal tonsils: see **tonsil.**

CCP superfamily: complement control protein superfamily.

CD antigens: a classification of cell surface proteins as 'clusters of differentiation' antigens based on their reactions with panels of monoclonal antibodies. The CD numbers are assigned to molecules by agreement at international workshops.

Cell-mediated immunity (CMI): specific immunity mediated by T lymphocytes which recognize major histocompatibility complex-bound antigens upon contact with the cells bearing them.

Cell-mediated immunity deficiency syndromes: syndromes characterized by failure to express reactions of cell-mediated immunity.

Cellular immunity: (1) term originated by Metchnikoff to refer to an increased ability of phagocytic cells to destroy or to digest parasitic organisms (see **phagocytosis**), and properly so used. Thus is a synonym for macrophage immunity. (2) Sometimes used to refer to cell-mediated immunity q.v.

Chediak–Higashi syndrome: disease of children inherited as autosomal recessive. The children show an increased susceptibility to severe pyogenic infection. There is a defect of granulopoiesis and the neutrophil leucocytes contain abnormally large lysosomal granules or phagolysosomes and are defective in microbicidal and chemotactic function. Similar syndromes are seen in a number of species, e.g. the beige mouse.

Chemokines: a family of molecules with cell-specific chemoattractant activity (see **chemotaxis**) and other activating properties for various cell types within the immune system.

Chemotaxis: reaction by which the direction of locomotion and the orientation of cells is determined by chemical substances. The cells become oriented and move towards (positive chemotaxis) or away from (negative chemotaxis) the source of a concentration gradient of the substance.

Chimerism: a state in which two or more genetically different propulations of cells coexist in an animal.

Class I MHC antigens: histocompatibility antigens composed of two non-covalently associated polypeptides; a type I transmembrane protein of MW 44 kDa heavy chain linked to $\beta 2$ microglobulin, MW 12 kDa. Class I antigens are expressed on the surface membranes of most nucleated cells, and their function is to present antigenic peptides to class I MHC-restricted T cells. The heavy chain consists of three extracellular domains (α_1, α_2 and α_3) of which the outer two (α_1, α_2) form an antigen-binding groove. This cleft in the surface of the protein has side walls consisting of α helical loops and a platform-like floor composed of β-pleated sheets. The groove accommodates peptides of 8–10 amino acids in length.

Class II MHC antigens: histocompatibility antigens composed of two non-covalently associated Type I transmembrane proteins: α chain, MW 32 kDa and β chain, MW 28 kDa. Class II MHC antigens are expressed predominantly on dendritic cells, B lymphocytes, macrophages and other accessory cells, but are inducible on other cells including epithelial T cells and vascular endothelium. Their function is to present antigenic peptides to class II MHC-restricted T cells, with the outer α_1 and β_1 domains forming an antigen-binding groove similar to, but larger than, that in class I MHC molecules.

Class switching: see **isotype switching**.

Classical complement pathway: the pathway of complement activation that commences with the binding of C1q to an antibody–antigen complex followed by the activation of C1, C4 and C2.

Clonal anergy: see **anergy**.

Clonal deletion: programmed cell death of inappropriately stimulated clones of antigen-reactive lymphocytes. See **immunological tolerance, negative selection.**

Colostrum: the first milk produced by the mother post partum. Viscid and yellow with high protein and high immunoglobulin content. Source of passive maternal immunity in newborn of many species.

Combined immunodeficiency: see **SCID**.

Complement: a system of at least 18 serum proteins and a group of membrane proteins which interact in a complex cascade reaction sequence. The components of the system are designated as numbers, e.g. C1, or as names, i.e. factor B. The classical complement pathway is activated primarily by immune complexes formed when antibody combines with antigen. The alternative pathway is activated by bacterial endotoxin, fungal and plant factors. Antigen-specific antibody is not required for initiation of this pathway. Both pathways allow the formation of the membrane attack complex, the insertion of which into cell membranes causes severe perturbation or holes resulting in an inability of the cell to survive. It should be noted that some species, notably pig, horse, dog and mouse, have complements that are not as haemolytically active as guinea pig or human complement.

Congenic strain (coisogenic strain): one of a number of separate strains of animals (e.g. mice) all constructed to possess identical genotypes except for a difference at a single gene locus. Although these strains are constructed to be genetically identical outside the single defined locus, the phenomena of mutation and genetic linkage ensure that mice within and between congenic strains will differ randomly at a minority of other loci.

Conglutinin: protein of the collectins protein family (q.v.) present in serum of *Bovidae* which can bind to complement (C3b)-bearing immune complexes in the presence of divalent cations. Not an antibody and not to be confused with immunoconglutinin q.v.

Constant region: the C-terminal portion of the heavy chain containing homology regions (immunoglobulin domains) C_H1, C_H2, C_H3, etc., or the C-terminal half of the light chain of an immunoglobulin molecule. So-called because the amino acid sequence in this region is constant from molecule to molecule except for amino acids at allotype marker sites. Many other members of the Ig superfamily contain domains homologous to the constant regions of immunoglobulin.

Costimulatory molecules: cell surface molecules other than the antigen receptor (TCR or membrane immunoglobulin) or its ligand (e.g. the major histocompatibility complex–antigenic peptide complex) that are required for an efficient response of lymphocytes to antigen.

Cryoglobulin: globulin, especially IgG or IgM, which precipitates spontaneously when serum is cooled below 37°C and redissolves on warming. Does not occur in normal serum. Cryoglobulinaemia may occur in association with myelomatosis, macroglobulinaemia, lymphoma and systemic lupus erythematosus (SLE). Characterized by peripheral vascular occlusion (Raynaud's phenomenon) and purpura of the extremities.

Cytokine receptor superfamily: a family of type I transmembrane proteins, many of which are receptors for cytokines or for haemopoietic growth factors or for hormones. Also known as the haemopoietic growth factor receptor superfamily or the haemopoietin superfamily.

Cytokines: generic name for proteins made and secreted by cells, which act as intercellular mediators with effects on growth, differentiation, activation, etc., of the same or other cells. Cytokines are important non-antigen-specific effector molecules in many immune and inflammatory responses. Lymphocytes are an important source of cytokines. It should be noted that cytokines have numerous functions outside the immune system, e.g. in developmental biology.

Cytotoxic T lymphocyte (CTL; cytolytic T lymphocyte; Tc; Terx): effector T lymphocyte subset (usually $CD8^+$, class I MHC antigen restricted) which directly lyses target cells. Cytotoxic T lymphocytes kill virus-infected cells provided that the latter carry syngeneic class I MHC antigens (see **MHC restriction**). Two major mechanisms are used for killing: (a) release of perforins, and (b) cytotoxic T lymphocyte-membrane Fas ligand binds to Fas on target cells thus inducing apoptosis in the latter.

D

Delayed-type hypersensitivity (DTH; delayed hypersensitivity type IV): hypersensitivity state mediated by primed T lymphocytes. The lesions, in which lymphocytes and macrophages are usually prominent, do not appear until about 24 hours after challenge of a primed subject with antigen, e.g. by intradermal inoculation.

Dendritic cell: the dendritic cells are a system of cells of stellate or dendritic morphology which are constitutively strongly class II MHC antigen positive and are important accessory cells, essential for primary immune responses. Originally derived from bone marrow, they are found throughout the body both in sites of contact with antigen (skin Langerhans cells, dendritic cells in gut, lung, etc.), and in peripheral lymphoid organs. Note that the follicular dendritic cells (q.v.) found in germinal centres are unrelated to the cells described above and are not bone marrow derived.

Di George's syndrome: failure of development of the parathyroids and thymus due to intrauterine damage to the third and fourth pharyngeal pouches. There is a defect manifest in infancy of cell-mediated immunity with low levels of circulating T lymphocytes, together with hypocalcaemia and tetany, and congenital heart defects.

Domain: sequence of a protein or peptide that forms a discrete structural unit, e.g. the constant regions and variable regions of immunoglobulins. Protein superfamilies comprise groups of proteins all of which contain domains with related tertiary structures, though the primary sequences within these domains may be different. The different superfamilies are defined on the basis of their content of such domains.

E

Effector cell: a cell that performs defined effector functions in immunity, either directly or as a result of signals from antigen-reactive lymphocytes.

Effector lymphocyte: lymphocyte which as a result of antigen-dependent differentiation has a direct functional role in the immune response, e.g. cytotoxic T lymphocyte, helper T lymphocyte, plasma cell.

Ellipsoids: fusiform stuctures that surround the capillaries at the termination of the penicillar arterioles of the spleen where these enter the red pulp. They consist of a sheath of high (or cuboidal) endothelial cells. Ellipsoids are prominent in the spleens of birds, pigs, horses and cats, but are difficult to distinguish in man and are absent in rodents.

Endoplasmic reticulum: a tubular cytoplasmic structure consisting of paired (parallel) membranes attached to the nuclear membrane. Present in all cells, it is most highly developed in protein-secreting cells where it is called rough-surfaced endoplasmic reticulum (RER) because of the numerous ribosomes attached to it. RER is prominent in protein-secreting cells such as plasma cells which secrete immunoglobulins.

Eosinophil leucocyte: a granulocyte found in normal blood (40–440 cells/mm^3 in man, i.e. up to 6% of total white cells), characterized by a bilobed nucleus and large, eosinophilic, cytoplasmic granules rich in cationic proteins.

Eosinophilia: increase in numbers of eosinophil leucocytes q.v. especially in blood, above physiological levels.

Particularly associated with immediate hypersensitivity reactions and responses to nematode worm infestations.

Epitope: the region on an antigen molecule to which antibody or the TCR (T cell receptor) binds specifically.

Extrinsic allergic alveolitis (EAA): a restrictive lung disease with constitutional fever and chills (influenza-like symptoms) caused by inhaling organic dusts. The most common syndromes are bird fancier's lung and farmer's lung.

F

F_1 hybrid: heterozygote belonging to the first generation derived from crossing genetically dissimilar parents.

Fab fragment: fragment obtained by papain hydrolysis of immunoglobulin molecules. The Fab fragment (MW ~ 45 kDa) consists of one light chain linked to the N-terminal half of the contiguous heavy chain (the Fd fragment). Two Fab fragments are obtained from each four chain molecule.

$F(ab')_2$ fragment: fragment obtained by pepsin digestion of immunoglobulin molecules (MW ~ 90 kDa). The $F(ab')_2$ fragment consists of that part of the immunoglobulin molecule which is on the N-terminal side of the site of pepsin digestion and therefore contains both Fab fragments plus the hinge region.

Farmer's lung: a syndrome of the extrinsic allergic alveolitis (q.v.) type. It is a disease mainly of farm-workers due in most instances to hypersensitivity to spores of thermophilic bacteria, mainly *Faenia rectivergula* (*Micropolyspora faeni*) and *Thermoactinomyces vulgaris*, organisms which occur in the dust of mouldy hay.

Fas (CD95): a type I transmembrane protein of the TNFR superfamily expressed on many cell types, including those of the myeloid cell series and lymphoid cell series. Cross-linking by anti-Fas antibody or by the natural Fas ligand induces apoptosis of the Fas-bearing cell. Fas is the murine equivalent of human Apo-1.

Fc fragment: the crystallizable fragment obtained by papain hydrolysis of immunoglobulin molecules. The Fc fragment of human IgG has a molecular weight of 50 kDa and consists of the C-terminal half of the two heavy chains linked by disulphide bonds. It has no antibody activity but contains the sites for complement and Fc receptor binding, placental transmission and the carbohydrate moiety of the molecule.

Fc′ fragment: a fragment produced in small amounts after papain hydrolysis of an immunoglobulin molecule, in addition to the Fc fragment. It is a non-covalently bonded dimer of the C3 homology region but without the terminal 13 amino acids, i.e. it is composed of the two C_H3 domains. The molecular weight of the dimer is 24 kDa (human IgG). Present in normal urine in small quantities.

Fc receptor: receptor, found on the plasma membrane of various cells, that binds the Fc fragment of immunoglobulin. A number of these receptors have been characterized.

Fcε receptor (IgE Fc receptor, FcεR): a molecular complex comprising an α, a β and two γ chains. The α chain binds to the Fc fragment of IgE with high affinity.

Fd fragment: the portion of the heavy chain of an immunoglobulin molecule N-terminal to the site of papain hydrolysis (cf. **Fc fragment**). It contains the variable region and part of the constant region.

Follicle: spherical accumulation of lymphocytes in lymphoid tissues.

Follicular dendritic cell (FDC): a cell with extensive dendritic processes found in the B cell areas of lymphoid tissue, i.e. in primary follicles and germinal centres. Follicular dendritic cells are not bone marrow-derived and are unrelated to the dendritic cells (q.v.) found in T cell areas and in other parts of the body (e.g. Langerhans cells, veiled cells, interdigitating cells).

Follicular hyperplasia: local or generalized enlargement of lymph nodes with increase in size and number of the follicles which typically contain active germinal centres (q.v.). This is a reactive change in the lymph nodes, usually following infection and is distinguishable from the lymphomas.

Forssman antigen: a glycolipid antigen present on tissue cells of many species, e.g. horse, sheep, mouse, dog and cat, but absent in man, rabbit, rat, pig and cow.

Freemartin: the female of twin bovine calves where the other twin is male and the two placentae have become fused *in utero*. Thus, the twins have exchanged cells before immunological maturity and are chimeras that do not reject grafts made from each other. In this situation the female calf is sterile due, amongst other things, to the influence of male hormones and can be recognized by physical examination.

G

GALT: see **gut associated lymphoid tissue**.

$\gamma\delta$ T cells: lineage of T lymphocyte possessing the $\gamma\delta$ form of the T cell receptor (TCR). Appears early in ontogeny, during thymus development, and also accounts for about 5% of mature T lymphocytes in peripheral lymphoid organs. In many species, $\gamma\delta$ T cells may be the predominant population of T lymphocyte at epithelial surfaces, such as the skin, intestine and genital tract. Some of these $\gamma\delta$ T cells may be extrathymically derived. $\gamma\delta$ T cells never express the $\alpha\beta$ TCR and they constitute a separate lineage of T cells.

Gamma (γ) globulin: the globulin fraction of serum that on electrophoresis shows the lowest anodic mobility at neutral pH. Contains mainly immunoglobulins.

Gene knockout mouse: a mouse in which a selected gene has been replaced by an inactive mutant and which therefore lacks the protein coded for by that gene.

Germ free: reared in the complete absence of bacteria and larger organisms. Freedom from all viruses is more difficult to achieve. Cf. **gnotobiotic, axenic**.

Germinal centre: a spherical aggregation of B lymphocytes which develops within the primary follicles of lymphoid tissues in response to stimulation by thymus-dependent antigens. Following antigen recognition in the T cell area, B cells migrate to the primary follicles where they proliferate massively to form a germinal centre, displacing the small recirculating B cells into the mantle zone (follicular mantle).

Germinal follicle: synonym for germinal centre.

Globulin: any serum protein whose anodic mobility on electrophoresis is less than that of albumin. Includes α, β and γ globulins; the latter fraction includes the immunoglobulins. Also commonly used to define those serum proteins which are precipitated by high concentrations of salts such as ammonium or sodium sulphate.

Glomerulonephritis: a term applied, with various qualifying prefixes, to a group of kidney diseases, in which the major lesion is in the glomeruli and is presumed to be immunologically mediated.

Gnotobiotic: descriptive of an environment in which all of the living organisms present are known, e.g. both a germ free mouse, and a mouse contaminated with a single known organism, may be described as gnotobiotic.

Goodpasture's syndrome: haemoptysis (coughing up blood) associated with proliferative glomerulonephritis. The glomerular basement membrane is thickened and IgG and, to a lesser extent, complement are deposited in a linear fashion along the basement membrane.

Graft rejection: destruction of tissue grafted into a genetically dissimilar recipient due to a specific immunological reaction against it by the recipient.

Graft-versus-host reaction: reaction of a graft containing immunologically competent T lymphocytes, against the tissues of a genetically non-identical recipient. The recipient must be unable to reject the graft either because of its immaturity (newborn animals, see **runt disease**), or its genetic constitution, or because it has been subjected to whole body irradiation or immunosuppression.

Granulocyte: one of a group of bone marrow-derived cells found in blood and tissue and characterized by the presence of numerous cytoplasmic granules. Three types of granulocyte can be differentiated by the morphology and staining properties of these granules (which give them their name); thus, neutrophil, eosinophil and basophil granulocytes (see under their synonyms **neutrophil leucocyte, eosinophil leucocyte** and **basophil leucocyte**).

Granuloma: the term is most frequently used to refer to a localized collection of macrophages and lymphocytes characteristic of chronic inflammatory lesions in which a delayed-type hypersensitivity (DTH) reaction is taking place.

Granzyme: granzymes are a family of serine proteases found in the granules of cytolytic lymphocytes (cytotoxic T lymphocytes, NK cells) and which are believed to play a part in lymphocyte-mediated cytotoxicity.

Gut-associated lymphoid tissue (GALT): lymphoid tissue closely associated with the gut, e.g. tonsils, Peyer's patches and appendix in man, sacculus rotundus in the rabbit, bursa of Fabricius in the chicken, etc.

GVH: see **graft-versus-host reaction**.

H

Haemolytic anaemia: anaemia due to an abnormal increase in the rate of destruction of circulating erythrocytes. Can result from metabolic abnormalities of the erythrocytes, from the development of antibodies to the erythrocytes, or from abnormalities of the mononuclear phagocyte system.

Haemolytic disease of the newborn: haemolytic anaemia in the fetus or newborn resulting from an excess of maternal anti-red cell antibody. In man, occurs due to antibody (usually Rhesus antibody) crossing the placenta. In the horse, pig and cattle, it follows the neonatal ingestion of colostrum, as, in these species, antibody does not cross the placenta.

Haplotype (haploid genotype): a cluster of genes inherited from one parent which, because of their close linkage on the same chromosome, are normally inherited together. In immunology usually refers to MHC genes.

Hassall's corpuscles: keratinized epithelial whorls or islands of cells found in the medulla of the thymus. These may be end-stage thymic epithelial cells and are associated with macrophages and apoptotic (see **apoptosis**) lymphocytes. Thus the Hassall's corpuscle may be a site of removal of dead cells.

Hay fever: acute nasal catarrh and conjunctivitis in atopic subjects caused by the inhalation of antigenic substances such as pollens (allergens q.v.) that are innocuous in normal persons. Due to immediate hypersensitivity (type I hypersensitivity reaction) following the reaction of cell-fixed IgE with the causative allergen. Often seasonal depending on concentration of the relevant antigen in air.

Heat shock proteins (HSP): ubiquitous intracellular proteins in all species, whose level increases when the organism is stressed. They were first identified following heat stress. They are involved in repair and re-folding of denatured proteins and they also assist in the folding and assembly of normal proteins, hence classified as 'chaperonins'. In immune cells, they assist in the assembly of immunoglobulin molecules and are believed to play an important role in antigen processing inasmuch as they assist the intracellular assembly of major histocompatibility complex (MHC) molecules.

Heavy chain: a polypeptide chain present in all immunoglobulin molecules. MW ~ 50 kDa in human IgG, 65 kDa in IgM. Each heavy chain is normally linked by disulphide bonds to a light chain and to another identical heavy chain. The heavy chain consists of a variable region (V_H) and a constant region composed of three or four domains (C_H1–C_H4), depending on immunoglobulin class. The amino acid sequence of the heavy chain constant region determines the class and immunoglobulin subclass, and the corresponding heavy chain is called the α chain, δ chain, ε chain, γ chain, or μ chain.

Heavy chain class: the group into which a heavy chain is placed by virtue of features of its primary or antigenic structure, common to all individuals of the same species, which distinguish it from heavy chains of other classes. These structural differences are found in the constant region. The heavy chain classes are α, δ, ε, γ and μ. See also **immunoglobulin class** and **isotype**.

Helper T lymphocyte (T_H lymphocyte, helper cell): a thymus-derived lymphocyte (usually $CD4^+$, class II MHC antigen-restricted, see **MHC restriction**) whose presence (help) is required for the production of normal levels of antibody by B lymphocytes and also for the normal development of cell-mediated immunity.

Heterospecific (heterologous): derived from or having specificity for a different species.

High endothelial venule (HEV): specialized venules found in the thymus-dependent area of the lymph node. Characterized by prominent, cuboidal, high endothelial lining cells. Recirculation of lymphocytes from blood to lymph takes place through the walls of these vessels. High endothelial cells in lymphoid tissue at different sites (lymph nodes, Peyer's patches, etc.) carry specific adhesion molecules (addressins) which are recognized by homing receptors on lymphocytes, thus allowing different populations of lymphocytes to home into specific lymphoid tissues.

Hinge region: a flexible proline-rich region of the heavy chain of the immunoglobulin molecule between the Fab fragment and the Fc fragment which acts as a hinge around which the Fab fragments can rotate. The angle between the Fab subunits may vary between 0° and 180°. The hinge region is adjacent to the sites of papain and pepsin hydrolysis.

Histamine: a vascular permeability factor, vasodilator and smooth muscle constrictor, widely distributed in biological tissues and found in high concentration in mast cells. Histamine and histamine-like substances are released when cell-bound IgE reacts with antigen. They cause the classical vascular lesions (weal and flare response) of immediate hypersensitivity. Anaphylatoxins also mediate release of histamine.

Histiocyte: a pathologist's term for a macrophage found within the tissues, in contrast to those found in the blood (monocytes) or serous cavities, etc. Some histiocytes appear to remain at the same site for long periods of time, e.g. those that retain dye particles in the skin after tattooing. They have a strong affinity for silver and other heavy metal stains.

Histocompatibility antigen: genetically determined alloantigen carried on the surface of nucleated cells of many tissues. Class I MHC antigens and Class II MHC antigens are of major importance in the recognition of antigens by T cells. C.f. **isoantigen**.

Holoxenic: conventionally reared animals. Cf. **axenic, gnotobiotic, germ free**.

Homing receptor: the receptor or receptors on leucocytes that specifically recognize addressins, i.e. adhesion molecules on vascular endothelial cells; thus allowing specific entry of a leucocyte into a particular tissue. Term used especially in the context of lymphocyte recirculation.

Homograft: an outmoded term for any graft made from one individual to another of the same species. Included allogeneic grafts (allografts), and syngeneic grafts. The latter terms are more informative and used by transplantation immunologists.

Humoral immunity: specific immunity mediated by antibodies, and thus ultimately by B lymphocytes.

Hypergammaglobulinaemia: raised serum gamma globulin level. Diffuse (i.e. not restricted to a single immunoglobulin class) increase in serum gamma globulin is associated with any condition where continued antigenic stimulation (see **antigen**) causes production of large amounts of antibody. In paraproteinaemias a sharp, high electrophoretic spike of immunoglobulin, which is monoclonal in origin, is seen.

Hypersensitivity: state of the previously immunized body in which tissue damage results from the immunological reaction to a further dose of antigen.

Hypervariable region: within the variable region of the heavy chains and light chains of immunoglobulin molecules, residues in certain positions show much higher variability from one molecule to another than do residues at other positions. They are partly, but not wholly, responsible for the specificity of the antigen-binding site and also the idiotypic variations between immunoglobulins secreted by different clones of cells. There are three hypervariable regions within the variable region of both heavy and light chain.

Hypocomplementaemia: any condition in which serum complement levels are low.

Hypogammaglobulinaemia: lowered serum immunoglobulin (gamma globulin) level.

I

ICAM-1, -2, -3 (intercellular adhesion molecules): a group of Type I transmembrane proteins within the Ig superfamily, having a variable number of constant region-like domains, and which bind to the $\beta2$ (leucocyte) integrins (see **integrin superfamily**). They are important mediators of adhesion of leucocytes to vascular endothelium and to one another, e.g. in clustering of lymphocytes around accessory cells in induction of an immune response.

Idiotope (idiotypic determinant): immunoglobulins are immunogenic, just like any other protein molecule, and antibodies can be made against any region of the molecule. An idiotope is an epitope in the variable region of an immunoglobulin that is characteristic of the immunoglobulin molecules produced by a single B cell clone, or a small number of clones. It may be present in the antigen-binding site or outside it.

Idiotype: set of one or more idiotopes (q.v.) by which a clone of immunoglobulin-forming cells can be distinguished from other clones. Some, known as individual, or private, idiotypes, appear to be unique to individuals. Others, known as inherited, public, or cross reacting idiotypes, are found in many members of the same animal species and even, in some cases, in more than one species.

IFN-α, -β (alpha-interferon and beta-interferon; type I interferons): two related proteins originally identified as being released by cells in response to viral infection. Their activity is non-specific in its spectrum of antiviral activity in that virtually all viruses are susceptible to their action. IFN-α was first known as leucocyte interferon since it is made by mononuclear phagocytes. IFN-β was known as fibroblast interferon.

IFN-γ (gamma-interferon): a cytokine produced by activated T cells and by NK cells which is a major macrophage activating factor. It enhances expression of class I MHC antigen and class II MHC antigen on many cells, including NK cells and B cells, for both of which it is a differentiation factor. IFN-γ release is particularly associated with the T_H1 cell subset of $CD4^+$ T cells and drives the cell-mediated immune response, but it is also made by $CD8^+$ T cells. IFN-γ has antiviral activity but has little sequence homology with IFN-α or IFN-β and these latter are more potent antiviral agents than IFN-γ.

Ig superfamily (immunoglobulin superfamily): a superfamily of cell membrane proteins which have in common the presence of extracellular domains (see **immunoglobulin domain**) with an imunoglobulin constant region or variable region-like structure, though most lack the polymorphism characteristic of immunoglobulin. Some proteins carry many such domains, others few. This is by far the largest of the membrane protein superfamilies. The diversity of immunoglobulin-like structures is assumed to result from evolution from a primordial immunoglobulin domain.

IgA: the major immunoglobulin of the external secretions (intestinal fluids, saliva, bronchial secretions, etc.) where it is found as a dimer linked to a secretory piece (transport piece) q.v. Also present in serum (concentration 1.5–4.0 mg/ml) as a monomer and in polymeric forms (dimer, trimer, tetramer). Does not cross human placenta. Present in human colostrum and milk, but is not major colostral immunoglobulin of cow or ewe.

IgD: immunoglobulin present in low concentrations in serum. Present as membrane immunoglobulin in the surface of B lymphocytes.

IgE: the main immunoglobulin associated with immediate hypersensitivity (type I hypersensitivity reaction). Present in serum in very low concentration (20–500 ng/ml) but elevated in atopic individuals. It shows high affinity for FcεRI on the surface of mast cells.

IgG: the major immunoglobulin in the serum of man. Homologous immunoglobulins are found in most species from amphibians upwards, but are not present in fish. There are four subclasses in man, IgG1, IgG2, IgG3 and IgG4, but the number varies in other species. All four subclasses are able to cross the human placenta.

IgM: high molecular weight (970 kDa) immunoglobulin. Phylogenetically the most primitive immunoglobulin, present in all vertebrates from lamprey upwards. In mammals it is mainly in the form of a cyclic pentamer of five basic four-chain units of two heavy chains and two light chains linked by disulphide bonds. In other species the predominant form may be a monomer (e.g., dogfish), tetramer (e.g., carp) or hexamer *(Xenopus)*. Heavy (μ) chain is larger than that of other immunoglobulins (MW 70 kDa). High carbohydrate content. Does not cross placenta in man but may in certain other species (e.g., rabbit).

Immediate hypersensitivity: IgE antibody-mediated hypersensitivity characterized by lesions resulting from release of histamine and other vasoactive substances (synonym type 1 hypersensitivity reaction). IgE antibody fixes to basophil leucocytes and, especially, to mast cells in the tissues.

Immune complex: a macromolecular complex of antigen and antibody molecules bound specifically together. May be present in soluble form especially in antigen excess. Complement components may be bound by immune complexes. Important in pathogenesis of certain hypersensitivity reactions.

Immune tolerance: see **immunological tolerance**.

Immunoconglutinin: autoantibody against fixed complement components, especially C3, also C4 (i.e. C3b and C4b). Serum immunoconglutinin levels reflect extent of complement fixation by *in vivo* immunological reactions and are raised in many bacterial, viral and parasitic infections and autoimmune diseases. Not to be confused with conglutinin q.v.

Immunodeficiency: any condition in which a deficiency of humoral immunity or cell-mediated immunity exists. Examples are severe combined immunodeficiency syndrome, cell-mediated immunity deficiency syndromes, X-linked agammaglobulinaemia, antibody deficiency syndrome, *inter alia*.

Immunoglobulin: member of a family of proteins each made up of light chains and heavy chains linked together by disulphide bonds. The members are divided into immunoglobulin classes and immunoglobulin subclasses (q.v.) determined by the amino acid sequence of their heavy chains. Most mammals have five immunoglobulin classes (IgM, IgG, IgA, IgD, IgE), although lower organisms have fewer, e.g. the cartilaginous fishes have only one immunoglobulin closely related to IgM. All antibodies are immunoglobulins; however, it is not certain that all immunoglobulin molecules function as antibodies. Present in serum and other body fluids. On electrophoresis show γ or β mobility relative to other serum proteins.

Immunoglobulin class (isotype): the group into which an immunoglobulin is placed by virtue of the amino acid sequence of the constant region of its heavy chain which distinguishes it from the other classes; the chains, being named with the Greek letter equivalent for each class. The γ chain of IgG for instance, differs in the amino acid sequence of its constant region from that of the heavy chains of the other immunoglobulin classes. IgA, IgD, IgE, IgG and IgM are immunoglobulin classes distinguishable respectively by their possession of α, δ, ε, γ and μ chains. It is the heavy chain constant region that determines the overall structure and properties of the immunoglobulin molecule.

Immunoglobulin domain: the three-dimensional structure formed by a single homology region of the heavy chain or light chain of an immunoglobulin, i.e. V_L region, C_L V_H region, C_H1–C_H4. Each homology region is folded into a similar three-dimensional shape which is believed to be shared by homology regions present in other members of the Ig superfamily. All members of the Ig superfamily are characterized by the presence of domains resembling the variable or constant domains of immunoglobulins.

Immunoglobulin subclass: subdivision within each immunoglobulin class, based on structural and antigenic differences in their heavy chains. Thus, human IgG has four subclasses: IgG1, IgG2, IgG3 and IgG4. The γ chains (γ1, γ2, γ3 and γ4) of the subclasses show closer sequence homology to each other than to the heavy chains of the other immunoglobulin classes. The subclasses differ structurally from one another, e.g. IgG1 and IgG4 have two inter-heavy chain disulphide bonds in the hinge region, IgG2 has four and IgG3 has 11. They also differ functionally, e.g. IgG1, 2 and 3 activate complement whereas IgG4 does not. Human IgA has two subclasses, IgA1 and IgA2. Immunoglobulins of other species also show subclass differences, e.g. mouse IgG1, IgG2a, IgG2b and IgG3.

Immunoglobulin superfamily: see **Ig superfamily**.

Immunological memory: concept formulated to explain the capacity of the immunological system to respond much faster and more powerfully to subsequent exposures to an antigen than it did at the first exposure.

Immunological rejection: destruction of foreign cells or tissues inoculated or grafted into a recipient due to a reaction of specific immunity against them.

Immunological tolerance: the induction of specific nonreactivity of the lymphoid tissues to an antigen capable in other circumstances of inducing active cell-mediated or humoral immunity. May follow contact with antigen in fetal or early post-natal life or, in adults, after administration of very high or very low doses of certain antigens. Immunological reactions to unrelated antigens are not affected by the induction of tolerance to any given antigen. Tolerance may be due to anergy, clonal deletion or active suppression of antigen-specific clones of T or B lymphocytes.

Inbred strain: experimental animals produced by sequential brother–sister matings. In immunology, the term usually refers to animals in the 20th and subsequent generations of such matings. Such animals are so homogeneous at histocompatibility loci that grafts can be freely exchanged between them without provoking graft rejection.

Incompatibility: antigenic non-identity between donor and recipient, e.g. in blood transfusion or tissue transplantation, such that harmful reactions may occur when donor material is introduced into the recipient. Examples of such reactions are transfusion reactions and immunological rejection.

Inflammatory cell: any cell present in an inflammatory lesion as part of the host response, e.g. neutrophil leucocytes, eosinophil leucocytes, macrophages, etc.

Integrin superfamily: a family of heterodimeric Type I transmembrane proteins each of which has an α and β chain. There are subfamilies determined by the β chain, thus β_1, β_2, etc., each of which may have multiple α chains. However, it is also clear that α chains can bind to more than one β chain, so the subfamily structure is not rigid. The β_2 integrins are also known as the leucocyte integrins since they are major mediators of leucocyte binding to vascular endothelium. This takes place in a first stage mediated by selectins (q.v.) which cause leucocytes to roll along the vessel wall, then a second stage in which integrins cause the leucocytes to stop, spread and transmigrate. Some of the β_1-integrins (VLA molecules, see **CD49**) are found on lymphocytes and perform the same function. Various other integrins have specific functions, e.g. $\alpha_4\beta_7$ is a homing receptor which allows lymphocytes to enter Peyer's patches. The cell surface ICAMs are ligands for integrins in leucocyte–endothelial interactions, but integrins also mediate binding of many cell types to extracellular matrix proteins such as fibronectin, laminin, vitronectin, etc.

Interdigitating cell: class II MHC antigen$^+$ cell of dendritic cell morphology found in the T cell areas of lymph nodes and other lymphoid tissues. Derived from Langerhans cells and other peripherally situated dendritic cells which, on contact with antigen, migrate in the afferent lymphatics to the lymph nodes. Important accessory cells because they present antigen to T cells and provide the first contact site with antigen from recirculating lymphocytes.

Interferons: see **IFN-α, -β; IFN-γ**.

Interleukins (IL) (inter-leucocytes): the name 'interleukin' is given to certain cytokines that act as intercellular signals. There is no logic to the interleukin designation: cytokines such as IFN-γ or TNF-α could as well be 'interleukins'. Nor is there any logic to the order in which the interleukins are numbered.

Intraepithelial lymphocyte: any lymphocyte found within an epithelium particularly the intestinal epithelium.

Isoantigen: antigen carried by several individuals belonging to the same group (usually blood groups) and which is often capable of eliciting an immune response in other individuals of the same species but belonging to other groups.

Isogeneic (isogenic): possessing absolutely identical genotypes, e.g. animals derived from the same egg, identical twins. Often used as a synonym for syngeneic as a descriptor of inbred strains. However, individuals of the latter are never absolutely identical in the sense implied by the term isogeneic.

Isograft: syngeneic grafts.

Isotype: classification of a molecule by comparison of its primary or antigenic structure with that of closely related molecules found within all members of the same species. Applied to the immunoglobulins, the isotype describes the immunoglobulin class and immunoglobulin subclass, light chain type and subtype and can also be applied to the variable region groups and subgroups.

Isotype switching: a process that occurs during an immune response in which a B cell switches from production of one immunoglobulin class to another class without loss of specificity.

Isotypic variation: structural variability of antigens common to all members of the same species, e.g. the antigenic differences which distinguish the immunoglobulin classes and types of immunoglobulin chains.

J

J chain: polypeptide chain (MW 15 kDa) with a high content of cysteine, found in the polymeric forms of IgA and IgM. Has been shown to link together two of the subunits in these immunoglobulins, thus maintaining the polymeric structure. The J chains from IgA and IgM are identical and only one J chain is present in each molecule.

K

Kallikreins (kininogenases): enzymes with indirect activity in increasing vascular permeability, vasodilatation and smooth muscle contraction. They are esterases which convert kininogens into pharmacologically active kinins.

κ (kappa) **chain**: one of the two types of light chain of immunoglobulins, the other being the λ (lambda) chain. An individual immunoglobulin molecule bears either two λ chains or two κ chains, never one of each. About 60% of human IgG molecules are of the κ type, 40% of the λ type, but in the mouse the ratio is 95% to 5%.

Kinins: peptides formed by the action of esterases known as kallikreins (kininogenases).

Knockout mouse: see **gene knockout mouse**.

Kupffer cell: a non-motile macrophage derived from blood monocytes and found lining the blood sinuses of the liver. As Kupffer cells are phagocytic and positioned in an area of high blood flow, they are highly active in removing foreign particles from the blood.

Kurloff cell: cell found in the peripheral blood, spleen and other organs of pregnant or oestrogen-treated guinea pigs. Contains a large inclusion body composed of mucoprotein and sulphated mucopolysaccharide and has been postulated to be a type of modified lymphocyte of unknown function.

L

λ (lambda) **chain**: one of the two types of light chain of immunoglobulins, the other being the κ (kappa) chain. An individual immunoglobulin molecule bears either two κ chains or two λ chains, never one of each. The ratio of κ to λ chains varies with species, e.g. about 60% of human IgG molecules are of the κ type, 40% of the λ type, whereas in the mouse the ratio is 95% κ chains to 5% λ chains.

Lamina propria: layer of connective tissue supporting the epithelium of the digestive tract and with it forming the mucosa. Contains the blood supply, lymphatic drainage and innervation of the mucosa and is the site of accumulation of lymphocytes, plasma cells, mast cells and macrophages in immunological reactions involving the gut.

Langerhans cell: an accessory cell of dendritic appearance found in the basal layers of the epidermis, derived from bone marrow and characterized by the presence of tennis racket-shaped cytoplasmic Birbeck granules. Strongly class II MHC antigen positive, weakly positive for Fcγ receptors and C3b receptors.

Late phase reaction: reaction which can begin about 5 hours after an immediate hypersensitivity reaction provoked by either skin testing or inhalation challenge by an allergen. The late phase reaction is characterized by inflammation, pruritus and a minimal cellular infiltration; its mechanism is probably due to late effects from the release of cytokines from mast cells or alveolar macrophages.

Leader peptide (signal peptide): a sequence of approximately 20, mainly hydrophobic, amino acids found at the N-termini of most nascent secreted proteins, e.g. the light chains and heavy chains of immunoglobulins, but absent from the secreted form. The peptide is rapidly cleaved from the nascent chains once they are released into the cisternal space of the endoplasmic reticulum and is responsible for vectorial release of the polypeptide chains and hence their secretion from the cell.

Leucocyte (leukocyte): the white cells of the blood and their precursors (see **myeloid cell series** and **lymphoid cell series**).

Leucocyte adhesion deficiency (leucocyte adhesion defect; LAD): a serious form of immunodeficiency due to a defect in the β_2 integrin (see **integrin superfamily**) chain (CD18) which results in defective function of leucocyte integrins (CD11/CD18) and failure of leucocytes to adhere to, and transmigrate through, vascular endothelium. The major defect is in neutrophil leucocyte mobilization.

Leukotrienes: pharmacologically active substances generated from arachidonic acid by the action of lipoxygenases.

Light chain: a polypeptide chain present in most immunoglobulin molecules. MW 22 kDa in man. Immunoglobulins, or their subunits if polymeric, are made up of two identical light chains linked to two heavy chains usually by disulphide bonds. Light chains are of two isotypes, κ and λ (see **κ (kappa) chain** and **λ (lambda) chain**) and a single immunoglobulin molecule or subunit always has two κ chains or two λ chains, never both. Light chain isotype is not related to immunoglobulin class differences.

Lupus erythematosus: skin disease with red scaly patches in exposed areas, especially butterfly-shaped area over the nose and cheeks. Not to be confused with systemic lupus erythematosus (SLE, q.v.) of which cutaneous lupus erythematosus is often a sign and to which it is related.

Lymph: the fluid that flows from all tissue cells of the body in lymphatic vessels, thus providing a medium for metabolic exchange and removal of waste products. Derived from the blood as an ultrafiltrate through the capillary walls, it returns to the blood stream, after passing through chains of lymph nodes (q.v.), by drainage from the thoracic duct into the vena cava.

Lymph node: small bean-shaped organ subdivided into a cortex and medulla and made up largely of lymphocytes and accessory cells, especially dendritic cells (interdigitating cells). The lymph node is a peripheral lymphoid organ and has both a lymphatic supply and a blood supply. The cortex is compartmentalized into T cell and B cell areas. Lymph nodes are distributed throughout the body, frequently in groups which drain lymph from a given area via afferent lymphatic vessels that pass into the node from peripheral tissue.

Lymphoblast: a blast cell of the lymphoid cell series with a nuclear pattern characterized by fine chromatin and basophilic nucleoli. Lymphoblasts are formed *in vivo* and *in vitro* following antigenic or mitogenic stimulation and divide to form populations of effector lymphocytes.

Lymphocyte: the cell type that carries receptors for, and recognizes, antigen and is therefore the mediator cell of specific immunity. There are two major forms of mature lymphocyte: the T lymphocyte, which generates the responses of cell-mediated immunity, and the B lymphocyte, which mediates humoral immunity and is the precursor of antibody-secreting cells. Unstimulated lymphocytes of both types are small cells (5–7 μm in diameter) with a large round or slightly indented nucleus and a narrow rim of cytoplasm.

Lymphocyte activation: the change seen when lymphocytes are cultured in the presence of a mitogen or of an antigen to which they are primed. The cells enter the cell cycle sequence, typically remaining in G1 for about 48 hours before entering the S phase and dividing. They increase in size, the cytoplasm becomes more extensive, and nucleoli are visible in the nucleus, which becomes less densely stained; after about 72 hours these cells resemble lymphoblasts. Activated lymphocytes differentiate into various functional forms, depending on the phenotype of the original cell, e.g. B cells to memory cells or plasma cells, T cells to cytotoxic T lymphocytes, helper T lymphocytes, cytokine-secreting cells, memory cells, etc.

Lymphocyte recirculation: the continuous passage of lymphocytes from blood to lymphoid tissues to lymph and thence back to blood. Recirculating cells are small resting cells.

Lymphocytosis: rise above normal of the number of lymphocytes, especially in blood.

Lymphoid cell series: cell series which includes lymphocytes, their precursors, e.g. thymocytes, and their progeny, e.g. plasma cells.

Lymphoid follicle (primary follicle): a tightly packed, spherical aggregation of cells in the cortex of a lymph node, the white pulp of the spleen or in other lymphoid tissues. Consists of a network of follicular dendritic cells, the spaces between which are packed with small recirculating B lymphocytes. Lymphoid follicles characterize the B cell areas of unstimulated lymphoid tissue, but during the secondary immune response to thymus dependent antigens, germinal centre develops within the follicle (secondary follicle).

Lymphoid tissues: body tissues in which the predominant cells are lymphocytes. They comprise the lymph, spleen, lymph nodes, thymus, Peyer's patches, pharyngeal tonsils, adenoids, and in birds the caecal tonsils (see **tonsils**) and bursa of Fabricius.

Lymphokine: any cytokine made and secreted by lymphocytes, commonly following antigen stimulation, e.g. IL-2, IL-4, etc.

Lymphoma: neoplastic disease of lymphoid tissue in which the abnormal cells are chiefly located in solid tissue, rather than found in the blood as in leukaemia, though there is overlap between the two.

Lymphoreticular tissue: synonym for lymphoid tissues.

Lysosome: a cytoplasmic organelle, limited by a membrane and containing hydrolytic enzymes (acid hydrolases). Present in many cells throughout the animal kingdom. Lysosomal enzymes are inert until released from the particle. These enzymes play an important part in intracellular digestion and are involved in many types of cell injury. They may also be released from the cell by exocytosis. For role of lysosomes in phagocytosis see **phagosome**.

Lysozyme: enzyme, first described by Fleming in 1922, present in and secreted by neutrophil leucocytes and macrophages and found in tears, nasal secretions, on the skin, and, in lower concentrations, in serum. Lyses certain bacteria, chiefly Gram positive cocci.

M

Mab (mAb): abbreviation for monoclonal antibody.

Macroglobulin: any globulin with a molecular weight above about 400 kDa. The best-studied serum macroglobulins are IgM.

Macroglobulinaemia: increase in level of macroglobulin in the serum. Usually refers to rise in IgM level. May follow antigenic challenge, e.g. in trypanosomiasis, or be due to paraproteinaemia, either primary, as in Waldenström's macroglobulinaemia, or secondary to diseases such as lymphoma or carcinoma, especially in the alimentary tract.

Macrophage: the mature cell of the mononuclear phagocyte system q.v. Macrophages are derived from blood monocytes which migrate into the tissues and differentiate there. They are strongly phagocytic of a wide variety of particulate materials, including microorganisms. They contain lysosomes, and possess microbicidal capacity. They are secretory cells which synthesize and release an enormous number of biologically active substances including cytokines, enzymes, inflammatory mediators, and microbicidal agents.

Macrophage activating factor: a generic term for any cytokine that activates macrophages, e.g. IFN-γ, TNF-α, GM-CSF.

Major histocompatibility complex (MHC): the collection of genes coding for the major histocompatibility antigens, see **class I MHC antigens, class II MHC antigens**.

MALT: mucosa associated lymhoid tissue.

Mantle zone (follicular mantle): a peripheral zone of small B lymphocytes found in a secondary follicle in which a germinal centre has developed. The mantle zone contains the small, recirculating, B cells which originally occupied the primary follicle but which have been pushed out as the germinal centre expands.

Marginal zone: a loosely packed area of T lymphocytes, B lymphocytes and macrophages which surrounds both the periarterial lymphatic sheath and the B cell follicles of the mammalian spleen q.v., particularly well-defined in rodents.

Mast cell: tissue cell (10–30 μm diameter) bearing high-affinity surface receptors for IgE (FcεRI) and strongly basophilic cytoplasmic granules, similar to, but smaller than, those of blood basophil leucocytes. The granules contain histamine, heparin, and tryptase, *inter alia*.

Masugi nephritis: experimental glomerulonephritis produced in one species (e.g. rat) by injection of antibody obtained from a second species (e.g. rabbit) that has been immunized with rat glomerular capillary basement membrane.

Maternal immunity (maternally transferred immunity): passive immunity (of humoral immunity type) acquired by the newborn animal from its mother. In man and other primates this is chiefly obtained before birth by the active transport of immunoglobulins across the placenta. The young of ungulates, in whom antibody is not transferred across the placenta, acquire it from the colostrum (q.v.), their intestines being permeable to immunoglobulins for a few days after birth. In mammals, secretory IgA in colostrum provides passive protection for the gut mucosa. The young of birds acquire maternal immunity from antibody in the egg yolk.

Medullary cord: area of the medulla of a lymph node close to the efferent lymphatic, composed largely of macrophages and, after antigenic stimulation, containing many plasma cells.

Medullary sinus: potential spaces in the medulla of a lymph node into which lymph drains before entering the efferent lymphatic.

Membrane attack complex: term used to denote the terminal complement components C5, C6, C7, C8 and C9, which associate to form the terminal attack complex ($\overline{C5b\text{-}9}$) on activation of either the classical complement pathway or the alternative pathway. $\overline{C5b\text{-}9}$ contains a hydrophobic region which allows it to insert into lipid bilayers and cause cell lysis.

Membrane immunoglobulin (B lymphocyte receptor; mIg; sIg): immunoglobulin on B cells in the form of a Type I transmembrane protein, thus synthesized by and acting as the antigen-specific receptor of the B cell. In unprimed cells, membrane immunoglobulin is of the IgM and IgD classes, but isotype switching (class switching) occurs during T cell-dependent immune responses, and these classes are replaced by IgG, IgA or IgE.

Memory cells: T lymphocytes or B lymphocytes which mediate immunological memory.

MHC: see **major histocompatibility complex**.

MHC restriction: the recognition by T lymphocytes of foreign antigen on the surface of a cell only in association with self-antigens of the major histocompatibility complex. $CD8^+$ T lymphocytes respond to foreign antigen in association with class I MHC antigens, whereas $CD4^+$ T lymphocytes respond to foreign antigen in association with class II antigens.

Microfilaments: long, fine strands about 5–8 nm wide composed of helically polymerized actin and found, usually as a network, in the cytoplasm or nucleus of eukaryotic cells, including all immunological cells.

Microfilaments are believed to mediate movement of the whole cell and of organelles within it. They form a contractile system which functions by the interaction of actin with actin-binding proteins and myosin.

Microglobulin: any globulin or globulin fragment of relatively low molecular weight (40 kDa or below). Has been used in case of low molecular weight proteins such as Bence–Jones protein in urine or similar proteins in serum. See **β_2 microglobulin**.

Microphage: Metchnikoff's term to describe polymorphonuclear leucocytes and other phagocytic cells of the myeloid cell series in contrast to macrophages q.v.

Microtubules: long hollow cylindrical structures (tubular in cross section) with an outer diameter of about 20–25 nm, found in the cytoplasm of eukaryotic cells including all cells of the immune system. Composed chiefly of tubulin.

Migration inhibition factor: the first immunologically important cytokine to be described was a factor released in delayed-type hypersensitivity reactions that inhibited macrophages from migrating out of capillary tubes. Still not unequivocally defined at the molecular level since several molecules with similar activities have been described.

Mixed leucocyte reaction (MLR): lymphocyte activation seen when mononuclear cells (mixtures of lymphocytes and accessory cells) from two genetically disparate individuals are cultured together *in vitro*. Due to a cell-mediated immune response of the cultured lymphocytes against foreign cell surface antigens. The strength of the reaction is directly related to the degree of incompatibility between the histocompatibility antigens of the two donors.

Monoclonal antibody: antibody produced by a single clone of cells or a clonally derived cell line, and therefore having a unique amino acid sequence. Commonly used to describe the antibody secreted by a hybridoma cell line, although strictly this is only monoclonal if one of the fusion partners is a non-producer. Very widely used as a specific reagent for the identification and study of antigens and particularly useful when the latter have not been purified.

Monocyte: a large, motile, amoeboid cell with an indented nucleus, precursor of the macrophage. Found in normal blood (200–800 cells per mm^3 or 2–10% of the total white cell count in humans). Derived from promonocytes in bone marrow and is the blood representative of the mononuclear phagocyte system. Monocytes remain in the blood for a short time (about 24 hours mean half-life) and then migrate into the tissues where they undergo further differentiation to become macrophages.

Monokine: generic term for secreted products of mononuclear phagocytes with regulatory effects on their own functions or those of others cells, e.g. IL-1, TNF-α.

Mononuclear cell: a vague term often used to refer to cells of the mononuclear, unlobulated phagocyte system or to lymphocytes, or to mixtures of the two, as seen in histological sections or fractionated blood leucocyte samples; in contrast to polymorphonuclear leucocytes.

Mononuclear phagocytes (mononuclear phagocyte system): a system of phagocytic cells of which the mature functioning form is the macrophage. The term mononuclear phagocyte system was introduced to replace the term 'reticuloendothelial system' which is now considered inaccurate. All mononuclear phagocytes are considered to share a common origin from the bone marrow promonocytes and to share a common function, i.e. phagocytosis and digestion of particulate material. If class II MHC antigen-positive they also act as accessory cells.

Mucosal mast cell: distinct population of mast cells found in mucosal tissues (especially intestine) characterized by production of specific serine protease and dependence on T lymphocyte-derived IL-3 for growth.

Myasthenia gravis: autoimmune disease characterized by progressive muscular weakness on exercise caused by faulty neuromuscular transmission. The patients' sera contain antibodies against the acetylcholine receptor on the post-synaptic membrane of the neuromuscular junction. These antibodies are the putative cause of the symptoms and can induce autoimmune myasthenia gravis in experimental animals.

Myeloid cell series: a series of bone marrow-derived cell lineages which include, as mature forms, the granular leucocytes (granulocytes) and the mononuclear phagocytes of blood.

Myeloma: a tumour of plasma cells, see **plasmacytoma, myelomatosis**.

Myeloma protein: immunoglobulin that is monoclonal produced by neoplastic plasma cells in myelomatosis in man, mouse and other species. Detected as an electrophoretically homogeneous para-protein.

Myelomatosis: disease characterized, in man, by neoplastic proliferation of plasma cells throughout the bone marrow. The neoplastic cells are monoclonal and produce large amounts of structurally identical immunoglobulin (paraprotein), usually of IgG or of IgA class though IgD and IgE have also been reported, forming a sharply localized band on serum electrophoresis and a characteristic monoclonal banding pattern on isoelectric focusing. Bence–Jones protein appears in the urine in a proportion of cases.

Myeloperoxidase: peroxidase found in the azurophil granules of the neutrophil leucocyte, which, together with hydrogen peroxidase and halide, forms a bactericidal system. Families with myeloperoxidase deficiency have been reported.

N

Naive lymphocyte: a lymphocyte that has not met antigen. An unprimed lymphocyte.

Native immunity: non-specific immunity resulting from

the genetic constitution of the host, e.g. immunity of man to canine distemper.

Natural antibody: antibody present in serum of normal individuals not known to have been immunized against the relevant antigen. Examples in man include the isohaemagglutinins of the ABO blood group system and antibodies to the Forssman antigen. They possibly result from an immune reaction against some closely related antigens of bacteria or food in the intestine of the infant.

Negative selection: clonal deletion of antigen-specific, receptor-bearing lymphocytes, resulting in permanent loss of the relevant cells and their progeny. Term usually applied to the deletion of self (autoantigen)-reactive T lymphocytes during development in the thymus, but may also refer to elimination of self-reactive B lymphocytes.

Neutrophil leucocyte: cell of myeloid cell series, the most numerous in normal peripheral blood (normal count in human blood 2500–7500 per mm^3 or 40–75% of the total white cell count). A motile, short-lived cell with multilobed nucleus and a cytoplasm filled with azurophil granules q.v. and specific granules q.v. which do not take up acidic or basic dyes strongly (hence name). Actively phagocytic with an efficient microbicidal capacity. Also reacts vigorously to chemotactic stimuli. Name usually abbreviated to neutrophil. Also known as polymorph or polymorphonuclear leucocyte or granulocyte (eosinophil leucocytes and basophil leucocytes also have mutilobed nuclei and cytoplasmic granules). It should be noted that in species other than man, the blood cells acting functionally as neutrophils may have granules whose staining properties are not 'neutrophil'. These are sometimes known as heterophil granulocytes.

Nitric oxide (NO): a highly reactive, colourless and odourless gas. In biological systems it has a half-life between 3 and 15 seconds and is derived from molecular oxygen and the guanidino nitrogen of L-arginine, in a reaction catalysed by NO synthase (NOS). Certain cells, e.g. vascular endothelium, generate NO through a constitutive NOS (cNOS, calcium-dependent). This mediates vascular relaxation. Functions such as neurotransmission and platelet aggregation are mediated through a similar constitutive pathway. Another form of NOS is inducible (iNOS, calcium-independent) by immunological stimuli such as IFN-γ, TNF-α and lipopolysaccharide to produce large amounts of NO which can be cytotoxic. This is an important microbicidal function of activated macrophages.

NK cell (natural killer cell): cytotoxic lymphocytes which lack the phenotypic markers of both T cells (TCR, CD3) and B cells (membrane immunoglobulin). Normally present as a minority population in blood. They contain prominent cytoplasmic granules and are morphologically distinguishable as large granular lymphocytes.

Non-specific immunity: mechanisms for the disposal of foreign and potentially harmful macromolecules, micro-oganisms or metazoa which do not involve the recognition of antigen and the mounting of a specific immune response. Such mechanisms include the action of lysozyme or anti-viral interferons, phagocytosis and chemical and physical barriers to infection. Protective immunity in invertebrates is of the non-specific type. Specific and non-specific immunity are so closely linked in vertebrates that it is often impossible to dissociate their actions.

NOS (nitric oxide synthase): see **nitric oxide**.

Nu nu mice: see **nude mice**.

Nude mice: mice with congenital absence of the thymus, and whose blood and thymus-dependent areas of the lymph nodes and spleen are depleted of T lymphocytes. These mice are homozygous for the gene 'nude' abbreviation nu, hence nu nu, and have no hair.

Nurse cell: see **thymic nurse cell**.

O

Opsonin: factor present in plasma and other body fluids which binds to particles, especially cells and microorganisms, and increases their susceptibility to phagocytosis.

Oral tolerance: the induction of specific immunological tolerance by oral administration of antigen. This is the usual result of feeding soluble antigens (e.g. food proteins) to a naive animal (non-immune animal).

Orthotopic graft: tissue or organ grafted to a site normally occupied by that tissue or organ.

Oxidative metabolic burst (respiratory burst): the rapid generation of reactive oxygen intermediates (q.v.) in neutrophil leucocytes and mononuclear phagocytes following phagocytosis and related stimuli (e.g. chemotactic factors).

P

Paracortex: the thymus-dependent area of a lymph node.

Paraprotein: any abnormal protein in serum but usually refers to immunoglobulin derived from an abnormally proliferating clone of neoplastic plasma cells. Paraproteins are normally monoclonal and appear as a sharply localized band on serum electrophoresis.

Paraproteinaemia: presence in serum of paraprotein. A diagnostic feature of myelomatosis and Waldenström's macroglobulinaemia.

Paratope: antigen-binding site of an antibody or TCR.

Passive immunity: immunity due, not to the production of a specific immune response by an individual, but to the presence in his tissues of antibody or primed lymphocytes derived from another immune individual. Examples are the immunity of the neonate against many infectious agents due to placentally or colostrally transferred maternal antibody (see **maternal immunity**) and the use of antitoxins to give protection against diphtheria or tetanus. In the case where lymphocytes are transferred (adoptive transfer), the term adoptive immunity is often used.

Passive transfer: transfer of immunity or hypersensitivity from an immune or primed donor to a previously non-immune animal by injection either of antibody or primed lymphocytes.

Perforin: protein found in granules of cytolytic cells such as cytotoxic T lymphocytes, in which it is formed during immune responses, and in NK cells, in which it is present constitutively. Normally found as a monomer that has no cytolytic activity, but on contact with target cells, the granules fuse with the lymphocyte plasma membrane, the perforins are released and, in the presence of Ca^{2+}, they form amphipathic ring polymers of 12–18 molecules which resemble the membrane attack complex of complement. These insert into the membrane of the target cell causing lysis of the latter.

Periarteriolar lymphatic sheath (PALS): the thymus-dependent area of the white pulp of the spleen q.v.

Peyer's patches: lymphoepithelial nodules in the submucosa of the small intestine, more prominent in the ileum (lower part) than in the jejunum. Contain lymphocytes including the precursors of IgA-producing B cells, germinal centres and thymus-dependent areas. These lymphoid tissues are separated from the lumen of the intestine by a single layer of columnar eptihelium (follicle-associated or dome epithelium). containing specialized cells (microfold-M-cells) which take up antigen from the lumen and transport it into the lymphoid areas. Peyer's patches are believed to be the principal site for induction of intestinal immune responses. There is distinct evidence that ileal Peyer's patches in ruminants could be a primary B cell organ.

Phagocyte: a cell that is able to ingest, and often to digest, large particles such as effete blood cells, bacteria, protozoa and dead tissue cells.

Phagocytosis: 'cell eating'. The ingestion of cells or particles by inclusion in a cytoplasmic phagosome q.v. In mammals, only cells of the mononuclear phagocyte system and neutrophil leucocytes are 'professional' phagocytes, although other cells may, on occasion, show facultative phagocytosis.

Phagolysosome: the product of the fusion of lysosomes with a phagosome. Materials included within it may be digested by hydrolysis. Following such digestion the vesicle may continue to function and is sometimes called a 'secondary lysosome'.

Phagosome: an intracellular vesicle in a phagocyte q.v. formed by invagination of the cell membrane and containing phagocytosed material. The latter is digested by lysosomal enzymes liberated into the vesicle following fusion of the phagosome with cytoplasmic lysosomes. The structure which results from this fusion is known as a phagolysosome.

Pharyngeal tonsil: see **tonsil**.

Pigeon fancier's lung: see **bird fancier's lung**.

Plasma: the fluid phase of blood in which the red and white blood cells are suspended.

Plasma cell: cell of B lymphocyte lineage with a major role in antibody synthesis and secretion. The cytoplasm is basophilic, rich in RNA and protein, and packed with rough-surfaced endoplasmic reticulum. The nucleus is often round and eccentrically placed, with clumped chromatin giving a 'clock face' or 'cartwheel' appearance. The plasma cell is the end cell of the B lymphocyte line. It is the major immunoglobulin secreting cell type and therefore the classical cell of humoral immunity. Present in lymphoid tissue and increased in numbers in the draining lymph node and at the site of entry of antigen following antigenic stimulation.

Plasmacytoma: a localized tumour of plasma cells in contrast to myelomatosis which is a diffuse plasma cell tumour in bone marrow and elsewhere.

Platelet (thrombocyte): a small non-nucleated 'cell' (3 μm diameter) found in mammalian blood, and derived from the megakaryocytes of bone marrow. It is important in blood coagulation as a generator of thromboplastin on contact with foreign surfaces and is essential for haemostasis and thrombosis.

PMN: see **polymorphonuclear leucocyte**.

Polymorphonuclear leucocyte: synonym for neutrophil leucocyte which, in its mature form, has a multilobed nucleus though other cell types, e.g. eosinophil leucocytes may also have multilobed nuclei.

Positive selection: the process which ensures that mature antigen-reactive T lymphocytes will only recognize foreign antigenic peptide when presented in association with a self major histocompatibility complex (MHC) molecule. This occurs during development in the thymus because T cells expressing T cell receptors (see **TCR**) capable of interacting with self MHC molecules are rescued from programmed cell death. Positive selection also occurs in germinal centre B cells which recognize their specific antigen in the form of immune complexes on follicular dendritic cells provided that accessory molecules are also present. High-affinity B cells are thus rescued from programmed cell death.

Post-capillary venules: small vessels through which blood flows after leaving the capillaries and before reaching the veins. It is between the endothelial cells of post-capillary venules (rather than capillaries) that most leucocytes migrate into inflammatory sites. The high endothelial venules (q.v.) of lymph nodes are specialized post-capillary venules through which the recirculating pool of lymphocytes pass from blood to lymph.

Pre-B lymphocyte (pre-B cell): B cell precursor which has developed from a pro-B lymphocyte, and which has begun to rearrange the heavy chain genes of immunoglobulin and to synthesize μ chains.

Pre-T lymphocyte (pre-T cell): bone marrow-derived precursor of T lymphocyte which has not yet undergone education or development in the thymus.

Primary follicle: see **lymphoid follicle.**

Primary immune response: the response of the animal body to an antigen on the first occasion that it encounters it.

Primary lymphoid organs: those lymphoid tissues (organs) that are essential to the ontogeny of the immune response, i.e. the thymus and, in birds, the bursa of Fabricius.

Primary lysosome: a lysosome prior to fusion with a phagosome.

Primary nodule: see **lymphoid follicle.**

Primed: (1) of a whole animal. Exposed to antigen in such a way that the antigen makes contact with the lymphoid tissue so that the appropriate responsive cells are activated. Further contact of a primed host with antigen usually results in a vigorous, rapid, secondary immune response. (2) Of lymphocytes. A primed lymphocyte is one that has been specifically activated in respect of a given antigen and can divide and give rise to effector lymphocytes and memory cells.

Primed lymphocyte: lymphocyte, primed (q.v.) specifically to an antigen.

Priming: (1) of immune responses. The events that follow initial contact with antigen. (2) Of neutrophil leucocytes. Addition of a stimulus at a dose too low to stimulate an oxidative metabolic burst, but which primes the cell so that on addition of a second stimulus, which may not be identical to the first, a much larger metabolic burst is obtained than in unprimed cells.

Privileged sites: sites in the body lacking normal lymphatic drainage and into which antigens, or tissue grafts, can be placed without stimulating an immune response, e.g. the central nervous system, the anterior chamber of the eye, and the cheek-pouch of the hamster.

Pro-B lymphocyte: the most immature identifiable precursor of the B lymphocyte.

Processing: see **antigen processing.**

Programmed cell death: death of cells due to activation of a genetic programme that instructs the cell to commit suicide. Requires new gene expression. Frequently takes the form of apoptosis, but the latter does not always involve gene expression by the dying cell. Important generally in developmental biology and, in immunology, in selection of lymphocytes during maturation.

Properdin (P; Factor P): protein of the alternative pathway of complement activation. Exists in the circulation as a mixture of polymers (monomer 53 kDa). Binds and stabilizes alternative pathway C3 convertase and C5 convertase preventing their spontaneous dissociation.

Prostaglandins: biologically active lipids generated by the action of cyclo-oxygenases on arachidonic acid. There is a large number of prostaglandins, which have a variety of activities as inflammatory mediators. The actions of different prostaglandins may be mutually antagonistic.

Proteasome: a proteolytic complex that degrades proteins within the cytosol (that is, the cytoplasm other than organelles and membranes) and nuclear proteins. Implicated in antigen processing for presentation by class I MHC.

Purine nucleoside phosphorylase deficiency (PNP deficiency): an autosomal recessive defect due to inheritance of a mutant form of PNP. Toxic metabolites accumulate in T cells and there is primarily a deficiency of cell-mediated immunity with normal numbers of B cells, though antibody production may be impaired as a secondary effect.

R

Receptor: macromolecule, usually a protein, which contains a site capable of selectively combining, with a varying degree of specificity, with complementary molecules known as ligands.

Recirculating pool: all the lymphocytes that continuously recirculate between blood and lymph, see **lymphocyte recirculation.**

Recirculation: see **lymphocyte recirculation.**

Recombinant strains: strains of animals (e.g. mice, sheep) in which the genes in the parental strains have either been reassorted (i.e. recombined) by breeding techniques (to give recombinant inbred strains) or altered by direct DNA manipulation. Recombinant inbred (RI) strains provide large numbers of virtually genetically uniform and homozygous mice in which the effects of reassorting various parental genes (e.g. heavy chain genes) can be studied.

Rejection: see **immunological rejection.**

Repertoire: (1) the number of different antibody or TCR (T cell receptor) variable region sequences produced by the immune system of a given species. (2) The number of different epitopes recognized by all the antibodies or T cell receptors produced by the immune system of a species. These two definitions are not synonymous since there is redundancy within the system, i.e. several antibody molecules or T cell receptors with different variable regions are able to recognize the same epitope. The number of different epitope-recognizing receptors on B cells or T cells produced by an individual animal or human at any one time is much less than the number that the immune system is capable of producing – the 'potential repertoire'.

Resident macrophage: macrophage present at a site in the absence of a known eliciting stimulus.

Respiratory burst: the increase in anaerobic glycolytic metabolism and oxygen consumption which occurs following activation of phagocytes, e.g. neutrophil leucocytes, by chemotactic factors, phagocytosis, etc., and which is accompanied by enhanced NADPH oxidase activity leading to generation of microbicidal reactive oxygen intermediates.

Reticulum cells: cells that, together with reticular fibres, make up the framework or stroma of lymphoid tissues such as the spleen and lymph nodes and of the bone marrow.

Reticular dysgenesis: the most complete form of severe combined immunodeficiency syndrome (SCID) in which there is a defect of maturation of all leucocytes, i.e. lymphocytes, granulocytes and monocytes.

Runt disease: disease which develops after injection of allogeneic lymhocytes into immunologically immature experimental animals. Characterized by loss of weight, failure to thrive, diarrhoea, splenomegaly and often death. An example of a graft-versus-host reaction.

S

Scavenger receptor: a type II transmembrane protein (trimeric) found on macrophages. There are two forms of different molecular weights. They bind a number of modified proteins such as acetyl-LDL and maleyl-HSA and may have a general role in scavenging modified proteins.

Scavenger receptor superfamily: a number of proteins other than the scavenger receptor have extracellular domains of the scavenger receptor type. They include CD5 and CD6.

SCID (severe combined immunodeficiency): a rare immunodeficiency state in infants presenting early with severe infections, diarrhoea and failure to thrive. Both humoral immunity and cell-mediated immunity are defective. There are various forms, e.g. (a) X-linked SCID in which there is a complete absence of T cells and T cell precursors but functional B cells are present; this form may be associated with mutations in the gene for the γ chain of IL-2R; (b) autosomal recessive forms including adenosine deaminase deficiency; and (c) in some cases, particular chains of the CD3 molecule are absent. The disease can be corrected by bone marrow transplantation, suggesting a stem cell defect (involving T cell precursors prior to migration to the thymus).

SCID mouse (severe combined immunodeficiency mouse): a mouse homozygous for a recessive mutation (SCID) on chromosome 16. Such mice have low numbers of both B cells and T cells, lack immunoglobulin in their serum and are deficient in both humoral immunity and cell-mediated immunity. Both T and B lymphocytes have a defective capacity to rearrange the genes coding for antigen receptors.

Secondary follicle: lymphoid follicle that contains a germinal centre. More frequently observed in secondary immune response than in primary immune response, hence name.

Secondary immune response: response of the immune system to an antigen to which it has already been primed and therefore has memory. There is very rapid production of large amounts of antibody over a few days followed by a slow exponential fall. The response of cell-mediated immunity follows a similar pattern.

Secondary lymphoid tissues: those lymphoid organs that are not essential to the ontogeny of the immune response, i.e. spleen, lymph nodes, tonsils, Peyer's patches.

Secondary lysosome: see **phagolysosome**.

Secretory IgA: the form of IgA found in external body secretions such as intestinal mucus, colostrum, milk, saliva, sweat and tears; mainly in the form of a dimer with a secretory piece bound to it. Thus distinct from that found in the serum, which has no secretory piece and, in humans, is predominantly monomeric.

Secretory piece: polypeptide of molecular weight 60 kDa found attached to dimers of the secretory form of IgA (see **secretory IgA**). Structurally unrelated to immunoglobulins and synthesized by epithelial cells in the gut, lung, mammary or other secretory tissues, not the plasma cells that synthesize the immunoglobulin. Has strong affinity for mucus thus prolonging retention of IgA on mucous surfaces. May also inhibit the destruction of IgA by enzymes in the digestive tract.

Selectins: a family of cell adhesion molecules. Type I transmembrane proteins which mediate the initial adhesion of leucocytes to vascular endothelium by low-affinity interactions and slow the cells up, causing them to roll along the side of the vessel. The leucocytes then become bound in a second step mediated by integrin superfamily proteins and migrate through the vessel wall.

Self tolerance: immunological tolerance to autoantigens. Such tolerance to self antigens accessible to the lymphoid tissues is thought to be acquired normally during fetal life.

Sequestered antigen: any antigen or epitope which is hidden from contact with immunologically competent cells and thus cannot stimulate an immune response. Antigens in certain privileged sites may be sequestered from the immune system.

Serotonin (5-hydroxytryptamine, 5HT): causes smooth muscle contraction, increased vascular permeability and vasoconstriction of larger vessels. Found in platelets and mast cells. It is also an important neurotransmitter.

Serum: fluid expressed from a blood clot as it contracts after coagulation of the blood. The essential difference between plasma and serum is that the latter does not contain fibrinogen.

Serum sickness: a hypersensitivity reaction to the injection of foreign antigens in large quantity, especially those contained in antisera used for passive immunization. The symptoms are due to the localization in the tissues of soluble immune complexes formed between antibody produced during the developing immune response and the large quantities of antigen still present (type III hypersensitivity reaction).

Severe combined immunodeficiency syndrome: see **SCID**.

Sezary syndrome: a T cell lymphoma with erythroderma

and other skin lesions. There are circulating T lymphoma cells with a characteristic irregular nuclear morphology. Associated with mycosis fungoides.

sIg: surface immunoglobulin. See **membrane immunoglobulin**. The prefix s is ambiguous and used for 'secretory' and 'soluble', etc., e.g. sIgA for secretory IgA found in external body secretions.

sIgA: see **secretory IgA**.

Signal peptide: see **leader peptide**.

SLE (systemic lupus erythematosus): autoimmune disease of man, characterized by widespread focal degeneration, 'fibrinoid necrosis' of connective tissue and disseminated lesions in many tissues including skin, joints, kidneys, pleura, peripheral vessels, peripheral nervous system, blood, etc. The glomerular lesions are particularly serious and have been shown to result from deposition of immune complexes in the glomerular capillaries. Cf. **lupus erythematosus**.

Small lymphocyte: cell 5–8 μm in diameter with a deeply staining nucleus and narrow rim of cytoplasm. The size simply indicates a resting (G0) cell and despite the apparent identity of morphology, small lymphocytes differ in origin and function. See **T lymphocyte**, **B lymphocyte**.

Somatic hypermutation: mutations that occur rapidly in the V-region (variable region) genes of the light chain and heavy chain during the formation of memory B cells, giving many more V-region sequences than are found in the germline genes from which they originated. Somatic hypermutation takes place in germinal centres.

Specific immunity: as a general concept, specific immunity is a non-susceptibility to re-infection by a pathogen, that develops in individuals who survive a first encounter with that same pathogen.

Specificity: a term defining selective reactivity between substances, e.g. of an antigen with its corresponding antibody or primed lymphocyte.

Spleen: a solid, encapsulated organ, deep red in colour, found in the upper abdomen. It is a peripheral lymphoid organ which has a major arterial supply and acts as a blood filter. The spleen comprises two fundamentally distinct types of tissue, the 'white pulp' (Malpighian body or corpuscle) which is a lymphoid tissue, and the 'red pulp'. (a) The white pulp forms a cuff or sheath of tissue round the arterioles which consists chiefly of lymphocytes together with a smaller number of splenic dendritic cells. The central zone of the white pulp (nearest the arterioles) is a thymus-dependent area and contains T cells. More peripherally are found B cells in spherical lymphoid follicles. Germinal centres (q.v.) develop in these follicles following antigenic stimulation. The arterioles terminate in venules between the white and the red pulp (the marginal zone) and it is from these venules that lymphocytes migrate into and populate the white pulp. The marginal zone itself is rich in B memory cells which are well placed to interact with antigens carried in the blood. (b) The red pulp consists of large numbers of blood-filled sinusoids in which phagocytosis of effete erythrocytes takes place. It also functions as a reserve site for haemopoiesis. It is rich in macrophages, and plasma cells are also found there.

Stem cell: the progenitor of all the cells of the immune system. A large cell with a rim of intensely RNA-rich (pyroninophilic) cytoplasm and pyroninophilic nucleoli within a leptochromatic nucleus (i.e. having narrow strands of chromatin) found in bone marrow and other haemopoietic tissues.

Stromal cells: see **reticulum cells**.

Subset: term used to classify functionally or structurally different populations of cells within a single cell type.

Superantigen: designation of a class of antigen that activates all T cells bearing one or more particular TCR Vβ sequences. Superantigens bind with high affinity to a region of class II MHC molecules that is outside the groove and to a region of the TCR Vβ chain away from the antigen-binding site, forming a direct link between MHC and TCR.

Superfamily: a group of proteins which show structural homology and putatively evolved from the same primordial gene.

Superoxide anion (O_2^-): oxygen molecule that carries an extra unpaired electron, and is therefore a free radical. Generated in neutrophil leucocytes and mononuclear phagocytes by one-electron-step reduction of molecular oxygen.

Supressor T lymphocyte: presently ill-defined subpopulation of T lymphocytes which directly suppresses the immune response.

Surface immunoglobulin: see **membrane immunoglobulin**.

Surrogate light chain: a protein found in precursors of B cells (pro-B lymphocytes and pre-B lymphocytes) but not in mature B lymphocytes. Derived from the genes V_{preB} and λ_5 which do not undergo translocation (see **immunoglobulin gene**) but are transcribed separately. The V_{preB} and λ_5 polypeptides form light chain-like structures which associate with the μ heavy chain (of IgM) made by B cell precursors before they start to make functional light chains.

Syngeneic (syngenic): genetically identical, usually applied to grafts made within an inbred strain.

Systemic lupus erythematosus: see SLE.

T

T–B cell cooperation: a process required for production of normal levels of antibody to thymus-dependent antigens, i.e. most antigens. Requires physical contact of helper T lymphocytes and B lymphocytes with accessory cells.

T cell: see **T lymphocyte**; the two names are interchangeable.

T cell-dependent antigen: see **thymus-dependent antigen**.

T cell-dependent area: see **thymus-dependent area**.

T cell-dependent immune response: the immune response of cell-mediated immunity. The latter term was introduced before T cells were discovered and is inexact but universally used.

T cell receptor: see **TCR.**

T cell subset: see **T lymphocyte subset.**

T-dependent antigen: see **thymus-dependent antigen.**

T-dependent area: see **thymus-dependent area.**

T helper cell: see **helper T lymphocyte.**

Thrombocyte: see **platelet.**

T-independent antigen: see **thymus independent antigen.**

T lymphocyte (also called a T cell; the two names are interchangeable): lymphocyte that is derived from the thymus. Carries an antigen-specific T cell receptor, and is $CD3^+$. T lymphocytes play a major role (a) as antigen reactive cells and effector cells in cell-mediated immunity; and (b) by cooperating with B lymphocytes in antibody production (humoral immunity) against thymus-dependent antigens.

T lymphocyte antigen receptor: see **TCR.**

T lymphocyte–B lymphocyte cooperation: see **T–B cell cooperation.**

T lymphocyte repertoire: the number of different epitopes to which the T lymphocytes of an individual animal are capable of responding. Cf. **B lymphocyte repertoire** and see **repertoire.**

T lymphocyte subset: group of T lymphocytes which are identified phenotypically by specific cell surface antigens and are characterized by a particular function.

T lymphocyte–T lymphocyte cooperation: postulated interaction between different T lymphocyte subsets in activation of cell-mediated immune responses. See also **T–B cell cooperation.**

TAP (transporter associated with antigen processing): proteins derived from MHC gene-associated *Tap* genes. TAP proteins transport the peptides generated from cytoplasmic cleavage of endogenously synthesized proteins across the membrane into the lumen of the endoplasmic reticulum, where they associate with, and stabilize, class I MHC antigen molecules. TAP proteins are essential for presentation of antigen with class I MHC.

TCR (T cell receptor; T lymphocyte receptor TcR): the molecule on the surface membrane of T lymphocytes capable of specifically binding antigen in association with MHC antigens. The molecule belongs to the Ig superfamily and consists of two Type I transmembrane protein disulphide-linked polypeptide chains of molecular weight 40–50 kDa. The majority ($\pm 90\%$) of human T lymphocytes express TCR composed of an $\alpha\beta$ heterodimer, while the remaining, mutually exclusive, population of T cells expresses a $\gamma\delta$ TCR. Each chain contains a variable (V) and constant (C) region, homologous to those found in immunoglobulin molecule variable regions and constant regions. In addition, the β and δ chains contain junctional (J) and diversity (D) regions which are analogous to those in immunoglobulin molecules. The α and γ chains also contain junctional regions. The antigen-binding site is formed primarily from the combination of the D and J regions, with a small contribution from the V region, and is responsible for interacting both with foreign peptide and self MHC molecules. The TCR is always non-covalently linked in the T cell membrane to the CD3 molecule, q.v.

TD antigen: see **thymus-dependent antigen.**

T_{DTH} lymphocyte: effector T lymphocyte in delayed-type hypersensitivity reaction. Usually $CD4^+$ and putatively may belong to the T_H1 subset of $CD4^+$ cells. Function depends on production of cytokines such as IFN-γ.

T_H lymphocyte (helper T lymphocyte): see T_H1 cells and T_H2 cells.

T_H1 cells: functional T lymphocyte subset of $CD4^+$ cells, characterized by their production of the cytokines IL-2, IFN-γ and TNF-α, -β and by failure to produce IL-4, IL-5 and IL-10. A polarization of the immune response towards T_H1 cell activity can occur in mice (and possibly man) and is associated with cell-mediated immunity *in vivo* including delayed-type hypersensitivity and T cell proliferation. The selective activation of T_H1 cells is favoured by the presence of IFN-γ and IL-12 and is inhibited by IL-4 and IL-10.

T_H2 cells: functional T lymphocyte subset of $CD4^+$ cells which, in mice (and possibly man) is distinct from the T_H1 subset, q.v. T_H2 cells are characterized by the production of IL-4, IL-5 and IL-10 and by the failure to produce IL-2 and IFN-γ. T_H2 cell activation is associated with helper activity for antibody production (see **helper T lymphocyte**) and the inhibition of T_H1-mediated responses. IL-4 is essential for the growth and differentiation of T_H2 cells.

Thoracic duct: duct that returns lymph to the blood.

Thymic cortex: see **thymus.**

Thymic epithelial cells: epithelial cells found in the thymic cortex and medulla (see **thymus**) and derived from the third branchial pouch. They are believed to control the education of T lymphocyte precursors by presenting self-peptides bound to MHC (major histocompatibility complex) molecules and by secreting cytokines which stimulate lymphocyte growth.

Thymic hypoplasia: congenital cell-mediated immunity deficiency syndrome in human infants. Often associated with hypoparathyroidism (see **Di George's syndrome**). The blood lymphocyte count is low, T lymphocytes being completely absent and the thymus-dependent areas of lymphoid tissues are depleted of lymphocytes. T-dependent B lymphocyte responses are also impaired. Recurrent opportunistic infections of the skin and respiratory tract may be present.

Thymic medullary hyperplasia: term used to signify the presence of germinal centres in the medulla of the thymus especially in myasthenia gravis. The term does not imply that thymic weight is increased.

Thymic nurse cell: a subset of large thymic epithelial cells with which thymocytes come into close contact.

Role in maturation and differentiation of T lymphocytes.

Thymocyte: any lymphocyte found within the thymus.

Thymus: the organ essential for the development of T lymphocytes. In mammals, it consists of two lobes situated in the anterior part of the thorax, ventral to the trachea and great vessels and is derived from the third and fourth branchial pouches. In birds, it is distributed along the neck as a series of lobes. Histologically, it consists mainly of lymphocytes distributed into distinct cortical and medullary areas on a network of reticular cells (stromal cells). In the cortex, these lymphocytes ('thymocytes') are rapidly dividing progeny of T lymphocyte precursors derived from haemopoietic tissues. During development in the cortex thymocytes are initially $CD4^-CD8^-$, but then become $CD4^+CD8^+$ and eventually develop into either $CD4^+CD8^-$ or $CD4^-CD8^+$, at which point they accumulate in the medulla before leaving to populate the peripheral lymphoid organs as mature T cells. $CD4^-CD8^-$ thymocytes begin to rearrange T cell receptor (TCR) genes randomly and, depending on the surface receptor expressed, undergo positive selection and negative selection. As a result 95% of the thymocytes produced never leave the thymus and die. In addition to lymphocytes, the thymus contains large numbers of epithelial cells, stromal cells and accessory dendritic cells which play important roles in positive and negative selection, both by presentation of self MHC (major histocompatibility complex) molecules and by production of growth factors for lymphocytes. The thymus functions mainly in the fetal and early neonatal period, and animals thymectomized at birth or congenitally athymic (see **nude mice, thymic hypoplasia**) have an absence of T lymphocytes. After puberty, the thymus atrophies, but may continue to produce new T lymphocytes and animals thymectomized during adult life gradually become deficient in T cells.

Thymus-dependent antigen (T-dependent antigen): an antigen that does not stimulate an antibody response in animals lacking a thymus. Cooperation with helper T lymphocytes is required in order for B lymphocytes to respond to such antigens by maturation into antibody-forming cells. Most proteins and other antigens which present a diversity of epitopes are thymus-dependent.

Thymus-dependent area: those areas of the peripheral lymphoid organs that appear selectively depleted of lymphocytes in neonatally thymectomized animals, and in babies and animals with congenital aplasia of the thymus (see **thymic hypoplasia** and **nude mice**). Anatomically, these areas are situated in the mid cortex (paracortical area) of lymph nodes, the centre of the Malpighian corpuscle of the spleen and in the internodular zone of Peyer's patches. In normal subjects they are mostly occupied by small lymphocytes of the recirculatory pool (see **lymphocyte recirculation**) which enter them by crossing high endothelial venules. They also contain various accessory cells including macrophages and dendritic cells (interdigitating cells).

Thymus-dependent cells: population of lymphocytes whose normal development depends on the presence of the thymus at birth, i.e. T lymphocytes q.v.

Thymus-derived cells: T lymphocytes. Most T lymphocytes, but perhaps not all, must mature in the thymus.

Thymus-independent antigen (T-independent antigen): an antigen that is able to stimulate B lymphocytes to produce antibody without the cooperation of T lymphocytes. An antibody response to such antigens can be stimulated in an animal lacking a thymus. T-independent antigens are, frequently, repeating polymers which present an array of identical epitopes to the lymphocyte.

Tingible body macrophage: macrophage within a germinal centre that has phagocytosed apoptotic B lymphocytes. The 'tingible bodies' (deeply staining debris) are remnants of ingested cells.

TNF (tumour necrosis factor): note that TNF-α and TNF-β are also frequently called TNF and LT (lymphotoxin) respectively. Thus, in many publications the designation TNF, without a following Greek letter, refers to the cytokine, TNF-α.

TNF family: a group of related proteins all of which usually exist as trimers and which bind to members of the TNFR superfamily. The family includes TNF-α and TNF-β.

TNFR superfamily (tumour necrosis factor receptor superfamily): a family of molecules named the TNFR superfamily by immunologists and the NGFR (nerve growth factor receptor) superfamily by others. It contains TNFR I and II, and NGFR, among others.

Tolerance: see **immunological tolerance**.

Tolerogenic: capable of inducing immunological tolerance.

Tonsil: accumulations of lymphoid tissue found in invaginations of the mucous membrane in the area between the mouth and the pharynx. Tonsils vary in extent in different species, being extensive in man and the horse and small in cattle. They are not found in mice. In man they form distinct organs. The most prominent of these are the two palatine tonsils; tubal, lingual and a single nasopharyngeal tonsil complete a ring round the area. The tonsil is a peripheral lymphoid organ. It contains a high proportion of B lymphocytes and is almost always rich in germinal centres. In birds, aggregations of lymphoid tissue with germinal centres found in each caecal wall near the point at which the twin caeca enter the junction of the large and small intestines are called 'caecal tonsils'.

Transfusion reaction: disease or physiological disturbance following transfusion of blood. Often due to a specific immune reaction of the recipient against antigens on the donor's red blood cells. In man, the most common and severe form of such reactions is due to ABO blood group system (q.v.) incompatibility as the recipient's plasma contains natural antibodies against the ABO antigens not present on his own cells. After repeated transfusions antibodies may be formed against other antigens on donor cells and cause reactions.

Transgene: a gene that has been transfected into the germ line to form a transgenic organism, e.g. a transgenic mouse.

Transgenic mouse: a mouse that carries a transgene that can be passed on to succeeding generations. The gene (DNA) is microinjected into fertilized eggs which are allowed to develop *in utero*. If the gene has been introduced into the germ line, the mice, as adults, will pass it on to their offspring. The mice that develop from transfected eggs are screened for the gene and selected mice are bred to achieve stable transmission of the gene.

Transplantation antigen: see **histocompatibility antigen**.

Transplantation immunology: the study of the immune response following transplantation of tissue from donor to recipient. Very largely, the study of cell-mediated immunity in this situation.

Type I hypersensitivity reaction: term used in Gell and Coombs' classification of hypersensitivity reactions. In the type I reaction, antigen combines with antibody which is fixed passively to the surfaces of cells, usually mast cells, and causes the release of vasoactive substances, thus synonymous with immediate hypersensitivity and anaphylaxis. The cell-bound antibody involved is usually IgE; the antigen is often known as an allergen. Typical diseases are hay fever and asthma.

Type II hypersensitivity reaction: term used in Gell and Coombs' classification of hypersensitivity reactions. In the type II reaction, antibody reacts either with a cell surface antigen or with an antigen or hapten which has become attached to the cell surface. Typical diseases include Goodpasture's syndrome and transfusion reactions. If the antibody is complement-fixing antibody, cell lysis occurs.

Type III hypersensitivity reaction: term used in Gell and Coombs' classification of hypersensitivity reactions. In this reaction the tissue damage is mediated by immune complexes, particularly soluble complexes formed in slight antigen excess. Typical diseases include serum sickness, extrinsic allergic alveolitis and systemic lupus erythematosus (SLE). In such diseases the complexes are deposited in the blood vessel walls and become surrounded by inflammatory cells, especially neutrophil leucocytes in the acute phase. Later, these are replaced by a mononuclear cell infiltrate.

Type IV hypersensitivity reaction: term used in Gell and Coombs's classification of hypersensitivity reactions. In the type IV reaction, primed lymphocytes react with antigen at the site of its deposition resulting in the formation of a lymphocyte–macrophage granuloma, e.g. in pulmonary tuberculosis. Circulating antibody is not involved in this reaction.

Type I transmembrane protein: a membrane protein in which the N-terminal region is extracellular, there is a single hydrophobic membrane-spanning region, and the C-terminal region is intracytoplasmic. A large majority of membrane-spanning proteins are of this type.

Type II transmembrane protein: a membrane protein in which the C-terminal region is extracellular, there is a single hydrophobic membrane-spanning region, and the N-terminal region is intracytoplasmic.

Type III transmembrane protein: a membrane protein which is folded so that the molecule crosses the membrane more than once. An important group which does this is the rhodopsin superfamily and the chemokine family. These have seven membrane-spanning domains and their N-terminal regions are outside, and the C-terminal regions inside, the cell. Another group, with four membrane-spanning domains (the TM4 superfamily), has both N- and C-termini within the cytoplasm. The β chain of the high affinity receptor for IgE (FcεRI) is of this type.

U

Unprimed: having never had contact with or responded to a given antigen (of animals, cells, etc.). The terms 'naive' and 'virgin' are also used to describe unprimed lymphocytes.

V

Variable region (V region): a sequence of approximately 115 amino acids at the N-terminus of the light chain (V_L) and the heavy chain (V_H) of immunoglobulin molecules and the α, β, γ and δ chain of the TCR (T cell receptor). The amino acid sequences in these regions are responsible for the great diversity of the antigen-binding site and T cell receptor repertoires, each clone of B cells or T cells producing molecules with different variable regions.

Vasoactive amines: substances containing amino groups, such as histamine and serotonin, which cause peripheral vasodilatation and increase the permeability of small vessels.

Veiled cell: a cell characterized by large veil-like processes; found in afferent lymph especially after priming with antigen. Veiled cells possess accessory cell function, and represent an intermediate stage between peripheral dendritic cells (e.g. Langerhans cells) and the dendritic cells of lymphoid tissues (e.g. interdigitating cells).

V_H region: the variable region of the heavy chain of immunoglobulin.

V_κ: the variable region of the κ (kappa) chain of immunoglobulin.

V_L region: the variable region of the light chain of immunoglobulin.

V_λ: the variable region of the λ (lambda) chain of immunoglobulin.

Virgin lymphocyte: a lymphocyte that has not met antigen. An unprimed lymphocyte.

W

Waldenström's macroglobulinaemia: disease occurring mainly in elderly males, characterized by proliferation of cells that make monoclonal IgM paraprotein. Serum IgM levels are high and there is lymph node enlargement, splenomegaly, a haemorrhagic tendency and hyperviscosity of the blood. Bone marrow and lymphoid tissues are infiltrated with pleomorphic lymphocytes and plasma cells.

White cells: the nucleated cells of the blood (i.e. granulocytes, lymphocytes and monocytes), so-called because they form a white layer over the erythrocytes when sedimented.

White pulp: see **spleen**.

X

X-linked agammaglobulinaemia (Bruton-type hypogammaglobulinaemia): antibody deficiency syndrome in boys which becomes manifest once maternally derived antibodies (see **maternal immunity**) have disappeared from the child's tissues. There are low numbers of circulating B cells and very low levels of all immunoglobulins. Pre-B lymphocytes are present in normal numbers in the bone marrow. Characterized by repeated, severe bacterial infections. There is a single defect in a gene on the X chromosome indentified as btk (Bruton Tyrosine Kinase). This codes for a protein tyrosine kinase presumptively required for B cell maturation beyond the pre-B lymphocyte stage.

Xenogeneic (xenogenic; heterogeneic): preferred term for grafted tissue that has been derived from a species different from the recipient.

Xenograft (heterograft): preferred term for a graft from a donor of dissimilar species.

Xenotype: structural or antigenic difference between molecules, e.g. immunoglobulins, cell membrane antigens, etc., derived from different species.

Z

Zymosan: cell wall fraction of yeast (*Saccharomyces cerevisiae*) which activates the alternative pathway to complement activation, and thus binds C3b. Frequently used for study of opsonic phagocytosis.

Index

Note: Page references in *italics* refer to Figures; those in **bold** refer to Tables

Actinobacillus pleuropneumoniae 379
acute-phase proteins (APPs)
 in the rat 187, **187**
 sheep **519**
acute-phase response (APR) proteins 273, **273**
Addison's disease in dogs 281
adhesion molecules
 in the cat 295
 in the pig 378, **378**
 in the rabbit 231–2, **231**
 in the rat 148, **148–9**
 in sheep 490
adult respiratory distress syndrome (ARDS), pig models for 401
Aeromonas salmonicida 29, 31
African swine fever 401, 406
agammaglobulinemia in horses 358
AIDS
 human 248, 289, 290, 353
 murine 594
Aleutian disease of mink 338, **338**, 340
Aleutian disease virus (ADV) 338
allergic asthma in cats 316
allergic breakthrough in humans 316
allergic bronchiolitis in cats 316
allergic colitis, in cats 316
allergic conjunctivitis in cats 316
allergic enteritis, in cats 316
allergic rhinitis 316, 324
alternative complement pathway in chicken (ACP) 100–1
Ambystoma mexicanum 63
Ambystoma tigrinum (tiger salamander) 69
Ambystomatidae family 63
amphibians
 immunobiology of T and B cells 68–9
 lymphocyte antigen-specific receptors 65–7
 lymphoid organs and lymphocytes 63–5
  B-cell receptors 66–7
  in vitro stimulation 65
  other sites of lymphocytopoiesis 64
  spleen 63–4
  surface lymphocyte markers 645
  T-cell receptors 65–6
  thymus 63
 ontogeny of immune response 69–70
 tumors of the immune system 70
anaphylactic shock in cats 315
angioneurotic edema in cats 315
anti-(2,4)-dinitrophenol (DNP) antibodies in amphibians 69
Anura 63
appendix, rabbit 226, *227*, *228*
arthritis
 adjuvant, in the rat 197
 associated with FIV infection in the cat 319
 collagen-induced, in the rat 183
 see also rheumatoid arthritis
asthma, allergic, in cats 316
Aujeszky's disease 401
autoimmune glomerulonephritis in the rat 196, 197
autoimmune hemolytic anemia (AIHA)
 in cats 317
 in dogs 280
autoimmune masticatory myopathy in dogs 281–2
autoimmune skin disease in dogs 280
autoimmune thrombocytopenia (AITP)
 in cats 317–18
 in dogs 280
 in pigs 406
avian immunology
 autoimmunity 111–13, **111**
 cell surface and secreted immunoglobulins in B-cell development 89–92
  bursa of Fabricius 89
  chicken Ig loci 89, *89*
  diversification of the primary immune repertoire 90
  Ig isotype switching 90
  Ig isotypes 91–2, **91**, 92
  rearrangement of chicken Ig genes 90
 complement 99
  avian complement components 99–100, **100**
  complement activation 100–1, **102**
  MHC and ontogeny 101
 cytokines 95–9, **96–7**
  genes **98**
  methods used to detect **99**
  receptors characterized **98**
 effect of viruses on immune functions 113
  DNA viruses **114**
  RNA viruses **115–16**
 immunocompetence of the embryo and newly hatched chick 104–5
  embryo vaccination 105
  maternal antibodies 105
  nonspecific immune defenses 105
  physiology 104
 leukocyte markers in the chicken 81–6
 lymphoid organs 73–80
  of the chicken **74–6**
  general description 73–6
 major histocompatibility complex (MHC) antigens 92–5
  genes 92, **93**
  regulation of immune response 94, **94**
  species-specific aspects of genetics or protein structure and function 95
  tissue distribution, structure and function of proteins and minor antigens 92–4

avian immunology *continued*
  mucosal gut immunity 105–8
    gut-associated immune system 106–8
  ontogeny of the immune system 101–4, **102**
    B-cell development 102–3
    early precursors 102
    other cell types 103
    T-cell development 103
  T-cell receptors, chicken 87–9
    distribution of $\alpha\beta$ and $\gamma\delta$ TCR populations 88
    species-specific aspects of avian T cells 89
      function of T-cell subsets 89
      TCR genes 89
    T-cell receptor repertoire ontogeny 87–8, **87**
      ontogeny of T-cell subsets 87
      ontogeny of TCR gene repertoires 88, **88**
  tumors of the immune system 108–11, **109**
    effects on immune function 111, **111**
    natural and experimental hosts 109, **109**
    pathogenesis of Marek's disease 110
    pathogenesis of retrovirus-induced tumors 110–11
    target cells for transformation by avian retroviruses 110
    target cells for transformation by Marek's disease virus 110
avian leukosis virus (ALV) 108

B-cell receptor complex (BCR) structure and receptors, mouse 577–80
B-cell receptors in amphibians 66–7
B cells
  in amphibians 64, 68–9
  in the cat 295
  in chicken 90–1, 102–3
  equine 347
  in fish 7–8
  rat 168–71
  sheep 511–12, 515
B lymphocytes
  in the cat 313
  in the chicken 86
    ChB1 86
    ChB3 86
    ChB2, ChB4 and ChB5 86
    ChB6 86
    ChB7, ChB8, ChB9, ChB10 and ChB11 86
    ChB12 86
  in fishes 27–9
  in pig 394–5
  in the rabbit **224**
  in sheep 490
B lymphoid tumors, rat 177–8
Babesia bovis 471
babesiosis 606
Bacillus cereus 432
bactenecins 399
badger, Eurasian (*Meles meles*) 337
balance hypothesis of autoimmunity 196, 197
Bartonella henselae 324
basophils in fishes 9, 31–3
big pad disease in cats 320
BLAD, 470
bone marrow
  in the cat 290
  mouse 564
  rat 139–40
  sheep 487
Bordetella bronchiseptica 324
bovine leukemia virus (BLV) 485, 530
Brachdanio rerio (zebrafish) 3, 24
bronchiolitis, allergic, in cats 316
bronchus-associated lymphoid tissue (BALT)
  chicken 73, 79, *81*
  in the pig 403
  in the rat 139, 141, 187–8, **187**
Brucella abortus
  in the pig 397
  in sheep 516
bullous pemphigoid
  in cats 318
  in dogs 280
Burkitt's lymphoma 178
bursa of Fabricius, chicken 73, 77–9, 89, 90–1, 104, 106

calcivirus in the cat 309
camelpox virus 426–7
camels and llamas
  blood groups 432
  complement system 432
  experimental immunizations 428–9
  immunodeficiencies 432
  immunoglobulins 423–8
    experimental immunizations 427–9
    heavy chains 426, **428**
    IgG subclasses 424–5
    light chains 425–6
    naturally occurring antibodies in infections 427–8
    other classes 425
  leukocytes and their markers 422–3
  lymphoid organs 422
  nonspecific immunity 432
  ontogeny of immune system 430
  passive transfer of immunity 430–2
  phylogeny of immune system 429–30
  tumors of the immune system 432
Camelus bactrianus (Bactrian camel) 421
Camelus dromedarius (dromedary camel) 421, 432
Candida albicans in horses 358
canine distemper virus 279
caprine arthritis encephalitis virus in the goat 502
caprine arthritis encephalomyelitis virus (CAEV) 528, 529
carcinoma of the cervix, human 242
Carrassius auratus (goldfish) 24
Castleman's disease 340
cat
  complement system 311–12, **312**
  cytokines 297–301
    IFN 300
    interleukins 297–300
    mast cell growth factor 301
    MHC 301–2
    nerve growth factor 301
    stem cell factor 301
    TNF-α 300
  immunodeficiency diseases 321–5
    adaptive immunity 322–3
    associated with innate immunity 323–4
    cell-mediated immunity 322
    deficiencies in adaptive immunity 324–5
    humoral immunity 322–3
    innate immunity 322
    innate versus adaptive immunity 322
    maternal immunity 323
  immunoglobulins 304–5, **306**
  immunological diseases 315–21
    allergic bowel disease 316–17
    allergies of the respiratory tract 316
    allergies of the skin 315–16

alloantibody diseases 317
autoantibody diseases 317–18
caused by cellular immunity 320–1
cell-mediated diseases of unknown etiology 320
chronic plasmacytic/lymphocytic enteritis 320
delayed-type hypersensitivity disease of the skin 320
granulomatous diseases of infectious etiology 320–1
eosinophilic granuloma complex 316
gammopathies (dysproteinemias or paraproteinemias) 321
monoclonal gammopathy 321
polyclonal gammopathy 321
immune complex diseases 318–20
mechanisms of allergy 315
systemic allergies 315
leukocyte differentiation antigens 294–7
adhesion molecules 295
B-cell antigens 295
NK cells 296
nonlineage 295–6
T-cell antigens 294–5
lymphoid organs 289–94
mucosal immunity 313–14
gastrointestinal immune system 313–14
cells of 314
functional studies of gut immune system 314
infection via the intestinal tract 314
secretory immunoglobulins 313–14
neonatal immune response 308–9
nonspecific immunity 309–11
eosinophils 310
globule leukocyte 310
mononuclear phagocytic system 311
natural barriers 309–10
natural killer cells 310–11
neutrophils 310
ontogeny of the immune system 313
B lymphocyte system 313
T lymphocyte system 313
passive transfer of maternal immunity 305–8
failure 308
immunity transferred by colostrum and milk 305–6
transplacental immunity 305
vaccination of kittens 306–7
red blood cell antigens 303–4, **303**
tumors of the immune system 325–7, **326**
cathelicidins 399
cattle
complement system 469–71
cell receptors for cleavage products of complement components 469
cf. complement system of other animals 471
complement components 469
control proteins 469
deficiencies 470
pathways of initiation 469
serum complement levels 470–1
cytokines 444–54, **448–51**
immunoglobulins 456–9
Ig gene mapping 459
Ig isotypes and physiochemical properties 456–7
Ig isotypes: functional properties 458
V genes 458–9
leukocytes and their markers 444–7
lymphoid organs 439–44
afferent lymph 442
efferent lymph 442
follicle-associated epithelium (FAE) and membranous cell (M cell) 443–4
gut-associated lymphoid tissue (GALT) 442–3
lymph and peripheral blood 442
lymph node 440, *441*
lymphocyte distribution 443–4
spleen 441–2, *441*
thymus 439–40
MHC complex antigens 459–62
class I antigens 459–60
class II antigens 460
disease associations **461**, 462
genetic organizations 460–2
methods of characterization 459–60
tissue distribution, structure and function of BoLA molecules 462
unique properties 462
nonspecific immunity 466–9
cellular components 467–9
molecular components 466–7
natural killer cells 469
physical and chemical barriers to infection 466
passive transfer of immunity 464–6
red blood cell antigens 462–4
applications and significance of red cell antigen diversity 464
blood group systems 462–3, **463**
identified by immune antisera 463
natural blood group system 463
structure of 463
T-cell receptor 454–6
distribution of $\alpha\beta$ and $\gamma\delta$ TCR populations 456
general features of TCR complexes 454, **454**
invariant components 455–6
molecular structure of TCR genes 454–5
$\alpha\beta$ chains 454–5
$\gamma\delta$ chains 455
Charcot–Marie–Tooth disease 198
Chediak–Higashi syndrome 339–40
chemokines
in fishes 11
in the rat 149, 186
chemotactic molecules in the dog **266**
chicken infectious anemia virus 104
Chlamydia psittaci 324
cholangiohepatitis in cats 320
chronic obstructive pulmonary disease (COPD) in the horse 352
chronic progressive polyarthritis in the cat 319
classical complement pathway (CCP) in chicken 101
Clostridium perfringens 432
cluster of differentiation (CD) antigens
in cattle 444–7, **445–6**
in the dog 263–4
equine *see* CD antigens in the horse
in fishes 7
in the mouse 586, **586–7**
rat 142, **143–7**
CD antigens in the horse 343–7, **344**
EqCD2 (EqWC3) 345
EqCD3 345
EqCD4 345
EqCD5 345
EqCD8–9 345–6
EqCD11a/18 346
EqCD13 346
EqCD44 346
EqMHC I and EqMHC II 346
EqWC1 346
EqWC2 346
EqWC4 (EqCD28) 346–7

CD2
  in the pig 374–5, 403
  in the rat 150
  in sheep 490
CD3, pig 375
CD4
  in the cat 294
  in the dog 264
  in the pig 375–6
  in the rabbit 226, **226**
  in the rat 150–1
CD5
  in the cat 295
  in the pig 376
  in the rabbit 226, 228, 235
  in the rat 142
  in sheep and goats 490, 530
CD6, pig 376
CD8
  in the cat 294
  in the rat 150–1
CD8, pig 376
CD8α in the dog 264
CD9 in the cat 295
CD28 in the dog 264
CD34 in the dog 264
CD38 in the dog 264
CD44, pig 376–7
CD45
  in the pig 377
  in the rat 142
CD86, pig 377
Coccidia 324
colony stimulating factors in fishes 13
complement
  in cattle 469–71
  in the chicken 99–102
  dog 274–5
  equine 355
  in fishes 10–11, 31, 36–8
  lagomorphs 243, **244–5**
  in mink 339
  in the horse 586, **586–7**
  pig 400–2
  rat 190–2
  sheep 522–3, **523**
C1
  in the cat 311
  in the dog 275
  in the rat 191
C2 in the rat 190, 191, 192
C3
  in the cat 311
  in chicken 100
  in the dog 275
  in fish 38
  in the rat 190, *190*, 191
C3 deficiency in the dog 277
C4
  in chicken 100
  in the dog 275
  in the rat 191
C5 in the cat 311
C5a in the rat 191
C6
  in chicken 100
  in the rat 190, 192
C6 deficiency in the pig 405
C9 in the rat 191
colitis, allergic, in cats 316
Concanavalin A 8
conglutinin 466–7, 471
conjunctival-associated lymphoid tissue (CALT) in chicken 73, 79, *80*
conjunctivitis, allergic, in cats 316
Cooperia curticei 427
Corynebacterium pyogenes in the camel 422
Cryptosporidia 324
cyclic hematopoiesis in the dog 278
Cyphotilapia frontosa 24
Cyprinus carpio (common carp) 24, 30
cytokine-induced neutrophil chemoattractant (CINC) in the rat 149
cytokines
  in the cat 297–301
  in cattle 444–54, **448–51**
  in chicken 95–9, **96–9**, 107, 117
  dog 267
  equine 350–3
  fishes 11–13, **12**, 43
  in lagomorphs 233
  in mink 338–9
  mouse 588–9, **588–9**
  pig 379–83, **380**
  rat 150, **152–9**, 186
  sheep 496–500, **497**

Daudi lymphoma in mice 606
defective leukemia virus (DLV) 108
dermatitis
  flea allergic, in cats 316
  miliary, in cats 315–16
dermatomycosis (ringworm) in cats 324
desmoglein-1 280
desmoplakins 280
diabetes mellitus, insulin dependent (IDDM)
  in dogs 281
  rat as model for 183, 195
Dirofilaria immitis 337
discoid lupus erythematosus (DLE)
  in the cat 318–19
  in dogs 280
DNA viruses, chicken **114**
dog
  autoimmunity 278–82
  chemotactic molecules **266**
  complement system 274–5
  cytokine receptors 268
  cytokines 267
  immunodeficiencies 277–8, **277**, **278**
  immunoglobulins 268, **268**
  leukocyte traffic and associated molecules 264–7
  leukocytes and their antigens 263–4
  lymphoid organs 261–3
  MHC antigens 268–9
  monoclonal antibodies 264, **264**. **265–6**
  mucosal immunity 275–7
    antigen transport 275–6
    intraepithelial lymphocytes 276
    Peyer's patches (Pps) and solitary lymphoid nodules 276–7
    secretory immunoglobulins 276
  neonatal immune responses 272–3
  nonspecific immunity 273–4
  ontogeny of the immune system 271
  passive transfer of immunity 271–2
  red blood cell antigens 270

T-cell receptor 268
tumors of the immune system 278, **279**
Dutonella vivax 432

E-selectin (ELAM-1)
in the dog 264
in the pig 378
in rabbit 232
ecthyma (of) virus in camels 426
eicosanoids fishes 11
Eimeria acervulina 107–8
Eimeria tenella 107, 108
elasmobranchs, major lymphoid tissues 3, *4*
endothelial adhesion molecules, sheep 490
enteritis, allergic, in cats 316
Enterococcus faecalis 398
eosinophilic enteritis in cats 317
eosinophils
in the cat 310
in fishes 9, 31–3
sheep 525–8
epigonal organ 6
Epstein-Barr virus 606, 610
equine immunoglobulin-secreting heterohybridomas 348
equine infectious anemia virus (EIAV) 347, 353, 528
equine influenza virus 353
Erysipelothrix rhusiopathiae 406
erythroblastosis fetalis in pigs 406
Escherichia coli 396, 398, 399, 432, 458, 465, 471
experimental allergic encephalomyelitis (EAE)
in rats 186
in sheep and goats 532
experimental allergic thyroiditis in sheep and goats 532
experimental autoimmune encephalomyelitis (EAE) 137, 183, 193, 195, 197

facial-conjunctival edema in cats 315
factor H deficiency in pigs 401, 406
feline AIDs 606
feline asthma 316
feline calcivirus infection 321
feline enteric coronavirus (FECV) 324
feline immunodeficiency virus (FIV) 289, 294, 297, 298, 299, 300, 306, 310, 311, 312, 313, 317, 325, 327
feline infectious peritonitis (FIP) 297, 310, 312
feline infectious peritonitis virus (FIPV) 321, 324
feline leukemia virus (FeLV) 289, 306, 307, 308, 310, 312, 317, 324, 325, 326
feline oncornavirus 289
feline panleukopenia virus (FPV) 306, 307, 308
feline retrovirus 290
feline rhinotracheitis virus 307
feline urologic syndrome (FUS) 323
ferret, domestic (*Mustela putorius*) 337
fishes 3–43
acquired immunodeficiencies 40–1
chemicals 40
stress 41
temperature 40
anatomy 3–6
complement system 31, 36–8
agnatha 37
chondrichythes 37
teleosts 37–8
cytokines 11–13, **12**
granulopoiesis and granulocytes in **34**
immunoglobulins 15–23, **16–19**
agnatha 15
antibody response 21
biochemical characterization of Ig 20–1
immunoglobulin genes 21–3
regulation of immunoglobulin genes 23
leukocyte traffic 9–11
chemotactic factors: *in vitro* migration 10–11, **10**
stress and pathological conditions affecting 10, **10**
within and between organs 9–10
leukocytes 6–9, 41
granulocytes 9
lymphocytes 7–8, 26–7, **27**
monocytes/macrophages 8–9
natural cytotoxic cells (NCC) 8
major histocompatibility complex (MHC) antigens 23–5
expression 25
functional studies 23–4
molecular cloning 24–5
polymorphism and genetic organization 25
major lymphoid organs 3–6, *4*, **4**
mucosal immunity 39–40
mucosal immune responses 39–40
mucosal immune system 39
oral tolerance 40
nonspecific immunity 31–4, **32–3**
leukocytes in **35–6**
ontogeny of the immune system 26–30
development of T and B lymphocytes 27–9
differentiation of lymphoid organs 26–7
differentiation of phagocytic system 27
ontogeny of B lymphocyte subpopulations 28–9, *28*
ontogeny of immune responsiveness 29–30, **30**
passive transfer of immunity 30–1
red blood cell antigens 25–6
taxa 3, *19*
T-cell receptor 13–15
tumors of the immune system 41, **42**
fowl pox 117
furunculosis 31

G-CSF in the pig 383
Gadus morhua (cod) 24
Giardia 324
Ginglymostoma cirratum (nurse shark) 24, 25, 37
glomerulonephritis
autoimmune, in the rat 196, 197
in the cat 319–20
membranoproliferative (MPGN) type II in the pig 401, 406
GM-CSF
in the dog 267
in the pig 383
in sheep 499
gombessa 23
graft-versus-host disease 406
granulocytes
in fishes 9, 31–3, **34**, 43
in the pig 399–400
in the rat 147, 184
GRO in rabbit 233
gut-associated lymphoid tissue (GALT)
in cattle 442–3
in chicken 73, 79, 106, 107–8
in fishes 6
in the rabbit 223, 224, *224*, 226, 245, 246, 247, 250
in the rat 139, 187–8, **187**
in sheep 488
Gymnophiona 63

haemobartonellosis 317

Haemobartonellosis felis 325
Harderian gland (HG) 79, *80*
Hashimoto's thyroiditis (lymphocytic thyroiditis) 111, 248, 281
head-associated lymphoid tissue (HALT), chicken 79
Helicobacter mustelae 337
Helicobacter pylori
  in the cat 313, 314
  in the ferret 337
hematopoiesis, cyclic, in the dog 278
hemolytic disease of the newborn, rabbit model of 223, 250
hemorrhagic enteritis (HE) virus 104, 105
herpes virus
  in the cat 309, 312, 324
  in chicken 117
  equine 347, 358
  in kittens 323
  turkey (HVT) 105
heterophils in fish 31–3
Hippoglossus hipposglossus (halibut), spleen in 6
hives in cats 315
horses and donkeys
  cytokines 350–3
    characterized equine cytokines 350–1
    role in equine disease 351
      possible models for T_H1 and T_H2 responses 352–3
      septicemia 351–2
  development of equine immunity 355–6
    immunocompetence in foals 356
    ontogeny of equine immune system 355–6
  disorders of the immune system 356–9
    allogeneic incompatibilities 358–9
    failure of passive transfer of immunity 358
    primary immunodeficiencies 356–8
    secondary immunodeficiencies 358
  immunoglobulins 347–9
  innate immunity 354–5
    complement 355
    natural killer and lymphokine activated killer cells 354–5
    neutrophil and macrophage function 355
  leukocytes and their antigens 343–7, **344**
    advances in equine immunology dependent on leukocyte markers 347
    equine T-cell receptor 347
    specific equine leukocyte antigens **344**
  MHC antigens
    equine MHC polymorphism 353
    genetic, biochemical and functional characteristics 353
    non-MHC lymphocyte alloantigens 354
    regulation of equine MHC expression 353–4
  reproductive immunology 359–61
  tumors of the immune system 359
HTLV1, rabbit model for 250
HTV1 in the rabbit 223, 250
human immunodeficiency virus (HIV) 289, 290, 294, 312, 321, 594
  rabbit model for 223, 250
  cf with rat immunodeficiency virus 194
hyperthyroidism 317
hypogammaglobulinemia , transient
  in horses 358
  of infancy in the dog 278

ICAM-1
  in the dog 264
  in the rat 187
ICAM-2 in the dog 264
ichthyophthiriasis 31
Ichthyopthiriasis multifillis 31
Ichthyostega 63
Ictalurus punctatus (channel catfish) 24, 30
idiopathic polyarthritis in the cat 319
idiopathic systemic vasculitis in the cat 319
Ig
  avian 86, 91–2, **91**, 92
  in camels and llamas 423–8
  in the cat 304–5, **306**
  in cattle 456–9
  in chicken 89–92
  dog 268, **268**
  equine 347–9
  in fishes 15–23, **16–19**
  in lagomorphs 235–8, **235**
  in mink 339, **339**
  mouse 570–86
  pig 384–7
  rat 171–3
  sheep 503–5
Ig heavy chains
  in camels and llamas 426, **428**
  in the cat 304–5
  in cattle **459**
  in fish 21, 22
  mouse 572–5
  rabbit **236, 237**
  rat 173–7, **174–6**
Ig light chain
  in camels and llamas 425–6
  in the cat 304
  equine 349
  in fish 22
  kappa, mouse 575
  lambda, mouse 575–6
  rabbit **237–8**
  rat 177, **178**
  in sheep 490
IgA
  in camels and llamas 423, 432
  in the cat 304, 305, 309, 313–14, 322, 323
  in cattle 457, 458, 464–5, 466
  in chicken 86, 92, 105, 107
  in the dog 272, 276
  equine 349
  in the mouse 572
  in the pig 384, 387, 395–6, 403
  rabbit 225, **226**, 233, 247
  in sheep 490, 514, 525
  in the rat 171, 172
IgA deficiency in the dog 277
IgD
  in camels and llamas 423
  in fish 21
  in the mouse 572
  in rabbit 235
  in the rat 172
  in sheep 490
IgE
  in the cat 305, 322
  in cattle 457, 458
  equine 349
  in the mouse 572
  in the rat 172
  in sheep 490
IgG
  in the cat 304, 305, 309, 322, 323
  in cattle 457, 458, 464, 466

in chicken 86, 92
in the dog 271–2, 276
in the horse 348–9, 355
in the pig 384–5, 387, 395–6, 403
in rabbit 233, 247
in the rat 171, 172
in sheep 490, 514, 525
IgM
in amphibians 68
in camels and llamas 423, 425, 432
in the cat 304, 305, 322, 323
in cattle 456–7, 458, 466
in chicken 86, 91–2, 105, 107
in the dog 272
in fish 20, 39–40
in the horse 349, 356
in the mouse 572
in the pig 384, 387, 395, 403
in rabbit **226**, 233, 247
in the rat 171, 172
in sheep 490, 525
IgM deficiency in horses 357–8
IgNAR in shark 20, 22
IgNARC in shark 20, 22
IgX in the rat 171
IgY
in amphibians 68–9, 70
in chicken 86
IL-1
in the cat 297
in the dog 267, 274
in fish 11–12
in the horse 350, 351, 352
in the pig 379, 403
in rabbit 233
IL-Iβ in fish 11
IL-2
in the cat 297, 314
in the dog 267, 268
fishes 12
in the horse 350, 354
in the pig 383, 403
in rabbit 233
in sheep 499
IL-3
in the pig 383
in sheep 499
IL-4
in the cat 297–8, 314
equine 350
in the pig 383, 403
in sheep 499
IL-5 in the cat 314
IL-6
in the cat 298, 314
in the dog 274
in the horse 350–1, 352
in the pig 379–82, 383
IL-7 in the dog 267
IL-8
in the dog 267
in the pig 382–3
in rabbit 233
IL-10
in the cat 298–9, 314
in the dog 267, 274
equine 351
in the pig 383
IL-11
in the dog 267
in the pig 383
IL-12
in the cat 299, **299**
in the pig 383
IL-15 in the pig 383
IL-16 in the cat 299
IL-18
in the cat 300
in the pig 383
immunocytomas, rat 177–8
immunoendocrinopathy syndrome 281
immunoglobulins *see under* Ig
immunopoiesis 563
infectious bronchitis virus 105
infectious bursal disease virus (IBDV) 104, 105
interferons 383
equine 351
fishes 13
IFN-α in the cat 300
IFN-β in the cat 300
IFN-γ
in the cat 300
in the dog 267
in sheep 499
interleukins in fishes 11–12 (*see also under* IL)

juvenile immunodeficiency syndrome in (JLIDS) camels 432

kidney, fishes 5, 26
Kupffer cells in chicken 103

Lactococcus lactis 432
lagomorphs 223–50
allelic exclusion 223
allotype suppression 223
autoimmune diseases 248–50, **250**
blood groups 242–3, **243**
complement 243, **244–5**
cytokines 233
idiotypes and network interactions in immune regualtion (latent allotypes) 223
immunodeficiencies 248
immunoglobulins 235–8, **235**
accessory molecule function 235
function 235
Ig gene loci 235
ontogeny of Ig repertoire 236
repertoire generation 236
species specific aspects of Ig structure 236–8
leukocyte markers 226–30, **229–30**
lymphocyte markers 228–30
specificity of reagents 227–8
leukocyte traffic 230–3
chemotactic molecules and populations recruited **232**, 233
evidence for organ-specific leukocyte recruitment 230–1
molecules involved in leukocyte–endothelial interactions 231–2
molecules involved in leukocyte migration 232, **232**
species-specific aspects 233
tissue-specific recruitment of leukocyte subpopulations 231
lymphoid organs 223–6
species-specific aspects 226
MHC antigens 238–42
disease association 242
expression and tissue distribution 242
species-specific aspect 242

lagomorphs, MHC antigens *continued*
as model for disease-related research 250
mucosal immunity 245–8
antigen transport and function of antigen-presenting cells 245–7
production and transport of mucosal Igs 247
structure/function of mucosal immune system 247–8
nonspecific immunity 243
passive transfer of immunity 243
T-cell receptor (TCR) 233–4
distribution of $\alpha\beta$ and $\gamma\delta$ TCR populations 234
genes 233–4, **233**
species-specific aspects of structure, function or development 234
TCR-associated proteins 234, **234**
TCR repertoire ontogeny 234
tumors of the immune system 248, **249**
V_H gene diversification mechanisms 223
Lama glama (domesticated llama) 421
Lama guanicoe (guanaco) 421
Lama pacos (alpaca) 421, 430
Lama vicugna (vicuña) 421
Lebistes reticulatus 30
lectin pathway in fish 36
Leishmania braziliensis 108
Leptospira pomona in the skunk 338
leukemia
lymphoid in the dog 278
mast cell, in the pig 405, 406
myeloid
in the dog 278
in the pig 405, 406
SCID mice as models for
acute lymphoblastic 609
acute myeloblastic 609
chronic lymphocytic 609, 610
chronic myelocytic 609
leukocyte adhesion deficiency in the dog 277
leukocytes
in the cat 294–7
in cattle 444–7
in the chicken *see* leukocytes in the chicken
in the dog 263–4
equine 343–7, **344**
in fishes 6–9, 31–3, 41, 43
pig 374–7, 403
sheep 489, **491–2**, **519**
leukocytes in the chicken 82–5
ChL2, ChL3, ChL4, ChL5, ChL7 and ChL10 82
ChL6 82–3
ChL9 83
ChL12 83
ChL13 83
CD45 83–4
CD57 84
HEMCAM 84–5
leukotriene B4 (LTB4) 11
Leydig's organ 6
lipopolysaccharide (LPS) 8
lipoxins 11
Lissamphibia 63
Listeria monocytogenes 522
lymph nodes
in camels *422*
in the cat 290–4, **291**, **292**, **293**, 308
in cattle 440, *441*
dog 261–3, **262**, *263*
mouse 565–7, *566*
in the pig 374, 402
rabbit 224–5
rat 138–9, 140–1
sheep 487
lymphadenitis in the camel 422, *423*
lymphoblasts in pig 379
lymphocytes
in cattle 443–4
in chicken 106–7
in fishes 7–8, 9–10, 26–7, **27**
in the rat 142
in sheep 490
lymphoid leukosis (LL) 108
lymphokine activated killer cells, equine 354–5
lymphoma in the dog 278
lymphopoeisis 563
lymphoproliferative disease virus (LPDV) 109
lymphosarcoma
in camels and llamas 434
in horses 358, 359

M cells
in cattle 443–4
in the dog 275–6
in rabbit 245–6
in the rat 188
in sheep and goats 524
macrophage activating factor (MAF) 13
macrophage inflammatory protein-2 (MIP-2) in the rat 149
macrophages
in cattle 468
in the dog 274, **274**
equine 347, 355
in fish 8–9, 10, 31–3, 43
in the rat 184, 186
Maedi-Visna virus (MVV) 485, 528, 529–30
major histocompatibility complex (MHC) antigens
in amphibians 67–8
in the cat 301–2
in cattle 459–62
chicken 92–5, 104, 113–17
dog 268–9
equine 353–4
fishes 23–5
in lagomorphs 238–42
in mink 339
pig 387–9
rabbit 226
malignant histiocytosis in the dog 278
mannan-binding lectin (MBL) in fish 36
mannose-binding protein (MBP) in the rat 185
Marek's disease 108, **109**
pathogenesis of 110, *110*
Marek's disease virus (MDV) 104, 105, 108, 110
marten (*Martes martes/foina/americana*) 337
mast cells
in the cat 301
in cattle 468
sheep **519**, 525–8
measles virus 394
melanomacrophages 5, 6, 9, 31
membranoproliferative glomerulonephritis (MPGN) type II in the pig 401, 406
Microsporum canis 324
migration inhibition factor (MIF) 11
mink (*Mustela vison*) 337, 338–40
complement 339
cytokines 338–9

  immunodeficiency 339–40
  immunoglobulins 339, **339**
  leukocytes and their markers 338
  MHC antigens 339
  nonspecific immunity 339
  ontogeny and passive transfer of immunoglobulins 339
mixed lymphoctyte reaction (MLR) 460
  in amphibians 65, 69
  in fish 23–4
  in rabbit 238
monoclonal antibodies (mAbs)
  in amphibians 64–5
  in the cat 296, **296**
  in cattle 444–7, **447**
  in the chicken 77, 81–2, **82–5**
  in the dog 264, **264**. **265–6**
  in fishes 7, 7, 8, 27–8
  in the horse **344**
  rabbit 226, 228
  in the rat 190
  in sheep 489, 490, **492**, 499
monocyte chemoattractant protein-1 (MCP-1)
  in rabbit 233
  in the rat 149
monocytes
  in the dog 274, **274**
  in fishes 8–9, 31–3
  in the rat 147, 184, 186
mononuclear phagocytes, chicken 86
Morone saxatilis (striped bass) 24
mouse
  cellular immunology 586–9
    CD markers 586, **586–7**
    cytokines 588–9, **588–9**
  immunoglobulins 570–86
    B-cell receptor complex (BCR) structure and receptors 577–80
      BCR structure 577–8
      co-receptors 578–80
    B-cell ontogeny and Ig repertoire expression 580–4
      B-cell repertoire in species 584
      establishment of V gene repertoire 580–3
      fetal/neonatal V gene repertoire vs adult V gene repertoire 583–4
    general structure 570–2
      biological properties 570–2, *571*
      physical and chemical aspects 570, **571**
    genetics 572–7
    heavy chains 572–5
      CH locus 573
      DH and JH gene segments 573
      V_H gene germline repertoire 573
      V_H segment organization 573
      V_H subgroup and family 572
    immunoglobulin gene polymorphisms 576–7
    kappa light chains 575
    lambda light chain 575–6
    surface light chain and pBCR 584–5
      B-cell development in mice carrying a deficient $\lambda 5$ gene 584–5
      B-cell subsets/lineages and natural antibodies 585–6
      expression in B-cell development 584
      functions of μH chains-SL complex pre-BCR 585
      gene and protein structure 584
  lymph organs 563–70
    central lymph organs 563–5
    peripheral lymph organs 565–70
  murine models of immunodeficiency 589–94, **589–93**
mucosa-associated lymph tissues (MALT)
  mouse 568–70
  in the pig 404
  in the rat 139, 141, 187
  in sheep 488
multicluster type Ig in Chondrichthyes 21
multiple sclerosis 195
mural nodules in the chicken 79
murine AIDS 594
myasthenia gravis
  in cats 318
  in dogs 281
Mycobacterium cheloni 321
Mycobacterium fortuitum 321
Mycobacterium lepraemurium 321
Mycobacterium smegmatis 321
Mycobacterium bovis
  in badgers 337
  in the ferret 337
Mycoplasma arthritidis 198
Mycoplasma hyorhinis 406
mycosis fungoides in the cat 327
myeloid markers, sheep 490
myeloma in the rat 177–8
myeloproliferative disease, equine 359
myxoma virus in the rabbit 223, 248, 250

nasal-associated lymph tissues (NALT) in the rat 139, 141, 187–8, **187**
Nanomonas congolense 432
natural antibodies in fishes 21
natural cytotoxic cells (NCC) in fishes 8
natural killer (NK) cells
  in the cat 296, 310–11
  in cattle 469
  in chicken 103, 105
  equine 354–5
neonatal isoerythrolysis
  in horses 358–9
  in kittens 317
nerve growth factor in the cat 301
neutrophils
  in the cat 310
  in cattle 467–8
  equine 355
  in fishes 9, 31–3
  in the rat 184
Newcastle disease virus (NDV) 105
Nocardia opaca 398

Onchorhynchus keta (chum salmon) 30
Onchorhynchus mykiss (rainbow trout) 24, 26, 31
Onchorhynchus nerka (coho salmon) 30
Oreochromis niloticus (tilapia) 24, 30
Ostertagia circumcincta 427
otter, river (*Lutra canadensis*) 337
otter, sea (*Enhydra lutris*) 337, 338

panleukopenia virus in the cat 309
papilloma in the rabbit 223, 242, 250
paroxysmal nocturnal hemoglobulinuria (PNH) 191
Pasteurella 529
Pasteurella haemolytica 520
Pasteurella multocida 379, 432
pemphigus erythematosus
  in cats 318
  in dogs 280

pemphigus foliaceous
  in cats 318
  in dogs 280
pemphigus vegetans in dogs 280
pemphigus vulgaris
  in cats 318
  in dogs 280
Peyer's patches
  in the cat 291
  in chicken 79, 104, 106
  in the dog 263, 275, 276, **276**
  in the pig 374, 402
  in the rabbit 224, 246
  in the rat 139, 141, 147–8
  sheep 525, **527**
pig
  autoimmunity 406
  complement system 400–2
    animal models of human disease 401–2
    complement deficiency states 401
    components 400–1, **400**
    infectious disease 401
    regulators of complement activation 401
    xenotransplantation 402
  cytokines and interferons 379–83, **380**
    detection **381**
    future prospects 383
    hematopoietic growth factors 383
    immune response cytokines 383
    inflammatory 379–83
    interferons 383
    oligonucleotide primer sequences **382**
  immunodeficiencies 405, **405**
  leukocyte traffic and associated molecules 377–9
    adhesion molecule detection 378
    adhesion molecules and experimental models 378
    differences in lymphocyte migration 378–9
    leukocyte migration 379
    lymphoblasts 379
  leukocytes 374–7
  lymphoid organs 373–4
    development, structure and function 373–4
  immunoglobulins 384–7
    ontogeny of Ig repertoire 386
    species-specific aspects 386–7
  MHC antigens 387–9
    class I antigens 387
    class II antigens 387
    class III genes 387
    distribution and function 387–9
    species-specific aspects 389
  mucosal immunity 402–5
    functional organization 404–5
    mammary glands 404
    oropharynx 402
    reproductive tract 404
    respiratory tract 403
    small intestine 402–3
      epithelium 403
      lamina propria 402–3
      mesenteric lymph node 402
      Peyer's patch 402
  neonatal immune responses 397–8
    neonate physiology 397
    nonspecific immune defense 397–8
    species-specific aspects 398
    specific immune defenses 398
  nonspecific immunity 398–400
    granulocytes 399–400
    natural and physical barriers 398–9
    soluble antimicrobial factors 399
  ontogeny of the immune system 394–5
    development of B lymphocytes 394–5
  passive transfer of immunity 395–7
    closure 396
    educational role of mammary secretions 397
    functions of mammary gland 395
    internal absorption of macromolecules in the neonate 396
    origins and composition of colostrum and milk 395–6
    placentation 395
    role of cells in mammary secretions 396
    transition from colostrum to milk 396
  red blood cell antigens 389–94
    serology 390–2
      EAA system 390
      EAB, EAD, EAG, EAI and EAO systems 390
      EAC, EAJ and EAP systems 391
      EAE system 391
      EAF system 391
      EAH system 391
      EAK system 391
      EAL system 392
      EAM system 392
      EAN system 392
      historical development 390
      reagents 390
      serological test 390
    significance and application 392–4
      heterozygosity and genetic distances 392
      mapping of blood group systems 392
      parentage testing 392
      relationship to production performance and diseases 392–3
    species-specific aspects 394
    terminology 389–94
    xenotransplantation 393–4
  T-cell receptor 383–4
  tumors of the immune system 405–6
Pipidae 63
plasma cell myeloma in horses 359
plasma cell tumors in the dog 278
plasmacytoma
  in the cat 327
  mouse 178
  in the pig 405
  in the rat 177–8
Pleurodeles 69
Pleurodeles waltl 63
Pleuronectes platessa (plaice) 9, 30
Pneumocystis carinii
  in the ferret 337
  in horses 358
pneumonia in cats 323–4
Poecilia formosa 24
polyarteritis nodosa in cats 319
polyarthritis 198
  idiopathic, in the cat 319
polychondritis of the ear in cats 320
polymyositis in dogs 281
proximal colonic lymphoid tissue (PCLT) in the rat 139, 141
Pseudomonas aeruginosa 402
pseudorabies in the pig 399, 401
Pteromyzoin marinus (sea lamprey) 31

Rana catesbeiana 63
Rana pipens 63

RANTES in the dog 267
rat
  autoimmunity 194–8, **194**
    additional rat models 198
    chemically induced 197–8
    induced by immune system modulation 196–7
      induced BUF rat model of thyroiditis 197
      induced models of IDDM 196
      syngeneic graft-vs-host (GvH) reactivity 196–7
    induced by immunization 197
    overview 194–5
    spontaneous models 195
      BBZ/Wor rat model of obese autoimmune non-insulin-dependent diabetes 195–6
      diabetes-prone bio-breeding (DP-BB) rat 195
      female BUF rat model of spontaneous autoimmune thyroiditis 196
    transgenic rat models 198
  B-cell development 168–71
    B-1 cells 170–2
    B-cell generation in adult bone marrow 168–9
    B-cell generation in fetal liver 169
    in vivo humoral immune response 171
    marginal zone B cells 170
    peripheral virgin B-cell differentiation 169–70
  B lymphoid tumors or immunocytomas 177–8
  complement system 190–2
    mechanisms of activation 190–1
    in the rat 191–2
  components of mucosal immune system 187–90
    passive immunity 189
    production and transport of mucosal Igs 189
    structure of the placenta 189–90
    transport of antigens 188–9
  cytokines and their receptors 150, **152–9**
  Ig variable region genes 173–7
    allotypic markers **173**
    complexity of the IgV_H locus 176–7, **176**
    development of Ig repertoire 177
    generation of Ig diversity 173
    Ig heavy-chain variable region sequences 173–6, **174–6**
    Ig light-chain sequences 177, **178**
  immunoglobulins 171–3
    nomenclature 171–3, **171**
  leukocyte traffic and associated molecules 142–9
    species-specific aspects 148–9
  lymphoid organs 137–41
    morphological and functional aspects 139–41
      blood 141, **142**
      central lymphoid organs 139–40
      lymph nodes 140–1
      mucosa-associated lymphoid tissues 141
      peripheral lymphoid organs 140–1
      spleen 140
    rat lymphoid system 137–9
      central part 137
      lymphatic system 138, *139*
      mucosa-draining lymph nodes 138–9
      organ/tissue-draining lymph nodes 139
      peripheral part 138–9
      skin-draining lymph nodes 138
  major histocompatibility complex (MHC) 179–83
    associations of RT1 with disease and/or immunological dysfunction 181–3, **183**
    history 179
    rat strains 180–1, **180**, **181–2**
    RT1 complex 179–80, **179**
  models of immunodeficiency 192–4, **192**
    brown Norway (BN) rat 193
    DP-BB rat 193
    gene-targeted genetic manipulation 194
    LEC rat 193
    Lewis rat 193–4
    nude rats 192–3
    SCID rat 194
    virus-induced/acquired immunodeficiency in 194
  nonspecific immunity 183–7
    infiltration of inflammatory cells 186–7
    local inflammation 184
    macrophages and granulocytes 184
    production of inflammatory mediators and cytokines 186
    recognition of microorganisms and their components 185–6
    skin and mucosa 183–4
    systemic effects of inflammation 187
  rat cluster of differentiation (CD) antigens 142, **143–7**
  T-cell development 150–61
    intra-thymic 150–1
    post-thymic 151–61
  T-cell receptor 162–8
    distribution and phenotype of $\alpha\beta$ and $\gamma\delta$ populations 168
      $\alpha\beta$ T cells 168
      $\gamma\delta$ T cells 168
    mAbs to rat TCR 164, **167**
    ontogeny of the rat TCR repertoire 164–8
    T-cell receptor sequences 162–4, **162–4**, **165–7**
    TCR associated proteins and accessory molecules involved in signalling 164, **167**
red blood cell antigens
  in the cat 303–4, **303**
  in cattle 462–4
  in the dog 270
  in the pig 389–94
  in sheep 509–10
Reiter's disease 319
respiratory burst activity (RBA) in fish 34
reticuloendotheliosis virus (REV) 109
rheumatoid arthritis (RA)
  in dogs 279, 280
  human 198, 607, 406
  in pigs 406
  rat as model for 148, 197
rhinitis, allergic 316, 324
Rhodococcus equi in the horse 355, 358
RNA viruses, chicken **115–16**
rotavirus in the pig 399
Rous sarcoma virus in chicken 95, **95**

sacculus rotundus, rabbit 226, 248
Salmo salar (Atlantic salmon) 24
Salmonella 488, 516
Salmonella dysenteriae 432
Salmonella typhimurium 432
Salvelinus leucomaenis 31
sarcoma in chicken 94
Schmidt's syndrome in humans 281
SCID mouse mutant
  Hu-PBL-SCID mouse 605–6
    description 605–6
    as model for immune diseases 606
    as model for infectious diseases 606
    xenogeneic engraftment 606
  models of human hematological disease 609–10
    Epstein-Barr virus lymphomas 610
    growth of human tumors of non-hematopoietic origin 610
    leukemia 609–10

SCID mouse mutant, models of human hematological disease *continued*
in testing leukemia therapy programs 610
RAG knock-out mice 608
SCI-Skin mouse model 606–7
CTL response 607
description 606–7
SCID and NK cell-deficient mice 608
SCID bone marrow engrafting 609
SCID-Hu/bone model 608–9
SCID-hu mouse model 604–7
chimera optimized by co-implanting various fetal human organs 605
description 604
as *in vivo* model for physiopathological analysis of human viral diseases 604–5
limitations 605
SCID 'leaky' phenotype and new immune-deficient murine strain 607–8
use without reconstitution 610–11
for immune system in reproduction 610–11
for infectious diseases 610
scleroderma in chicken 112
Semliki-Forest virus in sheep 499
Serpulina hyodysenteriae 379
serum amyloid A (SAA) in the dog 274
severe combined immunodeficiency (SCID)
in the horse 354–5, 356, 357
in mice 594
see also SCID mouse mutant
sheep and goats
complement 522–3, **523**
cytokines 496–500, **497**
expression and activity **498**
experimental investigation of autoimmunity 531–3
immunodeficiencies 528–30
immunoglobulins 503–5
concentration of Ig in biological fluids 503
expression and function 503
genes 503
Ig repertoire ontogeny 503–5
passive maternal transfer 505
recent reviews 505
repertoire diversification 505
structure of IgG 505
innate immune mechanisms 518, **518**
acute-phase proteins **519**
inducible nitric oxide synthase expression by goat macrophages 521–2
leukocytes and mast cells **519**
ovine-specific aspects of structure and function 520–1
reticuloendothelial system and immune defence 518–20
leukocyte migration 493–6
general considerations 493
leukocyte trafiic through stimulated tissues 494
leukocyte traffic through unstimulated tissues 493–4
lymphocyte migration/recirculation in the fetus 496
mediators of leukocyte migration 494–5
molecules involved in leukocyte migration 495–6
subset-specific and tissue-specific migration patterns of lymphocytes 495
leukocytes and their markers 489, **491–2**
endothelial adhesion molecules 490
lymphocyte markers 490
monoclonal antibody reagents 490, **492**
myeloid markers 490
lymphoid organs 485–9
bone marrow 487
fetal hemopoiesis 486
gut-associated lymphoid tissue 488
hemal nodes and mesenteric 'milk spots' 488
lymph nodes 487
spleen 486
stromal cells and accessory cells 488–9
thymus 486
mucosal immunity 523–8
antigen transport and function of antigen-presenting cells 524–5
components 523–4, **524**
intra-epithelial lymphocytes (Els) 525, **526**
Peyer's patches 525, **527**
production and transport of mucosal Ig 525
role of mast cells and eosinophils 525–8
neonatal immune system 515–18
immune system at birth 515
immune system during first month 516–17
precocious development of the fetal immune system 515
B-cell development 515
T-cell development 515
transition to postnatal life 515–16
ontogeny of the immune system 510–13
development of B cells 511–12
development of lymphoid organs 511
development of lymphatic system and lymphocyte recirculation 512
development of other cell lineages 512
development of T cells 511
ontogeny of immune responsiveness 512
recent review 513
ovine major HCA (Ovar) 505–9
genetics 506, **507–8**
methods used to characterize MHC antigens 505–6
ovine-specific aspects of MHC genetics or protein structure and function 509
tissue distribution, structure and function of ovine MHC antigens 506–8
red blood cell antigens 509–10
T-cell receptors 500–2
caprine T-cell receptor variable β chain (TCRVβ) repertoire 502
number and diversity of variable region genes 502
number, diversity and expression of C7 segments 502
prevalence of $\alpha\beta$ and $\gamma\delta$ T cells in different tissues 501
prevalence of sheep $\gamma\delta$ T cells 502
recent reviews 52
specific features of $\alpha\beta$ TCR 501
specific features of the $\gamma\delta$ 502
surface expression of CD3/TCR proteins 500
T-cell receptor genes 500
transmission of passive immunity 513–15
tumors of the immune system 530–1
clinical observations 530
histological, immunological and biochemical studies 530–1
viral expression 531
ex vivo studies 531
in vivo studies 531
Sjorgren's syndrome in dogs 280
skunk, striped (*Mephitis mephitis*) 337, 338
spleen
in amphibians 63–4
in the cat 290–1, 308
in cattle 441–2, *441*
chicken 79, *80*, 104
dog 261
ellipsoids 6
fishes 5–6, 41

melanomacrophage centers 6
mouse 567–8
rabbit 224
rat 140
sheep 486
Staphylococcus aureus 432, 458
stem cell factor in the cat 301
Streptococcus canis 324
Strongloides stercoralis in the ferret 337
subacute sclerosing panencephalitis (SSPE) 337
syphilis, rabbit model of 223, 250
systemic lupus erythematosus (SLE)
in the cat 317, 318
in the dog 278–9, 280
systemic vasculitis, idiopathic, in the cat 319

T cells
in amphibians 68–9
in the cat 294–5
in the chicken 103
in fish 8
in the rat 150–61
in sheep 500–2, 511, 515
T-cell receptor (TCR)
in amphibians 65–6
in cattle 454–6
in the chicken 87–9
in the dog 268
equine 347
in fishes 13–15
in lagomorphs 233–4
pig 383–4
rat 162–8
sheep 500–2
T lymphocytes
in the cat 313
in the chicken 85–6
CD3-TCR complex 85
CD4 and CD8 85
CD5 and CDw6 85
CD28 86
ChT1 85
ChT6 86
ChT11 86
in fishes 27–9
in the rabbit **224**
sheep 490
Taenia infection 526
teleosts, major lymphoid tissues 3, *4*
theileriosis 606
thymoma in the pig 405
thymopoietin 104
thymus
in amphibians 63
in the cat 290, 308
in cattle 439–40
in the chicken 73, 76–7, *78*, 104
in the dog 261
in fishes 5, 26–7, *27*, 41
in the mouse 564–5
in the rat 140
in sheep 486
thyroiditis
autoimmune
in chickens 111
in the rabbit 248–50
Hashimoto (lymphocytic)
in the chicken 111
in the dog 287
in the rabbit 248
spontaneous, in the rat 183
tonsils
in the cat 291
in the dog 263
toxic epidermal necrolysis (TEN) in cats 319
Toxoplasma gondii 299
transforming growth factor (TGF-β)
in chicken 107, 108
in fish 11, 13
transient hypogammaglobulinemia
in horses 358
of infancy in the dog 278
transition type of Ig in Actopterygii 21
transmissible gastroenteritis virus (TGEV) 396, 399
Triakis scyllia 25
Trichinella spiralis 189
Trichostrongylus colubriformis 427
trinitrophenol (TNP) 21
Triturus alpestris 63, 69
Triturus viridescens 63
Trypanosoma brucei 432
Trypanosoma brucei gambiense 432
Trypanosoma brucei rhodesiense 432
Trypanosoma cruzi 108
Trypanosoma evansi 432
trypanosomiasis 426, 432
tuberculosis, rabbit model for 223, 250
tumor necrosis factor (TNF-α)
in the cat 300
in fishes 11, 13
in the dog 267, 274
in the horse 351
in the pig 403
tumors of the immune system
in the cat 325–7, **326**
in the dog 278, **279**
in fishes 41, **42**
in horses 359
in the rabbit 248, **249**
in sheep 530–1

ulceroproliferative faucitis/periodontis 321
Urodela 63
urticaria in cats 315

VCAM in the rat 187
VCAM-1
in the dog 264
in the pig 378
Venezuelan equine encephalitis virus 356
Vibrio anguillarum vaccine 29, 39
viral plasmacytosis 338, **338**, 340
Visna infections in sheep 532
vitiligo in chicken 111, 113
Vogt–Koyanagi–Harada syndrome in dogs 281

weasel (*Mustela erminia/frenata/rixosa/sibirica*) 337

X-linked severe combined immunodeficiency in the dog 261, 277
Xenopus 38
Xenopus laevis 63, 66
MHC in 24
Xenopus ruwenzoriensis 66
Xenopus tropicalis 66
Xiphophorus coychianus (platyfish) 24

Yersinia ruckeri 29